Linear Models and Time-Series Analysis

Registered Offices
John Wiley & Sons, Inc., 111 River Street, Hoboken, NJ 07030, USA
John Wiley & Sons Ltd, The Atrium, Southern Gate, Chichester, West Sussex, PO19 8SQ, UK

Editorial Office
9600 Garsington Road, Oxford, OX4 2DQ, UK

For details of our global editorial offices, customer services, and more information about Wiley products visit us at www.wiley.com.

Wiley also publishes its books in a variety of electronic formats and by print-on-demand. Some content that appears in standard print versions of this book may not be available in other formats.

Library of Congress Cataloging-in-Publication Data

Names: Paolella, Marc S., author.
Title: Linear models and time-series analysis : regression, ANOVA, ARMA and GARCH / Dr. Marc S. Paolella.
Description: Hoboken, NJ : John Wiley & Sons, 2019. | Series: Wiley series in probability and statistics |
Identifiers: LCCN 2018023718 (print) | LCCN 2018032640 (ebook) | ISBN 9781119431855 (Adobe PDF) | ISBN 9781119431985 (ePub) | ISBN 9781119431909 (hardcover)
Subjects: LCSH: Time-series analysis. | Linear models (Statistics)
Classification: LCC QA280 (ebook) | LCC QA280 .P373 2018 (print) | DDC 515.5/5–dc23
LC record available at https://lccn.loc.gov/2018023718

Cover Design: Wiley
Cover Images: Images courtesy of Marc S. Paolella

Set in 10/12pt WarnockPro by SPi Global, Chennai, India
Printed in Singapore by C.O.S. Printers Pte Ltd

10 9 8 7 6 5 4 3 2 1

Contents

Preface

Cowards die many times before their deaths. The valiant never taste of death but once.
(William Shakespeare, Julius Caesar, Act II, Sc. 2)

The goal of this book project is to set a strong foundation, in terms of (usually small-sample) distribution theory, for the linear model (regression and ANOVA), univariate time-series analysis (ARMAX and GARCH), and some multivariate models associated primarily with modeling financial asset returns (copula-based structures and the discrete mixed normal and Laplace). The primary target audiences of this book are masters and beginning doctoral students in statistics, quantitative finance, and economics.

This book builds on the author's "Fundamental Statistical Inference: A Computational Approach", introducing the major concepts underlying statistical inference in the i.i.d. setting, and thus serves as an ideal prerequisite for this book. I hereafter denote it as book III, and likewise refer to my books on probability theory, Paolella (2006, 2007), as books I and II, respectively. For example, Listing III.4.7 refers to the Matlab code in Program Listing 4.7, chapter 4 of book III, and likewise for references to equations, examples, and pages.

As the emphasis herein is on relatively rigorous underlying distribution theory associated with a handful of core topics, as opposed to being a sweeping monograph on linear models and time series, I believe the book serves as a solid and highly useful prerequisite to larger-scope works. These include (and are highly recommended by the author), for time-series analysis, Priestley (1981), Brockwell and Davis (1991), Hamilton (1994), and Pollock (1999); for econometrics, Hayashi (2000), Pesaran (2015), and Greene (2017); for multivariate time-series analysis, Lütkepohl (2005) and Tsay (2014); for panel data methods, Wooldridge (2010), Baltagi (2013), and Pesaran (2015); for micro-econometrics, Cameron and Trivedi (2005); and, last but far from least, for quantitative risk management, McNeil et al. (2015). With respect to the linear model, numerous excellent books dedicated to the topic are mentioned below and throughout Part I.

Notably in statistics, but also in other quantitative fields that rely on statistical methodology, I believe this book serves as a strong foundation for subsequent courses in (besides more advanced courses in linear models and time-series analysis) multivariate statistical analysis, machine learning, modern inferential methods (such as those discussed in Efron and Hastie (2016), which I mention below), and also Bayesian statistical methods. As also stated in the preface to book III, the latter topic gets essentially no treatment there or in this book, the reasons being (i) to do the subject justice would require a substantial increase in the size of these already lengthy books and (ii) numerous excellent books dedicated to the Bayesian approach, in both statistics and econometrics, and at varying levels of sophistication, already exist. I believe a strong foundation in underlying distribution theory, likelihood-based inference, and prowess in computing are necessary prerequisites to appreciate Bayesian inferential methods.

The preface to book III contains a detailed discussion of my views on teaching, textbook presentation style, inclusion (or lack thereof) of end-of-chapter exercises, and the importance of computer programming literacy, all of which are applicable here and thus need not be repeated. Also, this book, like books I, II, and III, contains far more material than could be covered in a one-semester course.

This book can be nicely segmented into its three parts, with Part I (and Appendices A and B) addressing the linear (Gaussian) model and ANOVA, Part II detailing the ARMA and ARMAX univariate time-series paradigms (along with unit root testing and time-varying parameter regression models), and Part III dedicated to modern topics in (univariate and multivariate) financial time-series analysis, risk forecasting, and portfolio optimization. Noteworthy also is Appendix C on some multivariate distributional results, with Section C.1 dedicated to the characteristic function of the (univariate and multivariate) Student's t distribution, and Section C.2 providing a rather detailed discussion of, and derivation of major results associated with, the class of elliptic distributions.

A perusal of the table of contents serves to illustrate the many topics covered, and I forgo a detailed discussion of the contents of each chapter.

I now list some ways of (academically) using the book.[1] All suggested courses assume a strong command of calculus and probability theory at the level of book I, linear and matrix algebra, as well as the basics of moment generating and characteristic functions (Chapters 1 and 2 from book II). All courses *except the first* further assume a command of basic statistical inference at the level of book III. Measure theory and an understanding of the Lebesgue integral are *not* required for this book.

In what follows, "Core" refers to the core chapters recommended from this book, "Add" refers to additional chapters from this book to consider, and sometimes other books, depending on interest and course focus, and "Outside" refers to recommended sources to supplement the material herein with important, omitted topics.

1) One-semester beginning graduate course: Introduction to Statistics and Linear Models.
   - Core (not this book):
     Chapters 3, 5, and 10 from book II (multivariate normal, saddlepoint approximations, noncentral distributions).
     Chapters 1, 2, 3 (and parts of 7 and 8) from book III.
   - Core (this book):
     Chapters 1, 2, and 3, and Appendix A.
   - Add: Appendix D.
2) One-semester course: Linear Models.
   - Core (not this book):
     Chapters 3, 5, and 10 from book II (multivariate normal, saddlepoint approximations, noncentral distributions).
   - Core (this book):
     Chapters 1, 2, and 3, and Appendix A.
   - Add: Chapters 4 and 5, and Appendices B and D, select chapters from Efron and Hastie (2016).

1 Thanks to some creative students, other uses of the book include, besides a door stop and useless coffee-table centerpiece, a source of paper for lining the bottom of a bird cage and for mopping up oil spills in the garage.

- Outside (for regression): Select chapters from Chatterjee and Hadi (2012), Graybill and Iyer (1994), Harrell, Jr. (2015), Montgomery et al. (2012).[2]
- Outside (for ANOVA and mixed models): Select chapters from Galwey (2014), West et al. (2015), Searle and Gruber (2017).
- Outside (additional topics, such as generalized linear models, quantile regression, etc.): Select chapters from Khuri (2010), Fahrmeir et al. (2013), Agresti (2015).

3) One-semester course: Univariate Time-Series Analysis.
   - Core: Chapters 4, 5, 6, and 7, and Appendix A.
   - Add: Chapters 8, 9, and 10, and Appendix B.
   - Outside: Select chapters from Brockwell and Davis (2016), Pesaran (2015), Rachev et al. (2007).
4) Two-semester course: Time-Series Analysis.
   - Core: Chapters 4, 5, 6, 7, 8, 9, 10, and 11, and Appendices A and B.
   - Add: Chapters 12 and 13, and Appendix C.
   - Outside (for spectral analysis, VAR, and Kalman filtering): Select chapters from Hamilton (1994), Pollock (1999), Lütkepohl (2005), Tsay (2014), Brockwell and Davis (2016).
   - Outside (for econometric topics such as GMM, use of instruments, and simultaneous equations): Select chapters from Hayashi (2000), Pesaran (2015), Greene (2017).
5) One-semester course: Multivariate Financial Returns Modeling and Portfolio Optimization.
   - Core (not this book): Chapters 5 and 9 (univariate mixed normal, and tail estimation) from book III.
   - Core: Chapters 10, 11, 12, 13, and 14, and Appendix C.
   - Add: Chapter 5 (for TVP regression such as for the CAPM).
   - Outside: Select chapters from Alexander (2008), Jondeau et al. (2007), Rachev et al. (2007), Tsay (2010), Tsay (2012), and Zivot (2018).[3]
6) Mini-course on SAS.
   Appendix D is on data manipulation and basic usage of the SAS system. This is admittedly an oddity, as I use Matlab throughout (as a matrix-based prototyping language) as opposed to a primarily canned-procedure package, such as SAS, SPSS, Minitab, Eviews, Stata, etc.
   The appendix serves as a tutorial on the SAS system, written in a relaxed, informal way, walking the reader through numerous examples of data input, manipulation, and merging, and use of basic statistical analysis procedures. It is included as I believe SAS still has its strengths, as discussed in its opening section, and will be around for a long time. I demonstrate its use for ANOVA in Chapters 2 and 3. As with spoken languages, knowing more than one is often useful, and in this case being fluent in one of the prototyping languages, such as Matlab, R, Python, etc., and one of (if not the arguably most important) canned-routine/data processing languages, is a smart bet for aspiring data analysts and researchers.

In line with books I, II, and III, attention is explicitly paid to application and numeric computation, with examples of Matlab code throughout. The point of including code is to offer a framework for discussion and illustration of numerics, and to show the “mapping” from theory to computation,

2 All these books are excellent in scope and suitability for the numerous topics associated with applied regression analysis, including case studies with real data. It is part of the reason this author sees no good reason to attempt to improve upon them. Notable is Graybill and Iyer (1994) for their emphasis on prediction, and use of confidence intervals (for prediction and model parameters) as opposed to hypothesis tests; see my diatribe in Chapter III.2.8 supporting this view.

3 Jondeau et al. (2007) provides a toolbox of Matlab programs, while Tsay (2012) and Zivot (2018) do so for R.

in contrast to providing black-box programs for an applied user to run when analyzing a data set. Thus, the emphasis is on algorithmic development for implementations involving number crunching with vectors and matrices, as opposed to, say, linking to financial or other databases, string handling, text parsing and processing, generation of advanced graphics, machine learning, design of interfaces, use of object-oriented programming, etc.. As such, the choice of Matlab should not be a substantial hindrance to users of, say, R, Python, or (particularly) Julia, wishing to port the methods to their preferred platforms. A benefit of those latter languages, however, is that they are free. The reader without access to Matlab but wishing to use it could use GNU Octave, which is free, and has essentially the same format and syntax as Matlab.

The preface of book III contains acknowledgements to the handful of professors with whom I had the honor of working, and who were highly instrumental in "forging me" as an academic, as well as to the numerous fellow academics and students who kindly provided me with invaluable comments and corrections on earlier drafts of this book, and book III. Specific to this book, master's student (!!) Christian Frey gets the award for "most picky" (in a good sense), having read various chapters with a very fine-toothed comb, alerting me to numerous typos and unclarities, and also indicating numerous passages where "a typical master's student" might enjoy a bit more verbosity in explanation. Chris also assisted me in writing (the harder parts of) Sections 1.A and C.2. I would give him an honorary doctorate if I could. I am also highly thankful to the excellent Wiley staff who managed this project, as well as copy editor Lesley Montford, who checked every chapter and alerted me to typos, inconsistencies, and other aspects of the presentation, leading to a much better final product. I (grudgingly) take blame for any further errors.

Part I

Linear Models: Regression and ANOVA

1

The Linear Model

The application of econometrics requires more than mastering a collection of tricks. It also requires insight, intuition, and common sense.

(Jan R. Magnus, 2017, p. 31)

The natural starting point for learning about statistical data analysis is with a sample of independent and identically distributed (hereafter i.i.d.) data, say $\mathbf{Y} = (Y_1, \dots, Y_n)$, as was done in book III. The *linear regression model* relaxes both the identical and independent assumptions by (i) allowing the means of the Y_i to depend, in a linear way, on a set of other variables, (ii) allowing for the Y_i to have different variances, and (iii) allowing for correlation between the Y_i.

The linear regression model is not only of fundamental importance in a large variety of quantitative disciplines, but is also the basis of a large number of more complex models, such as those arising in panel data studies, time-series analysis, and generalized linear models (GLIM), the latter briefly introduced in Section 1.6. Numerous, more advanced data analysis techniques (often referred to now as algorithms) also have their roots in regression, such as the *least absolute shrinkage and selection operator* (LASSO), the *elastic net*, and *least angle regression* (LARS). Such methods are often now showcased under the heading of machine learning.

1.1 Regression, Correlation, and Causality

It is uncomfortably true, although rarely admitted in statistics texts, that many important areas of science are stubbornly impervious to experimental designs based on randomisation of treatments to experimental units. Historically, the response to this embarrassing problem has been to either ignore it or to banish the very notion of causality from the language and to claim that the shadows dancing on the screen are all that exists.

Ignoring the problem doesn't make it go away and defining a problem out of existence doesn't make it so. We need to know what we can safely infer about causes from their observational shadows, what we can't infer, and the degree of ambiguity that remains.

(Bill Shipley, 2016, p. 1)[1]

1 The metaphor to dancing shadows goes back a while, at least to Plato's Republic and the Allegory of the Cave. One can see it today in shadow theater, popular in Southeast Asia; see, e.g., Pigliucci and Kaplan (2006, p. 2).

Linear Models and Time-Series Analysis: Regression, ANOVA, ARMA and GARCH, First Edition. Marc S. Paolella.

The univariate linear regression model relates the scalar random variable Y to k other (possibly random) variables, or **regressors**, $x_1, \dots, x_k$ in a linear fashion,

$$Y = \beta_1 x_1 + \beta_2 x_2 + \cdots + \beta_k x_k + \epsilon, \tag{1.1}$$

where, typically, $\epsilon \sim \mathrm{N}(0, \sigma^2)$. Values $\beta_1, \dots, \beta_k$ and σ^2 are unknown, constant parameters to be estimated from the data. A more useful notation that also emphasizes that the means of the Y_i are not constant is

$$Y_i = \beta_1 x_{i,1} + \beta_2 x_{i,2} + \cdots + \beta_k x_{i,k} + \epsilon_i, \quad i = 1, 2, \dots, n, \tag{1.2}$$

where now a double subscript on the regressors is necessary. The ϵ_i represent the difference between the values of Y_i and the model used to represent them, $\sum_{j=1}^{k} \beta_j x_{i,j}$, and so are referred to as the **error terms**. It is important to emphasize that the error terms are i.i.d., but the Y_i are not. However, if we take $k = 1$ and $x_{i,1} \equiv 1$, then (1.2) reduces to $Y_i = \beta_1 + \epsilon_i$, which is indeed just the i.i.d. model with $Y_i \overset{\text{i.i.d.}}{\sim} \mathrm{N}(\beta_1, \sigma^2)$. In fact, it is usually the case that $x_{i,1} \equiv 1$ for any $k \geqslant 1$, in which case the model is said to **include a constant** or **have an intercept term**.

We refer to Y as the **dependent** (random) variable. In other contexts, Y is also called the **endogenous** variable, while the k regressors can also be referred to as the **explanatory**, **exogenous**, or **independent** variables, although the latter term should not be taken to imply that the regressors, when viewed as random variables, are necessarily independent from one another.

The linear structure of (1.1) is one way of building a relationship between the Y_i and a set of variables that "influence" or "explain" them. The usefulness of establishing such a relationship or **conditional** model for the Y_i can be seen in a simple example: Assume a demographer is interested in the income of people living and employed in Hamburg. A random sample of n individuals could be obtained using public records or a phone book, and (rather unrealistically) their incomes Y_i, $i = 1, \dots, n$, elicited. Assuming that income is approximately normally distributed, an **unconditional** model for income could be postulated as $\mathrm{N}(\mu_u, \sigma_u^2)$, where the subscript u denotes the unconditional model and the usual estimators for the mean and variance of a normal sample could be used.

(We emphasize that this example is just an excuse to discuss some concepts. While actual incomes for certain populations can be "reasonably" approximated as Gaussian, they are, of course, not: They are strictly positive, will thus have an extended right tail, and this tail might be heavy, in the sense of being Pareto—this naming being no coincidence, as Vilfredo Pareto worked on modeling incomes, and is also the source of what is now referred to in micro-economics as Pareto optimality. An alternative type of linear model, referred to as GLIM, that uses a non-Gaussian distribution instead of the normal, is briefly discussed below in Section 1.6. Furthermore, interest might not center on modeling the mean income—which is what regression does—but rather the median, or the lower or upper quantiles. This leads to quantile regression, also briefly discussed in Section 1.6.)

A potentially much more precise description of income can be obtained by taking certain factors into consideration that are highly related to income, such as age, level of education, number of years of experience, gender, whether he or she works part or full time, etc. Before continuing this simple example, it is imperative to discuss the three Cs: correlation, causality, and control.

Observe that (simplistically here, for demonstration) age and education might be positively correlated, simply because, as the years go by, people have opportunities to further their schooling and training. As such, if one were to claim that income tends to increase as a function of age, then one cannot conclude this arises out of "seniority" at work, but rather possibly because some of the older people have received more schooling. Another way of saying this is, while income and age are positively correlated, an increase in age is not necessarily **causal** for income; age and income may be **spuriously correlated**, meaning that their correlation is driven by other factors, such as education, which might indeed be causal for income. Likewise, if one were to claim that income tends to increase with educational levels, then one cannot claim this is due to education *per se*, but rather due simply to seniority at the workplace, possibly despite their enhanced education. Thus, it is important to include both of these variables in the regression.

In the former case, if a positive relationship is found between income and age *with education also in the regression*, then one can conclude a seniority effect. In the literature, one might say "Age appears to be a significant predictor of income, and this being concluded after having also **controlled for** education." Examples of controlling for the relevant factors when assessing causality are ubiquitous in empirical studies of all kinds, and are essential for reliable inference. As one example, in the field of "economics and religion" (which is now a fully established area in economics; see, e.g., McCleary, 2011), in the abstract of one of the highly influential papers in the field, Gruber (2005) states "Religion plays an important role in the lives of many Americans, but there is relatively little study by economists of the implications of religiosity for economic outcomes. This likely reflects the enormous difficulty inherent in separating the causal effects of religiosity from other factors that are correlated with outcomes." The paper is filled with the expression "having controlled for".

A famous example, in a famous paper, is Leamer (1983, Sec. V), showing how conclusions from a study of the factors influencing the murder rate are highly dependent on which set of variables are included in the regression. The notion of controlling for the right variables is often the vehicle for critiquing other studies in an attempt to correct potentially wrong conclusions. For example, Farkas and Vicknair (1996, p. 557) state "[Cancio et al.] claim that discrimination, measured as a residual from an earnings attainment regression, increased after 1976. Their claim depends crucially on which variables are controlled and which variables are omitted from the regression. We believe that the authors have omitted the key control variable—cognitive skill."

The concept of causality is fundamental in econometrics and other social sciences, and we have not even scratched the surface. The different ways it is addressed in popular econometrics textbooks is discussed in Chen and Pearl (2013), and debated in Swamy et al. (2015), Raunig (2017), and Swamy et al. (2017). These serve to indicate that the theoretical framework for understanding causality and its interface to statistical inference is still developing. The importance of causality for scientific inquiry cannot be overstated, and continues to grow in importance in light of artificial intelligence. As a simple example, humans understand that weather is (global warming aside) exogenous, and carrying an umbrella does not cause rain. How should a computer know this? Starting points for further reading include Pearl (2009), Shipley (2016), and the references therein.

Our development of the linear model in this chapter serves two purposes: First, it is the required theoretical statistical framework for understanding ANOVA models, as introduced in Chapters 2 and 3. As ANOVA involves designed experiments and randomization, as opposed to observational studies in the social sciences, we can avoid the delicate issues associated with assessing causality. Second, the linear model serves as the underlying structure of autoregressive time-series models as developed in Part II, and our emphasis is on statistical forecasting, as opposed to the development of structural economic models that explicitly need to address causality.

We now continue with our very simple illustration, just to introduce some terminology. Let $x_{i,2}$ denote the age of the ith person. A conditional model with a constant and age as a regressor is given by $Y_i = \beta_1 + \beta_2 x_{i,2} + \epsilon_i$, where $\epsilon_i \overset{\text{i.i.d.}}{\sim} \mathrm{N}(0, \sigma^2)$. The intercept is measured by β_1 and the slope of income

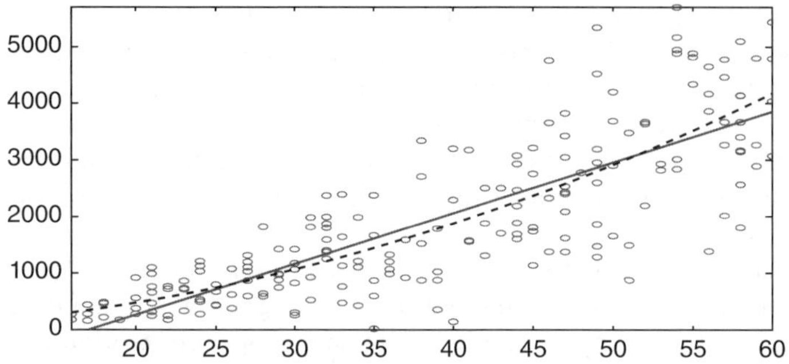

Figure 1.1 Scatterplot of age versus income overlaid with fitted regression curves.

is measured by β_2. Because age is expected to explain a considerable part of variability in income, we expect σ^2 to be significantly less than σ_u^2. A useful way of visualizing the model is with a scatterplot of $x_{i,2}$ and y_i. Figure 1.1 shows such a graph based on a fictitious set of data for 200 individuals between the ages of 16 and 60 and their monthly net income in euros. It is quite clear from the scatterplot that age and income are positively correlated. If age is neglected, then the i.i.d. normal model for income results in $\hat{\mu}_u = 1{,}797$ euros and $\hat{\sigma}_u = 1{,}320$ euros. Using the techniques discussed below, the regression model gives estimates $\hat{\beta}_1 = -1{,}465$, $\hat{\beta}_2 = 85.4$, and $\hat{\sigma} = 755$, the latter being about 43% smaller than $\hat{\sigma}_u$. The model implies that, conditional on the age x, the income Y is modeled as $\mathrm{N}(-1{,}465 + 85.4x, 755^2)$. This is valid only for $16 \leqslant x \leqslant 60$; because of the negative intercept, small values of age would erroneously imply a negative income. The fitted model $y = \hat{\beta}_1 + \hat{\beta}_2 x$ is overlaid in the figure as a solid line.

Notice in Figure 1.1 that the linear approximation underestimates income for both low and high age groups, i.e., income does not seem perfectly linear in age, but rather somewhat quadratic. To accommodate this, we can add another regressor, $x_{i,3} = x_{i,2}^2$, into the model, i.e., $Y_i = \beta_1 + \beta_2 x_{i,2} + \beta_3 x_{i,3} + \epsilon_i$, where $\epsilon_i \overset{\text{i.i.d.}}{\sim} \mathrm{N}(0, \sigma_q^2)$ and σ_q^2 denotes the conditional variance based on the quadratic model. It is important to realize that the model is still linear (in the constant, age, and age squared). The fitted model turns out to be $Y_i = 190 - 12.5x_{i,2} + 1.29x_{i,3}$, with $\hat{\sigma}_q = 733$, which is about 3% smaller than $\hat{\sigma}$. The fitted curve is shown in Figure 1.1 as a dashed line.

One caveat still remains with the model for income based on age: The variance of income appears to increase with age. This is a typical finding with income data and agrees with economic theory. It implies that both the mean and the variance of income are functions of age. In general, when the variance of the regression error term is not constant, it is said to be **heteroskedastic**, as opposed to **homoskedastic**. The generalized least squares extension of the linear regression model discussed below can be used to address this issue when the structure of the heteroskedasticity as a function of the **X** matrix is known.

In certain applications, the ordering of the dependent variable and the regressors is important because they are observed in time, usually equally spaced. Because of this, the notation Y_t will be used, $t = 1, \dots, T$. Thus, (1.2) becomes

$$Y_t = \beta_1 x_{t,1} + \beta_2 x_{t,2} + \cdots + \beta_k x_{t,k} + \epsilon_t, \quad t = 1, 2, \dots, T,$$

where $x_{t,i}$ indicates the tth observation of the ith explanatory variable, $i = 1, \dots, k$, and ϵ_t is the tth error term. In standard matrix notation, the model can be compactly expressed as

$$\mathbf{Y} = \mathbf{X}\boldsymbol{\beta} + \boldsymbol{\epsilon}, \tag{1.3}$$

where $[\mathbf{X}]_{t,i} = x_{t,i}$, i.e., with $\mathbf{x}_t = (x_{t,1}, \dots, x_{t,k})'$,

$$\mathbf{X} = \begin{bmatrix} \mathbf{x}_1' \\ \vdots \\ \mathbf{x}_T' \end{bmatrix} = \begin{bmatrix} x_{1,1} & x_{1,2} & \cdots & x_{1,k} \\ x_{2,1} & x_{2,2} & \cdots & x_{2,k} \\ \vdots & \vdots & & \vdots \\ x_{T,1} & x_{T,2} & & x_{T,k} \end{bmatrix}, \qquad \boldsymbol{\epsilon} \sim \mathrm{N}(\mathbf{0}, \sigma^2\mathbf{I}),$$

$\mathbf{Y}$ and $\boldsymbol{\epsilon}$ are $T \times 1$, $\mathbf{X}$ is $T \times k$ and $\boldsymbol{\beta}$ is $k \times 1$. The first column of $\mathbf{X}$ is usually $\mathbf{1}$, the column of ones. Observe that $\mathbf{Y} \sim \mathrm{N}(\mathbf{X}\boldsymbol{\beta}, \sigma^2\mathbf{I})$.

An important special case of (1.3) is with $k = 2$ and $x_{t,1} = 1$. Then $Y_t = \beta_1 + \beta_2 X_t + \epsilon_t$, $t = 1, \dots, T$, is referred to as the **simple linear regression model**. See Problems 1.1 and 1.2.

1.2 Ordinary and Generalized Least Squares

1.2.1 Ordinary Least Squares Estimation

The most popular way of estimating the k parameters in $\boldsymbol{\beta}$ is the **method of least squares**,[2] which takes $\widehat{\boldsymbol{\beta}} = \arg\min \mathsf{S}(\boldsymbol{\beta})$, where

$$\mathsf{S}(\boldsymbol{\beta}) = \mathsf{S}(\boldsymbol{\beta}; \mathbf{Y}, \mathbf{X}) = (\mathbf{Y} - \mathbf{X}\boldsymbol{\beta})'(\mathbf{Y} - \mathbf{X}\boldsymbol{\beta}) = \sum_{t=1}^{T} (Y_t - \mathbf{x}_t'\boldsymbol{\beta})^2, \tag{1.4}$$

and we suppress the dependency of S on $\mathbf{Y}$ and $\mathbf{X}$ when they are clear from the context.

Assume that $\mathbf{X}$ is of full rank k. One procedure to obtain the solution, commonly shown in most books on regression (see, e.g., Seber and Lee, 2003, p. 38), uses matrix calculus; it yields $\partial\, \mathsf{S}(\boldsymbol{\beta})/\partial\boldsymbol{\beta} = -2\mathbf{X}'(\mathbf{Y} - \mathbf{X}\boldsymbol{\beta})$, and setting this to zero gives the solution

$$\widehat{\boldsymbol{\beta}} = (\mathbf{X}'\mathbf{X})^{-1}\mathbf{X}'\mathbf{Y}. \tag{1.5}$$

This is referred to as the **ordinary least squares**, or o.l.s., estimator of $\boldsymbol{\beta}$. (The adjective "ordinary" is used to distinguish it from what is called generalized least squares, addressed in Section 1.2.3 below.) Notice that $\widehat{\boldsymbol{\beta}}$ is also the solution to what are referred to as the **normal equations**, given by

$$\mathbf{X}'\mathbf{X}\widehat{\boldsymbol{\beta}} = \mathbf{X}'\mathbf{Y}. \tag{1.6}$$

To verify that (1.5) indeed corresponds to the minimum of $\mathsf{S}(\boldsymbol{\beta})$, the second derivative is checked for positive definiteness, yielding $\partial^2\, \mathsf{S}(\boldsymbol{\beta})/\partial\boldsymbol{\beta}\partial\boldsymbol{\beta}' = 2\mathbf{X}'\mathbf{X}$, which is necessarily positive definite when $\mathbf{X}$ is full rank. Observe that, if $\mathbf{X}$ consists only of a column of ones, which we write as $\mathbf{X} = \mathbf{1}$, then $\widehat{\boldsymbol{\beta}}$ reduces to the mean, $\bar{Y}$, of the Y_t. Also, if $k = T$ (and $\mathbf{X}$ is full rank), then $\widehat{\boldsymbol{\beta}}$ reduces to $\mathbf{X}^{-1}\mathbf{Y}$, with $\mathsf{S}(\widehat{\boldsymbol{\beta}}) = 0$.

Observe that the derivation of $\widehat{\boldsymbol{\beta}}$ in (1.5) did not involve any explicit distributional assumptions. One consequence of this is that the estimator may not have any meaning if the maximally existing moment of the $\{\epsilon_t\}$ is too low. For example, take $\mathbf{X} = \mathbf{1}$ and $\{\epsilon_t\}$ to be i.i.d. Cauchy; then $\widehat{\beta} = \bar{Y}$ is a useless estimator. If we assume that the first moment of the $\{\epsilon_t\}$ exists and is zero, then, writing $\widehat{\boldsymbol{\beta}} = (\mathbf{X}'\mathbf{X})^{-1}\mathbf{X}'(\mathbf{X}\boldsymbol{\beta} + \boldsymbol{\epsilon}) = \boldsymbol{\beta} + (\mathbf{X}'\mathbf{X})^{-1}\mathbf{X}'\boldsymbol{\epsilon}$, we see that $\widehat{\boldsymbol{\beta}}$ is unbiased:

$$\mathbb{E}[\widehat{\boldsymbol{\beta}}] = \boldsymbol{\beta} + (\mathbf{X}'\mathbf{X})^{-1}\mathbf{X}'\mathbb{E}[\boldsymbol{\epsilon}] = \boldsymbol{\beta}. \tag{1.7}$$

2 This terminology dates back to Adrien-Marie Legendre (1752–1833), though the method is most associated in its origins with Carl Friedrich Gauss, (1777–1855). See Stigler (1981) for further details.

Next, if we have existence of second moments, and $\mathbb{V}(\epsilon) = \sigma^2\mathbf{I}$, then $\mathbb{V}(\hat{\boldsymbol{\beta}} \mid \sigma^2)$ is given by

$$\mathbb{E}[(\hat{\boldsymbol{\beta}} - \boldsymbol{\beta})(\hat{\boldsymbol{\beta}} - \boldsymbol{\beta})' \mid \sigma^2] = (\mathbf{X}'\mathbf{X})^{-1}\mathbf{X}'\mathbb{E}[\epsilon\epsilon']\mathbf{X}(\mathbf{X}'\mathbf{X})^{-1} = \sigma^2(\mathbf{X}'\mathbf{X})^{-1}. \tag{1.8}$$

It turns out that $\hat{\boldsymbol{\beta}}$ has the smallest variance among all linear unbiased estimators; this result is often referred to as the **Gauss–Markov Theorem**, and expressed as saying that $\hat{\boldsymbol{\beta}}$ is the best linear unbiased estimator, or BLUE. We outline the usual derivation, leaving the straightforward details to the reader. Let $\hat{\boldsymbol{\beta}}^* = \mathbf{A}'\mathbf{Y}$, where $\mathbf{A}'$ is a $k \times T$ nonstochastic matrix (it can involve $\mathbf{X}$, but not $\mathbf{Y}$). Let $\mathbf{D} = \mathbf{A} - \mathbf{X}(\mathbf{X}'\mathbf{X})^{-1}$. First calculate $\mathbb{E}[\hat{\boldsymbol{\beta}}^*]$ and show that the unbiased property implies that $\mathbf{D}'\mathbf{X} = \mathbf{0}$. Next, calculate $\mathbb{V}(\hat{\boldsymbol{\beta}}^* \mid \sigma^2)$ and show that $\mathbb{V}(\hat{\boldsymbol{\beta}}^* \mid \sigma^2) = \mathbb{V}(\hat{\boldsymbol{\beta}} \mid \sigma^2) + \sigma^2\mathbf{D}'\mathbf{D}$. The result follows because $\mathbf{D}'\mathbf{D}$ is obviously positive semi-definite and the variance is minimized when $\mathbf{D} = \mathbf{0}$.

In many situations, it is reasonable to assume normality for the $\{\epsilon_t\}$, in which case we may easily estimate the $k + 1$ unknown parameters σ^2 and β_i, $i = 1, \dots, k$, by maximum likelihood. In particular, with

$$f_{\mathbf{Y}}(\mathbf{y}) = (2\pi\sigma^2)^{-T/2} \exp\left\{-\frac{1}{2\sigma^2}(\mathbf{y} - \mathbf{X}\boldsymbol{\beta})'(\mathbf{y} - \mathbf{X}\boldsymbol{\beta})\right\}, \tag{1.9}$$

and log-likelihood

$$\ell(\boldsymbol{\beta}, \sigma^2; \mathbf{Y}) = -\frac{T}{2}\log(2\pi) - \frac{T}{2}\log(\sigma^2) - \frac{1}{2\sigma^2}\,\mathsf{S}(\boldsymbol{\beta}), \tag{1.10}$$

where $\mathsf{S}(\boldsymbol{\beta})$ is given in (1.4), setting

$$\frac{\partial\ell}{\partial\boldsymbol{\beta}} = -\frac{2}{2\sigma^2}\mathbf{X}'(\mathbf{Y} - \mathbf{X}\boldsymbol{\beta}) \quad \text{and} \quad \frac{\partial\ell}{\partial\sigma^2} = -\frac{T}{2\sigma^2} + \frac{1}{2\sigma^4}\,\mathsf{S}(\boldsymbol{\beta})$$

to zero yields the same estimator for $\boldsymbol{\beta}$ as given in (1.5) and $\tilde{\sigma}^2 = \mathsf{S}(\hat{\boldsymbol{\beta}})/T$. It will be shown in Section 1.3.2 that the maximum likelihood estimator (hereafter m.l.e.) of σ^2 is biased, while estimator

$$\hat{\sigma}^2 = \mathsf{S}(\hat{\boldsymbol{\beta}})/(T - k) \tag{1.11}$$

is unbiased.

As $\hat{\boldsymbol{\beta}}$ is a linear function of $\mathbf{Y}$, $(\hat{\boldsymbol{\beta}} \mid \sigma^2)$ is multivariate normally distributed, and thus characterized by its first two moments. From (1.7) and (1.8), it follows that $(\hat{\boldsymbol{\beta}} \mid \sigma^2) \sim \mathrm{N}(\boldsymbol{\beta}, \sigma^2(\mathbf{X}'\mathbf{X})^{-1})$.

1.2.2 Further Aspects of Regression and OLS

> The coefficient of multiple determination, R^2, is a measure many statisticians love to hate. This animosity exists primarily because the widespread use of R^2 inevitably leads to at least occasional misuse.
>
> (Richard Anderson-Sprecher, 1994)

In general, the quantity $\mathsf{S}(\hat{\boldsymbol{\beta}})$ is referred to as the **residual sum of squares**, abbreviated RSS. The **explained sum of squares**, abbreviated ESS, is defined to be $\sum_{t=1}^{T}(\hat{Y}_t - \bar{Y})^2$, where the *fitted value* of Y_t is $\hat{Y}_t := \mathbf{x}_t'\hat{\boldsymbol{\beta}}$, and the **total (corrected) sum of squares**, or TSS, is $\sum_{t=1}^{T}(Y_t - \bar{Y})^2$. (Annoyingly, both words "error" and "explained" start with an "e", and some presentations define SSE to be the error sum of squares, which is our RSS; see, e.g., Ravishanker and Dey, 2002, p. 101.)

The term *corrected* in the TSS refers to the adjustment of the Y_t for their mean. This is done because the mean is a "trivial" regressor that is not considered to do any real explaining of the dependent variable. Indeed, the total *uncorrected* sum of squares, $\sum_{t=1}^{T} Y_t^2$, could be made arbitrarily large just by adding a large enough constant value to the Y_t, and the model consisting of just the mean (i.e., an $\mathbf{X}$ matrix with just a column of ones) would have the appearance of explaining an arbitrarily large amount of the variation in the data.

While certainly $Y_t - \bar{Y} = (Y_t - \hat{Y}_t) + (\hat{Y}_t - \bar{Y})$, it is not immediately obvious that

$$\sum_{t=1}^{T} (Y_t - \bar{Y})^2 = \sum_{t=1}^{T} (Y_t - \hat{Y}_t)^2 + \sum_{t=1}^{T} (\hat{Y}_t - \bar{Y})^2,$$

i.e.,

$$\text{TSS} = \text{RSS} + \text{ESS}. \tag{1.12}$$

This fundamental identity is proven below in Section 1.3.2.

A popular statistic that measures the fraction of the variability of $\mathbf{Y}$ taken into account by a linear regression model that includes a constant, compared to use of just a constant (i.e., $\bar{Y}$), is the **coefficient of multiple determination**, designated as R^2, and defined as

$$R^2 = \frac{\text{ESS}}{\text{TSS}} = 1 - \frac{\text{RSS}}{\text{TSS}} = 1 - \frac{\text{S}(\hat{\boldsymbol{\beta}}, \mathbf{Y}, \mathbf{X})}{\text{S}(\bar{Y}, \mathbf{Y}, \mathbf{1})}, \tag{1.13}$$

where $\mathbf{1}$ is a T-length column of ones. The coefficient of multiple determination R^2 provides a measure of the extent to which the regressors "explain" the dependent variable over and above the contribution from just the constant term. It is important that $\mathbf{X}$ contain a constant or a set of variables whose linear combination yields a constant; see Becker and Kennedy (1992) and Anderson-Sprecher (1994) and the references therein for more detail on this point.

By construction, the observed R^2 is a number between zero and one. As with other quantities associated with regression (such as the nearly always reported "t-statistics" for assessing individual "significance" of the regressors), R^2 is a statistic (a function of the data but not of the unknown parameters) and thus *is a random variable.* In Section 1.4.4 we derive the F test for parameter restrictions. With J such linear restrictions, and $\hat{\boldsymbol{\gamma}}$ referring to the restricted estimator, we will show (1.88), repeated here, as

$$F = \frac{[\text{S}(\hat{\boldsymbol{\gamma}}) - \text{S}(\hat{\boldsymbol{\beta}})]/J}{\text{S}(\hat{\boldsymbol{\beta}})/(T-k)} \sim \text{F}(J, T-k), \tag{1.14}$$

under the null hypothesis H_0 that the J restrictions are true. Let $J = k - 1$ and $\hat{\boldsymbol{\gamma}} = \bar{Y}$, so that the restricted model is that all regressor coefficients, *except the constant* are zero. Then, comparing (1.13) and (1.14),

$$F = \frac{T-k}{k-1} \frac{R^2}{1-R^2}, \qquad \text{or} \qquad R^2 = \frac{(k-1)F}{(T-k)+(k-1)F}. \tag{1.15}$$

Dividing the numerator and denominator of the latter expression by $T - k$ and recalling the relationship between F and beta random variables (see, e.g., Problem I.7.20), we immediately have that

$$R^2 \sim \text{Beta}\left(\frac{k-1}{2}, \frac{T-k}{2}\right), \tag{1.16}$$

so that $\mathbb{E}[R^2] = (k-1)/(T-1)$ from, for example, (I.7.12). Its variance could similarly be stated. Recall that its distribution was derived under the null hypothesis that the $k-1$ regression coefficients are zero. This implies that R^2 is upward biased, and also shows that just adding superfluous regressors will always increase the expected value of R^2. As such, choosing a set of regressors such that R^2 is maximized is not appropriate for model selection.

However, the so-called **adjusted** R^2 can be used. It is defined as

$$R^2_{\text{adj}} = 1 - (1 - R^2)\frac{T-1}{T-k}. \tag{1.17}$$

Virtually all statistical software for regression will include this measure. Less well known is that it has (like so many things) its origin with Ronald Fisher; see Fisher (1925). Notice how, like the Akaike information criterion (hereafter AIC) and other penalty-based measures applied to the obtained log likelihood, when k is increased, the increase in R^2 is offset by a factor involving k in R^2_{adj}.

Measure (1.17) can be motivated in (at least) two ways. First, note that, under the null hypothesis,

$$\mathbb{E}[R^2_{\text{adj}}] = 1 - \left(1 - \frac{k-1}{T-1}\right)\frac{T-1}{T-k} = 0,$$

providing a perfect offset to R^2's expected value simply increasing in k under the null. A second way is to note that, while $R^2 = 1 - \text{RSS}/\text{TSS}$ from (1.13),

$$R^2_{\text{adj}} = 1 - \frac{\text{RSS}/(T-k)}{\text{TSS}/(T-1)} = 1 - \frac{\widehat{\mathbb{V}}(\hat{\boldsymbol{\epsilon}})}{\widehat{\mathbb{V}}(\mathbf{Y})},$$

the numerator and denominator being unbiased estimators of their respective variances, recalling (1.11). The use of R^2_{adj} for model selection is very similar to use of other measures, such as the (corrected) AIC and the so-called **Mallows'** C_k; see, e.g., Seber and Lee (2003, Ch. 12) for a very good discussion of these, and other criteria, and the relationships among them.

Section 1.2.3 extends the model to the case in which $\mathbf{Y} = \mathbf{X}\boldsymbol{\beta} + \boldsymbol{\epsilon}$ from (1.3), but $\boldsymbol{\epsilon} \sim \text{N}(\mathbf{0}, \sigma^2\boldsymbol{\Sigma})$, where $\boldsymbol{\Sigma}$ is a known, positive definite variance–covariance matrix. There, an appropriate expression for R^2 will be derived that generalizes (1.13). For now, the reader is encouraged to express R^2 in (1.13) as a ratio of quadratic forms, assuming $\boldsymbol{\epsilon} \sim \text{N}(\mathbf{0}, \sigma^2\boldsymbol{\Sigma})$, and compute and plot its density for a given $\mathbf{X}$ and $\boldsymbol{\Sigma}$, such as given in (1.31) for a given value of parameter a, as done in, e.g., Carrodus and Giles (1992). When $a = 0$, the density should coincide with that given by (1.16).

We end this section with an important remark, and an important example.

Remark It is often assumed that the elements of $\mathbf{X}$ are known constants. This is quite plausible in designed experiments, where $\mathbf{X}$ is chosen in such a way as to maximize the ability of the experiment to answer the questions of interest. In this case, $\mathbf{X}$ is often referred to as the **design matrix**. This will rarely hold in applications in the social sciences, where the $\mathbf{x}_t'$ reflect certain measurements and are better described as being observations of random variables from the multivariate distribution describing both $\mathbf{x}_t'$ and Y_t. Fortunately, under certain assumptions, one may ignore this issue and proceed as if $\mathbf{x}_t'$ were fixed constants and not realizations of a random variable.

Assume matrix $\mathbf{X}$ is no longer deterministic. Denote by $\mathbf{X}$ an outcome of random variable $\mathcal{X}$, with kT-variate probability density function (hereafter p.d.f.) $f_{\mathcal{X}}(\mathbf{X}\,;\boldsymbol{\theta})$, where $\boldsymbol{\theta}$ is a parameter vector. We require the following assumption:

0. The conditional distribution $\mathbf{Y} \mid (\mathcal{X} = \mathbf{X})$ depends only on $\mathbf{X}$ and parameters $\boldsymbol{\beta}$ and σ and such that $\mathbf{Y} \mid (\mathcal{X} = \mathbf{X})$ has mean $\mathbf{X}\boldsymbol{\beta}$ and finite variance $\sigma^2\mathbf{I}$.

For example, we could have $\mathbf{Y} \mid (\mathcal{X} = \mathbf{X}) \sim \mathrm{N}(\mathbf{X}\boldsymbol{\beta}, \sigma^2\mathbf{I})$. Under the stated assumption, the joint density of $\mathbf{Y}$ and $\mathcal{X}$ can be written as

$$f_{\mathbf{Y},\mathcal{X}}(\mathbf{y}, \mathbf{X} \mid \boldsymbol{\beta}, \sigma^2, \boldsymbol{\theta}) = f_{\mathbf{Y}\mid\mathcal{X}}(\mathbf{y} \mid \mathbf{X}; \boldsymbol{\beta}, \sigma^2) \cdot f_{\mathcal{X}}(\mathbf{X}; \boldsymbol{\beta}, \sigma^2, \boldsymbol{\theta}). \tag{1.18}$$

Now consider the following two additional assumptions:

1) The distribution of $\mathcal{X}$ does not depend on $\boldsymbol{\beta}$ or σ^2, so we can write $f_{\mathcal{X}}(\mathbf{X}; \boldsymbol{\beta}, \sigma^2, \boldsymbol{\theta}) = f_{\mathcal{X}}(\mathbf{X}; \boldsymbol{\theta})$.
2) The parameter space of $\boldsymbol{\theta}$ and that of $(\boldsymbol{\beta}, \sigma^2)$ are not related, that is, they are not restricted by one another in any way.

Then, with regard to $\boldsymbol{\beta}$ and σ^2, $f_{\mathcal{X}}$ is only a multiplicative constant and the log-likelihood corresponding to (1.18) is the same as (1.10) plus the additional term $\log f_{\mathcal{X}}(\mathbf{X}; \boldsymbol{\theta})$. As this term does not involve $\boldsymbol{\beta}$ or σ^2, the (generalized) least squares estimator still coincides with the m.l.e. When the above assumptions are satisfied, $\boldsymbol{\theta}$ and $(\boldsymbol{\beta}, \sigma^2)$ are said to be **functionally independent** (Graybill, 1976, p. 380), or **variation-free** (Poirier, 1995, p. 461). More common in the econometrics literature is to say that one assumes $\mathbf{X}$ to be **(weakly) exogenous** with respect to $\mathbf{Y}$.

The extent to which these assumptions are reasonable is open to debate. Clearly, without them, estimation of $\boldsymbol{\beta}$ and σ^2 is not so straightforward, as then $f_{\mathcal{X}}(\mathbf{X}; \boldsymbol{\beta}, \sigma^2, \boldsymbol{\theta})$ must be (fully, or at least partially) specified. If they hold, then

$$\mathbb{E}[\hat{\boldsymbol{\beta}}] = \mathbb{E}_{\mathcal{X}}[\mathbb{E}[\hat{\boldsymbol{\beta}} \mid \mathcal{X} = \mathbf{X}]] = \mathbb{E}_{\mathcal{X}}[\boldsymbol{\beta} + (\mathbf{X}'\mathbf{X})^{-1}\mathbf{X}'\mathbb{E}[\boldsymbol{\epsilon} \mid \mathcal{X}]] = \mathbb{E}_{\mathcal{X}}[\boldsymbol{\beta}] = \boldsymbol{\beta}$$

and

$$\mathbb{V}(\hat{\boldsymbol{\beta}} \mid \sigma^2) = \mathbb{E}_{\mathcal{X}}[\mathbb{E}[(\hat{\boldsymbol{\beta}} - \boldsymbol{\beta})(\hat{\boldsymbol{\beta}} - \boldsymbol{\beta})' \mid \mathcal{X} = \mathbf{X}, \sigma^2]] = \sigma^2\mathbb{E}_{\mathcal{X}}[(\mathcal{X}'\mathcal{X})^{-1}],$$

the latter being obtainable only when $f_{\mathcal{X}}(\mathbf{X}; \boldsymbol{\theta})$ is known.

A discussion of the implications of falsely assuming that $\mathbf{X}$ is not stochastic is provided by Binkley and Abbott (1987).[3] ■

Example 1.1 ***Frisch–Waugh–Lovell Theorem***
It is occasionally useful to express the o.l.s. estimator of each component of the partitioned vector $\boldsymbol{\beta} = (\boldsymbol{\beta}_1', \boldsymbol{\beta}_2')'$, where $\boldsymbol{\beta}_1$ is $k_1 \times 1$, $1 \leqslant k_1 < k$. With the appropriate corresponding partition of $\mathbf{X}$, model (1.3) is then expressed as

$$\mathbf{Y} = \begin{pmatrix} \mathbf{X}_1 & \mathbf{X}_2 \end{pmatrix} \begin{pmatrix} \boldsymbol{\beta}_1 \\ \boldsymbol{\beta}_2 \end{pmatrix} + \boldsymbol{\epsilon} = \mathbf{X}_1\boldsymbol{\beta}_1 + \mathbf{X}_2\boldsymbol{\beta}_2 + \boldsymbol{\epsilon}.$$

The normal equations (1.6) then read

$$\begin{pmatrix} \mathbf{X}_1' \\ \mathbf{X}_2' \end{pmatrix} \begin{pmatrix} \mathbf{X}_1 & \mathbf{X}_2 \end{pmatrix} \begin{pmatrix} \hat{\boldsymbol{\beta}}_1 \\ \hat{\boldsymbol{\beta}}_2 \end{pmatrix} = \begin{pmatrix} \mathbf{X}_1' \\ \mathbf{X}_2' \end{pmatrix} \mathbf{Y},$$

or

$$\mathbf{X}_1'\mathbf{X}_1\hat{\boldsymbol{\beta}}_1 + \mathbf{X}_1'\mathbf{X}_2\hat{\boldsymbol{\beta}}_2 = \mathbf{X}_1'\mathbf{Y} \quad \text{and} \quad \mathbf{X}_2'\mathbf{X}_1\hat{\boldsymbol{\beta}}_1 + \mathbf{X}_2'\mathbf{X}_2\hat{\boldsymbol{\beta}}_2 = \mathbf{X}_2'\mathbf{Y}, \tag{1.19}$$

3 We use the tombstone, QED, or halmos, symbol ■ to denote the end of proofs of theorems, as well as examples and remarks, acknowledging that it is traditionally only used for the former, as popularized by Paul Halmos.

so that

$$\widehat{\boldsymbol{\beta}}_1 = (\mathbf{X}_1'\mathbf{X}_1)^{-1}\mathbf{X}_1'(\mathbf{Y} - \mathbf{X}_2\widehat{\boldsymbol{\beta}}_2) \tag{1.20}$$

and $\widehat{\boldsymbol{\beta}}_2 = (\mathbf{X}_2'\mathbf{X}_2)^{-1}\mathbf{X}_2'(\mathbf{Y} - \mathbf{X}_1\widehat{\boldsymbol{\beta}}_1)$. To obtain an expression for $\widehat{\boldsymbol{\beta}}_2$ that does not depend on $\widehat{\boldsymbol{\beta}}_1$, let $\mathbf{M}_1 = \mathbf{I} - \mathbf{X}_1(\mathbf{X}_1'\mathbf{X}_1)^{-1}\mathbf{X}_1'$, premultiply (1.20) by $\mathbf{X}_1$, and substitute $\mathbf{X}_1\widehat{\boldsymbol{\beta}}_1$ into the second equation in (1.19) to get

$$\mathbf{X}_2'(\mathbf{I} - \mathbf{M}_1)(\mathbf{Y} - \mathbf{X}_2\widehat{\boldsymbol{\beta}}_2) + \mathbf{X}_2'\mathbf{X}_2\widehat{\boldsymbol{\beta}}_2 = \mathbf{X}_2'\mathbf{Y},$$

or, expanding and solving for $\widehat{\boldsymbol{\beta}}_2$,

$$\widehat{\boldsymbol{\beta}}_2 = (\mathbf{X}_2'\mathbf{M}_1\mathbf{X}_2)^{-1}\mathbf{X}_2'\mathbf{M}_1\mathbf{Y}. \tag{1.21}$$

A similar argument (or via symmetry) shows that

$$\widehat{\boldsymbol{\beta}}_1 = (\mathbf{X}_1'\mathbf{M}_2\mathbf{X}_1)^{-1}\mathbf{X}_1'\mathbf{M}_2\mathbf{Y}, \tag{1.22}$$

where $\mathbf{M}_2 = \mathbf{I} - \mathbf{X}_2(\mathbf{X}_2'\mathbf{X}_2)^{-1}\mathbf{X}_2'$.

An important special case of (1.21) discussed further in Chapter 4 is when $k_1 = k - 1$, so that $\mathbf{X}_2$ is $T \times 1$ and $\widehat{\boldsymbol{\beta}}_2$ in (1.21) reduces to the scalar

$$\widehat{\beta}_2 = \frac{\mathbf{X}_2'\mathbf{M}_1\mathbf{Y}}{\mathbf{X}_2'\mathbf{M}_1\mathbf{X}_2}. \tag{1.23}$$

This is a ratio of a bilinear form to a quadratic form, as discussed in Appendix A.

The Frisch–Waugh–Lovell theorem has both computational value (see, e.g., Ruud, 2000, p. 66, and Example 1.9 below) and theoretical value; see Ruud (2000), Davidson and MacKinnon (2004), and also Section 5.2. Extensions of the theorem are considered in Fiebig et al. (1996). ■

1.2.3 Generalized Least Squares

Now consider the more general assumption that $\boldsymbol{\epsilon} \sim \mathrm{N}(\mathbf{0}, \sigma^2\boldsymbol{\Sigma})$, where $\boldsymbol{\Sigma}$ is a known, positive definite variance–covariance matrix. The density of $\mathbf{Y}$ is now given by

$$f_{\mathbf{Y}}(\mathbf{y}) = (2\pi)^{-T/2}|\sigma^2\boldsymbol{\Sigma}|^{-1/2}\exp\left\{-\frac{1}{2\sigma^2}(\mathbf{y} - \mathbf{X}\boldsymbol{\beta})'\boldsymbol{\Sigma}^{-1}(\mathbf{y} - \mathbf{X}\boldsymbol{\beta})\right\}, \tag{1.24}$$

and one could use calculus to find the m.l.e. of $\boldsymbol{\beta}$. Alternatively, we could transform the model in such a way that the above results still apply. In particular, with $\boldsymbol{\Sigma}^{-1/2}$ the symmetric matrix such that $\boldsymbol{\Sigma}^{-1/2}\boldsymbol{\Sigma}^{-1/2} = \boldsymbol{\Sigma}^{-1}$, premultiply (1.3) by $\boldsymbol{\Sigma}^{-1/2}$ so that

$$\boldsymbol{\Sigma}^{-1/2}\mathbf{Y} = \boldsymbol{\Sigma}^{-1/2}\mathbf{X}\boldsymbol{\beta} + \boldsymbol{\Sigma}^{-1/2}\boldsymbol{\epsilon}, \qquad \boldsymbol{\Sigma}^{-1/2}\boldsymbol{\epsilon} \sim \mathrm{N}_T(\mathbf{0}, \sigma^2\mathbf{I}). \tag{1.25}$$

Then, using the previous maximum likelihood approach as in (1.10), with

$$\mathbf{Y}_* := \boldsymbol{\Sigma}^{-1/2}\mathbf{Y} \quad \text{and} \quad \mathbf{X}_* := \boldsymbol{\Sigma}^{-1/2}\mathbf{X} \tag{1.26}$$

in place of $\mathbf{Y}$ and $\mathbf{X}$ implies the normal equations

$$(\mathbf{X}'\boldsymbol{\Sigma}^{-1}\mathbf{X})\widehat{\boldsymbol{\beta}}_{\Sigma} = \mathbf{X}'\boldsymbol{\Sigma}^{-1}\mathbf{Y} \tag{1.27}$$

that generalize (1.6), and

$$\widehat{\boldsymbol{\beta}}_{\Sigma} = (\mathbf{X}_*'\mathbf{X}_*)^{-1}\mathbf{X}_*'\mathbf{Y}_* = (\mathbf{X}'\boldsymbol{\Sigma}^{-1}\mathbf{X})^{-1}\mathbf{X}'\boldsymbol{\Sigma}^{-1}\mathbf{Y}, \tag{1.28}$$

where the notation $\widehat{\boldsymbol{\beta}}_{\boldsymbol{\Sigma}}$ is used to indicate its dependence on knowledge of $\boldsymbol{\Sigma}$. This is known as the **generalized least squares** (g.l.s.) estimator, with variance given by

$$\mathbb{V}(\widehat{\boldsymbol{\beta}}_{\boldsymbol{\Sigma}} \mid \sigma^2) = \sigma^2(\mathbf{X}'\boldsymbol{\Sigma}^{-1}\mathbf{X})^{-1}. \tag{1.29}$$

It is attributed to A. C. Aitken from 1934. Of course, σ^2 is unknown. The usual estimator of $(T-k)\sigma^2$ is given by

$$\mathsf{S}(\boldsymbol{\beta}; \mathbf{Y}_*, \mathbf{X}_*) = (\mathbf{Y}_* - \mathbf{X}_*\widehat{\boldsymbol{\beta}}_{\boldsymbol{\Sigma}})'(\mathbf{Y}_* - \mathbf{X}_*\widehat{\boldsymbol{\beta}}_{\boldsymbol{\Sigma}}) = (\mathbf{Y} - \mathbf{X}\widehat{\boldsymbol{\beta}}_{\boldsymbol{\Sigma}})'\boldsymbol{\Sigma}^{-1}(\mathbf{Y} - \mathbf{X}\widehat{\boldsymbol{\beta}}_{\boldsymbol{\Sigma}}). \tag{1.30}$$

Example 1.2 Let $\epsilon_t \overset{\text{ind}}{\sim} \mathrm{N}(0, \sigma^2 k_t)$, where the k_t are known, positive constants, so that $\boldsymbol{\Sigma}^{-1} = \mathrm{diag}(k_1^{-1}, \dots, k_T^{-1})$. Then $\widehat{\boldsymbol{\beta}}_{\boldsymbol{\Sigma}}$ is referred to as the **weighted least squares** estimator. If in the Hamburg income example above, we take $k_t = x_t$, then observations $\{y_t, x_t\}$ receive weights proportional to x_t^{-1}. This has the effect of down-weighting observations with high ages, for which the uncertainty of the slope parameter is higher, and vice versa. ∎

Example 1.3 Let the model be given by $Y_t = \mu + \epsilon_t$, $t = 1, \dots, T$. With $\mathbf{X} = \mathbf{1}$, we have

$$(\mathbf{X}'\mathbf{X})^{-1}\mathbf{X}' = [T^{-1}, \dots, T^{-1}],$$

and the o.l.s. estimator of μ is just the simple average of the observations, $\bar{Y} = (\mathbf{X}'\mathbf{X})^{-1}\mathbf{X}'\mathbf{Y}$. Assume, however, that the ϵ_t are not i.i.d., but are given by the recursion $\epsilon_t = a\epsilon_{t-1} + U_t$, $|a| < 1$, and $U_t \overset{\text{i.i.d.}}{\sim} \mathrm{N}(0, \sigma^2)$. This is referred to as a *stationary first order autoregressive model*, abbreviated AR(1), and is the subject of Chapter 4. There, the covariance matrix of $\boldsymbol{\epsilon} = (\epsilon_1, \dots, \epsilon_T)'$ is shown to be $\mathrm{Cov}(\boldsymbol{\epsilon}) = \sigma^2\boldsymbol{\Sigma}$ with

$$\boldsymbol{\Sigma} = \frac{1}{1-a^2}\begin{bmatrix} 1 & a & a^2 & \cdots & a^{T-1} \\ a & 1 & a & \cdots & a^{T-2} \\ a^2 & a & 1 & \cdots & a^{T-3} \\ \vdots & \vdots & \vdots & \ddots & \vdots \\ a^{T-1} & a^{T-2} & a^{T-3} & \cdots & 1 \end{bmatrix}. \tag{1.31}$$

The g.l.s. estimator of μ is now a weighted average of the Y_t, where the weight vector is given by $\mathbf{w} = (\mathbf{X}'\boldsymbol{\Sigma}^{-1}\mathbf{X})^{-1}\mathbf{X}'\boldsymbol{\Sigma}^{-1}$. Straightforward calculation shows that, for $a = 0.5$, $(\mathbf{X}'\boldsymbol{\Sigma}^{-1}\mathbf{X})^{-1} = 4/(T+2)$ and

$$\mathbf{X}'\boldsymbol{\Sigma}^{-1} = \left[\frac{1}{2}, \frac{1}{4}, \frac{1}{4}, \dots, \frac{1}{4}, \frac{1}{2}\right]',$$

so that the first and last weights are $2/(T+2)$ and the middle $T-2$ are all $1/(T+2)$. Note that the weights sum to one. A similar pattern holds for all $|a| < 1$, with the ratio of the first and last weights to the center weights converging to $1/2$ as $a \to -1$ and to ∞ as $a \to 1$. Thus, we see that (i) for constant T, the difference between g.l.s. and o.l.s. grows as $a \to 1$ and (ii) for constant a, $|a| < 1$, the difference between g.l.s. and o.l.s. shrinks as $T \to \infty$. The latter is true because a finite number of observations, in this case only two, become negligible in the limit, and because the relative weights associated with these two values converges to a constant independent of T.

Now consider the model $Y_t = \mu + \epsilon_t$, $t = 1, \dots, T$, with $\epsilon_t = bU_{t-1} + U_t$, $|b| < 1$, $U_t \overset{\text{i.i.d.}}{\sim} \mathrm{N}(0, \sigma^2)$. This is referred to as an invertible *first-order moving average model*, or MA(1), and is discussed in detail in Chapter 6. There, it is shown that $\text{Cov}(\epsilon) = \sigma^2 \mathbf{\Sigma}$ with

$$\mathbf{\Sigma} = \begin{bmatrix} 1+b^2 & b & 0 & \cdots & 0 \\ b & 1+b^2 & \ddots & & \vdots \\ 0 & b & \ddots & & 0 \\ \vdots & 0 & \ddots & & b \\ 0 & \cdots & 0 & b & 1+b^2 \end{bmatrix}.$$

The weight vectors $\mathbf{w} = (\mathbf{X}'\mathbf{\Sigma}^{-1}\mathbf{X})^{-1}\mathbf{X}'\mathbf{\Sigma}^{-1}$ for the two values, $b = -0.9$ and $b = 0.9$, are plotted in Figure 1.2 for $T = 100$. This is clearly quite a different weighting structure than for the AR(1) model.

In the limiting case $b \to 1$, we have

$$Y_1 = \mu + U_0 + U_1, \quad Y_2 = \mu + U_1 + U_2, \quad \ldots, \quad Y_T = \mu + U_{T-1} + U_T$$

so that

$$\sum_{t=1}^{T} Y_t = T\mu + U_0 + U_T + 2\sum_{t=1}^{T-1} U_t,$$

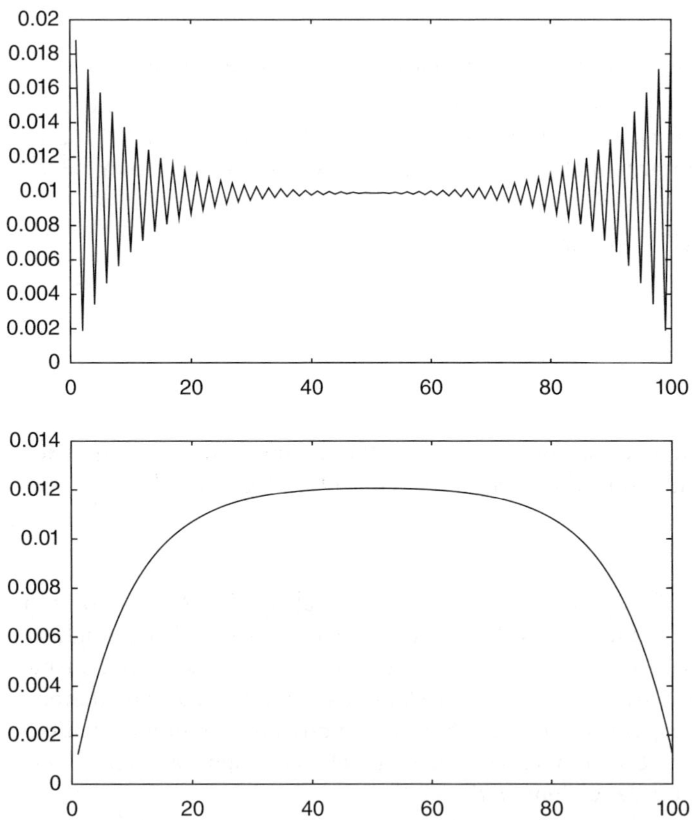

Figure 1.2 Weight vector for an MA(1) model with $T = 100$ and $b = 0.9$ (top) and $b = -0.9$ (bottom).

$\mathbb{E}[\bar{Y}] = \mu$ and

$$\mathbb{V}(\bar{Y}) = \frac{\sigma^2 + \sigma^2 + 4(T-1)\sigma^2}{T^2} = \frac{4\sigma^2}{T} - \frac{2\sigma^2}{T^2}.$$

For $T = 100$ and $\sigma^2 = 1$, $\mathbb{V}(\bar{Y} \mid b = 1) \approx 0.0398$. Similarly, for $b = -1$, $\sum_{t=1}^{T} Y_t = T\mu + U_0 + U_T$ and $\mathbb{V}(\bar{Y} \mid b = -1) = 2\sigma^2/T^2 = 0.0002$. ■

Consideration of the previous example might lead one to ponder if it is possible to specify conditions such that $\widehat{\boldsymbol{\beta}}_{\boldsymbol{\Sigma}}$ will equal $\widehat{\boldsymbol{\beta}}_{\mathbf{I}} = \widehat{\boldsymbol{\beta}}$ for $\boldsymbol{\Sigma} \neq \mathbf{I}$. A necessary and sufficient condition for $\widehat{\boldsymbol{\beta}}_{\boldsymbol{\Sigma}} = \widehat{\boldsymbol{\beta}}$ is if the k columns of $\mathbf{X}$ are linear combinations of k of the eigenvectors of $\boldsymbol{\Sigma}$, as first established by Anderson (1948); see, e.g., Anderson (1971, p. 19 and p. 561) for proof.

This question has generated a large amount of academic work, as illustrated in the survey of Puntanen and Styan (1989), which contains about 90 references (see also Krämer et al., 1996). There are several equivalent conditions for the result to hold, a rather useful and attractive one of which is that

$$\widehat{\boldsymbol{\beta}}_{\boldsymbol{\Sigma}} = \widehat{\boldsymbol{\beta}} \text{ if and only if } \mathbf{P}\boldsymbol{\Sigma} \text{ is symmetric,} \tag{1.32}$$

i.e., if and only if $\mathbf{P}\boldsymbol{\Sigma} = \boldsymbol{\Sigma}\mathbf{P}$, where $\mathbf{P} = \mathbf{X}(\mathbf{X}'\mathbf{X})^{-1}\mathbf{X}'$. Another is that there exists a matrix $\mathbf{F}$ satisfying $\mathbf{X}\mathbf{F} = \boldsymbol{\Sigma}^{-1}\mathbf{X}$, which is demonstrated in Example 1.5.

Example 1.4 With $\mathbf{X} = \mathbf{1}$ (a T-length column of ones), Anderson's condition implies that $\mathbf{1}$ needs to be an eigenvector of $\boldsymbol{\Sigma}$, or $\boldsymbol{\Sigma}1 = s\mathbf{1}$ for some nonzero scalar s. This means that the sum of each row of $\boldsymbol{\Sigma}$ must be the same value. This obviously holds when $\boldsymbol{\Sigma} = \mathbf{I}$, and clearly never holds when $\boldsymbol{\Sigma}$ is a diagonal weighting matrix with at least two weights differing.

To determine if $\widehat{\boldsymbol{\beta}}_{\boldsymbol{\Sigma}} = \widehat{\boldsymbol{\beta}}$ is possible for the AR(1) and MA(1) models from Example 1.3, we use a result of McElroy (1967), who showed that, if $\mathbf{X}$ is full rank and contains $\mathbf{1}$, then $\widehat{\boldsymbol{\beta}}_{\boldsymbol{\Sigma}} = \widehat{\boldsymbol{\beta}}$ if and only if $\boldsymbol{\Sigma}$ is full rank and can be expressed as $k_1\mathbf{I} + k_2\mathbf{1}\mathbf{1}'$, i.e., the equicorrelated case. We will see in Chapters 4 and 7 that this is never the case for AR(1) and MA(1) models or, more generally, for stationary and invertible ARMA(p, q) models. ■

Remark The previous discussion begets the question of how one could assess the extent to which o.l.s. will be inferior relative to g.l.s., notably because, in many applications, $\boldsymbol{\Sigma}$ will not be known. This turns out to be a complicated endeavor in general; see Puntanen and Styan (1989, p. 154) and the references therein for further details. Observe also how (1.28) and (1.29) assume the true $\boldsymbol{\Sigma}$. The determination of robust estimators for the variance of $\widehat{\boldsymbol{\beta}}$ for unknown $\boldsymbol{\Sigma}$ is an important and active research area in statistics and, particularly, econometrics (and for other model classes beyond the simple linear regression model studied here). The primary reference papers are White (1980, 1982), MacKinnon and White (1985), Newey and West (1987), and Andrews (1991), giving rise to the class of so-called **heteroskedastic and autocorrelation consistent** covariance matrix estimators, or HAC. With respect to computation of the HAC estimators, see Zeileis (2006), Heberle and Sattarhoff (2017), and the references therein. ■

It might come as a surprise that defining the coefficient of multiple determination R^2 in the g.l.s. context is not so trivial, and several suggestions exist. The problem stems from the definition in the o.l.s. case (1.13), with $R^2 = 1 - \mathsf{S}(\widehat{\boldsymbol{\beta}}, \mathbf{Y}, \mathbf{X})/\mathsf{S}(\bar{Y}, \mathbf{Y}, \mathbf{1})$, and observing that, if $\mathbf{1} \in \mathcal{C}(\mathbf{X})$ (the column space of $\mathbf{X}$, as defined below), then, via the transformation in (1.26), $\mathbf{1} \notin \mathcal{C}(\mathbf{X}_*)$.

To establish a meaningful definition, we first need the fact that, with $\hat{\mathbf{Y}} = \mathbf{X}\hat{\boldsymbol{\beta}}_{\boldsymbol{\Sigma}}$ and $\hat{\boldsymbol{\epsilon}} = \mathbf{Y} - \hat{\mathbf{Y}}$,

$$\mathbf{Y}'\boldsymbol{\Sigma}^{-1}\mathbf{Y} = \hat{\mathbf{Y}}'\boldsymbol{\Sigma}^{-1}\hat{\mathbf{Y}} + \hat{\boldsymbol{\epsilon}}'\boldsymbol{\Sigma}^{-1}\hat{\boldsymbol{\epsilon}}, \tag{1.33}$$

which is derived in (1.47). Next, from the normal equations (1.27) and letting $\mathbf{X}_i$ denote the ith column of $\mathbf{X}$, $i = 1, \dots, k$, we have a system of k equations, the ith of which is, with $\hat{\boldsymbol{\beta}}_{\boldsymbol{\Sigma}} = (\hat{\beta}_1, \dots, \hat{\beta}_k)'$,

$$(\mathbf{X}_i'\boldsymbol{\Sigma}^{-1}\mathbf{X}_1)\hat{\beta}_1 + (\mathbf{X}_i'\boldsymbol{\Sigma}^{-1}\mathbf{X}_2)\hat{\beta}_2 + \cdots + (\mathbf{X}_i'\boldsymbol{\Sigma}^{-1}\mathbf{X}_k)\hat{\beta}_k = \mathbf{X}_i'\boldsymbol{\Sigma}^{-1}\mathbf{Y}.$$

Similarly, premultiplying both sides of $\mathbf{X}\hat{\boldsymbol{\beta}}_{\boldsymbol{\Sigma}} = \hat{\mathbf{Y}}$ by $\mathbf{X}_i'\boldsymbol{\Sigma}^{-1}$ gives

$$(\mathbf{X}_i'\boldsymbol{\Sigma}^{-1}\mathbf{X}_1)\hat{\beta}_1 + (\mathbf{X}_i'\boldsymbol{\Sigma}^{-1}\mathbf{X}_2)\hat{\beta}_2 + \cdots + (\mathbf{X}_i'\boldsymbol{\Sigma}^{-1}\mathbf{X}_k)\hat{\beta}_k = \mathbf{X}_i'\boldsymbol{\Sigma}^{-1}\hat{\mathbf{Y}},$$

so that

$$\mathbf{X}_i'\boldsymbol{\Sigma}^{-1}(\mathbf{Y} - \hat{\mathbf{Y}}) = 0,$$

which we will see again below, in the context of projection, in (1.63). In particular, with $\mathbf{X}_1 = \mathbf{1} = (1, 1, \dots, 1)'$ the usual first regressor, $\mathbf{1}'\boldsymbol{\Sigma}^{-1}\hat{\mathbf{Y}} = \mathbf{1}'\boldsymbol{\Sigma}^{-1}\mathbf{Y}$. We now follow Buse (1973), and define the weighted mean to be

$$\bar{Y} := \bar{Y}_{\boldsymbol{\Sigma}} := \frac{\mathbf{1}'\boldsymbol{\Sigma}^{-1}\mathbf{Y}}{\mathbf{1}'\boldsymbol{\Sigma}^{-1}\mathbf{1}} \quad \left(= \frac{\mathbf{1}'\boldsymbol{\Sigma}^{-1}\hat{\mathbf{Y}}}{\mathbf{1}'\boldsymbol{\Sigma}^{-1}\mathbf{1}} \right), \tag{1.34}$$

which obviously reduces to the simple sample mean when $\boldsymbol{\Sigma} = \mathbf{I}$. The next step is to confirm by simply multiplying out that

$$(\mathbf{Y} - \bar{Y}\mathbf{1})'\boldsymbol{\Sigma}^{-1}(\mathbf{Y} - \bar{Y}\mathbf{1}) = \mathbf{Y}'\boldsymbol{\Sigma}^{-1}\mathbf{Y} - \frac{(\mathbf{1}'\boldsymbol{\Sigma}^{-1}\mathbf{Y})^2}{\mathbf{1}'\boldsymbol{\Sigma}^{-1}\mathbf{1}},$$

and, likewise,

$$(\hat{\mathbf{Y}} - \bar{Y}\mathbf{1})'\boldsymbol{\Sigma}^{-1}(\hat{\mathbf{Y}} - \bar{Y}\mathbf{1}) = \hat{\mathbf{Y}}'\boldsymbol{\Sigma}^{-1}\hat{\mathbf{Y}} - \frac{(\mathbf{1}'\boldsymbol{\Sigma}^{-1}\mathbf{Y})^2}{\mathbf{1}'\boldsymbol{\Sigma}^{-1}\mathbf{1}},$$

so that (1.33) can be expressed as

$$(\mathbf{Y} - \bar{Y}\mathbf{1})'\boldsymbol{\Sigma}^{-1}(\mathbf{Y} - \bar{Y}\mathbf{1}) = (\hat{\mathbf{Y}} - \bar{Y}\mathbf{1})'\boldsymbol{\Sigma}^{-1}(\hat{\mathbf{Y}} - \bar{Y}\mathbf{1}) + \hat{\boldsymbol{\epsilon}}'\boldsymbol{\Sigma}^{-1}\hat{\boldsymbol{\epsilon}}. \tag{1.35}$$

The definition of R^2 is now given by

$$R^2 = R^2_{\boldsymbol{\Sigma}} = 1 - \frac{\hat{\boldsymbol{\epsilon}}'\boldsymbol{\Sigma}^{-1}\hat{\boldsymbol{\epsilon}}}{(\mathbf{Y} - \bar{Y}\mathbf{1})'\boldsymbol{\Sigma}^{-1}(\mathbf{Y} - \bar{Y}\mathbf{1})}, \tag{1.36}$$

which is indeed analogous to (1.13) and reduces to it when $\boldsymbol{\Sigma} = \mathbf{I}$.

Along with examples of other, less desirable, definitions, Buse (1973) discusses the benefits of this definition, which include that it is interpretable as the proportion of the generalized sum of squares of the dependent variable that is attributable to the influence of the explanatory variables, and that it lies between zero and one. It is also zero when all the estimates coefficients (except the constant) are zero, and can be related to the F test as was done above in the ordinary least squares case.

1.3 The Geometric Approach to Least Squares

In spite of earnest prayer and the greatest desire to adhere to proper statistical behavior, I have not been able to say why the method of maximum likelihood is to be preferred over other methods, particularly the method of least squares.

(Joseph Berkson, 1944, p. 359)

The following sections analyze the linear regression model using the notion of projection. This complements the purely algebraic approach to regression analysis by providing a useful terminology and geometric intuition behind least squares. Most importantly, its use often simplifies the derivation and understanding of various quantities such as point estimators and test statistics. The reader is assumed to be comfortable with the notions of linear subspaces, span, dimension, rank, and orthogonality. See the references given at the beginning of Section B.5 for detailed presentations of these and other important topics associated with linear and matrix algebra.

1.3.1 Projection

The Euclidean **dot product** or **inner product** of two vectors $\mathbf{u} = (u_1, u_2, \dots, u_T)'$ and $\mathbf{v} = (v_1, v_2, \dots, v_T)'$ is denoted by $\langle \mathbf{u}, \mathbf{v} \rangle = \mathbf{u}'\mathbf{v} = \sum_{i=1}^{T} u_i v_i$. Observe that, for $\mathbf{y}, \mathbf{u}, \mathbf{w} \in \mathbb{R}^T$,

$$\langle \mathbf{y} - \mathbf{u}, \mathbf{w} \rangle = (\mathbf{y} - \mathbf{u})'\mathbf{w} = \mathbf{y}'\mathbf{w} - \mathbf{u}'\mathbf{w} = \langle \mathbf{y}, \mathbf{w} \rangle - \langle \mathbf{u}, \mathbf{w} \rangle. \tag{1.37}$$

The **norm** of vector $\mathbf{u}$ is $\|\mathbf{u}\| = \langle \mathbf{u}, \mathbf{u} \rangle^{1/2}$. The square matrix $\mathbf{U}$ with columns $\mathbf{u}_1, \dots, \mathbf{u}_T$ is **orthonormal** if $\mathbf{U}\mathbf{U}' = \mathbf{U}'\mathbf{U} = \mathbf{I}$, i.e., $\mathbf{U}' = \mathbf{U}^{-1}$, implying $\langle \mathbf{u}_i, \mathbf{u}_j \rangle = 1$ if $i = j$ and zero otherwise.

For a fixed $T \times k$ matrix $\mathbf{X}$, $k \leqslant T$ and usually such that $k \ll T$ ("is much less than"), the **column space** of $\mathbf{X}$, denoted $\mathcal{C}(\mathbf{X})$, or the **linear span** of the k columns $\mathbf{X}$, is the set of all vectors that can be generated as a linear sum of, or *spanned by*, the columns of $\mathbf{X}$, such that the coefficient of each vector is a real number, i.e.,

$$\mathcal{C}(\mathbf{X}) = \{\mathbf{y} : \mathbf{y} = \mathbf{X}\mathbf{b}, \mathbf{b} \in \mathbb{R}^k\}. \tag{1.38}$$

In words, if $\mathbf{y} \in \mathcal{C}(\mathbf{X})$, then there exists $\mathbf{b} \in \mathbb{R}^k$ such that $\mathbf{y} = \mathbf{X}\mathbf{b}$.

It is easy to verify that $\mathcal{C}(\mathbf{X})$ is a subspace of $\mathbb{R}^T$ with **dimension** $\dim(\mathcal{C}(\mathbf{X})) = \text{rank}(\mathbf{X}) \leqslant k$. If $\dim(\mathcal{C}(\mathbf{X})) = k$, then $\mathbf{X}$ is said to be a **basis matrix** (for $\mathcal{C}(\mathbf{X})$). Furthermore, if the columns of $\mathbf{X}$ are orthonormal, then $\mathbf{X}$ is an **orthonormal basis matrix** and $\mathbf{X}'\mathbf{X} = \mathbf{I}$.

Let $\mathbf{V}$ be a basis matrix with columns $\mathbf{v}_1, \dots, \mathbf{v}_k$. The method of **Gram–Schmidt** can be used to construct an orthonormal basis matrix $\mathbf{U} = [\mathbf{u}_1, \dots, \mathbf{u}_k]$ as follows. First set $\mathbf{u}_1 = \mathbf{v}_1 / \|\mathbf{v}_1\|$ so that $\langle \mathbf{u}_1, \mathbf{u}_1 \rangle = 1$. Next, let $\mathbf{u}_2^* = \mathbf{v}_2 - \langle \mathbf{v}_2, \mathbf{u}_1 \rangle \mathbf{u}_1$, so that

$$\langle \mathbf{u}_2^*, \mathbf{u}_1 \rangle = \langle \mathbf{v}_2, \mathbf{u}_1 \rangle - \langle \mathbf{v}_2, \mathbf{u}_1 \rangle \langle \mathbf{u}_1, \mathbf{u}_1 \rangle = \langle \mathbf{v}_2, \mathbf{u}_1 \rangle - \langle \mathbf{v}_2, \mathbf{u}_1 \rangle = 0, \tag{1.39}$$

and set $\mathbf{u}_2 = \mathbf{u}_2^* / \|\mathbf{u}_2^*\|$. By construction of $\mathbf{u}_2$, $\langle \mathbf{u}_2, \mathbf{u}_2 \rangle = 1$, and from (1.39), $\langle \mathbf{u}_2, \mathbf{u}_1 \rangle = 0$. Continue with $\mathbf{u}_3^* = \mathbf{v}_3 - \langle \mathbf{v}_3, \mathbf{u}_1 \rangle \mathbf{u}_1 - \langle \mathbf{v}_3, \mathbf{u}_2 \rangle \mathbf{u}_2$ and $\mathbf{u}_3 = \mathbf{u}_3^* / \|\mathbf{u}_3^*\|$, up to $\mathbf{u}_k^* = \mathbf{v}_k - \sum_{i=1}^{k-1} \langle \mathbf{v}_k, \mathbf{u}_i \rangle \mathbf{u}_i$ and $\mathbf{u}_k = \mathbf{u}_k^* / \|\mathbf{u}_k^*\|$. This renders $\mathbf{U}$ an orthonormal basis matrix for $\mathcal{C}(\mathbf{V})$.

Observe that, if $\mathbf{u} = \mathbf{P}_S\mathbf{y}$, then $\mathbf{P}_S\mathbf{u}$ must be equal to $\mathbf{u}$ because $\mathbf{u}$ is already in $\mathcal{S}$. This also follows algebraically from (1.42), i.e., $\mathbf{P}_S = \mathbf{T}\mathbf{T}'$ and $\mathbf{P}_S{}^2 = \mathbf{T}\mathbf{T}'\mathbf{T}\mathbf{T}' = \mathbf{T}\mathbf{T}' = \mathbf{P}_S$, showing that the matrix $\mathbf{P}_S$is **idempotent**, i.e., $\mathbf{P}_S\mathbf{P}_S = \mathbf{P}_S$. Therefore, if $\mathbf{w} = (\mathbf{I}_T - \mathbf{P}_S)\mathbf{y} \in \mathcal{S}^\perp$, then $\mathbf{P}_S\mathbf{w} = \mathbf{P}_S(\mathbf{I}_T - \mathbf{P}_S)\mathbf{y} = \mathbf{0}$. Another property of projection matrices is that they are symmetric, which follows directly from $\mathbf{P}_S = \mathbf{T}\mathbf{T}'$.

Example 1.6 Let $\mathbf{y}$ be a vector in $\mathbb{R}^T$ and $\mathcal{S}$ a subspace of $\mathbb{R}^T$ with corresponding projection matrix $\mathbf{P}_S$. Then, with $\mathbf{P}_{S^\perp} = \mathbf{I}_T - \mathbf{P}_S$ from (1.44),

$$\begin{aligned}\|\mathbf{P}_{S\perp}\mathbf{y}\|^2 &= \|\mathbf{y} - \mathbf{P}_S\mathbf{y}\|^2 = (\mathbf{y} - \mathbf{P}_S\mathbf{y})'(\mathbf{y} - \mathbf{P}_S\mathbf{y}) \\ &= \mathbf{y}'\mathbf{y} - \mathbf{y}'\mathbf{P}_S\mathbf{y} - \mathbf{y}'\mathbf{P}'_S\mathbf{y} + \mathbf{y}'\mathbf{P}'_S\mathbf{P}_S\mathbf{y} = \mathbf{y}'\mathbf{y} - \mathbf{y}'\mathbf{P}_S\mathbf{y} = \|\mathbf{y}\|^2 - \|\mathbf{P}_S\mathbf{y}\|^2,\end{aligned}$$

i.e.,

$$\|\mathbf{y}\|^2 = \|\mathbf{P}_S\mathbf{y}\|^2 + \|\mathbf{P}_{S^\perp}\mathbf{y}\|^2. \tag{1.45}$$

For $\mathbf{X}$ a full-rank $T \times k$ matrix and $\mathcal{S} = \mathcal{C}(\mathbf{X})$, this implies, for regression model (1.3) with $\widehat{\mathbf{Y}} = \mathbf{X}\widehat{\boldsymbol{\beta}}$ and $\widehat{\epsilon} = \mathbf{Y} - \mathbf{X}\widehat{\boldsymbol{\beta}}$,

$$\begin{aligned}\mathbf{Y}'\mathbf{Y} &= \widehat{\mathbf{Y}}'\widehat{\mathbf{Y}} + \widehat{\epsilon}'\widehat{\epsilon} \\ &= (\widehat{\mathbf{Y}} + \widehat{\epsilon})'(\widehat{\mathbf{Y}} + \widehat{\epsilon}).\end{aligned} \tag{1.46}$$

In the g.l.s. framework, use of (1.46) applied to the transformed model (1.25) and (1.26) yields, with $\widehat{\mathbf{Y}}_* = \mathbf{X}_*\widehat{\boldsymbol{\beta}}_\Sigma$ and $\widehat{\epsilon}_* = \mathbf{Y}_* - \widehat{\mathbf{Y}}_*$,

$$\mathbf{Y}'_*\mathbf{Y}_* = \widehat{\mathbf{Y}}'_*\widehat{\mathbf{Y}}_* + \widehat{\epsilon}'_*\widehat{\epsilon}_* = (\widehat{\mathbf{Y}}_* + \widehat{\epsilon}_*)'(\widehat{\mathbf{Y}}_* + \widehat{\epsilon}_*),$$

or, with $\widehat{\mathbf{Y}} = \mathbf{X}\widehat{\boldsymbol{\beta}}_\Sigma$ and $\widehat{\epsilon} = \mathbf{Y} - \widehat{\mathbf{Y}}$,

$$\begin{aligned}\mathbf{Y}'\boldsymbol{\Sigma}^{-1/2}\boldsymbol{\Sigma}^{-1/2}\mathbf{Y} &= \mathbf{Y}'_*\mathbf{Y}_* \\ &= (\widehat{\mathbf{Y}}_* + \widehat{\epsilon}_*)'(\widehat{\mathbf{Y}}_* + \widehat{\epsilon}_*) = (\widehat{\mathbf{Y}} + \widehat{\epsilon})'\boldsymbol{\Sigma}^{-1/2}\boldsymbol{\Sigma}^{-1/2}(\widehat{\mathbf{Y}} + \widehat{\epsilon}),\end{aligned}$$

or, finally,

$$\mathbf{Y}'\boldsymbol{\Sigma}^{-1}\mathbf{Y} = \widehat{\mathbf{Y}}'\boldsymbol{\Sigma}^{-1}\widehat{\mathbf{Y}} + \widehat{\epsilon}'\boldsymbol{\Sigma}^{-1}\widehat{\epsilon}, \tag{1.47}$$

which is (1.33), as was used for determining the R^2 measure in the g.l.s. case. ■

An equivalent definition of a projection matrix $\mathbf{P}$ onto $\mathcal{S}$ is when the following are satisfied:

$$\mathbf{v} \in \mathcal{S} \Rightarrow \mathbf{P}\mathbf{v} = \mathbf{v} \quad \text{(projection)} \tag{1.48}$$

$$\mathbf{w} \perp \mathcal{S} \Rightarrow \mathbf{P}\mathbf{w} = \mathbf{0} \quad \text{(perpendicularity)}. \tag{1.49}$$

The following result is both interesting and useful; it is proven in Problem 1.8, where further comments are given.

Theorem 1.2 If $\mathbf{P}$ is symmetric and idempotent with $\text{rank}(\mathbf{P}) = k$, then (i) k of the eigenvalues of $\mathbf{P}$ are unity and the remaining $T - k$ are zero, and (ii) $\text{tr}(\mathbf{P}) = k$.

This is understood as follows: If $T \times T$ matrix $\mathbf{P}$ is such that $\text{rank}(\mathbf{P}) = \text{tr}(\mathbf{P}) = k$ and k of the eigenvalues of $\mathbf{P}$ are unity and the remaining $T - k$ are zero, then it is not necessarily the case that $\mathbf{P}$ is symmetric and idempotent. However, if $\mathbf{P}$ is symmetric and idempotent, then $\text{tr}(\mathbf{P}) = k \Leftrightarrow \text{rank}(\mathbf{P}) = k$.

```
function G=makeG(X)        % G is such that M=G'G and I=GG'
k=size(X,2);               % could also use k = rank(X).
M=makeM(X);                % M=eye(T)-X*inv(X'*X)*X', where X is size TXk
[V,D]=eig(0.5*(M+M'));     % V are eigenvectors, D eigenvalues
e=diag(D);
[e,I]=sort(e);             % I is a permutation index of the sorting
G=V(:,I(k+1:end)); G=G';
```

Program Listing 1.2: Computes matrix $\mathbf{G}$ in Theorem 1.3. Function makeM is given in Listing B.2.

Let $\mathbf{M} = \mathbf{I}_T - \mathbf{P}_{\mathcal{S}}$ with $\dim(\mathcal{S}) = k$, $k \in \{1, 2, \dots, T-1\}$. As $\mathbf{M}$ is itself a projection matrix, then, similar to (1.42), it can be expressed as $\mathbf{VV}'$, where $\mathbf{V}$ is a $T \times (T-k)$ matrix with orthonormal columns. We state this obvious, but important, result as a theorem because it will be useful elsewhere (and it is slightly more convenient to use $\mathbf{V}'\mathbf{V}$ instead of $\mathbf{VV}'$).

Theorem 1.3 Let $\mathbf{X}$ be a full-rank $T \times k$ matrix, $k \in \{1, 2, \dots, T-1\}$, and $\mathcal{S} = \mathcal{C}(\mathbf{X})$ with $\dim(\mathcal{S}) = k$. Let $\mathbf{M} = \mathbf{I}_T - \mathbf{P}_{\mathcal{S}}$. The projection matrix $\mathbf{M}$ may be written as $\mathbf{M} = \mathbf{G}'\mathbf{G}$, where $\mathbf{G}$ is $(T-k) \times T$ and such that $\mathbf{GG}' = \mathbf{I}_{T-k}$ and $\mathbf{GX} = \mathbf{0}$.

A less direct, but instructive, method for proving Theorem 1.3 is given in Problem 1.5. Matrix $\mathbf{G}$ can be computed by taking its rows to be the $T-k$ eigenvectors of $\mathbf{M}$ that correspond to the unit eigenvalues. The small program in Listing 1.2 performs this computation. Alternatively, $\mathbf{G}$ can be computed by applying Gram–Schmidt orthogonalization to the columns of $\mathbf{M}$ and keeping the nonzero vectors.[5] Matrix $\mathbf{G}$ is not unique and the two methods just stated often result in different values.

It turns out that any symmetric, idempotent matrix is a projection matrix:

Theorem 1.4 The symmetry and idempotency of a matrix $\mathbf{P}$ are necessary and sufficient conditions for it to be the projection matrix onto the space spanned by its columns.

Proof: Sufficiency: We assume $\mathbf{P}$ is a symmetric and idempotent $T \times T$ matrix, and must show that (1.43) and (1.44) are satisfied for all $\mathbf{y} \in \mathbb{R}^T$. Let $\mathbf{y}$ be an element of $\mathbb{R}^T$ and let $\mathcal{S} = \mathcal{C}(\mathbf{P})$. By the definition of column space, $\mathbf{Py} \in \mathcal{S}$, which is (1.43). To see that (1.44) is satisfied, we must show that $(\mathbf{I} - \mathbf{P})\mathbf{y}$ is perpendicular to every vector in $\mathcal{S}$, or that $(\mathbf{I} - \mathbf{P})\mathbf{y} \perp \mathbf{Pw}$ for all $\mathbf{w} \in \mathbb{R}^T$. But

$$((\mathbf{I} - \mathbf{P})\mathbf{y})'\mathbf{Pw} = \mathbf{y}'\mathbf{Pw} - \mathbf{y}'\mathbf{P}'\mathbf{Pw} = \mathbf{0}$$

because, by assumption, $\mathbf{P}'\mathbf{P} = \mathbf{P}$.

For necessity, following Christensen (1987, p. 335), write $\mathbf{y} = \mathbf{y}_1 + \mathbf{y}_2$, where $\mathbf{y} \in \mathbb{R}^T$, $\mathbf{y}_1 \in \mathcal{S}$ and $\mathbf{y}_2 \in \mathcal{S}^{\perp}$. Then, using only (1.48) and (1.49), $\mathbf{Py} = \mathbf{Py}_1 + \mathbf{Py}_2 = \mathbf{Py}_1 = \mathbf{y}_1$ and

$$\mathbf{P}^2\mathbf{y} = \mathbf{P}^2\mathbf{y}_1 + \mathbf{P}^2\mathbf{y}_2 = \mathbf{Py}_1 = \mathbf{Py},$$

so that $\mathbf{P}$ is idempotent. Next, as $\mathbf{Py}_1 = \mathbf{y}_1$ and $(\mathbf{I} - \mathbf{P})\mathbf{y} = \mathbf{y}_2$,

$$\mathbf{y}'\mathbf{P}'(\mathbf{I} - \mathbf{P})\mathbf{y} = \mathbf{y}_1'\mathbf{y}_2 = 0,$$

5 In Matlab, the `orth` function can be used. The implementation uses the singular value decomposition (svd) and attempts to determine the number of nonzero singular values. Because of numerical imprecision, this latter step can choose too many. Instead, just use `[U,S,V]=svd(M); dim=sum(round(diag(S))==1); G=U(:,1:dim)';`, where `dim` will equal $T-k$ for full rank $\mathbf{X}$ matrices.

because $\mathbf{y}_1$ and $\mathbf{y}_2$ are orthogonal. As $\mathbf{y}$ is arbitrary, $\mathbf{P}'(\mathbf{I}-\mathbf{P})$ must be $\mathbf{0}$, or $\mathbf{P}' = \mathbf{P}'\mathbf{P}$. From this and the symmetry of $\mathbf{P}'\mathbf{P}$, it follows that $\mathbf{P}$ is also symmetric. ■

The following fact will be the key to obtaining the o.l.s. estimator in a linear regression model, as discussed in Section 1.3.2.

Theorem 1.5 Vector $\mathbf{u}$ in $\mathcal{S}$ is the closest to $\mathbf{y}$ in the sense that

$$\|\mathbf{y}-\mathbf{u}\|^2 = \min_{\tilde{u}\in\mathcal{S}} \|\mathbf{y}-\tilde{\boldsymbol{u}}\|^2.$$

Proof: Let $\mathbf{y} = \mathbf{u} + \mathbf{v}$, where $\mathbf{u} \in \mathcal{S}$ and $\mathbf{v} \in \mathcal{S}^{\perp}$. We have, for any $\tilde{\boldsymbol{u}} \in \mathcal{S}$,

$$\|\mathbf{y}-\tilde{\boldsymbol{u}}\|^2 = \|\mathbf{u}+\mathbf{v}-\tilde{\boldsymbol{u}}\|^2 = \|\mathbf{u}-\tilde{\boldsymbol{u}}\|^2 + \|\mathbf{v}\|^2 \geqslant \|\mathbf{v}\|^2 = \|\mathbf{y}-\mathbf{u}\|^2,$$

where the second equality holds because $\mathbf{v} \perp (\mathbf{u} - \tilde{\mathbf{u}})$. ■

The next theorem will be useful for testing whether the mean vector of a linear model lies in a subspace of $\mathcal{C}(\mathbf{X})$, as developed in Section 1.4.

Theorem 1.6 Let $\mathcal{S}_0 \subset \mathcal{S}$ be subspaces of $\mathbb{R}^T$ with respective integer dimensions r and s, such that $0 < r < s < T$. Further, let $\mathcal{S}\backslash\mathcal{S}_0$ denote the subspace $\mathcal{S} \cap \mathcal{S}_0^{\perp}$ with dimension $s-r$, i.e., $\mathcal{S}\backslash\mathcal{S}_0 = \{\mathbf{s} : \mathbf{s} \in \mathcal{S};\ \mathbf{s} \perp \mathcal{S}_0\}$. Then

a. $\mathbf{P}_{\mathcal{S}}\mathbf{P}_{\mathcal{S}_0} = \mathbf{P}_{\mathcal{S}_0}$ and $\mathbf{P}_{\mathcal{S}_0}\mathbf{P}_{\mathcal{S}} = \mathbf{P}_{\mathcal{S}_0}$.
b. $\mathbf{P}_{\mathcal{S}\backslash\mathcal{S}_0} = \mathbf{P}_{\mathcal{S}} - \mathbf{P}_{\mathcal{S}_0}$.
c. $\|\mathbf{P}_{\mathcal{S}\backslash\mathcal{S}_0}\mathbf{y}\|^2 = \|\mathbf{P}_{\mathcal{S}}\mathbf{y}\|^2 - \|\mathbf{P}_{\mathcal{S}_0}\mathbf{y}\|^2$.
d. $\mathbf{P}_{\mathcal{S}\backslash\mathcal{S}_0} = \mathbf{P}_{\mathcal{S}_0^{\perp}\backslash\mathcal{S}^{\perp}} = \mathbf{P}_{\mathcal{S}_0^{\perp}} - \mathbf{P}_{\mathcal{S}^{\perp}}$.
e. $\mathbf{P}_{\mathcal{S}}\mathbf{P}_{\mathcal{S}\backslash\mathcal{S}_0} = \mathbf{P}_{\mathcal{S}\backslash\mathcal{S}_0}\mathbf{P}_{\mathcal{S}} = \mathbf{P}_{\mathcal{S}\backslash\mathcal{S}_0}$.
f. $\|\mathbf{P}_{\mathcal{S}_0^{\perp}\backslash\mathcal{S}^{\perp}}\mathbf{y}\|^2 = \|\mathbf{P}_{\mathcal{S}_0^{\perp}}\mathbf{y}\|^2 - \|\mathbf{P}_{\mathcal{S}^{\perp}}\mathbf{y}^2\|$.

Proof: (part a) For all $\mathbf{y} \in \mathbb{R}^T$, as $\mathbf{P}_{\mathcal{S}_0}\mathbf{y} \in \mathcal{S}$, $\mathbf{P}_{\mathcal{S}}(\mathbf{P}_{\mathcal{S}_0}\mathbf{y}) = \mathbf{P}_{\mathcal{S}_0}\mathbf{y}$. Transposing yields the second result.

Another way of seeing this (and which is useful for proving the other results) is to partition $\mathbb{R}^T$ into subspaces $\mathcal{S}$ and $\mathcal{S}^{\perp}$, and then $\mathcal{S}$ into subspaces $\mathcal{S}_0$ and $\mathcal{S}\backslash\mathcal{S}_0$. Take as a basis for $\mathbb{R}^T$ the vectors

$$\underbrace{\overbrace{\mathbf{r}_1, \dots, \mathbf{r}_r}^{\mathcal{S}_0 \text{ basis}}, \overbrace{\mathbf{s}_{r+1}, \dots, \mathbf{s}_s}^{\mathcal{S}\backslash\mathcal{S}_0 \text{ basis}}}_{\mathcal{S} \text{ basis}}, \underbrace{\mathbf{z}_{s+1}, \dots, \mathbf{z}_T}_{\mathcal{S}^{\perp} \text{ basis}} \tag{1.50}$$

and let $\mathbf{y} = \mathbf{r} + \mathbf{s} + \mathbf{z}$, where $\mathbf{r} \in \mathcal{S}_0$, $\mathbf{s} \in \mathcal{S}\backslash\mathcal{S}_0$ and $\mathbf{z} \in \mathcal{S}^{\perp}$ are orthogonal. Clearly, $\mathbf{P}_{\mathcal{S}_0}\mathbf{y} = \mathbf{r}$ while $\mathbf{P}_{\mathcal{S}}\mathbf{y} = \mathbf{r} + \mathbf{s}$ and $\mathbf{P}_{\mathcal{S}_0}\mathbf{P}_{\mathcal{S}}\mathbf{y} = \mathbf{P}_{\mathcal{S}_0}(\mathbf{r}+\mathbf{s}) = \mathbf{r}$.

The remaining proofs are developed in Problem 1.9. ■

1.3.2 Implementation

For the linear regression model

$$\mathbf{Y}_{(T\times 1)} = \mathbf{X}_{(T\times k)}\boldsymbol{\beta}_{(k\times 1)} + \boldsymbol{\epsilon}_{(T\times 1)}, \tag{1.51}$$

with subscripts indicating the sizes and $\boldsymbol{\epsilon} \sim \mathrm{N}(\mathbf{0}, \sigma^2\mathbf{I}_T)$, we seek that $\widehat{\boldsymbol{\beta}}$ such that $\|\mathbf{Y} - \mathbf{X}\widehat{\boldsymbol{\beta}}\|^2$ is minimized. From Theorem 1.5, $\mathbf{X}\widehat{\boldsymbol{\beta}}$ is given by $\mathbf{P}_\mathbf{X}\mathbf{Y}$, where $\mathbf{P}_\mathbf{X} \equiv \mathbf{P}_{\mathcal{C}(\mathbf{X})}$ is an abbreviated notation for the projection matrix onto the space spanned by the columns of $\mathbf{X}$. We will assume that $\mathbf{X}$ is of full rank k, though this assumption can be relaxed in a more general treatment; see, e.g., Section 1.4.2.

If $\mathbf{X}$ happens to consist of k orthonormal column vectors, then $\mathbf{T} = \mathbf{X}$, where $\mathbf{T}$ is the orthonormal matrix given in (1.42), so that $\mathbf{P}_\mathbf{X} = \mathbf{T}\mathbf{T}'$. If (as usual), $\mathbf{X}$ is not orthonormal, with columns, say, $\mathbf{v}_1, \dots, \mathbf{v}_k$, then $\mathbf{T}$ could be constructed by applying the Gram–Schmidt procedure to $\mathbf{v}_1, \dots, \mathbf{v}_k$. Recall that, under our assumption that $\mathbf{X}$ is full rank, $\mathbf{v}_1, \dots, \mathbf{v}_k$ forms a basis (albeit not orthonormal) for $\mathcal{C}(\mathbf{X})$.

This can be more compactly expressed in the following way: From Theorem 1.1, vector $\mathbf{Y}$ can be decomposed as $\mathbf{Y} = \mathbf{P}_\mathbf{X}\mathbf{Y} + (\mathbf{I} - \mathbf{P}_\mathbf{X})\mathbf{Y}$, with $\mathbf{P}_\mathbf{X}\mathbf{Y} = \sum_{i=1}^k c_i\mathbf{v}_i$, where $\mathbf{c} = (c_1, \dots, c_k)'$ is the unique coefficient vector corresponding to the basis $\mathbf{v}_1, \dots, \mathbf{v}_k$ of $\mathcal{C}(\mathbf{X})$. Also from Theorem 1.1, $(\mathbf{I} - \mathbf{P}_\mathbf{X})\mathbf{Y}$ is perpendicular to $\mathcal{C}(\mathbf{X})$, i.e., $\langle(\mathbf{I} - \mathbf{P}_\mathbf{X})\mathbf{Y}, \mathbf{v}_i\rangle = 0$, $i = 1, \dots, k$. Thus,

$$\langle \mathbf{Y}, \mathbf{v}_j\rangle = \langle \mathbf{P}_\mathbf{X}\mathbf{Y} + (\mathbf{I} - \mathbf{P}_\mathbf{X})\mathbf{Y}, \mathbf{v}_j\rangle = \langle \mathbf{P}_\mathbf{X}\mathbf{Y}, \mathbf{v}_j\rangle = \left\langle \sum_{i=1}^k c_i\mathbf{v}_i, \mathbf{v}_j \right\rangle = \sum_{i=1}^k c_i\langle \mathbf{v}_i, \mathbf{v}_j\rangle,$$

$j = 1, \dots, k$, which can be written in matrix terms as

$$\begin{bmatrix} \langle \mathbf{Y}, \mathbf{v}_1\rangle \\ \langle \mathbf{Y}, \mathbf{v}_2\rangle \\ \vdots \\ \langle \mathbf{Y}, \mathbf{v}_k\rangle \end{bmatrix} = \begin{bmatrix} \langle \mathbf{v}_1, \mathbf{v}_1\rangle & \langle \mathbf{v}_1, \mathbf{v}_2\rangle & \cdots & \langle \mathbf{v}_1, \mathbf{v}_k\rangle \\ \langle \mathbf{v}_2, \mathbf{v}_1\rangle & \langle \mathbf{v}_2, \mathbf{v}_2\rangle & \cdots & \langle \mathbf{v}_2, \mathbf{v}_k\rangle \\ \vdots & \vdots & & \vdots \\ \langle \mathbf{v}_k, \mathbf{v}_1\rangle & \langle \mathbf{v}_k, \mathbf{v}_2\rangle & & \langle \mathbf{v}_k, \mathbf{v}_k\rangle \end{bmatrix} \begin{bmatrix} c_1 \\ c_2 \\ \vdots \\ c_k \end{bmatrix},$$

or, in terms of $\mathbf{X}$ and $\mathbf{c}$, as $\mathbf{X}'\mathbf{Y} = (\mathbf{X}'\mathbf{X})\mathbf{c}$. As $\mathbf{X}$ is full rank, so is $\mathbf{X}'\mathbf{X}$, showing that $\mathbf{c} = (\mathbf{X}'\mathbf{X})^{-1}\mathbf{X}'\mathbf{Y}$ is the coefficient vector for expressing $\mathbf{P}_\mathbf{X}\mathbf{Y}$ using the basis matrix $\mathbf{X}$. Thus, $\mathbf{P}_\mathbf{X}\mathbf{Y} = \mathbf{X}\mathbf{c} = \mathbf{X}(\mathbf{X}'\mathbf{X})^{-1}\mathbf{X}'\mathbf{Y}$, i.e.,

$$\mathbf{P}_\mathbf{X} = \mathbf{X}(\mathbf{X}'\mathbf{X})^{-1}\mathbf{X}'. \tag{1.52}$$

As $\mathbf{P}_\mathbf{X}\mathbf{Y}$ is unique from Theorem 1.1 (and from the full rank assumption on $\mathbf{X}$), it follows that the least squares estimator $\widehat{\boldsymbol{\beta}} = \mathbf{c}$. This agrees with the direct approach used in Section 1.2. Notice also that, if $\mathbf{X}$ is orthonormal, then $\mathbf{X}'\mathbf{X} = \mathbf{I}$ and $\mathbf{X}(\mathbf{X}'\mathbf{X})^{-1}\mathbf{X}'$ reduces to $\mathbf{X}\mathbf{X}'$, as in (1.42).

It is easy to see that $\mathbf{P}_\mathbf{X}$ is symmetric and idempotent, so that from Theorem 1.4 and the uniqueness of projection matrices (Problem 1.4), it is the projection matrix onto $\mathcal{S}$, the space spanned by its columns. To see that $\mathcal{S} = \mathcal{C}(\mathbf{X})$, we must show that, for all $\mathbf{Y} \in \mathbb{R}^T$, $\mathbf{P}_\mathbf{X}\mathbf{Y} \in \mathcal{C}(\mathbf{X})$ and $(\mathbf{I}_T - \mathbf{P}_\mathbf{X})\mathbf{Y} \perp \mathcal{C}(\mathbf{X})$. The former is easily verified by taking $\mathbf{b} = (\mathbf{X}'\mathbf{X})^{-1}\mathbf{X}'\mathbf{Y}$ in (1.38). The latter is equivalent to the statement that $(\mathbf{I}_T - \mathbf{P}_\mathbf{X})\mathbf{Y}$ is perpendicular to every column of $\mathbf{X}$. For this, defining the projection matrix

$$\mathbf{M} := \mathbf{I} - \mathbf{P}_\mathbf{X} = \mathbf{I}_T - \mathbf{X}(\mathbf{X}'\mathbf{X})^{-1}\mathbf{X}', \tag{1.53}$$

we have

$$\mathbf{X}'\mathbf{M}\mathbf{Y} = \mathbf{X}'(\mathbf{Y} - \mathbf{P}_\mathbf{X}\mathbf{Y}) = \mathbf{X}'\mathbf{Y} - \mathbf{X}'\mathbf{X}(\mathbf{X}'\mathbf{X})^{-1}\mathbf{X}'\mathbf{Y} = \mathbf{0}, \tag{1.54}$$

and the result is shown. Result (1.54) implies $\mathbf{M}\mathbf{X} = \mathbf{0}$. This follows from direct multiplication, but can also be seen as follows: Note that (1.54) holds for any $\mathbf{Y} \in \mathbb{R}^T$, and taking transposes yields $\mathbf{Y}'\mathbf{M}'\mathbf{X} = \mathbf{0}$, or, as $\mathbf{M}$ is symmetric, $\mathbf{M}\mathbf{X} = \mathbf{0}$.

Example 1.7 The method of Gram–Schmidt orthogonalization is quite naturally expressed in terms of projection matrices. Let $\mathbf{X}$ be a $T \times k$ matrix not necessarily of full rank, with columns $\mathbf{z}_1, \dots, \mathbf{z}_k$, $\mathbf{z}_1 \neq \mathbf{0}$. Define $\mathbf{w}_1 = \mathbf{z}_1/\|\mathbf{z}_1\|$ and

$$\mathbf{P}_1 = \mathbf{P}_{C(\mathbf{z}_1)} = \mathbf{P}_{C(\mathbf{w}_1)} = \mathbf{w}_1(\mathbf{w}_1'\mathbf{w}_1)^{-1}\mathbf{w}_1' = \mathbf{w}_1\mathbf{w}_1'.$$

Now let $\mathbf{r}_2 = (\mathbf{I} - \mathbf{P}_1)\mathbf{z}_2$, which is the component in $\mathbf{z}_2$ perpendicular to $\mathbf{z}_1$. If $\|\mathbf{r}_2\| > 0$, then set $\mathbf{w}_2 = \mathbf{r}_2/\|\mathbf{r}_2\|$ and $\mathbf{P}_2 = \mathbf{P}_{C(\mathbf{w}_1,\mathbf{w}_2)}$, otherwise set $\mathbf{w}_2 = \mathbf{0}$ and $\mathbf{P}_2 = \mathbf{P}_1$. This is then repeated for the remaining columns of $\mathbf{X}$. The matrix $\mathbf{W}$ with columns consisting of the j nonzero $\mathbf{w}_i$, $1 \leqslant j \leqslant k$, is then an orthonormal basis for $C(\mathbf{X})$. ∎

Example 1.8 Let $\mathbf{P}_\mathbf{X}$ be given in (1.52) with $\mathbf{1} \in C(\mathbf{X})$ and $\mathbf{P}_\mathbf{1} = \mathbf{1}\mathbf{1}'/T$ be the projection matrix onto $\mathbf{1}$, i.e., the line $(1, 1, \dots, 1)$ in $\mathbb{R}^T$. Then, from Theorem 1.6, $\mathbf{P}_\mathbf{X} - \mathbf{P}_\mathbf{1}$ is the projection matrix onto $C(\mathbf{X})\backslash C(\mathbf{1})$ and

$$\|(\mathbf{P}_\mathbf{X} - \mathbf{P}_\mathbf{1})\mathbf{Y}\|^2 = \|\mathbf{P}_\mathbf{X}\mathbf{Y}\|^2 - \|\mathbf{P}_\mathbf{1}\mathbf{Y}\|^2.$$

Also from Theorem 1.6, $\|\mathbf{P}_{\mathbf{X}\backslash\mathbf{1}}\mathbf{Y}\|^2 = \|\mathbf{P}_{\mathbf{1}^\perp\backslash\mathbf{X}^\perp}\mathbf{Y}\|^2 = \|\mathbf{P}_{\mathbf{1}^\perp}\mathbf{Y}\|^2 - \|\mathbf{P}_{\mathbf{X}^\perp}\mathbf{Y}\|^2$. As

$$\begin{aligned}
\|\mathbf{P}_{\mathbf{X}\backslash\mathbf{1}}\mathbf{Y}\|^2 &= \|(\mathbf{P}_\mathbf{X} - \mathbf{P}_\mathbf{1})\mathbf{Y}\|^2 = \sum(\hat{Y} - \bar{Y})^2,\\
\|\mathbf{P}_{\mathbf{1}^\perp}\mathbf{Y}\|^2 &= \|(\mathbf{I} - \mathbf{P}_\mathbf{1})\mathbf{Y}\|^2 = \sum(Y_t - \bar{Y})^2,\\
\|\mathbf{P}_{\mathbf{X}^\perp}\mathbf{Y}\|^2 &= \|(\mathbf{I} - \mathbf{P}_\mathbf{X})\mathbf{Y}\|^2 = \sum(Y_t - \hat{Y})^2,
\end{aligned}$$

we see that

$$\sum_{t=1}^{T}(Y_t - \bar{Y})^2 = \sum_{t=1}^{T}(Y_t - \hat{Y})^2 + \sum_{t=1}^{T}(\hat{Y} - \bar{Y})^2, \tag{1.55}$$

proving (1.12). ∎

Often it will be of interest to work with the estimated residuals of the regression (1.51), namely

$$\hat{\boldsymbol{\epsilon}} := \mathbf{Y} - \mathbf{X}\hat{\boldsymbol{\beta}} = (\mathbf{I}_T - \mathbf{P}_\mathbf{X})\mathbf{Y} = \mathbf{M}\mathbf{Y} = \mathbf{M}(\mathbf{X}\boldsymbol{\beta} + \boldsymbol{\epsilon}) = \mathbf{M}\boldsymbol{\epsilon}, \tag{1.56}$$

where $\mathbf{M}$ is the projection matrix onto the orthogonal complement of $\mathbf{X}$, given in (1.53), and the last equality in (1.56) follows because $\mathbf{M}\mathbf{X} = \mathbf{0}$, confirmed by direct multiplication or as shown in (1.54). From (1.4) and (1.56), the RSS can be expressed as

$$\text{RSS} = \mathsf{S}(\hat{\boldsymbol{\beta}}) = \hat{\boldsymbol{\epsilon}}'\hat{\boldsymbol{\epsilon}} = (\mathbf{M}\mathbf{Y})'\mathbf{M}\mathbf{Y} = \mathbf{Y}'\mathbf{M}\mathbf{Y} = \mathbf{Y}'(\mathbf{I} - \mathbf{P}_\mathbf{X})\mathbf{Y}. \tag{1.57}$$

Example 1.9 ***Example 1.1, the Frisch–Waugh–Lovell Theorem, cont.***
From the symmetry and idempotency of $\mathbf{M}_1$, the expression in (1.21) can also also be written as

$$\begin{aligned}
\hat{\boldsymbol{\beta}}_2 &= (\mathbf{X}_2'\mathbf{M}_1\mathbf{X}_2)^{-1}\mathbf{X}_2'\mathbf{M}_1\mathbf{Y} = (\mathbf{X}_2'\mathbf{M}_1'\mathbf{M}_1\mathbf{X}_2)^{-1}\mathbf{X}_2'\mathbf{M}_1'\mathbf{M}_1\mathbf{Y}\\
&= (\mathbf{Q}'\mathbf{Q})^{-1}\mathbf{Q}'\mathbf{Z},
\end{aligned}$$

where $\mathbf{Q} = \mathbf{M}_1\mathbf{X}_2$ and $\mathbf{Z} = \mathbf{M}_1\mathbf{Y}$. That is, $\hat{\boldsymbol{\beta}}_2$ can be computed *not* by regressing $\mathbf{Y}$ onto $\mathbf{X}_2$, but by regressing *the residuals of* $\mathbf{Y}$ onto *the residuals of* $\mathbf{X}_2$, where residuals refers to having removed the component spanned by $\mathbf{X}_1$. If $\mathbf{X}_1$ and $\mathbf{X}_2$ are orthogonal, then

$$\mathbf{Q} = \mathbf{M}_1\mathbf{X}_2 = \mathbf{X}_2 - \mathbf{X}_1(\mathbf{X}_1'\mathbf{X}_1)^{-1}\mathbf{X}_1'\mathbf{X}_2 = \mathbf{X}_2,$$

and, with $\mathbf{I} = \mathbf{M}_1 + \mathbf{P}_1$,

$$\begin{aligned}(\mathbf{X}_2'\mathbf{X}_2)^{-1}\mathbf{X}_2'\mathbf{Y} &= (\mathbf{X}_2'\mathbf{X}_2)^{-1}\mathbf{X}_2'(\mathbf{M}_1 + \mathbf{P}_1)\mathbf{Y} \\ &= (\mathbf{X}_2'\mathbf{X}_2)^{-1}\mathbf{X}_2'\mathbf{M}_1\mathbf{Y} = (\mathbf{Q}'\mathbf{Q})^{-1}\mathbf{Q}'\mathbf{Z},\end{aligned}$$

so that, under orthogonality, $\widehat{\boldsymbol{\beta}}_2$ can indeed be obtained by regressing $\mathbf{Y}$ onto $\mathbf{X}_2$. ■

It is clear that $\mathbf{M}$ should have rank $T - k$, or $T - k$ eigenvalues equal to one and k equal to zero. We can thus express $\hat{\sigma}^2$ given in (1.11) as

$$\hat{\sigma}^2 = \frac{S(\widehat{\boldsymbol{\beta}})}{T-k} = \frac{(\mathbf{MY})'\mathbf{MY}}{T-k} = \frac{\mathbf{Y}'\mathbf{MY}}{\text{rank}(\mathbf{M})} = \frac{\mathbf{Y}'(\mathbf{I}-\mathbf{P}_{\mathbf{X}})\mathbf{Y}}{\text{rank}(\mathbf{I}-\mathbf{P}_{\mathbf{X}})}. \tag{1.58}$$

Observe also that $\boldsymbol{\epsilon}'\mathbf{M}\boldsymbol{\epsilon} = \mathbf{Y}'\mathbf{MY}$.

It is now quite easy to show that $\hat{\sigma}^2$ is unbiased. Using properties of the trace operator and the fact $\mathbf{M}$ is a projection matrix (i.e., $\mathbf{M}'\mathbf{M} = \mathbf{MM} = \mathbf{M}$),

$$\begin{aligned}\mathbb{E}[\widehat{\boldsymbol{\epsilon}}'\widehat{\boldsymbol{\epsilon}}] &= \mathbb{E}[\boldsymbol{\epsilon}'\mathbf{M}'\mathbf{M}\boldsymbol{\epsilon}] = \mathbb{E}[\boldsymbol{\epsilon}'\mathbf{M}\boldsymbol{\epsilon}] = \text{tr}(\mathbb{E}[\boldsymbol{\epsilon}'\mathbf{M}\boldsymbol{\epsilon}]) = \mathbb{E}[\text{tr}(\boldsymbol{\epsilon}'\mathbf{M}\boldsymbol{\epsilon})] \\ &= \mathbb{E}[\text{tr}(\mathbf{M}\boldsymbol{\epsilon}\boldsymbol{\epsilon}')] = \text{tr}(\mathbf{M}\mathbb{E}[\boldsymbol{\epsilon}\boldsymbol{\epsilon}']) = \sigma^2\text{tr}(\mathbf{M}) = \sigma^2\text{rank}(\mathbf{M}) = \sigma^2(T-k),\end{aligned}$$

where the fact that $\text{tr}(\mathbf{M}) = \text{rank}(\mathbf{M})$ follows from Theorem 1.2. In fact, a similar derivation was used to obtain the general result (A.6), from which it directly follows that

$$\mathbb{E}[\boldsymbol{\epsilon}'\mathbf{M}\boldsymbol{\epsilon}] = \text{tr}(\sigma^2\mathbf{M}) + \mathbf{0}'\mathbf{M}\mathbf{0} = \sigma^2(T-k). \tag{1.59}$$

Theorem A.3 shows that, if $\mathbf{Y} \sim \text{N}(\boldsymbol{\mu}, \boldsymbol{\Sigma})$ with $\boldsymbol{\Sigma} > 0$, then the vector $\mathbf{CY}$ is independent of the quadratic form $\mathbf{Y}'\mathbf{AY}$ if $\mathbf{C}\boldsymbol{\Sigma}\mathbf{A} = 0$. Using this with $\boldsymbol{\Sigma} = \mathbf{I}$, $\mathbf{C} = \mathbf{P}$ and $\mathbf{A} = \mathbf{M} = \mathbf{I} - \mathbf{P}$, it follows that $\mathbf{X}\widehat{\boldsymbol{\beta}} = \mathbf{PY}$ and $(T-k)\hat{\sigma}^2 = \mathbf{Y}'\mathbf{MY}$ are independent. That is:

> Under the usual regression model assumptions (including that $\mathbf{X}$ is not stochastic, or is such that the model is variation-free), point estimators $\widehat{\boldsymbol{\beta}}$ and $\hat{\sigma}^2$ are independent.

This generalizes the well-known result in the i.i.d. case: Specifically, if $\mathbf{X}$ is just a column of ones, then $\mathbf{PY} = T^{-1}\mathbf{11}'\mathbf{Y} = (\bar{Y}, \bar{Y}, \dots, \bar{Y})'$ and $\mathbf{Y}'\mathbf{MY} = \mathbf{Y}'\mathbf{M}'\mathbf{MY} = \sum_{t=1}^{T}(Y_t - \bar{Y})^2 = (T-1)S^2$, so that $\bar{Y}$ and S^2 are independent.

As $\widehat{\boldsymbol{\epsilon}} = \mathbf{M}\boldsymbol{\epsilon}$ is a linear transformation of the normal random vector $\boldsymbol{\epsilon}$,

$$(\widehat{\boldsymbol{\epsilon}} \mid \sigma^2) \sim \text{N}(\mathbf{0}, \sigma^2\mathbf{M}), \tag{1.60}$$

though note that $\mathbf{M}$ is rank deficient (i.e., is less than full rank), with rank $T - k$, so that this is a degenerate normal distribution. In particular, by definition, $\widehat{\boldsymbol{\epsilon}}$ is in the column space of $\mathbf{M}$, so that $\widehat{\boldsymbol{\epsilon}}$ must be perpendicular to the column space of $\mathbf{X}$, or

$$\widehat{\boldsymbol{\epsilon}}'\mathbf{X} = \mathbf{0}. \tag{1.61}$$

If, as usual, $\mathbf{X}$ contains a column of ones, denoted $\mathbf{1}_T$, or, more generally, $\mathbf{1}_T \in \mathcal{C}(\mathbf{X})$, then (1.61) implies that $\sum_{t=1}^{T}\hat{\epsilon}_t = 0$.

We now turn to the generalized least squares case, with the model given by (1.3) and (1.24), and estimator (1.28). In this more general setting when $\boldsymbol{\epsilon} \sim \text{N}(\mathbf{0}, \sigma^2\boldsymbol{\Sigma})$, the residual vector is given by

$$\widehat{\boldsymbol{\epsilon}} = \mathbf{Y} - \mathbf{X}\widehat{\boldsymbol{\beta}}_{\boldsymbol{\Sigma}} = \mathbf{M}_{\boldsymbol{\Sigma}}\mathbf{Y}, \tag{1.62}$$

where $\mathbf{M}_{\boldsymbol{\Sigma}} = \mathbf{I}_T - \mathbf{X}(\mathbf{X}'\boldsymbol{\Sigma}^{-1}\mathbf{X})^{-1}\mathbf{X}'\boldsymbol{\Sigma}^{-1}$. Although $\mathbf{M}_{\boldsymbol{\Sigma}}$ is idempotent, it is not symmetric, and cannot be referred to as a projection matrix. Observe also that the estimated residual vector is no longer orthogonal to the columns of $\mathbf{X}$. Instead we have

$$\mathbf{X}'\boldsymbol{\Sigma}^{-1}(\mathbf{Y} - \mathbf{X}\widehat{\boldsymbol{\beta}}_{\boldsymbol{\Sigma}}) = \mathbf{0}, \tag{1.63}$$

so that the residuals do not necessarily sum to zero.

We now state a result from matrix algebra, and then use it to prove a theorem that will be useful for some hypothesis testing situations in Chapter 5.

Theorem 1.7 Let $\mathbf{V}$ be an $n \times n$ positive definite matrix, and let $\mathbf{U}$ and $\mathbf{T}$ be $n \times k$ and $n \times (n-k)$ matrices, respectively, such that, if $\mathbf{W} = [\mathbf{U}, \mathbf{T}]$, then $\mathbf{W}'\mathbf{W} = \mathbf{W}\mathbf{W}' = \mathbf{I}_n$. Then

$$\mathbf{V}^{-1} - \mathbf{V}^{-1}\mathbf{U}(\mathbf{U}'\mathbf{V}^{-1}\mathbf{U})^{-1}\mathbf{U}'\mathbf{V}^{-1} = \mathbf{T}(\mathbf{T}'\mathbf{V}\mathbf{T})^{-1}\mathbf{T}'. \tag{1.64}$$

Proof: See Rao (1973, p. 77). ■

Let $\mathbf{P} = \mathbf{P}_{\mathbf{X}}$ be the usual projection matrix on the column space of $\mathbf{X}$ from (1.52), let $\mathbf{M} = \mathbf{I}_T - \mathbf{P}$, and let $\mathbf{G}$ and $\mathbf{H}$ be matrices such that $\mathbf{M} = \mathbf{G}'\mathbf{G}$ and $\mathbf{P} = \mathbf{H}'\mathbf{H}$, in which case $\mathbf{W} = [\mathbf{H}', \mathbf{G}']$ satisfies $\mathbf{W}'\mathbf{W} = \mathbf{W}\mathbf{W}' = \mathbf{I}_T$.

Theorem 1.8 For the regression model given by (1.3) and (1.24), with $\widehat{\boldsymbol{\epsilon}} = \mathbf{M}_{\boldsymbol{\Sigma}}\mathbf{Y}$ from (1.62),

$$\widehat{\boldsymbol{\epsilon}}'\boldsymbol{\Sigma}^{-1}\widehat{\boldsymbol{\epsilon}} = \boldsymbol{\epsilon}'\mathbf{G}'(\mathbf{G}\boldsymbol{\Sigma}\mathbf{G}')^{-1}\mathbf{G}\boldsymbol{\epsilon}. \tag{1.65}$$

Proof: As in King (1980, p. 1268), using Theorem 1.7 with $\mathbf{T} = \mathbf{G}'$, $\mathbf{U} = \mathbf{H}'$, and $\mathbf{V} = \boldsymbol{\Sigma}$, and the fact that $\mathbf{H}'$ can be written as $\mathbf{X}\mathbf{K}$, where $\mathbf{K}$ is a $k \times k$ full rank transformation matrix, we have

$$\begin{aligned}
\boldsymbol{\epsilon}'\mathbf{G}'(\mathbf{G}\boldsymbol{\Sigma}\mathbf{G}')^{-1}\mathbf{G}\boldsymbol{\epsilon} &= \mathbf{U}'(\boldsymbol{\Sigma}^{-1} - \boldsymbol{\Sigma}^{-1}\mathbf{H}'(\mathbf{H}\boldsymbol{\Sigma}^{-1}\mathbf{H}')^{-1}\mathbf{H}\boldsymbol{\Sigma}^{-1})\mathbf{U} \\
&= \mathbf{U}'(\boldsymbol{\Sigma}^{-1} - \boldsymbol{\Sigma}^{-1}\mathbf{X}\mathbf{K}(\mathbf{K}'\mathbf{X}'\boldsymbol{\Sigma}^{-1}\mathbf{X}\mathbf{K})^{-1}\mathbf{K}'\mathbf{X}'\boldsymbol{\Sigma}^{-1})\mathbf{U} \\
&= \mathbf{U}'(\boldsymbol{\Sigma}^{-1} - \boldsymbol{\Sigma}^{-1}\mathbf{X}(\mathbf{X}'\boldsymbol{\Sigma}^{-1}\mathbf{X})^{-1}\mathbf{X}'\boldsymbol{\Sigma}^{-1})\mathbf{U} = \widehat{\boldsymbol{\epsilon}}'\boldsymbol{\Sigma}^{-1}\widehat{\boldsymbol{\epsilon}},
\end{aligned}$$

which is (1.65). ■

1.4 Linear Parameter Restrictions

> [D]eleting a small unimportant parameter from the model is generally a good idea, because we will incur a small bias but may gain much precision. This is true even if the estimated parameter happens to be highly 'significant', that is, have a large t-ratio. Significance indicates that we have managed to estimate the parameter rather precisely, possibly because we have many observations. It does not mean that the parameter is important.
>
> (Jan R. Magnus, 2017, p. 30)

In much applied regression analysis, the analyst will wish to know the extent to which certain linear restrictions on $\boldsymbol{\beta}$ hold. As the quote above by Magnus (2017) suggests, we recommend doing so via means more related to the purpose of the research, e.g., forecasting, and, particularly, in applications in the social sciences for which the notion of repeatability of the experiment does not apply, being aware of the pitfalls of the classic significance testing (use of p-values) and Neyman–Pearson hypothesis testing paradigm. This issue was discussed in some detail in Section III.2.8, where strong arguments were raised, and evidence presented, that significance and hypothesis testing might one day make it to the ash heap of statistical history. In addition to the numerous references provided in Section III.2.8, such as Ioannidis (2005), the interested reader is encouraged to read Ioannidis (2014), and a rebuttal to that paper in Leek and Jager (2017), as well as the very pertinent overview in Spiegelhalter (2017), addressing this issue and the more general theme of trustworthiness in statistical reports, amid concerns of reproducibility, fake news, and alternative facts.

1.4.1 Formulation and Estimation

A common goal in regression analysis is to test is whether an individual regression coefficient is "significantly" different than a given value, often zero. More general tests might involve testing whether the sum of certain coefficients is a particular value, or testing for the equality of two or more coefficients. These are all special cases of a general linear test that can be expressed as (regrettably with many Hs, but following standard terminology)

$$H_0 : \mathbf{H}\boldsymbol{\beta} = \mathbf{h}, \tag{1.66}$$

versus the alternative, H_1, corresponding to the unrestricted model. The matrix $\mathbf{H}$ is of dimension $J \times k$ and, without loss of generality, assumed to be of full rank J, so that $J \leqslant k$ and $\mathbf{h}$ is $J \times 1$. The null hypothesis can also be written

$$H_0 : \mathbf{Y} = \mathbf{X}\boldsymbol{\gamma} + \boldsymbol{\epsilon}, \quad \mathbf{X}\boldsymbol{\gamma} \in \mathcal{S}_H, \tag{1.67}$$

where

$$\mathcal{S}_H = \{\mathbf{z} : \mathbf{z} = \mathbf{X}\boldsymbol{\beta},\ \mathbf{H}\boldsymbol{\beta} = \mathbf{h},\ \boldsymbol{\beta} \in \mathbb{R}^k\}. \tag{1.68}$$

If $\mathbf{h} \neq \mathbf{0}$, then $\mathcal{S}_H$ is an **affine subspace** because it does not contain the zero element (provided both $\mathbf{X}$ and $\mathbf{H}$ are full rank, as is assumed).

As an important illustration, for testing if the last J regressors are not significant, i.e., if $\beta_{k-J+1} = \cdots = \beta_k = 0$, set $\mathbf{h} = \mathbf{0}$ and $\mathbf{H} = [\mathbf{0}_{J\times k-J} \mid \mathbf{I}_J]$. For example, if $k = 6$ and $J = 2$, then

$$\mathbf{H} = \begin{pmatrix} 0 & 0 & 0 & 0 & 1 & 0 \\ 0 & 0 & 0 & 0 & 0 & 1 \end{pmatrix}.$$

We next consider how $\boldsymbol{\gamma}$ in (1.67) can be estimated, followed by the distribution theory associated with the formal frequentist testing framework of the null hypothesis for assessing whether or not the data are in agreement with the proposed set of restrictions.

In many cases of interest, the reduced column space is easily identified. For example, if a set of coefficients are taken to be zero, then the nonzero elements of $\hat{\boldsymbol{\gamma}}$ are found by computing the o.l.s. estimator using an $\mathbf{X}$ matrix with the appropriate columns removed. In general, however, it will not always be clear how to identify the reduced column space, so that a more general method will be required. Theorem 1.9 gives a nonconstructive proof, i.e., we state the result and confirm it satisfies the requirements. We subsequently show two constructive proofs.

Theorem 1.9 Assuming $\mathbf{H}$ and $\mathbf{X}$ are full rank, the least squares estimator of $\boldsymbol{\gamma}$ in (1.67) is given by

$$\hat{\boldsymbol{\gamma}} = \hat{\boldsymbol{\beta}} + \mathbf{A}\mathbf{H}'[\mathbf{H}\mathbf{A}\mathbf{H}']^{-1}(\mathbf{h} - \mathbf{H}\hat{\boldsymbol{\beta}}), \tag{1.69}$$

where $\mathbf{A} = (\mathbf{X}'\mathbf{X})^{-1}$.

Proof: By definition, we require that $\hat{\boldsymbol{\gamma}}$ is the least squares estimator subject to the linear constraint. Thus, the proof entails showing that (1.69) satisfies the following two conditions:

1) $\mathbf{H}\hat{\boldsymbol{\gamma}} = \mathbf{h}$ and
2) $\|\mathbf{Y} - \mathbf{X}\hat{\boldsymbol{\gamma}}\|^2 \leqslant \|\mathbf{Y} - \mathbf{X}\mathbf{b}\|^2$ for all $\mathbf{b} \in \mathbb{R}^k$ such that $\mathbf{H}\mathbf{b} = \mathbf{h}$.

This is straightforward and detailed in Problem 1.6. ■

We will refer to $\hat{\boldsymbol{\gamma}}$ in (1.69) as the **restricted least squares**, or r.l.s., estimator. It can be derived in several ways, two important ones of which are now shown. A third way, using projection, is also straightforward and instructive; see, e.g., Ravishanker and Dey (2002, Sec. 4.6.2) or Seber and Lee (2003, p. 61).

Derivation of (1.69) Method I: This method makes use of the results for the generalized least squares estimator and does not explicitly require the use of calculus. We will need the following well-known matrix result: If matrices $\mathbf{A}, \mathbf{B}$ and $\mathbf{D}$ are such that $\mathbf{A} + \mathbf{B}\mathbf{D}\mathbf{B}'$ is a square matrix of full rank, then

$$(\mathbf{A} + \mathbf{B}\mathbf{D}\mathbf{B}')^{-1} = \mathbf{A}^{-1} - \mathbf{A}^{-1}\mathbf{B}(\mathbf{B}'\mathbf{A}^{-1}\mathbf{B} + \mathbf{D}^{-1})^{-1}\mathbf{B}'\mathbf{A}^{-1}. \tag{1.70}$$

See, e.g., Abadir and Magnus (2005, p. 107) for proof of the more general case of $(\mathbf{A} + \mathbf{B}\mathbf{D}\mathbf{C}')^{-1}$.

Let (uncharacteristically, using a lower case letter) $\mathbf{v}$ be a vector random variable with mean $\mathbf{0}$ and finite covariance matrix $\sigma_{\mathbf{v}}^2\mathbf{V}$, denoted $\mathbf{v} \sim (\mathbf{0}, \sigma_{\mathbf{v}}^2\mathbf{V})$. The constraint in (1.66) can be understood as the limiting case, as $\sigma_{\mathbf{v}}^2 \to 0$, of the *stochastic* set of extraneous information equations on $\boldsymbol{\beta}$,

$$\mathbf{H}\boldsymbol{\beta} + \mathbf{v} = \mathbf{h}. \tag{1.71}$$

The regression model $\mathbf{Y} = \mathbf{X}\boldsymbol{\beta} + \boldsymbol{\epsilon}$, $\mathbb{V}(\boldsymbol{\epsilon}) = \sigma^2\mathbf{I}_T$, can be combined with (1.71) via the so-called **mixed model** of Theil and Goldberger (1961) to give

$$\begin{pmatrix} \mathbf{Y} \\ \mathbf{h} \end{pmatrix} = \begin{pmatrix} \mathbf{X} \\ \mathbf{H} \end{pmatrix} \boldsymbol{\beta} + \begin{pmatrix} \boldsymbol{\epsilon} \\ \mathbf{v} \end{pmatrix}.$$

This can be expressed more compactly as

$$\mathbf{Y}_m = \mathbf{X}_m\boldsymbol{\beta}_m + \boldsymbol{\epsilon}_m, \quad \boldsymbol{\epsilon}_m \sim (\mathbf{0}, \boldsymbol{\Sigma}_m), \quad \boldsymbol{\Sigma}_m = \begin{pmatrix} \sigma^2\mathbf{I}_T & \mathbf{0} \\ \mathbf{0} & \sigma_{\mathbf{v}}^2\mathbf{V} \end{pmatrix},$$

where the subscript m denotes "mixed". Using generalized least squares,

$$\begin{aligned}
\hat{\boldsymbol{\beta}}_m &= (\mathbf{X}_m'\boldsymbol{\Sigma}_m^{-1}\mathbf{X}_m)^{-1}\mathbf{X}_m'\boldsymbol{\Sigma}_m^{-1}\mathbf{Y}_m \\
&= (\sigma^{-2}\mathbf{X}'\mathbf{X} + \sigma_{\mathbf{v}}^{-2}\mathbf{H}'\mathbf{V}^{-1}\mathbf{H})^{-1}(\sigma^{-2}\mathbf{X}'\mathbf{Y} + \sigma_{\mathbf{v}}^{-2}\mathbf{H}'\mathbf{V}^{-1}\mathbf{h}) \\
&= (\mathbf{X}'\mathbf{X} + \lambda\mathbf{H}'\mathbf{V}^{-1}\mathbf{H})^{-1}(\mathbf{X}'\mathbf{Y} + \lambda\mathbf{H}'\mathbf{V}^{-1}\mathbf{h}),
\end{aligned}$$

where $\lambda := \sigma^2/\sigma_{\mathrm{v}}^2$. Next, following Alvarez and Dolado (1994), use (1.70) with

$$\mathbf{A} := (\mathbf{X}'\mathbf{X})^{-1} \quad \text{and} \quad \mathbf{C}_\lambda := \mathbf{A}\mathbf{H}'(\mathbf{H}\mathbf{A}\mathbf{H}' + \lambda^{-1}\mathbf{V})^{-1}$$

to get

$$\begin{aligned}
\widehat{\boldsymbol{\beta}}_m &= [\mathbf{A} - \mathbf{C}_\lambda \mathbf{H}\mathbf{A}](\mathbf{X}'\mathbf{Y} + \mathbf{H}'(\lambda^{-1}\mathbf{V})^{-1}\mathbf{h}) \\
&= \mathbf{A}\mathbf{X}'\mathbf{Y} + \mathbf{A}\mathbf{H}'(\lambda^{-1}\mathbf{V})^{-1}\mathbf{h} - \mathbf{C}_\lambda\mathbf{H}\mathbf{A}\mathbf{X}'\mathbf{Y} - \mathbf{C}_\lambda\mathbf{H}\mathbf{A}\mathbf{H}'(\lambda^{-1}\mathbf{V})^{-1}\mathbf{h} \\
&= \widehat{\boldsymbol{\beta}} + \mathbf{C}_\lambda(\mathbf{H}\mathbf{A}\mathbf{H}' + \lambda^{-1}\mathbf{V})(\lambda^{-1}\mathbf{V})^{-1}\mathbf{h} - \mathbf{C}_\lambda\mathbf{H}\widehat{\boldsymbol{\beta}} - \mathbf{C}_\lambda\mathbf{H}\mathbf{A}\mathbf{H}'(\lambda^{-1}\mathbf{V})^{-1}\mathbf{h} \\
&= \widehat{\boldsymbol{\beta}} + \mathbf{C}_\lambda[\mathbf{H}\mathbf{A}\mathbf{H}'(\lambda^{-1}\mathbf{V})^{-1}\mathbf{h} + \mathbf{h} - \mathbf{H}\widehat{\boldsymbol{\beta}} - \mathbf{H}\mathbf{A}\mathbf{H}'(\lambda^{-1}\mathbf{V})^{-1}\mathbf{h}] \\
&= \widehat{\boldsymbol{\beta}} + \mathbf{C}_\lambda(\mathbf{h} - \mathbf{H}\widehat{\boldsymbol{\beta}}),
\end{aligned}$$

where $\widehat{\boldsymbol{\beta}}$ is the unrestricted least squares estimator. Letting $\sigma_{\mathrm{v}}^2 \to 0$ gives (1.69). Note that the inverse of $\mathbf{H}\mathbf{A}\mathbf{H}'$ exists because both $\mathbf{H}$ and $\mathbf{X}$ (and thus $\mathbf{A}$) are full rank. ∎

Remark The mixed model structure is useful in several regression modeling contexts, and is related to formal Bayesian methods, whereby model parameters are treated as random variables, though not requiring Bayesian methodology. For example, as stated by Lee and Griffiths (1979, pp. 4–5), "Thus, for stochastic prior information of the form given in [(1.71)], the mixed estimation procedure is more efficient, is distribution free, and does not involve a Bayesian argument."

It also provides the most straightforward derivation of the so-called Black–Litterman model for incorporating viewpoints into a statistical model for financial portfolio allocation; see, e.g., Kolm et al. (2008, p. 362), as well as Black and Litterman (1992), Meucci (2006), Giacometti et al. (2007), Brandt (2010, p. 313), and the references therein. ∎

Derivation of (1.69) Method II: The calculus technique of Lagrange multipliers is applicable in this setting.[6] Besides being of interest in itself for deriving $\widehat{\boldsymbol{\gamma}}$, we will subsequently need equation (1.72) derived along the way, in Section 1.4.2.

The method implies that the $k + J$ constraints

$$\frac{\partial}{\partial \widehat{\gamma}_i}\{\|\mathbf{Y} - \mathbf{X}\widehat{\boldsymbol{\gamma}}\|^2 + \boldsymbol{\lambda}'(\mathbf{H}\widehat{\boldsymbol{\gamma}} - \mathbf{h})\} = 0, \quad i = 1, \dots, k,$$
$$\mathbf{H}\widehat{\boldsymbol{\gamma}} - \mathbf{h} = \mathbf{0},$$

must be satisfied, where $\boldsymbol{\lambda} = (\lambda_1, \dots, \lambda_J)'$. The ith equation, $i = 1, \dots, k$, is easily seen to be

$$2\sum_{t=1}^{T}(Y_t - \mathbf{x}_t'\widehat{\boldsymbol{\gamma}})(-x_{it}) + (\text{the } i\text{th component of } \mathbf{H}'\boldsymbol{\lambda}) = 0,$$

so that the first k equations can be written together as $-2\mathbf{X}'(\mathbf{Y} - \mathbf{X}\widehat{\boldsymbol{\gamma}}) + \mathbf{H}'\boldsymbol{\lambda} = \mathbf{0}$. These, in turn, can be expressed together with constraint $\mathbf{H}\widehat{\boldsymbol{\gamma}} = \mathbf{h}$ as

$$\begin{bmatrix} 2\mathbf{X}'\mathbf{X} & \mathbf{H}' \\ \mathbf{H} & \mathbf{0} \end{bmatrix}\begin{bmatrix} \widehat{\boldsymbol{\gamma}} \\ \boldsymbol{\lambda} \end{bmatrix} = \begin{bmatrix} 2\mathbf{X}'\mathbf{Y} \\ \mathbf{h} \end{bmatrix}, \tag{1.72}$$

6 A particularly lucid discussion of Lagrange multipliers is provided by Hubbard and Hubbard (2002, Sec. 3.7).

```
X= [1 1 0 ; 1 1 0; 1 0 1; 1 0 1]; ell = [1 0 1];
kappaPRIME = pinv(X') * ell' % try to solve
% now check:
disc = ell' - X' * kappaPRIME; check = sum(abs(disc)) % should be zero if estimable
```

Program Listing 1.3: Attempts to solve $\boldsymbol{\ell}' = \mathbf{X}'\boldsymbol{\kappa}'$ for $\boldsymbol{\kappa}$ via use of the generalized inverse.

Clearly, $\mathbf{X}$ in (1.75) is rank deficient, with rank($\mathbf{X}$) = 2, also seen by deleting all redundant rows, to give

$$\mathbf{X}^* = \begin{bmatrix} 1 & 1 & 0 \\ 1 & 0 & 1 \end{bmatrix},$$

which is (full) rank 2. From (1.74),

$$\text{rank}\left(\begin{bmatrix} \mathbf{H} \\ \mathbf{X} \end{bmatrix}\right) = \text{rank}\left(\begin{bmatrix} \mathbf{H} \\ \mathbf{X}^* \end{bmatrix}\right) = \text{rank}\left(\begin{bmatrix} 0 & 1 & 1 \\ 1 & 1 & 0 \\ 1 & 0 & 1 \end{bmatrix}\right) = 3 = k,$$

so that estimator $\widehat{\boldsymbol{\gamma}}$ is unique, also seen from

$$\mathbf{W} = \begin{bmatrix} 2n & n & n & 0 \\ n & n & 0 & 1 \\ n & 0 & n & 1 \\ 0 & 1 & 1 & 0 \end{bmatrix},$$

which is (full) rank $k + J = 4$.

Without constraints on $\boldsymbol{\beta}$, for $\boldsymbol{\ell} = (1,\ 1,\ 1)$ and $\boldsymbol{\ell} = (0,\ 1,\ 1)$, $\boldsymbol{\ell}\boldsymbol{\beta}$ is not estimable because $\nexists\boldsymbol{\kappa}$ such that $\boldsymbol{\ell}' = \mathbf{X}'\boldsymbol{\kappa}'$, which the reader should confirm, and also should make intuitive sense. Likewise, $\boldsymbol{\ell}\boldsymbol{\beta}$ is estimable for $\boldsymbol{\ell} = (1,\ 0,\ 1)$ and $\boldsymbol{\ell} = (1,\ 1,\ 0)$ (both of which form the two unique rows of $\mathbf{X}$). These results can be checked using Matlab with the code given in Listing 1.3, taking $n = 2$. For example, running it with $\boldsymbol{\ell} = (1,\ 0,\ 1)$ yields solution $\boldsymbol{\kappa} = (0,\ 0,\ 1/2,\ 1/2)$. Inspection shows another solution to be $(1/2, -1/2,\ 1/2,\ 1/2)$, emphasizing that $\boldsymbol{\kappa}$ need not be unique, only that $\boldsymbol{\ell}' \in \mathcal{C}(\mathbf{X}')$.

A good discussion of estimability (and also its connection to their software) is provided in SAS/STAT 9.2 User's Guide (2008, Ch. 15), from which our notation was inspired (they use $\mathbf{L}$ and $\mathbf{K}$ in place of our $\boldsymbol{\ell}$ and $\boldsymbol{\kappa}$).

1.4.3 Moments and the Restricted GLS Estimator

Derivation of the first two moments of $\widehat{\boldsymbol{\gamma}}$ is straightforward: As $\widehat{\boldsymbol{\beta}}$ is unbiased, (1.69) implies

$$\mathbb{E}[\widehat{\boldsymbol{\gamma}}] = \boldsymbol{\beta} + \mathbf{AH}'(\mathbf{HAH}')^{-1}(\mathbf{h} - \mathbf{H}\boldsymbol{\beta}), \tag{1.76}$$

where, as usual, $\mathbf{A} = (\mathbf{X}'\mathbf{X})^{-1}$. It is then easy to verify that $\widehat{\boldsymbol{\gamma}} - \mathbb{E}[\widehat{\boldsymbol{\gamma}}] = (\mathbf{I} - \mathbf{B})(\widehat{\boldsymbol{\beta}} - \boldsymbol{\beta})$, where $\mathbf{B} = \mathbf{AH}'(\mathbf{HAH}')^{-1}\mathbf{H}$, and

$$(\mathbf{I} - \mathbf{B})\mathbf{A}(\mathbf{I} - \mathbf{B}') = \mathbf{A} - \mathbf{BA} - \mathbf{AB}' + \mathbf{BAB}' = \mathbf{A} - \mathbf{BA},$$

so that

$$\mathbb{V}(\hat{\gamma} \mid \sigma^2) = \mathbb{E}[(\hat{\gamma} - \mathbb{E}[\hat{\gamma}])(\hat{\gamma} - \mathbb{E}[\hat{\gamma}])' \mid \sigma^2] = (\mathbf{I} - \mathbf{B})\mathbb{V}(\hat{\boldsymbol{\beta}} \mid \sigma^2)(\mathbf{I} - \mathbf{B})'$$
$$= \sigma^2(\mathbf{I} - \mathbf{B})\mathbf{A}(\mathbf{I} - \mathbf{B})' = \sigma^2(\mathbf{I} - \mathbf{B})\mathbf{A} = \mathbb{V}(\hat{\boldsymbol{\beta}}) - \mathbf{K}, \tag{1.77}$$

where $\mathbf{K} = \sigma^2\mathbf{BA} = \sigma^2\mathbf{AH}'(\mathbf{HAH}')^{-1}\mathbf{HA}$ is positive semi-definite for $J < k$ (Problem 1.12), so that $\hat{\gamma}$ has a lower variance than $\hat{\boldsymbol{\beta}}$, assuming that the same estimate of σ^2 is used. Observe, however, that if the null hypothesis is wrong, then, via the bias evident in (1.76) with $\mathbf{h} \neq \mathbf{H}\boldsymbol{\beta}$, the mean squared error (hereafter m.s.e.) of $\hat{\gamma}$ could be higher than that of $\hat{\boldsymbol{\beta}}$. A good discussion of this and related issues is provided in Judge et al. (1985, pp. 52–62).

So far, the derivation of $\hat{\gamma}$ pertained to the linear regression model with i.i.d. normal errors. If the errors instead are of the form $\epsilon \sim \mathrm{N}(\mathbf{0}, \sigma^2\boldsymbol{\Sigma})$ for known positive definite matrix $\boldsymbol{\Sigma}$, then we can combine the methods of g.l.s. and r.l.s. In particular, just use (1.69) with $\boldsymbol{\Sigma}^{-1/2}\mathbf{Y}$ in place of $\mathbf{Y}$ and $\boldsymbol{\Sigma}^{-1/2}\mathbf{X}$ in place of $\mathbf{X}$. We will denote this estimator as $\hat{\gamma}_{\Sigma}$ and refer to it as the **restricted generalized least squares**, or r.g.l.s., estimator.

Example 1.10 We wish to compute by simulation the m.s.e. of $\hat{\boldsymbol{\beta}}$ based on the four estimators o.l.s., g.l.s., r.l.s. and r.g.l.s., using, for convenience, the scalar measure $M = \sum_{i=1}^{k} (\hat{\beta}_i - \beta)^2$. Let the model be

$$Y_t = \beta_1 + \beta_2 X_{t,2} + \beta_3 X_{t,3} + \beta_4 X_{t,4} + \epsilon_t, \quad t = 1, \dots, T = 20,$$

for $\epsilon = (\epsilon_1, \dots, \epsilon_T)' \sim \mathrm{N}(\mathbf{0}, \sigma^2\boldsymbol{\Sigma})$, where $\boldsymbol{\Sigma}$ is a known, full rank covariance matrix, and the regression parameters are constrained as $\beta_2 + \beta_3 + \beta_4 = 1$, for which we take $\beta_1 = 10$, $\beta_2 = 0.4$, $\beta_3 = -0.2$ and $\beta_4 = 1 - \beta_2 - \beta_3 = 0.8$. The choice of $\mathbf{X}$ matrix will determine the m.s.e., and so, for each of the 50,000 replications, we let $X_{t,i} \overset{\text{i.i.d.}}{\sim} \mathrm{N}(0, 1)$, $i = 2, 3, 4$, $t = 1, \dots, T$. Measure M is then approximated by its sample average.

Five models are used. The first takes $\epsilon \sim \mathrm{N}(0, \sigma^2 w_t)$, $w_t = \sqrt{t}$; the second is with $w_t = t$. The third and fourth models assume an AR(1) structure for ϵ_t (recall Example 1.3), with parameters $a = 0.25$ and

```
function compareRGLS
T=20; beta=[10 0.4 -0.2 0.8]'; H=[0 1 1 1]; h=1;
Sigma = diag( [(1:T)'].^(0.5)); Sigmainv=inv(Sigma);
[V,D]=eig(0.5*(Sigma+Sigma')); W=sqrt(D);
Sighalf = V*W*V'; Sighalfinv=inv(Sighalf);
sim=500; emat=zeros(sim,4);
for s=1:sim
  X=[ones(T,1),randn(T,3)]; y=X*beta+Sighalf*randn(T,1);
  OLS = inv(X'*X)*X'*y; GLS = inv(X'*Sigmainv*X)*X'*Sigmainv*y;
  RLS = OLSrestrict(y,X,H,h);
  RGLS = OLSrestrict(Sighalfinv*y,Sighalfinv*X,H,h);
  emat(s,:) = [sum((OLS-beta).^2) sum((GLS-beta).^2) ...
               sum((RLS-beta).^2) sum((RGLS-beta).^2)];
end
M=mean(emat)

function gamma = OLSrestrict(y,X,H,h)
  [J,k]=size(H); if nargin<4, h=zeros(J,1); end
  b=regress(y,X); A=inv(X'*X); gamma = b+A*H'*inv(H*A*H')*(h-H*b);
```

Program Listing 1.4: Compares performance of o.l.s., g.l.s., r.l.s., and r.g.l.s. for a specific model.

Table 1.1 Empirical mean squared error over the four regression parameters, based on 50,000 replications.

	Model				
Method	**1**	**2**	**3**	**4**	**5**
o.l.s.	0.80	2.73	0.30	0.44	0.36
g.l.s.	0.72	1.85	0.28	0.36	0.28
r.l.s.	0.56	1.90	0.22	0.35	0.27
r.g.l.s.	0.50	1.23	0.21	0.29	0.22

$a = 0.5$, respectively. The fifth model assumes an MA(1) structure for ϵ_t with $b = 0.5$. The program to compute M is given in Listing 1.4. The results are shown in Table 1.1.

We see that, for all the models, o.l.s. is the worst and r.g.l.s. is the best estimator. Model 2 stands out because the covariance matrix differs markedly from the identity matrix. As such, the difference between o.l.s. and g.l.s., and the difference between r.l.s. and r.g.l.s. is quite large. For the other models, these differences are less pronounced, particularly for model 3 (the AR(1) with $a = 0.25$). ■

1.4.4 Testing With $\mathbf{h} = \mathbf{0}$

> The source of all great mathematics is the special case, the concrete example. It is frequent in mathematics that every instance of a concept of seemingly great generality is in essence the same as a small and concrete special case.
>
> (Paul R. Halmos, 1985, p. 324)

The above quote from Halmos is not fully applicable here because the general case of $\mathbf{h} \neq \mathbf{0}$ *is* important. It is straightforward and subsequently detailed, but the derivation for the special case $\mathbf{h} = \mathbf{0}$ is both easier and more intuitive because it turns out that we can explicitly express the projection matrix corresponding to $\mathcal{S}_H$.

With $\mathcal{S} = \mathcal{C}(\mathbf{X})$ and $\mathcal{S}_H \subset \mathcal{S}$ as defined in (1.68), consider the hypothesis as given in (1.67), but with the additional normality assumption:

$$H_0 : \mathbf{Y} = \mathbf{X}\boldsymbol{\gamma} + \boldsymbol{\epsilon}, \ \boldsymbol{\epsilon} \sim \mathrm{N}(\mathbf{0}, \sigma^2\mathbf{I}), \ \mathbf{X}\boldsymbol{\gamma} \in \mathcal{S}_H$$
$$H_1 : \mathbf{Y} = \mathbf{X}\boldsymbol{\beta} + \boldsymbol{\epsilon}, \ \boldsymbol{\epsilon} \sim \mathrm{N}(\mathbf{0}, \sigma^2\mathbf{I}), \ \mathbf{X}\boldsymbol{\beta} \in \mathcal{S}.$$

For notational convenience, denote the projection matrix onto $\mathcal{C}(\mathbf{X})$ as simply $\mathbf{P}$ instead of $\mathbf{P}_\mathcal{S}$, let $\mathbf{M} = \mathbf{I} - \mathbf{P}$ and let $\mathbf{P}_\mathcal{H} = \mathbf{P}_{\mathcal{S}_H}$. With $\mathbf{h} = \mathbf{0}$, $\mathbf{X}\hat{\boldsymbol{\gamma}}$ from (1.69) can be expressed as

$$\begin{aligned} \mathbf{X}\hat{\boldsymbol{\gamma}} &= \mathbf{X}\hat{\boldsymbol{\beta}} - \mathbf{XAH}'[\mathbf{HAH}']^{-1}\mathbf{H}\hat{\boldsymbol{\beta}} \\ &= (\mathbf{XAX}' - \mathbf{XAH}'[\mathbf{HAH}']^{-1}\mathbf{HAX}')\mathbf{Y} \\ &= (\mathbf{P} - \mathbf{XAH}'[\mathbf{HAH}']^{-1}\mathbf{HAX}')\mathbf{Y} =: (\mathbf{P} - \mathbf{N})\mathbf{Y}, \end{aligned} \tag{1.78}$$

where $\mathbf{N}$ is so defined. Straightforward algebra verifies that $\mathbf{P} - \mathbf{N}$ is symmetric and idempotent, so that, from Theorem 1.4, it is the unique projection matrix onto the subspace

$$\{\mathbf{z} : \mathbf{z} = \mathbf{X}\boldsymbol{\beta}, \ \boldsymbol{\beta} \in \mathbb{R}^k, \ \mathbf{H}\boldsymbol{\beta} = \mathbf{0}\}.$$

Thus, for $\mathbf{h} = \mathbf{0}$, we can express $\mathbf{P}_{\mathcal{H}}$ explicitly as

$$\mathbf{P}_{\mathcal{H}} = \mathbf{P} - \mathbf{N} = \mathbf{I} - \mathbf{M} - \mathbf{N}, \quad \text{where} \quad \mathbf{P} - \mathbf{P}_{\mathcal{H}} = \mathbf{N} \tag{1.79}$$

is symmetric and idempotent. Then, from Theorem 1.2, rank($\mathbf{N}$) = tr($\mathbf{N}$), where

$$\text{tr}(\mathbf{N}) = \text{tr}(\mathbf{XAH}'[\mathbf{HAH}']^{-1}\mathbf{HAX}') = \text{tr}([\mathbf{HAH}']^{-1}\mathbf{HAX}'\ \mathbf{XAH}') = \text{tr}(\mathbf{I}_J) = J.$$

The constrained residual vector is then $\hat{\boldsymbol{\epsilon}}_H = \mathbf{Y} - \mathbf{X}\hat{\boldsymbol{\gamma}}$, or

$$(\mathbf{Y} - \mathbf{X}\hat{\boldsymbol{\beta}}) + (\mathbf{X}\hat{\boldsymbol{\beta}} - \mathbf{X}\hat{\boldsymbol{\gamma}}) = \mathbf{Y} - \mathbf{X}\hat{\boldsymbol{\gamma}} = (\mathbf{I} - \mathbf{P}_{\mathcal{H}})\mathbf{Y} = (\mathbf{M} + \mathbf{N})\mathbf{Y},$$

so that $(\mathbf{X}\hat{\boldsymbol{\beta}} - \mathbf{X}\hat{\boldsymbol{\gamma}}) = \mathbf{NY}$. The following result is (in light of previous results) simple, and very important:

From Theorem 1.6,

$$\mathbf{PP}_{\mathcal{H}} = \mathbf{P}_{\mathcal{H}}\mathbf{P} = \mathbf{P}_{\mathcal{H}}, \quad \text{and} \quad \mathbf{N} = \mathbf{P} - \mathbf{P}_{\mathcal{H}} = \mathbf{P}_{\mathcal{S}\setminus\mathcal{H}} \text{ is a projection matrix.} \tag{1.80}$$

In particular, note that $\mathbf{X}\hat{\boldsymbol{\gamma}} = \mathbf{P}_{\mathcal{H}}\mathbf{Y} = \mathbf{P}_{\mathcal{H}}\mathbf{PY} = \mathbf{P}_{\mathcal{H}}\mathbf{X}\hat{\boldsymbol{\beta}}$, so that $\mathbf{X}\hat{\boldsymbol{\gamma}}$ is the projection of $\mathbf{X}\hat{\boldsymbol{\beta}}$ onto $\mathcal{S}_H$.

If H_0 is true, then $\mathbf{PY}$ and $\mathbf{P}_{\mathcal{H}}\mathbf{Y}$ should be close, with the discrepancy arising only from sampling error. A natural measure[7] of the magnitude of the difference is the norm, $\|(\mathbf{P} - \mathbf{P}_{\mathcal{H}})\mathbf{Y}\|$, or its square, given by

$$[(\mathbf{P} - \mathbf{P}_{\mathcal{H}})\mathbf{Y}]'(\mathbf{P} - \mathbf{P}_{\mathcal{H}})\mathbf{Y} = \mathbf{Y}'(\mathbf{P} - \mathbf{P}_{\mathcal{H}})\mathbf{Y}.$$

From (A.6),

$$\mathbb{E}[\mathbf{Y}'(\mathbf{P} - \mathbf{P}_{\mathcal{H}})\mathbf{Y}] = \sigma^2\text{rank}(\mathbf{P} - \mathbf{P}_{\mathcal{H}}) + \boldsymbol{\beta}'\mathbf{X}'(\mathbf{P} - \mathbf{P}_{\mathcal{H}})\mathbf{X}\boldsymbol{\beta}, \tag{1.81}$$

where the latter term is, from (1.79), given by

$$\boldsymbol{\beta}'\mathbf{X}'(\mathbf{P} - \mathbf{P}_{\mathcal{H}})\mathbf{X}\boldsymbol{\beta} = \boldsymbol{\beta}'\mathbf{X}'\mathbf{NX}\boldsymbol{\beta} = \boldsymbol{\beta}'\mathbf{H}'[\mathbf{HAH}']^{-1}\mathbf{H}\boldsymbol{\beta}. \tag{1.82}$$

Under H_0, $\mathbf{X}\boldsymbol{\beta} = \mathbf{X}\boldsymbol{\gamma}$ so that

$$(\mathbf{P} - \mathbf{P}_{\mathcal{H}})\mathbf{X}\boldsymbol{\beta} = (\mathbf{P} - \mathbf{P}_{\mathcal{H}})\mathbf{X}\boldsymbol{\gamma} = \mathbf{0}, \tag{1.83}$$

and (1.81) reduces to

$$\mathbb{E}[\mathbf{Y}'(\mathbf{P} - \mathbf{P}_{\mathcal{H}})\mathbf{Y}] = \sigma^2\text{rank}(\mathbf{P} - \mathbf{P}_{\mathcal{H}}) = \sigma^2\text{rank}(\mathbf{N}) = J\sigma^2. \tag{1.84}$$

By using $\hat{\sigma}^2$ from the unrestricted model as an estimate for σ^2, as given in (1.58), and dividing $\mathbf{Y}'(\mathbf{P} - \mathbf{P}_{\mathcal{H}})\mathbf{Y}$ by $J\hat{\sigma}^2 = \text{rank}(\mathbf{P} - \mathbf{P}_{\mathcal{H}})\hat{\sigma}^2$, we expect the value

$$F = \frac{\mathbf{Y}'(\mathbf{P} - \mathbf{P}_{\mathcal{H}})\mathbf{Y} \ / \ \text{rank}(\mathbf{P} - \mathbf{P}_{\mathcal{H}})}{\hat{\sigma}^2} = \frac{\mathbf{Y}'(\mathbf{P} - \mathbf{P}_{\mathcal{H}})\mathbf{Y} \ / \ \text{rank}(\mathbf{P} - \mathbf{P}_{\mathcal{H}})}{\mathbf{Y}'(\mathbf{I} - \mathbf{P})\mathbf{Y} \ / \ \text{rank}(\mathbf{I} - \mathbf{P})} \tag{1.85}$$

7 Other measures, such as the sum or maximum of the vector of absolute values might also seem "natural". However, the sampling distribution of the chosen measure is tractable, and also leads to a UMPI test.

```
YX=[y,X]; fileID = fopen('coachingdata.txt','w');
fprintf(fileID,'%8.5g %1u %1u %1u %1u \r\n',YX'); fclose(fileID);
```

Program Listing 1.7: Outputs variables $\mathbf{y}$ and $\mathbf{X}$ generated in Listing 1.5 as a text file.

```
ods pdf file='Coaching Regression Output.pdf';
data coach;
  infile 'coachingdata.txt';
  input y X1-X4;
run;
proc reg data=coach;
  RestrictedModel: model y = X1-X4 / NOINT;
    restrict X1=X2, X3=X4;
  UnRestricted: model y = X1-X4 / NOINT;
    SameInterceptAndSlope: test X1=X2, X3=X4;
run;
ods _all_ close;
ods html;
```

SAS Program Listing 1.1: SAS statements for (i) reading the text data set produced from the Matlab output generated by the code in Listing 1.7, and (ii) performing a regression analysis of the restricted model and the unrestricted model, and, for the latter, conducting the F test for the restrictions in (1.97). The output is a report, as an Adobe portable document format (pdf), including several useful graphics.

Next, the code in SAS Listing 1.1 performs two regression analyses. The first is of the restricted model, where the `restrict` statement is used to indicate (in terms of the variable names associated with the $\mathbf{X}$ matrix, and not the β coefficients). The second is unrestricted, and performs the F test associated with the restriction we wish to test. Observe how the `NOINT` option is necessary to tell SAS not to include an intercept term (a column of ones) in the regression, which it otherwise does by default. The SAS output (not shown) for the test in (1.97) yields $F = 2.73$ with a p-value of 0.0787, agreeing with the values obtained above using manual calculations in Matlab. ■

A **time-series regression** is such that Y_t and $\mathbf{x}_t'$ correspond to time point t. For simplicity, assume that the time points for which observations are observed are equally spaced, so that $t = 1, \dots, T$. A simple case is the model $Y_t = \beta_1 + \beta_2 t + \epsilon_t$. Examples of dependent variables that could be modelled as a time-series regression include:

1) Quarterly sales of a certain product, using regressors such as quarterly "dummy" variables, price, and amount of advertising, as well as prices and amounts of advertising for similar products offered from various market competitors.
2) Monthly rate of:
   a) fatalities caused by car accidents, using as regressors monthly dummies and/or dummies for particular days, such as weekend days or holidays
   b) alcohol-related car accidents
   c) homicides caused by guns.

3) Blood pressure of a patient, measured at weekly intervals, with regressors such as weight, number of cigarettes smoked, etc.

There are occasions in which the (linear) relationship describing a variable over time undergoes a pronounced change, due perhaps to the occurrence of a relevant and major event at some time point t_0, $1 \leqslant t_0 \leqslant T$.[8] In this case, the model is said to undergo a **structural break** at time t_0. Referring to the above dependent variables, examples of events that might cause a structural break include:

1) Discovery of a significant positive (or negative) side-effect from consuming the product.
2) Introduction of a new law for:
   a) the mandatory wearing of seat belts,
   b) the legal threshold of blood alcohol levels deemed acceptable to drive,
   c) gun control.
3) Change in diet, medication, etc.

If a structural break occurs, then two coefficient vectors need to be estimated: the first, say $\boldsymbol{\beta}_{[1]}$, for the sample of data corresponding to time points $1, \dots, t_0$, and the second, say $\boldsymbol{\beta}_{[2]}$, corresponding to $t_0 + 1, \dots, T$. We assume that σ^2 in both segments of time is constant. Such a model is said to be a **piecewise (linear) regression** if we constrain the two regression lines to touch at t_0, i.e., if $\mathbf{x}'_{t_0}\boldsymbol{\beta}_{[1]} = \mathbf{x}'_{t_0}\boldsymbol{\beta}_{[2]}$ is imposed. Point t_0 is said to be a **knot** or **join point**. The extension to more than one knot should be clear.[9]

Example 1.13 Let $Y_t = a_1 + a_2 t + e_t$, $t = 1, \dots, t_0$, and $Y_t = b_1 + b_2 t + e_t$, $t = t_0 + 1, \dots, T$, with $e_t \overset{\text{i.i.d.}}{\sim} \mathrm{N}(0, \sigma^2)$, $t = 1, \dots, T$.[10] Then, for the regression function to be continuous over the whole range, it must be the case that $a_1 + a_2 t_0 = b_1 + b_2 t_0$, or

$$a_1 - b_1 + a_2 t_0 - b_2 t_0 = 0. \tag{1.98}$$

Another way of stating this model is

$$\mathbf{Y} = a_1 \mathbf{x}_1 + b_1 \mathbf{x}_2 + a_2 \mathbf{x}_3 + b_2 \mathbf{x}_4 + \mathbf{e} = \mathbf{X}\boldsymbol{\beta} + \mathbf{e},$$

where $\mathbf{X} = [\mathbf{x}_1\ \mathbf{x}_2\ \mathbf{x}_3\ \mathbf{x}_4]$, with

$$\mathbf{x}_1 = (\mathbf{1}'_{t_0}\ \mathbf{0}'_{T-t_0})', \quad \mathbf{x}_2 = (\mathbf{0}'_{t_0}\ \mathbf{1}'_{T-t_0})',$$
$$\mathbf{x}_3 = (1, 2, \dots, t_0, 0, \dots, 0)', \quad \mathbf{x}_4 = (0, \dots, 0, t_0 + 1, \dots, T)',$$

and parameter vector $\boldsymbol{\beta} = (a_1, b_1, a_2, b_2)'$ is subject to the constraint $\mathbf{H}\boldsymbol{\beta} = 0$ from (1.98), where $\mathbf{H} = [1 \quad -1 \quad t_0 \quad -t_0]$. From (1.69), the restricted parameter vector is

$$\hat{\boldsymbol{\gamma}} = \left(\mathbf{I}_4 - \frac{\mathbf{AH'H}}{\mathbf{HAH'}}\right)\hat{\boldsymbol{\beta}},$$

8 In fact, such a phenomenon can occur in any type of data for which the order of the observations is relevant. Another example would be for spatial data, e.g., weather measurements taken simultaneously at different locations.

9 Less obvious, however, is how to proceed if the locations of the knots are not known. See, for example, Judge et al. (1985, pp. 800-814) for discussion of this and other related issues.

10 If for $t = t_0 + 1, \dots, T$, we take $Y_t = b_1 + b_2(t - t_0) + e_t$, which is sometimes referred to as a **locally disjoint broken trend model**, its first usage being from Perron and Zhu (2005); see also Deng and Perron (2006), Sobreira and Nunes (2016), Chang and Perron (2016), and the references therein.

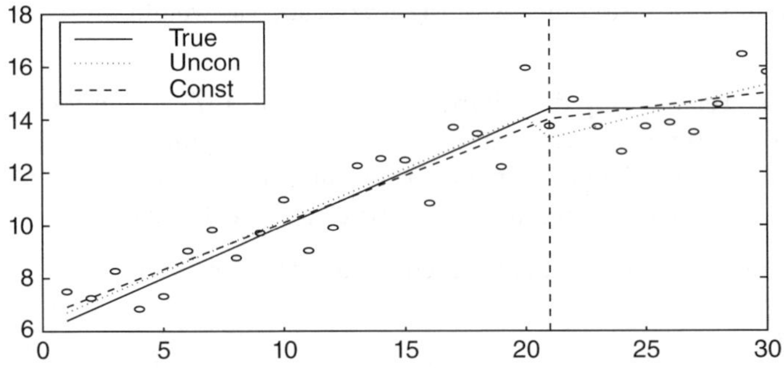

Figure 1.4 True and fitted piecewise regression.

where $\mathbf{A} = (\mathbf{X}'\mathbf{X})^{-1}$ and $\hat{\boldsymbol{\beta}}$ is the unrestricted estimated parameter vector. It is worth emphasizing that the value of the F test (1.88), and, hence, its p-value, depends only on $\mathbf{H}\boldsymbol{\beta}$ and is otherwise invariant to the choice of $\boldsymbol{\beta}$.

Figure 1.4 shows a simulated sample using $T = 30$, $t_0 = 21$, $\sigma^2 = 1$ and parameter values $a_1 = 6$, $a_2 = 0.4$, $b_2 = 0$ and $b_1 = a_1 + t_0 a_2 - t_0 b_2 = 14.4$, so that (1.98) is satisfied.[11] The p-value of the F test for constraint (1.98) is 0.130, so that the null hypothesis of a knot would not be rejected at conventional testing levels. In addition, the hypothesis that only one regression line is needed, i.e., that $a_1 = a_2$ and $b_1 = b_2$, was tested and resulted in a p-value of 0.0318. The data and plot were generated with the code in Listing 1.8.

Finally, to test whether the slope changes at the knot, let the unrestricted model be $Y_t = \alpha_1 + \alpha_2 t + \alpha_3(t - t_0)B_t + \epsilon_t$, $t = 1, \ldots, T$, where $\epsilon_t \overset{\text{i.i.d.}}{\sim} \text{N}(0, \sigma^2)$ and B_t is a boolean (or dummy) variable that is one if $t \geqslant t_0$ and zero otherwise, i.e., $B_t = \mathbb{I}_{\{t_0, t_0+1, \ldots\}}(t)$. The null hypothesis is that $\alpha_3 = 0$, for which the reduced column space is easy to express. For the data used, the p-value was 0.0310. As the true model is piecewise, it comes as no surprise that this p-value is quite close to the p-value given above for testing $a_1 = a_2$ and $b_1 = b_2$. ■

1.4.7 Confidence Intervals

Recall from (1.88) and (1.90) that, under the null hypothesis that $\mathbf{H}\boldsymbol{\beta} = \mathbf{h}$,

$$\frac{(\mathbf{H}\hat{\boldsymbol{\beta}} - \mathbf{h})'\mathbf{V}^{-1}(\mathbf{H}\hat{\boldsymbol{\beta}} - \mathbf{h})}{J\,\hat{\sigma}^2} \sim F_{J,T-k},$$

where $\mathbf{V} = \mathbf{H}(\mathbf{X}'\mathbf{X})^{-1}\mathbf{H}'$. This implies that

$$Q = \frac{(\mathbf{H}\hat{\boldsymbol{\beta}} - \mathbf{H}\boldsymbol{\beta})'\mathbf{V}^{-1}(\mathbf{H}\hat{\boldsymbol{\beta}} - \mathbf{H}\boldsymbol{\beta})}{J\,\hat{\sigma}^2} \sim F_{J,T-k} \tag{1.99}$$

11 The parameter values were chosen so that the data somewhat resemble actual data for rates of homicide in the USA, measured quarterly from 1985 to 1994, as shown in the Morbidity and Mortality Weekly Report from the Centers for Disease Control and Prevention (CDC), June 7, 1996, Vol. 45, No. 22, pp. 460–464. In their study, a piecewise linear regression was used to model the data.

```
function [pvalF1, pvalF2] = piecewise(seed,b2,doplot);
if nargin<2, b2=0.1; end, if nargin<3, doplot=1; end
t0=21; T=30; n=T-t0+1; x1=[ones(t0-1,1); zeros(n,1)]; x2=1-x1;
x3=[(1:t0-1)'; zeros(n,1)]; x4=[zeros(t0-1,1); (t0:T)']; X=[x1 x2 x3 x4];
a1=6; a2=0.4; b1=a1+a2*t0-b2*t0, beta=[a1 b1 a2 b2]'; sigma=1;
randn('state',seed); y=X*beta+sigma*randn(T,1); betahat=regress(y,X);
yfit=X*betahat; SSbeta=sum((y-yfit).^2); sigsqr_hat = SSbeta / (T-4);
% test the piecewise regression
H=[1 -1 t0 -t0]; J=1; gamma=OLSrestrict(y,X,H); yfitH=X*gamma;
SSgam=sum((y-yfitH).^2); F1 = (SSgam-SSbeta) / J / sigsqr_hat;
pvalF1 = 1-fcdf(F1,J,T-4);
if doplot==1
  true=X*beta;
  plot(1:T,true,'k-', 1:T,yfit,'g:', 1:T,yfitH,'r--', 1:T,y,'bo')
  set(gca,'fontsize',16), legend('True','Uncon','Const',2)
  ax=axis; h=line([t0 t0],[ax(3) ax(4)]); set(h,'linestyle','--')
end
% now test if both intercepts are equal and both slopes are equal
H=[1 -1 0 0; 0 0 1 -1]; J=2; gamma=OLSrestrict(y,X,H); yfitH=X*gamma;
SSgam=sum((y-yfitH).^2); F2 = (SSgam-SSbeta) / J / sigsqr_hat;
pvalF2=1-fcdf(F2,J,T-4);
```

Program Listing 1.8: Simulates and estimates a piecewise simple regression.

is a pivotal quantity for $\mathbf{H}\boldsymbol{\beta}$. In particular, letting $q = F^{-1}_{J,T-k}(1-\alpha)$ be the quantile such that $\Pr(Q \leqslant q) = 1-\alpha$, the ellipsoid $\{\mathbf{H}\boldsymbol{\beta} : Q \leqslant q\}$ is a $100(1-\alpha)\%$ confidence region for $\mathbf{H}\boldsymbol{\beta}$. If $J = 1$, then the region is just an interval.

Take, for example, the i.i.d. model: Let $Y_i \overset{\text{i.i.d.}}{\sim} \mathrm{N}(\mu, \sigma^2)$, $i = 1, \dots, n$, i.e., $\mathbf{X} = \mathbf{1}_n$ and $\boldsymbol{\beta} = \mu$, so that $\hat{\mu} = \bar{Y}$ and $Q = n(\hat{\mu} - \mu)^2/\hat{\sigma}^2 = (\hat{\mu} - \mu)^2/(S^2/n) \sim F_{1,n-1}$. Then, as $\sqrt{F^{-1}_{1,n-1}(1-\alpha)} = t^{-1}_{n-1}(1-\alpha/2)$, and from the symmetry of the Student's t distribution,

$$\{\mu : Q \leqslant q\} = \{\mu : |\hat{\mu} - \mu| \leqslant \sqrt{q}S/\sqrt{n}\} = (\hat{\mu} - \sqrt{q}S/\sqrt{n},\ \hat{\mu} + \sqrt{q}S/\sqrt{n})$$

is the usual confidence interval for μ. Similarly, for the general linear model with $J = 1$, $\mathbf{H}\boldsymbol{\beta}$ is a single linear combination of the elements in $\boldsymbol{\beta}$, which we denote $\boldsymbol{\ell}'\boldsymbol{\beta}$ for clarity, i.e., $\boldsymbol{\ell} = \mathbf{H}'$. Then $\mathbf{V} = \boldsymbol{\ell}'(\mathbf{X}'\mathbf{X})^{-1}\boldsymbol{\ell}$ is a scalar and, with $\mathbf{A} = (\mathbf{X}'\mathbf{X})^{-1}$,

$$\left\{ \boldsymbol{\ell}'\boldsymbol{\beta} : \frac{(\boldsymbol{\ell}'\hat{\boldsymbol{\beta}} - \boldsymbol{\ell}'\boldsymbol{\beta})^2}{\hat{\sigma}^2\, \boldsymbol{\ell}'\mathbf{A}\boldsymbol{\ell}} \leqslant q \right\} = \{\boldsymbol{\ell}'\boldsymbol{\beta} : |\boldsymbol{\ell}'\hat{\boldsymbol{\beta}} - \boldsymbol{\ell}'\boldsymbol{\beta}| \leqslant q^{1/2}\sqrt{\hat{\sigma}^2\, \boldsymbol{\ell}'\mathbf{A}\boldsymbol{\ell}}\} = \boldsymbol{\ell}'\hat{\boldsymbol{\beta}} \pm c\sqrt{\hat{\sigma}^2\, \boldsymbol{\ell}'\mathbf{A}\boldsymbol{\ell}}, \tag{1.100}$$

where $c = t^{-1}_{T-k}(1-\alpha/2)$. For $J \geqslant 2$, $\{\mathbf{H}\boldsymbol{\beta} : Q \leqslant q\}$ cannot be so easily "pivoted" to get intervals for the rows of $\mathbf{H}\boldsymbol{\beta}$, but, if $J = 2$ or $J = 3$, the region can be plotted.

Example 1.14 Let $Y_t = \beta_1 + \beta_2 t + e_t$, $t = 1, \dots, T$, $e_t \overset{\text{i.i.d.}}{\sim} \mathrm{N}(0, \sigma^2)$ and take $\mathbf{H} = \mathbf{I}_2$, so that the ellipsoid provides a confidence region for β_1 and β_2. For a simulated vector $\mathbf{Y}$ with $T = 10$, $\beta_1 = 1$, $\beta_2 = 2$, and $\sigma^2 = 1$, the region was computed with the program in Listing 1.9 and is shown in Figure 1.5 for the three common levels of significance $\alpha = 0.01$, 0.05, and 0.1. The relative size increase in going from $\alpha = 0.05$ to 0.01 is much larger than that from 0.1 to 0.05. ■

```
T=10; k=2; J=2; Y=1+2*(1:T)' + randn(T,1);
X=[ones(10,1),(1:10)'];
if 1==1, O=X; else O=orth(X); end
[betahat,BINT,R,RINT,STATS] = regress(Y,O,0.0001);
s2= sum(R.^2)/(T-k);
q90=finv(0.90,J,T-k); q95=finv(0.95,J,T-k); q99=finv(0.99,J,T-k);
Vi=(O'*O); % H is the 2X2 identity matrix
figure, h=plot(betahat(1),betahat(2),'k.'), set(h,'MarkerSize',30), hold on
inc=0.05;
for b1=BINT(1,1):inc:BINT(1,2)
  for b2=BINT(2,1):inc:BINT(2,2)
    beta=[b1 b2]'; Q=(betahat-beta)' * Vi * (betahat-beta) / (J*s2);
    if (Q <= q90), plot(b1,b2,'ro'), elseif (Q <= q95), plot(b1,b2,'gx')
    elseif (Q <= q99), plot(b1,b2,'b+'), end
  end
end, hold off
```

Program Listing 1.9: Generates ellipsoid for parameters of time-trend linear model. (Takes a relatively long to run; adjust `inc` accordingly.)

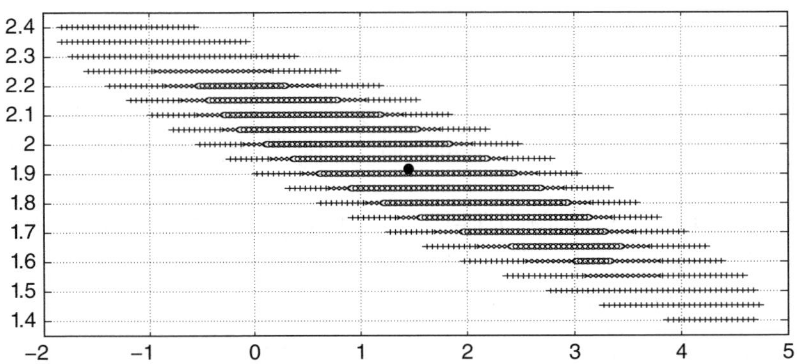

Figure 1.5 Ellipsoid for intercept β_1 (horizontal axis) and slope β_2 (vertical axis) for the model in Example 1.14, for $\alpha = 0.01$ (plus signs), $\alpha = 0.05$ (crosses) and $\alpha = 0.10$ (circles). The black dot is $\widehat{\boldsymbol{\beta}}$.

For $J = 3$, a three-dimensional plot of the region will be of limited use, while for $J \geqslant 4$, the whole region cannot be visualized as such, although one could plot it for two (or three) rows of $\mathbf{H}\boldsymbol{\beta}$ for fixed values of the remaining rows. This is clearly quite cumbersome and is essentially never done in practice. Instead, methods are used that yield simultaneous confidence intervals for each row of $\mathbf{H}\boldsymbol{\beta}$. One obvious way is to use Bonferroni's inequality as follows. Let $\hbar_i$ denote the ith row of $\mathbf{H}$, $i = 1, \dots, J$. Then the confidence region for $\hbar_i\boldsymbol{\beta}$ is precisely that in (1.100) with $\hbar_i$ instead of $\boldsymbol{\ell}'$. For simultaneous confidence intervals on the J values of $\hbar_i\boldsymbol{\beta}$, the **Bonferroni method** just takes $c = t_{T-k}^{-1}(1 - \alpha/(2J))$. The obvious disadvantage of this method is the inevitable large size of the intervals when J is large. An approach that makes explicit use of the normality assumption (and results in shorter confidence intervals) is based on the multivariate t distribution and referred to as **maximum modulus t intervals**; see Graybill (1976, Sec. 6.6) for further details.

We now consider another alternative to the Bonferroni intervals known as the S-method or **Scheffé's method**, from Scheffé (1953). We first need the following result: If $\mathbf{V} > 0$ (i.e., positive definite), and $\boldsymbol{\ell}$ and $\mathbf{b}$ are conformable vectors such that $\boldsymbol{\ell}'\mathbf{b}$ is a scalar, then

$$\max_{\boldsymbol{\ell}\neq\mathbf{0}} \frac{(\boldsymbol{\ell}'\mathbf{b})^2}{\boldsymbol{\ell}'\mathbf{V}\boldsymbol{\ell}} = \mathbf{b}'\mathbf{V}^{-1}\mathbf{b}. \tag{1.101}$$

Proof: First observe that, as matrix $\mathbf{V}$ enters only via a quadratic form, it can be assumed symmetric without loss of generality, and thus it makes sense to state that $\mathbf{V} > 0$, as all its eigenvalues are real. Take symmetric $\mathbf{V}^{1/2} > 0$ such that $\mathbf{V}^{1/2}\mathbf{V}^{1/2} = \mathbf{V}$ and define $\mathbf{u} = \mathbf{V}^{1/2}\boldsymbol{\ell}$ and $\mathbf{w} = \mathbf{V}^{-1/2}\mathbf{b}$, so that

$$\frac{(\boldsymbol{\ell}'\mathbf{b})^2}{\boldsymbol{\ell}'\mathbf{V}\boldsymbol{\ell}} = \frac{(\mathbf{u}'\mathbf{V}^{-1/2}\mathbf{V}^{1/2}\mathbf{w})^2}{\mathbf{u}'\mathbf{V}^{-1/2}\mathbf{V}\mathbf{V}^{-1/2}\mathbf{u}} = \frac{(\mathbf{u}'\mathbf{w})^2}{\mathbf{u}'\mathbf{u}} = \frac{\langle\mathbf{u},\mathbf{w}\rangle^2}{\|\mathbf{u}\|^2}.$$

From the Cauchy–Schwarz inequality (see Problem 1.7), $\langle\mathbf{u},\mathbf{w}\rangle^2 \leqslant \|\mathbf{u}\|^2\|\mathbf{w}\|^2$, with equality when $\mathbf{u} = \mathbf{w}$, i.e., $\mathbf{V}^{1/2}\boldsymbol{\ell} = \mathbf{V}^{-1/2}\mathbf{b}$ or $\boldsymbol{\ell} = \mathbf{V}^{-1}\mathbf{b}$. Thus, with $\boldsymbol{\ell} = \mathbf{V}^{-1}\mathbf{b}$,

$$\frac{\langle\mathbf{u},\mathbf{w}\rangle^2}{\|\mathbf{u}\|^2} = \|\mathbf{w}\|^2 = \|\mathbf{V}^{-1/2}\mathbf{b}\|^2 = \mathbf{b}'\mathbf{V}^{-1}\mathbf{b},$$

which is (1.101). See Graybill (1976, pp. 224–225) for an alternative proof. ■

Now, with $\mathbf{V} = \mathbf{H}(\mathbf{X}'\mathbf{X})^{-1}\mathbf{H}'$, $\boldsymbol{\theta} = \mathbf{H}\boldsymbol{\beta}$ and $\mathbf{b} = \widehat{\boldsymbol{\theta}} - \boldsymbol{\theta}$, (1.99) and (1.101) imply

$$\begin{aligned}
1-\alpha &= \Pr(Q \leqslant q) = \Pr((\widehat{\boldsymbol{\theta}} - \boldsymbol{\theta})'\mathbf{V}^{-1}(\widehat{\boldsymbol{\theta}} - \boldsymbol{\theta}) \leqslant Jq\hat{\sigma}^2) \\
&= \Pr\left(\max_{\boldsymbol{\ell}\neq\mathbf{0}} \frac{(\boldsymbol{\ell}'(\widehat{\boldsymbol{\theta}} - \boldsymbol{\theta}))^2}{\boldsymbol{\ell}'\mathbf{V}\boldsymbol{\ell}} \leqslant Jq\hat{\sigma}^2\right) = \Pr(|\boldsymbol{\ell}'(\widehat{\boldsymbol{\theta}} - \boldsymbol{\theta})| \leqslant \sqrt{Jq\hat{\sigma}^2\boldsymbol{\ell}'\mathbf{V}\boldsymbol{\ell}},\ \forall\boldsymbol{\ell}\neq\mathbf{0}),
\end{aligned}$$

where, as before, $q = F^{-1}_{J,T-k}(1-\alpha)$. That is, $\boldsymbol{\ell}'\widehat{\boldsymbol{\theta}} \pm \sqrt{Jq\hat{\sigma}^2\boldsymbol{\ell}'\mathbf{V}\boldsymbol{\ell}}$ simultaneously covers $\boldsymbol{\ell}'\boldsymbol{\theta}$ for an infinite set of vectors $\boldsymbol{\ell} \neq \mathbf{0}$ with level of significance $1-\alpha$. An alternative proof of this result using only basic calculus is given in Klotz (1969) and Roussas (1997, Sec. 17.4).

As only a finite number of such intervals will ever be constructed for a particular data set, the actual level exceeds $1-\alpha$. In particular, with $\boldsymbol{\ell}_i = (0,\dots,0,1,0\dots,0)'$ with the one in the ith position, $i = 1,\dots,J$, $\boldsymbol{\ell}_i'\widehat{\boldsymbol{\theta}} = \boldsymbol{\ell}_i'\mathbf{H}\widehat{\boldsymbol{\beta}} = \boldsymbol{\hbar}_i\widehat{\boldsymbol{\beta}}$, so that the J intervals $\boldsymbol{\hbar}_i\widehat{\boldsymbol{\beta}} \pm \sqrt{Jq\hat{\sigma}^2\boldsymbol{\ell}_i'\mathbf{V}\boldsymbol{\ell}_i}$ have simultaneous level of significance at least $1-\alpha$. As

$$\widehat{\mathbb{V}}(\boldsymbol{\hbar}_i\widehat{\boldsymbol{\beta}}) = \hat{\sigma}^2\boldsymbol{\hbar}_i\mathbf{A}\boldsymbol{\hbar}_i' = \hat{\sigma}^2\boldsymbol{\ell}_i'\mathbf{H}(\mathbf{X}'\mathbf{X})^{-1}\mathbf{H}'\boldsymbol{\ell}_i = \hat{\sigma}^2\boldsymbol{\ell}_i'\mathbf{V}\boldsymbol{\ell}_i,$$

these intervals are often written as $\boldsymbol{\hbar}_i\widehat{\boldsymbol{\beta}} \pm \sqrt{Jq\widehat{\mathbb{V}}(\boldsymbol{\hbar}_i\widehat{\boldsymbol{\beta}})}$, $i = 1,\dots,J$.

Example 1.15 Consider the same setup as in Example 1.14, with

$$\mathbf{A} = (\mathbf{X}'\mathbf{X})^{-1} = \frac{1}{15}\begin{bmatrix} 7 & -1 \\ -1 & 2/11 \end{bmatrix}$$

and $\mathbf{H} = \mathbf{I}_2$. Let $\boldsymbol{\ell}_1 = (1, 0)'$, $\boldsymbol{\ell}_2 = (0, 1)'$, $a_1 = \boldsymbol{\ell}_1'\mathbf{A}\boldsymbol{\ell}_1 = 7/15$ and $a_2 = \boldsymbol{\ell}_2'\mathbf{A}\boldsymbol{\ell}_2 = 2/165$. Then, with $J = 2$, $c = t_8^{-1}(1 - 0.05/4) \approx 2.7515$, the simultaneous 95% Bonferroni confidence intervals for β_1 and β_2 are $\hat{\beta}_i \pm c\hat{\sigma}\sqrt{a_i}$, $i = 1, 2$, with lengths $3.759\hat{\sigma}$ and $0.6059\hat{\sigma}$, respectively. With $J = k = 2$ and $q = F_{2,8}^{-1}(0.95) \approx 4.459$, the S-method confidence intervals are $\hat{\beta}_i \pm \hat{\sigma}\sqrt{2qa_i}$, $i = 1, 2$, with respective lengths $4.080\hat{\sigma}$ and $0.65776\hat{\sigma}$. The latter are about 8.5% longer than Bonferroni confidence intervals. ■

Remark In the previous example, the S-method intervals were longer than those from Bonferroni. To compare the lengths for other parameters, the top panel of Figure 1.6 plots the ratio of $t_{T-k}^{-1}(1 - \alpha/2J)$ to $\sqrt{JF_{J,T-k}^{-1}(1 - \alpha)}$ as a function of J, using $T - k = 40$ and three values of α. It would appear that the S-method is virtually useless compared to Bonferroni. This picture is misleading, however, because k or, more generally, the rank of $\mathbf{H}$ was not specified. In particular, with $\hbar_i$ the ith row of $\mathbf{H}$, assume $\hbar_1, \dots, \hbar_R$ are independent, $R \leqslant k$, and the remaining rows, $\hbar_{R+1}, \dots, \hbar_J$, are linear combinations of $\hbar_1, \dots, \hbar_R$. Let $\mathbf{H}^* = (\hbar_1', \dots, \hbar_R')'$ be the upper $R \times k$ portion of $\mathbf{H}$, so that

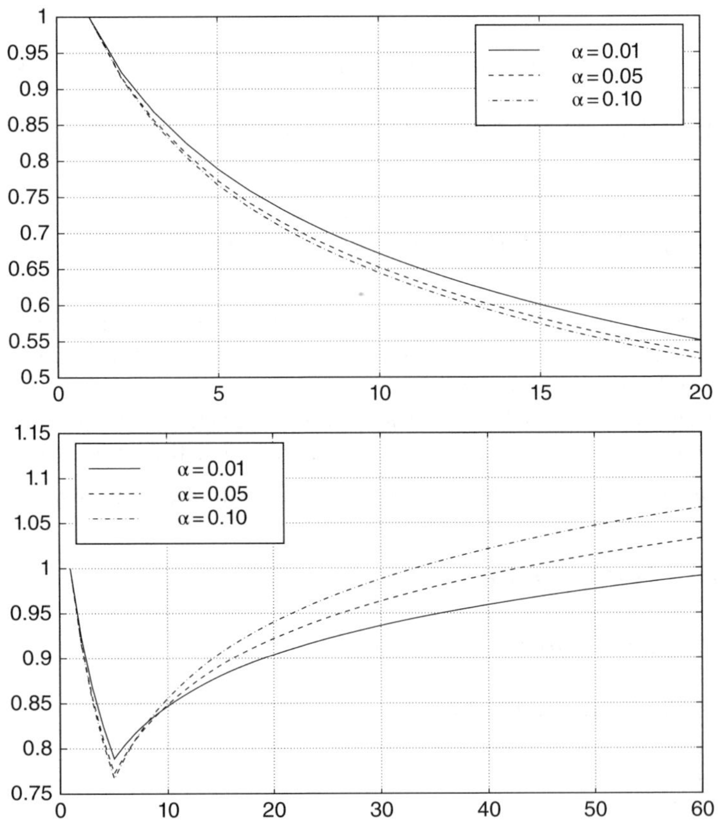

Figure 1.6 Ratio of lengths of Bonferroni to Scheffé confidence intervals. The top panel does not adjust for rank of **H**, while the bottom panel does adjust.

rank($\mathbf{H}$) = rank($\mathbf{H}^*$) = R. Then, with $\theta^* = (\theta_1^*, \dots, \theta_R^*)' = \mathbf{H}^*\boldsymbol{\beta}$, the S-method implies that

$$1 - \alpha = \Pr(|\boldsymbol{\ell}'(\hat{\theta}^* - \theta^*)| \leqslant \sqrt{Rq\hat{\sigma}^2\boldsymbol{\ell}'\mathbf{V}^*\boldsymbol{\ell}},\ \forall \boldsymbol{\ell} \in \mathbb{R}^R \setminus \mathbf{0}), \tag{1.102}$$

where $q = F^{-1}_{R,T-k}(1-\alpha)$ and $\mathbf{V}^* = \mathbf{H}^*(\mathbf{X}'\mathbf{X})^{-1}\mathbf{H}^{*\prime}$. But, by construction, each row $\hbar_i$ can be written as $\boldsymbol{\ell}_i'\mathbf{H}^*$ for some $\boldsymbol{\ell}_i \in \mathbb{R}^R \setminus \mathbf{0}$, so that (1.102) also includes the intervals for $\theta_{R+1}, \dots \theta_J$.[12] To see the effect this has, the right side of Figure 1.6 plots the ratio $t^{-1}_{T-k}(1-\alpha/2J)$ to $\sqrt{mF^{-1}_{m,T-k}(1-\alpha)}$ versus J, where $m = \min(J,k)$, $k = 5$ and, as before, $T - k = 40$. In this case, $\mathbf{H}^* = \mathbf{I}_k$. Indeed, if a relatively large number of intervals are to be computed, the S-method can be superior. ∎

In most realistic cases, the S-method gives rise to the longest intervals. Their additional length is the price to pay to be able to simultaneously construct infinitely many of them. In practice, their use allows a certain extent of "data mining", i.e., the researcher can keep computing intervals of interest until something "significant" is found, and still claim validity of the procedure. Preferably, however, one has a particular set of intervals in mind before the data are collected, to which the Bonferroni method (or others) can be applied.

Further details on confidence intervals can be found in numerous books on regression, including Ravishanker and Dey (2002, Sec. 7.3), Seber and Lee (2003, Ch. 5), and Khuri (2010, Ch. 7).

1.5 Alternative Residual Calculation

Recall from (1.60) that $\hat{\boldsymbol{\epsilon}} \sim \mathrm{N}(\mathbf{0}, \sigma^2\mathbf{M})$. Not only is $\mathbf{M}$ rank deficient, but the fact that the regression residuals are dependent on the $\mathbf{X}$ matrix implies that the distribution of common test statistics based on $\hat{\boldsymbol{\epsilon}}$, often ratios of quadratic forms, cannot be tabulated. This has historically been quite an inconvenience, though it should not be an issue now with modern computing power and the computational methods discussed in Section A.3. Perhaps the most popular example of a statistic whose use had been hampered by this fact (in the 1950s and 1960s) is the Durbin–Watson test D for detecting serial autocorrelation in the residuals; see Section 5.3.4. This was among the motivations for research on regression residuals that are independent of the regressor matrix.

Before proceeding, a comment on the relevance of this material is perhaps in order. In addition to being of historical importance for the reason just mentioned, we will also remark below that the recursive residuals are a special case of the ubiquitous and highly important Kalman filter. Next, as a theoretical curiosity, the derivation of the (below defined) BLUS and recursive residuals is instructive and, while arguably straightforward (especially after one sees the answer), is a great example of statistical mathematical ingenuity. Their practical relevance in some 21st century applications is admittedly less, such as in a machine-learning context and/or where large dimensional models are used, with mean terms being simply "regressed off" as part of a larger paradigm (see Section 11.2.2 for one such

---

12 Linear combinations of vectors are usually expressed in column form when using matrices. In this case,

$$\begin{pmatrix} | \\ \hbar_i' \\ | \end{pmatrix} = \ell_{i1}\begin{pmatrix} | \\ \hbar_1' \\ | \end{pmatrix} + \cdots + \ell_{iR}\begin{pmatrix} | \\ \hbar_R' \\ | \end{pmatrix} = \mathbf{H}^{*\prime}\begin{pmatrix} \ell_{i1} \\ \vdots \\ \ell_{iR} \end{pmatrix} = \mathbf{H}^{*\prime}\boldsymbol{\ell}_i, \quad i = 1, \dots, J,$$

or, taking transposes, $\hbar_i = \boldsymbol{\ell}_i'\mathbf{H}^*$.

example). As such, we illustrate the main concepts here, and place further details in Appendices 1.A and 1.B as optional reading for those interested in the proverbial "full Monty".

Several estimators of the regression residuals have been proposed, each sharing the three properties of linearity, unbiasedness, and a scalar covariance matrix; these are typically abbreviated with the acronym LUS. We denote such residuals by $\widehat{\epsilon}_{\text{LUS}} = \mathbf{CY}$, where $\mathbf{C}$ is a nonstochastic matrix (it can depend on $\mathbf{X}$, but not on $\mathbf{Y}$) satisfying $\mathbf{CX} = \mathbf{0}$ and $\mathbf{CC}' = \mathbf{I}$. Clearly, $\widehat{\epsilon}_{\text{LUS}} = \mathbf{CY}$ is linear in $\mathbf{Y}$, and as

$$\mathbb{E}[\widehat{\epsilon}_{\text{LUS}}] = \mathbb{E}[\mathbf{CY}] = \mathbb{E}[\mathbf{CX}\boldsymbol{\beta} + \mathbf{C}\boldsymbol{\epsilon}] = \mathbf{CX}\boldsymbol{\beta},$$

we see that the requirement $\mathbf{CX} = \mathbf{0}$ is necessary for unbiasedness. If $\mathbf{CC}' = \mathbf{I}$, then

$$\mathbb{E}[\widehat{\epsilon}_{\text{LUS}}\widehat{\epsilon}'_{\text{LUS}}] = \mathbb{E}[\mathbf{C}\boldsymbol{\epsilon}\boldsymbol{\epsilon}'\mathbf{C}'] = \sigma^2\mathbf{CC}' = \sigma^2\mathbf{I},$$

so that $\widehat{\epsilon}_{\text{LUS}}$ has a scalar covariance matrix.

Observe that the requirements $\mathbf{CX} = \mathbf{0}$ and $\mathbf{CC}' = \mathbf{I}$ (which is full rank) together imply that $\mathbf{C}$ cannot be $T \times T$, but rather $(T-k) \times T$, so that $\mathbf{CC}' = \mathbf{I}_{T-k}$ and $\widehat{\epsilon}_{\text{LUS}} \sim \mathrm{N}(\mathbf{0}, \sigma^2\mathbf{I}_{T-k})$. In particular, the rows of $\mathbf{C}$ are orthogonal to the columns of $\mathbf{X}$, i.e., they are contained in $\mathcal{C}(\mathbf{X})^{\perp}$, which has dimension $T-k$. Thus $\mathbf{CC}' = \mathbf{I} \Leftrightarrow$ the rows of $\mathbf{C}$ are orthogonal to one another $\Leftrightarrow$ there are at most $T-k$ rows in $\mathbf{C}$. Thus, only $T-k$ LUS residuals can be identified.

There are numerous matrices $\mathbf{C}$ that satisfy the LUS properties, and a "best" criteria was desired. This was pursued by Theil (1965, 1968) and Koerts (1967), and detailed in the books from Theil (1971) and Koerts and Abrahamse (1969). Consider the partition of the model

$$\begin{bmatrix} \mathbf{Y}_0 \\ \mathbf{Y}_1 \end{bmatrix} = \begin{bmatrix} \mathbf{X}_0 \\ \mathbf{X}_1 \end{bmatrix} \boldsymbol{\beta} + \begin{bmatrix} \boldsymbol{\epsilon}_0 \\ \boldsymbol{\epsilon}_1 \end{bmatrix} = \begin{bmatrix} \mathbf{X}_0 \\ \mathbf{X}_1 \end{bmatrix} \widehat{\boldsymbol{\beta}}_{\text{LS}} + \begin{bmatrix} \mathbf{e}_0 \\ \mathbf{e}_1 \end{bmatrix}, \tag{1.103}$$

where the quantities indexed with 0 have k rows and the quantities indexed with 1 contain the remaining $T-k$ rows. The vector $\boldsymbol{\epsilon}_0$ contains the k errors not represented in the LUS estimator. Given this partitioning, the best LUS, or **BLUS residuals**, denoted by $\widehat{\epsilon}_{\text{BLUS}}$, are defined as the vector of residuals among the class of LUS residuals that has the minimum expected sum of squared errors, i.e., the vector that minimizes

$$\mathbb{E}[(\widehat{\epsilon}_{\text{LUS}} - \boldsymbol{\epsilon}_1)'(\widehat{\epsilon}_{\text{LUS}} - \boldsymbol{\epsilon}_1)].$$

Some work is required to show that the vector of BLUS residuals can be expressed in the computationally attractive form

$$\widehat{\epsilon}_{\text{BLUS}} = \mathbf{e}_1 - \mathbf{X}_1\mathbf{X}_0^{-1} \left[\sum_{h=1}^{H} \frac{d_h}{1+d_h} \mathbf{q}_h\mathbf{q}_h' \right] \mathbf{e}_0, \tag{1.104}$$

where $d_1^2, \ldots, d_H^2$ are the eigenvalues of the matrix $\mathbf{X}_0(\mathbf{X}'\mathbf{X})^{-1}\mathbf{X}_0'$ that are less than one, $H \leqslant T-k$, and $\mathbf{q}_1, \ldots, \mathbf{q}_H$ are the corresponding eigenvectors. A detailed derivation is given in Appendix 1.A.

Furthermore, the $(T-k) \times T$ matrix $\mathbf{C}$ in this case is given by the partitioned matrix $\mathbf{C} = [\mathbf{C}_0\ \mathbf{C}_1]$ where the $(T-k) \times k$ matrix $\mathbf{C}_0$ and the $(T-k) \times (T-k)$ matrix $\mathbf{C}_1$ are derived by the following relationships:

$$\mathbf{C}_0 = -\mathbf{C}_1\mathbf{Z}, \quad \mathbf{C}_1 = \mathbf{PDP}',$$

where $\mathbf{Z} = \mathbf{X}_1\mathbf{X}_0^{-1}$, $\mathbf{D}$ is the $(T-k) \times (T-k)$ diagonal matrix whose first H successive diagonal elements are $d_1 \leqslant d_2 \leqslant \ldots \leqslant d_H < 1$ (the ds being the positive square roots of the d_k^2 defined in (1.104)),

```
function C = blusmat(X)
[T,k]=size(X); X_0 = X(1:k,:); X_1 = X(k+1:end,:);
Z = X_1*inv(X_0); D = eig(X_0*inv(X'*X)*X_0');
index1 = find(D<1 & D>0); H = size(index1,1);
D = [D(index1);ones(T-k-H,1)]; D = sort(D); D = diag(D);
[P tempD] = eig(eye(T-k) + Z*Z');
tempD = diag(tempD); [tempD index2] = sortrows(tempD);
P = P(:,index2(end:-1:1)); C_1 = P*D*P'; C = [-C_1*Z C_1];
```

Program Listing 1.10: Constructs the BLUS residual matrix **C**.

and $\mathbf{P}$ is the $(T-k)\times(T-k)$ orthogonal matrix with columns given by the eigenvectors of $\mathbf{I}+\mathbf{ZZ}'$ corresponding to the eigenvalues $1/d_1^2, \ldots, 1/d_H^2, 1, \ldots, 1$; see Appendix 1.A. The code in Listing 1.10 computes matrix $\mathbf{C}$.

One particular LUS residual estimator, the so-called **recursive residuals**, introduced by Hedayat and Robson (1970), Harvey and Phillips (1974), and Brown et al. (1975), is noteworthy. (Their use can be traced back all the way to Gauss; see Plackett, 1950; Stigler, 1981; and Young, 2011.) The procedure is computationally simple and turns out to be a special case of the Kalman filter; see the remarks in Section 5.6.

Phillips and Harvey (1974) show that the corresponding $\mathbf{C}$ matrix such that $\mathbf{V}=\mathbf{CY}$ and $\mathbf{V}\sim \mathrm{N}(\mathbf{0},\sigma^2\mathbf{I}_{T-k})$ can be expressed as

$$\mathbf{C}=\begin{pmatrix} \mathbf{a}_{k+1} & d_{k+1}^{-1/2} & 0 & \cdots & 0 \\ \mathbf{a}_{k+2} & & d_{k+2}^{-1/2} & & \vdots \\ \vdots & & & \ddots & \\ & & & & 0 \\ \mathbf{a}_T & & & & d_T^{-1/2} \end{pmatrix}, \tag{1.105}$$

of size $(T-k)\times T$, where, for $j=k+1,\ldots,T$,

$$\mathbf{a}_j = -d_j^{-1/2}\mathbf{x}_j'(\mathbf{X}_{j-1}'\mathbf{X}_{j-1})^{-1}\ \mathbf{X}_{j-1}', \quad d_j = 1+\mathbf{x}_j'(\mathbf{X}_{j-1}'\mathbf{X}_{j-1})^{-1}\mathbf{x}_j, \tag{1.106}$$

and $\mathbf{x}_j'$ is the jth row of $\mathbf{X}$. Note that $\mathbf{a}_j$ is a row vector with length $j-1$.

Direct multiplication verifies that $\mathbf{CX}=\mathbf{0}$ and $\mathbf{CC}'=\mathbf{I}_{T-k}$, and one may show (Theil, 1971, p. 209) that $\mathbf{C}'\mathbf{C}=\mathbf{M}$. Thus, in Theorem 1.3 above, one could take $\mathbf{G}$ to be $\mathbf{C}$. The program in Listing 1.11 computes (1.105). Appendix 1.B provides details on the derivation of the recursive residuals.

```
function C = recmat(X)
[T,k]=size(X); C=zeros(T,T);
for j=(k+1):T
  mid=inv (X(1:(j-1),:)' * X(1:(j-1),:));
  d=sqrt (1+X(j,:) * mid * X(j,:)');
  p2=mid * X(1:(j-1),:)'; v=-(X(j,:) * p2)/d;
  C(j,1:(j-1))=v; C(j,j)=1/d;
end
C=C((k+1):T,:);
```

Program Listing 1.11: Constructs the recursive residual matrix **C**.

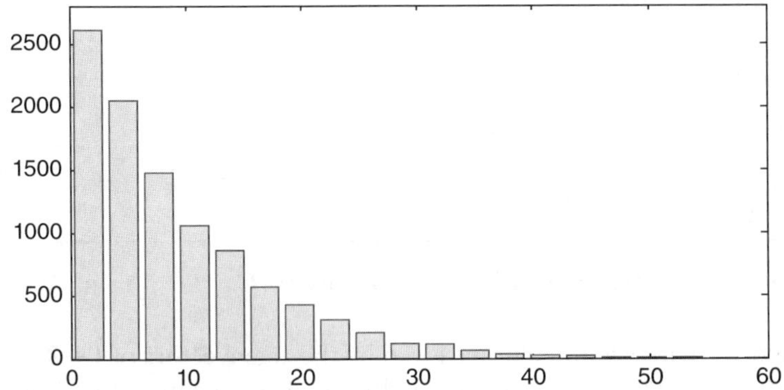

Figure 1.7 Simulated relative percentage change between the recursive and BLUS residuals for a model with intercept and time trend, and 20 observations.

Example 1.16 We wish to compare the magnitudes of the sum of squared BLUS and recursive residuals. Take the model to be $Y_j = 1 + 2j + e_j, j = 1, \ldots, 20$, with $e_j \overset{\text{i.i.d.}}{\sim} \text{N}(0, 1)$, so that the $\mathbf{X}$ matrix consists of a constant and a time vector. By using the code in Listings 1.10 and 1.11, it is a very simple Matlab exercise to simulate the model a large number of times and, for each, compute the relative percentage change between the recursive and BLUS residuals (i.e., $100 * (r - b)/r$, where r and b denote the sum of squares of the recursive and BLUS residuals, respectively).

Doing this for 10,000 replications and plotting the resulting histogram results in Figure 1.7. Note that, in every case, the sum of squared BLUS residuals is smaller than that for the recursive, as the theory dictates. Based on the simulation, there is more than a 35% chance that the relative percentage change will be more than 10%. ■

Remarks

a) Statistical tests common with the linear model using the BLUS residuals do not necessarily possess greater power than those using the "usual" o.l.s. residuals, or some other **C**. The use of BLUS residuals has faded considerably since the 1970s, although more recently Magnus and Sinha (2005) conducted studies comparing the power of BLUS against the recursive residuals when testing against heteroskedasticity (one of the original motivations for BLUS) and structural breaks (for which the recursive residuals are intuitively appealing). The reported simulation results lend mild support for the use of BLUS residuals over recursive residuals.
b) We will see later that the recursive residuals (or any LUS estimate) have other desirable properties that make their use valuable. In particular, in the context of time-series analysis, Chapter 8 will show that, for any $\mathbf{X}$ matrix, the coefficients of the sample autocorrelation function (SACF) based on the recursive residuals always have zero expectation and are symmetric, a property not shared by the SACF based on the usual o.l.s. residuals, even when $\mathbf{X}$ is only a column of ones. This is important because, in practice, the SACF coefficients are compared to their limiting distribution, which is normal (i.e., symmetric) with zero mean. For small samples and $\mathbf{X}$ matrices common in econometric applications, this can be an important factor. ■

1.6 Further Topics

> As it happens, the econometric modeling was done in the basement of the building and the econometric theory courses were taught on the top floor (the third). I was perplexed by the fact that the same language was used in both places. Even more amazing was the transmogrification of particular individuals who wantonly sinned in the basement and metamorphosed into the highest of high priests as they ascended to the third floor.
>
> (Edward Leamer, 1978, p. vi)

> With increasing interest in the stable distributions and their domains of attraction, the Cauchy distribution is found to occupy a less isolated position; indeed the normal distribution is extremal and rather special among stable distributions.
>
> (E. J. Pitman and E. J. Williams, 1967, p. 916)

> What the regression curve does is give a grand summary for the averages of the distributions corresponding to the set of x's. We could go further and compute several different regression curves corresponding to the various percentage points of the distributions and thus get a more complete picture of the set. Ordinarily this is not done, and so regression often gives a rather incomplete picture. Just as the mean gives an incomplete picture of a single distribution, so the regression curve gives a corresponding incomplete picture for a set of distributions.
>
> (Frederick Mosteller and John W. Tukey, 1977, p. 266)

An important special case of the linear model is the so-called analysis of variance, or ANOVA, for fixed and random effects, as introduced in Chapters 2 and 3, respectively. However, as these chapters are aimed at the underlying distribution theory of the core linear regression model and the ANOVA setting, numerous important topics associated with regression are regretfully not discussed. Two obvious ones are its extension to a multivariate framework, such as MANOVA and discriminant analysis (see, e.g., Huberty and Olejnik, 2006) and the use of Bayesian inferential methods (see, e.g., Christensen et al., 2011 and Gelman et al., 2013). Here, we mention several other omitted topics associated with regression analysis, albeit without much detail, so that the reader is at least aware of them, and provide useful references for further reading.

1) **Forecasting.**
   Based on regression model (1.3), interest might center on predicting the random variable Y_{T+1} for a given $\mathbf{x}_{T+1} = (x_{T+1,1}, \dots, x_{T+1,k})'$, so that $Y_{T+1} = \mathbf{x}'_{T+1}\boldsymbol{\beta} + \epsilon_{T+1}$, where $\epsilon_{T+1} \sim \mathrm{N}(0, \sigma^2)$. As $\widehat{\boldsymbol{\beta}}$ has the smallest variance among all linear unbiased estimators for $\boldsymbol{\beta}$, the minimum variance unbiased point estimator is $\widehat{Y}_{T+1} = \mathbf{x}'_{T+1}\widehat{\boldsymbol{\beta}}$, and, from (1.8),

$$\mathbb{V}(\widehat{Y}_{T+1} - Y_{T+1}) = \mathbb{V}(\widehat{Y}_{T+1}) + \mathbb{V}(Y_{T+1}) = \sigma^2\mathbf{x}'_{T+1}(\mathbf{X}'\mathbf{X})^{-1}\mathbf{x}_{T+1} + \sigma^2.$$

   Thus, an exact $100(1 - \alpha)\%$ confidence interval for Y_{T+1} is

$$\widehat{Y}_{T+1} \pm c\hat{\sigma}\sqrt{1 + \mathbf{x}'_{T+1}(\mathbf{X}'\mathbf{X})^{-1}\mathbf{x}_{T+1}}, \tag{1.107}$$

where $\hat{\sigma}^2$ is given in (1.11), and c is the $\alpha/2$ quantile of a Student's t random variable with $T-k$ degrees of freedom.

The reader is encouraged to set up the parametric and nonparametric bootstrap to generate confidence intervals for Y_{T+1} for both the Gaussian and non-Gaussian cases. Under the normality assumption, simulation can be used to confirm that the bootstrap results are comparable to the analytic method in (1.107). For a non-Gaussian, leptokurtic, and asymmetric distributional assumption, confidence intervals (hereafter c.i.s) based on (1.107) (i) will almost surely be such that the actual and nominal coverage probabilities are not equal, and (ii) restricted to being incorrectly symmetric. Bootstrap c.i.s are expected to be be more accurate, particularly as the level of non-Gaussianity increases.

Further details on multiple prediction intervals making use of the methods in Section 1.4.7 can be found in, e.g., Seber and Lee (2003, Sec. 5.3) and Rao et al. (2008, Ch. 6).

2) **Multicollinearity**.
Particularly in the social sciences, some regressors can be highly correlated with one another, and give rise to what is called multicollinearity. With very high correlation, the resulting standard errors on the coefficients are large, and thus the point estimates are rather imprecise. Several ways of dealing with this issue exist, including use of shrinkage (recall Section III.5.4), empirical Bayes estimators, **ridge regression** (which is related to the former two methods), and use of (generalized) cross validation.

Further methods that also relate more generally to model specification and estimation are the so-called garrote and LASSO estimators. The LASSO and ridge regression are generalized by the so-called **elastic net**. These tools are important for dimension reduction, variable selection, and improved predictive performance when modeling high-dimensional (big) data. Their respective Wikipedia entries are a good starting point and include original references, while further information can be found in textbook presentations such as Seber and Lee (2003, Sec. 12.5), Murphy (2012), Fahrmeir et al. (2013, Sec. 4.2), and Efron and Hastie (2016, Ch. 7, 12, 16). See also Lansangan and Barrios (2017) and the references therein for an introduction, further methods, and comparisons among them.

3) The choice of regressors, or, more generally, **model specification**.
Recall the reference to Leamer (1983) in Section 1.1, indicating the potentially severe implications resulting from the choice of variables to include in a regression. The tidy, impressive analytic results and distribution theory throughout this chapter are child's play (and arguably of secondary relevance) compared to the much thornier issue of model specification with real data, particularly from the social sciences. The quote by Magnus (2017) at the beginning of Section 1.4 serves to remind us that inspection of the "t-statistics" is not a viable method for model selection (in general agreement with the diatribe in Section III.2.8), and Magnus (2017, Sec. 2.14, 2.15) provides a very readable presentation of the bias/variance tradeoff associated with including a particular regressor into the model. The amusing quote by Leamer (1978) at the beginning of this section might be a reflection of the state of affairs during what might now appear to be a primordial age of econometrics, though it still contains more than just a grain of truth on the discrepancies between theory and practice.

As mentioned, model selection is related to multicollinearity—it might be preferred to simply omit regressors that are highly correlated with others. The inherent difficulty in establishing the "best" model is nicely stated in Seber and Lee (2003, p. 424): "The relative merits of ridge regression versus least squares and subset selection have been endlessly debated." Textbooks on regression analysis present many of the numerous ways that have been devised to select an optimal set (in some sense) from an available pool of regressors. See, e.g., the relevant chapters in Graybill and Iyer (1994), Ravishanker and Dey (2002), Seber and Lee (2003), Christensen (2011), Montgomery et al. (2012), Chatterjee and Hadi (2012), and Harrell, Jr. (2015).

Those books also cover numerous additional topics associated with applied regression analysis, and make use of real-data examples.

Particularly in econometrics, an influential body of work and methodology centers around the influential David F. Hendry, sometimes referred to general-to-specific (GETS) modeling, or the "LSE (London School of Economics) approach (to econometrics)" (see the same-titled Wikipedia entry). Good starting points include Hendry (1995, 2009), Castle et al. (2011), Hendry and Doornik (2014), and Castle et al. (2017).

4) **Missing values.**
It is not uncommon that one or more entries of the desired regressor matrix **X** are missing. A good starting point for methods of dealing with this important issue in the context of regression is Rao et al. (2008, Ch. 8). In a more general setting, analysis of data with missing values is addressed by so-called **multiple imputation**, often using simulation and, when applicable, an expectation-maximization (hereafter EM) algorithm. An internet search for books along the lines of "multiple imputation of missing data" will reveal numerous possible resources for addressing this common and pernicious issue when dealing with real data.
5) **Time-varying parameters**, such that one or more of the regression coefficients varies through time.
We deal with some aspects of this in Section 5.6. Consideration of such models leads naturally to the more general class of so-called state space models; see the references in Section 5.6.
6) One or more of the regression coefficients undergoes a **structural break**, i.e., a change in its value at some unknown point in time.
Estimation and testing in this case has been considered by numerous authors; see, e.g., Bai and Perron (1998, 2003), Qu and Perron (2007), Yamamoto and Perron (2013).[13] Another method is via **impulse indicator saturation**, as first investigated by Hendry (1999). It provides a general test for an unknown number of breaks, at unknown times, and is applicable in many model situations besides the linear regression model, such as vector autoregressions; see, e.g., Ericsson (2012), Castle et al. (2015), and the references therein for further development and application. It also has applications to testing for parameter constancy; see, e.g., Johansen and Nielsen (2009), Hendry and Doornik (2014), and the references therein. A package for R is available from Sucarrat et al. (2017) for automated GETS modeling of the mean and variance of a regression, and indicator saturation methods for detecting and testing for structural breaks in the mean.
7) Use of **robust estimators**.
In the presence of outliers, the least squares estimator is not optimal. Alternative estimation procedures have been developed to address this, e.g., Seber and Lee (2003, Sec. 3.13), Andersen (2008), and Huber and Ronchetti (2009, Ch. 7), as well as the note below on quantile regression.
8) **Partially adaptive estimation** for regression amid non-Gaussian disturbances.
This is related to the previous issue of robustness, but in that setting the assumption is that the disturbances are Gaussian, but such that one or more observations deviates substantially from

13 The authors conveniently provide Matlab codes for this last test, and others; see Perron's web page: http://people.bu.edu/perron/code.html.

the main group. Here, the assumption is not the presence of outliers *per se*, but rather that the underlying error distribution is non-Gaussian (and usually leptokurtic or heavy tailed, and possibly asymmetric), thus also giving rise to observations more extreme than the main cluster.

While general nonparametric methods are applicable in this setting, the method of partially adaptive estimation is very straightforward and still within the paradigm of parametric inference. It involves replacing the normality assumption with a flexible non-Gaussian distribution that embodies asymmetry and (semi-)heavy tails, and usually such that normality is a special or limiting case. General optimization routines will be required for computing the m.l.e., and bootstrap methods can be used for computing confidence intervals and other aspects of inference, such as forecasting.

The use of the Student's t distribution and its generalizations in regression analysis has been considered by McDonald and Newey (1988), Lange et al. (1989), and Butler et al. (1990). A less popular candidate, due to its historical complication regarding the evaluation of the p.d.f. (and thus the likelihood) is the (asymmetric) stable Paretian, as discussed in detail in Chapter II.8, and Sections III.9.4, III.9.5, and III.A.16. It also was the motivation for including the quote above by Pitman and Williams (1967).

The reason for its appeal, as compared to, say, use of (asymmetric) Student t variations, is the applicability of the generalized central limit theorem: One presumes that the standardized sum of all the neglected factors in the model (yielding the error term) converges to a stable distribution, of which normality is a special case. Note, however, that the non-Gaussian stable distribution does not possess a variance, and (as with any non-Gaussian distribution), the use of the bootstrap is recommended for inference on parameter and forecast uncertainty.

9) Use of **threshold regression**.
This is a type of **sample splitting model**, leading to far more general structures, such as cluster analysis and various multivariate methods in machine learning. As in Hansen (1999, 2000), under the assumption of two groups (referred to as classes, or regimes, in Hansen, 2000),

$$Y_t = \begin{cases} \mathbf{x}_t'\boldsymbol{\theta}_1 + \epsilon_t, & \text{if} \quad q_t \leqslant \gamma, \\ \mathbf{x}_t'\boldsymbol{\theta}_2 + \epsilon_t, & \text{if} \quad q_t > \gamma, \end{cases} \tag{1.108}$$

$t = 1, \ldots, T$, where $\mathbf{x}_t$ is a known $k \times 1$ vector; q_t is exogenous (not involving any Y_t) and is referred to as the threshold variable; and $\epsilon_t \overset{\text{i.i.d.}}{\sim} \mathrm{N}(0, \sigma^2)$. It can be an element of $\mathbf{x}_t$ and, for the asymptotic theory developed by Hansen (2000), is assumed to be continuous. Finally, γ is the **threshold parameter**. Let, as usual, the regressor matrix be $\mathbf{X} = [\mathbf{x}_1, \ldots, \mathbf{x}_T]'$, let $\mathbf{q} = [q_1, \ldots, q_T]'$ and $\mathbf{b} = \mathbb{I}\{\mathbf{q} \leqslant \gamma\}$, both $T \times 1$. Then, with $\mathbf{1}_k' = [1, 1, \ldots, 1]$ and selection matrix $\mathbf{S} = \mathbf{1}_k' \otimes \mathbf{b}$, define $\mathbf{X}_\gamma = \mathbf{S} \odot \mathbf{X}$, so that model (1.108) can be expressed as

$$\mathbf{Y} = \mathbf{X}\boldsymbol{\theta} + \mathbf{X}_\gamma\boldsymbol{\delta} + \boldsymbol{\epsilon} = \mathbf{Z}\boldsymbol{\beta} + \boldsymbol{\epsilon}, \tag{1.109}$$

where $\mathbf{Y}$ and $\boldsymbol{\epsilon}$ are defined in the usual way, $\boldsymbol{\theta} = \boldsymbol{\theta}_2$, $\mathbf{Z} = [\mathbf{X}, \ \mathbf{X}_\gamma]$ and $\boldsymbol{\beta} = [\boldsymbol{\theta}', \ \boldsymbol{\delta}']'$. Sample Matlab code to generate $\mathbf{X}_\gamma$ is given in Listing 1.12. For a given threshold γ, the usual least squares estimator (1.5) for $\boldsymbol{\beta}$ is used, and is also the m.l.e. under the usual Gaussian assumption on $\boldsymbol{\epsilon}$.

If γ were known, then the model reduces to the usual linear regression model, and the "significance" of $\boldsymbol{\delta}$ is assessed in the usual way, from Section 1.4. Matters are less clear when γ is to be elicited from the data. Let the **concentrated sum of squares** be given by (1.4), but as a function of γ, i.e.,

$$\mathsf{S}(\gamma) = \mathsf{S}(\gamma; \widehat{\boldsymbol{\beta}}; \mathbf{Y}, \mathbf{Z}) = \mathbf{Y}'\mathbf{M}_\gamma\mathbf{Y}, \quad \mathbf{M}_\gamma = \mathbf{I}_T - \mathbf{Z}(\mathbf{Z}'\mathbf{Z})^{-1}\mathbf{Z}'.$$

```
T=10; k=2; X=[ones(T,1), (1:T)']; b=rand(T,1)<0.5;
S = kron(ones(1,k),b); Xg = S.* X;
```

Program Listing 1.12: Example code for generating $\mathbf{X}_\gamma$ in (1.109).

(This is similar in concept to the **concentrated likelihood**, as will be used later in Section 5.6.3.1.) Assume $\gamma \in [\underline{\gamma}, \bar{\gamma}]$, and let

$$\hat{\gamma} = \underset{\gamma \in G}{\text{argmin}}\ \mathsf{S}(\gamma)$$

be the least squares estimator of γ, where $G = [\underline{\gamma}, \bar{\gamma}] \cap \{q_1, \dots, q_T\}$, noting that $\mathsf{S}(\gamma)$ takes on less than T distinct values. Hansen (2000) derives the asymptotic theory associated with estimator $\hat{\gamma}$, and approximate confidence intervals for γ based on a likelihood ratio statistic.
The case of model (1.108) with more than two groups is a straightforward generalization of this two-group setup. Examples of its use in macroeconomics include Rousseau and Wachtel (2002), Jude (2010), Stolbov (2013), Perri (2014), Pan et al. (2016), and the references therein.

10) **Quantile regression.**
The above quote by Mosteller and Tukey (1977) serves as a clear reminder of the limits of standard regression analysis and as one (of several) motivating factors for using quantile regression (QR). In particular, some contemplation reveals that, perhaps more often than not, it is not the mean that is of interest, but rather a particular quantile. For example, in income studies, interest might center on how the various exogenous factors influence not the mean income, but rather the lower 1, 5, and 10% quantiles, or their right-tail counterparts. Another benefit of QR compared to standard linear regression is that the median could be used instead of the mean as a type of robustified estimator, and/or its resulting implications (such as forecasts) compared to those based on the traditional use of the mean. Furthermore, QR allows for heteroskedasticity of the response function (recall the simple example in Figure 1.1) in a natural way, without requiring an explicit model for the error term that allows the exogenous variables to influence the estimates of σ_t (see, e.g., Fahrmeir et al., 2013, Ch. 10, for such an example and comparison to the use of QR).
A—clearly no longer relevant—disadvantage of QR is that closed-form solutions of the estimator no longer exist, and either linear programming techniques, or just general optimization algorithms, are required. One of the earliest survey articles on the topic is Koenker and Hallock (2001), while more detailed accounts can be found in the highly readable initial books of Koenker (2005) and Hao and Naiman (2007), as well as the newer Davino et al. (2014), which also provides code in R, SAS, and Stata.

11) **Generalized Linear Models.**
Above, we mentioned the use of robust estimators, or partially adaptive estimation, when the Gaussianity assumption is not applicable. However, these techniques are suitable when the unknown error distribution is "approximately Gaussian" in the sense of being unimodal, roughly bell-shaped, and having support over the whole real line. If the dependent variable is strictly positive and thus right-skewed, as occurs, for example, with lifetimes, waiting times, incomes, dividend payments, insurance claims, etc., then these aforementioned techniques are less applicable. Instead, one could model the expected value of a positive continuous random variable, such as the gamma, Pareto, (generalized) inverse Gaussian, etc., and the fitted regression coefficients would somehow need to be constrained such that $\mathbf{x}_t'\boldsymbol{\beta}$ is positive for all relevant $\mathbf{x}_t$.

Yet more complicated situations arise if the dependent variable is discrete, say, Bernoulli, binomial, multinomial, negative binomial, or Poisson. The above situation, as well as the discrete case, can all be elegantly handled by the use of what is referred to as the generalized linear model, or GLIM, whereby a transformation of the dependent variable is applied such that a regression can be used for modeling its mean. The assumed distribution of the dependent variable is usually taken to be a member of the exponential family, one example of which is the Gaussian, as studied in this chapter, in which case no transformation is required.

We briefly illustrate the mechanics assuming a Bernoulli distribution (with support zero and one) for the dependent variable Y. An example of this could be in so-called **credit scoring**, or **probability of default** models, whereby the credit-worthiness of a bank client (no or yes, i.e., 0 or 1) for receiving a loan is to be assessed, based on several exogenous factors (there are numerous books on this topic, e.g., Baesens et al. (2016) and Bluhm et al. (2010)). Let $\pi_i = \Pr(Y_i = 1) = \mathbb{E}[Y_i]$, and denote by η_i the linear predictor $\eta_i = \beta_1 x_{i,1} + \beta_2 x_{i,2} + \cdots + \beta_k x_{i,k} = \mathbf{x}_i'\boldsymbol{\beta}$, $i = 1, 2, \ldots, n$, as in (1.2). They are related via a **response function** h such that $\pi_i = h(\eta_i)$, where h is a strictly monotone increasing function that maps to the interval $(0, 1)$, such as the standard normal c.d.f. Φ, and inverse function $\eta_i = g(\pi_i)$, where function $g = h^{-1}$ is referred to as the **link function**. The so-called **logit** model takes

$$\pi_i = h(\eta_i) = \frac{\exp\{\eta_i\}}{1 + \exp\{\eta_i\}}, \quad g(\pi_i) = h^{-1}(\pi_i) = \log\left(\frac{\pi_i}{1 - \pi_i}\right) = \mathbf{x}_i'\boldsymbol{\beta},$$

while the **probit** model takes $\pi_i = h(\eta_i) = \Phi(\eta_i)$.

Good introductory accounts of GLIM (with the benefit of having books that cover numerous other aspects of linear and other models) can be found in Rao et al. (2008, Ch. 10), Khuri (2010, Ch. 13), Fahrmeir et al. (2013, Ch. 5), and Greene (2017), while several highly detailed books dedicated to the subject exist, such as Fahrmeir and Tutz (2001), Winkelmann (2008), and Agresti (2015).

1.7 Problems

Problem 1.1 Consider the simple linear regression model $Y_t = \beta_1 + \beta_2 X_t + \epsilon_t$, $t = 1, \ldots, T$.

a) By setting $\partial S(\boldsymbol{\beta})/\partial\beta_1$ to zero, show that $\widehat{\beta}_1 = \bar{Y} - \widehat{\beta}_2\bar{X}$. Using this with $0 = \partial S(\boldsymbol{\beta})/\partial\beta_2$, show that $\widehat{\beta}_2 = \hat{\sigma}_{X,Y}/\hat{\sigma}_X^2$, where $\hat{\sigma}_{X,Y}$ denotes the sample covariance between X and Y,

$$\hat{\sigma}_{X,Y} := \frac{1}{T-1}\sum_{t=1}^{T}(X_t - \bar{X})(Y_t - \bar{Y}),$$

and $\hat{\sigma}_X^2 := \hat{\sigma}_{X,X}$.

b) Show that $\widehat{Y}_t - \bar{Y} = \widehat{\beta}_2(X_t - \bar{X})$.

c) Define the standardized variables $x_t = (X_t - \bar{X})/\hat{\sigma}_X$ and $y_t = (Y_t - \bar{Y})/\hat{\sigma}_Y$, and consider the regression $y_t = \alpha_1 + \alpha_2 x_t + \varepsilon_t$. Show that $\hat{\alpha}_1 = 0$ and $\hat{\alpha}_2 = \hat{\rho}$, where $\hat{\rho} = \hat{\rho}_{X,Y}$ is the sample correlation between X and Y, with $|\hat{\rho}| \leqslant 1$. Thus, we can write

$$\widehat{Y}_t = \hat{\alpha}_1 + \hat{\alpha}_2 x_t = \hat{\rho} x_t,$$

and squaring and summing both sides yields $\hat{\rho}^2 = \sum \widehat{Y}_t^2 / \sum x_t^2$. Show that the R^2 statistics for the two regression models are the same, namely $\hat{\rho}^2$.

Problem 1.2 Show (1.12) directly (without use of Theorem 1.6) for the simple linear regression model $Y_t = \beta_1 + \beta_2 X_t + \epsilon_t$.

Problem 1.3 For nonsingular matrix $\mathbf{A}$, its **partitioned inverse** $\mathbf{A}^{-1}$ is

$$\begin{bmatrix} \mathbf{A}_{11} & \mathbf{A}_{12} \\ \mathbf{A}_{21} & \mathbf{A}_{22} \end{bmatrix}^{-1} = \begin{bmatrix} \mathbf{W}^{-1} & -\mathbf{W}^{-1}\mathbf{A}_{12}\mathbf{A}_{22}^{-1} \\ -\mathbf{A}_{22}^{-1}\mathbf{A}_{21}\mathbf{W}^{-1} & \mathbf{A}_{22}^{-1} + \mathbf{A}_{22}^{-1}\mathbf{A}_{21}\mathbf{W}^{-1}\mathbf{A}_{12}\mathbf{A}_{22}^{-1} \end{bmatrix} \tag{1.110}$$

$$= \begin{bmatrix} \mathbf{A}_{11}^{-1} + \mathbf{A}_{11}^{-1}\mathbf{A}_{12}\mathbf{Z}^{-1}\mathbf{A}_{21}\mathbf{A}_{11}^{-1} & -\mathbf{A}_{11}^{-1}\mathbf{A}_{12}\mathbf{Z}^{-1} \\ -\mathbf{Z}^{-1}\mathbf{A}_{21}\mathbf{A}_{11}^{-1} & \mathbf{Z} \end{bmatrix},$$

where $\mathbf{W} = \mathbf{A}_{11} - \mathbf{A}_{12}\mathbf{A}_{22}^{-1}\mathbf{A}_{21}$ and $\mathbf{Z} = \mathbf{A}_{22} - \mathbf{A}_{21}\mathbf{A}_{11}^{-1}\mathbf{A}_{12}$. This is a well-known result that can be found in numerous books on matrix algebra, and confirmed by computing $\mathbf{A}\mathbf{A}^{-1}$. Derive the Frisch–Waugh–Lovell theorem by applying the partitioned inverse (1.110) expression to (1.5).

Problem 1.4 Prove that the projection matrix $\mathbf{P}_{\mathcal{S}}$ in (1.42) is unique.
Hint: Let $\mathbf{H} = [\mathbf{h}_1 \ \mathbf{h}_2 \ \dots \ \mathbf{h}_k]$ be a different basis for $\mathcal{S}$. Justify that we can write $\mathbf{H} = \mathbf{T}\mathbf{A}$ for some $\mathbf{A}$.

Problem 1.5 This is a less direct, but instructive, method for proving Theorem 1.3. Let $\mathbf{M} = \mathbf{I}_T - \mathbf{P}_{\mathcal{S}}$ with $\dim(\mathcal{S}) = k$, $k \in \{1, 2, \dots, T-1\}$. Via the spectral decomposition, let $\mathbf{H}$ be an orthogonal matrix whose rows consist of the eigenvectors of $\mathbf{M}$. Partition $\mathbf{H}$ as

$$\mathbf{H} = \begin{bmatrix} \mathbf{H}_1 \\ \mathbf{H}_2 \end{bmatrix},$$

with "correct" sizes, and use Theorem 1.2 to write $\mathbf{H}\mathbf{M}\mathbf{H}'$ as a block matrix. Show that $\mathbf{M}\mathbf{H}_2' = \mathbf{0}$ and $\mathbf{H}_2' = \mathbf{P}_{\mathcal{S}}\mathbf{H}_2'$. This implies the rows of $\mathbf{H}_2$ are in $\mathcal{S}$. Use this to show $\mathbf{H}_1\mathbf{H}_2' = \mathbf{0} \Leftrightarrow \mathbf{H}_1\mathbf{M} = \mathbf{H}_1$. Postmultiply $\mathbf{H}'\mathbf{H} = \mathbf{I}_T$ by $\mathbf{M}$ to show $\mathbf{H}_1'\mathbf{H}_1 = \mathbf{M}$. Finally, show that $\mathbf{H}_1\mathbf{H}_1' = \mathbf{I}_{T-k}$.

Problem 1.6 Prove that the restricted least squares estimator $\hat{\gamma}$ given in (1.69) satisfies

1. $\mathbf{H}\hat{\gamma} = \mathbf{h}$ and
2. $\|\mathbf{Y} - \mathbf{X}\hat{\gamma}\|^2 \leqslant \|\mathbf{Y} - \mathbf{X}\mathbf{b}\|^2$ for all $\mathbf{b} \in \mathbb{R}^k$ such that $\mathbf{H}\mathbf{b} = \mathbf{h}$.

Hint: For 2, first show that, for every $\mathbf{b} \in \mathbb{R}^k$ such that $\mathbf{H}\mathbf{b} = \mathbf{h}$,

$$\|\mathbf{Y} - \mathbf{X}\mathbf{b}\|^2 = \|\mathbf{Y} - \mathbf{X}\hat{\beta}\|^2 + \|\mathbf{X}\hat{\beta} - \mathbf{X}\mathbf{b}\|^2,$$

and then argue it suffices to show that $\|\mathbf{X}\hat{\beta} - \mathbf{X}\hat{\gamma}\|^2 \leqslant \|\mathbf{X}\hat{\beta} - \mathbf{X}\mathbf{b}\|^2$. Add and subtract $\hat{\gamma}$ to the latter term, expand, and show the cross term is zero.

Problem 1.7 Let $\mathbf{u}, \mathbf{v} \in \mathbb{R}^n$. Prove the Cauchy–Schwarz inequality $\langle \mathbf{u}, \mathbf{v} \rangle \leqslant \|\mathbf{u}\| \, \|\mathbf{v}\|$ as follows.

1. Show that $0 \leqslant \langle \mathbf{u} - a\mathbf{v}, \mathbf{u} - a\mathbf{v} \rangle$ for all $a \in \mathbb{R}$.
2. Expand $\langle \mathbf{u} - a\mathbf{v}, \mathbf{u} - a\mathbf{v} \rangle$ and let $a = \langle \mathbf{u}, \mathbf{v} \rangle / \langle \mathbf{v}, \mathbf{v} \rangle$.

Problem 1.8 Prove Theorem 1.2, i.e., if $\mathbf{P}$ is symmetric and idempotent with $\text{rank}(\mathbf{P}) = k$, then (i) k of the eigenvalues of $\mathbf{P}$ are unity and the remaining $T - k$ are zero, and (ii) $\text{tr}(\mathbf{P}) = k$.
Hint: For (i), continue with the relation $\lambda\mathbf{x} = \mathbf{P}\mathbf{x} = \mathbf{P}\mathbf{P}\mathbf{x}$, and for (ii), let $\mathbf{P} = \mathbf{U}\mathbf{D}\mathbf{U}'$ and continue with the relation $k = \text{rank}(\mathbf{P}) = \text{tr}(\mathbf{D})$.

The converse of the result in Theorem 1.2 is, however, not true. For example, with $\mathbf{A} = \begin{bmatrix} 1 & 1 \\ 0 & 1 \end{bmatrix}$, rank($\mathbf{A}$) = tr($\mathbf{A}$) = 2 and a standard computation shows that the eigenvalues of $\mathbf{A}$ are both one. But $\mathbf{A}$ is neither symmetric nor idempotent.

Finally, there are related results without requiring symmetry. For example, the matrix

$$\mathbf{A} = \begin{bmatrix} 2 & -1/4 & -1/6 \\ 18 & -7/2 & -3 \\ -15 & 15/4 & 7/2 \end{bmatrix}$$

is not symmetric, but it is idempotent, with rank two, eigenvalues 0, 1 and 1, and tr($\mathbf{A}$) = 2. In general, if $\mathbf{A}$ is idempotent with k eigenvalues equal to one (and the rest zero), then rank($\mathbf{A}$) = tr($\mathbf{A}$) = k; see, e.g., Magnus and Neudecker (2007, p. 22).

Problem 1.9 Prove Theorem 1.6.

Problem 1.10 Partition the linear regression model (1.3) as

$$\mathbf{Y} = \begin{pmatrix} \mathbf{X}_1 & \mathbf{X}_2 \end{pmatrix} \begin{pmatrix} \boldsymbol{\beta}_1 \\ \boldsymbol{\beta}_2 \end{pmatrix} + \boldsymbol{\epsilon} = \mathbf{X}_1\boldsymbol{\beta}_1 + \mathbf{X}_2\boldsymbol{\beta}_2 + \boldsymbol{\epsilon}.$$

For convenience, let $\mathbf{M}_1 = \mathbf{M}_{\mathbf{X}_1} = \mathbf{I} - \mathbf{P}_{\mathbf{X}_1}$. Part (b) of Theorem 1.6 implies that $\mathbf{P}_{\mathbf{X}} = \mathbf{P}_{\mathbf{X}_1} + \mathbf{P}_{\mathbf{M}_1\mathbf{X}_2}$. Show this directly by using the projection and perpendicularity conditions (1.48) and (1.49).
Hint: Recall from the definition of column space (1.38) that, for an $\mathbf{x} \in \mathcal{C}(\mathbf{X})$, there exists a $\boldsymbol{\gamma}$ such that $\mathbf{x} = \mathbf{X}\boldsymbol{\gamma} = \mathbf{X}_1\boldsymbol{\gamma}_1 + \mathbf{X}_2\boldsymbol{\gamma}_2$, where $\boldsymbol{\gamma}$ is appropriately partitioned into $\boldsymbol{\gamma}_1$ and $\boldsymbol{\gamma}_2$.

Problem 1.11 Because $\mathbf{M}$ in (1.53) is a projection matrix onto $\mathcal{C}(\mathbf{X})^{\perp}$, it follows from Theorem 1.2 that rank($\mathbf{M}$) = $T - k$. Show this result using (B.67) and (B.68), i.e., if $\mathbf{A}$ and $\mathbf{B}$ are two matrices of the same size, then

$$\text{rank}(\mathbf{A} + \mathbf{B}) \leqslant \text{rank}(\mathbf{A}) + \text{rank}(\mathbf{B}),$$

and if $\mathbf{A}$ and $\mathbf{B}$ are $n \times n$ and $n \times k$ matrices, respectively, $k \geqslant 1$, then

$$\text{rank}(\mathbf{AB}) \geqslant \text{rank}(\mathbf{A}) + \text{rank}(\mathbf{B}) - n.$$

Problem 1.12 As in (1.66), let matrix $\mathbf{H}$ be of dimension $J \times k$ and full rank, with $J \leqslant k$. Show that $\mathbf{K} = \sigma^2 \mathbf{AH}'(\mathbf{HAH}')^{-1}\mathbf{HA}$ is positive semi-definite for $J < k$, where $\mathbf{A} = (\mathbf{X}'\mathbf{X})^{-1}$.
Hint: If you are not convinced of the following fact, then prove it first: If $\mathbf{A}$ is a real symmetric matrix of size n with full rank n, then so is $\mathbf{A}^{-1}$.
What happens when $J = k$?

Problem 1.13 Numerically find the minimum number of observations T required in Example 1.11 to achieve a given power, using $\alpha = 0.05$.

Problem 1.14 We had derived the restricted least squares estimator $\hat{\boldsymbol{\gamma}}$ for the model $\mathbf{Y} = \mathbf{X}\boldsymbol{\beta} + \boldsymbol{\epsilon}$ when the restriction $\mathbf{H}\boldsymbol{\beta} = \mathbf{h}$ holds, where $\mathbf{H}$ is $J \times k$ of full rank $J \leqslant k$. There is another way of doing this. It begins by expressing $\mathbf{H}\boldsymbol{\beta} = \mathbf{h}$ as $\boldsymbol{\beta} = \mathbf{S}\boldsymbol{\eta} + \mathbf{s}$, where the parameter vector $\boldsymbol{\eta}$ is of dimension $k - J$. That is, $\mathbf{Y} = \mathbf{X}\boldsymbol{\gamma} + \boldsymbol{\epsilon}$, where

$$\mathbf{X}\boldsymbol{\gamma} \in \mathcal{S}_H = \{\mathbf{y} : \mathbf{y} = \mathbf{X}\boldsymbol{\beta},\ \boldsymbol{\beta} = \mathbf{S}\boldsymbol{\eta} + \mathbf{s},\ \boldsymbol{\eta} \in \mathbb{R}^{k-J}\}.$$

An extensive treatment of the relation between these parameterizations is provided by Hirschberg and Slottje (1999).

For example, let $\boldsymbol{\beta} = (\beta_1, \dots, \beta_4)'$ and consider the constraint $\beta_2 = 2\beta_3$. Then we would take $\mathbf{H} = [0 \quad 1 \quad -2 \quad 0]$ and $\mathbf{h} = 0$. Alternatively, this can be expressed by

$$\boldsymbol{\beta} = \begin{pmatrix} \beta_1 \\ \beta_2 \\ \beta_2/2 \\ \beta_4 \end{pmatrix} = \begin{pmatrix} 1 & 0 & 0 \\ 0 & 1 & 0 \\ 0 & 1/2 & 0 \\ 0 & 0 & 1 \end{pmatrix} \begin{pmatrix} \beta_1 \\ \beta_2 \\ \beta_4 \end{pmatrix} + \mathbf{0}, \text{ i.e., } \mathbf{S} = \begin{pmatrix} 1 & 0 & 0 \\ 0 & 1 & 0 \\ 0 & 1/2 & 0 \\ 0 & 0 & 1 \end{pmatrix},$$

$\mathbf{s} = \mathbf{0}$ and $\eta = [\beta_1 \quad \beta_2 \quad \beta_4]'$.

a) Let $\boldsymbol{\beta} = (\beta_1, \dots, \beta_4)'$ but with the constraint that $\sum_{i=2}^4 \beta_i = 1$. Give the appropriate values of $\mathbf{H}$, $\mathbf{h}$, $\mathbf{S}$, $\boldsymbol{\eta}$ and $\mathbf{s}$.
b) For some given values of $\mathbf{S}$, $\boldsymbol{\eta}$ and $\mathbf{s}$, derive $\widehat{\boldsymbol{\gamma}}$.
   Hint: Plug in $\boldsymbol{\beta} = \mathbf{S}\boldsymbol{\eta} + \mathbf{s}$ into the regression model.
   (Ruud, 2000, pp. 79–80)
c) Express $\mathbf{X}\widehat{\boldsymbol{\gamma}}$ as $\mathbf{P_Z}\mathbf{Y} + (\mathbf{I} - \mathbf{P_Z})\mathbf{X}\mathbf{s}$, where $\mathbf{P_Z}$ is a projection matrix.
d) Show that the constraint $\mathbf{H}\boldsymbol{\beta} = \mathbf{h}$, where $\mathbf{H}$ is $J \times k$ and $\text{rank}(\mathbf{H}) = J \leqslant k$, can always be expressed as $\boldsymbol{\beta} = \mathbf{S}\boldsymbol{\eta} + \mathbf{s}$.
   (Ruud, 2000, p. 94(4.14a))

Problem 1.15 Recall the form of the generalized likelihood ratio statistic. For testing $H_0 : \mathbf{H}\boldsymbol{\beta} = \mathbf{h}$ in the linear model, it is given by

$$\text{LR} = \text{LR}(\mathbf{Y}, \mathbf{X}, \mathbf{H}, \mathbf{h}) = \frac{\max_{\sigma^2, \boldsymbol{\beta}: \mathbf{H}\boldsymbol{\beta} = \mathbf{h}} \mathcal{L}(\boldsymbol{\beta}, \sigma^2; \mathbf{Y})}{\max_{\sigma^2, \boldsymbol{\beta}} \mathcal{L}(\boldsymbol{\beta}, \sigma^2; \mathbf{Y})} = \frac{\mathcal{L}(\widehat{\boldsymbol{\gamma}}, \tilde{\sigma}^2_{\gamma}; \mathbf{Y})}{\mathcal{L}(\widehat{\boldsymbol{\beta}}, \tilde{\sigma}^2; \mathbf{Y})},$$

where $\widehat{\boldsymbol{\beta}} = (\mathbf{X}'\mathbf{X})^{-1}\mathbf{X}'\mathbf{Y}$ and $\tilde{\sigma}^2 = T^{-1}\mathsf{S}(\widehat{\boldsymbol{\beta}})$ refer to the unrestricted m.l.e. and $\widehat{\boldsymbol{\gamma}}$ and $\tilde{\sigma}^2_{\gamma} = T^{-1}\mathsf{S}(\widehat{\boldsymbol{\gamma}})$ refer to the restricted ones, where $\widehat{\boldsymbol{\gamma}}$ is given in (1.69). Show that a test of H_0 involving LR is equivalent to the F test given in (1.88).

Problem 1.16 This exercise will be of value in Section 2.5.2. Recall that, if $G \sim \text{Gam}(\alpha, \beta)$, $\alpha > 0$, $\beta > 0$, its p.d.f. is

$$f_G(x; \alpha, \beta) = \frac{\beta^\alpha}{\Gamma(\alpha)} x^{\alpha-1} \exp(-\beta x) \mathbb{I}(x > 0),$$

where

$$\Gamma(a) = \int_0^\infty x^{a-1} e^{-x}\, dx \quad \text{and} \quad \int_0^\infty x^{\alpha-1} \exp(-\beta x)\, dx = \frac{\Gamma(\alpha)}{\beta^\alpha}. \tag{1.111}$$

Let $G_i \overset{\text{ind}}{\sim} \text{Gam}(\alpha_i, 1)$ and let $R_1 = G_1/G_3$ and $R_2 = G_2/G_3$. It is clear that, conditional on G_3, R_1 and R_2 are independent. Show that without conditioning they are not, by confirming (omitting the obvious indicator functions)

$$f_{R_1,R_2}(r_1, r_2) = \frac{\Gamma(\alpha_1 + \alpha_2 + \alpha_3)}{\Gamma(\alpha_1)\Gamma(\alpha_2)\Gamma(\alpha_3)} \frac{r_1^{\alpha_1-1} r_2^{\alpha_2-1}}{(1 + r_1 + r_2)^{\alpha_1+\alpha_2+\alpha_3}},$$

which does not factor as $f_{R_1}(r_1) \times f_{R_2}(r_2)$. Further confirm that $f_{R_1,R_2}(r_1, r_2)$ integrates to one by using the function `dblquad` in Matlab.

1.A Appendix: Derivation of the BLUS Residual Vector

This appendix derives the BLUS residual vector (1.104). It is a detailed amalgam of the various proofs given in Theil (1965, 1968, 1971), Chow (1976), and Magnus and Sinha (2005), with the hope that the development shown here (that becomes visible and straightforward once atop the proverbial shoulders of giants, notably Henri Theil and Jan Magnus) serves as a clear, complete, and perhaps definitive derivation.[14]

Recall that we wish a residual estimator of the form $\widehat{\epsilon}_{\text{LUS}} = \mathbf{CY}$, where $\mathbf{C}$ is $(T-k) \times T$, and that the relevant minimization problem for the BLUS estimator is (writing just $\widehat{\epsilon}$ for $\widehat{\epsilon}_{\text{LUS}}$)

$$\widehat{\epsilon}_{\text{BLUS}} = \arg\min_{\widehat{\epsilon}} \mathbb{E}[(\widehat{\epsilon} - \epsilon_1)'(\widehat{\epsilon} - \epsilon_1)] \quad \text{subject to} \quad \mathbf{CX} = \mathbf{0},\ \mathbf{CC}' = \mathbf{I}, \tag{1.112}$$

where ϵ_1 is defined via the partition of the model in (1.103), repeated here as

$$\begin{bmatrix} \mathbf{Y}_0 \\ \mathbf{Y}_1 \end{bmatrix} = \begin{bmatrix} \mathbf{X}_0 \\ \mathbf{X}_1 \end{bmatrix} \beta + \begin{bmatrix} \epsilon_0 \\ \epsilon_1 \end{bmatrix} = \begin{bmatrix} \mathbf{X}_0 \\ \mathbf{X}_1 \end{bmatrix} \widehat{\beta}_{\text{LS}} + \begin{bmatrix} \mathbf{e}_0 \\ \mathbf{e}_1 \end{bmatrix}, \tag{1.113}$$

with ϵ_0 and $\mathbf{e}_0$ of size $k \times 1$, and ϵ_1 and $\mathbf{e}_1$ of size $(T-k) \times 1$.

We divide the derivation into several small parts.

Reduce the Two Constraints to One

The first part of the derivation consists in reducing the number of (matrix) constraints to one. The partition $\mathbf{C} = \begin{bmatrix} \mathbf{C}_0 & \mathbf{C}_1 \end{bmatrix}$ with $\mathbf{e} = \begin{bmatrix} \mathbf{e}_0 & \mathbf{e}_1 \end{bmatrix}'$, where $\mathbf{e}$ is of size $T \times 1$, yields

$$\mathbf{Ce} = \mathbf{C}_0\mathbf{e}_0 + \mathbf{C}_1\mathbf{e}_1, \tag{1.114}$$

where $\mathbf{C}_0$ is $(T-k) \times k$ and $\mathbf{C}_1$ is $(T-k) \times (T-k)$. Observe that the symmetry of $\mathbf{C}$ implies that of $\mathbf{C}_1$.

Using $\mathbf{CX} = \mathbf{0}$ and $\mathbf{X}'\mathbf{e} = \mathbf{0}$, we have

$$\mathbf{C}_0\mathbf{X}_0 + \mathbf{C}_1\mathbf{X}_1 = \mathbf{0}, \qquad \mathbf{X}'_0\mathbf{e}_0 + \mathbf{X}'_1\mathbf{e}_1 = \mathbf{0},$$

so that with

$$\mathbf{Z} = \mathbf{X}_1\mathbf{X}_0^{-1}, \tag{1.115}$$

we can write

$$\mathbf{e}_0 = -(\mathbf{X}_1\mathbf{X}_0^{-1})'\mathbf{e}_1 = -\mathbf{Z}'\mathbf{e}_1, \qquad \mathbf{C}_0 = -\mathbf{C}_1(\mathbf{X}_1\mathbf{X}_0^{-1}) = -\mathbf{C}_1\mathbf{Z}. \tag{1.116}$$

---

14 The author is grateful to my brilliant master's student Christian Frey for assembling this meticulous and detailed derivation from the original papers.

Further, using $\mathbf{CC}' = \mathbf{I}$, (1.116) yields

$$\mathbf{CC}' = \mathbf{C}_0\mathbf{C}_0' + \mathbf{C}_1\mathbf{C}_1' = \mathbf{C}_1\mathbf{ZZ}'\mathbf{C}_1' + \mathbf{C}_1\mathbf{C}_1' = \mathbf{C}_1[\mathbf{I} + \mathbf{ZZ}']\mathbf{C}_1' = \mathbf{I}, \tag{1.117}$$

so that both constraints $\mathbf{CX} = \mathbf{0}$ and $\mathbf{CC}' = \mathbf{I}$ are equivalent to (1.117). Moreover, by assumption $\mathbf{CX} = \mathbf{0}$, it follows that $\mathbf{CY} = \mathbf{C}\epsilon = \mathbf{Ce}$. As $\mathbf{CY} = (\mathbf{X}\beta + \epsilon) = \mathbf{C}\epsilon$ and $\mathbf{Ce} = \mathbf{C}(\mathbf{Y} - \hat{\beta}\mathbf{X}) = \mathbf{CY}$,

$$\hat{\epsilon} = \mathbf{CY} = \mathbf{C}\epsilon = \mathbf{C}_0\epsilon_0 + \mathbf{C}_1\epsilon_1 = -\mathbf{C}_1\mathbf{Z}\epsilon_0 + \mathbf{C}_1\epsilon_1,$$

and therefore

$$\begin{aligned}
&\mathrm{Cov}[(\hat{\epsilon} - \epsilon_1), (\hat{\epsilon} - \epsilon_1)] \\
&= \mathrm{Cov}[(-\mathbf{C}_1\mathbf{Z}\epsilon_0 + (\mathbf{C}_1 - \mathbf{I})\epsilon_1), (-\mathbf{C}_1\mathbf{Z}\epsilon_0 + (\mathbf{C}_1 - \mathbf{I})\epsilon_1)] \\
&= \sigma^2[\mathbf{C}_1(\mathbf{I} + \mathbf{ZZ}')\mathbf{C}_1' + \mathbf{I} - \mathbf{C}_1 - \mathbf{C}_1'].
\end{aligned} \tag{1.118}$$

The minimization problem for the BLUS estimator is then reduced to

$$\hat{\epsilon}_{\mathrm{BLUS}} = \arg\ \min_{\hat{\epsilon}} \mathbb{E}[(\hat{\epsilon} - \epsilon_1)'(\hat{\epsilon} - \epsilon_1)] \quad \text{subject to} \quad (1.117).$$

Solve with a Lagrangean Approach

Note that $\hat{\epsilon} = \mathbf{CY} = \mathbf{Ce}$, so that, with (1.118) and (1.117), the constrained minimization problem is equivalent to the Lagrangean

$$\begin{aligned}
L(\mathbf{C}_1, \lambda) = {} & \mathrm{tr}([\mathbf{C}_1(\mathbf{I} + \mathbf{ZZ}')\mathbf{C}_1' + \mathbf{I} - \mathbf{C}_1 - \mathbf{C}_1']) \\
& - \mathrm{tr}(\lambda[\mathbf{C}_1(\mathbf{I} + \mathbf{ZZ}')\mathbf{C}_1' - \mathbf{I}]),
\end{aligned} \tag{1.119}$$

where λ denotes the Lagrange multiplier matrix of dimension $(T - k) \times (T - k)$.

As $\partial \mathrm{tr}(\mathbf{AB})/\partial \mathbf{A} = \partial \mathrm{tr}(\mathbf{BA})/\partial \mathbf{A} = \mathbf{B}'$, the first-order condition with respect to $\mathbf{C}_1$ is

$$\frac{\partial L}{\partial \mathbf{C}_1} = 2\mathbf{C}_1(\mathbf{I} + \mathbf{ZZ}') - 2\mathbf{I} - 2\lambda\mathbf{C}_1(\mathbf{I} + \mathbf{ZZ}') = \mathbf{0}. \tag{1.120}$$

Symmetry of $\mathbf{C}_1$ Gives a Spectral Decomposition

To solve (1.120) for the two unknowns $\mathbf{C}_1$ and λ, postmultiply (1.120) by $\mathbf{C}_1'$ and use (1.117) to get

$$\lambda = \mathbf{I} - \mathbf{C}_1' = \mathbf{I} - \mathbf{C}_1, \tag{1.121}$$

which is obviously symmetric from the symmetry of $\mathbf{C}_1$. Substituting (1.121) in (1.120) yields

$$\mathbf{C}_1'\mathbf{C}_1(\mathbf{I} + \mathbf{ZZ}') = \mathbf{I}. \tag{1.122}$$

Thus, (1.122) and a spectral decomposition yield

$$\mathbf{C}_1^2 = (\mathbf{I} + \mathbf{ZZ}')^{-1} = \mathbf{PD}^2\mathbf{P}', \tag{1.123}$$

where, from the symmetry of $\mathbf{C}_1$, $\mathbf{D}^2$ is the $(T - k) \times (T - k)$ diagonal matrix with entries d_k^2 and $\mathbf{P}$ is the $(T - k) \times (T - k)$ orthogonal matrix ($\mathbf{PP}' = \mathbf{I}$) with columns given by the eigenvectors of $(\mathbf{I} + \mathbf{ZZ}')^{-1}$ corresponding to the eigenvalues $d_1^2, \ldots, d_{T-k}^2$. It is worth emphasizing that the symmetry of $\mathbf{C}_1$ ensures that the d_i are real.

Note that the notation $\mathbf{D}^2$ stands for the d_k^2 entries of matrix $\mathbf{D}^2$, just to avoid usage of the root symbol, while $\mathbf{D}$ is the diagonal matrix with entries d_k restricted to the positive square roots. The solution for (1.123) is then, say, $\mathbf{C}_1^* = (\mathbf{I} + \mathbf{ZZ}')^{-1/2} = \mathbf{PDP}'$. To simplify notation, we subsequently take $\mathbf{C}_1 \equiv \mathbf{C}_1^*$.

It is useful to introduce the partition

$$\mathbf{M} = \mathbf{I} - \mathbf{X}(\mathbf{X}'\mathbf{X})^{-1}\mathbf{X}' = \begin{bmatrix} \mathbf{M}_{00} & \mathbf{M}_{01} \\ \mathbf{M}_{10} & \mathbf{M}_{11} \end{bmatrix},$$

where $\mathbf{M}_{00} = \mathbf{I} - \mathbf{X}_0(\mathbf{X}'\mathbf{X})^{-1}\mathbf{X}_0'$, $\mathbf{M}_{01} = -\mathbf{X}_0(\mathbf{X}'\mathbf{X})^{-1}\mathbf{X}_1'$, $\mathbf{M}_{10} = \mathbf{M}_{01}'$, and $\mathbf{M}_{11} = \mathbf{I} - \mathbf{X}_1(\mathbf{X}'\mathbf{X})^{-1}\mathbf{X}_1'$, though we will make use only of $\mathbf{M}_{11}$. Direct multiplication shows that $\mathbf{M}_{11}^{-1} = \mathbf{I} + \mathbf{X}_1(\mathbf{X}_0'\mathbf{X}_0)^{-1}\mathbf{X}_1'$, i.e., using this latter claim, $\mathbf{M}_{11}\mathbf{M}_{11}^{-1}$ is

$$\begin{aligned}
&[\mathbf{I} - \mathbf{X}_1(\mathbf{X}'\mathbf{X})^{-1}\mathbf{X}_1'][\mathbf{I} + \mathbf{X}_1(\mathbf{X}_0'\mathbf{X}_0)^{-1}\mathbf{X}_1'] \\
&= \mathbf{I} - \mathbf{X}_1(\mathbf{X}'\mathbf{X})^{-1}\mathbf{X}_1' + \mathbf{X}_1(\mathbf{X}_0'\mathbf{X}_0)^{-1}\mathbf{X}_1' - \mathbf{X}_1(\mathbf{X}'\mathbf{X})^{-1}\mathbf{X}_1'\mathbf{X}_1(\mathbf{X}_0'\mathbf{X}_0)^{-1}\mathbf{X}_1' \\
&= \mathbf{I} - \mathbf{X}_1(\mathbf{X}'\mathbf{X})^{-1}\mathbf{X}_1' + \mathbf{X}_1(\mathbf{X}_0'\mathbf{X}_0)^{-1}\mathbf{X}_1' - \mathbf{X}_1(\mathbf{X}'\mathbf{X})^{-1}(\mathbf{X}'\mathbf{X} - \mathbf{X}_0'\mathbf{X}_0)(\mathbf{X}_0'\mathbf{X}_0)^{-1}\mathbf{X}_1' \\
&= \mathbf{I}.
\end{aligned}$$

Thus, with $\mathbf{Z} = \mathbf{X}_1\mathbf{X}_0^{-1}$ from (1.115),

$$\mathbf{M}_{11}^{-1} = \mathbf{I} + \mathbf{ZZ}', \tag{1.124}$$

from which it follows that $\mathbf{M}_{11} = (\mathbf{I} + \mathbf{ZZ}')^{-1}$. From (1.123) and (1.124), $\mathbf{M}_{11}^{-1} = (\mathbf{I} + \mathbf{ZZ}') = (\mathbf{C}_1^2)^{-1} = \mathbf{C}_1^{-2}$ so that, from (1.116),

$$\begin{aligned}
\hat{e}_{\text{BLUS}} &= \mathbf{CY} = \mathbf{Ce} = \mathbf{C}_0\mathbf{e}_0 + \mathbf{C}_1\mathbf{e}_1 = (-\mathbf{C}_1\mathbf{Z})(-\mathbf{Z}'\mathbf{e}_1) + \mathbf{C}_1\mathbf{e}_1 \\
&= \mathbf{C}_1(\mathbf{I} + \mathbf{ZZ}')\mathbf{e}_1 = \mathbf{C}_1\mathbf{M}_{11}^{-1}\mathbf{e}_1 = \mathbf{C}_1^{-1}\mathbf{e}_1 \\
&= \mathbf{e}_1 + (\mathbf{C}_1^{-1} - \mathbf{I})\mathbf{e}_1 \\
&= \mathbf{e}_1 + \sum_{k=1}^{T-k}(d_k^{-1} - 1)\mathbf{p}_k\mathbf{p}_k'\mathbf{e}_1,
\end{aligned} \tag{1.125}$$

where $\mathbf{p}_k$ are the eigenvectors and d_k^2 the eigenvalues of $\mathbf{M}_{11}$. The last equality follows by the existence of a spectral decomposition of $\mathbf{M}_{11} = \mathbf{C}_1^2 = \mathbf{PD}^2\mathbf{P}'$, so that

$$\mathbf{M}_{11}\mathbf{p}_k = [\mathbf{I} - \mathbf{X}_1(\mathbf{XX}')^{-1}\mathbf{X}_1']\mathbf{p}_k = d_k^2\mathbf{p}_k, \quad k = 1, \dots, T-k. \tag{1.126}$$

Premultiplying both sides of (1.126) by $\mathbf{X}_1'$ and using $\mathbf{X}_1'\mathbf{X}_1 = \mathbf{X}'\mathbf{X} - \mathbf{X}_0'\mathbf{X}_0$,

$$\begin{aligned}
\mathbf{X}_1'\mathbf{p}_k - (\mathbf{X}'\mathbf{X} - \mathbf{X}_0'\mathbf{X}_0)(\mathbf{X}'\mathbf{X})^{-1}\mathbf{X}_1'\mathbf{p}_k &= d_k^2\mathbf{X}_1'\mathbf{p}_k \\
\mathbf{X}_0'\mathbf{X}_0(\mathbf{X}'\mathbf{X})^{-1}\mathbf{X}_1'\mathbf{p}_k &= d_k^2\mathbf{X}_1'\mathbf{p}_k, \quad k = 1, \dots, T-k.
\end{aligned} \tag{1.127}$$

Now premultiplying both sides of (1.127) by $(\mathbf{X}_0')^{-1}$, using $\mathbf{Z} = \mathbf{X}_1\mathbf{X}_0^{-1}$, and rearranging,

$$[\mathbf{X}_0(\mathbf{X}'\mathbf{X})^{-1}\mathbf{X}_0' - d_k^2\mathbf{I}]\mathbf{Z}'\mathbf{p}_k = \mathbf{0}, \quad k = 1, \dots, T-k.$$

Use the Spectral Decomposition to Express the BLUS Estimator in terms of $\mathbf{e}_0$ and $\mathbf{e}_1$

Observe that d_k^2 is an eigenvalue of $\mathbf{X}_0(\mathbf{X}'\mathbf{X})^{-1}\mathbf{X}_0'$. As the eigenvectors $\mathbf{Z}'\mathbf{p}_k$ do not have unit length, we normalize by a scalar to get, for $d_k < 1$,

$$\mathbf{q}_k = \frac{d_k}{\sqrt{1 - d_k^2}}\mathbf{Z}'\mathbf{p}_k, \quad k = 1, \dots, T-k, \tag{1.128}$$

so that $\mathbf{q}_1, \ldots, \mathbf{q}_{T-k}$ have unit length and are pairwise orthogonal. As $\mathbf{P}$ is orthogonal, $\mathbf{P}^{-1} = \mathbf{P}'$, so that

$$\mathbf{ZZ}' = \mathbf{M}_{11}^{-1} - \mathbf{I} = (\mathbf{PD}^2\mathbf{P}')^{-1} - \mathbf{I} = (\mathbf{PD}^{-2}\mathbf{P}') - \mathbf{I},$$

and observe that

$$\mathbf{ZZ}'\mathbf{p}_k = \frac{1 - d_k^2}{d_k^2}\mathbf{p}_k, \quad k = 1, \ldots, T - k.$$

Thus, $\mathbf{q}_l'\mathbf{q}_k = 1$ if $l = k$ and zero otherwise for $k, l = 1, \ldots, T - k$. From

$$\frac{1 - d_k^2}{d_k^2}\mathbf{p}_k = \mathbf{Z}(\mathbf{Z}'\mathbf{p}_k) = \frac{\sqrt{1 - d_k^2}}{d_k}\mathbf{Zq}_k,$$

it follows that, if $d_k < 1$, $\mathbf{p}_k = \frac{d_k}{\sqrt{1-d_k^2}}\mathbf{Zq}_k$, $k = 1, \ldots T - k$, so that, with $\mathbf{e}_0 = -\mathbf{Z}'\mathbf{e}_1$ and $\mathbf{Z} = \mathbf{X}_1\mathbf{X}_0^{-1}$, the last line of (1.125) can be written as

$$\hat{\epsilon}_{\mathrm{BLUS}} = \mathbf{e}_1 + \sum_{k=1}^{T-k}\left(\frac{1}{d_k} - 1\right)\mathbf{p}_k\mathbf{p}_k'\mathbf{e}_1 \tag{1.129}$$

$$= \mathbf{e}_1 + \mathbf{Z}\sum_{k=1}^{T-k}\left(\frac{1}{d_k} - 1\right)\frac{d_k^2}{1 - d_k^2}\mathbf{q}_k\mathbf{q}_k'\mathbf{Z}'\mathbf{e}_1 \tag{1.130}$$

$$= \mathbf{e}_1 + \mathbf{X}_1\mathbf{X}_0^{-1}\sum_{k=1}^{T-k}\frac{d_k}{1 + d_k}\mathbf{q}_k\mathbf{q}_k'\mathbf{e}_0, \tag{1.131}$$

where in (1.129), the kth term in the sum is zero if $d_k = 1$. Thus, we can restrict the summation in (1.130) and (1.131) to $k = 1, \ldots, H$, where $d_k < 1$, for all $k = 1, \ldots, H$, with $H \leqslant T - k$. The result is sometimes expressed as a permutation of the elements d_h, $h = 1, \ldots, H$, say $d_1 \leqslant d_2 \leqslant \ldots \leqslant d_H < 1$, such that the d_h are nondecreasing. This yields (1.104), i.e.,

$$\hat{\epsilon}_{\mathrm{BLUS}} = \mathbf{e}_1 + \mathbf{X}_1\mathbf{X}_0^{-1}\sum_{h=1}^{H}\frac{d_h}{1 + d_h}\mathbf{q}_h\mathbf{q}_h'\mathbf{e}_0.$$

Observe that the BLUS estimator is represented as a deviation from the corresponding least squares errors.

Verification of Second-order Condition

As in Theil (1965), to verify that $\mathbf{C}^*$ or, equivalently, $\mathbf{C}_1^*$ is indeed a minimum of (1.123), consider an alternative estimator $\bar{\mathbf{C}}\mathbf{Y} = (\mathbf{C} + \mathbf{R})\mathbf{Y} = \left[\mathbf{C}_0' + \mathbf{R}_0' \quad \mathbf{C}_1' + \mathbf{R}_1'\right]\mathbf{Y}$, where $\mathbf{C}_1 = \mathbf{PDP}'$ is the optimal symmetric matrix $\mathbf{C}_1$ from the first-order condition (1.123) and, hence, $\mathbf{C}_0 = -\mathbf{C}_1\mathbf{Z} = -\mathbf{PDP}'\mathbf{Z}$ from (1.116). Note that, as before, $\mathbf{C}_1 \equiv \mathbf{C}_1^*$ and similarly $\mathbf{C} \equiv \mathbf{C}^*$. Recall that $\mathbf{D}$ is restricted to contain only positive diagonal entries (eigenvalues). We wish to show that $\mathbf{C}^* \leqslant \bar{\mathbf{C}}$ for all $\bar{\mathbf{C}}$.

From the assumption $\bar{\mathbf{C}}\mathbf{X} = \mathbf{0}$, it follows that $\mathbf{R}_0'\mathbf{X}_0 + \mathbf{R}_1'\mathbf{X}_1 = \mathbf{0}$, so that $\mathbf{R}_0' = -\mathbf{R}_1'\mathbf{Z}$, with $\mathbf{Z} = \mathbf{X}_1\mathbf{X}_0^{-1}$. Thus, the assumption $\bar{\mathbf{C}}\,\bar{\mathbf{C}}' = \mathbf{I}$, such that $\bar{\mathbf{C}}$ has a scalar covariance matrix, implies

$$(\mathbf{C} + \mathbf{R})'(\mathbf{C} + \mathbf{R}) = (\mathbf{C}_0 + \mathbf{R}_0)'(\mathbf{C}_0 + \mathbf{R}_0) + (\mathbf{C}_1 + \mathbf{R}_1)'(\mathbf{C}_1 + \mathbf{R}_1)$$

$$= (\mathbf{C}_1 + \mathbf{R}_1)'(\mathbf{I} + \mathbf{Z}\mathbf{Z}')(\mathbf{C}_1 + \mathbf{R}_1)$$
$$= (\mathbf{C}_1 + \mathbf{R}_1)'\mathbf{M}_{11}^{-1}(\mathbf{C}_1 + \mathbf{R}_1) = \mathbf{I},$$

where the last equality follows from (1.124). From (1.124) and (1.123), $\mathbf{M}_{11}^{-1} = \mathbf{C}_1^{-2}$, and

$$(\mathbf{I} + \mathbf{C}_1^{-1}\mathbf{R}_1)'(\mathbf{I} + \mathbf{C}_1^{-1}\mathbf{R}_1) = \mathbf{I}, \tag{1.132}$$

implying that $\mathbf{C}_1^{-1}\mathbf{R}_1 + (\mathbf{C}_1^{-1}\mathbf{R}_1)'$ is negative semi-definite. Indeed, with $\mathbf{N} := \mathbf{C}_1^{-1}\mathbf{R}_1$ and $\mathbf{v}' \in \mathbb{R}^{T-k}$ an arbitrary (real) nonzero row vector, premultiplying both sides of (1.132) with $\mathbf{v}'$ and postmultiplying by $\mathbf{v}$ gives

$$\mathbf{v}'(\mathbf{I} + \mathbf{N})'(\mathbf{I} + \mathbf{N})\mathbf{v} = \mathbf{v}'\mathbf{v}, \tag{1.133}$$

implying

$$\mathbf{v}'(\mathbf{N} + \mathbf{N}')\mathbf{v} = -\mathbf{v}'\mathbf{N}'\mathbf{N}\mathbf{v} \leqslant 0, \tag{1.134}$$

so that $\mathbf{N} + \mathbf{N}'$ is negative semi-definite.

Recall that the (unconstrained) objective function in (1.119) can be rewritten with $\mathbf{C}_1\mathbf{C}_1' = \mathbf{I}$. Also recall the properties of the trace operator, $\text{tr}(\mathbf{C}_1) = \text{tr}(\mathbf{C}_1')$, $\text{tr}(\mathbf{C}_1\mathbf{C}_1') = \text{tr}(\mathbf{C}_1'\mathbf{C}_1)$ and $\text{tr}(\mathbf{C}_1(\mathbf{Z}\mathbf{Z}')\mathbf{C}_1') = \text{tr}(\mathbf{C}_1\mathbf{C}_1'(\mathbf{Z}\mathbf{Z}'))$. Then the expectation in (1.112) is

$$\begin{aligned}\mathbb{E}[(\hat{\epsilon} - \epsilon_1)'(\hat{\epsilon} - \epsilon_1)] &= \text{tr}([\mathbf{C}_1(\mathbf{I} + \mathbf{Z}\mathbf{Z}')\mathbf{C}_1' + \mathbf{I} - \mathbf{C}_1 - \mathbf{C}_1']) \\ &= \text{tr}(\mathbf{C}_1\mathbf{C}_1') + \text{tr}(\mathbf{C}_1(\mathbf{Z}\mathbf{Z}')\mathbf{C}_1') + \text{tr}(\mathbf{I}) - 2\text{tr}(\mathbf{C}_1) \\ &= 2\text{tr}(\mathbf{I}) + \text{tr}(\mathbf{I}(\mathbf{Z}\mathbf{Z}')) - 2\text{tr}(\mathbf{C}_1).\end{aligned}$$

It follows that the unconstrained optimization problem as a function only of $\mathbf{C}_1$ is equal to

$$-\min_{\mathbf{C}_1} \text{tr}(\mathbf{C}_1) = \max_{\mathbf{C}_1} \text{tr}(\mathbf{C}_1) = \max_{\mathbf{C}_1} \text{tr}\left(\sum_{k=1}^{T-k} \frac{1}{d_k}\mathbf{p}_k\mathbf{p}_k'\right), \tag{1.135}$$

where the last equality follows from the spectral decomposition $\mathbf{C}_1 = \mathbf{P}\mathbf{D}\mathbf{P}'$; see (1.123). The objective function of the maximization problem (1.135) applied to $\mathbf{R}_1$ is then given as

$$\begin{aligned}\text{tr}(\mathbf{R}_1) = \text{tr}(\mathbf{C}_1\mathbf{N}) &= \text{tr}\left(\sum_{k=1}^{T-k} \frac{1}{d_k}\mathbf{p}_k\mathbf{p}_k'\mathbf{N}\right) = \text{tr}\left(\sum_{k=1}^{T-k} \frac{1}{d_k}\mathbf{p}_k'\mathbf{N}\mathbf{p}_k\right) \\ &= \frac{1}{2}\text{tr}\left(\sum_{k=1}^{T-k} \frac{1}{d_k}\mathbf{p}_k'(\mathbf{N} + \mathbf{N}')\mathbf{p}_k\right) \leqslant 0,\end{aligned}$$

so that, by the negative semi-definiteness of $(\mathbf{N} + \mathbf{N}')$, $\mathbf{N} = \mathbf{0}$, or, equivalently, $\mathbf{R} = \mathbf{0}$, are corresponding maxima of the objective function (1.135) given that the eigenvalues d_k, $k = 1, \ldots, T - k$, are positive. Therefore, $\mathbf{C}_1^*$ is a minimum of (1.119) and hence $\mathbf{C}^*$ is a minimum of (1.112).

1.B Appendix: The Recursive Residuals

Here we provide more detail on the recursive residuals in (1.105). Let $\hat{\boldsymbol{\beta}}_j = (\mathbf{X}_j'\mathbf{X}_j)^{-1}\mathbf{X}_j'\mathbf{Y}_j$ be the o.l.s. estimator obtained by using only the first j, $j \geqslant k$, observations, where $\mathbf{Y}_j$ is the $j \times 1$ vector of the first j

elements of $\mathbf{Y}$, and $\mathbf{X}_j$ is the $j \times k$ matrix of the first j rows of $\mathbf{X}$. As shown in Brown et al. (1975, p. 152), the $\hat{\boldsymbol{\beta}}_j$, $j = k+1, \ldots, T$, can be obtained recursively.

In particular, writing $\mathbf{X}_j'\mathbf{X}_j = \mathbf{X}_{j-1}'\mathbf{X}_{j-1} + \mathbf{x}_j\mathbf{x}_j'$, where $\mathbf{x}_j'$ is the jth row of $\mathbf{X}$, we can apply (1.70) with $\mathbf{A} = \mathbf{X}_{j-1}'\mathbf{X}_{j-1}$, $\mathbf{B} = \mathbf{x}_j$ and scalar $\mathbf{D} = 1$, to get

$$(\mathbf{X}_j'\mathbf{X}_j)^{-1} = (\mathbf{X}_{j-1}'\mathbf{X}_{j-1})^{-1} - \frac{(\mathbf{X}_{j-1}'\mathbf{X}_{j-1})^{-1}\mathbf{x}_j\mathbf{x}_j'(\mathbf{X}_{j-1}'\mathbf{X}_{j-1})^{-1}}{1 + \mathbf{x}_j'(\mathbf{X}_{j-1}'\mathbf{X}_{j-1})^{-1}\mathbf{x}_j}. \tag{1.136}$$

Postmultiplying (1.136) by $\mathbf{x}_j$ and simplifying easily yields

$$(\mathbf{X}_j'\mathbf{X}_j)^{-1}\mathbf{x}_j = \frac{(\mathbf{X}_{j-1}'\mathbf{X}_{j-1})^{-1}\mathbf{x}_j}{1 + \mathbf{x}_j'(\mathbf{X}_{j-1}'\mathbf{X}_{j-1})^{-1}\mathbf{x}_j}. \tag{1.137}$$

Next, from (1.6) and that $\hat{\boldsymbol{\beta}}_{j-1} = (\mathbf{X}_{j-1}'\mathbf{X}_{j-1})^{-1}\mathbf{X}_{j-1}'\mathbf{Y}_{j-1}$, write

$$\begin{aligned}
\mathbf{X}_j'\mathbf{X}_j\hat{\boldsymbol{\beta}}_j &= \mathbf{X}_j'\mathbf{Y}_j = \mathbf{X}_{j-1}'\mathbf{Y}_{j-1} + \mathbf{x}_jY_j = \mathbf{X}_{j-1}'\mathbf{X}_{j-1}\hat{\boldsymbol{\beta}}_{j-1} + \mathbf{x}_jY_j \\
&= (\mathbf{X}_{j-1}'\mathbf{X}_{j-1} + \mathbf{x}_j\mathbf{x}_j')\hat{\boldsymbol{\beta}}_{j-1} + \mathbf{x}_jY_j - \mathbf{x}_j\mathbf{x}_j'\hat{\boldsymbol{\beta}}_{j-1} \\
&= \mathbf{X}_j'\mathbf{X}_j\hat{\boldsymbol{\beta}}_{j-1} + \mathbf{x}_j(Y_j - \mathbf{x}_j'\hat{\boldsymbol{\beta}}_{j-1}),
\end{aligned}$$

premultiply with $(\mathbf{X}_j'\mathbf{X}_j)^{-1}$ and finally use (1.137) to get

$$\hat{\boldsymbol{\beta}}_j = \hat{\boldsymbol{\beta}}_{j-1} + \frac{(\mathbf{X}_{j-1}'\mathbf{X}_{j-1})^{-1}\mathbf{x}_j(Y_j - \mathbf{x}_j'\hat{\boldsymbol{\beta}}_{j-1})}{1 + \mathbf{x}_j'(\mathbf{X}_{j-1}'\mathbf{X}_{j-1})^{-1}\mathbf{x}_j}, \quad j = k+1, \ldots, T. \tag{1.138}$$

The standardized quantities

$$V_j = \frac{Y_j - \mathbf{x}_j'\hat{\boldsymbol{\beta}}_{j-1}}{\sqrt{1 + \mathbf{x}_j'(\mathbf{X}_{j-1}'\mathbf{X}_{j-1})^{-1}\mathbf{x}_j}}, \quad j = k+1, \ldots, T, \tag{1.139}$$

are defined to be the recursive residuals.

Let $\mathbf{V} = (V_{k+1}, \ldots, V_T)'$. We wish to derive the distribution of $\mathbf{V}$. Clearly, $\mathbb{E}[V_j] = 0$. For the variance, as Y_j and $\hat{\boldsymbol{\beta}}_{j-1}$ are independent for $j = k+1, \ldots, T$, and recalling (1.8),

$$\begin{aligned}
\mathbb{V}(V_j) &= \frac{1}{1 + \mathbf{x}_j'(\mathbf{X}_{j-1}'\mathbf{X}_{j-1})^{-1}\mathbf{x}_j}(\mathbb{V}(Y_j) + \mathbf{x}_j'\mathbb{V}(\hat{\boldsymbol{\beta}}_{j-1})\mathbf{x}_j) \\
&= \frac{1}{1 + \mathbf{x}_j'(\mathbf{X}_{j-1}'\mathbf{X}_{j-1})^{-1}\mathbf{x}_j}(\sigma^2 + \sigma^2\mathbf{x}_j'(\mathbf{X}_{j-1}'\mathbf{X}_{j-1})^{-1}\mathbf{x}_j) = \sigma^2.
\end{aligned}$$

Vector $\mathbf{V}$ has a normal distribution, because $\boldsymbol{\epsilon} \sim \mathrm{N}(\mathbf{0}, \sigma^2\mathbf{I})$, and each V_j can be expressed as

$$V_j = \frac{\epsilon_j - \mathbf{x}_j'(\mathbf{X}_{j-1}'\mathbf{X}_{j-1})^{-1}\sum_{k=1}^{j-1}\mathbf{x}_k\epsilon_k}{\sqrt{1 + \mathbf{x}_j'(\mathbf{X}_{j-1}'\mathbf{X}_{j-1})^{-1}\mathbf{x}_j}}. \tag{1.140}$$

To see this, note that $\mathbf{X}_{j-1}'(\mathbf{Y}_{j-1} - \mathbf{X}_{j-1}\boldsymbol{\beta}) = \sum_{k=1}^{j-1}\mathbf{x}_k\epsilon_k$ and hence for the numerator of V_j

$$Y_j - \mathbf{x}_j'\hat{\boldsymbol{\beta}}_{j-1} = \epsilon_j - \mathbf{x}_j'\hat{\boldsymbol{\beta}}_{j-1} + \mathbf{x}_j'\boldsymbol{\beta}$$

$$= \epsilon_j - \mathbf{x}_j'(\mathbf{X}_{j-1}'\mathbf{X}_{j-1})^{-1}\mathbf{X}_{j-1}'(\mathbf{Y}_{j-1} - \mathbf{X}_{j-1}\,\boldsymbol{\beta})$$

$$= \epsilon_j - \mathbf{x}_j'(\mathbf{X}_{j-1}'\mathbf{X}_{j-1})^{-1}\sum_{k=1}^{j-1}\mathbf{x}_k\epsilon_k.$$

For the covariances of $\mathbf{V}$, let N_j be the numerator in (1.140). For $j < i$, $\mathbb{E}[N_jN_i]$ is

$$\mathbb{E}(\epsilon_j\epsilon_i) - \mathbb{E}\left[\epsilon_j\mathbf{x}_i'(\mathbf{X}_{i-1}'\mathbf{X}_{i-1})^{-1}\sum_{k=1}^{i-1}\mathbf{x}_k\epsilon_k\right] - \mathbb{E}\left[\epsilon_i\mathbf{x}_j'(\mathbf{X}_{j-1}'\mathbf{X}_{j-1})^{-1}\sum_{k=1}^{j-1}\mathbf{x}_k\epsilon_k\right]$$
$$+\,\mathbb{E}\left[\mathbf{x}_j'(\mathbf{X}_{j-1}'\mathbf{X}_{j-1})^{-1}\left(\sum_{k=1}^{j-1}\mathbf{x}_k\epsilon_k\right)\mathbf{x}_i'(\mathbf{X}_{i-1}'\mathbf{X}_{i-1})^{-1}\left(\sum_{k=1}^{i-1}\mathbf{x}_k\epsilon_k\right)\right].$$

This, in turn, is

$$-\sigma^2\mathbf{x}_i'(\mathbf{X}_{i-1}'\mathbf{X}_{i-1})^{-1}\mathbf{x}_j + \sigma^2\sum_{k=1}^{j-1}[\mathbf{x}_j'(\mathbf{X}_{j-1}'\mathbf{X}_{j-1})^{-1}\mathbf{x}_k\mathbf{x}_i'(\mathbf{X}_{i-1}'\mathbf{X}_{i-1})^{-1}\mathbf{x}_k] \tag{1.141}$$

$$= -\sigma^2\mathbf{x}_i'(\mathbf{X}_{i-1}'\mathbf{X}_{i-1})^{-1}\mathbf{x}_j + \sigma^2\sum_{k=1}^{j-1}[\mathbf{x}_j'(\mathbf{X}_{j-1}'\mathbf{X}_{j-1})^{-1}\mathbf{x}_k\mathbf{x}_k'(\mathbf{X}_{i-1}'\mathbf{X}_{i-1})^{-1}\mathbf{x}_i] \tag{1.142}$$

$$= -\sigma^2\mathbf{x}_i'(\mathbf{X}_{i-1}'\mathbf{X}_{i-1})^{-1}\mathbf{x}_j + \sigma^2[\mathbf{x}_j'(\mathbf{X}_{j-1}'\mathbf{X}_{j-1})^{-1}(\mathbf{X}_{j-1}'\mathbf{X}_{j-1})(\mathbf{X}_{i-1}'\mathbf{X}_{i-1})^{-1}\mathbf{x}_i] = 0,$$

so that $\mathbf{V} \sim \mathrm{N}(\mathbf{0}, \sigma^2\mathbf{I}_{T-k})$.

1.C Appendix: Solutions

1) For the model $Y_t = \beta_1 + \beta_2X_t + \epsilon_t, t = 1, \dots, T$, with $\hat{\epsilon}_t = Y_t - \widehat{\beta}_1 - \widehat{\beta}_2X_t$, setting $\partial \mathrm{S}(\boldsymbol{\beta})/\partial\beta_1$ to zero gives $0 = -2\sum_{t=1}^T \hat{\epsilon}_t$ or

$$\widehat{\beta}_1 = \bar{Y} - \widehat{\beta}_2\bar{X}. \tag{1.143}$$

Using this in the equation $0 = \partial \mathrm{S}(\boldsymbol{\beta})/\partial\beta_2 = -2\sum_{t=1}^T X_t\widehat{\epsilon}_t$ and simplifying yields

$$\widehat{\beta}_2 = \frac{\sum_{t=1}^T X_tY_t - T\bar{X}\bar{Y}}{\sum_{t=1}^T X_t^2 - T\bar{X}^2} = \frac{\sum_{t=1}^T (X_t - \bar{X})(Y_t - \bar{Y})}{\sum_{t=1}^T (X_t - \bar{X})^2} = \frac{\hat{\sigma}_{X,Y}}{\hat{\sigma}_X^2}, \tag{1.144}$$

where $\hat{\sigma}_{X,Y}$ denotes the sample covariance between X and Y,

$$\hat{\sigma}_{X,Y} = \frac{1}{T-1}\sum_{t=1}^T (X_t - \bar{X})(Y_t - \bar{Y}),$$

and $\hat{\sigma}_X^2 = \hat{\sigma}_{X,X}$. From the first derivative equations, it follows that $\sum\hat{\epsilon}_t = \sum X_t\hat{\epsilon}_t = 0$. Also, as $\widehat{Y}_t = \widehat{\beta}_1 + \widehat{\beta}_2X_t$, it is easy to verify using (1.143) that

$$\widehat{Y}_t - \bar{Y} = \widehat{\beta}_2(X_t - \bar{X}). \tag{1.145}$$

Define the standardized variables $x_t = (X_t - \bar{X})/\hat{\sigma}_X$ and $y_t = (Y_t - \bar{Y})/\hat{\sigma}_Y$ (so that $\bar{x} = \bar{y} = 0$, $\hat{\sigma}_x^2 = \hat{\sigma}_y^2 = 1$ and $\sum x_t^2 = \sum y_t^2 = T - 1$) and consider the regression $y_t = \alpha_1 + \alpha_2 x_t + \varepsilon_t$. Then (1.143) implies $\hat{\alpha}_1 = 0$ and (1.144) implies

$$\hat{\alpha}_2 = \frac{\hat{\sigma}_{x,y}}{\hat{\sigma}_x \hat{\sigma}_y} = \hat{\sigma}_{x,y} = \frac{1}{T-1}\sum_{t=1}^{T} x_t y_t = \frac{(T-1)^{-1}}{\hat{\sigma}_X \hat{\sigma}_Y}\sum_{t=1}^{T}(X_t - \bar{X})(Y_t - \bar{Y}) = \frac{\hat{\sigma}_{X,Y}}{\hat{\sigma}_X \hat{\sigma}_Y} = \hat{\rho},$$

where $\hat{\rho} = \hat{\rho}_{X,Y}$ is the sample correlation between X and Y, with $|\hat{\rho}| \leqslant 1$. Thus, we can write

$$\hat{y}_t = \hat{\alpha}_1 + \hat{\alpha}_2 x_t = \hat{\rho} x_t,$$

and squaring and summing both sides yields $\hat{\rho}^2 = \sum \hat{y}_t^2 / \sum x_t^2$. The R^2 statistic is then

$$R^2 = \frac{\text{ESS}}{\text{TSS}} = \frac{\sum (\hat{y}_t - \bar{y})^2}{\sum (y_t - \bar{y})^2} = \frac{\hat{\rho}^2 \sum x_t^2}{\sum y_t^2} = \hat{\rho}^2.$$

Using (1.145) and (1.144), R^2 for the original model is

$$R^2 = \frac{\text{ESS}}{\text{TSS}} = \frac{\widehat{\beta}_2^2 \sum (X_t - \bar{X})^2}{\sum (Y_t - \bar{Y})^2} = \widehat{\beta}_2^2 \frac{\hat{\sigma}_X^2}{\hat{\sigma}_Y^2} = \frac{\hat{\sigma}_{X,Y}^2}{\hat{\sigma}_X^2 \hat{\sigma}_Y^2} = \hat{\rho}^2,$$

i.e., the same as for the regression with standardized components.

2) We need to show

$$\sum_{t=1}^{T} (Y_t - \bar{Y})^2 = \sum_{t=1}^{T} (Y_t - \widehat{Y}_t)^2 + \sum_{t=1}^{T} (\widehat{Y}_t - \bar{Y})^2.$$

From (1.143) and (1.145), we get

$$\widehat{Y}_t = \bar{Y} + \frac{\hat{\sigma}_{X,Y}}{\hat{\sigma}_X^2}(X_t - \bar{X}),$$

and using

$$\hat{\sigma}_{X,Y} = \frac{1}{T}\sum_{t=1}^{T}(X_t - \bar{X})(Y_t - \bar{Y}) = \frac{1}{T}\sum_{t=1}^{T} X_t Y_t - \bar{X}\bar{Y},$$

simple algebra shows that

$$\sum_{t=1}^{T} (Y_t - \bar{Y})^2 = \sum_{t=1}^{T} Y_t^2 - T\bar{Y}^2,$$

$$\sum_{t=1}^{T} (Y_t - \widehat{Y}_t)^2 = \sum_{t=1}^{T} Y_t^2 - T\bar{Y}^2 - T\frac{\hat{\sigma}_{X,Y}^2}{\hat{\sigma}_X^2},$$

$$\sum_{t=1}^{T} (\widehat{Y}_t - \bar{Y})^2 = T\frac{\hat{\sigma}_{X,Y}^2}{\hat{\sigma}_X^2},$$

proving the result.

3) From the appropriate partition

$$(\mathbf{X}'\mathbf{X}) = \begin{pmatrix} \mathbf{X}_1' \\ \mathbf{X}_2' \end{pmatrix} \begin{pmatrix} \mathbf{X}_1 & \mathbf{X}_2 \end{pmatrix} = \begin{pmatrix} \mathbf{X}_1'\mathbf{X}_1 & \mathbf{X}_1'\mathbf{X}_2 \\ \mathbf{X}_2'\mathbf{X}_1 & \mathbf{X}_2'\mathbf{X}_2 \end{pmatrix},$$

(1.110) implies that, with $\mathbf{U} = (\mathbf{X}_1'\mathbf{X}_1)^{-1}$ and $\mathbf{V} = (\mathbf{X}_2'\mathbf{X}_2)^{-1}$,

$$(\mathbf{X}'\mathbf{X})^{-1} = \begin{pmatrix} \mathbf{W}^{-1} & -\mathbf{W}^{-1}\mathbf{X}_1'\mathbf{X}_2\mathbf{V} \\ -\mathbf{V}\mathbf{X}_2'\mathbf{X}_1\mathbf{W}^{-1} & \mathbf{V} + \mathbf{V}\mathbf{X}_2'\mathbf{X}_1\mathbf{W}^{-1}\mathbf{X}_1'\mathbf{X}_2\mathbf{V} \end{pmatrix}$$

with $\mathbf{W} = \mathbf{X}_1'\mathbf{X}_1 - \mathbf{X}_1'\mathbf{X}_2\mathbf{V}\mathbf{X}_2'\mathbf{X}_1 = \mathbf{X}_1'\mathbf{M}_2\mathbf{X}_1$, where $\mathbf{M}_2 = \mathbf{I} - \mathbf{X}_2(\mathbf{X}_2'\mathbf{X}_2)^{-1}\mathbf{X}_2'$. Then

$$\hat{\boldsymbol{\beta}} = (\mathbf{X}'\mathbf{X})^{-1} \begin{pmatrix} \mathbf{X}_1' \\ \mathbf{X}_2' \end{pmatrix} \mathbf{Y}$$

gives

$$\hat{\boldsymbol{\beta}}_1 = (\mathbf{W}^{-1}\mathbf{X}_1' - \mathbf{W}^{-1}\mathbf{X}_1'\mathbf{X}_2\mathbf{V}\mathbf{X}_2')\mathbf{Y} = (\mathbf{X}_1'\mathbf{M}_2\mathbf{X}_1)^{-1}\mathbf{X}_1'\mathbf{M}_2\mathbf{Y},$$

as in (1.22), and

$$\begin{aligned}
\hat{\boldsymbol{\beta}}_2 &= (-\mathbf{V}\mathbf{X}_2'\mathbf{X}_1\mathbf{W}^{-1}\mathbf{X}_1' + (\mathbf{V} + \mathbf{V}\mathbf{X}_2'\mathbf{X}_1\mathbf{W}^{-1}\mathbf{X}_1'\mathbf{X}_2\mathbf{V})\mathbf{X}_2')\mathbf{Y} \\
&= (\mathbf{V}\mathbf{X}_2' + \mathbf{V}\mathbf{X}_2'\mathbf{X}_1\mathbf{W}^{-1}\mathbf{X}_1'(\mathbf{X}_2\mathbf{V}\mathbf{X}_2' - \mathbf{I}))\mathbf{Y} \\
&= \mathbf{V}\mathbf{X}_2'(\mathbf{Y} - \mathbf{X}_1(\mathbf{X}_1'\mathbf{M}_2\mathbf{X}_1)^{-1}\mathbf{X}_1'\mathbf{M}_2\mathbf{Y}) \\
&= (\mathbf{X}_2'\mathbf{X}_2)^{-1}\mathbf{X}_2'(\mathbf{Y} - \mathbf{X}_1\hat{\boldsymbol{\beta}}_1).
\end{aligned}$$

4) Observe that, as $\mathbf{T} = [\mathbf{w}_1, \mathbf{w}_2, \dots, \mathbf{w}_k]$ in (1.42) is an orthonormal basis for $\mathcal{S}$, all vectors in $\mathcal{S}$ can be represented by linear combinations of these $\mathbf{w}_i$. In particular, if $[\mathbf{h}_1, \mathbf{h}_2, \dots, \mathbf{h}_k]$ is a (different) basis for $\mathcal{S}$, then we can write $\mathbf{H} = \mathbf{T}\mathbf{A}$, where $\mathbf{H} = [\mathbf{h}_1\ \mathbf{h}_2\ \dots\ \mathbf{h}_k]$ and $\mathbf{A}$ is a full rank $k \times k$ matrix. As $\mathbf{T}'\mathbf{T} = \mathbf{I}$ and $\mathbf{H}'\mathbf{H} = \mathbf{I}$, we have $\mathbf{I} = \mathbf{H}'\mathbf{H} = \mathbf{A}'\mathbf{T}'\mathbf{T}\mathbf{A} = \mathbf{A}'\mathbf{A}$, so that $\mathbf{A}$ is orthogonal with $\mathbf{A}' = \mathbf{A}^{-1}$. Then $\mathbf{H}\mathbf{H}' = \mathbf{T}\mathbf{A}\mathbf{A}'\mathbf{T}' = \mathbf{T}\mathbf{T}'$, showing that $\mathbf{P}_{\mathcal{S}}$ is unique. Matrix $\mathbf{A}$ can be computed as $(\mathbf{T}'\mathbf{T})^{-1}\mathbf{T}'\mathbf{H}$. In Matlab, we can see this with the code in Listing 1.13.

5) Let $\mathbf{M} = \mathbf{I}_T - \mathbf{P}_{\mathcal{S}}$ with $\dim(\mathcal{S}) = k$, $k \in \{1, 2, \dots, T-1\}$. Via the spectral decomposition, let $\mathbf{H}$ be an orthogonal matrix whose rows consist of the eigenvectors of $\mathbf{M}$. From Theorem 1.2, $\mathbf{H}$ can be partitioned as

$$\mathbf{H} = \begin{bmatrix} \mathbf{H}_1 \\ \mathbf{H}_2 \end{bmatrix},$$

where $\mathbf{H}_1$ and $\mathbf{H}_2$ are of sizes $(T-k) \times T$ and $k \times T$, respectively, and such that

$$\mathbf{H}\mathbf{M}\mathbf{H}' = \begin{pmatrix} \mathbf{H}_1 \\ \mathbf{H}_2 \end{pmatrix} \mathbf{M} \begin{pmatrix} \mathbf{H}_1' & \mathbf{H}_2' \end{pmatrix} = \begin{pmatrix} \mathbf{H}_1\mathbf{M}\mathbf{H}_1' & \mathbf{H}_1\mathbf{M}\mathbf{H}_2' \\ \mathbf{H}_2\mathbf{M}\mathbf{H}_1' & \mathbf{H}_2\mathbf{M}\mathbf{H}_2' \end{pmatrix} = \begin{pmatrix} \mathbf{I}_{T-k} & \mathbf{0}_{(T-k)\times k} \\ \mathbf{0}_{k\times(T-k)} & \mathbf{0}_{k\times k} \end{pmatrix}.$$

Then $\mathbf{0} = \mathbf{H}_2\mathbf{M}\mathbf{H}_2' = \mathbf{H}_2\mathbf{M}'\mathbf{M}\mathbf{H}_2' = (\mathbf{M}\mathbf{H}_2')'\mathbf{M}\mathbf{H}_2'$ implies that $\mathbf{H}_2\mathbf{M} = \mathbf{M}\mathbf{H}_2' = \mathbf{0}$ or

$$\mathbf{0} = (\mathbf{I} - \mathbf{P}_{\mathcal{S}})\mathbf{H}_2' \Leftrightarrow \mathbf{H}_2' = \mathbf{P}_{\mathcal{S}}\mathbf{H}_2'.$$

```
T=rand(4,2); T=orth(T); Q=[1,2;3,4]; H=T*Q; H=orth(H);
A=inv(T'*T)*T'*H; H-T*A, A'*A
```

Program Listing 1.13: Computes $\mathbf{A} = (\mathbf{T}'\mathbf{T})^{-1}\mathbf{T}'\mathbf{H}$.

As $\mathbf{H}_2'$ is unchanged by projecting it onto $\mathcal{S}$, the rows of $\mathbf{H}_2$ are in $\mathcal{S}$. From this, and the fact that the rows of $\mathbf{H}$ are orthogonal,

$$\begin{aligned}\mathbf{H}_1\mathbf{H}_2' = \mathbf{0} &\Leftrightarrow \mathbf{H}_1\mathbf{P}_{\mathcal{S}}\mathbf{y} = \mathbf{0} \quad \forall\, \mathbf{y} \in \mathbb{R}^T \\ &\Leftrightarrow \mathbf{H}_1(\mathbf{I}\mathbf{y} - \mathbf{P}_{\mathcal{S}}\mathbf{y}) = \mathbf{H}_1\mathbf{y} \quad \forall\, \mathbf{y} \in \mathbb{R}^T \\ &\Leftrightarrow \mathbf{H}_1\mathbf{M}\mathbf{y} = \mathbf{H}_1\mathbf{y} \quad \forall\, \mathbf{y} \in \mathbb{R}^T \\ &\Leftrightarrow \mathbf{H}_1\mathbf{M} = \mathbf{H}_1.\end{aligned}$$

Postmultiplying $\mathbf{H}'\mathbf{H} = \mathbf{I}_T$ by $\mathbf{M}$ gives $\mathbf{H}_1'\mathbf{H}_1\mathbf{M} + \mathbf{H}_2'\mathbf{H}_2\mathbf{M} = \mathbf{M}$ or, as $\mathbf{H}_1\mathbf{M} = \mathbf{H}_1$ and $\mathbf{H}_2\mathbf{M} = \mathbf{0}$,

$$\mathbf{H}_1'\mathbf{H}_1 = \mathbf{M}. \tag{1.146}$$

Recall that the rows of $\mathbf{H}$ are orthonormal, so that

$$\mathbf{H}\mathbf{H}' = \begin{pmatrix}\mathbf{H}_1 \\ \mathbf{H}_2\end{pmatrix}(\mathbf{H}_1' \ \ \mathbf{H}_2') = \begin{pmatrix}\mathbf{H}_1\mathbf{H}_1' & \mathbf{H}_1\mathbf{H}_2' \\ \mathbf{H}_2\mathbf{H}_1' & \mathbf{H}_2\mathbf{H}_2'\end{pmatrix} = \mathbf{I}_T = \begin{pmatrix}\mathbf{I}_{T-k} & \mathbf{0}_{(T-k)\times k} \\ \mathbf{0}_{k\times(T-k)} & \mathbf{I}_{k\times k}\end{pmatrix}$$

and, in particular,

$$\mathbf{H}_1\mathbf{H}_1' = \mathbf{I}_{T-k}. \tag{1.147}$$

The result follows from (1.146) and (1.147).

6) Let $\mathbf{A} = (\mathbf{X}'\mathbf{X})^{-1}$. Direct substitution gives

$$\mathbf{H}\hat{\boldsymbol{\gamma}} = \mathbf{H}[\hat{\boldsymbol{\beta}} + \mathbf{A}\mathbf{H}'[\mathbf{H}\mathbf{A}\mathbf{H}']^{-1}(\mathbf{h} - \mathbf{H}\hat{\boldsymbol{\beta}})] = \mathbf{H}\hat{\boldsymbol{\beta}} + \mathbf{h} - \mathbf{H}\hat{\boldsymbol{\beta}},$$

so that the first condition is satisfied. To see the second, note that, for every $\mathbf{b} \in \mathbb{R}^k$ such that $\mathbf{H}\mathbf{b} = \mathbf{h}$, we can write

$$\|\mathbf{Y} - \mathbf{X}\mathbf{b}\|^2 = \|\mathbf{Y} - \mathbf{X}\hat{\boldsymbol{\beta}} + \mathbf{X}\hat{\boldsymbol{\beta}} - \mathbf{X}\mathbf{b}\|^2 = \|\mathbf{Y} - \mathbf{X}\hat{\boldsymbol{\beta}}\|^2 + \|\mathbf{X}\hat{\boldsymbol{\beta}} - \mathbf{X}\mathbf{b}\|^2, \tag{1.148}$$

because the cross term $(\mathbf{Y} - \mathbf{X}\hat{\boldsymbol{\beta}})'(\mathbf{X}\hat{\boldsymbol{\beta}} - \mathbf{X}\mathbf{b}) = \hat{\boldsymbol{\epsilon}}'\mathbf{X}(\hat{\boldsymbol{\beta}} - \mathbf{b}) = 0$ from (1.61). Because the first term in (1.148) does not depend on $\mathbf{b}$ or $\hat{\boldsymbol{\gamma}}$, it suffices to show that

$$\|\mathbf{X}\hat{\boldsymbol{\beta}} - \mathbf{X}\hat{\boldsymbol{\gamma}}\|^2 \leqslant \|\mathbf{X}\hat{\boldsymbol{\beta}} - \mathbf{X}\mathbf{b}\|^2. \tag{1.149}$$

First note that the cross term $(\mathbf{X}\hat{\boldsymbol{\beta}} - \mathbf{X}\hat{\boldsymbol{\gamma}})'(\mathbf{X}\hat{\boldsymbol{\gamma}} - \mathbf{X}\mathbf{b})$ vanishes because, from (1.69),

$$\begin{aligned}(\hat{\boldsymbol{\beta}} - \hat{\boldsymbol{\gamma}})'\mathbf{X}'\mathbf{X}(\hat{\boldsymbol{\gamma}} - \mathbf{b}) &= -(\mathbf{h} - \mathbf{H}\hat{\boldsymbol{\beta}})'[\mathbf{H}(\mathbf{X}'\mathbf{X})^{-1}\mathbf{H}']^{-1}\mathbf{H}(\mathbf{X}'\mathbf{X})^{-1}\mathbf{X}'\mathbf{X}(\hat{\boldsymbol{\gamma}} - \mathbf{b}) \\ &= -(\mathbf{h} - \mathbf{H}\hat{\boldsymbol{\beta}})'[\mathbf{H}(\mathbf{X}'\mathbf{X})^{-1}\mathbf{H}']^{-1}(\mathbf{H}\hat{\boldsymbol{\gamma}} - \mathbf{H}\mathbf{b}) = \mathbf{0},\end{aligned}$$

as $\mathbf{H}\hat{\boldsymbol{\gamma}} = \mathbf{h} = \mathbf{H}\mathbf{b}$. Thus, the right-hand side of (1.149) is

$$\|\mathbf{X}(\hat{\boldsymbol{\beta}} - \mathbf{b})\|^2 = \|\mathbf{X}(\hat{\boldsymbol{\beta}} - \hat{\boldsymbol{\gamma}} + \hat{\boldsymbol{\gamma}} - \mathbf{b})\|^2 = \|\mathbf{X}\hat{\boldsymbol{\beta}} - \mathbf{X}\hat{\boldsymbol{\gamma}}\|^2 + \|\mathbf{X}\hat{\boldsymbol{\gamma}} - \mathbf{X}\mathbf{b}\|^2,$$

and, as $\|\mathbf{X}\hat{\boldsymbol{\gamma}} - \mathbf{X}\mathbf{b}\|^2$ is non-negative, (1.149) is true. Strict equality holds when $\mathbf{X}\hat{\boldsymbol{\gamma}}$ equals $\mathbf{X}\mathbf{b}$, but as $\mathbf{X}$ is of full rank, this holds if and only if $\hat{\boldsymbol{\gamma}} = \mathbf{b}$.

7) From the definition of $\langle\cdot,\cdot\rangle$, for any $\mathbf{v} \in \mathbb{R}^n$, $\langle\mathbf{v},\mathbf{v}\rangle = \sum_{i=1}^n v_i^2 \geqslant 0$. For the second part,

$$\begin{aligned}\langle\mathbf{u} - a\mathbf{v}, \mathbf{u} - a\mathbf{v}\rangle &= \sum_{i=1}^n (u_i - av_i)^2 = \sum_{i=1}^n u_i^2 - 2a\sum_{i=1}^n u_iv_i + a^2\sum_{i=1}^n v_i^2 \\ &= \langle\mathbf{u},\mathbf{u}\rangle - 2a\langle\mathbf{u},\mathbf{v}\rangle + a^2\langle\mathbf{v},\mathbf{v}\rangle,\end{aligned}$$

so that, with $a = \langle \mathbf{u}, \mathbf{v} \rangle / \langle \mathbf{v}, \mathbf{v} \rangle$,

$$0 \leqslant \langle \mathbf{u}, \mathbf{u} \rangle - 2a\langle \mathbf{u}, \mathbf{v} \rangle + a^2 \langle \mathbf{v}, \mathbf{v} \rangle = \langle \mathbf{u}, \mathbf{u} \rangle - 2\frac{\langle \mathbf{u}, \mathbf{v} \rangle^2}{\langle \mathbf{v}, \mathbf{v} \rangle} + \frac{\langle \mathbf{u}, \mathbf{v} \rangle^2}{\langle \mathbf{v}, \mathbf{v} \rangle} = \langle \mathbf{u}, \mathbf{u} \rangle - \frac{\langle \mathbf{u}, \mathbf{v} \rangle^2}{\langle \mathbf{v}, \mathbf{v} \rangle},$$

or

$$\langle \mathbf{u}, \mathbf{v} \rangle^2 \leqslant \langle \mathbf{u}, \mathbf{u} \rangle \langle \mathbf{v}, \mathbf{v} \rangle.$$

As both sides are positive, taking square roots gives the inequality $\langle \mathbf{u}, \mathbf{v} \rangle \leqslant \|\mathbf{u}\| \; \|\mathbf{v}\|$, where $\|\mathbf{u}\|^2 = \langle \mathbf{u}, \mathbf{u} \rangle$.

8) (Theorem 1.2)
From idempotency, for any eigenvalue λ and corresponding eigenvector $\mathbf{x}$,

$$\lambda \mathbf{x} = \mathbf{P}\mathbf{x} = \mathbf{P}\mathbf{P}\mathbf{x} = \mathbf{P}\lambda\mathbf{x} = \lambda\mathbf{P}\mathbf{x} = \lambda^2 \mathbf{x},$$

which implies that $\lambda = \lambda^2$, so that the only solutions are $\lambda = 0$ or 1 (there are no complex solutions, though note that, from the assumption of symmetry, all eigenvalues are real anyway). Also from symmetry, the number of nonzero eigenvalues of $\mathbf{P}$ equals rank($\mathbf{P}$) $= k$, proving (i).
For (ii), form the spectral decomposition of $\mathbf{P}$ as $\mathbf{U}\mathbf{D}\mathbf{U}'$, where $\mathbf{U}$ is an orthogonal matrix and $\mathbf{D}$ is a diagonal matrix with k ones and $T - k$ zeros. Using the fact that (for conformable matrices) $\text{tr}(\mathbf{AB}) = \text{tr}(\mathbf{BA})$,

$$k = \text{rank}(\mathbf{P}) = \text{tr}(\mathbf{D}) = \text{tr}(\mathbf{U}\mathbf{D}\mathbf{U}') = \text{tr}(\mathbf{P}).$$

9) (Theorem 1.6)
   a) For convenience, we restate (1.50) from the proof in the text: Take as a basis for $\mathbb{R}^T$ the vectors

$$\underbrace{\overbrace{\mathbf{r}_1, \dots, \mathbf{r}_r}^{\mathcal{S}_0 \text{ basis}}, \overbrace{\mathbf{s}_{r+1}, \dots, \mathbf{s}_s}^{\mathcal{S}\backslash\mathcal{S}_0 \text{ basis}}}_{\mathcal{S} \text{ basis}}, \underbrace{\mathbf{z}_{s+1}, \dots, \mathbf{z}_T}_{\mathcal{S}^\perp \text{ basis}} \tag{1.150}$$

   and let $\mathbf{y} = \mathbf{r} + \mathbf{s} + \mathbf{z}$, where $\mathbf{r} \in \mathcal{S}_0$, $\mathbf{s} \in \mathcal{S}\backslash\mathcal{S}_0$ and $\mathbf{z} \in \mathcal{S}^\perp$ are orthogonal.
   b) Let $\mathbf{Q} = \mathbf{P}_{\mathcal{S}} - \mathbf{P}_{\mathcal{S}_0}$. From Theorem 1.4, if $\mathbf{Q}$ is symmetric and idempotent, then it is the projection matrix onto $\mathcal{C}(\mathbf{Q})$, but it is clearly symmetric and, from the first part of the theorem,

$$\mathbf{Q}\mathbf{Q} = \mathbf{P}_{\mathcal{S}}\mathbf{P}_{\mathcal{S}} - \mathbf{P}_{\mathcal{S}}\mathbf{P}_{\mathcal{S}_0} - \mathbf{P}_{\mathcal{S}_0}\mathbf{P}_{\mathcal{S}} + \mathbf{P}_{\mathcal{S}_0}\mathbf{P}_{\mathcal{S}_0} = \mathbf{P}_{\mathcal{S}} - \mathbf{P}_{\mathcal{S}_0}.$$

   For $\mathcal{C}(\mathbf{Q}) = \mathcal{S}\backslash\mathcal{S}_0$, it must be that, for $\mathbf{s} \in \mathcal{S}\backslash\mathcal{S}_0$ and $\mathbf{w} \in (\mathcal{S}\backslash\mathcal{S}_0)^\perp$, $\mathbf{Q}\mathbf{s} = \mathbf{s}$ and $\mathbf{Q}\mathbf{w} = \mathbf{0}$. As $\mathcal{S}\backslash\mathcal{S}_0 \subset \mathcal{S}$, $\mathbf{P}_{\mathcal{S}}\mathbf{s} = \mathbf{s}$ and, as $\mathbf{s} \perp \mathcal{S}_0$, $\mathbf{P}_{\mathcal{S}_0}\mathbf{s} = \mathbf{0}$, showing that $\mathbf{Q}\mathbf{s} = \mathbf{s}$. Next, from (1.150), $\mathbf{w}$ can be expressed as

$$\mathbf{w} = c_1\mathbf{r}_1 + \cdots + c_r\mathbf{r}_r + c_{s+1}\mathbf{z}_{s+1} + \cdots + c_T\mathbf{z}_T$$

   for some constants $c_i \in \mathbb{R}$. As $\mathbf{z}_i \perp \mathcal{S}$ (which implies $\mathbf{z}_i \perp \mathcal{S}_0 \subset \mathcal{S}$), $\mathbf{P}_{\mathcal{S}_0}\mathbf{w} = \mathbf{P}_{\mathcal{S}}\mathbf{w} = c_1\mathbf{r}_1 + \cdots + c_r\mathbf{r}_r$ so that $\mathbf{Q}\mathbf{w} = \mathbf{0}$. Thus, $\mathcal{C}(\mathbf{Q}) = \mathcal{S}\backslash\mathcal{S}_0$ and $\mathbf{P}_{\mathcal{S}\backslash\mathcal{S}_0} = \mathbf{Q} = \mathbf{P}_{\mathcal{S}} - \mathbf{P}_{\mathcal{S}_0}$. Note that this is a special case of the earlier result

$$\mathbf{P}_{\mathcal{S}^\perp} = \mathbf{P}_{\mathbb{R}^T\backslash\mathcal{S}} = \mathbf{P}_{\mathbb{R}^T} - \mathbf{P}_{\mathcal{S}} = \mathbf{I}_T - \mathbf{P}_{\mathcal{S}}.$$

c) As $\mathbf{P}_{S\backslash S_0} = \mathbf{P}_S - \mathbf{P}_{S_0}$,

$$\begin{aligned}\|\mathbf{P}_{S\backslash S_0}\mathbf{y}\|^2 &= \|\mathbf{P}_S\mathbf{y} - \mathbf{P}_{S_0}\mathbf{y}\|^2 = (\mathbf{P}_S\mathbf{y} - \mathbf{P}_{S_0}\mathbf{y})'(\mathbf{P}_S\mathbf{y} - \mathbf{P}_{S_0}\mathbf{y}) \\ &= \mathbf{y}'\mathbf{P}_S\mathbf{P}_S\mathbf{y} - \mathbf{y}'\mathbf{P}_{S_0}\mathbf{P}_S\mathbf{y} - \mathbf{y}'\mathbf{P}_S\mathbf{P}_{S_0}\mathbf{y} + \mathbf{y}'\mathbf{P}_{S_0}\mathbf{P}_{S_0}\mathbf{y} \\ &= \|\mathbf{P}_S\mathbf{y}\|^2 - \|\mathbf{P}_{S_0}\mathbf{y}\|^2\end{aligned}$$

using the results from part (a).

d) By expressing (1.150) as

$$\overbrace{\mathbf{r}_1, \dots, \mathbf{r}_r}^{S_0}, \overbrace{\underbrace{\mathbf{s}_{r+1}, \dots \mathbf{s}_s}_{S_0^\perp\backslash S^\perp}, \underbrace{\mathbf{z}_{s+1}, \dots \mathbf{z}_T}_{S_0^\perp \cap S^\perp = S^\perp}}^{S_0^\perp},$$

it is clear that $S\backslash S_0 = S_0^\perp\backslash S^\perp$. To verify the last equality,

$$\mathbf{P}_{S_0^\perp} - \mathbf{P}_{S^\perp} = (\mathbf{I} - \mathbf{P}_{S_0}) - (\mathbf{I} - \mathbf{P}_S) = \mathbf{P}_S - \mathbf{P}_{S_0} = \mathbf{P}_{S\backslash S_0}.$$

e) This follows easily from (1.150) because $\mathbf{P}_{S\backslash S_0}\mathbf{y} \in (S\backslash S_0) \subset S$, so that $\mathbf{P}_S(\mathbf{P}_{S\backslash S_0}\mathbf{y})$ remains $\mathbf{P}_{S\backslash S_0}\mathbf{y}$. Transposing gives the other equality.

10) For the projection condition, let $\mathbf{x} \in \mathcal{C}(\mathbf{X})$. We need to show that $(\mathbf{P}_{\mathbf{X}_1} + \mathbf{P}_{\mathbf{M}_1\mathbf{X}_2})\mathbf{x} = \mathbf{x}$. From the hint,

$$\begin{aligned}(\mathbf{P}_{\mathbf{X}_1} + \mathbf{P}_{\mathbf{M}_1\mathbf{X}_2})\mathbf{x} &= (\mathbf{P}_{\mathbf{X}_1} + \mathbf{P}_{\mathbf{M}_1\mathbf{X}_2})(\mathbf{X}_1\gamma_1 + \mathbf{X}_2\gamma_2) \\ &= \mathbf{P}_{\mathbf{X}_1}(\mathbf{X}_1\gamma_1 + \mathbf{X}_2\gamma_2) + \mathbf{P}_{\mathbf{M}_1\mathbf{X}_2}(\mathbf{X}_1\gamma_1 + \mathbf{X}_2\gamma_2).\end{aligned}$$

Clearly, $\mathbf{P}_{\mathbf{X}_1}\mathbf{X}_1\gamma_1 = \mathbf{X}_1\gamma_1$, and as $\mathbf{P}_{\mathbf{M}_1\mathbf{X}_2} = \mathbf{M}_1\mathbf{X}_2(\mathbf{X}_2'\mathbf{M}_1\mathbf{X}_2)^{-1}\mathbf{X}_2'\mathbf{M}_1$, we have $\mathbf{P}_{\mathbf{M}_1\mathbf{X}_2}\mathbf{X}_1 = \mathbf{0}$ (as $\mathbf{M}_1\mathbf{X}_1 = \mathbf{0}$) and $\mathbf{P}_{\mathbf{M}_1\mathbf{X}_2}\mathbf{X}_2 = \mathbf{M}_1\mathbf{X}_2$. Thus,

$$\begin{aligned}(\mathbf{P}_{\mathbf{X}_1} + \mathbf{P}_{\mathbf{M}_1\mathbf{X}_2})\mathbf{x} &= \mathbf{P}_{\mathbf{X}_1}(\mathbf{X}_1\gamma_1 + \mathbf{X}_2\gamma_2) + \mathbf{P}_{\mathbf{M}_1\mathbf{X}_2}(\mathbf{X}_1\gamma_1 + \mathbf{X}_2\gamma_2) \\ &= \mathbf{X}_1\gamma_1 + \mathbf{P}_{\mathbf{X}_1}\mathbf{X}_2\gamma_2 + \mathbf{M}_1\mathbf{X}_2\gamma_2 \\ &= \mathbf{X}_1\gamma_1 + (\mathbf{P}_{\mathbf{X}_1} + \mathbf{M}_1)\mathbf{X}_2\gamma_2 \\ &= \mathbf{X}_1\gamma_1 + \mathbf{X}_2\gamma_2 \\ &= \mathbf{X}\gamma = \mathbf{x},\end{aligned}$$

as $\mathbf{M}_1 = \mathbf{M}_{\mathbf{X}_1} = \mathbf{I} - \mathbf{P}_{\mathbf{X}_1}$.

For the perpendicularity condition, recall that the orthogonal complement of $\mathcal{C}(\mathbf{X})$ is

$$\mathcal{C}(\mathbf{X})^\perp = \{\mathbf{z} \in \mathbb{R}^T : \mathbf{X}'\mathbf{z} = \mathbf{0}\}. \tag{1.151}$$

Let $\mathbf{u} \in \mathcal{C}(\mathbf{X})^\perp$. We need to show that $(\mathbf{P}_{\mathbf{X}_1} + \mathbf{P}_{\mathbf{M}_1\mathbf{X}_2})\mathbf{u} = \mathbf{0}$. For the first term, note that, directly from (1.151), $\mathcal{C}(\mathbf{X})^\perp \subset \mathcal{C}(\mathbf{X}_1)^\perp$, i.e., if $\mathbf{u} \in \mathcal{C}(\mathbf{X})^\perp$, then $\mathbf{u} \in \mathcal{C}(\mathbf{X}_1)^\perp$, so that $\mathbf{P}_{\mathbf{X}_1}\mathbf{u} = \mathbf{0}$. For the second term, first note that, as $\mathcal{C}(\mathbf{X})^\perp \subset \mathcal{C}(\mathbf{X}_2)^\perp$, $\mathbf{X}_2'\mathbf{u} = \mathbf{0}$. As

$$\mathbf{P}_{\mathbf{M}_1\mathbf{X}_2} = \mathbf{M}_1\mathbf{X}_2(\mathbf{X}_2'\mathbf{M}_1\mathbf{X}_2)^{-1}\mathbf{X}_2'\mathbf{M}_1 = \mathbf{M}_1\mathbf{X}_2(\mathbf{X}_2'\mathbf{M}_1\mathbf{X}_2)^{-1}\mathbf{X}_2'(\mathbf{I} - \mathbf{P}_{\mathbf{X}_1}),$$

the condition $\mathbf{P}_{\mathbf{M}_1\mathbf{X}_2}\mathbf{u} = \mathbf{0}$ holds if both $\mathbf{X}_2'\mathbf{u} = \mathbf{0}$ and $\mathbf{P}_{\mathbf{X}_1}\mathbf{u} = \mathbf{0}$ hold, and we have just seen that these are both true, and we are done.

11) Write $\mathbf{I} = \mathbf{I} - \mathbf{P} + \mathbf{P}$, and use Theorem B.67 to get $T - \text{rank}(\mathbf{P}) \leqslant \text{rank}(\mathbf{I} - \mathbf{P})$. But, as $\mathbf{P}$ is idempotent, we have $\mathbf{0} = (\mathbf{I} - \mathbf{P})\mathbf{P}$, so from Theorem B.68, $T - \text{rank}(\mathbf{P}) \geqslant \text{rank}(\mathbf{I} - \mathbf{P})$. Together, they imply that $\text{rank}(\mathbf{I} - \mathbf{P}) = T - \text{rank}(\mathbf{P}) = k$.
12) For the statement in the hint, to see that $\mathbf{A}^{-1}$ is symmetric,

$$\mathbf{I} = \mathbf{A}\mathbf{A}^{-1} \Leftrightarrow \mathbf{I}' = \mathbf{I} = \mathbf{A}^{-1'}\mathbf{A} \Leftrightarrow \mathbf{I}\mathbf{A}^{-1} = \mathbf{A}^{-1'}\mathbf{A}\mathbf{A}^{-1} \Leftrightarrow \mathbf{A}^{-1} = \mathbf{A}^{-1'}.$$

As $\mathbf{A}$ is symmetric, all its eigenvalues are real, so that $\mathbf{A}$ has spectral decomposition $\mathbf{A} = \mathbf{U}\mathbf{D}\mathbf{U}'$ with $\mathbf{U}$ orthonormal and $\mathbf{D} = \text{diag}(d_1, \ldots, d_n)$ with each d_i real and positive. Then $\mathbf{A}^{-1} = \mathbf{U}\mathbf{D}^{-1}\mathbf{U}'$ (confirmed by calculating $\mathbf{A}\mathbf{A}^{-1}$) with $\mathbf{D}^{-1} = \text{diag}(d_1^{-1}, \ldots, d_n^{-1})$ with each $d_i^{-1} > 0$, implying that $\mathbf{A}^{-1}$ is also full rank.

To show that $\mathbf{K}$ is positive semi-definite: Let $\mathbf{x}$ be a $k \times 1$ real vector. We have to show that $\mathbf{x}'\mathbf{K}\mathbf{x} \geqslant 0$ for all $\mathbf{x}$ or, with $\mathbf{z} = \mathbf{H}\mathbf{A}\mathbf{x}$ and the fact that $\mathbf{A} = (\mathbf{X}'\mathbf{X})^{-1}$ is symmetric, that $\mathbf{z}'(\mathbf{H}\mathbf{A}\mathbf{H}')^{-1}\mathbf{z} \geqslant 0$. But this is true because $\mathbf{H}\mathbf{A}\mathbf{H}'$ (and, thus, $(\mathbf{H}\mathbf{A}\mathbf{H}')^{-1}$) is symmetric and full rank, i.e., $\mathbf{q}'\mathbf{H}\mathbf{A}\mathbf{H}'\mathbf{q} > 0$ for all nonzero $\mathbf{q}$.

Observe that $\mathbf{K}$ is not necessarily positive definite when $J < k$ because $\mathbf{z} = \mathbf{H}\mathbf{A}\mathbf{x}$ could be zero even for nonzero $\mathbf{x}$. This is the case, for example, with

$$\mathbf{H} = \begin{bmatrix} 1 & 0 & 0 \\ 0 & 1 & -1 \end{bmatrix}, \quad \mathbf{x} = \text{null}(\mathbf{H}) = \begin{bmatrix} 0 \\ \sqrt{2}/2 \\ \sqrt{2}/2 \end{bmatrix}.$$

If $J = k$ and, as always assumed, $\mathbf{H}$ is full rank, then $\mathbf{H}$ is a square matrix with unique inverse, and $\boldsymbol{\beta}$ is fully specified from the restrictions and the data have no influence on its estimate, i.e., the restriction $\mathbf{H}\boldsymbol{\beta} = \mathbf{h}$ implies that $\boldsymbol{\beta} = \mathbf{H}^{-1}\mathbf{h}$ and $\hat{\boldsymbol{\gamma}} = \mathbf{H}^{-1}\mathbf{h}$, which is not stochastic and, thus, has a zero covariance matrix. This agrees with the expression (1.77), because, with $J = k$,

$$\begin{aligned}\mathbf{K} &= \sigma^2\mathbf{A}\mathbf{H}'(\mathbf{H}\mathbf{A}\mathbf{H}')^{-1}\mathbf{H}\mathbf{A} \\ &= \sigma^2\mathbf{A}\mathbf{H}'\,\mathbf{H}'^{-1}\mathbf{A}^{-1}\mathbf{H}^{-1}\,\mathbf{H}\mathbf{A} = \sigma^2\mathbf{A} = \sigma^2(\mathbf{X}'\mathbf{X})^{-1} = \text{Var}(\hat{\boldsymbol{\beta}}).\end{aligned}$$

13) Using program `ncf.m` to compute the noncentral F c.d.f., the code in Listing 1.14 will do the job.

14)
   a) We take $\mathbf{H} = [0 \quad 1 \quad 1 \quad 1]$ and $\mathbf{h} = 1$. The constraint implies, for example, that $\beta_2 = 1 - \beta_3 - \beta_4$, so that $\mathbf{S}$, $\boldsymbol{\eta}$ and $\mathbf{s}$ are given via

$$\boldsymbol{\beta} = \begin{pmatrix} \beta_1 \\ 1 - \beta_3 - \beta_4 \\ \beta_3 \\ \beta_4 \end{pmatrix} = \begin{pmatrix} 1 & 0 & 0 \\ 0 & -1 & -1 \\ 0 & 1 & 0 \\ 0 & 0 & 1 \end{pmatrix} \begin{pmatrix} \beta_1 \\ \beta_3 \\ \beta_4 \end{pmatrix} + \begin{pmatrix} 0 \\ 1 \\ 0 \\ 0 \end{pmatrix}.$$

   b) The model is $\mathbf{Y} = \mathbf{X}\boldsymbol{\beta} + \boldsymbol{\epsilon} = \mathbf{X}\mathbf{S}\boldsymbol{\eta} + \mathbf{X}\mathbf{s} + \boldsymbol{\epsilon}$ or $\mathbf{Y} - \mathbf{X}\mathbf{s} = \mathbf{X}\mathbf{S}\boldsymbol{\eta} + \boldsymbol{\epsilon}$, so that, with $\mathbf{Y}^* = \mathbf{Y} - \mathbf{X}\mathbf{s}$ and $\mathbf{Z} = \mathbf{X}\mathbf{S}$,

$$\hat{\boldsymbol{\eta}} = (\mathbf{Z}'\mathbf{Z})^{-1}\mathbf{Z}'\mathbf{Y}^* = (\mathbf{S}'\mathbf{X}'\mathbf{X}\mathbf{S})^{-1}\mathbf{S}'\mathbf{X}'(\mathbf{Y} - \mathbf{X}\mathbf{s}).$$

```
powneed=0.90; beta=[0 -5 3 5]'; H=[1 -1 0 0; 0 0 1 -1]; sig2=9;
notenough=1; a=5;
while notenough
  a=a+1; n=4*a;
  dum1=[ones(n,1); zeros(n,1)]; dum2=1-dum1;
  time=kron((1:4)',ones(floor(n/4),1));
  c3=kron([1,0]',time); c4=kron([0,1]',time);
  X=[dum1 dum2 c3 c4]; A=inv(X'*X);
  theta=beta'*H'*inv(H*A*H')*H*beta/sig2;
  cutoff = finv(0.95,2,2*n-4); pow=1-ncf(cutoff,2,36,theta,0)
  if pow>=powneed, notenough=0; end
end
T=2*n
```

Program Listing 1.14: Finds minimum T for a given power `powneed` based on the setup in Example 1.11. Here, $T = 2n$, and n is incremented in steps of 4.

From the constraint $\boldsymbol{\beta} = \mathbf{S}\boldsymbol{\eta} + \mathbf{s}$,

$$\widehat{\boldsymbol{\gamma}} = \mathbf{S}\widehat{\boldsymbol{\eta}} + \mathbf{s} = \mathbf{S}(\mathbf{S}'\mathbf{X}'\mathbf{X}\mathbf{S})^{-1}\mathbf{S}'\mathbf{X}'(\mathbf{Y} - \mathbf{X}\mathbf{s}) + \mathbf{s}.$$

c) We have

$$\mathbf{X}\widehat{\boldsymbol{\gamma}} = \mathbf{X}\mathbf{S}(\mathbf{S}'\mathbf{X}'\mathbf{X}\mathbf{S})^{-1}\mathbf{S}'\mathbf{X}'(\mathbf{Y} - \mathbf{X}\mathbf{s}) + \mathbf{X}\mathbf{s} = \mathbf{P}_{\mathbf{Z}}\mathbf{Y} + (\mathbf{I} - \mathbf{P}_{\mathbf{Z}})\mathbf{X}\mathbf{s},$$

where $\mathbf{P}_{\mathbf{Z}} = \mathbf{Z}(\mathbf{Z}'\mathbf{Z})^{-1}\mathbf{Z}' = \mathbf{X}\mathbf{S}(\mathbf{S}'\mathbf{X}'\mathbf{X}\mathbf{S})^{-1}\mathbf{S}'\mathbf{X}'$ is clearly a projection matrix.

d) Choose $\mathbf{H}$ and $\boldsymbol{\beta}$ in such a way that the partition

$$\mathbf{H}\boldsymbol{\beta} = \begin{pmatrix} \mathbf{H}_1 & \mathbf{H}_2 \end{pmatrix} \begin{pmatrix} \boldsymbol{\beta}_{[1]} \\ \boldsymbol{\beta}_{[2]} \end{pmatrix} = \mathbf{H}_1\boldsymbol{\beta}_{[1]} + \mathbf{H}_2\boldsymbol{\beta}_{[2]} = \mathbf{h}$$

can be formed for which $\mathbf{H}_1$ is $J \times J$ and nonsingular. (This is always possible because $\mathbf{H}$ is full rank J.) Premultiplying by $\mathbf{H}_1^{-1}$ implies that $\boldsymbol{\beta}_{[1]} = \mathbf{H}_1^{-1}\mathbf{h} - \mathbf{H}_1^{-1}\mathbf{H}_2\boldsymbol{\beta}_{[2]}$ and

$$\boldsymbol{\beta} = \begin{pmatrix} \boldsymbol{\beta}_{[1]} \\ \boldsymbol{\beta}_{[2]} \end{pmatrix} = \begin{pmatrix} -\mathbf{H}_1^{-1}\mathbf{H}_2 \\ \mathbf{I}_{k-J} \end{pmatrix} \boldsymbol{\beta}_{[2]} + \begin{pmatrix} \mathbf{H}_1^{-1}\mathbf{h} \\ \mathbf{0}_{k-J} \end{pmatrix} = \mathbf{S}\boldsymbol{\eta} + \mathbf{s}.$$

15) From (1.9),

$$\mathcal{L}(\widehat{\boldsymbol{\beta}}, \tilde{\sigma}^2; \mathbf{Y}) = \frac{1}{(2\pi)^{T/2}\tilde{\sigma}^T} \exp\left\{-\frac{1}{2\tilde{\sigma}^2}(\mathbf{Y} - \mathbf{X}\widehat{\boldsymbol{\beta}})'(\mathbf{Y} - \mathbf{X}\widehat{\boldsymbol{\beta}})\right\}$$

$$= \frac{1}{(2\pi)^{T/2}\tilde{\sigma}^T} \exp\left\{-\frac{1}{2T^{-1}\mathrm{S}(\widehat{\boldsymbol{\beta}})}\mathrm{S}(\widehat{\boldsymbol{\beta}})\right\} = \frac{\mathrm{e}^{-T/2}}{(2\pi)^{T/2}\tilde{\sigma}^T},$$

and, similarly,

$$\mathcal{L}(\widehat{\boldsymbol{\gamma}}, \tilde{\sigma}_\gamma^2; \mathbf{Y}) = \frac{\mathrm{e}^{-T/2}}{(2\pi)^{T/2}\tilde{\sigma}_\gamma^T},$$

so that

$$R = \left(\frac{\tilde{\sigma}_\gamma}{\tilde{\sigma}}\right)^{-T} = \left(\frac{\tilde{\sigma}_\gamma^2}{\tilde{\sigma}^2}\right)^{-T/2}.$$

The GLRT rejects for small R, i.e., when $\tilde{\sigma}_\gamma^2/\tilde{\sigma}^2$ is large. In terms of sums of squares, R rejects when $\mathsf{S}(\hat{\gamma})/\mathsf{S}(\hat{\boldsymbol{\beta}})$ is large, or, equivalently, when

$$\frac{T-k}{J}\left(\frac{\mathsf{S}(\hat{\gamma})}{\mathsf{S}(\hat{\boldsymbol{\beta}})}-1\right)=\frac{[\mathsf{S}(\hat{\gamma})-\mathsf{S}(\hat{\boldsymbol{\beta}})]/J}{\mathsf{S}(\hat{\boldsymbol{\beta}})/(T-k)}=\frac{\mathsf{S}(\hat{\gamma})-\mathsf{S}(\hat{\boldsymbol{\beta}})}{J\hat{\sigma}^2}=F$$

is large. Thus, the F test and the GLRT are the same.

16) With $\mathbf{G}=(G_1,G_2,G_3)$, $R_3\equiv G_3$, and $\mathbf{R}=(R_1,R_2,R_3)$, the one-to-one transformation of $\mathbf{r}=(r_1,r_2,r_3)$ to $\mathbf{g}=(g_1,g_2,g_3)$ is $g_1=r_1r_3$, $g_2=r_2r_3$, and $g_3=r_3$. The Jacobian is

$$\mathbf{J}=\begin{bmatrix}\partial g_1/\partial r_1 & \partial g_2/\partial r_1 & \partial g_3/\partial r_1\\ \partial g_1/\partial r_2 & \partial g_2/\partial r_2 & \partial g_3/\partial r_2\\ \partial g_1/\partial r_3 & \partial g_2/\partial r_3 & \partial g_3/\partial r_3\end{bmatrix}=\begin{bmatrix}r_3 & 0 & 0\\ 0 & r_3 & 0\\ r_1 & r_2 & 1\end{bmatrix},\qquad \det(\mathbf{J})=r_3^2,$$

and, as

$$\begin{aligned}f_{\mathbf{G}}(\mathbf{g})=&\frac{1}{\Gamma(\alpha_1)}\frac{1}{\Gamma(\alpha_2)}\frac{1}{\Gamma(\alpha_3)}\mathbb{I}(g_1>0)\mathbb{I}(g_2>0)\mathbb{I}(g_3>0)\\ &\times g_1^{\alpha_1-1}g_2^{\alpha_2-1}g_3^{\alpha_3-1}\exp(-g_1-g_2-g_3),\end{aligned}$$

the joint density of $\mathbf{R}$ is

$$\begin{aligned}f_{\mathbf{R}}(\mathbf{r})&=f_{\mathbf{G}}(\mathbf{g})|\det(\mathbf{J})|\\ &=\frac{1}{\Gamma(\alpha_1)}\frac{1}{\Gamma(\alpha_2)}\frac{1}{\Gamma(\alpha_3)}r_3^{\alpha_1+\alpha_2+\alpha_3-1}r_1^{\alpha_1-1}r_2^{\alpha_2-1}\exp(-r_3(1+r_1+r_2)).\end{aligned}$$

As $g_3=r_3$, the margin $R_3\sim\text{Gam}(\alpha_3,1)$, and

$$\begin{aligned}f_{(R_1,R_2)|R_3}(r_1,r_2\mid r_3)&=\frac{f_{\mathbf{R}}(\mathbf{r})}{f_{R_3}(r_3)}\\ &\propto r_1^{\alpha_1-1}\exp(-r_3r_1)\times r_2^{\alpha_2-1}\exp(-r_3r_2)\times r_3^{\alpha_1+\alpha_2},\end{aligned}$$

so that, conditional on $R_3=r_3$, the density of R_1 and R_2 factors, and R_1 and R_2 are conditionally independent.

```
function I = gam3(a1,a2,a3)
up=20; I = dblquad(@RR,0,up,0,up);

  function A=RR(r1,r2)
    c = gamma(a1+a2+a3) / (gamma(a1)*gamma(a2)*gamma(a3));
    num = r1.^(a1-1).* r2.^(a2-1);
    den = (1+r1+r2).^(a1+a2+a3);
    A = c * num./den;
  end

end
```

Program Listing 1.15: Computes the integral in (1.152), confirming it is 1.000. The integral upper limit up would have to be chosen in a more intelligent manner to work for all values of input parameters a_1, a_2, and a_3.

For the joint density of R_1 and R_2, using (1.111), $f_{R_1,R_2}(r_1, r_2)$ is

$$\begin{aligned}
&\int_0^\infty f_{\mathbf{R}}(\mathbf{r})\,\mathrm{d}r_3 \\
&= \frac{1}{\Gamma(\alpha_1)}\frac{1}{\Gamma(\alpha_2)}\frac{1}{\Gamma(\alpha_3)} r_1^{\alpha_1-1} r_2^{\alpha_2-1} \int_0^\infty r_3^{\alpha_1+\alpha_2+\alpha_3-1} \exp(-r_3(1+r_1+r_2))\,\mathrm{d}r_3 \\
&= \frac{\Gamma(\alpha_1+\alpha_2+\alpha_3)}{\Gamma(\alpha_1)\Gamma(\alpha_2)\Gamma(\alpha_3)} \frac{r_1^{\alpha_1-1} r_2^{\alpha_2-1}}{(1+r_1+r_2)^{\alpha_1+\alpha_2+\alpha_3}}.
\end{aligned} \tag{1.152}$$

The program in Listing 1.15 shows how to use function `dblquad` within Matlab with what they call *nested functions* to perform the integration.

2

Fixed Effects ANOVA Models

Having established the basics of the linear model in Chapter 1, this chapter provides an introduction to one of the most important workhorses of applied statistics, the **analysis of variance**, or ANOVA, concentrating on the basics of fixed effects models. Section 2.1 explains the notions of fixed and random effects. Section 2.2 illustrates the analysis in the case of two groups, resulting in the usual t-test for significant differences between the means of two populations. This is extended in Section 2.3 to the case with two groups and ignored block effects, which is a special case of the two-way ANOVA. It also shows the relevance of the doubly noncentral F distribution and the usefulness of being able to calculate its c.d.f. quickly via a saddlepoint approximation.

A core part of this chapter is Section 2.4, providing the details of the (always Gaussian) one-way ANOVA model, and also the use of the SAS system for conducting the calculations with data. Section 2.5 extends this to the two-way ANOVA, with emphasis on rigorous derivation of the relevant distribution theory, and the use of (Matlab, but notably) SAS to perform the required calculations.

This chapter, and Chapter 3 on random effects models, are far from a complete treatment of ANOVA and designed experiments. References to textbooks that discuss higher-order models and other issues associated with ANOVA (such as the "messy" case for unbalanced designs, use of continuous covariates, checking model assumptions, and other practical issues with design of experiments and real data analysis, etc.) are given throughout, such as at the end of Section 2.4.6, the end of Section 2.5.4, and the beginning of Chapter 3.

2.1 Introduction: Fixed, Random, and Mixed Effects Models

> In general, practicing statisticians have tended to treat the distinction between fixed and random effects as an either-or affair, even while acknowledging that in many instances, the line between the two can be rather subtle.
>
> (W. W. Stroup and D. K. Mulitze, 1991, p. 195)

We begin by differentiating between so-called **fixed effects** and **random effects** models. The notion of fixed effects is nicely given by Searle et al. (1992, p. 3) as "the effects attributable to a finite set of levels of a factor that occur in the data and which are there because we are interested in them." As examples of levels associated with fixed effects, "smoker" and "non-smoker" are the two levels

Linear Models and Time-Series Analysis: Regression, ANOVA, ARMA and GARCH, First Edition. Marc S. Paolella.

associated with the factor smoking; "male" and "female" are the two (common) levels of gender; Austria, Belgium, Bulgaria, etc., are the 28 "levels" (member countries) of the European Union; and aripiprazole, fluoxetine, olanzapine, and ziprasidone are four psychopharmacological treatments for borderline personality disorder, etc.

Random effects can be described as those attributable to a very large or infinite set of levels of a factor, of which only a random sample occur in the data. For example, a random set of $a = 20$ public high schools are selected from a certain geographic area that contains hundreds of such schools. Interest centers not on the peculiarities of each of the (randomly chosen) 20 schools, but rather treating them as 20 random observations from a large population, in order to assess the variation attributable to differences in the schools. From each of these schools, $n = 15$ pupils in the same grade are randomly chosen to have their creative writing essays evaluated. This group of n, for each school, are the "cell replications", and are also random effects.

As another example, from a particular set of countries, say, the $b = 10$ countries in the Association of Southeast Asian Nations (ASEAN), a random set of $a = 20$ public high schools are selected from each country. Some countries will have thousands of such schools, and interest centers not on the peculiarities of each of the (randomly chosen) 20 schools, but rather treating them as 20 random observations from a large population. Within each of the 20 schools (in each of the countries), $n = 15$ pupils of the same age are chosen randomly. In this setting, school and pupil are random effects, while country is (decided to be) a fixed effect, "because we are interested in them." For each pupil, one records the gender: This is also a fixed effect. In this case, we have a so-called **mixed model**, as it contains both fixed and random effects (and possibly their interactions). In a pure fixed effects model, the observations in each cell are random replications (in our example, this is the $n = 15$ pupils), but this model is not referred to as a mixed model. Similarly, in a pure random effects model, there is (almost always) a grand mean, say μ, and this, being a fixed but unknown parameter, is a fixed effect, though the model in this case is not referred to as mixed. A mixed model will have both fixed and random factors besides the fixed grand mean and the random effect associated with the cell replications.

Further (usually continuous) variables that are known to have, or suspected of having, explanatory power, will often be included. These are called **covariates**. In our school performance example, these could include the parental income of each pupil and the Gini coefficient (for measuring economic inequality) of each country. Note that the former is different for each pupil, while the latter pertains only to the country. In this case, the analysis of such data is referred to as the **analysis of covariance**, or ANCOVA, and can be for fixed, random, or mixed models.

2.2 Two Sample t-Tests for Differences in Means

Every basic statistics course discusses the classic t-test for the null of equality of the means of two normal populations. This is done under the assumption of equal population variances, and usually also without the equality assumption. In both cases, the test decision is the same as that delivered by the binary result of zero being in or out of the corresponding confidence interval, the latter having been detailed in Section III.8.3.

Arguments in favor of the use of confidence intervals and the study of effect sizes, as opposed to the blind application of hypothesis tests, were discussed in Section III.2.8. There, it was also discussed how hypothesis testing can have a useful role in inference, notably in randomized studies that are repeatable. We now derive the distribution of the associated test statistic, under the equal variance assumption, using the linear model framework. This is an easy task, given the general results from Chapter 1.

Let $Y_{1j} \overset{\text{i.i.d.}}{\sim} \mathrm{N}(\mu_1, \sigma^2), j = 1, \dots, m$, independent of $Y_{2j} \overset{\text{i.i.d.}}{\sim} \mathrm{N}(\mu_2, \sigma^2), j = 1, \dots, n$, with $\sigma^2 > 0$. This can be expressed as the linear model $\mathbf{Y} = \mathbf{X}\boldsymbol{\beta} + \boldsymbol{\epsilon}$, where, in standard notation, $\boldsymbol{\epsilon} \sim \mathrm{N}(\mathbf{0}, \sigma^2\mathbf{I}_N)$, $N = m + n$,

$$\mathbf{X} = \begin{bmatrix} \mathbf{1}_m & \mathbf{0}_m \\ \mathbf{0}_n & \mathbf{1}_n \end{bmatrix}, \quad \boldsymbol{\beta} = \begin{bmatrix} \mu_1 \\ \mu_2 \end{bmatrix}, \quad \mathbf{Y} = \begin{bmatrix} \mathbf{Y}_1 \\ \mathbf{Y}_2 \end{bmatrix}, \tag{2.1}$$

and

$$\hat{\boldsymbol{\beta}} = (\mathbf{X}'\mathbf{X})^{-1}\mathbf{X}'\mathbf{Y} = \begin{bmatrix} m & 0 \\ 0 & n \end{bmatrix}^{-1} \begin{bmatrix} Y_{1\bullet} \\ Y_{2\bullet} \end{bmatrix} = \begin{bmatrix} \bar{Y}_{1\bullet} \\ \bar{Y}_{2\bullet} \end{bmatrix}, \tag{2.2}$$

where we define the notation

$$Y_{1\bullet} = \sum_{j=1}^{m} Y_{1j}, \quad \bar{Y}_{1\bullet} = \frac{Y_{1\bullet}}{m}, \quad \text{and likewise,} \quad Y_{2\bullet} = \sum_{j=1}^{n} Y_{2j}, \quad \bar{Y}_{2\bullet} = \frac{Y_{2\bullet}}{n}. \tag{2.3}$$

The residual sum of squares, $\mathrm{RSS} = \mathsf{S}(\hat{\boldsymbol{\beta}}) = \hat{\boldsymbol{\epsilon}}'\hat{\boldsymbol{\epsilon}}$, is immediately seen to be

$$\mathsf{S}(\hat{\boldsymbol{\beta}}) = \sum_{j=1}^{m} (Y_{1j} - \bar{Y}_{1\bullet})^2 + \sum_{j=1}^{n} (Y_{2j} - \bar{Y}_{2\bullet})^2 = (m-1)S_1^2 + (n-1)S_2^2, \tag{2.4}$$

where S_i^2 is the sample variance based on the data from group i, $i = 1, 2$. Thus, from (2.4) and (1.58), an unbiased estimator of σ^2 is

$$\hat{\sigma}^2 = \mathsf{S}(\hat{\boldsymbol{\beta}})/(m + n - 2). \tag{2.5}$$

In the case that $m = n$ (as we will consider below, with $a \geqslant 2$ groups instead of just two, for the balanced one-way fixed effects ANOVA model), (2.5) can be expressed as

$$(m = n), \ (a = 2), \qquad \hat{\sigma}^2 = \frac{1}{a(n-1)} \sum_{i=1}^{a} \sum_{j=1}^{n} (Y_{ij} - \bar{Y}_{i\bullet})^2. \tag{2.6}$$

Remark (1.57) states that $\mathrm{RSS} = \mathbf{Y}'(\mathbf{I} - \mathbf{P})\mathbf{Y} = \|\mathbf{Y}\|^2 - \|\mathbf{X}\hat{\boldsymbol{\beta}}\|^2$, where $\mathbf{P}$ is the usual projection matrix $\mathbf{P} = \mathbf{X}(\mathbf{X}'\mathbf{X})^{-1}\mathbf{X}'$. It is a useful exercise to confirm, in this simple setting, that this RSS formula also leads to (2.4). For clarity, let $\bar{Y}_{1\bullet}^2 = (\bar{Y}_{1\bullet})^2$. We have, from the definition of $\mathbf{X}$ and $\hat{\boldsymbol{\beta}}$ in (2.2),

$$\begin{aligned} \|\mathbf{Y}\|^2 - \|\mathbf{X}\hat{\boldsymbol{\beta}}\|^2 &= \sum_{j=1}^{m} Y_{1j}^2 + \sum_{j=1}^{n} Y_{2j}^2 - m\bar{Y}_{1\bullet}^2 - n\bar{Y}_{2\bullet}^2 \\ &= \sum_{j=1}^{m} (Y_{1j}^2 - \bar{Y}_{1\bullet}^2) + \sum_{j=1}^{n} (Y_{2j}^2 - \bar{Y}_{2\bullet}^2). \end{aligned} \tag{2.7}$$

But, as

$$\sum_{j=1}^{m} (Y_{1j} - \bar{Y}_{1\bullet})^2 = \sum_{j=1}^{m} Y_{1j}^2 - 2\sum_{j=1}^{m} Y_{1j}\bar{Y}_{1\bullet} + \sum_{j=1}^{m} \bar{Y}_{1\bullet}^2$$

$$= \sum_{j=1}^{m} Y_{1j}^2 - 2m\bar{Y}_{1\bullet}^2 + m\bar{Y}_{1\bullet}^2 = \sum_{j=1}^{m} Y_{1j}^2 - m\bar{Y}_{1\bullet}^2 \tag{2.8}$$

$$= \sum_{j=1}^{m} (Y_{1j}^2 - \bar{Y}_{1\bullet}^2),$$

and likewise for the second group, (2.7) is equivalent to (2.4). ■

The null hypothesis is that $\mu_1 = \mu_2$, and in the notation of Section 1.4, $\mathbf{H}\boldsymbol{\beta} = b$, with $J = 1$, $\mathbf{H} = [1, -1]$ and scalar $b = 0$. From (1.90) with $\mathbf{A} = (\mathbf{X}'\mathbf{X})^{-1}$, it follows that $\mathbf{HAH}' = m^{-1} + n^{-1}$. Thus, (1.87) is painlessly seen to be

$$\mathbf{Y}'(\mathbf{P} - \mathbf{P}_{\mathcal{H}})\mathbf{Y} = \mathsf{S}(\hat{\boldsymbol{\gamma}}) - \mathsf{S}(\hat{\boldsymbol{\beta}}) = (\mathbf{H}\hat{\boldsymbol{\beta}})'(\mathbf{HAH}')^{-1}\mathbf{H}\hat{\boldsymbol{\beta}} = \frac{(\bar{Y}_{1\bullet} - \bar{Y}_{2\bullet})^2}{m^{-1} + n^{-1}}. \tag{2.9}$$

Remark As we did above for (2.4), it is instructive to derive (2.9) by brute force, directly evaluating $\mathsf{S}(\hat{\boldsymbol{\gamma}}) - \mathsf{S}(\hat{\boldsymbol{\beta}})$. Here, it will be convenient to let $n_1 = m$ and $n_2 = n$, which would anyway be necessary in the general unbalanced case with $a \geqslant 2$ groups. Under the reduced model, $\mathbf{P}_{\mathcal{H}}\mathbf{Y} = \mathbf{X}\hat{\boldsymbol{\gamma}}$ with

$$\hat{\gamma} = \bar{Y}_{\bullet\bullet} = N^{-1} \sum_{i=1}^{2} \sum_{j=1}^{n_i} Y_{ij} = N^{-1} Y_{\bullet\bullet},$$

this being the mean of all the Y_{ij}, where $N = n_1 + n_2$. Then

$$\begin{aligned}
\mathsf{S}(\hat{\gamma}) &= \sum_{j=1}^{n_1} (Y_{1j} - \bar{Y}_{\bullet\bullet})^2 + \sum_{j=1}^{n_2} (Y_{2j} - \bar{Y}_{\bullet\bullet})^2 \\
&= (Y^2)_{1\bullet} - 2\bar{Y}_{\bullet\bullet}Y_{1\bullet} + n_1(\bar{Y}_{\bullet\bullet})^2 + (Y^2)_{2\bullet} - 2\bar{Y}_{\bullet\bullet}Y_{2\bullet} + n_2(\bar{Y}_{\bullet\bullet})^2 \\
&= (Y^2)_{1\bullet} + (Y^2)_{2\bullet} - N(\bar{Y}_{\bullet\bullet})^2,
\end{aligned}$$

which could have been more easily determined by realizing that, in this case, $\mathsf{S}(\hat{\gamma}) = (Y^2)_{\bullet\bullet} - N(\bar{Y}_{\bullet\bullet})^2$, and $(Y^2)_{\bullet\bullet} = (Y^2)_{1\bullet} + (Y^2)_{2\bullet}$. Observe that

$$\begin{aligned}
N(\bar{Y}_{\bullet\bullet})^2 &= N^{-1}(Y_{1\bullet} + Y_{2\bullet})^2 \\
&= N^{-1}(Y_{1\bullet})^2 + N^{-1}(Y_{2\bullet})^2 + 2N^{-1}Y_{1\bullet}Y_{2\bullet} \\
&= N^{-1}n_1^2\bar{Y}_{1\bullet}^2 + N^{-1}n_2^2\bar{Y}_{2\bullet}^2 + 2N^{-1}n_1n_2\bar{Y}_{1\bullet}\bar{Y}_{2\bullet}.
\end{aligned}$$

Next, from (2.4), and the latter expression in (2.8),

$$\mathsf{S}(\hat{\boldsymbol{\beta}}) = \sum_{j=1}^{n_1} Y_{1j}^2 - n_1\bar{Y}_{1\bullet}^2 + \sum_{j=1}^{n_2} Y_{2j}^2 - n_2\bar{Y}_{2\bullet}^2 = (Y^2)_{1\bullet} + (Y^2)_{2\bullet} - n_1\bar{Y}_{1\bullet}^2 - n_2\bar{Y}_{2\bullet}^2,$$

so that

$$\begin{aligned}
\mathsf{S}(\hat{\gamma}) - \mathsf{S}(\hat{\boldsymbol{\beta}}) &= n_1\bar{Y}_{1\bullet}^2\left(1 - \frac{n_1}{N}\right) + n_2\bar{Y}_{2\bullet}^2\left(1 - \frac{n_2}{N}\right) - 2\frac{n_1n_2}{N}\bar{Y}_{1\bullet}\bar{Y}_{2\bullet} \\
&= \frac{n_1n_2}{n_1 + n_2}(\bar{Y}_{1\bullet}^2 + \bar{Y}_{2\bullet}^2 - 2\bar{Y}_{1\bullet}\bar{Y}_{2\bullet})
\end{aligned}$$

$$= \frac{(\bar{Y}_{1\bullet} - \bar{Y}_{2\bullet})^2}{n_1^{-1} + n_2^{-1}},$$

which is the same as (2.9). ■

Based on (2.9), the F statistic (1.88) is

$$F = \frac{(\bar{Y}_{1\bullet} - \bar{Y}_{2\bullet})^2/(m^{-1} + n^{-1})}{((m-1)S_1^2 + (n-1)S_2^2)/(m+n-2)} = \frac{(\bar{Y}_{1\bullet} - \bar{Y}_{2\bullet})^2}{S_p^2(m^{-1} + n^{-1})} \sim F_{1,m+n-2}, \tag{2.10}$$

a central F distribution with 1 and $m + n - 2$ degrees of freedom, where

$$S_p^2 = \frac{(m-1)S_1^2 + (n-1)S_2^2}{m+n-2} \tag{2.11}$$

from (2.5) is referred to as the **pooled variance estimator** of σ^2. Observe that $F = T^2$, where

$$T = \frac{\bar{Y}_{1\bullet} - \bar{Y}_{2\bullet}}{S_p\sqrt{m^{-1} + n^{-1}}} \sim t_{m+n-2}$$

is the usual "t statistic" associated with the test. Thus, a two-sided t-test of size α, $0 < \alpha < 1$, would reject the null if $|T| > c_t$, where c_t is the quantile such that $\Pr(T > c_t) = \alpha/2$, or, equivalently, if $F > c$, where $\Pr(F > c) = \alpha$. Note that $c = c_t^2$.

Under the alternative, $F \sim F_{1,m+n-2}(\theta)$, where, from (1.82) with $\mathbf{A} = (\mathbf{X}'\mathbf{X})^{-1}$,

$$\theta = \frac{1}{\sigma^2}\boldsymbol{\beta}'\mathbf{H}'(\mathbf{HAH}')^{-1}\mathbf{H}\boldsymbol{\beta} = \frac{1}{\sigma^2}\frac{\delta^2}{m^{-1} + n^{-1}}, \qquad \delta = \mu_2 - \mu_1. \tag{2.12}$$

For a given value of θ, the power of the test is $\Pr(F > c)$. To demonstrate, let $m = n$ so that $\theta = n\delta^2/(2\sigma^2)$. In Matlab, we could use

```
n = 10; delta = 0.3; sig2=6; theta = n *delta^2 /2 /sig2;
c = finv(0.95,1,2*n-2); pow = 1 - spncf(c,1,2*n-2,theta);
```

where `spncf` refers to the saddlepoint c.d.f. approximation of the singly noncentral F distribution; see Section II.10.2. As an illustration, Figure 2.1 plots the power curve of the two-sided t-test as a function of δ, using $\sigma^2 = 1$, $\alpha = 0.05$, and three values of n. As expected, for a given δ, the power increases with n, and for a given n, the power increases with δ.

It is more useful, though not always possible, to first decide upon a size α and a power ρ, for given values of σ^2 and δ, and then calculate n. That requires solving for the smallest integer n such that

$$\Pr(F_{1,2n-2}(0) > c) \leqslant \alpha \quad \text{and} \quad \Pr(F_{1,2n-2}(n\delta^2/(2\sigma^2)) > c) \geqslant \rho.$$

Equivalently, and numerically easier, we find the smallest $n \in \mathbb{R}_{>0}$ such that

$$\Pr(F_{1,2n-2}(0) > c) = \alpha \quad \text{and} \quad \Pr(F_{1,2n-2}(n\delta^2/(2\sigma^2)) > c) = \rho, \tag{2.13}$$

and then round up to the nearest integer. A program to accomplish this is given in Listing 2.1. (It uses the saddlepoint approximation to the noncentral F distribution to save computing time.) This can then be used to find the required sample size n^* as a function of, say, σ^2. To illustrate, the top panel of Figure 2.2 plots n^* versus σ^2 for $\alpha = 0.05$, $\rho = 0.90$, and three values of δ. It appears that n^* is linear in σ^2, and this is now explained.

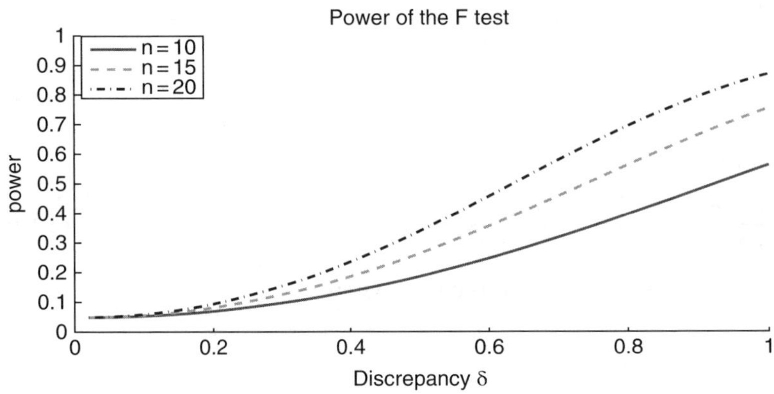

Figure 2.1 Power of the F test, given in (2.10) and (2.12), as a function of δ, using $\alpha = 0.05$ and $\sigma^2 = 1$.

Let $X_1, \ldots, X_n$ be an i.i.d. sample from a $\mathrm{N}(\mu, \sigma^2)$ population with σ^2 known. We wish to know the required sample size n for a one-sided hypothesis test of $H_0 : \mu = \mu_0$ versus $H_a : \mu = \mu_a$, for $\mu_a > \mu_0$, with size $\alpha \in (0, 1)$ and power $\rho \in (\alpha, 1)$. As $\bar{X}_n \sim \mathrm{N}(\mu_0, \sigma^2/n)$ under the null, let $Z = \sqrt{n}(\bar{X}_n - \mu_0)/\sigma \sim \mathrm{N}(0, 1)$, so that the required test cutoff value, c_α, is given by $\Pr(Z > c_\alpha \mid H_0) = \alpha$, or $c_\alpha = \Phi^{-1}(1 - \alpha)$. The power is

$$\begin{aligned}\rho &= \Pr(Z > c_\alpha \mid H_a) = \Pr(\bar{X}_n > \mu_0 + c_\alpha\sqrt{\sigma^2/n} \mid H_a) \\ &= \Pr\left(\frac{\bar{X}_n - \mu_a}{\sqrt{\sigma^2/n}} > \frac{\mu_0 - \mu_a + c_\alpha\sqrt{\sigma^2/n}}{\sqrt{\sigma^2/n}} \;\middle|\; H_a \right),\end{aligned}$$

or, simplifying, with $\delta = \mu_a - \mu_0$, the minimal sample size is $\lceil n \rceil$, where $\lceil \cdot \rceil$ denotes the ceiling function, i.e., $\lceil 2.3 \rceil = \lceil 2.8 \rceil = 3$, and

$$\begin{aligned}n &= \frac{\sigma^2}{\delta^2}(\Phi^{-1}(1 - \alpha) - \Phi^{-1}(1 - \rho))^2 \\ &= \frac{\sigma^2}{\delta^2}(\Phi^{-1}(1 - \alpha) + \Phi^{-1}(\rho))^2, \quad \rho \in (\alpha, 1). \qquad (2.14)\end{aligned}$$

Observe that (2.14) does not make sense for $\rho \in (0, \alpha)$. This formula is derived in most introductory statistics texts (see, e.g., Rosenkrantz, 1997, p. 299), and is easy because of the simplifying assumption that σ^2 is known, so that the t distribution (or F) is not required.

For the two-sided test, again assuming σ^2 known, it is straightforward to show that n is given by the solution to

$$\Phi(-z - k) + \Phi(-z + k) = \rho, \quad \text{where} \quad z = \Phi^{-1}(1 - \alpha/2) \quad \text{and} \quad k = \delta\sqrt{n}/\sigma, \qquad (2.15)$$

(see, e.g., Tamhane and Dunlop, 2000, pp. 248–249), which needs to be solved numerically. However, for $\delta > 0$, the term $\Phi(-z - k)$ will be relatively small, so that

$$n \approx \frac{\sigma^2}{\delta^2}\left(\Phi^{-1}\left(1 - \frac{\alpha}{2}\right) + \Phi^{-1}(\rho)\right)^2 \qquad (2.16)$$

should be highly accurate. These formulae all refer to testing with a single i.i.d. sample (and σ^2 known). These could, however, be applied to $D_i \overset{\text{i.i.d.}}{\sim} \mathrm{N}(\mu_D, \sigma_D^2)$, where $D_i = X_i - Y_i$ are computed from paired

```
function [n,c]=design1(delta,sigma2,alpha,power)
if nargin<4, power=0.90; end, if nargin<3, alpha=0.05; end
d2=delta^2; perc0=1-alpha; M=2; n=2;
c=ncf2cdfx(perc0,1,2*n-2,0,0); theta1=n*d2/(2*sigma2);
F=spncf(c,1,2*n-2,theta1,0);
while ( 1-F < power )
  n=n*M; c=ncf2cdfx(perc0,1,2*n-2,0,0); theta1=n*d2/(2*sigma2);
  F=spncf(c,1,2*n-2,theta1,0);
end
hib=n; lob= n/M; % this should bound n
% now use bisection:
versuch = (lob+hib)/2; valid=0; TOL=1e-8;
while (valid==0)
  z=betainv(alpha,(2*versuch-2)/2,1/2);
  c=((2*versuch-2)/z - (2*versuch-2))/1;
  theta1=versuch*d2/(2*sigma2); F=spncf(c,1,2*versuch-2,theta1,0);
  check=F-(1-power); valid= (abs( check ) < TOL);
  if (valid==0)
    if check<0, hib=versuch; else lob= versuch; end
    versuch= (lob+hib)/2;
  else n=versuch;
  end
end
n=ceil(n); z=betainv(alpha,(2*n-2)/2,1/2);
c=((2*n-2)/z - (2*n-2))/1;
% check the result
theta1=n*d2/(2*sigma2);
size_SPA=1-spncf(c,1,2*n-2,0,0) %#ok<NASGU,NOPRT>
size_exact=1-fcdf(c,1,2*n-2) %#ok<NASGU,NOPRT>
power_SPA = 1-spncf(c,1,2*n-2,theta1,0)  %#ok<NASGU,NOPRT>
power_exact = 1-ncf(c,1,2*n-2,theta1,0) %#ok<NASGU,NOPRT>

end % function
```

Program Listing 2.1: Computes $n^\star$ (and cutoff value c) for the given values δ, σ^2, α and ρ. The last part of the program takes into account that n is fractional. Round up n to get an integer and then recompute the cutoff value such that the size is exactly α. Functions `ncf2cdfx` and `spncf` use the saddlepoint approximation and are available in the set of programs associated with this book. The former is given in Listing 2.2. The word "Versuch" is a noun in German meaning "attempt" or "try", the latter being a reserved word in Matlab.

observations from a bivariate normal population. If the X_i and Y_i have the same variance σ^2 and the correlation between them is zero, then (2.14) and (2.16) can be applied with $\sigma_D^2 = \mathrm{Var}(D_i) = 2\sigma^2$. In particular, for the two-sided test,

$$n^* \approx 2\frac{\sigma^2}{\delta^2}\left(\Phi^{-1}\left(1-\frac{\alpha}{2}\right)+\Phi^{-1}(\rho)\right)^2. \tag{2.17}$$

Observe that (2.17) embodies two approximations: one is the nonzero term $\Phi(-z-k)$ in (2.15), the other is that σ^2 is known. It explains the linearity of n^* in Figure 2.2. To illustrate the accuracy, the bottom panel of Figure 2.2 is the same as the top panel, but using (2.17). We see that the approximation is excellent for the constellation of parameters under consideration.

```
function ncf2cdfx=ncf2cdfx(alpha,n1,n2,theta1,theta2)
% cutoff value of the (possibly doubly noncentral) F distribution using the SPA.
% Compare to Matlab's built in ncfinv and finv.

if (theta1>0) && (theta2>0), xval = 1.5*theta1/theta2; else xval=1; end
multip=1; cdf=2;
while (cdf>alpha)
  versuch= xval/multip;
  cdf = spncf(versuch,n1,n2,theta1,theta2); multip= multip*2;
end
lob= versuch;

multip= 1; cdf=-1;
while (cdf<alpha)
  versuch= xval*multip;
  cdf= spncf(versuch,n1,n2,theta1,theta2); multip= multip*2;
end
hib= versuch;

if 1==1 % Matlab's routine for minimization when bounds are known
  opt=optimset('TolX',1e-5,'Display','off');
  ncf2cdfx=fminbnd(@(x) spncf_(x,n1,n2,theta1,theta2,alpha),lob,hib,opt);
else % use bisection
  versuch = (lob+hib)/2; valid=0; TOL=1e-8;
  while (valid~=1)
    cdf= spncf(versuch,n1,n2,theta1,theta2);
    valid= (abs(cdf-alpha)<TOL);
    if (valid==1), ncf2cdfx= versuch;
    else
      if (cdf<alpha), lob= versuch; else hib= versuch; end
      versuch= (lob+hib)/2;
    end
  end
end
end % function

function disc=spncf_(x,n1,n2,theta1,theta2,alpha)
  disc=abs(spncf(x,n1,n2,theta1,theta2,2) - alpha);
end % function
```

Program Listing 2.2: Evaluation of the location-scale d-dimensional Student's t density. Continued from Listing 2.1.

2.3 The Two Sample *t*-Test with Ignored Block Effects

To consult the statistician after an experiment is finished is often merely to ask him to conduct a *post mortem* examination. He can perhaps say what the experiment died of.

(Sir Ronald A. Fisher, 1938, p. 17)

The following extension of the two-sample model is a special case of a two-way analysis of variance. We use it here to illustrate how easily the doubly noncentral F distribution can arise in an otherwise simple model. It also serves as an example emphasizing why experiments should be correctly planned,

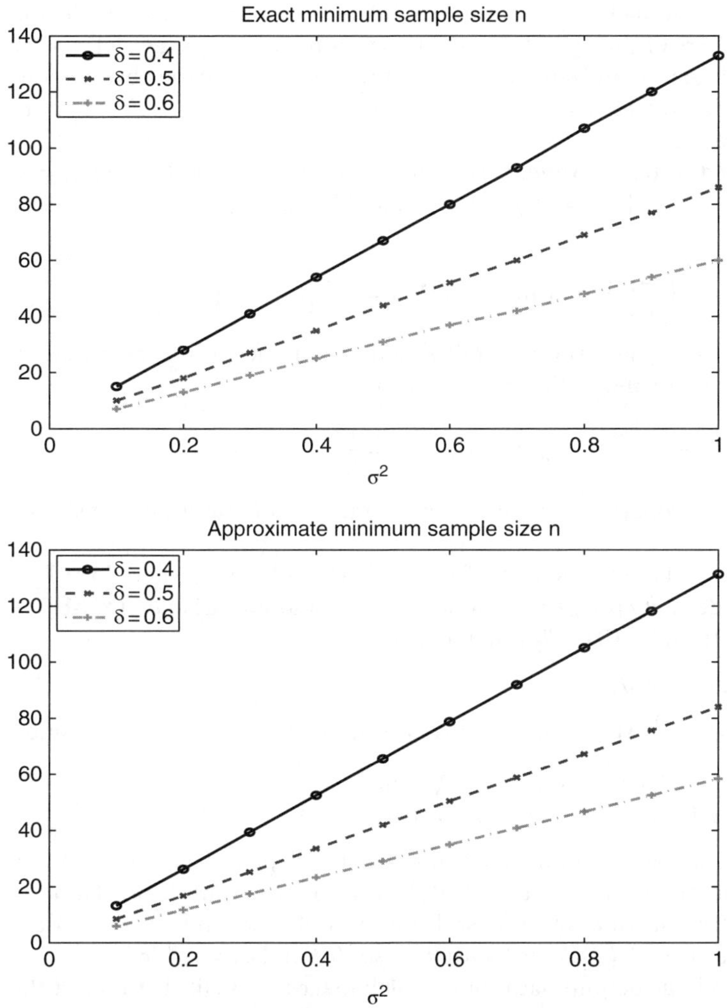

Figure 2.2 **Top:** Minimum required sample size as a function of σ^2, based on (2.13), using $\alpha = 0.05$ and $\rho = 0.90$. **Bottom::** Approximation to the sample size calculation computed using (2.17).

as emphasized in the above famous quote by Fisher: Particularly with modern statistical software, "analyzing" a data set is trivial (possibly with incorrect conclusions due to ignored effects or other problems, such as unaccounted for correlation, etc.), whereas designing an experiment often profits from input from a professional statistician.

For the sake of clarity, let Y_{1j} refer to a certain measurement of patient j when using treatment A, while Y_{2j} refers to use of treatment B. Under treatment A, the response to the treatment is δ on average, while under treatment B, it is zero. This is, so far, precisely the model described in the previous section.

Now suppose that the n values of Y_{1j} actually come from two different populations, say male and female. Assume there are equal numbers of each gender in the two treatment groups, i.e., let $n = 2s$, with s males and s females using treatment A, and similarly for treatment B. Arrange the observations with females first, i.e., $Y_{1,1}, \dots, Y_{1,s}$ refer to females, $Y_{1,s+1}, \dots, Y_{1,n}$ refer to males, $Y_{2,1}, \dots, Y_{2,s}$ refer to females, and $Y_{2,s+1}, \dots, Y_{2,n}$ refer to males.

The effect of being male is assumed to be **additive**, with quantity η. That means that $\mathbb{E}[Y_{1j}] = \delta$, $j = 1, \dots, s$, $\mathbb{E}[Y_{1j}] = \delta + \eta$, $j = s+1, \dots, n$, $\mathbb{E}[Y_{2j}] = 0$, $j = 1, \dots, s$, and $\mathbb{E}[Y_{2j}] = \eta$, $j = s+1, \dots, n$. In vector notation,

$$\boldsymbol{\mu}_{\mathbf{Y}_1} := \mathbb{E}[\mathbf{Y}_1] = \begin{pmatrix} \delta \mathbf{1}_s \\ (\delta + \eta)\mathbf{1}_s \end{pmatrix} =: \begin{pmatrix} \mathbf{u}_1 \\ \mathbf{u}_2 \end{pmatrix} \text{ and } \boldsymbol{\mu}_{\mathbf{Y}_2} := \mathbb{E}[\mathbf{Y}_2] = \begin{pmatrix} 0\mathbf{1}_s \\ \eta\mathbf{1}_s \end{pmatrix} =: \begin{pmatrix} \mathbf{v}_1 \\ \mathbf{v}_2 \end{pmatrix}.$$

In the case that both δ and η are not zero, the F statistic follows a doubly noncentral F distribution. To see this, first note that $\mathbb{E}[\bar{Y}_{1\bullet}] = \delta + \eta/2$ and $\mathbb{E}[\bar{Y}_{2\bullet}] = \eta/2$, implying

$$\bar{Y}_{1\bullet} - \bar{Y}_{2\bullet} \sim \mathrm{N}\left(\delta, \frac{2\sigma^2}{n}\right) \text{ and } \frac{(\bar{Y}_{1\bullet} - \bar{Y}_{2\bullet})^2}{2\sigma^2/n} \sim \chi^2(1, \theta_1), \quad \theta_1 = \frac{n\delta^2}{2\sigma^2},$$

recalling the definition and basic properties of the noncentral χ^2 distribution (see, e.g., Section II.10.1).

Using the pooled variance estimator (2.11), express $(2n-2)S_p^2$ as $\mathbf{Y}_1'\mathbf{M}\mathbf{Y}_1 + \mathbf{Y}_2'\mathbf{M}\mathbf{Y}_2$, where $\mathbf{M} = \mathbf{I}_n - \mathbf{1}_n\mathbf{1}_n'/n$. From Theorem A.1 and the additivity property of noncentral χ^2 random variables, $(\mathbf{Y}_1'\mathbf{M}\mathbf{Y}_1 + \mathbf{Y}_2'\mathbf{M}\mathbf{Y}_2)/\sigma^2 \sim \chi^2(2n-2, \theta_2)$, where, from (II.10.6), θ_2 is determined by

$$\mathbb{E}[(\mathbf{Y}_1'\mathbf{M}\mathbf{Y}_1 + \mathbf{Y}_2'\mathbf{M}\mathbf{Y}_2)/\sigma^2] = 2n - 2 + \theta_2. \tag{2.18}$$

But, from (A.6), $\mathbb{E}[\mathbf{Y}_1'\mathbf{M}\mathbf{Y}_1] = \mathrm{tr}(\sigma^2\mathbf{M}) + \boldsymbol{\mu}_{\mathbf{Y}_1}'\mathbf{M}\boldsymbol{\mu}_{\mathbf{Y}_1}$ with $\mathrm{tr}(\sigma^2\mathbf{M}) = \sigma^2(n-1)$ from Theorem 1.2, and

$$\boldsymbol{\mu}_{\mathbf{Y}_1}'\mathbf{M}\boldsymbol{\mu}_{\mathbf{Y}_1} = \begin{pmatrix} \mathbf{u}_1' & \mathbf{u}_2' \end{pmatrix} \begin{pmatrix} \mathbf{I}_s - \mathbf{1}_s\mathbf{1}_s'/n & -\mathbf{1}_s\mathbf{1}_s'/n \\ -\mathbf{1}_s\mathbf{1}_s'/n & \mathbf{I}_s - \mathbf{1}_s\mathbf{1}_s'/n \end{pmatrix} \begin{pmatrix} \mathbf{u}_1 \\ \mathbf{u}_2 \end{pmatrix} = \frac{s}{2}\eta^2,$$

after some simplification that the reader should confirm. As this does not depend on δ, $\mathbb{E}[\mathbf{Y}_2'\mathbf{M}\mathbf{Y}_2] = \mathbb{E}[\mathbf{Y}_1'\mathbf{M}\mathbf{Y}_1]$ because it is the same computation but with $\delta = 0$. It now follows from (2.18) that $\theta_2 = s\eta^2/\sigma^2$. The F statistic in (2.10) with nonzero δ and η is still the ratio of independent χ^2 random variables, but both are noncentral, i.e., $F \sim F_{1,4s-2}(\theta_1, \theta_2)$ with $\theta_1 = s\delta^2/\sigma^2$ and $\theta_2 = s\eta^2/\sigma^2$.

Via (2.18), large values of η^2 imply a large denominator of F, which makes it less likely to reject the null for a given value of θ_1. Thus, increasing η^2 will decrease the power of the test and also diminish its size. To see by how much, take $\alpha = 0.05$, $\rho = 0.90$ and $\delta = \sigma = 1$, for which (2.13) yields $n = 22$, i.e., $s = 11$. (Actually, the real value of n is 22.02, which should, technically speaking, be rounded up to 23. Using 22, the power is "only" 0.8997, while with 23, it is 0.9125.)

Figure 2.3 plots the size $\Pr(F > c; 0, \theta_2)$ (left axis) and power $\Pr(F > c; \theta_1, \theta_2)$ (right axis) of the test versus a grid of η-values between zero and three. The calculations were done using the saddlepoint approximation to the singly and doubly noncentral F distribution, as detailed in Butler and Paolella (2002a) (see also Section II.10.2). Its use explains why the size is not precisely 0.05 when $\eta = 0$. Use of the exact method takes over 100 times longer, with no appreciable gain in accuracy.

Such a graph is useful when designing an experiment, particularly when different opinions exist regarding η. Furthermore, the required sample size and cutoff value c can be computed with a nonzero η-value by solving

$$\Pr(F_{1,2n-2}(0, \theta_2) > c) = \alpha \quad \text{and} \quad \Pr(F_{1,2n-2}(\theta_1, \theta_2) > c) = \rho,$$

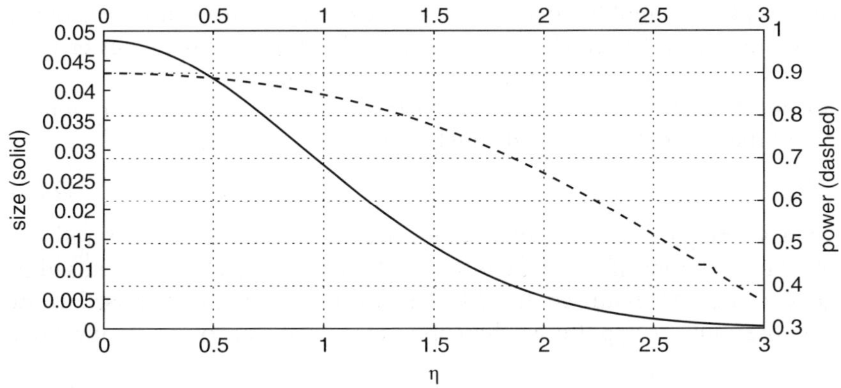

Figure 2.3 Size (solid, left axis) and power (dashed, right axis) for the two-way model ignoring the effect of gender, with $\alpha = 0.05$, $\rho = 0.90$ and $\delta = \sigma = 1$.

instead of (2.13). This was done for $\alpha = 0.05$, $\rho = 0.90$, $\delta = \sigma = 1$ and a grid of η-values between zero and one (using the saddlepoint approximation to save time). The sample size n^* stays constant at 22, while the cutoff value c smoothly drops from 4.01 down to 3.17. The reader is encouraged to confirm this result.

2.4 One-Way ANOVA with Fixed Effects

2.4.1 The Model

The one-way analysis of variance, or one-way ANOVA, extends the two-sample situation discussed above to $a \geqslant 2$ groups. For example, in an agricultural setting,[1] there might be $a \geqslant 2$ competing fertilizer mixtures available, the best one of which (in terms of weight of crop yield) is not theoretically obvious for a certain plant under certain conditions (soil, climate, amount of rain and sunshine, etc.). To help determine the efficacy of each fertilizer mixture, which ones are statistically the same, and, possibly, which one is best, an experiment could consist of forming na equally sized plots of land on which the plant is grown, such that all external conditions are the same for each plot (sunshine, rainfall, etc.), with n of the na plots, randomly chosen (to help account for any exogenous factor not considered), getting treated with the ith fertilizer mixture, $i = 1, \dots, a$. When the allocation of fertilizer treatments to the plots is done randomly, the crop yield of the na plots can be treated as independent realizations of random variables Y_{ij}, where i refers to the fertilizer used, $i = 1, \dots, a$, and j refers to which replication, $j = 1, \dots, n$.

The usual assumption is that the Y_{ij} are normally distributed with equal variance σ^2 and possibly different means μ_i, $i = 1, \dots, a$, so that the model is given by

$$Y_{ij} = \mu_i + \epsilon_{ij}, \quad \epsilon_{ij} \overset{\text{i.i.d.}}{\sim} \mathrm{N}(0, \sigma^2). \tag{2.19}$$

1 The techniques of ANOVA were actually founded for use in agriculture, which can be seen somewhat in the terminology that persists, such as "treatments", "plots", "split-plots" and "blocks". See Mahalanobis (1964) for a biography of Sir Ronald Fisher and others who played a role in the development of ANOVA. See also Plackett (1960) for a discussion of the major early developments in the field.

Observe that the normality assumption cannot be correct, as crop yield cannot be negative. However, it can be an excellent approximation if the probability is very small of a crop yield being lower than some small positive number, and is otherwise close to being Gaussian distributed.

The first, and often primary, question of interest is the extent to which the model can be more simply expressed as $Y_{ij} = \mu + \epsilon_{ij}$, i.e., all the μ_i are the same and equal μ. Formally, we wish to test

$$H_0 : \mu_1 = \mu_2 = \cdots = \mu_a \; (= \mu) \tag{2.20}$$

against the alternative that at least one pair of μ_i are different. It is worth emphasizing that, for $a > 2$, the alternative is not that all μ_i are different. If $a = 2$, then the method in Section 2.2 can be used to test H_0. For $a > 2$, a more general model is required. In addition, new questions can be posed, most notably: If we indeed can reject the equal-μ_i hypothesis, then precisely which pairs of μ_i actually differ from one another?

Instead of (2.19), it is sometimes convenient to work with the model parameterization given by

$$Y_{ij} = \mu + \alpha_i + \epsilon_{ij}, \quad i = 1, \dots, a, \quad j = 1, \dots, n. \tag{2.21}$$

i.e., $\mu_i = \mu + \alpha_i = \mathbb{E}[Y_{ij}]$, which can be interpreted as an overall mean μ plus a factor α_i for each of the a treatments. The $\mathbf{X}$ matrix is then similar to that given in (2.1), but with $a + 1$ columns, the first of which is a column of all ones, and thus such that $\mathbf{X}$ is rank deficient, with rank a.

In this form, we have $a + 1$ parameters for the a means, and the set of these $a + 1$ parameters is not identified, and only some of their linear combinations are estimable, recalling the discussion in Section 1.4.2. In this case, one linear restriction on the α_i is necessary in order for them to be estimable. A natural choice is $\sum_{i=1}^{a} \alpha_i = 0$, so that the α_i can be interpreted as deviations from the overall mean μ. The null hypothesis (2.20) can also be written $H_0 : \alpha_1 = \cdots = \alpha_a = 0$ versus H_a: at least one $\alpha_i \neq 0$.

2.4.2 Estimation and Testing

Based on the model assumptions of independence and normality, we would expect that the parameter estimators for model formulation (2.19) are given by $\hat{\mu}_i = \bar{Y}_{i\bullet}$, $i = 1, \dots, a$, and, recalling the notation of S_i^2 in (2.4), $\hat{\sigma}^2 = (n-1)\sum_{i=1}^{a} S_i^2/(na - a)$, the latter being a direct generalization of the pooled variance estimator of σ^2 in the two-sample case. This is indeed the case, and to verify these we cast the model in the general linear model framework by writing $\mathbf{Y} = \mathbf{X}\boldsymbol{\beta} + \boldsymbol{\epsilon}$, where

$$\mathbf{Y} = (Y_{11}, Y_{12}, \dots, Y_{1n}, \; Y_{21}, \dots, Y_{an})', \quad \mathbf{X} = \begin{pmatrix} \mathbf{1}_n & \mathbf{0}_n & \cdots & \mathbf{0}_n \\ \mathbf{0}_n & \mathbf{1}_n & \cdots & \vdots \\ \vdots & \vdots & \ddots & \\ \mathbf{0}_n & \mathbf{0}_n & \cdots & \mathbf{1}_n \end{pmatrix} = \mathbf{I}_a \otimes \mathbf{1}_n, \tag{2.22}$$

$\boldsymbol{\beta} = (\mu_1, \dots, \mu_a)'$, and $\boldsymbol{\epsilon}$ is similarly structured as $\mathbf{Y}$. Note that this generalizes the setup in (2.1) (but with the sample sizes in each group being the same) and is not the formulation in (2.21). Matrix $\mathbf{X}$ in (2.22) has full rank a.

As $\mathbf{X}$ is $na \times a$, there are $T = na$ total observations and $k = a$ regressors. The Kronecker product notation allows the design matrix to be expressed very compactly and is particularly helpful for representing $\mathbf{X}$ in more complicated models. It is, however, only possible when the number of replications is the same per treatment, which we assume here for simplicity of presentation. In this case, the model is said to be **balanced**. More generally, the ith group has n_i observations, $i = 1, \dots, a$, and if any two of the n_i are not equal, the model is **unbalanced**.

Using the basic facts that, for conformable matrices,

$$(\mathbf{A} \otimes \mathbf{B})' = (\mathbf{A}' \otimes \mathbf{B}') \quad \text{and} \quad (\mathbf{A} \otimes \mathbf{B})(\mathbf{C} \otimes \mathbf{D}) = (\mathbf{AC} \otimes \mathbf{BD}), \tag{2.23}$$

it is easy to verify that

$$(\mathbf{X}'\mathbf{X}) = (\mathbf{I}_a \otimes \mathbf{1}_n)'(\mathbf{I}_a \otimes \mathbf{1}_n) = n\mathbf{I}_a \quad \text{and} \quad \mathbf{X}'\mathbf{Y} = (\ Y_{1\bullet} \quad Y_{2\bullet} \quad \cdots \quad Y_{a\bullet}\ )', \tag{2.24}$$

yielding the least squares unbiased estimators

$$\hat{\boldsymbol{\beta}} = (\bar{Y}_{1\bullet}, \bar{Y}_{2\bullet}, \dots, \bar{Y}_{a\bullet})', \quad \mathsf{S}(\hat{\boldsymbol{\beta}}) = \sum_{i=1}^{a}\sum_{j=1}^{n}(Y_{ij} - \bar{Y}_{i\bullet})^2, \quad \hat{\sigma}^2 = \frac{\mathsf{S}(\hat{\boldsymbol{\beta}})}{a(n-1)}, \tag{2.25}$$

with $\hat{\sigma}^2$ generalizing that given in (2.6) for $a = 2$.

For the restricted model $\mathbf{Y} = \mathbf{X}\boldsymbol{\gamma} + \boldsymbol{\epsilon}$, i.e., the model under the null hypothesis of no treatment (fertilizer) effect, we could use (1.69) to compute $\hat{\boldsymbol{\gamma}}$ with the $J = a - 1$ restrictions represented as $\mathbf{H}\boldsymbol{\beta} = \mathbf{h}$ with, say,

$$\mathbf{H} = [\mathbf{I}_{a-1}, -\mathbf{1}_{a-1}] \quad \text{and} \quad \mathbf{h} = \mathbf{0}. \tag{2.26}$$

It should be clear for this model that

$$\hat{\gamma} = \hat{\mu} = \bar{Y}_{\bullet\bullet} \quad \text{and} \quad \mathsf{S}(\hat{\gamma}) = \sum_{i=1}^{a}\sum_{j=1}^{n}(Y_{ij} - \bar{Y}_{\bullet\bullet})^2, \tag{2.27}$$

so that the F statistic (1.88) associated with the null hypothesis (2.20) can be computed. Moreover, the conditions in Example 1.8 are fulfilled, so that (1.55) (with $\hat{Y} = \bar{Y}_{i\bullet}$ and $\bar{Y} = \bar{Y}_{\bullet\bullet}$) implies

$$\sum_{i=1}^{a}\sum_{j=1}^{n}(Y_{ij} - \bar{Y}_{\bullet\bullet})^2 = \sum_{i=1}^{a}\sum_{j=1}^{n}(Y_{ij} - \bar{Y}_{i\bullet})^2 + \sum_{i=1}^{a}\sum_{j=1}^{n}(\bar{Y}_{i\bullet} - \bar{Y}_{\bullet\bullet})^2, \tag{2.28}$$

and, in particular,

$$\mathsf{S}(\hat{\gamma}) - \mathsf{S}(\hat{\boldsymbol{\beta}}) = \sum_{i=1}^{a}\sum_{j=1}^{n}(\bar{Y}_{i\bullet} - \bar{Y}_{\bullet\bullet})^2 = n\sum_{i=1}^{a}(\bar{Y}_{i\bullet} - \bar{Y}_{\bullet\bullet})^2.$$

Thus, (1.88) gives

$$F = \frac{n\sum_{i=1}^{a}(\bar{Y}_{i\bullet} - \bar{Y}_{\bullet\bullet})^2/(a-1)}{\sum_{i=1}^{a}\sum_{j=1}^{n}(Y_{ij} - \bar{Y}_{i\bullet})^2/(na-a)} \sim F_{a-1,na-a}, \tag{2.29}$$

under H_0 from (2.20), which, for $a = 2$, agrees with (2.10) with $m = n$.

Remark The pitfalls associated with (and some alternatives to) the use of statistical tests for dichotomous model selection were discussed in Section III.2.8, where numerous references can be found, including recent ones such as McShane and Gal (2016) and Briggs (2016). We presume that the reader has got the message and realizes the ludicrousness of a procedure as simple as "if p-value is less than 0.05, the effect is significant", and "if p-value is greater than 0.05, there is no effect". We subsequently suppress this discussion and present the usual test statistics associated with ANOVA, and common to all statistical software, using the traditional language of "reject the null" and "not reject the null", hoping the reader understands that this nonfortuitous language is not a synonym for model selection. ∎

A test of size α "rejects" H_0 if $F > c$, where c is such that $\Pr(F > c) = \alpha$. We will sometimes write this as: The F test in (2.29) for H_0 rejects if $F > F^{\alpha}_{a-1,na-a}$, where $F^{\alpha}_{n,d}$ is the $100(1-\alpha)$th percent quantile of the $F_{n,d}$ distribution. As a bit of notational explanation to avoid any confusion, note how, as history has it, α is the standard notation for the significance level of a test, and how we use α_i in (2.21), this also being common notation for the fixed effects. Below, in (2.40), we will express F in matrix terms.

To determine the noncentrality parameter θ under the alternative hypothesis, we can use (1.82), i.e., $\theta = \boldsymbol{\beta}'\mathbf{H}'(\mathbf{HAH}')^{-1}\mathbf{H}\boldsymbol{\beta}/\sigma^2$, where $\mathbf{A} = (\mathbf{X}'\mathbf{X})^{-1}$. In particular, from (2.24) and (2.26), $\mathbf{HAH}' = n^{-1}\mathbf{HH}'$, and $\mathbf{HH}' = \mathbf{I}_{a-1} + \mathbf{1}_{a-1}\mathbf{1}'_{a-1}$. From (1.70), its inverse is

$$\mathbf{I}_{a-1} - \mathbf{1}_{a-1}(\mathbf{1}'_{a-1}\mathbf{I}_{a-1}\mathbf{1}_{a-1} + 1)^{-1}\mathbf{1}'_{a-1} = \mathbf{I}_{a-1} - a^{-1}\mathbf{1}_{a-1}\mathbf{1}'_{a-1},$$

so that

$$\boldsymbol{\beta}'\mathbf{H}'(n^{-1}\mathbf{HH}')^{-1}\mathbf{H}\boldsymbol{\beta} = n\boldsymbol{\beta}'\mathbf{H}'\mathbf{H}\boldsymbol{\beta} - na^{-1}\boldsymbol{\beta}'\mathbf{H}'\mathbf{1}_{a-1}\mathbf{1}'_{a-1}\mathbf{H}\boldsymbol{\beta}$$
$$= n\sum_{i=1}^{a-1}(\mu_i - \mu_a)^2 - \frac{n}{a}\left(\sum_{i=1}^{a-1}(\mu_i - \mu_a)\right)^2.$$

Notice that, when $a = 2$, this becomes n times

$$(\mu_1 - \mu_2)^2 - \frac{1}{2}(\mu_1 - \mu_2)^2 = \frac{1}{2}(\mu_1 - \mu_2)^2,$$

so that $\theta = n(\mu_1 - \mu_2)^2/(2\sigma^2)$, which agrees with (2.12) for $m = n$.

To simplify the expression for general $a \geqslant 2$, we switch to the alternative notation (2.21), i.e., $\mu_i = \mu + \alpha_i$ and $\sum_{i=1}^{a}\alpha_i = 0$. Then

$$\sum_{i=1}^{a-1}(\mu_i - \mu_a)^2 = \sum_{i=1}^{a-1}(\alpha_i - \alpha_a)^2 = \sum_{i=1}^{a}(\alpha_i - \alpha_a)^2 = \sum_{i=1}^{a}\alpha_i^2 - 2\alpha_a\sum_{i=1}^{a}\alpha_i + a\alpha_a^2 = \sum_{i=1}^{a}\alpha_i^2 + a\alpha_a^2$$

and

$$\frac{1}{a}\left(\sum_{i=1}^{a-1}(\mu_i - \mu_a)\right)^2 = \frac{1}{a}\left(\sum_{i=1}^{a-1}(\alpha_i - \alpha_a)\right)^2 = \frac{1}{a}(0 - \alpha_a - (a-1)\alpha_a)^2 = a\alpha_a^2,$$

so that

$$\theta = \frac{n}{\sigma^2}\sum_{i=1}^{a}\alpha_i^2. \tag{2.30}$$

Thus, with $F \sim F_{a-1,na-a}(\theta)$, the power of the test is $\Pr(F > c)$, where c is determined from (2.29) for a given probability α.

Remark Noncentrality parameter θ in (2.30) can be derived directly using model formulation (2.21), and the reader is encouraged to do so. Hint: We do so in the more general two-way ANOVA below; see (2.64). ∎

2.4.3 Determination of Sample Size

To determine n, the required number of replications in each of the a treatments, for a given significance α, power ρ, and value of σ^2, we solve

$$\Pr(F_{a,an-a}(0) > c) = \alpha \quad \text{and} \quad \Pr(F_{a,an-a}(\theta) > c) = \rho$$

for n and c, and then round up n to the nearest integer, giving, say, n^*. The program in Listing 2.1 is easily modified to compute this.

Remark It is worth emphasizing the luxury we have with the availability of cheap modern computing power. This makes such calculations virtually trivial. Use of the saddlepoint approximation to the noncentral F speeds things up further, allowing "what if" scenarios and plots of n as a function of various input variables to be made essentially instantaneously. To get an idea of how sample size determination was previously done and the effort put into construction of tabulated values, see Sahai and Ageel (2000, pp. 57–60). ■

Also similar to the sample size calculation in the two-sample case, the value of σ^2 must be specified. As σ^2 will almost always be unknown, an approximation needs to be used, for which there might be several (based on prior knowledge resulting from, perhaps, a pilot experiment, or previous, related experiments, or theoretical considerations, or, most likely, a combination of these). As n^* is an increasing function of σ^2, use of the largest "educated guess" for σ^2 would lead to a conservative choice of n^*. Arguably even more complicated is the specification of $\sum_{i=1}^{a} \alpha_i^2$, for which n^* is a decreasing function, i.e., to be conservative we need to choose the smallest relevant $\sum_{i=1}^{a} \alpha_i^2$.

One way to make such a choice is to choose a value δ that represents the smallest practically significant difference worth detecting between any two particular treatments, say 1 and 2. Then taking $|\alpha_1 - \alpha_2| = \delta$ and $\alpha_i = 0$, $i = 3, \dots, a$, together with the constraint $\sum_{i=1}^{a} \alpha_i = 0$ implies $\alpha_1 = \pm\delta/2$, $\alpha_2 = \mp\delta/2$ and $\sum_{i=1}^{a} \alpha_i^2 = \delta^2/2$. Specification of δ appears easier than $\sum_{i=1}^{a} \alpha_i^2$, although might lead to unnecessarily high choices of n^* if more specific information is available about the choice of the α_i.

In certain cases, an experiment is conducted in which the treatments are actually levels of a particular "input", the choice of which determines the amount of "output", which, say, is to be maximized. For example, the input might be the dosage of a drug, or the temperature of an industrial process, or the percentage of a chemical in a fertilizer, etc. Depending on the circumstances, the researcher might be free to choose the number of levels, a, as well as the replication number, n, but with the constraint that $na \leqslant N^*$. The optimal choice of a and n will depend not only on N^* and σ^2, but also on the approximate functional form (linear, quadratic, etc.) relating the level to the output variable; see, e.g., Montgomery (2000) for further details.

Alternatively, instead of different levels of some particular treatment, the situation might be comparing the performance of several different treatments (brands, methods, chemicals, medicines, etc.). In this case, there is often a **control group** that receives the "standard treatment", which might mean no treatment at all (or a placebo in medical studies involving humans), and interest centers on determining which, if any, treatments are better than the control, and, among those that are better, which is best. Common sense would suggest including only those treatments in the experiment that might possibly be better than the control. For example, imagine a study for comparing drugs that purport to increase the rate at which the human liver can remove alcohol from the bloodstream. The control group would consist of those individuals receiving no treatment (or, possibly, a placebo), while treatment with caffeine would not be included, as its (unfortunate) ineffectiveness is well-known.

Example 2.1 To see the effect on the power of the F test when superfluous treatments are included, let the first group correspond to the prevailing treatment and assume all other considered treatments do not have an effect. In terms of model formulation (2.21) with the natural constraint $\sum_{i=1}^{a} \alpha_i = 0$, we take $\alpha_1 = \delta$ and $\alpha_2 = \alpha_3 = \cdots = \alpha_a = -\delta/(a-1)$, so that $\sum_{i=1}^{a} \alpha_i^2 = \delta^2 a/(a-1)$. For $\sigma^2 = 1$, $n = 22$, test size $\alpha = 0.05$, and $\delta = 0.5$, the power is 0.90 for $a = 2$ and decreases as a increases. Figure 2.4

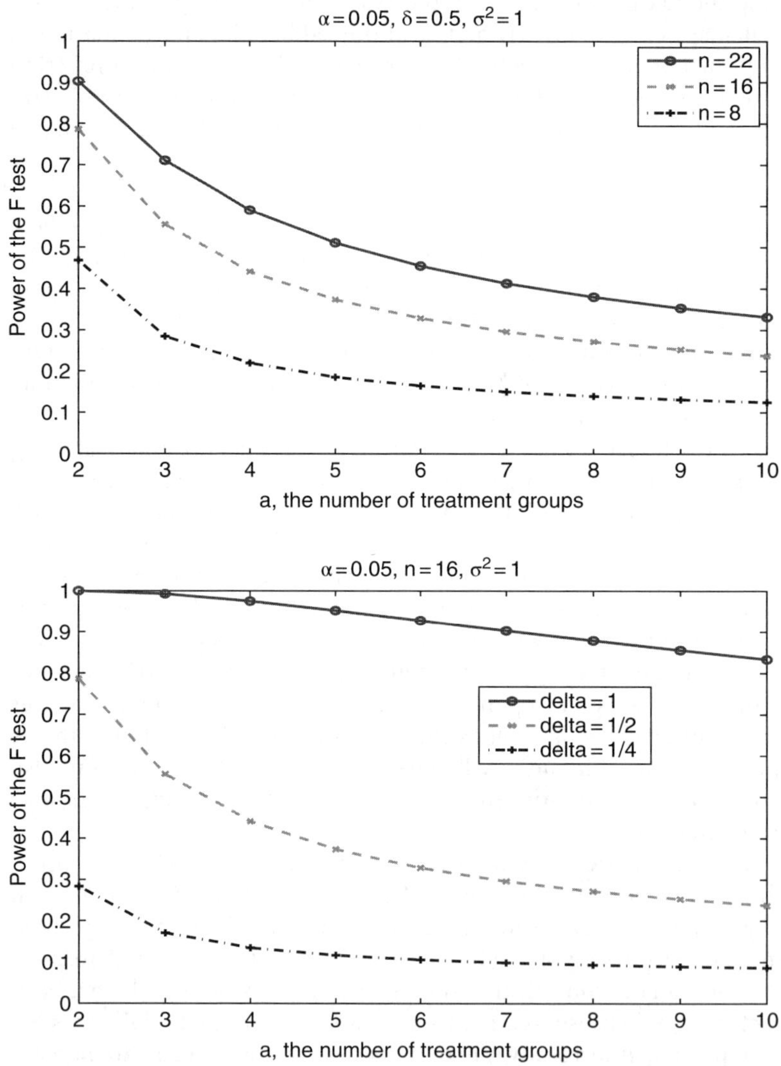

Figure 2.4 Top: Power of the F test as a function of a, for fixed α, δ, and σ^2, and three values of n. **Bottom:** Similar, but n is fixed at 16, and three values of δ are used. The middle dashed line is the same in both graphics.

plots the power, as a function of a, for various constellations of n and δ. Observe how the total sample size $N^* = na$ increases with n. This might not be realistic in practice, and instead N^* might be fixed, so that, as a increases, n decreases, and the power will drop far faster than shown in the plots. The reader is encouraged to reproduce the plots in Figure 2.4, as well as considering the case when N^* is fixed. ■

2.4.4 The ANOVA Table

We endow the various sums of squares arising in this model with particular names that are common (but not universal; see the Remark below) in the literature, as follows.

$$\sum_{i=1}^{a}\sum_{j=1}^{n}(Y_{ij} - \bar{Y}_{\bullet\bullet})^2 \text{ is called the } \underline{\textbf{t}}\textbf{otal (corrected) sum of squares} \text{ , abbreviated } SS_T;$$

$$\sum_{i=1}^{a}\sum_{j=1}^{n}(Y_{ij} - \bar{Y}_{i\bullet})^2 \text{ is the } \underline{\textbf{w}}\textbf{ithin (group) sum of squares} \text{ , abbreviated } SS_W,$$

also referred to as the **sum of squares due to error**; and, recalling (2.25) and (2.27),

$$\mathsf{S}(\hat{\gamma}) - \mathsf{S}(\hat{\beta}) \text{ is referred to as the } \underline{\textbf{b}}\textbf{etween (group) sum of squares}, \text{ or } SS_B.$$

That is, $SS_T = SS_W + SS_B$ from (2.28).

Remark It is important to emphasize that this notation, while common, is not universal. For example, in the two-way ANOVA in Section 2.5 below, there will be two factors, say A and B, and we will use SS_B to denote the latter. Likewise, in a three-factor model, the factors would be labeled A, B, and C.

In the two-way ANOVA case, some authors refer to the "more interesting" factor A as the "treatment", and the second one as a **block** (block here not in the sense of "preventing", but rather as "segmenting"), such as for "less interesting" things, such as gender, age group, smoker/non-smoker, etc. As the word block coincidentally also starts with a b, its associated sum of squares is denoted SS_B. ■

A more complete sum of squares decomposition is possible by starting with the **uncorrected total sum of squares,**

$$\begin{aligned}\sum_{i=1}^{a}\sum_{j=1}^{n} Y_{ij}^2 &= \sum_{i=1}^{a}\sum_{j=1}^{n}(Y_{ij} - \bar{Y}_{i\bullet} + \bar{Y}_{i\bullet} - \bar{Y}_{\bullet\bullet} + \bar{Y}_{\bullet\bullet})^2 \\ &= \sum_{i=1}^{a}\sum_{j=1}^{n}(Y_{ij} - \bar{Y}_{i\bullet})^2 + \sum_{i=1}^{a}\sum_{j=1}^{n}(\bar{Y}_{i\bullet} - \bar{Y}_{\bullet\bullet})^2 + \sum_{i=1}^{a}\sum_{j=1}^{n}\bar{Y}_{\bullet\bullet}^2,\end{aligned} \tag{2.31}$$

and verifying that all the cross terms are zero. For that latter task, let $\mathbf{P_X}$ be the projection matrix based on $\mathbf{X}$ in (2.22) and $\mathbf{P}_1$ the projection matrix based on a column of ones. Then the decomposition (2.31) follows directly from the algebraic identity

$$\mathbf{Y'IY} = \mathbf{Y'(I - P_X)Y} + \mathbf{Y'(P_X - P_1)Y} + \mathbf{Y'P_1Y}, \tag{2.32}$$

and the fact that

$$\mathsf{S}(\hat{\gamma}) - \mathsf{S}(\hat{\beta}) = \mathbf{Y'(P_X - P_1)Y}, \tag{2.33}$$

from (1.87). Recall from Theorem 1.6 that, if S_0 and S are subspaces of $\mathbb{R}^T$ such that $S_0 \subset S$, then $\mathbf{P}_S\mathbf{P}_{S_0} = \mathbf{P}_{S_0} = \mathbf{P}_{S_0}\mathbf{P}_S$. Thus, from (1.80) and that $\mathbf{1} \in \mathcal{C}(\mathbf{X})$, $\mathbf{P}_\mathbf{X} - \mathbf{P}_\mathbf{1}$ is a projection matrix.

Remark Anticipating the discussion of the two-way ANOVA in Section 2.5 below, we rewrite (2.32), expressing the single effect as A, and thus its projection matrix, as $\mathbf{P}_A$ instead of $\mathbf{P}_\mathbf{X}$:

$$\mathbf{Y}'\mathbf{Y} = \mathbf{Y}'\mathbf{P}_\mathbf{1}\mathbf{Y} + \mathbf{Y}'(\mathbf{P}_A - \mathbf{P}_\mathbf{1})\mathbf{Y} + \mathbf{Y}'(\mathbf{I} - \mathbf{P}_A)\mathbf{Y}, \tag{2.34}$$

where the three terms on the right-hand side are, respectively, the sums of squares with respect to the grand mean, the treatment effect, and the error term. Note that, in the latter term, $\mathbf{1} \in \mathcal{C}(A) = \mathcal{C}(\mathbf{X})$ (where A here refers to the columns of $\mathbf{X}$ associated with the factor A), and is why the term is *not* $\mathbf{Y}'(\mathbf{I} - \mathbf{P}_A - \mathbf{P}_\mathbf{1})\mathbf{Y}$.

Further, moving the last term in (2.32), namely $\mathbf{Y}'\mathbf{P}_\mathbf{1}\mathbf{Y}$, to the left-hand side of (2.34) gives the decomposition in terms of the **corrected total sum of squares**:

$$\mathbf{Y}'(\mathbf{I} - \mathbf{P}_\mathbf{1})\mathbf{Y} = \mathbf{Y}'(\mathbf{P}_A - \mathbf{P}_\mathbf{1})\mathbf{Y} + \mathbf{Y}'(\mathbf{I} - \mathbf{P}_A)\mathbf{Y}, \tag{2.35}$$

this being more commonly used. ■

Each of the sums of squares in (2.32) has an associated number of degrees of freedom that can be determined from Theorem A.1. In particular, for SS_T, $\text{rank}(\mathbf{I} - \mathbf{P}_\mathbf{1}) = na - 1$, for SS_W, $\text{rank}(\mathbf{I} - \mathbf{P}_\mathbf{X}) = na - a$, and for SS_B, as $\mathbf{P}_\mathbf{X} - \mathbf{P}_\mathbf{1}$ is a projection matrix,

$$\text{rank}(\mathbf{P}_\mathbf{X} - \mathbf{P}_\mathbf{1}) = \text{tr}(\mathbf{P}_\mathbf{X} - \mathbf{P}_\mathbf{1}) = \text{tr}(\mathbf{P}_\mathbf{X}) - \text{tr}(\mathbf{P}_\mathbf{1}) = \text{rank}(\mathbf{P}_\mathbf{X}) - \text{rank}(\mathbf{P}_\mathbf{1}) = a - 1, \tag{2.36}$$

from Theorem 1.2. Note also that $\mathbf{P}_\mathbf{X} = n^{-1}\mathbf{X}\mathbf{X}' = (\mathbf{I}_a \otimes \mathbf{J}_n)/n$, with trace $na/n = a$. Clearly, the sum of squares for the mean, $na\bar{Y}^2_{\bullet\bullet}$, and the uncorrected total sum of squares have one and na degrees of freedom, respectively.

From (2.33), the expected between (or treatment) sum of squares is

$$\mathbb{E}[SS_B] = \mathbb{E}[\mathbf{Y}'(\mathbf{P}_\mathbf{X} - \mathbf{P}_\mathbf{1})\mathbf{Y}] = \sigma^2\mathbb{E}[(\mathbf{Y}/\sigma)'(\mathbf{P}_\mathbf{X} - \mathbf{P}_\mathbf{1})(\mathbf{Y}/\sigma)], \tag{2.37}$$

so that, from (1.92), and recalling from (II.10.6) the expectation of a noncentral χ^2 random variable, i.e., if $Z \sim \chi^2(n, \theta)$, then $\mathbb{E}[Z] = n + \theta$, we have, with $J = a - 1$ and θ defined in (2.30),

$$\begin{aligned}\mathbb{E}[SS_B] &= \sigma^2(J + \boldsymbol{\beta}'\mathbf{X}'(\mathbf{P}_\mathbf{X} - \mathbf{P}_\mathbf{1})\mathbf{X}\boldsymbol{\beta}/\sigma^2) = \sigma^2(a - 1 + \theta)\\ &= \sigma^2(a-1) + n\sum_{i=1}^{a}\alpha_i^2.\end{aligned} \tag{2.38}$$

Similarly, from (1.93), $\mathbb{E}[SS_W] = \sigma^2(na - a)$.

Remark It is a useful exercise to derive (2.38) using the basic quadratic form result in (A.6), which states that, for $Y = \mathbf{X}'\mathbf{A}\mathbf{X}$ with $\mathbf{X} \sim \mathrm{N}_n(\boldsymbol{\mu}, \boldsymbol{\Sigma})$, $\mathbb{E}[Y] = \text{tr}(\mathbf{A}\boldsymbol{\Sigma}) + \boldsymbol{\mu}'\mathbf{A}\boldsymbol{\mu}$.

Before proceeding, the reader should confirm that, for $T = an$,

$$\mathbf{P}_\mathbf{1} = T^{-1}\mathbf{1}_T\mathbf{1}_T' = (na)^{-1}\mathbf{J}_a \otimes \mathbf{J}_n. \tag{2.39}$$

This is somewhat interesting in its own right, for it says that $\mathbf{P}_{\mathbf{1},an} = \mathbf{P}_{\mathbf{1},a} \otimes \mathbf{P}_{\mathbf{1},n}$, where $\mathbf{P}_{\mathbf{1},j}$ denotes the $j \times j$ projection matrix onto $\mathbf{1}_j$.

From (2.33), we have

$$\mathbb{E}[SS_B] = \mathbb{E}[S(\hat{\gamma}) - S(\hat{\beta})] = \mathbb{E}[\mathbf{Y}'(\mathbf{P_X} - \mathbf{P_1})\mathbf{Y}] = \mathbb{E}[\mathbf{Y}'\mathbf{P_X}\mathbf{Y}] - \mathbb{E}[\mathbf{Y}'\mathbf{P_1}\mathbf{Y}],$$

and, from (A.6) with

$$\mathbb{E}[\mathbf{Y}] = \boldsymbol{\mu} = \boldsymbol{\beta} \otimes \mathbf{1}_n \quad \text{and} \quad \boldsymbol{\beta} = (\mu_1, \dots, \mu_a)' = (\mu + \alpha_1, \dots, \mu + \alpha_a)',$$

we have

$$\begin{aligned}
\mathbb{E}[\mathbf{Y}'\mathbf{P_X}\mathbf{Y}] &= \sigma^2 \operatorname{tr}(\mathbf{P_X}) + \boldsymbol{\mu}'\mathbf{P_X}\boldsymbol{\mu} \\
&= a\sigma^2 + n^{-1}(\boldsymbol{\beta}' \otimes \mathbf{1}_n')(\mathbf{I}_a \otimes \mathbf{J}_n)(\boldsymbol{\beta} \otimes \mathbf{1}_n) \\
&= a\sigma^2 + n^{-1}(\boldsymbol{\beta}'\mathbf{I}_a\boldsymbol{\beta} \otimes \mathbf{1}_n'\mathbf{J}_n\mathbf{1}_n) \\
&= a\sigma^2 + n^{-1}(\boldsymbol{\beta}'\boldsymbol{\beta} \otimes n^2) \\
&= a\sigma^2 + n\sum_{i=1}^{a} \mu_i^2.
\end{aligned}$$

Similarly, with $\mathbf{P_1} = T^{-1}\mathbf{1}_T\mathbf{1}_T' = (na)^{-1}\mathbf{J}_a \otimes \mathbf{J}_n$,

$$\begin{aligned}
\mathbb{E}[\mathbf{Y}'\mathbf{P_1}\mathbf{Y}] &= \sigma^2 \operatorname{tr}(\mathbf{P_1}) + \boldsymbol{\mu}'\mathbf{P_1}\boldsymbol{\mu} = \sigma^2 + (na)^{-1}(\boldsymbol{\beta}' \otimes \mathbf{1}_n')(\mathbf{J}_a \otimes \mathbf{J}_n)(\boldsymbol{\beta} \otimes \mathbf{1}_n) \\
&= \sigma^2 + (na)^{-1}(\boldsymbol{\beta}'\mathbf{J}_a\boldsymbol{\beta} \otimes \mathbf{1}_n'\mathbf{J}_n\mathbf{1}_n) = \sigma^2 + (na)^{-1}\left(\left(\sum_{i=1}^{a} \mu_i\right)^2 \otimes n^2\right) \\
&= \sigma^2 + (na)^{-1}n^2(a\mu)^2 = \sigma^2 + na\mu^2.
\end{aligned}$$

Thus,

$$\mathbb{E}[\mathbf{Y}'(\mathbf{P_X} - \mathbf{P_1})\mathbf{Y}] = (a-1)\sigma^2 + n\left(\sum_{i=1}^{a} \mu_i^2 - a\mu^2\right),$$

but

$$\sum_{i=1}^{a} \mu_i^2 = a\mu^2 + \sum_{i=1}^{a} \alpha_i^2 + 2\mu\sum_{i=1}^{a} \alpha_i = a\mu^2 + \sum_{i=1}^{a} \alpha_i^2,$$

so that

$$\mathbb{E}[SS_B] = \mathbb{E}[\mathbf{Y}'(\mathbf{P_X} - \mathbf{P_1})\mathbf{Y}] = (a-1)\sigma^2 + n\sum_{i=1}^{a} \alpha_i^2,$$

as in (2.38). ■

For conducting statistical inference, it is usually more convenient to work with the **mean squares**, denoted MS, which are just the sums of squares divided by their associated degrees of freedom. For this model, the important ones are $MS_W = SS_W/(na - a)$ and $MS_B = SS_B/(a-1)$. Notice, in particular, that the F statistic in (2.29) can be written as

$$F = \frac{\mathbf{Y}'(\mathbf{P_X} - \mathbf{P_1})\mathbf{Y}/\operatorname{rank}(\mathbf{P_X} - \mathbf{P_1})}{\mathbf{Y}'(\mathbf{I} - \mathbf{P_X})\mathbf{Y}/\operatorname{rank}(\mathbf{I} - \mathbf{P_X})} = \frac{MS_B}{MS_W}. \tag{2.40}$$

The **expected mean squares** $\mathbb{E}[MS]$ are commonly reported in the analysis of variance. For this model, it follows from (2.36) and (2.38) that

$$\mathbb{E}[MS_B] = \sigma^2 + \frac{n}{a-1}\sum_{i=1}^{a}\alpha_i^2 = \sigma^2 + n\sigma_a^2, \tag{2.41}$$

where σ_a^2 is defined to be

$$\sigma_a^2 := (a-1)^{-1}\sum_{i=1}^{a}(\mu_i - \bar{\mu}_{\bullet})^2 = (a-1)^{-1}\sum_{i=1}^{a}(\alpha_i - \bar{\alpha}_{\bullet})^2 = (a-1)^{-1}\sum_{i=1}^{a}\alpha_i^2, \tag{2.42}$$

which follows because $\alpha_{\bullet} = \sum_{i=1}^{a}\alpha_i = 0$. Similarly, $\mathbb{E}[MS_W] = \sigma^2$.

Higher order moments of the mean squares, while not usually reported in this context, are straightforward to compute using the results in Section II.10.1.2. In particular, for $Z \sim \chi^2(n,\theta)$, along with $\mathbb{E}[Z] = n+\theta$, we have $\mathbb{V}(Z) = 2n + 4\theta$, and, most generally, for $s \in \mathbb{R}$ with $s > -n/2$,

$$\mathbb{E}[Z^s] = \frac{2^s}{e^{\theta/2}}\frac{\Gamma(n/2+s)}{\Gamma(n/2)}\,{}_1F_1(n/2+s, n/2; \theta/2), \quad s > -n/2,$$

as shown in (II.10.9). More useful for integer moments is, for $s \in \mathbb{N}$,

$$\mathbb{E}[Z^s] = 2^s\Gamma\left(s+\frac{n}{2}\right)\sum_{i=0}^{s}\binom{s}{i}\frac{(\theta/2)^i}{\Gamma(i+n/2)}, \quad s \in \mathbb{N}. \tag{2.43}$$

The various quantities associated with the sums of squares decomposition are typically expressed in tabular form, as shown in Table 2.1. Except for the expected mean squares, the output from statistical software will include the table using the values computed from the data set under examination. The last column contains the p-value p_B, which is the probability that a central F-distributed random variable with $a-1$ and $na-a$ degrees of freedom exceeds the value of the F statistic in (2.40). This number is often used for determining if there are differences between the treatments. Traditionally, a

Table 2.1 The ANOVA table for the balanced one-way ANOVA model. Mean squares denote the sums of squares divided by their associated degrees of freedom. Term σ_a^2 in the expected mean square corresponding to the treatment effect is defined in (2.42).

Source of variation	Degrees of freedom	Sum of squares	Mean square	Expected mean square	*F* statistic	*p*-value
Between (model)	$a-1$	SS_B	MS_B	$\sigma^2 + n\sigma_a^2$	MS_B/MS_W	p_B
Within (error)	$na-a$	SS_W	MS_W	σ^2		
Total (corrected)	$na-1$	SS_T				
Overall mean	1	$na\bar{Y}_{\bullet\bullet}^2$				
Total	na	$\mathbf{Y}'\mathbf{Y}$				

value under 0.1 (0.05, 0.01) is said to provide "modest" ("significant", "strong") evidence for differences in means, though recall the first Remark in Section 2.4.2, and the discussion in Section III.2.8.

If significant differences can be safely surmised, then the scientist would proceed with further inferential methods for ascertaining precisely which treatments differ from one another, as discussed below. Ideally, the experiment would be repeated several times, possibly with different designs and larger sample sizes, in line with Fisher's paradigm of using a "significant p-value" as (only) an indication that the experiment is worthy of repetition (as opposed to immediately declaring significance if p_B is less than some common threshold such as 0.05).

2.4.5 Computing Confidence Intervals

Section 1.4.7 discussed the Bonferroni and Scheffé methods of constructing simultaneous confidence intervals on linear combinations of the parameter vector $\boldsymbol{\beta}$. For the one-way ANOVA model, there are usually two sets of intervals that are of primary interest. The first is useful when one of the treatments, say the first, serves as a control, in which case interest centers on simultaneous c.i.s for $\mu_i - \mu_1$, $i = 2, \dots, a$. Whether or not one of the treatments is a control, the second set of simultaneous c.i.s is often computed, namely for all $a(a-1)/2$ differences $\mu_i - \mu_j$.

For the comparisons against a control, the Bonferroni procedure uses the cutoff value $c = t^{-1}_{na-a}(1 - \alpha/(2J))$, where we use the notation $t^{-1}_k(p)$ to denote the quantile of the Student's t distribution with k degrees of freedom, corresponding to probability p, $0 < p < 1$. Likewise, the Scheffé method takes $q = F^{-1}_{J,na-a}(1-\alpha)$, where $J = a - 1$. For all pairwise differences, Bonferroni uses $c = t^{-1}_{na-a}(1 - \alpha/(2D))$, $D = a(a-1)/2$, while the Scheffé cutoff value is still $q = F^{-1}_{a-1,na-a}(1-\alpha)$, recalling (1.102) and the fact that only $a - 1$ of the $a(a-1)/2$ differences are linearly independent.

Remark Methods are also available for deciding which population has the largest mean, most notably that from Bechhofer (1954). See also Bechhofer and Goldsman (1989), Fabian (2000), and the references therein. Detailed accounts of these and other methods can be found in Miller (1981), Hochberg and Tamhane (1987), and Hsu (1996). Miller (1985), Dudewicz and Mishra (1988, Sec. 11.2), Tamhane and Dunlop (2000), and Sahai and Ageel (2000) provide good introductory accounts. ∎

We illustrate the inferential consequences of the different intervals using simulated data. The Matlab function in Listing 2.3 generates data (based on a specified "seed" value) appropriate for a one-way ANOVA, using $n = 8$, $a = 5$, $\mu_1 = 12$, $\mu_2 = 11$, $\mu_3 = 10$, $\mu_4 = 10$, $\mu_5 = 9$ and $\sigma^2 = 4$. For seed value 1, the program produces the text file `anovadata.txt` with contents given in Listing 2.4 and which we will use shortly. A subsequent call to `p=anova1(x)` in Matlab yields the p-value 0.0017 and produces the ANOVA table (as a graphic) and a box plot of the treatments, as shown in Figure 2.5. While the p-value is indeed well under the harshest typical threshold of 0.01, the box plot shows that the true means are not well reflected in this data set, nor does the data appear to have homogeneous variances across treatments.

For computing the $a(a-1)/2 = 10$ simultaneous c.i.s of the differences of each pair of treatment means using $\alpha = 0.05$, the cutoff values $c = t^{-1}_{35}(1 - 0.05/20) = 2.9960$ and $q = F^{-1}_{4,35}(1 - 0.05) = 2.6415$ for the Bonferroni and Scheffé methods, respectively, are required. The appropriate value for the maximum modulus method is not readily computed, but could be obtained from tabulated sources for the standard values of $\alpha = 0.10$, 0.05 and 0.01.

```
1  function x=anovacreate(seed)
2  randn('state',seed); % this is now deprecated in Matlab,
3  %                      but still works in version R2010a
4  x=[]; n=8; sigma=2; mu=[12 11 10 10 9];
5  for i=1:5, x=[x sigma*randn(n,1)+mu(i)]; end
6  if exist('anovadata.txt'), delete('anovadata.txt'), end
7  pause(0.2), diary anovadata.txt
8  for i=1:5,
9    out=['T',num2str(i),' ',sprintf('%7.4f ',x(:,i))];
10   disp(out)
11 end
12 diary off
```

Program Listing 2.3: Matlab code to simulate data for one-way ANOVA, for a given seed value so it can be replicated and the data saved to a file for reading by SAS. The use of `diary` is very easy, but not ideal, as the output file will contain the Matlab lines of code at the beginning and end (and these need to be manually deleted). The use of function `fprintf` can be used instead; see Listing 1.7. Note that line 2 may not work in more recent versions of Matlab.

```
T1 13.7288 12.1884 10.2962 13.7470 11.1239 11.1407  9.7945 12.7925
T2  9.0701 11.3369  7.0693  9.5114  9.8954  9.3605 13.2183  9.7701
T3  9.4907  9.4603  6.6560  6.2479 11.1500  8.2677  5.7670  8.0711
T4 10.4255 10.9558 10.2013 10.5949 11.1403  6.7510 11.2869 11.3637
T5  9.0293  6.3969  6.4308 10.6244 10.6771 11.8406  7.0205  6.6335
```

Program Listing 2.4: Output from the program in Listing 2.3.

Instead of computing the various intervals "by hand" via Matlab (though that is not necessary; see their `multcompare` function), we use the SAS system, with the relevant code given in Listing 2.1 and output shown in several separate boxes below. (All the data processing commands used in Listing 2.1 are explained in Appendix D.) The same code can be used if the design is unbalanced. The SAS output we show is textual, though in more recent versions (as of this writing, version 9.4), the output is in very attractive hypertext markup language (HTML) format (and includes boxplots similar to the Matlab boxplot shown in Figure 2.5), and can easily be converted to both Adobe portable document format (pdf) and rich text format (rtf), the commands for which are illustrated in Listing 2.1.

ANOVA Table

Source	SS	df	MS	F
Columns	69.98	4	17.5	5.417
Error	113.1	35	3.23	
Total	183	39		

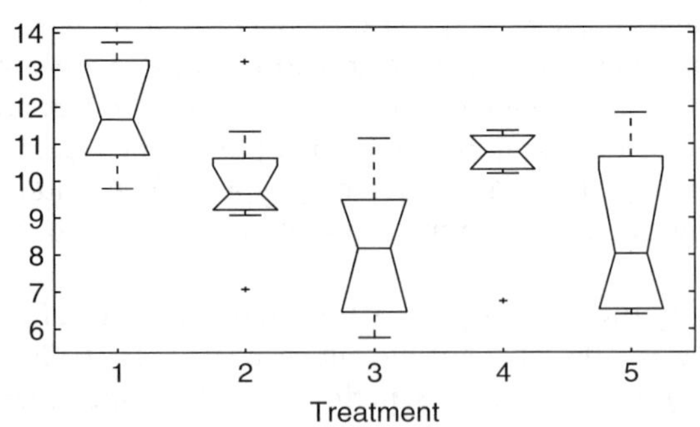

Figure 2.5 Matlab output for the ANOVA example.

```
                        The SAS System
                      The ANOVA Procedure
                    Class Level Information
            Class          Levels     Values
            treat               5     T1 T2 T3 T4 T5

                 Number of observations    40
```

SAS Output 2.1: First part of the output from `proc anova`.

```
                        The SAS System
                      The ANOVA Procedure
Dependent Variable: yield

                                        Sum of
 Source                     DF         Squares    Mean Square  F Value   Pr > F
 Model                       4      69.9846138     17.4961535     5.42   0.0017
 Error                      35     113.0527516      3.2300786
 Corrected Total            39     183.0373654

           R-Square     Coeff Var      Root MSE     yield Mean
           0.382352      18.40837      1.797242       9.763180
```

SAS Output 2.2: The ANOVA table is the second part of the output from `proc anova`.

SAS Outputs 2.1 and 2.2 accompany all calls to `proc anova`. The former, also in conjunction with the log output from SAS (not shown), assures the researcher that SAS is computing what he or she expects. The latter is, except for formatting (and that Matlab does not show the p-value in its table), the same as the Matlab output in the left of Figure 2.5 but contains four more statistics of potential interest. SAS Outputs 2.3, 2.4, and 2.5 show the simultaneous c.i.s using the three aforementioned methods. Each begins with a note regarding if the intervals are simultaneous (denoted as "controlling the experimentwise error rate" in SAS) or not, as well as information pertaining to the intervals, including the significance level α, the error degrees of freedom $an - a$, and the critical value.

We see that the Bonferroni intervals are considerably shorter than those using Scheffé, while the maximum modulus intervals are just slightly shorter than Bonferroni. To save space, two of the three outputs have been truncated, although for this data set each method yields the same conclusions regarding which differences contain zero, i.e., which treatment effects could be deemed to be the same, under the usual inferential paradigm of hypothesis testing (and thus subject to the same critique as discussed above). A shorter way of just showing which treatments are different (according to the computed 95% c.i.s) is graphically depicted by SAS and is shown in SAS Output 2.6.

There are several other methods of constructing simultaneous c.i.s for the $a(a-1)/2$ differences in treatment means. The most common method requires evaluation of the so-called **studentized range** distribution, which is not trivial, although critical values have been tabulated and, like the values associated with the maximum modulus method, are built in to SAS. This method is referred to as the

```
                        The SAS System

                     The ANOVA Procedure

             Bonferroni (Dunn) t Tests for yield

 NOTE: This test controls the Type I experimentwise error rate, but it
generally has a higher Type II error rate than Tukey's for all pairwise
                          comparisons.

          Alpha                                      0.05
          Error Degrees of Freedom                     35
          Error Mean Square                      3.230079
          Critical Value of t                     2.99605
          Minimum Significant Difference           2.6923

   Comparisons significant at the 0.05 level are indicated by ***.
                         Difference
         treat             Between      Simultaneous 95%
       Comparison            Means    Confidence Limits

       T1    - T4           1.5116     -1.1807   4.2039
       T1    - T2           1.9475     -0.7448   4.6398
       T1    - T5           3.2699      0.5776   5.9622  ***
       T1    - T3           3.7127      1.0204   6.4050  ***
       T4    - T1          -1.5116     -4.2039   1.1807
       T4    - T2           0.4359     -2.2564   3.1282
       T4    - T5           1.7583     -0.9340   4.4506
       T4    - T3           2.2011     -0.4912   4.8934
       T2    - T1          -1.9475     -4.6398   0.7448
       T2    - T4          -0.4359     -3.1282   2.2564
       T2    - T5           1.3224     -1.3699   4.0147
       T2    - T3           1.7652     -0.9271   4.4575
       T5    - T1          -3.2699     -5.9622  -0.5776  ***
       T5    - T4          -1.7583     -4.4506   0.9340
       T5    - T2          -1.3224     -4.0147   1.3699
       T5    - T3           0.4428     -2.2495   3.1351
       T3    - T1          -3.7127     -6.4050  -1.0204  ***
       T3    - T4          -2.2011     -4.8934   0.4912
       T3    - T2          -1.7652     -4.4575   0.9271
       T3    - T5          -0.4428     -3.1351   2.2495
```

SAS Output 2.3: Bonferroni simultaneous c.i.s from `proc anova` with the `BON` and `cldiff` options in the `means` statement. Notice the redundancy SAS provides by reporting the 10 intervals in two ways.

```
                    Scheffe's Test for yield

NOTE: This test controls the Type I experimentwise error rate, but it generally
has a higher Type II error rate than Tukey's for all pairwise comparisons.

                Alpha                                   0.05
                Error Degrees of Freedom                  35
                Error Mean Square                   3.230079
                Critical Value of F                  2.64147
                Minimum Significant Difference         2.921

    Comparisons significant at the 0.05 level are indicated by ***.

                          Difference
         treat             Between      Simultaneous 95%
      Comparison             Means    Confidence Limits

      T1    - T4            1.5116     -1.4094    4.4326
      T1    - T2            1.9475     -0.9735    4.8685
      T1    - T5            3.2699      0.3489    6.1908  ***
      T1    - T3            3.7127      0.7917    6.6336  ***
      T4    - T1           -1.5116     -4.4326    1.4094
      (etc.)
```

SAS Output 2.4: Similar to SAS Output 2.3 but for Scheffé simultaneous c.i.s. Abbreviated output.

```
            Studentized Maximum Modulus (GT2) Test for yield

NOTE: This test controls the Type I experimentwise error rate, but it generally
has a higher Type II error rate than Tukey's for all pairwise comparisons.

        Alpha                                                  0.05
        Error Degrees of Freedom                                 35
        Error Mean Square                                  3.230079
        Critical Value of Studentized Maximum Modulus       2.97460
        Minimum Significant Difference                        2.673

    Comparisons significant at the 0.05 level are indicated by ***.
                          Difference
         treat             Between       95% Confidence
      Comparison             Means           Limits
      T1    - T4            1.5116     -1.1615    4.1846
      T1    - T2            1.9475     -0.7255    4.6205
      T1    - T5            3.2699      0.5968    5.9429  ***
      T1    - T3            3.7127      1.0396    6.3857  ***
      T4    - T1           -1.5116     -4.1846    1.1615
      (etc.)
```

SAS Output 2.5: Similar to SAS Output 2.3 but for simultaneous c.i.s constructed using the maximum modulus method.

```
        Means with the same letter are not significantly different.

             Bon Grouping          Mean        N     treat
                        A       11.8515        8     T1
                        A
                   B    A       10.3399        8     T4
                   B    A
                   B    A        9.9040        8     T2
                   B
                   B             8.5816        8     T5
                   B
                   B             8.1388        8     T3
```

SAS Output 2.6: Depiction of which c.i.s contain zero using the Bonferroni method, as obtained from `proc anova` with the BON and `lines` options in the `means` statement. For this data set, the same grouping was obtained with Scheffé and maximum modulus.

```
              Tukey's Studentized Range (HSD) Test for yield

      NOTE: This test controls the Type I experimentwise error rate.

               Alpha                                       0.05
               Error Degrees of Freedom                      35
               Error Mean Square                       3.230079
               Critical Value of Studentized Range      4.06595
               Minimum Significant Difference            2.5836

      Comparisons significant at the 0.05 level are indicated by ***.

                              Difference
               treat           Between      Simultaneous 95%
            Comparison           Means     Confidence Limits
            T1     - T4         1.5116      -1.0720    4.0952
            T1     - T2         1.9475      -0.6361    4.5311
            T1     - T5         3.2699       0.6863    5.8535   ***
            T1     - T3         3.7127       1.1291    6.2963   ***
            T4     - T1        -1.5116      -4.0952    1.0720
            (etc.)
```

SAS Output 2.7: Similar to SAS Output 2.3 but for Tukey simultaneous c.i.s. Abbreviated output.

Tukey method, or just T-method. For a balanced design with the two main assumptions of normality and equal treatment variances satisfied, the Tukey c.i.s are the shortest.

Remark It is worth mentioning that Scheffé's method is more robust to violation of the latter two assumptions and can still be used for unbalanced data. In addition, while the cutoff value q in Scheffé's method is readily computed, that for the Tukey method is not, so that only those α-levels can be used for which its cutoff has been tabulated, namely 0.10, 0.05 and 0.01.

Scheffé (1959, Sec. 3.7) discusses further benefits of the S-method over the T-method; see also Sahai and Ageel (2000, p. 77) for a summary. ■

```
                    Dunnett's t Tests for yield

NOTE: This test controls the Type I experimentwise error for comparisons
                   of all treatments against a control.

                   Alpha                                   0.05
                   Error Degrees of Freedom                  35
                   Error Mean Square                   3.230079
                   Critical Value of Dunnett's t        2.55790
                   Minimum Significant Difference        2.2986


   Comparisons significant at the 0.05 level are indicated by ***.

                         Difference
          treat           Between      Simultaneous 95%
       Comparison           Means    Confidence Limits
       T4     - T1        -1.5116     -3.8102    0.7870
       T2     - T1        -1.9475     -4.2461    0.3511
       T5     - T1        -3.2699     -5.5684   -0.9713    ***
       T3     - T1        -3.7127     -6.0112   -1.4141    ***
```

SAS Output 2.8: Use of Dunnett's method, obtained with `means treat/DUNNETT('T1') cldiff;` in the `proc anova` statement.

SAS Output 2.7 shows the c.i.s using the T-method. They are indeed shorter than either the Bonferroni and Scheffé ones, although, in this case at least, the same conclusions would be drawn regarding which intervals contain zero or not.

If, in this experiment, one of the treatments is a control group, then simultaneous c.i.s can (and should) be produced by methods specifically designed for this purpose, such as **Dunnett's method**, which is also i mplemented in SAS's `anova` procedure. The output for the data set under study is shown in SAS Output 2.8. The resulting $a - 1$ intervals are indeed even shorter than those produced by the Tukey method. Again, however, inference regarding which treatments are different is the same for this data set.

It is instructive to repeat the previous exercise for several simulated data sets (not to mention the use of real data sets!) in order to get accustomed with the procedure. For example, running the program in Listing 2.3 with seed value 6 produced the boxplot in Figure 2.6 and a p-value for the F test of no treatment differences of 0.000467. The SAS program in Listing 2.1 was invoked again and produced the output that is abbreviated in SAS Output 2.9. Now we see quite a difference among the simultaneous c.i. methods.

2.4.6 A Word on Model Assumptions

Preliminary tests of $\sigma_1^2 = \sigma_2^2$ seem to be a fruitless pastime.

(Rupert G. Miller Jr., 1997, p. 58)

Up to this point, no attention has been paid to the plausibility of the model assumptions, methods of testing their validity, and consequences of their violation. These important issues are an integral part of the model-building process and cannot be overlooked in practice. The assumption of normality,

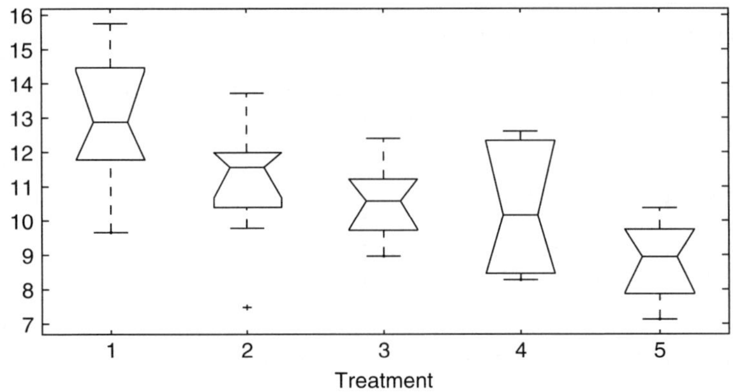

Figure 2.6 Matlab output from calling the function in Listing 2.3 as `x=anovacreate(6)`, and then running the built-in Matlab function `p=anova1(x)`.

```
options linesize=75 pagesize=65 nodate;
ods pdf file='ANOVA Output 1.pdf';
ods rtf file='ANOVA Output 1.rtf';

data test;
  infile 'anovadata.txt' flowover;
  retain treat; keep treat yield;
  input s $ @@;
  if substr(s,1,1) = 'T' then do;
    treat=s; delete; return;
  end;
  yield = input(s,7.5); if yield>.;
run;

proc anova;
  classes treat; model yield=treat;
  means treat / BON SCHEFFE lines cldiff;
run;
ods _all_ close;
ods html;
```

SAS Program Listing 2.1: SAS code for reading the text file of data, computing the ANOVA table, and constructing simultaneous c.i.s via the Bonferroni and Scheffé methods for the 10 pairs of mean differences using the SAS default of $\alpha = 0.05$. The term `ods` refers to the SAS' "Output Delivery System" and the commands here enable output to be generated as both pdf and rich text format files, both of which are automatically viewed in SAS.

for example, is partly justified by appealing to the central limit theorem, but is also preferred because of the tractability of the distribution of the F statistic under the null and alternative hypotheses. Certainly, not all real data will be from a normal population; an obvious example is lifetime data, which cannot be negative and which could exhibit extreme asymmetry. Another typical violation is when data exhibit more extreme observations than would be expected under normality. The distribution of

```
      Tukey Grouping          Mean       N     treat
                   A       12.9585       8     T1
                   A
              B    A       11.1345       8     T2
              B
              B            10.5498       8     T3
              B
              B            10.3519       8     T4
              B
              B             8.8256       8     T5

   Scheffe Grouping     Bon Grouping     SMM Grouping
                  A                A                A
                  A                A                A
             B    A           B    A           B    A
             B    A           B    A           B    A
             B    A           B    A           B    A
             B    A           B                B
             B    A           B                B
             B                B                B
             B                B                B
```

SAS Output 2.9: Partial results of `proc anova` for a different simulated data set. SMM refers to the (Studentized) maximum modulus method.

the F statistic and, more generally, the optimal way of assessing treatment differences with non-normal data are usually difficult to derive. Instead, nonparametric methods exist, and are often used if the normality assumption is not justified.

To get an idea of the consequences of non-normality, we simulate 10,000 times the p-value of the F test using i.i.d. Student's t data with location zero, scale one, and v degrees of freedom, $a = 4$ "treatments", and n observations per treatment. With normality, i.e., $v = \infty$, the simulated p-values should be uniformly distributed between zero and one. Figure 2.7 shows the resulting histograms for $n = 5$ and $v = 2$ (top) and $v = 4$ (bottom). We see that, for extreme data in which the variance does not exist, the behavior of the p-value (and, thus, the distribution of the F test statistic) deviates markedly from the behavior under normality, whereas for $v = 4$, which still implies quite heavy-tailed data, the behavior is not terribly far off. Table 2.2 shows the actual size of the F test with $\alpha = 0.05$, i.e., the fraction of p-values that were equal to or less than 0.05, for several further parameter constellations. Values less than $0.05 - 1.96\sqrt{0.05 \cdot 0.95/10000} = 0.0457$ are in bold face.

Similar calculations could be used to examine the (possibly size adjusted) power of the F test under the alternative hypothesis, or the effect of skewness on the size and power. As a typical asymmetric candidate, one could take $Y_{ij} \overset{\text{i.i.d.}}{\sim} \chi^2_v - v + \mu_i$, $i = 1, \ldots, a$. Asymmetric t distributions such as the noncentral t might also be entertained; they are easy to simulate from and allow control over both asymmetry and the thickness of the tails.

The other assumption that is often questioned is equal variances among the treatments. Graphical methods as well as formal tests exist for accessing the extent to which this and the normality assumption are violated. Textbooks dedicated to design of experiments, such as Gardiner and Gettinby (1998),

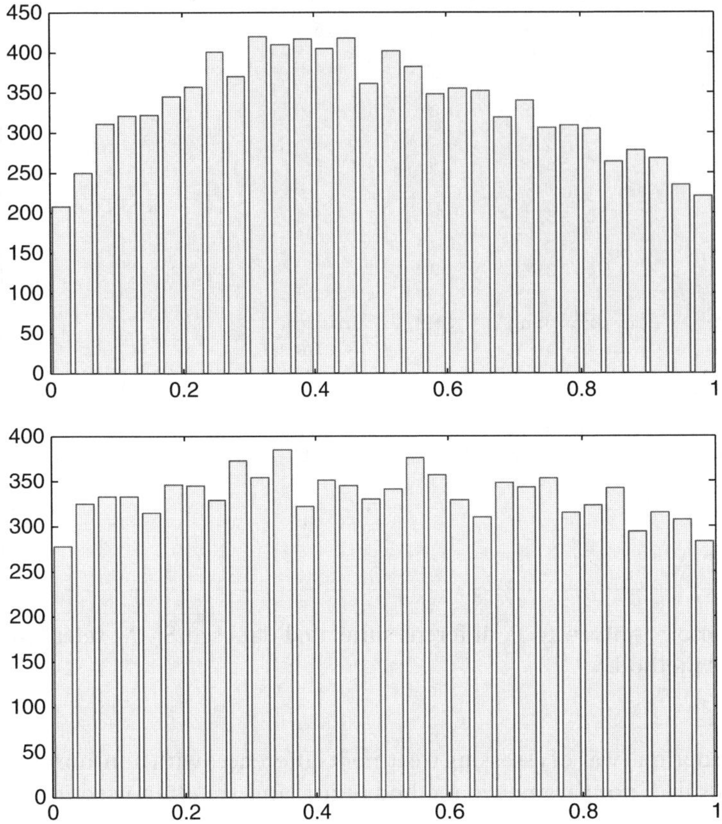

Figure 2.7 Histogram of 10,000 simulated p-values of the one-way ANOVA F test with $a = 4$ and $n = 5$ under the null hypothesis, but with i.i.d. Student's t data with 2 (top) and 4 (bottom) degrees of freedom.

Table 2.2 Empirical size of F test with $\alpha = 0.05$.

n\df	2	4	8	16	32	64	128
5	**0.033**	**0.043**	0.046	0.049	0.049	0.049	0.049
10	**0.034**	**0.042**	**0.044**	0.046	0.046	0.047	0.047
20	**0.036**	**0.045**	0.050	0.050	0.050	0.050	0.049
40	**0.038**	**0.045**	0.047	0.047	0.048	0.048	0.048

Dean and Voss (1999), and Montgomery (2000), provide ample discussion and examples of these and further issues. See also the excellent presentations of ANOVA and mixed models in Searle et al. (1992), MillerJr. (1997), Sahai and Ageel (2000), and Galwey (2014). For the analysis of covariance, as briefly mentioned in Section 2.1, an indispensable resource is Milliken and Johnson (2001).

2.5 Two-Way Balanced Fixed Effects ANOVA

The one-way fixed effects ANOVA model detailed in Section 2.4 is straightforwardly extended to support more than one factor. Here we consider the distribution theory of the balanced model with two factors. As a simple example to help visualize matters, consider again the agricultural example at the beginning of Section 2.4.1, and imagine an experiment in a greenhouse in which interest centers on $a \geqslant 2$ levels of a fertilizer and $b \geqslant 2$ levels of water. All ab combinations are set up, with n replications (plants) for each.

Once the ideas for the two-way ANOVA are laid out, the basic pattern for higher-order fixed effects ANOVA models with a balanced panel will be clear, and the reader should feel comfortable with conducting a data analysis in, say, SAS, or other software, and understand the output and how conclusions are (or should be) drawn.

After introducing the model in Section 2.5.1, Sections 2.5.2 and 2.5.3 present the basic theory of the cases without and with interaction, respectively, and the relevant ANOVA tables. Section 2.5.4 uses a simulated data set as an example to show the relevant coding in both Matlab and SAS.

2.5.1 The Model and Use of the Interaction Terms

For the two-way model, denote the first factor as A, with $a \geqslant 2$ treatments, and the second factor as B, with $b \geqslant 2$ treatments. The ordering of the two factors (i.e., which one is A and which one is B) is irrelevant, though, as mentioned in the Remark in Section 2.4.4, often A will refer to the factor associated with the scientific inquiry, while B is a **block**, accounting for differences in some attribute such as (for human studies) gender, age group, political affiliation, educational level, geographic region, time of day (see, e.g., Pope, 2016), etc., or, in industrial experiments, the factory line, etc.

The two-way fixed effect ANOVA model extends the forms in (2.19) and (2.21), and is expressed as

$$\begin{aligned} Y_{ijk} &= \mu_{ij} + \epsilon_{ijk}, \quad i = 1,2,\dots,a, \quad j = 1,2,\dots,b, \quad \epsilon_{ijk} \overset{\text{i.i.d.}}{\sim} \mathrm{N}(0,\sigma^2), \\ &= \mu + \alpha_i + \beta_j + (\alpha\beta)_{ij} + \epsilon_{ijk}, \end{aligned} \tag{2.44}$$

$k = 1,\dots,n$, subject to the constraints

$$\sum_{i=1}^{a} \alpha_i = 0, \quad \sum_{j=1}^{b} \beta_j = 0, \quad \sum_{i=1}^{a} (\alpha\beta)_{ij} = 0,\ \forall j, \quad \sum_{j=1}^{b} (\alpha\beta)_{ij} = 0, \forall i. \tag{2.45}$$

Terms $(\alpha\beta)_{ij}$ are referred to as the **interaction** factors (or effects, or terms). In general, the ijth group has n_{ij} observations, $i = 1,\dots,a$, $j = 1,\dots,b$, and if any of the n_{ij} are not equal, the model is unbalanced.

The usual ANOVA table will be shown below. It has in its output three F tests and their associated p-values, corresponding to the null hypotheses that $\sum_{i=1}^{a} \alpha_i = 0$ (no factor A effect), $\sum_{j=1}^{b} \beta_i = 0$ (no factor B effect), and $\sum_{i=1}^{a} \sum_{j=1}^{b} (\alpha\beta)_{ij} = 0$ (no interaction effect). One first inspects the latter; if the interaction effect can be deemed nonsignificant, then one proceeds to look at the former two. Violating our agreement in the Remark in Section 2.4.2 to subsequently suppress discussion of the dangers of use of p-values for model selection, we mention that an inspection of some published research studies, and even teaching notes on ANOVA, unfortunately use wording such as "As the p-value corresponding to the interaction effect is greater than 0.05, there is no interaction effect." A better choice of wording might be: "Based on the reported p-value, we will assume there is no significant interaction effect; and the subsequent analysis is conducted conditional on such, with the caveat that further experimental trials would be required to draw stronger conclusions on the presence of, and notably relevance of, interaction."

Observe that, if only the interaction factor is used (along with, of course, the grand mean), i.e., $Y_{ijk} = \mu + (\alpha\beta)_{ij} + \epsilon_{ijk}$, then this is equivalent to a one-way ANOVA with ab treatments. If the interaction effect is deemed significant, then the value of including the α_i and β_j effects is lowered and, possibly, rendered useless, depending on the nature of the interaction. In colloquial terms, one might describe the interaction effect as the presence of *synergy*, or the idea that a system is more than the sum of its parts. More specifically, assuming that the α_i and β_j are non-negative, the term **synergy** would be used if, due to the nonzero interaction effect $(\alpha\beta)_{ij}$, $\mathbb{E}[Y_{ijk}] > \mu + \alpha_i + \beta_j$, and the term **antagonism** would be used if $\mathbb{E}[Y_{ijk}] < \mu + \alpha_i + \beta_j$.

If there is no interaction effect (as one often hopes, as then nature is easier to describe), then the model reduces to $Y_{ijk} = \mu + \alpha_i + \beta_j + \epsilon_{ijk}$, and is such that the effect of the ith treatment from factor A does not depend on which treatment from factor B is used, and vice versa. In this case, the model is said to be **additive (in the main effects)**. This means, for example, that if one graphically plots, for a fixed j, $\hat{\mu}_{ij} = \hat{\mu} + \hat{\alpha}_i + \hat{\beta}_j + \widehat{(\alpha\beta)}_{ij}$ as a function of i, and overlays all j such plots, then the resulting lines will be approximately parallel (and vice versa). Such graphics are often produced by the ANOVA procedures in statistical software (see Figure 2.12 and, particularly, Figure 2.13 below) and typically accompany an empirical analysis. It should be obvious that, if the interaction terms are taken to be zero, then plots of $\hat{\mu}_{ij} = \hat{\mu} + \hat{\alpha}_i + \hat{\beta}_j$ will be, by construction, perfectly parallel.

2.5.2 Sums of Squares Decomposition Without Interaction

If one can assume there is no interaction effect, then the use of $n = 1$ is formally valid in (2.44), and otherwise not, though naturally the larger the cell sample size n, the more accurate the inference. As a concrete and simplified example to visualize things, imagine treatment A has three levels, referring to the percentage reduction in daily consumed calories (say, 75%, 50%, and 25%) for a dietary study measuring percentage weight loss. If factor B is gender (male or female), then one would not expect a significant interaction effect. Similarly, if factor B entails three levels of exercise, one might also expect that factors A and B influence Y_{ijk} linearly, without an interaction, or synergy, effect.

Model (2.44) without interaction is given by $Y_{ijk} = \mu + \alpha_i + \beta_j + \epsilon_{ijk}$, and when expressed as a linear model in matrix terms, it is $\mathbf{Y} = \mathbf{X}\boldsymbol{\beta} + \boldsymbol{\epsilon}$, where

$$\boldsymbol{\beta} = (\mu,\ \alpha_1, \dots, \alpha_a,\ \beta_1, \dots, \beta_b)'. \tag{2.46}$$

With $T = abn$, let $\mathbf{Y}$ be the $T \times 1$ vector formed by stacking the Y_{ijk} such that the last index, k, "moves quickest", in the sense of it changes on every row, followed by index j, which changes whenever k changes from n to 1, and finally index i changes slowest, whenever j changes from b to 1. The design matrix is then expressed as

$$\mathbf{X} = [\mathbf{X}_1 \mid \mathbf{X}_A \mid \mathbf{X}_B], \tag{2.47}$$

```
n=12; % n replications per cell
a=3;  % a treatment groups in the first factor
b=2;  % b treatment groups in the second factor
T=a*b*n; oa=ones(a,1); ob=ones(b,1); on=ones(n,1); obn=ones(b*n,1);
X1=ones(T,1); XA=kron(eye(a), obn); XB=kron( kron( oa, eye(b) ), on );
X=[X1, XA, XB];

% The three projection matrices
P1=X1*inv(X1'*X1)*X1'; PA=XA*inv(XA'*XA)*XA'; PB=XB*inv(XB'*XB)*XB'; %#ok<MINV>

% Claim: P1=PA*PB
diff = P1 - PA*PB; max(max(abs(diff)))
% Claim: PA-P1 is orthogonal to PB-P1
prod = (PA-P1)*(PB-P1); max(max(abs(prod)))
```

Program Listing 2.5: Generates the **X** matrix in (2.47) and (2.48).

where, denoting an n-length column of ones as $\underline{1}_n$ instead of $\mathbf{1}_n$ to help distinguish it from the identity matrix $\mathbf{I}_n$,

$$\begin{aligned}\mathbf{X}_1 &= \underline{1}_a \otimes \underline{1}_b \otimes \underline{1}_n = \underline{1}_T,\\ \mathbf{X}_A &= \mathbf{I}_a \otimes \underline{1}_b \otimes \underline{1}_n = \mathbf{I}_a \otimes \underline{1}_{bn},\\ \mathbf{X}_B &= \underline{1}_a \otimes \mathbf{I}_b \otimes \underline{1}_n. \end{aligned} \tag{2.48}$$

This is equivalent to first forming the **X** matrix corresponding to $n = 1$ and then post-Kronecker multiplying by $\underline{1}_n$, i.e.,

$$\mathbf{X}^{(1)} = [\underline{1}_a \otimes \underline{1}_b \mid \mathbf{I}_a \otimes \underline{1}_b \mid \underline{1}_a \otimes \mathbf{I}_b], \quad \mathbf{X} = \mathbf{X}^{(1)} \otimes \underline{1}_n. \tag{2.49}$$

It should be apparent that **X** is not full rank. The constraints $\sum_{i=1}^{a} \alpha_i = \sum_{j=1}^{b} \beta_j = 0$ need to be respected in order to produce the usual least squares estimator of $\boldsymbol{\beta}$ in (2.46).

Instead of using a whole page to write out an example of (2.47), the reader is encouraged to use the (top half of the) code in Listing 2.5 to understand the `kron` function in Matlab, and confirm that (2.47), (2.48), and (2.49) are correct.

Let $\mathbf{P}_1$, $\mathbf{P}_A$, and $\mathbf{P}_B$ be the respective projection matrices of $\mathbf{X}_1$, $\mathbf{X}_A$, and $\mathbf{X}_B$. In particular, letting $\mathbf{J}_m$ be the $m \times m$ matrix of ones,

$$\mathbf{P}_1 = (\underline{1}_T)(\underline{1}_T'\underline{1}_T)^{-1}(\underline{1}_T') = T^{-1}\mathbf{J}_T. \tag{2.50}$$

Likewise, using the Kronecker product facts from (2.23),

$$\begin{aligned}\mathbf{P}_A &= (\mathbf{I}_a \otimes \underline{1}_{bn})((\mathbf{I}_a \otimes \underline{1}_{bn}')(\mathbf{I}_a \otimes \underline{1}_{bn}))^{-1}(\mathbf{I}_a \otimes \underline{1}_{bn}')\\ &= (nb)^{-1}(\mathbf{I}_a \otimes \underline{1}_{bn})(\mathbf{I}_a \otimes \underline{1}_{bn}') = (nb)^{-1}(\mathbf{I}_a \otimes \mathbf{J}_{bn}). \end{aligned} \tag{2.51}$$

Observe that $\mathbf{P}_A$ is symmetric because of (2.23) and the symmetry of $\mathbf{I}_a$ and $\mathbf{J}_{bn}$, and is idempotent because

$$\mathbf{P}_A\mathbf{P}_A = (nb)^{-2}(\mathbf{I}_a \otimes \mathbf{J}_{bn})(\mathbf{I}_a \otimes \mathbf{J}_{bn}) = (nb)^{-2}(\mathbf{I}_a \otimes bn\mathbf{J}_{bn}) = \mathbf{P}_A.$$

Finally, for calculating $\mathbf{P}_B$, we need to extend the results in (2.23) to

$$(\mathbf{A} \otimes \mathbf{B} \otimes \mathbf{C})' = ((\mathbf{A} \otimes \mathbf{B}) \otimes \mathbf{C})' = ((\mathbf{A} \otimes \mathbf{B})' \otimes \mathbf{C}') = \mathbf{A}' \otimes \mathbf{B}' \otimes \mathbf{C}'$$

and

$$\begin{aligned}(\mathbf{A} \otimes \mathbf{B} \otimes \mathbf{C})(\mathbf{E} \otimes \mathbf{F} \otimes \mathbf{G}) &= ((\mathbf{A} \otimes \mathbf{B}) \otimes \mathbf{C})((\mathbf{E} \otimes \mathbf{F}) \otimes \mathbf{G}) \\ &= (\mathbf{A} \otimes \mathbf{B})(\mathbf{E} \otimes \mathbf{F}) \otimes \mathbf{CG} = (\mathbf{AE} \otimes \mathbf{BF}) \otimes \mathbf{CG} \\ &= \mathbf{AE} \otimes \mathbf{BF} \otimes \mathbf{CG}.\end{aligned}$$

Then

$$\begin{aligned}\mathbf{P}_B &= (\underline{1}_a \otimes \mathbf{I}_b \otimes \underline{1}_n)((\underline{1}_a' \otimes \mathbf{I}_b \otimes \underline{1}_n')(\underline{1}_a \otimes \mathbf{I}_b \otimes \underline{1}_n))^{-1}(\underline{1}_a' \otimes \mathbf{I}_b \otimes \underline{1}_n') \\ &= (\underline{1}_a \otimes \mathbf{I}_b \otimes \underline{1}_n)((\underline{1}_a'\underline{1}_a \otimes \mathbf{I}_b \otimes \underline{1}_n'\underline{1}_n))^{-1}(\underline{1}_a' \otimes \mathbf{I}_b \otimes \underline{1}_n') \\ &= (an)^{-1}(\underline{1}_a \otimes \mathbf{I}_b \otimes \underline{1}_n)(\underline{1}_a' \otimes \mathbf{I}_b \otimes \underline{1}_n') = (an)^{-1}(\mathbf{J}_a \otimes \mathbf{I}_b \otimes \mathbf{J}_n),\end{aligned} \tag{2.52}$$

which is also readily seen to be symmetric and idempotent.

Note that $\underline{1}_T \in \mathcal{C}(\mathbf{X}_A)$ and $\underline{1}_T \in \mathcal{C}(\mathbf{X}_B)$, and that the projection from $\mathbf{P}_1$ is "coarser" than that of $\mathbf{P}_A$ and $\mathbf{P}_B$, so that (and recalling that projection matrices are symmetric)

$$\mathbf{P}_A\mathbf{P}_1 = \mathbf{P}_1\mathbf{P}_A = \mathbf{P}_1, \quad \text{and} \quad \mathbf{P}_B\mathbf{P}_1 = \mathbf{P}_1\mathbf{P}_B = \mathbf{P}_1. \tag{2.53}$$

In light of $\underline{1}_T \in \mathcal{C}(\mathbf{X}_A)$ and $\underline{1}_T \in \mathcal{C}(\mathbf{X}_B)$, and also by way of thinking how to extend (2.35) from the one-way case, we are motivated to consider the matrices $\mathbf{P}_A - \mathbf{P}_1$ and $\mathbf{P}_B - \mathbf{P}_1$. From (2.53), it is trivial to confirm that $\mathbf{P}_A - \mathbf{P}_1$ and $\mathbf{P}_B - \mathbf{P}_1$ are (obviously symmetric and) idempotent, so that they are projection matrices. Thus,

$$\mathbf{P}_1(\mathbf{P}_A - \mathbf{P}_1) = \mathbf{0} = \mathbf{P}_1(\mathbf{P}_B - \mathbf{P}_1). \tag{2.54}$$

Also, $(\mathbf{P}_A - \mathbf{P}_1)(\mathbf{P}_B - \mathbf{P}_1) = \mathbf{P}_A\mathbf{P}_B - \mathbf{P}_A\mathbf{P}_1 - \mathbf{P}_1\mathbf{P}_B + \mathbf{P}_1\mathbf{P}_1 = \mathbf{P}_A\mathbf{P}_B - \mathbf{P}_1$. The second half of Listing 2.5 numerically confirms that $\mathbf{P}_1 = \mathbf{P}_A\mathbf{P}_B$, from which it follows that

$$\mathbf{0} = (\mathbf{P}_A - \mathbf{P}_1)(\mathbf{P}_B - \mathbf{P}_1), \tag{2.55}$$

as also confirmed numerically. The idea here is to illustrate the use of "proof by Matlab", which can be useful in more complicated settings when the algebra looks daunting. Of course, in this case, algebraically proving that

$$\mathbf{P}_1 = \mathbf{P}_A\mathbf{P}_B = \mathbf{P}_B\mathbf{P}_A \tag{2.56}$$

is very straightforward: Using (2.49) for simplicity, $\mathbf{P}_A\mathbf{P}_B$ is

$$\begin{aligned}&(\mathbf{I}_a \otimes \underline{1}_b)[(\mathbf{I}_a \otimes \underline{1}_b)'(\mathbf{I}_a \otimes \underline{1}_b)]^{-1}(\mathbf{I}_a \otimes \underline{1}_b)' \times (\underline{1}_a \otimes \mathbf{I}_b)[(\underline{1}_a \otimes \mathbf{I}_b)'(\underline{1}_a \otimes \mathbf{I}_b)]^{-1}(\underline{1}_a \otimes \mathbf{I}_b)' \\ &= (\mathbf{I}_a \otimes \underline{1}_b)(\mathbf{I}_a \otimes b)^{-1}(\mathbf{I}_a \otimes \underline{1}_b') \times (\underline{1}_a \otimes \mathbf{I}_b)(a \otimes \mathbf{I}_b)^{-1}(\underline{1}_a' \otimes \mathbf{I}_b) \\ &= b^{-1}(\mathbf{I}_a \otimes \underline{1}_b)(\mathbf{I}_a \otimes \underline{1}_b') \times a^{-1}(\underline{1}_a \otimes \mathbf{I}_b)(\underline{1}_a' \otimes \mathbf{I}_b) \\ &= b^{-1}(\mathbf{I}_a \otimes \mathbf{J}_b) \times a^{-1}(\mathbf{J}_a \otimes \mathbf{I}_b) = (ab)^{-1}(\mathbf{J}_a \otimes \mathbf{J}_b) = (ab)^{-1}\mathbf{J}_{ab},\end{aligned}$$

which is $\mathbf{P}_1$ of size $ab \times ab$. That $\mathbf{P}_A\mathbf{P}_B = \mathbf{P}_B\mathbf{P}_A$ follows from taking transposes and recalling that $\mathbf{P}_A$ and $\mathbf{P}_B$ are projection matrices and thus symmetric.

With (2.34) from the one-way case, and the previous projection matrices $\mathbf{P}_1$, $\mathbf{P}_A - \mathbf{P}_1$, and $\mathbf{P}_B - \mathbf{P}_1$ in mind, it suggests itself to inspect the algebraic identity

$$\mathbf{I} = \mathbf{P}_1 + (\mathbf{P}_A - \mathbf{P}_1) + (\mathbf{P}_B - \mathbf{P}_1) + (\mathbf{I} - (\mathbf{P}_A + \mathbf{P}_B - \mathbf{P}_1)), \tag{2.57}$$

where $\mathbf{I} = \mathbf{I}_T$, and $T = abn$. The orthogonality results (2.54), (2.55), and, as is easily confirmed using (2.56),

$$\begin{aligned}
\mathbf{P}_1(\mathbf{I} - (\mathbf{P}_A + \mathbf{P}_B - \mathbf{P}_1)) &= \mathbf{P}_1 - \mathbf{P}_1\mathbf{P}_A - \mathbf{P}_1\mathbf{P}_B + \mathbf{P}_1\mathbf{P}_1 = \mathbf{0},\\
(\mathbf{P}_A - \mathbf{P}_1)(\mathbf{I} - (\mathbf{P}_A + \mathbf{P}_B - \mathbf{P}_1)) &= \mathbf{P}_A(\mathbf{I} - (\mathbf{P}_A + \mathbf{P}_B - \mathbf{P}_1)) = \mathbf{0},\\
(\mathbf{P}_B - \mathbf{P}_1)(\mathbf{I} - (\mathbf{P}_A + \mathbf{P}_B - \mathbf{P}_1)) &= \mathbf{P}_B(\mathbf{I} - (\mathbf{P}_A + \mathbf{P}_B - \mathbf{P}_1)) = \mathbf{0},
\end{aligned}$$

imply that the terms on the right-hand side of (2.57) are orthogonal. Thus, similar to the decomposition in (2.32) and (2.35) for the one-way ANOVA, the corrected total sum of squares for the two-way ANOVA without interaction can be decomposed by subtracting $\mathbf{P}_1$ from both sides of (2.57) and writing

$$\mathbf{Y}'(\mathbf{I} - \mathbf{P}_1)\mathbf{Y} = \mathbf{Y}'(\mathbf{P}_A - \mathbf{P}_1)\mathbf{Y} + \mathbf{Y}'(\mathbf{P}_B - \mathbf{P}_1)\mathbf{Y} + \mathbf{Y}'(\mathbf{I} - (\mathbf{P}_A + \mathbf{P}_B - \mathbf{P}_1))\mathbf{Y}. \tag{2.58}$$

That is, $SS_T = SS_A + SS_B + SS_E$, where SS_T refers to the corrected total sum of squares.

Recall Theorem 1.2, which states that, if $\mathbf{P}$ is symmetric and idempotent, then $\text{rank}(\mathbf{P}) = k \Leftrightarrow \text{tr}(\mathbf{P}) = k$. This can be used precisely as in (2.36) above to determine the degrees of freedom associated with the various sum of squares, and construct the ANOVA Table 2.3. One could easily guess, and then confirm, that the degrees of freedom associated with SS_A and SS_B are $a - 1$ and $b - 1$, respectively, and that for SS_E is given by the (corrected) total $abn - 1$, minus those of SS_A and SS_B.

Next, recall:

1) Model (2.44) can be expressed as $\mathbf{Y} = \mathbf{X}\boldsymbol{\beta} + \boldsymbol{\epsilon}$, where $\boldsymbol{\beta}$ is given in (2.46) and $\boldsymbol{\epsilon} \sim \text{N}(\mathbf{0}, \sigma^2\mathbf{I}_T)$, $T = abn$, so that $\mathbf{Y} \sim \text{N}(\mathbf{X}\boldsymbol{\beta}, \sigma^2\mathbf{I}_T)$.
2) Theorem A.2, which states that, for $\mathbf{Y} \sim \text{N}(\boldsymbol{\mu}, \boldsymbol{\Sigma})$, $\boldsymbol{\Sigma} > 0$, the two quadratic forms $\mathbf{Y}'\mathbf{A}_1\mathbf{Y}$ and $\mathbf{Y}'\mathbf{A}_2\mathbf{Y}$ are independent if $\mathbf{A}_1\boldsymbol{\Sigma}\mathbf{A}_2 = \mathbf{A}_2\boldsymbol{\Sigma}\mathbf{A}_1 = \mathbf{0}$.

Table 2.3 The ANOVA table for the balanced two-way ANOVA model without interaction effect, where "error df" is $(abn - 1) - (a - 1) - (b - 1)$. Mean squares denote the sums of squares divided by their associated degrees of freedom. Table 2.4 is for the case with interaction, and also gives the expected mean squares.

Source of variation	Degrees of freedom	Sum of squares	Mean square	F statistic	p-value
Overall mean	1	$abn\bar{Y}^2_{\bullet\bullet}$			
Factor A	$a - 1$	SS_A	MS_A	MS_A/MS_E	p_A
Factor B	$b - 1$	SS_B	MS_B	MS_B/MS_E	p_B
Error	Error df	SS_E	MS_E		
Total (corrected)	$abn - 1$	SS_T			
Total	abn	$\mathbf{Y}'\mathbf{Y}$			

Thus, the orthogonality of the projection matrices in (2.58) and Theorem A.2 (with $\mathbf{\Sigma} = \sigma^2\mathbf{I}$) imply that MS_A, MS_B, and MS_E are all pairwise independent. As such, conditional on MS_E, ratios MS_A/MS_E and MS_B/MS_E are independent, and so must be functions of them. This implies that

$$\text{Conditional on } MS_E, p\text{-values } p_A \text{ and } p_B \text{ in Table 2.3 are independent.} \tag{2.59}$$

Unconditionally, ratios MS_A/MS_E and MS_B/MS_E, and thus their p-values, are not independent. This is also confirmed in Problem 1.16.

In our case here, we are working with projection matrices, so we can do a bit better. In particular, $SS_A = \mathbf{Y}'(\mathbf{P}_A - \mathbf{P}_1)'(\mathbf{P}_A - \mathbf{P}_1)\mathbf{Y}$, and

$$\mathbf{L}_A := (\mathbf{P}_A - \mathbf{P}_1)\mathbf{Y} \sim \mathrm{N}((\mathbf{P}_A - \mathbf{P}_1)\mathbf{X}\boldsymbol{\beta}, \sigma^2(\mathbf{P}_A - \mathbf{P}_1)).$$

Likewise defining $\mathbf{L}_B$ and $\mathbf{L}_E$, and letting $\mathbf{L} = [\mathbf{L}_A', \mathbf{L}_B', \mathbf{L}_E']'$, basic normal distribution theory implies that $\mathbf{L}$ follows a normal distribution with a block diagonal covariance matrix because of the orthogonality of the three projection matrices. As zero covariance implies independence under normality, it follows that $\mathbf{L}_A$, $\mathbf{L}_B$, and $\mathbf{L}_E$ are *completely* independent, not just pairwise.

Thus, separate functions of $\mathbf{L}_A$, $\mathbf{L}_B$, and $\mathbf{L}_E$, such as their sums of squares, are also completely independent, from which it follows that SS_A, SS_B, and SS_E (and thus MS_A, MS_B, and MS_E) are completely independent. This result is well known, referred to as Cochran's theorem, dating back to Cochran (1934), and usually proven via use of characteristic or moment generating functions; see, e.g., Khuri (2010, Sec. 5.5). Surveys of, and extensions to, Cochran's theorem can be found in Anderson and Styan (1982) and Semrl (1996). An admirable presentation in the context of elliptic distributions is given in Gupta and Varga (1993, Sec. 5.1).

Throughout the rest of this section on two-way ANOVA we will use a particular simulated data set for illustration, as detailed below, stored as variable `y` in Matlab. *The point right now is just to show the sums of squares in* (2.58) *computed in different ways.* In particular, they are computed (i) via SAS, (ii) via Matlab's canned function, and (iii) "by hand". The reason for the latter is to ensure a full understanding of what is being computed, as, realistically, one will not do these calculations manually, but just use canned routines in statistical software packages.

Based on our particular simulated data set introduced below, the SAS code for producing the two-way ANOVA table is given (a few pages) in SAS Program Listing 2.2. There, it is shown for the case when one wishes to include the interaction term. To omit the interaction term, as required now, simply change the `model` line to `model Happiness = Treatment Sport;`. The resulting ANOVA table is shown in SAS Output 2.10.

Matlab's `anovan` function can also compute this, and will be discussed below. The code to do so is given in Listing 2.10, using the first 25 lines, and changing line 25 to:

```
p=anovan(y,{fac1 fac2},'model','linear','varnames',{'Treatment A','Phy Act'})
```

The output is shown in Figure 2.8, and is the same as that from SAS.

Finally, to use Matlab for manually computing and confirming the output from the SAS `proc anova` and Matlab `anovan` functions, apply lines 1–9 from Listing 2.5, and then those in Listing 2.6, in conjunction with our simulated data set, to compute the sums of squares calculation in (2.58).

```
filename ein 'anova2prozac.txt';
ods pdf file='ANOVA Prozac Output.pdf';
ods rtf file='ANOVA Prozac Output.rtf';
data a;
  infile ein stopover;
  input Treatment $ Sport $ Happiness;
run;
proc anova;
  classes Treatment Sport;
  model Happiness = Treatment | Sport;
  means Treatment | Sport / SCHEFFE lines cldiff;
run;
ods _all_ close;
ods html;
```

SAS Program Listing 2.2: Runs the ANOVA procedure in SAS for the same data set used throughout this section. The notation `Treatment | Sport` is short for `Treatment Sport Treatment*Sport`.

```
Source          DF  Sum of Squares  Mean Square F Value Pr > F
Model            3    79.7993269     26.5997756  8.26    <.0001
Error           68  219.1019446      3.2220874
Corrected Total 71  298.9012715

Source       DF  Anova SS     Mean Square F Value Pr > F
Treatment    2   53.33396806 26.66698403 8.28    0.0006
Sport        1   26.46535881 26.46535881 8.21    0.0055
```

SAS Output 2.10: Analysis of the simulated data set that we will use throughout, and such that the model is $Y_{ijk} = \mu + \alpha_i + \beta_j + \epsilon_{ijk}$, i.e., does not use the interaction term. The same output for the two treatment effects sums of squares, and the error sums of squares, is given via Matlab in Figure 2.8.

2.5.3 Sums of Squares Decomposition With Interaction

We now develop the ANOVA table for the full model (2.44), with interaction. As mentioned above, in practice one starts with the full model in order to inspect the strength of the interaction term, usually hoping it is insignificant, as judged inevitably by comparing the p-value of the associated F test to the usual values of 0.10, 0.05, and 0.01. If the researcher decides it is insignificant and wishes to proceed without an interaction term, then, formally, all subsequent analysis, point estimates, and hypothesis test results are conditional on this decision, and one is in a pre-test estimation and pre-test testing framework.

If the interaction terms are strong enough, such that the model cannot be represented accurately without them, then the full two-way ANOVA model (2.44) can be expressed as $\mathbf{Y} = \mathbf{X}\boldsymbol{\beta} + \boldsymbol{\epsilon}$, with

$$\boldsymbol{\beta} = (\mu,\ \alpha_1, \dots, \alpha_a,\ \beta_1, \dots, \beta_b,\ (\alpha\beta)_{11}, (\alpha\beta)_{12}, \dots, (\alpha\beta)_{ab})', \tag{2.60}$$

Source	Sum Sq.	d.f.	Mean Sq.	F	Prob>F
Treatment A	53.334	2	26.6669	8.28	0.0006
Phy Act	26.465	1	26.4652	8.21	0.0055
Error	219.102	68	3.2221		
Total	298.901	71			

Figure 2.8 Same as SAS Output 2.10, but having used Matlab's function anovan. Note that in the fourth placed after the decimal, the mean square for treatment B ("Phy Act" in Matlab; "Sport" in SAS) differs among the two outputs (by one digit), presumably indicating that different numeric algorithms are used for their respective computations. This, in turn, is most surely irrelevant given the overstated precision of the Y measurements (they are not accurate to all 14 digits maintained in the computer), and that the F statistics and corresponding p-values are the same to all digits shown in the two tables.

```
% Decomposition using corrected total SS, for 2-way ANOVA, no interaction
SScT=y'*(eye(T)-P1)*y; SSA=y'*(PA-P1)*y;
SSB=y'*(PB-P1)*y; SSE=y'*(eye(T)-(PA+PB-P1))*y;
SSvec=[SScT, SSA, SSB, SSE]; disp(SSvec')
check=SScT-SSA-SSB-SSE; disp(check)
```

Program Listing 2.6: Computes the various sums of squares in (2.58), for the two-way ANOVA model without interaction, assuming that the simulated data set we use throughout (denoted y) is in memory (see below), and having executed lines 1–9 from Listing 2.5.

and

$$\mathbf{X} = [\mathbf{X}_1 \mid \mathbf{X}_A \mid \mathbf{X}_B \mid \mathbf{X}_{AB}], \tag{2.61}$$

where the first three terms are as in (2.48), and

$$\mathbf{X}_{AB} = \begin{pmatrix} \underline{1}_n & \mathbf{0}_n & \cdots & \mathbf{0}_n \\ \mathbf{0}_n & \underline{1}_n & \cdots & \vdots \\ \vdots & \vdots & \ddots & \\ \mathbf{0}_n & \mathbf{0}_n & \cdots & \underline{1}_n \end{pmatrix} = \mathbf{I}_a \otimes \mathbf{I}_b \otimes \underline{1}_n = \mathbf{I}_{ab} \otimes \underline{1}_n. \tag{2.62}$$

Note that (2.62) is the same as (2.22) for the one-way ANOVA model, but with ab different treatments instead of a.

The sum of squares decomposition (corrected for the grand mean) with interaction term is

$$\begin{aligned}\mathbf{Y}'(\mathbf{I} - \mathbf{P}_1)\mathbf{Y} &= \mathbf{Y}'(\mathbf{P}_A - \mathbf{P}_1)\mathbf{Y} + \mathbf{Y}'(\mathbf{P}_B - \mathbf{P}_1)\mathbf{Y} \\ &\quad + \mathbf{Y}'(\mathbf{P}_{AB} - \mathbf{P}_A - \mathbf{P}_B + \mathbf{P}_1)\mathbf{Y} + \mathbf{Y}'(\mathbf{I} - \mathbf{P}_{AB})\mathbf{Y},\end{aligned} \tag{2.63}$$

or $SS_T = SS_A + SS_B + SS_{AB} + SS_E$. As with (2.58), all terms in the center of the quadratic forms are orthogonal, e.g., recalling (2.56) and that otherwise the "more coarse" projection dominates,

$$\begin{aligned}&(\mathbf{P}_A - \mathbf{P}_1)(\mathbf{P}_{AB} - \mathbf{P}_A - \mathbf{P}_B + \mathbf{P}_1) \\ &= \mathbf{P}_A(\mathbf{P}_{AB} - \mathbf{P}_A - \mathbf{P}_B + \mathbf{P}_1) - \mathbf{P}_1(\mathbf{P}_{AB} - \mathbf{P}_A - \mathbf{P}_B + \mathbf{P}_1) \\ &= \mathbf{P}_A - \mathbf{P}_A - \mathbf{P}_1 + \mathbf{P}_1 - (\mathbf{P}_1 - \mathbf{P}_1 - \mathbf{P}_1 + \mathbf{P}_1) = \mathbf{0}.\end{aligned}$$

The reader is invited to quickly confirm the other cases.

It is of value to show (once) the sums of squares in (2.63) without matrix notation and contrast them with their analogous matrix expressions. As the reader should confirm,

$$
\begin{aligned}
SS_T &= \sum_{k=1}^{n}\sum_{i=1}^{a}\sum_{j=1}^{b} Y_{ijk}^2 - abn\bar{Y}_{\bullet\bullet\bullet}^2,\\
SS_A &= bn\sum_{i=1}^{a}(\bar{Y}_{i\bullet\bullet} - \bar{Y}_{\bullet\bullet\bullet})^2, \quad SS_B = an\sum_{j=1}^{b}(\bar{Y}_{\bullet j\bullet} - \bar{Y}_{\bullet\bullet\bullet})^2,\\
SS_{AB} &= n\sum_{i=1}^{a}\sum_{j=1}^{b}(\bar{Y}_{ij\bullet} - \bar{Y}_{i\bullet\bullet} - \bar{Y}_{\bullet j\bullet} + \bar{Y}_{\bullet\bullet\bullet})^2, \quad SS_E = \sum_{k=1}^{n}\sum_{i=1}^{a}\sum_{j=1}^{b}(Y_{ijk} - \bar{Y}_{ij\bullet})^2.
\end{aligned}
$$

Observe that $SS_{AB} + SS_E$ in (2.63) is precisely the SS_E term in (2.58). The reader is encouraged to construct code similar to that in Listings 2.5 and 2.6 to confirm the ANOVA sum of squares output shown in Figure 2.11 below for the two-way ANOVA with interaction. The relevant ANOVA table is given in Table 2.4.

From the facts that (i) MS_A and MS_E are independent and (ii) Theorem A.1 implies each is a χ^2 random variable divided by its respective degrees of freedom, we know that the distribution of $F_A := MS_A/MS_E$ is noncentral F, with $a-1$ numerator and $ab(n-1)$ denominator degrees of freedom, and numerator noncentrality

$$\theta_A = \frac{bn}{\sigma^2}\sum_{i=1}^{a}\alpha_i^2, \tag{2.64}$$

where (2.64) is a (correct) guess, based on the logical extension of (2.30), and subsequently derived. We first use it to obtain the expected mean square associated with treatment factor A. Again recalling that, for $Z \sim \chi^2(n,\theta)$, $\mathbb{E}[Z] = n + \theta$, we have, similar to (2.41), and recalling how σ^2 gets factored out

Table 2.4 The ANOVA table for the balanced two-way ANOVA model with interaction effect. Mean squares denote the sums of squares divided by their associated degrees of freedom. The expected mean squares are given in (2.65), (2.72), and (2.74)

Source of variation	Degrees of freedom	Sum of squares	Mean square	Expected mean square	*F* statistic	*p*-value
Overall mean	**1**	$abn\bar{Y}_{\bullet\bullet\bullet}^2$				
Factor A	$a-1$	SS_A	MS_A	$\mathbb{E}[MS_A]$	MS_A/MS_E	p_A
Factor B	$b-1$	SS_B	MS_B	$\mathbb{E}[MS_A]$	MS_B/MS_E	p_B
Factor A*B	$(a-1)(b-1)$	SS_{AB}	MS_{AB}	$\mathbb{E}[MS_{AB}]$	MS_{AB}/MS_E	p_{AB}
Error	$ab(n-1)$	SS_E	MS_E			
Total (corrected)	$abn-1$	SS_T				
Total	abn	$\mathbf{Y}'\mathbf{Y}$				

in front as in (2.37),

$$\mathbb{E}[MS_A] = \sigma^2 \frac{(a-1)+\theta_A}{a-1} = \sigma^2 + \frac{bn}{a-1}\sum_{i=1}^{a} \alpha_i^2. \tag{2.65}$$

Noncentrality term (2.64) can be formally derived by using (1.92), i.e.,

$$(\mathbf{Y}/\sigma)'(\mathbf{P}_A - \mathbf{P}_1)(\mathbf{Y}/\sigma) \sim \chi^2(a-1, \boldsymbol{\beta}'\mathbf{X}'(\mathbf{P}_A - \mathbf{P}_1)\mathbf{X}\boldsymbol{\beta}/\sigma^2), \tag{2.66}$$

and confirming that

$$\boldsymbol{\beta}'\mathbf{X}'(\mathbf{P}_A - \mathbf{P}_1)\mathbf{X}\boldsymbol{\beta} = \boldsymbol{\beta}'\mathbf{X}'(\mathbf{P}_A - \mathbf{P}_1)' \times (\mathbf{P}_A - \mathbf{P}_1)\mathbf{X}\boldsymbol{\beta} = bn\sum_{i=1}^{a} \alpha_i^2. \tag{2.67}$$

This would be very easy if, with $\boldsymbol{\alpha} = (\alpha_1, \dots, \alpha_a)'$, we can show

$$(\mathbf{P}_A - \mathbf{P}_1)\mathbf{X}\boldsymbol{\beta} = \mathbf{P}_A\mathbf{X}_A\boldsymbol{\alpha}. \tag{2.68}$$

If (2.68) is true, then note that, by the nature of projection, $\mathbf{P}_A\mathbf{X}_A = \mathbf{X}_A$, and $\mathbf{X}_A\boldsymbol{\alpha} = \boldsymbol{\alpha} \otimes \underline{1}_{bn}$, and the sum of the squares of the latter term is clearly $bn\sum_{i=1}^{a} \alpha_i^2$. To confirm (2.68), observe from (2.61) that

$$\begin{aligned}(\mathbf{P}_A - \mathbf{P}_1)\mathbf{X} &= (\mathbf{P}_A - \mathbf{P}_1)[\mathbf{X}_1 \mid \mathbf{X}_A \mid \mathbf{X}_B \mid \mathbf{X}_{AB}] \\ &= [\mathbf{0}_{T\times 1} \mid (\mathbf{P}_A - \mathbf{P}_1)\mathbf{X}_A \mid \mathbf{0}_{T\times b} \mid (\mathbf{P}_A - \mathbf{P}_1)\mathbf{X}_{AB}].\end{aligned} \tag{2.69}$$

The latter term $(\mathbf{P}_A - \mathbf{P}_1)\mathbf{X}_{AB} \neq \mathbf{0}$, but if we first assume the interaction terms $(\alpha\beta)_{ij}$ are all zero, then (2.69) implies

$$(\mathbf{P}_A - \mathbf{P}_1)\mathbf{X}\boldsymbol{\beta} = (\mathbf{P}_A - \mathbf{P}_1)\mathbf{X}_A\boldsymbol{\alpha}.$$

Now observe that $\mathbf{P}_1\mathbf{X}_A = T^{-1}\mathbf{J}_T(\mathbf{I}_a \otimes \underline{1}_{bn}) = a^{-1}\mathbf{J}_{T,a}$, where $\mathbf{J}_{T,a}$ is a $T \times a$ matrix of ones. This, and the fact that $\sum_{i=1}^{a} \alpha_i = 0$, implies $\mathbf{P}_1\mathbf{X}_A\boldsymbol{\alpha}$ is zero, and (2.68), and thus (2.64), are shown.

In the case with nonzero interaction terms, with

$$\boldsymbol{\gamma} = ((\alpha\beta)_{11}, (\alpha\beta)_{12}, \dots, (\alpha\beta)_{ab})', \tag{2.70}$$

we (cut corners and) confirm numerically that $(\mathbf{P}_A - \mathbf{P}_1)\mathbf{X}_{AB}\boldsymbol{\gamma} = \mathbf{0}$ (a T-length column of zeros), provided that the constraints on the interaction terms in (2.45) are met. It is *not* enough that all ab terms sum to zero. The reader is encouraged to also numerically confirm this, and, better, prove it algebraically.

Thus, $F_A \sim F_{a-1,ab(n-1)}(\theta_A)$, and the power of the test is $\Pr(F_A > c_A)$, where c_A is the cutoff value under the null (central) distribution for a given test significance level α. Based on the values we use below in an empirical example, namely $n = 12$, $a = 3$, $b = 2$, $\sigma = 2$, and $\sum_{i=1}^{a} \alpha_i^2 = 2/3$, (2.64) yields $\theta_A = 4$, so that the power of the test with significance level $\alpha = 0.05$ is 0.399, as computed with the code in Listing 2.7.

Analogous to (2.64), the test statistic associated with effect B is $F_B \sim F_{b-1,ab(n-1)}(\theta_B)$, where

$$\theta_B = \frac{an}{\sigma^2}\sum_{j=1}^{b} \beta_j^2, \tag{2.71}$$

which is $\theta_B = 81/8$ in our case, yielding a power of 0.880. Also analogously,

$$\mathbb{E}[MS_B] = \sigma^2 \frac{(b-1)+\theta_B}{b-1} = \sigma^2 + \frac{an}{b-1}\sum_{j=1}^{b} \beta_j^2. \tag{2.72}$$

```
n=12; a=3; b=2; sigma=2; dfA=a-1; dfB=b-1; dfErr=a*b*(n-1);
alpha=0.05; thetaA=4; thetaB=81/8;
cutA=finv(1-alpha,dfA,dfErr);
powerA = 1 - ncfcdf(cutA,dfA,dfErr,thetaA)
cutB=finv(1-alpha,dfB,dfErr);
powerB = 1 - ncfcdf(cutB,dfB,dfErr,thetaB)
```

Program Listing 2.7: Power calculations for the F tests in the two-way ANOVA with interaction.

Note that the distributions of the F_A and F_B tests in the case without interaction are similar, and use the denominator degrees of freedom taken from the SS_E in Table 2.3.

Now consider the interaction term. For convenience, let $\mathbf{R}_{AB} = (\mathbf{P}_{AB} - \mathbf{P}_A - \mathbf{P}_B + \mathbf{P}_1)$, and observe that $\mathbf{R}_{AB} = \mathbf{R}'_{AB}$ and $\mathbf{R}_{AB}\mathbf{R}_{AB} = \mathbf{R}_{AB}$. From (2.63), and similar to (2.66) and (2.67), we would need to prove that

$$\boldsymbol{\beta}'\mathbf{X}'\mathbf{R}_{AB}\mathbf{X}\boldsymbol{\beta} = n\sum_{i=1}^{a}\sum_{j=1}^{b}(\alpha\beta)_{ij}^2 \quad \text{or} \quad \mathbf{R}_{AB}\mathbf{X}\boldsymbol{\beta} = \boldsymbol{\gamma} \otimes \underline{1}_n, \tag{2.73}$$

where $\boldsymbol{\gamma}$ is defined in (2.70). It then follows from (2.73) that $\theta_{AB} = n\sigma^{-2}\sum_{i=1}^{a}\sum_{j=1}^{b}(\alpha\beta)_{ij}^2$, from which

$$\mathbb{E}[MS_{AB}] = \sigma^2 + \frac{n}{(a-1)(b-1)}\sum_{i=1}^{a}\sum_{j=1}^{b}(\alpha\beta)_{ij}^2. \tag{2.74}$$

To prove (2.73), we inspect $\mathbf{R}_{AB}\mathbf{X} = (\mathbf{P}_{AB} - \mathbf{P}_A - \mathbf{P}_B + \mathbf{P}_1)[\mathbf{X}_1 \mid \mathbf{X}_A \mid \mathbf{X}_B \mid \mathbf{X}_{AB}]$ and (as the reader is also welcome to) confirm

$$\begin{aligned} \underline{1}_T = \mathbf{P}_{AB}\mathbf{X}_1 = \mathbf{P}_A\mathbf{X}_1 = \mathbf{P}_B\mathbf{X}_1 = \mathbf{P}_1\mathbf{X}_1 \quad &\Rightarrow \quad \mathbf{R}_{AB}\mathbf{X}_1 = \mathbf{0}_T, \\ \mathbf{I}_a \otimes \underline{1}_{bn} = \mathbf{P}_{AB}\mathbf{X}_A = \mathbf{P}_A\mathbf{X}_A, \quad a^{-1}\mathbf{J}_{T,a} = \mathbf{P}_B\mathbf{X}_A = \mathbf{P}_1\mathbf{X}_A \quad &\Rightarrow \quad \mathbf{R}_{AB}\mathbf{X}_A = \mathbf{0}_T, \\ \underline{1}_a \otimes \mathbf{I}_b \otimes \underline{1}_n = \mathbf{P}_{AB}\mathbf{X}_B = \mathbf{P}_B\mathbf{X}_B, \quad b^{-1}\mathbf{J}_{T,b} = \mathbf{P}_A\mathbf{X}_B = \mathbf{P}_1\mathbf{X}_B \quad &\Rightarrow \quad \mathbf{R}_{AB}\mathbf{X}_B = \mathbf{0}_T, \end{aligned} \tag{2.75}$$

so that

$$\mathbf{R}_{AB}\mathbf{X}\boldsymbol{\beta} = \mathbf{R}_{AB}\mathbf{X}_{AB}\boldsymbol{\gamma} = (\mathbf{P}_{AB} - \mathbf{P}_A - \mathbf{P}_B + \mathbf{P}_1)\mathbf{X}_{AB}\boldsymbol{\gamma}. \tag{2.76}$$

Observe how in (2.75) the four terms generated from $\mathbf{R}_{AB}\mathbf{X}_1$ are all the same in magnitude (absolute value). Thus, by the nature of having two positive and two negative terms in $\mathbf{R}_{AB}$, their sum cancels. Increasing in complexity, for $\mathbf{R}_{AB}\mathbf{X}_A$ and $\mathbf{R}_{AB}\mathbf{X}_B$, observe that two terms are equal in magnitude, but have different signs, and the two other terms are equal in magnitude, but have different signs, and their sum cancels.

As perhaps then expected, $\mathbf{R}_{AB}\mathbf{X}_{AB}$ in (2.76) is the most complicated case, such that the four products generated by $\mathbf{R}_{AB}\mathbf{X}_{AB}$ are all different and cancellation does not occur. Some algebraic effort and practice with Kronecker products could then be invested to confirm that this indeed equals $\boldsymbol{\gamma} \otimes \underline{1}_n$, while a numerical confirmation is trivial in Matlab, and the reader is encouraged to at least do that.

2.5.4 Example and Codes

Imagine conducting an experiment to compare the effectiveness of various therapies for lowering anxiety, mitigating depression, or, more generally, "increasing happiness". For each patient, a progress measurement (say, some continuous measure such that zero implies no change from the initial state, and such that the larger it is, the higher is the improvement) is taken, once, after a fixed amount of time such that all treatments should have "kicked in", and doing so for a reasonably well-defined cohort, such as elderly people, people in a "mid-life crisis", or students attending university (the latter indeed being a high-risk group; see, e.g., Kitzrow, 2003). These example categories address the age of the patient, though other categories are possible, such as people with chronic pain and/or a particular disability or disease (e.g., Parkinson's). Let factor A describe the type of treatment, say, cognitive therapy (CT), meditation (MT), or use of Prozac (PZ), as discussed in Haidt (2006). (If multiple progress measurements are made through time, this gives rise to a type of **repeated measures** ANOVA.)

So far, this is a one-way design, though other factors might play a role. One possibility is gender, and another is if some form of physical activity is conducted that is reasonably appealing to the patient (or, less optimistically, the least unenjoyable), such as jogging, circuit training, weight lifting, or yoga (the latter having been investigated for its effectiveness; see, e.g., Kirkwood et al., 2005). Another possible set of factors are the subject's measurements associated with the so-called "big five (human) personality traits", namely openness to experience, conscientiousness, extraversion, agreeableness, and neuroticism. Further ideas might include levels indicating the extent of the person's religiosity, and also whether or not the person is a practicing Buddhist (see, e.g., Wright, 2017).

Use of treatment factor A, along with, say, gender and physical activity, gives rise to a three-way ANOVA. Omitting factors of relevance causes them to be "averaged over", and, if they do play a significant role, then their omission will cause the error variance to be unnecessarily high, possibly masking the differences in effects of the main factor under study. Worse, if there are ignored interaction effects, the analysis can be biased and possibly useless. Recall, in particular, the analysis in Section 2.3 when a block effect is erroneously ignored.

Let us assume for illustration that we use a balanced two-factor model, with therapy as factor A (with the $a = 3$ different treatments as mentioned above) and physical activity as factor B, with $b = 2$ categories "PA-NO" and "PA-YES". The data are fictitious, and not even loosely based on actual studies. For the cell means, we take

$$\begin{aligned} \mu_{11} &= 6+0, & \mu_{21} &= 6+0, & \mu_{31} &= 7+0, \\ \mu_{12} &= 6+1.5, & \mu_{22} &= 6+1.5, & \mu_{32} &= 7+1.5, \end{aligned}$$

and we need to figure out the values of μ, α_i, and β_j, respecting the constraints $\sum_{i=1}^{a} \alpha_i = \sum_{j=1}^{b} \beta_j = 0$. This can be done by solving the over-determined system of equations $\mathbf{Zc} = \mathbf{m}$, where

$$\mathbf{Z} = \begin{bmatrix} 1 & 1 & 0 & 0 & 1 & 0 \\ 1 & 0 & 1 & 0 & 1 & 0 \\ 1 & 0 & 0 & 1 & 1 & 0 \\ 1 & 1 & 0 & 0 & 0 & 1 \\ 1 & 0 & 1 & 0 & 0 & 1 \\ 1 & 0 & 0 & 1 & 0 & 1 \\ 0 & 1 & 1 & 1 & 0 & 0 \\ 0 & 0 & 0 & 0 & 1 & 1 \end{bmatrix}, \quad \mathbf{c} = \begin{bmatrix} \mu \\ \alpha_1 \\ \alpha_2 \\ \alpha_3 \\ \beta_1 \\ \beta_2 \end{bmatrix}, \quad \mathbf{m} = \begin{bmatrix} \mu_{11} \\ \mu_{21} \\ \mu_{31} \\ \mu_{12} \\ \mu_{22} \\ \mu_{32} \\ 0 \\ 0 \end{bmatrix}, \tag{2.77}$$

using the above values for the μ_{ij}. This, in turn, can be solved with $\mathbf{c} = (\mathbf{Z}'\mathbf{Z})^{-1}\mathbf{Z}'\mathbf{m}$.[2] It results in coefficients such that $\sum_{i=1}^{a} \alpha_i^2 = 2/3$ and $\sum_{j=1}^{b} \beta_j^2 = 9/8$, these being needed for power calculations.

2 In Matlab, with **Z** and **m** in memory, **c** can be computed as `c=Z\m`, which is shorthand for `mldivide(Z,m)`, and which, in this case, is `inv(Z'*Z)*Z'*m`.

```
n=12; % n replications per cell
a=3;  % a treatment groups in the first factor
b=2;  % b treatment groups in the second factor
sigma=2; % scale term of the errors

randn('state',1) % Deprecated in more recent versions of Matlab

% Put the data into a 3-dimensional array, as this seems the most logical
%  structure for storing data corresponding to a balanced, 2-way model
data3=zeros(a,b,n);
for i=1:a, for j=1:b %#ok<ALIGN>
  if i<=2, data3(i,j,:) = 6 + sigma*randn(n,1);
  else data3(i,j,:) = 7 + sigma*randn(n,1);
  end
  if j==2, data3(i,j,:) = data3(i,j,:) + 1.5; end
  % Use the following, with, say, n=2, a=4, b=3, to confirm that the
  %   conversion from data3 to data below is correct.
  % data3(i,j,:)=10*i+100*j+0.1*randn(n,1);
end, end
% Now convert such that it can be read into Matlab's anova2 procedure
% See the Matlab help file on anova2 for an illustration of
%    a data set and required format.
% Why they don't allow input as a 3D array?
% Their format, and this conversion, are a nuisance
tempa=zeros(n,b); data=[];
for i=1:a
  for j=1:b, tempa(:,j)=data3(i,j,:); end
  data=[data ; tempa]; %#ok<AGROW>
end
% In Matlab's anova2 procedure, "Rows" corresponds to the first level,
%    and "Columns" corresponds to the second level.
showoutput='on'; pvalues=anova2(data,n,showoutput);
```

Program Listing 2.8: Simulates two-way balanced fixed effects ANOVA data without interaction, and uses Matlab's function anova2 to perform the analysis. Note from the help file on anova2 the required data format. A different format is used with their more general function anovan, as used below.

Source	SS	df	MS	F	Prob>F
Columns	26.465	1	26.4652	8.11	0.0059
Rows	53.334	2	26.6669	8.17	0.0007
Interaction	3.781	2	1.8903	0.58	0.5631
Error	215.321	66	3.2624		
Total	298.901	71			

Figure 2.9 ANOVA table output from the Matlab code in Listing 2.8. Here, Columns refers to factor B, which is also clear because it has one degree of freedom, corresponding to use of $b = 2$. Similarly, Rows is for factor A, with $a = 3$ treatments (and, thus, two degrees of freedom).

The setup in (2.77) is yet more interesting. In the balanced case, we can replace the first six elements of the vector $\mathbf{m}$ with the respective cell means for each i, j combination, and the solution then yields the least squares estimates of the coefficients. We will verify this below in Listing 2.12. We can clearly permute the rows as we wish: Doing so such that $\mathbf{m} = [\mu_{11}, \ \mu_{12}, \ \mu_{21}, \ \mu_{22}, \ \mu_{31}, \ \mu_{32}]'$ instead, it should come as no surprise that the first $1 + a + b$ rows (all of them, except the last two, which embody the constraints) constitute precisely $\mathbf{X}^{(1)}$ given in (2.49). This can then be used to easily construct a program (in Matlab, etc.) to compute the least squares estimates in the balanced two-way case for any a and b, as the reader is encouraged to do.

The program in Listing 2.8 simulates the data with the above parameter values, noting that there is no interaction term $(\alpha\beta)_{ij}$ (see lines 11–14), and uses Matlab's function anova2 to perform the analysis. This results in the Matlab output in Figure 2.9, indicating that, at all conventional levels of significance (namely the smallest, $\alpha = 0.01$), the two main effects are significant, but the interaction effect is not.

Remark As we are no longer going to use Matlab to explicitly perform the numeric matrix ANOVA calculations, but rather calling their built-in functions anova2 and anovan, users of, say, R, should instead determine how to conduct this in R, with the correct ANOVA procedures, though still inspect how SAS is used, as shown below. This is all the more warranted, as most of the Matlab work in Listing 2.8 consists in putting the data into a suitable format for input into their anova2 function, and this has nothing to do, *per se*, with science and statistical inference. ■

This "data set" was obtained using a fixed seed value of one for the i.i.d. normal random errors, and we conveniently obtained correct inferential results at the usual significance levels with respect to the F tests for three factors. Repeating it with different seed values would surely sometimes result in erroneous conclusions. To confirm this, the program in Listing 2.9 repeats this exercise 100,000 times, and protocols the p-values of the three F tests, resulting in Figure 2.10.

The real point of this exercise is that such a simulation can be used to calculate the powers associated with the F tests in the ANOVA table, for a given (or several) value(s) of level of significance α, based

```
1  n=12; a=3; b=2; sigma=2;
2  sim=1e5; pvalA=zeros(sim,1); pvalB=pvalA; pvalAB=pvalA;
3  for rep=1:sim
4    % INCLUDE LINES 9-26 HERE
5    showoutput='off'; pvalues=anova2(data,n,showoutput);
6    pvalA(rep)=pvalues(2); pvalB(rep)=pvalues(1); pvalAB(rep)=pvalues(3);
7  end
8  corr(pvalA, pvalB)
9  boxes=50; figure, set(gca,'fontsize',16)
10 [histcount, histgrd] =hist(pvalA,boxes); h1=bar(histgrd,histcount);
11 set(h1,'facecolor',[0.94 0.94 0.94],'edgecolor',[0.9 0.7 1],'linewidth',1.8)
12 title('P-value of Effect A')
13 % ...similar for the other two graphics.
```

Program Listing 2.9: Performs a simulation of a balanced two-way ANOVA for assessing the distribution of the p-values associated with the F tests. In line 4, this means using lines 9–26 from Listing 2.8. Line 8 lends numerical support to (2.59) (though is not checking independence, but only correlation; note that the statistics p_A and p_B are not normally distributed).

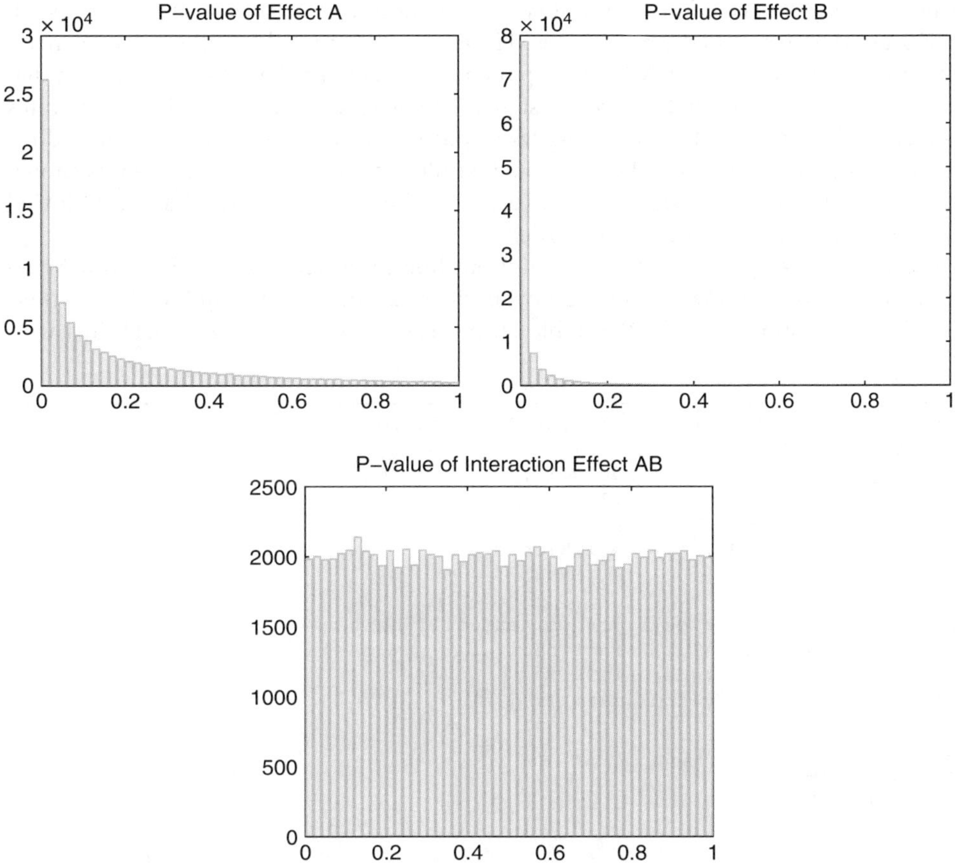

Figure 2.10 Histograms of p-values corresponding to the simulation from the code in Listing 2.9.

either on assumptions of the model parameters (as we do for convenience; see lines 11–14 in Listing 2.8) or, more practically, on an observed data set. In our setting here, with $\alpha = 0.05$, the empirical power associated with factor A (the fraction of p-values less than 0.05) is 0.40, while those for factors B and AB (the interaction) are 0.88 and 0.050, respectively. Note that those for factors A and B agree with the theoretical ones determined via the calculations in Listing 2.7, while the latter matches the level of significance α because, in our simulated model, there is no interaction effect.

In reality, instead of using the assumed model parameters, the simulated value, say $Y_{ijk}^{(s)}$, corresponding to the sth simulation, $s = 1, \ldots, S$, would be taken to be the mean of the n observations in the ijth cell of the actual data (obtained from, say, a **pilot study**, i.e., a small-scale preliminary study), plus an i.i.d. realization of an $\mathrm{N}(0, \hat{\sigma}^2)$ random variable, where $\hat{\sigma}^2$ is the estimate from the actual data. More generally (and perhaps more usefully, when analytic power calculations become more tricky), for an unbalanced design, the mean would be over the n_{ij} observations in the (ij)th cell, $i = 1, \ldots, a$, $j = 1, \ldots, b$. Such simulation can then be used to also determine the minimal cell sample size n to obtain the desired power of the F tests, so that a more accurate (and ideally balanced) subsequent study could be conducted.

We first show another way of performing the ANOVA calculations in Matlab that serves as a useful segue to the use of SAS. Matlab sports the more general function anovan, which allows for a general (not necessarily balanced) multi-way ANOVA. The input data also have a different structure than that used for anova2 (which is restricted to balanced designs). The program in Listing 2.10 (i) generates the same data set as in Listing 2.8, (ii) uses *cell arrays* to put the data into the format required for input to function anovan, and (iii) writes the data to a text file. The output from anovan is shown in Figure 2.11. It is, content-wise, the same as that in Figure 2.9, but now the factors can be endowed with useful names, and also appear in the desired order.

The generated text file anova2prozac.txt contains one line per observation, with the first being CT PA-NO 7.7288. The code in SAS Listing 2.2 now easily reads this text file and feeds it to their anova procedure. The corresponding ANOVA table output is the same as in Figure 2.11, just with

```
n=12; a=3; b=2; sigma=2; randn('state',1), data3=zeros(a,b,n);
for i=1:a, for j=1:b %#ok<ALIGN>
  if i<=2, data3(i,j,:) = 6 + sigma*randn(n,1);
  else data3(i,j,:) = 7 + sigma*randn(n,1);
  end
  if j==2, data3(i,j,:) = data3(i,j,:) + 1.5; end
end, end

% initialize the y-vector and the cell arrays, and fill them
T=a*b*n; y=zeros(T,1); fac1=cell(T,1); fac2=fac1;
for i=1:a, for j=1:b, for k=1:n %#ok<ALIGN>
  ind= n*b*(i-1) + n*(j-1) + k; y(ind)=data3(i,j,k);
  switch i
    case 1, fac1(ind)={'CT'};
    case 2, fac1(ind)={'MT'};
    case 3, fac1(ind)={'PZ'};
  end
  switch j
    case 1, fac2(ind)={'PA-NO'};
    case 2, fac2(ind)={'PA-YES'};
  end
end, end, end

% Use Matlab's most general ANOVA function, anovan
p=anovan(y,{fac1 fac2},'model','interaction', ...
         'varnames',{'Treatment A','Phy Act'});

% now output the data to a text file that can be read by SAS
fname='anova2prozac.txt'; if exist(fname,'file'), delete(fname), end
fileID = fopen(fname,'w');
for i=1:a*b*n
  str=[cell2mat(fac1(i)),' ',cell2mat(fac2(i)),' ',num2str(y(i),'%8.4f')];
  fprintf(fileID,'%s\r\n',str);
end
fclose(fileID);
```

Program Listing 2.10: First generates the same data set as in Listing 2.8, then puts the data into the format required for using function anovan (which involves cell arrays for the treatment names), and finally writes the data to a text file that can be read in by, for example, SAS.

Source	Sum Sq.	d.f.	Mean Sq.	F	Prob>F
Treatment A	53.334	2	26.6669	8.17	0.0007
Phy Act	26.465	1	26.4652	8.11	0.0059
Treatment A*Phy Act	3.781	2	1.8903	0.58	0.5631
Error	215.321	66	3.2624		
Total	298.901	71			

Figure 2.11 ANOVA table output from the Matlab code in Listing 2.10. Compare to Figure 2.9.

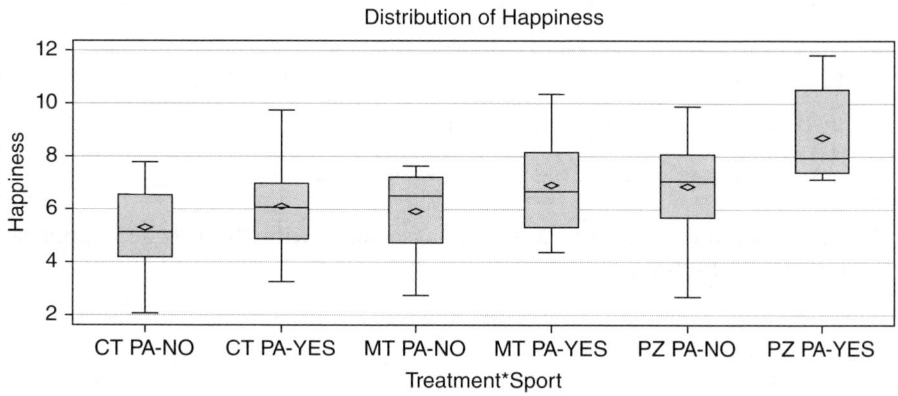

Figure 2.12 The default graphical output corresponding to the interaction effect `Treatment*Sport` from using the `means` statement in `proc anova` from SAS Listing 2.2.

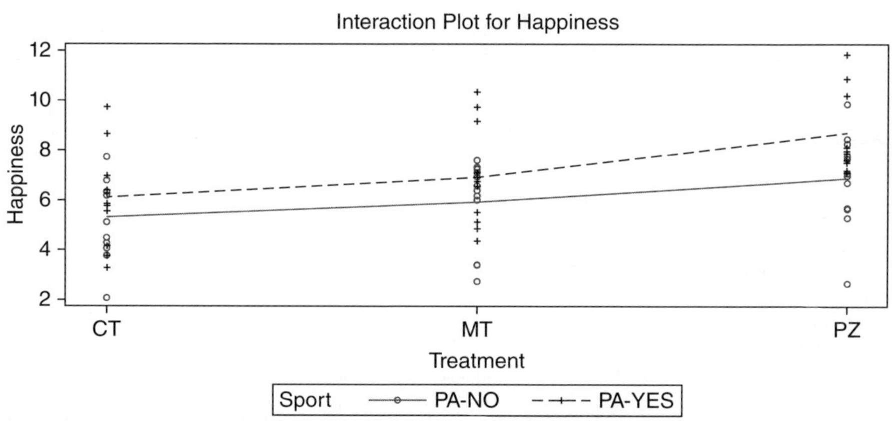

Figure 2.13 Default graphical output from SAS' `proc glm`, showing the same data as in Figure 2.12.

somewhat different formatting, and is omitted. Figure 2.12 is a set of boxplots for the *ab* treatments, and results from use of the `means` statement.

Using the same code as in SAS Listing 2.2, but with different pdf and rtf file output names, and changing the procedure call to `proc glm; classes Treatment Sport; model Happiness = Treatment | Sport; run;` produces the same ANOVA table, but a different graphic for the treatment means, as shown in Figure 2.13.

Remarks

a) SAS's `proc anova` (like Matlab's `anova2` function) requires balanced data. For unbalanced data, and other extras such as adding continuous covariates, use of random effects or mixed models, etc.,

```
n=12; a=3; b=2; T=a*b*n;
oa=ones(a,1); ob=ones(b,1); on=ones(n,1); obn=ones(b*n,1);
X1=ones(T,1); XA=kron(eye(a), obn); XB=kron( kron( oa, eye(b) ), on );
X=[X1, XA, XB];
fname='prozacX.txt'; if exist(fname,'file'), delete(fname), end
fileID = fopen(fname,'w');
fprintf(fileID,'%4u %4u %4u %4u %4u %4u\r\n',X'); fclose(fileID);
```

Program Listing 2.11: Generates and writes the **X** matrix associated with the Prozac happiness experiment, for the case with no interaction.

```
filename Yein 'anova2prozac.txt';
filename Xein 'prozacX.txt';
data Yvec;
  infile Yein stopover;
  input Treatment $ Sport $ Happiness;
run;
data Xmat;
  infile Xein stopover;
  input Int A1-A3 B1-B2;
run;
data YX;
  merge Yvec(keep=Happiness)
        Xmat;
run;
proc print data=YX; run;
proc reg;
  *model Happiness = Int A1-A3 B1-B2 / NOINT;
  model Happiness = A1-A3 B1-B2;
  restrict A1+A2+A3, B1+B2;
run;
```

SAS Program Listing 2.3: Reads in the ANOVA data, and also the relevant **X** regressor matrix as generated by Matlab from Listing 2.11, and runs `proc reg` to get the least squares coefficients. In the `restrict` statement, the various desired restrictions are listed one after another, separated by commas, and can specify to what they should be equal. In our setting, this is `A1+A2+A3=0, B1+B2=0`, but without the equals term, SAS understands this to mean equal to zero.

```
% First generate our usual data set used throughout
n=12; a=3; b=2; sigma=2; randn('state',1), data3=zeros(a,b,n);
for i=1:a, for j=1:b %#ok<ALIGN>
  if i<=2, data3(i,j,:) = 6 + sigma*randn(n,1);
  else data3(i,j,:) = 7 + sigma*randn(n,1);
  end
  if j==2, data3(i,j,:) = data3(i,j,:) + 1.5; end
end, end

% generate the Y vector
T=a*b*n; y=zeros(T,1);
for i=1:a, for j=1:b, for k=1:n %#ok<ALIGN>
  ind= n*b*(i-1) + n*(j-1) + k; y(ind)=data3(i,j,k);
end, end, end

% Now get the 6 cell means
mu=zeros(a,b);
for i=1:a, for j=1:b %#ok<ALIGN>
  mu(i,j)=mean(data3(i,j,:));
end, end

muvec=[mu(:) ; 0 ; 0]; % vectorize, and add the two zeros
Z=[1 1 0 0 1 0
   1 0 1 0 1 0
   1 0 0 1 1 0
   1 1 0 0 0 1
   1 0 1 0 0 1
   1 0 0 1 0 1
   0 1 1 1 0 0
   0 0 0 0 1 1];
c=Z\muvec % The least squares estimates of the model parameters
```

Program Listing 2.12: After generating our usual data set, as done in the beginning of Listing 2.10, use the over-identified system in (2.77) to generate the least squares estimates of the model parameters. They are identical to those given in the regression output from SAS (not shown here), based on the code in SAS Listing 2.3.

the SAS procedures `proc glm` and `proc mixed` are appropriate. These are their most general procedures under the Gaussianity assumption on the error term. While these could always be used, SAS maintains `proc anova` because it is computationally very efficient for a pure fixed effects ANOVA model with balanced data and, as mentioned in Appendix D, there was a time when one paid according to resources (time and memory) used. Similarly, SAS has `proc varcomp` and `proc nested`. The former supports mixed models (and both support unbalanced data), and both are computationally more efficient than use of their more advanced and subsuming `proc mixed` for pure random effects models. See Chapter 3 for examples of their use.

b) The simplicity of the decomposition of the sums of squares in the balanced ANOVA case, along with the elegance of using Kronecker products, is no longer available in the unbalanced case. Instead, different ways of computing the sums of squares are available, each with different interpretations. Conveniently at least, statistical software packages are set up to handle this case (such as Matlab's `anovan` function and SAS's `proc glm`), and produce the different sums of squares output and associated tests. The reader is encouraged to examine the output of these functions when using an unbalanced data set, as easily generated by modifying the above codes. Function `anovan` and `proc glm` are both called in the same way as with balanced data.

A detailed presentation of the unbalanced fixed effects ANOVA case, along with relevant SAS codes and discussion of output, is given in Khuri (2010, Ch. 10), while the highly regarded textbook by Milliken and Johnson (2009) is dedicated to "messy data" and contains a wealth of information, along with the use of SAS for conducting the analyses. ■

We end this section by using (2.77) to obtain the least squares estimates of the model parameters. The idea is to generate the relevant **X** matrix, write **X** (and the simulated **Y** data from the code in Listing 2.8) to text files, read them into SAS, merge them, and use their `proc reg` with the `restrict` statement to ensure $\sum_{i=1}^{a} \alpha_i = \sum_{j=1}^{b} \beta_j = 0$ to produce the least squares estimates of $\boldsymbol{\beta}$ in (2.46). The Matlab code in Listing 2.11 generates the **X** matrix and writes it to a text file. SAS Listing 2.3 then reads this in, merges it with the vector **Y** of happiness measurements, prints the data (as a check), and then executes `proc reg`. Finally, the Matlab code in Listing 2.12 uses the over-identified system in (2.77) to generate the least squares estimates of the model parameters. The reader (with access to Matlab and SAS) can easily conduct this, and confirm that the parameter estimates given in the SAS `proc reg` output (not shown here) are identical to those from the code in Listing 2.12.

3

Introduction to Random and Mixed Effects Models

Section 2.1, in the previous chapter on fixed effects ANOVA models, provided some introductory remarks on the distinction between fixed and random effects. This chapter is dedicated to random effects models, abbreviated as REMs, but also briefly touches on the mixed model case. In fixed effects ANOVA, interest is on least squares estimates associated with the treatments, testing their equality, and assessing which ones are statistically different. For example, with a two-way ANOVA without interaction, there are $2 + a + b$ parameters (the grand mean μ, the error variance, and the $a + b$ treatment parameters), albeit with sum restrictions. With REMs, there are only (besides the grand mean) **variance components**. For example, in a so-called two-way nested model with both effects random, no matter how many levels of factors A and B, there are three variance components—the error variance σ_e^2, and the variances from the two factors, σ_a^2 and σ_b^2—and thus only four model parameters to estimate.[1]

In Chapter 2, we referred to the levels of a particular fixed effect factor as the **levels**, or **treatments**. When using random effects, we will often use the term **classes**, this being common in the literature and also the statement in the various SAS procedures to designate a factor as being random.

We mostly assume throughout, as in the fixed effects ANOVA analysis in Chapter 2, that all models are balanced. This of course will not correspond to reality in all cases. The reason is tractability: With balance, the analysis is greatly simplified, allowing one to quickly get a handle on the majority of models of practical interest without getting lost in the (at times nontrivial) issues that arise with unbalanced data. We do address the unbalanced case from a heuristic/computational point of view in Sections 3.1.6 and 3.3.1.2, building on the elegant and easily computable results from the balanced case. This is not how the unbalanced case is usually handled (in fact, the author could not find any similar such presentation—and perhaps for good reason), though it is very instructive, and possibly of real use, replacing messy distribution theory with easy numerics such as simple optimization, simulation, and methods similar to the double bootstrap. The other assumption used throughout, without exception, is normality.

Substantially more detailed book-length compliments to this introductory chapter that we will frequently refer to are Sahai and Ojeda (2004) (dedicated to the balanced case) and Searle et al. (1992). Particularly for the unbalanced case, excellent resources, in addition to Searle et al. (1992), include

1 The term "components of variance" (nowadays variance components) seems to originate with Daniels (1939); see the discussion in Searle et al. (1992, p. 29) and further writings by Shayle Searle on the history of REMs and variance components analysis. It is an interesting coincidence that Henry Daniels pioneered the saddlepoint approximation, which is used in several contexts in this book, including Chapter 2, notably for the singly and doubly noncentral F distribution, and mentioned in this chapter, for confidence intervals for (functions of) variance components.

Linear Models and Time-Series Analysis: Regression, ANOVA, ARMA and GARCH, First Edition. Marc S. Paolella.

Khuri et al. (1998), Sahai and Ojeda (2005), and Milliken and Johnson (2009), along with the general linear model presentations in Graybill (1976), Khuri (2010), and Searle and Gruber (2017).

For all the models subsequently introduced, we assume, as usual, that the random effects are generated from mutually independent normally distributed random variables, and are denoted with lower-case Roman letters, e.g., a_i, b_j and b_{ij}. While we almost always use upper case to denote random variables, this convention appears more standard. Fixed effects, on the other hand, are, as in Chapter 2, denoted by lower-case Greek letters (α_i, β_j, etc.) and include the intercept μ. Thus, all unobserved quantities that have associated point and interval estimators are in Greek (fixed effects and the variance components σ_a^2, σ_e^2, etc.) while sampled values, whether observed or not (e.g., Y_{ij} and e_{ij}), are denoted with Roman letters. As such, to be consistent, we deviate from Chapters 1 and 2, and denote the linear model error term with an e, instead of ϵ, with variance σ_e^2 instead of σ_ϵ^2.

3.1 One-Factor Balanced Random Effects Model

The one-factor REM is the simplest case, and the obvious starting point. It is also an important model, serving to introduce the various concepts and procedures common to all REMs. As such, we go through the development slowly, with much detail, and also address the unbalanced case (albeit only partially, and in a non-conventional way). For the subsequent development of higher-order models, the pace is sped up, with some derivations and computational exercises given in the end of chapter exercises (answers are provided).

Recall the simple example mentioned in Section 2.1, where we sample $A = 20$ schools from a large population of schools belonging to some well-defined cohort (e.g., public high schools in a particular geographic area), and from each we sample $n = 15$ students in the same grade whose performance on some standardized test is to be evaluated. This is an example of a one-way REM. Examples abound in numerous fields of research: In agriculture (or forestry, animal studies, etc.), different plots of land can form the classes; in manufacturing, the classes can be from the factory lines of production, and/or the workers. In medicine, hospitals or clinics (or medical practitioners) can be the population of interest, etc. Returning to the school example, the variation due to the evaluators of the test might be the subject of interest.

3.1.1 Model and Maximum Likelihood Estimation

Let Y_{ij} denote the jth observation in the ith class, $i = 1, \ldots, A$, $j = 1, \ldots, n$. As with the fixed effects model, the n replications are random effects, but now the A values of the object under study (e.g., schools, hospitals, factory machines, sampled batches, segments of an ocean, galaxies, etc.) are also considered to be random realizations from a large population. Under the normality assumption, the model is represented as

$$Y_{ij} = \mu + a_i + e_{ij}, \quad a_i \overset{\text{i.i.d.}}{\sim} \mathrm{N}(0, \sigma_a^2), \quad e_{ij} \overset{\text{i.i.d.}}{\sim} \mathrm{N}(0, \sigma_e^2), \tag{3.1}$$

and such that a_i and e_{ij} are independent for all i and j. The three model parameters, which are assumed fixed but unknown, are μ, σ_a^2, and σ_e^2. From (3.1), the first two moments are

$$\mathbb{E}[Y_{ij}] = \mu, \quad \mathrm{Var}(Y_{ij}) = \sigma_a^2 + \sigma_e^2, \tag{3.2}$$

and

$$\mathrm{Cov}(Y_{ij}, Y_{ij'}) = \mathbb{E}[(a_i + e_{ij})(a_i + e_{ij'})] = \sigma_a^2, \quad j' \neq j. \tag{3.3}$$

(Notice here the use of the prime to denote "an alternative element", as opposed to a matrix transpose, or the first derivative.) In light of (3.3), σ_a^2 is denoted the **intra-class variance**, and we will denote σ_e^2 as the **error variance**.

In order to express the model in matrix notation, we first stack the Y_{ij} in "lexicon order", such that index j changes the fastest, giving

$$\mathbf{Y} = (Y_{11}, Y_{12}, \dots, Y_{1n}, Y_{21}, Y_{22}, \dots, Y_{2n}, \dots, Y_{11}, Y_{A2}, \dots, Y_{An})', \tag{3.4}$$

and define vector $\mathbf{e}$ similarly. Then, with $\mathbf{a} = (a_1, \dots, a_A)'$ and, similar to (2.22) using Kronecker product notation,

$$\begin{aligned} \mathbf{Y} &= (\underline{1}_A \otimes \underline{1}_n)\mu + (\mathbf{I}_A \otimes \underline{1}_n)\mathbf{a} + \mathbf{e} \\ &= \mathbf{X}\boldsymbol{\beta} + \boldsymbol{\epsilon}, \end{aligned} \tag{3.5}$$

where $\mathbf{X} = \underline{1}_{An}$, $\boldsymbol{\beta} = \mu$, and $\boldsymbol{\epsilon} = (\mathbf{I}_A \otimes \underline{1}_n)\mathbf{a} + \mathbf{e}$. We can thus express (3.1) and (3.5) as $\mathbf{Y} \sim \mathrm{N}_{An}(\boldsymbol{\mu}, \boldsymbol{\Sigma})$, where $\boldsymbol{\mu} = \mathbb{E}[\mathbf{Y}] = \mathbf{X}\boldsymbol{\beta}$ and, with $\mathbf{J}_n$ an $n \times n$ matrix of ones,

$$\begin{aligned} \boldsymbol{\Sigma} &= \mathrm{Var}(\mathbf{Y}) = \mathrm{Var}(\boldsymbol{\epsilon}) = (\mathbf{I}_A \otimes \underline{1}_n)\mathrm{Var}(\mathbf{a})(\mathbf{I}_A \otimes \underline{1}_n)' + \mathrm{Var}(\mathbf{e}) \\ &= (\mathbf{I}_A \otimes \underline{1}_n)\sigma_a^2 \mathbf{I}_A (\mathbf{I}_A \otimes \underline{1}_n)' + \sigma_e^2 \mathbf{I}_{An} = \sigma_a^2(\mathbf{I}_A \otimes \mathbf{J}_n) + \sigma_e^2(\mathbf{I}_A \otimes \mathbf{I}_n) \\ &= \mathbf{I}_A \otimes (\sigma_a^2 \mathbf{J}_n + \sigma_e^2 \mathbf{I}_n). \end{aligned} \tag{3.6}$$

Based on representation $\mathbf{Y} \sim \mathrm{N}_{An}(\boldsymbol{\mu}, \boldsymbol{\Sigma})$ and (3.6), it is straightforward to express the likelihood and numerically maximize it to obtain the m.l.e. The code for doing this is given in Listing 3.1. This exercise is beneficial for learning to "do things oneself" using basic principles, though (i) we will see below in Section 3.1.3 that there is a closed-form expression for the m.l.e., provided $\hat{\sigma}^2_{a,\mathrm{ML}}$ is positive, and (ii) for this model and, particularly, for more complicated models (such as one with mixed fixed and random effects, unbalanced data, continuous covariates, etc.), one would typically use canned reliable statistical software packages for the computations, as shown in Section 3.1.5.

Simulation is the easiest way of determining the small-sample performance of the m.l.e., and this was done for the constellation $A = 20$, $n = 15$, $\mu = 5$, $\sigma_a^2 = 0.4$, and $\sigma_e^2 = 0.8$, using $S = 10{,}000$ replications. Code for one such replication is shown in Listing 3.2, from which the reader can generate code to perform the simulation. (The code also computes the elements in the sums of squares decomposition given in (3.8), which we will need for other point and interval estimators.)

The simulation results are shown in Figure 3.1. The top panels show histograms of the estimated parameters, with the vertical dashed lines indicating the true parameters. (The m.l.e. is computed for μ, σ_a, and σ_e, and recall the invariance property of the m.l.e., such that $\hat{\sigma}_a^2$ is just the square of $\hat{\sigma}_a$.)

```
function [param, stderr, loglik, iters,bfgsok] = REM1wayMLE(y,A,n)
% param = [mu siga sige], sig is sigma, not sigma^2
ylen=length(y); if A*n ~= ylen, error('A and/or n wrong'), end
y=reshape(y,ylen,1); lo=1e-3; hi=2*std(y);
bound.lo=    [-1 lo lo]'; % mu, siga, sige
bound.hi=    [ 1 hi hi]';
bound.which=[ 0  1  1]';
initvec=[mean(y) std(y)/2 std(y)/2]';
opts=optimset('Display','None','TolX',1e-6,'MaxIter',200,...
  'MaxFunEval',600,'LargeScale','off'); bfgsok=1;
try
  [pout,fval,~,theoutput,~,hess]= fminunc(@(param) ...
    REM1_(param,y,A,n,bound),einschrk(initvec,bound),opts);
catch  %#ok<CTCH>
  disp('switching to use of simplex algorithm (fminsearch)')
  [pout,fval,~,theoutput]= fminsearch(@(param) ...
    REM1_(param,y,A,n,bound),einschrk(initvec,bound),opts);
  hess=eye(length(pout)); % just a place filler.
  bfgsok=0;
end
V=inv(hess); [param,V]=einschrk(pout,bound,V); param=param';
stderr=sqrt(diag(V))'; iters=theoutput.iterations; loglik=-fval;

function loglik=REM1_(param,y,A,n,bound)
if nargin<5, bound=0; end
if isstruct(bound), param=einschrk(real(param),bound,999); end
mu=param(1); siga=param(2); sige=param(3);
sigma2a=siga^2; sigma2e=sige^2;
muv=ones(A*n,1)*mu; J=ones(n,n); tmp=sigma2a*J+sigma2e*eye(n);
Sigma=kron(eye(A),tmp); loglik=-log(mvnpdf(y,muv,Sigma));
```

Program Listing 3.1: Maximum likelihood estimation of the three parameters of the one-way REM. Function `einschrk` is given in Listing III.4.7. An arbitrary positive lower bound is necessarily placed on the variance components. It was found that, as this bound gets closer to zero, numeric issues associated with the gradient/Hessian-based optimization method using the so-called BFGS algorithm (after the authors Charles George Broyden, Roger Fletcher, Donald Goldfarb, and David Shanno; see Section III.4.3.1) in Matlab version 2010 sometimes occur. To resolve this, if this happens, the program switches to use of the simplex method for optimization, which appears to never fail, though, for the same requested accuracy, requires far more function evaluations and thus takes longer. For the two constellations of parameters used in the simulations, and the imposed lower bound of 0.001 on σ_a and σ_e, the BFGS method never failed.

We see that, for this constellation, the m.l.e. appears close to unbiased and normally distributed, certainly for the fixed effect μ, but notably for σ_e^2 and reasonably so for σ_a^2.

The bottom panels show the histograms of the approximate standard errors (square roots of the variances) output from the BFGS algorithm (see, e.g., Section III.4.3 for details), with the vertical dashed lines being the best approximation of the truth: the sample standard error of the S m.l.e. point estimates of μ, $\hat{\sigma}_a$, and $\hat{\sigma}_e$, respectively. It thus appears that, for this constellation of parameters, inference on μ, σ_a^2, and σ_e^2 can safely be made using the asymptotic normal distribution and the

```
% desired parameters
A=20; n=15; mu=5; sigma2a=0.4; sigma2e=0.8;

% make Sigma matrix and generate a sample
muv=ones(A*n,1)*mu; J=ones(n,n); tmp=sigma2a*J+sigma2e*eye(n);
Sigma=kron(eye(A),tmp);
y=mvnrnd(muv,Sigma,1)'; % this is built into Matlab

% compute the various sums of squares
SST=sum(y'*y); Yddb=mean(y); SSu=A*n*Yddb^2; % Yddb is \bar{Y}_{dot dot}
H=kron(eye(A), ones(n,1)); Yidb=y'*H/n; % Yidb is \bar{Y}_{i dot}
SSa=n*sum( (Yidb-Yddb).^2 ); m=kron(Yidb', ones(n,1)); SSe=sum( (y-m).^2 );
check=SST-(SSu+SSa+SSe) % is zero

% MLE by brute force maximization
[param, stderr, loglik, iters,bfgsok] = REM1wayMLE(y,A,n);
AME=[param(1), param(2)^2, param(3)^2]

% MLE using closed form expression
mu_hat_MLE = mean(y); sigma2e_hat_MLE = SSe/A/(n-1);
sigma2a_hat_MLE = ( SSa/A - SSe/A/(n-1) )/n;
MLE = [mu_hat_MLE sigma2a_hat_MLE sigma2e_hat_MLE]
```

Program Listing 3.2: Generates a one-way REM data set, computes the sums of squares decomposition given in (3.8), calls the m.l.e. program in Listing 3.1, and computes the closed-form m.l.e. given in (3.21).

approximate standard errors output from use of the BFGS algorithm for computing the m.l.e. Another simple approximation to the standard errors is given in Section 3.1.3.

Figure 3.2 is similar to Figure 3.1, but based on $A = 7$ and $n = 5$. While the empirical distribution of $\hat{\sigma}_e^2$ is still close to Gaussian and the estimator appears virtually unbiased, its variance has increased markedly due to the reduction of $n = 15$ to $n = 5$. Having reduced $A = 20$ to $A = 7$, not only has the variation of $\hat{\sigma}_a^2$ increased, but it is no longer Gaussian, so that Wald confidence intervals based on the estimated standard error will not be particularly accurate. Below, we will discuss other ways of generating confidence intervals for $\hat{\sigma}_a^2$ that tend to be more accurate in such situations. While that seems beneficial for small sample sizes, the resulting intervals, however accurate, will be frustratingly wide.

3.1.2 Distribution Theory and ANOVA Table

In this and subsequent REMs, we will begin with the trivial "telescoping" identity

$$Y_{ij} = \bar{Y}_{\bullet\bullet} + (\bar{Y}_{i\bullet} - \bar{Y}_{\bullet\bullet}) + (Y_{ij} - \bar{Y}_{i\bullet}). \tag{3.7}$$

By squaring each term and summing, the reader is encouraged to confirm that the sums of all cross terms vanish, so that, similar to (2.28),

$$\begin{array}{llll} \sum_{i=1}^{A}\sum_{j=1}^{n} Y_{ij}^2 & = An\bar{Y}_{\bullet\bullet}^2 & + \; n\sum_{i=1}^{A}(\bar{Y}_{i\bullet} - \bar{Y}_{\bullet\bullet})^2 & + \; \sum_{i=1}^{A}\sum_{j=1}^{n}(Y_{ij} - \bar{Y}_{i\bullet})^2 \\ SST & = SS\mu & + \; SSa & + \; SSe, \end{array} \tag{3.8}$$

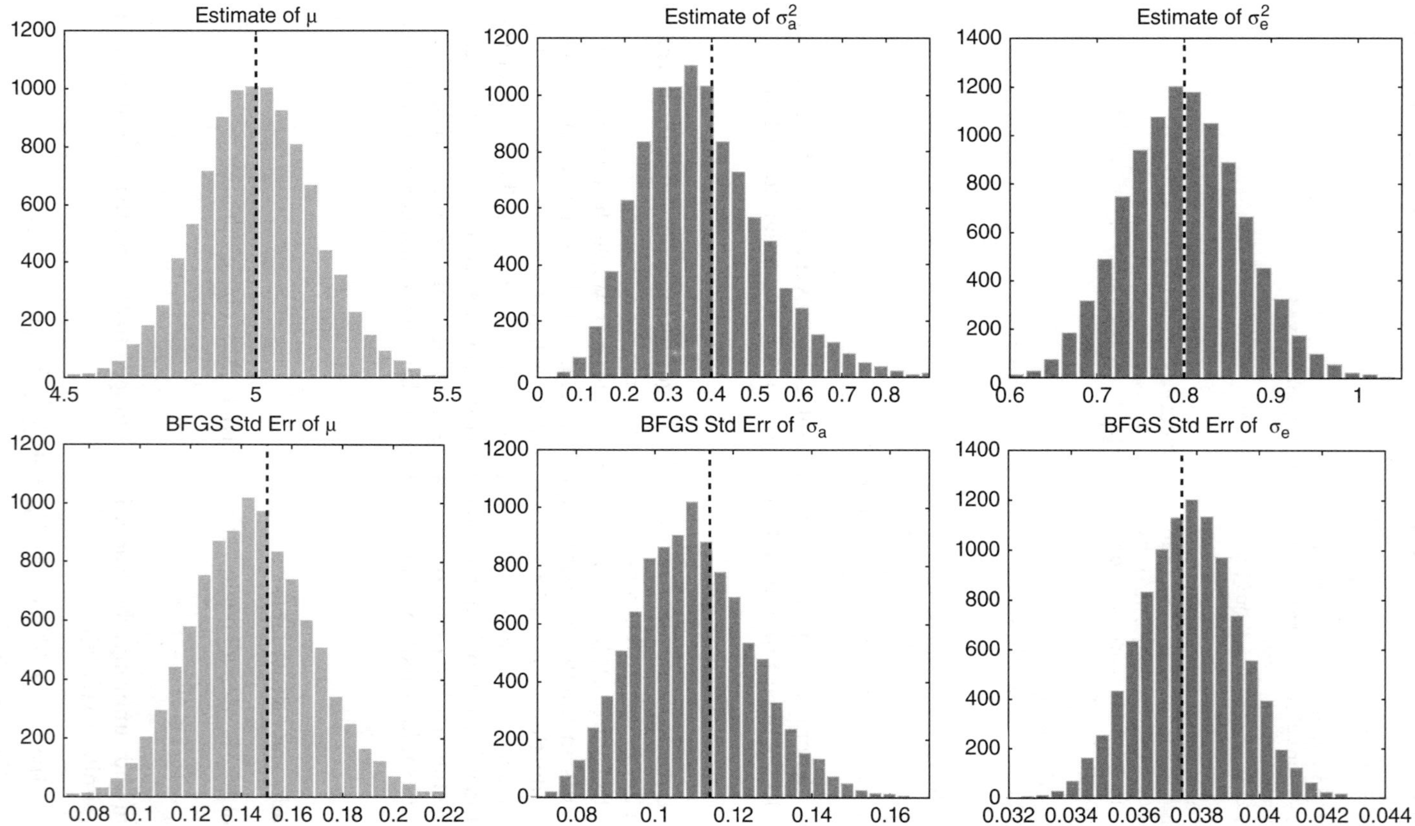

Figure 3.1 **Top:** Histograms of the m.l.e. of the three parameters, from left to right, μ, σ_a^2, and σ_e^2, of the one-way REM, based on $A = 20$, $n = 15$, and $S = 10{,}000$ replications. The vertical dashed line indicates the true value of the parameter in each graph. **Bottom:** Histograms of the approximate standard errors output from the BFGS algorithm, with the vertical dashed lines being the sample standard error of the S m.l.e. point estimates of μ, $\hat{\sigma}_a$, and $\hat{\sigma}_e$.

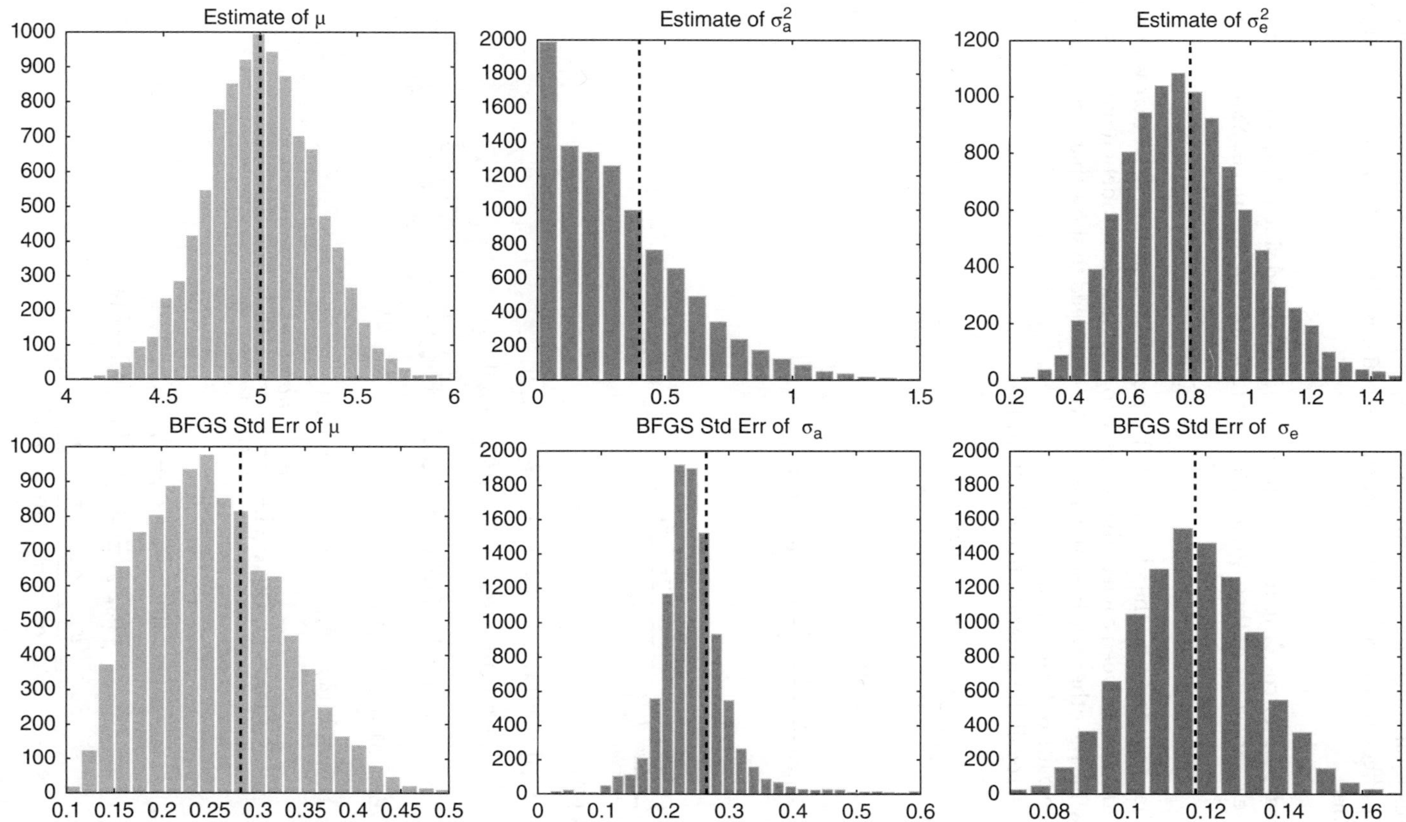

Figure 3.2 Same as Figure 3.1 but for $A = 7$ and $n = 5$.

where SST denotes total (uncorrected) sum of squares, $SS\mu$ is sum of squares for the mean, SSa is sum of squares for effect A, and SSe is the error sum of squares. Thus, SST is partitioned into the SS of the model factors.

Theorem 3.1 ***Independence*** The three terms on the right-hand side (r.h.s.) of (3.7) are independent, in which case so are sums of their squares (or any functions of them), i.e., $SS\mu$, SSa, and SSe are independent.

Proof: Observe that each term on the r.h.s. of (3.7) is normally distributed, so that we only need to verify that the covariance between each of them is zero to establish their independence. The first term, $\bar{Y}_{\bullet\bullet}$, has mean μ, while the other two have mean zero. We thus need to show that the expected product of each of the three pairs of terms is zero.

Before beginning, recall the notation from (2.3), and let

$$e_{i\bullet} = \sum_{j=1}^{n} e_{ij}, \quad \bar{e}_{i\bullet} = \frac{e_{i\bullet}}{n}, \qquad e_{\bullet\bullet} = \sum_{i=1}^{A}\sum_{j=1}^{n} e_{ij}, \quad \bar{e}_{\bullet\bullet} = \frac{e_{\bullet\bullet}}{An},$$

and similarly for $\bar{Y}_{i\bullet}$ and $\bar{Y}_{\bullet\bullet}$, so that

$$\bar{Y}_{i\bullet} = \frac{1}{n}\sum_{j=1}^{n} Y_{ij} = \frac{n\mu + na_i + e_{i\bullet}}{n} = \mu + a_i + \bar{e}_{i\bullet}, \tag{3.9}$$

and

$$\bar{Y}_{\bullet\bullet} = \frac{1}{An}\sum_{i=1}^{A}\sum_{j=1}^{n} Y_{ij} = \frac{An\mu + na_\bullet + e_{\bullet\bullet}}{An} = \mu + \bar{a}_\bullet + \bar{e}_{\bullet\bullet}.$$

Then, for the first pair,

$$\begin{aligned}
\mathbb{E}[\bar{Y}_{\bullet\bullet}(\bar{Y}_{i\bullet} - \bar{Y}_{\bullet\bullet})] &= \mathbb{E}[(\bar{a}_\bullet + \bar{e}_{\bullet\bullet})(a_i - \bar{a}_\bullet + \bar{e}_{i\bullet} - \bar{e}_{\bullet\bullet})] \\
&= \mathbb{E}[\bar{a}_\bullet(a_i - \bar{a}_\bullet)] + \mathbb{E}[\bar{e}_{\bullet\bullet}(\bar{e}_{i\bullet} - \bar{e}_{\bullet\bullet})] \\
&= \mathbb{E}[\bar{a}_\bullet a_i] - \mathbb{E}[\bar{a}_\bullet^2] + \mathbb{E}[\bar{e}_{\bullet\bullet}\bar{e}_{i\bullet}] - \mathbb{E}[\bar{e}_{\bullet\bullet}^2] = \frac{\sigma_a^2}{A} - \frac{\sigma_a^2}{A} + \frac{\sigma_e^2}{An} - \frac{\sigma_e^2}{An} = 0,
\end{aligned}$$

as in Graybill (1976, p. 610). Likewise,

$$\begin{aligned}
\mathbb{E}[\bar{Y}_{\bullet\bullet}(Y_{ij} - \bar{Y}_{i\bullet})] &= \mathbb{E}[(\bar{a}_\bullet + \bar{e}_{\bullet\bullet})(a_i - a_i + e_{ij} - \bar{e}_{i\bullet})] \\
&= \mathbb{E}[\bar{e}_{\bullet\bullet}e_{ij}] - \mathbb{E}[\bar{e}_{\bullet\bullet}\bar{e}_{i\bullet}] = \frac{\sigma_e^2}{An} - \frac{n\sigma_e^2}{An^2} = 0,
\end{aligned}$$

and

$$\begin{aligned}
\mathbb{E}[(\bar{Y}_{i\bullet} - \bar{Y}_{\bullet\bullet})(Y_{ij} - \bar{Y}_{i\bullet})] &= \mathbb{E}[(a_i - \bar{a}_\bullet + \bar{e}_{i\bullet} - \bar{e}_{\bullet\bullet})(e_{ij} - \bar{e}_{i\bullet})] \\
&= \mathbb{E}[\bar{e}_{i\bullet}e_{ij}] - \mathbb{E}[\bar{e}_{i\bullet}^2] - \mathbb{E}[\bar{e}_{\bullet\bullet}e_{ij}] + \mathbb{E}[\bar{e}_{\bullet\bullet}\bar{e}_{i\bullet}] \\
&= \frac{\sigma_e^2}{n} - \frac{\sigma_e^2}{n} - \frac{\sigma_e^2}{An} + \frac{n\sigma_e^2}{An^2} = 0,
\end{aligned}$$

confirming the result. ■

Theorem 3.2 ***Distribution***

$$\frac{SS\mu}{\gamma_a} \sim \chi_1^2\left(\frac{An\mu^2}{\gamma_a}\right), \qquad \frac{SSa}{\gamma_a} \sim \chi^2_{A-1}, \qquad \frac{SSe}{\sigma_e^2} \sim \chi^2_{A(n-1)}, \tag{3.10}$$

where $\gamma_a := n\sigma_a^2 + \sigma_e^2$.

Proof: It is not hard to show this directly (see, e.g., Graybill, 1976, p. 609), but a simple transformation can speed things up and is of great use when working with higher-order models. As in Stuart et al. (1999, p. 676), we define $H_i := a_i + \bar{e}_{i\bullet}$ and then verify that

$$\begin{aligned} Y_{ij} &= \quad \bar{Y}_{\bullet\bullet} \quad + (\bar{Y}_{i\bullet} - \bar{Y}_{\bullet\bullet}) + (Y_{ij} - \bar{Y}_{i\bullet}) \\ &= (\mu + \bar{H}_{\bullet}) + (H_i - \bar{H}_{\bullet}) + (e_{ij} - \bar{e}_{i\bullet}) \\ &= \mu + a_i + e_{ij}. \end{aligned} \tag{3.11}$$

Next, note that $\bar{H}_{\bullet} = \bar{a}_{\bullet} + \bar{e}_{\bullet\bullet}$ and $H_i \overset{\text{i.i.d.}}{\sim} \mathrm{N}(0, \sigma_a^2 + \sigma_e^2/n)$. Starting from the top right of (3.11), write

$$Y_{ij} - \bar{Y}_{i\bullet} = (\mu + a_i + e_{ij}) - (\mu + a_i + \bar{e}_{i\bullet}) = e_{ij} - \bar{e}_{i\bullet},$$

and, similarly,

$$\bar{Y}_{i\bullet} - \bar{Y}_{\bullet\bullet} = (\mu + a_i + \bar{e}_{i\bullet}) - (\mu + \bar{a}_{\bullet} + \bar{e}_{\bullet\bullet}) = H_i - \bar{H}_{\bullet},$$

and $\bar{Y}_{\bullet\bullet} = \mu + \bar{a}_{\bullet} + \bar{e}_{\bullet\bullet} = \mu + \bar{H}_{\bullet}$. Thus, for a given i, $\sigma_e^{-2}\sum_{j=1}^{n}(e_{ij} - \bar{e}_{i\bullet})^2 \sim \chi^2_{n-1}$ and

$$\sigma_e^{-2} SSe = \sigma_e^{-2} \sum_{i=1}^{A}\sum_{j=1}^{n}(e_{ij} - \bar{e}_{i\bullet})^2 \sim \chi^2_{A(n-1)}, \tag{3.12}$$

from the independence of the e_{ij} and the summability of independent chi-square random variables. Similarly, $\sigma_{H_i}^{-2}\sum_{i=1}^{A}(H_i - \bar{H}_{\bullet})^2 \sim \chi^2_{A-1}$. With $\gamma_a := n\sigma_a^2 + \sigma_e^2$,

$$\frac{SSa}{\gamma_a} = \frac{n\sum_{i=1}^{A}(H_i - \bar{H}_{\bullet})^2}{n\sigma_a^2 + \sigma_e^2} = \sigma_{H_i}^{-2}\sum_{i=1}^{A}(H_i - \bar{H}_{\bullet})^2 \sim \chi^2_{A-1}. \tag{3.13}$$

Finally, $(\mu + \bar{H}_{\bullet}) \sim \mathrm{N}(\mu, \sigma_a^2/A + \sigma_e^2/An)$ implies $\sqrt{An}(\mu + \bar{H}_{\bullet}) \sim \mathrm{N}(\sqrt{An}\,\mu, \gamma_a)$, so that $\sqrt{An/\gamma_a}(\mu + \bar{H}_{\bullet}) \sim \mathrm{N}(\sqrt{An/\gamma_a}\,\mu, 1)$, which, in turn, implies

$$\frac{An}{\gamma_a}(\mu + \bar{H}_{\bullet})^2 \sim \chi_1^2\left(\frac{An\mu^2}{\gamma_a}\right) \quad \text{or} \quad \frac{SS\mu}{\gamma_a} \sim \chi_1^2\left(\frac{An\mu^2}{\gamma_a}\right),$$

completing the proof. ∎

Remark Theorem 3.1 showed that $\bar{Y}_{\bullet\bullet}$, $\bar{Y}_{i\bullet} - \bar{Y}_{\bullet\bullet}$ and $Y_{ij} - \bar{Y}_{i\bullet}$ are independent for all i and j, from which it follows that sums of their squares (or any functions of them) are independent, so that $SS\mu$, SSa, and SSe are independent. We can also see this in the following way.

For a given i, we know from the independence property of $\bar{X}$ and S_X^2 for normal samples (see, e.g., Section II.3.7) that $\bar{e}_{i\bullet} \perp \sum(e_{ij} - \bar{e}_{i\bullet})^2$. This is the case for any i, i.e., also $\bar{e}_{i'\bullet} \perp \sum(e_{ij} - \bar{e}_{i\bullet})^2$, so that, from (3.12), $\bar{e}_{i\bullet} \perp SSe$. As SSa is a function only of $H_i = a_i + \bar{e}_{i\bullet}$, and SSe is not a function of a_i, we have $SSe \perp SSa$ (recalling $a_i \perp \bar{e}_{i\bullet}$). The same applies to $SS\mu$, being a function of $\bar{H}_{\bullet}$ and a fixed value

Table 3.1 ANOVA table for the balanced one-factor REM. The second column is specific to our model notation (3.1), and is not necessary, but shown for further clarity.

Source	Terms	df	*SS*	*EMS*
Mean	μ	1	$An\bar{Y}^2_{\bullet\bullet}$	$\sigma_e^2 + n\sigma_a^2 + An\mu^2$
A	$\{a_i\}$	$A-1$	$n\sum_{i=1}^{A}(\bar{Y}_{i\bullet} - \bar{Y}_{\bullet\bullet})^2$	$\sigma_e^2 + n\sigma_a^2$
Error	$\{e_{ij}\}$	$A(n-1)$	$\sum_{i=1}^{A}\sum_{j=1}^{n}(Y_{ij} - \bar{Y}_{i\bullet})^2$	σ_e^2
Total	$\{Y_{ij}\}$	An	$\sum_{i=1}^{A}\sum_{j=1}^{n}Y_{ij}^2$	

μ, i.e., $SSe \perp SS\mu$. Finally, as H_i are also normally distributed, $\bar{H}_\bullet \perp SSa$ and, as $SS\mu$ is a function of $\bar{H}_\bullet$ and a fixed value μ, $SS\mu \perp SSa$. ■

Dividing each *SS* term by its degrees of freedom and taking expected values yields the **expected mean squares**, or *EMS*. This gives, with $\gamma_a := n\sigma_a^2 + \sigma_e^2$,

$$\mathbb{E}[MS\mu] = \mathbb{E}[SS\mu] = \gamma_a \mathbb{E}\left[\chi_1^2\left(\frac{An\mu^2}{\gamma_a}\right)\right] = \gamma_a\left(1 + \frac{An\mu^2}{\gamma_a}\right) = \gamma_a + An\mu^2,$$
$$\mathbb{E}[MSa] = \mathbb{E}\left[\frac{SSa}{A-1}\right] = \frac{\gamma_a}{A-1}\mathbb{E}[\chi^2_{A-1}] = \gamma_a, \tag{3.14}$$

and

$$\mathbb{E}[MSe] = \mathbb{E}\left[\frac{SSe}{A(n-1)}\right] = \frac{\sigma_e^2}{A(n-1)}\mathbb{E}[\chi^2_{A(n-1)}] = \sigma_e^2, \tag{3.15}$$

recalling that $\mathbb{E}[\chi^2{}_\delta(\nu)] = \delta + \nu$. These are summarized in the ANOVA table (Table 3.1).
Recall the discussion of sufficiency and completeness in, e.g., Chapter III.7.

Theorem 3.3 ***Complete, Minimal Sufficient Statistics*** The set of complete, minimally sufficient statistics for μ, σ_a^2, and σ_e^2 is given by $SS\mu$, SSa, and SSe.

Proof: Sufficiency follows by expressing $\mathbf{Y} \sim \mathrm{N}_{An}(\boldsymbol{\mu}, \boldsymbol{\Sigma})$ with $\boldsymbol{\Sigma}$ given in (3.6) as

$$f_{\mathbf{Y}}(\mathbf{y}; \mu, \sigma_a^2, \sigma_e^2) = \frac{\exp\left\{-\frac{1}{2}\left[\frac{SSe}{\sigma_e^2} + \frac{SSa}{\gamma_a} + \frac{An}{\gamma_a}(\bar{Y}_{\bullet\bullet} - \mu)\right]\right\}}{(2\pi)^{An/2}(\sigma_e^2)^{A(n-1)/2}\gamma_a^{A/2}}, \tag{3.16}$$

(where $\gamma_a := n\sigma_a^2 + \sigma_e^2$), which the reader is encouraged to verify. The details are provided in, e.g., Searle et al. (1992, Sec. 3.7) and Sahai and Ojeda (2004, p. 26–27). For minimal sufficiency and completeness, see, e.g., Graybill (1976). ■

The reader can confirm (3.16) by using it for the calculation of the log-likelihood in the program in Listing (3.1). From this result, and recalling that the m.l.e. is a function of the sufficient statistics (see, e.g., Section III.7.1.2), one might expect that the m.l.e. can be algebraically expressed in terms of $SS\mu$, SSa, and SSe, which is indeed the case, as given below.

3.1.3 Point Estimation, Interval Estimation, and Significance Testing

Observe that, from the values of *EMS* in Table 3.1, comparing the magnitudes of *MSa* and *MSe* appears pertinent for assessing if $\sigma_a^2 > 0$. From the independence of the *SS*, the distribution of their ratio is tractable, and leads to

$$\frac{\frac{SSa}{\gamma_a}/(A-1)}{\frac{SSe}{\sigma_e^2}/A(n-1)} \sim F_{(A-1),A(n-1)} \quad \text{or} \quad F_a := \frac{MSa}{MSe} \sim \frac{\gamma_a}{\sigma_e^2} F_{A-1,A(n-1)}, \tag{3.17}$$

a scaled central F distribution, where, again, $\gamma_a := n\sigma_a^2 + \sigma_e^2$. If $\sigma_a^2 = 0$, then $\gamma_a = \sigma_e^2$ (and $\gamma_a/\sigma_e^2 = 1$), so that an α-level hypothesis test for $\sigma_a^2 = 0$ versus $\sigma_a^2 > 0$ rejects if $F_a > F^{\alpha}_{A-1,A(n-1)}$, where $F^{\alpha}_{n,d}$ is the $100(1-\alpha)$th percent quantile of the $F_{n,d}$ distribution.

There are several useful point estimators of σ_e^2 and σ_a^2, including the method of maximum likelihood, as shown in Section 3.1.1. Others include the "ANOVA method" (see below), restricted m.l.e. (denoted REML, the most recommended in practice and the default of software such as SAS), and Bayesian methods. Discussions and comparisons of these methods can be found in, e.g., Searle et al. (1992), Miller Jr. (1997), Sahai and Ojeda (2004), and Christensen (2011).

We demonstrate the easiest of these, which is also referred to as the "ANOVA method of estimation" (Searle et al., 1992, p. 59) and amounts to equating observed and expected sums of squares. From (3.15) and (3.14), $\mathbb{E}[MSe] = \sigma_e^2$ and $\mathbb{E}[MSa] = n\sigma_a^2 + \sigma_e^2$, so that

$$\hat{\sigma}_e^2 = MSe = \frac{SSe}{A(n-1)}, \quad \text{and} \quad \hat{\sigma}_a^2 = \frac{1}{n}(MSa - MSe) = \frac{1}{n}\left(\frac{SSa}{A-1} - \frac{SSe}{A(n-1)}\right) \tag{3.18}$$

yield unbiased estimators. Observe, however, that $\hat{\sigma}_a^2$ in (3.18) can be negative. That (3.18) is not the m.l.e. is then intuitively obvious because the likelihood is not defined for non-positive σ_a^2. We will see below in (3.21) that $\hat{\sigma}_e^2$ is indeed the m.l.e., and $\hat{\sigma}_a^2$ is nearly so. To calculate the probability that $\hat{\sigma}_a^2 < 0$, use (3.17) to obtain

$$\Pr(\hat{\sigma}_a^2 < 0) = \Pr(MSa < MSe) = \Pr\left(\frac{MSa}{MSe} < 1\right) = \Pr\left(F_{A-1,A(n-1)} < \frac{\sigma_e^2}{n\sigma_a^2 + \sigma_e^2}\right).$$

Searle et al. (1992, p. 66–69) and Lee and Khuri (2001) provide a detailed discussion of how the sample sizes and true values of the variance components influence $\Pr(\hat{\sigma}_a^2 < 0)$. In practice, in the case where $\hat{\sigma}_a^2 < 0$, one typically reports that $\sigma_a^2 = 0$, though formally the estimator ${}_+\sigma_a^2 = \max(0, \hat{\sigma}_a^2)$ is biased—an annoying fact for hardcore frequentists. Realistically, it serves as an indication that the model might be mis-specified, or a larger sample is required.

Note that, from Theorem 3.2 and (3.18),

$$\frac{SSe}{\sigma_e^2} \sim \chi^2_{A(n-1)} \quad \text{and} \quad \hat{\sigma}_e^2 = MSe = \frac{SSe}{A(n-1)},$$

from which it follows that

$$\text{Var}(\hat{\sigma}_e^2) = \frac{\text{Var}(SSe)}{A^2(n-1)^2} = \frac{1}{A^2(n-1)^2}\text{Var}\left(\frac{\sigma_e^2}{\sigma_e^2}SSe\right) = \frac{2A(n-1)}{A^2(n-1)^2}\sigma_e^4 = \frac{2\sigma_e^4}{A(n-1)}. \tag{3.19}$$

Similarly, with $\gamma_a = n\sigma_a^2 + \sigma_e^2$, as

$$\frac{SSa}{\gamma_a} \sim \chi^2_{A-1} \quad \text{and} \quad \hat{\sigma}_a^2 = \frac{MSa - MSe}{n} = \frac{1}{n}\left(\frac{SSa}{A-1} - \frac{SSe}{A(n-1)}\right),$$

and the independence of SSa and SSe, we have

$$\begin{aligned}\mathrm{Var}(\hat{\sigma}_a^2) &= \frac{1}{n^2}\left[\frac{\mathrm{Var}\,(SSa)}{(A-1)^2} + \frac{\mathrm{Var}(SSe)}{A^2(n-1)^2}\right] = \frac{1}{n^2}\left[\frac{\gamma_a^2\mathrm{Var}(SSa/\gamma_a)}{(A-1)^2} + \frac{\sigma_e^4\mathrm{Var}(SSe/\sigma_e^2)}{A^2(n-1)^2}\right] \\ &= \frac{1}{n^2}\left[\frac{\gamma_a^2 2(A-1)}{(A-1)^2} + \frac{\sigma_e^4 2A(n-1)}{A^2(n-1)^2}\right] = \frac{2}{n^2}\left[\frac{(n\sigma_a^2+\sigma_e^2)^2}{(A-1)} + \frac{\sigma_e^4}{A(n-1)}\right]. \end{aligned} \tag{3.20}$$

Replacing σ_a^2 and σ_e^2 by their point estimates and taking square roots, these expressions yield approximations to the standard error of $\hat{\sigma}_e^2$ and $\hat{\sigma}_a^2$, respectively (as were given in Scheffé, 1959, p. 228; see also Searle et al., 1992, p. 85), and can be used to form Wald confidence intervals for the parameters. These could be compared to the numerically obtained standard errors based on maximum likelihood estimation. The reader is invited to show that $\mathrm{Cov}(\hat{\sigma}_a^2, \hat{\sigma}_e^2) = -2\sigma_e^4/(An(n-1))$.

By equating the partial derivatives of the log-likelihood $\ell(\mu, \sigma_a^2, \sigma_e^2; \mathbf{y}) = \log f_{\mathbf{Y}}(\mathbf{y}; \mu, \sigma_a^2, \sigma_e^2)$ given in (3.16) to zero and solving, one obtains (see, e.g., Searle et al., 1992, p. 80; or Sahai and Ojeda, 2004, p. 35–36)

$$\hat{\mu}_{\mathrm{ML}} = \bar{Y}_{\bullet\bullet}, \quad \hat{\sigma}^2_{e,\mathrm{ML}} = \frac{SSe}{A(n-1)} = MSe, \quad \hat{\sigma}^2_{a,\mathrm{ML}} = \frac{1}{n}\left(\frac{SSa}{A} - \frac{SSe}{A(n-1)}\right), \tag{3.21}$$

provided $\hat{\sigma}^2_{a,\mathrm{ML}} > 0$. The reader is encouraged to numerically confirm this, which is very easy, using the codes in Listings 3.1 and 3.2.

Comparing (3.21) to (3.18), we see that the ANOVA method and the m.l.e. agree for $\hat{\sigma}_e^2$, and are nearly identical for $\hat{\sigma}_a^2$. The divisor of A in $\hat{\sigma}^2_{a,\mathrm{ML}}$ instead of $A-1$ from the ANOVA method implies a shrinkage towards zero. Recall in the i.i.d. setting for the estimators of variance σ^2, the m.l.e. has a divisor of (sample size) n, while the unbiased version uses $n-1$, and that the m.l.e. has a lower mean squared error. This also holds in the one-way REM setting here, i.e., $\mathrm{mse}(\hat{\sigma}^2_{a,\mathrm{ML}}) < \mathrm{mse}(\hat{\sigma}_a^2)$; see, e.g., Sahai and Ojeda (2004, Sec. 2.7) and the references therein.

We now turn to confidence intervals. Besides the Wald intervals, further interval estimators for the variance components (and various functions of them) are available. Recall (from, e.g., Chapter III.8) that a **pivotal quantity**, or **pivot**, is a function of the data and one or more (fixed but unknown model) parameters, but such that its distribution does not depend on any unknown model parameters. From (3.10),

$$Q(\mathbf{Y}, \sigma_e^2) = \frac{SSe}{\sigma_e^2} \sim \chi^2_{A(n-1)}$$

is a pivot, so that a $100(1-\alpha)\%$ confidence interval (c.i.) for the error variance σ_e^2 is

$$\Pr\left(l \leqslant \frac{SSe}{\sigma_e^2} \leqslant u\right) = \Pr\left(\frac{SSe}{u} \leqslant \sigma_e^2 \leqslant \frac{SSe}{l}\right), \tag{3.22}$$

where $\Pr(l \leqslant \chi^2_{A(n-1)} \leqslant u) = 1-\alpha$ and α is a chosen tail probability, typically 0.05.

Likewise, from (3.17) with $F_a = MSa/MSe$, $(\sigma_e^2/(n\sigma_a^2+\sigma_e^2))F_a \sim F_{A-1,A(n-1)}$, so that

$$\begin{aligned} 1-\alpha &= \Pr\left(\frac{L}{F_a} \leqslant \frac{\sigma_e^2}{n\sigma_a^2+\sigma_e^2} \leqslant \frac{U}{F_a}\right) \\ &= \Pr\left(\frac{F_a}{U} \leqslant 1 + n\frac{\sigma_a^2}{\sigma_e^2} \leqslant \frac{F_a}{L}\right) = \Pr\left(\frac{F_a/U-1}{n} \leqslant \frac{\sigma_a^2}{\sigma_e^2} \leqslant \frac{F_a/L-1}{n}\right), \end{aligned}$$

where L and U are given by $\Pr(L \leqslant F_{A-1,A(n-1)} \leqslant U) = 1-\alpha$.

Of particular interest is a confidence interval for the **intraclass correlation coefficient**, given by $\sigma_a^2/(\sigma_a^2+\sigma_e^2)$. Taking reciprocals in the c.i. for σ_a^2/σ_e^2 gives

$$\begin{aligned}\Pr\left(\frac{n}{F_a/L-1} \leqslant \frac{\sigma_e^2}{\sigma_a^2} \leqslant \frac{n}{F_a/U-1}\right) &= \Pr\left(1+\frac{n}{F_a/L-1} \leqslant \frac{\sigma_a^2+\sigma_e^2}{\sigma_a^2} \leqslant 1+\frac{n}{F_a/U-1}\right)\\ &= \Pr\left(\frac{1}{1+\frac{n}{F_a/U-1}} \leqslant \frac{\sigma_a^2}{\sigma_a^2+\sigma_e^2} \leqslant \frac{1}{1+\frac{n}{F_a/L-1}}\right) = 1-\alpha,\end{aligned}$$

or

$$1-\alpha = \Pr\left(\frac{F_a/U-1}{F_a/U-1+n} \leqslant \frac{\sigma_a^2}{\sigma_a^2+\sigma_e^2} \leqslant \frac{F_a/L-1}{F_a/L-1+n}\right), \tag{3.23}$$

where $F_a = MSa/MSe$ and L and U are given by $\Pr(L \leqslant F_{A-1,A(n-1)} \leqslant U) = 1-\alpha$.

It turns out that a pivot and, thus, an exact confidence interval for the intra-class covariance σ_a^2 is not available. One obvious approximation is to replace σ_e^2 with $\hat{\sigma}_e^2$ in the c.i. for σ_a^2/σ_e^2 to get

$$1-\alpha \approx \Pr\left(\hat{\sigma}_e^2\frac{F_a/U-1}{n} \leqslant \sigma_a^2 \leqslant \hat{\sigma}_e^2\frac{F_a/L-1}{n}\right), \tag{3.24}$$

which (perhaps obviously) performs well if An is large (Stapleton, 1995, p. 286), in which case $\hat{\sigma}_e^2 \to \sigma_e^2$. We saw in Section 3.1.1 that, when A is large, the Wald c.i. based on the m.l.e. will also be accurate. A more popular approximation than (3.24) due to Williams (1962) is

$$1-2\alpha \approx \Pr\left(\frac{SSa(1-U/F_a)}{nu^*} \leqslant \sigma_a^2 \leqslant \frac{SSa(1-L/F_a)}{nl^*}\right), \tag{3.25}$$

where u^* and l^* are such that $\Pr(l^* \leqslant \chi^2_{A-1} \leqslant u^*) = 1-\alpha$. See also Graybill (1976, p. 618–620) for derivation.

The reader is encouraged to compare the empirical coverage probabilities of these intervals to use of their asymptotically valid Wald counterparts from use of the m.l.e. and recalling that, for function $\boldsymbol{\tau}(\boldsymbol{\theta}) = (\tau_1(\boldsymbol{\theta}), \ldots, \tau_m(\boldsymbol{\theta}))'$ from $\mathbb{R}^k \to \mathbb{R}^m$,

$$\boldsymbol{\tau}(\widehat{\boldsymbol{\theta}}_{\mathrm{ML}}) \overset{\text{asy}}{\sim} \mathrm{N}(\boldsymbol{\tau}(\boldsymbol{\theta}),\ \dot{\boldsymbol{\tau}}\mathbf{J}^{-1}\dot{\boldsymbol{\tau}}'), \tag{3.26}$$

where $\dot{\boldsymbol{\tau}} = \dot{\boldsymbol{\tau}}(\boldsymbol{\theta})$ denotes the matrix with (i,j)th element $\partial\tau_i(\boldsymbol{\theta})/\partial\theta_j$ (see, e.g., Section III.3.1.4). In this case, the c.i. is formed using an **asymptotic pivot**.

The test for $\sigma_a^2 > 0$ is rather robust against leptokurtic or asymmetric alternatives, while the c.i.s for the variance components and their ratios are, unfortunately, quite sensitive to departures from normality. Miller Jr. (1997, p. 105–107) gives a discussion of the effects of non-normality on some of the hypothesis tests and confidence intervals.

3.1.4 Satterthwaite's Method

We have seen three ways of generating a c.i. for σ_a^2, namely via the generally applicable and asymptotically valid Wald interval based on the m.l.e. and its approximate standard error (resulting from either the approximate Hessian matrix output from the BFGS algorithm or use of (3.19) and (3.20)), and use of (3.24) and (3.25). A further approximate method makes use of a result due to Satterthwaite

(1946), and can also be applied quite generally for hypothesis testing and c.i.s in higher-order random and mixed models, often with better actual coverage probability than Wald. We now detail what is commonly referred to as **Satterthwaite's method**.

Throughout, we will let γ_i denote a weighted sum of variance components such that mean square M_i is an unbiased estimator of γ_i, i.e., $\mathbb{E}[M_i] = \gamma_i$. Interest in general centers on deriving an approximate c.i. for

$$\gamma = \sum_{i=1}^{k} h_i \gamma_i, \tag{3.27}$$

where the h_i, $i = 1, \ldots, k$, are a fixed set of coefficients. For the one-factor model of this section with $\gamma = \sigma_a^2$, and recalling (3.14) and (3.15), we let $\gamma_1 := \gamma_a = n\sigma_a^2 + \sigma_e^2$ and $\gamma_2 := \sigma_e^2$, and we want a c.i. for $\sigma_a^2 = h_1\gamma_1 + h_2\gamma_2$, with $h_1 = n^{-1}$ and $h_2 = -n^{-1}$.

Let $\{S_i\}$, $i = 1, \ldots, k$, denote a set of independent sum of squares values such that $S_i = d_i M_i$, where d_i and M_i are the corresponding degrees of freedom and mean squares, respectively. Then, with (3.13) serving as an example case, with $\hat{\gamma}_i := M_i = S_i/d_i$ and $\mathbb{E}[\hat{\gamma}_i] = \gamma_i$,

$$\frac{S_i}{\gamma_i} = \frac{d_i M_i}{\gamma_i} = \frac{d_i \hat{\gamma}_i}{\gamma_i} \sim \chi^2_{d_i}, \quad i = 1, \ldots, k.$$

The idea is that, as $d_i\hat{\gamma}_i/\gamma_i \sim \chi^2_{d_i}$, perhaps there is a value $d > 0$ such that the distribution of the weighted sum $d\hat{\gamma}/\gamma$ can be adequately approximated as χ^2_d, i.e.,

$$W := \frac{d\hat{\gamma}}{\gamma} \overset{\text{app}}{\sim} \chi^2_d, \quad \text{where} \quad \hat{\gamma} := \sum_{i=1}^{k} h_i \hat{\gamma}_i = \sum_{i=1}^{k} \frac{h_i S_i}{d_i}. \tag{3.28}$$

If the approximation is accurate, then, for l and u such that $1 - \alpha = \Pr(l \leqslant \chi^2_d \leqslant u)$,

$$1 - \alpha \approx \Pr(l \leqslant W \leqslant u) = \Pr\left(\frac{d\hat{\gamma}}{u} \leqslant \gamma \leqslant \frac{d\hat{\gamma}}{l}\right). \tag{3.29}$$

The first moment does not give information about the choice of d: As $\mathbb{E}[\hat{\gamma}_i] = \gamma_i$ and recalling (3.27), note that, for any $d > 0$,

$$\mathbb{E}[W] = \frac{d}{\gamma}\mathbb{E}[\hat{\gamma}] = \frac{d}{\gamma}\sum_{i=1}^{k}\mathbb{E}[h_i\hat{\gamma}_i] = \frac{d}{\gamma}\sum_{i=1}^{k} h_i\gamma_i = d = \mathbb{E}[\chi^2_d].$$

Using second moments, $\mathrm{Var}(S_i) = 2\gamma_i^2 d_i$ and

$$\mathrm{Var}(W) = 2\frac{d^2}{\gamma^2}\sum_{i=1}^{k}\frac{h_i^2\gamma_i^2}{d_i},$$

so equating $\mathrm{Var}(\chi^2_d) = 2d$ to $\mathrm{Var}(W)$ and solving for d yields

$$d = \frac{\gamma^2}{\sum_{i=1}^{k} h_i^2\gamma_i^2/d_i} = \frac{\left(\sum_{i=1}^{k} h_i\gamma_i\right)^2}{\sum_{i=1}^{k} h_i^2\gamma_i^2/d_i}, \tag{3.30}$$

which is clearly non-negative. To make (3.30) operational, one uses the observed mean square values, i.e.,

$$\hat{d} = \frac{\left(\sum_{i=1}^{k} h_i \hat{\gamma}_i\right)^2}{\sum_{i=1}^{k} h_i^2 \hat{\gamma}_i^2 / d_i} > 0. \tag{3.31}$$

For the approximate c.i. on σ_a^2, if $\gamma_1 := \gamma_a = n\sigma_a^2 + \sigma_e^2$ and $\gamma_2 := \sigma_e^2$, then $\hat{\gamma}_1 = S_1/d_1 = MSa$ and $\hat{\gamma}_2 = S_2/d_2 = MSe$, where $S_1 = SSa$, $d_1 = A - 1$, $S_2 = SSe$, and $d_2 = A(n-1)$. Notice that $S_1/\gamma_1 \sim \chi^2_{A-1}$ independent of $S_2/\gamma_2 \sim \chi^2_{A(n-1)}$ from (3.10), so that we have the general setup above with $k = 2$ and desire a c.i. for

$$\gamma = \sigma_a^2 = (\gamma_a - \sigma_e^2)/n = n^{-1}\gamma_1 - n^{-1}\gamma_2 = \sum_{i=1}^{2} h_i \gamma_i,$$

with $h_1 = n^{-1}$ and $h_2 = -n^{-1}$. Thus, from (3.29), replacing d with $\hat{d}$ from (3.31) as

$$\hat{d} = \frac{(h_1\hat{\gamma}_1 + h_2\hat{\gamma}_2)^2}{h_1^2\hat{\gamma}_1^2/d_1 + h_2^2\hat{\gamma}_2^2/d_2} = \frac{(\hat{\gamma}_1 - \hat{\gamma}_2)^2}{\hat{\gamma}_1^2/d_1 + \hat{\gamma}_2^2/d_2} = \frac{(MSa - MSe)^2}{\frac{(MSa)^2}{A-1} + \frac{(MSe)^2}{A(n-1)}} = \frac{n^2\hat{\sigma}_a^4}{\frac{(\hat{\sigma}_a^2+\hat{\sigma}_e^2)^2}{A-1} + \frac{\hat{\sigma}_e^4}{A(n-1)}},$$

an approximate $100(1-\alpha)\%$ c.i. for σ_a^2 is

$$\hat{d}\frac{(MSa - MSe)}{n\,u} \leqslant \sigma_a^2 \leqslant \hat{d}\frac{(MSa - MSe)}{n\,l}, \tag{3.32}$$

and $1 - \alpha = \Pr(l \leqslant \chi^2_{\hat{d}} \leqslant u)$. If $MSa \leqslant MSe$, the suggested interval is clearly of no use.

Recalling (3.18) and multiplying the terms in (3.32) by n, (3.32) can be written as

$$\hat{d}\frac{(MSa - MSe)}{u} \leqslant \mathbb{E}[MSa - MSe] \leqslant \hat{d}\frac{(MSa - MSe)}{l},$$

inspiring one to consider if, in general, with M_i denoting a mean square, an approximate interval for $\sum_{i=1}^{k} h_i M_i$ might be given by

$$\hat{d}\frac{\sum_{i=1}^{k} h_i M_i}{u} \leqslant \mathbb{E}\left[\sum_{i=1}^{k} h_i M_i\right] \leqslant \hat{d}\frac{\sum_{i=1}^{k} h_i M_i}{l}, \tag{3.33}$$

where l and u are given by $1 - \alpha = \Pr(l \leqslant \chi^2_{\hat{d}} \leqslant u)$ and

$$\hat{d} = \frac{\left(\sum_{i=1}^{k} h_i M_i\right)^2}{\sum_{i=1}^{k} h_i^2 M_i^2 / d_i}. \tag{3.34}$$

This is indeed the case when the M_i are mean squares such that $SSi/\mathbb{E}[MSi] = d_i MSi/\mathbb{E}[MSi] \sim \chi^2_{d_i}$, a central chi-square with d_i degrees of freedom, and the SSi are independent from one another. In the one factor REM, this is satisfied because $SSa/\mathbb{E}[MSa] \sim \chi^2_{d_a}$, independent of $SSe/\mathbb{E}[MSe] \sim \chi^2_{d_e}$, for $d_a = A - 1$ and $d_e = A(N-1)$. Under such conditions, $\sum h_i S_i/d_i = \sum h_i MSi$, a weighted sum of independent mean squares, and $\gamma = \sum h_i \gamma_i = \sum h_i \mathbb{E}[MSi]$, so that (3.29) can be written as (3.33). For $\hat{d}$, as $\mathbb{E}[MSi] = \gamma_i$ and $\hat{\gamma}_i = MSi$, (3.31) and (3.34) are also equivalent.

Finally, it can be shown that the conditions on the M_i are satisfied when they refer to the mean squares of random effects in balanced models. In mixed models, the SSi corresponding to the fixed effects (like $SS\mu$ in the one-factor REM, for instance) are distributed as multiples of noncentral chi-squares, while in unbalanced designs SSi is distributed as a weighted sum of chi-squares if the ith variance component σ_i is nonzero.

Remark It is important to note that negative values of one or more of the weights h_i imply that $\Pr(W < 0) > 0$ so that a chi-square (or any positive) approximation may be poor. Some of the Satterthwaite approximate intervals arising in practice are such that one or more of the h_i are, in fact, negative. This issue was addressed in Butler and Paolella (2002b) using (i) single bootstrap-based inference, (ii) a saddlepoint approximation to the relevant sums of χ^2 random variables arising in (3.28) based on the methods in Appendix A, and (iii) combining those two approaches to form a double bootstrap such that the inner bootstrap is replaced with the analytic (and thus far faster to calculate) saddlepoint approximation. The methods are both elegant and generally applicable, and can be compared to the variety of model-specific (and occasionally cumbersome) methods developed in Burdick and Graybill (1992).

Using two model classes (the three-way crossed model of Section 3.2.2 and the two-way nested model of Section 3.3.1.1) and a variety of parameter constellations, Butler and Paolella (2002b) demonstrate that, for small sample sizes, all three proposed methods result in more accurate actual confidence interval coverage of σ_a^2/σ_e^2 compared to the use of Satterthwaite, with the double bootstrap method being, unsurprisingly, the most accurate. ■

3.1.5 Use of SAS

Listings 1.7, 2.3, 2.10, and 2.11 showed various ways to output data generated in Matlab to a text file for subsequent reading into SAS. We do this again, building on the code in Listing 3.2, resulting in Listing 3.3 (and note that SAS could equally be used to generate data, as the interested student should pursue).

For one particular simulated data set, use of maximum likelihood via the code in Listing 3.1 yielded $\hat{\mu} = 4.8871$, $\hat{\sigma}_a^2 = 0.38457$, and $\hat{\sigma}_e^2 = 0.90377$, and produced a log-likelihood of -430.5. This data was then read into SAS and analyzed with their `proc varcomp`, as shown in SAS Listing 3.1. The output (not shown here) yields the same m.l.e. values to all shown significant digits. Using the ANOVA method of estimation (engaged in SAS using `method=type1`) yielded $\hat{\sigma}_a^2 = 0.40798$, and (the same as the m.l.e., as the theory suggests) $\hat{\sigma}_e^2 = 0.90377$. Using this method, SAS can also generate confidence intervals for the variance components with `proc varcomp`.

```
A=20; n=15; mu=5; sigma2a=0.4; sigma2e=0.8; muv=ones(A*n,1)*mu; J=ones(n,n);
tmp=sigma2a*J+sigma2e*eye(n); Sigma=kron(eye(A),tmp); y=mvnrnd(muv,Sigma,1);
school = kron( (1:A)' , ones(n,1) ); Out=[y' school];
fname='REM1A20n15.txt'; if exist(fname,'file'), delete(fname), end
fileID = fopen(fname,'w');
fprintf(fileID,'%8.5f %4u\r\n',Out'); fclose(fileID);
```

Program Listing 3.3: Generates and writes to a text file a one-way REM data set and the associated class variable, for input into SAS.

```
ods html close; ods html;
/* clear and close output window, open new */
filename ein 'REM1A20n15.txt';
data school;
  infile ein stopover;  input Y school;
run;
title 'REM 1 Way Example with A=20, n=15';
proc varcomp method=ml;
  class school;  model Y=school;
run;
proc varcomp method=type1;
  class school;  model Y=school / cl;
run;
```

SAS Program Listing 3.1: Reads in the data from the text file generated in Listing 3.3 and uses `proc varcomp` with maximum likelihood and the ANOVA method of estimation, the latter allowing for computation of confidence intervals.

```
proc mixed method=ml cl=wald nobound covtest;
  class school;
  model Y= / cl solution;
  random school;
run;
```

SAS Program Listing 3.2: Similar to Listing 3.1 but uses `proc mixed`. In the `model` statement, one lists only the fixed effects, and in this case there are none (besides the grand mean, which is used by default), while the `random` statement indicates the random effects.

Listing 3.2 shows how to conduct the same analysis using the more advanced and subsuming `proc mixed`. The latter also outputs (−2 times) the log-likelihood and the estimate of μ, and these agree with the Matlab output mentioned above. The point estimates of σ_a^2 and σ_e^2 are the same as those given above when using maximum likelihood and Wald confidence intervals are also output, ignoring the lower bound of zero by specifying the option `nobound`. In general, with mixed models (those containing both fixed and random effects besides the grand mean and the error term), `proc mixed` should be used instead of `proc glm` in SAS. See, e.g., Yang (2010) and the references therein for a clear discussion of the differences and the erroneous inference that could be obtained by using the latter.

3.1.6 Approximate Inference in the Unbalanced Case

With unbalanced data, the elegant model representation (3.5) and (3.6), and the subsequent simple distribution theory and point estimators, confidence intervals, and test statistics, are no longer applicable. To address this case, we take a simple, approximate approach, using "first principles" regarding the likelihood in the case that the extent of the unbalance is not large, e.g., the experiment was planned with balance, but a small number of cases could not be realized (exams got lost, test tubes broke, rats escaped, etc.). Sections 3.1.6.1 and 3.1.6.2 address point and interval estimation, respectively.

Our approximation (i) avoids having to construct the exact likelihood in the unbalanced case, (ii) is direct and easy to implement and applicable to all random effects models, (iii) leads to further insights, and (iv) is in line with the goals and scope of this book, namely to encourage the reader to think on his/her own, using existing first-principle skills. This is of course no replacement for a full, rigorous study of the unbalanced case, and the interested reader is directed to the references given in the introduction to this chapter for a detailed (but necessarily longer and more complicated) analysis.

3.1.6.1 Point Estimation in the Unbalanced Case

Notice that, whether balanced or not, the distribution of $\mathbf{Y}$ for any (Gaussian) random effects model is still multivariate normal. For the one-way REM, this is determined by (3.2) and (3.3), so that construction of $\boldsymbol{\Sigma}$ is not unwieldy, and the reader is encouraged to express the likelihood and design a program similar to that in Listing 3.1 to compute the m.l.e.

Our approach is to *treat the missing observations as parameters to be estimated jointly with the model parameters* μ, σ_a^2, and σ_e^2, and, when available, use the balanced-case closed-form m.l.e. expression of the latter. For the one-way REM case, the closed-form m.l.e. is given in (3.21). With closed-form m.l.e. expressions available in the balanced case, the likelihood is **concentrated**, such that numerical searching needs to take place only over the missing values. This procedure will not yield the true m.l.e. of μ, σ_a^2, and σ_e^2, as can easily be seen in a simpler case: Imagine data $X_i \overset{\text{i.i.d.}}{\sim} \mathrm{N}(\mu, \sigma^2)$, $i = 1, \dots, n$, such that X_i, $i = 1, \dots, k$, $1 \leqslant k < n$, are missing, and one applies this estimation method to obtain $\hat{\mu}$, $\hat{\sigma}^2$, and **imputations** $\{\widehat{X}_i\}_{i=1}^k$. Clearly, the latter and $\hat{\mu}$ will be equal to the mean of the available data. Using the closed-form m.l.e. solution to $\hat{\sigma}^2$ based on the data set augmented with the imputed values, not only is the sample size overstated, but also the imputed values are all constant and equal to the mean of the observed data, so that $\hat{\sigma}^2$ will be underestimated.

In the context of the one-way REM, we would thus expect that $\hat{\sigma}_e^2$ using this method will be smaller than the true m.l.e. Indeed, we will subsequently see that the estimates of μ and σ_a^2 are nearly the same as the true m.l.e. (often to four decimal places) in our experiments with $A = 20$, while the estimate of σ_e^2 is close to the m.l.e. and appears to be off by a scaling factor greater than one that can be approximated as a function of the cell sizes n_i, $i = 1, \dots, A$.

To generate the data, we start with a balanced panel and then replace some observations with Matlab's "not a number" designator NaN. Irrespective of where they are and how many there are, it turns out that it is very easy and elegant in Matlab to perform the likelihood maximization, as shown in the program in Listing 3.4. A simple and close starting value for all missing observations is the mean over the available observations. Notice also that no bounds need to be imposed on the missing values, making the code yet simpler. If more than one observation in a cell is missing (e.g., $Y_{1,1}$ and $Y_{1,2}$), we find, unsurprisingly, that the m.l.e.s of the missing values in that cell are always the same, but they do differ across cells. The m.l.e. is, unfortunately, not the simple mean of the observed cell values. However, a simple, closed-form expression is indeed available for the m.l.e. point estimators of the missing data, and is discussed and used below, but we first proceed naively, as often is the case as research unfolds.

The code in Listing 3.5 generates an unbalanced data set (this being equivalent to a balanced panel with missing values) and returns the point estimates of μ, σ_a^2, and σ_e^2 based on the approximate likelihood procedure. The code also writes the data to a text file, replacing NaN with a period, this being the designator for a missing value in SAS. Alternatively, one could simply omit the data lines corresponding to missing values—the analysis is the same in SAS. The code in SAS Listing 3.3 reads the text

```
function [param, V, stderr, loglik, iters] = REM1wayMLEMiss(y,A,n)
% param = {the set of missing data points}
ylen=length(y); if A*n ~= ylen, error('A and/or n wrong'), end
nmiss=sum(isnan(y)); inity=mean( y(~isnan(y)) );
initvec=ones(nmiss,1)*inity;
opts=optimset('Display','None','TolX',1e-5,'LargeScale','off');
[param,fval,~,theoutput,~,hess] = ...
   fminunc(@(param) REM1Miss_(param,y,A,n),initvec,opts);
V=inv(hess); param=param'; stderr=sqrt(diag(V))';
iters=theoutput.iterations; loglik=-fval;

function loglik=REM1Miss_(param,y,A,n)
% first fill in the missing values
loc=isnan(y); y(loc)=param; % that was perhaps easier than expected
% now compute SS based on the filled-in sample
SST=sum(y'*y); Yddb=mean(y);               SSu=A*n*Yddb^2;
H=kron(eye(A), ones(n,1)); Yidb=y'*H/n; SSa=n*sum( (Yidb-Yddb).^2 );
m=kron(Yidb', ones(n,1));                  SSe=sum((y-m).^2);
% Compute the MLE based on the filled-in sample
mu = mean(y); sigma2e = SSe/A/(n-1); sigma2a=(SSa/A-SSe/A/(n-1))/n;
% Finally, compute the log-likelihood
muv=ones(A*n,1)*mu; J=ones(n,n); tmp=sigma2a*J+sigma2e*eye(n);
Sigma=kron(eye(A),tmp); loglik=-log(mvnpdf(y,muv,Sigma));
```

Program Listing 3.4: Maximum likelihood estimation of the missing values (denoted in Matlab as NaN) causing the unbalance in a one-way REM, using the closed-form m.l.e. (3.21) of the model parameters based on the sums of squares in the imputed balanced model. Assumes $An \times 1$ vector $\mathbf{y}$, with entries for missing values as NaN, is in lexicon order (3.4). See Listing 3.5 for generating the data and calling `REM1wayMLEMiss`.

data file and applies `proc mixed` using the option to estimate the parameters using the (true) m.l.e. Observe how the SAS code is the same—no indication of balance or unbalance needs to be specified by the user.

Doing so with three missing values (via lines 11–13 in Listing 3.5) shows that $\hat{\mu}$ and $\hat{\sigma}_a^2$ are the same (to the digits shown by SAS), while the estimates for $\hat{\sigma}_e^2$ based on our approximate likelihood method, and SAS, differ slightly, with the latter being not only larger in all of runs attempted, but such that the ratio of the SAS value to ours was always about $1.0107 \approx An/(An-3) = 1.0101$, noting again that we use three missing values. Dividing n by the harmonic mean of the n_i of each cell produces $15/[20/(18/15 + 1/14 + 1/13)] = 1.0113$, and taking averages of these two yields, interestingly, 1.0107.

Repeating the exercise with 10 missing values (two from cells $i = 1$ and $i = 11$, and one from cells 2, 4, 8, 12, 14, and 18) reveals a similar pattern: The estimates $\hat{\mu}$ and $\hat{\sigma}_a^2$ from the two methods are the same, while that of $\hat{\sigma}_e^2$ from SAS is always about a factor 1.0368 higher. Indeed, $15/[20/(12/15 + 2/13 + 6/14)] = 1.0368$. As closed-form expressions for the true m.l.e. of the model parameters σ_a^2 and σ_e^2 are not available with unbalanced data (see, e.g., Searle et al., 1992, Ch. 6, for the general likelihood expression and the need for numeric optimization), it is not clear how this proportionality approximation can be justified or made rigorous. The interested reader is encouraged to investigate its viability for various A, n, and number of missing values.

```
1  % desired parameter constellation
2  A=20; n=15; mu=5; sigma2a=0.4; sigma2e=0.8;
3
4  % generate a balanced one-way REM
5  muv=ones(A*n,1)*mu; J=ones(n,n); tmp=sigma2a*J+sigma2e*eye(n);
6  Sigma=kron(eye(A),tmp); y=mvnrnd(muv,Sigma,1)';
7
8  % Now set some values to missing. The y vector is lexicon order.
9  Ymiss=[]; yoriginal=y; % save the original values if desired
10 % Take the following set of Y_{ij} entries as missing:
11 i=1; j=1; ind=n*(i-1)+j; Ymiss=[Ymiss y(ind)]; y(ind)=NaN;
12 i=1; j=2; ind=n*(i-1)+j; Ymiss=[Ymiss y(ind)]; y(ind)=NaN;
13 i=2; j=1; ind=n*(i-1)+j; Ymiss=[Ymiss y(ind)]; y(ind)=NaN;
14 % etc.
15 z=y; % keep the version with missing values
16
17 % estimate the missing values,
18 % using closed-form MLE for the model parameters
19 param = REM1wayMLEMiss(y,A,n);
20
21 % Replace the missing values by their imputed ones:
22 loc=isnan(y); y(loc)=param;
23
24 % compute the SS-values based on the imputed sample
25 SST=sum(y'*y); Yddb=mean(y); SSu=A*n*Yddb^2; % Yddb= \bar{Y}_{dot dot}
26 H=kron(eye(A), ones(n,1)); Yidb=y'*H/n; % Yidb= \bar{Y}_{i dot}
27 SSa=n*sum( (Yidb-Yddb).^2 );
28 m=kron(Yidb', ones(n,1)); SSe=sum( (y-m).^2 );
29
30 % compute the MLE based on the imputed sample
31 mu_hat_MLE = mean(y); sigma2e_hat_MLE = SSe/A/(n-1);
32 sigma2a_hat_MLE = ( SSa/A - SSe/A/(n-1) )/n;
33 MLE = [mu_hat_MLE sigma2a_hat_MLE sigma2e_hat_MLE]
34
35 % output the data to a text file for reading by SAS
36 school = kron( (1:A)' , ones(n,1) );
37 fname='REM1wayMissing.txt';
38 if exist(fname,'file'), delete(fname), end
39 fileID = fopen(fname,'w');
40 for i=1:A*n
41   yout=z(i); sout=school(i);
42   if isnan(yout), ystr='.     '; % dot and 5 spaces
43   else ystr=num2str(yout,'%8.4f');
44   end
45   sstr=num2str(sout,'%3u');
46   str=[ystr,' ',sstr]; fprintf(fileID,'%s\r\n',str);
47 end
48 fclose(fileID);
```

Program Listing 3.5: First generates a one-way REM data set and sets some values to missing (NaN in Matlab), recalling the lexicon ordering of the observation vector **y** given in (3.4). Then, via the program `REM1wayMLEMiss` in Listing 3.4, estimates the model, treating as unknown parameters the missing values and the actual model parameters μ, σ_a^2 and σ_e^2. The concentrated likelihood is used such that the latter set of parameters are algebraically given by the closed-form m.l.e. expression (3.21). Note that lines 4–6 and 23–31 are the same as those in Listing 3.2. Finally, the data are written to a text file using a "." instead of NaN for missing values, as used by SAS.

```
ods html close; ods html;
/* clear and close output window, open new */
filename ein 'REM1wayMissing.txt';
data school;
  infile ein stopover;  input Y school;
run;
title 'Unbalanced REM 1 Way Example';
proc mixed method=ml;
  class school;  model Y= / cl solution;  random school;
run;
```

SAS Program Listing 3.3: Reads in the unbalanced data from the text file generated in Listing 3.5 and uses `proc mixed` with maximum likelihood.

Figure 3.3 shows the small sample distribution of the estimators based on the approximate m.l.e. method, using $A = 20$, $n = 15$, $\sigma_a^2 = 0.4$, $\sigma_e^2 = 0.8$, and 10 missing values, and having applied the multiplicative factor 1.0368 to $\hat{\sigma}_e^2$, so that the histograms essentially reflect the distribution of the true m.l.e. The plots can be compared to those in Figure 3.1, which were based on the full, balanced panel for the same parameter constellation.

This approximate method could be applied to any random (or mixed) effects model such that the m.l.e. is available in closed form in the balanced case. This is also the case for the two-factor nested model discussed in Section 3.3.1. In the case that a closed-form expression for the m.l.e. is not available or unknown to the researcher, expressing the $\boldsymbol{\Sigma}$ matrix and the likelihood in the balanced case is very straightforward, as was seen in (3.6) for the one-way model, and as will be demonstrated below for crossed and nested models in Sections 3.2 and 3.3, respectively, so that one could easily numerically *maximize the likelihood with respect to the model parameters and the missing values, jointly*. Observe how, using the one-factor REM as an example, this just entails combining aspects of the programs in Listings 3.1 and 3.4. We emphasize again that this does *not* result in the m.l.e. of the model parameters, with at least that of σ_e^2 being off, though possibly to first order by a simple scaling factor that is a function of the cell sizes (n_i in the one-way case, n_{ij} in the two-way case, etc.).

As alluded to above, it turns out that we can also forgo the numeric determination of the point estimates of the missing values. From the definition of the model in (3.1) and using (3.9), for a given i, $a_i \sim \mathrm{N}(0, \sigma_a^2)$, $\bar{Y}_{i\bullet} = \mu + a_i + \bar{e}_{i\bullet} \sim \mathrm{N}(\mu, \sigma_a^2 + \sigma_e^2/n)$, $\mathrm{Cov}(a_i, \bar{Y}_{i\bullet}) = \sigma_a^2$, and a_i and $\bar{Y}_{i\bullet}$ are jointly normally distributed as

$$\begin{bmatrix} a_i \\ \bar{Y}_{i\bullet} \end{bmatrix} \sim \mathrm{N}\left(\begin{bmatrix} 0 \\ \mu \end{bmatrix}, \begin{bmatrix} \sigma_a^2 & \sigma_a^2 \\ \sigma_a^2 & \sigma_a^2 + \sigma_e^2/n \end{bmatrix} \right).$$

Thus, conditionally (see, e.g., Section II.3.22),

$$(a_i \mid \bar{Y}_{i\bullet} = \bar{y}_{i\bullet}) \sim \mathrm{N}(R(\bar{y}_{i\bullet} - \mu),\ \sigma_a^2(1 - R)), \quad R := \frac{\sigma_a^2}{\sigma_a^2 + \sigma_e^2/n}. \tag{3.35}$$

For a particular j such that Y_{ij} is missing from the panel, a suggested predictor for $Y_{ij} = \mu + a_i + e_{ij}$ is then $\mathbb{E}[\mu + a_i + e_{ij} \mid \bar{y}_{i\bullet}]$, or, replacing unknown parameters with estimators,

$$\hat{\mu} + \frac{\hat{\sigma}_a^2}{\hat{\sigma}_a^2 + \hat{\sigma}_e^2/n}(\bar{y}_{i\bullet} - \hat{\mu}), \tag{3.36}$$

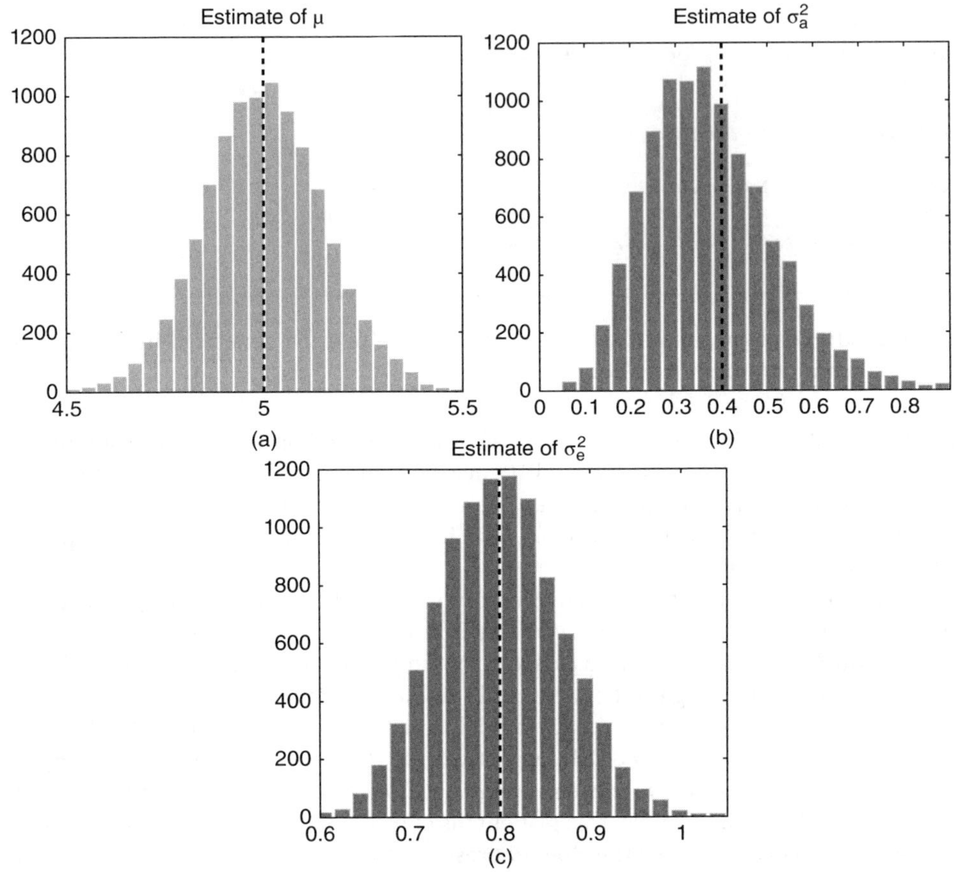

Figure 3.3 Similar to the top panels in Figure 3.1, namely histograms of the m.l.e. of the three parameters, from a) to c), μ, σ_a^2, and σ_e^2, of the one-way REM, based on $A = 20$, $n = 15$, and $S = 10{,}000$ replications, but such that 10 observations are missing, as shown in the code in Listing 3.5. These were obtained using the approximate m.l.e. method, and such that the obtained estimates for $\hat{\sigma}_e^2$ were multiplied by 1.0368. The vertical dashed line indicates the true value of the parameter in each graph.

where $\bar{y}_{i\bullet}$ is computed over the available $\{y_{ij}\}$ in the ith cell. This is referred to as the **best linear unbiased predictor**, or **BLUP**, a highly detailed discussion of which can be found in Searle et al. (1992, Ch. 7). When (3.35) is viewed as a likelihood, its maximum is at its expected value, explaining why the numerically determined optimal missing values coincide with (3.36).

Given an unbalanced data set such that the $An \times 1$ observation vector $\mathbf{Y}$ is in lexicon order (3.4), and the missing values causing the unbalance are indicated with NaN (such as simulated using lines 1–13 from Listing 3.5), code for computing the approximate m.l.e. of the one-way REM using BLUP for the missing values is given in Listing 3.6. Observe how we iterate between computing the BLUP imputed values and the parameter m.l.e. based on the balanced data, until convergence, and is thus similar to an expectation-maximization (EM) algorithm. Convergence occurs very quickly for the parameter constellations we used for demonstration, and is thus far faster than numeric optimization over the

```
function param=REM1wayMissMLEBLUP(y,A,n)
z=y(~isnan(y)); inity=mean(z); vv=var(z)/2;
mu=inity; sigma2e=vv; sigma2a=vv;
z=y; conv=0; tol=1e-5; maxit=20; iter=0; nivec=zeros(A,1);
while (~conv) && (iter<maxit), iter=iter+1; % disp(iter)
  for i=1:A % BLUP imputation
    indx=n*(i-1); icell=y((indx+1):(indx+n));  % the ith cell
    ii=~isnan(icell); cellbar=mean(icell(ii)); % ith cell mean
    ni=sum(ii); nivec(i)=ni; % keep track of the unbalance for later
    for j=1:n
      indx=n*(i-1)+j;
      blup = mu + (ni*sigma2a) / (sigma2e+ni*sigma2a) * (cellbar-mu);
      if isnan(y(indx)), z(indx)=blup; end
    end
  end
  % Update model parameters based on imputed vector z
  oldsigma2e=sigma2e; oldsigma2a=sigma2a; mu=mean(z);
  H=kron(eye(A), ones(n,1)); Yidb=z'*H/n; SSa=n*sum( (Yidb-mu).^2 );
  m=kron(Yidb', ones(n,1));                SSe=sum((z-m).^2);
  sigma2e = SSe/A/(n-1); sigma2a=(SSa/A-SSe/A/(n-1))/n;
  if sigma2a<=1e-3, sigma2a=1e-3; end
  conv=(abs(sigma2e-oldsigma2e)<tol) && (abs(sigma2a-oldsigma2a)<tol);
end
% apply harmonic adjustment factor to sigma^2_e
harm = A / sum(1./nivec); adj=n/harm; sigma2e=sigma2e*adj;
param=[mu sigma2a sigma2e];
```

Program Listing 3.6: Computes the approximate m.l.e. in the unbalanced case using the closed -orm balanced-case m.l.e. (3.21) and the BLUP for the missing values. Also, the *ad hoc* adjustment via the harmonic mean of the n_i is applied to $\hat{\sigma}_e^2$.

missing values, particularly as their number grows. Nevertheless, as elegant as this approach is, we repeat that it does not result in the true m.l.e.

Thus, this approximate method has some appeal, but unfortunately, even if the multiplicative correction factor for $\hat{\sigma}_e^2$ works for all sample size and parameter constellations, and also in other, higher-order models of interest, without a theory that dictates what it is (at least to first order), one requires the true m.l.e. to determine it (and determine if equality holds between the two estimation methods for the other variance components). Of course, simulation, as we did for Figure 3.3, could also be used, from which the multiplicative adjustment could be approximated based on minimizing the estimator's bias.

Realistically, in practice one uses canned statistical software packages that have the m.l.e. (and, more often used, the restricted m.l.e., or REML) reliably programmed for the general unbalanced case. The point of discussing the approximate m.l.e. method was to (i) illustrate what one could do for mildly unbalanced data with the availability of a closed-form m.l.e. for the balanced case, using basic likelihood principles and basic optimization in Matlab, and (ii) introduce the BLUP, and how it is not equal to simply the cell sample mean.

A good starting point for methods for confidence intervals of (functions of) variance components in REMss the book by Burdick and Graybill (1992), while Burch and Iyer (1997) and Lidong et al. (2008) derived further methods for certain models. A general treatment in random and mixed linear

Gaussian models, for unbalanced data, was developed in Cisewski and Hannig (2012). It is based on so-called **fiducial inference**, a concept dating back to Ronald Fisher (see the citations in Cisewski and Hannig, 2012), and such that a distribution on the parameter space is generated, as with Bayesian inference, but without requiring specification of a prior distribution.

3.1.6.2 Interval Estimation in the Unbalanced Case

We now turn to another use of the approximate likelihood method. Recall the exact confidence interval (3.23) for the intraclass correlation coefficient when the data are balanced. The codes in Listings 3.7 and 3.8 confirm that the actual coverage matches the nominal. In the unbalanced case, this no longer holds, though the interval could be computed by replacing the missing values by their m.l.e. estimates. Using the imputed data overstates the actual number of observations available, and so the resulting interval will tend to be too liberal, i.e., too short, having actual coverage less than the nominal. The reader is invited to check this, using the same programs, but such that in Listing 3.7, after line 5, a set of Y_{ij} values are set to NaN, say 20 of them, such that 0, 1, 2, 3, or 4 is removed from one of the $A = 20$ cells. With $n = 10$, this results in 10% missing values. Then, program `REM1wayMLEMiss` in Listing 3.4 can be used to estimate the missing values, say vector `miss`, and the **y** vector is then augmented with the imputed values with `loc=isnan(y); y(loc)=miss;` as in line 21 of Listing 3.5. Doing so with 10,000 replications and a nominal coverage of 0.90 resulted in an actual coverage of 0.860, which is indeed less than the nominal of 0.90, though not by much. Note that it is highly significantly below 0.90, based on the usual 95% Wald confidence interval for sums of Bernoulli trials. Repeating the exercise, again with 10,000 replications, but using $A = 10$, $n = 5$, and 10 missing values (20%), resulted in an actual coverage of only 0.762.

The method using the BLUP for imputation could also be used, and will be much faster. To do so, the program in Listing 3.6 would simply need to be augmented to output the estimates of the missing values. However, we chose to use the slower method of numerical searching for a reason: It also outputs the (approximate) variance-covariance matrix of the imputed values, and as their joint distribution is multivariate normal, the output fully describes their density. (The covariance matrix could be analytically determined: The variances are already given in (3.35). The numeric method saves us this effort, and is also applicable to higher-order models, where analytic calculations are less trivial.)

The idea is to simulate from this density, say $s_{\text{Miss}} = 1{,}000$ sets of missing values, and for each, impute the original data set to get a balanced panel, and compute the confidence interval with nominal coverage level $100(1-\alpha)\%$. This is very fast, as it just requires, for each of the s_{Miss} replications, simulating

```
1  A=20; n=10; mu=5; sigma2a=0.4; sigma2e=0.8;
2  ICC=sigma2a/(sigma2a+sigma2e);
3  muv=ones(A*n,1)*mu; J=ones(n,n);
4  tmp=sigma2a*J+sigma2e*eye(n); Sigma=kron(eye(A),tmp);
5  sim=1e3; cover=zeros(sim,1);
6  for loop=1:sim
7    y=mvnrnd(muv,Sigma,1)';
8    [lo,hi] = REM1wayCIforICC(y,A,n);
9    cover(loop) = (lo<ICC) && (ICC<hi);
10 end
11 empcov=mean(cover)
```

Program Listing 3.7: Confirms via simulation the equality of the actual and nominal coverage of the confidence interval (3.23) in the balanced case. Function `REM1wayCIforICC` is given in Listing 3.8.

```
function [lo,hi] = REM1wayCIforICC(y,A,n,alpha)
if nargin<4, alpha=0.10; end % 90% CI
Yddb=mean(y); H=kron(eye(A), ones(n,1)); Yidb=y'*H/n;
SSa=n*sum( (Yidb-Yddb).^2 ); m=kron(Yidb', ones(n,1)); SSe=sum((y-m).^2);
MSa=SSa/(A-1); MSe=SSe/A/(n-1); Fa=MSa/MSe;
L=finv(alpha/2, A-1, A*(n-1)); U=finv(1-alpha/2, A-1, A*(n-1));
lo=(Fa/U-1)/(Fa/U-1+n); hi=(Fa/L-1)/(Fa/L-1+n);
```

Program Listing 3.8: Computes the $100(1-\alpha)\%$ confidence interval (3.23) of the intraclass correlation coefficient for the balanced one-way REM model.

from a multivariate normal distribution and computing the simple function in Listing 3.8. Note that, if there were no missing values, then the resulting s_{Miss} intervals would all be identical. With unbalance, these intervals will all be different (and surely very close if the percentage of missing values is very small), and an idea for delivering a $100(1-\alpha)\%$ nominal c.i. is to take the q-quantile of the s_{Miss} lower confidence interval endpoints, and the $1-q$-quantile of the upper interval endpoints, where $0<q<1$ is a tuning parameter.

An obvious first guess for q is $q=\alpha/2$, though this turns out to not be optimal. Instead, we compute a multitude of confidence intervals using each value of q in a tight grid of q-values (this does not cost any appreciable computational time compared to use of just one value of q), and then, via simulation, can inspect the actual coverage as a function of q. Observe the similarity to the double bootstrap described in Chapter III.1, though note that the inner loop does not involve resampling, but just simple calibration. The code to perform the required calculations is surprisingly simple and shown in Listing 3.9.

Using this "quantile calibration heuristic" with 10,000 simulated data sets, nominal coverage 90%, and based on $A=20$, $n=10$, and 20 missing values, resulted in actual coverage probabilities depicted in the top panel of Figure 3.4, as a function of the quantile q. It thus appears that, for this choice of α, sample sizes, and constellation of true parameters, a choice of $q=0.355$ is optimal.

The bottom panel is similar, but having used $n=5$, implying 20% missing values. As expected, with a larger percentage of missing values, the slope of the line is larger: Recall that, if there were no missing values, then all the s_{Miss} intervals would be identical and the line would be flat. Also, the optimal q is different than the $n=10$ case, being about 0.235.

Remark Observe how the same heuristic can be applied to any confidence interval computable in the balanced case, such as the approximate one for σ_a^2 in (3.32) based on the Satterthwaite method. The reader is encouraged to design a program to use the above technique in the following way: Make a function that inputs a one-way REM unbalanced data set, and values A and n, and computes the parameter point estimates via the approximate m.l.e. method, with multiplicative adjustment for $\hat{\sigma}_e^2$.

Based on these point estimates, one then computes confidence intervals for the intraclass correlation coefficient, as well as σ_a^2 via the Satterthwaite method, using the above method but as a bootstrap: That is, a simulation based on, say, `Boot` = 1,000 parametric bootstrap replications of the data is performed, and, for each, the inner calibration loop with $s_{\text{Miss}}=1{,}000$ replications is computed over a grid of q-values to determine the optimal q. One could then simulate the performance of this method for, say, `sim` replications of data, to determine its actual coverage, for a given A, n, number of missing values, and constellation of true parameters.

```
A=20; n=10; mu=5; sigma2a=0.4; sigma2e=0.8;
ICC=sigma2a/(sigma2a+sigma2e);
muv=ones(A*n,1)*mu; J=ones(n,n);
tmp=sigma2a*J+sigma2e*eye(n); Sigma=kron(eye(A),tmp);
sim=1e4; qvec=0.01:0.005:0.7; qlen=length(qvec); cover=zeros(sim,qlen);
simCI=1000; lovec=zeros(simCI,1); hivec=zeros(simCI,1);
for loop=1:sim , if mod(loop,100)==0, disp(loop), end
  y=mvnrnd(muv,Sigma,1)';
  i=1; j=1; ind=n*(i-1)+j;  y(ind)=NaN;
  % etc. We used 20 missing values.
  loc=isnan(y); [mumiss,Vmiss] = REM1wayMLEMiss(y,A,n);
  % Now simulate from the distribution of missings
  for i=1:simCI
    mis = mvnrnd(mumiss,Vmiss,1)'; y(loc)=mis;
    [lo,hi] = REM1wayCIforICC(y,A,n);
    lovec(i)=lo; hivec(i)=hi;
  end
  for qloop=1:qlen
    quse=qvec(qloop); lo=quantile(lovec,quse);
    hi=quantile(hivec,1-quse);
    cover(loop,qloop) = (lo<ICC) && (ICC<hi);
  end
end
empcov=mean(cover);
figure, plot(qvec,empcov,'r-','linewidth',2)
```

Program Listing 3.9: Via simulation, computes and plots the mapping between quantile q and the actual coverage of the confidence interval for the intraclass correlation coefficient (ICC) with unbalanced data, using the method of simulating the missing observations from their computed distribution, and computing the c.i. based on the imputed, balanced sample.

We will implement this technique below in Section 3.3.1.2 in the context of the two-way nested REM. ∎

3.2 Crossed Random Effects Models

Once we move beyond one-factor models, each factor needs to be designated as either **crossed** (with some other factor) or **nested** (within another factor). This section examines the former case, such that all factors are crossed. In addition, with more than one factor, some could be fixed and some could be random, giving rise to a mixed model. In this section, we will restrict ourselves to all factors being random, and only mention that the two-factor crossed/mixed model is discussed in, e.g., Searle et al. (1992, p. 122). We will briefly look at an example of a mixed model later in Section 3.3.1.3, within the context of a nested model.

Recall Section 2.5 on the two-way ANOVA model with fixed effects. The setup in the two-factor crossed REM model is the same, but the classes are now random instead of fixed. As in the fixed effects case, the model can be additive in the two effects or include an interaction term. Continuing our high school student writing evaluation example, imagine now, in addition to the $A = 20$ schools chosen

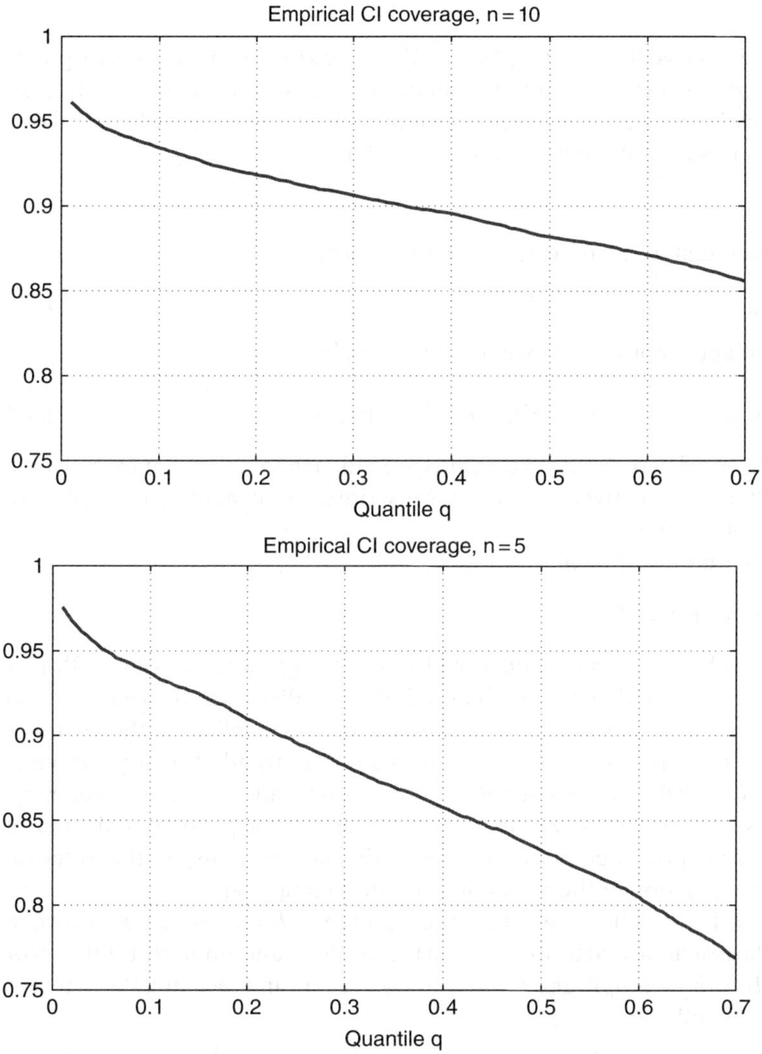

Figure 3.4 **Top:** Actual coverage probability as a function of tuning parameter q for the confidence interval of the intraclass correlation coefficient for the one-way unbalanced REM with $A = 20$, $n = 10$, 20 missing values, $\sigma_a^2 = 0.4$, and $\sigma_e^2 = 0.8$. **Bottom:** Same but having used $n = 5$.

randomly from a suitable population, B test evaluators are chosen randomly from a large population of candidates (such as undergraduate university admissions staff) and for each of the AB combinations, $n = 10$ high school pupils are randomly chosen. Observe how each class of factor A is **crossed** with each class of factor B. Interest centers on the variance components arising from the different schools (variance factor A), and the different evaluators (variance factor B), along with the error variance from the different pupils. This two-factor model is addressed in Section 3.2.1, while Section 3.2.2 considers the crossed model with three factors.

3.2.1 Two Factors

For the two-way crossed REM, we observe the set $\{Y_{ijk}\}$, where Y_{ijk} is the kth observation corresponding to the cross of the ith class from the first effect and the jth class of the second effect, $i = 1, \dots, A$, $j = 1, \dots, B$, $k = 1, \dots, n$, thus yielding a total of ABn observations. With two factors (whether fixed, random, or mixed), it is common to speak of the ith row and jth column.

3.2.1.1 With Interaction Term

This model with interaction is such that we assume Y_{ijk} can be represented as

$$Y_{ijk} = \mu + a_i + b_j + c_{ij} + e_{ijk}, \tag{3.37}$$

where the a_i, b_j, c_{ij}, and e_{ijk} are independent unobserved random variables with

$$a_i \overset{\text{i.i.d.}}{\sim} \mathrm{N}(0, \sigma_a^2), \quad b_j \overset{\text{i.i.d.}}{\sim} \mathrm{N}(0, \sigma_b^2), \quad c_{ij} \overset{\text{i.i.d.}}{\sim} \mathrm{N}(0, \sigma_c^2), \quad e_{ijk} \overset{\text{i.i.d.}}{\sim} \mathrm{N}(0, \sigma_e^2). \tag{3.38}$$

In other words, the particular observed A rows and B columns are independently drawn from a large population of row and column effects, respectively. Some authors write factor c_{ij} as $(ab)_{ij}$ to emphasize that it represents the interaction of a_i and b_j.

It follows from (3.37) and (3.38) that $\mathbb{E}[Y_{ijk}] = \mu$,

$$\mathrm{Var}(Y_{ijk}) = \sigma_Y^2 = \sigma_a^2 + \sigma_b^2 + \sigma_c^2 + \sigma_e^2,$$

$\mathrm{Cov}(Y_{ijk}, Y_{ijk'}) = \sigma_a^2 + \sigma_b^2 + \sigma_c^2$, $\mathrm{Cov}(Y_{ijk}, Y_{ij'k'}) = \sigma_a^2$, and $\mathrm{Cov}(Y_{ijk}, Y_{i'jk'}) = \sigma_b^2$, where, as in (3.3), i' is an element in $\{1, 2, \dots, A\} \backslash i$, etc. If $\sigma_c^2 = 0$, then the model is additive, as discussed below in Section 3.2.1.2, otherwise, there are interaction effects between the two classes. Unlike in the two-way fixed effects ANOVA, whereby inclusion of the interaction terms imply AB additional parameters (albeit subject to constraints), for the two-way crossed REM only a single additional parameter, σ_c^2, is required. Nevertheless, precise estimation of variance components is not possible with typical sample sizes (as seen from the often depressingly large width of confidence intervals), so that removal of σ_c^2, if justified, is beneficial for estimation of the remaining variance components.

As with the two-way ANOVA with fixed effects, we stack the Y_{ijk} in the $ABn \times 1$ vector $\mathbf{Y}$ in lexicon order such that index k changes fastest, followed by index j, and then index i, and similarly for the error vector $\mathbf{e}$. With $\mathbf{a} = (a_1, \dots, a_A)'$, $\mathbf{b} = (b_1, \dots, b_B)'$, and $\mathbf{c} = (c_{11}, c_{12}, \dots, c_{AB})'$, and recalling the matrices in the fixed effects case (2.48) and (2.62),

$$\begin{aligned}
\mathbf{Y} = \quad & (\underline{1}_A \otimes \underline{1}_B \otimes \underline{1}_n)\mu \\
& + (\mathbf{I}_A \otimes \underline{1}_B \otimes \underline{1}_n)\mathbf{a} \\
& + (\underline{1}_A \otimes \mathbf{I}_B \otimes \underline{1}_n)\mathbf{b} \\
& + (\mathbf{I}_A \otimes \mathbf{I}_B \otimes \underline{1}_n)\mathbf{c} \\
& + (\mathbf{I}_A \otimes \mathbf{I}_B \otimes \mathbf{I}_n)\mathbf{e},
\end{aligned} \tag{3.39}$$

from which the elegant structure reveals itself, and can be straightforwardly used to express $\mathbf{Y}$ for higher-order crossed models. Of course, (3.39) simplifies somewhat for computational purposes as

$$\begin{aligned}
\mathbf{Y} &= (\underline{1}_{ABn})\mu + (\mathbf{I}_A \otimes \underline{1}_{Bn})\mathbf{a} + (\underline{1}_A \otimes \mathbf{I}_B \otimes \underline{1}_n)\mathbf{b} + (\mathbf{I}_{AB} \otimes \underline{1}_n)\mathbf{c} + \mathbf{e} \\
&= \mathbf{X}\boldsymbol{\beta} + \boldsymbol{\epsilon},
\end{aligned} \tag{3.40}$$

where $\mathbf{X} = \underline{1}_{ABn}$, $\boldsymbol{\beta} = \mu$, and $\boldsymbol{\epsilon}$ is the rest of (3.40). Thus, $\mathbf{Y} \sim \mathrm{N}_{ABn}(\boldsymbol{\mu}, \boldsymbol{\Sigma})$, where $\boldsymbol{\mu} = \mathbf{X}\boldsymbol{\beta}$ and, from (3.40),

$$\begin{aligned}\boldsymbol{\Sigma} = \mathbb{V}(\mathbf{Y}) = \mathbb{V}(\boldsymbol{\epsilon}) &= (\mathbf{I}_A \otimes \underline{1}_{Bn})\mathrm{Var}(\mathbf{a})(\mathbf{I}_A \otimes \underline{1}_{Bn})' + \cdots + \mathrm{Var}(\mathbf{e}) \\ &= (\mathbf{I}_A \otimes \mathbf{J}_B \otimes \mathbf{J}_n)\sigma_a^2 \\ &\quad + (\mathbf{J}_A \otimes \mathbf{I}_B \otimes \mathbf{J}_n)\sigma_b^2 \\ &\quad + (\mathbf{I}_A \otimes \mathbf{I}_B \otimes \mathbf{J}_n)\sigma_c^2 \\ &\quad + (\mathbf{I}_A \otimes \mathbf{I}_B \otimes \mathbf{I}_n)\sigma_e^2,\end{aligned} \tag{3.41}$$

after some simplification similar to that used to obtain (3.6). Observe also the predictable pattern in (3.41), allowing for easy extension to higher-order (balanced, crossed, random effects) models.

Thus, the likelihood is easily expressed and, similar to the Matlab exercise in Section 3.1.1, the reader is invited to develop the code to compute the m.l.e. and approximate parameter standard errors. Note that (3.41) can be used for simulation, as was done in Listing 3.2, but it is perhaps easier to simulate **a**, **b**, and **c**, and use (3.37) directly, with a triple `for` loop, outputting also the classes. The generated data can be output to a text file and read in and analyzed by SAS, as in Section 3.1.5.

We now turn to the basic distribution theory associated with the model. As always, we start with the identity

$$Y_{ijk} = \bar{Y}_{\bullet\bullet\bullet} + (\bar{Y}_{i\bullet\bullet} - \bar{Y}_{\bullet\bullet\bullet}) + (\bar{Y}_{\bullet j\bullet} - \bar{Y}_{\bullet\bullet\bullet}) + (\bar{Y}_{ij\bullet} - \bar{Y}_{i\bullet\bullet} - \bar{Y}_{\bullet j\bullet} + \bar{Y}_{\bullet\bullet\bullet}) + (Y_{ijk} - \bar{Y}_{ij\bullet}). \tag{3.42}$$

Theorem 3.4 ***Independence and Distribution*** Squaring each term in (3.42) and summing over all subscripts results in all cross terms being zero, so that

$$SST = SS\mu + SSa + SSb + SSc + SSe, \tag{3.43}$$

where each *SS* term corresponds to its counterpart in (3.42) from left to right.

The r.h.s. *SS* values in (3.43) are mutually independent, and

$$\frac{SS\mu}{\gamma_\mu} \sim \chi_1^2\left(\frac{ABn\mu^2}{\gamma_\mu}\right), \quad \frac{SSa}{\gamma_a} \sim \chi^2_{A-1}, \quad \frac{SSb}{\gamma_b} \sim \chi^2_{B-1}, \quad \frac{SSc}{\gamma_c} \sim \chi^2_{(A-1)(B-1)}, \tag{3.44}$$

and $SSe/\sigma_e^2 \sim \chi^2_{AB(n-1)}$, where

$$\begin{aligned}\gamma_\mu &= Bn\sigma_a^2 + An\sigma_b^2 + n\sigma_c^2 + \sigma_e^2, & \gamma_a &= Bn\sigma_a^2 + n\sigma_c^2 + \sigma_e^2, \\ \gamma_b &= An\sigma_b^2 + n\sigma_c^2 + \sigma_e^2, & \gamma_c &= n\sigma_c^2 + \sigma_e^2.\end{aligned}$$

The corresponding ANOVA table is given in Table 3.2, along with the *EMS* values.

Proof: See Problem 3.1. ■

Inspecting *EMS* values in Table 3.2 immediately gives the ANOVA method point estimators

$$\hat{\sigma}_e^2 = MSe, \quad \hat{\sigma}_c^2 = \frac{MSc - MSe}{n}, \quad \hat{\sigma}_b^2 = \frac{MSb - MSc}{An}, \quad \hat{\sigma}_a^2 = \frac{MSa - MSc}{Bn}. \tag{3.45}$$

Table 3.2 ANOVA table for the balanced two-factor crossed REM.

Source	df	SS	EMS
Mean	1	$ABn\bar{Y}^2_{\bullet\bullet\bullet}$	$\sigma_e^2 + n\sigma_c^2 + An\sigma_b^2 + Bn\sigma_a^2 + ABn\mu^2$
A	$A-1$	$Bn\sum_{i=1}^{A}(\bar{Y}_{i\bullet\bullet} - \bar{Y}_{\bullet\bullet\bullet})^2$	$\sigma_e^2 + n\sigma_c^2 + \qquad + Bn\sigma_a^2$
B	$B-1$	$An\sum_{j=1}^{B}(\bar{Y}_{\bullet j\bullet} - \bar{Y}_{\bullet\bullet\bullet})^2$	$\sigma_e^2 + n\sigma_c^2 + An\sigma_b^2$
AB	$(A-1)(B-1)$	$n\sum_{i=1}^{A}\sum_{j=1}^{B}\begin{pmatrix}\bar{Y}_{ij\bullet} - \bar{Y}_{i\bullet\bullet}\\ -\bar{Y}_{\bullet j\bullet} + \bar{Y}_{\bullet\bullet\bullet}\end{pmatrix}^2$	$\sigma_e^2 + n\sigma_c^2$
Error	$AB(n-1)$	$\sum_{i=1}^{A}\sum_{j=1}^{B}\sum_{k=1}^{n}(Y_{ijk} - \bar{Y}_{ij\bullet})^2$	σ_e^2
Total	ABn	$\sum_{i=1}^{A}\sum_{j=1}^{B}\sum_{k=1}^{n}Y_{ijk}^2$	

Calculations similar to those in (3.19) and (3.20) lead to expressions for the sample variances as

$$\begin{aligned}
\mathrm{Var}(\hat{\sigma}_e^2) &= \frac{2\sigma_e^4}{AB(n-1)}, \quad \mathrm{Var}(\hat{\sigma}_c^2) = \frac{2}{n^2}\left[\frac{(\sigma_e^2+n\sigma_c^2)^2}{(A-1)(B-1)} + \frac{\sigma_e^4}{AB(n-1)}\right],\\
\mathrm{Var}(\hat{\sigma}_a^2) &= \frac{2}{B^2n^2}\left[\frac{(\sigma_e^2+n\sigma_c^2+Bn\sigma_a^2)^2}{A-1} + \frac{(\sigma_e^2+n\sigma_c^2)^2}{(A-1)(B-1)}\right],\\
\mathrm{Var}(\hat{\sigma}_b^2) &= \frac{2}{A^2n^2}\left[\frac{(\sigma_e^2+n\sigma_c^2+An\sigma_b^2)^2}{B-1} + \frac{(\sigma_e^2+n\sigma_c^2)^2}{(A-1)(B-1)}\right],
\end{aligned} \tag{3.46}$$

which can be used to form Wald confidence intervals for the variance components.

Unlike in the one-way case, this model is such that the m.l.e. does not have a closed-form solution (see, e.g., Sahai and Ojeda, 2004, Sec. 4.4.2), except for $\hat{\sigma}^2_{e,\mathrm{ML}}$, which is the same as in (3.45). No doubt (3.45) will be close to the m.l.e., and thus, if all estimates are positive, will serve as excellent starting values for numeric computation of the m.l.e. One can correctly speculate that, in all higher-order crossed models, the m.l.e. (or, more correctly, a solution to the set of log-likelihood derivative equations) is not expressible in closed form.

Similar to the motivation for the F test in (3.17) pertaining to the one-way REM case, the *EMS* values in Table 3.2, and the independence of the sums of squares, suggest the following F tests for σ_a^2, σ_b^2, and σ_c^2, respectively, with $P := (A-1)(B-1)$:

$$F_a = \frac{MSa}{MSc} \sim \frac{\gamma_a}{\gamma_c}F_{A-1,P}, \quad F_b = \frac{MSb}{MSc} \sim \frac{\gamma_b}{\gamma_c}F_{B-1,P}, \quad F_c = \frac{MSc}{MSe} \sim \frac{\gamma_c}{\sigma_e^2}F_{P,AB(n-1)}. \tag{3.47}$$

As an example of an exact confidence interval, from F_c in (3.47),

$$\begin{aligned}
1-\alpha &= \Pr\left(\frac{L}{F_c} < \frac{\sigma_e^2}{n\sigma_c^2+\sigma_e^2} < \frac{U}{F_c}\right) = \Pr\left(\frac{F_c/U-1}{n} < \frac{\sigma_c^2}{\sigma_e^2} < \frac{F_c/L-1}{n}\right)\\
&= \Pr\left(\frac{F_c-U}{nU+F_c-U} < \frac{\sigma_c^2}{\sigma_c^2+\sigma_e^2} < \frac{F_c-L}{nL+F_c-L}\right),
\end{aligned} \tag{3.48}$$

where L and U are such that $\Pr(L \leqslant F_{P,AB(n-1)} \leqslant U) = 1-\alpha$.

Not having pivots, exact confidence intervals for the individual variance components do not exist, though one can use asymptotic pivots via the Wald intervals formed from (3.46), as well as the easily derived ones using the Satterthwaite method from Section 3.1.4. In particular, the latter are

$$1-\alpha \approx \Pr\left(\hat{d}\frac{(MSc-MSe)}{n\,u} \leqslant \sigma_c^2 \leqslant \hat{d}\frac{(MSc-MSe)}{n\,l}\right), \quad \hat{d} = \frac{(MSc-MSe)^2}{\left(\frac{(MSc)^2}{(A-1)(B-1)} + \frac{(MSe)^2}{AB(n-1)}\right)},$$

$$1-\alpha \approx \Pr\left(\hat{d}\frac{(MSb-MSc)}{An\,u} \leqslant \sigma_b^2 \leqslant \hat{d}\frac{(MSb-MSc)}{An\,l}\right), \quad \hat{d} = \frac{(MSb-MSc)^2}{\left(\frac{(MSb)^2}{B-1} + \frac{(MSc)^2}{(A-1)(B-1)}\right)},$$

and

$$1-\alpha \approx \Pr\left(\hat{d}\frac{(MSa-MSc)}{Bn\,u} \leqslant \sigma_a^2 \leqslant \hat{d}\frac{(MSa-MSc)}{Bn\,l}\right), \quad \hat{d} = \frac{(MSa-MSc)^2}{\left(\frac{(MSa)^2}{(A-1)} + \frac{(MSc)^2}{(A-1)(B-1)}\right)},$$

for u and l such that $1-\alpha = \Pr(l \leqslant \chi^2_{\hat{d}} \leqslant u)$.

3.2.1.2 Without Interaction Term

If the analyst decides that the magnitude of σ_c^2 is negligible compared to the other variance components (typically as a result of failure to reject the null hypothesis that $\sigma_c^2 = 0$, based on the F_c test in (3.47), using a conventional test significance level, though possibly also coupled with theoretical knowledge of the process and/or results of previous, similar studies), then the model may be assumed additive. In this case, (3.45) becomes

$$\hat{\sigma}_e^2 = MSe, \qquad \hat{\sigma}_b^2 = \frac{MSb-MSe}{An}, \quad \hat{\sigma}_a^2 = \frac{MSa-MSe}{Bn}. \tag{3.49}$$

By squaring and summing identity (3.42) without the interaction term, i.e.,

$$Y_{ijk} = \bar{Y}_{\bullet\bullet\bullet} + (\bar{Y}_{i\bullet\bullet} - \bar{Y}_{\bullet\bullet\bullet}) + (\bar{Y}_{\bullet j\bullet} - \bar{Y}_{\bullet\bullet\bullet}) + (Y_{ijk} - \bar{Y}_{i\bullet\bullet} - \bar{Y}_{\bullet j\bullet} + \bar{Y}_{\bullet\bullet\bullet}), \tag{3.50}$$

it is straightforward to verify that all cross terms are zero, so that

$$SST = SS\mu + SSa + SSb + SSe,$$

where, as before, the r.h.s. SS values are mutually independent (see Problem 3.2). This results in ANOVA Table 3.3.

The distributions of $SS\mu/\gamma_\mu$, SSa/γ_a, SSb/γ_b, and SSe/σ_e^2 are the same as those in (3.44) but with γ_i values such that $\sigma_c^2 = 0$, i.e.,

$$\gamma_\mu = Bn\sigma_a^2 + An\sigma_b^2 + \sigma_e^2, \quad \gamma_a = Bn\sigma_a^2 + \sigma_e^2, \quad \text{and} \quad \gamma_b = An\sigma_b^2 + \sigma_e^2.$$

The first two F tests in (3.47) now become

$$F_a = \frac{MSa}{MSe} \sim \frac{\gamma_a}{\sigma_e^2} F_{A-1,d}, \quad F_b = \frac{MSb}{MSe} \sim \frac{\gamma_b}{\sigma_e^2} F_{B-1,d}, \tag{3.51}$$

where the denominator degrees of freedom in (3.51) is $d = ABn - A - B + 1$.

3.2.2 Three Factors

Now with three crossed factors, we observe Y_{ijkl}, the lth observation in the ith row, jth column, kth "pipe", $i = 1, \dots, A$, $j = 1, \dots, B$, $k = 1, \dots, C$, $l = 1, \dots, n$, and assume that Y_{ijkl} can be represented as

$$Y_{ijkl} = \mu + a_i + b_j + c_k + d_{ij} + f_{ik} + g_{jk} + h_{ijk} + e_{ijkl}, \tag{3.52}$$

Table 3.3 ANOVA table for the balanced two-factor crossed additive (no interaction effect) random effects model, where the error degrees of freedom is $d = ABn - A - B + 1$.

Source	df	SS	EMS
Mean	1	$ABn\bar{Y}^2_{\bullet\bullet\bullet}$	$\sigma_e^2 + An\sigma_b^2 + Bn\sigma_a^2 + ABn\mu^2$
A	$A-1$	$Bn\sum_{i=1}^{A}(\bar{Y}_{i\bullet\bullet} - \bar{Y}_{\bullet\bullet\bullet})^2$	$\sigma_e^2 + \qquad + Bn\sigma_a^2$
B	$B-1$	$An\sum_{j=1}^{B}(\bar{Y}_{\bullet j\bullet} - \bar{Y}_{\bullet\bullet\bullet})^2$	$\sigma_e^2 + An\sigma_b^2$
Error	d	$\sum_{i=1}^{A}\sum_{j=1}^{B}\sum_{k=1}^{n}\begin{pmatrix} Y_{ijk} - \bar{Y}_{i\bullet\bullet} \\ -\bar{Y}_{\bullet j\bullet} + \bar{Y}_{\bullet\bullet\bullet}\end{pmatrix}^2$	σ_e^2
Total	ABn	$\sum_{i=1}^{A}\sum_{j=1}^{B}\sum_{k=1}^{n} Y^2_{ijk}$	

where the a_i, b_j, c_k, d_{ij}, f_{ik}, g_{jk}, h_{ijk}, and e_{ijkl} are independent unobserved random variables with $a_i \overset{\text{i.i.d.}}{\sim} \mathrm{N}(0, \sigma_a^2)$, …, $e_{ijkl} \overset{\text{i.i.d.}}{\sim} \mathrm{N}(0, \sigma_e^2)$. As with the two-factor case, the particular observed A rows, B columns, and C pipes are independently drawn from a large population of row, column, and pipe effects, respectively. The model is additive if all interaction terms are zero.

As a logical extension of (3.42), the identity

$$\begin{aligned} Y_{ijkl} = \bar{Y}_{\bullet\bullet\bullet\bullet} &+ (\bar{Y}_{i\bullet\bullet\bullet} - \bar{Y}_{\bullet\bullet\bullet\bullet}) + (\bar{Y}_{\bullet j\bullet\bullet} - \bar{Y}_{\bullet\bullet\bullet\bullet}) + (\bar{Y}_{\bullet\bullet k\bullet} - \bar{Y}_{\bullet\bullet\bullet\bullet}) + (\bar{Y}_{ij\bullet\bullet} - \bar{Y}_{i\bullet\bullet\bullet} - \bar{Y}_{\bullet j\bullet\bullet} + \bar{Y}_{\bullet\bullet\bullet\bullet}) \\ &+ (\bar{Y}_{i\bullet k\bullet} - \bar{Y}_{i\bullet\bullet\bullet} - \bar{Y}_{\bullet\bullet k\bullet} + \bar{Y}_{\bullet\bullet\bullet\bullet}) + (\bar{Y}_{\bullet jk\bullet} - \bar{Y}_{\bullet j\bullet\bullet} - \bar{Y}_{\bullet\bullet k\bullet} + \bar{Y}_{\bullet\bullet\bullet\bullet}) \\ &+ (\bar{Y}_{ijk\bullet} - \bar{Y}_{ij\bullet\bullet} - \bar{Y}_{i\bullet k\bullet} - \bar{Y}_{\bullet jk\bullet} + \bar{Y}_{i\bullet\bullet\bullet} + \bar{Y}_{\bullet j\bullet\bullet} + \bar{Y}_{\bullet\bullet k\bullet} - \bar{Y}_{\bullet\bullet\bullet\bullet}) + (Y_{ijkl} - \bar{Y}_{ijk\bullet}) \end{aligned}$$

suggests itself as the correct one to use. Note how, omitting the $\bar{Y}_{\bullet\bullet\bullet\bullet}$ term throughout, the bracketed terms (except the last) are generated analogous to the structure of the inclusion-exclusion principle (Poincaré's theorem) in representing the union of events in basic probability. We will presume that, upon squaring and summing, calculations similar to (and correspondingly more tedious than) those of Problem 3.1 for the two-factor case give rise to the sums of squares decomposition

$$SST = SS\mu + SSa + SSb + SSc + SSd + SSf + SSg + SSh + SSe,$$

such that the r.h.s. SS values are mutually independent (see, e.g., Graybill, 1976, p. 641 for details). These are shown in Table 3.4 along with their corresponding EMS. Furthermore, letting $A' = A - 1$, $B' = B - 1$, and $C' = C - 1$, we state their distributions without proof as

$$\frac{SS\mu}{\gamma_\mu} \sim \chi_1^2\left(\frac{ABCn\mu^2}{\gamma_\mu}\right), \quad \frac{SSa}{\gamma_a} \sim \chi^2_{A'}, \quad \frac{SSb}{\gamma_b} \sim \chi^2_{B'}, \quad \frac{SSc}{\gamma_c} \sim \chi^2_{C'},$$

$$\frac{SSd}{\gamma_d} \sim \chi^2_{A'B'}, \quad \frac{SSf}{\gamma_f} \sim \chi^2_{A'C'}, \quad \frac{SSg}{\gamma_g} \sim \chi^2_{B'C'}, \quad \frac{SSh}{\gamma_h} \sim \chi^2_{A'B'C'},$$

and $SSe/\sigma_e^2 \sim \chi^2_{ABC(n-1)}$, where γ_μ, γ_a, etc., are given in Table 3.4.

Table 3.4 ANOVA table for the balanced three-factor crossed REM.

			EMS									
Effect	**df**	**SS**		***e***	***abc***	***bc***	***ac***	***ab***	***c***	***b***	***a***	***μ***
μ	1	$ABCn\bar{Y}^2_{\bullet\bullet\bullet}$	$\gamma_\mu =$	σ_e^2	$+\ n\sigma_h^2$	$+\ An\sigma_g^2$	$+\ Bn\sigma_f^2$	$+\ Cn\sigma_d^2$	$+\ ABn\sigma_c^2$	$+\ ACn\sigma_b^2$	$+\ BCn\sigma_a^2$	$+\ ABCn\mu^2$
a_i	$A-1$	$BCn\sum(\bar{Y}_{i\bullet\bullet\bullet}-\bar{Y}_{\bullet\bullet\bullet\bullet})^2$	$\gamma_a =$	σ_e^2	$+\ n\sigma_h^2$		$+\ Bn\sigma_f^2$	$+\ Cn\sigma_d^2$			$+\ BCn\sigma_a^2$	
b_j	$B-1$	$ACn\sum(\bar{Y}_{\bullet j\bullet\bullet}-\bar{Y}_{\bullet\bullet\bullet\bullet})^2$	$\gamma_b =$	σ_e^2	$+\ n\sigma_h^2$	$+\ An\sigma_g^2$		$+\ Cn\sigma_d^2$		$+\ ACn\sigma_b^2$		
c_k	$C-1$	$ABn\sum(\bar{Y}_{\bullet\bullet k\bullet}-\bar{Y}_{\bullet\bullet\bullet\bullet})^2$	$\gamma_c =$	σ_e^2	$+\ n\sigma_h^2$	$+\ An\sigma_g^2$	$+\ Bn\sigma_f^2$		$+\ ABn\sigma_c^2$			
d_{ij}	$(A-1)(B-1)$	$Cn\sum\sum\begin{pmatrix}\bar{Y}_{ij\bullet\bullet}-\bar{Y}_{i\bullet\bullet\bullet}\\-\bar{Y}_{\bullet j\bullet\bullet}+\bar{Y}_{\bullet\bullet\bullet\bullet}\end{pmatrix}^2$	$\gamma_d =$	σ_e^2	$+\ n\sigma_h^2$			$+\ Cn\sigma_d^2$				
f_{ik}	$(A-1)(C-1)$	$Bn\sum\sum\begin{pmatrix}\bar{Y}_{i\bullet k\bullet}-\bar{Y}_{i\bullet\bullet\bullet}\\-\bar{Y}_{\bullet\bullet k\bullet}+\bar{Y}_{\bullet\bullet\bullet\bullet}\end{pmatrix}^2$	$\gamma_f =$	σ_e^2	$+\ n\sigma_h^2$		$+\ Bn\sigma_f^2$					
g_{jk}	$(B-1)(C-1)$	$An\sum\sum\begin{pmatrix}\bar{Y}_{\bullet jk\bullet}-\bar{Y}_{\bullet j\bullet\bullet}\\-\bar{Y}_{\bullet\bullet k\bullet}+\bar{Y}_{\bullet\bullet\bullet\bullet}\end{pmatrix}^2$	$\gamma_g =$	σ_e^2	$+\ n\sigma_h^2$	$+\ An\sigma_g^2$						
h_{ijk}	$(A-1)(B-1)(C-1)$	$n\sum\sum\sum\begin{pmatrix}\bar{Y}_{ijk\bullet}-\bar{Y}_{ij\bullet\bullet}\\-\bar{Y}_{i\bullet k\bullet}-\bar{Y}_{\bullet jk\bullet}\\+\bar{Y}_{i\bullet\bullet\bullet}+\bar{Y}_{\bullet j\bullet\bullet}\\+\bar{Y}_{\bullet\bullet k\bullet}-\bar{Y}_{\bullet\bullet\bullet\bullet}\end{pmatrix}^2$	$\gamma_h =$	σ_e^2	$+\ n\sigma_h^2$							
e_{ijkl}	$ABC(n-1)$	$\sum\sum\sum\sum(Y_{ijkl}-\bar{Y}_{ijk\bullet})^2$		σ_e^2								
Total	$ABCn$	$\sum\sum\sum\sum Y_{ijkl}^2$										

Some staring at the *EMS* values in Table 3.4 yields the ANOVA method point estimators

$$\hat{\sigma}_e^2 = MSe, \quad \hat{\sigma}_h^2 = \frac{MSh - MSe}{n}, \tag{3.53}$$

and

$$\hat{\sigma}_g^2 = \frac{MSg - MSh}{An}, \quad \hat{\sigma}_f^2 = \frac{MSf - MSh}{Bn}, \quad \hat{\sigma}_d^2 = \frac{MSd - MSh}{Cn}. \tag{3.54}$$

For σ_c^2, it appears that use of the set $\{\gamma_c, \gamma_f, \gamma_g, \gamma_h\}$ will be fruitful, with

$$\begin{aligned}
\gamma_c &= \sigma_e^2 + n\sigma_h^2 + An\sigma_g^2 + Bn\sigma_f^2 + ABn\sigma_c^2,\\
\gamma_f &= \sigma_e^2 + n\sigma_h^2 + Bn\sigma_f^2,\\
\gamma_g &= \sigma_e^2 + n\sigma_h^2 + An\sigma_g^2,\\
\gamma_h &= \sigma_e^2 + n\sigma_h^2,
\end{aligned}$$

yielding

$$\hat{\sigma}_c^2 = \frac{MSc - MSf - MSg + MSh}{ABn}, \tag{3.55}$$

and, similarly,

$$\hat{\sigma}_b^2 = \frac{MSb - MSd - MSg + MSh}{ACn}, \quad \hat{\sigma}_a^2 = \frac{MSa - MSd - MSf + MSh}{BCn}. \tag{3.56}$$

Exact F tests for the second- and third-order interactions can be seen directly from the ANOVA table to be, with $P = A'B'C'$,

$$\begin{aligned}
F_d &= \frac{MSd}{MSh} \sim \frac{\gamma_d}{\gamma_h} F_{A'B',P}, \quad F_f = \frac{MSf}{MSh} \sim \frac{\gamma_f}{\gamma_h} F_{A'C',P},\\
F_g &= \frac{MSg}{MSh} \sim \frac{\gamma_g}{\gamma_h} F_{B'C',P}, \quad F_h = \frac{MSh}{MSe} \sim \frac{\gamma_h}{\sigma_e^2} F_{P,ABC(n-1)}.
\end{aligned}$$

As (3.55) and (3.56) suggest, from the *EMS* in Table 3.4, there does not exist exact F-ratios for testing σ_a^2, σ_b^2, and σ_c^2. For, say, σ_a^2, this would require there being a single *EMS* that is exactly equal to $\mathbb{E}[MSa] - BCn\sigma_a^2$. Notice though, from $\hat{\sigma}_a^2$ in (3.56), that

$$\begin{aligned}
\mathbb{E}[MSd] + \mathbb{E}[MSf] - \mathbb{E}[MSh] &= \sigma_e^2 + n\sigma_h^2 + Bn\sigma_f^2 + Cn\sigma_d^2\\
&= \mathbb{E}[MSa] - BCn\sigma_a^2,
\end{aligned} \tag{3.57}$$

so that the ratio

$$F_a' = \frac{MSa}{MSd + MSf - MSh} \tag{3.58}$$

is a test statistic such that large values would reject the null of $\sigma_a^2 = 0$, but it is not F distributed. Its distribution could be approximated by applying the Satterthwaite method to the denominator. A potentially problematic issue with F_a' is that it can be negative, and is why the next option is favored. Expressing (3.57) as

$$\mathbb{E}[MSa] + \mathbb{E}[MSh] = BCn\sigma_a^2 + \mathbb{E}[MSd] + \mathbb{E}[MSf] \tag{3.59}$$

yields the test statistic

$$F_a = \frac{MSa + MSh}{MSd + MSf}. \tag{3.60}$$

As $SSa/\mathbb{E}[MSa] = SSa/\gamma_a \sim \chi^2_{d_a}$ independent of $SSh/\mathbb{E}[MSh] = SSh/\gamma_h \sim \chi^2_{d_h}$ (for degrees of freedom $d_a = A'$ and $d_h = A'B'C'$), (3.28) from the Satterthwaite method suggests for the numerator of F_a that, for some constants h_1 and h_2,

$$W = \frac{d\hat{\gamma}}{\gamma} = d\,\frac{h_1(SSa/d_a) + h_2(SSh/d_h)}{h_1\mathbb{E}[MSa] + h_2\mathbb{E}[MSh]} \overset{\text{app}}{\sim} \chi^2_d, \tag{3.61}$$

where d is obtained from (3.34). But $h_1(SSa/d_a) = h_1 MSa$ and $h_2(SSh/d_h) = h_2 MSh$, so that (3.61) implies

$$h_1 MSa + h_2 MSh \overset{\text{app}}{\sim} (h_1\mathbb{E}[MSa] + h_2\mathbb{E}[MSh])\chi^2_d/d,$$

and likewise for the denominator of (3.60),

$$h_3 MSd + h_4 MSf \overset{\text{app}}{\sim} (h_3\mathbb{E}[MSd] + h_4\mathbb{E}[MSf])\chi^2_{d'}/d',$$

for estimated degrees of freedom value d'. Dividing each of the above two expressions by the scale term (and recalling that an F random variable is the ratio of two independent chi-squares divided by their respective degrees of freedom), it follows that

$$F_a = \frac{h_1 MSa + h_2 MSh}{h_3 MSd + h_4 MSf} \overset{\text{app}}{\sim} \frac{h_1\mathbb{E}[MSa] + h_2\mathbb{E}[MSh]}{h_3\mathbb{E}[MSd] + h_4\mathbb{E}[MSf]} F_{d,d'}, \tag{3.62}$$

a scaled F distribution with degrees of freedom d and d'. Furthermore, if

$$h_1\mathbb{E}[MSa] + h_2\mathbb{E}[MSh] = h_3\mathbb{E}[MSd] + h_4\mathbb{E}[MSf],$$

then $F_a \overset{\text{app}}{\sim} F_{d,d'}$. But, recalling (3.59), this is the case for $h_1 = h_2 = h_3 = h_4 = 1$ and $\sigma^2_a = 0$, so that, under the null hypothesis of $\sigma^2_a = 0$, $F_a \overset{\text{app}}{\sim} F_{d,d'}$, where, from (3.34),

$$d = \frac{(MSa + MSh)^2}{(MSa)^2/d_a + (MSh)^2/d_h} \quad \text{and} \quad d' = \frac{(MSd + MSf)^2}{(MSd)^2/d_d + (MSf)^2/d_f},$$

for $d_d = A'B'$, and $d_f = A'C'$. Under the alternative of $\sigma^2_a > 0$, the scale parameter in (3.62) is greater than one, implying that an approximate α-level test rejects the null of $\sigma^2_a = 0$ for large F_a, i.e., $F_a > F^{\alpha}_{d,d'}$, where, as always throughout, $F^{\alpha}_{d,d'}$ is the $100(1-\alpha)$th percent quantile of the $F_{d,d'}$ distribution.

The same analysis applies to F'_a, i.e., $MSa \overset{\text{exact}}{\sim} \mathbb{E}[MSa]\chi^2_{d_a}/d_a$ and, with the Satterthwaite approximation applied to the denominator,

$$F'_a = \frac{MSa}{MSd + MSf - MSh} \overset{\text{app}}{\sim} \frac{\mathbb{E}[MSa]}{\mathbb{E}[MSd] + \mathbb{E}[MSf] - \mathbb{E}[MSh]} F_{d_a,d''}$$

with

$$d'' = \frac{(MSd + MSf - MSh)^2}{(MSd)^2/d_d + (MSf)^2/d_f + (MSh)^2/d_h}.$$

The reader is encouraged to repeat this analysis to obtain approximate F tests for σ^2_b and σ^2_c.

Various confidence intervals of interest can be derived from the Satterthwaite method. For example, for σ_a^2/σ_e^2, Burdick and Graybill (1992, p. 136) show that (at the time of their writing), no procedure other than Satterthwaite is available. This case was investigated using the general procedures for the Satterthwaite class of ratios proposed in Butler and Paolella (2002b). For the three-way crossed model and confidence intervals for σ_a^2/σ_e^2, the bootstrap/saddlepoint-based method resulted in highly accurate actual coverage, substantially more than use of the Satterthwaite method, as A and/or σ_a^2 decrease. For large A and σ_a^2, the Satterthwaite method also performs well.

3.3 Nested Random Effects Models

An REM with two factors can be either crossed, as in Section 3.2.1, or **nested**, as studied now. Models with three or more factors can have aspects of both. It turns out that we have already seen an example of a nested REM: Recall the one-way model of Section 3.1 and observe how it can be envisioned as a two-stage design, whereby first, the A units, or classes, are randomly chosen from the relevant population and then, conditional on those chosen, from each a random sample of n units, or samples, are chosen. The factor corresponding to the samples is nested within the levels, or classes, of the treatment factor. While indeed a nested model, the adjective nested is typically used only when there are two or more factors (besides the error term), such that one is nested in another.

To see this hierarchy in the two-factor model, and how it differs from its crossed counterpart, let the first factor be school, with $A = 20$ schools being chosen from a large population of such, and, for each of the chosen schools, $B = 8$ teachers employed at the school are randomly chosen from the entire cohort of teachers. This implies that the factor "evaluator" (the teacher doing the grading) is nested within the factor "school", and there are not B classes of evaluators, but rather AB, grouped according to which school they are from. Each of the AB evaluators are asked to grade the writing assignment from n randomly chosen students such that the jth teacher in the ith school receives n exams from students in his or her school i, $i = 1, \dots, A$, $j = 1, \dots, B$.

3.3.1 Two Factors

For the two-way nested case, we will distinguish two cases. The first is such that both factors are random, and the second is such that the first factor is fixed, giving rise to (our first example of) a mixed model.

3.3.1.1 Both Effects Random: Model and Parameter Estimation

In this setup, we observe Y_{ijk}, the kth observation in the jth subclass of the ith class, $i = 1, \dots, A$, $j = 1, \dots, B$, $k = 1, \dots, n$, and assume that it can be represented as

$$Y_{ijk} = \mu + a_i + b_{ij} + e_{ijk}, \tag{3.63}$$

with $a_i \overset{\text{i.i.d.}}{\sim} \text{N}(0, \sigma_a^2)$, $b_{ij} \overset{\text{i.i.d.}}{\sim} \text{N}(0, \sigma_b^2)$, and $e_{ijk} \overset{\text{i.i.d.}}{\sim} \text{N}(0, \sigma_e^2)$. It is imperative to notice that there is no factor b_j: Recall that the nested factor has AB levels, and thus requires the double subscript. This can be compared to model (3.37), where the term with the double subscript refers to the interaction term resulting from crossing two factors. Some authors write factor b_{ij} as $b_{j(i)}$ to emphasize that the jth subclass is nested in the ith class.

From (3.63), it follows that

$$\mathbb{E}[Y_{ijk}] = \mu, \quad \text{Var}(Y_{ijk}) = \sigma_a^2 + \sigma_b^2 + \sigma_e^2, \quad \text{Cov}(Y_{ijk}, Y_{ij'k'}) = \sigma_a^2, \tag{3.64}$$

and

$$\text{Cov}(Y_{ijk}, Y_{ijk'}) = \mathbb{E}[(a_i + b_{ij} + e_{ijk})(a_i + b_{ij} + e_{ijk'})] = \sigma_a^2 + \sigma_b^2. \tag{3.65}$$

For representing the likelihood, we stack the Y_{ijk} in the $ABn \times 1$ vector $\mathbf{Y}$ in the usual lexicon ordering (and similar for the error vector $\mathbf{e}$), and let $\mathbf{a} = (a_1, \dots, a_A)'$ and $\mathbf{b} = (b_{11}, b_{12}, \dots, b_{AB})'$. Then, as in (3.39),

$$\mathbf{Y} = (\underline{1}_A \otimes \underline{1}_B \otimes \underline{1}_n)\mu + (\mathbf{I}_A \otimes \underline{1}_B \otimes \underline{1}_n)\mathbf{a} + (\mathbf{I}_A \otimes \mathbf{I}_B \otimes \underline{1}_n)\mathbf{b} + (\mathbf{I}_A \otimes \mathbf{I}_B \otimes \mathbf{I}_n)\mathbf{e}, \tag{3.66}$$

or, similar to (3.40),

$$\begin{aligned} \mathbf{Y} &= (\underline{1}_{ABn})\mu + (\mathbf{I}_A \otimes \underline{1}_{Bn})\mathbf{a} + (\mathbf{I}_{AB} \otimes \underline{1}_n)\mathbf{b} + \mathbf{e} \\ &= \mathbf{X}\boldsymbol{\beta} + \boldsymbol{\epsilon}, \end{aligned} \tag{3.67}$$

where $\mathbf{X} = \underline{1}_{ABn}$, $\boldsymbol{\beta} = \mu$, and $\boldsymbol{\epsilon}$ is the rest of (3.67). As usual, we let $\boldsymbol{\mu} = \mathbb{E}[\mathbf{Y}] = \mathbf{X}\boldsymbol{\beta}$. Then, (3.63) and (3.67) can be expressed as $\mathbf{Y} \sim \text{N}_{ABn}(\boldsymbol{\mu}, \boldsymbol{\Sigma})$, where, from (3.67),

$$\begin{aligned} \boldsymbol{\Sigma} &= \mathbb{V}(\mathbf{Y}) = \mathbb{V}(\boldsymbol{\epsilon}) \\ &= (\mathbf{I}_A \otimes \underline{1}_{Bn})\text{Var}(\mathbf{a})(\mathbf{I}_A \otimes \underline{1}_{Bn})' + (\mathbf{I}_{AB} \otimes \underline{1}_n)\text{Var}(\mathbf{b})(\mathbf{I}_{AB} \otimes \underline{1}_n)' + \text{Var}(\mathbf{e}) \\ &= (\mathbf{I}_A \otimes \mathbf{J}_B \otimes \mathbf{J}_n)\sigma_a^2 + (\mathbf{I}_A \otimes \mathbf{I}_B \otimes \mathbf{J}_n)\sigma_b^2 + (\mathbf{I}_A \otimes \mathbf{I}_B \otimes \mathbf{I}_n)\sigma_e^2. \\ &= (\mathbf{I}_A \otimes \mathbf{J}_{Bn})\sigma_a^2 + (\mathbf{I}_{AB} \otimes \mathbf{J}_n)\sigma_b^2 + \mathbf{I}_{ABn}\sigma_e^2. \end{aligned} \tag{3.68}$$

Simulating a vector $\mathbf{Y}$ for a given parameter constellation is easily done by first drawing $\mathbf{a}$, $\mathbf{b}$, and $\mathbf{e}$, and then computing (3.67). Maximum likelihood is also straightforwardly accomplished by modifying the program in Listing 3.1 to create, say, a Matlab function called REM2wayNestedMLE, which the reader is encouraged to do. Similar to the code in Listing 3.3, Listing 3.10 generates a data set and outputs it as a text file. This can be subsequently read into SAS and analyzed using several of their

```
1  A=10; B=6; n=8; mu=5; siga=1; sigb=0.4; sige=0.8;
2  a = siga*randn(A,1); b = sigb*randn(A*B,1); e = sige*randn(A*B*n,1);
3  y = ones(A*B*n,1)*mu ...
4      + kron(eye(A),ones(B*n,1))*a ...
5      + kron(eye(A*B),ones(n,1))*b + e;
6  school = kron( (1:A)' , ones(B*n,1) );
7  TeacherNestedInSchool = kron( (1:(A*B))' , ones(n,1) );
8  Out=[y school TeacherNestedInSchool];
9  fname='REM2nested.txt';
10 if exist(fname,'file'), delete(fname), end
11 fileID = fopen(fname,'w');
12 fprintf(fileID,'%8.5f %4u %4u\r\n',Out'); fclose(fileID);
13 % [param, stderr, loglik, iters, bfgsok] = REM2wayNestedMLE(y,A,B,n)
```

Program Listing 3.10: Generates and writes to a text file a two-way nested balanced REM data set, and the associated class variables, based on the parameter constellation given in line 1, for input into SAS. The last line, 9, if uncommented, calls the custom-made Matlab program to compute the m.l.e., though note that a closed-form solution exists; see (3.81).

```
ods html close; ods html;
filename ein 'REM2nested.txt';
data school; infile ein stopover; input Y school Evaluator; run;
title 'REM 2-Way Nested Example';
proc varcomp method=ml;
  class school Evaluator;
  *model Y=school Evaluator;
  model Y=school Evaluator(school);
run;
proc mixed method=ml cl ratio;
  class school Evaluator;
  model Y= / cl solution;
  *random school Evaluator;
  random school Evaluator(school);
run;
proc nested;
  class school Evaluator;
  var Y;
run;
```

SAS Program Listing 3.4: Reads in the data from the text file generated in Listing 3.3 and uses `proc varcomp` and `proc mixed` with maximum likelihood, and `proc nested` (which does not support maximum likelihood, and uses only the ANOVA method of estimation). `proc nested` does not support use of mixed models, i.e., inclusion of fixed effects (besides the grand mean), and also assumes the input data are sorted by the `class` variables, which is the case by virtue of how we generated and wrote the data. The `model` statements that are commented out in `proc varcomp` and `proc mixed` can also be used and deliver the same output.

procedures, as shown in SAS Program Listing 3.4. The m.l.e. obtained for a particular generated data set from Listing 3.10 and use of the custom Matlab function `REM2wayNestedMLE` was the same as those output from both SAS procedures, as was the obtained log-likelihood, as shown in the output of `proc mixed`. For this model, there is a closed-form expression for the m.l.e., provided the variance component estimates are positive, obviating the need for numeric calculations; see (3.81).

Theorem 3.5 ***Independence and Distribution*** By squaring and summing the expression

$$Y_{ijk} = \bar{Y}_{\bullet\bullet\bullet} + (\bar{Y}_{i\bullet\bullet} - \bar{Y}_{\bullet\bullet\bullet}) + (Y_{ij\bullet} - \bar{Y}_{i\bullet\bullet}) + (Y_{ijk} - \bar{Y}_{ij\bullet}), \tag{3.69}$$

and confirming that cross terms are zero, the *SS* decomposition is given by

$$SST = SS\mu + SSa + SSb + SSe, \tag{3.70}$$

where $SST = \sum_{i=1}^{A} \sum_{j=1}^{B} \sum_{j=1}^{n} Y_{ijk}^2$, and

$$\begin{aligned} SS\mu &= ABn\bar{Y}_{\bullet\bullet\bullet}^2, & SSa &= Bn \textstyle\sum_{i=1}^{A} (\bar{Y}_{i\bullet\bullet} - \bar{Y}_{\bullet\bullet\bullet})^2, \\ SSb &= n \textstyle\sum_{i=1}^{A} \sum_{j=1}^{B} (\bar{Y}_{ij\bullet} - \bar{Y}_{i\bullet\bullet})^2, & SSe &= \textstyle\sum_{i=1}^{A} \sum_{j=1}^{B} \sum_{j=1}^{n} (Y_{ijk} - \bar{Y}_{ij\bullet})^2. \end{aligned}$$

Proof: We wish show that $SS\mu$, SSa, SSb, and SSe are independent, and derive their distributions. Instead of directly showing the zero correlation between the $\binom{4}{2} = 6$ quantities, we generalize the method used in Section 3.1.2 by defining

$$G_{ij} = b_{ij} + \bar{e}_{ij\bullet} \quad \text{and} \quad H_i = a_i + \bar{b}_{i\bullet} + \bar{e}_{i\bullet\bullet} = a_i + \bar{G}_{i\bullet}\ .$$

Now write

$$\begin{aligned}
Y_{ijk} &= \bar{Y}_{\bullet\bullet\bullet} && + (\bar{Y}_{i\bullet\bullet} - \bar{Y}_{\bullet\bullet\bullet}) + (Y_{ij\bullet} - \bar{Y}_{i\bullet\bullet}) + (Y_{ijk} - \bar{Y}_{ij\bullet}) \\
&= (\mu + \bar{H}_\bullet) && + (H_i - \bar{H}_\bullet) \quad + (G_{ij} - \bar{G}_{i\bullet}) + (e_{ijk} - \bar{e}_{ij\bullet}) \\
&= \mu + a_i + \bar{G}_{i\bullet} && \qquad\qquad\qquad + (G_{ij} - \bar{G}_{i\bullet}) + (e_{ijk} - \bar{e}_{ij\bullet}) \\
&= \mu + a_i + b_{ij} && + \bar{e}_{ij\bullet} \qquad\qquad\qquad\qquad\quad + (e_{ijk} - \bar{e}_{ij\bullet}) \\
&= \mu + a_i + b_{ij} && + e_{ijk},
\end{aligned} \tag{3.71}$$

where the second row follows because, working from right to left,

$$\begin{aligned}
Y_{ijk} - \bar{Y}_{ij\bullet} &= (\mu + a_i + b_{ij} + e_{ijk}) - (\mu + a_i + b_{ij} + \bar{e}_{ij\bullet}) = e_{ijk} - \bar{e}_{ij\bullet}, \\
Y_{ij\bullet} - \bar{Y}_{i\bullet\bullet} &= (\mu + a_i + b_{ij} + \bar{e}_{ij\bullet}) - (\mu + a_i + b_{i\bullet} + \bar{e}_{i\bullet\bullet}) \\
&= b_{ij} - b_{i\bullet} + \bar{e}_{ij\bullet} - \bar{e}_{i\bullet\bullet} = G_{ij} - \bar{G}_{i\bullet}, \\
\bar{Y}_{i\bullet\bullet} - \bar{Y}_{\bullet\bullet\bullet} &= (\mu + a_i + b_{i\bullet} + \bar{e}_{i\bullet\bullet}) - (\mu + \bar{a}_\bullet + b_{\bullet\bullet} + \bar{e}_{\bullet\bullet\bullet}) \\
&= a_i - \bar{a}_\bullet + b_{i\bullet} - b_{\bullet\bullet} + \bar{e}_{i\bullet\bullet} - \bar{e}_{\bullet\bullet\bullet} = H_i - \bar{H}_\bullet,
\end{aligned}$$

and $\bar{Y}_{\bullet\bullet\bullet} = \mu + \bar{a}_\bullet + b_{\bullet\bullet} + \bar{e}_{\bullet\bullet\bullet} = \mu + \bar{H}_\bullet$.

Next observe that $SSe = \sum\sum\sum (e_{ijk} - \bar{e}_{ij\bullet})^2$, and

$$SSb = n\sum\sum (G_{ij} - \bar{G}_{i\bullet})^2, \quad SSa = Bn\sum (H_i - \bar{H}_\bullet)^2, \quad SS\mu = ABn(\mu + \bar{H}_\bullet)^2.$$

From the independence of $\bar{X}$ and S_X^2 for normal samples, $SSe \perp \bar{e}_{ij\bullet}$. As SSe is a function of only $\bar{e}_{ij\bullet}$, and G_{ij}, H_i and $\bar{H}_\bullet$ are functions of $\bar{e}_{ij\bullet}$ and other random variables independent of $\bar{e}_{ij\bullet}$, $SSe \perp SSb$, $SSe \perp SSa$, and $SSe \perp SS\mu$.

Similarly, $SSb \perp \bar{G}_{i\bullet}$ and, because SSb does not involve a_i, it is independent of functions of a_i and $\bar{G}_{i\bullet}$, i.e., of H_i, so that $SSb \perp SSa$ and $SSb \perp SS\mu$. Finally, $SSa \perp \bar{H}_\bullet$, so $SSa \perp SS\mu$.

For the distribution of SSe, as, for each given i, j pair, $\sigma_e^{-2}\sum_{k=1}^{n} (e_{ijk} - \bar{e}_{ij\bullet})^2 \sim \chi^2_{n-1}$, and, from the independence of all the e_{ijk}, $\sigma_e^{-2} SSe \sim \chi^2_{AB(n-1)}$.

Next, $G_{ij} \sim \mathrm{N}(0, \sigma_b^2 + \sigma_e^2/n)$ or $\sqrt{n}G_{ij} \sim \mathrm{N}(0, \gamma_b)$, where $\gamma_b = n\sigma_b^2 + \sigma_e^2$, and $SSb/\gamma_b \sim \chi^2_{A(B-1)}$. Similarly, $H_i \sim \mathrm{N}(0, \sigma_a^2 + \sigma_b^2/B + \sigma_e^2/Bn)$ or $\sqrt{Bn}H_i \sim \mathrm{N}(0, \gamma_a)$, where $\gamma_a = Bn\sigma_a^2 + n\sigma_b^2 + \sigma_e^2$, so that

$$\frac{SSa}{\gamma_a} = \frac{Bn}{Bn}\frac{\sum_{i=1}^{A} Bn(H_i - \bar{H}_\bullet)^2}{\gamma_a} = \frac{\sum_{i=1}^{A} (\sqrt{Bn}H_i - \sqrt{Bn}\bar{H}_\bullet)^2}{\gamma_a} \sim \chi^2_{A-1}. \tag{3.72}$$

Lastly, $\bar{H}_\bullet + \mu \sim \mathrm{N}(\mu, \sigma_a^2/A + \sigma_b^2/AB + \sigma_e^2/ABn)$ or $\sqrt{ABn}(\bar{H}_\bullet + \mu) \sim \mathrm{N}(\sqrt{ABn}\mu, \gamma_a)$, so that dividing by $\sqrt{\gamma_a}$ and squaring gives $(SS\mu/\gamma_a) \sim \chi_1^2(ABn\mu^2/\gamma_a)$. ■

Summarizing, $SS\mu$, SSa, SSb, and SSe are independent, and

$$\frac{SS\mu}{\gamma_a} \sim \chi_1^2\left(\frac{ABn\mu^2}{\gamma_a}\right), \quad \frac{SSa}{\gamma_a} \sim \chi^2_{A-1}, \quad \frac{SSb}{\gamma_b} \sim \chi^2_{A(B-1)}, \quad \frac{SSe}{\sigma_e^2} \sim \chi^2_{AB(n-1)}, \tag{3.73}$$

where $\gamma_a = Bn\sigma_a^2 + n\sigma_b^2 + \sigma_e^2$ and $\gamma_b = n\sigma_b^2 + \sigma_e^2$.

The *EMS* are given by

$$\mathbb{E}[MS\mu] = \gamma_a \mathbb{E}\left[\chi_1^2\left(\frac{ABn\mu^2}{\gamma_a}\right)\right] = \gamma_a\left(1 + \frac{ABn\mu^2}{\gamma_a}\right) = \gamma_a + ABn\mu^2, \tag{3.74}$$

$$\mathbb{E}[MSa] = \frac{\gamma_a}{A-1}\mathbb{E}[\chi_{A-1}^2] = \gamma_a, \quad \mathbb{E}[MSb] = \frac{\gamma_b}{A(B-1)}\mathbb{E}[\chi_{A(B-1)}^2] = \gamma_b, \tag{3.75}$$

and

$$\mathbb{E}[MSe] = \frac{\sigma_e^2}{AB(n-1)}\mathbb{E}[\chi_{AB(n-1)}^2] = \sigma_e^2. \tag{3.76}$$

These results are summarized in their standard fashion in Table 3.5.

It is valuable to explicitly consider how the sums of squares are computed, for an $ABn \times 1$ vector $\mathbf{Y}$ in the lexicon ordering (3.66), generated, say, from lines 1–3 in Listing 3.10. The key is to use the matrices in (3.67), as shown in lines 2 and 4, in the code given in Listing 3.11, which also computes the closed-form m.l.e. solution (3.81).

Table 3.5 ANOVA table for the balanced two-factor nested REM. Notation B(A) is short for "B within A", indicating the hierarchy of the nested factor.

Source	df	*SS*	*EMS*
Mean	1	$ABn\bar{Y}_{\bullet\bullet\bullet}^2$	$\sigma_e^2 + n\sigma_b^2 + Bn\sigma_a^2 + ABn\mu^2$
A	$A-1$	$Bn\sum_{i=1}^{A}(\bar{Y}_{i\bullet\bullet} - \bar{Y}_{\bullet\bullet\bullet})^2$	$\sigma_e^2 + n\sigma_b^2 + Bn\sigma_a^2$
B(A)	$A(B-1)$	$n\sum_{i=1}^{A}\sum_{j=1}^{B}(\bar{Y}_{ij\bullet} - \bar{Y}_{i\bullet\bullet})^2$	$\sigma_e^2 + n\sigma_b^2$
Error	$AB(n-1)$	$\sum_{i=1}^{A}\sum_{j=1}^{B}\sum_{k=1}^{n}(Y_{ijk} - \bar{Y}_{ij\bullet})^2$	σ_e^2
Total	ABn	$\sum_{i=1}^{A}\sum_{j=1}^{B}\sum_{k=1}^{n}Y_{ijk}^2$	

```
1  SST=sum(y'*y); Ydddb=mean(y); SSu=A*B*n*Ydddb^2; % Yddb= \bar{Y}_{dot dot dot}
2  H=kron(eye(A), ones(B*n,1)); Yiddb=y'*H/(B*n); % Yiddb= \bar{Y}_{i dot dot}
3  SSa=B*n*sum( (Yiddb-Ydddb).^2 );
4  H=kron(eye(A*B), ones(n,1)); Yijdb=y'*H/n; % Yijdb= \bar{Y}_{i j dot}
5  m = kron(Yiddb,ones(1,B)); SSb=n*sum( (Yijdb-m).^2 );
6  m=kron(Yijdb', ones(n,1)); SSe=sum( (y-m).^2 );
7  check=SST-(SSu+SSa+SSb+SSe) % is zero
8  % Now the MLE, if the MLE variance components are positive
9  MSe=SSe/A/B/(n-1); MSb=SSb/A/(B-1); MSa=SSa/(A-1);
10 sigma2eMLE=MSe; sigma2bMLE=(MSb-MSe)/n;
11 sigma2aMLE = ( (1-1/A)*MSa - MSb )/B/n;
12 muMLE=Ydddb; MLE=[muMLE sigma2aMLE sigma2bMLE sigma2eMLE]
```

Program Listing 3.11: Computes the *SS* values in (3.70) for a given vector $\mathbf{Y}$ in lexicon order, corresponding to a two-way nested, both factors random, balanced REM, and the closed-form m.l.e. solution (3.81).

With respect to hypothesis test statistics, a test for $\sigma_a^2 > 0$ will be based on MSa divided not by MSe (which would otherwise test $\sigma_a^2 = \sigma_b^2 = 0$) but MSa divided by MSb, i.e.,

$$\frac{\frac{SSa}{\gamma_a}/(A-1)}{\frac{SSb}{\gamma_b}/A(B-1)} \sim F_{(A-1),A(B-1)} \quad \text{or} \quad F_a = \frac{MSa}{MSb} \sim \frac{\gamma_a}{\gamma_b} F_{A-1,A(B-1)}, \tag{3.77}$$

a scaled central F distribution. If $\sigma_a^2 = 0$, then $\gamma_a = \gamma_b$, so that an α-level test for $\sigma_a^2 = 0$ versus $\sigma_a^2 > 0$ rejects if $F_a > F^{\alpha}_{A-1,A(B-1)}$, where $F^{\alpha}_{n,d}$ is the $100(1-\alpha)$th percent quantile of the $F_{n,d}$ distribution. Likewise,

$$\frac{\frac{SSb}{\gamma_b}/A(B-1)}{\frac{SSe}{\sigma_e^2}/AB(n-1)} \sim F_{A(B-1),AB(n-1)} \quad \text{or} \quad F_b = \frac{MSb}{MSe} \sim \frac{\gamma_b}{\sigma_e^2} F_{A(B-1),AB(n-1)} \tag{3.78}$$

is a scaled F distribution. For $\sigma_b^2 = 0$, $\gamma_b = \sigma_e^2$, so that an α-level test for $\sigma_b^2 = 0$ versus $\sigma_b^2 > 0$ rejects if $F_b > F^{\alpha}_{A(B-1),AB(n-1)}$.

Now turning to point estimators, from (3.75) and (3.76), $\mathbb{E}[MSe] = \sigma_e^2$, $\mathbb{E}[MSb] = n\sigma_b^2 + \sigma_e^2$, and $\mathbb{E}[MSa] = Bn\sigma_a^2 + n\sigma_b^2 + \sigma_e^2$, so that

$$\hat{\sigma}_e^2 = MSe, \quad \hat{\sigma}_b^2 = (MSb - MSe)/n, \quad \text{and} \quad \hat{\sigma}_a^2 = (MSa - MSb)/Bn \tag{3.79}$$

yield unbiased estimators using the ANOVA method of estimation. A closed-form solution to the set of equations that equate zero to the first derivatives of the log-likelihood is available, and is the m.l.e. if all variance component estimates are positive. For μ, the m.l.e. is

$$\hat{\mu}_{\text{ML}} = \bar{Y}_{\bullet\bullet\bullet}, \tag{3.80}$$

which turns out to be true for *all* pure random effects models, balanced or unbalanced; see, e.g., Searle et al. (1992, p. 146). For the variance components, if they are positive,

$$\hat{\sigma}^2_{e,\text{ML}} = MSe, \quad \hat{\sigma}^2_{b,\text{ML}} = (MSb - MSe)/n, \quad \hat{\sigma}^2_{a,\text{ML}} = ((1 - A^{-1})MSa - MSb)/Bn; \tag{3.81}$$

see, e.g., Searle et al. (1992, p. 148). Point estimators of other quantities of interest can be determined from the invariance property of the m.l.e. For example, the m.l.e. of $\rho := \sigma_a^2/\sigma_e^2$ is, comparing (3.79) and (3.81),

$$\hat{\rho}_{\text{ML}} = \frac{\hat{\sigma}^2_{a,\text{ML}}}{\hat{\sigma}^2_{e,\text{ML}}} \approx \frac{\hat{\sigma}_a^2}{\hat{\sigma}_e^2} = \frac{MSa - MSb}{Bn\, MSe} =: \hat{\rho}. \tag{3.82}$$

3.3.1.2 Both Effects Random: Exact and Approximate Confidence Intervals

For confidence intervals, the easiest (and usually of least relevance) is for the error variance. Similar to (3.22) for the one-factor case, from (3.73), SSe/σ_e^2 is a pivot, so that a $100(1-\alpha)\%$ confidence interval for σ_e^2 is given by $(SSe/u, SSe/l)$ because

$$1 - \alpha = \Pr\left(l \leqslant \frac{SSe}{\sigma_e^2} \leqslant u\right) = \Pr\left(\frac{SSe}{u} \leqslant \sigma_e^2 \leqslant \frac{SSe}{l}\right), \tag{3.83}$$

where l and u are given by $\Pr(l \leqslant \chi^2_{AB(n-1)} \leqslant u) = 1 - \alpha$, and $0 < \alpha < 1$ is a chosen tail probability, typically 0.05.

Exact intervals for some variance ratios of interest are available. From (3.78),

$$\frac{\sigma_e^2}{\gamma_b}F_b = \frac{\sigma_e^2}{n\sigma_b^2 + \sigma_e^2}F_b \sim F_{A(B-1),AB(n-1)}$$

is a pivot, and, using similar manipulations as in the one-factor case, we obtain the intervals

$$1-\alpha = \Pr\left(\frac{L}{F_b} < \frac{\sigma_e^2}{n\sigma_b^2+\sigma_e^2} < \frac{U}{F_b}\right) = \Pr\left(\frac{F_b/U-1}{n} < \frac{\sigma_b^2}{\sigma_e^2} < \frac{F_b/L-1}{n}\right)$$

$$= \Pr\left(\frac{F_b-U}{nU+F_b-U} < \frac{\sigma_b^2}{\sigma_e^2+\sigma_b^2} < \frac{F_b-L}{nL+F_b-L}\right), \tag{3.84}$$

where $\Pr(L \leqslant F_{A(B-1),AB(n-1)} \leqslant U) = 1-\alpha$.

Wald-based approximate confidence intervals for σ_a^2 and σ_b^2 can be computed in the usual way, and the Satterthwaite approximation is also available. In particular, for $\sigma_a^2 = (\gamma_a - \gamma_b)/(Bn)$, where $\gamma_a = Bn\sigma_a^2 + n\sigma_b^2 + \sigma_e^2$ and $\gamma_b = n\sigma_b^2 + \sigma_e^2$, with $h_1 = -h_2 = (Bn)^{-1}$ and $d_1 = A-1$, $d_2 = A(B-1)$, then either from (3.29) and (3.31), or (3.33) and (3.34),

$$\hat{d} = \frac{(h_1\hat{\gamma}_a + h_2\hat{\gamma}_b)^2}{(h_1^2\hat{\gamma}_a^2/d_1 + h_2^2\hat{\gamma}_b^2/d_2)} = \frac{(\hat{\gamma}_a - \hat{\gamma}_b)^2}{\hat{\gamma}_a^2/(A-1) + \hat{\gamma}_b^2/A(B-1)} = \frac{(MSa-MSb)^2}{\frac{(MSa)^2}{A-1} + \frac{(MSb)^2}{A(B-1)}}, \tag{3.85}$$

and, for $1-\alpha = \Pr(l \leqslant \chi^2_{\hat{d}} \leqslant u)$,

$$1-\alpha \approx \Pr\left(\hat{d}\frac{(MSa-MSb)}{Bn\,u} \leqslant \sigma_a^2 \leqslant \hat{d}\frac{(MSa-MSb)}{Bn\,l}\right).$$

Similarly, for $\sigma_b^2 = (\gamma_b - \sigma_e^2)/n = n^{-1}(\mathbb{E}[MSb] - \mathbb{E}[MSe])$,

$$\hat{d} = \frac{(MSb-MSe)^2}{\frac{(MSb)^2}{A(B-1)} + \frac{(MSe)^2}{AB(n-1)}}, \tag{3.86}$$

and

$$1-\alpha \approx \Pr\left(\hat{d}\frac{(MSb-MSe)}{n\,u} \leqslant \sigma_b^2 \leqslant \hat{d}\frac{(MSb-MSe)}{n\,l}\right), \tag{3.87}$$

for u and l such that $1-\alpha = \Pr(l \leqslant \chi^2_{\hat{d}} \leqslant u)$.

As is clear from (3.82), an exact interval for $\rho = \sigma_a^2/\sigma_e^2$ is not available because there is no exact pivot, but applying the Satterthwaite approximation using (3.85) results in

$$\frac{\hat{\rho}}{\rho} = \frac{(MSa-MSb)/\sigma_a^2}{Bn\,MSe/\sigma_e^2} \overset{\text{app}}{\sim} F_{\hat{d},AB(n-1)}$$

being an approximate one. Thus, with L and U given by $\Pr(L \leqslant F_{\hat{d},AB(n-1)} \leqslant U) = 1-\alpha$ for $0 < \alpha < 1$, an approximate c.i. for ρ is

$$1-\alpha \approx \Pr\left(\frac{\hat{\rho}}{U} < \rho < \frac{\hat{\rho}}{L}\right), \quad \hat{\rho} = \frac{MSa-MSb}{Bn\,MSe}. \tag{3.88}$$

The bootstrap/saddlepoint-based method of Butler and Paolella (2002b) is also applicable in this case and yields higher accuracy for small sample sizes.

Letting $V = \sigma_a^2 + \sigma_b^2 + \sigma_e^2$ be the total variance, other ratios, such as σ_a^2/V, σ_b^2/V, and $(\sigma_a^2 + \sigma_b^2)/V$, are also of potential interest, as well as $\sigma_a^2/(\sigma_a^2 + \sigma_b^2)$ and $\sigma_b^2/(\sigma_a^2 + \sigma_b^2)$. In the balanced setting, if exact intervals are not available, the Satterthwaite method and/or the bootstrap/saddlepoint-based method can be invoked. These could then, in turn, be used for the unbalanced case by the bootstrap/q-calibration exercise.

Similar to the idea in the remark at the end of Section 3.1.6.2, it is highly instructional (and potentially useful) to make a program that inputs an unbalanced panel for a two-way nested REM, and outputs (among other things, such as the approximate m.l.e. based on the method discussed in Section 3.1.6.1) a confidence interval for, say, $\rho = \sigma_a^2/\sigma_e^2$, based on (3.88) using the bootstrap/q-calibration exercise described in Section 3.1.6.2. Naturally, other confidence intervals, such as for the individual variance components or other ratios of interest, could also be incorporated.

In doing so, the first orders of business are to (i) write a program to compute the estimates of the missing values (via optimization to get also the approximate covariance matrix), using the closed-form expression for the m.l.e. of the model parameters μ, σ_a^2, σ_b^2, and σ_e^2, and (ii) confirm that the approximate m.l.e. for μ, σ_a^2, and σ_b^2 are essentially equal to the true m.l.e. (as computed, say, by SAS), and that of σ_e^2 is off by a multiplicative factor of 1.0735 for the constellation of parameters and number of (and constellation of) missing values used, namely $A = 10$, $B = 6$, $n = 8$, $\mu = 5$, $\sigma_a = 1$, $\sigma_b = 0.4$, $\sigma_e = 0.8$, and 30 missing observations.

The precise constellation of missing values we chose that gave rise to this multiplicative factor of 1.0735 is shown in Listing 3.12. This correction factor needs to be applied because otherwise, the bootstrap inference will be jeopardized. At this point, the reader might protest: How can this be done without access to the true m.l.e., in particular, without, say, SAS? As mentioned in Section 3.1.6.1 in the context of the one-way model, one could use simulation (and only Matlab), taking the multiplicative adjustment to be that value such that the estimator's (mean, or possibly median) bias is minimized.

Program `REM2wayNestedUnbalancedSatterforrho` (not shown, leaving it as a wonderful exercise for the reader) accomplishes this. Reasonably reliable assessment of the actual coverage would require use of at least $s = 1{,}000$ simulated data sets, in which case, with $s = 1{,}000$, a simple 95% binomial confidence interval of the actual coverage probability (assuming the true coverage, and the observed actual, is $p = 0.90$) is, to two digits, $0.90 \pm 1.96\sqrt{p(1-p)/s} = (0.88, 0.92)$.

Use of `Boot` $= 250$ bootstrap replications (and, for each, $s_{\text{Miss}} = 250$ replications of the missing data and computation of the balanced-case interval (3.88)) takes, for a single simulated data set, about 30 to 60 *minutes* on a typical PC at the time of writing (and use of one core only) to produce the confidence interval of ρ. Such a simulation with $s = 1{,}000$ was done (with 24 cores and 21 hours), and resulted in an actual coverage of 0.927, suggesting that the actual coverage might be slightly larger than the nominal. Use of `Boot` $= s_{\text{Miss}} = 1{,}000$ takes correspondingly longer and resulted in an actual coverage of 0.926, suggesting that use of 250 is adequate and such that the larger nominal coverage does not stem from too small a choice of `Boot` or s_{Miss}.

A histogram of the interval lengths (not shown) reveals that it is roughly Gaussian, with an elongated right tail. The average interval length was 4.0 and the sample standard deviation of the lengths was 1.9, indicating how much uncertainly is inherent in confidence intervals for (ratios of) variance components, even with a respectable sample size.

```
A=10; B=6; n=8; mu=5; siga=1; sigb=0.4; sige=0.8; bad=1;
while bad
  a=siga*randn(A,1); b=sigb*randn(A*B,1); e=sige*randn(A*B*n,1);
  y=ones(A*B*n,1)*mu ...
    + kron(eye(A),ones(B*n,1))*a ...
    + kron(eye(A*B),ones(n,1))*b + e;
  iset=1:2:9; % set some values to missing, here 30 of them
  for iloop=1:length(iset)
    i=iset(iloop);
    j=1; k=1; ind=B*n*(i-1)+n*(j-1)+k; y(ind)=NaN;
    j=1; k=2; ind=B*n*(i-1)+n*(j-1)+k; y(ind)=NaN;
    j=3; k=1; ind=B*n*(i-1)+n*(j-1)+k; y(ind)=NaN;
    j=5; k=1; ind=B*n*(i-1)+n*(j-1)+k; y(ind)=NaN;
    j=6; k=1; ind=B*n*(i-1)+n*(j-1)+k; y(ind)=NaN;
    j=6; k=2; ind=B*n*(i-1)+n*(j-1)+k; y(ind)=NaN;
  end
  try
    [mu_miss, V_miss]=REM2wayNestedMLEMiss(y,A,B,n);
    bad=min(eig(V_miss)) < 0.1;
  catch %#ok<CTCH>
    bad=1;
  end
end
```

Program Listing 3.12: Simulates a two-way nested, both factors random, balanced REM, with the indicated constellation of parameters, and then sets 30 of the values to missing, as indicated. The use of `while bad` is to ensure that the data set results in a valid approximate covariance matrix for the estimated missing values. It is very rare that this is problematic, but it is necessary when using the bootstrap procedure for approximate confidence intervals with unbalanced data. The use of `try` and `catch` is because, also rarely but possibly, the BFGS optimization algorithm, as used in program `REM2wayNestedMLEMiss` and based on Matlab version 2010, can fail. It is important to note that both of these selection mechanisms can induce a sample selection bias, and could affect the small sample properties of the point and interval estimators. We ignore this issue because, first, both mechanisms are rarely engaged, and, second, because interest here centers on development of concepts and teaching. A more rigorous analysis would have to address and resolve both issues.

Figure 3.5 shows the resulting point estimates of the variance components based on the approximate m.l.e. with multiplicative factor adjustment for $\hat{\sigma}_e^2$.

3.3.1.3 Mixed Model Case

Recall from Section 2.1 that a mixed effects model is one that contains both fixed and random effects, outside of the grand mean and the error term. We now describe the two-way nested mixed model, such that the first factor, A, is fixed, and the second factor, B, is nested in A, and is random. Using our perpetual example with schools and writing evaluations from the beginning of Section 3.3, the model is now similar to that of Section 3.3.1.1, where both factors are random, but now the schools are considered fixed ("because we are interested in them").

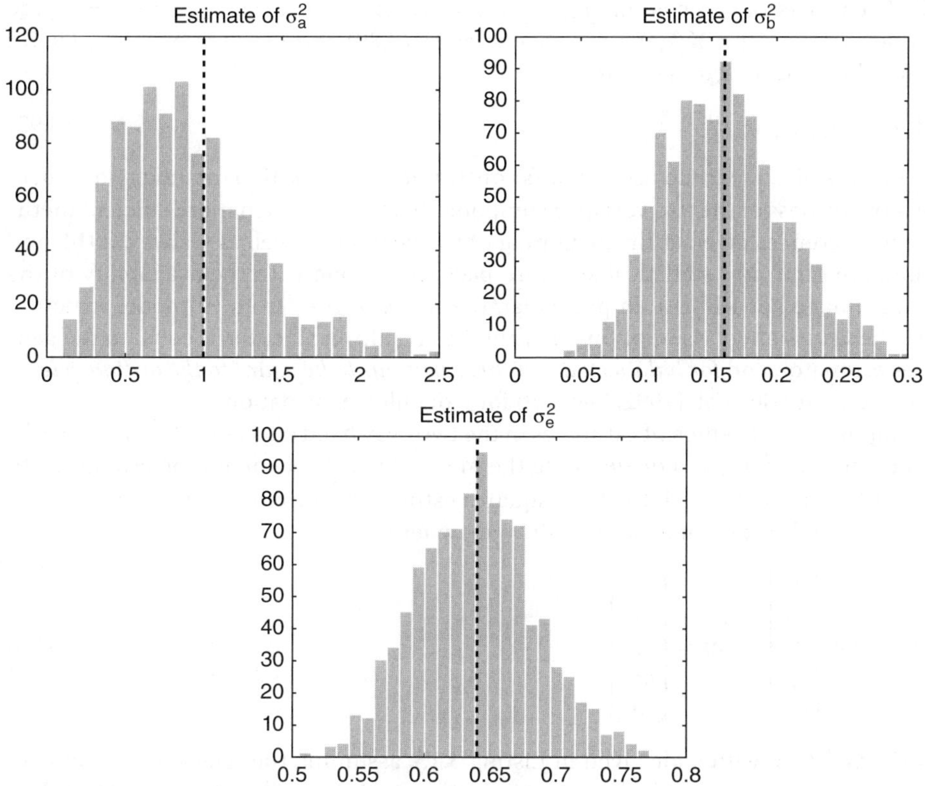

Figure 3.5 Point estimates of the three variance components for the two-way nested REM, based on the approximate m.l.e. with multiplicative factor adjustment for $\hat{\sigma}_e^2$ and use of 1,000 replications. True model parameters are those given in Listing 3.12, namely $\sigma_a^2 = 1^2$, $\sigma_b^2 = (0.4)^2$, $\sigma_e^2 = (0.8)^2$.

Exactly as in the case where both factors are assumed random, we observe Y_{ijk}, the kth observation in the jth subclass of the ith class, $i = 1, \dots, A$, $j = 1, \dots, B$, $k = 1, \dots, n$, but now assume that

$$Y_{ijk} = \mu + \alpha_i + b_{ij} + e_{ijk}, \quad \sum_{i=1}^{A} \alpha_i = 0, \quad b_{ij} \overset{\text{i.i.d.}}{\sim} \mathrm{N}(0, \sigma_b^2), \quad e_{ijk} \overset{\text{i.i.d.}}{\sim} \mathrm{N}(0, \sigma_e^2), \tag{3.89}$$

where the A classes are fixed levels of particular interest and, for each i, the B subclasses are randomly chosen. Differing from (3.65) and (3.65), first and second moments are

$$\mathbb{E}[Y_{ijk}] = \mu + \alpha_i, \quad \mathrm{Var}(Y_{ijk}) = \sigma_b^2 + \sigma_e^2, \quad \mathrm{Cov}(Y_{ijk}, Y_{ijk'}) = \sigma_b^2, \quad \mathrm{Cov}(Y_{ijk}, Y_{ij'k'}) = 0. \tag{3.90}$$

From expression (3.69), we again get decomposition (3.70).

Vector $\mathbf{Y}$ is expressed exactly the same as in (3.67), but using $\boldsymbol{\alpha} = (\alpha_1, \alpha_2, \dots, \alpha_A)'$ instead of $\mathbf{a}$, namely

$$\begin{aligned} \mathbf{Y} &= (\underline{1}_{ABn})\mu + (\mathbf{I}_A \otimes \underline{1}_{Bn})\boldsymbol{\alpha} + (\mathbf{I}_{AB} \otimes \underline{1}_n)\mathbf{b} + \mathbf{e} \\ &= \mathbf{X}\boldsymbol{\beta} + \boldsymbol{\epsilon}, \end{aligned} \tag{3.91}$$

where $\boldsymbol{\beta} = [\mu,\ \boldsymbol{\alpha}']'$, $\mathbf{X}$ consists of the column $\underline{1}_{ABn}$ followed by those of $\mathbf{I}_A \otimes \underline{1}_{Bn}$, and $\boldsymbol{\epsilon} = (\mathbf{I}_{AB} \otimes \underline{1}_n)\mathbf{b} + \mathbf{e}$. As always, let $\boldsymbol{\mu} := \mathbb{E}[\mathbf{Y}] = \mathbf{X}\boldsymbol{\beta}$. We can then express (3.89) and (3.91) as $\mathbf{Y} \sim \mathrm{N}_{ABn}(\boldsymbol{\mu}, \boldsymbol{\Sigma})$, where, similar to (3.68) but without component σ_a^2,

$$\boldsymbol{\Sigma} = (\mathbf{I}_{AB} \otimes \mathbf{J}_n)\sigma_b^2 + \mathbf{I}_{ABn}\sigma_e^2. \tag{3.92}$$

With the likelihood expressible, one could use Matlab's constrained optimization methods (to respect $\sum_{i=1}^{A} \alpha_i = 0$, and the positiveness of the two variance components), though much more efficient methods exist (see, e.g., the references given at the beginning of the chapter, as well as Galwey, 2014, and West et al., 2015) and are built into statistical software packages (along with the availability of the more popular restricted m.l.e., or REML). In particular, the m.l.e. of $\boldsymbol{\beta}$ is, from (i) model structure (3.91) and (3.92), (ii) the Gaussianity assumption on $\boldsymbol{\epsilon}$, and (iii) results in Chapter 1, equal to the generalized least squares estimator, and *for balanced data, this turns out to be equal to the ordinary least squares estimator*; see, e.g., Searle et al. (1992, Sec. 4.9) for a detailed explanation.

Recall (2.77) for computing the coefficient estimates in the two-way fixed effects ANOVA. Similarly, with $\mathbf{1}_A$ an A-length column of ones, and m_i denoting the mean of the Bn elements corresponding to the ith class of the first factor, $i = 1, \dots, A$, the least squares estimator $\widehat{\boldsymbol{\beta}}$ of $\boldsymbol{\beta}$ in (3.91) is given by the solution to the over-identified system of equations $\mathbf{Zc} = \mathbf{m}$, where

$$\mathbf{Z} = \begin{bmatrix} \mathbf{1}_A & \mathbf{I}_A \\ 0 & \mathbf{1}'_A \end{bmatrix}, \quad \mathbf{c} = \begin{bmatrix} \mu \\ \alpha_1 \\ \alpha_2 \\ \vdots \\ \alpha_A \end{bmatrix}, \quad \mathbf{m} = \begin{bmatrix} m_1 \\ m_2 \\ \vdots \\ m_A \\ 0 \end{bmatrix}, \tag{3.93}$$

The solution is $\mathbf{c} = (\mathbf{Z}'\mathbf{Z})^{-1}\mathbf{Z}'\mathbf{m}$, with code given in Listing 3.13, assuming the relevant variables are in computer memory, e.g., as constructed by lines 1–3 in Listing 3.10. This estimator is different than what is delivered by SAS' `proc mixed`, which sets $\hat{\alpha}_A = 0$ as the constraint. However, one can confirm that the estimates of the estimable functions $\hat{\mu} + \hat{\alpha}_i$ agree between SAS and use of (3.93).

Maximum likelihood estimates of the remaining parameters of the model, σ_b^2 and σ_e^2, can be obtained by numerically maximizing the log-likelihood of $(\mathbf{Y} - \mathbf{X}\widehat{\boldsymbol{\beta}}) \sim \mathrm{N}_{ABn}(\mathbf{0}, \boldsymbol{\Sigma})$, where $\boldsymbol{\Sigma}$ is given in (3.92). The reader is encouraged to construct a program, say `REM2wayNestedMixedMLE`, to accomplish this, and confirm that the point estimates of the two variance components are the same as those delivered by SAS. Regarding SAS code, first note that we can use the same data set as was generated by the code in Listing 3.10 for analysis in SAS, just treating the classes of factor A as fixed. The code in SAS Listing 3.5 then shows how the two-way mixed model is estimated with maximum likelihood.

As with previous models, we wish to show that $SS\mu$, $SS\alpha$, SSb, and SSe are independent, and derive their distributions and the corresponding EMS values. The independence of the SS follows the same

```
X=kron(eye(A),ones(B*n,1));
v=X*inv(X'*X)*X'*y; v=reshape(v,B*n,A)';
m=[v(:,1) ; 0]; Z=[ones(A,1), eye(A) ; 0 , ones(1,A)];
c=inv(Z'*Z)*Z'*m;
```

Program Listing 3.13: Computes the solution to (3.93).

```
ods html close; ods html;
/* close previous and open new */
filename ein 'REM2nested.txt';
data school;
  infile ein stopover;
  input Y school Evaluator;
run;
title 'Mixed REM 2-Way Nested Example';
proc varcomp method=ml;
  class school Evaluator;
  model Y=school Evaluator(school) / fixed=1;
run;
proc mixed method=ml cl=wald nobound covtest;
  class school Evaluator;
  model Y=school / cl solution;
  random Evaluator(school);
run;
```

SAS Program Listing 3.5: Reads in the data from the text file generated in Listing 3.3, treats the factor school as fixed, and uses `proc varcomp` and `proc mixed` with maximum likelihood.

argument as that for the two-factor nested REM with both effects random, in Section 3.3. The distributions of SSe/σ_e^2, SSb/γ, and $SS\mu/\gamma$ (noting $\bar{H}_\bullet = \bar{G}_{\bullet\bullet}$ because $\sum a_i = 0$) do not change, but where $\gamma = n\sigma_b^2 + \sigma_e^2$. Now, however, $H_i = (a_i + \bar{b}_{i\bullet} + \bar{e}_{i\bullet\bullet}) \sim \mathrm{N}(a_i, \sigma_b^2/B + \sigma_e^2/Bn)$ or $\sqrt{Bn}H_i \sim \mathrm{N}(\sqrt{Bn}a_i, \gamma)$ or $\sqrt{Bn/\gamma}H_i \sim \mathrm{N}(\kappa_i, 1)$, $\kappa_i = \sqrt{Bn/\gamma}a_i$, and $SSa/\gamma \sim \chi^2_{A-1}(v_a)$, where $v_a = \sum \kappa_i^2 = (Bn/\gamma)\sum a_i^2$.

Summarizing, with $\gamma_b = n\sigma_b^2 + \sigma_e^2$,

$$\frac{SS\mu}{\gamma_b} \sim \chi_1^2\left(\frac{ABn}{\gamma_b}\mu^2\right), \quad \frac{SS\alpha}{\gamma_b} \sim \chi^2_{A-1}\left(\frac{Bn}{\gamma_b}\sum \alpha_i^2\right), \quad \frac{SSb}{\gamma_b} \sim \chi^2_{A(B-1)}, \quad \frac{SSe}{\sigma_e^2} \sim \chi^2_{AB(n-1)}$$

are independent.

The EMS are derived just as in (3.74), (3.75), and (3.76) (but with $\sigma_a^2 = 0$), except for $\mathbb{E}[EM\alpha]$, given by

$$\mathbb{E}[EM\alpha] = \frac{\gamma_b}{A-1}\mathbb{E}[\chi^2_{A-1}(v_a)] = \frac{\gamma_b}{A-1}(A-1+v_a) = \gamma_b + \frac{Bn}{A-1}\sum a_i^2.$$

The results are organized in the ANOVA table (Table 3.6).

Inspection of the EMS in Table 3.6 shows that

$$F_\alpha = \frac{\dfrac{SS\alpha}{\gamma_b}/(A-1)}{\dfrac{SSb}{\gamma_b}/A(B-1)} = \frac{MS\alpha}{MSb} \sim F_{(A-1),A(B-1)}\left(\frac{Bn}{\gamma_b}\sum \alpha_i^2\right),$$

this being a (singly) noncentral F distribution. Under the null of no class effects (i.e., $\alpha_i = 0\ \forall i$), an α-level test rejects the null of no class effects if $F_\alpha > F^\alpha_{A-1,A(B-1)}$, where $F^\alpha_{n,d}$ is the $100(1-\alpha)$th percent quantile of the $F_{n,d}$ distribution. The test for the subclass random effect is the same as that in (3.78).

The ANOVA method estimators for the two variance components are the same as those for the all-random two-factor nested REM, namely $\hat{\sigma}_e^2 = MSe$ and $\hat{\sigma}_b^2 = (MSb - MSe)/n$. These calculations

Table 3.6 ANOVA table for balanced two-factor mixed nested REM.

Effect	df	SS	EMS
μ	1	$ABn\bar{Y}^2_{\bullet\bullet\bullet}$	$\sigma_e^2 + n\sigma_b^2 + ABn\mu^2$
α_i	$A-1$	$Bn\sum(\bar{Y}_{i\bullet\bullet} - \bar{Y}_{\bullet\bullet\bullet})^2$	$\sigma_e^2 + n\sigma_b^2 + \frac{Bn}{A-1}\sum\alpha_i^2$
b_{ij}	$A(B-1)$	$n\sum\sum(\bar{Y}_{ij\bullet} - \bar{Y}_{i\bullet\bullet})^2$	$\sigma_e^2 + n\sigma_b^2$
e_{ijk}	$AB(n-1)$	$\sum\sum\sum(Y_{ijk} - \bar{Y}_{ij\bullet})^2$	σ_e^2
Total	ABn	$\sum\sum\sum Y^2_{ijk}$	

can be done with the code in Listing 3.11 (and of course noting that they are not the m.l.e., as labeled there for the all-random two-factor nested model, and also ignoring the entry for `sigma2aMLE`). In SAS' `proc mixed`, one would change `method=ml` to `method=type1` in Listing 3.5 to produce the ANOVA method estimators. The estimates of the fixed effects are not affected by this choice.

The exact confidence interval for $\sigma_b^2/(\sigma_b^2 + \sigma_e^2)$ is the same as that given in (3.84), while the approximate one for σ_b^2 is given by (3.86) and (3.87).

3.3.2 Three Factors

We briefly outline now the three-factor nested case, along with the two mixed-model alternatives. The analysis follows the same general pattern as that in the two-factor case, so that even higher order models are straightforward to derive. Such models occur, not surprisingly, with far less frequency in practice. The ANOVA table and ANOVA point estimates for the four-factor nested REM can nevertheless be found in Graybill (1976, p. 639–640).

3.3.2.1 All Effects Random

We observe Y_{ijkl}, the lth observation in the kth subsubclass of the jth subclass of the ith class, $i = 1, \dots, A$, $j = 1, \dots, B$, $k = 1, \dots, C$, $l = 1, \dots, n$, and assume that it can be represented as

$$Y_{ijkl} = \mu + a_i + b_{ij} + c_{ijk} + e_{ijkl}, \tag{3.94}$$

where $a_i \overset{\text{i.i.d.}}{\sim} \mathrm{N}(0, \sigma_a^2)$, $b_{ij} \overset{\text{i.i.d.}}{\sim} \mathrm{N}(0, \sigma_b^2)$, $c_{ijk} \overset{\text{i.i.d.}}{\sim} \mathrm{N}(0, \sigma_c^2)$, and $e_{ijkl} \overset{\text{i.i.d.}}{\sim} \mathrm{N}(0, \sigma_e^2)$.

Theorem 3.6 ***Independence and Distribution*** By squaring and summing the expression

$$Y_{ijkl} = \bar{Y}_{\bullet\bullet\bullet\bullet} + (\bar{Y}_{i\bullet\bullet\bullet} - \bar{Y}_{\bullet\bullet\bullet\bullet}) + (Y_{ij\bullet\bullet} - \bar{Y}_{i\bullet\bullet\bullet}) + (Y_{ijk\bullet} - \bar{Y}_{ij\bullet\bullet}) + (Y_{ijkl} - \bar{Y}_{ijk\bullet}), \tag{3.95}$$

and confirming that cross terms are zero, the *SS* decomposition is given by

$$SST = SS\mu + SSa + SSb + SSc + SSe, \tag{3.96}$$

where each *SS* term corresponds to its counterpart in (3.95) from left to right. The values on the r.h.s. of (3.96) are independent, and

$$\frac{SS\mu}{\gamma_a} \sim \chi_1^2\left(\frac{ABCn\mu^2}{\gamma_a}\right), \quad \frac{SSa}{\gamma_a} \sim \chi^2_{A-1}, \quad \frac{SSb}{\gamma_b} \sim \chi^2_{A(B-1)},$$

Table 3.7 ANOVA table for the balanced three-factor nested REM.

Effect	df	SS	EMS
μ	1	$ABCn\bar{Y}^2_{\bullet\bullet\bullet\bullet}$	$\sigma_e^2 + n\sigma_c^2 + Cn\sigma_b^2 + BCn\sigma_a^2 + ABCn\mu^2$
a_i	$A-1$	$BCn\sum(\bar{Y}_{i\bullet\bullet\bullet} - \bar{Y}_{\bullet\bullet\bullet\bullet})^2$	$\sigma_e^2 + n\sigma_c^2 + Cn\sigma_b^2 + BCn\sigma_a^2$
b_{ij}	$A(B-1)$	$Cn\sum\sum(\bar{Y}_{ij\bullet\bullet} - \bar{Y}_{i\bullet\bullet\bullet})^2$	$\sigma_e^2 + n\sigma_c^2 + Cn\sigma_b^2$
c_{ijk}	$AB(C-1)$	$n\sum\sum\sum(\bar{Y}_{ijk\bullet} - \bar{Y}_{ij\bullet\bullet})^2$	$\sigma_e^2 + n\sigma_c^2$
e_{ijkl}	$ABC(n-1)$	$\sum\sum\sum\sum(Y_{ijkl} - \bar{Y}_{ijk\bullet})^2$	σ_e^2
Total	$ABCn$	$\sum\sum\sum\sum Y_{ijkl}^2$	

and

$$\frac{SSc}{\gamma_c} \sim \chi^2_{AB(C-1)}, \quad \frac{SSe}{\sigma_e^2} \sim \chi^2_{ABC(n-1)},$$

where $\gamma_a = BCn\sigma_a^2 + Cn\sigma_b^2 + n\sigma_c^2 + \sigma_e^2$, $\gamma_b = Cn\sigma_b^2 + n\sigma_c^2 + \sigma_e^2$, and $\gamma_c = n\sigma_c^2 + \sigma_e^2$.

Proof: See Problem 3.3. ■

The *EMS* are given in Table 3.7.

From Table 3.7, let

$$F_a = \frac{MSa}{MSb} \sim \frac{\gamma_a}{\gamma_b} F_{A-1,A(B-1)}, \qquad F_b = \frac{MSb}{MSc} \sim \frac{\gamma_b}{\gamma_c} F_{A(B-1),AB(C-1)},$$

and

$$F_c = \frac{MSc}{MSe} \sim \frac{\gamma_c}{\sigma_e^2} F_{AB(C-1),ABC(n-1)}.$$

Thus, if $\sigma_a^2 = 0$, then $\gamma_a = \gamma_b$, and an α-level test for $\sigma_a^2 = 0$ versus $\sigma_a^2 > 0$ rejects if $F_a > F^{\alpha}_{A-1,A(B-1)}$. If $\sigma_b^2 = 0$, then $\gamma_b = \gamma_c$, and an α-level test for $\sigma_b^2 = 0$ versus $\sigma_b^2 > 0$ rejects if $F_b > F^{\alpha}_{A(B-1),AB(C-1)}$. If $\sigma_c^2 = 0$, then $\gamma_c = \sigma_e^2$, so that an α-level test for $\sigma_c^2 = 0$ versus $\sigma_c^2 > 0$ rejects if $F_c > F^{\alpha}_{AB(C-1),ABC(n-1)}$.

From Table 3.7, the ANOVA method point estimators are

$$\hat{\sigma}_e^2 = MSe, \quad \hat{\sigma}_c^2 = \frac{MSc - MSe}{n}, \quad \hat{\sigma}_b^2 = \frac{MSb - MSc}{Cn}, \quad \hat{\sigma}_a^2 = \frac{MSa - MSb}{BCn}. \tag{3.97}$$

It is easy to generalize the Satterthwaite confidence intervals in the two-factor case to give

$$1 - \alpha \approx \Pr\left(\hat{d}\frac{(MSc - MSe)}{n\,u} \leqslant \sigma_c^2 \leqslant \hat{d}\frac{(MSc - MSe)}{n\,l}\right), \quad \hat{d} = \frac{(MSc - MSe)^2}{\left(\frac{(MSc)^2}{AB(C-1)} + \frac{(MSe)^2}{ABC(n-1)}\right)},$$

$$1 - \alpha \approx \Pr\left(\hat{d}\frac{(MSb - MSc)}{Cn\,u} \leqslant \sigma_b^2 \leqslant \hat{d}\frac{(MSb - MSc)}{Cn\,l}\right), \quad \hat{d} = \frac{(MSb - MSc)^2}{\left(\frac{(MSb)^2}{A(B-1)} + \frac{(MSc)^2}{AB(C-1)}\right)}$$

and

$$1-\alpha \approx \Pr\left(\hat{d}\frac{(MSa-MSb)}{BCn\ u} \leqslant \sigma_a^2 \leqslant \hat{d}\frac{(MSa-MSb)}{BCn\ l}\right), \quad \hat{d} = \frac{(MSa-MSb)^2}{\left(\frac{(MSa)^2}{(A-1)} + \frac{(MSb)^2}{A(B-1)}\right)},$$

for u and l such that $1-\alpha = \Pr(l \leqslant \chi^2_{\hat{d}} \leqslant u)$.

3.3.2.2 Mixed: Classes Fixed

We observe Y_{ijkl}, the lth observation in the kth subsubclass of the jth subclass of the ith class, $i = 1,\dots,A$, $j = 1,\dots,B$, $k = 1,\dots,C$, $l = 1,\dots,n$, and assume

$$Y_{ijkl} = \mu + \alpha_i + b_{ij} + c_{ijk} + e_{ijkl}, \quad \sum_{i=1}^{A} \alpha_i = 0, \quad b_{ij} \overset{\text{i.i.d.}}{\sim} \mathrm{N}(0,\sigma_b^2), \quad c_{ijk} \overset{\text{i.i.d.}}{\sim} \mathrm{N}(0,\sigma_c^2)$$

and $e_{ijkl} \overset{\text{i.i.d.}}{\sim} \mathrm{N}(0,\sigma_e^2)$. Problem 3.4 shows that

$$\frac{SS\mu}{\gamma_b} \sim \chi_1^2\left(\frac{ABCn\mu^2}{\gamma_b}\right), \quad \frac{SS\alpha}{\gamma_b} \sim \chi^2_{A-1}\left(\frac{BCn}{\gamma_b}\sum \alpha_i^2\right), \quad \frac{SSb}{\gamma_b} \sim \chi^2_{A(B-1)},$$

and, as in the all-random case, $SSc/\gamma_c \sim \chi^2_{AB(C-1)}$ and $SSe/\sigma_e^2 \sim \chi^2_{ABC(n-1)}$, where $\gamma_b = Cn\sigma_b^2 + n\sigma_c^2 + \sigma_e^2$ and $\gamma_c = n\sigma_c^2 + \sigma_e^2$. The EMS are given in Table 3.8.

From Table 3.8,

$$F_\alpha = \frac{\frac{SS\alpha}{\gamma_b}/(A-1)}{\frac{SSb}{\gamma_b}/A(B-1)} = \frac{MS\alpha}{MSb} \sim F_{(A-1),A(B-1)}\left(\frac{BCn}{\gamma_b}\sum \alpha_i^2\right)$$

follows a (singly) noncentral F distribution. Under the null of no class effects, an α-level test rejects the null of no class effects if $F_\alpha > F^\alpha_{A-1,A(B-1)}$. The tests for subclass and subsubclass randoms effects are identical to those in the all-random three-way effects model.

Point estimates for σ_e^2, σ_c^2, and σ_b^2 are given in (3.97), while approximate confidence intervals for σ_c^2 and σ_b^2 are the same as those in the all-random case.

Table 3.8 ANOVA table for the balanced three-factor mixed nested REM: Classes are fixed, sub- and subsubclasses are random.

Effect	df	SS	EMS
μ	1	$ABCn\bar{Y}^2_{\bullet\bullet\bullet\bullet}$	$\sigma_e^2 + n\sigma_c^2 + Cn\sigma_b^2 + ABCn\mu^2$
α_i	$A-1$	$BCn\sum(\bar{Y}_{i\bullet\bullet\bullet} - \bar{Y}_{\bullet\bullet\bullet\bullet})^2$	$\sigma_e^2 + n\sigma_c^2 + Cn\sigma_b^2 + \frac{BCn}{A-1}\sum\alpha_i^2$
b_{ij}	$A(B-1)$	$Cn\sum\sum(\bar{Y}_{ij\bullet\bullet} - \bar{Y}_{i\bullet\bullet\bullet})^2$	$\sigma_e^2 + n\sigma_c^2 + Cn\sigma_b^2$
c_{ijk}	$AB(C-1)$	$n\sum\sum\sum(\bar{Y}_{ijk\bullet} - \bar{Y}_{ij\bullet\bullet})^2$	$\sigma_e^2 + n\sigma_c^2$
e_{ijkl}	$ABC(n-1)$	$\sum\sum\sum\sum(Y_{ijkl} - \bar{Y}_{ijk\bullet})^2$	σ_e^2
Total	$ABCn$	$\sum\sum\sum\sum Y^2_{ijkl}$	

Table 3.9 ANOVA table for the balanced three-factor mixed nested REM: Classes and subclasses are fixed, subsubclasses are random.

Effect	df	SS	EMS
μ	1	$ABCn\bar{Y}^2_{\bullet\bullet\bullet\bullet}$	$\sigma_e^2 + n\sigma_c^2 + ABCn\mu^2$
α_i	$A-1$	$BCn\sum(\bar{Y}_{i\bullet\bullet\bullet} - \bar{Y}_{\bullet\bullet\bullet\bullet})^2$	$\sigma_e^2 + n\sigma_c^2 + \frac{BCn}{A-1}\sum\alpha_i^2$
β_{ij}	$A(B-1)$	$Cn\sum\sum(\bar{Y}_{ij\bullet\bullet} - \bar{Y}_{i\bullet\bullet\bullet})^2$	$\sigma_e^2 + n\sigma_c^2 + \frac{Cn}{A(B-1)}\sum\sum\beta_{ij}^2$
c_{ijk}	$AB(C-1)$	$n\sum\sum\sum(\bar{Y}_{ijk\bullet} - \bar{Y}_{ij\bullet\bullet})^2$	$\sigma_e^2 + n\sigma_c^2$
e_{ijkl}	$ABC(n-1)$	$\sum\sum\sum\sum(Y_{ijkl} - \bar{Y}_{ijk\bullet})^2$	σ_e^2
Total	$ABCn$	$\sum\sum\sum\sum Y_{ijkl}^2$	

3.3.2.3 Mixed: Classes and Subclasses Fixed

This is similar to the previous case but now $Y_{ijkl} = \mu + \alpha_i + \beta_{ij} + c_{ijk} + e_{ijkl}$, with

$$\sum_{i=1}^{A}\alpha_i = 0, \quad \sum_{j=1}^{B}\beta_{ij} = 0, \quad c_{ijk} \overset{\text{i.i.d.}}{\sim} \mathrm{N}(0,\sigma_c^2), \quad e_{ijkl} \overset{\text{i.i.d.}}{\sim} \mathrm{N}(0,\sigma_e^2),$$

i.e., both the α_i and the β_{ij} are fixed levels of particular interest and the β_{ij} are nested within the α_i (so that β_{ij} and $\beta_{i'j}$ are not related). Then, as shown in Problem 3.4,

$$\frac{SS\mu}{\gamma_c} \sim \chi_1^2(v_\mu), \quad \frac{SS\alpha}{\gamma_c} \sim \chi_{A-1}^2(v_\alpha), \quad \frac{SS\beta}{\gamma_c} \sim \chi_{A(B-1)}^2(v_\beta),$$
$$\frac{SSc}{\gamma_c} \sim \chi_{AB(C-1)}^2, \quad \text{and} \quad \frac{SSe}{\sigma_e^2} \sim \chi_{ABC(n-1)}^2,$$

where

$$v_\mu = \frac{ABCn\mu^2}{\gamma_c}, \quad v_\alpha = \frac{BCn}{\gamma_c}\sum_{i=1}^{A}\alpha_i^2, \quad v_\beta = \frac{Cn}{\gamma_c}\sum_{i=1}^{A}\sum_{j=1}^{B}\beta_{ij}^2,$$

and $\gamma_c = n\sigma_c^2 + \sigma_e^2$. The *EMS* are given in Table 3.9.

The test F_c for $\sigma_c^2 > 0$ is the same as before, while for testing class and subclass effects, respectively, the test statistics

$$F_\alpha = \frac{MS\alpha}{MSc} \sim F_{(A-1),AB(C-1)}(v_\alpha), \quad F_\beta = \frac{MSb}{MSc} \sim F_{A(B-1),AB(C-1)}(v_\beta),$$

should be used. Point estimates for σ_e^2 and σ_c^2 are given in (3.97). An approximate confidence interval for σ_c^2 is the same as in the all-random case.

3.4 Problems

Problem 3.1 For the two-factor crossed REM with interaction of Section 3.2.1, verify that (i) the cross terms from squaring and summing (3.42) indeed vanish, (ii) the r.h.s. terms in (3.43) are mutually independent, and (iii) the r.h.s. terms follow the stated distributions.

Problem 3.2 Similar to Problem 3.1 but now for the two-factor additive crossed REM of Section 3.2.1, verify the distributions and independence of the *SS*, along with the *E SS* values given in Table 3.3.

Problem 3.3 Extend the method used in Section 3.3.1.1 to the three-factor all-random model (3.94) in Section 3.3.2.1 to derive the distributions of $SS\mu$, SSa, SSb, SSc and SSe that lead to the ANOVA table (Table 3.7).

Problem 3.4 For both cases of the three-factor mixed nested REM, derive the distributions of the *SS* and the compute the *EMS* values.

Problem 3.5 Verify directly using basic principles that, for $G_{ij} \overset{\text{i.i.d.}}{\sim} \mathrm{N}(0,1)$,

$$\sum_{i=1}^{2}\sum_{j=1}^{2}(G_{ij} - \bar{G}_{i\bullet} - \bar{G}_{\bullet j} + \bar{G}_{\bullet\bullet})^2 \sim \chi^2_{(2-1)(2-1)}.$$

Try to extend this, also using only rudimentary distribution theory, to the more general case

$$\sum_{i=1}^{A}\sum_{j=1}^{B}(G_{ij} - \bar{G}_{i\bullet} - \bar{G}_{\bullet j} + \bar{G}_{\bullet\bullet})^2 \sim \chi^2_{(A-1)(B-1)},$$

with $A, B \in \{2, 3, \ldots\}$. Whether possible or not using basic principles, prove the latter using projection matrix results.

3.A Appendix: Solutions

1) i) Denote the five r.h.s. terms of (3.42) as $[1] = \bar{Y}_{\bullet\bullet\bullet}$, $[2] = \bar{Y}_{i\bullet\bullet} - \bar{Y}_{\bullet\bullet\bullet}$, $[3] = \bar{Y}_{\bullet j\bullet} - \bar{Y}_{\bullet\bullet\bullet}$, $[4] = \bar{Y}_{ij\bullet} - \bar{Y}_{i\bullet\bullet} - \bar{Y}_{\bullet j\bullet} + \bar{Y}_{\bullet\bullet\bullet}$ and $[5] = Y_{ijk} - \bar{Y}_{ij\bullet}$ and note that there are 10 cross terms. First observe that $\sum_i [4] = \sum_j [4] = 0$. It is easy to see that

$$[1]\sum_i [2] = [1]\sum_j [3] = [1]\sum_k [5] = [1]\sum_i [4] = 0.$$

Likewise,

$$[2]\sum_j [3] = 0, \quad [2]\sum_j [4] = 0, \quad [2]\sum_k [5] = 0$$

while

$$[3]\sum_i [4] = 0, \quad [3]\sum_k [5] = 0, \quad [4]\sum_k [5] = 0.$$

ii) As each term in (3.42) is normally distributed, independence of the *SS* can be directly shown by verifying that $\mathrm{Cov}([1],[2]) = \cdots = \mathrm{Cov}([4],[5]) = 0$ for each of the 10 pairs and all possible subscripts (Graybill, 1976, p. 630). For example, $\mathrm{Cov}([1],[4])$ is given by

$$\mathbb{E}\left[(\bar{a}_{\bullet} + \bar{b}_{\bullet} + \bar{c}_{\bullet\bullet} + \bar{e}_{\bullet\bullet\bullet}) \times \left(\begin{array}{c} (a_i + b_j + c_{ij} + \bar{e}_{ij\bullet}) - (a_i + \bar{b}_{\bullet} + \bar{c}_{i\bullet} + \bar{e}_{i\bullet\bullet}) \\ -(\bar{a}_{\bullet} + b_j + \bar{c}_{\bullet j} + \bar{e}_{\bullet j\bullet}) + (\bar{a}_{\bullet} + \bar{b}_{\bullet} + \bar{c}_{\bullet\bullet} + \bar{e}_{\bullet\bullet\bullet}) \end{array}\right)\right]$$

$$= \mathbb{E}[(\bar{c}_{\bullet\bullet} + \bar{e}_{\bullet\bullet\bullet}) \times (c_{ij} - \bar{c}_{i\bullet} - \bar{c}_{\bullet j} + \bar{c}_{\bullet\bullet} + \bar{e}_{ij\bullet} - \bar{e}_{i\bullet\bullet} - \bar{e}_{\bullet j\bullet} + \bar{e}_{\bullet\bullet\bullet})]$$
$$= \mathbb{E}[(\bar{c}_{\bullet\bullet})(c_{ij} - \bar{c}_{i\bullet} - \bar{c}_{\bullet j} + \bar{c}_{\bullet\bullet})] + \mathbb{E}[\bar{e}_{\bullet\bullet\bullet}(\bar{e}_{ij\bullet} - \bar{e}_{i\bullet\bullet} - \bar{e}_{\bullet j\bullet} + \bar{e}_{\bullet\bullet\bullet})]$$
$$= \frac{\sigma_c^2}{AB} - \frac{B\sigma_c^2}{AB^2} - \frac{A\sigma_c^2}{A^2B} + \frac{\sigma_c^2}{AB} + \text{(similar pattern)} = 0.$$

Alternatively, as with the nested models, a judiciously defined set of variables can simplify matters considerably. As in Stuart et al. (1999, p. 686), let

$$G_{ij} = \sqrt{n}(c_{ij} + \bar{e}_{ij\bullet}) \overset{\text{iid}}{\sim} \mathrm{N}(0, \gamma_c)$$
$$H_i = \sqrt{Bn}(a_i + \bar{c}_{i\bullet} + \bar{e}_{i\bullet\bullet}) = \sqrt{Bn}a_i + \sqrt{B}\bar{G}_{i\bullet}$$
$$K_j = \sqrt{An}(b_j + \bar{c}_{\bullet j} + \bar{e}_{\bullet j\bullet}) = \sqrt{An}b_j + \sqrt{A}\bar{G}_{\bullet j}$$
$$L = \sqrt{ABn}(\mu + \bar{a}_{\bullet} + \bar{b}_{\bullet} + \bar{c}_{\bullet\bullet} + \bar{e}_{\bullet\bullet\bullet}) \begin{array}{l} = \sqrt{ABn}(\mu + \bar{b}_{\bullet}) + \sqrt{A}\bar{H}_{\bullet} \\ = \sqrt{ABn}(\mu + \bar{a}_{\bullet}) + \sqrt{B}\bar{K}_{\bullet}, \end{array}$$

so that

$$\begin{aligned}
SS\mu &= ABn\bar{Y}^2_{\bullet\bullet\bullet} &&= L^2 \\
SSa &= Bn\sum(\bar{Y}_{i\bullet\bullet} - \bar{Y}_{\bullet\bullet\bullet})^2 &&= \sum(H_i - \bar{H}_{\bullet})^2 \\
SSb &= An\sum(\bar{Y}_{\bullet j\bullet} - \bar{Y}_{\bullet\bullet\bullet})^2 &&= \sum(K_i - \bar{K}_{\bullet})^2 \\
SSc &= n\sum\sum(\bar{Y}_{ij\bullet} - \bar{Y}_{i\bullet\bullet} - \bar{Y}_{\bullet j\bullet} + \bar{Y}_{\bullet\bullet\bullet})^2 &&= \sum\sum(G_{ij} - \bar{G}_{i\bullet} - \bar{G}_{\bullet j} + \bar{G}_{\bullet\bullet})^2 \\
SSe &= \sum\sum\sum(Y_{ijk} - \bar{Y}_{ij\bullet})^2 &&= \sum\sum\sum(e_{ijk} - \bar{e}_{ij\bullet})^2.
\end{aligned}$$

Then $\bar{e}_{ij\bullet} \perp SSe$ and, as the other SS are functions of $\bar{e}_{ij\bullet}$ (via G_{ij}, etc.) and random variables that do not arise in SSe (and are independent of $\bar{e}_{ij\bullet}$), it follows that SSe is independent of the other SS.

As

$$\mathrm{Cov}(\bar{G}_{i\bullet},\ G_{ij} - \bar{G}_{i\bullet} - \bar{G}_{\bullet j} + \bar{G}_{\bullet\bullet}) = \mathbb{E}[\bar{G}_{i\bullet}(G_{ij} - \bar{G}_{i\bullet} - \bar{G}_{\bullet j} + \bar{G}_{\bullet\bullet})]$$
$$= \mathbb{E}[\bar{G}_{i\bullet}\ G_{ij}] - \mathbb{E}[\bar{G}_{i\bullet}\ \bar{G}_{i\bullet}] - \mathbb{E}[\bar{G}_{i\bullet}\ \bar{G}_{\bullet j}] + \mathbb{E}[\bar{G}_{i\bullet}\ \bar{G}_{\bullet\bullet}]$$
$$= \frac{\gamma_c}{B} - \frac{B\gamma_c}{B^2} - \frac{\gamma_c}{AB} + \frac{B\gamma_c}{AB^2} = 0,$$

$\bar{G}_{i\bullet} \perp SSc$ and, as H_i is a function of $\bar{G}_{i\bullet}$ (and a_i), H_i is independent of SSc (SSc does *not* involve a_i) and thus $SSc \perp SSa$. Likewise, $SSc \perp SS\mu$ (as was also directly demonstrated above). From symmetry, $\mathrm{Cov}(\bar{G}_{\bullet j},\ G_{ij} - \bar{G}_{i\bullet} - \bar{G}_{\bullet j} + \bar{G}_{\bullet\bullet}) = 0$ so that $\bar{G}_{\bullet j} \perp SSc$ and, using the previous argument applied to the K_j, $SSc \perp SSb$.

For the remaining pairs,

$$\mathrm{Cov}(L,\ H_i - \bar{H}_{\bullet}) = \mathrm{Cov}((\bar{a}_{\bullet} + \bar{b}_{\bullet} + \bar{c}_{\bullet\bullet} + \bar{e}_{\bullet\bullet\bullet}),\ (a_i + \bar{c}_{i\bullet} + \bar{e}_{i\bullet\bullet}) - (\bar{a}_{\bullet} + \bar{c}_{\bullet\bullet} + \bar{e}_{\bullet\bullet\bullet}))$$
$$= \mathbb{E}[\bar{a}_{\bullet}(\ a_i - \bar{a}_{\bullet})] + \mathbb{E}[\bar{c}_{\bullet\bullet}(\bar{c}_{i\bullet} - \bar{c}_{\bullet\bullet})] + \mathbb{E}[\bar{e}_{\bullet\bullet\bullet}(\bar{e}_{i\bullet\bullet} - \bar{e}_{\bullet\bullet\bullet})]$$
$$= \frac{\sigma_a^2}{A} - \frac{\sigma_a^2}{A} + \frac{B\sigma_c^2}{AB^2} - \frac{\sigma_c^2}{AB} + \frac{Bn\sigma_e^2}{AB^2n^2} - \frac{\sigma_e^2}{ABn} = 0$$

so that $SS\mu \perp SSa$:

$$\mathrm{Cov}(L,\ K_i - \bar{K}_{\bullet}) = \mathrm{Cov}((\bar{a}_{\bullet} + \bar{b}_{\bullet} + \bar{c}_{\bullet\bullet} + \bar{e}_{\bullet\bullet\bullet}),\ (b_j + \bar{c}_{\bullet j} + \bar{e}_{\bullet j\bullet}) - (\bar{b}_{\bullet} + \bar{c}_{\bullet\bullet} + \bar{e}_{\bullet\bullet\bullet}))$$
$$= \mathbb{E}[\bar{b}_{\bullet}(b_j - \bar{b}_{\bullet})] + \mathbb{E}[\bar{c}_{\bullet\bullet}(\bar{c}_{\bullet j} - \bar{c}_{\bullet\bullet})] + \mathbb{E}[\bar{e}_{\bullet\bullet\bullet}(\bar{e}_{\bullet j\bullet} - \bar{e}_{\bullet\bullet\bullet})]$$
$$= 0 \quad \text{(from symmetry with above calculation)}$$

so that $SS\mu \perp SSb$; and $\mathrm{Cov}(H_i - \bar{H}_\bullet,\ K_i - \bar{K}_\bullet)$ is

$$\begin{aligned}
&\mathrm{Cov}((a_i + \bar{c}_{i\bullet} + \bar{e}_{i\bullet\bullet}) - (\bar{a}_\bullet + \bar{c}_{\bullet\bullet} + \bar{e}_{\bullet\bullet\bullet}),\ (b_j + \bar{c}_{\bullet j} + \bar{e}_{\bullet j\bullet}) - (b_\bullet + \bar{c}_{\bullet\bullet} + \bar{e}_{\bullet\bullet\bullet}))\\
&= \mathbb{E}[(\bar{c}_{i\bullet} - \bar{c}_{\bullet\bullet})(\bar{c}_{\bullet j} - \bar{c}_{\bullet\bullet})] + \mathbb{E}[(\bar{e}_{i\bullet\bullet} - \bar{e}_{\bullet\bullet\bullet})(\bar{e}_{\bullet j\bullet} - \bar{e}_{\bullet\bullet\bullet})]\\
&= \mathbb{E}[\bar{c}_{i\bullet}(\bar{c}_{\bullet j} - \bar{c}_{\bullet\bullet})] + \mathbb{E}[\bar{e}_{i\bullet\bullet}(\bar{e}_{\bullet j\bullet} - \bar{e}_{\bullet\bullet\bullet})] \quad \text{(from the previous calculations)}\\
&= \frac{\sigma_c^2}{AB} - \frac{B\sigma_c^2}{AB^2} + \frac{n\sigma_e^2}{BnAn} - \frac{Bn\sigma_e^2}{AB^2n^2} = 0,
\end{aligned}$$

showing that $SSa \perp SSb$.

iii) Recalling

$$\begin{aligned}
\gamma_\mu &= Bn\sigma_a^2 + An\sigma_b^2 + n\sigma_c^2 + \sigma_e^2, & \gamma_a &= Bn\sigma_a^2 + n\sigma_c^2 + \sigma_e^2,\\
\gamma_b &= An\sigma_b^2 + n\sigma_c^2 + \sigma_e^2, & \gamma_c &= n\sigma_c^2 + \sigma_e^2,
\end{aligned}$$

and observing

$$\begin{aligned}
H_i &= \sqrt{Bn}(a_i + \bar{c}_{i\bullet} + \bar{e}_{i\bullet\bullet}) \overset{\text{iid}}{\sim} \mathrm{N}(0, \gamma_a),\\
K_j &= \sqrt{An}(b_j + \bar{c}_{\bullet j} + \bar{e}_{\bullet j\bullet}) \overset{\text{iid}}{\sim} \mathrm{N}(0, \gamma_b),\\
L &= \sqrt{ABn}(\mu + \bar{a}_\bullet + \bar{b}_\bullet + \bar{c}_{\bullet\bullet} + \bar{e}_{\bullet\bullet\bullet}) \sim \mathrm{N}(\sqrt{ABn}\mu, \gamma_\mu),
\end{aligned}$$

the distributions of $SS\mu/\gamma_\mu$, SSa/γ_a, SSb/γ_b and SSe/γ_e are easily verified using the same derivation as in the one-way REM model.

A bit more work is required to confirm that $SSc/\gamma_c \sim \chi^2_{(A-1)(B-1)}$ or, for $G_{ij} \overset{\text{iid}}{\sim} \mathrm{N}(0, \gamma_c)$,

$$\gamma_c^{-1} \sum_i \sum_j (G_{ij} - \bar{G}_{i\bullet} - \bar{G}_{\bullet j} + \bar{G}_{\bullet\bullet})^2 \sim \chi^2_{(A-1)(B-1)},$$

though the result is intuitive, based on the two-factor fixed effects ANOVA model.

2) The independence and distributions of $SS\mu/\gamma_\mu$, SSa/γ_a and SSb/γ_b follow directly from the proof in the non-additive case (with $\sigma_c^2 = 0$). For the error term

$$(Y_{ijk} - \bar{Y}_{i\bullet\bullet} - \bar{Y}_{\bullet j\bullet} + \bar{Y}_{\bullet\bullet\bullet}) = e_{ijk} - \bar{e}_{i\bullet\bullet} - \bar{e}_{\bullet j\bullet} + \bar{e}_{\bullet\bullet\bullet},$$

its covariance with $\bar{Y}_{\bullet\bullet\bullet} = \mu + \bar{a}_\bullet + \bar{b}_\bullet + \bar{e}_{\bullet\bullet\bullet}$ is

$$\begin{aligned}
&\mathrm{Cov}(\bar{e}_{\bullet\bullet\bullet},\ e_{ijk} - \bar{e}_{i\bullet\bullet} - \bar{e}_{\bullet j\bullet} + \bar{e}_{\bullet\bullet\bullet})\\
&= \mathbb{E}[\bar{e}_{\bullet\bullet\bullet}\ e_{ijk}] - \mathbb{E}[\bar{e}_{\bullet\bullet\bullet}\ \bar{e}_{i\bullet\bullet}] - \mathbb{E}[\bar{e}_{\bullet\bullet\bullet}\ \bar{e}_{\bullet j\bullet}] + \mathbb{E}[\bar{e}^2_{\bullet\bullet\bullet}]\\
&= \frac{\sigma_e^2}{ABn} - \frac{Bn\sigma_e^2}{ABn\ Bn} - \frac{An\sigma_e^2}{ABn\ An} + \frac{\sigma_e^2}{ABn} = 0.
\end{aligned}$$

Its covariance with $\bar{Y}_{i\bullet\bullet} - \bar{Y}_{\bullet\bullet\bullet} = a_i - \bar{a}_\bullet + \bar{e}_{i\bullet\bullet} - \bar{e}_{\bullet\bullet\bullet}$ is, using the previous result,

$$\begin{aligned}
&\mathrm{Cov}(\bar{e}_{i\bullet\bullet},\ e_{ijk} - \bar{e}_{i\bullet\bullet} - \bar{e}_{\bullet j\bullet} + \bar{e}_{\bullet\bullet\bullet})\\
&= \mathbb{E}[\bar{e}_{i\bullet\bullet}\ e_{ijk}] - \mathbb{E}[\bar{e}_{i\bullet\bullet}\ \bar{e}_{i\bullet\bullet}] - \mathbb{E}[\bar{e}_{i\bullet\bullet}\ \bar{e}_{\bullet j\bullet}] + \mathbb{E}[\bar{e}_{i\bullet\bullet}\ \bar{e}_{\bullet\bullet\bullet}]\\
&= \frac{\sigma_e^2}{Bn} - \frac{Bn\sigma_e^2}{Bn\ Bn} - \frac{n\sigma_e^2}{Bn\ An} + \frac{Bn\sigma_e^2}{Bn\ ABn} = 0,
\end{aligned}$$

and similarly for $\bar{Y}_{\bullet j\bullet} - \bar{Y}_{\bullet\bullet\bullet}$, verifying independence of all SS terms.

3) Write

$$\begin{aligned}
Y_{ijkl} &= \bar{Y}_{\bullet\bullet\bullet\bullet} + (\bar{Y}_{i\bullet\bullet\bullet} - \bar{Y}_{\bullet\bullet\bullet\bullet}) + (Y_{ij\bullet\bullet} - \bar{Y}_{i\bullet\bullet\bullet}) + (Y_{ijk\bullet} - \bar{Y}_{ij\bullet\bullet}) + (Y_{ijkl} - \bar{Y}_{ijk\bullet}) \\
&= (\mu + \bar{I}_{\bullet}) + (I_i - \bar{I}_{\bullet}) + (H_{ij} - \bar{H}_{i\bullet}) + (G_{ijk} - \bar{G}_{ij\bullet}) + (e_{ijkl} - \bar{e}_{ijk\bullet}) \\
&\vdots \\
&= \mu + a_i + b_{ij} + c_{ijk} + e_{ijkl},
\end{aligned}$$

where

$$\begin{aligned}
G_{ijk} &= c_{ijk} + \bar{e}_{ijk\bullet}, & \bar{G}_{ij\bullet} &= \bar{c}_{ij\bullet} + \bar{e}_{ij\bullet\bullet}, \\
H_{ij} &= b_{ij} + \bar{G}_{ij\bullet}, & \bar{H}_{i\bullet} &= \bar{b}_{i\bullet} + \bar{G}_{i\bullet\bullet}, \\
I_i &= a_i + \bar{H}_{i\bullet}, & \bar{I}_{\bullet} &= \bar{a}_{\bullet} + \bar{H}_{\bullet\bullet},
\end{aligned}$$

so that

$$SSe = \sum\sum\sum\sum (e_{ijkl} - \bar{e}_{ijk\bullet})^2, \; SSc = n\sum\sum\sum (G_{ijk} - \bar{G}_{ij\bullet})^2,$$

$$SSb = Cn\sum\sum (H_{ij} - \bar{H}_{i\bullet})^2, \quad SSa = BCn\sum (I_i - \bar{I}_{\bullet})^2, \quad SS\mu = ABCn(\mu + \bar{I}_{\bullet})^2.$$

As in the lower order models,

$$\begin{aligned}
SSe \perp \bar{e}_{ijk\bullet} \quad &\Rightarrow \quad SSe \perp SSc, \quad SSe \perp SSb, \quad SSe \perp SSa, \quad SSe \perp SS\mu; \\
SSc \perp \bar{G}_{ij\bullet} \quad &\Rightarrow \quad SSc \perp SSb, \quad SSc \perp SSa, \quad SSc \perp SS\mu; \\
SSb \perp \bar{H}_{i\bullet} \quad &\Rightarrow \quad SSb \perp SSa, \quad SSb \perp SS\mu; \\
SSa \perp \bar{I}_{\bullet} \quad &\Rightarrow \quad SSa \perp SS\mu.
\end{aligned}$$

For the distributions,

$$\begin{aligned}
&\frac{SSe}{\sigma_e^2} \sim \chi^2_{ABC(n-1)} \\
G_{ijk} \sim \mathrm{N}\left(0, \sigma_c^2 + \frac{\sigma_e^2}{n}\right) \Rightarrow &\frac{SSc}{\gamma_c} \sim \chi^2_{AB(C-1)}, \quad \gamma_c = n\sigma_c^2 + \sigma_e^2 \\
H_{ij} \sim \mathrm{N}\left(0, \sigma_b^2 + \frac{\sigma_c^2}{C} + \frac{\sigma_e^2}{Cn}\right) \Rightarrow &\frac{SSb}{\gamma_b} \sim \chi^2_{A(B-1)}, \quad \gamma_b = Cn\sigma_b^2 + n\sigma_c^2 + \sigma_e^2 \\
I_i \sim \mathrm{N}\left(0, \sigma_a^2 + \frac{\sigma_b^2}{B} + \frac{\sigma_c^2}{BC} + \frac{\sigma_e^2}{BCn}\right) \Rightarrow &\frac{SSa}{\gamma_a} \sim \chi^2_{A-1}, \quad \gamma_a = BCn\sigma_a^2 + Cn\sigma_b^2 + n\sigma_c^2 + \sigma_e^2.
\end{aligned}$$

Finally,

$$(\mu + \bar{I}_{\bullet}) \sim \mathrm{N}(\mu, \sigma_a^2/A + \sigma_b^2/AB + \sigma_c^2/ABC + \sigma_e^2/ABCn),$$

so that

$$\sqrt{ABCn}(\mu + \bar{I}_{\bullet}) \sim \mathrm{N}(\mu\sqrt{ABCn}, \gamma_a), \quad \text{or} \quad \frac{SS\mu}{\gamma_a} \sim \chi_1^2\left(\frac{ABCn\mu^2}{\gamma_a}\right).$$

4) Recall the solution to Problem 3.3 in which

$$SSe = \sum\sum\sum (e_{ijk} - \bar{e}_{ij\bullet})^2, \quad SSb = n\sum\sum (G_{ij} - \bar{G}_{i\bullet})^2,$$

$$SSa = Bn\sum (H_i - \bar{H}_{\bullet})^2, \quad SS\mu = ABn(\mu + \bar{H}_{\bullet})^2.$$

The same argument shows that, in both mixed cases considered here, all the SS are independent and give rise to the same distributions for SS/σ_e^2 and SSc/γ_c. This also holds for SSb/γ_b in the first case.

Consider the first situation in which only the classes are fixed. For $SS\alpha$, observe that

$$I_i = (\alpha_i + \bar{H}_{i\bullet}) \sim \mathrm{N}(\alpha_i, \sigma_b^2/B + \sigma_c^2/BC + \sigma_e^2/BCn),$$

so that $\sqrt{BCn}I_i \sim \mathrm{N}(\sqrt{BCn}\alpha_i, \gamma_b)$, or $\sqrt{BCn/\gamma_b}I_i \sim \mathrm{N}(\kappa_i, 1)$, for $\kappa_i = \alpha_i\sqrt{BCn/\gamma_b}$ and $SS\alpha/\gamma_b \sim \chi^2_{A-1}(v_a)$, $v_a = \sum \kappa_i^2 = (BCn/\gamma_b)\sum \alpha_i^2$.

Similarly, as $\sum \alpha_i = 0$, $\bar{I}_\bullet = \bar{H}_{\bullet\bullet}$, and $(\mu + \bar{I}_\bullet) \sim \mathrm{N}(\mu, \sigma_b^2/AB + \sigma_c^2/ABC + \sigma_e^2/ABCn)$, or $\sqrt{ABCn}(\mu + \bar{I}_\bullet) \sim \mathrm{N}(\kappa_\mu, \gamma_b)$, where $\kappa_\mu = \mu\sqrt{ABCn}$ and $\gamma_b = Cn\sigma_b^2 + n\sigma_c^2 + \sigma_e^2$. Then $SS\mu/\gamma_b \sim \chi_1^2(v_\mu)$, for $v_\mu = \mu^2 ABCn/\gamma_b$. The EMS for b_{ij} and c_{ijk} do not change from the all-random model, while

$$\mathbb{E}[EM\mu] = \gamma_b\left(1 + \frac{ABCn}{\gamma_b}\mu^2\right) = \gamma_b + ABCn\mu^2,$$

$$\mathbb{E}[EM\alpha] = \frac{\gamma_b}{A-1}\left(A - 1 + \frac{BCn}{\gamma_b}\sum a_i^2\right) = \gamma_b + \frac{BCn}{A-1}\sum a_i^2.$$

Turning to the second case in which the classes and subclasses are fixed, $H_{ij} = (\beta_{ij} + \bar{G}_{ij\bullet}) \sim \mathrm{N}(\beta_{ij}, \sigma_c^2/C + \sigma_e^2/Cn) \Rightarrow \sqrt{Cn/\gamma_c} \sim \mathrm{N}(\kappa_{ij}, 1)$ leads to $SS\beta/\gamma_c \sim \chi^2_{A(B-1)}(v_\beta)$, $v_\beta = \sum_i\sum_j\kappa_{ij}^2 = (Cn/\gamma_c)\sum_i\sum_j\beta_{ij}^2$.

For $SS\alpha$, condition $\sum_j \beta_{ij} = 0$ yields

$$I_i = (\alpha_i + \bar{H}_{i\bullet}) = (\alpha_i + \bar{G}_{i\bullet\bullet}) = (\alpha_i + \bar{c}_{i\bullet\bullet} + \bar{e}_{i\bullet\bullet\bullet}) \sim \mathrm{N}(\alpha_i, \sigma_c^2/BC + \sigma_e^2/BCn),$$

or $SS\alpha/\gamma_c \sim \chi^2_{A-1}(v_\alpha)$, but where $v_\alpha = (BCn/\gamma_c)\sum \alpha_i^2$. Note that this is the same as in the first case above, except that γ_c replaces γ_b. Similarly, $SS\mu/\gamma_c \sim \chi_1^2(v_\mu)$, with now $v_\mu = \mu^2 ABCn/\gamma_c$. The EMS values follow the same calculation as above.

5) Let $G_{11}, G_{12}, G_{21}, G_{22}$ be i.i.d. N(0, 1). Then, with $A = B = 2$,

$$\bar{G}_{i\bullet} = B^{-1}(G_{i1} + \cdots G_{iB}) = (1/2)(G_{i1} + G_{i2}) \sim \mathrm{N}(0, 1/B) = N(0, 1/2),$$

similar for $\bar{G}_{\bullet j}$, and $\bar{G}_{\bullet\bullet} = (1/4)(G_{11} + G_{12} + G_{21} + G_{22})$,

$$\begin{aligned}
S &= \sum_{i=1}^{A}\sum_{j=1}^{B}(G_{ij} - \bar{G}_{i\bullet} - \bar{G}_{\bullet j} + \bar{G}_{\bullet\bullet})^2 \\
&= (G_{11} - \bar{G}_{1\bullet} - \bar{G}_{\bullet 1} + \bar{G}_{\bullet\bullet})^2 + (G_{12} - \bar{G}_{1\bullet} - \bar{G}_{\bullet 2} + \bar{G}_{\bullet\bullet})^2 \\
&\quad + (G_{21} - \bar{G}_{2\bullet} - \bar{G}_{\bullet 1} + \bar{G}_{\bullet\bullet})^2 + (G_{22} - \bar{G}_{2\bullet} - \bar{G}_{\bullet 2} + \bar{G}_{\bullet\bullet})^2 \\
&= (G_{11} - (1/2)(G_{11} + G_{12}) - (1/2)(G_{11} + G_{21}) + (1/4)(G_{11} + G_{12} + G_{21} + G_{22}))^2 \\
&\quad + (G_{12} - (1/2)(G_{11} + G_{12}) - (1/2)(G_{12} + G_{22}) + (1/4)(G_{11} + G_{12} + G_{21} + G_{22}))^2 \\
&\quad + (G_{21} - (1/2)(G_{21} + G_{22}) - (1/2)(G_{11} + G_{21}) + (1/4)(G_{11} + G_{12} + G_{21} + G_{22}))^2 \\
&\quad + (G_{22} - (1/2)(G_{21} + G_{22}) - (1/2)(G_{12} + G_{22}) + (1/4)(G_{11} + G_{12} + G_{21} + G_{22}))^2.
\end{aligned}$$

Using Maple, this reduces to 1/16 times

$$(G_{12} - G_{11} + G_{21} - G_{22})^2 + (G_{12} - G_{11} + G_{21} - G_{22})^2$$
$$+(G_{12} - G_{11} + G_{21} - G_{22})^2 + (G_{12} - G_{11} + G_{21} - G_{22})^2$$

or, with $C \sim \mathrm{N}(0, 4)$,

$$S = \frac{1}{4}(G_{11} - G_{12} - G_{21} + G_{22})^2 \sim \frac{1}{4}C^2 \sim \left(\frac{C}{2}\right)^2 \sim \chi^2_1 = \chi^2_{(2-1)(2-1)},$$

as was to be shown.

For the general case, the reader can try a proof by induction: Assume it holds for $A \geqslant 2$ and $B \geqslant 2$, and then attempt to confirm that it holds for $A + 1$ and B. (The proof then for increasing B is obviously symmetric with the case for increasing A).

Alternatively, one could proceed as follows. From

$$G_{ij} - \bar{G}_{i\bullet} - \bar{G}_{\bullet j} + \bar{G}_{\bullet\bullet} = (G_{ij} - \bar{G}_{\bullet\bullet}) - (\bar{G}_{i\bullet} - \bar{G}_{\bullet\bullet}) - (\bar{G}_{\bullet j} - \bar{G}_{\bullet\bullet})$$

and

$$\sum_i \sum_j (\bar{G}_{i\bullet} - \bar{G}_{\bullet\bullet})^2 = B \sum_i (\bar{G}_{i\bullet} - \bar{G}_{\bullet\bullet})^2,$$
$$\sum_i \sum_j (\bar{G}_{\bullet j} - \bar{G}_{\bullet\bullet})^2 = A \sum_j (\bar{G}_{\bullet j} - \bar{G}_{\bullet\bullet})^2,$$

and the three cross terms

$$\sum_i (\bar{G}_{i\bullet} - \bar{G}_{\bullet\bullet}) \sum_j (G_{ij} - \bar{G}_{\bullet\bullet}) = B \sum_i (\bar{G}_{i\bullet} - \bar{G}_{\bullet\bullet})^2$$
$$\sum_j (\bar{G}_{\bullet j} - \bar{G}_{\bullet\bullet}) \sum_i (G_{ij} - \bar{G}_{\bullet\bullet}) = A \sum_j (\bar{G}_{\bullet j} - \bar{G}_{\bullet\bullet})^2$$
$$\sum_i (\bar{G}_{i\bullet} - \bar{G}_{\bullet\bullet}) \sum_j (\bar{G}_{\bullet j} - \bar{G}_{\bullet\bullet}) = 0,$$

we have $\sum_i \sum_j (G_{ij} - \bar{G}_{i\bullet} - \bar{G}_{\bullet j} + \bar{G}_{\bullet\bullet})^2$

$$= \sum_i \sum_j (G_{ij} - \bar{G}_{\bullet\bullet})^2 - B \sum_i (\bar{G}_{i\bullet} - \bar{G}_{\bullet\bullet})^2 - A \sum_j (\bar{G}_{\bullet j} - \bar{G}_{\bullet\bullet})^2.$$

From basic normal theory results (see, e.g., (III.A.206)), the first term is clearly χ^2_{AB-1}. The second term is $\sum_i (\sqrt{B}\bar{G}_{i\bullet} - \sqrt{B}\bar{G}_{\bullet\bullet})^2$ and $\sqrt{B}\bar{G}_{i\bullet} \sim \mathrm{N}(0, 1)$, so that it is χ^2_{A-1}. Likewise, the third term is χ^2_{B-1}. Thus, if the three terms are independent, then their sum is χ^2 with $(AB - 1) - (A - 1) - (B - 1) = (A - 1)(B - 1)$.

Part II

Time Series Analysis: ARMAX Processes

4

The AR(1) Model

The auto-regressive assumption is often justified by the argument that omitted variables are subject to an auto-regressive process. This argument holds, however, only if all *omitted factors contributing to the additive disturbance are subject to auto-regressive processes with the* same *parameter. The widespread use of the auto-regressive correction in econometrics is explained by the fact that it accounts for serial correlation and is computationally efficient. The adaptive regression also explains serial correlation, is computationally efficient, and assumes an error structure which, in many situations, provides a better approximation of reality.*

(Thomas F. Cooley and Edward C. Prescott, 1973, p. 364)

In essentially all complex systems, whether in biology, economics, finance, medicine, meteorology, political science, sociology, etc., the actual mechanism that gives rise to the observed data is highly complicated and quite possibly changing over time. Moreover, it often involves a large number of (possibly interacting) factors, many of which will be difficult to measure and/or properly account for in a succinct model. As a result, it is of value to find simple approximations to reality that nevertheless capture some of the primary aspects of the process under study. This is the notion addressed in the above quote by Cooley and Prescott (1973), and for which allowing time-varying parameters might offer a better solution, as discussed in Section 5.6.

In this and subsequent chapters, interest centers on a finite set of univariate observations that form a time series, denoted $\{Y_t\}$, that are observed at equally spaced intervals of time, $t = 0, \dots, T$, and such that they are based on an underlying, unobserved, i.i.d. sequence of Gaussian random variables. The convention that time starts at zero (instead of one) is convenient when discussing the least squares estimator of the autoregressive parameter of an AR(1) model, and is often used (see, e.g., Hayashi, 2000, p. 573). Before proceeding, it is worth emphasizing the rather limited nature of our scope: There are often multiple time series of interest instead of just one, giving rise to multivariate time-series analysis (see, e.g., Hamilton, 1994, Lütkepohl, 2005, and Tsay, 2014), the data may not be equally spaced, and the Gaussianity assumption can be relaxed, not just to, say, continuous distributions exhibiting asymmetry and leptokurtosis, but also to discrete distributions (see, e.g., the numerous contributions in Davis et al., 2015, and the references therein).

Many processes that are observed through time exhibit **autocorrelation**, or the tendency for the observation in the current time period to be related, or correlated, to previous observations, usually in its very recent past. To explicitly capture this behavior, an autoregressive process can be used.

Linear Models and Time-Series Analysis: Regression, ANOVA, ARMA and GARCH, First Edition. Marc S. Paolella.

The simplest such model is the **first-order autoregressive**, or AR(1), process. Time series $\{Y_t\}$ follows a Gaussian AR(1) process if, for all t,

$$Y_t = aY_{t-1} + c + U_t, \qquad \text{with} \qquad U_t \overset{\text{i.i.d.}}{\sim} \mathrm{N}(0, \sigma^2), \tag{4.1}$$

and $a, c \in \mathbb{R}$. The same argument for using i.i.d. normal random variables for capturing the stochastic part of the linear model, namely an appeal to the central limit theorem, is applicable here as well. The U_t, provided they are i.i.d. but not necessarily Gaussian, are sometimes referred to as **white noise**, or the **innovation sequence** that "drives" the process.

Process (4.1) is a special case of the more general ARMA class of time-series models that will be considered in subsequent chapters. Nevertheless, (4.1) is quite useful in practice, and its properties are easier to derive than those of the more general ARMA process. As such, we dedicate a whole chapter to the AR(1) model.

4.1 Moments and Stationarity

We first derive the mean, variance, and covariances of the Y_t. Then, the concept of stationarity is introduced. With respect to a particular assumed data generating process, moments and stationarity conditions apply to $\{Y_t\}$, for all $t \in \mathbb{Z}$. Applications necessarily involve a finite amount of data, and, as mentioned above, we assume that the observed process starts at time $t = 0$.

By applying repeated substitution to (4.1), one can write, for any $0 < s \leqslant t$,

$$Y_t = a^s Y_{t-s} + c \sum_{i=0}^{s-1} a^i + \sum_{i=0}^{s-1} a^i U_{t-i}. \tag{4.2}$$

With $s = t$,

$$Y_t = a^t Y_0 + c \sum_{i=0}^{t-1} a^i + \sum_{i=0}^{t-1} a^i U_{t-i}, \tag{4.3}$$

so that, conditional on Y_0, Y_t is a weighted sum of i.i.d. normal random variables and a constant and is, hence, also normally distributed. Considering only the two interesting cases of $a = 1$ and $|a| < 1$, as $\mathbb{E}[U_t] = 0$, its conditional mean is easily seen to be

$$\mathbb{E}[Y_t \mid Y_0] = \begin{cases} ct + Y_0, & \text{if } a = 1, \\ c\left(\dfrac{1-a^t}{1-a}\right) + a^t Y_0, & \text{if } |a| < 1. \end{cases} \tag{4.4}$$

Remark l'Hôpital's rule is often useful for evaluating indeterminate forms or ratios: Let f and g, and their first derivatives, be continuous functions on (a, b). If

$$\lim_{x\to a^+} f(x) = \lim_{x\to a^+} g(x) = 0 \quad \text{and} \quad \lim_{x\to a^+} f'(x)/g'(x) = L,$$

then $\lim_{x\to a^+} f(x)/g(x) = L$. The result also holds for the limit as $x \to b^-$. Rigorous proofs can be found in virtually all real analysis textbooks. Most students remember this very handy result, but few can intuitively justify it. This can be done as follows: Assume f and g are continuous at a, so that $f(a) = g(a) = 0$. Recall that, for small $h > 0$,

$$f'(x) \approx \frac{f(x+h) - f(x)}{h} \quad \text{or} \quad f(x+h) \approx f(x) + hf'(x).$$

Using this gives

$$\lim_{x \to a^+} \frac{f(x)}{g(x)} = \lim_{h \to 0} \frac{f(a+h)}{g(a+h)}$$
$$\approx \lim_{h \to 0} \frac{f(a) + hf'(a)}{g(a) + hg'(a)} = \frac{f'(a)}{g'(a)} = \lim_{x \to a^+} \frac{f'(x)}{g'(x)},$$

which is the desired result. ∎

Notice that, from l'Hôpital's rule,

$$\lim_{a \to 1} \left(\frac{1 - a^t}{1 - a} \right) = \lim_{a \to 1} \left(\frac{-ta^{t-1}}{-1} \right) = t, \tag{4.5}$$

as would be expected from (4.4). Without l'Hôpital's rule, (4.5) can be seen directly as follows:

$$\begin{aligned}\lim_{a \to 1} \left(\frac{1 - a^t}{1 - a} \right) &= \lim_{a \to 1} \left(\frac{1}{1 - a} - \frac{a^t}{1 - a} \right) \\ &= \lim_{a \to 1} [1 + a + a^2 + a^3 + \cdots - (a^t + a^{t+1} + a^{t+2} + a^{t+3} + \cdots)] \\ &= \lim_{a \to 1} (1 + a + a^2 + a^3 + \cdots + a^{t-1}) = t.\end{aligned}$$

If $c = 0$, then $\mathbb{E}[Y_t \mid Y_0] = a^t Y_0$.

Using (4.3), the conditional variance of Y_t is $\mathbb{V}(Y_t \mid Y_0) = \mathbb{E}[(Y_t - \mathbb{E}[Y_t \mid Y_0])^2 \mid Y_0]$, or

$$\mathbb{E}[(U_t + aU_{t-1} + \cdots + a^{t-1}U_1)^2] = \sigma^2(1 + a^2 + a^4 + \cdots + a^{2t-2}),$$

i.e., for $t \geqslant 0$,

$$\mathbb{V}(Y_t \mid Y_0) = \begin{cases} \sigma^2 t, & \text{if} \quad |a| = 1, \\ \sigma^2 \left(\dfrac{1 - a^{2t}}{1 - a^2} \right), & \text{if} \quad |a| < 1, \end{cases} \tag{4.6}$$

as in Priestley (1981, p. 118). If $|a| < 1$, then $\mathbb{V}(Y_t)$ is finite in the limit as $t \to \infty$, and infinite otherwise. Furthermore, when $|a| < 1$, we define

$$\mu := \lim_{t \to \infty} \mathbb{E}[Y_t] = \frac{c}{1 - a}, \tag{4.7}$$

and

$$\gamma_0 := \lim_{t \to \infty} \mathbb{V}(Y_t) = \frac{\sigma^2}{1 - a^2}. \tag{4.8}$$

These are referred to as the **unconditional expected value** and **unconditional variance** of Y_t, respectively.

It is a very simple exercise for the reader to confirm that the model can be expressed as

$$Y_t - \mu = a(Y_{t-1} - \mu) + U_t, \quad |a| < 1, \tag{4.9}$$

which, upon rearranging, gives

$$Y_t = \mu(1 - a) + aY_{t-1} + U_t, \quad |a| < 1, \tag{4.10}$$

(and noting that $c = \mu(1 - a)$) showing that Y_t is a weighted average of its unconditional expectation, μ, and the previous period's value, Y_{t-1}, such that the weights sum to one, plus a Gaussian error term.

Figure 4.1 shows a simulated Gaussian AR(1) process with $a = 0.9, c = 0$, and $\sigma^2 = 1$. The first observation, Y_0, is drawn from the unconditional distribution $\mathrm{N}(\mu, \gamma_0)$, while the remaining are constructed via (4.1). The two dashed lines are $Y_0 a^t \pm 1.96V_t^{1/2}$, where $V_t = \mathbb{V}(Y_t \mid Y_0)$ from (4.6), and so provide 95% error bounds, individual for each $Y_t, t \geqslant 0$. One would thus expect about 5% of the observations to lie outside these bounds; for this series, 27 of the 500 are outside, which is indeed very close to 5%. Observe how it is only after about the 30th observation that the conditional distribution coincides with the unconditional $\mathrm{N}(\mu, \gamma_0)$ distribution. The rate at which this occurs depends on a; for $a = 0$, it is immediate, while as a approaches one, it occurs very slowly. Figure 4.2 is similar to 4.1 but with $a = 0.999$. Only at about $t = 2{,}500$ do the standard error lines begin to "flatten out".

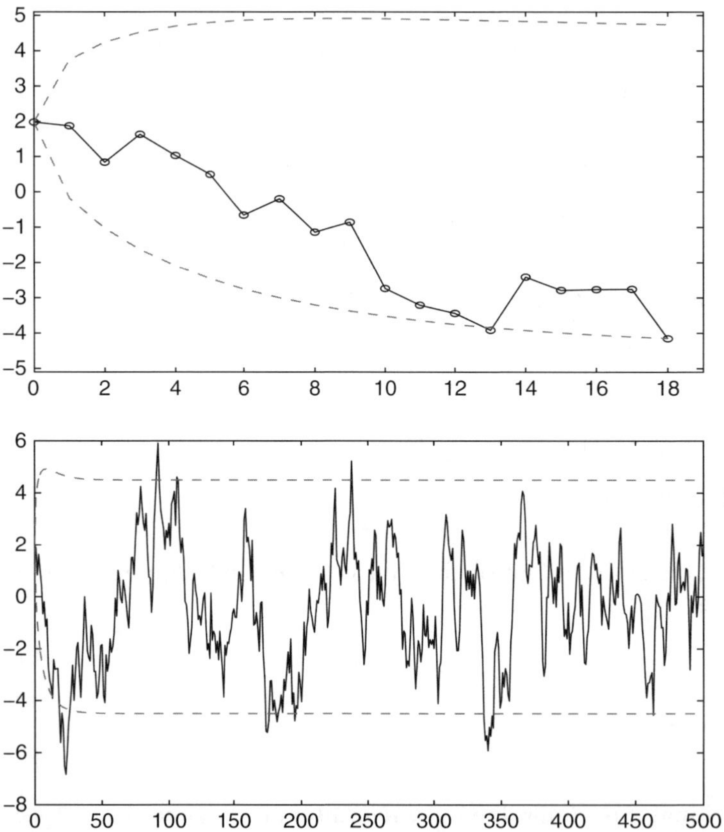

Figure 4.1 Example of a simulated AR(1) process with $a = 0.9$ and 95% error bounds. The top panel is just a magnified view of the beginning of the series. The figures are plotted such that the observations, indicated as dots, are connected to enhance visibility, though note that the process is not continuous.

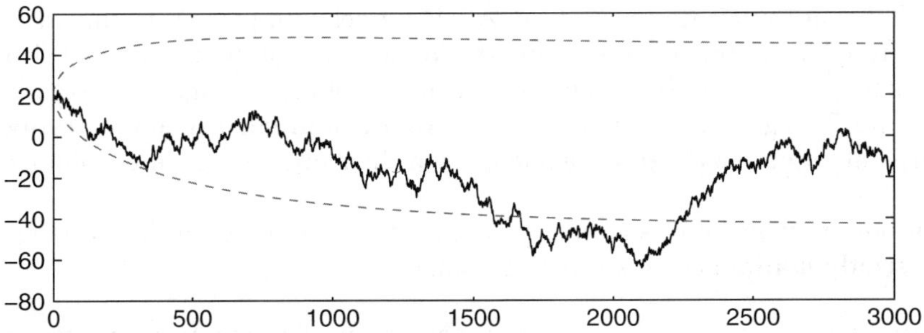

Figure 4.2 Example of AR(1) process with $a = 0.999$.

One of the most interesting and important aspects of time-series models is their covariance structure. For the AR(1) model with $|a| < 1$, using (4.2) and the fact that U_i is independent of Y_j for $i > j$, we have that, for $s \geqslant 0$ (and taking $c = 0$),

$$\mathrm{Cov}(Y_t, Y_{t-s}) = \mathrm{Cov}\left(a^s Y_{t-s} + \sum_{i=0}^{s-1} a^i U_{t-i},\ Y_{t-s} \right) = a^s \mathbb{E}[Y_{t-s}^2] = a^s \mathbb{V}(Y_{t-s}).$$

In particular, using (4.6), this is

$$\mathrm{Cov}(Y_t, Y_{t-s}) = \sigma^2 a^s \frac{1 - a^{2(t-s)}}{1 - a^2} = \sigma^2 \frac{a^s - a^{2t-s}}{1 - a^2}, \quad s \geqslant 0, \quad |a| < 1. \tag{4.11}$$

Somewhat more generally, for any r and t,

$$\mathrm{Cov}(Y_t, Y_r) = \sigma^2 \frac{a^{|t-r|} - a^{t+r}}{1 - a^2}, \tag{4.12}$$

as shown in Problem 4.3. Taking the limit of (4.11) as $t \to \infty$, we obtain the **unconditional covariance**,

$$\gamma_s := \lim_{t\to\infty} \mathrm{Cov}(Y_t, Y_{t-s}) = \frac{\sigma^2 a^s}{1 - a^2}, \quad s \geqslant 0, \quad |a| < 1.$$

For $s \geqslant 1$, this can be written as $\gamma_s = a\gamma_{s-1}$, for $|a| < 1$.

Now consider $\mathrm{Cov}(Y_t, Y_{t+s})$, for $s \geqslant 0$. Assume $c = 0$. From (4.2) with $t + s$ replacing t,

$$\mathrm{Cov}(Y_t, Y_{t+s}) = \mathrm{Cov}(Y_{t+s}, Y_t) = \mathrm{Cov}\left(a^s Y_t + \sum_{i=0}^{s-1} a^i U_{t+s-i}, Y_t \right) = a^s \mathbb{V}(Y_t).$$

Thus, for any $s \in \mathbb{Z}$,

$$\gamma_s = \frac{\sigma^2 a^{|s|}}{1 - a^2}, \quad |a| < 1. \tag{4.13}$$

It is important to note that, in contrast to (4.11), γ_s does not depend on the particular time point t, but rather only on the distance between two points of time. Furthermore, as $s \to \infty$, $\gamma_s \to 0$.

The Gaussian AR(1) process is **(weak-)stationary** if the condition $|a| < 1$ is fulfilled. More generally, a time series is weak-stationary if (i) the unconditional mean, unconditional variance, and unconditional covariances are finite and constant, and (ii) the unconditional covariances depend only on the time distance between two observations.

An example of a process that is not weak-stationary is an AR(1) model with $|a| \geqslant 1$. In this case, as the mean is not constant through time, the process is **non-stationary**. Non-stationarity could also arise if, say, parameter a is such that $|a| < 1$ changes (at one or more points) through time, or if c, or $\sigma^2 > 0$, changes through time. This gives rise to models with **structural breaks**, or **time-varying parameters**, for which a large literature has been developed, given their importance in economics (and surely other fields).

The opposite of weak-stationarity is not non-stationarity. A process could be such that it is not weak-stationary, but still is **strictly-stationary**, or strongly-stationary:

> A strictly-stationary process $\{Y_t\}$ is one whose joint probability distribution over k values in time does not change when the time index is shifted. That is, for a set of time indexes $t_1 < t_2 < \cdots < t_k$, $k \in \mathbb{N}$, the joint distribution of $Y_{t_1}, Y_{t_2}, \ldots, Y_{t_k}$ is the same as $Y_{t_1+s}, Y_{t_2+s}, \ldots, Y_{t_k+s}$, for all $s \in \mathbb{Z}$.

An example of a strictly-stationary process that is not weak-stationary is a sequence of i.i.d. Cauchy random variables: Observe that the joint distribution of any set of them is invariant to time shifts, but the mean (and variance) do not exist. If a strictly stationary process also has existence of second moments, then it is weak-stationary. A very important and useful fact is that, if process $\{Y_t\}$ is weakly stationary and such that the joint distribution of $Y_{t_1}, Y_{t_2}, \ldots, Y_{t_k}$, for a set of time indexes $t_1 < t_2 < \cdots < t_k$, and any $k \in \mathbb{N}$, is multivariate normal, then the process is also strictly stationary. Observe that the multivariate normal is characterized by its first two moments (the means, variances, and covariances); see, e.g., Chapter II.3, and Brockwell and Davis (1991, Sec. 1.6).

Remark The concept of weak-stationarity is a special case of **stationarity up to order** m. As in Priestley (1981, p. 105), time series $\{Y_t\}$ is stationary up to order m if, for a set of time indexes $t_1 < t_2 < \cdots < t_k$, $k \in \mathbb{N}$, and for all $s \in \mathbb{Z}$, all the joint moments up to order $m \geqslant 0$ of $\{Y(t_1), Y(t_2), \ldots, Y(t_k)\}$ exist, and equal the corresponding joint moments up to order m of $\{Y(t_1+s), Y(t_2+s), \ldots, Y(t_k+s)\}$. In particular,

$$\mathbb{E}[\{Y(t_1)\}^{m_1}\{Y(t_2)\}^{m_2}\cdots\{Y(t_k)\}^{m_k}] = \mathbb{E}[\{Y(t_1+s)\}^{m_1}\{Y(t_2+s)\}^{m_2}\cdots\{Y(t_k+s)\}^{m_k}],$$

for any $k \in \mathbb{N}$, and set of non-negative $\{m_i\}$ such that $m_1 + m_2 + \cdots + m_k \leqslant m$. Weak-stationarity is thus stationarity up to order two. ■

The AR(1) model with $|a| > 1$ is an example of what is referred to as an **explosive** process: Simulating and plotting such a process will quickly reveal why. If $a = 1$ and $c = 0$, then model (4.1) is referred to as a **random walk**, and is said to have a **unit root**: We will have much more to say about unit roots in Section 5.5.

If $a = 1$ and $c \neq 0$, then process (4.1) is said to be a **random walk with drift**. The name random walk attempts to describe the evolution of the Y_t through time: Assuming $c = 0$, the value of the process at time $t + 1$ is just the value in the previous time period t plus a random quantity that has equal probability of being positive or negative. As such, the process appears to randomly "walk" up or down through time. One of the distinguishing features of the random walk is that, recalling (4.6), its variance grows with t and is infinite as $t \to \infty$.

Remark Though sometimes attributed to George Pólya, the first use of the expression is usually accredited to Karl Pearson, from his article "The Problem of the Random Walk", in the July 1905 issue of *Nature* (Vol. LXXII). On page 294, it states "A man starts from a point O and walks 1 yard in a straight line; he then turns through any angle whatever and walks another 1 yard in a second straight line. He repeats this process n times. I require the probability that, after these stretches, he is at a distance between r and $r + \mathrm{d}r$ from his starting point O." ■

The top panel of Figure 4.3 illustrates a simulated random walk with $T = 200$ observations and $\sigma = 1$. The evolution of Y_t in the figure is very typical for small values of T in that artificial upward or downward trends appear over parts of the data. From the **data generating process**, or d.g.p., $Y_t = Y_{t-1} + U_t$, it is obvious that these trends are not genuine; they are referred to as **spurious trends**. Indeed, taking T larger makes this clearer: The bottom panel of Figure 4.3 shows the same random walk but with many more observations. One can already imagine the problems that will occur in the analysis of real data without (some) knowledge of the true underlying d.g.p.

Example 4.1 A simple model for stock prices is a random walk with drift, i.e., the price at time t, P_t, is the price at time $t-1$ plus a random quantity $U_t \overset{\text{i.i.d.}}{\sim} \mathrm{N}(0, \sigma^2)$, and possibly a small, positive,

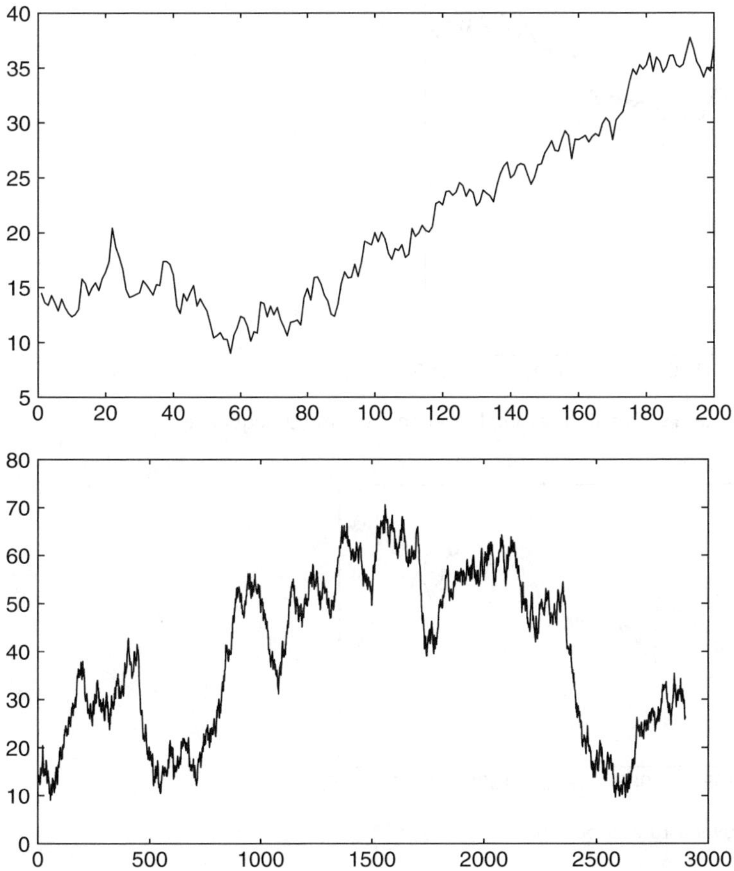

Figure 4.3 **Top:** Example of random walk. **Bottom:** Same random walk but showing more observations.

Figure 4.4 Random walk without drift (top) and with drift (bottom) based on the same U_t sequence.

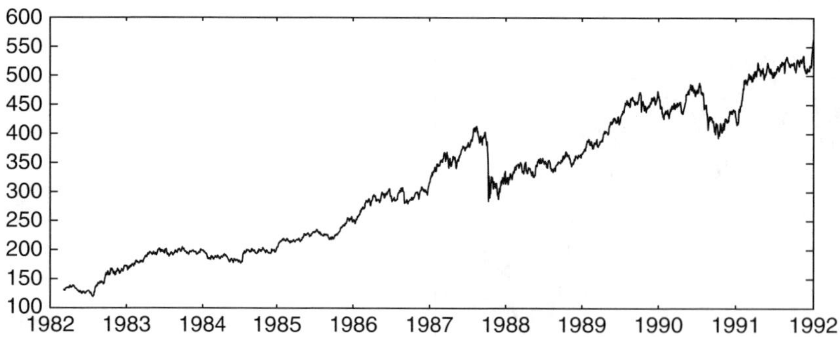

Figure 4.5 The daily S&P 500 stock index over a 10 year period.

constant value c that reflects the average rate of return over time for bearing the risk of the asset. That is, $P_t = P_{t-1} + c + U_t$. To illustrate, Figure 4.4 shows a random walk as before, i.e., without drift, and the same random walk but plus a constant drift term $c = 0.05$. This can be compared to Figure 4.5, which shows the S&P 500 stock index from 1982 to 1992. ■

The AR(1) process at the other stationarity border, $a = -1$, is less important than the $a = 1$ case, and behaves quite differently; see Problem 4.2.

4.2 Order of Integration and Long-Run Variance

Another concept related to stationarity is **order of integration**, denoted $I(d)$, where d is the order of integration and such that, first informally, the time series requires taking first differences d times in order to arrive at a stationary process. The definition from one of the seminal papers reads: "A series with no deterministic component which has a stationary, invertible ARMA representation after differencing d times is said to be integrated of order d..." (Engle and Granger, 1987, p. 252).

The order of integration plays a major role in the study of **co-integration** modeling of multivariate time series. While we will not need the concept in this book, it is worth pointing out that several definitions of $I(0)$ have been presented in the literature, and being $I(0)$ is not equivalent to being stationary. See Davidson (2009) for a detailed discussion of this issue, comparison of several definitions provided in the literature, and a suggestion for a formal definition in terms of an infinite stochastic sequence. Davidson (2009) shows that a necessary condition for a process to be $I(0)$ is if it admits a moving average (MA) representation (see Section 7.2) such that the MA coefficients are square-summable, i.e., $\sum_{j=0}^{\infty} b_j^2 < \infty$.

According to Hayashi (2000, p. 558), an $I(0)$ process is a strictly-stationary process whose **long-run variance** is finite and positive, where the long-run variance of process $\{Y_t\}$ is $\lim_{T\to\infty} \mathbb{V}(\sqrt{T}\bar{Y})$. This latter constraint purposely rules out the following case: Let $Y_t = e_t - e_{t-1}$, where $\{e_t\}$ are independent white noise, say $\{e_t\} \overset{\text{i.i.d.}}{\sim} \mathrm{N}(0,1)$. Observe that $Y_t \sim \mathrm{N}(0,2)$, and

$$\mathrm{Cov}(Y_t, Y_{t-1}) = \mathbb{E}[Y_t Y_{t-1}] = \mathbb{E}[(e_t - e_{t-1})(e_{t-1} - e_{t-2})] = -\mathbb{E}[e_{t-1}^2] = -1,$$

so that $\{Y_t\}$ is stationary. But $e_t - e_{t-1}$ being a first difference of i.i.d. random variables, in order for $\{Y_t\}$ to be $I(0)$, $\{e_t\}$ needs to be $I(1)$, but it is $I(0)$, thus giving rise to a definitional anomaly. We have

$$\sum_{t=1}^{T} Y_t = (e_1 - e_0) + (e_2 - e_1) + (e_3 - e_2) + \cdots = e_T - e_0,$$

and $\mathbb{V}(\sqrt{T}\bar{Y}) = \mathbb{V}\left(T^{-1/2} \sum_{t=1}^{T} Y_t\right) = \mathbb{V}(T^{-1/2}(e_T - e_0)) \xrightarrow{T\to\infty} 0$. Another way of seeing this is to let $V = \mathbb{V}(Y_t)$, and let e_0 be fixed. Then,

$$\begin{aligned}
Y_1 &= e_1 - e_0 \quad \Rightarrow e_1 = Y_1 + e_0\,, \\
Y_2 &= e_2 - e_1 = e_2 - e_0 - Y_1 \quad \Rightarrow e_2 = Y_2 + Y_1 + e_0\,, \\
Y_3 &= e_3 - e_2 = e_3 - e_0 - Y_2 - Y_1\,, \\
&\ \ \vdots \\
Y_t &= e_t - e_0 - \sum_{j=1}^{t-1} Y_t,
\end{aligned}$$

or

$$V = \mathbb{V}(e_t) - (t-1)V \quad \Rightarrow V = \frac{\mathbb{V}(e_t)}{t} \xrightarrow{t\to\infty} 0.$$

Hence the requirement that the long-run variance is finite and positive.

4.3 Least Squares and ML Estimation

4.3.1 OLS Estimator of a

Let $\mathbf{Y} = (Y_0, \ldots, Y_T)'$ be a sequence of $T+1$ observations from a stationary AR(1) process with autoregressive parameter a, additive term $c = 0$ and $U_t \overset{\text{i.i.d.}}{\sim} \text{N}(0, \sigma^2)$. (The situation with unknown c is a special case of the model class studied in Chapter 5). Define the $T \times (T+1)$ selection matrices $\mathbf{D}_T = [\mathbf{0} \mid \mathbf{I}_T]$ and $\mathbf{D}_{T-1} = [\mathbf{I}_T \mid \mathbf{0}]$, with $\mathbf{0}$ denoting a T-length column of zeros, and define

$$\mathbf{Y}_T := (Y_1, \ldots, Y_T)' = \mathbf{D}_T\mathbf{Y} \quad \text{and} \quad \mathbf{Y}_{T-1} := (Y_0, \ldots, Y_{T-1})' = \mathbf{D}_{T-1}\mathbf{Y}.$$

The AR(1) model has the form of a linear regression model, so the autoregressive parameter a can be estimated by regressing the last T observations onto the first T observations, and σ^2 can be estimated by the usual variance estimator in o.l.s. That is,

$$\widehat{a}_{\text{LS}} = \frac{\sum_{t=1}^{T} Y_t Y_{t-1}}{\sum_{t=0}^{T-1} Y_t^2} = \frac{\mathbf{Y}'_{T-1}\mathbf{Y}_T}{\mathbf{Y}'_{T-1}\mathbf{Y}_{T-1}} = \frac{\mathbf{Y}'\mathbf{D}'_{T-1}\mathbf{D}_T\mathbf{Y}}{\mathbf{Y}'\mathbf{D}'_{T-1}\mathbf{D}_{T-1}\mathbf{Y}}, \tag{4.14a}$$

and

$$\widehat{\sigma}^2_{\text{LS}} = \frac{\sum_{t=1}^{T} (Y_t - \widehat{a}_{\text{LS}} Y_{t-1})^2}{T-1}. \tag{4.14b}$$

These estimators are consistent and, except for very small sample sizes and/or cases in which $|a|$ is close to one, yield values that are very close to the m.l.e. Notice that, in $\widehat{\sigma}^2_{\text{LS}}$, the sum of the T squared residuals is divided by T minus the number of regressors, as is often done in least squares analysis to remove the bias of the estimator of σ^2; recall (1.59).

While there exist numerous methods of parameter estimation for the ARMA class of time-series models introduced later (the stationary AR(1) being a special case of which), the m.l.e. is usually preferred because it exhibits good small sample performance and possesses attractive asymptotic properties. We now illustrate three ways in which the likelihood can be determined.

4.3.2 Likelihood Derivation I

As $Y_0, U_1, U_2, \ldots, U_T$ are independent of one another, their joint p.d.f. is easily expressed. For Y_0, because observations previous to Y_0 are not available, we have to use its unconditional distribution: $Y_0 \sim \text{N}(0,\ \sigma^2/(1-a^2))$ or

$$f_{Y_0}(y) = \left(\frac{1-a^2}{2\pi\sigma^2}\right)^{1/2} \exp\left\{-\frac{1-a^2}{2\sigma^2}y^2\right\}. \tag{4.15}$$

Multiplying this by

$$f_{U_1,\ldots,U_T}(u_1,\ldots,u_T) = (2\pi\sigma^2)^{-T/2}\exp\left\{-\frac{1}{2\sigma^2}\sum_{t=1}^{T}u_t^2\right\}$$

yields the desired joint density. The transformation

$$\begin{array}{lcl} Y_0 & = & Y_0 \\ Y_1 & = & aY_0 + U_1 \\ Y_2 & = & aY_1 + U_2 \\ \vdots & & \\ Y_T & = & aY_{T-1} + U_T \end{array} \quad \Leftrightarrow \quad \begin{array}{lcl} Y_0 & = & Y_0 \\ U_1 & = & Y_1 - aY_0 \\ U_2 & = & Y_2 - aY_1 \\ \vdots & & \\ U_T & = & Y_T - aY_{T-1} \end{array}$$

has Jacobian

$$\mathbf{J} = \begin{bmatrix} 1 & 0 & 0 & \cdots & 0 \\ -a & 1 & 0 & & 0 \\ 0 & -a & 1 & \ddots & \vdots \\ \vdots & & \ddots & \ddots & 0 \\ 0 & \cdots & 0 & -a & 1 \end{bmatrix}, \qquad \det(\mathbf{J}) = 1,$$

so that, with $\mathbf{Y} = (Y_1,\ldots,Y_T)$,

$$f_{Y_0,\mathbf{Y}}(y_0,\mathbf{y}) = f_{Y_0}(y_0)\cdot(2\pi\sigma^2)^{-T/2}\exp\left\{-\frac{1}{2\sigma^2}\sum_{t=1}^{T}(y_t - ay_{t-1})^2\right\}, \tag{4.16}$$

where $f_{Y_0}(y_0)$ is given in (4.15).

A similar way of deriving (4.16) without the explicit use of the Jacobian is to express the joint density of $Y_0, Y_1,\ldots,Y_T$ as

$$f_{Y_0}f_{Y_1|Y_0}f_{Y_2|Y_1,Y_0}f_{Y_3|Y_2,Y_1,Y_0}\cdots f_{Y_T|Y_{T-1},\ldots,Y_1,Y_0}. \tag{4.17}$$

The density f_{Y_0} is given in (4.15). From (4.1), $Y_t = aY_{t-1} + U_t$, so that, conditional on Y_{t-1},

$$f_{Y_t|Y_{t-1},Y_{t-2},\ldots,Y_1,Y_0}(y_t \mid y_{t-1},y_{t-2},\ldots,y_1,y_0) = f_{Y_t|Y_{t-1}}(y_t \mid y_{t-1}),$$

and $(Y_t \mid Y_{t-1} = y_{t-1}) \sim \mathrm{N}(ay_{t-1},\sigma^2)$. Thus, the joint density of $Y_0, Y_1,\ldots,Y_T$ evaluated at $y_0, y_1,\ldots,y_T$ is

$$f_{Y_0}(y_0)\cdot\prod_{t=1}^{T}\frac{1}{\sigma\sqrt{2\pi}}\exp\left\{-\frac{1}{2\sigma^2}(y_t - ay_{t-1})^2\right\},$$

which is equivalent to (4.16).

The log-likelihood $\ell(a,\sigma;Y_0,\mathbf{Y}) = \ln f_{Y_0,\mathbf{Y}}(y_0,\mathbf{y};a,\sigma)$ can be straightforwardly programmed and maximized to obtain the m.l.e. and the (approximate) variance covariance matrix of a and σ. The resulting estimator is sometimes referred to as the **exact m.l.e.** because it is based on the exact likelihood, as opposed to an approximation, discussed next.

Alternatively, the **conditional m.l.e.** can be used, which, in this context, means conditioning on the first observation Y_0. Doing so implies that the conditional likelihood is just (4.17) but treating the term

f_{Y_0} as a constant (and, thus, omitting it). Importantly, inspection of (4.16) shows that the conditional m.l.e. is identical to the least squares estimator of a in (4.14) and also for σ^2 when $\hat{\sigma}^2_{\text{LS}}$ in (4.14) is divided by the number of observations instead of subtracting off the number of regressors. This also implies that, for known Y_0, $\sum_{t=1}^{T} Y_t Y_{t-1}$ and $\sum_{t=0}^{T-1} Y_t^2$ are **sufficient** for a, recalling the definition of sufficiency from, e.g., Section III.7.1. (See Forchini, 2000, for a study of the joint distribution of the minimal sufficient statistics for a and σ^2 in this setting with known Y_0.)

4.3.3 Likelihood Derivation II

It follows from (4.4), (4.12), and the normality assumption on the U_t, that $(\mathbf{Y} \mid Y_0 = y_0) \sim \mathrm{N}(\boldsymbol{\eta}, \sigma^2\boldsymbol{\Sigma}_0)$, where $\boldsymbol{\eta} = (ay_0, a^2y_0, \dots, a^T y_0)$ and the elements of $\boldsymbol{\Sigma}_0$ are given by (4.12) (without the σ^2). Thus, from the multivariate normal distribution,

$$f_{\mathbf{Y}|Y_0}(\mathbf{y} \mid y_0) = \frac{1}{|\sigma^2\boldsymbol{\Sigma}_0|^{1/2}(2\pi)^{T/2}} \exp\left\{-\frac{1}{2\sigma^2}(\mathbf{y}-\boldsymbol{\eta})'\boldsymbol{\Sigma}_0^{-1}(\mathbf{y}-\boldsymbol{\eta})\right\}. \tag{4.18}$$

The exact likelihood is then given by $f_{Y_0} \cdot f_{\mathbf{Y}|Y_0}$. It should be obvious that, computationally speaking, (4.16) is greatly preferred because (4.18) entails the construction and inverse of a $T \times T$ matrix.

4.3.4 Likelihood Derivation III

Instead of assuming that we observe the "start" of the time series, Y_0, we can envision having obtained a segment of a time series that extends infinitely far into the past. This implies that the unconditional expected value and covariances of the observations should be used. The former is zero from (4.7) and the latter are given in (4.13). In particular, the previous distinction between Y_0 and $\mathbf{Y}$ is no longer necessary and, with $\mathbf{y} = (y_0, y_1, \dots, y_T)'$,

$$f_{Y_0,\mathbf{Y}}(\mathbf{y}) = \frac{1}{|\sigma^2\boldsymbol{\Sigma}_{\text{unc}}|^{1/2}(2\pi)^{(T+1)/2}} \exp\left\{-\frac{1}{2\sigma^2}\mathbf{y}'\boldsymbol{\Sigma}_{\text{unc}}^{-1}\mathbf{y}\right\}, \tag{4.19}$$

where the (i,j)th element of $\boldsymbol{\Sigma}_{\text{unc}}$ is just γ_{i-j} (without the σ^2), i.e.,

$$\boldsymbol{\Sigma} = \boldsymbol{\Sigma}_{\text{unc}} = \frac{1}{1-a^2}\begin{bmatrix} 1 & a & a^2 & \cdots & a^T \\ a & 1 & a & \cdots & a^{T-1} \\ a^2 & a & 1 & \cdots & a^{T-2} \\ \vdots & \vdots & \vdots & \ddots & \vdots \\ a^T & a^{T-1} & a^{T-2} & \cdots & 1 \end{bmatrix}, \tag{4.20}$$

which is now a $(T+1)\times(T+1)$ matrix. In fact, (4.20) does not need to be computed and inverted because it can be shown that its inverse takes on the simple tri-diagonal band form

$$\boldsymbol{\Sigma}_{\text{unc}}^{-1} = \begin{bmatrix} 1 & -a & 0 & \cdots & 0 \\ -a & b & -a & & \vdots \\ 0 & -a & \ddots & & 0 \\ \vdots & & & b & -a \\ 0 & 0 & \cdots & -a & 1 \end{bmatrix}, \quad b = 1 + a^2, \tag{4.21}$$

as the reader should quickly confirm by direct multiplication. To make matters even more convenient, $|\boldsymbol{\Sigma}_{\text{unc}}| = 1/(1-a^2)$, independent of T. This determinant result turns out to be a special case of (6.21) for an AR(p) model.

Remark The general theory for inverses and determinants of patterned matrices such as (4.20) is well established. For this particular case, the result is given in Graybill (1983, p. 201), where more general, and other useful, interesting results can be found. Another impressive resource for results on structured matrices is Vandebril et al. (2008). ■

This form of the likelihood can be related to (4.16) by using the fact that $\boldsymbol{\Sigma}_{\text{unc}}^{-1}$ can be written as $\mathbf{C}'\mathbf{C}$, where

$$\mathbf{C} = \begin{bmatrix} \sqrt{1-a^2} & 0 & 0 & \cdots & 0 \\ -a & 1 & 0 & \cdots & 0 \\ 0 & -a & 1 & \ddots & \vdots \\ \vdots & 0 & \ddots & \ddots & 0 \\ 0 & \cdots & 0 & -a & 1 \end{bmatrix}.$$

Then

$$\mathbf{C}[y_0, y_1, \dots, y_T]' = [y_0\sqrt{1-a^2},\ y_1 - ay_0,\ \dots,\ y_T - ay_{T-1}]', \tag{4.22}$$

and

$$\mathbf{y}'\boldsymbol{\Sigma}_{\text{unc}}^{-1}\mathbf{y} = \mathbf{y}'\mathbf{C}'\mathbf{C}\mathbf{y} = y_0^2(1-a^2) + \sum_{t=1}^{T}(y_t - ay_{t-1})^2, \tag{4.23}$$

so that (4.19) and (4.16) are identical.

4.3.5 Asymptotic Distribution

If the data are generated by a stationary, mean-zero AR(1) process with i.i.d. $\mathrm{N}(0,\sigma^2)$ innovations, then the asymptotic distribution of the m.l.e. of a is given by

$$\sqrt{T}(\widehat{a}_{\text{ML}} - a) \overset{\text{asy}}{\sim} \mathrm{N}(0, 1-a^2), \tag{4.24}$$

i.e., for large enough samples, $\widehat{a}_{\text{ML}}$ is approximately normally distributed with mean a and variance $(1-a^2)/T$. The o.l.s. estimator has the same asymptotic distribution, i.e., $\sqrt{T}(\widehat{a}_{\text{LS}} - a) \overset{\text{asy}}{\sim} \mathrm{N}(0, 1-a^2)$. To informally motivate this latter result, use (4.14) to write

$$\widehat{a}_{\text{LS}} = \frac{\sum_{t=1}^{T} Y_t Y_{t-1}}{\sum_{t=0}^{T-1} Y_t^2} = \frac{\sum_{t=1}^{T}(aY_{t-1} + U_t)Y_{t-1}}{\sum_{t=1}^{T} Y_{t-1}^2} = a + \frac{\sum_{t=1}^{T} U_t Y_{t-1}}{\sum_{t=1}^{T} Y_{t-1}^2},$$

so that

$$\sqrt{T}(\widehat{a}_{\text{LS}} - a) = \frac{T^{-1/2}\sum_{t=1}^{T} U_t Y_{t-1}}{T^{-1}\sum_{t=1}^{T} Y_{t-1}^2} =: \frac{N}{D}.$$

As U_t is independent of Y_{t-1} and both have expected value zero, $\mathbb{E}[U_t Y_{t-1}] = 0$. Recall (see Section II.2.3) that, for independent random variables X and Y with means μ_X, μ_Y and finite variances σ_X^2, σ_Y^2,

$$\mathbb{E}[XY] = \mu_X\mu_Y \quad \text{and} \quad \mathbb{V}(XY) = \mu_Y^2\sigma_X^2 + \mu_X^2\sigma_Y^2 + \sigma_X^2\sigma_Y^2. \tag{4.25}$$

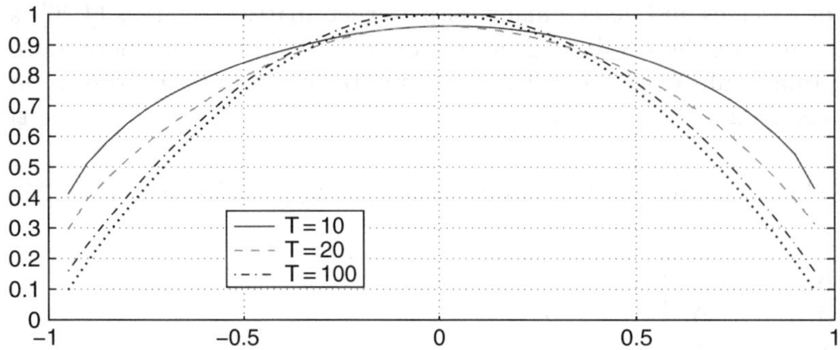

Figure 4.6 Variance of $\widehat{a}_{\text{LS}}$ times T as a function of a.

Thus, $\mathbb{V}(U_t Y_{t-1}) = \mathbb{V}(U_t)\mathbb{V}(Y_{t-1}) = \sigma^2\gamma_0$. As such, treating $\{U_t Y_{t-1}\}$ as an i.i.d. sequence, $\text{Var}(N) = T^{-1}T\mathbb{V}(U_t Y_{t-1}) = \sigma^2\gamma_0$ and, via the central limit theorem, $N \overset{\text{asy}}{\sim} \text{N}(0, \sigma^2\gamma_0)$. As denominator D is a consistent estimator of γ_0, we obtain

$$\sqrt{T}(\widehat{a}_{\text{LS}} - a) \overset{\text{asy}}{\sim} \frac{1}{\gamma_0}\text{N}(0, \sigma^2\gamma_0) = \text{N}\left(0, \frac{\sigma^2}{\gamma_0}\right) = \text{N}(0, 1 - a^2). \tag{4.26}$$

Observe that the asymptotic distribution does not involve σ^2. In fact, as shown in (4.37) below, $\widehat{a}_{\text{LS}}$ is independent of σ^2 for any sample size.

To illustrate the quality of the asymptotic expression for the variance, $(1 - a^2)/T$, Figure 4.6 shows T times the variance of $\widehat{a}_{\text{LS}}$ for three values of T, computed via simulation based on 10,000 replications, over a grid of a-values. The dotted line is $1 - a^2$, which is nearly reached for $T = 100$, while for smaller sample sizes, the variance curve is still essentially quadratic, but lower in a region around $a = 0$ and higher outside.

It is important to keep in mind that this asymptotic result relies on the normality of the innovations; see Problem 4.5 for the behavior of $\mathbb{V}(\widehat{a}_{\text{LS}})$ in some non-normal cases.

4.4 Forecasting

One of the most interesting and useful aspects of time-series analysis is extrapolating the model beyond the sample size T to obtain point and interval estimates of values that will be observed at a later date. This is referred to as **forecasting**, as a special case of the more general concept of **prediction**, though the two words are often used synonymously in the domain of statistical inference.[1]

Assume values of the process $\{Y_t\}$ are observed up to time T, which we refer to more generally as the **information set** up to time T, denoted Ω_T. Assume further that $c = 0$ in model (4.1), so that $Y_{T+1} = aY_T + U_{T+1}$. Then, a logical (and optimal) forecast of Y_{T+1} given Ω_T is aY_T, obtained by replacing the unobservable value of U_{T+1} by its expected value. We denote this as $Y_{T+1} \mid \Omega_T$ or, more commonly, as $Y_{T+1|T}$. As

$$Y_{T+1|T} - Y_{T+1} = aY_T - (aY_T + U_{T+1}) = -U_{T+1}\,,$$

1 At the risk of being pedantic, note that, in English, one might forecast the population size of the world, but we can only predict the future of humankind.

it follows that $\mathbb{E}[Y_{T+1|T} - Y_{T+1}] = 0$ and $\mathbb{V}(Y_{T+1|T} - Y_{T+1}) = \text{mse}(Y_{T+1|T}) = \sigma^2$. As a will almost always be unknown, it is replaced by an estimate, say $\hat{a} = \hat{a}_{\text{ML}}$, to get $\widehat{Y}_{T+1|T} := \hat{a}Y_T$ and

$$\begin{aligned}\text{mse}(\widehat{Y}_{T+1|T}) &= \mathbb{E}[(\widehat{Y}_{T+1|T} - Y_{T+1})^2] = \mathbb{E}[(\widehat{Y}_{T+1|T} - aY_T + aY_T - Y_{T+1})^2] \\ &= \mathbb{E}[(\widehat{Y}_{T+1|T} - aY_T)^2] + \mathbb{E}[(aY_T - Y_{T+1})^2] + \text{cross term} \\ &= \mathbb{E}[(\widehat{Y}_{T+1|T} - aY_T)^2] + \sigma^2, \end{aligned} \tag{4.27}$$

where the cross term

$$2\,\mathbb{E}[(\widehat{Y}_{T+1|T} - aY_T)(aY_T - Y_{T+1})] = 2\,\mathbb{E}[((\hat{a} - a)Y_T)(-U_{T+1})]$$

is zero because U_{T+1} is independent of $(\hat{a} - a)Y_T$. In this context, it is typical to speak of **mean square prediction error**, though we will not use this distinction here.

Observe from the nature of $\hat{a}$, from (4.14a) or (4.17), that the covariance between $\hat{a}$ and Y_T goes to zero as the sample size increases. Thus, (4.25), (4.8) and (4.24) imply that the first term in (4.27) is

$$\begin{aligned}\mathbb{E}[(\widehat{Y}_{T+1|T} - aY_T)^2] &= \mathbb{E}[((\hat{a} - a)Y_T)^2] \\ &\approx \mathbb{E}[Y_T^2]\mathbb{E}[(\hat{a} - a)^2] \approx \frac{\sigma^2}{1 - a^2}\frac{1 - a^2}{T} = \frac{\sigma^2}{T},\end{aligned} \tag{4.28}$$

so that

$$\text{mse}(\widehat{Y}_{T+1|T}) \approx (1 + T^{-1})\sigma^2. \tag{4.29}$$

Notice that this is independent of the true value of a and is asymptotically identical to $\text{mse}(Y_{T+1|T})$. In practice, as σ^2 is unknown, it is replaced by an estimate for computing (4.29).

Figure 4.7 shows the approximate m.s.e. of $\widehat{Y}_{T+1|T}$ for this model as a function of (positive values of) a, computed via simulation using 100,000 replications, based on $\sigma^2 = 4$ and $T = 10$. This was done for the three estimators: (i) the exact m.l.e. $\widehat{a}_{\text{ML}}$, (ii) the o.l.s. estimator $\widehat{a}_{\text{LS}}$, and (iii) the so-called **Yule–Walker estimator**, denoted $\widehat{a}_{\text{YW}}$. The latter is given by $\sum_{t=2}^{T} Y_t Y_{t-1} / \sum_{t=1}^{T} Y_t^2$ and discussed in Section 6.1.3.2. Notice that $\widehat{a}_{\text{LS}}$ and $\widehat{a}_{\text{YW}}$ are algebraically very close.

The first observation to be made from Figure 4.7 is that, with respect to m.s.e., the exact m.l.e. is superior to the conditional m.l.e. ($\widehat{a}_{\text{LS}}$) for all $0 < a < 1$, while for $0 < a < 0.6$, $\widehat{a}_{\text{YW}}$ is better than $\widehat{a}_{\text{ML}}$, but as a increases towards one, the m.s.e. of $\widehat{a}_{\text{YW}}$ grows significantly. Finally, for a near 0.5, the m.s.e.

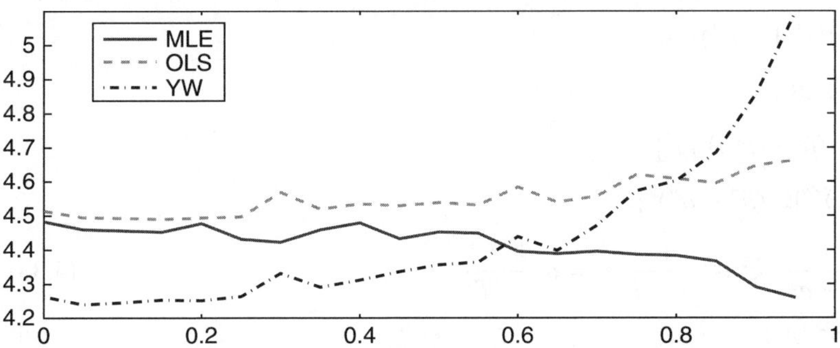

Figure 4.7 The m.s.e. of one-step ahead forecast $\widehat{Y}_{T+1|T} = \hat{a}Y_T$ for the AR(1) model as a function of autoregressive parameter a, with $\sigma = 2$. The solid line is the m.s.e. for $\widehat{a}_{\text{ML}}$, the dashed line is for $\widehat{a}_{\text{LS}}$, and the dash-dot line is for $\widehat{a}_{\text{YW}}$.

of $\widehat{a}_{\mathrm{ML}}$ is indeed very close to $\sigma^2(1+1/T) = 4.4$ from (4.29), but is higher for a near zero, and lower for a near one, indicating the approximate nature of (4.28).

A two-step ahead point forecast given Ω_T is denoted as $\widehat{Y}_{T+2|T}$. For a stationary AR(1) model with parameters a and σ^2, $\widehat{Y}_{T+2|T}$ is obtained by replacing all values in the equation for Y_{T+2} by their best estimates, i.e., as $Y_{T+2} = aY_{T+1} + U_{T+2}$, a is replaced by $\hat{a}$, Y_{T+1} by $\hat{Y}_{T+1|T}$, and U_{T+2} by zero. Thus, $\widehat{Y}_{T+2|T} = \hat{a}^2 Y_T$. Similarly, $\widehat{Y}_{T+h|T} = \hat{a}^h Y_T$, $h \geqslant 1$. Observe that $\lim_{h\to\infty} \hat{a}^h Y_T = 0$, so that "long term" forecasts converge to just the mean of the series. The m.s.e. of $\widehat{Y}_{T+h|T}$ can be obtained via a similar decomposition as (4.27), i.e., for $h \geqslant 1$,

$$\mathrm{mse}(\widehat{Y}_{T+h|T}) = \mathbb{E}[(\widehat{Y}_{T+h|T} - a^h Y_T)^2] + \mathbb{E}[(a^h Y_T - Y_{T+h})^2]. \tag{4.30}$$

For the latter term in (4.30), repeated substitution as in (4.2) yields

$$a^h Y_T - Y_{T+h} = a^h Y_T - \left(a^h Y_T + \sum_{i=1}^{h} a^{h-i} U_{T+i} \right),$$

from which it follows that

$$\mathbb{E}[(a^h Y_T - Y_{T+h})^2] = \sigma^2(1 + a^2 + a^4 + \cdots + a^{2(h-1)}). \tag{4.31}$$

For the first term on the r.h.s. of (4.30), with $\hat{a} = a + \epsilon$, applying the binomial theorem shows that

$$\hat{a}^h = (a+\epsilon)^h = \sum_{i=0}^{h} \binom{h}{i} a^i \epsilon^{h-i} = \epsilon^h + ha^1\epsilon^{h-1} + \cdots + ha^{h-1}\epsilon + a^h,$$

which, for large T (and, thus, small ϵ), is approximately $ha^{h-1}\epsilon + a^h$. Treating ϵ as a Gaussian random variable with mean zero, it follows that $\mathbb{E}[\hat{a}^h] = a^h$ and

$$\mathbb{V}(\hat{a}^h) \approx (ha^{h-1})^2 \mathbb{V}(\epsilon). \tag{4.32}$$

We can arrive at (4.32) in a different way, and also endow $\hat{a}^h$ with a distribution, as follows: As the asymptotic distribution of $\hat{a}$ is known from (4.24), that of $\hat{a}^h$ can be inferred. In particular, recall the **delta method** (see, e.g., Section III.3.1.4): For some differentiable function τ, $\tau(\hat{\theta}_{\mathrm{ML}}) \overset{\mathrm{asy}}{\sim} \mathrm{N}(\tau(\theta),\ \dot{\tau}^2 V)$, where V is the asymptotic variance of $\hat{\theta}_{\mathrm{ML}}$. Thus, with $\tau(\hat{a}) = \hat{a}^h$,

$$\sqrt{T}(\hat{a}^h - a^h) \overset{\mathrm{asy}}{\sim} \mathrm{N}(0, h^2 a^{2h-2}(1-a^2)).$$

Using this, we have, similar to (4.28),

$$\begin{aligned}
\mathbb{E}[(\widehat{Y}_{T+h|T} - a^h Y_T)^2] &= \mathbb{E}[((\hat{a}^h - a^h)Y_T)^2] \\
&\approx \mathbb{E}[Y_T^2]\mathbb{E}[(\hat{a}^h - a^h)^2] \\
&\approx \frac{\sigma^2}{1-a^2}\ h^2 a^{2h-2}\frac{1-a^2}{T} = \sigma^2 \frac{h^2 a^{2h-2}}{T}.
\end{aligned} \tag{4.33}$$

Thus, from (4.30), (4.31), and (4.33),

$$\mathrm{mse}(\widehat{Y}_{T+h|T}) \approx \sigma^2 \left(1 + a^2 + a^4 + \cdots + a^{2(h-1)} + \frac{h^2 a^{2h-2}}{T} \right). \tag{4.34}$$

As $\lim_{h\to\infty}(h^2a^{2h-2}) = 0$, $\lim_{h\to\infty}\text{mse}(\hat{Y}_{T+h|T}) = \sigma^2/(1-a^2) = \mathbb{V}(Y_t)$. In particular, the variance of the "long-term" forecast is just the variance of the series itself. In addition to the approximate nature of (4.34), observe that, for $h > 1$, (4.34) involves the two unknown parameters σ^2 and a. These need to be replaced by their respective estimates if (4.34) is to be computed in practice, adding further to its uncertainty.

To illustrate the nature of (4.34) in small samples, the top panel of Figure 4.8 shows the actual m.s.e. of h-step ahead forecasts, $h = 1, \ldots, 8$, of the AR(1) model with $a = 0.5$, $\sigma^2 = 4$, and $T = 10$ computed via simulation with 200,000 replications. These are compared to (4.34) using the true values of a and σ, and also (4.34) without the last term. Unsurprisingly, the m.s.e. based on the conditional m.l.e. is higher than that based on the exact m.l.e. More importantly, we see that the true m.s.e. based on the exact m.l.e. and (4.34) with all terms are reasonably close, particularly for $h \leqslant 3$, while use of (4.34) without the last term is relatively inaccurate, but converges to (4.34) as h increases. The limiting value as h increases is just the process variance, which in this case is $\sigma^2/(1-a^2) = 5.\bar{3}$.

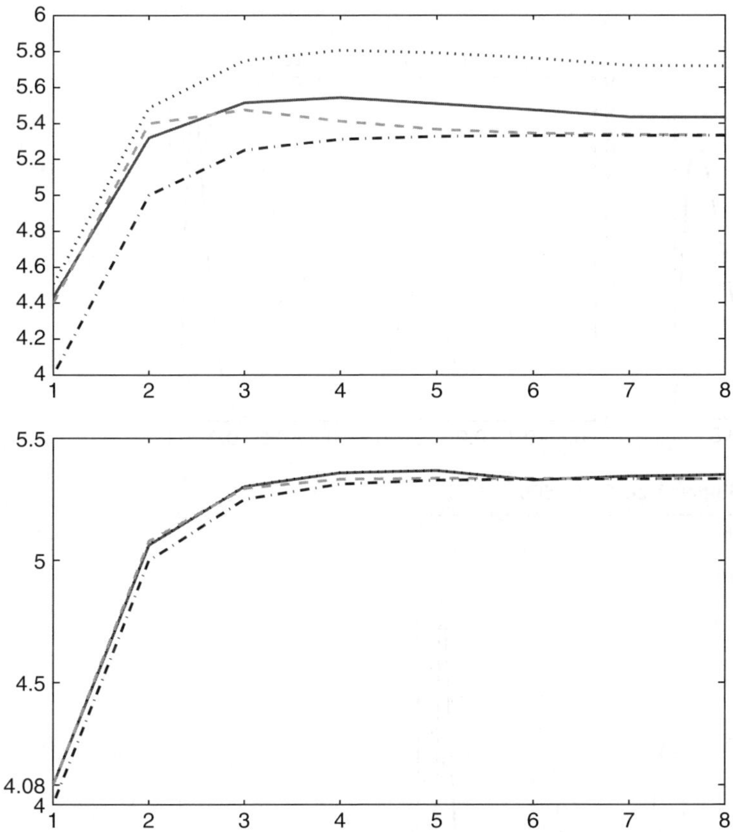

Figure 4.8 **Top:** True m.s.e. of h-step ahead forecasts, $h = 1, \ldots, 8$, for $a = 0.5$, $\sigma^2 = 4$, and $T = 10$ using exact m.l.e. (solid) and conditional m.l.e. (dotted), both restricted to yield stationary models. The dashed line is (4.34) using the true values of a and σ, while the dash-dot line is the same, but without the last term in (4.34). **Bottom:** Same, but with $T = 50$. Note the change of the y-axis.

4.5 Small Sample Distribution of the OLS and ML Point Estimators

In conclusion, we should be fully prepared for the serious bias of the least squares estimate, when we estimate the autoregressive model from a small sample typically dealt with in econometrics.

(Takamitsu Sawa, 1978, p. 169)

As for most models, the exact sampling distribution of the m.l.e. in the AR(1) model is not analytically tractable, but simulation offers an easy way of empirically approximating it. This was done for several values of a and T, using 1,000 replications. Kernel density estimates of the distribution of $\widehat{a}_{\text{ML}}$ are shown in Figure 4.9. With only $T + 1 = 20$ observations, the density of $\widehat{a}_{\text{ML}}$ is quite spread out, for $a = 0$, it appears symmetric around zero, while for $a = 0.5$, there is noticeable left skewness. This arises because $\widehat{a}_{\text{ML}}$ is constrained to lie between -1 and 1. The skewness is extreme for the $a = 0.9$ case, although the mode of the density is indeed quite close to 0.9.

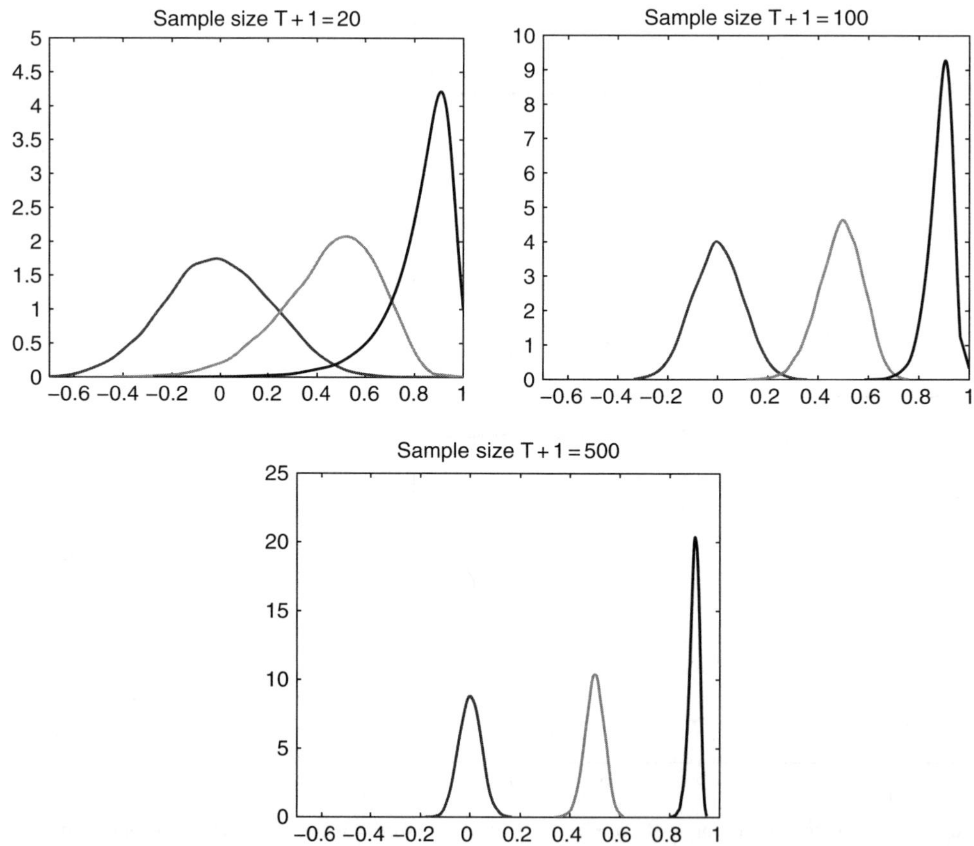

Figure 4.9 Kernel density estimates of the distribution of the m.l.e. of a in the AR(1) model using 10,000 replications (values $a = 0$, $a = 0.5$, and $a = 0.9$) and three sample sizes, as indicated.

Table 4.1 Small-sample behavior of AR(1) estimators $\widehat{a}_{LS}$ (o.l.s.) and $\widehat{a}_{ML}$ (m.l.e.) based on 1,000 simulated time series, near and on the stationarity border. The top panel gives the sample mean, the middle panel gives 1,000 times the variance, and the bottom panel gives the percentage of estimates exceeding 1.0.

	$T+1$	a = 0.90		0.95		0.99		1.00	
		o.l.s.	m.l.e.	o.l.s.	m.l.e.	o.l.s.	m.l.e.	o.l.s.	m.l.e.
Mean	20	0.831	0.833	0.887	0.889	0.950	0.950	0.919	0.832
	100	0.883	0.884	0.933	0.934	0.976	0.978	0.982	0.970
	500	0.896	0.896	0.946	0.946	0.986	0.987	0.997	0.995
$10^3 \cdot$ Var	20	20.5	17.3	15.8	12.5	8.98	6.62	21.6	21.2
	100	2.44	2.29	1.56	1.41	0.692	0.520	0.986	0.955
	500	0.406	0.399	0.223	0.216	0.0661	0.0601	0.0404	0.0327
% > 1	20	3.89	0	10.9	0	26.6	0	32.9	0
	100	0.0	0	0.15	0	9.67	0	31.9	0
	500	0.0	0	0.0	0	0.14	0	32.1	0

We also observe that, as a increases from 0 to 1, the variance of $\widehat{a}_{ML}$ decreases, agreeing with (4.24). As expected, as T grows, the density becomes less skew and more Gaussian in appearance, centered on the true value of a.

Now consider what happens as a approaches one. The exact likelihood cannot be evaluated at $a = 1$ because $f_{Y_0}(\cdot) = 0$. Thus, when computing the m.l.e., the optimization algorithm must be prevented from trying values of $\widehat{a} \geqslant 1$. This motivates use of the conditional m.l.e. $\widehat{a}_{LS}$ from (4.14a), which does not require f_{Y_0}. For the three sample sizes $T = 20$, $T = 100$, and $T = 500$, and the four values of a, 0.90, 0.95, 0.99, and 1.0, the mean and variance of $\widehat{a}_{LS}$ and $\widehat{a}_{ML}$, based on 10,000 simulated time series, are shown in Table 4.1, along with the percentage of estimates that equal or exceed unity.

Inspection of the table reveals several facts:

- As the sample size increases, both estimators improve in terms of bias and variance.
- For all values of $a \geqslant 0.90$, the variance of $\widehat{a}_{ML}$ is smaller than $\widehat{a}_{LS}$.
- For the stationary models, the variance decreases as a increases towards one, as was also seen in Figure 4.9. But when $a = 1$ and T is small, the variance of both estimators jumps up considerably.
- Both the o.l.s. and m.l.e. are extremely downward biased for $T = 20$ and moderately so for $T = 100$; for $T = 500$, the bias is zero when measured with two significant digits.
- For the stationary models, the bias of the m.l.e. is slightly less than that of the o.l.s. estimator, with their difference being more pronounced for smaller sample sizes.
- For $a = 1$, the m.l.e. exhibits a much greater bias than the o.l.s. estimator, particularly for small T. This is due to the fact that both estimators have a relatively high variance when $a = 1$ but the m.l.e. cannot equal or exceed one.
- A value of $\widehat{a} > 1$ is of little practical value if it can be assumed that the process is not explosive, in which case it will most likely be truncated to one. For the random walk with $T = 20$, if all

occurrences of $\widehat{a}_{\mathrm{LS}} > 1$ are set to one, then the mean is 0.9045 (not shown in the table), so that, with respect to bias, the truncated o.l.s. estimator is preferred to the m.l.e. Similar results hold for the larger sample sizes: For $T = 100$ (500), the mean of the truncated o.l.s. values is 0.9800 (0.9960).

The previous analysis is useful because many time series of interest, notably in macro-economics, have small T, and resemble a random walk, so that $\hat{a}$ for the AR(1) model will be close to one. In order to make useful inferences in such cases (such as a valid confidence interval for a), we require the p.d.f. and c.d.f. of $\widehat{a}_{\mathrm{LS}}$.

For model (4.1) with $c = 0$, i.e., $Y_t = aY_{t-1} + U_t$, $U_t \overset{\text{i.i.d.}}{\sim} \mathrm{N}(0, \sigma^2)$, let as before $\mathbf{Y} = (Y_0, \dots, Y_T)'$ and $\mathbf{U} = (U_0, \dots, U_T) \sim \mathrm{N}(\mathbf{0}, \sigma^2 \mathbf{I}_{T+1})$. From (4.15), Y_0 can be expressed as $Y_0 = bU_0$, where

$$b = \begin{cases} (1 - a^2)^{-1/2}, & \text{if } a \in (-1, 1), \\ 0, & \text{otherwise,} \end{cases} \tag{4.35}$$

and

$$\begin{aligned} Y_1 &= aY_0 + U_1 = abU_0 + U_1, \\ Y_2 &= aY_1 + U_2 = a^2 bU_0 + aU_1 + U_2, \end{aligned}$$

etc., so that $\mathbf{Y} = \mathbf{RU}$, where

$$\mathbf{R} = \mathbf{R}(a) = \begin{bmatrix} b & 0 & 0 & \cdots & 0 & 0 \\ ba & 1 & 0 & \cdots & 0 & 0 \\ ba^2 & a & 1 & \cdots & 0 & 0 \\ \vdots & \vdots & \vdots & \ddots & \vdots & \vdots \\ & & & \ddots & 1 & 0 \\ ba^T & a^{T-1} & a^{T-2} & \cdots & a & 1 \end{bmatrix}. \tag{4.36}$$

The distribution of the o.l.s. estimator $\widehat{a}_{\mathrm{LS}}$ in (4.14) is thus that of the quadratic form

$$\frac{\mathbf{U}'\mathbf{R}'\mathbf{D}'_{T-1}\mathbf{D}_T\mathbf{RU}}{\mathbf{U}'\mathbf{R}'\mathbf{D}'_{T-1}\mathbf{D}_{T-1}\mathbf{RU}} = \frac{\mathbf{U}'\mathbf{AU}}{\mathbf{U}'\mathbf{BU}}, \tag{4.37}$$

where $\mathbf{A} = (\mathbf{R}'\mathbf{D}'_{T-1}\mathbf{D}_T\mathbf{R} + \mathbf{R}'\mathbf{D}'_T\mathbf{D}_{T-1}\mathbf{R})/2$ (so that it is symmetric, recalling the beginning of Chapter A) and $\mathbf{B} = \mathbf{R}'\mathbf{D}'_{T-1}\mathbf{D}_{T-1}\mathbf{R}$. Note that σ^2 cancels from the numerator and denominator, showing that $\widehat{a}_{\mathrm{LS}}$ is invariant to its value.

The numeric methods of Appendix A.1 can be used to compute its p.d.f. and c.d.f. Figure 4.10 shows the density of $\widehat{a}_{\mathrm{LS}}$ for several a and T calculated from (i) simulation and (ii) using the p.d.f. saddlepoint approximation of (4.37). The latter is highly accurate even for (the impractically small sample size of) $T + 1 = 10$, and improves as T increases.

The probability that $\widehat{a}_{\mathrm{LS}}$ is greater than one can be calculated from the c.d.f. Using the exact calculation for $T = 20$ yields 4.09%, 10.9%, 26.8%, and 33.0%, for $a = 0.90, 0.95, 0.99$, and 1.0, respectively. These are very close to the values obtained via simulation, as shown in Table 4.1.[2]

The low order moments of $\widehat{a}_{\mathrm{LS}}$ can be computed using the results in Appendix B.2 or obtained by numerically integrating $x^n\bar{f}(x)$, where $\bar{f} = \hat{f} / \int \hat{f}(x)\,dx$ is the normalized saddlepoint density of $\widehat{a}_{\mathrm{LS}}$.

2 For comparison, the second-order c.d.f. s.p.a. gives values 3.93%, 10.3%, 25.1%, and 32.7%; note the relative inaccuracy for $a = 0.99$ and this sample size. Alternatively, numerically integrating the normalized second-order p.d.f. gives 26.6% for this case, which does compare well with the exact value of 26.8%.

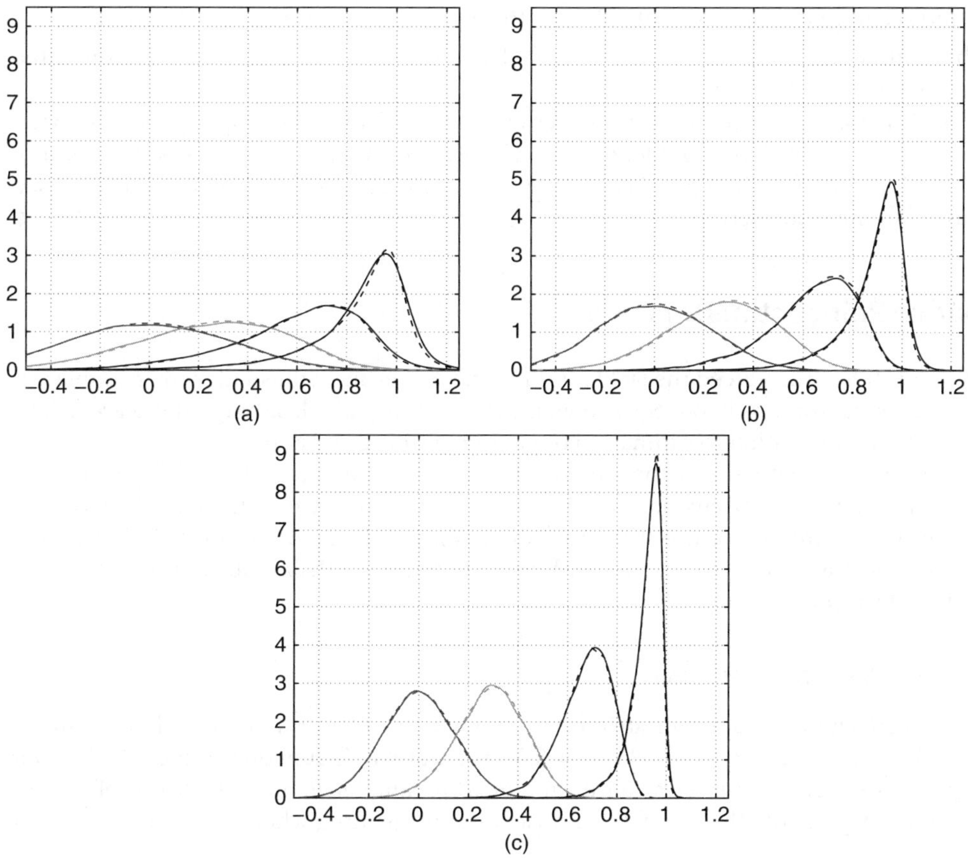

Figure 4.10 Density of $\hat{a}_{LS}$ based on $a = 0$, $a = 0.3$, $a = 0.7$, and $a = 0.95$, for $T + 1 = 10$ (a), $T + 1 = 20$ (b), and $T + 1 = 50$ (c) using simulation and kernel density estimation based on 10,000 replications (solid) and the saddlepoint approximation of (4.37) (dashed).

For example, with $T + 1 = 20$, the means corresponding to $a = 0.90$, 0.95, 0.99, and 1.0 are 0.830, 0.885, 0.950, and 0.916, respectively, while the variances (times 1,000) are 20.2, 15.9, 9.10, and 22.3. Observe that these are very close to the values in Table 4.1 obtained from simulation.

Remark A common measure of persistence in an autoregressive time-series model is the **half life**, defined to be the time required for a unit shock to dissipate by 50%, and, for the AR(1) model, computed as $\hat{h} = \ln(1/2)/\ln(\hat{\alpha})$, for $0 < \hat{\alpha} < 1$. Examples of its use can be found in the economic literature on **purchasing power parity**; see, e.g., the references in Chen and Giles (2011), as well as a discussion on its extension to the AR(p) case.

For the AR(1) model, the density of $\hat{h}$ is, via univariate transformation,

$$f_{\hat{h}|C}(h \mid C) = \frac{(1/2)^{1/h} \ln 2}{h^2 \Pr(C)} f_{\hat{a}_{LS}}((1/2)^{1/h}), \quad C = \{0 < \hat{\alpha} < 1\}. \tag{4.38}$$

$f_{\hat{a}_{\mathrm{LS}}}$ can quickly and accurately be approximated by $\hat{f}_{\hat{a}_{\mathrm{LS}}}$, the density saddlepoint approximation (s.p.a.), and Pr(C) can be computed from the s.p.a. of the c.d.f. of $\hat{a}_{\mathrm{LS}}$. The c.d.f. of $\hat{h} \mid C$ at h can be computed from the s.p.a. for $\Pr(\hat{a}_{\mathrm{LS}} \leqslant (1/2)^{1/h})/C$.

Chen and Giles (2011) show that no positive integer moments of $\hat{h}$ exist, lending explanation to the difficulty in the literature of ascribing confidence intervals to h. Given the bias of the o.l.s. estimator of α, the half life h is often computed using a bias-adjusted estimator of α, such as the median-unbiased estimator, discussed in the next section. ■

4.6 Alternative Point Estimators of *a*

Sections III.7.4.4 and III.8.4 introduced the jackknife, and the mean-bias-adjusted, median-unbiased, and mode-unbiased estimators, respectively, showing examples in an i.i.d. setting. These are briefly reviewed here, and their application to estimation of parameter a is discussed.

The latter group of estimators rely on the analytic expressions for the p.d.f., c.d.f., and moments of the o.l.s. estimator discussed in Chapters A and B. In particular, the calculation of the mean-bias-adjusted estimator requires the mean of (4.37), which can be computed from (B.36), the median unbiased estimator requires the c.d.f., while computation of the mode unbiased estimator requires evaluation of the p.d.f.

4.6.1 Use of the Jackknife for Bias Reduction

The basic jackknife, using notation appropriate for time series, is as follows: Assume we have a sample of T observations $\mathbf{Y} = (Y_1, Y_2, \dots, Y_T)'$ and $\hat{\theta} = S = S(\mathbf{Y})$ is a statistic (a function of the data but not of the unknown parameters) that serves as an estimator of parameter θ, and which we refer to as the base estimator. Let $\mathbf{Y}_{(t)}$ denote the set of $T-1$ observations resulting when observation Y_t is not included, i.e.,

$$\mathbf{Y}_{(t)} = (Y_1, Y_2, \dots, Y_{t-1}, Y_{t+1}, \dots, Y_T)', \quad t = 1, 2, \dots, T, \tag{4.39}$$

and let $S_{(t)} = S(\mathbf{Y}_{(t)})$, $t = 1, \dots, T$. The delete-1 jackknife estimator of θ based on $\hat{\theta}$ is given by

$$\hat{\theta}^* = TS - (T-1)\bar{S}_{\bullet}, \qquad \bar{S}_{\bullet} = T^{-1} \sum_{t=1}^{T} S_{(t)}, \tag{4.40}$$

where $\bar{S}_{\bullet}$ is the average of the $S_{(t)}$. Assume the expansion

$$\text{bias}(S) = \mathbb{E}[S] - \theta = \frac{a_1}{T} + \frac{a_2}{T^2} + \cdots \tag{4.41}$$

holds, for constants a_i that can depend on θ but not on sample size T. Then

$$\begin{aligned}
\mathbb{E}[\hat{\theta}^*] &= T\mathbb{E}[S] - (T-1)\mathbb{E}[S_1] \\
&= T\left(\theta + \frac{a_1}{T} + \frac{a_2}{T^2} + \frac{a_3}{T^3} + \cdots\right) - (T-1)\left(\theta + \frac{a_1}{T-1} + \frac{a_2}{(T-1)^2} + \frac{a_3}{(T-1)^3} + \cdots\right) \\
&= \theta + a_2\left(\frac{1}{T} - \frac{1}{T-1}\right) + a_3\left(\frac{1}{T^2} - \frac{1}{(T-1)^2}\right) + \cdots = \theta - \frac{a_2}{T(T-1)} + O(T^{-3}),
\end{aligned}$$

showing that the first-order term a_1/T drops out and the second-order term is only slightly larger than a_2/T^2 in (4.41). If, for all θ, $\hat{\theta}$ itself is unbiased (so that $a_1 = a_2 = \cdots = 0$), then, clearly, $\hat{\theta}^*$ is also unbiased.

This is applicable to an i.i.d. sample, with some simple examples shown in Section III.7.4.4 and far more detail provided in Shao and Tu (1995). In a time series (or spacial data) context with ordered data $Y_1, Y_2, \ldots, Y_T$, observe how removing an observation and computing, say, the estimator of the AR(1) parameter assuming model (4.1) is no longer valid, as the time series structure is disturbed. Instead, non-overlapping or moving-block sub-samples are used. The former was developed for the AR(1) model by Phillips and Yu (2005), while both are considered in Chambers (2013), where the method is also extended to the AR(p) case, building on results developed in Shaman and Stine (1988) and Patterson (2000b).

Observe how there is overlap in the sub-samples in (4.40). For the jackknife with non-overlapping sub-samples, let ℓ denote the length of the sub-sample and m denote the number of sub-samples, such that (perhaps overly optimistically) $T = m \times \ell$; define $\mathbf{Y}_{(i)}$ to be a sub-sample of $\mathbf{Y}$ given by

$$\mathbf{Y}_{(i)} = (Y_{(i-1)\ell+1}, \ldots, Y_{i\ell})', \quad i = 1, \ldots, m; \tag{4.42}$$

and let $S_{(i)} = S(\mathbf{Y}_{(i)})$, $i = 1, \ldots, m$. The jackknife estimator for non-overlapping sub-samples is defined as in Phillips and Yu (2005) to be

$$\hat{\theta}^{(m)} = \left(\frac{T}{T-\ell}\right) S - \left(\frac{\ell}{T-\ell}\right) \bar{S}_{\bullet}, \qquad \bar{S}_{\bullet} = m^{-1} \sum_{i=1}^{m} S_{(i)}, \tag{4.43}$$

where, as before, $\bar{S}_{\bullet}$ is the average of the $S_{(i)}$. As $\ell = T/m$ and $T - \ell = (T/m)(m-1)$, (4.43) can be written as

$$\hat{\theta}^{(m)} = \left(\frac{m}{m-1}\right) S - \left(\frac{1}{m-1}\right) \bar{S}_{\bullet}, \qquad \bar{S}_{\bullet} = m^{-1} \sum_{i=1}^{m} S_{(i)}. \tag{4.44}$$

Observe the special case of (4.43) with $\ell = T - 1$ (so that $T - \ell = 1$), $m = T$, and use of definition (4.39) instead of (4.42) for $\mathbf{Y}_{(i)}$, $i = 1, \ldots, m = T$, yields $\hat{\theta}^*$ in (4.40).

The (more realistic) situation when the sub-sample lengths are not all equal, as well as the use of moving block sub-samples (such that the sub-samples have overlap), are addressed in Chambers (2013).

We illustrate the jackknife procedure for non-overlapping sub-samples (4.44) in the AR(1) case, with the o.l.s. estimator as the base, $T = 20$, and values $m \in \{2, 4, 5\}$. Figure 4.11 shows the results, over a grid of a-values. The top panels indicate the mean-bias and, true to the theory, the jackknife reduces it, compared to the o.l.s. estimator, with the most reduction for $m = 2$. The middle panels show that the median-bias is also reduced, while the lower panels indicate, unfortunately, that the m.s.e. is, for $0 < a < 0.85$, lower with the o.l.s. estimator, while for $a > 0.85$, the reduction in m.s.e. from the jackknife, for any m, is not appreciable. This graphical analysis can be compared to Figure 4.13, showing the performance of other bias-adjusted estimators, as well as that of the exact m.l.e.

4.6.2 Use of the Bootstrap for Bias Reduction

While the bootstrap was showcased in Chapter III.1 as a reliable and generally applicable means of computing an interval estimator, it can also be deployed for bias reduction. The jackknife can be

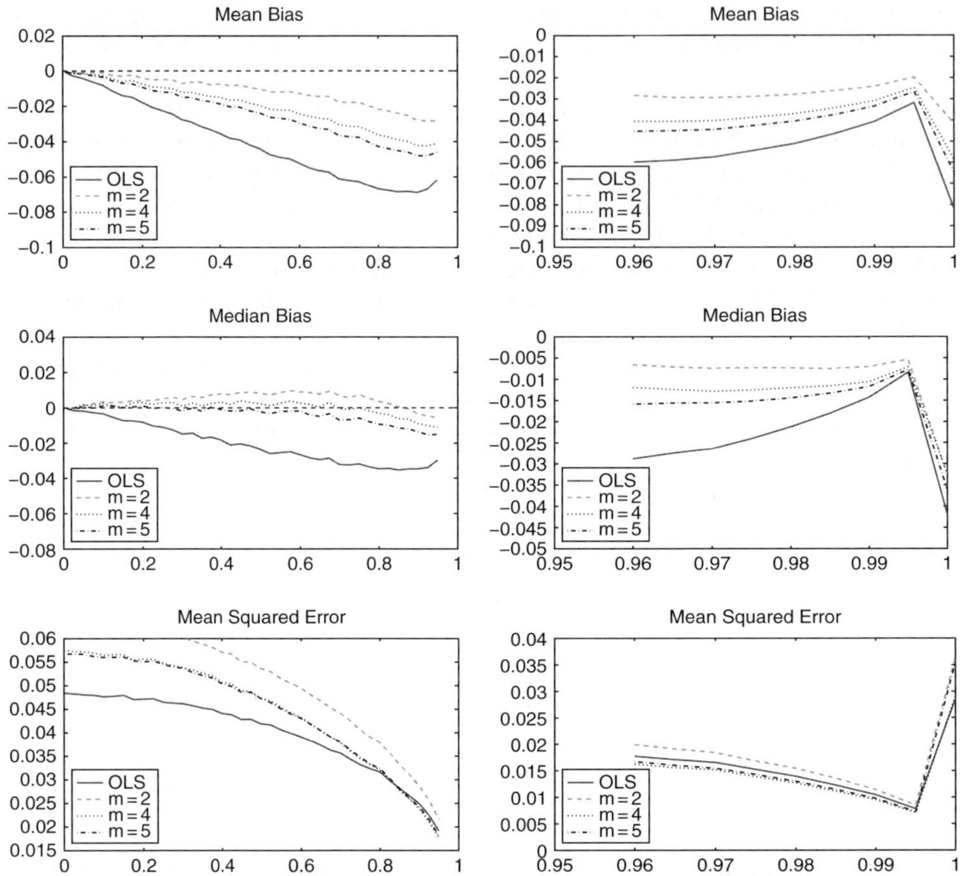

Figure 4.11 **Left:** Performance comparison as a function of AR(1) parameter a, of the least squares estimator, denoted as OLS, along with jackknife bias-adjusted estimator (4.44) for several values of m, as indicated, and based on 100,000 replications. The top graphs show the mean-bias, the middle graphs show the median-bias, and the bottom graphs show the m.s.e. **Right:** Same but concentrating on the area near the unit root.

viewed as an approximation of the bootstrap; see, e.g., Efron (1979) and Shao and Tu (1995). Like with the jackknife and the other methods discussed below for bias-adjusted estimators, its use can result in the final estimator possessing a higher mean squared error. The procedure works as follows for $|a| < 1$:

1) Denote by B the number of bootstrap replications, chosen large enough such that the inferential result (here, a bias-adjusted estimate) does not change appreciably as B is increased.
2) For sample $\mathbf{Y} = (Y_1, Y_2, \ldots, Y_T)'$, compute the base estimator $\hat{\theta} = S = S(\mathbf{Y})$, which can be either $\hat{a}_{\mathrm{ML}}$ or $\hat{a}_{\mathrm{LS}}$ in our context, as well as the associated model residuals (the filtered time series innovations) $\hat{\mathbf{U}} = (\hat{U}_1, \ldots, \hat{U}_T)'$, which are presumed to be i.i.d.
3) For the nonparametric bootstrap, let $\mathbf{U}^{(b)}$ be the bth bootstrap replication of $\hat{\mathbf{U}}$, formed by sampling with replacement. For each $\mathbf{U}^{(b)}$, $b = 1, \ldots, B$, construct time series $\mathbf{Y}^{(b)}$ by taking $\mathbf{Y}^{(b)} = \mathbf{\Sigma}^{1/2}\mathbf{U}^{(b)}$, where $\mathbf{\Sigma}$ is given in (4.20), conditional on $\hat{\theta}$ (either $\hat{a}_{\mathrm{ML}}$ or $\hat{a}_{\mathrm{LS}}$).

4) For each bootstrap data set $\mathbf{Y}^{(b)}$, calculate $\hat{\theta}^{(b)} = S(\mathbf{Y}^{(b)})$, $b = 1, \ldots, B$. In the AR(1) context, this is $\hat{a}_{\mathrm{ML}}^{(b)}$ or $\hat{a}_{\mathrm{LS}}^{(b)}$. The parametric bootstrap is similar, but draws $\hat{\mathbf{U}}^{(b)}$ from the assumed parametric distribution and conditional on its estimated parameters from step 2, such as $\hat{\sigma}^2$.
5) Denote the arithmetic average of the $\hat{\theta}^{(b)}$ as $\bar{\hat{\theta}}$. The bootstrap estimated bias of $\hat{\theta}$ is given by $\bar{\hat{\theta}} - \hat{\theta}$, and the bias-adjusted estimator is then $\hat{\theta} - (\bar{\hat{\theta}} - \hat{\theta}) = 2\hat{\theta} - \bar{\hat{\theta}}$.

The reader is encouraged to implement this and, via simulation, generate results paralleling those in Figure 4.11, as well as reproducing the jackknife results in Figure 4.11.

Bias-adjusted estimators (based on the bootstrap) for the vector autoregressive model, in both the stationary and non-stationary (unit root or explosive) cases, is pursued in Engsted and Pedersen (2014).

4.6.3 Median-Unbiased Estimator

By definition, an estimator $\hat{\theta}$ is **median-unbiased** for θ if, for each value θ in the parameter space, θ is a median of $\hat{\theta}$. The median-unbiased estimator was first proposed in the context of the AR(1) model by Andrews (1993), with further developments in Carstensen and Paolella (2003).

The following bias correction procedure then makes $\hat{a}_{\mathrm{Med}}$ a median-unbiased estimator: $\hat{a}_{\mathrm{Med}}$ takes that value of a that yields the o.l.s. estimator to have a median equal to the o.l.s. estimate obtained from the data. More formally, let $\mathrm{Med}(\hat{a}_{\mathrm{LS}} \mid a) = m(a)$ denote the median function of $\hat{a}_{\mathrm{LS}}$ when a is the true parameter, and let $m^{-1} : (m(-1), m(1)] \to (-1, 1]$ denote its inverse which, as $m(a)$ is strictly increasing, is properly defined. The median-unbiased estimator $\hat{a}_{\mathrm{Med}}$ is then given by

$$\hat{a}_{\mathrm{Med}} = \begin{cases} 1, & \text{if } \hat{a}_{\mathrm{LS}} > m(1), \\ m^{-1}(\hat{a}_{\mathrm{LS}}), & \text{if } m(-1) < \hat{a}_{\mathrm{LS}} \leqslant m(1), \\ -1, & \text{if } \hat{a}_{\mathrm{LS}} \leqslant m(-1). \end{cases} \tag{4.45}$$

Given the observed value of the o.l.s. estimator, say $\hat{a}_{\mathrm{LS}}^{\mathrm{O}}$, the estimator can be expressed for $m(-1) < \hat{a}_{\mathrm{LS}}^{\mathrm{O}} \leqslant m(1)$ as

$$\hat{a}_{\mathrm{Med}} = m^{-1}(\hat{a}_{\mathrm{LS}}) = \mathrm{argmin}_a |\mathrm{Med}(\hat{a}_{\mathrm{LS}} \mid a) - \hat{a}_{\mathrm{LS}}^{\mathrm{O}}|, \tag{4.46}$$

or, equivalently, with $F_{\hat{a}_{\mathrm{LS}}}$ denoting the c.d.f. of $\hat{a}_{\mathrm{LS}}$,

$$\hat{a}_{\mathrm{Med}} = m^{-1}(\hat{a}_{\mathrm{LS}}) = \mathrm{argmin}_a |F_{\hat{a}_{\mathrm{LS}}}(\hat{a}_{\mathrm{LS}}^{\mathrm{O}} \mid a) - 0.5|, \tag{4.47}$$

which lends itself to computation.

4.6.4 Mean-Bias Adjusted Estimator

The ability to quickly and accurately evaluate $F_{\hat{a}_{\mathrm{LS}}}$ facilitates modification of the median unbiasedness procedure to obtain an approximately **mean-unbiased** estimator. This approach of bias correction is not new; it has, for example, been pursued in MacKinnon and Smith, Jr. (1998) in a general context, and in Tanizaki (2000) for this setting. Analytic results on the mean-bias are derived in Bao (2007) and Bao and Ullah (2007), and could be used with the method discussed in Section III.7.4.1 for bias reduction.

The mean-bias reducing method amounts to interpreting $m(\cdot)$ as the analogously defined mean function in (4.45), i.e., let $m(a) = \mathbb{E}[\widehat{a}_{\text{LS}} \mid a]$. Like the median function, it is strictly increasing for $-1 < a < 1$, so that its inverse exists. In particular, for $m(-1) < \widehat{a}^{O}_{\text{LS}} \leqslant m(1)$,

$$\widehat{a}_{\text{Mean}} = m^{-1}(\widehat{a}_{\text{LS}}) = \operatorname{argmin}_a |\mathbb{E}[\widehat{a}_{\text{LS}} \mid a] - \widehat{a}^{O}_{\text{LS}}|. \tag{4.48}$$

Estimator $\widehat{a}_{\text{Mean}}$ is not exactly mean-unbiased, not only because of the truncation at -1 and 1, but also because of the nonlinearity of the mean function, i.e., $\mathbb{E}[m^{-1}(\widehat{a}_{\text{LS}})] \neq m^{-1}(\mathbb{E}[\widehat{a}_{\text{LS}}]) = a$.

A reliable method for evaluating $\mathbb{E}[\widehat{a}_{\text{LS}}]$ is required in (4.48), which is now briefly discussed. Contrary to Tanizaki (2000), who used simulation to obtain the mean function, the computation of $\mathbb{E}[\widehat{a}_{\text{LS}}]$ can be more accurately achieved by means of the relation

$$\mathbb{E}[\widehat{a}_{\text{LS}}] = \int_0^\infty [1 - F_{\widehat{a}_{\text{LS}}}(t)]\,\mathrm{d}t - \int_{-\infty}^0 F_{\widehat{a}_{\text{LS}}}(t)\,\mathrm{d}t \tag{4.49}$$

(see, e.g., (I.7.71)). In this context, the use of the s.p.a. for evaluating (4.49) allows for substantial time savings without sacrificing accuracy relevant for empirical work. Alternatively, and potentially faster for larger sample sizes, is to use the expressions for the first and second moments of a ratio of quadratic forms in normal variables as given in Appendix B.2.

4.6.5 Mode-Adjusted Estimator

Use of bias adjustment methods based on the mean and median (as measures of central tendency), leads naturally to consideration of the third such measure: the mode, as introduced in Broda et al. (2007). Following (4.46) and (4.48), it is natural to define the **mode-adjusted** estimator as

$$\widehat{a}_{\text{Mode}} = m^{-1}(\widehat{a}_{\text{LS}}) = \operatorname{argmin}_a |\text{Mode}(\widehat{a}_{\text{LS}} \mid a) - \widehat{a}^{O}_{\text{LS}}|, \tag{4.50}$$

where $m(\cdot)$ is now interpreted in (4.45) as the mode function. In comparison to $\widehat{a}_{\text{Med}}$ and $\widehat{a}_{\text{Mean}}$, which are well-defined and unique for continuous distributions with finite first moment, use of $\widehat{a}_{\text{Mode}}$ only makes sense if the relevant distribution is unimodal. Indeed, inspection shows that, for sample sizes greater than five, the p.d.f. of $\widehat{a}_{\text{LS}}$ is unimodal and, paralleling the requirements of the median and mean, the mode function of $\widehat{a}_{\text{LS}}$ is strictly increasing for $|a| < 1$, thus guaranteeing that $\widehat{a}_{\text{Mode}}$ is uniquely defined.

Let $f_{\widehat{a}_{\text{LS}}}(x; a)$ denote the p.d.f. of $\widehat{a}_{\text{LS}}$ at x when the true parameter is a. From the definition of the mode, it follows that (4.50) is equivalent to choosing $\widehat{a}$ such that the density $f_{\widehat{a}_{\text{LS}}}(x; \widehat{a})$ attains its maximum at the observed value of $\widehat{a}_{\text{LS}}$. That is, we can write

$$\widehat{a}^{O}_{\text{LS}} = \operatorname{argmax}_x f_{\widehat{a}_{\text{LS}}}(x; \widehat{a}_{\text{Mode}}), \tag{4.51}$$

i.e., $\widehat{a}_{\text{Mode}}$ is the (unique) value of a such that the observed value is a mode of $f_{\widehat{a}_{\text{LS}}}(x; a)$.

Both (4.50) and (4.51) can be operationalized to construct an algorithm for the computation of $\widehat{a}_{\text{Mode}}$. However, both involve a nested root search (the outer one being for the objective function, the inner one to obtain the mode), and so are relatively much slower than computation of $\widehat{a}_{\text{Med}}$ or $\widehat{a}_{\text{Mean}}$. As such, we instead compute (4.51) as

$$\left.\frac{\partial f_{\widehat{a}_{\text{LS}}}(x; \widehat{a}_{\text{Mode}})}{\partial x}\right|_{x=\widehat{a}^{O}_{\text{LS}}} = 0. \tag{4.52}$$

This is justified under the stated assumptions of unimodality and monotonicity of the mode of $\widehat{a}_{\rm LS}$ as a function of a (for $|a| < 1$). As only a single univariate root search is required in (4.52), its use with numerical differentiation will be much faster than (4.50) or (4.51) and, by avoiding the otherwise necessary nested root search, it is also numerically more reliable.[3]

4.6.6 Comparison

To see that the different point estimators will exhibit differing small-sample properties, Figure 4.12 plots the mean, median, and mode, minus a, of (4.37), based on a sample size of $T + 1 = 20$ observations, over a grid of a-values.

Figure 4.13 shows the bias and m.s.e. results based on a simulation with 10,000 replications, whereby OLS and MLE refer to the least squares and maximum likelihood estimators, respectively, while MBA, MED and MOD refer to the mean-bias-adjusted, median-unbiased, and mode-unbiased estimators, respectively. The top panels of Figure 4.13 show the mean-bias, computed as the average of the 10,000 observations for each value of a, minus a. As expected, the mean-bias for the MBA estimator is very close to zero, though it does deviate somewhat from zero as a increases. This is possible because the mean of (4.37) as a function of a is not linear, deviating from linearity more so as a approaches one. All the other estimators exhibit a negative mean-bias that grows in magnitude as a moves from zero to about 0.85, with the most bias from the mode-unbiased estimator MOD. The middle panel shows the median-bias, computed analogously as the mean-bias. True to the theory, the median-unbiased estimator MED indeed exhibits virtually no median-bias for any value of a, while MBA has positive median-bias and the remaining have negative median-bias, the most being from MOD.

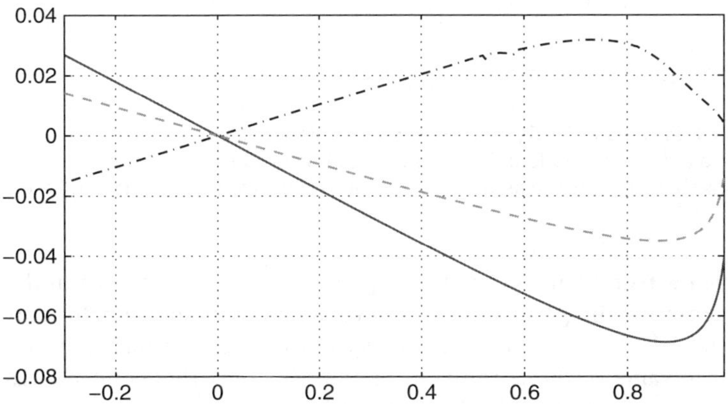

Figure 4.12 The mean (solid), median (dashed), and mode (dash-dot) of (4.37), minus a, versus a, for $T = 19$.

3 The only possible caveat to use of the mode-unbiased estimator is the approximate nature of the density via the s.p.a. To check this, we used the numerical second derivative of the exact c.d.f. of $\widehat{a}_{\rm LS}$, which was found to be numerically quite reliable, but extremely time-consuming compared to use of the s.p.a. Using the very small sample size of $T = 25$, we found that the differences in $\widehat{a}_{\rm Mode}$ based on the s.p.a. and use of the exact c.d.f. were affected in only the third or fourth decimal place, thus confirming that use of the s.p.a. in this context will not jeopardize the accuracy of the method by any appreciable amount. (As T increases, so does the accuracy of the s.p.a. because the distribution of $\widehat{a}_{\rm LS}$ approaches the normal, for which the s.p.a. is exact.)

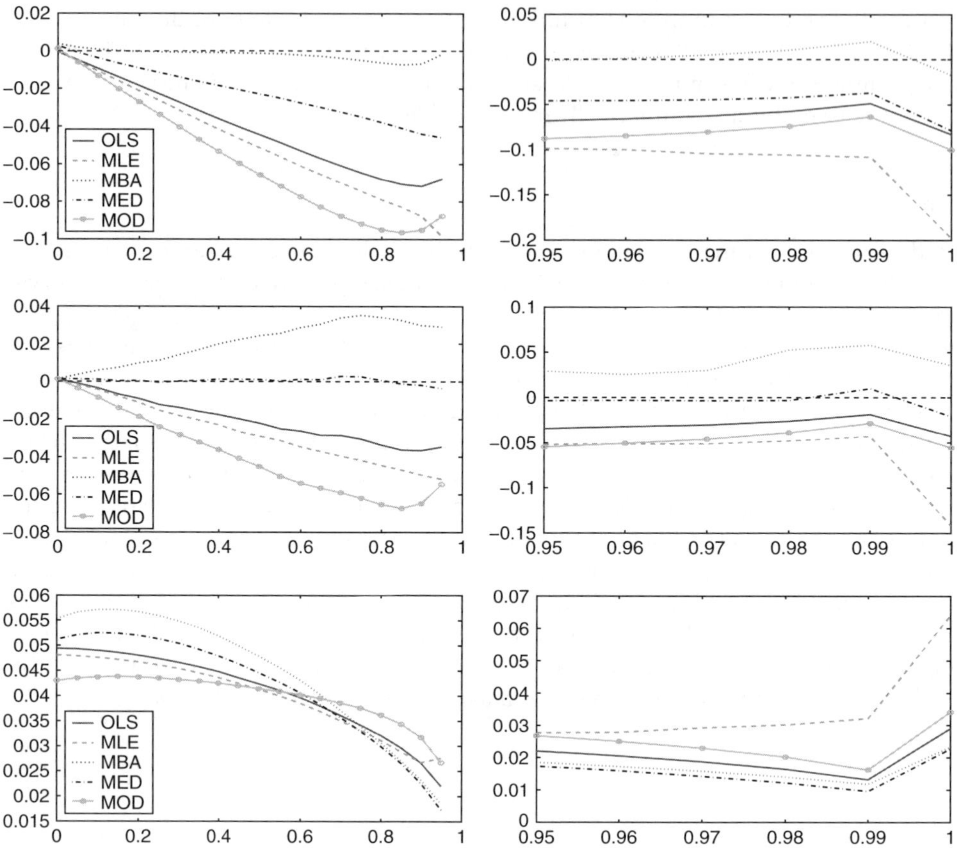

Figure 4.13 Performance comparison of the least squares (OLS), exact maximum likelihood (MLE), mean-bias-adjusted (MBA), median-unbiased (MED), and mode-unbiased (MOD) estimators for parameter a in the AR(1) model with $T + 1 = 20$. The top graphs show the mean-bias, the middle graphs show the median-bias, and the bottom graphs show the m.s.e.

Arguably the most important single measurement is the mean squared error, m.s.e., shown in the bottom panels. As is typical when examining the performance of several competitive estimators for parameters in more complicated models, the best estimator (with respect to m.s.e. or similar criteria) will, unfortunately, depend on the true value of the parameter being estimated. This concept was emphasized in Section III.1.1.2, in the context of a simpler, i.i.d. model.

In this case, o.l.s. and m.l.e. are relatively close in performance for all $a < 0.9$, with the latter being the preferred of the two, while for $0.9 < a < 1$, o.l.s. is better. The other estimators give rise to rather different behavior depending on a. For values of a between 0 and 0.4, the MOD is preferred by a considerable margin, while the MBA performs worst. For $0.75 < a < 1$, the MBA and MED dominate, with the latter being slightly better. For $0.7 < a < 0.95$, MOD has the highest m.s.e., while for $a > 0.95$, the m.l.e. is uniformly the worst performer.

We will revisit these estimators in Chapter 5, in the context of the AR(1) model with exogenous regressors.

4.7 Confidence Intervals for *a*

A simple way of computing confidence intervals (c.i.s) for a and σ when $|a| < 1$ is to use the asymptotic normality of the m.l.e. (valid for $|a| < 1$) in conjunction with the numerically computed Hessian matrix. However, the c.i. for a will become problematic for small sample sizes and values of a near one because (i) the upper bound of the c.i. could exceed one, (ii) the density of the m.l.e. is highly left skewed (and thus not Gaussian), and (iii) the m.l.e. of a is downwards biased, so that the actual coverage probability will be lower than the nominal.

The first two problems can be corrected by using the bootstrap, as described in Section 4.6.2. Let $\widehat{a}_{\rm ML}^{(i)}$ denote the ith bootstrap estimate, $i = 1, \ldots, B$. Then, for a particular choice of significance level α, the appropriate sample quantiles from the $\widehat{a}_{\rm ML}^{(i)}$ are calculated to determine the bootstrap interval. Note that (i) the downward bias of $\widehat{a}_{\rm ML}$ will jeopardize the performance of the bootstrap for a near one and (ii) the smaller the choice of α, the larger B should be, to ensure adequate sampling in the tails.

For this model class, an analytic method is available for computing an exact c.i. for a, based on the o.l.s. estimator. Similar to the construction of the median-unbiased estimator of a, let $q_p(a)$ be the p-quantile of random variable $\widehat{a}_{\rm LS}$, based on T observations, when the true value of the parameter is a. That is, $q_p(a)$ is a function of T, p, and a, and is implicitly given by

$$p = \int_{-\infty}^{q_p(a)} f_{\widehat{a}_{\rm LS}}(x; a)\,\mathrm{d}x = F_{\widehat{a}_{\rm LS}}(q; a), \quad 0 < p < 1.$$

For fixed T and p, $0 < p < 1$, q_p is a function of a. Assuming that $q_p(a)$ is monotone for all $a \in (-1, 1]$, the inverse function q_p^{-1} is well-defined. Let $\hat{a}_{\rm LS}^{\rm O}$ be the observed value of $\hat{a}_{\rm LS}$, and let $c = q_p^{-1}(\widehat{a}_{\rm LS}^{\rm O}) \Leftrightarrow q_p(c) = \widehat{a}_{\rm LS}^{\rm O}$. As in Andrews (1993), an exact $100(1-\alpha)\%$ c.i. for a is then given by

$$(\hat{c}_L, \hat{c}_U), \tag{4.53}$$

where, for given values $\alpha_1 \geqslant 0$, $\alpha_2 \geqslant 0$, and $\alpha_2 > \alpha_1$ such that $\alpha = \alpha_1 + (1-\alpha_2)$ for $0 < \alpha < 1$, the lower and upper bounds are given by

$$\hat{c}_L = \begin{cases} 1, & \text{if } \widehat{a}_{\rm LS}^{\rm O} > q_{\alpha_2}(1), \\ q_{\alpha_2}^{-1}(\widehat{a}_{\rm LS}^{\rm O}), & \text{if } q_{\alpha_2}(-1) < \widehat{a}_{\rm LS}^{\rm O} \leqslant q_{\alpha_2}(1), \\ -1, & \text{if } \widehat{a}_{\rm LS}^{\rm O} \leqslant q_{\alpha_2}(-1), \end{cases} \tag{4.54}$$

and

$$\hat{c}_U = \begin{cases} 1, & \text{if } \widehat{a}_{\rm LS}^{\rm O} > q_{\alpha_1}(1), \\ q_{\alpha_1}^{-1}(\widehat{a}_{\rm LS}^{\rm O}), & \text{if } q_{\alpha_1}(-1) < \widehat{a}_{\rm LS}^{\rm O} \leqslant q_{\alpha_1}(1), \\ -1, & \text{if } \widehat{a}_{\rm LS}^{\rm O} \leqslant q_{\alpha_1}(-1), \end{cases} \tag{4.55}$$

respectively. For example, if $\alpha = 0.10$, we can take $\alpha_2 = 0.95$ and $\alpha_1 = 0.05$.

To see that interval (4.53) has correct coverage, ignore the truncation and observe that

$$\Pr(\hat{c}_L \leqslant a \leqslant \hat{c}_U) = \Pr(\hat{c}_L \leqslant a) - \Pr(\hat{c}_U \leqslant a) = \alpha_2 - \alpha_1 = 1 - \alpha,$$

which follows because

$$\Pr(\hat{c}_L \leqslant a) = \Pr(q_{\alpha_2}^{-1}(\widehat{a}_{\rm LS}) \leqslant a) = \Pr(\widehat{a}_{\rm LS} \leqslant q_{\alpha_2}(a)) = \alpha_2,$$

```
1 function [lo,hi]=ar1andrewsCI(y,X,alpha)
2 T=length(y)-1; % Assumes y is of length T+1, Y_0,Y_1,...,Y_T
3 if nargin<3, alpha=0.10; end
4 if nargin<2, X=[]; end
5 if numel(X)==1 % signal to use a particular X matrix as follows:
6   type=X;
7   if type==1, X=ones(T,1);
8   elseif type==2, X=[ones(T,1) (1:T)'];
9   else error('Type of X matrix not defined')
10   end
11 end
12 if isempty(X), M=eye(T);
13 %else XT=X(:,2:end); XT1=X(:,1:(end-1)); Z=[XT XT1]; M=makeM(Z);
14 else M=makeM(X);
15 end
16
17 DT=[zeros(T,1) eye(T)]; D1=[eye(T) zeros(T,1)];
18 ahat=(y'*D1'*M*DT*y)/(y'*D1'*M*D1*y);
19 p1=alpha/2; p2=1-alpha/2;
20 tol=1e-5; opt=optimset('Display','none','TolFun',tol,'TolX',tol);
21 if 1==2 % the lower bound, as defined in Andrews
22   q1=fzero(@(a) ff(a,1,M,p2),0.99,opt); qm1=fzero(@(a) ff(a,-1,M,p2),-0.99,opt);
23   if q1<ahat, lo=1; elseif ahat<qm1, lo=-1;
24   else lo=fzero(@(a) ff(a,ahat,M,p2),0,opt);
25   end
26 else % faster but not formally correct
27   lo=fzero(@(a) ff(a,ahat,M,p2),0,opt);
28 end
29 if 1==2
30   q1=fzero(@(a) ff(a,1,M,p1),0.99,opt); qm1=fzero(@(a) ff(a,-1,M,p1),-0.99,opt);
31   if q1<ahat, hi=1; elseif ahat<qm1, hi=-1;
32   else hi=fzero(@(a) ff(a,ahat,M,p1),0,opt);
33   end
34 else
35   hi=fzero(@(a) ff(a,ahat,M,p1),0,opt);
36 end
37
38 function M=makeM(X), [T,~]=size(X); M=eye(T)-X*pinv(X'*X)*X';
39 function d=ff(a,ahat,M,p), cdf=ar1olsdist(a,ahat,M); d=cdf-p;
40 function cdf=ar1olsdist(a,ahat,M,imhof) % cdf of OLS a-hat in ARX(1) model
41 if nargin<4, imhof=0; end
42 [T,~]=size(M); if a>=1, b=0; else b=1/sqrt(1-a^2); end
43 aa=a.^(0:T)'; R=toeplitz(aa,[1 zeros(1,T)]); R(:,1)=R(:,1)*b;
44 DT=[zeros(T,1) eye(T)]; D1=[eye(T) zeros(T,1)];
45 A=R'*D1'*M*DT*R; A=(A+A')/2; B=R'*D1'*M*D1*R;
46 if imhof, cdf=imhofratio(ahat,A,B,1,0);
47 else
48   SPAorder=1; [~,cdf]=sparatio(ahat,A,B,1,0,SPAorder,0);
49   if cdf<0 % flag that SPA failed, try Imhof
50     disp('SPA failed, trying Imhof'), [~,cdf]=sparatio(ahat,A,B,1,0,SPAorder,1);
51   end
52 end
```

Program Listing 4.1: Calculates the c.i. (4.53), using the s.p.a. for the c.d.f. of the ratio of quadratic forms. Anticipating use of regressors, as shown in Chapter 5, the code in lines 4–15 handles the **X** matrix. In the setting in this chapter, there is no **X** matrix and matrix **M** is just the identity matrix. The function makeM, as was given in Listing B.2, is replaced by the version shown in line 39, using a generalized inverse, because the **M** matrix in (5.11) in Chapter 5 could be singular. However, in our examples it suffices to use **X** instead of **Z**, as discussed in Section 5.2, thus line 13 is commented out and line 14 is used.

and similarly for $\hat{c}_U$.[4] Values of α_1 and α_2 may be chosen to give, for example, equal-tailed or one-sided intervals. For a 90% equal-tailed c.i., one takes $\alpha_1 = (1 - 0.90)/2 = 0.05$ and $\alpha_2 = 1 - \alpha_1 = 0.95$. (Note that $\alpha_1 + \alpha_2 = 1$ when equal tail intervals are chosen, so that $\alpha_1 = \alpha/2$, otherwise they do not sum to one.)

Values $\hat{c}_L$ and $\hat{c}_U$ are computationally straightforward to obtain using the methods and programs developed in Appendix A.3. The program in Listing 4.1 computes (4.53), optionally (as we have) just using the middle terms in (4.54) and (4.55). It is thus easy to confirm the actual coverage probability and determine the average interval length for a given T, a, and α. To illustrate the 90% equal-tail

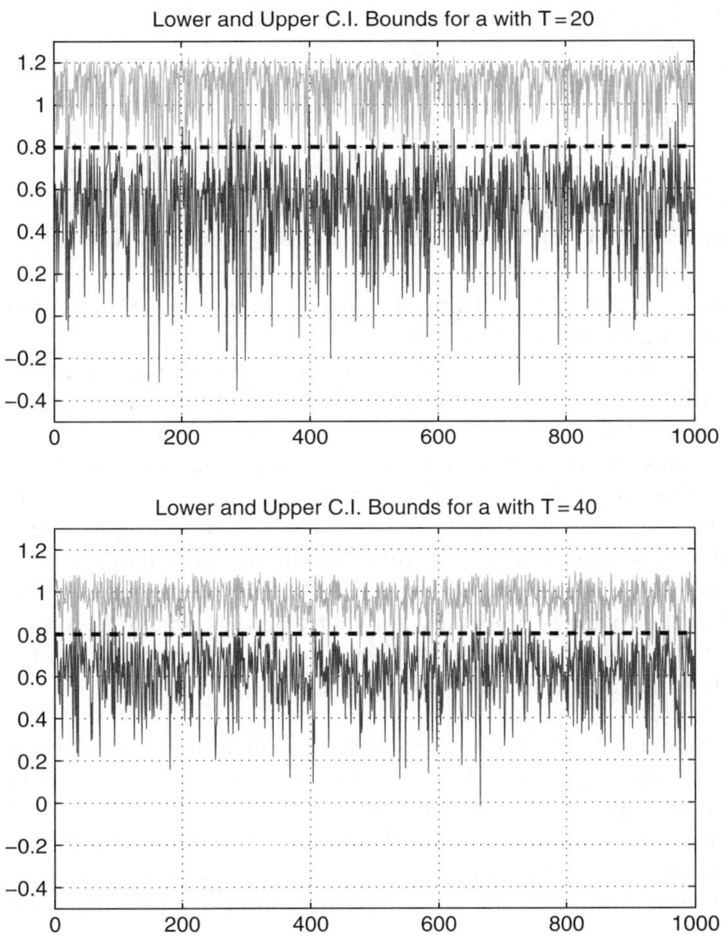

Figure 4.14 Confidence intervals over 1,000 replications when $a = 0.8$, for the indicated sample sizes.

4 Andrews (1993) notes that, in order to maintain exact coverage when $a = 1$, $\hat{c}_L$ should rather be defined as $\hat{c}_L > 1$ if $\hat{a}_{\text{LS}} > q_{\alpha_2}(1)$, resulting in an empty set for the confidence interval. Otherwise, the coverage is $1 - \alpha_1 > 1 - \alpha$ when $a = 1$, as the true parameter cannot lie to the left of the interval.

```
1  T=40; a=0.80; sim=1000; lovec=zeros(sim,1); hivec=lovec; cover=lovec;
2  for i=1:sim, disp(i)
3    % generate an AR(1) process, use a warm-up of 40 observations
4    U=randn(T+40,1); y=zeros(T+40,1);
5    for t=2:T+40, y(t)=a*y(t-1)+U(t); end, y=y((end-T+1):end);
6    % get confidence interval and keep track of coverage
7    X=[]; % no X matrix
8    %X=1; % a constant
9    %X=2; % a constant and time trend
10   [lo,hi]=ar1andrewsCI(y,X); lovec(i)=lo; hivec(i)=hi;
11   cover(i)=(a>lo) && (a<hi);
12 actual_coverage = mean(cover) %#ok<NOPTS>
13 figure, plot(1:sim,lovec,'r-', 1:sim,hivec,'g-'), grid
14 set(gca,'fontsize',16), ylim([-0.5 1.3])
15 title(['Lower and Upper C.I. Bounds for a with T=',int2str(T)])
16 line([0 1000], [a a],'color','k','linestyle','-','linewidth',3)
17 lowside = mean(a>lovec), hiside = mean(a<hivec) %#ok<NOPTS>
18 end
```

Program Listing 4.2: Generates the graphics in Figure 4.14. Anticipating use of regressors, as shown in Chapter 5, the code in lines 7–9 allows a choice of **X** matrix. In our setting, there is no **X** matrix.

confidence intervals for a particular value of a, say $a = 0.8$, Figure 4.14 shows them for 1,000 simulated values, and for two sample sizes, having used the code in Listing 4.2.

To illustrate the coverage over a range of a, the top panel of Figure 4.15 shows the actual coverage of the nominal 90% c.i. for a grid of a-values between zero and 0.95, based on a simulation with 20,000 replications, using $\sigma = 1$. As expected, interval (4.53) exhibits the correct coverage,[5] while the coverage of the asymptotic-based c.i. drops off markedly as a increases, as expected from the bias of the point estimator and the deviation from normality of its distribution. The right panel shows the lengths of the intervals, along with the length of the asymptotic-based c.i. but truncating its upper end at one. Observe how the length decreases almost four-fold when moving from $a = 0$ to $a = 0.95$.

Figure 4.16 is similar, but shows the results for the asymptotic-based 90% c.i.s for σ and its truncation at zero. From the top panel, we see that the coverage probability for values of a less than 0.8 is about 0.865, but then drops off as a approaches 1. The lengths of the intervals are, in comparison to c.i.s for a, virtually constant for $0 \leqslant a \leqslant 0.8$, but then decrease as a approaches one.

Remark In the case of an AR(p) model (as discussed in Section 6.1), an exact method for a median-unbiased point estimator analogous to (4.45), and its extension for confidence intervals in (4.53), appears to not be possible, but an approximate method is developed in Andrews and Chen (1994). In the AR(1) case but allowing for an unknown form of heteroskedasticity, Romano and Wolf (2001) and Andrews and Guggenberger (2014) present a method for constructing a confidence interval with asymptotically correct size, the latter not requiring any tuning (sub-sample size) parameters. ∎

5 It deviates slightly as a approaches one because of numerical inaccuracies in calculating the quantities in $\hat{c}_L$ and $\hat{c}_U$.

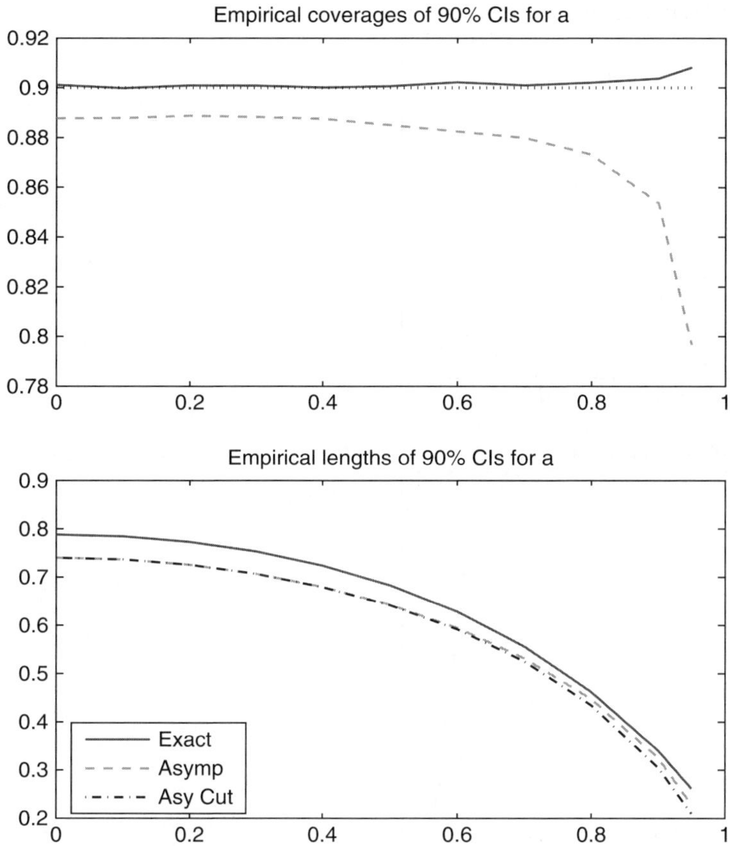

Figure 4.15 Comparison of actual confidence interval coverage (top panel) and length (bottom panel) for parameter a in the AR(1) model with $T = 19$, based on simulation with 20,000 replications and level of significance $\alpha = 0.90$. "Exact" (solid) refers to (4.53), "Asymp" (dashed) refers to c.i.s based on the m.l.e. and its asymptotic normal distribution. and "Asy Cut" (dash dot) is the same as "Asymp" but truncating the c.i.s to lie between −1 and 1.

4.8 Problems

Problem 4.1 Construct programs to simulate an AR(1) process and to estimate it via least squares. Use them to make code that replicates Figure 4.6.

Problem 4.2 Consider the AR(1) model with parameter $a = -1$. As y_t is just the negative of its predecessor y_{t-1} plus an error term, the process will tend to oscillate back and forth around zero. Use recursive substitution to show that $Y_0, Y_2, Y_4, \ldots$, is a random walk with i.i.d. $\mathrm{N}(0, 2\sigma^2)$ innovations. Then simulate and plot the process for a large value of T to see how the variance appears to change through time.

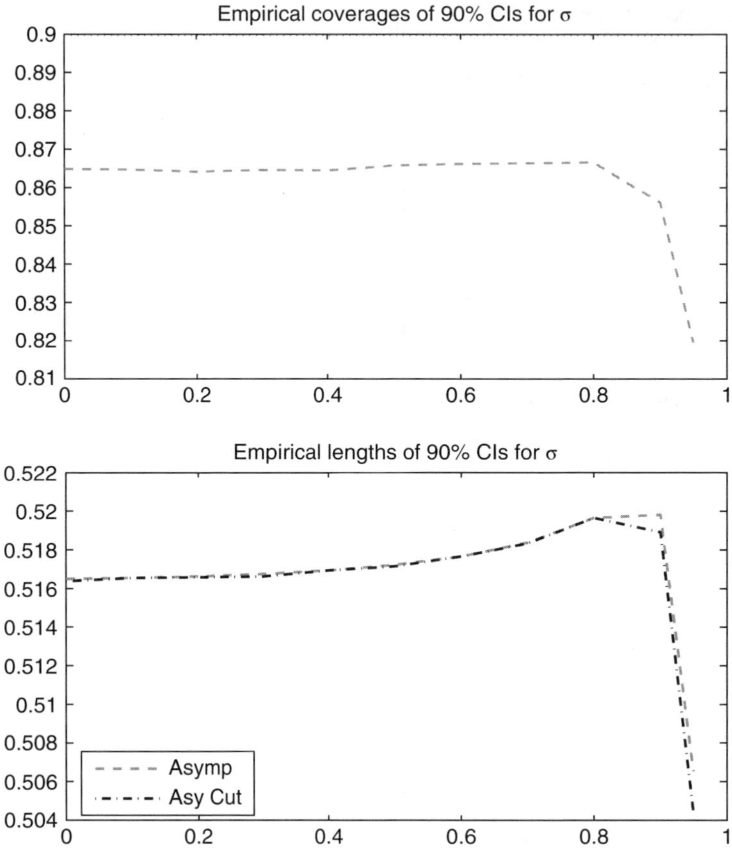

Figure 4.16 Similar to Figure 4.15 but for parameter σ. "Asy Cut" (dash dot) is the same as "Asymp" but truncating the c.i.s to lie above zero.

Problem 4.3 Show (4.12), i.e., that, for any r and t,

$$\mathrm{Cov}(Y_t, Y_r) = \sigma^2 \frac{a^{|t-r|} - a^{t+r}}{1 - a^2}. \tag{4.56}$$

Then verify the following special and limiting cases:

- If $r = t$, then (4.12) reduces to the expression for $\mathbb{V}(Y_t)$ given in the top of (4.6).
- If $r = t - s$ for $s > 0$, then (4.12) reduces to the expression for the covariance given in (4.11).
- If the time index is shifted forward in time by v units, i.e., instead of observing Y_1 through Y_T we observe Y_{1+v} through Y_{T+v}, then (4.12) becomes, for $1 \leqslant t, r \leqslant T$,

$$\mathrm{Cov}(Y_{t+v}, Y_{r+v}) = \sigma^2 \frac{a^{|t-r|} - a^{t+r+2v}}{1 - a^2}$$

and, in the limit as $v \to \infty$, this approaches $\sigma^2 a^{|t-r|}/(1 - a^2)$, which is γ_{t-r} from (4.13).

Problem 4.4 Recall Figure 4.8, in which the top panel shows the m.s.e. of h-step ahead forecasts for $a = 0.5$, $\sigma^2 = 4$, and $T = 10$. Notice that (for the chosen values of a and T), the empirically observed m.s.e. decreases after $h = 4$. Show algebraically that the expression in (4.34) decreases after $h = 3$ by computing the difference $\text{mse}(\widehat{Y}_{T+4|T}) - \text{mse}(\widehat{Y}_{T+3|T})$ and determining when this is negative. Under what conditions is (4.34) nondecreasing in h?

Problem 4.5 Recall the discussion in Section 4.3.5 regarding the asymptotic variance of $\widehat{a}_{\text{LS}}$ and its small sample behavior. In practice, the normality assumption might not be valid, with the usual violation being leptokurtosis. The effect of non-normality can be investigated in this case by repeating the simulation exercise in Section 4.3.5 with a heavy-tailed distribution such as Student's $t(v)$ instead of the normal. Do so for different values of v and examine the behavior as v changes. What do you expect to see?
Also use a contaminated normal distribution such as 80% N(0,1) and 20% N(0,16). Do you expect the results to be similar to the Student's t case?

5

Regression Extensions: AR(1) Errors and Time-varying Parameters

The place of econometrics at the centre of economics is now confirmed.

(The Economist, Oct. 11, 2003, p. 84)

A popular univariate time-series model that is useful in itself and also serving as a baseline for advanced models in econometrics is the regression framework of Chapter 1 combined with the AR(1) model for the regression error term from Chapter 4. This chapter considers this model in detail.

After discussing the likelihood in Section 5.1, we develop point and interval estimators for the AR(1) parameter amid regressor covariates in Section 5.2. Section 5.3 discusses methods for testing the null hypothesis of the AR(1) coefficient being zero. Section 5.4 builds on the methods from Section 4.6 for bias-adjusted point estimation of the AR(1) parameter. Section 5.5 details some basic methods for unit-root testing. Finally, we turn to the regression model with time-varying $\boldsymbol{\beta}$ coefficients in Section 5.6.

5.1 The AR(1) Regression Model and the Likelihood

Let $\mathbf{x}_t'$, $t = 0, 1, \ldots, T$, be a set of $1 \times k$ vectors of non-stochastic, known constants, such that $\mathbf{X} = [\mathbf{x}_0, \ldots, \mathbf{x}_T]'$ is a full rank $(T + 1) \times k$ matrix. The model is now given by two equations; the first is the **observation equation**,

$$Y_t = \mathbf{x}_t'\boldsymbol{\beta} + \epsilon_t, \tag{5.1}$$

which is identical to the linear model (1.3). The difference between (5.1) and (1.3) is the assumptions on the ϵ_t, brought out in the second model equation,

$$\epsilon_t = a\epsilon_{t-1} + U_t, \qquad U_t \overset{\text{i.i.d.}}{\sim} \mathrm{N}(0, \sigma^2), \tag{5.2}$$

referred to as the **latent equation**. The names indicate that the Y_t are observed, but not the ϵ_t. The ϵ_t are also referred to as latent variables.

There are now $k + 2$ parameters: $\boldsymbol{\beta} \in \mathbb{R}^k$, $a \in (-1, 1]$, and $\sigma^2 > 0$. If $a = 0$, then the AR(1) structure in (5.2) renders the ϵ_t i.i.d., and (5.1) becomes just the linear model (1.3). If, instead, there are no regressors, then $Y_t = \epsilon_t$ and the model reduces to the pure AR(1) process (4.1). If $a \in (-1, 1)$, then ϵ_0 is assumed to be a realization from its unconditional distribution, $\mathrm{N}(0, \sigma^2/(1 - a^2))$, while, if $a = 1$, then ϵ_0 is taken to be an arbitrary constant.

Linear Models and Time-Series Analysis: Regression, ANOVA, ARMA and GARCH, First Edition. Marc S. Paolella.

The latent equation (5.2) can also be written as $(1 - aL)\epsilon_t = U_t$, where $L\epsilon_t = \epsilon_{t-1}$ and L is referred to as the **lag operator**. (We will make extensive use of the lag operator in Chapter 6.) Treating L as a variable and multiplying (5.1) by the polynomial $(1 - aL)$ gives

$$(1 - aL)Y_t = (1 - aL)\mathbf{x}_t'\boldsymbol{\beta} + (1 - aL)\epsilon_t,$$

or

$$Y_t = aY_{t-1} + (1 - aL)\mathbf{x}_t'\boldsymbol{\beta} + U_t. \tag{5.3}$$

Example 5.1 Let the $\mathbf{x}_t'$ consist of a constant and time trend, denoted $\mathbf{x}_t' = (1, t)$, so that the second column of $\mathbf{X}$ is $(0, 1, 2, \ldots, T)'$ and $\mathbf{x}_t'\boldsymbol{\beta} = \beta_1 + \beta_2 t$. Then, as $L1 = 1$ and $Lt = t - 1$,

$$\begin{aligned}(1 - aL)\mathbf{x}_t'\boldsymbol{\beta} &= (1 - aL)(\beta_1 + \beta_2 t)\\ &= (\beta_1 + \beta_2 t) - a(\beta_1 + \beta_2(t - 1))\\ &= (1 - a)\beta_1 + a\beta_2 + (1 - a)\beta_2 t,\end{aligned}$$

so that (5.3) can be written as

$$Y_t = aY_{t-1} + \beta_1^* + \beta_2^* t + U_t, \tag{5.4}$$

where $\beta_1^* = (1 - a)\beta_1 + a\beta_2$ and $\beta_2^* = (1 - a)\beta_2$. Notice that, if $a = 1$, then $\beta_1^* = \beta_2$, $\beta_2^* = 0$, and the model reduces to a random walk with drift, given by $Y_t = \beta_2 + Y_{t-1} + U_t$. Going the other way, for $|a| < 1$,

$$\beta_1 = \frac{\beta_1^*(1 - a) - a\beta_2^*}{(1 - a)^2}, \quad \beta_2 = \frac{\beta_2^*}{1 - a},$$

these being referred to as **common factor restrictions**. ■

Let $\mathbf{Y} = (Y_0, Y_1, \ldots, Y_T)'$ and assume $|a| < 1$. From (5.1), $\mathbf{Y} \sim \mathrm{N}(\mathbf{X}\boldsymbol{\beta}, \sigma^2\boldsymbol{\Sigma})$, where $\sigma^2\boldsymbol{\Sigma}$ is the $(T + 1) \times (T + 1)$ covariance matrix of $\mathbf{Y} - \mathbf{X}\boldsymbol{\beta} = \boldsymbol{\epsilon} = (\epsilon_0, \epsilon_1, \ldots, \epsilon_T)'$. From (5.2), $\boldsymbol{\Sigma}$ is given by (4.20). Thus, $f_{\mathbf{Y}}(\mathbf{y})$ is the same as in (4.19) but with $\mathbf{y} - \mathbf{X}\boldsymbol{\beta}$ replacing $\mathbf{y}$ on the r.h.s., and, recalling the decomposition of $\boldsymbol{\Sigma}$ in (4.23),

$$(\mathbf{Y} - \mathbf{X}\boldsymbol{\beta})'\boldsymbol{\Sigma}^{-1}(\mathbf{Y} - \mathbf{X}\boldsymbol{\beta}) = (1 - a^2)\epsilon_0^2 + \sum_{t=1}^{T}(\epsilon_t - a\epsilon_{t-1})^2, \tag{5.5}$$

giving the expression for the exact likelihood

$$\mathcal{L}(\boldsymbol{\beta}, \sigma^2, a; \mathbf{Y}) = K_a \exp\left\{-\frac{1}{2\sigma^2}\left[(1 - a^2)\epsilon_0^2 + \sum_{t=1}^{T}(\epsilon_t - a\epsilon_{t-1})^2\right]\right\}, \tag{5.6}$$

where $K_a = \sqrt{1 - a^2}\,/(2\pi\sigma^2)^{(T+1)/2}$. This is straightforward to evaluate and maximize to obtain point estimates of a, $\boldsymbol{\beta}$, and σ^2.

An equivalent expression for the likelihood can be obtained by using the model representation in (5.3): Let $\theta_t = (1 - aL)\mathbf{x}_t'\boldsymbol{\beta}$, $t = 1, \ldots, T$. As before, we assume that ϵ_0 is drawn from its unconditional distribution $\mathrm{N}(0, \gamma_0)$, where $\gamma_0 = \sigma^2/(1 - a^2)$ from (4.8). Then $Y_0 = \mathbf{x}_0'\boldsymbol{\beta} + \epsilon_0 \sim \mathrm{N}(\mathbf{x}_0'\boldsymbol{\beta}, \gamma_0)$. Conditionally, $Y_t \mid Y_{t-1} \sim \mathrm{N}(aY_{t-1} + \theta_t, \sigma^2)$, $t = 1, \ldots, T$, i.e.,

$$f_{Y_t \mid Y_{t-1}}(y_t \mid y_{t-1}) = \frac{1}{\sqrt{2\pi}\sigma}\exp\left\{-\frac{1}{2\sigma^2}[y_t - (ay_{t-1} + \theta_t)]^2\right\}, \tag{5.7}$$

so that the joint density of $\mathbf{Y}$ can be expressed as

$$f_{Y_0}(y_0) \times \prod_{i=1}^{T} f_{Y_t \mid Y_{t-1}}(y_t \mid y_{t-1}). \tag{5.8}$$

The equivalence of (5.6) and (5.8) follows because both representations use the same unconditional density of Y_0, and the term in the exponent of (5.7) can be written as

$$\begin{aligned} y_t - (ay_{t-1} + \theta_t) &= y_t - ay_{t-1} - (1 - aL)\mathbf{x}_t'\boldsymbol{\beta} \\ &= (y_t - \mathbf{x}_t'\boldsymbol{\beta}) - a(y_{t-1} - \mathbf{x}_{t-1}'\boldsymbol{\beta}) = z_t - az_{t-1}. \end{aligned}$$

From (5.6), some simple algebra (that the reader should confirm) shows that the likelihood function can also be expressed as follows:

$$\mathcal{L}(\boldsymbol{\beta}, \sigma^2, a; \mathbf{Y}) = K_a \exp\left\{ -\frac{1}{2\sigma^2} \left[(1 + a^2)\boldsymbol{\epsilon}'\boldsymbol{\epsilon} - a^2(\epsilon_0^2 + \epsilon_T^2) - 2a \sum_{t=1}^{T} \epsilon_t \epsilon_{t-1} \right] \right\}. \tag{5.9}$$

Accurate starting values (or even as final values) can be obtained by iterating between simple estimators for $\boldsymbol{\beta}$, σ^2, and a. In particular, let $\widehat{\boldsymbol{\beta}}^{(1)} = \widehat{\boldsymbol{\beta}}_{\mathrm{LS}} = (\mathbf{X}'\mathbf{X})^{-1}\mathbf{X}'\mathbf{Y}$ be the standard o.l.s. estimator appropriate in a linear regression model assuming an i.i.d. Gaussian error process, with residuals $\widehat{\boldsymbol{\epsilon}}^{(1)} = \mathbf{Y} - \mathbf{X}\widehat{\boldsymbol{\beta}}^{(1)}$. Then let $\widehat{a}^{(1)} = \widehat{a}_{\mathrm{LS}}$ from (4.14) based on the residuals $\widehat{\boldsymbol{\epsilon}}^{(1)}$. Next, let $\widehat{\boldsymbol{\beta}}^{(2)} = \widehat{\boldsymbol{\beta}}_{\mathrm{GLS}} = (\mathbf{X}'\boldsymbol{\Sigma}^{-1}\mathbf{X})^{-1}\mathbf{X}'\boldsymbol{\Sigma}^{-1}\mathbf{Y}$, where $\boldsymbol{\Sigma}$ is (4.20) based on $\widehat{a}^{(1)}$, from which $\widehat{\boldsymbol{\epsilon}}^{(2)} = \mathbf{Y} - \mathbf{X}\widehat{\boldsymbol{\beta}}^{(2)}$ are computed. From $\widehat{\boldsymbol{\epsilon}}^{(2)}$, let $\widehat{a}^{(2)} = \widehat{a}_{\mathrm{LS}}$ from (4.14). The process can be repeated until convergence for a given tolerance. The estimator for σ^2 can be taken from either the $\widehat{\boldsymbol{\beta}}_{\mathrm{LS}}$ step or the $\widehat{a}_{\mathrm{LS}}$ step. Similar to the pure AR(1) case, this procedure will not perform well for small sample sizes and a near one.

Interval estimates of the parameters can easily be computed using the asymptotically valid Wald intervals, based on the approximate standard errors generated as a by-product from Hessian-based optimization routines, as discussed at length in Chapter III.4. More accurate intervals can be obtained via use of the parametric or nonparametric bootstrap. In both cases, sampling involves the U_t. For the ith parametric bootstrap draw, $U_t^{(i)} \overset{\text{i.i.d.}}{\sim} \mathrm{N}(0, \widehat{\sigma}^2)$, from which $\epsilon_t^{(i)} = \widehat{a}\epsilon_{t-1}^{(i)} + U_t^{(i)}$ is constructed, as in (5.2), with $\epsilon_0^{(i)} \sim \mathrm{N}(0, \hat{\sigma}^2/(1 - \hat{a}^2))$, and then $Y_t^{(i)} = \mathbf{x}_t'\widehat{\boldsymbol{\beta}} + \epsilon_t^{(i)}$, from (5.1), $t = 0, 1, \dots, T$. The nonparametric bootstrap is similar, but the $U_t^{(i)}$ are sampled, with replacement, from the filtered innovation sequence $\widehat{U}_t$, $t = 1, \dots, T$.

5.2 OLS Point and Interval Estimation of *a*

As above, let $\mathbf{Y} = (Y_0, Y_1, \dots, Y_T)'$ and define, as in Section 4.3, the $T \times (T+1)$ matrices $\mathbf{D}_T = [\mathbf{0} \mid \mathbf{I}_T]$ and $\mathbf{D}_{T-1} = [\mathbf{I}_T \mid \mathbf{0}]$, so that

$$\mathbf{Y}_T = (Y_1, \dots, Y_T)' = \mathbf{D}_T\mathbf{Y} \quad \text{and} \quad \mathbf{Y}_{T-1} = (Y_0, \dots, Y_{T-1})' = \mathbf{D}_{T-1}\mathbf{Y}.$$

Then, generalizing Example 5.1 to the case with k regressors,

$$Y_t = aY_{t-1} + \mathbf{x}_t'\boldsymbol{\beta} - a\mathbf{x}_{t-1}'\boldsymbol{\beta} + U_t, \quad t = 1, \dots, T,$$

or, in matrix form,

$$\mathbf{Y}_T = a\mathbf{Y}_{T-1} + \mathbf{Z}\boldsymbol{\gamma} + \mathbf{U}_T, \tag{5.10}$$

where $\boldsymbol{\gamma} = [\boldsymbol{\beta}', -a\boldsymbol{\beta}']'$, $\mathbf{Z} = [\mathbf{X}_T, \mathbf{X}_{T-1}]$, $\mathbf{X}_T = [\mathbf{x}_1, \dots, \mathbf{x}_T]'$, $\mathbf{X}_{T-1} = [\mathbf{x}_0, \dots, \mathbf{x}_{T-1}]'$, and $\mathbf{U}_T = (U_1, \dots, U_T)'$. Applying the Frisch–Waugh–Lovell theorem from Example 1.1 shows that, with[1]

$$\mathbf{M} = \mathbf{I}_T - \mathbf{Z}(\mathbf{Z}'\mathbf{Z})^{-}\mathbf{Z}', \tag{5.11}$$

the o.l.s. estimator of a can be expressed as

$$\widehat{a}_{\text{LS}} = \frac{\mathbf{Y}'_{T-1}\mathbf{M}\mathbf{Y}_T}{\mathbf{Y}'_{T-1}\mathbf{M}\mathbf{Y}_{T-1}}. \tag{5.12}$$

Observe from Example 5.1 that, for a model with intercept ($\mathbf{X}$ being a column of ones) or an intercept and time-trend model ($\mathbf{X} = [\mathbf{1}, \ \mathbf{t}]$), the column spaces of $\mathbf{X}$ and $\mathbf{Z}$ are equal, and (5.11) can be replaced with $\mathbf{M} = \mathbf{I}_T - \mathbf{X}(\mathbf{X}'\mathbf{X})^{-1}\mathbf{X}'$.

We now show that $\widehat{a}_{\text{LS}}$ is independent of $\boldsymbol{\beta}$ for any exogenous regressor matrix $\mathbf{X}$, generalizing the result in Andrews (1993), which is restricted to $\mathbf{X}$ corresponding to a constant and time trend. This can be done in two ways. One is based on the observation that the o.l.s. estimator is a function of $\mathbf{Y}$ only through a so-called *maximal invariant* whose distribution is free of $\boldsymbol{\beta}$, see, e.g., Dufour and King (1991). A second way is to use a *singular value decomposition*, hereafter SVD (see, e.g., Harville, 1997, Sec. 21.12; Gentle, 2007, Sec. 7.7; Lay et al., 2015, p. 435) of $\mathbf{X}$ (which allows for linearly dependent regressors, as we require), and is now demonstrated.

Theorem 5.1 Estimator $\widehat{a}_{\text{LS}}$ is independent of $\boldsymbol{\beta}$ for any exogenous regressor matrix $\mathbf{X}$.

Proof: First note that the computation of (5.12) requires the $T \times 2k$ matrix $\mathbf{Z}$ to be of full column rank. This is not always satisfied, e.g., if the regressors include a constant, a constant and a linear time trend, or a specific combination of impulse and step dummies. For a particular $\mathbf{X}$ such as the constant and time-trend model, the linearly dependent columns can be "calculated out by hand", as in Andrews (1993).

In order that the proof is valid for all $r = \text{rank}(\mathbf{Z}) \leqslant 2k$, let the SVD of $\mathbf{Z}$ be $\mathbf{Z} = \mathbf{Q}\mathbf{W}\mathbf{V}'$, where $\mathbf{Q}$ and $\mathbf{V}$ are $T \times r$ and $2k \times r$ matrices, respectively, of full column rank r, and $\mathbf{W}$ is an $r \times r$ diagonal matrix of full rank. Moreover, $\mathbf{Q}'\mathbf{Q} = \mathbf{V}'\mathbf{V} = \mathbf{I}_r$. Clearly, only r different parameters in $\boldsymbol{\gamma}$ are identified. Defining $\tilde{\mathbf{Z}} = \mathbf{Q}\mathbf{W}$ and $\tilde{\boldsymbol{\gamma}} = \mathbf{V}'\boldsymbol{\gamma}$, we can rewrite (5.10) as

$$\mathbf{Y}_T = a\mathbf{Y}_{T-1} + \tilde{\mathbf{Z}}\tilde{\boldsymbol{\gamma}} + \mathbf{U}_T. \tag{5.13}$$

The linearly dependent columns of $\mathbf{Z}$ are effectively removed by means of the SVD of $\mathbf{Z}$. The o.l.s. estimator for a can now be obtained as in (5.12) with $\mathbf{M}$ replaced by $\mathbf{M} = \mathbf{I}_T - \tilde{\mathbf{Z}}(\tilde{\mathbf{Z}}'\tilde{\mathbf{Z}})^{-}\tilde{\mathbf{Z}}' = \mathbf{I}_T - \mathbf{Q}\mathbf{Q}'$.

As in Andrews (1993), to show that $\widehat{a}_{\text{LS}}$ does not depend on the value of $\boldsymbol{\beta}$, it suffices to show that the residuals $\mathbf{M}\mathbf{Y}_T$ and $\mathbf{M}\mathbf{Y}_{T-1}$ do not depend on $\boldsymbol{\beta}$. As

$$\mathbf{M}\mathbf{Y}_T = \mathbf{M}\mathbf{X}_T\boldsymbol{\beta} + \mathbf{M}\boldsymbol{\epsilon}_T \quad \text{and} \quad \mathbf{M}\mathbf{Y}_{T-1} = \mathbf{M}\mathbf{X}_{T-1}\boldsymbol{\beta} + \mathbf{M}\boldsymbol{\epsilon}_{T-1},$$

1 The notation $\mathbf{A}^{-}$ denotes a **generalized inverse** of $\mathbf{A}$. It satisfies $\mathbf{A}\mathbf{A}^{-}\mathbf{A} = \mathbf{A}$. A generalized inverse always exists for $\mathbf{A}$ symmetric, and also satisfies $\mathbf{A}^{-}\mathbf{A}\mathbf{A}^{-} = \mathbf{A}^{-}$, known as the reflexive property, and is symmetric. Finally, and crucially, $\mathbf{Z}(\mathbf{Z}'\mathbf{Z})^{-}\mathbf{Z}'$ is the perpendicular projection operator onto $\mathcal{C}(\mathbf{Z})$; see, e.g., Seber and Lee (2003, p. 476) or Christensen (2011, p. 430). Detailed presentations of generalized inverses can be found in several books; in addition to the aforementioned references; see, e.g., Ravishanker and Dey (2002, Ch. 3) and Dhrymes (2013, Sec. 3.3). See also Remark (a) in Section C.2.3.

where $\boldsymbol{\epsilon}_T$ and $\boldsymbol{\epsilon}_{T-1}$ are defined analogously to $\mathbf{Y}_T$ and $\mathbf{Y}_{T-1}$, this amounts to showing that $\mathbf{MX}_T = \mathbf{0}$ and $\mathbf{MX}_{T-1} = \mathbf{0}$ because neither $\mathbf{M}\boldsymbol{\epsilon}_T$ nor $\mathbf{M}\boldsymbol{\epsilon}_{T-1}$ depend on $\boldsymbol{\beta}$. Partitioning $\mathbf{V}'$ into the first k and the last k columns $\mathbf{V}_1'$ and $\mathbf{V}_2'$, respectively, we obtain $\mathbf{Z} = [\mathbf{X}_T, \mathbf{X}_{T-1}] = [\mathbf{QWV}_1', \mathbf{QWV}_2']$ and, thus, $\mathbf{X}_T = \mathbf{QWV}_1'$ and $\mathbf{X}_{T-1} = \mathbf{QWV}_2'$. Now,

$$\mathbf{MX}_T = (\mathbf{I}_T - \mathbf{QQ}')\mathbf{QWV}_1' = (\mathbf{Q} - \mathbf{Q})\mathbf{WV}_1' = \mathbf{0}$$

and

$$\mathbf{MX}_T = (\mathbf{I}_T - \mathbf{QQ}')\mathbf{QWV}_2' = (\mathbf{Q} - \mathbf{Q})\mathbf{WV}_2' = \mathbf{0},$$

confirming that $\widehat{a}_{\mathrm{LS}}$ is independent of $\boldsymbol{\beta}$. ■

The invariance of $\widehat{a}_{\mathrm{LS}}$ with respect to σ^2 follows from the structure of (5.12) as a ratio, such that σ^2 cancels. Thus, we can assume $\boldsymbol{\beta} = \mathbf{0}$ and $\sigma^2 = 1$ in the following, without loss of generality.

We now have $\mathbf{MY}_T = \mathbf{MD}_T\boldsymbol{\epsilon}$ and $\mathbf{MY}_{T-1} = \mathbf{MD}_{T-1}\boldsymbol{\epsilon}$, where $\boldsymbol{\epsilon} = [\epsilon_0, \boldsymbol{\epsilon}_T']'$. Substituting this into (5.12) shows that $\widehat{a}_{\mathrm{LS}}$ has the same distribution as

$$\frac{\boldsymbol{\epsilon}'\mathbf{D}_{T-1}'\mathbf{MD}_T'\boldsymbol{\epsilon}}{\boldsymbol{\epsilon}'\mathbf{D}_{T-1}'\mathbf{MD}_{T-1}'\boldsymbol{\epsilon}} = \frac{\mathbf{U}'\mathbf{R}'\mathbf{D}_{T-1}'\mathbf{MD}_T\mathbf{RU}}{\mathbf{U}'\mathbf{R}'\mathbf{D}_{T-1}'\mathbf{MD}_{T-1}\mathbf{RU}} = \frac{\mathbf{U}'\mathbf{AU}}{\mathbf{U}'\mathbf{BU}}, \tag{5.14}$$

where $\mathbf{A}$ and $\mathbf{B}$ are so defined (and taking $\mathbf{A}$ to be the symmetric version, i.e., $(\mathbf{A}' + \mathbf{A})/2$) and $\boldsymbol{\epsilon} = \mathbf{RU}$ for $\mathbf{U} = (U_0, \dots, U_T)' \sim \mathrm{N}(\mathbf{0}, \mathbf{I}_{T+1})$, with $\mathbf{R} = \mathbf{R}(a)$ given in (4.35) and (4.36). In case there are no exogenous regressors, set $\mathbf{M} = \mathbf{I}$ and all conditioning on $\mathbf{X}$ is replaced by conditioning on T. The methods detailed in Chapters A and B can be used to compute the distribution and moments of $\widehat{a}_{\mathrm{LS}}$, respectively.

An exact confidence interval (meaning that the nominal and actual coverage coincide) for a can be computed precisely as in Section 4.7. Note that the code in Listing 4.1 was already designed to support the use of a (possibly rank deficient) regressor matrix.

As discussed in Chapter III.2, much intellectual effort has been invested into the Neyman–Pearson hypothesis testing framework in statistical and econometric modeling for investigating various parametric assumptions, and the resulting tests can often be used to construct valid confidence intervals for parameters. The **Durbin–Watson test** is discussed in Appendix B, and detailed further below in Section 5.3.2, with test statistic $D = \widehat{\boldsymbol{\epsilon}}'\mathbf{A}\widehat{\boldsymbol{\epsilon}}/\widehat{\boldsymbol{\epsilon}}'\widehat{\boldsymbol{\epsilon}}$, where $\widehat{\boldsymbol{\epsilon}} = \mathbf{MY}$ and $\mathbf{A}$ is given in (B.8). Let d denote the observed test statistic for a given data set $\mathbf{Y}$ and regression matrix $\mathbf{X}$, where $\mathbf{Y} \sim \mathrm{N}(\mathbf{X}\boldsymbol{\beta}, \sigma^2\boldsymbol{\Sigma}(a))$, with a, $|a| < 1$, the unknown AR(1) parameter. For a one-sided c.i. of a of the form $(\underline{a}, 1)$ with significance level α, consider taking $\underline{a}$ to be that value such that

$$\alpha = \Pr(D_{\underline{a}} \leqslant d) := \Pr\left(\frac{\mathbf{Z}'\mathbf{MAMZ}}{\mathbf{Z}'\mathbf{MZ}} \leqslant d\right), \quad \mathbf{Z} \sim \mathrm{N}(\mathbf{0}, \boldsymbol{\Sigma}(\underline{a})), \tag{5.15}$$

where we can ignore $\mathbf{X}\boldsymbol{\beta}$ because $\mathbf{MX} = \mathbf{0}$ and σ^2 cancels from the ratio. The idea is to find the point $\underline{a}$ such that, for $-1 < a < \underline{a}$ we would reject the null, while for $\underline{a} < a < 1$ we do not reject.

This is perhaps best explained graphically: For a simulated data set of $T = 50$ observations from an AR(1) regression model with $\mathbf{X} = [\mathbf{1}, \mathbf{t}]$ and true $a = 0.5$, Figure 5.1 shows the density of the Durbin–Watson statistic D under the null of $a = 0$, with the vertical dashed line indicating the observed test statistic d, overlaid with the density of D corresponding to $\boldsymbol{\Sigma}(\underline{a})$. The area under the density to the left of the vertical dashed line is $\alpha = 0.05$. The code used to generate Figure 5.1 is shown in Listing 5.1.

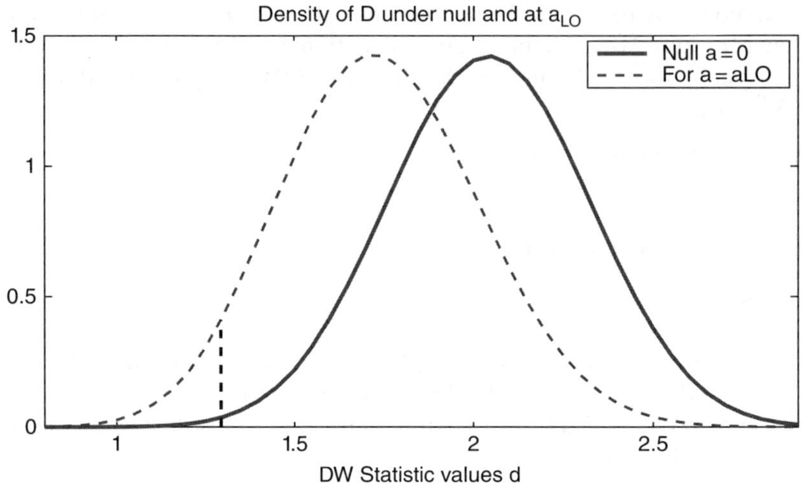

Figure 5.1 Demonstration of computation of $\underline{a}$ in (5.15).

```
1  T=50; X=[ones(T,1) (1:T)']; M=makeM(X); A=makeDW(T);
2  a=0.5; Si=leeuwAR(a,T); [V,D]=eig(Si); S12=V*D^(-1/2)*V'; S=V*D^(-1)*V';
3  e=S12*randn(T,1); eh=M*e; d=(eh'*A*eh)/(eh'*eh);
4  aLO=fzero(@(aa) cdfratio(d,M*A*M,M,inv(leeuwAR(aa,T)),[],1)-0.05, 1.5);
5
6  rvec=0:0.05:4;
7  a=0; Si=leeuwAR(a,T); [V,D]=eig(Si); S12=V*D^(-1/2)*V';
8  pdf0=ROQpdfgeary(rvec,S12*M*A*M*S12,S12*M*S12);
9
10 a=aLO; Si=leeuwAR(a,T); [V,D]=eig(Si); S12=V*D^(-1/2)*V';
11 pdfLO=ROQpdfgeary(rvec,S12*M*A*M*S12,S12*M*S12);
12 pdfatd=interp1(rvec,pdfLO,d);
13
14 figure, plot(rvec,pdf0,'r-','linewidth',3), hold on
15 plot(rvec,pdfLO,'b--','linewidth',2), hold off
16 set(gca,'fontsize',16), xlabel('DW Statistic values d')
17 title('Density of D under null and at a_{LO}')
18 yy=ylim; yy=yy(2); ylim([0 yy]), xlim([0.8 2.9])
19 legend('Null a=0','For a=aLO')
20 line([d d],[0 pdfatd],'linewidth',2,'linestyle','--','color','k')
```

Program Listing 5.1: Simulates an AR(1) regression model, computes $\underline{a}$ (impressively, in the single line 4), and generates Figure 5.1. Program `ROQpdfgeary` for computing the p.d.f. of a ratio of quadratic forms is available in the book's associated collection of programs. Program `leeuwAR` is given in Listing 7.5.

The reader is encouraged to set up a simulation for a grid of true values of $a \geqslant 0$ and several sample sizes T, and confirm that the actual coverage of the generated c.i. is indeed the nominal of $1 - \alpha$.

5.3 Testing $a = 0$ in the ARX(1) Model

There is always a well-known solution to every human problem—neat, plausible, and wrong.
(Henry Louis Mencken, 1920)[2]

The goal of this section is to test the null hypothesis that the autoregressive parameter a in (5.2) is zero—an endeavor that was deemed very important starting in the late 1940s, with the seminal work being from Durbin and Watson (1950).

5.3.1 Use of Confidence Intervals

As a dual to use of a statistical hypothesis testing framework to generate a confidence interval for a, as in Section 5.2 above, we show how to construct a two-sided equal-tail test based on the confidence interval of $\widehat{a}_{\text{LS}}$ from Section 5.2. Very simply, the test rejects if zero is not in the confidence interval. For illustration, Figure 5.2 shows the power for two sample sizes, based on $\alpha = 0.10$ and several regressor matrices. The bottom right panel uses an $\mathbf{X}$ matrix consisting of a constant, time trend, a vector with the first half zeros and the second half ones, and a vector with the first half zeros and the second half a time trend squared, as a model to capture a break in the intercept and trend (recall the regression examples in Section 1.4.6). We denote the associated matrix as $\mathbf{X} = [\mathbf{1}, \mathbf{t}, D1, Dt^2]$.[3] As expected, the power is higher as the sample size T increases. We also see that the power gets lower as the $\mathbf{X}$ matrix increases in complexity and number of regressors.

5.3.2 The Durbin–Watson Test

We next consider statistics that were designed specifically for testing $a = 0$ and explain in what sense these tests are optimal, starting with the Durbin–Watson test. Recall Chapter III.2, in which the basic concepts of hypothesis testing were discussed, such as simple and composite, unbiasedness, consistency, and uniformly most powerful (UMP) and uniformly most powerful unbiased (UMPU). We will meet other concepts below, such as point optimal tests and invariance.

---

2 Mencken was better known for his disdain of the 1920 Republican Presidential candidate, Warren Harding, and the folly of many voters, resulting in a now-well-cited quote. On July 26, 1920, he published a column in the Baltimore newspaper *The Evening Sun*, writing "The larger the mob, the harder the test. In small areas, before small electorates, a first-rate man occasionally fights his way through, carrying even the mob with him by the force of his personality. But when the field is nationwide, and the fight must be waged chiefly at second and third hand, and the force of personality cannot so readily make itself felt, then all the odds are on the man who is, intrinsically, the most devious and mediocre—the man who can most adeptly disperse the notion that his mind is a virtual vacuum." He went on to say, now famously, "On some great and glorious day the plain folks of the land will reach their heart's desire at last, and the White House will be adorned by a downright moron."

3 For this matrix, $\mathbf{Z}'\mathbf{Z}$ in (5.11) is not full rank, and the generalized inverse is required. However, in this case, using full rank $\mathbf{X}$ in place of $\mathbf{Z}$ yields the same results.

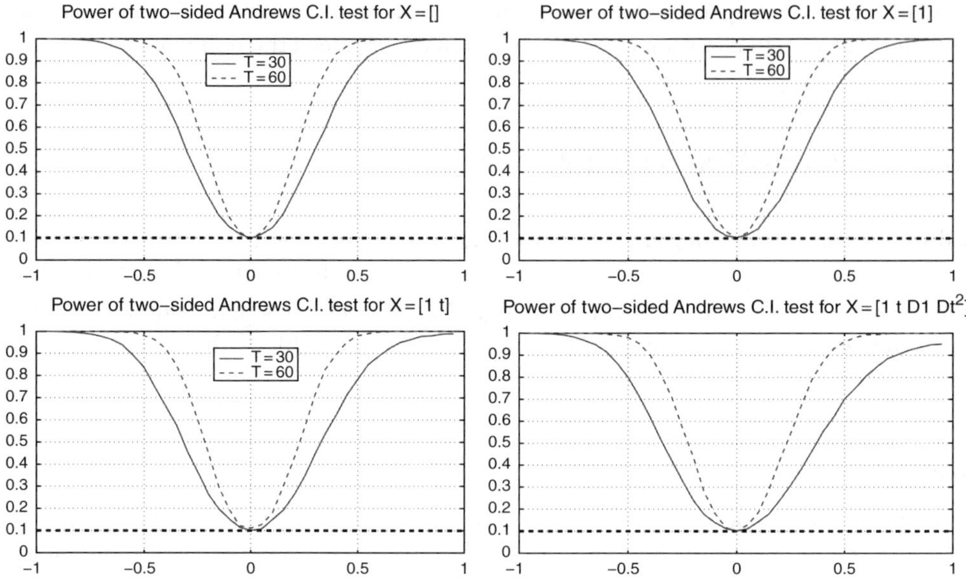

Figure 5.2 Power of the test for the null of $a = 0$ based on the two-sided equal-tail confidence interval for a, for significance level $\alpha = 0.10$ and $T = 30$ and $T = 60$, for various $\mathbf{X}$ matrices, as indicated, and based on 5,000 simulated replications.

Observe that likelihood expression (5.9) is approximately (note the middle term in the exponent, changing a^2 to a)

$$\begin{aligned}
\mathcal{L}(\boldsymbol{\beta}, \sigma^2, a; \mathbf{y}) &\approx K_a \exp\left\{-\frac{1}{2\sigma^2}\left[(1+a^2)\boldsymbol{\epsilon}'\boldsymbol{\epsilon} - a(\epsilon_0^2 + \epsilon_T^2) - 2a\sum_{t=1}^{T}\epsilon_t\epsilon_{t-1}\right]\right\} \\
&= K_a \exp\left\{-\frac{1}{2\sigma^2}[(1+a^2)\boldsymbol{\epsilon}'\boldsymbol{\epsilon} - 2a\boldsymbol{\epsilon}'\boldsymbol{\Theta}\boldsymbol{\epsilon}]\right\}, \quad \boldsymbol{\Theta} = \mathbf{I} - \frac{1}{2}\mathbf{A}, \\
&= K_a \exp\left\{-\frac{1}{2\sigma^2}[\boldsymbol{\epsilon}'((1-a)^2\mathbf{I} + a\mathbf{A})\boldsymbol{\epsilon}]\right\},
\end{aligned} \tag{5.16}$$

as the reader should quickly confirm (algebraically and/or numerically), where $\mathbf{A}$ is the matrix associated with the Durbin–Watson test, given in (B.8), and the test statistic is, repeating from (B.16), with $\hat{\boldsymbol{\epsilon}} = \mathbf{MY} = \mathbf{M}\boldsymbol{\epsilon}$ the o.l.s. regression residuals, and $\mathbf{M} = \mathbf{I} - \mathbf{X}(\mathbf{X}'\mathbf{X})^{-1}\mathbf{X}'$, as given in (1.53),

$$D = \frac{\sum_{t=2}^{T}(\hat{\epsilon}_t - \hat{\epsilon}_{t-1})^2}{\sum_{t=1}^{T}\hat{\epsilon}_t^2} = \frac{\hat{\boldsymbol{\epsilon}}'\mathbf{A}\hat{\boldsymbol{\epsilon}}}{\hat{\boldsymbol{\epsilon}}'\hat{\boldsymbol{\epsilon}}} = \frac{\boldsymbol{\epsilon}'\mathbf{M}'\mathbf{A}\mathbf{M}\boldsymbol{\epsilon}}{\boldsymbol{\epsilon}'\mathbf{M}'\mathbf{M}\boldsymbol{\epsilon}} = \frac{\boldsymbol{\epsilon}'\mathbf{M}\mathbf{A}\mathbf{M}\boldsymbol{\epsilon}}{\boldsymbol{\epsilon}'\mathbf{M}\boldsymbol{\epsilon}}. \tag{5.17}$$

Anderson (1948) (see also Anderson, 1971) showed that, if K_a is neglected in (5.16) and the k columns of $\mathbf{X}$ can be expressed as linear combinations of k of the eigenvectors of $\mathbf{A}$, then the statistic D in (5.17) provides a UMP one-sided test for $H_0 : a = 0$ vs. $H_1 : a > 0$ or $H_1 : a < 0$.

Remark See also Durbin and Watson (1971), Kariya (1977), Kariya and Eaton (1977), and King (1980) regarding this optimality in the more general elliptic distribution setting. As the eigenvector condition will not be fulfilled precisely in general, the Durbin–Watson test (B.16) is said to be approximately UMP. Cassing and White (1983) consider the impact of this eigenvector assumption on the power of the D test. Another aspect that influences the performance of D (and other tests for first-order autocorrelation) is when the observation equation (5.1) is mis-specified. The case when relevant explanatory variables are missing from the regressor matrix was addressed in Examples B.6 and B.7. ■

More generally, for the regression model (5.1) such that $\boldsymbol{\epsilon} \sim \mathrm{N}(\mathbf{0}, \sigma^2\boldsymbol{\Sigma}(\lambda))$, where $\boldsymbol{\Sigma} > 0$ is of any form (and not only the AR(1) model), King and Hillier (1985) show via application of the generalized Neyman–Pearson lemma (see, e.g., Lehmann, 1986, p. 96; Ferguson, 1967, p. 235) that the test of $\lambda = 0$ versus $\lambda > 0$ that rejects for (depending on the application) small or large values of

$$\frac{\widehat{\boldsymbol{\epsilon}}'\mathbf{L}(0)\widehat{\boldsymbol{\epsilon}}}{\widehat{\boldsymbol{\epsilon}}'\widehat{\boldsymbol{\epsilon}}}, \quad \mathbf{L}(\boldsymbol{\lambda}) = \frac{\partial[\boldsymbol{\Sigma}(\lambda)]^{-1}}{\partial\lambda}, \tag{5.18}$$

is **locally best invariant** (LBI) or **point optimal invariant** (POI), where $\widehat{\boldsymbol{\epsilon}} = \mathbf{MY}$ are the ordinary least squares residuals. See also Cox (1983), Chester (1984), and McCabe and Leybourne (2000) regarding this derivation.

The term "locally best" means that the power function has maximal slope as $\lambda \to 0$, while for invariance, observe that this statistic is invariant to changes in the scale of $\mathbf{Y}$ (because σ^2 cancels from the numerator and denominator) and also invariant to $\boldsymbol{\beta}$ (because $\mathbf{MX} = \mathbf{0}$). This is often written in the literature as saying that the test statistic is invariant to translations of the form $\mathbf{Y}^* = \gamma_0\mathbf{Y} + \mathbf{X}\boldsymbol{\gamma}$ for γ_0 a *positive* scalar and $\boldsymbol{\gamma}$ is any real $k \times 1$ vector; see, e.g., King (1980, Eq. (3.2)).

For example, in the AR(1) testing case with error covariance matrix $\boldsymbol{\Sigma}(a)$, we see from (5.16) that $[\boldsymbol{\Sigma}(a)]^{-1} \approx (1-a)^2\mathbf{I} + a\mathbf{A}$, so that $\mathbf{L}(0) = \mathbf{A} - 2\mathbf{I}$, and (5.18) immediately yields the Durbin–Watson test statistic.

5.3.3 Other Tests for First-order Autocorrelation

Since 1970, other tests for first-order autocorrelation have been proposed, also expressible as a ratio of quadratic forms. These include that from King (1981), given by

$$D' = \frac{\widehat{\boldsymbol{\epsilon}}'\tilde{\mathbf{A}}\widehat{\boldsymbol{\epsilon}}}{\widehat{\boldsymbol{\epsilon}}'\widehat{\boldsymbol{\epsilon}}} = D + \frac{\hat{\epsilon}_1^2 + \hat{\epsilon}_T^2}{\sum_{t=1}^T \hat{\epsilon}_t^2}, \tag{5.19}$$

where D is given in (B.16) and (5.17), and $\tilde{\mathbf{A}}$ is $\mathbf{A}$ as given in (B.8) but taking the top left and bottom right entries to be two instead of one. Clearly, as T grows, the two tests will be equivalent. King (1981) showed that D' has higher power than D for $a < 0$.

Building on the work of Kadiyala (1970) and Durbin and Watson (1971), Berenblut and Webb (1973) (hereafter B-W) proposed a test for first-order autocorrelation that has higher power than D for large values of a, i.e., as $a \to 1$. Their test statistic is a modification of a generalized likelihood ratio, with the numerator taking the fixed value $a = 1$ and the denominator taking $a = 0$ (but $\boldsymbol{\beta}$ is estimated in both, hence "generalized"). Similar to the approximation in (5.16) and assumptions on the $\mathbf{X}$ matrix, B-W showed that the test is approximately UMP in a neighborhood of $a = 1$, i.e., it is an (approximate) LBI test.

Recall from (4.21) the form of $\boldsymbol{\Sigma}^{-1}$ in (5.5). B-W define the distribution of the first observation in a different way, so that this matrix, denoted $\mathbf{V}^{-1}(a)$ to distinguish it, is (4.21) but such that the $(1, 1)$

element is $b = 1 + a^2$ instead of 1. Berenblut and Webb (1973, p. 38) mention possible use of the likelihood ratio

$$\Lambda = \frac{\sup_{\Theta} \mathcal{L}(\theta; \mathbf{Y}, \mathbf{X})}{\sup_{\Theta^0} \mathcal{L}(\theta; \mathbf{Y}, \mathbf{X})} \approx \frac{(\mathbf{Y} - \mathbf{X}\tilde{\boldsymbol{\beta}})'\mathbf{V}^{-1}(\hat{a})(\mathbf{Y} - \mathbf{X}\tilde{\boldsymbol{\beta}})}{(\mathbf{Y} - \mathbf{X}\hat{\boldsymbol{\beta}})'(\mathbf{Y} - \mathbf{X}\hat{\boldsymbol{\beta}})}, \tag{5.20}$$

(with the r.h.s. being an approximation because it omits some terms in the likelihood) where $\Theta = \{\boldsymbol{\beta} \in \mathbb{R}^k,\ a \in (-1, 1]\}$, $\Theta^0 = \{\boldsymbol{\beta} \in \mathbb{R}^k,\ a = 0\}$, $\tilde{\boldsymbol{\beta}} = (\mathbf{X}'\mathbf{V}^{-1}(\hat{a})\mathbf{X})^{-1}\mathbf{X}'\mathbf{V}^{-1}(\hat{a})\mathbf{Y}$ is the generalized least squares estimator (and m.l.e.) of $\boldsymbol{\beta}$, and $\hat{a}$ is the m.l.e. of a. As Λ is a (generalized) likelihood ratio test, under certain conditions it will be UMP asymptotically, though this does not imply it will have good power properties in small samples.

B-W did not pursue it, for computational reasons that will be made clear below. Instead, they took $\mathbf{B} = \mathbf{V}^{-1}(1)$, which is the same as the Durbin–Watson matrix $\mathbf{A}$ in (B.8), but with the $(1, 1)$ element being 2 instead of 1, and $\tilde{\boldsymbol{\beta}} = (\mathbf{X}'\mathbf{B}\mathbf{X})^{-1}\mathbf{X}'\mathbf{B}\mathbf{Y}$, and propose the test statistic

$$G = \frac{(\mathbf{Y} - \mathbf{X}\tilde{\boldsymbol{\beta}})'\mathbf{B}(\mathbf{Y} - \mathbf{X}\tilde{\boldsymbol{\beta}})}{(\mathbf{Y} - \mathbf{X}\hat{\boldsymbol{\beta}})'(\mathbf{Y} - \mathbf{X}\hat{\boldsymbol{\beta}})} = \frac{\mathbf{Y}'\mathbf{W}\mathbf{Y}}{\mathbf{Y}'\mathbf{M}\mathbf{Y}}, \quad \mathbf{W} = \mathbf{B} - \mathbf{B}\mathbf{X}(\mathbf{X}'\mathbf{B}\mathbf{X})^{-1}\mathbf{X}'\mathbf{B}, \tag{5.21}$$

where $\mathbf{M}$ is the usual o.l.s. residual projection matrix given in (B.15) and (1.53). Similar to the canonical reduction of D in (B.18), G can be expressed as (when there is a total of T observations) $\sum_{i=1}^{T-k} \lambda_i \chi_i^2 / \sum_{i=1}^{T-k} \chi_i^2$, where the λ_i are the eigenvalues of $\mathbf{MW}$.

The program in Listing 5.2 computes the relevant cutoff values for the three test statistics D, G and Λ. Those for the first two use the methodology and programs developed in Section A.3.1, and require less than a second, while that for Λ requires simulation, involving exact maximum likelihood estimation of the ARX(1) model (using the more general code we develop for the so-called ARMAX model in Chapter 7). This is clearly the most time-consuming part of the computation (and would have been prohibitive in the 1970s). The program then conducts a simulation, for a given value of parameter a, to determine the power. Again, because of the need to compute the m.l.e. (as opposed to just the D and G statistics), this calculation is rather time-consuming.

Figure 5.3 shows the power of the three tests, for the same two sample sizes as used in Figure 5.2, and based on the $\mathbf{X}$ matrices used in the bottom panels of that figure. Observe that the size is correct. For $\mathbf{X} = [\mathbf{1},\ \mathbf{t}]$ in the top panel, the D and G tests have the same power, while in the bottom panel, G (slightly) dominates for $a > 0.7$, while D has (slightly) higher power for $0 < a < 0.5$. The test based on Λ has much lower power for $a > 0$, and also power less than the size for a substantial part of the parameter space of a, showing that it is a **biased test**. Comparing the graphs in Figure 5.3 to the bottom panels of Figure 5.2, we see that the D and G tests are more powerful, as expected, given their approximate UMP nature.

It is noteworthy that the Λ test has good power properties for $a < 0$ and appears unbiased. The reader is invited to investigate the power of the D, D' and G tests for the $a < 0$ case (which involves rejecting for large values of the statistics), and compare with that of Λ.

Another test statistic for $a = 0$ expressible as a ratio of quadratic forms is that of King (1985a), given by

$$K(a^*) = \frac{\mathbf{Y}'\mathbf{G}'[\mathbf{G}\boldsymbol{\Sigma}(a^*)\mathbf{G}']^{-1}\mathbf{G}\mathbf{Y}}{\mathbf{Y}'\mathbf{M}\mathbf{Y}}, \quad |a^*| < 1, \tag{5.22}$$

```
1 function [DWp,BWp,LRTp]=DWBWLRTsim(a,T,X,alpha)
2 % determine power for Durbin Watson, Berenblut-Webb, and LRT
3 % for ARX model with sample size T, AR(1) param a, X matrix and
4 %   significance level alpha (default 0.05)
5 persistent DWc BWc LRTc
6 sim=1e4;
7 if nargin<3 || isempty(X), X=0; end
8 if nargin<4, alpha=0.05; end
9
10 if numel(X)==1 % signal to use a particular X matrix as follows:
11   type=X;
12   if type==0, X=[];
13   elseif type==1, X=ones(T,1);
14   elseif type==2, X=[ones(T,1) (1:T)'];
15   elseif type==3
16     c=round(T/2);
17     D1=[zeros(c,1) ; ones(c,1)]; if length(D1)>T, D1=D1(1:(end-1)); end
18     Dt=[zeros(c,1) ; ((c+1):T)'];if length(Dt)>T, Dt=Dt(1:(end-1)); end
19     X=[ones(T,1), (1:T)', D1, Dt.^2];
20   else error('Type of X matrix not defined')
21   end
22 end
23 [Tchk,k]=size(X); if Tchk ~= T, error('T and X incompatible'), end
24 M=makeM(X); A=makeDW(T); B=A; B(1,1)=2;
25 W=B-B*X*inv(X'*B*X)*X'*B; %#ok<MINV>
26 if isempty(DWc)
27   disp('Calculating cutoff values')
28   useimhof=1;
29   DWc=fzero(@(r) cdfratio(r,M*A*M,M,eye(T),[],useimhof)-alpha, 1.45)
30   BWc=fzero(@(r) cdfratio(r,W,     M,eye(T),[],useimhof)-alpha, 1.45)
31   LRT=zeros(sim,1);
32   for i=1:sim
33     e=randn(T,1);
34     [~, ~, ~, ~, llfull]=armareg(e,X,1,0,1);
35     S=e'*M*e; % Residual sum of squares for OLS
36     s2=S/(T-k); llols = -T/2*log(2*pi) - T/2*log(s2) - S/2/s2;
37     LRT(i)=2*(llols-llfull);
38   end
39   LRTc=quantile(LRT,alpha) %#ok<NOPRT>
40 end
```

Program Listing 5.2: Computes the power of the D, G, and Λ tests for model (5.1)–(5.2) with passed AR(1) parameter a, sample size T, regressor matrix $\mathbf{X}$, and significance level α (default 0.05). Passing $\mathbf{X}$ as scalar is used to generate the typical regressor matrices of none, constant, and time trend, as well as the one with a trend break, $[\mathbf{1}, \mathbf{t}, D1, Dt^2]$. Use `clear DWBWLRTsim` to remove the cutoff values (defined as `persistent` variables). Function `armareg` is given in Listing 7.7 in Chapter 7. The log-likelihood corresponding to the o.l.s. model calculated in lines 35–36 uses (1.4), (1.10), and (1.56). The program is continued in Listing 5.3.

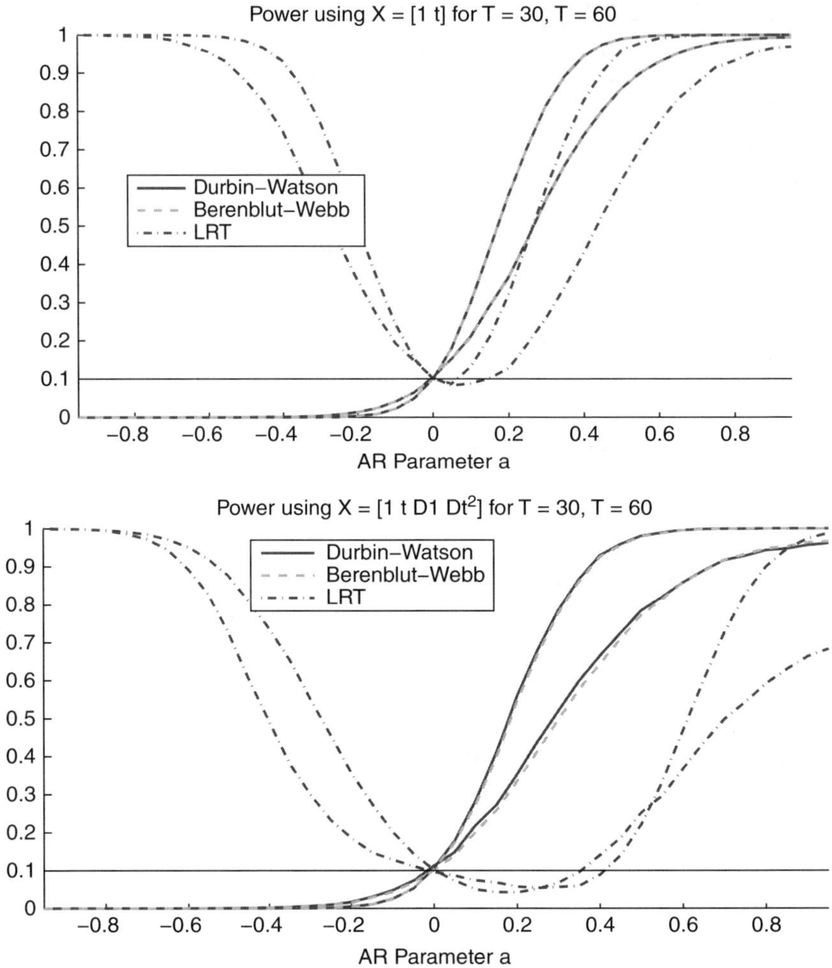

Figure 5.3 Power of the *D*, *G*, and Λ tests, for significance level $\alpha = 0.10$, two sample sizes $T = 30$ and $T = 60$, and two **X** matrices, as indicated. The lines corresponding to *D* and *G* are graphically indistinguishable in the top plot.

rejecting for small values of $K(a^*)$, and where **G** is from Theorem 1.3. As with (5.18), this test statistic is invariant to translations of the form $\mathbf{Y}^* = \gamma_0 \mathbf{Y} + \mathbf{X}\boldsymbol{\gamma}$ for γ_0 a positive scalar and $\boldsymbol{\gamma}$ is any real $k \times 1$ vector. The test based on (5.22) is POI because it is the most powerful (invariant) test at the point $a = a^*$, as shown in King (1980).

Using (1.65), we see that (5.22) is a type of likelihood ratio test, but such that the m.l.e. estimate of a is not used, but rather a fixed value of a in the alternative space. As discussed in King (1985a), it is similar to the B-W test in that it is point optimal, but such that the chosen point, say a^*, about which power is optimized, is not equal to one. (More precisely, the B-W test can be viewed as an approximation of (5.22) as $a^* \to 1$. Likewise, the Durbin–Watson test, which is of the form (5.18) and thus designed to have maximal power for a close to zero, approximates (5.22) as $a^* \to 0$.)

```
1  DWr=zeros(sim,1); BWr=DWr; LRTr=DWr;
2  if a==0, S12=eye(T); Si=eye(T); S=eye(T);
3  else
4    b=1+a^2; r=[b -a zeros(1,T-2)];
5    Si=toeplitz(r); Si(1,1)=1; Si(T,T)=1;
6    [V,D]=eig(Si); S12=V*D^(-1/2)*V'; S=inv(Si);
7  end
8  for i=1:sim
9    e=S12*randn(T,1); % normal innovations. Next line uses stable
10   %staba=1.5; stabb=0.8; e=S12*stabgen(T,staba,stabb)';
11   eh=M*e; D=(eh'*A*eh)/(eh'*eh); G=(e'*W*e)/(e'*M*e);
12   if 1==2 % The LRT using true value of a
13     BetaGLS=inv(X'*Si*X)*X'*Si*e; s2=1; % little sigma^2
14     Term=(e-X*BetaGLS)'*Si*(e-X*BetaGLS);
15     llfull=-(T/2)*log(2*pi) -0.5*log(det(s2*S))-Term/2/s2;
16   else % LRT with full MLE estimation
17     [~, ~, ~, ~, llfull]=armareg(e,X,1,0,1);
18   end
19   S=e'*M*e; s2=S/(T-k); llols = -T/2*log(2*pi) - T/2*log(s2) - S/2/s2;
20   LR=2*(llols-llfull);
21   DWr(i)=D<DWc; BWr(i)=G<BWc; LRTr(i)=LR<LRTc;
22 end
23 DWp=mean(DWr); BWp=mean(BWr); LRTp=mean(LRTr);
```

Program Listing 5.3: Continued from Listing 5.2. Program `stabgen` in line 9 (commented out) is for generating stable Paretian variates, and is given in Listing III.A.5.

We will encounter related point optimal tests below, namely, in Section 5.5.1 for a unit root, and in Section 5.6.3.2 for regression parameter constancy. As emphasized in King (1985a, p. 29), the observation vector $\mathbf{Y}$ should not be used to choose the value of a^*. Observe how this differs from the Λ test in (5.20), which we saw has relatively lower power.

Remarks

a) It is of interest to investigate the size and power properties of the tests amid non-Gaussian innovations. We conducted this for the D and G tests using the asymmetric stable Paretian distribution (accomplished by activating line 9 in Listing 5.3). For (approximately) $1.4 < \alpha < 2$, where here α denotes the stable tail index, there was almost no change in the size or power of the test. (The graphics are not shown, as they look virtually identical to those in Figure 5.3, and, not having computed the Λ statistic, are computed within seconds.) As tail index α decreases, the size starts to drop, though not by much, and the results seem essentially invariant to the choice of asymmetry parameter β. As such, while the tests are not invariant to use of innovation distributions exhibiting even rather heavy tails and asymmetry, they are highly robust to them. This agrees with the findings of Ali and Sharma (1993), who examined this robustness in more detail with distributions other than the stable Paretian.

b) Further tests, such as the nonparametric ones by Geary (1970) and Bartels (1982, 1984), do not require the relatively more complicated distribution theory associated with quadratic forms, but (unsurprisingly, given the approximate UMP result noted above) tend to have lower power than D and related tests; see, e.g., Dubbelman et al. (1978). ■

5.3.4 Further Details on the Durbin–Watson Test

> There is nothing more frightful than ignorance in action.
>
> (Johann von Goethe)

Section 5.3.4.1 outlines some historical aspects of the famous Durbin–Watson test D, including the traditionally used bounds test that readers might have seen in a first course in econometrics. It also bolsters the arguments outlined in Section III.2.8 against use of significance and hypothesis testing for model selection. Section 5.3.4.2 studies the limiting power properties of D, and, in doing so, demonstrates why alternative tests with higher power for more extreme autocorrelation, such as the B-W test (5.21), might be preferred in some contexts.

5.3.4.1 The Bounds Test, and Critique of Use of p-Values

Before approximately 1970, there were no precise algorithmic methods to calculate the c.d.f. of the D statistic (5.17) in order to get a p-value. Just obtaining the eigenvalues of the $T \times T$ matrix $\mathbf{MA}$ was quite a chore, given the limited access to computers and the necessary computational algorithms.

Henshaw, Jr. (1966) proposed use of the four-parameter beta distribution (the two shape parameters, as well as the two endpoints), such that the moments coincide with the first four integer moments of D. This has been shown by several authors (see, e.g., Harrison, 1972) to yield a highly accurate approximation that can be used in practice instead of an exact or saddlepoint-based approximation. A similar approximation was developed by Ali (1983), who used a four-parameter Pearson distribution. These approximations can be obtained without explicitly calculating the eigenvalues—recall (B.5). (Interestingly, though of less relevance in modern computing environments, the exact distribution can also be obtained without calculating eigenvalues; see Farebrother, 1985, 1994, and the references therein.)

This idea of using the beta distribution was known to Durbin and Watson in 1950: They suggested its use, based on matching the first two moments and assuming the support of D to be $(0, 4)$, if the bounds test is inconclusive. (This is not as accurate as the approximation of Henshaw, Jr. (1966) using four moments.) However, at the time they had little confidence in the beta, as the necessary computing power to check it was simply not available. Evaluating the beta approximation still involves non-trivial matrix calculations, as well as evaluation of the incomplete beta function for the c.d.f., so that, in the 1950s, it would have been impractical to ask an applied researcher, or anyone for that matter, to calculate this routinely.

Their solution, which to this day is still used (though is arguably now superfluous; see below), is referred to as the **bounds test**. It consists of constructing lower and upper bounding random variables, say d_L and d_U, whose distributions are independent of $\mathbf{X}$, and tabulating their values for various sample sizes T and number of regressors k, at the popular cutoff significance levels of 0.05 and 0.01. In this way, for a test with significance level α (either 0.05 or 0.01), if the calculated d statistic is less than d_L, then one would reject the null hypothesis. Similarly, if d is greater than d_U, then one would not reject it. The test is inconclusive when $d_L < d < d_U$, which is, appropriately, called the **inconclusive region**.

This method is described in numerous econometrics textbooks, including recent ones (see, e.g., Wooldridge, 2009, p. 415; Baum, 2006, p. 157), as well as the Wikipedia entry on the Durbin–Watson statistic, though the methodology for computing the c.d.f., and thus the p-value, is well-known now, and tables of the bounds (and the dreaded result of winding up in the inconclusive region) can be done away with.

To derive these bounding random variables for fixed T and k, we use a method that is now common, and is more general than that used by Durbin and Watson, namely the **Poincaré Separation Theorem**: Let $\nu_1 \leqslant \nu_2 \leqslant \cdots \leqslant \nu_T$ be the ordered eigenvalues of $\mathbf{A}$, as given in (B.9), and let $\lambda_1 \leqslant \lambda_2 \leqslant \cdots \leqslant \lambda_{T-k}$ be the ordered eigenvalues of $\mathbf{MA}$. Corollary 1 of Theorem B.5 in Section B.5 implies

$$\nu_i \leqslant \lambda_i \leqslant \nu_{i+k}, \quad i = 1, \dots, T-k,$$

so that

$$d_L := \frac{\sum_{i=1}^{T-k} \nu_i \chi_i^2}{\sum_{i=1}^{T-k} \chi_i^2} \leqslant \frac{\sum_{i=1}^{T-k} \lambda_i \chi_i^2}{\sum_{i=1}^{T-k} \chi_i^2} \leqslant \frac{\sum_{i=1}^{T-k} \nu_{i+k} \chi_i^2}{\sum_{i=1}^{T-k} \chi_i^2} := d_U, \tag{5.23}$$

as the χ_i^2 are positive.

Remarks

a) Although the problem had been reduced to determining specific quantiles of just two bounding random variables for several combinations of T and k, this was still not trivial in 1950. In order to calculate these values, Durbin and Watson used a technique involving an expansion in Jacobi polynomials. In doing so, they used further approximations, reducing the problem to one involving simply a large number of elementary calculations, suitable for a hand calculator.

   In his interview with P. C. B. Phillips (1988), James Durbin recalled how the calculations for their published tables were actually constructed. They had at that time, "a room with perhaps eight or ten young ladies operating desk calculators, supervised by an older lady of forbidding demeanor. They did the computing." Although they claimed two-digit accuracy, they were not sure for quite some time.

b) In the late 1960s, Durbin and Watson were planning their third paper and wanted to recalculate their published tables with the help of "modern" computing. Johan Koerts and Adriaan P. J. Abrahamse had independently decided to investigate the same question, and informed Durbin that, to two digits, their initial tables were correct. By this point in time, far more precise methods had been developed to evaluate the c.d.f. of D, such as by Pan Jie-Jian (1964) and Koerts and Abrahamse (1969), the latter giving a Fortran program to compute it to a user-specified degree of accuracy, based on the work of Imhof (1961), which uses the inversion formula method for calculating the c.d.f., as discussed in Chapter A. ■

We state some disadvantages of using the bounds test, as compared to computing the p-value.

- For fixed k, the size of the inconclusive region is inversely proportional to the sample size T, and can be quite large for moderate T , as is common in economic data sets.
- Even when the test is conclusive, the researcher is prevented from reporting the corresponding p value, which conveys much more information than just the binary result of the hypothesis test.
- The tables have entries for a limited number of values of the sample size T, so that more often than not, one must (linearly) interpolate. This requires not only more time, but introduces a small amount of error.
- Use of tables is outdated and inconvenient. Econometric packages now have the tables built in (with all their aforementioned limitations), while several modern statistical software packages compute the exact or approximate p-value.

- The bounds test without modification is inappropriate for detecting autocorrelation at higher lags, which might accompany seasonal data. As an example, monthly sales data might not only have first-order serial correlation, but also 12th order. The **generalized Durbin–Watson test** naturally suggests itself, generalizing (B.16) to

$$D_j = \frac{\sum_{t=j+1}^{T} (\hat{\epsilon}_t - \hat{\epsilon}_{t-j})^2}{\sum_{t=1}^{T} \hat{\epsilon}_t^2} = \frac{\boldsymbol{\epsilon}'\mathbf{M}\mathbf{A}_j\mathbf{M}\boldsymbol{\epsilon}}{\boldsymbol{\epsilon}'\mathbf{M}\boldsymbol{\epsilon}}, \tag{5.24}$$

  $j = 1, 2, \ldots$, where $\mathbf{A}_j = \mathbf{D}_j'\mathbf{D}_j$ generalizes (B.8), with $\mathbf{D}$ the $(T-j) \times T$ Toeplitz matrix with first column $[-1, 0, \ldots, 0]'$ as before, and first row $[-1, 0, \ldots, 0, 1, 0, \ldots, 0]$, where there are $j-1$ zeros between -1 and 1 (see Vinod, 1973; and Ali, 1987). The SAS programming language, for example, computes these (and the associated exact p-values), while Matlab's function `dwtest` also returns a p-value, but is only for $j = 1$.
- The same methods that are used to compute the p-value for D and related tests (such as that from Berenblut and Webb, 1973) can also evaluate their corresponding power against AR(1) alternatives, as discussed in Section 5.3.4.2. Tabulating the power is impractical, as it would need to be done for various T k, a and α, but is anyway futile, as the power of these tests strongly depends on $\mathbf{X}$.

In addition to the above disadvantages of using the bounds test instead of direct calculation of the p-value, one might question the use of such tests, even if the p-value is delivered. This was elaborated upon in more generality in Section III.2.8. In particular, the Fisher significance testing framework (use of p-values) was designed for situations involving assessment of treatment effects in designed studies, with the intention of deciding if the experiment should be repeated (several times) to confirm and measure efficacy. In our setting here, a single data set from a (vastly) more complicated data generating process (hereafter d.g.p.) is under study, and interest centers on the choice of model: Regression with or without an AR(1) term for the residual process. While the Durbin–Watson (and related) test statistics have value in indicating deviations from a regression model with i.i.d. error terms, it is no longer clear how one should use the p-value for deciding on the appropriate model.

While earlier researchers may not have questioned the use of the general hypothesis testing paradigm in econometrics, some explicitly considered the effect of basing the choice of model on the result of the Durbin–Watson test. For example, Nakamura and Nakamura (1978) investigated the pretest estimator of β_2 in (B.17) when the choice of model is determined by the outcome of D for a given significance level α. Nakamura and Nakamura (1978, p. 207) conclude: "Our results so far suggest that tests of significance for autocorrelation might best be dispensed with in estimating [regression relationships] in favor of a practice of always transforming." (Here, "transform" refers to estimating the regression model with AR(1) disturbance term, which, at the time, was not so trivial. They used the so-called Cochrane–Orcutt method of estimation, which is not the same as, and inferior to, the m.l.e., as well as the other methods we will explore in Section 5.4.)

Their results were independently corroborated by Fomby and Guilkey (1978), who showed that, if a pretest estimator for $\boldsymbol{\beta}$ is used based on the Durbin–Watson statistic, then the optimal significance level α *is far greater than 0.05*, and more like 0.50, when measuring the performance of $\widehat{\boldsymbol{\beta}}$ based on m.s.e. Notice that this implies yet another reason why the tabulated bounds for the D test, using significance levels 0.01 and 0.05, are of little use.

Similar findings regarding the inappropriateness of the traditional significance levels have been shown to be the case in the unit root testing framework; see Kim and Choi (2017).

5.3.4.2 Limiting Power as $a \to \pm 1$

Use of the Durbin–Watson test statistic D in (5.17), or, equivalently, D_1 in (5.24), is expected to result in among the most powerful tests for first-order serial correlation in the Gaussian linear regression model (5.1)–(5.2) because it is approximately UMP, in the sense detailed in Section 5.3.2. With the ability to quickly and reliably compute the distribution of ratios of quadratic forms in normal variables (with any positive definite covariance matrix), the power of the D and related tests are easily determined, and can be compared to each other, as in Figure 5.3 above, for a particular $\mathbf{X}$ matrix, as a function of autoregressive parameter a. This section looks at some analytic results regarding the power of D. In particular, we can characterize the limiting power as $a \to 1$ (with similar results holding for $a \to -1$), and see how the choice of $\mathbf{X}$ affects it.

Let α be the tail probability corresponding to the test significance level, and let

$$\pi_D(a_+) = \Pr(D \leqslant d_\alpha^+ \mid a = a_+), \quad 0 \leqslant a_+ < 1, \tag{5.25}$$

denote the power of the D test for testing against positive autocorrelation (suppressing the dependence on the $\mathbf{X}$ matrix), with d_α^+ the test cutoff value corresponding to α. Here, a_+ is a specific value of the autocorrelation parameter, and the null hypothesis is $H_0 : a = 0$. By construction, $\pi_D(0) = \Pr(D \leqslant d_\alpha^+ \mid a = 0) = \alpha$. The power function for testing against negative autocorrelation is defined similarly, namely

$$\pi_D(a_-) = \Pr(D \geqslant d_\alpha^- \mid a = a_-), \quad -1 < a_- \leqslant 0, \tag{5.26}$$

though interest is usually on (5.25).

Assume a sample of size T. From (5.17), and similar to the rearrangement in (A.32),

$$\pi_D(a_+) = \Pr\left(\frac{\boldsymbol{\epsilon}'\mathbf{MAM}\boldsymbol{\epsilon}}{\boldsymbol{\epsilon}'\mathbf{M}\boldsymbol{\epsilon}} \leqslant d_\alpha^+ \mid a = a_+\right) = \Pr(\boldsymbol{\epsilon}'(\mathbf{Q} - d_\alpha^+\mathbf{M})\boldsymbol{\epsilon} \leqslant 0 \mid a = a_+), \tag{5.27}$$

where $\mathbf{Q} = \mathbf{MAM}$. Assume $\boldsymbol{\epsilon} \sim \mathrm{N}(\mathbf{0}, \sigma^2\boldsymbol{\Sigma})$, with $\boldsymbol{\Sigma}$ (of size $T \times T$) given in (4.20) (and observe that we restrict $0 \leqslant a_+ < 1$ so that it is well-defined). Let $\boldsymbol{\eta} = (\sigma^2\boldsymbol{\Sigma})^{-1/2}\boldsymbol{\epsilon} \sim \mathrm{N}(\mathbf{0}, \mathbf{I}_T)$, so that $\boldsymbol{\epsilon} = (\sigma^2\boldsymbol{\Sigma})^{1/2}\boldsymbol{\eta}$, and substitute this into (5.27) to get

$$\pi_D(a_+) = \Pr(\sigma^2\boldsymbol{\eta}'\boldsymbol{\Sigma}^{1/2}(\mathbf{Q} - d_\alpha^+\mathbf{M})\boldsymbol{\Sigma}^{1/2}\boldsymbol{\eta} \leqslant 0 \mid a = a_+).$$

As the right-hand side of the inequality is zero, we may cancel the positive constants σ^2 and $(1 - a^2)^{-1}$ from $\boldsymbol{\Sigma}$ to get the slightly simpler

$$\pi_D(a_+) = \Pr(\boldsymbol{\eta}'\mathbf{V}^{1/2}(\mathbf{Q} - d_\alpha^+\mathbf{M})\mathbf{V}^{1/2}\boldsymbol{\eta} \leqslant 0 \mid a = a_+),$$

where $[\mathbf{V}]_{i,j]=a^{|i-j|}}$, i.e.,

$$\mathbf{V} = \begin{bmatrix} 1 & a & a^2 & \cdots & a^{T-1} \\ a & 1 & a & \cdots & a^{T-2} \\ a^2 & a & 1 & \cdots & a^{T-3} \\ \vdots & \vdots & \vdots & \ddots & \vdots \\ a^{T-1} & a^{T-2} & a^{T-3} & \cdots & 1 \end{bmatrix}.$$

As $\mathbf{V}^{1/2}(\mathbf{Q} - d_\alpha^+\mathbf{M})\mathbf{V}^{1/2}$ is (real and) symmetric, it can be expressed as $\mathbf{L}\boldsymbol{\Lambda}\mathbf{L}'$, with $\boldsymbol{\Lambda} = \mathrm{diag}([\lambda_1, \dots, \lambda_T])$ the diagonal matrix of eigenvalues, and $\mathbf{L}$ the corresponding orthogonal matrix, as was similarly done in (A.2)–(A.4). Thus,

$$\pi_D(a_+) = \Pr(\boldsymbol{\eta}'\mathbf{L}\boldsymbol{\Lambda}\mathbf{L}'\boldsymbol{\eta} \leqslant 0 \mid a = a_+) = \Pr\left(\sum_{i=1}^{T} \lambda_i \eta_i^2 \leqslant 0 \mid a = a_+\right), \tag{5.28}$$

where $\eta_i^2 \overset{\text{i.i.d.}}{\sim} \chi^2(1)$, $i = 1, \dots, T$, because $\mathbf{L}'\boldsymbol{\eta} \sim \mathrm{N}(\mathbf{0}, \mathbf{I}_T)$. Also observe that, from Theorem B.4, the nonzero λ_i are also the eigenvalues of $(\mathbf{Q} - d_\alpha^+\mathbf{M})\mathbf{V} = \mathbf{M}(\mathbf{A} - d_\alpha^+\mathbf{I})\mathbf{M}\mathbf{V}$ or, equivalently, those of $\mathbf{V}(\mathbf{Q} - d_\alpha^+\mathbf{M}) = \mathbf{V}\mathbf{M}(\mathbf{A} - d_\alpha^+\mathbf{I})\mathbf{M}$ (where the latter expression is just the transpose of $\mathbf{M}(\mathbf{A} - d_\alpha^+\mathbf{I})\mathbf{M}\mathbf{V}$, and thus has the same eigenvalues, as the reader should confirm in general). For testing against negative autocorrelation, $\pi_D(a_-)$ is the same as (5.28) except that we use d_α^- instead of d_α^+ in constructing the λ_i, and the inequality sign is reversed.

From Figure 5.3, one might think that, as $|a| \to 1$, $\pi_D(a) \to 1$. Unfortunately, this is not usually true, with the limiting power, which we denote by $\pi_D(1^-)$, depending on the $\mathbf{X}$ matrix (and thus also the sample size). It turns out that $\pi_D(1^-)$ can be below the size of the test, or even zero, rendering D (and similar autocorrelation tests expressible as ratios of quadratic forms) biased. Denote the T-vector of ones as $\mathbf{1} = (1, 1, \dots, 1)'$. As in Krämer (1985), observe that, as $a \to +1$, $\mathbf{V} \to \mathbf{1}\mathbf{1}' =: \mathbf{V}^+$, the matrix consisting of all ones, so that $\mathbf{V}^+$ is of rank one. Hence, via (B.66), $(\mathbf{Q} - d_\alpha^+\mathbf{M})\mathbf{V}^+$ is at most rank one, and there is at most one nonzero eigenvalue in the sum in (5.28). Assume there is one such value, say λ^*. Then, as the η^2 are strictly positive, when this single eigenvalue is negative (positive), the limiting power is one (zero).

Recall the definition of column space from (1.38). For determining the power as $|a| \to 1$, it is useful to differentiate between the two exclusive and exhaustive cases: Either $\mathbf{1} \in \mathcal{C}(\mathbf{X})$ or $\mathbf{1} \notin \mathcal{C}(\mathbf{X})$. The latter case amounts to a regression without a constant term or appropriate set of dummy variables, and implies that $\lambda^* \neq 0$ because if $\mathbf{1} \notin \mathcal{C}(\mathbf{X})$, then $\mathbf{M}\mathbf{V}^+ = \mathbf{M}\mathbf{1}\mathbf{1}' \neq \mathbf{0}$, from the Projection Theorem 1.1, and (1.54). Thus, if $\mathbf{1} \notin \mathcal{C}(\mathbf{X})$, then there is exactly one nonzero eigenvalue λ^*, and, recalling that the trace of a matrix is equal to the sum of its eigenvalues, λ^* is given by the trace of $(\mathbf{Q} - d_\alpha^+\mathbf{M})\mathbf{V}^+$. The limiting power in this case is either zero or one. Krämer (1985) gives a simple example of an $\mathbf{X}$ matrix such that $\pi_D(a_+)$ initially increases in a_+, but then begins to decrease as $a_+ \to 1$, with $\pi_D(1^-) = 0$. Thus, in this case, $\pi_D(a_+)$ is not monotonic in a_+, and D is biased. This undesirable feature of D was also noted and studied in Tillman (1975) and King (1985a).

An illustration is provided in the (rather uneventful) graphic in the top panel of Figure 5.4, demonstrating that, for each of the three $\mathbf{X}$ matrices considered and all values of T shown, the limiting power is one. The reader should confirm this theoretical finding by plotting, for a fixed sample size T and each of the used $\mathbf{X}$ matrices, the power of the D test for a grid of a-values such as $0.900, 0.901, \dots, 0.999$, as computed from (part of) the program in Listing 5.2.

If the limiting power is zero, then one could add the regressor $\mathbf{1}$, and thereby increase the power for suitably large a, which appears rather dubious, as adding a regressor the researcher believes to be incorrect should not enhance a statistical procedure. As discussed in Krämer (1985), the resolution to this contradiction is to note that, in this setting, the Durbin–Watson test is perhaps not optimal, nor possibly the o.l.s. procedure for estimation.[4] Use of an alternative test, such as the B-W test (5.21) or that of King (1985a) in (5.22), is suggested. Note that, as shown by Krämer (1985) and Zeisel (1989), there exist $\mathbf{X}$ matrices with $\mathbf{1} \notin \mathcal{C}(\mathbf{X})$ such that $\pi_D(1^-)$ remains at zero for any T.

4 As further stated in Krämer and Zeisel (1990, p. 371), "[This] does not imply that by adding or removing an intercept, one can control the power of the test. Whether or not the design matrix should contain an intercept is no matter of choice but rather dictated by the underlying data generating process."

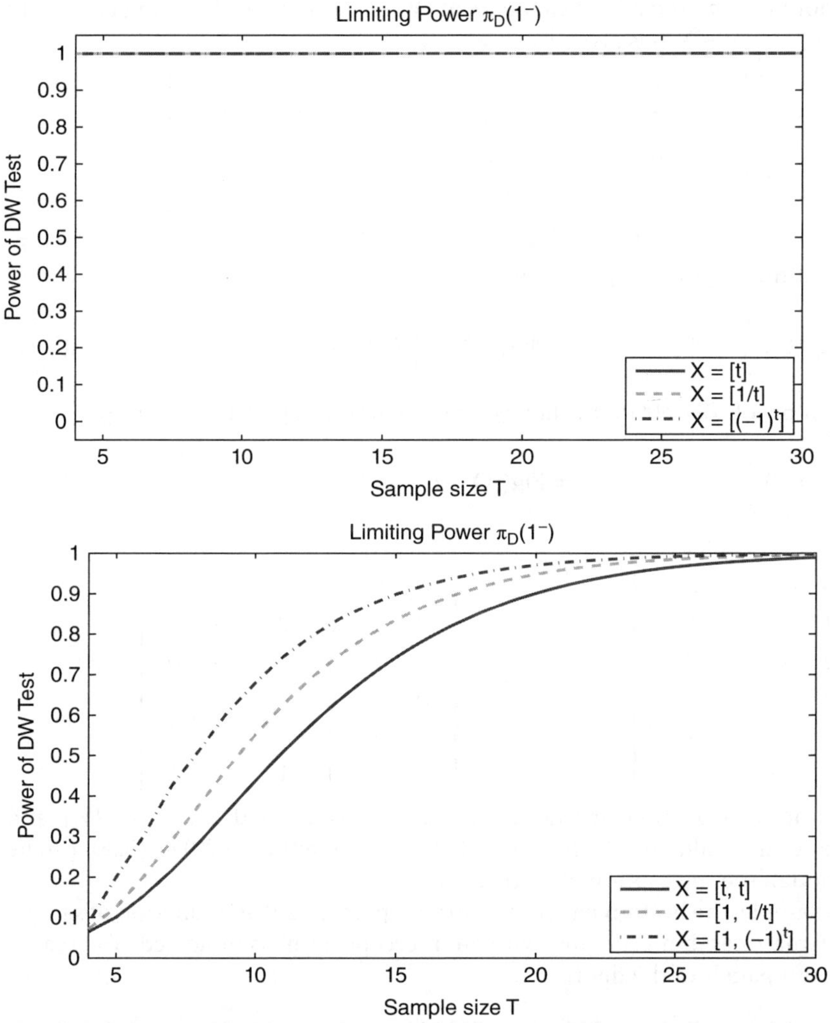

Figure 5.4 Top: Limiting power corresponding to a significance level of $\alpha = 0.05$, for several **X** matrices, none of which contain a column of ones, as a function of sample size T. **Bottom:** Same but showing the limiting power (5.29), such that the **X** matrices contain a column of ones.

Otherwise, as is more common, $\mathbf{1} \in \mathcal{C}(\mathbf{X})$, and $\pi_D(1^-)$ lies strictly between zero and one, as rigorously shown by Zeisel (1989) and Krämer and Zeisel (1990), building on the work of Tillman (1975). Following the elegant derivation in Krämer and Zeisel (1990), replace λ_i by $\tilde{\lambda}_i = (1-a)^{-1}\lambda_i$ in (5.28) and notice that the power is unchanged. The $\{\tilde{\lambda}_i\}$ are the eigenvalues of $\mathbf{V}(1-a)^{-1}(\mathbf{Q} - d_\alpha^+\mathbf{M})$. We know in this case that $\mathbf{V}^+(\mathbf{Q} - d_\alpha^+\mathbf{M}) = \mathbf{0}$, so that

$$\lim_{a\to 1}(1-a)^{-1}\mathbf{V}(\mathbf{Q} - d_\alpha^+\mathbf{M}) = \lim_{a\to 1}(1-a)^{-1}(\mathbf{V} - \mathbf{V}^+)(\mathbf{Q} - d_\alpha^+\mathbf{M}).$$

Observe that, in the limit, both $(1-a)$ and all the elements of $(\mathbf{V}-\mathbf{V}^{+})$ are zero. Applying l'Hopital's rule elementwise, $\lim_{a\to 1}(1-a)^{-1}(\mathbf{V}-\mathbf{V}^{+})$ is given by

$$\mathbf{W} := -(|i-j|)_{i,j} = -\lim_{a\to 1}\begin{bmatrix} 0 & 1 & \cdots & (T-1)a^{T-2} \\ 1 & 0 & \cdots & (T-2)a^{T-3} \\ 2a & 1 & \cdots & \\ \vdots & \vdots & \ddots & \vdots \\ (T-1)a^{T-2} & (T-2)a^{T-3} & \cdots & 0 \end{bmatrix},$$

which is full rank. Thus, the limiting power $\pi_D(1^-)$ is given by

$$\pi_D(1^-) = \Pr\left(\sum_{i=1}^{T} \lambda_i \eta_i^2 \leqslant 0\right), \quad (\lambda_1, \dots, \lambda_T) = \text{Eig}((\mathbf{Q} - d_\alpha^{+}\mathbf{M})\mathbf{W}). \tag{5.29}$$

Using a less direct method of proof, Zeisel (1989) showed that the limiting power is given by

$$\pi_D(1^-) = \Pr\left(\sum_{i=1}^{T} \gamma_i \eta_i^2 \leqslant 0\right), \quad (\gamma_1, \dots, \gamma_T) = \text{Eig}((\mathbf{Q} - d_\alpha^{+}\mathbf{M})\mathbf{Z}), \tag{5.30}$$

where

$$\mathbf{Z} = \begin{bmatrix} 0 & 0 & 0 & 0 & \cdots & 0 \\ 0 & 1 & 1 & 1 & \cdots & 1 \\ 0 & 1 & 2 & 2 & \cdots & 2 \\ 0 & 1 & 2 & 3 & \cdots & 3 \\ \vdots & \vdots & \vdots & \vdots & \ddots & \vdots \\ 0 & 1 & 2 & 3 & \cdots & T-1 \end{bmatrix} = \mathbf{U}\mathbf{U}', \quad \mathbf{U} = \begin{bmatrix} 0 & 0 & 0 & 0 & \cdots & 0 \\ 0 & 1 & 0 & 0 & \cdots & 0 \\ 0 & 1 & 1 & 0 & \cdots & 0 \\ 0 & 1 & 1 & 1 & \cdots & 0 \\ \vdots & \vdots & \vdots & \vdots & \ddots & \vdots \\ 0 & 1 & 1 & 1 & \cdots & 1 \end{bmatrix},$$

i.e., $\mathbf{Z} = [\min(i,j)-1]_{i,j}$. Numeric inspection via the code in Listing 5.4 shows that $\{\lambda_i\} = 2\{\gamma_i\}$, and note that the factor of two does not influence the power in (5.30), so that either (5.29) or (5.30) can be used to compute $\pi_D(1^-)$ (though see the caption of Listing 5.4).

As a contrast to the first example, the bottom panel of Figure 5.4 plots (5.29) as a function of sample size T, for the same three $\mathbf{X}$ matrices, but each now with an intercept term, as indicated. The reader is encouraged to replicate both panels of the figure.

```
1 T=200; X=[ones(T,1), (1:T)'];
2 %T=25; k=2; X=[ones(T,1), randn(T,k)];
3 A=makeDW(T); M=makeM(X); Q=M*A*M;
4 W=zeros(T,T); for i=1:T, for j=1:T, W(i,j)=-abs(i-j); end, end
5 Z=zeros(T,T); for i=1:T, for j=1:T, Z(i,j)=min(i,j)-1; end, end
6 alpha=0.05; d=fzero(@(r) cdfratio(r,M*A*M,M,eye(T),[],1)-alpha, 1.45);
7 lamW=sort(eig((Q-d*M)*W)); lamZ=sort(eig(Z*(Q-d*M)));
8 disp(max(abs(lamW - 2*lamZ)))
```

Program Listing 5.4: Code for confirming equality of (5.29) and (5.30) when $\mathbf{1} \in \mathcal{C}(\mathbf{X})$. Use of the commented out line 2 also works, though as T and k increase, the equality often does not hold, even if $\mathbf{X}$ is full rank, presumably because of round-off error. In the event they are not equal, calculating the power via the program in Listing 5.2 for $a = 0.999$ reveals that use of $\mathbf{W}$ and (5.29) leads to more accurate limiting power than use of $\mathbf{Z}$ and (5.30), as the reader should confirm.

Krämer and Zeisel (1990) and Small (1993) showed that similar findings regarding the limiting power apply to the B-W test (5.21) and the POI test (5.22), i.e., for the latter $\pi_K(1^-)$ is zero or one when $\mathbf{1} \notin \mathcal{C}(\mathbf{X})$, and is otherwise strictly between zero and one. The study of $\pi_D(1^-)$ is augmented in Bartels (1992) to include the random walk case, i.e., $a = 1$. Wan et al. (2007) investigated the limiting power of the Durbin–Watson and related tests in the presence of correct and, notably, mis-specified linear restrictions on the regression coefficients (see also examples B.6 and B.7, and Section 1.4).

5.4 Bias-Adjusted Point Estimation

This section builds on the methods in Section 4.6 for bias-adjusted estimators of a, now allowing for a set of regressors. It is well-known that $\widehat{a}_{\text{LS}}$ is downward biased, increasingly so for a near one; recall the results in Section 4.5. While various procedures exist to partially correct for this, no operational method has so far been devised that is exactly mean-unbiased. It is, however, straightforward to construct a median-unbiased estimator, hereafter denoted $\widehat{a}_{\text{Med}}$, and first pursued in this context by Andrews (1993).

Perhaps unsurprisingly, it turns out that none of the bias-corrected methods is uniformly better with respect to all criteria, though it appears that their relative behavior is virtually invariant to the choice of sample size and set of regressors, as detailed in Broda et al. (2007). Moreover, this fortuitous behavior remains (approximately) constant for a variety of non-normal innovation distribution assumptions commonly entertained in practice. The optimal choice of estimator depends (essentially) only on the true value of the autoregressive parameter a, but in virtually the same way for any model design and distributional assumption. For example, as demonstrated in Figure 5.6, the mean-adjusted estimator has the lowest mean squared error for all a between about 0.7 and 1.0—a result of interest given that many (macroeconomic) series exhibit high persistence or even near unit-root behavior. It is interesting to note that, while this work has its origins in that of Andrews (1993), who proposed and studied the median-unbiased estimator in this context, *for no range of a is the median unbiased estimator optimal in terms of mean squared error.*

Before turning to the results, we first look at the effects and magnitudes induced by the different corrections. Figure 5.5a plots values of $\widehat{a}_{\text{LS}}$ on the ordinate (y-axis) versus the corresponding quantity that should be added to $\widehat{a}_{\text{LS}}$ to arrive at $\widehat{a}_{\text{Med}}$ on the abscissa (x-axis). For example, with $T = 10$ and an $\mathbf{X}$ matrix consisting of an intercept and time trend, if $\widehat{a}_{\text{LS}} = 0.2$, then $\widehat{a}_{\text{Med}} \approx 0.68$. As expected, the amount of correction decreases as the sample size increases. One also sees that, particularly for smaller sample sizes, the amount of correction increases substantially when the $\mathbf{X}$ matrix changes from $\mathbf{1}$ to $[\mathbf{1},\ \mathbf{t}]$.

Figure 5.5b is similar, but shows the correction appropriate for $\widehat{a}_{\text{Mean}}$. Observe that it differs significantly from Figure 5.5b only for values of $\widehat{a}_{\text{LS}}$ less than -0.4. Figure 5.5c shows the correction appropriate for $\widehat{a}_{\text{Mode}}$, and is quite unlike its two counterparts. This implies that its small-sample properties will differ markedly from those of $\widehat{a}_{\text{Med}}$ and $\widehat{a}_{\text{Mean}}$, as detailed next.

A simulation was conducted with 10,000 replications, based on $T = 19$ observations, using the two regression matrices $\mathbf{X} = \mathbf{1}$ and $\mathbf{X} = [\mathbf{1},\ \mathbf{t}]$. Figure 5.6 shows the results just for the mean squared error (m.s.e.), and is thus similar to the bottom two panels in Figure 4.13 (which correspond to no regression matrix), except that the m.l.e. was not computed. The first characteristic one sees is that, as the complexity of the regression matrix increases, i.e., when going from no $\mathbf{X}$ matrix to $\mathbf{X} = \mathbf{1}$ to $\mathbf{X} = [\mathbf{1},\ \mathbf{t}]$, the least squares estimator becomes less attractive compared to the adjusted estimators, particularly

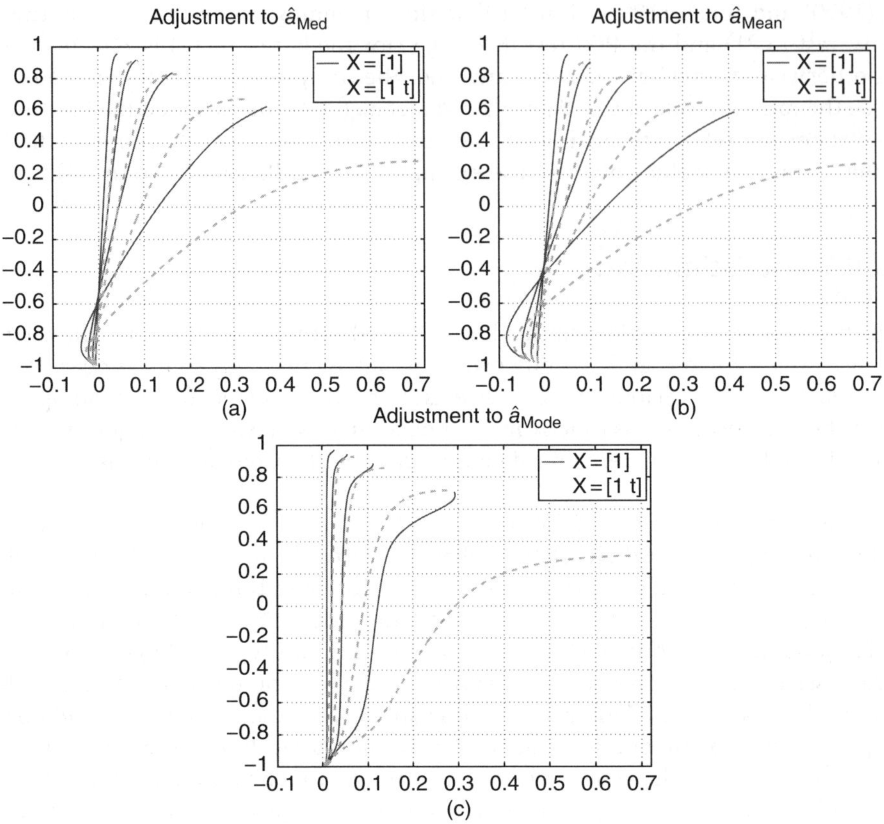

Figure 5.5 Adjustment to $\hat{a}_{LS}$ corresponding to $\hat{a}_{Med}$ (a), $\hat{a}_{Mean}$ (b), and $\hat{a}_{Mode}$ (c), shown for the two indicated **X** matrices (solid and dashed lines). The four sample sizes used are $T = 10$, $T = 25$, $T = 50$, and $T = 100$, moving from right to left within each plot.

as a increases towards one. Secondly, the magnitude of the m.s.e. increases for all estimators as **X** increases in complexity.

Approximately speaking, in the case with no **X** matrix, the m.s.e. is 0.035 for all of the estimators at $a = 0.7$, for $\mathbf{X} = \mathbf{1}$ and $a = 0.7$, the m.s.e. is 0.052 for all adjusted estimators (but not for the o.l.s. estimator), and for $\mathbf{X} = [\mathbf{1}, \ \mathbf{t}]$ and $a = 0.7$, the m.s.e. is 0.075 for all adjusted estimators, which is more than double the value corresponding to the no **X** matrix case.

Otherwise, the results are similar when comparing the ordering of the estimators, with $\hat{a}_{LS}$ being best for $a < -0.1$, the mode-unbiased estimator being preferred for $-0.1 < a < 0.5$ for no **X** matrix and $-0.1 < a < 0.7$ for the other two matrices, and for $0.7 < a < 1$, the mean-bias-adjusted estimator is preferred (with the median-unbiased estimator relatively close in performance). Rather conveniently, it turns out that these results are qualitatively very similar for different sample sizes, and numerous **X** matrices tried.

Remarks

a) An interesting and useful property of the three bias-corrected estimators is that they are one-to-one transformations of the least squares estimator, say $\hat{a}_{BC} = m_{BC}^{-1}(\hat{a}_{LS})$, where BC denotes

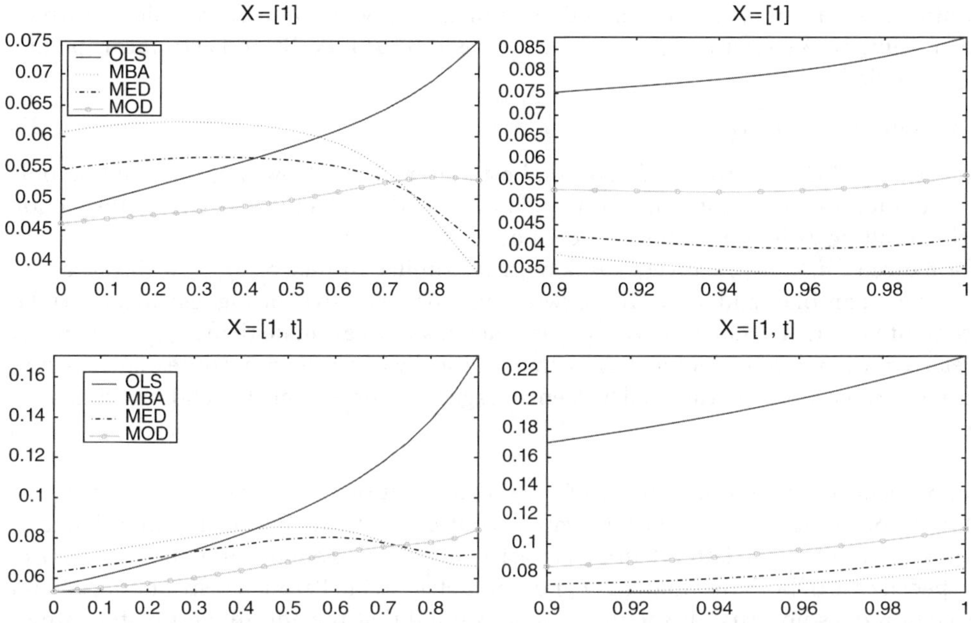

Figure 5.6 Comparison of the m.s.e. for the least squares (OLS), mean bias-adjusted (MBA), median unbiased (MED), and mode unbiased (MOD) estimators for parameter a in the AR(1) model with $T = 19$. The top panels refer to use of $\mathbf{X} = \mathbf{1}$, while the bottom panels are for $\mathbf{X} = [\mathbf{1}\ \mathbf{t}]$. The right panels just focus on the range $0.9 < a < 1$.

the respective method of bias correction, i.e., BC $\in$ {Mean, Med, Mode}, and $m_{\mathrm{BC}}^{-1}(\widehat{a}_{\mathrm{LS}})$ is the inverse mean, median, and mode function, respectively. Of course, $m_{\mathrm{BC}}^{-1}(\widehat{a}_{\mathrm{LS}})$ is not available analytically, and is computed by numerical methods. For a given sample size and $\mathbf{X}$ matrix, the adjusted estimators can be calculated for a tight grid of $\widehat{a}_{\mathrm{LS}}$ values, from which properties of interest such as the median and moments (for the bias and m.s.e.) could be obtained by numeric integration.

Alternatively, as each of the three estimators takes under a second to compute on a modern PC, a direct, brute-force simulation exercise is also feasible. We use a combination of these two methods, which involves simulation, but capitalizes on the one-to-one transformation of the estimators. This results in a ten-fold decrease in computation time compared to direct simulation, and is now discussed.

The first step is to compute $m_{\mathrm{BC}}^{-1}(\widehat{a}_{\mathrm{LS}})$ over a judiciously chosen set of $\widehat{a}_{\mathrm{LS}}$ values. In particular, for any given parameter constellation, an unequally spaced grid of points is dynamically constructed using a recursive algorithm that ensures a specified accuracy, as required in the next step. This avoids redundant calculation and saves considerable time.

For the second step, (i) simulate a time series from model (5.1)–(5.2), (ii) calculate $\widehat{a}_{\mathrm{LS}}$, and (iii) from $\widehat{a}_{\mathrm{LS}}$ use linear interpolation from the grid obtained in the first step to obtain the corresponding bias-corrected estimators. As parts (i), (ii) and (iii) are all numerically cheap, this second step is extremely fast, enabling use of a very large number of replications (we used 10,000), so that the inherent variation arising from simulation can be effectively eliminated. Thus, for any given $\mathbf{X}$ matrix, the mean- and median-bias, and the m.s.e., of the three bias-adjusted estimators can be routinely computed over a grid of a-values.

b) Consider treating the p.d.f. of $\widehat{a}_{\mathrm{LS}}$ as a likelihood function and choosing a to be the value for which it obtains its maximum. We refer to this as the **pseudo maximum likelihood estimator** (p.m.l.), denoted $\widehat{a}_{\mathrm{PML}}$, and defined as

$$\widehat{a}_{\mathrm{PML}} = \mathrm{argmax}_a \, f_{\widehat{a}_{\mathrm{LS}}}(a; \widehat{a}_{\mathrm{LS}}). \tag{5.31}$$

Note the similarity of (5.31) to the mode-adjusted estimator (4.51). However, this method only requires computation of the maximum of one p.d.f., and is thus significantly faster. Inspection shows that the resulting estimator is uniquely defined.[5]

The performance of $\widehat{a}_{\mathrm{PML}}$ is, relatively speaking, very similar to that of $\widehat{a}_{\mathrm{Mode}}$ with respect to both mean- and median-bias, and m.s.e. These two estimators also have the highest linear correlation when computed from a simulation. As $\widehat{a}_{\mathrm{PML}}$ is much faster to evaluate than $\widehat{a}_{\mathrm{Mode}}$, the former might be preferred when used in conjunction with numerically intensive procedures, such as simulation and bootstrap exercises. The reader is encouraged to implement this and investigate its performance. ∎

The small-sample behavior with respect to non-Gaussian innovations is now examined. We use the constant and time-trend model with $T = 25$ observations, but take the distributional assumption to be Cauchy, which possesses tails much fatter than usually arises in empirical applications in econometrics and serves as a special case of both the Student's t and symmetric stable Paretian distribution, and for which the expectation does not exist. It should be kept in mind that the adjusted estimators are all based on the normal assumption used in the calculation of the distribution of $\widehat{a}_{\mathrm{LS}}$ in (5.14).

The results are shown in Figure 5.7. The overall behavior of the estimators is still similar to the normal case. For example, $\widehat{a}_{\mathrm{Mean}}$ is still approximately unbiased over most of the parameter space, exhibiting an increase in bias as $|a|$ approaches one, as in the normal case. Estimator $\widehat{a}_{\mathrm{Med}}$ is no longer median-unbiased, but is approximately so for $a < 0.8$. Interestingly, $\widehat{a}_{\mathrm{Mode}}$ is also approximately median-unbiased. The differences in m.s.e. among the adjusted estimators are somewhat less pronounced, although, qualitatively speaking, the envelope of minimum m.s.e. is virtually the same as before, i.e., $\widehat{a}_{\mathrm{LS}}$ is recommended over most of the negative a range, $\widehat{a}_{\mathrm{Mode}}$ for $-0.1 < a < 0.7$ and $\widehat{a}_{\mathrm{Mean}}$ for $a > 0.7$. Similar results were found based on Laplace, as well as Student's t and asymmetric stable Paretian distributional assumptions for a variety of tail thickness parameters. Thus, it appears that, particularly for small sample sizes, the choice of $\mathbf{X}$ has more of an impact than does—even considerable—deviation from normality.

5.5 Unit Root Testing in the ARX(1) Model

> How to link a theoretical model with empirical evidence in a scientifically valid way is a tremendously difficult task that has been debated as long as economics and econometrics have existed. The dilemma facing an empirical economist/econometrician is that there are many economic models but only one economic reality: which of the models should be chosen?
>
> (Katarina Juselius, 2018, p. 1)

5 Note that the uni-modality of $f_{\widehat{a}_{\mathrm{LS}}}(t \mid a, \mathbf{X})$ with respect to t does not automatically imply a unique maximum with respect to a.

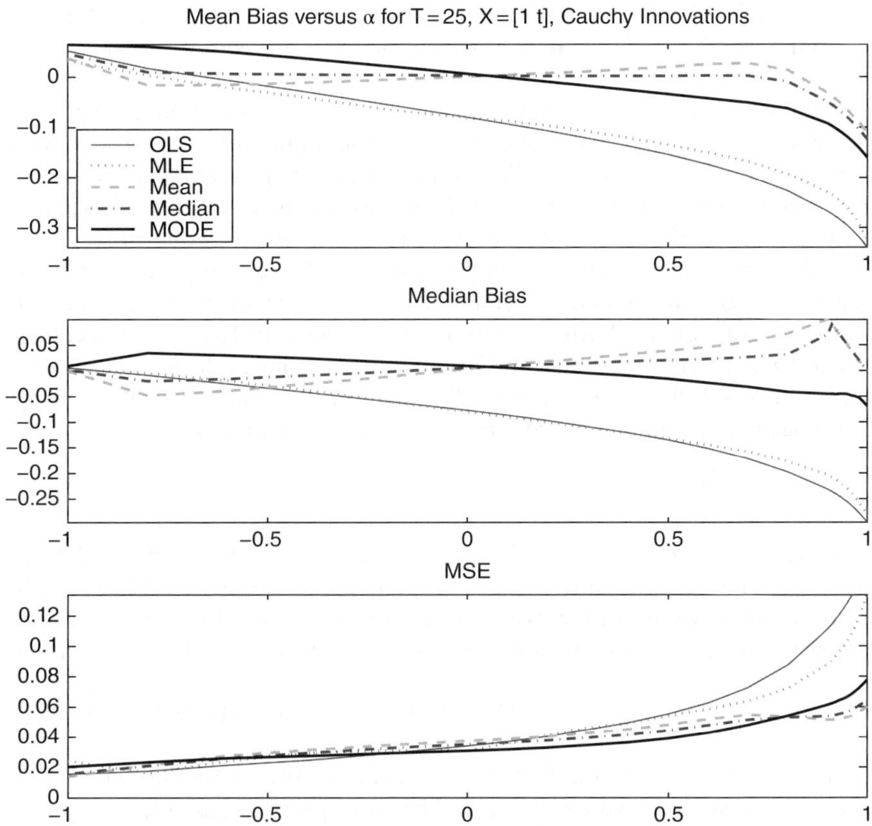

Figure 5.7 The performance of the various adjusted estimators, having used Cauchy innovations.

Starting with the seminal work of Dickey and Fuller (1979), unit root testing has become the focus of a large number of research papers and is now a routine part of applied econometrics. Part of the reason is that a unit root implies that the effect of the innovation sequence never dies out and there is no mean to which the process reverts. In the context of economics, this has interpretations to "shocks" (the innovations) to the economy and policy implications; see, e.g., Campbell and Mankiw (1987), Cochrane (1988), Cribari-Neto (1996), and the references therein.

This can be contrasted with a **trend-stationary model**, i.e., model (5.1)–(5.2) with mean dictated by $\mathbf{x}_t'\boldsymbol{\beta}$ (often with a time trend in the $\mathbf{X}$ matrix) and $|a| < 1$. For the latter, shocks die out over time (the speed of which is dictated by a) and are **mean-reverting** (or have an **attractor**) towards $\mathbf{x}_t'\boldsymbol{\beta}$. Another reason for the popularity of unit root tests is because the decision between $|a| < 1$ and $a = 1$ has implications for so-called **co-integration models**. Pre-testing for unit roots is usually the first step in building such models; see, e.g., Watson (1994), Hamilton (1994), Maddala and Kim (1998), Patterson (2000a), Hayashi (2000), Lütkepohl (2005), Zivot and Wang (2006), and the references therein on unit root testing and co-integration modeling. The unit root literature is extensive: Starting points for deeper study include the previous references, as well as Stock (1994), Hatanaka (1996), Perron (2006), Patterson (2011, 2012), and Choi (2015). See also Abadir (1998), Larsson (1995, 1998), and the references therein for details on approximations of the small-sample distribution of unit root tests and their asymptotics.

Much of the literature on unit root testing appeals to the use of asymptotic distribution theory, often involving (functionals of) Brownian motion and use of tabulated cutoff-values for finite sample sizes based on simulation. In certain model contexts, this will be unavoidable, while in simpler ones exact distribution theory for finite samples via ratios of quadratic forms, as outlined in Appendices A and B, is available, and, for general (non-stochastic) $\mathbf{X}$ and sample size, a p-value can be delivered.

Section 5.5.1 begins with the first generation of tests, for which the null is a unit root ($a = 1$), followed by Section 5.5.2, addressing the subsequent development of tests for which the null is $a < 1$. This situation in which tests are available for both forms of the null is very useful. In practice, tests from both types are conducted when analyzing real data. We will present some model paradigms later in Section 7.7 when *neither* of these nulls is true, such as allowing parameter a to be time-varying, giving rise to a so-called stochastic unit root model, and the use of fractional integration.

5.5.1 Null is $a = 1$

> If you apply unit root tests to an hour of second by second temperature data from 9 to 10 AM you will think it has both a linear trend and a unit root. Millisecond data will not help you to detect climate change. That's why unit root tests are a problem. You have to think, and consider the span of data you have and the frequency of mean reversion that makes economic sense in your data.
>
> (John H. Cochrane, April 2015, internet blog)

The starting point is the o.l.s. estimator of a in model (5.1)–(5.2), and its distribution under the null hypothesis of $a = 1$. The most basic setup presumes the model has no regressors and is given by $Y_t = aY_{t-1} + U_t$, $U_t \overset{\text{i.i.d.}}{\sim} \mathrm{N}(0, \sigma^2)$, and the null hypothesis is $H_0 : a = 1$ versus $H_1 : a < 1$. Dickey and Fuller (1979) propose to use $\tau = (\widehat{a}_{\mathrm{LS}} - 1)/\mathrm{std}(\widehat{a}_{\mathrm{LS}})$ as the test statistic. Under the null of $a = 1$, $\widehat{a}_{\mathrm{LS}}$ does not

```
sim=1e4; reject=zeros(sim,1); alpha=0.05; T=25; % 25 or 50
avec=0:0.02:1; alen=length(avec); pow=zeros(alen,1);
model='AR'; % AR or ARD or TS, for, respectively, X=[], X[1], and X[1 t]
for aloop=1:alen, a=avec(aloop); disp(a)
  parfor i=1:sim
    U=randn(T+40,1); y=zeros(T+40,1);
    for t=2:T+40, y(t)=a*y(t-1)+U(t); end
    y=y((end-T+1):end);
    reject(i)=adftest(y,'model',model,'alpha',alpha);
  end
  pow(aloop) = mean(reject);
end
figure, plot(avec,pow,'r-','linewidth',3)
```

Program Listing 5.5: Simulates the power of the Dickey–Fuller unit root test using the built-in Matlab function `adftest`, which supports use of the three $\mathbf{X}$ matrices used in Figure 5.8. When simulating the Y_t, a "burn in" period of 40 observations is used to remove the effect of the initial value of Y, here zero.

have the asymptotic distribution given in (4.26), and simulation is required to obtain the cutoff values associated with the usual levels of significance, for a given sample size.

The Dickey–Fuller test is built into many high-level programming languages and statistical software packages. In Matlab, the relevant function is `adftest` (where the "a" stands for "augmented", referring to the test in the AR(p) case). Cutoff values obtained via simulation for the usual significance levels have been tabulated for the aforementioned no-regressor model, as well as model (5.1)–(5.2) with $\mathbf{X} = [\mathbf{1}]$ and $\mathbf{X} = [\mathbf{1}, \mathbf{t}]$. The code in Listing 5.5 calculates the power of the test based on simulation, for these three $\mathbf{X}$ matrices, and was used to produce the top panels in Figure 5.8.

Now consider the use of a one-sided test derived from the c.i. of parameter a based on $\widehat{a}_{\text{LS}}$ as in Section 4.7, whereby appeals to asymptotics or simulation for obtaining cutoff values are not required. As it is based on $\widehat{a}_{\text{LS}}$, one might expect it to have similar power properties to the classic Dickey–Fuller test. The test rejects if the value of one is not in the confidence interval. The bottom panels of Figure 5.8 show the power curves corresponding to this test with significance $\alpha = 0.05$, and demonstrate that they are higher than those for the Dickey–Fuller test, particularly in the case with regressors. The reader is encouraged to set up the code to replicate these results.

Hisamatsu and Maekawa (1994) suggest use of the Durbin–Watson statistic for testing for a unit root in the pure AR(1) case (no regressors), assuming the first observation is zero. Their simulations show that it has nearly the same power properties as the Dickey and Fuller (1979) tests. The test can be augmented to the case with regressors, and is expressible as, and amenable to the analytic results for, ratios of quadratic forms, provided that a modification is made to the assumption on the first observation, as in Hisamatsu and Maekawa (1994). Under the null that $a = 1$, (4.20) is rank deficient, recalling its determinant is $1/(1 - a^2)$. However, recalling Section 5.3, we can take $\boldsymbol{\Sigma}^{-1}$ with $a = 1$ to be $\mathbf{B}$ in (5.21), using the modification from Berenblut and Webb (1973) on the first observation so that it is full rank.

Then, for (first observation modified) model (5.1)–(5.2), with $\mathbf{Z} = \mathbf{B}^{1/2}\boldsymbol{\epsilon} \sim \text{N}(\mathbf{0}, \sigma^2\mathbf{I}_T)$, $\mathbf{A}$ the Durbin–Watson matrix as given in (B.8), $\mathbf{H} = \mathbf{B}^{-1/2}\mathbf{MAMB}^{-1/2}$, $\mathbf{K} = \mathbf{B}^{-1/2}\mathbf{MB}^{-1/2}$, and $\mathbf{M} = \mathbf{I}_T - \mathbf{X}(\mathbf{X}'\mathbf{X})^{-1}\mathbf{X}'$ being the usual projection matrix onto the orthogonal complement of regressor matrix $\mathbf{X}$, as given in (1.53), the test statistic and its distribution under the null of a unit root and the use of the modified model is

$$R = \frac{\boldsymbol{\epsilon}'\mathbf{MAM}\boldsymbol{\epsilon}}{\boldsymbol{\epsilon}'\mathbf{M}\boldsymbol{\epsilon}} = \frac{\boldsymbol{\epsilon}'\mathbf{B}^{1/2}\mathbf{B}^{-1/2}\mathbf{MAMB}^{-1/2}\mathbf{B}^{1/2}\boldsymbol{\epsilon}}{\boldsymbol{\epsilon}'\mathbf{B}^{1/2}\mathbf{B}^{-1/2}\mathbf{MB}^{-1/2}\mathbf{B}^{1/2}\boldsymbol{\epsilon}} = \frac{\mathbf{Z}'\mathbf{HZ}}{\mathbf{Z}'\mathbf{KZ}}, \tag{5.32}$$

and we reject the null hypothesis of a unit root for large R. Observe that R is invariant to scale term $\sigma > 0$ in (5.2). Its c.d.f. can be evaluated using program `cdfratio` in Listing A.3, and the α-quantile (cutoff value for the test) can be evaluated using the `fzero` command as in line 29 of Listing 5.2.

The short code in Listing 5.6 calculates the cutoff $c = c(1 - \alpha, T, \mathbf{X})$ for $T = 50$ and significance level 0.05 (corresponding to the 95% quantile, as we reject for large R), for an $\mathbf{X}$ matrix consisting of a constant and time trend, and confirms the level of the test via simulation with (an indulgence of) ten million replications, yielding 0.0500.

It might appear that the power for $|a| < 1$ and cutoff $c = c(1 - \alpha, T, \mathbf{X})$, using the usual matrix $\boldsymbol{\Sigma}$ for the covariance matrix for $\boldsymbol{\epsilon}$, is given by

$$\begin{aligned}\text{(wrong)}\quad \Pr(R > c) &= \Pr\left(\frac{\boldsymbol{\epsilon}'\mathbf{B}^{1/2}\mathbf{B}^{-1/2}\mathbf{MAMB}^{-1/2}\mathbf{B}^{1/2}\boldsymbol{\epsilon}}{\boldsymbol{\epsilon}'\mathbf{B}^{1/2}\mathbf{B}^{-1/2}\mathbf{MB}^{-1/2}\mathbf{B}^{1/2}\boldsymbol{\epsilon}} > c\right)\\ &= \Pr\left(\frac{\mathbf{Z}'\mathbf{HZ}}{\mathbf{Z}'\mathbf{KZ}} > c\right), \quad \mathbf{Z} = \mathbf{B}^{1/2}\boldsymbol{\epsilon} \sim \text{N}(\mathbf{0}, \sigma^2\mathbf{B}^{1/2}\boldsymbol{\Sigma}\mathbf{B}^{1/2}).\end{aligned} \tag{5.33}$$

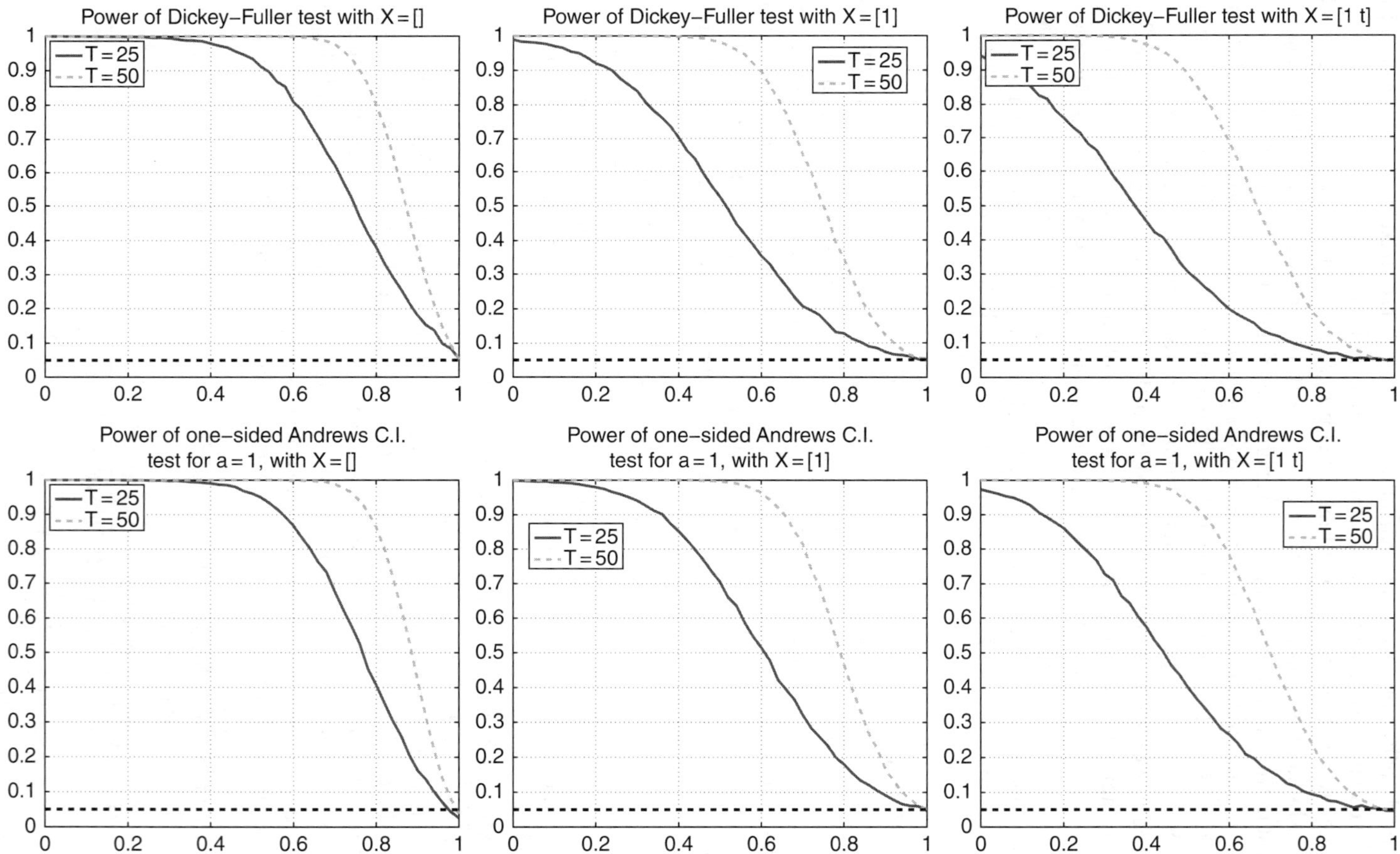

Figure 5.8 **Top:** The power of the Dickey–Fuller unit root test, for two sample sizes and three **X** matrices, as indicated. **Bottom:** Same, but based on the one-sided c.i. of $\hat{a}_{LS}$ as in Section 4.7.

```
alpha=0.95; T=50; X=[ones(T,1), (1:T)'];
if isempty(X), M=eye(T); k=0; else M=makeM(X); [~,k]=size(X); end
A=makeDW(T); B=A; B(1,1)=2; [V,D]=eig(B); Bm12=V*sqrt(inv(D))*V';
H=Bm12*M*A*M*Bm12; K=Bm12*M*Bm12;
Rc=fzero(@(r) cdfratio(r,H,K,eye(T),[],1)-alpha, 0.5);
sim=1e7; rej=zeros(sim,1); % now simulate the size
for i=1:sim
  U=[0 ; randn(T-1,1)]; e=cumsum(U);
  R=(e'*M*A*M*e)/(e'*M*e); rej(i)=R>Rc;
end
thesize=mean(rej) %#ok<NOPTS>
```

Program Listing 5.6: Calculates the cutoff value for the unit root test (5.32) and simulates the size.

However, observe that, because the model assumption in the test statistic under the null was modified, as $a \to 1$, (5.33) is incorrect (though will be close as the sample size increases, as the reader can confirm). Similarly, with $\mathbf{G}$ as given in Theorem 1.3,

$$\text{(wrong)} \quad \Pr(R > c) = \Pr\left(\frac{\boldsymbol{\epsilon}'\mathbf{MAM}\boldsymbol{\epsilon}}{\boldsymbol{\epsilon}'\mathbf{M}\boldsymbol{\epsilon}} > c\right) = \Pr\left(\frac{\boldsymbol{\epsilon}'\mathbf{G}'\mathbf{GAG}'\mathbf{G}\boldsymbol{\epsilon}}{\boldsymbol{\epsilon}'\mathbf{G}'\mathbf{G}\boldsymbol{\epsilon}} > c\right)$$
$$= \Pr\left(\frac{\mathbf{Z}'\tilde{\mathbf{A}}\mathbf{Z}}{\mathbf{Z}'\mathbf{Z}} > c\right), \quad \mathbf{Z} = \mathbf{B}^{1/2}\mathbf{Z} = \mathbf{G}\boldsymbol{\epsilon} \sim \mathrm{N}_{T-k}(\mathbf{0}, \sigma^2\mathbf{G}\boldsymbol{\Sigma}\mathbf{G}'), \tag{5.34}$$

where $\tilde{\mathbf{A}} = \mathbf{GAG}'$ is $(T-k) \times (T-k)$, is also incorrect, and results in very similar values as (5.33). However, if we replace $\boldsymbol{\Sigma}$ in (5.34) with $\tilde{\boldsymbol{\Sigma}}$, where the latter is "almost" the variance covariance matrix of a stationary AR(1) process with parameter a, but computed as the inverse of $\boldsymbol{\Sigma}^{-1}(a)$ such that the (1,1) element is $b = 1 + a^2$ instead of one, then the power is nearly exact, though it appears in this setting that (because of the change of the assumption on the first observation of the model) simulation is the only way to get exact values.

The (nearly exact) power can be very quickly computed using our usual program `cdfratio` applied to (5.34) but, as mentioned, using $\tilde{\boldsymbol{\Sigma}}$ instead of $\boldsymbol{\Sigma}$. Figure 5.9 illustrates the power for three values of sample size T and two $\mathbf{X}$ matrices, based on the analytic calculation and simulation. The slight difference in the powers between the analytic and simulated calculations for sample size $T = 25$ is apparent. Note that this is not due to s.p.a. error, having performed the analytic calculation using both the exact and s.p.a. methods.

These figures can be compared to the middle and right panels of Figure 5.8, for $T = 25$ and $T = 50$: We see that the powers are virtually identical to those of the test derived from the one-sided c.i. of parameter a based on $\widehat{a}_{\mathrm{LS}}$.

We now turn to the test proposed by Shively (2001), which builds upon related work by Sargan and Bhargava (1983), Bhargava (1986), and Dufour and King (1991). In order to facilitate optimal comparison to the bottom panel in Figure 5.9, Figure 5.10 shows the resulting power. We immediately see that, particularly for $T = 25$, it has lower power than the R test (5.32). We detail it because its construction is interesting and has some relations to the test statistics developed in Section 5.6 for time-varying regression coefficients.

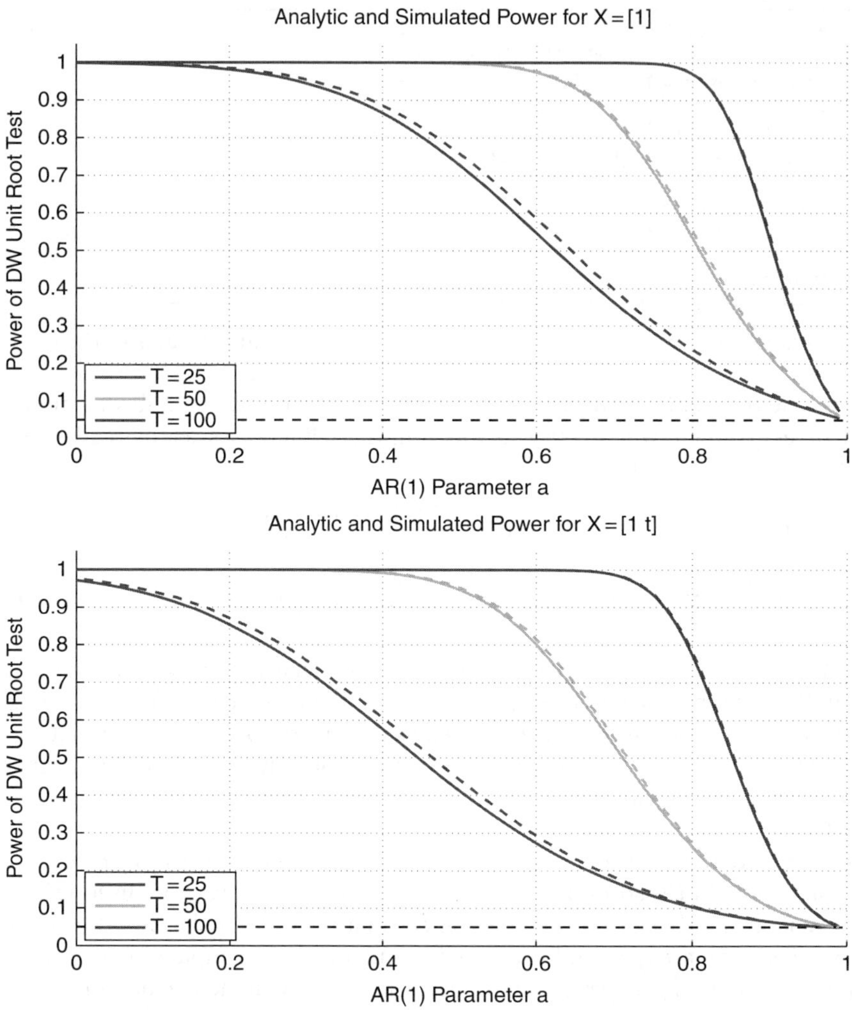

Figure 5.9 **Top:** Approximate power of the R test (5.32) computed analytically (solid line) and from simulation (dashed line) with significance level 0.05, for $T = 25$, $T = 50$, and $T = 100$, using $\mathbf{X} = [\mathbf{1}]$. **Bottom:** Same, but based on $\mathbf{X} = [\mathbf{1},\ \mathbf{t}]$.

In this setting, the null hypothesis is the random walk with drift, $Y_t = \delta + Y_{t-1} + U_t$, $U_t \overset{\text{i.i.d.}}{\sim} \mathrm{N}(0, \sigma^2)$, and the alternative is the stationary model with intercept and time trend given in (5.1)–(5.2) for $\mathbf{X} = [\mathbf{1}, \mathbf{t}]$, or (5.4). Recursive substitution of the null model gives

$$\begin{aligned} Y_t &= \delta + Y_{t-1} + U_t \\ &= \delta + \delta + Y_{t-2} + U_{t-1} + U_t \\ &= \delta + \delta + \delta + Y_{t-3} + U_{t-2} + U_{t-1} + U_t \\ &\ \ \vdots \\ &= Y_0 + \delta t + U_t^*, \quad U_t^* = U_1 + \cdots + U_t, \end{aligned}$$

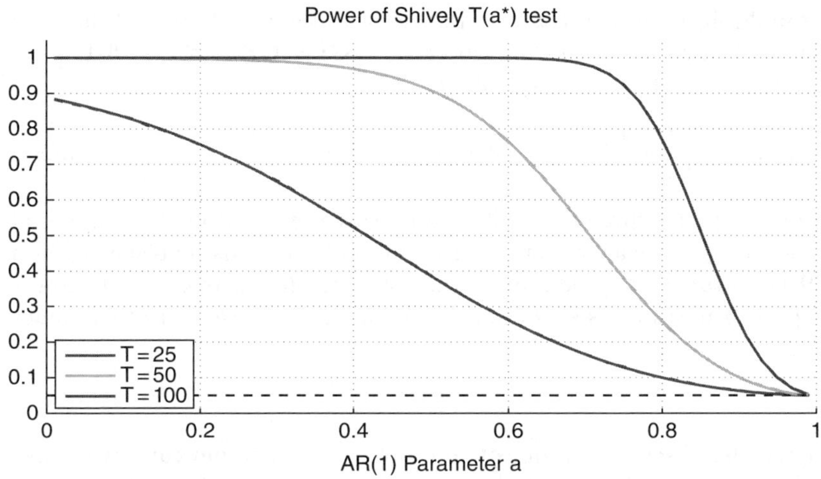

Figure 5.10 Similar to the right panel of Figure 5.9, but based on the test (5.35) from Shively (2001).

where Y_0 is the unknown starting value of the random walk. This can be expressed in matrix notation as $\mathbf{Y} = \mathbf{X}\boldsymbol{\beta} + \mathbf{U}^*$, $\mathbf{X} = [\mathbf{1}, \mathbf{t}]$, and $\mathbf{Y} \sim \mathrm{N}(\mathbf{X}\boldsymbol{\beta}, \sigma^2\tilde{\boldsymbol{\Sigma}})$, with the (i, j)th element of $\tilde{\boldsymbol{\Sigma}}$ given by $\min(i, j)$, or

$$\tilde{\mathbf{Y}} := \tilde{\boldsymbol{\Sigma}}^{-1/2}\mathbf{Y} = \tilde{\boldsymbol{\Sigma}}^{-1/2}\mathbf{X}\boldsymbol{\beta} + \tilde{\boldsymbol{\Sigma}}^{-1/2}\mathbf{U}^* =: \tilde{\mathbf{X}}\boldsymbol{\beta} + \mathbf{V}, \quad \mathbf{V} \sim \mathrm{N}(\mathbf{0}, \sigma^2\mathbf{I}),$$

where $:=$ $(=:)$ indicates that the element(s) on the left (right) hand side are so-defined.

The alternative model, denoted $H(a)$, is $\mathbf{Y} = \mathbf{X}\boldsymbol{\beta} + \boldsymbol{\epsilon}$ with $\boldsymbol{\epsilon} \sim \mathrm{N}(\mathbf{0}, \sigma^2\boldsymbol{\Sigma}(a))$, where $\boldsymbol{\Sigma}(a)$ is the covariance matrix of a stationary AR(1) process with parameter a. Multiplying this by $\tilde{\boldsymbol{\Sigma}}^{-1/2}$ gives

$$\tilde{\mathbf{Y}} = \tilde{\mathbf{X}}\boldsymbol{\beta} + \boldsymbol{\eta}, \quad \boldsymbol{\eta} = \tilde{\boldsymbol{\Sigma}}^{-1/2}\boldsymbol{\epsilon} \sim \mathrm{N}(\mathbf{0}, \sigma^2\boldsymbol{\Omega}(a)), \quad \boldsymbol{\Omega}(a) = \tilde{\boldsymbol{\Sigma}}^{-1/2}\boldsymbol{\Sigma}(a)\tilde{\boldsymbol{\Sigma}}^{-1/2}.$$

The proposed test is then precisely the same form as (5.22), i.e., we reject for small values of test statistic

$$T(a^*) = \frac{\mathbf{Y}'\tilde{\boldsymbol{\Sigma}}^{-1/2}\tilde{\mathbf{G}}'\mathbf{N}^{-1}(a^*)\tilde{\mathbf{G}}\tilde{\boldsymbol{\Sigma}}^{-1/2}\mathbf{Y}}{\mathbf{Y}'\tilde{\boldsymbol{\Sigma}}^{-1/2}\tilde{\mathbf{M}}\tilde{\boldsymbol{\Sigma}}^{-1/2}\mathbf{Y}}, \quad \mathbf{N}(a) = \tilde{\mathbf{G}}\boldsymbol{\Omega}(a)\tilde{\mathbf{G}}', \tag{5.35}$$

where $\tilde{\mathbf{M}} = \mathbf{I}_T - \tilde{\mathbf{X}}(\tilde{\mathbf{X}}'\tilde{\mathbf{X}})^{-1}\tilde{\mathbf{X}}'$, $\tilde{\mathbf{G}}$ from Theorem 1.3 is $(T-2) \times T$ and such that $\tilde{\mathbf{M}} = \tilde{\mathbf{G}}'\tilde{\mathbf{G}}$, $\tilde{\mathbf{G}}\tilde{\mathbf{G}}' = \mathbf{I}_{T-k}$, and $\tilde{\mathbf{G}}\tilde{\mathbf{X}} = \mathbf{0}$. Value a^* is a point in the stationary support of a and, crucially, such that it does not depend on the data but rather is chosen in advance.[6] In this model setting, the $\mathbf{X}$ matrix is fixed, and Shively (2001) recommends use of the value $a^* = 0.9$. As the vector of ones is included in the regression, the invariance property implies that $T(a^*)$ does not depend on specification of Y_0, unlike the test of Dufour and King (1991); see also Remark (b) below.

Our usual methods for ratios of quadratic forms in normal variables can be used to determine the appropriate cutoff value, $c = c(\alpha, T, a^*)$, of test statistic (5.35), for a given T and significance level α, for which we use 0.05. In particular, the distribution of (5.35) under the null is the stated ratio of quadratic forms in $\mathbf{Y} \sim \mathrm{N}(\mathbf{X}\boldsymbol{\beta}, \sigma^2\tilde{\boldsymbol{\Sigma}})$, and one ignores the σ^2 as it cancels from the ratio, and ignores

6 This is emphasized in a similar context in King (1985a, p. 29), stating "$\mathbf{Y}$ should not be used in any way to choose the value of $[a^*]$. Its use would mean that the test is no longer [most powerful invariant] at $[a = a^*]$…". A similar statement is given in King (1985b, p. 213).

$\mathbf{X}\boldsymbol{\beta}$ as it is removed from pre-multiplication with $\tilde{\mathbf{G}}\tilde{\boldsymbol{\Sigma}}^{-1/2}$. Alternatively, the distribution of the ratio under the null is the same as that of $(\mathbf{Z}'\mathbf{N}^{-1}(a^*)\mathbf{Z})/(\mathbf{Z}'\mathbf{Z})$, where $\mathbf{Z} := \tilde{\mathbf{G}}\tilde{\mathbf{Y}} = \tilde{\mathbf{G}}\tilde{\boldsymbol{\Sigma}}^{-1/2}\mathbf{Y} \sim \mathrm{N}(\mathbf{0}, \mathbf{I}_{T-2})$.

The power of the test, as a function of a (and α, T, and a^*) is[7]

$$\Pr(T(a^*) \leqslant c \mid H(a)) = \Pr\left(\frac{\mathbf{Z}'\mathbf{N}^{-1}(a^*)\mathbf{Z}}{\mathbf{Z}'\mathbf{Z}} \leqslant c\right), \quad \mathbf{Z} \sim \mathrm{N}(\mathbf{0}, \sigma^2\mathbf{N}(a)). \tag{5.36}$$

Figure 5.10 shows the power of the test for three different sample sizes as a function of a, computed analytically from (5.36) (via the s.p.a.) and via simulation (they are optically indistinguishable), and can be compared to Figure 5.9. Unfortunately, as mentioned, the power of $T(a^*)$ with $a^* = 0.9$ is lower over most of the parameter space. The reader is encouraged to compute its power for other values of a^*.

Remarks

a) Cochrane (1991) and Blough (1992) discuss a critique of unit root tests in that they can have power against any stationary alternative only if they also have excessive probability of false rejection for some unit root processes. With respect to the low power of unit root tests and the choice of significance level, see Kim and Choi (2017).
b) There are several other unit root test statistics that can be expressed as a ratio of quadratic forms in normal variables, though there does not exist a test that is uniformly most powerful (King, 1987b). A relevant issue in this testing framework is the initial regression error term—assumptions on it affect the power of the test; see, e.g., Müller and Elliott (2003), Choi (2015, Sec. 2.4.10), and the references therein. Tests with high power irrespective of the initial error term, nesting several seemingly disparate tests in the literature and which are amenable to exact inference (under normality), are developed in Broda et al. (2009). Unit root tests can also be used to generate near-optimal confidence intervals for parameter a when it is close to unity; see, e.g., Elliott and Stock (2001).
c) The exact small-sample distribution theory no longer holds when leaving the Gaussian framework. We wish to investigate the size and power of the test when we still use this assumption but take the error term to be non-Gaussian. Figure 5.11 shows the results, only for $T = 25$, having used the same, analytically determined cutoff value under normality, but taking the innovation sequence to be (symmetric) stable Paretian for several values of stable tail index α (and using simulation to determine the actual size and power). As α decreases, the actual size indeed decreases, but to such a small extent that, even for very heavy-tailed innovation processes, the test is still highly accurate, and with very similar power curves. Further information on unit root and other tests with stable Paretian innovations can be found in Rachev and Mittnik (2000, Ch. 14–15).
d) Particularly in economics, there is ample reason to suspect that the true d.g.p. is not as simple as dictated by a linear regression model. In particular, the $\mathbf{X}$ matrix could well be mis-specified because of structural breaks in the parameters of the constant and time trend, and this will have a negative impact on the ability of unit root tests to reject the null of $a = 1$; see, e.g., Perron (1989), and the references given at the beginning of this section. When such breaks are ignored, the size of standard unit root tests tends to zero as the magnitude of the break(s) increase. Work on unit root testing in the presence of structural breaks includes that by Breitung (2002), Kurozumi (2002),

7 The expressions for the power given in the middle of page 541 of Shively (2001) are incorrect.

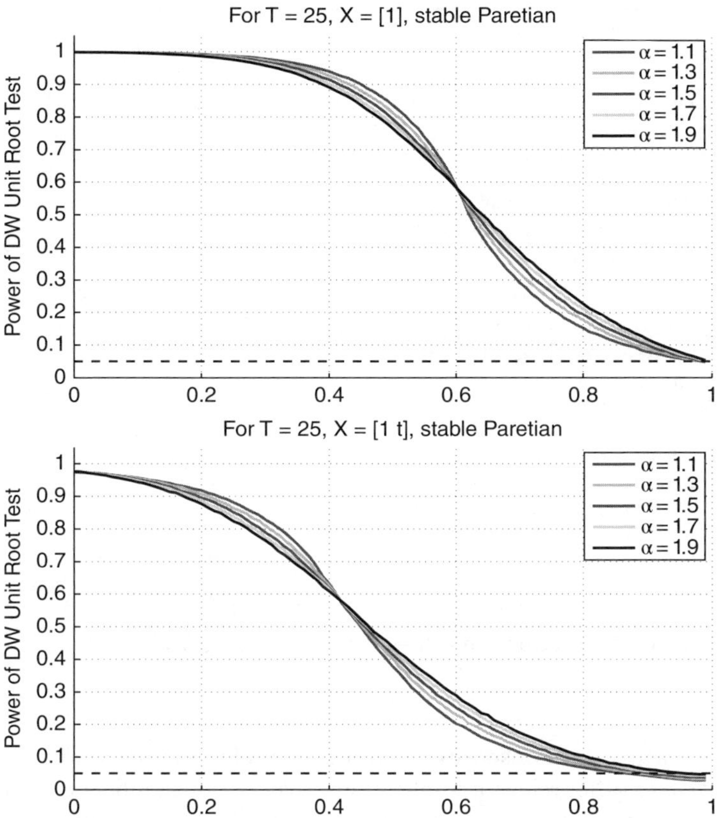

Figure 5.11 **Top:** Power of the R test (5.32) computed from simulation from 10,000 replications, using $S_{\alpha,0}$ innovations, with significance level 0.05, $T = 25$, and $\mathbf{X} = [\mathbf{1}]$. **Bottom:** Same, but for $\mathbf{X} = [\mathbf{1},\ \mathbf{t}]$.

Saikkonen and Lütkepohl (2002), Kim and Perron (2009), and the references therein, as well as innovation variance shifts; see, e.g., Kim et al. (2002).

To demonstrate this, we use the test statistic in (5.32), based falsely on $\mathbf{X} = [\mathbf{1},\ \mathbf{t}]$, with the actual $\mathbf{X}$ given by $[\mathbf{1}, \mathbf{t}, D1, Dt]$, similar to the matrix described at the beginning of Section 5.3 (but with Dt and not Dt^2). Figure 5.12 shows the actual size, based on simulation, of the unit root test, as a function of β_3 (top) and β_4 (bottom), where these are the coefficients for the latter two columns of the actual $\mathbf{X}$ matrix. As they increase in magnitude, the size drops towards zero. The code in Listing 5.7 (omitting the graphics commands) was used to generate the plots. The test developed in Kim and Perron (2009) can remedy this situation, and is such that the time of the break point is unknown.[8]

e) Recall Section 4.6.1 on the use of the jackknife for bias reduction in the stationary AR(1) model. Their use in the unit root setting is developed in Chambers and Kyriacou (2013) and Chen and Yu (2015), while Chambers and Kyriacou (2018) consider the near-unit root case. ■

8 Matlab code for that test is available from those authors.

```
alpha=0.95; T=25; X=[ones(T,1), (1:T)'];
M=makeM(X); G=makeG(X); [~,k]=size(X);
A=makeDW(T); B=A; B(1,1)=2;
[V,D]=eig(B); Bm12=V*sqrt(inv(D))*V';
H=Bm12*M*A*M*Bm12; K=Bm12*M*Bm12;
Rc=fzero(@(r) cdfratio(r,H,K,eye(T),[],1)-alpha, 0.5);
c=round(T/2); % now generate actual X matrix
D1=[zeros(c,1) ; ones(c,1)]; if length(D1)>T, D1=D1(1:(end-1)); end
Dt=[zeros(c,1) ; ((c+1):T)'];if length(Dt)>T, Dt=Dt(1:(end-1)); end
X2=[ones(T,1), (1:T)', D1, Dt]; b1=5; b2=2;

a=1; sim=1e5; power1=zeros(sim,1);
%bvec=0:0.1:10; blen=length(bvec); power=zeros(blen,1);
bvec=0:0.05:0.8; blen=length(bvec); power=zeros(blen,1);
for bloop=1:blen,  b=bvec(bloop); disp(b)
  beta=[b1 b2 0 b]'; % or beta=[b1 b2 b 0]';
  for i=1:sim
    U=[0 ; randn(T-1,1)]; e=zeros(T,1);
    for t=2:T, e(t)=a*e(t-1)+U(t); end
    Y=X2*beta+e; R=(Y'*M*A*M*Y)/(Y'*M*Y); power1(i)=R>Rc;
  end
  power(bloop)=mean(power1);
end
```

Program Listing 5.7: Code for generating the values shown in Figure 5.12.

5.5.2 Null is $a < 1$

As emphasized in Section III.2.8, there is a growing consensus regarding the preference of use of confidence intervals and study of effect sizes (or other relevant implications) over use of significance and hypothesis testing. Besides the arbitrary choice of the significance level, a crucial issue concerns what one does in light of the binary result of a hypothesis test for a unit root: Either the test does not reject the null of a unit root, or it rejects, and subsequently, one usually conditions on the result, i.e., proceeds as if it is the case.[9] It is more sensible (though often far more difficult, in terms of distribution theory) to invoke a pre-test testing or pre-test estimation framework in which one explicitly accounts for the conditioning on the result of the pre-conducted test in subsequent testing or estimation exercises.

To help temper this issue in the unit root testing context, Kwiatkowski et al. (1992), hereafter KPSS, investigated the hypothesis test such that the null is stationarity and the alternative is a unit root. In doing so, they and many subsequent researchers have found that, for many economic data sets of interest, the (usually Dickey–Fuller) test with null of a unit root, and also the KPSS test, do not reject their respective nulls, implying the lack of strong evidence in favor of, or against, a unit root.

Notice that this conclusion can also be drawn from the use of (correct) confidence intervals, computed via the method discussed above in Section 5.2: If the interval includes unity, and is short enough (a subjective decision that *cannot* be outsourced to, and objectified by, the use of a binary hypothesis test), then one can be relatively certain that (to the extent that the model and choice of regressors is reasonable) the assumption of a unit root is tenable; whereas, if the interval includes unity but

9 As an example, applying the augmented Dickey–Fuller unit root test to monthly observations from the NYSE Composite Index, Narayan (2006, p. 105) reports the test statistic and the 5% cutoff value, and concludes "…[W]e are unable to reject the unit root null hypothesis. This implies that US stock price has a unit root."

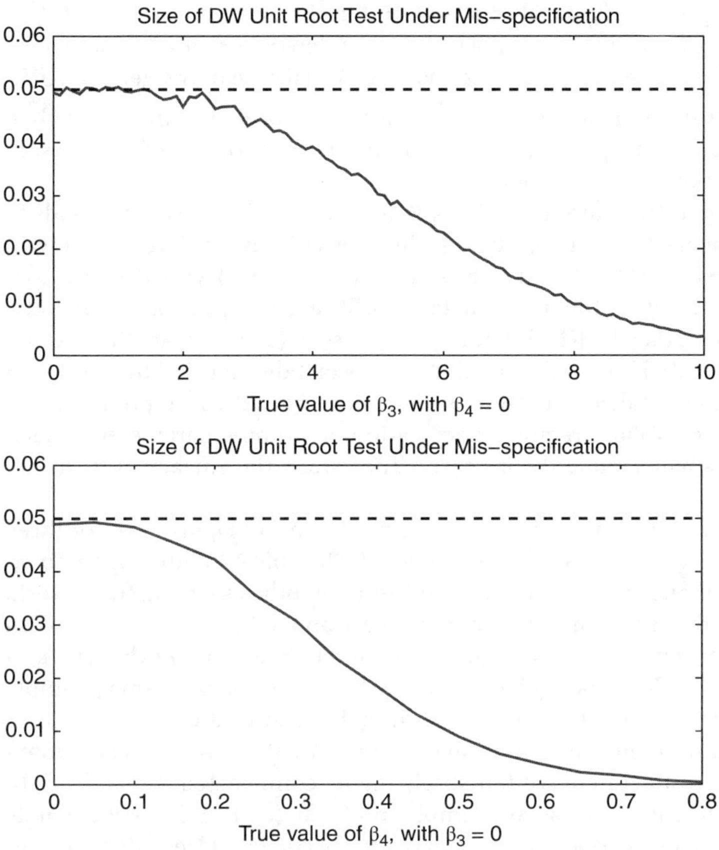

Figure 5.12 Top: Actual size of the R test (5.32), for $T = 25$ and nominal size $\alpha = 0.05$, computed via simulation with 100,000 replications, as a function of β_3, the coefficient referring to the regressor capturing the break in the constant. **Bottom:** Same, but as a function of β_4, the coefficient referring to the regressor capturing the break in the trend.

is long enough, then one cannot be so sure, and could proceed by investigating inference (such as forecasting or assessing the relevance of particular regressors) in both the stationary and unit root settings. Finally, if the confidence interval does not contain unity, then one has more assurance that a unit root may not be tenable, though, of course, the choice of the confidence level associated with the interval influences this.

We now turn to the KPSS test. Repeating the model (5.1)–(5.2) here for convenience, the observation and latent equations, respectively, are given by

$$Y_t = \mathbf{x}_t'\boldsymbol{\beta} + \epsilon_t, \quad \epsilon_t = a\epsilon_{t-1} + U_t, \qquad U_t \overset{\text{i.i.d.}}{\sim} \mathrm{N}(0, \sigma^2), \tag{5.37}$$

and interest centers on knowing if $a = 1$ or $|a| < 1$. The model can be expressed somewhat differently, with two error terms, as

$$Y_t = \alpha_t + \mathbf{z}_t'\boldsymbol{\beta} + \epsilon_t, \quad \alpha_t = \alpha_{t-1} + U_t, \tag{5.38}$$

where $\mathbf{z}_t$ embodies a set of known regressors (typically a time trend in the unit root literature), $\{\epsilon_t\}$ denotes a stationary time-series process (such as an AR(1) model), independent of U_t, assumed to be an i.i.d. white noise (not necessarily Gaussian) process. If, as in a special case of the general KPSS framework, we assume that $\epsilon_t \overset{\text{i.i.d.}}{\sim} \mathrm{N}(0, \sigma^2)$, independent of $U_t \overset{\text{i.i.d.}}{\sim} \mathrm{N}(0, \lambda\sigma^2)$, for $\sigma_2 > 0$ and $\lambda \geqslant 0$, then this is exactly model (5.58) given below (with $x_t = 1$ in (5.58)), proposed in the context of a regression model with α_t being a time-varying regression coefficient.

We wish to test the null of $\lambda = 0$ versus the alternative of $\lambda > 0$. Notice how the null corresponds to the desired null hypothesis of a stationary time series, whereas the alternative is a unit root. Below, in Section 5.6.3, as in Nyblom and Mäkeläinen (1983) and Nabeya and Tanaka (1988), we will develop an exact (meaning, the small-sample distribution theory is tractable) LBI test, using ratios of quadratic forms. This is in fact precisely the test studied by KPSS in the special case of (5.38)—a fact they explicitly state (Kwiatkowski et al., 1992, Sec. 2). However, instead of using exact distribution theory, KPSS derive the asymptotic distribution under weaker assumptions on $\{\epsilon_t\}$ (as the i.i.d. assumption will not be tenable for many time series of interest) and require specification of a tuning parameter ℓ (such that $\ell = 0$ corresponds to the exact small-sample theory case). They study the efficacy of its use in small samples via simulation.

The power of the test in the case of $z_t = t$ is shown below, in the right panel of Figure 5.21, for three sample sizes, and agree with the values given in Kwiatkowski et al. (1992, Table 4, column 6) obtained via simulation. We recommend that, if one wishes to use the unit root hypothesis testing framework, both a test with null of a unit root and a test with a stationary null are applied. If, as is the case with many unit root tests in the former group, and for the more general KPSS and the Leybourne and McCabe (1994, 1999) tests (discussed in Remark (b) below) in the latter group, exact small-sample distribution theory is not available for assessing the power, one should use simulation.

Above, we mentioned that all such inference is conditional on the extent to which the assumed model is a reasonable approximation to the unknown but surely highly complicated actual d.g.p. In general, one might be skeptical of the efficacy of such a simple model as (5.1)–(5.2) to adequately describe phenomena as complex as major economic measures. We partially address this later, in Section 7.7, where we discuss some alternative models that nest the unit root process as a limiting special case.

Remarks

a) Augmenting the previous comment on possible alternative models, one needs to keep in mind the idea that the complexity of the model (when used for highly complex phenomena) will be to a large extent dictated by the available number of data points, as discussed in Section III.3.3. For example, one could argue that, if both the (say) Dickey–Fuller and (say) KPSS tests do not reject their respective nulls (and tests with higher power are not available), then one should attempt to obtain more data, so that the power of both tests is higher, and more definitive conclusions can be drawn. This argument is flawed in the sense that (besides the obvious fact that more data might not be available), if the true d.g.p. is not equal to the one assumed (this being almost surely the case), then the availability of more data might be better used in conjunction with a richer model that more adequately describes the d.g.p., instead of one so simple as a regression model (with constant regressors) and either a stationary or unit root AR(p) error structure.
b) Leybourne and McCabe (1994, 1999) argue that economic time series are "often best" represented as ARIMA processes instead of pure AR or random walk models, and consider the null hypothesis of a stationary ARMA(p, 1) process (possibly with regressors, or ARMAX) versus the alternative

of an ARIMA$(p, 1, 1)$ process. In particular, using notation that we will detail in Chapter 6, the model is

$$\phi(L)Y_t = \alpha_t + \beta t + \epsilon_t, \quad \alpha_t = \alpha_{t-1} + U_t, \tag{5.39}$$

where $\phi(L) = 1 - \phi_1 L - \cdots - \phi_p L^p$, i.e.,

$$Y_t = \alpha_t + \beta t + \phi_1 Y_{t-1} + \cdots \phi_p Y_{t-p} + \epsilon_t, \quad \alpha_t = \alpha_{t-1} + U_t,$$

the ϕ_i are such that $Y_t = \phi_1 Y_{t-1} + \cdots + \phi_p Y_{t-p} + \epsilon_t$ is stationary (see Section 6.1.1), $\epsilon_t \overset{\text{i.i.d.}}{\sim} \mathrm{N}(0, \sigma_\epsilon^2)$, independent of $U_t \overset{\text{i.i.d.}}{\sim} \mathrm{N}(0, \sigma_U^2)$, $\sigma_\epsilon^2 > 0$, and $\sigma_U^2 \geqslant 0$. Observe how this generalizes (5.38). The null is that (5.39) is trend stationary, i.e., $\sigma_U^2 = 0$, with all $\alpha_t = \alpha_0 =: \alpha$. The alternative is that $\sigma_U^2 > 0$ and is such that it is a "local departure" resembling the ARIMA$(p, 1, 1)$ process

$$\phi(L)(1 - L)Y_t = \beta + (1 - \theta L)\xi_t, \quad 0 < \theta < 1, \quad \xi_t \overset{\text{i.i.d.}}{\sim} \mathrm{N}(0, \sigma_\xi^2),$$

where the (necessary in this context) assumption is made that there is no zero pole cancellation, i.e., $(1 - \theta L)$ is not a factor of the polynomial $\phi(L)$ (see Chapter 7). The relationship of σ_ξ^2 to σ_ϵ^2, θ and σ_U^2 is detailed in Leybourne and McCabe (1994, p. 158).

Exact distribution theory is not available for the general model (5.39). In this setting, p is a tuning parameter that needs to be specified: For economic time series, Leybourne and McCabe (1994) argue that it should be greater than zero, and conveniently show via simulation that choosing p too large is not costly in terms of actual size and power. The case with $p = 0$ results in tractable small-sample theory, as discussed above. The two tests of Leybourne and McCabe (1994, 1999) differentiate themselves by how the estimate of σ_ϵ^2 is computed. Matlab implements both tests in their function `lmctest`, along with the KPSS test as `kpsstest`. ■

5.6 Time-Varying Parameter Regression

In some problems it seems reasonable to assume that the regression coefficients are not constants but chance variables.

(Abraham Wald, 1947, p. 586)

The potential pitfalls confronting empirical research include inadequate theory, data inaccuracy, hidden dependencies, invalid conditioning, inappropriate functional form, non-identification, parameter non-constancy, dependent, heteroskedastic errors, wrong expectations formation, mis-estimation and incorrect model selection.

(David F. Hendry, 2009, p. 3)

In class there is much discussion of the assumptions of exogeneity, homoskedasticity, and serial correlation. However, in practice it may be unstable regression coefficients that are most troubling. Rarely is there a credible economic rationale for the assumption that the slope coefficients are time invariant.

(Robert F. Engle, 2016, p. 643)

5.6.1 Motivation and Introductory Remarks

The above three quotes, as well as that from Cooley and Prescott (1973) at the beginning of Chapter 4, should serve as indicators of the relevance and popularity of (regression) models that allow for some form of time variation in one or more parameters. A strong critique of the usual, fixed-coefficient linear model, in favor of one with random coefficients, is given in Swamy et al. (1988).

Starting with Wald (1947) and particularly since the late 1960s, an enormous amount of research has been published on time-varying parameter (TVP) regression models; so much, that already by the mid 1970s, an annotated bibliography was deemed appropriate; see Johnson (1977). More recent overviews are provided by Dziechciarz (1989), Freimann (1991), and Swamy and Tavlas (1995, 2001). Their use can found use in numerous settings, including testing the capital asset pricing model (CAPM) in finance; see, e.g., Bos and Newbold (1984), as one of the earliest such references, and, more recently, Engle (2016) and Bali et al. (2016a,b).

We detail three basic types of models for time-varying regression parameters, and their associated statistical tests, in Sections 5.6.2, 5.6.3, and 5.6.4, respectively. Before commencing, we provide some remarks.

Remarks

a) This is a large, important, and ever-growing field of research, and we only cover some fundamental, albeit still relevant, structures and statistical tests. A more general modeling framework is discussed in Creal et al. (2013), while a related, but conceptually different (and more modern) regression-type model with TVPs was introduced in Hastie and Tibshirani (1993), with a recent survey of the field by Park et al. (2015). Different, more general, tests of parameter constancy are developed in Nyblom (1989) and Hansen (1992). Likelihood-based methods for detecting model constancy of parameters and mis-specification in general are discussed in McCabe and Leybourne (2000), Golden et al. (2016), and the references therein.
b) We will see below that parameter estimation is straightforward in the first two classes of models considered. However, a far more general framework applicable to all the models, and others not considered herein, is to cast the model into the so-called **state space representation** and use the methods of **Kalman filtering**. This is now a very well-studied area, with estimation techniques, inferential methods, and computational algorithms that were not available "in the earlier days". As in Durbin (2000), the linear Gaussian state space model is given by

$$\mathbf{Y}_t = \mathbf{X}_t'\boldsymbol{\alpha}_t + \boldsymbol{\epsilon}_t, \quad \boldsymbol{\alpha}_t = \mathbf{T}_t\boldsymbol{\alpha}_{t-1} + \mathbf{U}_t,$$

with $\boldsymbol{\epsilon}_t \sim \mathrm{N}(\mathbf{0}, \mathbf{H}_t)$ independent of $\mathbf{U}_t \sim \mathrm{N}(\mathbf{0}, \mathbf{Q}_t)$. Notice here that the observed time series $\mathbf{Y}_t$ can be multivariate, and $\boldsymbol{\alpha}_t$, referred to as the *state vector at time t*, can, but need not, evolve as a random walk. Moreover, the covariance matrices of $\boldsymbol{\epsilon}_t$ and $\mathbf{U}_t$ can also vary with time.

It can be shown that generation of the recursive residuals from Section 1.5 is a special case of the Kalman filter, see, e.g., Harvey (1993, p. 99) and Durbin and Koopman (2012, p. 150). An early and very accessible reference on use of Kalman filtering for the linear regression model with TVPs is Morrison and Pike (1977).

Book-length treatments on state space methods aimed at statisticians and econometricians include West and Harrison (1997)[10] and Durbin and Koopman (2012), while Chui and Chen

10 The title of the book by West and Harrison, *Bayesian Forecasting and Dynamic Models* makes their modeling slant and intended audience rather clear. On page 35 of the first edition, they write "It is now well-known that, in normal [dynamic

(1999) is aimed more at engineers. An implementation augmenting Matlab's tools for state space modeling is provided by Peng and Aston (2011). See also the relevant chapters in Hamilton (1994), Brockwell and Davis (1991, 2016), and Shumway and Stoffer (2000), and the filtering method discussed in Rao (2000).

This framework is also necessary for incorporating time-varying linear constraints into the time-varying regression model, as briefly discussed in the Remark at the end of Section 1.4.1. ■

5.6.2 The Hildreth–Houck Random Coefficient Model

The two influential papers of Rao (1965) and Hildreth and Houck (1968) studied estimation of, and inference based on, the regression model with random coefficients. They were not the first to propose the model, nor methods for estimation (see the references below, and in their papers), though their work has become associated with this random coefficient structure, and is often referred to as the Hildreth–Houck random coefficient (HHRC) model. It serves as an excellent starting point for more general structures, such as the Rosenberg formulation discussed in Section 5.6.4, and the much more general state space framework, mentioned above.

The HHRC model is given by

$$\begin{aligned} Y_t &= X_{t,1}(\beta_1 + V_{t,1}) + \cdots + X_{t,k}(\beta_k + V_{t,k}), \quad V_{t,i} \sim \mathrm{N}(0, \sigma_i^2) \\ &= \mathbf{x}_t'\boldsymbol{\beta}_t, \quad \boldsymbol{\beta}_t = \boldsymbol{\beta} + \mathbf{V}_t, \quad \mathbf{V}_t \overset{\text{i.i.d.}}{\sim} \mathrm{N}(\mathbf{0}, \boldsymbol{\Lambda}) \qquad (5.40) \\ &= \mathbf{x}_t'\boldsymbol{\beta} + U_t, \quad U_t = \sum_{i=1}^{k} X_{t,i}V_{t,i} = \mathbf{x}_t'\mathbf{V}_t, \qquad (5.41) \end{aligned}$$

or $\mathbf{Y} = \mathbf{X}\boldsymbol{\beta} + \mathbf{U}$, where, in our usual notation, $\mathbf{Y} = (Y_1, \ldots, Y_T)'$, $\mathbf{U} = (U_1, \ldots, U_T)'$, $\mathbf{V}_t = [V_{t,1}, \ldots, V_{t,k}]'$, and (also as usual) $\mathbf{x}_t = [X_{t,1}, \ldots, X_{t,k}]'$ is assumed fixed (or weakly exogenous), with $\mathbf{X} = [\mathbf{x}_1, \ldots, \mathbf{x}_T]'$ $T \times k$ of full rank. The standard HHRC model assumes that the $V_{t,i}$ are independent, $i = 1, \ldots, k$, $t = 1, \ldots, T$, so that $\boldsymbol{\Lambda}$ is diagonal, i.e.,

$$\boldsymbol{\Lambda} = \mathrm{diag}(\boldsymbol{\sigma}^{(2)}), \quad \boldsymbol{\sigma}^{(2)} = (\sigma_1^2, \ldots, \sigma_k^2)'.$$

Observe how there is no regression equation error term ϵ_t in (5.41), as in the usual regression model (5.1), because it is assumed that $X_{t,1} = 1$, in which case $V_{t,1}$ serves this purpose. Adding an ϵ_t with variance σ^2 would render parameters σ^2 and σ_1^2 to be unidentifiable.

As an illustration of data following the HHRC model, Figure 5.13 depicts simulated realizations from (5.41) based on a regression with intercept and time trend.

We now turn to the covariance structure and estimation of the HHRC model. Observe that $\mathbb{E}[\mathbf{U}] = \mathbf{0}$ and

$$\mathbf{H}(\boldsymbol{\Lambda}) := \mathbb{V}(\mathbf{U}) = \mathbb{E}[\mathbf{U}\mathbf{U}'] = \mathrm{diag}(\mathbf{h}) = \mathbf{X}\boldsymbol{\Lambda}\mathbf{X}' \odot \mathbf{I}_T, \qquad (5.42)$$

linear models] with known variances, the recurrence relationships for sequential updating of posterior distributions are essentially equivalent to the Kalman filter [...]. It was clearly not, as many people appear to believe, that Bayesian Forecasting is founded upon Kalman Filtering [...]. To say that 'Bayesian Forecasting is Kalman Filtering' is akin to saying that statistical inference is regression!"

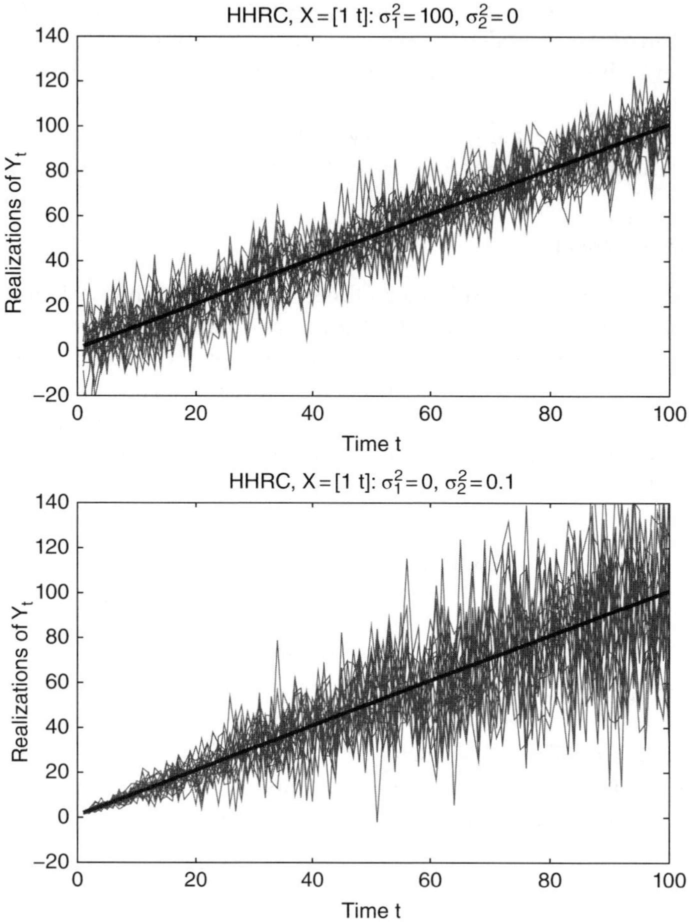

Figure 5.13 Twenty realizations of model (5.41) with $\mathbf{X} = [\mathbf{1},\ \mathbf{t}]$, $\boldsymbol{\beta} = (1, 1)'$, and σ_1^2 and σ_2^2 as indicated in the titles. The thick solid line is the true mean of Y_t, obtained by setting $\sigma_1^2 = \sigma_2^2 = 0$.

where $\mathbf{h} := (H_1, \dots, H_T)'$, because, for $t \neq s$, $\mathbb{E}[U_t U_s] = 0$, while, as $V_{t,i} \perp V_{t,j}$ for $i \neq j$,

$$H_t := \mathbb{E}[U_t^2] = \mathbb{E}\left[\left(\sum_{i=1}^{k} X_{t,i} V_{t,i}\right)\left(\sum_{j=1}^{k} X_{t,j} V_{t,j}\right)\right] = \mathbb{E}\left[\sum_{i=1}^{k} X_{t,i}^2 V_{t,i}^2\right]$$

$$= \sum_{i=1}^{k} X_{t,i}^2 \mathbb{E}[V_{t,i}^2] = \sum_{i=1}^{k} X_{t,i}^2 \sigma_i^2 = \mathbf{x}_t' \boldsymbol{\Lambda} \mathbf{x}_t. \tag{5.43}$$

A natural generalization is to let $\boldsymbol{\Lambda}$ in (5.41) be any positive semi-definite covariance matrix, as studied in Nelder (1968) and Swamy (1971). Observe how, while (5.41) and the usual linear regression model (5.1) have the same conditional means, the latter has constant variance for all t, whereas the variance for model (5.41) depends on t and $\mathbf{x}_t$, and is thus a **heteroskedastic** regression model. This distinction is sometimes reflected in referring to (5.1) and (5.41) as **regression models of the first and second kind**, respectively; see Fisk (1967) and Nelder (1968).

Remark Nelder (1968, p. 304) also provides a nice motivation for the use of a regression model of the second kind by considering (in line with much of the work of Ronald Fisher) an example from agricultural statistics: "Consider, for example, an agricultural experiment with fertilizers; if x is the amount of fertilizer applied and y the yield, one can think of the field plot as a black box converting input x into output y, and assert that, whereas we know fairly exactly how much fertilizer we put in and how much yield we got out from each plot, what we do not know are the parameters of the individual black boxes (plots) that did the conversion. Thus, assuming a linear relation for simplicity, we are led to a model of the second kind with $y_i = b_{0i} + b_{1i}x_i$, where b_{0i} and b_{1i} define the conversion process over the plots with means β_0 and β_1 and a variance matrix [in our notation] $\boldsymbol{\Lambda}$." ■

If $\boldsymbol{\Lambda}$ is known, then the usual generalized least squares solution (1.28) is applicable to determine $\widehat{\boldsymbol{\beta}}$, as

$$\widehat{\boldsymbol{\beta}}_{\Lambda} = \widehat{\boldsymbol{\beta}}_{\text{GLS}}(\boldsymbol{\Lambda}) = (\mathbf{X}'\mathbf{H}(\boldsymbol{\Lambda})\mathbf{X})^{-1}\mathbf{X}'\mathbf{H}(\boldsymbol{\Lambda})\mathbf{Y}. \tag{5.44}$$

If, far more likely, $\boldsymbol{\Lambda}$ is not known, then the exact likelihood is easily expressed, given that $\mathbf{Y} \sim \text{N}(\mathbf{X}\boldsymbol{\beta}, \mathbf{H})$. In this case, as (5.44) is the m.l.e. of $\boldsymbol{\beta}$ given $\boldsymbol{\Lambda}$, we can use (5.44) to form the **concentrated likelihood**, as first noted by Rubin (1950) in this context, and given by

$$\mathcal{L}(\boldsymbol{\Lambda}; \mathbf{Y}) = \frac{1}{|\mathbf{H}|^{1/2}(2\pi)^{T/2}} \exp\left\{-\frac{1}{2}((\mathbf{Y} - \mathbf{X}\widehat{\boldsymbol{\beta}}_{\Lambda})'\mathbf{H}^{-1}(\mathbf{Y} - \mathbf{X}\widehat{\boldsymbol{\beta}}_{\Lambda}))\right\}. \tag{5.45}$$

Thus, numeric maximization needs to be applied only over the $k(k+1)/2$ unique terms in $\boldsymbol{\Lambda}$ in the general case, or the k variance terms in $\boldsymbol{\Lambda}$, for the diagonal HHRC case.

Remark Unfortunately, as discussed in Zaman (2002), for the case with general positive semi-definite $\boldsymbol{\Lambda}$, the likelihood suffers from the same issue as with discrete mixtures of normals (recall Section III.5.1.3) in that the likelihood can tend to infinity. For the (more typical) case of diagonal $\boldsymbol{\Lambda}$, if the elements of $\mathbf{X}$ are from a continuous distribution such that the probability of any element being zero is zero, then, as the sample size increases, the probability of encountering one of the singularities during numeric estimation decreases. In general with this model, maximum likelihood estimation can behave poorly, as reported in Froehlich (1973) and Dent and Hildreth (1977); see also the simulation results below. ■

We wish to first develop a least squares estimator for $\boldsymbol{\Lambda}$, as in Thiel and Mennes (1959) (and Hildreth and Houck, 1968; Froehlich, 1973; and Crockett, 1985). To this end, let $\mathbf{R} = \mathbf{MY}$ be the o.l.s. residual vector, where $\mathbf{M} = \mathbf{I}_T - \mathbf{X}(\mathbf{X}'\mathbf{X})^{-1}\mathbf{X}'$ from (1.53). As $\mathbf{R} = \mathbf{MY} = \mathbf{MU}$, $\mathbb{E}[\mathbf{R}] = \mathbf{0}$ and $\mathbb{V}(\mathbf{R}) = \mathbf{MHM}$. Denote by $\dot{\mathbf{Z}}$ the elementwise squares of each element of the matrix, i.e., $\dot{\mathbf{Z}} = \mathbf{Z} \odot \mathbf{Z}$, where $\odot$ denotes the Hadamard, or elementwise product.[11] To proceed, we will require the following two basic results:

Theorem 5.2 For $\mathbf{A}$ and $\mathbf{B}$ $T \times T$ matrices, and $\mathbf{U} \sim (\mathbf{0}, \mathbf{H})$ of length T,

$$\mathbb{E}[\mathbf{AU} \odot \mathbf{BU}] = \text{diag}(\mathbf{A}\mathbb{E}[\mathbf{UU}']\mathbf{B}'). \tag{5.46}$$

11 This is the notation used in several research papers on the HHRC model, though more formal, and general, notation for Hadamard multiplication would be $\mathbf{Z}^{\circ 2}$, as suggested by Reams (1999).

Proof: This is the elementary observation that, for $T \times 1$ vectors $\mathbf{x}$ and $\mathbf{y}$, $\mathbf{x} \odot \mathbf{y} = \text{diag}(\mathbf{x}\mathbf{y}')$. With $\mathbf{U}$ a $T \times 1$ vector, $\mathbf{AU} \odot \mathbf{BU} = \text{diag}((\mathbf{AU})(\mathbf{BU})')$. Thus, (5.46) follows because $(\mathbf{BU})' = \mathbf{U}'\mathbf{B}'$ and the linearity of expectation. ■

Theorem 5.3 Let $\mathbf{A}$ and $\mathbf{B}$ be $m \times n$ matrices, and let $\mathbf{H}$ be an $n \times n$ diagonal matrix with diagonal entries given by vector $\mathbf{h}$, i.e., $\mathbf{H} = \text{diag}(\mathbf{h})$. Then

$$\text{diag}(\mathbf{AHB}') = (\mathbf{A} \odot \mathbf{B})\mathbf{h}. \tag{5.47}$$

Proof: Writing out both sides confirms the result. See Horn (1994, p. 305) for details and further related results. ■

We now have, from (5.46) and (5.47),

$$\mathbb{E}[\dot{\mathbf{R}}] = \mathbb{E}[\mathbf{MU} \odot \mathbf{MU}] = \text{diag}(\mathbf{M}\mathbb{E}[\mathbf{UU}']\mathbf{M}') = \text{diag}(\mathbf{MHM}) = (\mathbf{M} \odot \mathbf{M})\mathbf{h} = \dot{\mathbf{M}}\mathbf{h}, \tag{5.48}$$

as stated in Hildreth and Houck (1968) without proof.

Remark Observe from (5.42) that we can write $\mathbf{h} = \mathbb{E}[\mathbf{U} \odot \mathbf{U}]$. One might thus wonder if we can obtain (5.48) directly, from the conjecture that, for $T \times T$ matrices $\mathbf{A}$ and $\mathbf{B}$,

$$\mathbb{E}[\mathbf{AU} \odot \mathbf{BU}] \stackrel{?}{=} \mathbb{E}[(\mathbf{A} \odot \mathbf{B})(\mathbf{U} \odot \mathbf{U})] = (\mathbf{A} \odot \mathbf{B})\mathbb{E}[\mathbf{U} \odot \mathbf{U}] = (\mathbf{A} \odot \mathbf{B})\mathbf{h},$$

which would be the case if the elegant-looking result $\mathbf{AU} \odot \mathbf{BU} \stackrel{?}{=} (\mathbf{A} \odot \mathbf{B})(\mathbf{U} \odot \mathbf{U})$ were true. The reader can confirm numerically that this is not the case in general, and also not when taking $\mathbf{A}$ and $\mathbf{B}$ both to be a projection matrix $\mathbf{M}$. ■

From (5.43) and (5.48),

$$\dot{\mathbf{R}} = \dot{\mathbf{M}}\mathbf{h} + \boldsymbol{\epsilon} = \dot{\mathbf{M}}\dot{\mathbf{X}}\boldsymbol{\sigma}^{(2)} + \boldsymbol{\epsilon}, \tag{5.49}$$

where the error term $\boldsymbol{\epsilon} = (\epsilon_1, \dots, \epsilon_T)'$ denotes the discrepancy between $\dot{\mathbf{R}}$ and $\mathbb{E}[\dot{\mathbf{R}}]$. Thus, o.l.s. can be applied to (5.49) to obtain estimator

$$\tilde{\boldsymbol{\sigma}}^{(2)}_{\text{OLS}} = (\dot{\mathbf{X}}'\dot{\mathbf{M}}^2\dot{\mathbf{X}})^{-1}\dot{\mathbf{X}}'\dot{\mathbf{M}}\dot{\mathbf{R}}. \tag{5.50}$$

An alternative estimator, from Rao (1968), though also proposed in Hildreth and Houck (1968), and derived in detail in Froehlich (1973), is

$$\tilde{\boldsymbol{\sigma}}^{(2)}_{\text{MQ}} = (\dot{\mathbf{X}}'\dot{\mathbf{M}}\dot{\mathbf{X}})^{-1}\dot{\mathbf{X}}'\dot{\mathbf{R}}. \tag{5.51}$$

This is the so-called *minimum norm quadratic unbiased estimation* estimator, or MINQUE, as coined by Rao (1968). Both (5.50) and (5.51) are consistent estimators, as shown by Hildreth and Houck (1968), while their asymptotic normality is proven in Crockett (1985) and Anh (1988). As (5.50) or (5.51) could contain negative elements, we take $\hat{\boldsymbol{\sigma}}^{(2)} = \max(\tilde{\boldsymbol{\sigma}}^{(2)}, \mathbf{0}_k)$. Constrained optimization could also be used to avoid the latter construct, as discussed in Hildreth and Houck (1968) and Froehlich (1973).

The following iterated least-squares estimation procedure (for diagonal $\mathbf{\Lambda}$) then suggests itself: Compute the o.l.s. residuals $\mathbf{R} = \mathbf{MY}$ and then $\widehat{\boldsymbol{\sigma}}^{(2)}$. Next, with $\widehat{\mathbf{\Lambda}} = \text{diag}(\widehat{\boldsymbol{\sigma}}^{(2)})$, take

$$\widehat{\boldsymbol{\beta}}(\widehat{\mathbf{\Lambda}}) = (\mathbf{X}'\mathbf{H}(\widehat{\mathbf{\Lambda}})\mathbf{X})^{-1}\mathbf{X}'\mathbf{H}(\widehat{\mathbf{\Lambda}})\mathbf{Y} \tag{5.52}$$

from the g.l.s. estimator (5.44). Observe that one could then iterate on $\widehat{\boldsymbol{\beta}}(\widehat{\mathbf{\Lambda}})$ and $\widehat{\boldsymbol{\sigma}}^{(2)}$ until convergence, though in our code and simulation below we perform only two iterations. By the nature of least squares and projection, the distribution of $\widehat{\boldsymbol{\beta}} - \boldsymbol{\beta}$ is invariant to the choice of $\boldsymbol{\beta}$.

As shown in Griffiths (1972) and Lee and Griffiths (1979) (see also Judge et al., 1985, p. 807, and the references therein), for known $\mathbf{\Lambda}$, the minimum variance unbiased estimator of $\boldsymbol{\beta}_t$ in (5.40) is *not* $\widehat{\boldsymbol{\beta}}_{\mathbf{\Lambda}}$ in (5.44), but rather given by, with $H_t = \mathbf{x}_t'\mathbf{\Lambda}\mathbf{x}_t$ from (5.43),

$$\widehat{\boldsymbol{\beta}}_t = \widehat{\boldsymbol{\beta}}_{\mathbf{\Lambda}} + \frac{Y_t - \mathbf{x}_t'\widehat{\boldsymbol{\beta}}_{\mathbf{\Lambda}}}{H_t}\mathbf{\Lambda}\mathbf{x}_t.$$

In practice, $\mathbf{\Lambda}$ would be replaced by $\widehat{\mathbf{\Lambda}}$.

We turn now to the small-sample distribution of the estimators, obtained by simulation. Figure 5.14 shows histograms of the least squares estimates of the parameters based on $T = 100$, $\mathbf{X} = [\mathbf{1},\ \mathbf{t}]$, $\boldsymbol{\beta} = (0, 0)'$, $\sigma_1^2 = 10$, $\sigma_2^2 = 0.1$, and 10,000 replications. We see that, except for $\hat{\sigma}_1^2$, they are close to unbiased and reasonably Gaussian in shape, though the right tail of $\hat{\sigma}_2^2$ is somewhat elongated. The least squares estimator can be used to obtain starting values for the m.l.e., computed based on (5.44) and (5.45). Figure 5.15 shows the resulting histograms based on the m.l.e. The distribution of $\hat{\beta}_1$ exhibits an (unexplained) bimodality. On a more positive note, $\hat{\sigma}_1^2$ has much less pile-up at zero, and its mode is

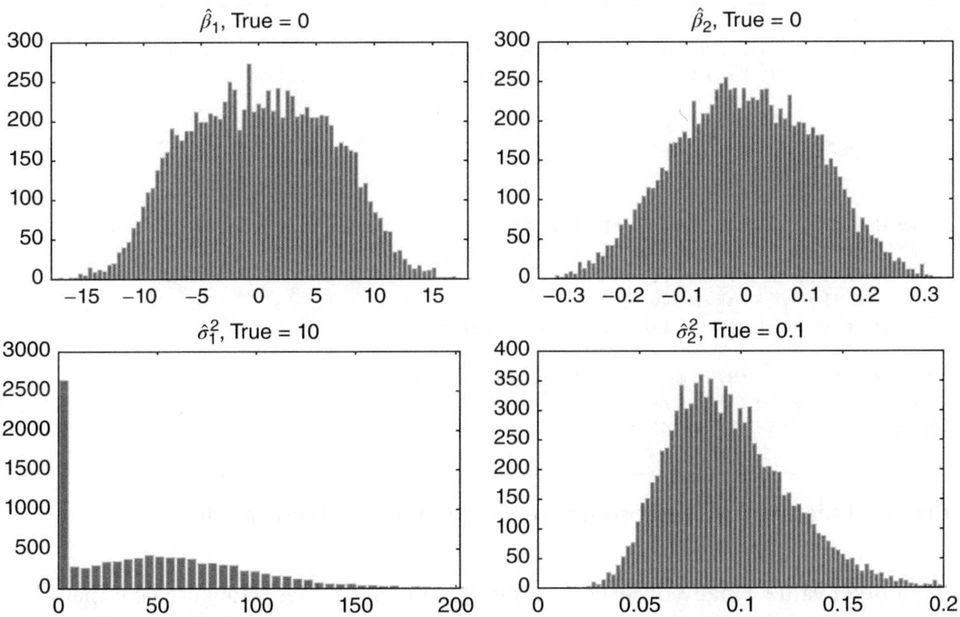

Figure 5.14 Histograms of the least squares estimators for the HHRC model based on $T = 100$, $\mathbf{X} = [\mathbf{1},\ \mathbf{t}]$, $\boldsymbol{\beta} = (0, 0)'$, $\sigma_1^2 = 10$, $\sigma_2^2 = 0.1$, and 10,000 replications. For $\hat{\sigma}_1^2$, about 25% of the estimates were zero.

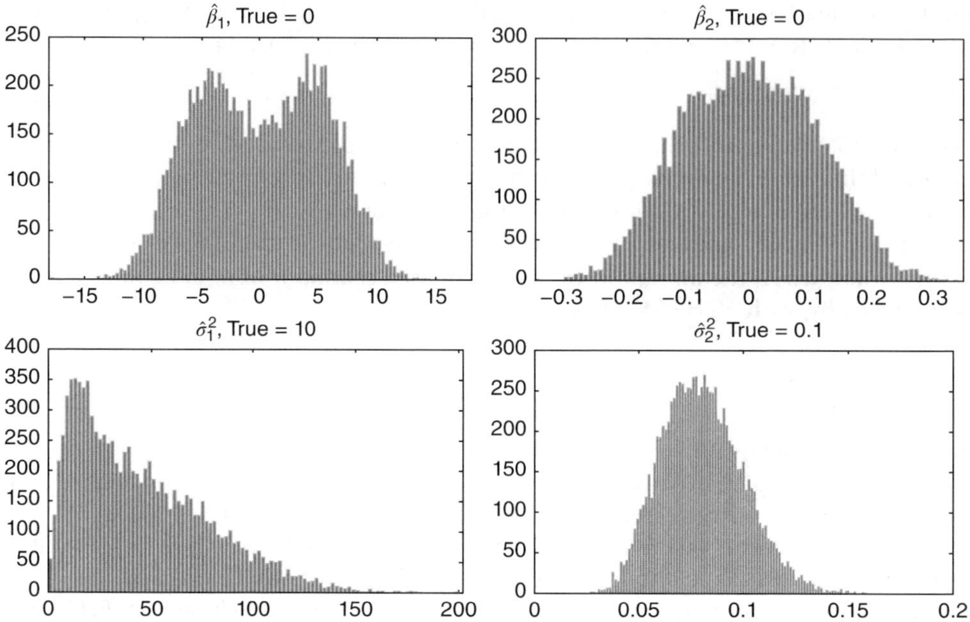

Figure 5.15 Same as Figure 5.14 but based on the m.l.e.

```
function [betahat,Sighat]=HHRCOLS(Y,X)
if nargin<1 % simulate data of an intercept-trend model
  T=100; X=[ones(T,1), (1:T)']; s1=10; s2=0.1; % sigma^2_1 and sigma^2_2
  beta1true=0; beta2true=0;
  beta1=beta1true+sqrt(s1)*randn(T,1);
  beta2=beta2true+sqrt(s2)*randn(T,1);
  Y = sum( X .* [beta1, beta2] , 2);
end
[T,k]=size(X); M=makeM(X); R=M*Y; Z=(M.*M)*(X.*X);
Sighat=max(inv(Z'*Z)*Z'*(R.*R), zeros(k,1)); %#ok<MINV>
Lam=diag(Sighat); H=(X*Lam*X').*eye(T);
betahat=inv(X'*H*X)*X'*H*Y; %#ok<MINV>
if 1==1 % could stop there, but do one more iteration
  R=Y-X*betahat;
  Sighat=max(inv(Z'*Z)*Z'*(R.*R), zeros(k,1)); %#ok<MINV>
  Lam=diag(Sighat); H=(X*Lam*X').*eye(T);
  betahat=inv(X'*H*X)*X'*H*Y; %#ok<MINV>
end
```

Program Listing 5.8: Least squares estimation of the HHRC model.

close to the true parameter value. Finally, observe that the right tail of $\hat{\sigma}_2^2$ is less elongated compared to the least squares estimator, resulting in the m.l.e. having lower variance and being closer to Gaussian.

The program in Listing 5.8 computes the least squares estimates (and optionally simulates the process), while that in Listing 5.9 is for the m.l.e.

```
function [betahat,sighat]=HHRCMLE(Y,X)
[~,SighatOLS]=HHRCOLS(Y,X);
if SighatOLS(1) <= 1e-6, SighatOLS(1)=1; end
if SighatOLS(2) <= 1e-6, SighatOLS(2)=0.01; end
initvec = [min(SighatOLS(1), 190) min(SighatOLS(2), 0.28)];
bound.lo=[0 0]; bound.hi=[200 0.3]; bound.which=[1 1];
maxiter=200; tol=1e-5;
opts=optimset('Display','off', 'Maxiter',maxiter, ...
      'TolFun',tol,'TolX',tol,'LargeScale','Off');
[pout,~,~,~,~,hess] = fminunc(@(param) ...
    HHRCLik(param,Y,X,bound), einschrk(initvec,bound), opts);
V=inv(hess); [param,V]=einschrk(pout,bound,V);
sighat=param'; Lam=diag(sighat); H=(X*Lam*X').*eye(T);
betahat=inv(X'*H*X)*X'*H*Y; %#ok<MINV>

function ll=HHRCLik(param,Y,X,bound)
if nargin<4, bound=0; end
if any(isinf(param)) || any(isnan(param)) || any(~isreal(param))
  paramvec=[1 0.01];
else
  if isstruct(bound)
    paramvec=einschrk(param,bound,999);
  else
    paramvec=param;
  end
end
s1=paramvec(1); s2=paramvec(2); sigv=[s1,s2];
Lam=diag(sigv); [T,~]=size(X); H=(X*Lam*X').*eye(T);
beta=inv(X'*H*X)*X'*H*Y; %#ok<MINV>
h=diag(H); h=max(1e-12, h); H=diag(h);
if 1==2
  f=mvnpdf(Y,X*beta,H); ll=log(f);
else
  Z=Y-X*beta; Hinv=diag(1./h);
  ll=-0.5*sum(log(h))-0.5*Z'*Hinv*Z;
end
ll = -sum(ll);
```

Program Listing 5.9: Maximum likelihood estimation of the HHRC model.

Repeating the exercise with $T = 1{,}000$ (results not shown) yielded estimates for β_2 and σ_2^2 around their true values, with lower dispersion than for $T = 100$ and, for the least squares estimator, the empirical distribution of $\hat{\sigma}_2^2$ was highly Gaussian in appearance. However, for both estimators, the dispersion of $\hat{\beta}_1$ *increased*, and $\hat{\sigma}_1^2$ often took on very large values. The qualitatively same results were obtained for the m.l.e. when using the true parameter values as starting values, instead of the least squares estimates. This finding appears to contradict the consistency results for the least squares and maximum likelihood estimators. Note that this model, with $\mathbf{X} = [\mathbf{1},\ \mathbf{t}]$, has $\mathbb{V}(Y_t) = \mathbb{V}(U_t) = H_t = \sigma_1^2 + t^2\sigma_2^2$, from (5.43), as depicted in the bottom panel of Figure 5.13. Recalling that $X_{t,1} = 1$ and $V_{t,1} \sim \mathrm{N}(0, \sigma_1^2)$ takes the role of the usual error term in the regression model, it appears that, with the ever-increasing variance of U_t as t grows, estimation of intercept coefficient β_1 might not become more accurate as t grows.

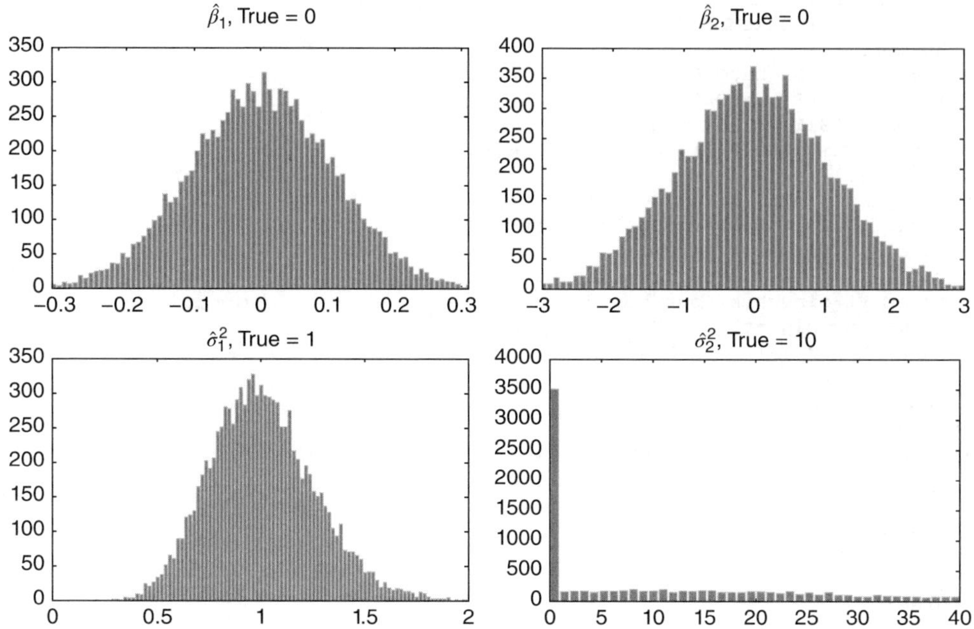

Figure 5.16 Histograms of the least squares estimators for the HHRC model based on $T = 100$, $\mathbf{X} = [\mathbf{1},\ \mathbf{v}]$, where $\mathbf{v}$ is the eigenvector $\mathbf{v}_i$ in (B.10) with $i = \text{round}(T/3)$, $\boldsymbol{\beta} = (0, 0)'$, $\sigma_1^2 = 1$, $\sigma_2^2 = 10$, and 10,000 replications. For $\hat{\sigma}_2^2$, about 33% of the estimates were zero.

To help corroborate this, the experiment was repeated, but using $\mathbf{X} = [\mathbf{1},\ \mathbf{v}]$, where $\mathbf{v}$ is the eigenvector $\mathbf{v}_i$ in (B.10) with $i = \text{round}(T/3)$; see Example B.5. As the magnitude of $\mathbf{v}$ does not grow with t, one might expect different results. This is indeed the case: Based on the least squares estimator, histograms of the point estimates are shown in Figure 5.16. Compared to the case with $\mathbf{X} = [\mathbf{1},\ \mathbf{t}]$, we now see that σ_1^2 is estimated very accurately and has a near-Gaussian distribution, while σ_2^2 has a large pile-up at zero, and is otherwise far too large.

To study the case with $\mathbf{X} = [\mathbf{1},\ \mathbf{t}]$ further, assume β_1 and β_2 are known and, without loss of generality, let $\beta_1 = \beta_2 = 0$, so that $Y_t = U_t$, with $\mathbb{V}(U_t) = \sigma_1^2 + t^2\sigma_2^2$. The top panel of Figure 5.17 shows a histogram of the computed m.l.e.s of σ_1^2, assuming both $\boldsymbol{\beta} = (\beta_1,\ \beta_2)'$ and σ_2^2 are known, based on $T = 1{,}000$ and 1,000 replications. Observe in this case, that

$$\ell(\sigma_1^2) = -\frac{1}{2}\sum_{t=1}^{T} \ln(\sigma_1^2 + t\sigma_2^2) - \frac{1}{2}\sum_{t=1}^{T} \frac{Y_t^2}{\sigma_1^2 + t\sigma_2^2},$$

and, with $H_t = \sigma_1^2 + t\sigma_2^2 = \mathbb{E}[Y_t^2]$,

$$\dot{\ell}_1 := \frac{\mathrm{d}}{\mathrm{d}\sigma_1^2}\ell(\sigma_1^2) = -\frac{1}{2}\sum_{t=1}^{T} \frac{1}{\sigma_1^2 + t\sigma_2^2} + \frac{1}{2}\sum_{t=1}^{T} \frac{Y_t^2}{(\sigma_1^2 + t\sigma_2^2)^2} = \frac{1}{2}\sum_{t=1}^{T}\left(\frac{Y_t^2 - H_t}{H_t^2}\right),$$

so that $\hat{\sigma}_1^2$ could be determined by (numerically) solving $\dot{\ell}_1 = 0$.

The bottom panels of Figure 5.17 similarly show histograms of $\hat{\sigma}_1^2$ and $\hat{\sigma}_2^2$ assuming known $\boldsymbol{\beta} = (0,\ 0)'$. The behavior of $\hat{\sigma}_1^2$ is, in both cases, demonstrably better, suggesting that, for this model

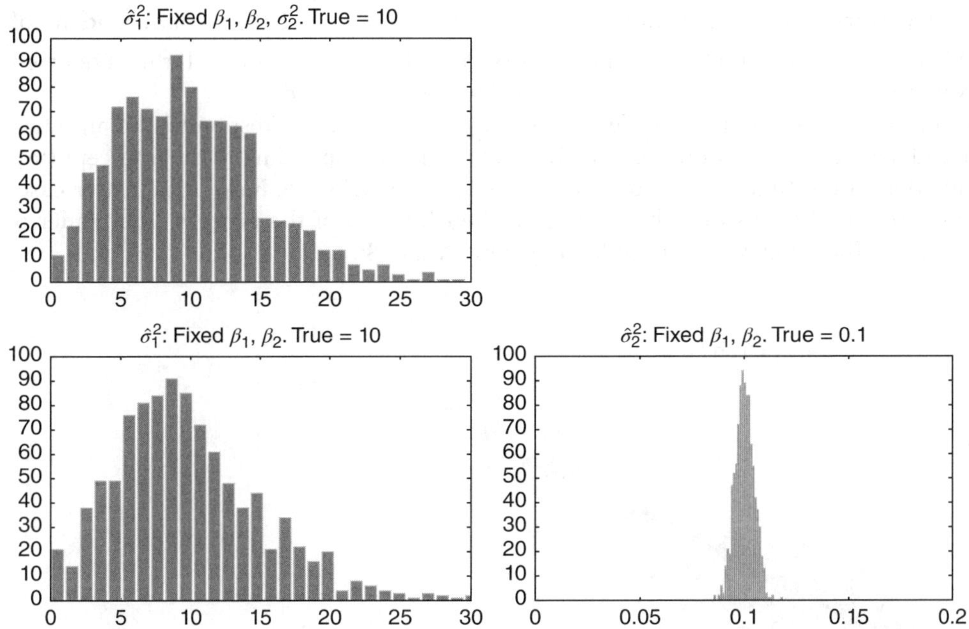

Figure 5.17 Histograms of the m.l.e. for the HHRC model for $\mathbf{X} = [\mathbf{1},\ \mathbf{t}]$, $\boldsymbol{\beta} = (0, 0)'$, $\sigma_1^2 = 10$, $\sigma_2^2 = 0.1$, but now based on $T = 1{,}000$, and 1,000 replications, and assuming known β_1 and β_2. The top panel further assumes σ_2^2 is also known.

and choice of $\mathbf{X}$ matrix, there could be an identifiability issue with β_1 and σ_1^2. This finding behooves consideration of constrained optimization, such as a restriction on the sum of the variance terms, as done, for example, by Zaman (2002) in his simulation study. Another potential approach is use of shrinkage estimation, possibly along the lines of Hamilton (1991), as also noted by Zaman (2002). A more formal Bayesian approach also suggests itself; see, e.g., Liu and Hanssens (1981). Finally, one could entertain a different model that has better understood estimation properties, such as the random walk coefficient model, as considered in the next section.

Testing the null of the usual regression model versus the variance structure of the HHRC is a special case of heteroskedasticity tests in "classic" regression analysis; see, e.g., the excellent (and at the time, state of the art) presentation in Judge et al. (1985), as well as Evans and King (1985, 1988), who provide point optimal tests expressible as ratios of quadratic forms, as will be detailed below in Sections 5.6.3.2 and 5.6.4 in the context of different TVP regression models.

5.6.3 The TVP Random Walk Model

Generalizing the baseline regression model (1.3), the linear model with coefficients that are time-varying and evolve as a random walk is given by

$$Y_t = \mathbf{x}_t'\boldsymbol{\alpha}_t + \epsilon_t, \quad \boldsymbol{\alpha}_t = \boldsymbol{\alpha}_{t-1} + \mathbf{U}_t, \tag{5.53}$$

for known set of vectors $\mathbf{x}_t \in \mathbb{R}^k$, and unknown $\boldsymbol{\alpha}_t \in \mathbb{R}^k$, where $\epsilon_t \overset{\text{i.i.d.}}{\sim} \text{N}(0, \sigma^2)$ independent of $\mathbf{U}_t \overset{\text{i.i.d.}}{\sim} \text{N}(\mathbf{0}, \lambda\sigma^2\mathbf{\Psi})$, for $\sigma_2 > 0$, $\lambda \geqslant 0$, and covariance matrix $\mathbf{\Psi}$. When $k = 1$ and $x_t = 1$, this is referred to as a **local level model**, and a specification of the distribution of α_0 is made.

Figure 5.18 shows sample realizations of the model using an $\mathbf{X}$ matrix corresponding to an intercept and time trend, i.e., $\mathbf{X} = [\mathbf{1}, \ \mathbf{t}]$, allowing only the coefficient corresponding to the intercept (top graphic) and only that of the time trend (bottom graphic) to vary. Observe how, in the latter case, use of only $\lambda = 0.001$ can induce so much variation in the evolution of the process. The reader is encouraged to replicate these figures, using code for general $\mathbf{X}$ and $\mathbf{\Psi}$.

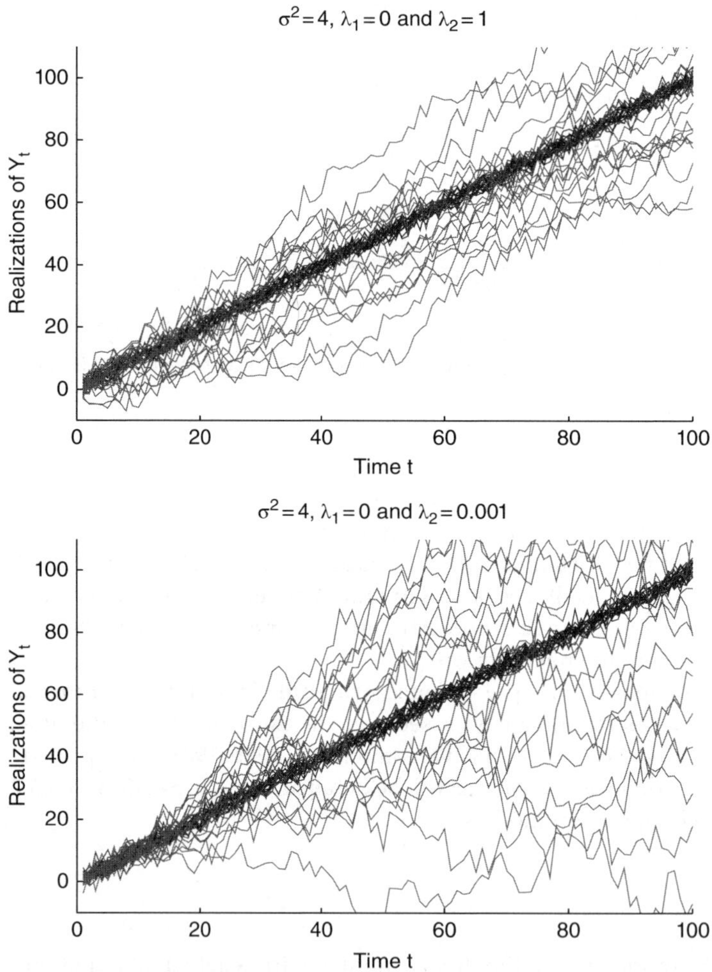

Figure 5.18 **Top:** Twenty realizations of model (5.53) with $\mathbf{X} = [\mathbf{1}, \ \mathbf{t}]$, $\boldsymbol{\alpha}_0 = [0, \ 1]'$, $\sigma^2 = 4$, $\mathbf{\Psi} = \text{diag}([1, 0])$, and two values of λ. **Bottom:** Same, but having used $\mathbf{\Psi} = \text{diag}([0, 1])$.

We first detail the covariance structure and estimation of the model in Section 5.6.3.1, followed by methods for testing for this form of time variation in Section 5.6.3.2.

5.6.3.1 Covariance Structure and Estimation

Let $\tilde{\boldsymbol{\alpha}}_t := \boldsymbol{\alpha}_t - \boldsymbol{\alpha}_0$, so that $\tilde{\boldsymbol{\alpha}}_t = \tilde{\boldsymbol{\alpha}}_{t-1} + \mathbf{U}_t$. Observe that $\tilde{\boldsymbol{\alpha}}_0 = \mathbf{0}$ and $\mathbb{E}[\boldsymbol{\alpha}_t] = \mathbb{E}[\tilde{\boldsymbol{\alpha}}_t] = \mathbf{0}$. Setting $V_t := \mathbf{x}_t'\tilde{\boldsymbol{\alpha}}_t + \epsilon_t$, the observation equation can be expressed as $Y_t = \mathbf{x}_t'\boldsymbol{\alpha}_0 + V_t$ or, with $\mathbf{Y} = (Y_1, \dots, Y_T)'$, $\mathbf{X} = (\mathbf{x}_1, \dots, \mathbf{x}_T)'$, and $\mathbf{V} = (V_1, \dots, V_T)'$, as

$$\mathbf{Y} = \mathbf{X}\boldsymbol{\alpha}_0 + \mathbf{V}, \quad \mathbb{V}(\mathbf{V}) = \sigma^2\boldsymbol{\Omega}(\lambda), \quad \boldsymbol{\Omega}(\lambda) = \lambda\mathbf{W} \odot (\mathbf{X}\boldsymbol{\Psi}\mathbf{X}') + \mathbf{I}, \tag{5.54}$$

where $\boldsymbol{\alpha}_0$ is an unknown constant, and $[\mathbf{W}]_{s,t} = \min(s,t)$. This follows because $\mathbb{E}[V_t] = 0$ and, for $t < s$, using (II.3.6) gives

$$\begin{aligned}\mathrm{Cov}(V_t, V_s) &= \mathbb{E}[(\mathbf{x}_t'\tilde{\boldsymbol{\alpha}}_t + \epsilon_t)(\mathbf{x}_s'\tilde{\boldsymbol{\alpha}}_s + \epsilon_s)] = \mathbb{E}[(\mathbf{x}_t'\tilde{\boldsymbol{\alpha}}_t)(\mathbf{x}_s'\tilde{\boldsymbol{\alpha}}_s)'] = \mathbf{x}_t'\mathbb{E}[\tilde{\boldsymbol{\alpha}}_t\tilde{\boldsymbol{\alpha}}_s']\mathbf{x}_s \\ &= \mathbf{x}_t'\mathbb{E}[\tilde{\boldsymbol{\alpha}}_t(\tilde{\boldsymbol{\alpha}}_t + \mathbf{U}_{t+1} + \cdots + \mathbf{U}_s)']\mathbf{x}_s = \mathbf{x}_t'\mathbb{V}(\tilde{\boldsymbol{\alpha}}_t)\mathbf{x}_s = t\lambda\sigma^2\mathbf{x}_t'\boldsymbol{\Psi}\mathbf{x}_s.\end{aligned}$$

Similarly, for $s < t$, $\mathrm{Cov}(V_t, V_s) = s\lambda\sigma^2\mathbf{x}_t'\boldsymbol{\Psi}\mathbf{x}_s$, and

$$\mathbb{V}(V_t) = \mathbb{E}[(\mathbf{x}_t'\tilde{\boldsymbol{\alpha}}_t + \epsilon_t)(\mathbf{x}_t'\tilde{\boldsymbol{\alpha}}_t + \epsilon_t)] = \mathbf{x}_t'\mathbb{E}[\tilde{\boldsymbol{\alpha}}_t\tilde{\boldsymbol{\alpha}}_t']\mathbf{x}_t + \mathbb{V}(\epsilon_t) = \sigma^2 + t\lambda\sigma^2\mathbf{x}_t'\boldsymbol{\Psi}\mathbf{x}_s.$$

Remark We might attempt to use features of Matlab to compute matrix $\mathbf{W}$ as fast as possible. Letting $[\mathbf{E}]_{t,s} = \mathbb{I}(t \geqslant s)$ be the lower triangular matrix with all nonzero entries equal to one, $\mathbf{W}$ can be expressed as $\mathbf{E}\mathbf{E}'$. Matrix $\mathbf{E}$ can be computed in several ways, such as, with T defined:

a) `E=zeros(T,T); for t=1:T, for s=1:T, E(t,s)=(t>=s); end, end`
b) `E=zeros(T,T); for i=1:T, for j=1:i, E(i,j)=1; end, end`
c) `E=toeplitz(ones(T,1),[1 zeros(1,T-1)]);`
d) `E=toeplitz(ones(T,1),zeros(1,T));` % correct but generates a warning message and then, $\mathbf{W}$ is computed as `W=E*E'`. Of these four, option (b) is the fastest. Alternatively, $\mathbf{W}$ can be computed directly via
e) `W=diag(1:T)/2; for t=1:T, for s=(t+1):T, W(t,s)=min(t,s); end, end, W=W+W';`
f) `W=zeros(T,T); for t=1:T, for s=1:T, W(t,s)=min(t,s); end, end`

where (e) attempts to capitalize on the symmetry of $\mathbf{W}$. It turns out that, of all the methods, the simple "brute force" way (f) is the fastest. ■

As $\mathbf{Y} \sim \mathrm{N}(\mathbf{X}\boldsymbol{\alpha}_0, \sigma^2\boldsymbol{\Omega}(\lambda))$, the log-likelihood for a given regressor $\mathbf{X}$ matrix and set of data $(Y_1, \dots, Y_T)'$ is easily expressed and numeric techniques can be used to compute the m.l.e. As in Cooley and Prescott (1973), and as done for the HHRC model in Section 5.6.2 above, it suggests itself to use the known m.l.e.s of $\boldsymbol{\alpha}_0$ and σ^2 from the generalized least squares solutions (1.28) and (1.30),

$$\hat{\boldsymbol{\alpha}}_0(\lambda) = (\mathbf{X}'\boldsymbol{\Omega}(\lambda)\mathbf{X})^{-1}\mathbf{X}'\boldsymbol{\Omega}(\lambda)\mathbf{Y}, \quad \hat{\sigma}^2(\lambda) = T^{-1}(\mathbf{Y} - \mathbf{X}\hat{\boldsymbol{\alpha}}_0(\lambda))'\boldsymbol{\Omega}(\lambda)(\mathbf{Y} - \mathbf{X}\hat{\boldsymbol{\alpha}}_0(\lambda)),$$

to form the concentrated likelihood, given by (suppressing the notational dependence of $\hat{\boldsymbol{\alpha}}_0$, $\hat{\sigma}^2$, and $\boldsymbol{\Omega}$ on λ)

$$\mathcal{L}(\lambda; \mathbf{Y}) = \frac{1}{(\hat{\sigma}^2)^{T/2}|\boldsymbol{\Omega}|^{1/2}(2\pi)^{T/2}} \exp\left\{-\frac{1}{2}((\mathbf{Y} - \mathbf{X}\hat{\boldsymbol{\alpha}}_0)'\boldsymbol{\Omega}^{-1}(\mathbf{Y} - \mathbf{X}\hat{\boldsymbol{\alpha}}_0))\right\},$$

or, simplifying, taking logs, and omitting the (2π) term as it does not depend on λ,

$$\ell(\lambda; \mathbf{Y}) = -\frac{T}{2} \ln \hat{\sigma}^2(\lambda) - \frac{1}{2} \ln|\mathbf{\Omega}(\lambda)|. \tag{5.55}$$

Thus, numeric maximization needs to be applied only over λ.[12]

To illustrate, Figure 5.19 shows kernel density plots of $\hat{\lambda}_{\mathrm{ML}}$ based on $T = 100$, $\mathbf{X} = [\mathbf{1},\ \mathbf{t}]$, and $\mathbf{\Psi} = \mathrm{diag}([1, 0])$. For each of the two values of λ, the simulation was run for three values of σ^2, confirming that $\hat{\lambda}_{\mathrm{ML}}$ is independent of σ^2. An upper bound of 30 was imposed on $\hat{\lambda}_{\mathrm{ML}}$. While the mode of $\hat{\lambda}_{\mathrm{ML}}$ is close to the true value, we see that, as λ increases, so does the probability that $\hat{\lambda}_{\mathrm{ML}}$ can assume large

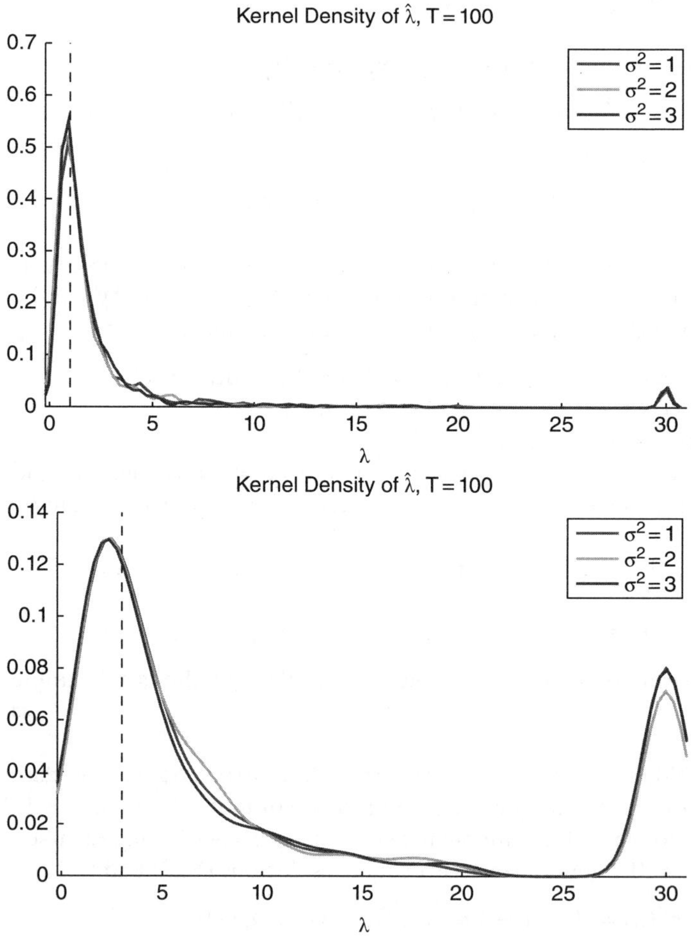

Figure 5.19 Kernel density plots of $\hat{\lambda}_{\mathrm{ML}}$ based on 1,000 replications, sample size $T = 100$, $\mathbf{X} = [\mathbf{1},\ \mathbf{t}]$, $\mathbf{\Psi} = \mathrm{diag}([1, 0])$, and $\lambda = 1$ (top) and $\lambda = 3$ (bottom).

12 In case the elements of $\mathbf{\Psi}$ are not specified, they can also be estimated, but observe that they are not all identified because of the multiplicative factor λ. One could omit λ and let $\mathrm{diag}(\mathbf{\Psi}) = (\lambda_1, \ldots, \lambda_k)$.

values far into the right tail. More specifically, for $\lambda = 3$, 775 of the $\hat{\lambda}_{ML}$-values lie between 0 and 20, while 225 piled up at the imposed border of 30. Thus, there is over a 20% chance of m.l.e. failure. Inspection based on one of the "failed" data sets shows that the log-likelihood increases quickly as λ increases from 0 to about 10, and then is relatively flat (for values far beyond the imposed estimation upper limit of 30). It thus appears that, for some data sets, $\ell(\lambda; \mathbf{Y})$ can increase in λ without bound (though it is nearly flat).

If we restrict attention to the cases such that the $\hat{\lambda}_{ML}$ did not hit the imposed upper bound of 30, it is of interest to see the behavior of the approximate standard errors of $\hat{\lambda}_{ML}$, as they could be used to form approximate (Wald) confidence intervals for λ that would presumably improve as T increases and the distribution of $\hat{\lambda}_{ML}$ is closer to Gaussian (observe the large skewness in Figure 5.19). Scatterplots of the $\hat{\lambda}_{ML}$-values versus their corresponding approximate standard errors returned from the Hessian-based optimization algorithm, overlaid with a fitted regression line, are shown in Figure 5.20 for the two cases $\lambda = 1$ and $\lambda = 3$. For the latter, the fitted regression of the standard error (SE) is

$$\text{SE} \approx -0.2611 + 0.5834\hat{\lambda}_{ML} + 0.0908\hat{\lambda}_{ML}^2, \tag{5.56}$$

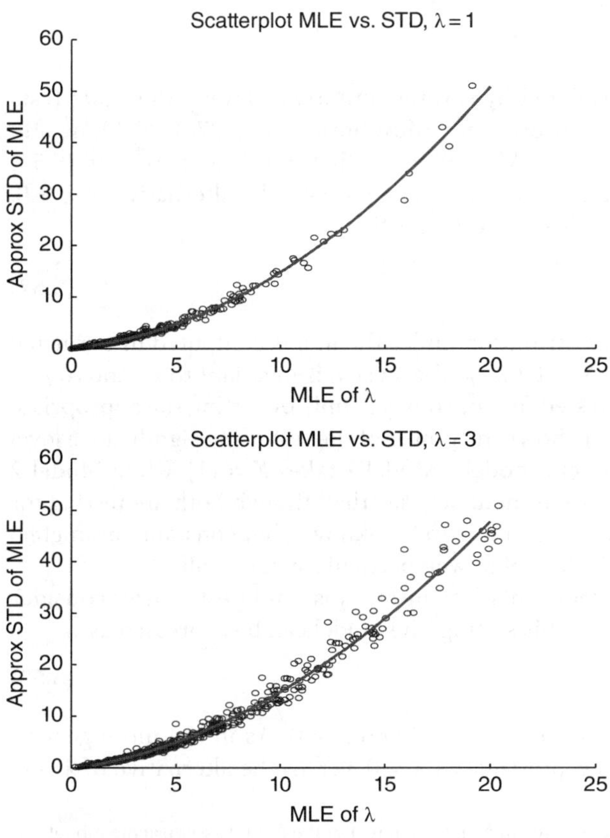

Figure 5.20 Scatterplots showing the approximate standard error as a function of $\hat{\lambda}_{ML}$, where the values used correspond to the $\hat{\lambda}_{ML}$ used in Figure 5.19 but having been truncated such that $\hat{\lambda}_{ML} < 25$.

with $R^2 = 0.98$. Thus, when $\hat{\lambda}_{\rm ML}$ is close to (the true value of) 3, the approximate standard error is about 2.3, implying that zero will be in the 90% (and obviously 95%) Wald confidence interval. More generally, this is the case for $\hat{\lambda}_{\rm ML} \leqslant 11$ (and $\leqslant 15$ for a 95% c.i.). The parametric bootstrap in this case will not be of much help: Even if we (i) get lucky enough that $\hat{\lambda}_{\rm ML} \approx \lambda$ and (ii) we reject the bth bootstrap replication if $\hat{\lambda}_{\rm ML}^{(b)}$ hits the upper bound of some reasonably imposed constraint, we see that the sampling variation of the $\hat{\lambda}_{\rm ML}^{(b)}$ is still very high. In this case (for $\lambda = 3$), the sample standard error of the 775 "valid" replications is 3.9, which is almost double that suggested by (5.56). As mentioned, the Wald c.i.s are anyway a poor choice in this context because of the strong asymmetry of the distribution of $\hat{\lambda}_{\rm ML}^{(b)}$.

The method of generating a confidence interval from use of a hypothesis test statistic (shown at the end of Section 5.2) using the test given in (5.57) below, was not successful in this case, as the reader is invited to confirm. A possible solution to this issue is developed in Stock and Watson (1998), who develop asymptotically valid confidence intervals and median-unbiased point estimators for λ in model (5.53).

A point forecast of Y_{t+h} requires knowledge of $\hat{\boldsymbol{\alpha}}_T$, and this is best obtained by use of filtering methods, as discussed in Remark (**b**) in the beginning of Section 5.6.

5.6.3.2 Testing for Parameter Constancy

Testing for TVP in model (5.53) has been considered by several authors, including the exact tests (known small-sample distribution theory) of LaMotte and McWhorter, Jr. (1978), Nyblom and Mäkeläinen (1983), and Nabeya and Tanaka (1988).[13] We will detail that of the latter two. In (5.53) and (5.54), the null hypothesis of constant $\boldsymbol{\alpha}_t$ is expressible as $\lambda = 0$, versus the alternative of $\lambda > 0$. Then, *for* $\boldsymbol{\Psi}$ *known*, direct application of (5.18) yields the test statistic

$$L = \frac{\mathbf{Y}'\mathbf{M}[\mathbf{W} \odot (\mathbf{X}\boldsymbol{\Psi}\mathbf{X}')]\mathbf{M}\mathbf{Y}}{\mathbf{Y}'\mathbf{M}\mathbf{Y}}, \tag{5.57}$$

and the null is rejected for large values of L. Its distribution under the null is computed in our usual way for ratios of quadratic forms, and such that $\mathbf{Y} \sim \mathrm{N}(\mathbf{0}, \mathbf{I}_T)$ (because L is invariant to σ^2 and $\boldsymbol{\alpha}_0$).

Observe that a subset of the vector $\boldsymbol{\alpha}_t$ can be tested for constancy simply by setting the appropriate elements of $\boldsymbol{\Psi}$ to zero. As an example, Figure 5.21 shows the power of the test using significance level $\alpha = 0.05$ and three sample sizes T, for two different models. Model 1 takes $\mathbf{X} = [\mathbf{1}]$, while Model 2 refers to use of $\mathbf{X} = [\mathbf{1},\ \mathbf{t}]$ and $\boldsymbol{\Psi} = \mathrm{diag}([1, 0])$. We immediately see that, though both are testing for parameter variation in the intercept of the regression, adding the (unknown, but constant parameter) time trend to the regression induces a substantial loss of power, particularly for small T.

The case of testing only one of the regression coefficients for constancy is of interest, and we consider it in more detail, showing a more powerful test. In this setting, the model can be expressed as

$$Y_t = x_t\alpha_t + \mathbf{z}_t'\boldsymbol{\beta} + \epsilon_t, \quad \alpha_t = \alpha_{t-1} + U_t, \tag{5.58}$$

where $\epsilon_t \overset{\text{i.i.d.}}{\sim} \mathrm{N}(0, \sigma^2)$ independent of $U_t \overset{\text{i.i.d.}}{\sim} \mathrm{N}(0, \lambda\sigma^2)$, for $\sigma_2 > 0$ and $\lambda \geqslant 0$. As in the more general case (5.53), the null hypothesis of constant α_t is expressible as $\lambda = 0$, versus the alternative of $\lambda > 0$.

---

13 A discussion and illustration of various tests for parameter constancy in the context of the capital asset pricing model (CAPM) is given in Wells (1996, Ch. 2). Dangl and Halling (2012) investigate the out-of-sample predictive performance of regression models with random walk coefficients for the monthly returns on the S&P 500 index, demonstrating its efficacy in this context.

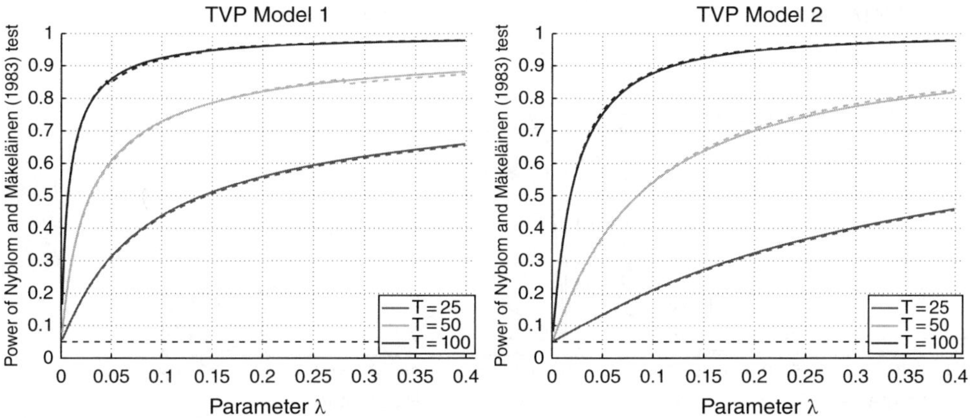

Figure 5.21 The power of the Nyblom and Mäkeläinen (1983) test (5.57) for three sample sizes and two models, as described in the text. Solid (dashed) lines were computed using the exact (s.p.a.) method; they are essentially indistinguishable, with use of the s.p.a. being about three times faster.

We illustrate the test from Shively (1988a) for model (5.58), which was shown to have higher power than other tests in this context. The setup is the same as that of Nyblom and Mäkeläinen (1983), namely use of a locally most powerful test, but such that a different point for the optimality is used.

As before, with $\tilde{\alpha}_t = \alpha_t - \alpha_0$ and $V_t = x_t\tilde{\alpha}_t + \epsilon_t$, the model can be expressed as $\tilde{\alpha}_t = \tilde{\alpha}_{t-1} + U_t$ and $Y_t = x_t\alpha_0 + \mathbf{z}_t'\boldsymbol{\beta} + V_t$, or, in obvious matrix notation,

$$\mathbf{Y} = \begin{bmatrix} \mathbf{x} & \mathbf{Z} \end{bmatrix} \begin{bmatrix} \alpha_0 \\ \boldsymbol{\beta} \end{bmatrix} + \mathbf{V}, \quad \mathbb{V}(\mathbf{V}) = \sigma^2\boldsymbol{\Omega}(\lambda), \quad \boldsymbol{\Omega}(\lambda) = \lambda\mathbf{DWD} + \mathbf{I}, \tag{5.59}$$

where $\mathbf{D} = \text{diag}(\mathbf{x})$. The derivation of $\mathbb{V}(\mathbf{V})$ is similar to that of the more general case above in (5.54), but can be seen directly as follows: As $\mathbb{E}[\alpha_t] = \mathbb{E}[\tilde{\alpha}_t] = 0$,

$$\mathbb{V}(V_t) = x_t^2\mathbb{V}(\tilde{\alpha}_t) + \mathbb{V}(\epsilon_t) = \sigma^2(x_t^2\lambda t + 1),$$

and, for $t < s$,

$$\begin{aligned} \text{Cov}(V_t, V_s) &= \mathbb{E}[(x_t\tilde{\alpha}_t + \epsilon_t)(x_s\tilde{\alpha}_s + \epsilon_s)] = x_t x_s\mathbb{E}[\tilde{\alpha}_t\tilde{\alpha}_s] \\ &= x_t x_s\mathbb{E}[\tilde{\alpha}_t(\tilde{\alpha}_t + U_{t+1} + \cdots + U_s)] = x_t x_s\mathbb{V}(\tilde{\alpha}_t) = \sigma^2 x_t x_s\lambda t, \end{aligned}$$

or, for all t and s, $\text{Cov}(V_t, V_s) = \sigma^2(x_t x_s\lambda\min(t, s) + \mathbb{I}\{t = s\})$, which is (5.59).

The null and alternative are $\mathbf{V} \sim \text{N}(\mathbf{0}, \sigma^2\mathbf{I})$ and $\mathbf{V} \sim \text{N}(\mathbf{0}, \sigma^2\boldsymbol{\Omega}(\lambda))$, respectively. Then, similar to the construction of tests (5.22) and (5.35), a POI test statistic for some particular value of λ, say, λ^*, is given by

$$T(\lambda^*) = \frac{\mathbf{Y}'\mathbf{G}'[\mathbf{G}\boldsymbol{\Omega}(\lambda^*)\mathbf{G}']^{-1}\mathbf{GY}}{\mathbf{Y}'\mathbf{MY}}, \tag{5.60}$$

where $\mathbf{M}$ and $\mathbf{G}$ are computed in the usual way, based on matrix $[\mathbf{x} \;\; \mathbf{Z}]$, and we reject the null for small values of $T(\lambda^*)$. Via use of (1.65), this is stated in Nyblom and Mäkeläinen (1983, Eq. (2.5)), though they chose to use the test statistic (5.57), for which power is maximized around $\lambda = 0$. Similar to proposals

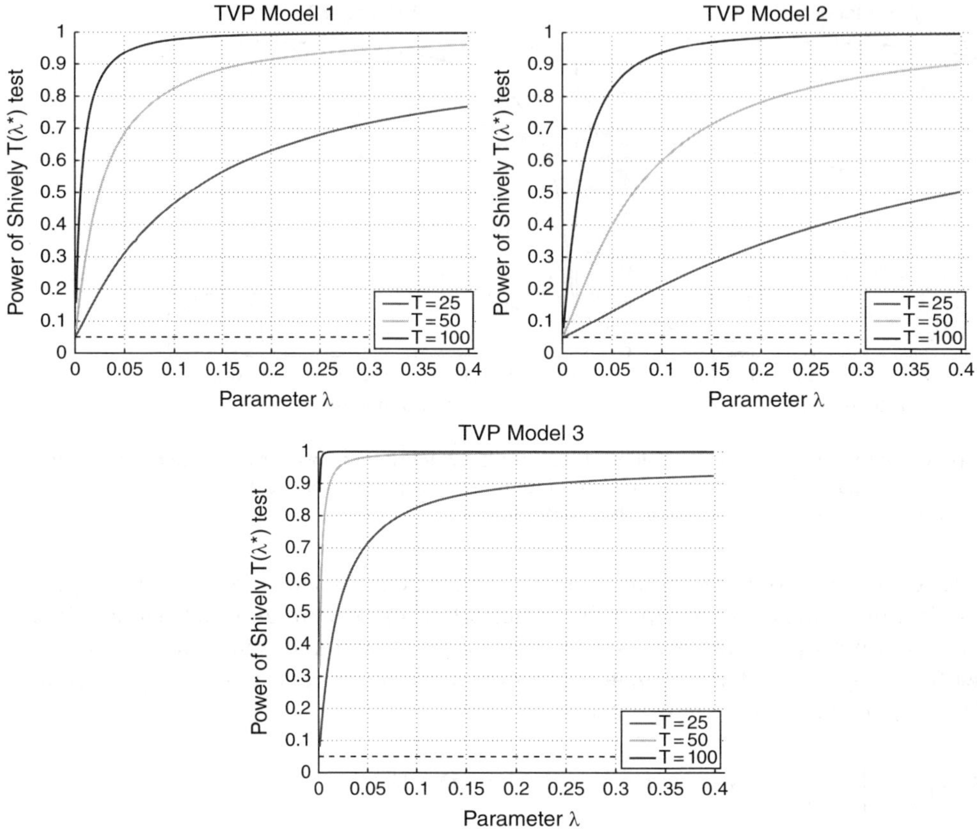

Figure 5.22 The power of the Shively (1988a) test (5.60) and the value of λ^* such that the power of the test based on $T(\lambda^*)$ is 0.5 when the true λ equals λ^*, for three sample sizes and three models, as described in the text.

in King (1985a) and Franzini and Harvey (1983), Shively (1988a) suggests choosing the value of λ^* in (5.60) such that the power of the test based on $T(\lambda^*)$ when the true λ equals λ^* is 0.5. Observe that, as usual, $T(\lambda^*)$ is invariant to scale changes in $\mathbf{Y}$, i.e., σ^2 cancels from the numerator and denominator.

Figure 5.22 shows the power for three sample sizes and three design matrices. Model 1 is as above, with $x_t = 1$ and no $\mathbf{Z}$ matrix, as also studied in Shively (1988a). Model 2 is the same as above, with $x_t = 1$ and $z_t = t$. Thus, the first two panels of Figure 5.22 can be compared to those in Figure 5.21. We see that the test based on (5.60) is indeed more powerful.

Finally, model 3 takes $x_t = \sqrt{t}$ and $z_t = 1$. An attempt with $x_t = t$ and $z_t = 1$ did not work (for any sample size) because of numeric problems obtaining λ^*, though this is not too surprising in light of Figure 5.18, which shows that relatively very small values of λ need to be used. The reader is encouraged to implement this test and replicate our shown results: We suggest use of Matlab's `fminbnd` function for use in determining λ^*.

For model 1, computation of λ^* for a grid of values between sample sizes $T = 20$ and $T = 120$ reveals that λ^* is smoothly decreasing in T, and can be approximated as

$$\text{Model 1:} \qquad \lambda^* \approx -0.0469 - \frac{6.4190}{T} + \frac{1.0371}{T^{1/2}} + \frac{130.2943}{T^2},$$

yielding a regression R^2 of just over 0.9999. Its use results in the power being graphically identical to the top panel in Figure 5.22. A similar exercise yields

$$\text{Model 2:} \qquad \lambda^* \approx -0.3972 - \frac{51.8997}{T} + \frac{8.6150}{T^{1/2}} + \frac{714.3107}{T^2}.$$

Returning to the problematic case of the desired model 3 with $x_t = t$ and $z_t = 1$ but using either of the previous approximations to λ^* reveals that the power is relatively very high, reaching unity for $\lambda \approx 0.03$ ($\lambda \approx 0.002$) for $T = 25$ ($T = 50$).

5.6.4 Rosenberg Return to Normalcy Model

Rosenberg (1973) proposed a model that nests both (5.41) and (a limiting case of) (5.53), such that $\boldsymbol{\alpha}_t$ is not a random walk under the null, but rather a stationary (vector) AR(1) process. For the case with only one possibly TVP, the model can be expressed as

$$Y_t = x_t\alpha_t + \mathbf{z}_t'\boldsymbol{\beta} + \epsilon_t, \quad \alpha_t - \mu = \phi(\alpha_{t-1} - \mu) + U_t, \quad |\phi| < 1, \tag{5.61}$$

where $\epsilon_t \overset{\text{i.i.d.}}{\sim} \text{N}(0, \sigma^2)$ independent of $U_t \overset{\text{i.i.d.}}{\sim} \text{N}(0, \lambda\sigma^2)$, for $\sigma_2 > 0$ and $\lambda \geqslant 0$. This is now often referred to as the **(Rosenberg) return-to-normalcy model**, or just the Rosenberg model. With $\phi = 0$, the model reduces to the HHRC model (5.41), while with $\phi = 1$, it reduces to the random walk model (5.53). Recalling (4.10), the law of motion for α_t can be expressed as $\alpha_t = (1 - \phi)\mu + \phi\alpha_{t-1} + U_t$, showing that α_t is a weighted average of μ and α_{t-1}, such that the weights sum to one. As (5.61) nests both the HHRC and the baseline (constant regressor) regression models, there are two possible null hypotheses.

We concentrate on the first null hypothesis of interest, $H_0 : \lambda = 0$, in which case, along with $\alpha_0 = \mu$, (5.61) reduces to the usual baseline linear regression model. We wish to test the null of $\lambda = 0$ versus the stochastic coefficient case of $\{\lambda > 0, |\phi| < 1\}$. This testing situation is somewhat more challenging because parameter ϕ is not identified under the null, but only under the alternative, so that the usual likelihood ratio test cannot be applied. A test was operationalized in Watson and Engle (1985) by applying the method proposed by Davies (1977, 1987), which builds on the work of Roy (1953). Its small sample distribution is not tractable, but Watson and Engle (1985) provide a method for calculating the critical value of the test such that the size of the test is asymptotically bounded. Alternative approaches were developed in King (1987a) and Shively (1988b), yielding tests expressible as ratios of quadratic forms (in normal variables) and thus can make use of exact small-sample inference. We detail that of Shively (1988b), as he demonstrates that his test has higher power for many models and alternatives of interest.

As before, the model can be expressed as $Y_t = x_t\mu + \mathbf{z}_t'\boldsymbol{\beta} + V_t$, $V_t = x_t(\alpha_t - \mu) + \epsilon_t$, $\alpha_t - \mu = \phi(\alpha_{t-1} - \mu) + U_t$, or, in matrix notation, and recalling (4.13),

$$\mathbf{Y} = \begin{bmatrix} \mathbf{x} & \mathbf{Z} \end{bmatrix} \begin{bmatrix} \mu \\ \boldsymbol{\beta} \end{bmatrix} + \mathbf{V}, \quad \mathbb{V}(\mathbf{V}) = \sigma^2\boldsymbol{\Omega}(\lambda, \phi), \quad \boldsymbol{\Omega}(\lambda, \phi) = \lambda\mathbf{D}\boldsymbol{\Sigma}(\phi)\mathbf{D} + \mathbf{I},$$

where $\mathbf{D} = \text{diag}(\mathbf{x})$ and $\mathbf{\Sigma}(\phi)$ is the covariance matrix of a stationary AR(1) model, as in (4.13), i.e., $[\mathbf{\Sigma}(\phi)]_{s,t} = \phi^{|s-t|}/(1-\phi^2)$.

Analogous to the test statistic (5.22), Shively (1988b) proposes a POI test for a specified alternative $\{\lambda^*, \phi^*\}$. He suggests use of $\phi^* = 0.7$ and, similar to the method of Shively (1988a) for model (5.58), choosing λ^* such that the power of the test based on statistic $T(\lambda^*, \phi^*)$ at $\{\lambda^*, 0.7\}$ is 0.5, where

$$T(\lambda^*, \phi^*) = \frac{\mathbf{Y}'\mathbf{G}'[\mathbf{G}\mathbf{\Omega}(\lambda^*, \phi^*)\mathbf{G}']^{-1}\mathbf{G}\mathbf{Y}}{\mathbf{Y}'\mathbf{M}\mathbf{Y}}, \tag{5.62}$$

and we reject for small values of $T(\lambda^*, \phi^*)$. This can be set up precisely as with test statistic (5.60), and the reader is encouraged to do so.

Figure 5.23 is similar to Figure 5.22, showing the power of the $T(\lambda^*, \phi^*)$ test for the same three type of models, but in the form (5.61), e.g., Model 1 is given by

$$Y_t = \alpha_t + \epsilon_t, \quad \alpha_t - \mu = \phi(\alpha_{t-1} - \mu) + U_t, \quad |\phi| < 1, \tag{5.63}$$

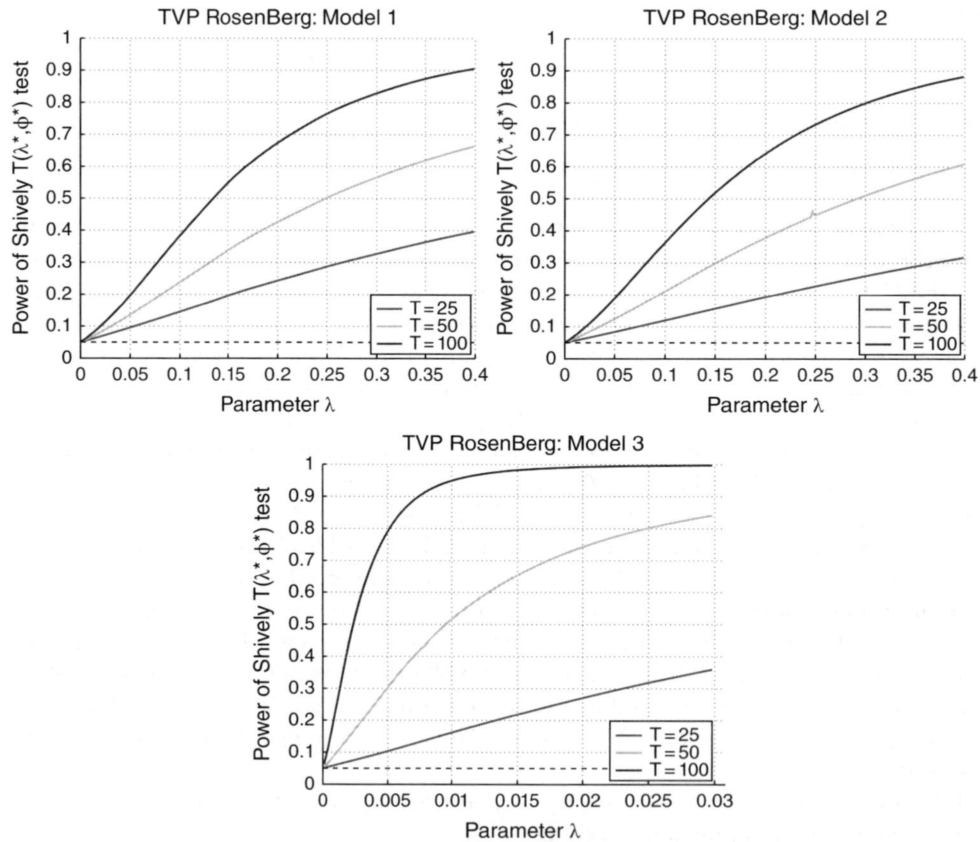

Figure 5.23 The power of the Shively (1988b) test (5.62) and the value of λ^* such that the power of the test based on $T(\lambda^*, \phi^*)$ is 0.5 when the true λ equals λ^*, for three sample sizes and three models, as described in the text.

$t = 1, \dots, T$.[14] Comparing Figures 5.22 and 5.23, we see that the power associated with the use of the return-to-normalcy alternative model is much lower than with the random walk alternative model.

Remarks

a) Brooks (1993) proposes an alternative method of selecting $\{\lambda^*, \phi^*\}$ in the Shively (1988b) test, based on the idea of Cox and Hinkley (1974, p. 102) to maximize some weighted average of powers.
b) Shively (1988b) also provides a highly accurate approximation to the small-sample distribution of the test statistic used in Watson and Engle (1985), and develops a test similar to (5.62) but for the ARIMA(1,1,0) alternative model

$$Y_t = x_t \alpha_t + \mathbf{z}_t' \boldsymbol{\beta} + \epsilon_t, \quad \alpha_t = \alpha_{t-1} + \phi(\alpha_{t-1} - \alpha_{t-2}) + U_t, \quad |\phi| < 1. \tag{5.64}$$

c) The second null hypothesis is the HHRC model, versus the more general formulation (5.61). Brooks and King (1994) (see also Brooks, 1995, 1997) propose a test statistic similar to the above ones, expressible as a ratio of quadratic forms, so that our usual computational machinery is applicable. The decision between the two models has been considered by Bos and Newbold (1984) and Brooks et al. (1994) in the context of the capital asset pricing model.
d) The more general alternative hypothesis involving all, or a subset of, the regression coefficients evolving according to a stationary **vector autoregressive process**, denoted VAR(p), and such that p is known, has been investigated by Lin and Teräsvirta (1999). ■

14 The test using this model was also demonstrated in Shively (1988b) for $T = 31$. For this sample size, we obtain the same value, $\lambda^* = 0.436$, and the same power for the select values he reports in his Table 1.

6

Autoregressive and Moving Average Processes

Statements about parameter values have been discussed as if parameters have a clearly-defined tangible existence, whereas in most cases, they are at best mathematical artifacts introduced only in order to provide the most useful approximation available to the behaviour of the underlying reality. It is all too easy to lose sight of the fact that the real purpose of the analysis is to make statements about this reality rather than about the models that approximate it.

(James Durbin, 1987, p. 179)

There are many extensions of the AR(1) model, a very natural one of which is to include more lagged terms, yielding the AR(p) model. Another important one is to consider lags of the error term U_t, giving rise to moving average, or MA(q), models. In this chapter, these two models will be introduced and methods for their estimation discussed.

6.1 AR(p) Processes

A natural generalization of the AR(1) model (4.1) is to allow more past values of Y_t into the equation; this is called the **autoregressive model of order** p, or AR(p) model, given by

$$Y_t = c + a_1 Y_{t-1} + a_2 Y_{t-2} + \cdots + a_p Y_{t-p} + U_t, \tag{6.1}$$

where here and throughout the chapter,

$$U_t \overset{\text{i.i.d.}}{\sim} \mathrm{N}(0, \sigma^2), \tag{6.2}$$

as in (4.1). Model (6.1) can be more compactly expressed as $a(L)Y_t = c + U_t$, where

$$a(L) = 1 - a_1 L - \cdots - a_p L^p \tag{6.3}$$

is a polynomial in L, and L is the **lag operator** such that $LY_t = Y_{t-1}$. As one might imagine, the AR(p) parameterization allows a much richer class of dynamic behavior than the AR(1) case.

There are two popular conventions used for denoting the observations of an AR(p) time-series model. The first extends that used in the AR(1) case in Chapter 4: We label the observations $Y_{1-p}, Y_{2-p}, \ldots, Y_0, Y_1, \ldots, Y_T$, so that one has a total of $T + p$ observations available. For example, if $p = 2$, the sample is $Y_{-1}, Y_0, \ldots, Y_T$. This has the advantage that the o.l.s. estimator of the autoregressive terms uses T observations. The second notation just labels the sequence as $Y_1, \ldots, Y_T$, for a total

Linear Models and Time-Series Analysis: Regression, ANOVA, ARMA and GARCH, First Edition. Marc S. Paolella.

of T observations. Both notations have advantages and disadvantages. We primarily use the former, but in general choose the one that is more convenient for the task at hand.

6.1.1 Stationarity and Unit Root Processes

In the AR(1) case, the polynomial (6.3) is just $a(L) = 1 - a_1 L$. When L is treated as a variable, the solution to the equation $a(L) = 0$ is $1/a_1$, so that the stationarity condition can be stated as requiring that the root of the AR(1) polynomial is greater than one in absolute value. This carries over to the AR(p) case: The model is stationary when the moduli of all p (possibly complex) roots of the polynomial $a(L)$ are greater than one. If the complex numbers are plotted in the usual fashion, then this is equivalent to requiring that the roots lie outside of the complex unit circle.

For the special case with $p = 2$, the simple quadratic formula can be used: Treating L as a variable gives

$$a(L) = 1 - a_1 L - a_2 L^2 = (1 - \lambda_1 L)(1 - \lambda_2 L),$$

where λ_1 and λ_2 are so defined, and the roots of $a(L)$ are thus λ_1^{-1} and λ_2^{-1}. Multiplying this by L^{-2} and setting $\lambda = L^{-1}$ gives $\lambda^2 - a_1\lambda - a_2 = (\lambda - \lambda_1)(\lambda - \lambda_2)$, with solution

$$\lambda_{1,2} = \frac{1}{2}a_1 \pm \frac{1}{2}\sqrt{a_1^2 + 4a_2}. \tag{6.4}$$

The roots are complex if $a_1^2 + 4a_2 < 0$, and real otherwise. In general, express $a(L)$ as

$$a(L) = 1 - a_1 L - \cdots - a_p L^p = (1 - \lambda_1 L) \cdot \cdots \cdot (1 - \lambda_p L), \tag{6.5}$$

with roots $\lambda_1^{-1}, \dots, \lambda_p^{-1}$. These can be computed numerically. The model is stationary when $|\lambda_i^{-1}| > 1$ or, equivalently, when $|\lambda_i| < 1$, $i = 1, \dots, p$, where $|\lambda_i|$ is the modulus of λ_i.

Remark The roots are easily computed in Matlab using the built-in `roots` function. For example, if the model is an AR(2) with parameters $a_1 = 1.2$ and $a_2 = -0.8$, then executing `rr=roots([0.8 -1.2 1])` returns the two complex roots $0.75 \pm 0.8292i$. The modulus is computed as `abs(rr)`, giving in this case 1.1180 for both roots, so that the model is stationary. In general, if $a = (a_1, \dots, a_p)$ is the autoregressive parameter vector, then executing `rr=roots([-a(end:-1:1) 1])` returns the p roots. ■

If $|\lambda_i| < 1$, $i = 1, \dots, p-1$, and $\lambda_p = 1$, then the process $a(L)Y_t = c + U_t$ has a unit root (recall Section 5.5) and will resemble a random walk, with drift if $c \neq 0$. The process can be written as $(1 - \lambda_1 L) \cdot \cdots \cdot (1 - \lambda_{p-1} L)(1 - L)Y_t = c + U_t$ or, with $X_t := (1 - L)Y_t = Y_t - Y_{t-1}$, as $(1 - \lambda_1 L) \cdot \cdots \cdot (1 - \lambda_{p-1} L)X_t = c + U_t$. That is, the first difference of an AR(p) process with a unit root is a stationary AR($p - 1$) process. The practical implication of this fact is that, when faced with time-series data that resemble a random walk, the first difference $X_t = Y_t - Y_{t-1}$ can be computed and subsequently analyzed to infer a guess for p (or for p and q in the case of a mixed ARMA(p, q) model), this being the topic of Chapter 9, and conduct parameter estimation, as discussed below and in Chapter 7.

While functions for the computation of polynomial roots are standard in numerical toolboxes and computing languages, there is a computationally less sophisticated method for determining if the AR polynomial corresponds to a stationary process, due originally to Schur (1917) and Cohn (1922):

A necessary and sufficient condition that all the roots of polynomial $\alpha(z) = \alpha_0 + \alpha_1 z + \cdots + \alpha_p z^p$ lie outside the unit circle is that

$$S_\alpha := \mathbf{T}_1\mathbf{T}_1' - \mathbf{T}_2'\mathbf{T}_2 > 0, \tag{6.6}$$

i.e., that S_α is positive definite, where $\mathbf{T}_1$ and $\mathbf{T}_2$ are the Toeplitz matrices given by

$$\mathbf{T}_1 = \begin{bmatrix} \alpha_0 & 0 & \cdots & 0 \\ \alpha_1 & \alpha_0 & \cdots & 0 \\ \vdots & \vdots & \ddots & \vdots \\ \alpha_{p-1} & \alpha_{p-2} & \cdots & \alpha_0 \end{bmatrix} \quad \text{and} \quad \mathbf{T}_2 = \begin{bmatrix} \alpha_p & \alpha_{p-1} & \cdots & \alpha_1 \\ 0 & \alpha_p & \cdots & \alpha_2 \\ \vdots & \vdots & \ddots & \vdots \\ 0 & 0 & \cdots & \alpha_p \end{bmatrix}. \tag{6.7}$$

We will refer to (6.6) as the Schur condition. By taking $\alpha_0 = 1$ and $\alpha_i = -a_i$, $i = 1, \ldots, p$, (6.6) can be computed for the AR polynomial.

Remark It seems little is gained by this, as positive definiteness of a symmetric matrix is equivalent to all the (necessarily real) eigenvalues being positive, and computation of the latter is essentially equivalent to polynomial root solving. However, another way to verify that S_α is positive definite is to check that all the **leading principle minors** of S_α are positive. Recall that, if $\mathbf{A}$ is an $n \times n$ matrix, the leading principal minor of $\mathbf{A}$ of order k, $1 \leqslant k \leqslant n$, is the determinant of the matrix obtained by deleting the last $n - k$ rows and columns of $\mathbf{A}$. Matrix $\mathbf{A}$ is positive definite if and only if all the leading principal minors are positive (see, e.g., Abadir and Magnus, 2005, p. 223).

Pagano (1973, p. 541) notes that, in the context of the Schur condition, the calculation of the leading principal minors can be numerically unstable. See also Pollock (1999, pp. 157–158) for more details on the Schur condition. ■

Example 6.1 For every $p \in \mathbb{N}$, the $(1, 1)$ element of S_α is easily seen to be given by $1 - a_p^2$, so that it is necessary that $|a_p| < 1$ for the AR(p) model to be stationary. ■

Example 6.2 In the $p = 1$ case, the Schur condition reduces to $1 - a_1^2 > 0$, or $|a_1| < 1$, as noted in Chapter 4. ■

Example 6.3 For $p = 2$, S_α is

$$\begin{bmatrix} 1 & 0 \\ -a_1 & 1 \end{bmatrix}\begin{bmatrix} 1 & -a_1 \\ 0 & 1 \end{bmatrix} - \begin{bmatrix} -a_2 & 0 \\ -a_1 & -a_2 \end{bmatrix}\begin{bmatrix} -a_2 & -a_1 \\ 0 & -a_2 \end{bmatrix} = \begin{bmatrix} 1 - a_2^2 & -a_1 - a_2 a_1 \\ -a_1 - a_2 a_1 & 1 - a_2^2 \end{bmatrix},$$

from which the two leading principle minors are

$$1 - a_2^2 \quad \text{and} \quad (a_2 + 1)^2(a_1 + a_2 - 1)(a_2 - a_1 - 1),$$

giving conditions $-1 < a_2 < 1$ and $(a_1 + a_2 - 1)(a_2 - a_1 - 1) > 0$. Expanding the latter, simplifying, and using the former, these can be written as

$$-1 < a_2 < 1, \quad |a_1| < 1 - a_2. \tag{6.8}$$

From the second of these,

$$a_2 - 1 < a_1 < 1 - a_2, \tag{6.9}$$

and adding a_2 shows that $2a_2 - 1 < a_1 + a_2 < 1$; in particular, that

$$a_1 + a_2 < 1 \tag{6.10}$$

is a necessary condition for stationarity.

From conditions $-1 < a_2 < 1$ and (6.9), it is necessary that $a_1 < 2$ for stationarity.

To see that the model has a unit root when $a_1 + a_2 = 1$ and $a_1 < 2$, let $a_2 = 1 - a_1$ and use (6.4) to compute the roots

$$\frac{1}{2}a_1 \pm \frac{1}{2}\sqrt{a_1^2 + 4(1 - a_1)} = \frac{1}{2}(a_1 \pm |2 - a_1|).$$

If $a_1 < 2$, then

$$\lambda_1 = \frac{1}{2}(a_1 + (2 - a_1)) = 1 \quad \text{and} \quad \lambda_2 = \frac{1}{2}(a_1 - (2 - a_1)) = a_1 - 1 < 1,$$

showing that there is exactly one unit root. ∎

Example 6.4 For the AR(3) process $Y_t = 1.2Y_{t-1} - 0.8Y_{t-2} + 0.59Y_{t-3} + U_t$, we construct the matrices

$$\mathbf{T}_1 = \begin{bmatrix} 1 & 0 & 0 \\ -1.2 & 1 & 0 \\ 0.8 & -1.2 & 1 \end{bmatrix}, \quad \mathbf{T}_2 = \begin{bmatrix} -0.59 & 0.80 & -1.20 \\ 0 & -0.59 & 0.80 \\ 0 & 0 & -0.59 \end{bmatrix},$$

and compute

$$S_\alpha = \begin{bmatrix} 0.652 & -0.728 & 0.092 \\ -0.728 & 1.452 & -0.728 \\ 0.092 & -0.728 & 0.652 \end{bmatrix},$$

with leading principle minors 0.6519, 0.4165, and 0.0113, showing that the process is stationary. In this case, the minimum eigenvalue of S_α is 0.0092 and the minimum of the modulus of the roots of the polynomial $1 - 1.2L + 0.8L^2 - 0.59L^3$ is 1.0073, also confirming stationarity.

If we instead take $a_3 = 0.61$, the leading principle minors are 0.6279, 0.3896, and −0.0113, showing that the process is not stationary. The minimum eigenvalue of S_α is −0.0095 and the minimum of the modulus of the roots of the corresponding polynomial is 0.9930, also showing that the model is not stationary. For $a_3 = 0.6$, the moduli of the AR polynomial roots are 1.2910, 1.2910, and 1.000, so that the model has exactly one unit root. Observe in this case that $a_1 + a_2 + a_3 = 1$, which can be compared to (6.10) in the $p = 2$ case. ∎

6.1.2 Moments

Instead of proceeding as we did in Section 4.1 to compute μ, the mean of process (6.1) in the limit as $t \to \infty$, we instead assume that the process is stationary and calculate the expected value of (6.1). With $\mu = \lim_{t\to\infty} \mathbb{E}[Y_t]$, this gives

$$\begin{aligned} \mu &= c + a_1\mu + a_2\mu + \cdots + a_p\mu \\ &= \frac{c}{1 - a_1 - a_2 - \cdots - a_p} = \frac{c}{a(1)}, \end{aligned} \tag{6.11}$$

where $a(\cdot)$ is the polynomial in (6.3). Recall that, for the AR(1) model to be stationary, it is necessary that $a_1 < 1$, while for the AR(2) model, from (6.10), it is necessary that $a_1 + a_2 < 1$. We found a similar result for the AR(3) model in Example 6.4. From the form of (6.11), one might conjecture that, in order for the AR(p) process to be stationary, it should be necessary that $\sum_{i=1}^{p} a_i < 1$, and, if $\sum_{i=1}^{p} a_i = 1$, then there is one unit root. This is indeed true, although we do not formally prove it.

Example 6.5 The AR(4) model

$$Y_t = -0.2Y_{t-1} + 1.1Y_{t-2} + 0.4Y_{t-3} - 0.3Y_{t-4} + U_t \tag{6.12}$$

is such that $a_1 + a_2 + a_3 + a_4 = 1$. The model can also be written as

$$(1 - \lambda_1 L)(1 - \lambda_2 L)(1 - \lambda_3 L)(1 - L)Y_t = U_t,$$

with $|\lambda_1^{-1}| = 2.4839$ and $|\lambda_{2,3}^{-1}| = 1.1584$ (to 4 decimal places). Denote the first difference of Y_t as $X_t = (1 - L)Y_t$, so that $(1 - \lambda_1 L)(1 - \lambda_2 L)(1 - \lambda_3 L)X_t = U_t$ is a stationary AR(3) process. Multiplying out, the model for X_t can be written as

$$X_t = -1.2X_{t-1} - 0.1X_{t-2} + 0.3X_{t-3} + U_t. \tag{6.13}$$

A simulated realization of (6.12) is shown in the top panel of Figure 6.1. The "overall movement" indeed has the usual characteristic of a random walk, but there is also short-term autocorrelation because of the stationary AR(3) component.

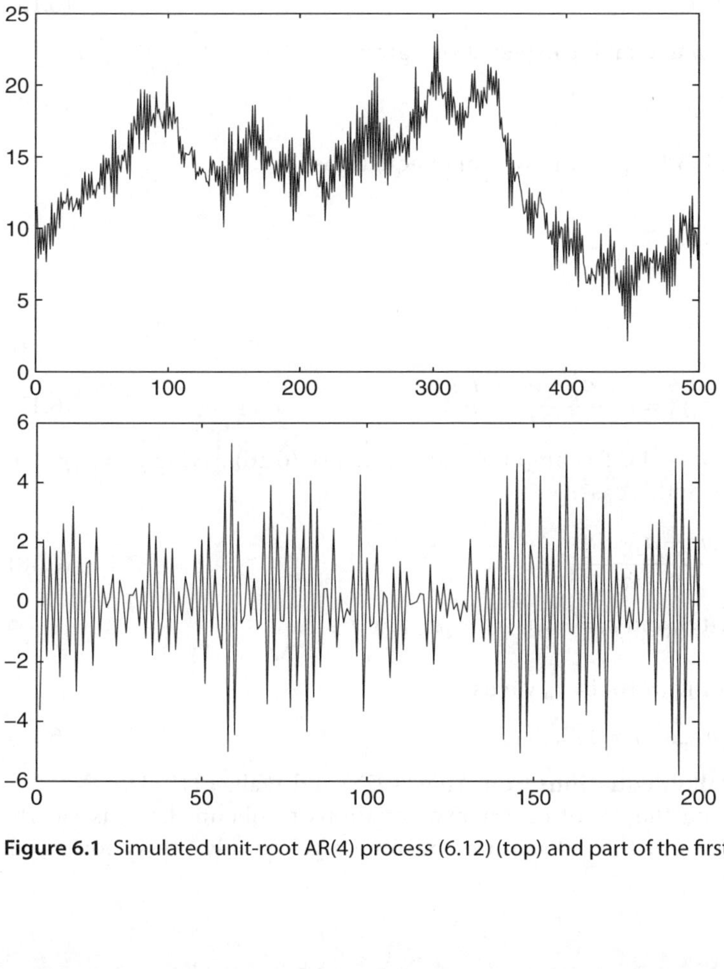

Figure 6.1 Simulated unit-root AR(4) process (6.12) (top) and part of the first difference series (bottom).

Notice the abrupt and persistent "change of direction" of the series around observation 350. This is purely an artifact of chance, though when faced with similar-looking real data (say, a stock price or exchange rate), a natural inclination might be to consider the change to be evidence of a structural break in the model purported to describe the evolution of the series. The right panel shows the first 200 observations of $X_t = (1 - L)Y_t$. While this series appears mean stationary (around zero), one might question the constancy of the variance, which appears to change with time. The reason for the seeming "volatility clustering" of model (6.13) is the large negative coefficient on X_{t-1} and the proximity of the polynomial roots to the unit circle. ■

We now turn to the unconditional variance, $\gamma_0 = \lim_{t\to\infty} \mathbb{V}(Y_t)$, and the unconditional covariances, $\gamma_s = \lim_{t\to\infty} \text{Cov}(Y_t, Y_{t-s})$. Similar to the AR(1) case in (4.13), $\gamma_s = \gamma_{-s}$.

To derive the $\{\gamma_s\}$, it is advantageous to use the expression for μ in (6.11), and quickly confirm that (6.1) can be written as

$$Y_t - \mu = a_1(Y_{t-1} - \mu) + a_2(Y_{t-2} - \mu) + \cdots + a_p(Y_{t-p} - \mu) + U_t, \tag{6.14}$$

generalizing expression (4.9) in the stationary AR(1) case. Now, multiplying both sides of (6.14) by $Y_t - \mu$, taking expectations, and using the relation $\gamma_s = \gamma_{-s}$, we have

$$\gamma_0 = a_1\gamma_1 + a_2\gamma_2 + \cdots + a_p\gamma_p + \sigma^2. \tag{6.15}$$

Similarly, multiplying (6.14) by $Y_{t-j} - \mu$ and taking expectations gives

$$\gamma_j = a_1\gamma_{j-1} + a_2\gamma_{j-2} + \cdots + a_p\gamma_{j-p}, \quad j = 1, 2, \ldots. \tag{6.16}$$

Example 6.6 With $p = 2$, (6.15) and (6.16) give the system of equations

$$\begin{aligned}\gamma_0 &= a_1\gamma_1 + a_2\gamma_2 + \sigma^2,\\ \gamma_1 &= a_1\gamma_0 + a_2\gamma_1,\\ \gamma_2 &= a_1\gamma_1 + a_2\gamma_0,\end{aligned}$$

which can be solved to yield

$$\gamma_0 = \sigma^2\frac{(1 - a_2)}{D}, \quad \gamma_1 = \sigma^2\frac{a_1}{D}, \quad \gamma_2 = \sigma^2\frac{(a_1^2 + a_2 - a_2^2)}{D}, \tag{6.17}$$

where $D = (a_2 + 1)(a_1 + a_2 - 1)(a_2 - a_1 - 1)$. To compute γ_j for $j \geqslant 3$, use (6.16), giving $\gamma_j = a_1\gamma_{j-1} + a_2\gamma_{j-2}$. The correlations $\rho_j = \gamma_j/\gamma_0$ are, in this case,

$$\rho_0 = 1, \quad \rho_1 = \frac{a_1}{1 - a_2}, \quad \rho_2 = \frac{a_1^2 + a_2 - a_2^2}{1 - a_2}, \tag{6.18}$$

valid for $a_2 < 1$ (true under stationarity), and $\rho_j = a_1\rho_{j-1} + a_2\rho_{j-2}$ for $j \geqslant 3$. ■

For a stationary AR(p) model, dividing (6.16) by γ_0 yields

$$\rho_j = a_1\rho_{j-1} + a_2\rho_{j-2} + \cdots + a_p\rho_{j-p}, \quad j = 1, 2, \ldots, \tag{6.19}$$

which are referred to as the **Yule–Walker equations** from Yule (1927) and Walker (1931).

As in Example 6.6 for $p = 2$, solving the set of covariance equations to obtain the γ_s is clearly theoretically feasible for any p. However, the calculations become algebraically messy for $p \geqslant 4$

```
function [Vi,detVi]=leeuwAR(a,T);
% a is [a1 a2 ... ap] of a stationary AR(p) model
p=length(a); a=-a; if nargin < 2, T=p+1; end
firrowP = [1 zeros(1,T-1)]; fircolP = [1 a zeros(1,T-p-1)];
P = toeplitz(fircolP,firrowP); P1 = P(1:p,1:p);
firrowQ1 = a(p:-1:1); fircolQ1 = [a(p) zeros(1,p-1)];
Q1 = toeplitz(fircolQ1,firrowQ1); Q = [Q1; zeros(T-p,p)];
Vi = P'*P - Q*Q';
if nargout>=2,  detVi = det ( P1'*P1 - Q1*Q1' ); end
```

Program Listing 6.1: Computes (6.20) and its determinant.

and numerically prohibitive as p grows. Instead, several other methods have been developed for calculating the γ_s that are far more expedient; see Galbraith and Galbraith (1974), McLeod (1975), De Gooijer (1978), and Mittnik (1988). We use the expression by van der Leeuw (1994), which is convenient in matrix-based software and, for the AR(p) case, delivers the inverse of the variance-covariance matrix.

Let $\sigma^2\boldsymbol{\Sigma}$ denote the $T \times T$ unconditional covariance matrix of $Y_1, \dots, Y_T$, and denote the (ij)th element of $\boldsymbol{\Sigma}$ as γ_{i-j}. For $p < T$,

$$\boldsymbol{\Sigma}^{-1} = \mathbf{P}'\mathbf{P} - \mathbf{Q}\mathbf{Q}', \tag{6.20}$$

where $\mathbf{P}$ is the $T \times T$ band matrix with $(\alpha_0, \alpha_1, \dots, \alpha_p, 0, \dots, 0)'$ as the first column, $\alpha_0 = 1$, $\alpha_i = -a_i$, $i = 1, \dots, p$, and

$$\mathbf{Q} = \begin{bmatrix} \mathbf{T}_2 \\ \mathbf{0} \end{bmatrix},$$

of size $T \times p$, where $\mathbf{T}_2$ is given in (6.7). Observe that the upper $p \times p$ portion of $\mathbf{P}$ is just $\mathbf{T}_1$ from (6.7). The covariances $\gamma_0, \gamma_1, \dots, \gamma_{T-1}$ are then given by the first row or column of the inverse of (6.20). Useful also is that

$$|\boldsymbol{\Sigma}^{-1}| = |\mathbf{T}_1'\mathbf{T}_1 - \mathbf{T}_2\mathbf{T}_2'|, \quad T > p, \tag{6.21}$$

with the special case of $p = 1$ easily seen to be $1 - a_1^2$. A program to compute $\boldsymbol{\Sigma}^{-1}$ and its determinant is given in Listing 6.1.

6.1.3 Estimation

6.1.3.1 Without Mean Term

We first consider the case with $c = 0$. Because we have a method of computing the covariance matrix $\boldsymbol{\Sigma}$ of the Y_t, under the assumption that $U_t \overset{\text{i.i.d.}}{\sim} \mathrm{N}(0, \sigma^2)$, the likelihood of the Y_t is just the multivariate normal density with mean $\mathbf{0}$ and covariance matrix $\sigma^2\boldsymbol{\Sigma}$, identical to the expression in (4.19), but where $\boldsymbol{\Sigma}$ corresponds to the AR(p) model with parameters $\mathbf{a} = (a_1, a_2, \dots, a_p)'$. Observe from (4.19) that what is required is not $\boldsymbol{\Sigma}$, but rather $\boldsymbol{\Sigma}^{-1}$ and its determinant, which is what (6.20) conveniently delivers. Nevertheless, if T is large, the evaluation of (6.20) and (4.19) can still be costly. Instead, it makes sense to partition the likelihood as in (4.16),

$$f_{\mathbf{Y}_0,\mathbf{Y}}(\mathbf{y}_0, \mathbf{y}) = f_{\mathbf{Y}_0}(\mathbf{y}_0) f_{\mathbf{Y}|\mathbf{Y}_0}(\mathbf{y} \mid \mathbf{y}_0), \tag{6.22}$$

where $\mathbf{Y}_0 = (Y_{1-p}, Y_{2-p}, \dots, Y_0)'$ and $\mathbf{Y} = (Y_1, Y_2, \dots, Y_T)'$.

For example, with the AR(2) model and using (6.17), the inverse of

$$\boldsymbol{\Sigma} = \begin{bmatrix} \gamma_0 & \gamma_1 \\ \gamma_1 & \gamma_0 \end{bmatrix} \quad \text{is} \quad \boldsymbol{\Sigma}^{-1} = \begin{bmatrix} 1 - a_2^2 & -a_1(1+a_2) \\ -a_1(1+a_2) & 1 - a_2^2 \end{bmatrix}, \tag{6.23}$$

with determinant

$$|\boldsymbol{\Sigma}^{-1}| = (a_2+1)^2((1-a_2)^2 - a_1^2). \tag{6.24}$$

Then

$$\begin{bmatrix} Y_{-1} \\ Y_0 \end{bmatrix} \sim \mathrm{N}\left(\begin{bmatrix} 0 \\ 0 \end{bmatrix}, \sigma^2 \begin{bmatrix} \gamma_0 & \gamma_1 \\ \gamma_1 & \gamma_0 \end{bmatrix} \right), \tag{6.25}$$

or

$$f_{Y_{-1},Y_0}(\mathbf{y}_0) = \frac{|\boldsymbol{\Sigma}^{-1}|^{1/2}}{(2\pi\sigma^2)^{p/2}} \exp\left\{ -\frac{1}{2\sigma^2} \mathbf{y}_0' \boldsymbol{\Sigma}^{-1} \mathbf{y}_0 \right\}, \quad p = 2,$$

and

$$f_{\mathbf{Y} \mid (Y_{-1},Y_0)}(\mathbf{y} \mid \mathbf{y}_0) = \prod_{t=1}^{T} \frac{1}{(2\pi\sigma^2)^{1/2}} \exp\left\{ -\frac{1}{2\sigma^2} (y_t - a_1 y_{t-1} - a_2 y_{t-2})^2 \right\}.$$

In the AR(2) case, via (6.17) we were able to derive $\boldsymbol{\Sigma}^{-1}$ of size 2×2. If we use (6.20) instead, then T must be at least $p+1$, giving the 3×3 matrix

$$\begin{aligned} \boldsymbol{\Sigma}^{-1} &= \begin{bmatrix} 1 & -a_1 & -a_2 \\ 0 & 1 & -a_1 \\ 0 & 0 & 1 \end{bmatrix} \begin{bmatrix} 1 & 0 & 0 \\ -a_1 & 1 & 0 \\ -a_2 & -a_1 & 1 \end{bmatrix} - \begin{bmatrix} -a_2 & -a_1 \\ 0 & -a_2 \\ 0 & 0 \end{bmatrix} \begin{bmatrix} -a_2 & 0 & 0 \\ -a_1 & -a_2 & 0 \end{bmatrix} \\ &= \begin{bmatrix} 1 & -a_1 & -a_2 \\ -a_1 & 1 + a_1^2 - a_2^2 & -a_1 \\ -a_2 & -a_1 & 1 \end{bmatrix}, \end{aligned} \tag{6.26}$$

also with determinant (6.24). We would then compute the likelihood as

$$\begin{aligned} \mathcal{L}(\mathbf{a}, \sigma; \mathbf{y}_0, \mathbf{y}) &= f_{Y_{-1},Y_0,Y_1}(\mathbf{y}_0) f_{(Y_2,\ldots,Y_T) \mid (Y_{-1},Y_0,Y_1)}(\mathbf{y}) \\ &= \frac{|\boldsymbol{\Sigma}^{-1}|^{1/2}}{(2\pi\sigma^2)^{(p+1)/2}} \exp\left\{ -\frac{1}{2\sigma^2} \mathbf{y}_0' \boldsymbol{\Sigma}^{-1} \mathbf{y}_0 \right\} \\ &\quad \times \prod_{t=2}^{T} \frac{1}{(2\pi\sigma^2)^{1/2}} \exp\left\{ -\frac{1}{2\sigma^2} (y_t - a_1 y_{t-1} - a_2 y_{t-2})^2 \right\}, \end{aligned} \tag{6.27}$$

where here, $p = 2$, $\mathbf{a} = (a_1, a_2)$, $\mathbf{y}_0 = (y_{-1}, y_0, y_1)'$ and $\mathbf{y} = (y_2, \ldots, y_T)'$. As a check, inverting (6.26), taking its upper 2×2 submatrix and inverting it indeed yields $\boldsymbol{\Sigma}^{-1}$, as given in (6.23).

For general p, use (6.20) to obtain the $(p+1) \times (p+1)$ matrix $\boldsymbol{\Sigma}^{-1}$ and calculate the likelihood as

$$\begin{aligned} \mathcal{L}(\mathbf{a}, \sigma; \mathbf{y}_0, \mathbf{y}) &= f_{Y_{1-p},Y_{2-p},\ldots,Y_0,Y_1}(\mathbf{y}_0) f_{(Y_2,\ldots,Y_T) \mid (Y_{1-p},Y_{2-p},\ldots,Y_0,Y_1)}(\mathbf{y}) \\ &= \frac{|\boldsymbol{\Sigma}^{-1}|^{1/2}}{(2\pi\sigma^2)^{(p+1)/2}} \exp\left\{ -\frac{1}{2\sigma^2} \mathbf{y}_0' \boldsymbol{\Sigma}^{-1} \mathbf{y}_0 \right\} \\ &\quad \times \prod_{t=2}^{T} \frac{1}{(2\pi\sigma^2)^{1/2}} \exp\left\{ -\frac{1}{2\sigma^2} \left(y_t - \sum_{i=1}^{p} a_i y_{t-i} \right)^2 \right\}, \end{aligned} \tag{6.28}$$

```
function [MLE, stderr]=exactarp(y,p)
n=length(y); y=reshape(y,n,1); initvec=[yw(y,p); std(y)];
tol=1e-5; maxiter=200; show='none'; % 'iter','notify', or 'final'.
options = optimset('Display',show,'TolX',tol,'Tolfun',tol, ...
    'MaxIter',maxiter,'LargeScale','off');
[MLE,loglik,exitflag]=fminunc(@exactarp_,initvec,options,y,p);
if nargout>1
  H = -hessian(@exactarp_,MLE,y,p);
  stderr=real(sqrt(diag(inv(H))));
end

function loglik=exactarp_(param,y,p)
a=param(1:(end-1)); rr=roots([-a(end:-1:1); 1]); rootcheck=min(abs(rr));
if rootcheck<=1, loglik=abs(0.999-rootcheck)*1e8; return, end
sig=abs(param(end)); % this is not sigma^2, but just (little) sigma.
n=length(y); [Vi,detVi]=leeuwAR(a'); start=y(1:(p+1));
lik0=0.5*log(detVi)-((n)/2)*log(2*pi*sig^2)-(0.5/sig^2)*start'*Vi*start;
res=y((p+2):end); for i=1:p,  res=res-a(i)*y(p+2-i:end-i); end
ll=lik0-sum(res.^2)/2/sig^2; loglik = -ll;
```

Program Listing 6.2: Computes the exact m.l.e. of an AR(p) model with known mean zero and Gaussian innovations. Programs `yw` and `leeuwAR` are given in Listings 6.3 and 6.1, respectively, while program `hessian` is given in Listing III.5.4.

where $\mathbf{y}_0$ and $\mathbf{y}$ are appropriately defined. The m.l.e.s, denoted $\widehat{\mathbf{a}}_{\mathrm{ML}}$ and $\hat{\sigma}^2_{\mathrm{ML}}$, are those values of $\mathbf{a}$ and σ^2 that maximize (the log of) (6.28) under the constraint of stationarity.

A program to compute the exact m.l.e. of a stationary AR(p) model with known mean zero is given in Listing 6.2. It calls the program in Listing 6.3 below to compute starting values and uses the program in Listing 6.1 above to compute the (inverse of the) covariance matrix of the first $p+1$ values. To enforce stationarity, a simple penalty term for parameter values corresponding to non-stationary models is used.

The left panel of Figure 6.2 shows a contour plot of the likelihood for a simulated AR(2) time series with 15 observations, known mean zero and known scale parameter $\sigma = 1$, as a function of a_1 (horizontal axis) and a_2 (vertical axis). The inscribed triangle indicates the region of stationarity, and the dark dot indicates the true parameters, which in this case are $a_1 = a_2 = 0$. The middle panel is similar,

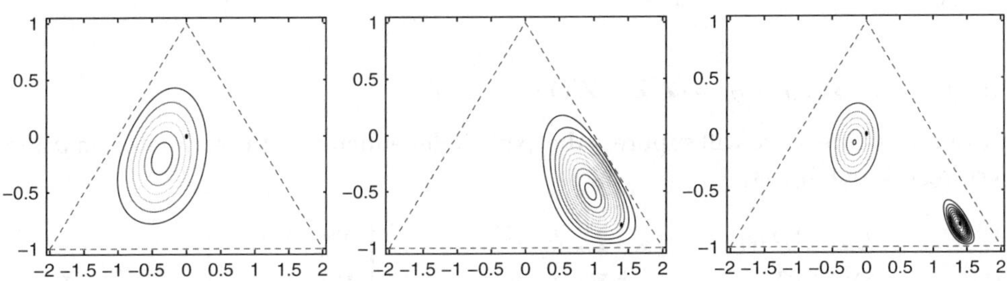

Figure 6.2 Contour plots of likelihoods of simulated AR(2) time series as functions of a_1 (horizontal axis) and a_2 (vertical axis). The left panel takes $a_1 = a_2 = 0$ and $T = 15$ observations, the middle panel takes $a_1 = 1.4$, $a_2 = -0.8$, and $T = 15$, and the right panel shows both these cases, but based on simulated series with $T = 30$ observations.

but based on an AR(2) model with $a_1 = 1.4$ and $a_2 = -0.8$. Observe how the likelihood behaves as the parameters approach the stationarity border, and also how far the true parameter is from the m.l.e., which, of course, is at the center of the concentric circles. The last panel overlays the likelihood of the previous two models, but based on simulated series with 30 observations. The reader is encouraged to construct a program to replicate Figure 6.2.

Asymptotically, under certain assumptions on the innovation sequence (that include as a special case being i.i.d. Gaussian), for a stationary AR(p) process,

$$\sqrt{T}(\hat{\mathbf{a}}_{\text{ML}} - \mathbf{a}) \overset{\text{asy}}{\sim} \text{N}(\mathbf{0}, \sigma^2\mathbf{\Gamma}^{-1}), \tag{6.29}$$

where $\mathbf{a} = (a_1, \dots, a_p)'$ and $\hat{\mathbf{a}}_{\text{ML}}$ are both p-dimensional, and $\mathbf{\Gamma}$ is just $\mathbf{\Sigma}$, but of size $p \times p$, i.e., the (ij)th element of $\mathbf{\Gamma}$ is γ_{i-j}. See, e.g., Brockwell and Davis (1991, Sec. 8.8, 10.8) for the required conditions and proof. For $p = 1$, this reduces to $\sqrt{T}(\hat{a}_{\text{ML}} - a) \overset{\text{asy}}{\sim} \text{N}(0, 1 - a^2)$, while for $p = 2$, from (6.23),

$$\sqrt{T}(\hat{\mathbf{a}}_{\text{ML}} - \mathbf{a}) \overset{\text{asy}}{\sim} \text{N}\left(\begin{bmatrix} 0 \\ 0 \end{bmatrix}, \begin{bmatrix} 1 - a_2^2 & -a_1(1 + a_2) \\ \cdot & 1 - a_2^2 \end{bmatrix}\right). \tag{6.30}$$

Observe how the asymptotic variance of both a_1 and a_2 only depends on a_2. Problem 6.8 asks the reader to check this via simulation.

6.1.3.2 Starting Values

An important issue that arises in the numeric maximization of the log of likelihood (6.28) is the starting values of the $p + 1$ parameters $\mathbf{a} = (a_1, \dots, a_p)'$ and σ^2. We present two easily computed estimators.

Least Squares As in the AR(1) case, the o.l.s. estimator is applicable. This is obtained by taking the dependent variable to be $(Y_1, \dots, Y_T)'$ and using the $T \times p$ design matrix

$$\mathbf{Z} = \begin{bmatrix} Y_0 & Y_{-1} & \cdots & Y_{1-p} \\ Y_1 & Y_0 & \cdots & Y_{2-p} \\ \vdots & \vdots & & \vdots \\ \vdots & \vdots & & \vdots \\ Y_{T-2} & Y_{T-3} & \cdots & Y_{T-p-1} \\ Y_{T-1} & Y_{T-2} & \cdots & Y_{T-p} \end{bmatrix}. \tag{6.31}$$

Then

$$\hat{\mathbf{a}}_{\text{LS}}(p) = (\hat{a}_{\text{LS}}(1, p), \dots, \hat{a}_{\text{LS}}(p, p))' = (\mathbf{Z}'\mathbf{Z})^{-1}\mathbf{Z}'(Y_1, \dots, Y_T)'.$$

When it is clear from the context, we will suppress the explicit dependence of the estimator on p and just write $\hat{\mathbf{a}}_{\text{LS}}$. Writing this out, $\hat{\mathbf{a}}_{\text{LS}}$ is

$$\begin{bmatrix} \sum_{i=0}^{T-1} Y_i^2 & \sum_{i=-1}^{T-2} Y_i Y_{i+1} & \cdots & \sum_{i=1-p}^{T-p} Y_i Y_{i-1+p} \\ \sum_{i=0}^{T-1} Y_i Y_{i-1} & \sum_{i=-1}^{T-2} Y_i^2 & \cdots & \sum_{i=1-p}^{T-p} Y_i Y_{i-2+p} \\ \vdots & \vdots & & \vdots \\ \sum_{i=0}^{T-1} Y_i Y_{i+1-p} & \sum_{i=-1}^{T-2} Y_i Y_{2+i-p} & \cdots & \sum_{i=1-p}^{T-p} Y_i^2 \end{bmatrix}^{-1} \begin{bmatrix} \sum_{i=1}^{T} Y_i Y_{i-1} \\ \sum_{i=1}^{T} Y_i Y_{i-2} \\ \vdots \\ \sum_{i=1}^{T} Y_i Y_{i-p} \end{bmatrix}. \tag{6.32}$$

For clarity of structure, with $p = 3$ this is

$$\begin{bmatrix} \sum_{i=0}^{T-1} Y_i^2 & \sum_{i=-1}^{T-2} Y_iY_{i+1} & \sum_{i=-2}^{T-3} Y_iY_{i+2} \\ \sum_{i=0}^{T-1} Y_iY_{i-1} & \sum_{i=-1}^{T-2} Y_i^2 & \sum_{i=-2}^{T-3} Y_iY_{i+1} \\ \sum_{i=0}^{T-1} Y_iY_{i-2} & \sum_{i=-1}^{T-2} Y_iY_{i-1} & \sum_{i=-2}^{T-3} Y_i^2 \end{bmatrix}^{-1} \begin{bmatrix} \sum_{i=1}^{T} Y_iY_{i-1} \\ \sum_{i=1}^{T} Y_iY_{i-2} \\ \sum_{i=1}^{T} Y_iY_{i-3} \end{bmatrix}.$$

By its matrix construction, $\mathbf{Z}'\mathbf{Z}$ is symmetric, which is also seen by carefully looking at the individual elements in (6.32). Observe also that the elements along any chosen diagonal of $\mathbf{Z}'\mathbf{Z}$ are close, but not identical.

The estimate of σ^2 is computed as usual, namely the sum of squared residuals divided by either T (for the conditional m.l.e.) or $T - p$ to adjust for bias. As in the AR(1) case, the o.l.s. estimator is the value that maximizes the conditional likelihood $f_{(Y_1,\ldots,Y_T)\mid(Y_{1-p},Y_{2-p},\ldots,Y_0)}(\mathbf{y};\mathbf{a},\sigma^2)$. It will yield estimates that are reasonably close to the exact m.l.e. values, except for small sample sizes and/or cases for which the process is close to the stationarity border.

Thus, the o.l.s. estimator is equivalent to the conditional m.l.e., and assuming (correctly) that, asymptotically, the first p observations become negligible when the model is stationary, one would expect that $\hat{\mathbf{a}}_{\text{LS}}$ has the same asymptotic distribution as the m.l.e. This is true, and was shown by Mann and Wold (1943), i.e., $\sqrt{T}(\hat{\mathbf{a}}_{\text{LS}} - \mathbf{a}) \overset{\text{asy}}{\sim} \text{N}(\mathbf{0}, \sigma^2\mathbf{\Gamma}^{-1})$ (see also Hamilton, 1994, pp. 215–216; and Fuller, 1996, Sec. 8.2.1).

Yule–Walker The Yule–Walker equations (6.19) can also be used to derive estimates of the model parameters. We denote them as $\hat{a}_{\text{YW}}(i,p)$, $i = 1,\ldots,p$, for $p < T$. Recalling that $\rho_s = \rho_{-s}$ and using the sample counterparts (detailed in Section 8.1.1) for observed time series $Y_1,\ldots,Y_T$,

$$\hat{\rho}_s = \frac{\hat{\gamma}_s}{\hat{\gamma}_0}, \qquad \hat{\gamma}_s = T^{-1}\sum_{t=s+1}^{T} Y_tY_{t-s}, \tag{6.33}$$

we arrive at the set of equations

$$\hat{\mathbf{r}} = \begin{bmatrix} \hat{\rho}_1 \\ \hat{\rho}_2 \\ \vdots \\ \hat{\rho}_p \end{bmatrix} = \begin{bmatrix} 1 & \hat{\rho}_1 & \cdots & \hat{\rho}_{p-1} \\ \hat{\rho}_1 & \ddots & \ddots & \vdots \\ \vdots & & & \hat{\rho}_1 \\ \hat{\rho}_{p-1} & \hat{\rho}_{p-2} & \cdots & 1 \end{bmatrix} \begin{bmatrix} a_1 \\ a_2 \\ \vdots \\ a_p \end{bmatrix} = \widehat{\mathbf{R}}\mathbf{a}, \tag{6.34}$$

where $\hat{\mathbf{r}}$ and $\widehat{\mathbf{R}}$ are so defined. Note that $\widehat{\mathbf{R}} = \widehat{\mathbf{\Gamma}}/\hat{\gamma}_0$. This can be solved for

$$\hat{\mathbf{a}}_{\text{YW}}(p) = (\hat{a}_{\text{YW}}(1,p),\ldots \hat{a}_{\text{YW}}(p,p))'$$

as $\widehat{\mathbf{R}}^{-1}\hat{\mathbf{r}}$, for any $p < T$. Based on (6.32) and (6.33), notice the striking similarity to the least squares estimator. As with $\hat{\mathbf{a}}_{\text{LS}}$, we suppress the argument p and just write $\hat{\mathbf{a}}_{\text{YW}}$.

For σ^2, (6.15) can be used to obtain

$$\hat{\sigma}^2 = \hat{\gamma}_0 - \sum_{i=1}^{p} \hat{a}_i\hat{\gamma}_i = \hat{\gamma}_0 - \hat{\mathbf{a}}'(\hat{\gamma}_0\hat{\mathbf{r}}) = \hat{\gamma}_0(1 - \hat{\mathbf{r}}'\widehat{\mathbf{R}}^{-1}\hat{\mathbf{r}}). \tag{6.35}$$

For $p = 1$, the solution is just $\hat{a}_{\text{YW}} = \hat{\rho}_1$. This will always be less in absolute value than the least squares estimator (4.14) because

$$\hat{a}_{\text{YW}} = \hat{\rho}_1 = \frac{\sum_{t=2}^{T} Y_tY_{t-1}}{\sum_{t=1}^{T} Y_t^2} \lessdot \frac{\sum_{t=2}^{T} Y_tY_{t-1}}{\sum_{t=1}^{T-1} Y_t^2} = \hat{a}_{\text{LS}}, \tag{6.36}$$

```
function [ayw,aols]=yw(y,p)
y=reshape(y,length(y),1); r=localsacf(y,p);
v=[1; r(1:end-1)]; R=toeplitz(v,v); ayw=inv(R)*r;
if nargout>1 % the o.l.s. estimator
  z=y(p+1:end); zl=length(z); Z=[];
  for i=1:p, Z=[Z y(p-i+1:p-i+zl)]; end
  R=inv(Z'*Z)*Z'; aols=R*z;
end

function acf=localsacf(x,imax) % computes the estimates of gamma_i
T=length(x); a=zeros(imax,1);
for i=1:imax, a(i)= sum(x(i+1:T) .* x(1:T-i) ); end
acf=a./sum(x.^2);
```

Program Listing 6.3: Computes the Yule–Walker and the least squares estimator for an AR(p) model.

where $a \lessdot b$ means $|a| < |b|$. These expressions are clearly asymptotically equivalent. For $p = 2$, and writing $r_i = \hat{\rho}_i$, $i = 1, 2$,

$$\hat{\mathbf{a}}_{\mathrm{YW}} = \begin{bmatrix} \hat{a}_{\mathrm{YW}}(1,2) \\ \hat{a}_{\mathrm{YW}}(2,2) \end{bmatrix} = \begin{bmatrix} 1 & r_1 \\ r_1 & 1 \end{bmatrix}^{-1} \begin{bmatrix} r_1 \\ r_2 \end{bmatrix} = \frac{1}{(r_1-1)(r_1+1)} \begin{bmatrix} r_1(r_2-1) \\ r_1^2 - r_2 \end{bmatrix}. \tag{6.37}$$

Although it is now somewhat more difficult to algebraically relate $\hat{\mathbf{a}}_{\mathrm{YW}}$ to $\hat{\mathbf{a}}_{\mathrm{LS}}$, one can surmise from the validity of the Yule–Walker equations and the consistency of the $\hat{\rho}_i$ that $\hat{\mathbf{a}}_{\mathrm{YW}}(2)$ is consistent for a_1 and a_2 when the model is AR(2) or, more generally, that $\hat{\mathbf{a}}_{\mathrm{YW}}(p)$ is consistent for $a_1, \dots, a_p$ when the model is stationary AR(p). Unsurprisingly, under the appropriate assumptions, $\hat{\mathbf{a}}_{\mathrm{YW}}$ has the same asymptotic distribution as $\hat{\mathbf{a}}_{\mathrm{ML}}$ and $\hat{\mathbf{a}}_{\mathrm{LS}}$. A proof can be found in Brockwell and Davis (1991, Sec. 8.10).

Listing 6.3 gives a program to compute the Yule–Walker and least squares estimates.

6.1.3.3 With Mean Term

Usually, parameter c will not be known, though in some cases, such as with stock returns, economic theory can provide a value (in this case, zero). Consider model (6.1), but now with all $p + 2$ parameters, c, σ, and $a_1, \dots, a_p$, unknown. The extension of (6.28) to handle the case with unknown c is quite straightforward. Recall from (6.11) that

$$\mathbb{E}[Y_t] = \mu = c/(1 - a_1 - \cdots - a_p), \tag{6.38}$$

so that, similar to (6.25), $\mathbf{Y}_0 \sim \mathrm{N}(\boldsymbol{\mu}, \boldsymbol{\Sigma})$, where $\mathbf{Y}_0 = (Y_{1-p}, Y_{2-p}, \dots, Y_0, Y_1)'$ is of length $p+1$, $\boldsymbol{\mu} = (\mu, \dots, \mu)'$ and (6.20) can be used to calculate $\boldsymbol{\Sigma}^{-1}$. The likelihood is just the product of $f_{\mathbf{Y}_0}$ and $f_{\mathbf{Y}}$, where $\mathbf{Y} = (Y_2, \dots, Y_T)'$ and

$$f_{\mathbf{Y}}(\mathbf{y}) = \prod_{t=2}^{T} \frac{1}{(2\pi\sigma^2)^{1/2}} \exp\left\{ -\frac{1}{2\sigma^2} \left(y_t - c - \sum_{i=1}^{p} a_i y_{t-i} \right)^2 \right\}.$$

One simple way of obtaining starting values for c and the a_i is to take $\hat{\mu} = \bar{Y}$ and then use either Yule–Walker or least squares based on $\dot{Y}_t = Y_t - \hat{\mu}$ to obtain estimates of the a_i. An estimate for c is then obtained from (6.38). One could also iterate on this procedure.

Estimation in the more general case in which the mean of the Y_t involves a set of k regression coefficients will be dealt with in Section 7.4 in the more general setting of an ARMA model.

6.1.3.4 Approximate Standard Errors

It is desirable to have (an approximation of) the standard errors of the parameters, from which asymptotically valid confidence intervals can be constructed. Often in statistical software a p-value for each a_i coefficient is reported, corresponding to the test that $a_i = 0$. These can be computed based on the asymptotic result (6.29). The variance covariance matrix $\mathbf{\Gamma}^{-1}$ can be approximated for a finite-length observed time series in an obvious way by replacing the a_i used to construct $\mathbf{\Sigma}$ in (6.20) with their respective m.l.e.s. An alternative way is to replace the γ_i in $\mathbf{\Gamma}$ with their sample counterparts $\hat{\gamma}_i = T^{-1}\sum_{t=i+1}^{T} Y_t Y_{t-i}$ for observed time series $Y_1, \dots, Y_T$. An estimate of σ^2 can be obtained from (6.35).

Another way to get standard errors of the parameters, but which involves more computation, is to approximate the Hessian matrix via numerical differentiation using the estimated m.l.e.[1] Lastly, and most computationally expensive of the methods stated so far, is to use the bootstrap.

Example 6.7 To investigate the performance of some of the aforementioned methods for getting standard errors of the AR parameters, a simulation was done with 10,000 replications, using an AR(1) model with the three values $a = 0$, 0.5, and 0.9, and two sample sizes $T = 10$ and 50 (and constant value $\sigma^2 = 4$). For each model and method, the average of the 10,000 standard errors of $\hat{a}_{\text{ML}}$ is reported, which we refer to just as SE. The exact m.l.e. was used for estimation. In addition, it would be expected to make a difference if we know the mean of the series, or—as is far more common in practice—if we do not, in which case it needs to be estimated along with a and σ. Both ways are considered. The results are shown in Table 6.1.

As expected, we see that, for all methods of SE construction, (i) as $|a|$ increases, SE decreases, (ii) as T increases, SE decreases, and (iii) for a given T and a, SE is higher if the mean is assumed unknown. Relatively speaking, the method based on use of $\hat{\gamma}_i$ is the best in almost all cases, with second best being the use of the Hessian matrix. The worst performer is using $\hat{a}_{\text{ML}}$ to construct the asymptotic variance–covariance matrix. This result is somewhat surprising, as one would expect a function of the $\hat{\gamma}_i$ to have a higher sampling variance than a function of $\hat{a}_{\text{ML}}$. Also, particularly for $T = 10$ and $a = 0.9$, one would expect the true and asymptotic distribution of $\hat{\gamma}_i$ to deviate considerably, and so favoring use of the SE based on the numeric Hessian matrix.

It is interesting to consider what happens when the estimated model is mis-specified. Taking the true model to be an AR(2) with $a_1 = 1.2$ and $a_2 = -0.8$ and repeating the above exercise, just for $T = 10$ and c known, *and (wrongly) using an AR(1) model* results in an empirical (sample standard deviation of the 10,000 $\hat{a}_{\text{ML}}$ values) standard error for $\hat{a}_{\text{ML}}$ of 0.042, while all three methods discussed above resulted in about the same SE of 0.106.

It is not surprising that use of the variance covariance matrix in (6.29) with any method of its estimation is relatively inaccurate because (6.29) is only valid when the value of p used for the model is at least as large as the true value of p. ■

---

1 Recall that this method has the drawback that numerical derivatives are functions of the tuning parameter h dictating the perturbation in the function, though for well-behaved functions there will usually be a reasonable range of h-values such that the elements of the Hessian will be approximately constant. Alternatively, the selection of h can be avoided by using the final value of the Hessian matrix that is built up when using the quasi-Newton numerical maximization methods. However, it is also subject to variation because it is a function of the convergence criteria imposed on the estimation. It is not at all clear which method will be more accurate on average, or for a particular data set.

Table 6.1 Comparison of estimated standard errors for the AR(1) model. "Emp" is the empirically observed standard error of $\hat{a}_{\mathrm{ML}}$, calculated as the sample standard deviation of the $\hat{a}_{\mathrm{ML}}$ based on simulation with 10,000 replications; $(\hat{\gamma})$ is short for $[\hat{\sigma}^2\boldsymbol{\Gamma}^{-1}](\hat{\gamma})$, which refers to use of the sample covariances to form the asymptotic variance–covariance matrix in (6.29); $(\hat{\mathbf{a}}_{\mathrm{ML}})$ is short for $[\hat{\sigma}^2\boldsymbol{\Gamma}^{-1}](\hat{\mathbf{a}}_{\mathrm{ML}})$ and refers to use of $\hat{a}_{\mathrm{ML}}$ to form the asymptotic variance–covariance matrix in (6.29); and "Hess" refers to use of the estimated Hessian matrix constructed from the quasi-Newton method used to numerically maximize the likelihood. The values in parentheses indicate the sample standard error of the 10,000 estimates. Entries under "Known $c = 0$" assume the process has zero mean, so that only a and σ are estimated, and the $\hat{\gamma}_i$ are formed as in (8.6), i.e., without subtracting the mean from the data. For "c jointly estimated", the model is extended to include an **X** matrix consisting of a column of ones, and the $\hat{\gamma}_i$ are formed with mean subtraction, as in (8.10). Boldface entries indicate being closest to the empirically observed standard error of $\hat{a}_{\mathrm{ML}}$.

		Known $c = 0$				c jointly estimated			
T	a	Emp	$(\hat{\gamma})$	$(\hat{\mathbf{a}}_{\mathrm{ML}})$	Hess	Emp	$(\hat{\gamma})$	$(\hat{\mathbf{a}}_{\mathrm{ML}})$	Hess
10	0	0.310	**0.304** (0.017)	0.300 (0.022)	0.319 (0.051)	0.310	**0.302** (0.018)	0.297 (0.025)	0.319 (0.057)
	0.5	0.279	0.280 (0.034)	0.270 (0.042)	**0.279** (0.065)	0.316	0.300 (0.022)	0.288 (0.032)	**0.312** (0.060)
	0.9	0.207	**0.212** (0.053)	0.170 (0.068)	0.153 (0.091)	0.311	**0.277** (0.032)	0.249 (0.055)	0.267 (0.086)
50	0	0.140	**0.140** (0.0019)	0.140 (0.0020)	0.143 (0.044)	0.140	**0.140** (0.0019)	0.140 (0.0020)	0.143 (0.029)
	0.5	0.123	**0.123** (0.0091)	0.122 (0.0095)	0.125 (0.050)	0.128	0.125 (0.0087)	0.125 (0.0091)	**0.126** (0.015)
	0.9	0.0713	**0.0721** (0.016)	0.0669 (0.017)	0.0638 (0.021)	0.0906	**0.0834** (0.016)	0.0774 (0.018)	0.0760 (0.020)

6.2 Moving Average Processes

6.2.1 MA(1) Process

Observe that p parameters are required to model the dynamic portion of the AR(p) model and, if the true process has in fact a "long memory", then p could be rather large. In the spirit of what is called **parsimonious model building**, whereby we recognize that we will never find the underlying true model and try instead to find the simplest model that "adequately" describes the process, we might consider the above AR(p) structure, but with the restriction that $a_i = f_i(a)$, where f_i is some known function that depends only on the single parameter a, $i = 1, \dots, p$. Because we can choose the f_i, the model is more flexible than the AR(1) model, but has the same number of parameters. The greater flexibility comes from allowing observations from more than one time period in the past to affect the current realization, albeit in a restricted way, dictated by the specification of the f_i.

Let the model be $Y_t = aY_{t-1} - a^2Y_{t-2} + a^3Y_{t-3} - \cdots + U_t$, for $|a| < 1$, and let $p \to \infty$ so that the model can be expressed as

$$(1 - aL + a^2L^2 - a^3L^3 + \cdots)Y_t = U_t. \tag{6.39}$$

Because

$$1 - aL + a^2L^2 - a^3L^3 + \cdots = \sum_{j=0}^{\infty}(-aL)^j = \frac{1}{1 + aL}, \tag{6.40}$$

the model can be written as

$$\frac{1}{1+aL}Y_t = U_t \qquad \text{or} \qquad Y_t = (1+aL)U_t = U_t + aU_{t-1}.$$

This is referred to as a **moving average model of order 1**, or MA(1), with parameter a. The usual convention is to write the model as

$$Y_t = c + U_t + bU_{t-1}. \tag{6.41}$$

The name moving average refers to the fact that Y_t is a (weighted) average of U_t and U_{t-1}, and this average "moves" along the U_t as t changes.

One could imagine having started with formulation (6.41), and seeing what this implies for a finite data set. Simple recursive substitution yields $Y_1 = U_1 + bU_0$, or $U_1 = Y_1 - bU_0$ and

$$Y_2 = U_2 + bU_1 = U_2 + b(Y_1 - bU_0) = U_2 + bY_1 - b^2U_0$$

or $U_2 = Y_2 - bY_1 + b^2U_0$ and $U_3 = Y_3 - bU_2 = Y_3 - bY_2 + b^2Y_1 - b^3U_0$. Clearly, for general $t \geqslant 1$,[2]

$$U_t = Y_t - bY_{t-1} + b^2Y_{t-2} - b^3Y_{t-3} + \cdots + (-b)^tU_0 = \sum_{k=0}^{t-1}(-bL)^kY_t + (-b)^tU_0. \tag{6.42}$$

This is just the finite version of the infinite series in (6.40). Thus, dividing the MA(1) model $Y_t = (1+bL)U_t$ by $(1+bL)$ is justified for $|b| < 1$, so we can write

$$\frac{1}{1+bL}Y_t = U_t,$$

and $1/(1+bL)$ is understood to be the above finite sequence for a finite set of data, while in the limit, we arrive at the geometric infinite sequence because $|b| < 1$ implies $(-b)^tU_0 \to 0$.

As $U_t \overset{\text{i.i.d.}}{\sim} \mathrm{N}(0,\sigma^2)$, taking expectations of both sides of (6.41) reveals that $\mathbb{E}[Y_t] = c$,

$$\gamma_0 = \mathbb{V}(Y_t) = \mathbb{E}[(Y_t - c)^2] = \mathbb{E}[(U_t + bU_{t-1})^2] = (1+b^2)\sigma^2, \tag{6.43}$$

$\gamma_1 = b\sigma^2$, and $\gamma_s = 0$ for $s \geqslant 2$. The correlation structure is thus

$$\rho_1 = \frac{b}{1+b^2}, \qquad \rho_2 = \rho_3 = \cdots = 0, \tag{6.44}$$

and ρ_1 has a maximum of 1/2 at $|b| = 1$.

Simulating an MA(1) process is straightforward; it just involves evaluating the recursion $Y_t = U_t + bU_{t-1}$ with $U_t \overset{\text{i.i.d.}}{\sim} \mathrm{N}(0,\sigma^2)$, $t = 0, \ldots, T$. Listing 6.4 implements this method.

Let $c = 0$ and $\sigma^2 = 1$, and consider the model $Y_t = U_t + b^{-1}U_{t-1}$ for $b \neq 0$. From (6.43), $\gamma_0 = (1 + b^{-2})$, $\gamma_1 = b^{-1}$ and

$$\rho_1 = \frac{b^{-1}}{1+b^{-2}} = \frac{b^2}{b^2}\frac{b^{-1}}{1+b^{-2}} = \frac{b}{1+b^2},$$

2 Some students set up the sum of the Y_t terms differently, by working backwards, thus getting

$$U_t = \sum_{j=1}^{t}(-bL)^{t-j}Y_t,$$

and these are of course equivalent, seen by letting $k = t - j$ so that when $j = t$, $k = 0$, etc.

```
function y=ma1sim(nobs,sigma,b)
u=sigma*randn(nobs,1); y=zeros(nobs,1); y(1)=u(1)+b*sigma*randn(1,1);
for i=2:nobs, y(i) = u(i) + b*u(i-1); end
```

Program Listing 6.4: Simulates an MA(1) series with Gaussian innovations.

so that the processes $Y_t = U_t + b^{-1}U_{t-1}$ and $Y_t = U_t + bU_{t-1}$ have exactly the same correlation structure and either model could be used. From (6.39), the model $Y_t = U_t + bU_{t-1}$ can be written as $(1 + bL)^{-1}Y_t = U_t$ or

$$Y_t = bY_{t-1} - b^2Y_{t-2} + b^3Y_{t-3} - \cdots + U_t, \tag{6.45}$$

and the alternating sign geometric series of these coefficients converges for $|b| < 1$ and diverges for $|b| \geqslant 1$. Because it is more convenient to work with convergent series, the MA(1) model with $|b| < 1$ is preferred to $|b| > 1$. If $|b| < 1$, then the model is said to be **invertible**, and is otherwise **non-invertible**. For $b = 1$, there is only one representation of the model, but as the sum of coefficients in (6.45) diverges, the borderline case of $b = 1$ is also deemed non-invertible.

From the simple covariance structure, matrix $\mathbf{\Sigma}$ and its inverse are easily constructed, so that the likelihood and, thus, the m.l.e. can be computed. Unfortunately, an expression resembling (6.28) that partitions the likelihood is not available with the moving average model. One way of proceeding is just to compute $\mathbf{\Sigma}$ corresponding to the whole T-length sample, along with its inverse and determinant, and then evaluate the likelihood function as

$$\mathcal{L}(\mathbf{b}, \sigma; \mathbf{y}) = \frac{|\mathbf{\Sigma}^{-1}|^{1/2}}{(2\pi\sigma^2)^{T/2}} \exp\left\{-\frac{1}{2\sigma^2}\mathbf{y}'\mathbf{\Sigma}^{-1}\mathbf{y}\right\}, \qquad \mathbf{y} = (y_1, \dots, y_T)'. \tag{6.46}$$

The first problem associated with the numeric maximization of the log of (6.46) is that $\hat{b}$ should be restricted to lie in $(-1, 1]$ so that the resulting model (except at the borderline) is invertible. This can be achieved in the following way: Let σ^2 be any positive value, let $0 < |b| < 1$, and consider the two models $Y_t = U_t + bU_{t-1}$, $U_t \overset{\text{i.i.d.}}{\sim} \mathrm{N}(0, \sigma_U^2)$, and $Z_t = V_t + b^{-1}V_{t-1}$, $V_t \overset{\text{i.i.d.}}{\sim} \mathrm{N}(0, \sigma_V^2)$, such that $\mathbb{V}(Y_t) = \mathbb{V}(Z_t)$. This implies

$$\mathbb{V}(Y_t) = (1 + b^2)\sigma_U^2 = (1 + b^{-2})\sigma_V^2 = \mathbb{V}(Z_t),$$

or

$$\sigma_U^2 = \sigma_V^2 \frac{1 + b^{-2}}{1 + b^2} = \frac{\sigma_V^2}{b^2}.$$

This can be used during the iterative maximization of the likelihood as follows: If the numeric function maximizer attempts a value of $|b| > 1$, set $b \leftarrow 1/b$ and then $\sigma^2 \leftarrow \sigma^2/b^2$.

The second problem with numeric maximization of (6.46) is that it will be computationally slow when T is over, say, 100, and will be essentially impossible to compute for data sets with thousands of observations (as arise, e.g., in the analysis of daily, or even higher frequency financial asset returns).[3] There exists a method for computing the inverse of this patterned matrix that is (much) faster than $O(T^3)$, as shown in Uppuluri and Carpenter (1969), as well as a closed-form approximation, from Durbin (1959).

3 This concern, in turn, is mitigated by reality: It is unlikely that the moving average parameter is constant over such long time periods. In reality, shorter moving windows for estimation are used to account for this issue. See also Chapter 13 in this regard.

The method we suggest is to use an approximation to the likelihood function in place of the exact expression. Its use not only speeds up the calculations enormously, but, precisely for larger sample sizes, its approximate nature becomes negligible compared to use of the exact m.l.e. If we condition on $U_0 = u_0 := 0 = \mathbb{E}[U_0]$, then the simple recursion $U_t = Y_t - bU_{t-1}$, $t = 1, \dots, T$ can be used to get the "filtered" values $\widehat{U}_t = Y_t - b\widehat{U}_{t-1}$. The **conditional likelihood** is then just the product of T normal densities evaluated at the $\widehat{U}_t$, i.e.,

$$\mathcal{L}^{\text{cond}}(b, \sigma^2; \mathbf{Y}, U_0) = \prod_{t=1}^{T} \phi(\widehat{U}_t; 0, \sigma^2), \quad \phi(u; 0, \sigma^2) = \frac{1}{\sqrt{2\pi}\sigma} \exp\left\{-\frac{1}{2}\frac{u^2}{\sigma^2}\right\}, \tag{6.47}$$

and its log would be maximized over b and σ. The programs in Listings 6.5 and 6.6 implement both the exact and conditional m.l.e. calculation.

Asymptotically, the two methods of likelihood calculation are equivalent when the model is invertible. However, there will be differences when working with small sample sizes. Figure 6.3 shows the results of a simulation study, based on 10,000 replications, of the behavior of the exact and conditional m.l.e. of an MA(1) model for two values of b, $\sigma = 10$, and sample size $T = 15$. The sampling variation for $\hat{b}$ is large for both methods of estimation because of the small sample size, with the exact m.l.e. having a tendency to pile up at the border more so than for the conditional m.l.e. For $\hat{\sigma}$, the bias of the exact m.l.e. is small, but apparent. The same simulation but with $b = 0$ reveals that conditional and exact m.l.e.s result in virtually identical small-sample distributions. The reader is encouraged to replicate these results.

```
function [param, stderr, resid]=ma1(y,exact)
ylen=length(y); y=reshape(y,ylen,1); initvec=[0 std(y)]';
opt=optimset('Display','iter','TolX',1e-3,'MaxIter',100,'LargeScale','off');
if exact==1, [param,loglik,exitflag]=fminunc(@exactma1_,initvec,opt,y);
else         [param,loglik,exitflag]=fminunc(@condma1_,initvec,opt,y);
end
b=param(1); littlesig=abs(param(2));
if abs(b)>1, b=1/b; littlesig=littlesig/abs(b); end
param=[b littlesig]';
if nargout>1
  if exact==1, H = -hessian(@exactma1_,param,y); stderr=sqrt(diag(inv(H)));
  else         H = -hessian(@condma1_,param,y); stderr=sqrt(diag(inv(H)));
  end
end
if nargout>2
  if exact==1
    Sigma=ma1Sigma(b,ylen); SigInv=inv(Sigma);
    [V,D]=eig(0.5*(SigInv+SigInv')); W=sqrt(D); SigInvhalf = V*W*V';
    resid = SigInvhalf*y/littlesig;
  else
    [garb,uvec]=condma1_(param,y); resid=uvec/littlesig;
  end
end
```

Program Listing 6.5: Computes the m.l.e. (exact or conditional) of an MA(1) model. Set `exact` to 1 to compute the exact m.l.e., otherwise the conditional m.l.e. is computed. The program is continued in Listing 6.6.

```
function [loglik,uvec]=condma1_(param,y)
ylen=length(y); uvec=zeros(ylen,1); pastu=0;
b=param(1); sig=abs(param(2)); % this is NOT sigma^2, but just (little) sigma.
if abs(b)>1, b=1/b; sig=sig/abs(b); end
for t=1:ylen,  u=y(t)-b*pastu; uvec(t)=u; pastu=u; end
ll = - ylen * log(sig) - sum(uvec.^2)/(2*sig.^2); loglik = -ll;

function loglik=exactma1_(param,y)
ylen=length(y); b=param(1); sig=abs(param(2));
if abs(b)>1, b=1/b; sig=sig/abs(b); end
Sigma=ma1Sigma(b,ylen); % varcov matrix, but not scaled by little sigma.
Vi=inv(Sigma); detVi=det(Vi);
if detVi<=0, loglik=abs(detVi+0.01)*1e4; return, end
ll = - ylen * log(sig) + 0.5*log(detVi) - y'*Vi*y/(2*sig^2); loglik = -ll;

function Sigma=ma1Sigma(b,ylen); % construct the varcov in a primitive way:
Sigma=zeros(ylen,ylen); v=(1+b^2); for i=1:ylen, Sigma(i,i)=v; end
for i=1:(ylen-1), Sigma(i,i+1)=b; Sigma(i+1,i)=b; end
```

Program Listing 6.6: Continuation of Listing 6.5.

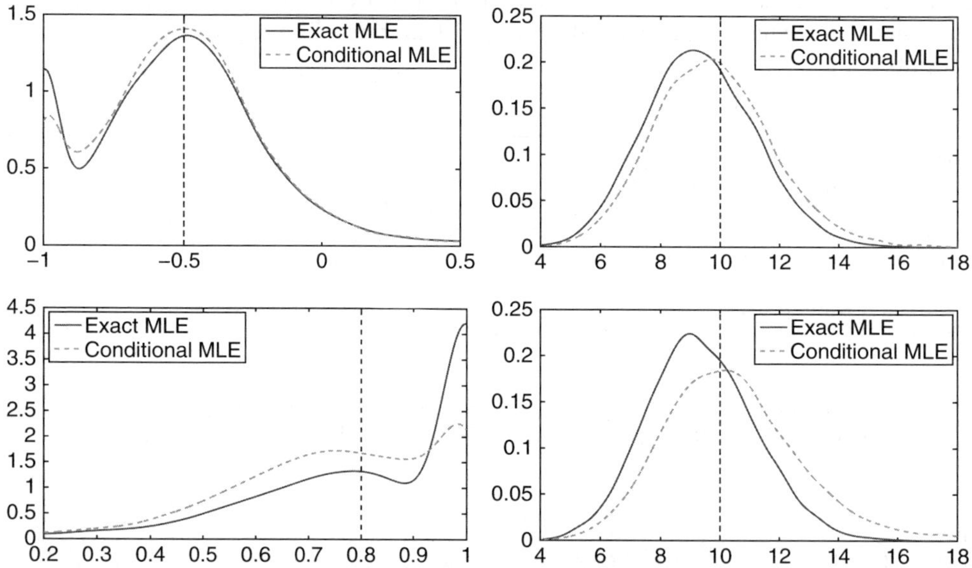

Figure 6.3 Left: Density of the exact m.l.e. (solid) and conditional m.l.e. (dashed) of parameter b in the MA(1) model based on $T = 15$ observations, for true $b = -0.5$ (top), $b = 0.8$ (bottom), and $\sigma = 10$. **Right:** Same, but for $\hat{\sigma}$.

The MA model (and, more generally, the ARMA model) is also amenable to the state space representation and use of the Kalman filter for computing the likelihood, and has the advantage of not requiring inversion of $T \times T$ matrices; see the references in Section 5.6 for details.

As the MA(1) model has only two unknown parameters, the choice of starting values is not overly important: When using just the naive value of zero for b and the sample variance for σ^2 (as done in the program in Listing 6.5), one or two iterations of the numeric function maximization algorithm are enough to pull the values into a region very close to the final values. Nevertheless, it is, in general, better practice to use more intelligent starting values if easily computed ones are available. See Problem 6.3 for details.

6.2.2 MA(q) Processes

The MA(1) model (6.41) can be extended in a natural way to the MA(q) model, given by

$$Y_t = c + U_t + b_1 U_{t-1} + b_2 U_{t-2} + \cdots + b_q U_{t-q} = b(L)U_t, \tag{6.48}$$

where

$$b(L) = 1 + b_1 L + \cdots + b_q L^q. \tag{6.49}$$

The mean of Y_t is clearly c. Using the i.i.d. property of the U_t,

$$\begin{aligned}\mathrm{Cov}(Y_t, Y_{t+s}) &= \mathbb{E}[(U_t + b_1 U_{t-1} \cdots + b_q U_{t-q})(U_{t+s} + b_1 U_{t+s-1} + \cdots + b_q U_{t+s-q})] \\ &= \sigma^2 \sum_{i=0}^{q-|s|} b_i b_{i+|s|},\end{aligned}$$

where $b_0 \equiv 1$. Thus, $\gamma_0 = \mathbb{V}(Y_t) = \sigma^2 \sum_{i=0}^{q} b_i^2$ and

$$\gamma_s = \mathrm{Cov}(Y_t, Y_{t+s}) = \begin{cases} \sigma^2 \sum_{i=0}^{q-|s|} b_i b_{i+|s|}, & |s| \leqslant q, \\ 0, & |s| > q, \end{cases} \tag{6.50}$$

from which the $\rho_s = \gamma_s/\gamma_0$ can be calculated. Van der Leeuw (1994) shows that the covariance matrix of an MA(q) model with $\sigma = 1$ can be expressed as

$$\mathbf{\Sigma} = \mathbf{M}\mathbf{M}' + \mathbf{N}\mathbf{N}', \tag{6.51}$$

where $\mathbf{M}$ is the $T \times T$ lower triangular band matrix with first column $(1, b_1, \dots, b_q, 0, \dots, 0)$ and

$$\mathbf{N} = \begin{bmatrix} \mathbf{N}_1 \\ \mathbf{0} \end{bmatrix}, \qquad \mathbf{N}_1 = \begin{bmatrix} b_q & b_{q-1} & \cdots & b_1 \\ 0 & b_q & \cdots & b_2 \\ \vdots & \vdots & \ddots & \vdots \\ 0 & 0 & \cdots & b_q \end{bmatrix}, \tag{6.52}$$

where $\mathbf{N}$ is of size $T \times q$. Note that matrices $\mathbf{M}$ and $\mathbf{N}$ parallel those of $\mathbf{P}$ and $\mathbf{Q}$ for the AR(p) case, and that for the AR(p) model, $\mathbf{\Sigma}^{-1}$ was directly constructed, whereas it is $\mathbf{\Sigma}$ for the MA(q) model.

For example, with $q = 2$, $\sigma^2 = 1$, and $T = 4$,

$$\begin{aligned}\mathbf{\Sigma} &= \begin{bmatrix} 1 & 0 & 0 & 0 \\ b_1 & 1 & 0 & 0 \\ b_2 & b_1 & 1 & 0 \\ 0 & b_2 & b_1 & 1 \end{bmatrix} \begin{bmatrix} 1 & b_1 & b_2 & 0 \\ 0 & 1 & b_1 & b_2 \\ 0 & 0 & 1 & b_1 \\ 0 & 0 & 0 & 1 \end{bmatrix} + \begin{bmatrix} b_2 & b_1 \\ 0 & b_2 \\ 0 & 0 \\ 0 & 0 \end{bmatrix} \begin{bmatrix} b_2 & 0 & 0 & 0 \\ b_1 & b_2 & 0 & 0 \end{bmatrix} \\ &= \begin{bmatrix} 1 + b_1^2 + b_2^2 & b_1 + b_1 b_2 & b_2 & 0 \\ b_1 + b_1 b_2 & 1 + b_1^2 + b_2^2 & b_1 + b_1 b_2 & b_2 \\ b_2 & b_1 + b_1 b_2 & 1 + b_1^2 + b_2^2 & b_1 + b_1 b_2 \\ 0 & b_2 & b_1 + b_1 b_2 & 1 + b_1^2 + b_2^2 \end{bmatrix},\end{aligned}$$

which agrees with (6.50).

The concept of invertibility is also extended to the MA(q) case. Paralleling the development of the stationarity condition for AR(p) models, express polynomial (6.49) as

$$b(L) = 1 + b_1 L + \cdots + b_q L^q = (1 - \eta_1 L) \cdot \cdots \cdot (1 - \eta_q L), \tag{6.53}$$

so that the roots[4] of $b(L)$ are given by $\eta_1^{-1}, \ldots, \eta_q^{-1}$. The model is invertible when $|\eta_i^{-1}| > 1$ or, equivalently, when $|\eta_i| < 1$, $i = 1, \ldots, q$, where $|\eta_i|$ is the modulus of η_i. Unlike the AR model, we can "flip" any set of the η_i and the correlation structure remains the same. For example, let $b_1 = -0.5$ and $b_2 = -0.24$. As

$$1 - 0.5L - 0.24L^2 = (1 + 0.3L)(1 - 0.8L),$$

the roots are $\eta_1^{-1} = -10/3$ and $\eta_2^{-1} = 10/8$, so that, as $|\eta_i^{-1}| > 1$, $i = 1, 2$, the model is invertible. Flipping η_2 gives

$$(1 + 0.3L)\left(1 - \frac{1}{0.8}L\right) = 1 - 0.95L - 0.375\ L^2 =: 1 + b_1^* L + b_2^* L^2 =: b^*(L),$$

which is obviously not invertible, but the MA(2) models based on $b(L)$ and $b^*(L)$ have exactly the same correlation (but not covariance) structure, namely, to four digits, $\rho_1 = -0.2906$, $\rho_2 = -0.1835$, and $\rho_i = 0$, $i \geqslant 3$. Interestingly, the parameters b_i need not correspond to an invertible MA process in order for (6.51) to be valid. The Schur condition can also be checked for the MA polynomial via (6.6) and (6.7), with q in place of p, $\alpha_0 = 1$, and $\alpha_i = b_i$, $i = 1, \ldots, q$.

Using (6.51), the exact m.l.e. can be obtained by extending the program in Listing 6.5 in an obvious way. Also, the conditional m.l.e. is straightforwardly computed by setting $U_0, \ldots, U_{1-q}$ to zero. This is illustrated in the context of the more general ARMA setting in Section 7.4.

Asymptotically, under certain assumptions on the innovation sequence, for both the exact and conditional m.l.e. of $\mathbf{b} = (b_1, \ldots, b_q)'$,

$$\sqrt{T}(\widehat{\mathbf{b}}_{\mathrm{ML}} - \mathbf{b}) \overset{\text{asy}}{\sim} \mathrm{N}(\mathbf{0}, \sigma^2 \boldsymbol{\Gamma}_*^{-1}), \tag{6.54}$$

where $\boldsymbol{\Gamma}_*$ is the covariance matrix of $\mathbf{V} = (V_t, V_{t-1}, \ldots, V_{t-q+1})'$, the V_t follow the AR(q) process $b(L)V_t = U_t$, and $b(L) = 1 + b_1 L + \cdots + b_q L^q$. Thus, the result parallels that for $\widehat{\mathbf{a}}_{\mathrm{ML}}$ in the AR(p) model, but replacing a_i with $-b_i$. For the MA(1) case,

$$\sqrt{T}(\widehat{b}_{\mathrm{ML}} - b) \overset{\text{asy}}{\sim} \mathrm{N}(0, 1 - b^2), \tag{6.55}$$

while for $q = 2$, replacing a_i with $-b_i$ in the covariance matrix of (6.30),

$$\sqrt{T}(\widehat{\mathbf{b}}_{\mathrm{ML}} - \mathbf{b}) \overset{\text{asy}}{\sim} \mathrm{N}\left(\begin{bmatrix} 0 \\ 0 \end{bmatrix}, \begin{bmatrix} 1 - b_2^2 & b_1(1 - b_2) \\ \cdot & 1 - b_2^2 \end{bmatrix}\right). \tag{6.56}$$

The result for $q = 3$ is given in (6.62) below; see Problem 6.5. Proofs of (6.54) can be found in several textbooks, including the 1970 monograph by Box and Jenkins and subsequent editions (Box et al., 2008), Fuller (1996, Sec. 8.4), and Brockwell and Davis (1991, Sec. 8.8, 10.8).

The covariance matrix in (6.54) can be approximated by using the values of $\widehat{\mathbf{b}}_{\mathrm{ML}}$ in place of their theoretical counterparts and/or the Hessian matrix can be approximated from the likelihood function, as discussed in Section 6.1.3.4. The reader is encouraged to investigate via simulation which method delivers better estimates of the standard error.

Forecasting MA(q) processes is dealt with in Section 7.5.

4 Similar to the expression for the AR polynomial, if $b = (b_1, \ldots, b_q)$ is the MA parameter vector, then executing `rr=roots([b(end:-1:1) 1])` computes the roots in Matlab.

6.3 Problems

Problem 6.1 Use Matlab to graphically determine the ranges of a_1 and a_2 for the AR(3) process to be stationary. Algebraically verify the limiting case $a_3 \downarrow -1$.

Problem 6.2 Recall the discussion of conditional and exact estimation of the MA(1) model. To investigate this, design a simulation to compare the bias and m.s.e. of the exact and conditional m.l.e. for b over a grid of b-values for $T = 15$, $\sigma = 1$.

Problem 6.3 We wish to develop simple estimators for the two parameters of an MA(1) model that can be used as starting values for computing the m.l.e. We consider two. First, use (6.44) to derive a method of moments estimator for b, and, based on this, use (6.43) to get an estimate of σ.

The second way is to use the fact that an MA(1) model can be represented as an infinite AR model, which suggests estimating an AR(p) model via least squares and setting

$$\hat{b} = \hat{a}_1. \tag{6.57}$$

The choice of p will of course influence the quality of the estimator: If p is chosen too small, then the AR(p) model is "very" mis-specified, so that $\hat{b}$ will be quite biased, while if p is chosen too large, then the variance of $\hat{b}$ will be large. This tradeoff becomes acute as $|b|$ approaches one. Koreisha and Pukkila (1990) recommend taking p to be $\sqrt{T}$, rounded off to the nearest integer. Make a program to compute this.

Now compare via simulation the performance (in terms of bias and m.s.e.) of the two estimators for b. Use a grid of b-values, $T = 15$ observations, $\sigma = 1$, and 10,000 replications.

Problem 6.4 For the MA(2) polynomial $b(L) = 1 + b_1L + b_2L^2$ with $b_2 > 0$, there is a range of values of b_1 such that the moduli of the two roots are equal and constant. For example, with $b_2 = 0.8$, the moduli are 1.118 for $-1.78 < b_1 < 1.78$. Explain.

Problem 6.5 Using (6.20) and (6.54) (and a symbolic software package such as Maple), show (6.55) and (6.56), and derive the expression for the $q = 3$ case. Assume throughout that $\sigma^2 = 1$.

Problem 6.6 In Example 6.6, the ρ_i were derived for the AR(2) model and given in (6.18). Observe that, if $a_2 = 0$, then $\rho_1 = a_1$ and $\rho_2 = a_1^2 = \rho_1^2$. It is of interest to know if ρ_2 can equal ρ_1^2 if $a_2 \neq 0$. Show that it cannot.

Problem 6.7 Write a program that produces graphs like those in Figure 6.3. Do so for different values of b.

Problem 6.8 For a stationary AR(2) process, the asymptotic distribution of the AR(2) parameters is given in (6.30) to be

$$\sqrt{T}(\hat{\mathbf{a}}_{\text{ML}} - \mathbf{a}) \overset{\text{asy}}{\sim} \text{N}\left(\begin{bmatrix} 0 \\ 0 \end{bmatrix}, \begin{bmatrix} 1 - a_2^2 & -a_1(1 + a_2) \\ \cdot & 1 - a_2^2 \end{bmatrix}\right).$$

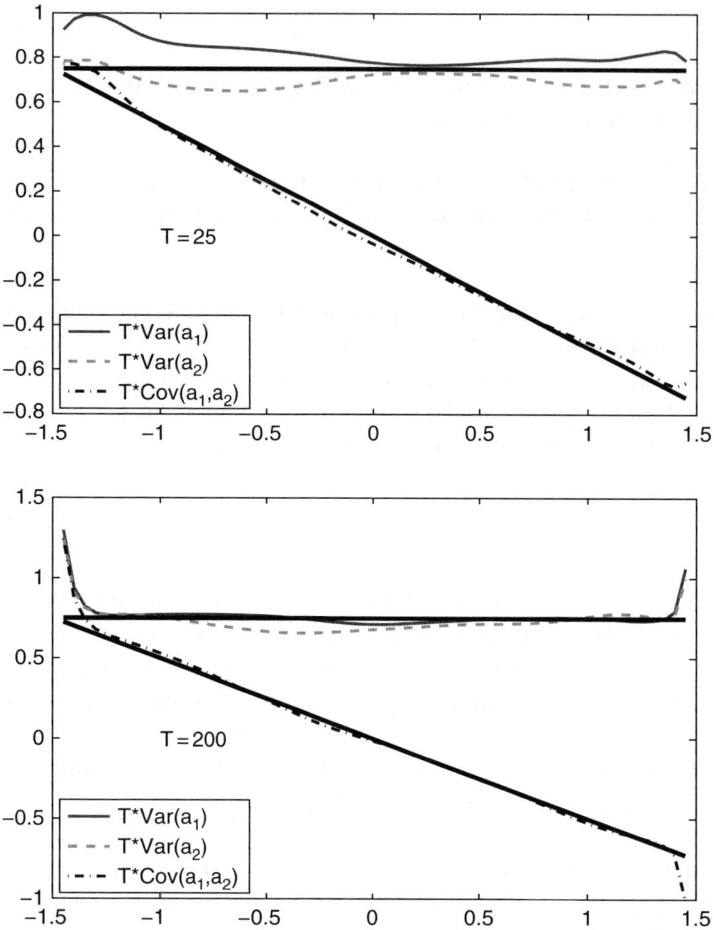

Figure 6.4 **Top:** Variance and covariance of estimated AR(2) parameters, as a function of a_1, for constant $a_2 = -0.5$ and sample size $T = 25$. Solid lines show the asymptotic values from (6.30). **Bottom:** Same, but for $T = 200$.

Write a program to investigate this via simulation for $a_2 = -0.5$ and a grid of a_1-values over the stationarity region (6.8), and produce graphs like those in Figure 6.4, which correspond to $T = 25$ and $T = 200$. Use either Yule–Walker or least squares for estimation.

6.A Appendix: Solutions

Solution to Problem 6.1 Figure 6.5 shows two views of the stationarity region of an AR(3) model. The panels were created with the code in Listing 6.7.

Figure 6.6 is similar, but each graph uses a constant value of a_3. It was created with the code in Listing 6.8.

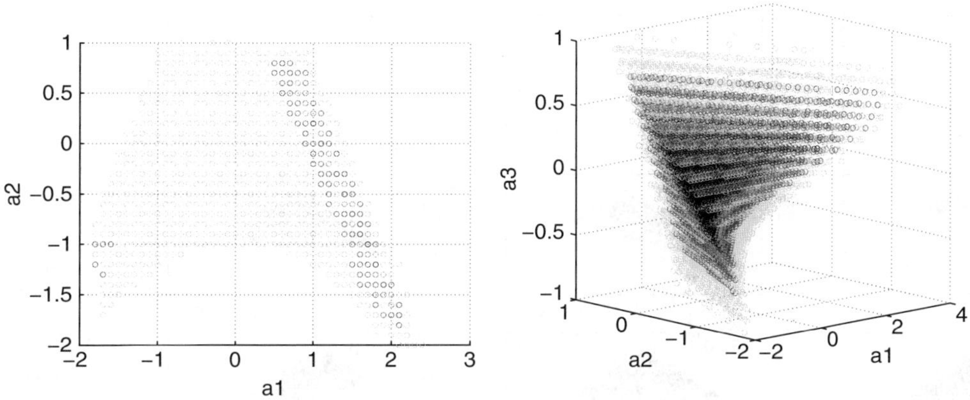

Figure 6.5 Two views of the allowed range of parameters a_1, a_2, a_3 of the AR(3) model such that the process is stationary.

```
%mm=[];
figure
for a1=-3:0.1:3
  for a2=-2:0.1:2
    for a3=-1:0.1:1
      a=[a1 a2 a3]; m=min( abs(roots([-a(end:-1:1) 1])) );
      if m>1
        if     m<1.1, plot3(a1,a2,a3,'yo')
        elseif m<1.2, plot3(a1,a2,a3,'ro')
        elseif m<1.3, plot3(a1,a2,a3,'go')
        elseif m<1.4, plot3(a1,a2,a3,'bo')
        else          plot3(a1,a2,a3,'ko')
        hold on, %mm=[mm m];
        end
      end
    end
  end
end
hold off
set(gca,'fontsize',16), grid, xlabel('a1'), ylabel('a2'), zlabel('a3')
view(-0.5,90)  % two-dimensional view
```

Program Listing 6.7: Produces Figure 6.5. The commented out pieces of the code construct a vector mm. A histogram of it can be made to see the relative size of the minimum modulus of stationary models, which was useful in determining the ranges for the colors used in the plot.

To verify what happens as a_3 approaches -1, use the Schur condition to get

$$S_a = \begin{bmatrix} 1 & 0 & 0 \\ -a_1 & 1 & 0 \\ -a_2 & -a_1 & 1 \end{bmatrix} \begin{bmatrix} 1 & -a_1 & -a_2 \\ 0 & 1 & -a_1 \\ 0 & 0 & 1 \end{bmatrix} - \begin{bmatrix} -a_3 & 0 & 0 \\ -a_2 & -a_3 & 0 \\ -a_1 & -a_2 & -a_3 \end{bmatrix} \begin{bmatrix} -a_3 & -a_2 & -a_1 \\ 0 & -a_3 & -a_2 \\ 0 & 0 & -a_3 \end{bmatrix}$$

$$= \begin{bmatrix} -a_3^2+1 & -a_1 - a_2 a_3 & -a_2 - a_1 a_3 \\ -a_1 - a_2 a_3 & a_1^2 - a_2^2 - a_3^2 + 1 & -a_1 - a_2 a_3 \\ -a_2 - a_1 a_3 & -a_1 - a_2 a_3 & -a_3^2 + 1 \end{bmatrix}.$$

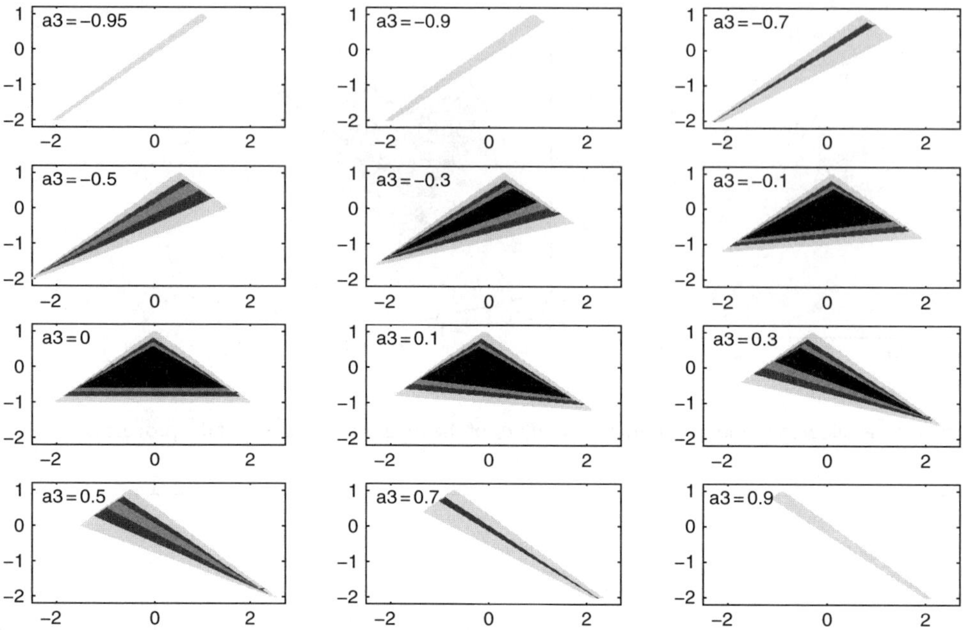

Figure 6.6 Allowed range of a_1 and a_2 in the AR(3) model for a fixed value of a_3, for several values of a_3.

```
a3vec=[-0.95 -0.9:0.2:-0.1 0 0.1:0.2:0.9]; lw=1.3;
for lp=1:length(a3vec)
  a3=a3vec(lp), subplot(4,3,lp)
  for a1=-3:0.02:3
    for a2=-2:0.02:2
      a=[a1 a2 a3]; m=min( abs(roots([-a(end:-1:1) 1])) );
      if m>1
        if     m<1.1, h=plot(a1,a2,'y.');
        elseif m<1.2, h=plot(a1,a2,'r.');
        elseif m<1.3, h=plot(a1,a2,'g.');
        elseif m<1.4, h=plot(a1,a2,'b.');
        else          h=plot(a1,a2,'k.');
        end
        set(h,'linewidth',lw), hold on
      end
    end
  end
  hold off, set(gca,'fontsize',12), axis([-2.5 2.7 -2.2 1.2])
  str=['a3=',num2str(a3)]; text(-2.3,1.0,str,'fontsize',16)
end, orient tall
```

Program Listing 6.8: Produces Figure 6.6.

Thus, $1 - a_3^2 > 0$ and, with $A_1 = a_2 + a_1 a_3$ and $A_2 = 1 - a_3^2 > 0$,

$$\det \begin{bmatrix} -a_3^2 + 1 & -a_1 - a_2 a_3 \\ -a_1 - a_2 a_3 & a_1^2 - a_2^2 - a_3^2 + 1 \end{bmatrix} = -(A_1 - A_2)(A_1 + A_2) > 0. \quad (6.58)$$

Now let $a_3 \downarrow -1$ so that $A_2 \downarrow 0$. Then (6.58) reduces to $A_1^2 < 0$ or $(a_2 - a_1)^2 < 0$, i.e., that $a_1 = a_2$.

Solution to Problem 6.2 The code in Listing 6.9 was used to perform the computation, based on 10,000 replications, and produce the graphs in Figure 6.7. We see, somewhat surprisingly, that the conditional m.l.e. is actually preferred in terms of both bias and m.s.e. for most of the parameter space. For $|b| > 0.7$, the exact m.l.e. exhibits lower bias, while for $|b| > 0.8$ it also has smaller m.s.e.

Solution to Problem 6.3 Using (6.44), its solutions are $b = 0$, if $\rho_1 = 0$, and

$$b = \frac{1}{2\rho_1}\left(1 \pm \sqrt{1 - 4\rho_1^2}\ \right), \quad \rho_1 \neq 0. \quad (6.59)$$

This is valid because $|\rho_1| < 1/2$ for an MA(1) model. To ensure $|\hat{b}| < 1$, i.e., an invertible model, the solution in (6.59) with the negative sign is taken, i.e.,

$$\hat{b} = \begin{cases} 0, & \text{if } \hat{\rho}_1 = 0, \\ \dfrac{1 - \sqrt{(1 - 2\hat{\rho}_1)(1 + 2\hat{\rho}_1)}}{2\hat{\rho}_1}, & \text{if } 0 < |\hat{\rho}_1| < \frac{1}{2}, \\ \operatorname{sgn}(\hat{\rho}_1)\ 0.95 & \text{if } |\hat{\rho}_1| \geqslant \frac{1}{2}, \end{cases} \quad (6.60)$$

where the value 0.95 is arbitrary. Of course, $\Pr(\hat{\rho}_1 = 0) = 0$, and if $|\hat{\rho}_1| \geqslant 1/2$, this might be a signal that an MA(1) model is not appropriate for the data.

```
bvec=-0.95:0.05:0.95; sig=1; T=15; sim=10000;
b1=zeros(sim,length(bvec)); b2=b1; true=kron(ones(sim,1),bvec);
for bloop=1:length(bvec), b=bvec(bloop);
  for i=1:sim
    y=ma1sim(T,sig,b,i); if mod(i,100)==0, disp([i, b]), end
    param=ma1(y,1); b1(i,bloop)=param(1); % exact
    param=ma1(y,2); b2(i,bloop)=param(1); % conditional
  end
end
figure, set(gca,'fontsize',16)
plot(bvec,mean(b1)-bvec,'r-',bvec,mean(b2)-bvec,'g--','linewidth',3)
title('Bias'), legend('Exact MLE','Conditional MLE')
figure, set(gca,'fontsize',16)
plot(bvec, mean((b1-true).^2), 'r-', ...
     bvec, mean((b2-true).^2), 'g--', 'linewidth',3)
title('MSE'), legend('Exact MLE','Conditional MLE')
```

Program Listing 6.9: Simulates the exact and conditional MLE of an MA(1) model and compares their bias and m.s.e. for parameter b. Function `ma1sim` is given in Listing 6.4, while the function for estimation of the MA(1) model is given in Listings 6.5 and 6.6.

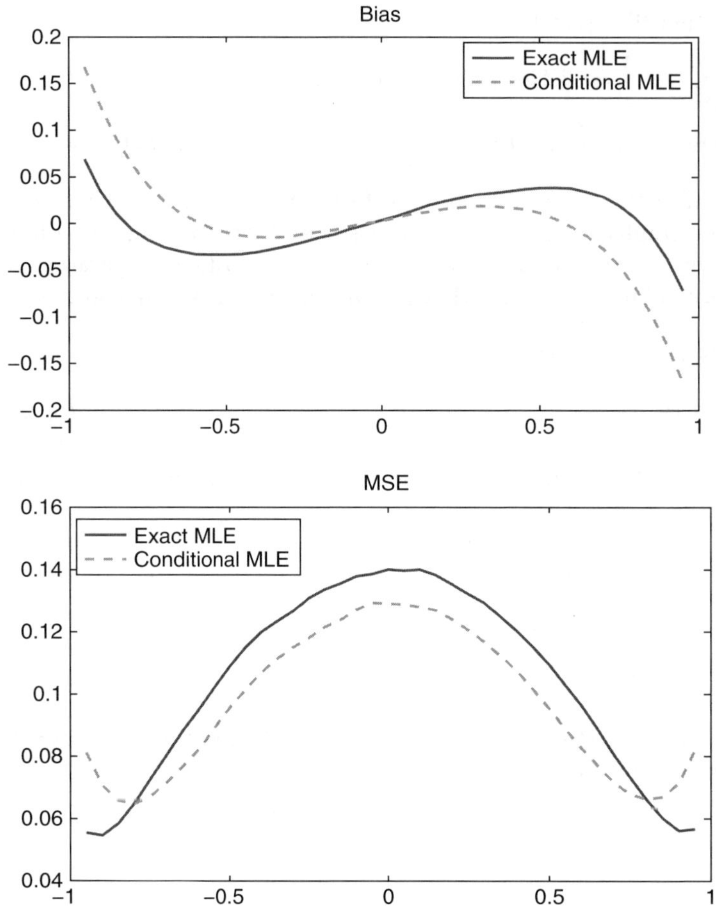

Figure 6.7 The bias (top) and m.s.e. (bottom) of the exact m.l.e. (solid line) and conditional m.l.e. (dashed line) of $\hat{b}$ in the MA(1) model as a function of parameter b, based on $T = 15$ observations, $\sigma = 1$, and 10,000 replications. The graphics were produced from the code in Listing 6.9.

```
T=length(y); p=round(sqrt(T)); z=y(p+1:end); zl=length(z); Z=[];
for i=1:p, Z=[Z y(p-i+1:p-i+zl)]; end,   a=inv(Z'*Z)*Z'*z; b=a(1)
```

Program Listing 6.10: Code for computing (6.57).

An estimate of σ^2 follows from (6.43) as $\hat{\sigma}^2 = \hat{\sigma}_Y^2/(1 + \hat{b}^2)$, where $\hat{\sigma}_Y^2 = S_Y^2$ is the sample variance of the series $Y_1, \dots, Y_T$.

For the approach based on an estimated AR(p) model, we set $\hat{b} = \hat{a}_1$ as in (6.57), where a_1 is the first of p AR terms estimated via least squares. The code in Listing 6.10 can be used to compute (6.57).

For the comparison, Figure 6.8 shows the resulting bias and m.s.e. Compared to Figure 6.7, we see that the bias and m.s.e. of the initial estimators are considerably larger than for the m.l.e.s as $|b|$

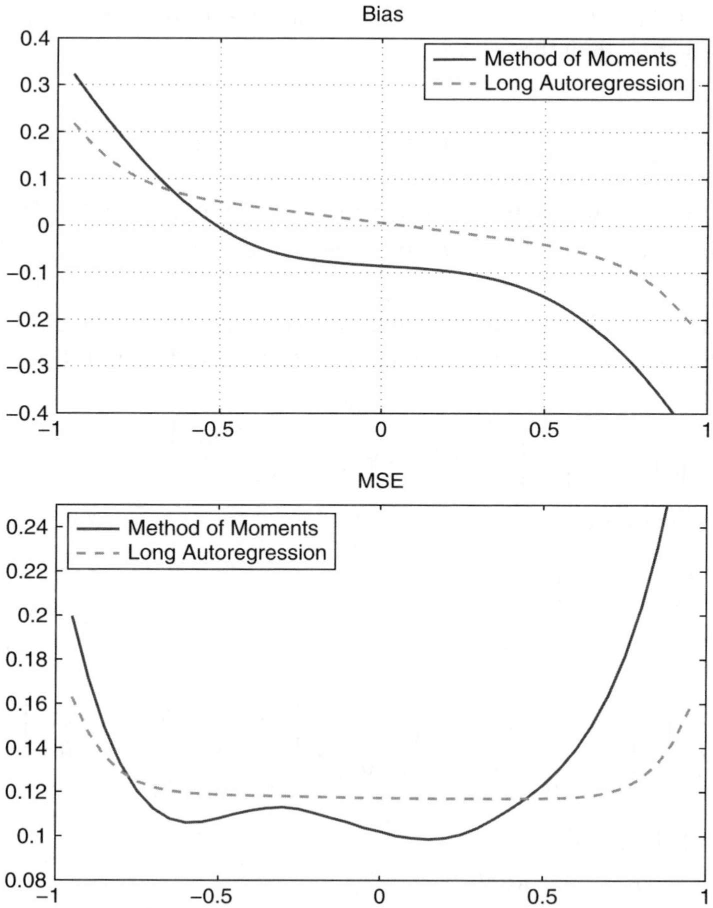

Figure 6.8 The bias (top) and m.s.e. (bottom) of the moments-method (6.60) (solid line) and long autoregression method (6.57) (dashed line) of estimating $\hat{b}$ with closed-form expressions, for the MA(1) model as a function of parameter b, based on $T = 15$ observations, $\sigma = 1$, and 10,000 replications.

approaches one, but are comparable for b near zero. Method (6.57) has smaller bias than (6.60) for the entire parameter space except for a small region around $b = -0.5$. Regarding the m.s.e., (6.60) is slightly better than (6.57) for $-0.8 < b < 0.5$, but (6.57) exhibits a much lower m.s.e. than (6.60) as b increases towards 1.0. Taken all together, (6.57) would be preferred, but (6.60) is far cheaper numerically.

Solution to Problem 6.4 Write $b(L) = 1 + b_1 L + b_2 L^2 = (1 - \eta_1 L)(1 - \eta_2 L)$, multiply by L^{-2} and set $\eta = L^{-1}$ to get $\eta^2 + b_1\eta + b_2 = (\eta - \eta_1)(\eta - \eta_2)$ with roots

$$\eta_{1,2} = -\frac{1}{2}b_1 \pm \frac{1}{2}\sqrt{b_1^2 - 4b_2}. \tag{6.61}$$

Values $\eta_{1,2}^{-1}$ are the two roots of $b(L)$. If $b_2 > 0$ and $b_1^2 - 4b_2 < 0$, then the latter term in (6.61) is a purely imaginary number, and the the moduli of the $\eta_{1,2}$ are

$$\sqrt{\left(-\frac{1}{2}b_1\right)^2 + \left(\frac{1}{2}\sqrt{4b_2 - b_1^2}\right)^2} = \sqrt{\frac{1}{4}b_1^2 + \frac{1}{4}(4b_2 - b_1^2)} = \sqrt{b_2},$$

which is obviously constant in b_1 and the moduli of the roots of $b(L)$ are $b_2^{-1/2}$ for $|b_1| < 2b_2^{1/2}$. For $b_2 = 0.8$, $b_2^{-1/2} = \sqrt{5}/2$, and the range for b_1 is $\pm 4/\sqrt{5}$ or, approximately, ± 1.788854382.

Solution to Problem 6.5 For the MA(1) model, $\mathbf{\Gamma}_*$ can be computed as the inverse of the 1×1 matrix of $\mathbf{\Sigma}$ from (6.20). We cannot compute $\mathbf{\Gamma}_*^{-1}$ directly from (6.20) because (6.20) requires the use of at least $p + 1$ elements. Thus,

$$\mathbf{\Sigma}^{-1} = \mathbf{P}'\mathbf{P} - \mathbf{Q}\mathbf{Q}' = \begin{bmatrix} 1 & b_1 \\ 0 & 1 \end{bmatrix}\begin{bmatrix} 1 & 0 \\ b_1 & 1 \end{bmatrix} - \begin{bmatrix} b_1 \\ 0 \end{bmatrix}\begin{bmatrix} b_1 & 0 \end{bmatrix} = \begin{bmatrix} 1 & b_1 \\ b_1 & 1 \end{bmatrix},$$

and the inverse of the upper 1×1 submatrix of

$$\mathbf{\Sigma} = \begin{bmatrix} 1 & b_1 \\ b_1 & 1 \end{bmatrix}^{-1} = \frac{1}{1 - b_1^2}\begin{bmatrix} 1 & -b_1 \\ -b_1 & 1 \end{bmatrix}$$

is $1 - b_1^2$. For $q = 2$,

$$\mathbf{\Sigma}^{-1} = \begin{bmatrix} 1 & b_1 & b_2 \\ 0 & 1 & b_1 \\ 0 & 0 & 1 \end{bmatrix}\begin{bmatrix} 1 & 0 & 0 \\ b_1 & 1 & 0 \\ b_2 & b_1 & 1 \end{bmatrix} - \begin{bmatrix} b_2 & b_1 \\ 0 & b_2 \\ 0 & 0 \end{bmatrix}\begin{bmatrix} b_2 & 0 & 0 \\ b_1 & b_2 & 0 \end{bmatrix} = \begin{bmatrix} 1 & b_1 & b_2 \\ b_1 & b_1^2 - b_2^2 + 1 & b_1 \\ b_2 & b_1 & 1 \end{bmatrix}$$

and, with $K = (b_2 - 1)(b_1 - b_2 - 1)(b_1 + b_2 + 1)$,

$$\mathbf{\Sigma} = \frac{1}{K}\begin{bmatrix} 1 + b_2 & -b_1 & -b_2 + b_1^2 - b_2^2 \\ -b_1 & 1 + b_2 & -b_1 \\ -b_2 + b_1^2 - b_2^2 & -b_1 & 1 + b_2 \end{bmatrix},$$

so that the inverse of the upper 2×2 submatrix of $\mathbf{\Sigma}$ is

$$\mathbf{\Gamma}_*^{-1} = \begin{bmatrix} 1 - b_2^2 & b_1 - b_1 b_2 \\ b_1 - b_1 b_2 & 1 - b_2^2 \end{bmatrix}.$$

Similarly for the $q = 3$ case,

$$\mathbf{\Sigma}^{-1} = \begin{bmatrix} 1 & b_1 & b_2 & b_3 \\ 0 & 1 & b_1 & b_2 \\ 0 & 0 & 1 & b_1 \\ 0 & 0 & 0 & 1 \end{bmatrix}\begin{bmatrix} 1 & 0 & 0 & 0 \\ b_1 & 1 & 0 & 0 \\ b_2 & b_1 & 1 & 0 \\ b_3 & b_2 & b_1 & 1 \end{bmatrix} - \begin{bmatrix} b_3 & b_2 & b_1 \\ 0 & b_3 & b_2 \\ 0 & 0 & b_3 \\ 0 & 0 & 0 \end{bmatrix}\begin{bmatrix} b_3 & 0 & 0 & 0 \\ b_2 & b_3 & 0 & 0 \\ b_1 & b_2 & b_3 & 0 \end{bmatrix}$$

$$= \begin{bmatrix} 1 & b_1 & b_2 & b_3 \\ b_1 & b_1^2 - b_3^2 + 1 & b_1 + b_1 b_2 - b_2 b_3 & b_2 \\ b_2 & b_1 + b_1 b_2 - b_2 b_3 & b_1^2 - b_3^2 + 1 & b_1 \\ b_3 & b_2 & b_1 & 1 \end{bmatrix},$$

with inverse

$$\mathbf{\Sigma} = \frac{1}{K}\begin{bmatrix} -b_2+b_1b_3+b_3^2-1 & b_1-b_2b_3 & b_2-b_1b_3-b_1^2+b_2^2 & J \\ b_1-b_2b_3 & -b_2+b_1b_3+b_3^2-1 & b_1-b_2b_3 & b_2-b_1b_3-b_1^2+b_2^2 \\ b_2-b_1b_3-b_1^2+b_2^2 & b_1-b_2b_3 & -b_2+b_1b_3+b_3^2-1 & b_1-b_2b_3 \\ J & b_2-b_1b_3-b_1^2+b_2^2 & b_1-b_2b_3 & -b_2+b_1b_3+b_3^2-1 \end{bmatrix},$$

where

$$J = b_3 - 2b_1b_2 + b_2b_3 + b_1^3 - b_3^3 - b_1b_2^2 - b_1b_3^2 + b_1^2b_3 + b_2^2b_3\,,$$
$$K = (b_1+b_2+b_3+1)(b_2-b_1b_3+b_3^2-1)(b_2-b_1-b_3+1).$$

The inverse of the upper 3×3 submatrix of this is

$$\mathbf{\Gamma}_*^{-1} = \begin{bmatrix} 1-b_3^2 & b_1-b_2b_3 & b_2-b_1b_3 \\ b_1-b_2b_3 & b_1^2-b_2^2-b_3^2+1 & b_1-b_2b_3 \\ b_2-b_1b_3 & b_1-b_2b_3 & 1-b_3^2 \end{bmatrix}, \tag{6.62}$$

and

$$\operatorname{tr}(\mathbf{\Gamma}_*^{-1}) = b_1^2 - b_2^2 - 3(b_3^2 - 1).$$

Solution to Problem 6.6 Such values of a_2 are given by the solutions to

$$\frac{a_1^2 + a_2 - a_2^2}{1 - a_2} = \frac{a_1^2}{(1-a_2)^2}, \tag{6.63}$$

which are 0, $1 - a_1$, and $a_1 + 1$. But the latter condition in (6.9), namely $a_1 < 1 - a_2$, is violated for the solution $a_2 = 1 - a_1$, and the former condition, $a_2 - 1 < a_1$, is violated for the solution $a_2 = 1 + a_1$, so that there are no stationary AR(2) models with nonzero a_2 and such that $\rho_2 = \rho_1^2$.

```
b=-0.5; sig=10; T=15; sim=1000;
b1=zeros(sim,1); sig1=b1; bstd1=b1; b2=b1; sig2=b1; bstd2=b1;
for i=1:sim
  if mod(i,100)==0, i, end
  y=ma1sim(T,sig,b,i);
  [param, stderr]=ma1(y,1);
  b1(i)=param(1); sig1(i)=param(2); bstd1(i)=stderr(1);
  [param, stderr]=ma1(y,0);
  b2(i)=param(1); sig2(i)=param(2); bstd2(i)=stderr(1);
end
figure
[f,g]=kerngau(b1); plot(g,f,'r-'), hold on
[f,g]=kerngau(b2); plot(g,f,'g--'), hold off
figure
[f,g]=kerngau(sig1); plot(g,f,'r-'), hold on
[f,g]=kerngau(sig2); plot(g,f,'g--'), hold off
mean(bstd1), std(b1)
mean(bstd2), std(b2)
```

Program Listing 6.11: Code used to produce Figure 6.3. Change the first line, b, to investigate the behavior for different b.

```
T=25; % T=200; Need to also change text command below.
sig2=1; a2=-0.5; sim=500; lb=-(1-a2); ub=1-a2;
a1v=(lb+0.05):0.05:(ub-0.05);
var1=zeros(length(a1v),1); var2=var1; cov12=var1;
asyvar=var1; asycov=var1;
for aloop=1:length(a1v)
  a1=a1v(aloop), a1est=zeros(sim,1); a2est=a1est;
  for s=1:sim
    y=armasim(T,sig2,[a1 a2],[],s,500); ayw=yw(y,2);
    a1est(s)=ayw(1); a2est(s)=ayw(2);
  end
  varcov=T*cov(a1est,a2est);
  var1(aloop)=varcov(1,1); var2(aloop)=varcov(2,2);
  cov12(aloop)=varcov(1,2);
  asyvar(aloop)=1-a2^2; asycov(aloop)=-a1*(1+a2);
end
h=plot(a1v,var1,'r-',a1v,var2,'g--',a1v,cov12,'b-.')
for i=1:3, set(h(i),'linewidth',2), end
hold on
h=plot(a1v,asyvar,'k-',a1v,asycov,'k-')
for i=1:2, set(h(i),'linewidth',3), end
hold off
set(gca,'fontsize',16), text(-1,0,'T=25','fontsize',20)
legend('T*Var(a_1)', 'T*Var(a_2)', 'T*Cov(a_1,a_2)', ...
        'Location','Southwest')
```

Program Listing 6.12: Produces Figure 6.4.

Alternatively, values of a_2 such that $\rho_2 = a_1^2$ are given by the solutions to

$$\frac{a_1^2 + a_2 - a_2^2}{1 - a_2} = a_1^2,$$

which are zero if $a_1 = 0$, and zero and $1 + a_1^2$ if $a_1 \neq 0$. But from the first condition in (6.8), $a_2 < 1$, which is clearly not fulfilled by $1 + a_1^2$, so that, for stationary AR(2) models, $\rho_2 \neq a_1^2$.

Solution to Problem 6.7 The code in Listing 6.11 can be used.

Solution to Problem 6.8 The code in Listing 6.12 was used to produce Figure 6.4.

7

ARMA Processes

Parsimoniously parameterized time-series models were developed as aids to short-term forecasting, where the fiction that the analyst has discovered the 'true' model is innocuous. Such fiction, however, is far from innocuous when attempting to base inference about long-run behavior on these fitted models.

(Paul Newbold *et al.*, 1993)

The AR(p) and MA(q) time-series models seen in the previous chapter are straightforward to combine, resulting in the so-called ARMA(p, q) model—a very flexible model class capable of producing a variety of autocorrelation structures. The infinite AR and MA expansions of this model, as developed in Section 7.2, will be seen to play an important role in estimation and forecasting, these being discussed in Sections 7.3, 7.4, and 7.5. This is done within the ARMAX model, which augments the ARMA error structure with a set of regressors, as in Chapter 5. Section 7.6 builds on the material in Section 5.4 for obtaining an improved estimator of the AR(1) parameter. Finally, Section 7.7 briefly introduces ARMA-type models that embody certain forms of nonlinearity, and/or can serve as an alternative to a near or exact unit root process.

7.1 Basics of ARMA Models

7.1.1 The Model

Similar to (6.39), consider the infinite AR model whose coefficients are functions of a single parameter a, given by

$$Y_t = aY_{t-1} + a^2Y_{t-2} + a^3Y_{t-3} + \cdots + U_t, \quad |a| < 1/2, \tag{7.1}$$

where here and throughout the chapter,

$$U_t \overset{\text{i.i.d.}}{\sim} \mathrm{N}(0, \sigma^2), \tag{7.2}$$

as in (6.2). Model (7.1) can be written as $(1 - aL - a^2L^2 - \cdots)Y_t = U_t$, and, with $z = aL$, the polynomial is

$$1 - z - z^2 - \cdots = 2 - (1 + z + z^2 + \cdots) = 2 - \frac{1}{1-z} = \frac{1-2z}{1-z},$$

Linear Models and Time-Series Analysis: Regression, ANOVA, ARMA and GARCH, First Edition. Marc S. Paolella.

so that

$$\frac{1-2z}{1-z}Y_t = U_t,$$

or $Y_t - 2aY_{t-1} = U_t - aU_{t-1}$ or

$$Y_t = 2aY_{t-1} + U_t - aU_{t-1}. \tag{7.3}$$

Model (7.3) is stationary for $|a| < 1/2$. This model combines an AR(1) with an MA(1) structure, though in a restricted way. Relaxing the constraint gives a model with the form $Y_t = aY_{t-1} + U_t + bU_{t-1}$, which is, appropriately, referred to as an ARMA(1,1) model (and is stationary if $|a| < 1$).

More generally, by combining the AR(p) and MA(q) structures and introducing the constant c as in (6.1) and (6.48), the ARMA(p, q) model

$$a(L)Y_t = c + b(L)U_t \tag{7.4}$$

is obtained, where

$$a(L) = 1 - a_1L - \cdots - a_pL^p \quad \text{and} \quad b(L) = 1 + b_1L + \cdots + b_qL^q. \tag{7.5}$$

The same exercise that led to (6.11) shows that the mean is

$$\mathbb{E}[Y_t] = \mu = c/(1 - a_1 - \cdots - a_p). \tag{7.6}$$

Using this value of μ, it is easy to verify that model (7.4) can be written as

$$a(L)(Y_t - \mu) = b(L)U_t.$$

Thus, as in the pure MA case, the mean does not depend on the MA parameters. For the time being, we consider the known mean case, so that, without loss of generality, let $\mu = 0$. The regression case, which generalizes the use of c in (7.4), is dealt with in Section 7.4.2.

7.1.2 Zero Pole Cancellation

Recalling the notation in (6.5) and (6.53), we can write $a(L) = \prod_{i=1}^{p}(1 - \lambda_iL)$ and, similarly for the MA polynomial, $b(L) = \prod_{j=1}^{q}(1 - \eta_jL)$, so that

$$Y_t = \frac{(1-\eta_1L)(1-\eta_2L)\cdots(1-\eta_qL)}{(1-\lambda_1L)(1-\lambda_2L)\cdots(1-\lambda_pL)}U_t. \tag{7.7}$$

Assume that the roots of both the AR and MA polynomials are all outside the unit circle. If there is a pair (i, j) such that $\eta_i = \lambda_j$, then the two factors $(1 - \lambda_jL)$ and $(1 - \eta_iL)$ cancel in (7.7), and the model is expressible as an ARMA($p - 1, q - 1$). This is sometimes referred to as **zero pole cancellation**. This can be continued until there are no more terms to cancel, giving the most parsimonious expression for the model. This is also important for model estimation: If there is a term $(1 - \lambda_jL)$ common to both polynomials, then λ_j, and thus the ARMA model, is not identified. For this reason, we only entertain stationary, invertible ARMA models such that the AR and MA polynomials do not have any roots in common.

Remark In practice, with a finite data set, if the true model is a stationary and invertible ARMA(p, q) process, then the likelihood of an ARMA($p + k, q + k$) process can have a unique maximum. However, the ARMA parameters will have rather large standard errors, and computation of the roots of the AR and MA polynomials will reveal that k roots are (approximately) shared. Often, with modern computing power, one just fits the model for numerous p and q, and uses the AIC and BIC criteria to determine the best model; see Chapter 9. ∎

One might wonder why we cannot cancel roots $(1 - \lambda L)$ and $(1 - \eta L)$ when $\lambda = \eta$ and such that $|\lambda| \leqslant 1$. This is because the process with the cancelled roots, and the one without, may not be the same: Recall the discussion around (6.42) that motivated the meaning of dividing by a term of the form $(1 - \lambda L)$. We informally concluded that this makes sense only when the root is outside the unit circle, i.e., if modulus $|\lambda| < 1$. This is true, and more formally shown in functional analysis, where L is the **left shift operator on a sequence**. A useful discussion is provided in Dhrymes (2013, Sec. 6.2). The following example illustrates the point.

Example 7.1 Consider the model

$$(1 + bL)Y_t = (1 + bL)U_t. \tag{7.8}$$

Writing $Y_t = -bY_{t-1} + bU_{t-1} + U_t$, recursive substitution easily gives

$$Y_t = U_t + (-1)^t b^t (Y_0 - U_0) \quad .$$

For $|b| < 1$, in the limit (in the sense of, for any fixed t, the time series extends infinitely into the past), $Y_t = U_t$ and, thus, we can cancel the roots in (7.8). For $b = 1$, $Y_t = U_t$ plus the term $(Y_0 - U_0)$, with alternating signs—not a very intuitive model, and certainly not equivalent to a white noise process. For $b > 1$, the process explodes. One might argue that taking $Y_0 = U_0 = E[U_t] = 0$ will resolve the problem. If one *defines* the model in that way, then it works. Otherwise, the argument is wrong: In terms of simulation, we should be able to sample values of U_0 and Y_0 from their respective unconditional distributions (or choose any real numbers in fact) and, with a long enough burn-in period, the process is virtually independent of the actual values chosen for U_0 and Y_0. Clearly, for $b \geqslant 1$, this will not be the case. Furthermore, for $b \geqslant 1$, it is not at all clear what the unconditional distribution of Y_0 is, assuming it even exists.

Thus, for $|b| < 1$, we can cancel the common factor, but for $|b| \geqslant 1$, we cannot. ∎

7.1.3 Simulation

The simulation of an ARMA process is quite straightforward. One way is to compute the square root of the variance–covariance matrix corresponding to the specified ARMA model (given in Section 7.4) and then right-multiply it by a vector of i.i.d. standard normal random variables. For large T, this will clearly be time-consuming. Instead, we just use a loop and set Y_t equal to the weighted sum of the p past values of Y and the q past values of U dictated by the parameters of the ARMA(p, q) model. Important then is the choice of starting values $Y_{1-p}, \dots, Y_0$, which could be determined from the aforementioned method based on the exact covariance matrix. Though that is arguably the best way, we just set values $Y_{1-p}, \dots, Y_0$ to zero, simulate $500 + T$ observations, and deliver the final T values, where 500 is obviously arbitrary, and referred to as the **burn-in period**. (However, recall Figure 4.2, which demonstrates the relevance of the initial observation for an AR(1) model with a close to unity, or, in general, when the process is close to the stationarity border, in which case the burn-in period might need to be larger.) This is implemented in Listing 7.1.

```
function y=armasim(nobs,sig2,pv,qv)
if nargin<4, qv=[]; end
p=length(pv); q=length(qv); pv=reshape(pv,1,p); qv=reshape(qv,1,q);
warmup=500; e=sqrt(sig2)*randn(nobs+warmup,1); init=0;
evec=zeros(q,1); yvec=zeros(p,1); y=zeros(nobs+warmup,1);
for i=1:nobs+warmup
  if p>0, y(i) = y(i) + pv*yvec; end
  if q>0, y(i) = y(i) + qv*evec; end
  y(i) = y(i) + e(i);
  if p>1, yvec(2:p)=yvec(1:p-1); end, yvec(1)=y(i);
  if q>1, evec(2:q)=evec(1:q-1); end, evec(1)=e(i);
end
y=y(warmup+1:end);
```

Program Listing 7.1: Simulates `nobs` observations of an ARMA process with innovation variance `sig2`, AR parameters passed as vector `pv`, MA parameters as vector `qv`, and uses a burn-in period of 500. For example, to generate a series with 1,000 observations from the ARMA(2,1) model with $a_1 = 1.2, a_2 = -0.8, b_1 = -0.5$, and $\sigma^2 = 1$, use `y=armasim(1000,1,[1.2 -0.8],-0.5,1);`.

7.1.4 The ARIMA(p, d, q) Model

An important extension of the ARMA model (7.4)–(7.5) is when the data generating process is such that it needs to be differenced d times to be a stationary ARMA(p, q) model, referred to as an ARIMA(p, d, q) process. Most often, d is either zero or one, though the case of $d = 2$ does arise in practice. The process $\{Z_t\}$ is then expressed as

$$a(L)(1 - L)^d Z_t = c + b(L)U_t, \qquad d \in \mathbb{N}, \tag{7.9}$$

and the polynomials $a(L)$ and $b(L)$ are given in (7.5). Note that $Y_t := (1 - L)^d Z_t$ is ARMA(p, q). If $d = 1$, then the process is said to have a unit root, the testing of which is detailed in Section 5.5. The constant c can be replaced by a more general structure, such as a regression equation $\mathbf{x}_t'\boldsymbol{\beta}$, as done throughout Chapter 5, and dealt with below in Section 7.4.2. Throughout the remainder of this chapter, we assume that d is known, and concern centers on working with process (7.4)–(7.5). Forecasting an ARIMA($p, 1, q$) model is dealt with in Section 7.5.3.

Remarks

a) A further extension of the ARIMA model class is to allow for a structure addressing **seasonality**, such as when working with quarterly or monthly data. Then both the non-seasonal and seasonal parts are endowed with an ARIMA structure, denoted ARIMA$(p, d, q) \times (P, D, Q)_s$ or ARIMA$(p, d, q)(P, D, Q)_s$, where s denotes the periodicity of the seasonality, such as four or twelve. The model is then

$$a(L)A(L)(1 - L)^d(1 - L^s)^D Z_t = c + b(L)B(L)U_t, \tag{7.10}$$

where, similar to (7.5),

$$A(L) = 1 - A_1 L^s - \cdots - A_P L^{Ps} \quad \text{and} \quad B(L) = 1 + B_1 L^s + \cdots + B_Q L^{Qs}. \tag{7.11}$$

For example, the zero-mean ARIMA$(1, 0, 0)(1, 0, 0)_4$ model is given by

$$1 - aL - AL^4 + aAL^5 = (1 - aL)(1 - AL^4)Y_t = U_t,$$

and can be viewed as an AR(5) model such that the coefficients for lags 2 and 3 are zero (a **subset** AR model), and that of lag 5 is constrained to be $a \times A$. The reader is encouraged to develop a program to estimate a seasonal ARMA model for general a, A, b, B, with the polynomial multiplication function `conv` being of great use. (Note that packages such as R and SAS have procedures for this model, and the reader can compare his/her results to those from canned routines.)

b) A conceptually different, relatively new, and potentially very useful technique suitable for modeling seasonal time series is so-called **singular spectrum analysis**, or SSA. See Zhigljavsky (2010) for an overview and, among others, Hassani and Thomakos (2010), Hassani et al. (2013a,b), Silva and Hassani (2015), de Carvalho and Rua (2017), and the references therein for further methodological details and applications to economic time series, and Arteche and García-Enríquez (2017) for use with stochastic volatility models applied to financial data. ■

7.2 Infinite AR and MA Representations

We have already seen from (6.45) that an invertible MA(1) model can be represented by an infinite AR model. Similarly, a stationary AR(1) model can be expressed as an infinite MA. In particular, if $Y_t = aY_{t-1} + U_t$, or $(1 - aL)Y_t = U_t$, $|a| < 1$, then

$$Y_t = (1 - aL)^{-1}U_t = (1 + aL + a^2L^2 + \cdots)U_t = U_t + aU_{t-1} + a^2U_{t-2} + \cdots.$$

These results generalize: An invertible MA(q) process can be represented as an infinite AR, and a stationary AR(p) can be represented as an infinite MA. To illustrate the latter, consider the stationary AR(p) process $a(L)Y_t = U_t$ with $a(L) = 1 - a_1L - \cdots - a_pL^p$. The infinite MA representation is given by

$$Y_t = a^{-1}(L)U_t = \psi(L)U_t = (1 + \psi_1L + \psi_2L^2 + \cdots)U_t.$$

The coefficients in $\psi(L)$ can be obtained by treating L as the variable in a polynomial and multiplying both sides of $a^{-1}(L) = \psi(L)$ by $a(L)$, i.e., $1 = a(L)\psi(L)$, and then equating coefficients of L^j. That is,

$$\begin{aligned} a(L)\psi(L) &= (1 - a_1L - \cdots - a_pL^p)(1 + \psi_1L + \psi_2L^2 + \cdots) \\ &= 1 + (\psi_1L - a_1L) + (\psi_2 - a_1\psi_1 - a_2)L^2 + (\psi_3 - a_1\psi_2 - a_2\psi_1 - a_3)L^3 + \cdots \end{aligned}$$

so that

$$\begin{aligned} \psi_1 - a_1 = 0 \quad &\Rightarrow \quad \psi_1 = a_1 \\ \psi_2 - a_1\psi_1 - a_2 = 0 \quad &\Rightarrow \quad \psi_2 = a_1\psi_1 + a_2 \\ \psi_3 - a_1\psi_2 - a_2\psi_1 - a_3 = 0 \quad &\Rightarrow \quad \psi_3 = a_1\psi_2 + a_2\psi_1 + a_3 \\ &\vdots \end{aligned}$$

or

$$\psi_0 = 1, \quad \psi_j = \sum_{k=1}^{j} a_k\psi_{j-k}, \quad j = 1, 2, \ldots.$$

Note that, for an AR(1) process, $a_i = 0$ for $i \geqslant 2$, and $\psi_i = a_1^i$, $i = 0, 1, 2, \ldots$.

Besides being of theoretical interest and also required for forecasting (see Section 7.5), the infinite MA representation shows that AR and MA models are only "interchangeable" when infinite numbers of terms are used. By combining the two structures, very flexible correlation structures can be realized

```
function arcoef=infAR(a,b,n)
% AR coef a=(a_1,...,a_p) and MA coef b=(b_1,...,b_q)
% Call -infAR(-b,-a,n) to compute the infinite MA representation
q=length(b); p=length(a); a=reshape(a,p,1);
d=zeros(n,1); a=[a ; zeros(n-p,1)];
for j=1:n, s=0;
  for k=1:min(j,q),  if j-k==0, s=s-b(k); else s=s+b(k)*d(j-k); end, end
  d(j)=a(j) - s;
end
arcoef=d;
```

Program Listing 7.2: Computes (7.15).

with far fewer parameters than would be required with pure AR or MA models, in line with the idea of parsimonious model building.

The previous derivation of the infinite MA expression for an AR(p) model can be extended to the mixed (stationary and invertible) ARMA(p, q) case in a straightforward way: The model is now

$$Y_t = a^{-1}(L)b(L)U_t = \psi(L)U_t, \tag{7.12}$$

with $\{U_t\}$ as in (7.2), so that we equate coefficients of L^j in

$$b(L) = a(L)\psi(L)$$

to get the ψ_i. With $\psi_0 = 1$, it is straightforward to verify that this leads to the recursive expression $\psi_j = b_j + \sum_{k=1}^{j} a_k \psi_{j-k}$ or, using the finiteness of p and q,

$$\psi_0 = 1, \qquad \psi_j = b_j \mathbb{I}(j \leqslant q) + \sum_{k=1}^{\min(j,p)} a_k \psi_{j-k}, \quad j \geqslant 1. \tag{7.13}$$

Similarly, we can express the model as $\pi(L)Y_t = b^{-1}(L)a(L)Y_t = U_t$, where

$$\pi(L) = 1 - \pi_1 L - \pi_2 L^2 + \cdots$$

is the infinite AR polynomial, the terms of which are computed by equating coefficients of L^j in $b(L)\pi(L) = a(L)$ to get

$$\pi_0 = -1, \qquad \pi_j = a_j \mathbb{I}(j \leqslant p) - \sum_{k=1}^{\min(j,q)} b_k \pi_{j-k}, \qquad j \geqslant 1. \tag{7.14}$$

The infinite AR representation is then

$$Y_t = \pi_1 Y_{t-1} + \pi_2 Y_{t-2} + \cdots. \tag{7.15}$$

Program `infAR(a,b,n)` in Listing 7.2 computes (7.15). If in (7.14) we replace π_i with $-\psi_i$, $i \geqslant 0$, a_i with $-b_i$, and b_i with $-a_i$, $i = 1, \dots, \max(p, q)$, we obtain $\psi_0 = 1$ and, for $j \geqslant 1$,

$$-\psi_j = -b_j \mathbb{I}(j \leqslant q) - \sum_{k=1}^{\min(j,p)} (-a_k)(-\psi_{j-k}), \tag{7.16}$$

which is precisely (7.13). Thus, calling `-infAR(-b,-a,n)` delivers the ψ_i.

Let $a_i = 0$ for $i > p$; $b_i = 0$ for $i > q$; and $\pi_i = 0$ for $i < 0$. Then (7.14) is

$$\pi_0 = -1, \qquad \pi_j = a_j - b_1\pi_{j-1} - \cdots - b_{j-1}\pi_1 + b_j, \quad j = 1, 2, \ldots,$$

which can be expressed in matrix terms as

$$\tilde{\mathbf{a}} = (\mathbf{I}_{r+1} + \mathbf{B})\boldsymbol{\pi}, \tag{7.17}$$

where

$$\boldsymbol{\pi} = \begin{bmatrix} -1 \\ \pi_1 \\ \pi_2 \\ \vdots \\ \pi_r \end{bmatrix}, \quad \tilde{\mathbf{a}} = \begin{bmatrix} -1 \\ a_1 \\ a_2 \\ \vdots \\ a_r \end{bmatrix} \quad \text{and} \quad \mathbf{B} = \begin{bmatrix} 0 & 0 & \cdots & 0 & 0 \\ b_1 & 0 & & 0 & 0 \\ b_2 & b_1 & \ddots & \vdots & \vdots \\ \vdots & \vdots & & 0 & 0 \\ b_r & b_{r-1} & \cdots & b_1 & 0 \end{bmatrix}.$$

This could be used to get the matrix expression $\boldsymbol{\pi} = (\mathbf{I}_{r+1} + \mathbf{B})^{-1}\tilde{\mathbf{a}}$, where $(\mathbf{I}_{r+1} + \mathbf{B})^{-1}$ always exists because $|\mathbf{I}_{r+1} + \mathbf{B}| = 1$. This could be computed for $r = p$ and then, for $j > p$, the recursion

$$\pi_j = -b_1\pi_{j-1} - b_2\pi_{j-2} - \cdots - b_q\pi_{j-q}$$

would be used. In Matlab, with vectors a and b as the r-length, zero-padded AR and MA parameter row vectors, this is just

```
B=toeplitz([0 b]',zeros(r+1,1)); at=[-1 a]'; pi=inv(eye(r+1)+B)*at;
```

(7.18)

with $\pi_1, \ldots, \pi_r$ given as `pi(2:end)`. Similarly, $\psi_1, \ldots, \psi_r$ would be computed by executing `temp=b; b=-a; a=-temp;` running (7.18) and delivering `-pi(2:end)`. As an example, for the ARMA(1,3) model with $a_1 = 0.5$, $b_1 = -1.1$, $b_2 = 0.7$ and $b_3 = -0.3$, $\psi_1 = -0.6$, $\psi_2 = 0.4$, $\psi_3 = -0.1$, and, from (7.13), $\psi_i = 2^{3-i}\psi_3$, $i \geqslant 4$.

Another value of this exercise is that the latter equation can be used to express the b_i in terms of the π_i by building a system of equations for $j = p+1, \ldots, p+q$, i.e., with $\pi_0 = 1$ and $\pi_i = 0$ for $i < 0$,

$$\begin{bmatrix} \pi_{p+1} \\ \pi_{p+2} \\ \vdots \\ \pi_{p+q} \end{bmatrix} = - \begin{bmatrix} \pi_p & \pi_{p-1} & \cdots & \pi_{p+1-q} \\ \pi_{p+1} & \pi_p & \cdots & \pi_{p+2-q} \\ \vdots & & \ddots & \\ \pi_{p+q-1} & \pi_{p+q-2} & \cdots & \pi_p \end{bmatrix} \begin{bmatrix} b_1 \\ b_2 \\ \vdots \\ b_q \end{bmatrix} =: -\boldsymbol{\Pi}\mathbf{b}, \tag{7.19}$$

which can be solved in an obvious way for $\mathbf{b} = (b_1, \ldots, b_q)'$, provided matrix $\boldsymbol{\Pi}$ is of full rank. With $\mathbf{b}$ known, (7.17) can be computed with $r = p$ to obtain $\tilde{\mathbf{a}}$ and, thus, $\mathbf{a} = (a_1, \ldots, a_p)'$. In other words, the parameter vectors $\mathbf{a}$ and $\mathbf{b}$ can be easily recovered from $\pi_1, \ldots, \pi_{p+q}$. A program to compute this is given in Listing 7.3.

7.3 Initial Parameter Estimation

Without doubt, the m.l.e. is among the most important estimators for the parameters of an ARMA model, and Section 7.4 below develops the likelihood function. However, when $q > 0$, there is no

```
function [a,b,detP]=pitoab(d,p,q)
% d is vector of infinite AR coefficients pi_1,...,pi_{p+q+1}
%   corresponding to the ARMA(p,q) process with parameter vectors
%   a=(a_1,...,a_p) and b=(b_1,...,b_q)
r=max(p,q); d=reshape(d,length(d),1); row=zeros(r,1);
if p==0, col=[-1 ; d(p+1:p+r-1)]; else col=d(p:p+r-1); end
for i=0:r-1
  if p-i==0; row(i+1)=-1; elseif p-i<0, row(i+1)=0; else row(i+1)=d(p-i); end
end
Pimat=toeplitz(col,row); pivec=d(p+1:p+r); detP=det(Pimat);
if abs(detP)<1e-7
  %disp(['determinant is ',num2str(detP),'. Try another p and/or q'])
  a=0; b=0; return
else
  b=-inv(Pimat)*pivec; B=toeplitz([0; b],zeros(r+1,1));
  atilde=(eye(r+1)+B)*[-1 ; d(1:r)]; a=atilde(2:p+1); b=b(1:q);
end
```

Program Listing 7.3: Recovers the a_i and b_i from the π_i via (7.19) and (7.17).

closed-form expression for the maximum of the conditional (let alone exact) likelihood of an ARMA process, so that the likelihood needs to be numerically maximized. This is, of course, no longer a hindrance, but starting values will still be required, with poor ones potentially leading to a local inferior maximum of the likelihood. For this purpose, computationally cheap estimators are required, and we present two such methods.

7.3.1 Via the Infinite AR Representation

The ability to recover the a_i and b_i from the π_i lends itself immediately to a way of obtaining simple, closed-form estimators for the parameters of an ARMA model: Estimate via Yule–Walker or least squares (see Listing 6.3) an AR(p^*) model, where p^* is a function of T (say, $\sqrt{T}$, rounded) to get AR coefficients $\hat{\pi}_i$, $i = 1, \dots, p^*, p^* \geqslant p + q + 1$. Then use (7.19) and (7.17) with $\hat{\pi}_i$ in place of the unknown π_i. The sample variance of the residuals from the AR(p^*) model serves as an estimator of σ^2. The choice of p^* will play a role in the accuracy of the results in the usual way: Small values will induce a bias into the $\hat{\pi}_i$ and large values will increase their variance.

To illustrate, consider the zero-mean ARMA(2,1) model with $a_1 = 1.2$, $a_2 = -0.8$, $b_1 = -0.3$, and $\sigma = 2$. Figure 7.1 shows the resulting kernel density estimates of the estimated parameters, for sample size $T = 1{,}000$, based on the conditional m.l.e., as discussed below in Section 7.4, and this infinite AR method. The latter is clearly not as efficient as the m.l.e., though, for the three ARMA parameters, it appears, at least for this parameter constellation and sample size, unbiased.

The reader is encouraged to design a short program, say `armaviainfAR(y,p,q)`, that inputs a time series, p and q, and outputs the vector of estimated parameters (hint: all the hard work is done in Program Listing 7.3), and replicate this study and try other parameter constellations.

7.3.2 Via Infinite AR and Ordinary Least Squares

As in Section 7.3.1, we begin by estimating an AR(p^*) model, where p^* is chosen as a function of the series length T. Now, it is not the estimated AR parameters that are of use, but the $T - p^*$ filtered residuals of this AR(p^*) model, $\hat{U}_t$, $t = p^* + 1, \dots, T$, which serve as estimates of the true innovation

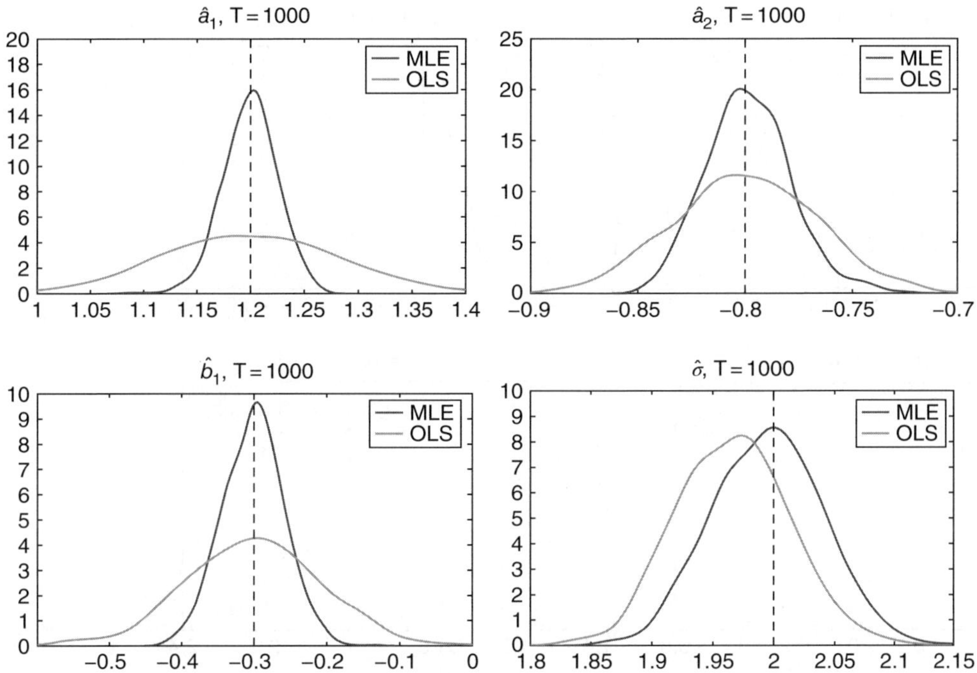

Figure 7.1 Comparison of the conditional m.l.e. and the infinite AR representation method of Section 7.3.1 for estimation of ARMA(2,1) parameters (shown as vertical lines in the plots) for $T = 1{,}000$ observations, based on 10,000 replications.

counterparts. These can be used as regressors in the model

$$Y_t - \hat{U}_t = \sum_{i=1}^{p} a_i Y_{t-i} + \sum_{j=1}^{q} b_j \hat{U}_{t-j} + \xi_t, \quad t = p^* + 1 + q, \dots, T. \tag{7.20}$$

This is implemented in Listing 7.4. As a first example, we use the same ARMA(2,1) model as used above, with $a_1 = 1.2$, $a_2 = -0.8$, $b_1 = -0.3$, and $\sigma = 2$. Figure 7.2 shows the kernel density results for

```
function param = armaols(y,p,q)
% assumes zero mean stationary invertible ARMA(p,q)
% param = [AR terms, MA terms, sigma]
L=ceil(sqrt(length(y))); z=y(L+1:end);
Z=toeplitz(y(L:end-1),y(L:-1:1));
uhat=(eye(length(z)) - Z*inv(Z'*Z)*Z') * z; %#ok<*MINV>
sigmahat = std(uhat); yy=z-uhat; X=[]; m=max(p,q);
for i=1:p, X=[X z(m-i+1 : length(z)-i)]; end %#ok<*AGROW>
for i=1:q, X=[X uhat(m-i+1 : length(uhat)-i)]; end
yuse=yy(m+1:end); ARMAparam=inv(X'*X)*X'*yuse;
param=[ARMAparam ; sigmahat];
```

Program Listing 7.4: Computes 7.20.

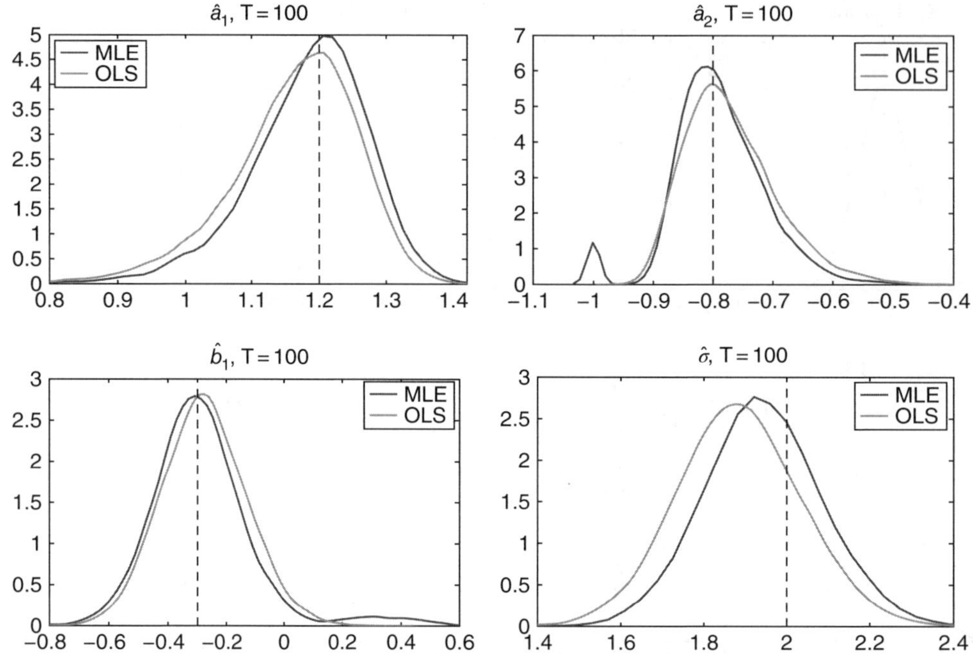

Figure 7.2 Comparison of m.l.e. and o.l.s.-based methods for estimation of ARMA(2,1) parameters (shown as vertical lines in the plots) for $T = 100$ observations, based on 10,000 replications.

sample size $T = 100$. It is noteworthy that method (7.20) is far faster than (certainly exact, but also conditional) maximum likelihood. It clearly performs quite well relative to the m.l.e. benchmark, and has the apparent advantage of avoiding the inferior local likelihood maximum indicated in the plots. Figure 7.3 is similar, but having used $T = 1{,}000$ observations, and can thus be compared to Figure 7.1.

Next, we use the zero-mean ARMA(1,3) model with $a_1 = 0.5$, $b_1 = -1.1$, $b_2 = 0.7$, and $b_3 = -0.3$. From (7.16) and calling `-infAR(-b,-a,6)`, the first six terms of the infinite MA expansion are -0.6000, 0.4000, -0.1000, -0.0500, -0.0250, and -0.0125. Figure 7.4 shows the results of method (7.20) and the conditional m.l.e. for the four ARMA parameters. Now, the results are highly discrepant, with both methods consistently drawn to two different optima, with that of the m.l.e. being correct. The medians of the 10,000 point estimates for method (7.20) are $\hat{a}_1 = -0.1643$, $\hat{b}_1 = -0.4383$, $\hat{b}_2 = 0.2996$, and $\hat{b}_3 = -0.0367$, which result in an infinite MA expansion of -0.6026, 0.3986, -0.1022, 0.0168, -0.0028, and 0.0005, these being indeed close to the true values. (The results for $\hat{\sigma}$ are not shown: The m.l.e. resulted in a Gaussian-looking distribution centered at the true value of 2.0, while that of method (7.20) was similar, but centered at 1.98.)

Remarks

a) An idea for further practice is the following: Presumably, for a given sample size T, known p and q, and true ARMA parameters corresponding to a stationary, invertible process, there exists an optimal p^*, in the sense that the resulting, say, m.s.e., is the smallest in aggregate across the $p + q$ parameters. Thus, an idea to improve the method is to first take, say, $p^* = p^*_{(1)} := \lfloor \sqrt{T} \rfloor$, and obtain

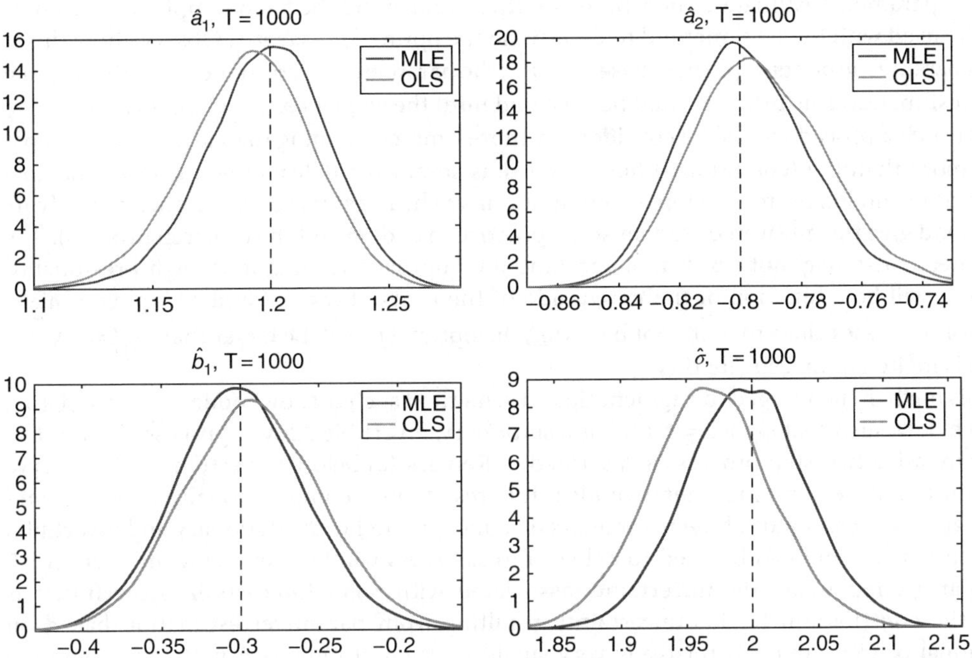

Figure 7.3 Same as Figure 7.2 but using $T = 1{,}000$ observations.

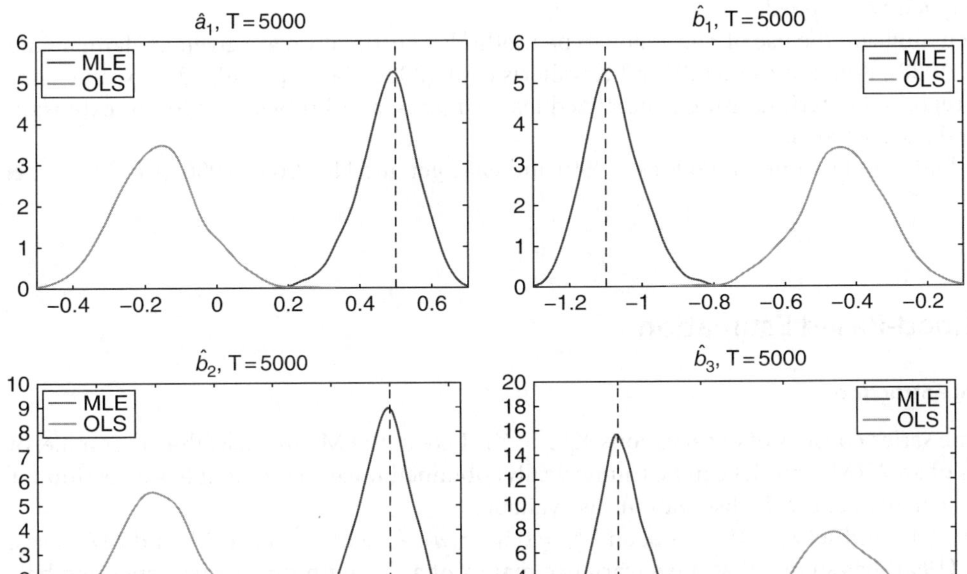

Figure 7.4 Similar to Figures 7.2 and 7.3, but for the zero-mean ARMA(1,3) model with $a_1 = 0.5$, $b_1 = -1.1$, $b_2 = 0.7$, $b_3 = -0.3$, and $T = 5{,}000$ observations.

the $p + q + 1$ parameter estimates. Then, based on these values and the fixed sample size T, simulation can be used with the o.l.s. method to determine the optimal p^*, say $p^*_{(2)}$. Observe how this is similar to a parametric bootstrap. Then, based on $p^*_{(2)}$, the parameters corresponding to the original data are re-estimated. This process could be repeated until the sequence $\{p^*_{(i)}\}_{i=1}^{\infty}$ converges.

Note that such a procedure will be considerably more time-consuming than just using $p^*_{(1)}$, possibly even more than the (conditional) m.l.e., and thus somewhat defeats the purpose. The goal of this exercise is no longer to develop a computationally cheap estimator, but rather to study its properties (and give the motivated student some practice in coding and the research process). The iterative scheme also may not result in a substantially improved estimator, though presumably, improvement will be a function of the magnitude of the roots of the AR and MA polynomials, such that, for processes close to unit-root behavior, the optimal p^* will be larger than $p^*_{(1)} = \lfloor \sqrt{T} \rfloor$. The reader is invited to investigate this.

b) The reader is further encouraged to augment this o.l.s. method such that the model is $Y_t = \mathbf{x}_t'\boldsymbol{\beta} + \epsilon_t$, as in (7.25) below, such that $a(L)\epsilon_t = b(L)U_t$ is a stationary, invertible ARMA process. This can be done in an iterative two-step process as described in Remark (a) below in Section 7.4.2. Observe that, without the aforementioned iterative idea, this results in a computationally very fast, and reasonably accurate, estimator of the parameters of a linear model with (stationary and invertible) ARMA disturbances, and can be used to deliver forecasts. As will be seen below in Section 7.5 (in particular, Example 7.5), the uncertainty associated with point forecasts into the future is large, and the contribution to this uncertainty resulting from parameter estimation based on the conditional or exact m.l.e. is relatively very small. As such, if interest centers on generating forecasts, particularly for a large number of time series, then this o.ls.-based method will be attractive, given its reasonable accuracy compared to the m.l.e., and its enormous advantage in terms of simplicity and speed.

c) Further discussions on the use of regression-type methods of estimation are given in Hannan and McDougall (1988), Kapetanios (2003), and Kavalieris et al. (2003). See Appendix 7.1 for an extension of the regression method using generalized least squares, and Problem 7.2 for an extension using iterated least squares.

Further methods can be found in Pollock (1999) and Granger and Newbold (1986, p. 87). ■

7.4 Likelihood-Based Estimation

7.4.1 Covariance Structure

Assume the time series consists of observations $Y_1, \dots, Y_T$. Like in the MA(q) case, the exact m.l.e. of the parameters of an ARMA model can be (numerically) obtained once a computable expression for $\boldsymbol{\Sigma}$, the covariance matrix of the T observations, is available.

Let $m = \max(p, q)$ (and set $a_i = 0$ if $i > p$, and $b_i = 0$ if $i > q$). As with the pure AR and MA cases, van der Leeuw (1994) has shown that the covariance matrix of a T-length time series generated by a stationary and invertible ARMA(p, q) process can be expressed as

$$\boldsymbol{\Sigma} = [\mathbf{N}\ \mathbf{M}][\bar{\mathbf{P}}'\bar{\mathbf{P}} - \bar{\mathbf{Q}}\bar{\mathbf{Q}}']^{-1}[\mathbf{N}\ \mathbf{M}]', \tag{7.21}$$

```
function V=leeuwARMA(a,b,T);
% a=(a_1,...,a_p), where the sign convention on the a_i is such that
%                  Y_t = a_1 Y_{t-1} + a_2  Y_{t-2} + ... a_p  Y_{t-p}
% b=(b_1,...,b_q), and T is desired size of the covariance matrix
% pass a as [] for a pure MA model, and b as [] for a pure AR model

p=length(a); q=length(b); a=reshape(a,1,p); b=reshape(b,1,q);
m=max(p,q); a=[a zeros(1,m-p)]; b=[b zeros(1,m-q)]; p=m; q=m; % zero pad

a=-a; Tuse=T+p;
firrowP = [1 zeros(1,Tuse-1)]; fircolP = [1 a zeros(1,Tuse-p-1)];
P = toeplitz(fircolP,firrowP);
firrowQ1 = a(p:-1:1); fircolQ1 = [a(p) zeros(1,p-1)];
Q1 = toeplitz(fircolQ1,firrowQ1); Q = [Q1; zeros(Tuse-p,p)];

firrowM = [1 zeros(1,T-1)]; fircolM = [1 b zeros(1,T-q-1)];
M = toeplitz(fircolM,firrowM);
firrowN1 = b(q:-1:1); fircolN1 = [b(q) zeros(1,q-1)];
N1 = toeplitz(fircolN1,firrowN1); N = [N1; zeros(T-q,q)];

middle=P'*P - Q*Q'; outer=[N M]; V = outer * inv(middle) * outer';
```

Program Listing 7.5: Computes (7.21).

where $\bar{\mathbf{P}}$ and $\bar{\mathbf{Q}}$ have the same structure as matrices $\mathbf{P}$ and $\mathbf{Q}$ in (6.20), but are of order $(T+m)\times(T+m)$ and $(T+m)\times m$, respectively, and $\mathbf{N}$ and $\mathbf{M}$ are given in (6.51). Listing 7.5 shows the code to compute this.

Example 7.2 For the ARMA(1,1) model $Y_t = aY_{t-1} + U_t + bU_{t-1}$ with $T = 2$,

$$\bar{\mathbf{P}} = \begin{bmatrix} 1 & 0 & 0 \\ -a & 1 & 0 \\ 0 & -a & 1 \end{bmatrix}, \quad \bar{\mathbf{Q}} = \begin{bmatrix} a \\ 0 \\ 0 \end{bmatrix}, \quad \mathbf{M} = \begin{bmatrix} 1 & 0 \\ b & 1 \end{bmatrix}, \quad \mathbf{N} = \begin{bmatrix} b \\ 0 \end{bmatrix}$$

and (7.21) gives

$$\boldsymbol{\Sigma} = \begin{bmatrix} b & 1 & 0 \\ 0 & b & 1 \end{bmatrix} \begin{bmatrix} 1 & -a & 0 \\ -a & a^2+1 & -a \\ 0 & -a & 1 \end{bmatrix}^{-1} \begin{bmatrix} b & 0 \\ 1 & b \\ 0 & 1 \end{bmatrix} = \begin{bmatrix} \gamma_0 & \gamma_1 \\ \gamma_1 & \gamma_0 \end{bmatrix},$$

where

$$\gamma_0 = \frac{1+2ab+b^2}{1-a^2}, \quad \gamma_1 = \frac{(1+ab)(a+b)}{1-a^2}, \tag{7.22}$$

and higher-order covariances are computed as $\gamma_i = a\gamma_{i-1}$, $i \geqslant 2$. ■

While both elegant and easily programmed, (7.21) will be exceedingly slow as T grows. This problem can be eliminated by calculating (7.21) for $T^* = m+1$, $m = \max(p,q)$, and using the following recursion for the remaining elements:

$$\gamma_k = \sum_{i=1}^{p} a_i \gamma_{k-i}, \quad k = m+2, \dots, T. \tag{7.23}$$

```
function Sigma=acvf(a,b,T);
p=length(a); q=length(b); m=max(p,q)+1; %+1 because of the way leeuwARMA works
V=leeuwARMA(a,b,m); gamma=[V(:,1) ; zeros(T-m,1) ];
for k=m+1:T,  s=0; for i=1:p, s=s+a(i)*gamma(k-i); end, gamma(k)=s; end
Sigma=toeplitz(gamma);
```

Program Listing 7.6: A faster way of computing $\mathbf{\Sigma}$ when T is large. The method assumes $T > \max(p, q) + 1$ without explicitly checking for it.

To see the validity of (7.23), first zero-pad the AR or MA polynomial so that $p = q = m$, then multiply the equation for Y_t by Y_{t-k} (assuming $\mathbb{E}[Y_t] = 0$ without loss of generality) to give

$$Y_t Y_{t-k} = a_1 Y_{t-k} Y_{t-1} + \cdots + a_m Y_{t-k} Y_{t-m} + Y_{t-k} U_t + b_1 Y_{t-k} U_{t-1} + \cdots + b_m Y_{t-k} U_{t-m},$$

and take expectations to get (using the fact that $\gamma_i = \gamma_{-i}$)

$$\gamma_k = a_1 \gamma_{k-1} + \cdots + a_m \gamma_{k-m} + \sum_{i=0}^{m} \mathbb{E}[Y_{t-k} U_{t-i}]. \tag{7.24}$$

As $\mathbb{E}[Y_{t-k} U_{t-i}] = 0$ if $t - i > t - k$, or $k > i$, the latter sum in (7.24) is zero if $k > m$, which justifies (7.23). The only reason k starts from $m + 2$ instead of $m + 1$ in (7.23) is that (7.21) requires $T > m$. This method is implemented in Listing (7.6) and the reader can verify its large speed advantage for large T.

Remark The first explicit computer-programmable methods for calculating $\boldsymbol{\gamma}_m = (\gamma_0, \dots, \gamma_m)$ for an ARMA model appear to be given by McLeod (1975) and Tunnicliffe Wilson (1979), although, as McLeod also states, the method was used for some special ARMA cases in the first edition (1970) of the seminal Box and Jenkins monograph. A closed-form matrix expression for $\boldsymbol{\gamma}_m$ appears to have been first given by Mittnik (1988), while Zinde-Walsh (1988) and Karanasos (2000) derive expressions for γ_i based on the b_i and *the roots* of the AR polynomial, with Karanasos' result restricted to the case with distinct (real or complex) roots. ■

7.4.2 Point Estimation

Once $\mathbf{\Sigma}$ is numerically available, the likelihood is straightforward (in principle) to calculate and maximize. The drawback, however, of *any* method for calculating $\mathbf{\Sigma}$, whatever its speed, is that a $T \times T$ matrix *inverse* needs to be calculated at each likelihood evaluation. Keep in mind that this problem evaporates when working with pure AR(p) models: From (6.28), the exact likelihood is partitioned so that only $\mathbf{\Sigma}^{-1}$ of size $p + 1$ needs to be calculated—and $\mathbf{\Sigma}^{-1}$ can be directly calculated via (6.20), thus even avoiding the small matrix inversion.

With MA or ARMA processes, this luxury is no longer available. As T gets into the hundreds, the calculation of $\mathbf{\Sigma}^{-1}$ for MA or ARMA processes becomes prohibitive. The method involving use of the Kalman filter would be preferred for computing the exact m.l.e., as it involves matrices only on the order of $\max(p, q + 1)$. The startup conditions on the filter for calculating the exact likelihood need to be addressed; see, e.g., Jones (1980) and Harvey and Pierse (1984). With large sample sizes, the conditional m.l.e. will result in nearly the same results as use of the exact m.l.e., and is trivial to program, as discussed next.

The conditional m.l.e. simply combines the conditioning arguments used in the separate AR and MA cases. In particular, the first p values of Y_t are assumed fixed, and all q unobservable values of U_t are taken to be zero. The conditional likelihood still needs to be numerically maximized, but as there are no $T \times T$ matrices to invert, the method is very fast for large T and, unless the AR and/or MA polynomials are close to the stationarity (invertibility) borders, there will not be much difference in the conditional and exact m.l.e. values.

Similar to the development in Chapter 5, we can introduce a regression term into the model via the observation equation

$$Y_t = \mathbf{x}_t' \boldsymbol{\beta} + \epsilon_t, \tag{7.25}$$

but with the latent equation being given by the ARMA process $a(L)\epsilon_t = b(L)U_t$, termed ARMAX.

Observe that the joint distribution of $\mathbf{Y} = (Y_1, Y_2, \dots, Y_T)'$ is $\mathrm{N}(\mathbf{X}\boldsymbol{\beta}, \sigma^2\boldsymbol{\Sigma})$, where $\mathbf{X} = [\mathbf{x}_1, \dots, \mathbf{x}_T]'$ and is assumed to be full rank, of size $T \times k$, and $\sigma^2\boldsymbol{\Sigma}$ is the $T \times T$ covariance matrix of $\mathbf{Y} - \mathbf{X}\boldsymbol{\beta} =: \epsilon = (\epsilon_1, \epsilon_2, \dots, \epsilon_T)'$. As $\boldsymbol{\Sigma}$ is readily computable via (7.21) and (7.23), the exact likelihood of $\mathbf{Y}$ can be straightforwardly computed and, thus, the m.l.e. of the parameter vector

$$\boldsymbol{\theta} = (\boldsymbol{\beta}', \ \mathbf{a}', \ \mathbf{b}', \ \sigma)' \tag{7.26}$$

can be obtained, where $\mathbf{a} = (a_1, \dots, a_p)$ and $\mathbf{b} = (b_1, \dots, b_q)$. If $\mathbf{X}$ is just a column of ones, i.e., $\mathbf{X}\boldsymbol{\beta}$ is just β_1, then the model is equivalent to model (7.4), but the way of introducing the constant term into the model is different. In particular, with (7.4), the mean is given by (7.6), whereas with (7.25), the mean is β_1.

A program to compute the conditional and exact m.l.e. of the parameters of model (7.25) is given in Listings 7.7 and 7.8. Initial estimates for $\boldsymbol{\beta}$ are obtained by o.l.s., and those for the ARMA parameters are just zeros, though one of the methods discussed in Section 7.3 could easily be used instead. For $p = 1$ and/or $q = 1$, the parameters are constrained to lie between -1 and 1. In the general ARMA case, the program incorporates a simple, "brute force" method of imposing stationarity and invertibility: Illustrating for stationarity, the roots of the autoregressive polynomial are computed at each evaluation of the likelihood. If any are on or inside the unit circle, then the likelihood of the model is not computed, and a very small value is returned in its place. It is chosen proportional to the extent of the violation, so as to give the optimization routine a chance to "find its way back".

Remarks

a) If there are regressors in the model, as in (7.25), then, given the ARMA parameters, the covariance matrix $\boldsymbol{\Sigma}$ of the ϵ_t can be constructed (up to a scale constant σ) and the g.l.s. estimator (1.28) can be used to obtain the m.l.e. of $\boldsymbol{\beta}$. This is attractive because it is a closed-form solution, but not obtainable, as $\boldsymbol{\Sigma}$ is not known. As such, the simple iterative method suggests itself: Starting with the o.l.s. estimate of $\boldsymbol{\beta}$, compute the (say, conditional) m.l.e. of the ARMA parameters using the o.l.s. residuals. Based on these, compute $\widehat{\boldsymbol{\Sigma}}$ and use it to compute the g.l.s. estimator of $\boldsymbol{\beta}$. This can be repeated until convergence.

   The benefit of such a method is that numerical optimization is necessary only for a subset of the model parameters, thus providing a speed advantage, similar in principle to use of the EM algorithm. Observe, however, that an approximate joint covariance matrix is not available from this method. If confidence intervals for the parameters are desired, or confidence regions for a set of them, or the distribution of forecasts, then the bootstrap (single or double, parametric or nonparametric) can be applied, as discussed in Chapter III.1.3.

```
function [param, stderr, resid, varcov, loglik]=armareg(y,X,p,q,exact)
% Set exact=1 for exact ML, otherwise conditional ML is used.
% param=[B ; ar terms ; ma terms ; sigma]
% stderr is same shape as param and gives approximate standard errors
% resid is the estimated white noise series
% varcov is the entire (estimated) variance covariance matrix
% Pass X as [] if there is no constant term.
% If X is a scalar, it is set to a vector of ones
ylen=length(y); y=reshape(y,ylen,1);  if length(X)==1, X=ones(ylen,1); end
if isempty(X), res=y; beta=[]; nrow=ylen; ncol=0;
else [nrow,ncol]=size(X); beta=inv(X'*X)*X'*y; res=y-X*beta;
end
if p+q==0,  sigma=sqrt(res'*res/ylen); param=[beta' sigma]'; return, end
initvec=[beta' zeros(1,p+q) std(y)]';
if (p+q)==1 % for an AR(1) or MA(1) model.
  %%%%%%%%        beta        a or b  scale
  bound.lo=    [-ones(1,ncol)    -1       0          ]';
  bound.hi=    [ ones(1,ncol)     1       2*std(y) ]';
  bound.which=[zeros(1,ncol)     1       1          ]';
elseif (p==1) & (q==1)
  %%%%%%%%        beta         a   b     scale
  bound.lo=    [-ones(1,ncol)    -1  -1     0          ]';
  bound.hi=    [ ones(1,ncol)     1   1     2*std(y) ]';
  bound.which=[zeros(1,ncol)     1   1     1          ]';
else
  bound.which=zeros(1,length(initvec)); % no bounds at all.
end

mletol=1e-4; MaxIter=100; MaxFunEval=MaxIter*length(initvec);
opt=optimset('Display','None','TolX',mletol,'MaxIter',MaxIter,...
  'MaxFunEval',MaxFunEval,'LargeScale','off');
[pout,negloglik,exitflag,theoutput,grad,hess]= ...
  fminunc(@arma_,einschrk(initvec,bound),opt,y,X,p,q,exact,bound);
loglik=-negloglik; varcov=inv(hess);
[param,varcov]=einschrk(pout,bound,varcov);
if nargout>1 % get varcov and standard errors
  if 1==1 % direct Hessian calc instead of bfgs output
    H = -hessian(@arma_,param,y,X,p,q,exact); varcov=inv(H);
  end
  stderr=sqrt(diag(varcov));
end
if nargout>2 % get residuals
  littlesig=param(end);
  if exact==1
    if isempty(X), z=y; else beta = param(1:ncol); z=y-X*beta; end
    a=param(ncol+1:ncol+p); b=param(ncol+p+1:end-1);
    Sigma = acvf(a,b,nrow); SigInv=inv(Sigma);
    [V,D]=eig(0.5*(SigInv+SigInv')); W=sqrt(D); SigInvhalf = V*W*V';
    resid = SigInvhalf*z/littlesig;
  else
    [garb,uvec]=arma_(param,y,X,p,q,0); resid=uvec/littlesig;
  end
end
```

Program Listing 7.7: Computes the exact and conditional m.l.e. of the parameters in the linear regressionmodel with ARMA disturbances. The program is continued in Listing 7.8.

```
function [loglik,uvec]=arma_(param,y,X,p,q,exact,bound)
if nargin<7, bound=0; end
if isstruct(bound), param=einschrk(real(param),bound,999); end
if any(isinf(param)) | any(isnan(param))
  param=zeros(length(param),1); param(end)=1;
end
if isempty(X), nrow=length(y); ncol=0;
else, [nrow,ncol]=size(X); regbeta=param(1:ncol); end
a=param(ncol+1:ncol+p); b=param(ncol+p+1:end-1);
sig=abs(param(end)); % this is NOT sigma^2, but just (little) sigma.
if p>0 % enforce stationarity
  rootcheck=min(abs(roots([ -a(end:-1:1); 1 ])));
  if rootcheck<=1.0001, loglik=abs(1.01-rootcheck)*1e6; return, end
end
if q>0 % enforce invertibility
  rootcheck=min(abs(roots([ b(end:-1:1); 1 ])));
  if rootcheck<=1.0001, loglik=abs(1.01-rootcheck)*1e6; return, end
end
if isempty(X), z=y; else, z=y-X*regbeta; end

if exact==1 % get the exact likelihood
  uvec=0;
  if (p==1) & (q==0) % speed this case up considerably.
    K=(-nrow/2) * log(2*pi); s2=sig^2;
    e=z(1)^2*(1-a^2) + sum( (z(2:end) - a*z(1:end-1)).^2 );
    % True ll is: ll = 0.5*log(1-a^2) + K - nrow * log(sig) - e/2/s2;
    % If you include K, this is not compatible with the general case below.
    % ll = 0.5*log(1-a^2) - nrow * log(sig) - e/2/s2;
    ll = K + 0.5*log(1-a^2) - nrow * log(sig) - e/2/s2;
  else
    Sigma=acvf(a,b,nrow); Vi=inv(Sigma); detVi=det(Vi);
    if detVi<=0, loglik=abs(detVi+0.01)*1e4; return, end
    ll = -nrow * log(sig) + 0.5*log(detVi) - z'*Vi*z/(2*sig^2);
  end
else % conditional likelihood
  reversearvec= a(p:-1:1); % avoid reversing the part of z each time
  uroll=zeros(q,1); % a rolling window of U_t hat values
  uvec=zeros(nrow-p,1); % all the T-p U_t hat values
  for t=p+1:nrow
    u=z(t);
    if p>0, u=u-sum( z((t-p):t-1).*reversearvec ); end
    if q>0, u=u-sum(uroll.*b); uroll=[u ; uroll(1:q-1)]; end
    uvec(t-p)=u;
  end
  ll = - nrow * log(sig) - sum(uvec.^2)/(2*sig.^2);
end
loglik = -ll;
```

Program Listing 7.8: Continued from Listing 7.7.

b) An advantage of using the state space representation and Kalman filtering techniques is that it can be set up to allow for **missing values**. This is explicitly dealt with in Jones (1980) and Harvey and Pierse (1984), and, in more generality, throughout the monograph by Durbin and Koopman (2012). Here, we just mention one possible way of proceeding for an ARMAX(p, q) model when faced with missing values in the time series, based on the method of **multiple imputation**.

Assume a time series from $t = 1$ to $t = T$ and observation t is missing, $t \in \{p+1, p+2, \dots, T-1\}$. Using the current value of $\widehat{\theta}$ in (7.26) and the current set of filtered innovations $\widehat{U}_1$, $\widehat{U}_2$, …, $\widehat{U}_{t-1}$ (where unavailable values are replaced by their expected value of zero), compute a point estimate of Y_t, say $\tilde{Y}_t$, based on its optimal forecast as in (7.47) given below, and add to it a value $\tilde{U}_t$ that is, for a nonparametric bootstrap type of imputation, drawn from the current set of $\{\widehat{U}_t\}$, or, for a parametric bootstrap type of imputation, drawn from a $\mathrm{N}(0, \hat{\sigma}^2)$ distribution. The likelihood can then be computed in the usual way, and the m.l.e. determined. This is repeated B times, and a set of B point estimates $\widehat{\theta}_1, \dots, \widehat{\theta}_B$ is obtained, from which the mean or median could be taken as the single point estimate for each parameter. The empirical distribution of the set of B point estimates can be plotted as histograms or kernel density estimates, indicating the distribution of the point estimators taking into account the variation induced by the unknown values of Y_t.

This method is easily extended to the case in which multiple observations are missing, including the case for which two or more adjacent observations are missing, by computing $\tilde{Y}_{t+1}$ conditional on $\tilde{Y}_t$, etc. The reader is encouraged to augment the program for the conditional and exact m.l.e. given in Listings 7.7 and 7.8 to implement this technique, passing an additional vector boolean argument indicating which observations are present and missing. Observe how this method can be combined with the traditional bootstrap in order to obtain bootstrap distributions of the parameters that account for their uncertainty from the missing values as well as their sampling error, and similarly used to generate prediction intervals associated with point forecasts.

c) For the conditional m.l.e., conditioning on the first p-values of the series gives rise to the arguably unattractive property that, for a given time series and set of ARMA(p, q) parameters, the likelihood will not be the same when using an ARMA($p+1, q$) model with the same parameters, and ($p+1$)th autoregressive coefficient $\hat{a}_{p+1} = 0$.

d) With respect to ensuring stationarity and invertibility of the ARMA model during estimation, more sophisticated techniques of constrained optimization could be used (in conjunction with the polynomial roots or the Schur condition), as allowed for in Matlab's function `fmincon`.

e) Matlab (in its system identification toolbox) has a built-in routine for ARMA estimation that runs extremely fast. For a pure ARMA model, one would execute `m=armax(y,[p,q]); ahat=-m.a(2:end); bhat=m.c(2:end);`. The performance of this method can be compared to those in Section 7.3. ■

7.4.3 Interval Estimation

Asymptotically, for both the exact and conditional m.l.e. of $\theta = (a_1, \dots, a_p, b_1, \dots, b_q)'$ corresponding to a stationary and invertible ARMA process,

$$\sqrt{T}(\hat{\theta}_{\mathrm{ML}} - \theta) \overset{\mathrm{asy}}{\sim} \mathrm{N}(\mathbf{0}, \mathbf{C}), \quad \mathbf{C} = \sigma^2 \begin{bmatrix} \mathbb{E}[\mathbf{X}\mathbf{X}'] & \mathbb{E}[\mathbf{X}\mathbf{V}'] \\ \mathbb{E}[\mathbf{V}\mathbf{X}'] & \mathbb{E}[\mathbf{V}\mathbf{V}'] \end{bmatrix}^{-1}, \tag{7.27}$$

where $\mathbf{X} = (X_t, X_{t-1}, \dots, X_{t-p+1})$, with $a(L)X_t = U_t$ and $\mathbf{V} = (V_t, V_{t-1}, \dots, V_{t-q+1})'$ with $b(L)V_t = U_t$. This clearly generalizes the AR(p) result in (6.29) and the MA(q) result in (6.54). Proofs of (7.27) and the precise conditions under which it holds can be found in the references stated at the end of Section 6.2.2.

Example 7.3 For $p = q = 1$ and using the infinite MA representations for the AR polynomials,

$$\mathbb{E}[X_t V_t] = \mathbb{E}\left[\left(\sum_{i=0}^{\infty} a^i U_{t-i}\right)\left(\sum_{j=0}^{\infty} (-b)^j U_{t-j}\right)\right] = \sum_{i=0}^{\infty} (-ab)^i = \frac{1}{1+ab},$$

so that

$$\mathbf{C} = \sigma^2 \begin{bmatrix} \frac{\sigma^2}{1-a^2} & \frac{\sigma^2}{1+ab} \\ \frac{\sigma^2}{1+ab} & \frac{\sigma^2}{1-b^2} \end{bmatrix}^{-1}$$

$$= \frac{1+ab}{(a+b)^2} \begin{bmatrix} (1-a^2)(1+ab) & -(1-a^2)(1-b^2) \\ -(1-a^2)(1-b^2) & (1-b^2)(1+ab) \end{bmatrix}. \tag{7.28}$$

If $b = 0$, then $[\mathbf{C}]_{1,1} = (1-a^2)/a^2$ and $[\mathbf{C}]_{2,2} = 1/a^2$ from (7.28). Thus, wrongly estimating an AR(1) process as an ARMA(1, 1) results in the asymptotic variance of $\hat{a}_{\text{ML}}$ and $\hat{b}_{\text{ML}}$ increasing without bound as a approaches zero. This makes sense, as there is zero pole cancellation and the parameters are no longer identified. More generally, this holds if $a = -b$. ■

Godolphin and Unwin (1983) developed an efficient algorithm to alleviate the otherwise tedious evaluation of matrix $\mathbf{C}$ in (7.27) when $p + q$ is not very small. With it, $\mathbf{C}$ can be numerically computed with the m.l.e. values $\hat{\boldsymbol{\theta}}_{\text{ML}}$ replacing $\boldsymbol{\theta}$. Potentially easier (and possibly more accurate) is to use the approximate Hessian matrix from the likelihood function, as discussed in Section 6.1.3.4. Based on it and the asymptotic normality of the estimators, approximate one-at-a-time confidence intervals (c.i.s) for each of the parameters can be constructed.

A way of obtaining more accurate interval estimators for the parameters is to use the bootstrap, as was demonstrated for the AR(1) case in Section 4.7. In particular, given the estimated residuals $\hat{\mathbf{U}} = (\hat{U}_1, \dots, \hat{U}_T)'$ (which are approximately i.i.d. normal if the true data generating process is a stationary, invertible ARMA(p, q) model with normal innovations and p and q are correctly specified), let $\mathbf{U}^{(i)}$ be the ith bootstrap replication of the $\hat{\mathbf{U}}$, formed by sampling from the $\hat{U}_t$ with replacement. For each $\mathbf{U}^{(i)}$, generate time series $\mathbf{Y}^{(i)} = \boldsymbol{\Sigma}^{1/2}\mathbf{U}^{(i)}$, where $\boldsymbol{\Sigma} = \boldsymbol{\Sigma}(\hat{\boldsymbol{\theta}}_{\text{ML}})$ is based on the m.l.e. of the original data, and compute the corresponding m.l.e. $\hat{\boldsymbol{\theta}}^{(i)}_{\text{ML}}$. This is conducted B times and the usual method of obtaining c.i.s for each of the parameters is used.

Example 7.4 For $p = q = 1$, $a = -0.3$, $b = 0.7$ and using $B = 2{,}000$ bootstrap replications, 90% c.i.s were constructed for a and b for $s = 1{,}000$ simulated time series with $T = 30$ and $\sigma^2 = 16$. The length of each interval and whether or not it covered the true parameter were recorded. This was also done for intervals based on the approximate Hessian matrix returned with the m.l.e. values and the asymptotic normal distribution. For parameter a, the actual coverage of the asymptotic c.i. was 0.77, with average length 0.96. The bootstrap c.i. had actual coverage 0.91 and mean length 1.3.

Thus, for this relatively small sample size, the bootstrap interval appears far superior to the use of the asymptotic result. For parameter b, however, coverage of the asymptotic c.i. was 0.72 with length

0.96, while the bootstrap coverage was 0.98 with length 1.5. While still closer in coverage than the asymptotic c.i., the bootstrap interval is apparently too large. The reader is encouraged to investigate the performance based on the double bootstrap. In addition, there are more accurate methods for constructing (single) bootstrap c.i.s without much more computational effort; see Efron (2003), Davison et al. (2003), Efron and Hastie (2016, Ch. 11), and the references therein. ∎

7.4.4 Model Mis-specification

Assume the true model is a stationary, invertible ARMA(p, q) process with i.i.d. normal innovations, and, based on a "reasonably sized" data set, one estimates an ARMA(p, q) model. Then, the estimated residuals should be close to i.i.d. normal. Under the normality assumption, independence can be informally checked by the use of correlograms (sample autocorrelations) applied to the residuals, as detailed in Chapter 8.

The model is said to be over-specified if one fits an ARMA(p^*, q^*), where either $p^* = p$ and $q^* > q$, or $p^* > p$ and $q^* = q$, or $p^* > p$ and $q^* > q$. One might think that, in all of these over-specified cases, as the sample size tends to infinity, the parameter vector will be asymptotically unbiased and normally distributed, such that the parameters that are nonzero attain their true values, and the zero ones will be zero. This is, however, only true in the first two cases, i.e., $p^* = p$ and $q^* > q$, or $p^* > p$ and $q^* = q$. In the last case, with $p^* > p$ and $q^* > q$, we have the zero pole cancellation problem, discussed after (7.7). Assume $p^* = p + 1$ and $q^* = q + 1$. Then the extraneous factor, say $(1 - \lambda L)$, in both the AR and MA polynomial is not identified, meaning that the equivalent model arises for any $|\lambda| < 1$. Because of cancellation, the likelihood has no information about λ, and the value of zero is just as likely as any other value for $|\lambda| < 1$. This implies that the coefficients in the AR and MA polynomial are not identified.

As already mentioned above, one of the common modern strategies of deciding on p and q is to fit numerous ARMA models (possibly all p, q combinations such that, say, $0 \leqslant p \leqslant 5$ and $0 \leqslant q \leqslant 5$ for a non-seasonal model) and then take the one that yields the smallest AIC or BIC value; see Chapter 9. The aforementioned issue with zero pole cancellation implies that this strategy could lead to numeric estimation problems for some models. Observe, however, that this will not be the case if we restrict ourselves to only AR(p) processes (which are anyway extremely fast to estimate when using the conditional likelihood approach). Of course, such an approach will not be optimal if the true model is genuinely from an ARMA(p, q) with $q > 0$. Yet, in practice:

> It is often understood that the true data generating process is not ARMA, and an AR(p) model for a suitably chosen p offers a good approximation.

Methods for choosing p include the same usage of the information criteria, and also ones that are based on small sample distribution theory; see Section 9.5.

Now consider the under-specified case: One fits an ARMA(p^*, q^*), where $p^* < p$ and/or $q^* < q$. (Note this includes the case such that, say, $p^* < p$ and $q^* > q$; it is still under-specified.) For illustration, take $p = 2$, $q = 1$, and $p^* = q^* = 1$. Expressing the true AR(2) polynomial as $(1 - \lambda_1 L)(1 - \lambda_2 L)$, and denoting the fitted stationary invertible ARMA(1,1) coefficients as $\hat{a}^*$ and $\hat{b}^*$, the residuals $\{\hat{\epsilon}_t^*\}$ are given by

$$\hat{\epsilon}_t^* = \frac{(1 - \hat{a}^* L)}{(1 - \hat{b}^* L)} Y_t = \frac{(1 - \hat{a}^* L)}{(1 - \hat{b}^* L)} \frac{(1 + bL)}{(1 - \lambda_1 L)(1 - \lambda_2 L)} \epsilon_t, \tag{7.29}$$

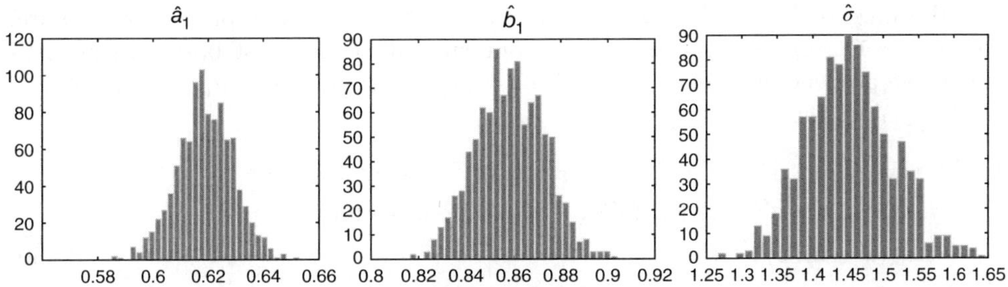

Figure 7.5 Histograms of fitted ARMA(1,1) parameters when the true model is ARMA(2,1) with $a_1 = 1.2$, $a_2 = -0.8$, $b_1 = 0.5$, and $\sigma = 1$, for $T = 1{,}000$ and 1,000 replications.

which is (depending if factors cancel) either an ARMA(3,2), an ARMA(2,1), or AR(1). The under-specified model will not necessarily result in an AR(1) residual process, in the sense that the roots cancel. More specifically, given continuity, $\Pr(\hat{a}^* = \lambda_1) = \Pr(\hat{a}^* = \lambda_2) = 0$, so that the roots will never precisely cancel, but in principle, they might be "close enough" that, for practical purposes, they cancel. However, recalling (6.4) and the condition on a_1 and a_2 such that the roots will be complex, note that, if the AR(2) polynomial is such that the roots are complex pairs, then cancellation could never occur in this example.

In general, the under-specified model maximizes the restricted likelihood, capturing part of the autocorrelation structure of the data, and its residuals are "closer" to i.i.d. than the original time series. As an example, consider what happens if we fit an ARMA(1,1) model when the true model is ARMA(2,1) with parameters $a_1 = 1.2$, $a_2 = -0.8$, $b_1 = 0.5$, and $\sigma = 1$. In this case, as $a_1^2 + 4a_2 < 0$, the roots of the AR polynomial form a complex pair, and cancellation is not possible. Figure 7.5 shows histograms of the three estimated parameters of the under-specified ARMA(1,1) model based on simulation and estimation with the conditional m.l.e., using sample size $T = 1{,}000$ and $s = 1{,}000$ replications. The mean of the $\hat{\sigma}$ is 1.45, which is lower than $\sqrt{\gamma_0}$ of the ARMA(2,1) process, 3.09, computed from (7.21), but higher than the true value of σ. The mean of $\hat{a}_1$ is 0.62, and that of $\hat{b}_1$ is 0.86.

As another example, let the true process be an AR(2) with $a_1 = 1.0$ and $a_2 = -0.2$, which is stationary, with AR polynomial roots 3.618 and 1.382. When we fit an under-specified AR(1) model, the mean of $\hat{a}_1$ based on $s = 1{,}000$ replications is 0.83, which can be seen as the "best compromise value", and clearly does not induce any cancellation of roots in (7.29). The reader is encouraged to replicate the simulations done here.

7.5 Forecasting

Forecasting is like trying to drive a car blindfolded and following directions given by a person who is looking out of the back window.

(Anonymous)

7.5.1 AR(p) Model

For point prediction of an AR(p) process, the extension of the AR(1) case developed in Section 4.4 is straightforward. For $h = 1$, point forecast $\hat{Y}_{T+1|T}$ is formed by substituting estimates in place of unknowns into the r.h.s. of $Y_{T+1} = \sum_{i=1}^{p} a_i Y_{T+1-i} + U_{T+1}$, so that U_{T+1} is replaced by zero and a_i is replaced by $\hat{a}_i$, $i = 1, \ldots, p$. For the m.s.e. of a one-step ahead forecast based on an estimated AR(p) model, we proceed as in (4.27), but with $\mathbf{a} = (a_1, a_2, \ldots, a_p)'$, $\hat{\mathbf{a}} = (\hat{a}_1, \hat{a}_2, \ldots, \hat{a}_p)'$, and $\mathbf{Y}_t = (Y_t, Y_{t-1}, \ldots, Y_{t-p+1})'$, to get

$$\begin{aligned} \text{mse}(\hat{Y}_{T+1|T}) &= \mathbb{E}[(\hat{Y}_{T+1|T} - Y_{T+1})^2] = \mathbb{E}[(\hat{Y}_{T+1|T} - \mathbf{a}'\mathbf{Y}_T + \mathbf{a}'\mathbf{Y}_T - Y_{T+1})^2] \\ &= \mathbb{E}[(\hat{Y}_{T+1|T} - \mathbf{a}'\mathbf{Y}_T)^2] + \mathbb{E}[(\mathbf{a}'\mathbf{Y}_T - Y_{T+1})^2] + \text{cross term} \\ &= \mathbb{E}[((\hat{\mathbf{a}}' - \mathbf{a}')\mathbf{Y}_T)^2] + \sigma^2. \end{aligned} \tag{7.30}$$

It is intuitive that, for a stationary AR(p) model, the dependence between $\hat{a}_i$ and Y_t, for all $i = 1, \ldots, p$ and $t = 1, \ldots, T$, weakens as $T \to \infty$. Let $d_i = \hat{a}_i - a_i$. From the identity $\left(\sum_{i=1}^{r} x_i\right)^2 = \sum_{i=1}^{r} \sum_{j=1}^{r} x_i x_j$ and treating $\hat{a}_i$ and Y_t as if they were independent,

$$\begin{aligned} \mathbb{E}[((\hat{\mathbf{a}}' - \mathbf{a}')\mathbf{Y}_T)^2] &= \mathbb{E}[(d_1 Y_T + d_2 Y_{T-1} + \cdots + d_i Y_{T-i+1} + \cdots + d_p Y_{T-p+1})^2] \\ &= \sum_{i=1}^{p} \sum_{j=1}^{p} \mathbb{E}[d_i Y_{T-i+1}\, d_j Y_{T-j+1}] \qquad (7.31) \\ &\approx \sum_{i=1}^{p} \sum_{j=1}^{p} \text{Cov}(\hat{a}_i, \hat{a}_j)\gamma_{i-j}. \qquad (7.32) \end{aligned}$$

From (6.29), $\sqrt{T}(\hat{\mathbf{a}}_{\text{ML}} - \mathbf{a}) \overset{\text{asy}}{\sim} \text{N}(\mathbf{0}, \sigma^2\boldsymbol{\Gamma}^{-1})$, where $\boldsymbol{\Gamma}$ is the $p \times p$ unconditional covariance matrix of $Y_1, \ldots, Y_p$, with (ij) th element γ_{i-j}. Thus, taking $\text{Cov}(\hat{a}_i, \hat{a}_j)$ to be its large-sample approximation $(\sigma^2/T)\boldsymbol{\Gamma}^{-1}$, the product $\text{Cov}(\hat{a}_i, \hat{a}_j)\gamma_{i-j}$ in (7.32) is the (ij)th term in the matrix $(\sigma^2/T)\boldsymbol{\Gamma}^{-1} \odot \boldsymbol{\Gamma}$, where $\odot$ is the Hadamard, or elementwise, product. As the sum of all the elements of a $p \times p$ matrix $\mathbf{A}$ can be written as $\mathbf{1}'\mathbf{A}\mathbf{1}$, where $\mathbf{1}$ is a column of p ones, (7.32) is $(\sigma^2/T)\mathbf{1}'(\boldsymbol{\Gamma}^{-1} \odot \boldsymbol{\Gamma})\mathbf{1}$.

It turns out that, for any full rank symmetric matrix $\mathbf{K}$ of size m,

$$\mathbf{1}'(\mathbf{K} \odot \mathbf{K}^{-1})\mathbf{1} = m, \tag{7.33}$$

so that (7.32) reduces to $\sigma^2 p/T$. To prove (7.33), we use the following result.

Theorem 7.1 Let $\mathbf{A}$ and $\mathbf{B}$ be $m \times n$ (real) matrices, $\mathbf{x}$ any $n \times 1$ (real) vector, and $\mathbf{D} = \text{diag}(\mathbf{x})$. Then, the ith diagonal entry of matrix $\mathbf{ADB}'$ coincides with the ith entry of vector $(\mathbf{A} \odot \mathbf{B})\mathbf{x}$ for all $i = 1, \ldots m$, i.e.,

$$[\mathbf{ADB}']_{ii} = [(\mathbf{A} \odot \mathbf{B})\mathbf{x}]_i, \quad \forall\ 1 \leqslant i \leqslant m. \tag{7.34}$$

Proof: See, e.g., Horn (1994, p. 305), or Schott (2005, p. 296). ∎

Then, (7.33) follows because, for $\mathbf{x} = \mathbf{1}_m = \mathbf{1}$, $\mathbf{D} = \text{diag}(\mathbf{x}) = \mathbf{I}_m = \mathbf{I}$, and $\mathbf{K}$ a symmetric $m \times m$ matrix of full rank, (7.34) with $\mathbf{A} = \mathbf{K}$ and $\mathbf{B} = \mathbf{K}^{-1}$ implies

$$\mathbf{1}'(\mathbf{K} \odot \mathbf{K}^{-1})\mathbf{1} = \sum_{i=1}^{m} [(\mathbf{K} \odot \mathbf{K}^{-1})\mathbf{1}]_i = \sum_{i=1}^{m} [\mathbf{KDK}^{-1}]_{ii} = \sum_{i=1}^{m} [\mathbf{I}]_{ii} = m.$$

Problem 7.5 shows this in other ways. The result is the pleasantly simple expression

$$\text{mse}(\hat{Y}_{T+1|T}) \approx \sigma^2 \left(1 + \frac{p}{T}\right), \tag{7.35}$$

which generalizes (4.29) in the $p = 1$ case. Result (7.35) has been given by Bloomfield (1972, p. 505), derived in the context of the spectral analysis of time series.

The easiest way of determining the quality of approximation (7.35) is via simulation. For the AR(2) model with $a_1 = 1.2, a_2 = -0.8$, and $\sigma^2 = 4$, the true $\text{mse}(\widehat{Y}_{T+1|T})$ based on the exact m.l.e. and 100,000 replications is 4.99, 4.44, and 4.30 for $T = 10$, 20, and 30, respectively, which can be compared to the values from (7.35) of 4.80, 4.40, and 4.27. Similarly, for $a_1 = -0.6, a_2 = 0.2$, simulation resulted in 4.91, 4.45, and 4.29 for $T = 10$, 20, and 30.

Though further simulation would be required before making general statements, it appears that (7.35) is almost exact for $T = 30$ and is still vastly better for $10 < T < 30$ than use of just its limiting expression as $T \to \infty$, i.e., σ^2.[1]

The method of computing an h-step ahead point forecast of an AR(p) process is the same as for $h = 1$: we replace unknown values on the r.h.s. of $Y_{T+h} = \sum_{i=1}^{p} a_i Y_{T+h-i} + U_{T+h}$ with estimates. In particular, U_{T+h} is replaced by its expected value of zero, the a_i are replaced by their estimates, while for the Y_t, if $t \leqslant T$, then the observed value Y_t is used, otherwise its forecast $\widehat{Y}_{t|T}$ is used. For example, with $p = 3$,

$$\begin{aligned}
\widehat{Y}_{T+1|T} &= \hat{a}_1 Y_T + \hat{a}_2 Y_{T-1} + \hat{a}_3 Y_{T-2}, \\
\widehat{Y}_{T+2|T} &= \hat{a}_1 \widehat{Y}_{T+1|T} + \hat{a}_2 Y_T + \hat{a}_3 Y_{T-1}, \\
\widehat{Y}_{T+3|T} &= \hat{a}_1 \widehat{Y}_{T+2|T} + \hat{a}_2 \widehat{Y}_{T+1|T} + \hat{a}_3 Y_T, \quad \text{and} \\
\widehat{Y}_{T+h|T} &= \hat{a}_1 \widehat{Y}_{T+h-1|T} + \hat{a}_2 \widehat{Y}_{T+h-2|T} + \hat{a}_3 \widehat{Y}_{T+h-3|T}, \quad h > p.
\end{aligned}$$

Of course, $\widehat{Y}_{T+h|T}$ can be expressed as a linear combination of $Y_T, Y_{T-1}, \dots, Y_{T-p+1}$; for the AR(3) case with $h = 2$,

$$\begin{aligned}
\widehat{Y}_{T+2|T} &= \hat{a}_1 \widehat{Y}_{T+1|T} + \hat{a}_2 Y_T + \hat{a}_3 Y_{T-1} \\
&= \hat{a}_1(\hat{a}_1 Y_T + \hat{a}_2 Y_{T-1} + \hat{a}_3 Y_{T-2}) + \hat{a}_2 Y_T + \hat{a}_3 Y_{T-1} \\
&= (\hat{a}_1^2 + \hat{a}_2) Y_T + (\hat{a}_1 \hat{a}_2 + \hat{a}_3) Y_{T-1} + \hat{a}_1 \hat{a}_3 Y_{T-2}.
\end{aligned} \tag{7.36}$$

Letting the coefficient of Y_{T-i} be designated by $\hat{a}_i^{(2)}$, we can, in general, express $\widehat{Y}_{T+h|T}$ as

$$\widehat{Y}_{T+h|T} = \sum_{i=1}^{p} \hat{a}_i^{(h)} Y_{T-i+1}. \tag{7.37}$$

---

1 It is interesting, though perhaps pure coincidence, that use of

$$\text{mse}(\hat{Y}_{T+1|T}) \overset{?}{\approx} \sigma^2 \left(1 + \frac{p}{T} + \frac{p^2}{T^2}\right)$$

yields values of 4.96, 4.44, and 4.30 for $T = 10$, 20, and 30, which are remarkably close to the true values. Further simulations, with a variety of T, p, and $\mathbf{a}$, would be required to assess its validity. Taking this conjecture one step further, it might be the case that

$$\text{mse}(\hat{Y}_{T+1|T}) \overset{?}{\approx} \sigma^2 \sum_{j=0}^{\infty} \left(\frac{p}{T}\right)^j = \frac{T}{T-p}\sigma^2.$$

Unfortunately, even if this is true, it will have little value for small T because then the independence assumption of $\hat{a}_i$ and Y_t will not be tenable. To illustrate, for $T = 5, p = 2$, and $\sigma^2 = 4$, (7.35) and its two extensions are, respectively, 5.6, 6.24, and 6.67; simulation (assuming known σ^2) resulted in 6.02, indicating either that this "generalization" is wrong, or that the independence assumption becomes crucial for very small T, or both.

A simple way of calculating the $\hat{a}_i^{(h)}$ in (7.37) for any h and p is to express the model as the **vector AR(1) process**

$$\mathbf{Y}_t = \mathbf{A}\mathbf{Y}_{t-1} + \mathbf{U}_t, \tag{7.38}$$

where $\mathbf{Y}_t = (Y_t, Y_{t-1}, \dots, Y_{t-p+1})'$ and $\mathbf{U}_t = (U_t, 0, \dots, 0)'$ are $p \times 1$ vectors, and $\mathbf{A}$ is the $p \times p$ matrix

$$\mathbf{A} = \begin{bmatrix} a_1 & a_2 & \cdots & a_{p-1} & a_p \\ 1 & 0 & & 0 & 0 \\ 0 & 1 & & 0 & 0 \\ \vdots & \vdots & \ddots & \vdots & \vdots \\ 0 & 0 & \cdots & 1 & 0 \end{bmatrix}. \tag{7.39}$$

Let $\widehat{\mathbf{A}}$ denote $\mathbf{A}$ with the estimated values of the a_i. As $\mathbb{E}[\mathbf{U}_{T+h}] = \mathbf{0}$, (7.38) implies that $\mathbf{Y}_{T+h|T} = \mathbf{A}^h\mathbf{Y}_T$. Thus, the $\hat{a}_i^{(h)}$ are the elements in the first row of $\widehat{\mathbf{A}}^h$ and $\widehat{Y}_{T+h|T}$ is the first element of $\widehat{\mathbf{Y}}_{T+h|T} = \widehat{\mathbf{A}}^h\mathbf{Y}_T$. Defining $\mathbf{e}_1 = (1, 0, \dots, 0)'$, we can express this as

$$\widehat{Y}_{T+h|T} = \mathbf{e}_1'\widehat{\mathbf{A}}^h\mathbf{Y}_T. \tag{7.40}$$

For example, the coefficients in (7.36) are given by the first row in

$$\widehat{\mathbf{A}}^2 = \begin{bmatrix} \hat{a}_1 & \hat{a}_2 & \hat{a}_3 \\ 1 & 0 & 0 \\ 0 & 1 & 0 \end{bmatrix} \begin{bmatrix} \hat{a}_1 & \hat{a}_2 & \hat{a}_3 \\ 1 & 0 & 0 \\ 0 & 1 & 0 \end{bmatrix} = \begin{bmatrix} \hat{a}_1^2 + \hat{a}_2 & \hat{a}_1\hat{a}_2 + \hat{a}_3 & \hat{a}_1\hat{a}_3 \\ \hat{a}_1 & \hat{a}_2 & \hat{a}_3 \\ 1 & 0 & 0 \end{bmatrix}.$$

Also, simple substitution in (7.38) as in the scalar AR(1) case shows that

$$\mathbf{Y}_{T+h} = \mathbf{A}^h\mathbf{Y}_T + \sum_{i=0}^{h-1} \mathbf{A}^i\mathbf{U}_{T+h-i}, \tag{7.41}$$

and

$$Y_{T+h} = \mathbf{e}_1'\mathbf{Y}_{T+h}. \tag{7.42}$$

For $p = 1$, the first term in (7.41) is just $a_1^h Y_T$ and, if $|a_1| < 1$, then $a_1^h \to 0$ as $h \to \infty$. More generally, if the AR(p) model is stationary, then $\mathbf{A}^h \to 0$ as $h \to \infty$. To see this, write $\mathbf{A} = \mathbf{U}\boldsymbol{\Lambda}\mathbf{U}'$, where $\mathbf{U}$ is orthogonal and $\boldsymbol{\Lambda} = \mathrm{diag}(\lambda_1, \dots, \lambda_p)$ are the eigenvalues of $\mathbf{A}$. Then, $\mathbf{A}^h \to 0$ if each $|\lambda_i| < 1$. The result follows because the λ_i are the values that satisfy

$$1 - a_1\lambda^{-1} - \cdots - a_{p-1}\lambda^{1-p} - a_p\lambda^{-p} = 0,$$

as shown, for example, in Hamilton (1994, p. 21).

If the model is stationary, then, as $h \to \infty$, $\mathbf{Y}_t = \sum_{i=0}^{\infty} \mathbf{A}^i\mathbf{U}_{t-i}$ from (7.41). This is an infinite MA representation with $\mathbf{e}_1'\mathbf{A}^i\mathbf{e}_1 = \psi_i$, i.e., the ith term in the infinite moving average expression for Y_t. Thus, from (7.41) and (7.42),

$$\mathrm{mse}(\widehat{Y}_{T+h|T}) = \mathbb{E}[(\widehat{Y}_{T+h|T} - \mathbf{e}_1'\mathbf{A}^h\mathbf{Y}_T + \mathbf{e}_1'\mathbf{A}^h\mathbf{Y}_T - Y_{T+h})^2]$$

$$= \mathbb{E}[(\mathbf{e}_1'\widehat{\mathbf{A}}^h\mathbf{Y}_T - \mathbf{e}_1'\mathbf{A}^h\mathbf{Y}_T)^2] + \mathbb{E}\left[\left(\mathbf{e}_1'\mathbf{A}^h\mathbf{Y}_T - \mathbf{e}_1'\left(\mathbf{A}^h\mathbf{Y}_T + \sum_{i=0}^{h-1}\mathbf{A}^i\mathbf{U}_{T+h-i}\right)\right)^2\right] + 0$$

$$= \mathbb{E}[(\mathbf{e}_1'(\widehat{\mathbf{A}}^h - \mathbf{A}^h)\mathbf{Y}_T)^2] + \mathbb{E}\left[\left(-\mathbf{e}_1' \sum_{i=0}^{h-1} \mathbf{A}^i \mathbf{U}_{T+h-i}\right)^2\right], \tag{7.43}$$

and the latter term is just $\sigma^2 \sum_{i=0}^{h-1} \psi_i^2$.

This decomposition makes clear that the m.s.e. consists of a term reflecting estimated parameter uncertainty, and a term involving the future unknown innovations. The former term is significantly more difficult to handle for general h than it was with $h = 1$. Using a critical result in Neudecker (1969), Yamamoto (1976) showed that

$$\mathbb{E}[(\mathbf{e}_1'(\widehat{\mathbf{A}}^h - \mathbf{A}^h)\mathbf{Y}_T)^2] \approx \frac{\sigma^2}{T}\text{tr}[\mathbf{M}_h'\boldsymbol{\Gamma}^{-1}\mathbf{M}_h\boldsymbol{\Gamma}], \tag{7.44}$$

where, as above, $\boldsymbol{\Gamma}$ is the $p \times p$ unconditional covariance matrix of $Y_1, \dots, Y_p$, the matrix $\mathbf{M}_h$ is the upper left $p \times p$ submatrix of $\sum_{i=0}^{h-1}(\mathbf{A}'^{\,i} \otimes \mathbf{A}^{h-1-i})$, $\otimes$ is the Kronecker matrix product, and $\mathbf{A}$ is given in (7.39). It is easy to see that, for general p and $h = 1$, (7.44) reduces to $\sigma^2 p/T$ as derived above. Also, for general h and $p = 1$, it reduces to the last term given in (4.34).

In practice, one often just computes the latter term in (7.43), i.e.,

$$\text{mse}(\widehat{Y}_{T+h|T}) \approx \sigma^2 \sum_{i=0}^{h-1} \psi_i^2, \tag{7.45}$$

which obviously underestimates $\text{mse}(\widehat{Y}_{T+h|T})$ because it neglects the forecast error arising from the parameter uncertainty. Matters increase in complexity when dealing with ARMA(p, q) processes, compounded even further with an unknown mean term.

7.5.2 MA(q) and ARMA(p, q) Models

For the MA(q) model, the same principles as above are applied. For point estimates, with $\widehat{\mathbf{b}} = (\hat{b}_1, \hat{b}_2, \dots, \hat{b}_q)'$ and $\widehat{\mathbf{U}}_t = (\widehat{U}_t, \widehat{U}_{t-1}, \dots, \widehat{U}_{t-q+1})'$,

$$\widehat{Y}_{T+h|T} = \sum_{j=1}^{q} \hat{b}_j \widehat{U}_{T+h-j} = \widehat{\mathbf{b}}'\widehat{\mathbf{U}}_{T+h-1}.$$

As none of the $\widehat{U}_t$ are observed, the filtered values, i.e., the model residuals $\widehat{U}_t$, $t = 1, \dots, T$, are used in their place. If $t > T$, then $\mathbb{E}[U_t] = 0$ is used. For example, with $q = 2$,

$$\begin{aligned}
\widehat{Y}_{T+1|T} &= \hat{b}_1\widehat{U}_T + \hat{b}_2\widehat{U}_{T-1},\\
\widehat{Y}_{T+2|T} &= 0 + \hat{b}_2\widehat{U}_T, \quad \text{and}\\
\widehat{Y}_{T+h|T} &= 0, \quad h > q.
\end{aligned}$$

For the m.s.e., we ignore the part in its decomposition that accounts for the discrepancy between $\mathbf{b}$ and $\widehat{\mathbf{b}}$; this can be most effectively dealt with via use of the bootstrap as discussed below. What remains is

$$\mathbb{E}\left[\left(\sum_{j=1}^{q} b_j\widehat{U}_{T+h-j} - \left(U_{T+h} + \sum_{j=1}^{q} b_j U_{T+h-j}\right)\right)^2\right] = \mathbb{E}\left[\left(-U_{T+h} + \sum_{j=1}^{q} b_j(\widehat{U}_{T+h-j} - U_{T+h-j})\right)^2\right].$$

Further assuming that $\widehat{U}_{T+h-j} = U_{T+h-j}$ for $T + h - j \leqslant T$ (or $h \leqslant j$), and with $\widehat{U}_{T+h-j} = 0$ for $T + h - j > T$ (or $h > j$), this reduces to (with $b_0 = 1$),

$$\text{mse}(\widehat{Y}_{T+h|T}) \approx \mathbb{E}\left[\left(-\sum_{j=0}^{h-1} b_j U_{T+h-j}\right)^2\right] = \sigma^2(1 + b_1^2 + b_2^2 + \cdots + b_{h-1}^2), \tag{7.46}$$

where $b_j = 0$ if $j > q$. Observe that this is the same as (7.45) because the ψ_i are just the MA parameters. It is important to keep in mind that, in both (7.45) and (7.46), parameter uncertainty is not taken into account and, in the latter, also the error incurred by the assumption that $\widehat{U}_{T+h-j} = U_{T+h-j}$ for $h \leqslant j$. Both of these sources of error are accounted for via the bootstrap.

Point estimates for the ARMA(p, q) case follow analogously by combining the techniques in the AR and MA special cases. For example, with $p = 3$ and $q = 2$,

$$\widehat{Y}_{T+1|T} = \hat{a}_1 Y_T + \hat{a}_2 Y_{T-1} + \hat{a}_3 Y_{T-2} + \hat{b}_1 \widehat{U}_T + \hat{b}_2 \widehat{U}_{T-1}, \tag{7.47a}$$

$$\widehat{Y}_{T+2|T} = \hat{a}_1 \widehat{Y}_{T+1|T} + \hat{a}_2 Y_T + \hat{a}_3 Y_{T-1} + 0 + \hat{b}_2 \widehat{U}_T, \tag{7.47b}$$

$$\widehat{Y}_{T+3|T} = \hat{a}_1 \widehat{Y}_{T+2|T} + \hat{a}_2 \widehat{Y}_{T+1|T} + \hat{a}_3 Y_T, \quad \text{and} \tag{7.47c}$$

$$\widehat{Y}_{T+h|T} = \hat{a}_1 \widehat{Y}_{T+h-1|T} + \hat{a}_2 \widehat{Y}_{T+h-2|T} + \hat{a}_3 \widehat{Y}_{T+h-3|T}, \quad h > \max(p, q). \tag{7.47d}$$

An alternative way of computing the point forecasts is to use the infinite AR representation (7.15). For a pure AR(p) model, the forecasting methods are equivalent; otherwise, they will numerically differ. The program in Listing 7.9 implements both methods to compute $1, 2, \ldots, h$ step ahead forecasts corresponding to an (estimated) ARMA model. If $\mathbb{E}[Y_t] = \mathbf{x}_t'\boldsymbol{\beta}$, then the $\mathbf{X}$ matrix corresponding to Y_1 through Y_T is passed to the routine as was done in Listing 7.7. One would add $\mathbf{x}_{T+i}'\widehat{\boldsymbol{\beta}}$ to the computed output from the program for $i = 1, \ldots, h$. The program can also be easily augmented to output approximate forecast standard errors based on (7.49) given below, using the estimated parameters in place of σ and the ψ_i.

In light of (7.45) and (7.46), one might expect for the ARMA case that $\text{mse}(\widehat{Y}_{T+h|T}) \approx \sigma^2 \sum_{i=0}^{h-1} \psi_i^2$, when parameter uncertainty is ignored. This is indeed the case. We have

$$\begin{aligned}\text{mse}(\widehat{Y}_{T+h|T}) &= \mathbb{E}[(\widehat{Y}_{T+h|T} - Y_{T+h})^2] = \mathbb{E}[(\widehat{Y}_{T+h|T} - Y_{T+h|T} + Y_{T+h|T} - Y_{T+h})^2] \\ &= \mathbb{E}[(\widehat{Y}_{T+h|T} - Y_{T+h|T})^2] + \mathbb{E}[(Y_{T+h|T} - Y_{T+h})^2] + \begin{matrix}\text{cross}\\ \text{term}\end{matrix}\end{aligned} \tag{7.48}$$

where, as in all previous cases, the cross term is zero. Using its infinite MA representation, $Y_{T+h} = \sum_{i=0}^{\infty} \psi_i U_{T+h-i}$ while that for $Y_{T+h|T}$ is the same, except that $U_t = 0$ for $t > T$, so $Y_{T+h|T} = \sum_{i=h}^{\infty} \psi_i U_{T+h-i}$. Thus, further assuming $\widehat{U}_t = U_t$ for $t \leqslant T$, the middle term in (7.48) is

$$\mathbb{E}[(Y_{T+h|T} - Y_{T+h})^2] = \mathbb{E}\left[\left(\sum_{i=0}^{h-1} \psi_i U_{T+h-i}\right)^2\right],$$

and ignoring the uncertainty of the estimated parameters as given by the first term in (7.48), we have

$$\text{mse}(\widehat{Y}_{T+h|T}) \approx \sigma^2 \sum_{i=0}^{h-1} \psi_i^2. \tag{7.49}$$

```
function [fore, param, stderr, resid, varcov]=armafore(y,X,p,q,exact,h,useinfAR)
if nargin<7, useinfAR=0; end
fore=zeros(h,1); [param, stderr, resid, varcov]=armareg(y,X,p,q,exact);
if h==0, return, end
[nrow,ncol]=size(X); avec=param(ncol+1:ncol+p)'; bvec=param(ncol+p+1:end-1)';
if useinfAR==0
 if p>0, yvec=y(end:-1:end+1-p); end, if q>0, uvec=resid(end:-1:end+1-q); end
 for i=1:h
   if p>0, fore(i)=fore(i)+avec*yvec; end
   if q>0, fore(i)=fore(i)+bvec*uvec; uvec=[0 ; uvec(1:end-1)]; end
   if p>0, yvec=[fore(i) ; yvec(1:end-1)]; end
 end
else %infinite AR method. Identical for pure AR models.
  arinfupperlim=100; % arbitrary!
  n=min(length(y),arinfupperlim);
  arcoef=infAR(avec,bvec,n); yvec=y(end:-1:end-n+1);
  for i=1:h, fore(i)=sum(arcoef.*yvec); yvec=[fore(i) ; yvec(1:end-1)]; end
end
```

Program Listing 7.9: Returns same quantities as program `armareg.m` in Listing 7.7 but also returns 1 through h step ahead point forecasts for the ARMA part of the model; the regressor term still needs to be added to the ARMA forecasts. The code that uses the infinite AR representation chooses an arbitrary AR length. Instead, one could use the length of the time series, or (if that is excessively large) choose the cutoff value such that the last AR coefficient (and all subsequent ones) are smaller in magnitude than some specified value, like 10^{-6}.

An analysis of the (downward) bias inherent in (7.49) is analyzed in detail by Ansley and Newbold (1981). They find, among other things, that the bias becomes more extreme as the model gets closer to the stationarity and/or the invertibility border.

The bootstrap could be used to compute more accurate c.i.s for the point forecasts. It would be used just as described earlier, but instead of (or in addition to) keeping the bootstrap parameter estimates, for each of B resampled series, an h-step ahead forecast would be made, say $\widehat{Y}^{(B)}_{T+h|T}$, with the U_{T+i} chosen from the $\widehat{U}_t$, $t = 1, \dots, T$, with replacement instead of set to zero. In doing so, the error arising both from the future U_t and from the estimated parameter uncertainty are taken into account. Based on the appropriate quantiles, a c.i. can be constructed. Also, a kernel density estimate of these values provides a nonparametric forecast of the entire density of $Y_{T+h|T}$. A parametric density forecast approximation can be obtained by fitting, say, a noncentral t, a normal inverse Gaussian, a mixture of normals, etc., to the set of $\widehat{Y}^{(B)}_{T+h|T}$.

Example 7.5 A sample series of length $T = 100$ based on the ARMA(2, 1) model with $a_2 = 1.2$, $a_2 = -0.8$, $b_1 = 0.5$, $\sigma = 1$ and intercept and trend regressor term $10 + 0.04t$ was simulated, its parameters estimated and the first 30 out-of-sample forecasts constructed, based on both the true and estimated parameters. The exact m.l.e.s (with standard errors in parentheses) were $\widehat{\beta}_1 = 10.71(0.48)$, $\widehat{\beta}_2 = 0.0278(0.0081)$, $\hat{a}_1 = 1.125(0.075)$, $\hat{a}_2 = -0.747(0.072)$, $\hat{b}_1 = 0.451(0.11)$, and $\hat{\sigma} = 0.983(0.070)$. The series and the forecasts are shown in Figure 7.6, along with 90% c.i.s based on use of (7.49) with the estimated values, i.e., $\hat{\sigma}^2 \sum_{i=0}^{h-1} \widehat{\psi}_i^2$. (The estimated regression line was added to the lower and upper values for the ARMA c.i.s.)

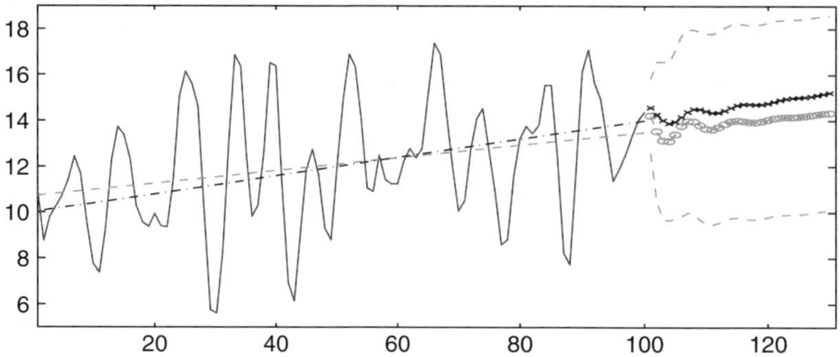

Figure 7.6 Simulated time series (solid line) with out-of-sample forecasts based on estimated parameters (circles) and based on the true parameters (crosses). The straight dashed (dash-dot) line is the estimated (exact) regression term.

We see that a considerable portion of the difference between the point forecasts based on the estimated and true parameters can be attributed to the trend line, to which they converge quickly. Also, after about the 5-step ahead forecast, the size of the c.i.s are approximately that of the (detrended) time series itself, rendering accurate forecasts much further than, say, two steps ahead, almost impossible. The size of the c.i. for the one-step ahead forecast is however considerably smaller than that for the detrended time series itself, but recall that it does not take parameter uncertainty into account.

Finally, the conditional m.l.e. was computed and the largest relative percentage difference from the exact m.l.e. was for β_2, which changed to 0.0269 (just over -3%); the other parameters changed by less than 1%. If one were to overlay the forecasts based on the conditional m.l.e. onto the plot, they would be virtually indistinguishable from the point forecasts based on the exact m.l.e. The difference between their point forecasts is thus relative to the difference between point forecasts based on the estimated and true parameters, essentially zero.

This minute difference becomes completely negligible when considering the width of the c.i.s, i.e., taking the uncertainty of the future U_t into account. As such, if the primary goal of the analysis is forecasting, then use of the conditional m.l.e. instead of the exact m.l.e. appears acceptable, and even use of the methods in Section 7.3 might be adequate, especially if numerous time series require (automated) predicting. ■

Adding a bit of insult to injury to the analysis in the previous example, one should keep in mind that the exercise used the knowledge that the true data generating process is a covariance stationary ARMA(2,1) model with normal innovations. For real time series, not only will p and q not be known, but the ARMA class itself may not be appropriate. Chapter 9 will have more to say about the selection of p and q.

Finally, the assumption that the i.i.d. innovation sequence is normally distributed may not be tenable, with the most common deviations being fatter tails and asymmetry. This point was investigated in detail by Harvey and Newbold (2003) using forecast errors based on macroeconomic time series. They conclude that "… the frequently made assumption of forecast error normality is untenable, its use resulting in overly narrow prediction intervals", and that "… evidence of skewness was also displayed for the vast majority of variables and horizons".

They recommend replacing the normal assumption with an asymmetric t, such as the noncentral t. Observe that exact maximum likelihood estimation with such a distributional assumption is not straightforward, lending even more support for use of the conditional m.l.e. The reader is encouraged to adapt the programs in this chapter to support estimation of an ARMAX model with innovations from an asymmetric, heavy-tailed distribution whose shape parameters are jointly estimated with the remaining ARMAX model parameters.

7.5.3 ARIMA(p, d, q) Models

Let $\{Z_t\}$ follow an ARIMA(p, d, q) process (7.9), i.e., $a(L)(1-L)^d Z_t = c + b(L)U_t$, with $U_t \overset{\text{i.i.d.}}{\sim} \text{N}(0, \sigma^2)$ as in (7.2), such that $Y_t = (1-L)^d Z_t$ is a stationary, invertible ARMA(p, q) process. For data $\{Z_t\}_{t=1}^T$, the dth difference is taken to obtain $Y_{1+d}, Y_{2+d}, \dots, Y_T$. For example, if $d = 1$, then $Y_2 = Z_2 - Z_1$, $Y_3 = Z_3 - Z_2, \dots, Y_T = Z_T - Z_{T-1}$, while if $d = 2$, then, as $(1-L)^2 = 1 - 2L + L^2$,

$$\begin{aligned} Y_3 &= Z_3 - 2Z_2 + Z_1, \\ Y_4 &= Z_4 - 2Z_3 + Z_2, \\ &\vdots \\ Y_T &= Z_T - 2Z_{T-1} + Z_{T-2}. \end{aligned}$$

Given estimates of the ARMA parameters of $\{Y_t\}$, say $\hat{a}(L)$ and $\hat{b}(L)$, such as obtained from the methods in Sections 7.3 or 7.4, forecasts of $Y_{T+h|T}$ are constructed as before; see, e.g., (7.47). Then, to form forecasts $\widehat{Z}_{T+h|T}$, $h = 1, 2, \dots$, the differencing operation needs to be reversed: For example, with $d = 1$, as $Y_{T+1} = Z_{T+1} - Z_T$, we have $Z_{T+1} = Z_T + Y_{T+1}$, or $\widehat{Z}_{T+1|T} = Z_T + \widehat{Y}_{T+1|T}$, $\widehat{Z}_{T+2|T} = \widehat{Z}_{T+1|T} + \widehat{Y}_{T+2|T} = Z_T + \widehat{Y}_{T+1|T} + \widehat{Y}_{T+2|T}$, etc. Prediction intervals can be straightforwardly and reliably computed using the parametric or nonparametric bootstrap. The reader is encouraged to devise a program that inputs a time series assumed to be ARIMA($p, 1, q$), along with p, q, and h, and outputs the h forecast point forecasts and associated 95% prediction intervals based on a nonparametric bootstrap, along with a time-series plot showing the original series, the forecasts, and their prediction intervals.

7.6 Bias-Adjusted Point Estimation: Extension to the ARMAX($1, q$) model

This section was written with Simon Broda and Kai Carstensen

Recall Section 5.4, in which we examined methods for obtaining improved estimators of the AR(1) parameter. Although the first-order autoregressive model is undoubtedly one of the most important models in practice and continues to be the focus of many theoretical contributions in econometrics, it may fail to adequately capture the autocorrelation structure inherent in the data generating process. The ARMA class of models is one type of generalization. In the AR(p) case, an approximately median-unbiased estimator was proposed in Andrews and Chen (1994). However, in the spirit of parsimonious model building, in some situations introducing a moving average component may be more appropriate, thus leading to the ARMAX($1, q$) model.

MA terms may arise due to aggregation (Chambers, 2004) or other data transformations (Galbraith and Zinde-Walsh, 1999). As an empirical example, Ng and Perron (2001) find strongly negative MA(1) parameters for the inflation series of the G7 countries. MA components have received particular attention in the study of unit roots because the "usual" tests are severely hampered by their presence; see Phillips and Perron (1988) and Schwert (1989a). Studies such as Ng and Perron (1995, 2001) and Galbraith and Zinde-Walsh (1999) have addressed this issue by constructing tests that take account of the presence of an MA term by either augmenting the test equation with higher-order AR components or by directly estimating the MA parameters.

The bias-adjusted estimators developed in Section 5.4 can be extended to the ARMAX$(1, q)$ model. For illustration, we restrict ourselves to the $q = 1$ case, with the extension to higher order q being clear. Extending (5.1)–(5.2), the ARMAX(1,1) model can be written as

$$Y_t = \mathbf{x}_t'\boldsymbol{\beta} + \epsilon_t, \qquad t = 0, \dots, T,$$

where

$$\epsilon_t = a\epsilon_{t-1} + bU_{t-1} + U_t, \qquad t = 1, \dots, T, \qquad U_t \overset{\text{iid}}{\sim} \mathrm{N}(0, \sigma^2), \tag{7.50}$$

and, from (7.22),

$$\epsilon_0 \sim \mathrm{N}(0, \sigma^2 d), \qquad d = \frac{b^2 + 2ab + 1}{1 - a^2}. \tag{7.51}$$

If the MA(1) parameter b is known, then the estimators developed in Section 5.4 can be computed as before, with $\mathbf{R}$ in (5.14) replaced by the symmetric square root of

$$\boldsymbol{\Sigma} \equiv \mathbb{E}[\boldsymbol{\epsilon}\boldsymbol{\epsilon}'] = \mathrm{T}(\boldsymbol{c}, \boldsymbol{c}'), \tag{7.52}$$

where

$$\boldsymbol{\epsilon} = (\epsilon_0, \epsilon_1, \dots, \epsilon_T)', \quad \boldsymbol{c} = \sigma^2(d \ ea \ ea^2 \ \dots \ ea^T)', \quad e = \frac{a(1 + b^2) + b(1 + a^2)}{1 - a^2}, \tag{7.53}$$

and $\mathrm{T}(\boldsymbol{c}, \boldsymbol{r})$ denotes a Toeplitz matrix with $\boldsymbol{c}$ as its first column and $\boldsymbol{r}$ as its first row. Extensions for the $q > 1$ case could use the convenient matrix results of Mittnik (1988) or van der Leeuw (1994) for computing (7.52).

For this procedure to be valid, it is required that, for every value of b, the mean, median, and mode functions are strictly increasing in a. Inspection for several sample sizes and choices of $\mathbf{X}$ shows that this is indeed the case, thus guaranteeing the existence of the respective inverse function. Figure 7.7 shows the inverse mean function of $\widehat{a}_{\mathrm{LS}}$ for a model with $\mathbf{X} = \mathbf{1}$ and $T = 50$. Observe how, for each value of b, a different inverse mean function is obtained.

Typically, of course, the MA(1) parameter b is not known. A natural idea is to replace nuisance parameters in the covariance matrix by an estimator, which is also the approach taken by Phillips and Sul (2003) to allow for cross-sectional dependence in their proposed median-unbiased estimator for panel data. We suggest use of the following iterative scheme:

1) Let $\widehat{a}_{\mathrm{LS}}^{O}$ be the observed value of the o.l.s. estimator of a given in (5.12).
2) Obtain an initial estimate $\widehat{b}_0$ for b and set $i = 0$.
3) Compute the bias-corrected estimator

$$\widehat{a}^i := m_{\widehat{b}_i}^{-1}(\widehat{a}_{\mathrm{LS}}^{O}), \tag{7.54}$$

where m denotes the median, mode, and mean functions, respectively, with b in (7.52) replaced by $\widehat{b}_i$.

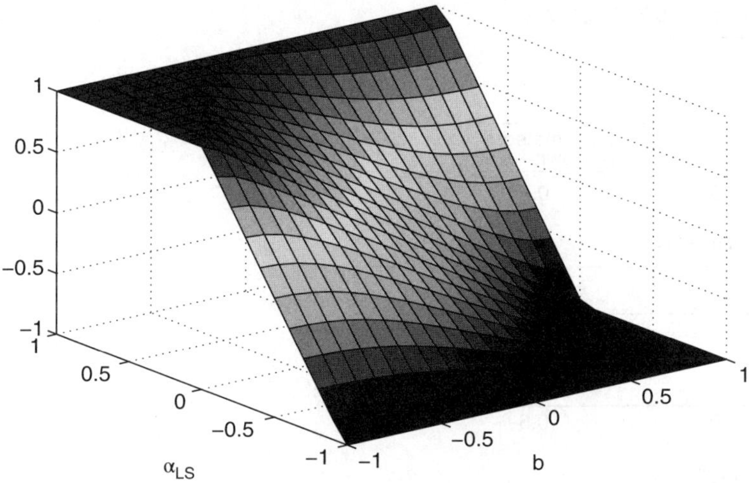

Figure 7.7 Inverse mean function for various values of $\widehat{a}^{O}_{\mathrm{LS}}$ (here denoted as α_{LS}) and b for the ARMAX(1,1) model with $\mathbf{X} = \mathbf{1}$ and $T = 50$ observations.

4) Set $i = i + 1$. Obtain a new estimate $\widehat{b}_i$ for b by exact maximum likelihood, conditional on $a = \widehat{a}_{i-1}$. If $|\widehat{b}_i - \widehat{b}_{i-1}|$ exceeds a given tolerance, then repeat from step 3.

In the below experiments, the exact m.l.e. as an initial estimator for b was used, along with a convergence tolerance of 10^{-3} in step 4. If this scheme converges (which it did in all trials), we denote the final value of $\widehat{a}$ as the MA(q)-*modified* mean- (median-, mode-, respectively) adjusted estimator.

With the additional complexity of the MA term, a simulation scheme as was used for the pure AR(1) case, which makes use of the one-to-one nature of the estimators to $f_{\widehat{a}_{\mathrm{LS}}}$, is no longer practical for assessing the small sample properties of the MA(1)-modified estimator. Instead, direct simulation is used. Because of the increased computation time required by the MA-modified estimator and the fact that its performance needs to be assessed over the two-dimensional support of parameters a and b, we investigate only a single model, based on 1,000 replications. We chose the representative sample size $T = 50$ and $\mathbf{X} = \mathbf{1}$, and restricted ourselves to nonnegative values of a.

Three estimation situations are studied. First, the MA(1)-modified estimators are used, as stated above; second, the bias-adjusted estimators *ignoring* the MA component, i.e., wrongly assuming $b = 0$; and third, the bias-adjusted estimators, *given* the MA(1) component, i.e., using the ARMA(1,1) covariance matrix with b known. The latter situation is obviously unrealistic in practice, but serves as a theoretical benchmark for the former two.

We first concentrate on the median-unbiased estimator because, under the assumption that b is known, its theoretical property of exact median-unbiasedness offers a simple check on the validity of the procedure. Furthermore, we illustrate the results using $a = 0.7$, which, recalling the results in Section 5.4, is (approximately) the single point at which $\widehat{a}_{\mathrm{Med}}$ is optimal in terms of m.s.e. when $b = 0$.

For this model, the first plot in Figure 7.8 shows the mean bias, as a function of b, of (i) the exact m.l.e., (ii) the unmodified median-unbiased estimator, and (iii) its MA(1)-modified counterpart, for both b known and estimated. The middle and lower graphs are similar, but show the median-bias and m.s.e., respectively. From the middle graph, we see that, when b is known, the MA(1)-modified

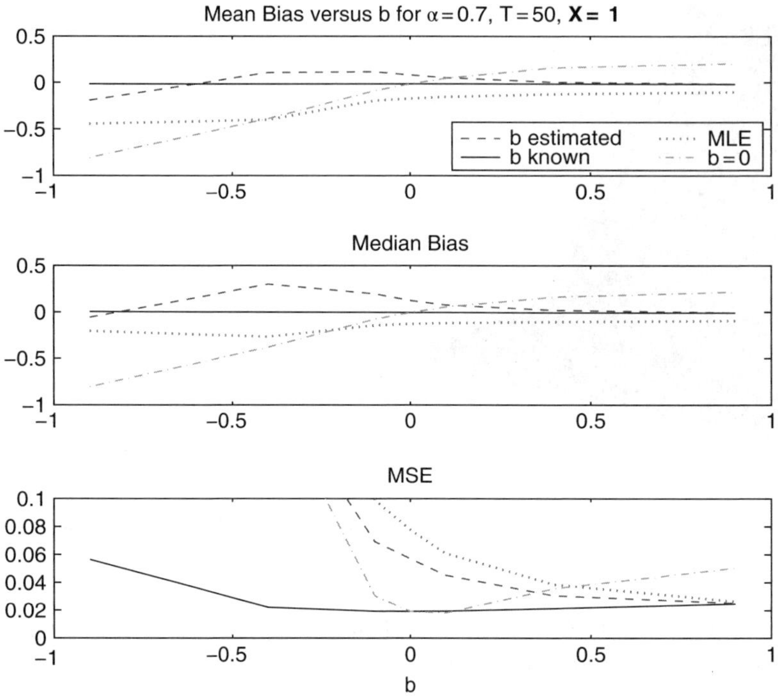

Figure 7.8 Mean bias (top), median bias (middle), and m.s.e. (bottom) of the exact m.l.e. (dotted line) and the MA(1)-modified median-unbiased estimator with b known (solid line), b estimated (dashedline), and $b = 0$ (dash-dot line), based on a model with a constant, $T = 50$ observations and normal innovations. The bottom graph is truncated.

median-adjusted estimator is exactly median-unbiased, in agreement with the theory. With respect to mean bias, we see from the top graph that the modified median-unbiased estimator is less biased than the exact m.l.e.

Of arguably most relevance is the m.s.e., as shown in the bottom graph of Figure 7.8. Unsurprisingly, comparing the bias-adjusted estimators with their MA(1)-modified counterparts, we see that it is advantageous (in terms of m.s.e.) to neglect the MA term for "moderate" values of b, as the reduction in bias is outweighed by the additional variation induced by estimating the MA term. For this model and sample size, one should neglect the MA term for, approximately, $|b| < 0.4$, although this cutoff will of course also depend on the true values of a to some extent. In particular, as a approaches unity, this cutoff increases, i.e., as we approach the unit root model, the MA term should be ignored unless $|b|$ is rather high. These conclusions can be expected to be valid for different sample sizes and design matrices as well, so that some prior knowledge on the part of the researcher can help decide whether or not to take a potential moving average component into account.

As expected, the MA(1)-modified estimator with b known performs best in terms of m.s.e. But, more relevantly, when b is estimated, the modified median-unbiased estimator still outperforms the exact m.l.e. Again, the relative performance depends on the magnitudes of both a and b. The dependence on a can be seen from Figure 7.9, which shows the relative improvement in m.s.e. achieved by the proposed procedure over the exact m.l.e. for the same model with a constant and $T = 50$

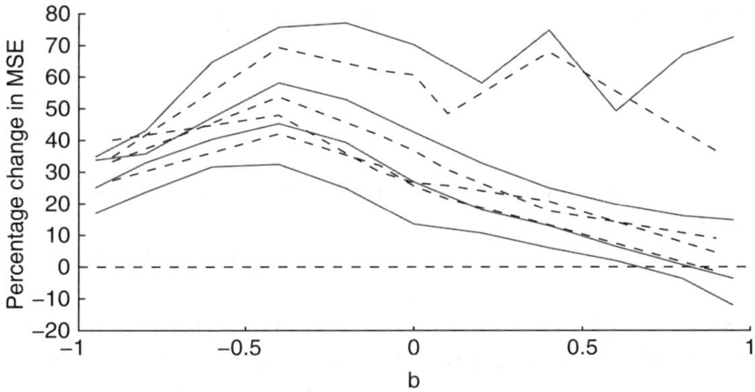

Figure 7.9 Percentage reduction in m.s.e. compared to the exact m.l.e. of the modified mean-adjusted (solid line) and median-unbiased (dashed line) estimators: From bottom to top, $a = 0.7, 0.8, 0.9, 1$.

observations, versus b, for $a = 0.7$, 0.8, 0.9, and 1.0. The results for the pure AR case carry over to the ARMAX(1,1) setting considered here, in the sense that, for this range of the autoregressive parameter, the MA(1)-modified median-unbiased estimator exhibits a lower m.s.e., while otherwise the exact m.l.e. performs better.

The previous results do not tell the whole story because (recalling the results in Section 5.4) $\hat{a}_{\text{Med}}$ is not optimal for $a > 0.7$. Instead, it is $\hat{a}_{\text{Mean}}$ which is of interest in this parameter range. As such, we also include the performance of the MA(1)-modified mean-adjusted estimator in Figure 7.9. Consistent with the results for $b = 0$, we see that, for the range of the autoregressive parameter under investigation, it outperforms the MA(1)-modified median-unbiased estimator. For $0 < a < 0.7$, the performance of both are reasonably similar and not shown.

As such, we make the following recommendations. For high-persistence models (with $a > 0.7$), if a moderate to strong MA(1) component is presumed, then use the MA(1)-modified mean-adjusted estimator, otherwise $\hat{a}_{\text{Mean}}$ (without MA modification) is preferred. For models with less persistence ($-0.1 < a < 0.7$), one should consider use of the MA(1)-modified median and mode-adjusted estimators, unless there is only a weak MA component, in which case, $\hat{a}_{\text{Mode}}$ should be used.

7.7 Some ARIMAX Model Extensions

> The new era of *practical* non-linear time series modelling is, without doubt, long overdue.
>
> (Howell Tong and K. S. Lim, 1980, p. 245)

There is substantial evidence for "nonlinearities" in a variety of economic data, and an associated large number of proposed models and methods that deviate in some fashion from the use of linear ARMAX and related structures applied to possibly first-differenced data; see, e.g., the monographs by Tong (1990), Granger and Teräsvirta (1993), Franses and van Dijk (2000), Fan and Yao (2003), Teräsvirta et al. (2010), and Haldrup et al. (2014), as well as Guidolin et al. (2008), Scholz et al. (2012, 2015) and, notably, the numerous references therein.

Though our primary concern in Parts I and II of this book is establishing a strong foundation for linear inference, this section briefly outlines some interesting and useful time-series structures that provide viable alternatives to the strictly linear ARMAX model with a single innovation sequence, applied to a data set, or its first difference.

7.7.1 Stochastic Unit Root

> For reasons that are probably obvious, stock market prices have been the most analysed economic data during the past forty years or so.
>
> (Clive W. J. Granger, 1992, p. 3)

One might easily imagine that, in systems as complex as economies, regression models with a possibly unit root process for the error term are too simplistic, and a better (albeit surely still mis-specified) model for the actual d.g.p. allows the autoregressive parameter to vary over time. This idea can be motivated by considering stock returns. Let the time series $\{P_t\}$, $t = 1, \ldots, T$, denote a sequence of prices of a financial asset observed at equally spaced intervals, such as the daily closing price (and ignoring weekends if the asset is not traded, such as stocks). Prices approximately follow a random walk, and so are not covariance stationary. The returns, being formed from first differences, would then be stationary, and are thus the objects primarily used for modeling and prediction.

Simple returns are defined as $R_t = (P_t - P_{t-1})/P_{t-1}$.[2] Let $\mathbb{E}_s[X]$ denote the expected return of random variable X based on information available up to and including time s. Then, to motivate consideration of a stochastic unit root using the simple returns, we have $\mathbb{E}_{t-1}[R_t] = \mathbb{E}_{t-1}[P_t]/P_{t-1} - 1$, or $\mathbb{E}_{t-1}[P_t] = (1 + \mathbb{E}_{t-1}[R_t])P_{t-1}$. With $\epsilon_t := P_t - \mathbb{E}_{t-1}[P_t]$, $\delta_t = \mathbb{E}_{t-1}[R_t]$, and $\alpha_t = 1 + \delta_t$, the price process can be expressed as

$$P_t = \alpha_t P_{t-1} + \epsilon_t. \tag{7.55}$$

This is an AR(1) model with random autoregressive coefficient. Recursive substitution n times yields

$$\begin{aligned}
P_t &= \alpha_t P_{t-1} + \epsilon_t \\
&= \alpha_t\{\alpha_{t-1}P_{t-2} + \epsilon_{t-1}\} + \epsilon_t \\
&= \alpha_t\{\alpha_{t-1}[\alpha_{t-2}P_{t-3} + \epsilon_{t-2}] + \epsilon_{t-1}\} + \epsilon_t \\
&= \alpha_t\{\alpha_{t-1}[\alpha_{t-2}(\alpha_{t-3}P_{t-4} + \epsilon_{t-3}) + \epsilon_{t-2}] + \epsilon_{t-1}\} + \epsilon_t \\
&= \alpha_t\alpha_{t-1}\alpha_{t-2}\alpha_{t-3}P_{t-4} + \alpha_t\epsilon_{t-1} + \alpha_t\alpha_{t-1}\epsilon_{t-2} + \alpha_t\alpha_{t-1}\alpha_{t-2}\epsilon_{t-3} + \epsilon_t \\
&=: \vdots \\
&= P_{t-n}\left(\prod_{i=0}^{n-1}\alpha_{t-i}\right) + \sum_{j=1}^{n-1}\left(\prod_{i=0}^{j-1}\alpha_{t-i}\right)\epsilon_{t-j} + \epsilon_t.
\end{aligned} \tag{7.56}$$

Conditions on $\{(\alpha_t, \epsilon_t)\}$ such that $\{P_t\}$ is strictly and/or covariance stationary are given in Gonzalo and Montesinos (2002), and depend on the convergence of infinite sequence $\{\psi_{t,j}\}_{j=0}^{\infty}$, where $\psi_{t,j} = \prod_{i=0}^{j-1}\alpha_{t-i}$. See also Vervaat (1979), Nicholls and Quinn (1982), Tjøstheim (1986), Brandt (1986), Pourahmadi (1986, 1988), and Karlsen (1990) on stationarity conditions for this model.

2 Later, in Part III, we will use the continuously compounded percentage returns associated with the price process, given by $R_t = 100(\ln P_t - \ln P_{t-1})$, where their relation to simple returns is discussed in Section I.4.4.3. If, for example, the prices are measured daily, then the R_t are called the daily **percentage log returns**, or just the daily returns.

One natural candidate takes model (7.55) with

$$\delta_t \overset{\text{i.i.d.}}{\sim} \mathrm{N}(0, \sigma_\delta^2) \quad \text{indep. of} \quad \epsilon_t \overset{\text{i.i.d.}}{\sim} \mathrm{N}(0, \sigma_\epsilon^2). \tag{7.57}$$

This is a **stochastic unit root**, or STUR process, as considered in Leybourne et al. (1996a,b), Granger and Swanson (1997), and the references therein. We will see in Section 10.1 that this simple process is such that $\{R_t\}$ has no autocorrelation, but gives rise to volatility clustering and autocorrelations in the squared returns, just like genuine financial asset returns data.

Another candidate, as analyzed in Leybourne et al. (1996a), takes (7.55) with $\delta_0 = 0$, and otherwise

$$\delta_t = \rho\delta_{t-1} + \eta_t, \quad |\rho| \leqslant 1, \quad \epsilon_t \overset{\text{i.i.d.}}{\sim} \mathrm{N}(0, \sigma_\epsilon^2) \quad \text{indep. of} \quad \eta_t \overset{\text{i.i.d.}}{\sim} \mathrm{N}(0, \sigma_\eta^2). \tag{7.58}$$

For most realistic economic time series, there is a need for further lags of P_t and also possibly exogenous regressors. As such, (7.58) is augmented in Leybourne et al. (1996a) to $P_t^* = \alpha_t P_{t-1}^* + \epsilon_t$, where $P_t^* = P_t - \mathbf{x}_t'\boldsymbol{\beta} - \sum_{i=1}^p \phi_i P_{t-i}$, where the ϕ_i, $i = 1, \dots, p$, are such that the corresponding AR(p) model is stationary; see Section 6.1.1. Leybourne et al. (1996a) illustrate how the parameters of the model can be estimated via a state space representation and Kalman filtering; see Remark (b) in the beginning of Section 5.6. From their application to six major stock indexes, Sollis et al. (2000) conclude that "Evidence supporting the stochastic unit root hypothesis is found. However, the implementation of this model generally leads to only very minuscule gains in the prediction of daily prices…"

Another possible STUR structure is

$$P_t = \exp(a_t)P_{t-1} + \epsilon_t, \quad a_t = \phi_0 + \phi_1 a_{t-1} + \eta_t, \tag{7.59}$$

where $\epsilon_t \overset{\text{i.i.d.}}{\sim} \mathrm{N}(0, \sigma_\epsilon^2)$, independent of $\eta_t \overset{\text{i.i.d.}}{\sim} \mathrm{N}(0, \sigma_\eta^2)$. The first equation in (7.59) is referred to as the **measurement equation**, while the second is the **transition equation**. Two estimation methods for (7.59) were proposed in Granger and Swanson (1997), though they gave rise to "wild estimates" that are "fairly imprecise", while a Bayesian estimation approach from Jones and Marriott (1999) appears more reliable. Tests for changes in the persistence of the process have been considered by Busetti and Taylor (2004) and Harvey et al. (2006).

Model (7.59) and its estimation via Bayesian techniques has been generalized in Yang and Leon-Gonzalez (2010) to the GSTUR model

$$P_t = v_t + \delta t + \gamma, \tag{7.60a}$$

$$v_t = \exp(a_t)v_{t-1} + \sum_{i=1}^{l} \lambda_i \triangle v_{t-i} + \epsilon_t, \tag{7.60b}$$

$$a_t = \phi_0 + \phi_1 a_{t-1} + \cdots + \phi_p a_p + \eta_t, \tag{7.60c}$$

where $\triangle v_{t-i} = v_{t-i} - v_{t-i-1}$ in the measurement equation are lagged first differences. The structure in (7.60c) is an AR(p) process, as studied at length in Section 6.1. In particular, as in (6.11), (7.60c) is assumed to be stationary, with unconditional mean given by

$$\mu_a = \lim_{t\to\infty} \mathbb{E}[a_t] = \phi_0 + \phi_1\mu_a + \phi_2\mu_a + \cdots + \phi_p\mu_a = \frac{\phi_0}{1 - \phi_1 - \phi_2 - \cdots - \phi_p}. \tag{7.61}$$

The GSTUR model reduces to a random walk with $\delta = \gamma = l = p = \phi_0 = \sigma_\eta^2 = 0$, while for $\gamma = l = p = \sigma_\eta^2 = 0$ and $\phi_0 = -\infty$, it reduces the linear regression $P_t = \gamma + \delta t + \epsilon_t$, $\epsilon_t \overset{\text{i.i.d.}}{\sim} \mathrm{N}(0, \sigma_\epsilon^2)$. Yang and

Leon-Gonzalez (2010) demonstrate with several financial and economic time series that there is very strong evidence for a stochastic unit root compared to a random walk model.

7.7.2 Threshold Autoregressive Models

This short section introduces a particular nonlinear autoregressive structure, namely the so-called **threshold autoregressive** model, or TAR, as developed and discussed in Tong (1978, 1983, 1990, 2007, 2011), Tong and Lim (1980), and Hansen (1997); see also the survey paper by Chen et al. (2011b), and the festschrift in Tong's honor, Chan (2009). The TAR model is a piecewise linear autoregression, similar to the threshold regression briefly discussed in Section 1.6. With two regimes and one autoregressive term, it is given by

$$Y_t = \begin{cases} a_0 + a_1^{(1)} Y_{t-1} + \epsilon_t, & \text{if} \quad q_t \leqslant \gamma, \\ a_0 + a_1^{(2)} Y_{t-1} + \epsilon_t, & \text{if} \quad q_t > \gamma, \end{cases} \tag{7.62}$$

$t = 1, \dots, T$, where threshold variable q_t is either exogenous (not involving any Y_t) or is based on Y_{t-d} for some $d \geqslant 1$, γ is the threshold, or threshold parameter, and $\epsilon_t \overset{\text{i.i.d.}}{\sim} \mathrm{N}(0, \sigma^2)$. The parameters to be estimated are a_0, $a_1^{(1)}$, $a_1^{(2)}$, σ, and γ. If q_t is taken to be Y_{t-d}, then it is referred to as a **self-exciting TAR** (SETAR) model, and $d \geqslant 1$ is the **delay parameter**. In many applications, a range of values of d will be tried, such as $\{1, 2, \dots, d_{\max}\}$, and most often, d is taken to be one, whether for daily data (e.g., Shively, 2003) or quarterly data (e.g., Kapetanios and Shin, 2006).

More generally, with $k + 1$ threshold values, p autoregressive lags, and different intercept and scale terms in each regime, the TAR(k, p) model is given by

$$Y_t = \mathbf{x}_t' \boldsymbol{\theta}^{(j)} + \sigma^{(j)} \epsilon_t, \quad \text{if} \quad \gamma_{j-1} < q_t \leqslant \gamma_j, \quad j = 1, \dots, k, \tag{7.63}$$

where $\mathbf{x}_t' = (1,\ Y_{t-1}, \dots, Y_{t-p})$ and $-\infty = \gamma_0 < \gamma_1 < \cdots < \gamma_k = \infty$. The parameters to be estimated are $\boldsymbol{\theta}^{(j)}$, $\sigma^{(j)}$, and γ_i, $i = 1, \dots, k-1$.

Yadav et al. (1994) provide a natural setting for the use of a threshold model in the context of the price differences of equivalent assets. Caner and Hansen (2001),[3] Gonzalo and Montesinos (2002), and Kapetanios and Shin (2006) address testing and the associated distribution theory under various situations such that the process initially appears $I(1)$, i.e., contains a unit root. The idea is that, while the Dickey–Fuller and related unit root tests may not reject the null of a unit root, tests that allow for a TAR or SETAR alternative (with, typically, two or three regimes) often will reject the null. For example, Gonzalo and Montesinos (2002) propose a so-called **threshold autoregressive stochastic unit root model**, or TARSUR, whereby the largest root of the AR polynomial is less than one in some regimes, larger than one in others, and in such a way that, on average, it is equal to one. The resulting process is strictly stationary, and such that one regime is stationary and the other is (mildly) explosive.

Several extensions of model (7.63) have been considered. A multivariate threshold model is developed in Tsay (1998). The (theoretically more challenging) threshold MA and ARMA cases have been studied; see Tong (1990), Brockwell et al. (1992), Ling (1999), Ling and Tong (2005), Amendola et al. (2006), Ling et al. (2007), and the references therein. The use of thresholds in conjunction with GARCH-type structures for modeling the conditional heteroskedasticity in financial asset returns (see Chapter 10) has been addressed by numerous authors; see, e.g., Li and Li (1996), Brooks (2001), Chen et al. (2005), Chen et al. (2008a,b), So and Choi (2009), and the references therein.

---

3 Programs for model estimation and inference are provided (in Gauss, Matlab, and R) by Bruce Hansen on his web page.

It is also possible to enable a continuous transition between the threshold regimes, giving rise to the **smooth transition autoregressive** (STAR) model. The idea goes back to Bacon and Watts (1971), and was subsequently pursued in the econometric literature; see, e.g., Chan and Tong (1986), Teräsvirta (1994, 1998), van Dijk et al. (2002), and the references therein. Another extension is from Astatkie et al. (1997), who examine a nested variation of TAR. Finally, Chen et al. (2012) argue that, in many applications of interest, there will be more than one threshold variable, and provide details on the case with two such variables, each with a single partition, thus giving rise to a model with four regimes.[4]

7.7.3 Fractionally Integrated ARMA (ARFIMA)

> We believe that our work, along with the other recent size and power studies, provide a joint condemnation of the widespread mechanical application of unit root tests.
>
> (Francis X. Diebold and Glenn D. Rudebusch, 1991, p. 160)

In most applications, the ARIMA(p, d, q) process (7.9) will be postulated with $d = 0$ or $d = 1$, and this behooves the question if there is anything "in between" these two extremes. This can be achieved by taking d to be a real number, and the lag polynomial $(1 - L)^d$ is viewed as an operator whose inverse can be computed using its Taylor series expansion, as shown below. As the midpoint of zero and one is $1/2$, one might guess that the resulting process is stationary if $0 \leqslant d < 1/2$, which indeed turns out to the case, and is in fact more generally true for $-1/2 < d < 1/2$. The model can be expressed exactly as in (7.9), except that the range of d changes, i.e.,

$$a(L)(1 - L)^d Z_t = c + b(L)U_t, \quad d \in \mathbb{R}, \tag{7.64}$$

and is referred to as a (Gaussian) ARFIMA(p, d, q) process where, as usual, $U_t \overset{\text{i.i.d.}}{\sim} \mathrm{N}(0, \sigma^2)$, and the F stands for fractional. The process is stationary and invertible if the usual conditions given above on polynomials $a(L)$ and $b(L)$ are satisfied, and $|d| < 1/2$.

Recalling the basic random walk process from Section 4.1 and how the effect of innovation U_t never dies out—which we might call "infinite memory", model (7.64) is referred to as a **long memory process**, and is such that the effect of U_t is not as transient as in a stationary ARMA model, but also not permanent. More specifically, an ARFIMA(p, d, q) process exhibits long memory for $d \in (0, 1/2)$; so-called **intermediate memory** for $d \in (-1/2, 0)$, and is non-stationary for $d \in (1/2, 1)$, but such that the process is **level reverting** in the sense that the impact of U_t is not permanent; see, e.g., Granger and Joyeux (1980), Hosking (1981), Cheung and Lai (1993), Wu and Crato (1995), and the references therein for further details.

Once one is accustomed to thinking of d as a real number instead of being restricted to taking on a measure-zero quantity such as zero or one, model (7.64) seems very natural, and the restriction of $d \in \{0, 1\}$ seems quite unnatural. In particular, for real time-series data that appear to exhibit long memory (because it somewhat resembles a random walk and the sample autocorrelations are large

---

4 For this last model, the intrigued reader can work on developing a computer program that estimates the parameters associated with the general structure, where the inputs include, along with the observed time series $\mathbf{Y} = (Y_1, \dots, Y_T)'$, autoregressive lag order p, and any relevant exogenous variables as, say $T \times k$ matrix $\mathbf{X}$, a set of s threshold variables (as, say, $T \times s$ matrix $\mathbf{Q}$), the endpoints of the closed set indicating the support of each (as, say, an $s \times 2$ matrix $\boldsymbol{\Gamma}$ with the ith row containing $\underline{\gamma}_i$ and $\bar{\gamma}_i$, $i = 1, \dots, s$), and, for each of the s threshold variables, the number of desired thresholds (say, as $\mathbf{k} = (k_1, \dots, k_s)'$). There are thus $R = (k_1 + 1)(k_2 + 1) \cdots (k_s + 1)$ regimes. The fixed value of p can be relaxed, as in Chen et al. (2012), such that each of the R regimes has its own autoregressive lag length.

and die off slowly), the use of $d = 1$ seems rather extreme, as it induces "infinite memory"—the effect of a shock at period t on future observations *never* dies off. Chapter 8 is dedicated to studying the autocorrelation function of stationary, short-memory series, and the reader can have a peak at Figure 8.5 to see the behavior of the sample autocorrelation function of a random walk. The autocorrelation function of a stationary ARFIMA process dies off very slowly, at a hyperbolic (instead of geometric) rate depending on d.

If the observed time series is such that use of $d = 2$ might suggest itself, or if it resembles a unit-root type process with $1/2 \leqslant d$, one can compute first differences and apply the fractionally differenced model (7.64). That is, the postulated model for $\{Z_t\}$ would be $a(L)(1-L)(1-L)^d Z_t = c + b(L)U_t$ with $d \in \mathbb{R} \cap (-1/2, 1/2)$.

Long memory models for addressing the hyperbolic decay of sample autocorrelations of certain data sets have been in use since at least 1950 in fields such as hydrology, meteorology, geophysics, and climatology. Granger (1980) demonstrates that series with long memory can arise from aggregation of short-memory processes, lending some theoretical support to their use in modeling economic data. Lo (1991) demonstrates lack of long memory in (daily and monthly) stock returns, once the effects of short-range dependence are accounted for. Long memory in the absolute or squared returns, however, is very prominent in asset returns; see Section 10.6.3, in particular, Figure 10.15.

A large review of long memory processes (up to the mid 1990s) can be found in Baillie (1996). Diebold and Rudebusch (1991) demonstrate, somewhat expectedly, that application of the Dickey–Fuller unit root tests have very low power against fractional alternatives. The reader is encouraged confirm this, by simulation, and plotting the power of the various unit root tests from Section 5.5 over a grid of d-values, for several sample sizes.

Various methods exist for parameter estimation. Exact calculation of the covariance matrix is tractable, and is used for computing the m.l.e.; see Yajima (1985), Sowell (1992), Chung (1994), and the references therein. Naturally, the reader is encouraged to implement this and perform simulations to investigate the small-sample behavior of the m.l.e. Results can be compared to existing software packages, such as R, which has routines for simulation and various methods of estimation of ARFIMA models. The Gaussianity assumption can also be relaxed; see, e.g., Scherrer et al. (2007) and Kwan et al. (2012).

The infinite MA representation $Z_t = (1-L)^{-d} a^{-1}(L) b(L) U_t$ of (zero-mean) model (7.64) shows that, if we can calculate the expansion $(1-L)^{-d}$, we obtain a way to simulate an ARFIMA process and also calculate the covariance matrix, and thus express the likelihood. To this end, let $f(z) = (1-z)^{-d}$ for $|d| < 1/2$, so that $f'(z) = d \cdot (1-z)^{-d-1}$ and, more generally,

$$f^{(j)}(z) = (d+j-1) \cdot (d+j-2) \cdots (d+1) \cdot (d) \cdot (1-z)^{-d-j}.$$

The Taylor series of f around $z = 0$ is thus

$$\begin{aligned}(1-z)^{-d} &= f(0) + \left.\frac{\partial f}{\partial z}\right|_{z=0} \cdot z + \frac{1}{2!}\left.\frac{\partial^2 f}{\partial z^2}\right|_{z=0} \cdot z^2 + \frac{1}{3!}\left.\frac{\partial^3 f}{\partial z^3}\right|_{z=0} \cdot z^3 + \cdots \\ &= 1 + dz + (1/2!)(d+1)dz^2 + (1/3!)(d+2)(d+1)dz^3 + \cdots,\end{aligned}$$

suggesting to represent the operator $(1-L)^{-d}$ as

$$1 + dL + \frac{(d+1)dL^2}{2!} + \frac{(d+2)(d+1)dL^3}{3!} + \cdots = \sum_{j=0}^{\infty} h_j L^j, \tag{7.65}$$

where $h_0 = 1$ and

$$h_j = \frac{(d+j-1)(d+j-2)(d+j-3)\cdots(d+1)(d)}{j!}. \tag{7.66}$$

It can be shown (see, e.g., Hamilton, 1994, Sec. 15.A for a basic and detailed derivation) that, for $d < 1$, as $j \to \infty$, $h_j \approx (j+1)^{d-1}$. This hyperbolic (as opposed to geometric) behavior gives rise to the long memory property of an ARFIMA model. Sequence $\sum_{j=0}^{\infty} h_j^2 < \infty$ (referred to as **square summability**) for $|d| < 1/2$, but not for $d \geqslant 1/2$, and is the reason for the covariance stationarity of an ARFIMA$(0, d, 0)$ model for $|d| < 1/2$.

7.8 Problems

Problem 7.1 Verify (7.3) by simulation, i.e., simulate the infinite AR model and estimate an ARMA(1,1).

Problem 7.2 Recall the use of o.l.s. for estimating the parameters of an ARMA(p, q) model in Section 7.3.2. Kapetanios (2003) suggested extending this method so that it is computed in an iterative fashion until convergence, referred to as iterative ordinary least squares, or i.o.l.s. The method begins by estimating an AR(p^*) model and computing the residuals $\widehat{U}_t^{(0)}$, $t = p^* + 1, \ldots, T$. In the second step, these residuals are used as proxies for the true innovation sequence, and the regression

$$Y_t - \widehat{U}_t^{(0)} = \sum_{i=1}^{p} a_i Y_{t-i} + \sum_{j=1}^{q} b_j \widehat{U}_{t-j}^{(0)} + \xi_t \tag{7.67}$$

is estimated. Unlike the standard o.l.s. approach, i.o.l.s. makes use of the residuals $\widehat{\xi}_t$, $t = p^* + \max(p, q) + 1, \ldots, T$. In particular, they are fed back into the regression (7.67) as the new proxies for the innovation sequence. This procedure can be iterated until the change in residuals is negligible. At the ith iteration, we set $\widehat{U}_t^{(i+1)} = \widehat{\xi}_t^{(i)}$, and the model is estimated anew.

Program this method,[5] and conduct simulations as in Section 7.3.2, comparing with kernel density plots the estimators from the o.l.s. and i.o.l.s. Figure 7.10 shows the results corresponding to those in Figure 7.2. We see that the iterated method indeed conveys a small advantage.

Problem 7.3 This exercise considers a different strategy for a time-series regression model that puts more emphasis on the ARMA parameters.
Recall that the ordinary least squares residual vector can be expressed as $\widehat{\epsilon}_{\text{LS}} = \mathbf{MY}$, where $\mathbf{M} = \mathbf{I} - \mathbf{X}(\mathbf{X}'\mathbf{X})^{-1}\mathbf{X}'$. In the iterative method discussed in the chapter, one would begin by computing $\widehat{\epsilon}_{\text{LS}}$ and then treating it as a mean-zero ARMA process. Notice, however, that $\mathbb{V}(\widehat{\epsilon}_{\text{LS}}) = \sigma^2 \mathbf{M\Sigma M}$, which is neither full rank nor the covariance matrix of the desired ARMA process. The full rank condition can be alleviated as follows:
Recall from Theorem 1.3 that $\mathbf{M}$ may be written as $\mathbf{M} = \mathbf{G}'\mathbf{G}$, where $\mathbf{G}$ is $(T-k) \times T$ such that $\mathbf{GG}' = \mathbf{I}_{T-k}$ and $\mathbf{GX} = \mathbf{0}$. As $\mathbf{Y} \sim \text{N}(\mathbf{X}\boldsymbol{\beta}, \sigma^2\boldsymbol{\Sigma})$, we have $\mathbf{W} := \mathbf{GY} \sim \text{N}(\mathbf{0}, \sigma^2\mathbf{G\Sigma G}')$, where $\mathbf{W}$ has

5 The author is grateful to master's students Jan Krepl, Antonio Polino, Anna Stepuk, and Michal Svatoň for researching, programming, and investigating this method, as well as several other (not so good) ideas from the author.

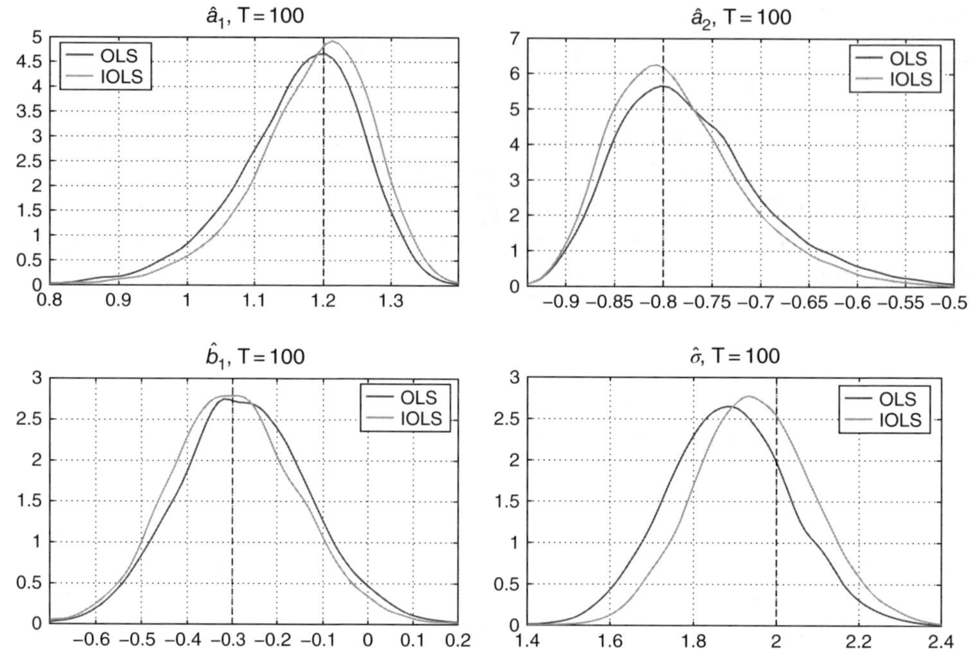

Figure 7.10 Similar to Figure 7.2, comparing the o.l.s. and i.o.l.s. methods for estimation of ARMA(2,1) parameters for $T = 100$ observations, based on 10,000 replications.

length $T - k$ and $\mathbf{G\Sigma G}'$ is a $(T-k) \times (T-k)$ full rank matrix (provided, of course, that $\mathbf{\Sigma}$ is full rank). The exact likelihood of $\mathbf{W}$ can then be maximized, as a function of the $p + q$ ARMA parameters and σ. This avoids joint estimation of $\boldsymbol{\beta}$ and the ARMA coefficients.

To further eliminate σ from the maximization, we could maximize the likelihood of $\mathbf{W}^* := \mathbf{W}/\hat{\sigma}$, where, approximately, $\mathbf{W}^* \sim \mathrm{N}(\mathbf{0}, \mathbf{G\Sigma G}')$. An estimate of σ could be obtained by estimating the full model in the usual fashion, say by conditional or exact maximum likelihood. This might seem to defeat the purpose because one then presumably has the "best" obtainable estimator, but this is not certain, as demonstrated via simulation:

First consider the AR(1) model with unknown mean. Compare via simulation (with 1,000 replications and $T = 40$) the bias and m.s.e. of the estimated autoregressive parameter $\hat{a}$ based on three estimators: (i) the exact m.l.e., yielding the three parameter estimates $\widehat{\beta}_{\mathrm{ML}}$ for the intercept, $\widehat{a}_{\mathrm{ML}}$ and $\widehat{\sigma}_{\mathrm{ML}}$, (ii) the method that maximizes the likelihood of $\mathbf{W}$, assuming knowledge of σ (so that only parameter a needs to be estimated), and (iii) maximizing the likelihood of $\mathbf{W}^* = \mathbf{W}/\widehat{\sigma}_{\mathrm{ML}}$. The latter also has only one parameter to estimate, but note that σ is *not* assumed known. Interest centers on the difference in performance of the first and third estimators, while the second one (which is not realistic) serves as a benchmark for the third.

Repeat the simulation, with the same time series, but using a regressor matrix consisting of a column of ones and a time vector.

Lastly, consider the model $Y_t = 10 + 0.5t + \epsilon_t$, $t = 1, \ldots, T$, with $T = 30$ and ϵ_t an MA(2) process with $b_1 = -1.2$, $b_2 = 0.8$, and $\sigma^2 = 10$. Use 1,000 replications and compare the m.s.e. of $\widehat{b}_1$ and $\widehat{b}_2$ based on $\mathbf{W}^*$ and the exact m.l.e.

Problem 7.4 Via simulation, compute the m.s.e. of $\widehat{Y}_{T+1|T}$ for an MA(1) model. Use $T = 20$, $\sigma^2 = 4$, and a grid of values of b. Do so using both the conditional and exact m.l.e. Also compute the m.s.e. based on $\widehat{b}$ from the exact m.l.e., but using $\widehat{U}_t$ based on use of the conditional m.l.e. What do you find? Repeat for $T = 100$ just using the conditional m.l.e.

Problem 7.5 Let $\mathbf{K}$ be any full rank symmetric matrix of size m. Show that

$$\mathbf{1}'(\mathbf{K} \odot \mathbf{K}^{-1})\mathbf{1} = m, \tag{7.68}$$

using the following two methods:

a) Let $\mathbf{A} = \{a_{ij}\}$ and $\mathbf{B} = \{b_{ij}\}$ be $m \times m$ matrices, with $\mathbf{B}$ symmetric, and show that $\mathrm{tr}(\mathbf{AB}) = \mathbf{1}'(\mathbf{A} \odot \mathbf{B})\mathbf{1}$.

(Contributed by David Harville)

b) Write

$$\mathbf{1}'(\mathbf{K} \odot \mathbf{K}^{-1})\mathbf{1} = \sum_{i=1}^{m} \sum_{j=1}^{m} [K]_{ij} \left[\frac{1}{|\mathbf{K}|} \mathbf{K}^{\mathrm{adj}} \right]_{ij},$$

where $\mathbf{K}^{\mathrm{adj}}$ is the adjoint matrix. Now use the facts that $\mathbf{K} = \mathbf{K}'$ and $|\mathbf{K}| = |\mathbf{K}'|$.

(Contributed by Ronald Butler)

7.A Appendix: Generalized Least Squares for ARMA Estimation

Recall the ordinary least squares (o.l.s.) method for ARMA estimation in Section 7.3.2. A related, quickly computed, and potentially more accurate parameter estimator corresponding to the general stationary and invertible ARMA(p,q) model

$$Y_t = \sum_{i=1}^{p} a_i Y_{t-i} + \sum_{j=1}^{q} b_j U_{t-j} + U_t \tag{7.69}$$

is presented in Koreisha and Pukkila (1990).[6] As with the o.l.s. method, an AR(p^*) model with p^* chosen as, say, $\lfloor \sqrt{T} \rfloor$ is fit to the data. (Paralleling the discussion in Section 7.3.2, a simulation-based iterative procedure could be developed to determine a better choice of p^*.) The residuals $\{\widehat{U}_t\}$ from the AR(p^*) autoregression may then serve as an estimator of the innovation series $\{U_t\}$ and can be used as regressors in the model

$$Y_t - \widehat{U}_t = \sum_{i=1}^{p} a_i Y_{t-i} + \sum_{j=1}^{q} b_j \widehat{U}_{t-j} + \xi_t.$$

It is then assumed that the true innovation series can be written as

$$U_t = \widehat{U}_t + \epsilon_t, \qquad \epsilon_t \overset{\text{i.i.d.}}{\sim} \mathrm{N}(0, 1), \tag{7.70}$$

6 The author is grateful to master's students Patrick Aschermayr, Dmitrii Dmitriev, Christian Frey, and Shuo Yang, who cleaned up, improved, augmented, and implemented my initial set of notes on this topic.

i.e., the residuals equal the true innovation series plus an i.i.d. zero-mean (and known variance) Gaussian error term. Combining (7.69) and (7.70) yields

$$Y_t - \hat{U}_t = \sum_{i=1}^{p} a_i Y_{t-i} + \sum_{j=1}^{q} b_j \hat{U}_{t-j} + \epsilon_t + \sum_{j=1}^{q} b_j \epsilon_{t-j}, \tag{7.71}$$

which is a regression with moving average errors that can be efficiently estimated by generalized least squares (g.l.s.). Let $m = \max(p, q)$ and $\tau = p^* + m + 1$, and define

$$Z_t = Y_t - \hat{U}_t, \quad \boldsymbol{x}_t = (Y_{t-1}, \ldots, Y_{t-p}, \hat{U}_{t-1}, \ldots, \hat{U}_{t-q}), \quad \xi_t = \epsilon_t + \sum_{j=1}^{q} b_j \epsilon_{t-j},$$

for $t = \tau, \ldots, T$. To enable a matrix representation of (7.71) as $\mathbf{Y} = \mathbf{X}\boldsymbol{\beta} + \boldsymbol{\xi}$, let

$$\mathbf{Y} = (Z_\tau, \ldots, Z_T)', \quad \mathbf{X} = (\boldsymbol{x}_\tau, \ldots, \boldsymbol{x}_T)',$$
$$\boldsymbol{\xi} = (\xi_\tau, \ldots, \xi_T)', \quad \boldsymbol{\beta} = (a_1, \ldots, a_p, b_1, \ldots, b_q)',$$

noting that $\mathbf{X}$ is $(T - p^* - m) \times (p + q)$, $\mathbf{Y}$ is $(T - p^* - m) \times 1$, $\boldsymbol{\xi}$ is $(T - p^* - m) \times 1$, and $\boldsymbol{\beta}$ is $(p + q) \times 1$. Thus,

$$\hat{\boldsymbol{\beta}}_{\text{GLS}} = (\mathbf{X}'\hat{\boldsymbol{\Sigma}}^{-1}\mathbf{X})^{-1}\mathbf{X}'\hat{\boldsymbol{\Sigma}}^{-1}\mathbf{Y}, \quad \hat{\boldsymbol{\Sigma}} = \boldsymbol{\Sigma}(\hat{\boldsymbol{b}}_0), \tag{7.72}$$

and $\hat{\boldsymbol{b}}_0$ are the MA parameter estimates from the regression of (7.72) with $\boldsymbol{\Sigma} = \mathbf{I}$. This is implemented in Listing 7.10, and uses methods for computing $\boldsymbol{\Sigma}^{-1}$ that are discussed below. (Note that this g.l.s. estimator can be used to update $\hat{\boldsymbol{\Sigma}}$, so that the procedure can be iterated until convergence. Based on a few simulation experiments, it was found that this added little value, and is not implemented in the code.)

To illustrate, Figure 7.11 is similar to Figure 7.10, comparing the performance of the baseline o.l.s. method and the g.l.s. estimator (7.72). As with the i.o.l.s. estimator, use of g.l.s. also conveys a (modest) advantage for this parameter constellation and sample size, though, given its extra complexity, it may not be worth the effort. The reader is naturally encouraged to experiment with other ARMA models and sample sizes, hopefully finding cases for which its advantage is more substantial.

Computation of $\boldsymbol{\Sigma}^{-1}$

Computing the inverse of a $T \times T$ matrix is an $O(T^3)$ operation, and is thus the bottleneck when evaluating (7.72). The special structure of $\boldsymbol{\Sigma}$ can be capitalized upon to decrease the required computational time, and some ways of doing so are now detailed.

Let $\boldsymbol{\epsilon} = (\epsilon_{\tau-q}, \epsilon_{\tau-q+1}, \ldots, \epsilon_T)'$ be $(T - p^*) \times 1$, so that $\boldsymbol{\xi} = \mathbf{M}_1\boldsymbol{\epsilon}$, where $\mathbf{M}_1 = \mathbf{M}_1(\mathbf{b})$ is a $(T - p^* - m) \times (T - p^*)$ Toeplitz matrix with first row $(b_q, b_{q-1}, \ldots, 1, 0, \ldots, 0)$ and first column $(b_q, 0, \ldots, 0)'$, i.e.,

$$\mathbf{M}_1 = \begin{bmatrix} b_q & b_{q-1} & \cdots & \cdots & \cdots & 1 & 0 & \cdots & 0 \\ 0 & b_q & b_{q-1} & \cdots & \vdots & \vdots & 1 & 0 & 0 \\ 0 & 0 & b_q & b_{q-1} & \cdots & \vdots & \vdots & 1 & 0 \\ 0 & 0 & 0 & b_q & b_{q-1} & \cdots & \cdots & \cdots & 1 \end{bmatrix}. \tag{7.73}$$

Then $\boldsymbol{\Sigma} = \mathbb{E}[\mathbf{M}_1\boldsymbol{\epsilon}(\mathbf{M}_1\boldsymbol{\epsilon})'] = \mathbf{M}_1\mathbf{M}_1$, as $\boldsymbol{\epsilon} \sim \mathrm{N}(\mathbf{0}, \mathbf{I})$.

```
function param = armaglsSTUDENTS(y,p,q,sigma_inv_method)
if nargin<4, sigma_inv_method = 'other'; end
T=length(y); L=ceil(sqrt(length(y)));
z=y(L+1:end); Z=toeplitz(y(L:end-1),y(L:-1:1));
uhat=(eye(length(z))-Z*inv(Z'*Z)*Z') * z; yy=z-uhat;
Y=yy(m+1:end); X=[]; m=max(p,q);
for i=1:p, X=[X z(m-i+1 : length(z)-i)]; end %#ok<*AGROW>
for i=1:q, X=[X uhat(m-i+1 : length(uhat)-i)]; end
beta=inv(X'*X)*(X'*Y); b=[1 beta(p+1:end)'];
switch sigma_inv_method
  case 'Cholesky'
    M1=toeplitz([b(q+1) (zeros(T-L-m-1,1))'], [fliplr(b) (zeros(T-L-m-1,1))']);
    Sigma = M1*M1'; P=chol(Sigma, 'lower'); invP=inv(P); Xhat=invP*X; %#ok<*MINV>
    param = inv(Xhat'*Xhat)*(Xhat'*(invP*Y));
  case 'Recurrent'
    M = toeplitz([b (zeros(T-L-q-1,1))'], [1 (zeros(T-L-1,1))']);
    invSigma = inverseM1M1(M,m);
    param = inv(X'*invSigma*X)*(X'*invSigma*Y);
  case 'Direct'
    M1=toeplitz([b(q+1) (zeros(T-L-m-1,1))'], [fliplr(b) (zeros(T-L-m-1,1))']);
    Sigma = M1*M1'; invSigma = inv(Sigma);
    param = inv(X'*invSigma*X)*(X'*invSigma*Y);
  otherwise
    M1=toeplitz([b(q+1) (zeros(T-L-m-1,1))'], [fliplr(b) (zeros(T-L-m-1,1))']);
    Sigma = M1*M1'; invSigma = inv(Sigma);
    param = inv(X'*invSigma*X)*(X'*invSigma*Y);
end

function [M1M1inv]=inverseM1M1(M,r)
%tf=istril(M); % Introduced in Version R2014a
%assert((tf==1), 'Matrix is not square/lower triangular');
[~, n]=size(M);
for i=0:-1:(-n+1)
  d=diag(M,i); assert(all(d == d(1)), 'Matrix not Toeplitz');
end
MMinv=(invtoeplitz(M))'*invtoeplitz(M); E=MMinv(1:r,1:r);
F=MMinv(1:r,(r+1):n);
G=MMinv((r+1):n,1:r); H=MMinv((r+1):n,(r+1):n);
M1M1inv=H-G*inv(E)*F;

function [invM]=invtoeplitz(M)
c_0=1/M(1,1); c(1)=c_0; [~, n]=size(M);
for j=2:n
  sum=0; for h=1:(j-1), sum=sum+(M(j+1-h,1)*c(h)); end
  c(j)=-(1/M(1,1))*sum;
end
b=[M(1,1), zeros(1,n-1)]; invM=toeplitz(c,b);
```

Program Listing 7.10: Computes the g.l.s. estimator (7.72) for a stationary, invertible ARMA(p, q) process with choice of method for the inversion of $\mathbf{\Sigma}$.

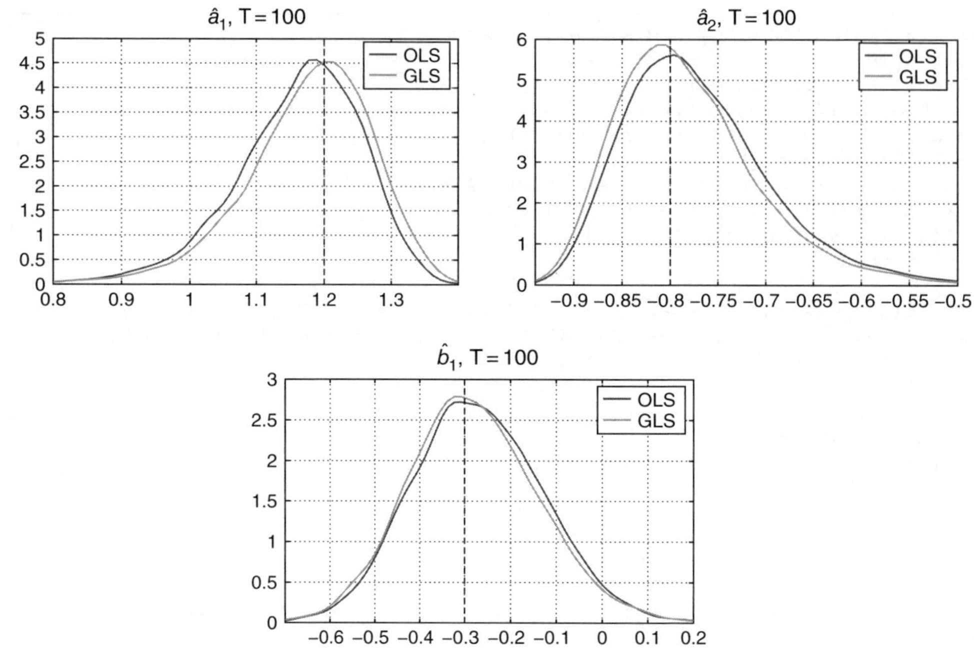

Figure 7.11 Similar to Figure 7.10, but comparing the o.l.s. and g.l.s. methods for estimation of the ARMA(2,1) parameters for $T = 100$ observations, based on 10,000 replications.

In order to apply the recurrence relation (7.76) below, let $\mathbf{M}$ be the square $(T-p^*)\times(T-p^*)$ matrix

$$\mathbf{M}=\begin{bmatrix}\mathbf{D}\\ \mathbf{M}_{11}\end{bmatrix},\qquad \mathbf{D}=\begin{bmatrix}1 & 0 & \dots & \dots & 0 & \dots & \dots & \dots & 0\\ b_1 & 1 & 0 & \dots & 0 & \dots & \dots & \dots & 0\\ b_2 & b_1 & 1 & \dots & 0 & \dots & \dots & \dots & 0\\ \vdots & \vdots & \ddots & 1 & \vdots & \vdots & \vdots & \vdots & 0\\ b_{q-1} & b_{q-2} & \dots & b_1 & 1 & 0 & 0 & 0 & 0\end{bmatrix}, \tag{7.74}$$

where $\mathbf{D}$ is $q\times(T-p^*)$ Toeplitz, $\mathbf{M}_{11}$ is $(T-p^*-q)\times(T-p^*)$ Toeplitz, the same as $\mathbf{M}_1$, but with $(m-q)$ less rows. Note that, if $m=\max(p,q)=q$, then $\mathbf{M}_1=\mathbf{M}_{11}$. Then,

$$\mathbf{M}=\begin{bmatrix}1 & 0 & \dots & \dots & 0 & \dots & \dots & \dots & 0\\ b_1 & 1 & 0 & \dots & 0 & \dots & \dots & \dots & 0\\ b_2 & b_1 & 1 & \dots & 0 & \dots & \dots & \dots & 0\\ \vdots & \vdots & \ddots & 1 & \vdots & \vdots & \vdots & \vdots & 0\\ b_{q-1} & b_{q-2} & \dots & b_1 & 1 & 0 & 0 & 0 & 0\\ b_q & b_{q-1} & \dots & \dots & \dots & 1 & 0 & \dots & 0\\ 0 & b_q & b_{q-1} & \dots & \vdots & \vdots & 1 & 0 & 0\\ 0 & 0 & b_q & b_{q-1} & \dots & \vdots & \vdots & 1 & 0\\ 0 & 0 & 0 & b_q & b_{q-1} & \dots & b_2 & b_1 & 1\end{bmatrix}. \tag{7.75}$$

Interest centers on $(\mathbf{MM}')^{-1} = (\mathbf{M}')^{-1}\mathbf{M}^{-1} = (\mathbf{M}^{-1})'\mathbf{M}^{-1}$. By noting that, if $\mathbf{M}$ is lower diagonal Toeplitz, its inverse is also a $(T-p^*)\times(T-p^*)$ lower diagonal Toeplitz, there exists a simple explicit expression for the inverse of $\mathbf{M}^{-1}$, with first row $(1, 0, \dots, 0)$ and first column $(c_0, c_1, \dots, c_{T-p^*})'$. Define for simplicity $\bar{n} = T - p^*$. Then, as detailed in Vecchio (2003),

$$c_0 = \frac{1}{b_0}, \quad c_k = -\frac{1}{b_0}\sum_{i=0}^{k-1} b_{k-i}c_i, \quad 1 \leqslant k \leqslant \bar{n}, \tag{7.76}$$

where $b_i = 0$ for $i > q$.

To see why this holds, note that $\mathbf{M}^{-1}$ is lower diagonal Toeplitz, given by

$$\mathbf{M}^{-1} = \begin{bmatrix} 1 & 0 & \dots & \dots & 0 & \dots & \dots & \dots & 0 \\ c_1 & 1 & 0 & \dots & 0 & \dots & \dots & \dots & 0 \\ c_2 & c_1 & 1 & \dots & 0 & \dots & \dots & \dots & 0 \\ \vdots & \vdots & \ddots & 1 & \vdots & \vdots & \vdots & \vdots & 0 \\ c_{q-1} & c_{q-2} & \dots & c_1 & 1 & 0 & 0 & 0 & 0 \\ c_q & c_{q-1} & \dots & \dots & \dots & 1 & 0 & \dots & 0 \\ 0 & c_q & c_{q-1} & \dots & \vdots & \vdots & 1 & 0 & 0 \\ 0 & 0 & c_q & c_{q-1} & \dots & \vdots & \vdots & 1 & 0 \\ 0 & 0 & 0 & c_q & c_{q-1} & \dots & c_2 & c_1 & 1 \end{bmatrix}, \tag{7.77}$$

so that, as $\mathbf{MM}^{-1} = \mathbf{I}_{\bar{n}}$, the c_k can be obtained by solving

$$\begin{aligned} b_0c_0 &= 1 \\ b_1c_0 + b_0c_1 &= 0 \\ b_2c_0 + b_1c_1 + b_0c_2 &= 0 \\ &\vdots \\ b_{\bar{n}}c_0 + b_{\bar{n}-1}c_1 + \dots + b_0c_{\bar{n}} &= 0. \end{aligned} \tag{7.78}$$

As $b_0 = 1$, it follows that $c_0 = 1$. Solving the second equation for c_1 and noting that b_1 is given by the first entry in the second row of matrix $\mathbf{M}$, c_1 is obtained. This holds similarly for the subsequent terms, and recurrence formula (7.76) follows.

Having obtained the inverse of $\mathbf{M}$, $(\mathbf{MM}')^{-1}$ follows by $(\mathbf{MM}')^{-1} = (\mathbf{M}^{-1})'\mathbf{M}^{-1}$. As an important remark, note that the product of two Toeplitz matrices is in general not (lower or upper triangular) Toeplitz. This is easily confirmed by multiplying a lower triangular Toeplitz matrix by a upper triangular Toeplitz matrix such as its transpose. However, the product of two lower (upper) triangular Toeplitz matrices is Toeplitz. To obtain $\mathbf{\Sigma}^{-1}$, write

$$\begin{aligned} \mathbf{MM}' &= \begin{bmatrix} \mathbf{D} \\ \mathbf{M}_1 \end{bmatrix} \begin{bmatrix} \mathbf{D} \\ \mathbf{M}_1 \end{bmatrix}' = \begin{bmatrix} \mathbf{DD}' & \mathbf{DM'}_1 \\ \mathbf{M}_1\mathbf{D}' & \mathbf{M}_1\mathbf{M}_1' \end{bmatrix} \\ &=: \begin{bmatrix} \mathbf{A}^{11}_{m\times m} & \mathbf{A}^{12}_{m\times(T-p^*-m)} \\ \mathbf{A}^{21}_{(T-p^*-m)\times m} & \mathbf{A}^{22}_{(T-p^*-m)\times(T-p^*-m)} \end{bmatrix}, \end{aligned} \tag{7.79}$$

and let

$$(\mathbf{MM}')^{-1} =: \begin{bmatrix} \mathbf{E}_{m\times m} & \mathbf{F}_{m\times(T-p^*-m)} \\ \mathbf{G}_{(T-p^*-m)\times m} & \mathbf{H}_{(T-p^*-m)\times(T-p^*-m)} \end{bmatrix}. \tag{7.80}$$

Then

$$\begin{aligned} \mathbf{MM}'(\mathbf{MM}')^{-1} &= \begin{bmatrix} \mathbf{A}^{11} & \mathbf{A}^{12} \\ \mathbf{A}^{21} & \mathbf{A}^{22} \end{bmatrix} \begin{bmatrix} \mathbf{E} & \mathbf{F} \\ \mathbf{G} & \mathbf{H} \end{bmatrix} \\ &= \begin{bmatrix} \mathbf{I}_{m\times m} & \mathbf{0}_{m\times(T-p^*-m)} \\ \mathbf{0}_{(T-p^*-m)\times m} & \mathbf{I}_{(T-p^*-m)\times(T-p^*-m)} \end{bmatrix}, \end{aligned} \tag{7.81}$$

implying

$$\begin{aligned} &\mathbf{A}^{11}\mathbf{E} + \mathbf{A}^{12}\mathbf{G} = \mathbf{I}_{m\times m}, \\ &\mathbf{A}^{11}\mathbf{F} + \mathbf{A}^{12}\mathbf{H} = \mathbf{0}_{m\times(T-p^*-m)}, \\ &\mathbf{A}^{21}\mathbf{E} + \mathbf{A}^{22}\mathbf{G} = \mathbf{0}_{(T-p^*-m)\times m}, \end{aligned} \tag{7.82}$$

$$\mathbf{A}^{21}\mathbf{F} + \mathbf{A}^{22}\mathbf{H} = \mathbf{I}_{(T-p^*-m)\times(T-p^*-m)}. \tag{7.83}$$

From (7.82), it follows that $\mathbf{A}^{21} = -\mathbf{A}^{22}\mathbf{GE}^{-1}$ so that, from (7.83),

$$(-\mathbf{A}^{22}\mathbf{GE}^{-1})\mathbf{F} + \mathbf{A}^{22}\mathbf{H} = \mathbf{I}_{(T-p^*-m)\times(T-p^*-m)},$$

and

$$\mathbf{A}^{22}(\mathbf{H} - \mathbf{GE}^{-1}\mathbf{F}) = \mathbf{I}_{(T-p^*-m)\times(T-p^*-m)}. \tag{7.84}$$

From (7.84), $\mathbf{H} - \mathbf{GE}^{-1}\mathbf{F} = (\mathbf{A}^{22})^{-1}$, so that, with $(\mathbf{A}^{22})^{-1} = (\mathbf{M}_1\mathbf{M}_1')^{-1}$,

$$\mathbf{\Sigma}^{-1} = (\mathbf{M}_1\mathbf{M}_1')^{-1} = \mathbf{H} - \mathbf{GE}^{-1}\mathbf{F}, \tag{7.85}$$

which corresponds to the **Schur complement** of the block matrix $\mathbf{H}$. Note that expression (7.85) still contains the computation of the inverse matrix $\mathbf{E}^{-1}$, which is $O(m^3)$, and thus far smaller than $O(T^3)$.

Another approach to speed up the computation is to make use of the Cholesky decomposition

$$\mathbf{\Sigma} = \mathbf{PP}', \tag{7.86}$$

which is applicable as $\mathbf{\Sigma}$ is a symmetric, positive semi-definite matrix. Moreover, as $\mathbf{\Sigma}$ is a band matrix, its Cholesky decomposition can be obtained by a fast computational algorithm. For example, the Matlab function `chol(A)` delivers the Cholesky decomposition of matrix $\mathbf{A}$. The resulting matrix $\mathbf{P}$ is lower triangular and therefore easy to invert, so that, using (7.86), we can rewrite (7.72) as

$$\begin{aligned} \beta_{\text{GLS}} &= (\tilde{\mathbf{X}}'\tilde{\mathbf{X}})^{-1}\tilde{\mathbf{X}}'\tilde{\mathbf{Y}} \\ &= ((\mathbf{P}^{-1}\mathbf{X})'(\mathbf{P}^{-1}\mathbf{X}))^{-1}(\mathbf{P}^{-1}\mathbf{X})'(\mathbf{P}^{-1}\mathbf{Y}) \\ &= ((\mathbf{X}'(\mathbf{P}')^{-1}\mathbf{P}^{-1}\mathbf{X}))^{-1}\mathbf{X}'(\mathbf{P}')^{-1}\mathbf{P}^{-1}\mathbf{Y} \\ &= (\mathbf{X}'\mathbf{\Sigma}^{-1}\mathbf{X})^{-1}\mathbf{X}'\mathbf{\Sigma}^{-1}\mathbf{Y}. \end{aligned}$$

In this way, the computational burden is considerably lowered.

The latter approach is adopted by Koreisha and Pukkila (1990). A fact overlooked by those authors is that computation of the Cholesky factors can be entirely avoided. To see this, recall from (6.51) and (6.52) that $\mathbf{\Sigma}$ can be written as

$$\mathbf{\Sigma} = \mathbf{M}_2\mathbf{M}_2' + \mathbf{N}\mathbf{N}', \tag{7.87}$$

where $\mathbf{N}$ is $(T - p^* - m) \times m$ matrix and $\mathbf{M}_2$ is a $(T - p^* - m) \times (T - p^* - m)$ lower triangular matrix and thus, by definition, the Cholesky factor of $\mathbf{M}_2\mathbf{M}_2'$. The Cholesky decomposition of $\mathbf{\Sigma}$ can therefore be obtained by updating $\mathbf{M}_2$, this being an $O(T^2)$ operation, as opposed to computing the Cholesky factors from scratch, which is $O(T^3)$. A description of the algorithm for $\text{rank}(\mathbf{M}_2) = 1$ is given in Gill et al. (1974), while for $\text{rank}(\mathbf{M}_2) = k$, $k \in \mathbb{N}$, see Davis (2006), where also a specialized algorithm for sparse matrices is presented that would allow for additional time savings. For MA(1) models, i.e., $q = \text{rank}(\mathbf{N}) = 1$, the Matlab function `cholupdate` can be used, which unfortunately does not accept sparse matrices as input arguments. The general case with arbitrary q requires custom programming.

From a practical point of view, approaches that make direct use of the Cholesky decomposition for matrix inversion (even if the computation of the Cholesky factors can be avoided, as given in (7.87)) will often not outperform standard, optimally-coded inversion approaches (such as the function `inv` in Matlab) in modern computational software packages. The reason for this is that standard inversion algorithms already use the most efficient inversion approach for symmetric positive semi-definite (or even lower triangular) and sparse matrices. However, comparisons show that making use of the lower triangular Toeplitz structure (or so-called recurrent block matrix approach) are faster than use of the Matlab function `inv` for sufficiently large matrices.

7.B Appendix: Multivariate AR(p) Processes and Stationarity, and General Block Toeplitz Matrix Inversion

In order to emphasize the importance of having fast inversion algorithms for Toeplitz matrices, we briefly discuss the case of stationary multivariate AR(p) processes. Consider a model where the output $\mathbf{X}_t$ of a system connected with the input $\mathbf{Y}_t$ is given, as in Akaike (1973), by

$$\mathbf{X}_t = \sum_{i=1}^{p} \mathbf{A}_i\mathbf{Y}_{t-i} + \mathbf{U}_t, \quad \mathbb{E}[\mathbf{U}_t] = \mathbf{0}, \tag{7.88}$$

where $\mathbf{X}_t$ is $e \times 1$, $\mathbf{A}_i$ is $e \times d$, $\mathbf{Y}_t$ is $d \times 1$, and $\mathbf{U}_t$ is $e \times 1$. Vectors $\mathbf{X}_t$, $\mathbf{Y}_t$, and $\mathbf{U}_t$ are assumed to be jointly (weak-)stationary stochastic processes.

To be clear, recall that, if it exists, the covariance of two vector random variables $\mathbf{X} = (X_1, \dots, X_n)'$ and $\mathbf{Y} = (Y_1, \dots, Y_m)'$, with expectations $\boldsymbol{\mu}_\mathbf{X}$ and $\boldsymbol{\mu}_\mathbf{Y}$, respectively, is given by $\text{Cov}(\mathbf{X}, \mathbf{Y}) := \mathbb{E}[(\mathbf{X} - \boldsymbol{\mu}_\mathbf{X})(\mathbf{Y} - \boldsymbol{\mu}_\mathbf{Y})']$, an $n \times m$ matrix with (ij)th element $\sigma_{X_i,Y_j} = \text{Cov}(X_i, Y_j)$. From symmetry, $\text{Cov}(\mathbf{X}, \mathbf{Y}) = \text{Cov}(\mathbf{Y}, \mathbf{X})'$. (See, e.g., page II.99, though this definition is standard and appears in numerous book presentations.)

Now consider another stochastic process $\mathbf{V}_t$ of dimension $d \times 1$ with finite second moments such that $\mathbb{E}[\mathbf{V}_t] = \mathbf{0}$, and let $\boldsymbol{\gamma}_{j,\mathbf{X}} = \text{Cov}(\mathbf{X}_t, \mathbf{V}_{t-j})$, of size $e \times d$, $\boldsymbol{\gamma}_{j-i,\mathbf{Y}} = \text{Cov}(\mathbf{Y}_{t-i}, \mathbf{V}_{t-j})$ of size $d \times d$, and $\text{Cov}(\mathbf{U}_t, \mathbf{V}_{t-j}) = \mathbf{0}$, where $\mathbf{0}$ is an $e \times d$ matrix of zeros, $j = 1, \dots, p$. Note that all these covariance matrices are not functions of time t, but only of lags i and j.

From (7.88) and that, by assumption, $\mathbb{E}[\mathbf{U}_t \mathbf{V}'_{t-j}] = \mathbf{0}$,

$$\mathbb{E}[\mathbf{X}_t \mathbf{V}'_{t-j}] = \sum_{i=1}^{p} \mathbf{A}_i \mathbb{E}[\mathbf{Y}_{t-i} \mathbf{V}'_{t-j}], \quad j = 1, \ldots, p. \tag{7.89}$$

As $\mathbb{E}[\mathbf{U}_t] = \mathbf{0}_e$ and $\mathbb{E}[\mathbf{V}_t] = \mathbf{0}_d$, the (weak-)stationarity assumption implies that $\gamma_{j,\mathrm{X}} = \mathbb{E}[\mathbf{X}_t \mathbf{V}'_{t-j}]$ and $\gamma_{j-i,\mathrm{Y}} = \mathbb{E}[\mathbf{Y}_{t-i} \mathbf{V}'_{t-j}]$ are functions only of j and $j - i$, respectively. Hence, (7.89) yields

$$\left[\gamma_{1,\mathrm{X}} \gamma_{2,\mathrm{X}} \cdots \gamma_{p,\mathrm{X}}\right] = \left[\mathbf{A}_1 \mathbf{A}_2 \ldots \mathbf{A}_p\right] \mathbf{T}, \text{ where} \tag{7.90}$$

$$\mathbf{T} := \begin{bmatrix} \gamma_{0,\mathrm{Y}} & \gamma_{1,\mathrm{Y}} & \gamma_{2,\mathrm{Y}} & \cdots & \gamma_{p-1,\mathrm{Y}} \\ \gamma_{-1,\mathrm{Y}} & \gamma_{0,\mathrm{Y}} & \gamma_{1,\mathrm{Y}} & \cdots & \gamma_{p-2,\mathrm{Y}} \\ \gamma_{-2,\mathrm{Y}} & \gamma_{-1,\mathrm{Y}} & \gamma_{0,\mathrm{Y}} & \cdots & \gamma_{p-3,\mathrm{Y}} \\ \vdots & \vdots & \vdots & \ddots & \vdots \\ \gamma_{-p+1,\mathrm{Y}} & \gamma_{-p+2,\mathrm{Y}} & \gamma_{-p+3,\mathrm{Y}} & \cdots & \gamma_{0,\mathrm{Y}} \end{bmatrix}. \tag{7.91}$$

If $\gamma_{j-i,\mathrm{Y}}$ and $\gamma_{j,\mathrm{X}}$ are given and the inverse of the block Toeplitz matrix $\mathbf{T}$ exists, then (7.90) can be solved for $\mathbf{A}_i$. Setting $\mathbf{X}_t = \mathbf{Y}_{t+k}$, $k = 1, 2, \ldots$, and $\mathbf{V}_{t-i} = \mathbf{Y}_{t-i}$, the solution to $\mathbf{A}_i$ of (7.90) gives the least mean square error $(k + 1)$-step ahead linear predictor based on the past p observations of $\mathbf{Y}_t$. Note that this solution can be considered as the solution to the filtering problem, where $\mathbf{X}_t = \mathbf{Y}_{t+k} + \mathbf{W}_{t+k}$ and the process $\mathbf{W}_t$ is uncorrelated with the process $\mathbf{Y}_t$. Alternatively, if one takes $\mathbf{X}_t = \mathbf{Y}_t$ and $\mathbf{V}_{t-j} = \mathbf{Y}_{t-j-q}$, (7.90) corresponds to the set of Yule–Walker equations for the d-dimensional mixed autoregressive moving average process of order q and p.

In order to invert the block Toeplitz matrix $\mathbf{T}$, which in this general setting is not necessarily a symmetric matrix, efficient inversion algorithms are needed (see Akaike, 1973). A well-known approach is the **Trench–Durbin** algorithm for the inversion of symmetric positive definite Toeplitz matrices in $O(n^2)$ flops. An excellent introductory treatment of this and other algorithms related to Toeplitz matrices is given in Golub and Loan (2012).

8

Correlograms

The correlogram is probably the most useful tool in time-series analysis after the time plot.
(Chris Chatfield, 2001, p. 30)

Interpreting a correlogram is one of the hardest tasks in time-series analysis…
(Chris Chatfield, 2001, p. 31)

Among the major tools traditionally associated with univariate time-series analysis are two sample correlograms that provide information about the correlation structure and, within the ARMA(p, q) model class, about possible candidates for p and q. These are studied in detail in this chapter.

8.1 Theoretical and Sample Autocorrelation Function

8.1.1 Definitions

Recall the calculation of the autocovariances γ_s, or $\gamma(s)$, $s = 0, 1, 2, \ldots$, of a stationary, invertible ARMA(p, q) process, as discussed in Section 7.4.1. It is more common in applications with real data and the assessment of suitable values of p and q to work with the standardized version, namely the **autocorrelations**. They are given by

$$\rho_s = \text{Corr}(Y_t, Y_{t-s}) = \frac{\text{Cov}(Y_t, Y_{t-s})}{\sqrt{\mathbb{V}(Y_t)\mathbb{V}(Y_{t-s})}} = \frac{\gamma_s}{\gamma_0}. \tag{8.1}$$

The set of values $\rho_1, \rho_2, \ldots$ is referred to as the **(theoretical) autocorrelation function**, abbreviated TACF (or just ACF). For example, recalling the autocovariances of the AR(1) process, as given in (4.13), we have

$$\rho_s = \frac{\gamma_s}{\gamma_0} = a^{|s|}, \quad s \in \mathbb{Z}. \tag{8.2}$$

Very common in time-series analysis is to plot ρ_s for $s = 1$ up to some arbitrary value (that rarely exceeds 30 for non-seasonal data). This is referred to as a **correlogram**. Two examples for an AR(1) process are shown in Figure 8.1.[1] Indeed, for the AR(1) model, the shape of the ACF is quite predictable, given the very simple form of ρ_s in (8.2).

1 To produce such graphs in Matlab, use the `stem` function. The correlograms displayed in this chapter were generated by modifying the `stem` function to make the lines thicker and remove the circle at the top of each spike.

Linear Models and Time-Series Analysis: Regression, ANOVA, ARMA and GARCH, First Edition. Marc S. Paolella.

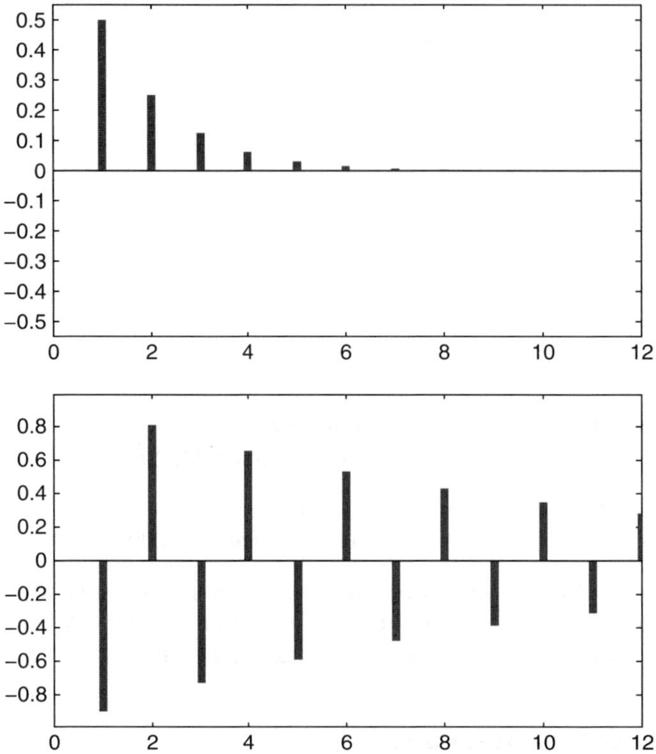

Figure 8.1 TACF of the AR(1) process with $a = 0.5$ (top) and $a = -0.9$ (bottom).

Figure 8.2 shows the ACF for several stationary AR(3) models, illustrating the variety of shapes that are possible for the modest value of $p = 3$.

We state one more definition and a characterization result that are relevant for defining the sample counterpart to the TACF, and will play an important role when we examine the joint distribution of the sample autocorrelations in Section 8.1.3.

A function $\kappa : \mathbb{Z} \to \mathbb{R}$ is said to be positive semi-definite if, for all $n \in \mathbb{N}$

$$\sum_{r=1}^{n} \sum_{s=1}^{n} \kappa(t_r - t_s) z_r z_s \geqslant 0, \tag{8.3}$$

for all (sets of time points) $\mathbf{t} = (t_1, \dots, t_n)' \in \mathbb{Z}^n$ and all $\mathbf{z} = (z_1, \dots, z_n)' \in \mathbb{R}^n$.

The result we now need is that a function $\gamma : \mathbb{Z} \to \mathbb{R}$ is the autocovariance function of a weakly stationary time series if and only if γ is even, i.e., $\gamma(h) = \gamma(-h)$ for all $h \in \mathbb{Z}$, and is positive semi-definite. To show $\Rightarrow$, let $\mathbf{t} = (t_1, \dots, t_n)' \in \mathbb{Z}^n$ and $\mathbf{z} = (z_1, \dots, z_n)' \in \mathbb{R}^n$, and let $\mathbf{Y_t} = (Y_{t_1}, \dots, Y_{t_n})'$ be a set of random variables such that $\mathbb{E}[\mathbf{Y_t}] = \mathbf{0}$ and having finite second moments. Then, with

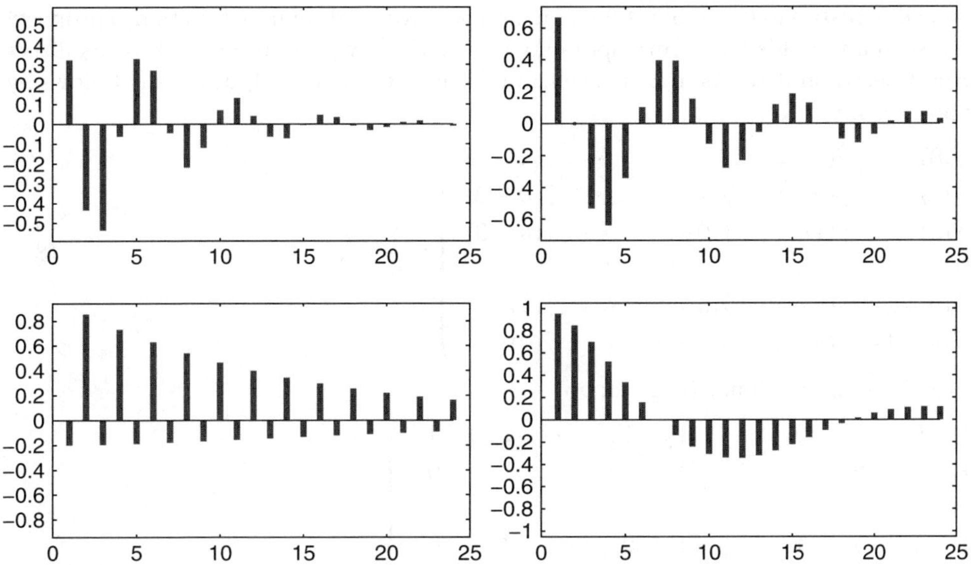

Figure 8.2 TACF of the stationary AR(3) model with parameters $\mathbf{a} = (a_1, a_2, a_3) = (0.4, -0.5, -0.2)$ (top left), $\mathbf{a} = (1.2, -0.8, 0)$ (top right), $\mathbf{a} = (-0.03, 0.85, 0)$ (bottom left) and $\mathbf{a} = (1.4, -0.2, -0.3)$ (bottom right).

$\boldsymbol{\Gamma}_n = [\gamma(t_r - t_s)]_{r,s=1}^n$ the covariance matrix of $\mathbf{Y_t}$, the symmetry of $\boldsymbol{\Gamma}_n$ implies γ is even, and

$$0 \leqslant \mathbb{V}(\mathbf{a}'\mathbf{Y_t}) = \mathbf{a}'\mathbb{E}[\mathbf{Y_t'}\mathbf{Y_t'}]\mathbf{a} = \mathbf{a}'\boldsymbol{\Gamma}_n\mathbf{a} = \sum_{r=1}^{n}\sum_{s=1}^{n}\gamma(t_r - t_s)a_r a_s, \tag{8.4}$$

thus satisfying (8.3). The proof of $\Leftarrow$ is more advanced, and can be found, e.g., in Brockwell and Davis (1991, p. 27).

Dividing (8.4) by $\gamma(0)$ shows that the autocorrelation function (8.1) corresponding to a stationary time series is also positive semi-definite. In particular, with $\mathbf{t} = (1, \dots, n)'$, $\mathbf{Y_t} = (Y_1, \dots, Y_n)'$, and

$$\mathbf{R}_n = \frac{1}{\gamma(0)}\boldsymbol{\Gamma}_n = \begin{pmatrix} 1 & \rho_1 & \cdots & \rho_{n-1} \\ \rho_1 & 1 & \cdots & \rho_{n-2} \\ \vdots & & \ddots & \vdots \\ \rho_{n-1} & \cdots & \rho_1 & 1 \end{pmatrix}, \tag{8.5}$$

we require that $\mathbf{R}_n \geqslant 0$.

Assume we have T equally spaced observations from a time series, say $Y_1, \dots, Y_T$, generated by a stationary, mean-zero model. The obvious "plug-in estimator", or natural sample counterpart of γ_s is $(T-s)^{-1}\sum_{t=s+1}^{T} Y_t Y_{t-s}$, but it is advantageous to use a divisor of T instead of $T - s$, i.e.,

$$\widehat{\gamma}_s = T^{-1}\sum_{t=s+1}^{T} Y_t Y_{t-s}, \tag{8.6}$$

which is a form of shrinkage towards zero. As is typical with such shrinkage estimators, (8.6) is biased, but has a lower mean squared error than its unbiased counterpart; see Priestley (1981, p. 323–324).

A further compelling reason to use (8.6) is that it yields a positive semi-definite function, a property that we have just seen also holds for γ_s corresponding to a stationary process, but not for its direct sample counterpart based on data. As in Brockwell and Davis (1991, Sec. 7.2), this easily follows by expressing, for any $1 \leqslant n \leqslant T$,

$$\widehat{\boldsymbol{\Gamma}}_n = \begin{pmatrix} \hat{\gamma}(0) & \hat{\gamma}(1) & \hat{\gamma}(2) & \cdots & \hat{\gamma}(n-1) \\ \hat{\gamma}(1) & \hat{\gamma}(0) & \hat{\gamma}(1) & \cdots & \hat{\gamma}(n-2) \\ \hat{\gamma}(2) & \hat{\gamma}(1) & \hat{\gamma}(0) & \cdots & \hat{\gamma}(n-3) \\ \vdots & \cdots & \cdots & \ddots & \\ \hat{\gamma}(n-2) & \hat{\gamma}(n-3) & \hat{\gamma}(n-4) & \cdots & \hat{\gamma}(1) \\ \hat{\gamma}(n-1) & \hat{\gamma}(n-2) & \hat{\gamma}(n-3) & \cdots & \hat{\gamma}(0) \end{pmatrix} = \frac{1}{n}\mathbf{L}\mathbf{L}', \tag{8.7}$$

where $\mathbf{L}$ is the $n \times (2n-1)$ "band matrix" given by

$$\mathbf{L} = \begin{pmatrix} 0 & 0 & \cdots & 0 & Y_1 & Y_2 & Y_3 & \cdots & Y_{n-1} & Y_n \\ 0 & \cdots & 0 & Y_1 & Y_2 & Y_3 & \cdots & \cdots & Y_n & 0 \\ \vdots & & & & & & & & & \vdots \\ 0 & Y_1 & Y_2 & Y_3 & \cdots & Y_n & 0 & \cdots & 0 & 0 \\ Y_1 & Y_2 & Y_3 & \cdots & Y_n & 0 & \cdots & 0 & 0 & 0 \end{pmatrix}.$$

Thus, for any $\mathbf{z} = (z_1, \dots, z_n)' \in \mathbb{R}^n$, $\mathbf{z}'\widehat{\boldsymbol{\Gamma}}_n\mathbf{z} = n^{-1}(\mathbf{z}'\mathbf{L})(\mathbf{L}'\mathbf{z}) \geqslant 0$.

It is noteworthy that the matrices $\mathbf{R}_n$ in (8.5) and $\widehat{\boldsymbol{\Gamma}}_n$ in (8.7) are symmetric and **persymmetric**, where the latter means a square matrix that is symmetric with respect to the northeast-to-southwest diagonal.

It will be subsequently convenient to express $\hat{\rho}_s$ as a ratio of quadratic forms. Use of (8.6) implies that the sample estimate of ρ_s is given by

$$\hat{\rho}_s = R_s := \frac{\hat{\gamma}_s}{\hat{\gamma}_0} = \frac{\sum_{t=s+1}^{T} Y_t Y_{t-s}}{\sum_{t=1}^{T} Y_t^2} = \frac{\mathbf{Y}'\mathbf{A}_s\mathbf{Y}}{\mathbf{Y}'\mathbf{Y}}, \tag{8.8}$$

where $\mathbf{Y} = (Y_1, \dots, Y_T)'$ and the (i,j)th element of $\mathbf{A}_s$ is given by $\mathbb{I}\{|i-j| = s\}/2$, $i,j = 1, \dots, T$. For example, with $T = 5$,

$$\mathbf{A}_1 = \begin{bmatrix} 0 & \frac{1}{2} & 0 & 0 & 0 \\ \frac{1}{2} & 0 & \frac{1}{2} & 0 & 0 \\ 0 & \frac{1}{2} & 0 & \frac{1}{2} & 0 \\ 0 & 0 & \frac{1}{2} & 0 & \frac{1}{2} \\ 0 & 0 & 0 & \frac{1}{2} & 0 \end{bmatrix} \quad \text{and} \quad \mathbf{A}_2 = \begin{bmatrix} 0 & 0 & \frac{1}{2} & 0 & 0 \\ 0 & 0 & 0 & \frac{1}{2} & 0 \\ \frac{1}{2} & 0 & 0 & 0 & \frac{1}{2} \\ 0 & \frac{1}{2} & 0 & 0 & 0 \\ 0 & 0 & \frac{1}{2} & 0 & 0 \end{bmatrix}, \tag{8.9}$$

etc. A program to compute the $\mathbf{A}$ matrices is given in Listing 8.1.

```
function A=makeA(T,m) % A = 0.5 * 1( |i-j| = m)
v=zeros(T,1); v(m+1)=1; A=0.5*toeplitz(v,v');
```

Program Listing 8.1: Computes $\mathbf{A}_m$ of size $T \times T$.

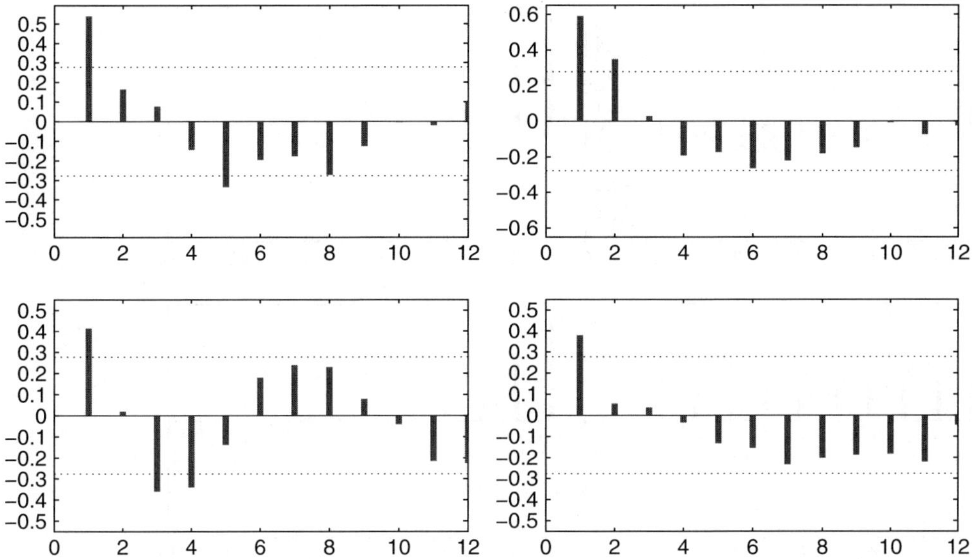

Figure 8.3 The SACFs of four simulated AR(1) time series with $a = 0.5$ and $T = 50$.

The **sample ACF**, abbreviated SACF, is given by the random variable $\mathbf{R}_m = (R_1, \dots, R_m)'$. The upper limit m can be as high as $T - 1$ but, practically speaking, can be taken to be, say, $\min(T/2, 30)$.

The *observed* values of the SACF, say $\mathbf{r} = (r_1, \dots, r_m)'$, based on a stationary and invertible ARMA time-series process, will obviously not exactly resemble the corresponding TACF, but they will be close for large enough T. To illustrate, Figure 8.3 shows the SACFs of four simulated AR(1) time series, each with $a = 0.5$ and $T = 50$. The two horizontal dashed lines are given by $\pm 1.96/\sqrt{T}$ and provide an asymptotically valid 95% c.i. for each individual r_s, as will be discussed in Section 8.1.3.2.

For now, it suffices to observe that, at least for sample sizes around $T = 50$, the SACF does not strongly resemble its theoretical counterpart. Figure 8.4 is similar but uses $T = 500$ observations instead. The SACF is now far closer to the TACF, but can still take on patterns that noticeably differ from the true values.

In practice, $\mathbb{E}[Y_t]$ is unknown and, assuming stationarity, is constant for all t and estimated as the sample mean, say $\hat{\mu}$. Then, the sample covariance in (8.6) is computed as

$$\hat{\gamma}_s = T^{-1} \sum_{t=s+1}^{T} (Y_t - \hat{\mu})(Y_{t-s} - \hat{\mu}) = T^{-1} \sum_{t=s+1}^{T} \hat{\epsilon}_t \hat{\epsilon}_{t-s}, \tag{8.10}$$

and

$$R_s = \frac{\hat{\boldsymbol{\epsilon}}' \mathbf{A}_s \hat{\boldsymbol{\epsilon}}}{\hat{\boldsymbol{\epsilon}}' \hat{\boldsymbol{\epsilon}}}, \tag{8.11}$$

where $\hat{\boldsymbol{\epsilon}} = (\hat{\epsilon}_1, \dots, \hat{\epsilon}_T)' = \mathbf{Y} - \hat{\mu}$. The plotted SACF $\mathbf{R}_m = (R_1, \dots, R_m)'$ based on (8.11) is one of the primary graphical tools used in time-series analysis.

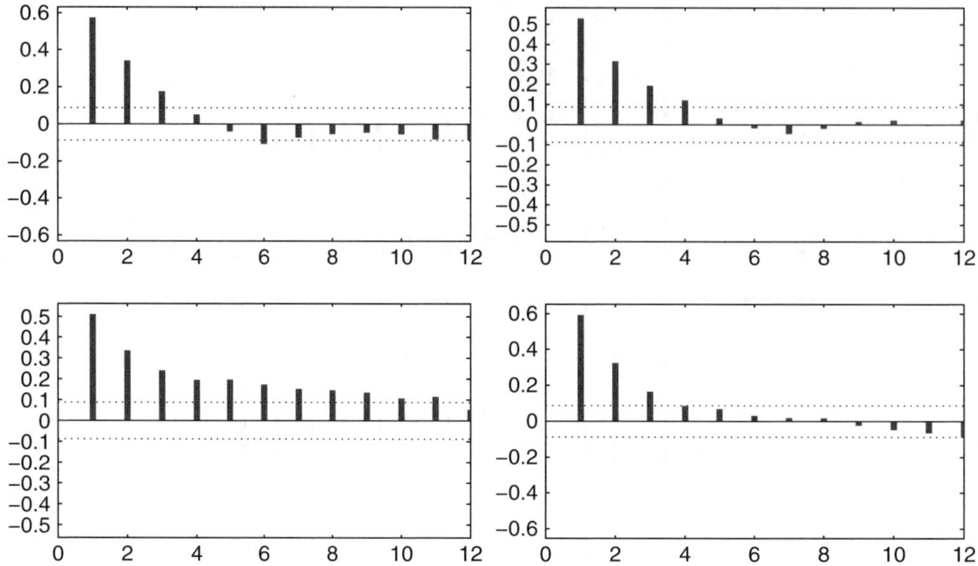

Figure 8.4 The SACFs of four simulated AR(1) time series with $a = 0.5$ and $T = 500$.

The statistics $\{\hat{\gamma}_s\}$ in (8.10) have the interesting property that

$$\sum_{s=-(T-1)}^{T-1} \hat{\gamma}_s = 0. \tag{8.12}$$

To prove (8.12), following Percival (1993), we first construct the $T \times T$ symmetric matrix

$$\mathbf{S} = \begin{bmatrix} (Y_1 - \bar{Y})(Y_1 - \bar{Y}) & (Y_1 - \bar{Y})(Y_2 - \bar{Y}) & \cdots & (Y_1 - \bar{Y})(Y_T - \bar{Y}) \\ (Y_2 - \bar{Y})(Y_1 - \bar{Y}) & (Y_2 - \bar{Y})(Y_2 - \bar{Y}) & & (Y_2 - \bar{Y})(Y_T - \bar{Y}) \\ \vdots & & \ddots & \vdots \\ (Y_T - \bar{Y})(Y_1 - \bar{Y}) & \cdots & & (Y_T - \bar{Y})(Y_T - \bar{Y}) \end{bmatrix}.$$

The sum of the diagonal elements of $\mathbf{S}$ is $T\hat{\gamma}_0$, while the sum of the elements along the sth sub- or super-diagonal is $T\hat{\gamma}_s$, $s = 1, 2, \ldots, T-1$. As $\hat{\gamma}_s = \hat{\gamma}_{-s}$, the sum of all T^2 elements in $\mathbf{S}$ is $T\sum_{s=-(T-1)}^{T-1} \hat{\gamma}_s$. However, each row (and column) sum is easily seen to be zero, so that the sum of all T^2 elements in $\mathbf{S}$ is zero, showing (8.12).

Dividing by $\hat{\gamma}_0$ and using the fact that $\hat{\gamma}_s = \hat{\gamma}_{-s}$, (8.12) can also be written as

$$\sum_{s=1}^{T-1} R_s = -\frac{1}{2}, \tag{8.13}$$

which implies that $R_s < 0$ for at least one value of $s \in \{1, 2, \ldots, T-1\}$. This helps to explain why, in each of the four SACF plots in Figure 8.3, several of the spikes are negative, even though the theoretical ACF (shown in the top panel of Figure 8.1) is strictly positive. As a more extreme case, Figure 8.5 shows the SACF for a simulated random walk with 200 observations.

A program to compute the sample ACF for a given time series is shown in Listing 8.2.

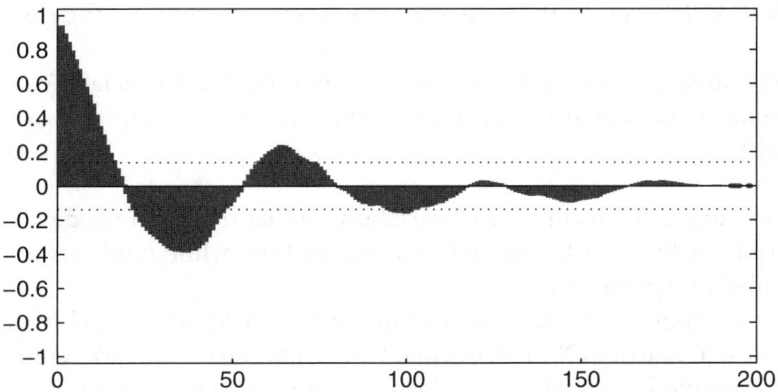

Figure 8.5 Sample ACF for a simulated random walk with 200 observations.

```
function sacf=sampleacf(Y,imax,removemean,doplot)
if nargin<3, removemean=1; end, if nargin<4, doplot=0; end
if removemean, Y=Y-mean(Y); end, T=length(Y); a=zeros(imax,1);
for i=1:imax, a(i)= sum(Y(i+1:T) .* Y(1:T-i) ); end, sacf=a./ sum(Y.^2);
if doplot,  stem(sacf), se=1.96/sqrt(T);
  line([0,imax],[0,0],'linestyle','-','linewidth',1)
  line([0,imax],[-se,-se],'linestyle',':','linewidth',1)
  line([0,imax],[ se, se],'linestyle',':','linewidth',1)
  rm=max(abs(sacf)); rm=1.1*max(rm,0.5); ax=axis; axis([ax(1) ax(2)  -rm rm])
end
```

Program Listing 8.2: Computes and plots the sample autocorrelation function for time series `Y` up to lag `imax`.

8.1.2 Marginal Distributions

Generalizing the use of the sample mean, assume now that

$$\mathbf{Y} \sim \mathrm{N}(\mathbf{X}\boldsymbol{\beta}, \boldsymbol{\Psi}^{-1}), \tag{8.14}$$

where $\mathbf{X}$ is a known, full rank $T \times k$ matrix of exogenous variables, and $\boldsymbol{\beta}$ and $\boldsymbol{\Psi}^{-1}$ are fixed but unknown. (In Section 8.1.3.3 it will be a bit more notationally convenient to work with $\boldsymbol{\Psi}^{-1}$ rather than $\boldsymbol{\Psi}$ as the variance covariance matrix.) Then, $\hat{\boldsymbol{\epsilon}}$, the o.l.s. regression residuals based on $\mathbf{Y}$ and $\mathbf{X}$, can be computed and used to construct the SACF via (8.11). From Chapter 1, the o.l.s. residual vector can be expressed as $\hat{\boldsymbol{\epsilon}} = \mathbf{M}\mathbf{Y} = \mathbf{M}\boldsymbol{\epsilon}$, where $\mathbf{M} = \mathbf{I}_T - \mathbf{X}(\mathbf{X}'\mathbf{X})^{-1}\mathbf{X}'$. If $\boldsymbol{\epsilon} \sim \mathrm{N}(\mathbf{0}, \boldsymbol{\Psi}^{-1})$, then $\mathbf{M}\boldsymbol{\epsilon} \sim \mathrm{N}(\mathbf{0}, \mathbf{M}\boldsymbol{\Psi}^{-1}\mathbf{M})$. In the "null hypothesis case" of i.i.d. error terms, $\boldsymbol{\Psi} = \mathbf{I}$ and $\mathbf{M}\boldsymbol{\epsilon} \sim \mathrm{N}(\mathbf{0}, \mathbf{M})$.

From (1.61), if $\mathbf{1}_T \in \mathcal{C}(\mathbf{X})$, then $\sum_{i=1}^{T} \hat{\epsilon}_i = 0$, and (8.13) also holds.

Remark An alternative to the usual o.l.s. residuals is to use the recursive residuals, as discussed in Section 1.5, to compute the elements of the SACF. They are computed as in (8.11), except that matrix $\mathbf{A}_s$ is of size $T - k$ and $\hat{\boldsymbol{\epsilon}}$ refers to the recursive residual vector. To distinguish them from the usual SACF, we denote the elements of the SACF based on the recursive residuals as $\breve{R}_s$. Recall that, if the true regression model error terms are i.i.d., then so are the $T - k$ recursive residuals. This has the advantage that each component of the SACF based on them also has mean zero, which is not the case for the usual SACF.

Furthermore, their distribution is symmetric (about zero), which is also not true for the usual SACF components. This fact can be useful when using the SACF for *model identification*; see Chapter 9. See Problem 8.6 for more details and proof. ■

From (8.11), R_s (or $\breve{R}_s$) is a ratio of quadratic forms in normal variables, so that the methods developed in Appendix A can be straightforwardly used to compute its distribution, while methods for computation of its moments are detailed in Appendix B.

To illustrate, Figure 8.6 shows the correlogram of $\mathbb{E}[\mathbf{R}_m]$ corresponding to an MA(1) model with parameter $b = 0.5$ for three different design matrices $\mathbf{X}$ in (8.14), and two sample sizes. This was computed using the program `sacfmom` developed in Problem 8.1. Recalling the covariance structure of an MA(q) model from (6.50), the first spike in the TACF of an MA(1) model is nonzero, while the remaining are zero. In this case, $\rho_1 = b/(1 + b^2) = 0.4$. The top left panel of Figure 8.6 shows the case for which $\mathbb{E}[Y_t]$ is known: We see that $\mathbb{E}[\mathbf{R}_m]$ is indeed very close to the *shape* of the TACF, but the first spike is only 0.314 instead of 0.4.[2]

The case of known $\mathbb{E}[Y_t]$ is not practical, and serves as a benchmark, with the use of an intercept model ($\mathbf{X} = \mathbf{1}$) being more realistic for typical univariate time-series analysis. This case is shown in the middle panel, while the right panel shows the situation with an intercept and time-trend model. In the top panels, they both deviate markedly from the correct shape, this being due to the unrealistic small sample size of $T = 10$. The bottom panels are the same, but with $T = 50$.

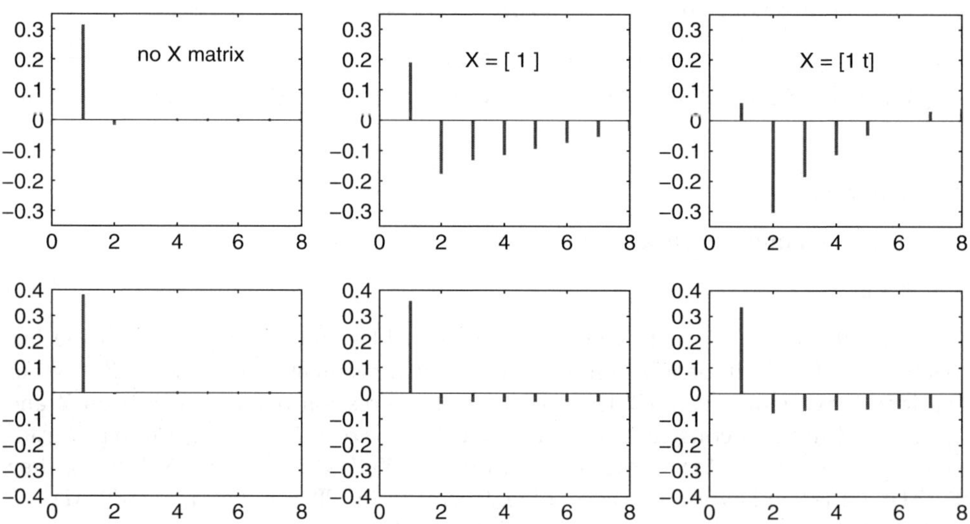

Figure 8.6 Top: The mean SACF for an MA(1) model with $b = 0.5$ and $T = 10$, i.e., the mean of each R_s is shown. Left is with known mean, middle is for an intercept term in the model, i.e., with an **X** matrix consisting of a column of ones, and right is for an intercept and trend model, i.e., with an **X** matrix consisting of a column of ones and the time-trend vector $1, 2, \ldots, T$. **Bottom:** Same but with $T = 50$.

2 Had we used the divisor of $T - s$ for the sample analog of γ_s, the value would be multiplied by 10/9, i.e., 0.349, which is indeed less biased. However, the second spike would also increase in magnitude, by a factor of 10/8.

Figure 8.6 is useful for knowing the expected shape of the SACF for a particular ARMA model. It would be useful to augment the correlogram of $\mathbb{E}[\mathbf{R}_m]$ with intervals reflecting the variance.

Based on $\mathbb{E}[R_s]$ and $\mathbb{E}[R_s^2]$, the variance and standard deviation (std) can be calculated. The top panel of Figure 8.7 plots the std of R_1 through R_{10} for the MA(1) model with $b = 0.5$ and intercept, using sample sizes $T = 20, 30, 40$ and 50, while the bottom panel is similar but for $T = 30$ and several values of b. Although we can calculate the exact distribution of R_s, it is easier to use the fact that the asymptotic distribution of each R_s is normal (see Section 8.1.3.2), and plot, for each s, a line extending from, say, $\mathbb{E}[R_s] - 1.645\sqrt{\mathbb{V}(R_s)}$ to $\mathbb{E}[R_s] + 1.645\sqrt{\mathbb{V}(R_s)}$, thus providing intervals with approximate 90% coverage probability for each R_s. This is shown in Figure 8.8 for an AR(2) model with $a_1 = 1.2$, $a_2 = -0.8$ and an intercept (i.e., with the $\mathbf{X}$ matrix in (8.14) consisting of a column of ones).

This idea could be used with real data by examining the extent to which the $\widehat{R}_s$ fall within their respective intervals. This is shown in Figure 8.9 for $\widehat{R}_1, \dots, \widehat{R}_{10}$, computed based on $\mathbf{X} = \mathbf{1}$, using 90%

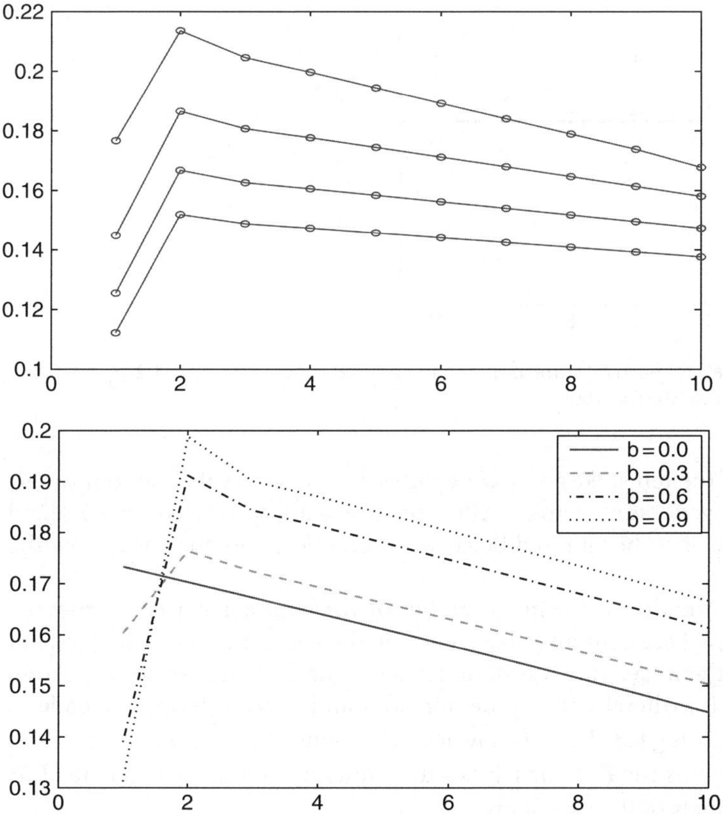

Figure 8.7 Top panel is the standard deviation of R_s, $s = 1, \dots, 10$, for an MA(1) process with $b = 0.5$, intercept term, and four sample sizes $T = 20, 30, 40$, and 50 (top to bottom). The bottom panel is similar but for $T = 30$ and $b = 0, 0.3, 0.6$, and 0.9.

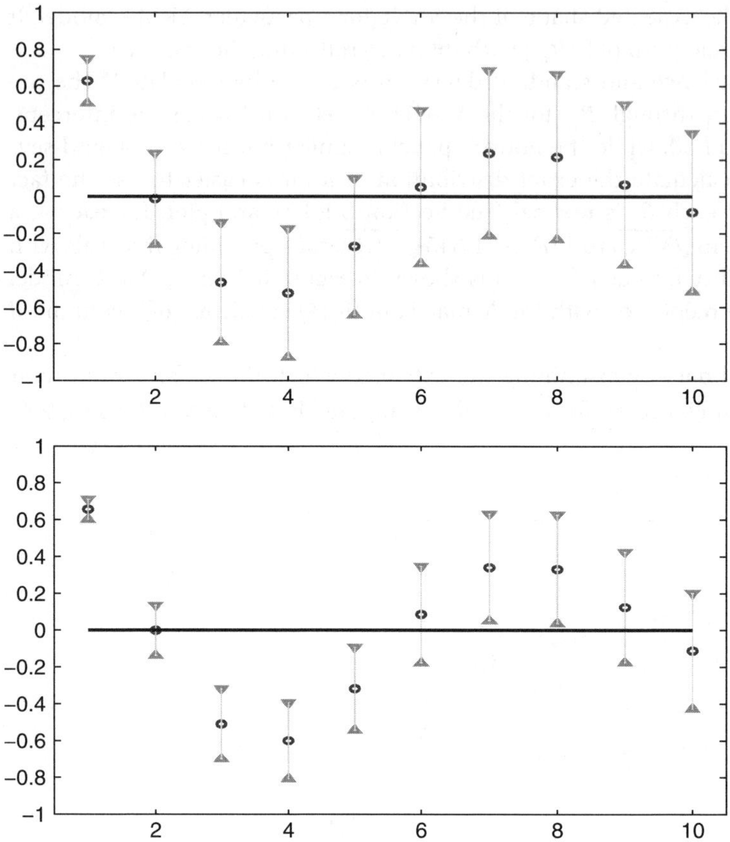

Figure 8.8 $\mathbb{E}[R_s]$ and individual 90% intervals for each R_s, corresponding to an AR(2) model with $a_1 = 1.2$, $a_2 = -0.8$, and $\mathbf{X} = \mathbf{1}$. Top (bottom) panel is based on $T = 30$ ($T = 100$).

intervals (so that, on average, one of the ten spikes *should* be outside of its interval), and four different data sets, each of length $T = 30$, generated from an AR(2) process with $a_1 = 1.2$, $a_2 = -0.8$, and $\sigma = 1$. The usefulness of this exercise is rather limited because, in practice, one does not know the true parameters!

The obvious thing to do is use an estimate, say the m.l.e., of the ARMA parameters to construct the intervals. That is, values $\mathbb{E}[R_s]$ and $\mathbb{V}[R_s]$ are computed based not on the true ARMA parameters, but on the point estimates obtained via the m.l.e. This was done for the same four time series as used in Figure 8.9 and is shown in Figure 8.10, whereby the model includes an intercept term. The code to compute one such graph is given in Listing 8.3. There is now less of a chance that $\widehat{R}_s$ will lie outside of the interval because the intervals are—as the $\widehat{R}_s$ themselves—determined from the actual data. This is apparent for $\widehat{R}_8$ in the lower left panels of the two figures.

More often than not, the "wrong" choices for p and q will be made in practice—almost with certainty in fact, because the true data generating process is unlikely to be precisely given by a stationary ARMA model. To see what happens when the wrong model is chosen, the previous exercise can be repeated

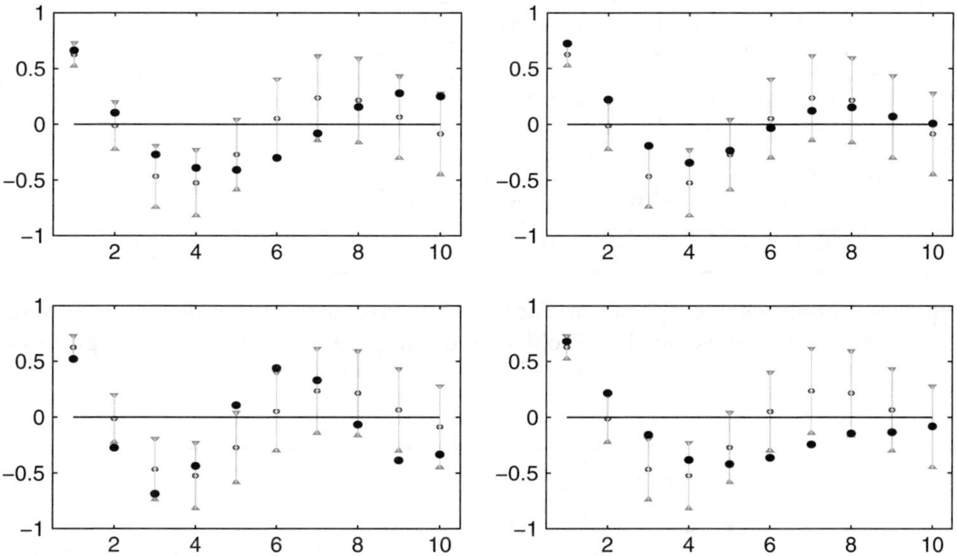

Figure 8.9 SACF of four simulated AR(2) time series with $T = 30$, $a_1 = 1.2$, and $a_2 = -0.8$ (big solid circles) with overlaid 90% bounds for the SACF based on the true parameters.

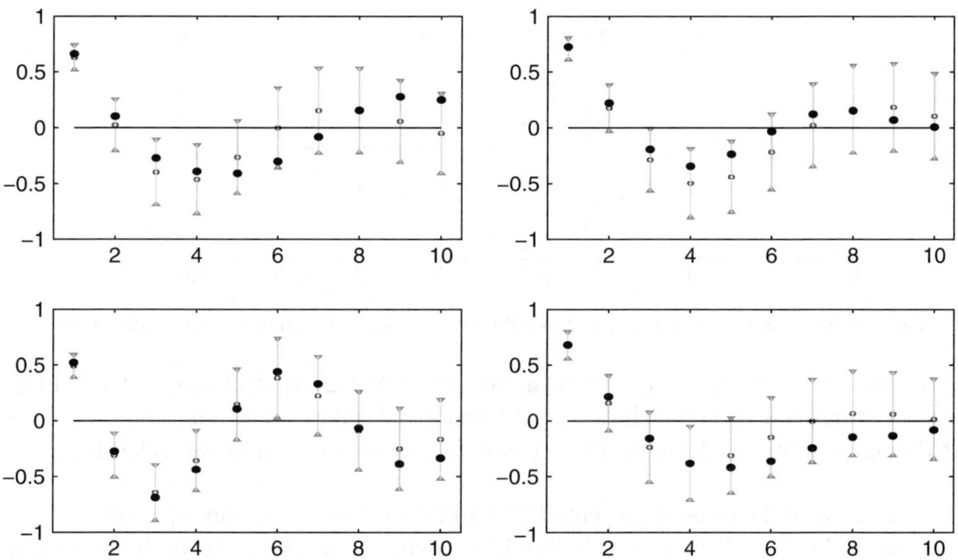

Figure 8.10 SACF of four simulated AR(2) time series with $T = 30$, $a_1 = 1.2$, and $a_2 = -0.8$ (big solid circles) with overlaid 90% bounds for the SACF based on the m.l.e. of the parameters.

```
a=[1.2 -0.8]; T=30; X=[ones(T,1)]; up=10; seed=1;
y=armasim(T,1,a,0,seed); param=armareg(y,X,2,0,1); ahat=param(2:3)
[mu,m2]=sacfmom(X,leeuwARMA(ahat,0,T),1:up); std=sqrt(m2-mu.^2);
lo=mu-norminv(0.95)*std; hi=mu+norminv(0.95)*std;
plot(1:up,mu,'ro',1:up,lo,'g^',1:up,hi,'gv','linewidth',2)
axis([0.5 up+0.5 -1 1])
ln=line([1:up ; 1:up],[lo hi]'); set(ln,'color',[1 1 0])
ln=line([1 up],[0 0]); set(ln,'linewidth',2), set(gca,'fontsize',20)
ss=sampleacf(y,up); hold on, hh=plot(1:up,ss,'ko');
set(hh,'linewidth',5), hold off
```

Program Listing 8.3: Computes a graph as in Figure 8.10. Program `leeuwARMA` is given in Listing 7.5, program `sacfmom` is developed in Problem 8.1., andprogram `sampleacf` is given in Listing 8.2.

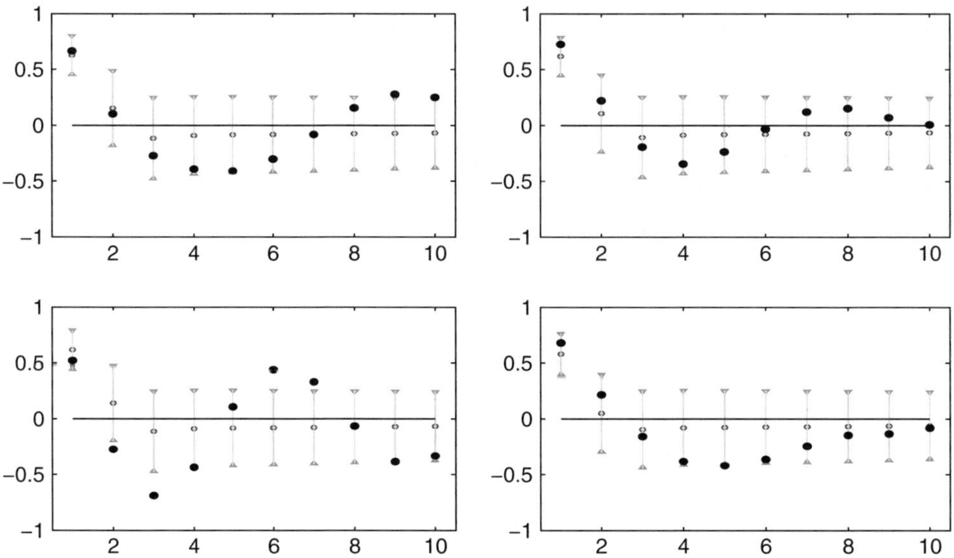

Figure 8.11 Same as Figure 8.10, using the same four simulated time series, but based on the *wrong model* MA(2).

with the same four series as before, but (incorrectly) assuming an MA(2) model instead of an AR(2). The results are shown in Figure 8.11. Notice how only the first $q = 2$ spikes are flexible, and the others are condemned to hover around zero. Now, the chance of $\widehat{R}_s$ falling outside its interval is clearly higher than 10%.

Of course, as an AR(2) model can be represented as an infinite MA, if we choose q larger, then the discrepancy between the $\widehat{R}_s$ and the intervals should decrease. Figure 8.12 shows the result but having used an MA(6) model. In this case, the intervals indeed "track" the $\widehat{R}_s$ better than with an MA(2) model, but outliers are still to be found.

Remark Moments of the R_s in the more general setting such that **Y** follows an elliptic distribution (as discussed in Section C.2) are derived in Kan and Wang (2010). ∎

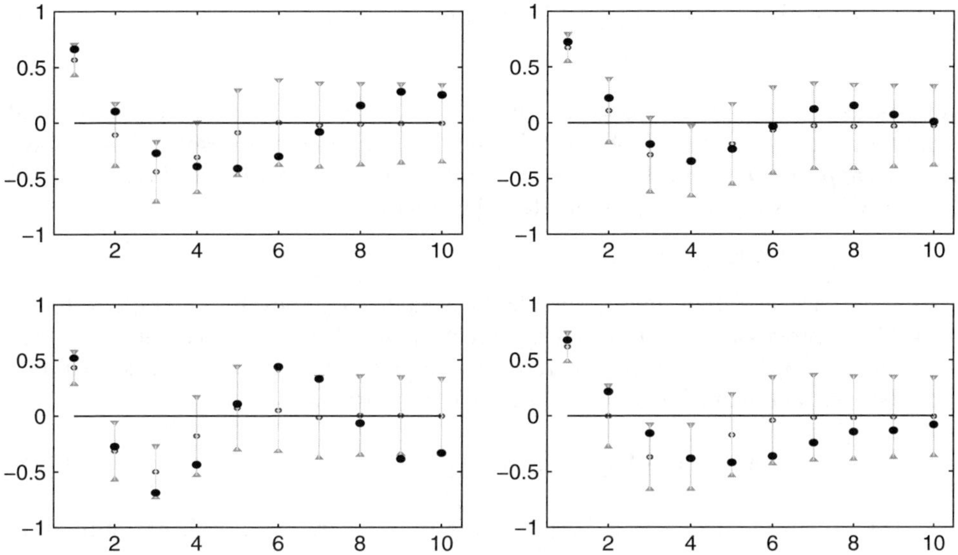

Figure 8.12 Same as Figures 8.10 and 8.11, using the same four simulated time series, but based on the *wrong model* MA(6).

8.1.3 Joint Distribution

This section is more advanced, and contains several results that are not fully derived here (albeit with references). The reader can skim or skip it. The results, however, are very useful, as shown in Section 9.5.

Considerably more challenging than computation of the distribution of the individual R_s is the *joint* distribution of the R_s, $s = 1, \dots, m$. For $\mathbf{R}_m = (R_1, \dots, R_m)'$, we detail its support, its asymptotic distribution, a saddlepoint approximation, and, based on the latter, an approximation to the conditional distribution of R_m given $(R_1, \dots, R_{m-1})'$. While interesting in its own right from a theoretical point of view, this conditional distribution can be used for identification of the AR lag length p; see Section 9.5.

8.1.3.1 Support

Recall from (8.5) that the autocorrelation function corresponding to a stationary time series is positive semi-definite. Recall also from the Remark in Section 6.1.1 on leading principle minors how use of their determinants can confirm the positive definiteness of a matrix. These imply that we can characterize the support of $\mathbf{R}_m = (R_1, \dots, R_m)'$ corresponding to a stationary time series as

$$\mathfrak{S}_m = \{\mathbf{r}_i = (r_1, r_2, \dots, r_i) : |\mathbf{S}_i| > 0, \quad i = 1, \dots, m\}, \tag{8.15}$$

where $\mathbf{S}_i$ is the $(i+1) \times (i+1)$ symmetric and persymmetric band matrix given by

$$\mathbf{S}_i = \begin{pmatrix} 1 & r_1 & \cdots & r_i \\ r_1 & 1 & \cdots & r_{i-1} \\ \vdots & & \ddots & \vdots \\ r_i & \cdots & r_1 & 1 \end{pmatrix}. \tag{8.16}$$

For $m = 1$, $|\mathbf{S}_1| = 1 - r_1^2$, so that $\mathfrak{F}_1 = \{r : |r_1| < 1\}$. For $m = 2$,

$$\mathfrak{F}_2 = \{(r_1, r_2) : -1 < r_1 < 1,\ 2r_1^2 - 1 < r_2 < 1\}. \tag{8.17}$$

A drawback of expression (8.15) is that it does not immediately provide bounds on the range of permissible values of a given R_m for $m > 1$. We use the following approach to resolve this. Let $\mathfrak{F}_m$ denote the *conditional support* of R_m *given* $\mathbf{R}_{m-1} = \mathbf{r}_{m-1}$. In other words, given observed values $\mathbf{r}_{m-1} = (r_1, \dots, r_{m-1})$ for the SACF, the support of R_m is an interval given by

$$\mathfrak{F}_m = \{r : -1 \leqslant r_{\min} < r < r_{\max} \leqslant 1\},$$

where values $r_{\min}$ and $r_{\max}$ are such that, for r_m outside these values, $\mathbf{r}_m$ does not correspond to the ACF of a stationary AR(m) process. Values $r_{\min}$ and $r_{\max}$ are now straightforward to determine based on the inequality constraint $|\mathbf{S}_m| > 0$ from (8.15). To do so, we need the following well-known fact from matrix algebra (see, e.g., Graybill, 1983, pp. 184–185): For square matrix

$$\mathbf{A} = \begin{bmatrix} \mathbf{A}_{11} & \mathbf{A}_{12} \\ \mathbf{A}_{21} & \mathbf{A}_{22} \end{bmatrix}, \qquad |\mathbf{A}| = |\mathbf{A}_{11}|\ |\mathbf{A}_{22} - \mathbf{A}_{21}\mathbf{A}_{11}^{-1}\mathbf{A}_{12}|. \tag{8.18}$$

From (8.16) and (8.18), $|\mathbf{S}_m| = |\mathbf{S}_{m-1}|(1 - \mathbf{v}_m'\mathbf{S}_{m-1}^{-1}\mathbf{v}_m)$, where $\mathbf{v}_m = (r_m, r_{m-1}, \dots, r_1)'$. Assuming that $\mathbf{r}_{m-1}$ lies in the support of the distribution of $\mathbf{S}_{m-1}$, $|\mathbf{S}_{m-1}| > 0$, and r_m ranges over

$$\{r_m : 1 - \mathbf{v}_m'\mathbf{S}_{m-1}^{-1}\mathbf{v}_m > 0\}. \tag{8.19}$$

Letting $\mathbf{S}_{m-1}^{-1} = \mathbf{W} = [w_{ij}]$, we have, with $v_1 = r_m$,

$$\mathbf{v}_m'\mathbf{W}\mathbf{v}_m = \sum_{i=1}^{m}\sum_{j=1}^{m} w_{ij}v_iv_j = r_m\left(w_{11}r_m + \sum_{j=2}^{m} w_{1j}v_j\right) + r_m\sum_{i=2}^{m} w_{i1}v_i + \sum_{i=2}^{m}\sum_{j=2}^{m} w_{ij}v_iv_j,$$

i.e., the conditional support of R_m is given by the solution to $Ar_m^2 + Br_m + C = 0$, where, using the symmetry of $\mathbf{W}$,

$$A = -w_{11}, \quad B = -2\sum_{j=2}^{m} w_{1j}v_j, \quad C = 1 - \sum_{i=2}^{m}\sum_{j=2}^{m} w_{ij}v_iv_j.$$

Problem 8.2 shows that this yields

$$2r_1^2 - 1 < r_2 < 1 \tag{8.20}$$

and

$$\frac{(r_1 + r_2)^2}{r_1 + 1} - 1 < r_3 < \frac{(r_1 - r_2)^2}{r_1 - 1} + 1. \tag{8.21}$$

A program to compute the interval of support of R_m given SACF values $(r_1, \dots, r_{m-1})'$ is given in Listing 8.4.

8.1.3.2 Asymptotic Distribution

Consider the SACF corresponding to a stationary, invertible ARMA(p, q) model with known mean (say zero). With $\boldsymbol{\rho} = (\rho_1, \dots, \rho_m)'$, Bartlett (1946) showed that, under certain conditions,

$$\sqrt{T}(\mathbf{R}_m - \boldsymbol{\rho}) \overset{\text{asy}}{\sim} \mathrm{N}(\mathbf{0}, \mathbf{W}), \quad \text{or} \quad \mathbf{R}_m \overset{\text{asy}}{\sim} \mathrm{N}(\boldsymbol{\rho}, T^{-1}\mathbf{W}), \tag{8.22}$$

```
function range=sacfrange(robs)
robs=reshape(robs,length(robs),1); m=length(robs)+1;
if m==1, lo=-1; hi=1;
elseif m==2, lo=2*robs^2-1; hi=1;
elseif m==3, r1=robs(1); r2=robs(2); lo=(r1+r2)^2/(r1+1)-1; hi=(r1-r2)^2/(r1-1)+1;
else
  W=inv(toeplitz([1 ; robs])); v=[-99 ; robs(end:-1:1)];
  A=-W(1,1); B=-2*W(1,2:end)*v(2:end); C=1-v(2:end)'*W(2:end,2:end)*v(2:end);
  wurzel=sort(roots([A B C])); lo=wurzel(1); hi=wurzel(2);
end
range=[lo hi];
```

Program Listing 8.4: Computes the conditional support of $\mathbf{R}_m$ given $R_1 = r_1, \ldots, R_{m-1} = r_{m-1}$, and robs is the vector of observed SACF values $\mathbf{r}_{m-1} = (r_1, r_2, \ldots, r_{m-1})'$.

(the latter expression being informal notation), where the (i,j) component of the matrix $\mathbf{W}$ is given by what is usually referred to as **Bartlett's formula,**

$$w_{ij} = \sum_{k=1}^{\infty} \{\rho_{k+i} + \rho_{k-i} - 2\rho_i\rho_k\} \times \{\rho_{k+j} + \rho_{k-j} - 2\rho_j\rho_k\}. \tag{8.23}$$

Proofs of Bartlett's formula are given in several texts; Priestley (1981, pp. 324–326) and Pollock (1999, pp. 670–671) offer particularly straightforward and instructive derivations, while Brockwell and Davis (1991) provide rigorous derivations under two sets of conditions. See also Anderson (1992) for further discussion of the required conditions for Bartlett's formula.

If the process is just white noise, then $\rho_i = 0$, $i = 1, 2, \ldots$, and, assuming existence of fourth moments, $\mathbf{W}$ reduces to the identity matrix. That is, in finite samples of white noise (with fourth moments), the R_s are approximately i.i.d. $\mathrm{N}(0, T^{-1})$. As white noise is the usual initial null hypothesis when analyzing a time series, it is common to plot individual 95% confidence intervals for each R_s. These are the dotted lines in Figures 8.4 and 8.5.

Example 8.1 Let ϵ_t follow the AR(1) process $\epsilon_t = a\epsilon_{t-1} + U_t$ with $|a| < 1$. Then ρ_i is given by $a^{|i|}$ and w_{ij} can be written as

$$\begin{aligned} w_{ij} &= \sum_{k=1}^{\infty} (a^{|k-i|} - a^{|k+i|})(a^{|k-j|} - a^{|k+j|}) \\ &= \sum_{k=1}^{m} (a^{|k-i|} - a^{|k+i|})(a^{|k-j|} - a^{|k+j|}) + \sum_{k=m+1}^{\infty} (a^{k-i} - a^{k+i})(a^{k-j} - a^{k+j}), \end{aligned}$$

where $m = \max(i,j)$. For R_1, $i = j = 1$ and

$$\begin{aligned} w_{11} &= (1-a^2)^2 + \sum_{k=2}^{\infty} (a^{k-1} - a^{k+1})^2 = (1-a^2)^2 + (a^{-2} - 2 + a^2) \sum_{k=2}^{\infty} a^{2k} \\ &= (1-a^2)^2 + (a^{-2} - 2 + a^2)\frac{a^4}{1-a^2} = (1-a^2)^2 + a^2(1-a^2) = 1 - a^2. \end{aligned}$$

Higher-order terms are similarly computed. For $(R_1, R_2, R_3)'$,

$$\mathbf{W} = (1-a^2)\begin{bmatrix} 1 & 2a & 3a^2 \\ 2a & 3a^2+1 & 2a(2a^2+1) \\ 3a^2 & 2a(2a^2+1) & 5a^4+3a^2+1 \end{bmatrix}.$$

When evaluated at the modest value of $a = 0.25$,

$$\mathbf{W} = \begin{bmatrix} 0.9375 & \cdot & \cdot \\ 0.4688 & 1.1133 & \cdot \\ 0.1758 & 0.5273 & 1.1316 \end{bmatrix} \quad \text{and} \quad \mathbf{W}_{\text{corr}} = \begin{bmatrix} 1 & \cdot & \cdot \\ 0.4588 & 1 & \cdot \\ 0.1707 & 0.4698 & 1 \end{bmatrix},$$

where $\mathbf{W}_{\text{corr}}$ denotes the correlation matrix. Observe that the correlation between adjacent spikes is quite high, and is independent of T. For the moderate value $a = 0.5$, the variances of the first three R_s based on its asymptotic distribution are $0.75/T$, $1.31/T$, and $1.55/T$, respectively, and the correlation between R_1 and R_2 is 0.76. ■

One way of evaluating Bartlett's formula (8.23) is by numerically summing terms until the desired degree of accuracy is obtained. More convenient methods for computing Bartlett's formula corresponding to a zero-mean stationary ARMA(p, q) process have been given by Kanto (1988) and Boshnakov (1996); we now summarize the results from Kanto (1988).[3] First buffer with zeros either polynomial $a(L)$ or $b(L)$ so that the model $a(L)y_t = b(L)\epsilon_t$ is an ARMA(m, m) with $m = \max(p, q)$, i.e., if $m > p$, then $a_i = 0$ for $i = p+1, p+2, \ldots, m$. Define $\mathbf{a} = (1, -a_1, \ldots, -a_m)'$, $\mathbf{b} = (1, b_1, \ldots, b_m)'$ and let the band-matrix operator Toep$(\mathbf{c}, \mathbf{r})$ denote the Toeplitz matrix with first column $\mathbf{c}$ and first row $\mathbf{r}$, and Hank$(\mathbf{c})$ denote the Hankel matrix with first row and column $\mathbf{c}$, for example,

$$\text{Toep}\left(\begin{bmatrix}1\\2\\3\end{bmatrix}, [4\ 5]\right) = \begin{bmatrix} 1 & 4 & 5 \\ 2 & 1 & 4 \\ 3 & 2 & 1 \end{bmatrix}, \qquad \text{Hank}\left(\begin{bmatrix}1\\2\\3\end{bmatrix}\right) = \begin{bmatrix} 1 & 2 & 3 \\ 2 & 3 & 0 \\ 3 & 0 & 0 \end{bmatrix}.$$

Then, similar to the results in Mittnik (1988), the first $m+1$ autocovariances can be expressed as

$$\boldsymbol{\gamma} = (\gamma_0, \gamma_1, \ldots, \gamma_m)' = \mathbf{C}^{-1}\mathbf{N}\mathbf{A}^{-1}\mathbf{b}\sigma^2, \tag{8.24}$$

where $\mathbf{A} = \text{Toep}(\mathbf{a}, \mathbf{0}_{1\times m})$, $\mathbf{N} = \text{Hank}(\mathbf{b})$, and $\mathbf{C}$ is given by $\mathbf{A} + \text{Hank}(\mathbf{a})$ but with the first column replaced by the first column of $\mathbf{A}$; higher-order autocovariances can be computed as $\gamma_l = \sum_{i=1}^{p} a_i \gamma_{l-i}$, $l \geqslant m+1$. Now define $\tilde{\boldsymbol{\gamma}} = \mathbf{C}_{\tilde{\mathbf{a}}}^{-1}\mathbf{N}_{\tilde{\mathbf{b}}}\mathbf{A}_{\tilde{\mathbf{a}}}^{-1}\tilde{\mathbf{b}}$, where $\tilde{\mathbf{a}} = \mathbf{D}\mathbf{a}$, $\tilde{\mathbf{b}} = \mathbf{E}\mathbf{b}$,

$$\mathbf{D} = \text{Toep}\left(\begin{bmatrix} -\mathbf{a} \\ \mathbf{0}_{m\times 1} \end{bmatrix}, \mathbf{0}_{1\times m}\right) = \begin{bmatrix} 1 & 0 & \cdots & & 0 \\ -a_1 & 1 & & & \vdots \\ \vdots & -a_1 & \ddots & & \\ & \vdots & \ddots & & 0 \\ -a_{m-1} & & & & 1 \\ -a_m & -a_{m-1} & & & -a_1 \\ 0 & -a_m & \ddots & & \vdots \\ \vdots & & \ddots & & \\ & & & & -a_{m-1} \\ 0 & & & 0 & -a_m \end{bmatrix},$$

3 In Kanto (1988), vector $\mathbf{a}$ should be defined as given herein, T in his equation (7) should be Γ, and matrices $\mathbf{D}$ and $\mathbf{E}$ are both of size $(2m+1) \times (m+1)$.

$\mathbf{E} = \text{Toep}\left(\begin{bmatrix}\mathbf{b}\\ \mathbf{0}_{m\times 1}\end{bmatrix}, \mathbf{0}_{1\times m}\right)$, $\mathbf{A}_{\tilde{\mathbf{a}}} = \text{Toep}(\tilde{\mathbf{a}}, \mathbf{0}_{1\times 2m})$, $\mathbf{N}_{\tilde{\mathbf{b}}} = \text{Hank}(\tilde{\mathbf{b}})$, and $\mathbf{C}_{\tilde{\mathbf{a}}}$ is given by $\mathbf{A}_{\tilde{\mathbf{a}}} + \text{Hank}(\tilde{\mathbf{a}})$, but with the first column replaced by the first column of $\mathbf{A}_{\tilde{\mathbf{a}}}$. The (ij)th element of $\mathbf{W}$ in (8.23) is then given by

$$w_{ij} = \frac{\tilde{\gamma}_{i-j} + \tilde{\gamma}_{i+j} - 2\rho_j\tilde{\gamma}_i - 2\rho_i\tilde{\gamma}_j + 2\rho_i\rho_j\tilde{\gamma}_0}{\gamma_0^2}, \tag{8.25}$$

where $\rho_i = \gamma_i/\gamma_0$ and $\tilde{\gamma}_{-i} = \tilde{\gamma}_i$. A program to compute $\boldsymbol{\rho}$ and $\mathbf{W}$ is given in Listing 8.5.

8.1.3.3 Small-Sample Joint Distribution Approximation

Assume for the moment that there are no regression effects and let $\epsilon \sim \text{N}(\mathbf{0}, \boldsymbol{\Omega}^{-1})$ with $\boldsymbol{\Omega}^{-1} > 0$. While no tractable exact expression for the p.d.f. of $\mathbf{R}_m$ appears to exist, a saddlepoint approximation is shown in Butler and Paolella (1998) to be given by

$$\hat{f}_{\mathbf{R}_m}(\mathbf{r}) = (2\pi)^{-\frac{m}{2}}|\boldsymbol{\Omega}|^{\frac{1}{2}}|\widehat{\mathbf{H}}_{\boldsymbol{\Omega}}|^{-\frac{1}{2}}|\widehat{\mathbf{P}}_{\boldsymbol{\Omega}}|^{-\frac{1}{2}}(\text{tr}\{\widehat{\mathbf{P}}_{\boldsymbol{\Omega}}^{-1}\})^m, \tag{8.26}$$

where $\mathbf{r} = (r_1, \dots, r_m)$,

$$\widehat{\mathbf{P}}_{\boldsymbol{\Omega}} = \widehat{\mathbf{P}}_{\boldsymbol{\Omega}}(\widehat{\mathbf{s}}) = \boldsymbol{\Omega} + 2\mathbf{r}'\widehat{\mathbf{s}}\ \mathbf{I}_T - 2\sum_{i=1}^{m}\widehat{s}_i\mathbf{A}_i, \tag{8.27}$$

and $\widehat{\mathbf{H}}_{\boldsymbol{\Omega}} = \widehat{\mathbf{H}}_{\boldsymbol{\Omega}}(\widehat{\mathbf{s}})$ with (ij)th element given by

$$\widehat{h}_{ij} = -\frac{1}{2}\frac{\partial^2}{\partial\widehat{s}_i\partial\widehat{s}_j}\log(|\widehat{\mathbf{P}}_{\boldsymbol{\Omega}}|) = 2\ \text{tr}\{\widehat{\mathbf{P}}_{\boldsymbol{\Omega}}^{-1}(\mathbf{A}_i - r_i\mathbf{I}_T)\widehat{\mathbf{P}}_{\boldsymbol{\Omega}}^{-1}(\mathbf{A}_j - r_j\mathbf{I}_T)\}, \tag{8.28}$$

$i, j = 1, \dots, m$. Saddlepoint vector $\widehat{\mathbf{s}} = (\widehat{s}_1, \dots, \widehat{s}_m)$ solves

$$0 = -\frac{1}{2}\frac{\partial}{\partial\widehat{s}_i}\log|\widehat{\mathbf{P}}_{\boldsymbol{\Omega}}| = \text{tr}\{\widehat{\mathbf{P}}_{\boldsymbol{\Omega}}^{-1}(\mathbf{A}_i - r_i\mathbf{I}_T)\}, \qquad i = 1, \dots, m, \tag{8.29}$$

```
function [rho,W]=kanto(a,b,dim)
% a=(a1,...,ap), b=(b1,...,bq), dim is size of W requested.
% EXAMPLE: for model y(t) = 1.2 y(t-1) -0.8 y(t-2) + e(t),  a=[1.2 -0.8];
p=length(a); q=length(b); a=-a;  % Kanto uses other sign convention for AR terms
m=max(dim,max(length(a),length(b))); aa=zeros(m,1); bb=zeros(m,1); aa(1:p)=a;
bb(1:q)=b; a=[1; aa]; b=[1; bb];
A=toeplitz(a,[1 zeros(1,m)]); psi=inv(A)*b; B=hankel(b); C=A+hankel(a);
C(:,1)=A(:,1); gamma=inv(C)*B*psi; rho=gamma/gamma(1);
D=toeplitz([a; zeros(m,1)] , [1 zeros(1,m)] );
E=toeplitz([b; zeros(m,1)] , [1 zeros(1,m)] );
atil=D*a; btil=E*b; A=toeplitz(atil,[atil(1) zeros(1,2*m)]);
psi=inv(A)*btil; B=hankel(btil); C=A+hankel(atil); C(:,1)=A(:,1);
gtil=inv(C)*B*psi;
for k=1:dim
  for l=1:dim
    W(k,l) = gtil(abs(k-l)+1) + gtil(k+l+1) - 2*rho(l+1)*gtil(k+1) ...
             - 2*rho(k+1)*gtil(l+1) + 2*rho(k+1)*rho(l+1)*gtil(1);
    W(k,l) = W(k,l) / (gamma(1))^2;
  end
end
```

Program Listing 8.5: Computes $\boldsymbol{\rho}$ via (8.24) and $\mathbf{W}$ via (8.25).

and, in general, needs to be numerically obtained. In the null setting for which $\mathbf{\Omega} = \mathbf{I}_T$, $\text{tr}\{\widehat{\mathbf{P}}_{\mathbf{I}}^{-1}\} = T$ so that the last factor in (8.26) is just T^m.

The extension of (8.26) for use with regression residuals based on (8.14), i.e., $\mathbf{Y} \sim \text{N}(\mathbf{X}\boldsymbol{\beta}, \mathbf{\Psi}^{-1})$, is not immediately possible because the covariance matrix of $\widehat{\boldsymbol{\epsilon}}$ is not full rank and a canonical reduction of the residual vector is required. As $\mathbf{M}$ is an orthogonal projection matrix, Theorem 1.3 showed that it can be expressed as $\mathbf{M} = \mathbf{G}'\mathbf{G}$, where $\mathbf{G}$ is $(T-k) \times T$ and such that $\mathbf{G}\mathbf{G}' = \mathbf{I}_{T-k}$ and $\mathbf{G}\mathbf{X} = \mathbf{0}$. Then

$$R_s = \frac{\widehat{\boldsymbol{\epsilon}}'\mathbf{A}_s\widehat{\boldsymbol{\epsilon}}}{\widehat{\boldsymbol{\epsilon}}'\widehat{\boldsymbol{\epsilon}}} = \frac{\boldsymbol{\epsilon}'\mathbf{M}\mathbf{A}_s\mathbf{M}\boldsymbol{\epsilon}}{\boldsymbol{\epsilon}'\mathbf{M}\boldsymbol{\epsilon}} = \frac{\boldsymbol{\epsilon}'\mathbf{G}'\mathbf{G}\mathbf{A}_s\mathbf{G}'\mathbf{G}\boldsymbol{\epsilon}}{\boldsymbol{\epsilon}'\mathbf{G}'\mathbf{G}\boldsymbol{\epsilon}} = \frac{\mathbf{w}'\tilde{\mathbf{A}}_s\mathbf{w}}{\mathbf{w}'\mathbf{w}}, \tag{8.30}$$

where $\mathbf{w} = \mathbf{G}\boldsymbol{\epsilon}$ and $\tilde{\mathbf{A}}_s = \mathbf{G}\mathbf{A}_s\mathbf{G}'$ is a $(T-k) \times (T-k)$ symmetric matrix.

By setting $\mathbf{\Omega}^{-1} = \mathbf{G}\mathbf{\Psi}^{-1}\mathbf{G}'$, approximation (8.26) becomes valid using $\mathbf{w} \sim \text{N}(\mathbf{0}, \mathbf{\Omega}^{-1})$ and $\mathbf{G}\mathbf{A}_s\mathbf{G}'$ in place of $\boldsymbol{\epsilon}$ and $\mathbf{A}_s$, respectively. Note that, in the null case with $\mathbf{Y} \sim \text{N}(\mathbf{X}\boldsymbol{\beta}, \sigma^2\mathbf{I}_T)$, $\mathbf{\Omega}^{-1} = \sigma^2\mathbf{I}_{T-k}$. A program to compute (8.26) based on regression residuals corresponding to regressor matrix $\mathbf{X}$ is given in Listing 8.6.

```
function [f,s,Pi,H]=sacfpdf(rvec,X,Psiinv,sstart)
global amat Omega r m n
r=rvec; m=length(r); r=reshape(r,m,1);
[T,k]=size(X); if k==0, T=length(Psiinv); G=eye(T); else, G=makeG(X); end
Omega=inv( G*Psiinv*(G') ); n=length(Omega);
amat=zeros(n,n,m); for i=1:m, amat(:,:,i) = G*makeA(T,i)*(G'); end
if nargin<4, sstart=zeros(m,1); end
options = optimset('Display','iter','Tolfun',1e-6,'MaxIter',10,...
          'LevenbergMarquardt','off','LargeScale','on');
[s,fval,exitflag]=fsolve(@spe,sstart,options);
if exitflag<=0, f=0; Pi=-1; return, end % det(Pi)<0. Signal failure to the caller
Pi=poi(s); H=makeH(Pi);
f = (2*pi)^(-m/2) * sqrt(det(Omega)) / sqrt(det(H)) * sqrt(det(Pi)) * (trace(Pi))^m;

function f=spe(s)
global amat r m n
for i=1:m, tt=poi(s) * (amat(:,:,i) - r(i)*eye(n)); f(i)=trace(tt); end

function Pi = poi(s) % Pi is inverse of matrix P
global amat Omega r m n
S=zeros(n,n); for i=1:m, S=S+s(i)*amat(:,:,i); end
Pi = inv( Omega + 2*r'*s*eye(n) - 2*S );

function H = makeH(Pi)
global amat r m n
for i=1:m, for j=i:m
  tt=Pi*(amat(:,:,i) - r(i)*eye(n))*Pi*(amat(:,:,j) - r(j)*eye(n));
  H(i,j) = 2*trace(tt); H(j,i) = H(i,j);
end, end
```

Program Listing 8.6: Computes the saddlepoint joint density $\hat{f}_{\mathbf{R}_m}(\mathbf{r})$, where $\mathbf{r}$ is passed as `rvec`, based on o.l.s. regression residuals from model (8.14), i.e., $\mathbf{Y} \sim \text{N}\left(\mathbf{X}\boldsymbol{\beta}, \mathbf{\Psi}^{-1}\right)$, with $\mathbf{\Psi}^{-1}$ passed as `Psiinv`. Function `makeG` is given in Listing 1.2 and `makeA` is given above in Listing 8.1.

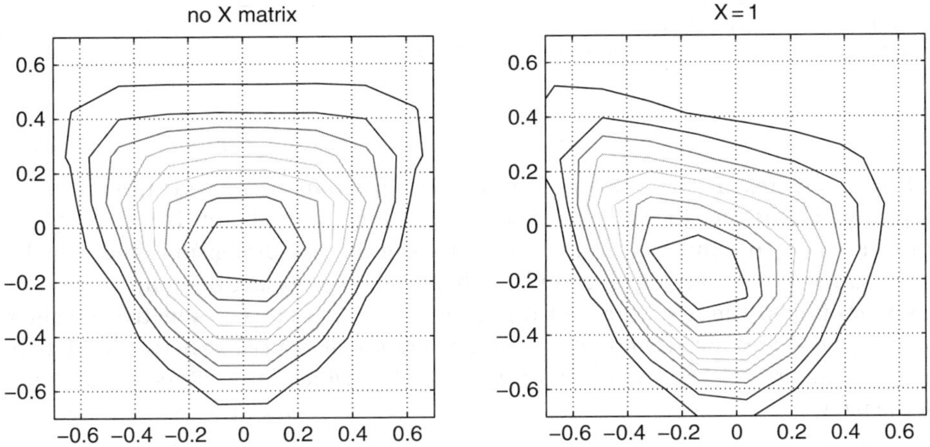

Figure 8.13 Simulated joint density (based on 500,000 replications) of R_1 (x-axis) and R_2 (y-axis) with $T = 10$ and $\Omega = \mathbf{I}_T$.

Remark The matrix $\mathbf{G}$ used in (8.30) is not unique and can be replaced, for example, by $\mathbf{GK}$, where $\mathbf{K}$ is a real $T \times T$ matrix such that $\mathbf{KK}' = \mathbf{I}_T$, $\mathbf{K}'\mathbf{MK} = \mathbf{M}$ and $\mathbf{GKX} = \mathbf{0}$. Another choice is the $(T - k) \times T$ matrix $\mathbf{C}$ associated with the recursive regression residuals, as discussed in Section 1.5. Recall that $\mathbf{CX} = \mathbf{0}$, $\mathbf{CC}' = \mathbf{I}_{T-k}$, and $\mathbf{C}'\mathbf{C} = \mathbf{M}$.

Certainly, the true distribution of $\mathbf{R}_m$ does not depend on whether $\mathbf{G}$ or $\mathbf{C}$ is used, but the saddlepoint density approximation might when applied in this context through the values of $\{\mathbf{A}_l\}$, $l = 1, \ldots, m$, and also through Ω in the non-null setting. Conveniently, use of $\mathbf{G}$ or $\mathbf{C}$ yields exactly the same density calculations, as was verified numerically for numerous cases but remains to be algebraically proven. ■

Example 8.2 To examine the accuracy of the approximation under the null setting of white noise, we compute $f_{\mathbf{R}_2}$ and $\widehat{f}_{\mathbf{R}_2}$, the joint distribution of R_1 and R_2, for $\mathbb{E}[Y_t]$ known (no regressor case) and $\mathbb{E}[Y_t]$ constant but unknown (so $\mathbf{X} = \mathbf{1}$). The very small sample size of $T = 10$ is used to ensure the non-normality of the distribution and thus provide a challenge to the quality of the approximation (recall that both $f_{\mathbf{R}_m}$ and $\widehat{f}_{\mathbf{R}_m}$ are asymptotically normal). Figure 8.13 shows a contour plot of the true joint density, obtained via simulation, with and without mean removal, using the Matlab code given in Listing 8.7.

In both cases, but particularly for the (almost always more relevant) case with mean removal, we see that the joint density deviates greatly from the asymptotic bivariate normal density. The joint saddlepoint densities (8.26) for these two cases are shown in the top panel of Figure 8.14, produced using the code in Listing 8.8. Comparing these to the plots in Figure 8.13, it is clear that the saddlepoint approximation is quite accurate. Figure 8.14 also shows the case with $T = 30$. It is closer to, though still deviates from, its asymptotic $T^{-1/2}\mathrm{N}(\mathbf{0}, \mathbf{I}_2)$ distribution. ■

Example 8.3 To illustrate the non-null setting, Figure 8.15 shows the s.p.a. (8.26) and the simulated p.d.f. $f_{\mathbf{R}_2}$ corresponding to an AR(1) process with $a_1 = 0.5$, sample size $T = 10$, and constant, unknown

```
up=500000; T=10; pair=zeros(up,2);
for i=1:up,   e=randn(T,1); pair(i,:)=sampleacf(e,2,1)'; end
[ heights, xycoord ] = hist3(pair,[10,10]);
contour(xycoord{1}, xycoord{2}, heights', 9)
grid, set(gca,'fontsize',14), axis([-0.7 0.7 -0.7 0.7])
```

Program Listing 8.7: Generates the left plot in Figure 8.13. Program `sampleacf` is given in Listing 8.2. Function `hist3` generates the values of a bivariate histogram (and plots it if no output is specified). The optional second argument is the number of bins of the two dimensions and defaults to 10 for both. The accuracy of the contour plot can be enhanced by increasing the number of simulations and the number of bins. The first output of `hist3` given by `heights` is a matrix of values proportional to the joint p.d.f., and the second output, `xycoord`, is called a *cell structure* in Matlab. In this case, it has two elements, `xycoord{1}` and `xycoord{2}`, which are vectors giving the r_1 and r_2 ordinates of the p.d.f.

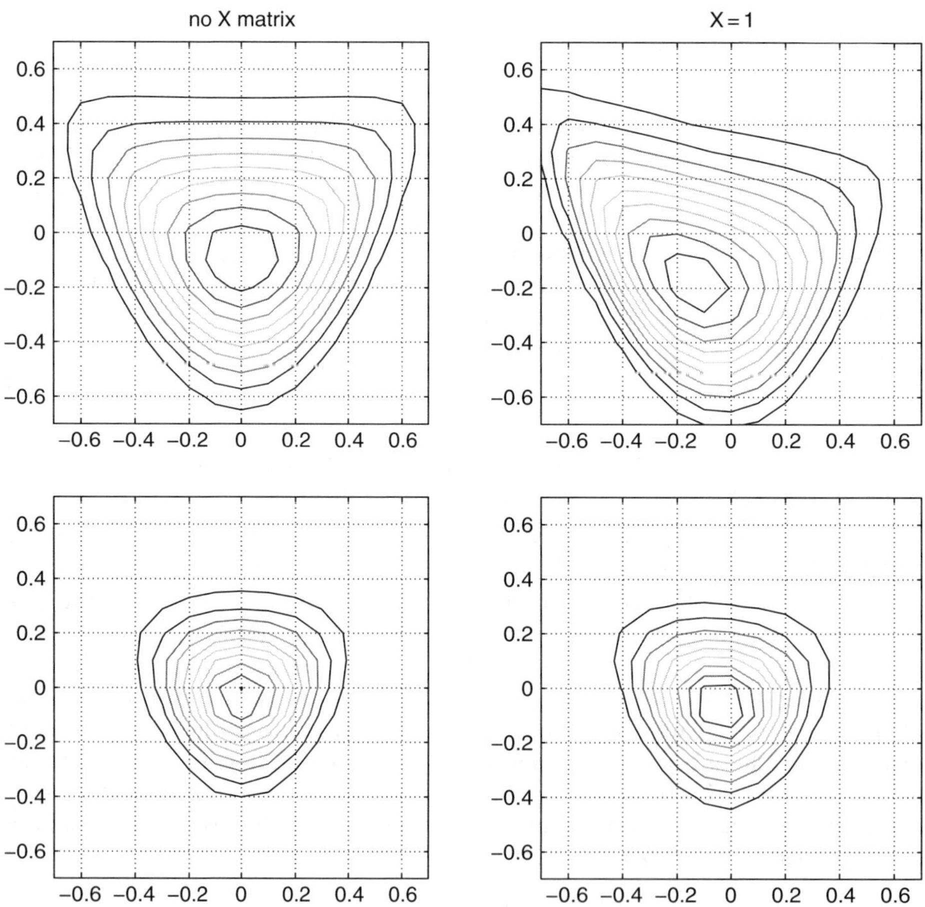

Figure 8.14 Saddlepoint density (8.26) of R_1 (x-axis) and R_2 (y-axis) for $\mathbf{\Omega} = \mathbf{I}_T$ and sample sizes $T = 10$ (top) and $T = 30$ (bottom).

```
T=30; X=ones(T,1); a1=0; a2=0; Psiinv=inv(leeuwAR([a1 a2],T));
c1=1; for r1=-0.7:0.1:0.7
  c2=1; for r2=-0.7:0.1:0.7
    rvec=[r1 r2]; c1c2=[c1 c2], f(c1,c2)=sacfpdf(rvec,X,Psiinv);
    c2=c2+1;
  end
  c1=c1+1;
end
contour([-0.7:0.1:0.7],[-0.7:0.1:0.7],f')
grid, set(gca,'fontsize',14), axis([-0.7 0.7 -0.7 0.7])
```

Program Listing 8.8: Code used to compute the plots in Figures 8.14 and 8.15. Program `sacfpdf` is given above in Listing 8.6 and `leeuwAR` is given in Listing 6.1 (recall that `leeuwAR` returns the *inverse* of the variance covariance matrix).

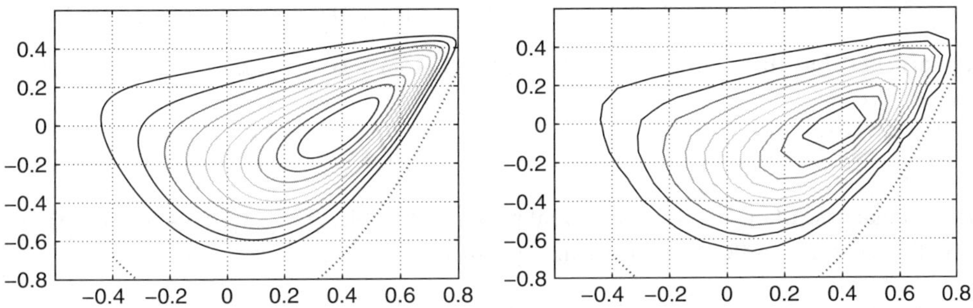

Figure 8.15 The left graph is saddlepoint density (8.26) of R_1 (x-axis) and R_2 (y-axis) for $T = 10$, $\mathbf{X} = \mathbf{1}$ and $\mathbf{\Omega}$ corresponding to an AR(1) model with $a_1 = 0.5$. The right graph is the corresponding density based on 1,000,000 simulations of the SACF (R_1, R_2). The code in Listing 8.7 was used for the simulation, but with simulated AR(1) time series instead of i.i.d. normal sequences, produced from the program `armasim` given in Listing 7.1. In both graphs, the dotted line to the right of the contour plot is the lower endpoint of the support of R_2, based on (8.20). It might happen in the simulated p.d.f. that the density appears to go beyond the allowed support, but this is just an artifact of Matlab's plotting procedure; executing `r1=pair(:,1); r2=pair(:,2); sum(r2 <= (2*r1.*r1-1))` yields precisely zero, i.e., all SACF pairs are in the support $\mathfrak{S}_2$ given in (8.17).

mean ($\mathbf{X} = \mathbf{1}$). As with the null setting, the s.p.a. density appears quite accurate, even for this very small sample size. ■

Example 8.4 Another way of judging the accuracy of the s.p.a. density approximation is via the Box and Pierce (1970) Q-statistic, given by $Q_m = TS$, where T is the sample size and $S = \sum_{i=1}^{m} R_i^2$. It can be shown that $Q_m \overset{\text{asy}}{\sim} \chi^2_m$ under the null of white noise.[4] The transformation from $\widehat{f}_{\mathbf{R}_{(m)}}$ to f_{Q_m} will involve an $(m-1)$-dimensional integration, which prohibits use of the calculation for even moderate m. For $m = 2$, we can proceed as follows.

First let X_1 and X_2 be continuous random variables with joint pdf f_{X_1,X_2}. To derive an expression for the density of $S = X_1^2 + X_2^2$, start by letting $S = X_1^2 + X_2^2$, $Z = X_1^2$, $X_1 = \pm\sqrt{Z}$, and $X_2 = \pm\sqrt{S - Z}$.

4 A small-sample improvement to the distribution of Q_m that is commonly used in practice was given by Ljung and Box (1978), known as as the Ljung–Box statistic.

Then, considering each of the four possible sign configurations on the x_i,

$$f_{S,Z}(s,z) = |\det \mathbf{J}| f_{X_1,X_2}(x_1,x_2) \mathbb{I}_{(-\infty,0)}(x_1) \mathbb{I}_{(-\infty,0)}(x_2) + \cdots$$

with

$$\mathbf{J} = \begin{pmatrix} \partial x_1/\partial z & \partial x_1/\partial s \\ \partial x_2/\partial z & \partial x_2/\partial s \end{pmatrix} = \begin{pmatrix} \pm\frac{1}{2}z^{-1/2} & \cdot \\ 0 & \pm\frac{1}{2}(s-z)^{-1/2} \end{pmatrix}$$

and, as z and $s-z$ are both positive, all $|\det \mathbf{J}|$ are the same, namely

$$|\det \mathbf{J}| = \frac{1}{4} z^{-1/2}(s-z)^{-1/2}.$$

Thus

$$f_{S,Z}(s,z) = \frac{1}{4} z^{-1/2}(s-z)^{-1/2} \times [f_{X_1,X_2}(-\sqrt{z}, -\sqrt{s-z}) + \cdots] \mathbb{I}_{(0,s)}(z), \tag{8.31}$$

where the term in brackets has the form $f(-,-) + f(-,+) + f(+,-) + f(+,+)$ and

$$f_S(s) = \int_0^s f_{S,Z}(s,z)\, \mathrm{d}z.$$

Remark For the special case of i.i.d. standard normal random variables, $f_{X_1,X_2}(x_1,x_2) = \exp\{-\frac{1}{2}(x_1^2 + x_2^2)\}/(2\pi)$, so that all four terms in (8.31) are the same, yielding

$$f_{S,Z}(s,z) = \frac{1}{4} z^{-1/2}(s-z)^{-1/2} \frac{4}{2\pi} \mathrm{e}^{-\frac{1}{2}(z+(s-z))} = \frac{1}{2\pi} z^{-1/2}(s-z)^{-1/2} \mathrm{e}^{-\frac{1}{2}s} \mathbb{I}_{(0,s)}(z).$$

Observe that, to compute

$$I = \int_0^s x^a (s-x)^b \, \mathrm{d}x, \qquad s \in (0,1), \quad a,b > 0,$$

use $u = 1 - x/s$ (so that $x = (1-u)s$ and $\mathrm{d}x = -s\,\mathrm{d}u$) to get

$$\begin{aligned} I &= \int_0^s x^a (s-x)^b \, \mathrm{d}x \\ &= -s \int_1^0 ((1-u)s)^a (s-(1-u)s)^b \, \mathrm{d}u = s^{a+b+1} \int_0^1 (1-u)^a u^b \, \mathrm{d}u \\ &= s^{a+b+1} B(b+1, a+1), \end{aligned} \tag{8.32}$$

Then,

$$f_S(s) = \frac{1}{2\pi} \mathrm{e}^{-\frac{1}{2}s} \int_0^s z^{-1/2}(s-z)^{-1/2} \, \mathrm{d}z = \frac{1}{2\pi} \mathrm{e}^{-\frac{1}{2}s} B\left(\frac{1}{2}, \frac{1}{2}\right) = \frac{1}{2} \mathrm{e}^{-\frac{1}{2}s} \mathbb{I}_{(0,\infty)}(s),$$

so that $S \sim \mathrm{Exp}(1/2)$ and also $S \sim \chi^2_2$. ■

In our case here, $f_S(s) = \int_0^{\min(s,1)} f_{S,Y_1}(s, y_1)\, \mathrm{d}y_1$, where

$$\widehat{f}_{S,Y_1}(s,y_1) = \frac{1}{4} y_1^{-1/2}(s-y_1)^{-1/2} [\widehat{f}_{R_1,R_2}(-\sqrt{y_1}, -\sqrt{s-y_1}) + \cdots], \tag{8.33}$$

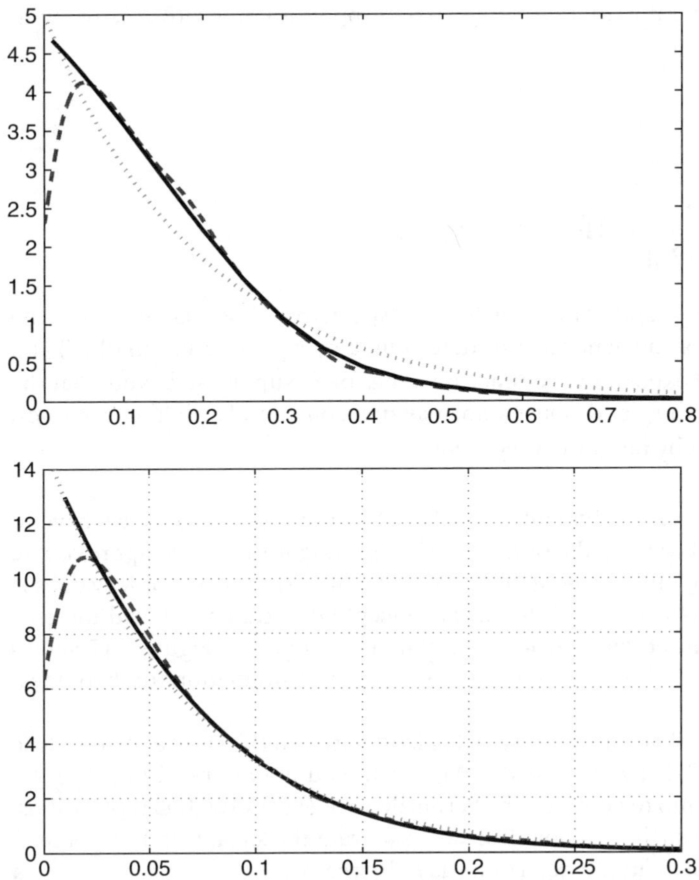

Figure 8.16 Density of $S = R_1^2 + R_2^2$ based on (8.33) (solid), simulation (dashed), and asymptotic (dotted). Top panel is for $T = 10$; bottom is for $T = 30$. The decline of the kernel density in the left tail is an artifact of the method; further simulation reveals that the true density indeed increases upwards.

and the term in square brackets is the sum of the f_{R_1,R_2} evaluated at the four sign combinations of the arguments.

Figure 8.16 shows the density in (8.33) for $\mathbf{X} = \mathbf{1}, \mathbf{\Omega} = \mathbf{I}$ and the two sample sizes $T = 10$ and $T = 30$, overlaid with a kernel density estimate computed from 5,000 simulated replications, as well as the asymptotic distribution. Even for $T = 10$, the agreement between the saddlepoint approximation and the simulated density is extremely high, further indicating the accuracy of (8.26). Nevertheless, for the more practical sample size of $T = 30$, the asymptotic distribution is quite accurate as well. ■

8.1.4 Conditional Distribution Approximation

Interest in this section centers on the distribution of the scalar random variable R_m given $\mathbf{R}_{m-1} = \mathbf{r}_{m-1}$, where $\mathbf{R}_{m-1} = (R_1, \ldots, R_{m-1})'$ and $\mathbf{r}_{m-1} = (r_1, \ldots, r_{m-1})'$. Following the methodology developed in Barndorff-Nielsen and Cox (1979), Butler and Paolella (1998) derive a conditional double saddlepoint density computed as the ratio of two single p.d.f. approximations (8.26), or, with $\mathbf{r}_m = [\mathbf{r}'_{m-1}\ r_m]'$,

$$\hat{f}_{R_m|\mathbf{R}_{m-1}}(r_m \mid \mathbf{r}_{m-1}) = \frac{\hat{f}_{\mathbf{R}_m}(\mathbf{r}_m)}{\hat{f}_{\mathbf{R}_{m-1}}(\mathbf{r}_{m-1})} = \sqrt{\frac{|\hat{\mathbf{H}}_{m-1}|\ |\hat{\mathbf{P}}_{m-1}|}{2\pi|\hat{\mathbf{H}}_m|\ |\hat{\mathbf{P}}_m|}}(\mathrm{tr}\{\hat{\mathbf{P}}_m^{-1}\})^m(\mathrm{tr}\{\hat{\mathbf{P}}_{m-1}^{-1}\})^{1-m}, \tag{8.34}$$

where $\hat{\mathbf{P}}_{m-1}$ and $\hat{\mathbf{H}}_{m-1}$ are the $\hat{\mathbf{P}}_{\Omega}$ and $\hat{\mathbf{H}}_{\Omega}$ matrices, respectively, associated with the $(m-1)$-dimensional saddlepoint $\hat{\mathbf{s}}_{m-1}$ of the denominator determined by $\mathbf{r}_{m-1}$, as given in (8.27) and (8.28), and likewise for $\hat{\mathbf{P}}_m$ and $\hat{\mathbf{H}}_m$, and explicit dependence on $\boldsymbol{\Omega}$ has been suppressed. Note that the denominator in (8.34) does not vary with r_m and is just a normalizing constant. For higher accuracy, it could be replaced by $\int \hat{f}_{\mathbf{R}_m}$, computed by numeric integration.

Example 8.5 The accuracy of (8.34) is more difficult to verify with simulation because we have to condition on a measure-zero quantity. For the p.d.f. of $R_2 \mid R_1$, this is straightforward to approximate by simulating pairs (R_1, R_2) and keeping only those such that $R_1 \approx r_1$; the resulting set of R_2 values is a sample from the distribution of $R_2 \mid (R_1 = r_1)$. While this is conceptually straightforward for general R_m, the amount of simulation required will become prohibitive as m grows. Figure 8.17 shows $f_{R_2|R_1}(r_2 \mid r_1)$ for $r_1 = 2/3$ (top) and $r_1 = 1/3$ (bottom) using this method of simulation (dashed lines) and the s.p.a. (8.34) (solid lines).

Observe how the simulated conditional density involves a tradeoff between the choice of how close $R_1 \approx r_1$ (we used the interval $(2/3 - \epsilon, 2/3 + \epsilon)$ for $\epsilon = 0.0005$) and the number of resulting observations of R_2 that are input into the algorithm to compute the kernel density. (From 100,000 replications, 325 were contained in the interval). As such, it is not clear how accurate the s.p.a. is in this context, and, as mentioned, it will be more problematic to determine as m increases. ■

We now turn to the conditional c.d.f. for which we define

$$\tau_1 = \widehat{\Pr}(R_1 < r_1) \quad \text{and} \quad \tau_m = \widehat{\Pr}(R_m < r_m \mid \mathbf{R}_{m-1} = \mathbf{r}_{m-1}), \quad m > 1, \tag{8.35}$$

where $\widehat{\Pr}$ anticipates that these quantities can only be approximated. When $m = 1$, τ_1 can be computed as discussed in Section 8.1.2. For $m > 1$, the p.d.f. (8.34) can be integrated (and normalized for higher accuracy) to give the approximation

$$\tau_m = \frac{\int_{\mathfrak{I}'_m} \hat{f}_{R_m|\mathbf{R}_{m-1}}(x \mid \mathbf{r}_{m-1})\, dx}{\int_{\mathfrak{I}_m} \hat{f}_{R_m|\mathbf{R}_{m-1}}(x \mid \mathbf{r}_{m-1})\, dx},$$

where $\mathfrak{I}'_m = \mathfrak{I}_m \cap (-1, r_m]$ and $\mathfrak{I}_m$ is the conditional support of R_m given $\mathbf{R}_{m-1} = \mathbf{r}_{m-1}$, as detailed in Section 8.1.3.1. Using this, the denominator in (8.34) cancels, so that

$$\tau_m = \frac{\int_{\mathfrak{I}'_m} \hat{f}_{\mathbf{R}_m}(\mathbf{r}_m)\, dx}{\int_{\mathfrak{I}_m} \hat{f}_{\mathbf{R}_m}(\mathbf{r}_m)\, dx}, \quad \mathbf{r}_m = [\mathbf{r}'_{m-1}\ x]'. \tag{8.36}$$

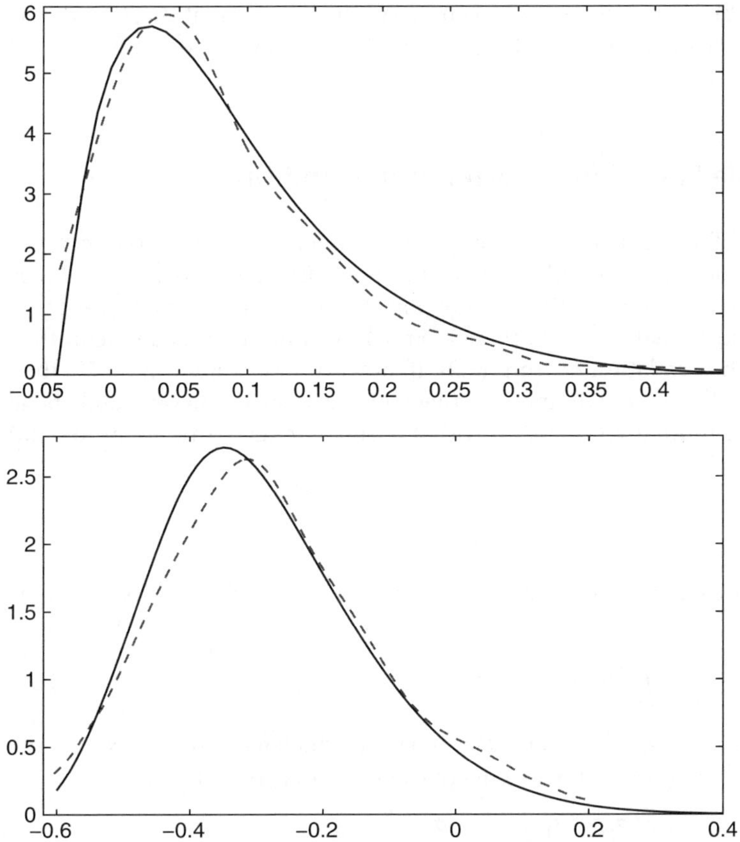

Figure 8.17 The conditional p.d.f. of R_2 given $R_1 = 2/3$ (top) and $R_1 = 1/3$ (bottom) for an AR(2) model with constant, unknown mean ($\mathbf{X} = \mathbf{1}$) and parameters $a_1 = 1.2$ and $a_2 = -0.8$. The solid line is the s.p.a. (8.34) and the dashed line is based on simulation. The Matlab code to generate the plots is developed in Problem 8.4.

While computation of (8.36) is straightforward, it is possible to derive an approximation to the integral similar in spirit to the Lugannani and Rice (1980) saddlepoint approximation to the c.d.f. of a univariate random variable. Using the method of Temme (1982) and Barndorff-Nielsen and Cox (1989, Sec. 3.9), a double saddlepoint c.d.f. approximation is shown in Butler and Paolella (1998) to be

$$\tau_m = \Phi(w_0) + \phi(w_0)\left(\frac{1}{w_0} - \frac{1}{v_0}\right), \tag{8.37}$$

for $\hat{s}_m \neq 0$, where, as usual, Φ and ϕ denote the c.d.f. and p.d.f. of the standard normal distribution, respectively, and

$$\begin{aligned} w_0 &= \operatorname{sgn}(\hat{s}_m)\sqrt{\log(|\hat{\mathbf{P}}_m|/|\hat{\mathbf{P}}_{m-1}|)}, \\ v_0 &= \hat{s}_m(|\hat{\mathbf{H}}_m| \;/\; |\hat{\mathbf{H}}_{m-1}|)^{\frac{1}{2}}[\operatorname{tr}(\hat{\mathbf{P}}_{m-1}^{-1}) \;/\; \operatorname{tr}(\hat{\mathbf{P}}_m^{-1})]^{m-1}. \end{aligned} \tag{8.38}$$

See Butler (2007, Sec. 2.3.2) for further discussion on the accuracy of this approximation. This will be used in Section 9.5 to develop a powerful method of selecting the autoregressive lag order for a given time series and **X** matrix.

8.2 Theoretical and Sample Partial Autocorrelation Function

Recall that the *s*th element of the ACF, ρ_s, measures the correlation between Y_t and Y_{t-s}. The measure is unconditional in the sense that it does not condition on any random variables, in particular, those lying between Y_t and Y_{t-s}. Take the zero mean AR(1) model, for example, with $Y_t = aY_{t-1} + U_t$: It is clear from the construction of the Y_t that, if $a \neq 0$, then Y_1 and Y_3 will not be independent; they are jointly normally distributed with correlation a^2 from (8.2). If, however, we condition on Y_2, then the *conditional* correlation between Y_1 and Y_3 will indicate their (linear) association over and above the association resulting from their mutual relationship with Y_2. This is referred to as the **partial autocorrelation** at lag $s = 2$.

8.2.1 Partial Correlation

Recall that **Y** is an (*n*-variate, non-degenerate) multivariate normal random variable if its density is given by

$$f_{\mathbf{Y}}(\mathbf{y}) = \frac{1}{|\boldsymbol{\Sigma}|^{1/2}(2\pi)^{n/2}} \exp\left\{-\frac{1}{2}((\mathbf{y}-\boldsymbol{\mu})'\boldsymbol{\Sigma}^{-1}(\mathbf{y}-\boldsymbol{\mu}))\right\}, \tag{8.39}$$

denoted $\mathbf{Y} \sim \mathrm{N}(\boldsymbol{\mu}, \boldsymbol{\Sigma})$, where $\boldsymbol{\mu} = (\mu_1, \dots, \mu_n)' \in \mathbb{R}^n$ and $\boldsymbol{\Sigma} > 0$ with (ij)th element σ_{ij}, $\sigma_i^2 := \sigma_{ii}$. The mean is $\boldsymbol{\mu} = \mathbb{E}[\mathbf{X}] := \mathbb{E}[(X_1, \dots, X_n)']$, and the variance covariance matrix is given by

$$\boldsymbol{\Sigma} = \mathbb{V}(\mathbf{X}) := \mathbb{E}[(\mathbf{X}-\boldsymbol{\mu})(\mathbf{X}-\boldsymbol{\mu})'] = \begin{bmatrix} \sigma_1^2 & \sigma_{12} & \cdots & \sigma_{1n} \\ \sigma_{21} & \sigma_2^2 & & \sigma_{2n} \\ \vdots & & \ddots & \vdots \\ \sigma_{n1} & \sigma_{n2} & & \sigma_n^2 \end{bmatrix}.$$

With $\mathbf{A} \in \mathbb{R}^{m \times n}$ a full rank matrix with $m \leqslant n$, the set of linear combinations

$$\mathbf{L} = (L_1, \dots, L_m)' = \mathbf{A}\mathbf{Y} \sim \mathrm{N}(\mathbf{A}\boldsymbol{\mu}, \mathbf{A}\boldsymbol{\Sigma}\mathbf{A}'),$$

using the fact that $\mathbb{V}(\mathbf{A}\mathbf{X} + \mathbf{b}) = \mathbf{A}\boldsymbol{\Sigma}\mathbf{A}'$, and $\mathbf{A}\boldsymbol{\Sigma}\mathbf{A}' > 0$.

Now suppose that $\mathbf{Y} = (Y_1, \dots, Y_n)' \sim \mathrm{N}(\boldsymbol{\mu}, \boldsymbol{\Sigma})$ is partitioned into two subvectors $\mathbf{Y} = (\mathbf{Y}'_{(1)}, \mathbf{Y}'_{(2)})'$, where $\mathbf{Y}_{(1)} = (Y_1, \dots, Y_p)'$ and $\mathbf{Y}_{(2)} = (Y_{p+1}, \dots, Y_n)'$ with $\boldsymbol{\mu}$ and $\boldsymbol{\Sigma}$ partitioned accordingly such that $\mathbb{E}[\mathbf{Y}_{(i)}] = \boldsymbol{\mu}_{(i)}$, $\mathbb{V}(\mathbf{Y}_{(i)}) = \boldsymbol{\Sigma}_{ii}$, $i = 1, 2$, and $\mathrm{Cov}(\mathbf{Y}_{(1)}, \mathbf{Y}_{(2)}) = \boldsymbol{\Sigma}_{12}$, i.e., $\boldsymbol{\mu} = (\boldsymbol{\mu}'_{(1)}, \boldsymbol{\mu}'_{(2)})'$ and

$$\boldsymbol{\Sigma} = \begin{bmatrix} \boldsymbol{\Sigma}_{11} & \vdots & \boldsymbol{\Sigma}_{12} \\ \cdots & \cdots & \cdots \\ \boldsymbol{\Sigma}_{21} & \vdots & \boldsymbol{\Sigma}_{22} \end{bmatrix}, \quad \boldsymbol{\Sigma}_{21} = \boldsymbol{\Sigma}'_{12}.$$

Using the previous partition notation, two very important properties of the multivariate normal distribution are as follows.

1. $\mathbf{Y}_{(1)}$ and $\mathbf{Y}_{(2)}$ are independent iff $\boldsymbol{\Sigma}_{12} = \mathbf{0}$, i.e., zero correlation implies independence.

2. The conditional distribution of $\mathbf{Y}_{(1)}$ given $\mathbf{Y}_{(2)}$ is normal. In particular, if $\mathbf{\Sigma}_{22} > 0$ (which is true if $\mathbf{\Sigma} > 0$), then

$$(\mathbf{Y}_{(1)} \mid \mathbf{Y}_{(2)} = \mathbf{y}_{(2)}) \sim \mathrm{N}\left(\boldsymbol{\mu}_{(1)} + \mathbf{\Sigma}_{12}\mathbf{\Sigma}_{22}^{-1}(\mathbf{y}_{(2)} - \boldsymbol{\mu}_{(2)}),\ \mathbf{\Sigma}_{11} - \mathbf{\Sigma}_{12}\mathbf{\Sigma}_{22}^{-1}\mathbf{\Sigma}_{21}\right). \tag{8.40}$$

Example 8.6 Let

$$\mathbf{Y} = \begin{bmatrix} Y_1 \\ Y_2 \\ Y_3 \end{bmatrix} \sim \mathrm{N}(\boldsymbol{\mu}, \mathbf{\Sigma}), \quad \boldsymbol{\mu} = \begin{bmatrix} 2 \\ 1 \\ 0 \end{bmatrix}, \quad \mathbf{\Sigma} = \begin{bmatrix} 2 & 1 & 1 \\ 1 & 3 & 0 \\ 1 & 0 & 1 \end{bmatrix}.$$

Because $\det(\mathbf{\Sigma}) = 2 \neq 0$, $\mathbf{Y}$ is not degenerate. To derive the distribution of $Y_2 \mid (Y_1, Y_3)$, first rewrite the density as

$$\begin{bmatrix} Y_2 \\ Y_1 \\ Y_3 \end{bmatrix} \sim \mathrm{N}\left(\begin{bmatrix} \mu_{(1)} \\ \boldsymbol{\mu}_{(2)} \end{bmatrix}, \begin{bmatrix} \Sigma_{11} & \mathbf{\Sigma}_{12} \\ \mathbf{\Sigma}_{21} & \mathbf{\Sigma}_{22} \end{bmatrix} \right),$$

where $\mu_{(1)}$ and Σ_{11} are scalars, with

$$\begin{bmatrix} \mu_{(1)} \\ \boldsymbol{\mu}_{(2)} \end{bmatrix} = \begin{bmatrix} 1 \\ \cdots \\ 2 \\ 0 \end{bmatrix}, \quad \begin{bmatrix} \Sigma_{11} & \mathbf{\Sigma}_{12} \\ \mathbf{\Sigma}_{21} & \mathbf{\Sigma}_{22} \end{bmatrix} = \begin{bmatrix} 3 & 1 & 0 \\ 1 & 2 & 1 \\ 0 & 1 & 1 \end{bmatrix}.$$

Then, from (8.40),

$$Y_2 \mid (Y_1, Y_3) \sim \mathrm{N}\left(\mu_{(1)} + \mathbf{\Sigma}_{12}\mathbf{\Sigma}_{22}^{-1}(\mathbf{y}_{(2)} - \boldsymbol{\mu}_{(2)}),\ \Sigma_{11} - \mathbf{\Sigma}_{12}\mathbf{\Sigma}_{22}^{-1}\mathbf{\Sigma}_{21}\right),$$

i.e., substituting and simplifying,

$$\begin{aligned} \mathbb{E}[Y_2 \mid (Y_1, Y_3)] &= \mu_{(1)} + \mathbf{\Sigma}_{12}\mathbf{\Sigma}_{22}^{-1}(\mathbf{y}_{(2)} - \boldsymbol{\mu}_{(2)}) \\ &= 1 + \begin{bmatrix} 1 & 0 \end{bmatrix} \begin{bmatrix} 2 & 1 \\ 1 & 1 \end{bmatrix}^{-1} \left(\begin{bmatrix} y_1 \\ y_3 \end{bmatrix} - \begin{bmatrix} 2 \\ 0 \end{bmatrix} \right) = y_1 - y_3 - 1 \end{aligned}$$

and

$$\begin{aligned} \mathbb{V}(Y_2 \mid (Y_1, Y_3)) &= \Sigma_{11} - \mathbf{\Sigma}_{12}\mathbf{\Sigma}_{22}^{-1}\mathbf{\Sigma}_{21} \\ &= 3 - \begin{bmatrix} 1 & 0 \end{bmatrix} \begin{bmatrix} 2 & 1 \\ 1 & 1 \end{bmatrix}^{-1} \begin{bmatrix} 1 \\ 0 \end{bmatrix} = 2, \end{aligned}$$

so that $Y_2 \mid (Y_1, Y_3) \sim \mathrm{N}(y_1 - y_3 - 1,\ 2)$. ■

Let $\mathbf{Y} = (Y_1, Y_2, \ldots, Y_n)' \sim \mathrm{N}(\boldsymbol{\mu}, \mathbf{\Sigma})$, with $\mathbf{\Sigma} > 0$ and, as usual, the (ij)th element of $\mathbf{\Sigma}$ denoted by σ_{ij}. Let indices i and j be such that $1 \leqslant i < j \leqslant n$. Let $\mathbf{Y}_{(1)} = (Y_i, Y_j)'$ and $\mathbf{Y}_{(2)} = \mathbf{Y} \backslash \mathbf{Y}_{(1)}$, i.e., $\mathbf{Y}_{(2)}$ is $\mathbf{Y}$ but with the elements Y_i and Y_j removed. Let $\mathbf{\Sigma}_{11} = \mathbb{V}(\mathbf{Y}_{(1)})$, $\mathbf{\Sigma}_{22} = \mathbb{V}(\mathbf{Y}_{(2)})$, and $\mathbf{\Sigma}_{12} = \mathbf{\Sigma}_{21}' = \mathrm{Cov}(\mathbf{Y}_{(1)}, \mathbf{Y}_{(2)})$, so that, with $\mathbf{Y}^* = (\mathbf{Y}_{(1)}', \mathbf{Y}_{(2)}')' = (Y_i, Y_j, \mathbf{Y}_{(2)}')'$,

$$\mathbb{V}(\mathbf{Y}^*) = \begin{bmatrix} \mathbf{\Sigma}_{11} & \mathbf{\Sigma}_{12} \\ \mathbf{\Sigma}_{21} & \mathbf{\Sigma}_{22} \end{bmatrix}.$$

Using (8.40), let $\mathbf{C}$ be the 2×2 conditional covariance matrix given by

$$\mathbf{C} = \begin{bmatrix} \sigma_{ii|\mathbf{Y}_{(2)}} & \sigma_{ij|\mathbf{Y}_{(2)}} \\ \sigma_{ji|\mathbf{Y}_{(2)}} & \sigma_{jj|\mathbf{Y}_{(2)}} \end{bmatrix} = \mathbf{\Sigma}_{11} - \mathbf{\Sigma}_{12}\mathbf{\Sigma}_{22}^{-1}\mathbf{\Sigma}_{21} \, .$$

The *partial correlation of Y_i and Y_j, given* $\mathbf{Y}_{(2)}$, is defined by

$$\rho_{ij|\mathbf{Y}_{(2)}} = \rho_{ij|(\{1,2,\dots,n\}\backslash\{i,j\})} = \frac{\sigma_{ij|\mathbf{Y}_{(2)}}}{\sqrt{\sigma_{ii|\mathbf{Y}_{(2)}}\sigma_{jj|\mathbf{Y}_{(2)}}}} = \frac{\sigma_{ij|\mathbf{Y}_{(2)}}}{\sqrt{\sigma^2_{i|\mathbf{Y}_{(2)}}\sigma^2_{j|\mathbf{Y}_{(2)}}}}. \tag{8.41}$$

Example 8.7 (Example 8.6 cont.) To compute $\rho_{13|2}$, first write

$$\begin{bmatrix} Y_1 \\ Y_3 \\ Y_2 \end{bmatrix} \sim \mathrm{N}\left(\begin{bmatrix} \boldsymbol{\mu}_{(1)} \\ \mu_{(2)} \end{bmatrix}, \begin{bmatrix} \mathbf{\Sigma}_{11} & \mathbf{\Sigma}_{12} \\ \mathbf{\Sigma}_{21} & \Sigma_{22} \end{bmatrix} \right),$$

where

$$\begin{bmatrix} \boldsymbol{\mu}_{(1)} \\ \mu_{(2)} \end{bmatrix} := \mathbb{E}\begin{bmatrix} Y_1 \\ Y_3 \\ \cdots \\ Y_2 \end{bmatrix} = \begin{bmatrix} \mu_1 \\ \mu_3 \\ \cdots \\ \mu_2 \end{bmatrix} = \begin{bmatrix} 2 \\ 0 \\ \cdots \\ 1 \end{bmatrix}$$

and

$$\begin{bmatrix} \mathbf{\Sigma}_{11} & \mathbf{\Sigma}_{12} \\ \mathbf{\Sigma}_{21} & \Sigma_{22} \end{bmatrix} := \mathbb{V}\left(\begin{bmatrix} Y_1 \\ Y_3 \\ \cdots \\ Y_2 \end{bmatrix} \right) = \begin{bmatrix} \sigma_{11} & \sigma_{13} & \sigma_{12} \\ \sigma_{31} & \sigma_{33} & \sigma_{32} \\ \sigma_{21} & \sigma_{23} & \sigma_{22} \end{bmatrix} = \begin{bmatrix} 2 & 1 & 1 \\ 1 & 1 & 0 \\ 1 & 0 & 3 \end{bmatrix},$$

so that

$$\mathbf{C} = \mathbf{\Sigma}_{11} - \mathbf{\Sigma}_{12}\mathbf{\Sigma}_{22}^{-1}\mathbf{\Sigma}_{21} = \begin{bmatrix} 2 & 1 \\ 1 & 1 \end{bmatrix} - \begin{bmatrix} 1 \\ 0 \end{bmatrix} [\,3\,]^{-1} [\,1 \quad 0\,] = \begin{bmatrix} 5/3 & 1 \\ 1 & 1 \end{bmatrix}$$

and

$$\rho_{13|(2)} = \frac{1}{\sqrt{5/3 \cdot 1}} = \sqrt{3/5}.$$

In general terms,

$$\begin{aligned} \mathbf{C} &= \mathbf{\Sigma}_{11} - \mathbf{\Sigma}_{12}\mathbf{\Sigma}_{22}^{-1}\mathbf{\Sigma}_{21} \\ &= \begin{bmatrix} \sigma_{11} & \sigma_{13} \\ \sigma_{31} & \sigma_{33} \end{bmatrix} - \begin{bmatrix} \sigma_{12} \\ \sigma_{32} \end{bmatrix} [\,\sigma_{22}\,]^{-1} [\,\sigma_{12} \quad \sigma_{32}\,] \\ &= \begin{bmatrix} \sigma_{11} - \sigma_{12}^2/\sigma_{22} & \sigma_{13} - \sigma_{12}\sigma_{32}/\sigma_{22} \\ \sigma_{31} - \sigma_{32}\sigma_{12}/\sigma_{22} & \sigma_{33} - \sigma_{32}^2/\sigma_{22} \end{bmatrix} \end{aligned}$$

and

$$\rho_{13|(2)} = \frac{\sigma_{13} - \sigma_{12}\sigma_{32}/\sigma_{22}}{\sqrt{(\sigma_{11} - \sigma_{12}^2/\sigma_{22})\,(\sigma_{33} - \sigma_{32}^2/\sigma_{22})}} = \frac{\sigma_{22}\sigma_{13} - \sigma_{12}\sigma_{32}}{\sqrt{\sigma_{22}\sigma_{11} - \sigma_{12}^2}\sqrt{\sigma_{22}\sigma_{33} - \sigma_{32}^2}}$$
$$= \frac{\sigma_{22}\sigma_{13} - \sigma_{12}\sigma_{32}}{\sqrt{\sigma_{22}\sigma_{11}\left(1 - \dfrac{\sigma_{12}^2}{\sigma_{22}\sigma_{11}}\right)}\sqrt{\sigma_{22}\sigma_{33}\left(1 - \dfrac{\sigma_{32}^2}{\sigma_{22}\sigma_{33}}\right)}},$$

or

$$\rho_{13|(2)} = \frac{\sigma_{22}\sigma_{13} - \sigma_{12}\sigma_{32}}{\sqrt{\sigma_{22}\sigma_{11}\sigma_{22}\sigma_{33}}\sqrt{(1-\rho_{12}^2)(1-\rho_{23}^2)}} = \frac{\dfrac{\sigma_{13}}{\sqrt{\sigma_{11}\sigma_{33}}} - \dfrac{\sigma_{12}}{\sqrt{\sigma_{22}\sigma_{11}}}\dfrac{\sigma_{32}}{\sqrt{\sigma_{22}\sigma_{33}}}}{\sqrt{(1-\rho_{12}^2)(1-\rho_{23}^2)}}$$
$$= \frac{\rho_{13} - \rho_{12}\rho_{23}}{\sqrt{(1-\rho_{12}^2)(1-\rho_{23}^2)}}. \tag{8.42}$$

Using the previous numbers, $\rho_{13} = 1/\sqrt{2}$, $\rho_{12} = 1/\sqrt{6}$, and $\rho_{23} = 0$, so that (8.42) gives

$$\rho_{13|(2)} = \frac{\rho_{13} - \rho_{12}\rho_{23}}{\sqrt{(1-\rho_{12}^2)(1-\rho_{23}^2)}} = \frac{1/\sqrt{2}}{\sqrt{(1-(1/\sqrt{6})^2)}} = \sqrt{\frac{3}{5}},$$

as before. ■

Example 8.8 Let $\mathbf{Y} = (Y_1, \dots, Y_4)' \sim \mathrm{N}(\mathbf{0}, \mathbf{\Sigma})$ with

$$\mathbf{\Sigma} = \frac{1}{1-a^2}\begin{bmatrix} 1 & a & a^2 & a^3 \\ a & 1 & a & a^2 \\ a^2 & a & 1 & a \\ a^3 & a^2 & a & 1 \end{bmatrix}$$

for a value of a such that $|a| < 1$, so that

$$\begin{bmatrix} Y_1 \\ Y_3 \\ Y_4 \\ Y_2 \end{bmatrix} \sim \mathrm{N}(\mathbf{0}, \mathbf{\Omega}), \qquad \mathbf{\Omega} = \frac{1}{1-a^2}\begin{bmatrix} 1 & a^2 & a^3 & a \\ a^2 & 1 & a & a \\ a^3 & a & 1 & a^2 \\ a & a & a^2 & 1 \end{bmatrix}.$$

Then, with the appropriate partitions for $\boldsymbol{\mu}$ and $\mathbf{\Omega}$,

$$(Y_1, Y_3, Y_4 \mid Y_2)' \sim \mathrm{N}(\boldsymbol{\nu},\ \mathbf{C}),$$

where

$$\boldsymbol{\nu} = \boldsymbol{\mu}_{(1)} + \mathbf{\Omega}_{12}\mathbf{\Omega}_{22}^{-1}(\mathbf{y}_{(2)} - \boldsymbol{\mu}_{(2)}) = \begin{bmatrix} 0 \\ 0 \\ 0 \end{bmatrix} + \begin{bmatrix} a \\ a \\ a^2 \end{bmatrix}[1]^{-1}(y_2 - 0) = \begin{bmatrix} ay_2 \\ ay_2 \\ a^2y_2 \end{bmatrix}$$

and

$$\begin{aligned}
\mathbf{C} &= \mathbf{\Omega}_{11} - \mathbf{\Omega}_{12}\mathbf{\Omega}_{22}^{-1}\mathbf{\Omega}_{21} \\
&= \frac{1}{1-a^2}\begin{bmatrix} 1 & a^2 & a^3 \\ a^2 & 1 & a \\ a^3 & a & 1 \end{bmatrix} - \frac{1}{1-a^2}\begin{bmatrix} a \\ a \\ a^2 \end{bmatrix}[1]^{-1}\begin{bmatrix} a & a & a^2 \end{bmatrix} \\
&= \frac{1}{1-a^2}\left(\begin{bmatrix} 1 & a^2 & a^3 \\ a^2 & 1 & a \\ a^3 & a & 1 \end{bmatrix} - \begin{bmatrix} a^2 & a^2 & a^3 \\ a^2 & a^2 & a^3 \\ a^3 & a^3 & a^4 \end{bmatrix}\right) \\
&= \frac{1}{1-a^2}\left(\begin{bmatrix} 1-a^2 & 0 & 0 \\ 0 & 1-a^2 & a-a^3 \\ 0 & a-a^3 & 1-a^4 \end{bmatrix}\right) = \begin{bmatrix} 1 & 0 & 0 \\ 0 & 1 & a \\ 0 & a & a^2+1 \end{bmatrix}.
\end{aligned} \tag{8.43}$$

It follows that

$$\rho_{13|(2)} = \frac{\sigma_{13|(2)}}{\sqrt{\sigma_{11|(2)}\ \sigma_{33|(2)}}} = \frac{0}{1} = 0, \qquad \rho_{14|(2)} = \frac{\sigma_{14|(2)}}{\sqrt{\sigma_{11|(2)}\ \sigma_{44|(2)}}} = \frac{0}{\sqrt{1+a^2}} = 0$$

and

$$\rho_{34|(2)} = \frac{\sigma_{34|(2)}}{\sqrt{\sigma_{33|(2)}\ \sigma_{44|(2)}}} = \frac{a}{\sqrt{1+a^2}}. \tag{8.44}$$

Equivalently, from (8.42),

$$\rho_{13|(2)} = \frac{\rho_{13} - \rho_{12}\rho_{23}}{\sqrt{(1-\rho_{12}^2)(1-\rho_{23}^2)}} = \frac{\rho_2 - \rho_1\rho_1}{\sqrt{(1-\rho_1^2)(1-\rho_1^2)}} = \frac{a^2-a^2}{1-a^2} = 0,$$

because $\rho_{13} = \mathrm{Corr}(Y_t, Y_{t-2}) = \rho_2 = a^2$ and $\rho_{12} = \rho_{23} = \mathrm{Corr}(Y_t, Y_{t-1}) = \rho_1 = a$. That is, Y_t and Y_{t-2} are conditionally uncorrelated after having taken into account their correlation with Y_{t-1}. Observe how it was critical to condition on the observation(s) between the two random variables of interest. Conditional on Y_2, $Y_3 = (aY_2 + U_3) \sim \mathrm{N}(aY_2, \sigma^2)$ and

$$Y_4 = aY_3 + U_4 = a(aY_2 + U_3) + U_4 = (a^2Y_2 + aU_3 + U_4) \sim \mathrm{N}(a^2Y_2, \sigma^2(a^2+1)).$$

The covariance between Y_3 and Y_4 conditional on Y_2 is then, from basic principles,

$$\begin{aligned}
\sigma_{34|(2)} &= \mathrm{Cov}(Y_3, Y_4 \mid Y_2) = \mathbb{E}[(Y_3 - \mathbb{E}[Y_3 \mid Y_2])(Y_4 - \mathbb{E}[Y_4 \mid Y_2]) \mid Y_2] \\
&= \mathbb{E}[(Y_3 - aY_2)(Y_4 - a^2Y_2) \mid Y_2] = \mathbb{E}[Y_3Y_4 - Y_3a^2Y_2 - aY_4Y_2 + a^3Y_2^2 \mid Y_2] \\
&= \mathbb{E}[(aY_2 + U_3)(a^2Y_2 + aU_3 + U_4) \mid Y_2] \\
&\quad - \mathbb{E}[Y_3a^2Y_2 \mid Y_2] - \mathbb{E}[aY_4Y_2 \mid Y_2] + \mathbb{E}[a^3Y_2^2 \mid Y_2] \\
&= a^3Y_2^2 + a\sigma^2 - a^2Y_2aY_2 - aY_2a^2Y_2 + a^3Y_2^2 = a\sigma^2,
\end{aligned}$$

so that the conditional correlation is given by

$$\rho_{34|(2)} = \text{Corr}(Y_3, Y_4 \mid Y_2) = \frac{\text{Cov}(Y_3, Y_4 \mid Y_2)}{\sqrt{\mathbb{V}(Y_3 \mid Y_2)\mathbb{V}(Y_4 \mid Y_2)}} = \frac{a\sigma^2}{\sqrt{\sigma^2\sigma^2(a^2+1)}} = \frac{a}{\sqrt{1+a^2}},$$

which is only zero if $a = 0$ (in which case all the observations are i.i.d.). Note that this expression for $\rho_{34|(2)}$ agrees with the derivation in (8.44). ■

8.2.2 Partial Autocorrelation Function

Above we said that the partial autocorrelation at $s = 2$ is the conditional correlation between Y_1 and Y_3 "over and above the association resulting from their mutual relationship with Y_2." This informal statement is now made more precise, so that we can define the theoretical partial autocorrelation function, or TPACF.

8.2.2.1 TPACF: First Definition

Let $\mathbf{X} = (X_1, \dots, X_n)'$ have zero mean and full rank covariance matrix $\mathbf{\Sigma}$. For a constant integer p, $1 < p < n$, define $\mathbf{X}^{(1)} = (X_1, \dots, X_p)'$ and $\mathbf{X}^{(2)} = (X_{p+1}, \dots, X_n)'$. Let $\mathcal{S}$ be the subspace of all linear combinations of the subset $\mathbf{X}^{(2)}$. (For our purposes here, $\mathbf{X}$ will be a rearrangement of a subset of time series $\mathbf{Y} = (Y_1, \dots, Y_T)$, which has a joint multivariate normal distribution, such as was done in Examples 8.7 and 8.8.) By the Projection Theorem 1.1, each X_i, $i = 1, \dots, p$, can be expressed as

$$X_i = X_{i,1} + X_{i,2}\,, \quad \text{where} \quad X_{i,2} \in \mathcal{S}, \quad X_{i,1} \in \mathcal{S}^{\perp}.$$

In particular, there exists a real vector of coefficients $\mathbf{a}_i'$ for each $X_{i,2}$ such that $X_{i,2} = \mathbf{a}_i'\mathbf{X}^{(2)}$, i.e.,

$$\begin{pmatrix} X_{1,2} \\ \vdots \\ X_{p,2} \end{pmatrix} = \begin{pmatrix} \mathbf{a}_1' \\ \vdots \\ \mathbf{a}_p' \end{pmatrix} \begin{pmatrix} X_{p+1} \\ \vdots \\ X_n \end{pmatrix} =: \mathbf{A}\mathbf{X}^{(2)}. \tag{8.45}$$

Because of the orthogonality,

$$\mathbf{\Sigma}_{12} = \text{Cov}(\mathbf{X}^{(1)}, \mathbf{X}^{(2)}) = \text{Cov}\left(\begin{bmatrix} X_{1,1} \\ \vdots \\ X_{p,1} \end{bmatrix} + \begin{bmatrix} X_{1,2} \\ \vdots \\ X_{p,2} \end{bmatrix}, \begin{bmatrix} X_{p+1} \\ \vdots \\ X_n \end{bmatrix} \right)$$

$$= \text{Cov}\left(\begin{bmatrix} X_{1,2} \\ \vdots \\ X_{p,2} \end{bmatrix}, \begin{bmatrix} X_{p+1} \\ \vdots \\ X_n \end{bmatrix} \right) = \text{Cov}(\mathbf{A}\mathbf{X}^{(2)}, \mathbf{X}^{(2)}) = \mathbf{A}\mathbf{\Sigma}_{22}\,.$$

Thus, $\mathbf{A} = \mathbf{\Sigma}_{12}\mathbf{\Sigma}_{22}^{-1}$ and, from (8.45),

$$\mathbb{V}\begin{pmatrix} X_{1,2} \\ \vdots \\ X_{p,2} \end{pmatrix} = \mathbb{V}(\mathbf{A}\mathbf{X}^{(2)}) = \mathbf{A}\mathbf{\Sigma}_{22}\mathbf{A}' = \mathbf{\Sigma}_{12}\mathbf{\Sigma}_{22}^{-1}\mathbf{\Sigma}_{22}\mathbf{\Sigma}_{22}^{-1}\mathbf{\Sigma}_{12}' = \mathbf{\Sigma}_{12}\mathbf{\Sigma}_{22}^{-1}\mathbf{\Sigma}_{21}\,.$$

Problem 8.5 verifies that

$$\boldsymbol{\Sigma}_{11} = \mathbb{V}\left(\begin{bmatrix} X_{1,1} \\ \vdots \\ X_{p,1} \end{bmatrix} + \begin{bmatrix} X_{1,2} \\ \vdots \\ X_{p,2} \end{bmatrix}\right) = \mathbb{V}\left(\begin{bmatrix} X_{1,1} \\ \vdots \\ X_{p,1} \end{bmatrix}\right) + \mathbb{V}\left(\begin{bmatrix} X_{1,2} \\ \vdots \\ X_{p,2} \end{bmatrix}\right), \tag{8.46}$$

from which it follows that

$$\mathbb{V}\left(\begin{bmatrix} X_{1,1} \\ \vdots \\ X_{p,1} \end{bmatrix}\right) = \boldsymbol{\Sigma}_{11} - \mathbb{V}\left(\begin{bmatrix} X_{1,2} \\ \vdots \\ X_{p,2} \end{bmatrix}\right) = \boldsymbol{\Sigma}_{11} - \boldsymbol{\Sigma}_{12}\boldsymbol{\Sigma}_{22}^{-1}\boldsymbol{\Sigma}_{21} .$$

Recalling the definition of partial correlation in (8.41), this shows that, for $1 \leqslant i < j \leqslant p$,

$$\mathrm{Cov}(X_{i,1}, X_{j,1}) = \rho_{ij|(p+1,\ldots,n)}, \tag{8.47}$$

i.e., that $\rho_{ij|(p+1,\ldots,n)}$ is the correlation coefficient of the residuals of X_i and X_j after removing the parts of X_i and X_j that lie in $\mathcal{S}$.

The **theoretical partial autocorrelation function**, or TPACF, is given by the set of coefficients $(\alpha_{11}, \alpha_{22}, \ldots, \alpha_{mm})$, where typical element α_{ss} is defined to be the partial correlation between Y_t and Y_{t-s} conditional on the Y_i between the two, i.e., $\alpha_{11} = \rho_1$ and

$$\alpha_{ss} = \rho_{t,t-s|(t-1,\ldots,t-s+1)} = \rho_{1,1+s|(2,\ldots,s)}, \qquad s > 1. \tag{8.48}$$

8.2.2.2 TPACF: Second Definition

In light of the projection theory result (8.47) and the implications of Example 1.9, an equivalent definition of element α_{ss} is the last coefficient in a linear projection of Y_t on its most recent s values, i.e.,

$$\widehat{Y}_t = \alpha_{s1}Y_{t-1} + \alpha_{s2}Y_{t-2} + \cdots + \alpha_{ss}Y_{t-s}. \tag{8.49}$$

This definition explains the use of the double subscript on α.

For the AR(1) model, this implies that $\alpha_{11} = \rho_1 = a$ and $\alpha_{ss} = 0$, $s > 1$. When this PACF is plotted as correlogram, it will look like those in Figure 8.1 but with only the first spike; the others are zero. For the AR(p) model, this implies that $\alpha_{ss} = 0$ for $s > p$. Figure 8.18 is the PACF counterpart to Figure 8.2. Notice how the value of the last nonzero "spike" is always equal to the value of the last nonzero autocorrelation coefficient.

The definition (8.49) can also be viewed as resulting from the computation of regression (6.32) for an infinite sample size. But asymptotically, the matrices in (6.32) approach the theoretical counterparts illustrated in the Yule–Walker equations (6.34). Thus, α_{ss} can be computed by solving the system of equations

$$\begin{bmatrix} \rho_1 \\ \rho_2 \\ \vdots \\ \rho_s \end{bmatrix} = \begin{bmatrix} \rho_0 & \rho_1 & \cdots & \rho_{s-1} \\ \rho_1 & \ddots & \ddots & \vdots \\ \vdots & & & \rho_1 \\ \rho_{s-1} & \rho_{s-2} & \cdots & \rho_0 \end{bmatrix} \begin{bmatrix} \alpha_{s1} \\ \alpha_{s2} \\ \vdots \\ \alpha_{ss} \end{bmatrix} =: \mathbf{C}_s \begin{bmatrix} \alpha_{s1} \\ \alpha_{s2} \\ \vdots \\ \alpha_{ss} \end{bmatrix}, \tag{8.50}$$

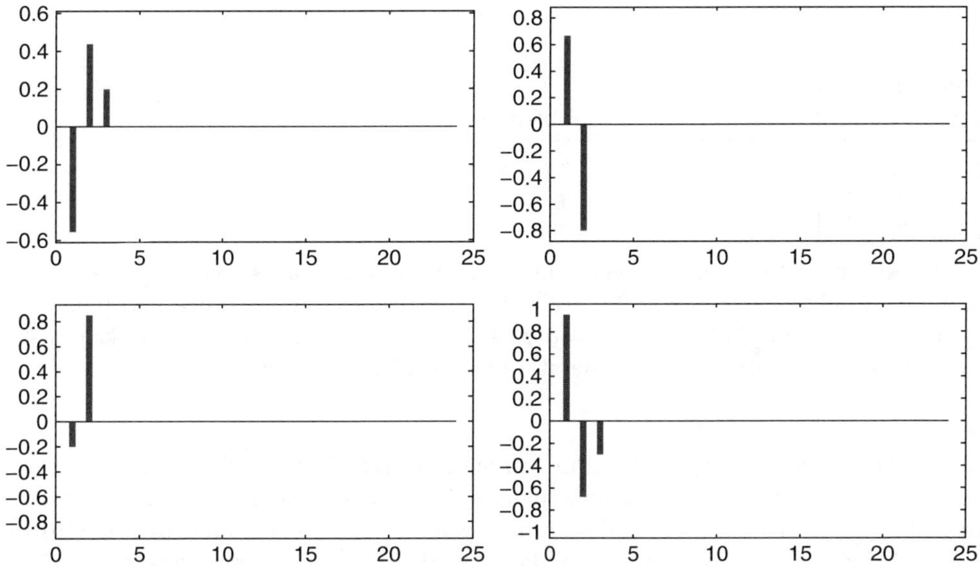

Figure 8.18 TPACF of the stationary AR(3) model with parameters $\mathbf{a} = (a_1, a_2, a_3) = (0.4, -0.5, -0.2)$ (top left), $\mathbf{a} = (1.2, -0.8, 0)$ (top right), $\mathbf{a} = (-0.03, 0.85, 0)$ (bottom left), and $\mathbf{a} = (1.4, -0.2, -0.3)$ (bottom right).

where $\mathbf{C}_s$ is so defined. The ρ_i could be obtained from (6.20) for a pure AR process, or from (7.21) and (7.23) for an ARMA process. In fact, because only value α_{ss} is required from (8.50), Cramer's rule (see, e.g., Trench, 2003, p. 374; or Munkres, 1991, p. 21) can be used, i.e.,

$$\alpha_{ss} = \frac{|\,\mathbf{C}_s^*\,|}{|\,\mathbf{C}_s\,|}, \qquad s = 1, 2, \ldots, \tag{8.51}$$

where matrix $\mathbf{C}_s^*$ is obtained by replacing the last column of matrix $\mathbf{C}_s$ by the column vector $(\rho_1, \rho_2, \ldots, \rho_s)'$, i.e.,

$$\mathbf{C}_s^* = \left[\begin{array}{cccc|c} 1 & \rho_1 & \cdots & \rho_{s-2} & \rho_1 \\ \rho_1 & 1 & & \rho_{s-3} & \rho_2 \\ \rho_2 & \rho_1 & & \rho_{s-4} & \rho_3 \\ \vdots & & & \vdots & \vdots \\ \rho_{s-2} & & & 1 & \rho_{s-1} \\ \rho_{s-1} & \rho_{s-2} & \cdots & \rho_1 & \rho_s \end{array}\right].$$

Applying (8.51), the first three terms of the PACF are given by

$$\alpha_{11} = \frac{|\,\rho_1\,|}{|1|} = \rho_1, \qquad \alpha_{22} = \frac{\begin{vmatrix} 1 & \rho_1 \\ \rho_1 & \rho_2 \end{vmatrix}}{\begin{vmatrix} 1 & \rho_1 \\ \rho_1 & 1 \end{vmatrix}} = \frac{\rho_2 - \rho_1^2}{1 - \rho_1^2}, \tag{8.52}$$

and

$$\alpha_{33} = \frac{\begin{vmatrix} 1 & \rho_1 & \rho_1 \\ \rho_1 & 1 & \rho_2 \\ \rho_2 & \rho_1 & \rho_3 \end{vmatrix}}{\begin{vmatrix} 1 & \rho_1 & \rho_2 \\ \rho_1 & 1 & \rho_1 \\ \rho_2 & \rho_1 & 1 \end{vmatrix}} = \frac{\rho_3 + \rho_1\rho_2(\rho_2 - 2) - \rho_1^2(\rho_3 - \rho_1)}{(1 - \rho_2) - (1 - \rho_2 - 2\rho_1^2)}. \tag{8.53}$$

Notice that, for an AR(1) model with parameter a, the numerator of the expression for α_{22} is zero, and for α_{33} the numerator simplifies to $a^3 + a^5 - 2a^3 - a^3(a^2 - 1) = 0$. For an AR(2) process with parameters a_1 and a_2, the ρ_i are given in (6.18), with $\rho_3 = a_1\rho_2 + a_2\rho_1$. A symbolic computing package such as Maple can then be used to verify that the numerator of α_{33} is identically zero.

8.2.2.3 Sample Partial Autocorrelation Function

The **sample partial ACF**, or SPACF, is just the finite sample counterpart of the theoretical PACF. For its computation, (8.51) can be used with the sample values $\hat{\rho}_i$, though a computationally more efficient method of computing the α_{ss} from a set of correlations is given by the so-called **Durbin–Levinson algorithm**; see, e.g., Brockwell and Davis (1991) and Pollock (1999) for clear derivations and original references. A matrix-based implementation of this is given in Listing 8.9.

Alternatively (but not equivalent numerically for finite samples), the regression method based on (8.49) and fitting the coefficients with least squares can be used. Matlab's function `parcorr` computes it this way. Recall Examples 1.1 and 1.9 on the Frisch–Waugh–Lovell theorem. In particular, as we are interested in only one of the coefficients, it can be expressed as the ratio of quadratic forms in (1.23), and is thus amenable to eliciting its small-sample distribution. The small-sample distribution of the *joint* density of the SPACF can be obtained by transforming the density of the SACF; see Butler and Paolella (1998) and the references therein for details on the required Jacobian. It can be shown that, for i.i.d. normal data (and other uncorrelated processes that relax the normality assumption), $T^{1/2}\hat{\alpha}_{ii}$ is asymptotically standard normal; see, e.g., Priestley (1981) and Brockwell and Davis (1991).

The SPACF for the time series that were used in generating the SACFs in Figure 8.3 are shown in Figure 8.19. The dashed lines indicate asymptotic 95% c.i.s for the individual spikes assuming a white-noise model.

```
function pacf = pacfcomp(acf)
n=length(acf); acf1 = acf(2:n); n=n-1;
[t,p] = chol(toeplitz(acf(1:n)));
if p>0, q=p-1; else q=n; end
r = acf1(1:q);
for k=1:q
  r(k) = r(k)/t(k,k);
  if k<q, r((k+1):q) = r((k+1):q) - t(k,(k+1):q)'*r(k); end
end
pacf = r./diag(t);
if p>0, pacf((q+1):n) = zeros((n-q),1); end
```

Program Listing 8.9: Computes the SPACF based on the SACF, using the Cholesky decomposition of the sample correlation matrix; see, e.g., Pourahmadi (2001, Ch. 7).

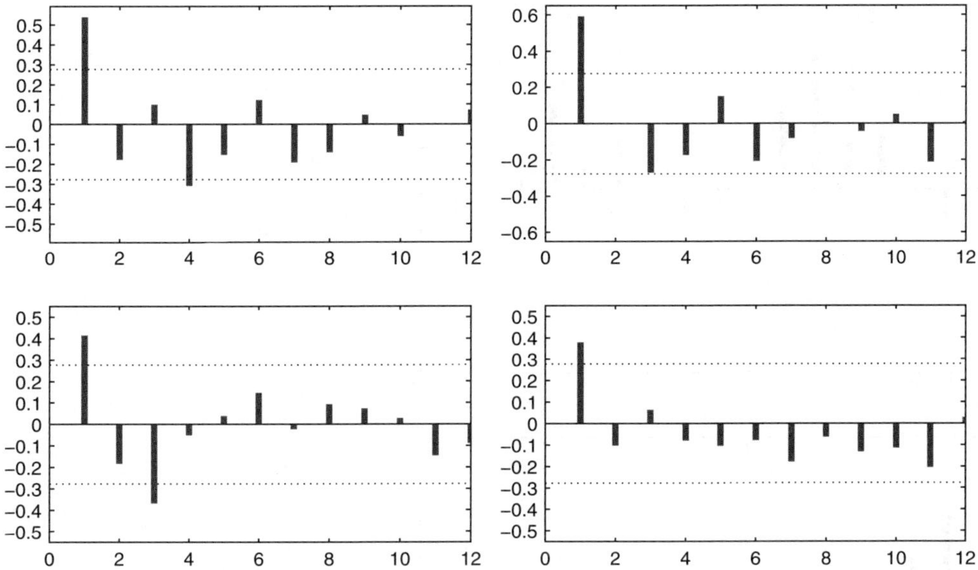

Figure 8.19 The SPACFs of the four simulated AR(1) time series with $a = 0.5$ and $T = 50$ as were used in Figure 8.3.

Example 8.9 Consider the AR(5) process with parameters $a_1 = 1.1$, $a_2 = 0$, $a_3 = -0.6$, $a_4 = 0$, and $a_5 = 0.4$. In this case, some of the coefficients are zero; this is referred to as a **subset autoregressive model**. The ACF and PACF are shown in Figure 8.20. Observe that it is *not* the case that the second and fourth spikes in the PACF are zero, but only that it cuts off after lag 5. ■

Recall from (6.44) that ρ_1 for an MA(1) model has a supremum of 1/2. As such, the ACF of an MA(1) process will have a single spike at lag 1 bounded by 1/2, and all other spikes are zero. Knowing that the ACF spikes corresponding to a (stationary) AR(p) model are nonzero and exponentially decay after lag p, it might appear strange that the MA(1) model, which, from (6.39), can be represented as an infinite AR, has correlations (6.44). This apparent paradox is explained by noting the special structure of the AR coefficients in (6.39). When chosen exactly in such a way, the correlation structure (6.44) arises.

Regarding the PACF, one might think that, on the one hand, as the MA(1) can be expressed as an AR(∞), the PACF should never cut off; or that, on the other hand, because the unconditional correlation is zero after lag 1, the PACF should also be zero after lag one. To directly see that the latter is wrong, note that, either from the partial correlation in (8.42) with $\rho_{ij} = \text{Corr}(Y_i, Y_j)$ or directly from the latter expression in (8.52),

$$\alpha_{22} = \rho_{13|(2)} = \frac{\rho_{13} - \rho_{12}\rho_{23}}{\sqrt{(1-\rho_{12}^2)(1-\rho_{23}^2)}} = \frac{\rho_2 - \rho_1\rho_1}{\sqrt{(1-\rho_1^2)(1-\rho_1^2)}} = \frac{\rho_2 - \rho_1^2}{1-\rho_1^2}$$

or, recalling (6.44),

$$\alpha_{22} = \frac{-\left(\frac{b}{1+b^2}\right)^2}{1-\left(\frac{b}{1+b^2}\right)^2} = \frac{-b^2}{1+b^2+b^4}.$$

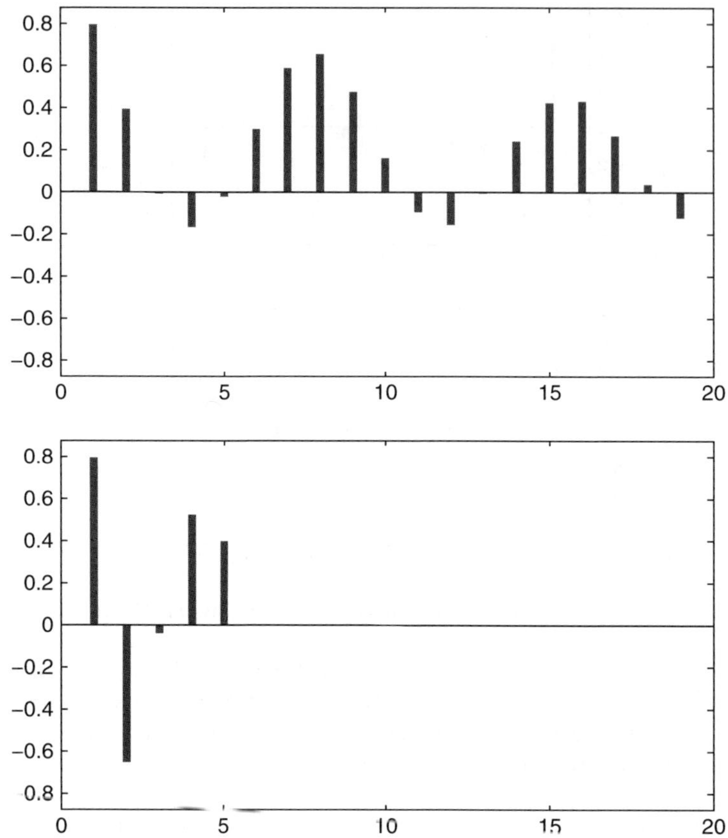

Figure 8.20 The theoretical ACF (top) and PACF (bottom) of the subset AR(5) model with $a_1 = 1.1$, $a_3 = -0.6$, and $a_5 = 0.4$.

From (8.52), $\alpha_{33} = b^3/(1 + b^2 + b^4 + b^6)$ which, for $b = 0.5$, is 8/85. The pattern for α_{kk} suggests that

$$\alpha_{kk} = \frac{(-b)^k(b^2 - 1)}{1 - b^{2(k+1)}},$$

which is indeed true; see, e.g., Brockwell and Davis (1991, Sec. 3.4). Figure 8.21 plots the theoretical ACF and PACF corresponding to the MA(1) model with $b = 0.5$, so that $\rho_1 = 0.4$, showing that, while the ACF cuts off after lag 1, the PACF dies out exponentially. Similarly, for an MA(q) process, (6.50) implies that the TACF cuts off after lag q. But, just as in the MA(1) case, the TPACF for an MA(q) process with $q \geqslant 1$ does not cut off and dies off exponentially instead.

Remark Students sometimes have difficulty explaining exactly why the ACF cuts off but the PACF does not for MA(q) models. To help understand this more fully, note that a regression of $Y_t - c$ onto $Y_{t-2} - c$ (for an infinite sample size) would yield a value of zero because Y_t and Y_{t-2} are uncorrelated. However, now taking $c = 0$ for simplicity, in the regression of Y_t on Y_{t-1} and Y_{t-2}, it is *not* the case that

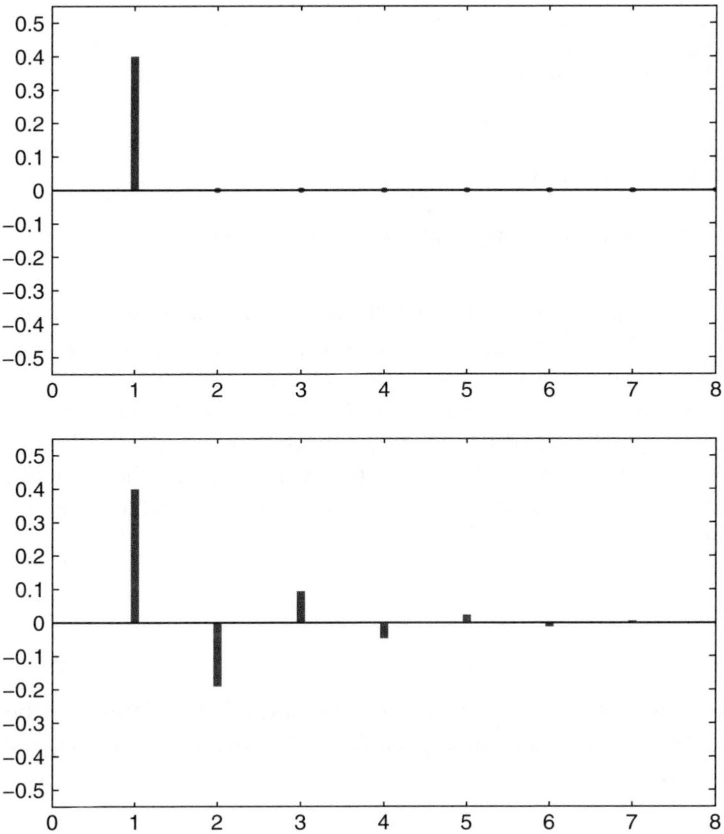

Figure 8.21 The theoretical ACF (top) and PACF (bottom) for the MA(1) model with $b = 0.5$.

the coefficient of Y_{t-2} will be zero because regressors Y_{t-1} and Y_{t-2} are correlated and Y_t is correlated with Y_{t-1}.

To verify this outside of a time series context, let

$$\begin{pmatrix} Y \\ X_1 \\ X_2 \end{pmatrix} \sim \mathrm{N}(\mathbf{0},\ \boldsymbol{\Sigma}), \quad \boldsymbol{\Sigma} = \begin{bmatrix} \sigma_1^2 & \sigma_{12} & 0 \\ \sigma_{12} & \sigma_2^2 & \sigma_{23} \\ 0 & \sigma_{23} & \sigma_3^2 \end{bmatrix}, \tag{8.54}$$

for σ_{12} and σ_{23} nonzero (and, of course, such that $\boldsymbol{\Sigma} > 0$). Then $\mathrm{Cov}(Y, X_1) \neq 0$, $\mathrm{Cov}(X_1, X_2) \neq 0$, but $\mathrm{Cov}(Y, X_2) = 0$. By simulating a large sample from this multivariate normal density, we can (i) regress the values of Y onto those of X_1 and should find a nonzero coefficient, (ii) regress Y onto X_2 and should find a zero coefficient, and (iii) regress Y onto both X_1 and X_2 and should find both regression coefficients to be nonzero. The code in Listing 8.10 performs this with a “data set” of one million observations, and confirms the claim. ■

```
s1=1; s2=1; s3=1; s12=0.4; s23=-0.7; S=[s1^2 s12 0; s12 s2^2 s23; 0 s23 s3^2]
[V,D]=eig(S); C=V*sqrt(D)*V'; T=1e6; z=randn(3,T); N=(C*z)';
cov(N) % just as a check: It should be approx S
Y=N(:,1); X1=N(:,2); X2=N(:,3);
inv(X1'*X1)*X1'*Y % should be nonzero
inv(X2'*X2)*X2'*Y % should be zero
inv(X'*X)*X'*Y % both coefficients should be nonzero!
```

Program Listing 8.10: Simulates from (8.54) and computes regression coefficients.

Summing up, for a (stationary) AR(p) process, $p \geqslant 1$, the TACF decays exponentially,[5] but the TPACF "cuts off" after the pth spike, i.e., it is zero. Matters are reversed for an MA(q) process: The TACF cuts off after the qth spike, while the TPACF is nonzero and decays exponentially.

For an ARMA(p, q) process with $p > 0$ and $q > 0$, *neither* the TACF or the TPACF cut off. The practical side of this result is that the sample ACF and PACF can be inspected and, if one of them appears to "cut off", then p and q can be guessed. Of course, in practice, with real data that may not even be from a stationary model, let alone a stationary ARMA model, let alone a pure AR or pure MA model, matters are not always so clear cut.

8.3 Problems

Problem 8.1 Write a program that computes the values in Figures 8.6 and 8.7, i.e., computes the first two raw moments of R_s based on o.l.s. residuals for a given $\mathbf{X}$ matrix, and such that the true residuals have a given ARMA covariance matrix.

Problem 8.2 Show (8.20) and (8.21).

Problem 8.3 Let Y_t follow the AR(2) model $Y_t = 1.2Y_{t-1} - 0.8Y_{t-2} + U_t$, where $U_t \overset{\text{i.i.d.}}{\sim} \text{N}(0, \sigma^2)$ and $\sigma > 0$. For the SACF, show that

$$\mathbf{W}_{\text{corr}} = \begin{bmatrix} 1 & \cdot & \cdot \\ 0.873 & 1 & \cdot \\ 0.531 & 0.865 & 1 \end{bmatrix}$$

and, asymptotically,

$$\begin{pmatrix} R_1 - 2/3 \\ R_2 - 0 \end{pmatrix} \sim \text{N}\left[\begin{pmatrix} 0 \\ 0 \end{pmatrix}, T^{-1}\begin{pmatrix} 0.0617 & 0.1481 \\ & 0.4667 \end{pmatrix}\right].$$

Similar to Example 8.3, calculate and plot contour plots of the simulated and saddlepoint densities for $T = 20$ and $\mathbf{X} = \mathbf{1}$, and also the asymptotic distribution.

Problem 8.4 Write a Matlab program to generate the plots in Figure 8.17.

---

5 It is, however, not true that all the spikes are nonzero. As an example, for the AR(2) model with $a_1 = 1.2$ and $a_2 = -0.8$, $\rho_2 = 0$. See Figure 8.2.

Problem 8.5 Verify (8.46). It is enough to take $p = 2$.

Problem 8.6 Assume that $\mathbf{Y} = \mathbf{X}\boldsymbol{\beta} + \boldsymbol{\epsilon}$, with $\boldsymbol{\epsilon} \sim \mathrm{N}(\mathbf{0}, \sigma^2\mathbf{I})$, and $\mathbf{X}$ full rank, of size $T \times k$. Let $\breve{R}_1, \ldots, \breve{R}_m$ be the elements of the SACF computed from the regression residuals based on $\mathbf{X}$, but using the recursive residuals, as discussed in Section 1.5. That is,

$$\breve{R}_s = \frac{\widehat{\boldsymbol{\epsilon}}'\mathbf{A}_s\widehat{\boldsymbol{\epsilon}}}{\widehat{\boldsymbol{\epsilon}}'\widehat{\boldsymbol{\epsilon}}},$$

where $\mathbf{A}_s$ is given in (8.9), but is of size $(T-k) \times (T-k)$, and $\widehat{\boldsymbol{\epsilon}} = \mathbf{C}\mathbf{Y} = \mathbf{C}\boldsymbol{\epsilon}$ are the $T-k$ recursive residuals. Show analytically that the density of $\breve{R}_s$ is symmetric about zero. Verify this computationally by comparing $F_{\breve{R}_s}(z)$ with $1 - F_{\breve{R}_s}(-z)$ over a grid of z values. Similarly, show computationally that the R_s, i.e., the SACF elements based on the usual regression residuals, are not symmetric about their mean (which is nonzero).

8.A Appendix: Solutions

Solution to Problem 8.1 There are two ways of setting up the ratio of quadratic forms pertaining to R_s such that program `sawa` in in Listing B.4 can be used. Common to both is to write

$$R_s = \frac{\widehat{\boldsymbol{\epsilon}}'\mathbf{A}_s\widehat{\boldsymbol{\epsilon}}}{\widehat{\boldsymbol{\epsilon}}'\widehat{\boldsymbol{\epsilon}}} = \frac{\boldsymbol{\epsilon}'\mathbf{M}'\mathbf{A}_s\mathbf{M}\boldsymbol{\epsilon}}{\boldsymbol{\epsilon}'\mathbf{M}'\mathbf{M}\boldsymbol{\epsilon}} = \frac{\boldsymbol{\epsilon}'\mathbf{M}\mathbf{A}_s\mathbf{M}\boldsymbol{\epsilon}}{\boldsymbol{\epsilon}'\mathbf{M}\boldsymbol{\epsilon}},$$

as in (B.22). The first way uses the Cholesky decomposition of the variance–covariance matrix to get

$$R_s = \frac{\boldsymbol{\epsilon}'\boldsymbol{\Sigma}^{-1/2}\boldsymbol{\Sigma}^{1/2}\mathbf{M}\mathbf{A}_s\mathbf{M}\boldsymbol{\Sigma}^{1/2}\boldsymbol{\Sigma}^{-1/2}\boldsymbol{\epsilon}}{\boldsymbol{\epsilon}'\boldsymbol{\Sigma}^{-1/2}\boldsymbol{\Sigma}^{1/2}\mathbf{M}\boldsymbol{\Sigma}^{1/2}\boldsymbol{\Sigma}^{-1/2}\boldsymbol{\epsilon}} = \frac{\mathbf{W}'\mathbf{A}\mathbf{W}}{\mathbf{W}'\mathbf{B}\mathbf{W}},$$

where $\mathbf{W} = \boldsymbol{\Sigma}^{-1/2}\boldsymbol{\epsilon} \sim \mathrm{N}_T(\mathbf{0}, \mathbf{I})$, $\mathbf{A} = \boldsymbol{\Sigma}^{1/2}\mathbf{M}\mathbf{A}_s\mathbf{M}\boldsymbol{\Sigma}^{1/2}$, and $\mathbf{B} = \boldsymbol{\Sigma}^{1/2}\mathbf{M}\boldsymbol{\Sigma}^{1/2}$. The program in Listing 8.11 implements this when the value `the_method` is set to 1.

The second way is the rest of (B.22), i.e.,

$$R_s = \frac{\boldsymbol{\epsilon}'\mathbf{G}'\mathbf{G}\mathbf{A}_s\mathbf{G}'\mathbf{G}\boldsymbol{\epsilon}}{\boldsymbol{\epsilon}'\mathbf{G}'\mathbf{G}\boldsymbol{\epsilon}} = \frac{\mathbf{Z}'\tilde{\mathbf{A}}_s\mathbf{Z}}{\mathbf{Z}'\mathbf{Z}},$$

where $\tilde{\mathbf{A}}_s = \mathbf{G}\mathbf{A}_s\mathbf{G}'$ is $(T-k) \times (T-k)$, but now $\mathbf{Z} = \mathbf{G}\boldsymbol{\epsilon} \sim \mathrm{N}_{T-k}(\mathbf{0}, \mathbf{G}\boldsymbol{\Sigma}\mathbf{G}')$. Listing 8.11 implements this when the value `the_method` is set to any other value than 1. In this case, the Matlab variable `A` refers to the numerator matrix of the quadratic form, and is the matrix $\tilde{\mathbf{A}}_s$, while the denominator matrix is just the $T-k$ identity matrix.

The reader should verify that these two methods indeed yield identical numerical values for the moments of R_s.

Solution to Problem 8.2 For $m = 2$,

$$\mathbf{W} = \begin{bmatrix} 1 & r_1 \\ r_1 & 1 \end{bmatrix}^{-1} = \frac{1}{1 - r_1^2}\begin{bmatrix} 1 & -r_1 \\ -r_1 & 1 \end{bmatrix}, \qquad v = \begin{bmatrix} r_2 \\ r_1 \end{bmatrix},$$

```
function [mu,mom2]=sacfmom(X,Sigma,svec)
mu=zeros(length(svec),1); mom2=mu; T=length(Sigma);
if isempty(X)
  M=eye(T); k=0;
else
  M=eye(T)-X*inv(X'*X)*X'; [garb,k]=size(X);
end;
the_method=1; % set either to 1, or something else.
if the_method==1
  [V,D]=eig(0.5*(Sigma+Sigma')); Q = V*sqrt(D)*V';
  for j=1:length(svec)
    s=svec(j); A=Q'*M*makeA(T,s)*M*Q; B=Q'*M*Q;
    if nargout==1, mu(j)=sawa(A,B,eye(T));
    else, [s1,s2]=sawa(A,B,eye(T)); mu(j)=s1; mom2(j)=s2;
    end
  end
else
  if isempty(X), G=eye(T); else G=makeG(X); end
  varcov=G*Sigma*G';
  for j=1:length(svec)
    s=svec(j); A=G*makeA(T,s)*G';
    if nargout==1, mu(j)=sawa(A,eye(T-k),varcov);
    else, [s1,s2]=sawa(A,eye(T-k),varcov); mu(j)=s1; mom2(j)=s2;
    end
  end
end

function A=makeA(T,m) % A = 0.5 * 1( |i-j| = m)
v=zeros(T,1); v(m+1)=1; A=0.5*toeplitz(v,v');
```

Program Listing 8.11: Returns the first two raw moments of R_s based on regression residuals with full rank $T \times k$ design matrix **X**. For no **X** matrix, pass the empty matrix `[ ]`. The function parameter `svec` is a vector of s values and `Sigma` is the covariance matrix of the true residuals, usually from an ARMA process, as computed via program `leeuwARMA` in Listing 7.5. Program `sawa` is given in Listing B.4.

so that

$$A = -\frac{1}{1-r_1^2}, \quad B = -2\frac{-r_1}{1-r_1^2}r_1, \quad C = 1 - \frac{1}{1-r_1^2}r_1 r_1,$$

and the roots of $Ar_2^2 + Br_2 + C$ are $(-B \pm \sqrt{B^2 - 4AC})/(2A)$, or, as $1 - r_1^2 > 0$,

$$\frac{-\frac{2r_1^2}{1-r_1^2} \pm \sqrt{\left(\frac{2r_1^2}{1-r_1^2}\right)^2 + \frac{4}{1-r_1^2}\frac{1-2r_1^2}{1-r_1^2}}}{-\frac{2}{1-r_1^2}} = \frac{-2r_1^2 \pm \sqrt{(2r_1^2)^2 + 4(1-2r_1^2)}}{-2} = (2r_1^2 - 1), \quad 1.$$

That is, given $R_1 = r_1$,

$$2r_1^2 - 1 < R_2 < 1. \tag{8.55}$$

For $m = 3$, $v = \begin{bmatrix} r_3 & r_2 & r_1 \end{bmatrix}'$,

$$\mathbf{W} = \begin{bmatrix} 1 & r_1 & r_2 \\ r_1 & 1 & r_1 \\ r_2 & r_1 & 1 \end{bmatrix}^{-1}$$

$$= \frac{1}{(1-r_2)(r_2-2r_1^2+1)} \begin{bmatrix} 1-r_1^2 & -r_1+r_1r_2 & -r_2+r_1^2 \\ -r_1+r_1r_2 & 1-r_2^2 & -r_1+r_1r_2 \\ -r_2+r_1^2 & -r_1+r_1r_2 & 1-r_1^2 \end{bmatrix},$$

$$A = -w_{11} = -\frac{1-r_1^2}{(1-r_2)(r_2-2r_1^2+1)},$$

$$B = -2(w_{12}v_2 + w_{13}v_3) = -\frac{2r_1(r_1^2-2r_2+r_2^2)}{(1-r_2)(r_2-2r_1^2+1)},$$

and

$$C = 1 - \sum_{i=2}^{m}\sum_{j=2}^{m} w_{ij}v_iv_j = 1 - (w_{22}v_2^2 + 2w_{23}v_2v_3 + w_{33}v_3^2)$$

$$= 1 - \frac{r_1^2(1-r_1)(1+r_1) + r_2(1-r_2)(r_2^2-2r_1^2+r_2)}{(1-r_2)(r_2-2r_1^2+1)}$$

$$= \frac{r_1^4 - 3r_1^2 + 4r_1^2r_2 - 2r_1^2r_2^2 + 1 - 2r_2^2 + r_2^4}{(1-r_2)(r_2-2r_1^2+1)}.$$

With

$$W = (2r_1(r_1^2-2r_2+r_2^2))^2 + 4(1-r_1^2)(r_1^4 - 3r_1^2 + 4r_1^2r_2 - 2r_1^2r_2^2 + 1 - 2r_2^2 + r_2^4)$$
$$= 4(1-r_2)^2(r_2-2r_1^2+1)^2$$

and the facts that $1 - r_2 > 0$ and

$$r_2 - 2r_1^2 + 1 > 0 \quad \Leftrightarrow \quad r_2 > 2r_1^2 - 1,$$

where $r_2 > 2r_1^2 - 1$ is the constraint obtained for the $m = 2$ case from (8.55), we have

$$W^{1/2} = 2(1-r_2)(r_2-2r_1^2+1).$$

Then $(-B \pm \sqrt{B^2 - 4AC})/(2A)$ simplifies to (noting that A, B, C all have the same denominators)

$$\frac{2r_1(r_1^2-2r_2+r_2^2) \pm \sqrt{W}}{-2(1-r_1^2)}$$

$$= \frac{r_1(r_1^2-2r_2+r_2^2) \pm (1-r_2)(r_2-2r_1^2+1)}{-(1+r_1)(1-r_1)}$$

$$= \frac{2r_1r_2 - r_1 + r_1^2 + r_2^2 - 1}{1+r_1}, \quad \frac{-(r_1 - 2r_1r_2 + r_1^2 + r_2^2 - 1)}{1-r_1}$$

$$= \frac{2r_1r_2 + r_1^2 + r_2^2}{1+r_1} - 1, \quad \frac{-(-2r_1r_2 + r_1^2 + r_2^2)}{1-r_1} + 1 = \frac{(r_1+r_2)^2}{1+r_1} - 1, \quad \frac{(r_2-r_1)^2}{r_1-1} + 1,$$

and computation with some values of r_1 and r_2 shows that the ordering is

$$\frac{(r_1+r_2)^2}{r_1+1} - 1 < r_3 < \frac{(r_1-r_2)^2}{r_1-1} + 1.$$

Solution to Problem 8.3 For the AR(2) model $Y_t = 1.2Y_{t-1} - 0.8Y_{t-2} + U_t$, the methods in Section 6.1.2 lead to $\rho_1 = 2/3$, $\rho_2 = 0$, and $\rho_3 = -8/15$, and (8.25) yields

$$\mathbf{W} = \begin{bmatrix} 0.0617 & \cdot & \cdot \\ 0.1481 & 0.4667 & \cdot \\ 0.1284 & 0.5748 & 0.9471 \end{bmatrix} \quad \text{or} \quad \mathbf{W}_{\text{corr}} = \begin{bmatrix} 1 & \cdot & \cdot \\ 0.873 & 1 & \cdot \\ 0.531 & 0.865 & 1 \end{bmatrix},$$

i.e., asymptotically,

$$\begin{pmatrix} R_1 - 2/3 \\ R_2 - 0 \end{pmatrix} \sim \mathrm{N}\left[\begin{pmatrix} 0 \\ 0 \end{pmatrix}, T^{-1}\begin{pmatrix} 0.0617 & 0.1481 \\ & 0.4667 \end{pmatrix}\right]. \tag{8.56}$$

The code in Listing 8.12 produces the graphs shown in Figure 8.22.

Solution to Problem 8.4 The program is given in Listing 8.13.

Solution to Problem 8.5 With $p = 2$, to simplify matters, we eliminate the double subscript and let X_1, X_2, Y_1 and Y_2 be mean zero, finite variance random variables such that $X_i \perp Y_j$, for all combinations of $i, j \in \{1, 2\}$.

Thus, we wish to show that

$$\mathbb{V}\left(\begin{bmatrix} X_1 + Y_1 \\ X_2 + Y_2 \end{bmatrix}\right) = \mathbb{V}\left(\begin{bmatrix} X_1 \\ X_2 \end{bmatrix}\right) + \mathbb{V}\left(\begin{bmatrix} Y_1 \\ Y_2 \end{bmatrix}\right).$$

Let $Z_i = X_i + Y_i$, so that

$$\mathbb{V}\left(\begin{bmatrix} Z_1 \\ Z_2 \end{bmatrix}\right) = \mathbb{E}\left[\begin{bmatrix} Z_1 \\ Z_2 \end{bmatrix}\begin{bmatrix} Z_1 & Z_2 \end{bmatrix}\right] = \begin{bmatrix} \mathbb{E}[Z_1^2] & \mathbb{E}[Z_1Z_2] \\ \cdot & \mathbb{E}[Z_2^2] \end{bmatrix}$$

or

$$\mathbb{V}\left(\begin{bmatrix} X_1 + Y_1 \\ X_2 + Y_2 \end{bmatrix}\right) = \begin{bmatrix} \mathbb{E}[(X_1+Y_1)^2] & \mathbb{E}[(X_1+Y_1)(X_2+Y_2)] \\ \cdot & \mathbb{E}[(X_2+Y_2)^2] \end{bmatrix}$$

$$= \begin{bmatrix} \mathbb{E}[X_1^2 + 2X_1Y_1 + Y_1^2] & \mathbb{E}[X_1X_2 + X_1Y_2 + Y_1X_2 + Y_1Y_2] \\ \cdot & \mathbb{E}[X_2^2 + 2X_2Y_2 + Y_2^2] \end{bmatrix}.$$

```
% The simulated pdf
% Note use of the larger number of bins in hist3. Using the
%    the default leads to a poor looking contour plot.
clear all, up=1000000; T=20; pair=zeros(up,2);
X=[ones(T,1)]; M=makeM(X); % this is the general setup
for i=1:up
  if mod(i,1000)==0, i, end
  y=armasim(T,1,[1.2 -0.8],[],i); resid=M*y;
  pair(i,:)=sampleacf(resid,2)';
end
[ heights, xycoord ]=hist3(pair,[20,20]);
contour(xycoord{1},xycoord{2},heights',7)
grid, set(gca,'fontsize',14), axis([0.47 0.77 -0.3 0.3])

r2b=[]; for r1=0.47:0.01:0.77, r2b=[r2b 2*r1^2-1]; end
hold on, h=plot(0.47:0.01:0.77,r2b,'r:'), set(h,'linewidth',2), hold off
% These two lines get repeated in each segment below too

% The SPA pdf
clear all, T=20; X=[ones(T,1)]; a1=1.2; a2=-0.8;
Psiinv=inv(leeuwAR([a1 a2],T));
c1=1; for r1=0.47:0.01:0.77
  c2=1; for r2=-0.3:0.01:0.3
    rvec=[r1 r2]; c1c2=[c1 c2], f(c1,c2)=sacfpdf(rvec,X,Psiinv);
    c2=c2+1;
  end
  c1=c1+1;
end
contour([0.47:0.01:0.77],[-0.3:0.01:0.3],f')
grid, set(gca,'fontsize',14), axis([0.47 0.77 -0.3 0.3])

% Asymptotic pdf
clear all, T=20; a1=1.2; a2=-0.8;
mu=[2/3 0]'; Sigma=[0.0617 0.1481; 0.1481 0.4667] / T;
c1=1; for r1=0.47:0.01:0.77
  c2=1; for r2=-0.3:0.01:0.3
    rvec=[r1 r2]; fasy(c1,c2)=mvnpdf(rvec',mu,Sigma); c2=c2+1;
  end, c1=c1+1;
end
contour([0.47:0.01:0.77],[-0.3:0.01:0.3],fasy')
grid, set(gca,'fontsize',14), axis([0.47 0.77 -0.3 0.3])
```

Program Listing 8.12: Generates the graphs in Figure 8.22.

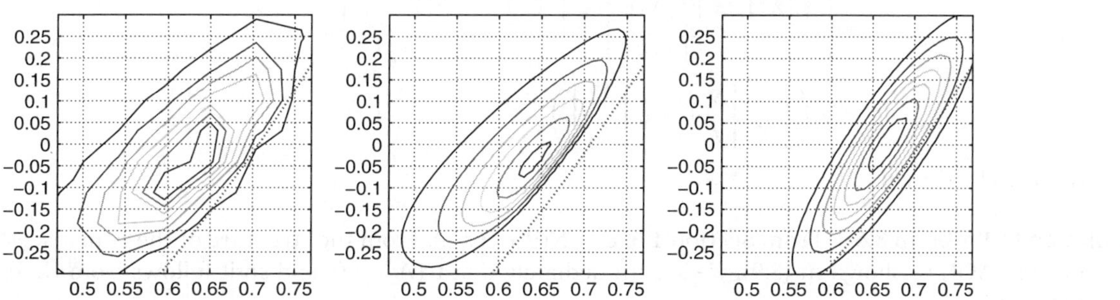

Figure 8.22 Similar to Figure 8.15 but based on $T = 20$ and an AR(2) model with $a_1 = 1.2$ and $a_2 = -0.8$. The left graph is based on simulation, the middle graph is the SPA, and the right graph is the asymptotic distribution given in (8.56).

```
% Simulation
clear all, up=100000; T=10; a1=1.2; a2=-0.8; pair=zeros(up,2);
X=[ones(T,1)]; % M=makeM(X);
for i=1:up
  if mod(i,1000)==0, i, end
  y=armasim(T,1,[a1 a2],[],i); resid=y; % resid=M*y;
  pair(i,:)=sampleacf(resid,2)';
end
eps=0.0005; targ=2/3; lo=targ-eps; hi=targ+eps;
pp=pair(:,1); bool = find((pp<hi) & (pp>lo));
use=pair(bool,2); length(use) % do we have enough data?
[simpdf,grd] = kerngau(use);

% SPA
Psiinv=inv(leeuwAR([a1 a2],T));
r1=targ; r2vec=-0.05:0.01:0.45; f=zeros(length(r2vec),1);
for i=1:length(r2vec)
  r2=r2vec(i); rvec=[r1 r2];
  f(i)=sacfpdf(rvec,X,Psiinv); % not yet normalized.
end
denom1 = sacfpdf(r1,X,Psiinv);
denom2 = sum(f)*0.01; % approximate the area under the pdf.
f = f / denom2;

plot(grd,simpdf,'r--', r2vec,f,'b-')
set(gca,'fontsize',14), axis([-0.05 0.45 0 6.1])
```

Program Listing 8.13: The first segment of code simulates the SACF (R_1, R_2) for an AR(2) model with unknown mean and parameters $a_1 = 1.2$ and $a_2 = -0.8$, and produces a set of R_2 realizations such that $R_1 \approx 2/3$. Accuracy can be enhanced by increasing the number of replications (parameter `up`) and decreasing the width of the interval for R_1 (parameter `eps`). The second segment computes the SPA (8.34) but uses simple numeric integration to get the integration constant instead of the denominator in (8.34). In this case, with $R_1 = 2/3$, `denom1` is 2.70 and `denom2` is 3.06. For $R_1 = 1/3$, `denom1` is 0.845 and `denom2` is 0.850.

From the orthogonality, $\mathbb{E}[X_1Y_1] = \mathbb{E}[X_1]\mathbb{E}[Y_1] = 0 \cdot 0 = 0$, and, similarly, $\mathbb{E}[X_1Y_2] = \mathbb{E}[X_2Y_1] = 0$, so that

$$\begin{aligned}\mathbb{V}\left(\begin{bmatrix} X_1 + Y_1 \\ X_2 + Y_2 \end{bmatrix}\right) &= \begin{bmatrix} \mathbb{E}[X_1^2] + \mathbb{E}[Y_1^2] & \mathbb{E}[X_1X_2] + \mathbb{E}[Y_1Y_2] \\ \cdot & \mathbb{E}[X_2^2] + \mathbb{E}[Y_2^2] \end{bmatrix} \\ &= \begin{bmatrix} \mathbb{E}[X_1^2] & \mathbb{E}[X_1X_2] \\ \cdot & \mathbb{E}[X_2^2] \end{bmatrix} + \begin{bmatrix} \mathbb{E}[Y_1^2] & \mathbb{E}[Y_1Y_2] \\ \cdot & \mathbb{E}[Y_2^2] \end{bmatrix} \\ &= \mathbb{V}\left(\begin{bmatrix} X_1 \\ X_2 \end{bmatrix}\right) + \mathbb{V}\left(\begin{bmatrix} Y_1 \\ Y_2 \end{bmatrix}\right),\end{aligned}$$

as was to be shown.

Solution to Problem 8.6 From Section 1.5, $\hat{\epsilon} \sim \mathrm{N}(\mathbf{0}, \sigma^2\mathbf{I}_{T-k})$. From the structure of the $\mathbf{A}$ matrices for the SACF, as shown in (8.9), we see immediately that $\mathrm{tr}(\mathbf{A}_s) = 0$, and so it follows from (B.5) that $\mathbb{E}[\breve{R}_s] = 0$.

To show symmetry, note that the numerator of $\breve{R}_s$ is

$$\widehat{\boldsymbol{\epsilon}}'\mathbf{A}_s\widehat{\boldsymbol{\epsilon}} = \sum_{t=s+1}^{T-k} \widehat{\epsilon}_t\widehat{\epsilon}_{t-s}. \tag{8.57}$$

This has expectation zero, from (A.6). Observe that, from (B.22), the structure in (8.57) is not preserved for the R_s, i.e., for the elements of the SACF based on the usual regression residuals. It is the structure in (8.57) that implies symmetry. To illustrate, take $s = 2$ as an example. Then the numerator of $\breve{R}_2$ is

$$\widehat{\boldsymbol{\epsilon}}'\mathbf{A}_2\widehat{\boldsymbol{\epsilon}} = \sum_{t=3}^{T-k} \widehat{\epsilon}_t\widehat{\epsilon}_{t-2} = \widehat{\epsilon}_3\widehat{\epsilon}_1 + \widehat{\epsilon}_4\widehat{\epsilon}_2 + \widehat{\epsilon}_5\widehat{\epsilon}_3 + \cdots + \widehat{\epsilon}_{T-k}\widehat{\epsilon}_{T-k-2}$$

and

$$\begin{aligned} -\widehat{\boldsymbol{\epsilon}}'\mathbf{A}_2\widehat{\boldsymbol{\epsilon}} &= (-\widehat{\epsilon}_3)\widehat{\epsilon}_1 + (-\widehat{\epsilon}_4)\widehat{\epsilon}_2 + \widehat{\epsilon}_5(-\widehat{\epsilon}_3) + \widehat{\epsilon}_6(-\widehat{\epsilon}_4) + \cdots \\ &= (\mathbf{S}_2\widehat{\boldsymbol{\epsilon}})'\mathbf{A}_2(\mathbf{S}_2\widehat{\boldsymbol{\epsilon}}), \end{aligned}$$

where $\mathbf{S}_2 := \text{diag}(1, 1, -1, -1, 1, 1, \ldots)$. As $\mathbf{S}_2\mathbf{S}_2' = \mathbf{I}_{T-k}$, $\mathbf{S}_2\widehat{\boldsymbol{\epsilon}} \sim \text{N}(\mathbf{0}, \sigma^2\mathbf{I}_{T-k})$, showing that $\widehat{\boldsymbol{\epsilon}}'\mathbf{A}_2\widehat{\boldsymbol{\epsilon}}$ and $-\widehat{\boldsymbol{\epsilon}}'\mathbf{A}_2\widehat{\boldsymbol{\epsilon}}$ have the same distribution, i.e., the distribution of $\widehat{\boldsymbol{\epsilon}}'\mathbf{A}_2\widehat{\boldsymbol{\epsilon}}$ is symmetric about zero. Using the facts that $\widehat{\boldsymbol{\epsilon}}'\widehat{\boldsymbol{\epsilon}}$ is always positive and that $\breve{R}_s$ has mean zero, it follows that the distribution of $\breve{R}_2$ is also symmetric about zero. A similar argument can be applied to each $\breve{R}_s$, $s = 1, 2, \ldots$.

As a numerical illustration, let $\mathbf{X}$ have an intercept and trend, and take the sample size to be $T = 8$. The following code uses programs `makeA` from Listing 8.1 and `cdfratio` from Listing A.3 to compute the cdf of $\breve{R}_1$ through $\breve{R}_4$ at zero.

```
T=8; X=[ones(T,1) (1:T)']; k=2;
for s=1:4, A=makeA(T-k,s); cdfratio(0,A,eye(T-k),1), end
```

We indeed get 0.5000 as the answer for each s. To check for symmetry, use the following code, which compares $F_{\breve{R}_s}(z)$ with $1 - F_{\breve{R}_s}(-z)$ over a grid of z values.

```
s=2; A=makeA(T-k,s); z=0:0.01:1; f1=  cdfratio( z,A,eye(T-2),1);
f2=1-cdfratio(-z,A,eye(T-2),1); plot(z,f1-f2)
```

The difference `f1-f2` is zero for all values of z.

This will not be the case for the SACF based on the usual regression residuals. This can be seen numerically by comparing $F_{R_s}(m + z)$ with $1 - F_{R_s}(m - z)$ over a grid of z values, where $m = \mathbb{E}[R_s]$, which we know is not zero in general. The setup is that in (B.22), and the following code is used for computation:

```
G=makeG(X); s=2; Atilde = G*makeA(T,s)*G'; m=mean(diag(Atilde))
f1=  cdfratio(m+z,Atilde,eye(T-2),1);
f2=1-cdfratio(m-z,Atilde,eye(T-2),1);
plot (z,f1-f2)
```

This results in a plot clearly showing that the density is not symmetric.

9

ARMA Model Identification

There are two things you are better off not watching in the making: sausages and econometric estimates.

(Edward Leamer, 1983, p. 37)

Establishing plausible values of p and q associated with an ARMA(p, q) model corresponding to a given set of time-series data constitutes an important part of what is referred to as (univariate time series) **model identification**, a term and procedure popularized by the highly influential book on time-series analysis by the prolific George Box and Gwilym Jenkins, the first of which appeared in 1970; see Box et al. (2008).[1] Other aspects of the Box and Jenkins paradigm include parameter estimation and out-of-sample forecasting, which were covered in previous chapters.

9.1 Introduction

> One reason why Akaike does not accept the problem of ARMA order selection as that of estimating an unknown true order, (m_0, h_0), say, is that there is no fundamental reason why a time series need necessarily follow a 'true' ARMA model.
>
> (Raj J. Bhansali, 1993, p. 51)

Before proceeding with methods for choosing p and q, it is important to emphasize that the term "model identification" includes a former, and important, step concerned with deciding if and what data transformations are required to induce stationarity, such as removing a time trend or other regressor effects, taking logs, or first differences, or even difference of logs, etc. Pankratz (1983, Ch. 7), Lütkepohl and Krätzig (2004, Ch. 2), and Box et al. (2008), among others, discuss appropriate data transformations for inducing stationarity. In what follows, we will assume that the initial series has been appropriately transformed, and the resulting series is not only (weak) stationary, but also a realized sample path from a stationary, invertible ARMA(p, q) model.

1 As for a bit of historical trivia, Box' doctoral thesis advisor was Egon Pearson, son of Karl Pearson. Pearson senior and Ronald Fisher had a longstanding rivalry that ultimately prevented Fisher from ever formally having an academic chair in statistics. In 1978, George Box married Joan Fisher, one of Fisher's (five) daughters.

Linear Models and Time-Series Analysis: Regression, ANOVA, ARMA and GARCH, First Edition. Marc S. Paolella.

Emphasizing the message in the above quote from Bhansali (1993), it is essential to realize that an ARMA(p, q) model is nothing but an approximation to the actual, unknown data generating process, and there are no "true" values of p and q that need to be determined. Instead, values of p and q are selected (and the corresponding parameters are then estimated) that provide an acceptable approximation to the true, but unknown (and almost always unknowable) data generating process. The ARMA class of models is quite rich in the sense that, even with $p + q$ relatively small, a very wide variety of correlation structures are possible. The goal of identification is to select the most appropriate choice (or choices) of p and q.

Given the flexibility of the autocorrelation structure possible with ARMA models, it might seem tempting to just pick large enough values of p and q, perhaps as a function of the available sample size, so as to ensure that the autocorrelation structure of the given data set is arbitrarily closely replicated by the ARMA model. We learned via the demonstration in Figure 8.12 that this is possible just by fitting an MA(q) model with large enough q, even if the data are not generated by an MA model. (Of course, a high order AR model will also work, and is easier to estimate.) The problem with such a strategy is that the parameters need to be estimated, and the more there are, the lower will be their accuracy.

Furthermore, when such a model is used to make forecasts, it tends to perform inadequately, if not disastrously. Such a model is said to be **overfitted**. A better model will embody the principle of parsimony, recalling the discussion at the beginning of Section 6.2.1: The goal is to find the smallest values of p and q that capture "an adequate amount" or the "primary features of", the correlation structure. The reader is correct in having the feeling that there is a considerable amount of subjectivity involved in this activity! Fortunately, *some* of this subjectivity is removable. The remainder of this chapter discusses several ways of model identification; they need not be used exclusively, but can (and usually are) combined.

As mentioned, it is important to keep in mind that the d.g.p. of most real phenomena are complicated, and a stationary, Gaussian ARMA process is just an approximation. Numerous variations of this model class have been proposed that involve adding nonlinearity aspects to the baseline ARMA model, though their efficacy for forecasting has been questioned. As is forcefully and elegantly argued in Zellner (2001) in a general econometric modeling context, it is worthwhile having an ordering of possible models in terms of **complexity** (a term only informally defined), with higher probabilities assigned to simpler models. Moreover, Zellner (2001, Sec. 3) illustrates the concept with the choice of ARMA models, discouraging the use of MA components in favor of pure AR processes, even if it entails more parameters, because "counting parameters" is not necessarily a measure of complexity (see also Keuzenkamp and McAleer, 1997, p. 554). This agrees precisely with the general findings of Makridakis and Hibon (2000, p. 458), who state that "statistically sophisticated or complex models do not necessarily produce more accurate forecasts than simpler ones".

We begin in Section 9.2 by discussing the classic method, which, like reading palms of hands, or tea leaves at the bottom of the cup, involves visual inspection on behalf of the modeler and "analysis" of the sample correlograms. Section 9.3 considers the standard frequentist paradigm of significance testing. Section 9.4 presents the use of penalty criteria, this being the most used, and arguably most useful method in terms of general applicability, ease of implementation, and effectiveness. Section 9.5 considers the aforementioned aspect of complexity, restricting the model class (initially at least) to just AR(p), and develops a near-exact testing paradigm that explicitly supports the use of exogenous regressors. It is shown to outperform the penalty criteria in several cases. Section 9.6 shows a simple, fast method for selecting p for an AR(p) model. Finally, Section 9.7 briefly discusses more sophisticated pattern recognition methods of determining p and q in the ARMA modeling framework.

9.2 Visual Correlogram Analysis

> The list of individuals and firms that have been badly hurt financially by inadequate "reading of the tea leaves" is daunting, including Sir Isaac Newton, and more recently, Long-Term Capital Management…
>
> (Steve Pincus and Rudolf E. Kalman, p. 13713, 2004)

Computation and visual inspection of the sample correlograms was popularized in the 1960s, and showcased in the pioneering 1970 monograph by George Box and Gwilym Jenkins (see the subsequent fourth edition, Box et al., 2008). In light of the lack of computing power that is now ubiquitously available, the technique had its merits, and is still instructional and taught in the ARMA model building framework. It involves examining the sample ACF (SACF) and sample PACF (SPACF) to get candidate values for p and q. As discussed at the end of Section 8.2, if the SACF appears to "cut off", then one can postulate that the model is an MA(q), where q is taken to be the number of spikes before "cutoff". Similarly, if the SPACF cuts off, then an AR(p) model would be declared. Numerous examples of this, with real data, are provided in (the arguably now outdated, but well-written and, at the time, useful) Abraham and Ledolter (1983). Both authors were doctoral students of George Box.

The idea is illustrated in Figures 8.3 and 8.19, which show the SACF and SPACF of four simulated AR(1) time series with parameter $a = 0.5$ and based on $T = 50$ observations. In particular, note from Figure 8.19 that the SPACF is not exactly zero after the first spike, but most of them are indeed within the asymptotic one-at-a-time 95% confidence interval band. Keep in mind the nature of these bands: They are only asymptotically valid, so that in small samples their accuracy is jeopardized. Furthermore, if the time series under investigation consists of regression residuals, then, as was illustrated in Figure 8.6 and those in Section 8.1.3.3, the $\mathbf{X}$ matrix can play a major role in the actual distribution of the elements of the SACF and SPACF, particularly for sample sizes under, say, $T = 100$. Secondly, as these are one-at-a-time 95% intervals, one *expects* one spike in 20, on average, to fall outside the interval when the null hypothesis of no autocorrelation is true.

What one typically does in practice (and is one of the reasons giving rise to the famous quote by Ed Leamer above) is add some personal, subjective, a priori beliefs into the decision of which spikes to deem significant (based presumably on the culmination of experience on behalf of the modeler). These beliefs typically include considering low-order spikes to be more important (for non-seasonal data of course), so that, for example, in the bottom left panel of Figure 8.19, one might well entertain an AR(3) model. If, however, a "lone spike" appears, of high order (say, larger than 8) and of length not greatly exceeding the edge of the confidence band, then it would be dismissed as "probably arising just from sampling error". Further complicating matters is the correlation of the spikes in the correlograms, so that "significant" spikes tend to arise in clusters.

Of course, if the true process comes from a mixed ARMA model, then neither correlogram cuts off. Figure 9.1 provides an example with artificial data, consisting of 100 points generated from an ARMA(2,2) model with parameters $a_1 = 1.1$, $a_2 = -0.4$, $b_1 = -0.5$, $b_2 = 0.7$, $c = 0$, and $\sigma^2 = 1$. The top two panels show the theoretical ACF (TACF) and theoretical PACF (TPACF) corresponding to the process. The second row shows a time plot of the actual data.

This particular realization of the process is interesting (and not unlikely), in that certain segments of the data appear to be from a different process. A researcher confronted with this data might be inclined to find out what (say, macroeconomic) event occurred near observation 35 that reversed a downward trend to an upward one, amidst clear periodic behavior, only to change again to a

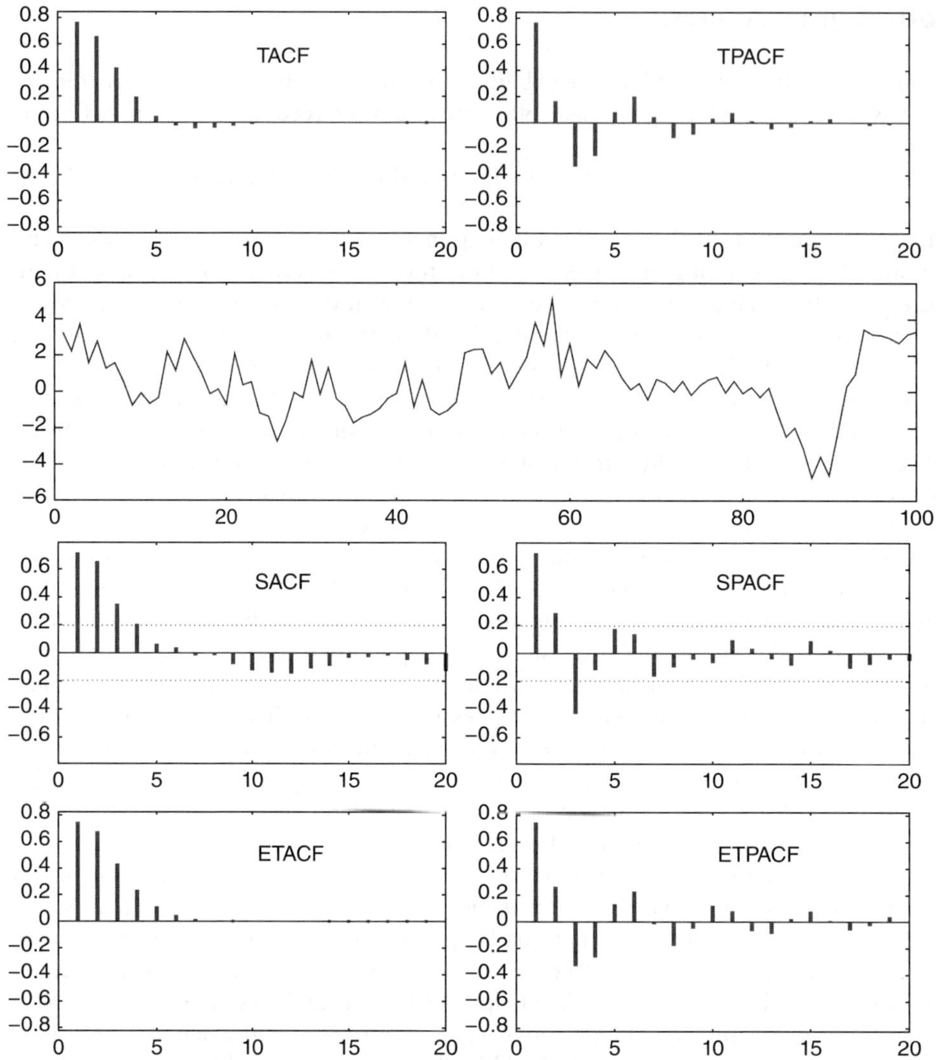

Figure 9.1 Top panels are the theoretical correlograms corresponding to a stationary and invertible ARMA(2,2) model with parameters $a_1 = 1.1, a_2 = -0.4, b_1 = -0.5, b_2 = 0.7, c = 0$, and $\sigma^2 = 1$. The second row shows a realization of the process, with its sample correlograms plotted in the third row. The last row shows the theoretical ACF and PACF but based on the *estimated* ARMA(2,2) model of the data.

downward trend without periodic behavior, and, finally, "crash" near observation 90, but bounce back abruptly in a rallying trend. Of course, having generated the process ourselves, we know it is indeed stationary and what appear to be anomalies in the data are just artifacts of chance. This illustrates the benefit of parsimonious modeling: If we were to introduce dummy exogenous variables to account for the handful of "outliers" in the data, and/or use more sophisticated structures to capture the apparent changes in the model, etc., it would all be for nought: The model arrived at after hours or days of serious academic contemplation and work would be utterly wrong, and while able to fit the observed data well, would produce unreliable forecasts, not to mention a false understanding of causal economic relationships.

The reader should not get the impression that most, if not all, data sets are actually stationary; on the contrary, most real data sets are most likely *not* stationary! But the nature of the non-stationarities is so difficult to guess at that simple, parsimonious models are often preferred, as mentioned above in Section 9.1, with respect to forecasting prowess.

Returning to the identification step, the third row of Figure 9.1 shows the sample correlograms, which do indeed somewhat resemble the theoretical ones. Based on the decay of the sample ACF and the cutoff of the PACF at lag 3, it would seem that an AR(3) model would be appropriate. The last row shows the theoretical correlograms that correspond to the *estimated* ARMA(2,2) model (assuming a known mean of zero). The m.l.e. values (and approximate standard errors in parentheses) are $\hat{a}_1 = 0.946(0.12)$, $\hat{a}_2 = -0.245(0.12)$, $\hat{b}_1 = -0.364(0.076)$, $\hat{b}_2 = 0.817(0.078)$, and $\hat{\sigma} = 0.966(0.069)$. Notice that these correlograms are closer to the true ones than are the sample correlograms. This is quite reasonable because more information is used in their construction (in particular, knowledge of the parametric model being an ARMA(2,2) and maximum likelihood estimation, as opposed simply to sample moments). Of course, this knowledge of p and q is not realistic in practical settings.

In practice, it is also a good idea to compute the correlograms for different segments of the data, the number of segments depending on the available sample size. If the data are from a stationary process, then the SACFs and SPACFs for the different segments should be similar in appearance. Figure 9.2 shows the sample correlograms corresponding to the two halves of the data under investigation. While they clearly have certain similarities, notice that the SACF from the first half appears to cut off after two large spikes (suggesting an MA(2) model), while the SACF for the second half dies out gradually,

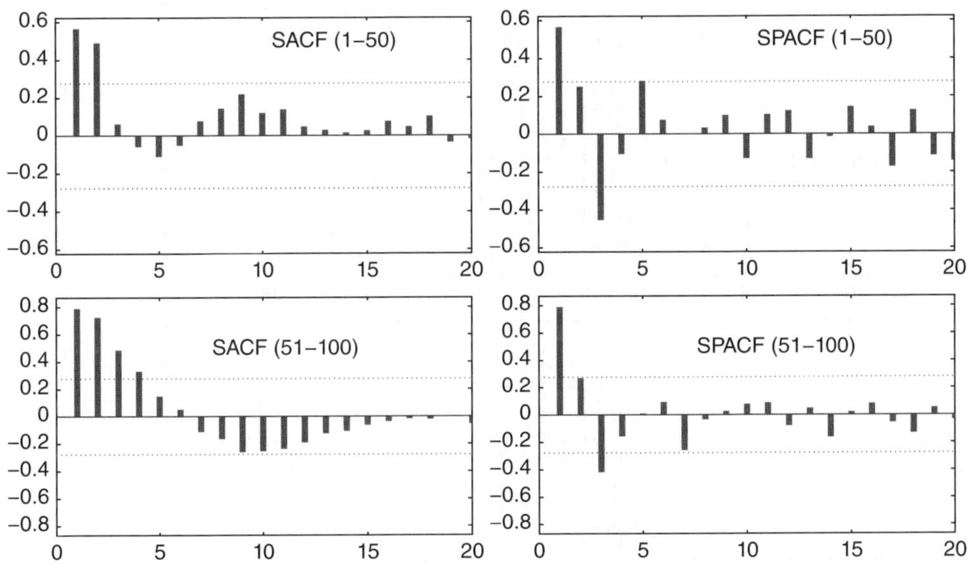

Figure 9.2 An informal graphical test for covariance stationarity is to compute the sample correlograms for non-overlapping segments of the data.

indicative of an AR or ARMA process with $p > 0$. Assuming stationarity, we add to our collection of tentative models an MA(2) and an ARMA(1,1).

Once a handful of candidate p, q values are decided upon, the models are estimated and the residuals are computed. Then the SACF and SPACF correlograms of the residuals can be inspected, which we denote as RSACF and RSPACF, respectively. Ideally, we would find the smallest values of p and q such that the RSACF and RSPACF appear to correspond to white noise. The true sampling distributions of the RSACF and RSPACF are far more difficult than those of the SACF and RPACF—which are themselves intractable and can only be approximated, recalling the discussion in Section 8.1.3. As such, we only consider their asymptotic distribution: Assuming an ARMA(p, q) model was fit to the data, using consistent estimators, and such that the true data generating process is indeed an ARMA(p, q), the asymptotic distributions of the RSACF and RSPACF are the same as those of the SACF and SPACF under the null hypothesis of white noise. Thus, the usual bounds corresponding to asymptotically valid one-at-a-time 95% confidence intervals can be overlaid onto the RSACF and RSPACF correlograms.

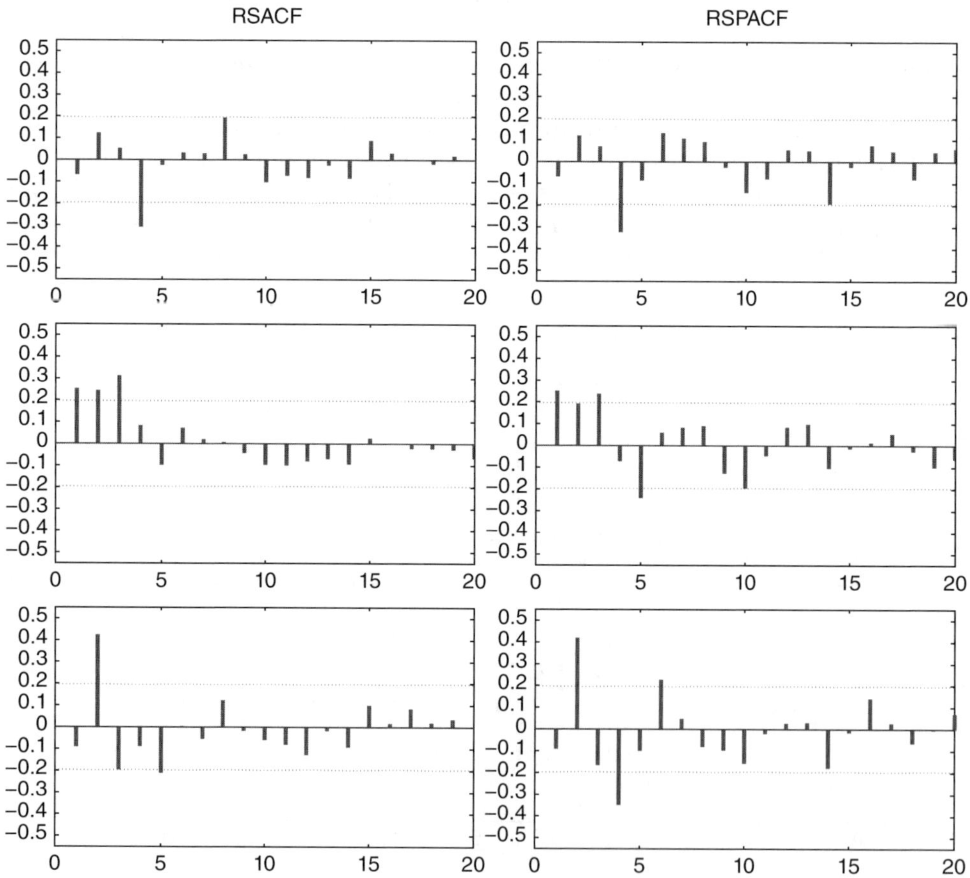

Figure 9.3 The RSACF (left) and RSPACF (right) for models AR(3) (top), MA(2) (middle), and ARMA(1,1) (bottom).

Figure 9.3 shows these plots for the three candidate models (based on residuals from the exact m.l.e., and estimated without any regressors, not even an intercept). Each of the three entertained models considerably violates the white noise hypothesis (each for different reasons) and so must be deemed inappropriate. One could ponder these further and come up with a second round of candidate values of p and q that attempts to take into account the deficiencies brought out here. Based on their RSACF and RSPACF plots, this process could be iterated until "convergence". (And lending more ammunition to Leamer's above quote.)

Below, we will introduce the method of penalty criteria for the determination of p and q. Their use suggests either an ARMA(1,2) or an ARMA(2,2). So, Figure 9.4 is similar to Figure 9.3, but corresponds to these two mixed models. In addition, because of the significant spikes at lags 4 and 5 of the previous RSPACFs, we also consider an AR(5) model. Of course, we know that the true model is ARMA(2,2), but the AR(5) could indeed be competitive because (i) it contains only one more parameter than the true model, (ii) the infinite AR representation of the true model might be adequately approximated by an AR(5), and (iii) AR models are, in general, easier to estimate and have lower "complexity" than MA or ARMA models, recalling the discussion in Section 9.1. Inspection of the plots shows that all of the

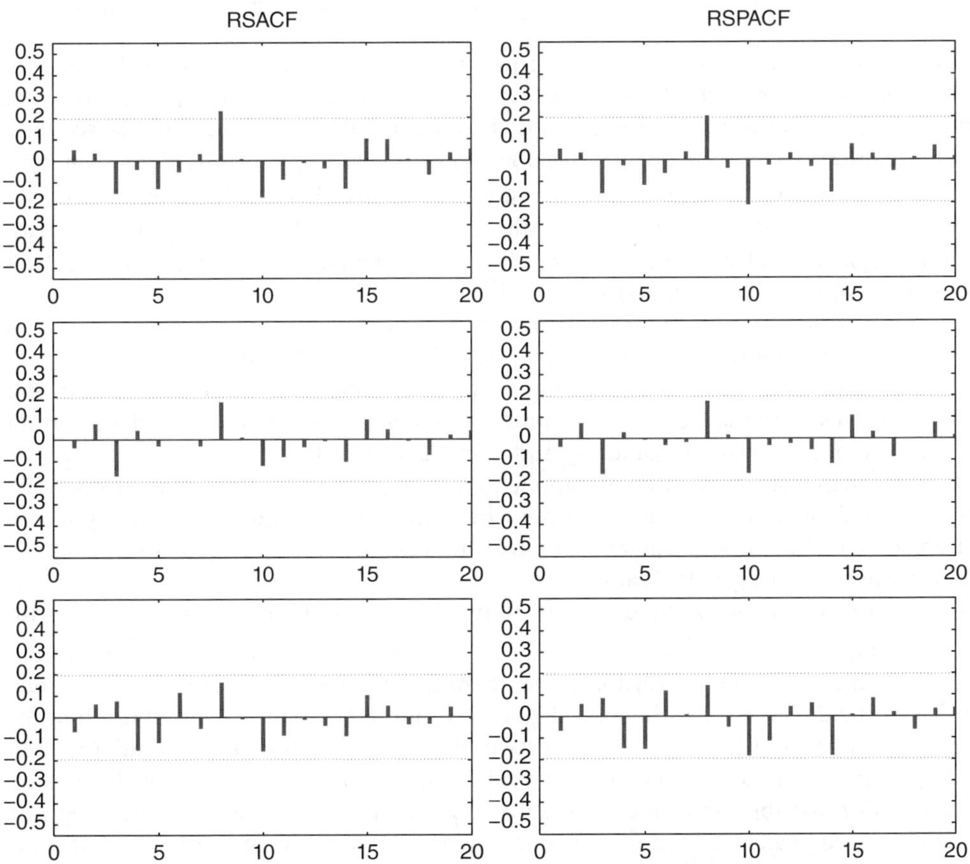

Figure 9.4 The RSACF (left) and RSPACF (right) for models ARMA(1,2) (top), ARMA(2,2) (middle), and AR(5) (bottom).

models appear adequate. The "lone spike" at lag 8 that is slightly significant for the ARMA(1,2) model should not be any cause for alarm, recalling that we expect one out of 20 spikes to be statistically significant at the 95% level when using one-at-a-time confidence intervals.

9.3 Significance Tests

> The three golden rules of econometrics are test, test, and test.
>
> (David Hendry, 1980, p. 403)

> [E]conometric testing, as against estimation, is not worth anything at all. Its marginal product is zero. It is a slack variable.
>
> (Deirde [formerly Donald] McCloskey, 2000, p. 18)

> I had been the author of unalterable evils; and I lived in daily fear, lest the monster whom I had created should perpetrate some new wickedness.
>
> (Mary Shelley's scientist, Victor Frankenstein)

In this context, and within the Neyman–Pearson hypothesis testing framework for model selection, it would seem appropriate to conduct a hypothesis test on a parameter in question, where the null hypothesis is that it is zero and the alternative is that it is nonzero. This is very straightforward when assuming the validity of the asymptotic normal distribution in small samples. In particular, let $\widehat{T}_i = \hat{\theta}_i/\widehat{\mathrm{SE}}(\hat{\theta}_i)$, $i = 1, \dots, p+q$, be the ith standardized parameter estimate. Then the hypothesis test $H_0 : \theta_i = 0$ versus $H_1 : \theta_i \neq 0$ with significance level α would reject the null if the p-value associated with $\widehat{T}_i$, as computed based on a standard normal distribution, is less than α. Equivalently, one checks if zero is contained in the $100(1-\alpha)\%$ confidence interval of θ_i.

The problem is how, if possible, to link the choice of α to the purpose of the analysis, not to mention that significance testing was not initially proposed for model selection; see the discussion in Section III.2.8 for more detail. This is also brought out in the above first two fully conflicting quotes (from two highly respected scientists). The blind use of hypothesis testing for model selection arose out of a historical quirk, misunderstanding, and convenience and strength of precedence. Its Shelleyian wickedness manifests itself in giving applied data researchers a false sense of scientific integrity and a childish algorithm for model creation: Check the p-value, and make a dichotomous decision based on $\mathbb{I}(p < 0.05)$, without any concern for the lack of ability for replication, and the connection to the purpose of studying the data.

To assess the performance of this method, for 500 simulated series, the model $Y_t = \mu + \epsilon_t$, with $\epsilon_t = a_1\epsilon_{t-1} + a_2\epsilon_{t-2} + a_3\epsilon_{t-3} + U_t$, $U_t \overset{\text{iid}}{\sim} \mathrm{N}(0, \sigma^2)$, $t = 1, \dots, T$, and $\mu = a_1 = a_2 = a_3 = 0$ was estimated using exact maximum likelihood with approximate standard errors of the parameters obtained by numerically evaluating the Hessian at the m.l.e. Figure 9.5 shows the empirical distribution of the $\tau_i = F(t_i)$, where t_i is the ratio of the m.l.e. of a_i to its corresponding approximate standard error, $i = 1, 2, 3$, and $F(\cdot)$ refers to the c.d.f. of the Student's t distribution with $T-4$ degrees of freedom. The use of the Student's t distribution is of course not exact, but can be motivated by recalling that the conditional m.l.e. is equivalent to the use of least squares, in which case the t distribution is correct. Indeed, the use of the Student's t was found to be slightly better than the standard normal for the smaller sample sizes.

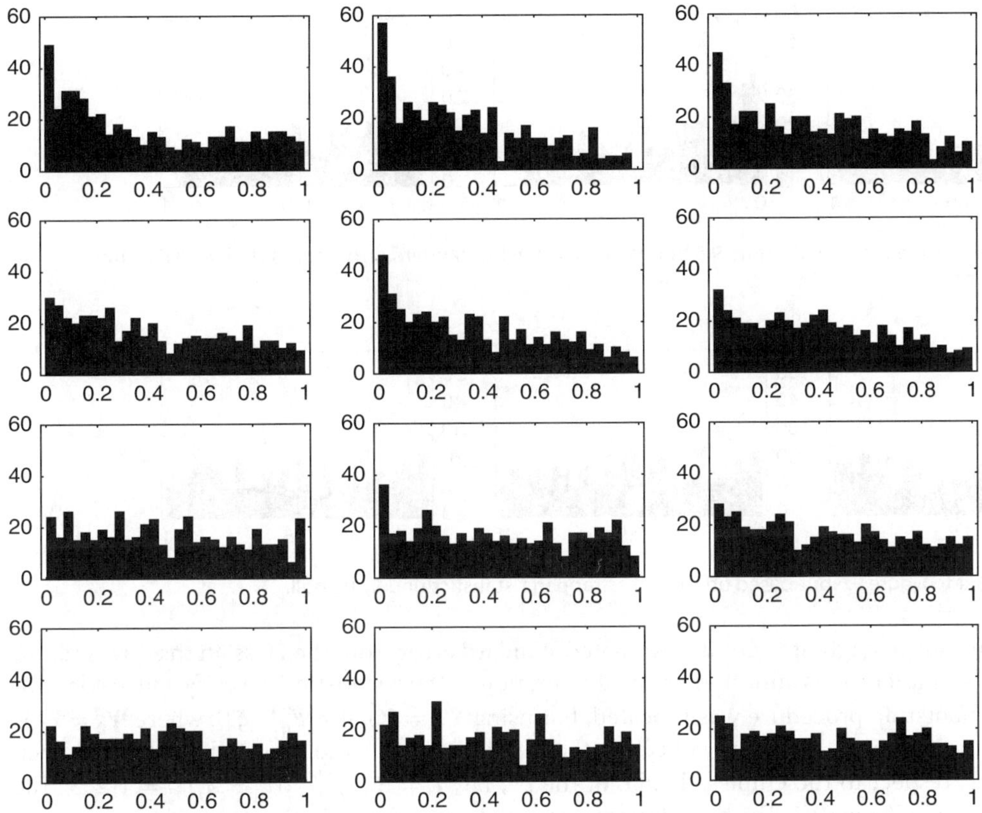

Figure 9.5 Empirical distribution of $F(t_i)$, $i = 1, 2, 3$ (left, middle, and right panels), where t_i is the ratio of the m.l.e. of a_i to its corresponding approximate standard error and $F(\cdot)$ is the Student's t cdf with $T - 4$ degrees of freedom. Rows from top to bottom correspond to $T = 15$, $T = 30$, $T = 100$, and $T = 200$, respectively.

The top two rows correspond to $T = 15$ and $T = 30$. While indeed somewhat better for $T = 30$, it is clear that the usual distributional assumption (normality, or use of a t distribution) does not hold. The last two rows correspond to $T = 100$ and $T = 200$, for which the asymptotic distribution is adequate.

Remark The most numerically sensitive part associated with the m.l.e. is the computation of the approximate standard errors. The single bootstrap can be deployed for improving their quality, as discussed in Section 7.4.3. This was done for each of the simulated time series in the $T = 15$ case using $B = 100$ bootstrap replications, the jth of which being formed as $\mathbf{Y}^{(j)} = \mathbf{X}\widehat{\boldsymbol{\beta}} + \widehat{\sigma}\widehat{\boldsymbol{\Psi}}^{-1/2}\boldsymbol{\xi}^{(j)}$, where the hatted terms denote the m.l.e. values and $\boldsymbol{\xi}^{(j)}$ was formed by with-replacement sampling from the residual vector $\widehat{\sigma}^{-1}\widehat{\boldsymbol{\Psi}}^{1/2}(\mathbf{y} - \mathbf{X}\widehat{\boldsymbol{\beta}})$. The standard error of the m.l.e. is then taken to be the sample standard deviation of the B bootstrap m.l.e. values. The performance of the resulting t-statistics are shown in Figure 9.6. Compared to the top row of Figure 9.5, there is indeed some improvement, but they are still quite far from being uniformly distributed. Qualitatively similar results were obtained by use of the parametric bootstrap, taking $\boldsymbol{\xi}^{(j)}$ to be i.i.d. standard normal draws.

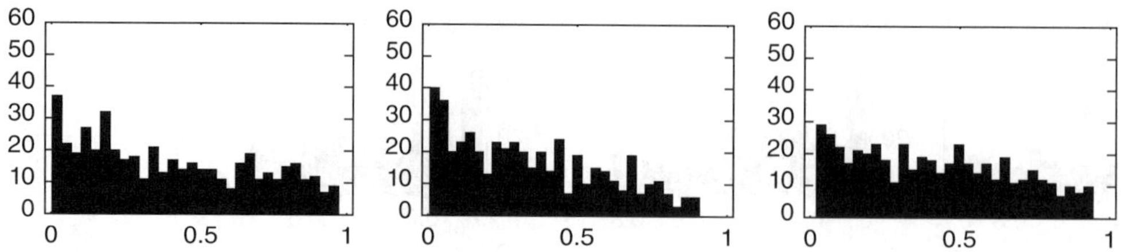

Figure 9.6 Similar to the top row of Figure 9.5 but having used m.l.e. standard errors computed from $B = 100$ bootstrap iterations.

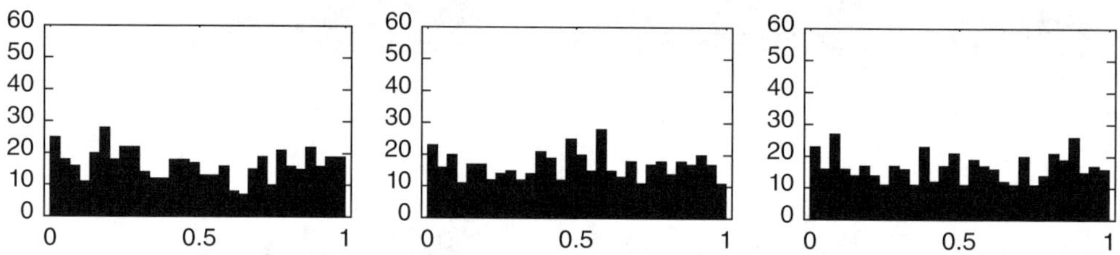

Figure 9.7 Similar to Figure 9.6 but based on the bootstrapped t-statistics under the null.

Thus, and not surprisingly, it is not the estimated standard error from the Hessian that gives rise to the problem, but rather the assumption on the distribution of the t-statistic. To verify and accommodate this, the bootstrap procedure was repeated, but using $\mathbf{Y}^{(j)} = \mathbf{X}\hat{\boldsymbol{\beta}} + \hat{\sigma}\boldsymbol{\Psi}_0^{-1/2}\xi^{(j)}$, where $\boldsymbol{\Psi}_0 = \mathbf{I}_T$ is the null assumption, $j = 1, \dots, B$, and collecting the B t-statistics $t^{(j)}$. The reported p value τ_i is then computed with respect to the empirical c.d.f. of the $t^{(j)}$, i.e., $\tau_i = B^{-1}\sum_{j=1}^{B} \mathbb{I}(t_i^{(j)} < t_i)$, $i = 1, 2, 3$. The results are shown in Figure 9.7. The bootstrap method is clearly reliable in this context.

Its drawback is the time required for computation, especially as B should be considerably larger than 100. In particular, the CACF testing paradigm, as discussed below in Section 9.5, is far faster. ■

Continuing with the data set shown in Figure 9.1, the 95% confidence intervals for the estimated ARMA(2,2) model based on the asymptotic normality of the m.l.e. are $(0.705 < a_1 < 1.187)$, $(-0.487 < a_2 < -0.0024)$, $(-0.513 < b_1 < -0.216)$, $(0.665 < b_2 < 0.969)$, and $(0.832 < \sigma < 1.100)$. It is imperative to keep in mind that these are one-at-a-time intervals, and not simultaneous. From these, there is "some evidence" that a_2 might not differ from zero. This is also in agreement with the results from Figure 9.4, which suggest that an ARMA(1,2) is adequate, compared to an ARMA(2,2). Of course, the intervals presented are based on asymptotic theory, which is not always reliable in small samples. The bootstrap can be used, as discussed directly above, to obtain more accurate intervals. Doing so with $B = 2{,}000$ replications yielded $(0.309 < a_1 < 1.158)$, $(-0.441 < a_2 < 0.423)$, $(-0.521 < b_1 < 0.346)$, $(0.308 < b_2 < 0.999)$, and $(0.824 < \sigma < 1.158)$. Some of these intervals are much larger in size and could well be too large (recall the results in Example 7.4). Nevertheless, the evidence that a_2 could be zero is now quite large (and the significance of b_1 could also be drawn into question).

Another statistic that can be used to assess if parameter θ_i is significantly different from zero is the likelihood ratio, or

$$r_i^2 = -2(\ell_{\text{res}} - \ell_{\text{unr}}) \overset{\text{asy}}{\sim} \chi_1^2, \tag{9.1}$$

where ℓ_{res} refers to the log-likelihood of the restricted model (i.e., with the parameter of interest, θ_i, restricted to zero) evaluated at the m.l.e., and ℓ_{unr} is that for the unrestricted model. We can also use the **signed likelihood ratio statistic**, given by

$$r_i = \mathrm{sgn}(\widehat{\theta}_i - \theta_{i0})\sqrt{r_i^2} = \mathrm{sgn}(\widehat{\theta}_i)\sqrt{-2(\ell_{\mathrm{res}} - \ell_{\mathrm{unr}})} \overset{\mathrm{asy}}{\sim} \mathrm{N}(0,1), \tag{9.2}$$

where θ_{i0} is the value of θ_i under the null hypothesis, which in this case is just zero. Use of (9.1) or (9.2) has two advantages over the use of a confidence interval (or p-value). First, (9.1) is easily extendable for testing the significance of a set of coefficients, with the degrees of freedom equal to the number of imposed restrictions. Second, it will usually be more accurate in the sense that (9.2) will be closer to normally distributed than $\widehat{T}_i$. This is intuitively plausible because more information goes into the calculation of r_i than $\widehat{T}_i$ via the estimation of two models instead of one. Furthermore, $\widehat{T}_i$ makes use of the approximate standard error of $\widehat{\theta}_i$, which is difficult to estimate accurately, whereas finding the maximum of the likelihood of the restricted model to a high degree of accuracy is usually quite straightforward.

Their differences in accuracy can be quickly assessed via simulation. For illustration purposes, we consider the ARMA(1,1) model with $a_1 = 0.7$ and $b_1 = -0.2$. For simulated processes from this ARMA(1,1) model, we calculate $\widehat{T} = \widehat{\theta}/\widehat{\mathrm{SE}}(\widehat{\theta})$ and the signed likelihood ratio statistic r from (9.2) corresponding to a_2, i.e., we estimate an ARMA(1,1) and an ARMA(2,1). Similarly, $\widehat{T}$ and r are calculated for b_2 by additionally estimating an ARMA(1,2). Figure 9.8 shows the results in the form of a normal qqplot using a sample size of $T = 40$ and based on 300 replications. We see immediately

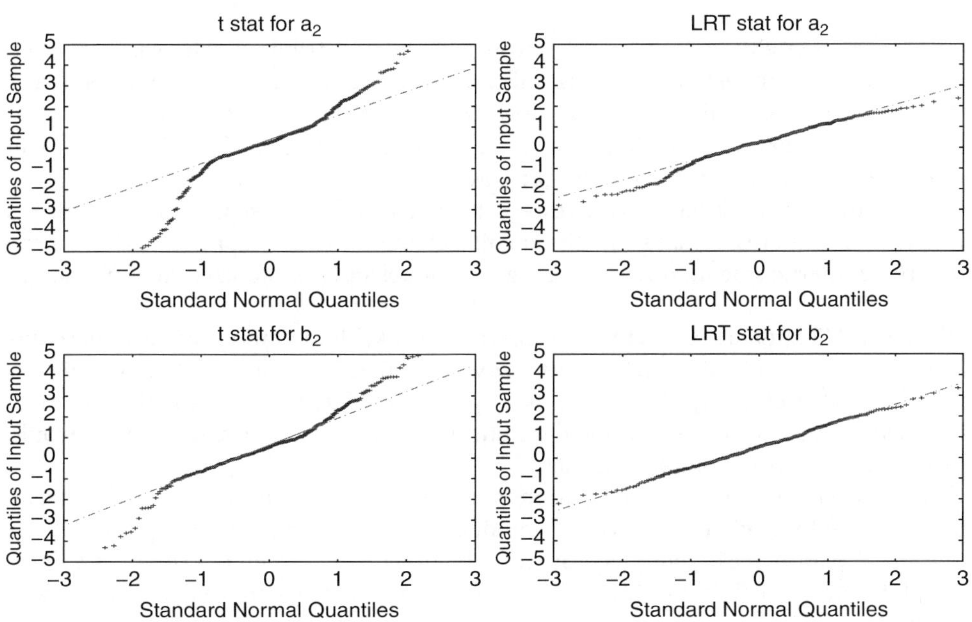

Figure 9.8 QQ plot of 300 simulated values of the statistic $\widehat{T} = \widehat{\theta}/\widehat{\mathrm{SE}}(\widehat{\theta})$ (denoted t stat) and the signed likelihood ratio statistic r (denoted LRT stat) for testing ARMA(2,1) and ARMA(1,2) when the true model is an ARMA(1,1) with $a = 0.7$ and $b = -0.2$, based on $T = 40$ observations.

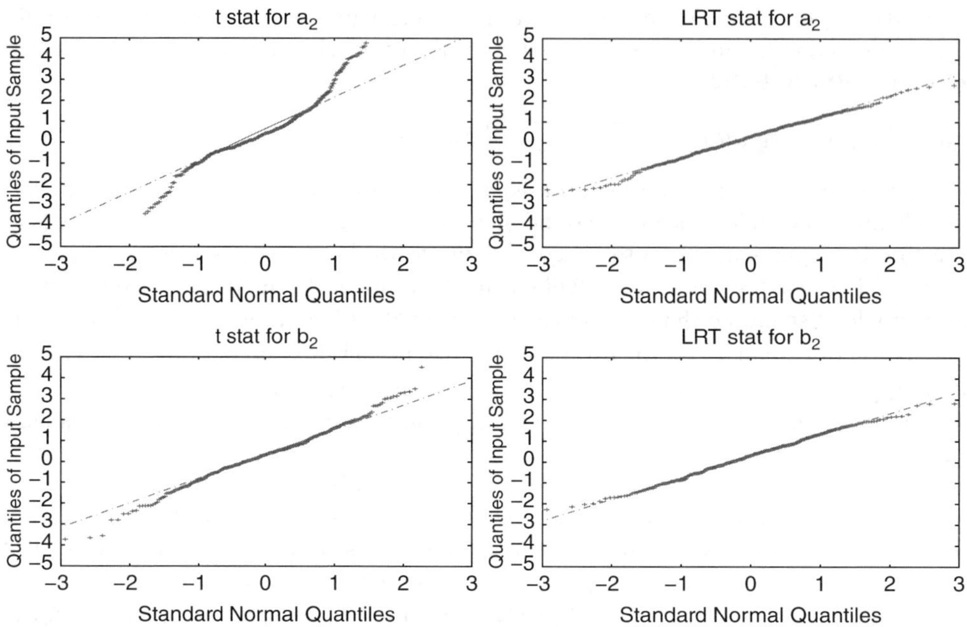

Figure 9.9 Same as Figure 9.8 but using $T = 100$ observations.

that the statistic $\widehat{T}$ for testing either a_2 or b_2 is far from normally distributed, whereas r is much closer. Furthermore, while r for testing $a_2 = 0$ is still not accurate enough for inferential use, r corresponding to testing $b_2 = 0$ is almost exactly normally distributed. Figure 9.9 is similar, but uses 100 observations. All four measures improve in terms of normality, though $\widehat{T}$ for a_2 is still unacceptable and r for a_2 is now almost exactly normally distributed. In summary, (i) for a given sample size, r appears to be more reliable with respect to its asymptotic distribution, (ii) matters improve as the sample size increases, and (iii) the quality of the normality approximation to the distribution of r depends on the true model, and, for a given model and sample size, can differ across parameters.

Returning to the ARMA(2,2) data set we are working with, the likelihood ratio statistic r^2 for comparing an ARMA(1,2) to an ARMA(2,2) is 3.218, with a p-value of 0.927. As this is just under 0.95, we would (just barely) "accept" (better: not reject) the null hypothesis of $a_2 = 0$, whereas the 95% confidence interval (based on the asymptotic normal distribution and not the bootstrap analysis) would have led us to (just barely) reject the null hypothesis.

In general, when the coefficient under investigation is not the pth AR term or the qth MA term, setting it to zero gives rise to a **subset** ARMA(p, q) model, which will have less than $p + q$ ARMA coefficients. If the "true" parameter is genuinely (or close enough to) zero, then restricting it to zero and re-estimation of the other coefficients will result in different and more accurate values, and a more parsimonious model.

9.4 Penalty Criteria

> Unthinking approaches have been the common modus operandi and using "all possible models" are frequently seen in the literature. "Let the computer find out" is a poor strategy and usually reflects the fact that the researcher did not bother to think clearly about the problem of interest and its scientific setting.
>
> The hard part, and the one where training has been so poor, is the a priori thinking about the science of the matter before data analysis—even before data collection.
>
> (Kenneth P. Burnham and David R. Anderson, 2002, p. 147 and p. 144)

We turn now to the method of order selection based on penalty functions. While there are several, the most popular penalty methods are (i) the **Akaike information criterion**, or AIC, (ii) the **corrected AIC**, or AICC, and (iii) the **(Schwarz's) Bayesian information criterion**, or BIC (or SBC), given, respectively, by

$$\mathrm{AIC} = \ln \hat{\sigma}^2 + \frac{2z}{T}, \quad \mathrm{AICC} = \ln \hat{\sigma}^2 + \frac{T+z}{T-z-2}, \quad \mathrm{BIC} = \ln \hat{\sigma}^2 + \frac{z \ln T}{T}, \tag{9.3}$$

where $z = p + q + k$, with k being the number of regressors in the mean equation, and $\hat{\sigma}^2$ is the (preferably exact) m.l.e. of σ^2. Other methods include the **final prediction error** (FPE) and the **Hannan–Quinn** (HQ) criterion,

$$\mathrm{FPE} = \hat{\sigma}^2 \frac{T+z}{T-z}, \quad \mathrm{HQ} = \ln \hat{\sigma}^2 + 2\frac{z \ln \ln T}{T}. \tag{9.4}$$

Details on the origins, justification, derivation, and asymptotic properties of these and other criteria, as well as original references, can be found in Konishi and Kitagawa (2008), Brockwell and Davis (1991), Choi (1992, Ch. 3), and McQuarrie and Tsai (1998). Lütkepohl (2005) discusses their use for identification with multivariate time series. An excellent source of information on these measures is Burnham and Anderson (2002), which, in addition to covering the technicalities of the penalty criteria, is mostly concerned with the underpinnings of model selection and their realistic use in data analysis.

One chooses the model that gives rise to the smallest criterion value. Observe the tradeoff between the first term in each criteria, $\ln \hat{\sigma}^2$, which can only get smaller as successively more terms are added to the model (much like the R^2 statistic in regression analysis, which increases even when noise vectors are included in the exogenous variable set), and the second term, which is increasing in z but tempered by the sample size. The decision of which models to include in the "contest" is, of course, a subjective one.

While not strictly necessary, calculation of these measures typically involves maximum likelihood estimation of numerous ARMA models and is, thus, somewhat computationally intensive. For example, one might consider all 36 ARMA(p, q) constellations for $0 \leqslant p \leqslant 5$ and $0 \leqslant q \leqslant 5$. This computational burden is no longer a relevant issue with modern computing power, but was not routinely feasible before, say, 1980.

Very briefly and informally, the AIC tends to overfit, while the AICC corrects this problem and has better small-sample and asymptotic properties than the AIC. The BIC also enjoys good asymptotic properties and has the tendency to select fewer parameters than the AICC.

Penalty function methods have at least three advantages over other methods, such as the significance testing paradigm of Section 9.3, the informal assessment of the sample ACFs of Section 9.2, and various pattern identification methods, as discussed below in Section 9.7. First, they are considerably simpler to understand, at least with respect to the tradeoff argument just discussed. Second, they are easily used in modeling contexts for which correlogram inspection or pattern identification methods are either far more complicated or not applicable, such as seasonal ARMA, subset ARMA, periodic ARMA, fractional integrated ARMA, time-varying parameter ARMA, multivariate ARMA, as well as other nonlinear time-series models such as threshold, bilinear, GARCH and Markov switching models. Third, they are more easily implemented in a computer algorithm to choose the best model automatically (notwithstanding the above quote by Burnham and Anderson, 2002). A final and compelling reason to prefer penalty function methods is that they work well; see the above references and also Koreisha and Yoshimoto (1991), Choi (1992, Ch. 3), and Koreisha and Pukkila (1995).

For the data set shown in Figure 9.1, when based on all 15 possible ARMA(p, q) models with $0 \leqslant p \leqslant 3$ and $0 \leqslant q \leqslant 3$, the AIC and AICC chose an ARMA(2,2) (the correct specification), while the BIC chose an ARMA(1,2). This agrees with the known behavior of BIC to prefer more parsimonious models, and also coincides with the results from the significance test on $\hat{a}_2$. This exercise also emphasizes the point that, for the model chosen by a particular criterion (in this case, the AIC or AICC), not all

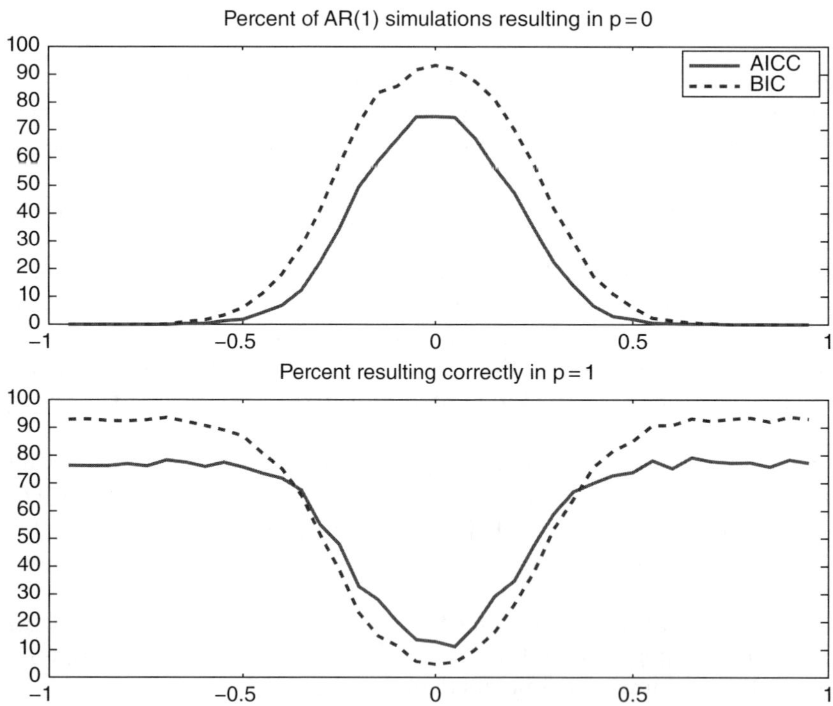

Figure 9.10 Simulation-based performance of the AICC and BIC criteria (9.3) for an AR(1) model as a function of parameter a, with $T = 50$ and $p_{\max} = 5$.

estimated parameter values from that selected model will necessarily be significantly different from zero when using the common significance level of $\alpha = 0.05$.

To further illustrate the performance of the AICC and BIC, Figure 9.10 shows the results of a simulation study of an AR(1) model with known mean over a grid of values of parameter a, using $T = 50$, exact maximum likelihood estimation, and based on 1,000 replications. The second panel shows the percentage of correct selections. We see that the AICC is better for the range $|a| < 0.4$. This is because the BIC has a higher probability of under-selection than the AICC, which becomes acute near $a = 0$. This is made clearer in the top panel, which shows the percentage of $p = 0$ selections. The reader is encouraged to replicate this study and also consider the other criteria in (9.4).

We now turn to the AR(2) model, and first use $T = 25$. With two parameters, the performance of the selection criteria cannot be as easily plotted as for the AR(1) case. To accommodate this, Figure 9.11 plots, for various a_1 and a_2 combinations spanning their support given in (6.8), one of seven symbols,

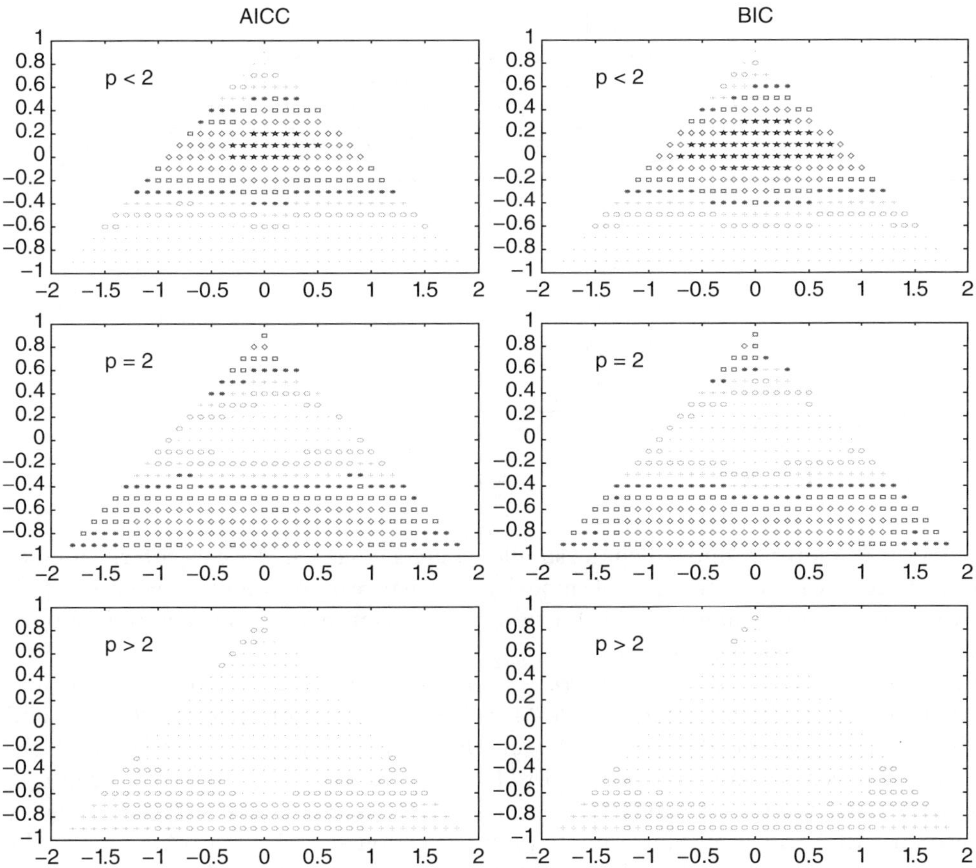

Figure 9.11 Simulation-based performance of AICC (left) and BIC (right) criteria in terms of percentage of under-selection (top), correct selection (middle), and over-selection (bottom) for an AR(2) model as a function of parameters a_1 (y–axis) and a_2 (x–axis), with $T = 25$ and $p_{max} = 4$. Legend is, for $k = 100/7$, dots 0–k%, circles k–$2k$%, plus $2k$–$3k$%, star $3k$–$4k$%, square $4k$–$5k$%, diamond $5k$–$6k$%, pentagram $6k$–$7k$%.

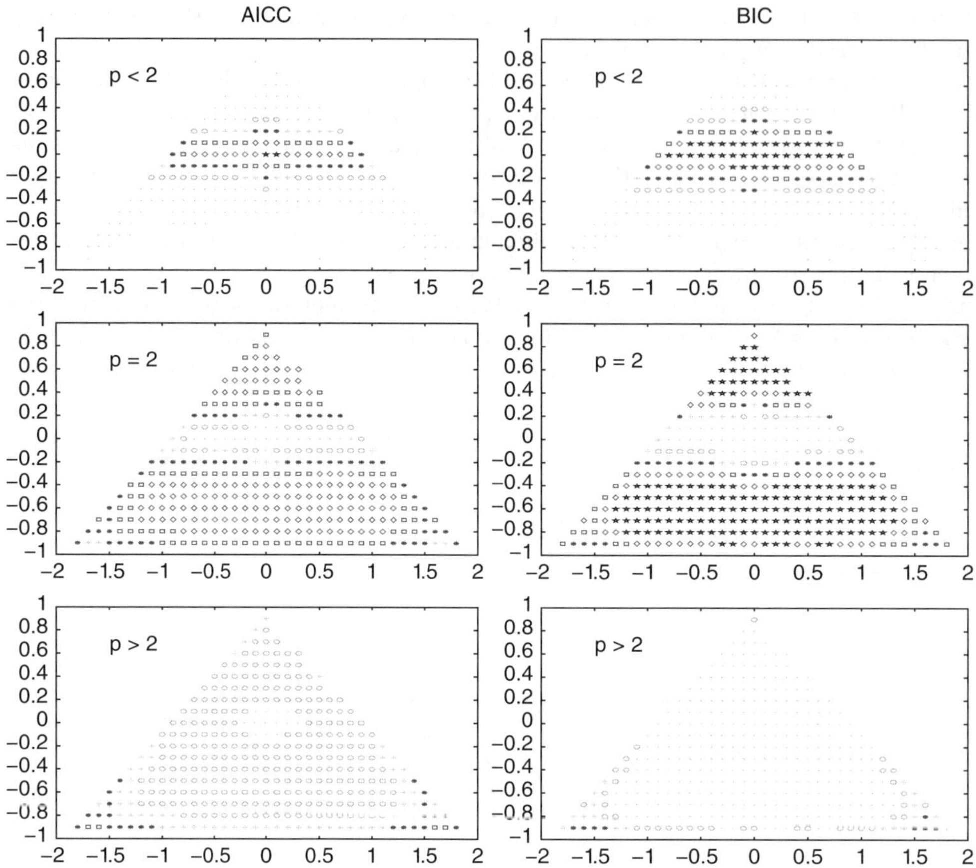

Figure 9.12 Same as Figure 9.11 but based on $T = 100$ and $p_{max} = 8$.

each of which indicates an interval into which the simulated percentage fell. A clear pattern emerges that is in agreement with the results for the AR(1) model: Performance is poorest near the origin ($a_1 = a_2 = 0$), improves as a_1 and/or a_2 move away from zero, and worsens near the edge of the support, where the probability of over-selection increases.

Also, as with the AR(1) case, the BIC is more conservative and has a lower rate of over-selection (and higher rate of under-selection) compared to the AICC. Figure 9.12 is similar, but for $T = 100$ and $p_{max} = 8$. For the AICC, it appears that the probability of selecting $p = 2$ has not changed remarkably, so that the benefit of increased sample size is cancelled by the increase in p_{max}. This increase in p_{max} has, however, clearly increased the probability of over-selection. Quite different is the performance of the BIC: Unlike the AICC, the probability of both under-selection and over-selection has actually gone down for some sets of parameters, and the probability of correct selection has increased by a large margin.

9.5 Use of the Conditional SACF for Sequential Testing

> It can be said that, like most problems of statistical inference, the choice of the order of the autoregressive model to be fitted to the time series data has been basically formulated until now as that of estimation or as that of testing of hypotheses. Neither of these two formulations suit the objectives of the experimenter in many situations when it is recognized that no unique model can describe satisfactorily the true underlying process and that more than one model should be retained for further consideration.
>
> (Quang Phuc Duong, 1984)

Several sequential testing procedures for ARMA model order selection have been proposed in the time-series academic literature. For example, Jenkins and Alevi (1981) and Tiao and Box (1981) consider methods based on the asymptotic distribution of the SPACF under the null of white noise. More generally, Pötscher (1983) considers determination of optimal values of p and q by a sequence of Lagrange multiplier tests. In particular, for a given choice of maximal orders, P and Q, and a chain of (p, q)-values $(p_0, q_0) = (0, 0), (p_1, q_1), \dots , (p_K, q_K) = (P, Q)$, such that either $p_{i+1} = p_i$ and $q_{i+1} = q_i + 1$ or $p_{i+1} = p_i + 1$ and $q_{i+1} = q_i$, $i = 0, 1, \dots , K = P + Q$, a sequence of Lagrange-multiplier tests are performed, and this for each possible chain. The optimal orders are obtained when the test does not reject for the first time. As noted by Pötscher (1983, p. 876), "strong consistency of the estimators is achieved if the significance levels of all the tests involved tend to zero with increasing size…"

This forward search procedure is superficially similar to the method proposed herein, and also requires specification of a sequence of significance levels. Our method differs in two important regards. First, near-exact small-sample distribution theory is employed by use of conditional saddlepoint approximations. Second, we explicitly allow for, and account for, a mean term in the form of a regression $\mathbf{X}\boldsymbol{\beta}$.

There are two crucial results that allow for the development of this method. The first is the following: Anderson (1971, Sec. 6.3.2) has shown for the regression model with circular AR(m) errors (so $\epsilon_1 \equiv \epsilon_T$) and the columns of $\mathbf{X}$ restricted to Fourier regressors, i.e.,

$$Y_t = \beta_1 + \sum_{s=1}^{(k-1)/2} \left\{ \beta_{2s} \cos\left(\frac{2\pi st}{T}\right) + \beta_{2s+1} \sin\left(\frac{2\pi st}{T}\right) \right\} + \epsilon_t, \tag{9.5}$$

that the uniformly most powerful unbiased (UMPU) test of AR($m-1$) versus AR(m) disturbances rejects for values of r_m falling sufficiently far out in either tail of the conditional density

$$f_{R_m \mid R_{(m-1)}}(r_m \mid \mathbf{r}_{(m-1)}), \tag{9.6}$$

where $\mathbf{r}_{(m-1)} = (r_1, \dots , r_{m-1})'$ denotes the observed value of the vector of random variables $R_{(m-1)}$. A p-value can be computed as $\min\{\tau_m, 1 - \tau_m\}$, where, as in (8.35),

$$\tau_1 = \Pr(R_1 < r_1) \quad \text{and} \quad \tau_m = \Pr(R_m < r_m \mid R_{(m-1)} = \mathbf{r}_{(m-1)}), \quad m > 1. \tag{9.7}$$

The $m = 1$ case was discussed in detail in Section 5.3. The optimality of the test breaks down in either the non-circular model and/or with arbitrary exogenous $\mathbf{X}$, but does provide strong motivation for an approximately UMPU test in the general setting considered here. This is particularly so for economic time series, as they typically exhibit seasonal (i.e., cyclical) behavior that can mimic the Fourier regressors in (9.5) (see, e.g., Dubbelman et al., 1978; King, 1985a, p. 32).

The second crucial result involves the tractability of the small-sample distribution via a conditional saddlepoint approximation. Recall Section 8.1.4 on approximating the distribution of the scalar random variable R_m given $\mathbf{R}_{m-1} = \mathbf{r}_{m-1}$, where $\mathbf{R}_{m-1} = (R_1, \dots, R_{m-1})'$ and $\mathbf{r}_{m-1} = (r_1, \dots, r_{m-1})'$. The conditional p.d.f. $f_{R_m|\mathbf{R}_{m-1}}(r_m \mid \mathbf{r}_{m-1})$ is given in (8.34), while the conditional c.d.f. (9.7) is given in (8.37) and (8.38). With the ability to calculate these distributions, this model selection strategy was operationalized and studied in Butler and Paolella (2017). In particular, the sequential series of tests

$$H_m : a_m = 0, \quad H_{m-1} : a_m = a_{m-1} = 0, \quad \dots, \quad H_1 : a_m = \cdots = a_1 = 0 \tag{9.8}$$

is performed. Testing stops when the first hypothesis is rejected (and all remaining are then also rejected). A natural way of implementing the sequence of p-values for selecting the autoregressive lag order p is to take the largest value $j \in \{1, \dots, m\}$ such that $\tau_j < c$ or $\tau_j > 1 - c$, or set it to zero if no such extreme τ_j occurs. We refer hereafter to this as the **conditional ACF testing method**, or CACF. The CACF method (9.8) is implemented in the program in Listings 9.1 and 9.2.

```
function [pvaluevec, phat]=ButPao(Y,X,c)
% INPUT
%   Time series column vector Y
%   Regression matrix X, if not passed, defaults to a column of ones.
%   Pass [] for no X matrix
%   c is an maxp-length vector of significance levels, with default
%     maxp=7 and c=[c_1,..,c_maxp]=[0.175 0.15 0.10 0.075 0.05 0.025]
% OUTPUT:
%   pvaluevec is the vector of p-values, starting with AR(1).
%   phat is the estimated AR(p) order, based on the p-values, and c.
global Omega G T k r maxp
if nargin<3
  maxp=7; c=[0.175 0.15 0.125 0.10 0.075 0.05 0.025];
else
  maxp=length(c);
end
T=length(Y); if nargin<2, X=ones(T,1); end
if isempty(X), k=0; else [~,k]=size(X); end
pvaluevec=NaN(maxp,1); r=NaN(maxp,1);
if isempty(X), M=eye(T); else M=makeM(X); end
if isempty(X), G=eye(T); else G=makeG(X); end % G is such that M=G'G and I=GG'
e=M*Y; for i=1:maxp, r(i)=(e'*makeA(T,i)*e)/(e'*e); end
pvaluevec(1)=cdfratio(r(1),G*makeA(T,1)*G',eye(T-k),eye(T-k),[],2);
Omega=eye(T-k); % For pure AR(p) case, null is Identity.
% Omegai=G*Psii*G'; Omega=inv(Omegai); % More general varcov
options = optimset('Display','off');
phat=-1; m=1; sinit=0; s=fsolve(@spe,sinit,options);
P=makeP(s); [V,D]=eig(P); D=diag(D); tol=1e-6;
if any(D)<tol, disp('P<=0'), D(D<tol)=tol; P=V*diag(D)*V'; end
Pi=inv(P); H=makeH(m,Pi); [V,D]=eig(H); D=diag(D); tol=1e-6;
if any(D)<tol, disp('H<=0'), D(D<tol)=tol; H=V*diag(D)*V'; end
```

Program Listing 9.1: The CACF sequential testing methodology (9.7) and (9.8) using the saddlepoint approximation. Functions makeM, makeG, and makeA are given in Listings B.2, 1.2, and 8.1, respectively. Continued in Listing 9.2.

```
for m=2:maxp
  Pm1=P; Pim1=Pi; Hm1=H;
  sinit=zeros(m,1); s=fsolve(@spe,sinit,options);
  P=makeP(s); [V,D]=eig(P); D=diag(D); tol=1e-6;
  if any(D)<tol, disp('P<=0'), D(D<tol)=tol; P=V*diag(D)*V'; end
  Pi=inv(P); H=makeH(m,Pi); [V,D]=eig(H); D=diag(D); tol=1e-6;
  if any(D)<tol, disp('H<=0'), D(D<tol)=tol; H=V*diag(D)*V'; end
  sm=s(end); w0=sign(sm)*sqrt( log( det(P)/det(Pm1) ) );
  v0=sm*sqrt(det(H)/det(Hm1)); v0=v0*( tr(Pim1)/tr(Pi) )^(m-1);
  if (isreal(w0) && isreal(v0))
    pvaluevec(m)=normcdf(w0) + normpdf(w0)*(1/w0 - 1/v0);
  end
end
if all(isreal(pvaluevec))
  phat=0;
  for i=1:maxp
    if (pvaluevec(i)<c(i)) || (pvaluevec(i)>(1-c(i))), phat=i; end
  end
end

function f=spe(s),  global G k T r
m=length(s); f=zeros(m,1);
for i=1:m
  Pi=inv(makeP(s)); GAG=G*makeA(T,i)*G'; f(i)=tr(Pi*(GAG-r(i)*eye(T-k)));
end

function P = makeP(s),  global Omega G k T r
m=length(s); Sum=zeros(T-k,T-k);
for i=1:m, GAG=G*makeA(T,i)*G'; Sum=Sum+s(i)*GAG; end
rr=r(1:m); P = Omega + 2*(rr'*s)*eye(T-k) - 2*Sum;

function H = makeH(m,Pinv),  global G T k r
H=zeros(m,m); I=eye(T-k);
for i=1:m,  Ai=G*makeA(T,i)*G';
  for j=1:m,  Aj=G*makeA(T,j)*G';
    H(i,j)=2*tr( Pinv*(Ai-r(i)*I)*Pinv*(Aj-r(j)*I) );
  end
end

function t=tr(Z), t=sum(diag(Z));
```

Program Listing 9.2: Continued from Listing 9.1.

Observe how this hypothesis-test-driven strategy, similar to the use of model estimation and the penalty-based criteria, is subject to the critique spelled out in the above quote by Duong (1984). We will see, however, that this method does allow some subjectivity from the modeler to be incorporated into the selection, as well as being able to generate a set of candidate models, as opposed to the binary result of a typical hypothesis test, or the mechanical procedure of choosing the model with the smallest penalty-based criteria. This is done via allowing c to be a vector, $\mathbf{c}$, instead of a scalar, as will be subsequently discussed, chosen to incorporate prior knowledge on behalf of the modeler, and this $\mathbf{c}$ can be changed to possibly result in different order selections.

The effectiveness of this strategy will clearly be quite dependent on the choices of $m = p_{\max}$ and (so far, scalar) c, as well as on the accuracy of the conditional saddlepoint approximation used. Note

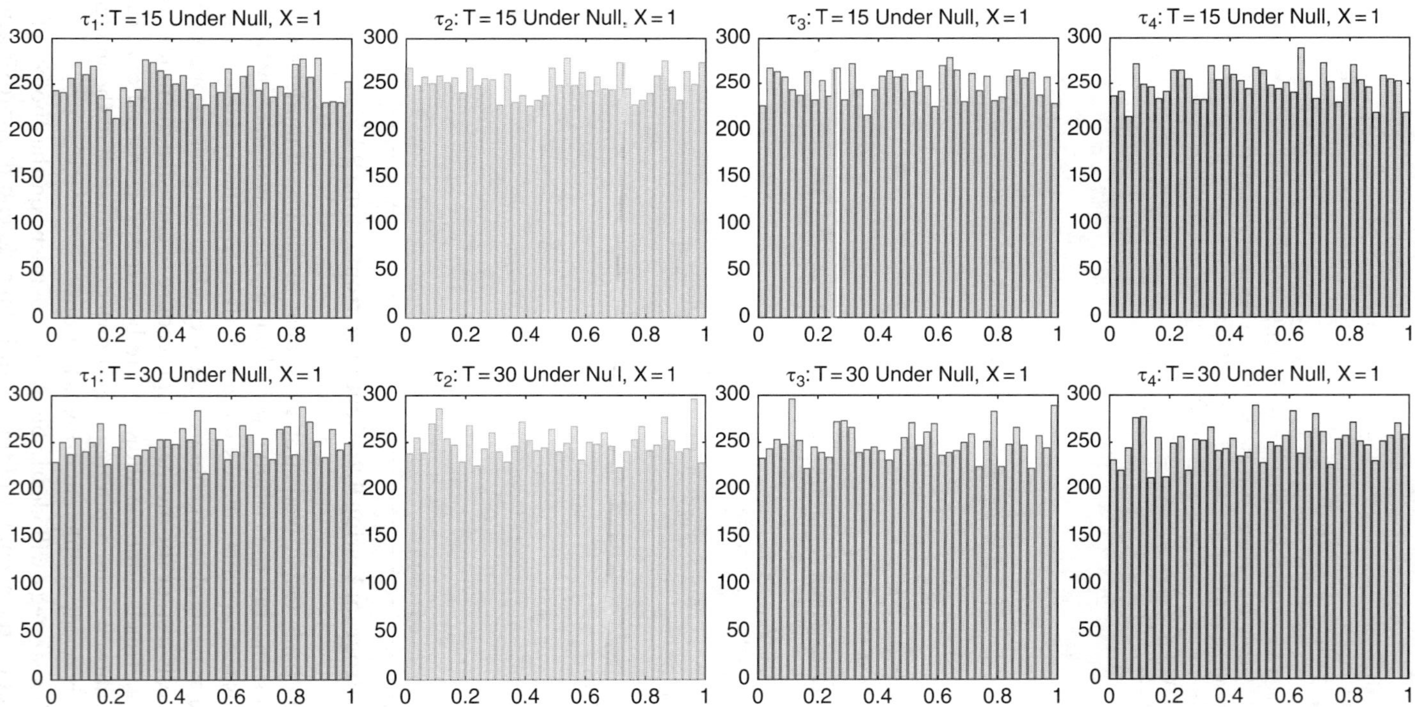

Figure 9.13 Histograms of $\tau_1, \dots, \tau_4$, based on 10,000 replications with true data being iid normal, and taking $\mathbf{X} = \mathbf{1}$. Top (bottom) panels are for $T = 15$ ($T = 30$).

that, while penalty function methods also require an upper limit m, the CACF has the extra "tuning parameter" c that can be seen as either a blessing (for a more Bayesian-oriented researcher interested in inference for a specific data set) or a curse (for a hardcore frequentist who assumes the d.g.p. is actually correctly specified, and interested strictly in asymptotic consistency and behavior in fictitious infinite repeated trials). A natural value might be $c = 0.025$, so that, under the null of zero autocorrelation, p assumes a particular wrong value with approximate probability 0.05, and $p = 0$ is chosen approximately with probability $1 - 0.05m$, i.e.,

$$\begin{aligned}\Pr\{\text{choose } p = 0 \mid \text{white noise}\} &= \Pr\{\tau_j \in (0.025, 0.975), \quad j = 1, \ldots, m\} \\ &= (1 - 0.05)^m \approx 1 - 0.05m, \end{aligned} \tag{9.9}$$

from the binomial expansion.

To assess the accuracy of the conditional saddlepoint approximation to (9.6), consider its distribution under the null of no autocorrelation. That is, when $\boldsymbol{\Psi}^{-1} = \mathbf{I}_T$ in (1.3) for the linear model $\mathbf{Y} = \mathbf{X}\boldsymbol{\beta} + \boldsymbol{\epsilon}$. We expect $\tau_i \overset{\text{iid}}{\sim} \text{Unif}(0, 1)$. This was empirically tested by computing $\tau_1, \ldots, \tau_4$ in (9.7) for 10,000 time-series replications, each consisting of T independent standard normal simulated random variables, for $T = 15$ and $T = 30$, but with mean removal, i.e., taking $\mathbf{X} = \mathbf{1}$. Histograms of the resulting τ_i, as shown in Figure 9.13, are in agreement with the uniform assumption. Furthermore, the absolute sample correlations between each pair of the τ_i were all less than 0.02 for $T = 15$ and less than 0.013 for $T = 30$. The program to generate these plots is given in Listing 9.3. These results are in stark contrast to the empirical distribution of the "t-statistics" and the associated p-values, shown above in Figure 9.5.

As a first case for illustration, the optimal AR lag orders among the choices $p = 0$ through $p = 4$ (i.e., $m = p_{\max} = 4$) were determined via the CACF method for each of 1000 simulated AR(1) series of length $T = 30$ and AR parameter a, using $a = 0, 0.1, 0.3, \ldots, 0.9$,[2] and the regression model based on $\mathbf{X} = \mathbf{1}$. The cutoff value $c = 0.025$ was used. The results are shown in Figure 9.14. Observe that, for $a = 0$, about 5% select either $p = 1$ or $p = 2$ or $p = 3$ or $p = 4$, exactly as the theory suggests. The code to create the plots is given in Listing 9.4.

Next, consider comparing the performance of the CACF method to the various penalty criteria in (9.3) and (9.4), using the same setup except ($m = 4, c = 0.025$) with the constant and time-trend model $\mathbf{X} = [\mathbf{1}, \ \mathbf{t}]$. Figure 9.15 show the results. The CACF method dominates in the null model, while for

```
T=15; sim=1e4; pmat=zeros(sim,4);
for i=1:sim, disp(i), Y=randn(T,1); p=ButPao(Y)'; pmat(i,:)=p; end
for i=1:4
  if i==1, ec=[1 0 0]; elseif i==2, ec=[0 1 0];
  elseif i==3, ec=[0 0 1]; else ec=[0.5 0.3 0]; end
  use=pmat(:,i); ok=(use>0 & use<1); use=use(ok); length(use)
  figure, [histcount, histgrd] =hist(use,40); h1=bar(histgrd,histcount);
  set(h1,'facecolor',[0.9 0.9 0.9],'edgecolor',ec,'linewidth',1.2)
  set(gca,'fontsize',16)
  title(['\tau_',int2str(i),': T=',int2str(T),' Under Null, X=1'])
end
```

Program Listing 9.3: Code for generating the bar plots in Figure 9.13.

2 Negative values of a were also considered; the results essentially paralleled their positive counterparts.

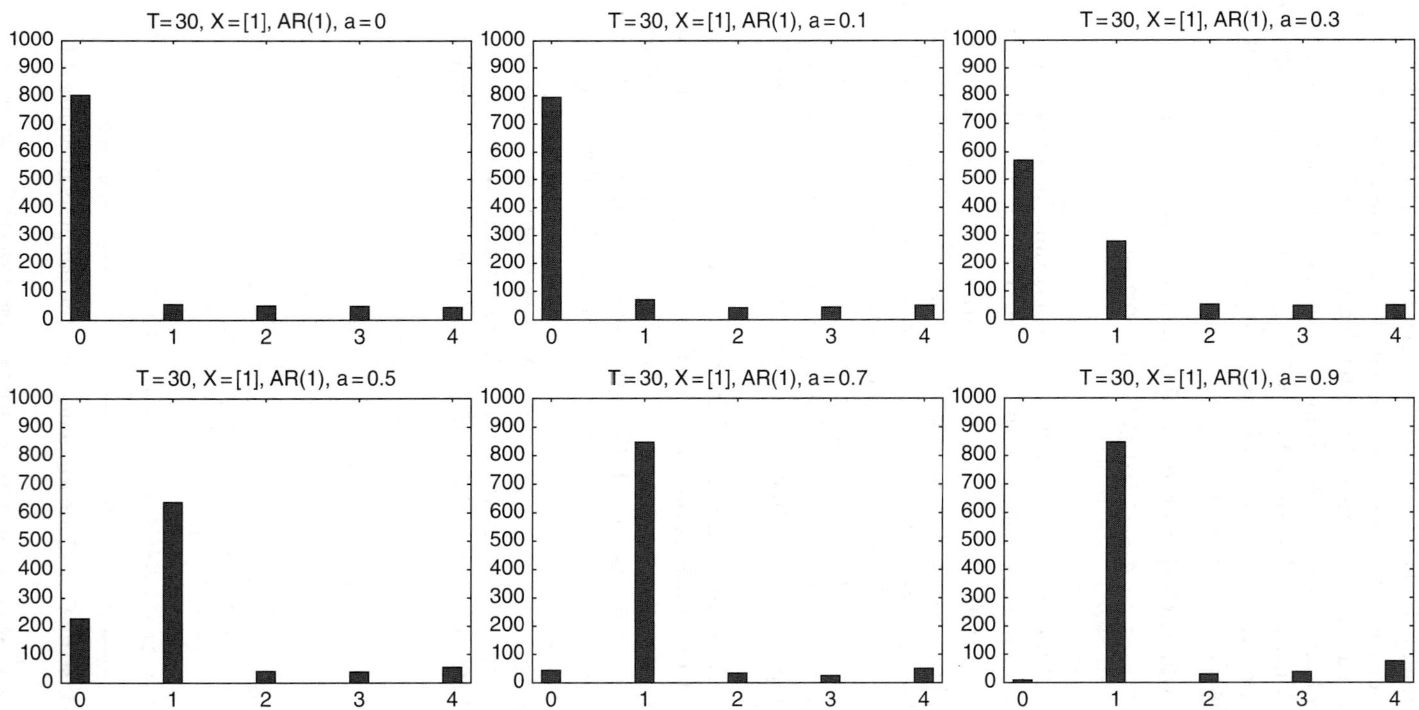

Figure 9.14 Bar plots corresponding to the chosen value of AR lag length p, based on the CACF method, a true AR(1) process with parameter a, $T = 30$, and $\mathbf{X} = \mathbf{1}$, using $c = 0.025$ and $m = 4$.

```
T=30; sim=1e3; phivec=[0 0.1 0.3 0.5 0.7 0.9]; pvec=zeros(sim,1);
for philoop=1:length(phivec)
  phi=phivec(philoop);
  for i=1:sim
    disp(i), Y=armasim(T,1,phi); X=ones(T,1); c=ones(4,1)*0.025;
    [~, phat]=ButPao(Y,X,c); pvec(i)=phat;
  end
  tt=tabulate(pvec); figure, set(gca,'fontsize',16)
  h=bar(tt(:,1),tt(:,2)); set(h,'barwidth',0.2,'facecolor','r')
  title(['T=',int2str(T),', X=[1], AR(1), a=',num2str(phi)])
  xlim([-0.2 4.2]), ylim([0 sim])
end
```

Program Listing 9.4: Code for generating the bar plots in Figure 9.14.

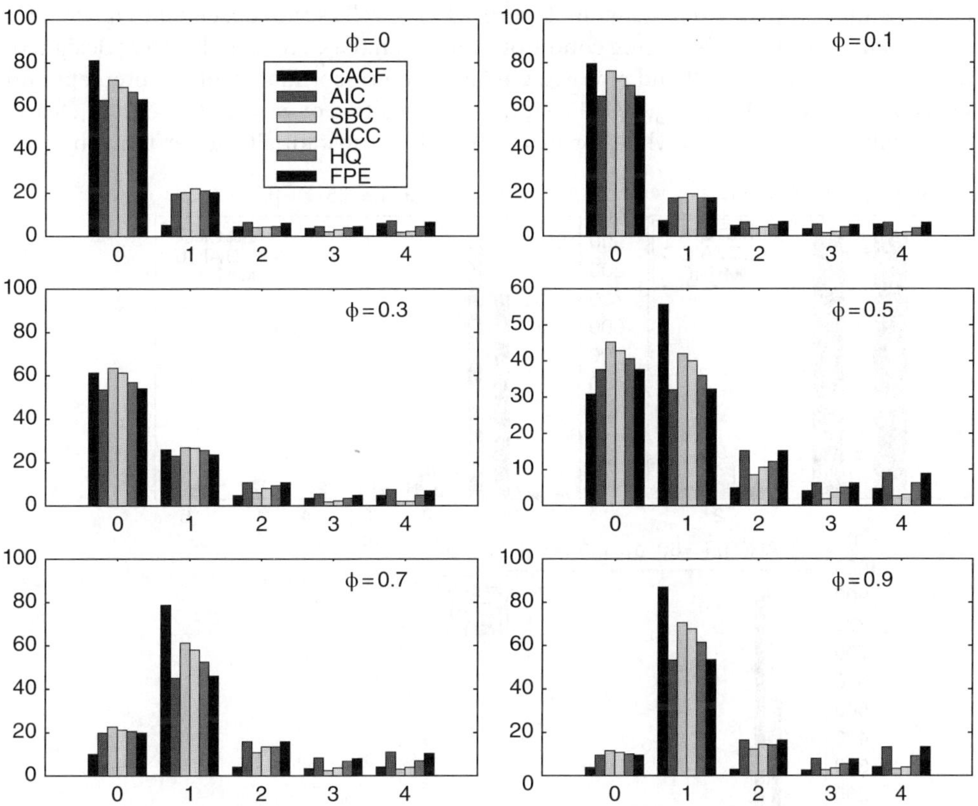

Figure 9.15 Performance of the various methods in the AR(1) case using $\mathbf{X} = [\mathbf{1},\ \mathbf{t}]$, $T = 30$, $m = 4$, $c = 0.025$, and true AR(1) parameter denoted as ϕ in the graphics.

small absolute values of a the CACF under-selects more than the penalty-based criteria. For $a \geqslant 0.5$, the CACF is blatantly superior. This emphasizes the benefit of explicitly taking the regressors into account when computing (9.7).

A potential concern with the CACF method is what happens if m is much larger than the true p. To investigate this, we stay with the AR(1) example, but consider only the case with $a = 0.5$, and use three choices of m, namely 2, 6, and 10. We first do this with a larger sample size of $T = 60$ and no $\mathbf{X}$ matrix, which should convey an advantage to the penalty-based measures relative to CACF. For the former, we use only the AICC and BIC. Figure 9.16 shows the results, based on 1,000 replications. With $m = 2$, all three methods are very accurate, with CACF and BIC being about equal with respect to the probability of choosing the correct p of one, and slightly beating AICC. With $m = 6$, the BIC dominates. The nature of the CACF methodology is such that, when m is much larger than p, the probability of overfitting (choosing p too high) will increase, according to the choice of c. With $m = 10$, this is apparent. In this case, the BIC is superior, and also substantially stronger than the more liberal AICC. The code in Listing 9.5 was used to generate Figure 9.16, as well as the subsequent Figure 9.17.

We now conduct a similar exercise, but using conditions for which the CACF method was designed, namely a smaller sample size of $T = 30$ and a more substantial regressor matrix of an intercept and time-trend regression, i.e., $\mathbf{X} = [\mathbf{1}, \mathbf{t}]$. Figure 9.17 shows the results. For $m = 2$, the CACF clearly outperforms the penalty-based criteria, while for $m = 6$, which is substantially larger than the true

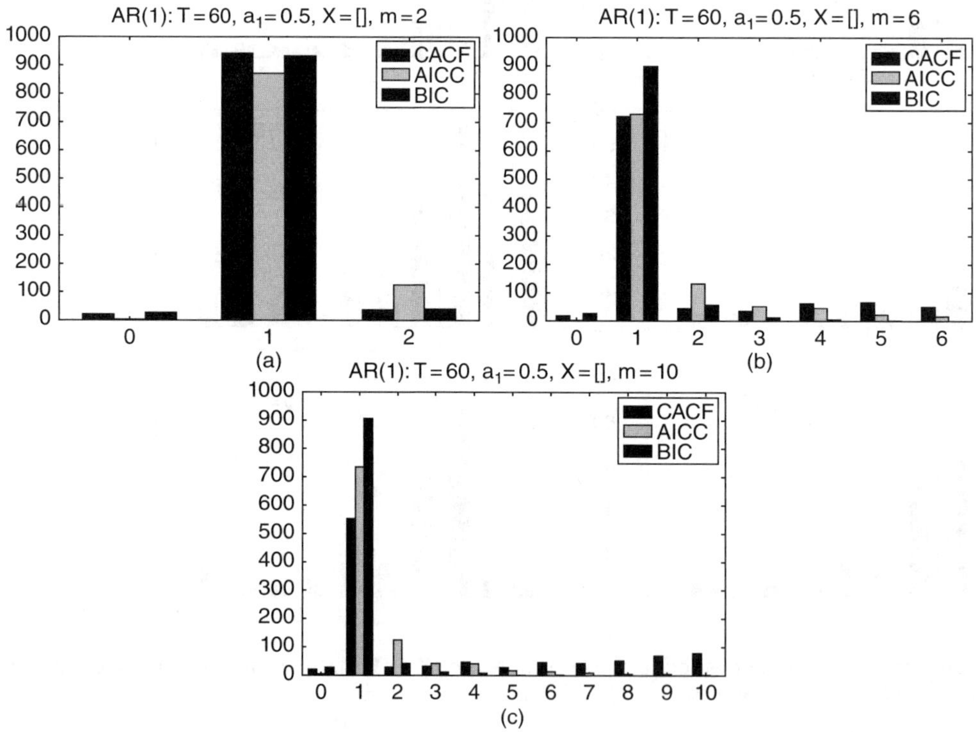

Figure 9.16 Histograms corresponding to the chosen value of AR lag length p, based on the CACF, AICC, and BIC, using three values of tuning parameter m (2, 6, and 10, from a to c), and 1,000 replications. True model is Gaussian AR(1) with parameter $a = 0.5$, sample size $T = 60$, and known mean (no $\mathbf{X}$ matrix). The CACF method uses $c = 0.025$.

```
T=30; phi=0.5;  % sample size and AR(1) parameter
% X=[]; k=0;
X=[ones(T,1) , (1:T)']; [~,k]=size(X);
pmaxvec=2:4:10; mleexact=1; q=0;
sim=1e3; pvecCACF=zeros(sim,1); pvecAICC=zeros(sim,1); pvecBIC=zeros(sim,1);
for pmaxloop=1:length(pmaxvec)
  pmax=pmaxvec(pmaxloop);
  c=ones(pmax,1)*0.025; % use for scalar c
  % vv=0:(pmax-1); c=pmax/80 - 0.0125*vv; % use for vector c
  for i=1:sim, disp([pmax, i])
    Y=armasim(T,1,phi); [~, phat]=ButPao(Y,X,c); pvecCACF(i)=phat;
    AICCmin=1e20; BICmin=1e20;
    for p=0:pmax
      param=armareg(Y,X,p,q,mleexact); lsig2=log(param(end)^2);
      K=p+k; AICC=lsig2+(T+K)/(T-K-2); BIC=lsig2 + K*log(T)/T;
      if AICC<AICCmin, AICCmin=AICC; pvecAICC(i)=p; end
      if BIC<BICmin, BICmin=BIC; pvecBIC(i)=p; end
    end
  end
  pmat=[pvecCACF pvecAICC pvecBIC];
  figure, hist(pmat,0:pmax), set(gca,'fontsize',16), legend('CACF','AICC','BIC')
  str=['AR(1): T=',int2str(T),', a_1=0.5, X=[1,t], m=',int2str(pmax)]; title(str)
  xlim([-0.5 pmax+0.5]), ylim([0 sim]), drawnow
end
```

Program Listing 9.5: Code for generating the histograms in Figures 9.16 and 9.17.

$p = 1$, the CACF chooses the correct p with the highest probability of the three selection methods, though the AICC is very close. For the very large $m = 10$ (which, for $T = 30$, might be deemed inappropriate), CACF and BIC perform about the same with respect to the probability of choosing the correct p of one, while AICC dominates. Thus, in this somewhat extreme case (with $T = 30$ and $m = 10$), the CACF still performs competitively, due to its nearly exact small-sample distribution theory and the presence of an **X** matrix.

An aspect of the CACF method that greatly enhances its ability and is not applicable with penalty-based model selection methods is the use of different significance levels for the sequential tests. This allows an objective way of incorporating prior notions of preferring low order, parsimoniously parameterized models. For non-seasonal AR models, Butler and Paolella (2017) suggest the use of a simple linear sequence of significance level values, such as, for $m = 7$,

$$\mathbf{c} = [0.175, \ 0.15, \ 0.125, 0.10, \ 0.075, \ 0.05, \ 0.025], \tag{9.10}$$

This was found to work remarkably well in many situations, compared to the penalty-based methods. As an illustration, consider an AR(4) model with parameters $a_1 = 0.4$, $a_2 = -0.3$, $a_3 = 0.2$, and a_4 takes on the six values -0.1 through -0.6. In an attempt to use a more complicated regression matrix that is typical in econometric applications, an **X** matrix corresponding to an intercept and time-trend model with structural break is used, i.e., for $T = 30$,

$$\mathbf{X}' = \begin{bmatrix} 1 & 1 & 1 & \cdots & 1 & \cdots & 1 \\ 1 & 2 & 3 & \cdots & 16 & \cdots & 30 \\ 0 & 0 & 0 & \cdots & 1 & \cdots & 1 \\ 0 & 0 & 0 & \cdots & 1 & \cdots & 15 \end{bmatrix}. \tag{9.11}$$

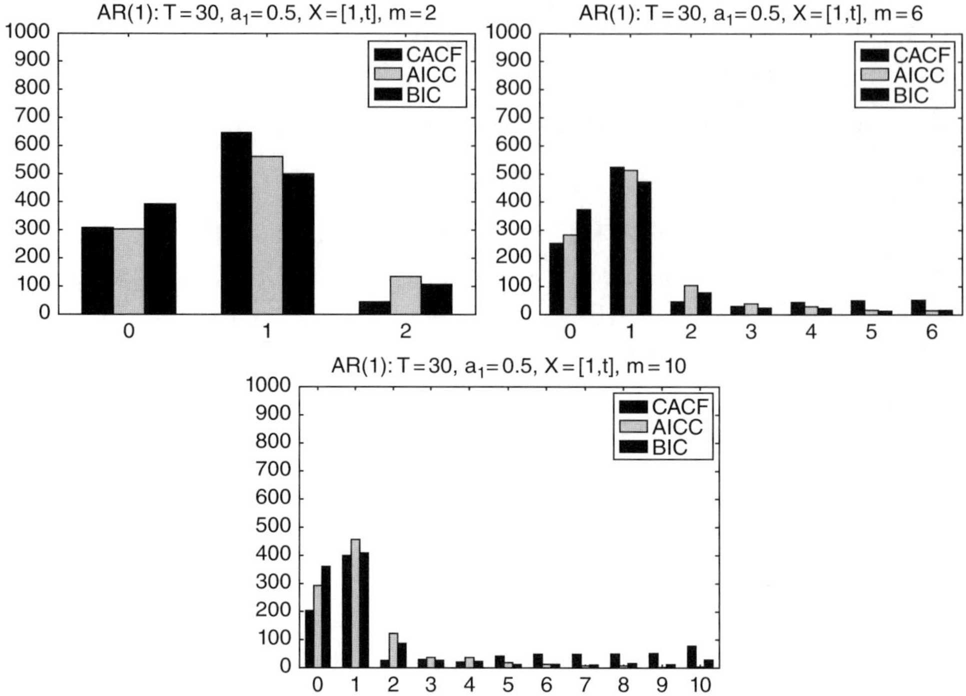

Figure 9.17 Similar to Figure 9.16, but for sample size $T = 30$ and $\mathbf{X} = [\mathbf{1},\ \mathbf{t}]$.

Figure 9.18 shows the resulting selection of p, based on 1,000 replications and a sample size of $T = 30$. Figure 9.19 is similar, but having used $T = 100$. The reader is encouraged to compare these results to those obtained using the various penalty criteria.

Based on further simulations, this model selection paradigm was demonstrated in Butler and Paolella (2017) to be competitive with, and often superior to, the use of penalty criteria, with the reasons being that (i) it is based on a sequence of tests that are (not exactly, but as close as possible to) UMP and (ii) the small-sample distribution theory explicitly accounts for the $\mathbf{X}$ matrix, and this is enabled by use of the highly accurate conditional saddlepoint approximation. Those findings are based on the assumption that the true model is known to be a linear regression with correctly specified $\mathbf{X}$ matrix, error terms are from a Gaussian AR(p) process, tuning parameter m is chosen such that $m \geqslant p$, and parameters $p, a_1, \ldots, a_p$, $\boldsymbol{\beta}$, and σ^2 are fixed but unknown. It is appropriate to challenge these heroic assumptions somewhat.

We modify this setup by assuming, similarly, that the true data generating process is $\mathbf{Y} = \mathbf{X}\boldsymbol{\beta} + \boldsymbol{\epsilon}$, with $\epsilon_t = a_1\epsilon_{t-1} + U_t$ a stationary AR(1) process, but now such that $U_t \overset{\text{iid}}{\sim} t_v(0, \sigma)$, $t = 1, \ldots, T$, i.e., Student's t with v degrees of freedom, location zero, and scale $\sigma > 0$.

Figures 9.20 and 9.21 are similar to Figure 9.10, again based on the true $p = 1$, a sample size of $T = 50$ (and taking $m = 5$ and $c = 0.025$), but using four different values of degrees of freedom parameter v, and *for all methods* falsely assuming Gaussianity. Figure 9.20 is for the known mean case, while Figure 9.21 assumes the constant and time-trend model $\mathbf{X} = [\mathbf{1},\ \mathbf{t}]$. We see that none of the methods are substantially affected by use of even very heavy-tailed innovation sequences, notably the CACF,

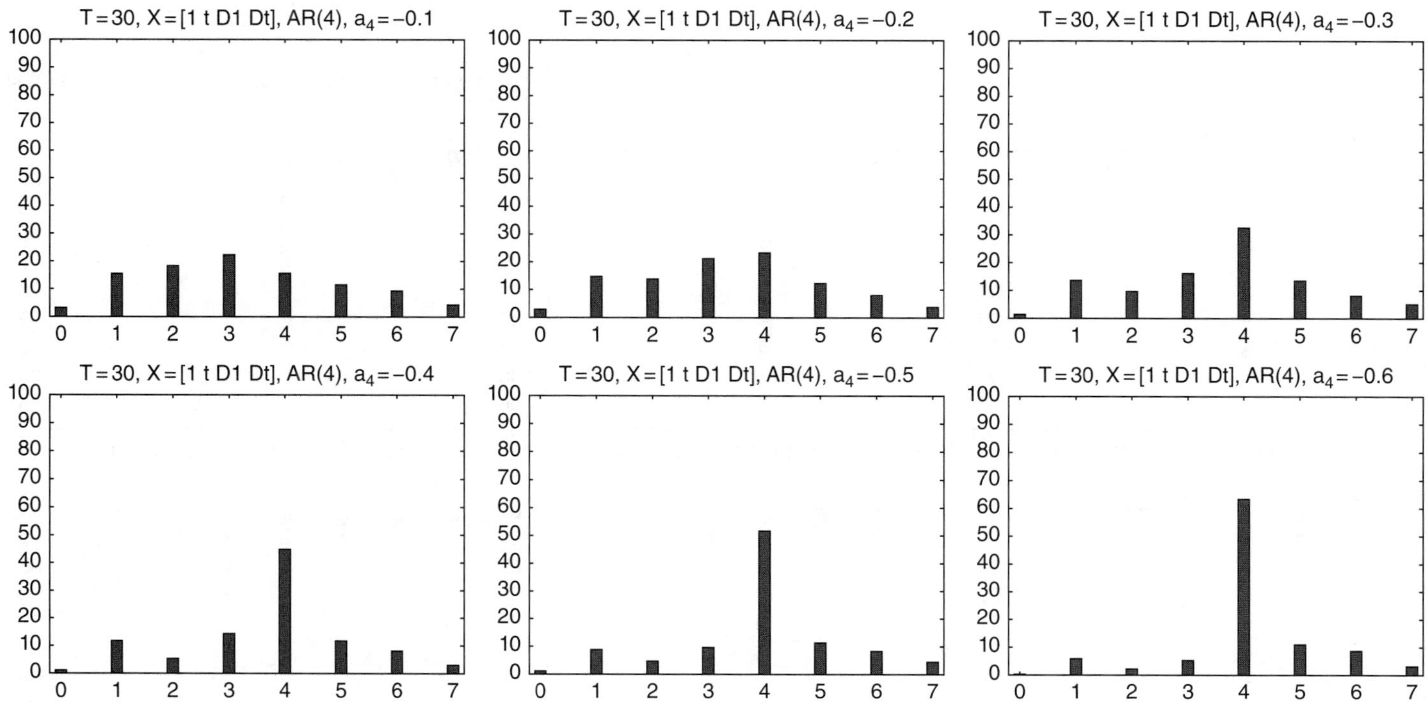

Figure 9.18 Based on the CACF method, bar plots corresponding to the chosen value of AR lag length p, in percent, based on 1,000 replications, using a true AR(4) process with parameters $a_1 = 0.4$, $a_2 = -0.3$, $a_3 = 0.2$, and a_4 taking on the six values −0.1 through −0.6, as indicated in the titles of the plots. The sample size is $T = 30$ and $\mathbf{X}$ is given in (9.11). The CACF tuning parameters are $m = 7$ and $\mathbf{c}$ as given in (9.10).

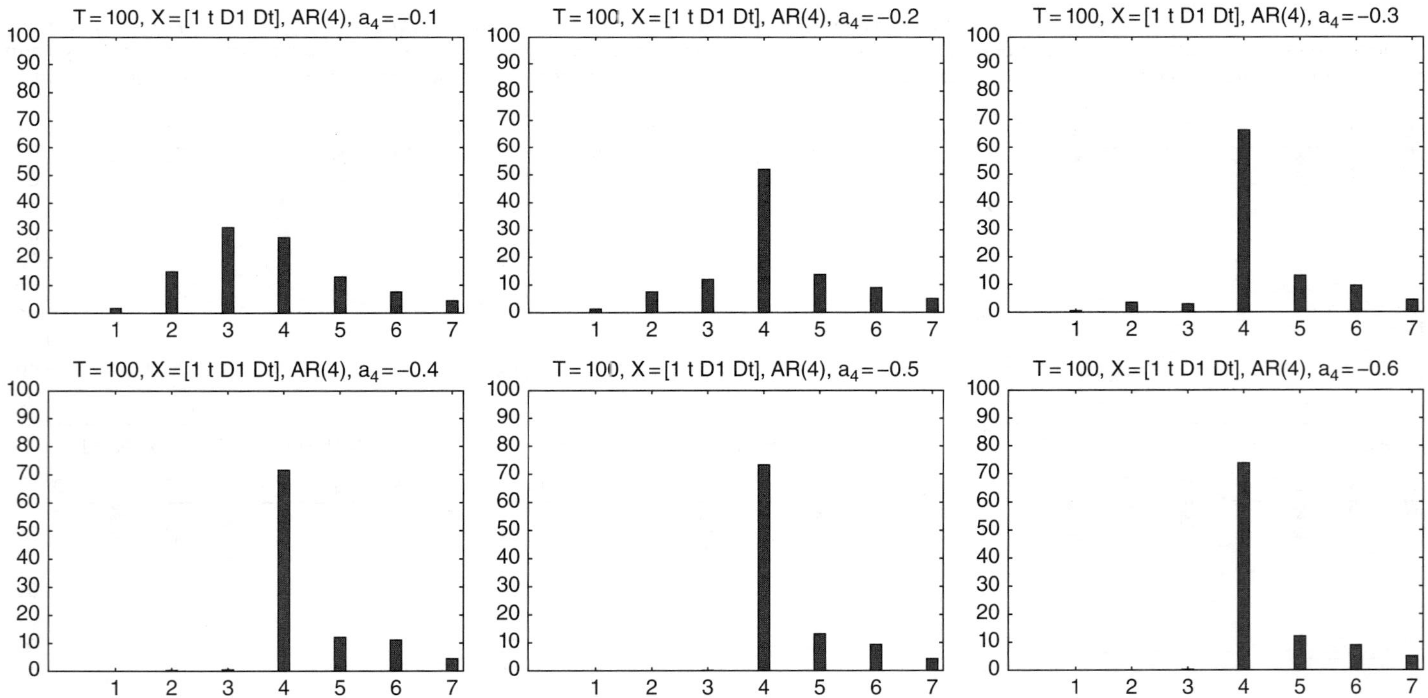

Figure 9.19 Similar to Figure 9.18 but based on $T = 100$.

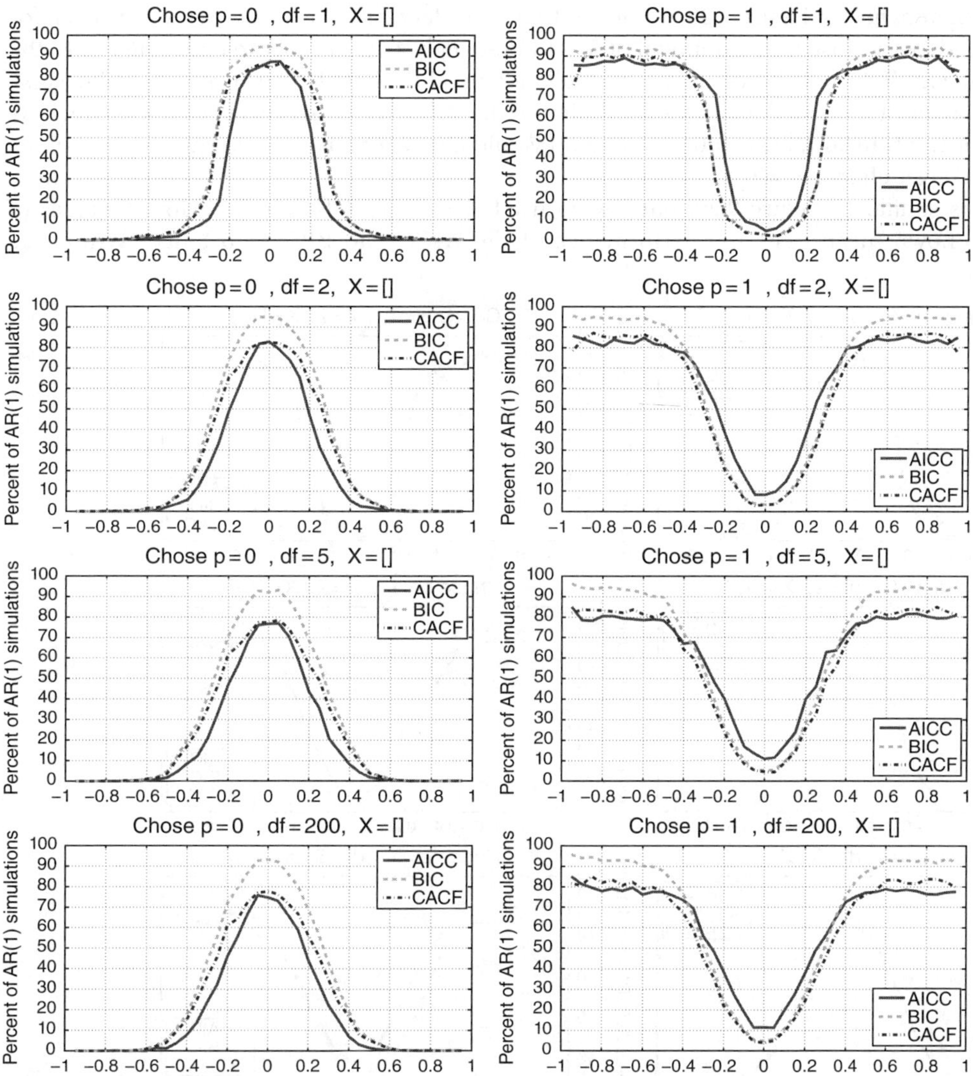

Figure 9.20 Similar to Figure 9.10: Performance of the three indicated AR order selection methods as a function of autoregressive parameter a, for sample size $T = 50$, known mean (denoted by $\mathbf{X} = [\ ]$), and $p_{\max} = m = 5$, when the true AR order is $p = 1$ and (falsely) assuming Gaussianity. The true innovation sequence consists of i.i.d. Student's $t(\nu)$ realizations, with df $= \nu$ indicated in the titles (from top to bottom, $\nu = 1$, $\nu = 2$, $\nu = 5$, and $\nu = 200$). Left (right) panels indicate the percentage of the 1,000 replications that resulted in choosing $p = 0$ ($p = 1$).

which explicitly uses the normality assumption in the small-sample distribution theory. However, note from the top left panel of Figure 9.20 (which corresponds to $\nu = 1$, or Cauchy innovations) that, for $p = a = 0$, instead of $0.774 = (1 - 0.05)^m \approx 1 - 0.05m = 0.75$ from (9.9), the null of $p = 0$ is chosen about 86% of the time, while from the third and fourth rows (for $\nu = 5$ and $\nu = 200$), it is about 78%. Similarly, the choice of $p = 1$ when $p = a = 0$ should occur about 5% of the time for the CACF method, but, from the right panels of Figure 9.20, it is lower than this, decreasing as v decreases. However, for $v = 5$, it is already very close to the nominal of 5%. Interestingly, with respect to choosing $p = 0$, the behavior of the CACF for all choices of v is virtually identical to the AICC near $a = 0$, while as $|a|$ grows, the behavior of the CACF coincides with that of the BIC. (Note that this behavior is precisely what we do *not* want: Ideally, for $a = 0$, the method would always choose $p = 0$, while for $a \neq 0$, the method would never choose $p = 0$.)

Figure 9.21 is similar to Figure 9.20, but having used $\mathbf{X} = [\mathbf{1}, \ \mathbf{t}]$. Observe how, for all values of v, unlike the known mean case, the performance of the AICC and BIC is no longer symmetric

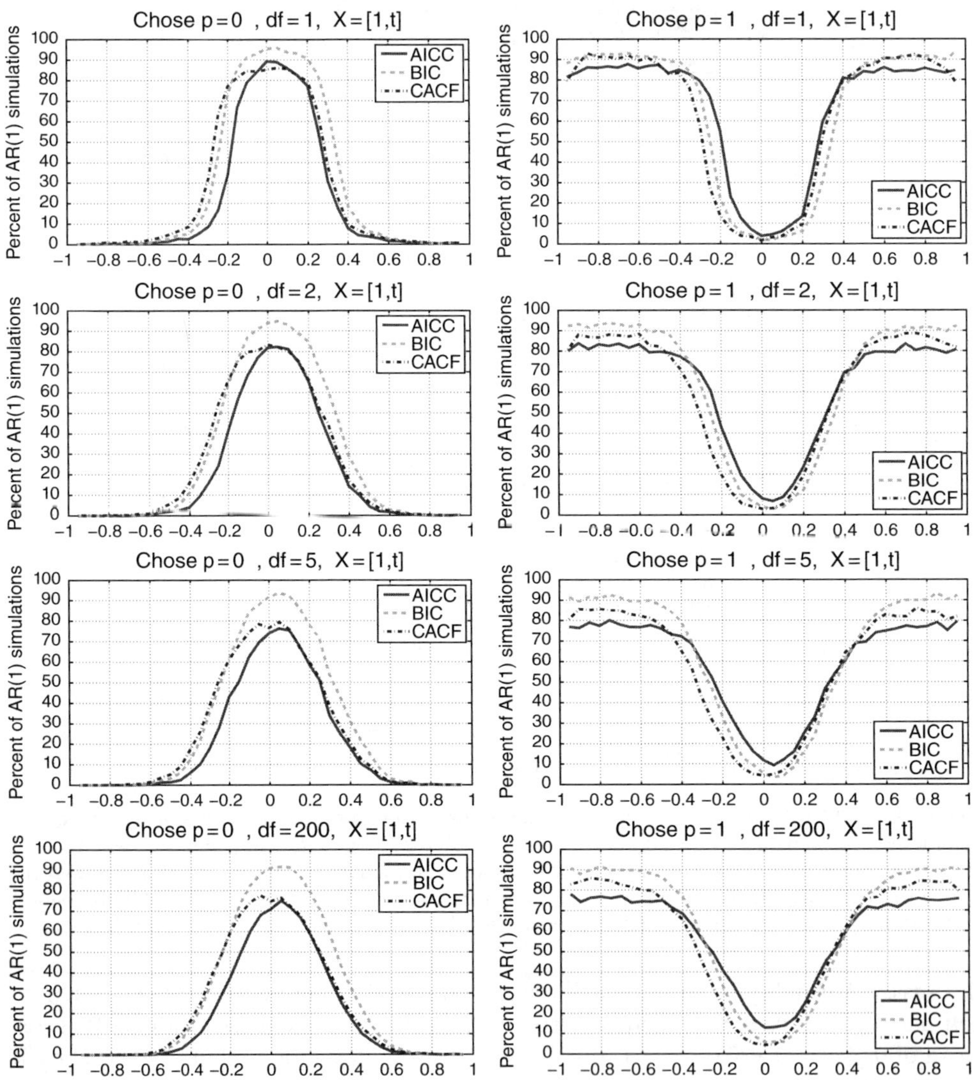

Figure 9.21 Same as Figure 9.20 but having used $\mathbf{X} = [\mathbf{1}, \ \mathbf{t}]$.

about $\phi = 0$, but the CACF is still virtually symmetric. Fascinatingly, we see from the left panels of Figure 9.21 that the CACF probability of choosing $p = 0$ virtually coincides with that of the AICC for $\phi \geqslant 0$, while for $\phi < -0.2$, it virtually coincides with that of the BIC.

The reader is encouraged to replicate this study of the behavior of the methods amid non-Gaussianity, and also consider asymmetric innovations, such as via use of the asymmetric stable Paretian or noncentral t. Different values of p and a variety of AR(p) coefficients could also be investigated. Finally, it also makes sense to allow the penalty-based selection criteria to use the information about the true innovation process, i.e., model parameter estimation is conducted using the true d.g.p., this being a feature that the CACF method does not (currently) have as the theory was developed under the assumption of Gaussianity. In this case, exact maximum likelihood is not straightforward, but the conditional m.l.e. based on a non-Gaussian distributional assumption is easily programmed, and serves as a great exercise for the motivated student. (Far more challenging would be an attempt at deriving the conditional saddlepoint approximation for the sample autocorrelation function for innovations coming from an elliptic class of distributions such as symmetric stable Paretian.)

Remarks

a) Note that the CACF method can be used for determining p, but does not indicate which of the p AR parameters (except the last) are "significant" and which are not, as might arise in a subset AR model. If the data analyst believes that some of the p coefficients could be zero, then one could use (9.8) for determining p, and then, for example, use the information criteria AIC and/or BIC applied to all 2^{p-1} subset AR(p) models, as well as possibly the signed likelihood ratio statistic (9.2) for assessing which AR coefficients can be set to zero. In general, this author frowns upon such procedures (recall, again, the opening quote by Ed Leamer), viewing instead an AR(p) model as a simple approximation to a much more complicated underlying reality, though the idea of a subset model could be of use if, say, p is relatively high, but many a_i coefficients might realistically be zero. This would occur, for example, in a seasonal autoregressive model.

b) A partial extension to the ARMA(p, q) model is possible. We assume that the regression error terms follow a stationary, invertible ARMA(p, q) process *with q known*, but the MA parameters $\boldsymbol{\theta} = (\theta_1, \dots, \theta_q)$ and p are not known. A possible method for eliciting p using the SACF method is as follows. Iterate the two steps starting with $p_1 = 0$:

   1) Estimate the ARMAX(p_i, q) model to obtain $\widehat{\boldsymbol{\theta}}$,
   2) Compute $\tau_1, \dots, \tau_m$ with $\boldsymbol{\Psi}^{-1}$ corresponding to the MA(q) model with parameters $\widehat{\boldsymbol{\theta}}$, from which p_{i+1} is determined.

   Iteration stops when $p_{i+1} = p_i$ (or i exceeds some preset value), and p_{i+1} is set as before for a given value or vector of c. Butler and Paolella (2017) provide simulation studies showing the effectiveness of this strategy (assuming Gaussianity), and that the CACF results in being often among the best methods, along with AICC and BIC.

c) What emerges from the various simulation studies is the known (but—we believe—not *well*-known) fact that the small-sample performance of all the penalty-based criteria is highly dependent on the actual autoregressive model parameters; see, e.g., Rahman and King (1999) and the references therein. The same result turns out (unfortunately) to also be true for the CACF method. ■

9.6 Use of the Singular Value Decomposition

This section was written with Patrick Walker

Another method to determine the optimal order of an AR(p) process (without covariates) is based on the singular value decomposition (SVD) of either a data matrix or an autocorrelation matrix of the observed process. SVD is a matrix factorization that generalizes the eigenvalue decomposition of symmetric positive semi-definite matrices to any real or complex matrix. It has many applications in statistics and signal processing, and can generally be used for determining the rank of a matrix. More formally, let $\mathbf{M}$ be any real or complex $m \times n$ matrix. There exists a factorization $\mathbf{M} = \mathbf{USV}^*$, where $\mathbf{U}$ is a unitary $m \times m$ and $\mathbf{V}$ a unitary $n \times n$ matrix, $\mathbf{V}^*$ is the conjugate transpose of $\mathbf{V}$, and $\mathbf{S}$ is an $m \times n$ rectangular diagonal matrix (though note it is not necessarily a square matrix) with real diagonal entries $\alpha_1 \geqslant \ldots \geqslant \alpha_r \geqslant 0$, where r is the rank of matrix $\mathbf{M}$; see the references given just above Theorem 5.1.

A computationally efficient iterative algorithm to determine the order of an autoregressive process based on the SVD is presented in Konstantinides (1991); see also Dickie and Nandi (1994). Let n be larger than the order of the process, p. Then, for some $m \geqslant n$, an $m \times n$ matrix $\mathbf{X}_n$ is formed from the data sample as follows. (For this method, we use the notation more common in the engineering literature.) Given an observed data sequence $(x(1), x(2), \ldots, x(N))$ of length N, construct the data matrix $\mathbf{X}_n$ as

$$\mathbf{X}_n = [X(N-1), \quad X(N-2), \quad \ldots \quad X(N-n)],$$

where $X(N-k)$, $k = 1, 2, \ldots, n$, is an m-dimensional column vector defined as

$$X(N-k) = [x(N-k), \quad x(N-k-1), \quad \ldots, \quad x(N-k-m+1)]'.$$

Analogously define the vector $X(N)$ as

$$X(N) = [x(N), \quad x(N-1), \quad \ldots, \quad x(N-m+1)]'.$$

The optimal order of the autoregressive process is then determined as the so-called **effective rank** of matrix $\mathbf{X}_n$, defined as the number of singular values that are larger than a certain threshold. The calculation of this threshold requires one to distinguish between significantly small and insignificantly large singular values. For this reason, the iterative algorithm proposed in Konstantinides (1991) makes use of the confidence bounds for perturbed singular values of noisy data matrices derived in Konstantinides and Yao (1988). The algorithm of Konstantinides (1991) for order determination of autoregressive processes is given as follows.

1) Choose n larger than an initial guess of the process order. Choose $m \geqslant n$, for example $m = 2n + 1$. Build matrix $\mathbf{X}_n$ and vector $X(N)$.
2) Perform the SVD of data matrix $\mathbf{X}_n$ to get singular values $\alpha_1 \geqslant \cdots \geqslant \alpha_n$.
3) Compute the least squares solution $a(n) = [a_1, \quad \ldots, \quad a_n]'$ of the problem $\mathbf{X}_n a(n) = X(N)$. Estimate the noise variance $\sigma_e^2 = \text{MSE}_x/m = ||X(N) - \mathbf{X}_n a(n)||_2^2/m$.
4) Compute the threshold $\delta = \sqrt{c(m)}\sigma_e$, where $c(m) = \chi^2_{1-\alpha}(m)$ is the $(1-\alpha)$-percentile of the χ^2-distribution with m degrees of freedom. Determine the rank k such that $\alpha_k > \delta \geqslant \alpha_{k+1}$.
5) Repeat steps 1–4 with $n = k + 1$ to find the rank l of the new data matrix $\mathbf{X}_{k+1}$. If $1 < l < k$, then set $k = l$ and repeat step 5; else stop and return l as the optimal order.

```
function order = ARorderKonst(x,n,siglevel)
%   x = time series of observed data (vector of size N x 1)
%   n = largest value of p to try.

if nargin<3, siglevel= 0.05; end;
N = length(x); m = 2*n+1; X = zeros(N,m);
X(N,:) = x(N:-1:N-m+1);
for i = 1:n, X(N-i,:) = x(N-i:-1:N-i-m+1); end
boldX = flipud(X(N-n:1:N-1,:))';

% Iterative algorithm: initial step
% 2.) Singular value decomposition of data matrix
alpha = svd(boldX);

% 3.) Calculate error variance
an = (boldX' * boldX) \ (boldX' * X(N,:)'); % OLS solution of X_n * a(n) = X(N)
sigmaE = sqrt(norm(X(N,:)' - boldX*an)^2/m); % error variance

% 4.) Calculate threshold epsilon_L and find effective rank of data matrix
c = chi2inv(1-siglevel,m);
epsilonL= sqrt(c)*sigmaE;
k = length(find(alpha > epsilonL));

% 5.) Iterate the following steps until break criteria fulfilled
while 1
  n= k+1; m = 2*n+1;
  clear X
  X = zeros(N,m); X(N,:) = x(N:-1:N-m+1);
  for i = 1:n, X(N-i,:) = x(N-i:-1:N-i-m+1); end
  clear boldX
  boldX = flipud(X(N-n:1:N-1,:))';
  % 5.2.)
  alpha = svd(boldX);
  % 5.3.)
  an = (boldX' * boldX) \ (boldX' * X(N,:)'); % OLS solution
  sigmaE = sqrt(norm(X(N,:)' - boldX*an)^2/m);
  % 5.4.)
  c = chi2inv(1-siglevel,m); epsilonL= sqrt(c)*sigmaE;
  k2= length(find(alpha > epsilonL));
  if (k2<k && k2>1)
    k = k2;
  else
    order = k2;
    break
  end
end
```

Program Listing 9.6: AR order selection algorithm based on the singular value decomposition from Konstantinides (1991).

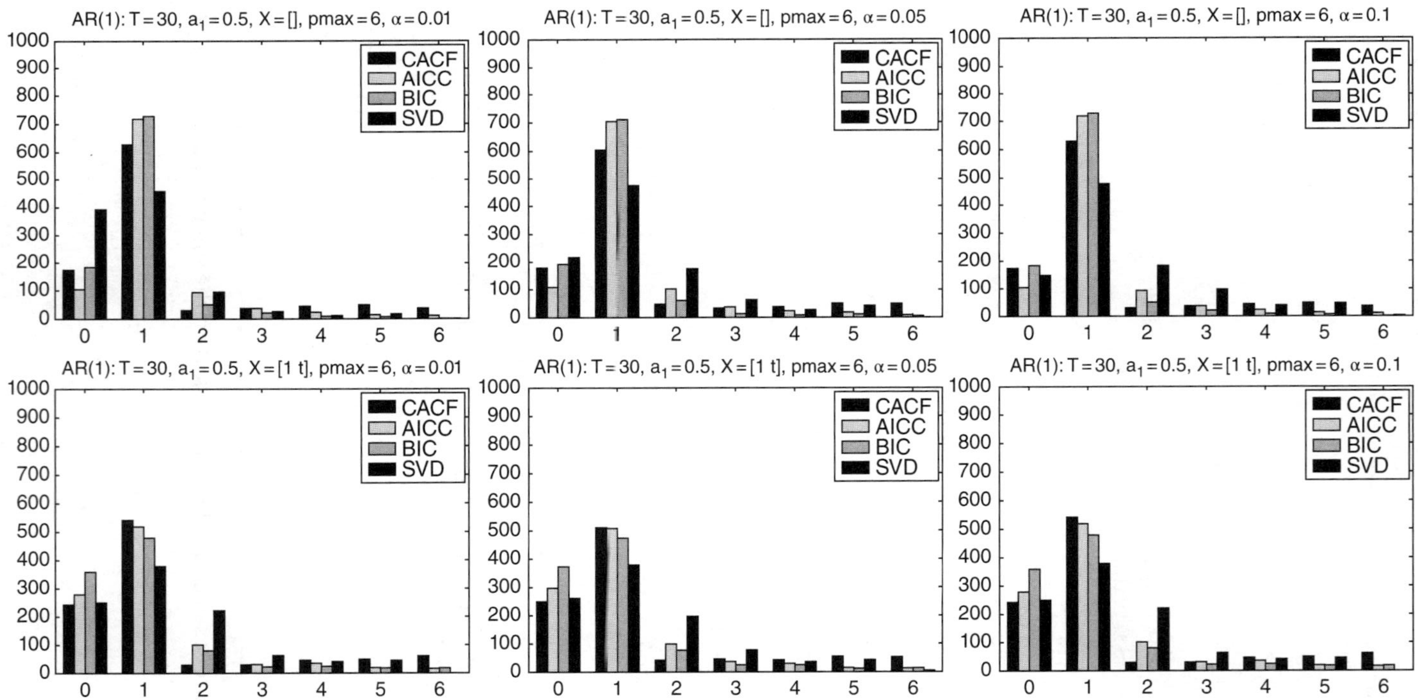

Figure 9.22 Histograms, based on 1,000 replications and sample size $T = 30$, corresponding to the chosen value of AR lag length p, for an AR(1) model with parameter $a = 0.5$, for the CACF, AICC, BIC, and SVD methods, using $p_{\max} = 6$. The top panels correspond to a pure AR(1) model with no regressor matrix, while the bottom panels use $\mathbf{X} = [\mathbf{1},\ \mathbf{t}]$, and the SVD method is applied to the ordinary least squares residuals based on this $\mathbf{X}$ matrix. From left to right, the three values $\alpha = 0.01$, $\alpha = 0.05$, $\alpha = 0.10$, as the tuning parameter of the SVD method, are used.

The choice of $m = 2n + 1$ is arbitrary and is a tuning parameter of the model. One needs $N \geqslant n + m$ observations for the algorithm to work. For n between 4 and 8, the minimum number of observations is thus between 13 and 25 under this choice of m. While other iterative order determination algorithms evaluate the testing criterion for all values $n = 1, \ldots, p + 1$, the algorithm of Konstantinides (1991) does not need this and is thus much faster. The program in Listing 9.6 implements the method.

The value of $\alpha = 0.05$ imposed in Konstantinides (1991) is certainly a natural choice in light of the ubiquitous nature of its use as the type I error rate in hypothesis testing. Nevertheless, it is also a tuning parameter of the method, and the value of 0.05 may not be optimal with respect to a more typical "loss function" regarding the choice of p.

The first simulation experiment uses an AR(1) model with parameter $a = 0.5$. Figure 9.22 is similar to Figures 9.16 and 9.17, but includes the use of the SVD method, and only considers $p_{\max} = 6$. Three values of tuning parameter α are used, 0.01, 0.05, and 0.10. It appears that, for this model and sample size, in the case with no $\mathbf{X}$ matrix, the use of $\alpha = 0.01$ is inferior to use of $\alpha = 0.05$, with the latter having a much lower chance of choosing $p = 0$, and a slightly higher chance of choosing the correct $p = 1$. The AICC and BIC are clearly superior in the no $\mathbf{X}$ case, while the CACF method is (very slightly) preferred with $\mathbf{X} = [\mathbf{1},\ \mathbf{t}]$, as in Figure 9.17.

The next experiment uses an AR(4) model with $a_1 = 0.4$, $a_2 = -0.3$, $a_3 = 0.2$, and two values of a_4, namely $a_4 = -0.1$ and $a_4 = -0.6$ (which is stationary, with maximum modulus of the AR polynomial roots being 0.91), sample size $T = 90$, and no $\mathbf{X}$ matrix. The upper lag order is $p_{\max} = 7$. Two sets of tuning parameters for the CACF and SVD methods are used. Figure 9.23 shows the results. For the case with $a_4 = -0.1$ (upper panels), no method performs well, noting that the BIC and the SVD with $\alpha = 0.05$ have a high preponderance of choosing a lag length of 0, 1, or 2. With $a_4 = -0.6$ (bottom panels), matters look somewhat better, with the BIC, followed by AICC, being blatantly superior to the other methods.

9.7 Further Methods: Pattern Identification

> Doing econometrics is like trying to learn the laws of electricity by playing the radio.
>
> (Guy Orcutt, cited by Leamer, 1983, p. 31)

Optical inspection of the SACF and SPACF, as discussed in Section 9.2, is a special case of more general "pattern recognition" methods. There are other correlograms available that exhibit different properties than the SACF and SPACF and so provide a fuller picture of the correlation structure. The most popular of these are the two **inverse correlograms**, named the sample inverse ACF, or SIACF, and the sample inverse partial ACF, or SIPACF. These mimic the SACF and SPACF but essentially reverse the roles of p and q, so that, for example, both the SPACF and SIACF should "cut off" for an AR(p) model.[3]

Another set of complementary correlograms to the SACF and SPACF are the **modified SACF and SPACF**, which have the convenient property that they cut off even for stationary and invertible ARMA(p, q) models such that both p and q are nonzero. See, e.g., Choi (1992) and the references

3 The SIACF was introduced by Cleveland (1972) in the context of the spectral analysis of time series, and Chatfield (1979) provided the time-domain definition. Abraham and Ledolter (1984) have demonstrated that, for full-order AR processes (all coefficients are non-zero), the SPACF is more powerful than the SIACF, while Etuk (2000) showed via simulation that, for identification of *subset autoregressive models* (at least one of the $a_i = 0$, $i < p$), the SIACF is often better.

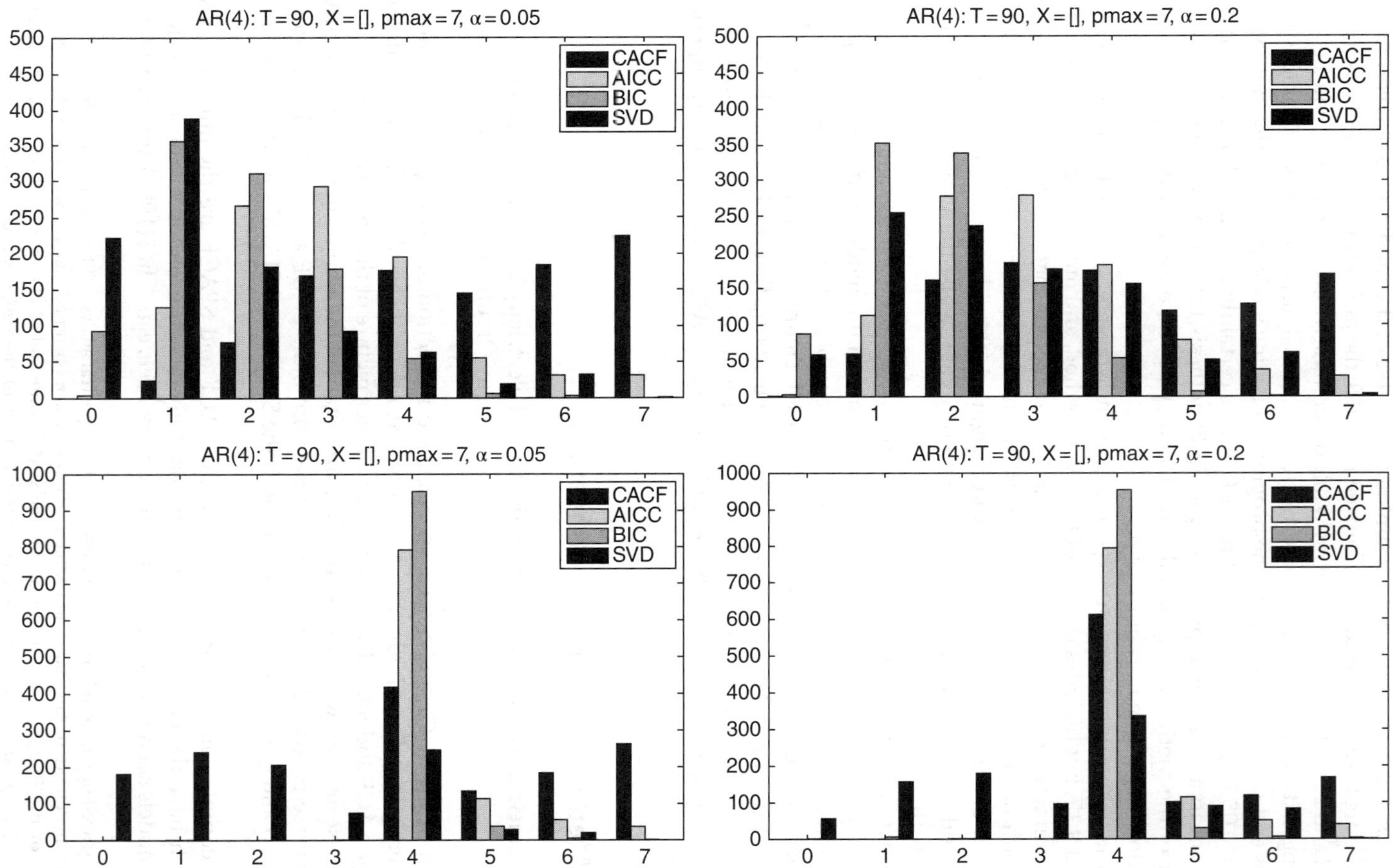

Figure 9.23 **Top:** Similar to Figure 9.22, but based on an AR(4) model with $a_1 = 0.4$, $a_2 = -0.3$, $a_3 = 0.2$, and $a_4 = -0.1$, no **X** matrix, sample size $T = 90$, $p_{\max} = 7$, and two sets of tuning parameters. The left panels use $\alpha = 0.05$ for the SVD and $c = 0.125$ for the CACF, while right panels use $\alpha = 0.2$ and $c = 0.075$. **Bottom:** Same but $a_4 = -0.6$.

therein. While appealing, the sampling distributions are complicated, and they tend to not work well in practice.

There are also other procedures that produce a set of data that is patterned in some way depending on p and q. As a brief demonstration of a possible pattern identification-type method that could be used to elicit p and q, recall matrix $\mathbf{\Pi}$ defined in (7.19). There it was used with the true values of p and q and is full rank. Let us see what happens when p and q are arbitrary. We compute the determinant of $\mathbf{\Pi}$ for all combinations of $p = 0, 1, \dots, p_{\max}$ and $q = 0, 1, \dots, q_{\max}$, based on the true π_i, and record in a $p_{\max} \times q_{\max}$ matrix, say $\mathbf{D}$, a zero if the absolute value of the determinant is less than 10^{-7} and a one otherwise. Using $p_{\max} = q_{\max} = 7$, the left matrix in (9.12) corresponds to an ARMA(1, 3) model (with $a_1 = 0.5$, $b_1 = -1.1$, $b_2 = 0.7$, and $b_3 = -0.3$) and the right matrix to an ARMA(2, 5) model with $\mathbf{a} = [0.3 \ \ 0.2]$ and $\mathbf{b} = [-0.4 \ \ 0.4 \ \ -0.3 \ \ -0.1 \ \ -0.3]$.

$$\mathbf{D} = \begin{bmatrix} 1 & 1 & 1 & 1 & 1 & 1 & 1 \\ 1 & 1 & 1 & 0 & 0 & 0 & 0 \\ 1 & 1 & 1 & 0 & 0 & 0 & 0 \\ 0 & 0 & 0 & 0 & 0 & 0 & 0 \\ 0 & 0 & 0 & 0 & 0 & 0 & 0 \\ 0 & 0 & 0 & 0 & 0 & 0 & 0 \\ 0 & 0 & 0 & 0 & 0 & 0 & 0 \end{bmatrix} \qquad \mathbf{D} = \begin{bmatrix} 1 & 1 & 1 & 1 & 1 & 1 & 1 \\ 1 & 1 & 1 & 1 & 1 & 1 & 1 \\ 1 & 1 & 1 & 1 & 1 & 0 & 0 \\ 1 & 1 & 1 & 1 & 1 & 0 & 0 \\ 1 & 1 & 1 & 1 & 1 & 0 & 0 \\ 0 & 0 & 0 & 0 & 0 & 0 & 0 \\ 0 & 0 & 0 & 0 & 0 & 0 & 0 \end{bmatrix} \tag{9.12}$$

A clear pattern emerges: The value of p is the number of rows with *all* ones, and q is the number of rows with *any* ones. As long as $p_{\max}$ and $q_{\max}$ are chosen large enough, trial and error appears to confirm that this rule holds for all stationary and invertible ARMA models, but it would need to be (and probably has been; see Beguin et al., 1980) algebraically proven.[4] A method for determining p and q from a sample time series immediately suggests itself: Compute estimates $\hat{\pi}_i$ using o.l.s., $i = 1, \dots, p_{\max} + q_{\max}$, compute $\mathbf{D}$, and look (hope?) for a clear pattern. Tuning parameters include $p_{\max}$, $q_{\max}$ and the threshold for declaring if the determinant is zero.

For the data set in Figure 9.1, the "determinant matrix" $\mathbf{D}$, computed using $p_{\max} = q_{\max} = 5$, takes the form

$$\mathbf{D} = \begin{bmatrix} 1 & 1 & 1 & 1 & 1 \\ 1 & 1 & 1 & 1 & 0 \\ 0 & 0 & 0 & 0 & 0 \\ 0 & 0 & 0 & 0 & 0 \\ 0 & 0 & 0 & 0 & 0 \end{bmatrix}, \tag{9.13}$$

which suggests an ARMA(1,2) model. If the second row ended in a one instead of a zero, we could have claimed an ARMA(2,2) model instead. These model choices are in agreement with those from the other methods. However, matrix $\mathbf{D}$ was obtained using a threshold (for deciding if the determinant is zero) around 0.2, which is quite far from zero in this context, illustrating the sensitivity of the method to the choice of tuning parameters.

4 See Choi (1992), Chan (1999), and the references therein for related methods and discussion of other pattern methods for the mixed ARMA case.

The book by Choi (1992) is dedicated to the topic of ARMA model selection and provides the best starting place for further information and research in this area. In particular, Choi (1992, Ch. 5) provides detail on various pattern identification methods. Anderson (1994) is well worth looking at as well. He writes in his introduction "[Choi's chapter 5] presents about a dozen methods proposed by several statisticians. The purpose of the present chapter is to organize these methods coherently and relate them to each other." Further, good reading about ARMA modeling in general is provided by (a different) Anderson (1995).

Part III

Modeling Financial Asset Returns

10

Univariate GARCH Modeling

The goal of this chapter is to develop the primary topics associated with the class of univariate GARCH models, as well as some less common but highly useful methods for estimation. One of their primary applications is for risk prediction of financial portfolios of assets, and this will be detailed in Chapter 11. This basic univariate GARCH framework is not as limited as it might seem: It can be used to form *multivariate* models of financial asset returns and as an important application in the context of portfolio optimization. We save that discussion also for Chapter 11, concentrating herein on several core aspects of the univariate case.

The outline of this chapter is as follows. After some introductory remarks in Section 10.1, Section 10.2 presents the fundamental properties of the baseline Gaussian GARCH model and details its estimation. Section 10.3 builds on this by discussing some simple but important extensions. Section 10.4 is concerned with estimation of GARCH models when the underlying i.i.d. process is specifically noncentral Student's t, denoted NCT-GARCH. Section 10.5 is dedicated to the GARCH model with a stable Paretian distributional assumption, denoted $S_{\alpha,\beta}$-GARCH, and discusses testing the stability and i.i.d. assumptions of the filtered innovations process. Section 10.6 details a GARCH-type model that does not fit into the class of extensions from Section 10.3, but embodies a richer dynamic structure based on a discrete normal mixture distribution that leads to improved out-of-sample forecasts.

10.1 Introduction

> People in the academic world often do well, not because they are smarter than others, but because they have chosen somehow sexier fields to research.
>
> (Emanuel Parzen, in Newton, 2002)

Volatility clustering is one of several so-called **stylized facts**—typically observed empirical characteristics and regularities of financial asset price changes, or returns. This clustering is perhaps best defined as stated in Mandelbrot (1963): "large changes tend to be followed by large changes, of either sign, and small changes tend to be followed by small changes", where "changes" refer to differences in the underlying price. Instead of price changes, we will model the returns, as were defined in Section 7.7.1. The returns on many, if not virtually all, financial assets measured at the weekly, daily, and higher frequency levels exhibit volatility clustering, such as currency exchange

Linear Models and Time-Series Analysis: Regression, ANOVA, ARMA and GARCH, First Edition. Marc S. Paolella.

rates, stock and futures prices, crude oil (see, e.g., Baumeister and Peersman, 2013), convertible bonds (see, e.g., Wang and Li, 2011), fixed-income exchange traded funds (ETFs),[1] etc.

The other primary stylized facts of asset returns are high leptokurtosis and mild asymmetry. See Granger et al. (2000), Cont (2001), Teräsvirta and Zhao (2011), and the references therein for discussion of these and further stylized facts.

The **autoregressive conditional heteroskedasticity** (ARCH) model, and particularly its generalized form, hereafter **GARCH**, are now cornerstone structures for addressing the volatility clustering in financial time series, with their origins in the works of Engle (1982), McCulloch (1985a), Bollerslev (1986, 1987), and Taylor (1986), followed by an enormous subsequent growth in model extensions, applications, and theoretical underpinnings. There are now many variations of the GARCH idea—arguably too many—culminating in an amusing "alphabet soup" of associated acronyms, including the tongue-in-cheek YAARCH (Yet Another ARCH). A (much-needed) overview can be found in Bollerslev (2010). Aspiring academics may wish to admonish the remark from Moosa (2017, Sec. 1.6): "There have been more sequels to ARCH than to *Jaws*, *Rocky*, *Rambo*, and *Die Hard* put together. As for 'better' models, it is not obvious to me in what ways the extensions and alternative are better—it has been an extravaganza that served no purpose whatsoever, apart from providing the means whereby students get their PhDs and academics get their promotions."

The econometric term "heteroskedasticity" just means that the variance is not constant.[2] In this context, "volatility" is typically defined as the square root of the variance, which, in the Gaussian case, is the scale term of the normal distribution, σ. We *define* volatility as the (time-varying) scale term, and thus it will make sense to speak of (conditional) volatility also for models in which the variance of the underlying i.i.d. random variables, also referred to as the **innovation process** or innovation sequence, does not exist, such as in the stable Paretian case, or Student's t with degrees of freedom less than or equal to two.

A clear illustration of pronounced volatility clustering is given in Figure 10.1, showing the daily returns on the exchange rate of various currencies compared to the U.S. dollar around the Asian Financial Crisis (AFC) commencing in Thailand in July 1997, as analyzed by Mittnik and Paolella (2000).

Remark The massive devaluation in July 1997 of the Thai bhat is considered to have triggered the financial crisis that led to the collapse of foreign exchange and equity prices in many East Asian countries. During the 1980s, Thailand's capital markets became increasingly liberalized and opened up for foreign investors. As in other countries in the region, the Thailand economy experienced high economic growth and large inflows of private capital throughout the 1990s, while maintaining an effectively pegged nominal exchange rate. The capital inflow led to an appreciation of the real exchange rate that, in turn, negatively affected firms' exports and profit margins and, ultimately, led to cash flow shortages. From 1990 to 1997, Thai companies increased the U.S. dollar-denominated foreign debt outstanding from international bond market activities by a factor of about 10, to $12.9 billion (Harvey and Roper, 1999).

---

1 Investing in the fixed-income (bond) market is quite different from investing in (liquid) stocks because of the over-the-counter (OTC) nature of the markets, lack of liquidity, high markups, bond maturities, etc. An ETF eliminates most of these problems, as well as providing diversification, so that the instrument can be bought and sold similar to a liquid stock.

2 While simple as a concept, the word's spelling was, for quite some time, not, with the decision of using a "k" or a "c" becoming a serious and divisive issue among otherwise pleasant econometricians, culminating in the argument in McCulloch (1985b) for use of "k". Research continues on this fascinating topic; see, e.g., Paloyo (2011).

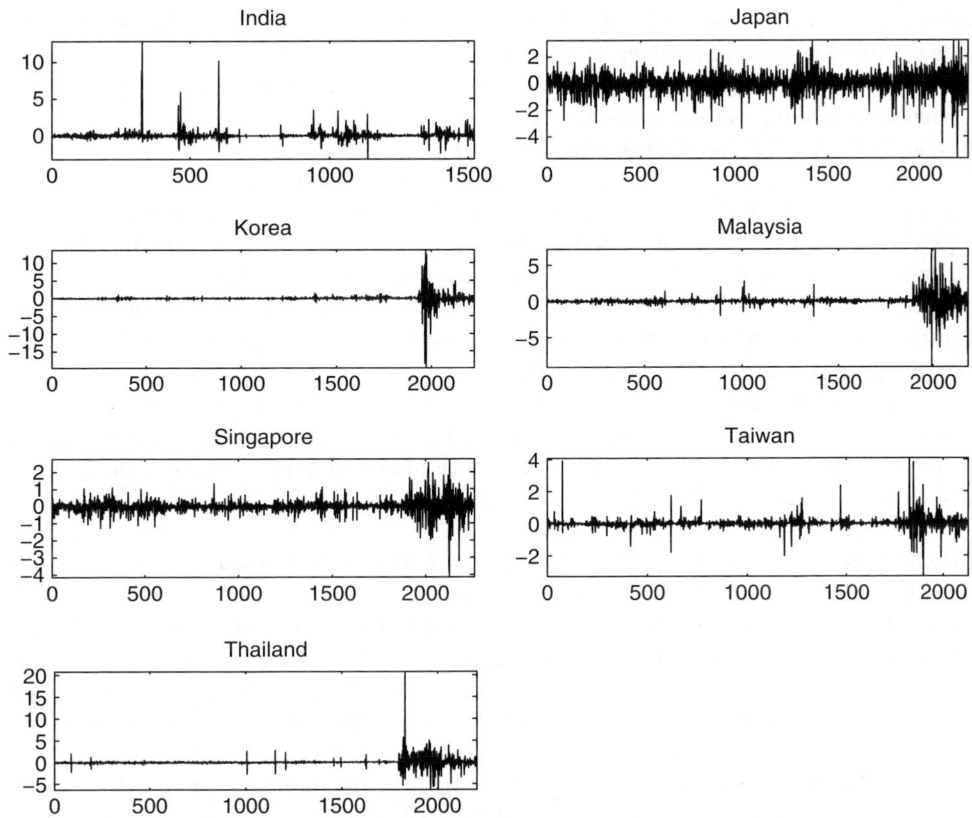

Figure 10.1 The returns $R_t = 100 \times (\ln P_t - \ln P_{t-1})$, where P_t is the exchange rate at time t, of various Asian currencies versus the USD, from 1 January 1990 to 31 December 1998 (with the exception of the Indian rupee, which extends only until 2 July 1998).

In February 1997, Somprasong became the first Thai company that failed to serve foreign debt. In May of that year, the Thai bhat came under heavy attacks from speculators acting on the economic slowdown as well as political instability. The largest finance company, Finance One, failed shortly thereafter. In June, Thailand's finance minister, who had strongly resisted a devaluation of the bhat, resigned. On July 2, the Bank of Thailand announced a managed float of the bhat and requested "technical assistance" from the International Monetary Fund. The resulting devaluation from 24.5 to 30.2 bhat per U.S. dollar is considered to have triggered the East Asian crisis. Subsequent to that, negative news from the Thai economy as well as from other countries in the region led to violent swings for the bhat, reaching a low point on January 12, 1998 with an exchange rate of 56.1 bhat per U.S. dollar.

For a more detailed account of the AFC, see, e.g., Corsetti et al. (1999a,b) and Harvey and Roper (1999). ∎

Almost precisely a decade later, enormous increases and fluctuations in volatility arose again, due to the U.S. banking and liquidity crisis, sometimes referred to as the sub-prime crisis, or the Global

Financial Crisis (GFC) starting around mid 2007.[3] For a study of the volatility during this period, see, e.g., Banulescu et al. (2016) and the references therein, notably also their graphics of the (annualized, so-called realized) volatility annotated with some of the major events that occurred during the GFC, one graphic of which, using a type of robust GARCH estimation based on methodology from Creal et al. (2011, 2013) and Harvey (2013a), is shown in Figure 10.2.

We will see below in Example 10.1 the rather remarkable result that a simple GARCH model (albeit driven by a heavy-tailed innovation process) can mimic the behavior of the returns on currency, stock, and other asset prices even during extreme market periods and crisis conditions, such as the AFC and GFC. In particular, this is accomplished without the need for more elaborate models, such as those incorporating "structural breaks" or "regime switching", these being two examples of models with time-varying parameters. This is not to be misunderstood as implying that simple GARCH-type models cannot be improved upon in terms of forecasting ability by accounting for structural breaks and/or regime switching. Quite on the contrary, they can, and this should not come as a surprise, given that most GARCH-type models used are very simplistic time-series structures that have between three and five parameters for describing what is in reality a complicated phenomenon. See, for example, the discussion of the autocorrelation function below in Section 10.6.3 regarding structural breaks, and Section 10.6.5 regarding the use of regime switching models.

The basic GARCH model is a simple, ARMA-type structure as was developed in Part II, applied not to the mean, but to the variance (or, more generally, the scale term, as discussed above, or its square) of the innovation sequence. Like ARMA models in classic time-series analysis, GARCH and all its variations are stochastic processes that do not purport to address the reason for conditional heteroskedasticity or claim to be the true data generating process (d.g.p.) of what is most surely a very complicated underlying phenomenon. However, due to the nature of financial asset returns and their highly persistent volatility, simple GARCH models turn out to be very effective for modeling and predicting the scale terms. Possible economic explanations as to why financial return series exhibit volatility clustering and heavy tails are discussed in Kirchler and Huber (2007) and the references therein.

Remarks

a) GARCH-type processes are not the only stochastic models that can mimic the volatility clustering and other stylized facts of asset returns. For example, recall the stochastic unit root model (7.57), namely

$$P_t = (1 + \delta_t)P_{t-1} + \epsilon_t, \quad \delta_t \overset{\text{i.i.d.}}{\sim} \text{N}(0, \sigma_\delta^2) \quad \text{indep. of} \quad \epsilon_t \overset{\text{i.i.d.}}{\sim} \text{N}(0, \sigma_\epsilon^2). \tag{10.1}$$

It can generate data that strongly resemble a financial price process and whose returns exhibit "GARCH effects"; see, e.g., Yoon (2003). For example, Figure 10.3 shows a simulated (price) process $\{P_t\}$ from (10.1), their log percentage returns, and the SACF of the returns and the squared returns, using the code in Listing 10.1.

b) Another stochastic process that gives rise to GARCH effects and has numerous other appealing properties is the **multifractal model**. Its use often results in superior forecasts compared to GARCH and is gaining in popularity. See Lux (2008), Wang et al. (2016), Lux et al. (2016), Segnon et al. (2017), Lux and Segnon (2018), and the references therein.

3 The starting date for the sub-prime crisis is often taken to be early August, 2007; see Covitz et al. (2013) and the references therein.

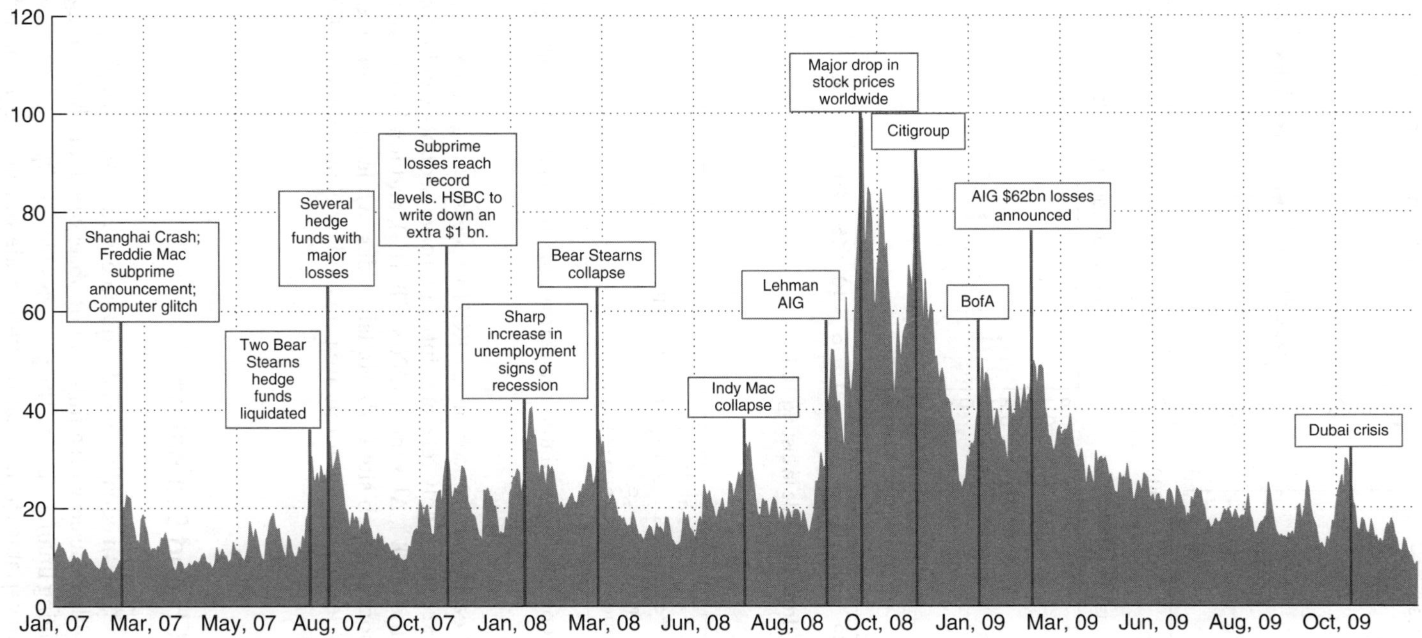

Figure 10.2 Annualized volatility during the GFC based on a robustified GARCH model and marked with occurrences of several major events. This graphic, courtesy of Peter Hansen, is Figure 4 in Banulescu et al. (2016).

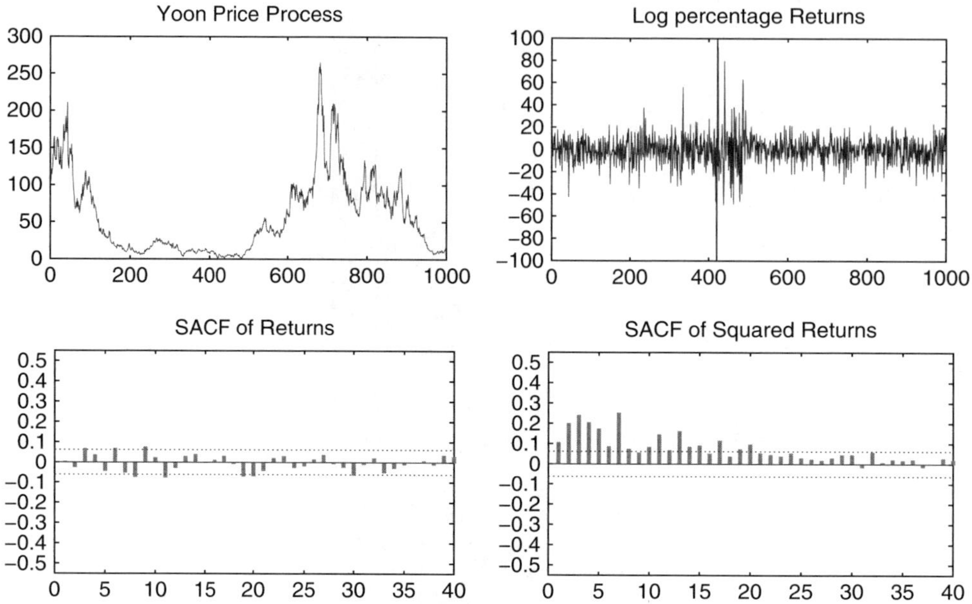

Figure 10.3 Realization of (10.1) from the code in Listing 10.1.

```
T=1e3; a=randn(T,1)/10; e=0.005+randn(T,1)/1;
P=zeros(T,1); P(1)=100; for t=2:T, P(t)=(1+a(t))*P(t-1)+e(t); end
if any(P<=1e-3), P=P+abs(min(P))+1e-2; end
lP=log(P); R=100*(lP(2:end)-lP(1:(end-1)));
```

Program Listing 10.1: Generates a simulated realization of (10.1) and the corresponding returns.

c) For further reading and earlier references on GARCH processes, the surveys by Bollerslev et al. (1994) and Palm (1997) are still highly relevant; introductory accounts can be found in Patterson (2000a, Ch. 16), Morimune (2007), Alexander (2008, Ch. II.4), and Jondeau et al. (2007, Ch. 4), the latter also having an associated web site with Matlab codes, while more technical and overarching (pun intended) book-length presentations are provided by Gourieroux (1997) and Francq and Zakoïan (2010). ■

10.2 Gaussian GARCH and Estimation

Let the time series $\{R_t\}$, $t \in \mathbb{Z}$, be an equally spaced sequence of random variables, forming a stochastic process. This will invariably be used with a finite set of observed data, these being the (percentage log) returns on some underlying financial asset. The process given by $R_t = \epsilon_t = Z_t\sigma_t$, where $\{Z_t\}$ refers to a sequence of i.i.d. random variables from location-zero, scale-one density $f_Z(\cdot)$, and scale parameter

σ_t being computed from the recursion

$$\sigma_t^2 = c_0 + \sum_{i=1}^{r} c_i \epsilon_{t-i}^2 + \sum_{j=1}^{s} d_j \sigma_{t-j}^2, \tag{10.2}$$

is referred to as a (discrete time) GARCH(r,s) process.

While we will use (10.2) throughout this chapter, the reader should know that different notations will be found in the (large) GARCH literature. For example, a common one is to denote the scale term as h_t, and the parameters in (10.2) as ω, α_i, and β_i, respectively, i.e.,

$$h_t = \omega + \sum_{i=1}^{r} \alpha_i \epsilon_{t-i}^2 + \sum_{j=1}^{s} \beta_j h_{t-j}. \tag{10.3}$$

If $Z_t \overset{\text{i.i.d.}}{\sim} \mathrm{N}(0, 1)$, then we can be more specific and say that $\{R_t\}$ follows a GARCH(r,s) process with normal innovations, or, in short, a normal (or Gaussian) GARCH(r,s) model. In this case, $\mathbb{E}[Z_t^2] = 1$. The reason for defining the seemingly superfluous ϵ_t instead of just using R_t is because the model could be augmented with a location term, such as $R_t = \mu + \epsilon_t$, with $\epsilon_t = Z_t \sigma_t$, or a time-series structure for the mean, such as an ARMA model, as will be done in (10.18) below.

10.2.1 Basic Properties

Let Ω_t denote the **information set at time** t, or the set of observed random variables available about the process generating the returns up to time t. For the discrete time GARCH process (10.2), Ω_t is taken to be $(\dots, R_{t-1}, R_t)$, i.e., infinitely many past returns, from which the GARCH model parameters $\boldsymbol{\theta} = (\mathbf{c}', \mathbf{d}')'$, where $\mathbf{c} = (c_0, c_1, \dots, c_r)'$ and $\mathbf{d} = (d_1, d_2, \dots, d_s)'$, and past scale terms $(\sigma_t, \sigma_{t-1}, \dots)$, could (in principle) be exactly elicited. The following aspect is one of the defining features of GARCH-type models and differentiates them from a competing class referred to as **stochastic volatility**, or SV, models:

> Given Ω_{t-1}, σ_t is a known constant, computed from (10.2). It is not stochastic.

The calculation of the one-step ahead forecast $\sigma_t \mid \Omega_{t-1}$ is discussed in Section 10.3.3 in a somewhat more general context. Obviously, in practice, one does not have an infinite number of past observations, so that Ω_T is just $(R_1, R_2, \dots, R_{T-1}, R_T)$, and $(r + s + 1)$-length parameter vector $\boldsymbol{\theta}$ and the s past scale terms $(\sigma_0, \sigma_{-1}, \dots, \sigma_{1-s})$ need to be estimated from the available data. This will be dealt with below in detail.

For the moment, let $\Omega_t = (R_t, R_{t-1}, \dots)$, so that, with normal innovations, the conditional variance of R_t, given Ω_{t-1}, is

$$\mathbb{E}[R_t^2 \mid \Omega_{t-1}] = \mathbb{E}[\sigma_t^2 Z_t^2 \mid \Omega_{t-1}] = \sigma_t^2 \mathbb{E}[Z_t^2 \mid \Omega_{t-1}] = \sigma_t^2 \mathbb{E}[Z_t^2] = \sigma_t^2. \tag{10.4}$$

Now consider the *unconditional* variance of R_t. First, let $\mathbb{E}[\sigma_{\blacksquare}^2]$ denote the unconditional expectation of σ_t^2. Then, from (10.2), and that the $\{Z_t\}$ are i.i.d.,

$$\mathbb{E}[\sigma_{\blacksquare}^2] = c_0 + \mathbb{E}\left[\sum_{i=1}^{r} c_i \epsilon_{t-i}^2\right] + \mathbb{E}\left[\sum_{j=1}^{s} d_j \sigma_{\blacksquare}^2\right]$$

$$= c_0 + \sum_{i=1}^{r} c_i \mathbb{E}[Z_{t-i}^2]\mathbb{E}[\sigma_{\blacksquare}^2] + \sum_{j=1}^{s} d_j \mathbb{E}[\sigma_{\blacksquare}^2] \quad (Z_t \text{ is independent of } \sigma_t^2)$$

$$= c_0 + \mathbb{E}[Z_t^2] \sum_{i=1}^{r} c_i \mathbb{E}[\sigma_{\blacksquare}^2] + \sum_{j=1}^{s} d_j \mathbb{E}[\sigma_{\blacksquare}^2] = c_0 + \mathbb{E}[\sigma_{\blacksquare}^2] \left(\mathbb{E}[Z_t^2] \sum_{i=1}^{r} c_i + \sum_{j=1}^{s} d_j \right)$$

$$= \frac{c_0}{1 - \mathbb{E}[Z_t^2] \sum_{i=1}^{r} c_i - \sum_{j=1}^{s} d_j}. \tag{10.5}$$

Thus, unconditionally, $\mathbb{V}(R_t) = \mathbb{V}(Z_t)\mathbb{E}[\sigma_{\blacksquare}^2]$, which, for $Z_t \overset{\text{i.i.d.}}{\sim} \mathrm{N}(0, 1)$, is

$$\mathbb{V}(R_t) = \mathbb{E}[\sigma_{\blacksquare}^2] = \frac{c_0}{1 - V_{\mathrm{N}}}, \quad V_{\mathrm{N}} := \sum_{i=1}^{r} c_i + \sum_{j=1}^{s} d_j. \tag{10.6}$$

Observe that $\mathbb{V}(R_t)$ does not exist if $V_{\mathrm{N}} \geqslant 1$ (where the N denotes normality) and is such that, the closer V_{N} is to one, the larger is the unconditional variance (or volatility) of R_t. This is sometimes referred to as the "sum to one" condition. In practice, once the model is estimated, the statistic $\widehat{V}_{\mathrm{N}} = \sum_{i=1}^{r} \hat{c}_i + \sum_{j=1}^{s} \hat{d}_j$ gives a measure of the extent of persistence of shocks to the volatility of the series.

Formulae for higher moments of R_t can also be obtained. For example, in the GARCH(1,1) case (with normal innovations), the kurtosis of $R_t = \epsilon_t = Z_t\sigma_t$, and the autocorrelation function of R_t^2, if they exist, are given, respectively, by

$$\mathrm{kurt}(R_t) = 3\frac{1 - (c_1 + d_1)^2}{1 - (3c_1^2 + 2c_1 d_1 + d_1^2)} > 3, \quad r(\tau) = (c_1 + d_1)^{\tau-1} \frac{c_1(1 - d_1^2 - c_1 d_1)}{1 - d_1^2 - 2c_1 d_1}, \tag{10.7}$$

see e.g., Bollerslev (1988) and He and Teräsvirta (1999a,b). These are generalized (10.24) and (10.25).

As scale terms, all σ_t need to be positive, so that constraints on the parameters $\boldsymbol{\theta}$ are required. It is easy to see that it is sufficient if $c_i > 0$, $i = 0, \ldots, r$, and $d_j \geqslant 0$, $j = 1, \ldots, s$. In the very important $r = s = 1$ case, this is also a necessary condition (that needs to be imposed during estimation). Necessary and sufficient conditions in the general r, s case are given in Nelson and Cao (1992) and are such that some coefficients can be negative. These conditions can be done away with by expressing the model in terms of $\ln \sigma^2$ instead of σ^2, giving rise to the so-called EGARCH model by Nelson (1991), and the log-GARCH model; see Francq et al. (2013) and the references therein.

We next consider the choice of r and s. In the classic time-series literature on ARMA(p, q) models, much intellectual effort has been expended on how best to determine the orders p and q of the AR and MA components, respectively; recall Chapter 9. Fortunately, when modeling financial return series, it is almost always adequate simply to take $r = s = 1$. Notice one could use the AIC and BIC information criteria to determine the optimal r and s, though most empirical work just assumes $r = s = 1$. In this case, (10.2) reduces to $\sigma_t^2 = c_0 + c_1\epsilon_{t-1}^2 + d_1\sigma_{t-1}^2$, which can be interpreted as the variance at time t being decomposed into three simple pieces: A baseline value c_0, a fraction, d_1, of the previous period's variance, and a fraction, c_1, of the squared magnitude of the previous period's return.

10.2.2 Integrated GARCH

It would appear from (10.6) that, if $V_{\mathrm{N}} = 1$, the model would give rise to an explosive process, such that the magnitude of the R_t increases on average without bound. This is not the case: The normal

GARCH model (10.2) with $V_{\mathrm{N}} = 1$ is called the **integrated GARCH**, or IGARCH, and yields a strictly stationary process; see Nelson (1990) and the book references mentioned above. This also serves as an example of a process that is strictly stationary but not covariance stationary. Besides being a common process for modeling actual data (and such that there is one less parameter to estimate), the IGARCH construction is useful for emphasizing an important empirical aspect of GARCH modeling applied to financial returns or any (economic or other) time series whose d.g.p. is, as mentioned, undoubtedly not actually given by such a simple model:

> For genuine financial returns data, as the sample size is allowed to increase, the estimated (normal) GARCH parameters tend towards the IGARCH border (or exceed it, if not constrained), this being typically attributed to model mis-specification.

For further discussion of this point, and empirical evidence, see Diebold and Lopez (1996), Hillebrand (2005), Chavez-Demoulin et al. (2014), and the references therein. An analogy can be made to a more classical aspect of econometric modeling, namely unit root testing, from Section 5.5. In particular, as the true underlying d.g.p. is surely more complicated than the simple structure (5.1)–(5.2) assumed there, possibly because of structural breaks, time-varying parameters, a stochastic unit root, or other forms of model mis-specification, it is the case that, for a stationary process (in particular, without a unit root) and, as the sample size T increases, the power of unit root tests can drop to zero. Recall the example in Figure 5.12.

10.2.3 Maximum Likelihood Estimation

We now turn to parameter estimation of the normal GARCH model (10.2), but augment it with a location term as $R_t = \mu + \epsilon_t$, $\epsilon_t = Z_t\sigma_t$, $t = 1, \dots, T$. While moment-based methods are available (see, e.g., Baillie and Chung, 2001; Prono, 2016; and the references therein), we concentrate on use of the likelihood. Observe that, by construction, the likelihood is just the product of the location- and scale-transformed random variables $\sigma_t^{-1} f_Z((R_t - \mu)/\sigma_t)$. Basic code for computing the m.l.e. is given in Listing 10.2. For actual application, in order to help avoid inferior local maxima of the log-likelihood, we suggest using the method of profile likelihood for estimation discussed below.

Francq and Zakoïan (2004) prove the consistency and asymptotic normality of the maximum likelihood estimator of the GARCH model parameters. To assess the small-sample properties of the m.l.e., we use simulation. Such a study obviously requires being able to simulate a GARCH process, and the reader is encouraged to write a simple function to accomplish this.[4] Figure 10.4 shows the results of such a simulation, based on 1,000 replications, via histograms of the estimated parameters, for generated series with $T = 1{,}000$ observations, and using $c_0 = 0.04$, $c_1 = 0.05$, and $d_1 = 0.8$. Also shown are histograms of $\hat{c}_1 + \hat{d}_1$, as a measure of the persistence, as well as the sample variance of the filtered innovations, say $\widehat{\mathbb{V}}(\widehat{Z}_t)$. The reader should reproduce the graphics in Figure 10.4 using own code for simulation and estimation, as well as those via the built-in Matlab routines (and similarly for users of R, Python, etc.). More ambitious readers can conduct a simulation involving use of various estimation procedures, including those in canned packages such as EViews, SAS, etc., and compare their performance (in terms of bias, variance, m.s.e., etc.), and differences of the resulting estimates.

---

4 Matlab as of version R2012a has functions to do this, aptly named `garch` and `simulate`, as well as `estimate` and `forecast`; see the respective help files. Users of R will have a variety of such packages available, notably fGarch, from Diethelm Wuertz and Yohan Chalabi.

```
function [param,stderr,loglik,zvec] = babygarch(y)
% normal-GARCH(1,1) with power=2. y is vector of log percentage returns
initvec=          [0    0.04    0.05 0.8];
%                 mu     c_0     c_1  d_1
   bound.lo=     [-4    0       0     0  ];
   bound.hi=     [ 4    0.5     1     1  ];
   bound.which=[ 1      1       1     1  ];
opt=optimset('Display','None', 'Maxiter',500, 'TolFun',1e-6, ...
     'TolX',1e-6,'LargeScale','off');
init=einschrk(initvec,bound);
[pout,~,~,~,~,hess] = fminunc(@(param) like(param,y,bound),init,opt);
[loglik,zvec]=like(pout,y,bound);
V=pinv(hess)/length(y);
[param,V]=einschrk(pout,bound,V); stderr=sqrt(diag(V));

function [loglik,zvec]=like(param,y,bound)
param=einschrk(real(param),bound,999);
meanterm=param(1); c0=param(2); c1=param(3); d1=param(4);
e=y-meanterm; [zvec,sigvec]=ungarch(e,c0,c1,d1);
K=sqrt(2*pi); ll = -0.5 * zvec.^2 - log(K) - log(sigvec);
loglik = -mean(ll);

function [eout,sigvec]=ungarch(e,c0,c1,d1)
sigvec=zeros(length(e),1); e2=e.^2;
denom=1-c1-d1; if denom>0.001, sinit=c0/denom; else sinit=mean(e2); end
einit=sinit;
% do the recursion in sigvec^delta because it is faster
sigvec(1)=c0+c1*einit+d1*sinit;
for t=2:length(e), sigvec(t)=c0 + c1 *e2(t-1) + d1*sigvec(t-1); end
sigvec=sigvec.^(1/2); eout=e./sigvec;
```

Program Listing 10.2: Calculates the m.l.e. of the normal-GARCH(1,1) model (10.2) with added location term for the returns. Function `einschrk`, for imposing simple box constraints on the parameters, is given and discussed in Section III.4.3.2.

When the model is correctly specified, $\widehat{\mathbb{V}}(\widehat{Z}_t)$ should be unity, and this is the case. Except for estimated location term $\hat{\mu}$, the distributions of the parameters are not Gaussian, even for the relatively large sample size used. Further simulations with larger T confirm that they all appear to approach Gaussianity and are unbiased.

The choice of starting values for $\hat{c}_0$, $\hat{c}_1$, and $\hat{d}_1$ are important, as the log-likelihood can exhibit more than one local maxima. In the previous simulation, we used the true values as starting values—a luxury obviously not available in real life. This issue of multiple maxima has been noted by Ma et al. (2006), Winker and Maringer (2009), and Paolella and Polak (2015a), though seems to be often ignored, and can lead to inferior forecasts and jeopardize results in applied work.[5] This unfortunate observation might help explain the results of Brooks et al. (2001) in their extensive comparison of econometric software. In particular, they find that, with respect to estimating just the simple normal

5 When the SAS system, hailed as the industry standard and absolute benchmark in the 1980s, introduced GARCH estimation, they obviously used what would appear to be intelligent starting values: If there are no "GARCH effects", then in the normal-GARCH model, c_0 will be the unconditional variance, estimated as the sample variance, and c_1 and d_1 are zero. This can often be a local inferior maximum of the likelihood, and the method failed.

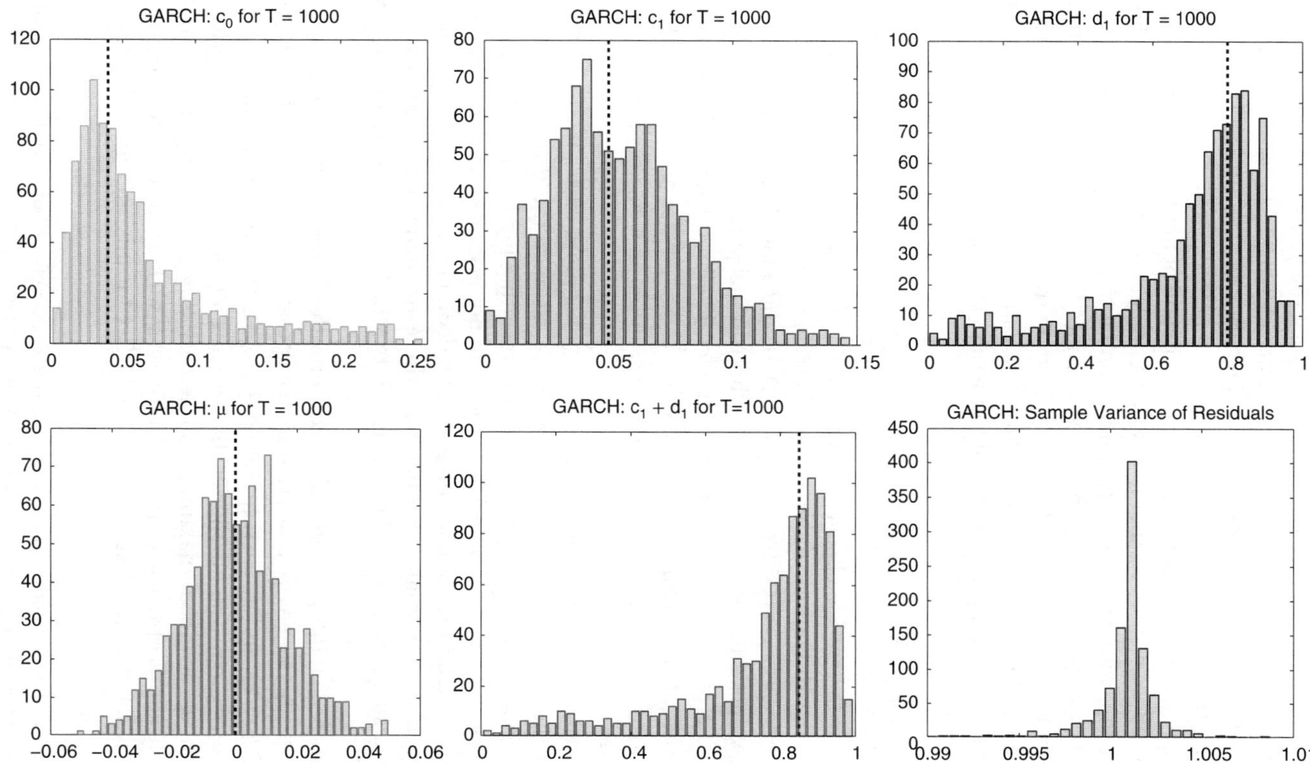

Figure 10.4 Results of m.l.e. estimation of a normal GARCH(1,1) process with $T = 1{,}000$ observations, using 1,000 replications. True parameters are indicated by vertical dashed lines.

GARCH model, "the results produced using a default application of several of the most popular econometrics packages differ considerably from one another" (Brooks et al., 2001, p. 54). Another reason for discrepant results is the choice of ϵ_0 and σ_0 to start the GARCH(1,1) recursion, for which several suggestions exist in the literature. For the Gaussian-GARCH(1,1) model, we take $\hat{\sigma}_0^2$ to be the sample unconditional variance of the R_t, and $\hat{\epsilon}_0^2 = \kappa\hat{\sigma}_0^2$, where κ is given in (10.13) in the context of the more general APARCH model.

As in Paolella and Polak (2015a), we illustrate the phenomenon of multiple maxima with a real (and typical) data set, and propose a solution that is simple to implement. We use the 250, 500, 750, and 1,000 daily (percentage log) returns on AT&T, starting on December 16, 1996, and estimate model (10.2) and also model (10.9) given below (which just changes the exponent in (10.2) from $\delta = 2$ to $\delta = 1$). The solid lines in Figure 10.5 (with the y-axis given on the *right* of the figures) show the profile log-likelihood (p.l.) obtained by fixing the value of c_0, and based on a grid of 100 points of c_0 between zero and 1.1 times the sample variance of the series. That is, for a fixed value of c_0, and with $\mathbf{R} = (R_1, \dots, R_T)$, we compute

$$\widehat{\boldsymbol{\theta}}_{\text{p.l.}}(c_0) = \arg\max_{\boldsymbol{\theta}_{\text{p.l.}}} \ell(\boldsymbol{\theta}_{\text{p.l.}}; \mathbf{R}), \qquad \boldsymbol{\theta}_{\text{p.l.}} = (c_1, d_1)'. \tag{10.8}$$

For example, in the case with $T = 500$ observations and $\delta = 1$, the maximum occurs for the fixed value of c_0 (this being one of the 100 points of the grid) of 0.6338, with the estimated values of the other model parameters obtained as $\hat{c}_1 = 0.1308$ and $\hat{d}_1 = 0.2174$, and a log-likelihood value of −671.0185. Using these as starting values and optimizing over the three model parameters yields $\hat{c}_0 = 0.6345$, the same value of $\hat{c}_1$, $\hat{d}_1 = 0.2166$, and (to 7 digits) the same log-likelihood value. The dashed line (with the y-axis given on the *left* of each figure) shows the value $\widehat{V}_{\text{N}} = \hat{c}_1 + \hat{d}_1$ corresponding to each of the 100 $\widehat{\boldsymbol{\theta}}_{\text{p.l.}}(c_0)$ values. Its essential linearity for the $\delta = 1$ case, and piecewise linearity for the $\delta = 2$ case, is noteworthy.

Our goal is to demonstrate that multiple maxima of the likelihood exist. To show this, we do the following. For each of the 100 values of c_0 in the grid, we estimate the three GARCH parameters *jointly*, using as starting values $[c_0, \boldsymbol{\theta}_{\text{p.l.}}(c_0)]$, and plot the resulting log-likelihood as a circle. In each of the figures there are clearly far fewer than 100 circles, but not just one—each one corresponds to a local maximum of the log-likelihood. Worse, some of them do not lie on the profile log-likelihood line. Thus, we see that, particularly for smaller sample sizes, there are many local maxima, and the choice of starting value plays a substantial role in determining to which local maximum the optimization routine will converge.[6]

To obtain (with high probability) the global maximum, the following procedure suggests itself: (i) Based on a set of n equally spaced c_0-values that span its possible range, compute (10.8), (ii) take the value of c_0 from the set, say c_0^*, and its corresponding $\widehat{\boldsymbol{\theta}}_{\text{p.l.}}(c_0^*)$ that results in the largest log-likelihood as starting values, to (iii) estimate the full model. The larger is n (hence, the finer the grid), the higher the probability of reaching the global maximum; some trials suggest that a grid of length $n = 10$ is adequate for most applications. The use of more parameters, such as for the mean equation (10.18), or more elaborate GARCH structures such as the APARCH formulation given next, or additional

6 It might be thought that use of different optimization algorithms could be beneficial, but this is not necessarily the case if they are given a single starting value. For example, the qualitatively same graphical result was obtained when using the usual quasi-Newton, simplex (Matlab's function `fminsearch`), and other heuristic optimization algorithms discussed in Section III.4.4.

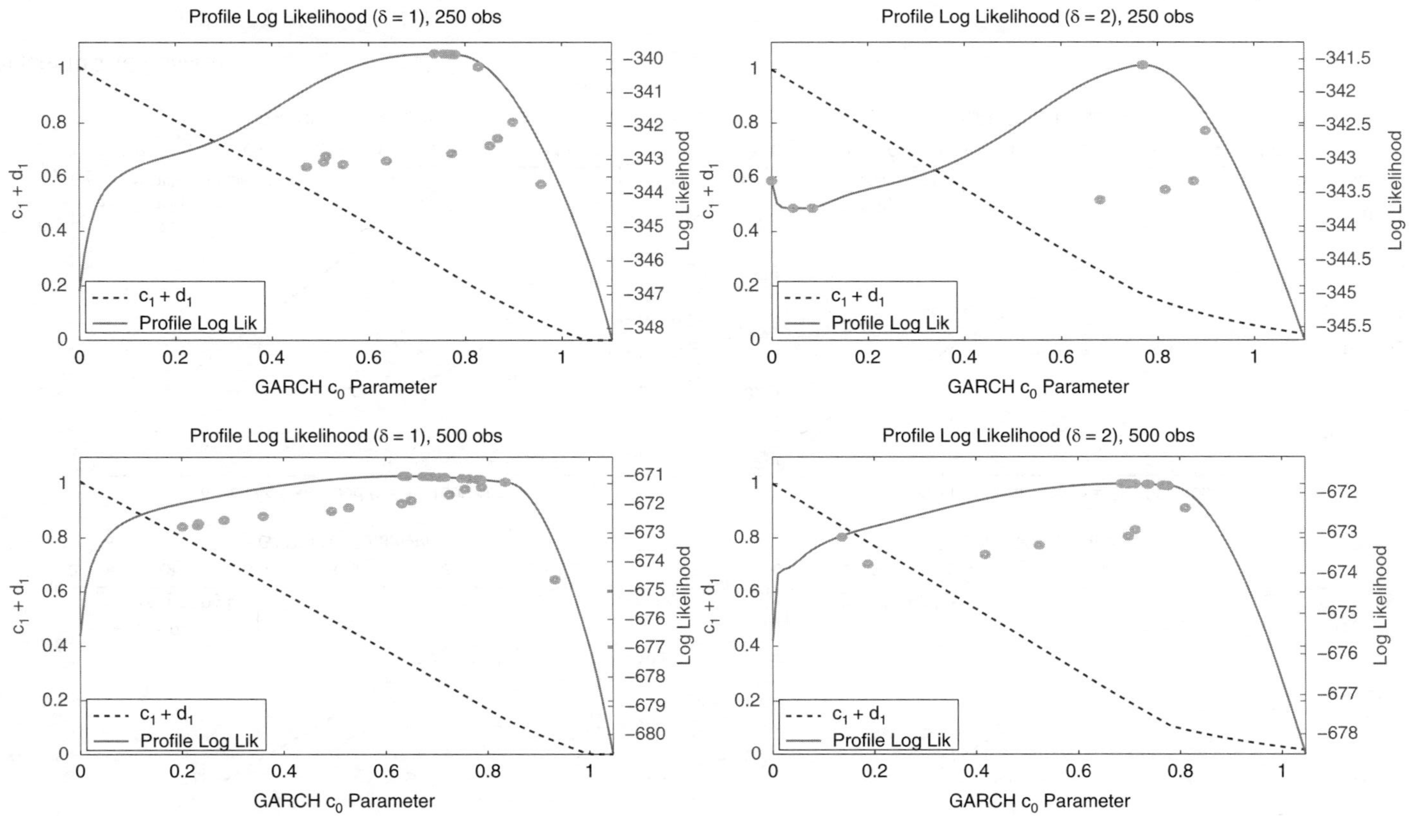

Figure 10.5 The profile log-likelihood (solid) in c_0, for the returns on AT&T, and the local maxima of the log-likelihood (circles). Left panels correspond to model (10.9) given below (and indicated with $\delta = 1$), while the right panels correspond to model (10.2) (indicated with $\delta = 2$).

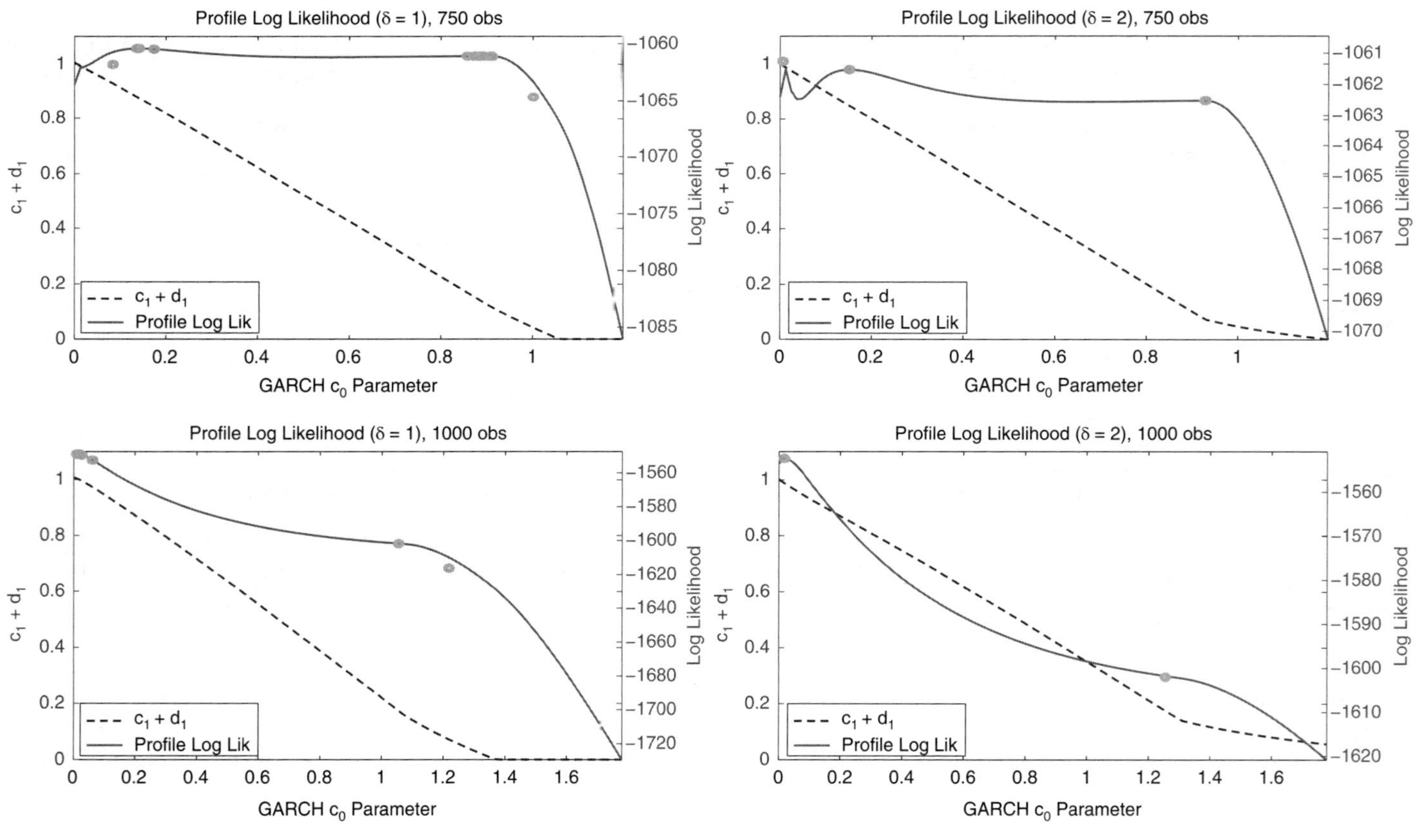

Figure 10.5 (*Continued*)

shape parameter(s) of a non-Gaussian distribution (e.g., Student's t, GAt, NCT, stable Paretian, NIG, etc.), also subsequently discussed, can further exacerbate the problem of multiple local maxima of the likelihood.

10.2.4 Variance Targeting Estimator

Recall that the unconditional variance, $\mathbb{V}(R_t)$, when $\{R_t\}$ follows a normal-GARCH(r, s) process, is given by (10.6), if it exists. When it is computed with the GARCH parameter point estimates, it results in what is referred to as the **implied variance** (or implied volatility, though the latter term is more often associated with the volatility of an underlying financial derivative instrument, notably an option), though the term "parametric fitted implied variance" is arguably more precise. Assuming $\mathbb{V}(R_t) < \infty$, the usual sample variance S_T^2, where T denotes the sample size, is a consistent (albeit not necessary efficient) estimator of it, and both variance estimators are asymptotically normal. The asymptotic distribution of the difference of the two estimators is developed in Horváth et al. (2006).

In the typical case of $r = s = 1$, (10.6) reduces to $\mathbb{V}(R_t) = c_0/(1 - c_1 - d_1)$, so that an estimator for c_0 is $\hat{c}_0^* = (1 - \hat{c}_1 - \hat{d}_1)S_T^2$. As S_T^2 is a trivially computed, closed-form, non-parametric estimator, maximum likelihood needs to be conducted only over two, instead of three, dimensions, to obtain, say, $\hat{c}_1^*$ and $\hat{d}_1^*$, and this, besides being faster, can also help avoid the local likelihood maxima situation discussed above. The set $\{\hat{c}_0^*, \hat{c}_1^*, \hat{d}_1^*\}$ is referred to as the **variance targeting estimator**, or VTE. The idea appears to have been first explicitly used and discussed in Engle and Mezrich (1996). Observe that the VTE is not applicable in the IGARCH setting, which also has only two free parameters, though if one is willing to impose that, say, $\hat{c}_1^* + \hat{d}_1^* = 0.99$, then there is only one free parameter to be estimated.

Theoretical properties and simulation-based results for the finite-sample performance of the VTE versus the full m.l.e. can be found in Francq et al. (2011), Hill and Renault (2012), Vaynman and Beare (2014), and Anatolyev and Khrapov (2015). In particular, Francq et al. (2011) give conditions under which the VTE is asymptotically normal, which include that fourth moments of R_t exist. They show, unsurprisingly, that "variance targeting may result in a serious deterioration of the asymptotic precision when the moment condition is close to be violated". Anatolyev and Khrapov (2015) find that parameter estimates and associated model forecasts are less accurate with the VTE, notably so when the innovations process is heavy-tailed; see Section 10.3.1. They conclude that "if computational costs are not prohibitive, variance targeting should probably be avoided." As a contrast, Francq et al. (2011) show that, when the model is mis-specified, the VTE can be better than the use of the QMLE, i.e., the use of the normal distribution when the actual innovations process is not normal; see Section 10.3.2 for more discussion of the QMLE.

10.3 Non-Gaussian ARMA-APARCH, QMLE, and Forecasting

10.3.1 Extending the Volatility, Distribution, and Mean Equations

As a first alternative to the law of motion for the volatility (10.2), we can use an exponent of one instead of two, i.e.,

$$\sigma_t = c_0 + \sum_{i=1}^{r} c_i |\epsilon_{t-i}| + \sum_{j=1}^{s} d_j \sigma_{t-j}, \tag{10.9}$$

as advocated by Taylor (1986) and Schwert (1989b). This formulation tends to provide a better fit, in line with the results of Nelson and Foster (1994), who demonstrate that this model is a more efficient filter of the unconditional variance in the presence of leptokurtic error distributions. A further benefit of (10.9) is that it is applicable if the innovation sequence does not possess second moments, such as for some of the non-Gaussian extensions considered below.

A formulation that generalizes (10.2) and (10.9) is the **asymmetric power ARCH**, abbreviated APARCH(r, s), from Ding et al. (1993), given by

$$\sigma_t^\delta = c_0 + \sum_{i=1}^{r} c_i(|\epsilon_{t-i}| - \gamma_i\epsilon_{t-i})^\delta + \sum_{j=1}^{s} d_j\sigma_{t-j}^\delta, \quad \delta > 0, \ |\gamma_i| < 1. \tag{10.10}$$

Its benefit is that it nests, with only two additional parameters over the usual Bollerslev (1986) GARCH$(1, 1)$ model (10.2), at least five previously proposed GARCH extensions at that time (Ding et al., 1993, p. 98), notably the popular so-called GJR-GARCH model of Glosten et al. (1993). In the $r = s = 1$ case, it is necessary and sufficient that $c_0 > 0$, $c_1 > 0$, $d_1 \geqslant 0$, along with $\delta > 0$, $|\gamma_1| < 1$. Parameter $\gamma_i \neq 0$ allows σ_t to respond asymmetrically to positive and negative shocks ϵ_{t-i}, with its use often leading to improved forecasts. This extension is also useful for describing the so-called **news impact curve**; see Engle and Ng (1993). It turns out that, for daily financial asset returns data, the likelihood is often relatively flat in parameter δ, with its maximum between one and two. We advocate just setting it equal to one. The APARCH model has been well-studied; see, e.g., He and Teräsvirta (1999b,a), Ling and McAleer (2002), Karanasos and Kim (2006), and Francq and Zakoïan (2010, Ch. 10).

Another extension of (10.2) is the **(generalized) quadratic ARCH**, or Q-GARCH, structure, introduced in Sentana (1995). The idea of Q-ARCH is to view the law of motion as a Taylor series of the true d.g.p. in terms of the past ϵ_t, and include a second-order term. By (i) augmenting the Q-ARCH structure with the past σ_t terms as in the GARCH extension of ARCH and (ii) similar to (10.10) relaxing the usual exponent of two, the (power) Q-GARCH(r, s) model is given by

$$\sigma_t^\delta = c_0 + \sum_{i=1}^{r} c_i\epsilon_{t-i} + \sum_{i=1}^{r} c_{ii}|\epsilon_{t-i}|^\delta + 2\sum_{i=1}^{r}\sum_{j=i+1}^{r} c_{ij}\epsilon_{t-i}\epsilon_{t-j} + \sum_{j=1}^{s} d_j\sigma_{t-j}^\delta. \tag{10.11}$$

See also Mittnik et al. (2000) and Park et al. (2011) for applications of the Q-GARCH model.

Other such formulations exist, notably the class of fractionally integrated, or **FIGARCH** models, to account for the very slow decay in the absolute or squared autocorrelations; see, e.g., Bollerslev and Mikkelsen (1996), Baillie et al. (1996), Conrad and Haag (2006), Caporin (2003), Tayefi and Ramanathan (2012), and the references therein. See Harvey (2013b) for a detailed account and applications of **dynamic conditional score**, or DCS, models, often in conjunction with skew-t type distributions. DCS models involve a modification of traditional GARCH models and their estimation. Their application is highlighted in Gao and Zhou (2016). The survey in Teräsvirta (2009) discusses yet more GARCH-type constructions.

The next aspect of the Gaussian GARCH model that is often extended is the distribution of the i.i.d. innovation sequence. A natural starting point for the innovations distribution assumption (and this having been the assumption in the original ARCH and GARCH formulations) is to take the Z_t to be i.i.d. Gaussian, though this was soon realized to be highly inadequate. The computed residuals (filtered innovations) of a normal-GARCH model applied to daily or higher-frequency data tend to be non-Gaussian, notably with substantial leptokurtosis and mild asymmetry, similar to the asset returns themselves. This will have implications for risk prediction and asset allocation; see Sections 11.1 and 11.3, respectively. In fact, it turns out that the choice of distributional assumption is of more importance with respect to risk and density prediction than the functional form of the GARCH recursion; see, e.g., Mittnik and Paolella (2000), Bao et al. (2006, 2007), and Kuester et al. (2006). Many candidate distributions suggest themselves; see Palm (1996), McDonald (1997), and Broda and Paolella (2011) for overviews of relevant distributions.

As examples of non-Gaussian distributions used in conjunction with GARCH-type models, Bollerslev (1987), Nelson (1991), and Granger and Ding (1995) proposed the use of the Student's t, the GED (see Example II.7.2), and the Laplace, respectively. The normal inverse Gaussian (NIG) distribution (see Section II.9.4) in conjunction with univariate GARCH modeling is studied in Jensen and Lunde (2001), and in the multivariate setting by Aas et al. (2005) and Broda and Paolella (2009a). The use of the (asymmetric) stable Paretian in the univariate case was investigated by several authors, including Liu and Brorsen (1995), Rachev and Mittnik (2000), Mittnik et al. (2002), Mittnik and Paolella (2003), and Paolella (2016); see Broda et al. (2013) and the references therein for extensions to the multivariate setting.

Perhaps the most used distributional assumption in conjunction with GARCH models for forecasting VaR and ES is an asymmetric Student's t, for which there are several variations, including the GAt (Mittnik and Paolella, 2000; Kuester et al., 2006; Problem II.7.7 and Section III.A.8) and related constructions (Theodossiou, 1998; Giot and Laurent, 2003; Zhu and Galbraith, 2010, 2011; Diamandis et al., 2011; Harvey and Sucarrat, 2014), a limiting asymmetric case of the generalized hyperbolic distribution (Aas and Haff, 2006; Paolella and Polak, 2015b), and the noncentral Student's t (Harvey and Siddique, 1999; Krause and Paolella, 2014). In Section 10.4 we will detail the APARCH model with noncentral t innovations.

Analogous to (10.5) and (10.6), the APARCH model (10.10) gives rise to a strictly stationary series for parameter values such that

$$V := \sum_{i=1}^{r} \kappa_i c_i + \sum_{j=1}^{s} d_j \leqslant 1, \tag{10.12}$$

where

$$\kappa_i := \mathbb{E}[(|Z| - \gamma_i Z)^{\delta}] \tag{10.13}$$

depends on the density specification $f_Z(\cdot)$ of the i.i.d. innovation sequence. The integrated APARCH model results for parameter constellations satisfying $V = 1$, while, for $V > 1$, the model will be explosive.

The (power) Q-GARCH model can also be endowed with an i.i.d. non-Gaussian innovation sequence. Taking iterated expectations of (10.11) shows that $\mathbb{E}[\epsilon_{t-i}\epsilon_{t-j}] = 0$ for $i \neq j$, so that $\mathbb{E}[\sigma^{\delta}] = c_0/(1 - V_Q)$, where

$$V_Q = \sum_{i=1}^{r} c_{ii}\mathbb{E}[|Z|^{\delta}] + \sum_{j=1}^{s} d_j. \tag{10.14}$$

Note that $\mathbb{E}[|Z|^{\delta}]$ in (10.14) is just the κ defined in (10.13) with $\gamma_i = 0$.

Besides being necessary for calculating the volatility persistence measure V, the κ_i are also required for generating correct starting values for recursions (10.10) and (10.11) when evaluating the likelihood. For $Z \sim \mathrm{N}(0,1)$, the reader should confirm that

$$\mathbb{E}[(|Z| - \gamma Z)^\delta] = \frac{1}{\sqrt{2\pi}}[(1+\gamma)^\delta + (1-\gamma)^\delta]2^{(\delta-1)/2}\Gamma\left(\frac{\delta+1}{2}\right).$$

Note that, with $\delta = 2$ and $\gamma = 0$, this reduces to $\kappa = \mathbb{E}[|Z|^2] = 1$, so that (10.12) reduces to the stationarity condition (10.6).

For $Z \sim t_\nu$, i.e., Student's t with ν degrees of freedom and $\nu > \delta$, a straightforward integral calculation (that the reader should also verify) yields

$$\mathbb{E}[(|Z| - \gamma Z)^\delta] = \nu^{\delta/2}\frac{1}{2\sqrt{\pi}}[(1+\gamma)^\delta + (1-\gamma)^\delta]\Gamma\left(\frac{\delta+1}{2}\right)\Gamma\left(\frac{\nu-\delta}{2}\right)\Big/\Gamma\left(\frac{\nu}{2}\right).$$

For $\delta = 2$ and $\gamma = 0$, this reduces to $\kappa = \mathbb{E}[|Z|^2] = \mathbb{V}(Z) = \nu/(\nu-2)$.

For the asymmetric stable Paretian distribution, the expression for (10.13) in the $\gamma = 0$ case is shown via some lengthy calculations in Section II.8.3 to be

$$\mathbb{E}[|X|^r] = \kappa^{-1}\Gamma\left(1 - \frac{r}{\alpha}\right)(1+\tau^2)^{r/2\alpha}\cos\left(\frac{r}{\alpha}\arctan\tau\right), \quad -1 < r < \alpha, \tag{10.15}$$

where, noting that, for ϵ not a negative integer, $\Gamma(-1+\epsilon) = \Gamma(\epsilon)/(-1+\epsilon)$,

$$\tau = \beta\tan(\pi\alpha/2), \qquad \kappa = \begin{cases} \Gamma(1-r)\cos(\pi r/2), & \text{if } r \neq 1, \\ \pi/2, & \text{if } r = 1. \end{cases}$$

For $Z \sim t'(k,\theta)$, i.e., the noncentral t distribution with degrees of freedom $k > \delta > 0$, noncentrality θ, and $\gamma = 0$ in (10.13),

$$\mathbb{E}[|Z|^\delta] = k^{\delta/2}\frac{\Gamma((k-\delta)/2)\Gamma((1+\delta)/2)}{\Gamma(k/2)\sqrt{\pi}}\,{}_1F_1\left(-\frac{\delta}{2}, \frac{1}{2}; \frac{\theta^2}{2}\right); \tag{10.16}$$

see Chapter II.10. For $k = 4$, $\theta = 0$, and $\delta = 1$, $\mathbb{E}[|Z|] = 1$. As $k = 4$ is considered to be the best general compromise value for many financial return series under a Student's t distributional assumption (see, e.g., Platen and Rendek, 2008), and also in light of the large confidence intervals associated with its estimation, we can, without any substantial loss of accuracy, just use the value of one in place of computing $\mathbb{E}[|Z|^\delta]$ for $Z \sim t'(k,\theta)$.

Finally, for the GAt distribution, Problem II.7.7(b) shows that, for $\nu d > \delta > 0$,

$$\mathbb{E}[(|Z| - \gamma Z)^\delta] = \frac{(1+\gamma)^\delta\theta^{-(\delta+1)} + \theta^{\delta+1}(1-\gamma)^\delta}{\theta^{-1}+\theta}\frac{B\left(\frac{\delta+1}{d}, \nu - \frac{\delta}{d}\right)}{B\left(\frac{1}{d}, \nu\right)}\nu^{\delta/d}. \tag{10.17}$$

Example 10.1 We mentioned above that a simple GARCH model can give rise to realized processes that resemble the seemingly rather non-stationary returns in Figure 10.1.[7] Figure 10.6 shows three

---

7 Whether or not the returns corresponding to the various Asian currencies before, during, and after the crisis are actually from a strictly stationary process is neither definitively answerable, nor necessary. What is relevant in our setting is to find a model that can deliver accurate forecasts of volatility and tail risk measures. What one could reasonably say is that, over long periods of time, of course the process is not stationary, given the highly complex nature of the global financial system, changing economic and political conditions, changing technology and legal structures in the financial industry, etc.

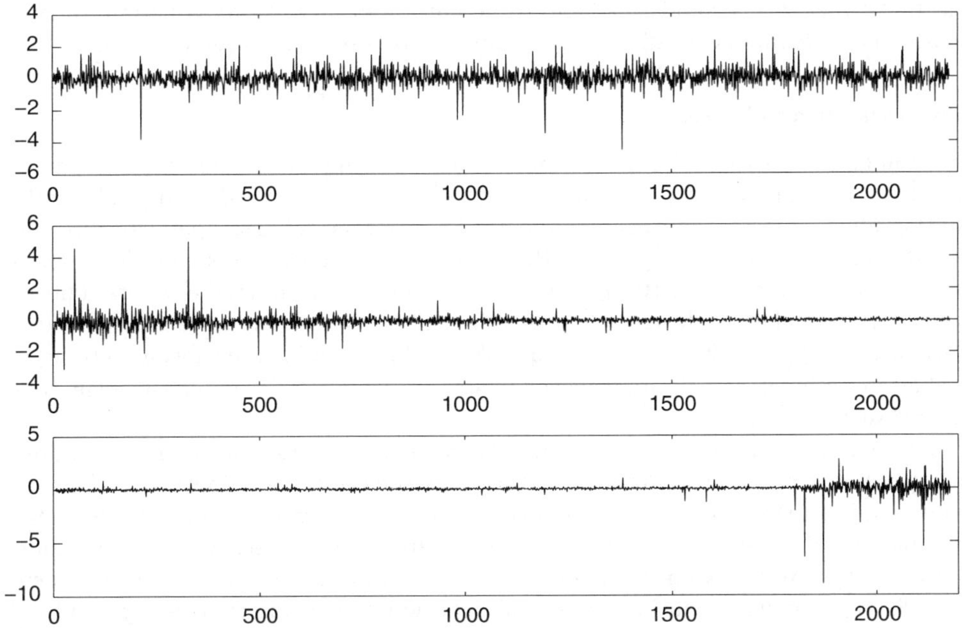

Figure 10.6 Three realizations of a simulated t_4-IGARCH(1, 1) process with $c_0 = 0.01$ and $c_1 = 0.02$.

sample path realizations of a simulated IGARCH model with Student's t_4 innovations. The first one looks like "business as usual", the second appears to have a consistently decreasing conditional and unconditional variance, while the third appears, like the genuine Asian currency returns, to suddenly change in structure.

The reason for these seemingly disparate behaviors is that the sample size is finite: Longer runs would eventually begin to resemble each other in terms of the actual process behavior. The use of IGARCH ensures that the volatility is maximally persistent without being explosive, while use of the heavy-tailed innovations causes the process to exhibit possibly sudden changes in volatility and very heavy tails of the unconditional process. ■

The third augmentation of the baseline GARCH model is to endow the mean term with a statistical time-series structure, such as a (stationary and invertible) ARMA model, in which case we replace $R_t = \epsilon_t = Z_t\sigma_t$ with

$$R_t - a_0 = \sum_{i=1}^{p} a_i(R_{t-i} - a_0) + \epsilon_t + \sum_{j=1}^{q} b_j\epsilon_{t-j}, \qquad \epsilon_t = Z_t\sigma_t, \tag{10.18}$$

yielding what is often referred to as an ARMA(p, q)-GARCH(r, s) model. The concept of efficient markets, combined with modern ease of trading and low transaction costs, high liquidity, and extensive use of trading algorithms by large financial institutions, render attempts at finding exploitable signals in the mean of daily returns via (10.18) nearly a waste of time, though for short windows of data the parameter a_0 could be estimated instead of taking it to be zero because of the so-called

momentum effect.[8] Further terms could be added to the mean and/or volatility equation, such as exogenous variables (macroeconomic, trading volume, intraday volatility measurements, etc.).

10.3.2 Model Mis-specification and QMLE

Recall that estimation of a mis-specified GARCH model, such as using (10.2) when the true model has, for example, time-varying parameters or omits relevant exogenous variables, often results in the estimated GARCH model parameters approaching the IGARCH border as the sample size increases. This also occurs if the distributional assumption on the innovation sequence is mis-specified. By way of illustration based on simulation, Figure 10.7 is similar to Figure 10.4, and shows the results of fitting the normal-GARCH(1,1) model when the true d.g.p. is t_3-GARCH(1,1) (with the same μ, c_0, c_1, and d_1 as used previously, and $V = [3/(3-2)]c_1 + d_1 = 0.95$). We see that $\hat{c}_0$ and $\hat{c}_1$ are substantially larger than their true values, resulting in $\hat{c}_1 + \hat{d}_1$ often exceeding its true value of 0.85, with an average value of 0.931, this being close to the true V.

This finding might entice one to always use a GARCH model with a non-Gaussian innovations distribution assumption that nests (or yields as a limiting case) the normal distribution, such as the Student's t, in order to mitigate this effect. However, this issue opens a wider discussion. The use of the **quasi-maximum likelihood** estimator, or QMLE (see the remark below), whereby a GARCH model is fit under the so-called **synthetic assumption** of Gaussian innovations, can still result in consistent estimates of the GARCH parameters if the innovations distribution is not Gaussian (see, e.g., Hall and Yao, 2003; Berkes et al., 2003a; Francq and Zakoïan, 2004). This requires that the process is strictly stationary. More restrictively, in order for consistency and that the asymptotic distribution of the QMLE is Gaussian, the existence of fourth moments of the true innovations distribution is required, which may not be fulfilled and, as seen in Sections III.9.1 and III.9.2, is anyway difficult to verify. (The situation when the tail index, i.e., the supremum of the maximally existing moment, say α, of the true innovations distribution is such that $\alpha \in (2, 4)$ has been considered in Mikosch and Straumann, 2006, Thm. 4.4. They show that, in addition to being consistent, the estimator possesses a non-Gaussian stable Paretian limiting distribution.)

To shed some light on the behavior of the estimator in finite samples, we repeat the previous exercise that generated Figure 10.7, but using t_5 instead of t_3 innovations, so that fourth moments exist, and doing so for a sample size of $T = 10{,}000$, with the hope that the consistency and asymptotic normality of the estimator is sufficiently engaged. As seen from Figure 10.8, this appears to not be the case. Similar to the simulation based on t_3 innovations, the mean (over the 1,000 replications) of $\hat{c}_1 + \hat{d}_1$ is 0.880, which is very close to $V = [5/(5-2)]c_1 + d_1 = 0.883$.

Further complicating matters is that, unless the distribution (up to the unknown parameters) is correctly specified, the GARCH parameters can be inconsistent. See Fan et al. (2014) and the references therein for further details and a possible resolution to this issue, and Anatolyev and Khrapov (2015) for further empirical confirmation of this.

One could further argue that the *entire* model is mis-specified: Surely the true d.g.p. of financial asset returns is not abiding truly by a (possibly non-Gaussian) strictly stationary GARCH-type process.

---

8 This is the observed tendency, across time and markets, for falling (rising) asset prices to fall (rise) further. An internet search will reveal the large literature on this issue, with a good starting point being Bali et al. (2016a, Ch. 11) and the references therein. A possible explanation of this violation of market efficiency is that it is a self-fulfilling prophesy: Uninformed traders (gamblers) following a spurious trend literally induce and propagate it, thus giving rise to the momentum effect. Put colloquially, "the trend is your friend".

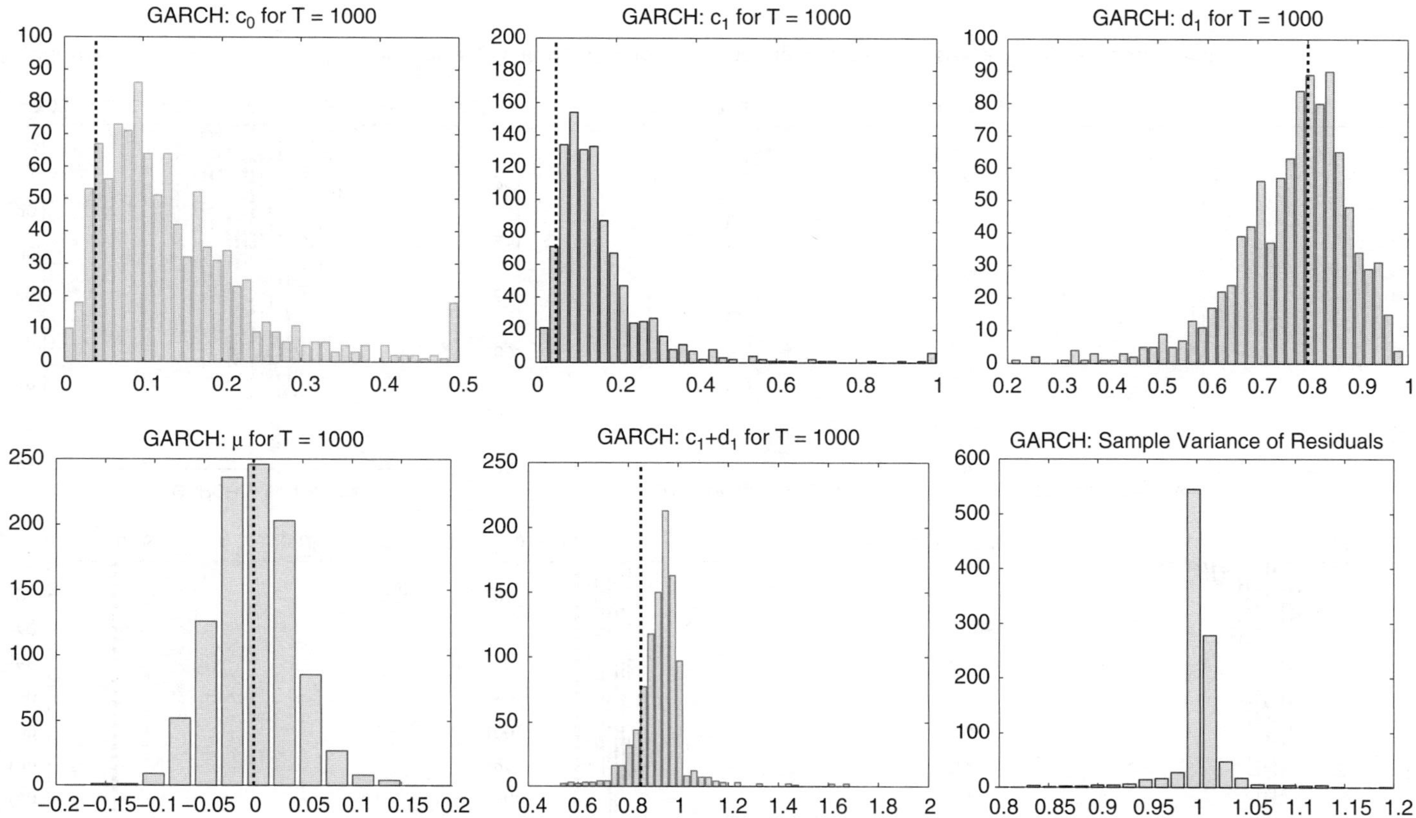

Figure 10.7 Similar to Figure 10.4, but showing the m.l.e. of a fitted (mis-specified) normal-GARCH(1,1), when using a t_3-GARCH(1,1) process as the d.g.p., with $T = 1{,}000$ observations, using 1,000 replications.

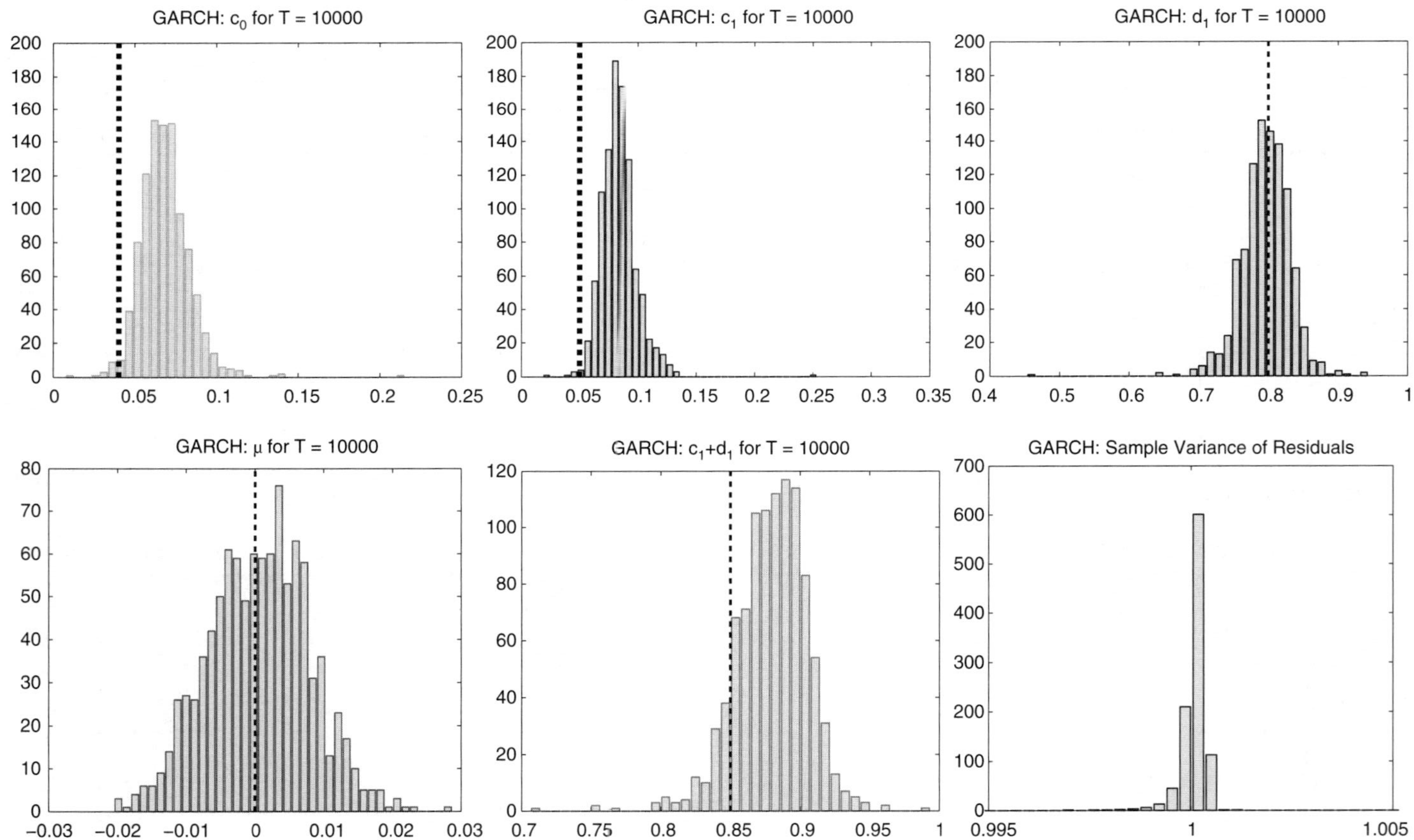

Figure 10.8 Similar to Figure 10.7, but based on a t_5-GARCH(1,1) process, and having used a sample size of $T = 10{,}000$ observations.

Notice this implies, among other things, that the innovation distribution parameters and the GARCH parameters are constant through time, which does not seem realistic for many years of data. We take the viewpoint that all models, in most endeavors, particularly empirical finance, are mis-specified, and the best judge of a model in this context is its ability to deliver accurate (say, risk, or density) forecasts, or portfolio allocations, better than all competing models.

Remark In general, the QMLE maximizes a simpler, usually Gaussian, likelihood, in place of the genuine, and most likely unknown one. See White (1982), Bollerslev and Wooldridge (1992), White (1994), and the references therein for theoretical discussions. As stated by Bollerslev and Wooldridge (1992, p. 144), "Taken literally, the assumption of conditional normality can be quite restrictive. The symmetry imposed under normality is difficult to justify in general, and the tails of even conditional distributions often seem to be fatter than that of the normal distribution. The extensive use of maximum likelihood under the assumption of normality is almost certainly due to its relative simplicity and the widespread familiarity with its properties under ideal conditions. Because maximum likelihood under normality is so widely used, it is important to investigate its properties in a setting general enough to include most cases of interest to applied researchers."

It is not clear where the first usage of the term QMLE comes from. One idea is, in his response to the discussion of his paper, Nelder (1968, p. 328) writes "I take Mr. Fisk's point that Gaussian estimation is free of distributional assumptions, and perhaps should have used a term such as 'quasi-likelihood' for the quantity to be maximized."

QMLE is related to *pseudo-likelihood estimation*, which omits certain dependency structures from the likelihood that are not necessarily important to the analysis; see Cox and Reid (2004) and the references therein. ■

10.3.3 Forecasting

For a given specification of the innovation p.d.f. $f_Z(\cdot)$, say GAt, along with estimates of the distributional parameters, in this case $\hat{\nu}$, $\hat{d}$, and $\hat{\theta}$, and APARCH parameters $\widehat{\boldsymbol{\theta}}$, the h-step ahead density forecast $\hat{f}_{t+h|t}(\cdot)$ conditional on Ω_t is the p.d.f. of the random variable given by $\hat{\sigma}_{t+h|t}\tilde{Z}$, for $\tilde{Z} \sim \text{GA}t(\hat{\nu}, \hat{d}, \hat{\theta})$. The volatility term $\hat{\sigma}_{t+h|t}$ is recursively evaluated from (10.10) with

$$\hat{\sigma}^{\delta}_{t+\ell|t} = \hat{c}_0 + \sum_{i=1}^{r} \hat{c}_i E^{(i)}_{t+\ell-i} + \sum_{j=1}^{s} \hat{d}_j \hat{\sigma}^{\delta}_{t+\ell-j|t}, \qquad \ell = 1, \dots, h,$$

where $\hat{\sigma}_{k|t}$, $k = 1, \dots, t$, are the filtered in-sample volatilities, $\hat{\sigma}_{k|t}$, $k > t$, denote recursively computed out-of-sample forecasts,

$$E^{(i)}_k = \begin{cases} (|\tilde{\epsilon}_k| - \hat{\gamma}_i\tilde{\epsilon}_k)^{\hat{\delta}}, & \text{if } k \leqslant t, \\ \mathbb{E}(|\epsilon_k| - \hat{\gamma}_i\epsilon_k)^{\hat{\delta}} = \hat{\sigma}^{\hat{\delta}}_k \kappa_i, & \text{if } k > t, \end{cases}$$

$\tilde{\epsilon}_k$, $k \leqslant t$, denote the filtered ϵ-values, and κ_i is given by (10.13) evaluated with the appropriate estimated parameters. For example, with GAt innovations, $\kappa_i = \kappa(\hat{\nu}, \hat{d}, \hat{\theta}, \hat{\gamma}_i, \hat{\delta})$. Density forecasts using the Gaussian-GARCH(1,1) model (10.2) are clearly a special case. Similar calculations apply to the Q-GARCH construction.

For the more general location-scale model, the forecast distribution is that of $\hat{\mu}_{t+h|t} + \hat{\sigma}_{t+h|t}\tilde{Z}$, where $\hat{\mu}_{t+h|t}$ might be given by, say, an AR(p) model as in (10.18).

10.4 Near-Instantaneous Estimation of NCT-APARCH(1,1)

Here we showcase use of the APARCH(1,1) model (10.10) coupled with an i.i.d. noncentral t (NCT) innovations process. This is a flexible model that can capture the leptokurtosis inherent in the filtered Gaussian-GARCH residuals, as well as asymmetry. The APARCH(1,1) formulation allows a different type of asymmetry, namely in volatility at time t, depending on the sign of (filtered) ϵ_{t-1} and the value of parameter γ_1.

It is preferred to have a mean-zero innovations process driving the GARCH-type model, so we express it, with a location term, as (using here a_0 as the location term, anticipating possible usage of autoregressive parameters)

$$R_t = a_0 + Z_t^* \sigma_t, \quad Z_t^* = Z_t - \mu, \quad Z_t \overset{\text{iid}}{\sim} \text{NCT}(k, \theta), \tag{10.19}$$

where

$$\mu = \mathbb{E}[Z_t] = \theta \left(\frac{k}{2} \right)^{1/2} \frac{\Gamma(k/2 - 1/2)}{\Gamma(k/2)}, \quad k > 1,$$

with $k \in \mathbb{R}_{>1}$ being the degrees of freedom parameter (and such that we require the mean of the process to exist) and $\theta \in \mathbb{R}$ being the NCT noncentrality parameter. Joint maximum likelihood estimation of all seven model parameters (assuming δ in (10.10) is fixed) is straightforward, and the reader is encouraged to program this, though observe that estimation will be rather slow, due to the nature of the p.d.f. of the NCT; see Sections III.A.14 and II.10.4.

Our goal is to develop a method to deliver accurate parameter estimates of the NCT-APARCH(1,1) model (for typical daily stock return data) requiring as little computational time as possible, so that, among other things, it can be used for portfolio optimization via the method in Paolella (2017), which requires its computation thousands of times. The method, as developed in Krause and Paolella (2014) and now described, involves fixing the APARCH parameters to judiciously chosen values, and using an estimation procedure for the remaining parameters that does not require computing the likelihood.

As $\mathbb{E}[R_t] = a_0$ in (10.19), assume for the moment that we can estimate the intercept parameter a_0 by taking the median (as a robust estimator) of the R_t. Next, with a fixed set of values for the APARCH(1,1) parameters (the choice of which is subsequently discussed), the APARCH filter is applied to the location-adjusted returns $R_t - \hat{a}_0$, yielding a (presumed) set of i.i.d. location-zero, scale-one NCT residuals. For the third step, to elicit $\hat{k}$ and $\hat{\theta}$, we use the fast table lookup method developed in Section III.9.3.3. Observe that this procedure requires computing a median of a vector of data, one run of the APARCH filter, and then applying the quantile-based table lookup procedure to obtain the NCT shape parameters. It is thus conducted on the order of microseconds.

However, the estimator of a_0 can be improved with little effort, and it is worthwhile in the context of portfolio optimization, as its performance is very sensitive to the predictive mean of the returns. In particular, instead of using the median, we use the trimmed mean procedure discussed in Example III.4.3 for Student's t data (and ignoring the NCT asymmetry). This requires knowing k, as elicited in the third step of the procedure, and so an iterative procedure suggests itself, as follows:

1) Take a_0 to be the sample median of the returns data, say $\hat{a}_0^{(1)} = \text{median}(\{R_t\})$.
2) Apply the fixed APARCH filter (its choice of parameters being discussed below) to the location-adjusted returns $R_t - \hat{a}_0^{(1)}$. This results in a set of data, say $\mathbf{Z}^{(1)} = (Z_1^{(1)}, \dots, Z_T^{(1)})$, which are (presumed close to) i.i.d., with unit scale term.

3) Based on $\mathbf{Z}^{(1)}$, compute the estimators of k and θ via the table lookup procedure from Section III.9.3.3.
4) Based on $\hat{k}$, determine the optimal trimming value α, say $\alpha_*^{(1)}$, as discussed in Example III.4.3.
5) Let $\hat{a}_0^{(2)} = \text{trim}(\{R_t\}, \alpha_*^{(1)})$.

Steps 2 through 5 can be repeated, applying the APARCH filter to $R_t - \hat{a}_0^{(2)}$ to get $\mathbf{Z}^{(2)}$. Then obtain $\hat{k}$ and $\hat{\theta}$ from the table lookup, get $\alpha_*^{(2)}$, and set $\hat{a}_0^{(3)} = \text{trim}(\{R_t\}, \alpha_*^{(2)})$, etc. The method converged after three or at most four iterations in numerous studies conducted with real and simulated data, and two iterations are usually enough. If speed is critical, then only one pass could be conducted, using, instead of median($\{R_t\}$), the trimmed mean for a typical value of k for conditional (i.e., APARCH scale-adjusted) returns, such as $k = 6$.

One of the three core ideas in this procedure is the fixing of the APARCH parameters instead of estimating them. This is done by choosing a typical set of values associated with daily stock returns data. This works because the fitted parameters tend to be similar across various asset returns at the daily level. This idea is not new, and goes back at least to the RiskMetrics™ technical document from J.P. Morgan/Reuters (1996, pp. 80–81), in which an integrated GARCH(1,1) model with $d_1 = 0.94$ (and $c_0 = 0$) is proposed. A similar idea was discovered in Lux and Kaizoji (2007) for FIGARCH models, such that volatility predictions improve across a group of financial time-series returns when the individually estimated time-series parameters are averaged.

With maximum likelihood estimation straightforward to conduct and available in many econometric software packages, most academic researchers tended to "let the data speak for themselves", and did not embrace the idea of just fixing the parameters, though the forecasting results between the two models are virtually the same for short horizons (and slightly better for long-term horizons, with the fixed parameters); see, e.g., Neely and Weller (2002) in the context of currency exchange rates. One can view both the fitted GARCH model and the RiskMetrics™ fixed-parameter suggestion as special cases of shrinkage-based maximum likelihood estimation, with the optimal amount of shrinkage most likely not at either of these two extremes.

We now provide a different argument that supports the use of fixed GARCH/APARCH coefficients instead of estimation, and also shows how to determine the optimal fixed values. We compare (i) the variation of the m.l.e. of the NCT-GARCH parameters based on typical financial returns data with (ii) the variation of the m.l.e. from simulation of the NCT-GARCH process using a typical parameter vector as the true values. If the variation in (i) is smaller than that in (ii), then it stands to reason that estimation of the GARCH parameters can be forgone without great loss of accuracy and replaced by typical values obtained in (i) (for which we choose $c_0 = 0.04$, $c_1 = 0.05$ and $d_1 = 0.90$, based approximately on the median values shown in Figure 10.9).[9]

We have already seen in Figure 10.4 that the dispersion of the estimated parameters is, even for a sample size of $T = 1{,}000$, rather high, indicating that this conjecture might have merit. To add verification, we consider the daily percentage log returns of the 30 components of the Dow Jones Industrial Index (DJIA) from Wharton/CRSP (as used in April 2013), from January 1, 1993 until

9 This choice of parameters assumes a certain "uniformity" of the estimates. In particular, focusing on one of the three parameters, say c_0, if its mean or median value from Figure 10.9 (approximately 0.05, though we found use of 0.04 to be better) is used, then it could be the case that, for many of the data sets used to construct the figure, the values of $\hat{c}_1$ and $\hat{d}_1$ corresponding to $\hat{c}_0 \approx 0.04$ are *not* close to their respective median values in the figure. Some investigation and use of different fixed sets of $\{c_0, c_1, d_1\}$ revealed that this is conveniently not the case, and the set of values obtained as the medians of the parameters, individually, are close to optimal.

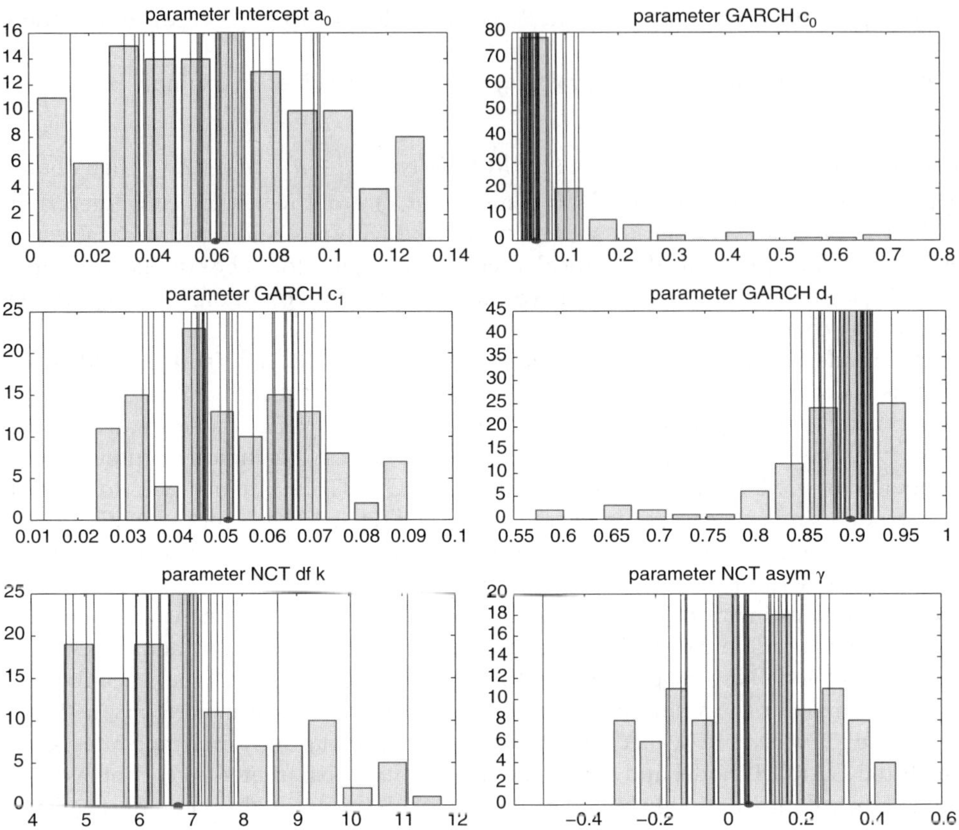

Figure 10.9 The m.l.e. parameter estimates corresponding to the NCT-GARCH model for the 30 assets of the DJIA, from January 1, 1993 until December 31, 2012, using non-overlapping windows of length $T = 1{,}000$. The circle on the x-axis indicates the median of the data. The thin vertical lines refer to the average parameter value for each of the 30 assets. (In the lower right panel, the NCT noncentrality parameter is denoted as γ.)

December 31, 2012. For (i), we use non-overlapping windows of length 1,000 (yielding 150 sets of parameter estimates), and the results are shown in Figure 10.9. (Although we are only concerned with the GARCH parameters, we show all six parameters for completeness.)

This can be compared to Figure 10.10, which shows the computation corresponding to (ii), i.e., the m.l.e. simulation results based on series of length $T = 1{,}000$, generated from an NCT-GARCH process with parameters $a_0 = 0.06, c_0 = 0.04, c_1 = 0.05, d_1 = 0.90$, degrees of freedom $k = 7$, and noncentrality parameter $\theta = 0.05$. We see that the variation of (i) is indeed smaller than that of (ii), for GARCH parameters c_0 and c_1, though it is not quite the case for parameter d_1 because of the elongated left tail in the distribution for d_1 in Figure 10.9. However, most of the mass is indeed centered around the value 0.90.

We now address the asymmetry parameter γ_1 in the APARCH(1,1) formulation. When estimating all the parameters of the NCT-APARCH model for sample sizes of $T = 1{,}000$, and particularly so for

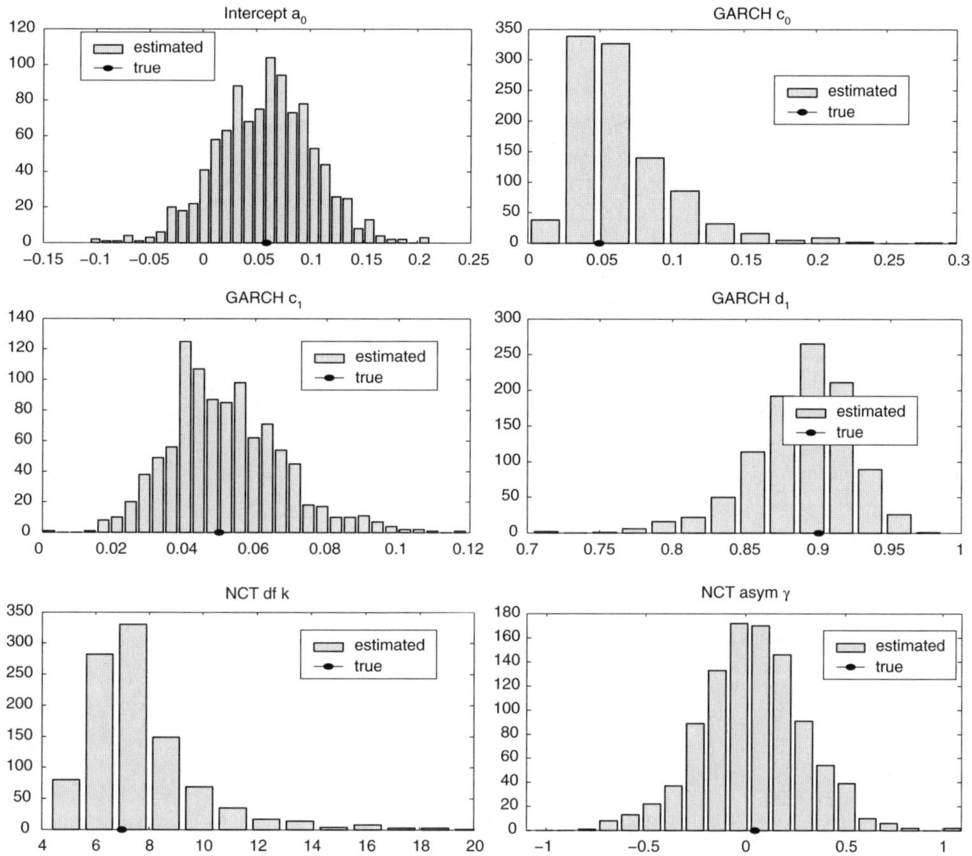

Figure 10.10 The m.l.e. parameter estimates corresponding to the NCT-GARCH model for simulated NCT GARCH data, using length $T = 1{,}000$ and 1,000 replications. The circle on the x-axis indicates the median of the data. (In the lower right panel, the NCT noncentrality parameter is denoted as γ.)

$T = 250$, it was found that $\hat{\gamma}_1$ is rather erratic, when viewed over moving windows through time, and often approaches (and touches) its upper boundary value of one.[10]

Given the problematic asymmetry parameter in small sample sizes, Krause and Paolella (2014) fix the three parameters associated with the traditional GARCH model (as stated above), and then, conditional on those, choose the optimal value of γ_1 with respect to out-of-sample value-at-risk (VaR)

10 Many applied research papers use, and advocate, larger sample sizes when fitting GARCH-type models, justified from simulation results such as those shown in Figure 10.4. However, if, as we claim, the d.g.p. is changing through time, then use of shorter windows of data makes sense (to optimally address the bias-variance tradeoff) for forecasting applications, or use of weighted likelihood, as discussed in Chapter 13. With respect to the APARCH asymmetry parameter γ_1, to rule out any computational errors, simulations were conducted. When using a sample size of 250, it was found that the final m.l.e.s are very sensitive to the choice of starting values, and appear to result in biased estimates of γ_1. However, when using very large sample sizes (e.g., 25,000), the estimator looks as one would expect, namely virtually Gaussian and centered around the true parameter values, and this having been achieved using any reasonable set of starting values, not just the true ones.

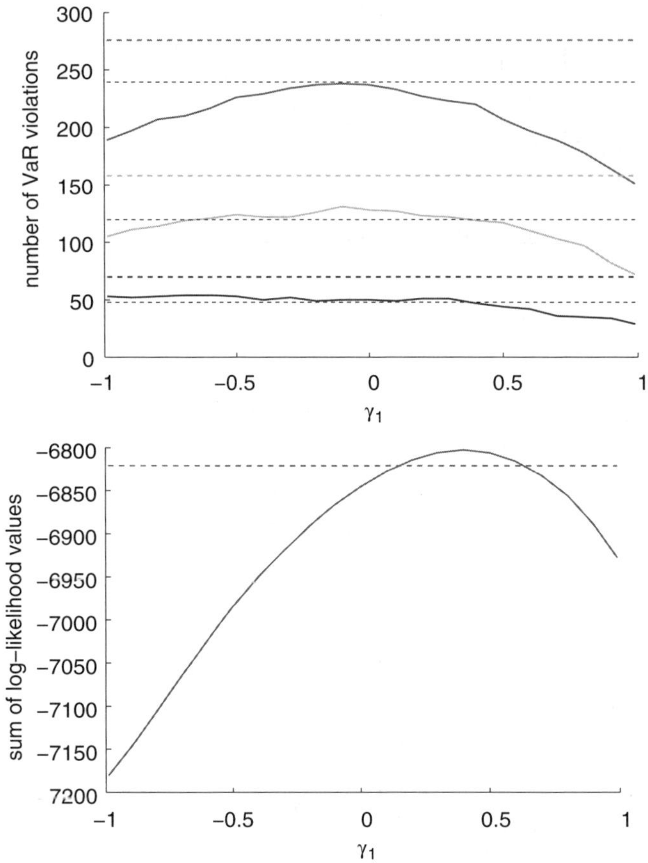

Figure 10.11 Illustration of the effect of varying the APARCH asymmetry parameter γ_1 on the number of VaR violations (top) and the sum of the predicted log-likelihood values (bottom). Results are out-of-sample (4,787 forecasts) for the period January 4, 1993, to December 31, 2012, obtained from a rolling window exercise with window size $T = 250$. The data set under study is the 20-year sequence of daily (percentage log) returns of the equally weighted portfolio of DJIA-30 components (as of April 2013). The dashed lines refer to the NCT-GARCH(1,1) model and the solid lines to the NCT-APARCH(1,1) model with γ_1 being varied. The dotted lines in the left panel depict the expected number of VaR violations at the 1% (lower lines; blue), 2.5% (middle lines; green) and 5% (upper lines; red) significance level, respectively.

and density forecast quality, the latter measured by evaluating the log predicted NCT density at the realized returns, as done in Paolella (2015) and Paolella and Polak (2015a). Figure 10.11 shows the results of this exercise for $T = 250$. (The results for $T = 1{,}000$ were also computed and were qualitatively very similar.)

We see that the NCT-APARCH model (with fixed GARCH parameters) performs in a superior way for most choices of γ_1, with the number of VaR violations being closer to its expected value. A similar result is observed in terms of density prediction: For $0.25 \leqslant \gamma_1 \leqslant 0.55$ the NCT-APARCH model outperforms the GARCH case. From these plots, it appears that taking $\gamma_1 = 0.4$ is the optimal choice to improve the forecast quality.

Thus, we advocate fixing the APARCH parameters as

$$c_0 = 0.04, \quad c_1 = 0.05, \quad d_1 = 0.90, \quad \text{and } \gamma_1 = 0.4, \tag{10.20}$$

to be used in the iterative (or one-pass) scheme for estimation of the NCT-APARCH(1,1) process (10.19).

10.5 $S_{\alpha,\beta}$-APARCH and Testing the IID Stable Hypothesis

The method described for fast estimation of the NCT-APARCH(1,1) model in Section 10.4 can be adapted for use with an $S_{\alpha,\beta}$-APARCH(1,1), i.e., use of asymmetric stable Paretian innovations instead of NCT. It requires constructing a table lookup estimator for the two shape parameters α and β for location-zero, scale-one i.i.d. stable data, as well as developing the trimmed-mean estimator for a_0 based on the assumption of (symmetric) stable data and a fixed value of α.

This was done in Paolella (2016), with two goals. First, for developing a fast and numerically reliable method for estimating the $S_{\alpha,\beta}$-APARCH(1,1) model not requiring use of the stable likelihood and, second, to assess the appropriateness of the conditional stable assumption for asset returns data—an issue that has been debated and researched since the 1960s. The idea is to obtain the $S_{\alpha,\beta}$-APARCH(1,1) filtered innovations, and then apply the (also numerically fast) tests for stability developed in Section III.9.5, namely the so-called ALHADI, τ_{20}, and $A + \tau_{20}$ tests. All these tests use particular features of the stable Paretian distribution, and have correct size and good power properties (with the latter having the best). It appears more difficult to design a powerful test for the null of i.i.d. NCT.

Detailed analysis for several individual (percentage log) stock return series is presented in Paolella (2016), along with summary histograms of the p-values from three tests for stability, formed by using numerous data sets. The latter are reproduced in Figures 10.12 and 10.13, based, respectively, on 957 and 4100 time series of stock returns, from the DJIA and S&P500 stock indexes. Note that the ability to very quickly compute parameter estimates of the $S_{\alpha,\beta}$-APARCH(1,1) model, as well as statistics and their respective p-values for testing the i.i.d. stable Paretian distribution assumption, is what allows these histograms to be rapidly produced.

Under the null of stability of the innovations (and, implicitly, that the APARCH(1,1) filter is adequate for inducing a near-i.i.d. process), the p-values should be uniformly distributed on $(0, 1)$. We see this

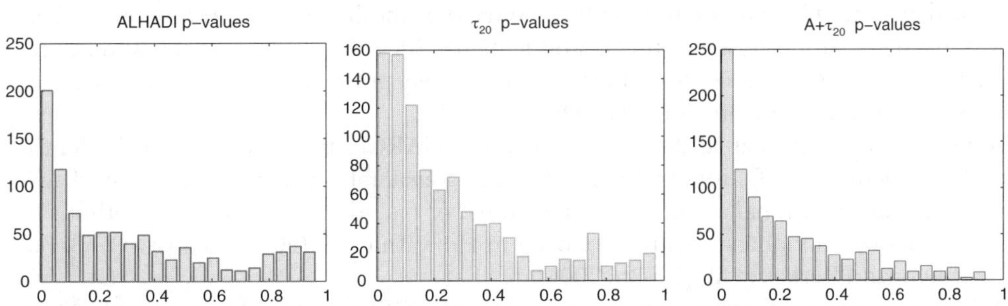

Figure 10.12 Histograms of the p-values of the three tests for stability, based on 957 data sets formed from 33 windows of data using $T = 500$, and 29 stocks comprising the DJIA.

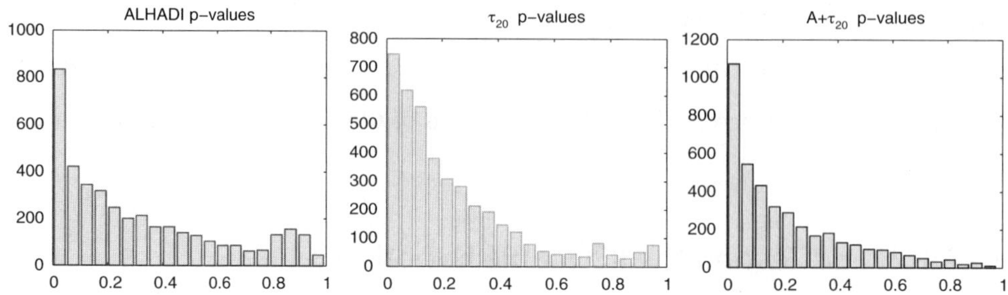

Figure 10.13 Same as Figure 10.12 but having used the 100 largest market-cap stocks from the S&P500 index, resulting in 4100 data sets and *p*-values.

is not the case, providing strong evidence against the hypothesis that all, or most, stock returns are conditionally stable Paretian.

Remarks

a) There have been numerous critiques of the use of the stable Paretian distribution for modeling real phenomena, notably financial asset returns, most of which can be dismissed as false or no longer applicable; see the discussion in Section III.9.2.3. The above test results provide evidence that, more often than not, the (conditional) stable distribution for stock returns is not true. In particular, the τ_{20} test is based on the summability property of the stable Paretian distribution, so that, from the middle panels of Figures 10.12 and 10.13, the large pileup of p-values on the left (and clear violation of uniformity) indicates that the filtered $S_{\alpha,\beta}$-APARCH(1,1) residuals do not obey summability: As consecutive observations are summed, the tail tends to becomes thinner. This is sometimes taken to be one of the stylized facts of asset returns, thus ruling out the stable Paretian being the actual distribution. This does not, however, detract from deploying it as an excellent and useful approximation for risk forecasting purposes (see the next remark). Models that can account for the (often but not always observed) stylized fact of decreasing tail thickness as the frequency of the measured returns decreases (from, say, daily, to weekly, to monthly) are considered in Grabchak and Samorodnitsky (2010).
b) One should differentiate in this context between the use of formal testing procedures for a distributional assumption and the appropriateness, or lack thereof, of using that distribution. In particular, use of the stable distribution in conjunction with GARCH-type models for risk prediction has been shown to be quite successful (see, e.g., Mittnik and Paolella, 2003; Broda et al., 2013; and the references therein). This lends evidence that, at least from a purely practical point of view, the model has merit. This point is also made in Nolan (1999).
c) A very large number of empirical studies exist that use a GARCH-type model with Student's t or generalized exponential (GED) innovations, these being available in many software packages. Rarely, if ever, is the distributional assumption questioned or tested in a formal way. Methods for GARCH residual distributional testing are proposed in Horváth et al. (2001), Bai (2003), Berkes and Horváth (2003), and Berkes et al. (2003b, 2004).

   If powerful tests existed for distributions such as the t and GED, as they do now for the stable Paretian, it might be the case that they are also rejected for the majority of stock return series. Instead of formal testing, some studies will (correctly) compare the forecasting performance of several competitive models. The usual finding is that the use of a GARCH-type model allowing for asymmetric shocks to volatility, in conjunction with (almost any) leptokurtic, asymmetric distribution, such as the NCT or stable, but also one with semi-heavy tails, such as the NIG (for which the m.g.f. exists), and even a thin-tailed distribution, such as the mixed normal, delivers competitive risk and density forecasts.

It is important keep in mind that all models and distributions employed for modeling non-trivial real data are nothing but approximations and are necessarily wrong, and, in the context of financial asset returns and other heavy-tailed phenomena, *without an infinite amount of i.i.d. data, tail measurements will always be inaccurate*; see Sections III.9.1 and III.9.2 for discussion of this issue. This is the reason why non-Gaussian (asymmetric and) leptokurtic distributions with differing tail behaviors (heavy or thin) can lead to good forecasting performance. As such, at least in finance, while distributional testing is an important diagnostic, a crucial measure of the utility of a model is in the application to forecasting, such as downside risk, or portfolio optimization—for which different (non-Gaussian) models can be compared and ranked. In the context of portfolio construction and out-of-sample performance, see Section 11.3.

d) Both in-sample and out-of-sample diagnostics and (some) testing procedures can have value for assessing the appropriateness of a model, but the purpose of the model must always be considered. For example, if a forecast of tomorrow's VaR is required on 100,000 client portfolios, then speed, numeric reliability, and practicality will play a prominent role. In the complicated game of financial risk forecasting and asset allocation, it is highly unlikely that a single model will be found to consistently outperform all others in every regard, but the appropriate use of distributional testing, out-of-sample performance diagnostics, and some "informed common sense" can lead to a model that reliably fulfils its intended purpose. ∎

The other "half" of the null hypothesis of i.i.d. stable innovations of stable-GARCH filtered data is that they are i.i.d. We discuss two tests in this regard, and investigate the performance of one of them.

A popular test for the i.i.d. hypothesis of a univariate time series was proposed in Brock et al. (1996), usually referred to as BDS, from the first three authors (and the associated working paper version from them in 1987). Under the null, the test statistic is asymptotically standard normal, though deviates from this in small samples; see Kanzler (1998) for tabulated cutoffs and a publicly available Matlab implementation of the BDS test. According to Brock et al. (1996), their test does not require existence of "higher moments". It was extended to the multivariate case in Baek and Brock (1992). Its effectiveness for GARCH-type effects has been investigated by several authors; see Brooks and Heravi (1999), Chen and Kuan (2002), Caporale et al. (2005), Zivot and Wang (2006, Ch. 18), and the references therein. The BDS test requires a choice of the so-called embedding dimension, m, and different choices can lead to different test outcomes. This issue is addressed in Matilla-García et al. (2014).

Based on Kanzler's (1998) implementation and default settings, we confirm via simulation that the distribution of the BDS test statistic appears non-Gaussian and heavy-tailed for i.i.d. stable data with $\alpha < 2$. One possibility is to apply the non-parametric bootstrap for a given set of data to determine the p-value of the test, as implemented in the Eviews software package, though we do not pursue this. The interested reader is encouraged to investigate the extension by Matilla-García et al. (2014) and, possibly in conjunction with the bootstrap, construct a BDS-based test with correct size, and investigate its performance under the null of heavy-tailed i.i.d. data and the power against alternatives of interest, such as (stable-)GARCH.

We next consider the two tests for the i.i.d. hypothesis, based on generalized runs distributions, by Cho and White (2011), denoted GR-L1 and GR-sup.[11] These tests do not have any moment conditions, and so are appropriate for testing the i.i.d. assumption for stable data.

As a demonstration, Figure 10.14 shows p-values of the two tests for sets of i.i.d. normal and stable data, and normal GARCH processes. We see that, under the null of normal and stable i.i.d. data (first two rows), or their absolute values (third row), the p-values of both tests deviate somewhat from the uniform distribution, so that, for normal data, or a given value of α for stable data, and sample size T, simulation is required to determine the appropriate cutoff values for tests with the usual levels of significance. From the fourth row, we see that, with normal-GARCH data, both tests have little power,

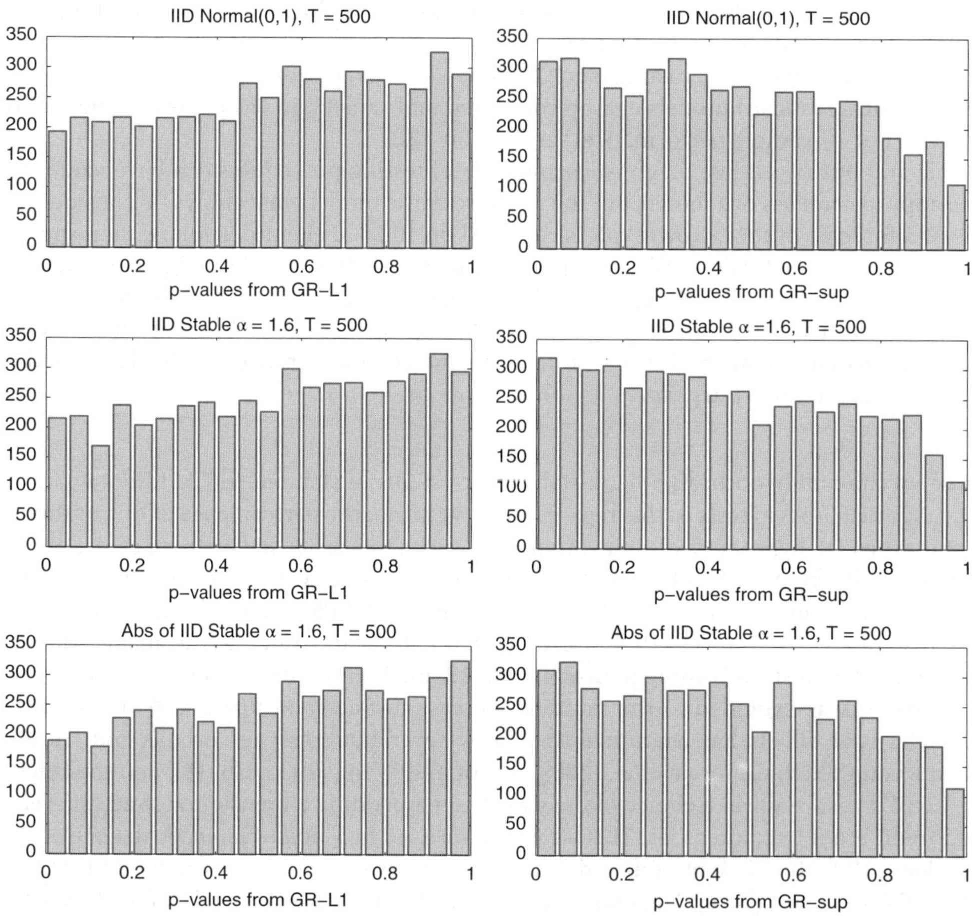

Figure 10.14 Histograms of p-values of the GR-L1 (left) and GR-sup (right) tests, based on 5000 replications for sample size $T = 500$, of i.i.d. normal data (top row); $S_{1.6,0}(0, 1)$ data (second row); absolute value of i.i.d. $S_{1.6,0}(0, 1)$ data (third row); normal-GARCH data with $c_0 = 0.05$, $c_1 = 0.05$, and $d_1 = 0.90$ (fourth row); and absolute value of same normal-GARCH data (last row).

11 Matlab code has been kindly provided to the author by Jin Seo Cho, which delivers also p-values of both tests.

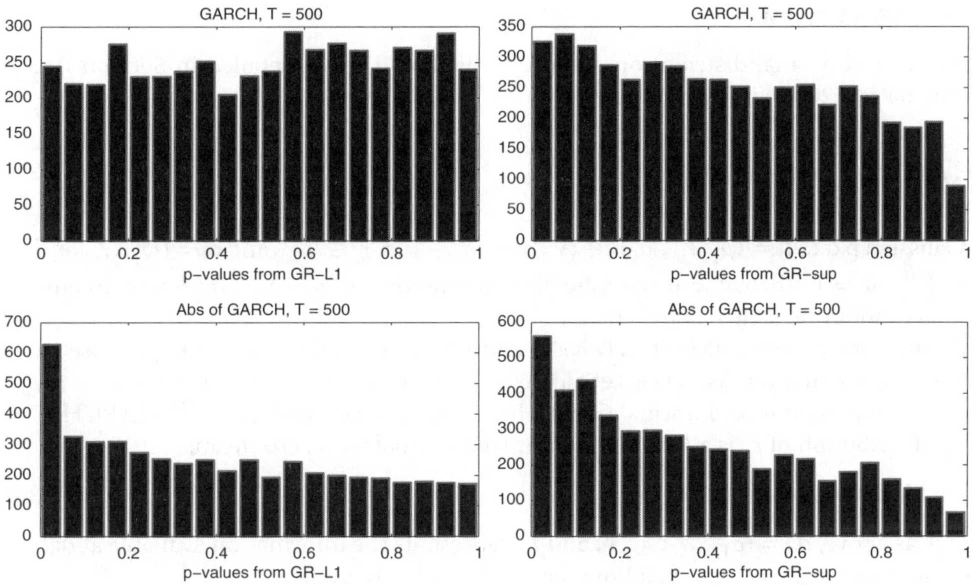

Figure 10.14 (*Continued*)

though taking absolute values of the simulated GARCH data does lead to the tests having some power, as seen from the pile-up of the p-values towards zero in the bottom panels.

Based on this short analysis, we can conclude that testing for i.i.d.-ness appears difficult, and further investigations will be necessary to obtain tests with high power.

10.6 Mixed Normal GARCH

10.6.1 Introduction

As we have seen above, different generalizations of the baseline GARCH model (10.2), such as (10.10) and (10.11), can be devised, serving as better filters for the unknown actual law of motion underlying the scale term through time, and allowing for asymmetric effects of shocks on volatility. Additionally, and more importantly for risk and density prediction, the distribution of the i.i.d. innovation sequence $\{Z_t\}$ can be changed to a non-Gaussian one. Both of these enhancements to the original Gaussian-GARCH formulation are valuable, but are still limited in the extent to which they can capture additional dynamics in the underlying process.

We now discuss a class of GARCH-type models, referred to as **mixed normal GARCH**, that deviates in an important way from the previous structures, though still nests (10.2) as a special case. This class of models is beneficial because it gives rise to more complicated dynamics that result in better in-sample fits and, crucially, better risk and density forecasts. What this mixed normal GARCH class is *not* is the usual GARCH equation (10.2) driven by an innovation process from a k-component discrete mixture of normals distribution, but rather a matrix-based structure that essentially embodies k Gaussian GARCH equations, allowing for "interactions" between them.

10.6.2 The MixN(k)-GARCH(r, s) Model

The k-component mixed normal distribution, denoted MixN($\boldsymbol{\omega}, \boldsymbol{\mu}, \boldsymbol{\sigma}$), is detailed in Section III.5.1 and Frühwirth-Schnatter (2006); its p.d.f. is given by

$$f_{\text{MN}}(y; \boldsymbol{\omega}, \boldsymbol{\mu}, \boldsymbol{\sigma}) = \sum_{j=1}^{k} \omega_j \phi(y; \mu_j, \sigma_{jt}^2), \tag{10.21}$$

where ϕ is the Gaussian p.d.f., $\boldsymbol{\mu} = (\mu_1, \dots, \mu_k)' \in \mathbb{R}^k$, $\boldsymbol{\sigma} = (\sigma_1, \dots, \sigma_k)' \in \mathbb{R}^k_{>0}$, and $\boldsymbol{\omega} = (\omega_1, \dots, \omega_k)' \in (0, 1)^k$ such that $\sum_{j=1}^k \omega_j = 1$. Throughout the following, we impose $\mu_k = -\sum_{j=1}^{k-1}(\omega_j/\omega_k)\mu_j$ to ensure that the associated random variable has zero mean.

Independently and concurrently, Haas et al. (2004a) and Alexander and Lazar (2006) proposed and studied the following construction. As in Haas et al. (2004a), we say that time series $\boldsymbol{\varepsilon} = (\varepsilon_1, \varepsilon_2, \dots, \varepsilon_T)$ is generated by a k-component mixed normal GARCH(r, s) process, denoted MixN(k)-GARCH(r, s), if the conditional distribution of ε_t is a k-component mixed normal with zero mean,

$$\varepsilon_t \mid \mathcal{F}_{t-1} \sim \text{MixN}(\boldsymbol{\omega}, \boldsymbol{\mu}, \boldsymbol{\sigma}_t), \tag{10.22}$$

where $\boldsymbol{\omega}$ and $\boldsymbol{\mu}$ are as above, $\boldsymbol{\sigma}_t = (\sigma_{1,t}, \dots, \sigma_{k,t})'$, and $\mathcal{F}_t$ represents the information available at date t. The component variances $\sigma_{i,t}^2$, $i = 1, \dots, k$, follow the GARCH-like structure

$$\boldsymbol{\sigma}_t^{(2)} = \boldsymbol{\gamma}_0 + \sum_{i=1}^{r} \boldsymbol{\gamma}_i \varepsilon_{t-i}^2 + \sum_{j=1}^{s} \boldsymbol{\Psi}_j \boldsymbol{\sigma}_{t-j}^{(2)}, \tag{10.23}$$

where $\boldsymbol{\gamma}_i = (\gamma_{i,1}, \gamma_{i,2}, \dots, \gamma_{i,k})'$, $i = 0, \dots, r$, are $k \times 1$ vectors, $\boldsymbol{\Psi}_j$, $j = 1, \dots, s$, are $k \times k$ matrices, and $\boldsymbol{\sigma}_t^{(\delta)} = (\sigma_{1,t}^\delta, \sigma_{2,t}^\delta, \dots, \sigma_{k,t}^\delta)'$, for $\delta \in \mathbb{R}_{>0}$. The parameters of the model need to be such that $\boldsymbol{\sigma}_t^{(\delta)} > 0$, where, in case of non-scalars, $>$ indicates element-wise inequality. As above, a (possibly time-varying) mean term can be incorporated, so that the model becomes (using c_t instead of μ_t as above for the mean) $R_t = c_t + \epsilon_t$.

Special cases of model (10.22)–(10.23) had been proposed earlier, and include the formulations by Vlaar and Palm (1993) and Bauwens et al. (1999), with their relationships to the general model detailed in Haas et al. (2004a). Parameter conditions for the non-negativity of all elements of $\boldsymbol{\sigma}_t^{(2)}$ could be derived by writing the model in ARCH(∞) form and applying the results in Nelson and Cao (1992) and Conrad and Karanasos (2009), but in the case of the (practically most relevant) first-order diagonal model discussed next, it follows from the GARCH(1, 1) analogy that the conditions $\boldsymbol{\gamma}_0 > 0$, $\boldsymbol{\gamma}_1 \geqslant 0$, and $\boldsymbol{\Psi}_1 \geqslant 0$ are required.

As with the simple GARCH model (10.2), in most applications $r = s = 1$ suffices, so in the following we take $r = s = 1$ and denote the (i, j)th element of $\boldsymbol{\Psi}_1$ by $\psi_{i,j}$. As discussed and demonstrated in Haas et al. (2004a), it is reasonable to restrict the $\boldsymbol{\Psi}_j$ to be diagonal, so in this case we refer to the diagonal elements of $\boldsymbol{\Psi}_1$ as ψ_i, $i = 1, \dots, k$. Even with $r = s = 1$ and diagonal $\boldsymbol{\Psi}_1$ (but with $k > 1$), the model is able to generate pseudo-long memory in the autocorrelation function of the squares of the returns, as illustrated below.

In this special but very useful case of $r = s = 1$, k unrestricted, and diagonal $\boldsymbol{\Psi}_1$, which we then denote as $\boldsymbol{\beta}$, we obtain the following reasonably palatable expressions for the moments and autocorrelation function (see Haas et al., 2004a, App. B). Defining

$$C_{11} = \boldsymbol{\gamma}_1 \boldsymbol{\omega}' + \boldsymbol{\beta},$$

$$
\begin{aligned}
C_{21} &= (\gamma_1\omega') \otimes \gamma_0 + \gamma_0 \otimes (\gamma_1\omega') + \gamma_0 \otimes \beta + \beta \otimes \gamma_0 + (\beta \otimes \gamma_1)\omega'\mu^{(2)} \\
&\quad + (\gamma_1 \otimes \beta)\omega'\mu^{(2)} + 6(\gamma_1 \otimes \gamma_1)(\omega \odot \mu^{(2)})', \\
C_{22} &= 3(\gamma_1 \otimes \gamma_1)\text{vec}\,[\text{diag}(\omega)]' + \beta \otimes (\gamma_1\omega') + (\gamma_1\omega') \otimes \beta + \beta \otimes \beta, \\
d_1 &= \gamma_0 + \gamma_1\omega'\mu^{(2)}, \\
d_2 &= \gamma_0 \otimes \gamma_0 + (\gamma_0 \otimes \gamma_1 + \gamma_1 \otimes \gamma_0)\omega'\mu^{(2)} + (\gamma_1 \otimes \gamma_1)\omega'\mu^{(4)},
\end{aligned}
$$

we have

$$
\begin{aligned}
\mathbb{E}[\sigma_t^{(2)}] &= (I_k - C_{11})^{-1} d_1, \\
\mathbb{E}[\sigma_t^{(2)}\sigma_t^{(2)'}] &= [(I_{k^2} - C_{22})^{-1} C_{21}(I_k - C_{11})^{-1},\ (I_{k^2} - C_{22})^{-1}] \begin{bmatrix} d_1 \\ d_2 \end{bmatrix},
\end{aligned}
$$

and

$$
\mathbb{E}[\varepsilon_t^2] = \omega'\mathbb{E}[\sigma_t^2] + \omega'\mu^{(2)}, \tag{10.24a}
$$

$$
\mathbb{E}[\varepsilon_t^4] = 3\text{vec}\,[\text{diag}(\omega)]'\ \text{vec}(\mathbb{E}[\sigma_t^{(2)}\sigma_t^{(2)'}]) + 6\omega'(\mu^{(2)} \odot \mathbb{E}[\sigma_t^{(2)}]) + \omega'\mu^{(4)}, \tag{10.24b}
$$

from which the kurtosis can be computed. Further defining

$$
v = \frac{\gamma_0\mathbb{E}[\varepsilon_t^2] + \gamma_1\mathbb{E}[\varepsilon_t^4] + \beta(\mathbb{E}[\sigma_t^{(2)}\sigma_t^{(2)'}] + \mathbb{E}[\sigma_t^{(2)}]\mu^{(2)'})\omega - \mathbb{E}[\sigma_t^{(2)}]\mathbb{E}[\varepsilon_t^2]}{\mathbb{E}[\varepsilon_t^4] - \mathbb{E}^2[\varepsilon_t^2]},
$$

the autocorrelation function of ε_t^2 is given by

$$
r(\tau) = \frac{\text{Cov}(\varepsilon_{t-\tau}^2, \varepsilon_t^2)}{\mathbb{V}(\varepsilon_t^2)} = \frac{\mathbb{E}[\varepsilon_t^2\varepsilon_{t-\tau}^2] - \mathbb{E}^2[\varepsilon_t^2]}{\mathbb{E}[\varepsilon_t^4] - \mathbb{E}^2[\varepsilon_t^2]} = \begin{cases} \omega' v, & \tau = 1, \\ \omega'(\gamma_1\omega' + \beta)^{\tau-1} v, & \tau > 1, \end{cases} \tag{10.25}
$$

generalizing the results in (10.7).

10.6.3 Parameter Estimation and Model Features

The price to pay for having such an elaborate structure and the enhanced dynamics is, of course, more parameters to estimate. For example, with $k = 3$ components, there are 14 free parameters to estimate. This behooves discussion. From a numeric point of view, the larger number of parameters is not necessarily problematic, as the likelihood is easily expressed and quickly evaluated, and numerous applications with real and simulated data indicate that generic Hessian-based optimization routines are usually effective in locating a plausible maximum of the likelihood. There are two related caveats to be mentioned here. The first is that, as with the i.i.d. mixed normal model as discussed in Section III.5.1, and with the plain Gaussian-GARCH model (10.2) as discussed in Section 10.2, there can exist several *plausible* maxima of the likelihood function, and only via use of various starting values can one obtain what is likely to be the global maximum.

The second caveat involves avoiding maxima that are not plausible but, due to the nature of mixture models, can arise and plague estimation. This mixture degeneracy problem is unfortunately germane to all mixture models. In the i.i.d. case, several ways of dealing with this problem are discussed in Section III.5.1, including use of simple box constraints and quasi-Bayesian shrinkage priors. Within the mixed normal GARCH framework, such solutions are not as straightforward, with its elaborate structure and high parameterization. A solution has been proposed in Broda et al. (2013) that deals with this problem elegantly, generally, and effectively, and is also applicable in the i.i.d. case. The method works by appending to the log-likelihood two judiciously chosen penalty and shrinkage terms, and thus does not entail any more numeric work than is involved in directly maximizing the likelihood.

While significantly more complicated and numerically more intensive, a full Bayesian approach is also possible, as pursued by Bauwens and Rombouts (2007b). Similar to the use of the quasi-Bayesian prior of Hamilton (1991) in the i.i.d. normal mixture, by specifying an "informative prior" (even if very weak), the degeneracy issue evaporates. Another method for estimation is pursued by Lee and Lee (2009), but that approach does not solve the mixture degeneracy problem.

Another feature of mixture GARCH models is their ability to allow for conditional as well as unconditional component models. With (say, daily) financial returns data, the component of the mixture that captures the most volatile observations can often be adequately modeled with a relatively high, but constant, variance—it does not require a GARCH structure. This is notably the case with the MixN(3)-GARCH(1,1) model: the third component (the one associated with the highest volatility) is capturing a few scattered outlying observations, and no GARCH structure is required for it, whereas the dynamics in the volatility are being driven by the first two components of the model. We label this construction as MixN(k, g)-GARCH(r, s) or, in short, for $r = s = 1$, just MixN(k, g), to indicate that only g, $g \leqslant k$, components follow a GARCH(1,1) process dictated by (10.23).

Any reasonably intelligent model with enough parameters will generate an excellent in-sample fit. We are thus concerned with the possibility that the mixed normal GARCH model is over-fitting the data. This appears to not be the case. With $g = k = 3$ (and 14 parameters), the conservative likelihood penalty measure, BIC (and certainly the less conservative AIC) favors it over the use of competitive models in the location-scale class (10.2), in almost all cases, with real data. Between the MixN(3,3) and the MixN(3,2), the latter is favored. Moreover, and perhaps the most important reason for its use in modeling asset returns, the mixed normal GARCH model delivers excellent out-of-sample forecasts (quantile and density) relative to genuine (as opposed to, and of course also against, straw man) competitors, as demonstrated in Kuester et al. (2006), Paolella and Steude (2008), Paolella and Taschini (2008), and Broda et al. (2013).

Part of the reason for this strong performance is that normal mixture distributions have been found in numerous studies to fit the distribution of asset returns under an i.i.d. assumption very well; see the discussion and references in Haas et al. (2004a) and Paolella (2015). Another benefit involves skewness: A feature of the model not possible with the formulation in (10.2) and its APARCH and Q-GARCH extensions is **time-varying skewness**. A growing amount of literature indicates that not only are there asymmetries in the reaction of the volatility of the process and the innovations distribution, but that this asymmetry is time varying, and is relevant for asset allocation; see Rockinger and Jondeau (2002), Jondeau and Rockinger (2003, 2009, 2012), Haas et al. (2004a), and the references therein. With more than one component, time-varying skewness is automatically accommodated in the mixed normal GARCH model, i.e., it is inherent in the model without requiring an explicit, *ad hoc* specification of a conditional skewness process appended to (10.2). The relevant equations and graphical illustrations of the time-varying skewness are given in Haas et al. (2004a). One could also use asymmetric densities instead of the normal, with the skew-normal of Azzalini (1985) being a natural suggestion; see the discussion of the skew-normal distribution in Section III.A.8, and Haas (2010) and Geweke and Amisano (2011) for its use in the context of (mixture and regime switching) GARCH models.

Another reason for the strength of the model involves the autocorrelation function: One of the stylized facts of asset returns is a slow, hyperbolic decline of the autocorrelation function of the absolute or squared returns. The FIGARCH model construct, as mentioned in Section 10.3.1, was proposed precisely to capture this feature, though open issues remain regarding stationarity and parameter estimation; see Caporin (2003) and Tayefi and Ramanathan (2012). Their use, in turn, has been criticized because the aforementioned apparent hyperbolic behavior can be induced by ignoring model or parameter structural breaks; see, e.g., Lamoureux and Lastrapes (1990), Lux and Kaizoji (2007), Lux (2008), Lee et al. (2010), Hillebrand and Medeiros (2016), Wang et al. (2016), and the references therein.

A benefit of the mixed normal GARCH model is that it can give rise to a *pseudo long-memory process*, by which we mean that the autocorrelation function of the squared returns exhibits a decay that mimics long-memory behavior, though the process is not a genuine long-memory one, and eventually exhibits exponential decay. To illustrate, Figure 10.15 shows the improvement in the match between theoretical and sample autocorrelations, based on the squared returns, for the (diagonal) MixN(k)-GARCH(1, 1) model over the plain GARCH(1,1) model, where the theoretical values are computed from (10.25), based on the fitted parameters. The advantage of the model (with $k > 1$) arises from the structure of the roots of the ARMA representation of the ϵ_t^2 process (see Haas et al., 2004a), and is not shared by traditional GARCH(1,1) models or the special, "linear" cases of the mixed normal GARCH model with $r = s = 1$, diagonal $\boldsymbol{\Psi}_1$ and $k = 2$, as proposed by Vlaar and Palm (1993), Bauwens et al. (1999), and Bai et al. (2003).

Yet another advantage of the mixed normal GARCH framework, as compared to, say, a Student's t GARCH model, is in its applicability to (univariate and multivariate) option pricing, as discussed in Alexander and Lazar (2006), and developed further in Badescu et al. (2008), Rombouts and Stentoft (2009, 2011), and the references therein. This also holds for continuous mixtures of normals in the multivariate setting; see Paolella and Polak (2015b).

Several (still univariate) generalizations of the MixN(k)-GARCH(r, s) model (10.22)–(10.23) exist. First, one can replace the normal distribution with a more general one, such as the aforementioned

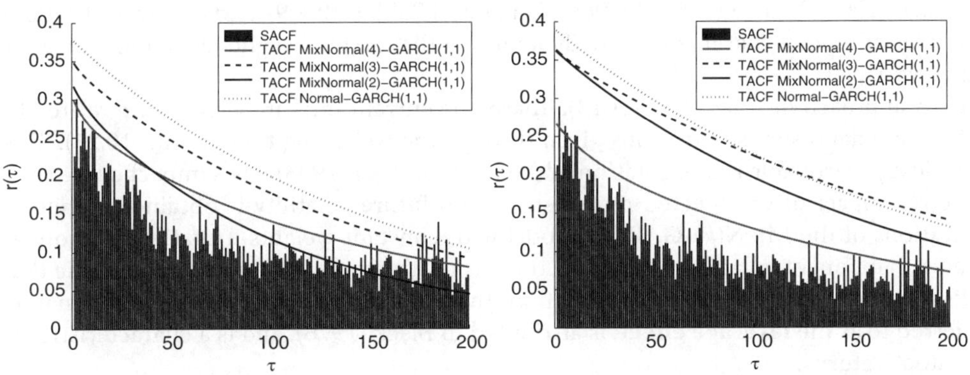

Figure 10.15 **Left:** Sample autocorrelation function (SACF) in lag τ, $\tau = 1, 2, \ldots, 200$, of the squared returns from the 20 years of daily NASDAQ Composite returns from January 1, 1990 to December 31, 2010, overlaid with their theoretical counterparts, given in (10.25), for the fitted plain GARCH(1,1) and various fitted MixN(k)-GARCH(1,1) models. **Right:** Same but based on the 4000 daily NASDAQ Composite returns until March 16, 2011.

skew-normal. The benefits of using the (symmetric and asymmetric) stable Paretian is detailed in Broda et al. (2013). Second, in light of the **leverage effect** (see below), a fruitful method of introducing asymmetries into the structure of the model is developed in Alexander and Lazar (2004). Third, a method that also addresses the leverage effect, but in a conceptually very different way, is to allow the weights of the components to vary with time, as illustrated next in Section 10.6.4. A fourth way is the extension to a Markov-switching framework, as briefly discussed in Section 10.6.5.

10.6.4 Time-Varying Weights

An extension of the class of MixN(k)-GARCH(r, s) models that results in further improved forecast accuracy is to allow for time-varying mixing weights. These are related, though not equivalent, to Markov-switching models, such as that of Hamilton (1989), which have found many applications in macroeconomics and finance (see Section 10.6.5). The conditional densities of such mixture models are endowed with great flexibility. As illustrated in Haas et al. (2006b), the predictive density may even become bimodal, depending on the expected jump size.

Mixture models with mixing weights depending on lagged process values and/or exogenous variables have been employed quite successfully throughout the literature. An example is the modeling of exchange rate behavior in target zones, where a jump component reflects the probability of realignments, and the probability of a jump depends on interest differentials and, possibly, further explanatory variables incorporating market expectations; see, e.g., Vlaar and Palm (1993), Bekaert and Gray (1998), Neely (1999), Klaster and Knot (2002), and Haas et al. (2006a). Another example is modeling the nonlinear relation between hedge fund returns and market risk factors; see Tashman and Frey (2009) and Tashman (2010).

Some of the models used in the previously mentioned references have some similarities to the class of smooth transition GARCH (STGARCH) models (see Teräsvirta, 2009; Medeiros and Veiga, 2009) and the component GARCH specification of Bauwens and Storti (2009). The difference is that, in the latter models, the weighting applies to the volatility parameters directly (as compared to the densities as in (10.21)), so that the conditional distribution is not a mixture. See also Wong and Li (2001), Alexander and Lazar (2005), Bauwens et al. (2006b), Lange and Rahbek (2009), and Cheng et al. (2009) for further examples of models that incorporate dynamics of the mixing weights depending on a set of predetermined variables.

We illustrate the approach in Haas et al. (2013). There, the current mixture weights are related to past returns via sigmoidal response functions given through the weighting function in (10.26). This leads to an empirically reasonable representation of the Engle and Ng (1993) news impact curve such that an asymmetric impact of unexpected return shocks on future volatility is obtained. While in empirical applications of the MixN(k)-GARCH model with constant weights, negative component means, and higher component volatilities coincide, there is no *dynamic* asymmetry in the sense that negative shocks tend to increase *future* volatility more than positive shocks. This type of dynamic asymmetry, referred to as the **leverage effect**, is attributed to Black (1976), and is a characteristic, or stylized fact, of stock returns.

The general model structure in Haas et al. (2013) takes the form

$$\omega_{j,t} = \frac{\lambda_j}{1 + \sum_{i=1}^{k-1} \lambda_i}, \quad j = 1, \dots, k-1, \quad \omega_{k,t} = 1 - \sum_{i=1}^{k-1} \omega_{i,t}, \tag{10.26}$$

where, mimicking the structure of an asymmetric GARCH-type model,

$$\lambda_j = \exp\left(\gamma_{0,j} + \sum_{i=1}^{u} \gamma_{i,j}\varepsilon_{t-i} + \sum_{i=1}^{v} \kappa_{i,j}\omega_{j,t-i} + \sum_{i=1}^{w} \delta_{i,j}|\varepsilon_{t-i}|^d \right), \tag{10.27}$$

denoted, say, TV(u, v, w)-MixN(k)-GARCH(r, s), where u, v and w are the orders of the lagged ε_t, lagged ω_j, and lagged $|\varepsilon_t|^d$, $d > 0$, respectively. A more parsimonious version of this general structure still capable of capturing the time variation in the mixing weights is obtained by setting $v = w = 0$, along with $r = s = 1$, and we abbreviate this as TV(u)-MixN(k) or, recalling the above restricted (k, g) formulation such that $k - g$ components are not endowed with a GARCH structure, as TV(u)-MixN(k, g).

The choice of sigmoidal weighting functions in (10.26) was inspired from the resulting shape of the fitted *non-parametric* weighting function, given as follows. Let $m > 0$ be the number of intervals, and let $\boldsymbol{\theta} = (\theta_1, \dots, \theta_{m-1})$ denote the sorted vector of interval bounds, where $\theta_i < \theta_{i+1}$. Then, for $k = 2$ components, the non-parametric weighting function is given by

$$\omega_{1,t}(\varepsilon_{t-1}) = \begin{cases} b_1, & \text{if } \varepsilon_{t-1} < \theta_1, \\ b_2, & \text{if } \theta_1 \leqslant \epsilon_{t-1} < \theta_2, \\ b_3, & \text{if } \theta_2 \leqslant \epsilon_{t-1} < \theta_3, \\ & \vdots \\ b_{m-1}, & \text{if } \theta_{m-2} \leqslant \epsilon_{t-1} < \theta_{m-1}, \\ b_m, & \text{if } \varepsilon_{t-1} \geqslant \theta_{m-1}, \end{cases} \tag{10.28}$$

where $\omega_{2,t}(\varepsilon_{t-1}) = 1 - \omega_{1,t}$ and $b_i \in (0, 1)$, $i = 1, \dots, m$, are parameters to be estimated. The choice of m involves the usual bias-variance tradeoff and should be chosen at least as a function of the length of the return series. For choosing the intervals in (10.28) defined by $\boldsymbol{\theta} = (\theta_1, \dots, \theta_{m-1})$, we recommend simply an equally spaced grid. The two intercepts b_1 and b_m, corresponding to the ends of the innovation range, may not be measured accurately due to the sparseness of observations in this area.

To illustrate, the left and middle panels of Figure 10.16 show the estimated weight intercepts, $\hat{b}_i$, for the TV-MixN(2)-GARCH(1,1) model with non-parametric weighting function (10.28),

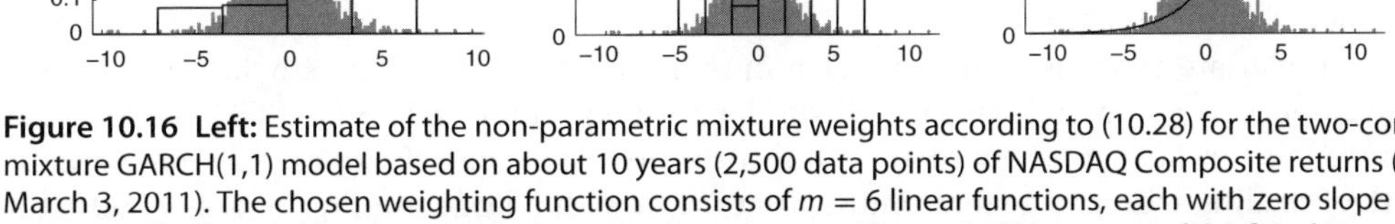

Figure 10.16 Left: Estimate of the non-parametric mixture weights according to (10.28) for the two-component mixture GARCH(1,1) model based on about 10 years (2,500 data points) of NASDAQ Composite returns (April 6, 2011 to March 3, 2011). The chosen weighting function consists of $m = 6$ linear functions, each with zero slope and estimated intercept. The first and last are very close to zero. Superimposed is a scaled histogram of the fitted innovations. **Middle:** Same but for $m = 10$ (some of estimated weights are essentially zero); ignore the y axis label with π. **Right:** The fitted sigmoidal mixing weights based on (10.26) and (10.27) with $v = w = 0$, $u = 1$, and $k = 2$, i.e., $\omega_{1t} = (1 + \exp\{-\gamma_0 - \gamma_1\epsilon_{t-1}\})^{-1}$.

for two choices of m. Indeed, the "staircase formation", as the theory predicts, is obtained and encourages the use of the sigmoidal structures. The right panel shows the estimated weight function based on (10.26) and (10.27) with $v = w = 0$, $u = 1$, and $k = 2$, i.e., TV(1)-MixN(2), with $\omega_{1t} = (1 + \exp\{-\gamma_0 - \gamma_1 \epsilon_{t-1}\})^{-1}$. As the first mixture component represents the low-volatility regime, this reveals that negative and positive shocks have an asymmetric impact on future volatility in the sense that negative news surprises increase volatility more than positive news surprises. Note that, for the original, constant-weight MixN(k) model, the graph would be a straight line.

By allowing for the asymmetric news impact curve, it is expected that the time-varying weights extension should convey significant out-of-sample forecast improvements over and above the already admirable ones delivered by the MixN(k) model. This indeed turns out to be the case, as detailed in Haas et al. (2006b, 2013), highlighting the importance of incorporating the leverage effect into the mixed normal GARCH framework.

10.6.5 Markov Switching Extension

Since the work of Hamilton (1989), the use and development of models in economics and finance based on a Markov switching structure continues to grow. See, notably, the survey of Hamilton (2008) and the various articles in the special issue dedicated to the topic in Dufrénot and Jawadi (2017).

The two major formulations in the GARCH context such that there is a fixed and finite number of states are due to Gray (1996) and Haas et al. (2004b). The latter has several advantages over the former, including (i) ease of estimation, (ii) tractability of stationarity conditions, and (iii) improved out-of-sample forecast performance. A survey of, further empirical evidence on, and details regarding the stationarity conditions and moments of the Markov-switching GARCH extension are given in Haas and Paolella (2012). This model has been implemented as the `MSGtool` toolbox for Matlab by Chuffart (2017), and as the `MSGARCH` package for R by Ardia et al. (2017b). See also Ardia et al. (2017a) on the forecasting performance of Markov-switching GARCH models using a variety of asset classes.

A yet further richer dynamic structure is possible by allowing for an infinite number of states; see the detailed developments and encouraging results in Dufays (2016) and Shi and Song (2016).

10.6.6 Multivariate Extensions

Another possible generalization is to extend the model to the multivariate framework. This was done by Bauwens et al. (2007) and Haas et al. (2009), the latter model allowing for asymmetries. Both show that the multivariate model clearly identifies the existence of two components with distinctly different volatility dynamics, and that the low (high) volatility component is associated with positive (negative) means, implying that the low and high volatility components can be interpreted as bull and bear markets, respectively. In the asymmetric multivariate mixture model, a leverage effect is shown to be present in the high-volatility component.

While these multivariate extensions are very rich in their ability to model the stochastic behavior of multivariate asset returns, their drawback is, similar to many multivariate GARCH models, the curse of dimensionality, rendering them inapplicable for more than a handful of assets. There are (at least) two ways of dealing with the general multivariate case with a large number of assets, but still having the benefit of discrete mixture structures. The first is to use the so-called technique of **independent components analysis** (ICA) with each *univariate* component modeled by a mixed normal (or mixed stable) GARCH model (see Broda et al., 2013) or other non-Gaussian processes, whether i.i.d. or with GARCH; see Broda and Paolella (2009a), Chen et al. (2015), Ghalanos et al. (2015), and the references therein. The second way is to forgo GARCH structures in favor of a multivariate mixed normal (or Laplace) i.i.d. model, and use short windows of estimation, and parameter shrinkage, as detailed in Paolella (2015), Gambacciani and Paolella (2017), and Chapter 14, possibly in conjunction with weighted likelihood; see Chapter 13.

A third, somewhat related, way is using **principle components analysis** (PCA) in the context of the so-called COMFORT model discussed in Section 11.2.4, which involves use of a continuous normal mixture distribution; see Paolella et al. (2018b) for details.

11

Risk Prediction and Portfolio Optimization

Building on the framework from Chapter 10, we now consider some applications of univariate GARCH modeling when working with weekly, daily, or higher frequency financial asset returns data. Section 11.1 overviews their use in conjunction with prediction of value at risk (VaR) and expected shortfall (ES), along with a description of other methods designed for that purpose. Section 11.2 scratches the surface of multivariate GARCH modeling by presenting four such methods, all of which are such that estimation is primarily based on *univariate* GARCH, thus avoiding the curse of the dimensionality issue in estimation and other problems associated with some high-dimensional (and highly parameterized) multivariate GARCH models that have been proposed. The most basic one is the constant-conditional-correlation GARCH, referred to as CCC, and its popular extension, dynamic CC, or DCC. These are used in Section 11.3 to introduce the basics of portfolio optimization, where also the so-called univariate collapsing method for portfolio allocation is discussed, along with the concept of ES span.

11.1 Value at Risk and Expected Shortfall Prediction

The value-at-risk (VaR) and expected shortfall (ES) are among the most popular tail risk measures used in quantitative risk management. For continuous random variable X with finite expected value, the ξ-level ES of X, denoted $\text{ES}(X, \xi)$, can be expressed as the tail conditional expectation

$$\text{ES}(X, \xi) = \frac{1}{\xi} \int_{-\infty}^{q_{X,\xi}} u\, f_X(u)\, du = \mathbb{E}[X \mid X \leqslant q_{X,\xi}], \quad \xi \in (0, 1), \tag{11.1}$$

where the ξ-quantile of X is denoted $q_{X,\xi}$ and is such that $\text{VaR}(X, \xi) = q_{X,\xi}$ is the ξ-level value-at-risk corresponding to one unit of investment. In some presentations, VaR and ES are the negatives of the definitions above, so that the risk measures are positive numbers. Section III.A.8 provides a discussion of several important issues concerning VaR and ES, derives the ES for several common distributions used in empirical finance, and gives a large number of references to the literature.

One of the primary uses of GARCH modeling is for generating accurate short-term predictions of tail risk measures, based often on daily (or higher frequency) financial asset returns data. The NCT-APARCH(1,1) model from Section 10.4, and the MixN(k), MixN(k,g), and TV(u)-MixN(k) models from Section 10.6, perform very well in this regard. The first of these belongs to the class of models in which a parametric non-Gaussian distribution is coupled with a GARCH-type law of motion for the scale term, for which many variations exist. That class can be extended by further

Linear Models and Time-Series Analysis: Regression, ANOVA, ARMA and GARCH, First Edition. Marc S. Paolella.

allowing for dynamics in the shape parameters of the distribution; see Hansen (1994), Gerlach et al. (2013), and Gabrielsen et al. (2015). In addition to these fully parametric specifications, there are several other methods of VaR and ES prediction for univariate financial return series that were explicitly designed for this purpose. Some of these include:

1) The weighting method of Boudoukh et al. (1998).
   This treats the returns as i.i.d., thus ignoring, among other things, the volatility clustering, but places more weight on recent returns than ones further in the past. Boudoukh et al. (1998) do this by assigning weights that sum to one and decay with a geometric rate. The VaR forecast is determined from the empirical c.d.f. of the weighted returns, i.e., the appropriate sample quantile. This method is appealing because one can argue that the recent past is more important than events further back in time for generating a forecast for one, or a small number of, periods in the future, and thus there is a preference for shorter windows. When done without weighting or accounting for GARCH effects, resulting in what is called the method of **(simple) historical simulation**, past crisis and high-volatility periods are possibly not included in the window for estimation, resulting in the risk for the next period being highly under-estimated. See the following discussion on FHS.
   The idea of weighting observations through time, or, more generally, the assumed underlying i.i.d. sequence of innovations in the likelihood of the data, to account for the fact that the observed data are not i.i.d., or, more generally, that the assumed model is mis-specified, is elaborated upon in Chapter 13.
2) The use of **filtered historical simulation**, or FHS, from Hull and White (1998) and Barone-Adesi et al. (1999, 2002).[1]
   This method fits a GARCH model to (as with all models, a past window of specified length of) the time series of returns, such as (10.2) or (10.10), to generate the deterministic GARCH forecast of the scale term, $\hat{\sigma}_{t+1}$, and also the filtered innovation sequence $\{\widehat{Z}_t\}$. A VaR forecast is then given by $\hat{\sigma}_{t+1}$ times the relevant sample quantile, say q, from the $\{\widehat{Z}_t\}$, and an ES forecast can also be generated based on the $\{\widehat{Z}_t\}$ that exceed q. Pritsker (2006) and Kuester et al. (2006) demonstrate the viability of the method, notably compared to use of (simple) historical simulation.
   While the performance of *all* empirical methods for generating a VaR forecast are dependent on the choice of window size, this is particularly acute for (simple) historical simulation because it ignores the stochastic and highly changing nature of the volatility. In particular, if the recent past is "calm" and the chosen window length is such that the most recent high volatility period is not included, then the VaR forecast will tend to be too liberal, underestimating the risk. Or, imagine the window length is such that it just covers a highly volatile period in the past. As the window is progressed, the crisis period will exit the window, and the VaR prediction drops (in magnitude) severely from one day to the next. FHS is less sensitive to this issue because of the use of a GARCH filter.
   Observe how the non-parametric bootstrap can be applied to the $\{\widehat{Z}_t\}$ and thus used to generate confidence intervals of the VaR and ES. See also Gao and Song (2008), the textbook presentations in Dowd (2005) and Christoffersen (2011), and the references therein, for further information on FHS.

1 See the associated web site http://filteredhistoricalsimulation.com/. Use of FHS is also detailed on the Matlab help page for the topic: Using Bootstrapping and Filtered Historical Simulation to Evaluate Market Risk.

3) The **EVT-GARCH** approach from McNeil and Frey (2000) and the related considerations in Chavez-Demoulin et al. (2014).
This method is similar to FHS in that it first uses a GARCH filter to (i) obtain the filtered innovations and (ii) generate a prediction of the scale σ_{t+1} based on the information set up to time t (this prediction being deterministic, recalling the discussion near the beginning of Section 10.2), and then fits a generalized Pareto distribution (GPD) to the tails of the filtered innovations (this being motivated by extreme value theory and the so-called **peaks-over-threshold** method, or POT), from which a VaR and ES forecast can be computed. See also Rocco (2014).
As with FHS, the choice of GARCH filter and also the innovations assumption for the chosen GARCH model play a role in the accuracy of the forecasts, as detailed in Kuester et al. (2006). The fact that the GPD is fit to the filtered innovations to obtain the predictive quantile, as opposed to using the fully specified parametric structure of the GARCH model with a non-Gaussian innovations assumption, would seem to imply that the choice of innovation distribution used in the GARCH filter should not play a role. In fact, according to quasi maximum likelihood theory, the choice should be Gaussian: Recall the discussion in Section 10.3.2 and the findings of Fan et al. (2014) and Anatolyev and Khrapov (2015).
However, if the (non-Gaussian)-GARCH model is just viewed as an approximate, mis-specified filter to the underlying d.g.p., then use of a flexible one accounting for all the stylized facts of the data should result in the filtered innovation sequence being closer to i.i.d. This appears to be the case, as shown in Kuester et al. (2006). The use of an GAt-APARCH(1,1) model in conjunction with the EVT method of McNeil and Frey (2000) results in excellent out-of-sample performance of VaR forecasts.
4) The robustified semiparametric GARCH method of Mancini and Trojani (2011).
This method is related to FHS and EVT-GARCH, but employs robust statistical methods for estimation of the filtered scale terms from the GARCH equation, as well as a robustified resampling scheme for the GARCH residuals that controls bootstrap instability due to outlying observations. This leads to improved VaR forecasts and also smoother prediction intervals for VaR over time.
5) Quantile regression methods, namely the so-called **CAViaR** method, initiated in Engle and Manganelli (2004).
This method is notable because it directly models the quantity of interest, using various functional forms for the VaR. One of its strengths is that use of a GARCH filter is not required, though this method does not fair as well as other methods in horse-race comparisons. Some variations of the method are proposed in Kuester et al. (2006), and an extension allowing for incorporation of implied volatility estimates is considered in Jeon and Taylor (2013). The CAViaR framework has been extended to the multivariate setting in White et al. (2015).
6) Use of conditional autoregressive logit (CARL) models, from Taylor and Yu (2016).
Several variations of the proposed CARL model class are used in Taylor and Yu (2016) for modeling and forecasting the exceedance probability, i.e., the probability that the realization at time $t+1$ exceeds a specified value, in either the left or right tail. This is the opposite of VaR prediction, which is a quantile, for a given probability. Taylor and Yu (2016) also propose a time-varying POT method building on the CARL model for VaR and ES prediction, and demonstrate its strong forecasting performance.[2]

2 The authors provide code in GAUSS; see http://onlinelibrary.wiley.com/journal/10.1111/(ISSN)1467-985X/homepage/179_4.htm.

7) Use of so-called **expectiles** and the resulting conditional autoregressive expectile (CARE) models; see Taylor (2008), Kuan et al. (2009), Gerlach and Chen (2016), Bellini and Di Bernardino (2017), and the references therein.
8) The use of the Gaussian-GARCH model (10.2) in conjunction with the bootstrap and a bias correction adjustment for improved VaR prediction; see Christoffersen and Gonçalves (2005), Giamouridis (2006), Pascual et al. (2006), and Hartz et al. (2006).
   The benefit of this method is its simplicity, given the very low number of parameters and ease of estimation of the Gaussian-GARCH model compared to more elaborate formulations, as discussed in Section 10.2. See also Chen et al. (2011a).
9) Further non-parametric methods; see Chen and Tang (2005), Cai and Wang (2008), Martins-Filho et al. (2016), Wang and Zhao (2016), and the references therein.
10) Use of realized volatility.
   Based on high-frequency intra-day data, when available, daily realized volatility can be "observed" (i.e., independent of a model and essentially error free) and then used for daily prediction purposes; see Martens (2001), Giot and Laurent (2004), Galbraith and Kisinbay (2005), Koopman et al. (2005), and the references therein. Giot and Laurent (2004) demonstrate with a variety of data sets that the method does *not* lead to improvements in forecast quality when compared to use of a skewed-t A-PARCH model for daily returns.
11) Use of implied volatility induced from option prices.
   A detailed account of volatility prediction based on option prices is given in Poon and Granger (2003). From their review, there is favorable evidence that this model class produces competitive volatility forecasts. See also Cesarone and Colucci (2016), Barone-Adesi (2016), and the references therein.

While the modeling techniques in the above list have been demonstrated to yield competitive VaR forecasts, they do not deliver an entire parametric density forecast for the future portfolio return. Having this density is of value for at least two reasons. First, interest might center not just on prediction of a particular tail risk measure, but rather on the *entire* distribution. Density forecasting has grown in importance in finance and other areas of econometrics because of its added value when working with asymmetric loss functions and non-Gaussian data; see Timmermann (2000) and Tay and Wallis (2000) for surveys, and Amisano and Giacomini (2007) for some associated tests. The second reason for preferring models that deliver an entire (parametric) density forecast is that univariate density predictions for (what turns out to be linear combinations of) individual assets can be analytically combined to yield the density of a *portfolio* of such assets, thus allowing portfolio optimization; see the discussion below in Section 11.3.

Typically, when **backtesting** a model for VaR prediction, i.e., estimating it over moving windows of a large time-series sample and computing, for each window, an h-step-ahead VaR prediction, one computes the resulting sequence of indicator functions (0 or 1) representing whether or not the actual return at time $t + h$ exceeded the forecasted VaR based on the model and the information set up to and including time t. For VaR backtesting, the nonzero components of this sequence are sometimes referred to as **(VaR) violations** or **hits**. If a nominal probability for the VaR quantile of, say, $\xi = 0.01$ is chosen, then, based on a set of w moving windows, one hopes to obtain $w/100$ hits. The resulting

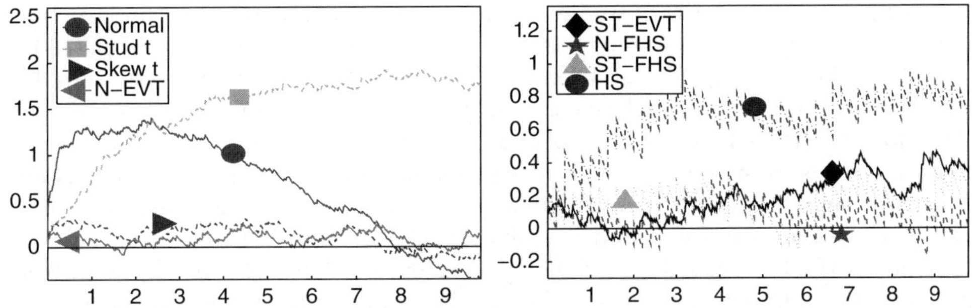

Figure 11.1 Examples of deviation plots for illustrating the unconditional coverage of VaR predictions. The *x*-axis is the VaR level (the tail probability) in percent, with 1, 2.5, and 5 being commonly checked values. The *y*-axis shows the deviation, so that a value of zero is ideal. Instead of showing tables of results for several VaR levels, such a graphic is more appealing and contains more information. The graphics are taken from Kuester et al. (2006), and pertain to VaR forecasts based on moving windows of 500 observations, from the log percentage returns of the daily closing prices of the NASDAQ composite index, from its inception on February 8, 1971, to June 22, 2001, yielding a total of 7,681 observations. The index itself is a market value-weighted portfolio of more than 5,000 stocks. The various models depicted are described in detail in Kuester et al. (2006).

number is obviously a random variable that follows a binomial distribution with parameters w and ξ under the null hypothesis that the model is correct.

The deviation of the actual number of hits compared to the expected number of $w\xi$ is a measure that, based on the $\text{Bin}(w, \xi)$ distribution, is used to assess the quality of the *unconditional* coverage probability associated with the model. While the proportion of hits can be tabulated for various models and probability levels, the use of a graphic is far more appealing and revealing. For example, Figure 11.1 shows the "deviation plots" for several VaR models and a large range of probabilities (from 0.001 to 0.10), as were initiated and used in Kuester et al. (2006). The VaR levels can be read off the horizontal axis, while the vertical axis depicts, for each VaR level, the excess of percentage violations over the VaR level. The goal here is not to compare the models for that particular data set (NASDAQ returns), but to illustrate the use of the deviation plot, and also demonstrate the varying performance of different models. The actual models used include some of the ones discussed above, and are described in detail in Kuester et al. (2006).

While the unconditional coverage is clearly very important, also of strong relevance is the *conditional* coverage, taking into account that the hits should be i.i.d. Bernoulli. For example, if a backtest based on a certain model and number of moving windows w results in precisely $w\xi$ hits, then the unconditional coverage is perfect, but if the hits tend to cluster together, then the model is clearly not generating i.i.d. realizations, and there would be predictability in them (and a sequence of severe losses close in time, which is highly undesirable for financial institutions and investors). For tests that address both unconditional and conditional coverage, see Christoffersen (1998, 2009), Haas (2005, 2009), Francioni and Herzog (2012), Abad et al. (2014), Pelletier and Wei (2016), and the numerous references therein.

Such tests are used in numerous empirical studies, as well as in comparisons of new and existing univariate and multivariate models for VaR prediction; see, e.g., Kuester et al. (2006), Bao et al. (2006, 2007), Santos et al. (2013), Slim et al. (2016), and Paolella and Polak (2017). As an example from the latter paper, Figure 11.2 shows the returns on the equally weighted portfolio through time based on the 30 stocks of the Dow Jones Industrial Index (DJIA), overlaid with the one-step-ahead VaR predictions and the realized hit sequences from several models. Use of an i.i.d. but non-Gaussian model results in both too many hits (though less than obtained with an i.i.d. Gaussian model) and also clustering of the hits, while use of the Gaussian DCC model results in too many hits, but less clustering. The use of the non-Gaussian GARCH so-called COMFORT model, discussed in 11.2.4, results in the best unconditional performance (number of hits) as well as the best conditional performance, i.e., less clustering of realized hits compared to the other models.

Backtesting the performance of the predicted ES is less straightforward, and is an ongoing research topic at the time of writing. See Section III.A.8 for references on backtesting ES amid the fact that it is not **elicitable**.

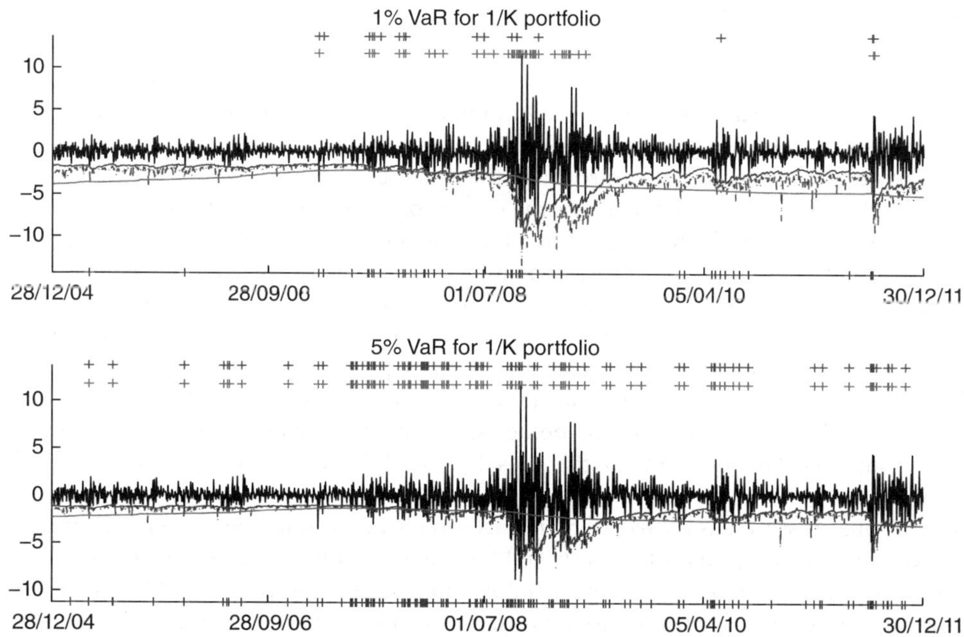

Figure 11.2 Center black lines are the returns on the equally weighted portfolio constructed from the 2,767 daily returns of $K = 30$ components of the DJIA from January 2, 2001, to December 30, 2011 (based on the index composition as of June 8, 2009). Overlaid as colored lines are the associated one-day-ahead 1% (top) and 5% (bottom) VaR forecasts, using: (i) one of the non-Gaussian GARCH COMFORT models (dashed red line), (ii) a non-Gaussian but i.i.d. model (solid magenta line), and the Gaussian DCC model (solid blue line). Further overlaid are the VaR violations, depicted by + signs on the top and bottom of the graphs, using the same color as corresponds to the lines for the VaR predictions.

11.2 MGARCH Constructs Via Univariate GARCH

11.2.1 Introduction

While direct extensions of (10.2) are possible, giving rise to various types of multivariate GARCH (hereafter MGARCH) models, the proliferation of parameters and thus the ensuing estimation problems, for even modest number of assets d, renders many such constructions virtually useless for applications in risk assessment or portfolio management (asset allocation). Several alternative formulations for MGARCH have been proposed that either substantially reduce the number of parameters that require numeric optimization, or, possibly while embodying a potentially large number of parameters, are such that the number of parameters to be *simultaneously* estimated by a generic optimization routine is very small. One fruitful and popular avenue in this latter direction is to build an MGARCH model by use of univariate GARCH models applied to each of the constituent series, sometimes referred to as **equation by equation** modeling, followed by a subsequent step that models the joint correlation structure. This estimation framework is explicitly considered in Francq and Zakoïan (2016),who prove its strong consistency and asymptotic normality in a general framework, including DCC-type models. The subsequent sections illustrate several methods of doing so, though before proceeding, we provide some important remarks.

Remarks

a) Other popular multivariate models include the so-called VEC model of Bollerslev et al. (1988); the BEKK model of Engle and Kroner (1995), so named after the authors of an earlier version, namely Baba, Engle, Kraft, and Kroner (see also Caporin and McAleer, 2008, 2012); the model of Kroner and Ng (1998), which is a weighted average of the CCC and (diagonal) BEKK models; the GARCC random coefficient model of McAleer et al. (2008), which generalizes the BEKK; the factor-GARCH models of Engle et al. (1990), Alexander and Chibumba (1996), Chan et al. (1999), Alexander (2001), Vrontos et al. (2003), and Santos and Moura (2014); and the generalized orthogonal GARCH, or GO-GARCH, models of van der Weide (2002), Lanne and Saikkonen (2007), Zhang and Chan (2009), Broda and Paolella (2009a), Boswijk and van der Weide (2011), and Ghalanos et al. (2015). The GO-GARCH construction is related to the so-called class of rotated ARCH models of Noureldin et al. (2014), which include a variant of the DCC model discussed below, and such that there are only $2d$ or even only $d + 1$ parameters requiring numeric optimization.
   A multivariate extension of the Q-GARCH model (10.11) is given in Sentana (1995), while, as mentioned above, multivariate generalizations of the univariate model in Section 10.6 have been proposed and investigated by Bauwens et al. (2007) and Haas et al. (2009). While these yield models that allow for a very rich dynamic structure, because of parameter proliferation, they are useful for only a small number of assets (though they could be used to drive the factors in a factor-GARCH setup). See the survey articles of Bauwens et al. (2006a) and Silvennoinen and Teräsvirta (2009) for discussions of many of these, and further multivariate model constructions.
b) Asymptotic properties of the variance targeting estimator (VTE) in the multivariate setting have been studied by Pedersen and Rahbek (2014) for the BEKK-GARCH model, and in Francq et al. (2016) for the CCC-GARCH model, while Burda (2015) uses covariance targeting in the general BEKK-GARCH model.

c) All multivariate time-series models share the problem that historical prices for some of the current assets of interest may not be available in the past, such as bonds with particular maturities, private equity, new public companies, merger companies, etc.; see Andersen et al. (2007, p. 515) for discussion and some resolutions to this issue. As our concern herein is on the statistical methodology, we skirt this important issue by considering only equities from major indexes (such as the components of the DJIA). We also ignore the issue of **survivorship bias**, whereby, based on the current date, we obtain past stock prices of the firms in the index, ignoring the fact that, in the past, some companies exited the index (and possibly went bankrupt), and new ones entered. See, e.g., Shumway (1997) and the references therein. This is a form of hindsight bias, and can result in analyses of model performance being exaggerated. When used for actual investment purposes, forecasting applications should attempt to incorporate the probability of bankruptcy. ∎

11.2.2 The Gaussian CCC and DCC Models

> …, joint distributions estimated over periods without panics will mis-estimate the degree of correlation between asset returns during panics. Under these circumstances, fear and disengagement by investors often result in simultaneous declines in the values of private obligations, as investors no longer realistically differentiate among degrees of risk and liquidity, and increases in the values of riskless government securities. Consequently, the benefits of portfolio diversification will tend to be overestimated when the rare panic periods are not taken into account.
>
> (Alan Greenspan, 1999)

Arguably the most popular method of generating an MGARCH model via univariate GARCH is the **constant conditional correlation**, or CCC, model of Bollerslev (1990). For each of the component series, a univariate GARCH model is fit and the filtered innovations are ordered as columns in a matrix. The sample correlation of this matrix is used to estimate the correlations. Thus, this MGARCH model fulfils the desired aspect of ease of estimation in two ways. First, with respect to the univariate GARCH models, these require only joint estimation of two (in the Gaussian IGARCH case) to five (Gaussian APARCH) parameters each, and the optimization could be parallelized across assets, for further time savings. Second, via use of the sample correlation estimator applied to the matrix of filtered innovations, this large set of correlation parameters is trivially and nearly instantaneously estimated. This assumes, however, that the correlations are constant through time. Observe that, via the time-varying volatility from the individual fitted GARCH recursions, the covariance matrix itself is changing over time. The **dynamic conditional correlation**, or DCC, model of Engle (2002, 2009), and the **varying correlation**, or VC, model of Tse and Tsui (2002), augments this basic structure with a simple, two-parameter addition that allows for motion also in the correlations, as will be shown below.

We limit our discussion herein to the Gaussian DCC model (of which CCC is a special case) and a semi-parametric variant of it. See Remark (b) below and the subsequent subsections for some discussion of the non-Gaussian CCC, DCC, and other GARCH-type model settings. Let $\mathbf{Y}_t = (Y_{t,1}, Y_{t,2}, \dots, Y_{t,d})'$ be a d-dimensional vector of asset returns, equally spaced through time (where we use the letter $\mathbf{Y}$ instead of $\mathbf{R}$ for the asset returns because $\mathbf{R}$ will be used to designate a correlation matrix, mimicking the notation in Engle, 2002). The ith univariate series, $i = 1, \dots, d$, is assumed to follow the Gaussian GARCH(1,1) model (10.2) with possibly unknown mean, given by

$$Y_{t,i} - \mu_i = Z_{t,i}\sigma_{t,i}, \quad \sigma_{t,i}^2 = c_{0,i} + c_{1,i}(Y_{t-1,i} - \mu_i)^2 + d_{1,i}\sigma_{t-1,i}^2, \quad Z_t \overset{\text{i.i.d.}}{\sim} \mathrm{N}(0,1). \tag{11.2}$$

Differing from the notation in (10.2) (and to be consistent with that used in Engle, 2002), let $\epsilon_t = Z_t$, with $\epsilon_t = (\epsilon_{t,1}, \dots, \epsilon_{t,d})'$.

We abbreviate $\mathbf{Y}_{t|\Omega_{t-1}}$, where Ω_t is, as in Section 10.2.1, the information set at time t, as just $\mathbf{Y}_{t|t-1}$. The DCC model can then be expressed as

$$\mathbf{Y}_{t|t-1} \sim \mathrm{N}_d(\boldsymbol{\mu}, \mathbf{H}_t), \quad \mathbf{H}_t = \mathbf{D}_t\mathbf{R}_t\mathbf{D}_t, \tag{11.3}$$

in conjunction with (11.2), where $\boldsymbol{\mu} = (\mu_1, \dots, \mu_d)'$, $\mathbf{D}_t^2 = \mathrm{diag}([\sigma_{t,1}^2, \dots, \sigma_{t,d}^2])$, and $\{\mathbf{R}_t\}$ the set of $d \times d$ matrices of time-varying conditional correlations with dynamics specified by

$$\mathbf{R}_t := \mathbb{E}[\epsilon_t\epsilon_t' \mid \Omega_{t-1}] = \mathrm{diag}\,(\mathbf{Q}_t)^{-1/2}\mathbf{Q}_t\,\mathrm{diag}\,(\mathbf{Q}_t)^{-1/2}, \tag{11.4}$$

$t = 1, \dots, T$. Observe that

$$\epsilon_t = \mathbf{D}_t^{-1}(\mathbf{Y}_t - \boldsymbol{\mu}). \tag{11.5}$$

The $\{\mathbf{Q}_t\}$ form a sequence of conditional matrices parameterized by

$$\mathbf{Q}_t = \mathbf{S}(1 - a - b) + a(\epsilon_{t-1}\epsilon_{t-1}') + b\mathbf{Q}_{t-1}, \tag{11.6}$$

with $\mathbf{S}$ the $d \times d$ unconditional correlation matrix (Engle, 2002, p. 341) of the ϵ_t, and parameters a and b are estimated via maximum likelihood conditional on estimates of all other parameters, as discussed next. Matrices $\mathbf{S}$ and $\mathbf{Q}_0$ can be estimated with the usual plug-in sample correlation based on the filtered ϵ_t; see also Bali and Engle (2010) and Engle and Kelly (2012) on estimation of the DCC model. Observe that the resulting $\mathbf{Q}_t$ from the update in (11.6) will not necessarily be precisely a correlation matrix; this is the reason for the standardization in (11.4). The CCC model is a special case of (11.3), with $a = b = 0$ in (11.6).

The mean vector, $\boldsymbol{\mu}$, can be set to zero (and considered to be an extreme shrinkage estimator, with target determined from the economic theory of efficient markets) as done, e.g., in Kroner and Ng (1998, Sec. 5), or estimated using the sample mean of the returns, as in Engle and Sheppard (2001) and McAleer et al. (2008). If estimation is to be used, then, in a more general non-Gaussian context, it is best estimated jointly with the other parameters associated with each univariate return series. This is particularly important amid heavy-tails, in which case the sample mean has relatively low efficiency compared to the m.l.e.; see Paolella and Polak (2017) for some details in this regard.

Let $\mathbf{Y} = [\mathbf{Y}_1, \dots, \mathbf{Y}_T]'$, and denote the set of parameters as $\boldsymbol{\theta}$. The log-likelihood of the remaining parameters, conditional on $\boldsymbol{\mu}$, is given by

$$\begin{aligned} \ell(\boldsymbol{\theta}; \mathbf{Y}, \boldsymbol{\mu}) &= -\frac{1}{2}\sum_{t=1}^{T}(d\ln(2\pi) + \ln(|\mathbf{H}_t|) + (\mathbf{Y}_t - \boldsymbol{\mu})'\mathbf{H}_t^{-1}(\mathbf{Y}_t - \boldsymbol{\mu})) \\ &= -\frac{1}{2}\sum_{t=1}^{T}(d\ln(2\pi) + 2\ln(|\mathbf{D}_t|) + \ln(|\mathbf{R}_t|) + \epsilon_t'\mathbf{R}_t^{-1}\epsilon_t). \end{aligned} \tag{11.7}$$

Then, as in Engle (2002), adding and subtracting $\epsilon_t'\epsilon_t$, $\ell = \ell(\boldsymbol{\theta}; \mathbf{Y}, \boldsymbol{\mu})$ can be decomposed as the sum of volatility and correlation terms, say $\ell = \ell_V + \ell_C$, where

$$\ell_V = -\frac{1}{2}\sum_{t=1}^{T}(d\ln(2\pi) + 2\ln(|\mathbf{D}_t|) + \epsilon_t'\epsilon_t), \tag{11.8}$$

and

$$\ell_C = -\frac{1}{2}\sum_{t=1}^{T}(\ln(|\mathbf{R}_t|) + \epsilon_t'\mathbf{R}_t^{-1}\epsilon_t - \epsilon_t'\epsilon_t). \tag{11.9}$$

In this way, a two-step maximum likelihood estimation procedure can be applied. First, estimate the GARCH model parameters for each univariate returns series as discussed in Section 10.2.3, and construct the standardized residuals. Second, maximize the conditional likelihood with respect to parameters a and b in (11.6) based on the filtered residuals from the previous step. Note that, in the CCC model, the correlation matrix is assumed to be constant over time, with $\mathbf{R}_t = \mathbf{R}$, and the standardization step in (11.4) is not necessary.

Remarks

a) See Caporin and McAleer (2013) for several critiques of the DCC construction (including the standardization step (11.4)), and Aielli (2013) for a modified DCC model, termed cDCC, with potentially better small-sample properties. Fermanian and Malongo (2017) provide conditions for the existence and uniqueness of strictly stationary solutions of the DCC model. An interesting alternative to the DCC model is discussed in Section 11.2.3.
b) One might argue that having only two parameters for modeling the evolution of an entire correlation matrix will not be adequate. While this is certainly true, the models of Engle (2002) and Tse and Tsui (2002) have two strong points: First, their use is perhaps better than no parameters (as in the CCC model) and, second, it allows for easy implementation and estimation. Matrix generalizations of the simple DCC structure that allow the number of parameters to be a function of d, and also introducing asymmetric extensions of the DCC idea, are considered in Engle (2002) and Cappiello et al. (2006), though with a potentially very large number of parameters, the usual estimation and inferential problems arise.
   Bauwens and Rombouts (2007a) consider an approach in which similar series are pooled into one of a small number of clusters, such that their GARCH parameters are the same within a cluster. A related idea is to group series with respect to their correlations, generalizing the DCC model; see, e.g., Vargas (2006), Billio et al. (2006), Zhou and Chan (2008), Billio and Caporin (2009), Engle and Kelly (2012), So and Yip (2012), Aielli and Caporin (2013), and the references therein.
   An alternative approach is to assume a Markov switching structure between two (or more) regimes, each of which has a CCC structure, as first proposed in Pelletier (2006), and augmented to the non-Gaussian case in Paolella et al. (2018a). Such a construction implies many additional parameters, but their estimation makes use of the usual sample correlation estimator, thus avoiding the curse of dimensionality, and shrinkage estimation can be straightforwardly invoked to improve performance. The idea is that, for a given time segment, the correlations are constant, and take on one set (of usually two, or at most three sets) of values. This appears to be better than attempting to construct a model that allows for their variation at every point in time. The latter, notably with the aforementioned matrix asymmetric DCC extensions, might be "asking too much of the data" and inundated with too many parameters requiring joint numeric optimization. Paolella et al. (2018a) demonstrate strong out-of-sample performance of their non-Gaussian Markov switching CCC model with two regimes, compared to the Gaussian CCC case, the Gaussian CCC switching case, the Gaussian DCC model, and the non-Gaussian single component CCC of Paolella and Polak (2015a).

c) CCC- and DCC-type MGARCH models that support non-Gaussian innovation processes have been proposed by various researchers. These include Aas et al. (2005), using the multivariate normal inverse Gaussian (NIG); Jondeau et al. (2007, Sec. 6.2) and Wu et al. (2015), using the multivariate skew-Student density; Santos et al. (2013) using a multivariate Student's t; Virbickaite et al. (2016) using a Dirichlet location-scale mixture of multivariate normals; and Paolella and Polak (2015b,c, 2017) using the multivariate generalized hyperbolic, the latter in a full maximum-likelihood framework applicable for large d because of the availability of an EM algorithm; see Section 11.2.4 below. ∎

11.2.3 Morana Semi-Parametric DCC Model

Morana (2015) proposes a variation of the DCC model that incorporates a semi-parametric aspect, and denotes it SP-DCC. See also Morana (2017) and Morana and Sbrana (2017) for further details, applications, and simulation results. Similar to (11.3), let

$$\mathbf{Y}_t = \boldsymbol{\mu} + \boldsymbol{\varepsilon}_t, \quad \text{and} \quad \boldsymbol{\varepsilon}_t = \mathbf{H}_t^{1/2}\mathbf{Z}_t, \tag{11.10}$$

where $\boldsymbol{\varepsilon}_t = (\varepsilon_{t,1}, \ldots, \varepsilon_{t,d})'$ and $\mathbf{Z}_t$ is i.i.d. with first two moments $\mathbb{E}[\mathbf{Z}_t] = \mathbf{0}_d$ and $\mathbb{V}(\mathbf{Z}_t) = \mathbf{I}_d$. Observe the difference between $\boldsymbol{\epsilon}_t$ in (11.5) as used for the DCC construction, and $\boldsymbol{\varepsilon}_t$ as used here, this arising as it was deemed desirable to keep the notations used in the original works. They are related by $\boldsymbol{\epsilon}_t = \mathbf{D}_t^{-1}\boldsymbol{\varepsilon}_t$. The mean term $\boldsymbol{\mu}$ is estimated as in the DCC model, namely via the sample averages of the return series.

Denote the (ij)th element of $\mathbf{H}_t$ by $h_{t,ij}$, $i,j = 1, \ldots, d$, and assume that the conditional variances $h_{t,i} := h_{t,ii} = \mathbb{V}(Y_{t,i} \mid \Omega_{t-1})$ respectively follow the strictly stationary GARCH(1,1) process (11.2) (using the notation of Morana, 2015, which is the same as shown in (10.3), but now for the ith series)

$$h_{t,i} = \omega_i + \alpha_i \varepsilon_{t-1,i}^2 + \beta_i h_{t-1,i}, \quad i = 1, \ldots, d. \tag{11.11}$$

Differing from the usual DCC construction, the conditional covariances are determined by use of the **polarization identity**

$$4 \cdot \text{Cov}(A, B) = \mathbb{V}(A + B) - \mathbb{V}(A - B), \tag{11.12}$$

arising from the simple fact given in (III.A.62) that $\mathbb{V}(A \pm B) = \mathbb{V}(A) + \mathbb{V}(B) \pm 2\text{Cov}(A, B)$, for any two random variables A and B with existing second moments. The off-diagonal elements of $\mathbf{H}_t$, $h_{t,ij} = \text{Cov}(Y_{t,i}, Y_{t,j} \mid \Omega_{t-1})$, can then be represented as

$$4 \cdot h_{t,ij} = \mathbb{V}_{t-1}(Y_{t,i} + Y_{t,j}) - \mathbb{V}_{t-1}(Y_{t,i} - Y_{t,j}), \quad i,j = 1, \ldots, d, \quad i \neq j, \tag{11.13}$$

where $\mathbb{V}_{t-1}(Y_{t,i})$ is shorthand for $\mathbb{V}(Y_{t,i} \mid \Omega_{t-1})$. Next, define the aggregate variables

$$Y_{t,ij}^+ := Y_{t,i} + Y_{t,j}, \quad Y_{t,ij}^- := Y_{t,i} - Y_{t,j}, \quad \varepsilon_{t,ij}^+ := \varepsilon_{t,i} + \varepsilon_{t,j}, \quad \varepsilon_{t,ij}^- := \varepsilon_{t,i} - \varepsilon_{t,j}, \tag{11.14}$$

and assume the conditional variance processes $h_{t,ij}^+ := \mathbb{V}_{t-1}(Y_{t,ij}^+)$ and $h_{t,ij}^- := \mathbb{V}_{t-1}(Y_{t,ij}^-)$ are given, respectively, by the GARCH(1,1) specifications

$$h_{t,ij}^+ = \omega_{ij}^+ + \alpha_{ij}^+ \varepsilon_{t-1,ij}^{+2} + \beta_{ij}^+ h_{t-1,ij}^+, \quad i,j = 1, \ldots d, \quad i \neq j, \tag{11.15}$$

and

$$h_{t,ij}^- = \omega_{ij}^- + \alpha_{ij}^- \varepsilon_{t-1,ij}^{-2} + \beta_{ij}^- h_{t-1,ij}^-, \quad i,j = 1, \ldots, d, \quad i \neq j. \tag{11.16}$$

By substituting (11.15) and (11.16) into (11.13), the implied parametric structure for the conditional covariance $h_{ij,t}$ can be expressed as

$$\begin{aligned} 4 \cdot h_{t,ij} &= \omega_{ij}^{+} + \alpha_{ij}^{+}\varepsilon_{t-1,ij}^{+2} + \beta_{ij}^{+}h_{t-1,ij}^{+} - \omega_{ij}^{-} - \alpha_{ij}^{-}\varepsilon_{t-1,ij}^{-2} - \beta_{ij}^{-}h_{t-1,ij}^{-} \\ &= \omega_{ij}^{+} - \omega_{ij}^{-} + \alpha_{ij}^{+}(\varepsilon_{t-1,i} + \varepsilon_{t-1,j})^2 - \alpha_{ij}^{-}(\varepsilon_{t-1,i} - \varepsilon_{t-1,j})^2 \\ &\quad + \beta_{ij}^{+}h_{t-1,ij}^{+} - \beta_{ij}^{-}h_{t-1,ij}^{-}. \end{aligned} \tag{11.17}$$

Note that, by assuming constant GARCH parameters across aggregate series, i.e., $\alpha_{ij}^{+} = \alpha_{ij}^{-} =: \alpha$ and $\beta_{ij}^{+} = \beta_{ij}^{-} =: \beta$, and rearranging (11.17) with $\omega_{ij} := (\omega_{ij}^{+} - \omega_{ij}^{-})/4$,

$$h_{t,ij} = \omega_{ij} + \alpha\varepsilon_{t-1,i}\varepsilon_{t-1,j} + \beta h_{t-1,ij},$$

showing how the SP-DCC model is more flexible than the usual DCC construct.

The log-likelihood can be expressed as in (11.7), decomposed similarly, as $\ell = \ell_V + \ell_C$, and a two-step procedure can be used. The volatility part of the likelihood is the same as in (11.8), namely (and recalling $\boldsymbol{\epsilon}_t = \mathbf{D}_t^{-1}\boldsymbol{\varepsilon}_t$)

$$\ell_V = -\frac{1}{2}\sum_{t=1}^{T}(d\ln(2\pi) + 2\ln(|\mathbf{D}_t|) + \boldsymbol{\varepsilon}_t'\mathbf{D}_t^{-1}\mathbf{D}_t^{-1}\boldsymbol{\varepsilon}_t). \tag{11.18}$$

Differing from the DCC model, SP-DCC does not maximize (11.9), but rather the sum of individual GARCH likelihoods for the aggregate series $Y_{t,ij}^{+}$ and $Y_{t,ij}^{-}$, i.e., $\ell_{\rm SP} = \ell_{\rm SP}^{+} + \ell_{\rm SP}^{-}$, where

$$\ell_{\rm SP}^{+} = -\frac{1}{2}\sum_{t=1}^{T} 2\sum_{i=1}^{d}\sum_{j>i}^{d}\left(\ln(2\pi) + \ln h_{t,ij}^{+} + \frac{\varepsilon_{t,ij}^{+2}}{h_{t,ij}^{+}}\right),$$

and similarly for $\ell_{\rm SP}^{-}$. This is jointly maximized by separately maximizing each term. Hence, the conditional variances for the aggregates $h_{t,ij}^{+}$ and $h_{t,ij}^{-}$, $i,j = 1,\dots,d$, $i \neq j$, are estimated equation by equation by means of quasi-maximum likelihood using the aggregated conditional mean residuals $\varepsilon_{t,ij}^{+}$ and $\varepsilon_{t,ij}^{-}$ from (11.14).

Through the polarization identity, the $h_{t,ij}$, $i,j = 1,\dots,d$, $i \neq j$, are estimated non-parametrically via $4 \cdot \hat{h}_{t,ij} = \hat{h}_{t,ij}^{+} - \hat{h}_{t,ij}^{-}$, $i,j = 1,\dots,d$, $i \neq j$. Finally, the estimator of the conditional correlation matrix $\mathbf{R}_t$ is given by $\widehat{\mathbf{R}}_t = \widehat{\mathbf{D}}_t^{-1}\widehat{\mathbf{H}}_t\widehat{\mathbf{D}}_t^{-1}$, where $\widehat{\mathbf{D}}_t^2 = \text{diag}([\hat{h}_{1,t},\dots,\hat{h}_{d,t}])$.

As in Morana (2015), an ex-post correction to ensure that $\widehat{\mathbf{R}}_t$ is positive definite at each point in time can be implemented as follows. First, if required, the estimated conditional correlations in $\widehat{\mathbf{R}}_t$, $\hat{\rho}_{ij,t}$, $i \neq j$, are bounded to lie within the range $-1 \leqslant \hat{\rho}_{t,ij} \leqslant 1$ by applying the so-called **sign-preserving bounding transformation**

$$\hat{\rho}_{t,ij}^{*} = \hat{\rho}_{t,ij}(1 + \hat{\rho}_{t,ij}^{k})^{-1/k}, \quad k \in \{2,4,\dots\}, \tag{11.19}$$

where k is selected optimally by minimizing the sum of squared Frobenious norms over the temporal sample

$$\arg\min_{k}\sum_{t=1}^{T}\|\widehat{\mathbf{R}}_t - \widehat{\mathbf{R}}_t^{*}\|_F^2 = \arg\min_{k}\sum_{t=1}^{T}\sum_{i=1}^{d}\sum_{j=1}^{d}|\hat{\rho}_{ij,t} - \hat{\rho}_{ij,t}^{*}|^2. \tag{11.20}$$

Second, if required, positive definiteness is enforced by means of nonlinear shrinkage of the negative eigenvalues of the $\widehat{\mathbf{R}}_t^{*}$ matrix toward their corresponding positive average values over the temporal sequence in which they are positive. Denote the spectral decomposition as $\widehat{\mathbf{R}}_t^* = \widehat{\mathbf{E}}_t \widehat{\mathbf{V}}_t \widehat{\mathbf{E}}_t'$, where $\widehat{\mathbf{V}}_t$ is the diagonal matrix of sorted eigenvalues and the columns of $\widehat{\mathbf{E}}_t$ are the associated orthogonal eigenvectors, and let $\widehat{\mathbf{V}}_t^*$ be the diagonal matrix with adjusted eigenvalues. The adjusted estimators are then

$$\widehat{\mathbf{R}}_t^{**} = \widehat{\mathbf{E}}_t \widehat{\mathbf{V}}_t^* \widehat{\mathbf{E}}_t', \quad \text{and} \quad \widehat{\mathbf{H}}_t^{**} = \widehat{\mathbf{D}}_t \widehat{\mathbf{R}}_t^{**} \widehat{\mathbf{D}}_t. \tag{11.21}$$

An implementation of the SP-DCC method is available as part of the set of programs associated with the book.[3] Function `SPDCC1step` is such that, for an input set of returns data, the mean vector and variance-covariance matrix corresponding to the one-step-ahead predictive density are output, and thus can be used for portfolio optimization, as described below in Section 11.3.2.

11.2.4 The COMFORT Class

Recall from Section 10.3.1 that, in the univariate GARCH setting, when modeling daily (or higher frequency) financial asset returns with interest centering on density or VaR forecasting, the assumption on the innovations distribution almost always plays a more important role than does the functional form of the law of motion for the scale term. It is, unsurprisingly, also the case in the multivariate setting, notably amid non-ellipticity of the returns, and with portfolio allocation applications in mind. It thus suggests itself to use a CCC or DCC structure with a non-Gaussian distribution, though this is not so trivial in terms of estimation.

A common, simple way of attempting to address this has been to employ a two-step procedure, whereby first, via an appeal to quasi maximum likelihood (recall Section 10.3.2), a Gaussian CCC or DCC model is fit to the data, and, based on the ensuing residuals, a non-Gaussian distribution, such as the multivariate Student's *t*, is fit (see, e.g., Santos et al., 2013, and the references therein). Conveniently for applied researchers, both steps are available in numerous canned econometrics packages such as Eviews. However, use of this *ad hoc* method is certainly inferior to full m.l.e., and is not obvious if the resulting parameters are consistent. Its use is also compounded by the possibility of incorrectly accounting for how the dispersion matrix in the assumed non-Gaussian multivariate distribution is estimated; see Paolella and Polak (2017) for details.

The latter authors also show that use of joint maximum likelihood estimation, enabled by use of an EM algorithm developed in Paolella and Polak (2015b) for a CCC model with a multivariate generalized hyperbolic distribution (of which Student's t is a limiting case), results in superior out-of-sample density and value-at-risk forecasting performance. It also delivers impressive portfolio performance—far better than use of Gaussian DCC; see Paolella and Polak (2015c). The price to pay for using this **common market factor non-Gaussian returns model**, or COMFORT, is having to understand a more complicated stochastic process and the required estimation technique.

The starting point of the model is the **multivariate normal mean-variance mixture distribution** or MNMVM. The d-dimensional random vector $\mathbf{Y}$ is said to have such a distribution if $\mathbf{Y} = \mathbf{m}(G) + \mathbf{H}^{1/2}\sqrt{G}\mathbf{Z}$, where $\mathbf{Z} \sim \mathrm{N}(\mathbf{0}, \mathbf{I}_d)$, where $G \geqslant 0$ is a non-negative, univariate random variable, independent of $\mathbf{Z}$, $\mathbf{H}$ is a $d \times d$ symmetric and positive definite matrix, and $\mathbf{m} : [0, \infty) \to \mathbb{R}^d$ is a measurable function. The name MNMVM comes from the fact that $\mathbf{Y} \mid (G = g) \sim \mathrm{N}(\mathbf{m}(g), g\mathbf{H})$. The multivariate generalized hyperbolic (MGHyp) distribution, as introduced by Barndorff-Nielsen (1977),

3 The author is grateful to Claudio Morana and Matthias Hartmann for supplying their original codes, and to Marco Gambacciani for adapting them for use with the profile likelihood method of univariate GARCH estimation from Section 10.2.3 and generating the one-step-ahead forecasts, as required for the predictive density.

is a special case of MNMVM with $\mathbf{m}(G) = \mu + \gamma G$, for $d \times 1$ vector $\boldsymbol{\gamma} \in \mathbb{R}^d$ and $G \sim \text{GIG}\,(\lambda, \chi, \psi)$, i.e., generalized inverse Gaussian. A highly detailed presentation in the univariate case, with links to the many special cases, and details on the GIG distribution, is given in Chapter II.9. (We will see this form again, and go into more detail, in Section 12.2, for the special case of the multivariate noncentral Student's t distribution, and further generalize the structure in Section 12.6.)

One of the benefits of use of the MGHyp for applications to portfolio optimization in finance is that, if $\mathbf{Y} \sim \text{MGHyp}(\boldsymbol{\mu}, \boldsymbol{\gamma}, \mathbf{H}, \lambda, \chi, \psi)$, where $\boldsymbol{\mu}$ is a location vector and $\mathbf{H}$ is a dispersion matrix, then the weighted sums of margins (the portfolio distribution), say $\mathbf{w}'\mathbf{Y}$, is univariate GHyp, i.e., $\mathbf{w}'\mathbf{Y} \sim \text{GHyp}(\mathbf{w}'\boldsymbol{\mu}, \mathbf{w}'\boldsymbol{\gamma}, \mathbf{w}'\mathbf{H}\mathbf{w}, \lambda, \chi, \psi)$. See, e.g., McNeil et al. (2015) for a proof. Different choices of shape parameters λ, χ, and ψ give rise to different tail behavior, from thin tails (the Gaussian and Laplace are limiting and special cases), to so-called semi-heavy tails such that the distribution is leptokurtic but still possesses a moment generating function, to genuinely heavy tailed (the Student's t being a limiting case). While the parameters of the MGHyp, notably the shape parameters λ, χ, and ψ, are identified, certain parameter restrictions are required; see McNeil et al. (2015) and Paolella and Polak (2015b) for details. Furthermore, use of all three shape parameters with typically sized data sets results in a rather flat likelihood, so one usually restricts one or two of them, giving rise to the numerous known special cases of the distribution.

The COMFORT model uses the MGHyp distribution with a CCC or DCC augmentation of the dispersion matrix. That is, for a set of d financial assets, with associated (percentage log) return vector $\mathbf{Y}_t = (Y_{t,1}, Y_{t,2}, \ldots, Y_{t,d})'$, for time $t = 1, \ldots, T$, the model is given by

$$\mathbf{Y}_t = \boldsymbol{\mu} + \boldsymbol{\gamma} G_t + \boldsymbol{\varepsilon}_t, \quad \boldsymbol{\varepsilon}_t = \mathbf{H}_t^{1/2} \sqrt{G_t} \mathbf{Z}_t, \tag{11.22}$$

where $\mathbf{H}_t = \mathbf{S}_t \boldsymbol{\Gamma}_t \mathbf{S}_t$, such that $\mathbf{S}_t = \text{diag}(s_{1,t}, \ldots, s_{d,t})$ is a scale matrix, and $s_{k,t} > 0$, $k = 1, \ldots, d$, are the scale terms driven by the modified GARCH equation dynamics

$$s_{k,t}^2 = \omega_k + \alpha_k (y_{t-1,k} - \mu_k - \gamma_k G_{t-1})^2 + \beta_k s_{k,t-1}^2. \tag{11.23}$$

The dependency matrix $\boldsymbol{\Gamma}_t$ can be assumed time invariant, $\boldsymbol{\Gamma}_t = \boldsymbol{\Gamma}$, as in the CCC model, or structured analogous to the DCC model.

As mentioned above, despite the large parameterization as d increases, estimation by full (joint with all parameters) maximum likelihood estimation is feasible and fast via use of an EM algorithm; see Paolella and Polak (2015b) for details.

This modeling paradigm turns out to yield some remarkable results and insights:

1) The required incorporation of a sequence of univariate latent (positive, continuous) random variables, denoted $\{G_t\}$, can be endowed with the interpretation as a **common market factor**, and is able to account for information arrivals and jumps in such a way that, conditional on it, the returns distribution is Gaussian. This allows for two, essentially orthogonal, structures to model the data: A univariate "jump process" for modeling aberrations and news arrivals, and a GARCH structure for modeling the persistence in volatility. Even for very large d, as is typical for portfolios of major financial institutions, all model parameters are quickly and simultaneously estimated via joint maximum likelihood, enabled by an EM algorithm. This results in the $\{G_t\}$ sequence being imputed (filtered), and it can be plotted.

   As an example, the COMFORT model was fit to 11 years of daily data consisting of the $d = 30$ stocks that comprise the DJIA index. The top panel of Figure 11.3 plots the returns for Merck & Co. Inc., with the second and third rows showing the filtered $\{G_t\}$ sequence and the filtered scaled

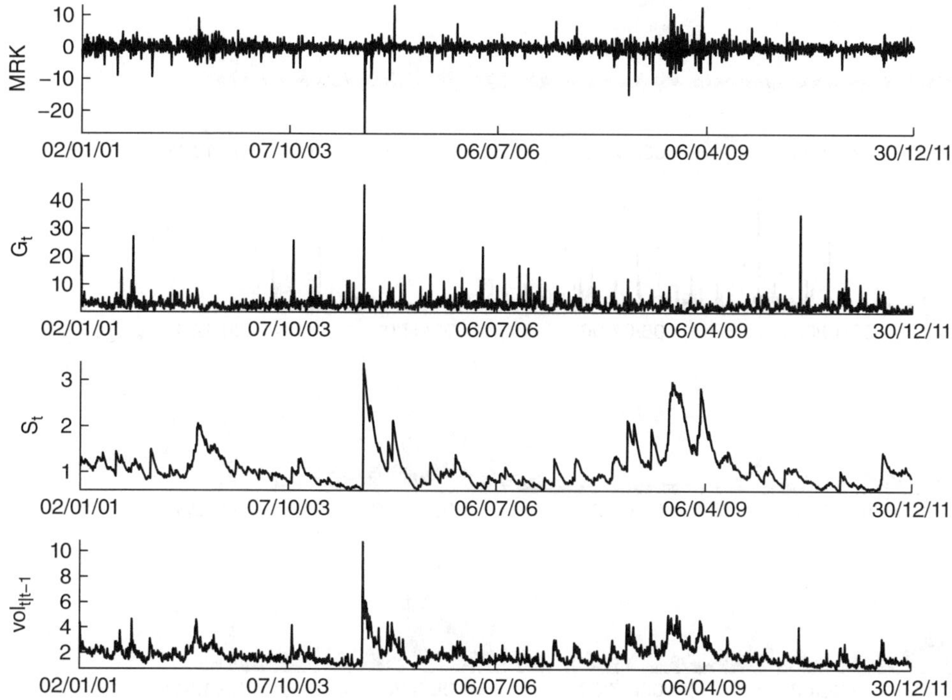

Figure 11.3 The (log percentage) returns on Merck & Co. Inc. for the dates indicated (top) and several filtered time series associated with the COMFORT model.

terms, denoted S_t, from the conditional Gaussian GARCH equation associated with Merck. The fourth panel shows the volatilities, as computed by appropriately combining the $\{G_t\}$ and $\{S_t\}$ (see Paolella and Polak, 2015b, for details). Observe how there is little relation between $\{G_t\}$ and $\{S_t\}$: As a first example, at the return below −20%, the filtered G_t value "picks this up", though (because of the bad news arrival) it was also the onset of a high volatility period, as seen in $\{S_t\}$.

As a second example, around the time of the global financial crisis in 2008 and 2009, the volatility of Merck, as seen from the returns in the top panel, is clearly relatively very high, as is, correspondingly, the $\{S_t\}$ around that period, *but the* $\{G_t\}$ *sequence is rather quiet* because, while the volatility is persistent and being adequately modeled by GARCH, there were no "major surprises" that needed to be caught by G_t. There are a handful of very large G_t "spikes" outside of the one associated with the over −20% drop, and these are not associated with any particular increase in the filtered S_t, but do influence the volatility via the combination of $\{G_t\}$ and $\{S_t\}$. The idea is that the latter two quantities are somewhat orthogonal and each is "doing its job". Without G_t, all there would be is the GARCH-induced volatility, and, from the visible enormous variation in $\{G_t\}$, it is clear that without $\{G_t\}$, the model would be rather mis-specified.

Figure 11.4 is similar, but shows all 30 series overlaid. The graphic emphasizes that $\{G_t\}$ is a univariate sequence, and also shows that the various $\{S_t\}$ are highly correlated through time, as are the COMFORT volatilities in the last panel, though they exhibit more variation than just the $\{S_t\}$ because of the influence of $\{G_t\}$.

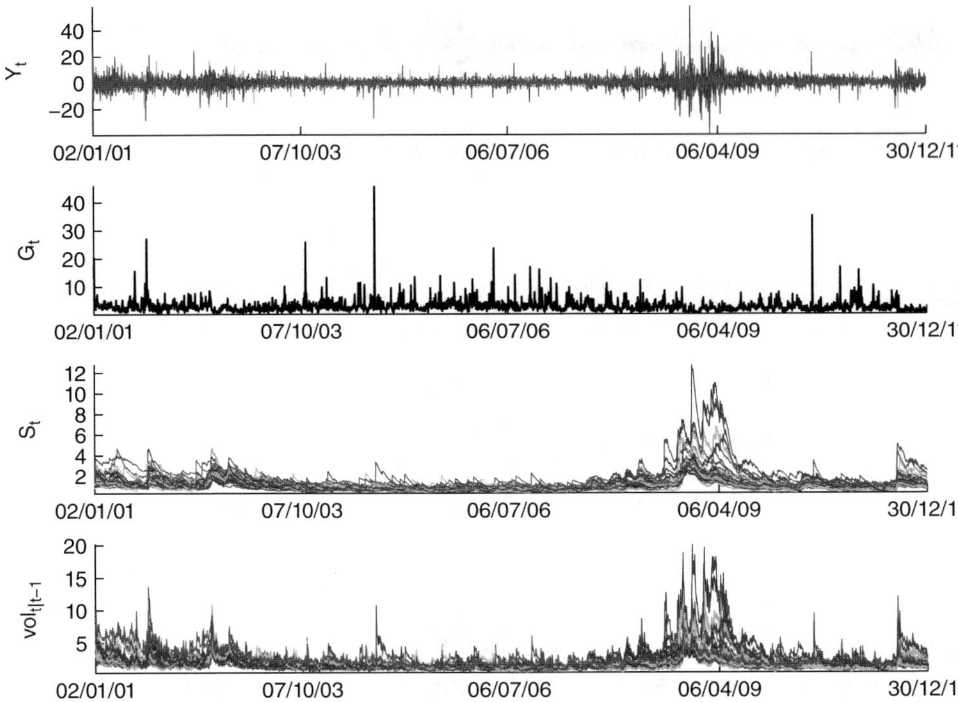

Figure 11.4 Similar to Figure 11.3, but overlaying the results for all 30 series associated with the DJIA.

2) It is worth contrasting the aforementioned *ad hoc* method of using the Gaussian CCC or DCC residuals to fit an i.i.d. multivariate Student's t distribution, with the COMFORT class of models: The former estimates the univariate series as Gaussian-GARCH, and then fits the degrees of freedom parameter of a multivariate Student's t, while COMFORT also fits univariate Gaussian-GARCH models to each margin, but *conditional* on the filtered $\{G_t\}$ sequence, in an iterative fashion, via the EM algorithm. This implies that there is feedback during estimation between the filtered $\{G_t\}$ sequence and the conditional Gaussian GARCH parameters. There is obviously no such feedback in the *ad hoc* method.
Somewhat fascinatingly, by disentangling these two effects, the estimated, conditionally Gaussian univariate GARCH processes from (11.23) yield essentially the *same parameters* across assets, i.e., the $\hat{\omega}_{0,i}$, when freely estimated for each of d assets, $i = 1, \ldots, d$, are virtually the same (and this having been confirmed by using numerous starting values in the estimation, and conducted also on numerous data sets), and similarly for $\hat{\alpha}_{1,i}$ and $\hat{\beta}_{1,i}$. This is far from the case in the Gaussian CCC or DCC setting and, thus, in the *ad hoc* Student's t DCC model. In light of the COMFORT model, this variation in GARCH parameters can be interpreted as the result of model mis-specification: The GARCH(1,1) structure is inadequate for capturing all the features of the individual series, as was discussed earlier, in Section 10.2.2, with the GARCH parameters moving towards the IGARCH border as the sample size increases.
This implies that, conditional on the latent sequence describing the common market factor, persistence in volatility, as captured by a GARCH(1,1) process, *is the same across all assets.* While of

great theoretical interest, this feature has the useful practical implication that the estimation of the univariate GARCH models (11.23) can be foregone (this being the most time-consuming part of the estimation process), replaced either by one joint estimation, or just fixing the three GARCH parameters to values that consistently arise for various sets of daily stock returns data. We term this Fast ReducEd Estimation, or FREE-COMFORT.

3) A third benefit of the COMFORT model class is that an extension to a pseudo type of **stochastic volatility** (SV) model is possible. Recalling the brief discussion at the beginning of Section 10.2.1, SV models are considered more realistic, as the volatility at time t is not simply a deterministic function of information up to time $t-1$. The price to pay for allowing a further source of randomness to enter into the volatility at time t is intractability of the likelihood and the requirement of more sophisticated methods for parameter estimation. In the COMFORT context, it was found that there is some predictability in the (otherwise i.i.d.) univariate latent sequence $\{G_t\}$, and a model extension was proposed that has some elements of an SV model, but such that the likelihood is still accessible, allowing for straightforward parameter estimation; see Paolella and Polak (2015b) for details on the formal connection to SV models, along with an application to multivariate option pricing. A different model in the univariate case that builds on the classic GARCH structure but allows for elements of an SV model and is such that it still is amenable to traditional likelihood optimization is proposed and studied in Smetanina (2017).
4) The incorporation of the univariate $\{G_t\}$ sequence allows one to move from the Gaussian CCC-GARCH model to the non-Gaussian COMFORT model, and its ensuing enhanced ability for risk assessment and portfolio allocation. The former can be thought of as the COMFORT model with constant G_t for all time periods. However, this single $\{G_t\}$ sequence is latent to all d (say) stocks, and that may not be realistic. One might argue that each industry sector should be endowed with its own such sequence. Such a construction no longer enjoys the convenient distribution theory associated with the MGHyp, and simulation from the predictive distribution would be required. Models that make use of this idea and reap benefits in terms of forecasting and portfolio construction ability are pursued in Näf et al. (2018b,a), with an introduction to the latter given in Section 12.6.

11.2.5 Copula Constructions

The development and use of copula-based models for various phenomena in finance continues to grow unabated. Part of the reason for their appeal is that they allow the (possibly time-varying, such as via GARCH) univariate series to be modeled separately, and independently of the copula structure that links them into a multivariate distribution.

Basic knowledge of copula theory, and some experience with the methodology, is now considered essential for financial econometricians. Informally, a copula is a multivariate distribution such that the univariate margins are Unif $(0, 1)$, and fully describes the dependence among the margins. The copula (as its name suggests, for those skilled in Latin) can be viewed as a structure to tie, or join, a set of univariate marginal distributions. Their use and applicability have grown in various fields, particularly quantitative risk management and empirical finance. Detailed accounts can be found in Bradley and Taqqu (2003), Nelsen (2006), McNeil et al. (2015), Joe (2015), and Ibragimov and Prokhorov (2017), while Fermanian (2017) provides (at the time of writing) a recent overview of their use in financial econometrics. Surveys of the use of copula-based models for financial time series and volatility models are given by Patton (2009), Genest et al. (2009), and Heinen and Valdesogo (2012).

We will detail the use of one particular, and very straightforward, copula construction for modeling and predicting asset returns in Chapter 12.

11.3 Introducing Portfolio Optimization

11.3.1 Some Trivial Accounting

Assume a universe of financial assets, such as currencies, commodities, or stocks, that can be traded (i.e., have the required liquidity) at the desired frequency, such as monthly, weekly, daily, intra-day, etc. For illustration, we assume daily stock trading on companies $1, \dots, d$. Further, assume there is an investment amount of capital at time t, say C_t. We want to detail the evolution of the wealth through time.

It is useful to first do so without incorporating transaction costs; these are dealt with below. Further, we assume no short-selling, which is typical in many contexts and mandatory in others. Let

$$\mathbf{w}_t = (w_{1,t}, \dots, w_{d,t})' \in [0,1]^d, \quad \sum_{k=1}^{d} w_{k,t} = 1, \tag{11.24}$$

denote the non-negative weight vector summing to one that describes a portfolio for the d stocks at time t. Observe that this is another assumption, namely that we attempt to invest all the money available. More realistic settings could allow for a "risk fear strategy" such that, based on some calculated signal at time t, the investor wishes to exit the market (anticipating perhaps very high risk or a market downturn), and thus allowing for $\sum_{k=1}^{d} w_{k,t} = 0$.

Denote the price of stock k at time t as $P_{k,t} > 0$. In what follows, we assume time is measured in some discrete unit such as hours, days, weeks, etc. For our purposes, assume daily, and when we speak of time t, one can fix this to mean, say, at 4:00PM on day t.

Based on some investment strategy (such as use of an assumed stochastic process with parameters fit from a window of past data up to time t and a prediction for time $t+1$), we decide to hold the portfolio with weights $\mathbf{w}_t$, such that $C_t w_{k,t}$ is invested in the kth asset, $k = 1, \dots, d$. This will often not be possible because of discreteness, and thus entails buying

$$\alpha_{k,t} := \left\lfloor \frac{C_t w_{k,t}}{P_{k,t}} \right\rfloor$$

stocks of company k, where $\lfloor \cdot \rfloor$ denotes the floor function, i.e., $\lfloor 2.8 \rfloor = \lfloor 2.1 \rfloor = 2$. Thus, the amount of money invested in stock for company k is $\alpha_{k,t} P_{k,t}$, which has an upper bound of $C_t w_{k,t}$ and will be close to this bound when $C_t / P_{k,t}$ is large. We denote the total amount invested (wealth) as $\mathcal{W}$; in particular, at the beginning of the investment process, this is $\mathcal{W}_1 := \mathcal{W}_{1|1}$, where we define

$$\mathcal{W}_{s|t} := \sum_{k=1}^{d} \alpha_{k,t} P_{k,s},$$

i.e., $\mathcal{W}_{s|t}$ is the *portfolio wealth* at time s, given the prices at time s and the number of shares at time t. Observe that, as the weights sum to one,

$$\mathcal{W}_{t|t} = \sum_{k=1}^{d} \alpha_{k,t} P_{k,t} \approx \sum_{k=1}^{d} C_t w_{k,t} = C_t,$$

with C_t being the upper bound on the amount invested, $\mathcal{W}_{t|t} \leqslant C_t$. When $\mathcal{W}_{t|t} = C_t$, we will call this the *full investment approximation*.

At time $t+1$, we know the prices $P_{k,t+1}$, $k = 1, \dots, d$, and the portfolio is worth

$$\mathcal{W}_{t+1|t} := \sum_{k=1}^{d} \alpha_{k,t} P_{k,t+1} = \sum_{k=1}^{d} \alpha_{k,t} \frac{P_{k,t}}{P_{k,t}} P_{k,t+1} \approx C_t \sum_{k=1}^{d} w_{k,t} \frac{P_{k,t+1}}{P_{k,t}}. \tag{11.25}$$

From the last expression in (11.25), it is clear that we wish to have chosen a zero weight in stocks such that the price change from time t to $t+1$ is negative, and ideally a weight of one in the stock whose relative price increase is the largest.

Remark Observe that (11.25) is not valid for negative weights. This is because a negative weight implies short selling, which means the stocks are borrowed at time t (for a fee, just like you pay interest when you borrow money from a bank), immediately sold, and purchased at a future date (for which you hope the price has gone down) to return to the lender. Expression (11.25) could be augmented to support short selling by taking

$$\mathcal{W}_{t+1|t} \approx C_t \sum_{k=1}^{d} |w_{k,t}| \left(\frac{P_{k,t+1}}{P_{k,t}} \right)^{\mathrm{sgn}(w_{k,t})},$$

though this will not be necessary to compute the returns below in (11.28) because there the relative price *difference* is used. ■

The percentage return of the portfolio at time $t+1$ based on starting time t, denoted $R_{P,t+1|t}$, is given by, with the full investment approximation,

$$R_{P,t+1|t} := 100 \frac{\mathcal{W}_{t+1|t} - \mathcal{W}_{t|t}}{\mathcal{W}_{t|t}} \approx 100 \frac{C_t \sum_{k=1}^{d} w_{k,t} \frac{P_{k,t+1}}{P_{k,t}} - C_t}{C_t} \tag{11.26}$$

$$= 100 \left(\sum_{k=1}^{d} w_{k,t} \frac{P_{k,t+1}}{P_{k,t}} - \frac{1}{d} \sum_{k=1}^{d} \frac{P_{k,t}}{P_{k,t}} \right) = 100 \sum_{k=1}^{d} \frac{1}{P_{k,t}} \left(w_{k,t} P_{k,t+1} - \frac{P_{k,t}}{d} \right). \tag{11.27}$$

Observe in (11.27) how, if the weights $w_{k,t}$ were chosen to be equal, i.e., the equally weighted portfolio, then (11.27) reduces to (with the full investment approximation)

$$\text{(equal weights) } R_{P,t+1|t} \approx \frac{1}{d} \sum_{k=1}^{d} 100 \left(\frac{P_{k,t+1} - P_{k,t}}{P_{k,t}} \right) = \frac{1}{d} \sum_{k=1}^{d} R_{k,t+1|t},$$

where $R_{k,t+1|t} := 100(P_{k,t+1} - P_{k,t})/P_{k,t}$ is the (simple) percentage return on asset k at time $t+1$ calculated with respect to its price at time t. As the portfolio weights sum to one, we can also write (11.26) as

$$R_{P,t+1|t} \approx 100 \left(\sum_{k=1}^{d} w_{k,t} \frac{P_{k,t+1}}{P_{k,t}} - \sum_{k=1}^{d} w_{k,t} \right) = 100 \sum_{k=1}^{d} w_{k,t} \frac{P_{k,t+1} - P_{k,t}}{P_{k,t}}$$

$$= \sum_{k=1}^{d} w_{k,t} R_{k,t+1|t} = \mathbf{w}_t' \mathbf{R}_{t+1|t}, \tag{11.28}$$

where $\mathbf{R}_{t+1|t} := (R_{1,t+1|t}, \dots, R_{d,t+1|t})'$, generalizing the equally weighted case.

Now consider the multi-step returns. We first illustrate the returns with the full investment approximation. With the new price information at time $t+1$, the weights are updated to vector $\mathbf{w}_{t+1}$, as calculated by the investment method, and the wealth that can be invested is $\mathcal{W}_{t+1|t} = C_{t+1}$. We thus buy and sell shares of the d assets such that we have $\alpha_{k,t+1} = \lfloor C_{t+1} w_{k,t+1} / P_{k,t+1} \rfloor$ shares in company k, which, under the full investment approximation, implies a wealth in company k of $\alpha_{k,t+1} P_{k,t+1} = C_{t+1} w_{k,t+1}$. At time $t+2$, the prices $P_{k,t+2}$ are realized,

$$\mathcal{W}_{t+2|t+1} = \sum_{k=1}^{d} \alpha_{k,t+1} P_{k,t+2} = \sum_{k=1}^{d} \alpha_{k,t+1} \frac{P_{k,t+1}}{P_{k,t+1}} P_{k,t+2} \approx C_{t+1} \sum_{k=1}^{d} w_{k,t+1} \frac{P_{k,t+2}}{P_{k,t+1}},$$

and

$$R_{P,t+2|t} = 100 \frac{\mathcal{W}_{t+2|t+1} - \mathcal{W}_{t|t}}{\mathcal{W}_{t|t}}. \tag{11.29}$$

Consider now the return for h periods ahead, $h \geqslant 1$, i.e., based on starting time t, we want the return at time $t+h$. For the initial purchase at time t, it is algebraically convenient to let $\mathcal{W}_{t|t-1} := \mathcal{W}_{t|t}$. From (11.29) for time $t+h$,

$$\begin{aligned}\frac{1}{100} R_{P,t+h|t} + 1 &= \frac{\mathcal{W}_{t+h|t+h-1}}{\mathcal{W}_{t|t}} = \prod_{\tau=t+1}^{t+h} \left(\frac{\mathcal{W}_{\tau|\tau-1}}{\mathcal{W}_{\tau-1|\tau-2}} \right) \\ &= \prod_{\tau=t+1}^{t+h} \left(1 + \frac{\mathcal{W}_{\tau|\tau-1} - \mathcal{W}_{\tau-1|\tau-2}}{\mathcal{W}_{\tau-1|\tau-2}} \right) = \prod_{\tau=t+1}^{t+h} \left(1 + \frac{1}{100} R_{P,\tau|\tau-1} \right),\end{aligned}$$

or

$$R_{P,t+h|t} = 100 \left(\prod_{\tau=t+1}^{t+h} \left(1 + \frac{1}{100} R_{P,\tau|\tau-1} \right) - 1 \right). \tag{11.30}$$

Recall the Taylor series $\log(1+x) = \sum_{i=1}^{\infty} (-1)^{i+1} x^i / i$ for $|x| < 1$, with first-order approximation $\log(1+x) \approx x$. Thus, return $R_{P,t+2|t}$ satisfies $R_{P,t+2|t} \approx 100 \log(\mathcal{W}_{t+2|t+1} / \mathcal{W}_{t|t})$ (see, e.g., page I.152), and

$$\begin{aligned}R_{P,t+2|t} &\approx 100 \log \left(\frac{\mathcal{W}_{t+2|t+1}}{\mathcal{W}_{t|t}} \right) \\ &= 100 \log \left(\frac{\mathcal{W}_{t+2|t+1}}{\mathcal{W}_{t+1|t}} \frac{\mathcal{W}_{t+1|t}}{\mathcal{W}_{t|t}} \right) = 100 \log \left(\frac{\mathcal{W}_{t+2|t+1}}{\mathcal{W}_{t+1|t}} \right) + 100 \log \left(\frac{\mathcal{W}_{t+1|t}}{\mathcal{W}_{t|t}} \right) \\ &= R_{P,t+2|t+1} + R_{P,t+1|t} = \mathbf{w}'_{t+1} \mathbf{R}_{t+2|t+1} + \mathbf{w}'_t \mathbf{R}_{t+1|t}.\end{aligned}$$

In general, at time $t+h$, using both the log and full investment approximations,

$$R_{P,t+h|t} \approx \sum_{i=1}^{h} \mathbf{w}'_{t+i-1} \mathbf{R}_{t+i|t+i-1}. \tag{11.31}$$

This is the commonly used approximation for illustrating and comparing the performance of investment methods. It is conservative compared to (11.30) because $\log(1+x) \leqslant x$ for $|x| < 1$ (see, e.g., Lang, 1997, p. 87). The difference between (11.30) and (11.31) can also be exemplified as follows. For $h = 3$, (11.30) is

$$\frac{1}{100} R_{P,t+3|t} + 1 = \left(1 + \frac{1}{100} R_{P,t+1|t} \right) \left(1 + \frac{1}{100} R_{P,t+2|t+1} \right) \left(1 + \frac{1}{100} R_{P,t+3|t+2} \right)$$

or, multiplying out,

$$\begin{aligned} R_{P,t+3|t} &= R_{P,t+1|t} + R_{P,t+2|t+1} + R_{P,t+3|t+2} \\ &\quad + \frac{1}{100}(R_{P,t+1|t}R_{P,t+2|t+1} + R_{P,t+1|t}R_{P,t+3|t+2} + R_{P,t+2|t+1}R_{P,t+3|t+2}) \\ &\quad + \frac{1}{100^2}R_{P,t+1|t}R_{P,t+2|t+1}R_{P,t+3|t+2}. \end{aligned} \tag{11.32}$$

Thus, (11.31) ignores the higher-order terms in (11.32), which are clearly very small, but become visible over long investment horizons.

Now consider relaxing the full investment approximation. Assume the investor starts with capital C_t and invests in portfolio $\sum_{k=1}^{d} \alpha_{k,t}P_{k,t} \leqslant C_t$, and let the "savings" be that amount that cannot be invested because of the discreteness, i.e.,

$$S_t := C_t - \sum_{k=1}^{d} \alpha_{k,t}P_{k,t}.$$

At time $t+1$, the portfolio is worth $\mathcal{W}_{t+1|t} = \sum_{k=1}^{d} \alpha_{k,t}P_{k,t+1}$ and imagine the investor sells everything, obtaining $C_{t+1} = S_t + \mathcal{W}_{t+1|t}$. (We also assume the interest on S_t from time period t to $t+1$ is zero, which currently is not so unrealistic, though is easily accommodated.) She then purchases the portfolio with $\alpha_{k,t+1}$ shares from company k, at price $P_{k,t+1}$, $k = 1, \dots, d$, where $\alpha_{k,t+1} := \lfloor C_{t+1}w_{k,t+1}/P_{k,t+1} \rfloor$. Naturally, in practice, one does not sell everything and then repurchase the new portfolio because of transaction costs and the fact that there will be overlap between the two portfolios. Instead, one buys or sells the shares of company k to adjust $\alpha_{k,t}$ to $\alpha_{k,t+1}$, $k = 1, \dots, d$. Without transaction costs and assuming a zero bid-ask spread, this is equivalent to imagining selling everything and then purchasing the updated portfolio anew.

It follows that

$$S_{t+1} = C_{t+1} - \sum_{k=1}^{d} \alpha_{k,t+1}P_{k,t+1}.$$

At time $t+2$, the portfolio is worth $\mathcal{W}_{t+2|t+1} = \sum_{k=1}^{d} \alpha_{k,t+1}P_{k,t+2}$, and selling gives

$$\begin{aligned} C_{t+2} &= S_{t+1} + \mathcal{W}_{t+2|t+1} = C_{t+1} - \sum_{k=1}^{d} \alpha_{k,t+1}P_{k,t+1} + \sum_{k=1}^{d} \alpha_{k,t+1}P_{k,t+2} \\ &= C_{t+1} + \sum_{k=1}^{d} \alpha_{k,t+1}(P_{k,t+2} - P_{k,t+1}) = S_t + \sum_{k=1}^{d} \alpha_{k,t}P_{k,t+1} + \sum_{k=1}^{d} \alpha_{k,t+1}(P_{k,t+2} - P_{k,t+1}) \\ &= C_t + \sum_{k=1}^{d} \alpha_{k,t}(P_{k,t+1} - P_{k,t}) + \sum_{k=1}^{d} \alpha_{k,t+1}(P_{k,t+2} - P_{k,t+1}). \end{aligned}$$

Continuing, she purchases $\alpha_{k,t+2}$ shares from company k, at price $P_{k,t+2}$, $k = 1, \dots, d$, and $S_{t+2} = C_{t+2} - \sum_{k=1}^{d} \alpha_{k,t+2}P_{k,t+2}$. At time $t+3$, the portfolio is worth

$$\mathcal{W}_{t+3|t+2} = \sum_{k=1}^{d} \alpha_{k,t+2}P_{k,t+3}, \quad \text{and selling gives} \quad C_{t+3} = S_{t+2} + \mathcal{W}_{t+3|t+2},$$

which reduces to

$$C_t + \sum_{k=1}^{d} \alpha_{k,t}(P_{k,t+1} - P_{k,t}) + \sum_{k=1}^{d} \alpha_{k,t+1}(P_{k,t+2} - P_{k,t+1}) + \sum_{k=1}^{d} \alpha_{k,t+2}(P_{k,t+3} - P_{k,t+2}).$$

The pattern should now be clear, so that, at period $t + h$,

$$C_{t+h} = C_t + \sum_{\tau=t}^{t+h-1} \sum_{k=1}^{d} \alpha_{k,\tau}(P_{k,\tau+1} - P_{k,\tau}), \tag{11.33}$$

and the percentage return is

$$R_{P,t+h|t} = 100\frac{C_{t+h} - C_t}{C_t}, \quad \text{or} \quad C_{t+h} = C_t\left(1 + \frac{R_{P,t+h|t}}{100}\right). \tag{11.34}$$

We now address how to account for transaction costs. To do so, we could adjust C_t at each t, but it is easier to imagine having a separate account (without interest) to pay the costs, and these costs are removed from the return calculated at time $t + i$, $i = 1, \dots, h$. To this end, let T_1 be the initial startup cost for buying the $\sum_{k=1}^{d} \alpha_{k,1}$ shares. This can be seen as imagining the existing portfolio at time $t = 0$ to consist of $\alpha_{k,0} = 0$, $k = 1, \dots, d$, and thus we take $T_{1|0} := T_1$. Let $T_{t+1|t}$ be the total induced transaction cost for buying or selling $\alpha_{k,t+1} - \alpha_{k,t}$ shares on company k, $k = 1, \dots, d$, at the price at time $t + 1$, and similarly for $T_{t+2|t+1}, \dots, T_{t+h-1|t+h-2}$.

A typical approximation uses so-called **proportional transaction costs**, and does not account for the bid-ask spread, i.e., the different buying (ask) and selling (bid) prices. This method will be subsequently discussed. Assuming the investment procedure stops at time $t + h$, all shares are sold at time $t + h$, at cost T_{t+h}, and we define $T_{t+h|t+h-1} := T_{t+h}$. The *gross* percentage return, i.e., before paying transaction costs, is (11.34), which we now denote as

$$R^G_{P,t+h|t} = 100\frac{C_{t+h} - C_t}{C_t}. \tag{11.35}$$

The *net* percentage return, meaning after transaction costs, is then

$$R^N_{P,t+h|t} = 100\frac{C_{t+h} - \sum_{i=0}^{h} T_{t+i|t+i-1} - C_t}{C_t}. \tag{11.36}$$

The *proportional transaction cost* approximation, as in DeMiguel et al. (2013), assumes that transaction costs lead to a deduction of the total portfolio return proportional to the amount of portfolio rebalancing. It is defined as

$$(100 + R^N_{P,t+h|t}) = \left(1 - \kappa \sum_{i=0}^{h} \sum_{k=1}^{d} |w_{k,t+1+i} - w_{k,(t+i)^+}|\right)(100 + R^G_{P,t+h|t}), \tag{11.37}$$

where

1) $w_{k,t}$ from (11.24) is the portfolio weight of asset k, computed at time t, held from time t to $t + 1$.
2) w_{k,t^+} is the proportion of wealth that is held in asset k at time $t + 1$ just before rebalancing the portfolio, i.e.,

$$w_{k,t^+} = \frac{\alpha_{k,t} P_{k,t+1}}{\sum_{k=1}^{d} \alpha_{k,t} P_{k,t+1}}. \tag{11.38}$$

3) $\kappa > 0$ quantifies the level of proportional transaction cost, with values of 0.005 and 0.010 (five and ten basis points, respectively) being typical.

Observe how (11.37) and (11.38) account for the change in the proportion of wealth invested in asset k due to a change in asset prices from time t to $t + 1$.

An important aspect of this method is how the equally weighted portfolio is treated. This is characterized by $w_{k,t} = 1/d$ for all assets $k = 1, \ldots, d$ and all time periods $t = 1, 2, \ldots$. When the relative prices of assets change from time t to $t + 1$, $w_{k,t^+} \neq 1/d$, and the portfolio needs to be rebalanced (and incurs transaction costs).

Observe that (11.37) utilizes the total returns, as opposed to the log returns (11.31), to guarantee proportionality of the transaction costs to the portfolio value. To see this, rewrite (11.37) using $C_{t+h} = C_t(1 + R^N_{P,t+h|t}/100)$ from the right-hand side of (11.34) to get

$$C_{t+h} = C_t \left(1 - \kappa \sum_{i=0}^{h} \sum_{k=1}^{d} |w_{k,t+1+i} - w_{k,(t+i)^+}| \right) \left(1 + \frac{R^G_{P,t+h|t}}{100} \right). \tag{11.39}$$

For $h = 1$, (11.39) appears in, among others, DeMiguel et al. (2009b). From this, the total transaction cost amount can be expressed as a fraction of the final wealth, proportional to the amount of rebalancing, i.e.,

$$C_{t+h} \kappa \sum_{i=0}^{h} \sum_{k=1}^{d} |w_{k,t+1+i} - w_{k,(t+i)^+}| = \sum_{i=0}^{h} T_{t+i|t+i-1}. \tag{11.40}$$

This approximation is implemented in the program in Listing 11.1.

```
function [pandl_net,pandl_gross] = netreturns(wmat,rmat,pmat,kap)
% Computes the portfolio returns net of transactions costs as
%    r_t =( 1 - kap sum_j=1^N |w_j,t - w_j,(t-1)+| ) * ( w_t .* r_t) where
%    w_j,(t-1)+ is the portfolio weight in asset j at time t before rebalancing;
%    w_j,t is the desired portfolio weight at time t after rebalancing;
%    kap is the proportional transaction cost;
%    w_t is the vector of portfolio weights; and
%    r_t is the vector of returns.
%    p_t is the vector of prices
pandl_gross = sum( wmat .* rmat ,2 ); pandl_net = zeros(size(pandl_gross));
pandl_net(1) = pandl_gross(1);
wmatplus =  zeros(size(wmat));
alpha = zeros(size(wmat));
for t=2:length(pandl_gross)
    alpha(t-1,:) = wmat(t-1,:)./ pmat(t-1,:);
    % without loss of generality the total wealth invested is 1
    wmatplus(t-1,:) = (alpha(t-1,:).*pmat(t,:)) ./ (sum(alpha(t-1,:).*pmat(t,:),2));
    pandl_net(t) =  ( ( 1 -  kap * sum( abs( wmat(t,:) - wmatplus(t-1,:) ) ,2 ) ) ...
        * ( 100 + pandl_gross(t) ) ) - 100;
end
```

Program Listing 11.1: Computes the returns net of transaction costs.

A further approximation involves use of only the returns on each asset. This is convenient and will often be enough. The implementation is given in Listing 11.2.

```
function [pandl_net,pandl_gross] = netreturns(wmat,rmat,kap)
pandl_gross = sum( wmat .* rmat ,2 );
pandl_net=zeros(size(pandl_gross));
pandl_net(1)=pandl_gross(1);
for t=2:length(pandl_gross)
  pandl_net(t) = ( ( ( 1 -  kap * sum( abs( wmat(t,:) - wmat(t-1,:) ) ,2 ) ) ...
    * (100 + pandl_gross(t))) - 100 );
end
```

Program Listing 11.2: Further approximation of accounting for transaction costs, requiring only the returns.

Remarks

a) To help reduce transaction costs without an appreciable effect on performance, one approach is to impose some form of constraint on the rebalancing of the portfolio weights; see DeMiguel et al. (2009a, 2014), Fan et al. (2012), Fastrich et al. (2015), and the references therein.
Another method is to "tame" the evolution of the covariance matrix, allowing for some dynamic variation, but not as much as induced by use of traditional multivariate GARCH models. One way of accomplishing this is by using principle components analysis (PCA), as investigated in Paolella et al. (2018b) (in a non-Gaussian context).
The use of PCA in this context is not new, with the seminal works being Ding (1994) and Alexander and Chibumba (1996), with subsequent embellishments by Alexander (2001, 2002, 2008). The idea is conceptually very simple: The spectral decomposition of the covariance matrix is computed and, instead of using univariate GARCH processes for all margin processes, only a small number of leading principle components of the covariance matrix are endowed with a GARCH structure (and the remaining eigenvalues are set to zero). Finally, the reader should know that the general methodology of PCA goes back to Pearson (1901) and Hotelling (1933); a good textbook starting point is Jolliffe (2002), though PCA now gains even more popularity via its applicability in machine learning, and is discussed in many such textbooks.
Other methods include shrinking the ex-post portfolio weights towards a constant target portfolio, as demonstrated in Suh (2016), and use of averaging of covariance matrix forecasts over subsequent rolling windows to stabilize portfolio weights and thus reduce transaction costs; see, e.g., Hautsch et al. (2015).
b) We are still not done—we have not accounted for paid dividends on the stocks. These can only increase returns (even after adjusting for the fact that dividends might be taxed as income), so a conservative measure of returns can ignore them. Often, one uses returns that are (split and) *dividend adjusted*, such that the dividend is added to the return, and one can proceed as above, with the returns automatically adjusted for dividends. In the case of nonzero-coupon bonds, one would need to incorporate coupon payments. ■

11.3.2 Markowitz and DCC

Consider a set of d financial assets, such as highly liquid stocks on major exchanges, for which returns are observed over a specified period of time and frequency (e.g., daily, ignoring the weekend effect for stocks), and assume (as is common in numerous real contexts) that short-selling will not be used. The set of valid portfolio weights is thus

$$\mathcal{A} = \{\mathbf{a} \in [0,1]^d \;:\; \mathbf{1}_d' \, \mathbf{a} = 1\}. \tag{11.41}$$

In the classic portfolio optimization framework going back to the seminal work of Markowitz (1952), the returns are assumed to be an i.i.d. multivariate normal process with mean $\boldsymbol{\mu}$ and variance $\boldsymbol{\Sigma}$. One wishes to determine the portfolio weight vector, say $\mathbf{a}^*$, that yields the lowest variance of the predictive portfolio percentage return at time $t+1$, given information up to time t, say $P_{t+1|t}$, conditional on its expected value being greater than some positive threshold τ_{daily}. That is,

$$\mathbf{a}^* = \arg\min_{\mathbf{a}\in\mathcal{A}} \mathbb{V}(P_{t+1|t,\mathbf{a}}) \quad \text{such that} \quad \mathbb{E}[P_{t+1|t,\mathbf{a}}] \geqslant \tau_{\text{daily}}, \tag{11.42}$$

where $\mathcal{A}$ is given in (11.41), and, with discrete compounding,

$$\tau_{\text{daily}} = 100\left(\left(1+\frac{\tau}{100}\right)^{1/250} - 1\right), \quad \tau = 100\left(\left(1+\frac{\tau_{\text{daily}}}{100}\right)^{250} - 1\right), \tag{11.43}$$

for $\tau = \tau_{\text{annual}}$ the desired annual percentage return, here calculated assuming 250 business days per year. In this i.i.d. Gaussian Markowitz setting, note that

$$\widehat{\mathbb{E}}[P_{t+1|t,\mathbf{a}}] = \mathbf{a}'\widehat{\boldsymbol{\mu}}, \qquad \widehat{\mathbb{V}}(P_{t+1|t,\mathbf{a}}) = \mathbf{a}'\widehat{\boldsymbol{\Sigma}}\mathbf{a}, \tag{11.44}$$

where $\widehat{\boldsymbol{\mu}}$ and $\widehat{\boldsymbol{\Sigma}}$ refer to the usual plug-in unbiased estimators of the Gaussian distribution parameters. When short selling is allowed in the i.i.d. Markowitz framework, there is a well-known closed-form solution to (11.42); see, e.g., Ruppert (2004, Sec. 5.5) for a clear exposition and also Matlab codes for calculating and plotting the ubiquitous efficiency frontier and calculation of the tangency portfolio. For the long-only case, numerical methods are required to determine it. This is very easy to set up in Matlab using their constrained optimization program `fmincon`, with bare-bones code given in Listing 11.3 for this i.i.d. case, as well as with using the predicted variance-covariance matrix from the DCC model.

```
function w = PortMNS(data, tau, DCC)
if nargin<3, DCC=0; end
if DCC, [mu,Sigma] = DCC1step(data); else mu = mean(data); Sigma = cov(data); end
DEDR=100*((tau/100 + 1)^(1/250)-1); feas=max(mu) <= DEDR;
if feas, w=meanvar(mu,Sigma,DEDR)'; else w=zeros(length(mu),1); end

function w = meanvar(mu, Sigma, tau)
opt=optimset('Algorithm','active-set','LargeScale', 'off','Display','off');
d=length(mu); A = -mu; B = -tau; LB = zeros(1,d); UB = ones(1,d); w0=UB/d;
Aeq = ones(1,d); Beq = 1; % sum(w) = 1
w = fmincon(@fv, w0, A, B,  Aeq, Beq, LB, UB, [], opt, Sigma);

function f = fv(w, Sigma), f = w * Sigma * w';
```

Program Listing 11.3: MNS stands for Markowitz No Short (selling). Delivers the long-only mean-variance pure Markowitz (i.i.d. model with plug-in estimators for mean and variance–covariance) portfolio weight vector or the long-only mean-variance portfolio weight vector based on the DCC density prediction of the mean and covariance matrix. It is based on a set of returns passed as `data` and for a given desired expected yearly return τ. Function `DCC1step` is from the author, and computes the predictive distribution (here, Gaussian, and thus characterized by the mean vector and covariance matrix) corresponding to the fitted DCC model of Section 11.2.2, having used the profile likelihood method of univariate Gaussian GARCH estimation discussed at the end of Section 10.2.

Remarks

a) In the more general elliptic setting, which nests the Gaussian distribution and can allow for heavy tails such as via a multivariate t distribution, (11.42) is still valid, provided that second moments exist.

b) The success of the method depends crucially on the estimates $\widehat{\boldsymbol{\mu}}$ and $\widehat{\boldsymbol{\Sigma}}$. For a particular length of data, say T, the number of parameters, and thus the estimation error, grow with the number of assets d, and is such that, for typical T, even a modest choice of d will lead to highly unreliable estimates and, thus, poor performance. This was emphasized in DeMiguel et al. (2009b), who provide an example showing that, in order to outperform the **equally weighted portfolio** (equal weights for each asset; see Section 11.3.3 below) in the case of monthly updating with 25 assets, about 3000 months (250 years) of past historical returns are required. Not only is that completely unrealistic, but this is all the more problematic if the data generating process of the returns is changing over time.

   Many studies have shown the deleterious effect of estimation error, and developed suggestions for mitigating the problem; see, e.g., Frankfurter et al. (1971), Kalymon (1971), Klein and Bawa (1976), Frost and Savarino (1986), Britten-Jones (1999), and Kolm et al. (2014), as well as Brandt (2010) for an overview. Shrinkage estimation is a key methodology in this pursuit. See, e.g., Jorion (1986), Ledoit and Wolf (2004, 2012), and Kan and Zhou (2007) for model parameter shrinkage, and DeMiguel et al. (2009a,b), Brown et al. (2013), and the references therein for portfolio weight shrinkage.

   Chen and Yuan (2016) propose to restrict the portfolio weight vector to a subspace determined by using only a subset of the spectral decomposition of the estimated covariance matrix. The idea is to offset the loss of efficiency from restricting the investment set by reduced estimation error.

   Another approach for determining the portfolio weights that avoids the pitfalls inherent in the estimation of the large number of parameters associated with first and second moments (let alone possible use of higher-order moments) is by Brandt et al. (2009), and briefly discussed in a remark below in Section 11.3.4. There, another method for avoiding the "curse of dimensionality" suited for daily (or higher frequency) data, called the univariate collapsing method, is presented.

c) One would think that use of a multivariate GARCH-type model such as CCC or DCC should result in superior portfolio performance at the daily level, given the blatant non-i.i.d. nature of the data. This is true if investment and portfolio updating can take place without transaction costs (see Section 11.3.1). As reality dictates, transaction costs are significant, particularly for individual investors, but also for financial institutions. When using GARCH-type models, the **turnover**, i.e., the sum of the absolute changes of the portfolio weights induced when re-balancing, tends to be vastly higher compared to use of an i.i.d. model. This is because of the far greater changes from period to period of the estimated covariance matrix. As such, it is often the case that an i.i.d. model can outperform the use of DCC with even modest transaction costs. ■

In the non-elliptic setting (elliptical distributions, and examples of non-elliptic ones being discussed in Section C.2), given the asymmetry of the portfolio distribution, the variance as the risk measure is not as desirable. Instead, left-tail risk measures such as value-at-risk (VaR) and expected shortfall (ES) are commonly used, and indeed lead to different allocations than with use of the variance; see, e.g., Embrechts et al. (2002) and Campbell and Kräussl (2007). In this case, the portfolio optimization problem can be expressed as follows: We want the portfolio weight vector $\mathbf{a}^*$ that yields the least expected shortfall (in magnitude—recall the ES will be negative, so, formally, we want the largest ES) of the predictive portfolio return random variable $P_{t+1|t}$, conditional on its expected value being greater than τ_{daily}, i.e.,

$$\mathbf{a}^* = \arg\min_{\mathbf{a}\in\mathcal{A}} |\text{ES}(P_{t+1|t,\mathbf{a}}, \xi)| \quad \text{such that} \quad \mathbb{E}[P_{t+1|t,\mathbf{a}}] \geqslant \tau_{\text{daily}}, \tag{11.45}$$

where ξ is a pre-specified probability associated with the ES (for which we take 0.01). This will be used in the model discussed below in Section 11.3.4 and also in Chapter 12.

11.3.3 Portfolio Optimization Using Simulation

It is useful to think about what one can do if the function `fmincon`, as invoked, for example, by the code in Listing 11.3, or *any* black-box constrained optimizer, were not available. The first reason to entertain this idea is that, in more advanced model settings, particularly for large d, the numeric optimizer could encounter an inferior local minimum of (11.42) or (11.45), as well as exhibit other numeric problems, as discussed at length in Paolella (2014, Sec. 4.2). The second reason is that, when the objective function is not smooth in the parameters (as occurs when using a model such as the one discussed below in Section 11.3.4), `fmincon` will tend to fail, as it attempts to use gradient and Hessian information; see Paolella (2014) for a demonstration. In these cases, the evolutionary optimization algorithms discussed in Section III.4.4 would seem to be appropriate, though they are still subject to the possibility of returning an inferior local minimum, as well as having relatively slow convergence properties.[4]

The issue of avoiding potential inferior local minima (whether with use of gradient/Hessian-based methods or with evolutionary algorithms) can be mitigated by ensuring that the starting value passed to the optimization method is such that the associated objective function is "close enough" to the global minimum. This can be done with simulation, and is, in fact, a viable method in and of itself for locating a suitable portfolio vector, obviating any need for (traditional or evolutionary) optimization algorithms.

The method of simulation is extraordinarily simple: We randomly draw s portfolio weights (how to do so being subsequently discussed), where s will be a function of d and the desired accuracy (and possibly also depending on features of the data and the nature of the optimum; see below). Those draws that do not match all the required constraints (such as the mean constraint in (11.42), but potentially many others, as is typical in pension funds and financial institutional investors) are deleted. From the remaining, choose the portfolio vector that most closely satisfies (11.42) or (11.45). This vector could then be used as a starting value for the optimization methods, or, if s is high enough, the optimal portfolio vector (up to simulation discrepancy) is obtained, and use of optimization algorithms can be forgone.

Note that, as $s \to \infty$, the probability of locating $\mathbf{a}^*$, if it exists (its existence depending on the choice of τ), goes to one. Observe that, by the nature of simulation-based estimation with finite s, (11.42) or (11.45) will not be exactly obtained, but only approximated. We argue that this is not a drawback: All

4 Moreover, those heuristic optimization methods, as presented in Section III.4.4, were not designed to handle general constraints. If the only constraints are box constraints, i.e., fixed bounds on one or more parameters, then a straightforward transformation, as was done in Section III.4.3.2, can be employed. In our setting, we do have the simple fixed bound of [0, 1] on each of the portfolio weights, but additionally we require that their sum is equal to one, and also that the constraint on the minimum expected return is met. The reader interested in the CMAES optimization algorithm is encouraged to explore how constraints can be embedded, possibly by appending the objective function with penalty terms to respect the desired constraints.

models are wrong w.p.1, are anyway subject to estimation error, and the portfolio delivered will depend on the chosen data set, in particular, how much past data to use and which assets to include, and, in the case of non-ellipticity, also depends on the choice of ξ (see, e.g., Rockafellar and Uryasev, 2000; Embrechts et al., 2002). As such, the method should be judged not on how well (11.42) or (11.45) can be evaluated *per se*, but rather on the out-of-sample portfolio performance, for a given model, given universe of assets, and conditional on all tuning parameters (such as τ and s).

The primary starting point for sampling portfolio weight vectors is to obtain values that are uniform on the simplex (11.41). This is achieved by taking $\mathbf{a} = (a_1, \dots, a_d)'$ to be

$$\mathbf{a} = \mathbf{U}^{(\log)}/\mathbf{1}_d'\mathbf{U}^{(\log)}, \quad \mathbf{U}^{(\log)} = (\log U_1, \dots, \log U_d)', \quad U_i \overset{\text{iid}}{\sim} \text{Unif}(0,1), \tag{11.46}$$

see, e.g., Devroye (1986). However, use of

$$\mathbf{a} = \mathbf{U}^{(q)}/\mathbf{1}_d'\mathbf{U}^{(q)}, \quad \mathbf{U}^{(q)} = (U_1^q, \dots, U_d^q)', \quad U_i \overset{\text{iid}}{\sim} \text{Unif}(0,1), \tag{11.47}$$

is valuable for exploring other parts of the parameter space. In particular, the non-uniformity corresponding to $q = 1$ is such that there is a disproportionate number of values close to the **equally weighted portfolio**. As $q \to 0$, (11.47) collapses to equal weights. This is useful, given the well-studied ability of the seemingly naive equally weighted portfolio to outperform other allocation methods, as discussed in the subsequent remark.

Remark The equally weighted (or, commonly, "$1/N$") portfolio simply takes the portfolio weights to be equal. As the weights need to sum to one, the weight of each asset is, in our notation, $1/d$. This can be seen as an extreme form of shrinkage such that the choice of portfolio weights does not depend on the data itself, but only on the number of assets.

Studies of the high performance of the equally weighted portfolio relative to classic Markowitz allocation goes back at least to Bloomfield et al. (1977), with further analysis provided by DeMiguel et al. (2009b) and Brown et al. (2013). Fugazza et al. (2015) confirm that the "startling" performance by such a naive strategy indeed holds at the monthly level, but fails to extend to longer-term horizons, when asset return predictability is taken into account. This finding is thus very relevant for mutual- and pension-fund managers, and motivates the search for techniques to improve upon the $1/N$ strategy, particularly for short-term horizons such as monthly, weekly or daily.

See also Stivers (2018) for a possible explanation of why the $1/N$ portfolio can outperform traditional mean-variance approaches for asset allocation. ■

As $q \to \infty$ in (11.47), $\mathbf{a}$ will approach a vector of all zeroes, except for a one at the position corresponding to the largest U_i. Thus, large values of q can be used for exploring what we will refer to as **corner solutions**, or allocations such that only a small number of stocks have appreciable weight, and the remaining ones have weights close to or equal to zero. Figure 11.5 illustrates these sampling methods via scatterplots of a_1 versus a_2 for $d = 3$, using $s = 1{,}000$ points.

The sampling methods can be mixed: A data-driven heuristic is developed in Paolella (2017) for determining the proportions of portfolio vectors to be generated via uniform sampling (11.46) and from (11.47) for $q = 1$ and $q = 8$, resulting in a lower total number of replications required to reach an acceptable minimum of (11.42) or (11.45).

A natural way of checking the efficacy of the simulation method is to use it when the optimal portfolio can be easily obtained, such as with the i.i.d. Markowitz setting and use of (11.42). As our sampling

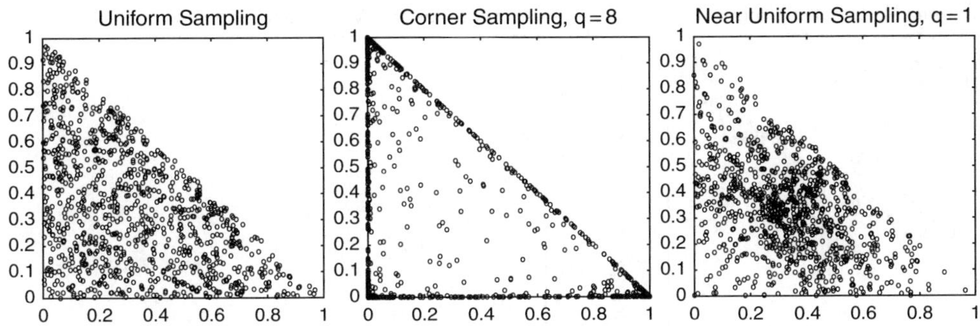

Figure 11.5 Scatterplot of the first two out of three portfolio weights, for different sampling schemes.

schemes are defined so far with only non-negative portfolio weights from (11.41) and (11.46), we use the long-only i.i.d. Markowitz setting.[5] The goal is to conduct a backtesting exercise over moving windows of stock return data, computing for each window the optimal portfolio corresponding to the i.i.d. long-only Markowitz framework, using the program in Listing 11.3, and the optimal portfolio obtained via simulation with s replications. For each of the s replications, a random portfolio vector is selected, and its mean and variance are computed from (11.44). Of these s results, those not fulfilling the mean criterion are discarded, and from the remaining, the one with the smallest variance is returned.

The idea is to repeat this exercise with ever-increasing numbers of simulated portfolios s, and confirm that, as s increases, so does the closeness between the numeric-optimized portfolio and the portfolio obtained via the simulation method. The optimal s is then chosen as the smallest value such that the results of the sampling method are "adequately close" to those obtained from the optimized Markowitz portfolio solution. An obvious measure is the average (over the moving windows) Euclidean distance between the analytical-optimized and sampling-based optimal portfolio.

For this exercise, we use the $T = 3{,}923$ daily (percentage log) returns on 29 stocks from the DJIA-30, from June 1, 1999 to December 31, 2014. (The DJIA-30 index consists of 30 stocks, but for the dates we use, the Visa company is excluded due to its late IPO in 2008.) With $\tau = 10\%$, observe that, for some time periods (for which we use moving windows of length $w = 250$), there might be no solution to the portfolio problem, and the portfolio vector consisting of all zeros is returned, i.e., trading is not conducted on that day. The left panel of Figure 11.6 shows the resulting cumulative returns, including those for the equally weighted portfolio, giving our first demonstration that the simple $1/N$ strategy can outperform the Markowitz allocation framework.

(The extent of this outperformance is understated: When transaction costs are accounted for, the $1/N$ strategy will be relatively even better, as it induces far lower turnover than other strategies.) The interested reader is encouraged to replicate these findings, as it serves as a good warmup to more advanced methods.

We see from the figure that, even for $s = 10{,}000$, the uniform sampling-based method is not able to fully reproduce the optimized portfolio vector results. They differ substantially only during periods for which the sampling method is not able to find a portfolio that satisfies the mean constraint.

5 To allow for negative weights in the sampling scheme, one could simply randomly assign a positive or negative sign to the simulated vector **a**, and then renormalize. However, in this case, each weight is no longer restricted to lie in $[0, 1]$, and some constraints would need to be imposed on the lower and upper limits.

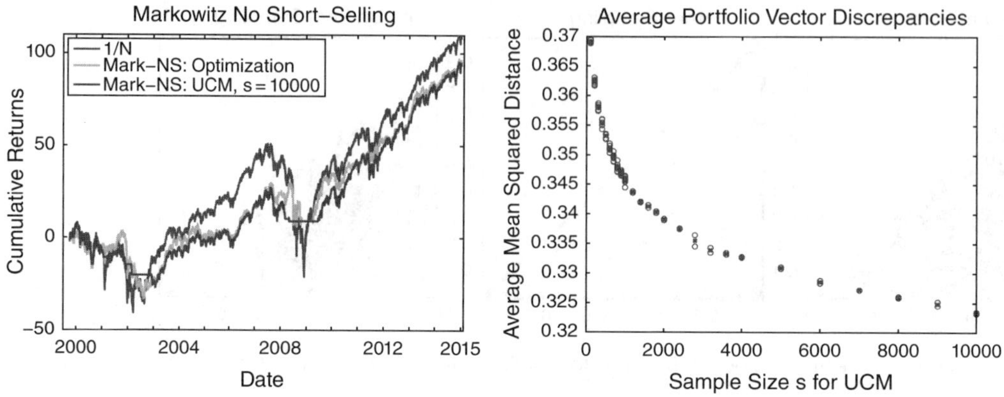

Figure 11.6 Left: Cumulative return sequences of the DJIA data using the equally weighted allocation and the Markowitz iid long-only framework (denoted Mark-NS), based on moving windows of $w = 250$ returns. **Right:** Circles indicate the average, over all the windows, of $\| \mathbf{w}^A - \mathbf{w}^U \|_2$, where $\mathbf{w}^A$ and $\mathbf{w}^U$ refer to the analytic (optimized) and UCM-based portfolio vectors, respectively. This was conducted $h = 8$ times per sample size s for $s \leqslant 1000$, and otherwise $h = 2$ times. Crosses indicate the average over the h-values.

The average portfolio vector discrepancies, as a function of s, are plotted in the right panel. The trajectory indicates that the concept works, but even $s = 10{,}000$ is not yet adequate, and that the primary issue arises during periods for which relatively few random portfolios will obtain the desired mean constraint. Based on this analysis, it is clear that brute-force sampling will not be appropriate, and more clever sampling strategies are required. Section 11.3.4 addresses this issue.

11.3.4 The Univariate Collapsing Method

With more sophisticated models and distributional assumptions, simple formulae such as those in (11.44) are often not available. We discuss a simple and general alternative method of calculating the mean, variance, and, in particular, the ES. Consider a set of d assets for which returns are observed at a particular frequency (such as daily) over a specified period of time. For a particular set of portfolio weights $\mathbf{a} = (a_1, a_2, \dots, a_d)'$ (chosen either from the simulation-based methodology or by a numeric optimizer), a univariate time series, say $\mathbf{R}_P = (R_{P,1}, R_{P,2}, \dots, R_{P,T})'$, which we will call the **constructed portfolio return series**, is computed from the d time series of past observed asset returns, $\mathbf{R}_1, \dots, \mathbf{R}_d$, as $\mathbf{R}_P = a_1\mathbf{R}_1 + \cdots + a_d\mathbf{R}_d$. In our toy example under the i.i.d. assumption, the sample mean and variance of $\mathbf{R}_P$ replaces the analytic calculation in (11.44).

The idea of the constructed portfolio return series is to look at the past returns of the portfolio dictated by weight vector **a**. These returns are “fictitious” in the sense that the particular portfolio designated by vector **a** was most likely not held by the active portfolio manager over the specified time period. It shows the returns that would have occurred if those portfolio weights were used and not changed over time. The use of the constructed portfolio series for risk *assessment*, whereby the portfolio weights are known, and only the risk of the position is required (typically VaR or ES), is within the scope of univariate GARCH modeling, and has been well-studied; see, e.g., Kuester et al. (2006) and the references therein. Its use for risk *management*, whereby active portfolio trading is engaged, is less common; see Manganelli (2004), Bauwens et al. (2006a, p. 143), Andersen et al. (2007, p. 541), Christoffersen (2009, Sec. 3), Paolella (2014), and the references therein. Our goal is to use

$\mathbf{R}_P$, in conjunction with a modeling technique and method for portfolio optimization, for (active) risk management.

Generalizing the toy example above, the—for daily returns data, rather untenable—i.i.d. assumption is replaced by assuming a GARCH-type time-series model for $\mathbf{R}_P$, and this is fit to $\{R_{P,t}\}_{t=1}^{T}$. Then, an h-step-ahead (univariate) density prediction is formed, from which any measurable quantity of interest, such as the mean, variance, VaR, and ES, can be (analytically or empirically) computed. We refer to this as the **univariate collapsing method**, or UCM. While very straightforward conceptually, the problem is the computational time required. In an MGARCH model such as DCC, estimation is performed once, the predictive mean and covariance matrix are determined, and then portfolio optimization is conducted based on the multivariate predictive density. With UCM, for every entertained portfolio vector (by simulation or an optimization algorithm), the constructed portfolio return series needs to be computed (that being computationally trivial) and, in particular, a univariate GARCH-type model needs to be estimated and, from its h-step-ahead density prediction, the mean and a risk measure (variance or VaR or ES) needs to be computed. The latter steps of GARCH model estimation and analytic computation of the ES are the severe bottlenecks of the otherwise useful and straightforward method, and partially explain why, compared to other methods, it has not received much (academic at least) attention.

One solution to this computational issue, as proposed in Paolella (2014), is to use the NCT-APARCH(1,1) model and the fast estimation technique discussed in Section 10.4. Moreover, as the predictive distribution is then NCT (with scale being determined from the APARCH update), the VaR (the left tail quantile) and the ES are also delivered instantaneously via the pre-tabulation method. This enormous gain in speed allows the performance of UCM to be investigated in backtest exercises and further developed.

Observe that, by using an asymmetric, heavy-tailed distribution as the innovation sequence of a flexible GARCH-type model that allows for asymmetric responses to the sign of the returns, UCM respects all the major *univariate* stylized facts of asset returns, as well as a *multivariate* aspect that many models do not address, namely non-ellipticity (see Section C.2), as induced, for example, by differing tail thicknesses and asymmetries across assets. This latter feature is accomplished *in an indirect way* by assuming that the conditional portfolio distribution can be adequately approximated by a noncentral Student's t distribution (NCT). If the underlying assets were to actually follow a location-scale multivariate noncentral t distribution, then their weighted convolution is also noncentral t. This motivation is unfortunately highly tempered, first by the fact that the scale terms are not constant across time, but rather exhibit strong GARCH-like behavior—and it is known that GARCH processes are not closed under summation; see, e.g., Nijman and Sentana (1996). Second, the multivariate NCT necessitates that each asset has the same tail thickness (degrees of freedom), this being precisely an assumption we wish to avoid, in light of evidence against it. Third, in addition to the fact that the underlying process generating the returns is surely not precisely a multivariate noncentral t-GARCH process, even if this were a reasonable approximation locally, it is highly debatable if the process is stationary, particularly over several years.

As such, and as also remarked by Manganelli (2004), UCM (i) relies on the fact that the pseudo-historical time series corresponding to a particular portfolio weight vector can be very well approximated by (as our choice) an NCT-APARCH process and (ii) uses shorter windows of estimation (say, 250 observations, or about one year of daily trading data) to account for the non-stationarity of the underlying process.

The primary benefit of the UCM is that it avoids the ever-increasing complexity, implementation, numerical issues, and parameter estimation inaccuracy associated with multivariate models, particularly those that support differing tail thicknesses of the assets and embody a multivariate GARCH-type structure.

With the UCM there is no need to employ formal numeric optimization methods to obtain the desired portfolio, nor optimization of model parameters associated with an elaborate multivariate model for the return process. This avoids all their associated problems, such as initial values, local maxima, convergence issues, and specification of tolerance parameters. Moreover, while a multivariate model explicitly captures features such as the (possibly time-varying) covariance matrix, this often necessitates estimation of many parameters, and the curse of model mis-specification can be magnified, as well as the curse of dimensionality, in the sense that, the more parameters there are to estimate, the larger is the magnitude of estimation error. Of course, for the latter, shrinkage estimation is a notably useful method for error reduction. However, not only is the ideal method of shrinkage not known, but even if it were, the combined effect of the two curses can be detrimental to the multivariate density forecast.

Finally, note that, with the UCM, the objective function will not be differentiable in the portfolio weights. This, however, is irrelevant when used with the simulation-based methodology for determining the optimal portfolio.

Remark Observe how portfolio optimization usually first involves obtaining the multivariate predictive distribution of the returns at the future date of interest, and then, in a second step, based on that predictive distribution, the optimal portfolio weight vector is determined. The UCM method does not make use of this two-step approach, but rather uses only univariate information of (potentially thousands) of candidate portfolio distributions to determine the optimal portfolio. The idea of avoiding the usual two-step approach is not new. For example, Brandt et al. (2009) propose a straightforward and successful method that directly models the portfolio weight of each asset as a function of the asset's characteristics, such as market capitalization, book-to-market ratio, and lagged return (momentum), as in Fama and French (1993, 1996). In doing so, and as emphasized by those authors, they avoid the large dimensionality issue of having to model first- and, notably, second-order moments (let alone third-order moments to capture higher-order effects and asymmetries, in which case, the dimensionality explodes).

Based on their suggestion of factors, the method of Brandt et al. (2009) is particularly well-suited to monthly (as they used) or lower-frequency re-balancing, such as bi-monthly, quarterly, or yearly. Our goal is higher frequency re-balancing, such as daily, in which case GARCH-type effects become highly relevant. Fletcher (2017) has independently confirmed the efficacy of the Brandt et al. (2009) approach, using the largest 350 stocks in the United Kingdom. However, he finds that (i) the performance benefits are concentrated in the earlier part of the sample period and have disappeared in recent years, and (ii) there are no performance benefits from use of the methodology based on random subsets of those 350 largest stocks. ■

With respect to the UCM, it is shown in Paolella (2017) that use of naive sampling, as described above, is problematic in the sense that s needs to be very large, and even then performance is not much better than use of the simple $1/N$ strategy. Large improvements are gained by (i) using a data-driven heuristic for mixing the sampling schemes of (11.46) and (11.47), (ii) avoiding trading if a certain fraction of sampled portfolios do not meet the mean-constraint requirement, (iii) augmenting the search for the optimal portfolio by accounting for characteristics of the individual stock returns, referred to there as the *performance ratio of individual time forecasts* or, amusingly, the PROFITS measure, and (iv) invoking a cutoff mechanism such that one of two portfolios is chosen based on the ES.

The interested reader is referred to Paolella (2017) for a detailed account of this so-called UCM(τ, s):DDS+DONT(k_C)+PROFIT(k_S, k_{CS})+τ^*(k_{ES}) method or, far more appealing, just UCM2 (perhaps standing for "You See Money Too"). The key to developing such a strategy is to avoid the pernicious trap of **backtest overfitting**, as discussed in Section 13.3 and detailed in Paolella (2017).

To avoid too many lines on the graph (especially without color), we first limit ourselves here to showing the performance of the UCM2 methodology compared to the use of the i.i.d. Markowitz, DCC-Markowitz, and equally weighted strategies, along with the method denoted MCD, based on an i.i.d. discrete two-component multivariate normal mixture model (MixN), as detailed in Chapter 14. Figure 11.7 shows the results. As the UCM is stochastic in nature, eight runs are depicted. We see that the overall best performer is the MCD method in terms of cumulative returns, and it is noteworthy that this model does not use any type of GARCH filter, but rather an i.i.d. framework based on short windows of estimation (to account for the time-varying scale) and use of shrinkage estimation.

Arguably of more interest than the cumulative returns themselves is a risk-adjusted performance measure. The most important (or at least the most common) is the **Sharpe ratio** (assuming a risk-free interest rate of zero), here computed simply as

$$\text{SR} = \frac{250\ \bar{r}}{\sqrt{250}\ \text{std}(\boldsymbol{r})}, \tag{11.48}$$

where $\boldsymbol{r} = (r_1, \dots, r_T)$ denotes the collection of observed one-step-ahead portfolio results.

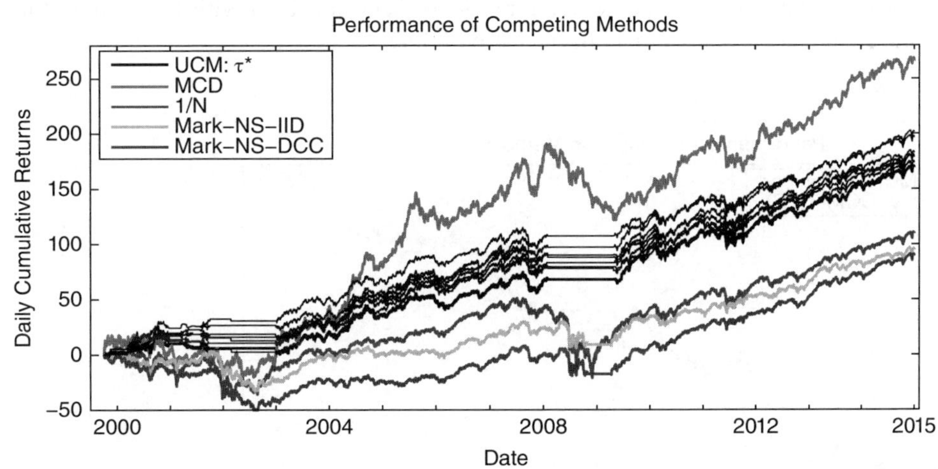

Figure 11.7 The same as the left panel of Figure 11.6, i.e., comparison of cumulative returns for the stocks listed on the DJIA, but with other methods. From top to bottom, the first is the i.i.d. two-component mixed normal distribution with parameters estimated via the MCD methodology, from Gambacciani and Paolella (2017) (in the color version, in purple). This is followed by eight runs of the UCM2 method based on 900 replications (black lines), the equally weighted method (red line), Markowitz (no short selling) based on the i.i.d. assumption (green line), and Markowitz (no short selling) but using the Gaussian DCC-GARCH model for computing the expected returns and their covariance matrix (blue line).

Remarks

a) Other, more recent measures of downside risk-adjusted return that have some advantages over the Sharpe ratio are the so-called **stable tail adjusted return ratio**, or STARR (Martin et al., 2003), and the **Sortino ratio** (Sortino and van der Meer, 1991). Gambacciani and Paolella (2017) show that these measures also favor the MCD model.

b) It is imperative to note that transaction costs were not accounted for in the comparisons illustrated in Figures 11.7 and 11.8. Besides necessarily lowering all the plotted returns (except the equally weighted, which is not affected by use of the simple proportional transaction cost approximations), it could change their relative ranking. For example, while the two Markowitz cases of i.i.d. and DCC-GARCH result in similar performance, inspection of the actual portfolio weights over time reveals that they are much more volatile for the latter, as is typical when GARCH-type filters are used, and presumably would thus induce greater transaction costs, making the DCC approach yet less competitive. ■

For $1/N$, Markowitz-IID, and Markowitz-DCC, we obtain Sharpe ratios of 0.38, 0.47, and 0.43, respectively. The mean Sharpe ratio over the eight UCM runs is 0.95, while that for the MCD method is 0.66. The reason for the superior (and quite good) performance of the UCM method in terms of Sharpe ratio is because the UCM method is unique among the methods shown in its ability to avoid trading during, and the subsequent losses associated with, crisis periods.

As such, it recommends itself to develop a similar methodology for incorporation into the MCD method. This results in the graph shown in Figure 11.8, with label "MixN-MCD-Enhanced" and results in a Sharpe ratio of 0.91. Overlaid is the analogous performance graphic using the method developed in Näf et al. (2018b), which augments the COMFORT model of Section 11.2.4 with more than one latent random variable sequence. It results in a Sharpe ratio of 0.62.

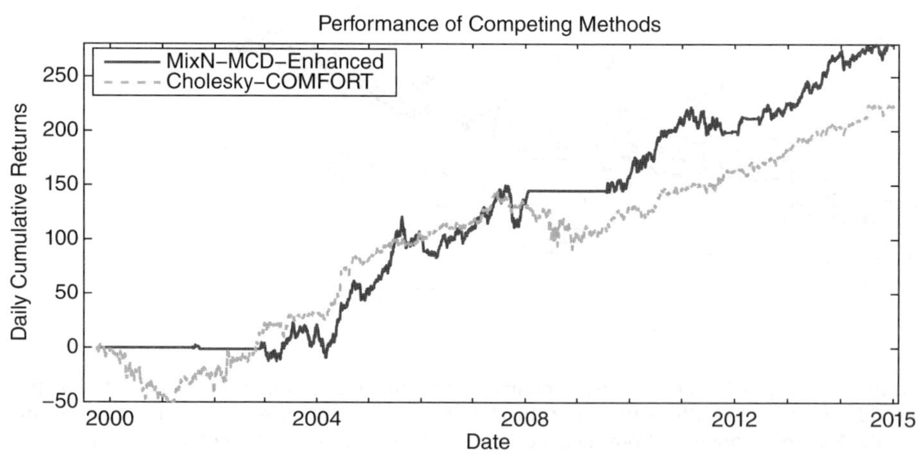

Figure 11.8 Similar to Figure 11.7 but using (i) a modified version of the mixed normal MCD method such that a signal, based on information up to time t, is used to determine if trading should take place at time $t+1$ or not, and (ii) a new variant of the COMFORT method discussed in Section 11.2.4.

11.3.5 The ES Span

A benefit of the simulation-based approach to determining the optimal portfolio weight vector, as compared to use of direct (gradient/Hessian-based, or evolutionary) optimization algorithms, is that one obtains as a by-product the so-called **ES span**, as introduced in Paolella (2014). Based on a particular time segment of returns data consisting of d assets, denoted as $\mathbf{D}$, and a specified tail probability ξ, we define the distribution of possible values that the ES can take on, over the set of all $\mathbf{a}$, when $\mathbf{a}$ is uniformly chosen over the simplex (11.41), and conditional on a chosen model $\mathcal{M}$, to be $\text{span}_{\text{ES}}(\mathbf{D}, \mathcal{M}, \xi)$. The values obtained from simulation can be plotted as a histogram, and convey knowledge of the distribution of the ES corresponding to $\mathbf{D}$ (and $\mathcal{M}$ and ξ). Observe that use of optimization algorithms gives no such information—they just return a single value (also dependent on $\mathbf{D}$, $\mathcal{M}$ and ξ), which hopefully is the global optimum.

The spread of the ES values, measured as, say, the (sample) variance or interquartile range of $\text{span}_{\text{ES}}(\mathbf{D}, \mathcal{M}, \xi)$, or other measures, such as the distance from the minimal ES value to, say, the ES corresponding to the equally weighted portfolio (possibly scaled by the sample variance or interquartile range), contain information about the relative sensitivity of risk to changes in the portfolio vector, and might be of use in a trading strategy. As an example from Paolella (2014), Figure 11.9 shows approximations of the ES span based on two models. In particular, the $\text{span}_{\text{ES}}(\mathbf{D}, \mathcal{M}, 0.01)$ is depicted as a histogram based on 100,000 replications, drawn uniformly from the simplex, for $\mathbf{D}$ being the matrix of 252 log percentage returns of the 30 stocks in the DJIA, corresponding to years 2005 (left panels) and 2008 (right panels), based on the UCM (top panels) and the i.i.d. two-component multivariate normal mixture model, fit via m.l.e. with shrinkage, as detailed in Chapter 14 (bottom panels).

By contrasting the ES span for the same data set $\mathbf{D}$ based on two models, we see the dependency of the span on the choice of model, and by how much they can differ. Unexpectedly, for both models, the ES span differs remarkably between the two years 2005 and 2008, with 2005 having been chosen because it was a relatively quiet, low-volatility, low-risk period, and 2008 being in the midst of the liquidity crisis. Indicated on the plots as long vertical lines are the minimum ES based on 100,000 simulated portfolios, and the ES corresponding to use of constrained numeric optimization.[6] Also in the plots are short vertical lines, indicating the ES corresponding to putting all the weight on a single stock. (The x-axis was truncated for readability, so that not all 30 are shown.)

Also shown is the ES corresponding to the equally weighted portfolio, denoted "$1/N$". (Further shown is the ES corresponding to the Markowitz minimum variance portfolio. This is just for curiosity: it is not directly comparable to the rest of the graphic because it allows for short selling.) Observe that the ES of the equally weighted portfolio is very close to the center of the span for 2008, for both models, while for 2005 it is much more left of center. This seems logical: The data in 2005 are much less volatile than those in 2008 (during the unfolding of the liquidity crisis) and, more relevantly, also have much thinner tails. As such, the equally weighted portfolio for 2005 should, via the central limit theorem, yield a more Gaussian-like distribution (and, thus, a lower ES) than that in 2008.

We now wish to provide more detail on the claim that the conditional tails in 2008 are thicker than those in 2005. Before proceeding, we explicitly remind the reader what we mean by the

6 The constrained minimization was based on Matlab's `fmincon` function. For the UCM, the ES, being the objective function to minimize, is "close to", but not precisely continuous in the portfolio weights, so that the reported `fmincon` result (whose algorithm requires differentiability and, thus, continuity) is based on the best result obtained from 1,000 runs using different starting values, drawn randomly and uniformly from the simplex. This obviously takes a large amount of time, and was just done for illustration. It is not feasible in practice.

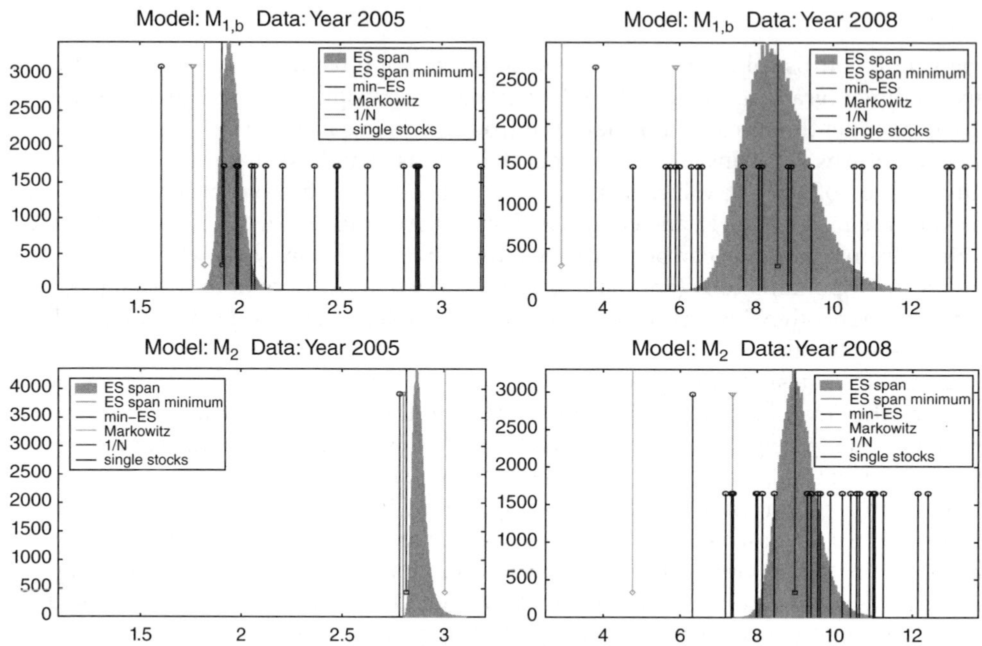

Figure 11.9 **Top:** The ES span, $\text{span}_{\text{ES}}(\mathbf{D}, \mathcal{M}, 0.01)$, based on the UCM NCT-APARCH(1,1) model of Section 11.3.4 (denoted in the title as $\mathcal{M}_{1,b}$), shown as a histogram from 100,000 random portfolio replications drawn uniformly from the simplex via (11.46), for **D** corresponding to the 252 trading days of years 2005 (left) and 2008 (right). Its minimum value is denoted by the (orange) line "ES span minimum", while the minimum ES obtained by constrained optimization (fmincon; repeated 1,000 times because of non-differentiability of the objective function) is indicated by the (red) line "min-ES". The (green) line "Markowitz" indicates the ES corresponding to the minimum variance portfolio allowing for short selling (i.e., negative portfolio weights). The short vertical (black) lines indicate the ES corresponding to putting a weight of one on a single asset (and the rest zero). The x-axis was truncated on the right to improve readability, so that some (or all, in the case of the lower left panel) of the ES values corresponding to individual assets are not shown. **Bottom:** Same as top, but based on the i.i.d. discrete two-component multivariate normal mixture model, as discussed in Chapter 14, fit via maximum likelihood with shrinkage (denoted in the title as $\mathcal{M}_2$).

just-mentioned "conditional" tails. This is conditioning on the changing volatility, captured by using a GARCH model for the time-varying scale term. The conditional distribution of asset returns is of far more relevance when interest centers on short-term forecasting and asset allocation. We estimated the NCT-APARCH model for each of the 30 series in 2005. This yields an *average* estimated degrees of freedom parameter of 15.6. The same exercise applied to the 30 stocks for the year 2008 resulted in (the shockingly low value of) 4.0. Now recall the discussions in Section III.9.1 regarding fallacious inference about the tail index when basing it on a fully specified parametric distribution. In particular, if the true underlying process is i.i.d. stable Paretian with tail index (obtained after a trivial bit of trial and error) 1.78, then the estimated tail index *under the erroneous location scale Student's t assumption* is, on average, 4.0 (and between 3.5% and 4.5 90% of the time when conducted using series of length $T = 5{,}000$). The point is: Not knowing the true distribution of the returns, it is very difficult to make reliable inference about the tail index, and the rather low values of the fitted degrees of freedom parameters in the year 2008 under the conditional (accounting for GARCH)

Student's t assumption leads us to question the existence of certainly fourth, but also third, and even second moments in many of the stocks (recall the *average* was 4.0, so approximately half the stocks have an estimated degrees of freedom below this value). It is important to emphasize that the aforementioned average degrees of freedom parameter of 4.0 is *not* referring to the unconditional estimated parametric tail indexes (which would be influenced by conditional heteroskedasticity), but rather the conditional (parametric) tail index (via the degrees of freedom parameter of the NCT) for an NCT-APARCH model, i.e., the varying volatility is accounted for.

Next, we look at the short vertical lines corresponding to investment in a single stock. For 2005, most such investments deliver a much higher ES than the minimum ES portfolio, whereas for 2008, a small number of stocks are such that their ES values are not relatively far from the minimum ES. This at first might appear to be a perverse anomaly (or a mistake), but it makes sense when we juxtapose the facts in the previous discussion about the conditional tail index, the fallacy of parametric-based tail index estimation, and remind ourselves that we cannot assume the conditional returns are literally Student's t. In particular, if one imagines a case in which most stocks have a *genuine* tail index below two (and we again emphasize that its determination is difficult and cannot be based on a parametric assumption such as Student's t or stable Paretian; recall Sections III.9.1 and III.9.2), then, as d increases and (obviously erroneously assuming they are independent), their convolution will be in the domain of attraction of a non-Gaussian stable law, and diversification may not be any better than use of a particular individual stock. Of course, stock returns are not independent; they are, unfortunately, usually all positively dependent. (One can speak of positively *correlated* if second moments exist.) If some stock returns were negatively correlated, then, clearly, diversification would help lower risk. As this is not the case, and given their very heavy tails in 2008, this explains how it can be that holding a single asset may not be much riskier than holding an equally weighted portfolio of all of them.

12

Multivariate *t* Distributions

The multivariate normal distribution is the right starting point for modeling many phenomena, though it will prove inadequate in several contexts. For daily stock returns, we are motivated to consider distributions that exhibit leptokurtosis in the univariate margins, as well as possible asymmetry. Distributions that nest the multivariate normal, or yield it as a limiting case, and possess desirable features enabling them to be of use in a variety of applications (in particular, but not only, empirical finance) include the multivariate generalized hyperbolic distribution, or MGHyp (Section 11.2.4), discrete mixtures of multivariate normals (Chapter 14), and the multivariate noncentral t, or MVNCT (Section 12.2 below). While very flexible, the MGHyp and the MVNCT are such that each univariate margin has the same tail thickness parameter. This is not appropriate in some situations, and we present distributions that allow for heterogeneous tail behavior.

We proceed as follows. The short Section 12.1, on the multivariate Student's t distribution, serves as a warmup for Section 12.2, on its noncentral extension. Sections 12.3 and 12.4 are dedicated to a group of bivariate distributions that generalize the t such that they can exhibit differing tail thickness parameters. Section 12.5 details a copula construction, referred to as the meta-elliptical t, of which we will make extensive use. Section 12.6 discusses a class of distributions formed as a so-called heterogenous multivariate normal mean-variance mixture. Some summary comments are given in Section 12.7, while Sections 12.A and 12.B are appendices that detail some further aspects of the copula construction from Section 12.5.

12.1 Multivariate Student's *t*

Our natural starting point is the d-dimensional multivariate Student's t distribution, some details of which are given in Appendix C.1. When endowed with a correlation structure, it is sometimes expressed in what we will refer to as its **canonical form**, without location and scale parameters, namely

$$f_{\mathrm{T}}(\mathbf{t};\mathbf{R},v)=\frac{\Gamma\left(\frac{v+d}{2}\right)}{\Gamma\left(\frac{v}{2}\right)(v\pi)^{d/2}|\mathbf{R}|^{1/2}}\left(1+\frac{\mathbf{t}'\mathbf{R}^{-1}\mathbf{t}}{v}\right)^{-(v+d)/2}, \tag{12.1}$$

Linear Models and Time-Series Analysis: Regression, ANOVA, ARMA and GARCH, First Edition. Marc S. Paolella.

where v is the degrees of freedom parameter, and $\mathbf{R}$ is a positive definite correlation matrix such that its (ij)th element satisfies:

$$\{\rho_{ij} : \rho_{ii} = 1, \quad -1 < \rho_{ij} < 1, \quad i \neq j, \quad \rho_{ji} = \rho_{ij}, \quad i,j = 1, \ldots, d\}. \tag{12.2}$$

To obtain the density of $\mathbf{X} \sim t_v(\boldsymbol{\mu}, \boldsymbol{\Sigma})$, let scale vector $\boldsymbol{\sigma} = (\sigma_1, \ldots, \sigma_d)' \in \mathbb{R}^d_{>0}$, and let $\boldsymbol{\Sigma} = \mathbf{SRS}$, with $\mathbf{S} = \text{diag}(\boldsymbol{\sigma})$. Observe that $\boldsymbol{\Sigma} > 0$. Then, for $\boldsymbol{\mu} = (\mu_1, \ldots, \mu_d)' \in \mathbb{R}^d$, the usual location-scale transform implies, for $\mathbf{t} = (t_1, \ldots, t_d)'$ and $t_i = (x_i - \mu_i)/\sigma_i$, $i = 1, \ldots, d$, $\mathbf{X} = \boldsymbol{\mu} + \mathbf{ST}$, with

$$f_{\mathbf{X}}(\mathbf{x}; \boldsymbol{\mu}, \boldsymbol{\Sigma}, v) = \frac{f_{\mathbf{T}}(\mathbf{t}; \mathbf{R}, v)}{\sigma_1 \sigma_2 \cdots \sigma_d}, \qquad \mathbf{R} = \mathbf{S}^{-1} \boldsymbol{\Sigma} \mathbf{S}^{-1}. \tag{12.3}$$

From (C.23), $\mathbb{E}[\mathbf{X}] = \boldsymbol{\mu}$, if $v > 1$, and $\mathbb{V}(\mathbf{X}) = \{v/(v-2)\}\boldsymbol{\Sigma}$, if $v > 2$.

It is of value to recall the mixture representation of the Student's t from (C.20). In particular, let $\mathbf{Z} \sim \mathrm{N}_d(\mathbf{0}, \mathbf{I})$, $\boldsymbol{\Sigma}$ a $d \times d$ symmetric positive definite matrix, and $G \sim \text{IGam}(v/2, v/2)$, $v > 0$. Then $\boldsymbol{\Sigma}^{1/2}\mathbf{Z} \sim \mathrm{N}_d(\mathbf{0}, \boldsymbol{\Sigma})$,

$$\mathbf{X} = \boldsymbol{\mu} + \sqrt{G}\boldsymbol{\Sigma}^{1/2}\mathbf{Z} \sim t_v(\boldsymbol{\mu}, \boldsymbol{\Sigma}), \tag{12.4}$$

and $(\mathbf{X} \mid G = g) \sim \mathrm{N}(\boldsymbol{\mu}, g\boldsymbol{\Sigma})$. For simulating from (12.1) (with extension to the location-scale case obtained in the usual way), representation (12.4) can be used. Matlab conveniently has a simulation method built in, for example in the bivariate case with the indicated parameters,

```
T=500; C = [1 0.4; 0.4 1]; df = 3; x = mvtrnd(C, df, T);
```

For general dimension d, the best way of estimating the parameters associated with (12.3) is via the EM algorithm, as described, for example, in McLachlan and Krishnan (2008). In the bivariate case, there are only six parameters, and we can use direct likelihood optimization, resulting in the program in Listing 12.1. It calls the density evaluation in Listing 12.2, which directly works with the location-scale form (12.3), though the reader is encouraged to make the minor modification to use the canonical form (12.1), which is how Matlab has it implemented in their function `mvtpdf` (and apply the location-scale transformation). The benefit of their implementation is that it accepts a $T \times d$ matrix of data, with the rows being the entire set of d-variate observations, instead of requiring T calls to the program in Listing 12.2 for each likelihood evaluation.

Example 12.1 We begin by fitting the *univariate* location-scale *noncentral* Student's t distribution, or NCT (detailed in Section III.9.3), denoted $T \sim t'(k, \theta)$, where parameters k and θ are the degrees of freedom and noncentrality (asymmetry), respectively, to each of the 30 individual series comprising the DJIA index, using the 5,037 daily returns from January 4, 1993, to December 31, 2012. The left panel of Figure 12.1 shows the estimated values of k, and their associated confidence intervals, obtained via the non-parametric bootstrap.[1] As the multivariate t assumes an equal degrees of freedom parameter for each margin, we see that this assumption is not applicable to all 30 assets, though it might be reasonable for certain subsets of them. The right panel is similar, but corresponds to parameter θ. ■

1 Not shown is the same plot but having used the confidence intervals more easily obtained as a by-product from the Hessian-based quasi-Newton optimization method and the asymptotic normality of the m.l.e. They were all shorter than their bootstrap counterparts and, of course, symmetric.

```
function [param,stderr,iters,loglik,Varcov] = MVTestimation(x)
% param:  (k, mu1, mu2, Sigma_11, Sigma_12, Sigma_22)
[nobs d]=size(x); if d~=2, error('not done yet, use EM'), end
if d==2
  %%%%%%%        k     mu1  mu2   s11    s12  s22
  bound.lo=    [  0.2 -1    -1     0.01 -90 0.01];
  bound.hi=    [ 20    1     1    90     90   90];
  bound.which= [  1    0     0     1      1    1];
  initvec      =[2    -0.8 -0.2  20      2   10];
end
maxiter=300; tol=1e-7; MaxFunEvals=length(initvec)*maxiter;
opts=optimset('Display','iter','Maxiter',maxiter,'TolFun',tol,'TolX',tol,...
    'MaxFunEvals',MaxFunEvals,'LargeScale','Off');
[pout,fval,~,theoutput,~,hess]= ...
  fminunc(@(param) MVTloglik(param,x,bound),einschrk(initvec,bound),opts);
V=inv(hess)/nobs; % Don't negate because we work with the negative of the loglik
[param,V]=einschrk(pout,bound,V); % transform and apply delta method to get V
param=param'; Varcov=V; stderr=sqrt(diag(V)); % Approximate standard errors
loglik=-fval*nobs; iters=theoutput.iterations;

function ll=MVTloglik(param,x,bound)
if nargin<3, bound=0; end
if isstruct(bound), param=einschrk(real(param),bound,999); end
[nobs d]=size(x); Sig=zeros(d,d); k=param(1); mu=param(2:3); % Assume d=2
Sig(1,1)=param(4); Sig(2,2)=param(6); Sig(1,2)=param(5); Sig(2,1)=Sig(1,2);
if min(eig(Sig))<1e-10, ll=1e5;
else
  pdf=zeros(nobs,1);
  for i=1:nobs, pdf(i) = mvtpdfmine(x(i,:),k,mu,Sig); end
  llvec=log(pdf); ll=-mean(llvec); if isinf(ll), ll=1e5; end
end
```

Program Listing 12.1: Estimates the six parameters of the bivariate location-scale Student's t.

```
function y = mvtpdfmine(x,df,mu,Sigma)
% x is a d X 1 vector. Unlike Matlab's version, cannot pass a matrix.
% Matlab's routine accepts a correlation (not dispersion) matrix.
% So, just need to do the usual scale transform. For example:
%   x=[0.2 0.3]'; C = [1 .4; .4 1]; df = 2;
%   scalevec=[1 2]'; xx=x./scalevec; mvtpdf(xx,C,df)/prod(scalevec)
% Same as:
%   Sigma = diag(scalevec) * C * diag(scalevec); mvtpdfmine(x,df,[],Sigma)
d=length(x);
if nargin<3, mu = []; end, if isempty(mu), mu = zeros(d,1); end
if nargin<4, Sigma = eye(d); end
x = reshape(x,d,1); mu = reshape(mu,d,1); term = (x-mu)' * inv(Sigma) * (x-mu);
logN=-((df+d)/2)*log(1+term/df); logD=0.5*log(det(Sigma))+(d/2)*log(df*pi);
y = exp(gammaln((df+d)/2) - gammaln(df/2) + logN - logD);
```

Program Listing 12.2: Evaluation of the location-scale d-dimensional Student's t density.

Remark Although we are focussing on the i.i.d. setting in this chapter, one might be curious about how Figure 12.1 would change if, instead of examining the shape parameters of the NCT for the unconditional returns, we do so conditioning on the GARCH filtered scale terms. As discussed in Chapter 10, once GARCH effects are removed by jointly estimating the parameters of an NCT-GARCH(1,1) model, we would expect that the estimated tail index corresponding to the NCT should increase (thinner tails). This is indeed the case, as shown in Figure 12.2. Interestingly, the asymmetry parameters are more clearly negative when based on the NCT-GARCH-filtered residuals.

It is important to emphasize that the unconditional returns are surely not i.i.d. NCT distributed—they are obviously not i.i.d., and the NCT is just an approximation. As such, the estimated degrees of freedom parameter from the NCT is *not* a measure of the supremum of the maximally existing moment of the returns; see the discussion in Section III.9.1 for further details on this important point. The NCT-GARCH(1,1) model is certainly "less mis-specified" than its i.i.d. counterpart, but is also

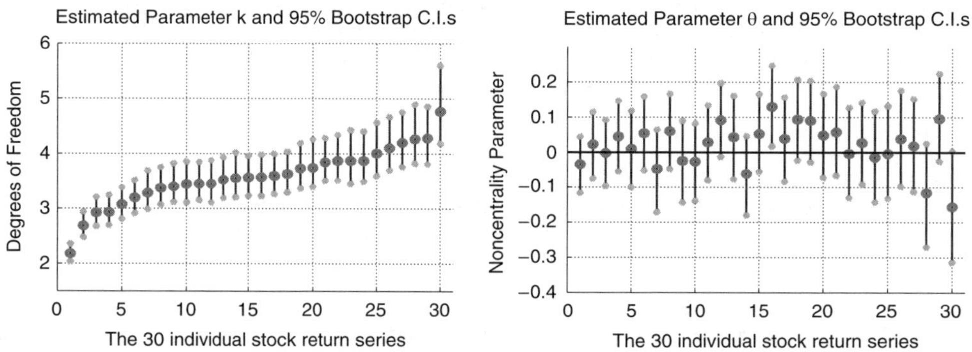

Figure 12.1 **Left:** The sorted values of the estimated tail thickness parameters (degrees of freedom) $\hat{k}$ of the noncentral Student's t distribution, for the 30 daily stock return series comprising the DJIA index, along with approximate 95% confidence intervals obtained via the non-parametric bootstrap with $B = 1{,}000$ replications. **Right:** The same as the left panel, but for the noncentrality parameter θ. *The ordering is the same as in the left panel*, thus allowing a comparison.

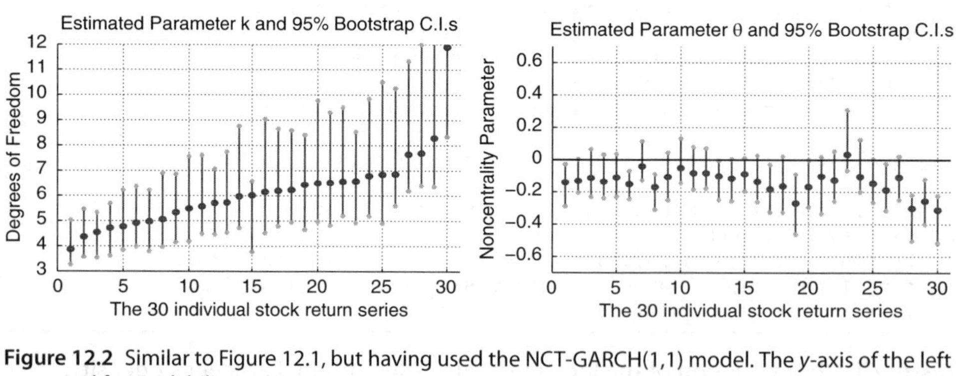

Figure 12.2 Similar to Figure 12.1, but having used the NCT-GARCH(1,1) model. The y-axis of the left graphic is truncated for readability.

wrong w.p.1, in terms of both the assumed GARCH innovations distribution and the GARCH model itself, which is just a simple mechanism to address the non-i.i.d. nature of the data.

Valid conclusions from juxtaposing the two figures include:

i) Incorporation of a GARCH filter helps address some of the heavy-tailed nature of the data. (One could speak of addressing the leptokurtosis, but this would then assume existence of fourth moments.)
ii) Conditional on use of the NCT distribution, the tail index is not constant across all assets. However, based on the confidence intervals, it appears that it can be deemed the same for many assets, particularly in the GARCH case.
iii) The lengths of the bootstrap confidence intervals are roughly proportional to the magnitude of the point estimates, so that, in the GARCH case, there is, relatively speaking, larger sampling error associated with the tail index k, as compared to the asymmetry parameter.
iv) When conditioning on the scale term via a GARCH model, the point estimates of asymmetry parameter θ are negative for 29 out of the 30 assets, and have a lower sampling error than in the i.i.d. case.

■

Example 12.2 ***(Example 12.1 cont.)***
Now consider the bivariate data set consisting of the 1,945 daily returns, from the shorter period of June 2001 to March 2009, of the two stocks with the most extreme individually estimated values of k, these being Bank of America (with $\hat{k} = 1.5$ for this period) and Wal-Mart Stores ($\hat{k} = 4.4$). Their returns are plotted in Figure 12.3, and it is apparent that the global financial crisis impacted the former relatively much more than the latter. A scatterplot of the two series is given in the top panel of Figure 12.4, while the bottom panel shows the fitted multivariate t distribution (12.3), overlaid with the scatterplot of the data, but such that, for optical clarity, the points for which both returns were less than three in absolute value were omitted.

The point estimates (and approximate standard errors) are $\hat{v} = 2.01(0.09)$, $\mu_1 = 0.011(0.029)$, $\mu_2 = -0.011(0.027)$, $\sigma_1^2 = 1.09(0.06)$, $\sigma_{12} = 0.442(0.033)$, and $\sigma_2^2 = 0.884(0.04)$, with an obtained log-likelihood value of −7199.3. Observe how the value of $\hat{v}$ lies between the two individually estimated values of $\hat{k}$ (though it is closer to the smaller of the two), thus underestimating the risk (or probability of extreme occurrences) of Bank of America and significantly overestimating that of Wal-Mart. ■

The previous two examples serve as motivation for use of multivariate distributions such that the margins can have individual tail shape parameters. In addition, stock (and other asset) returns typically exhibit asymmetry. This was seen above in the right panel of Figure 12.1, showing that individual assets tend to have positive skewness (yet from Figure 12.2, their GARCH-filtered counterparts are nearly all negative). Interestingly, aggregate stock returns, such as market indexes and their associated exchange traded funds, tend to exhibit negative skewness. A resolution to this seeming puzzle is provided in Albuquerque (2012).

From a risk management point of view, if, say, the left tail is somewhat heavier than the right (i.e., negative asymmetry) and a symmetric model is fit to the data, then the, say, 1% quantile from the fitted distribution (a negative value, in the left tail) will be smaller in magnitude (i.e., closer to zero)

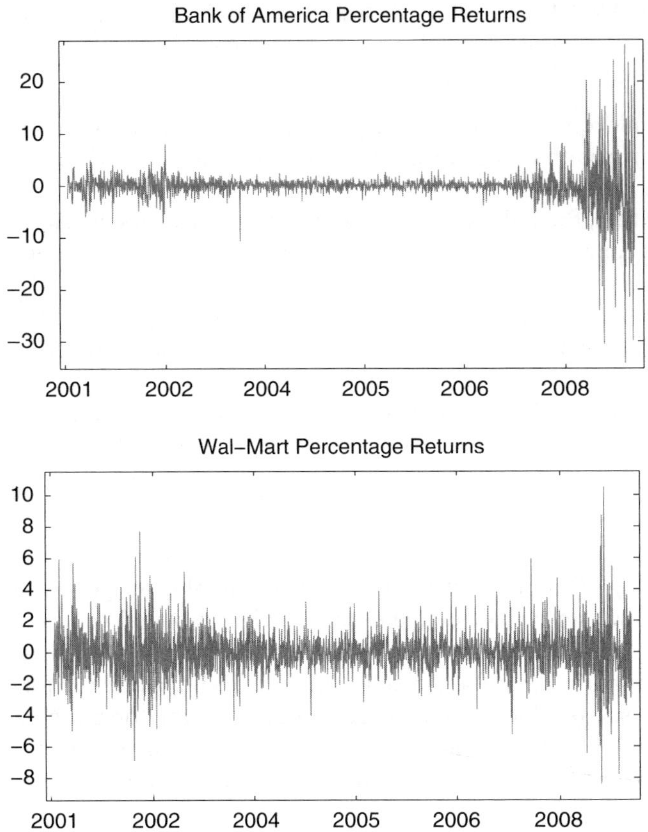

Figure 12.3 Daily percentage log returns on Bank of America (top) and Wal-Mart (bottom).

than it should, i.e., it will underestimate the loss associated with an extreme negative event. We first consider an important asymmetric extension of (12.3) in Section 12.2, and then turn to the harder task of how to extend the distribution such that each margin is endowed with its own degrees of freedom parameter.

12.2 Multivariate Noncentral Student's *t*

There are several asymmetric extensions to the multivariate Student's t available; see Kotz and Nadarajah (2004), Genton (2004), Nadarajah and Dey (2005), Arellano-Valle and Genton (2010), and the references therein for an overview. One of the earliest, from Kshirsagar (1961), is a direct extension of the univariate noncentral t (NCT) and, being a continuous mixture of normals, has some useful properties. We refer to it in short as MVNCT: Let $\boldsymbol{\gamma} = (\gamma_1, \dots, \gamma_d)' \in \mathbb{R}^d$, $\mathbf{Z} \sim \mathrm{N}_d(\mathbf{0}, \mathbf{I})$, $\mathbf{R}$ a $d \times d$ correlation matrix (12.2), and $G \sim \mathrm{IGam}(v/2, v/2)$, $v > 0$, independent of $\mathbf{Z}$. Then, similar to (12.4),

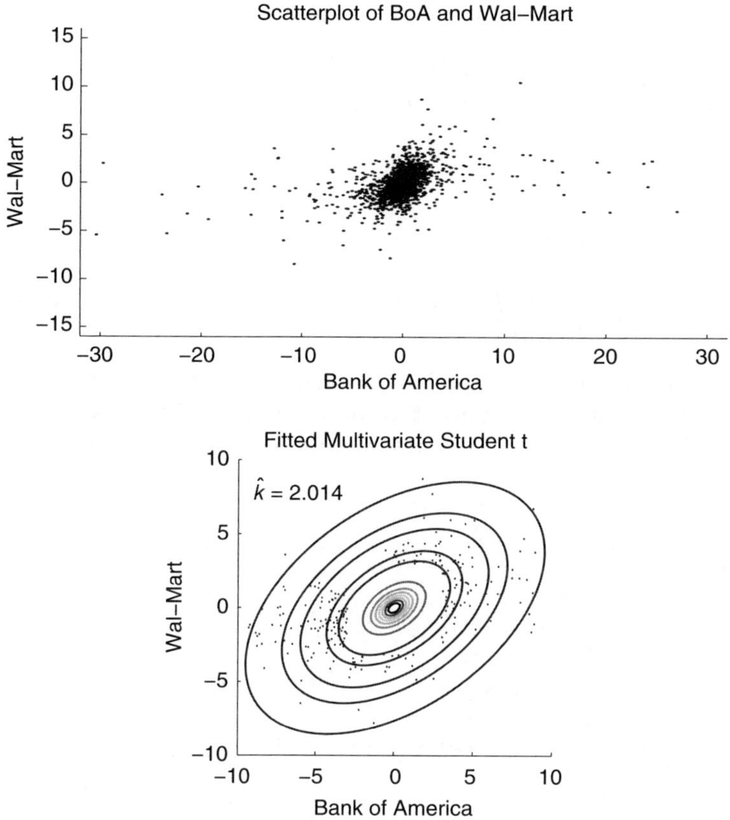

Figure 12.4 Top: Scatterplot of the returns on Bank of America and Wal-Mart for the $T = 1{,}945$ observations. **Bottom:** Scatterplot, now with truncated and equal axes, and omitting points near the center, with an overlaid contour plot of the fitted multivariate Student's t density.

$\mathbf{Y} = (\boldsymbol{\gamma} + \mathbf{R}^{1/2}\mathbf{Z}) \sim \mathrm{N}_d(\boldsymbol{\gamma}, \mathbf{R})$, and

$$\mathbf{T} = \sqrt{G}\mathbf{Y} = \sqrt{G}\boldsymbol{\gamma} + \sqrt{G}\mathbf{R}^{1/2}\mathbf{Z} \sim \mathrm{MVNCT}(\mathbf{0}, \boldsymbol{\gamma}, \mathbf{R}, v) \tag{12.5}$$

is said to follow a *Kshirsagar* (1961) *d-dimensional multivariate noncentral t distribution* (in short, noncentral t) with degrees of freedom v, noncentrality vector $\boldsymbol{\gamma}$, and correlation matrix $\mathbf{R}$. Recalling the relation between the gamma, inverse gamma, and χ^2 distributions, an equivalent representation sometimes seen in the literature is the following: Let $\mathbf{Z} \sim \mathrm{N}_d(\boldsymbol{\gamma}, \mathbf{R})$, independent of $C \sim \chi^2(v)$. Then $\mathbf{T} = \mathbf{Z}/\sqrt{C/v} \sim \mathrm{MVNCT}(\mathbf{0}, \boldsymbol{\gamma}, \mathbf{R}, v)$.

Note that, from the construction in (12.5),

$$(\mathbf{T} \mid G = g) \sim \mathrm{N}(g\boldsymbol{\gamma}, g\mathbf{R}), \tag{12.6}$$

implying that $\mathbf{T}$ is, analogous to the usual multivariate Student's t, a continuous mixture of normals, and, also from (12.5), all the univariate margins are noncentral t. If $\boldsymbol{\gamma} = \mathbf{0}$, then $\mathbf{T}$ is elliptic (in this case, spherical), and otherwise is non-elliptic; see the discussion in Section C.2.

The p.d.f. of $\mathbf{T} \sim \text{MVNCT}(\mathbf{0}, \boldsymbol{\gamma}, \mathbf{R}, v)$, denoted $f_{\mathbf{T}} = f_{\mathbf{T}}(\mathbf{x}; \mathbf{0}, \boldsymbol{\gamma}, \mathbf{R}, v)$, is given by

$$f_{\mathbf{T}} = \frac{\Gamma((v+d)/2)}{(v\pi)^{d/2}\Gamma(v/2)|\mathbf{R}|^{1/2}} \exp\left\{-\frac{1}{2}\boldsymbol{\gamma}'\mathbf{R}^{-1}\boldsymbol{\gamma}\right\}\left(1 + \frac{\mathbf{x}'\mathbf{R}^{-1}\mathbf{x}}{v}\right)^{-(v+d)/2} \times \sum_{k=0}^{\infty} g_k(\mathbf{x}; \boldsymbol{\mu}, \boldsymbol{\gamma}, \mathbf{R}, v), \tag{12.7}$$

where

$$g_k(\mathbf{x}; \boldsymbol{\mu}, \boldsymbol{\gamma}, \mathbf{R}, v) = \frac{2^{k/2}\Gamma((v+d+k)/2)}{k!\Gamma((v+d)/2)}\left(\frac{\mathbf{x}'\mathbf{R}^{-1}\boldsymbol{\gamma}}{\sqrt{v + \mathbf{x}'\mathbf{R}^{-1}\mathbf{x}}}\right)^k. \tag{12.8}$$

The derivation is similar to that in the univariate case; see Section II.10.4.1.1. Recall Section III.10.3.2 on a fast, accurate approximation to the p.d.f. of the univariate noncentral t. With very minor modification, this can also be used in the MVNCT case. The program in Listing 12.3 accomplishes this.

```
function pdfln = mvnctpdfln(x, mu, gam, v, Sigma)
% x         d X T matrix of evaluation points
% mu, gam  d-length location and noncentrality vector
% v is df; Sigma is the dispersion matrix.
[d,t] = size(x); C=Sigma; [R, err] = cholcov(C, 0);
assert(err == 0, 'C is not (semi) positive definite');
mu=reshape(mu,length(mu),1); gam=reshape(gam,length(gam),1);
vn2 = (v + d) / 2; xm = x - repmat(mu,1,t); rho = sum((R'\xm).^2,1);
pdfln = gammaln(vn2) - d/2*log(pi*v) - gammaln(v/2) - ...
        sum(slog(diag(R))) - vn2*log1p(rho/v);
if (all(gam == 0)), return; end
idx = (pdfln >= -37); maxiter=1e4; k=0;
if (any(idx))
  gcg = sum((R'\gam).^2); pdfln = pdfln - 0.5*gcg; xcg = xm' * (C \ gam);
  term = 0.5*log(2) + log(xcg) - 0.5*slog(v+rho');
  term(term == -inf) = log(realmin); term(term == +inf) = log(realmax);
  logterms = gammaln((v+d+k)/2) - gammaln(k+1) - gammaln(vn2) + k*term;
  ff = real(exp(logterms)); logsumk = log(ff);
  while (k < maxiter)
    k=k+1;
    logterms = gammaln((v+d+k)/2) - gammaln(k+1) - gammaln(vn2) + k*term(idx);
    ff = real(exp(logterms-logsumk(idx))); logsumk(idx)=logsumk(idx)+log1p(ff);
    idx(idx) = (abs(ff) > 1e-4); if (all(idx == false)), break, end
  end
  pdfln = real(pdfln+logsumk');
end

function y = slog(x) % Truncated log. No -Inf or +Inf.
y = log(max(realmin, min(realmax, x)));
```

Program Listing 12.3: The direct density approximation (d.d.a.) to the (log of the) d-variate canonical MVNCT density.

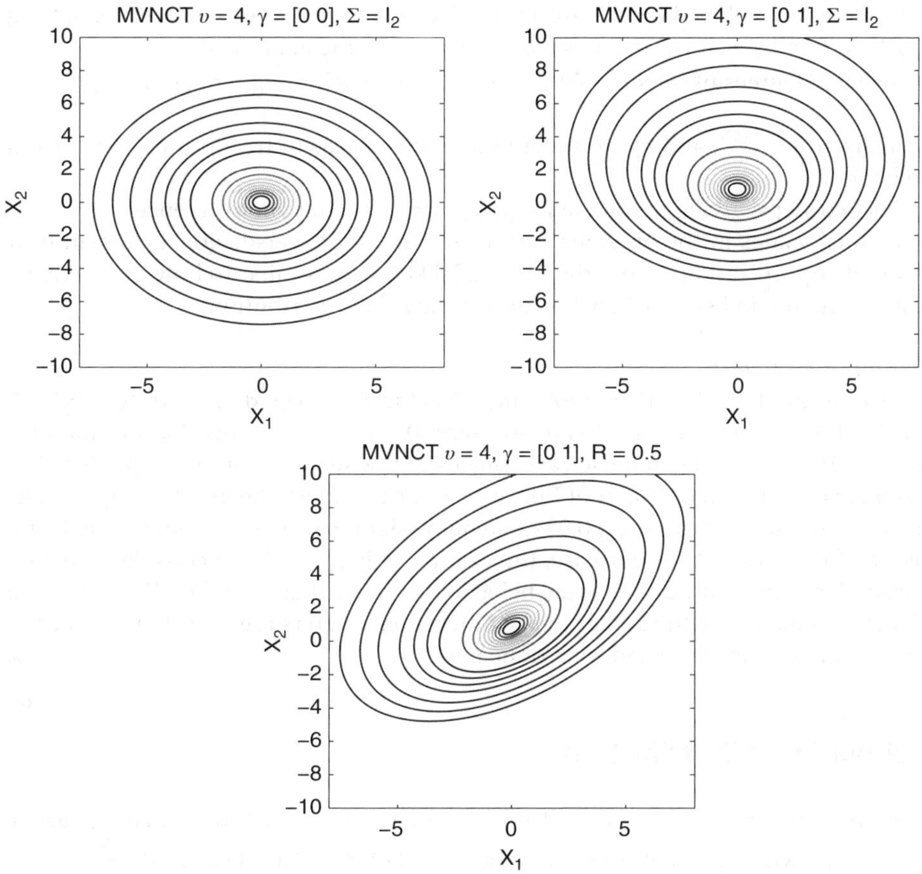

Figure 12.5 Bivariate contour plots of three MVNCT densities.

Figure 12.5 shows contour plots of the bivariate MVNCT density, using $v = 4, \gamma_1 = 0$, and two different values for γ_2 (and correlation zero and 0.5). The code to produce these plots is given in Listing 12.4. It is instructive, as it shows two ways of generating the plots: First, with basic principles and FOR loops, giving a program that is easy to understand and portable to all languages, and, second, using the vectorized capabilities and specific commands of Matlab (namely `meshgrid` and `reshape`). The latter is far faster because the evaluation of the (log) density is done "all at once" in a vectorized fashion, but also because the double FOR loop just to generate the large matrix of coordinates is surprisingly slow.

Location vector $\boldsymbol{\mu}$ and scale vector $\boldsymbol{\sigma} = (\sigma_1, \dots, \sigma_d)'$ can be introduced precisely as in (12.3) to give $\mathbf{X} = \boldsymbol{\mu} + \mathbf{ST}$, $\mathbf{S} = \text{diag}(\boldsymbol{\sigma})$, and we write $\mathbf{X} \sim \text{MVNCT}(\boldsymbol{\mu}, \boldsymbol{\gamma}, \boldsymbol{\Sigma}, v)$, where $\mathbf{R} = \mathbf{S}^{-1}\boldsymbol{\Sigma}\mathbf{S}^{-1}$. Estimation of the location-scale MVNCT in the bivariate case can be done with a simple modification to the program in Listing 12.1. It is given in Listing 12.5, and is used below in Example 12.3. For the general d-variate case, a two-step method can be used that avoids having to estimate all the parameters simultaneously. It works as follows (and the reader is encouraged...).

1) Recalling that the margins of the MVNCT are univariate noncentral t, estimate each, getting parameters $\hat{\mu}_i^{[1]}, \hat{\gamma}_i^{[1]}, \hat{\sigma}_i^{[1]}, \hat{v}_i^{[1]}, i = 1, \ldots, d$, and set $\hat{v}^{[1]}$ equal to the mean of the $\hat{v}_i^{[1]}$.
2) Conditional on the fixed degrees of freedom $\hat{v}^{[1]}$, estimate again each margin to get $\hat{\mu}_i^{[2]}, \hat{\gamma}_i^{[2]}, \hat{\sigma}_i^{[2]}$, $i = 1, \ldots, d$.
3) Conditional on the fixed $\hat{\mu}_i^{[2]}, \hat{\gamma}_i^{[2]}$, and $\hat{\sigma}_i^{[2]}$, estimate the single degree of freedom value $\hat{v}^{[2]}$ from the MVNCT likelihood.
4) Repeat the previous two steps, giving the sequence $\hat{\mu}_i^{[j]}, \hat{\gamma}_i^{[j]}, \hat{\sigma}_i^{[j]}, \hat{v}^{[j]}$, until convergence.
5) Conditional on the final values in the previous step, estimate each lower-diagonal element of $\mathbf{R}$ individually (univariate optimizations) from the MVNCT likelihood, similar to how the elements in the $\mathbf{R}$ matrix are estimated in Section 12.5.4 below for the AFaK distribution.

Example 12.3 ***(Example 12.1 cont.)***
We fit the MVNCT to the Bank of America (BoA) and Wal-Mart returns data, getting $\hat{v} = 2.02$, $\hat{\gamma}_1 = -0.157$ (for BoA), while $\hat{\gamma}_2$ (for Wal-Mart) is nearly zero, 0.036. The obtained log-likelihood is −7194.5, compared to −7199.3 in the symmetric case, suggesting (when compared to a χ^2_2 distribution) "parameter significance". Clearly, only that of BoA is "significant" (with the usual asymptotically determined estimate of its standard error being 0.044). Genuine significance is best determined with respect to the measure of real interest: In our case, this is forecasting, as will be considered below. The fact that the margins have markedly different tail behaviors indicate that even the MVNCT is still "too mis-specified", and conclusions with respect to asymmetry parameters are best drawn once the heterogeneous tail behavior issue is addressed, as is done next. ■

12.3 Jones Multivariate *t* Distribution

Jones (2002) proposed two constructions for a multivariate t distribution such that each univariate margin is endowed with its own degrees of freedom parameter. Let $Z_i \overset{\text{iid}}{\sim} \mathrm{N}(0, 1)$ and $W_i \overset{\text{indep}}{\sim} \chi^2(n_i)$, $i = 1, \ldots, d$, such that they are all mutually independent, and taking $W_i = 0$ w.p.1, when $n_i = 0$. For values $0 = v_0 < v_1 \leqslant \cdots \leqslant v_d$, let $n_i = v_i - v_{i-1}$, $i = 1, \ldots, d$. The first is to take

$$T_1 = \frac{\sqrt{v_1} Z_1}{\sqrt{W_1}}, \quad T_2 = \frac{\sqrt{v_2} Z_2}{\sqrt{W_1 + W_2}}, \ldots, T_d = \frac{\sqrt{v_d} Z_d}{\sqrt{W_1 + \cdots + W_d}}, \tag{12.9}$$

which, from the additivity of independent χ^2 random variables, is such that T_i is Student's t with v_i degrees of freedom. Note that construction (12.9) with $v_1 = v_2 = \cdots = v_d$ is equivalent to the usual multivariate t distribution (12.3), with zero mean vector and $\mathbf{\Sigma}$ the identity matrix.

The second construction takes

$$T_1 = \sqrt{v_1} Z_1 / \sqrt{W_1}, \quad T_2 = \sqrt{v_2} Z_2 / \sqrt{W_1 + U_2}, \ldots, T_d = \sqrt{v_d} Z_d / \sqrt{W_1 + U_d}, \tag{12.10}$$

where $U_i \sim \chi^2(v_i - v_1)$, $i = 2, \ldots, d$, with the Z_i, W_1, and U_i all mutually independent. In the $d = 2$ case, both of these constructions are equivalent. In this case, Jones (2002) shows that, for r_1 and r_2 both nonnegative integers, $r_1 < v_1$ and $r_1 + r_2 < v_2$,

$$\mathbb{E}[T_1^{r_1} T_2^{r_2}] = \frac{v_1^{r_1/2} v_2^{r_2/2} \Gamma\{(r_1+1)/2)\} \Gamma\{(r_2+1)/2\} \Gamma\{(v_1 - r_1)/2\} \Gamma\{(v_2 - r_1 - r_2)/2\}}{\pi \Gamma(v_1/2) \Gamma\{(v_2 - r_1)/2\}}, \tag{12.11}$$

```
v=4; gam=[0 1]'; R12=0.5; R=[1 R12; R12 1]; Xvec=-8:0.02:8; Yvec=-10:0.02:10;
if 1==2 % Manual (slow)
  % XY=zeros(2,length(Yvec)*length(Xvec)); % don't need.
  Z=zeros(length(Xvec),length(Yvec));
  for xl=1:length(Xvec),  x=Xvec(xl);
    for yl=1:length(Yvec),  y=Yvec(yl); use=[x ; y];
      % pos=(xl-1)*length(Yvec)+yl; XY(:,pos)=use; % don't need
      Z(xl,yl)= exp( mvnctpdfln(use, gam, R, v) );
    end
  end
else % Vectorized (fast)
  [X,Y]=meshgrid(Xvec,Yvec); XY=[X(:)' ; Y(:)'];
  Z = exp( mvnctpdfln(XY, gam, R, v) );
  Z = reshape(Z',length(Yvec),length(Xvec))'; % note the end transpose!
end
figure, contour(Xvec,Yvec,Z',9,'linewidth',2), hold on
levvec=[40 20 10 5 2 1 0.5]/10000;
for z=1:length(levvec), lev=levvec(z);
  contour(Xvec,Yvec,Z',[lev lev],'linewidth',2)
end
set(gca,'fontsize',16), xlabel('X_1'), ylabel('X_2')
str1=['MVNCT v=',int2str(v),', \gamma=[',int2str(gam(1)),' ', ...
    num2str(gam(2)),'], '];
if R12==0, str2='\Sigma = I_2'; else str2=['R = ',num2str(R12)]; end
title([str1 str2]), axis equal, xlim([-8 8]), ylim([-10 10])
```

Program Listing 12.4: Generates the plots in Figure 12.5. The lines commented out with "don't need" were there just to confirm that the set of pair coordinates for the density are the same in both the slow and fast way of computing.

```
function [param,stderr,iters,loglik,Varcov] = MVNCT2estimation(x)
[d T]=size(x); if d~=2, error('not done yet, use 2-step'), end
%%%%%%%%        k      mu1    mu2   scale1   scale2   R12  gam1   gam2
bound.lo=    [ 1.1     -1     -1    0.01     0.01    -1    -4     -4 ];
bound.hi=    [  20      1      1     100      100     1     4      4 ];
bound.which= [  1       0      0       1        1     1     1      1 ];
initvec     =[  3       0      0       2        2   0.5     0      0 ];
maxiter=300; tol=1e-6; MaxFunEvals=length(initvec)*maxiter;
opts=optimset('Display','iter','Maxiter',maxiter,'TolFun',tol,'TolX',tol,...
    'MaxFunEvals',MaxFunEvals,'LargeScale','Off');
[pout,fval,~,theoutput,~,hess] = ...
  fminunc(@(param) MVNCTloglik(param,x,bound),einschrk(initvec,bound),opts);
V=inv(hess)/T; [param,V]=einschrk(pout,bound,V); param=param';
Varcov=V; stderr=sqrt(diag(V)); loglik=-fval*T; iters=theoutput.iterations;

function ll=MVNCTloglik(param,x,bound)
if nargin<3, bound=0; end
if isstruct(bound), param=einschrk(real(param),bound,999); end
k=param(1); mu=param(2:3); scale=param(4:5); gam=param(7:8);
R12=param(6); R=[1 R12; R12 1]; if min(eig(R))<1e-4, ll=1e5;
else
  xx=x; for i=1:2, xx(i,:)=(x(i,:)-mu(i))/scale(i); end
  llvec = mvnctpdfln(xx, gam, R, k) - log(prod(scale));
  ll=-mean(llvec); if isinf(ll), ll=1e5; end
end
```

Program Listing 12.5: Maximum likelihood estimation of the location-scale MVNCT for $d = 2$ dimensions.

if r_1 and r_2 are even, and zero otherwise. For example, with $r_1 = r_2 = 1$, this yields $\mathrm{Cov}(T_1, T_2) = \mathbb{E}[T_1 T_2] = 0$. Likewise, for $v_1 > 2$ and $v_2 > 2$, it is easy to confirm that, respectively for $r_1 = 2$ and $r_2 = 0$, and $r_1 = 0$ and $r_2 = 2$,

$$\mathbb{V}(T_1) = \mathbb{E}[T_1^2] = \frac{v_1}{v_1 - 2} \quad \text{and} \quad \mathbb{V}(T_2) = \frac{v_2}{v_2 - 2}, \tag{12.12}$$

which agree with the variance expression for (12.3) when $v_1 = v_2$.

Jones (2002) also derives the density expression

$$f_{T_1,T_2}(t_1, t_2; \mathbf{v}) = \frac{1}{\pi\sqrt{v_1 v_2}} \frac{\Gamma((v_1+1)/2)}{\Gamma((v_2+1)/2)} \frac{\Gamma(v_2/2+1)}{\Gamma(v_1/2)} \times \frac{{}_2F_1(v_2/2+1,\ (v_2-v_1)/2;\ (v_2+1)/2;\ z)}{m^{v_2/2+1}}, \tag{12.13}$$

where $z = (t_1^2/v_1)/m$ and $m = 1 + t_1^2/v_1 + t_2^2/v_2$. See, e.g., Section II.5.3 for the definition of, and methods of computation for, the ${}_2F_1$ function. The reader is encouraged to algebraically (or numerically) confirm that (12.13) agrees with (12.1) when $v_1 = v_2$.

One arguable drawback of constructions (12.9) and (12.10) is that the T_i can never be independent (except in the limit as the v_i tend to infinity), a characteristic shared by the usual multivariate t distribution. Moreover, if we wish to endow (12.9) with correlation between the T_i, and/or noncentrality terms, then we can expect the derivation, and the final form, of a closed-form or single integral expression for the joint distribution to be far more complicated in the $d = 2$ case, let alone the general d-variate case.

Initiating this, we can extend Jones' construction (12.9) to support a **dispersion matrix** (which is related to, but not necessarily equal to, a covariance matrix) $\boldsymbol{\Sigma}$ and noncentrality parameters $\boldsymbol{\beta}$ as follows. We take $\mathbf{X} = (X_1, X_2)' \sim \mathrm{N}(\boldsymbol{\beta}, \boldsymbol{\Sigma})$, with $\boldsymbol{\beta} = (\beta_1, \beta_2)' \in \mathbb{R}^2$ and $\boldsymbol{\Sigma}$ a 2×2 symmetric, positive definite matrix, independent of $W_i \overset{\text{indep}}{\sim} \chi^2(n_i)$, $i = 1, 2$, where $n_1 = v_1$ and $n_2 = v_2 - v_1$, and $0 < v_1 \leqslant v_2 < \infty$.

Then, defining in addition $T_3 = \sqrt{W_1}$ and $T_4 = \sqrt{W_1 + W_2}$, so that

$$W_1 = T_3^2, \quad W_2 = T_4^2 - T_3^2, \quad X_1 = T_3 T_1/\sqrt{v_1}, \quad X_2 = T_4 T_2/\sqrt{v_2},$$

the Jacobian is

$$\mathbf{J} = \begin{bmatrix} \partial X_1/\partial T_1 & \partial X_1/\partial T_2 & \partial X_1/\partial T_3 & \partial X_1/\partial T_4 \\ \partial X_2/\partial T_1 & \partial X_2/\partial T_2 & \partial X_2/\partial T_3 & \partial X_2/\partial T_4 \\ \partial W_1/\partial T_1 & \partial W_1/\partial T_2 & \partial W_1/\partial T_3 & \partial W_1/\partial T_4 \\ \partial W_2/\partial T_1 & \partial W_2/\partial T_2 & \partial W_2/\partial T_3 & \partial W_2/\partial T_4 \end{bmatrix} = \begin{bmatrix} T_3/\sqrt{v_1} & 0 & T_1/\sqrt{v_1} & 0 \\ 0 & T_4/\sqrt{v_2} & 0 & T_2/\sqrt{v_2} \\ 0 & 0 & 2T_3 & 0 \\ 0 & 0 & -2T_3 & 2T_4 \end{bmatrix},$$

and $|\det \mathbf{J}| = 4T_3^2 T_4^2 / \sqrt{v_1 v_2}$. Thus

$$f_{T_1,T_2,T_3,T_4}(t_1, t_2, t_3, t_4) = \frac{4t_3^2 t_4^2}{\sqrt{v_1 v_2}} f_{X_1,X_2,W_1,W_2}(t_3 t_1/\sqrt{v_1},\ t_4 t_2/\sqrt{v_2},\ t_3^2,\ t_4^2 - t_3^2),$$

and $f_{T_1,T_2}(t_1, t_2) = \int\int f_{T_1,T_2,T_3,T_4}(t_1, t_2, t_3, t_4)\, \mathrm{d}t_3\, \mathrm{d}t_4$. Clearly,

$$f_{X_1,X_2}(x_1, x_2) = \frac{1}{|\boldsymbol{\Sigma}|^{1/2}(2\pi)} \exp\left\{-\frac{1}{2}((\mathbf{x} - \boldsymbol{\beta})'\boldsymbol{\Sigma}^{-1}(\mathbf{x} - \boldsymbol{\beta}))\right\},$$

while, for $v_2 > v_1$,

$$f_{W_1,W_2}(w_1, w_2; v_1, v_2) = \frac{1}{2^{v_1/2}\Gamma(v_1/2)} w_1^{v_1/2-1} e^{-w_1/2} \mathbb{I}_{(0,\infty)}(w_1) \times \frac{1}{2^{(v_2-v_1)/2}\Gamma((v_2 - v_1)/2)} w_2^{(v_2-v_1)/2-1} e^{-w_2/2} \mathbb{I}_{(0,\infty)}(w_2),$$

and, for $v_2 = v_1$,

$$f_{W_1,W_2}(w_1, w_2; v_1) = \frac{1}{2^{v_1/2}\Gamma(v_1/2)} w_1^{v_1/2-1} e^{-w_1/2} \mathbb{I}_{(0,\infty)}(w_1) \times \mathbb{I}_{[0]}(w_2).$$

Thus,

$$f_{T_1,T_2}(t_1, t_2; \mathbf{v}, \boldsymbol{\beta}, \boldsymbol{\Sigma}) = \int\int_{0<t_3<t_4<\infty} \frac{4t_3^2 t_4^2}{\sqrt{v_1 v_2}} \frac{2^{-v_1/2}}{\Gamma(v_1/2)} \frac{(2\pi)^{-1}|\boldsymbol{\Sigma}|^{-1/2}}{2^{(v_2-v_1)/2}\Gamma((v_2 - v_1)/2)} \times t_3^{v_1-2} e^{-t_3^2/2} (t_4^2 - t_3^2)^{(v_2-v_1)/2-1} e^{-(t_4^2-t_3^2)/2} \times \exp\left\{-\frac{1}{2}((\mathbf{h} - \boldsymbol{\beta})'\boldsymbol{\Sigma}^{-1}(\mathbf{h} - \boldsymbol{\beta}))\right\} \mathrm{d}t_3\, \mathrm{d}t_4, \tag{12.14}$$

where $\mathbf{v} = (v_1, v_2)$, with $0 < v_1 < v_2 < \infty$, and $\mathbf{h} = [t_3 t_1/\sqrt{v_1}, \quad t_4 t_2/\sqrt{v_2}]'$, to ease the notation. A location term, say $\boldsymbol{\mu} = (\mu_1, \mu_2)' \in \mathbb{R}^2$, could be trivially added.

One could pursue a similar expression for general d, in which case there are $2d$ random variables involved in (12.9), and the joint p.d.f. of $\mathbf{T} = (T_1, \ldots, T_d)'$ will involve a d-dimensional integral. It should be obvious that such an exercise will be of limited value because of the curse of dimensionality aspect as d grows. Another, less obvious reason is the associated numeric problems arising with its evaluation when any pair v_i and v_{i+1} (recall they are ordered) are close in value; see the closing remarks in Section 12.7.

Clearly, if $\boldsymbol{\beta} = \mathbf{0}$ and all the v_i are equal (to, say, v), then $\mathbf{T} \sim T_v(\mathbf{0}, \boldsymbol{\Sigma})$. Thus, in light of (12.12), one might hope that, for $\boldsymbol{\beta} = \mathbf{0}$ and $2 < v_1 < \cdots < v_p$, $\mathbb{V}(\mathbf{T})$ is simply given by $\mathbf{K\Sigma K}$, where $\mathbf{K}$ is the diagonal matrix with iith element $\kappa_i = \sqrt{v_i/(v_i - 2)}$, $i = 1, \ldots, k$. Simulation quickly indicates that this is *not* the case. An indication of the complexity of $\mathbb{V}(\mathbf{T})$ when off-diagonal elements are present can be gleamed from (12.17), which gives the correlation between T_1 and T_2 in the bivariate case, with no asymmetry parameters, when the linkage between T_1 and T_2 is expressed via an angle θ.

Figure 12.6 shows (12.14) for $v_1 = 2$, $v_2 = 8$, and several constellations of the other parameters. Computation of p.d.f. (12.14) via numeric integration in these cases was numerically unproblematic because v_1 and v_2 are very well-separated. The reader is encouraged to program (12.14) and replicate Figure 12.6.

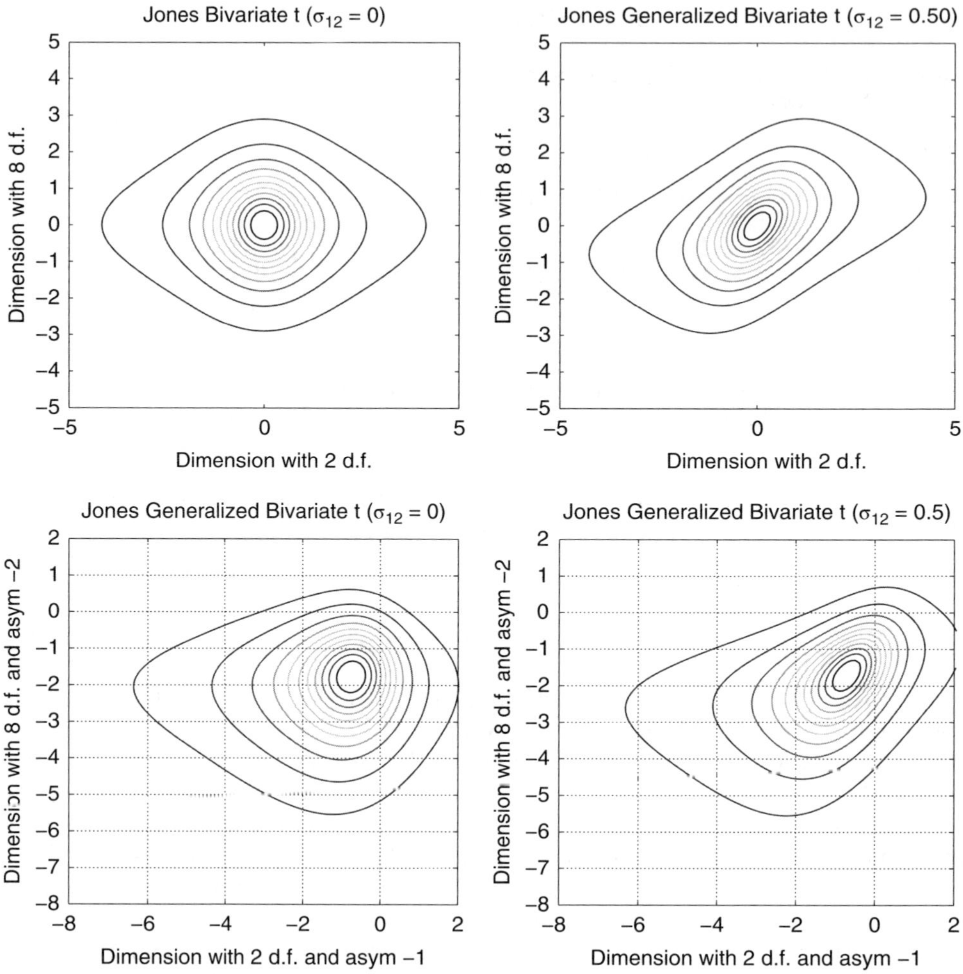

Figure 12.6 Top left: The bivariate Jones (2002) distribution (12.9) for $v_1 = 2$ and $v_2 = 8$ degrees of freedom. **Top right:** Same, but its generalization (12.14) with $\sigma_1^2 = \sigma_2^2 = 1$ and $\sigma_{12} = 0.5$ (and no noncentrality). **Bottom:** Similar to top, but additionally introduce asymmetry via noncentrality parameters $\beta_1 = -1$ and $\beta_2 = -2$.

12.4 Shaw and Lee Multivariate *t* Distributions

Several ideas in the bivariate case are developed in Shaw and Lee (2008), such as a bivariate t with a closed-form expression (involving the ${}_2F_1$ function) for the density and possibly independent marginals, albeit the same degrees of freedom, and a bivariate distribution with a closed-form expression (involving the ${}_1F_1$ function) for the density, but such that one marginal is normal. They also provide a relation to, and generalization of, the Jones (2002) structure (12.9) by proposing to take, in our notation,

$$T_1 = \sqrt{\frac{v_1}{W_1}} Z_1, \quad T_2 = \sqrt{\frac{v_2}{W_1 + W_2}} (Z_1 \sin(\theta) + Z_2 \cos(\theta)), \tag{12.15}$$

where $W_1 \sim \chi^2(v_1)$, $W_2 \sim \chi^2(v_2 - v_1)$, $Z_1, Z_2 \sim \mathrm{N}(0, 1)$, all four completely independent, $0 < v_1 \leqslant v_2 < \infty$, and $\theta \in [0, 2\pi)$. Simulation from (12.15) is trivial, and Shaw and Lee (2008) derive the density expression

$$f_{T_1,T_2}(t_1, t_2; v_1, v_2, \theta) = \frac{2\Gamma\left(\frac{a+b+2}{2}\right) / \cos(\theta)}{\Gamma\left(\frac{a}{2}\right)\Gamma\left(\frac{b}{2}\right)\pi\sqrt{a(a+b)}} \int_0^1 \frac{u^a (1-u^2)^{(b-2)/2}}{(\alpha_2 + (\alpha_1 - \alpha_2)u^2 - \beta u)^{(a+b+2)/2}} \, du \,, \tag{12.16}$$

where $a = v_1$, $b = v_2 - v_1$, and

$$\alpha_2 = 1 + \frac{t_2^2}{(a+b)\cos^2(\theta)}, \quad \alpha_1 = \alpha_2 + \frac{t_1^2}{a \cos^2(\theta)}, \quad \beta = \frac{2 \sin(\theta) t_1 t_2}{\sqrt{a(a+b)}\cos^2(\theta)}.$$

This reduces to the closed-form expression given by Jones (2002) when $\theta = 0$. Shaw and Lee (2008) also show that the correlation between T_1 and T_2 is given by

$$\rho(T_1, T_2; v_1, v_2, \theta) = \sin(\theta) \frac{\Gamma\left(\frac{a-1}{2}\right)\Gamma\left(\frac{a+b-2}{2}\right)}{\Gamma\left(\frac{a}{2}\right)\Gamma\left(\frac{a+b-1}{2}\right)} \sqrt{\frac{a}{2} - 1} \sqrt{\frac{a+b}{2} - 1}, \tag{12.17}$$

for $a = v_1 > 2$ (and recalling $b = v_2 - v_1$ and that $v_2 \geqslant v_1$).

One drawback of (12.16) is that the margins are not independent when $\theta = 0$, in which case it coincides with (12.9). To devise a construction that yields independent marginals when $\theta = 0$, Shaw and Lee (2008) replace $W_1 + W_2$ in (12.15) by a $\chi^2(v_2)$ random variable that does not depend on W_1, i.e.,

$$T_1 = \sqrt{\frac{v_1}{W_1}} Z_1, \quad T_2 = \sqrt{\frac{v_2}{W_2}} (Z_1 \sin(\theta) + Z_2 \cos(\theta)), \tag{12.18}$$

where $W_i \overset{\text{ind}}{\sim} \chi^2(v_i)$, $i = 1, 2$, independent of, as before, $Z_1, Z_2 \overset{\text{ind}}{\sim} \mathrm{N}(0, 1)$. (Another structure for accomplishing this is given below in Section 12.6). They show that the p.d.f. $f_{T_1,T_2}(t_1, t_2; v_1, v_2, \theta)$ can be expressed as

$$\frac{\Gamma(r_{12})\Gamma(r_{22})\,{}_2F_1\left(r_{12}, r_{22}; \frac{1}{2}; \frac{\gamma^2}{4\alpha_1\alpha_2}\right)\sqrt{\alpha_1\alpha_2} + \gamma\Gamma(r_{11})\Gamma(r_{21})\,{}_2F_1\left(r_{11}, r_{21}; \frac{3}{2}; \frac{\gamma^2}{4\alpha_1\alpha_2}\right)}{\alpha_1^{r_{11}}\alpha_2^{r_{21}} \cos(\theta)\pi\sqrt{v_1 v_2}\Gamma(v_1/2)\Gamma(v_2/2)}, \tag{12.19}$$

where, to ease the notation, we let $r_{ij} = v_i/2 + 1/j$, $i = 1, 2$, $j = 1, 2$, and

$$\alpha_i = 1 + \frac{t_i^2}{v_i \cos^2(\theta)}, \quad i = 1, 2, \quad \text{and} \quad \gamma = \frac{2 \sin(\theta) t_1 t_2}{\sqrt{v_1 v_2}\cos^2(\theta)}.$$

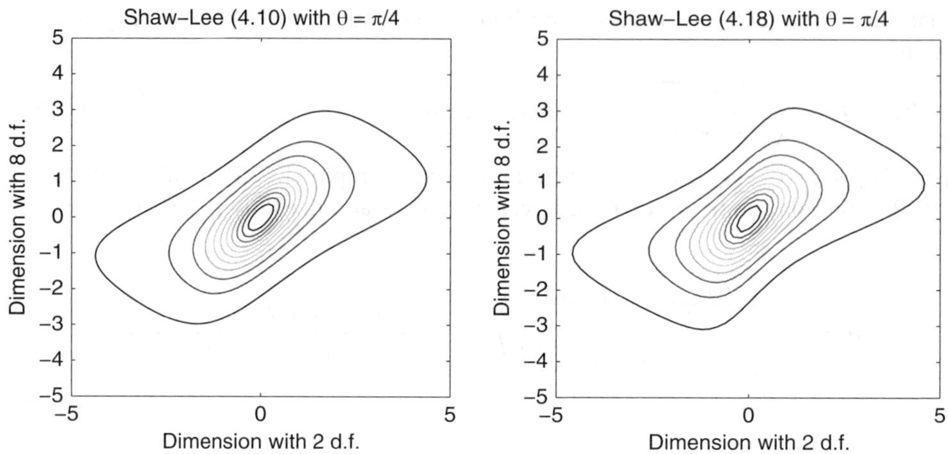

Figure 12.7 Contour plots of (12.16) (Shaw and Lee, 2008, Eq. 4.10) and (12.19) (Shaw and Lee, 2008, Eq. 4.18) for $v_1 = 2$, $v_2 = 8$, and $\theta = \pi/4$. Compare to the top right panel of Figure 12.6.

One can confirm, both algebraically and numerically, that the density factors into the product of the univariate Student's t marginals when $\theta = 0$. They also show that the correlation between T_1 and T_2 is

$$\rho(T_1, T_2; v_1, v_2, \theta) = \sin(\theta)\frac{\Gamma\left(\frac{v_1-1}{2}\right)\Gamma\left(\frac{v_2-1}{2}\right)}{\Gamma\left(\frac{v_1}{2}\right)\Gamma\left(\frac{v_2}{2}\right)}\sqrt{\frac{v_1}{2}-1}\sqrt{\frac{v_2}{2}-1}, \tag{12.20}$$

for $v_1 > 2$ and $v_2 > 2$.

To illustrate the difference between the two distributions when T_1 and T_2 are correlated, contour plots of (12.16) and (12.19) are shown in the left and right panels, respectively, of Figure 12.7, using $v_1 = 2$ and $v_2 = 8$ degrees of freedom, and a dependence parameter of $\theta = \pi/4$. These can be compared to the top right panel of Figure 12.6 (and the bottom right panel of Figure 12.12). Unfortunately, extension of (12.15) or (12.18) to the d-dimensional case, or incorporation of noncentrality (asymmetry) parameters (even in the bivariate case), is presumably intractable.

12.5 The Meta-Elliptical *t* Distribution

A rather general way of producing a d-variate distribution such that the ith univariate margin is, say, Student's $t(v_i)$, is to use a **copula construction**. Continuous copula-based distributions are very general in that the margins can be taken to be essentially any continuous distribution, while the dependency structure, via the copula, is also very flexible, though in reality there are only a handful of choices that are typically used. When the distribution is such that the margins are $t(v_i)$ and the copula is based on a multivariate t distribution, it is often referred to as a meta-elliptical t, as discussed in this section. More advanced aspects of t copula constructions can be found in Demarta and McNeil (2005) and Nikoloulopoulos et al. (2009).

12.5.1 The FaK Distribution

The canonical p.d.f. of the **meta-elliptical t distribution** proposed in Fang et al. (2002) is given by[2]

$$f_{\mathbf{X}}(\mathbf{x};\mathbf{v},\mathbf{R}) = \psi(\Phi_{v_0}^{-1}(\Phi_{v_1}(x_1)),\dots,\Phi_{v_0}^{-1}(\Phi_{v_d}(x_d));\mathbf{R},v_0)\prod_{i=1}^{d}\phi_{v_i}(x_i), \tag{12.21}$$

where $\mathbf{x} = (x_1,\dots,x_d)' \in \mathbb{R}^d$, $\mathbf{v} = (v_0, v_1,\dots,v_d)' \in \mathbb{R}^{d+1}_{>0}$, $\phi_v(x)$, and $\Phi_v(x)$ denote, respectively, the univariate Student's t p.d.f. and c.d.f., with v degrees of freedom, evaluated at x, $\mathbf{R}$ is a d-dimensional correlation matrix (12.2), and, with $\mathbf{z} = (z_1, z_2,\dots,z_d)'$, $\psi(\cdot;\cdot) = \psi(z_1, z_2,\dots,z_d;\mathbf{R}, v)$ is the **density weighting function** given by

$$\psi(\cdot;\cdot) = \frac{\Gamma\{(v+d)/2\}\{\Gamma(v/2)\}^{d-1}}{[\Gamma\{(v+1)/2\}]^d|\mathbf{R}|^{1/2}}\left(1+\frac{\mathbf{z}'\mathbf{R}^{-1}\mathbf{z}}{v}\right)^{-(v+d)/2}\prod_{i=1}^{d}\left(1+\frac{z_i^2}{v}\right)^{(v+1)/2}. \tag{12.22}$$

If $v_0 = v_i$, $i = 1, 2,\dots,d$, then $x_i = \Phi_{v_0}^{-1}(\Phi_{v_i}(x_i))$, $i = 1,\dots,d$, and $\mathbf{T} \sim t_v(\mathbf{0},\mathbf{R})$, where we set $v = v_0$.

Fang et al. (2002) refer to this as a multivariate asymmetric t distribution and write $\mathbf{T} \sim \mathrm{AM}t_d(\cdot)$, where d is the dimension of $\mathbf{T}$, but we choose not to use this notation because, while the multivariate density is indeed asymmetric,[3] the univariate margins are not. We express a random variable $\mathbf{T}$ with location vector $\boldsymbol{\mu} = (\mu_1,\dots,\mu_d)'$, scale terms $\sigma_i > 0$, $i = 1,\dots,d$, and positive definite **dispersion matrix** (not covariance matrix) $\boldsymbol{\Sigma} = \mathbf{DRD}$, where $\mathbf{D} = \mathrm{diag}([\sigma_1,\dots,\sigma_d])$ and $\mathbf{R}$ is a correlation matrix (12.2), as $\mathbf{T} \sim \mathrm{FaK}(\mathbf{v},\boldsymbol{\mu},\boldsymbol{\Sigma})$, with FaK a reminder of the involved authors, and density

$$f_{\mathbf{T}}(\mathbf{y};\mathbf{v},\boldsymbol{\mu},\boldsymbol{\Sigma}) = \frac{f_{\mathbf{X}}(\mathbf{x};\mathbf{v},\mathbf{R})}{\sigma_1\sigma_2\cdots\sigma_d},\quad \mathbf{x} = \left(\frac{y_1-\mu_1}{\sigma_1},\dots,\frac{y_d-\mu_d}{\sigma_d}\right)',\quad \mathbf{R} = \mathbf{D}^{-1}\boldsymbol{\Sigma}\mathbf{D}^{-1}, \tag{12.23}$$

with $f_{\mathbf{X}}$ given in (12.21). The margin $(T_i - \mu_i)/\sigma_i$ is standard Student's t with v_i degrees of freedom, irrespective of v_0. Thus, $\mathbb{E}[T_i] = \mu_i$, if $v_i > 1$, and $\mathbb{V}(T_i) = \sigma_i^2 v_i/(v_i - 2)$, if $v_i > 2$.

Simulation of $\mathbf{T} = (T_1,\dots,T_d)' \sim \mathrm{FaK}(\mathbf{v},\boldsymbol{\mu},\boldsymbol{\Sigma})$ can be done as follows. With $\mathbf{R}$ as given in (12.23), simulate $\mathbf{Y} = (Y_1,\dots,Y_d)' \sim t_{v_0}(\mathbf{0},\mathbf{R})$ and set

$$T_i = \mu_i + \sigma_i\Phi_{v_i}^{-1}(\Phi_{v_0}(Y_i)),\quad i = 1,\dots,d. \tag{12.24}$$

This is implemented in Listing 12.6 (for the more general AFaK setting discussed below). The margin $Y_i \sim t_{v_0}$, so that in (12.24), $\Phi_{v_0}(Y_i) \sim \mathrm{Unif}(0, 1)$. Thus, from the **probability integral transform,** $\Phi_{v_i}^{-1}(\Phi_{v_0}(Y_i)) \sim t_{v_i}$. The T_i are not independent because the Y_i are not independent.

Figure 12.8 shows a selection of examples from the bivariate FaK distribution, all with zero location and unit scales.

The parameter v_0 influences the dependency structure of the distribution. To illustrate this, Figure 12.9 shows (12.21) with $v_1 = 2$, $v_2 = 4$, and six different values of v_0, and with $\mathbf{R} = \mathbf{I}$, so that all the X_i are uncorrelated. Overlaid onto each plot is a scatterplot of 100,000 simulated realizations of the density, but such that, for clarity, the points in the middle of the density are not shown. When compared to scatterplots of financial returns data, it appears that only values of $v_0 \geqslant \max_i v_i$,

2 There is a minor typographical error in the p.d.f. as given in Fang et al. (2002, Eq. 4.1) that is not mentioned in the corrigendum in Fang et al. (2005), but which is fixed in the p.d.f. as given in the monograph of Kotz and Nadarajah (2004, Eq. 5.16), but which itself introduces a new typographical error.

3 A multivariate cumulative distribution function is said to be symmetric if $F_{\mathbf{X}}(X_1, X_2,\dots,X_d) = F_{\mathbf{X}}(X_{i_1}, X_{i_2},\dots,X_{i_d})$, for any permutation $\{i_1, 1_2,\dots,i_d\}$ of $\{1, 2,\dots,d\}$. This condition is equivalent to exchangeability; see Section I.5.2.3.

```
function T=FaKrnd(sim,v,mu,scales,R,noncen)
v0=v(1); v=v(2:end); p=length(v);
if nargin<6, noncen=[]; end
if isempty(noncen), noncen=zeros(p+1,1); end
a0=noncen(1); a=noncen(2:end);
T=zeros(sim,p);
for j=1:sim
  r=mvtrnd(R,v0,1);
  for i=1:p
    if a0==0, term1=tcdf(r(i),v0); else term1=nctcdf(r(i),v0,a0); end
    if a(i)==0
      term2 = tinv(term1,v(i));
    else
      term2 = nctinv(term1,v(i),a(i));
    end
    T(j,i)=mu(i)+scales(i)*term2;
  end
end
```

Program Listing 12.6: Simulates realizations from FaK distribution (12.23) for `noncen` all zeros, otherwise is for the asymmetric FaK (AFaK) introduced below.

$i = 1, \ldots, d$, are of interest, and one could entertain just setting $v_0 = \max_i v_i$. Based on the empirical forecasting exercises for financial returns data in Paolella and Polak (2015a), $\hat{v}_0$ is very close to $\max(\hat{v}_1, \hat{v}_2)$ when it is freely estimated and suggests forgoing its estimation.

Let $\mathbf{V} = \mathbb{V}(\mathbf{T})$ denote the covariance matrix of $\mathbf{T} = (T_1, \ldots, T_d)' \sim \text{FaK}(\mathbf{v}, \boldsymbol{\mu}, \boldsymbol{\Sigma})$. Unfortunately, Fang et al. (2002) and Abdous et al. (2005) are silent on the off-diagonal elements of $\mathbf{V}$. This is addressed with approximations in Section 12.B.

12.5.2 The AFaK Distribution

While flexible in terms of allowing for differing degrees of freedom, each marginal is still restricted to being symmetric about its location parameter. A simple idea is to replace (12.21) with a structure that augments each Student's t margin with a noncentrality parameter, say $\theta_i \in \mathbb{R}$, $i = 0, 1, \ldots, d$.

Thus, for $\mathbf{X} = (X_1, \ldots, X_d)'$, $\mathbf{x} = (x_1, \ldots, x_d)'$, and with $\phi_{v,\theta}(x)$ and $\Phi_{v,\theta}(x)$ denoting the p.d.f. and c.d.f., respectively, of the noncentral t distribution at $x \in \mathbb{R}$,

$$f_{\mathbf{X}}(\mathbf{x}; \mathbf{v}, \boldsymbol{\theta}, \mathbf{R}) = \psi(\Phi^{-1}_{v_0,\theta_0}(\Phi_{v_1,\theta_1}(x_1)), \ldots, \Phi^{-1}_{v_0,\theta_0}(\Phi_{v_d,\theta_d}(x_d)); \mathbf{R}, v_0) \prod_{i=1}^{d} \phi_{v_i,\theta_i}(x_i), \quad \theta_0 = 0, \tag{12.25}$$

still in conjunction with (12.22). The location-scale variant $f_{\mathbf{T}}(\mathbf{y}; \mathbf{v}, \boldsymbol{\theta}, \boldsymbol{\mu}, \boldsymbol{\Sigma})$ is analogous to (12.23),

$$f_{\mathbf{T}}(\mathbf{y}; \mathbf{v}, \boldsymbol{\theta}, \boldsymbol{\mu}, \boldsymbol{\Sigma}) = \frac{f_{\mathbf{X}}(\mathbf{x}; \mathbf{v}, \boldsymbol{\theta}, \mathbf{R})}{\sigma_1 \sigma_2 \cdots \sigma_d}, \quad \mathbf{x} = \left(\frac{y_1 - \mu_1}{\sigma_1}, \ldots, \frac{y_d - \mu_d}{\sigma_d} \right)', \quad \mathbf{R} = \mathbf{D}^{-1} \boldsymbol{\Sigma} \mathbf{D}^{-1}, \tag{12.26}$$

where $f_{\mathbf{X}}$ is given in (12.25) and, as before, $\mathbf{D} = \text{diag}([\sigma_1, \ldots, \sigma_d])$. For a random variable with density (12.26), we will write $\mathbf{T} \sim \text{AFaK}(\mathbf{v}, \boldsymbol{\theta}, \boldsymbol{\mu}, \boldsymbol{\Sigma})$, for **asymmetric FaK**. Drawing sample AFaK realizations is the same as that for FaK, but replacing the Φ_{v_i} in (12.24) by Φ_{v_i,θ_i}, $i = 0, 1, \ldots, d$.

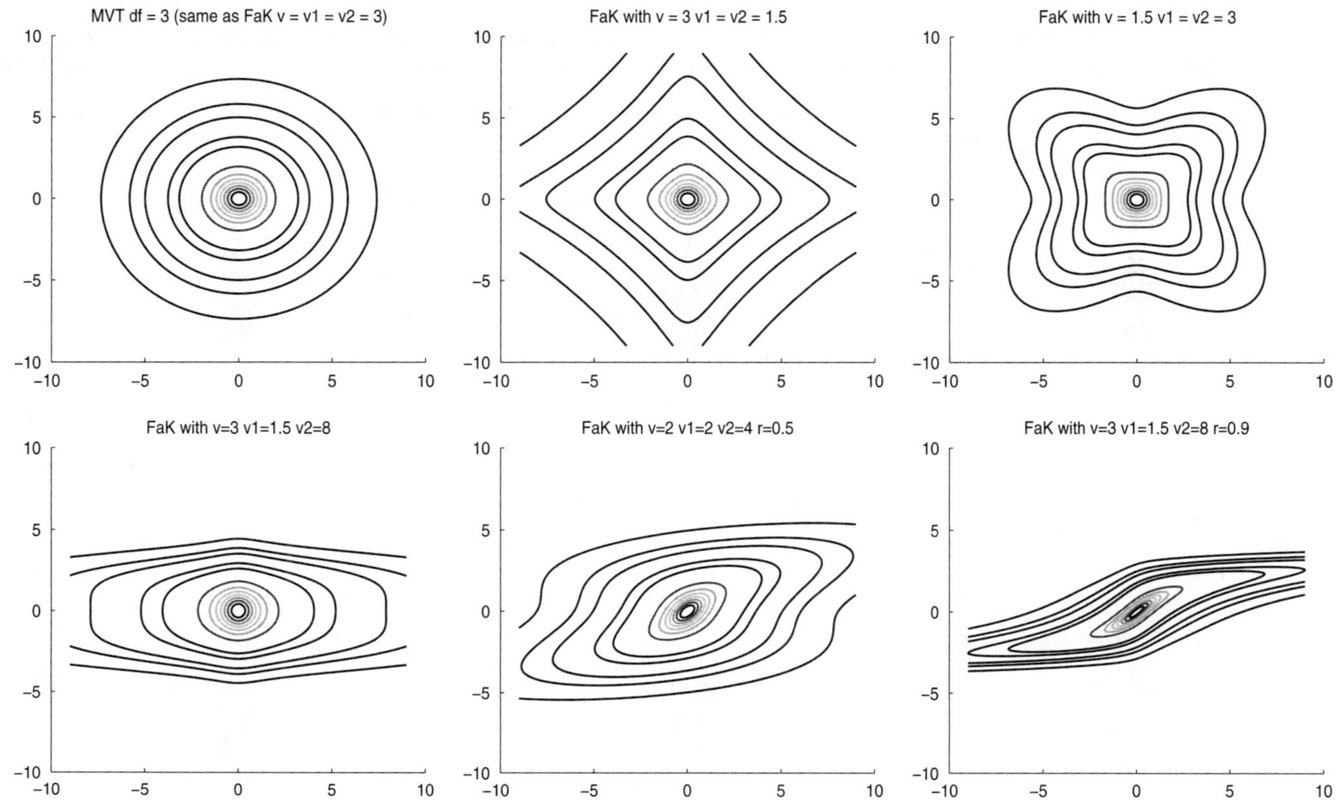

Figure 12.8 Examples of the bivariate FaK distribution (12.23), each with zero location and unit scale, and degrees of freedom parameters given in the title (writing v for parameter v_0).

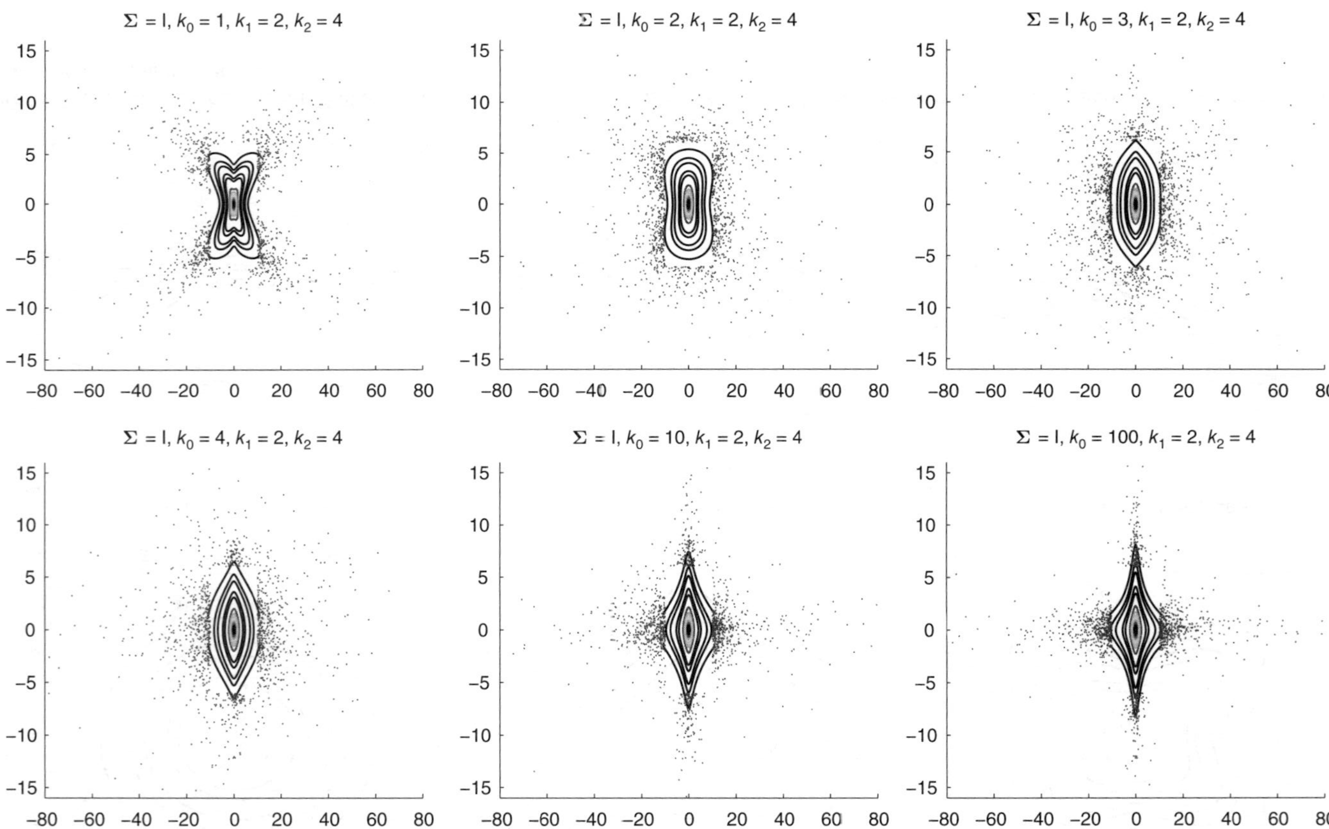

Figure 12.9 Examples of the bivariate FaK distribution (12.23) based on simulation, each with zero location, unit scale, and zero correlation. The case with $v_0 = \max_i v_i$ in the bottom left (in the graphics titles, having used k instead of v) appears the most appropriate for financial returns data.

```
function f=FFKpdfvec(ymat,v,theta,mu,scale,R)
[nobs,p]=size(ymat);
if length(theta)== p, theta=[0 ; theta(:)]; end
if length(theta)~= (p+1), error('bad theta'), end
if length(v)~= (p+1), error('bad v'), end
xmat=zeros(nobs,p); Zmat=zeros(nobs,p);
for i=1:p, xmat(:,i) = ( ymat(:,i)-mu(i) ) / scale(i); end
v0=v(1); v=v(2:end); % degrees of freedom parameters
a0=theta(1); a=theta(2:end); % a for asymmetry parameters
tol=1e-6; pdfprod=ones(nobs,1);
if sum(abs(theta))<1e-10 % close enough to zero
  for i=1:p
    x=xmat(:,i); pdfprod=pdfprod.*tpdf(x,v(i));
    cdf=tcdf(x,v(i)); cdf=min(cdf,1-tol); cdf=max(cdf,tol);
    Zmat(:,i)=tinv(cdf,v0);
  end
else
  for i=1:p
    x=xmat(:,i); pdfprod=pdfprod.*nctpdf(x,v(i),a(i));
    %x=xmat(:,i); pdfprod=pdfprod .* exp(  stdnctpdfln_j(x,v(i),a(i)) );
    % previous line is slightly faster
    cdf=nctcdf(x,v(i),a(i)); cdf=min(cdf,1-tol); cdf=max(cdf,tol);
    if abs(a0)<1e-10
      Zmat(:,i)=tinv(cdf,v0);
    else
      Zmat(:,i)=nctinv(cdf,v0,a0);
    end
  end
end
term1 = exp( gammaln((v0+p)/2) + (p-1)*gammaln(v0/2) - p*gammaln((v0+1)/2) );
term1 = term1 / sqrt(det(R)); Rinv=inv(R); temp=zeros(nobs,1);
for i=1:nobs, z=Zmat(i,:)'; temp(i)=z'*Rinv*z; end %#ok<MINV>
term2=(1+temp/v0).^(-(v0+p)/2); term3=prod( (1+Zmat.^2/v0).^((v0+1)/2) ,2);
f = (term1.*term2.*term3.*pdfprod) / prod(scale);
```

Program Listing 12.7: Computes the p-dimensional (A)FaK density for a set of nobs points.

The program in Listing 12.7 computes the density of the (A)FaK for a set of points. Useful also is to be able to compute the c.d.f., at least in the bivariate case (higher dimensions will be very slow). The program in Listing 12.8 shows how this is elegantly accomplished with nested integration.

It should be clear from the construction (and simulations confirm) that, for $\theta_0 = 0$, and irrespective of v_0, the univariate margins are $T_i \sim t'(v_i, \theta_i, \mu_i, \sigma_i)$, where $t'(v, \theta, \mu, \sigma)$ denotes the (singly) noncentral Student's t distribution with v degrees of freedom, noncentrality parameter θ, location μ, and scale σ. (This fact is crucial to the first step of our two-step estimation method discussed below.) It is less obvious what happens when $\theta_0 \neq 0$, as ψ in (12.25) may not be a copula. Simulations confirm that the margins are in fact *not* $t'(v_i, \theta_i, \mu_i, \sigma_i)$ for $\theta_0 \neq 0$, and we do not further entertain this case.

Thus, using the expectation of the univariate NCT distribution, if $\min_i(v_i) > 1$, $\mathbb{E}[\mathbf{T}] = (\mathbb{E}[T_1], \mathbb{E}[T_2], \ldots, \mathbb{E}[T_d])'$, where

$$\mathbb{E}[T_i] = \mu_i + \theta_i \left(\frac{v_i}{2}\right)^{1/2} \frac{\Gamma(v_i/2 - 1/2)}{\Gamma(v_i/2)}, \qquad i = 1, \ldots, d. \tag{12.27}$$

```
function cdf = AFaKcdf(df,noncen,mu,scale,R,xup,yup)
% pass df as [v0 v1 v2] and noncen as [0 gam1 gam2]
% As a check: (pick any v)
%  v=3; df=[v v v]; noncen=[0 0 0]; scale=[1 1]; mu=[0 0]; R12=0; R=[1 R12; R12 1];
%  AFaKcdf(df,noncen,mu,scale,R,0,0)    % Should return 0.250000. It does.

ATOL=1e-10; RTOL=1e-6;  % 10 and 6 are the defaults
cdf = quadgk(@(yvec) int1(yvec,df,noncen,mu,scale,R,xup), ...
      -Inf,yup,'AbsTol',ATOL,'RelTol',RTOL);

function Int=int1(yvec,df,noncen,mu,scale,R,xup)
Int=zeros(size(yvec)); ATOL=1e-10; RTOL=1e-6;
for i=1:length(yvec),  y=yvec(i);
  Int(i) = quadgk(@(x) int2(x,y,df,noncen,mu,scale,R), ...
           -Inf,xup,'AbsTol',ATOL,'RelTol',RTOL);
end

function f = int2(x,y,df,noncen,mu,scale,R)
yy=y*ones(1, length(x)); pass=[x ; yy];
f=FFKpdfvec(pass',df,noncen,mu,scale,R)';
```

Program Listing 12.8: Computes the bivariate (A)FaK c.d.f. at `xup`, `yup`. See also Listing 12.21, for the covariance.

To illustrate the new construction (12.25), Figure 12.10 shows several contour plots in the bivariate case. Numeric integration confirms that (12.25) yields a proper density.

12.5.3 FaK and AFaK Estimation: Direct Likelihood Optimization

We wish to develop a program for computing the m.l.e. based on direct evaluation of the log-likelihood. It is useful, as it shows how to deal with estimation in the general d-dimensional case, notably the off-diagonal elements of $\mathbf{R}$. This requires use of the so-called **vech operator** and its inverse procedure, where vech returns as a vector the elements on and below the diagonal of a matrix. This is already built into Matlab, as function `tril` (lower triangular), but to invert it requires a bit of coding. The reader should confirm that a symmetric $m \times m$ matrix has $u = (m+1)m/2$ unique elements, and that, in terms of u, $m = (\sqrt{8u+1} - 1)/2$. The program in Listing 12.9 computes vech and its inverse.

The program in Listings 12.10–12.12 computes the m.l.e. for the FaK or AFaK model of arbitrary dimension d. As it is easy to simulate from the AFaK, as given in Listing 12.6, we do the obvious thing: Based on a sample size of $n = 10{,}000$ and for several parameter constellations, for $d = 2$ and $d = 3$, the m.l.e. was computed (using starting values relatively far off from the true values) and seen to agree with the true parameter values to about three significant digits for the degrees of freedom parameters, and three or more for the remaining parameters. The program also outputs (as `param.Sig`) the covariance matrix based on the estimated parameters, using approximations (12.52) and (12.56) developed below in Section 12.B. This can be compared to the usual sample covariance matrix of the data.

The reader is encouraged to expand the program to incorporate the parametric and nonparametric bootstrap for determining confidence intervals of the parameters, as used in the next example.

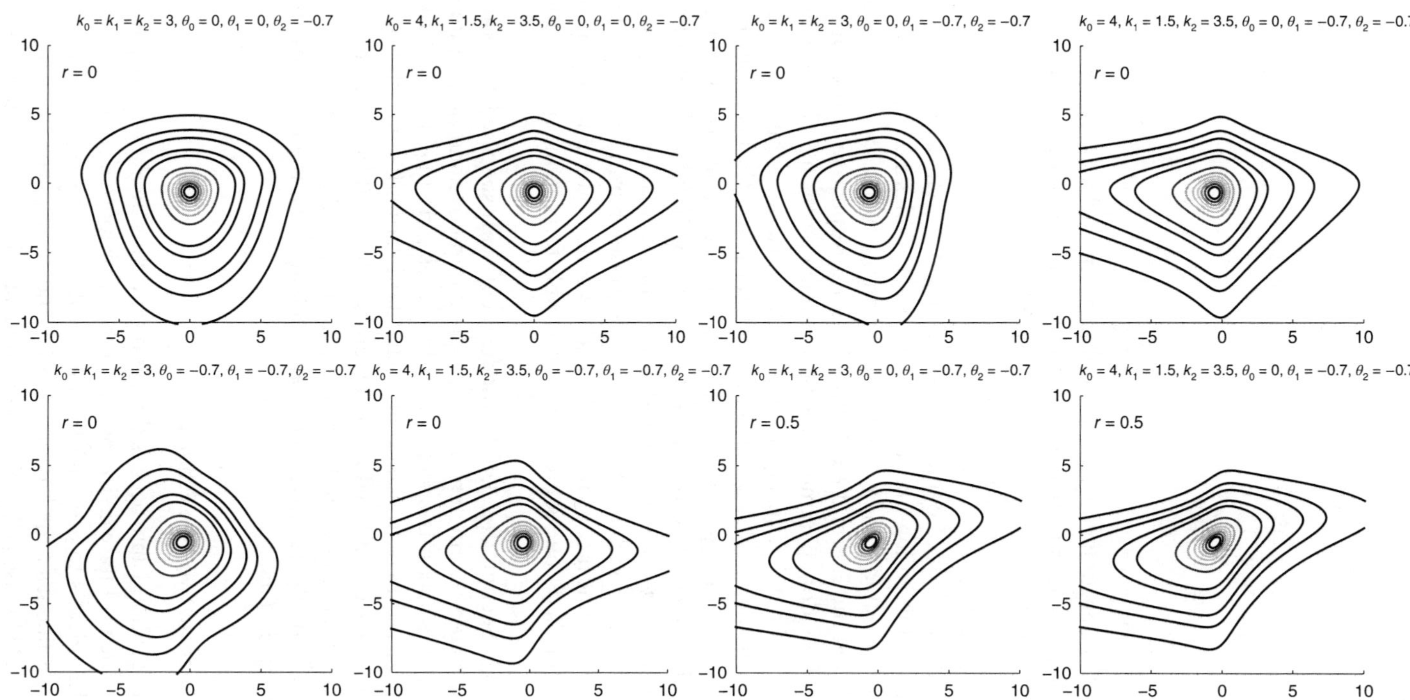

Figure 12.10 Examples of the bivariate AFaK (asymmetric FaK) distribution (12.25), each with zero location and unit scale (and using k instead of v).

```
function v=vech(M,invert)
if nargin>1 % M is a vector, and we return a matrix
  vec=M; n=(sqrt(8*length(vec)+1)-1)/2;
  V=zeros(n,n);
  for i=1:n
    take=(n-i+1); comp=vec(1:take); vec=vec(take+1:end);
    V(i:n,i)=comp; V(i,i:n)=comp';
  end, v=V;
else  % M is a matrix, and we return vech(M)
  tt=tril(M); v=tt(tt~=0);
end
```

Program Listing 12.9: If one argument is passed, computes the vech of a matrix. If the second argument is passed, constructs the matrix from a passed vector of values to invert the vech operator. That is, for symmetric matrix M, we have `M == vech(vech(M),1).`

Example 12.4 ***(Example 12.1 cont.)***
Recall the bivariate fit to the returns of Bank of America and Wal-Mart stock prices, using the multivariate t, resulting in an obtained log-likelihood of −7199.3. Tables 12.1 and 12.2 show the m.l.e. parameter estimates and associated standard errors based on the two Shaw and Lee (2008) constructions (12.15) and (12.18), and the FaK and AFaK distributions, respectively. The ability to allow each margin its own degrees of freedom parameter enables a much better fit, as seen from the obtained log-likelihood values. The FaK obtains the highest log-likelihood among the three symmetric models, while the AFaK results in a further increase (this being necessarily so—the likelihood can only increase when parameters are added), resulting in a likelihood ratio test p-value (based on its asymptotic χ^2_2 distribution under the null) of 0.030. As was emphasized in Section III.2.8 and will be further discussed in Chapter 13, the fact that this p-value is below 0.05 does not persuade us in any way to choose the AFaK model over the FaK or to conclude that asymmetry in the margins is "important" or "significant", and this, especially when one factors in that the i.i.d. (A)FaK model is anyway mis-specified with probability one. What counts is (in our context) out-of-sample forecasting performance among a set of practical models.

Recalling the discussion in Section III.2.3, if the model is correctly specified, then one would expect that inference on the estimated parameters based on the parametric and nonparametric bootstrap will be similar, while as the model mis-specification increases, one might expect these to diverge. For the four models considered here, the estimated standard errors from both bootstrap methods are reasonably close for the degrees of freedom parameters (which will generally have relatively wide confidence intervals, given their estimation difficulty), while those of the other parameters are all rather close, suggesting (informally at least) that these models are flexible enough to capture the salient features of the (unconditional) data. ■

12.5.4 FaK and AFaK Estimation: Two-Step Estimation

A drawback of the direct likelihood approach is that, as d (and thus the number of parameters) grows, estimation time will become prohibitive. To counter this, we propose a somewhat obvious two-step estimation procedure. The idea is to estimate the univariate margins separately, these being either a location-scale (central) Student's t, or location-scale noncentral t. For the latter, while the d.d.a. program in Listing 12.3 could be used (it is obviously applicable in the univariate case as well), we use

```
function [param,stderr,iters,loglik,Varcov] = AFaKestimation(data, AFaK, initvec)
% data is the usual T X p matrix of financial asset returns
% Set AFaK to 0 (default) for (symmetric) FaK
%             1 for AFaK (with noncentrality terms), theta_0= 0

if nargin<2, AFaK=0; end
if nargin<3, initvec=[]; else initvec=reshape(initvec,1,length(initvec)); end
[nobs p]=size(data);
O=ones(1,p); Odf=ones(1,p+1); ORmat=ones(1,p*(p-1)/2);
switch AFaK
  case 0
    %             df       mu    scale   R (off diagonals)
    bound.lo=    [1.0*Odf, -1*O, 0.01*O, -0.99*ORmat ];
    bound.hi=    [ 20*Odf,  1*O, 1000*O,  0.99*ORmat ];
    bound.which=[Odf     ,  0*O, O,       ORmat ];

    if isempty(initvec)
      initvec=[3*Odf, 0*O, O, 0.01*ORmat];
    end
  case 1 % here, there are an additional p noncen parameters
    %             df       noncen    mu   scale    R (off diagonals)
    bound.lo=    [1.0*Odf, -6*O,     -1*O, 0.01*O, -0.99*ORmat ];
    bound.hi=    [ 15*Odf,  6*O,      1*O, 1000*O,  0.99*ORmat ];
    bound.which=[Odf     , O,        0*O, O,       ORmat ];

    if isempty(initvec)
      initvec=[3*Odf, 0*O, 0*O, O, 0.01*ORmat];
    end
  otherwise
    error('Not valid AFaK Model Option')
end

maxiter=300; tol=1e-5; MaxFunEvals=25*maxiter;
opts=optimset('Display','iter','Maxiter',maxiter,'TolFun',tol,'TolX',tol, ...
    'MaxFunEvals',MaxFunEvals,'LargeScale','Off');

if 1==1
  [pout,fval,~,theoutput,~,hess]= fminunc(@(param) ...
      FFKloglik(param,data,AFaK,bound),einschrk(initvec,bound),opts);
else
  [pout,fval,~,theoutput]= fminsearch(@(param) ...
      FFKloglik(param,data,AFaK,bound),einschrk(initvec,bound),opts);
  hess=eye(length(pout));
end
V=inv(hess)/nobs; [param,V]=einschrk(pout,bound,V);
param=param'; Varcov=V; stderr=sqrt(diag(V));
loglik=-fval*nobs; iters=theoutput.iterations;
```

Program Listing 12.10: Computes the m.l.e. based on direct evaluation of the likelihood of the FaK or AFaK distribution (12.26). Continued in Listing 12.11.

```
if AFaK==0
  PP=param; clear param
  param.df=PP(1:p+1); param.mu=PP(p+2:2*p+1); param.scale=PP(2*p+2:3*p+1);
  Rterms=PP(3*p+2:end); param.Rterms=Rterms;
elseif AFaK==1
  PP=param; clear param
  param.df=PP(1:p+1); param.noncen=PP(p+2:2*p+1);
  param.mu=PP(2*p+2:3*p+1); param.scale=PP(3*p+2:4*p+1);
  Rterms=PP(4*p+2:end); param.Rterms=Rterms;
end

% now augment so I can vech it:
Rt=Rterms'; RR=[1 , Rt(1:p-1)]; Rt=Rt(p:end);
for i=2:p-1 % have to add a total of p ones into Rterms
  RR=[RR , 1 , Rt(1:p-i)]; Rt=Rt(p-i+1:end);  %#ok<AGROW>
end
RR=[RR , 1]; param.R=vech(RR,1);

kvec=param.df(2:end);
if AFaK==0
  kappa = sqrt(kvec./(kvec-2)); M = diag(param.scale .* kappa);
else
  theta=param.noncen;
  m1=sqrt(kvec/2) .* gamma(kvec/2-1/2) ./ gamma(kvec/2) .* theta;
  m2=kvec./(kvec-2) .* (1+theta.^2); varterm = m2-m1.^2;
  M=diag(param.scale .* sqrt(varterm));
end
param.Sig = M*param.R*M;
```

Program Listing 12.11: Continued from Listing 12.10.

```
function ll=FFKloglik(param,data,AFaK,bound)
if nargin<4, bound=0; end
if isstruct(bound)
  paramvec=einschrk(real(param),bound,999);
else
  paramvec=param;
end
[~, p]=size(data);
if AFaK==0 % symmetric case
  dfvec=paramvec(1:p+1); muvec=paramvec(p+2:2*p+1);
  scalevec = paramvec(2*p+2:3*p+1);
  Rterms=paramvec(3*p+2:end);
  noncenvec=zeros(1,p+1);
elseif AFaK==1 % asymmetric case but with v0==0
  dfvec=paramvec(1:p+1); noncenvec=[0 paramvec(p+2:2*p+1)];
  muvec=paramvec(2*p+2:3*p+1);
  scalevec = paramvec(3*p+2:4*p+1); Rterms=paramvec(4*p+2:end);
end
Rt=Rterms; RR=[1 , Rt(1:p-1)]; Rt=Rt(p:end);
for i=2:p-1, RR=[RR , 1 , Rt(1:p-i)]; Rt=Rt(p-i+1:end); end
RR=[RR , 1]; R=vech(RR,1);  if min(eig(R))<1e-5, ll=1e5; return, end
pdf = FFKpdfvec(data,dfvec,noncenvec,muvec,scalevec,R);
llvec=log(pdf); ll=-mean(llvec); if isinf(ll), ll=1e5; end
```

Program Listing 12.12: Continued from Listing 12.11.

Table 12.1 The estimated parameters using the set of 1,945 daily (log percentage) returns of Bank of America and Wal-Mart Stores, as depicted in Figures 12.3 and 12.4. "S-L" refers to Shaw and Lee, "loglik" is the log-likelihood evaluated at the obtained m.l.e., "std err Hess" refers to the approximate (asymptotic normal-based) standard errors obtained as output from the optimization, and "std err NPB" and "std err PB" refer to use of the nonparametric and parametric bootstrap, respectively.

S–L (12.15)	**Loglik**	v_1	v_2	μ_1	μ_2	**Scale 1**	θ	**Scale 2**
MLE	−7092	1.620	3.705	0.0274	−0.0070	0.923	0.545	1.081
Std err NPB		(0.078)	(0.317)	(0.026)	(0.029)	(0.027)	(0.026)	(0.030)
Std err PB		(0.082)	(0.337)	(0.035)	(0.036)	(0.033)	(0.031)	(0.035)
S–L (12.18)	**Loglik**	v_1	v_2	μ_1	μ_2	**Scale 1**	θ	**Scale 2**
MLE	−7142	1.601	4.813	0.0313	−0.0057	0.926	0.587	1.160
Std err Hess		(0.077)	(0.491)	(0.027)	(0.030)	(0.030)	(0.023)	(0.031)
Std err NPB		(0.072)	(0.508)	(0.027)	(0.028)	(0.027)	(0.023)	(0.029)
Std err PB		(0.067)	(0.498)	(0.024)	(0.024)	(0.029)	(0.023)	(0.032)

Table 12.2 Similar to Table 12.1 but for the FaK and AFaK distributions, as well as the MESTI, discussed in Section 12.6.2.

FaK (12.23)	**Loglik**	v_0	v_1	v_2			μ_1	μ_2	**Scale 1**	R_{12}	**Scale 2**
MLE	−7086	3.975	1.464	3.873			0.0331	0.0027	0.857	0.492	1.106
Std err Hess		(0.50)	(0.067)	(0.34)			(0.026)	(0.028)	(0.028)	(0.020)	(0.030)
Std err NPB		(0.56)	(0.058)	(0.38)			(0.025)	(0.031)	(0.024)	(0.020)	(0.030)
Std err PB		(0.53)	(0.068)	(0.35)			(0.026)	(0.028)	(0.029)	(0.020)	(0.033)
AFaK (12.26)	**Loglik**	v_0	v_1	v_2	θ_1	θ_2	μ_1	μ_2	**Scale 1**	R_{12}	**Scale 2**
MLE	−7079	3.849	1.473	3.887	−0.165	0.136	0.191	−0.192	0.857	0.492	1.106
Std err Hess		(0.48)	(0.068)	(0.35)	(0.055)	(0.094)	(0.057)	(0.12)	(0.028)	(0.020)	(0.030)
Std err NPB		(0.55)	(0.060)	(0.37)	(0.060)	(0.094)	(0.062)	(0.12)	(0.024)	(0.020)	(0.030)
Std err PB		(0.53)	(0.066)	(0.35)	(0.064)	(0.093)	(0.061)	(0.12)	(0.027)	(0.021)	(0.031)
MESTI (12.35)	**Loglik**		k_1	k_2	β_1	β_2	μ_1	μ_2	**Scale 1**	R_{12}	**Scale 2**
MLE	−7144		1.445	4.357	−0.159	0.147	0.186	−0.205	0.857	0.525	1.127

the saddlepoint approximation (s.p.a.), as discussed in Section III.10.3.1, given its speed and accuracy over a very large portion of the parameter space. This determines the location μ_i, the scale σ_i, and the one or two shape parameters v_i and θ_i, $i = 1, \dots, d$. For the copula shape parameter v_0, we take $\hat{v}_0 = \max\{\hat{v}_i\}$, though one could do a univariate optimization over this parameter, conditional on all others, very quickly.

```
function param = AFaK2stepSIMPLE(data, AFaK)
% Two-step for (A)FaK. Uses SPA for noncentral t.
% data is a T X p matrix of asset returns.  AFaK=0 for FaK. Else AFaK
if nargin<2, AFaK=0; end
[~, p]=size(data); dfvec=zeros(p,1); muvec=zeros(p,1); scalevec=zeros(p,1);
if AFaK==0
  for i=1:p
    pp = Studentstestimation(data(:,i)); % df, location and scale of Student t
    dfvec(i)=pp(1); muvec(i)=pp(2); scalevec(i)=pp(3);
  end
  theta=zeros(p,1); noncenvec=theta;
else
  noncenvec=zeros(p,1);
  for i=1:p
    pp = Noncentraltestimation(data(:,i),1,1); % Uses the SPA
    noncenvec(i)=pp(1); dfvec(i)=pp(2); %pp = [noncen df mu scale]
    muvec(i)=pp(3); scalevec(i)=pp(4);
  end
end
param.df=[max(dfvec) ; dfvec]; param.noncen=[0 ; noncenvec];
param.mu=muvec; param.scale = scalevec;
param.R=corr(data); v=vech(param.R); param.Rterms=v(v<0.999999);
```

Program Listing 12.13: Two-step estimation of the FaK and AFaK models. Program `Studentstestimation` is the same as `tlikmax` in Listing III.4.6, just augmented for weighted likelihood (not used here), as discussed in Chapter 13. Program `Noncentraltestimation` estimates the parameters of the location-scale NCT using the saddlepoint approximation, as discussed in Section III.9.3.

What remains is parameter **R**, which can be quickly estimated by the sample correlation of the location-scale adjusted data. The program in Listing 12.13 shows the simple code to accomplish this. Note that, like the direct likelihood method in Section 12.5.3, the two-step method can also be augmented to deliver confidence intervals and/or standard errors of the parameters with the parametric and nonparametric bootstrap. As guessed, the reader is encouraged to do so. With (i) the s.p.a. for the noncentral t, (ii) just setting $\hat{v}_0 = \max\{\hat{v}_i\}$, and (iii) using the sample correlation, this two-step procedure is extremely fast.

Simulations confirm that, for FaK and AFaK, the location, scale, degrees of freedom, and, for AFaK, the noncentrality parameters, are estimated with virtually the same accuracy as the direct method. We wish to examine the accuracy of the correlation term $\widehat{R}_{12}$, using the bivariate case. We perform a simulation in the FaK case (as it is faster to simulate and estimate) using 500 replications, a sample size of 10,000, two values of R_{12}, fixed $v_2 = 4$, and three values of v_1, namely 1.8, 2.0, and 2.5. For the former two cases, second moments do not exist, and we would expect that the estimated correlation term in the two-step procedure will not perform well.

This is confirmed in Figure 12.11. The top two panels show kernel density estimates of the resulting $\widehat{R}_{12}$, based on the two-step method. As v_1 decreases below two, the sample-correlation-based estimator of R_{12} becomes highly biased and with a high variability. The bottom two panels are similar, based on the same simulated data, but having used full maximum likelihood. In this case, the kernel densities corresponding to each of the three choices of v_1 *are identical* (up to three significant digits)

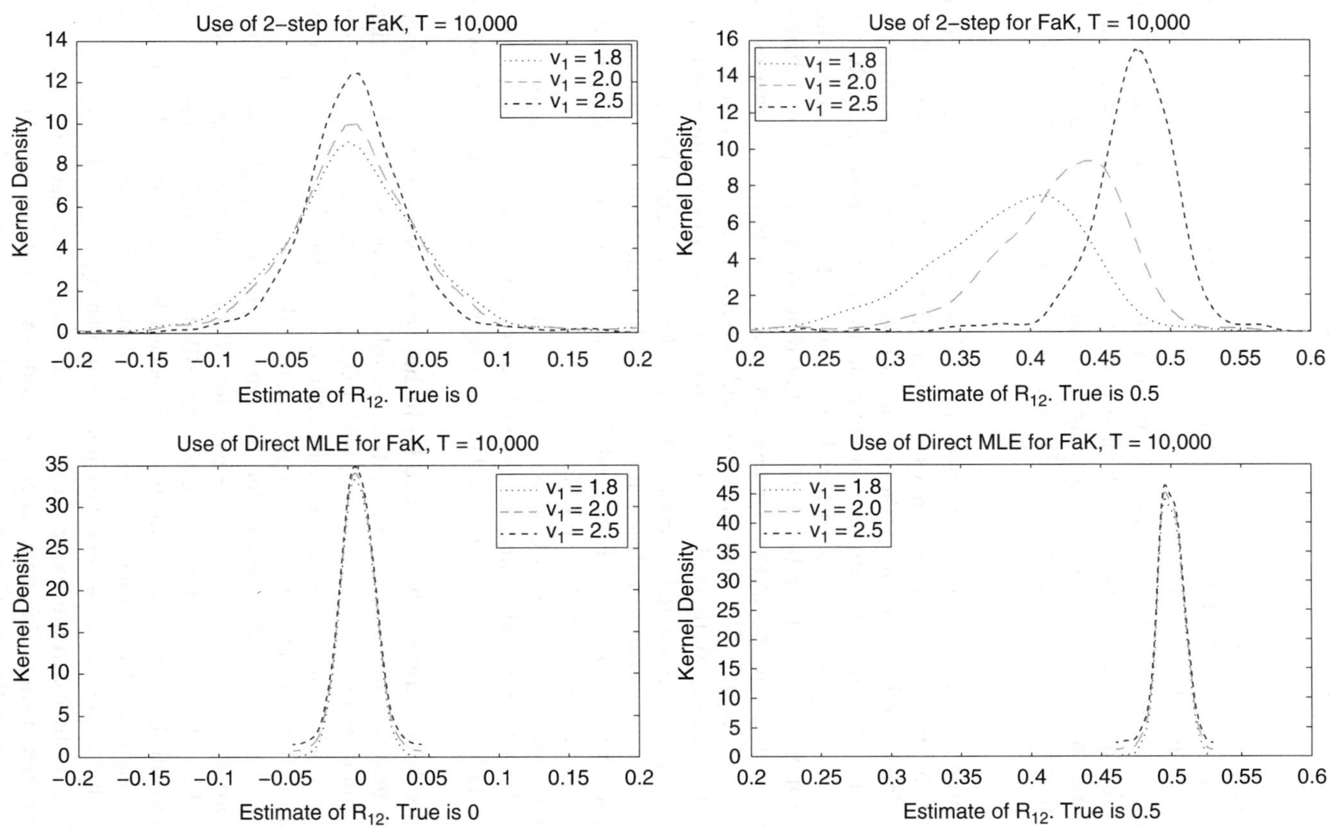

Figure 12.11 **Top:** Kernel density plots of $\widehat{R}_{12}$ based on 500 replications and $T = 10{,}000$ observations using the two-step method of estimating the parameters of the FaK model, with $\upsilon_2 = 4$, υ_1 indicated in the graphs, and two choices of R_{12}, zero and 0.5. **Bottom:** Same, but having used full maximum likelihood estimation via the direct method, and plotted with a vertical offset because *they are otherwise graphically identical*; see the text for explanation.

and are thus plotted with a vertical offset, just for visualization purposes. This is not a mistake: It happens because, for each of the three v_i-values, the 500 simulated series are based on seed values $1, 2, \dots, 500$.

Example 12.5 ***(Example 12.2 cont.)***
For the same data as used in Table 12.2, the point estimates for the AFaK are $\hat{v}_1 = 1.48$, $\hat{v}_2 = 4.36$ (and $\hat{v}_0 = \hat{v}_2$); $\hat{\theta}_1 = -0.16$, $\hat{\theta}_2 = 0.15$ (and $\hat{\theta}_0 \equiv 0$); $\hat{\mu}_1 = 0.19$, $\hat{\mu}_2 = -0.21$; $\hat{\sigma}_1 = 0.87$, $\hat{\sigma}_2 = 1.13$, and $\widehat{R}_{12} = 0.34$. These compare well with the results based on the full m.l.e. given in Table 12.2, though the estimates of R_{12} differ somewhat, as might have been expected, given that the existence of second moments of the first time series is questionable, rendering the sample correlation matrix an inconsistent and possibly highly unreliable estimator, particularly, as seen from Figure 12.11, as $|R_{12}|$ grows away from zero. ■

There are (at least) three ways of dealing with the estimation of the correlation matrix when second moments do not exist. The first is to use a different estimator for the R_{ij}, as is common in the copula literature, such as Kendall's τ and Spearman's ρ (these being conveniently built into Matlab).

The second way to address the problem, common for financial data, is to impose a GARCH process on the scale terms of each margin. The unconditional distribution of a GARCH process (or more general stochastic volatility structures) can be very heavy tailed, while the conditional distribution, having addressed the changing scale term, results in less heavy tails of the innovation process, besides the fact that the model itself might be "less mis-specified" and (often, but not always) better suited for short-horizon density forecasting. With Student's $t(v)$ as the assumed process, unconditional daily returns data typically exhibit an estimated v between two and five (see Figure 12.1), whereas, after controlling for GARCH effects, $\hat{v}$ is almost always at least four. Thus, the traditional estimator of the correlation matrix $\mathbf{R}$, when based on GARCH-filtered data, should be unproblematic, albeit clearly not as efficient as the m.l.e., as seen from Figure 12.11.

Note that the problem persists if the non-Gaussian stable distribution is the correct innovation process in a conditional GARCH model. In practice, one of course never knows the correct distribution, and so, in a heavy-tailed context, we recommend (with or without GARCH), the following third solution.

A third way, given its demonstrated efficiency and far higher speed, is to use the first part of the two-step procedure to get all the margin parameter estimates, followed by sequential univariate optimizations of the likelihood, as in the direct method, with fixed parameters from the first step, just over each of the R_{ij} for which $\hat{v}_i$ or $\hat{v}_j$ is less than (or just above) two. This latter step is given in the function in Listing 12.14. Thus, even for large d, this procedure will still be relatively fast.

Inspection of Figure 12.11 suggests that, under the true d.g.p. of (A)FaK, particularly with low degrees of freedom, this procedure should be used for *all* of the (lower diagonal) R_{ij} terms, given (i) the much higher efficiency of the m.l.e. estimator, (ii) its applicability irrespective of the values of v_i and v_j, and (iii) the fact that the likelihood is accessible and only $d(d-1)/2$ univariate optimizations are required. The reader is encouraged to make the program to do this, say `FangFangKotzestimation2step`—a very easy job, given Listings 12.10, 12.13, and 12.14. To ensure the resulting $\widehat{\mathbf{R}}$ is positive definite, we can repeatedly shrink its off-diagonal elements to their average until this is obtained, using (13.5) given later, with code shown in Listing 12.15.

Related (advanced) results on the behavior of sample covariance matrices in a heavy-tailed setting can be found in Davis et al. (2016a,b) and the references therein.

```
1  function [param,stderr,iters,loglik,Varcov] = AFaK_MLE_R_biv(Y,paramfix,rho)
2  if nargin<7, rho=1; end
3  nobs=length(Y);
4  bound.lo=-0.999; bound.hi=0.999; bound.which=1;
5  maxiter=300; tol=1e-5; MaxFunEvals=25*maxiter;
6  opts=optimset('Display','none','Maxiter',maxiter,'TolFun',tol,'TolX',tol, ...
7      'MaxFunEvals',MaxFunEvals,'LargeScale','Off');
8  Rsamp=corr(Y); initvec=Rsamp(1,2);
9  [pout,fval,~,theoutput,~,hess]= fminunc(@(param) ...
10     AFaK_R_loglik(param,Y,paramfix,rho,bound), ...
11     einschrk(initvec,bound),opts);
12 V=inv(hess)/nobs; [param,V]=einschrk(pout,bound,V);
13 Varcov=V; stderr=sqrt(diag(V));
14 loglik=-fval*nobs; iters=theoutput.iterations;
15
16 function ll = AFaK_R_loglik(param,Y,paramfix,rho,bound)
17 if isstruct(bound), R12=einschrk(real(param),bound,999); else R12=param; end
18 dfvec=paramfix.df; noncenvec=paramfix.noncen;
19 muvec=paramfix.locvec; scalevec=paramfix.scalevec;
20 R=[1 R12; R12 1]; T=length(Y); tvec=(1:T);
21 omega=(T-tvec+1).^(rho-1); w=T*omega/sum(omega);
22 pdf = FFKpdfvec(Y,dfvec,noncenvec,muvec,scalevec,R);
23 llvec=log(pdf) .* w'; ll=-mean(llvec); if isinf(ll), ll=1e5; end
```

Program Listing 12.14: Maximum likelihood estimation of the correlation term (off-diagonal term R_{12}) of the bivariate AFaK model, conditional on all the other model parameters, as passed in the structure `paramfix`, with entries given in lines 18 and 19. Scalar `rho` passed to the function is for weighted likelihood, as discussed in Section 13.

```
low=1e-6; bad=any(eig(param.R)<low); sR2=0.02;
while bad
  a=U'*(param.R-eye(p))*U/p/(p-1);
  disp('Shrinking R to enforce positive definite')
  param.R = (1-sR2) * param.R + sR2 *( (1-a)*eye(p)+a*J );
  bad=any(eig(param.R)<low);
end
```

Program Listing 12.15: Shrink the estimated correlation matrix via (13.5) until it is positive definite.

12.5.5 Sums of Margins of the AFaK

In the i.i.d. setting, based on a time series of (A)FaK data $\mathbf{Y}_1, \ldots, \mathbf{Y}_T$, for $\mathbf{Y}_t = (Y_{1,t}, \ldots, Y_{d,t})'$, $t = 1, \ldots, T$, the predictive densities for time periods $T+1, T+2, \ldots$, are all the same, namely, from (12.26),

$$f_{\mathbf{Y}}(\mathbf{y}; \widehat{\mathbf{v}}, \widehat{\boldsymbol{\theta}}, \widehat{\boldsymbol{\mu}}, \widehat{\boldsymbol{\Sigma}}) = \frac{f_{\mathbf{X}}(\mathbf{x}; \widehat{\mathbf{v}}, \widehat{\boldsymbol{\theta}}, \widehat{\mathbf{R}})}{\hat{\sigma}_1 \hat{\sigma}_2 \cdots \hat{\sigma}_d}, \quad \mathbf{x} = \left(\frac{y_1 - \hat{\mu}_1}{\hat{\sigma}_1}, \ldots, \frac{y_d - \hat{\mu}_d}{\hat{\sigma}_d} \right)', \tag{12.28}$$

where $f_{\mathbf{X}}$ is given in (12.25), and hatted values indicate parameter estimates. We are concerned with the distribution of $P = \mathbf{w}'\mathbf{Y}$, where $\mathbf{w} = (w_1, \ldots, w_d)' \in \mathbb{R}^d$. This arises, for example, in the study of the portfolio distribution, when $\mathbf{Y}$ is the random variable of, say, tomorrow's returns on a specified set of d

assets. In that case, $\sum_{i=1}^{d} w_i = 1$ and is often restricted such that w_i is non-negative (no short-selling), so that $\mathbf{w} \in \mathcal{A}$, where $\mathcal{A}$ is given in (11.41).

The mean of P is just $\mathbb{E}[P] = \mathbf{w}'\mathbb{E}[\mathbf{Y}]$, and using (12.27). However, unlike with the distribution of the sum of (weighted) margins from the multivariate (noncentral) t distribution, that of the FaK or AFaK is analytically intractable when the v_i are not all the same. An idea is to use simulation of P, from which any measurable quantity of interest can be elicited; for portfolio analysis, this is typically the mean (which we already have analytically), and one or more measures of risk, such as the variance, value-at-risk (VaR), or the expected shortfall (ES).

The distribution of P is empirically generated by drawing s_1 replications from (12.28), stored in a $d \times s_1$ matrix, say $\mathbf{M}$, and then computing the $s_1 \times 1$ vector

$$\widetilde{\mathbf{P}} = \widetilde{\mathbf{P}}_{\mathbf{w}} = \mathbf{w}'\mathbf{M}. \tag{12.29}$$

The problem with this idea in the aforementioned context is that a relatively large number of replications will be necessary for accurate tail risk measures. In particular, if only one vector $\mathbf{w}$ is of interest (such as for risk *assessment* of a given portfolio), this method will be relatively slow, most notably for the AFaK case, and a faster method based on a parametric approximation is developed in Section 12.A.

If, however, interest centers on potentially thousands of candidate $\mathbf{w}$, as would be required in portfolio optimization (or risk *management*), then it makes sense to simulate $\mathbf{M}$ once, based on a large number of replications, say $s_1 = 10{,}000$, and then use (12.29) for each $\mathbf{w}$, from which the desired risk measure can be obtained empirically. For example, assuming $\mathbf{M}$ has been generated and $\mathbf{w}$ is a candidate portfolio vector, the VaR and ES can be empirically computed as discussed in Section III.1.7 via the code:

```
P=w'*M; VaR=quantile(P,0.01); Plo=P(P<=VaR); ES=mean(Plo);
```

The estimates of the desired risk quantities for a given $\mathbf{w}$, and the resulting optimal portfolio vector, say $\mathbf{w}^*$, are dependent on $\mathbf{M}$, with the dependence weakening as $s_1 \to \infty$. Thus, the choice of s_1 should be taken as large as possible, given the nature of the application. We will make use of this model and method for obtaining the ES for portfolio optimization in Section 13.4.

12.6 MEST: Marginally Endowed Student's *t*

Recall the construction of the multivariate Student's t in (12.4), and the MVNCT in (12.5) via the latent univariate random variable $G \sim \text{IGam}(v/2, v/2)$, $v > 0$. These are special cases of the multivariate normal mean-variance mixture distribution, or MNMVM, introduced in Section 11.2.4. We wish to generalize this to the case such that each margin is endowed with its own G_i, $i = 1, \ldots, d$. Using the canonical form (the reason for which is discussed below), we define the vector random variable $\mathbf{X}$ to be a **heterogenous multivariate normal mean-variance mixture distribution** if

$$\mathbf{X} = \mathbf{m}(\mathbf{G}) + \mathbf{D}\mathbf{R}^{1/2}\mathbf{Z}, \tag{12.30}$$

where $\mathbf{Z} \sim \text{N}_d(\mathbf{0}, \mathbf{I})$, $\mathbf{D} = \text{diag}([G_1^{1/2}, G_2^{1/2}, \ldots, G_d^{1/2}])$ and $\mathbf{G} = (G_1, G_2, \ldots, G_d)'$ is a vector of possibly dependent non-negative continuous scalar-valued random variables (but still independent of $\mathbf{Z}$), $\mathbf{R}$ is a positive definite correlation matrix (12.2), and $\mathbf{m} : \mathbb{R}^d \to \mathbb{R}^d$ is a measurable mean function. From (12.30), it follows that $(\mathbf{X} \mid \mathbf{G} = \mathbf{g}) \sim \mathrm{N}(\mathbf{m}(\mathbf{g}), \mathrm{diag}(\mathbf{g})^{1/2}\mathbf{R}\,\mathrm{diag}(\mathbf{g})^{1/2})$.

12.6.1 SMESTI Distribution

As a first special case of (12.30), consider taking $G_i \overset{\text{indep}}{\sim} \mathrm{IGam}(k_i/2, k_i/2)$, $i = 1, \ldots, d$, with $\mathbf{k} = (k_1, k_2, \ldots, k_d)' \in \mathbb{R}^d_{>0}$ and $\mathbf{m}(\mathbf{G}) = \boldsymbol{\mu}$. Then $\mathbf{X} = \boldsymbol{\mu} + \mathbf{D}\mathbf{R}^{1/2}\mathbf{Z}$, implying $(\mathbf{X} \mid \mathbf{G} = \mathbf{g}) \sim \mathrm{N}(\boldsymbol{\mu}, \mathbf{DRD})$. Observe that $\boldsymbol{\mu}$ is a location term. Multiplying each margin of $\mathbf{X} - \boldsymbol{\mu}$ by a scale term $\sigma_i > 0$ gives $\mathbf{X} = \boldsymbol{\mu} + \mathbf{D}\boldsymbol{\Sigma}^{1/2}\mathbf{Z}$, with $\boldsymbol{\Sigma} = \mathbf{SRS}$, with $\mathbf{S} = \mathrm{diag}(\boldsymbol{\sigma})$ and $\boldsymbol{\sigma} = (\sigma_1, \ldots, \sigma_d)'$, implying $(\mathbf{X} \mid \mathbf{G} = \mathbf{g}) \sim \mathrm{N}(\boldsymbol{\mu}, \mathbf{D}\boldsymbol{\Sigma}\mathbf{D})$.

The p.d.f. of $\mathbf{X}$ is

$$f_{\mathbf{X}}(\mathbf{x}; \mathbf{k}, \boldsymbol{\mu}, \boldsymbol{\Sigma}) = \int_0^\infty \int_0^\infty \cdots \int_0^\infty f_{\mathbf{X}\mid\mathbf{G}}(\mathbf{x}; \mathbf{g}) \prod_{i=1}^d f_{G_i}(g_i; k_i/2, k_i/2)\, dg_1\, dg_2 \cdots dg_d. \tag{12.31}$$

We will denote this as $\mathbf{X} \sim \mathrm{SMESTI}(\mathbf{k}, \boldsymbol{\mu}, \boldsymbol{\Sigma})$, where SMESTI stands for *symmetric marginally endowed Student's t: independent case*.

A reason this construction might be seemingly uninteresting is that, except for extraordinarily low dimensions, (12.31) involves a nested d-dimensional integral, nullifying any possibility of computing the likelihood of a given data set. However, a two-step approach similar to the AFaK can be used: First estimate the location, scale, and degrees of freedom (and, for the asymmetric case discussed below, the noncentrality parameter) for each margin, and then, in a second step, deal with the off-diagonal terms of the dispersion matrix. This will be done in Section 12.6.3. In addition, by construction, simulating realizations of $\mathbf{X}$ is trivial; see below for a program to do this.

In the bivariate case, the p.d.f. of $\mathbf{X} = (X_1, X_2)'$ is given by

$$\int_0^\infty \int_0^\infty f_{\mathbf{X}\mid\mathbf{G}}(\mathbf{x}; \mathbf{g}) f_{G_1}(g_1; k_1/2, k_1/2) f_{G_2}(g_2; k_2/2, k_2/2)\, dg_1\, dg_2, \tag{12.32}$$

where, with $\phi(\cdot; \boldsymbol{\mu}, \boldsymbol{\Sigma})$ denoting the multivariate normal p.d.f. with mean $\boldsymbol{\mu}$ and variance covariance matrix $\boldsymbol{\Sigma}$,

$$f_{\mathbf{X}\mid\mathbf{G}}(\mathbf{x}; \mathbf{g}) = \phi(\mathbf{x}; \boldsymbol{\mu}, \mathbf{D}\boldsymbol{\Sigma}\mathbf{D}), \quad \boldsymbol{\mu} = \begin{bmatrix} \mu_1 \\ \mu_2 \end{bmatrix}, \quad \mathbf{D}\boldsymbol{\Sigma}\mathbf{D} = \begin{bmatrix} g_1\sigma_{11}^2 & \sqrt{g_1 g_2}\sigma_{12} \\ \sqrt{g_1 g_2}\sigma_{12} & g_2\sigma_{22}^2 \end{bmatrix}.$$

If we restrict $G_1 = G_2 =: G$, then $g_1 = g_2$, $k_1 = k_2 =: k$, and, with $G \sim \mathrm{IGam}(k/2, k/2)$ and $\mathbf{1}_d$ a d-length column of ones, (12.32) simplifies to

$$\begin{aligned} f_{\mathbf{X}}(\mathbf{x}; k\mathbf{1}_2, \boldsymbol{\mu}, \boldsymbol{\Sigma}) &= \int_0^\infty \int_0^\infty \phi(\mathbf{x}; \boldsymbol{\mu}, g_1\boldsymbol{\Sigma}) f_G(g_1; k/2, k/2) f_G(g_2; k/2, k/2)\, dg_1\, dg_2 \\ &= \int_0^\infty \phi(\mathbf{x}; \boldsymbol{\mu}, g_1\boldsymbol{\Sigma}) f_G(g_1; k/2, k/2)\, dg_1 \times \int_0^\infty f_G(g_2; k/2, k/2)\, dg_2, \end{aligned}$$

which is (12.3) for $d = 2$. The generalization to $d > 2$ is obvious.

Now consider the restriction that $k_1 = k_2 = k$, but not that $G_1 = G_2$. Then (12.32) is

$$f_{\mathbf{X}}(\mathbf{x}; k\mathbf{1}_2, \boldsymbol{\mu}, \boldsymbol{\Sigma}) = \int_0^\infty \int_0^\infty \phi(\mathbf{x}; \boldsymbol{\mu}, \mathbf{D}\boldsymbol{\Sigma}\mathbf{D}) f_{G_1}(g_1; k/2, k/2) f_{G_2}(g_2; k/2, k/2)\, dg_1\, dg_2, \tag{12.33}$$

while if we further take $\boldsymbol{\Sigma} = \text{diag}([\sigma_1^2, \sigma_2^2])$, then $\mathbf{D}\boldsymbol{\Sigma}\mathbf{D} = \text{diag}([g_1\sigma_1^2, \quad g_2\sigma_2^2])$, and (12.33) reduces to

$$\int_0^\infty \phi(x_1; \mu_1, g_1\sigma_1^2) f_{G_1}(g_1; k/2, k/2)\, \text{d}g_1 \times \int_0^\infty \phi(x_2; \mu_2, g_2\sigma_2^2) f_{G_2}(g_2; k/2, k/2)\, \text{d}g_2,$$

which is the product of the margins, each being Student's t with k degrees of freedom, showing that a type of multivariate t, with a single degree of freedom *and independent margins*, is a special case of the proposed SMESTI distribution. If $\boldsymbol{\Sigma}$ is endowed with off-diagonal elements, then, like the multivariate normal distribution, the dependence among the X_i is strictly via $\boldsymbol{\Sigma}$. Observe that the same decomposition goes through without restricting the k_i to be equal—all that is required is that $\boldsymbol{\Sigma}$ is diagonal, so that a multivariate t type of distribution, with independent margins and with different degrees of freedom for each marginal, is a special case of the proposed distribution.

Computationally speaking, density expression (12.31) can, in principle, be evaluated for any d using an algorithm that recursively calls a univariate numerical integration routine, until the inner integral is reached, in which case the integrand is delivered. This will of course be maddeningly slow for d larger than, say, three. The code to do this is given (for the more general MESTI case) in Listing 12.17.

As a check, when $\boldsymbol{\Sigma}$ is diagonal, the density can (and should) be evaluated as the product of location-scale univariate (noncentral, if asymmetric; seen below) Student's t p.d.f.s. Their equality was confirmed for $d = 2$ and 3. To illustrate, Figure 12.12 contrasts the usual MVT (12.3) and the SMESTI distribution for $d = 2$, as given in (12.32), with the highlight being the lower right panel, showing a case with two different degrees of freedom and non-diagonal covariance matrix.

It follows from the mixture construction that, for $\mathbf{X} \sim \text{SMESTI}(\mathbf{k}, \boldsymbol{\mu}, \boldsymbol{\Sigma})$, $\mathbb{E}[\mathbf{X}] = \boldsymbol{\mu}$, if $\min\{k_i\} > 1$, and does not exist otherwise. From the independence property of the components when $\boldsymbol{\Sigma}$ is diagonal, it immediately follows (even for non-diagonal $\boldsymbol{\Sigma}$) that $\mathbb{V}(X_i) = [k_i/(k_i - 2)]\sigma_i^2$ if $k_i > 2$, and does not otherwise exist, $i = 1, \ldots, d$. Now, the idea that $\mathbb{V}(\mathbf{X})$ is possibly given by $\mathbf{K}\boldsymbol{\Sigma}\mathbf{K}$, where $\mathbf{K}$ is the diagonal matrix with iith element $\sqrt{k_i/(k_i - 2)}$, $i = 1, \ldots, d$, is easily dismissed, for the following reason: If all the k_i are equal, then this yields the same covariance matrix as that for (12.3), but these matrices must be different, owing to the different dependency structure of their elements arising from using either a single latent variable G, in (12.3), or a set of d of them, as in (12.31). It turns out that the exact expression for $\text{Cov}(X_i, X_j)$ is tractable, and is given in (12.40).

12.6.2 AMESTI Distribution

We wish to extend the SMESTI structure such that the margins can exhibit asymmetry. To this end, let $\boldsymbol{\beta} = (\beta_1, \ldots, \beta_d)' \in \mathbb{R}^d$ and $\mathbf{m}(\mathbf{G}) = \boldsymbol{\mu} + \mathbf{D}\boldsymbol{\beta}$. Then

$$\mathbf{X} = \boldsymbol{\mu} + \mathbf{D}\boldsymbol{\beta} + \mathbf{D}\mathbf{R}^{1/2}\mathbf{Z}, \tag{12.34}$$

where $\mathbf{D}$ (and the G_i, $\boldsymbol{\mu}$ and $\mathbf{Z}$) are defined as before. Then $(\mathbf{X} \mid \mathbf{G} = \mathbf{g}) \sim \text{N}(\boldsymbol{\mu} + \mathbf{D}\boldsymbol{\beta}, \mathbf{DRD})$, generalizing the MVNCT (12.5). The resulting p.d.f. of $\mathbf{X}$ is given by the same integral expression in (12.31), denoted $f_{\mathbf{X}}(\mathbf{x}; \mathbf{k}, \boldsymbol{\beta}, \boldsymbol{\mu}, \mathbf{R})$, and we write either $\mathbf{X} \sim \text{MESTI}(\mathbf{k}, \boldsymbol{\beta}, \boldsymbol{\mu}, \mathbf{R})$ or, to emphasize its asymmetric property, $\mathbf{X} \sim \text{AMESTI}(\mathbf{k}, \boldsymbol{\beta}, \boldsymbol{\mu}, \mathbf{R})$. Observe that $X_i = \mu_i + G_i^{1/2}\beta_i + G_i^{1/2}Z_i$, where $Z_i \sim \text{N}(0, 1)$, so that the margins of $\mathbf{X}$ are each location-μ_i, scale-one noncentral t.

If we had instead defined $\mathbf{X}$ as $\boldsymbol{\mu} + \mathbf{D}\boldsymbol{\beta} + \mathbf{D}\boldsymbol{\Sigma}^{1/2}\mathbf{Z}$, for $\boldsymbol{\Sigma} = \mathbf{SRS}$ as in the SMESTI case, then this implies that $X_i = \mu_i + G_i^{1/2}\beta_i + G_i^{1/2}\sigma_i Z_i$, and this is *not* the construction of the (univariate) noncentral t (which assumes unit scale instead of $\sigma_i Z_i$). While μ_i is indeed a location parameter, it is not the case that $(X_i - \mu_i)$ is multiplied by σ_i, so that, in this construction, σ_i is not a scale parameter. (This issue does not arise in the SMESTI case, as $\boldsymbol{\beta} = \mathbf{0}$.)

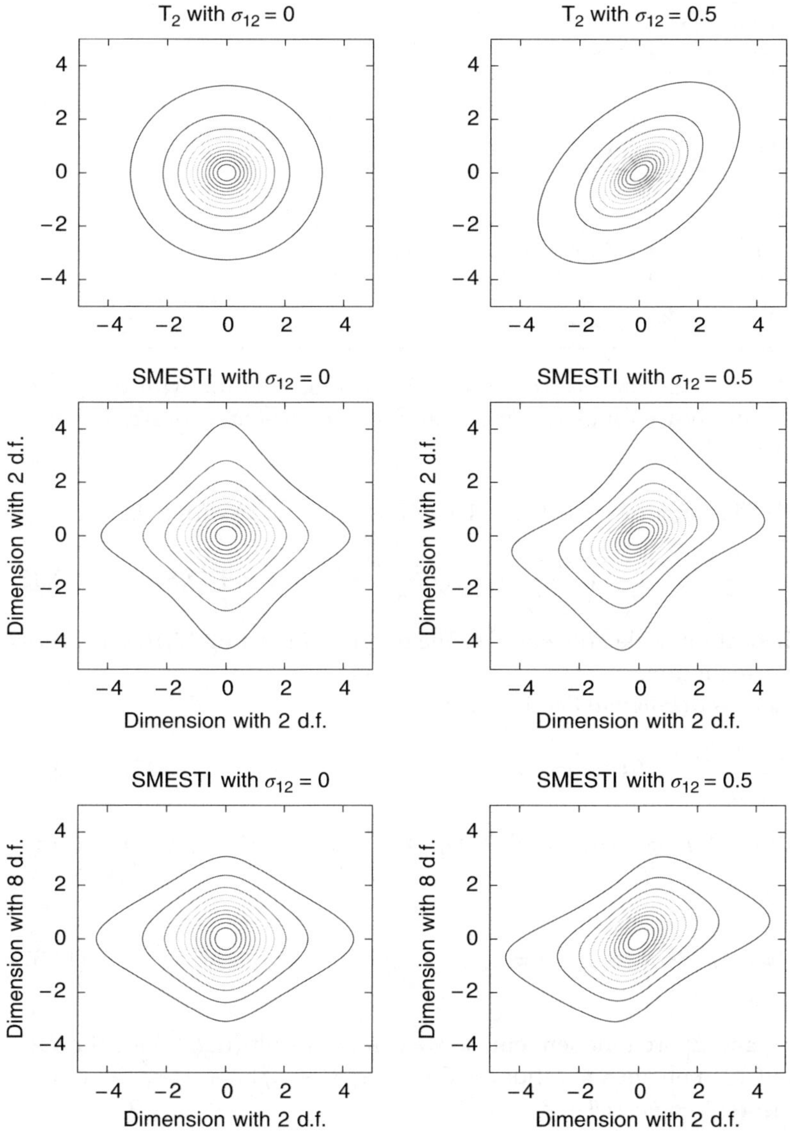

Figure 12.12 Top row shows the usual MVT (12.3) with $k = 2$ degrees of freedom, zero mean vector, $\sigma_1^2 = \sigma_2^2 = 1$, and two values of σ_{12}, zero (left) and 0.5 (right). The middle and last rows show the SMESTI distribution with $k_1 = k_2 = 2$ and $k_1 = 2, k_2 = 8$, respectively (same $\boldsymbol{\mu}$ and $\boldsymbol{\Sigma}$ as first row).

```
function M = MESTIsim(k,beta,mu,scale,R,T)
d=length(k); beta=reshape(beta,1,d); D=eye(d); M=zeros(T,d);
%[Vv,Dd] = eig(R); R12=Vv*sqrt(Dd)*Vv; % for Way 2 below
for t=1:T
  for i=1:d,  ki=k(i);
    V=gamrnd(ki/2,1,[1 1])/(ki/2); G=1./V; % either this...
    %chi2=random('chi2',ki,1,1); G = 1./(chi2/ki); % or this.
    D(i,i)=sqrt(G);
  end
  muN=(D*beta')'; VN=D*R*D; M(t,:)=mvnrnd(muN,VN,1); % Way 1
  %Z = mvnrnd(zeros(1,d),eye(d))'; M(t,:)=D*beta'+D*R12*Z; % Way 2
end
for i=1:d, M(:,i)=scale(i)*M(:,i)+mu(i); end
```

Program Listing 12.16: Simulates T realizations from the (S)MESTI distribution with the passed parameters. Two equivalent ways are shown for generating G, and two equivalent ways are shown for generating the MESTI random variable.

We denote a location-scale MESTI random variable as $\mathbf{M} \sim \text{MESTI}(\mathbf{k}, \boldsymbol{\beta}, \boldsymbol{\mu}, \boldsymbol{\Sigma})$ with p.d.f.

$$f_{\mathbf{M}}(\mathbf{y}; \mathbf{k}, \boldsymbol{\beta}, \boldsymbol{\mu}, \boldsymbol{\Sigma}) = \frac{f_{\mathbf{X}}(\mathbf{x}; \mathbf{k}, \boldsymbol{\beta}, \mathbf{0}, \mathbf{R})}{\sigma_1 \sigma_2 \cdots \sigma_d}, \quad \mathbf{x} = \left(\frac{y_1 - \mu_1}{\sigma_1}, \ldots, \frac{y_d - \mu_d}{\sigma_d} \right)', \quad \mathbf{R} = \mathbf{S}^{-1} \boldsymbol{\Sigma} \mathbf{S}^{-1}. \tag{12.35}$$

The univariate margins are each location-scale noncentral t. The program in Listing 12.16 shows how to simulate from the (S)MESTI distribution.

Let $\mathbf{X} \sim \text{MESTI}(\mathbf{k}, \boldsymbol{\beta}, \mathbf{0}, \mathbf{R})$. Then, as detailed in Section II.10.4.3,

$$\mathbb{E}[X_i] = \beta_i \left(\frac{k_i}{2} \right)^{1/2} \frac{\Gamma(k_i/2 - 1/2)}{\Gamma(k_i/2)}, \quad \text{if } k_i > 1, \quad i = 1, \ldots, d. \tag{12.36}$$

For the variance of X_i, from (III.A.124) and (III.A.125), $\mathbb{V}(G_i^{1/2}) = \mathbb{E}[G_i] - (\mathbb{E}[G_i^{1/2}])^2$, with $\mathbb{E}[G_i] = k_i/(k_i - 2)$ and

$$\mathbb{E}[G_i^{1/2}] = \sqrt{\frac{k_i}{2}} \frac{\Gamma\left(\frac{k_i - 1}{2}\right)}{\Gamma\left(\frac{k_i}{2}\right)} =: A_i, \quad \text{if } k_i > 1, \quad i = 1, \ldots, d. \tag{12.37}$$

By construction from (12.34), G_i and Z_i are independent, so we can use result (II.2.36) for the variance of a product: For r.v.s G and Y, in obvious notation, $\mathbb{V}(GY) = \mu_Y^2 \sigma_G^2 + \mu_G^2 \sigma_Y^2 + \sigma_G^2 \sigma_Y^2$. Now, with $Y = \beta + Z$ and (dropping subscripts) $X = G^{1/2}(\beta + Z) = G^{1/2} Y$,

$$\begin{aligned} \mathbb{V}(X) &= \beta^2 \mathbb{V}(G_i^{1/2}) + (\mathbb{E}[G_i^{1/2}])^2 \cdot 1 + \mathbb{V}(G^{1/2}) \cdot 1 = (1 + \beta^2) \mathbb{V}(G^{1/2}) + A^2 \\ &= (1 + \beta^2) \left[\frac{k}{k - 2} - A^2 \right] + A^2, \end{aligned}$$

i.e.,

$$\mathbb{V}(X_i) = \left(\frac{k_i}{k_i - 2} \right) + \beta_i^2 \left[\frac{k_i}{k_i - 2} - \frac{k_i}{2} \left(\frac{\Gamma\left(\frac{k_i - 1}{2}\right)}{\Gamma\left(\frac{k_i}{2}\right)} \right)^2 \right], \quad \text{if } k_i > 2, \quad i = 1, \ldots, d. \tag{12.38}$$

For the covariance, from (III.A.86),

$$\mathbb{V}(\mathbf{X}) = \mathbb{E}_{\mathbf{G}}[\mathbb{V}(\mathbf{X} \mid \mathbf{G})] + \mathbb{V}_{\mathbf{G}}(\mathbb{E}[\mathbf{X} \mid \mathbf{G}]) = \mathbb{E}_{\mathbf{G}}[\mathbf{DRD}] + \mathbb{V}(\mathbf{D}\boldsymbol{\beta}, \mathbf{D}\boldsymbol{\beta}). \tag{12.39}$$

As G_i and G_j are independent for $i \neq j$, $\text{Cov}(\mathbf{D}\boldsymbol{\beta}, \mathbf{D}\boldsymbol{\beta}) = \text{diag}(\beta_1^2 \mathbb{V}(G_1^{1/2}), \dots, \beta_d^2 \mathbb{V}(G_d^{1/2}))$, so that $\text{Cov}(X_i, X_j)$ for $i \neq j$ does not depend on $\boldsymbol{\beta}$. With this and again using independence, it follows from (12.39) that $\text{Cov}(X_i, X_j) = \mathbb{E}[G_i^{1/2}]\mathbb{E}[G_j^{1/2}] = A_i A_j$ from (12.37), $i \neq j$, i.e.,

$$\text{Cov}(X_i, X_j) = \sigma_{ij} \sqrt{\frac{k_i}{2}\frac{k_j}{2}} \frac{\Gamma\left(\frac{k_i - 1}{2}\right)\Gamma\left(\frac{k_j - 1}{2}\right)}{\Gamma\left(\frac{k_i}{2}\right)\Gamma\left(\frac{k_j}{2}\right)}, \quad i \neq j, \quad k_i, k_j > 1. \tag{12.40}$$

The program in Listing 12.17 computes the MESTI density at a given point `xvec` for any dimension d, though it is rather slow for $d = 3$, and for any $d \geqslant 4$ becomes prohibitive, given the curse of dimensionality. Its value is that it illustrates the useful technique of general d-dimensional numeric integration conducted recursively. As a test case for $d = 2$ and $d = 3$ with a diagonal $\mathbf{R}$ matrix (so that the margins are independent), first set `prodtogg=0` in line 3 of Listing 12.17 and run the following code:

```
%x=[0 1]; k=[2 7]; beta=[-0.5 1]; mu=[1 2]; scale=[1 2];
x=[0 1 0]; k=[2 7 3]; beta=[-0.5 1 2]; mu=[1 2 3]; scale=[1 2 3];
MESTIpdf(x,k,beta,mu,scale)
```

Then, do the same but with `prodtogg=1` to see that they are equal to machine precision.

Figure 12.13 shows the bivariate MESTI density, as computed using the aforementioned program, for the same parameter constellations as were used in the bottom four panels of Figure 12.12, but having used $\beta_i = -i$, $i = 1, 2$. With tail thickness and asymmetry parameters for each marginal, and a covariance matrix to account for dependence, the MESTI distribution is quite flexible. However, it does not have the feature of **tail dependence**, this being a recognized stylized fact of asset returns. To allow for tail dependence, we need to drop the independence assumption on the G_i, as discussed in Section 12.6.5.

12.6.3 MESTI Estimation

With the p.d.f. available, full maximum likelihood estimation is trivial to set up, using our usual code for such things. The program in Listing 12.18 is given for completeness, though without large parallel processing for line 32, it is essentially useless, even for $d = 2$. It could serve as a base for developing the code for estimating, via full maximum likelihood, the MEST extension in Section 12.6.5, though with the aforementioned caveat in mind about the necessity of parallel computing.

Estimation of the (S)MESTI model for general d and large sample sizes can be conducted very fast using the aforementioned two-step procedure, similar to use with the (A)FaK, where here the univariate Student's t (or NCT) is estimated to obtain the $\hat{\mu}_i$, $\hat{k}_i$ (and $\hat{\beta}_i$), $i = 1, \dots, d$, and in a second step the σ_{ii} and σ_{ij} are obtained via the method of moments, equating the usual sample estimates of them with (12.38) and (12.40), conditioning on $k_i = \hat{k}_i$ (and $\beta_i = \hat{\beta}_i$).

The short code in Listing 12.20 confirms the estimation (and the simulation) procedures work correctly.

```
function f = MESTIpdf(x,k,beta,mu,scale,R)
% pdf of the MESTI asymmetric marginally endowed students t
prodtogg=1; % set to 1 to use product, when indep.
d=length(x); x=reshape(x,1,d);
if nargin<6 || isempty(R), R=eye(d); end
if nargin<5 || isempty(scale), scale=ones(1, d); end, scale=reshape(scale,1,d);
if nargin<4 || isempty(mu), mu=zeros(1, d); end,      mu=reshape(mu,1,d);
if nargin<3 || isempty(beta), beta=zeros(1, d); end, beta=reshape(beta,1,d);
xx=(x-mu)./scale; dd=R-eye(d);
if norm(dd,1)<1e-14 && prodtogg % Are they independent?
  f=1; for i=1:d, f=f*nctpdf(xx(i), k(i), beta(i)); end
else
  gfixed=[]; f=quadgk(@(gi) recursint(gi,gfixed,xx,k,beta,R), 1e-8, Inf);
end
f=f/prod(scale);

function Int=recursint(gi,gfixed,x,k,beta,R)
d=length(x); gilen=length(gi); Int=zeros(size(gi));
if length(gfixed)==(d-1) % last, inner integral
  for gg=1:gilen
    gp=gi(gg); D=eye(d); D(1,1)=sqrt(gp);
    for i=1:(d-1), D(i+1,i+1)=sqrt(gfixed(i)); end
    P=1; normteil=mvnpdf(x,beta*D,D*R*D);
    for i=1:d
      guse=D(i,i)^2; kuse=k(i); P=P*IGampdf(guse,kuse/2,kuse/2);
    end
    Int(gg)=normteil*P;
  end
else % continue the recursion
  for gg=1:gilen
    gfixednew=[gfixed gi(gg)];
    Int(gg) = quadgk(@(gi) recursint(gi,gfixednew,x,k,beta,R), 1e-8, Inf);
  end
end

function f=IGampdf(x,a,b), f = b^a / gamma(a) * x^(-(a+1)) * exp(-b/x);
```

Program Listing 12.17: Computes the p.d.f. of the MESTI distribution for any d via recursive numeric integration, for the single d-length point `xvec`.

Example 12.6 ***(Example 12.1 cont.)***
We continue with the Bank of America and Wal-Mart data, now fitting the MESTI distribution. Estimation is done using the two-step method, and then the log-likelihood is computed from the program in Listing 12.17, though observe that this is not the log-likelihood corresponding to the true m.l.e., but rather the result of the two-step estimation method. The obtained log-likelihood is −7144, showing the expected large improvement over the usual MVT, but pales compared to the AFaK result in Table 12.2, of −7079. The latter has one additional parameter, but a crucial one, as it allows for tail dependence, whereas MESTI does not. The parameter estimates are given in the last line of Table 12.2 and are very similar to the corresponding ones from the AFaK. ■

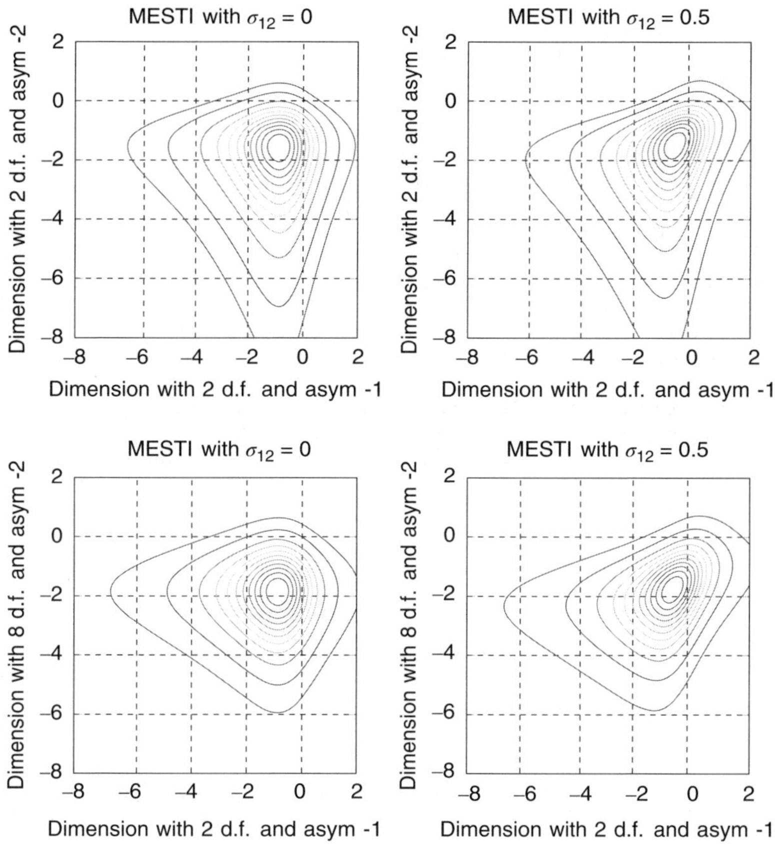

Figure 12.13 Similar to the bottom four panels of Figure 12.12, but for the MESTI distribution, with $\beta_1 = -1$ and $\beta_2 = -2$.

Remarks

a) In light of the relatively poor performance of the sample-based estimator of the elements of **R** in the (A)FaK model when the relevant degrees of freedom are small (recall Figure 12.11), we can anticipate the same poor performance in the (S)MESTI context. An idea is to use the method of indirect inference (Chapter III.10), with the auxiliary function being the (A)FaK likelihood for the individual correlations, to determine the σ_{ij} for the (S)MESTI. This appears reasonable, as (i) the AFaK is a similar distribution in many regards to MESTI, with nearly the same number of parameters and a clear "mapping" between them, and (ii) simulating from the MESTI is trivial and fast. What a fantastic take-home exam idea!

b) In a portfolio optimization context, the sums of (weighted) margins is required; this can be approximated exactly the same as was done with the (A)FaK, given that the (S)MESTI margins are also (noncentral) Student's t and that simulation from MESTI is fast, faster in fact than with the AFaK. ■

```
1  function [param,stderr,iters,loglik,Varcov] = MESTIest(x,initvec)
2  % Marginally Endowed Student's t, Independent G_i case
3  % Full MLE for the d=2 dimension case.
4  [nobs d]=size(x); if d~=2, error('Full MLE only for d=2'), end
5  if nargin<2, initvec=[]; end
6  %%%%%%%        k1   k2    beta1 beta2 mu1  mu2    scale1   scale2 R12
7  bound.lo=    [0.3 0.3  -10    -10     -1    -1  0.01      0.01  -20];
8  bound.hi=    [ 20  20   10     10      1     1   20        20     20];
9  bound.which=[  1   1     1      1      0     0    1         1      1];
10 %if isempty(initvec), initvec=[2 4 0 0 0 0 2 2 0]; end
11 if isempty(initvec), initvec=[1.5 3.9 -0.17 0.14 0.2 -0.2 0.9 1.1 0.5]; end
12 maxiter=300; tol=1e-5; MaxFunEvals=25*maxiter;
13 opts=optimset('Display','iter','Maxiter',maxiter,'TolFun',tol,'TolX',tol, ...
14     'MaxFunEvals',MaxFunEvals,'LargeScale','Off');
15 [pout,fval,~,theoutput,~,hess]= ...
16     fminunc(@(param) MESTIloglik(param,x,bound),einschrk(initvec,bound),opts);
17 V=inv(hess)/nobs; [param,V]=einschrk(pout,bound,V); param=param';
18 Varcov=V; stderr=sqrt(diag(V)); loglik=-fval*nobs; iters=theoutput.iterations;
19 
20 function ll=MESTIloglik(param,x,bound)
21 if nargin<3, bound=0; end
22 if isstruct(bound)
23   paramvec=einschrk(real(param),bound,999);
24 else
25   paramvec=param;
26 end
27 [nobs d]=size(x); R=eye(d);
28 k=paramvec(1:2); beta=paramvec(3:4); mu=paramvec(5:6);
29 scale=paramvec(7:8); R(1,2)=paramvec(9); R(2,1)=R(1,2);
30 pdf=zeros(nobs,1); hand=plot(1:nobs,pdf,'g-');
31 for i=1:nobs
32   pdf(i) = MESTIpdf(x(i,:),k,beta,mu,scale,R);
33   if i==1, delete(hand), end
34   hand=plot(1:nobs, pdf, 'g-',1:i,pdf(1:i),'r-o');
35   title(['df = ',num2str(k)]), drawnow % Watch the slow show...
36 end
37 llvec=log(pdf); ll=-mean(llvec); if isinf(ll), ll=1e5; end
```

Program Listing 12.18: Computes the m.l.e. of the MESTI distribution in the $d = 2$ case. Calls program MESTIpdf from Listing 12.17. Assuming use of only one core (no parallel processing), the graphics commands in lines 33 and 35 allow the p.d.f. evaluation to be seen unfolding at each t.

12.6.4 AoN$_m$-MEST

Let $\mathbf{X}_{\bullet,\bullet} = [\,\mathbf{X}_{1,\bullet} \mid \mathbf{X}_{2,\bullet} \mid \cdots \mid \mathbf{X}_{d,\bullet}\,]$ be the $T \times d$ matrix of the return series under study, where $\mathbf{X}_{i,\bullet} = (X_{i,1}, \dots, X_{i,T})'$, $i = 1, \dots, d$, and $\mathbf{X}_{\bullet,t} = (X_{1,t}, X_{2,t}, \dots, X_{d,t})'$, $t = 1, \dots, T$. We impose a structure such that each asset belongs to one of m groups, $1 \leqslant m \leqslant d$. Let n_j be the number of components in the jth group, so that $\sum_{j=1}^{m} n_j = d$. After reordering and renumbering indices, we have m groups, such that we can partition the components in $\mathbf{X}$ as $\mathbf{X}_{\bullet,\bullet} = [\,\mathbf{X}^1_{\bullet,\bullet} \mid \mathbf{X}^2_{\bullet,\bullet} \mid \cdots \mid \mathbf{X}^m_{\bullet,\bullet}\,]$, where

$$
\begin{aligned}
\mathbf{X}^1_{\bullet,\bullet} &= [\mathbf{X}^1_{1,\bullet} \mid \mathbf{X}^1_{2,\bullet} \mid \cdots \mid \mathbf{X}^1_{n_1,\bullet}],\\
\mathbf{X}^2_{\bullet,\bullet} &= [\mathbf{X}^2_{1,\bullet} \mid \mathbf{X}^2_{2,\bullet} \mid \cdots \mid \mathbf{X}^2_{n_2,\bullet}],\\
&\ \ \vdots\\
\mathbf{X}^m_{\bullet,\bullet} &= [\mathbf{X}^m_{1,\bullet} \mid \mathbf{X}^m_{2,\bullet} \mid \cdots \mid \mathbf{X}^m_{n_m,\bullet}].
\end{aligned}
$$

```
function param = MESTIest2(data)
[~, d]=size(data); X=data; R=eye(d);
beta=zeros(d,1); k=zeros(d,1); mu=zeros(d,1); scale=zeros(d,1);
for i=1:d
  pNCT = Noncentraltestimation(data(:,i),1,1);
  beta(i)=pNCT(1); k(i)=pNCT(2); mu(i)=pNCT(3); scale(i)=pNCT(4);
end
param.beta=beta; param.df=k; param.mu=mu; param.scale=scale;
for i=1:d, X(:,i) = ( X(:,i)-mu(i) ) / scale(i); end
SC=cov(X); % Sample Covariance
for i=1:d % Method of moments for the off-diagonal elements
  for j=(i+1):d,  ki=k(i); kj=k(j);
    t1=sqrt(ki*kj/4)*gamma((ki-1)/2)*gamma((kj-1)/2)/gamma(ki/2)/gamma(kj/2);
    R(i,j)=SC(i,j)/t1; R(j,i)=R(i,j);
  end
end
param.R=R;
```

Program Listing 12.19: Estimates the parameters of the (S)MESTI distribution via the two-step method.

```
k=[4 5]; beta=[2 -2]; mu=[1 2]; scale=[4 3]; r=0.6; R=[1 r; r 1];
T=1e5; M = MESTIsim(k,beta,mu,scale,R,T);
param = MESTIest2(M); param.df, param.beta, param.mu, param.scale, param.R
```

Program Listing 12.20: MESTI simulation and estimation.

The components within a group are jointly MVNCT, i.e., for $j \in \{1, \dots, m\}$,

$$\mathbf{X}^{j}_{\bullet,t} = \boldsymbol{\mu}_j + \sqrt{G_j}\boldsymbol{\beta}_j + \sqrt{G_j}\boldsymbol{\Gamma}_{jj}\mathbf{Z}, \tag{12.41}$$

where $G_j \sim \text{IGam}(k_j/2, k_j/2)$ and $\mathbf{Z} \sim \text{N}(\mathbf{0}, \mathbf{I})$. The model can then be stated as

$$\mathbf{X}_{\bullet,t} \overset{\text{iid}}{\sim} \text{AoN}_m\text{-MEST}(\mathbf{k}, \boldsymbol{\beta}, \boldsymbol{\mu}, \boldsymbol{\Gamma}), \quad t = 1, \dots, T, \tag{12.42}$$

where $\mathbf{k}$ is the $m \times 1$ vector of degrees of freedom parameters, $\boldsymbol{\beta}$ is the $m \times 1$ vector of asymmetry (noncentrality) parameters, $\boldsymbol{\mu}$ is the location term, of size $d \times 1$, and $\boldsymbol{\Gamma}$ is the $d \times d$ dispersion matrix, partitioned as

$$\boldsymbol{\Gamma} = \begin{pmatrix} \boldsymbol{\Gamma}_{11} & \boldsymbol{\Gamma}_{12} & \cdots & \boldsymbol{\Gamma}_{1m} \\ \boldsymbol{\Gamma}_{21} & \boldsymbol{\Gamma}_{22} & \cdots & \\ \vdots & & \ddots & \vdots \\ \boldsymbol{\Gamma}_{m1} & \cdots & & \boldsymbol{\Gamma}_{mm} \end{pmatrix}. \tag{12.43}$$

We now turn to estimation of the model parameters. Given a certain grouping, parameter estimation is conducted in two steps.

Step 1: For each of the univariate return series, the degrees of freedom k_i, noncentrality β_i, location μ_i, and scale parameters σ_i, $i = 1, \dots, d$, are estimated via maximum likelihood, but using the closed-form, vectorized saddlepoint approximation to the NCT density, as in Broda and Paolella (2007), for speed reasons. (In the GARCH case, if the model in (12.41) is endowed with the fixed APARCH structure (10.20), then the method in Section 10.4 can be used, setting $\hat{\mu}_i = 0$ for all i.)

Thus, in the first step, estimates of $\mathbf{k}$, $\boldsymbol{\beta}$, $\boldsymbol{\mu}$, and some of the elements of $\widehat{\boldsymbol{\Gamma}}$ are obtained. Note that the model requires the k_i to be equal for all components in a group, i.e., $\mathbf{k}$ now has only m distinct components. Denote these values as $k_{[1]}, k_{[2]}, \dots, k_{[m]}$. One rather rudimentary way to operationalize this equality restriction is to take the arithmetic mean or median over all k_i in a certain group, i.e., using the former, for group j,

$$\hat{k}_{[j]} = n_j^{-1} \sum_{l=1}^{n_j} \hat{k}_l. \tag{12.44}$$

Step 2: Estimate $\boldsymbol{\Gamma}$. The diagonal matrices $\boldsymbol{\Gamma}_{jj}$ are the dispersion matrices for the components in group j, $j = 1, \dots, m$, while the off-diagonal matrices $\boldsymbol{\Gamma}_{ij}$ contain dispersion terms from series in the different groups i and j. Note that these need not be square, nor of the same size, due to the different number of components in the groups. As we assume in this construction that dependency of returns in different groups is only via a covariance term, the estimates of the (n, l)th element of $\boldsymbol{\Gamma}_{ij}$, σ_{nl}^{ij}, is calculated from the expression of the covariance of the MESTI distribution, i.e.,

$$\mathrm{Cov}(X_{n,t}^i, X_{l,t}^j) = \sigma_{nl}^{ij} \sqrt{\frac{k_i}{2}\frac{k_j}{2}} \frac{\Gamma\left(\frac{k_i-1}{2}\right)\Gamma\left(\frac{k_j-1}{2}\right)}{\Gamma\left(\frac{k_i}{2}\right)\Gamma\left(\frac{k_j}{2}\right)}, \quad i \neq j, \quad n = 1, \dots, n_i, \quad l = 1, \dots, n_j. \tag{12.45}$$

Estimates $\hat{\sigma}_{nl}^{ij}$ are obtained by replacing the left-hand side of (12.45) by its usual plug-in estimator for the covariance between the nth element of the ith group and the lth element of the jth group at time t. With assets in the same group j sharing a single latent variable, the MEST structure simplifies to the MVNCT, so that estimates of the off-diagonals of $\boldsymbol{\Gamma}_{jj}$, $j = 1, \dots, m$, are calculated from

$$\mathrm{Cov}(X_{n,t}^j, X_{l,t}^j) = \sigma_{nl}^{jj} \frac{k_i}{k_i - 2} + \beta_n^j \beta_l^j \left(\frac{k_j}{k_j - 2} - \frac{k_j}{2} \left(\frac{\Gamma\left(\frac{k_j-1}{2}\right)}{\Gamma\left(\frac{k_j}{2}\right)} \right)^2 \right), \quad l \neq n. \tag{12.46}$$

Example 12.7 The AoN model requires that the assets are grouped, such that the components in each group share a common latent variable, this being the G_t sequence discussed in Section 11.2.4. A purely data-driven way could be to estimate the COMFORT model on each univariate asset to get the imputed G_t sequences, and attempt to group them via some measure such as correlation. Instead, and driven more by financial economic theory, we use the grouping dictated by the official sector to which the asset belongs.

For our example, we use daily data for the 416 stocks listed in the in the S&P500 index that are available from January 5, 2004, to May 16, 2014. (The data are from Bloomberg, and are dividend and split adjusted.) For each stock, we estimate the parameters of the NCT-APARCH model using the method discussed in Section 10.4, and thus, for all 416 stocks, taking just a few seconds. (This is the reason for the granularity of the estimates—which is irrelevant given their sampling error—and the upper limit of 30 for the estimated degrees of freedom parameter and the restriction to lie between −1 and 1 for the estimated noncentrality parameter.) Figure 12.14 plots the estimated degrees of freedom

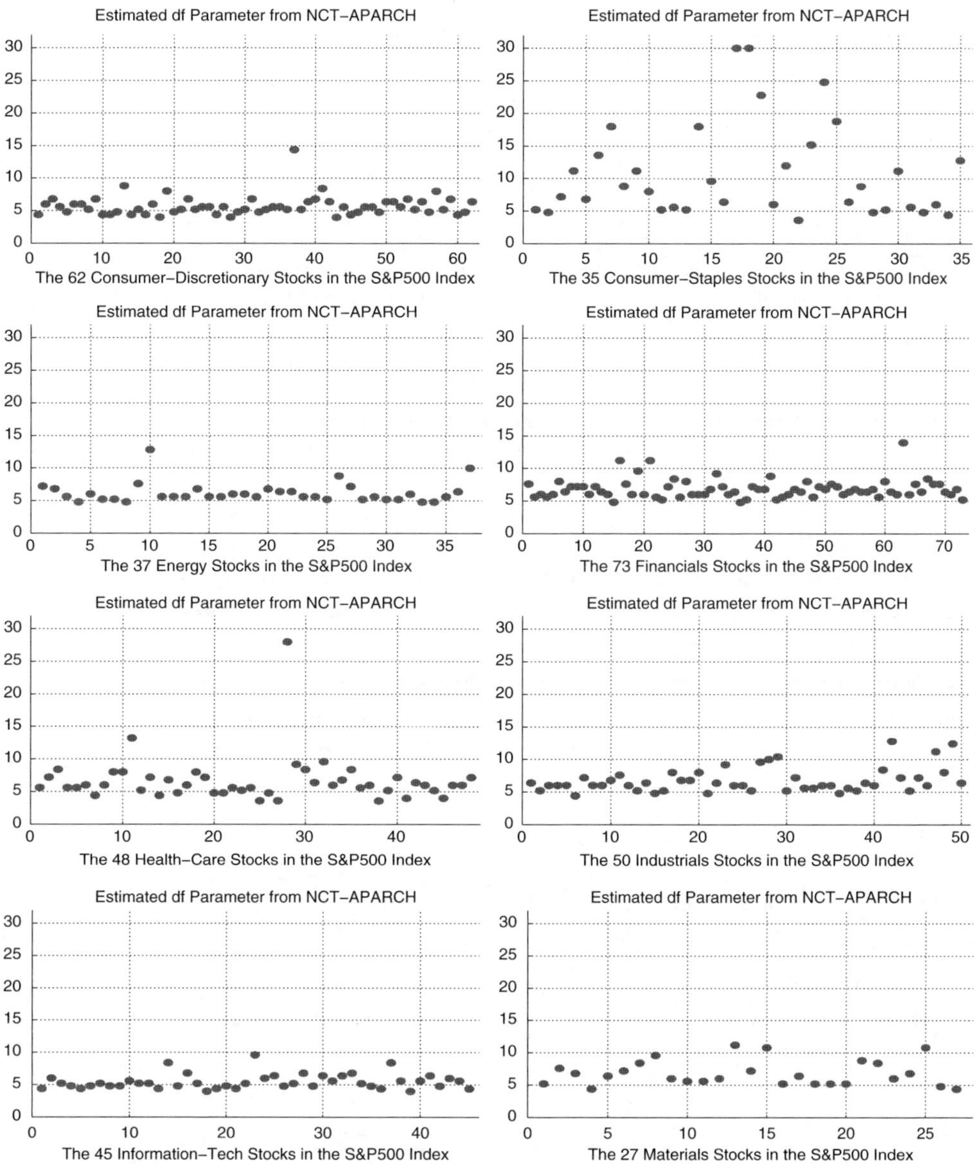

Figure 12.14 Estimated degrees of freedom parameter from a fitted NCT-APARCH model using the fast estimation method discussed in Section 10.4, and based on the 10 years of daily returns data of the S&P500 from January 2004 to May 2014.

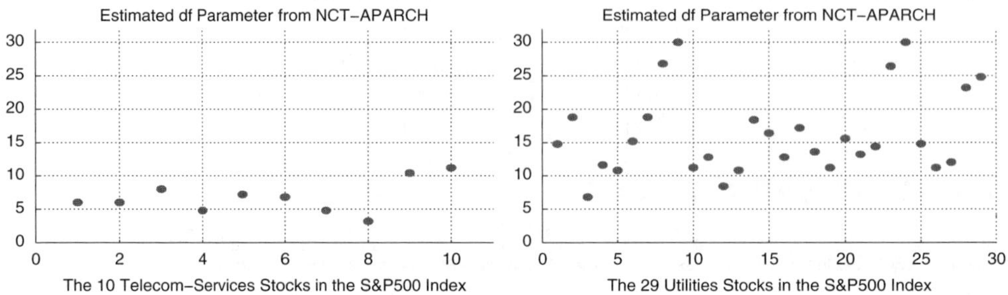

Figure 12.14 (*Continued*)

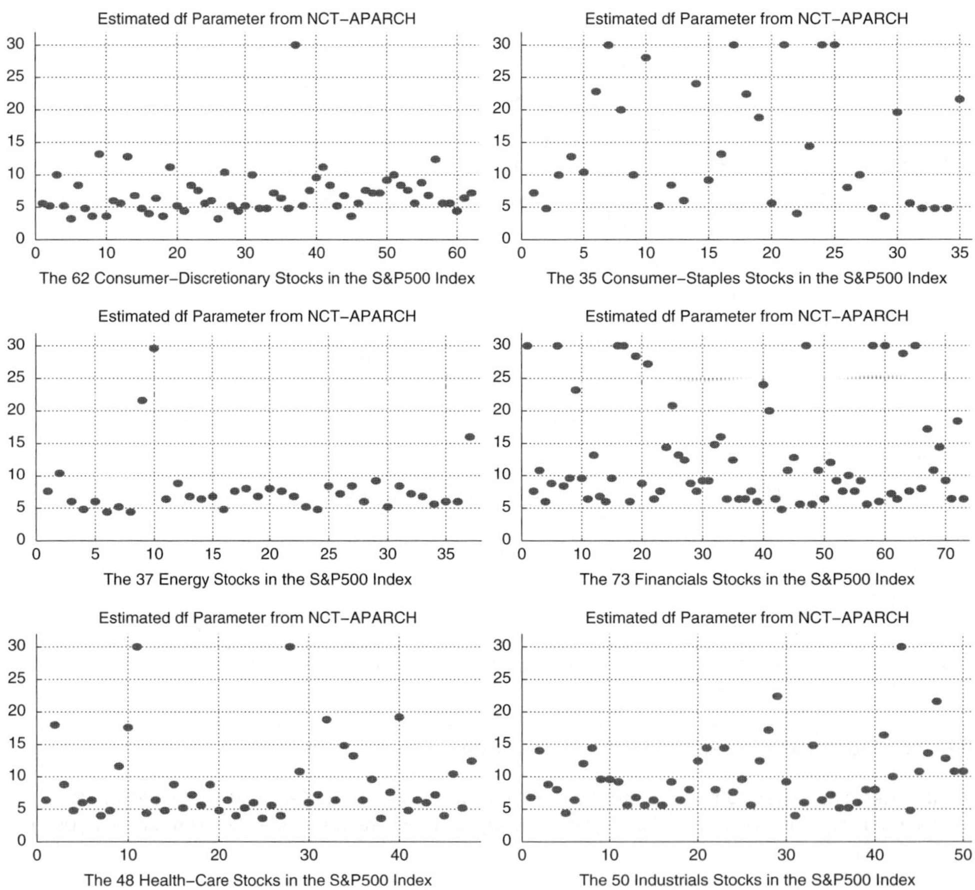

Figure 12.15 Same as Figure 12.14 but based on only the last 500 observations (two years of data).

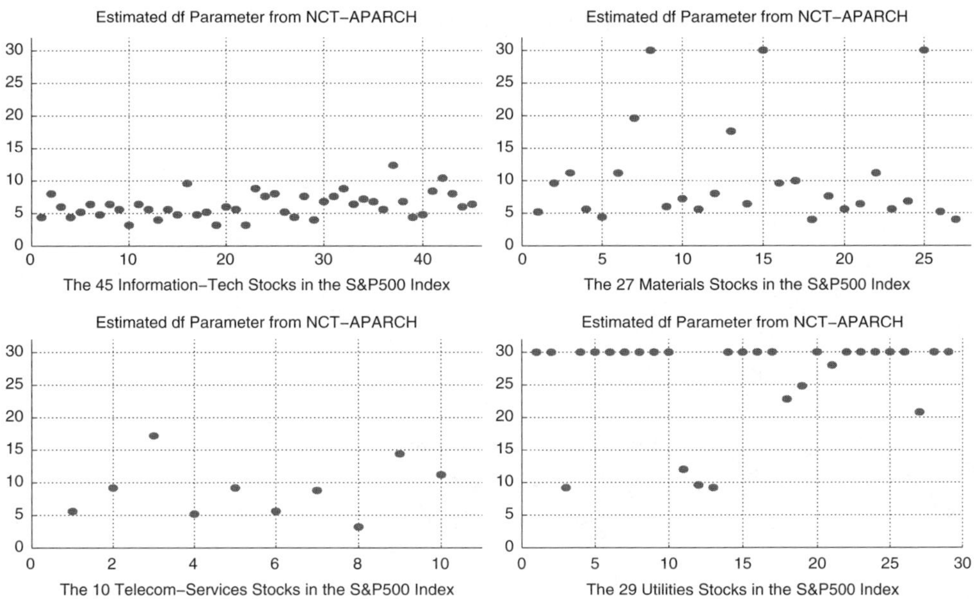

Figure 12.15 (*Continued*)

parameter for each stock, grouped by sector, computed over the whole available 10-year time period. Figure 12.15 is similar, but just uses the last 500 observations (about two years of trading data). These are summarized in Figure 12.16, which forms boxplots of the estimates based on sector. It appears that the sectors form reasonable, albeit not perfect, groups with respect to the tail behavior of the assets, and motivates the use of sector splitting for the AoN method.

Figures 12.17 and 12.18 are analogous to Figures 12.14 and 12.15, but show the estimated noncentrality parameter of the fitted NCT-APARCH. There is a strong association between the asymmetry parameters within a sector, with several such that one best assumes it is zero (given the imprecision of the estimator), while for consumer discretionary stocks (materials and utilities), the point estimates are predominantly positive (negative) and, for each stock in the sector, could be taken to be the mean or median of the individual point estimates as a form of shrinkage estimation.

With respect to inspecting the estimated (parametric) tail index of the sectors to motivate this method of grouping, it is important to note that, while the stochastic process generating a latent variable G_t sequence necessarily dictates the tail behavior and asymmetry, it is not the case that two series with even identical tail behaviors come from the same G_t sequence. Thus, this method is only a proxy. It is nevertheless appealing from the point of view that shocks to the economy generated within a sector will affect other assets in that sector, and possibly less so the stocks outside of it.

As with all financial econometric models, the proof of the pudding is in the eating: The only way to judge the efficacy of the idea is with out of sample forecasting—either risk, density, or, most usefully, portfolio weights. For this model, there is no analytic tractable form of the portfolio density, and simulation is required. The same method as in Section 12.5.5 could be used. See Näf et al. (2018a) for details on the portfolio performance of the AoN-MEST model. ■

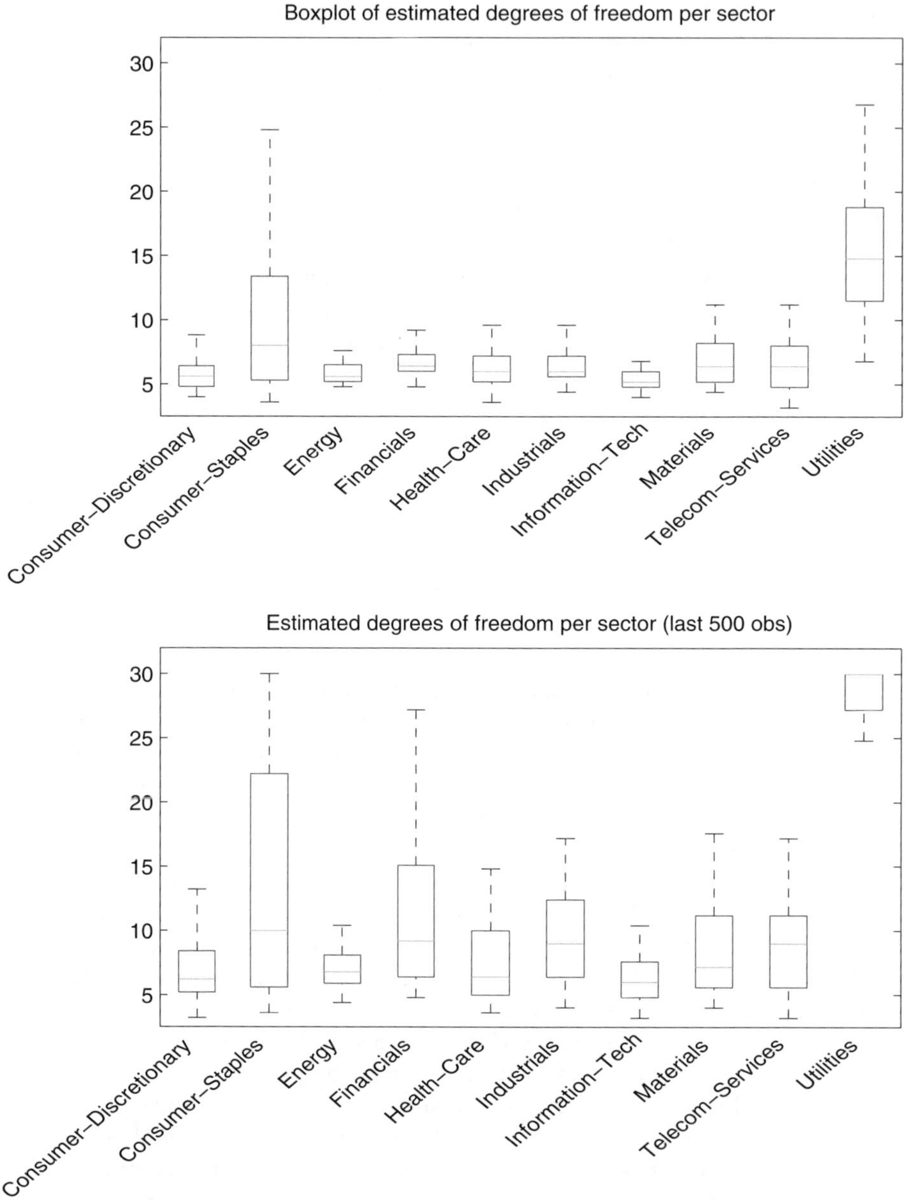

Figure 12.16 Top: Boxplots of the estimated degrees of freedom parameters based on division by financial sector, having used the entire 10 year data period. **Bottom:** Same, but based on only the last 500 observations (two years of data).

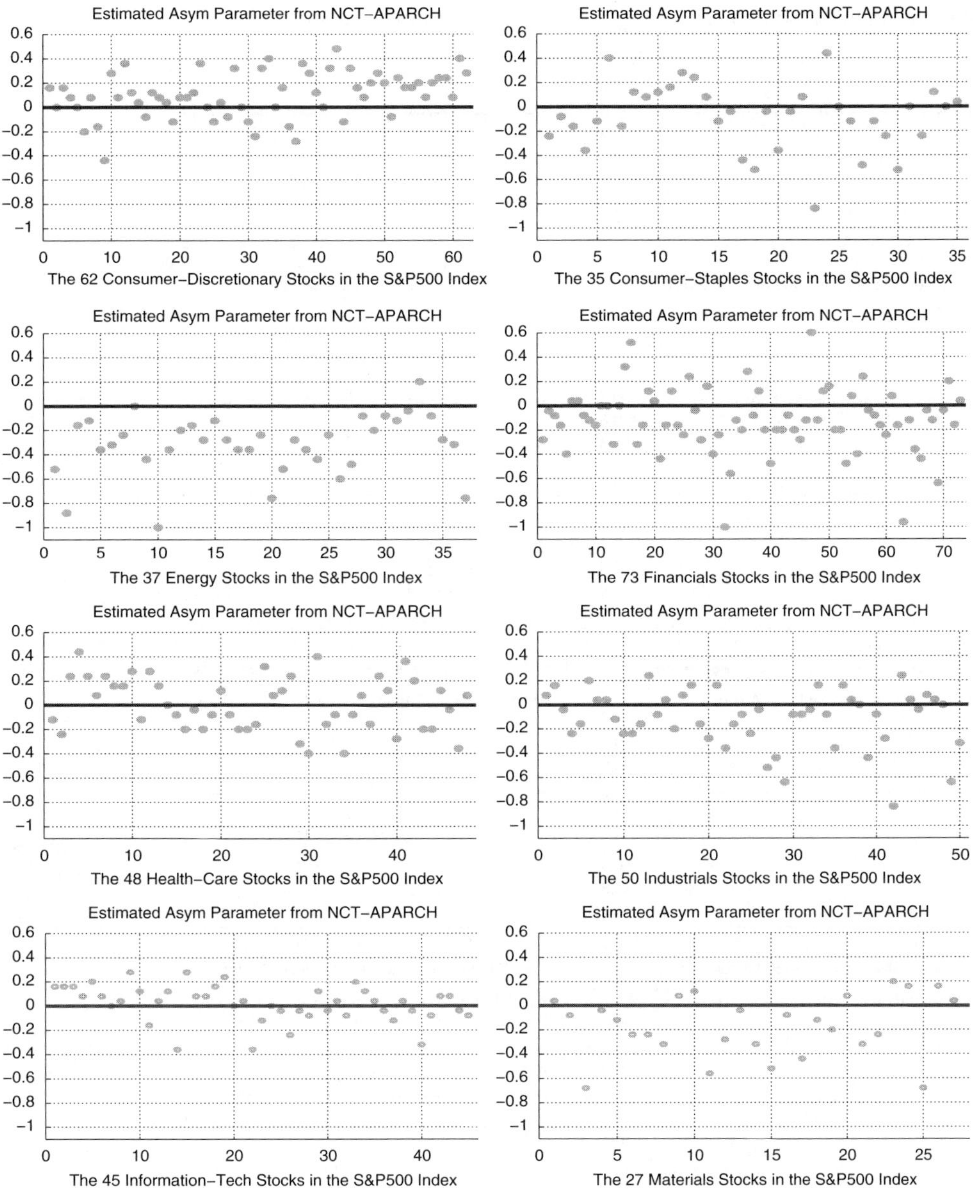

Figure 12.17 Similar to Figure 12.14, but showing the estimated noncentrality (asymmetry) parameter from a fitted NCT-APARCH model, based on the 10 years of daily returns data.

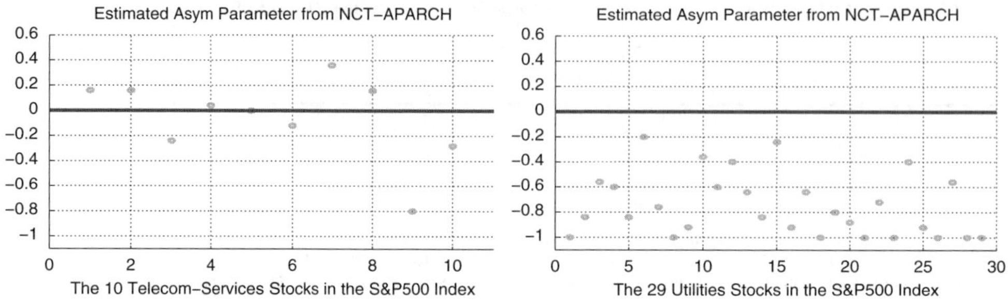

Figure 12.17 *(Continued)*

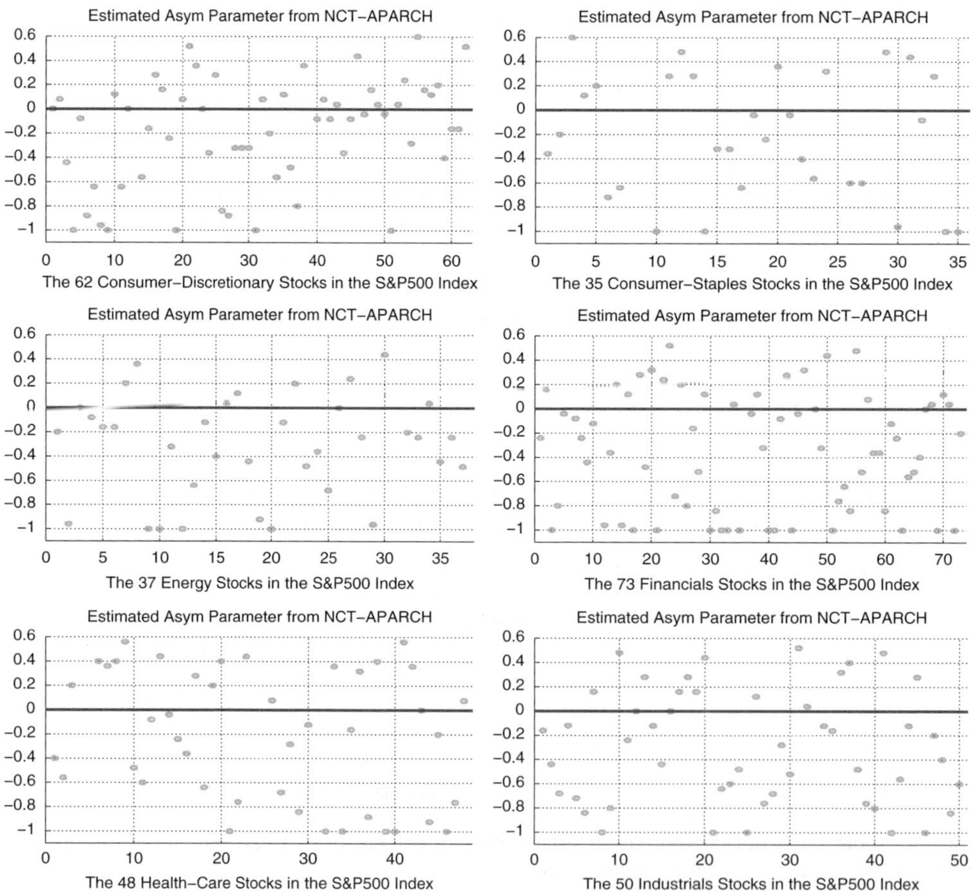

Figure 12.18 Same as Figure 12.17 but based on only the last 500 observations (two years of data). Note that some points might be missing due to the size of the *y*-axis, chosen to be the same as in Figure 12.17.

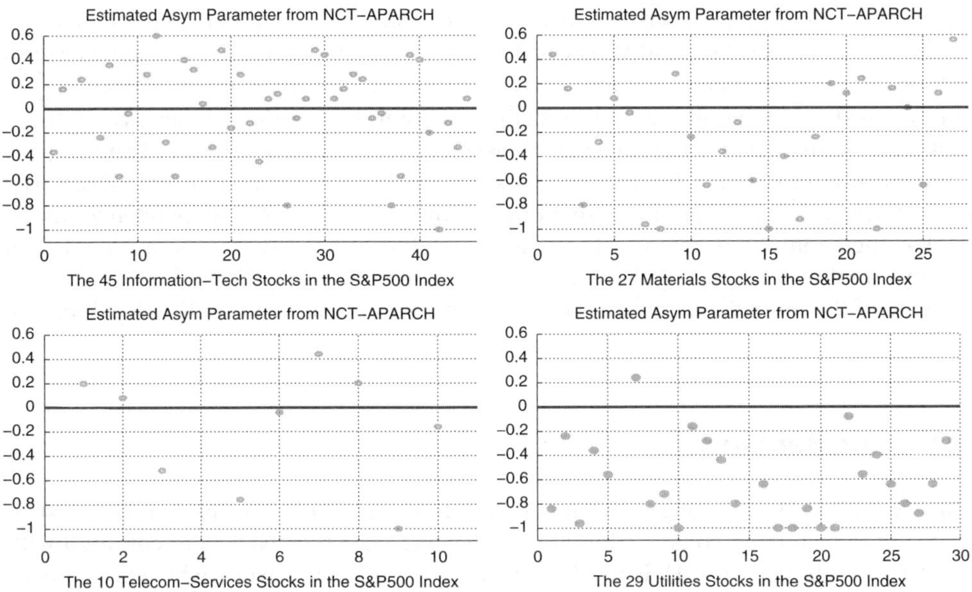

Figure 12.18 *(Continued)*

12.6.5 MEST Distribution

A potential limitation of the MESTI structure is that dependence between margins is strictly through matrix $\boldsymbol{\Sigma}$. This has the significant disadvantage of not allowing for tail dependence. An elegant way of introducing dependency among the margins beyond covariance is to allow dependence among the G_i by endowing them with a multivariate structure, yielding the MEST distribution. Observe the special case in which G_i and G_j are perfectly correlated: If, say, $G_i = G_j$ w.p.1, $i, j = 1, \ldots, d$, then this is precisely the structure of the MVNCT, which exhibits tail dependence (except in the limiting Gaussian case); see, e.g., McNeil et al. (2005, p. 211), and the discussion in Jondeau (2016).

Multivariate generalizations of the generalized inverse Gaussian (MGIG) suggest themselves as natural structures for models of $\mathbf{G}_t = (G_{1,t}, \ldots, G_{d,t})'$, where, anticipating applications to time series, we endow each G_i with a further subscript, t, indexing time. We mention four possibilities. One of the earliest approaches to construct an MGIG was done in Barndorff-Nielsen et al. (1992), using random effect models and yielding a multivariate distribution with GIG margins. Their construction is quite simple, which turns out to be both its merit and weakness: While it allows the specification of several properties of the distribution, most importantly its m.g.f., an expression for the density cannot be derived.

Another candidate is from Minami (2003), who proposed a sophisticated multivariate extension to the inverse Gaussian distribution. Unfortunately, some initial attempts at simulating it did not result in success. A third option is, instead of taking $\mathbf{D}_t = \mathrm{diag}(G_{1,t}^{1/2}, G_{2,t}^{1/2}, \ldots, G_{d,t}^{1/2})$, to entertain a distribution that models the entire matrix $\mathbf{D}_t$ in (12.30). A candidate in this case would be the matrix generalized inverse Gaussian distribution. It was originally proposed in Barndorff-Nielsen et al. (1982), while Butler (1998) derives various properties and its relation to other distributions. The resulting model is surely highly flexible, but might be over-parameterized in typical applications.

The last option we mention is to use a copula construction to join the G_i, each being IGam (or, more generally, GIG). We favor this latter idea, as it is straightforward to construct, clearly has the desired margin distributions, and is such that evaluation of $\mathbb{E}[G_{i,t}^{1/2}]$ and, more challenging, $\eta_{ij} = \mathbb{E}[G_{i,t}^{1/2} G_{j,t}^{1/2}]$ is possible. In this context, note that we are not modeling the joint distribution of, say, asset returns with a copula, but rather with a very rich normal mean-variance mixture distribution, and only require inducing dependency among the $G_{i,t}$, $i = 1, \ldots, d$. As such, we have the luxury of not necessarily requiring sophisticated copula structures, but rather only a simple way of inducing dependence among them and such that their margins remain in the IGam (or GIG) class. This can be accomplished using even the very simple Gaussian copula. Even if the G_i are linked only by traditional correlation, the components of the *financial assets* (or whatever the components represent) will exhibit tail dependence.

12.7 Some Closing Remarks

The Jones (2002) distribution (12.9), along with our simple extension (12.14), the Shaw and Lee (2008) extension (12.16), and their related construction (12.19), have at least two aspects in common with our proposed MESTI distribution (12.31). First, they are all easy to simulate from, which is useful for studying the small-sample behavior of the parameter estimates, and also crucial if the method of indirect inference is to be used (Chapter III.10). Second, for $d > 2$, (12.9) also has no known closed-form solution (see Jones, 2002, page 170). In the $d = 2$ case, the density of (12.9) has a closed-form expression, but this comes at the price of not having a dispersion matrix or noncentrality terms, as added in (12.14). Once these are added, the computation of the density involves, for all $d > 1$, a d-dimensional integral, just like the MESTI.

The constructions from Shaw and Lee (2008) are, unfortunately, only limited to the bivariate case, so that what remains is a comparison between the Jones (2002) extension (12.14), the AFaK distribution (12.25), and the MESTI distribution (12.31).

1) The univariate margins of (12.31) are independent (Student's t) for $\boldsymbol{\Sigma}$ diagonal. With a diagonal dispersion matrix, the marginals of (12.14), similar to those of its special case (12.3), are uncorrelated, but not independent. The same holds for (12.25); see Abdous et al. (2005, p. 11), who refer to this as a "serious limitation".
2) During estimation of (12.9) or (12.14), the degrees of freedom k_i, $i = 1, \ldots, d$, have to be in ascending order, thus causing a label-switching problem. The v_i are fully decoupled in (12.25) and (12.31), completely obviating this problem.
3) Both evaluation of the generalization of (12.14) to the d-dimensional case and evaluation of (12.31) suffer from the curse of dimensionality as d grows. However, we have confirmed that (12.31) is, numerically speaking, completely unproblematic for any set of $v_1, \ldots, v_d$, whereas (12.14) reveals itself to be quite numerically problematic when (as tested for the $d = 2$ case) v_2 is close to v_1, as would be expected by the construction of the density. However, and quite disappointingly, this region of "closeness" is not some small epsilon-distance, but rather large enough to obviate its use in practice (as revealed by the potential closeness of the v_i from Figure 12.1).
For example, numerical problems occurred for the case when $v_1 = 3$ and $v_1 < v_2 < 3.8$ when very high accuracy is demanded from the numeric integration, worsening and eventually failing as $v_2 \to v_1$ no matter what the desired accuracy, and presumably will only be exacerbated for larger d.

For well-separated values of v_1 and v_2 (and no $\boldsymbol{\Sigma}$ matrix or noncentrality parameters), we recover, to virtually machine precision, the values obtained from the closed-form density expression in Jones (2002).

In contrast to the statement in Shaw and Lee (2008, p. 1285) that (12.16) "is a good basis for numerical computation provided $b > 0$", (12.16) also suffers numerically for v_2 near v_1, the critical distance depending on the desired accuracy of the numeric integration. When values of v_1 and v_2 are used such that no numerical problems occur, and $\theta = 0$, the obtained values agree, to machine precision, with those obtained for the bivariate density of (12.9).

12.A ES of Convolution of AFaK Margins

Here we propose an approximate method for computing the VaR and ES of a (weighted) convolution of the margins of an AFaK distribution that is faster than the simulation method discussed in Section 12.5.5 if only one $\mathbf{w}$ is relevant. Consider using a relatively small simulation sample size s_1 and approximating the distribution of $\widetilde{\mathbf{P}}$ in (12.29) with a flexible, fat-tailed, asymmetric parametric distribution whose parameters are estimated from the set of s_1 i.i.d. simulated observations. From this fitted distribution, the ES (and any measure of interest) can be analytically calculated. To this end, we entertain use of four such location-scale distributions: the GAt, a mixture of two GAt distributions (MixGAt), the noncentral t, and the asymmetric stable Paretian distribution.

The c.d.f. of the two-component MixGAt is no more complicated to evaluate than the single GAt. In particular, the c.d.f. of the K-component MixGAt is given by

$$F_P(z) = \sum_{j=1}^{K} \lambda_j F_{Z_j}\left(\frac{z-u_j}{c_j}; d_j, v_j, \theta_j\right), \tag{12.47}$$

where the jth GAt c.d.f. mixture component is given as the closed-form expression in (III.A.123), so that a quantile can be found by simple one-dimensional root searching. Similar to calculations for the ES of mixture distributions in Section III.A.8 and Broda and Paolella (2011), the ES of the mixture is given by

$$\begin{aligned}\mathrm{ES}(P;\xi) &= \frac{1}{\xi}\int_{-\infty}^{q_{P,\xi}} x f_P(x)\mathrm{d}x = \frac{1}{\xi}\sum_{j=1}^{K}\lambda_j \int_{-\infty}^{q_{P,\xi}} x c_j^{-1} f_{Z_j}\left(\frac{x-u_j}{c_j}\right)\mathrm{d}x \\ &= \frac{1}{\xi}\sum_{j=1}^{K}\lambda_j \int_{-\infty}^{\frac{q_{P,\xi}-u_j}{c_j}} (c_j z + u_j) c_j^{-1} f_{Z_j}(z) c_j \mathrm{d}z,\end{aligned}$$

giving

$$\begin{aligned}\mathrm{ES}(P;\xi) &= \frac{1}{\xi}\sum_{j=1}^{K}\lambda_j \left[c_j \int_{-\infty}^{\frac{q_{P,\xi}-u_j}{c_j}} z f_{Z_j}(z)\mathrm{d}z + u_j \int_{-\infty}^{\frac{q_{P,\xi}-u_j}{c_j}} f_{Z_j}(z)\mathrm{d}z\right] \\ &= \frac{1}{\xi}\sum_{j=1}^{K}\lambda_j \left[c_j S_{1,Z_j}\left(\frac{q_{P,\xi}-u_j}{c_j}\right) + u_j F_{Z_j}\left(\frac{q_{P,\xi}-u_j}{c_j}\right)\right],\end{aligned} \tag{12.48}$$

where $q_{P,\xi}$ is the ξ-quantile of P, S_{1,Z_j} is given in (III.A.95), and F_{Z_j} is the c.d.f. of the GAt distribution.

Remark While estimation of the two-component MixGAt is straightforward using standard maximum likelihood estimation, it was found that this occasionally resulted in an inferior, possibly bi-modal fit that optically did not agree well with a kernel density estimate. This artefact arises from the nature of mixture distributions and the problems associated with the likelihood, as discussed in detail in Chapter III.6. The methods discussed there can be used to rectify this issue, though we mention here an alternative, general method based on a so-called augmented likelihood procedure. The technique was first presented in Broda et al. (2013) in the context of discrete mixtures of stable Paretian, and is adapted for the mixture GAt as follows.

Let $f(x;\boldsymbol{\theta}) = \sum_{i=1}^{K} \lambda_i f_i(x;\boldsymbol{\theta}_i)$ be the univariate p.d.f. corresponding to a K-component mixture distribution with component weights $\lambda_1, \dots, \lambda_K$ positive and summing to one. The likelihood function is

$$\ell^*(\boldsymbol{\theta};\mathbf{x}) = \sum_{t=1}^{T} \log \sum_{i=1}^{K} \lambda_i f_i(x_t;\boldsymbol{\theta}_i), \tag{12.49}$$

where $\mathbf{x} = (x_1, \dots, x_T)'$ is the sequence of evaluation points, and $\boldsymbol{\theta} = (\boldsymbol{\lambda}, \boldsymbol{\theta}_1, \dots, \boldsymbol{\theta}_K)'$ is the vector of all model parameters. Assuming that the $\boldsymbol{\theta}_i$ include location and scale parameters, ℓ^* will be plagued with "spikes" (degenerate maxima).

To avoid these, we remove unbounded states from the likelihood function by introducing a smoothing (or shrinkage) term κ that, as $\kappa \to \infty$, drives all components to be the same (irrespective of their assigned mixing weight), implying that the mixture loses its otherwise inherently large flexibility. The suggested augmented likelihood function is given by

$$\tilde{\ell}(\boldsymbol{\theta};\mathbf{x},\kappa) = \ell^*(\boldsymbol{\theta};\mathbf{x}) + \kappa \sum_{i=1}^{K} \frac{1}{T} \sum_{t=1}^{T} \log f_i(x_t;\boldsymbol{\theta}_i), \tag{12.50}$$

where $\kappa \geqslant 0$ controls the shrinkage strength. If all component densities f_i are of the same type, then larger values of κ lead to more similar parameter estimates across components, with identical estimates in the limit, as $\kappa \to \infty$. At $\kappa = 0$, (12.50) reduces to (12.49). We term

$$\hat{\boldsymbol{\theta}}_{\mathrm{ALE}}(\kappa) = \arg\max_{\boldsymbol{\theta}} \tilde{\ell}(\boldsymbol{\theta};\mathbf{X},\kappa)$$

the **augmented likelihood estimator**, or ALE, and it is, for fixed κ or κ growing at a slow enough rate compared to the sample size T, consistent. By changing κ, smooth density estimates can be enforced, even for small sample sizes. For MixGAt with $K = 2$ and 250 observations, we find that $\kappa = 10$ works well. ■

We now turn to the use of the location-scale noncentral t (NCT). Estimation of the four parameters is done via maximum likelihood, using the saddlepoint approximation to the density. There is no closed-form expression for the VaR (quantiles) and the ES (note the latter needs the former), but numeric methods are straightforward and fast, relative to the time cost of parameter estimation; see the program and discussion in Section III.A.14.[4]

4 Another idea that is far faster is to use the table-lookup method for the VaR and ES, based on the two shape parameters, as developed in Krause and Paolella (2014), and then incorporate the location and scale terms from the fact that VaR and ES preserve location-scale transformations; see (III.A.128). However, that table-lookup method was designed for a shifted NCT based on its expectation (12.57), such that, when location parameter μ is zero, its mean is zero, as is desirable when working with GARCH-type processes.

For the stable Paretian distribution, estimation of the four parameters is done using the empirical c.f. estimator (see Section III.9.4.5), given its enormous speed advantage and comparable efficiency to the m.l.e. For VaR and ES, a table construction was made such that, for given α and β, the VaR and ES are delivered based on table-lookup. Once the table is constructed, this method is essentially instantaneous.[5] An alternative fast method based on a saddlepoint approximation is developed in Broda et al. (2017).

The motivation for using the stable is the conservative nature of the delivered ES. In particular, the GAt, MixGAt, and NCT are all asymmetric variations of the Student's t distribution which, while clearly heavy-tailed, still potentially possess a variance; as opposed to the stable, except in the measure-zero case of $\alpha = 2$. As such, the ES delivered from the stable will almost always be larger than those from the t-based models. This might be desirable when more conservative estimates of tail risk are desired, and, in the context of portfolio optimization, will affect the optimized portfolio vectors and the out-of-sample performance.

Thus, for these four parametric models, they are all straightforward to estimate and, as just discussed, computation of the VaR and ES is very fast and numerically reliable. We compare their performance against the pure simulation method based on 100,000 replications, and use for vector $\mathbf{w}$ (as a neutral and typical choice) the equally weighted portfolio, i.e., $w_i = 1/d$, $i = 1, \ldots, d$.

Remark The tail behavior associated with the $\widetilde{\mathbf{P}} = \widetilde{\mathbf{P}}_{\mathbf{w}}$ in (12.29), given the AFaK model and the parameters, is not subject to debate: Its distribution is analytically intractable, but involves convolutions of (dependent) random variables with power tails, and, as such, will also have power tails—unless all the $v_i > 2$ and as $d \to \infty$, so that the Gaussian central limit theorem is applicable. If the v_i are less than two, then the resulting convolution is expected to be in the domain of attraction of a non-Gaussian stable law.

It is fallacious to argue that, as our model involves use of the (noncentral) Student's t, with estimated degrees of freedom parameters above two, the convolution will have a finite variance, and so the stable distribution cannot be considered. It is crucial to realize first that *the model we employ is wrong w.p.1* (and also subject to estimation error) and, second, recalling that, if an i.i.d. set of stable data with, say, $\alpha = 1.7$ is estimated as a Student's t model, the resulting estimated degrees of freedom will not be below two, but rather closer to four (see Section III.9.1). ■

With respect to computational time for estimating the AFaK model and evaluating the ES for each of the four parametric distributions based on a sample size of $s_1 = 1{,}000$, the noncentral t required 0.20 seconds (on a 4.3GHz PC), the GAt and MixGAt required 0.23 and 1.96 seconds, respectively, and the stable required 0.00064 seconds. Generation of the s_1 samples required 2.9 seconds. (Generating the $s_1 = 1e5$ samples for the purely simulation-based method required a bit under 100 times that, or about 4.5 minutes.) The empirical calculation of the ES based on $s_1 = 1e5$ samples required approximately 0.35 seconds. The bottleneck is clearly the generation of samples, and this is because of the evaluation

5 Note, again, from (III.A.128) that VaR and ES preserve location-scale transformations, so that the table needs only be constructed for the two shape parameters α and β. If $\alpha > 1$ and the location parameter is zero, then the mean is zero; see Section II.8.3.1. Given the extremely heavy tails of the stable, and the fact that the density is difficult to evaluate far into the tails, direct evaluation of the ES (or construction of the aforementioned table) will be problematic if numeric integration using the p.d.f. is employed. However, a definite integral formula for the ES that is straightforward to numerically evaluate has been derived by Stoyanov et al. (2006), and this was used to build the tables; see Section III.1.16 for details.

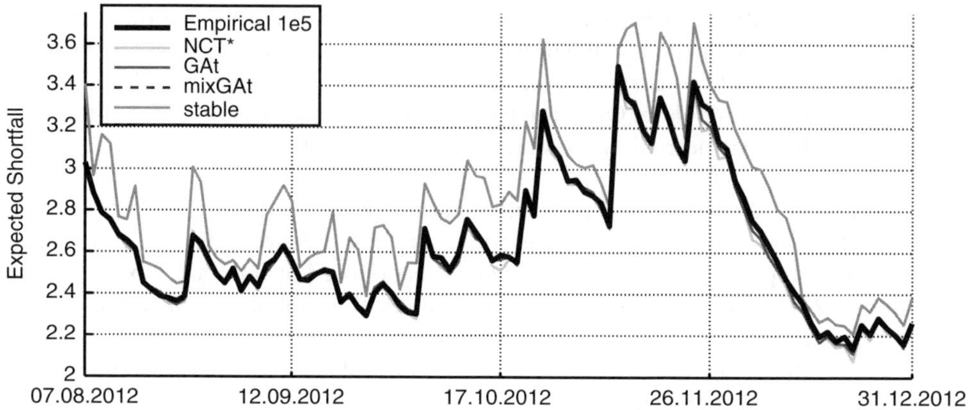

Figure 12.19 Comparison of five methods of estimating ES for a sequence of 100 rolling windows, using the equally weighted portfolio based on the $d = 30$ constituents (as of April 2013) of the Dow Jones Industrial Average index (Wharton/CRSP), with starting dates August 8, 2012 to December 31, 2012. These are plotted as a function of time, and based on the equally weighted portfolio.

of the NCT quantile function in (12.24). In summary, it is fastest to use $s_1 = 1e3$ samples and one of the parametric methods to obtain the ES.

With respect to accuracy, for practical purposes, the ES delivered by the GAt, MixGAt, and NCT are all very close to the empirical one, while the stable, as predicted, is considerably more conservative; see Figure 12.19. Closer inspection shows that the GAt and NCT are very close in performance, while the MixGAt is virtually unbiased and has the smallest deviation from the empirical values. However, because of the relatively high estimation time of the MixGAt (and the fact that the model is wrong anyway), we advocate use of the NCT.

Our next and final goal is to determine an approximation to the smallest value of s_1, say s_1^*, such that the sampling variance of the ES determined from the parametric methods is less than some threshold. This value, s_1^*, is then linked to the tail thickness of the various predictive returns distributions over the non-overlapping windows of data used above. The logic is that, as the tail thickness (degrees of freedom) parameter of the fitted NCT decreases, more samples will be required to estimate it accurately, as its determination is primarily driven by tail observations.

To compute s_1^* for a particular data set, the ES is calculated $n = 50$ times for a fixed s_1, based on n simulations of $\widetilde{\mathbf{P}}$ in (12.29) (and having used the NCT for its approximation). This is conducted for a range of s_1 values, and s_1^* is taken to be the smallest number such that the sample variance of the n ES values is less than a threshold value. Figure 12.20 shows the results for selected values of s_1. As expected, the spread of the ES values across rolling windows decreases as s_1 increases. As can be seen from the middle right panel, a roughly linear relationship is obtained for the logarithm of ES variance.

A simple regression approach suggests itself. For a variance threshold of $\exp\{-2\} \approx 0.14$, some trial and error results in

$$s_1(\hat{v}_P) = \lceil 100 + (49.5 - 3.8\ \hat{v}_P + 100.5\ \hat{v}_P^{-1})^2 \rceil \mathbb{I}\{\hat{v}_P \leqslant 15\} + 100\mathbb{I}\{\hat{v}_P > 15\}. \tag{12.51}$$

The procedure is then: From an initial set of 300 AFaK samples, the ES is evaluated, s_1 is computed from (12.51), and, if $s_1 > 300$ (or $\hat{v}_P < 11.58$), an additional $s_1 - 300$ samples are drawn.

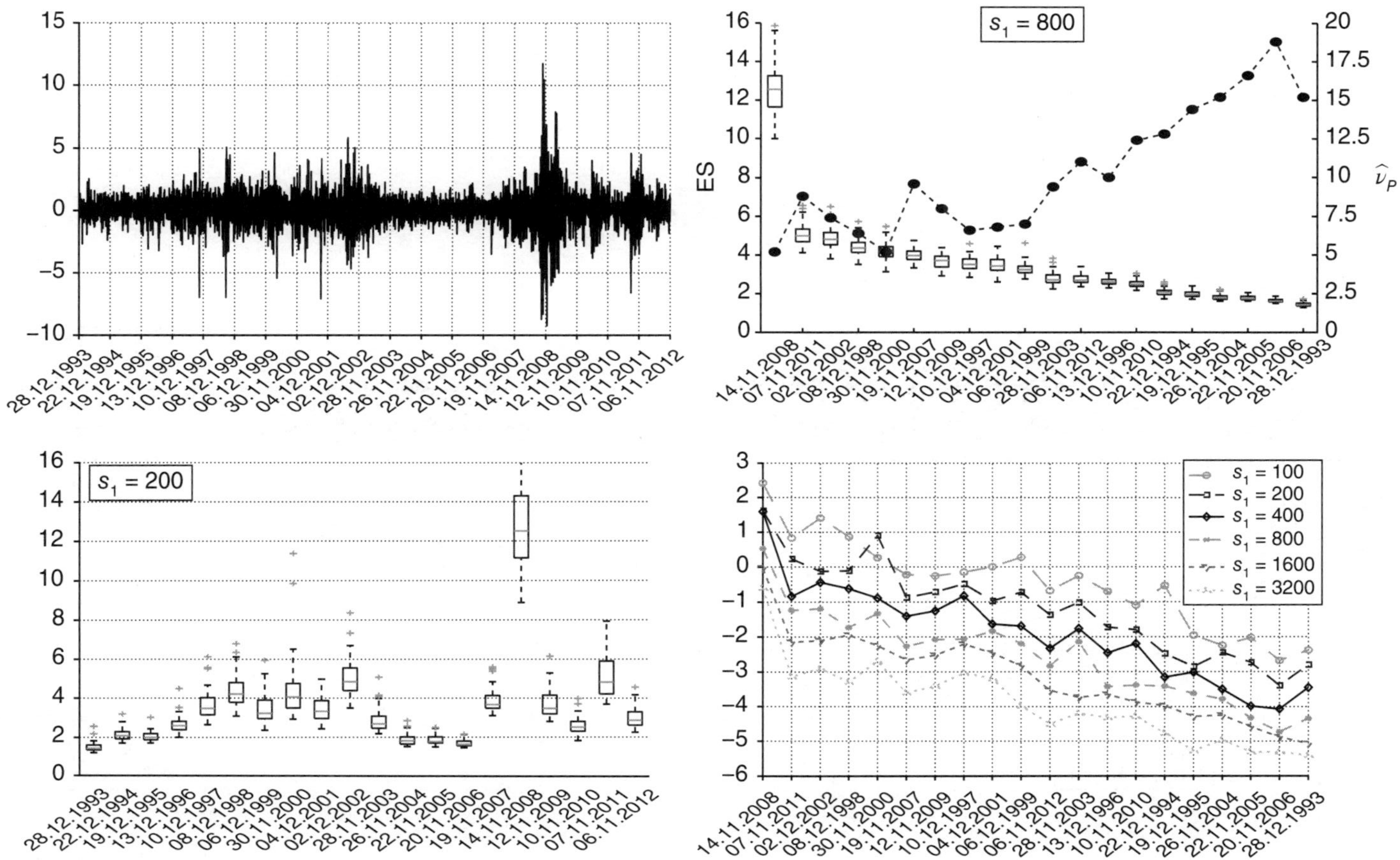

Figure 12.20 **Upper left:** Percentage log returns of the equally weighted portfolio. **Mid and lower left:** Boxplots of 1% ES values obtained from 50 simulations based on s_1 draws from the fitted copula for different non-overlapping rolling windows of size 250, spanning January 4, 1993, to December 31, 2012. Timestamps denote the most recent date included in a data window. All values are obtained via the NCT estimator. **Upper right:** Boxplots of 1% ES values sorted in descending order by the average ES value, overlayed by the average of the estimated degrees of freedom parameters. **Mid right:** ES variances in log scale across rolling windows for different samples sizes s_1, sorted by the average ES value per window. **Lower right:** Linear approximation of the above panel, overlayed by the linear approximation of the estimated degrees of freedom, based on $s_1 = 3{,}200$.

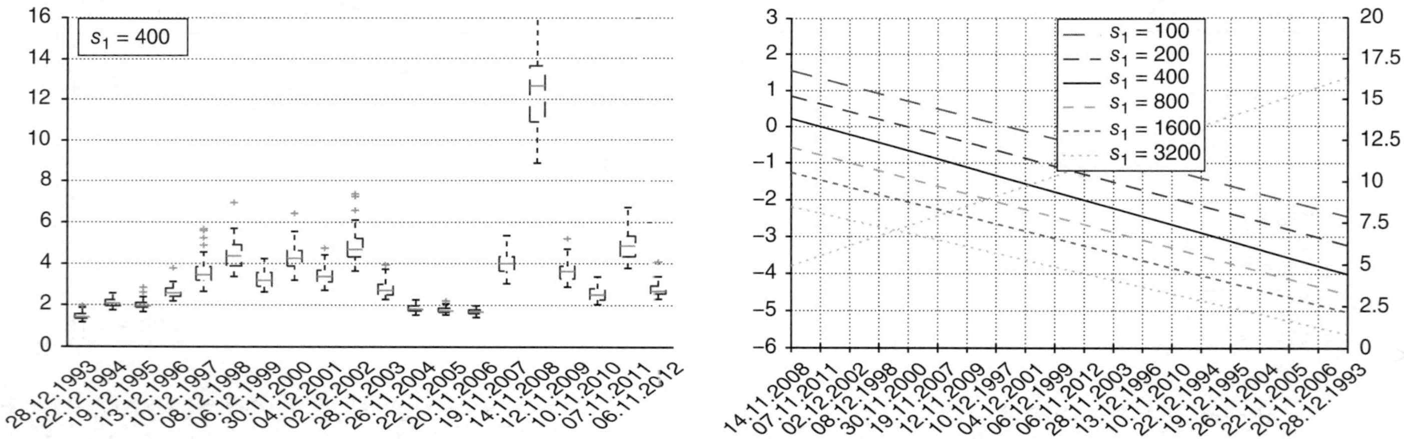

Figure 12.20 (*Continued*)

12.B Covariance Matrix for the FaK

At the end of Section 12.5.1 we remarked that a closed-form expression for the covariance matrix $\mathbf{V}$ of $\mathbf{T} = (T_1, \dots, T_d)' \sim \text{FaK}(\mathbf{v}, \boldsymbol{\mu}, \boldsymbol{\Sigma})$ appears elusive. An attempt based on representation (12.24) appears fruitless. We might guess that, if $v_i > 2$, $i = 1, \dots, d$, then

$$\mathbf{V} \stackrel{?}{=} \mathbf{K}\boldsymbol{\Sigma}\mathbf{K}, \qquad \mathbf{K} = \text{diag}([\kappa_1, \kappa_2, \dots, \kappa_d]), \qquad \kappa_i = \sqrt{v_i/(v_i - 2)}. \tag{12.52}$$

By computing the covariance via bivariate numeric integration, conducted using the program in Listing 12.21, we can confirm that (12.52) is at least a reasonable approximation for the range of values considered. For example, the code in Listing 12.22 can be used to compare the approximate and (numerically computed) exact values.

```
function covval = AFaKcovint(df,noncen,mu,scale,R)
ATOL=1e-10; RTOL=1e-6;  % 10 and 6 are the defaults
covval = quadgk(@(yvec) int(yvec,df,noncen,mu,scale,R), ...
         -Inf,Inf,'AbsTol',ATOL,'RelTol',RTOL);

function Int=int(yvec,df,noncen,mu,scale,R)
Int=zeros(size(yvec));
ATOL=1e-10; RTOL=1e-6; % 10 and 6 are the defaults
for i=1:length(yvec)
  y=yvec(i);
  Int(i) = quadgk(@(x) AFaKcov(x,y,df,noncen,mu,scale,R), ...
           -Inf,Inf,'AbsTol',ATOL,'RelTol',RTOL);
end

function f = AFaKcov(x,y,df,noncen,mu,scale,R)
dfvec=df(2:end); theta=noncen(2:end);
m1=sqrt(dfvec/2) .* gamma(dfvec/2-1/2) ./ gamma(dfvec/2) .* theta;
yy=y*ones(1, length(x)); tx=x-m1(1)-mu(1); ty=yy-m1(2)-mu(2);
pass=[x ; yy]; f=tx.*ty.*FFKpdfvec(pass',df,noncen,mu,scale,R)';
```

Program Listing 12.21: Computes the covariance of the bivariate (A)FaK. Nested univariate numeric integration based on adaptive Gauss–Kronrod quadrature is used, as implemented in Matlab's quadgk routine. Their implementation conveniently supports integration over infinite intervals, and is more accurate than use of their other numeric integration routines, notably the canned routine for bivariate integration, dblquad, even in conjunction with an extreme error tolerance. The cases in the graphs for which $v_0 = v_1 = v_2$ were also computed with numeric integration, and as they are exact (the discrepancy being on the order of less than 1×10^{-8} for all σ_{12} between -0.9 and 0.9), we can be rather confident that the values for the $v_1 \neq v_2$ cases are quite accurate.

```
df=[4 3 4]; noncen=[0 0 0]; scale=[3 0.1];
mu=[3 -7]; R12=0.5; R=[1 R12; R12 1];
K = sqrt(diag( [df(2)/(df(2)-2) , df(3)/(df(3)-2)] ));
ApproxSigma = K*diag(scale)*R*diag(scale)*K
TrueCov = AFaKcovint(df,noncen,mu,scale,R)
```

Program Listing 12.22: Approximate and exact covariance of the bivariate (A)FaK.

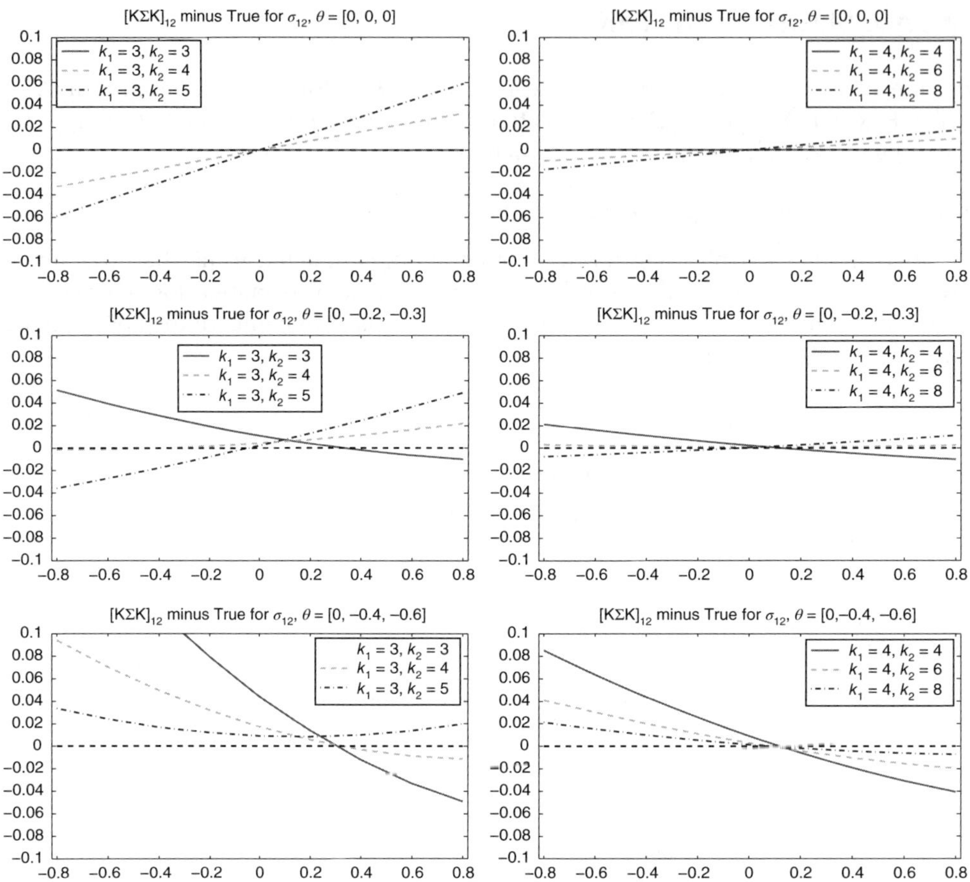

Figure 12.21 **Top:** Illustration of the discrepancy between the approximation of $V_{12} = \text{Cov}(T_1, T_2)$ (obtained as the off-diagonal term $\mathbf{K\Sigma K}$) and the true value (obtained by bivariate numeric integration), as a function of σ_{12}, where $\mathbf{T} = (T_1, T_2)' \sim \text{FaK}(\mathbf{v}, \boldsymbol{\mu}, \boldsymbol{\Sigma})$, with $\boldsymbol{\mu} = \mathbf{0}$, $\sigma_1 = \sigma_2 = 1$, σ_{12} varies along the x-axis, and $\mathbf{v} = (v_0,\ v_1,\ v_2)'\ v_0 = \max(v_1, v_2)$, with v_1 and v_2 specified in the legend of the plots. **Middle and bottom:** Same, but with using the AFaK distribution with $\theta_0 = 0$ but nonzero θ_1 and θ_2.

To illustrate, the top two panels of Figure 12.21 show the discrepancy between the single covariance term in the 2×2 matrix $\mathbf{K\Sigma K}$ from (12.52) and the true covariance between T_1 and T_2, obtained via numeric integration, over a grid of σ_{12} values, where v_0 is always taken to be $\max(v_1, v_2)$. Notice that, for the cases with $v_1 = v_2$ (and, thus, $v_0 = v_1 = v_2$), the FaK coincides with (12.3), with covariance precisely $\mathbf{K\Sigma K}$. This is also seen in the graphs.

The nonzero discrepancy visible from the plots appears to increase monotonically in $|\sigma_{12}|$ (for fixed v_i), in $|v_2 - v_1|$, and in $\min(v_1, v_2)$. It also appears linear and symmetric about $\sigma_{12} = 0$, suggesting that we take, with $V_{ij} := [\mathbf{V}]_{ij}$ the ijth element $\mathbf{V}$, $i, j = 1, \ldots, d$,

$$V_{12} = \text{Cov}(T_1, T_2) = [\mathbf{K\Sigma K}]_{12} - b_{12}\sigma_{12} = (\kappa_1\kappa_2 - b_{12})\sigma_{12}, \tag{12.53}$$

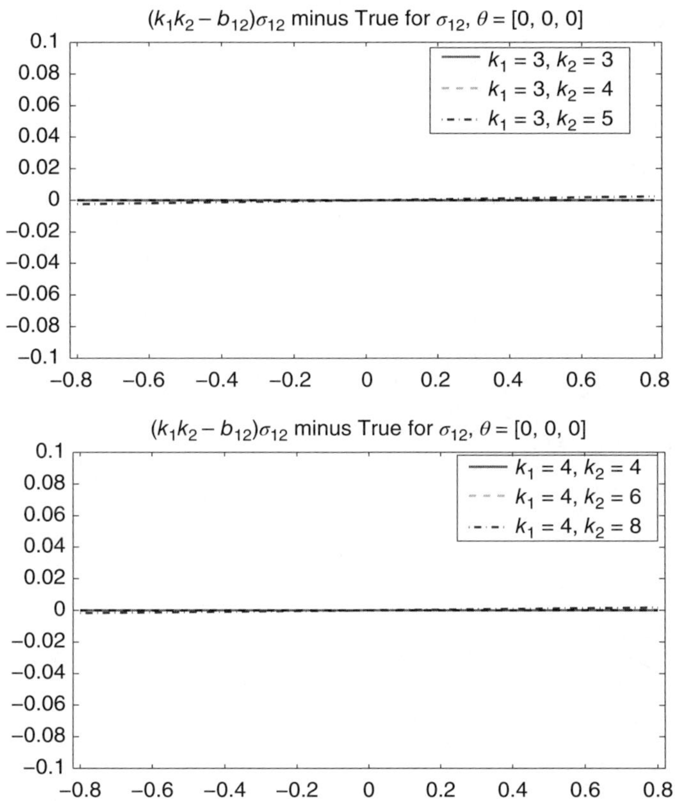

Figure 12.22 Same as top two panels of Figure 12.21 but based on (12.53) and (12.54) (and using k instead of v in the legend).

where $b_{12} = b(v_1, v_2)$ is the the slope of the line depicted in the top panels, κ_1 and κ_2 are given in (12.52), and assuming that $v_0 = \max(v_1, v_2)$. Some trial and error based on simulation for a range of values of v_1 and v_2 yielded the approximation

$$0.6 \cdot b(v_1, v_2) = c_1 g + c_2 \delta + c_3 \delta m + c_4 \delta m^{1/2} \tag{12.54}$$

for b, resulting in an R^2 regression coefficient of 0.985, where

$$m := \min(v_1, v_2), \quad \delta := |v_2 - v_1|, \quad g := \Gamma(\delta + 1/2),$$

and, with enough significant digits to maintain the accuracy,

$$c_1 = -0.00043, \quad c_2 = 0.2276, \quad c_3 = 0.0424, \quad c_4 = -0.1963.$$

Figure 12.22 shows the same top two panels of Figure 12.21 but based on (12.53) and (12.54). The additional terms from (12.53) and (12.54) result in further accuracy, presumably enough for practical applications. The results are, however, limited to the bivariate case, with $v_0 = \max(v_1, v_2)$. If this latter constraint on v_0 is adopted, then the result might hold in the general d-variate case: Observe that, for $\mathbf{T} = (T_1, \dots, T_d)' \sim \text{FaK}(\mathbf{v}, \boldsymbol{\mu}, \boldsymbol{\Sigma})$, if the bivariate marginal distribution of $(T_i, T_j)'$

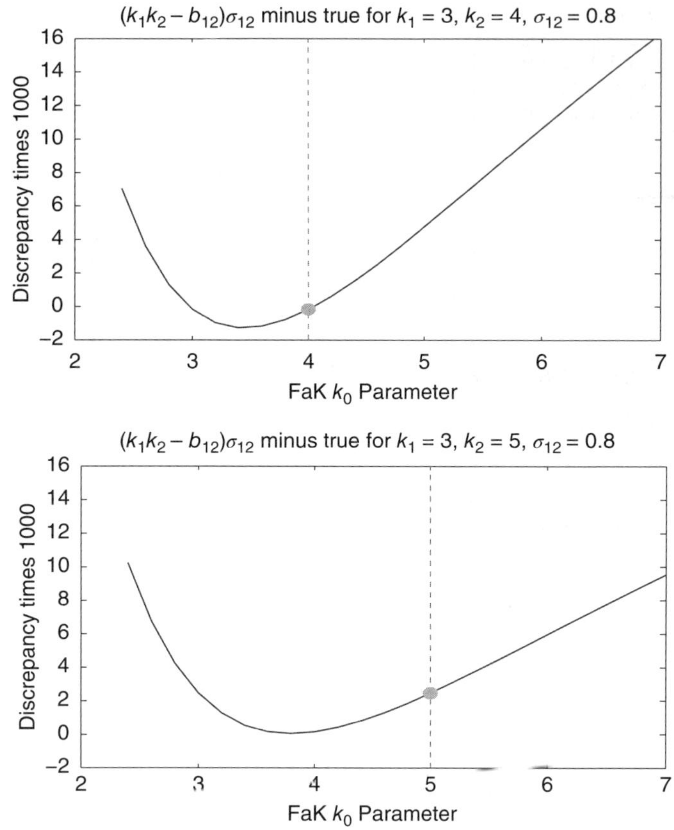

Figure 12.23 **Top:** Similar to Figure 12.22 but for $v_1 = 3$, $v_2 = 4$, a fixed value of σ_{12} of 0.8, and as a function of v_0. The vertical dashed line indicates the case with $v_0 = \max(v_1, v_2)$, which agrees with the corresponding point in the bottom panel of Figure 12.22 (right-most point of the dashed line). **Bottom:** Same, but for $v_1 = 3$, $v_2 = 5$. (Note that notation k instead of v is used in the titles, sparing the lazy author a re-computation of the graphics.)

is $\text{FaK}([v_0, v_i, v_j]', [\mu_i, \mu_j]', \mathbf{\Sigma}_{i,j})$, where $\mathbf{\Sigma}_{i,j}$ is the 2×2 dispersion matrix of the ith and jth entries of $\mathbf{\Sigma}$, then clearly $\text{Cov}(T_1, T_2)$ is unaffected by $v_3, v_4, \dots, v_d$. Given that the univariate margins are Student's t, this conjecture seems likely, and simulation (by estimating the FaK parameters of the three bivariate distributions formed from a simulated FaK with $d = 3$ with 100,000 observations) essentially confirms this.

For example, taking the tri-variate case with $v_1 = 3$, $v_2 = 4$, $v_3 = 5$, and $v_0 = \max(v_i) = 5$, the bivariate margins (T_1, T_3) and (T_2, T_3) are such that v_0 is the maximum of the two v_i-values, but not for (T_1, T_2). The top panel of Figure 12.23 investigates this case in more detail, showing the error incurred by the approximation (12.53) as a function of v_0, for a fixed value of σ_{12} of 0.8, with $v_1 = 3$ and $v_2 = 4$. The case with $v_0 = \max(v_1, v_2) = 4$ corresponds very close to the minimal error obtained for all v_0.

As such, use of the structure of (12.53) for all pairs,

$$V_{ij} \approx (\kappa_i \kappa_j - b_{ij})\sigma_{ij}, \qquad b_{ij} = b(v_i, v_j), \tag{12.55}$$

appears to be a reasonable approximation, keeping in mind that the slope term b_{ij} is not exact, and b_{ij} was determined only for the bivariate case with $v_0 = \max(v_i, v_j)$. The reader is encouraged to investigate this in more detail, replicating and augmenting the findings and graphs shown here.

We now turn to the AFaK case. Let $\mathbf{T} \sim \text{AFaK}(\mathbf{t}; \mathbf{v}, \boldsymbol{\theta}, \boldsymbol{\mu}, \boldsymbol{\Sigma})$, with $\theta_0 = 0$ and $\min_i(v_i) > 2$. As with the symmetric case, we hope that, to first order, $\mathbb{V}(\mathbf{T})$ is reasonably approximated by $\mathbf{K\Sigma K}$, where $\mathbf{K}$ is now

$$\mathbf{K} = \text{diag}([\kappa_1, \kappa_2, \dots, \kappa_d]), \qquad \kappa_i = \sqrt{\mathbb{V}(S_i)}, \qquad S_i = (T_i - \mu_i)/\sigma_i, \tag{12.56}$$

i.e., the diagonal matrix with iith element given by the square root of the variance of the singly noncentral t random variable $S_i \sim t'(v_i, \theta_i, 0, 1)$, computed from the expression for the mean,

$$\mathbb{E}[S] = \theta\left(\frac{v}{2}\right)^{1/2} \frac{\Gamma(v/2 - 1/2)}{\Gamma(v/2)}, \tag{12.57}$$

as in (12.36), and $\mathbb{E}[S^2] = [v/(v-2)](1 + \theta^2)$.

The middle and bottom panels in Figure 12.21 show that the linearity and symmetry about $\sigma_{12} = 0$ no longer hold when noncentrality parameters are introduced, although the discrepancy between the true $\text{Cov}(T_i, T_j)$ and that given by the corresponding element of approximation $\mathbf{K\Sigma K}$ remains small. This will break down as the asymmetry increases and/or as $\min(v_i) \to 2$.

The reader interested in this model is encouraged to develop an approximation to $\text{Cov}(T_i, T_j)$ improving upon $\mathbf{K\Sigma K}$, similar to (12.53) and (12.54). Such an approximate mapping is a type of **response surface**. The program in Listing 12.21 using bivariate numeric integration would be used to generate a set of exact covariance values over a four-dimensional grid of values in v_1, v_2, θ_1, and θ_2, and then trial and error is required for finding an accurate response surface based on polynomials and other terms involving $\mathbf{v}$ and $\boldsymbol{\theta}$ for $\text{Cov}(T_1, T_2)$. A final program would input, for any dimension d, $\mathbf{v}$, $\boldsymbol{\theta}$, $\boldsymbol{\mu}$, and $\boldsymbol{\Sigma}$, and output the (approximation to the) covariance matrix of $\mathbf{T} \sim \text{AFaK}(\mathbf{t}; \mathbf{v}, \boldsymbol{\theta}, \boldsymbol{\mu}, \boldsymbol{\Sigma})$. As the resulting response surface is evaluated very fast, one could use it to estimate the model parameters with the method of moments, i.e., choose the parameters to minimize the difference between the sample mean vector and sample variance covariance matrix, and their theoretical counterparts. The result is a type of robust estimator, in the sense that the likelihood was not used, which could well be mis-specified.

13

Weighted Likelihood

[I]t is worth asking why do we continue to study non-linear time series models, if they are analytically difficult, can rarely be given economic interpretation, and are very hard to use for practical tasks such as forecasting.

(Clive W. J. Granger, 2008, p. 2)

13.1 Concept

The goal of this chapter is not to present another model for asset returns, but rather a way of augmenting virtually any time-series model with very little effort that results in improved forecasts. There are, in fact, two ways, and we have already discussed one of them (notably in book III, though also in this chapter below, and in Chapter 14), namely use of shrinkage. From a conceptual point of view, shrinkage helps to address the annoying fact that there is a finite amount of data available for estimation, and uses the tradeoff of bias and variance to deliver estimators with, in aggregate, lower mean squared error. The second inferential augmentation, use of **weighted likelihood**, works from a different angle, and addresses the fact that, in essentially all realistic applications in time-series modeling, the proposed model is wrong w.p.1, and is in some (often unknown) way mis-specified. Note that, as shrinkage and weighted likelihood are addressing different aspects of the estimation problem, they can (and should) be used in conjunction. Conveniently and importantly, neither entails a more complicated estimation procedure, though both require the use of tuning parameters that need to be optimized for the desired purpose of the model (which is, in our setting, forecasting).

Weighted likelihood can be used in conjunction with what the researcher deems to be the "best" model in the sense of being "least mis-specified", but also in models that are blatantly mis-specified. The reason one would use the latter is because of ease of estimation—the "best" model might be relatively sophisticated and entail a complicated estimation paradigm, and/or is such that simulation is trickier (as would be used for checking the small-sample behavior of estimators) and/or whose stationarity conditions are more elaborate or unknown. In particular, we have in mind to use an i.i.d. setting for modeling financial asset returns, which, as was emphasized in Chapter 10 on GARCH structures, is quite obviously mis-specified. Such an approach, when used with weighted likelihood, still requires respecting the time-series nature of the data (i.e., the natural ordering through time), as

Linear Models and Time-Series Analysis: Regression, ANOVA, ARMA and GARCH, First Edition. Marc S. Paolella.

opposed to allowing for the data to enter an inferential procedure permuted in some way. That is, if the data are truly i.i.d., then their ordering has no relevance; they are **exchangeable**.[1]

The idea is to recognize that, in all traditional likelihood-based inference for i.i.d. data, each observation (or, in the non-i.i.d. case, possibly the i.i.d. innovation, or error term, associated with that observation) is implicitly equally weighted in the likelihood. This is optimal if the data generating process (d.g.p.) is correctly specified. In a time-series context, notably for modeling financial asset returns, it is essentially understood that the underlying d.g.p. is quite complicated, and *any* postulated model is wrong w.p.1. All models will be mis-specified to some extent, some models more than others, though it is not obvious what a correct metric is for "degree of mis-specification". Use of penalized in-sample-fit measures such as AIC and BIC can help choose among competing models, as well as (and preferably) out-of-sample performance. As all models will be wrong in this context, we can envision our chosen model to be a reasonable approximation when used on short time intervals, though as the size of the data window increases, its "degree of mis-specification" is expected to increase.

While it might, at first blush, appear that the extent of a model's mis-specification is an analytic concept that has nothing whatsoever to do with the amount of data available for estimation, the demands of reality when working with nontrivial d.g.p.s suggest that the two are indeed intimately linked. Essentially, the amount of available data decisively dictates the possible complexity of the model.

If we somehow knew the true parametric form of the d.g.p., and estimation of its parameters were computationally feasible, then it would be optimal to use all the available data, and possibly the correct d.g.p.,[2] and equally weight the data in the likelihood, as we have so far implicitly done. As this is not the case, we are left with fitting a mis-specified model that serves as a reasonable local approximation to the d.g.p., so the question becomes: How much data should be used? A small window of observations leads to less bias but very high variance, and vice versa for a large window. As the complexity of the model increases, more data could be used.

Moreover, for time-series data, if the goal is to construct a density forecast at the future time $T+1$, then it stands to reason that, amid a mis-specified model that does not account for how, say, the parameters change through time, more recent observations contain relatively more information about the distribution at time $T+1$ than do observations much further in the past. The same concept could, for example, be applied to spatial data, such that observations closer to the target area to be forecasted to receive more weight than more distant observations.

In general, the idea that parameters change over (say) time was strongly embraced in the latter half of the 20th century with regression modeling (and continues unabated), as discussed in Section 5.6. With respect to the Hildreth–Houck random coefficient model, it was pointed out early on that, along with endowing the coefficients with randomness, they should also be augmented by making them (usually simple linear) functions of other observable random variables that change through time. As stated by Singh et al. (1976, p. 341), "... we assume that the typical regression coefficient $\beta_i(t)$ is subject to two influences that cause it to deviate from its average value $\bar{\beta}_i$. The first of these, following [Hildreth–Houck], is a random disturbance that possesses certain distributional properties.

1 See Section I.5.2.3 for a formal definition. This is also the assumption used for the non-parametric bootstrap. Note that exchangeability does not imply independence.

2 Recall the quote by Magnus (2017) at the beginning of Section 1.4 regarding possibly omitting estimation of some parameters, even if they correspond to genuine effects in the true model, because of having a finite amount of data, and such that precision of more relevant parameters can be gained (at the cost of bias). An example is the asymmetry parameter in the APARCH model from Section 10.3.1: When this asymmetric effect is mild, it is better off in terms of out-of-sample forecasting ability to just set it to zero. Observe how this is a form of shrinkage estimation.

The second is due to the influence of factors that may vary systematically with time". This concept ultimately gave rise to the more general time-varying regression model structures surveyed in Park et al. (2015). The use of weighted likelihood can thus be seen as a "poor man's" way of improving the forecasts from a model for which it is speculated or known that the parameters (such as regression coefficients) are changing through time and, possibly, depend on certain variables that are not easily obtained.

As already mentioned, the most notable stylized fact of asset returns measured at a daily frequency is volatility clustering. Indeed, one of the reasons the class of GARCH models is successful is because it models the volatility essentially as a weighted average of past volatilities, with more weight on recent observations. This is well-captured in the following quote by the originator of the ARCH model:

> The assumption of equal weights seems unattractive, as one would think that the more recent events would be more relevant and therefore should have higher weights. Furthermore the assumption of zero weights for observations more than one month old is also unattractive.
>
> (Robert Engle, 2001, p. 159)

While Engle's statement is taken somewhat out of context (it refers to the use of a GARCH filter for modeling a time-varying volatility), it embodies precisely the more general idea that, when the true d.g.p. of a time series is complicated, a mis-specified model to be used for forecasting the future can be improved by conducting its estimation such that more recent observations receive relatively more weight than those further in the past. Such a method adds considerably to the forecasting power of a model for financial asset returns—even when time-varying volatility via a GARCH-type model is used.

To add some intuition to the use of weighted likelihood, let the true d.g.p. be nonlinear (as will be the case in almost all nontrivial phenomena of interest). As suggested in the quote above from Granger (2008), specification of the nonlinear model can be rather difficult, and the ensuing problems associated with forecasting (and possibly estimation) can preclude its practical use. Consider using a local linear approximation (as could be obtained from a first-order Taylor series, for example, from the true d.g.p., if it were somehow available). This will be useful on smaller windows of data, but the parameters associated with the linear approximation will need to change through time (or, for spatial models, through space). This is in fact the content of Granger (2008), appealing to a result he derived with Halbert White, referred to as White's theorem. It states that, for time series $\{Y_t\}$ with finite mean and $\Pr(Y_t = 0) = 0$, there exists sequences $\{p_t\}$ and $\{e_t\}$ such that $Y_t = p_t Y_{t-1} + e_t$, i.e., the model can be expressed linearly, with time-varying coefficients.

The specification of the law of motion for sequence $\{p_t\}$ might be challenging, and instead one can consider assuming it is constant on short windows of data. This, in turn, can be approximated by using the entire data set, but such that observations at time t are given more weight than those at time $t - s$, $s > 0$. Thus, the nonlinearities associated with the true d.g.p. can be implicitly modeled via use of a linear model with time-varying parameters or weighted likelihood.

Another notable feature of this strategy, also emerging in Engle's quote, is that the researcher is relieved of having to choose an arbitrary cutoff for the data window, and can, in principle, use all relevant data instead of just an arbitrarily chosen amount, such as one or four years. Indeed, in light of the above reasoning, it should actually seem quite odd that, in the difficult game of time-series prediction, there should exist a precise point of time in the past such that the data previous to that point are of absolutely no relevance to the analysis, while the data that do get included are implicitly equally weighted.

To implement the weighting scheme for a set of υ observations, a vector of weights $\boldsymbol{\varpi} = (\varpi_1, \dots, \varpi_\upsilon)$ is used such that it is standardized to sum to a constant, such as υ (as with the conventional m.l.e.), or to one, which is what we choose. The model parameters are then estimated by maximizing the weighted likelihood, whereby the log-likelihood component associated with period t is multiplied by ϖ_t, $t = 1, 2, \dots, \upsilon$. We use the simple hyperbolic weighting scheme given by

$$\varpi_t \propto (\upsilon - t + 1)^{\rho - 1}, \qquad \sum_{t=1}^{\upsilon} \varpi_t = 1, \tag{13.1}$$

where the single parameter ρ dictates the shape of the weighting function. Values of $\rho < 1$ ($\rho > 1$) cause more recent observations to be given relatively more (less) weight than those values further in the past, while $\rho = 1$ corresponds to the standard, equally weighted likelihood.

Example 13.1 Let $\mathbf{Y}_t \overset{\text{ind}}{\sim} \mathrm{N}_2(\mathbf{0}, \mathbf{R}_t)$, $t = 1, \dots, T$, where $\mathbf{R}_t$ is a correlation matrix with the single parameter being its (1, 2) element $R_{t,12} = 0.5t/T$. Notice that $R_{1,12}$ starts at (almost) zero and changes linearly such that, for $t = T$, is 0.5. This is an example of a time-varying parameter model. If we estimate R_{12} with the usual, equally weighted likelihood and assuming an i.i.d. sequence, then we expect $\widehat{R}_{12}$ to be close to 0.25, though the density prediction at time $T + 1$ ideally would use an $\widehat{R}_{12}$ close to 0.5.

We imagine that we don't know the true time-varying nature of R_{12}, and consider the use of weighted likelihood. We do this using values of weighting parameter $\rho = 0.01, 0.02, \dots, 1$, with $\rho = 1$ corresponding to traditional, equally weighted estimation, applying (13.1) to the usual plug-in estimator of correlation. Weighted estimation of correlation matrices based on the usual plug-in sample estimator is considered in detail in Pozzi et al. (2012), who also provide Matlab code as function `weightedcorrs`. We use that routine for our results, though the reader is encouraged to construct the basic weighted sample correlation estimator him- or herself.

The results, for two sample sizes T and 40 replications, are shown in Figure 13.1a. As ρ decreases towards 0.01, the "effective sample size" is decreasing, and the variance of $\widehat{R}_{12}$ increases, though is also becoming less biased. As (also) expected, the variance of $\widehat{R}_{12}$ is larger for the smaller sample size of $T = 1{,}000$.

Now let $\mathbf{Y}_t \overset{\text{ind}}{\sim} \mathrm{FaK}_2(\mathbf{v}, \mathbf{0}, \mathbf{R}_t)$, as introduced in Section 12.5.1, where the subscript 2 denotes the dimension d, $\mathbf{v} = (\upsilon_0, \upsilon_1, \upsilon_2) = (4, 4, 4)'$, and the same structure as for the normal case is used for $\mathbf{R}_t$. Figure 13.1b shows the resulting $\widehat{R}_{12}$ when the estimator is the weighted sample correlation, while Figure 13.1c uses the weighted m.l.e., conditional on the *true* parameters $\mathbf{v}$, $\boldsymbol{\mu} = (0, 0)'$, and scales $\boldsymbol{\sigma} = (1, 1)'$. (One could also inspect $\widehat{R}_{12}$ when all parameters are jointly estimated, though besides taking longer to compute in this exercise, the point is to compare the m.l.e. of R_{12} versus the sample correlation estimator, which itself does not make use of the other parameters.)

As with the normal case, the variance of $\widehat{R}_{12}$ increases as ρ decreases, its bias decreases, and is lower for the larger of the two sample sizes. Moreover, its variance when using the (weighted) sample correlation is substantially higher when using FaK instead of the normal, and, most crucially, for FaK, $\widehat{R}_{12}$ has much lower variance when the (weighted) m.l.e. is used, in agreement with the results in Figure 12.11.

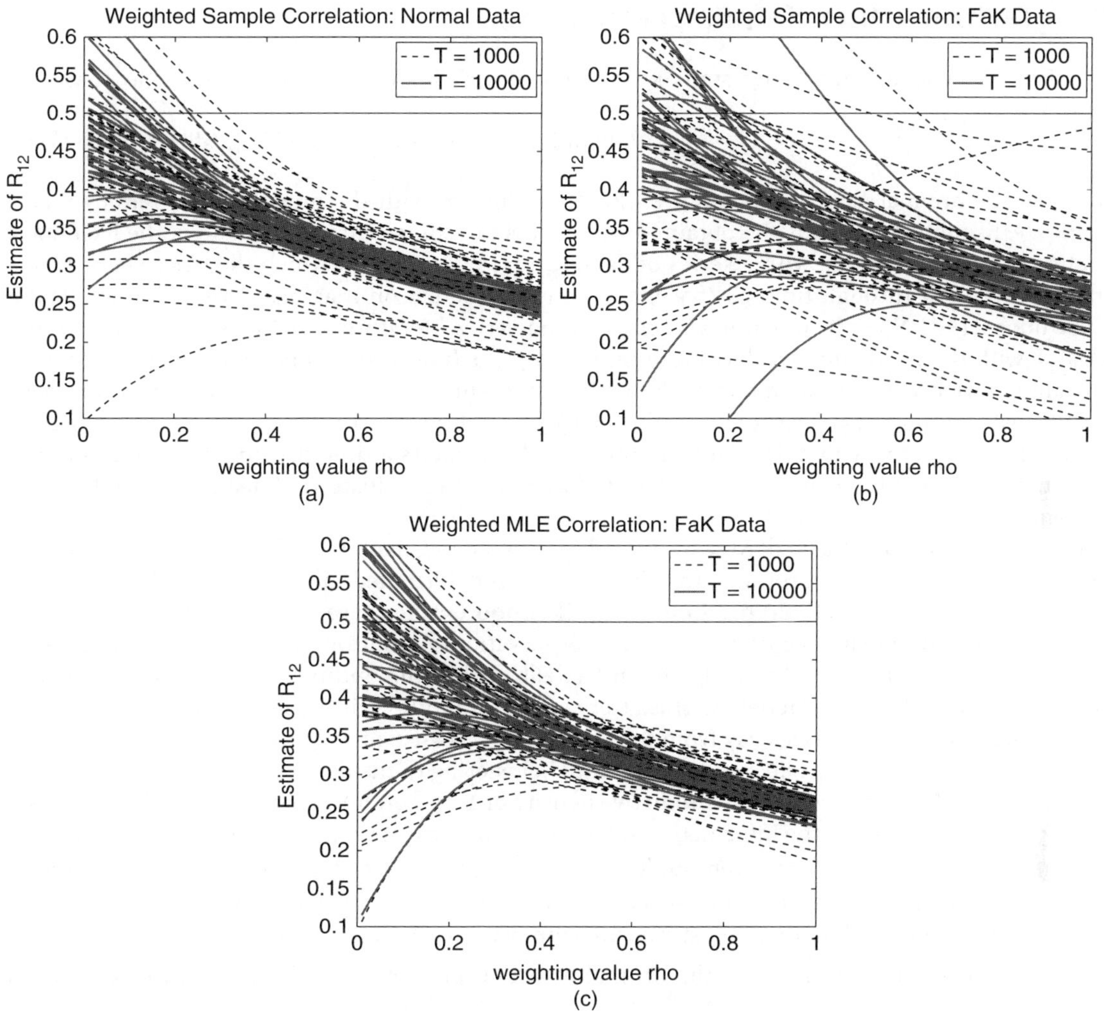

Figure 13.1 (a) Estimates of $\widehat{R}_{12}$ using the weighted sample correlation estimator, as a function of weighting parameter ρ, for bivariate normal data generated as $\mathbf{Y}_t \overset{\text{ind}}{\sim} \mathrm{N}_2(\mathbf{0}, \mathbf{R}_t)$, $t = 1, \dots, T$, where $\mathbf{R}_t$ is a correlation matrix with single parameter $R_{t,12} = 0.5t/T$, so that the correlation is varying linearly through time, from zero to 0.5. (b) Same as (a), again using the weighted sample correlation estimator, but for bivariate FaK data with $v_0 = v_1 = v_2 = 4$ and the same correlation structure. (c) Same as (b), but estimation is based on the weighted m.l.e.

The reader is invited to construct the code to reproduce the graphs in Figure 13.1 and, at least for the case for bivariate normal data generated as $\mathbf{Y}_t \overset{\text{ind}}{\sim} \mathrm{N}_2(\mathbf{0}, \mathbf{R}_t)$, perform far more than 40 replications and compute and plot the mean squared error as a function of ρ. Presumably, it will be an approximately quadratic function such that its minimum is reached for a value of ρ somewhere between zero and one. ■

13.2 Determination of Optimal Weighting

> The only useful function of a statistician is to make predictions, and thus to provide a basis for action.
>
> (William Edwards Deming; quoted in Wallis, 1980, p. 321).

Note that the optimal value of ρ cannot be jointly estimated with the model parameters by maximizing the likelihood, but must be obtained with respect to some criterion outside of the likelihood function. Consider working with a series of data, say $\mathbf{Y}_1, \mathbf{Y}_2, \dots, \mathbf{Y}_T$, in which the observations have a natural ordering through time or space, and concern centers on making a prediction about the as-yet-unobserved $\mathbf{Y}_{T+1}$. In much of the classic time-series literature, emphasis was on predicting $\mathbb{E}[\mathbf{Y}_{T+1}]$, with the nearly universal assumption that $\mathbf{Y}_{T+1}$ is (multivariate) normally distributed with constant covariance matrix $\boldsymbol{\Sigma}$. In numerous applications, empirical finance being a highly prominent one, the distribution of $\mathbf{Y}_{T+1}$, either unconditionally, or conditional on $\mathbf{Y}_1, \mathbf{Y}_2, \dots, \mathbf{Y}_T$, is (often highly) non-Gaussian and, for daily financial asset returns, also exhibits a time-varying covariance matrix. As such, there is merit in constructing a density forecast for $\mathbf{Y}_{T+1}$, instead of just its first and second moments.

In the context of using the i.i.d. Mix_2N_d model (presented in Chapter 14) for obtaining a predictive density for the DJIA-30 asset returns, the choice of ρ was determined in Paolella (2015) by using the average of the so-called **realized predictive log-likelihood** values based on one-step-ahead predictions, formed from moving windows of $\upsilon = 250$ observations (about one year of daily trading data). For the particular data set and model used in Paolella (2015), the optimal ρ was found to be about 0.7, with respect to (13.3) given below. The latter changes smoothly with ρ, and is monotone decreasing as ρ increases or decreases away from 0.7. Notice that, via use of weighted likelihood, the model implicitly addresses the non-i.i.d. aspect of volatility clustering of the data.

To be more precise, we require the predictive density of $\mathbf{Y}_{t+1}$, conditional on I_t, the information set up to time t. In our context, the information set is just the past υ observations of daily data. We denote the predictive density based on given model $\mathcal{M}$ as $f^{\mathcal{M}}_{t+1|I_t}(\cdot;\widehat{\boldsymbol{\theta}})$, where $\widehat{\boldsymbol{\theta}}$ is an estimator of parameter set $\boldsymbol{\theta}$. As a simpler example, take model $\mathcal{M}$ to be the univariate, mean-zero, stationary first-order autoregressive model with homoskedastic normal innovations (4.1), i.e., $Y_t = aY_{t-1} + U_t$, $|a| < 1$, and $U_t \overset{\text{i.i.d.}}{\sim} \text{N}(0, \sigma^2)$. Then $f_{t|I_{t-1}}(y; \hat{a}, \widehat{\sigma}^2)$ is the normal distribution with mean $\hat{a}y_{t-1}$ and variance $\widehat{\sigma}^2$, i.e., $\phi(y; \hat{a}y_{t-1}, \widehat{\sigma}^2)$.

Observe that either a growing or moving window can be used; we use the latter, given our premise that the local linear approximation to the d.g.p. is not stationary over large segments of time, and also exhibits volatility clustering, so that use of relatively smaller, fixed-size windows is preferred. For judging the quality of the density forecasts, we use the average (over all the moving windows) of logs of the values of the forecast density itself, evaluated at next period's actual realization. In particular, based on model $\mathcal{M}$, the realized predictive log-likelihood at time $t+1$ is given by

$$\pi_{t+1}(\mathcal{M}, \upsilon) = \log\, f^{\mathcal{M}}_{t+1|I_t}(\mathbf{y}_{t+1}; \widehat{\boldsymbol{\theta}}). \tag{13.2}$$

This $\pi_{t+1}(\mathcal{M}, \upsilon)$ is computed for each $t = \upsilon, \dots, T-1$, where T is the length of the entire time series under study, and their average,

$$S_T(\mathcal{M}_i, \upsilon) = \frac{1}{(T-\upsilon)} \sum_{t=\upsilon+1}^{T} \pi_t(\mathcal{M}_i, \upsilon), \tag{13.3}$$

is reported, for the models $\mathcal{M}_1, \dots, \mathcal{M}_m$ under consideration. We refer to this as the **normalized sum of the realized predictive log-likelihood**. In this way, the choice and calibration of the model is tied to directly what is of interest—its ability to forecast.

This method has gained in prominence, compared to inspection of, say, point estimates of forecasts, particularly for non-Gaussian models and models that are less concerned with mean prediction, but rather volatility; both of these conditions being precisely the case in financial time-series analysis. See Dawid (1984, 1985a,b, 1986), Diebold and Mariano (1995), Diebold et al. (1998), Christoffersen (1998), Timmermann (2000), Tay and Wallis (2000), Corradi and Swanson (2006), Gneiting et al. (2007), Geweke and Amisano (2010), Maneesoonthorn et al. (2012), and Paolella and Polak (2015a) for methodological developments and applications in financial forecasting.

To reiterate, if the actual d.g.p. were known and feasibly estimated, then no weighting should be employed (i.e., $\rho = 1$). Observe that this idea could be used as a metric to rank and judge the efficacy of various models and the "degree of mis-specification": The smaller the optimal ρ is for a given model (i.e., larger weighting is required), the more it deviates from the actual d.g.p.

Remarks

a) We use the convention that, with no subscript on $\hat{\theta}$ in $f^{\mathcal{M}}_{t|I_{t-1}}(\cdot;\hat{\theta})$, this implies that $\hat{\theta}$ has been estimated based on I_{t-1}. However, this need not be the case, and in many models, $\hat{\theta}$ is not updated for every t. For example, in the the AR(1) model, we could estimate a and σ^2 only every, say, $o = 20$ observations, but the forecast for time t is still $\hat{a}y_{t-1}$, because $y_{t-1} \in I_{t-1}$, but $\hat{a}$ may not have been "refreshed" with y_{t-1}. We denote the predictive density of $\mathbf{Y}_t$ conditional on I_{t-1} but using only the parameter estimate based on I_ζ as $f^{\mathcal{M}}_{t|I_{t-1}}(\cdot;\hat{\theta}_\zeta)$, $\zeta \leqslant t-1$. If we wish to re-estimate θ only every o observations, then in a computer program a FOR loop is used to traverse from $t = \tau_0 + 1$ up to $t = T$, where τ_0 indicates where the forecasting exercise starts (and usually equals v), and the parameters would be re-estimated if $\text{rem}(t - \tau_0 - 1, o) = 0$, where rem is the remainder function with $\text{rem}(a, b) = a - nb$ for $n = \lfloor a/b \rfloor$. We can then express the tth density forecast as

$$f^{\mathcal{M}}_{t|I_{t-1}}(\cdot;\hat{\theta}_\zeta), \qquad \zeta = t - 1 - \text{rem}(t - \tau_0 - 1, o). \tag{13.4}$$

Note that, for $o = 1$, this reduces to $\zeta = t - 1$. We take $o = 1$ when estimation is fast, while for models such that estimation is relatively time consuming, a value $o > 1$ should be considered.

b) Lest the reader get the impression that weighted likelihood is only a technique to augment an i.i.d. model as a substitute for (in our context) a (possibly, but not necessarily) more appropriate time-series model, we wish to emphasize that, quite on the contrary, it can also be applied with the latter. The idea is that, w.p.1, even the time-series model employed for inference is mis-specified, and so weighting recent observations more than those in the past will lead to better predictions. This was shown to be the case by Mittnik and Paolella (2000) in the context of VaR prediction for financial time series modeled with GARCH-type processes, and also by Paolella and Steude (2008). In the latter paper, several models, ranging in complexity from very simple to rather sophisticated, were used, and the very intuitive and confirming result emerged that, as the GARCH-type model employed increased in complexity and (crucially) effectiveness for prediction with traditional, un-weighted maximum likelihood, its optimal value of weighting parameter increases towards one, i.e., less weighting is required.

Weighted likelihood can also be used in conjunction with the bootstrap to compute confidence intervals for value at risk (VaR) and expected shortfall (ES); see Broda and Paolella (2011). ∎

13.3 Density Forecasting and Backtest Overfitting

It is worth reflecting again on the earlier quote by Granger (2008) regarding the use of complicated nonlinear time-series models. The AFaK model from Section 12.5.1, like the Mix_2N_d in Chapter 14, is analytically simple, fast, and straightforward to estimate and forecast. While the underlying true d.g.p. of a (particularly multivariate) sequence of financial asset returns is surely highly complicated and nonlinear, via the links between nonlinear models, linear models with time-varying parameters, and weighted likelihood, we can argue that use of a linear model (in our setting, actually an i.i.d. model with no relationship, linear or otherwise, between time points) with weighted likelihood offers a potentially reasonable approximation to the true d.g.p. for forecasting purposes that also exhibits the practically useful aforementioned benefits (ease of estimation and forecasting, etc.). However, the ultimate judge of a time-series (or spatial) model is almost always its ability to forecast, as considered below.

Weighted likelihood can be used in conjunction with shrinkage. An essentially perfect context for the use of shrinkage is the correlation matrix $\mathbf{R}$, given that their off-diagonal elements are measuring a common phenomenon and their numbers grow on the order of d^2. It is known that covariance- and, thus, correlation-matrices can be subject to large estimation error, particularly as the ratio d/T grows, and shrinkage estimation becomes crucial; see, e.g., Jorion (1986), Kan and Zhou (2007), Levina et al. (2008), Ledoit and Wolf (2004, 2012), and the references therein.

Denote by $\widehat{\mathbf{R}}$ an estimator of $\mathbf{R}$, such as the sample correlation estimator or the m.l.e. Shrinkage towards zero can be applied to its off-diagonal elements by taking the estimator to be $\tilde{\mathbf{R}} = (1 - s_{\mathbf{R}})\widehat{\mathbf{R}} + s_{\mathbf{R}}\mathbf{I}$, for some $0 \leqslant s_{\mathbf{R}} \leqslant 1$. Alternatively, shrinkage towards the average of the correlation coefficients can be used. A bit of thought reveals that this can be algebraically expressed as, with $a = \mathbf{1}'(\widehat{\mathbf{R}} - \mathbf{I})\mathbf{1}/(d(d-1))$ and $\mathbf{1}$ a d-vector of ones,

$$\tilde{\mathbf{R}} = (1 - s_{\mathbf{R}})\widehat{\mathbf{R}} + s_{\mathbf{R}}((1-a)\mathbf{I} + a\mathbf{1}\mathbf{1}'). \tag{13.5}$$

Use of (13.5), with $s_{\mathbf{R}} = 0.2$, was demonstrated in Paolella and Polak (2015a) (in the context of the AFaK model with a GARCH structure for the scale terms) to be most effective, in terms of out-of-sample density forecasting, using the $d = 30$ components of the DJIA index from January 1993 to December 2012. We will demonstrate below a similar result for the AFaK model but using the i.i.d. setting.

Remark In a financial context, correlations among asset returns tend to be positive, and change over time. These two characteristics further support use of simple shrinkage constructs such as (13.5), while the latter suggests that smaller windows of estimation—as we anyway advocate in light of an unknown and complex d.g.p.—are beneficial, thus increasing ratio d/T and the necessity of employing shrinkage to reduce the m.s.e. of $\widehat{\mathbf{R}}$.

Relevant to a discussion of correlations between financial assets changing over time is the concept of **financial contagion**. Researchers define the term in different ways. For example, Forbes and Rigobon (2002) argue that financial contagion is an increase in cross-market comovement after a sudden shock to one market (or country), while Dungey and Martin (2007) separate contagion from **spillover**. The difference between these two types of linkages is related to the timing of transmission, with contagion referring to a shock that occurs contemporaneously in two markets, while a spillover involves a time lag. Dungey and Martin (2007) demonstrate that spillover effects are larger than contagion effects. More detail can be found in Dungey et al. (2018) and Caporin et al. (2017), who recently study contagion via use of high-frequency data. Dungey et al. (2018) examine contagion through the episodes of flight-to-safety (moving assets from stocks to gold) and flight-to-quality (stocks to bonds), whereas Caporin et al. (2017) explore so-called systemic co-jumps.

Regarding why correlations among assets (in a single market or in multiple markets) tend to change over time and "move together", one possible, and surely partial, explanation is the following, which we will also refer to as contagion: As markets drop and investors begin to sell not just the distressed stocks, but everything, out of fear and desire for liquidity, more assets begin to fall; other nervous investors follow suit, markets drop further, and the correlations among assets begin to increase. As put by Ilmanen (2011, p. 14), "Sharp liquidations tend to occur amidst tightening financial conditions, and these in turn reinforce price and liquidity declines. These forces contribute to the short-term momentum and long-term reversal patterns observed for many investments."

One can view this as a form of violation of "efficient markets", and the potential for so-called behavioral finance models for assisting in explaining human behavior and the rationality of decisions amid irrational market participants. It also serves as an example of why traditional hedging strategies—designed to deal with dropping stock prices by offsetting with other instruments or low-correlated stocks, fail, precisely when they are required, and thus the need for more advanced financial engineering and econometric tools. See, e.g., Solnik and Longin (2001), Pesaran and Pick (2007), and the references therein for more substantial, detailed discussions and explanations for this effect. ■

What we require is a program to estimate the i.i.d. AFaK model, using the two-step method with the correlation terms determined optionally via maximum likelihood or the sample correlation, both with the weighting procedure, and such that, for the margins and **R**, weighted likelihood can be used, along with shrinkage via (13.5). We name it `FangFangKotzestimation2step`. In the FaK case, it calls program `Studentstestimation`, as also called in Listing 12.13, and is the same as program `tlikmax` in program Listing III.4.6, except that `Studentstestimation` supports weighted likelihood. The only changes required are to additionally pass to it the scalar `rho`, and augment the evaluation of the log-likelihood with the code in Listing 13.1. For the AFaK case, program `Noncentraltestimation` is similar, but using the s.p.a. density approximation to the NCT.

```
T=length(x); tvec=(1:T); omega=(T-tvec+1).^(rho-1); w=T*omega'/sum(omega);
ll = -mean(w.*llvec);
```

Program Listing 13.1: Required addition to program `Studentstestimation` to support weighted likelihood, which is otherwise the same as program `tlikmax` in Listing III.4.6.

Next, we need a program that computes (13.3) over a grid of $s_{\mathbf{R}}$ values from (13.5) and ρ-values from (13.1) and plots the resulting 3D performance graphic. This is given in Listing 13.2. The reader is encouraged to expand this, performing the parameter estimation only every, say, $o = 10$ trading days to save time without great loss of applicability, via (13.4).[3]

3 Another idea is to augment the code such that the previous window's final parameter estimates are used as the starting values for the next window, as the parameters are not expected to change very much. This task is not so crucial with this model, as it appears that the final estimates are not dependent on the choice of starting values, nor is much time wasted using inferior starting values. Also, it could be a bit tricky when using the `parfor` statement, enabling parallel processing.

Based on the daily (percentage log) returns of the 30 stocks comprising the DJIA index, from June 2001 to March 2009 (yielding 1,945 vector observations), the resulting graphic for the case of the i.i.d. FaK model and using sample correlations is shown in Figure 13.2 (and having taken about 15 hours of computing, using four cores). We see the appealing result that performance is close to quadratic in both ρ and $s_{\mathbf{R}}$, with the maximum occurring at $\tilde{\rho} = 0.45$ and $\tilde{s}_{\mathbf{R}} = 0.30$ (with respect to the coarse grid chosen).

The previous exercise was conducted for several models, and the results are collected in Table 13.1. For the FaK model but using the m.l.e. for the off-diagonal elements of $\mathbf{R}$, the optimal values are $\tilde{\rho} = 0.50$ and $\tilde{s}_{\mathbf{R}} = 0.30$, yielding a slightly higher achieved maximum of (13.3) of 45.3986. (The resulting 3D figure is very similar in appearance to that in Figure 13.2, and is omitted.) This demonstrates that use of the m.l.e. does add to forecasting performance (for the FaK and this data set), but it is far from obvious if this relatively small gain (also considering the additional computational cost) is significant in a meaningful sense, such as with respect to applications such as hedging or portfolio optimization investment strategies.

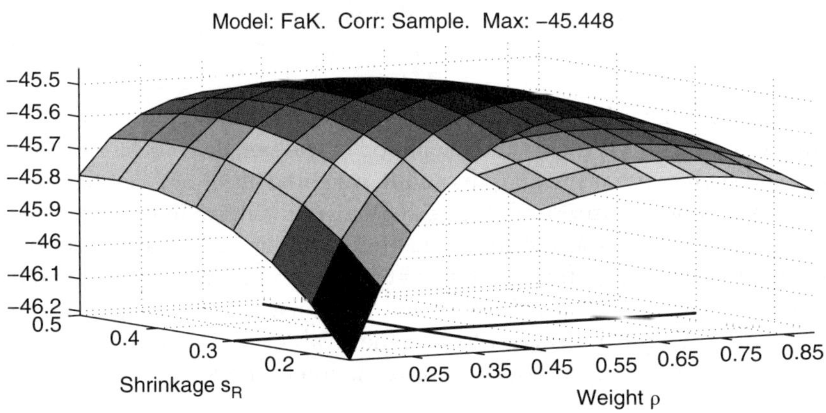

Figure 13.2 Density forecast measure (13.3) over a grid of $s_{\mathbf{R}}$ values from (13.5) and ρ-values from (13.1), for the FaK model, using sample correlation.

Table 13.1 The obtained average realized predictive log-likelihood (13.3) (to four significant digits) for various models and weighted likelihood (ρ) and correlation shrinkage ($s_{\mathbf{R}}$) parameter settings. Last column is the difference of (13.3) from that of the first entry.

Model	Type	Correlations	ρ	$s_{\mathbf{R}}$	(13.3)	Diff
FaK	i.i.d.	Sample	0.45	0.30	−45.45	0.00
FaK	i.i.d.	Sample	1.00	0.00	−46.05	0.60
FaK	i.i.d.	m.l.e.	0.50	0.30	−45.40	−0.05
AFaK	i.i.d.	Sample	0.45	0.30	−45.53	0.08
AFaK	i.i.d.	m.l.e.	0.45	0.30	−45.83	0.38

```
if matlabpool('size') == 0
  matlabpool open
  matlabpool size
end % use 4 cores. Commands are for Matlab version 10.
% parpool % higher Matlab versions
% data is the T X d matrix of (daily log percentage) returns
[T,d]=size(data); wsize=500;
MLEforR=0; AFaK=0; % choose. Use of MLE and AFaK is much slower.
rhovec=0.05:0.05:0.9; s_Rvec=0.0:0.05:0.5; % choose grid
rholen=length(rhovec); sRlen=length(s_Rvec); rpllmat=zeros(rholen,sRlen);
tic
  for rholoop=1:rholen, rho=rhovec(rholoop);
    parfor sRloop=1:sRlen, s_R=s_Rvec(sRloop);
      rpllvec=zeros(T-wsize-1+1,1); % put here because parfor
      for t=(wsize+1):T % make density prediction for time t
        disp([rho, s_R, t])
        Y=data((t-wsize):(t-1),:); % previous wsize returns
        Yt=data(t,:); % actual realized return at time t
        param = FangFangKotzestimation2step(Y,AFaK,rho,MLEforR,s_R);
        v=param.df; theta=param.noncen; mu=param.mu; scale=param.scale;
        rpllvec(t-wsize)=FFKpdfvec(Yt,v,theta,mu,scale,param.R);
      end
      rpllmat(rholoop,sRloop)=mean(log(rpllvec));
    end
  end
toc
% chop off some
rhovec=0.05:0.05:0.9; rhovec=rhovec(3:end);
s_Rvec=0.0:0.05:0.5;  s_Rvec=s_Rvec(4:end); use=rpllmat(3:end,4:end);

% surface plot
surf(rhovec,s_Rvec,use')
set(gca,'fontsize',16), zlim([min(use(:)) max(use(:))])
xlabel('Weight \rho'), ylabel('Shrinkage s_R')
xlim([rhovec(1) rhovec(end)]), ylim([s_Rvec(1) s_Rvec(end)])
set(gca,'XTick',0.15:0.1:0.85), set(gca,'YTick',0.2:0.1:0.5)

% get values of rho and s_R at function maximum
[TheMax idx] = max(use(:)); [rhomaxi sRmaxi] = ind2sub(size(use),idx);
rhomax=rhovec(rhomaxi); s_Rmax=s_Rvec(sRmaxi);
disp(['Achieved Max: ',num2str(TheMax), ...
     ' for rho=',num2str(rhomax),', s_R=',num2str(s_Rmax)])
title(['Model: FaK. Corr: Sample.  Max: ',num2str(TheMax)])

% plot lines showing coordinates at maximum
zz=zlim; zz=zz(1); xx=rhomax; yy=s_Rmax;
line([xx xx],ylim,[zz zz],'color','k','linewidth',3)
line(xlim,[yy yy],[zz zz],'color','k','linewidth',3)
```

Program Listing 13.2: Constructs (13.3) for a given data set `data`, over a double grid of $s_{\mathbf{R}}$ and ρ-values.

The i.i.d. AFaK model, using the sample correlation for estimating $\mathbf{R}$, resulted in a maximum of 45.5323, occurring at $\tilde{\rho} = 0.45$ and $\tilde{s}_{\mathbf{R}} = 0.30$. Note, perhaps surprisingly, that the obtained maximum of (13.3) based on the FaK with sample correlation is (slightly) higher than that of AFaK, even though it appears that most assets have "significant" asymmetry. This is not an inconsistency or a paradox, but rather the nature of statistical inference, and worth emphasizing.

> Classic assessment of parameter significance (notably at the conventional levels) does not imply improvements in forecasts, particularly, but not only, in highly parameterized time-series models.

This issue touches upon many important topics in statistics, such as use of p-values for assessing "parameter significance" (recall Section III.3.8), multiple hypothesis testing, backtest overfitting (discussed below), shrinkage estimation, and model selection procedures such as the lasso, elastic net, and related methods. Indeed, as noted in Rapach and Zhou (2013, p. 328) in their extensive survey of stock return prediction methods, "Some studies argue that, despite extensive in-sample evidence of equity premium predictability, popular predictors from the literature fail to outperform the simple historical average benchmark forecast in out-of-sample tests. Recent studies, however, provide improved forecasting strategies that deliver statistically and economically significant out-of-sample gains relative to the historical average benchmark." Further recent examples include Harvey and Liu (2016) and Harvey et al. (2016), who discuss the inadequacy of the usual t-test procedure for determining the factors driving the cross-section of expected financial asset returns, while Diebold and Yilmaz (2015) illustrate use of the elastic net in conjunction with highly parameterized vector autoregressive models for multivariate financial variable prediction and assessment of "connectedness".

Remark In general, with such data-driven ideas, one needs to exercise some caution. In our case, Figure 13.2 shows that the forecasting quality is very smooth in ρ and $s_{\mathbf{R}}$, and such that it is monotonically decreasing to the left and right of the optimal ρ and $s_{\mathbf{R}}$ values. If it were instead the case that the plot were somewhat erratic (jittery) in behavior or, worse, jittery and no visible approximate quadratic shape, then choosing the optimal value where the erratic graphic happens to obtain its maximum would be highly suspect and almost surely unreliable for genuine increases in forecasting performance: In such a case, we would be modeling in-sample noise, and not capturing a genuine "signal" useful for forecasting.

This touches upon the topic of **backtest overfitting**, in which numerous (possibly dozens of) tuning parameters are optimized in-sample or, more often, on an out-of-sample exercise as we have done, and result in impressive performance. However, it is fictitious and does not lead to gains (and actually often leads to losses) when used in a genuine (true future) out-of-sample prediction framework. Good starting points to this literature include Bailey et al. (2014) and Zhu et al. (2017). That both of these papers are in the context of finance should not surprise: The lure of finding signals in stock price data is very enticing to many people (as well as brokers and electronic platforms happy to make commissions on naive gamblers or—more politically correct—"noise traders", though the latter do help to provide liquidity), and with easy access to past stock returns and powerful computing, one can try literally thousands of models quickly, and then "pick the best one", thinking one has used his or her intelligence and expertise gained from an introductory statistics course to become wealthy.

The "models" often used in this context are typically rather simplistic, moving-average-based calculations on the price process (as opposed to the returns process) with a large variety of window lengths, and under an implicit assumption of a local mean-reversion in stock prices, not to mention the so-called "technical trading" rules, in which one believes in certain patterns recognizable to the human eye, such as "head and shoulders" and "cup and handle", etc. Appeals to wishful thinkers less inclined to study mathematics and stochastic processes with arguments how a simple (and unpredictable) random walk (recall Figure 4.3) or a stationary ARMA model realization (recall Figure 9.1 and the subsequent discussion), particularly with leptokurtic innovations, easily gives rise to such patterns, as well as fictitious "trends", usually fall on deaf ears.

The ever-poular mantra "Past performance is not an indicator of future performance" is an understatement: Strong backtest performance might literally be an indicator of negative future results, with the optimized "strategy" easily beaten by trivial allocation methods such as putting an equal amount of money in all available assets, commonly referred to as the equally weighted or "$1/N$" strategy. This equally weighted strategy is nothing more than an extreme form of shrinkage, and has been shown in numerous studies to work shockingly well, to the disgruntlement of "talented fund managers with many years of experience and MBAs"; recall the discussion in Section III.2.1.1. As our method involves only two parameters that result in well-behaved performance, one can be cautiously optimistic, but the real proof comes only in genuine out-of-sample performance.

To further illustrate the concept, the common story goes as follows. A well-dressed businessman enters a nice bar and enters conversation with some apparently well-off gentlemen (usually assessed by the wristwatch) who regularly frequent the establishment. After the usual pleasantries, he explains that he is a highly successful investor, and goes so far as to say that, based on his advanced statistical models, the stock market will increase on each of the next four business days. He then politely exits, and repeats the exercise at another posh bar, but says that the stock market will increase over the next three business days, and then drop on the fourth. He continues this at different bars (presumably not drinking too much), exhausting all the 16 possible permutations of performance over the next four business days.

The next business day, markets will go up or down (with essentially the same probability as a flip of a fair coin), and, assuming the market went up, he returns that evening to the eight bars in which his stated prediction was correct, and, casually, in the midst of advanced-sounding statistical talk, reminds them of his success. The subsequent day, he returns to the four bars in which his prediction was "correct" twice in a row, etc., until after the fourth day, he returns to the single bar for which his streak of success held true. "Gentlemen, you are surely now convinced. Who wants to invest?"

The point is, besides the "method of prediction" being completely random, that he does not disclose all the failed methods considered. This is the concept underlying backtest overfitting. ■

Having shown the benefits of shrinking the off-diagonals of $\widehat{\mathbf{R}}$, and the use of weighted likelihood, we now entertain shrinking the estimates of the v_i. Juxtaposing the usual multivariate Student's t with the (A)FaK, these can be seen as the two extremes of a continuum: One has the degrees of freedom parameter equal across all margins, while the other has them all (and w.p.1 when estimating) *unequal*. We have already demonstrated that the former is too inflexible for the DJIA data set, but the latter might be *too flexible*, in the following sense:

> While as tail thickness measures, none of the v_i are precisely equal, the amount of data being used to estimate them (limited also by the fact that we believe the v_i are changing, hopefully slowly, over time) is too small to obtain the desired accuracy. We are asking too much of the data, in terms of the amount of data available, and the parameterization of the model.

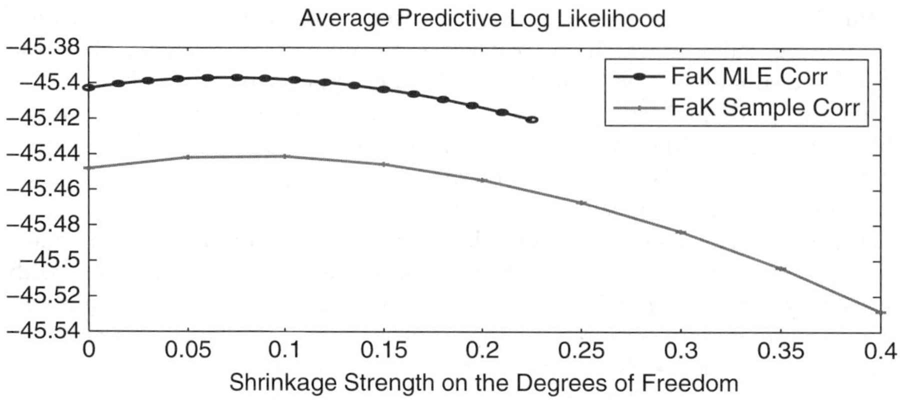

Figure 13.3 Density forecasting performance measure (13.3) as a function of degrees of freedom shrinkage parameter s_v, for the i.i.d. FaK model applied to the usual DJIA data, using the two forms of estimating correlation matrix $\mathbf{R}$, and with fixed $\tilde{\rho} = 0.45$ and $\tilde{s}_{\mathbf{R}} = 0.30$.

This is precisely where shrinkage can play a useful role. Also observe that the v_i are estimated with relatively much more uncertainty than the location and scale parameters, as they are tail measures, so that shrinkage could be expected to reduce their overall m.s.e. Finally, recall the left panel of Figure 12.1, which suggests that, for many assets, the degrees of freedom might be very similar across assets, so that shrinkage will be beneficial as a way of pooling information across assets. (One could also entertain forming, say, three clusters, such that each $\hat{v}_i$ takes on only one of three possible values, this also being a form of shrinkage. This is clearly more difficult to implement, and is considered in Section 12.6.4.) Estimates of the noncentrality parameters θ_i could be subjected to shrinkage in a similar way (with the right panel of Figure 12.1 suggesting zero as the natural target).

Let s_v be the shrinkage strength on the v_i, and the mean of the $\hat{v}_i$, denoted $\bar{v}$, be the target, so that $\tilde{v}_i = s_v \bar{v} + (1 - s_v)\hat{v}_i$ are the resulting shrinkage estimators for the v_i, $i = 1, \dots, d$. An ideal setup would be one such that s_v, $s_{\mathbf{R}}$, and ρ are each endowed with a tight grid of values, and their optima are determined by computing (13.3) over all the combinations induced by the three grids. While feasible, it could take many weeks or even months to run. This is an example of the **curse of dimensionality**, such that each additional dimension increases the computational time by a large multiplicative factor. Instead, we "cut corners" (nearly literally), and fix the values of ρ and $s_{\mathbf{R}}$ to $\tilde{\rho} = 0.45$ and $\tilde{s}_{\mathbf{R}} = 0.30$, respectively. Thus, for a grid of s_v-values, we have a one-dimensional search problem. The results are shown in Figure 13.3. The performance is smooth in s_v, with its optimal value (for this data set, choice of window length 500, and chosen grid coarseness) being approximately $\tilde{s}_v = 0.075$ for $\mathbf{R}$ estimated both via sample correlations and m.l.e. The rather low value of $\tilde{s}_v$ indicates that this idea may not be very fruitful and the aforementioned idea of use of clusters might be better.

13.4 Portfolio Optimization Using (A)FaK

Before applying the AFaK model to real data, we investigate its performance using simulated data and based on the true model parameters. This obviously unrealistic setting serves as a check on the methodology and also (assuming the method is programmed correctly), will illustrate the large variation in the performance of the methods due strictly to the nature of statistical sampling.

We simulate first from the multivariate t distribution (hereafter MVT), using $d = 10$ dimensions, $v = 4$ degrees of freedom, each component of the mean vector being i.i.d. $\mathrm{N}(0, 0.1^2)$ realizations, and each of the scale terms being i.i.d. $\mathrm{Exp}(1, 1)$ (i.e., scale one, and location one) realizations. The off-diagonal elements of correlation matrix $\mathbf{R}$ are taken to be i.i.d. $\mathrm{Beta}(4, 9)$, with mean $4/13$ and such that the resulting matrix is positive definite. The code in Listing 13.3 performs this, using our `FaKrnd` routine instead of `mvtrnd`, as we will generalize this exercise subsequently to the FaK case.

```
1  d=10; sim=3e3; bad=1;
2  while bad
3    R=eye(d);
4    for i=1:d, for j=(i+1):d, R(i,j)=betarnd(4,9); R(j,i)=R(i,j); end, end
5    bad=any(eig(R)<1e-4);
6  end
7  dffix=4; df=dffix*ones(d+1,1); % all the same. Is the MVT
8  mu=0+0.1*randn(d,1); scales=1+exprnd(1,d,1);
9  Sigma=diag(scales) * R * diag(scales);
10 data=FaKrnd(sim,df,mu,scales,R,zeros(d+1,1)); % MVT or FaK iid sequence
```

Program Listing 13.3: Generates `sim` i.i.d. realizations of the MVT with random parameters. By changing line 7, different margin degrees of freedom can be specified, and the output are then FaK realizations, of which MVT is a special case.

The next step is to compute the optimal portfolio vectors and the associated realized returns. This is conducted over moving windows of (arbitrarily chosen) length 1,000 and with a desired expected annual return of $\tau = 10\%$ from (11.43), using (i) the long-only Markowitz procedure based on the usual sample estimators of the mean and covariance, and (ii) the long-only allocation method based on simulation, from (11.46), *and knowledge of the true MVT parameters.* The simple code for the former is given in Listing 13.4, while that for the latter is given in Listing 13.5.

```
DEAR=10; winsize=1000; [T,~]=size(data); Ret=zeros(T-winsize,1);
for t=(winsize+1):T,  if mod(t,100)==0, disp(t), end
  Y=data((t-winsize):(t-1),:); P = PortMNS(Y, DEAR); Yt=data(t,:); RR=P'*Yt';
  if ~isnan(RR), Ret(t-winsize)=RR; end
end
CSMarkRet=cumsum(Ret); SharpeMarkowitz=mean(Ret)/std(Ret);
```

Program Listing 13.4: For given data set `data`, computes the long-only Markowitz portfolio vectors for a desired expected annual return (DEAR) of $\tau = 10\%$, over windows of length winsize $= 1,000$, their realized returns, and then the cumulative returns and the Sharpe ratio (with risk-free rate of zero). Program `PortMNS` is given in Listing 11.3.

This exercise runs quickly, and was conducted several times, with six representative results shown in Figure 13.4. The cumulative realized returns are plotted from (i) the long-only Markowitz method and (ii) the simulation-based method using the known MVT distribution and $s = 10,000$ replications (`asim` in Listing 13.5). Also plotted is the performance of the equally weighted ("$1/N$") portfolio. Further overlaid are 100 cumulative returns based on randomly selecting the long-only portfolio weights

```
1  asim=1e4; agiveup=asim/10;
2  xi=0.01; DEDR= 100*((DEAR/100 + 1)^(1/250) - 1);
3  [T,d]=size(data); muvec=mu; PortMat=zeros(T-winsize,d);
4  for t=(winsize+1):T,  if mod(t,100)==0, disp(t), end
5    bestES=-1e9; foundDEDRokay=0;
6    for i=1:asim
7      if foundDEDRokay || (i <= agiveup)
8        a=-log(rand(d,1)); a=a/sum(a); ER = a'*muvec; VaR = tinv(xi,dffix);
9        ES = -tpdf(VaR,dffix) / xi * ( (dffix+VaR^2)/(dffix-1) );
10       ES = ER + a'*Sigma*a * ES;
11       if (ER>=DEDR) && (ES>bestES), besta=a; bestES=ES; foundDEDRokay=1; end
12     end
13   end
14   if foundDEDRokay, PortMat(t-winsize,:)=besta; end
15 end
16 Ret=zeros(T-winsize,1); % compute the returns from FaK
17 for t=(winsize+1):T
18   Yt=data(t,:); P = PortMat(t-winsize,:); RR=P*Yt';
19   if ~isnan(RR), Ret(t-winsize)=RR; end
20 end
21 CSFaKRet=cumsum(Ret); SharpeFaK=mean(Ret)/std(Ret);
```

Program Listing 13.5: Similar to Listing 13.4 but using simulation from (11.46) based on `asim` = 10,000 replications, and using the knowledge that the true d.g.p. is i.i.d. MVT, and the true parameters—observe how `dffix` comes from line 7 in Listing 13.3. The formula for the ES of a standard Student's t in line 9 is given in (III.A.121), while that for the portfolio (weighted sums of margins) is computed using (III.A.126) and (C.28). If after `agiveup` samples, no portfolio is found that satisfies the mean constraint, we give up (to save time), and the portfolio vector is taken to be all zeros, i.e., no investment is made (and implicitly, existing assets would be sold). Adding to the unrealistic setting with fully known d.g.p., we also do not account for transaction costs.

at each point in time. This provides a guide for assessing if the allocation methods are genuinely outperforming a "pure luck strategy". The code to generate such a plot is given in Listing 13.6.

In each of the six cases, the true MVT parameters are different, having been generated from the code in Listing 13.3, but come from the same underlying distribution, as discussed above. The fact that the MVT case is using the true parameter values gives it an edge in terms of total returns, as seen in the middle- and lower-left panels, though in other cases it does not perform better in finite-time experiments, such as in the middle right panel.

> The take-away message is that, even over a period using 2,000 days of trading, allocation based on the true model and true parameters may not outperform the somewhat naive Markowitz approach (at least in terms of total return), and that the latter can even be beaten by the very naive $1/N$ strategy.

Thus, one should be extraordinarily cautious when claims are made about the viability of various trading strategies. The reader is encouraged to repeat this exercise and also plot the cumulative returns corresponding to the MVT model, but using estimated instead of the true parameters.

We take one next step (of several) towards reality and leave the elliptic world, using instead the FaK model with heterogeneous degrees of freedom, but still (ludicrously, for academic purposes) assuming

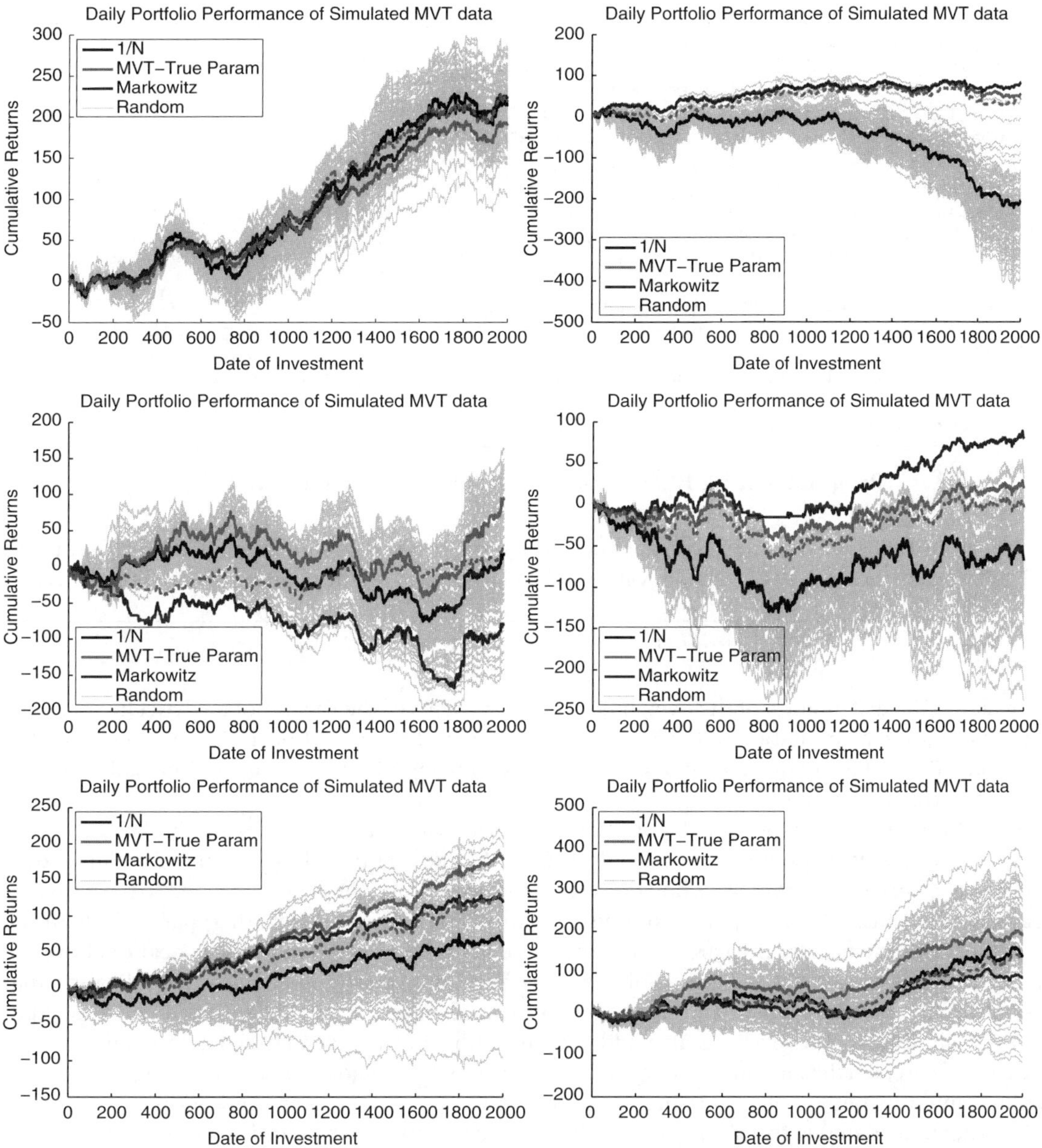

Figure 13.4 Cumulative returns of the equally weighted, Markowitz, and MVT models, the latter using the true parameter values and simulation based on s samples to obtain the optimal portfolio. The thinner, dashed (red) line uses $s = 1{,}000$ instead of $s = 10{,}000$ (thicker, solid, red line). In all but the top left case, use of $s = 10{,}000$ is at least as good as $s = 1{,}000$ and in some cases, such as the last four panels, leads to substantially better results.

```
a=ones(d,1)/d; rep=100;
for t=(winsize+1):T, Yt=data(t,:); Ret(t-winsize)=a'*Yt'; end
CSRet1d=cumsum(Ret); Sharpe1N=mean(Ret)/std(Ret);
xData=1:(T-winsize); figure, hold on % need plot to make the legend
plot(xData,CSRet1d,'k-',xData,CSFaKRet,'r-',xData,CSMarkRet,'b-','linewidth',3)
for i=1:rep  % now the random portfolios
  a=-log(rand(d,1)); a=a/sum(a);
  for t=(winsize+1):T, Yt=data(t,:); Ret(t-winsize)=a'*Yt'; end
  CSRet=cumsum(Ret); plot(xData,CSRet,'g-')
  if i==1, legend('1/N','FaK','Markowitz','Random','Location','NorthWest'), end
end
% Plot them again to see the lines more clearly.
plot(xData,CSRet1d,'k-',xData,CSFaKRet,'r-',xData,CSMarkRet,'b-','linewidth',3)
hold off
title('Daily Portfolio Performance of Simulated MVT data','fontsize',14)
xlabel('Date of Investment','fontsize',16)
ylabel('Cumulative Returns','fontsize',16)
```

Program Listing 13.6: Generates the plots shown in Figure 13.4.

the model, and the true parameters, are known. The same method of simulation is used as in Listing 13.3, but we take the degrees of freedom values to be i.i.d. Unif(2, 7), and change line 7 in Listing 13.3 to

```
df=2+(7-2)*rand(d,1); df=[max(df); df];
```

Recall that the distribution of the (weighted) sum of margins of the FaK is not analytically tractable, requiring that the computation of the ES is done via the method in Section 12.5.5, namely using the empirical VaR and ES, obtained from $s_1 = 10{,}000$ draws. Listing 13.7 shows the required code to determine the optimal portfolio.

Results for four runs are shown in Figure 13.5, with other runs (not shown) being similar. We obtain our hoped-for result that the FaK model outperforms Markowitz (which is designed for elliptic data with existence of second moments), and does so particularly when the set of v_i tended to have smaller (heavier-tail) values. The $1/N$ portfolio is also seen to be inferior in this setting, particularly in the last of the four shown runs. The FaK graphs are also such that they systematically lie near or above the top of the cloud of cumulative returns obtained from random portfolio allocations, indicating that accounting for the heavy-tailed and heterogeneous-tailed nature of the data indeed leads to superior asset allocation. This exercise also adds confirmation to the fact that allocations differ in the non-elliptic case, particularly amid heavy tails, and also that the algorithm for obtaining the optimal portfolio, and the method of calculating the ES for a given portfolio vector, are working.

The crucial next step is to still use knowledge that the d.g.p. is FaK, but use parameter estimates instead of the true values, based on the two-step estimator with the conditional m.l.e. for the elements of $\mathbf{R}$, and this along with shrinkage for $\mathbf{R}$ with $s_{\mathbf{R}} = 0$ and $s_{\mathbf{R}} = 0.30$, as developed in Section 13.3. For the code, just replace line 6 in Listing 13.7 with the code in Listing 13.8.

Figure 13.6 is similar to Figure 13.5, and uses the same generated data, so that the two figures can be directly compared. The degradation in performance of the FaK model is apparent: The realistic necessity of parameter estimation when using parametric models takes a strong toll for all of the four runs shown, and also shrinkage of $\widehat{\mathbf{R}}$ does not help, but rather, at least for the cases shown and

```
1  asim=1e3; agiveup=asim/10; xi=0.01; DEDR= 100*((DEAR/100 + 1)^(1/250) - 1);
2  [T,d]=size(data); muvec=mu; PortMat=zeros(T-winsize,d); s1=1e4;
3  for t=(winsize+1):T
4    bestES=-1e9; foundDEDRokay=0;
5    %%%%%%%%%%%%%%%%%% now because is FaK and not MVT. Use true parameters.
6    param.df=df; param.noncen=nc; param.R=R; param.mu=mu; param.scale=scales;
7    M=FaKrnd(s1,param.df,muvec,param.scale,param.R,param.noncen)';
8    %%%%%%%%%%%%%%%%%%
9    for i=1:asim
10     if foundDEDRokay || (i <= agiveup)
11       a=-log(rand(d,1)); a=a/sum(a); ER = a'*muvec;
12       %%%%%%%%%%%%%%%%%%
13       % now because is FaK and not MVT.
14       P=a'*M;
15       if 1==2 % Just for comparison. It is much slower. Based on NCT fit.
16         paramNCT = Noncentraltestimation(P,1,1);
17         theta=paramNCT(1); v=paramNCT(2); mu=paramNCT(3); scale=paramNCT(4);
18         [ES, VaR] = nctES(xi,v,theta); ES=mu+scale*ES;
19       else % USE THIS with FaK: empirical VaR and ES. Much faster.
20         VaR=quantile(P,0.01); Plo=P(P<=VaR); ES=mean(Plo);
21         % slower: P=sort(P); lowP=P(1:round(s1*xi)); ES=mean(lowP);
22       end
23       %%%%%%%%%%%%%%%%%%
24       if (ER>=DEDR) && (ES>bestES), besta=a; bestES=ES; foundDEDRokay=1; end
25     end
26   end
27   if foundDEDRokay, PortMat(t-winsize,:)=besta; end
28 end
```

Program Listing 13.7: Similar to Listing 13.5, but for the FaK distribution, again using known parameters.

```
  MLEforR=1; AFaK=0; rho=0; s_R=0; % or s_R=0.30
  Y=data((t-winsize):(t-1),:);
  param = FangFangKotzestimation2step(Y,AFaK,MLEforR,rho,s_R);
  muvec = param.mu;
```

Program Listing 13.8: Replace line 6 in Listing 13.7 with this to conduct parameter estimation instead of using the true values.

the choice of $s_{\mathbf{R}} = 0.30$, predominantly hurts. Admittedly, the choice of $s_{\mathbf{R}} = 0.30$, as determined in Section 13.3, was obtained with respect to density forecasting, for a real financial returns data set with $d = 30$ and a window size of 250, as opposed to our context here, which is portfolio optimization, for simulated FaK data, with $d = 10$ and a window size of 1000. The reader is invited to determine the optimal shrinkage in this setting, though it is doubtful that much will be obtained in terms of cumulative return performance.

It is worth emphasizing that, in general, the quality of density forecasts and portfolio performance are not necessarily "comonotonic" with respect to tuning parameters, in the sense that the best, say, tuning parameters for shrinkage and weighted likelihood for density forecasts are not necessarily

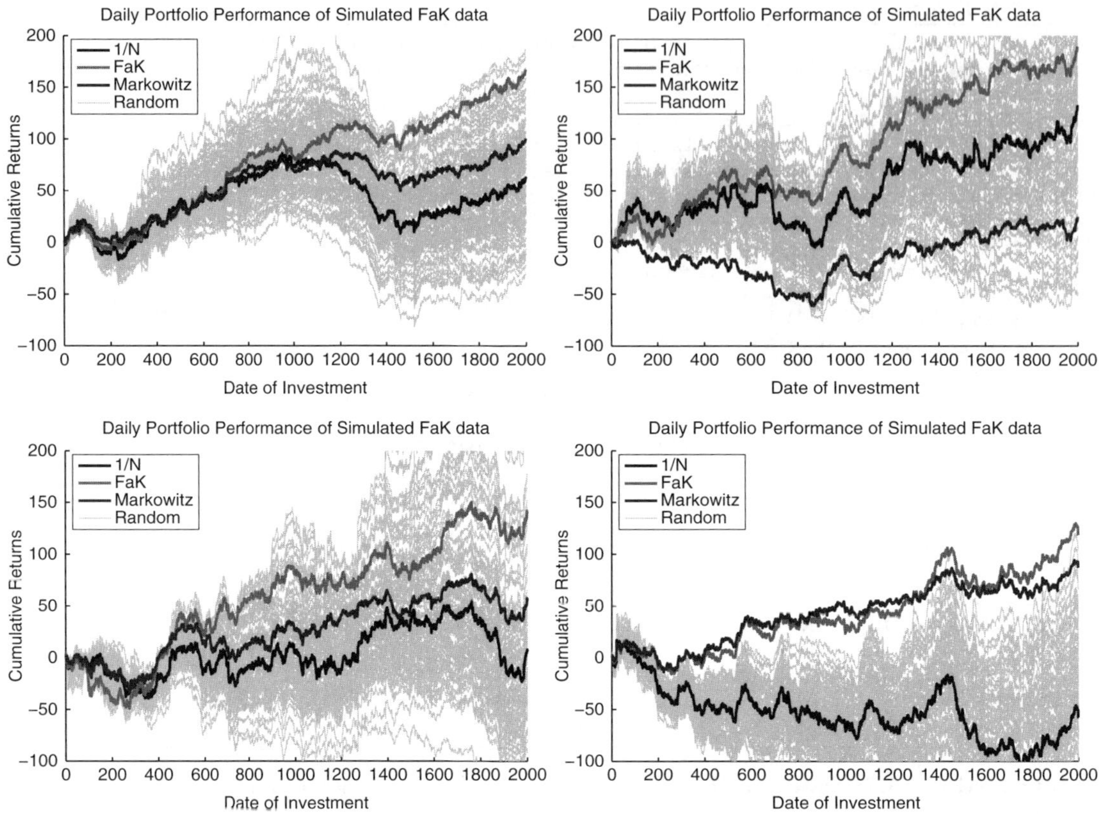

Figure 13.5 Similar to Figure 13.4, but based on the FaK model, using the true parameter values. All plots were truncated in order to have the same y-axis.

the best values for portfolio performance. Note that, if the *true* multivariate predictive density were somehow available, then the optimal portfolio (as defined by some measure accounting for risk and return) can be elicited from it. However, there is still an important caveat here that we wish to emphasize:

> Actual performance, even with the true model, is probabilistic, and thus only with repeated investment over very many time periods would it be the case that, on average, the desired return is achieved with respect to the specified risk. As (i) the true predictive density is clearly not attainable (because the specified model is wrong w.p.1, along with the associated estimation error) and (ii) backtest exercises necessarily involve a finite amount of data (so that the real long-term performance cannot be assessed with great accuracy), there will be a difference between inference based on density forecast and portfolio performance.

This exercise serves to illustrate a case in which the estimation error associated with highly parameterized models—even in the unrealistic setting in which the parametric model (here, i.i.d. FaK) is known—induces a dramatic loss in out-of-sample performance. This underscores the point made in Section III.2.8 regarding use of classic inferential methods, such as inspecting the t-statistics

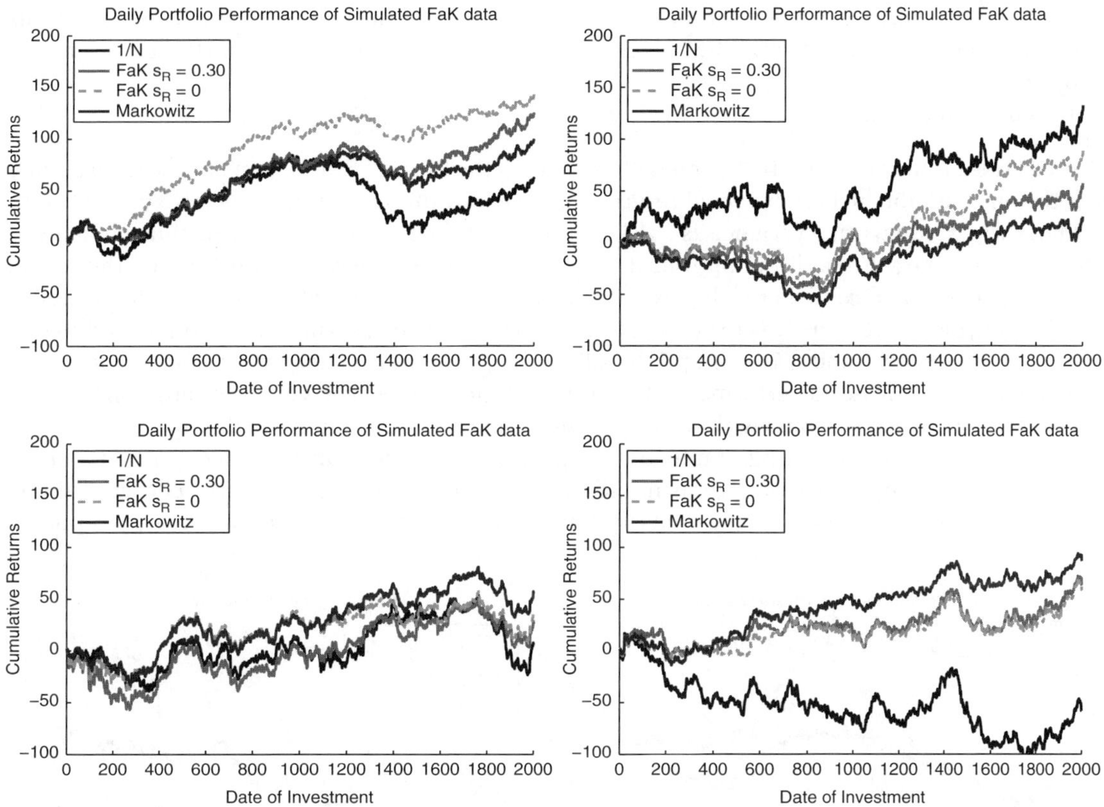

Figure 13.6 Performance comparison using the same four data sets as in Figure 13.5, and having estimated the FaK parameters.

associated with estimated parameters, when interest centers on forecasting—particularly, but not only, in highly parameterized time-series models.

Not yet willing to give up, we consider an alternative investment strategy that capitalizes on the nature of how the optimal portfolio is determined. In particular, as we use random sampling instead of a black-box optimization algorithm to determine optimal portfolio (11.45), we have access to s (in our case, $s = 10{,}000$) portfolios. We attempt to use these in a simple, creative way, and apply the following algorithm for a given desired expected annual return τ, for which we use 10%:

1) For a given data set of dimension d, window length, and τ, estimate the FaK model, possibly with shrinkage. (In the cases shown, we use $s_{\mathbf{R}} = 0$.)
2) Attempt s random portfolios (we use $s = 10{,}000$ for $d = 10$), and if after $s/10$ generations no portfolio reaches the desired expected annual return (the τ-constraint), give up (and trading does not occur).
3) Assuming the exit in step 2 is not engaged, from the s portfolios, store those that meet the τ-constraint, amassing a total of υ valid portfolios.
4) If $\upsilon < s/100$, then do not trade. The idea is that, if so few portfolios meet the τ-constraint, then, taking the portfolio parameter uncertainty into account, it is perhaps unlikely that the expected return will actually be met.

5) Assuming $v \geqslant s/100$, keep the subset consisting of the (at most) $s/10$ with the lowest ES. (This requires that the stored ES, and the associated stored expected returns and portfolio vectors, are sorted).
6) From this remaining subset, choose that portfolio with the highest expected return.

The core idea is to collect the 10% of portfolios yielding the lowest ES, and then choose from among them the one with the highest expected return. Observe how this algorithm could also be applied to the Markowitz setting, using variance as a risk measure, but then the sampling algorithm would need to be used, as opposed to a direct optimization algorithm, as is applicable with (11.45). The reader can investigate this and confirm to what extent similar results hold. This alternative method contains several tuning parameters, such as the choice of τ, the window size, s, shrinkage $s_{\mathbf{R}}$, and the (arbitrary) values of $s/100$ in step 4, and $s/10$ in step 5. Recalling the discussion of backtest overfitting above, one is behooved to investigate its performance for a range of such values (and data sets), and confirm that the results are reasonably robust with respect to their choices around an optimal range.

Figure 13.7 shows the resulting graphs based again on the same four simulated data sets. There now appears to be some space for optimism, and tweaking the—somewhat arbitrarily chosen—tuning parameters surely will lead to enhanced performance. The intrigued reader is encouraged to pursue

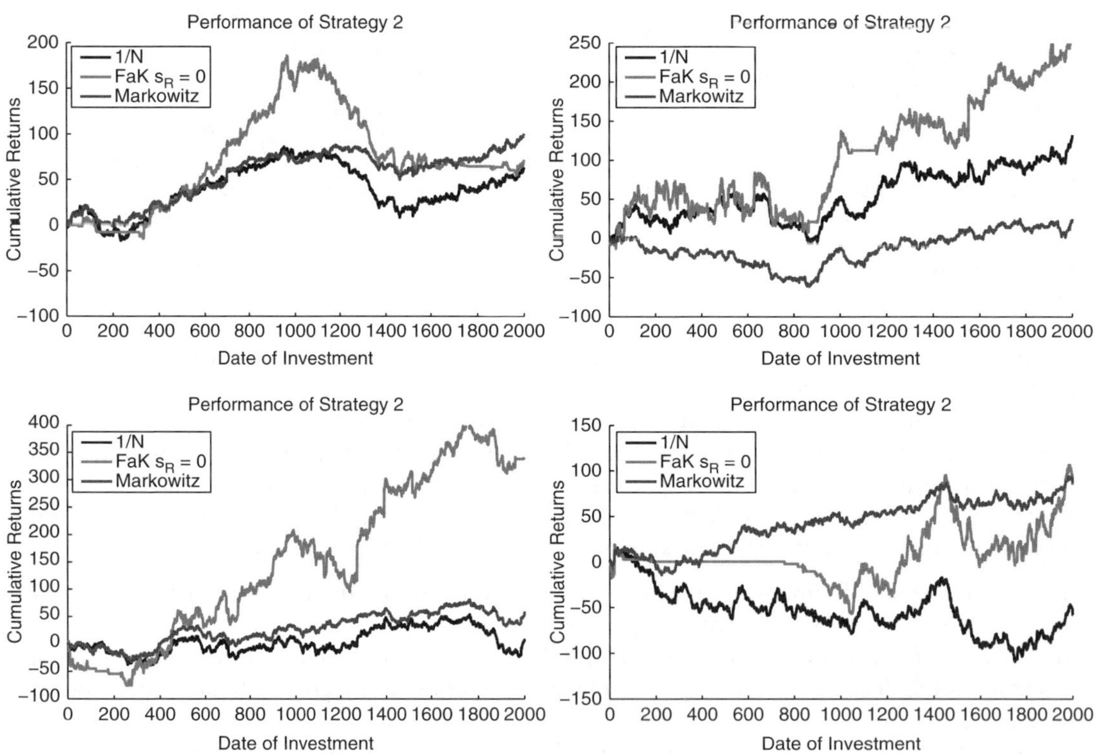

Figure 13.7 Similar to Figure 13.6, with estimated parameters and using $s_R = 0$, but having used the alternative investment strategy based on choosing among the 10% of generated portfolios with the lowest ES the one with the highest expected return.

this, and investigate it with simulated and real data, ideally computing additional performance measures such as the Sharpe and related ratios, and taking into account transaction costs, as discussed in Section 11.3.1. The AFaK model (with simulated AFaK and real data) could also be used, though note that it is computationally slower because of estimation and the determination of the ES. Finally, for real daily returns data, incorporation of a GARCH-type filter applied to each of the margins could be beneficial, given the clear conditional heteroskedasticity, though such results are often tempered via incorporation of transaction costs, given that models that use a GARCH-type structure have much higher turnover than their i.i.d. counterparts.

As a final step and "progression to the next level", we use real data, namely the closing prices on the 30 stocks on the DJIA, but instead of the daily data from June 2001 to March 2009 as used in Section 13.3, we use an updated data set, from January 2, 2001 to July 21, 2016 (conveniently including the market turmoil associated with the Brexit event). However, we still refrain from accounting for transaction costs. Figure 13.8 shows the obtained cumulative returns based on the equally weighted portfolio, the FaK model (obviously, estimating the parameters) in conjunction with the alternative investment strategy outlined above, Markowitz (the latter two restricted to no short-selling), and randomly generated portfolios with non-negative weights. The only merit one can ascribe to the FaK/alternative investment strategy is that it avoids trading during the financial crisis period, though as time goes on its performance is overshadowed by both Markowitz and $1/N$, and *none* of the methods used in this study do particularly better than the average of the random portfolios after about the middle of 2013.

One can compare these results to the better performances shown in Figures 11.7 and 11.8. While general conclusions are difficult to draw, it appears safe to say that naive application of simple

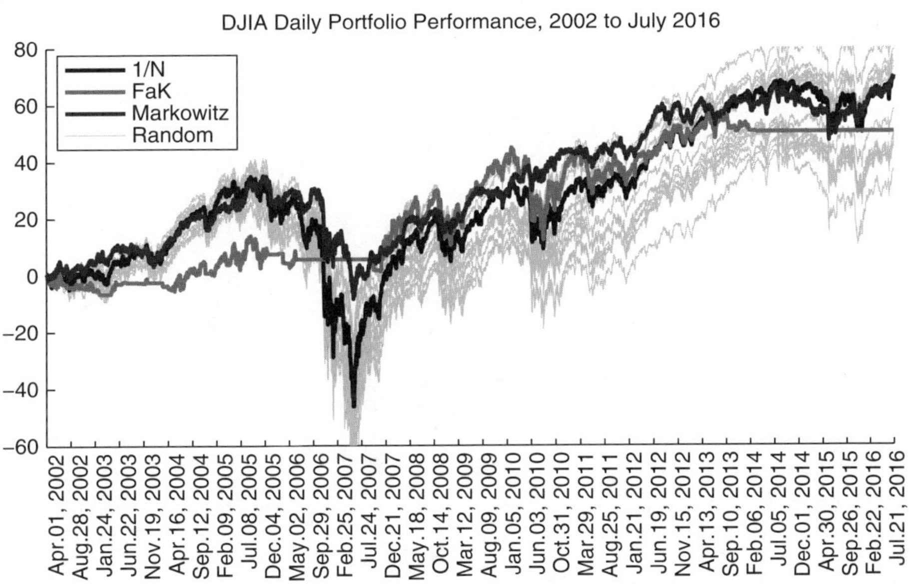

Figure 13.8 Cumulative returns on portfolios of the 30 stocks in the DJIA index, using the FaK model with the alternative investment strategy, the $1/N$ allocation, Markowitz (no short selling), and 400 random portfolios (showing only the most extreme ones to enhance graphic readability).

copula-based models, while straightforward (due to the ability to separately specify and estimate the margins and the copula) and appealing (because the margins are easily endowed with heterogeneous tail behavior), may not deliver as much bang for the buck as different non-Gaussian stochastic processes such as the COMFORT-based paradigm and the mixture distribution paradigm. A further disadvantage of the copula methodology not shared by the latter two frameworks is that simulation is required to obtain the necessary characteristics of the predictive portfolio distribution.

14

Multivariate Mixture Distributions

Occasionally, papers are published suggesting how returns can be forecast using a simple statistical model, and presumably these techniques are the basis of the decisions of some financial analysts. More likely the results are fragile: once you try to use them, they go away.

(Clive W. J. Granger, 2005, p. 36)

The next obvious step is towards using predictive, or conditional, distributions. Major problems remain, particularly with parametric forms and in the multivariate case. For the center of the distribution a mixture of Gaussians appears to work well but these do not represent tail probabilities in a satisfactory fashion.

(Clive W. J. Granger, 2005, p. 37)

Use of the i.i.d. univariate discrete mixture of normals distribution, or MixN, as detailed in Chapter III.5.1, allows for great enrichment in modeling flexibility compared to the Gaussian. Here, we extend this to the multivariate case. We also develop the methodology for mixtures of (multivariate) Laplace, this distribution having the same tail behavior (short, or thin tails) as the normal, but such that it is leptokurtic. This is advantageous for modeling heavier-tailed data, such as financial asset returns. We will also see other important concepts such as mixture diagnostics and an alternative estimation paradigm for multivariate mixtures.

14.1 The Mix_kN_d Distribution

Like its univariate counterpart, use of the multivariate mixed normal distribution has a long history in statistics, and the scope of applications to which it is applied continues to expand, notably in biology, medicine, finance and, somewhat more recently, machine learning; see McLachlan and Peel (2000), Frühwirth-Schnatter (2006), Bishop (2006, Ch. 9), Schlattmann (2009), and Murphy (2012, Ch. 11).

Linear Models and Time-Series Analysis: Regression, ANOVA, ARMA and GARCH, First Edition. Marc S. Paolella.

```
function y = mixMVNsim(mu1,mu2,Sig1,Sig2,lam,n)
[V,D]=eig(Sig1); C1=V*sqrt(D)*V'; [V,D]=eig(Sig2); C2=V*sqrt(D)*V';
d=length(mu1); y=zeros(n,d);
for i=1:n
  z=randn(d,1);
  if rand<lam, y(i,:)=mu1+C1*z; else y(i,:)=mu2+C2*z; end
end
```

Program Listing 14.1: Simulates n i.i.d. realizations from a Mix_2N_d distribution.

14.1.1 Density and Simulation

Let $\mathbf{Y}_t = (Y_{t,1}, Y_{t,2}, \dots, Y_{t,d})' \overset{\text{i.i.d.}}{\sim} \text{Mix}_k\text{N}_d(\boldsymbol{M}, \boldsymbol{\Psi}, \boldsymbol{\lambda})$, $t = 1, \dots, T$, where Mix_kN_d denotes the k-component, non-singular d-dimensional multivariate mixed normal distribution, with

$$\begin{aligned} \boldsymbol{M} &= \left[\boldsymbol{\mu}_1 \,|\, \boldsymbol{\mu}_2 \,|\cdots|\, \boldsymbol{\mu}_k \right], \quad \boldsymbol{\mu}_j = (\mu_{1j}, \mu_{2j}, \dots, \mu_{dj})', \\ \boldsymbol{\Psi} &= \left[\boldsymbol{\Sigma}_1 \,|\, \boldsymbol{\Sigma}_2 \,|\cdots|\, \boldsymbol{\Sigma}_k \right], \quad \boldsymbol{\lambda} = (\lambda_1, \dots, \lambda_k), \end{aligned} \tag{14.1}$$

$\boldsymbol{\Sigma}_j > 0$ (i.e., positive definite), $j = 1, \dots, k$, and

$$f_{\text{Mix}_k\text{N}_d}(\mathbf{y}; \boldsymbol{M}, \boldsymbol{\Psi}, \boldsymbol{\lambda}) = \sum_{j=1}^{k} \lambda_j f_{\text{N}}(\mathbf{y}; \boldsymbol{\mu}_j, \boldsymbol{\Sigma}_j), \quad \lambda_j \in (0,1), \quad \sum_{j=1}^{k} \lambda_j = 1, \tag{14.2}$$

with f_{N} denoting the d-variate normal distribution. Yakowitz and Spragins (1968) have proven that the class of Mix_kN_d distributions is **identified** (see Section III.5.1.1).

Simulating realizations from the $\text{Mix}_k\text{N}_d(\boldsymbol{M}, \boldsymbol{\Psi}, \boldsymbol{\lambda})$ distribution is straightforward; the short program in Listing 14.1 shows this for $k = 2$.

14.1.2 Motivation for Use of Mixtures

As for the univariate case, there are many multivariate distributions that nest (or yield as a limiting case) the normal, and otherwise allow for thicker tails, such as the multivariate Student's t or, more generally, the multivariate generalized hyperbolic (MGHyp) and multivariate noncentral t (MVNCT) distributions, the latter two also allowing for asymmetry. These assist in addressing some of the common stylized facts of financial asset returns. However, a discrete mixture distribution is of particular relevance for financial returns data because of its ability to capture the following two additional stylized facts associated with *multivariate* asset returns:

1) The so-called **leverage** or **down-market effect**, or the negative correlation between volatility and asset returns. A popular explanation for this phenomenon is attributed to Black (1976), who noted that a falling stock price implies a higher leverage on the firm's capital structure (debt to equity ratio), and thus a higher probability of default. This increase in risk is then reflected in a higher stock price volatility.[1]

1 While the effect is empirically visible for falling stock prices, it is less apparent, or missing, for rising prices, calling into question Black's explanation. The empirical effect of negative correlation between volatility and returns also appears in other asset classes (such as exchange rates and commodities) for which Black's explanation is not applicable. See, e.g., Figlewski and Wang (2000), Hens and Steude (2009), Hasanhodzic and Lo (2011), and the references therein for further details.

2) The so-called **contagion effect**, or the tendency of the correlation between asset returns to increase during pronounced market downturns, as well as during periods of higher volatility. See the remark in Section 13.3 for more discussion of this.

The stylized facts of heavy tails, asymmetry, and volatility clustering in the univariate returns distribution, along with changing correlations among the assets, such as the contagion effect, are sometimes referred to as the proverbial **four horsemen** in multivariate financial asset returns; see, e.g., Allen and Satchell (2014) and Bianchi et al. (2016).

We will show some empirical evidence for these effects below, and discuss how a mixture distribution is well-suited for capturing them. A third stylized fact that the Mix_kN_d for $k > 1$ (and also the MGHyp and MVNCT) can capture, but not the (usual, central) multivariate t, hereafter MVT, is non-ellipticity; see Section C.2. Evidence against ellipticity for financial asset returns, driven in part from the two aforementioned stylized facts, as well as so-called time-varying tail-dependence and heterogeneous tail indexes, is provided in McNeil et al. (2005), Chicheportiche and Bouchaud (2012), Paolella and Polak (2015a), and the references therein.

Remark A stylized fact of multivariate financial asset returns that the mixed normal does not formally capture is **tail dependence**, or the dependency (or co-movement) between returns falling in the tails of the distributions (see, e.g., McNeil et al., 2005, Sec. 5.2.3 and the references therein). This is because more extreme market conditions are being modeled essentially by one of the two (in our case, the second; see below) components of the Mix_2N_d distribution, which, being Gaussian, does not have tail dependence.

However, observe that, if there really were just two "states of nature", say "business as usual" and "crisis", then the Mix_2N_d model does allow for this effect, as the covariance matrix in the second component will be different than that of the first component (and the contagion effect is captured). To formally have a tail dependence structure, the Gaussian assumption would need to be replaced with a distribution that has tail dependence, such as a (noncentral) multivariate Student's t, a multivariate generalized hyperbolic, or a copula structure, though observe that, as the number of components k increases, the Mix_kN_d distribution can arbitrarily accurately approximate the tail behavior of such distributions.

This latter statement should not be interpreted as an argument to choose k "as large as possible". As we have seen many times here and in book III, the choice of k involves a tradeoff, with large k inducing many more parameters and, thus, decreased precision of the parameter estimates. The optimal choice should depend on the desired application, such as, in empirical finance, risk prediction, density forecasting, portfolio optimization, etc. ■

With a Mix_2N_d model, we would expect to have the higher-weighted, or primary, component, say the first, capturing the more typical, "business as usual" stock return behavior, with a near-zero mean vector $\boldsymbol{\mu}_1$, and the second component capturing the more volatile, "crisis" behavior, with

- (much) higher variances in $\boldsymbol{\Sigma}_2$ than in $\boldsymbol{\Sigma}_1$,
- significantly larger correlations, reflecting the contagion effect,
- and a predominantly negative $\boldsymbol{\mu}_2$, reflecting the down-market effect.

A distribution with only a single mean vector and covariance matrix (such as the MVT, MVNCT, and MGHyp) cannot capture this behavior, no matter how many additional shape parameters for the tail thickness and asymmetry the distribution possesses. We will subsequently see that these three features are germane to the DJIA-30 data set.

To get some feeling for the data, Figure 14.1 shows three sets of bivariate scatterplots and the corresponding contour plots of the fitted Mix_2N_2 model. It might be of interest to know which assets are the least correlated during turbulent market periods in which contagion effects can be strong. The first column of panels shows the result for the two stocks with the lowest correlation in the estimated covariance matrix $\hat{\mathbf{\Sigma}}_2$ of all $\binom{30}{2}$ pairs, this being for Hewlett-Packard and Kraft Foods, with a correlation in component 2 of 0.27 (and 0.17 in component 1).[2] The middle column shows the pair for which the correlations change the most between components one and two, these being Chevron and Walt Disney. The first component correlation is 0.25, while the second is 0.63. The last column shows the pair for which the correlation in the second component was largest. Unsurprisingly, it is between Chevron and Exxon Mobil, both in the same sector of energy, oil, and gas. The correlations between these two are 0.79 and 0.88, in the first and second components, respectively. The program in Listing 14.2 shows how to locate these pairs.

14.1.3 Quasi-Bayesian Estimation and Choice of Prior

With multivariate distributions, the number of parameters requiring estimation can be large, even for a modest number of dimensions d, and often grows quadratically with d, so that direct likelihood maximization via generic optimization routines will be impractical. For the multivariate normal distribution, the closed-form solution for the m.l.e. is very straightforward. When working with mixtures of normals (univariate or multivariate), no such closed-form solution exists. However, the univariate EM algorithm can be extended easily to the multivariate MixN case. Just as with the univariate mixed normal distribution, we will see that use of shrinkage estimation is of enormous value in the multivariate setting.

Anticipating use of the EM algorithm, denote the latent, or hidden, variable associated with the tth observation $\mathbf{Y}_t$ as $\mathbf{H}_t = (H_{t,1}, \dots, H_{t,k})'$, $t = 1, \dots, T$, where $H_{t,j} = 1$ if $\mathbf{Y}_t$ came from the jth component, and zero otherwise, $j = 1, \dots, k$. The joint density of $\mathbf{Y}_t$ and $\mathbf{H}_t$ is, with $\boldsymbol{\theta} = \{\boldsymbol{M}, \boldsymbol{\Psi}, \boldsymbol{\lambda}\}$ and $\mathbf{h} = (h_1, \dots, h_k)$,

$$f_{\mathbf{Y}_t|\mathbf{H}_t}(\mathbf{y} \mid \mathbf{h}; \boldsymbol{\theta}) f_{\mathbf{H}_t}(\mathbf{h}; \boldsymbol{\theta}) = \mathbb{I}\left(\sum_{j=1}^{k} h_j = 1\right) \prod_{j=1}^{k} [\lambda_j f_{\mathrm{N}}(\mathbf{y}; \boldsymbol{\mu}_j, \mathbf{\Sigma}_j)]^{h_j} \mathbb{I}_{\{0,1\}}(h_j). \tag{14.3}$$

With $\mathbf{Y} = (\mathbf{Y}_1, \dots, \mathbf{Y}_T)'$ and $\mathbf{H} = (\mathbf{H}_1, \dots, \mathbf{H}_T)'$, the complete data log-likelihood is

$$\ell_c(\boldsymbol{\theta}; \mathbf{Y}, \mathbf{H}) = \sum_{t=1}^{T} \sum_{j=1}^{k} H_{t,j} \log \lambda_j + \sum_{t=1}^{T} \sum_{j=1}^{k} H_{t,j} \log f_{\mathrm{N}}(\mathbf{Y}_t; \boldsymbol{\mu}_j, \mathbf{\Sigma}_j). \tag{14.4}$$

2 It is actually the second lowest correlation; the first is between General Motors (GM) and Merck, but this is primarily due to the massive losses GM suffered, so that its second component correlations with other series are among the lowest anyway. As GM is no longer in the DJIA index, we chose not to use it. Further observe that we just picked the pair with the numerically lowest correlation in $\hat{\mathbf{\Sigma}}_2$, and it might be that this value is not statistically different from the second lowest value or the third, etc. Given the i.i.d. assumption, this could be straightforwardly assessed by the parametric or nonparametric bootstrap, from which, e.g., one-at-a-time confidence intervals on the correlation parameters could be computed.

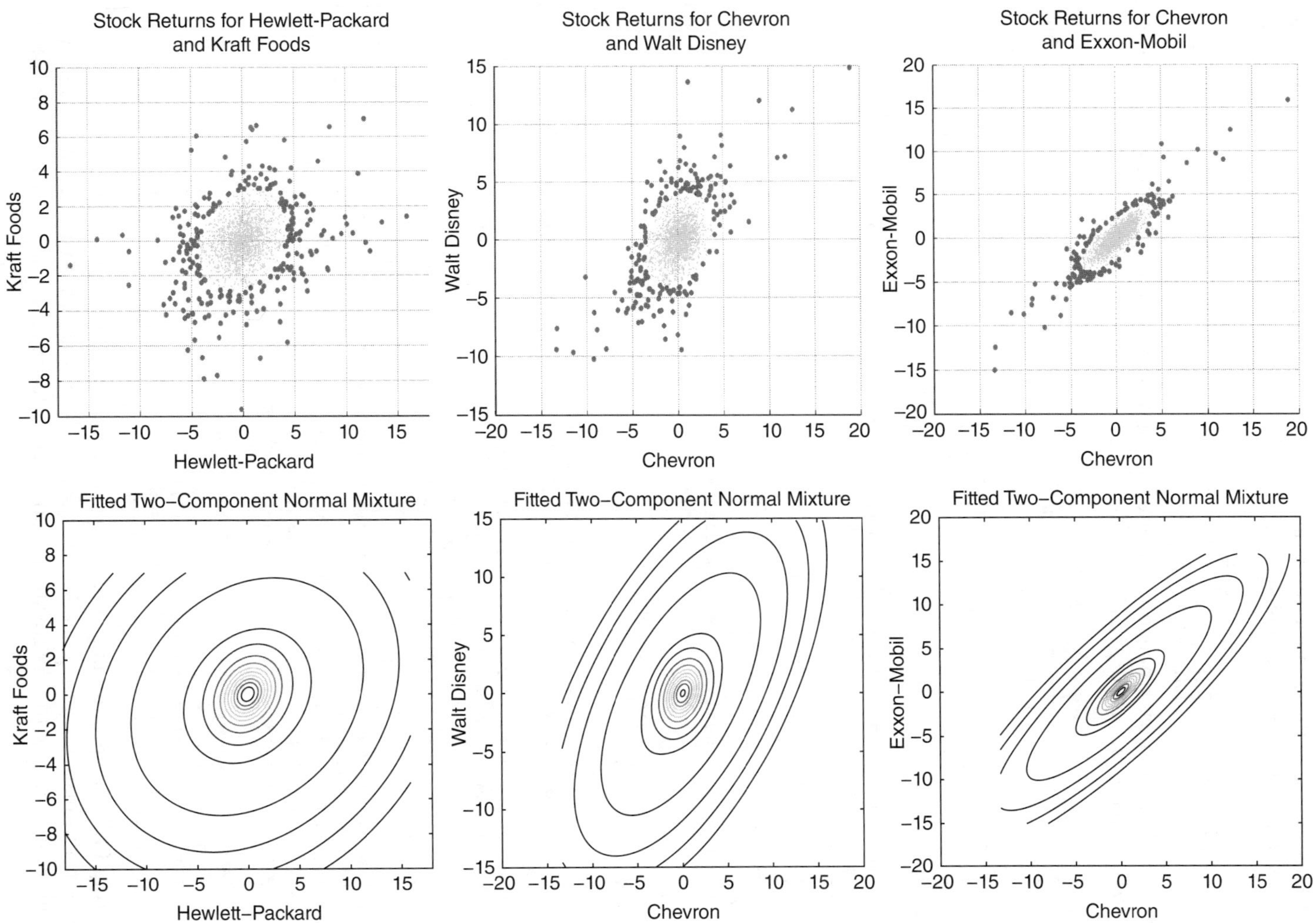

Figure 14.1 Examples of scatterplots between pairs of stock return series (top) and their corresponding contour plots of the fitted Mix_2N_2 distribution (bottom). In the scatterplots, the smaller (larger) dots correspond to the points assigned to the first (second) component, as determined by the approximate split discussed in Section 14.2.1.

Then, calculations similar to those in the univariate case yield the EM algorithm. In particular, the conditional expectation of the $H_{t,j}$ is calculated from

$$\Pr(H_{t,j} = 1 \mid \mathbf{Y}_t = \mathbf{y}; \theta) = \frac{\lambda_j f_{\mathrm{N}}(\mathbf{y}; \boldsymbol{\mu}_j, \boldsymbol{\Sigma}_j)}{\sum_{j=1}^{k} \lambda_j f_{\mathrm{N}}(\mathbf{y}; \boldsymbol{\mu}_j, \boldsymbol{\Sigma}_j)}, \quad j = 1, \dots, k. \tag{14.5}$$

We state the resulting parameter updating equations, augmented by the quasi-Bayesian prior of Hamilton (1991), as in the univariate MixN case. They are

$$\hat{\lambda}_j = \frac{1}{T} \sum_{t=1}^{T} H_{t,j}, \quad \hat{\boldsymbol{\mu}}_j = \frac{c_j \mathbf{m}_j + \sum_{t=1}^{T} H_{t,j} \mathbf{Y}_t}{c_j + \sum_{t=1}^{T} H_{t,j}}, \quad j = 1, \dots, k, \tag{14.6}$$

```
[T,d]=size(data); [mu1,mu2,Sig1,Sig2] = mixnormEMm(data,50);
[~,C1]=cov2corr(Sig1); C1=(C1+C1')/2;
[~,C2]=cov2corr(Sig2); C2=(C2+C2')/2;

% Locate the two assets that exhibit the lowest 2nd-component correlation
%%%%%%%%%%%%%%%%%%%%%%%%%
% WAY 1: brute force code:
cmin=1; asset1=1; asset2=1;
for row=1:(d-1)
  for col=(row+1):d
    c=C2(row,col);
    if c<cmin, cmin=c; asset1=row; asset2=col; end
  end
end
mincorr = cmin, asset1, asset2 %#ok<NOPTS>
%%%%%%%%%%%%%%%%%%%%%%%%%
% WAY 2: elegant and fast:
Use=C2(:); minC=min(Use), loc=find(Use==minC); loc=loc(1); %#ok<NOPTS>
asset1=ceil(loc/d), asset2=mod(loc,d); if asset2==0, asset2=d; end, asset2
%%%%%%%%%%%%%%%%%%%%%%%%%
% WAY 3: If the ceil and mod functions were not available:
Use=C2(:); minC=min(Use), loc=find(Use==minC); loc=loc(1); %#ok<NOPTS>
garb=zeros(d^2,1); garb(loc)=1; garb=reshape(garb,d,d);
asset1=find(sum(garb,1)==1), asset2=find(sum(garb,2)==1)
%%%%%%%%%%%%%%%%%%%%%%%%%

% Locate the two assets that exhibit the maximal difference in correlations
%   between the two components
%  (Note: correlations in this context are almost always >0)
Use=abs(C1-C2); Use=Use(:); loc=find(Use==max(Use)); loc=loc(1);
asset1=ceil(loc/d), asset2=mod(loc,d); if asset2==0, asset2=d; end, asset2

% Locate the largest 2nd-component correlation
Use=C2; for i=1:d, Use(i,i)=0; end
Use=Use(:); loc=find(Use==max(Use)); loc=loc(1);
asset1=ceil(loc/d), asset2=mod(loc,d); if asset2==0, asset2=d; end, asset2
```

Program Listing 14.2: Code for finding interesting pairs of data from the DJIA-30 dataset. It assumes thereturns are in the matrix `data`. Function `cov2corr` is in Matlab's finance toolbox, and just converts a covariance matrix to a correlation matrix. Function `mixnormEMm` is given below in Listing 14.6.

and

$$\widehat{\boldsymbol{\Sigma}}_j = \frac{\mathbf{B}_j + \sum_{t=1}^{T} H_{t,j}(\mathbf{Y}_t - \widehat{\boldsymbol{\mu}}_j)(\mathbf{Y}_t - \widehat{\boldsymbol{\mu}}_j)' + c_j(\mathbf{m}_j - \widehat{\boldsymbol{\mu}}_j)(\mathbf{m}_j - \widehat{\boldsymbol{\mu}}_j)'}{a_j + \sum_{t=1}^{T} H_{t,j}}, \tag{14.7}$$

$j = 1, \ldots, k$. Fixed quantities $\mathbf{m}_j \in \mathbb{R}^d$, $a_j \geqslant 0$, $\mathbf{B}_j$ a $d \times d$ positive definite matrix, and $c_j \geqslant 0$, indicate the prior information, with interpretations analogous to the univariate case.

Thus, genuine maximum likelihood (possibly with shrinkage via the quasi-Bayesian prior) can be conducted extremely fast, even for a large number of parameters. For the application we consider, with $k = 2$ components and $d = 30$ assets, there are $d^2 + 3d + 1 = 991$ parameters to estimate. In a general optimization setting with so many parameters this would be essentially infeasible, even with modern computing power, while the EM algorithm is very simple to implement and, using a 3-GHz desktop PC, takes about one tenth of a second, using $T = 1{,}000$ observations. The program is given in Listing 14.6 and incorporates the use of weighted likelihood, as discussed in Chapter 13.

When using the Gaussian framework (i.e., single-component multivariate normal) for financial portfolio optimization, the use of shrinkage applied to the sample means, variances, and covariances of the returns to improve performance is well-known; see, e.g., Jorion (1986), Jagannathan and Ma (2003), Kan and Zhou (2007), Bickel and Levina (2008), Fan et al. (2008), and the references therein. Here, we extend this idea to the Mix_kN_d case. A natural candidate for the prior would be to take $\mathbf{m}_j$ to be a d-vector of zeros, and $\mathbf{B}_j$ the d-dimensional identity matrix, corresponding to shrinkage to the standard normal. Our choice will be similar to this, but altered in such a way to be more meaningful in the context of modeling daily equity returns in general, as subsequently explained. The precise values are obtained based on "loose calibration" to the DJIA-30 data (explained below), and thus form a data-driven prior, further distancing it from a traditional Bayesian approach, though it is similar in principle to the use of so-called empirical Bayes procedures. The relationship between the empirical Bayes approach and shrinkage estimation are discussed in Berger (1985, Sec. 4.5), Lehmann and Casella (1998, Sec. 4.6), Robert (2007, Sec. 2.8.2, 10.5), and the references therein.

The top two panels in Figure 14.2 show the 30 values of $\widehat{\boldsymbol{\mu}}_1$ and $\widehat{\boldsymbol{\mu}}_2$, obtained from fitting the $\text{Mix}_2\text{N}_{30}$ model to the DJIA-30 data set via the EM algorithm, but using only a very weak prior (enough such that the singularities are avoided). These values are in accordance with our aforementioned discussion of the two regimes at work in the financial market. While the means in $\widehat{\boldsymbol{\mu}}_1$ are closely centered around zero, those from $\widehat{\boldsymbol{\mu}}_2$ are nearly all negative, and with a much higher magnitude than those from $\widehat{\boldsymbol{\mu}}_1$. From the middle row of panels, we see that the variances from $\widehat{\boldsymbol{\Sigma}}_2$ of the 30 components are about 10 times the size of those from $\widehat{\boldsymbol{\Sigma}}_1$. Thus, the second component indeed captures the high volatility "regime" of the returns, and is associated with a relatively strong negative mean term. Finally, we see from the bottom panels that the correlations between the 30 assets are also higher in the second component, reflecting the contagion effect. As already mentioned, while being leptokurtic and asymmetric, distributions such as the MGHyp (and its special or limiting cases) and MVNCT have only one location vector and dispersion matrix, and so cannot capture these two separate types of market behavior.

Based on these findings, and in line with the usual motivation for the James–Stein estimator for the mean vector of a multivariate normal population with independent components (see Section III.5.4), our prior is one that shrinks the means, variances, and covariances from each of the two components towards their average values over the $d = 30$ series, as shown in Figure 14.2. Thus, $\mathbf{m}_1$ is a vector of all zeros, $\mathbf{m}_2$ is a vector with all elements equal to -0.1; $\mathbf{B}_1$ is the prior strength, ω, times the matrix with

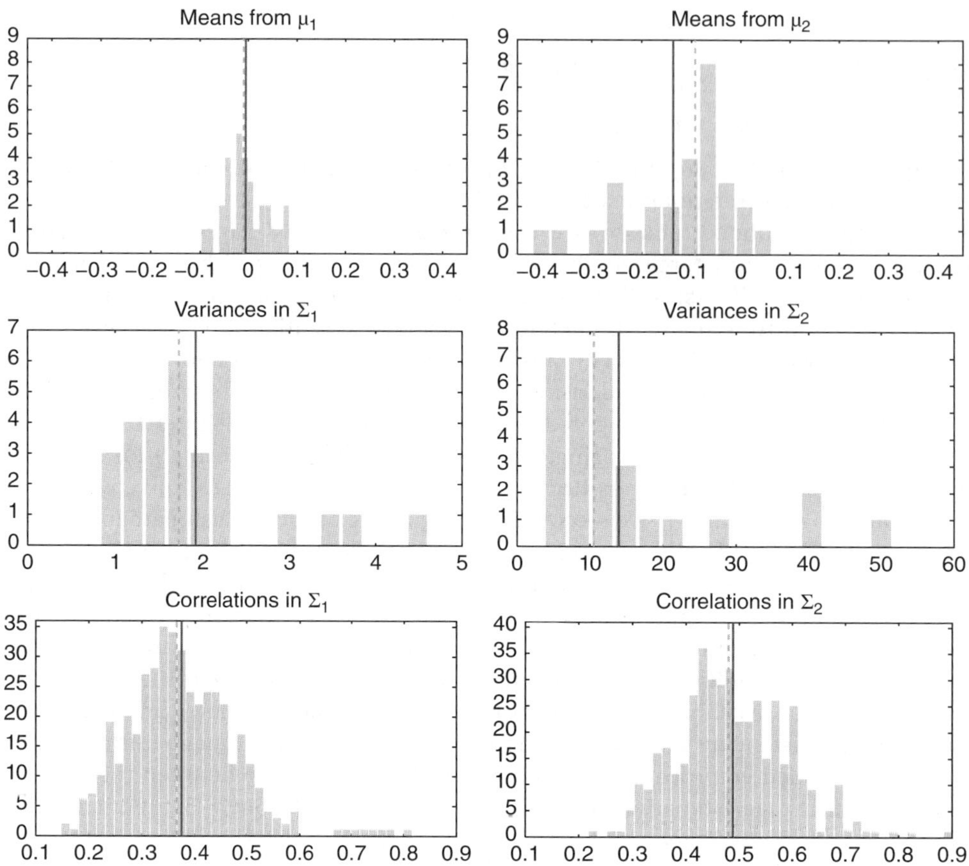

Figure 14.2 The estimated $d = 30$ means (top), 30 variances (middle), and 435 correlations (bottom) for the first (left) and second (right) components of the normal mixture corresponding to the DJIA-30 data set under study. Solid (dashed) vertical lines show the mean (median).

its $d = 30$ diagonal elements equal to 1.5, and off-diagonal elements equal to 0.6; for $\mathbf{B}_2$, the variance and covariance are 10 and 4.6, respectively. While several shrinkage targets for the covariance matrix have been proposed, the one with constant correlations is the easiest, and also does not suffer from a potential criticism of adding too much information via the prior. Use of constant correlation as a shrinkage target was advocated by Ledoit and Wolf (2004), who show that it yields comparable performance to other choices.

Weight a_j reflects our strength in the prior of the variance–covariance matrix $\mathbf{\Sigma}_j$, $j = 1, 2$. We take $a_1 = 2\omega$ and $a_2 = \omega/2$ because $\mathbf{\Sigma}_2$ is far more variable than $\mathbf{\Sigma}_1$, though the value of 2 is arbitrary and could be viewed as a further tuning parameter, along with ω. Weight c_j reflects our strength in the prior of mean vector $\boldsymbol{\mu}_j$, $j = 1, 2$. These should be higher than the a_j for two reasons. First, an appeal to the efficient market hypothesis provides some justification for shrinking the means in the first, primary component of the mixture towards zero, while the blatant down-market effect in financial crises lends support for shrinking the mean in the second component of the mixture towards a negative value.

The second reason is that errors in the estimated mean vector are considered more consequential in asset allocation and portfolio management (see, e.g., Best and Grauer, 1991, 1992; Chopra and Ziemba, 1993), so that the benefits of shrinkage could be quite substantial.[3] In light of this, we take $c_j = 20\omega, j = 1, 2$. (The large factor of 20 was determined by some trial and error based on the simulation exercise discussed next. It could also serve as a further tuning parameter.) The shrinkage prior, as a function of the scalar hyper-parameter ω, is then

$$\begin{aligned} a_1 = 2\omega, \quad a_2 = \omega/2, \quad c_1 = c_2 = 20\omega, \quad \mathbf{m}_1 = \mathbf{0}_d, \quad \mathbf{m}_2 = (-0.1)\mathbf{1}_d, \\ \mathbf{B}_1 = a_1[(1.5 - 0.6)\mathbf{I}_d + 0.6\mathbf{J}_d], \quad \mathbf{B}_2 = a_2[(10 - 4.6)\mathbf{I}_d + 4.6\mathbf{J}_d], \end{aligned} \tag{14.8}$$

where $\mathbf{1}_d$ and $\mathbf{J}_d$ are the $d \times 1$ column vector and $d \times d$ matrix of ones, respectively. As the numerical values in (14.8) were obtained by calibration to a typical set of financial stock returns, but only loosely, in the sense that each margin receives the same prior structure and the correlations are constant in each of the two prior dispersion matrices, we expect this prior to be useful for *any* such set of financial data that exhibits the usual stylized facts of (daily) asset returns.

The only tuning parameter that remains to be chosen is ω. The effect of different choices of ω is easily demonstrated with a simulation study, using the $\text{Mix}_2\text{N}_{30}$ model, with parameters given by the m.l.e. of the 30 return series (whose parameter values are depicted in Figure 14.2). We used $T = 250$ observations (which is roughly the number of trading days in one year), a choice of 11 different values of ω, and 10,000 replications for each ω. All 110,000 estimations were successful, at least in the sense that the program in Listing 14.6 never failed, with the computation of *all* of them requiring about 20 minutes (on a single core, 3.2 GHz PC).[4]

For assessing the quality of the estimates, we use the same technique as in the univariate MixN case, namely, the log sum of squares as the summary measure, noting that, as with the univariate case, we have to convert the estimated parameter vector if the component labels are switched. That is,

$$M^*(\widehat{\boldsymbol{\theta}}, \boldsymbol{\theta}) = \min\{M(\widehat{\boldsymbol{\theta}}, \boldsymbol{\theta}), M(\widehat{\boldsymbol{\theta}}, \boldsymbol{\theta}^{=})\}, \tag{14.9}$$

for

$$\boldsymbol{\theta} = (\boldsymbol{\mu}_1', \ \boldsymbol{\mu}_2', \ (\text{vech}(\boldsymbol{\Sigma}_1))', \ (\text{vech}(\boldsymbol{\Sigma}_2))', \ \lambda_1)', \tag{14.10}$$

where the vech operator of a matrix forms a column vector consisting of the elements on and below the main diagonal (see the beginning of Section 12.5.3),

$$M(\widehat{\boldsymbol{\theta}}, \boldsymbol{\theta}) := \log(\widehat{\boldsymbol{\theta}} - \boldsymbol{\theta})'(\widehat{\boldsymbol{\theta}} - \boldsymbol{\theta}), \tag{14.11}$$

3 While this result is virtually conventional wisdom now, it has been challenged by Bengtsson (2003), who shows that the presumed deleterious impact of the estimation errors of the mean vector might be exaggerated, and that errors in the covariance matrix can be equally detrimental. As such, shrinkage of both the mean vector and covariance matrix should be beneficial.

4 One might inquire about the potential for multiple local plausible maxima of the log likelihood. To (very partially) address this, for each of the 110,000 replications in the simulation study, the model was estimated twice, based on different starting values, these being (i) the true parameter values and (ii) the default starting values, which we take to be the prior values from (14.8). Interestingly, for all 10,000 data sets and each value of ω, without a single exception, the final likelihood values obtained based on the two different starting values were identical up to the tolerance requested of the EM estimation algorithm, namely 10^{-6}. While we did not compare the parameter values, this is quite strong evidence that the two starting values led to the same maximum each time. (As an "idiot check", estimating each model twice, but using the same starting values, yields genuinely identical likelihood values, up to full machine precision.)

and $\boldsymbol{\theta}^{=}$ refers to the parameter vector obtained by switching the labels of the two components, i.e., $\boldsymbol{\theta}^{=} = (\boldsymbol{\mu}_2', \ \boldsymbol{\mu}_1', \ (\text{vech}(\boldsymbol{\Sigma}_2))', \ (\text{vech}(\boldsymbol{\Sigma}_1))', \ (1-\lambda_1))'$.

The boxplots in Figure 14.3 show, for each value of ω, the discrepancy measure M^* from (14.9), but decomposed into four components consisting of the aggregate of the elements in $\widehat{\boldsymbol{\mu}}_1$, $\widehat{\boldsymbol{\mu}}_2$, $\widehat{\boldsymbol{\Sigma}}_1$, and $\widehat{\boldsymbol{\Sigma}}_2$, respectively (and ignoring λ_1). This is valuable because, in addition to being able to assess their estimation uncertainly separately, we can also see the impact of the choices of the a_j and c_j, $j = 1, 2$. (If we were to pool all 991 parameters, then those from $\widehat{\boldsymbol{\Sigma}}_2$ would dominate the measure.) The improvement to both mean vectors is quite substantial, with the last boxplot in each graph, labeled $\omega = \infty$, having been based on $\omega = 10^5$, illustrating the case when the prior is allowed to dominate. The improvement from the shrinkage is less dramatic for the covariance matrices, with larger values of ω eventually leading to an increase in average estimation error. A reasonable choice of ω appears to be 20, though we will see below in the context of density forecasting with the DJIA-30 data (which are certainly not generated from an i.i.d. $\text{Mix}_2\text{N}_{30}$ process, as used in the previous simulation) that higher values of ω are desirable.

14.1.4 Portfolio Distribution and Expected Shortfall

In financial applications with portfolio optimization, interest centers on weighted sums of the univariate margins of the joint Mix_kN_d distribution. This is the random variable describing the portfolio returns, which, at time t and for portfolio weight vector $\mathbf{a}$ and parameter vector $\boldsymbol{\theta}$, we will denote as $P_t(\mathbf{a}, \boldsymbol{\theta})$.

Theorem 14.1 Let $\mathbf{Y}_t \sim \text{Mix}_k\text{N}_d(\boldsymbol{M}, \boldsymbol{\Psi}, \boldsymbol{\lambda})$, with $\boldsymbol{\theta} = \{\boldsymbol{M}, \boldsymbol{\Psi}, \boldsymbol{\lambda}\}$ as given in (14.1), and portfolio return $P = P_t(\mathbf{a}, \boldsymbol{\theta}) = \mathbf{a}'\mathbf{Y}_t$. For any $\mathbf{a} \in \mathbb{R}^d$,

$$f_P(x; \boldsymbol{\theta}) = \sum_{j=1}^{k} \lambda_j \phi(x; \mu_j, \sigma_j^2), \tag{14.12}$$

where $\phi(x; \mu, \sigma^2)$ denotes the univariate normal distribution with mean μ and variance σ^2, evaluated at x, $\mu_j = \mathbf{a}'\boldsymbol{\mu}_j$, and $\sigma_j^2 = \mathbf{a}'\boldsymbol{\Sigma}_j\mathbf{a}$, $j = 1, \ldots, k$.

Proof: Let $\mathbf{X} \sim \text{N}_d(\boldsymbol{\mu}, \boldsymbol{\Sigma})$, with characteristic function

$$\varphi_{\mathbf{X}}(\mathbf{t}) = \mathbb{E}[\exp(i\mathbf{t}'\mathbf{X})] = \exp\left(i\mathbf{t}'\boldsymbol{\mu} - \frac{1}{2}\mathbf{t}' \ \boldsymbol{\Sigma} \ \mathbf{t}\right) =: \varphi(\mathbf{t}; \boldsymbol{\mu}, \boldsymbol{\Sigma}), \tag{14.13}$$

for $\mathbf{t} \in \mathbb{R}^d$. As scalar $S = \mathbf{a}'\mathbf{X} \sim \text{N}(\mathbf{a}'\boldsymbol{\mu}, \mathbf{a}'\boldsymbol{\Sigma}\mathbf{a})$ for $\mathbf{a} = (a_1, \ldots, a_d)' \in \mathbb{R}^d$, (14.13) implies that

$$\varphi_S(t) = \varphi(t; \mathbf{a}'\boldsymbol{\mu}, \mathbf{a}'\boldsymbol{\Sigma}\mathbf{a}) = \mathbb{E}[\exp(it\mathbf{a}'\mathbf{X})] = \int_{\mathbb{R}^d} \exp(it\mathbf{a}'\mathbf{x}) \, \mathrm{d}F_{\text{N}}(\mathbf{x}; \boldsymbol{\mu}, \boldsymbol{\Sigma}). \tag{14.14}$$

Let $\mathbf{Y} \sim \text{Mix}_k\text{N}_d(\boldsymbol{M}, \boldsymbol{\Psi}, \boldsymbol{\lambda})$. With discrete random variable C such that $f_C(c) = \lambda_c$, $\lambda_c \in (0, 1)$, $\sum_{c=1}^{k} \lambda_c = 1$, we can express the mixed normal density as

$$f_{\mathbf{Y}}(\mathbf{y}) = \int f_{\mathbf{Y}|C}(\mathbf{y} \mid c) \, \mathrm{d}F_C(c) = \sum_{c=1}^{k} \lambda_c f_{\text{N}}(\mathbf{y}; \boldsymbol{\mu}_c, \boldsymbol{\Sigma}_c). \tag{14.15}$$

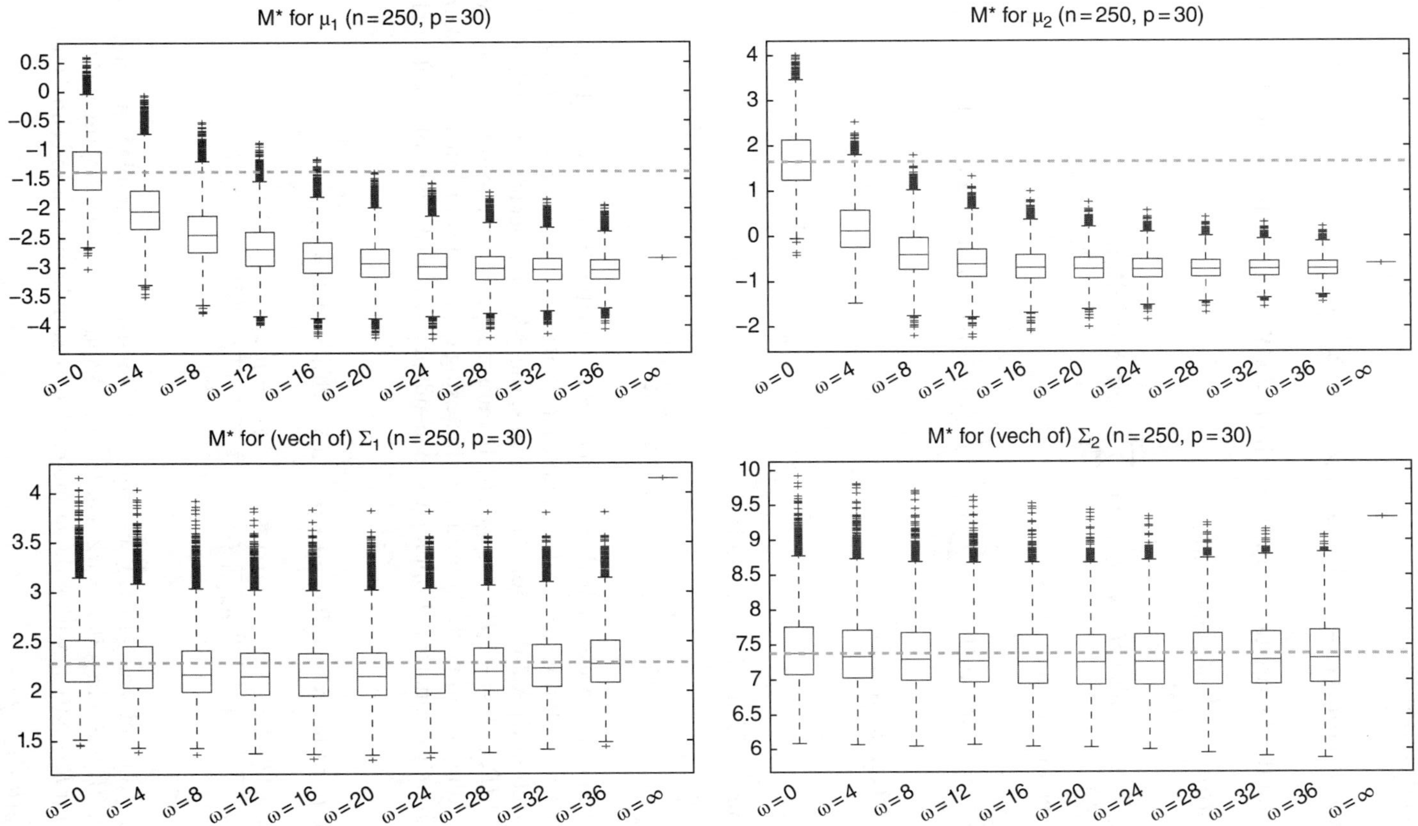

Figure 14.3 Estimation accuracy, as a function of prior strength parameter ω, measured as four divisions of M^* from (14.9) (λ_1 is ignored), based on simulation with 10,000 replications and $T = 250$, of the parameters of the $\text{Mix}_2\text{N}_{30}$ model, using as true parameters the m.l.e. of the DJIA-30 data set.

Then, from (14.13) and (14.15),

$$\varphi_{\mathbf{Y}}(\mathbf{t}) = \int_{\mathbb{R}^d} \exp(i\mathbf{t}'\mathbf{y})\, \mathrm{d}F_{\mathbf{Y}}(\mathbf{y}) = \sum_{c=1}^{k} \lambda_c \exp\left(i\mathbf{t}'\boldsymbol{\mu}_c - \frac{1}{2}\mathbf{t}'\,\boldsymbol{\Sigma}_c\,\mathbf{t}\right),$$

and interest centers on the distribution of the portfolio $P = \mathbf{a}'\mathbf{Y}$. Its c.f. is, from (14.14),

$$\begin{aligned}\varphi_P(t) &= \mathbb{E}[\exp(itP)] = \int_{\mathbb{R}^d} \exp(it\mathbf{a}'\mathbf{y})\, \mathrm{d}F_{\mathbf{Y}}(\mathbf{y}) \\ &= \sum_{c=1}^{k} \lambda_c \int_{\mathbb{R}^d} \exp(it\mathbf{a}'\mathbf{y})\, \mathrm{d}F_{\mathrm{N}}(\mathbf{y}; \boldsymbol{\mu}_c, \boldsymbol{\Sigma}_c) = \sum_{c=1}^{k} \lambda_c \varphi(t; \mathbf{a}'\boldsymbol{\mu}_c, \mathbf{a}'\boldsymbol{\Sigma}_c\mathbf{a}),\end{aligned}$$

and applying the inversion theorem gives

$$\begin{aligned}f_P(x) &= \frac{1}{2\pi}\int_{-\infty}^{\infty} \exp(-itx)\varphi_P(t)\, \mathrm{d}t = \sum_{c=1}^{k} \lambda_c \frac{1}{2\pi}\int_{-\infty}^{\infty} \exp(-itx)\varphi(t; \mathbf{a}'\boldsymbol{\mu}_c, \mathbf{a}'\boldsymbol{\Sigma}_c\mathbf{a})\, \mathrm{d}t \\ &= \sum_{c=1}^{k} \lambda_c f_{\mathrm{N}}(x; \mathbf{a}'\boldsymbol{\mu}_c, \mathbf{a}'\boldsymbol{\Sigma}_c\mathbf{a}),\end{aligned}$$

which is (14.12). ■

The first two moments of P are (see Example II.7.14)

$$\mathbb{E}_{\theta}[P] = \sum_{c=1}^{k} \lambda_c \mu_c, \qquad \mathbb{V}_{\theta}(P) = \sum_{c=1}^{k} \lambda_c(\sigma_c^2 + \mu_c^2) - (\mathbb{E}_{\theta}[P])^2, \tag{14.16}$$

using standard notation to express their dependence on parameter $\boldsymbol{\theta}$. The c.d.f. of P is $F_P(x; \boldsymbol{\theta}) = \sum_{c=1}^{k} \lambda_c \Phi((x - \mu_c)/\sigma_c)$, with Φ the standard normal c.d.f. Denote the ξ-quantile of P as $q_{P,\xi}$, for $0 < \xi < 1$. Recall from Section III.A.7 that, for P continuous, the ξ-level expected shortfall is (using the minus convention) $\mathrm{ES}\,(P, \xi; \boldsymbol{\theta}) = -\mathbb{E}[P \mid P \leqslant q_{P,\xi}; \boldsymbol{\theta}]$. In our setting, an analytic expression is available, so that the objective function in portfolio optimization using expected shortfall as the risk measure is instantly and accurately evaluated. Dropping the dependency of the ES on $\boldsymbol{\theta}$ for notational convenience, we have

Theorem 14.2 For portfolio return $P = P_t(\mathbf{a}, \boldsymbol{\theta}) = \mathbf{a}'\mathbf{Y}_t$ with p.d.f. (14.12),

$$\mathrm{ES}(P, \xi) = \sum_{j=1}^{k} \frac{\lambda_j \Phi(c_j)}{\xi}\left\{\mu_j - \sigma_j \frac{\phi(c_j)}{\Phi(c_j)}\right\}, \quad c_j = \frac{q_{P,\xi} - \mu_j}{\sigma_j}, \quad j = 1, \ldots, k. \tag{14.17}$$

Proof: With $P \sim \mathrm{Mix}_k\mathrm{N}_1$ and p.d.f. (14.12), we require the following two simple facts, both of which are shown in Section III.A.8. First, if $Y = \sigma Z + \mu$ for $\sigma > 0$ and $\mathrm{ES}(Z; \xi)$ exists, then $\mathrm{ES}(Y, \xi) = \mu + \sigma\,\mathrm{ES}\,(Z, \xi)$. Second, for $R \sim \mathrm{N}(0, 1)$ with p.d.f. ϕ and c.d.f. Φ, a simple integration shows that

$$\mathrm{ES}(R, \xi) = -\phi\{\Phi^{-1}(\xi)\}/\xi. \tag{14.18}$$

Let $q_{P,\xi}$ be the ξ-quantile of P, $X_j \sim \mathrm{N}(\mu_j, \sigma_j^2)$, $c_j := (q_{P,\xi} - \mu_j)/\sigma_j$, and $Z \sim \mathrm{N}(0,1)$. Based on the substitution $z = (x - \mu_j)/\sigma_j$,

$$\begin{aligned}
\mathrm{ES}(P,\xi) &= \frac{1}{\xi}\int_{-\infty}^{q_{P,\xi}} x f_P(x)\,\mathrm{d}x = \frac{1}{\xi}\sum_{j=1}^{k}\lambda_j \int_{-\infty}^{q_{P,\xi}} x\,\sigma_j^{-1} f_Z\left(\frac{x-\mu_j}{\sigma_j}\right)\mathrm{d}x \\
&= \frac{1}{\xi}\sum_{j=1}^{k}\lambda_j \int_{-\infty}^{\frac{q_{P,\xi}-\mu_j}{\sigma_j}} (\sigma_j z + \mu_j)\sigma_j^{-1} f_Z(z)\,\sigma_j\,\mathrm{d}z \\
&= \frac{1}{\xi}\sum_{j=1}^{k}\lambda_j \left[\sigma_j \int_{-\infty}^{c_j} z f_Z(z)\,\mathrm{d}z + \mu_j \int_{-\infty}^{c_j} f_Z(z)\,\mathrm{d}z\right]. \qquad (14.19)
\end{aligned}$$

Using (14.18) and (14.19), we obtain (14.17). ■

14.2 Model Diagnostics and Forecasting

All models are wrong, but some are useful.

(George Edward Pelham Box, 1979)

14.2.1 Assessing Presence of a Mixture

Recall that the filtered $H_{t,j}$ values from (14.5) have support [0, 1] and can be referred to as the **posterior probabilities** that observation $\mathbf{Y}_t$ came from component j, $t = 1, \ldots, T$, $j = 1, 2$, conditional on all the $\mathbf{Y}_t$ and the estimated parameters. It is natural to plot the values of $H_{t,1}$, versus the time ordering t, $t = 1, \ldots, 1{,}945$. These are shown in the left panel of Figure 14.4, as returned from the EM algorithm after it converged. The right panel is the same, but just showing the last 250 values. It appears that the two components are well separated, with most values being very close to either zero or one.

While this would appear to add even more support to our claim that there exist two reasonably distinct "regimes", this is actually not the case: The same effect occurs if the data come from a (single component) leptokurtic multivariate distribution such as Student's t or Laplace. To illustrate,

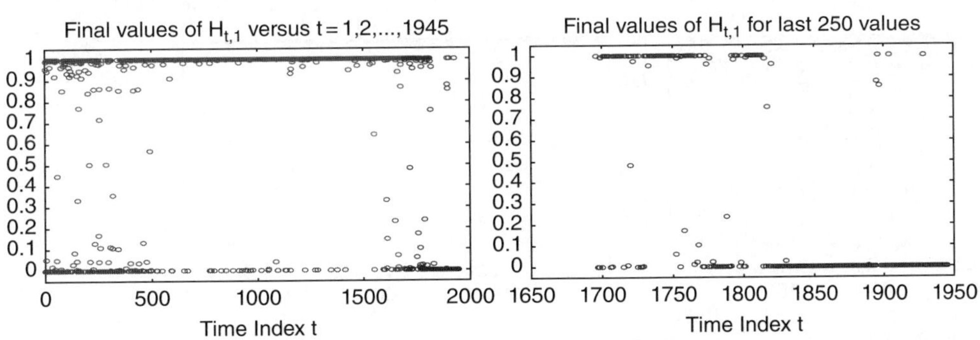

Figure 14.4 Final values of $\widehat{H}_{t,1}$ returned from the EM algorithm based on the $\mathrm{Mix}_2\mathrm{N}_{30}$ model, applied to the DJIA-30 data set.

we first simulate a set of $T = 1{,}945$ return vectors, each an i.i.d. draw from the multivariate normal distribution with the mean and covariance matrix chosen as the sample mean and covariance from the DJIA-30 stock return data, and attempt to fit the $\text{Mix}_2\text{N}_{30}$ model. The code for this is given in Listing 14.3.

Of course, there is only one component, and the parameters of the mixture model are not identified. As perhaps expected, the EM algorithm converges slowly (over 1,000 iterations in this case). The likelihood is (except for the singularities) relatively flat in λ, with true value zero or one, and thus not in the interior of the parameter space. For this simulated data set, the m.l.e. was $\hat{\lambda} = 0.55$. Further simulations resulted in similar behavior. The left panel of Figure 14.5 shows the final values of $\widehat{H}_{t,1}$. There is clearly far less separation than with the actual DJIA-30 data. The sum of the diagonal of the sample covariance matrix of the DJIA-30 data is 146, while the sums for the two "mixture components" $\widehat{\boldsymbol{\Sigma}}_1$ and $\widehat{\boldsymbol{\Sigma}}_2$, based on the simulated multivariate normal data, were 136 and 155, respectively, showing that there is hardly any difference in the two components.

The right panel of Figure 14.5 is similar, but based on $T = 1{,}945$ samples of i.i.d. data generated from the multivariate Laplace distribution given below in (14.31). A realization from this distribution is very simple to generate using its mixture representation: For $b > 0$, $G \sim \text{Gam}(b, 1)$ and $(\mathbf{Y} \mid G = g) \sim \text{N}(\mathbf{0}, g\boldsymbol{\Sigma})$. The code for the plot is given in Listing 14.4.

Now, the sum of the diagonals of $\widehat{\boldsymbol{\Sigma}}_1$ and $\widehat{\boldsymbol{\Sigma}}_2$ are 34 and 218, respectively, with a clear separation of the two normal components, even though *the data were not generated from a two-component mixture of normals.* As the shape parameter b decreases towards one, the univariate marginal distributions become very peaked and leptokurtic, allowing a clear separation of the data (under the incorrect assumption of a MixN). As $b \to \infty$, the distribution becomes Gaussian, so that the resulting plot of the $\widehat{H}_{t,1}$ begins to look like the left panel of Figure 14.5.

```
T=1945; Y=mvnrnd(mean(data),cov(data),T);
[mu1,mu2,Sig1,Sig2,lam,ll,H1] = mixnormEMm (Y,0.1,[]);
figure, plot(1:1945,H1,'ro')
sum(diag(cov(data))), sum(diag(Sig1)), sum(diag(Sig2))
```

Program Listing 14.3: Simulates T i.i.d. realizations from the d-dimensional multivariate normal distribution and estimates the Mix_2N_d model. `data` is the $T \times d$ DJIA-30 daily returns matrix.

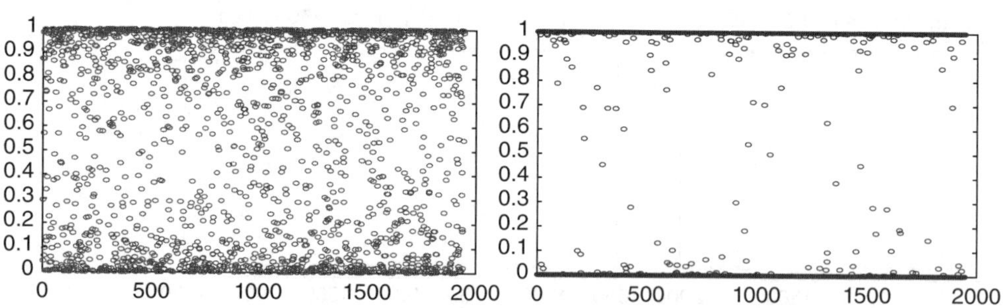

Figure 14.5 Left: Final values of $\widehat{H}_{t,1}$ returned from the EM algorithm based on the $\text{Mix}_2\text{N}_{30}$ model for a simulated set of multivariate normal data with $T = 1{,}945$, $d = 30$, using a mean and covariance equal to the sample mean and covariance from the DJIA-30 data set. **Right:** Same, but having used a multivariate Laplace distribution with $b = 1$.

```
d=30; b=1; mu=mean(data); Sig=cov(data)/b; T=1945; Y=zeros(T,d);
for i=1:T
  theta=gamrnd(b,1,[1,1]); Y(i,:)=mvnrnd(zeros(1,d),theta*Sig,1) + mu;
end
[mu1,mu2,Sig1,Sig2,lam,ll,H1] = mixnormEMm (Y,0.1,[]);
figure, plot(1:1945,H1,'bo')
sum(diag(cov(data))), sum(diag(Sig1)), sum(diag(Sig2))
```

Program Listing 14.4: Simulates T i.i.d. realizations from the d-dimensional multivariate Laplace distribution (14.31) and estimates the Mix_2N_d model. The expression for `Sig` in the first line comes from the variance of the Laplace distribution, see (14.32).

Thus, we see that our empirical justification for using a mixture distribution with two components stems from the results in Figure 14.2, namely that the means of the first and second component differ markedly, with the latter being primarily negative, and that the correlations in the second component are on average higher than those associated with component 1. As we have seen by considering Figures 14.4 and 14.5, the larger variances associated with the second component arise if the data are actually generated from a two-component mixture of normals, *but also if the data are generated from a (single-component) multivariate Laplace.*

> The separation apparent in Figure 14.4 is necessary, but not sufficient, to support the hypothesis that the data were generated by a mixture distribution.

14.2.2 Component Separation and Univariate Normality

Returning now to the DJIA-30 data, the separation apparent from Figure 14.4 is also highly advantageous because it allows us to assign each $\mathbf{Y}_t$ to one of the two components, in most cases with what appears to be rather high confidence. Once done, we can assess how well each of the two estimated multivariate normal distributions fits the observations assigned to its component. While we could use the rule to assign the tth observation $\mathbf{Y}_t$ to component 1 if $\widehat{H}_{t,1} > 0.5$, and to component 2 otherwise (which would result in 1,490 observations assigned to component 1, or 76.6%, which is nearly the same as $\hat{\lambda}_1 = 0.763$), we instead use the criteria $\widehat{H}_{t,1} > 0.99$, choosing to place those $\mathbf{Y}_t$ whose corresponding values of $\widehat{H}_{t,1}$ suggest even a slight influence from component 2 into this more volatile component. This results in 1,373 observations assigned to component 1, or 70.6% of the observations, and 572 to the second component.

Once the data are (inevitably imperfectly) split, we wish to assess the normality of each of the two components. There are many tests for composite *univariate* normality; see Chapter III.6 for some of these. Unfortunately, testing composite *multivariate* normality is not trivial; see Thode, Jr. (2002, Chap. 9) and the survey article from Mecklin and Mundfrom (2004). Part of the reason for the complexity of testing multivariate normality is that there are many ways a distribution can depart from it, so that no single test will be optimal. Examining only the univariate margins (as was illustrated in Section III.6.5.1) is not ideal because they do not uniquely determine the joint distribution. (Example II.3.2 shows a distribution for which all marginals, univariate and multivariate, are normal, but the joint distribution is not.) Nevertheless, we proceed first by inspecting the behavior of the univariate margins, so that we can possibly suggest a more suitable multivariate distribution that at least accounts for the univariate empirical behavior of the data.

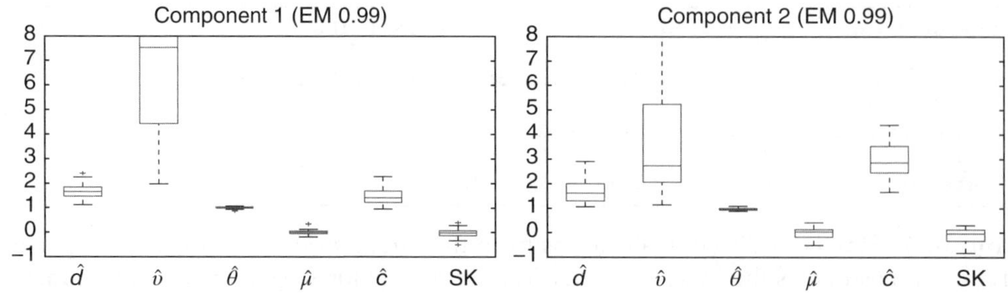

Figure 14.6 Truncated boxplots of the fitted GAt parameters of the $d = 30$ return series in the first (left) and second (right) component. Parameter d has nothing to do with our use of d for the dimension of the data, 30 in our case.

Based on the split into the two components, we will estimate, for each of the $d = 30$ univariate series in each of the two components, a flexible, asymmetric, fat-tailed distribution (that nests the normal as a limiting case), and inspect the parameters to learn about the univariate margins. For this, we use the GAt distribution (III.A.124). Figure 14.6 shows the (truncated) boxplots of the five GAt estimated parameters over the $d = 30$ time series, along with the sample skewness.

For the first component, the sample skewness is virtually centered around zero and has a much lower variation than those for the second component, indicating that we can assume symmetry in the marginals for the first component. For both components, the estimated value of the asymmetry parameter θ barely deviates from unity, lending support that the asymmetry exhibited in the asset returns is well-explained by using two symmetric components in a mixture distribution.

The scale terms for the first component are, as expected, much lower than those in the second component. In addition, while the values of $\hat{\upsilon}$ (the tail thickness parameter, with $\upsilon = \infty$ corresponding to exponential tails as with the normal and GED distributions) in the first component are, on average, quite high, and far higher than $\hat{\upsilon}$ for the second component, some of those 30 values are still rather small, the smallest, corresponding to the stock returns of McDonald's corporation, being 1.98.[5] This fact adds considerable weight against the multivariate normality hypothesis for each of the two components, though there are very few stocks such as McDonald's that have such aberrant behavior, and so ending the story here would be premature.

To investigate this further, consider the following heuristic procedure. For each of the d series, but *not* separating them into the two components, we fit the GAt, first with no parameter restrictions (other than those required by the parameter space of the distribution), and second, with the restriction that $90 < \hat{\upsilon} < 100$, which essentially forces normality if GAt distribution parameter $d = 2$ and $\theta = 1$, or Laplace if $d = 1$ and $\theta = 1$, though it is important to emphasize that $\hat{d}$ and $\hat{\theta}$ were not constrained in this way.[6] Then, we compute the asymptotically valid p-value of the likelihood ratio test. If that value

---

5 The maximally existing moment of the GAt is bound above by υd. In this case, $\hat{d}$ is 2.41, so that $\hat{\upsilon}\hat{d} = 4.8$, and this is also the stock with the lowest such product. Recall from Chapter III.9 that this does *not* imply an estimate for the supremum of the maximally existing moment of 4.8 because of the flawed nature of using a parametric model for determining the maximally existing moment.

6 For each estimation, several different starting values were used to help ensure the global maximum was found. In particular, we used as starting values $\hat{d} = 1.4$, $\hat{\theta} = 0.98$, $\hat{\mu} = 0$, $\hat{c} = 3$, and that for $\hat{\upsilon}$ was chosen from an equally spaced grid of 10 points from its lowest possible value (we used 0.5 in the unrestricted, and 90 in the restricted) to its highest allowed value of 100. Doing this made a difference in about 10% of the entries in the table, confirming that multiple maxima of the likelihood are possible for this model.

```
cut=0.05; comp0outliers=zeros(30,1);
for stock=1:30, stock
  y=data(:,stock); pval=0; remove=-1;
  while pval<cut
    use=y; remove=remove+1;
    for i=1:remove
      loc=find(abs(use)==max(abs(use))); use=[use(1:loc-1) ; use(loc+1:end)];
    end
    [param,stderr,iters,loglikUNR] = GAtestimation(use,0.5);
    [param,stderr,iters,loglikRES] = GAtestimation(use,90);
    stat=2*(loglikUNR-loglikRES); pval = 1-chi2cdf(stat,1);
  end
  comp0outliers(stock)=remove;
end
```

Program Listing 14.5: Removes "outliers" until the p-value from the likelihood ratio test exceeds 0.05.

is less than 0.05, we remove the largest value (in absolute terms) from the series, and re-compute the estimates and the p-value. This is repeated until the p-value exceeds 0.05, and we report the smallest number of observations required to be removed in order to achieve this. The Matlab code to perform this calculation is shown in Listing 14.5, assuming the returns are stored in a $T \times d$ matrix `data`. (The second parameter passed to `GAtestimation` is the lower limit on $\hat{v}$.)

The results are given in Table 14.1, in the rows labeled "All". The other rows are the same, but having used the observations allocated to components 1 and 2. Thus, for example, stock number 5 (Bank of America) is such that the 65 most extreme values had to be removed from the series to get the p-value above 0.05, but *no observations* from component 1 needed to be removed, and only three from component 2. Except for stock numbers 6 (Boeing) and 22 (McDonald's), either zero or one outlier, or two (in two cases), had to be removed from the first component.

While this is a heuristic method with unknown theoretical performance and arbitrarily chosen significance level 0.05, it does provide some evidence that a mixture of two normal distributions can

Table 14.1 Number of observations required to be removed until the likelihood ratio test comparing GA*t* and normality does not reject at the 0.05 level.

Stock #	1	2	3	4	5	6	7	8	9	10	11	12	13	14	15
All	19	19	5	2	65	21	11	27	60	10	30	19	28	31	8
Comp1	0	0	0	0	0	3	0	0	1	1	1	0	0	0	0
Comp2	0	1	0	0	3	1	1	6	8	2	0	8	0	1	1
Stock #	**16**	**17**	**18**	**19**	**20**	**21**	**22**	**23**	**24**	**25**	**26**	**27**	**28**	**29**	**30**
All	7	12	8	5	19	7	15	28	3	14	10	11	6	8	10
Comp1	0	0	2	2	0	0	7	0	0	0	0	0	0	0	0
Comp2	0	0	0	1	0	0	1	2	0	2	1	3	0	2	1

account for nearly, but not all, of the leptokurtic behavior in the returns, as well as the asymmetry, and, as already mentioned, has the advantage over other asymmetric, fat-tailed multivariate distributions in that the two components can capture highly distinct behavior that would otherwise have to be averaged over when using only a single component.

Remark There is another advantage of a mixture of normals compared to use of a single-component multivariate distribution with one or several additional shape parameters, such as the MGHyp and all its special cases, the MVNCT, etc. As discussed in Chapter 12, for these distributions the shape parameters (such as the degrees of freedom parameter in the Student's t) dictate the thickness of the tails, common to all d univariate dimensions. Based on the results in the previous table, this would be too restrictive. For example, with stock number 4 (AT&T), only two extreme values needed to be removed to induce thin-tailed behavior, while Bank of America required 65.

To further substantiate this, recall Figure 12.1, showing the (sorted) values of the estimated degrees of freedom parameter corresponding to the univariate location-scale Student's t distribution, for each of the $d = 30$ return series (the entire series, and not split into two components), along with approximate 95% confidence intervals computed via the nonparametric bootstrap. Of interest is if the v_i, $i = 1, 2, \dots, 30$, can be deemed equal, as would be required for the multivariate Student's t distribution. This is clearly untenable.

As mentioned above, the Mix_2N_d distribution does not formally exhibit tail dependence and does not have a tail index: It is a thin, or short-tailed distribution, so that the tail behavior of each of the d dimensions is also the same. However, via the mixture and its two sets of location and dispersion parameters, the margins are leptokurtic and can "mimic" heterogeneous tail behavior, *but only up to a point*, usually adequate for the actual range of the data, but eventually, the tail behaviors of the margins are all the same, and are thin-tailed, Gaussian.

Another way of addressing this issue, and allowing for each margin to have its own tail thickness parameter is via use of a copula structure, as with the (A)FaK distribution in Chapter 12. ∎

We now consider the McDonald's results in more detail, this being the worst-fit case. Figure 14.7 shows the unrestricted and restricted fitted densities (top panels) corresponding to the first and second components (left and right, respectively), and the fitted densities after having removed the seven (one) most extreme values from the first (second) component (bottom panels). The unrestricted and restricted densities are surprisingly close, with their differences only observable in the tails of the distribution. Once the extreme values are removed, the unrestricted and restricted densities are nearly indistinguishable.

In all cases, but particularly the top panels (for which no extreme values were removed), the value of GAt shape parameter $\hat{d}$ (not to be confused with the dimension of the multivariate data set, d) decreases substantially when going from the unrestricted to the restricted model. In particular, $\hat{d} = 1.4$ (1.2) for the first (second) component. This is because a lower value of d implies a higher kurtosis, and so it is able to offset the restriction that $\hat{v}$ is constrained to lie above 90. Recalling that $d = 1$ in the GAt (with $\theta = 1$ and $v \to \infty$) corresponds to the Laplace distribution, this motivates considering the use of a mixture of two multivariate Laplace distributions instead of multivariate normal. We return to this idea in Section 14.5.2.

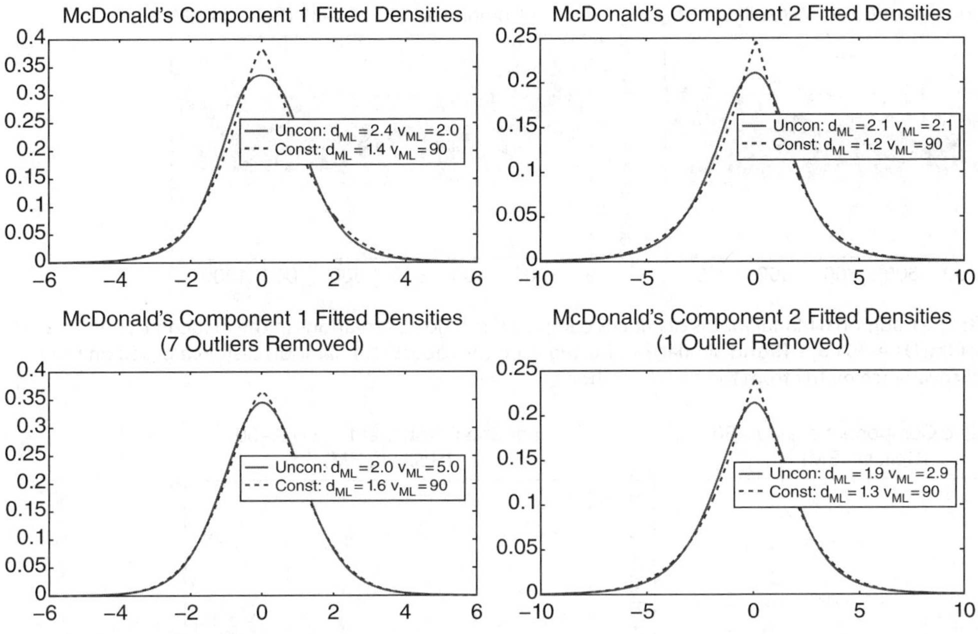

Figure 14.7 The first (left) and second (right) components of the McDonald's stock returns, with unrestricted and restricted GA*t* densities, without and with outlier removal.

14.2.3 Component Separation and Multivariate Normality

We now turn to one way of assessing the multivariate normality of the two components. The left panel of Figure 14.8 shows the Mahalanobis distances corresponding to the multivariate normal fit of the DJIA returns. (This is the same as Figure III.3.4.) With the ability to partition the data into the two components, we can construct the same graphic, but applied to each component. The left panel of Figure 14.9 shows this for the first component, in which case 14% of the observations lie

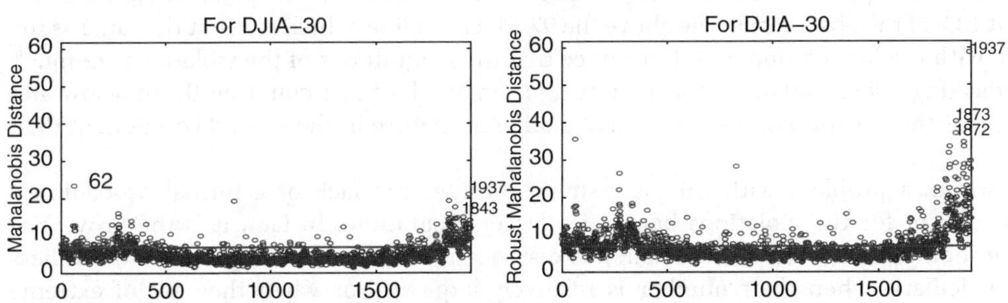

Figure 14.8 **Left:** The traditional Mahalanobis distances computed for the DJIA-30 returns, with 15% of the observations above the cutoff line. **Right:** Similar, but having used the robust Mahalanobis distance based on the mean vector and covariance matrix from the m.c.d. method, resulting in 33% above.

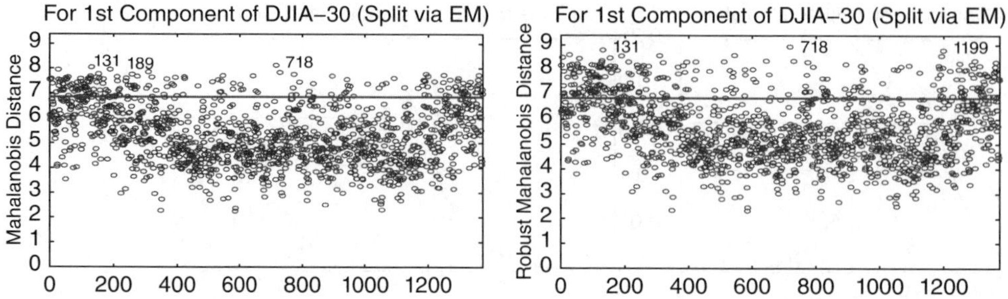

Figure 14.9 Left: The traditional Mahalanobis distances computed for the observations in the first component, based on the EM-split of the DJIA-30 data. **Right:** Similar, but having used the robust Mahalanobis distance based on the mean vector and covariance matrix from the m.c.d. method.

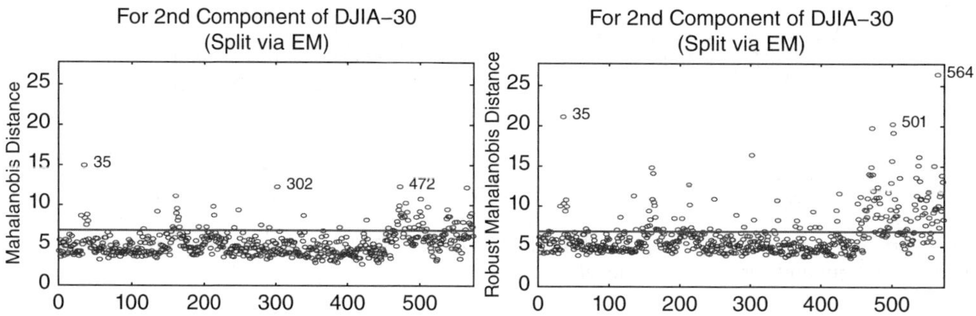

Figure 14.10 Similar to Figure 14.9 but for the second component.

above the 97.5% cutoff line, instead of 2.5%, as would be expected under the null hypothesis of i.i.d. normality. Thus, as also determined above via the previous univariate analysis, the first component is not multivariate normal, though it is noteworthy that the magnitudes of the Mahalanobis distances that exceed the cutoff are rather modest.

The same graphic but applied to the second component is shown in the left panel of Figure 14.10, and is such that 13% of the observations lie above the 97.5% cutoff line, which is about the same as the 14% associated with the first component. But notice that the magnitudes of the violations are much larger, thus indicating (albeit outside a formal testing framework of just counting those below and above the line) that the violation of normality is of a different nature in the second component than that in the first.

There is, however, a problem with this assessment, besides the lack of a formal probabilistic framework to account for the violations being of a larger magnitude. In fact, it is the proverbial elephant in the room: The use of the Mahalanobis distance is sensitive to the presence of outliers in the data, particularly when their number is relatively large and/or when they are of extreme magnitudes, because they have a strong and deleterious effect on the estimates of the mean and covariance matrix, with the pernicious effect of allowing the genuine outliers to mask themselves. To address this, we use the minimum covariance determinant, or m.c.d. method, discussed in Section III.3.1.3.

Similar to the Mahalanobis distance, outliers are identified in the m.c.d. method via the so-called **robust Mahalanobis distance** (hereafter, r.M.d.), given by

$$\mathrm{RMD}(\mathbf{y};\alpha) = \sqrt{(\mathbf{y}-\hat{\boldsymbol{\mu}}_r)'\hat{\mathbf{S}}_r^{-1}(\mathbf{y}-\hat{\boldsymbol{\mu}}_r)}, \tag{14.20}$$

where $\mathbf{y}$ is a d-dimensional column vector, and $\hat{\boldsymbol{\mu}}_r$ and $\hat{\mathbf{S}}_r$ are robust estimators of the location vector and scatter matrix, respectively.

The right panel of Figure 14.10 shows the result for the second component when using the m.c.d. robust estimators of the mean vector and covariance matrix. It indicates that 26% of the observations lie above the cutoff line, but also shows that the violations occur predominantly at the end of the time period. This period corresponds precisely with the massive drops and high volatility of the financial markets during the Global Financial Crisis. This robust estimator is clearly a superior tool in situations such as this, in which a substantial number of observations are present that deviate from the overall typical behavior and cause genuine outliers to be masked.

The right panel in Figure 14.9 shows the corresponding plot for the first component. The differences between the usual and robust Mahalanobis distances are rather small in comparison to those for the second component, though it results in 20% of the values exceeding the cutoff line. The right panel of Figure 14.8 shows the robust Mahalanobis distances (14.20) when computed for the returns themselves (and not split into two components), indicating that the extent of non-normality of the returns data is much stronger than what the traditional Mahalanobis distance indicates.

Taken altogether, we have considerable evidence that neither component, but particularly the second, is adequately modeled with a Gaussian distribution. One way of addressing this is to use a mixture of distributions whose components allow for leptokurtic behavior, such as the Laplace. This is done in Section 14.5.

14.2.4 Mixed Normal Weighted Likelihood and Density Forecasting

To apply the weighted likelihood scheme of Chapter 13 in this context, the tth term entering into the log-likelihood gets multiplied by its corresponding weight ϖ_t, $t = 1, \dots, v$. When using the EM algorithm for the $\mathrm{Mix}_k\mathrm{N}_d$ model, this direct implementation is not available. To accommodate this, several options were considered in Paolella (2015) and the following was found to be the best choice: Multiply each $\mathbf{Y}_t$ appearing in the mean updating equation (14.6) by ϖ_t, multiply each $\mathbf{Y}_t$ appearing in the variance updating equation (14.7) by ϖ_t, and multiply each $H_{t,j}$ in the component weight updating equation (14.6) by ϖ_t.

It turns out that most of the improvement comes from applying the weights to the $\hat{\boldsymbol{\Sigma}}_j$, whereas virtually no improvement is obtained from weighting the $\hat{\boldsymbol{\mu}}_j$. Interestingly, this is virtually the opposite result compared to the gains in forecasting performance attributable to the use of shrinkage, which improves the mean forecast significantly, but hardly affects the variance and covariance estimates; recall Figure 14.3. Thus, the use of shrinkage and weighted likelihood contribute to forecasting improvement in a nearly orthogonal fashion. That weighted likelihood in this context improves the estimates of $\hat{\boldsymbol{\Sigma}}_j$ (with respect to forecasting) relatively the most was to be expected, given the volatility clustering inherent in the data, and the fact that we do not account for it by, say, a GARCH-type law of motion for the volatility.

The Matlab implementation of the quasi-Bayesian EM algorithm for estimating the Mix_2N_d distribution with weighted likelihood is given in Listing 14.6. The default setup is to apply weighted likelihood only to the $\widehat{\boldsymbol{\Sigma}}_j$. The extension to the k-component case is straightforward, and the reader is encouraged to set it up for $k = 3$. Compared to the univariate case, the code is more involved.

```
function [mu1,mu2,Sig1,Sig2,lam,loglik,H1,crit,iter] ...
       = mixnormEMm (y,omega,init,rho,tol,maxit)
if nargin < 6, maxit=1e4; end, if nargin < 5, tol=1e-6; end
if nargin < 4, rho=1; end,      if nargin < 3, init=[]; end
if nargin < 2, omega=0; end
if length(rho)==1
  weightmeans=0; weightsigmas=1; weightlam=0;
else
  weightmeans=rho(2); weightsigmas=rho(3); weightlam=rho(4);
end
[n,p]=size(y);

% weighted likelihood
tvec=(1:n)'; likew=(n-tvec+1).^(rho(1)-1); likew=n*likew/sum(likew);

if 1==1 % based on typical financial data
  s1=1.5; cov1=0.6; s2=10; cov2=4.6; m1=zeros(p,1); m2=-0.1*ones(p,1);
else % arbitrary
  s1=1; cov1=0.0; s2=1; cov2=0.0; m1=zeros(p,1); m2=zeros(p,1);
end
psig1=zeros(p,p); psig2=zeros(p,p);
for i=1:p, for j=1:p %#ok<ALIGN>
  if i==j, psig1(i,j)=s1; else psig1(i,j)=cov1; end
  if i==j, psig2(i,j)=s2; else psig2(i,j)=cov2; end
end, end
a1=2*omega; a2=omega/2; c1=20*omega; c2=20*omega; B1=a1*psig1; B2=a2*psig2;

if isempty(init)
  mu1=m1; mu2=m2; Sig1=psig1; Sig2=5*psig2; lam=0.8;
else
  mu1=init.mu1; mu2=init.mu2; Sig1=init.Sig1; Sig2=init.Sig2; lam=init.lam;
end
```

Program Listing 14.6: Estimates the parameters of the Mix_2N_d distribution using the EM algorithm with quasi-Bayesian prior. (The code uses p instead of d for the dimension.) Input `y` is the $n \times d$ matrix of data. `omega` is the prior strength; pass 0 (default) for standard m.l.e., i.e., no prior information, or pass a positive value as the strength. `init` contains initial values as as structure, i.e., `init.mu1`, `init.mu2`, `init.Sig1`, `init.Sig2`, and `init.lam`. The default is []. `rho` indicates the weight for hyperbolic weighted likelihood, with a weight of 1 yielding equally weighted (usual) likelihood, and values less than 1 putting more weight on recent observations. Or pass vector `[rho weightmeans weightsigmas weightlam]` where the latter three are boolean values and dictate which of the parameters receive the weights. Default is to use only the $\boldsymbol{\Sigma}_i$. `tol` is the required tolerance for each parameter to assume convergence. `maxit` is the maximum allowed number of iterations before giving up. Continued in Listing 14.7.

```
wscheme=2; % See below how this is used.
iter = 0; crit=0; pdftol=1e-200; eigtol=1e-12;
new = [mu1 ; mu2 ; Sig1(:) ; Sig2(:) ; lam];
while 1
  iter=iter+1; old=new;
  if iter==1000, iter, end
  Sig1=(Sig1+Sig1')/2; Sig2=(Sig2+Sig2')/2; % sometimes off by a tiny amount

  [V,D] = eig(Sig1); dd=diag(D);
  if any(dd<eigtol), dd=max(dd,eigtol); D=diag(dd); Sig1=V*D*V'; end
  [V,D] = eig(Sig2); dd=diag(D);
  if any(dd<eigtol), dd=max(dd,eigtol); D=diag(dd); Sig2=V*D*V'; end

  Comp1=mvnpdf(y,mu1',Sig1); Comp1=max(Comp1,pdftol);
  Comp2=mvnpdf(y,mu2',Sig2); Comp2=max(Comp2,pdftol);
  mixn =lam*Comp1+(1-lam)*Comp2; H1=lam*Comp1./mixn; H2=1-H1;

  if weightmeans, G1=H1.*likew; G2=H2.*likew; else G1=H1; G2=H2; end
  if wscheme==1, N1=sum(G1); N2=sum(G2); else N1=sum(H1); N2=sum(H2); end
  rep1 = repmat(G1,1,p); rep2 = repmat(G2,1,p);
  mu1 = ( c1*m1 + sum( rep1 .* y )' ) / (c1 + N1);
  mu2 = ( c2*m2 + sum( rep2 .* y )' ) / (c2 + N2);

  if weightsigmas, G1=H1.*likew; G2=H2.*likew; else G1=H1; G2=H2; end
  if wscheme==1, N1=sum(G1); N2=sum(G2); else N1=sum(H1); N2=sum(H2); end
  rep1 = repmat(G1,1,p); rep2 = repmat(G2,1,p);
  ymm = y - repmat(mu1',n,1); ymmH = rep1 .* ymm; outsum1=ymmH'*ymm;
  Sig1 = (B1 + c1*(m1-mu1)*(m1-mu1)' + outsum1 ) / (a1+N1);
  ymm = y - repmat(mu2',n,1); ymmH = rep2 .* ymm; outsum2=ymmH'*ymm;
  Sig2 = (B2 + c2*(m2-mu2)*(m2-mu2)' + outsum2 ) / (a2+N2);

  if weightlam, G1=H1.*likew; else G1=H1; end
  lam = mean(G1);  new = [mu1 ; mu2 ; Sig1(:) ; Sig2(:) ; lam];
  crit = max (abs (old-new)); if (crit < tol) || (iter >= maxit), break, end
end
loglik=sum(log(mixn));
```

Program Listing 14.7: Continuation of Listing 14.6.

In particular, to avoid FOR loops (being much slower than using the internal vectorized Matlab functions), we make judicious use of the `repmat` function. This is also done in Matlab's `mvnpdf` function for the multivariate normal p.d.f., which is far faster for a large number of observations than computing the density in a direct, less sophisticated way.

14.2.5 Density Forecasting: Optimal Shrinkage

We can compute $\pi_t(\mathcal{M}_i, v)$ from (13.2), for a set of $\text{Mix}_2\text{N}_{30}$ models, estimated with shrinkage prior (14.8) with a given value of ω, denoting this by $\mathcal{M}_\omega$. We do this using a moving window of size $v = 250$, starting at observation $\tau_0 = v = 250$, and updating parameter vector $\widehat{\theta}$ at each time increment ($o = 1$; $\zeta = t - 1$).

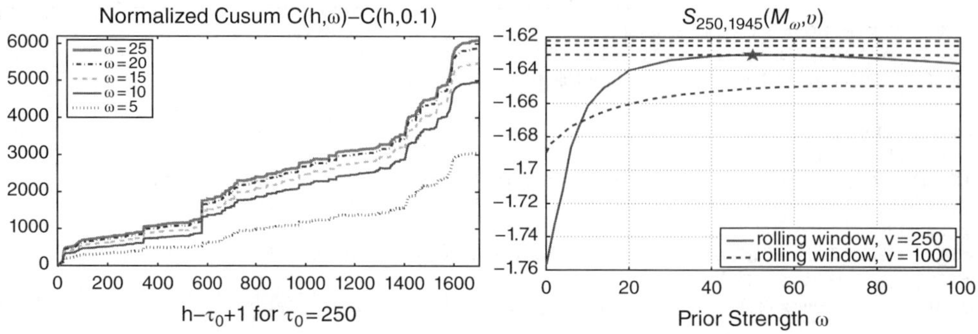

Figure 14.11 **Left:** Plot of the standardized cusum $C(h, \omega) - C(h, 0.1)$ where C is given in (14.21), versus $h - \tau_0 + 1$, for $\tau_0 = 250$ and several $\omega = 5, 10, 15, 20, 25$. **Right:** The normalized sum of the realized predictive log likelihood (14.22) as a function of prior strength hyper-parameter ω, and based on estimation with a moving window of length $\upsilon = 250$ (solid line) and $\upsilon = 1{,}000$ (dashed line). For the latter, τ_0 is still 250, and we use the convention in (14.23). The star shows the best obtained value, corresponding to a prior weight of $\omega = 50$, and is the same star in both panels of Figure 14.16, while the top-most horizontal line is the same line in the right panel of Figure 14.16, showing the additional improvement from the methods discussed in Section 14.2.4.

The left panel of Figure 14.11 plots the cumulative sum (cusum) of the $\pi_t(\mathcal{M}_\omega, 250)$, normalized by subtracting the cusum of $\pi_t(\mathcal{M}_{0.1}, 250)$, that is, we plot $C(h, \omega) - C(h, 0.1)$, where

$$C(h, \omega) = \sum_{t=\tau_0+1}^{h} \pi_t(\mathcal{M}_\omega, 250), \quad h = \tau_0 + 1, \dots, T, \tag{14.21}$$

versus $h - \tau_0 + 1$, for $\omega = 5, 10, 15, 20, 25$ and $T = 1{,}945$.

The code to construct the plot is given in Listing 14.8. It is possible, if not likely, that there exist multiple plausible maxima of the likelihood function. Ideally, via use of many starting values, a local optimum would be located that is, with high probability, the global one. We instead use just two starting values, as follows: For a given window of observations, the first starting value is simply the final value obtained from the previous window. As these two data sets just differ by two observations, we expect the m.l.e.s from both of them to be close, so that this should be a very reasonable starting value. Nevertheless, it is possible (and occurs with nonnegligible frequency; use the `cnta` and `cntb` variables in the program in Listing 14.8 to see) that this leads precisely to an inferior local maximum. As such, our second starting value is the simple, default one used in the estimation program given in Listing 14.6.

The improvement in forecast accuracy is virtually monotonically increasing with both increasing h and increasing ω, providing very strong evidence that shrinkage estimation vastly outperforms the use of the m.l.e. in this context. In fact, the gains from using the shrinkage estimator compared to the m.l.e. are higher than indicated in Figure 14.11 because we used the benchmark model $\mathcal{M}_{0.1}$ instead of the m.l.e. $\mathcal{M}_0$. This was done because the former is numerically far more reliable than use of no prior information, which occasionally settles on a singularity in the mixture likelihood. What is not clear is if the overall gain, $C(T, 25) - C(T, 0.1) = 6{,}098$ (the top of the graph), is "significant" in some sense. The answer depends on the application and how the forecasts are to be used. If, for example,

```
[n,p]=size(data); omegavec=[0.1 5:25]; omegalen=length(omegavec);
win=250; up=n-win; logpdf=zeros(up,1);
for oloop=1:omegalen
  omega=omegavec(oloop); init=[]; cnta=0; cntb=0;
  for start=1:up, if mod(start,100)==0, start, end
    use=data(start:(start+win-1),:); % data for estimation
    y1=data(start+win,:); % what actually happens tomorrow
    %%%%%%%%%%%%%%%%%%%%%%%%ESTIMATE %%%%%%%%%%%%%%%%%%%%%%%%%%%%%%%%%%%%
    [mu1a,mu2a,Sig1a,Sig2a,lama,lla] = mixnormEMm (use,omega,init);
    [mu1b,mu2b,Sig1b,Sig2b,lamb,llb] = mixnormEMm (use,omega,[]);
    if lla>llb % keep track of how often this happened
      cnta=cnta+1; mu1=mu1a; mu2=mu2a; Sig1=Sig1a; Sig2=Sig2a; lam=lama;
    else
      cntb=cntb+1; mu1=mu1b; mu2=mu2b; Sig1=Sig1b; Sig2=Sig2b; lam=lamb;
    end
    Sig1=(Sig1+Sig1')/2; Sig2=(Sig2+Sig2')/2; % occasionally needed!
    % use these parameter estimates as starts for the next window
    init.mu1=mu1; init.mu2=mu2; init.Sig1=Sig1; init.Sig2=Sig2; init.lam=lam;
    % compute the realized predictive log likelihood
    Comp1=mvnpdf(y1,mu1',Sig1); Comp2=mvnpdf(y1,mu2',Sig2);
    mixn =lam*Comp1+(1-lam)*Comp2;
    %%%%%%%%%%%%%%%%%%%%%%%%%%%%%%%%%%%%%%%%%%%%%%%%%%%%%%%%%%%%%%%%%%%%%%%
    logpdf(start)=log(mixn); % store the realized predictive log likelihood
  end
  eval(['logpdf',int2str(omega),'=logpdf;'])
end
vv=1:(length(data)-win); figure, hold on
plot(vv,cumsum(logpdf25)-cumsum(logpdf0),'m-','linewidth',3)
% further, similar plot commands
hold off
```

Program Listing 14.8: Constructs the left panel of Figure 14.11, assuming the returns are collected inmatrix `data`. Code between the big comment lines is also used in Listing 14.9.

they are used for determining the weights in a financial portfolio, then one natural measure would be the increase in average return for a given level of (some measure of) risk.[7]

For now, we will normalize the value $C(T, \omega)$ by dividing it by the number of time points used in the sum, $T - \tau_0$ (in this case 1,695), as was done in (13.3), and also by the dimension of the random variable under study, in this case, $d = 30$. This facilitates comparison for different values of τ_0 and d. That is, for a given model $\mathcal{M}$ and window size v, we take the **normalized sum of the realized predictive log-likelihood** to be

$$S_{\tau_0,T}(\mathcal{M}, v) = \frac{1}{(T - \tau_0)d} \sum_{t=\tau_0+1}^{T} \pi_t(\mathcal{M}, v), \tag{14.22}$$

where d is the dimension of the data. It is thus the average realized predictive log-likelihood, averaged over the number of time points used and the dimension of the random variable under study.

7 The clear bottom line in finance is, from the viewpoint of statistics, quite welcome because it provides a very explicit objective function and method for comparison that most everybody agrees upon. Of course, this might also be deemed distasteful: The expression sometimes used in the financial industry, CIMITYM, is a good case in point. It stands for: Cash Is More Important Than Your Mother.

The right panel of Figure 14.11 plots $S_{250,1945}(\mathcal{M}_\omega, 250)$ (solid line) as a function of ω, the weight of the shrinkage prior, from which we can see that the optimal amount of shrinkage based on $v = 250$ is, say, $\omega^*(250) = 50$. The plot also shows $S_{250,1945}(\mathcal{M}_\omega, 1000)$, i.e., having used a moving window of size $v = 1{,}000$, and $\omega^*(1000) \approx 65$. Observe that $\tau_0 < v$, which, formally, does not make sense. We use the convention that, for computing $\pi_t(\mathcal{M}, v)$, we use the previous

$$\min(v, t-1) \tag{14.23}$$

observations. So, for example, with $\tau_0 = 250$ and $v = 1{,}000$, at observation $t = 251$, we use the past 250 observations, at $t = 252$, we use the past 251 observations, etc., up to $t = 1{,}000$, for which we use the past 999 observations; for $t \geqslant 1{,}001$, we use the previous $v = 1{,}000$ observations in the information set I_{t-1}. If we instead had used $\tau_0 = v = 250$ and $\tau_0 = v = 1000$ for the two cases shown in the right panel of Figure 14.11, then comparison would still be possible because they are standardized by dividing by $T - \tau_0$, but it would be less desirable because the realized predictive log likelihoods for $t = 251, \dots, 1000$ would be omitted from the $v = 1{,}000$ case. If the d.g.p. is changing through time, and that period was, say, relatively more difficult to estimate than later periods, then the comparison would be biased.

The code to implement this is given in Listing 14.9; note that, to save space, the actual lines of estimation code that are identical to those in Listing 14.8 are omitted.

From the right panel of Figure 14.11, we immediately see three facts, the first two of which are well-known and intuitive, the third less so:

1) When using the m.l.e. ($\omega = 0$ in the plot), use of a larger sample size v for estimation (in this case 1,000 versus 250) leads to improvement in the density forecasts.
2) The effect of shrinkage (or the prior in a Bayesian method) decreases as the sample size v increases.
3) When using $\omega^*(v)$, the optimal amount of shrinkage for a given sample size v, the quality of the density forecasts are not necessarily better as v increases.

When we take these three observed facts together, it might seem somewhat puzzling, if not disturbing, that forecast accuracy improves so much by using a shrinkage prior, and, when using it, the accuracy gets *worse* as the window size is increased. *This is because the assumed d.g.p. is wrong.* One

```
1  [n,p]=size(data); omegavec=[0.1 1 3:3:75 80:4:120]; omegalen=length(omegavec);
2  spll=zeros(omegalen,1); winstart=250; winsize=1000; logpdf=zeros(n-winstart,1);
3  for oloop=1:omegalen
4    omega=omegavec(oloop); init=[]; cnta=0; cntb=0;
5    for track=(winstart+1):n
6      firstpoint=max(1,track-winsize);
7      use=data(firstpoint:(track-1),:); y1=data(track,:);
8      %%%%%% CODE FOR ESTIMATION SAME AS IN PREVIOUS PROGRAM %%%%%
9      logpdf(track-winstart)=log(mixn);
10   end
11   spll(oloop)=sum(logpdf);
12 end
13 norm1000=spll/(n-winstart)/p;
```

Program Listing 14.9: Similar to Listing 14.8 but allows for differing values of τ_0 (`winstart`) and size of rolling window v (`winsize`). The coded referred to in line 8 is from Listing 14.8.

blatant reason it is wrong is because we assume an i.i.d. model, but the volatility is time-varying and somewhat persistent ("mild" and "wild" observations tend to be followed by another mild or wild observation, respectively). Moreover, from the right panel of Figure 14.10 we also see that long "crisis" periods can occur. Both of these observations have in common the feature of persistence, so that use of an i.i.d. model with a small window could be reasonably accurate.

> The problem with using too short a window is that the estimated parameter vector $\widehat{\theta}$ will have a high variance, while using too long a window will cause $\widehat{\theta}$ to be biased. Thus, we are in the all–too-common situation of facing a **bias-variance tradeoff**, and seek the window length that finds the optimal tradeoff between them.

In light of this, we could entertain finding the optimal window size, say v^*, given by

$$v^* = \arg\max_{v} S_{\tau_0,T}(\mathcal{M}_{\omega^*(v)}, v), \tag{14.24}$$

which, in this case, would be somewhere between $v = 1$ and $v = 1{,}000$. (In general, we require at least $k(d+1)$ observations to estimate a Mix_kN_d model, but because we use the shrinkage prior (14.8) with $\omega > 0$, we can use less than this number of observations and still not land on a singularity in the likelihood.) Figure 14.12 illustrates the idea by showing the normalized sum of the realized predictive log-likelihood (14.22) versus window size v, for three values of hyper-parameter ω and two values of τ_0, 100 (left) and 600 (right). From the right panel, we see that use of $\omega = 50$ dominates the other two values of ω for all v, and that, irrespective of ω (at least for the three values considered), the optimal choice of v is between 250 and 275. Comparison with the left panel shows that the effect of the choice of prior strength has a far greater impact on the performance when using a smaller τ_0, even though the density forecasts for observations beyond $t = 600$ in the time series are the same for both values of τ_0.

Comparing the two graphs corresponding to $\omega = 50$ shows that (14.22) is lower for the $\tau_0 = 100$ case for all v, in particular, $v = 100$. Recalling that (14.22) standardizes by the number of π_t in the sum, this indicates that the density prediction of observations $t = 101, \ldots, 600$ was, relative to the remaining observations, less successful.

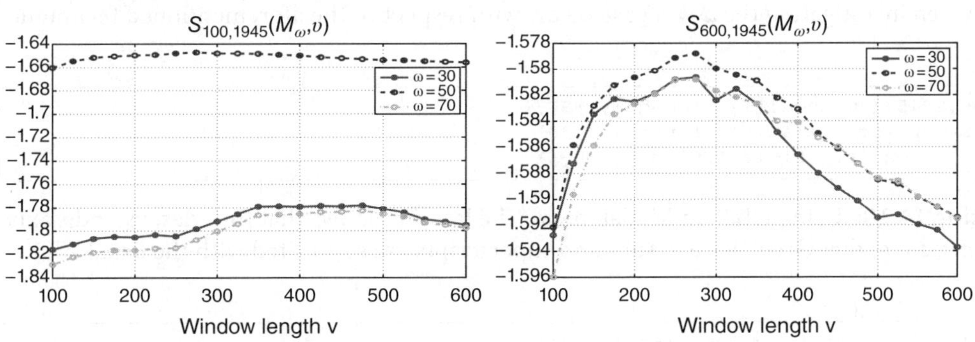

Figure 14.12 Both panels show the normalized sum of the realized predictive log-likelihood (14.22) as a function of moving window size v, $v = 100, 200, \ldots, 600$, for three values of prior strength hyper-parameter ω. The left uses $\tau_0 = 100$, while the right uses $\tau_0 = 600$. In the left panel, the plot for $\omega = 50$ (the one at the top) has the same shape as the corresponding one in the right panel when the plot is magnified, with its maximum also at $v = 275$.

Remarks

a) It is a curious coincidence that use of one year of daily trading data is essentially optimal. Even more interestingly, this is precisely the quantity of data often used in financial institutions, partly because of regulatory requirements; see, e.g., the Bank for International Settlements: Basel Committee on Banking Supervision (2009). Perhaps there is indeed more of an enlightened basis to their decision than just the natural human appeal of using precisely one calendar year, or perhaps it is *because* so many people instinctively prefer the use of one year that it becomes a self-fulfilling prophecy, though we will not concern ourselves with this speculation. Either way, it is only optimal after we apply (a sizeable amount of) shrinkage; without it, as we saw from Figure 14.11, it is not true, and using more data is better.
b) Somewhat less intellectually intriguing, to get full LaTeX equations into the title of the plot in Matlab, use the code in Listing 14.10. For the legend in the plot, the `interpreter` option is not supported by Matlab's legend command, but the statements in Listing 14.11 will do the trick. ■

In all the empirical analysis up to this point, we used all $d = 30$ series that constitute the DJIA. However, it is of obvious interest to investigate the performance using different values of d. With $d = 4$ and using the first four series in the DJIA, namely 3M Company (MMM), Alcoa Inc. (AA), American Express Company (AXP), and AT&T Inc. (T), the left panel of Figure 14.13 shows that a window size of about $\upsilon = 150$ is optimal, and for which $\omega^*(150) \approx 10$. Similarly, we take $p = 8$, adding the next four stocks to the set (Bank of America Corporation (BAC), Boeing Company (BA), Caterpillar Inc. (CAT), and Chevron Corporation (CVX)). Again, $\upsilon = 150$ is preferred, with $\omega^*(150) \approx 20$.

This would appear to significantly temper our previous comment regarding the optimality of using approximately one year of daily trading data. However, in Section 14.2.6 when we use weighted likelihood and moving averages of λ, we will find that use of $\upsilon = 250$ is still better than $\upsilon = 150$; see Figure 14.17. The reason for this is very appealing: While there *is* more information in a window of 250 observations than there is with 150, emphasis (via weighted likelihood and moving averages of λ) needs to be placed on more recent observations. Without this, the $\upsilon = 150$ case will outperform the $\upsilon = 250$ case simply because relatively less valuable observations have been removed from the window.

From the plots in Figure 14.13, $\omega^*(250) \approx 12$ for $d = 4$, while for $d = 8$, $\omega^*(250) \approx 22$. We will use the former result when investigating the $d = 4$ case below with respect to the aforementioned technique

```
str ='$S_{250,1945}(\mathcal{M}_{\omega},v)$';
title(str,'interpreter','latex','fontsize',20)
% basic LaTeX is easier: title('\Sigma_{\mu}')
```

Program Listing 14.10: Using LaTeX in Matlab plots. The last line shows that (in titles, legends, axis labels, text commands) that Greek letters, sub- and superscripts are supported with the usual LaTeX.

```
h=legend('','','','','','Location','SouthEast');
str={'$v=250$','$v=200$','$v=150$','$v=100$','$v=50$'};
set(h,'String',str,'interpreter','latex','fontsize',20)
```

Program Listing 14.11: Using LaTeX in Matlab legends in graphics.

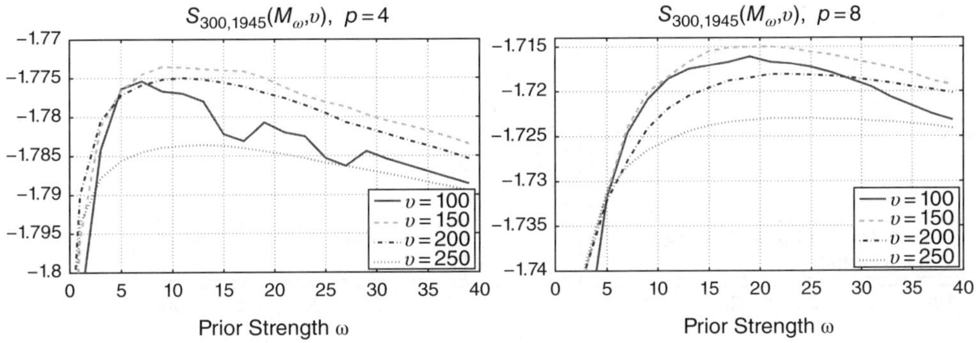

Figure 14.13 Similar to the right panel of Figure 14.11, except that here the left panel shows the results for only $d = 4$ assets (and four values of υ), while the right is for $d = 8$. Observe that $\tau_0 = 300$ for all four window sizes, so that the density predictions were based on precisely the same data points, namely $t = \tau_0 + 1 = 301$ to $t = T = 1{,}945$.

of using weighted likelihood and moving averages of λ. However, these optimal values of ω for different values of d contain further information that behooves us to investigate the situation more closely. In particular, we might conjecture that $\omega^*(250)$ is increasing in d. To investigate this, we conducted this exercise with all seven non-overlapping subsets of $d = 4$ series (omitting series 29 and 30). Even better, we do this for all 15 subsets of $d = 2$, all 10 subsets of $d = 3$, etc., for $d = 2, 3, \dots, 10$, as well as $d = 15$, and also $d = 20$ (using series 1 through 20, and 11 through 30, so that, in this case, there is overlap), $d = 25$ (series 1 through 25), and all $d = 30$ series. For each data set, values $\omega = 1, 2, \dots, 60$ were tried.

The results are shown in Figure 14.14. The left panel shows the value of $\omega \in \{1, 2, \dots, 60\}$ that yielded the highest values of the attained normalized sum of the realized predictive log-likelihood (14.22), denoted $\omega^*(250)$, as well as the fitted regression line (in d and d^2). The right panel shows the corresponding maximal values of the predictive log-likelihoods (14.22). Recalling that (14.22) divides by d, the values are comparable; we see that, as d increases, the quality of the forecasts tends to increase, and also the variability decreases. This indicates that, at least with respect to predictive log-likelihood, more assets are better than less, which is intuitively what we would expect (given that they are all correlated), but is diametrically opposed to what is found in practice, with respect to portfolio construction, using conventional models; see, e.g., DeMiguel et al. (2009b) and the references

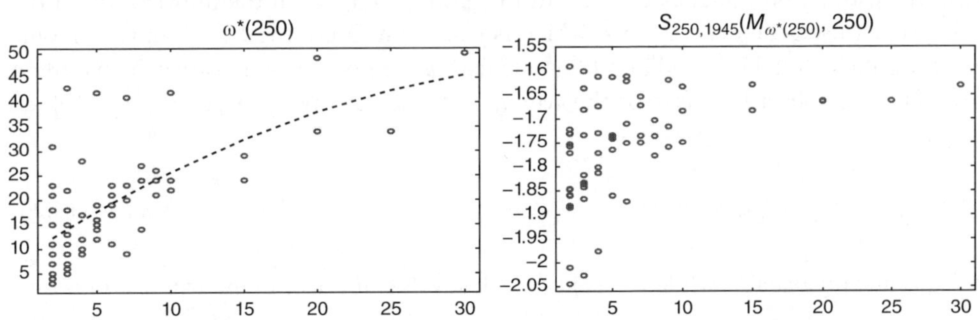

Figure 14.14 The optimal value of ω (left) and the corresponding values of the attained normalized sum of the realized predictive log-likelihood (14.22) (right), for various subsets of the DJ-30 assets under study.

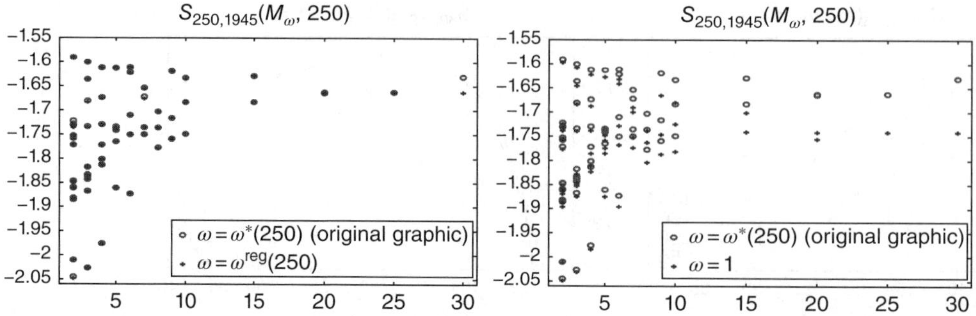

Figure 14.15 **Left:** Overlays same plot in the right panel of Figure 14.14, and additionally shows, as crosses, the result when taking ω to be from the regression line depicted in the left panel of Figure 14.14, i.e., round($\mathbf{X}\hat{\boldsymbol{\beta}}$), where $\mathbf{X} = [1, d, d^2]$ and $\hat{\boldsymbol{\beta}} = [8.4272, 1.9604, -0.0240]'$. The resulting values based on $\omega^*(250)$ and $\omega^{reg}(250)$ are virtually identical, except for the $d = 30$ case. **Right:** Same as left, but based on a fixed value of $\omega = 1$.

therein. With respect to portfolio performance, the i.i.d. Mix_2N_d model has the potential to do well; recall Figure 11.7, which shows results based on use of the m.c.d. estimator.

The left panel of Figure 14.15 is the same as the right one in Figure 14.14, but uses the value of ω obtained from the aforementioned fitted regression line, denoted $\omega^{reg}(250)$. There is virtually no difference in quality, except for the $d = 30$ case. As a comparison, the right panel of Figure 14.15 is similar to the left, but uses the fixed value $\omega = 1$ for all the data sets. In this case, the differences are far more pronounced and they increase in d, as expected, given that $\omega^{reg}(250)$ increases in d.

14.2.6 Moving Averages of λ

Keep in mind the three most important aspects of real data analysis: compromise, compromise, and compromise.

(Ed Leamer, 1997, p. 552)

Recall Figure 14.4, showing the final values of $\hat{H}_{t,1}$, $t = 1, \ldots, 1{,}945$, output from the EM algorithm, based on the $\text{Mix}_2\text{N}_{30}$ model for the DJIA-30 dataset. Particularly from the right panel, it is apparent that the $\hat{H}_{t,1}$ are highly correlated, indicating that today's value of $\hat{H}_{t,1}$ might be a good predictor of tomorrow's. In the estimation schemes used up to this point, we ignored the information in the sequence $\{\hat{H}_{t,1}\}$ and just used the m.l.e. $\hat{\lambda}_1 = \hat{\lambda}$, which is just the mean of the $\hat{H}_{t,1}$ from (14.6). One natural suggestion would be to take $\hat{\lambda}$ used in $\mathcal{M}$ for calculating the predictive density of $\mathbf{Y}_t$ based on a rolling window of $v = 250$ observations to be the average of the latest, say, v_λ values of $\{\hat{H}_{\cdot,1}\}$, which we denote as $\hat{\lambda}_{v,v_\lambda}$, i.e.,

$$\hat{\lambda}_{v,v_\lambda} = v_\lambda^{-1} \sum_{t=v-v_\lambda+1}^{v} \hat{H}_{t,1}, \quad 1 \leqslant v_\lambda \leqslant v. \tag{14.25}$$

If $v_\lambda = v$, then $\hat{\lambda}_{v,v}$ is just the usual $\hat{\lambda}$, while $\hat{\lambda}_{v,1}$ is just the last value of $\{\hat{H}_{\cdot,1}\}$ in the window of observations.

The solid line in the right panel of Figure 14.16 shows $S_{\tau_0,T}(\mathcal{M}_{v_\lambda}, v)$, for $v = \tau_0 = 250$, as a function of v_λ, based on weighted likelihood parameter $\rho = 0.7$ (applied only to the $\hat{\boldsymbol{\Sigma}}_j$), and where $\mathcal{M}_{v_\lambda}$ is the

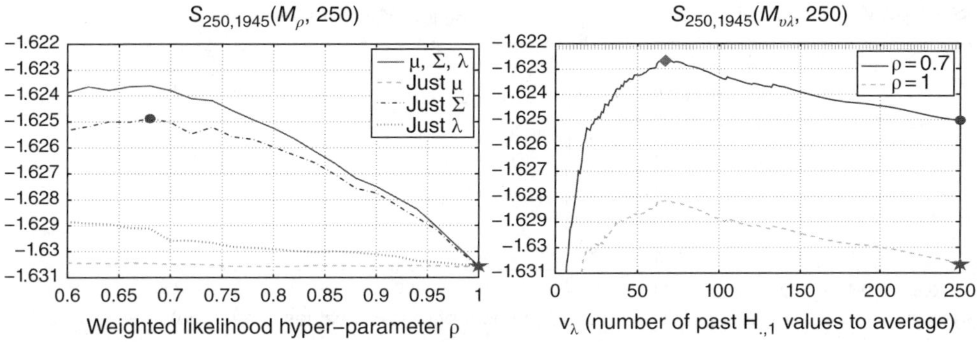

Figure 14.16 Left: Normalized sum of the realized predictive log-likelihood, for $\upsilon = 250$ and shrinkage hyper-parameter $\omega = 50$, as a function of hyper-parameter ρ, which controls the shape of the weights used in the weighted likelihood calculation. The star at the bottom right of the left and right panels is the same star shown in the right panel of Figure 14.11. **Right:** Normalized sum of the realized predictive log-likelihood, for $\upsilon = 250$ and shrinkage hyper-parameter $\omega = 50$, as a function of υ_λ, which dictates how many of the latest υ_λ values of $\{\widehat{H}_{\cdot,1}\}$ are averaged to form the value of $\hat{\lambda}$. The big circle in both plots represents the same value. Both plots have the same *y*-axis range, so that it is easy to see the improvement in using $\upsilon_\lambda = 70$ with weighted likelihood for $\rho = 0.7$ applied just to the $\widehat{\boldsymbol{\Sigma}}_j$, compared to using $\upsilon_\lambda = 250$ with weighted likelihood applied to all model parameters, including λ. Finally, the horizontal line at the top of the graph is the result of taking $\hat{\lambda}$ to be $0.75\hat{\lambda}_{250,70} + 0.25\hat{\lambda}_{250,1}$.

$\mathrm{Mix}_2\mathrm{N}_{30}$ model but such that $\hat{\lambda}$ is replaced by $\hat{\lambda}_{\upsilon,\upsilon_\lambda}$ in (14.25). The dashed line is similar, but corresponds to $\rho = 1$, showing that virtually the same amount of improvement is gained with this method, irrespective of ρ, and suggesting that the optimal value of υ_λ is practically independent of ρ. Indeed, we see that there are nearly monotone gains in forecast accuracy obtained as υ_λ is decreased, and a maximum is reached at about $\upsilon_\lambda = 70$, after which the quality drops off rather abruptly. As $\upsilon_\lambda \to 1$, i.e., as we approach the strategy of taking $\hat{\lambda}$ to be the last value of $\{\widehat{H}_{\cdot,1}\}$, the performance turns abysmal, and the graph was truncated. Thus, it appears that taking υ_λ corresponding to about 14 weeks of daily data is superior to use of the whole year.

The strong correlation among the $\widehat{H}_{t,1}$ apparent in Figure 14.4 would suggest that the previous day's value of $\widehat{H}_{t,1}$ should still somehow be of value. Some trial and error (and arguable possible indulgence in backtest overfitting) reveals that taking $\hat{\lambda}$ to be

$$\hat{\lambda}_{\mathrm{mix}} := 0.75\hat{\lambda}_{250,70} + 0.25\hat{\lambda}_{250,1} \tag{14.26}$$

results in further improvement, shown as the horizontal dotted line near the top of the graph in the right panel of Figure 14.16.

While the graphics in Figure 14.16 impressively display the increase in density forecast performance from using weighted likelihood with $\rho = 0.7$ and from using $\upsilon_\lambda = 70$, the right panel of Figure 14.11 shows that these gains are relatively small compared to what is achieved from shrinkage. In that plot, the horizontal dashed lines at the top show the incremental gains from the weighted likelihood for the $\widehat{\boldsymbol{\Sigma}}_j$ with $\rho = 0.7$, and the additional gain obtained by further taking $\hat{\lambda}$ to be (14.26). Thus, these additional tools provide only modest improvements relative to what is achieved with shrinkage. A more substantial improvement can be made by improving upon the normality assumption, as detailed in Section 14.5.2.

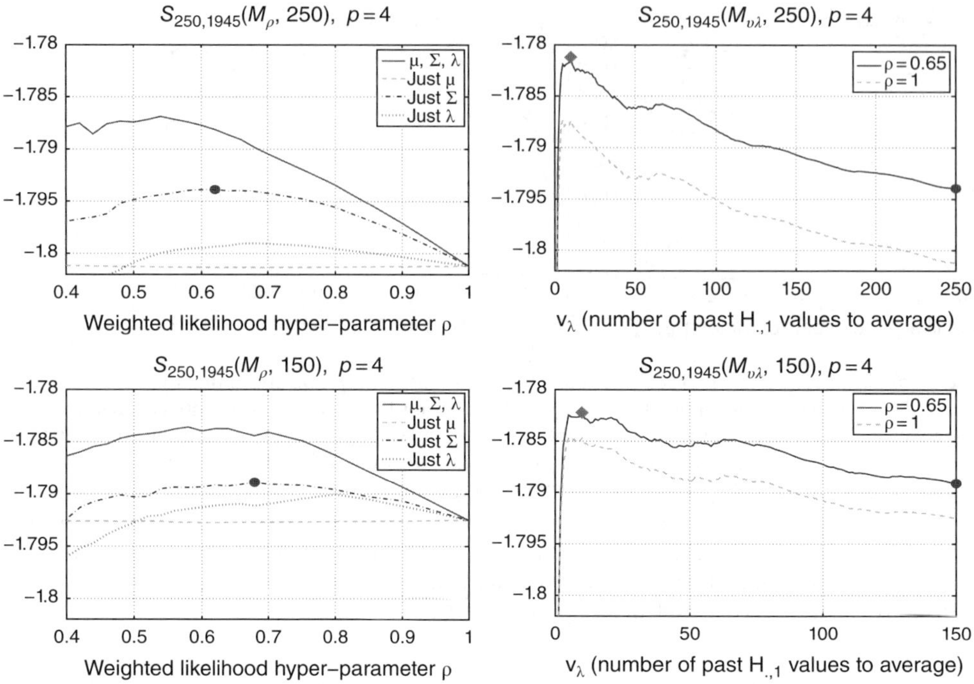

Figure 14.17 Similar to Figure 14.16 but using $d = 4$ assets instead of 30. Top is window size $\upsilon = 250$ and $\omega = 12$ (as ascertained from the left panel of Figure 14.13); bottom is window size $\upsilon = 150$ and $\omega = 10$. For both window sizes, $\tau_0 = 250$ so that the results are directly comparable.

At the end of Section 14.2.5, we also considered the case with $d = 4$ and $d = 8$ assets. Figure 14.17 is similar to Figure 14.16, but is for the $d = 4$ case, and having used two window sizes: $\upsilon = 250$ (with shrinkage parameter $\omega = 12$, in accordance with the results from the left panel of Figure 14.13) and $\upsilon = 150$ (with $\omega = 10$). The results are very similar qualitatively to the $d = 30$ case, particularly the optimal value of ρ, which we take to be 0.65. However, observe that the optimal value of υ_λ from (14.25) is 10, though interestingly we see from the right panels of Figure 14.17 that there is a local maximum at $\upsilon_\lambda = 70$, precisely the optimal value for the $d = 30$ case. Using values $0.75\hat{\lambda}_{\upsilon,70} + 0.25\hat{\lambda}_{\upsilon,1}$ and $0.75\hat{\lambda}_{\upsilon,10} + 0.25\hat{\lambda}_{\upsilon,1}$ gave results (not shown) that were very close to (and below) the optimal value shown in the right panels as the large diamond. The rapid decline of the quality as υ_λ decreases from 10 to 1 is alarming, so that use of $\upsilon_\lambda = 70$ might be a safer choice in practice. (A similar result was found using a window size of $\upsilon = 500$ observations—a global, and sharp, peak at $\upsilon_\lambda = 10$ and a local maximum at $\upsilon_\lambda = 70$.)

Finally, comparing the two cases of $\upsilon = 250$ and $\upsilon = 150$ (note that the scaling of the y-axis is the same in all four plots), we see that, while the $\upsilon = 150$ case is superior without weighted likelihood and just using the default of $\upsilon_\lambda = \upsilon$, the $\upsilon = 250$ case at its optimal value, with weighted likelihood and $\upsilon_\lambda = 10$, is better than the $\upsilon = 150$ case at its optimal value (albeit not by much). It is less a matter that $\upsilon = 250$ with weighted likelihood and $\upsilon_\lambda = 10$ is *better* than $\upsilon = 150$, it is enough that they are close: This indicates, as already briefly discussed in the previous section, that there is indeed more information about the model parameters in the last 250 observations than in the last 150, but because

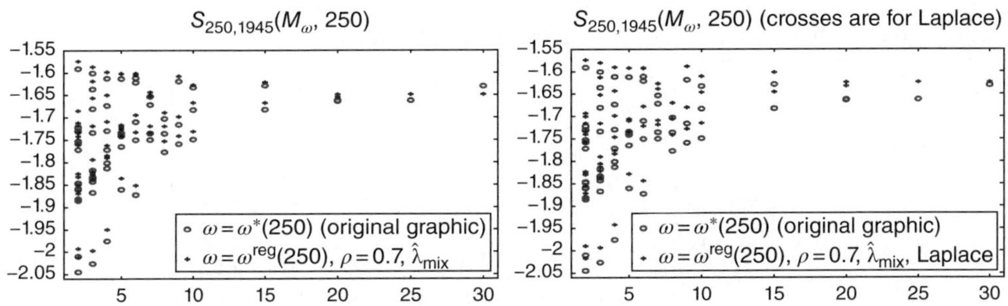

Figure 14.18 Left: Overlays same plot in the right panel of Figure 14.14, and additionally shows, as crosses, the result when taking ω to be from the regression line from the left panel of Figure 14.14 and additionally with (i) weighted likelihood, with $\rho = 0.7$ and just applied to the $\widehat{\boldsymbol{\Sigma}}_j$, and (ii) a moving average of λ from (14.26), i.e., $0.75\hat{\lambda}_{250,70} + 0.25\hat{\lambda}_{250,1}$. **Right:** Same as left, except that instead of the Mix_2N_d distribution, we use the Mix_2Lap_d distribution (14.41), introduced in Section 14.5.

the model is (without question in nearly every application) mis-specified, we need to weight more recent observations and information relatively heavier than those further in the past. This is precisely what weighted likelihood and use of (14.25) are accomplishing.

The left panel of Figure 14.18 illustrates the improvement from weighted likelihood and use of $\hat{\lambda}_{\text{mix}}$ in (14.26) for the same groups of assets used in Figures 14.14 and 14.15. The improvement is considerable for smaller d, and it appears the only case for which there is a decrease in performance is for $d = 30$, though this is not because of the weighted likelihood and use of $\hat{\lambda}_{\text{mix}}$, but rather because of the substantial difference with $d = 30$ between using $\omega^*(250)$ and $\omega^{\text{reg}}(250)$, as shown in the left panel of Figure 14.15. The right panel of Figure 14.18 shows the results when using the two-component mixture of Laplace distribution, instead of the normal mixture, as introduced below in Section 14.5.2. Its use bestows an improvement in forecast quality, particularly as d increases.

Example 14.1 Up to this point, we have only illustrated use of the Mix_2N_d model on the stock returns associated with the components of the DJIA-30. This was done using a prior (14.8) that was "loosely calibrated" to that data, and we found optimal choices from (i) weighted likelihood (13.1) with $\rho \approx 0.7$, (ii) use of window length $\upsilon \approx 250$, (iii) prior strength ω as a function of d given in Figure 14.14, and (iv) use of $\hat{\lambda}_{\upsilon,\upsilon_\lambda}$ in (14.25) with $\upsilon_\lambda \approx 70$ when $d = 30$.

Of interest is to see what happens when using a different data set of related type, namely stock returns from a highly developed, liquid stock market. Because such markets have common stylized facts, the hope is that the Mix_2N_d model is still applicable and the aforementioned tuning parameters are nearly optimal. For the prior (14.8), one could study the performance characteristics as a function of the set of associated fixed numeric values, though observe we did not do this with the DJIA-30 data: Presumably, one could improve the density forecasting performance by finding more suitable values, but this would be more of an exercise in backtest overfitting, as discussed in Section 13.3. As such, we just adopt prior (14.8) with the values stated there.

In this exercise, we use the returns on the $d = 20$ stocks associated with the Swiss Market Index (SMI-20), from November 10, 2000 to August 31, 2009, and investigate density forecasting performance. Figure 14.19 assesses this as a function of ω, ρ, and use of $\hat{\lambda}_{\upsilon,\upsilon_\lambda}$ for the SMI-20. We see that the optimal value of weighted likelihood parameter ρ (applied just to the $\widehat{\boldsymbol{\Sigma}}_j$) is between 0.6 and 0.7,

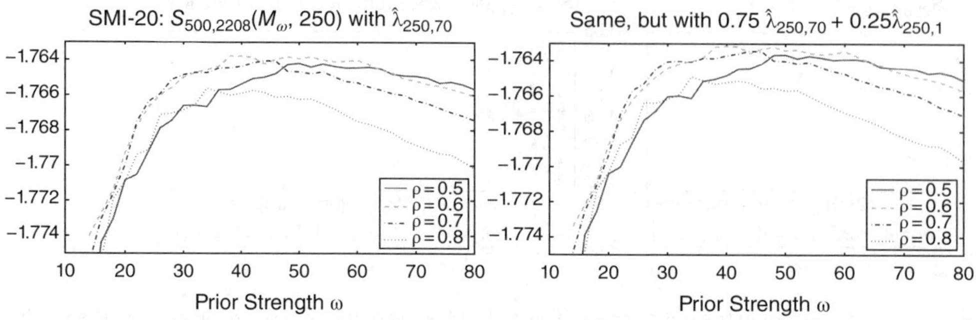

Figure 14.19 Left: The normalized sum of the realized predictive log-likelihood (14.22) for the $\text{Mix}_2\text{N}_{20}$ based on the 2,208 daily returns for $d = 20$ returns on the SMI stocks as a function of shrinkage hyper-parameter ω, for four values of weighted likelihood parameter ρ (applied just to the $\hat{\boldsymbol{\Sigma}}_j$), and based on moving windows of length $\upsilon = 250$, with $\tau_0 = 500$. The left uses a moving average of the estimated component weight λ $\hat{\lambda}_{\upsilon,\upsilon_\lambda}$ from (14.25) with $\upsilon_\lambda = 70$. **Right:** Similar, but uses (14.26).

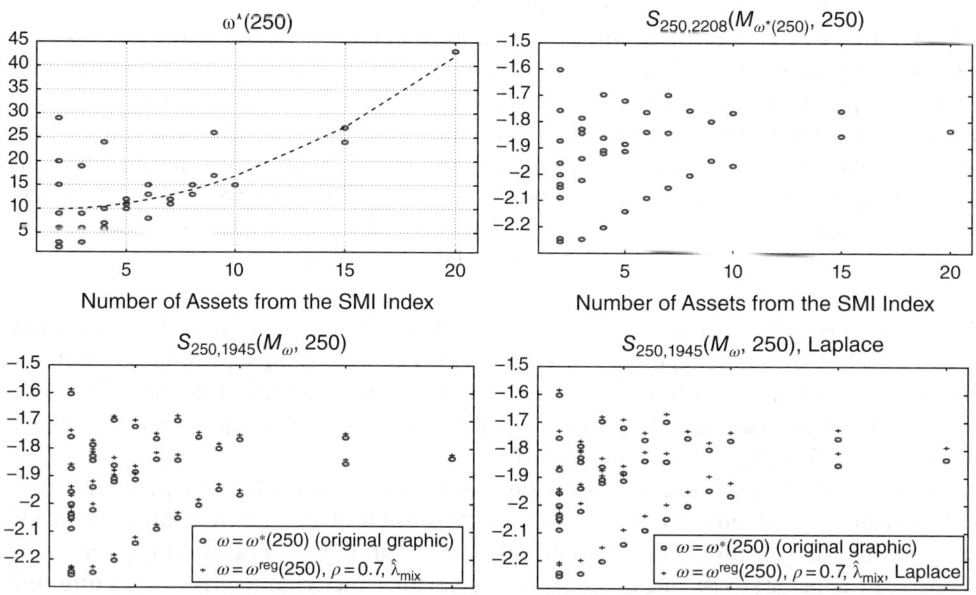

Figure 14.20 Top panels parallel those in Figure 14.14, but using the SMI-20 data (with the two points corresponding to 15 assets refer to stocks 1 through 15, and 6 to 20, so that they do contain overlap). The bottom panels are similar to those in Figure 14.18, showing the incremental improvement by using weighted likelihood and moving averages of λ (left) and by using the mixture Laplace distribution (right).

precisely in agreement with the corresponding value for the DJIA-30 data, as shown in Figure 14.16, and that use of the moving average (14.26) is superior to use of $\hat{\lambda}_{v,v_\lambda}$ from (14.25) with $v_\lambda = 70$. Finally, $\omega^*(20) = 38$, which, interestingly, perhaps coincidentally, is *precisely* the value obtained from the regression line shown in the left panel of Figure 14.14, referring to the DJIA-30 data.

Figure 14.20 shows the performance based on subsets of the SMI-20, and is similar to Figures 14.14 and 14.18. The results essentially mirror those obtained using the DJIA-30.

The reader is encouraged to explore other such markets (Nikkei-225, FTSE-100, DAX-30, S&P500, etc.), but also others such as less liquid emerging stock markets, commodities, foreign exchange, fixed income (via exchange traded funds), etc., with the hope that the model and the associated prior and tuning parameters found for the DJIA-30 are roughly optimal. ■

One should keep in mind a critique of our approach that we ignore, namely **survivorship bias**, as already briefly discussed in Section 11.2.2. We have used the definition of the DJIA constituents in 2009, and similarly for the SMI-20, and then obtained the daily (closing, dividend and split-adjusted) price data of those stocks. But, had we really been in the year 2000, the list will change as some companies potentially go bankrupt and exit (and new ones enter). The reason we allow ourselves to ignore this issue is that we are concentrating on the development of statistical methodology, though for real applications, survivorship bias needs to be addressed in genuine backtesting studies, as well as other issues such as transaction costs and related practicalities of trading.

14.3 MCD for Robustness and Mix_2N_d Estimation

The traditional purpose of the minimum covariance determinant method (m.c.d.) is to deliver estimates of $\boldsymbol{\mu}$ and $\boldsymbol{\Sigma}$ for a set of data that are (purported to be) i.i.d. from a multivariate normal distribution such that they are resistant to outliers, i.e., **robust**. We have already seen an application of m.c.d. above in Section 14.2.3, with Figures 14.9 and 14.10 showing its use and benefit for the robust Mahalanobis distance. Our goal is to consider using m.c.d. in a different way than initially intended, namely for estimation of the normal mixture model.

From Section 14.2.1, recall how we used the values $\{\hat{H}_{t,1}\}$, output from the EM algorithm, to split the $T = 1{,}945$ observations into two groups, which we then analyzed separately for assessing normality. It is of interest to see what happens if we use that method to separate the data, and then, to each of the two groups, fit a (single-component) multivariate normal distribution. What has to be decided is the cutoff value; recall above we used the criteria $\hat{H}_{t,1} > 0.99$. Below, we use $\hat{H}_{t,1} > c_{\text{EM}}$ for $c_{\text{EM}} = 0.99$, as well as 0.5 and 0.10.

Estimation of the multivariate normal distribution with just one component, using our usual shrinkage prior, just requires computing (14.6) and (14.7), with the H_{ij} being either zero or one. There is no iteration and the parameters are thus essentially instantaneously computed. Schematically,

$$\mathbf{Y} \to \boxed{\text{EM alg}} \to \boxed{H_{t,1} < c_{\text{EM}}?} \to \{\mathbf{Y}_t \mapsto \mathbf{Y}^{(1)}\}, \tag{14.27}$$

$t = 1, \ldots, T$, where $\mathbf{Y}^{(1)}$ denotes the collection of $\mathbf{Y}_t$ assigned to the first component, and likewise for $\mathbf{Y}^{(2)}$. Then, the estimate of λ is just the fraction of observations assigned to the first component (the mean of the Boolean r.v.s $\mathbb{I}(\hat{H}_{t,1} \geqslant 0.99)$, $t = 1, \ldots, T$), and estimates of the two location vectors and

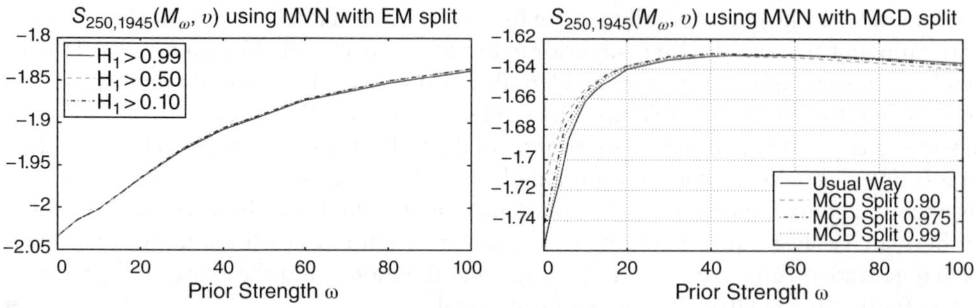

Figure 14.21 **Left:** Similar to the right panel of Figure 14.11, but such that the density forecasts were formed via (14.27). Note the difference in the y-axis compared to the right panel of Figure 14.11: Use of (14.27) and fitting two separate MVN distributions performs comparatively poorly. **Right:** Same, but based on the m.c.d. method of separation (14.29), for different values of tuning parameter α, with overlaid graph from the right panel of Figure 14.11, showing that use of the (prior-augmented) m.l.e. via the EM algorithm results in nearly the same as with use of the m.c.d. split method. In fact, the latter, at all three cutoff values, is slightly better for $\omega \leqslant 40$.

dispersion matrices are via (14.6) and (14.7). Schematically,

$$\{\mathbf{Y}^{(1)}, \mathbf{Y}^{(2)}, \hat{\lambda}\} \rightarrow \{\hat{\boldsymbol{\mu}}_i, \hat{\boldsymbol{\Sigma}}_i; \quad i = 1, 2\}. \tag{14.28}$$

The left panel of Figure 14.21 is similar to the right panel of Figure 14.11 (the normalized sum of the realized predictive log-likelihood (14.22) as a function of prior strength ω, based on a moving window of length $\upsilon = 250$) but, for each moving window of observations, having used the split (14.27) and estimation (14.28). Note that the scales of these two plots are quite different, with the method of individual component estimation based on splitting via the EM algorithm being vastly inferior to the original method, shown in Figure 14.11, which involved joint estimation, via the EM algorithm, of all the model parameters, $\boldsymbol{\mu}_1$, $\boldsymbol{\mu}_2$, $\boldsymbol{\Sigma}_1$, $\boldsymbol{\Sigma}_2$, and λ. This result conflicts with our otherwise quite favorable separation results indicated in Figure 14.4.

Recall the right panel of Figure 14.8, showing the robust Mahalanobis distance of each $\mathbf{y}_t$, based on the m.c.d. estimator of the covariance matrix. It offers another way of splitting the data.

The only tuning parameter required is α, for the cutoff value $c(\alpha, d) = F^{-1}_{\chi^2}(\alpha, d)$, the 100α percentile of the χ^2_d distribution, with d the dimension of each $\mathbf{Y}_t$ (in our case, 30). In the right panels of Figures 14.9, and 14.10, the cutoff value was based on $\alpha = 0.975$. Schematically,

$$\mathbf{Y} \rightarrow \boxed{\text{MCD alg}} \rightarrow \boxed{\text{RMD } (\mathbf{Y}_t; \alpha) < c(\alpha, d)?} \rightarrow \{\mathbf{Y}_t \mapsto \mathbf{Y}^{(1)}\}, \tag{14.29}$$

$t = 1, \ldots, T$, followed by (14.28), where the estimate of λ is the fraction of observations assigned to the first component by the m.c.d. algorithm (and those of $\hat{\boldsymbol{\mu}}_i$, $\hat{\boldsymbol{\Sigma}}_i$, $i = 1, 2$, are based on (14.6) and (14.7), respectively, as with the EM split).

The right panel of Figure 14.21 is similar to the left panel, but having used the m.c.d. method for separation, and also overlays the same solid line as shown in the right panel of Figure 14.11. It is important to emphasize that the method of estimation (single-component multivariate normal, applied to each of the two components, and with our usual shrinkage prior) is identical: all that has changed is the method of separating the two components. Now we see that the results are essentially identical to (and actually slightly better than) those obtained from jointly estimating all the model parameters with (prior-augmented) maximum likelihood via the EM algorithm. This might appear difficult to justify, as the m.l.e. is expected to be the best estimator as the sample size grows. The reason that the m.c.d. method can outperform it is that the actual d.g.p. is not Mix_2N_d.

This discovery would appear to imply that we can forego the joint estimation of the model parameters, with its more time-consuming iterative computational method, and can instead use the closed-form estimators, which also means a full avoidance of inferior local maxima of the likelihood. While this is essentially true, the price to pay for this luxury is the computational complexity of the m.c.d., being about 20 times slower. This, in turn, was addressed by Gambacciani and Paolella (2017), who propose a variation of the m.c.d. method, called StepMCD, suitable for computing a large number of moving windows of time-series data (as required for assessing model performance). It deviates from repeated m.c.d. estimation by using the final optimal h-subset from the previous window's m.c.d. estimation as the starting values for the new h-subset. In this way, a substantial reduction in total computational time is obtained, while still keeping the robustness and the efficiency of the m.c.d. estimator. This idea, besides the large increase in speed, actually leads to a (slight) *increase* in out-of-sample performance compared to direct use of FASTMCD applied to each moving window.

14.4 Some Thoughts on Model Assumptions and Estimation

> Econometric textbooks reveal a pronounced lack of concern for the foundations of probability in regard to economic phenomena, while focusing on myopic accounts of estimation and inference in some well-specified abstract models.
>
> (Omar Hamouda and Robin Rowley, 1996, p. 133)

> Some of our assumptions are so closely held that we will cling to them, even in the face of overwhelming evidence.
>
> (Rory Miller, 2008, p. 21)

> We all have a tendency to think that the world must conform to our prejudices. The opposite view involves some effort of thought, and most people would die sooner than think—in fact they do so.
>
> (Bertrand Russell, 1925, p. 166)

It is noteworthy that the proposed m.c.d. methodology for fitting a Mix_2N_d distribution presented in Section 14.3 deviates from the mainstream literature regarding parameter estimation. The overwhelming majority of statistical modeling, as presented in both research and textbooks, and in both the frequentist and Bayesian paradigms, makes a separation between deciding on the model to be used for the data and its method of estimation. This separation is not discussed, but rather appears to be *understood*. Indeed, there exist applications for which the d.g.p. might be relatively easy to characterize, so that this separation is valid, at least as a starting point. However, for complex data, particularly multivariate, and certainly for (univariate and particularly multivariate) financial asset returns data, this will almost surely not be the case.

Traditionally, a model is assumed, possibly a blatantly simple approximation to reality (particularly in the classic time-series literature such as the ARMA class of models from Part II), and then extensive efforts are dedicated to devising methods of parametric inference, most notably point estimation. (The author is guilty of this too; see, e.g., Section 7.6.) The statistics and econometrics literature is dominated by methodology for point estimation; though its inappropriateness for actual data analysis has been noted by numerous academics; see, e.g., the excellent discussion and references in Berkson (1980) and Hampel (1996). The quote from Hamouda and Rowley (1996) at the beginning of this section embodies this idea precisely.

A different way of proceeding is to view the model, and its method of estimation, *as a single unit*. It is crucial to acknowledge that a two- (or more) component multivariate MixN is surely not the actual d.g.p. of financial returns data, rendering discussions about optimal unbiased estimators, biased estimators that minimize the mean squared error, or estimators with ideal asymptotic properties somewhat superfluous. Instead, recognizing that multivariate financial asset returns data are quite complex, and surely not strictly stationary over long periods of time, one should choose the parametric functional form based on its practicality (e.g., the distribution of convolutions of the MixN margins is straightforward; see Section 14.1.4), its simplicity (particularly compared to numerous multivariate GARCH models), and its ability to capture important stylized facts of asset returns, and, *jointly*, choose a method of parameter estimation that may not enjoy classic properties of unbiasedness, consistency, asymptotic normality, etc., under the purported but wrong model, but rather *leads to statistically verifiable and blatantly improved forecasts* compared to alternative estimation methods.

> The unit of {model, estimation method} is judged on (i) its ease of computability, (ii) its forecasting ability, and (iii) conveniently in finance, on a nearly universal ideal of being able to generate improved asset allocation—which can be objectively ranked in terms of its risk/return performance.

As discussed, this notion is contrary to much (but not all of) of mainstream methodology, and the quotes by Russell (2009) and Miller (2008) (both stated in completely different contexts) seem rather appropriate here. The idea is certainly not new, for example many methods and techniques in machine learning and analysis of big data embrace this idea as a core principle.

Risking beating the proverbial dead horse, one might envision the thoughts of a representative employee concerned about the quality of her pension fund (particularly in light of the difficult demographic issues faced by pension funds in Europe and elsewhere, and the numerous contributions in the media discussing their inadequate performance), and faced with the choice of her pension fund manager using an investment strategy based on a model that uses an unbiased, asymptotically efficient estimator (such as the usual plug-in estimators for the mean and variance-covariance matrix under an i.i.d. assumption on the returns, in the classic—but still used, see Allen et al. (2016)—Markowitz framework), or one that is based on a different model that is (also) wrong w.p.1, uses a possibly inconsistent estimator (this being anyway ill-defined if the d.g.p. is not stationary), but that *consistently delivers better returns at lower risk*. The bottom line desire should be quite clear.

> All models (certainly in finance) are wrong; some can be useful, but might require leaving the comforting, yet fictitious, assumptions from normative economics (how things *should* behave) and the assumption that the d.g.p. is stationary through time.

In addition to the benefit conveyed from the m.c.d. method for the Mix_2N_d in this context, having the possibility to split the data via m.c.d. is valuable because we might wish to replace the multivariate normal distribution in either or both components with a different one, for which joint estimation might be complicated, but is straightforward for a single component.

14.5 The Multivariate Laplace and Mix_kLap_d Distributions

Recall from Figure 14.6 how the estimated GAt parameters $\hat{d}$ (not to be confused with the number of dimensions, d) computed for each asset in the two components lie roughly between one and two, and also how, when estimated tail parameter $\hat{\upsilon}$ was constrained to lie above 90, $\hat{d}$ moved towards one. (If the data were truly Laplace, then, as the sample size grows, we expect $\hat{d}$ to be close one, and $\hat{\upsilon}$, without constraint, to be large.) This provides a modicum of motivation for us to consider a multivariate Laplace distribution instead of the multivariate normal for each of the two components of the mixture.

Our motivation is tempered by the fact that the same effect could occur for other heavy-tailed distributions—by limiting $\hat{\upsilon}$ in the GAt, its only recourse to account for the heavier tails is via parameter d. As such, a natural choice would be to entertain a mixture of multivariate Student's t distributions, say Mix_2T_d. This was done, but the Mix_2Lap_d was more successful (in terms of ability to numerically compute the m.l.e.) because (i) the first component is close enough to normal, such that a leptokurtic distribution with finite positive moments of all orders fits better than the Student's t (being genuinely fat-tailed and does not possess a moment generating function), and (ii) the distribution of the second component, by itself, can be reasonably approximated by the Student's t, but, when jointly conducted for both components, the inappropriateness of the Mix_2T_d manifests itself with difficulties in convergence and relatively flat likelihoods of the two degrees of freedom parameters, most notably for the first component.

> Essentially, the mixture aspect of the model is already addressing much of the seeming heavy-tailed nature of asset returns, so that what remains can be accommodated by a non-Gaussian, leptokurtic, but thin-tailed distribution.

Remark It is worth emphasizing that the actual distribution of financial asset returns, while surely not Gaussian, is *not* necessarily heavy-tailed; the demonstrations and arguments in Heyde and Kou (2004) and Sections III.9.1 and III.9.2 should settle this point. This explains why models with different tail behaviors (heavy, semi-heavy, and thin) can be used for successful VaR prediction, as discussed in Sections 10.3.1, 10.6, and 11.1. Recall also that the determination of the tail behavior of a process based on a finite amount of data is very difficult because, by definition, there are very few observations in the tails (and it is not even clear where the tail "starts"; it is a limiting concept). ∎

The subsequent subsections are organized as follows. Sections 14.5.1 and 14.5.2 present the (single component) multivariate Laplace distribution and the Mix_kLap_d, respectively, as well as the EM algorithms associated with prior-augmented maximum likelihood estimation of their parameters. Section 14.5.3 considers estimation and forecasting performance of the Mix_kLap_d when using the m.c.d. split and separate component estimation. Section 14.5.4 discusses how parameter vector $\mathbf{b}$ can be estimated, while Section 14.5.5 gives the portfolio distribution and the expected shortfall associated with the Mix_kLap_d. Finally, Section 14.5.6 presents a method of fast evaluation of the required Bessel function under restrictions appropriate when using the Mix_kLap_d distribution for modeling financial asset returns data.

14.5.1 The Multivariate Laplace and EM Algorithm

Example II.7.19 showed that the basic, univariate Laplace distribution results from a continuous normal variance mixture construction. We can extend this to the multivariate case; it will require the integral form of the modified Bessel function of the third kind,

$$K_{\upsilon}(x) = \frac{1}{2}\int_0^{\infty} u^{\upsilon-1}\exp\left[-\frac{x}{2}\left(\frac{1}{u}+u\right)\right]\,du. \tag{14.30}$$

While the various Bessel functions are conveniently available in Matlab and other computational packages, and evaluated relatively quickly and accurately, it turns out to be the bottleneck (the slowest element) in the estimation procedure. We will see below that we will only require computing (14.30) for a select set of υ-values, and for which it can be evaluated exactly via a finite expansion, so that, after having pre-computed the relevant coefficients in that expansion, its calculation is rendered nearly instantaneous.

Let $(\mathbf{Y} \mid G = g) \sim \mathrm{N}_d\ (\boldsymbol{\mu}, g\boldsymbol{\Sigma})$ for $\boldsymbol{\mu} \in \mathbb{R}^d$ and $\boldsymbol{\Sigma} > 0$, i.e., positive definite, and let $G \sim \mathrm{Gam}\ (b, 1)$, and set $m = (\mathbf{y} - \boldsymbol{\mu})'\boldsymbol{\Sigma}^{-1}(\mathbf{y} - \boldsymbol{\mu})$. Recall that $|g\boldsymbol{\Sigma}| = g^d|\boldsymbol{\Sigma}|$. Then, with substitution $u = \sqrt{m/2}/g$ (and using the fact that $m > 0$ and $g > 0$ with probability one), the p.d.f. of $\mathbf{Y}$ can be expressed via (III.A.79) as

$$f_{\mathbf{Y}}(\mathbf{y}, \boldsymbol{\mu}, \boldsymbol{\Sigma}, b) = \int_0^{\infty} f_{\mathbf{Y}|G}(\mathbf{y}; g) f_G(g)\,dg,$$

or

$$\int_0^{\infty} \frac{1}{|g\boldsymbol{\Sigma}|^{1/2}(2\pi)^{d/2}}\exp\left\{-\frac{1}{2}(\mathbf{y}-\boldsymbol{\mu})'(g\boldsymbol{\Sigma})^{-1}(\mathbf{y}-\boldsymbol{\mu})\right\}\frac{1}{\Gamma(b)}g^{b-1}\exp(-g)\,dg$$

$$= \frac{1}{|\boldsymbol{\Sigma}|^{1/2}(2\pi)^{d/2}}\frac{1}{\Gamma(b)}\int_0^{\infty} g^{-d/2+b-1}\exp\left\{-\frac{m}{2g}-g\right\}\,dg$$

$$= \frac{1}{|\boldsymbol{\Sigma}|^{1/2}(2\pi)^{d/2}}\frac{2}{\Gamma(b)}\left(\frac{m}{2}\right)^{b/2-d/4}\frac{1}{2}\int_0^{\infty} u^{d/2-b-1}\exp\left\{-\frac{\sqrt{2m}}{2}(u+u^{-1})\right\}\,du$$

$$= \frac{1}{|\boldsymbol{\Sigma}|^{1/2}(2\pi)^{d/2}}\frac{2}{\Gamma(b)}\left(\frac{m}{2}\right)^{b/2-d/4}K_{b-d/2}(\sqrt{2m}), \tag{14.31}$$

using the easily verified fact that $K_{\upsilon}(x) = K_{-\upsilon}(x)$. We write $\mathbf{Y} \sim \mathrm{Lap}(\boldsymbol{\mu}, \boldsymbol{\Sigma}, b)$.

From the law of the iterated expectation, $\mathbb{E}[Y] = \mathbb{E}[\mathbb{E}[\mathbf{Y} \mid G]] = \mathbb{E}[\boldsymbol{\mu}] = \boldsymbol{\mu}$, while use of the conditional variance formula yields

$$\mathbb{V}(\mathbf{Y}) = \mathbb{E}[\mathbb{V}(\mathbf{Y} \mid G)] + \mathbb{V}(\mathbb{E}[\mathbf{Y} \mid G]) = \mathbb{E}[G]\boldsymbol{\Sigma} = b\boldsymbol{\Sigma}, \tag{14.32}$$

recalling from (I.7.9) that $\mathbb{E}[G] = b$. Further properties of this distribution are given in Podgórski and Kozubowski (2001).

Remarks

a) Krzanowski and Marriott (1994, Eq. 2.38) state (without derivation) the density but using $b = d/2$, which yields a simplification such that the $(m/2)$ term in the density is no longer present, and refer to that as the multivariate Laplace. Note, however, that the Bessel function $K_0(\cdot)$ still remains, and does not simplify to a less complicated expression.

b) Using the fact that $K_{1/2}(x) = \sqrt{\pi/(2x)}\ e^{-x}$ for $x > 0$, we can choose values for b in (14.31) such that the Bessel function is no longer present. In particular, these are $b = (d-1)/2$ or $(d+1)/2$, with the former only valid for $d > 1$. Using the latter, we get, after simplifying (and again with a location parameter),

$$f_{\mathbf{Y}}(\mathbf{y},\boldsymbol{\mu},\boldsymbol{\Sigma}) = \frac{|\boldsymbol{\Sigma}|^{-1/2}}{\sqrt{2}(2\pi)^{(d-1)/2}} \frac{\exp(-\sqrt{2m})}{\Gamma((d+1)/2)}, \tag{14.33}$$

where $m = (\mathbf{y}-\boldsymbol{\mu})'\boldsymbol{\Sigma}^{-1}(\mathbf{y}-\boldsymbol{\mu})$. As a check, for $d = 1$ (and $\boldsymbol{\mu} = \mathbf{0}$), (14.33) reduces to

$$f_Y(y;\sigma) = \frac{1}{\sqrt{2}\sigma}\exp\left(-\sqrt{2}\frac{|y|}{\sigma}\right) = \frac{1}{2}\frac{1}{\sigma/\sqrt{2}}\exp\left(-\frac{|y|}{\sigma/\sqrt{2}}\right),$$

a univariate Laplace density with scale term $\sigma/\sqrt{2}$.

c) Density (14.31) with $b = 1$ is a special case of the asymmetric multivariate Laplace distribution studied in Kotz et al. (2000) and Kozubowski and Podgórski (2001). In particular, with $\mathbf{a} = (a_1,\dots,a_d)'$ a vector of asymmetry parameters, the density $f_{\mathbf{Y}}(\mathbf{y})$ is, again with $m = (\mathbf{y}-\boldsymbol{\mu})'\boldsymbol{\Sigma}^{-1}(\mathbf{y}-\boldsymbol{\mu})$, given by

$$\frac{2\exp\{(\mathbf{y}-\boldsymbol{\mu})'\boldsymbol{\Sigma}^{-1}\mathbf{a}\}}{|\boldsymbol{\Sigma}|^{1/2}(2\pi)^{d/2}}\left(\frac{m}{2+\mathbf{a}'\boldsymbol{\Sigma}^{-1}\mathbf{a}}\right)^{\upsilon/2} K_{\upsilon}\left(\sqrt{(2+\mathbf{a}'\boldsymbol{\Sigma}^{-1}\mathbf{a})m}\right), \tag{14.34}$$

where $\upsilon = b - d/2 = 1 - d/2$. They also provide a useful overview and references of the numerous distributions that have been coined multivariate Laplace. See also the monograph dedicated to the topic by Kotz et al. (2001). ■

To get a sense of the leptokurtic nature of the distribution, Figure 14.22 shows a contour plot in the bivariate case of the multivariate normal and the multivariate Laplace for three values of b, $\boldsymbol{\mu} = \mathbf{0}$, and identity dispersion matrix scaled by b, so that $\mathbb{V}(\mathbf{Y})$ is the same for each. The case with $b = 90$ is nearly identical to the bivariate normal, suggesting that, as $b \to \infty$, it converges in distribution to a normal. For the case with $\boldsymbol{\Sigma} = \mathrm{diag}([\sigma_1^2,\dots,\sigma_d^2])$, (14.33) reduces to

$$\frac{\exp\left\{-\sqrt{2\sum_{i=1}^{d}(y_i-\mu_i)^2/\sigma_i^2}\right\}}{\Gamma((d+1)/2)\ \pi^{(d-1)/2}}\prod_{i=1}^{d}\frac{1}{\sqrt{2}\sigma_i},$$

from which it is clear that the distribution does not factor into a product of marginals, so that there is dependency among the elements of $\mathbf{Y}$ even if $\boldsymbol{\Sigma}$ is diagonal (in which case the components are uncorrelated). Figure 14.23 shows the joint density of two independent univariate Laplace random variables for three values of b. For small b, these look quite different from the joint distributions shown in Figure 14.22, though as $b \to \infty$, it appears that the distribution approaches the normal. This is indeed the case, and the reader is encouraged to prove it: From Section II.9.5.2 (and Table II.9.2), the univariate Laplace converges to the normal distribution as $b \to \infty$, provided a scale term δ is introduced, and such that δ/b converges to a constant. This result also holds in the multivariate case.

We now develop an EM algorithm for estimating (14.31). Besides being of value in its own right, it will help set the stage for the more complicated EM algorithm for the discrete mixture case given below. Conditional on G_i, the distribution of $\mathbf{Y}_i$ is normal, so let $G_1, G_2, \dots, G_T$ be the latent, unobserved, i.i.d. $\mathrm{Gam}(b, 1)$ random variables, and where we assume that b is known. Assuming known b initially simplifies matters; in Section 14.5.4, after the more general context of the $\mathrm{Mix}_k\mathrm{Lap}_d$ case is presented, we address the estimation of parameter b.

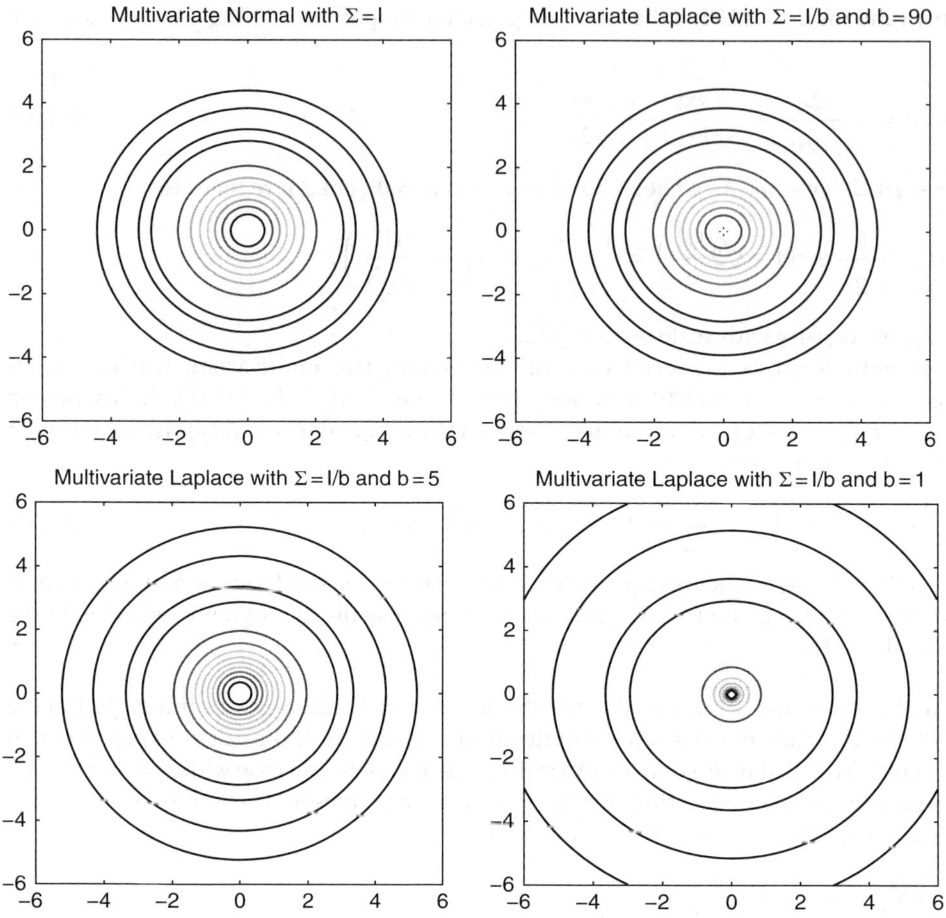

Figure 14.22 Examples of the bivariate Laplace distribution; top left is normal, for comparison. The contour lines have the same height across plots.

Omitting constants that do not depend on $\boldsymbol{\theta} = [\boldsymbol{\mu}', \text{ vech } (\boldsymbol{\Sigma})']'$, the log joint density of $\mathbf{Y}_t$ and G_t is (after multiplying by two),

$$\begin{aligned}\log f_{\mathbf{Y}_t,G_t}(\mathbf{y}_t, g_t; \boldsymbol{\mu}, \boldsymbol{\Sigma}) &= \log f_{\mathbf{Y}_t|G_t}(\mathbf{y}_t; g_t) + \log f_{G_t}(g_t)\\ &\propto -\log|\boldsymbol{\Sigma}| - g_t^{-1}(\mathbf{y}_t - \boldsymbol{\mu})'\boldsymbol{\Sigma}^{-1}(\mathbf{y}_t - \boldsymbol{\mu}),\end{aligned}$$

so that the complete data log-likelihood is, with $\mathbf{Y} = (\mathbf{Y}_1, \dots, \mathbf{Y}_T)$, $\mathbf{G} = (G_1, \dots, G_T)$, and using the fact that $\text{tr}(\mathbf{AB}) = \text{tr}(\mathbf{BA})$,

$$\ell_c(\boldsymbol{\theta}; \mathbf{Y}, \mathbf{G}) = -T\log|\boldsymbol{\Sigma}| - \sum_{t=1}^{T} G_t^{-1}\boldsymbol{\Sigma}^{-1}(\mathbf{Y}_t - \boldsymbol{\mu})(\mathbf{Y}_t - \boldsymbol{\mu})'. \tag{14.35}$$

The derivation of the m.l.e. of $\boldsymbol{\mu}$ and $\boldsymbol{\Sigma}$ follows similarly to that in Example III.3.8, and yields

$$\hat{\boldsymbol{\mu}} = \frac{\sum_{t=1}^{T} G_t^{-1}\mathbf{Y}_t}{\sum_{t=1}^{T} G_t^{-1}}, \quad \hat{\boldsymbol{\Sigma}} = T^{-1}\sum_{t=1}^{T} G_t^{-1}(\mathbf{Y}_t - \hat{\boldsymbol{\mu}})(\mathbf{Y}_t - \hat{\boldsymbol{\mu}})'. \tag{14.36}$$

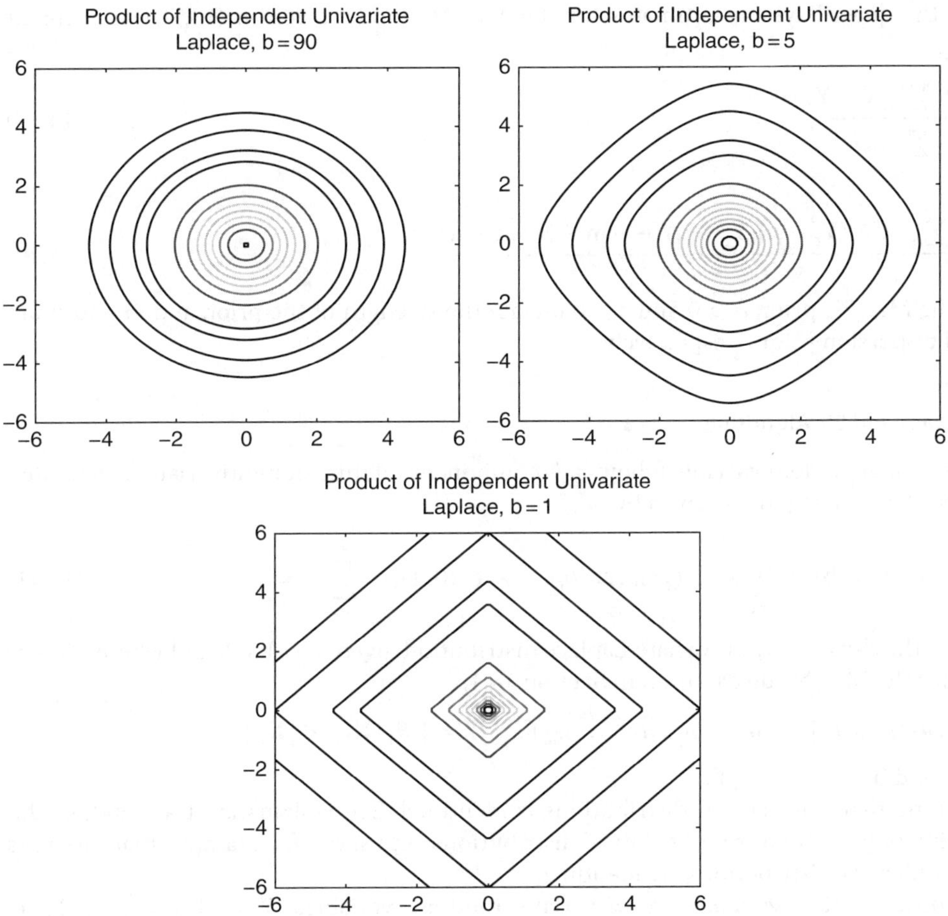

Figure 14.23 Bivariate distributions as products of i.i.d. univariate Laplace with scale $1/b$.

Next, with $m_t = (\mathbf{y}_t - \boldsymbol{\mu})'\boldsymbol{\Sigma}^{-1}(\mathbf{y}_t - \boldsymbol{\mu})$,

$$f_{G_t|\mathbf{Y}_t}(g_t; \mathbf{y}_t) = \frac{f_{\mathbf{Y}_t,G_t}(\mathbf{y}_t, g_t; \boldsymbol{\mu}, \boldsymbol{\Sigma})}{f_{\mathbf{Y}_t}(\mathbf{y}_t; \boldsymbol{\mu}, \boldsymbol{\Sigma})} = \frac{g_t^{-d/2+b-1} \exp\{-m_t/(2g_t) - g_t\}}{2(m_t/2)^{b/2-d/4} K_{b-d/2}(\sqrt{2m_t})} \mathbb{I}_{(0,\infty)}(g_t). \tag{14.37}$$

This is the generalized inverse Gaussian (GIG) distribution (see Section II.9.4.1), with $(G_t \mid \mathbf{Y}_t) \sim \mathrm{GIG}(b - d/2, m_t, 2)$ and, using (II.9.18),

$$\mathbb{E}[G_t^{-1} \mid \mathbf{y}_t] = \frac{K_{b-d/2-1}(\sqrt{2m_t})}{(m_t/2)^{1/2} K_{b-d/2}(\sqrt{2m_t})}, \tag{14.38}$$

which, for $b = (d+1)/2$, simplifies to $(m_t/2)^{-1/2}$.

As g_t^{-1} enters linearly in (14.35), the conditional expectation of the complete data log-likelihood, $\mathbb{E}_{\theta^{(s)}}[\ell_c(\boldsymbol{\theta}; \mathbf{Y}, \mathbf{G}) \mid \mathbf{Y}]$, with respect to the hidden variables $\mathbf{G}$, given the observed data $\mathbf{Y}$, and the value of parameter $\boldsymbol{\theta}$ at the sth iteration just involves substituting (14.38) in place of g_t^{-1} in (14.35). The EM algorithm then consists of iterating between (14.36) and (14.38) until convergence.

Analogous to the quasi-Bayesian estimator for the Mix_kN_d model given in (14.7), we augment (14.36) by taking

$$\hat{\boldsymbol{\mu}} = \frac{c\mathbf{m} + \sum_{t=1}^{T} G_t^{-1}\mathbf{Y}_t}{c + \sum_{t=1}^{T} G_t^{-1}} \tag{14.39}$$

and

$$\hat{\boldsymbol{\Sigma}} = \frac{\mathbf{B} + \sum_{t=1}^{T} G_t^{-1}(\mathbf{Y}_t - \boldsymbol{\mu})(\mathbf{Y}_t - \boldsymbol{\mu})' + c(\mathbf{m} - \hat{\boldsymbol{\mu}})(\mathbf{m} - \hat{\boldsymbol{\mu}})'}{a + T}, \tag{14.40}$$

where, as with the Mix_kN_d prior, $a \geqslant 0$ and $c \geqslant 0$ dictate the strength of the prior, and $\mathbf{m}$ and $\mathbf{B}$ are the location and dispersion priors, respectively.

14.5.2 The Mix_kLap_d and EM Algorithm

We say a d-dimensional random variable follows a k-component mixture of multivariate Laplace distributions, or Mix_kLap_d, if its p.d.f. is given by

$$f_{\text{Mix}_k\text{Lap}_d}(\mathbf{y}; \boldsymbol{M}, \boldsymbol{\Psi}, \boldsymbol{\lambda}, \mathbf{b}) = \sum_{j=1}^{k} \lambda_j f_{\text{Lap}}(\mathbf{y}; \boldsymbol{\mu}_j, \boldsymbol{\Sigma}_j, b_j), \quad \lambda_j \in (0,1), \quad \sum_{j=1}^{k} \lambda_j = 1, \tag{14.41}$$

with f_{Lap} denoting the d-variate multivariate Laplace distribution given in (14.31), and where, similar to our notation for the Mix_kN_d distribution in Section 14.1,

$$\boldsymbol{M} = \left[\, \boldsymbol{\mu}_1 \,|\, \boldsymbol{\mu}_2 \,|\cdots|\, \boldsymbol{\mu}_k \,\right], \quad \boldsymbol{\mu}_j = (\mu_{1j}, \mu_{2j}, \ldots, \mu_{dj})', \quad \boldsymbol{\Psi} = \left[\, \boldsymbol{\Sigma}_1 \,|\, \boldsymbol{\Sigma}_2 \,|\cdots|\, \boldsymbol{\Sigma}_k \,\right],$$

$\boldsymbol{\lambda} = (\lambda_1, \ldots, \lambda_k)'$, and $\mathbf{b} = (b_1, \ldots, b_k)'$.

We tacitly assume that this class of distributions is identified. See Holzmann et al. (2006), who study identifiability of finite mixtures of elliptical distributions, and show, for example, that mixtures of multivariate Student's t distributions are identifiable.

Assume we observe the sequence of d-variate random variables $\mathbf{Y}_t = (Y_{t,1}, Y_{t,2}, \ldots, Y_{t,d})'$, $t = 1, \ldots, T$, with $\mathbf{Y}_t \overset{\text{i.i.d.}}{\sim} \text{Mix}_k\text{Lap}_d(\boldsymbol{M}, \boldsymbol{\Psi}, \boldsymbol{\lambda}, \mathbf{b})$, and, for now, take vector $\mathbf{b}$ to be a set of known constants (this will be relaxed below). Interest centers on estimation of the remaining parameters,

$$\boldsymbol{\theta} = [\text{vec}(\boldsymbol{M}),\ \text{vech}\,(\boldsymbol{\Sigma}_1)',\ \text{vech}\,(\boldsymbol{\Sigma}_2)',\ \ldots\ \text{vech}\,(\boldsymbol{\Sigma}_k)',\ \boldsymbol{\lambda}']', \tag{14.42}$$

similar to (14.10), where the vech operator is defined.

As with the development of the EM algorithm for the Mix_kN_d distribution, denote the hidden variable associated with the tth observation $\mathbf{Y}_t$ as $\mathbf{H}_t = (H_{t,1}, \ldots, H_{t,k})$, where $H_{t,j} = 1$ if $\mathbf{Y}_t$ came from the jth component, and zero otherwise, so that, with $\mathbf{h} = (h_1, \ldots, h_k)$,

$$f_{\mathbf{H}_t}(\mathbf{h}) = \prod_{j=1}^{k} \lambda_j^{h_j} \mathbb{I}_{\{0,1\}}(h_j) \mathbb{I}\left(\sum_{j=1}^{k} h_j = 1\right). \tag{14.43}$$

Then, analogous to the Mix_kN_d case, the joint density of $\mathbf{Y}_t$ and $\mathbf{H}_t$ is

$$f_{\mathbf{Y}_t|\mathbf{H}_t}(\mathbf{y}, \mathbf{h}; \boldsymbol{\theta}) f_{\mathbf{H}_t}(\mathbf{h}; \boldsymbol{\theta}) = \prod_{j=1}^{k} [\lambda_j f_{\text{Lap}}(\mathbf{y}; \boldsymbol{\mu}_j, \boldsymbol{\Sigma}_j, b_j)]^{h_j} \mathbb{I}_{\{0,1\}}(h_j) \mathbb{I}\left(\sum_{j=1}^{k} h_j = 1\right),$$

where $\boldsymbol{\theta}$ is given in (14.42), and $\mathbb{E}[H_{t,j} \mid \mathbf{Y}_t] = \Pr(H_{t,j} = 1 \mid \mathbf{Y}_t = \mathbf{y}_t)$ is

$$\hbar_{t,j} := \frac{\lambda_j f_{\text{Lap}}(\mathbf{y}_t; \boldsymbol{\mu}_j, \boldsymbol{\Sigma}_j, b_j)}{\sum_{j=1}^{k} \lambda_j f_{\text{Lap}}(\mathbf{y}_t; \boldsymbol{\mu}_j, \boldsymbol{\Sigma}_j, b_j)}, \quad t = 1, \dots, T, \quad j = 1, \dots, k. \tag{14.44}$$

Recall that, from the construction of the single-component multivariate Laplace distribution, $(\mathbf{Y}_t \mid G_t = g_t) \sim \mathrm{N}(0, g_t\boldsymbol{\Sigma})$, where $G_t \sim \mathrm{Gam}(b, 1)$, and $\mathbf{G} = (G_1, \dots, G_T)$ was part of the complete data log-likelihood. In the mixture context, G_t can come from one of k distributions, $\mathrm{Gam}(b_j, 1)$, with p.d.f. $f_{\text{Gam}}(g_t; b_j) = g_t^{b_j} \exp(-g_t)/\Gamma(b_j)\mathbb{I}_{(0,\infty)}(g_t)$, $j = 1, \dots, k$, so its specification requires conditioning on $\mathbf{H}_t$. That is,

$$(G_t \mid \mathbf{H}_t) = (G_t \mid H_{t,j} = 1) \sim \mathrm{Gam}(b_j, 1) \quad \text{and} \quad (\mathbf{Y}_t \mid H_{t,j} = 1, G_t = g_t) \sim \mathrm{N}(0, g_t\boldsymbol{\Sigma}).$$

Then $f_{\mathbf{Y}_t,G_t,\mathbf{H}_t}(\mathbf{y}, g, \mathbf{h}; \boldsymbol{\theta}) = f_{\mathbf{Y}_t|G_t,\mathbf{H}_t}(\mathbf{y}; g, \mathbf{h}, \boldsymbol{\theta}) f_{G_t|\mathbf{H}_t}(g; \mathbf{h}, \boldsymbol{\theta}) f_{\mathbf{H}_t}(\mathbf{h}; \boldsymbol{\theta})$, where, noting that, conditional on $\mathbf{H}_t$ being $\mathbf{h}_t$ with jth element one,

$$f_{\mathbf{Y}_t|G_t,\mathbf{H}_t}(\mathbf{y}, g, \mathbf{h}; \boldsymbol{\theta}) = \prod_{j=1}^{k} [f_{\mathrm{N}}(\mathbf{y}; \boldsymbol{\mu}_j, g\boldsymbol{\Sigma}_j)]^{h_j}. \tag{14.45}$$

Similarly,

$$f_{G_t|\mathbf{H}_t}(g; \mathbf{h}) = \prod_{j=1}^{k} [f_{\text{Gam}}(g; b_j)]^{h_j}, \tag{14.46}$$

so that, from (14.43), (14.45) and (14.46),

$$f_{\mathbf{Y}_t,G_t,\mathbf{H}_t}(\mathbf{y}, g, \mathbf{h}; \boldsymbol{\theta}) = \prod_{j=1}^{k} [\lambda_j f_{\mathrm{N}}(\mathbf{y}; \boldsymbol{\mu}_j, g\boldsymbol{\Sigma}_j) f_{\text{Gam}}(g; b_j)]^{h_j} \mathbb{I}_{\{0,1\}}(h_j) \mathbb{I}\left(\sum_{j=1}^{k} h_j = 1\right).$$

The complete data log-likelihood is

$$\ell_c(\boldsymbol{\theta}; \mathbf{Y}, \mathbf{G}, \mathbf{H}) = \sum_{t=1}^{T} \log f_{\mathbf{Y}_t,\mathbf{G}_t,\mathbf{H}_t}(\mathbf{y}, g, \mathbf{h}; \boldsymbol{\theta}),$$

where $\mathbf{Y} = (\mathbf{Y}_1, \dots, \mathbf{Y}_T)$, $\mathbf{G} = (G_1, \dots, G_T)$, and $\mathbf{H} = (\mathbf{H}_1, \dots, \mathbf{H}_T)$. Then, recalling that $|g\boldsymbol{\Sigma}| = g^d|\boldsymbol{\Sigma}|$ for $g \in \mathbb{R}_{>0}$, with d the dimension, $\ell_c(\boldsymbol{\theta}; \mathbf{Y}, \mathbf{G}, \mathbf{H})$ is given by

$$\begin{aligned}
&\sum_{t=1}^{T} \log f_{\mathbf{Y}_t|\mathbf{G}_t,\mathbf{H}_t}(\mathbf{y}; g, \mathbf{h}, \boldsymbol{\theta}) + \sum_{t=1}^{T} \log f_{\mathbf{G}_t|\mathbf{H}_t}(g; \mathbf{h}, \boldsymbol{\theta}) + \sum_{t=1}^{T} \log f_{\mathbf{H}_t}(\mathbf{h}; \boldsymbol{\theta}) \\
&= \sum_{t=1}^{T} \sum_{j=1}^{k} h_j \left\{ -\frac{d}{2}\log(2\pi) - \frac{d}{2}\log(g) - \frac{1}{2}\log|\boldsymbol{\Sigma}_j| - \frac{1}{2}(\mathbf{y} - \boldsymbol{\mu}_j)'(g\boldsymbol{\Sigma}_j)^{-1}(\mathbf{y} - \boldsymbol{\mu}_j) \right\} \\
&\quad + \sum_{t=1}^{T} h_j\{b_j \log(g) - g - \log\Gamma(b_j)\} + \sum_{t=1}^{T} \sum_{j=1}^{k} h_j \log \lambda_j,
\end{aligned}$$

where, in the last expression after the equals sign, note that the elements of $\mathbf{h} = (h_1, \dots, h_k)$ and g are changing at each time point t.

Omitting terms that do not depend on $\boldsymbol{\theta}$ from (14.42), keeping in mind that the b_j are known, multiplying by two, and using capital letters for the $\mathbf{Y}_t$, $\mathbf{G}_t$, and $\mathbf{H}_t$ to indicate them as random variables, we can write

$$\ell_c(\boldsymbol{\theta}; \mathbf{Y}, \mathbf{G}, \mathbf{H}) = \sum_{t=1}^{T} \sum_{j=1}^{k} H_{t,j} \{ -\log|\boldsymbol{\Sigma}_j| - G_t^{-1} (\mathbf{Y}_t - \boldsymbol{\mu}_j)' \boldsymbol{\Sigma}_j^{-1} (\mathbf{Y}_t - \boldsymbol{\mu}_j) \}$$
$$+ 2 \sum_{t=1}^{T} \sum_{j=1}^{k} H_{t,j} \log \lambda_j. \tag{14.47}$$

As in the case for $\mathrm{Mix}_k\mathrm{N}_d$, the λ_j are disjoint from the other parameters in $\boldsymbol{\theta}$, so that their m.l.e.s can be determined separately, and, intuitively, yield the same result as (14.6), namely

$$\hat{\lambda}_j = \frac{1}{T} \sum_{t=1}^{T} H_{t,j}, \quad j = 1, \dots, k. \tag{14.48}$$

Estimators for $\boldsymbol{\mu}$ and $\boldsymbol{\Sigma}$ from the complete data log-likelihood follow easily because of the binary nature of the $H_{t,j}$ and that $\sum_{j=1}^{k} H_{t,j} = 1$, so that we have k independent multivariate Laplace populations, each with $\sum_{t=1}^{T} H_{t,j}$ observations. Thus, we can apply (14.36), i.e.,

$$\widehat{\boldsymbol{\mu}}_j = \frac{\sum_{t=1}^{T} H_{t,j} G_t^{-1} \mathbf{Y}_t}{\sum_{t=1}^{T} H_{t,j} G_t^{-1}}, \quad \widehat{\boldsymbol{\Sigma}}_j = \frac{\sum_{t=1}^{T} H_{t,j} G_t^{-1} (\mathbf{Y}_t - \widehat{\boldsymbol{\mu}}_j)(\mathbf{Y}_t - \widehat{\boldsymbol{\mu}}_j)'}{\sum_{t=1}^{T} H_{t,j}}, \tag{14.49}$$

$j = 1, \dots, k$. As a direct generalization of the quasi-Bayesian estimator (14.7) for the $\mathrm{Mix}_k\mathrm{N}_d$ model, and (14.39) and (14.40) for the single-component multivariate Laplace, we take

$$\widehat{\boldsymbol{\mu}}_j = \frac{c_j \mathbf{m}_j + \sum_{t=1}^{T} H_{t,j} G_t^{-1} \mathbf{Y}_t}{c_j + \sum_{t=1}^{T} H_{t,j} G_t^{-1}}, \tag{14.50}$$

and

$$\boldsymbol{\Sigma}_j = \frac{\mathbf{B}_j + \sum_{t=1}^{T} H_{t,j} G_t^{-1} (\mathbf{Y}_t - \widehat{\boldsymbol{\mu}}_j)(\mathbf{Y}_t - \widehat{\boldsymbol{\mu}}_j)' + c_j (\mathbf{m}_j - \widehat{\boldsymbol{\mu}}_j)(\mathbf{m}_j - \widehat{\boldsymbol{\mu}}_j)'}{a_j + \sum_{t=1}^{T} H_{t,j}}, \tag{14.51}$$

$j = 1, \dots, k$, and use the prior values given below in (14.54).

As usual in our EM algorithm derivations, computation of (14.48) and (14.49) is not feasible because $\mathbf{G}$ and $\mathbf{H}$ are not observed. Hence, the E-step: Conditional on the observed $\mathbf{Y}_t$ and the value of $\widehat{\boldsymbol{\theta}}$, say $\boldsymbol{\theta}^{(s)}$, in the sth step of the iterative scheme, we compute

$$Q(\boldsymbol{\theta}; \boldsymbol{\theta}^{(s)}) = \mathbb{E}_{\boldsymbol{\theta}^{(s)}} [\ell_c(\boldsymbol{\theta}; \mathbf{Y}, \mathbf{G}, \mathbf{H}) \mid \mathbf{Y} = \mathbf{y}],$$

the conditional expectation of the complete data log-likelihood with respect to the hidden random variables, given the observed data $\mathbf{y}$, and using as parameter $\boldsymbol{\theta}$ the current value $\boldsymbol{\theta}^{(s)}$. For this, we need $\mathbb{E}_{\boldsymbol{\theta}^{(s)}}[H_{t,j} \mid \mathbf{Y}_t]$, which is just (14.44).

Next, for the tth observation but using the parameters of the jth component, let

$$m_{t,j} = (\mathbf{y}_t - \boldsymbol{\mu}_j)' \boldsymbol{\Sigma}_j^{-1} (\mathbf{y}_t - \boldsymbol{\mu}_j), \quad t = 1, \dots, T, \quad j = 1, \dots, k,$$

and recall that, when conditioning on $\mathbf{H}_t = \mathbf{h}$, the jth element of $\mathbf{h}$ is one (and the rest are zero). Then, to compute $\mathbb{E}_{\theta^{(s)}}[G_t^{-1} \mid \mathbf{Y}_t = \mathbf{y}, \mathbf{H}_t = \mathbf{h}]$, we require $f_{G_t|\mathbf{Y}_t,\mathbf{H}_t}(g; \mathbf{y}, \mathbf{h})$, or

$$\frac{f_{\mathbf{Y}_t,\mathbf{G}_t,\mathbf{H}_t}(\mathbf{y}, g, \mathbf{h}; \theta)}{f_{\mathbf{Y}_t,\mathbf{H}_t}(\mathbf{y}, \mathbf{h}; \theta)} = \frac{f_{\mathbf{Y}_t|\mathbf{G}_t,\mathbf{H}_t}(\mathbf{y}; g, \mathbf{h}, \theta) f_{\mathbf{G}_t|\mathbf{H}_t}(g; \mathbf{h}, \theta)}{f_{\mathbf{Y}_t|\mathbf{H}_t}(\mathbf{y}, \mathbf{h}, \theta)} \frac{f_{\mathbf{H}_t}(\mathbf{h}; \theta)}{f_{\mathbf{H}_t}(\mathbf{h}; \theta)}$$
$$= \frac{\prod_{j=1}^k [f_{\mathrm{N}}(\mathbf{y}; \boldsymbol{\mu}_j, g\boldsymbol{\Sigma}_j) f_{\mathrm{Gam}}(g; b_j)]^{h_j}}{\prod_{j=1}^k [f_{\mathrm{Lap}}(\mathbf{y}; \boldsymbol{\mu}_j, \boldsymbol{\Sigma}_j, b_j)]^{h_j}} = \frac{f_{\mathrm{N}}(\mathbf{y}; \boldsymbol{\mu}_j, g\boldsymbol{\Sigma}_j) f_{\mathrm{Gam}}(g; b_j)}{f_{\mathrm{Lap}}(\mathbf{y}; \boldsymbol{\mu}_j, \boldsymbol{\Sigma}_j, b_j)},$$

but this is the same density as given in (14.37), so that we can use the result from (14.38), just changing b to b_j and m_t to $m_{t,j}$, to get

$$\zeta_{t,j} := \mathbb{E}_{\theta^{(s)}}[G_t^{-1} \mid \mathbf{Y}_t = \mathbf{y}, \mathbf{H}_t = \mathbf{h}] = \frac{K_{b_j-d/2-1}(\sqrt{2m_{t,j}})}{(m_{t,j}/2)^{1/2} K_{b_j-d/2}(\sqrt{2m_{t,j}})}, \tag{14.52}$$

$t = 1, \ldots, T$, $j = 1, \ldots, k$. Then, with $W = \ell_c(\theta; \mathbf{Y}, \mathbf{G}, \mathbf{H}) \mid (\mathbf{Y} = \mathbf{y})$, we have $\mathbb{E}[W] = \mathbb{E}[\mathbb{E}[W \mid \mathbf{H}_t]]$, or, using (14.47) and (14.52),

$$\begin{aligned}
&\mathbb{E}_{\theta^{(s)}}[\ell_c(\theta; \mathbf{Y}, \mathbf{G}, \mathbf{H}) \mid \mathbf{Y} = \mathbf{y}] \\
&= \mathbb{E}_{\theta^{(s)}}[\mathbb{E}_{\theta^{(s)}}[\ell_c(\theta; \mathbf{Y}, \mathbf{G}, \mathbf{H}) \mid \mathbf{Y} = \mathbf{y}; \mathbf{H}_t = \mathbf{h}] \mid \mathbf{Y} = \mathbf{y}] \\
&= \mathbb{E}_{\theta^{(s)}}\left[\sum_{t=1}^T \sum_{j=1}^k H_{t,j}\{-\log|\boldsymbol{\Sigma}_j| - \zeta_{t,j}(\mathbf{Y}_t - \boldsymbol{\mu}_j)'\boldsymbol{\Sigma}_j^{-1}(\mathbf{Y}_t - \boldsymbol{\mu}_j)\}\right. \\
&\quad \left. +2\sum_{t=1}^T \sum_{j=1}^k H_{t,j} \log \lambda_j \mid \mathbf{Y} = \mathbf{y}\right],
\end{aligned} \tag{14.53}$$

which follows because $\mathbb{E}_{\theta^{(s)}}[H_{t,j}G_t^{-1} \mid \mathbf{Y}_t, \mathbf{H}_t] = H_{t,j}\mathbb{E}_{\theta^{(s)}}[G_t^{-1} \mid \mathbf{Y}_t, \mathbf{H}_t]$. As (14.53) is linear in the $H_{t,j}$, we need only the expectation of the $H_{t,j}$, conditional on $\mathbf{Y} = \mathbf{y}$ and $\theta^{(s)}$, as given by (14.44).

Similar to (14.8), and in light of the variance, as given in (14.32), we take

$$\begin{aligned}
&a_1 = 2\omega, \quad a_2 = \omega/2, \quad c_1 = c_2 = 20\omega, \quad \mathbf{m}_1 = \mathbf{0}_d, \quad \mathbf{m}_2 = -0.1\mathbf{1}_d, \\
&\mathbf{B}_1 = \frac{a_1}{b_1}[(1.5 - 0.6)\mathbf{I}_d + 0.6\mathbf{J}_d], \quad \mathbf{B}_2 = \frac{a_2}{b_2}[(10 - 4.6)\mathbf{I}_d + 4.6\mathbf{J}_d],
\end{aligned} \tag{14.54}$$

where hyper-parameter ω controls the strength of the prior.

Thus, with a starting value $\theta^{(0)}$, the EM algorithm iterates between (14.44), (14.48), and (14.49), where it is understood that, in (14.44), λ_j, $\boldsymbol{\mu}_j$ and $\boldsymbol{\Sigma}_j$ are the most current values from $\boldsymbol{\theta}^{(s)}$, and in (14.48) and (14.49), the expectations of $H_{t,j}$ and G_t^{-1} are used, these being $\hbar_{t,j}$ in (14.44) and $\zeta_{t,j}$ in (14.52), respectively.

Remarks There are some issues related to computation that apply to both the normal and Laplace mixture cases. During the EM iterations, roundoff error can induce $\widehat{\boldsymbol{\Sigma}}_1$ or $\widehat{\boldsymbol{\Sigma}}_2$ to deviate slightly from symmetry, which would invalidate their status as covariance or dispersion matrices. A simple and effective solution is just to set $\widehat{\boldsymbol{\Sigma}}_j = (\widehat{\boldsymbol{\Sigma}}_j + \widehat{\boldsymbol{\Sigma}}_j')/2$, $j = 1, \ldots, k$. Less trivial is the possibility that one or more of the $\widehat{\boldsymbol{\Sigma}}_j$ are rank deficient, which is analogous to one or more of the scale terms σ_j in the univariate case approaching zero. It was found that simply setting eigenvalues of $\widehat{\boldsymbol{\Sigma}}_j$ lower than some threshold to that threshold during the iteration was enough to prevent the algorithm from ceasing, and allowing it to either find a more plausible solution or, on occasion, to return a near-rank-deficient $\hat{\boldsymbol{\Sigma}}_j$.

This latter case is preventable with the use of $a_j > 0$; just taking $\mathbf{B}_j = \mathbf{I}_d$ and $a_j = 0.1$ in (14.7) is adequate, whereby an extremely small amount of prior information is enough to prevent the optimizer from landing on a singularity point of the likelihood. (Another approach for ensuring full rank of the $\hat{\boldsymbol{\Sigma}}_j$ during estimation in the normal mixture case is proposed in Ingrassia and Rocci, 2007.)

As with the mixture of normals (both univariate and multivariate), we can safely presume that the likelihood surface of the Mix_kLap_d model (with its 991 parameters for $k = 2$, $d = 30$ and fixed b_1 and b_2) has more than one local maxima. In a time series context in which moving windows of data are used for estimation, we suggest the following strategy. Perform two estimations, such that their starting values are (i) the final values of the previous window and (ii) the prior values as given in (14.54), i.e., $\boldsymbol{\mu}_i = \mathbf{m}_i$, $\boldsymbol{\Sigma}_i = \mathbf{B}_i/a_i$, $i = 1, 2$. The one returning the higher likelihood value is used. ■

14.5.3 Estimation via MCD Split and Forecasting

We have seen in Section 14.3 that the DJIA-30 stock returns data can be successfully decomposed into two sets using the m.c.d. method. We can attempt the same idea for obtaining parameter estimates of the $\text{Mix}_2\text{Lap}_{30}$ distribution: We apply the m.c.d. split, and then the two location vectors and dispersion matrices associated with the $\text{Mix}_2\text{Lap}_{30}$ are separately estimated using the EM algorithm developed in Section 14.5.1. The mixture weight of the first component, $\lambda = \lambda_1$, is, as in the Mix_2N_d case, just the fraction of observations assigned to the first component by the m.c.d. procedure. The shape parameters b_1 and b_2 are determined as in Section 14.5.4.

We wish to determine the optimal value of ω for the shrinkage prior for the $\text{Mix}_2\text{Lap}_{30}$, applied to the DJIA-30 data, using the normalized sum of the realized predictive log-likelihood (14.22) as a function of ω, for a moving window of length $v = 250$, and estimating the model via the use of the m.c.d. split followed by separate estimation of each Laplace component. We do this for two sets of fixed values of b_1 and b_2; the first are $b_1 = 7.5$ and $b_2 = 3.5$, which were found to be optimal when using the entire data set for estimation, and $b_1 = b_2 = 5.5$, these being optimal when using only the last 500 observations (see Section 14.5.4). This two-year period of observations (March 2007 to March 2009) occurs during a full unfolding of the Global Financial Crisis, massive market downturns and relatively high market volatility, and so it is not surprising that even the first component is picking up non-normality of the data. This also suggests that it is not ideal to use "as much data as possible", but rather a sample size T such that performance is maximized. Thus, the sample size becomes a tuning parameter, though better is to use weighted likelihood.

We conduct the same density forecasting exercise used to obtain the right panel of Figure 14.11, based on the $\text{Mix}_2\text{N}_{30}$, and so, for comparison purposes, we replicate this in the right panel of Figure 14.24, having only changed the scaling on the y-axis. The left panel shows the new results as the solid and dashed lines. The use of the Laplace instead of the normal clearly leads to better density forecasts, with the optimal value of ω being about 40. There is very little difference when using either $b_1 = 7.5$ and $b_2 = 3.5$ or $b_1 = b_2 = 5.5$, though the latter gives the larger result. This is comforting because joint estimation of all the parameters, including the b_i, while straightforward, is more time-consuming.

We also observe that the improvement as ω increases from 0.1 to 40 is far less than obtained with the normal distribution. This makes good sense: The mis-specified normal distribution is such that its tails are too thin for the data, and so its m.l.e. without shrinkage attempts to compensate by increasing

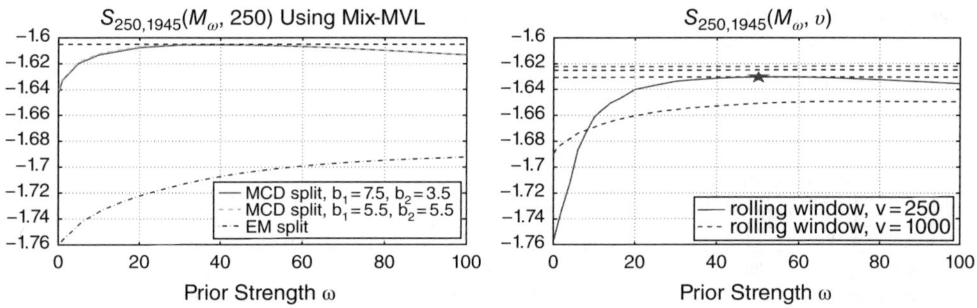

Figure 14.24 **Left:** The normalized sum of the realized predictive log-likelihood versus ω, based on the $\text{Mix}_2\text{Lap}_{30}$ estimated via m.c.d. split and prior-augmented m.l.e. via the EM algorithm for each separate Laplace component, using a moving window of length $v = 250$, and two sets of fixed (b_1, b_2) Laplace parameters (solid and dashed lines). The dash-dot line is the result when having separated the components based on the EM algorithm output of $\widehat{H}_{t,1}$ for the $\text{Mix}_2\text{N}_{30}$ distribution. **Right:** Same as the right panel of Figure 14.11 for $v = 250$, except having changed the scaling of the y-axis; it is just for comparison to the left panel graphic.

the variance parameters. Furthermore, under a Gaussian assumption, the m.l.e. of the mean is the sample average, but we know this is not optimal for heavier-tailed distributions, so that it will tend to be unduly influenced by outliers—which shrinkage is admirably able to counteract. The Laplace distribution is surely also mis-specified, but significantly less so than the normal, so that less "tinkering" via shrinkage is required.

Superimposed on the plot with the dash-dot line are the results based on the same calculations, but having used the EM algorithm (for the normal mixture) for component separation, as in (14.27) and (14.28), instead of via the m.c.d. from (14.29). Analogous to the results from Section 14.3, separation based on m.c.d. is far more successful.

The left panel of Figure 14.25 reproduces the graph of the normalized sum of the realized predictive log-likelihood, based on the m.c.d. split, with $b_1 = b_2 = 5.5$, from the left panel of Figure 14.24 as the dotted line, and overlays it with the analogous results based on jointly estimating all the model parameters via the EM algorithm (hereafter, just "joint estimation"), and having used different priors. In particular, the m.c.d. split and the solid line in the figure both use what we will call prior 1, as given in (14.54). The use of joint estimation is superior to use of the m.c.d. split, though the latter gives a smoother realized predictive log-likelihood as a function of ω than does joint estimation.

The second and third priors we consider are of the same form as (14.54), but with different weights on the covariance matrices; they are given by

$$\begin{aligned}
&\text{prior 1:} \quad a_1 = 2\omega, \quad a_2 = \omega/2, \quad c_1 = c_2 = 20\omega, \\
&\text{prior 2:} \quad a_1 = 1\omega, \quad a_2 = 1\omega, \quad c_1 = c_2 = 20\omega, \\
&\text{prior 3:} \quad a_1 = 1.5\omega, \quad a_2 = 1.5\omega, \quad c_1 = c_2 = 20\omega.
\end{aligned} \tag{14.55}$$

From the figure, we see that priors 2 and 3 not only yield better forecasts for all ω in the relevant range, but, particularly for prior 3, their plots are also much smoother, indicating more stable and reliable estimation. The optimal amount of shrinkage for priors 2 and 3 are $\omega^*(250) = 55$ and $\omega^*(250) = 40$, respectively, and yield nearly the same normalized sum of the realized predictive log-likelihood. In what follows, we use prior 3 and $\omega = 40$.

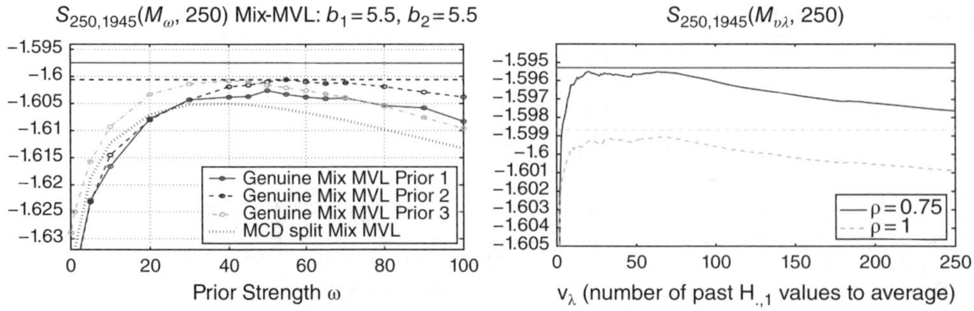

Figure 14.25 **Left:** The normalized sum of the realized predictive log-likelihood versus ω, based on the two-component multivariate Laplace mixture distribution, using a moving window of length $\upsilon = 250$, for $b_1 = b_2 = 5.5$. The line labeled "MCD split Mix MVL" is the same graph as shown in Figure 14.24 with the label "MCD split, $b_1 = 5.5$, $b_2 = 5.5$". The others are the result of joint estimation, using the priors given in (14.55). The horizontal solid line at the top of the graph shows the value obtained based on prior 3, $\omega = 40$, when using weighted likelihood, with $\rho = 0.75$ and applied to just the $\mathbf{\Sigma}_j$. **Right:** This is an analog to the right panel of Figure 14.16: With $\mathcal{M}_{\upsilon_\lambda}$ the $\text{Mix}_2\text{Lap}_{30}$ model with $\hat{\lambda}$ replaced by $\hat{\lambda}_{\upsilon,\upsilon_\lambda}$ the figure shows the normalized sum of the realized predictive log-likelihood versus υ_λ, based on prior 3 for $\omega = 40$, with two weighted likelihood values $\rho = 1$ (corresponding to equal weights) and $\rho = 0.75$, applied to just the $\mathbf{\Sigma}_j$. The two horizontal lines are the result of taking $\hat{\lambda}$ to be $0.75\hat{\lambda}_{250,70} + 0.25\hat{\lambda}_{250,1}$, for each of the two values of ρ.

The method of weighted likelihood via (13.1) is again applicable. The solid horizontal line at the top of the left panel of Figure 14.25 shows the result; it is based on weighted likelihood applied just to the $\hat{\mathbf{\Sigma}}_j$, with parameter $\rho = 0.75$ (and using prior 3 from (14.55) with $\omega = 40$). The right panel of Figure 14.25 is similar to that in Figure 14.16, showing the normalized sum of the realized predictive log likelihood $S_{\tau_0,T}(\mathcal{M}_{\upsilon_\lambda}, \upsilon)$, for $\upsilon = \tau_0 = 250$, as a function of υ_λ, where $\mathcal{M}_{\upsilon_\lambda}$ is the $\text{Mix}_2\text{Lap}_{30}$ model with $\hat{\lambda}$ replaced by $\hat{\lambda}_{\upsilon,\upsilon_\lambda}$ (and having used prior 3 given in (14.55), with $\omega = 40$). Interestingly, just as with the $\text{Mix}_2\text{N}_{30}$ model, we again obtain nearly monotone gains in forecast accuracy as υ_λ is decreased, and a maximum is reached at about $\upsilon_\lambda = 70$. Also, again taking $\hat{\lambda}$ to be $0.75\hat{\lambda}_{250,70} + 0.25\hat{\lambda}_{250,1}$ results in further improvement, as indicated by the horizontal lines.

Figure 14.26 shows the analogous results for the $d = 4$ assets case. From the left panel, we see that the optimal prior strength is $\omega = 15$, and the improvement over and above this when using weighted likelihood with prior 3 and $\rho = 0.65$ is substantial. The right panel shows that υ_λ between 5 and 7 is optimal, though we again observe the massive cutoff in performance at $\upsilon_\lambda = 4$ and below, so that use of this optimal value seems risky—perhaps 10 is a better compromise that is nearly optimal, and further away from the slippery slope. Also, there are local maxima between $\upsilon_\lambda = 50$ and $\upsilon_\lambda = 70$, as with the Mix_2N_4 model for $d = 4$, and coinciding with the global maximum when using $d = 30$ assets.

14.5.4 Estimation of Parameter b

The EM algorithm presented above for obtaining the (prior-augmented) m.l.e. of parameter vector $\boldsymbol{\theta}$, as given in (14.42), for the Mix_kLap_d model so far assumed known $\mathbf{b}$. We discuss here three ways of estimating $\mathbf{b}$, but implement only the third one.

The first way is to augment the EM algorithm such that, at each iteration, a new step maximizes the likelihood with respect to the k unknown values in $\mathbf{b}$, conditional on the other parameters, similar

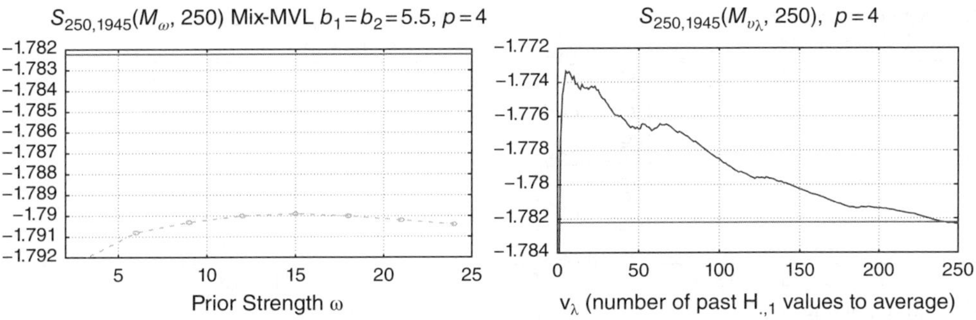

Figure 14.26 **Left:** Similar to the left panel of Figure 14.25, expect for $d = 4$ assets, and just showing the normalized sum of the realized predictive log-likelihood based on joint estimation via the EM algorithm, based on prior 3 in (14.55), and taking $b_1 = b_2 = 5.5$ (as with the $d = 30$ case). The horizontal solid line at the top of the graph shows the value obtained when using prior 3, $\omega = 15$ and weighted likelihood, with $\rho = 0.65$ and applied to just the $\boldsymbol{\Sigma}_j$. **Right:** Similar to the right panels of Figures 14.17 and 14.25, except based on the Mix_2Lap_4 model, $d = 4$, and just for the weighted likelihood hyperparameter $\rho = 0.65$ (as usual, applied just to the $\boldsymbol{\Sigma}_j$). The horizontal solid line is precisely the same as the one in the left panel.

to the way the EM algorithm for a multivariate Student's t, or mixtures of them, can be augmented to estimate the unknown degrees of freedom parameter associated with each component (see, e.g., McLachlan and Peel, 2000, Sec. 7.5). While possible, this step requires use of generic optimization routines, and will thus substantially increase the relative estimation time.

The second way is to use the EM algorithm as we present it, namely with a fixed value of $\mathbf{b}$, and do so for various $\mathbf{b}$ until the likelihood is maximized. A generic optimizer can be applied to this k-dimensional problem, noting that k is typically only two or three, but for every one of its function evaluations the above EM algorithm needs to be run. When tracing out the maximum log-likelihood as a function of $\mathbf{b}$, this results in the profile log-likelihood in $\mathbf{b}$; recall Section 10.2. The resulting choice of $(\widehat{\mathbf{b}}' \, \widehat{\boldsymbol{\theta}}')'$ is the joint m.l.e., and is equivalent to the first method mentioned—and approximately equally time-consuming, though it might exhibit different convergence properties. We do not pursue either of these methods, instead using the third way, which is much faster, but is *not* equivalent to the previous two methods, and is most certainly inferior from an estimation efficiency point of view, though it turns out to be adequate in this model and application context; recall the discussion in Section 14.3.

This third way entails splitting the data into two groups (either by m.c.d. or the EM algorithm) and then examining the profile log-likelihood of the *single-component* multivariate Laplace distribution in b_i, $i = 1, 2$. The benefit of this is that the EM algorithm for the single-component multivariate Laplace distribution with known b is very fast, and only a univariate optimum needs to be located. The downside is that the split into two data sets is imperfect—there is loss of information and introduction of bias, and the resulting chosen $\hat{b}_i$ values will not be the m.l.e.s, nor necessarily share the asymptotic properties of the m.l.e. However, as the b_i dictate the tail behavior of the distribution, the uncertainty (statistical error) associated with their true m.l.e.s is substantial, and so the impact of using this much faster method turns out to be minimal. Moreover, forecasting exercises conducted in Paolella (2015) using a range of $\hat{b}_i$ values indicate that this method is adequate—no gain in forecast quality is achieved by choosing alternative values of the $\hat{b}_i$, while choosing values that are far from those selected by this method indeed results in inferior forecasts.

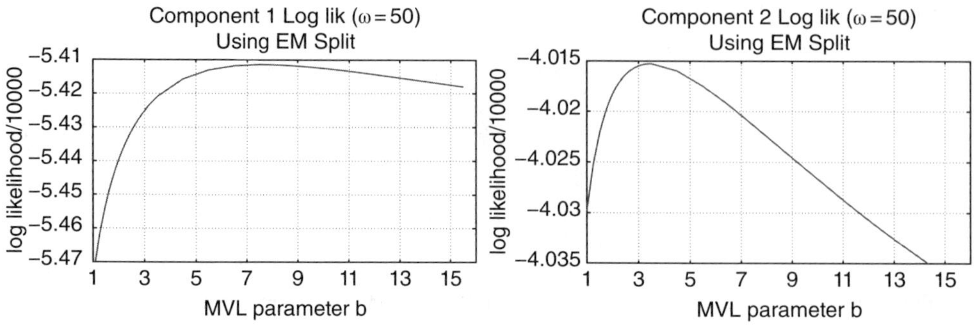

Figure 14.27 Profile log-likelihood as a function of the parameter b of the single-component multivariate Laplace distribution fit to each of the two components of the $T = 1{,}945$ daily returns of the $d = 30$ stocks composing the Dow Jones Industrial Index from June 13, 2001, to March 11, 2009. The data were decomposed into the two components using the EM algorithm for the two-component multivariate Laplace distribution.

To conduct the split, we proceed as follows: (i) arbitrary values of b_1 and b_2 are chosen, (ii) the EM algorithm for the two-component multivariate mixture Laplace distribution is run, and (iii) using (14.44), those observations for which $\widehat{H}_{t,1} = \hbar_{t,j} > 0.99$ are assigned to component 1, otherwise to component 2. (The value of 0.99 of course represents a tuning parameter that will affect the final results. Several reasonable values were tried, and the results were not sensitive to its choice.) Step (iv) then estimates single-component multivariate Laplace distributions, for each of the two components, over a grid of b_1 and b_2 values (the choice of grid being discussed below) and in (v) the optimal values of the b_i are obtained from the two profile log-likelihoods. These optimal values of the b_i are then used as the fixed values instead of those chosen in step (i), and the iterative method is conducted again, starting with step (ii). This could be repeated "until convergence", but we use just two iterations.

Figure 14.27 shows the results when this method is applied to the DJIA data. It depicts the profile log-likelihood (divided by 10,000) of b_1 and b_2, for each of the two components (which have, respectively, 1,312 and 633 observations). The estimations also use a shrinkage prior with weight $\omega = 50$. The maximum is approximately $\widehat{\mathbf{b}} = (7.5, 3.5)$. From the plots, it is obvious that the sampling error associated with b_1 is higher than that of b_2. The profile log-likelihood of b_2 is far more peaked, with a maximum around three, so that that the second component has higher kurtosis than the normal. Thus, in the context of modeling daily multivariate financial asset returns data, the Mix_2Lap_d will have a clear advantage over the Mix_2N_d.

Also observe that use of the value $b = (d + 1)/2 = 15.5$ is, particularly for component 2, highly untenable, so that the simplified multivariate Laplace density without the Bessel function in (14.33), and the simplification in (14.38), are unfortunately of no use.

14.5.5 Portfolio Distribution and Expected Shortfall

As with the Mix_kN_d case, of interest is the distribution of $P = P_t(\mathbf{a}, \boldsymbol{\theta}) = \mathbf{a}'\mathbf{Y}_t$ when $\mathbf{Y}_t \sim \text{Mix}_k\text{Lap}_d(\boldsymbol{M}, \boldsymbol{\Psi}, \boldsymbol{\lambda}, \mathbf{b})$. First consider the case with $\mathbf{L} \sim \text{Lap}(\boldsymbol{\mu}, \boldsymbol{\Sigma}, b)$ with density (14.31). Then, for $\mathbf{a} \in \mathbb{R}^d$, $P = \mathbf{a}'\mathbf{L} \sim \text{Lap}(\mathbf{a}'\boldsymbol{\mu}, \mathbf{a}'\boldsymbol{\Sigma}\mathbf{a}', b)$, where this is (14.31) with $d = 1$. The result is a special case for normal mean–variance mixture distributions (in particular, the multivariate generalized hyperbolic, or MGHyp, but also the multivariate noncentral t), as shown in, e.g., McNeil et al. (2005, p. 76). Now let $\mathbf{Y} \sim \text{Mix}_k\text{Lap}_d(\boldsymbol{M}, \boldsymbol{\Psi}, \boldsymbol{\lambda}, \mathbf{b})$, with density (14.41), for $\boldsymbol{M} = \begin{bmatrix}\boldsymbol{\mu}_1 & \boldsymbol{\mu}_2 & \cdots & \boldsymbol{\mu}_k\end{bmatrix}$,

$\boldsymbol{\Psi} = \left[\boldsymbol{\Sigma}_1 \ \boldsymbol{\Sigma}_2 \ \cdots \ \boldsymbol{\Sigma}_k\right]$, $\mathbf{b} = (b_1, \ldots, b_k)'$, and $\boldsymbol{\lambda} = (\lambda_1, \ldots, \lambda_k)$. Let $P = \mathbf{a}'\mathbf{Y}$. Then, analogous to result (14.12) and using the same format of proof, we find that

$$f_P(x) = \sum_{c=1}^{k} \lambda_c \ \mathrm{Lap}(x; \mathbf{a}'\boldsymbol{\mu}_c, \mathbf{a}'\boldsymbol{\Sigma}_c\mathbf{a}, b_c). \tag{14.56}$$

Notice that the univariate Laplace distributions in (14.56) are of the form (14.31), with $d = 1$. Similar to (14.16), and in light of (14.32), we have, with $\mu_c = \mathbf{a}'\boldsymbol{\mu}_c$ and $\sigma_c^2 = \mathbf{a}'\boldsymbol{\Sigma}_c\mathbf{a}$, $c = 1, \ldots, k$,

$$\mu_P = \mathbb{E}[P] = \sum_{c=1}^{k} \lambda_c \mu_c, \qquad \sigma_P^2 = \mathbb{V}(P) = \sum_{c=1}^{k} \lambda_c (b_c \sigma_c^2 + \mu_c^2) - \mu_P^2. \tag{14.57}$$

Also, analogous to (14.17), and denoting the density and distribution function of the univariate Laplace distribution in (14.56) as f and F, respectively, calculation shows that

$$\mathrm{ES}(P, \xi) = \sum_{j=1}^{k} \frac{\lambda_j F(c_j, 0, 1, b_j)}{\xi} \left(\mu_j - \sigma_j b_j \frac{f(c_j, 0, 1, b_j + 1)}{F(c_j, 0, 1, b_j)} \right), \quad c_j = \frac{q_{P,\xi} - \mu_j}{\sigma_j}. \tag{14.58}$$

The c.d.f. F can be calculated with numeric integration.

14.5.6 Fast Evaluation of the Bessel Function

We finally return to the issue of evaluating the Bessel function (14.30). An asymptotic expansion of $K_v(z)$ as given in Watson (1922, p. 202) is

$$K_v(z) = \sqrt{\frac{\pi}{2z}} e^{-z} \times E(v, z), \quad \text{where}$$

$$E(v, z) = 1 + \frac{4v^2 - 1^2}{1!8z} + \frac{(4v^2 - 1^2)(4v^2 - 3^2)}{2!(8z)^2} + \frac{(4v^2 - 1^2)(4v^2 - 3^2)(4v^2 - 5^2)}{3!(8z)^3} + \cdots .$$

Thus, when v is of the form $v = n - 1/2$, $n = 1, 2, \ldots$, we have the exact relations

$$\begin{aligned} E(1/2, z) &= 1, \\ E(3/2, z) &= (1 + 1/z), \\ E(5/2, z) &= (1 + 3/z + 3/z^2), \\ E(7/2, z) &= 1 + 6/z + 15/z^2 + 15/z^3, \end{aligned}$$

and, in general, we find

$$E(v, z) = 1 + \sum_{i=1}^{k} \frac{(k+i)!}{2^i (k-i)! i!} \frac{1}{z^i}, \quad \text{for } v = n - \frac{1}{2}, \quad n \in \mathbb{N}. \tag{14.59}$$

Based on Figure 14.27, it appears as though we can safely suffice ourselves with evaluating the log-likelihood on a grid of half-integer b-values.

Remark For the disconcerted reader who objects to the use of a grid of half-integer b-values and prefers to have the m.l.e. evaluated to near machine precision, we offer the gentle reminder that there is always a trade-off in real applications between estimation speed and accuracy, and the purpose of

```
function K=quickbesselk(v,z)
v=abs(v); k=v-0.5;
if (k-floor(k)<1e-8) && (k<=12.01)
  S=1; k=floor(k);
  cmat=[
    1    0     0       0        0         0          0          0           0            0            0            0
    3    3     0       0        0         0          0          0           0            0            0            0
    6   15    15       0        0         0          0          0           0            0            0            0
   10   45   105     105        0         0          0          0           0            0            0            0
   15  105   420     945      945         0          0          0           0            0            0            0
   21  210  1260    4725    10395     10395          0          0           0            0            0            0
   28  378  3150   17325    62370    135135     135135          0           0            0            0            0
   36  630  6930   51975   270270    945945    2027025    2027025           0            0            0            0
   45  990 13860  135135   945945   4729725   16216200   34459425    34459425            0            0            0
   55 1485 25740  315315  2837835  18918900   91891800  310134825   654729075    654729075            0            0
   66 2145 45045  675675  7567560  64324260  413513100 1964187225  6547290750  13749310575  13749310575            0
   78 3003 75075 1351350 18378360 192972780 1571349780 9820936125 45831035250 151242416325 316234143225 316234143225
  ];
  for i=1:k
    % First way: quicker than Matlab's besselk only for very small k
    %coef = exp( gammaln(k+i+1)-gammaln(k-i+1)-gammaln(i+1) - i*log(2) );

    % Second way: also not faster for large k, but delivers more accurate coefficients of the matrix above
    % t1=fact(k+i); t2=fact(k-i); t3=fact(i); coef = t1/t2/t3/2^i

    coef = cmat(k,i); % The fastest way!
    S=S+ coef / z^i;
  end
  K=sqrt(pi/2/z) * exp(-z) * S;
else
  K=besselk(v,z);
end
```

Program Listing 14.12: Fast computation of the Bessel function $K_v(z)$, based on (14.59) and pre-computation of the relevant coefficients.

the application should help dictate this. Recall the discussion in Section 14.3. If the goal is density forecasting and financial portfolio optimization, then it turns out to make no appreciable difference if the $\hat{b}_i$-values are restricted to the half-integer grid. Moreover, by changing the (often arbitrarily chosen) length, T, of the data set (or the window size, when used in a backtesting exercise), the point estimates of the $\hat{b}_i$ can vary considerably more than 0.5 or 1.0, thus rendering the concept of "the exact m.l.e." meaningless. ■

Use of (14.59) is about 10 times faster to evaluate than Matlab's implementation, but only when having pre-computed the coefficients in (14.59). This is done for the function in Program listing 14.12, and was constructed such that, for $d = 30$, we can deliver the values of the required Bessel function for $b = 0.5, 1.5, \dots, 27.5$. It requires the following short program to evaluate the factorial function without recourse to the gamma function:

```
function p=fact(n), if n-1<1e-14, p=1; else p=n*fact(n-1); end
```

Part IV

Appendices

Appendix A

Distribution of Quadratic Forms

Quadratic forms in normal variables play a key role in the distribution theory associated with linear regression and time-series models. This appendix develops the tools necessary for understanding the material in Chapters 1 to 5, as well as the more specialized material in Appendix B.

Function $\mathbb{R}^n \to \mathbb{R}$ is a **quadratic form** if it can be expressed as $\mathbf{x} \mapsto \mathbf{x}'\mathbf{A}\mathbf{x}$, where $\mathbf{A}$ is an $n \times n$ real symmetric matrix. Let $\mathbf{A}$ be such a matrix, and let $\mathbf{X} \sim \mathrm{N}_n(\boldsymbol{\mu}, \boldsymbol{\Sigma})$ with $\boldsymbol{\Sigma} > 0$. The scalar random variable (hereafter abbreviated r.v.) $Y = \mathbf{X}'\mathbf{A}\mathbf{X}$ is referred to as a quadratic form (in normal variables).

If $Y = \mathbf{X}'\mathbf{B}\mathbf{X}$ with $\mathbf{B}$ not symmetric, observe that, as a scalar, $Y' = Y$, i.e., $(\mathbf{X}'\mathbf{B}\mathbf{X})' = \mathbf{X}'\mathbf{B}'\mathbf{X}$ so that $Y = \mathbf{X}'(\mathbf{B}+\mathbf{B}')\mathbf{X}/2 = \mathbf{X}'\mathbf{A}\mathbf{X}$ with $\mathbf{A} = (\mathbf{B}+\mathbf{B}')/2$. Matrix $\mathbf{A}$ is symmetric, so there is no loss in generality in working with symmetric matrices.

A more general structure is a **bilinear quadratic form**, $\mathbf{X}'\mathbf{A}\mathbf{Y}$, where $\mathbf{X}$ is $n \times 1$, $\mathbf{Y}$ is $m \times 1$ and $\mathbf{A}$ is $n \times m$. If $n = m$ and $\mathbf{A}$ is symmetric, then $\mathbf{X}'\mathbf{A}\mathbf{Y}$ can be written as a quadratic form via

$$\mathbf{X}'\mathbf{A}\mathbf{Y} = \mathbf{Z}'\mathbf{B}\mathbf{Z} \quad \text{for} \quad \mathbf{Z} = \begin{pmatrix} \mathbf{X} \\ \mathbf{Y} \end{pmatrix}, \ \mathbf{B} = \frac{1}{2}\begin{pmatrix} \mathbf{0} & \mathbf{A} \\ \mathbf{A} & \mathbf{0} \end{pmatrix}.$$

A.1 Distribution and Moments

Let $Y = \mathbf{X}'\mathbf{A}\mathbf{X}$ with $\mathbf{X} \sim \mathrm{N}_n(\boldsymbol{\mu}, \boldsymbol{\Sigma})$, $\boldsymbol{\Sigma} > 0$. While the density of $\mathbf{X}$ is tractable, Y is a very complicated function of $\mathbf{X}$, though it possesses a structure that lends itself to expression in simpler terms; this is the key to studying its distribution.

A.1.1 Probability Density and Cumulative Distribution Functions

First, let $\boldsymbol{\Sigma}^{\frac{1}{2}}$ be a matrix such that $\boldsymbol{\Sigma}^{\frac{1}{2}}\boldsymbol{\Sigma}^{\frac{1}{2}} = \boldsymbol{\Sigma}$. Recall that $\boldsymbol{\Sigma}^{\frac{1}{2}}$ is easily computed using the spectral decomposition and is symmetric and positive-definite. Then $\boldsymbol{\Sigma}^{-\frac{1}{2}}\mathbf{X} \sim \mathrm{N}_n\left(\boldsymbol{\Sigma}^{-\frac{1}{2}}\boldsymbol{\mu}, \mathbf{I}\right)$. Next, write

$$Y = \mathbf{X}'\mathbf{A}\mathbf{X} = \mathbf{X}'\mathbf{I}\mathbf{A}\mathbf{I}\mathbf{X} = \mathbf{X}'\boldsymbol{\Sigma}^{-\frac{1}{2}}\boldsymbol{\Sigma}^{\frac{1}{2}}\mathbf{A}\boldsymbol{\Sigma}^{\frac{1}{2}}\boldsymbol{\Sigma}^{-\frac{1}{2}}\mathbf{X}, \tag{A.1}$$

and let the spectral decomposition of $\boldsymbol{\Sigma}^{\frac{1}{2}}\mathbf{A}\boldsymbol{\Sigma}^{\frac{1}{2}}$ be given by $\mathbf{P}\boldsymbol{\Lambda}\mathbf{P}'$, where $\mathbf{P}$ is an orthogonal matrix and $\boldsymbol{\Lambda} = \mathrm{diag}([\lambda_1, \dots, \lambda_n]) = \mathrm{Eig}(\boldsymbol{\Sigma}^{\frac{1}{2}}\mathbf{A}\boldsymbol{\Sigma}^{\frac{1}{2}}) = \mathrm{Eig}(\boldsymbol{\Sigma}\mathbf{A}) = \mathrm{Eig}(\mathbf{A}\boldsymbol{\Sigma})$. Then, from (A.1),

$$F_Y(y) = \Pr\left(\mathbf{X}'\boldsymbol{\Sigma}^{-\frac{1}{2}}\mathbf{P}\boldsymbol{\Lambda}\mathbf{P}'\boldsymbol{\Sigma}^{-\frac{1}{2}}\mathbf{X} \leqslant y\right) = \Pr(\mathbf{W}'\boldsymbol{\Lambda}\mathbf{W} \leqslant y), \tag{A.2}$$

Linear Models and Time-Series Analysis: Regression, ANOVA, ARMA and GARCH, First Edition. Marc S. Paolella.

where

$$\mathbf{W} = \mathbf{P}'\boldsymbol{\Sigma}^{-\frac{1}{2}}\mathbf{X} \sim \mathrm{N}(\boldsymbol{\nu}, \mathbf{I}_n), \qquad \boldsymbol{\nu} = \mathbf{P}'\boldsymbol{\Sigma}^{-\frac{1}{2}}\boldsymbol{\mu} = (\nu_1, \dots, \nu_n)'. \tag{A.3}$$

This decomposition is sometimes referred to as the **principle axis theorem**; see Scheffé (1959, p. 397).

Recall the definition of a noncentral χ^2 random variable: If $(X_1, \dots, X_n) \sim \mathrm{N}_n(\boldsymbol{\mu}, \mathbf{I})$, with $\boldsymbol{\mu} = (\mu_1, \dots, \mu_n)'$, then $X = \sum_{i=1}^{n} X_i^2$ follows a noncentral χ^2 distribution with n degrees of freedom and noncentrality parameter $\theta = \sum_{i=1}^{n} \mu_i^2$. We write $X \sim \chi^2(n, \theta)$.

From (A.2) and (A.3),

$$\mathbf{W}'\boldsymbol{\Lambda}\mathbf{W} = \sum_{i=1}^{\mathrm{rank}(\mathbf{A})} \lambda_i W_i^2, \qquad W_i^2 \overset{\mathrm{ind}}{\sim} \chi^2(1, \nu_i^2), \tag{A.4}$$

is a weighted sum of rank $(\mathbf{A})$ independent noncentral χ^2 random variables, each with one degree of freedom. Methods and programs for computing this distribution are detailed in Section II.10.1.5, including inversion of the characteristic function and the saddlepoint approximation (s.p.a.). The program in Listing A.1 implements this decomposition to compute the p.d.f. and c.d.f. using both ways.

Example A.1 Let $\mathbf{X} = (X_1, \dots, X_n)' \sim \mathrm{N}_n(\boldsymbol{\mu}, \boldsymbol{\Sigma})$ with $\boldsymbol{\Sigma} > 0$ and $\boldsymbol{\mu} = (\mu_1, \dots, \mu_n)'$. Consider the sample variance of $X_1, \dots, X_n$, denoted S^2. Let $\mathbf{1}_n$ denote an n-length column vector of ones, $\mathbf{J}_n$ an $n \times n$ matrix of ones, and

$$\mathbf{M} = \mathbf{I}_n - \mathbf{1}_n(\mathbf{1}_n'\mathbf{1}_n)^{-1}\mathbf{1}_n' = \mathbf{I}_n - n^{-1}\mathbf{J}_n, \tag{A.5}$$

so that $\mathbf{MX} = \mathbf{X} - \bar{X}$. As detailed in Chapter 1, $\mathbf{M}$ is a rank $m = n - 1$ matrix with one eigenvalue equal to zero and $n - 1$ eigenvalues equal to one. It is easy to confirm that $\mathbf{M}' = \mathbf{M}$ and $\mathbf{MM} = \mathbf{M}$, so that

$$Y = \sum_{i=1}^{n} (X_i - \bar{X})^2 = (\mathbf{MX})'(\mathbf{MX}) = \mathbf{X}'\mathbf{M}'\mathbf{MX} = \mathbf{X}'\mathbf{MX}$$

```
1  function [f,F,svec]=XAXdistribution(xvec,mu,Sigma,A,spa)
2  if nargin<5, spa=1; end
3  svec=[]; lx = length(xvec); f=zeros(lx,1); F=f;
4  [V,D]=eig(0.5*(Sigma+Sigma')); W=sqrt(D); Sighalf = V * W * V';
5  R=Sighalf*A*Sighalf; [P,Lam]=eig(R); lam=diag(Lam); % R = P Lam P'
6  v = P' * inv(Sighalf) * mu; nonc = v.^2;
7  ok=abs(lam)>1e-7; nonc=nonc(ok); lam=lam(ok); dfvec=ones(length(lam),1);
8  if spa==1, [f,F,svec] = spaweightedsumofchisquare(2,xvec,lam,dfvec,nonc);
9  else       [f,F]=weightedsumofchisquare(xvec,lam,dfvec,nonc);
10 end
```

Program Listing A.1: The p.d.f. and c.d.f. of $\mathbf{X}'\mathbf{AX}$ evaluated at each element of `xvec`, where $\mathbf{X} \sim \mathrm{N}_n(\boldsymbol{\mu}, \boldsymbol{\Sigma})$, using either the inversion formulae and numeric integration or the saddlepoint approximation. Programs `weightedsumofchisquare.m` and `spaweightedsumofchisquare.m` use the methods developed in Section II.10.1 and are available in the collection of programs. The reason for using $(\boldsymbol{\Sigma} + \boldsymbol{\Sigma}')/2$ instead of simply $\boldsymbol{\Sigma}$ in the fourth line of the program is that Matlab's eigenvalue/vector routine `eig` is apparently quite sensitive to numerically small deviations from symmetry. For symmetric matrix $\mathbf{A}$, it should be the case that calling `[V,D]=eig(A)` yields orthogonal $\mathbf{V}$ and real diagonal $\mathbf{D}$ such that $\mathbf{A} = \mathbf{VDV}'$. Perturbing $\mathbf{A}$ slightly can render $\mathbf{V}$ non-orthogonal and $\mathbf{A} \neq \mathbf{VDV}'$.

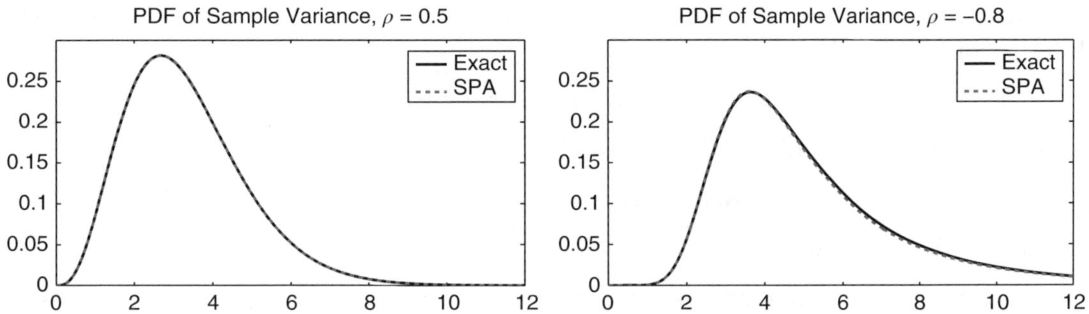

Figure A.1 True (via inversion formula) and second-order s.p.a. density of the sample variance S^2, for a sample of size 10 for $\mathbf{X} \sim \mathrm{N}(\boldsymbol{\mu}, \boldsymbol{\Sigma})$ with $\boldsymbol{\mu} = (-2,-1,0,1,2,2,1,0,-1,-2)'$ and $\boldsymbol{\Sigma}$ corresponding to an AR(1) process with parameter ρ. In the left panel, for $\rho = 0.5$, the two graphs are optically indistinguishable. The s.p.a. is about 14 times faster to compute.

```
function [f,F]=samplevariancedistribution(xvec,mu,Sigma,spa)
if nargin<4, spa=1; end
n = length(mu); X = ones(n,1); A = eye(n) - X * inv(X'*X) * X';
m=length(mu)-1; [f0,F0]=XAXdistribution(m*xvec,mu,Sigma,A,spa);
f=m*f0; F=F0;
```

Program Listing A.2: The p.d.f. and c.d.f. at values in `xvec` of the sample variance S^2 for data $\mathbf{X} \sim \mathrm{N}(\boldsymbol{\mu}, \boldsymbol{\Sigma})$ using either numeric inversion of the characteristic function or the saddlepoint approximation.

is a quadratic form. The eigenvalues $\{\lambda_i\}$ of $\boldsymbol{\Sigma}^{\frac{1}{2}}\mathbf{M}\boldsymbol{\Sigma}^{\frac{1}{2}}$ are nonnegative because $\mathbf{M}$ is positive semi-definite and $\boldsymbol{\Sigma}$ is positive definite.[1]

A scale transformation yields the density $f_{S^2}(s) = mf_Y(ms)$, $m = n-1$, while the c.d.f. is given by $F_{S^2}(s) = \Pr(Y \leqslant ms)$. These can be computed via the program in Listing A.2.

To illustrate, let $n = 10$, $\boldsymbol{\mu} = (-2,-1,0,1,2,2,1,0,-1,-2)'$, and $\boldsymbol{\Sigma}$ correspond to a first-order autoregressive, or AR(1), process with parameter ρ, for which the (i,j)th element of $\boldsymbol{\Sigma}$ is given by $\rho^{|i-j|}/(1-\rho^2)$, as detailed in Chapter 4. Figure A.1 plots the density f_{S^2} for two values of ρ and demonstrates the accuracy of the s.p.a. See also Problem A.1.[2] ■

A.1.2 Positive Integer Moments

Although the raw moments of $Y = \mathbf{X}'\mathbf{A}\mathbf{X}$ for $\mathbf{X} \sim \mathrm{N}_n(\boldsymbol{\mu}, \boldsymbol{\Sigma})$ are straightforward to compute using the decomposition in (A.4) and the moments for a noncentral χ^2, direct calculations are also possible.

1 To see this, first observe that, if $\boldsymbol{\Sigma}$ positive definite, then $\boldsymbol{\Sigma}^{\frac{1}{2}}$ can be constructed and is also positive definite, so that for any vector $\mathbf{w} \in \mathbb{R}^n \backslash \mathbf{0}$, $\mathbf{z} := \boldsymbol{\Sigma}^{\frac{1}{2}}\mathbf{w} \neq \mathbf{0}$. Next, $\mathbf{w}'\boldsymbol{\Sigma}^{\frac{1}{2}}\mathbf{M}\boldsymbol{\Sigma}^{\frac{1}{2}}\mathbf{w} = \mathbf{z}'\mathbf{M}\mathbf{z} \geqslant 0$ because $\mathbf{M}$ is positive semi-definite. (Of course, $\mathbf{z}'\mathbf{M}\mathbf{z} \geqslant 0$ also follows simply because $Y = (n-1)S^2 = \mathbf{X}'\mathbf{M}\mathbf{X}$ cannot be negative.) Thus, the eigenvalues of $\boldsymbol{\Sigma}^{\frac{1}{2}}\mathbf{M}\boldsymbol{\Sigma}^{\frac{1}{2}}$ are nonnegative.

2 As is also mentioned in Chapter 1, we use the tombstone, QED, or halmos, symbol ■ to denote the end of proofs of theorems, as well as examples and remarks, acknowledging that it is traditionally only used for the former, as popularized by Paul Halmos.

For the expected value of Y, begin with the obvious relationship $Y = \text{tr}(Y)$ and use the fact that, for conformable matrices, $\text{tr}(\mathbf{AB}) = \text{tr}(\mathbf{BA})$ to get

$$\mathbb{E}[Y] = \mathbb{E}[\text{tr}(\mathbf{X}'\mathbf{AX})] = \mathbb{E}[\text{tr}(\mathbf{AXX}')] = \text{tr}\mathbb{E}[\mathbf{AXX}'] = \text{tr}(\mathbf{A}\ \mathbb{E}[\mathbf{XX}']).$$

Next, as $\boldsymbol{\Sigma} = \mathbb{E}[(\mathbf{X} - \boldsymbol{\mu})(\mathbf{X} - \boldsymbol{\mu})'] = \mathbb{E}[\mathbf{XX}'] - \boldsymbol{\mu}\boldsymbol{\mu}'$, it follows that

$$\mathbb{E}[Y] = \text{tr}(\mathbf{A}(\boldsymbol{\Sigma} + \boldsymbol{\mu}\boldsymbol{\mu}')) = \text{tr}(\mathbf{A}\boldsymbol{\Sigma}) + \boldsymbol{\mu}'\mathbf{A}\boldsymbol{\mu}. \tag{A.6}$$

Some more effort is required to show that

$$\text{Cov}(\mathbf{X}'\mathbf{AX}, \mathbf{X}'\mathbf{BX}) = 2\text{tr}(\mathbf{A}\boldsymbol{\Sigma}\mathbf{B}\boldsymbol{\Sigma}) + 4\boldsymbol{\mu}'\mathbf{A}\boldsymbol{\Sigma}\mathbf{B}\boldsymbol{\mu}, \tag{A.7}$$

from which

$$\mathbb{V}(Y) = 2\text{tr}[(\mathbf{A}\boldsymbol{\Sigma})^2] + 4\boldsymbol{\mu}'\mathbf{A}\boldsymbol{\Sigma}\mathbf{A}\boldsymbol{\mu} \tag{A.8}$$

follows. Proofs of (A.7) can be found in, e.g., Searle (1971, Sec. 2.5) (where original references can be found), Graybill (1983, p. 367), Mathai and Provost (1992, pp. 53, 76), and Schott (2005, p. 418). Also of interest is that $\text{Cov}(\mathbf{CX}, \mathbf{X}'\mathbf{AX}) = 2\mathbf{C}\boldsymbol{\Sigma}\mathbf{A}\boldsymbol{\mu}$, the proof of which can be found in the previous references.

The positive integer moments of Y are tractable when $\mathbf{X} \sim \text{N}(\mathbf{0}, \mathbf{I})$. From (A.4), we can write $Y = \sum_{i=1}^{n} \lambda_i W_i^2$, where $W_i^2 \overset{\text{i.i.d.}}{\sim} \chi^2(1)$ and $\{\lambda_i\} = \text{Eig}(\mathbf{A})$. As the moment generating function (m.g.f.) of W_i^2 is $(1 - 2s)^{-1/2}$ for $s < 1/2$,

$$\mathbb{M}_Y(s) = \prod_{i=1}^{n} (1 - 2\lambda_i s)^{-1/2}, \quad \mathbb{K}_Y(s) = \log \mathbb{M}_Y(s) = -\frac{1}{2} \sum_{i=1}^{n} \log(1 - 2\lambda_i s), \tag{A.9}$$

with rth derivative, $r = 1, 2, \ldots$, easily verified to be

$$\frac{\text{d}^r \mathbb{K}_Y(s)}{\text{d}s^r} = \alpha_r \sum_{i=1}^{n} \frac{\lambda_i^r}{(1 - 2\lambda_i s)^r}, \tag{A.10}$$

where $\alpha_r = 2(r-1)\alpha_{r-1} = (r-1)!2^{r-1}$, in particular, $\alpha_1 = 1, \alpha_2 = 2, \alpha_3 = 8$, and $\alpha_4 = 48$. Recalling that the rth derivative of $\mathbb{K}_Y(s)$ evaluated at $s = 0$ is the rth cumulant κ_r, and the relationship between the trace of a matrix and its eigenvalues, we have

$$\kappa_r = \alpha_r \sum_{i=1}^{n} \lambda_i^r = \alpha_r t_r, \quad t_r := \sum_{i=1}^{n} \lambda_i^r = \text{tr}(\mathbf{A}^r).$$

Next, using the relationship between cumulants and moments given by

$$\mathbb{E}[Y^r] = \sum_{j=0}^{r-1} \binom{r-1}{j} \kappa_{j+1} \mathbb{E}[Y^{r-1-j}], \quad r = 0, 1, \ldots, \tag{A.11}$$

(see, e.g., Severini, 2005, p. 114), we have, inserting the above expressions for κ_{j+1} and α_r, and reversing the sum,

$$\mathbb{E}[Y^r] = (r-1)! \sum_{j=0}^{r-1} 2^j \frac{t_{j+1}\mathbb{E}[Y^{r-1-j}]}{(r-1-j)!} = (r-1)! \sum_{i=0}^{r-1} \frac{2^{r-i-1}}{i!} t_{r-i} \mathbb{E}[Y^i]. \tag{A.12}$$

For $r = 1$, $\mathbb{E}[Y] = \text{tr}(\mathbf{A})$, while for $r = 2$,

$$\mathbb{E}[Y^2] = 2\,\text{tr}(\mathbf{A}^2) + [\text{tr}(\mathbf{A})]^2, \tag{A.13}$$

so that $\mathbb{V}(Y) = 2\,\text{tr}(\mathbf{A}^2)$, which agrees with (A.8). Lastly, for $r = 3$, (A.12) simplifies to

$$\mathbb{E}[Y^3] = 8\,\text{tr}(\mathbf{A}^3) + 6\text{tr}(\mathbf{A}^2)\,\text{tr}(\mathbf{A}) + [\text{tr}(\mathbf{A})]^3. \tag{A.14}$$

Example A.2 For $n = 10$ and $\boldsymbol{\mu}$ as given in Example A.1, with $\rho = 0.5$, $\mathbb{E}[Y]$ using decomposition (A.4) is $\sum_{i=1}^{9} \lambda_i(1 + \nu_i^2) = 29.9$. Use of (A.6) yields the same. This gives $\mathbb{E}[S^2] = 3.32$. Similarly, $\mathbb{V}(Y) = \sum_{i=1}^{9} \lambda_i^2(2 + 4\nu_i^2) = 186.3$, agreeing with that from (A.8). ■

Remark The use of quadratic forms and their moments arises in so-called **delta-gamma hedging** in financial risk management, in particular calculation of the value at risk (VaR) of non-linear portfolios. See, e.g., Rounvinez (1997), Britten-Jones and Schaefer (1999), Castellacci and Siclari (2003), Jondeau et al. (2007, Sec. 8.5.2), and the references therein. ■

A.1.3 Moment Generating Functions

This section derives some m.g.f.s that are useful in various contexts. We begin with a different derivation than that used in (A.9) for the m.g.f. of $Y = \mathbf{X}'\mathbf{A}\mathbf{X}$ for $\mathbf{X} \sim \text{N}_n(\mathbf{0}, \mathbf{I})$, then consider the m.g.f. of $Y = \mathbf{X}'\mathbf{A}\mathbf{X}$ for $\mathbf{X} \sim \text{N}_n(\boldsymbol{\mu}, \mathbf{I})$, and finally derive the joint m.g.f. of $N = \mathbf{X}'\mathbf{A}\mathbf{X}$ and $D = \mathbf{X}'\mathbf{B}\mathbf{X}$ for $\mathbf{X} \sim \text{N}_n(\boldsymbol{\mu}, \mathbf{I})$.

First observe that, from the (non-degenerate) n-dimensional multivariate normal distribution, we immediately have the identity

$$|\boldsymbol{\Sigma}|^{1/2} = \int_{\mathbb{R}^n} (2\pi)^{-n/2} \exp\left\{-\frac{1}{2}\mathbf{x}'\boldsymbol{\Sigma}^{-1}\mathbf{x}\right\} d\mathbf{x}. \tag{A.15}$$

Let $\mathbf{X} \sim \text{N}_n(\mathbf{0}, \mathbf{I})$ and let $\mathbf{A}$ be a symmetric $n \times n$ matrix with spectral decomposition $\mathbf{A} = \mathbf{O}\mathbf{D}\mathbf{O}'$ with $\mathbf{D} = \text{diag}([\lambda_1, \dots, \lambda_n])$ the eigenvalues of $\mathbf{A}$. The m.g.f. of $Y = \mathbf{X}'\mathbf{A}\mathbf{X}$ is

$$\begin{aligned}
\mathbb{M}_Y(s) = \mathbb{E}\left[e^{s\mathbf{X}'\mathbf{A}\mathbf{X}}\right] &= \int_{\mathbb{R}^n} (2\pi)^{-n/2} \exp\left\{s\mathbf{x}'\mathbf{A}\mathbf{x} - \frac{1}{2}\mathbf{x}'\mathbf{x}\right\} d\mathbf{x} \\
&= \int_{\mathbb{R}^n} (2\pi)^{-n/2} \exp\left\{-\frac{1}{2}\mathbf{x}'(\mathbf{I} - 2s\mathbf{A})\mathbf{x}\right\} d\mathbf{x} \\
&\overset{(A.15)}{=} |\mathbf{I} - 2s\mathbf{A}|^{-1/2}.
\end{aligned} \tag{A.16}$$

As $1 = |\mathbf{I}| = |\mathbf{O}'\mathbf{O}|$ and the determinant of a product is the product of determinants,

$$|\mathbf{O}'||\mathbf{I} - 2s\mathbf{A}||\mathbf{O}| = |\mathbf{O}'(\mathbf{I} - 2s\mathbf{A})\mathbf{O}| = |\mathbf{I} - 2s\mathbf{D}| = \prod_{i=1}^{n}(1 - 2s\lambda_i), \tag{A.17}$$

so that (A.16) can also be expressed as in (A.9), with convergence strip determined as follows. We want $1 - 2s\lambda_i > 0, i = 1, \dots, n$, and if $\lambda_i > 0$, then $1 - 2s\lambda_i > 0 \Leftrightarrow s < 1/(2\lambda_i)$, and if $\lambda_i < 0$, then $1 - 2s\lambda_i > 0 \Leftrightarrow s > 1/(2\lambda_i)$. Let $\underline{\lambda} = 2\min \lambda_i$ and $\bar{\lambda} = 2\max \lambda_i$. If $\underline{\lambda} > 0$ (so that all λ_i are positive), then $\mathbb{M}_Y(s)$ is finite for $s < \bar{\lambda}^{-1}$. If $\bar{\lambda} < 0$ (so that all λ_i are negative), then $\mathbb{M}_Y(s)$ is finite for $s > \underline{\lambda}^{-1}$. Otherwise, $\mathbb{M}_Y(s)$ exists for $\underline{\lambda}^{-1} < s < \bar{\lambda}^{-1}$.

Turning now to the case with nonzero mean, the first fact we need is that the m.g.f. of $W^2 \sim \chi^2(n,\theta)$ is given by

$$\mathbb{M}_{W^2}(s) = (1-2s)^{-n/2} \exp\left\{\frac{s\theta}{1-2s}\right\}, \quad s < 1/2, \tag{A.18}$$

as was shown in two ways in Problem II.10.6. Let $W_i^2 \overset{\text{ind}}{\sim} \chi^2(n_i, \nu_i^2)$, $i = 1, \dots, n$, and let $S = \sum_{i=1}^n \lambda_i W_i^2$. It follows from (A.18) and the independence of the W_i^2 that

$$\mathbb{M}_S(s) = \prod_{i=1}^n \mathbb{M}_{W_i^2}(\lambda_i s) = \prod_{i=1}^n (1-2\lambda_i s)^{-n_i/2} \exp\left\{\frac{\lambda_i s \nu_i^2}{1-2\lambda_i s}\right\}, \tag{A.19}$$

with convergence strip determined exactly the same as was done after (A.17). The case with $n_i = 1$, $i = 1, \dots, n$ is often of most interest, as in (A.4).

Let $Y = \mathbf{X}'\mathbf{A}\mathbf{X}$, $\mathbf{X} \sim \mathrm{N}_n(\boldsymbol{\mu}, \mathbf{I})$, spectral decomposition $\mathbf{A} = \mathbf{P}\boldsymbol{\Lambda}\mathbf{P}'$, $\boldsymbol{\Lambda} = \operatorname{diag}([\lambda_1, \dots, \lambda_n])$. Then, as in (A.2) to (A.4), the m.g.f. of Y is the same as that of $S = \sum_{i=1}^n \lambda_i W_i^2$, where $W_i^2 \overset{\text{ind}}{\sim} \chi^2(1, \nu_i^2)$ and $\boldsymbol{\nu} = (\nu_1, \dots, \nu_n)' = \mathbf{P}'\boldsymbol{\mu}$. That is, $\mathbb{M}_Y(s)$ is given in (A.19) with $n_i = 1$, $i = 1, \dots, n$. This can be directly written in matrix terms as

$$\mathbb{M}_Y(s) = |\mathbf{I}_n - 2s\boldsymbol{\Lambda}|^{-1/2} \exp\{s\boldsymbol{\nu}'\boldsymbol{\Lambda}(\mathbf{I}_n - 2s\boldsymbol{\Lambda})^{-1}\boldsymbol{\nu}\} \tag{A.20}$$

or, after a bit of algebra (Problem A.3),

$$\mathbb{M}_Y(s) = |\boldsymbol{\Omega}|^{-\frac{1}{2}} \exp\left\{-\frac{1}{2}\boldsymbol{\mu}'(\mathbf{I}_n - \boldsymbol{\Omega}^{-1})\boldsymbol{\mu}\right\}, \quad \boldsymbol{\Omega} = \mathbf{I}_n - 2s\mathbf{A}, \tag{A.21}$$

which generalizes (A.16) to the noncentral case.

Expression (A.21) can also be obtained directly as a special case of the last result we need in this section: the joint m.g.f. of $N = \mathbf{X}'\mathbf{A}\mathbf{X}$ and $D = \mathbf{X}'\mathbf{B}\mathbf{X}$, where, as before, $\mathbf{X} \sim \mathrm{N}_n(\boldsymbol{\mu}, \mathbf{I})$. The result is just an application of the following fundamental result: Let $A(\mathbf{x}) = \mathbf{x}'\mathbf{A}\mathbf{x} + \mathbf{x}'\mathbf{a} + a$ and $B(\mathbf{x}) = \mathbf{x}'\mathbf{B}\mathbf{x} + \mathbf{x}'\mathbf{b} + b$ be functions of $\mathbf{x}$, where $a, b \in \mathbb{R}$, $\mathbf{x}, \mathbf{a}, \mathbf{b} \in \mathbb{R}^n$, and $\mathbf{A}$ and $\mathbf{B}$ are symmetric $n \times n$ matrices with $\mathbf{B}$ positive definite. Then

$$\begin{aligned}\int_{\mathbb{R}^n} A(\mathbf{x}) e^{-B(\mathbf{x})} d\mathbf{x} &= \frac{1}{2}\pi^{n/2} |\mathbf{B}|^{-1/2} \exp\left\{\frac{1}{4}(\mathbf{b}'\mathbf{B}^{-1}\mathbf{b}) - b\right\} \\ &\quad \times \left[\operatorname{tr}(\mathbf{A}\mathbf{B}^{-1}) - \mathbf{b}'\mathbf{B}^{-1}\mathbf{a} + \frac{1}{2}\mathbf{b}'\mathbf{B}^{-1}\mathbf{A}\mathbf{B}^{-1}\mathbf{b} + 2a\right],\end{aligned} \tag{A.22}$$

as shown in, e.g., Graybill (1976, p. 48) and Ravishanker and Dey (2002, p. 142). We have

$$\mathbb{M}_{N,D}(s,t) = \int_{\mathbb{R}^n} \exp(s\mathbf{x}'\mathbf{A}\mathbf{x} + t\mathbf{X}'\mathbf{B}\mathbf{X}) f_{\mathbf{X}}(\mathbf{x}; \boldsymbol{\mu}, \mathbf{I}) d\mathbf{x},$$

where $f_{\mathbf{X}}(\mathbf{x}; \boldsymbol{\mu}, \mathbf{I}) = (2\pi)^{-T/2} \exp\left(-\frac{1}{2}(\mathbf{x} - \boldsymbol{\mu})'(\mathbf{x} - \boldsymbol{\mu})\right)$. Expanding and combining the two terms in the exponent gives

$$\mathbb{M}_{N,D}(s,t) = (2\pi)^{-T/2} \int_{\mathbb{R}^n} \exp\left(-\frac{1}{2}\mathbf{x}'\mathbf{S}\mathbf{x} + \mathbf{x}'\mathbf{s} + s_0\right) d\mathbf{x},$$

where

$$\mathbf{S} = \mathbf{S}(s,t) = \mathbf{I} - 2s\mathbf{A} - 2t\mathbf{B}, \quad \mathbf{s} = -\boldsymbol{\mu}, \quad \text{and} \quad s_0 = -\frac{1}{2}\boldsymbol{\mu}'\boldsymbol{\mu}.$$

This integral is a special case of (A.22), with solution

$$\mathbb{M}_{N,D}(s,t) = (2\pi)^{-T/2} \cdot \frac{1}{2}\pi^{T/2}\left|\frac{1}{2}\mathbf{S}\right|^{-1/2} \exp\left(\frac{1}{4}\mathbf{s}'\left(\frac{1}{2}\mathbf{S}\right)^{-1}\mathbf{s} - s_0\right) \cdot 2$$

or

$$\mathbb{M}_{N,D}(s,t) = |\mathbf{S}|^{-1/2} \exp\left(-\frac{1}{2}\boldsymbol{\mu}'(\mathbf{I}-\mathbf{S}^{-1})\boldsymbol{\mu}\right). \tag{A.23}$$

Note that, when $t = 0$, $\mathbb{M}_{N,D}(s,0) = \mathbb{M}_N(s)$ reduces to (A.21).

The next example offers some practice with matrix algebra and the results developed so far, and proves a more general result. It can be skipped upon first reading.

Example A.3 A natural generalization of the quadratic form $\mathbf{X}'\mathbf{A}\mathbf{X}$ is

$$Z = \mathbf{X}'\mathbf{A}\mathbf{X} + \mathbf{a}'\mathbf{X} + d, \qquad \mathbf{X} \sim \mathrm{N}_n(\boldsymbol{\mu}, \boldsymbol{\Sigma}), \tag{A.24}$$

where $\mathbf{a}$ is an $n \times 1$ vector and d is a scalar. We wish to show that the m.g.f. is

$$\mathbb{M}_Z(s) = \exp\left\{s(d + \boldsymbol{\mu}'\mathbf{A}\boldsymbol{\mu} + \mathbf{a}'\boldsymbol{\mu}) + s^2 \sum_{i=1}^{n} \frac{c_i^2}{1-2s\lambda_i}\right\} \prod_{i=1}^{n} (1-2s\lambda_i)^{-1/2}, \tag{A.25}$$

where $\boldsymbol{\Sigma}^{1/2}\mathbf{A}\boldsymbol{\Sigma}^{1/2} = \mathbf{P}\boldsymbol{\Lambda}\mathbf{P}'$ with $\mathbf{P}$ orthogonal, $\boldsymbol{\Lambda} = \mathrm{diag}([\lambda_1, \dots, \lambda_n])$, and

$$(c_1, \dots, c_n)' = \mathbf{P}'(\boldsymbol{\Sigma}^{1/2}\mathbf{a}/2 + \boldsymbol{\Sigma}^{1/2}\mathbf{A}\boldsymbol{\mu}).$$

To see this, from the multivariate normal p.d.f., the m.g.f. of Z is

$$\mathbb{E}[\mathrm{e}^{sZ}] = \frac{1}{(2\pi)^{n/2}|\boldsymbol{\Sigma}|^{1/2}} \int_{\mathbb{R}^n} \exp\left\{s\mathbf{x}'\mathbf{A}\mathbf{x} + s\mathbf{a}'\mathbf{x} + sd - \frac{1}{2}(\mathbf{x}-\boldsymbol{\mu})'\boldsymbol{\Sigma}^{-1}(\mathbf{x}-\boldsymbol{\mu})\right\} \mathrm{d}\mathbf{x},$$

and the exponent can be rearranged as

$$\begin{aligned} & s\mathbf{x}'\mathbf{A}\mathbf{x} + s\mathbf{a}'\mathbf{x} + sd - \frac{1}{2}(\mathbf{x}-\boldsymbol{\mu})'\boldsymbol{\Sigma}^{-1}(\mathbf{x}-\boldsymbol{\mu}) \\ &\quad = -\frac{1}{2}(\boldsymbol{\mu}'\boldsymbol{\Sigma}^{-1}\boldsymbol{\mu} - 2sd) + \frac{1}{2}(\boldsymbol{\mu} + s\boldsymbol{\Sigma}\mathbf{a})'(\mathbf{I} - 2s\mathbf{A}\boldsymbol{\Sigma})^{-1}\boldsymbol{\Sigma}^{-1}(\boldsymbol{\mu} + s\boldsymbol{\Sigma}\mathbf{a}) \\ &\quad\quad - \frac{1}{2}(\mathbf{x}-\mathbf{m})'(\boldsymbol{\Sigma}^{-1} - 2s\mathbf{A})(\mathbf{x}-\mathbf{m}), \end{aligned}$$

where $\mathbf{m} = (\boldsymbol{\Sigma}^{-1} - 2s\mathbf{A})^{-1}\boldsymbol{\Sigma}^{-1}(\boldsymbol{\mu} + s\boldsymbol{\Sigma}\mathbf{a})$. As $\boldsymbol{\Sigma} > 0$ and $\mathbf{A}$ is finite, there exists a neighborhood N_0 around zero such that, for $s \in N_0$, $\boldsymbol{\Sigma}^{-1} - 2s\mathbf{A} > 0$. Recognizing the kernel of the multivariate normal distribution,

$$\int_{\mathbb{R}^n} \exp\left[-\frac{1}{2}(\mathbf{x}-\mathbf{m})'(\boldsymbol{\Sigma}^{-1} - 2s\mathbf{A})(\mathbf{x}-\mathbf{m})\right] \mathrm{d}\mathbf{x} = (2\pi)^{n/2}|(\boldsymbol{\Sigma}^{-1} - 2s\mathbf{A})|^{-1/2},$$

the integral becomes

$$\mathbb{M}_Z(s) = |\mathbf{I} - 2s\mathbf{A}\boldsymbol{\Sigma}|^{-1/2} \times \exp\{E\}, \tag{A.26}$$

where

$$E := -\frac{1}{2}(\boldsymbol{\mu}'\boldsymbol{\Sigma}^{-1}\boldsymbol{\mu} - 2sd) + \frac{1}{2}(\boldsymbol{\mu} + s\boldsymbol{\Sigma}\mathbf{a})'(\mathbf{I} - 2s\mathbf{A}\boldsymbol{\Sigma})^{-1}\boldsymbol{\Sigma}^{-1}(\boldsymbol{\mu} + s\boldsymbol{\Sigma}\mathbf{a}). \tag{A.27}$$

Now let $\boldsymbol{\Sigma}^{1/2}$ be the symmetric square root of $\boldsymbol{\Sigma}$ and set $\boldsymbol{\Sigma}^{1/2}\mathbf{A}\boldsymbol{\Sigma}^{1/2} = \mathbf{P}\boldsymbol{\Lambda}\mathbf{P}'$ with $\mathbf{P}$ orthogonal, and $\boldsymbol{\Lambda} = \text{diag}([\lambda_1, \ldots, \lambda_n])$ the eigenvalues of $\boldsymbol{\Sigma}^{1/2}\mathbf{A}\boldsymbol{\Sigma}^{1/2}$, the nonzero ones of which are the same as those of $\mathbf{A}\boldsymbol{\Sigma}$. Then, with $|\mathbf{P}'\mathbf{P}| = |\mathbf{I}| = 1$ and recalling that the determinant of a product is the product of the determinants,

$$\begin{aligned}
|\mathbf{I} - 2s\mathbf{A}\boldsymbol{\Sigma}| &= |\boldsymbol{\Sigma}^{-1/2}\boldsymbol{\Sigma}^{1/2}|\ |\mathbf{I} - 2s\mathbf{A}\boldsymbol{\Sigma}| = |\boldsymbol{\Sigma}^{-1/2}|\ |\boldsymbol{\Sigma}^{1/2}|\ |\mathbf{I} - 2s\mathbf{A}\boldsymbol{\Sigma}| \\
&= |\boldsymbol{\Sigma}^{1/2}|\ |\mathbf{I} - 2s\mathbf{A}\boldsymbol{\Sigma}|\ |\boldsymbol{\Sigma}^{-1/2}| = |\boldsymbol{\Sigma}^{1/2}\boldsymbol{\Sigma}^{-1/2} - 2s\boldsymbol{\Sigma}^{1/2}\mathbf{A}\boldsymbol{\Sigma}\boldsymbol{\Sigma}^{-1/2}| \\
&= |\mathbf{I} - 2s\boldsymbol{\Sigma}^{1/2}\mathbf{A}\boldsymbol{\Sigma}^{1/2}| = |\mathbf{I} - 2s\mathbf{P}\boldsymbol{\Lambda}\mathbf{P}'| = |\mathbf{P}\mathbf{P}' - 2s\mathbf{P}\boldsymbol{\Lambda}\mathbf{P}'| \\
&= |\mathbf{P}|\ |\mathbf{I} - 2s\boldsymbol{\Lambda}|\ |\mathbf{P}'| = |\mathbf{P}'|\ |\mathbf{P}|\ |\mathbf{I} - 2s\boldsymbol{\Lambda}| = |\mathbf{P}'\mathbf{P}||\mathbf{I} - 2s\boldsymbol{\Lambda}| \\
&= |\mathbf{I} - 2s\boldsymbol{\Lambda}| = \prod_{i=1}^{n}(1 - 2s\lambda_i),
\end{aligned}$$

so that

$$|\mathbf{I} - 2s\mathbf{A}\boldsymbol{\Sigma}|^{-1/2} = \prod_{i=1}^{n}(1 - 2s\lambda_i)^{-1/2}. \tag{A.28}$$

Next, we simplify E in (A.27). First recall that $(\mathbf{AB})^{-1} = \mathbf{B}^{-1}\mathbf{A}^{-1}$, so that

$$\begin{aligned}
(\mathbf{I} - 2s\mathbf{A}\boldsymbol{\Sigma})^{-1}\boldsymbol{\Sigma}^{-1} &= [\boldsymbol{\Sigma}(\mathbf{I} - 2s\mathbf{A}\boldsymbol{\Sigma})]^{-1} = (\boldsymbol{\Sigma} - 2s\boldsymbol{\Sigma}\mathbf{A}\boldsymbol{\Sigma})^{-1} \\
&= [\boldsymbol{\Sigma}^{1/2}(\mathbf{I} - 2s\boldsymbol{\Sigma}^{1/2}\mathbf{A}\boldsymbol{\Sigma}^{1/2})\boldsymbol{\Sigma}^{1/2}]^{-1} \\
&= \boldsymbol{\Sigma}^{-1/2}(\mathbf{I} - 2s\boldsymbol{\Sigma}^{1/2}\mathbf{A}\boldsymbol{\Sigma}^{1/2})^{-1}\boldsymbol{\Sigma}^{-1/2}.
\end{aligned}$$

Then

$$\begin{aligned}
E &= -\frac{1}{2}[(\boldsymbol{\Sigma}^{-1/2}\boldsymbol{\mu})'(\boldsymbol{\mu}\boldsymbol{\Sigma}^{-1/2}) - 2sd] \\
&\quad + \frac{1}{2}(\boldsymbol{\Sigma}^{-1/2}\boldsymbol{\mu} + s\boldsymbol{\Sigma}^{1/2}\mathbf{a})'(\mathbf{I} - 2s\boldsymbol{\Sigma}^{1/2}\mathbf{A}\boldsymbol{\Sigma}^{1/2})^{-1}(\boldsymbol{\Sigma}^{-1/2}\boldsymbol{\mu} + s\boldsymbol{\Sigma}^{1/2}\mathbf{a}) \\
&= s(d + \boldsymbol{\mu}'\mathbf{A}\boldsymbol{\mu} + \mathbf{a}'\boldsymbol{\mu}) \\
&\quad + (s^2/2)(\boldsymbol{\Sigma}^{1/2}\mathbf{a} + 2\boldsymbol{\Sigma}^{1/2}\mathbf{A}\boldsymbol{\mu})'(\mathbf{I} - 2s\boldsymbol{\Sigma}^{1/2}\mathbf{A}\boldsymbol{\Sigma}^{1/2})^{-1}(\boldsymbol{\Sigma}^{1/2}\mathbf{a} + 2\boldsymbol{\Sigma}^{1/2}\mathbf{A}\boldsymbol{\mu}),
\end{aligned}$$

or

$$\begin{aligned}
E &= s(d + \boldsymbol{\mu}'\mathbf{A}\boldsymbol{\mu} + \mathbf{a}'\boldsymbol{\mu}) \\
&\quad + (s^2/2)(\boldsymbol{\Sigma}^{1/2}\mathbf{a} + 2\boldsymbol{\Sigma}^{1/2}\mathbf{A}\boldsymbol{\mu})'\mathbf{P}\mathbf{P}'(\mathbf{I} - 2s\boldsymbol{\Sigma}^{1/2}\mathbf{A}\boldsymbol{\Sigma}^{1/2})^{-1}\mathbf{P}\mathbf{P}'(\boldsymbol{\Sigma}^{1/2}\mathbf{a} + 2\boldsymbol{\Sigma}^{1/2}\mathbf{A}\boldsymbol{\mu}) \\
&= s(d + \boldsymbol{\mu}'\mathbf{A}\boldsymbol{\mu} + \mathbf{a}'\boldsymbol{\mu}) \\
&\quad + (s^2/2)(\boldsymbol{\Sigma}^{1/2}\mathbf{a} + 2\boldsymbol{\Sigma}^{1/2}\mathbf{A}\boldsymbol{\mu})'\mathbf{P}(\mathbf{P}'\mathbf{P} - 2s\mathbf{P}'\boldsymbol{\Sigma}^{1/2}\mathbf{A}\boldsymbol{\Sigma}^{1/2}\mathbf{P})^{-1}\mathbf{P}'(\boldsymbol{\Sigma}^{1/2}\mathbf{a} + 2\boldsymbol{\Sigma}^{1/2}\mathbf{A}\boldsymbol{\mu}),
\end{aligned}$$

or, with $\mathbf{c} = (c_1, \ldots, c_n)' = \mathbf{P}'(\boldsymbol{\Sigma}^{1/2}\mathbf{a} + 2\boldsymbol{\Sigma}^{1/2}\mathbf{A}\boldsymbol{\mu})$,

$$E = s(d + \boldsymbol{\mu}'\mathbf{A}\boldsymbol{\mu} + \mathbf{a}'\boldsymbol{\mu}) + \frac{s^2}{2}\mathbf{c}'(\mathbf{I} - 2s\boldsymbol{\Lambda})^{-1}\mathbf{c}.$$

Putting this together with (A.26), (A.27), and (A.28) gives (A.25). ∎

A.2 Basic Distributional Results

Let $\mathbf{X} \sim \mathrm{N}_n(\boldsymbol{\mu}, \boldsymbol{\Sigma})$ with $\boldsymbol{\Sigma} > 0$, so that $\mathbf{Z} = \boldsymbol{\Sigma}^{-1/2}\mathbf{X} \sim \mathrm{N}_n(\boldsymbol{\Sigma}^{-1/2}\boldsymbol{\mu}, \mathbf{I})$. Recalling the definition of the noncentral χ^2 distribution, it follows that $\mathbf{Z}'\mathbf{Z} = \mathbf{X}'\boldsymbol{\Sigma}^{-1}\mathbf{X} \sim \chi^2(n, \theta)$, where the noncentrality term is $\theta = (\boldsymbol{\Sigma}^{-1/2}\boldsymbol{\mu})'(\boldsymbol{\Sigma}^{-1/2}\boldsymbol{\mu}) = \boldsymbol{\mu}'\boldsymbol{\Sigma}^{-1}\boldsymbol{\mu}$. Important special cases include:

$$\text{If} \quad \mathbf{X} \sim \mathrm{N}_n(\boldsymbol{\mu}, \sigma^2\mathbf{I}_n), \quad \text{then} \quad \mathbf{X}'\mathbf{X}/\sigma^2 \sim \chi^2(n, \theta), \quad \theta = \boldsymbol{\mu}'\boldsymbol{\mu}/\sigma^2.$$

$$\text{If} \quad \mathbf{X} \sim \mathrm{N}_n(\mathbf{0}, \boldsymbol{\Sigma}), \quad \text{then} \quad \mathbf{Z}'\mathbf{Z} = \mathbf{X}'\boldsymbol{\Sigma}^{-1}\mathbf{X} \sim \chi^2(n).$$

It is of both theoretical and practical interest to know the general conditions for matrix $\mathbf{A}$ such that $\mathbf{X}'\mathbf{A}\mathbf{X} \sim \chi^2(r, \theta)$ for some r, $0 < r \leqslant n$; in particular, if there are other $\mathbf{A}$ besides $\boldsymbol{\Sigma}^{-1}$. There are: it turns out to be necessary and sufficient that rank $(\mathbf{A}\boldsymbol{\Sigma}) = r$ and $\mathbf{A}\boldsymbol{\Sigma}$ is idempotent, i.e., that $\mathbf{A}\boldsymbol{\Sigma} = \mathbf{A}\boldsymbol{\Sigma}\mathbf{A}\boldsymbol{\Sigma}$, in which case $\theta = \boldsymbol{\mu}'\mathbf{A}\boldsymbol{\mu}$. To show this, we first prove the following three results.

1) Let $\mathbf{P}$ be an $n \times n$ symmetric matrix. Then $\mathbf{P}$ is idempotent with rank r if and only if $\mathbf{P}$ has r unit and $n - r$ zero eigenvalues.

   Proof:

   a) $(\Rightarrow)$ For any eigenvalue λ and corresponding eigenvector $\mathbf{x}$ of $\mathbf{P}$, idempotency implies $\lambda\mathbf{x} = \mathbf{P}\mathbf{x} = \mathbf{P}\mathbf{P}\mathbf{x} = \mathbf{P}\lambda\mathbf{x} = \lambda\mathbf{P}\mathbf{x} = \lambda^2\mathbf{x}$, i.e., $\lambda = \lambda^2$. The roots of the equation $\lambda^2 - \lambda = 0$ are zero and one. From the symmetry of $\mathbf{P}$, the number of nonzero eigenvalues of $\mathbf{P}$ equals rank $(\mathbf{P}) = r$.[3]
   b) $(\Leftarrow)$ Let $\mathbf{P} = \mathbf{U}\mathbf{D}\mathbf{U}'$ with $\mathbf{U}$ orthogonal and $\mathbf{D} = \mathrm{diag}(\lambda_i)$, $\lambda_1 = \cdots = \lambda_r = 1$ and $\lambda_{r+1} = \cdots = \lambda_n = 0$. From symmetry, $\mathrm{rank}(\mathbf{P}) = r$. Also, $\mathbf{P}^2 = \mathbf{U}\mathbf{D}\mathbf{U}'\mathbf{U}\mathbf{D}\mathbf{U}' = \mathbf{U}\mathbf{D}\mathbf{D}\mathbf{U}' = \mathbf{U}\mathbf{D}\mathbf{U}' = \mathbf{P}$. ■

2) Let $\mathbf{X} \sim \mathrm{N}_n(\mathbf{0}, \mathbf{I})$ and $Y = \mathbf{X}'\mathbf{A}\mathbf{X}$, for $\mathbf{A}$ symmetric. Then $Y \sim \chi^2(r, 0)$ if and only if $\mathbf{A} = \mathbf{A}\mathbf{A}$ with rank $(\mathbf{A}) = r$.

   Proof:

   a) $(\Leftarrow)$ From 1(a), $\mathbf{A}$ can be written as $\mathbf{U}\mathbf{D}\mathbf{U}'$ with $\mathbf{D} = \mathrm{diag}(\lambda_i)$, $\lambda_1 = \cdots = \lambda_r = 1$ and $\lambda_{r+1} = \cdots = \lambda_n = 0$. With $\mathbf{Z} = \mathbf{U}'\mathbf{X} \sim \mathrm{N}(\mathbf{0}, \mathbf{I})$, $Y = \mathbf{X}'\mathbf{U}\mathbf{D}\mathbf{U}'\mathbf{X} = \mathbf{Z}'\mathbf{D}\mathbf{Z} = \sum_{i=1}^r Z_i^2 \sim \chi^2(r)$.
   b) $(\Rightarrow)$ Let $\{\lambda_i\}$ be the eigenvalues of $\mathbf{A}$. Equating the m.g.f. of $\mathbf{X}'\mathbf{A}\mathbf{X}$ from (A.9) and that of a $\chi^2(r, 0)$ r.v. from (A.18) implies

   $$\prod_{i=1}^{n} (1 - 2s\lambda_i)^{-1/2} = (1 - 2s)^{-r/2},$$

   whose square is a polynomial in s in a neighborhood of zero. As such, the two must have the same degree and roots, implying that $\lambda_1 = \cdots = \lambda_r = 1$ and $\lambda_{r+1} = \cdots = \lambda_n = 0$. The result now follows from 1(b). ■

3) Let $\mathbf{X} \sim \mathrm{N}_n(\boldsymbol{\mu}, \mathbf{I})$ and $Y = \mathbf{X}'\mathbf{A}\mathbf{X}$. Then $Y \sim \chi^2(r, \theta)$, $\theta = \boldsymbol{\mu}'\mathbf{A}\boldsymbol{\mu}$, if and only if $\mathbf{A}$ is idempotent with $\mathrm{rank}(\mathbf{A}) = r$.

---

3 Recall that, in general, if matrix $\mathbf{A}$ (possibly asymmetric) has r nonzero eigenvalues, then $\mathrm{rank}(\mathbf{A}) \geqslant r$, while if $\mathbf{A}$ is symmetric and has r nonzero eigenvalues, then $\mathrm{rank}(\mathbf{A}) = r$; see, e.g., Magnus and Neudecker (2007).

Proof:

a) (⇐) Similar to 2(a), but with $\mathbf{Z} = \mathbf{U}'\mathbf{X} \sim \mathrm{N}(\mathbf{v}, \mathbf{I})$, where $\mathbf{v} = \mathbf{U}'\boldsymbol{\mu}$, so that $Y = \mathbf{Z}'\mathbf{D}\mathbf{Z} = \sum_{i=1}^{r} Z_i^2 \sim \chi^2(r, \theta)$, where θ is determined by

$$\theta = \sum_{i=1}^{r} v_i^2 = \mathbf{v}'\mathbf{D}\mathbf{v} = \boldsymbol{\mu}'\mathbf{U}\mathbf{D}\mathbf{U}'\boldsymbol{\mu} = \boldsymbol{\mu}'\mathbf{A}\boldsymbol{\mu}.$$

b) (⇒) As $\mathbf{A}$ is symmetric, we can express it as $\mathbf{A} = \mathbf{O}\boldsymbol{\Lambda}\mathbf{O}'$ with $\mathbf{O}$ orthogonal and $\boldsymbol{\Lambda} = \mathrm{diag}(\lambda_i)$ the eigenvalues of $\mathbf{A}$. Let $\boldsymbol{v} = (v_1, \dots, v_n)' = \mathbf{O}'\boldsymbol{\mu}$. By equating the m.g.f. of $\mathbf{X}'\mathbf{A}\mathbf{X}$, as given in (A.19) (with $n_i = 1$) with that of a $\chi^2(r, \theta)$ r.v., as given in (A.18), we see that

$$\prod_{i=1}^{n} (1 - 2\lambda_i s)^{-1/2} \exp\left\{ \sum_{i=1}^{n} \frac{\lambda_i s v_i^2}{1 - 2s\lambda_i} \right\} = (1 - 2s)^{-r/2} \exp\left\{ \frac{s}{1 - 2s}\theta \right\}$$

must hold for all s in a neighborhood of zero. It can be shown[4] that this implies the desired condition on the λ_i, and the result follows from 1(b). ■

The following two theorems, A.1 and A.2, are of great relevance for working with the Gaussian linear model, notably in ANOVA. Original references, some history of their (at times faulty) development, and references to alternative "accessible" proofs in the noncentral case, are provided in Khuri (2010, Sec. 1.6).

Theorem A.1 ***Distribution of Quadratic Form*** Let $\mathbf{X} \sim \mathrm{N}_n(\boldsymbol{\mu}, \boldsymbol{\Sigma})$ with $\boldsymbol{\Sigma}$ positive definite. The quadratic form $\mathbf{X}'\mathbf{A}\mathbf{X}$ follows a $\chi^2(r, \theta)$ distribution, where $r = \mathrm{rank}(\mathbf{A}\boldsymbol{\Sigma})$, $\mathbf{A}$ symmetric, and $\theta = \boldsymbol{\mu}'\mathbf{A}\boldsymbol{\mu}$, if and only if $\mathbf{A}\boldsymbol{\Sigma}$ is idempotent.

Proof: Let $\mathbf{Z} = \boldsymbol{\Sigma}^{-1/2}(\mathbf{X} - \boldsymbol{\mu}) \sim \mathrm{N}_n(\mathbf{0}, \mathbf{I}_n)$ with $\boldsymbol{\Sigma}^{1/2}\boldsymbol{\Sigma}^{1/2} = \boldsymbol{\Sigma}$, so that

$$\begin{aligned}\mathbf{X}'\mathbf{A}\mathbf{X} &= (\boldsymbol{\Sigma}^{1/2}\mathbf{Z} + \boldsymbol{\mu})'\mathbf{A}(\boldsymbol{\Sigma}^{1/2}\mathbf{Z} + \boldsymbol{\mu}) = (\boldsymbol{\Sigma}^{1/2}(\mathbf{Z} + \boldsymbol{\Sigma}^{-1/2}\boldsymbol{\mu}))'\mathbf{A}\boldsymbol{\Sigma}^{1/2}(\mathbf{Z} + \boldsymbol{\Sigma}^{-1/2}\boldsymbol{\mu}) \\ &= (\mathbf{Z} + \boldsymbol{\Sigma}^{-1/2}\boldsymbol{\mu})'\boldsymbol{\Sigma}^{1/2}\mathbf{A}\boldsymbol{\Sigma}^{1/2}(\mathbf{Z} + \boldsymbol{\Sigma}^{-1/2}\boldsymbol{\mu}) = \mathbf{V}'\mathbf{B}\mathbf{V},\end{aligned}$$

where $\mathbf{V} = (\mathbf{Z} + \boldsymbol{\Sigma}^{-1/2}\boldsymbol{\mu}) \sim \mathrm{N}(\boldsymbol{\Sigma}^{-1/2}\boldsymbol{\mu}, \mathbf{I}_n)$ and $\mathbf{B} = \boldsymbol{\Sigma}^{1/2}\mathbf{A}\boldsymbol{\Sigma}^{1/2}$. Let

$$\theta = (\boldsymbol{\Sigma}^{-1/2}\boldsymbol{\mu})'\mathbf{B}(\boldsymbol{\Sigma}^{-1/2}\boldsymbol{\mu}) = \boldsymbol{\mu}'\mathbf{A}\boldsymbol{\mu}.$$

From result 3 above, and that $\boldsymbol{\Sigma}^{-1/2}$ and $\boldsymbol{\Sigma}^{-1}$ are full rank,[5]

$$\begin{aligned}\mathbf{V}'\mathbf{B}\mathbf{V} \sim \chi^2(r, \theta) &\Leftrightarrow \mathbf{B}\mathbf{B} = \mathbf{B}, \ \mathrm{rank}(\mathbf{B}) = r \\ &\Leftrightarrow \boldsymbol{\Sigma}^{1/2}\mathbf{A}\boldsymbol{\Sigma}^{1/2}\boldsymbol{\Sigma}^{1/2}\mathbf{A}\boldsymbol{\Sigma}^{1/2} = \boldsymbol{\Sigma}^{1/2}\mathbf{A}\boldsymbol{\Sigma}^{1/2}, \ \mathrm{rank}(\mathbf{A}) = r \\ &\Leftrightarrow \mathbf{A}\boldsymbol{\Sigma}\mathbf{A} = \mathbf{A}, \ \mathrm{rank}(\mathbf{A}) = r \\ &\Leftrightarrow \mathbf{A}\boldsymbol{\Sigma}\mathbf{A}\boldsymbol{\Sigma} = \mathbf{A}\boldsymbol{\Sigma}, \ \mathrm{rank}(\mathbf{A}\boldsymbol{\Sigma}) = r.\end{aligned}$$

The last condition is also equivalent to $\boldsymbol{\Sigma}\mathbf{A}\boldsymbol{\Sigma}\mathbf{A} = \boldsymbol{\Sigma}\mathbf{A}$, seen by transposing both sides (and recalling that both $\mathbf{A}$ and $\boldsymbol{\Sigma}$ are symmetric). ■

---

4 See, e.g., Ravishanker and Dey (2002, p. 175) and the references stated therein.

5 Recall that, if $\mathbf{A}$ is an $m \times n$ matrix, $\mathbf{B}$ an $m \times m$ matrix, and $\mathbf{C}$ an $n \times n$ matrix, and if $\mathbf{B}$ and $\mathbf{C}$ are nonsingular, then $\mathrm{rank}(\mathbf{A}) = \mathrm{rank}(\mathbf{B}\mathbf{A}\mathbf{C})$. See, e.g., Schott (2005, p. 13).

Theorem A.2 ***Independence of Two Quadratic Forms*** Let $\mathbf{X} \sim \mathrm{N}_n(\boldsymbol{\mu}, \boldsymbol{\Sigma})$, $\boldsymbol{\Sigma} > 0$. The two quadratic forms $\mathbf{X}'\mathbf{A}_1\mathbf{X}$ and $\mathbf{X}'\mathbf{A}_2\mathbf{X}$ are independent if $\mathbf{A}_1\boldsymbol{\Sigma}\mathbf{A}_2 = \mathbf{A}_2\boldsymbol{\Sigma}\mathbf{A}_1 = \mathbf{0}$.

Proof: Let $\mathbf{Z} = \boldsymbol{\Sigma}^{-1/2}\mathbf{X} \sim \mathrm{N}_n(\boldsymbol{\Sigma}^{-1/2}\boldsymbol{\mu}, \mathbf{I}_n)$ and $\mathbf{A}_i^* = \boldsymbol{\Sigma}^{1/2}\mathbf{A}_i\boldsymbol{\Sigma}^{1/2}$, $i = 1, 2$, so that $\mathbf{X}'\mathbf{A}_i\mathbf{X} = \mathbf{Z}'\mathbf{A}_i^*\mathbf{Z}$ and $\mathbf{A}_1^*\mathbf{A}_2^* = \boldsymbol{\Sigma}^{1/2}\mathbf{A}_1\boldsymbol{\Sigma}\mathbf{A}_2\boldsymbol{\Sigma}^{1/2} = \mathbf{0}$. Let $k = \mathrm{rank}(\mathbf{A}_1)$, $0 < k \leqslant n$, and take $\mathbf{A}_1^* = \mathbf{UDU}'$ for $\mathbf{U}$ orthogonal and $\mathbf{D} = \mathrm{diag}(\lambda_i)$ with $\lambda_{k+1} = \cdots = \lambda_n = 0$. With $\mathbf{W} = (W_1, \ldots, W_n)' = \mathbf{U}'\mathbf{Z} \sim \mathrm{N}_n(\mathbf{U}'\boldsymbol{\Sigma}^{-1/2}\boldsymbol{\mu}, \mathbf{I}_n)$,

$$\mathbf{X}'\mathbf{A}_1\mathbf{X} = \mathbf{Z}'\mathbf{A}_1^*\mathbf{Z} = \mathbf{Z}'\mathbf{UDU}'\mathbf{Z} = \mathbf{W}'\mathbf{DW} = \sum_{i=1}^{k} \lambda_i W_i^2, \tag{A.29}$$

and, with $\mathbf{B} = \mathbf{U}'\mathbf{A}_2^*\mathbf{U}$, $\mathbf{X}'\mathbf{A}_2\mathbf{X} = \mathbf{Z}'\mathbf{A}_2^*\mathbf{Z} = \mathbf{W}'\mathbf{U}'\mathbf{A}_2^*\mathbf{UW} = \mathbf{W}'\mathbf{BW}$. As

$$\mathbf{DB} = \mathbf{U}'\mathbf{A}_1^*\mathbf{U}\ \mathbf{U}'\mathbf{A}_2^*\mathbf{U} = \mathbf{U}'\mathbf{A}_1^*\mathbf{A}_2^*\mathbf{U} = \mathbf{0},$$

recalling the structure of $\mathbf{D}$, it must be the case that $\mathbf{B}$ can be partitioned as, say, $\mathbf{B} = \begin{pmatrix} \mathbf{0}_{k\times n} \\ \tilde{\mathbf{B}}_{\ell\times n} \end{pmatrix}$; $\ell = n - k$, but the symmetry of $\mathbf{B}$ then implies that $\mathbf{B} = \begin{pmatrix} \mathbf{0}_{k\times k} & \mathbf{0}_{k\times \ell} \\ \mathbf{0}_{\ell\times k} & \check{\mathbf{B}}_{\ell\times\ell} \end{pmatrix}$, i.e., $\mathbf{W}'\mathbf{BW} = \mathbf{X}'\mathbf{A}_2\mathbf{X}$ involves only $W_{k+1}, \ldots, W_n$. From (A.29), the result follows. ∎

Example A.4 As a partial converse of Theorem A.2, let $\mathbf{X} \sim \mathrm{N}_n(\mathbf{0}, \mathbf{I}_n)$ and assume the two quadratic forms $\mathbf{X}'\mathbf{A}_1\mathbf{X}$ and $\mathbf{X}'\mathbf{A}_2\mathbf{X}$ are independent, each following a central χ^2 distribution. As the sum of independent central χ^2 r.v.s is also χ^2, $\mathbf{X}'(\mathbf{A}_1 + \mathbf{A}_2)\mathbf{X}$ is χ^2 and Theorem A.1 implies that $\mathbf{A}_1 + \mathbf{A}_2$ is idempotent. Thus, as both $\mathbf{A}_1$ and $\mathbf{A}_2$ must also be idempotent,

$$\mathbf{A}_1 + \mathbf{A}_2 = (\mathbf{A}_1 + \mathbf{A}_2)^2 = \mathbf{A}_1 + \mathbf{A}_1\mathbf{A}_2 + \mathbf{A}_2\mathbf{A}_1 + \mathbf{A}_2,$$

so that $\mathbf{A}_1\mathbf{A}_2 + \mathbf{A}_2\mathbf{A}_1 = \mathbf{0}$. Pre-multiplying this with $\mathbf{A}_1$, and then post-multiplying it by $\mathbf{A}_1$, yields $\mathbf{A}_1\mathbf{A}_2 + \mathbf{A}_1\mathbf{A}_2\mathbf{A}_1 = \mathbf{0}$ and $\mathbf{A}_1\mathbf{A}_2\mathbf{A}_1 + \mathbf{A}_2\mathbf{A}_1 = \mathbf{0}$, respectively. Thus

$$\mathbf{0} = \mathbf{0} - \mathbf{0} = \mathbf{A}_1\mathbf{A}_2 + \mathbf{A}_1\mathbf{A}_2\mathbf{A}_1 - (\mathbf{A}_1\mathbf{A}_2\mathbf{A}_1 + \mathbf{A}_2\mathbf{A}_1) = \mathbf{A}_1\mathbf{A}_2 - \mathbf{A}_2\mathbf{A}_1,$$

and $\mathbf{A}_1\mathbf{A}_2 = \mathbf{A}_2\mathbf{A}_1 = \mathbf{0}$. ∎

Problems A.6 and A.7 give some practice using Theorems A.1 and A.2, while Problem A.8 asks the reader to prove the following result.

Theorem A.3 ***Independence of Vector and Quadratic Form*** Let $\mathbf{Y} \sim \mathrm{N}_n(\boldsymbol{\mu}, \boldsymbol{\Sigma})$, $\boldsymbol{\Sigma} > 0$. Vector $\mathbf{BY}$, with $\mathbf{B}$ a real $q \times n$ matrix, is independent of $\mathbf{Y}'\mathbf{AY}$ if $\mathbf{B}\boldsymbol{\Sigma}\mathbf{A} = \mathbf{0}$.

Proof: See Problem A.8. ∎

A.3 Ratios of Quadratic Forms in Normal Variables

For symmetric matrices $\mathbf{A}$ and $\mathbf{B}$, the ratio given by

$$R = \frac{\mathbf{X}'\mathbf{AX}}{\mathbf{X}'\mathbf{BX}}, \quad \mathbf{X} \sim \mathrm{N}_n(\boldsymbol{\mu}, \boldsymbol{\Sigma}), \quad \mathbf{B} \neq \mathbf{0}, \quad \mathbf{B} \geqslant 0, \quad \boldsymbol{\Sigma} > 0, \tag{A.30}$$

arises in many contexts in which quadratic forms appear. The restriction that $\mathbf{B}$ is positive semi-definite but nonzero ensures that the denominator is positive with probability one.[6] Let $\boldsymbol{\Sigma}^{\frac{1}{2}}$ be such that $\boldsymbol{\Sigma}^{\frac{1}{2}}\boldsymbol{\Sigma}^{\frac{1}{2}} = \boldsymbol{\Sigma}$, and let $\mathbf{A}^* = \boldsymbol{\Sigma}^{\frac{1}{2}}\mathbf{A}\boldsymbol{\Sigma}^{\frac{1}{2}}$, $\mathbf{B}^* = \boldsymbol{\Sigma}^{\frac{1}{2}}\mathbf{B}\boldsymbol{\Sigma}^{\frac{1}{2}}$, and $\mathbf{Z} = \boldsymbol{\Sigma}^{-1/2}\mathbf{X} \sim \mathrm{N}(\boldsymbol{\Sigma}^{-1/2}\boldsymbol{\mu}, \mathbf{I})$. Then

$$R = \frac{\mathbf{X}'\mathbf{A}\mathbf{X}}{\mathbf{X}'\mathbf{B}\mathbf{X}} = \frac{\mathbf{Z}'\mathbf{A}^*\mathbf{Z}}{\mathbf{Z}'\mathbf{B}^*\mathbf{Z}},$$

so that we may assume $\boldsymbol{\Sigma} = \mathbf{I}$ without loss of generality. Observe that, if $\mathbf{X} \sim \mathrm{N}(\mathbf{0}, \sigma^2\mathbf{I})$, then σ^2 can be factored out of the numerator and denominator, so that R does not depend on σ^2.

A.3.1 Calculation of the CDF

Let $\mathbf{X} \sim \mathrm{N}_n(\boldsymbol{\mu}, \mathbf{I})$. For computing the c.d.f. of ratio R in (A.30) at a given value r, construct the spectral decomposition

$$\mathbf{A} - r\mathbf{B} = \mathbf{P}\boldsymbol{\Lambda}\mathbf{P}', \tag{A.31}$$

$\boldsymbol{\Lambda} = \mathrm{diag}([\lambda_1, \dots, \lambda_n])$, and let $\mathbf{W} = \mathbf{P}'\mathbf{X} \sim \mathrm{N}_n(\boldsymbol{\nu}, \mathbf{I})$, where $\boldsymbol{\nu} = \mathbf{P}'\boldsymbol{\mu} = (\nu_1, \dots, \nu_n)'$. Then

$$\begin{aligned}\Pr(R \leqslant r) &= \Pr(\mathbf{X}'\mathbf{A}\mathbf{X} \leqslant r\,\mathbf{X}'\mathbf{B}\mathbf{X}) = \Pr(\mathbf{X}'(\mathbf{A} - r\mathbf{B})\mathbf{X} \leqslant 0) \\ &= \Pr(\mathbf{X}'\mathbf{P}\boldsymbol{\Lambda}\mathbf{P}'\mathbf{X} \leqslant 0) = \Pr(\mathbf{W}'\boldsymbol{\Lambda}\mathbf{W} \leqslant 0) = F_S(0),\end{aligned} \tag{A.32}$$

where $S = \sum_{i=1}^n \lambda_i W_i^2$ and $W_i^2 \overset{\text{ind}}{\sim} \chi^2(1, \nu_i^2)$, so that S is a weighted sum of noncentral χ^2 random variables, each with one degree of freedom and noncentrality parameter ν_i^2, $i = 1, \dots, n$. The λ_i are the eigenvalues of $\mathbf{A} - r\mathbf{B}$, some of which, depending on $\mathbf{A}$ and $\mathbf{B}$, might be zero.

If $\mathbf{B} > 0$, then both $\mathbf{B}^{1/2}$ and $\mathbf{B}^{-1/2}$ exist, and R can be written as

$$R = \frac{\mathbf{X}'\mathbf{A}\mathbf{X}}{\mathbf{X}'\mathbf{B}\mathbf{X}} = \frac{\mathbf{X}'\mathbf{B}^{\frac{1}{2}}\mathbf{B}^{-\frac{1}{2}}\mathbf{A}\mathbf{B}^{-\frac{1}{2}}\mathbf{B}^{\frac{1}{2}}\mathbf{X}}{\mathbf{X}'\mathbf{B}^{\frac{1}{2}}\mathbf{B}^{\frac{1}{2}}\mathbf{X}} = \frac{\mathbf{Y}'\mathbf{C}\mathbf{Y}}{\mathbf{Y}'\mathbf{Y}}, \tag{A.33}$$

where $\mathbf{Y} = \mathbf{B}^{1/2}\mathbf{X}$ and $\mathbf{C} = \mathbf{B}^{-1/2}\mathbf{A}\mathbf{B}^{-1/2}$. The support of R is given by the following result.

Theorem A.4 Let $\mathbf{x} \in \mathbb{R}^T \setminus \mathbf{0}$ so that $\mathbf{x}'\mathbf{x} > 0$, and $\mathbf{A}$ be a symmetric real $T \times T$ matrix. Then

$$\lambda_{\min} \leqslant \frac{\mathbf{x}'\mathbf{A}\mathbf{x}}{\mathbf{x}'\mathbf{x}} \leqslant \lambda_{\max}, \tag{A.34}$$

where $\lambda_{\min}$ and $\lambda_{\max}$ are the (necessarily real) minimum and maximum eigenvalues of $\mathbf{A}$, respectively.

Proof: Order the T eigenvalues of $\mathbf{A}$ as

$$\lambda_{\min} = \lambda_1 \leqslant \lambda_2 \leqslant \dots \leqslant \lambda_T = \lambda_{\max},$$

and let $\mathbf{S}$ be an orthogonal $T \times T$ matrix such that $\mathbf{S}'\mathbf{A}\mathbf{S} = \boldsymbol{\Lambda} := \mathrm{diag}([\lambda_1, \lambda_2, \dots, \lambda_T])$. Define $\mathbf{y} = \mathbf{S}'\mathbf{x}$. Then $\mathbf{x}'\mathbf{A}\mathbf{x} = \mathbf{y}'\mathbf{S}'\mathbf{A}\mathbf{S}\mathbf{y} = \mathbf{y}'\boldsymbol{\Lambda}\mathbf{y}$ and $\mathbf{x}'\mathbf{x} = \mathbf{y}'\mathbf{S}'\mathbf{S}\mathbf{y} = \mathbf{y}'\mathbf{y}$. As a sum of squares, $\mathbf{y}'\mathbf{y} \geqslant 0$, so that $\mathbf{y}'(\lambda_{\min}\mathbf{I})\mathbf{y} \leqslant \mathbf{y}'\boldsymbol{\Lambda}\mathbf{y} \leqslant \mathbf{y}'(\lambda_{\max}\mathbf{I})\mathbf{y}$, i.e., $\lambda_{\min}\mathbf{y}'\mathbf{y} \leqslant \mathbf{y}'\boldsymbol{\Lambda}\mathbf{y} \leqslant \lambda_{\max}\mathbf{y}'\mathbf{y}$. Substituting the previous two

---

6 If $\mathbf{B}$ has z zero eigenvalues, $0 < z < n$, then there exists a z-dimensional hyperplane $\mathcal{Z}$ in $\mathbb{R}^n$ (e.g., a line for $z = 1$, etc.) such that, for $\mathbf{X} \in \mathcal{Z}$, $\mathbf{X}'\mathbf{B}\mathbf{X} = 0$. However, $\mathcal{Z}$ has measure zero in $\mathbb{R}^n$ so that, with probability one, $\mathbf{X}'\mathbf{B}\mathbf{X} > 0$.

```
function F=cdfratio(rvec,A,B,Sigma,mu,method)
if nargin<6, method=1; end
if nargin<5 || isempty(mu), mu=zeros(length(A),1); end
[V,D]=eig(0.5*(Sigma+Sigma')); W=sqrt(D); Sighalf = V*W*V'; SI=inv(Sighalf);
A=Sighalf*A*Sighalf; B=Sighalf*B*Sighalf; mu=SI*mu;
rl=length(rvec); F=zeros(rl,1); n=length(A);
for rloop=1:rl
  r=rvec(rloop); [P,Lam] = eig((A-r*B)); Lam=real(diag(Lam)); v=P'*mu; nc=v.^2;
  if method==1
    F(rloop)=myimhof(0,Lam,ones(n,1),nc);
  else
    [~,cdfval] = spaweightedsumofchisquare(2,0,Lam,ones(n,1),nc);
    F(rloop)=cdfval;
  end
end
```

Program Listing A.3: Computes the c.d.f. (A.32). Program `myimhof.m` is given in Listing II.10.6, though we recommend changing the integration function `quadl` to `quadgk`. Programs `myimhof.m` (updated version) and `spaweightedsumofchisquare.m` use the methods developed in Section II.10.1 and are available in the collection of programs.

equalities and dividing by $\mathbf{x}'\mathbf{x}$, the result follows. A different proof of this fundamental result, using calculus, and not using the spectral decomposition, is given in Ravishanker and Dey (2002, pp. 45–46). ■

It then follows from (A.34) and the fact that $\Pr(\mathbf{Y}'\mathbf{Y} \leqslant 0) = 0$ that the support of R is the interval between the smallest and largest eigenvalues of matrix $\mathbf{C}$ in (A.33), or, equivalently, those of $\mathbf{B}^{-1}\mathbf{A}$. For values of r in the interior of the support, i.e., r such that $0 < F_R(r) < 1$, (A.32) states that

$$0 < \Pr(R \leqslant r) = F_S(0) = \Pr\left(\sum_{i=1}^{n} \lambda_i \chi_i^2(1, \nu_i^2) \leqslant 0\right) < 1,$$

(where, to be clear, the λ_i are the eigenvalues of $\mathbf{A} - r\mathbf{B}$), implying that at least one of the λ_i must be negative and at least one positive, i.e., $\lambda_{\min} < 0$ and $\lambda_{\max} > 0$.

As the characteristic function and m.g.f. of S are tractable, the c.d.f. inversion formula or the s.p.a. can be applied to compute $F_R(r) = F_S(0)$.[7] A program for computing (A.32) using either of these is given in Listing A.3. Note that, for each value of r, (A.31) needs to be computed to obtain the ν_i and λ_i. For large n, this will be the most time-consuming part of the computation.

A.3.2 Calculation of the PDF

The characteristic function of R is not tractable in general, so that direct application of the inversion formula is not possible. We consider three ways for its computation.

7 The application of the inversion formula in this context, and the s.p.a., are detailed in book II. See also Tanaka (1996, Sec. 6.6), Helstrom (1996), Problem II.10.14, and Marsh (1998) for further aspects of the s.p.a. in this context.

A.3.2.1 Numeric Differentiation

One obvious possibility is to numerically differentiate the c.d.f. in (A.32) as

$$f_R(r) \approx \frac{F_R(r+\delta) - F_R(r-\delta)}{2\delta}, \tag{A.35}$$

for an appropriate choice of $\delta > 0$. This will certainly be adequate for graphical purposes, but to achieve very high accuracy F_R would need to be evaluated to virtually machine precision and higher-order numeric derivative expressions should be used.

Example A.5 Let $\mathbf{X} = (X_1, X_2)' \sim \mathrm{N}(\boldsymbol{\mu}, \mathbf{I}_2)$ for $\boldsymbol{\mu} = (\mu_1, \mu_2)'$ and observe that

$$R = \frac{\mathbf{X}'\mathbf{A}\mathbf{X}}{\mathbf{X}'\mathbf{B}\mathbf{X}} = \frac{X_1X_2}{X_1^2} = \frac{X_2}{X_1}, \quad \mathbf{A} = \begin{pmatrix} 0 & 1/2 \\ 1/2 & 0 \end{pmatrix}, \quad \mathbf{B} = \begin{pmatrix} 1 & 0 \\ 0 & 0 \end{pmatrix}, \tag{A.36}$$

so that R is a Cauchy-like ratio. From Example II.2.18, its exact density is given by

$$f_{X_2/X_1}(r) = \exp\left(\frac{k^2 - c}{2}\right)\frac{1}{2\pi}\left(\frac{b}{a}\sqrt{\frac{2\pi}{a}}(1 - 2\Phi(-k)) + 2a^{-1}\exp\left(-\frac{k^2}{2}\right)\right),$$

where $a = 1 + r^2$, $b = \mu_1 + r\mu_2$, $c = \mu_1^2 + \mu_2^2$, $k = b/\sqrt{a}$, and Φ is the standard normal c.d.f. For $\boldsymbol{\mu} = (0, 0)'$, the reduces the standard Cauchy density $\pi^{-1}(1 + r^2)^{-1}$.

We consider the case with $\boldsymbol{\mu} = (0.1, 2)'$. The density is plotted in Figure A.4. The left panel of Figure A.2 shows the relative percentage error (r.p.e.), defined as r.p.e. $= 100(\text{Approx} - \text{True})/\text{True}$, of the approximation in (A.35) using the equally spaced grid of values $r = -20, -19.9, \ldots, 19.9, 20$, $\delta = 10^{-7}$, and c.d.f. values computed by the inversion formula (see the figure caption for details). The mean of the absolute r.p.e. values is about 10^{-5} and their median is about 4×10^{-6}. ■

A.3.2.2 Use of Geary's formula

Let N and D be continuous r.v.s with joint c.f. $\varphi_{N,D}$ and such that $\Pr(D > 0) = 1$ and $\mathbb{E}[D] < \infty$. Example II.1.24 details what we refer to as **Geary's formula**, from Geary (1944), which shows that

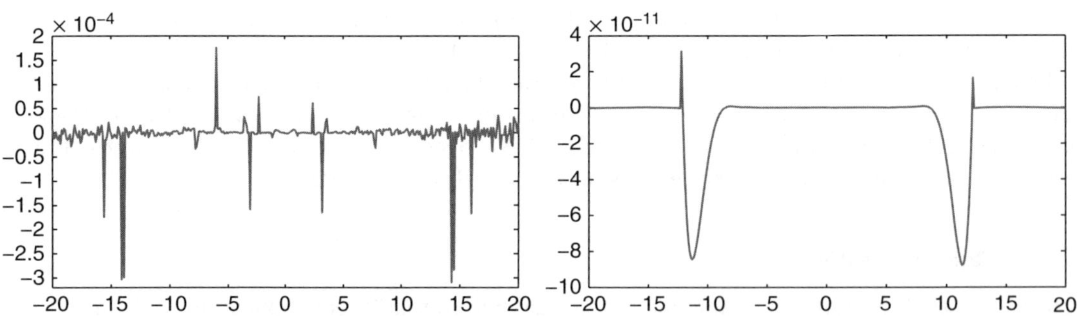

Figure A.2 **Left:** relative percentage error (r.p.e.) incurred when using (A.35) with $\delta = 10^{-7}$ for the p.d.f. of (A.36) with $\boldsymbol{\mu} = (0.1, 2)'$. **Right:** r.p.e. using the exact expression (A.39). For both the c.d.f. in (A.35) and the p.d.f. in (A.39), numeric integration using Matlab's integration routine `quadgk` was used with default tolerance parameters for the absolute and relative error. It is well-suited to these integrands because, paraphrasing from their documentation, "[it] may be most efficient for oscillatory integrands and any smooth integrand at high accuracies. It supports infinite intervals and can handle moderate singularities at the endpoints."

the density of $R = N/D$ can be written as

$$f_R(r) = \frac{1}{2\pi i}\int_{-\infty}^{\infty}\left[\frac{\partial\varphi_{N,D}(s,t)}{\partial t}\right]_{t=-rs} \mathrm{d}s. \tag{A.37}$$

With $N = \mathbf{X}'\mathbf{A}\mathbf{X}$ and $D = \mathbf{X}'\mathbf{B}\mathbf{X}$, and from (A.23) we have

$$\mathbb{M}_{N,D}(s,t) = |\boldsymbol{\Omega}|^{-\frac{1}{2}}\exp\left\{-\frac{1}{2}\boldsymbol{\mu}'(\mathbf{I}-\boldsymbol{\Omega}^{-1})\boldsymbol{\mu}\right\}, \quad \boldsymbol{\Omega} = \mathbf{I} - 2(s\mathbf{A} + t\mathbf{B}),$$

and we assume $\varphi_{N,D}(s,t) = \mathbb{M}_{N,D}(is,it)$. Then, we can rewrite (A.37) as

$$f_R(r) = \frac{1}{\pi}\int_0^{\infty}\mathrm{Re}[\mathbb{M}^*(is)]\,\mathrm{d}s, \tag{A.38}$$

where $\mathbb{M}^*(s) := [\partial\mathbb{M}_{N,D}(s,t)/\partial t]_{t=-rs}$ is given in Butler and Paolella (2008) as

$$\mathbb{M}^*(s) = \left[\prod_{i=1}^{n}(1-2s\lambda_i)^{-1/2}\right]\exp\left\{s\sum_{i=1}^{n}\frac{\lambda_i\nu_i^2}{1-2s\lambda_i}\right\}\left[\mathrm{tr}\ \mathbf{D}^{-1}\mathbf{H} + \boldsymbol{\nu}'\mathbf{D}^{-1}\mathbf{H}\mathbf{D}^{-1}\boldsymbol{\nu}\right],$$

with spectral decomposition $\mathbf{A} - r\mathbf{B} = \mathbf{P}\boldsymbol{\Lambda}\mathbf{P}'$ as in (A.31), $\boldsymbol{\Lambda} = \mathrm{diag}([\lambda_1,\ldots,\lambda_n])$, $\mathbf{P} = [\mathbf{p}_1,\ldots,\mathbf{p}_n]$, $\mathbf{D} = \mathbf{I} - 2s\boldsymbol{\Lambda}$, $\mathbf{H} = \mathbf{P}'\mathbf{B}\mathbf{P}$, $\boldsymbol{\nu} = \mathbf{P}'\boldsymbol{\mu}$, and we have exploited the fact that $\mathrm{Re}[\mathbb{M}^*(is)]$ is an even function of s.

The following result is derived in Broda and Paolella (2009b). Let $R = \mathbf{X}'\mathbf{A}\mathbf{X}/\mathbf{X}'\mathbf{B}\mathbf{X}$, where $\mathbf{X} \sim \mathrm{N}_n(\boldsymbol{\mu},\mathbf{I})$, $\mathbf{B} \neq 0$, $\mathbf{B} \geqslant 0$. Then the density of R is

$$f_R(r) = \frac{1}{\pi}\int_0^{\infty}\frac{\rho(u)\cos\beta(u) - u\delta(u)\sin\beta(u)}{2\gamma(u)}\,\mathrm{d}u, \tag{A.39}$$

where

$$\beta(u) = \frac{1}{2}\sum_{i=1}^{n}\arctan a_i + \frac{\theta_i a_i}{c_i}, \qquad \gamma(u) = \exp\left\{\frac{1}{2}\sum_{i=1}^{n}\frac{\theta_i b_i}{c_i} + \frac{1}{4}\ln c_i\right\},$$

$$\rho(u) = \mathrm{tr}\ \mathbf{H}\mathbf{F}^{-1} + \boldsymbol{\nu}'\mathbf{F}^{-1}(\mathbf{H} - u^2\boldsymbol{\Lambda}\mathbf{H}\boldsymbol{\Lambda})\mathbf{F}^{-1}\boldsymbol{\nu}, \qquad \delta(u) = \mathrm{tr}\ \mathbf{H}\boldsymbol{\Lambda}\mathbf{F}^{-1} + 2\boldsymbol{\nu}'\mathbf{F}^{-1}\mathbf{H}\boldsymbol{\Lambda}\mathbf{F}^{-1}\boldsymbol{\nu},$$

$a_i = \lambda_i u$, $b_i = a_i^2$, $c_i = 1 + b_i$, $\theta_i = \nu_i^2 = (\mathbf{p}_i'\boldsymbol{\mu})^2$, and $\mathbf{F} = \mathbf{I} + u^2\boldsymbol{\Lambda}^2$.

Calculation of (A.39) is implemented in program `ROQpdfgeary.m`, which is not shown here but included with the collection associated with the book. The right panel of Figure A.2 shows the r.p.e. based on (A.39) for the same distribution as used in Example A.5; we see it delivers values virtually identical to the analytic solution available in this special case.

A.3.2.3 Use of Pan's Formula

Based on contour integration and building on the work of Grad and Solomon (1955), Pan Jie-Jian (1964) developed what became a popular algorithm for the distribution of a weighted sum of independent central χ_1^2 random variables, such that the weights are distinct. See also Durbin and Watson (1971), Farebrother (1980, 1984, 1990, 1994), and the discussion in Section II.10.1.4. It can be reduced to calculating a simple truncated infinite sum, with number of terms, say N, such that the accuracy is a function of N. This is implemented in program `ROQpdfpan.m`, not shown here, but available in the collection of programs.

Example A.6 Let $R = 1/(C + C^{-1})$, where $C \sim \mathrm{Cau}(0,1)$, with density $f_C(c) = \pi^{-1}(1+c^2)^{-1}$. We wish to calculate the exact p.d.f. of R via transformation, and, by expressing it as a ratio of quadratic forms, compare f_R with the numeric methods using Geary's formula, and Pan's method, via programs `ROQpdfgeary.m` and `ROQpdfpan.m`, respectively.

Differentiating $r = 1/(c + c^{-1})$ shows that r has its maximum at $c = 1$, so that $|R| < 1/2$. Solving for c is the same as solving for $c^2 - cr^{-1} + 1 = 0$, or, for $r > 0$ and $c > 0$,

$$c = \frac{1 - (1 - 4r^2)^{1/2}}{2r}, \qquad \frac{\mathrm{d}c}{\mathrm{d}r} = \frac{1 - \sqrt{1 - 4r^2}}{2r^2\sqrt{1 - 4r^2}} = \frac{1}{2r^2}\left(\frac{1}{\sqrt{1 - 4r^2}} - 1\right).$$

Note that, for $0 < r < 1/2$, $0 < \sqrt{1 - 4r^2} < 1$, $\mathrm{d}c/\mathrm{d}r > 0$. Thus, from symmetry, and simplifying,

$$\begin{aligned} f_R(r) &= f_C(c)\frac{\mathrm{d}c}{\mathrm{d}r}\mathbb{I}(r > 0) + f_C(c)\frac{\mathrm{d}c}{\mathrm{d}r}\mathbb{I}(r < 0) \\ &= \frac{4r^2(1 - s_r)}{\pi r^2 s_r(4r^2 + (1 - s_r)^2)}\mathbb{I}\left(|r| < \frac{1}{2}\right), \quad s_r = \sqrt{1 - 4r^2}. \end{aligned} \tag{A.40}$$

In terms of quadratic forms, recalling that a Cauchy random variable can be expressed as a ratio of independent standard normal r.v.s, say X_1 and X_2, we have

$$R = \frac{1}{\frac{X_1}{X_2} + \frac{X_2}{X_1}} = \frac{X_1 X_2}{X_1^2 + X_2^2} = \frac{\mathbf{X}'\mathbf{A}\mathbf{X}}{\mathbf{X}'\mathbf{X}}, \quad \mathbf{A} = \begin{bmatrix} 0 & 1/2 \\ 1/2 & 0 \end{bmatrix}.$$

The eigenvalues of $\mathbf{A}$ are $-1/2$ and $1/2$, so that, from (A.34), $|R| < 1/2$, in agreement with the previous result.

The p.d.f. is plotted in the left panel of Figure A.3, based on a grid of 96 points. Its calculation over the grid using (A.39) via program `ROQpdfgeary.m` takes about 0.5 seconds (on a typical PC at the time of writing). Use of Pan's method, based on $N = 100$ points, requires 0.2 seconds, but from the right panel of Figure A.3, this is seen to be relatively inaccurate. Much higher accuracy is achieved using $N = 1{,}000$, though it is still not as good as the use of (A.39), and requires over 190 seconds. Thus, this example demonstrates a case when the Pan method is much slower than the Geary method. The program in Listing A.4 was used for the calculation. ■

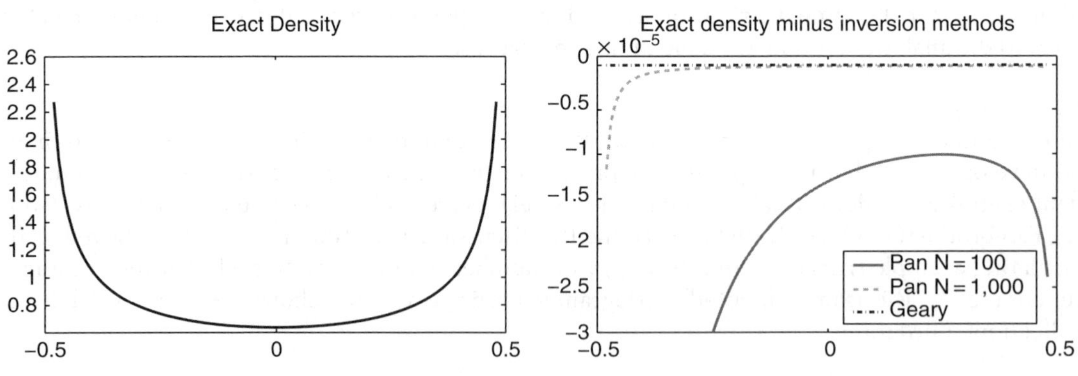

Figure A.3 **Left:** Exact density (A.40). **Right:** Discrepancy between exact density and use of the Geary and Pan methods. The graph is truncated from below.

```
r=[-0.48:0.01:-0.01 , 0.01:0.01:0.48]'; r2=r.^2; s=sqrt(1-4*r2); d=1-s;
num = 4 * r2 .* d; den = pi * r2 .* s .* ( 4*r2 + d.^2 ); f = num./den;
figure, set(gca,'fontsize',16)
plot(r,f,'b-','linewidth',2), title('Exact Density')
A=[0 1/2 ; 1/2 0]; B=eye(2);
tic, fPan2 = ROQpdfpan(r,A,B,100); toc
tic, fPan3 = ROQpdfpan(r,A,B,1000); toc
tic, fGeary=ROQpdfgeary(r,A,B,zeros(2,1)); toc
figure, set(gca,'fontsize',16)
plot(r,f-fPan2,'r-', r,f-fPan3,'g--', r,f-fGeary,'b-.','linewidth',2)
legend('Pan N=100','Pan N=1000','Geary','location','SouthEast')
title('Exact density minus inversion methods')
```

Program Listing A.4: Generates the graphics in Figure A.3. Programs `ROQpdfpan` and `ROQpdfgeary` are available in the collection of programs associated with the book.

A.3.2.4 Saddlepoint Approximation

As before, let $R = N/D$ with $N = \mathbf{X}'\mathbf{A}\mathbf{X}$ and $D = \mathbf{X}'\mathbf{B}\mathbf{X}$. The same derivation that leads to (A.37) also shows that, with C the (constructed) random variable associated with m.g.f.

$$\mathbb{M}_C(s) = \frac{1}{\mathbb{E}[D]} \frac{\partial}{\partial t} \mathbb{M}_{N,D}(s,t)|_{t=-rs}, \tag{A.41}$$

the density of R is

$$f_R(r) = \mathbb{E}[D] f_C(0), \tag{A.42}$$

where f_C is the density of C. We approximate f_C with $\hat{f}_C$, the s.p.a. applied to $\mathbb{M}_C$, so that the s.p.a. of the density of R is, from (A.42), $\hat{f}_R(r) = \mathbb{E}[D]\,\hat{f}_C(0)$, as was done in Daniels (1954, Sec. 9). As shown in Butler and Paolella (2008), this leads to

$$\hat{f}_R(r) = \frac{J(\hat{s})}{\sqrt{2\pi K_S''(\hat{s})}} \exp\{\mathbb{K}_S(\hat{s})\}, \tag{A.43}$$

where $\mathbb{K}_S$ is the cumulant generating function (c.g.f.) of S in (A.32) and $\hat{s}$ solves $\mathbb{K}_S'(\hat{s}) = 0$. Quantity $J(\hat{s})$ is computed from

$$J(s) = \mathrm{tr}(\mathbf{UH}) + \mathbf{v}'\mathbf{UHU}\mathbf{v}, \tag{A.44}$$

with $\mathbf{U} = (\mathbf{I} - 2s\boldsymbol{\Lambda})^{-1}$, and $\mathbf{H}$, $\mathbf{P}$, $\boldsymbol{\Lambda}$ and $\mathbf{v}$ as in Section A.3.2.2.

A second-order saddlepoint density approximation for the general case can be derived. In particular, from Butler (2007, p. 383),

$$\tilde{f}_R(r) = \hat{f}_R(r)(1 + O), \tag{A.45}$$

where $\hat{f}_R(r)$ is given in (A.43),

$$O = \left(\frac{\hat{\kappa}_4}{8} - \frac{5}{24}\hat{\kappa}_3^2\right) + \frac{J_r'(\hat{s})\hat{\kappa}_3}{2J_r(\hat{s})\sqrt{K_S''(\hat{s})}} - \frac{J_r''(\hat{s})}{2J_r(\hat{s})K_S''(\hat{s})}, \tag{A.46}$$

$\hat{\kappa}_i = K_S^{(i)}(\hat{s})/K_S''(\hat{s})^{i/2}$, and $J_r'(\hat{s}) = 2\mathrm{tr}(\mathbf{U}\boldsymbol{\Lambda}\mathbf{UH}) + 4\mathbf{v}'\mathbf{U}\boldsymbol{\Lambda}\mathbf{UHU}\mathbf{v}$.

```
function [pdf,cdf,svec]=sparatio(rvec,A,B,Sigma,mu,DANIELS)
if nargin<6, DANIELS=2; end
pdf=zeros(length(rvec),1); cdf=pdf; svec=pdf;
n=length(A);
if length(Sigma)==1
  Sighalf=eye(n);
  nullcase = (max(max(abs(B-eye(n)))) < 1e-12) & (DANIELS==1);
  if nargin>=5, if length(mu)>1, nullcase=0; end, end
else
  nullcase=0;
  [V,D]=eig(makesym(Sigma)); % see end of program
  W=sqrt(D); Sighalf = V*W*V';
end
SI=inv(Sighalf); rl=length(rvec); s=0;
%%%%%%%%%%%%%%%%%%%%%%%%%%%%%%%%%%%%%%%%%%%%%%%%%%%%%%%%%%%%
% nullcase is 1 (i.e., true) if
%   1. B=Identity          2. Sigma = Identity
%   3. mu is zero          4. DANIELS=1 (i.e., 1st order SPA)
% If so, this runs faster, particularly for large n
%%%%%%%%%%%%%%%%%%%%%%%%%%%%%%%%%%%%%%%%%%%%%%%%%%%%%%%%%%%%
if nullcase==1
  zi=eig(A);
else
  A=Sighalf * A * Sighalf; B=Sighalf * B * Sighalf;
end
for rloop=1:rl
  r=rvec(rloop);
  if nullcase==1, nc=zeros(n,1); Lam=zi-r;
  else
    [P,Lam] = eig(makesym((A-r*B))); Lam=diag(Lam);
    if nargin<5          % mu not passed, so set it to 0
      nu=zeros(n,1); nc=nu;
    elseif length(mu)==1  % mu passed as scalar. Set to 0
      nu=zeros(n,1); nc=nu;
    else
      nu=P'*SI*mu; nc=nu.^2;
    end
  end
```

Program Listing A.5: Computes the saddlepoint p.d.f. and c.d.f. approximations of (A.30). To use this program, we need to modify the header of program `spaweightedsum` in Listing II.10.3 to `[pdf,cdf,svec,K,Kpp,kap3,forboth] = spaweightedsumofchisquare(...` in order to pass back the necessary quantities for computing the p.d.f. Pass `Sigma` as a scalar to take $\boldsymbol{\Sigma} = \mathbf{I}$. If mu is not passed, or if mu is passed as a scalar, then $\boldsymbol{\mu}$ is taken to be the zero vector. Set DANIELS=2 (default value if not passed) to use the higher order term in (A.46). The program is continued in Listing A.6.

```
    [garbage,cdf_,s,K,Kpp,kap3,forboth] = ...
       spaweightedsumofchisquare(DANIELS,0,Lam,ones(n,1),nc,s);

    cdf(rloop)=cdf_; svec(rloop)=s;
    if nullcase==1
      ubiq=1-2*Lam*s;
      pdf(rloop) = n*exp(sum(-0.5*log(ubiq))) / ...
          sqrt(4*pi*sum(Lam.^2 .* ubiq.^(-2)));
    else
      H = P'*B*P; L=diag(Lam); I=eye(length(Lam)); U = inv(I-2*s*L);
      UH=U*H; tt = UH*U*nu; J = sum(diag(UH)) + nu'*tt;
      pdf(rloop) = real( J * exp(K) / sqrt(2*pi*Kpp) );
      if DANIELS==2
        Jp = 2*sum(diag(U*L*U*H)) + 4*nu'*U*L*tt;
        % Now numerically calculate Jpp, the 2nd deriv of J(s)
        stol=1e-6; sm=s-stol; su=s+stol;
        Um = inv(I-2*sm*L); Uu = inv(I-2*su*L);
        ttm = Um*H*Um*nu;    ttu = Uu*H*Uu*nu;
        Jpm = 2*sum(diag(Um*L*Um*H)) + 4*nu'*Um*L*ttm;
        Jpu = 2*sum(diag(Uu*L*Uu*H)) + 4*nu'*Uu*L*ttu;
        Jpp = (Jpu-Jpm)/(2*stol);
        term3=Jp*kap3 / 2 / J / sqrt(Kpp); term4=Jpp / 2 / J / Kpp;
        O=real( forboth + term3 - term4 );
        pdf(rloop) = pdf(rloop)*(1+O);
      end
    end
end

function X=makesym(Z), X = 0.5*(Z+Z');
```

Program Listing A.6: Continuation of Listing A.5.

The second derivative of J_r, as required in (A.46), could also be algebraically formulated, but it is easily and accurately numerically obtained. The program in Listing A.5 computes (A.43) and (A.45).

Remarks

a) An important special case of the general ratio R is when $\boldsymbol{\mu} = \mathbf{0}$. Then $\boldsymbol{\nu} = \mathbf{0}$, and (A.43) easily reduces to

$$\hat{f}_R(r) = \frac{\text{tr}\,(\mathbf{I} - 2\hat{s}\boldsymbol{\Lambda})^{-1}\mathbf{H}}{\sqrt{4\pi \sum_{i=1}^{n} \lambda_i^2 (1 - 2\lambda_i \hat{s})^{-2}}} \prod_{i=1}^{n} (1 - 2\lambda_i \hat{s})^{-1/2}, \tag{A.47}$$

which was first derived by Lieberman (1994a,b).

b) If, further, $\boldsymbol{\mu} = \mathbf{0}$, $\boldsymbol{\Sigma} = \sigma^2\mathbf{I}$, and $\mathbf{B} = \mathbf{I}$, then matters simplify considerably. First, $\text{tr}\,(\mathbf{I} - 2\hat{s}\boldsymbol{\Lambda})^{-1}\mathbf{H} = n$, seen by noting that $\mathbf{H} = \mathbf{P}'\mathbf{P} = \mathbf{I}$, implying

$$\text{tr}\,(\mathbf{I}_n - 2\hat{s}\boldsymbol{\Lambda})^{-1} = \sum_{i=1}^{n} (1 - 2\hat{s}\lambda_i)^{-1}.$$

Next, as the saddlepoint equation solves $0 = \mathbb{K}_S'(\hat{s}) = \sum_{i=1}^n \lambda_i(1-2\hat{s}\lambda_i)^{-1}$, it follows that $\sum_{i=1}^n (1-2\hat{s}\lambda_i)^{-1} = n$, because

$$n = \sum_{i=1}^n \frac{1-2\hat{s}\lambda_i}{1-2\hat{s}\lambda_i} = \sum_{i=1}^n \frac{1}{1-2\hat{s}\lambda_i} - 2\hat{s}\sum_{i=1}^n \frac{\lambda_i}{1-2\hat{s}\lambda_i}.$$

Thus, from (A.47), $\hat{f}_R(r)$ can be expressed as

$$\hat{f}_R(r) = \frac{n\prod_{i=1}^n (1-2\lambda_i\hat{s})^{-1/2}}{\sqrt{4\pi\sum_{i=1}^n \lambda_i^2(1-2\lambda_i\hat{s})^{-2}}}. \tag{A.48}$$

It is easy to confirm that the eigenvalues of $(\mathbf{A} - r\mathbf{I})$ are given by

$$\lambda_i = \zeta_i - r, \qquad \zeta = \mathrm{Eig}(\mathbf{A}). \tag{A.49}$$

These need to be calculated only once, so that density (A.48) is easily computed as a function of r. This seemingly very special case actually arises in various applications, typically as the distribution of a statistic under a null hypothesis. ■

Example A.7 Continuing with the Cauchy-like ratio in Example A.5 with $\boldsymbol{\mu}' = (\mu_1, \mu_2) = (0.1, 2)$, Figure A.4 shows the exact density and s.p.a. (A.43). We see that the first-order s.p.a. does not pick up the bimodality of the true p.d.f., and evaluating the r.p.e. for ever larger values of r indicates that it converges to 21.62 in the right tail.

This corresponds to a limiting ratio of $\lim_{r\to\infty} f_R(r)/\hat{f}_R(r)$, of 0.8222. It is possible to analytically determine this limiting ratio for the general ratio of quadratic forms under various conditions on **A** and **B**, as detailed in Butler and Paolella (2008).

For this example, the relevant expression is

$$\lim_{r\to\infty}\frac{f_R(r)}{\hat{f}_R(r)} = \frac{\sqrt{2\pi(1-2t_0)}(2t_0)^{\frac{n-1}{2}}u_0e^{-\eta_2}}{B\left(\frac{1}{2},\frac{n+1}{2}\right)\frac{n}{2}}\,{}_1F_1\left(\frac{n}{2};\frac{1}{2};\frac{\nu_0^2}{2}\right), \tag{A.50}$$

where

$$t_0 = \frac{1}{4n}\{2n-1+\nu_0^2-\sqrt{(\nu_0^2+2n-1)^2-(2n-1)^2+1}\},$$

$$u_0 = \sqrt{\frac{n-1}{2}+\frac{2t_0^2}{(1-2t_0)^2}+\frac{4\nu_0^2t_0^2}{(1-2t_0)^3}}, \qquad \eta_2 = \frac{\nu_0^2}{2(1-2t_0)},$$

and, for this case, $n = 2$ and $\nu_0 = \mu_2 = 2$.[8] Computing (A.50) indeed yields 0.8222. Expression (A.50) also applies to the ratio of tail areas, $\lim_{r\to\infty} \Pr(R > r)/\widehat{\Pr}(R > r)$, where $\widehat{\Pr}$ refers to use of the first-order c.d.f. saddlepoint approximation.

Figure A.4 also shows the second-order s.p.a. (A.45). Observe that it can pick up the bimodality of R. The r.p.e. of the second-order approximation is still high near the two peaks and, far into the tail, it stays constant at about 2.1%, which is far better than the first-order approximation r.p.e. of 21.62.

8 In general, ν_0 is a more complicated function of **A**, **B** and $\boldsymbol{\mu}$; see Butler and Paolella (2008, Lemma 7).

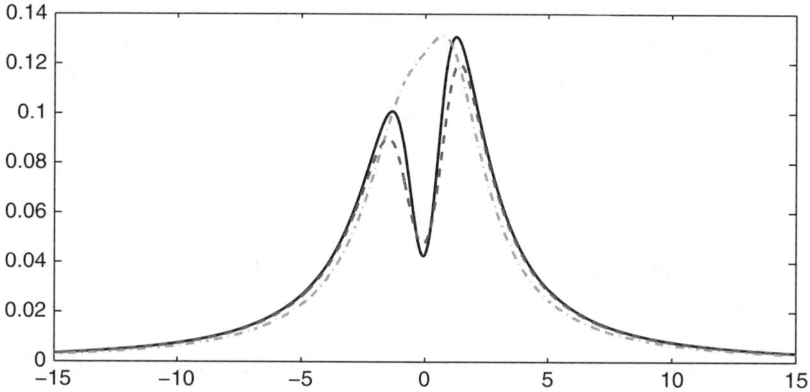

Figure A.4 Exact density (solid), second-order s.p.a. (A.45) (dashed) and normalized first-order s.p.a. (A.43) (dash-dot) of R in (A.36) using $\boldsymbol{\mu} = (0.1, 2)'$.

Of course, in most statistical contexts, vector $\mathbf{X}$ in ratio R will be of length equal or close to the sample size (usually far greater than two), which has the effect of making the distribution of R somewhat more like that of a normal random variable, with correspondingly higher accuracy of the s.p.a.s. ■

A.4 Problems

> It's quite satisfying—hell, it's incredibly satisfying—to face some important problem and find a solution that works.
>
> (Jack Kilby)[9]

> I'm old enough to know you can't close your mind to new ideas. You have to test out every possibility if you want something new.
>
> (Dr. John Goodenough)[10]

Problem 1.1 Recall Example A.1, which detailed the calculation of the distribution of the sample variance. Construct a program that plots the density s.p.a. for a given value of $\boldsymbol{\mu}$ and AR(1) parameter ρ of the $\boldsymbol{\Sigma}$ matrix. How would you expect the density to behave as a function of ρ for, say, $\boldsymbol{\mu} = \mathbf{0}$? Plot it for several ρ. Also make the program such that it can also plot a kernel density estimate of the p.d.f. based on simulation.

9 Unlike Thomas Edison, Alexander Graham Bell, Henry Ford, and the Wright Brothers, the Kansas engineer Jack St. Clair Kilby (1923–2005) is not an American household name, though he deserves to be. His invention of the integrated circuit has permanently changed human civilization. It has won him numerous awards, including the National Medal of Science in 1970 and the Nobel Prize in Physics in 2000.

10 Quoted in Kennedy (2017). Goodenough started university studies at the age of 23, and was informed by a physics professor that he was already too old to succeed in the field (and thus not living up to his last name). In 1980, at the age of 57, he co-invented the lithium-ion battery. At the age of 94, with his team at the University of Texas at Austin, he filed a patent for a new type of solid-state battery that is much safer, lighter, cheaper, more durable, and lasts longer than anything of its kind, and is poised to revolutionize the electric car industry.

Problem 1.2 In the proof of Theorem A.1, the last line states $\mathbf{A\Sigma A} = \mathbf{A} \Leftrightarrow \mathbf{A\Sigma A\Sigma} = \mathbf{A\Sigma}$. This exercise just details this a bit, and practices some basic linear algebra. Let $\mathbf{B}$ and $\mathbf{C}$ be symmetric $n \times n$ matrices, not necessarily of full rank, and let $\mathbf{F}$ be $n \times n$. Trivially, $\mathbf{B} = \mathbf{C} \Rightarrow \mathbf{BF} = \mathbf{CF}$. If $\mathbf{BF} = \mathbf{CF}$ and $\mathbf{F}$ is full rank, then $\mathbf{F}^{-1}$ exists, and $\mathbf{BF} = \mathbf{CF} \Rightarrow \mathbf{BFF}^{-1} = \mathbf{CFF}^{-1} \Rightarrow \mathbf{B} = \mathbf{C}$. For $n = 2$, construct an $\mathbf{F}$, $\mathbf{B}$ and $\mathbf{C}$ such that $\mathbf{BF} = \mathbf{CF}$ but $\mathbf{B} \neq \mathbf{C}$.

Problem 1.3 Derive (A.21) from (A.20).

Problem 1.4 For $\mathbf{X} \sim \mathrm{N}_n(\mathbf{0}, \mathbf{\Sigma})$, derive the joint m.g.f. of vector $\mathbf{N}$ and scalar D, where $D = \mathbf{X}'\mathbf{X}$ and $\mathbf{N} = (\mathbf{X}'\mathbf{A}_1\mathbf{X}, \mathbf{X}'\mathbf{A}_2\mathbf{X}, \dots, \mathbf{X}'\mathbf{A}_m\mathbf{X})'$.

Problem 1.5 Construct a Matlab program that simulates the ratio of normal random variables in (A.36) and produces a kernel density estimate. Do so for

$$\boldsymbol{\mu} = \begin{bmatrix} 1 \\ 2 \end{bmatrix}, \quad \mathbf{\Sigma} = \begin{bmatrix} 1 & -0.9 \\ -0.9 & 1 \end{bmatrix},$$

and overlay the resulting graph with the s.p.a.

Problem 1.6 Let $\mathbf{A}$ and $\mathbf{B}$ be symmetric matrices such that $\mathbf{A} + \mathbf{B} = \mathbf{I}$. Let $\mathbf{Z} \sim \mathrm{N}_n(\mathbf{0}, \mathbf{I})$.

a) Prove: If $\mathbf{A} = \mathbf{A}^2$, then $\mathbf{B} = \mathbf{B}^2$ and $\mathbf{AB} = \mathbf{0}$.
b) Prove: If $\mathbf{Z}'\mathbf{AZ} \sim \chi^2_k$, then $\mathbf{Z}'\mathbf{BZ} \sim \chi^2_{n-k}$ and $\mathbf{Z}'\mathbf{AZ} \perp \mathbf{Z}'\mathbf{BZ}$.

Problem 1.7 Let $\mathbf{A}$, $\mathbf{B}$ and $\mathbf{C}$ be $n \times n$ symmetric matrices such that $\mathbf{A} + \mathbf{B} + \mathbf{C} = \mathbf{I}$ and $\mathbf{C} \geqslant 0$. Let $\mathbf{Z} \sim \mathrm{N}_n(\mathbf{0}, \mathbf{I})$.

a) Prove: If $\mathbf{A} = \mathbf{A}^2$ and $\mathbf{B} = \mathbf{B}^2$, then $\mathbf{AB} = \mathbf{AC} = \mathbf{BC} = \mathbf{0}$ and $\mathbf{C} = \mathbf{C}^2$. Hint: For a $T \times k$ matrix $\mathbf{X}$, the **column space** of $\mathbf{X}$, denoted $\mathcal{C}(\mathbf{X})$ is the set of all vectors that can be generated as a linear sum of the columns of $\mathbf{X}$, such that the coefficient of each vector is a real number, i.e.,

$$\mathcal{C}(\mathbf{X}) = \{\mathbf{y} : \mathbf{y} = \mathbf{Xb}, \mathbf{b} \in \mathbb{R}^k\}. \tag{A.51}$$

In words, if $\mathbf{y} \in \mathcal{C}(\mathbf{X})$, then there exists $\mathbf{b} \in \mathbb{R}^k$ such that $\mathbf{y} = \mathbf{Xb}$. This is used extensively in Chapter 1, and (A.51) is the same as (1.38). The hint now consists of letting $\mathbf{a} \in \mathcal{C}(\mathbf{A})$.
b) Prove: If $\mathbf{Z}'\mathbf{AZ} \sim \chi^2_k$ and $\mathbf{Z}'\mathbf{BZ} \sim \chi^2_m$, then $\mathbf{Z}'\mathbf{CZ} \sim \chi^2_{n-k-m}$ and $\mathbf{Z}'\mathbf{AZ}$, $\mathbf{Z}'\mathbf{BZ}$ and $\mathbf{Z}'\mathbf{CZ}$ are mutually independent.

Problem 1.8 Prove Theorem A.3: Let $\mathbf{Y} \sim \mathrm{N}_n(\boldsymbol{\mu}, \mathbf{\Sigma})$, $\mathbf{\Sigma} > 0$. Vector $\mathbf{BY}$, with $\mathbf{B}$ a real $q \times n$ matrix, is independent of $\mathbf{Y}'\mathbf{AY}$ if $\mathbf{B\Sigma A} = \mathbf{0}$.

a) First prove it in the case for which $\mathbf{\Sigma} = \mathbf{I}$ by setting $\mathbf{A} = \mathbf{UDU}'$, where $\mathbf{U}$ is orthogonal, $\mathbf{D} = \mathrm{diag}([\lambda_1, \dots, \lambda_k, 0, \dots, 0])$ and $k = \mathrm{rank}(\mathbf{A})$. Let $\mathbf{Z} = \mathbf{U}'\mathbf{Y}$ and $\mathbf{B}^* = \mathbf{BU}$, and show that $\mathbf{B}^*\mathbf{D} = \mathbf{0}$.
b) Prove the general case by setting $\mathbf{Z} = \mathbf{\Sigma}^{-1/2}\mathbf{Y}$, $\mathbf{B}^* = \mathbf{B\Sigma}^{1/2}$, and using the previous special case.

A.A Appendix: Solutions

1) The program in Listing A.7 serves as a basis. One can add more "features".

```
function s2sim(sim,n,rho,xhi)
mu=zeros(n,1); % mu = ([1 2 3 4 5 5 4 3 2 1] - 3)';
Sigma = makevarcovAR1(n,rho); % see below
xvec=0.05:0.05:xhi; fspa=samplevariancedistribution(xvec,mu,Sigma,1);
figure, plot(xvec,fspa,'b-','linewidth',2)
if sim>0
  S2 = zeros(sim,1); [V,D]=eig(0.5*(Sigma+Sigma'));
  W=sqrt(D); Sighalf = V * W * V';
  for i=1:sim, X = mu + Sighalf * normrnd(0,1,n,1); S2(i) = var(X); end
  pdf = ksdensity(S2,xvec); hold on
  plot(xvec,pdf,'r--','linewidth',2), hold off
  legend('saddlepoint','Simulated')
end
set(gca,'fontsize',16)

function S = makevarcovAR1(n,rho);
S=zeros(n,n);
for i=1:n
  for j=i:n, v=rho^(j-i); S(i,j)=v; S(j,i)=v; end
end
S=S/(1-rho^2);
```

Program Listing A.7: Simulates sample variance, S^2, and plots kernel density (computed with Matlab's function `ksdensity` which is part of its statistics toolbox). Compare to Figure A.1.

2) Take $\mathbf{F}$ to be a matrix of rank one, $\mathbf{B}$ and $\mathbf{C}$ symmetric, say

$$\mathbf{F}=\begin{bmatrix}1 & 2\\ 1 & 2\end{bmatrix},\quad \mathbf{B}=\begin{bmatrix}1 & 2\\ 2 & 3\end{bmatrix},\quad \mathbf{C}=\begin{bmatrix}c_1 & c_2\\ c_2 & c_3\end{bmatrix}.$$

Then

$$\begin{bmatrix}3 & 6\\ 5 & 10\end{bmatrix}=\mathbf{BF}=\mathbf{CF}=\begin{bmatrix}c_1+c_2 & 2c_1+2c_2\\ c_2+c_3 & 2c_2+2c_3\end{bmatrix},$$

and taking $c_1=2$ implies $c_2=1$, which implies $c_3=4$, or

$$\mathbf{C}=\begin{bmatrix}2 & 1\\ 1 & 4\end{bmatrix},\quad \text{and } \mathbf{By}=\mathbf{Cy}\qquad \forall\, \mathbf{y}\in \mathcal{C}(\mathbf{F})=y\begin{bmatrix}1\\ 1\end{bmatrix},\quad y\in\mathbb{R},$$

but clearly, $\mathbf{By}\neq\mathbf{Cy}\ \forall \mathbf{y}\in\mathbb{R}^2$. If $\mathcal{C}(\mathbf{F})=\mathbb{R}^2$, then $\mathbf{By}=\mathbf{Cy}\qquad \forall\,\mathbf{y}\in\mathcal{C}(\mathbf{F})=\mathbb{R}^2$, and $\mathbf{B}=\mathbf{C}$.

3) To see the first term (the determinant), note that $\mathbf{P}'\mathbf{P}=\mathbf{I}$ and $|\mathbf{P}'\mathbf{P}|=|\mathbf{P}'||\mathbf{P}|$ and $\mathbf{\Lambda}=\mathbf{P}'\mathbf{AP}$, so that

$$\begin{aligned}|\mathbf{\Omega}|^{-\frac{1}{2}} &= |\mathbf{P}'\mathbf{P}|^{-\frac{1}{2}}|\mathbf{\Omega}|^{-\frac{1}{2}} = |\mathbf{P}'|^{-\frac{1}{2}}|\mathbf{\Omega}|^{-\frac{1}{2}}|\mathbf{P}|^{-\frac{1}{2}} = |\mathbf{P}'\mathbf{\Omega}\mathbf{P}|^{-\frac{1}{2}}\\ &= |\mathbf{P}'(\mathbf{I}_n-2s\mathbf{A})\mathbf{P}|^{-\frac{1}{2}} = |(\mathbf{P}'\mathbf{I}_n\mathbf{P}-2s\mathbf{P}'\mathbf{AP})|^{-\frac{1}{2}} = |(\mathbf{I}_n-2s\mathbf{\Lambda})|^{-\frac{1}{2}}.\end{aligned}$$

For the term in the exponent, with $\boldsymbol{\nu}=\mathbf{P}'\boldsymbol{\mu}$, $\mathbf{I}=\mathbf{P}'\mathbf{P}$ and $\mathbf{\Lambda}=\mathbf{P}'\mathbf{AP}$,

$$\begin{aligned}s\boldsymbol{\nu}'\mathbf{\Lambda}(\mathbf{I}-2s\mathbf{\Lambda})^{-1}\boldsymbol{\nu} &= s(\boldsymbol{\mu}'\mathbf{P})\mathbf{P}'\mathbf{AP}(\mathbf{P}'\mathbf{P}-2s\mathbf{P}'\mathbf{AP})^{-1}(\mathbf{P}'\boldsymbol{\mu})\\ &= s(\boldsymbol{\mu}'\mathbf{P})\mathbf{P}'\mathbf{AP}(\mathbf{P}'(\mathbf{I}-2s\mathbf{A})\mathbf{P})^{-1}(\mathbf{P}'\boldsymbol{\mu})\\ &= s(\boldsymbol{\mu}'\mathbf{P})\mathbf{P}'\mathbf{APP}'(\mathbf{I}-2s\mathbf{A})^{-1}\mathbf{P}(\mathbf{P}'\boldsymbol{\mu})\\ &= s\boldsymbol{\mu}'\mathbf{A}(\mathbf{I}-2s\mathbf{A})^{-1}\boldsymbol{\mu}.\end{aligned}$$

It remains to show that this is equal to

$$-\frac{1}{2}\boldsymbol{\mu}'(\mathbf{I}-(\mathbf{I}-2s\mathbf{A})^{-1})\boldsymbol{\mu}.$$

Observe that

$$\mathbf{I}=(\mathbf{I}-\mathbf{A})(\mathbf{I}-\mathbf{A})^{-1}=(\mathbf{I}-\mathbf{A})^{-1}-\mathbf{A}(\mathbf{I}-\mathbf{A})^{-1}$$

or, rearranging and replacing $\mathbf{A}$ by $k\mathbf{A}$,

$$\mathbf{A}(\mathbf{I}-k\mathbf{A})^{-1}=-\frac{1}{k}(\mathbf{I}-(\mathbf{I}-k\mathbf{A})^{-1}),$$

from which the result follows using $k=2s$.

4) This is the same as the m.g.f. of $\mathbf{B}=(\mathbf{X}'\mathbf{A}_1\mathbf{X},\dots,\mathbf{X}'\mathbf{A}_m\mathbf{X},\mathbf{X}'\mathbf{I}\mathbf{X})$, which, with $\mathbf{s}=(s_1,\dots,s_{m+1})$, is given by

$$\mathbb{M}_{\mathbf{B}}(\mathbf{s})=\mathbb{E}\left[\exp\left\{\sum_{i=1}^{m+1}s_i\mathbf{X}'\mathbf{A}_i\mathbf{X}\right\}\right],$$

or

$$\mathbb{M}_{\mathbf{B}}(\mathbf{s})=\int_{\mathbb{R}^n}(2\pi)^{-n/2}|\boldsymbol{\Sigma}|^{-1/2}\exp\left\{-\frac{1}{2}\left[\mathbf{x}'\left(\boldsymbol{\Sigma}^{-1}-2\sum_{i=1}^{m+1}s_i\mathbf{A}_i\right)\mathbf{x}\right]\right\}\,d\mathbf{x}.$$

Letting $\mathbf{z}=\left(\boldsymbol{\Sigma}^{-1}-2\sum_{i=1}^{m+1}s_i\mathbf{A}_i\right)^{1/2}\mathbf{x}$, and recalling the method of multivariate transformation using the Jacobian (see, e.g., Sec. I.9.1), it follows that

$$d\mathbf{x}=\left|\boldsymbol{\Sigma}^{-1}-2\sum_{i=1}^{m+1}s_i\mathbf{A}_i\right|^{-1/2}d\mathbf{z},$$

and

$$\begin{aligned}\mathbb{M}_{\mathbf{B}}(\mathbf{s})&=|\boldsymbol{\Sigma}|^{-1/2}\left|\boldsymbol{\Sigma}^{-1}-2\sum_{i=1}^{m+1}s_i\mathbf{A}_i\right|^{-1/2}\int_{\mathbb{R}^n}(2\pi)^{-T/2}e^{-\frac{1}{2}\mathbf{z}'\mathbf{z}}\,d\mathbf{z}\\&=\left|\mathbf{I}-2\sum_{i=1}^{m+1}s_i\mathbf{A}_i\boldsymbol{\Sigma}\right|^{-1/2}.\end{aligned}$$

Thus,

$$\mathbb{M}_{\mathbf{N},D}(\mathbf{s},t)=\left|\mathbf{I}_n-2\sum_{i=1}^{m}s_i\mathbf{A}_i\boldsymbol{\Sigma}-2t\boldsymbol{\Sigma}\right|^{-1/2}.$$

5) The program in Listing A.8 performs the required calculations and Figure A.5 shows the desired plots.

6) a) Squaring $\mathbf{B}=\mathbf{I}-\mathbf{A}$ gives

$$\mathbf{B}^2=(\mathbf{I}-\mathbf{A})(\mathbf{I}-\mathbf{A})=\mathbf{I}-\mathbf{A}-\mathbf{A}+\mathbf{A}^2=\mathbf{I}-\mathbf{A}-\mathbf{A}+\mathbf{A}=\mathbf{I}-\mathbf{A}=\mathbf{B}$$

and

$$\mathbf{A}\mathbf{B}=\mathbf{A}(\mathbf{I}-\mathbf{A})=\mathbf{A}-\mathbf{A}^2=\mathbf{A}-\mathbf{A}=\mathbf{0}.$$

```
A=[0 1/2 ; 1/2 0]; B=[1 0 ; 0 0]; rho=-0.9; Sig=[1 rho; rho 1]; mu=[1 2]';
r=-12:0.025:12; [pdf,cdf,svec]=sparatio(r,A,B,Sig,mu,2);
sim=15000; rr = zeros(sim,1); [V,D]=eig(0.5*(Sig+Sig'));
W=sqrt(D); Sighalf = V * W * V';
for i=1:sim, X = mu + Sighalf * randn(2,1); rr(i) = X(2)/X(1); end
pdfrr = ksdensity(rr,r);
figure, plot(r,pdfrr,'r--',r,pdf,'b-','linewidth',2)
set(gca,'fontsize',16), grid, axis([-12 12 0 0.32])
```

Program Listing A.8: Simulates ratio of independent normal random variables, computes and plots kernel density estimate and compares it with the saddlepoint approximation.

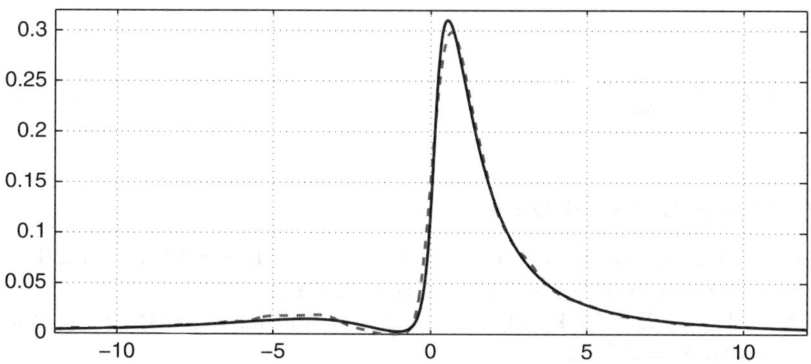

Figure A.5 Saddlepoint (solid) and kernel density estimate (dashed) of X_2/X_1, where $(X_1, X_2)' \sim \mathrm{N}(\boldsymbol{\mu}, \boldsymbol{\Sigma})$, $\boldsymbol{\mu} = (1, 2)'$ and $\boldsymbol{\Sigma}$ such that $\mathbb{V}(X_1) = \mathbb{V}(X_2) = 1$ and $\mathrm{Corr}(X_1, X_2) = -0.9$.

b) Theorem A.1 with $\boldsymbol{\Sigma} = \mathbf{I}$ implies that $\mathbf{A} = \mathbf{A}^2$. The previous part then implies that $\mathbf{B} = \mathbf{B}^2$ and $\mathbf{AB} = \mathbf{0}$, so that, again from Theorem A.1, $\mathbf{Z}'\mathbf{BZ} \sim \chi^2_{n-k}$. As

$$\mathbf{0} = \mathbf{0}' = (\mathbf{AB})' = \mathbf{B}'\mathbf{A}' = \mathbf{BA}, \tag{A.52}$$

Theorem A.2 implies that $\mathbf{Z}'\mathbf{AZ} \perp \mathbf{Z}'\mathbf{BZ}$.

7) a) From the hint, let $\mathbf{a} \in \mathcal{C}(\mathbf{A})$, so that $\exists \mathbf{b}$ such that $\mathbf{a} = \mathbf{Ab}$, and this implies $\mathbf{Aa} = \mathbf{AAb} = \mathbf{Ab} = \mathbf{a}$. Next, $\mathbf{a} = \mathbf{Ia} = (\mathbf{A} + \mathbf{B} + \mathbf{C})\mathbf{a} = \mathbf{a} + \mathbf{Ba} + \mathbf{Ca}$, so that $(\mathbf{B} + \mathbf{C})\mathbf{a} = \mathbf{0}$ or $\mathbf{a}'\mathbf{Ba} + \mathbf{a}'\mathbf{Ca} = 0$. As $\mathbf{B}' = \mathbf{B}$ and $\mathbf{B} = \mathbf{B}^2$, we have

$$\mathbf{a}'\mathbf{Ba} = \mathbf{a}'\mathbf{BBa} = \mathbf{a}'\mathbf{B}'\mathbf{Ba} = (\mathbf{Ba})'(\mathbf{Ba}) \geqslant 0. \tag{A.53}$$

It was given that $\mathbf{C} \geqslant 0$, so that $\mathbf{a}'\mathbf{Ba} + \mathbf{a}'\mathbf{Ca} = 0$ implies that $\mathbf{a}'\mathbf{Ba} = \mathbf{a}'\mathbf{Ca} = 0$. Now, (A.53) implies that, if $0 = \mathbf{a}'\mathbf{Ba}$, then $\mathbf{Ba} = \mathbf{0}$. But, as $\mathbf{Aa} = \mathbf{a}$, this means that $\mathbf{BAa} = \mathbf{0}$, or that $\mathbf{BA} = \mathbf{0}$. It then follows from (A.52) that $\mathbf{AB} = \mathbf{0}$. Next, the condition

$$\mathbf{A} + \mathbf{B} + \mathbf{C} = \mathbf{I} \tag{A.54}$$

implies that $\mathbf{A}^2 + \mathbf{AB} + \mathbf{AC} = \mathbf{A}$, or $\mathbf{A}^2 + \mathbf{AC} = \mathbf{A}$, or $\mathbf{A} + \mathbf{AC} = \mathbf{A}$, so that $\mathbf{AC} = \mathbf{0}$. Similarly, postmultiplying (A.54) by $\mathbf{B}$ gives $\mathbf{AB} + \mathbf{B}^2 + \mathbf{CB} = \mathbf{B}$, or $\mathbf{B} + \mathbf{CB} = \mathbf{B}$, so that $\mathbf{CB} = \mathbf{0}$ and, from the symmetry of $\mathbf{C}$ and $\mathbf{B}$, (A.52) can be used to see that $\mathbf{BC} = \mathbf{0}$. Finally, postmultiplying (A.54) by $\mathbf{C}$ gives $\mathbf{AC} + \mathbf{BC} + \mathbf{C}^2 = \mathbf{C}$, or $\mathbf{C}^2 = \mathbf{C}$.

b) Theorem A.1 with $\mathbf{\Sigma} = \mathbf{I}$ implies that $\mathbf{A} = \mathbf{A}^2$ and $\mathbf{B} = \mathbf{B}^2$, so that the previous part implies that $\mathbf{AB} = \mathbf{AC} = \mathbf{BC} = \mathbf{0}$ and $\mathbf{C} = \mathbf{C}^2$. As $\mathbf{AB} = \mathbf{0}$, Theorem A.2 implies that $\mathbf{Z}'\mathbf{AZ} \perp \mathbf{Z}'\mathbf{BZ}$. As a sum of independent χ^2 r.v.s is also χ^2 (see Example II.2.3), $\mathbf{Z}'\mathbf{AZ} + \mathbf{Z}'\mathbf{BZ} = \mathbf{Z}'(\mathbf{A} + \mathbf{B})\mathbf{Z} \sim \chi^2_{k+m}$, so that, from Theorem A.1, $\mathbf{A} + \mathbf{B}$ is idempotent. This also easily follows, as

$$(\mathbf{A} + \mathbf{B})(\mathbf{A} + \mathbf{B}) = \mathbf{A}^2 + \mathbf{B}^2 + \mathbf{AB} + \mathbf{BA} = \mathbf{A} + \mathbf{B}.$$

Then, as $(\mathbf{A} + \mathbf{B}) + \mathbf{C} = \mathbf{I}$, the results from Problem A.6 imply that $\mathbf{Z}'\mathbf{CZ} \sim \chi^2_{n-k-m}$. Finally, as $\mathbf{AC} = \mathbf{BC} = \mathbf{0}$, Theorem A.2 implies that $\mathbf{Z}'\mathbf{AZ} \perp \mathbf{Z}'\mathbf{CZ}$ and $\mathbf{Z}'\mathbf{BZ} \perp \mathbf{Z}'\mathbf{CZ}$.

8) a) From the hint, with $\mathbf{A} = \mathbf{UDU}'$, $\mathbf{U}$ orthogonal and $\mathbf{D} = \text{diag}([\lambda_1, \dots, \lambda_k])$, where $k = \text{rank}(\mathbf{A})$, let $\mathbf{Z} = \mathbf{U}'\mathbf{Y} \sim \text{N}(\mathbf{U}'\boldsymbol{\mu}, \mathbf{I})$, so that $\mathbf{Y} = \mathbf{UZ}$, and let $\mathbf{B}^* = \mathbf{BU}$ (where $\mathbf{B}$ and $\mathbf{B}^*$ are $q \times n$) so that $\mathbf{B} = \mathbf{B}^*\mathbf{U}'$. Then

$$\begin{aligned} \mathbf{BY} &= \mathbf{BUZ} = \mathbf{B}^*\mathbf{Z}, \\ \mathbf{Y}'\mathbf{AY} &= \mathbf{Y}'\mathbf{UDU}'\mathbf{Y} = \mathbf{Z}'\mathbf{DZ} = \sum_{i=1}^{k} \lambda_i Z_i^2, \end{aligned} \tag{A.55}$$

and, as $\mathbf{U}'$ is full rank,

$$\mathbf{0} = \mathbf{BA} = \mathbf{B}^*\mathbf{U}'\mathbf{UDU}' = \mathbf{B}^*\mathbf{DU}' \Leftrightarrow \mathbf{B}^*\mathbf{D} = \mathbf{0}.$$

From the structure of $\mathbf{D}$, the first k columns of $\mathbf{B}^*$ must be zero, so that $\mathbf{B}^*\mathbf{Z} = \mathbf{BY}$ is a function only of $Z_{k+1}, \dots, Z_n$. This and (A.55) show the independence of $\mathbf{Y}'\mathbf{AY}$ and $\mathbf{BY}$.

b) From the hint, let $\mathbf{Z} = \mathbf{\Sigma}^{-1/2}\mathbf{Y} \sim \text{N}(\mathbf{\Sigma}^{-1/2}\boldsymbol{\mu}, \mathbf{I})$, so that $\mathbf{Y} = \mathbf{\Sigma}^{1/2}\mathbf{Z}$, and let $\mathbf{B}^* = \mathbf{B\Sigma}^{1/2}$, so that $\mathbf{BY} = \mathbf{B\Sigma}^{1/2}\mathbf{Z} = \mathbf{B}^*\mathbf{Z}$. Then, with $\mathbf{A}^* = \mathbf{\Sigma}^{1/2}\mathbf{A\Sigma}^{1/2}$,

$$\mathbf{Y}'\mathbf{AY} = \mathbf{Z}'\mathbf{\Sigma}^{1/2}\mathbf{A\Sigma}^{1/2}\mathbf{Z} = \mathbf{Z}'\mathbf{A}^*\mathbf{Z},$$

and $\mathbf{B}^*\mathbf{A}^* = \mathbf{B\Sigma}^{1/2}\mathbf{\Sigma}^{1/2}\mathbf{A\Sigma}^{1/2} = (\mathbf{B\Sigma A})\mathbf{\Sigma}^{1/2} = \mathbf{0}$, because of the assumption that $\mathbf{B\Sigma A} = \mathbf{0}$. From the previous part, $\mathbf{B}^*\mathbf{A}^* = \mathbf{0}$ means that $\mathbf{Z}'\mathbf{A}^*\mathbf{Z} = \mathbf{Y}'\mathbf{AY}$ is independent of $\mathbf{B}^*\mathbf{Z} = \mathbf{BY}$, as was to be shown.

Appendix B

Moments of Ratios of Quadratic Forms

Appendix A presented methods for the calculation of the density and cumulative distribution function of (ratios of) quadratic forms. This appendix considers their moments and some applications.

Relatively straightforward expressions are available for the mean and variance of a quadratic form in normal variables, and also for higher moments via recursion (A.12) in the special case with $\mathbf{X} \sim \mathrm{N}(\mathbf{0}, \sigma^2\mathbf{I})$. Matters are not as nice when working with ratios of such forms, but results are available. Unsurprisingly, these increase in complexity as we move from $\mathbf{X} \sim \mathrm{N}(\mathbf{0}, \mathbf{I})$ to $\mathbf{X} \sim \mathrm{N}(\boldsymbol{\mu}, \boldsymbol{\Sigma})$. We consider these in turn. Throughout, let $R = \mathbf{X}'\mathbf{A}\mathbf{X}/\mathbf{X}'\mathbf{B}\mathbf{X}$.

Note: In an effort to use interesting, illustrative examples throughout this appendix, some basic notions of the linear regression, AR(1), and ARX(1) models, as developed in Chapters 1, 4, *and* 5, *respectively, are required.*

B.1 For $\mathbf{X} \sim \mathrm{N}_n(\mathbf{0}, \sigma^2\mathbf{I})$ and $\mathbf{B} = \mathbf{I}$

First note that $\sigma^2 > 0$ can be set to one, without loss of generality, as it can be factored out of $\mathbf{X}$ and it cancels from the numerator and denominator. Let the spectral decomposition of $\mathbf{A}$ be given by $\mathbf{A} = \mathbf{P}\boldsymbol{\Lambda}\mathbf{P}'$, with $\boldsymbol{\Lambda} = \mathrm{diag}([\lambda_1, \dots, \lambda_n])$ the eigenvalues of $\mathbf{A}$. Then

$$R = \frac{\mathbf{X}'\mathbf{A}\mathbf{X}}{\mathbf{X}'\mathbf{X}} = \frac{\mathbf{X}'\mathbf{P}\boldsymbol{\Lambda}\mathbf{P}'\mathbf{X}}{\mathbf{X}'\mathbf{P}\mathbf{P}'\mathbf{X}} = \frac{\mathbf{Y}'\boldsymbol{\Lambda}\mathbf{Y}}{\mathbf{Y}'\mathbf{Y}}, \tag{B.1}$$

where $\mathbf{Y} = \mathbf{P}'\mathbf{X} \sim \mathrm{N}_n(\mathbf{0}, \mathbf{I})$. Thus, R can be expressed as

$$R = \frac{\sum_{i=1}^n \lambda_i \chi_i^2}{\sum_{i=1}^n \chi_i^2} =: \frac{U}{V}, \tag{B.2}$$

where U and V are defined to be the numerator and denominator, respectively, and the χ_i^2 are i.i.d. central χ_1^2 random variables. From (A.34), $\lambda_{\min} \leqslant R \leqslant \lambda_{\max}$, where $\lambda_{\min}$ and $\lambda_{\max}$ refer respectively to the smallest and largest eigenvalues of $\mathbf{A}$. Thus, R has finite support, and all positive integer moments exist.

Linear Models and Time-Series Analysis: Regression, ANOVA, ARMA and GARCH, First Edition. Marc S. Paolella.

As in Example II.2.22, let X_i be a set of i.i.d. r.v.s with finite mean and such that $\Pr(X_i = 0) = 0$, and define $S := \sum_{i=1}^n X_i$ and $R_i := X_i/S$, $i = 1, \dots, n$. As $\sum_{i=1}^n R_i$ is not stochastic, it equals its expected value, i.e.,

$$1 = \sum_{i=1}^n R_i = \mathbb{E}\left[\sum_{i=1}^n R_i\right] = n\mathbb{E}[R_1],$$

and (as was intuitively obvious), $\mathbb{E}[R_i] = n^{-1}$. Note that, if the X_i are positive r.v.s (or negative r.v.s), then $0 < R_i < 1$, and the expectation exists. Now let the X_i be i.i.d. positive r.v.s and let λ_i, $i = 1, \dots, n$, be a set of known constants. The expectation of

$$R := \frac{\sum_{i=1}^n \lambda_i X_i}{\sum_{i=1}^n X_i} = \frac{\sum_{i=1}^n \lambda_i X_i}{S} = \sum_{i=1}^n \lambda_i R_i$$

is

$$\mathbb{E}[R] = \sum_{i=1}^n \lambda_i \mathbb{E}[R_i] = n^{-1} \sum_{i=1}^n \lambda_i.$$

To connect to our setting, let $X_i \overset{\text{i.i.d.}}{\sim} \chi_1^2$. As $\mathbb{E}[X_i] = 1$,

$$\mathbb{E}[R] = \frac{\sum_{i=1}^n \lambda_i}{n} = \frac{\mathbb{E}\left[\sum_{i=1}^n \lambda_i X_i\right]}{\mathbb{E}\left[\sum_{i=1}^n X_i\right]}.$$

Thus, we see an example for which the expectation of a nonlinear function is, uncharacteristically, the function of the expectations. The reason is because it is a ratio, and this result holds somewhat more generally; see Heijmans (1999) for discussion. The result hinges on work from Basu in 1955, and is referred to as **Basu's lemma** or theorem.[1] In this case, Basu's lemma can be used to elegantly show that, in (B.2), R is independent of V, in which case $\mathbb{E}[RV] = \mathbb{E}[R]\mathbb{E}[V]$, so that $U = RV$ implies $\mathbb{E}[U] = \mathbb{E}[R]\mathbb{E}[V]$.

We show now an alternative, older proof of this independence that was discovered by Pitman, in 1937; see Stuart and Ord (1994, p. 529).

Theorem B.1 ***Independence of Ratio and Denominator*** Ratio R in (B.1) is independent of its denominator.

Proof: Let $Q = q(\mathbf{X}) = \mathbf{X}'\mathbf{X}$ and let $H = h(\mathbf{X})$ be any scale-free function of $\mathbf{X}$ (such as R). As $f_{\mathbf{X}}(\mathbf{x}) = (2\pi)^{-n/2} \exp(-\mathbf{x}'\mathbf{x}/2)$, with $\theta_k = it_k$, $k = 1, 2$, the joint c.f. of H and Q is

$$\begin{aligned}\varphi_{H,Q}(t_1, t_2) &= \mathbb{E}[\exp(\theta_1 H + \theta_2 Q)] \\ &= \int_{-\infty}^{\infty} \cdots \int_{-\infty}^{\infty} \exp[it_1 h(\mathbf{x}) + it_2 q(\mathbf{x})] f_{\mathbf{X}}(\mathbf{x})\, \mathrm{d}x_1 \dots \mathrm{d}x_n \\ &\propto \int_{-\infty}^{\infty} \cdots \int_{-\infty}^{\infty} \exp[\theta_1 h(\mathbf{x})] \exp\left[-\frac{1}{2}(1 - 2\theta_2) q(\mathbf{x})\right] \mathrm{d}x_1 \dots \mathrm{d}x_n.\end{aligned}$$

1 The development of Basu's lemma requires the important concept of **ancillarity** in mathematical statistics. Accessible and detailed discussions of these issues can be found in, e.g., Boos and Hughes-Oliver (1998), Ghosh (2002), Casella and Berger (2002), and Davison (2003). A basic development of Basu's lemma with correct proof of necessary and sufficient conditions is given in Koehn and Thomas (1975).

Let $y_i = (1-2\theta_2)^{1/2}x_i$, so $dx_i = (1-2\theta_2)^{-1/2}\,dy_i$, $q(\mathbf{x}) = \mathbf{x}'\mathbf{x} = (1-2\theta_2)^{-1}\mathbf{y}'\mathbf{y}$ and $h(\mathbf{x}) = h(\mathbf{y})$ (because h is scale-free). Then

$$\begin{aligned}\varphi_{H,Q}(t_1,t_2) &\propto \int_{-\infty}^{\infty}\cdots\int_{-\infty}^{\infty}\exp[\theta_1 h(\mathbf{y})]\exp\left[-\frac{1}{2}\mathbf{y}'\mathbf{y}\right](1-2\theta_2)^{-n/2}\,dy_1\dots dy_n\\ &= (1-2\theta_2)^{-n/2}\mathbb{E}[\exp(\theta_1 h(\mathbf{y}))],\\ &\propto \mathbb{E}[\exp(\theta_1 H)],\end{aligned}$$

which does not involve Q, showing that H and Q are independent. ∎

The consequence of Theorem B.1 is that $U = RV$ implies $\mathbb{E}[U] = \mathbb{E}[R]\mathbb{E}[V]$, and, more generally, $U^p = (RV)^p$ implies $\mathbb{E}[U^p] = \mathbb{E}[R^p]\mathbb{E}[V^p]$ for all p such that the expectations exist, i.e.,

$$\mathbb{E}\,[R^p] = \frac{\mathbb{E}\,[U^p]}{\mathbb{E}\,[V^p]}. \tag{B.3}$$

It is this latter fact that is critical for the easy evaluation of the moments of R. Calculating the raw moments of R has been reduced to deriving the raw moments of both U and V.

From (A.6) with $\boldsymbol{\mu} = \mathbf{0}$ and $\boldsymbol{\Sigma} = \mathbf{I}$, the numerator expected value is $\mathbb{E}[U] = \mathbb{E}[\mathbf{Y}'\boldsymbol{\Lambda}\mathbf{Y}] = \text{tr}(\mathbf{A})$, so that, with $t_p = \sum_i \lambda_i^p = \text{tr}(\mathbf{A}^p)$,

$$\mathbb{E}[R] = \frac{\mathbb{E}[U]}{\mathbb{E}[V]} = \frac{\text{tr}(\mathbf{A})}{n} = \frac{t_1}{n}.$$

Likewise, for the variance, (A.8) implies $\mathbb{V}(U) = 2\,\text{tr}(\mathbf{A})^2$, so that

$$\mathbb{E}[U^2] = \mathbb{V}(U) + (\mathbb{E}[U])^2 = 2\,\text{tr}(\mathbf{A})^2 + [\text{tr}(\mathbf{A})]^2,$$

and

$$\begin{aligned}\mathbb{V}(R) = \mathbb{E}[R^2] - (\mathbb{E}[R])^2 &= \frac{\mathbb{E}[U^2]}{\mathbb{E}[V^2]} - \frac{t_1^2}{n^2} = \frac{2\,\text{tr}(\mathbf{A})^2 + [\text{tr}(\mathbf{A})]^2}{\mathbb{V}(\chi_n^2) + (\mathbb{E}[(\chi_n^2)])^2} - \frac{t_1^2}{n^2}\\ &= \frac{2t_2 + t_1^2}{2n + n^2} - \frac{t_1^2}{n^2} = 2\frac{nt_2 - t_1^2}{n^2(n+2)}.\end{aligned}$$

More generally, as V is a central χ^2 with n degrees of freedom, it is straightforward to show (see Example I.7.5) that its pth raw moment, $p = 1, 2, \dots$, is given by

$$\mathbb{E}[V] = n, \quad \mathbb{E}[V^p] = n(n+2)(n+4)\cdots(n+2(p-1)). \tag{B.4}$$

The positive integer moments of U are given in (A.12), and recalling from (I.4.47) that raw and central moments are related by

$$\mu_p = \mathbb{E}[(R-\mu)^p] = \sum_{i=0}^{p}(-1)^i\binom{p}{i}\mu'_{p-i}\mu^i,$$

some basic algebra gives μ_3 and μ_4, so that skewness and kurtosis can be computed. Summarizing,

$$\mu = \frac{t_1}{n}, \quad \mu_2 = 2\frac{nt_2 - t_1^2}{n^2(n+2)}, \quad \mu_3 = 8\frac{n^2t_3 - 3nt_1t_2 + 2t_1^3}{n^3(n+2)(n+4)}, \tag{B.5}$$

and

$$\mu_4 = 12\frac{n^3(4t_4 + t_2^2) - 2n^2(8t_1t_3 + t_2t_1^2) + n(24t_1^2t_2 + t_1^4) - 12t_1^4}{n^4(n+2)(n+4)(n+6)},$$

where $t_p = \sum_i \lambda_i^p = \mathrm{tr}(\mathbf{A}^p)$. With $\eta_p := \sum_{i=1}^n (\lambda_i - \bar{\lambda})^p$, these can also be expressed as

$$\mu = \mathbb{E}[R] = \bar{\lambda}, \qquad \mu_2 = \frac{2\eta_2}{n(n+2)},$$
$$\mu_3 = \frac{8\eta_3}{n(n+2)(n+4)}, \qquad \mu_4 = \frac{48\eta_4 + 12\eta_2^2}{n(n+2)(n+4)(n+6)}. \tag{B.6}$$

The next example involves an application based on the famous and still-popular Durbin and Watson (1950, 1971) autocorrelation test statistic. It will be used in subsequent examples, and is discussed further in Section 5.3.4.

Example B.1 ***Durbin–Watson, no regressors***
The Durbin–Watson test can be used to test if a time series of observations exhibits first-order serial autocorrelation. This is the topic of Chapter 4, though for now we just require the model, which is $Y_t = aY_{t-1} + U_t, t = 1, \ldots, T$, where $U_t \overset{\text{i.i.d.}}{\sim} \mathrm{N}(0, \sigma^2)$, and the observations are consecutive, having been observed at equally spaced time points. For $\mathbf{Y} = (Y_1, \ldots, Y_T)'$, the test statistic is given by

$$D = \frac{\sum_{t=2}^T (Y_t - Y_{t-1})^2}{\sum_{t=1}^T Y_t^2} = \frac{\mathbf{Y}'\mathbf{A}\mathbf{Y}}{\mathbf{Y}'\mathbf{Y}}, \tag{B.7}$$

where $\mathbf{A}$ is the tri-diagonal Toeplitz (diagonal-constant) matrix given by

$$\mathbf{A} = \mathbf{D}'\mathbf{D} = \begin{bmatrix} 1 & -1 & & & & \\ -1 & 2 & -1 & & & \\ & -1 & 2 & -1 & & \\ & & \ddots & \ddots & \ddots & \\ & & & -1 & 2 & -1 \\ & & & & -1 & 1 \end{bmatrix}, \tag{B.8}$$

and $\mathbf{D}$ is the $(T-1) \times T$ Toeplitz matrix with first rows and columns given by $[-1, 1, 0, \ldots, 0]$ and $[-1, 0, \ldots, 0]'$, respectively. As we will need to compute this matrix, we give the code for it in Listing B.1.

The null hypothesis is that $a = 0$, so that $\mathbf{Y} \sim \mathrm{N}_T(\mathbf{0}, \sigma^2\mathbf{I})$. Under this assumption, the moments of D can be computed using (B.6), though simplified expressions are given below. Conveniently, von

```
function A=makeDW(T)
A=2*eye(T); A(1,1)=1; A(T,T)=1;
for i=1:(T-1), A(i,i+1)=-1; A(i+1,i)=-1; end
```

Program Listing B.1: Computes matrix **A** in (B.8).

Neumann (1941) showed that the eigenvalues of **A** are given by

$$\lambda_h = 2 - 2\cos\left(\frac{\pi(h-1)}{T}\right) = 4\sin^2\left(\frac{\pi(h-1)}{2T}\right), \quad h = 1, \dots, T, \tag{B.9}$$

which are clearly positive for $h \geqslant 2$ and zero for $h = 1$, so that matrix **A** is positive semi-definite with rank $T - 1$. The corresponding eigenvectors are given by

$$\begin{aligned} \mathbf{v}_1 &= T^{-1/2}(1, 1, \dots, 1)', \\ \mathbf{v}_i &= \sqrt{\frac{2}{T}}\,(\cos k_i, \cos 3k_i, \dots \cos k_i(2T-1))', \quad i = 2, \dots, T, \end{aligned} \tag{B.10}$$

where $k_i := \pi(i-1)/(2T)$. From (A.34) and (B.9), the support of D is

$$\mathcal{S}_D = [0, \lambda_{\max}], \quad \text{with} \quad \lambda_{\max} = 4\sin^2\left(\frac{\pi(T-1)}{2T}\right) < 4. \tag{B.11}$$

The p.d.f. of D can be calculated by the methods discussed in Section A.3.2. As an illustration, for two (unrealistically small) values of T, the p.d.f. under the null hypothesis of $a = 0$ is shown in Figure B.1, along with histograms based on simulated values. For $T = 4$, the p.d.f. has quite a non-Gaussian shape that even the second-order s.p.a. is not able to capture. Matters change already for $T = 6$.

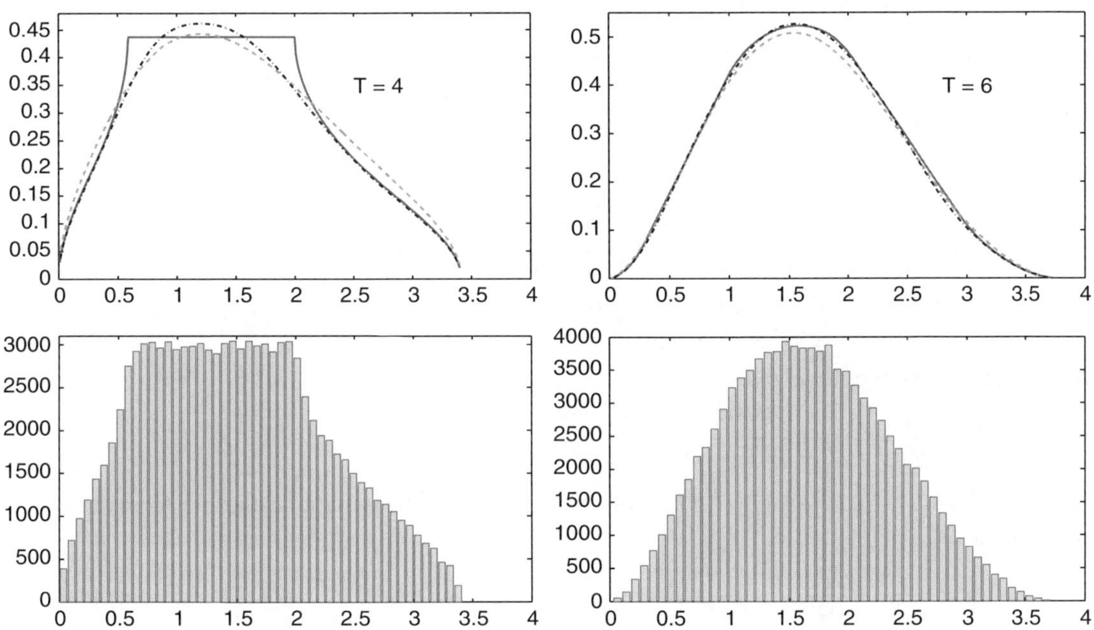

Figure B.1 **Top:** The exact density of D in (B.7) for $\mathbf{Y} \sim \mathrm{N}_T(\mathbf{0}, \sigma^2\mathbf{I})$ (solid) and via the first-order (dashed) and second-order (dash-dot) s.p.a. **Bottom:** Histograms of 100,000 simulated values.

From (B.5) and $\mathbf{A}$ in (B.8),

$$\mathbb{E}[D] = \frac{\text{tr}(\mathbf{A})}{T} = \frac{2(T-2)+2}{T} = \frac{2(T-1)}{T}.$$

For the variance, it is easy to verify that $\text{diag}(\mathbf{A}^2) = (2, 6, 6, \ldots, 6, 2)$, so that $t_2 = \text{tr}(\mathbf{A}^2) = 6(T-2) + 4 = 6T - 8$ and, from (B.5),

$$\mu_2 = \mathbb{V}(D) = 2\frac{nt_2 - t_1^2}{n^2(n+2)} = 2\frac{T(6T-8) - (2(T-1))^2}{T^2(T+2)} = 4\frac{T^2-2}{T^2(T+2)}, \qquad \text{(B.12)}$$

which approaches $4/T$ in the limit as $T \to \infty$.

Some computation reveals that $\text{diag}(\mathbf{A}^3) = (5, 19, 20, 20, \ldots, 20, 19, 5)$, so that $t_3 = 10 + 38 + 20(T-4) = 20T - 32$ and

$$\mu_3 = 8\frac{T^2 t_3 - 3Tt_1t_2 + 2t_1^3}{T^3(T+2)(T+4)} = 32\frac{T-2}{T^3(T+4)}.$$

Thus,

$$\text{skew}(D) = \frac{\mu_3}{\mu_2^{3/2}} = 4\frac{T-2}{T+4}\left(\frac{T+2}{T^2-2}\right)^{3/2} = \frac{4}{T^{3/2}}\left(1 + O\left(\frac{1}{T}\right)\right) \xrightarrow{T\to\infty} 0.$$

Further computation shows $\text{diag}(\mathbf{A}^4) = (14, 62, 70, 70, \ldots, 70, 62, 14)$, so that

$$t_4 = 28 + 124 + 70(T-4) = 70T - 128.$$

Using a symbolic computing package such as Maple, define

$$t_1 = 2T - 2, \qquad t_2 = 6T - 8,$$
$$t_3 = 20T - 32, \qquad t_4 = 70T - 128,$$

and simplify the expression for μ_4 to get

$$\begin{aligned}\mu_4 &= 12\frac{T^3(4t_4 + t_2^2) - 2T^2(8t_1t_3 + t_2t_1^2) + T(24t_1^2t_2 + t_1^4) - 12t_1^4}{T^4(T+2)(T+4)(T+6)} \\ &= 48\frac{T^5 + 6T^4 - 4T^3 - 16T^2 + 4T - 48}{T^4(T+2)(T+4)(T+6)}.\end{aligned}$$

Computing (with Maple) then gives

$$\text{kurt}(D) = \frac{\mu_4}{\mu_2^2} = 3\frac{Q(T+2)}{(T+4)(T+6)(T^2-2)^2},$$

where $Q = T^5 + 6T^4 - 4T^3 - 16T^2 + 4T - 48$. Deleting all but the two highest-order terms in Q and simplifying gives

$$\text{kurt}(D) \approx 3\frac{T^6 + 6T^5}{T^6 + 10T^5},$$

which converges to 3 as $T \to \infty$.

This suggests that D is asymptotically normally distributed, i.e., using our informal asymptotic notation, for large T, $D \overset{\text{app}}{\sim} \text{N}(2, 4/T)$. This can be rigorously proven; see, e.g., Srivastava (1987) and the references therein. ■

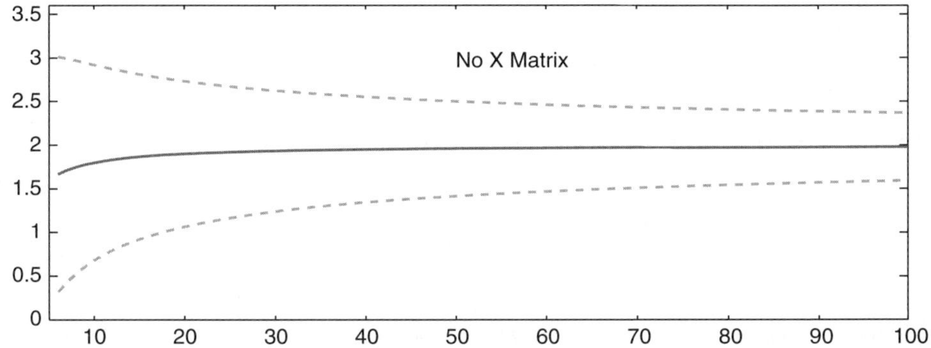

Figure B.2 Exact mean (solid), and the mean plus and minus 1.96 times the exact standard deviation (dashed) of the Durbin–Watson test statistic (B.7) under the null hypothesis of no autocorrelation, versus sample size, starting at $T = 6$.

Figure B.2 plots $\mathbb{E}[D]$, along with $\mathbb{E}[D]$ plus and minus 1.96 times the square root of the variance, as a function of sample size T. This could form the basis of a trivially computed, approximate test of the null hypothesis of zero correlation with significance level 0.05 that should be very accurate for "reasonable" sample sizes. This luxury is lost in the next example, which illustrates the more popular, but more complicated, application of the Durbin–Watson test. The ability to compute the c.d.f. via the inversion formula or saddlepoint methods will then be of great use.

Example B.2 ***Durbin–Watson, with regressors***
The Durbin–Watson test was actually designed to test for first order autocorrelation in the ordinary least squares (o.l.s.) regression residuals. (The reader not familiar with regression might wish to have a look at the beginnings of Chapters 1, 4, and 5. We put the example here, as the main emphasis is on working with quadratic forms.) The model is

$$Y_t = \mathbf{x}_t'\boldsymbol{\beta} + \epsilon_t, \tag{B.13}$$

where $\mathbf{x}_t'$, $t = 0, 1, \dots, T$, is a set of $1 \times k$ known constants such that $\mathbf{X} = [\mathbf{x}_1, \dots, \mathbf{x}_T]'$ is a full rank $T \times k$ matrix, $\mathbf{Y} = (Y_1, \dots, Y_T)'$ is the observed random variable, and

$$\epsilon_t = a\epsilon_{t-1} + U_t, \qquad U_t \overset{\text{i.i.d.}}{\sim} \mathrm{N}(0, \sigma^2). \tag{B.14}$$

Vector $\boldsymbol{\beta}$, along with a and σ, are the unknown parameters of the model. Under the null hypothesis of no autocorrelation, a in (B.14) is zero and $\epsilon_t \overset{\text{i.i.d.}}{\sim} \mathrm{N}(0, \sigma^2)$.

The o.l.s. estimator is $\widehat{\boldsymbol{\beta}} = (\mathbf{X}'\mathbf{X})^{-1}\mathbf{X}'\mathbf{Y}$ and the residual vector $\widehat{\epsilon} := \mathbf{Y} - \mathbf{X}\widehat{\boldsymbol{\beta}}$ can be expressed as $\mathbf{MY}$, where

$$\mathbf{M} = \mathbf{I} - \mathbf{X}(\mathbf{X}'\mathbf{X})^{-1}\mathbf{X}'. \tag{B.15}$$

As $\mathbf{MX} = \mathbf{0}$, the residuals are $\widehat{\epsilon} = \mathbf{MY} = \mathbf{M}\epsilon$. We give the trivial code for computing $\mathbf{M}$ in Listing B.2 because will make use of it often.

```
function M=makeM(X); [T,col]=size(X); M=eye(T)-X*inv(X'*X)*X';
```

Program Listing B.2: Computes matrix **M**.

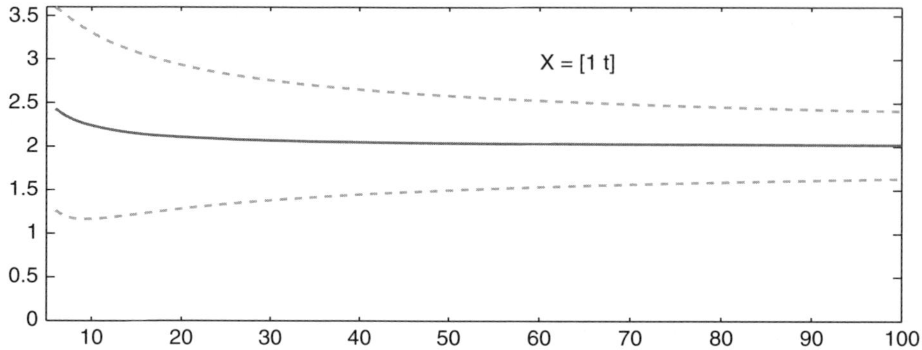

Figure B.3 Exact mean (solid), and the mean ±1.96 times the square root of the exact variance (dashed) of the Durbin–Watson test statistic (B.18) under the null hypothesis of no autocorrelation, versus sample size, starting at $T = 6$. The **X** matrix consists of a column of ones and a time-trend vector, $1, 2, \dots, T$.

The test statistic is

$$D = \frac{\sum_{t=2}^{T} (\hat{\epsilon}_t - \hat{\epsilon}_{t-1})^2}{\sum_{t=1}^{T} \hat{\epsilon}_t^2} = \frac{\hat{\epsilon}' \mathbf{A} \hat{\epsilon}}{\hat{\epsilon}' \hat{\epsilon}} = \frac{\epsilon' \mathbf{M}' \mathbf{A} \mathbf{M} \epsilon}{\epsilon' \mathbf{M}' \mathbf{M} \epsilon} = \frac{\epsilon' \mathbf{M} \mathbf{A} \mathbf{M} \epsilon}{\epsilon' \mathbf{M} \epsilon}. \tag{B.16}$$

Observe that (B.16) is not of the form (B.1) because the denominator matrix $\mathbf{M} \neq \mathbf{I}$. However, under the null hypothesis, D can be expressed as $\mathbf{Z}'\mathbf{\Lambda}\mathbf{Z}/\mathbf{Z}'\mathbf{Z}$, where $\mathbf{Z} \sim \mathrm{N}_{T-k}(\mathbf{0}, \sigma^2\mathbf{I})$. Crucially, observe that the dimension is $T - k$, and not T. This is referred to as **canonical reduction**, and can be shown in (at least) three ways, as detailed directly below.

Thus, D in (B.16) under the null hypothesis can be expressed as in (B.1), with $T - k$ components, and as such, the moments can be computed as usual. To illustrate, Figure B.3 shows the exact mean of D as a function of the sample size, along with lines indicating the range of the 95% confidence interval, based on model

$$Y_t = \beta_1 + \beta_2 t + \epsilon_t, \tag{B.17}$$

so that the **X** matrix consists of a constant and linear time trend, which we denote by $\mathbf{X} = [\mathbf{1},\ \mathbf{t}]$. This can be compared to Figure B.2, showing the case without a regressor matrix. ■

We now show three ways to conduct the canonical reduction of (B.16) (and related statistics).

1) The first method is the easiest, when using Theorem 1.3 from Chapter 1, which states that **M** can be expressed as $\mathbf{M} = \mathbf{G}'\mathbf{G}$, where **G** is $(T - k) \times T$ and such that $\mathbf{G}\mathbf{G}' = \mathbf{I}_{T-k}$ and $\mathbf{G}\mathbf{X} = \mathbf{0}$. Then,

$$D = \frac{\epsilon' \mathbf{M} \mathbf{A} \mathbf{M} \epsilon}{\epsilon' \mathbf{M} \epsilon} = \frac{\epsilon' \mathbf{G}' \mathbf{G} \mathbf{A} \mathbf{G}' \mathbf{G} \epsilon}{\epsilon' \mathbf{G}' \mathbf{G} \epsilon} = \frac{\mathbf{Z}' \tilde{\mathbf{A}} \mathbf{Z}}{\mathbf{Z}' \mathbf{Z}} = \frac{\sum_{i=1}^{T-k} \lambda_i \chi_i^2}{\sum_{i=1}^{T-k} \chi_i^2}, \tag{B.18}$$

where $\tilde{\mathbf{A}} = \mathbf{G}\mathbf{A}\mathbf{G}'$ is $(T - k) \times (T - k)$, and $\mathbf{Z} = \mathbf{G}\epsilon \sim \mathrm{N}_{T-k}(\mathbf{0}, \sigma^2\mathbf{I})$. Note that **G** does not need to be computed, as the nonzero eigenvalues of $\tilde{\mathbf{A}} = \mathbf{G}\mathbf{A}\mathbf{G}'$ are the same as the nonzero eigenvalues of $\mathbf{G}'\mathbf{G}\mathbf{A} = \mathbf{M}\mathbf{A}$. Nevertheless, code for computing **G** is given in Listing 1.2 in Chapter 1, as it will be used elsewhere.

```
T=10; k=2; X=randn(10,2); % Just pick a random X matrix
C=ones(T,1); M1=makeM(C); % M1 is a "centering matrix".
Z=M1*X; sum(Z) % Z are the centered columns of X. Check that they sum to zero
M=makeM(Z); A=makeDW(T); rank(M*A) % rank is 7, not 10-2=8.
```

Program Listing B.3: Inspection of rank(**MA**).

2) If full rank **X** contains a constant term, or if all the columns of **X** do not have zero mean, then one can verify that rank(**MA**) $= T - k$. The code in Listing B.3 verifies that, when **X** is centered and still full rank, rank(**MA**) $= T - k - 1$.

When rank(**MA**) $= T - k$, the distribution of D follows directly from Theorems B.2 and B.3, detailed in Section B.5 below. That is, as rank(**M**) = rank(**MAM**) $= T - k$ and $\mathbf{M} \cdot \mathbf{MAM} = \mathbf{MAM} \cdot \mathbf{M}$ we can take **P** to be an orthogonal matrix that simultaneously diagonalizes **MAM** and **M**, say $\mathbf{P}'\mathbf{MAMP} = \mathbf{D}_1$ and $\mathbf{P}'\mathbf{MP} = \mathbf{D}_2$ with $\mathbf{D}_1 = \mathrm{diag}([\lambda_1, \dots, \lambda_{T-k}, 0, \dots, 0])$ and $\mathbf{D}_2 = \mathrm{diag}([1, \dots, 1, 0, \dots, 0])$, where $\mathbf{D}_2$ contains $T - k$ ones. Thus

$$D = \frac{\epsilon'\mathbf{MAM}\epsilon}{\epsilon'\mathbf{M}\epsilon} = \frac{\epsilon'\mathbf{PD}_1\mathbf{P}'\epsilon}{\epsilon'\mathbf{PD}_2\mathbf{P}'\epsilon} = \frac{\mathbf{Z}'\mathbf{D}_1\mathbf{Z}}{\mathbf{Z}'\mathbf{D}_2\mathbf{Z}} = \frac{\sum_{i=1}^{T-k} \lambda_i \chi_i^2}{\sum_{i=1}^{T-k} \chi_i^2},$$

where $\mathbf{Z} = \mathbf{P}'\epsilon \sim \mathrm{N}(\mathbf{0}, \sigma^2\mathbf{I}_T)$.

3) A more tedious (but still quite instructional) method for canonical reduction was used by Durbin and Watson in their seminal 1950 paper. With $\epsilon \sim \mathrm{N}_T(\mathbf{0}, \boldsymbol{\sigma}^2\mathbf{I})$ and assuming that the regression matrix **X** is full rank k, let **L** be the orthogonal matrix such that

$$\mathbf{L}'\mathbf{ML} = \begin{bmatrix} \mathbf{I}_{T-k} & \mathbf{0} \\ \mathbf{0} & \mathbf{0} \end{bmatrix},$$

and define $\xi = \mathbf{L}'\epsilon$. As $\mathbf{L}'\mathbf{L} = \mathbf{LL}' = \mathbf{I}$, we see that $\xi \sim \mathrm{N}(\mathbf{0}, \boldsymbol{\sigma}^2\mathbf{I})$ and that

$$D = \frac{\epsilon'\mathbf{MAM}\epsilon}{\epsilon'\mathbf{M}\epsilon} = \frac{\epsilon'\mathbf{LL}'\mathbf{MLL}'\mathbf{ALL}'\mathbf{MLL}'\epsilon}{\epsilon'\mathbf{LL}'\mathbf{MLL}'\epsilon} = \frac{\xi' \begin{bmatrix} \mathbf{I}_{T-k} & \mathbf{0} \\ \mathbf{0} & \mathbf{0} \end{bmatrix} \mathbf{L}'\mathbf{AL} \begin{bmatrix} \mathbf{I}_{T-k} & \mathbf{0} \\ \mathbf{0} & \mathbf{0} \end{bmatrix} \xi}{\xi' \begin{bmatrix} \mathbf{I}_{T-k} & \mathbf{0} \\ \mathbf{0} & \mathbf{0} \end{bmatrix} \xi}. \tag{B.19}$$

If we define **H** to be the upper left $(T - k) \times (T - k)$ matrix of $\mathbf{L}'\mathbf{AL}$, and define **K** to be the orthogonal matrix such that $\mathbf{K}'\mathbf{HK} = \boldsymbol{\Lambda} = \mathrm{diag}([\lambda_1, \dots, \lambda_{T-k}])$ (noting that $\mathbf{K}'\mathbf{K} = \mathbf{KK}' = \mathbf{I}$) and define

$$\zeta = \begin{bmatrix} \mathbf{K} & \mathbf{0} \\ \mathbf{0} & \mathbf{I}_k \end{bmatrix}' \xi,$$

so that $\zeta \sim \mathrm{N}(\mathbf{0}, \sigma^2\mathbf{I})$, we get, continuing (B.19),

$$D = \frac{\xi' \begin{bmatrix} \mathbf{H} & \mathbf{0} \\ \mathbf{0} & \mathbf{0} \end{bmatrix} \xi}{\xi' \begin{bmatrix} \mathbf{I}_{T-k} & \mathbf{0} \\ \mathbf{0} & \mathbf{0} \end{bmatrix} \xi} = \frac{\xi' \begin{bmatrix} \mathbf{K} & \mathbf{0} \\ \mathbf{0} & \mathbf{I}_k \end{bmatrix} \begin{bmatrix} \boldsymbol{\Lambda} & \mathbf{0} \\ \mathbf{0} & \mathbf{0} \end{bmatrix} \begin{bmatrix} \mathbf{K} & \mathbf{0} \\ \mathbf{0} & \mathbf{I}_k \end{bmatrix}' \xi}{\xi' \begin{bmatrix} \mathbf{K} & \mathbf{0} \\ \mathbf{0} & \mathbf{I}_k \end{bmatrix} \begin{bmatrix} \mathbf{I}_{T-k} & \mathbf{0} \\ \mathbf{0} & \mathbf{0} \end{bmatrix} \begin{bmatrix} \mathbf{K} & \mathbf{0} \\ \mathbf{0} & \mathbf{I}_k \end{bmatrix}' \xi} = \frac{\zeta' \begin{bmatrix} \boldsymbol{\Lambda} & \mathbf{0} \\ \mathbf{0} & \mathbf{0} \end{bmatrix} \zeta}{\zeta' \begin{bmatrix} \mathbf{I}_{T-k} & \mathbf{0} \\ \mathbf{0} & \mathbf{0} \end{bmatrix} \zeta}.$$

We see the last equation in the above line is just a ratio of weighted χ^2 random variables, namely that given at the end of (B.18), with the λ_i the nonzero eigenvalues of

$$\begin{bmatrix} \mathbf{I}_{T-k} & \mathbf{0} \\ \mathbf{0} & \mathbf{0} \end{bmatrix} \mathbf{L}'\mathbf{A}\mathbf{L} \begin{bmatrix} \mathbf{I}_{T-k} & \mathbf{0} \\ \mathbf{0} & \mathbf{0} \end{bmatrix} = \mathbf{L}'\mathbf{M}\mathbf{L}\mathbf{L}'\mathbf{A}\mathbf{L}\mathbf{L}'\mathbf{M}\mathbf{L} = \mathbf{L}'\mathbf{M}\mathbf{A}\mathbf{M}\mathbf{L},$$

which are also the eigenvalues of $\mathbf{L}\mathbf{L}'\mathbf{M}\mathbf{A}\mathbf{M} = \mathbf{M}\mathbf{A}\mathbf{M}$ or those of $\mathbf{M}\mathbf{M}\mathbf{A} = \mathbf{M}\mathbf{A}$.

Example B.3 ***Durbin–Watson, constant regressor***
An important special case of the regression model used in the previous example is $Y_t = \beta + \epsilon_t$, so that all the Y_t have the same expected value. This corresponds to an $\mathbf{X}$ matrix consisting of just a column of ones, in which case $\mathbf{M} = \mathbf{I}_n - n^{-1}\mathbf{J}_T$ and the least squares estimator reduces to the sample mean, $T^{-1}\sum_{t=1}^{T} Y_t$.

From the structure of $\mathbf{A}$ in (B.8), it is clear that $\mathbf{J}\mathbf{A} = \mathbf{0}$, so that $\mathbf{M}\mathbf{A} = \mathbf{A}$. Similarly, it is easy to see that $\mathbf{M}\mathbf{A}\mathbf{M} = \mathbf{A}$; this also follows because $\mathbf{M}$ and $\mathbf{A}$ are symmetric, so that taking transposes of $\mathbf{M}\mathbf{A} = \mathbf{A}$ gives $\mathbf{A}\mathbf{M} = \mathbf{A}$, so that $(\mathbf{M}\mathbf{A})\mathbf{M} = \mathbf{A}\mathbf{M} = \mathbf{A}$. A bit of intuition can be added to this: Without mean-adjusting (centering) the data, D is given by

$$D_0 = \frac{\sum_{t=2}^{T}(Y_t - Y_{t-1})^2}{\sum_{t=1}^{T} Y_t^2} = \frac{\mathbf{Y}'\mathbf{A}\mathbf{Y}}{\mathbf{Y}'\mathbf{Y}},$$

while centering results in

$$D_1 = \frac{\sum_{t=2}^{T}((Y_t - \bar{Y}) - (Y_{t-1} - \bar{Y}))^2}{\sum_{t=1}^{T}(Y_t - \bar{Y})^2} = \frac{\sum_{t=2}^{T}(Y_t - Y_{t-1})^2}{\sum_{t=1}^{T}(Y_t - \bar{Y})^2} = \frac{\mathbf{Y}'\mathbf{A}\mathbf{Y}}{\mathbf{Y}'\mathbf{M}\mathbf{Y}},$$

which has the same numerator as D_0. This would not work if the regression terms $\bar{Y}$ were replaced with their more general counterparts $\mathbf{x}_t'\hat{\boldsymbol{\beta}}$ and $\mathbf{x}_{t-1}'\hat{\boldsymbol{\beta}}$.

From the discussion just after (B.18), and the fact that $\mathbf{A}$ always has one zero eigenvalue, the $T-1$ values of λ_i are the nonzero eigenvalues of the $T \times T$ matrix $\mathbf{A}$. Thus, for example, the support will not extend to zero, as was the case without regressors in Example B.1 above. From (B.9), observe that, for any given $T \geqslant 2$, $\cos(\pi/T) = -\cos(\pi(T-1)/T)$ (draw the unit circle to see this), so that $\lambda_2 + \lambda_T = 4$. Similarly, $\lambda_3 + \lambda_{T-1} = 4$, etc., so that the set $\{\lambda_i : i = 1, \ldots, T\} = \{4 - \lambda_i : i = 1, \ldots, T\}$. Thus, for $\lambda_{\min} < d < 2$,

$$\begin{aligned} \Pr(D \leqslant d) &= \Pr\left(\sum_{i=1}^{T-1} \lambda_i \chi_i^2 \leqslant d \sum_{i=1}^{T-1} \chi_i^2\right) = \Pr\left(\sum_{i=1}^{T-1}(4-\lambda_i)\chi_i^2 \leqslant d \sum_{i=1}^{T-1} \chi_i^2\right) \\ &= \Pr\left(\sum_{i=1}^{T-1} \lambda_i \chi_i^2 \geqslant (4-d) \sum_{i=1}^{T-1} \chi_i^2\right) = \Pr(D \geqslant 4-d), \end{aligned}$$

and the p.d.f. is symmetric about two. Figure B.4 illustrates this. While indeed symmetric, the p.d.f.s for very small T are quite far from that of the normal distribution. In these cases, the second-order s.p.a. captures the tails reasonably well, but not the sharp peak or flat top of the true p.d.f. for $T < 7$.

For the moments of D in this case, we require the $T-1$ eigenvalues of $\tilde{\mathbf{A}}_{T-1}$, where the subscript denotes the size of the matrix and $\tilde{\mathbf{A}} = \mathbf{G}\mathbf{A}\mathbf{G}'$ as in (B.18). But, as $\mathbf{M}\mathbf{A} = \mathbf{A}$, these are just the $T-1$

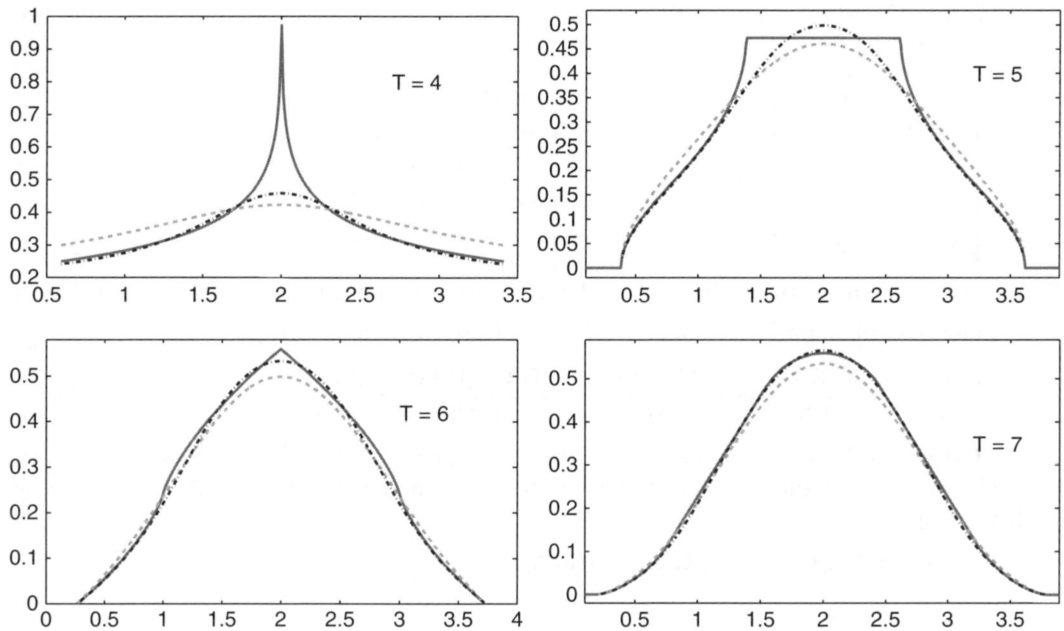

Figure B.4 The exact density of D in (B.16) for the model $Y_t = \beta + \epsilon_t$ (solid) and the first-order (dashed) and second-order (dash-dot) s.p.a.

positive eigenvalues of $\mathbf{A}_T$, say $\lambda_1, \ldots, \lambda_{T-1}$ (take $\lambda_T = 0$). Using (B.5), and noting that $t_p = \text{tr}(\mathbf{A}_T^p) = \sum_{i=1}^{T} \lambda_i^p = \sum_{i=1}^{T-1} \lambda_i^p$,

$$\mathbb{E}[D] = \frac{\text{tr}(\mathbf{A}_T)}{T-1} = \frac{2(T-2)+2}{T-1} = 2,$$

as we knew, given the symmetry of D about 2. For the variance, first observe that $\text{diag}(\mathbf{A}^2) = (2, 6, 6, \ldots, 6, 2)$, implying $\text{tr}(\mathbf{A}_T^2) = 6(T-2)+4$. Then, from (B.5),

$$\mathbb{V}(D) = 2\frac{(T-1)\text{tr}(\mathbf{A}_T^2) - (\text{tr}(\mathbf{A}_T))^2}{(T-1)^2(T-1+2)} = 4\frac{(T-2)}{(T-1)(T+1)}, \tag{B.20}$$

after simplifying, which obviously is approximately $4/T$ for large T. Higher moments could be similarly computed. ∎

We now examine a different statistic that is of great importance in time-series analysis. As in Example B.2, consider the regression model $Y_t = \mathbf{x}_t'\boldsymbol{\beta} + \epsilon_t$, where $\mathbf{x}_t'$, $t = 0, 1, \ldots, T$, is a set of $1 \times k$ known constants such that $\mathbf{X} = [\mathbf{x}_1, \ldots, \mathbf{x}_T]'$ is a full rank $T \times k$ matrix, $\mathbf{Y} = (Y_1, \ldots, Y_T)'$, and the residuals are $\widehat{\boldsymbol{\epsilon}} = \mathbf{MY} = \mathbf{M}\boldsymbol{\epsilon}$. The sth **sample autocorrelation** is given by

$$R_s = \frac{\sum_{t=s+1}^{T} \hat{\epsilon}_t \hat{\epsilon}_{t-s}}{\sum_{t=1}^{T} \hat{\epsilon}_t^2} = \frac{\widehat{\boldsymbol{\epsilon}}'\mathbf{A}_s\widehat{\boldsymbol{\epsilon}}}{\widehat{\boldsymbol{\epsilon}}'\widehat{\boldsymbol{\epsilon}}}, \tag{B.21}$$

where $s \in \{1, 2, \ldots, T-1\}$ and the (i,j)th element of $\mathbf{A}_s$ is given by $\mathbb{I}\{|i-j| = s\}/2$, $i,j = 1, \ldots, T$. For example, with $T = 5$,

$$\mathbf{A}_1 = \begin{bmatrix} 0 & \frac{1}{2} & 0 & 0 & 0 \\ \frac{1}{2} & 0 & \frac{1}{2} & 0 & 0 \\ 0 & \frac{1}{2} & 0 & \frac{1}{2} & 0 \\ 0 & 0 & \frac{1}{2} & 0 & \frac{1}{2} \\ 0 & 0 & 0 & \frac{1}{2} & 0 \end{bmatrix} \quad \text{and} \quad \mathbf{A}_2 = \begin{bmatrix} 0 & 0 & \frac{1}{2} & 0 & 0 \\ 0 & 0 & 0 & \frac{1}{2} & 0 \\ \frac{1}{2} & 0 & 0 & 0 & \frac{1}{2} \\ 0 & \frac{1}{2} & 0 & 0 & 0 \\ 0 & 0 & \frac{1}{2} & 0 & 0 \end{bmatrix}.$$

The R_s are discussed in detail in Chapter 8. Here, we are only concerned about their low-order moments under the null hypothesis that $\epsilon_t \overset{\text{i.i.d.}}{\sim} \mathrm{N}(0, \sigma^2)$. As in (B.16) and (B.18),

$$R_s = \frac{\hat{\epsilon}'\mathbf{A}_s\hat{\epsilon}}{\hat{\epsilon}'\hat{\epsilon}} = \frac{\epsilon'\mathbf{M}'\mathbf{A}_s\mathbf{M}\epsilon}{\epsilon'\mathbf{M}'\mathbf{M}\epsilon} = \frac{\epsilon'\mathbf{M}'\mathbf{A}_s\mathbf{M}\epsilon}{\epsilon'\mathbf{M}\epsilon} = \frac{\epsilon'\mathbf{G}'\mathbf{G}\mathbf{A}_s\mathbf{G}'\mathbf{G}\epsilon}{\epsilon'\mathbf{G}'\mathbf{G}\epsilon} = \frac{\mathbf{Z}'\tilde{\mathbf{A}}_s\mathbf{Z}}{\mathbf{Z}'\mathbf{Z}}, \tag{B.22}$$

where $\tilde{\mathbf{A}}_s = \mathbf{G}\mathbf{A}_s\mathbf{G}'$ is $(T-k) \times (T-k)$ and $\mathbf{Z} = \mathbf{G}\epsilon \sim \mathrm{N}_{T-k}(\mathbf{0}, \sigma^2\mathbf{I})$. Thus, the first two moments are given by (B.5), where, using the fact that $\mathrm{tr}(\mathbf{AB}) = \mathrm{tr}(\mathbf{BA})$ for conformable matrices $\mathbf{A}$ and $\mathbf{B}$ such that $\mathbf{AB}$ and $\mathbf{BA}$ are square,

$$t_1 = \mathrm{tr}(\tilde{\mathbf{A}}_s) = \mathrm{tr}(\mathbf{G}\mathbf{A}_s\mathbf{G}') = \mathrm{tr}(\mathbf{G}'\mathbf{G}\mathbf{A}_s) = \mathrm{tr}(\mathbf{M}\mathbf{A}_s),$$

and

$$t_2 = \mathrm{tr}(\tilde{\mathbf{A}}_s^2) = \mathrm{tr}(\mathbf{G}\mathbf{A}_s\mathbf{G}'\mathbf{G}\mathbf{A}_s\mathbf{G}') = \mathrm{tr}(\mathbf{G}'\mathbf{G}\mathbf{A}_s\mathbf{G}'\mathbf{G}\mathbf{A}_s) = \mathrm{tr}(\mathbf{M}\mathbf{A}_s\mathbf{M}\mathbf{A}_s).$$

That is,

$$\mathbb{E}[R_s] = \frac{\mathrm{tr}(\mathbf{M}\mathbf{A}_s)}{T-k} \quad \text{and} \quad \mathbb{V}(R_s) = 2\frac{(T-k)\mathrm{tr}(\mathbf{M}\mathbf{A}_s)^2 - \mathrm{tr}^2(\mathbf{M}\mathbf{A}_s)}{(T-k)^2(T-k+2)}, \tag{B.23}$$

where $\mathrm{tr}(\mathbf{A})^2 = \mathrm{tr}(\mathbf{A}^2)$ and $\mathrm{tr}^2(\mathbf{A}) = [\mathrm{tr}(\mathbf{A})]^2$. Expressions for the third and fourth moments follow similarly.

Now consider the important special case when $k = 1$ and $\mathbf{X} = \mathbf{1}$. As in Paolella (2003),

$$\mathbf{M}\mathbf{A}_s = \left(\mathbf{I}_T - \frac{1}{T}\mathbf{1}\mathbf{1}'\right)\mathbf{A}_s = \mathbf{A}_s - \frac{1}{T}\mathbf{B},$$

where

$$\mathbf{B} = \begin{cases} \left[\frac{1}{2}\mathbf{1}_{T\times s} \mid \mathbf{1}_{T\times(T-2s)} \mid \frac{1}{2}\mathbf{1}_{T\times s}\right], & \text{if} \quad s \leqslant T/2, \\ \left[\frac{1}{2}\mathbf{1}_{T\times(T-s)} \mid \mathbf{0}_{T\times(2s-T)} \mid \frac{1}{2}\mathbf{1}_{T\times(T-s)}\right], & \text{if} \quad s > T/2, \end{cases}$$

and $\mathbf{0}_{r\times s}$ $(\mathbf{1}_{r\times s})$ denotes the $r \times s$ matrix of zeros (ones). Perhaps more clearly, for $s \leqslant T/2$, $\mathbf{B}$ can be expressed as

$$\mathbf{B} = \begin{bmatrix} \underbrace{\begin{matrix} \frac{1}{2} & & \frac{1}{2} \\ \vdots & \cdots & \vdots \\ \frac{1}{2} & & \frac{1}{2} \end{matrix}}_{s} & \underbrace{\begin{matrix} 1 & & 1 \\ \vdots & \cdots & \vdots \\ 1 & & 1 \end{matrix}}_{T-2s} & \underbrace{\begin{matrix} \frac{1}{2} & & \frac{1}{2} \\ \vdots & \cdots & \vdots \\ \frac{1}{2} & & \frac{1}{2} \end{matrix}}_{s} \end{bmatrix}.$$

Denote the (ij)th element of $\mathbf{M}$ by m_{ij} and the (ij)th element of $\mathbf{A}_s$ by a_{ij}. Then, from the structure of $\mathbf{M}$ and $\mathbf{A}_s$,

$$\begin{aligned}\operatorname{tr}(\mathbf{MA}_s) &= \sum_{i=1}^{T}\left(\sum_{j=1}^{T} m_{ij}a_{ji}\right) = \sum_{i=1}^{T} m_{ii}a_{ii} + \sum_{i\neq j}^{T}\sum^{T} m_{ij}a_{ji}\\ &= -\frac{1}{T}\sum_{i\neq j}^{T}\sum^{T} a_{ji} = -\frac{T-s}{T}.\end{aligned}$$

Thus, from (B.23), when $\mathbf{X} = \mathbf{1}$,

$$\mathbb{E}[R_s] = -\frac{T-s}{T(T-1)}, \qquad s = 1, 2, \ldots, T-1. \tag{B.24}$$

The sign of $\mathbb{E}[R_s]$ was to be expected because the residuals sum to zero, so that a small amount of negative correlation is induced.

For the variance, denote the (ij)th element of $\mathbf{B}$ as b_{ij}, and observe that

$$\mathbf{B}^2 = \begin{cases} [c\mathbf{1}_{T\times s} \mid d\mathbf{1}_{T\times(T-2s)} \mid c\mathbf{1}_{T\times s}], & \text{if} \quad s \leqslant T/2,\\ [c\mathbf{1}_{T\times(T-s)} \mid \mathbf{0}_{T\times(2s-T)} \mid c\mathbf{1}_{T\times(T-s)}], & \text{if} \quad s > T/2,\end{cases}$$

where $c = \frac{1}{2}\left(\frac{1}{2}\cdot s\cdot 2 + (T-2s)\right) = (T-s)/2$ and $d = T-s$. It follows from the symmetry of $\mathbf{A}_s$ that

$$\operatorname{tr}(\mathbf{A}_s^2) = \sum_{i=1}^{T}\sum_{j=1}^{T} a_{ij}^2 = \left(\frac{1}{2}\right)^2 \cdot 2\cdot(T-s) = \frac{T-s}{2},$$

$$\begin{aligned}\operatorname{tr}(\mathbf{BA}_s) &= \sum_{i=1}^{T}\sum_{j=1}^{T} a_{ji}b_{ji} = 2\cdot\frac{1}{2}\begin{cases}\frac{s}{2} + (T-2s), & \text{if} \quad s \leqslant T/2,\\ \frac{s}{2}(T-s), & \text{if} \quad s > T/2,\end{cases}\\ &= \begin{cases} T - \frac{3}{2}s, & \text{if} \quad s \leqslant T/2,\\ \frac{s}{2}(T-s), & \text{if} \quad s > T/2,\end{cases}\end{aligned}$$

and

$$\begin{aligned}\operatorname{tr}(\mathbf{B}^2) &= \begin{cases} 2sc + d(T-2s), & \text{if} \quad s \leqslant T/2,\\ 2(T-s)c, & \text{if} \quad s > T/2,\end{cases}\\ &= (T-s)^2,\end{aligned}$$

having used that fact that, for matrix $\mathbf{H}$ with (ij)th element h_{ij}, $\operatorname{tr}(\mathbf{H}'\mathbf{H}) = \sum_{i=1}^{T}\sum_{j=1}^{T} h_{ij}^2$ (see, e.g., Graybill, 1983, p. 300). Combining terms,

$$\begin{aligned}\operatorname{tr}(\mathbf{MA}_s)^2 &= \operatorname{tr}\left(\mathbf{A}_s^2 - \frac{1}{T}\mathbf{BA}_s - \frac{1}{T}\mathbf{A}_s\mathbf{B} + \frac{1}{T^2}\mathbf{B}^2\right)\\ &= \frac{1}{2T^2}\begin{cases} T^3 - sT^2 - 2T^2 + 2sT + 2s^2, & \text{if} \quad s \leqslant T/2,\\ T^3 - sT^2 - 2sT + 2s^2, & \text{if} \quad s > T/2,\end{cases}\end{aligned}$$

which, from (B.23), yields an expression for $\mathbb{V}(R_s)$ when $\mathbf{X} = \mathbf{1}$ as

$$\begin{cases} \dfrac{T^4 - (s+3)T^3 + 3sT^2 + 2s(s+1)T - 4s^2}{(T-1)^2(T+1)T^2}, & \text{if} \quad 0 < s \leqslant T/2,\\[2ex] \dfrac{T^4 - (s+1)T^3 - (s+2)T^2 + 2s(s+3)T - 4s^2}{(T-1)^2(T+1)T^2}, & \text{if} \quad T/2 < s < T.\end{cases} \tag{B.25}$$

This expression was also derived by Dufour and Roy (1985) and Anderson (1993) using different methods of proof.[2] For $1 \leqslant s \leqslant T/2$ (essentially the only relevant part in practice), the reader can verify that the approximation

$$\mathbb{V}(R_s) \approx \frac{T-2}{T^2} - \frac{T-3}{T^3}s ,$$

derived from (B.25), is extremely accurate (and is not poor on the other half), though offers no substantial computational benefit over the exact calculation. However, it prominently shows that the variance is practically (affine) linear in s, with a negative slope, implying that low values of s (precisely the ones required in practice) have the highest variance.

Remark There is another interesting consequence of Theorem B.1. Again with $R = U/V$ for $U = \mathbf{X}'\mathbf{A}\mathbf{X}$, $V = \mathbf{X}'\mathbf{X}$, $\mathbf{X} \sim \mathrm{N}_n(\mathbf{0}, \sigma^2\mathbf{I})$, and $\boldsymbol{\zeta} = \mathrm{Eig}(\mathbf{A})$, the independence of R and V implies, for r such that $\min \zeta_i < r < \max \zeta_i$,

$$F_R(r) = \Pr(R \leqslant r) = \Pr(R \leqslant r \mid V = 1) = \Pr(U \leqslant r \mid V = 1).$$

The joint m.g.f. of U and V is

$$\begin{aligned}\mathbb{M}_{U,V}(s,t) &= \mathbb{E}[\exp\{sU + tV\}] \\ &= \mathbb{E}\left[\exp\left\{s\sum_{i=1}^{n}\zeta_i\chi_i^2 + t\sum_{i=1}^{n}\chi_i^2\right\}\right] = \mathbb{E}\left[\exp\left\{\sum_{i=1}^{n}(s\zeta_i + t)\chi_i^2\right\}\right] \\ &= \prod_{i=1}^{n}[1 - 2(s\zeta_i + t)]^{-1/2}. \end{aligned} \tag{B.26}$$

Based on this, the conditional saddlepoint approximation discussed in Section II.5.2.1 is applicable, and it would be of interest to compare the accuracy of the c.d.f. approximation from its use with the one discussed in Section A.3.1. However, it turns out that they are identical, as proven in Butler and Paolella (1998) in a more general setting, and shown for this case in Appendix B.6. ∎

B.2 For $\mathbf{X} \sim \mathrm{N}(\mathbf{0}, \boldsymbol{\Sigma})$

Let $\mathbf{X} \sim \mathrm{N}(\mathbf{0}, \boldsymbol{\Sigma})$, with $\boldsymbol{\Sigma} > 0$.[3] As such, we can compute a matrix $\boldsymbol{\Sigma}^{-1/2} > 0$ such that $\boldsymbol{\Sigma}^{-1/2}\boldsymbol{\Sigma}^{-1/2} = \boldsymbol{\Sigma}^{-1}$. First write

$$R = \frac{\mathbf{X}'\mathbf{A}\mathbf{X}}{\mathbf{X}'\mathbf{B}\mathbf{X}} = \frac{\mathbf{X}'\boldsymbol{\Sigma}^{-1/2}\boldsymbol{\Sigma}^{1/2}\mathbf{A}\boldsymbol{\Sigma}^{1/2}\boldsymbol{\Sigma}^{-1/2}\mathbf{X}}{\mathbf{X}'\boldsymbol{\Sigma}^{-1/2}\boldsymbol{\Sigma}^{1/2}\mathbf{B}\boldsymbol{\Sigma}^{1/2}\boldsymbol{\Sigma}^{-1/2}\mathbf{X}} = \frac{\mathbf{Y}'\mathbf{A}^*\mathbf{Y}}{\mathbf{Y}'\mathbf{B}^*\mathbf{Y}}, \tag{B.27}$$

where $\mathbf{A}^* = \boldsymbol{\Sigma}^{1/2}\mathbf{A}\boldsymbol{\Sigma}^{1/2}$, $\mathbf{B}^* = \boldsymbol{\Sigma}^{1/2}\mathbf{B}\boldsymbol{\Sigma}^{1/2}$, and $\mathbf{Y} = \boldsymbol{\Sigma}^{-1/2}\mathbf{X} \sim \mathrm{N}(\mathbf{0}, \mathbf{I})$. Observe that, if $\mathbf{B} = \boldsymbol{\Sigma}^{-1}$, then the analysis in Section B.1 is still valid.

We now proceed as in Sawa (1978). From (A.23), the joint m.g.f. of $A = \mathbf{Y}'\mathbf{A}^*\mathbf{Y}$ and $B = \mathbf{Y}'\mathbf{B}^*\mathbf{Y}$ is given by $\mathbb{M}_{A,B}(t_1, t_2) = |\mathbf{I}_T - 2t_1\mathbf{A}^* - 2t_2\mathbf{B}^*|^{-1/2}$. Let the spectral decomposition of $\mathbf{B}^*$ be $\mathbf{P}'\boldsymbol{\Lambda}\mathbf{P}$, where

2 Dufour and Roy (1985) falsely state the top expression (B.25) for all s.

3 In most applications, it is useful to take $\mathbf{X} \sim \mathrm{N}(\mathbf{0}, \sigma^2\boldsymbol{\Sigma})$, where $\sigma > 0$ is a scale term. Observe that such a scaling factor cancels out in the ratio, so we can take it to be unity without loss of generality.

$\mathbf{\Lambda} = \text{diag}([\lambda_1, \dots, \lambda_n])$ are the eigenvalues of $\mathbf{B}^*$ and $\mathbf{P}'\mathbf{P} = \mathbf{I}_T$. Then

$$\begin{aligned}
\mathbb{M}_{A,B}(t_1, t_2) &= |\mathbf{P}'\mathbf{P}|^{-1/2} |\mathbf{I}_T - 2t_1\mathbf{A}^* - 2t_2\mathbf{B}^*|^{-1/2} \\
&= |\mathbf{P}'|^{-1/2} |\mathbf{I}_T - 2t_1\mathbf{A}^* - 2t_2\mathbf{B}^*|^{-1/2} |\mathbf{P}|^{-1/2} \\
&= |\mathbf{P}'\mathbf{P} - 2t_1\mathbf{P}'\mathbf{A}^*\mathbf{P} - 2t_2\mathbf{P}'\mathbf{B}^*\mathbf{P}|^{-1/2} \\
&= |\mathbf{I}_T - 2t_1\mathbf{C} - 2t_2\mathbf{\Lambda}|^{-1/2} = |\mathbf{R}(t_1, t_2)|^{-1/2},
\end{aligned}$$

where $\mathbf{C} = \mathbf{P}'\mathbf{A}^*\mathbf{P}$, with (i,j)th element c_{ij}, and $\mathbf{R}(t_1, t_2) = \mathbf{I}_T - 2t_1\mathbf{C} - 2t_2\mathbf{\Lambda}$. For convenience, we subsequently write $\mathbf{R} = \mathbf{R}(t_1, t_2)$, though the dependence on t_1 and t_2 must be kept in mind.

The pth moment, $p \in \mathbb{N}$, is now obtainable using the Sawa (1972) result derived in (II.1.24),

$$\mathbb{E}[R^p] = \mathbb{E}\left[\left(\frac{A}{B}\right)^p\right] = \frac{1}{\Gamma(p)} \int_0^\infty (t_2)^{p-1} \left[\frac{\partial^p}{\partial t_1^p} \mathbb{M}_{A,B}(t_1, -t_2)\right]_{t_1=0} \mathrm{d}t_2. \tag{B.28}$$

Observe that

$$\frac{\partial \mathbf{R}}{\partial t_1} = -2\mathbf{C}. \tag{B.29}$$

For $p = 1$, (B.29) and (B.72) from Section B.5 below imply

$$\frac{\partial}{\partial t_1} |\mathbf{R}| = -2|\mathbf{R}| \cdot \text{tr}(\mathbf{R}^{-1}\mathbf{C}), \tag{B.30}$$

so that

$$\frac{\partial}{\partial t_1} \mathbb{M}_{A,B}(t_1, t_2) = -\frac{1}{2} |\mathbf{R}|^{-\frac{3}{2}} \left(\frac{\partial}{\partial t_1} |\mathbf{R}|\right) = |\mathbf{R}|^{-\frac{1}{2}} \text{tr}\,(\mathbf{R}^{-1}\mathbf{C}). \tag{B.31}$$

Thus,

$$\frac{\partial}{\partial t_1} \mathbb{M}_{A,B}(t_1, -t_2) = |\mathbf{I}_T - 2t_1\mathbf{C} + 2t_2\mathbf{\Lambda}|^{-\frac{1}{2}} \;\text{tr}[(\mathbf{I}_T - 2t_1\mathbf{C} + 2t_2\mathbf{\Lambda})^{-1}\mathbf{C}],$$

and, evaluated at $t_1 = 0$,

$$\begin{aligned}
\left.\frac{\partial}{\partial t_1} \mathbb{M}_{A,B}(t_1, -t_2)\right|_{t_1=0} &= |\mathbf{I}_T + 2t_2\mathbf{\Lambda}|^{-\frac{1}{2}} \;\text{tr}[(\mathbf{I}_T + 2t_2\mathbf{\Lambda})^{-1}\mathbf{C}] \\
&= \prod_{i=1}^{T} (1 + 2\lambda_i t_2)^{-\frac{1}{2}} \;\sum_{j=1}^{T} \frac{c_{jj}}{1 + 2\lambda_j t_2}.
\end{aligned} \tag{B.32}$$

For $p = 2$, it is convenient to first define

$$\begin{aligned}
S_1 &= S_1(t_1, t_2) := \text{tr}^2(\mathbf{R}^{-1}\mathbf{C}) = (\text{tr}\,\mathbf{R}^{-1}\mathbf{C})^2 \quad \text{and} \\
S_2 &= S_2(t_1, t_2) := \text{tr}(\mathbf{R}^{-1}\mathbf{C})^2 = \text{tr}(\mathbf{R}^{-1}\mathbf{C}\mathbf{R}^{-1}\mathbf{C}).
\end{aligned}$$

Then,

$$\begin{aligned}
\frac{\partial^2 |\mathbf{R}|}{\partial t_1^2} &\overset{(B.30)}{=} \frac{\partial}{\partial t_1}(-2|\mathbf{R}| \cdot \text{tr}\,\mathbf{R}^{-1}\mathbf{C}) \\
&= \frac{\partial}{\partial t_1}(-2|\mathbf{R}|) \cdot \text{tr}\,\mathbf{R}^{-1}\mathbf{C} - 2|\mathbf{R}| \frac{\partial}{\partial t_1}(\text{tr}\,\mathbf{R}^{-1}\mathbf{C})
\end{aligned}$$

$$\overset{(B.70)}{\underset{(B.71)}{=}} 4|\mathbf{R}|S_1 - 2|\mathbf{R}|\mathrm{tr}\left(\mathbf{R}^{-1}\frac{\partial \mathbf{C}}{\partial t_1} + \frac{\partial \mathbf{R}^{-1}}{\partial t_1}\mathbf{C}\right)$$

$$\overset{(B.73)}{=} 4|\mathbf{R}|S_1 - 2|\mathbf{R}|\mathrm{tr}\left(\mathbf{0} - \mathbf{R}^{-1}\frac{\partial \mathbf{R}}{\partial t_1}\mathbf{R}^{-1}\mathbf{C}\right)$$

$$\overset{(B.29)}{=} 4|\mathbf{R}|(S_1 - S_2), \tag{B.33}$$

and

$$\frac{\partial^2 \mathbb{M}_{A,B}(t_1,t_2)}{\partial t_1^2} \overset{(B.31)}{=} \frac{\partial}{\partial t_1}\left(-\frac{1}{2}|\mathbf{R}|^{-\frac{3}{2}}\frac{\partial |\mathbf{R}|}{\partial t_1}\right)$$

$$= \frac{3}{4}|\mathbf{R}|^{-\frac{5}{2}}\left(\frac{\partial |\mathbf{R}|}{\partial t_1}\right)^2 - \frac{1}{2}|\mathbf{R}|^{-\frac{3}{2}}\frac{\partial^2 |\mathbf{R}|}{\partial t_1^2}$$

$$\overset{(B.30)}{\underset{(B.33)}{=}} \frac{3}{4}|\mathbf{R}|^{-\frac{5}{2}}(4|\mathbf{R}|^2 S_1) - \frac{1}{2}|\mathbf{R}|^{-\frac{3}{2}}4|\mathbf{R}|(S_1 - S_2)$$

$$= |\mathbf{R}|^{-\frac{1}{2}}(S_1 + 2S_2). \tag{B.34}$$

Finally,

$$S_1(0,-t_2) = \mathrm{tr}^2[(\mathbf{I}_T + 2t_2\mathbf{\Lambda})^{-1}\mathbf{C}]$$

$$= \sum_{i=1}^{T}\frac{c_{ii}}{1+2\lambda_i t_2}\sum_{j=1}^{T}\frac{c_{jj}}{1+2\lambda_j t_2} = \sum_{i=1}^{T}\sum_{j=1}^{T}\frac{c_{ii}c_{jj}}{(1+2\lambda_i t_2)(1+2\lambda_j t_2)}.$$

For $S_2(0,-t_2)$, a small example is helpful: With $T = 3$ and $k_i := (1+2\lambda_i t_2)^{-1}$,

$$\mathbf{R}^{-1}\mathbf{C} = \begin{bmatrix} k_1 & & \\ & k_2 & \\ & & k_3 \end{bmatrix}\begin{bmatrix} c_{11} & c_{12} & c_{13} \\ c_{21} & c_{22} & c_{23} \\ c_{31} & c_{32} & c_{33} \end{bmatrix} = \begin{bmatrix} k_1c_{11} & k_1c_{12} & k_1c_{13} \\ k_2c_{21} & k_2c_{22} & k_2c_{23} \\ k_3c_{31} & k_3c_{32} & k_3c_{33} \end{bmatrix}$$

so that the (ii)th element of $\mathbf{R}^{-1}\mathbf{C}\mathbf{R}^{-1}\mathbf{C}$ is $k_i\sum_j k_j c_{ij}c_{ji} = k_i\sum_j k_j c_{ij}^2$, because $\mathbf{C}$ is symmetric. Summing over i to form the trace gives the result, which is, for general T,

$$S_2(0,-t_2) = \sum_{i=1}^{T}\sum_{j=1}^{T}\frac{c_{ij}^2}{(1+2\lambda_i t_2)(1+2\lambda_j t_2)}.$$

Thus,

$$\left.\frac{\partial^2 \mathbb{M}_{A,B}(t_1,-t_2)}{\partial t_1^2}\right|_{t_1=0} = \prod_{i=1}^{T}(1+2\lambda_i t_2)^{-\frac{1}{2}} \cdot \sum_{i=1}^{T}\sum_{j=1}^{T}\frac{c_{ii}c_{jj}+2c_{ij}^2}{(1+2\lambda_i t_2)(1+2\lambda_j t_2)}. \tag{B.35}$$

Substituting (B.32) and (B.35) into (B.28) and changing the variable of integration from t_2 to t gives expressions for the first two moments, summarized as follows: For $R = \mathbf{X}'\mathbf{A}\mathbf{X}/\mathbf{X}'\mathbf{B}\mathbf{X}$ with $\mathbf{X} \sim \mathrm{N}(\mathbf{0},\mathbf{\Sigma})$, let $\mathbf{B}^* = \mathbf{\Sigma}^{1/2}\mathbf{B}\mathbf{\Sigma}^{1/2} = \mathbf{P}'\mathbf{\Lambda}\mathbf{P}$ for $\mathbf{\Lambda} = \mathrm{diag}([\lambda_1,\dots,\lambda_n])$ and $c_{ij} = [\mathbf{P}'\mathbf{A}^*\mathbf{P}]_{ij}$, with $\mathbf{A}^* = \mathbf{\Sigma}^{1/2}\mathbf{A}\mathbf{\Sigma}^{1/2}$. Then, with

$$\zeta_i = 1 + 2\lambda_i t \quad \text{and} \quad \bar{\lambda}(t) = \prod_{i=1}^{T}\zeta_i^{-1/2},$$

$$\mathbb{E}[R] = \int_0^\infty \sum_{j=1}^{n} \zeta_i^{-1} \bar{\lambda}(t)\, c_{jj}\, \mathrm{d}t, \tag{B.36}$$

and

$$\mathbb{E}[R_l^2] = \int_0^\infty \sum_{i=1}^{n} \sum_{j=1}^{n} \zeta_i^{-1} \zeta_j^{-1} t\, \bar{\lambda}(t)\, (c_{ii} c_{jj} + 2c_{ij}^2)\, \mathrm{d}t\ . \tag{B.37}$$

A program to compute (B.36) and (B.37) is given in Listing B.4.

Remarks

a) The integrals can be approximated by numerical integration over a finite range, say 0 to t^*. De Gooijer (1980) gave approximate expressions for the roundoff error of the finitely evaluated integrals for the first and second moments, as well as formulae for their truncation errors. Paolella (2003) studied the behavior of the upper limit t^*, for a given accuracy level, as a function of n, $\boldsymbol{\Sigma}$, and regressor matrices. It was found that t^* was less than (the unexpectedly small value of) two

```
function [mom1,mom2]=sawa(A,B,Sigma)
[V,D]=eig(0.5*(Sigma+Sigma')); S12=V*sqrt(D)*V';
Astar=S12*A*S12; Bstar=S12*B*S12;
tol=1e-8; [P,lambda]=eig(0.5*(Bstar+Bstar'));
lambda=diag(lambda); C=P'*Astar*P; c=diag(C);
upper=1e-3;
while abs(sawaint1(upper,c,lambda))>tol, upper=upper*2; end;
mom1=quadl(@sawaint1,0,upper,tol,0,c,lambda);
if nargout>1
  upper=1e-3;
  while abs(sawaint2(upper,C,lambda))>tol, upper=upper*2; end;
  mom2=quadl(@sawaint2,0,upper,tol,0,C,lambda);
end

function I=sawaint1(uvec,c,lambda)
I=zeros(size(uvec));
for loop=1:length(uvec)
  t=uvec(loop); zeta=1+2*lambda*t; lambar=prod(zeta.^(-1/2));
  I(loop)=lambar*sum( c./zeta );
end;

function I=sawaint2(uvec,C,lambda)
I=zeros(size(uvec)); c=diag(C);
for loop=1:length(uvec)
  t=uvec(loop); zeta=1+2*lambda*t; lambar=prod(zeta.^(-1/2));
  K=zeros(size(C));
  for i=1:length(C), for j=1:length(C)
      K(i,j)=(c(i)*c(j)+2*C(i,j)^2)/zeta(i)/zeta(j);
  end, end
  K=K*lambar*t; I(loop)=sum(sum(K));
end;
```

Program Listing B.4: Computes the mean, mom1, and, if nargout $= 2$, the second raw moment, mom2, of the ratio of quadratic forms $\mathbf{Y}'\mathbf{A}\mathbf{Y}/\mathbf{Y}'\mathbf{B}\mathbf{Y}$ where $\mathbf{Y} \sim \mathrm{N}(\mathbf{0}, \boldsymbol{\Sigma})$.

for all cases considered, and that, of the three factors, the sample size n appears to exerts the most influence on t^*.

b) Extensions to the third and fourth moments were given by De Gooijer (1980) as

$$\mathbb{E}[R_l^3] = \frac{1}{2}\int_0^\infty \sum_{i=1}^n \sum_{j=1}^n \sum_{r=1}^n \zeta_i^{-1}\zeta_j^{-1}\zeta_r^{-1} t^2 \bar{\lambda}(t) \times (c_{ii}c_{jj}c_{rr} + 6c_{ij}^2 c_{rr} + 8c_{ij}c_{jr}c_{ri})\, \mathrm{d}t$$

and

$$\mathbb{E}[R_l^4] = \frac{1}{6}\int_0^\infty \sum_{i=1}^n \sum_{j=1}^n \sum_{r=1}^n \sum_{s=1}^n \zeta_i^{-1}\zeta_j^{-1}\zeta_r^{-1}\zeta_s^{-1} t^3\, \bar{\lambda}(t) \times (c_{ii}c_{jj}c_{rr}c_{ss} + 32c_{ij}c_{jr}c_{ri}c_{ss} + 12c_{ij}^2 c_{rs}c_{sr} + 48c_{ij}c_{jr}c_{rs}c_{si})\, \mathrm{d}t.$$

Ali (1984) presented a simplification of the formulae that leads to a decrease in computation time for the higher moments. ■

Example B.4 Morin-Wahhab (1985) gave analytic expressions (in terms of hypergeometric functions of many variables) for the positive integer moments of

$$\frac{\sum_{i=1}^{p_1} a_i X_i + \sum_{j=1}^{p_3} c_j Z_j}{\sum_{i=1}^{p_2} b_i Y_i + \sum_{j=1}^{p_3} d_j Z_j}, \tag{B.38}$$

where $X_i, i = 1, \dots, p_1, Y_j, j = 1, \dots, p_2$, and $Z_k, k = 1, \dots, p_3$, are independent central χ^2 random variables with ℓ_i, m_j, and n_k integer degrees of freedom, respectively. Using the fact that $C \sim \chi_n^2$ can be expressed as the sum of n i.i.d. χ_1^2 r.v.s, (B.38) can be expressed as the ratio $\mathbf{Z}'\mathbf{C}\mathbf{Z}/\mathbf{Z}'\mathbf{D}\mathbf{Z}$, where

$$\mathbf{C} = \text{diag}\,([a_1\mathbf{J}_{\ell_1}, \dots, a_{p_1}\mathbf{J}_{\ell_{p_1}}, \mathbf{0}_{m_\bullet}, c_1\mathbf{J}_{n_1}, \dots, c_{p_3}\mathbf{J}_{n_{p_3}}]),$$
$$\mathbf{D} = \text{diag}\,([\mathbf{0}_{\ell_\bullet}, b_1\mathbf{J}_{m_1}, \dots, b_{p_2}\mathbf{J}_{m_{p_2}}, d_1\mathbf{J}_{n_1}, \dots, d_{p_3}\mathbf{J}_{n_{p_3}}]),$$

$\mathbf{J}_h$ $(\mathbf{0}_h)$ denotes an h-length vector of ones (zeros), $\ell_\bullet = \sum_{i=1}^{p_1} \ell_i$, $m_\bullet = \sum_{j=1}^{p_2} m_j$, $n_\bullet = \sum_{k=1}^{p_3} n_k$, $\mathbf{Z} \sim \mathrm{N}_n(\mathbf{0}, \mathbf{I})$, and $n = \ell_\bullet + m_\bullet + n_\bullet$. As the analytic expressions are not readily evaluated numerically, it is more expedient to use (B.36) and (B.37) and the Ali (1984) results for higher-order moments. ■

Example B.5 ***Examples B.1–B.3 cont.***

The first two moments of the Durbin–Watson statistic (B.16), but now under the alternative hypothesis that a in (B.14) is not zero, can be determined via (B.36) and (B.37), as programmed in Listing B.4. For this time-series model, the (i, j)th element of $\boldsymbol{\Sigma}$ is given by

$$\frac{a^{|i-j|}}{1 - a^2}, \tag{B.39}$$

as was also used in Example A.1, and derived in (4.13).

Simulation can also be used to compute the first two (and, conveniently, higher) moments, and also serves as a check on the derivation, programming, and numeric accuracy of the integral formulae. The reader should now be quite comfortable with such programming tasks, but, recognizing repetition as our didactic friend, we give a program for such in Listing B.5.

```
T=10; a=0.5; Sigma = toeplitz((a).^(0:(T-1)))/(1-a^2);
[V,D]=eig(0.5*(Sigma+Sigma')); S=V*sqrt(D)*V';
X=[ones(T,1) (1:T)']; M=makeM(X); B=M; A=M*makeDW(T)*M;
sim=1e7; D=zeros(sim,1);
for i=1:sim, r=S*randn(T,1); D(i)=(r'*A*r)/(r'*B*r); end
simulated_mean_var = [mean(D), var(D)]
[mom1,mom2]=sawa(A,B,Sigma); true_mean_var = [mom1 mom2-mom1^2]
```

Program Listing B.5: Simulation and exact calculation for obtaining the mean and variance of the Durbin–Watson statistic under the alternative hypothesis (in this case,with $\mathbf{X} = [\mathbf{1} \;\; \mathbf{t}]$). Programs `makeDW` and `makeM` are given in Example B.2.

Figure B.5 shows the mean and T times the variance of the Durbin–Watson statistic as a function of a, where we multiply by T in light of (B.12) and (B.20). This is done for three sample sizes, $T = 10$, 20, and 40, and three $\mathbf{X}$ matrices, the intercept model considered in Example B.3, intercept and time trend, denoted $\mathbf{X} = [\mathbf{1}\ \mathbf{t}]$, and $\mathbf{X} = [\mathbf{1}\ \mathbf{t}\ \mathbf{v}]$, where $\mathbf{v}$ is the eigenvector $\boldsymbol{v}_i$ in (B.10) with $i = \text{round}(T/3)$. This latter choice might seem strange, but the cyclical nature of the $\boldsymbol{v}_i$ is also a common feature in economic data, so that use of $\boldsymbol{v}_i$, along with an intercept and time trend, yields an $\mathbf{X}$ matrix that is somewhat typical in econometrics (see, e.g., Dubbelman et al., 1978 and King, 1985a, p. 32).

As $\mathbf{X}$ increases in complexity (moving from the top to the bottom panels in Figure B.5), we see that, for small sample sizes, the mean and, particularly, the variance, deviate greatly from what appears to be their asymptotic values. ■

B.3 For $\mathbf{X} \sim \mathrm{N}(\boldsymbol{\mu}, \mathbf{I})$

Expressions for the moments of the ratio

$$R = \frac{\mathbf{X}'\mathbf{H}\mathbf{X}}{\mathbf{X}'\mathbf{K}\mathbf{X}} =: \frac{N}{D}, \quad \mathbf{X} \sim \mathrm{N}_T(\boldsymbol{\mu}, \mathbf{I}), \tag{B.40}$$

are still tractable when certain restrictions on the (without loss of generality, symmetric) matrices $\mathbf{H}$ and $\mathbf{K}$ are fulfilled. In particular, (i) $\mathbf{K}$ is idempotent, (ii) $\mathbf{H}$ and $\mathbf{K}$ commute, i.e., $\mathbf{HK} = \mathbf{KH}$, and (iii) $r := \text{rank}(\mathbf{H}) = \text{rank}(\mathbf{K}) = \text{rank}(\mathbf{HK})$ for $1 \leqslant r \leqslant T$. We also require $\mathbf{K} \geqslant 0$ so that $\Pr(D > 0) = 1$, but this is automatically fulfilled if $\mathbf{K}$ is symmetric and idempotent (in which case its eigenvalues are either zero or one). There are important applications in which these conditions are fulfilled, so that they are not as restrictive as they perhaps appear. We show the derivation from Ghazal (1994). Before commencing, note that, if $\mathbf{X} \sim \mathrm{N}(\boldsymbol{\mu}, \sigma^2\mathbf{I})$, then

$$R = \frac{\mathbf{X}'\mathbf{H}\mathbf{X}}{\mathbf{X}'\mathbf{K}\mathbf{X}} \frac{\sigma^{-2}}{\sigma^{-2}} = \frac{(\mathbf{X}/\sigma)'\mathbf{H}(\mathbf{X}/\sigma)}{(\mathbf{X}/\sigma)'\mathbf{K}(\mathbf{X}/\sigma)} = \frac{\mathbf{Y}'\mathbf{H}\mathbf{Y}}{\mathbf{Y}'\mathbf{K}\mathbf{Y}},$$

where $\mathbf{Y} = \mathbf{Y}/\sigma \sim \mathrm{N}(\boldsymbol{\mu}/\sigma, \mathbf{I})$, i.e., we can take $\sigma = 1$ without loss of generality.

The joint moment generating function of $D = \mathbf{X}'\mathbf{K}\mathbf{X}$ and $N = \mathbf{X}'\mathbf{H}\mathbf{X}$ follows directly from (A.23), with the only difference being that we use $\mathbb{M}_{D,N}$ instead of $\mathbb{M}_{N,D}$ because it is slightly more convenient. With

$$\mathbf{S} = \mathbf{S}(t_1, t_2) = \mathbf{I} - 2t_1\mathbf{K} - 2t_2\mathbf{H}, \quad \mathbf{s} = -\boldsymbol{\mu}, \quad \text{and} \quad s_0 = -\frac{1}{2}\boldsymbol{\mu}'\boldsymbol{\mu},$$

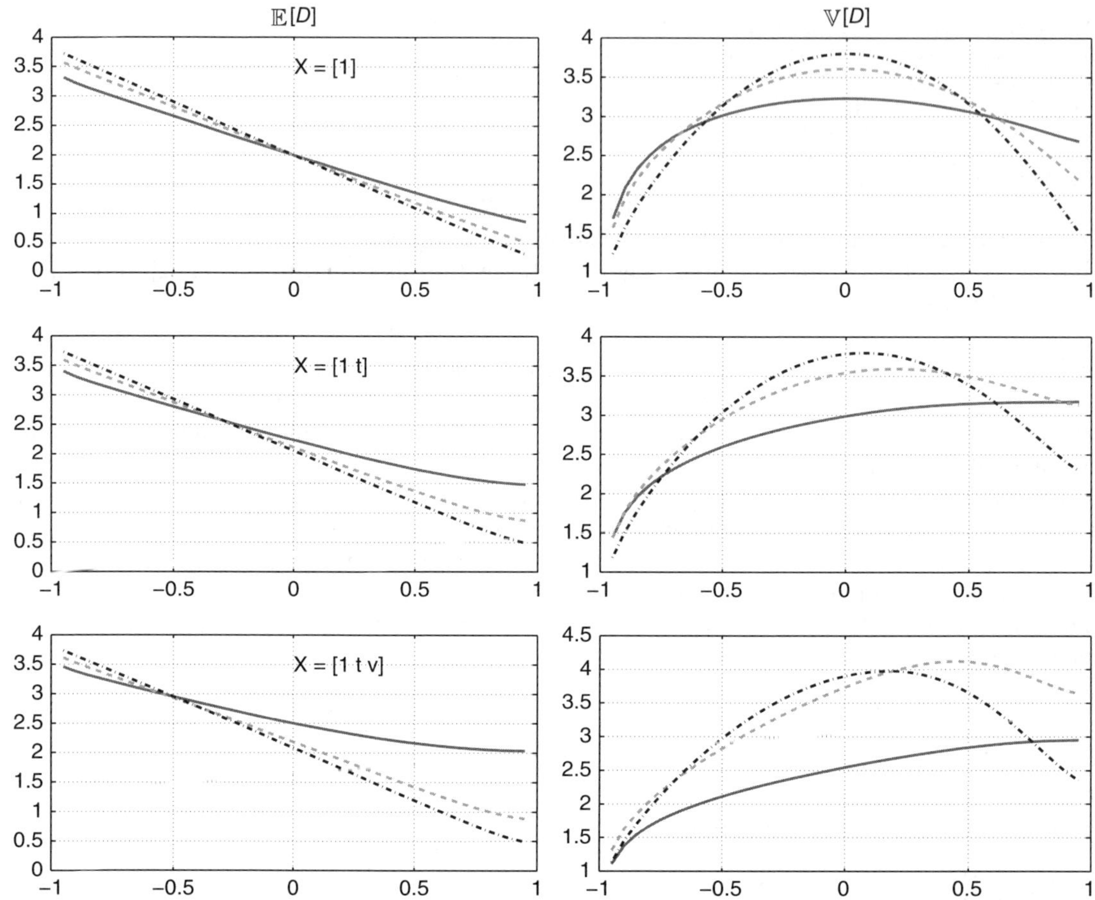

Figure B.5 For the three sample sizes $T = 10$ (solid), $T = 20$ (dashed) and $T = 40$ (dash-dot), the mean (left) and T times the variance (right) of the Durbin–Watson statistic (B.16), as a function of the autoregressive parameter a in (B.14), for the intercept model $\mathbf{X} = [\mathbf{1}]$ (top panels), intercept and time trend $\mathbf{X} = [\mathbf{1}\ \mathbf{t}]$ (middle panels), and intercept, time trend and cyclical, $\mathbf{X} = [\mathbf{1}\ \mathbf{t}\ \mathbf{v}]$ (bottom panels), where $\mathbf{v}$ is the eigenvector v_i in (B.10) with $i = \text{round}(T/3)$.

we have

$$\mathbb{M}_{D,N}(t_1, t_2) = |\mathbf{S}|^{-1/2} \exp\left(-\frac{1}{2}\boldsymbol{\mu}'(\mathbf{I} - \mathbf{S}^{-1})\boldsymbol{\mu}\right). \tag{B.41}$$

The necessity of the conditions on $\mathbf{H}$ and $\mathbf{K}$ stated above will now become clear. If they are satisfied, then Theorem B.3 in Section B.5 can be employed as follows. There exists a matrix $\mathbf{Q}$ such that

$$\mathbf{Q}'\mathbf{Q} = \mathbf{Q}\mathbf{Q}' = \mathbf{I}_T, \tag{B.42}$$

$$\mathbf{Q}'\mathbf{K}\mathbf{Q} =: \boldsymbol{\Omega} = \begin{bmatrix} \mathbf{I}_r & \mathbf{0} \\ \mathbf{0} & \mathbf{0} \end{bmatrix} = \text{diag}([\omega_1, \omega_2, \dots, \omega_T]), \tag{B.43}$$

where $\omega_1 = \omega_2 = \cdots = \omega_r = 1, \omega_{r+1} = \omega_{r+2} = \cdots = \omega_T = 0$, and

$$\mathbf{Q}'\mathbf{H}\mathbf{Q} =: \mathbf{\Lambda} = \begin{bmatrix} \mathbf{\Lambda}_r & \mathbf{0} \\ \mathbf{0} & \mathbf{0} \end{bmatrix} = \operatorname{diag}([\lambda_1, \lambda_2, \dots, \lambda_T]), \tag{B.44}$$

where $\lambda_1, \dots, \lambda_r$ are the r nonzero eigenvalues of $\mathbf{H}$ and $\lambda_{r+1} = \cdots = \lambda_T = 0$. Together, (B.42), (B.43), and (B.44) imply

$$\mathbf{S} = \mathbf{Q}(\mathbf{I} - 2t_1\mathbf{\Omega} - 2t_2\mathbf{\Lambda})\mathbf{Q}'. \tag{B.45}$$

From (B.42) and (B.45), and recalling that the determinant of a product is the product of the determinants, $|\mathbf{S}| = |\mathbf{I} - 2t_1\mathbf{\Omega} - 2t_2\mathbf{\Lambda}| = \prod_{i=1}^{T}(1 - 2t_1 - 2t_2\lambda_i)$, and (B.43) and (B.44) then imply that

$$|\mathbf{S}| = \prod_{i=1}^{r}(1 - 2t_1 - 2t_2\lambda_i). \tag{B.46}$$

Now let $[\mu_1^*, \mu_2^*, \dots, \mu_T^*]' := \boldsymbol{\mu}^* := \mathbf{Q}'\boldsymbol{\mu}$. Clearly, $\boldsymbol{\mu}^{*'}\boldsymbol{\mu}^* = \boldsymbol{\mu}'\mathbf{Q}\mathbf{Q}'\boldsymbol{\mu} = \boldsymbol{\mu}'\boldsymbol{\mu}$. Also

$$\sum_{i=1}^{r} \mu_i^{*2} = \boldsymbol{\mu}^{*'}\mathbf{\Omega}\boldsymbol{\mu}^* = \boldsymbol{\mu}'\mathbf{Q}\mathbf{\Omega}\mathbf{Q}'\boldsymbol{\mu} = \boldsymbol{\mu}'\mathbf{K}\boldsymbol{\mu} =: 2\delta, \tag{B.47}$$

implying $\sum_{i=r+1}^{T} \mu_i^{*2} = \boldsymbol{\mu}'\boldsymbol{\mu} - 2\delta$.

From (B.42) and (B.45), $\mathbf{S}^{-1} = \mathbf{Q}(\mathbf{I} - 2t_1\mathbf{\Omega} - 2t_2\mathbf{\Lambda})^{-1}\mathbf{Q}'$, so that

$$\begin{aligned}
-\frac{1}{2}\boldsymbol{\mu}'(\mathbf{I} - \mathbf{S}^{-1})\boldsymbol{\mu} &= -\frac{1}{2}\boldsymbol{\mu}'\boldsymbol{\mu} + \frac{1}{2}\boldsymbol{\mu}'\mathbf{S}^{-1}\boldsymbol{\mu} = -\frac{1}{2}\boldsymbol{\mu}^{*\,\prime}\boldsymbol{\mu}^* + \frac{1}{2}\boldsymbol{\mu}^{*\,\prime}(\mathbf{I} - 2t_1\mathbf{\Omega} - 2t_2\mathbf{\Lambda})^{-1}\boldsymbol{\mu}^* \\
&= -\frac{1}{2}\sum_{i=1}^{T}\mu_i^{*2} + \frac{1}{2}\sum_{i=1}^{T}\frac{\mu_i^{*2}}{1 - 2\omega_i t_1 - 2t_2\lambda_i} \\
&= -\frac{1}{2}\sum_{i=1}^{T}\mu_i^{*2} + \frac{1}{2}\sum_{i=1}^{r}\frac{\mu_i^{*2}}{1 - 2t_1 - 2t_2\lambda_i} + \frac{1}{2}\sum_{i=r+1}^{T}\frac{\mu_i^{*2}}{1} \\
&= -\frac{1}{2}\sum_{i=1}^{r}\mu_i^{*2} + \frac{1}{2}\sum_{i=1}^{r}\frac{\mu_i^{*2}}{1 - 2t_1 - 2t_2\lambda_i} \\
&\overset{(B.47)}{=} -\delta + \sum_{i=1}^{r}\frac{\mu_i^{*2}/2}{1 - 2t_1 - 2t_2\lambda_i}.
\end{aligned} \tag{B.48}$$

Now, (B.46) and (B.48) allow writing (B.41) as

$$\mathbb{M}_{D,N}(t_1, t_2) = \prod_{i=1}^{r}(1 - 2t_1 - 2t_2\lambda_i)^{-\frac{1}{2}} \exp\left(-\delta + \sum_{i=1}^{r}\frac{\mu_i^{*2}/2}{1 - 2t_1 - 2t_2\lambda_i}\right), \tag{B.49}$$

which lends itself to differentiation. From Sawa (1972) (see p. II.15–16 for derivation):

(Sawa, 1972) Let X_1 and X_2 be r.v.s such that $\Pr(X_1 > 0) = 1$, with joint m.g.f. $\mathbb{M}_{X_1,X_2}(t_1, t_2)$, which exists for $t_1 < \epsilon$ and $|t_2| < \epsilon$, for $\epsilon > 0$. Then the kth order moment, $k \in \mathbb{N}$, of X_2/X_1, if it exists, is given by

$$\mathbb{E}\left[\left(\frac{X_2}{X_1}\right)^k\right] = \frac{1}{\Gamma(k)}\int_{-\infty}^{0}(-t_1)^{k-1}\left[\frac{\partial^k}{\partial t_2^k}\mathbb{M}X_1, X_2(t_1, t_2)\right]_{t_2=0} \mathrm{d}t_1. \tag{B.50}$$

Differentiating (B.49) is simplified by using the fact that, for any positive differentiable function $f(x)$,

$$\frac{\mathrm{d}f(x)}{\mathrm{d}x} = \frac{\mathrm{d}\exp(\ln(f(x)))}{\mathrm{d}x} = f(x)\cdot\frac{\mathrm{d}}{\mathrm{d}x}\ln(f(x)).$$

The first derivative of (B.49) is now

$$\frac{\partial \mathbb{M}_{D,N}(t_1,t_2)}{\partial t_2} = \mathbb{M}_{D,N}(t_1,t_2)\cdot\frac{\partial}{\partial t_2}\ln \mathbb{M}_{D,N}(t_1,t_2), \tag{B.51}$$

and

$$\frac{\partial \ln \mathbb{M}_{D,N}(t_1,t_2)}{\partial t_2} = \frac{\partial}{\partial t_2}\left[-\frac{1}{2}\sum_{i=1}^{r}\ln(1-2t_1-2t_2\lambda_i) - \delta + \sum_{i=1}^{r}\frac{\mu_i^{*2}/2}{1-2t_1-2t_2\lambda_i}\right]$$

$$= \sum_{i=1}^{r}\frac{\lambda_i}{1-2t_1-2t_2\lambda_i} + \sum_{i=1}^{r}\frac{\lambda_i\mu_i^{*2}}{(1-2t_1-2t_2\lambda_i)^2}, \tag{B.52}$$

so that

$$\left.\frac{\partial \mathbb{M}_{D,N}(t_1,t_2)}{\partial t_2}\right|_{t_2=0} = \prod_{i=1}^{r}(1-2t_1)^{-\frac{1}{2}}\exp\left(-\delta + \sum_{i=1}^{r}\frac{\mu_i^{*2}/2}{1-2t_1}\right) \times\left[\sum_{i=1}^{r}\frac{\lambda_i}{1-2t_1} + \sum_{i=1}^{r}\frac{\lambda_i\mu_i^{*2}}{(1-2t_1)^2}\right].$$

For convenience, define

$$\gamma_m := \boldsymbol{\mu}'\mathbf{H}^m\boldsymbol{\mu} = \boldsymbol{\mu}'\mathbf{Q}\boldsymbol{\Lambda}^m\mathbf{Q}'\boldsymbol{\mu} = \boldsymbol{\mu}^{*\prime}\boldsymbol{\Lambda}^m\boldsymbol{\mu}^* = \sum_{i=1}^{r}\lambda_i^m\mu_i^{*2}, \quad m\in\mathbb{N}, \tag{B.53}$$

and

$$\alpha_m := \mathrm{tr}(\mathbf{H}^m) = \sum_{i=1}^{r}\lambda_i^m, \quad m\in\mathbb{N}. \tag{B.54}$$

Using these and that $\delta = \frac{1}{2}\sum_{i=1}^{r}\mu_i^{*2}$ from (B.47), we get[4]

$$\left.\frac{\partial \mathbb{M}_{D,N}(t_1,t_2)}{\partial t_2}\right|_{t_2=0} = \left[\frac{\alpha_1}{(1-2t_1)^{r/2+1}} + \frac{\gamma_1}{(1-2t_1)^{r/2+2}}\right]\times\exp\left(-\delta + \frac{\delta}{1-2t_1}\right).$$

In order to simplify this, define

$$h(x,c,\delta,m) := \frac{1}{(1-2x)^{c+m}}\exp\left\{-\delta + \frac{\delta}{1-2x}\right\}, \tag{B.55}$$

so that

$$\left.\frac{\partial \mathbb{M}_{D,N}(t_1,t_2)}{\partial t_2}\right|_{t_2=0} = \alpha_1 h\left(t_1,\frac{r}{2},\delta,1\right) + \gamma_1 h\left(t_1,\frac{r}{2},\delta,2\right).$$

4 This differs from Ghazal (1994, Eq. 2.24) because of a minor error in that presentation.

From (B.50), we require

$$H(c,\delta,m,n) := \frac{1}{\Gamma(n)} \int_{-\infty}^{0} (-x)^{n-1} h(x,c,\delta,m)\,\mathrm{d}x.$$

Substituting $t = 1/(1-2x)$ leads to

$$H(c,\delta,m,n) = \frac{\mathrm{e}^{-\delta}}{2^n\Gamma(n)} \int_0^1 t^{c+m-n-1}(1-t)^{n-1}\mathrm{e}^{\delta t}\,\mathrm{d}t.$$

Now, using the integral expression for the confluent hypergeometric function (II.5.27),

$${}_1F_1(a,b;z) = \frac{1}{B(a,b-a)} \int_0^1 y^{a-1}(1-y)^{b-a-1}\mathrm{e}^{zy}\,\mathrm{d}y,$$

and Kummer's transformation (II.5.29), ${}_1F_1\,(a,b,x) = \mathrm{e}^x\ {}_1F_1\,(b-a,b,-x)$, gives

$$H(c,\delta,m,n) = \frac{\Gamma(c+m-n)}{\Gamma(c+m)}\frac{1}{2^n}\ {}_1F_1(n,c+m;-\delta), \quad c+m>n. \tag{B.56}$$

Thus, we finally arrive at the pleasantly compact expression

$$\mathbb{E}[R] = \alpha_1 f(1,1) + \gamma_1 f(2,1), \tag{B.57}$$

where $f(m,n) := H\left(\frac{r}{2},\delta,m,n\right)$. A similar calculation verifies that

$$\mathbb{E}[R^2] = (2\alpha_2 + \alpha_1^2)f(2,2) + 2(2\gamma_2 + \alpha_1\gamma_1)f(3,2) + \gamma_1^2 f(4,2), \tag{B.58}$$

and further tedious work, as done by Ghazal (1994), shows that

$$\begin{aligned}\mathbb{E}[R^3] = {} & [8\alpha_3 + 6\alpha_2\alpha_1 + \alpha_1^3]f(3,3) + 3[8\gamma_3 + 4\alpha_1\gamma_2 + \gamma_1(2\alpha_2+\alpha_1^2)]f(4,3)\\ & + 3\gamma_1(4\gamma_2 + \alpha_1\gamma_1)f(5,3) + \gamma_1^3 f(6,3),\end{aligned}$$

and

$$\begin{aligned}\mathbb{E}[R^4] = {} & [48\alpha_4 + 32\alpha_3\alpha_1 + 12\alpha_2\alpha_1^2 + 12\alpha_2^2 + \alpha_1^4]f(4,4)\\ & + 4[48\gamma_4 + 24\alpha_1\gamma_3 + 6\gamma_2(2\alpha_2+\alpha_1^2) + \gamma_1(8\alpha_3 + 6\alpha_2\alpha_1 + \alpha_1^3)]f(5,4)\\ & + 6[16\gamma_3\gamma_1 + 8\gamma_2(\gamma_2 + \alpha_1\gamma_1) + \gamma_1^2(2\alpha_2 + \alpha_1^2)]f(6,4)\\ & + 4\gamma_1^2(6\gamma_2 + \alpha_1\gamma_1)f(7,4) + \gamma_1^4 f(8,4).\end{aligned}$$

A program to compute (B.57) and (B.58) is given in Listing B.6.

Example B.6 ***Example B.2 cont.***
We wish to see if the Durbin–Watson statistic D in (B.16) is a suitable candidate for (B.40). This requires setting $\mathbf{H} = \mathbf{MAM}$ and $\mathbf{K} = \mathbf{M}$. As $\mathbf{M}$ is idempotent, condition (i) (after (B.40)) is satisfied and so is condition (ii), because $\mathbf{HK} = \mathbf{MAMM} = \mathbf{MAM} = \mathbf{MMAM} = \mathbf{KH}$. For condition (iii) to hold, recall that, if the $T \times k$ full-rank matrix $\mathbf{X}$ contains a constant term (as is usual), or if all the columns of $\mathbf{X}$ do not have zero mean, then rank($\mathbf{MA}$) $= T - k =$ rank($\mathbf{M}$) (see the second canonical reduction argument on page 703). So, we require this condition on $\mathbf{X}$ in order for rank($\mathbf{MA}$) = rank($\mathbf{M}$) to hold, and condition (iii) would follow if rank($\mathbf{MA}$) = rank($\mathbf{MAM}$), but this is true, as $\mathbf{MA} = \mathbf{MMA}$ and $\mathbf{MAM}$ have the same nonzero eigenvalues.

```
function [mom1,mom2] = ghazal(H,K,mu)
if matdif(H,H')   > 1e-12, error('H not symmetric'), end
if matdif(K,K')   > 1e-12, error('K not symmetric'), end
if matdif(K,K*K)   > 1e-12, error('K not idempotent'), end
if matdif(H*K,K*H) > 1e-12, error('H and K do not commute'), end
rh=rank(H); rk=rank(K); rhk=rank(H*K);
if (rh~=rk) | (rk~=rhk), error('ranks do not agree') else r=rh; end
if r==0, error('rank is zero'), end

delta = 0.5 * mu' * K * mu; gam1 = mu' * H * mu; gam2 = mu' * H^2 * mu;
alf1 = trace(H);      alf2 = trace(H^2);
mom1 = alf1*HH(r/2,delta,1,1) + gam1*HH(r/2,delta,2,1);
mom2 = (2*alf2+alf1^2)*HH(r/2,delta,2,2) ...
       + 2*(2*gam2+alf1*gam1)*HH(r/2,delta,3,2) ...
       + gam1^2*HH(r/2,delta,4,2);

function d = matdif(A,B), d = max(max(abs(A-B)));
function v = HH(c,delta,m,n)
k = gamma(c+m-n) / gamma(c+m) / 2^n;
if exist('hypergeom','file')
  v = k * hypergeom(n,c+m,-delta); % in the symbolic toolbox in Matlab
else
  v = k * f11(n,c+m,-delta); % the Laplace approximation from page II.197
end
```

Program Listing B.6: Computes (B.57) and (B.58).

Before proceeding, we can verify that (B.57) and (B.58) indeed reduce to the expression given in Section B.1 when $\boldsymbol{\mu} = \mathbf{0}$. In this case, $\delta = \gamma_m = 0$, and, as

$$ {}_1F_1(a,b,0) = \frac{\Gamma(b)}{\Gamma(a)\,\Gamma(b-a)} \int_0^1 t^{a-1}(1-t)^{b-a-1}\,\mathrm{d}t = 1, $$

it follows that

$$ H\left(\frac{r}{2}, 0, m, n\right) = \frac{1}{2^n} \frac{\Gamma\left(\frac{r}{2}+m-n\right)}{\Gamma\left(\frac{r}{2}+m\right)}. $$

Thus, for the mean,

$$ \mathbb{E}[R] = \alpha_1 f(1,1) = \mathrm{tr}(\mathbf{H}) \frac{1}{2} \frac{\Gamma\left(\frac{r}{2}+1-1\right)}{\Gamma\left(\frac{r}{2}+1\right)} = \frac{\mathrm{tr}(\mathbf{H})}{r} = \frac{\mathrm{tr}(\mathbf{MAM})}{T-k} = \frac{\mathrm{tr}(\mathbf{MA})}{T-k}, $$

which, in conjunction with (B.18), agrees with (B.5). The second moment is

$$ \mathbb{E}[R^2] = (2\alpha_2 + \alpha_1^2) f(2,2) = (2\,\mathrm{tr}(\mathbf{H}^2) + \mathrm{tr}^2(\mathbf{H})) \frac{1}{2^2} \frac{\Gamma\left(\frac{r}{2}+2-2\right)}{\Gamma\left(\frac{r}{2}+2\right)} $$

$$ = \frac{2\,\mathrm{tr}(\mathbf{H}^2) + \mathrm{tr}^2(\mathbf{H})}{r(r+2)}, $$

so that

$$\mathbb{V}(R) = \frac{2\,\mathrm{tr}(\mathbf{H}^2) + \mathrm{tr}^2(\mathbf{H})}{r(r+2)} - \frac{\mathrm{tr}^2(\mathbf{H})}{r^2} = 2\frac{r\,\mathrm{tr}(\mathbf{H}^2) - \mathrm{tr}^2(\mathbf{H})}{r^2(r+2)},$$

which is precisely as given in (B.5).

Now consider an example for which the expectation of the regression error term is not zero. In particular, let the true model be

$$\mathbf{Y} = \mathbf{X}\boldsymbol{\beta} + \mathbf{Z}\boldsymbol{\alpha} + \boldsymbol{\eta}, \quad \boldsymbol{\eta} \sim \mathrm{N}(\mathbf{0}, \sigma^2\mathbf{I}). \tag{B.59}$$

We estimate the under-specified model $\mathbf{Y} = \mathbf{X}\boldsymbol{\beta} + \boldsymbol{\epsilon}$, where $\boldsymbol{\epsilon} = \mathbf{Z}\boldsymbol{\alpha} + \boldsymbol{\eta} \sim \mathrm{N}(\boldsymbol{\mu}, \sigma^2\mathbf{I})$, with $\boldsymbol{\mu} = \mathbf{Z}\boldsymbol{\alpha}$, which is unknown (and wrongly assumed to be zero). Then, with $\mathbf{M} = \mathbf{I} - \mathbf{X}(\mathbf{X}'\mathbf{X})^{-1}\mathbf{X}'$ and $\widehat{\boldsymbol{\epsilon}} = \mathbf{M}\mathbf{Y} = \mathbf{M}\boldsymbol{\epsilon} = \mathbf{M}(\mathbf{Z}\boldsymbol{\alpha} + \boldsymbol{\eta})$, and using the symmetry and idempotency of $\mathbf{M}$,

$$D = \frac{\widehat{\boldsymbol{\epsilon}}'\mathbf{A}\widehat{\boldsymbol{\epsilon}}}{\widehat{\boldsymbol{\epsilon}}'\widehat{\boldsymbol{\epsilon}}} = \frac{(\mathbf{Z}\boldsymbol{\alpha} + \boldsymbol{\eta})'\mathbf{MAM}(\mathbf{Z}\boldsymbol{\alpha} + \boldsymbol{\eta})}{(\mathbf{Z}\boldsymbol{\alpha} + \boldsymbol{\eta})'\mathbf{M}(\mathbf{Z}\boldsymbol{\alpha} + \boldsymbol{\eta})}.$$

We consider a special case for illustration. Assume that a particular time series is generated by the model $Y_t = \beta_1 + \beta_2 t + \alpha v_t + \eta_t$, $t = 1, \dots, T$, where $\eta_t \overset{\text{i.i.d.}}{\sim} \mathrm{N}(0, \sigma^2)$, β_1, β_2 and α are unknown coefficients, and $\mathbf{v} = (v_1, \dots, v_T)'$ is the vector used in Example B.5.

We are interested in the mean and variance of the Durbin–Watson statistic, D, if the model is incorrectly specified by omitting vector $\mathbf{v}$ from the regression, which, with its sinusoidal form, represents a variable that describes the cyclical nature of the Y_t series. With $\sigma = 1$, Figure B.6 plots the mean of D corresponding to the under-specified model $Y_t = \beta_1 + \beta_2 t + \epsilon_t$, along with 1.96 times the square root of its variance, as a function of α. (The reader should confirm that the values of β_1 and β_2 are irrelevant).

We see that the mean of D decreases as α moves away from zero (the same values result using negative values of α), and the variance decreases. The horizontal dashed lines serve to indicate where D would lie, with 95% probability, if α were truly zero. For α larger than about six, the Durbin–Watson test would tend to reject its null hypothesis of zero autocorrelation in the residuals, *even though there is no autocorrelation in the true residuals.* What is happening is that, because the omitted regressor

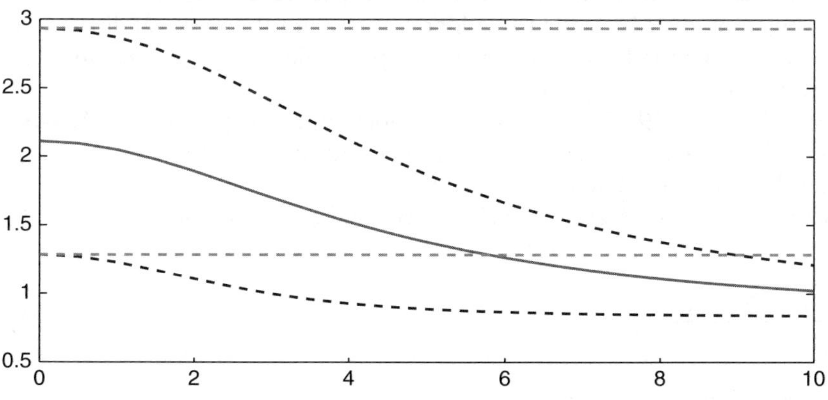

Figure B.6 The mean of D, and the mean plus and minus 1.96 times its standard deviation, as a function of α, when using the mis-specified model that erroneously assumes $\alpha = 0$.

has sinusoidal behavior and is part of the residual of the under-specified model, it mimics the behavior of an autocorrelated series, so that, as its presence increases (via increasing $|\alpha|$), the distribution of D deviates further from the null case.

This effect is well-known in econometrics, so that significance of the Durbin–Watson test can be interpreted as evidence that the model is mis-specified. Given that (reliable) data on certain economic variables are sometimes not available, such an occurrence is more of the rule than the exception in econometrics. Because the desired data are not available, the correct model cannot be estimated, but one can instead estimate the under-specified model together with an autoregressive process like (B.14), which could lead to more accurate estimation of $\boldsymbol{\beta}$ and produce better forecasts of Y_t. ■

Remark The more general case of $\mathbb{E}[N^p/D^q]$ from (B.40), for $p, q \geqslant 0$ and p not necessarily an integer, and with less restrictions than we used, is addressed in Bao and Kan (2013). See also Roberts (1995) and Ullah et al. (1995). ■

B.4 For X ~ N(μ, Σ)

This is the most general case we consider, and is naturally the most difficult. First observe that we can always write

$$R = \frac{\mathbf{X}'\mathbf{H}\mathbf{X}}{\mathbf{X}'\mathbf{K}\mathbf{X}} = \frac{\mathbf{X}'\boldsymbol{\Sigma}^{-1/2}\boldsymbol{\Sigma}^{1/2}\mathbf{H}\boldsymbol{\Sigma}^{1/2}\boldsymbol{\Sigma}^{-1/2}\mathbf{X}}{\mathbf{X}'\boldsymbol{\Sigma}^{-1/2}\boldsymbol{\Sigma}^{1/2}\mathbf{K}\boldsymbol{\Sigma}^{1/2}\boldsymbol{\Sigma}^{-1/2}\mathbf{X}} = \frac{\mathbf{Z}'\mathbf{L}\mathbf{Z}}{\mathbf{Z}'\mathbf{N}\mathbf{Z}},$$

where $\mathbf{Z} = \boldsymbol{\Sigma}^{-1/2}\mathbf{X} \sim \mathrm{N}(\boldsymbol{\Sigma}^{-1/2}\boldsymbol{\mu}, \mathbf{I}_T)$, $\mathbf{L} = \boldsymbol{\Sigma}^{1/2}\mathbf{H}\boldsymbol{\Sigma}^{1/2}$, and $\mathbf{N} = \boldsymbol{\Sigma}^{1/2}\mathbf{K}\boldsymbol{\Sigma}^{1/2}$. If the condition $\mathbf{K}\boldsymbol{\Sigma}\mathbf{H} = \mathbf{H}\boldsymbol{\Sigma}\mathbf{K}$ holds, then

$$\begin{aligned}\mathbf{L}\mathbf{N} = \boldsymbol{\Sigma}^{1/2}\mathbf{H}\boldsymbol{\Sigma}^{1/2}\boldsymbol{\Sigma}^{1/2}\mathbf{K}\boldsymbol{\Sigma}^{1/2} &= \boldsymbol{\Sigma}^{1/2}\mathbf{H}\boldsymbol{\Sigma}\mathbf{K}\boldsymbol{\Sigma}^{1/2}\\ &= \boldsymbol{\Sigma}^{1/2}\mathbf{K}\boldsymbol{\Sigma}\mathbf{H}\boldsymbol{\Sigma}^{1/2} = \boldsymbol{\Sigma}^{1/2}\mathbf{K}\boldsymbol{\Sigma}^{1/2}\boldsymbol{\Sigma}^{1/2}\mathbf{H}\boldsymbol{\Sigma}^{1/2} = \mathbf{N}\mathbf{L},\end{aligned}$$

and the commutative property necessary in Section B.3 holds. Condition $\mathbf{K}\boldsymbol{\Sigma}\mathbf{H} = \mathbf{H}\boldsymbol{\Sigma}\mathbf{K}$ might not be fulfilled in real applications, but moreover $\mathbf{N}$ needs to be idempotent, and the rank condition also needs to be met. Thus, the results of Section B.3 are not generally applicable when $\boldsymbol{\Sigma} \neq \sigma^2\mathbf{I}$, and other methods will have to be entertained.

Analytic results are available: Magnus (1986) derives a computable integral expression for $\mathbf{H}$ symmetric and $\mathbf{K}$ positive semi-definite. See also Bao and Kan (2013) and the references therein. We discuss two alternative, somewhat easier methods.

The first simply uses a Taylor series approximation: Let $X = \mathbf{X}'\mathbf{H}\mathbf{X}$ and $Y = \mathbf{X}'\mathbf{K}\mathbf{X}$, so that $R = X/Y$. From (II.2.32) and (II.2.33),

$$\mathbb{E}[R] \approx \frac{\mu_X}{\mu_Y}\left(1 + \frac{\mathbb{V}(Y)}{\mu_Y^2} - \frac{\mathrm{Cov}(X,Y)}{\mu_X\mu_Y}\right), \tag{B.60}$$

$$\mathbb{V}(R) \approx \frac{\mu_X^2}{\mu_Y^2}\left(\frac{\mathbb{V}(X)}{\mu_X^2} + \frac{\mathbb{V}(Y)}{\mu_Y^2} - \frac{2\,\mathrm{Cov}(X,Y)}{\mu_X\mu_Y}\right), \tag{B.61}$$

where $\mu_X = \mathbb{E}[X]$, $\mu_Y = \mathbb{E}[Y]$ and, from (A.6), (A.7), and (A.8),

$$\mu_X = \text{tr}(\mathbf{H\Sigma}) + \boldsymbol{\mu}'\mathbf{H}\boldsymbol{\mu}, \quad \mathbb{V}(X) = 2\,\text{tr}(\mathbf{H\Sigma H\Sigma}) + 4\boldsymbol{\mu}'\mathbf{H\Sigma H}\boldsymbol{\mu},$$
$$\mu_Y = \text{tr}(\mathbf{K\Sigma}) + \boldsymbol{\mu}'\mathbf{K}\boldsymbol{\mu}, \quad \mathbb{V}(Y) = 2\,\text{tr}(\mathbf{K\Sigma K\Sigma}) + 4\boldsymbol{\mu}'\mathbf{K\Sigma K}\boldsymbol{\mu},$$

and $\text{Cov}(X, Y) = 2\,\text{tr}(\mathbf{H\Sigma K\Sigma}) + 4\boldsymbol{\mu}'\mathbf{H\Sigma K}\boldsymbol{\mu}$.

It is difficult to state when these expressions will be accurate, as they depend on many variables, though we can be confident that (B.60) will usually be more accurate than (B.61). Often, R will be a statistic associated with a particular model, and as the sample size increases, so will the accuracy of (B.60) and (B.61), all other things being equal. This is best demonstrated with an example.

Example B.7 ***Example B.6 cont.***
We continue to examine the behavior of the Durbin–Watson statistic when the model is mis-specified and there are regressors missing from the observation equation (B.13). As in Example B.6, the true regression model is $Y_t = \beta_1 + \beta_2 t + \beta_3 v_t + \epsilon_t$, $t = 1, \dots, T$, along with an autoregressive error term $\epsilon_t = a\epsilon_{t-1} + U_t$, $U_t \overset{\text{i.i.d.}}{\sim} \text{N}(0, \sigma^2)$.

We compute the mean and variance of D as a function of the autoregressive parameter a when the model is under-specified by omitting the regressor $\mathbf{v}$, so that the error term has mean $\boldsymbol{\mu} = \beta_3\mathbf{v}$ (and covariance matrix $\boldsymbol{\Sigma}$ from the AR model). This is done for $T = 20$ and $\beta_3 = 6$, and shown in Figure B.7. The solid line shows the exact values (computed using the method described below), while the dashed lines were computed with (B.60) and (B.61). We see that the mean is approximated very well with (B.60), while (B.61) breaks down as $a \to -1$.

For comparison, the mean and variance of D when $\beta_3 = 0$ are also plotted as dash-dot lines. The inscribed arrows show that, if the true autoregressive parameter a is near -0.74, then the expected value of D computed under the mis-specified model will be near the value that one expects under the null hypothesis of no autocorrelation! This sinister fact should be kept in mind when confronted with the results of an econometric regression analysis that claims non-significance of the Durbin–Watson

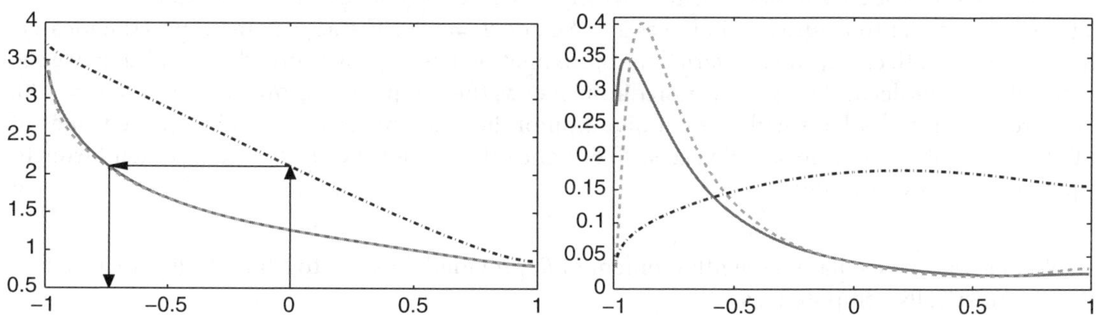

Figure B.7 The mean (left) and variance (right) (not multiplied by T), of the Durbin–Watson statistic, as a function of the autoregressive parameter a. The true model is $Y_t = \beta_1 + \beta_2 t + \beta_3 v_t + \epsilon_t$, $t = 1, \dots, T$, $\epsilon_t = a\epsilon_{t-1} + U_t$, $U_t \overset{\text{i.i.d.}}{\sim} \text{N}(0, \sigma^2)$, with $\beta_3 = 6$, $T = 20$, and vector $\mathbf{v} = (v_1, \dots, v_T)'$ is the same as used in Example B.6, but the regression model is mis-specified as $Y_t = \beta_1 + \beta_2 t + \epsilon_t$. The solid lines are the exact values; the dashed lines were computed from (B.60) and (B.61). The dash-dot lines show the exact mean and variance when $\beta_3 = 0$ (so that the model would not be under-specified). The arrows in the left plot indicate how to determine that value of a such that the mean of D would be precisely the same value if there were no autocorrelation ($a = 0$) and if the model were not mis-specified; it is $a = -0.74$.

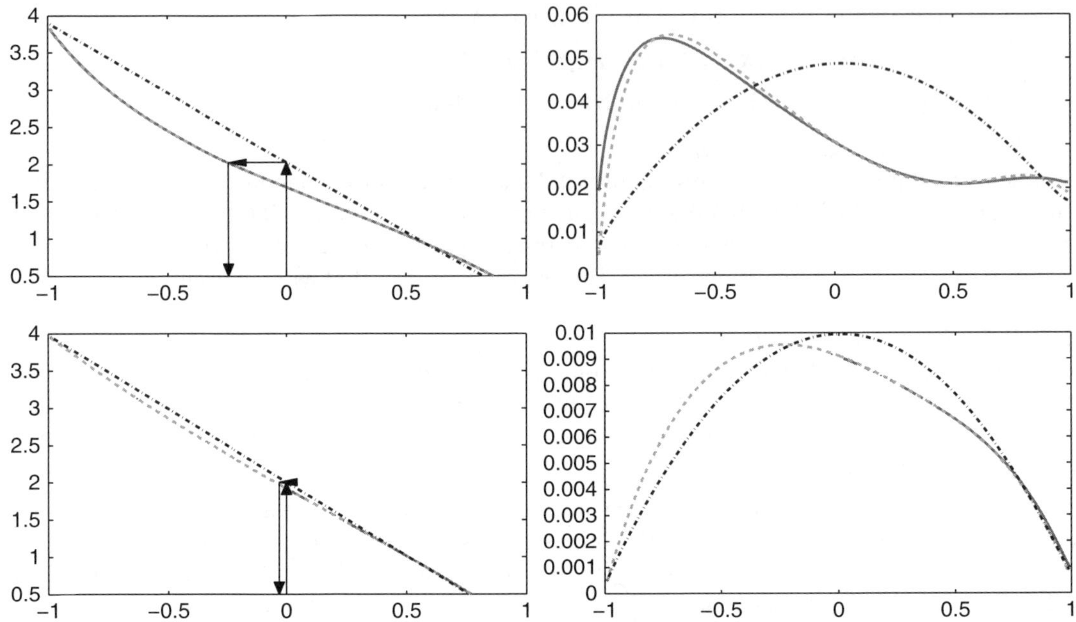

Figure B.8 Same as Figure B.7 but using $T = 80$ (top) and $T = 400$ (bottom).

statistic. It would render the parameter estimates biased, jeopardizing conclusions drawn from them (and possibly answering the embarrassing question as to why the coefficients sometimes "have the wrong sign"). There is some good news: The under-specified model can potentially still be used to produce forecasts: they will obviously not be as good as ones produced with a correctly specified model, and their confidence intervals will be wrong, because they depend on the $\mathbf{X}$ matrix.

Figure B.8 is similar to Figure B.7, but uses $T = 80$ and $T = 400$. The approximation to the mean is virtually exact in both cases, and the variance approximation has improved greatly. In addition, we see that the effect of model under-specification diminishes as the sample size grows. Finally, for $T = 400$, there were numeric problems with the computation of the exact moments for most of the values of $a < 0$. Conveniently, for large sample sizes, (B.60) and (B.61) are very accurate and much faster to compute than the exact values. ■

An obvious way of calculating the nth moment of R, provided it exists, to a high degree of accuracy is just to numerically compute it as

$$\mathbb{E}[R^n] = \int r^n f_R(r)\,\mathrm{d}r, \tag{B.62}$$

using the exact or saddlepoint methods for the p.d.f. Alternatively, one can use the c.d.f. of R via the expression

$$\mathbb{E}[R^n] = \int_0^\infty n r^{n-1}(1 - F_R(r))\,\mathrm{d}r - \int_{-\infty}^0 n r^{n-1} F_R(r)\,\mathrm{d}r, \tag{B.63}$$

which was derived in Problem I.7.13. As an example with the Durbin–Watson test, it is clear from (B.16) that $D \geqslant 0$, so that (B.63) simplifies to

$$\mathbb{E}[D^n] = \int_0^\infty n r^{n-1}(1 - F_D(r))\,\mathrm{d}r.$$

This can be further refined by recalling from (B.18) that D can be expressed as $\mathbf{Z}'\tilde{\mathbf{A}}\mathbf{Z}/\mathbf{Z}'\mathbf{Z}$, where the nonzero eigenvalues of $\tilde{\mathbf{A}}$ are the same as those of $\mathbf{MA}$, so that (A.34) implies $0 \leqslant d_{\min} < D < d_{\max}$, where $d_{\min}$ and $d_{\max}$ are the minimum and maximum eigenvalues of $\mathbf{MA}$. Thus,

$$\begin{aligned}\mathbb{E}[D^n] &= \int_{d_{\min}}^{d_{\max}} n r^{n-1}(1 - F_D(r))\,\mathrm{d}r = \int_{d_{\min}}^{d_{\max}} n r^{n-1}\,\mathrm{d}r - \int_{d_{\min}}^{d_{\max}} n r^{n-1} F_D(r)\,\mathrm{d}r \\ &= d_{\max}^n - d_{\min}^n - n \int_{d_{\min}}^{d_{\max}} r^{n-1} F_D(r)\,\mathrm{d}r. \end{aligned} \tag{B.64}$$

Because $F_D(r) = 0$ for $r < d_{\min}$, it is easy to verify that $d_{\min}$ could be replaced by a non-negative number less than $d_{\min}$, in particular, zero, the lower bound of D. Similarly, as $F_D(r) = 1$ for $r > d_{\max}$, the value $d_{\max}$ could be replaced by any number larger than it, such as 4, which is the upper bound of D: This follows from (A.34) and (B.16), with $D = \hat{\epsilon}'\mathbf{A}\hat{\epsilon}/\hat{\epsilon}'\hat{\epsilon}$, and (B.11). Thus, we can also write the pleasantly simple looking

$$E[D^n] = 4^n - n \int_0^4 r^{n-1} F_D(r)\,\mathrm{d}r.$$

As an aside, another way to see that $0 < D < 4$ is by comparing it to the first sample autocorrelation, R_1, given in (B.21). Because it is a sample correlation, $|R_1| < 1$, and the reader can check that

$$D = 2(1 - R_1) - \frac{\hat{\epsilon}_1^2 + \hat{\epsilon}_T^2}{\sum_{t=1}^T \hat{\epsilon}_t^2},$$

which implies that $0 < D < 4$.

The program in Listing B.7 computes (B.64) for a given $\mathbf{X}$ matrix, choice of n, and values for $\boldsymbol{\mu}$ and $\boldsymbol{\Sigma}$. The parameter `method` allows the choice between using the exact c.d.f. or the s.p.a. The program

```
function m=dwmoment(n,X,method,Sigma,mu)
[T,k]=size(X); M=makeM(X);
if nargin<3, method=1; end
if nargin<4, Sigma=eye(T); end
if nargin<5, mu=zeros(T,1); end
A=M*makeDW(T)*M; B=M; ee=eig(A); dmin=min(ee); dmax=max(ee);
m = dmax^n - dmin^n - n*quadgk(@(rvec)fun(rvec,n,A,B,Sigma,mu,method),dmin,dmax);
m=real(m); % remove possible roundoff error

function I=fun(rvec,n,A,B,Sigma,mu,method)
I = rvec.^(n-1) .* cdfratio(rvec,A,B,Sigma,mu,method)';
```

Program Listing B.7: Computes the nth moment of the Durbin–Watson statistic for regression model $\mathbf{Y} = \mathbf{X}\boldsymbol{\beta} + \boldsymbol{\epsilon}$, and $\boldsymbol{\epsilon} \sim \mathrm{N}_T(\boldsymbol{\mu}, \boldsymbol{\Sigma})$. Matrix $\boldsymbol{\Sigma}$ corresponds in our applications in this chapter to the AR(1) model, and is given by (B.39). Pass `method` as 1 (default) to use exact c.d.f. calculation (numeric inversion of the c.f.) or pass 2 to use the s.p.a.

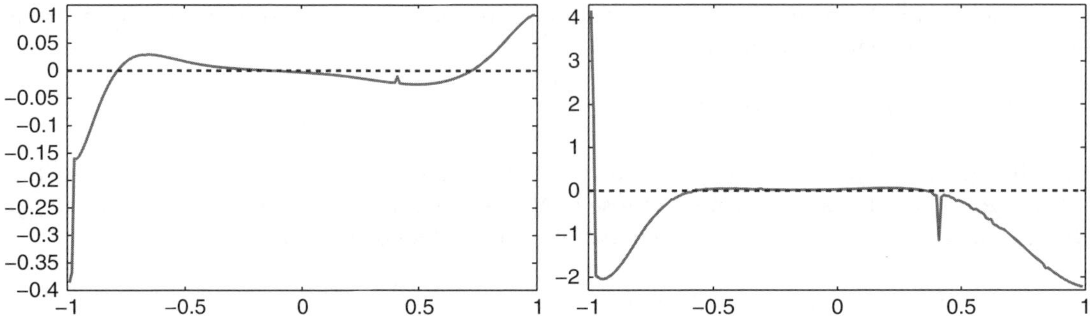

Figure B.9 The relative percentage error 100(Approx − Exact)/Exact, as a function of parameter a, based on the s.p.a. for the mean (left) and variance (right) of the Durbin–Watson statistic using the same parameter values as those in Figure B.7.

is quite short, because all the real work is done by calling function `cdfratio`, given in Listing A.3, which itself calls routines for the inversion formula for the c.d.f., or the saddlepoint approximation. As a check on its implementation, it can be used to compute the mean and variance shown in Figure B.6; use of (B.57) and (B.58) takes about 2/3 of the computing time compared to (B.64).

To assess the error introduced via use of the s.p.a., Figure B.9 shows the relative percentage error, defined as r.p.e. = 100(Approx − Exact)/Exact between them, as a function of parameter a. We see that use of the s.p.a. for computing (B.64) for the mean yields less than one tenth of 1% error for almost the entire parameter space of a, though it begins to break down for values of a extremely close to −1. This is the same behavior that the Taylor series approximation exhibited, though on a much smaller scale. For the variance, the percentage error is under a tenth of 1% for $-0.6 < a < 0.4$, and under 2% for $|a| < 0.98$, and begins to break down as $a \to -1$. Again, this behavior is similar to that of the Taylor series approximation, but with a vastly higher accuracy.

Example B.8 Recall the sample autocorrelation R_s from (B.21), with $|R_s| \leqslant 1$.[5] Thus, from (B.63), the mean of R_s can be expressed as

$$\mathbb{E}[R_s] = \int_0^1 (1 - F_{R_s}(r))\,\mathrm{d}r - \int_{-1}^0 F_{R_s}(r)\,\mathrm{d}r = 1 - \int_{-1}^1 F_{R_s}(r)\,\mathrm{d}r,$$

and, in general,

$$\mathbb{E}[R_s^n] = 1 - n \int_{-1}^1 r^{n-1} F_{R_s}(r)\,\mathrm{d}r. \tag{B.65}$$

This is readily computed using an obvious modification of the code in Listing B.7. ■

---

5 The support of R_s could be even smaller, however, depending on the eigenvalues of $\mathbf{A}_s$ via relation (A.34). With $T = 10$ and $s = 1$, the eigenvalues range between ± 0.9595. If the mean of the data is subtracted before computing R_1 (corresponding to regression residuals with an $\mathbf{X}$ matrix equal to a column of ones), then, along the lines of (B.18), the support of R_1 is bound by the smallest and largest (nonzero) eigenvalues of $\mathbf{MA}_1$, which, for $T = 10$, are -0.9595 and 0.8413. From (B.24), the mean in this case is $-1/10$.

B.5 Useful Matrix Algebra Results

The following small collection of matrix results are standard and can be found in numerous books dedicated to the subject, e.g., Searle (1982), Graybill (1983), Horn (1994), Harville (1997), Schott (2005), Abadir and Magnus (2005), Magnus and Neudecker (2007), Gentle (2007), and Khuri and Searle (2017). More succinct accounts emphasizing the required results for studying linear models (with proofs) can be found in Ravishanker and Dey (2002, Ch. 1–3), and Seber and Lee (2003, App. A), both of which also serve as excellent compliments to this book for augmenting Chapters 1, 2, and 3.

If **A** and **B** are two conformable matrices, then

$$\text{rank}(\mathbf{AB}) \leqslant \min(\text{rank}(\mathbf{A}), \text{rank}(\mathbf{B})). \tag{B.66}$$

If **A** and **B** are two matrices of the same size, then

$$\text{rank}(\mathbf{A}+\mathbf{B}) \leqslant \text{rank}(\mathbf{A}) + \text{rank}(\mathbf{B}). \tag{B.67}$$

If **A** and **B** are $n \times n$ and $n \times k$ matrices, respectively, $k \geqslant 1$, then

$$\text{rank}(\mathbf{AB}) \geqslant \text{rank}(\mathbf{A}) + \text{rank}(\mathbf{B}) - n. \tag{B.68}$$

If **A**, **B**, and **C** are three matrices such that the following products are defined, then

$$\text{rank}(\mathbf{AB}) + \text{rank}(\mathbf{BC}) \leqslant \text{rank}(\mathbf{B}) + \text{rank}(\mathbf{ABC}). \tag{B.69}$$

Note that if $\mathbf{B} = \mathbf{I}$, (B.69) reduces to (B.68).

If **A** is a square matrix whose diagonal elements are differentiable functions of x, then

$$\frac{\partial\,\text{tr}(\mathbf{A})}{\partial x} = \text{tr}\left(\frac{\partial \mathbf{A}}{\partial x}\right). \tag{B.70}$$

If **A** and **B** are matrices whose elements are differentiable functions of x and such that the product **AB** is defined, then

$$\frac{\partial \mathbf{AB}}{\partial x} = \frac{\partial \mathbf{A}}{\partial x}\mathbf{B} + \mathbf{A}\frac{\partial \mathbf{B}}{\partial x}. \tag{B.71}$$

If **A** is a symmetric matrix whose elements are differentiable functions of x, then

$$\frac{\partial |\mathbf{A}(x)|}{\partial x} = |\mathbf{A}|\,\text{tr}\left(\mathbf{A}^{-1}\frac{\partial \mathbf{A}}{\partial x}\right), \tag{B.72}$$

(see Searle, 1982, p. 337, eq. (38) and (39)) and

$$\frac{\partial \mathbf{A}^{-1}(x)}{\partial x} = -\mathbf{A}^{-1}\left(\frac{\partial \mathbf{A}}{\partial x}\right)\mathbf{A}^{-1}, \tag{B.73}$$

(see Searle, 1982, p. 335, eq. (22)).

Theorem B.2 If **A** and **B** are two $T \times T$ symmetric matrices, then a necessary and sufficient condition for an orthogonal matrix **P** to exist such that $\mathbf{P}'\mathbf{AP} = \mathbf{D}_{\text{A}}$ and $\mathbf{P}'\mathbf{BP} = \mathbf{D}_{\text{B}}$ where both $\mathbf{D}_{\text{A}}$ and $\mathbf{D}_{\text{B}}$ are diagonal and the elements of $\mathbf{D}_{\text{A}}$ are the eigenvalues of **A**, is that **A** and **B** commute, i.e., $\mathbf{AB} = \mathbf{BA}$.

Proof:

($\Leftarrow$) Assume $\mathbf{AB} = \mathbf{BA}$. Following Searle (1982, p. 312) and Graybill (1983, p. 406), we take $\mathbf{R}$ to be the orthogonal matrix, and $\mathbf{D}_\mathrm{A}$ to be the diagonal matrix, such that

$$\mathbf{R}'\mathbf{A}\mathbf{R} = \mathbf{D}_\mathrm{A} = \mathrm{diag}(\lambda_i \mathbf{I}_{m_i}) = \begin{bmatrix} \lambda_1 \mathbf{I}_{m_1} & \mathbf{0} & \cdots & \mathbf{0} \\ \mathbf{0} & \lambda_2 \mathbf{I}_{m_2} & \cdots & \mathbf{0} \\ \vdots & \vdots & \ddots & \vdots \\ \mathbf{0} & \mathbf{0} & \cdots & \lambda_s \mathbf{I}_{m_s} \end{bmatrix},$$

where λ_i is one of the s distinct eigenvalues of $\mathbf{A}$ of multiplicity m_i, and $\sum_{i=1}^s m_i = T$. Define

$$\mathbf{C} = \mathbf{R}'\mathbf{B}\mathbf{R} = (\mathbf{C}_{i,j})_{i,j} = \begin{bmatrix} \mathbf{C}_{11} & \mathbf{C}_{12} & \cdots & \mathbf{C}_{1s} \\ \mathbf{C}_{21} & \mathbf{C}_{22} & \cdots & \\ \vdots & \vdots & \ddots & \vdots \\ \mathbf{C}_{s1} & & \cdots & \mathbf{C}_{ss} \end{bmatrix}, \tag{B.74}$$

where the partition of $\mathbf{C}$ is the same as that of $\mathbf{D}_\mathrm{A}$. As $\mathbf{R}$ is full rank and orthogonal, and $\mathbf{AB} = \mathbf{BA}$,

$$\mathbf{D}_\mathrm{A}\mathbf{C} = \mathbf{R}'\mathbf{A}\mathbf{R}\mathbf{R}'\mathbf{B}\mathbf{R} = \mathbf{R}'\mathbf{A}\mathbf{B}\mathbf{R} = \mathbf{R}'\mathbf{B}\mathbf{A}\mathbf{R} = \mathbf{R}'\mathbf{B}\mathbf{R}\mathbf{R}'\mathbf{A}\mathbf{R} = \mathbf{C}\mathbf{D}_\mathrm{A}.$$

Equating $\mathbf{C}\mathbf{D}_\mathrm{A}$ and $\mathbf{D}_\mathrm{A}\mathbf{C}$ easily shows that $\mathbf{C}_{i,j}\lambda_j = \lambda_i \mathbf{C}_{i,j}\ \forall\ (i,j)$. However, as $\lambda_i \neq \lambda_j$ for $i \neq j$, $\mathbf{C}_{i,j} = \mathbf{0}$ for $i \neq j$. This shows that $\mathbf{C}$ is block diagonal, $\mathrm{diag}(\mathbf{C}_{i,i})$, and, as $\mathbf{B}$ is symmetric, so is $\mathbf{C}$, implying that the $\mathbf{C}_{i,i}$ are also symmetric. Thus, there exists orthogonal matrices $\mathbf{Q}_i$ and diagonal matrices $\mathbf{D}_i$, $i = 1, \ldots, s$, such that $\mathbf{Q}_i'\mathbf{C}_{i,i}\mathbf{Q}_i = \mathbf{D}_i$.

Define $\mathbf{Q} := \mathrm{diag}(\mathbf{Q}_{i,i})$ and $\mathbf{D}_\mathrm{B} := \mathrm{diag}(\mathbf{D}_i)$, so that $\mathbf{Q}'\mathbf{C}\mathbf{Q} = \mathbf{D}_\mathrm{B}$, or

$$\mathbf{Q}'\mathbf{R}'\mathbf{B}\mathbf{R}\mathbf{Q} = \mathbf{D}_\mathrm{B},$$

with $\mathbf{Q}$ orthogonal and block diagonal. Observe that the block structures of $\mathbf{Q}$ and $\mathbf{D}_\mathrm{A}$ are the same, and as $\mathbf{D}_\mathrm{A} = \mathrm{diag}(\lambda_i \mathbf{I}_{m_i})$, $\mathbf{Q}$ and $\mathbf{D}_\mathrm{A}$ commute, implying $\mathbf{Q}'\mathbf{D}_\mathrm{A}\mathbf{Q} = \mathbf{D}_\mathrm{A}$, or

$$\mathbf{Q}'\mathbf{R}'\mathbf{A}\mathbf{R}\mathbf{Q} = \mathbf{D}_\mathrm{A}.$$

Let $\mathbf{P} = \mathbf{R}\mathbf{Q}$. As both $\mathbf{R}$ and $\mathbf{Q}$ are orthogonal, multiplying shows that $\mathbf{P}\mathbf{P}' = \mathbf{P}'\mathbf{P} = \mathbf{I}$, i.e., $\mathbf{P}$ is an orthogonal matrix diagonalizing both $\mathbf{A}$ and $\mathbf{B}$.

($\Rightarrow$) To prove that $\mathbf{AB} = \mathbf{BA}$ is necessary, assume an orthogonal matrix $\mathbf{P}$ exists such that $\mathbf{P}'\mathbf{A}\mathbf{P} = \mathbf{D}_\mathrm{A}$ and $\mathbf{P}'\mathbf{B}\mathbf{P} = \mathbf{D}_\mathrm{B}$ with $\mathbf{D}_\mathrm{A}$ and $\mathbf{D}_\mathrm{B}$ diagonal. As $\mathbf{D}_\mathrm{A}\mathbf{D}_\mathrm{B}$ is diagonal (and thus equal to its transpose), $\mathbf{D}_\mathrm{A}\mathbf{D}_\mathrm{B} = \mathbf{D}_\mathrm{B}\mathbf{D}_\mathrm{A}$, so that

$$\mathbf{AB} = \mathbf{P}\mathbf{D}_\mathrm{A}\mathbf{P}'\mathbf{P}\mathbf{D}_\mathrm{B}\mathbf{P}' = \mathbf{P}\mathbf{D}_\mathrm{A}\mathbf{D}_\mathrm{B}\mathbf{P}' = \mathbf{P}\mathbf{D}_\mathrm{B}\mathbf{D}_\mathrm{A}\mathbf{P}' = \mathbf{P}\mathbf{D}_\mathrm{B}\mathbf{P}'\mathbf{P}\mathbf{D}_\mathrm{A}\mathbf{P}' = \mathbf{BA},$$

proving the theorem. ■

The following extension is used in Example B.2 and is also fundamental to the results of Section B.3.

Theorem B.3 Let $\mathbf{A}, \mathbf{B} \in \mathbb{R}^{T \times T}$ be symmetric matrices that commute. If $\mathbf{B}$ is also idempotent and of rank $r \leqslant T$, then $\mathbf{D}_\mathrm{B}$ is in fact the diagonal matrix of eigenvalues of $\mathbf{B}$, with r ones and $T - r$ zeros. If in addition

$$\mathrm{rank}(\mathbf{B}) = \mathrm{rank}(\mathbf{A}) = \mathrm{rank}(\mathbf{BA}) = r,$$

then $\mathbf{D}_\mathrm{B}$ has its ones in the same i, i position in $\mathbf{D}_\mathrm{B}$ as the nonzero λ_i in $\mathbf{D}_\mathrm{A}$.

Proof: Using the same notation as in the proof of Theorem B.2, for $\mathbf{D}_B$ to have the required structure, each $\mathbf{D}_i$ must consist of nothing but zeros and ones on its diagonal. This holds iff $\mathbf{C}_{i,i}$ is both symmetric and idempotent $\forall i$, or, recalling that $\mathbf{C} = \text{diag}(\mathbf{C}_{i,i})$, iff $\mathbf{C}$ is symmetric and idempotent. From (B.74) and the fact that $\mathbf{R}$ is orthogonal, this is equivalent to $\mathbf{B}$ being symmetric and idempotent, i.e., a projection matrix. As $\mathbf{P}$ is orthogonal, $\text{rank}(\mathbf{D}_B) = \text{rank}(\mathbf{B}) = r$, so that $\mathbf{D}_B$ consists of exactly r ones and $T - r$ zeros. For the second statement in the theorem, observe that $\text{rank}(\mathbf{BA}) = \text{rank}(\mathbf{PD}_B\mathbf{D}_A\mathbf{P}')$. As $\mathbf{PD}_B\mathbf{D}_A\mathbf{P}'$ is symmetric, its rank is equal to the number of its nonzero eigenvalues, which is equal to the number of nonzero eigenvalues of $\mathbf{D}_B\mathbf{D}_A$, which is a diagonal matrix with rank r only when $\mathbf{D}_B$ has its ones in the same i,i position in $\mathbf{D}_B$ as the nonzero λ_i in $\mathbf{D}_A$. ■

We now turn to the Poincaré separation theorem, as needed to establish a result given in (5.23). The following presentation is based on Magnus and Neudecker (2007, p. 232–237). We start with a theorem and two lemmas needed to prove the theorem.

Theorem B.4 If $\mathbf{A}$ is an $m \times n$ matrix and $\mathbf{B}$ is an $n \times m$ matrix, $n \geqslant m$, then the nonzero eigenvalues of $\mathbf{AB}$ are the same as those of $\mathbf{BA}$, and the latter has an additional $n - m$ zero eigenvalues.

Proof: See, e.g., Magnus and Neudecker (2007, p. 16) or Rao et al. (2008, p. 498). ■

Lemma B.1 Let $\mathbf{A}$ be a real, symmetric, $T \times T$ matrix with ordered eigenvalues $\lambda_1 \leqslant \lambda_2 \leqslant \ldots \leqslant \lambda_T$, and $\mathbf{S} = (\mathbf{s}_1, \mathbf{s}_2, \ldots, \mathbf{s}_T)$ orthogonal such that $\mathbf{S}'\mathbf{AS} = \text{diag}(\lambda_1, \lambda_2, \ldots, \lambda_T)$. Let $\mathbf{x}$ be any $T \times 1$ real vector, and let $\mathbf{R}_k := (\mathbf{s}_1, \mathbf{s}_2, \ldots, \mathbf{s}_k)$, and $\mathbf{T}_k := (\mathbf{s}_k, \mathbf{s}_{k+1}, \ldots, \mathbf{s}_T)$. Then

$$\lambda_k = \min_{\mathbf{x}:\mathbf{R}'_{k-1}\mathbf{x}=\mathbf{0}} \frac{\mathbf{x}'\mathbf{Ax}}{\mathbf{x}'\mathbf{x}}, \quad k = 2, \ldots, T, \quad \lambda_k = \max_{\mathbf{x}:\mathbf{T}'_{k+1}\mathbf{x}=\mathbf{0}} \frac{\mathbf{x}'\mathbf{Ax}}{\mathbf{x}'\mathbf{x}}, \quad k = 1, \ldots, T-1. \tag{B.75}$$

Proof: Let $\mathbf{y} = \mathbf{S}'\mathbf{x}$. Partition $\mathbf{S}$ and $\mathbf{y}$ as $\mathbf{S} = (\mathbf{R}_{k-1}, \mathbf{T}_k)$ and $\mathbf{y} = \begin{pmatrix} \mathbf{y}_1 \\ \mathbf{y}_2 \end{pmatrix}$, respectively, so that $\mathbf{x} = \mathbf{Sy} = \mathbf{R}_{k-1}\mathbf{y}_1 + \mathbf{T}_k\mathbf{y}_2$. As $\begin{pmatrix} \mathbf{y}_1 \\ \mathbf{y}_2 \end{pmatrix} = \begin{pmatrix} \mathbf{R}'_{k-1} \\ \mathbf{T}'_k \end{pmatrix}\mathbf{x}$,

$$\mathbf{R}'_{k-1}\mathbf{x} = \mathbf{0} \Leftrightarrow \mathbf{y}_1 = \mathbf{0} \Leftrightarrow \mathbf{x} = \mathbf{T}_k\,\mathbf{y}_2.$$

As $\mathbf{S}$ is an orthogonal matrix that diagonalizes $\mathbf{A}$,

$$\mathbf{s}'_i\mathbf{s}_j = \begin{cases} 1, & \text{if } i = j, \\ 0, & \text{if } i \neq j, \end{cases} \qquad \mathbf{s}'_i\mathbf{A}\mathbf{s}_j = \begin{cases} \lambda_j, & \text{if } i = j, \\ 0, & \text{if } i \neq j, \end{cases}$$

and, in particular, $\mathbf{T}'_k\mathbf{A}\mathbf{T}'_k = \text{diag}([\lambda_k, \lambda_{k+1}, \ldots, \lambda_T])$. This gives

$$\min_{\mathbf{R}'_{k-1}\mathbf{x}=\mathbf{0}} \frac{\mathbf{x}'\mathbf{Ax}}{\mathbf{x}'\mathbf{x}} = \min_{\mathbf{x}=\mathbf{T}_k\mathbf{y}_2} \frac{\mathbf{x}'\mathbf{Ax}}{\mathbf{x}'\mathbf{x}} = \min \frac{\mathbf{y}'_2(\mathbf{T}'_k\mathbf{A}\mathbf{T}'_k)\mathbf{y}_2}{\mathbf{y}'_2\mathbf{y}_2} = \lambda_k,$$

where the last equality follows from Theorem A.4. The second equality in (B.75) is proved similarly, partitioning $\mathbf{S}$ as $(\mathbf{R}_k, \mathbf{T}_{k+1})$. ■

Lemma B.2 Let $\mathbf{A}$ be defined as in Lemma B.1 above. For every $T \times (k-1)$ matrix $\mathbf{B}$, and every $T \times (T-k)$ matrix $\mathbf{C}$,

$$\min_{\mathbf{x}:\mathbf{B}'\mathbf{x}=\mathbf{0}} \frac{\mathbf{x}'\mathbf{Ax}}{\mathbf{x}'\mathbf{x}} \leqslant \lambda_k \leqslant \max_{\mathbf{x}:\mathbf{C}'\mathbf{x}=\mathbf{0}} \frac{\mathbf{x}'\mathbf{Ax}}{\mathbf{x}'\mathbf{x}}, \qquad 1 \leqslant k \leqslant T,$$

with only the first (second) equality applying when $k = T$ $(k = 1)$.

Proof: Let $\mathbf{R} := \mathbf{R}_k$ and $\mathbf{S}$ be defined as in Lemma B.1, so that $\mathbf{R}'\mathbf{R} = \mathbf{I}_k$, and

$$\mathbf{R}'\mathbf{A}\mathbf{R} = \operatorname{diag}([\lambda_1, \lambda_2, \dots, \lambda_k]).$$

As there can be at most $k-1$ independent columns in the $(k-1) \times k$ matrix $\mathbf{B}'\mathbf{R}$, there exists a $\mathbf{p} \neq \mathbf{0}$ such that

$$\mathbf{B}'\mathbf{R}\mathbf{p} = \mathbf{0}. \tag{B.76}$$

Thus, the restriction $\mathbf{B}'\mathbf{x} = \mathbf{0}$ is equivalent to setting $\mathbf{x} = \mathbf{R}\mathbf{p}$ when $\mathbf{p}$ satisfies (B.76). It follows that

$$\min_{\mathbf{B}'\mathbf{x}=0} \frac{\mathbf{x}'\mathbf{A}\mathbf{x}}{\mathbf{x}'\mathbf{x}} = \min_{\mathbf{x}=\mathbf{R}_\mathbf{p}} \frac{\mathbf{x}'\mathbf{A}\mathbf{x}}{\mathbf{x}'\mathbf{x}} = \min_{\mathbf{p}:\mathbf{B}'\mathbf{R}_\mathbf{p}=0} \frac{\mathbf{p}'(\mathbf{R}'\mathbf{A}\mathbf{R})\mathbf{p}}{\mathbf{p}'\mathbf{p}} \leqslant \max_{\mathbf{p}:\mathbf{B}'\mathbf{R}_\mathbf{p}=0} \frac{\mathbf{p}'(\mathbf{R}'\mathbf{A}\mathbf{R})\mathbf{p}}{\mathbf{p}'\mathbf{p}} = \lambda_k,$$

where the last equality follows from Theorem A.4.

The proof of the second inequality in the lemma follows similarly, using $\mathbf{T} := \mathbf{T}_k$ defined in the previous lemma in place of $\mathbf{R}$. ■

Theorem B.5 ***Poincaré Separation Theorem*** Let $\mathbf{A}$ be defined as in Lemma B.1 above, with eigenvalues $\lambda_1 \leqslant \lambda_2 \leqslant \dots \leqslant \lambda_T$. Let $\mathbf{G}$ be a $T \times k$ matrix, $1 \leqslant k \leqslant T$, such that $\mathbf{G}'\mathbf{G} = \mathbf{I}_k$, and denote the sorted eigenvalues of $\mathbf{G}'\mathbf{A}\mathbf{G}$ by $\mu_1 \leqslant \mu_2 \leqslant \dots \leqslant \mu_k$. Then $\lambda_i \leqslant \mu_i \leqslant \lambda_{n-k+i}$, $i = 1, 2, \dots, k$.

Proof: For $k = 1$, $\mathbf{G}$ is a $T \times 1$ vector. Renaming it $\mathbf{x}$, and noting that $\mu_1 = \mathbf{x}'\mathbf{A}\mathbf{x}$, we see this is a restatement of Theorem A.4. For $k = T$, the theorem states that $\mathbf{A}$ and $\mathbf{G}'\mathbf{A}\mathbf{G}$ have the same eigenvalues, which follows from Theorem B.4.

For $2 \leqslant k \leqslant T-1$, let $\mathbf{S}, \mathbf{R} := \mathbf{R}_{i-1}$ and $\mathbf{T} := \mathbf{T}_{T-j}$ be defined as in Lemma B.1. For $1 \leqslant i \leqslant k$,

$$\lambda_i = \min_{\mathbf{x}:\mathbf{R}'\mathbf{x}=0} \frac{\mathbf{x}'\mathbf{A}\mathbf{x}}{\mathbf{x}'\mathbf{x}} \leqslant \min_{\substack{\mathbf{R}'\mathbf{x}=0\\ \mathbf{x}=\mathbf{G}\mathbf{y}}} \frac{\mathbf{x}'\mathbf{A}\mathbf{x}}{\mathbf{x}'\mathbf{x}} = \min_{\mathbf{y}:\mathbf{R}'\mathbf{G}\mathbf{y}=0} \frac{\mathbf{y}'\mathbf{G}'\mathbf{A}\mathbf{G}\mathbf{y}}{\mathbf{y}'\mathbf{y}} \leqslant \mu_i\,.$$

The first equality follows from Lemma B.1 for $i > 1$, and from Theorem A.4 for $i = 1$, whereby $\mathbf{x}$ is not restricted. The second equality holds because of the added restriction that $\mathbf{x}$ lies in the k-dimensional subspace spanned by the columns of $\mathbf{G}$. The last inequality follows from Lemma B.2, taking $\mathbf{B}$ (which was arbitrary) to be $\mathbf{R}'\mathbf{G}$.

Next, let $T - k + 1 \leqslant j \leqslant T - 1$. By a similar argument,

$$\lambda_j = \max_{\mathbf{x}:\mathbf{T}'\mathbf{x}=0} \frac{\mathbf{x}'\mathbf{A}\mathbf{x}}{\mathbf{x}'\mathbf{x}} \geqslant \max_{\substack{\mathbf{T}'\mathbf{x}=0\\ \mathbf{x}=\mathbf{G}\mathbf{y}}} \frac{\mathbf{x}'\mathbf{A}\mathbf{x}}{\mathbf{x}'\mathbf{x}} = \max_{\mathbf{y}:\mathbf{T}'\mathbf{G}\mathbf{y}=0} \frac{\mathbf{y}'\mathbf{G}'\mathbf{A}\mathbf{G}\mathbf{y}}{\mathbf{y}'\mathbf{y}} \geqslant \mu_{k-T+j}\,.$$

The result follows by taking $j = T - k + i$. ■

Corollary 1 Let $\mathbf{A}$ be defined as in Lemma B.1 above, with eigenvalues $\lambda_1 \leqslant \lambda_2 \leqslant \dots \leqslant \lambda_T$, and let $\mathbf{M}$ be a $T \times T$ projection matrix of rank k, $1 \leqslant k \leqslant T$. Denoting the nonzero ordered eigenvalues of $\mathbf{MAM}$ as $\mu_1 \leqslant \mu_2 \leqslant \dots \leqslant \mu_k$, we have, for $i = 1, 2, \dots, k$,

$$\lambda_i \leqslant \mu_i \leqslant \lambda_{n-k+i}\,.$$

Proof: Theorem 1.3 shows that $\mathbf{M}$ can be expressed as $\mathbf{M} = \mathbf{G}\mathbf{G}'$, where $\mathbf{G}'\mathbf{G} = \mathbf{I}_k$. The result now follows from Theorem B.4, i.e., the positive eigenvalues of $\mathbf{G}\mathbf{G}'\mathbf{A}\mathbf{G}\mathbf{G}'$ are the same as those of $\mathbf{G}'\mathbf{A}\mathbf{G}$. ■

B.6 Saddlepoint Equivalence Result

We prove the result stated after (B.26), this being a special case of the results in Butler and Paolella (1998). In order to do so, we first repeat the relevant saddlepoint formulae from Chapter II.5. The Lugannani and Rice (1980) expression for the s.p.a. to the c.d.f. of continuous r.v. X is given by

$$\hat{F}_X(x) = \Phi(\hat{w}) + \phi(\hat{w}) \left\{ \frac{1}{\hat{w}} - \frac{1}{\hat{u}} \right\}, \qquad x \neq \mathbb{E}[X], \tag{B.77}$$

where Φ and ϕ are the c.d.f. and p.d.f. of the standard normal distribution, respectively, and

$$\hat{w} = \text{sgn}(\hat{s})\sqrt{2\hat{s}x - 2\,\mathbb{K}_X(\hat{s})}, \quad \hat{u} = \hat{s}\sqrt{\mathbb{K}_X''(\hat{s})}. \tag{B.78}$$

We refer to this as the single-saddlepoint.

The Skovgaard (1987) double-saddlepoint result is as follows. Let $\mathbb{K}(s,t)$ denote the joint c.g.f. for continuous random variables X and Y, assumed convergent over $\mathcal{S}$, an open neighborhood of $(0,0)$. The gradient of $\mathbb{K}$ is $\mathbb{K}'(s,t) = (\mathbb{K}_s'(s,t), \mathbb{K}_t'(s,t))'$, where

$$\mathbb{K}_s'(s,t) := \frac{\partial}{\partial s}\mathbb{K}(s,t), \quad \mathbb{K}_t'(s,t) := \frac{\partial}{\partial t}\mathbb{K}(s,t),$$

and

$$\mathbb{K}''(s,t) := \begin{bmatrix} \mathbb{K}_{ss}''(s,t) & \mathbb{K}_{st}''(s,t) \\ \mathbb{K}_{ts}''(s,t) & \mathbb{K}_{tt}''(s,t) \end{bmatrix}, \quad \mathbb{K}_{ss}''(s,t) = \frac{\partial^2}{\partial s^2}\mathbb{K}(s,t), \quad \text{etc.}, \tag{B.79}$$

is the Hessian.

Let $\mathcal{X}$ be the interior of the convex hull of the joint support of (X,Y). Skovgaard (1987) derived a double-saddlepoint approximation for the conditional c.d.f. of X at x given $Y = y$, for $(x,y) \in \mathcal{X}$. In this case, the gradient is a one-to-one mapping from the convergence strip $\mathcal{S}$ onto $\mathcal{X}$. There are two saddlepoints to compute when using this approximation; the first is the unique pre-image of (x,y) in $\mathcal{S}$, denoted $(\tilde{s},\tilde{t})$, computed as the solutions to

$$\mathbb{K}_s'(\tilde{s},\tilde{t}) = x, \qquad \mathbb{K}_t'(\tilde{s},\tilde{t}) = y. \tag{B.80}$$

This is the numerator saddlepoint in the approximation. The second is found by fixing $s = 0$ and solving $\mathbb{K}_t'(0,\tilde{t}_0) = y$ for the unique value of $\tilde{t}_0$ in $\{t : (0,t) \in \mathcal{S}\}$. This is the denominator saddlepoint. The c.d.f. approximation is then given by

$$\Pr(X \leqslant x \mid Y = y) \approx \Phi(\tilde{w}) + \phi(\tilde{w})\{\tilde{w}^{-1} - \tilde{u}^{-1}\}, \qquad \tilde{s} \neq 0, \tag{B.81}$$

where

$$\tilde{w} = \text{sgn}(\tilde{s})\sqrt{2}\sqrt{\tilde{s}x + \tilde{t}y - \mathbb{K}(\tilde{s},\tilde{t}) - \tilde{t}_0 y + \mathbb{K}(0,\tilde{t}_0)}, \tag{B.82}$$

$$\tilde{u} = \tilde{s}\sqrt{|\mathbb{K}''(\tilde{s},\tilde{t})| \;/\; \mathbb{K}_{tt}''(0,\tilde{t}_0)}. \tag{B.83}$$

Because the forms of the Lugannani–Rice (B.77) and Skovgaard (B.81) approximations are the same, this amounts to showing that $\hat{w} = \tilde{w}$ and $\hat{u} = \tilde{u}$, where hats and tildes indicate single- and double-saddlepoint quantities, respectively.

In particular, for the (first-order) single-s.p.a., from (A.32) and (A.49), $\hat{F}_R(r) = \hat{F}_S(0)$, where $S = \sum_{i=1}^{n}(\zeta_i - r)\chi_i^2$, with $\chi_i^2 \overset{\text{i.i.d.}}{\sim} \chi^2(1)$, $i = 1, \dots, n$, ζ_i are the eigenvalues of $\mathbf{A}$, and $\min\zeta_i < r < \max\zeta_i$. For

$$\mathbb{K}_S\,(s) = -(1/2)\sum_{i=1}^{n} \ln(1 - 2s(\zeta_i - r)),$$

(B.78) is

$$\hat{w} = \operatorname{sgn}(\hat{s})\sqrt{-2\,\mathbb{K}_S\,(\hat{s})}, \quad \hat{u} = \hat{s}\sqrt{\mathbb{K}_S''(\hat{s})},$$

where saddlepoint $\hat{s}$ is the unique root of

$$\sum_{i=1}^{n} \frac{\zeta_i - r}{1 - 2\hat{s}(\zeta_i - r)} = \mathbb{K}_S'\,(\hat{s}) = 0 \tag{B.84}$$

in

$$\mathcal{S}_0 = \{s : 1 - 2s(\zeta_i - r) > 0, \quad i = 1, \dots, n\}.$$

For the double-s.p.a., from (B.26),

$$\mathbb{K}\,(s, t) = \mathbb{K}_{U,V}\,(s, t) = -(1/2)\sum_{i=1}^{n} \ln(1 - 2(s\zeta_i + t)),$$

and $\tilde{s}, \tilde{t}$ are the unique values in

$$\begin{aligned}\mathcal{S}_1 &= \{(s,t) : t < (1 - 2s\zeta_i)/2, \quad i = 1, \dots, n\}\\ &= \left\{(s,t) : t < \frac{1}{2}\min(1 - 2s\min_i\zeta_i,\, 1 - 2s\max_i\zeta_i)\right\}\end{aligned}$$

that satisfy

$$\sum_{i=1}^{n} \frac{\zeta_i}{1 - 2(\tilde{s}\zeta_i + \tilde{t})} = \mathbb{K}_s'(\tilde{s}, \tilde{t}) = r, \tag{B.85a}$$

$$\sum_{i=1}^{n} \frac{1}{1 - 2(\tilde{s}\zeta_i + \tilde{t})} = \mathbb{K}_t'(\tilde{s}, \tilde{t}) = 1. \tag{B.85b}$$

From (B.82) and (B.83) with $x = r$ and $y = 1$,

$$\tilde{w} = \operatorname{sgn}(\tilde{s})\sqrt{2}\sqrt{\tilde{s}r + \tilde{t} - \mathbb{K}(\tilde{s}, \tilde{t}) - \tilde{t}_0 + \mathbb{K}(0, \tilde{t}_0)}, \quad \tilde{u} = \tilde{s}\sqrt{\frac{|\mathbb{K}''(\tilde{s}, \tilde{t})|}{\mathbb{K}_{tt}''(0, \tilde{t}_0)}},$$

where $\mathbb{K}''(s, t)$ denotes the Hessian (B.79). Observe that $\mathbb{K}_S$ takes one argument, and refers to the c.g.f. for the single-s.p.a., while $\mathbb{K}$ with two arguments, e.g., $\mathbb{K}(s, t)$, refers to the c.g.f. for the double-s.p.a., so there should be no source of confusion.

We first show that $(\tilde{s}, \tilde{t})$ is related to $\hat{s}$ according to

$$\tilde{s} = n\hat{s}, \tag{B.86a}$$

$$\tilde{t} = \frac{1 - n(1 + 2\hat{s}r)}{2}, \tag{B.86b}$$

a relationship ascertained by trial and error. The true value of $(\tilde{s}, \tilde{t})$ is characterized as the unique solution to (B.85) in region $\mathcal{S}_1$. To show that the right side of (B.86) solves these equations, simply substitute for $(\tilde{s}, \tilde{t})$ in terms of $\hat{s}$; each equation then reduces to the single-saddlepoint equation. To illustrate, for (B.85a), this is

$$0 = \sum_{i=1}^{n} \frac{\zeta_i}{1 - 2(\tilde{s}\zeta_i + \tilde{t})} - r = \sum_{i=1}^{n} \frac{\zeta_i}{1 - 2(\tilde{s}\zeta_i + \tilde{t})} - \frac{r}{n} \sum_{i=1}^{n} \frac{1 - 2(\tilde{s}\zeta_i + \tilde{t})}{1 - 2(\tilde{s}\zeta_i + \tilde{t})}$$

$$= \sum_{i=1}^{n} \frac{\zeta_i - \frac{r}{n} + 2\frac{r}{n}(\tilde{s}\zeta_i + \tilde{t})}{1 - 2(\tilde{s}\zeta_i + \tilde{t})} = \sum_{i=1}^{n} \frac{\zeta_i - \frac{r}{n} + 2\frac{r}{n}\left(n\hat{s}\zeta_i + \frac{1-n(1+2\hat{s}r)}{2}\right)}{1 - 2\left(n\hat{s}\zeta_i + \frac{1-n(1+2\hat{s}r)}{2}\right)}$$

$$= \frac{2r\hat{s} + 1}{n} \sum_{i=1}^{n} \frac{\zeta_i - r}{1 - 2\hat{s}(\zeta_i - r)},$$

which is equivalent to (B.84). A similar calculation using (B.85b) yields the same result. As $\hat{s}$ is unique in $\mathcal{S}_0$, the right-hand side of (B.86) indeed solves (B.85). To be the unique root in $\mathcal{S}_1$ and, hence, the true value of $(\tilde{s}, \tilde{t})$, it must lie in $\mathcal{S}_1$. To see that this is true, note that

$$(\tilde{s}, \tilde{t}) \in \mathcal{S}_1 \Leftrightarrow \tilde{t} < \frac{1 - 2\tilde{s}\zeta_i}{2} \, \forall i,$$

and substituting, this is

$$\frac{1 - n(1 + 2\hat{s}r)}{2} < \frac{1 - 2n\hat{s}\zeta_i}{2} \, \forall i \Leftrightarrow 1 + 2\hat{s}(\zeta_i - r) > 0 \, \forall i \Leftrightarrow \hat{s} \in \mathcal{S}_0.$$

The value of $\tilde{t}_0$ for the denominator saddlepoint is the solution to (B.85b) with $\tilde{s}$ fixed to 0, i.e., the solution to $\sum_{i=1}^{n}(1 - 2\tilde{t}_0)^{-1} = 1$, or $\tilde{t}_0 = (1 - n)/2$.

Thus, substituting into the expression for $\tilde{w}$,

$$\tilde{w} = \operatorname{sgn}(\tilde{s})\sqrt{2}\sqrt{\tilde{s}r + \tilde{t} - \mathbb{K}(\tilde{s}, \tilde{t}) - \tilde{t}_0 + \mathbb{K}(0, \tilde{t}_0)}$$

$$= \operatorname{sgn}(n\hat{s})\sqrt{2}\sqrt{\begin{array}{c} n\hat{s}r + \frac{1-n(1+2\hat{s}r)}{2} + \frac{1}{2}\sum_{i=1}^{n} \ln\left(1 - 2\left(n\hat{s}\zeta_i + \frac{1-n(1+2\hat{s}r)}{2}\right)\right) \\ -\frac{1-n}{2} - \frac{1}{2}\sum_{i=1}^{n} \ln\left(1 - 2\left(\frac{1-n}{2}\right)\right) \end{array}}$$

$$= \operatorname{sgn}(\hat{s})\sqrt{2}\sqrt{\begin{array}{c} \frac{1-n}{2} + \frac{1}{2}\sum_{i=1}^{n} \ln(n(1 - 2\hat{s}(\zeta_i - r))) \\ -\frac{1-n}{2} - \frac{n \ln n}{2} \end{array}}$$

$$= \operatorname{sgn}(\hat{s})\sqrt{\sum_{i=1}^{n} \ln(1 - 2\hat{s}(\zeta_i - r))} = \hat{w}.$$

Showing $\hat{u} = \tilde{u}$ is more difficult, with a direct comparison being of no use. Instead, differentiating both sides of equality $\tilde{s} = n\hat{s}$ with respect to r to yields

$$|\mathbb{K}''(\tilde{s}, \tilde{t})| = 2n^{-3}\mathbb{K}_S''(\hat{s}), \tag{B.87}$$

which is shown below. Then, using

$$\mathbb{K}''_{tt}(s,t) = 2\sum_{i=1}^{n}(1-2(s\zeta_i+t))^{-2} \qquad \text{so that} \qquad \mathbb{K}''_{tt}(0,\tilde{t}_0) = \frac{2}{n},$$

(B.87) implies the final result needed for the proof:

$$\tilde{u} = \tilde{s}\sqrt{\frac{|\mathbb{K}''(\tilde{s},\tilde{t})|}{\mathbb{K}''_{tt}(0,\tilde{t}_0)}} = n\hat{s}\sqrt{\frac{2n^{-3}\mathbb{K}''_S(\hat{s})}{2/n}} = \hat{s}\sqrt{\mathbb{K}''_S(\hat{s})} = \hat{u}.$$

For the derivation of (B.87), compute the derivative of the double saddlepoint equations (B.85) with respect to r, using the chain rule (I.A.147), to get

$$\frac{\partial \mathbb{K}'_s(\tilde{s},\tilde{t})}{\partial r} = \mathbb{K}''_{ss}(\tilde{s},\tilde{t})\frac{\partial \tilde{s}}{\partial r} + \mathbb{K}''_{st}(\tilde{s},\tilde{t})\frac{\partial \tilde{t}}{\partial r} = \frac{\partial r}{\partial r} = 1 \quad \text{and}$$
$$\frac{\partial \mathbb{K}'_t(\tilde{s},\tilde{t})}{\partial r} = \mathbb{K}''_{ts}(\tilde{s},\tilde{t})\frac{\partial \tilde{s}}{\partial r} + \mathbb{K}''_{tt}(\tilde{s},\tilde{t})\frac{\partial \tilde{t}}{\partial r} = \frac{\partial 0}{\partial r} = 0,$$

and write this as the system of equations

$$\mathbb{K}''(\tilde{s},\tilde{t})\begin{bmatrix}\partial\tilde{s}/\partial r\\ \partial\tilde{t}/\partial r\end{bmatrix} = \begin{bmatrix}1\\0\end{bmatrix}.$$

Use of Cramer's rule (see, e.g., Trench, 2003, p. 374; or Munkres, 1991, p. 21) then gives

$$\frac{\partial \tilde{s}}{\partial r} = \frac{\begin{vmatrix}1 & \mathbb{K}''_{st}(\tilde{s},\tilde{t})\\ 0 & \mathbb{K}''_{tt}(\tilde{s},\tilde{t})\end{vmatrix}}{\begin{vmatrix}\mathbb{K}''_{ss}(\tilde{s},\tilde{t}) & \mathbb{K}''_{st}(\tilde{s},\tilde{t})\\ \mathbb{K}''_{st}(\tilde{s},\tilde{t}) & \mathbb{K}''_{tt}(\tilde{s},\tilde{t})\end{vmatrix}} = \frac{\mathbb{K}''_{tt}(\tilde{s},\tilde{t})}{|\mathbb{K}''(\tilde{s},\tilde{t})|}, \tag{B.88}$$

and, substituting for $\tilde{s}$ and $\tilde{t}$,

$$\mathbb{K}''_{tt}(\tilde{s},\tilde{t}) = 2\sum_{i=1}^{n}(1-2(\tilde{s}\zeta_i+\tilde{t}))^{-2} = \frac{2}{n^2}\sum_{i=1}^{n}(1-2\hat{s}(\zeta_i-r))^{-2}. \tag{B.89}$$

To determine $\partial\hat{s}/\partial r$, differentiate the single-saddlepoint equation $\mathbb{K}'_S(\hat{s}) = 0$ in (B.84) using the chain rule to get

$$0 = \sum_{i=1}^{n}\frac{\partial \mathbb{K}'_S(\hat{s})}{\partial(\zeta_i - r)}\frac{\partial(\zeta_i-r)}{\partial r} + \frac{\partial \mathbb{K}'_S(\hat{s})}{\partial \hat{s}}\frac{\partial \hat{s}}{\partial r} = -\sum_{i=1}^{n}\frac{1}{(1-2\hat{s}(\zeta_i-r))^2} + \mathbb{K}''_S(\hat{s})\frac{\partial \hat{s}}{\partial r},$$

which, from (B.89), implies

$$\frac{n^2}{2}\frac{\mathbb{K}''_{tt}(\tilde{s},\tilde{t})}{\mathbb{K}''_S(\hat{s})} = \frac{\partial \hat{s}}{\partial r}.$$

But, as $\tilde{s} = n\hat{s}$, this and (B.88) implies

$$\frac{1}{n}\frac{\mathbb{K}''_{tt}(\tilde{s},\tilde{t})}{|\mathbb{K}''(\tilde{s},\tilde{t})|} = \frac{1}{n}\frac{\partial \tilde{s}}{\partial r} = \frac{\partial \hat{s}}{\partial r} = \frac{n^2}{2}\frac{\mathbb{K}''_{tt}(\tilde{s},\tilde{t})}{\mathbb{K}''_S(\hat{s})},$$

and simplifying yields $2n^{-3}\mathbb{K}''_S(\hat{s}) = |\mathbb{K}''(\tilde{s},\tilde{t})|$, which is (B.87).

Appendix C

Some Useful Multivariate Distribution Theory

This appendix serves to collect some useful results that fall in the domain of probability and distribution theory associated with multivariate random variables. We begin in Section C.1 with a simple derivation of the characteristic function of the Student's t distribution, along with that for weighted linear sums of the univariate margins from the multivariate case. Besides being of general interest, this material will be of use in Chapter 12.

Section C.2 provides a rather detailed introduction to the important concept of ellipticity. This class of distributions arises in a wide variety of statistical applications, and, in particular, is highly relevant for quantitative risk management; see, e.g., McNeil et al. (2015). Throughout Part III, we refer to the notion, and show evidence, that the unconditional distribution of daily stock returns tends to be non-elliptic. As such, it is important to understand what ellipticity entails. This is all the more important because *conditional* models for the data generating process of financial asset returns that yield the best performance in terms of (multivariate) density forecasts and portfolio construction often are such that the predictive distribution is indeed *elliptic*; see, in particular, the evidence and discussions in Paolella et al. (2018a,b).

C.1 Student's *t* Characteristic Function

Recall that the moment generating function (m.g.f.) of r.v. X is given by $\mathbb{M}_X(t) = \mathbb{E}[e^{tX}]$. It exists if it is finite on a neighborhood of zero, i.e., if there exists an $h > 0$ such that, $\forall t \in (-h, h)$, $\mathbb{M}_X(t) < \infty$. If the m.g.f. of random variable X exists, then the largest (open) interval $\mathcal{I}$ around zero such that $\mathbb{M}_X(t) < \infty$ for $t \in \mathcal{I}$ is referred to as the convergence strip (of the m.g.f. of X). When it exists, the m.g.f. uniquely determines, or characterizes, the distribution, i.e., for a given m.g.f. there is a unique corresponding c.d.f. (up to sets of measure zero). This fact is useful when the m.g.f. of a random variable is known, but not its p.d.f. or c.d.f. In addition, if the m.g.f. of random variable X exists, then its mean is given by $\mu_X = \mathbb{M}'_X(0)$, and higher-order raw moments $\mu'_j(X)$ can be computed as $\mathbb{M}^{(j)}_X(0)$, $j = 2, 3, \ldots$, where $\mathbb{M}^{(j)}(t)$ denotes the jth derivative with respect to t.

Trivially but usefully, if $\mathbb{M}_Z(t)$ is the m.g.f. of r.v. Z and $X = \mu + \sigma Z$ for $\sigma > 0$, then

$$\mathbb{M}_X(t) = \mathbb{E}[e^{tX}] = \mathbb{E}[e^{t(\mu+\sigma Z)}] = e^{t\mu}\mathbb{M}_Z(t\sigma). \tag{C.1}$$

The m.g.f. of vector $\mathbf{X} = (X_1, \ldots, X_d)'$ is defined to be $\mathbb{M}_\mathbf{X}(\mathbf{t}) = \mathbb{E}[e^{\mathbf{t}'\mathbf{X}}]$, where $\mathbf{t} = (t_1, \ldots, t_d)'$. As in the univariate case, this characterizes the distribution of $\mathbf{X}$ and, thus, all the (univariate and multivariate)

Linear Models and Time-Series Analysis: Regression, ANOVA, ARMA and GARCH, First Edition. Marc S. Paolella.

margins as well. In particular, observe that

$$\mathbb{M}_{\mathbf{X}}((0,\ldots,0,t_j,0,\ldots,0)') = \mathbb{E}[e^{t_jX_j}] = \mathbb{M}_{X_j}(t_j), \quad j=1,\ldots,d, \tag{C.2}$$

so that knowledge of $\mathbb{M}_{\mathbf{X}}$ implies knowledge of $\mathbb{M}_{X_j}$, $j=1,\ldots,d$, similar to knowledge of $f_{\mathbf{X}}$ implies knowledge of all the d univariate marginal p.d.f.s f_{X_j}, but knowing all the f_{X_j} (or all the $\mathbb{M}_{X_j}$) does not convey a full characterization of $f_{\mathbf{X}}$ (or $\mathbb{M}_{\mathbf{X}}$).

For r.v. $\mathbf{Z}=(Z_1,\ldots,Z_d)'$ with m.g.f. $\mathbb{M}_{\mathbf{Z}}$, let $\mathbf{X}=\boldsymbol{\mu}+\boldsymbol{\Sigma}^{1/2}\mathbf{Z}$, for vector $\boldsymbol{\mu}=(\mu_1,\ldots,\mu_d)'\in\mathbb{R}^d$ and $d\times d$ positive definite matrix $\boldsymbol{\Sigma}$, which, in general, we denote with the shorthand $\boldsymbol{\Sigma}>0$, with typical entry denoted σ_{ij} and diagonal elements denoted σ_j^2, $j=1,\ldots,d$. Then, the extension of (C.1) to the multivariate case takes the form

$$\mathbb{M}_{\mathbf{X}}(\mathbf{t}) = e^{\mathbf{t}'\boldsymbol{\mu}}\mathbb{M}_{\mathbf{Z}}(\boldsymbol{\Sigma}^{1/2}\mathbf{t}). \tag{C.3}$$

The characteristic function (c.f.) of r.v. X is given by $\varphi_X(t)=\mathbb{E}[e^{itX}]$ for $t\in\mathbb{R}$, and where i denotes the imaginary unit such that $i^2=-1$. Unlike the m.g.f., it exists for all random variables, though in some cases obtaining an analytic expression can be difficult. The uniqueness theorem states that a distribution is uniquely determined by its c.f., i.e., if random variables X and Y have c.d.f.s F_X and F_Y, and c.f.s φ_X and φ_Y, respectively, and $\varphi_X(t)=\varphi_Y(t)$ for all $t\in\mathbb{R}$, then $F_X=F_Y$ "almost everywhere", meaning, they can differ only on sets of measure zero. This is what is meant when we say "the unique c.f. of r.v. X is φ_X". This is written as

$$\mathbf{X}\overset{d}{=}\mathbf{Y} \Leftrightarrow \varphi_{\mathbf{X}}=\varphi_{\mathbf{Y}}. \tag{C.4}$$

If the m.g.f. $\mathbb{M}_X(t)$ exists, then, under further conditions that are satisfied in most cases used in statistical inference (see the discussion and references in Section II.1.2.4), $\varphi_X(t)$ can be determined by simply evaluating $\mathbb{M}_X(it)$, though observe that the latter is not formally defined, as the m.g.f. is defined as a mapping from the real line.

Similar to the vector m.g.f., the c.f. of random vector $\mathbf{X}=(X_1,\ldots,X_d)'$ is uniquely given by

$$\varphi_{\mathbf{X}}(\mathbf{t}) = \mathbb{E}[e^{i\mathbf{t}'\mathbf{X}}], \quad \text{where} \quad \mathbf{t}=(t_1,\ldots,t_d)'\in\mathbb{R}^d. \tag{C.5}$$

We will also need the modified Bessel function of the third kind. It is given by

$$K_z(x) = \frac{1}{2}\int_0^\infty u^{z-1}\exp\left[-\frac{x}{2}\left(\frac{1}{u}+u\right)\right]\,du, \quad z\in\mathbb{R}, \quad x\in\mathbb{R}_+. \tag{C.6}$$

An introduction is provided in Section II.9.2, where further references can be found. In Matlab, $K_z(x)$ can be computed with the built-in function `besselk(z,x)`.

We now turn to the derivation of the c.f. of a Student's t random variable with v degrees of freedom (denoted, in short, as $t(v)$ or t_v). The proper generalized hyperbolic, or proper GHyp (where proper refers to the parameters being in an open subset of the parameter space, so that the m.g.f. exists), has p.d.f. given by, with $y_x=\sqrt{\delta^2+(x-\mu)^2}$,

$$f_{\mathrm{GHyp}}(x;\lambda,\alpha,\beta,\delta,\mu) = \frac{(\alpha^2-\beta^2)^{\frac{\lambda}{2}}y_x^{\lambda-\frac{1}{2}}}{\sqrt{2\pi}\alpha^{\lambda-\frac{1}{2}}\delta^\lambda K_\lambda(\delta\sqrt{\alpha^2-\beta^2})}K_{\lambda-\frac{1}{2}}(\alpha y_x)e^{\beta(x-\mu)}, \tag{C.7}$$

for $\lambda\in\mathbb{R}$, $\alpha>0$, $\beta\in(-\alpha,\alpha)$, $\delta>0$, $\mu\in\mathbb{R}$, and $K_z(x)$ defined in (C.6). The m.g.f. is

$$\mathbb{M}_X(t) = e^{\mu t}\frac{K_\lambda(\delta\sqrt{\alpha^2-(\beta+t)^2})}{K_\lambda(\delta\sqrt{\alpha^2-\beta^2})\left(\frac{\alpha^2-(\beta+t)^2}{\alpha^2-\beta^2}\right)^{\lambda/2}} = e^{\mu t}\frac{K_\lambda(\sqrt{\chi\psi_t})}{K_\lambda(\sqrt{\chi\psi})(\psi_t/\psi)^{\lambda/2}}, \tag{C.8}$$

where $\psi_t = \alpha^2 - (\beta + t)^2$, $\chi = \delta^2$ and $\psi = \alpha^2 - \beta^2 > 0$. The convergence strip is given by $-\alpha - \beta < t < \alpha - \beta$. See Section II.9.5.2.1 for derivations of (C.7) and (C.8).

For $\lambda < 0$ and, in the limit as $\alpha = |\beta| \to 0$, the proper GHyp converges to a $t(v)$ distribution and $v = \delta^2 = -2\lambda$ (see Section II.9.5.2.3). As the supremum of the maximally existing moment of $X \sim t(v)$ is v, the m.g.f. cannot exist on an open neighborhood of zero, but the c.f. can be obtained by taking the limit of $\varphi_X(t) = \mathbb{M}_X(it)$ in (C.8). Setting location and asymmetry terms μ and β to zero, and fixing $\lambda = -v/2$,

$$\sqrt{\chi\psi} = v^{1/2}\alpha, \qquad \lim_{\alpha\to 0} \sqrt{\chi\psi_{it}} = v^{1/2}|t|,$$

$$\lim_{\alpha\to 0}\left(\frac{\psi_{it}}{\psi}\right)^{\lambda/2} = \lim_{\alpha\to 0}\left(\frac{\alpha^2+t^2}{\alpha^2}\right)^{-v/4} = \lim_{\alpha\to 0}\left(\sqrt{\frac{\alpha^2+t^2}{\alpha^2}}\right)^{-v/2} = |t|^{-v/2}\lim_{\alpha\to 0}\alpha^{v/2}.$$

Then, using the well-known results $K_z(x) = K_{-z}(x)$ and $K_z(x) \simeq \Gamma(z)2^{z-1}x^{-z}$, for $x \downarrow 0$, $z > 0$ (where a derivation of the latter can be found in Song et al., 2014), we have, with $\lambda = -v/2$,

$$\begin{aligned}\varphi_X(t; v) &= \lim_{\alpha\to 0}\frac{K_{v/2}(\sqrt{\chi\psi_{it}})}{K_{v/2}(\sqrt{\chi\psi})(\psi_{it}/\psi)^{\lambda/2}}\\ &= \lim_{\alpha\to 0}\frac{K_{v/2}(v^{1/2}|t|)|t|^{v/2}\alpha^{v/2}}{\Gamma(v/2)2^{v/2-1}v^{-v/4}\alpha^{v/2}} = \frac{K_{v/2}(v^{1/2}|t|)(v^{1/2}|t|)^{v/2}}{\Gamma(v/2)2^{v/2-1}},\end{aligned} \tag{C.9}$$

as was given by Hurst (1995), based on a different method of derivation. Note that $\varphi_X(t; v)$ is real because X is symmetric about zero.

Remark According to Dreier and Kotz (2002), the derivation of (C.9) "has been a topic of some controversy and difficulties in statistical literature for the last 30 years. Several approaches were suggested involving incomplete, complicated and sometimes convoluted proofs." Result (C.9) is stated (without reference or derivation) in the reference work of Kotz and Nadarajah (2004, p. 40), while Platen and Heath (2006, p. 37) and Seneta (2004, p. 186) attribute it to Hurst (1995). Seneta (2004) also remarks that the result was essentially given in a different context and is related "by a simple duality argument" in Madan and Seneta (1990).

Special cases were known before the more elegant general result (C.9). In particular, for odd degrees of freedom,

$$\varphi_X(t; 3) = (1 + |t\sqrt{3}|)\exp(-|t\sqrt{3}|), \tag{C.10}$$

which was used in Example II.4.27, and

$$\varphi_X(t; 5) = \left(1 + |t\sqrt{5}| + \frac{5}{3}t^2\right)\exp(-|t\sqrt{5}|). \tag{C.11}$$

To confirm (C.10), use (C.9) and

$$K_v(z) = \sqrt{\frac{\pi}{2z}}e^{-z} \times E(v, z), \tag{C.12}$$

where

$$E(v, z) = 1 + \frac{4v^2 - 1^2}{1!8z} + \frac{(4v^2-1^2)(4v^2-3^2)}{2!(8z)^2} + \frac{(4v^2-1^2)(4v^2-3^2)(4v^2-5^2)}{3!(8z)^3} + \cdots \tag{C.13}$$

from Watson (1922, p. 202) to write, with $\Gamma(3/2) = \sqrt{\pi}/2$,

$$\varphi_X(t;3) = \frac{3^{3/4}|t|^{3/2}}{2^{3/2-1}\Gamma(3/2)} K_{3/2}\left(|t|\sqrt{3}\right)$$
$$= \frac{3^{3/4}|t|^{3/2}}{2^{3/2-2}\sqrt{\pi}} \sqrt{\frac{\pi}{2|t|\sqrt{3}}} \mathrm{e}^{-(|t|\sqrt{3})} \left(1 + \frac{1}{|t|\sqrt{3}}\right) = \mathrm{e}^{-(|t|\sqrt{3})}\left(1 + |t|\sqrt{3}\right).$$

See Johnson et al. (1995, p. 367) for further special cases, and Dreier and Kotz (2002) for further references on methods of derivation. ■

The c.f. of the multivariate t distribution is similar in form to (C.9), as discussed next. The d-dimensional, zero-location (vector), identity-scale (matrix), multivariate Student's t distribution with $v > 0$ degrees of freedom has density

$$f_{\mathbf{X}}(\mathbf{x}; v) = \frac{\Gamma\left(\frac{v+d}{2}\right)}{\Gamma\left(\frac{v}{2}\right)(v\pi)^{d/2}} \left(1 + \frac{\mathbf{x}'\mathbf{x}}{v}\right)^{-(v+d)/2}, \tag{C.14}$$

for $\mathbf{x} = (x_1, \dots, x_d)'$, and we typically write $\mathbf{X} \sim t_v$ or $\mathbf{X} \sim t_v(\mathbf{0}, \mathbf{I})$, the latter anticipating the location-scale case given below. The c.f. corresponding to (C.14) was first (correctly) given by Sutradhar (1986) (without use of the Bessel function, but with different expressions for when v is odd, even, and fractional), while Song et al. (2014) derive it (and that of a type of generalized multivariate t) by extending the method for the univariate case from Hurst (1995), resulting in a much more compact expression in terms of the Bessel function. With $\mathbf{t} = (t_1, \dots, t_d)' \in \mathbb{R}^d$, it is given by

$$\varphi_{\mathbf{X}}(\mathbf{t}; v) = \frac{K_{v/2}(\|\sqrt{v}\mathbf{t}\|)(\|\sqrt{v}\mathbf{t}\|)^{v/2}}{\Gamma(v/2)2^{v/2-1}}, \qquad \|\mathbf{t}\| = \sqrt{\mathbf{t}'\mathbf{t}}, \tag{C.15}$$

which strongly parallels the univariate case (C.9). See Kotz and Nadarajah (2004, Ch. 2) and the references therein for further discussion.

For vector $\boldsymbol{\mu} = (\mu_1, \dots, \mu_d)' \in \mathbb{R}^d$ and $d \times d$ dispersion matrix $\boldsymbol{\Sigma} > 0$ with typical entry denoted σ_{ij} and diagonal elements denoted σ_j^2, $j = 1, \dots, d$, the location-scale version of (C.14) is given by

$$f_{\mathbf{X}}(\mathbf{x}; \boldsymbol{\mu}, \boldsymbol{\Sigma}, v) = \frac{\Gamma\left(\frac{v+d}{2}\right)}{\Gamma\left(\frac{v}{2}\right)(v\pi)^{d/2}|\boldsymbol{\Sigma}|^{1/2}} \left(1 + \frac{(\mathbf{x}-\boldsymbol{\mu})'\boldsymbol{\Sigma}^{-1}(\mathbf{x}-\boldsymbol{\mu})}{v}\right)^{-(v+d)/2}, \tag{C.16}$$

and we typically write $\mathbf{X} \sim t_v(\boldsymbol{\mu}, \boldsymbol{\Sigma})$. This distribution arises as follows. First recall that r.v. X is said to follow an inverse gamma (IGam) distribution if its p.d.f. is given by

$$f_X(x; \alpha, \beta) = [\beta^\alpha/\Gamma(\alpha)]x^{-(\alpha+1)} \exp\{-\beta/x\}\mathbb{I}_{(0,\infty)}(x), \qquad \alpha > 0, \beta > 0. \tag{C.17}$$

As in Problem I.7.9 (and as the reader should quickly confirm),

$$\mathbb{E}[X^r] = \frac{\Gamma(\alpha - r)}{\Gamma(\alpha)}\beta^r, \qquad \alpha > r, \tag{C.18}$$

so that, for example,

$$\mathbb{E}[X] = \frac{\beta}{\alpha - 1}, \quad \mathbb{V}(X) = \frac{\beta^2}{(\alpha - 1)^2(\alpha - 2)}, \tag{C.19}$$

if $\alpha > 1$ and $\alpha > 2$, respectively. Let $G \sim \text{IGam}(v/2, v/2)$, $v \in \mathbb{R}_{>0}$. Realizations of G can be simulated by use of the following code in Matlab:

```
v=4; T=1000; Y=gamrnd(v/2,1,[T 1]) / (v/2); G=1./Y;
% Or this: chi2=random('chi2',v,T,1); G = 1./(chi2/v);
```

Now let $\mathbf{Z} = (Z_1, Z_2, \ldots, Z_d)' \sim \text{N}_d(\mathbf{0}, \mathbf{\Sigma})$. Then

$$\mathbf{X} = (X_1, X_2, \ldots, X_d)' = \boldsymbol{\mu} + \sqrt{G}\mathbf{Z} \tag{C.20}$$

follows a d-variate multivariate Student's t distribution with v degrees of freedom, location parameter $\boldsymbol{\mu}$, and dispersion matrix $\mathbf{\Sigma}$. From (C.18),

$$\mathbb{E}[G^{1/2}] = \sqrt{v/2}\frac{\Gamma\left(\frac{v-1}{2}\right)}{\Gamma\left(\frac{v}{2}\right)}, \quad v > 1, \tag{C.21}$$

so that, for $v > 1$, $\mathbb{E}[\mathbf{X}]$ exists, and, from (C.20), $\mathbb{E}[\mathbf{X}] = \boldsymbol{\mu} + \mathbb{E}[G^{1/2}]\mathbb{E}[\mathbf{Z}] = \boldsymbol{\mu}$. From (C.19),

$$\mathbb{E}[G] = v/(v-2), \text{ if } v > 2, \tag{C.22}$$

implying

$$\mathbb{E}[\mathbf{X}] = \boldsymbol{\mu}, \text{ if } v > 1, \qquad \mathbb{V}(\mathbf{X}) = \frac{v}{v-2}\mathbf{\Sigma}, \text{ if } v > 2. \tag{C.23}$$

Expression (C.20) is equivalent to saying that $(\mathbf{X} \mid G = g) \sim \text{N}(\boldsymbol{\mu}, g\mathbf{\Sigma})$, in which case we can write, similar to Example II.7.21,

$$f_{\mathbf{X}}(\mathbf{x}; \boldsymbol{\mu}, \mathbf{\Sigma}, v) = \int_0^\infty f_{\mathbf{X}|G}(\mathbf{x}; g) f_G(g; v/2, v/2)\, \mathrm{d}g. \tag{C.24}$$

From the c.f. analog of (C.3), the c.f. corresponding to (C.16) is

$$\varphi_{\mathbf{X}}(\mathbf{t}; \boldsymbol{\mu}, \mathbf{\Sigma}, v) = \mathbb{E}[\mathrm{e}^{i\mathbf{t}'\mathbf{X}}] = \mathrm{e}^{i\mathbf{t}'\boldsymbol{\mu}}\frac{K_{v/2}(\|\sqrt{v}\mathbf{\Sigma}^{1/2}\mathbf{t}\|)(\|\sqrt{v}\mathbf{\Sigma}^{1/2}\mathbf{t}\|)^{v/2}}{\Gamma(v/2)2^{v/2-1}}. \tag{C.25}$$

Let $j \in \{1, 2, \ldots, n\}$ and define $\mathbf{t} = (0, \ldots, 0, t, 0, \ldots, 0)'$, where t appears in the jth position. Similar to (C.2), we can compute the marginal c.f. corresponding to X_j. Observe that, with $\mathbf{\Sigma}^{1/2}$ symmetric (as can be obtained via the spectral decomposition method of calculating it), $\|\sqrt{v}\mathbf{\Sigma}^{1/2}\mathbf{t}\| = \sqrt{v}\sqrt{\mathbf{t}'\mathbf{\Sigma}\mathbf{t}} = \sqrt{v}|t|\sigma_j$. Thus,

$$\varphi_{X_j}(t) = \mathrm{e}^{it_j\mu_j}\frac{K_{v/2}(v^{1/2}|t|\sigma_j)(v^{1/2}|t|\sigma_j)^{v/2}}{\Gamma(v/2)2^{v/2-1}}, \tag{C.26}$$

so that, from the uniqueness theorem and (C.9), $X_j \sim t_v(\mu_j, \sigma_j)$. See Ding (2016) for a simple derivation of (and corrections to mistakes in previous literature) of the conditional distribution of subsets of $\mathbf{X}$ given a different subset, paralleling the result for the multivariate normal in (8.40).

Let $\mathbf{X} = (X_1, \ldots, X_d)' \sim t_v(\boldsymbol{\mu}, \boldsymbol{\Sigma})$ with p.d.f. (C.16), and define $S = \sum_{j=1}^{d} a_i X_i = \mathbf{a}'\mathbf{X}$ for $\mathbf{a} = (a_1, \ldots, a_d)' \neq \mathbf{0}$, i.e., S is a non-zero weighted sum of the univariate margins. Then

$$\begin{aligned}
\varphi_S(t) &= \mathbb{E}_S[\mathrm{e}^{itS}] = \mathbb{E}_{\mathbf{X}}[\mathrm{e}^{it\mathbf{a}'\mathbf{X}}] = \mathbb{E}_{\mathbf{X}}[\mathrm{e}^{i(t\mathbf{a})'\mathbf{X}}] = \varphi_{\mathbf{X}}(t\mathbf{a}; \boldsymbol{\mu}, \boldsymbol{\Sigma}, v) \\
&= \mathrm{e}^{it\mathbf{a}'\boldsymbol{\mu}} \frac{K_{v/2}(\| \sqrt{vt}\boldsymbol{\Sigma}^{1/2}\mathbf{a}\|)(\| \sqrt{vt}\boldsymbol{\Sigma}^{1/2}\mathbf{a}\|)^{v/2}}{\Gamma(v/2)2^{v/2-1}} \\
&= \mathrm{e}^{it\mu_S} \frac{K_{v/2}(v^{1/2}|t|\kappa)(v^{1/2}|t|\kappa)^{v/2}}{\Gamma(v/2)2^{v/2-1}} = \mathrm{e}^{it\mu_S}\varphi_T(\kappa t; v),
\end{aligned} \tag{C.27}$$

where $T \sim t_v$, $\mu_S = \mathbf{a}'\boldsymbol{\mu}$ and $\kappa = \|\boldsymbol{\Sigma}^{1/2}\mathbf{a}\| = \sqrt{\mathbf{a}'\boldsymbol{\Sigma}\mathbf{a}} > 0$. Thus, from the c.f. analog of (C.1), $S \stackrel{d}{=} \mu_S + \kappa T$, a location-scale Student's t with v degrees of freedom, where $\stackrel{d}{=}$ means equality in distribution (and is not to be confused with the dimension d of $\mathbf{X}$).

This method of proof can be extended to show a more general result encompassing (C.26) and (C.27). Let $\mathbf{X} \sim t_v(\boldsymbol{\mu}, \boldsymbol{\Sigma})$. For $1 \leqslant k \leqslant d$, $\mathbf{c} \in \mathbb{R}^k$, and $\mathbf{B}$ a $k \times d$ real matrix,

$$\mathbf{c} + \mathbf{B}\mathbf{X} \sim t_v(\mathbf{c} + \mathbf{B}\boldsymbol{\mu}, \mathbf{B}\boldsymbol{\Sigma}\mathbf{B}'). \tag{C.28}$$

This result also follows from the more general statement for elliptic random variables given in Theorem C.14 of Section C.2 below.

It is important to remember that the (weighted) sum of two or more *independent* Student's t r.v.s is not Student's t. In particular, it is *not* the case that, if $X_i \stackrel{\text{ind}}{\sim} t_{v_i}(\mu_i, \sigma_i)$, $i = 1, \ldots, d$, then $\sum_{i=1}^{d} X_i$ follows some t distribution. One might wonder if this is possible when the v_i are all equal: It is still not the case. This can be confirmed by applying the convolution formula for $d = 2$ and confirming that the resulting expression does not agree with the p.d.f. of a t r.v. This implies that, unless $v \to \infty$, one cannot construct density (C.16) such that the X_i are independent. Indeed, for $d = 2$, $\mu_1 = \mu_2 = 0$, $\sigma_1 = \sigma_2 = 1$, $0 < v < \infty$, and $X_i \stackrel{\text{ind}}{\sim} t_v(0, 1)$, $i = 1, 2$,

$$\begin{aligned}
f_{X_1}(x_1) f_{X_2}(x_2) &= \frac{\Gamma\left(\frac{v+1}{2}\right) v^{\frac{v}{2}}}{\sqrt{\pi}\,\Gamma\left(\frac{v}{2}\right)} (v + x_1^2)^{-\frac{v+1}{2}} \times \frac{\Gamma\left(\frac{v+1}{2}\right) v^{\frac{v}{2}}}{\sqrt{\pi}\,\Gamma\left(\frac{v}{2}\right)} (v + x_2^2)^{-\frac{v+1}{2}} \\
&\neq \frac{\Gamma\left(\frac{v+2}{2}\right)}{\Gamma\left(\frac{v}{2}\right)(v\pi)} \left(1 + \frac{\mathbf{x}'\mathbf{x}}{v}\right)^{-(v+2)/2} = f_{\mathbf{X}}(\mathbf{x}; \boldsymbol{\mu}, \boldsymbol{\Sigma}, v),
\end{aligned} \tag{C.29}$$

where the latter density is the joint distribution of $\mathbf{X} = (X_1, X_2)' \sim t_v(\boldsymbol{\mu}, \boldsymbol{\Sigma})$ from (C.16), for $\boldsymbol{\mu} = (\mu_1, \mu_2)'$ and $\boldsymbol{\Sigma} = \mathbf{I}_2$.

This result can be confirmed from the formulation of the multivariate Student's t as a continuous scale mixture of normals in (C.24). Note that each univariate marginal is affected by the mixing random variable g, so that they can never be independent (unless $v \to \infty$, resulting in the normal distribution).

C.2 Sphericity and Ellipticity

This section was written together with Christian Frey.

C.2.1 Introduction

In the multivariate setting, it is useful to distinguish between elliptic and non-elliptic distributions. Informally, an elliptic distribution preserves a type of rotational symmetry. For univariate random variables, the concept of ellipticity reduces to symmetry of the p.d.f. The multivariate normal and Student's t are among the canonical examples of elliptic distributions. The latter, with zero location and identity matrix for the dispersion, is given in (C.14), and is also **spherical**, this being a special case of elliptic distributions, as discussed below.

To gain an appreciation for their study, let $Z_i \overset{\text{i.i.d.}}{\sim} \text{N}(0,1)$, $i = 1, 2$. This could also be stated as $\mathbf{Z} = (Z_1, Z_2)' \sim \text{N}(\mathbf{0}, \mathbf{I}_2)$, and, as we will see below, the distribution of $\mathbf{Z}$ is spherical. Recall that $C = Z_1/Z_2 \sim \text{Cau}(0,1)$. Similarly, if $\mathbf{Z} \sim t_v(\mathbf{0}, \mathbf{I}_2)$ with density (C.14), then it is still the case that $C = Z_1/Z_2 \sim \text{Cau}(0,1)$, for any $v > 0$. This "Cauchy ratio property" holds whenever $\mathbf{Z}$ comes from a spherical distribution (and $\Pr(Z_2 = 0) = 0$).

It is important to emphasize that this result does *not* hold when $Z_i \overset{\text{i.i.d.}}{\sim} t_v(0,1)$, $i = 1, 2$; recall (C.29). In this latter case, with $v = 1$, $Z_i \overset{\text{i.i.d.}}{\sim} \text{Cau}(0,1)$, $i = 1, 2$, and the distribution of $C = Z_1/Z_2$ is given in (III.A.149); it is not Cauchy. It is only in the limit as $v \to \infty$ for $Z_i \overset{\text{i.i.d.}}{\sim} t_v(0,1)$, $i = 1, 2$, in which case Z_1 and Z_2 are (independent and) Gaussian. The code in Listing C.1 informally demonstrates the Cauchy ratio result for the case with $v = 1$ and $\mathbf{Z} \sim t_v(\mathbf{0}, \mathbf{I}_2)$, comparing it to the histogram generated by taking ratios of i.i.d. standard normals.

The reason this result holds is because a multivariate t random vector $\mathbf{X} = (X_1, X_2)'$ can be expressed as in (C.20), namely, as a continuous univariate random variable, say R, independent of $\mathbf{Z} = (Z_1, Z_2)' \sim \text{N}(\mathbf{0}, \mathbf{I}_2)$, such that $\Pr(R > 0) = 1$, multiplied by $\mathbf{Z}$. Thus, $X_1/X_2 = Z_1/Z_2$. This is, in fact, the inherent structure of spherical r.v.s, as will be seen below in Theorem C.1(**d**) (as well as Theorem C.2 and Example C.8).

As another example, let $\mathbf{Z} = (Z_1, \dots, Z_n)' \sim \text{N}(\mathbf{0}, \sigma^2 \mathbf{I}_n)$ and R be a continuous random variable, independent of $\mathbf{Z}$, such that $\Pr(R > 0) = 1$. Then, with

$$T_Z = \frac{\bar{Z}_n}{S_n/\sqrt{n}}, \quad \text{where} \quad \bar{Z}_n = n^{-1}\sum_{i=1}^{n} Z_i, \quad S_n^2 = (n-1)^{-1}\sum_{i=1}^{n}(Z_i - \bar{Z}_n)^2,$$

```
sim=1e5; df=1; T=mvtrnd(eye(2),df,sim); C=T(:,1)./T(:,2);
C=C(abs(C)<14); figure, hist(C,100)
% Now using ratio of indep standard normals
C=randn(sim,1)./randn(sim,1); C=C(abs(C)<14); figure, hist(C,100)
```

Program Listing C.1: Graphically compares the distribution of a ratio of independent standard normals, which is Cauchy, to the ratio of two random variables from the multivariate Student's t distribution with identity dispersion matrix and arbitrary (positive) degrees of freedom. Function `mvtrnd` is built into Matlab and (as the name suggests) generates i.i.d. realizations of a multivariate t distribution with the specified dispersion matrix and degrees of freedom.

we know that $T_Z \sim t_{n-1}$, i.e., Student's t with $n-1$ degrees of freedom; see Section II.3.7 for a detailed derivation. Now let $\mathbf{X} = R\mathbf{Z}$. It is simple to see that the above T statistic, computed based on $\mathbf{X}$, algebraically reduces to T_Z. Thus, all spherical random vectors have this "t-statistic property".

We now mention some prominent distributions that are not elliptic. The proper MGHyp (C.7) with asymmetry parameter vector $\boldsymbol{\gamma} \neq \mathbf{0}$, the non-degenerate multivariate discrete mixture of normals from Chapter 14 with location parameters $\boldsymbol{\mu}_i$ not all equal, and the multivariate noncentral t (MVNCT) given in (12.5), with finite degrees of freedom and asymmetry parameter vector $\boldsymbol{\gamma} \neq \mathbf{0}$, are all examples of **non-elliptic** distributions.

Remark Observe that, within the MVNCT setting, for example, the point $\boldsymbol{\gamma} = \mathbf{0}$ has measure zero in $\mathbb{R}^d$, so that, if $\boldsymbol{\gamma}$ were sampled from a continuous multivariate distribution (an idea comfortable for Bayesians) with support being some open subset of $\mathbb{R}^d$ including the origin, then the distribution of the resulting random vector will be elliptic with probability zero (w.p. 0). This line of reasoning holds for any non-elliptical distribution that nests an elliptic one as a special case such that the associated parameter (vector) assumes a measure-zero value.

As such, the formally correct answer to the question "Is the unknown distribution that generated the data elliptic?" is "No, w.p.1". What might be meant, however, is "(Based on inspection of the data,) is the distribution that generated them close enough to being elliptic that we can, for application purposes, assume so?" This is a more reasonable question, and can be decided based upon a likelihood ratio test if one assumes a parametric framework and (heroically) the class of distribution is known. Without such a distributional assumption, testing for ellipticity is more challenging; some tests, and a demonstration, are given in Section C.2.4.

One might indeed consider use of a parametric distributional class because one requires, say, a predictive distribution for some application, e.g., financial risk measurement or portfolio construction. In that case, instead of being concerned with a Neyman–Pearson hypothesis testing and dichotomous decision framework, it is more useful to base the decision on a comparison of criteria related to the purpose of the modeling exercise, using both the unrestricted (non-elliptic) and restricted (elliptic) cases. Continuing with the finance example, such criteria might include the performance of out-of-sample density or risk measurement forecasting or exercises, or (risk-adjusted) portfolio performance. ■

Our goal here is to highlight the main concepts and results associated with elliptic distributions. The key insight concerning elliptic (and as a special case, spherical) distributions, formulated in Theorem C.1(**d**), is due to Schoenberg (1938) and marks the starting point for building a theory around elliptical distributions. In addition to several references given throughout the discussion, Schoenberg (1938), Kelker (1970), Cambanis et al. (1981), Fang et al. (1989), Gupta and Varga (1993), and Frahm (2004) are primary sources for proofs of the stated results.

C.2.2 Sphericity

Recall (from, say, the beginning of Section 1.3.1) that an n-dimensional orthonormal matrix is a real $n \times n$ matrix whose rows, or columns, constitute an orthonormal basis for $\mathbb{R}^n$. Equivalently, a real $n \times n$ matrix $\mathbf{U}$ with ith column $\mathbf{u}_i$ is orthonormal if $\mathbf{U}'\mathbf{U} = \mathbf{I}_n$, or $\mathbf{u}_i'\mathbf{u}_j = 1$ if $i = j$, and zero otherwise. Hereafter, we will use the terms orthonormal and orthogonal interchangeably, as use of the latter is common in the literature. Notice that this orthogonality implies $\mathbf{U}^{-1} = \mathbf{U}'$, so that $\mathbf{U}\mathbf{U}' = \mathbf{I}_n$ as well.

Remark Multiplication by an orthogonal matrix does not change the length of a column vector $\mathbf{x} \in \mathbb{R}^n$, as

$$||\mathbf{Ux}||^2 = \mathbf{x}'\mathbf{U}'\mathbf{Ux} = \mathbf{x}'\mathbf{x} = ||\mathbf{x}||^2, \tag{C.30}$$

but induces only a rotation or reflection. Result (C.30) turns out to be a necessary and sufficient condition, and thus can also be used as a definition. In particular, if $\mathbf{U}$ is orthogonal, then (C.30) shows that $||\mathbf{Ux}||^2 = ||\mathbf{x}||^2$. Now only assume that $||\mathbf{Ux}||^2 = ||\mathbf{x}||^2$ for all $\mathbf{x} \in \mathbb{R}^n$. Then, with $\mathbf{x} = \mathbf{e}_i = (0, 0, \ldots, 0, 1, 0, \ldots, 0)'$ the column vector of all zeros except in the ith position having a one, $||\mathbf{Ux}||^2 = ||\mathbf{x}||^2$ implies that

$$1 = ||\mathbf{e}_i||^2 = ||\mathbf{x}||^2 = ||\mathbf{Ux}||^2 = ||\mathbf{u}_i||^2 = \mathbf{u}_i'\mathbf{u}_i,$$

i.e., $\mathbf{u}_i'\mathbf{u}_i = 1$, $i = 1, \ldots, n$. Now let $\mathbf{x} = \mathbf{e}_i + \mathbf{e}_j$, $i, j \in \{1, 2, \ldots, n\}$, with $i \neq j$. Then

$$2 = ||\mathbf{x}||^2 = ||\mathbf{Ux}||^2 = ||\mathbf{u}_i + \mathbf{u}_j||^2 = ||\mathbf{u}_i||^2 + ||\mathbf{u}_j||^2 + 2\mathbf{u}_i'\mathbf{u}_j = 2 + 2\mathbf{u}_i'\mathbf{u}_j,$$

so that $\mathbf{u}_i'\mathbf{u}_j = 0$. That is, if $||\mathbf{Ux}||^2 = ||\mathbf{x}||^2$ holds for all $\mathbf{x} \in \mathbb{R}^n$, then $\mathbf{U}$ is orthogonal (orthonormal). ■

The n-dimensional column random vector $\mathbf{X}$ is said to have a **spherically symmetric** distribution if, for every $\mathbf{H} \in \mathcal{O}(n)$, $\mathbf{X} \overset{d}{=} \mathbf{HX}$, where $\mathcal{O}(n)$ is the set of all $n \times n$ orthogonal matrices, and $\overset{d}{=}$ means equality in distribution.

Example C.1 Denote by $\mathbf{U}_n$ a column random vector distributed uniformly on the unit sphere in $\mathbb{R}^n$. For every orthogonal matrix $\mathbf{H}$, $\mathbf{HU}_n \overset{d}{=} \mathbf{U}_n$ because multiplying by $\mathbf{H}$ preserves lengths and induces only rotations on the sphere. Thus, $\mathbf{U}_n$ has a spherically symmetric distribution. ■

If $\mathbf{X}$ has density $f_{\mathbf{X}}$ and is spherically symmetric, then $f_{\mathbf{X}}(\mathbf{x})$ is a function of $\mathbf{x}$ only via $\mathbf{x}'\mathbf{x}$. For example, the multivariate normal and multivariate Student's t (C.16) in the uncorrelated, equal scales, zero mean case are spherically symmetric. Observe that this cannot serve as a definition, as not all r.v.s possess density functions, e.g., the stable Paretian.

Example C.2 From Chapter II.8 and Section III.A.16, recall that the c.f. of a scaled univariate symmetric stable r.v. is $\varphi_X(t; \alpha, c) = \exp\{-c^\alpha |t|^\alpha\}$, for $0 < \alpha \leqslant 2$ and $c > 0$. If the elements of $\mathbf{X} = (X_1, X_2, \ldots, X_n)'$ are i.i.d. with this c.f., then their joint c.f. is of the form $\exp\left(-c^\alpha \sum_{i=1}^n |t_i|^\alpha\right)$ (and that of their sum S is $\varphi_S(t; \alpha, c) = \exp\{-nc^\alpha |t|^\alpha\}$). This motivates the following definition. Random vector $\mathbf{X}$ is said to have an ***α*-symmetric** distribution if its c.f. is of the form

$$\varphi_{\mathbf{X}}(\mathbf{t}) = g(|t_1|^\alpha + |t_2|^\alpha + \ldots + |t_n|^\alpha), \quad \mathbf{t} = (t_1, t_2, \ldots, t_n)', \tag{C.31}$$

for some function g; see Cambanis et al. (1983). Use of $\alpha = 2$ results in the spherically symmetric distributions. ■

Example C.3 Let $\mathbf{X} \sim \mathrm{N}(\mathbf{0}, \mathbf{I}_n)$. The c.f. of a univariate standard normal distribution is $\varphi_X(t) = \exp(-t^2/2)$, and that of $\mathbf{X}$ is, with $\mathbf{t} = (t_1, \ldots, t_n)'$,

$$\varphi_{\mathbf{X}}(\mathbf{t}) = \exp\left(-\frac{1}{2}\mathbf{t}'\mathbf{t}\right) = \exp\left(-\frac{1}{2}(t_1^2 + t_2^2 + \ldots + t_n^2)\right), \tag{C.32}$$

so that, from (C.31), the multivariate standard normal distribution is spherically symmetric, or α-symmetric, with $\alpha = 2$.

This also follows because, for every $\mathbf{H} \in \mathcal{O}(n)$, as $\mathbf{H}$ is orthonormal, $\mathbf{HH}' = \mathbf{I}_n$, and as $\mathbb{V}(\mathbf{HX}) = \mathbf{HI}_n\mathbf{H}' = \mathbf{I}_n$, $\mathbf{HX} \sim \mathrm{N}(\mathbf{0}, \mathbf{I}_n)$, so that $\mathbf{X} \overset{d}{=} \mathbf{HX}$, and thus by definition, $\mathbf{X}$ is spherically symmetric. ■

Random vector $\mathbf{X} = (X_1, X_2, \dots, X_n)'$ is called an n **-dimensional version** of the univariate r.v. Y if $\boldsymbol{\lambda}'\mathbf{X} \overset{d}{=} c(\boldsymbol{\lambda})Y$ for all $\boldsymbol{\lambda} \in \mathbb{R}^n$ for some function $c(\cdot)$ such that $c(\boldsymbol{\lambda}) > 0$ if $\boldsymbol{\lambda} \neq \mathbf{0}$. In particular, taking $c(\boldsymbol{\lambda}) = (\boldsymbol{\lambda}'\boldsymbol{\lambda})^{1/2}$ results in a spherical distribution, while taking $c(\boldsymbol{\lambda}) = c(\boldsymbol{\lambda}; \alpha) = \left(\sum_{1=1}^{n} |\lambda_i|^{\alpha}\right)^{1/\alpha}$, $0 < \alpha \leqslant 2$, yields an α-symmetric distribution.

Example C.4 Again with $\mathbf{X} \sim \mathrm{N}_n(\mathbf{0}, \mathbf{I}_n)$ and letting $Y \sim \mathrm{N}(0, 1)$, we know (see, e.g., Chapter II.3) that, for all $\boldsymbol{\lambda} = (\lambda_1, \dots, \lambda_n)' \in \mathbb{R}^n \backslash \{\mathbf{0}\}$ and $c(\boldsymbol{\lambda}) = (\boldsymbol{\lambda}'\boldsymbol{\lambda})^{1/2}$, $\boldsymbol{\lambda}'\mathbf{X} \sim \mathrm{N}(0, \boldsymbol{\lambda}'\boldsymbol{\lambda})$, which has the same distribution as $c(\boldsymbol{\lambda})Y$. Thus, the usual multivariate normal distribution is an n-dimensional version of the univariate case. ■

The next theorem introduces some relevant notation, and characterizes spherical symmetry. Recall the definition of the multivariate characteristic function of random vector $\mathbf{X}$ in (C.5), and the uniqueness theorem regarding c.f.s and the distribution of $\mathbf{X}$.

Theorem C.1 For random vector $\mathbf{X} = (X_1, X_2, \dots, X_n)'$, statements (**a**) through (**d**) are equivalent:

a) $\mathbf{X} \overset{d}{=} \mathbf{HX}$ for every $\mathbf{H} \in \mathcal{O}(n)$.
b) The unique c.f. of $\mathbf{X}$, $\varphi_{\mathbf{X}}(\mathbf{t})$, $\mathbf{t} = (t_1, \dots, t_n)'$, can be expressed in the form $\varphi(\mathbf{t}'\mathbf{t})$, for some $\varphi \in \boldsymbol{\Phi}_n$, where φ is the **characteristic generator** of the spherical distribution $\mathbf{X}$ and
$$\boldsymbol{\Phi}_n = \{\varphi(\mathbf{t}) : \varphi(t_1^2 + t_2^2 + \dots + t_n^2) \text{ is an } n\text{-dimensional c.f.}\} \tag{C.33}$$
is the family of all possible characteristic generators for dimension n. We write $\mathbf{X} \sim S_n(\varphi)$. Observe the notational distinction between $\varphi_{\mathbf{X}}$ and φ. For the latter, one could use, say, $\varphi_{[\mathbf{X}]}$ or $\tilde{\varphi}_{\mathbf{X}}$ or $\phi_{\mathbf{X}}$ or $\psi_{\mathbf{X}}$ or ψ, for more clarity.
c) For any $\mathbf{a} \in \mathbb{R}^n$, $\mathbf{a}'\mathbf{X} \overset{d}{=} \|\mathbf{a}\|X_1$.
d) One can express $\mathbf{X}$ as
$$\mathbf{X} \overset{d}{=} R\mathbf{U}_n \tag{C.34}$$
for a continuous univariate random variable R, such that $\Pr(R \geqslant 0) = 1$, independent of $\mathbf{U}_n$, where $\mathbf{U}_n$ is uniformly distributed on the unit sphere surface in $\mathbb{R}^n$. Random variable R is referred to as the **generating variate** or **radial random variable**, F_R the unique **generating c.d.f.**, and $\mathbf{U}_n$ the **uniform base** of the spherical distribution.
e) For $\mathbf{U}_n$ in part (**d**), $\mathbb{E}[\mathbf{U}_n] = \mathbf{0}$ and $\mathbb{V}(\mathbf{U}_n) = \mathbf{I}_n/n$.

Proof of (a) ⇒ (b) ⇒ (c) ⇒ (a):

(a) ⇒ (b) For any $\mathbf{H} \in \mathcal{O}(n)$,
$$\varphi_{\mathbf{X}}(\mathbf{t}) \equiv \varphi_{\mathbf{HX}}(\mathbf{t}) = \mathbb{E}[e^{i\mathbf{t}'\mathbf{HX}}] = \mathbb{E}[e^{i(\mathbf{H}'\mathbf{t})'\mathbf{X}}] = \varphi_{\mathbf{X}}(\mathbf{H}'\mathbf{t}),$$
where use of $\equiv$ denotes invocation of the assumption made, here (**a**). For these characteristic functions to be equal for any $\mathbf{H} \in \mathcal{O}(n)$, it must be the case that $\varphi_{\mathbf{X}}(\mathbf{t})$ depends only on the length of $\mathbf{t}$, i.e., if $\varphi_{\mathbf{X}}(\mathbf{t})$ can be expressed in the form $\varphi(\mathbf{t}'\mathbf{t})$, for some $\varphi \in \boldsymbol{\Phi}_n$.

(b) $\Rightarrow$ **(c)** For any $\mathbf{a} \in \mathbb{R}^n$, with t scalar, $\varphi_{\mathbf{a}'\mathbf{X}}(t)$ is

$$\mathbb{E}[\mathrm{e}^{it(\mathbf{a}'\mathbf{X})}] = \mathbb{E}[\mathrm{e}^{i(t\mathbf{a})'\mathbf{X}}] = \varphi_{\mathbf{X}}(t\mathbf{a}) \equiv \varphi(t^2\mathbf{a}'\mathbf{a}) = \mathbb{E}[\mathrm{e}^{it^2(a_1^2X_1+\cdots+a_n^2X_n)}].$$

Likewise,

$$\varphi_{\|\mathbf{a}\|\mathbf{X}_1}(t) = \mathbb{E}[\mathrm{e}^{it\|\mathbf{a}\|\mathbf{X}_1}] \equiv \varphi(t^2\|\mathbf{a}\|^2) = \mathbb{E}[\mathrm{e}^{it^2(a_1^2X_1+\cdots+a_n^2X_1)}].$$

Recalling that $X_1, \dots, X_n$ have the same law, $\varphi_{\mathbf{a}'\mathbf{X}}(t) = \varphi_{\|\mathbf{a}\|\mathbf{X}_1}(t)$, so that, by the uniqueness theorem (C.4), (**c**) follows.

(c) $\Rightarrow$ **(a)** For any $\mathbf{H} \in \mathcal{O}(n)$,

$$\varphi_{\mathbf{HX}}(\mathbf{t}) = \mathbb{E}[\mathrm{e}^{i\mathbf{t}'\mathbf{HX}}] = \mathbb{E}[\mathrm{e}^{i(\mathbf{H}'\mathbf{t})'\mathbf{X}}] \equiv \mathbb{E}[\mathrm{e}^{i\|\mathbf{H}'\mathbf{t}\|X_1}] = \mathbb{E}[\mathrm{e}^{i\|\mathbf{t}\|X_1}] \equiv \mathbb{E}[\mathrm{e}^{i\mathbf{t}'\mathbf{X}}] = \varphi_{\mathbf{X}}(\mathbf{t}).$$

The proof of part (**d**) is more involved than the simple proofs for (**a**) to (**c**), and given separately below, along with that of part (**e**). ∎

Example C.5 ***Example C.3 cont.***
By direct inspection of (C.32), it immediately follows that the characteristic generator of $\mathbf{X} \sim \mathrm{N}(\mathbf{0}, \mathbf{I}_n)$ is $\varphi(u) = \exp(-u^2/2)$, where $u = \|\mathbf{t}\|^2$. ∎

Theorem C.1: Proof of (a) $\Leftrightarrow$ **(d):** We prove a slightly extended version of the equivalence between (**a**), (**b**), and (**d**). Recalling (C.33), a function $\varphi(\cdot)$ is contained in the set $\mathbf{\Phi}_n$ if and only if

$$\varphi(\|\mathbf{t}\|^2) = \int_0^\infty \Omega_n(\|\mathbf{t}\|^2 r^2)\,\mathrm{d}F_R(r), \tag{C.35}$$

where $\Omega_n(\|\mathbf{t}\|^2)$ is the c.f. of the uniformly distributed random variable $\mathbf{U}_n$ on the unit sphere, and $F_R(r)$ is the c.d.f. of the univariate random variable R with support $[0, \infty)$.

Let

$$S_1^{n-1} = \{\mathbf{s} \in \mathbb{R}^n : \mathbf{s}'\mathbf{s} = \|\mathbf{s}\|^2 = 1\}, \quad n \geqslant 2,$$

be the **unit sphere** in $\mathbb{R}^n$, and observe that S_1^{n-1} has $n-1$ dimensions because of the norm constraint. Next, let

$$\overline{S_1^{n-1}} = \int_{\mathbf{s}\in S_1^{n-1}} \mathrm{d}S(\mathbf{s}) = \frac{2\pi^{n/2}}{\Gamma(n/2)}, \tag{C.36}$$

which is derived in Theorem C.6 below. Note that, for $n = 2$ and radius r, this is the familiar length of a circle, i.e., $2\pi r = \frac{\mathrm{d}}{\mathrm{d}r}\pi r^2$. Here, $\mathrm{d}S(\cdot)$ is the area element of the unit sphere, $\overline{S_1^{n-1}}$ is the surface integral (surface area) of the unit sphere in $\mathbb{R}^n$, and we set $\mathrm{d}S := \mathrm{d}S_1^{n-1}$ for notational simplicity. (Introductions to surface integrals can be found in Apostol, 1969, Ch. 12, and Trench, 2003, p. 452). Then

$$\Omega_n(\|\mathbf{t}\|^2) = \frac{1}{\overline{S_1^{n-1}}} \int_{\mathbf{s}\in S_1^{n-1}} \exp\{i\mathbf{t}'\mathbf{s}\}\,\mathrm{d}S(\mathbf{s}) = \int_{\mathbf{u}\in S_1^{n-1}} \exp\{i\mathbf{t}'\mathbf{u}\}\,\mathrm{d}F_{\mathbf{U}}(\mathbf{u}), \tag{C.37}$$

where

$$\mathrm{d}F_{\mathbf{U}}(\cdot) = \frac{1}{\overline{S_1^{n-1}}}\,\mathrm{d}S(\cdot) = \frac{\Gamma(n/2)}{2\pi^{n/2}}\,\mathrm{d}S(\cdot).$$

Note that

$$\frac{\mathrm{d}F_{\mathbf{U}}(\cdot)}{\mathrm{d}S(\cdot)} = f_{\mathbf{U}}(\cdot) = \frac{1}{\overline{S_1^{n-1}}} = \frac{\Gamma(n/2)}{2\pi^{n/2}} \tag{C.38}$$

is the density of the uniformly distributed r.v. $\mathbf{U}_n$ on the unit sphere, as is derived in Theorem C.6 below, where, as already defined, $\overline{S_1^{n-1}}$ is its total surface and $\Omega_n(\|\mathbf{t}\|^2)$ is the c.f. of the uniformly distributed r.v. $\mathbf{U}_n$ on the unit sphere.

We wish to show that the stochastic representation

$$\mathbf{X} \stackrel{d}{=} R\mathbf{U}_n$$

holds, where $\mathbf{U}_n$ is uniformly distributed on the surface of the unit sphere S_1^{n-1}, random variable R is independent of $\mathbf{U}_n$ and such that $\Pr(R \geqslant 0) = 1$, and $R \sim F_R$ is related to φ by (C.35).

Necessity: Assume that $\varphi(\cdot)$ can be expressed as in (C.35). Let r.v. R be independent of $\mathbf{U}_n$ and have c.d.f. F_R such that $\Pr(R \geqslant 0) = 1$. The c.f. of $R\mathbf{U}_n$ is

$$\begin{aligned}\varphi_{R\mathbf{U}_n}(\mathbf{t}) &= \mathbb{E}[\exp\{i\mathbf{t}'R\mathbf{U}_n\}] = \mathbb{E}[\ \mathbb{E}[\exp\{i\mathbf{t}'R\mathbf{U}_n\} \mid R]\ ] \\ &= \int_0^\infty \Omega_n(r^2\|\mathbf{t}\|^2)\,\mathrm{d}F_R(r) = \varphi(\|\mathbf{t}\|^2),\end{aligned} \tag{C.39}$$

where the third equality follows by noting that $\mathbf{U}_n \in S_n(\varphi)$ and using C.1(b). Hence, $\varphi(\cdot) \in \mathbf{\Phi}_n$, and, by C.1(**b**) $\Rightarrow$ C.1(**a**), $\mathbf{X} = R\mathbf{U}_n \in S_n(\varphi)$.

Sufficiency: Assume $\varphi(\cdot) \in \mathbf{\Phi}_n$. Then, $g(t_1, \dots, t_n) \equiv \varphi(\mathbf{t}'\mathbf{t})$ is, by (**a**) $\Leftrightarrow$ (**b**) in Theorem C.1, a c.f. of some r.v. $\mathbf{X} \in S_n(\varphi)$ with c.d.f. $F_{\mathbf{X}}$. Hence, $g(t_1, \dots, t_n)$ is a rotational/radial symmetric function of $t_1, \dots, t_n$. As $g(\|\mathbf{t}\|) = g(\|\mathbf{t}\|s_1, \dots, \|\mathbf{t}\|s_n) = g(\|\mathbf{t}\|\mathbf{s})$ for every $\mathbf{s} \in S_1^{n-1}$, setting again $\mathrm{d}S := \mathrm{d}S_1^{n-1}$ for notational simplicity and using (C.38), we have

$$\begin{aligned}\varphi(\mathbf{t}'\mathbf{t}) &= \frac{1}{\overline{S_1^{n-1}}} \cdot g(\|\mathbf{t}\|) \cdot \overline{S_1^{n-1}} \\ &= \frac{1}{\overline{S_1^{n-1}}} \int_{\mathbf{s}\in S_1^{n-1}} g(\|\mathbf{t}\|) \cdot \mathrm{d}S(\mathbf{s}) \\ &= \frac{1}{\overline{S_1^{n-1}}} \int_{\mathbf{s}\in S_1^{n-1}} g(\|\mathbf{t}\|\mathbf{s})\,\mathrm{d}S(\mathbf{s}) \\ &= \int_{\mathbf{u}\in S_1^{n-1}} \left[\int_{\mathbb{R}^n} \mathrm{e}^{i\|\mathbf{t}\|\mathbf{u}'\mathbf{x}}\,\mathrm{d}F_{\mathbf{X}}(\mathbf{x})\right] \mathrm{d}F_{\mathbf{U}}(\mathbf{u}) \\ &= \int_{\mathbb{R}^n} \left[\int_{\mathbf{u}\in S_1^{n-1}} \mathrm{e}^{i\|\mathbf{t}\|\mathbf{u}'\mathbf{x}}\,\mathrm{d}F_{\mathbf{U}}(\mathbf{u})\right] \mathrm{d}F_{\mathbf{X}}(\mathbf{x}) \\ &= \int_{\mathbb{R}^n} \Omega_n(\|\mathbf{t}\|^2\|\mathbf{x}\|^2)\,\mathrm{d}F_{\mathbf{X}}(\mathbf{x}) = \int_0^\infty \Omega_n(\|\mathbf{t}\|^2 r^2)\,\mathrm{d}F_{\|\mathbf{X}\|}(r),\end{aligned} \tag{C.40}$$

where the interchange of integrals in the fourth line can be justified by Fubini's theorem because the integrand $\exp\{i\|\mathbf{t}\|\mathbf{u}'\mathbf{x}\}$ is non-negative, and Ω_n is the usual notation for the characteristic generator of a r.v. distributed on the unit sphere. The first equality in the last line follows because $\mathbf{U}_n \in S_n(\varphi)$ and by C.1(**a**) $\Rightarrow$ C.1(**b**) (as in the necessity direction). The last equality in the last line follows by defining

$$F_{\|\mathbf{X}\|}(r) := \int_{\|\mathbf{x}\|\leqslant r} \mathrm{d}F_{\mathbf{X}}(\mathbf{x}) = \Pr(\|\mathbf{X}\| \leqslant r),$$

so that $F_{\|\mathbf{X}\|}(\cdot)$ is a c.d.f. over $[0, \infty)$. This is possible because $\|\mathbf{X}\|$ takes only values on the non-negative half line. Further, (C.40) and (C.39) are equivalent, implying

$$\varphi(\|\mathbf{t}\|^2) = \int_0^\infty \Omega_n(\|\mathbf{t}\|^2 r^2)\,\mathrm{d}F_{\|\mathbf{X}\|}(r) = \int_0^\infty \Omega_n(\|\mathbf{t}\|^2 r^2)\,\mathrm{d}F_R(r),$$

and observing that a probability measure is, similarly to the uniqueness theorem (C.4) for the c.f., uniquely defined by its Laplace transform (see, e.g., Kallenberg, 2002, Thm. 4.3, for a proof, or Paolella, 2007, Eq. 1.17, 1.18, for the statement). Result $F_{\|\mathbf{X}\|}(r) = F_R(r)$ follows and therefore (C.35). Comparing (C.40) with (C.39), we see that (C.40) is the c.f. of $R\mathbf{U}_n$, where R is a r.v. with c.d.f. $F_{\|\mathbf{X}\|}$. The representation $\mathbf{X} \stackrel{d}{=} R\mathbf{U}_n$ thus follows. ■

Remark Result (C.34) is due to Schoenberg (1938). His main motivation was to investigate the connection between Fourier–Stieltjes integrals and the class of Laplace–Stieltjes integrals. The former are positive definite functions. See also Bochner's Theorem (mentioned in, e.g., Paolella, 2007, p. 24) for the relation between positive definite functions and characteristic functions. The latter are completely monotone functions.

Schoenberg expected a certain relation between the two, as in both cases (i) the defining kernel is the exponential function and (ii) both classes are convex, multiplicative and closed; see Schoenberg (1938, p. 813) for details. A subclass of positive definite functions of particular interest in the current context are characteristic functions that are of rotational (or radial) symmetry as in C.1(b). He shows in his Theorem C.1 that these r.v.s can be completely described by (C.35).

In words, (C.35) says that the c.f. of rotational (or radial) symmetric functions equals the one-sided Laplace–Stieltjes transform of the c.f. of a uniformly distributed random variable on the unit sphere. This can immediately be translated to the intuitive statement $\mathbf{X} \stackrel{d}{=} R\mathbf{U}_n$, which says that a spherical r.v. $\mathbf{X}$ (forming an n-dimensional sphere) can be decomposed in distribution into a random radius R and uniformly distributed random points on the unit sphere. ■

The following theorem helps to understand the nature of R given the distribution of $\mathbf{X}$, and has been used in forming a test for (conditional) elliptical symmetry (Zhu and Neuhaus, 2003).

Theorem C.2 Suppose $\mathbf{X} \stackrel{d}{=} R\mathbf{U}_n \sim S_n(\varphi)$ and $\Pr(\mathbf{X} = \mathbf{0}) = 0$. Consider additionally the factorization of $\mathbf{U}_n = (\mathbf{U}_1, \mathbf{U}_2)$, where $\mathbf{U}_1$ is $m \times 1$ and $\mathbf{U}_2$ is $(n-m) \times 1$ for integers $1 \leqslant m < n$.

a) $\|\mathbf{X}\|$ and $\mathbf{X}/\|\mathbf{X}\|$ are independent, and

$$\|\mathbf{X}\| \stackrel{d}{=} R \quad \text{and} \quad \mathbf{X}/\|\mathbf{X}\| \stackrel{d}{=} \mathbf{U}_n. \tag{C.41}$$

b) $(\mathbf{U}_1, \mathbf{U}_2) \stackrel{d}{=} (B\mathbf{U}_1, (1-B^2)^{1/2}\mathbf{U}_2)$, where $B \geqslant 0$, $\mathbf{U}_1$, and $\mathbf{U}_2$ are independent and $B^2 \sim \mathrm{Beta}(m/2, (n-m)/2)$.

Proof: (This follows Cambanis et al., 1981)

a) This result relies on Schoenberg's key insight in Theorem C.1(d), namely that every spherical random variable $\mathbf{X}$ can be decomposed in distribution as $\mathbf{X} \stackrel{d}{=} R\mathbf{U}_n$. Recall that a necessary and sufficient condition for two random variables $\mathbf{Y}$ and $\mathbf{Z}$ to have the same distribution is

$\mathbb{E}[f(\mathbf{Y})] = \mathbb{E}[f(\mathbf{Z})]$ for every non-negative (or bounded) Borel-measurable function f; see, e.g., Problem III.A.17(24). Thus,

$$\mathbf{X}'\mathbf{X} \stackrel{d}{=} R\mathbf{U}_n'\mathbf{U}_n R \stackrel{d}{=} R^2, \tag{C.42}$$

which follows by computing $\mathbb{E}[f(R\mathbf{U}_n'\mathbf{U}_n R)] = \mathbb{E}[f(R^2)]$ giving the last equality. Hence, we have $(\|\mathbf{X}\|, \mathbf{X}/\|\mathbf{X}\|) \stackrel{d}{=} (R, \mathbf{U}_n)$, as the mapping $x \mapsto (\|x\|, x/\|x\|)$ is (Borel-)measurable on $\mathbb{R}\backslash\{0\}$. This result will be often used below.

b) Without loss of generality, take $\mathbf{X} = (\mathbf{X}_1, \mathbf{X}_2) \sim \mathrm{N}(\mathbf{0}, \mathbf{I}_n)$, where $\mathbf{X}_1$ is $m \times 1$ and $\mathbf{X}_2$ is $(n-m) \times 1$. As $\mathbf{X}_1$ and $\mathbf{X}_2$ are independent, it follows from Theorem C.2(**a**) that

$$\mathbf{X}_1/\|\mathbf{X}_1\| = \mathbf{U}_1, \quad \mathbf{X}_2/\|\mathbf{X}_2\| = \mathbf{U}_2, \quad \|\mathbf{X}_1\|, \quad \|\mathbf{X}_2\|,$$

are jointly independent. Define $B := \mathbf{X}_1/\|\mathbf{X}\|$. Then B, $\mathbf{U}_1$, and $\mathbf{U}_2$ are independent. Now

$$B^2 = \frac{\|\mathbf{X}_1\|^2}{\|\mathbf{X}\|^2} = \frac{\|\mathbf{X}_1\|^2}{\|\mathbf{X}_1\|^2 + \|\mathbf{X}_2\|^2} \sim \mathrm{Beta}(m/2, (n-m)/2),$$

because $\|\mathbf{X}_1\|^2 \sim \chi^2_m$, $\|\mathbf{X}_2\|^2 \sim \chi^2_{n-m}$, and both are independent. As

$$B\mathbf{U}_1 = \frac{\|\mathbf{X}_1\|}{\|\mathbf{X}\|}\frac{\|\mathbf{X}_1\|}{\|\mathbf{X}_1\|} = B, \quad 1 - B^2 = \left(1 - \frac{\|\mathbf{X}_1\|}{\|\mathbf{X}\|}\right) = \frac{\|\mathbf{X}\| - \|\mathbf{X}_1\|}{\|\mathbf{X}\|} = \frac{\|\mathbf{X}_2\|}{\|\mathbf{X}\|},$$

Theorem C.2(**a**) implies

$$(\mathbf{U}_1, \mathbf{U}_2) = \mathbf{U}_n \stackrel{d}{=} \frac{\mathbf{X}}{\|\mathbf{X}\|} = \left(\frac{\mathbf{X}_1}{\|\mathbf{X}\|}, \frac{\mathbf{X}_2}{\|\mathbf{X}\|}\right) = (B\mathbf{U}_1, (1 - B^2)^{1/2}\mathbf{U}_2),$$

showing the result. ■

We can now complete the proof of Theorem C.1.

Theorem C.1: Proof of (e): We wish to show that $\mathbb{E}[\mathbf{U}_n] = \mathbf{0}$ and $\mathbb{V}(\mathbf{U}_n) = \mathbf{I}_n/n$. Without loss of generality, take $\mathbf{X} \sim \mathrm{N}_n(\mathbf{0}, \mathbf{I}_n)$. From Theorem C.2(**a**), $\mathbf{X} \stackrel{d}{=} \|\mathbf{X}\|\mathbf{U}_n$, where $\|\mathbf{X}\|$ is independent of $\mathbf{U}_n$. Note that $\mathbb{E}[\mathbf{X}] = \mathbf{0}$ and $\mathbb{V}(\mathbf{X}) = \mathbf{I}_n$. As $\|\mathbf{X}\|^2 \sim \chi^2_n$, $\mathbb{E}[\|\mathbf{X}\|] > 0$ and $\mathbb{E}[\|\mathbf{X}\|^2] = n$, the statement immediately follows. ■

Theorem C.3 Let $X \stackrel{d}{=} R\mathbf{U}_n \sim S_n(\varphi)$ and $\mathbf{X} = (\mathbf{X}_1, \mathbf{X}_2)$, where $\mathbf{X}_1$ is $m \times 1$ and $\mathbf{X}_2$ is $(n-m) \times 1$. Provided that the conditional random variable $\mathbf{X}_2 \mid (\mathbf{X}_1 = \mathbf{x}_1)$ exists, it is also spherically distributed and can be represented stochastically by

$$\mathbf{X}_2 \mid (\mathbf{X}_1 = \mathbf{x}_1) \stackrel{d}{=} R^*\mathbf{U}_2, \tag{C.43}$$

where $\mathbf{U}_2$ is $(n-m) \times 1$ and uniformly distributed on S^{n-m-1}, the generating variate R^* is

$$R^* = R(1 - B^2)^{1/2} \mid (RB\mathbf{U}_1 = \mathbf{x}_1), \tag{C.44}$$

$\mathbf{U}_1$ is $m \times 1$ and uniformly distributed on S^{m-1}, $B^2 \sim \mathrm{Beta}(m/2, (n-m)/2)$, and R, B^2, $\mathbf{U}_1$, and $\mathbf{U}_2$ are mutually independent.

Proof: (As in Fang et al., 1989) From Theorem C.2(b),

$$(\mathbf{U}_1, \mathbf{U}_2) \stackrel{d}{=} (B\mathbf{U}_1, (1-B^2)^{1/2}\mathbf{U}_2),$$

hence

$$(\mathbf{X}_1, \mathbf{X}_2) \stackrel{d}{=} R(B\mathbf{U}_1, (1-B^2)^{1/2}\mathbf{U}_2).$$

Note that the random variables R, B^2, U_1, and U_2 are mutually independent. ■

Example C.6 Another use of Theorem C.2 is that it yields an immediate method for simulating $\mathbf{U}_n$ by taking $\mathbf{X} \sim \mathrm{N}(\mathbf{0}, \mathbf{I}_n)$ in (C.41). This method of generation turns out to be among the most efficient as n grows; see Harman and Lacko (2010) and the references therein. Figure C.1 depicts the $n = 3$ case. ■

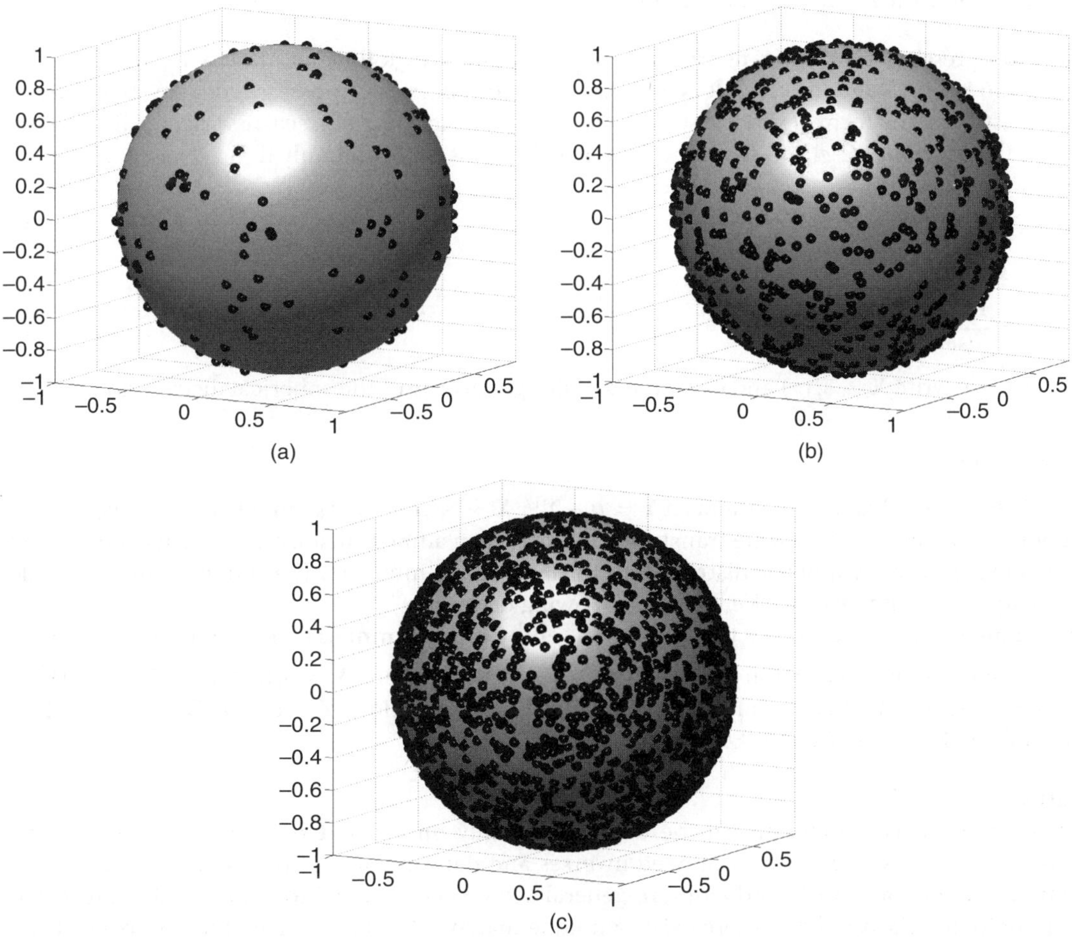

Figure C.1 The unit sphere for $n = 3$ with 200 (a), 1000 (b), and 3000 (c) random uniformly distributed points.

Example C.7 For $\mathbf{X} \sim \mathrm{N}(\mathbf{0}, \mathbf{I}_n)$ with $\varphi(u) = \exp(-u/2)$, we know from Example C.3 that $\mathbf{X}$ is spherical, so that, from Theorem C.1, it has the representation $\mathbf{X} \stackrel{d}{=} R\mathbf{U}_n$. From Theorem C.2, $R \stackrel{d}{=} \|\mathbf{X}\|$ and $\|\mathbf{X}\|^2 \sim \chi^2_n$. ■

Example C.8 Let $\mathbf{Z} \sim \mathrm{N}(\mathbf{0}, \mathbf{I}_n)$, independent of $S^2 \sim \chi^2_k$, and let $\mathbf{X} = \sqrt{k}\mathbf{Z}/S$, noting that $\Pr(S = 0) = 0$. Then, paralleling the univariate case, $\mathbf{X}$ has a multivariate Student's t distribution with k degrees of freedom, or $\mathbf{X} \sim t_k$. From Example C.7, we can also write $\mathbf{X} = \sqrt{k}R\mathbf{U}_n/S$ for $R^2 \sim \chi^2_n$. As R, $\mathbf{U}_n$, and S are independent, $\mathbf{X}$ has a spherical distribution. Let $R_* = \sqrt{k}R/S$, so that $\mathbf{X} = R_*\mathbf{U}_n$. Then $R^2_* = R^2/(S^2/k)$ and $R^2_*/n = (R^2/n)/(S^2/k) \sim F(n, k)$. ■

Example C.9 If the characteristic generator of $\mathbf{X}$ is given by $\varphi(\mathbf{x}'\mathbf{x}) = \exp(-c(\mathbf{x}'\mathbf{x})^\alpha)$, where $c \in \mathbb{R}_{>0}$ and $0 < \alpha \leqslant 2$, we say that it follows a **symmetric multivariate stable law**. Setting $\alpha = 1$ results in the family of multivariate Cauchy distributions. ■

Before progressing to elliptic distributions, we remark the following, which is subsumed in Theorem C.6 below. Random variable $\mathbf{X} \sim S_n(\varphi)$ does not necessarily possess a density, but when it does, it must be of the form $g(\mathbf{x}'\mathbf{x})$ for some non-negative function g of a scalar variable. This can be used to define a density $C_n g(\mathbf{x}'\mathbf{x})$ for some spherical distribution, if and only if

$$h_n = \int_0^\infty t^{n/2-1} g(t)\,\mathrm{d}t < \infty, \tag{C.45}$$

where $t = \mathbf{x}'\mathbf{x}$, and

$$C_n = \frac{\Gamma(n/2)}{\pi^{n/2}} \frac{1}{h_n}. \tag{C.46}$$

In this case, we write $\mathbf{X} \sim S_n(g)$ and call g the **density generator** of the spherical distribution.

C.2.3 Ellipticity

Let $\mathbf{X} \sim \mathrm{N}(\mathbf{0}, \mathbf{I}_n)$ as in Example C.3, and let $\mathbf{Y} = \boldsymbol{\mu} + \boldsymbol{\Sigma}^{1/2}\mathbf{X} \sim \mathrm{N}(\boldsymbol{\mu}, \boldsymbol{\Sigma})$, for symmetric $\boldsymbol{\Sigma} > 0$, as detailed in Section II.3.3. This location scale transform, also called an affine transformation, is used to extend the class of spherically symmetric distributions to elliptically symmetric. The condition that the scale matrix is square can be relaxed. We have the following definition.

An n-dimensional random vector $\mathbf{Y}$ has an **elliptically symmetric** distribution with location parameter $\boldsymbol{\mu}$ and dispersion matrix $\boldsymbol{\Sigma}$ if $\mathbf{Y} \stackrel{d}{=} \boldsymbol{\mu} + \mathbf{A}'\mathbf{X}$, where $\mathbf{X} \sim S_k(\varphi)$, $\boldsymbol{\mu} \in \mathbb{R}^n$, $\mathbf{A} \in \mathbb{R}^{k\times n}$, $\mathrm{rank}(\mathbf{A}) = k$, $\mathbf{A}'\mathbf{A} = \boldsymbol{\Sigma}$, hence $\mathrm{rank}(\boldsymbol{\Sigma}) = k$. We use the notation $\mathbf{Y} \sim \mathrm{EC}_n(\boldsymbol{\mu}, \boldsymbol{\Sigma}, \varphi)$. Note that $S_n(\varphi) \subseteq \mathrm{EC}_n(\boldsymbol{\mu}, \boldsymbol{\Sigma}, \varphi)$ and $S_n(\varphi) = \mathrm{EC}_n(\mathbf{0}, \mathbf{I}, \varphi)$.

Remarks

a) An n-variate density function can be defined not only on $\mathbb{R}^n$, but on the lower dimensional subspace $\mathbb{R}^k$, $0 < k < n$. As we will need inverses and determinants of non-square (rectangular) matrices, we give a short introduction to generalized inverses related to the current context. For every finite matrix $\mathbf{A} \in \mathbb{R}^{k\times n}$, there exists a unique matrix $\mathbf{A}^- \in \mathbb{R}^{n\times k}$, called the Moore–Penrose (or pseudo-/generalized inverse), satisfying the following (Penrose) equations

$$\mathbf{A}\mathbf{A}^-\mathbf{A} = \mathbf{A}, \qquad \mathbf{A}^-\mathbf{A}\mathbf{A}^- = \mathbf{A}^-, \tag{C.47}$$

$$(\mathbf{A}\mathbf{A}^-)' = \mathbf{A}\mathbf{A}^-, \qquad (\mathbf{A}^-\mathbf{A})' = \mathbf{A}^-\mathbf{A}, \tag{C.48}$$

where $\mathbf{A}'$ denotes the usual transpose. If $\mathbf{A}$ is square and non-singular, then $\mathbf{A}^- = \mathbf{A}^{-1}$ satisfies the four Penrose equations. It can be shown that the Moore–Penrose inverse is unique, see, e.g., Ben-Israel and Greville (2003, Sec. 1.2). As $\mathbf{A} \in \mathbb{R}^{k\times n}$ has column rank k, $(\mathbf{AA}')$ is invertible. This leads to $\mathbf{A}^-$ being given by $\mathbf{A}^- = \mathbf{A}'(\mathbf{AA}')^{-1}$, because

$$\mathbf{AA}^-\mathbf{A} = \mathbf{A}[\mathbf{A}'(\mathbf{AA}')^{-1}]\mathbf{A} = (\mathbf{AA}')(\mathbf{AA}')^{-1}\mathbf{A} = \mathbf{A}.$$

$\mathbf{A}^-$ is said to be a **right-inverse**, because $\mathbf{AA}^- = \mathbf{I}_k$. Moreover, as the reader should quickly confirm, $(\mathbf{A}^-)'\mathbf{A}' = \mathbf{I}_k$. As $(\mathbf{AA}')^{-1}$ is a symmetric square matrix, the spectral decomposition theorem implies that we can write

$$(\mathbf{AA}')^{-1} = \mathbf{OD}^{-1}\mathbf{O}',$$

where $\mathbf{O}$ is an orthonormal $n \times n$ square matrix and $\mathbf{D}$ is an $n \times n$ diagonal matrix containing the eigenvalues of $\mathbf{A}'\mathbf{A}$. Further, $\mathbf{D}^{-1}$ is diagonal, with reciprocals of the positive main diagonal elements of $\mathbf{D}$ whereas all zero elements of $\mathbf{D}$ are retained unchanged. We can then define the absolute value of the pseudo-determinant of a rectangular matrix $\mathbf{A} \in \mathbb{R}^{k\times n}$ in a natural way as

$$|\mathbf{A}| \equiv |\det(\mathbf{A})| := \prod_{i=1}^{n} \sqrt{D_{ii}}, \tag{C.49}$$

where D_{ii} is the ith diagonal element of $\mathbf{D}$, $i = 1, \dots, n$. Note that both the Moore–Penrose inverse and the absolute pseudo-determinant are generalizations of their standard function counterparts. In the following, we interchangeably use $|\mathbf{A}| \equiv |\det(\mathbf{A})|$ for the standard absolute determinant (i.e., for non-singular matrices) and for the pseudo-determinant for rectangular matrices. It will always be clear from the dimensionality of the matrix which one is meant.

b) In the probability literature, and in particular on elliptical distributions, one frequently encounters the terminology of an "absolutely continuous" distribution. The following explanations serve as a short introduction to this measure theoretic topic. A random variable X defined as a (Borel-)measurable function from the probability space $(\mathbb{R}, \mathcal{B}(\mathbb{R}), \Pr(\cdot))$ to the measurable space $(\mathbb{R}, \mathcal{B}(\mathbb{R}))$ is said to be **absolutely continuous** if and only if there is a non-negative Borel-measurable function f on $\mathbb{R}$ such that

$$F(x) = \int_{-\infty}^{x} f(t)\,\mathrm{d}t, \quad x \in \mathbb{R},$$

where f is the density function of X (because $F(x) \to 1$ as $x \to \infty$ and $\int_{-\infty}^{\infty} f(x)\,\mathrm{d}x = 1$). If X is absolutely continuous with density f, then it follows that

$$\Pr(X(B)) = \int_B f(x)\,\mathrm{d}x \quad \text{for each } B \in \mathcal{B}(\mathbb{R}).$$

The measure λ defined by $\lambda(B) := \int_B f(x)\,\mathrm{d}x$, $B \in \mathcal{B}(\mathbb{R})$ satisfies $\lambda(a, b] = F(b) - F(a)$, $a < b$, where λ is the Lebesgue–Stieltjes measure corresponding to F, implying $\Pr(X) = \lambda$. Hence, absolute continuity of X means equivalently

$$\Pr(X(A)) = 0 \Rightarrow \lambda(A) = 0, \quad \forall A \in \mathcal{B}(\mathbb{R}).$$

In words, the Lebesgue–Stieltjes measure λ has the same null sets as the probability measure $\Pr(X)$. If $\Pr(X)$ has additionally the same null sets as the Lebesgue–Stieltjes measure, then both measures are called equivalent. A concrete example where this does not hold is if X is a **Dirac delta function**, i.e., $\Pr(X = c) = 1$ but $\lambda(X = c) = 0$. Note that the considerations above hold equivalently for the multivariate case where all marginal random variables are independent from each other, as then $\Pr(\mathbf{X} = \mathbf{c}) = 0$. For more details on this topic, we refer to a measure-theoretic probability book such as Ash and Doléans-Dade (2000, p. 175). ■

The next theorem characterizes elliptic random variables.

Theorem C.4 $\mathbf{Y} \sim \mathrm{EC}_n(\boldsymbol{\mu}, \boldsymbol{\Sigma}, \varphi)$ if and only if

a) the c.f. of $\mathbf{Y}$ is $\varphi_{\mathbf{Y}}(\mathbf{t}) = \exp\{i\mathbf{t}'\boldsymbol{\mu}\}\varphi(\mathbf{t}'\boldsymbol{\Sigma}\mathbf{t})$ for some scalar function φ
b) there exists a continuous, non-negative univariate random variable R, independent of $\mathbf{U}_k$, such that

$$\mathbf{Y} \stackrel{d}{=} \boldsymbol{\mu} + R\mathbf{A}'\mathbf{U}_k, \qquad \mathbf{A}'\mathbf{A} = \boldsymbol{\Sigma}. \tag{C.50}$$

Proof: (Motivated from Fang et al., 1989)

a) **Sufficiency:** The c.f. of $\mathbf{Y}$ is

$$\begin{aligned}\varphi_{\mathbf{Y}}(\mathbf{t}) &= \mathbb{E}[\exp(i\mathbf{t}'\mathbf{Y})] = \mathbb{E}[\exp(i\mathbf{t}'(\boldsymbol{\mu} + \mathbf{A}'\mathbf{X}))] = \exp(i\mathbf{t}'\boldsymbol{\mu})\mathbb{E}[\exp(i\mathbf{t}'\mathbf{A}'\mathbf{X})] \\ &= \exp(i\mathbf{t}'\boldsymbol{\mu})\mathbb{E}[\exp(i\mathbf{t}'\mathbf{A}'\mathbf{A}\mathbf{t})] = \exp(i\mathbf{t}'\boldsymbol{\mu})\varphi(\mathbf{t}'\boldsymbol{\Sigma}\mathbf{t}),\end{aligned} \tag{C.51}$$

where the second to last equality follows from Theorem C.1(**b**).
Necessity: Define $\mathbf{X} := (\mathbf{Y} - \boldsymbol{\mu})'\mathbf{A}^-$, where, as before, $\mathbf{A}^-$ denotes the Moore–Penrose pseudo inverse such that $\mathbf{A} = \mathbf{A}\mathbf{A}^-\mathbf{A}$. As $\mathbf{A}$ has column rank k, $(\mathbf{A}\mathbf{A}')$ is invertible, and $\mathbf{A}^-$ is given as $\mathbf{A}^- = \mathbf{A}'(\mathbf{A}\mathbf{A}')^{-1}$, where $\mathbf{A}^-$ is $n \times k$. Moreover, we need $(\mathbf{A}')^-$ which is given by $(\mathbf{A}')^- = (\mathbf{A}\mathbf{A}')^{-1}\mathbf{A}$, where $(\mathbf{A}')^-$ is $k \times n$. This can be verified by computing $\mathbf{A}'(\mathbf{A}')^-\mathbf{A}' = \mathbf{A}'$, i.e. $\mathbf{A}'(\mathbf{A}\mathbf{A}')^{-1}\mathbf{A}\mathbf{A}' = \mathbf{A}'$. Thus, with the if direction in Theorem C.4(**a**), the c.f. of $\mathbf{X}$ is

$$\begin{aligned}\varphi_{\mathbf{X}}(\mathbf{t}) &= \varphi_{\mathbf{Y}-\boldsymbol{\mu}}((\mathbf{A}^-)\mathbf{t}) = \varphi(\mathbf{t}'(\mathbf{A}')^-\boldsymbol{\Sigma}(\mathbf{A}^-)\mathbf{t}) \\ &= \varphi(\mathbf{t}'(\mathbf{A}')^-(\mathbf{A}'\mathbf{A})(\mathbf{A}^-)\mathbf{t}) \\ &= \varphi(\mathbf{t}'((\mathbf{A}\mathbf{A}')^{-1}\mathbf{A})(\mathbf{A}'\mathbf{A})(\mathbf{A}'(\mathbf{A}\mathbf{A}')^{-1})\mathbf{t}) \\ &= \varphi(\mathbf{t}'(\mathbf{A}\mathbf{A}')^{-1}(\mathbf{A}\mathbf{A}')(\mathbf{A}\mathbf{A}')(\mathbf{A}\mathbf{A}')^{-1}\mathbf{t}) = \varphi(\mathbf{t}'\mathbf{t}), \quad \mathbf{t} \in \mathbb{R}^k.\end{aligned}$$

Note that the last equality can also be derived from the third equality by noting that $(\mathbf{A}')^-\mathbf{A}' = \mathbf{I}_k$ and $\mathbf{A}\mathbf{A}^- = \mathbf{I}_k$. Then, by theorem C.1(**b**) ⇒ (**a**), $\mathbf{X} \sim S_n(\varphi)$, and, hence, $\mathbf{Y} \sim \mathrm{EC}_n(\boldsymbol{\mu}, \boldsymbol{\Sigma}, \varphi)$.

b) **Sufficiency:** This is an immediate consequence of Theorem C.1(c).
Necessity: Again with $\mathbf{X} := (\mathbf{Y} - \boldsymbol{\mu})'\mathbf{A}^-$, the c.f. of $\mathbf{X}$ is given by $\varphi_{\mathbf{X}}(\mathbf{t}) = \varphi(\mathbf{t}'\mathbf{t})$, $\mathbf{t} \in \mathbb{R}^k$, so that, by Theorem C.1(**b**) ⇒ (**d**), $\mathbf{X}$ can be written as $\mathbf{X} \stackrel{d}{=} R\mathbf{U}_k$. Hence, with $\mathbf{X} = R\mathbf{U}_k = (\mathbf{Y} - \boldsymbol{\mu})'\mathbf{A}^-$, we have $\mathbf{Y} \stackrel{d}{=} \boldsymbol{\mu} + R\mathbf{A}'\mathbf{U}_k \stackrel{d}{=} \boldsymbol{\mu} + \mathbf{A}'\mathbf{X} \sim \mathrm{EC}_n(\boldsymbol{\mu}, \boldsymbol{\Sigma}, \varphi)$. ■

Theorem C.5 Let $\mathbf{Y} \sim \mathrm{EC}_n(\boldsymbol{\mu}, \boldsymbol{\Sigma}, \varphi)$. Then $Q(\mathbf{Y}) = (\mathbf{Y} - \boldsymbol{\mu})'\boldsymbol{\Sigma}^-(\mathbf{Y} - \boldsymbol{\mu}) \stackrel{d}{=} R^2$, where $\boldsymbol{\Sigma}^-$ is the generalized inverse of $\boldsymbol{\Sigma}$. For $k = n$ and $\boldsymbol{\Sigma}$ full rank, $\boldsymbol{\Sigma}^- = \boldsymbol{\Sigma}^{-1}$.

Proof: (From Cambanis et al., 1981) First note that

$$\begin{aligned}\boldsymbol{\Sigma}\mathbf{A}^-(\mathbf{A}^-)'\boldsymbol{\Sigma} &= (\mathbf{A}'\mathbf{A})\mathbf{A}^-(\mathbf{A}^-)'(\mathbf{A}'\mathbf{A}) = \mathbf{A}'(\mathbf{A}\mathbf{A}^-)((\mathbf{A}^-)'\mathbf{A}')\mathbf{A} \\ &= \mathbf{A}'(\mathbf{A}\mathbf{A}^-)(\mathbf{A}\mathbf{A}^-)'\mathbf{A} = \mathbf{A}'\mathbf{I}_k\mathbf{I}_k\mathbf{A} = \mathbf{A}'\mathbf{A} = \boldsymbol{\Sigma},\end{aligned}$$

where the first equality in the second line follows because, as before, $\mathbf{AA}^- = \mathbf{AA}'(\mathbf{AA}')^{-1} = \mathbf{I}_k$. Hence,

$$\boldsymbol{\Sigma}^- = \mathbf{A}^-(\mathbf{A}^-)' \tag{C.52}$$

by the definition of the generalized inverse. From the full rank representation (C.50),

$$Q(\mathbf{Y}) = R^2\mathbf{U}_k\mathbf{A}(\mathbf{A}^-(\mathbf{A}^-)')\mathbf{A}'\mathbf{U}_k \overset{d}{=} R^2\mathbf{U}_k(\mathbf{AA}^-)(\mathbf{AA}^-)'\mathbf{U}_k \overset{d}{=} R^2\mathbf{U}_k\mathbf{I}_k\mathbf{I}_k\mathbf{U}_k \overset{d}{=} R^2,$$

as claimed. ■

Example C.10 ***Example C.7 cont.***
Let $\mathbf{X} \sim \mathrm{N}(\mathbf{0}, \mathbf{I}_n)$ and $\mathbf{Y} = \boldsymbol{\mu} + \boldsymbol{\Sigma}^{1/2}\mathbf{X} \sim \mathrm{N}(\boldsymbol{\mu}, \boldsymbol{\Sigma})$, for symmetric $\boldsymbol{\Sigma} > 0$. Then, from Theorem C.4(**c**),

$$Q(\mathbf{Y}) = (\mathbf{Y} - \boldsymbol{\mu})'\boldsymbol{\Sigma}^{-1}(\mathbf{Y} - \boldsymbol{\mu}) = (\mathbf{Y} - \boldsymbol{\mu})'\boldsymbol{\Sigma}^{-1/2'}\boldsymbol{\Sigma}^{-1/2}(\mathbf{Y} - \boldsymbol{\mu}) \overset{d}{=} \mathbf{X}'\mathbf{X} = \|\mathbf{X}\|^2 \sim \chi^2_n,$$

which is indeed the distribution of R^2 in this case. ■

The next theorem provides a complete characterization of density functions for elliptically contoured distributions.

Theorem C.6 Let $\mathbf{Y} \sim \mathrm{EC}_n(\boldsymbol{\mu}, \boldsymbol{\Sigma}, \varphi)$ and assume that $\mathbf{A}'\mathbf{A} = \boldsymbol{\Sigma}$ and $\mathrm{rank}(\boldsymbol{\Sigma}) = k$. Further, assume that the density of $\mathbf{Y}, f_{\mathbf{Y}}(\mathbf{y})$, exists and

$$h_k = \int_0^\infty t^{k/2-1} g_k(t)\,\mathrm{d}t < \infty, \quad t \in [0, \infty). \tag{C.53}$$

Then

$$f_{\mathbf{Y}}(\mathbf{y}) = \frac{C_k}{|\mathbf{A}|} \cdot g_k((\mathbf{y} - \boldsymbol{\mu})'\boldsymbol{\Sigma}^-(\mathbf{y} - \boldsymbol{\mu})), \quad \text{where} \quad C_k = \frac{\Gamma(k/2)}{\pi^{k/2}}\frac{1}{h_k}. \tag{C.54}$$

If, additionally, $k = n$, so that $\boldsymbol{\Sigma}$ has full rank, then

$$f_{\mathbf{Y}}(\mathbf{y}) = \frac{C_n}{|\boldsymbol{\Sigma}|^{1/2}} \cdot g_n((\mathbf{y} - \boldsymbol{\mu})'\boldsymbol{\Sigma}^{-1}(\mathbf{y} - \boldsymbol{\mu})). \tag{C.55}$$

Further, $g(\cdot)$ is the density generator of $\mathbf{Y}$ and given by

$$g_k(t) = \frac{\Gamma(k/2)}{2\pi^{k/2}} \cdot t^{-1/2(k-1)} \cdot f_R(t^{1/2}), \quad t > 0. \tag{C.56}$$

Note that, for an elliptical r.v. $\mathbf{Y}$,

$$t = ((\mathbf{y} - \boldsymbol{\mu})'\mathbf{A}^-)((\mathbf{y} - \boldsymbol{\mu})'\mathbf{A}^-)' = (\mathbf{y} - \boldsymbol{\mu})'\boldsymbol{\Sigma}^-(\mathbf{y} - \boldsymbol{\mu}). \tag{C.57}$$

where $\boldsymbol{\Sigma}^- = \boldsymbol{\Sigma}^{-1}$ if $\mathrm{rank}(\boldsymbol{\Sigma}) = n$. For $\mathbf{X}$ spherical, $t = \mathbf{x}'\mathbf{x}$, and, hence, $g_n(\mathbf{x}'\mathbf{x})$ follows.

The first term of the density generator of $\mathbf{Y}$ is the density of a r.v. distributed uniformly on the unit sphere S_1^{k-1}, denoted (as before) by $f_{\mathbf{U}(\cdot)}$, and f_R is the p.d.f. of R. In this case, we write $\mathbf{Y} \sim \mathrm{EC}_n(\boldsymbol{\mu}, \boldsymbol{\Sigma}, \varphi)$ and call $g(\cdot)$ the **density generator** of the elliptical distribution.

Proof: (From Huber, 1982, Anderson, 2003, and Frahm, 2004) We show first the functional form of $f_{\mathbf{Y}}(\mathbf{y})$. Assume that $\mathbf{Y} \sim \mathrm{EC}_n(\boldsymbol{\mu}, \boldsymbol{\Sigma}, \varphi)$ so that, by Theorem C.4(**b**), $\mathbf{Y} \overset{d}{=} \boldsymbol{\mu} + R\mathbf{A}'\mathbf{U}_k$, $\mathbf{A}'\mathbf{A} = \boldsymbol{\Sigma}$. As (i) R is independent of $\mathbf{U}_k$, (ii) the p.d.f. of R is absolutely continuous on $(0, \infty)$, and (iii) $\Pr(R = 0) = F_R(0) = 0$, the joint density of $\mathbf{X} = R\mathbf{U}_k$ is given by

$$f_{R,\mathbf{U}_k}(r, \mathbf{u}) = f_R(r) \cdot f_{\mathbf{U}_k}(\mathbf{u}), \quad r > 0, \quad \mathbf{u} \in S_1^{k-1}. \tag{C.58}$$

To derive the density of $\mathbf{X}$, define the transformation $h : (r, \mathbf{u}) \mapsto r\mathbf{u} = \mathbf{x}$ and note that h is injective. The p.d.f. of $\mathbf{X}$ is then given by

$$f_{\mathbf{X}}(\mathbf{x}) = f_{R,\mathbf{U}_k}(h^{-1}(\mathbf{x})) \cdot |\mathbf{J}_h|^{-1}, \quad \mathbf{x} \neq \mathbf{0}, \tag{C.59}$$

where $\mathbf{J}_h$ is the Jacobian determinant of $\partial(r\mathbf{u})/\partial(r, \mathbf{u})$. Denote by

$$S_r^{k-1} = \{\mathbf{x} \in \mathbb{R}^k : \mathbf{x}'\mathbf{x} = ||\mathbf{x}||^2 = r, r > 0\}, \quad n \geqslant 2, \tag{C.60}$$

the unit sphere with radius r.

We now show two methods for determining the Jacobian in (C.59).

(**First approach**) The Jacobian matrix of the transformation $h : (r, \mathbf{u}) \mapsto r\mathbf{u} = \mathbf{x}$ is not lower triangular, so a direct computation results in mixed terms for $|\mathbf{J}_h|$ and no tractable expression is available. Instead we parameterize the transformation as $h : \mathbf{u} \mapsto r\mathbf{u} = \mathbf{x}$, where r is taken as the parameter because we can always identify r^{k-1} by the fundamental relationship in Theorem (C.2)(**a**) as $r^{k-1} = ||\mathbf{x}||^{k-1}$, recalling that the unit sphere has $k - 1$ dimensions. The Jacobian determinant of the parameterized transformation is then given by

$$|\mathbf{J}_h| = \begin{bmatrix} r & 0 & 0 & \dots & 0 \\ 0 & r & \ddots & 0 & 0 \\ \vdots & \ddots & \ddots & 0 & 0 \\ 0 & 0 & \dots & \ddots & 0 \\ 0 & 0 & \dots & 0 & r \end{bmatrix} = r^{k-1} = ||\mathbf{x}||^{k-1}, \quad \mathbf{x} \neq 0.$$

(**Second approach**) As $\partial r\mathbf{u}/\partial r$ has unit length and is orthogonal to each tangent plane $\partial r\mathbf{u}/\partial \mathbf{u}$ of the surface of the sphere S_r^{k-1}, we can write

$$\begin{bmatrix} \partial(r\mathbf{u})/\partial r \\ \partial(r\mathbf{u})/\partial \mathbf{u} \end{bmatrix} \begin{bmatrix} \partial(r\mathbf{u})/\partial r, & \partial(r\mathbf{u})/\partial \mathbf{u} \end{bmatrix} = \begin{bmatrix} \mathbf{1} & \mathbf{0}' \\ \mathbf{0} & r^2\mathbf{I}_{k-1} \end{bmatrix}.$$

The Jacobian $\partial(r\mathbf{u})/\partial(r, \mathbf{u})$ is then given by

$$\frac{\partial(r\mathbf{u})}{\partial(r, \mathbf{u})} = \begin{bmatrix} \mathbf{1} & \mathbf{0}' \\ \mathbf{0} & r\mathbf{I}_{k-1} \end{bmatrix},$$

and the absolute value of the determinant is

$$|\mathbf{J}_h| = \det\left(\begin{bmatrix} \mathbf{1} & \mathbf{0}' \\ \mathbf{0} & r\mathbf{I}_{k-1} \end{bmatrix} \right) = r^{k-1} = ||\mathbf{x}||^{k-1}, \quad \mathbf{x} \neq \mathbf{0}.$$

Moreover, $h^{-1}(\mathbf{x}) = (r, \mathbf{u}) = (||\mathbf{x}||, \mathbf{x}/||\mathbf{x}||)$ by Theorem C.2(**a**), so that the p.d.f. of $\mathbf{X}$ is

$$\begin{aligned} f_{\mathbf{X}}(\mathbf{x}) &= f_{R,\mathbf{U}_k}(||\mathbf{x}||, \mathbf{x}/||\mathbf{x}||) \cdot ||\mathbf{x}||^{-(k-1)} \\ &= f_{\mathbf{U}_k}(\mathbf{u}) \cdot ||\mathbf{x}||^{-(k-1)} \cdot f_R(||\mathbf{x}||). \end{aligned}$$

Define now the mapping $q : \mathbf{x} \mapsto \boldsymbol{\mu} + \mathbf{A}'\mathbf{x} = \mathbf{y}$ and note that q is injective as $\mathbf{A}\mathbf{A}^- = \mathbf{I}_k$. The absolute value of the Jacobian determinant of $\partial(\boldsymbol{\mu} + \mathbf{A}'\mathbf{x})/\partial\mathbf{x}$ is equal to $|\mathbf{J}_q| = |\det(\mathbf{A})|$, where $|\det(\mathbf{A})|$

corresponds to the pseudo-determinant defined in (C.49), and thus the p.d.f. of $\mathbf{Y} \stackrel{d}{=} \boldsymbol{\mu} + \mathbf{A}'\mathbf{X}$ is given by

$$\mathbf{y} \mapsto f_{\mathbf{Y}}(\mathbf{y}) = f_{\mathbf{X}}(q^{-1}(\mathbf{y})) = f_{\mathbf{X}}((\mathbf{y} - \boldsymbol{\mu})'\mathbf{A}^{-}) \cdot |\det(\mathbf{A})|^{-1}.$$

Hence, the p.d.f. of $\mathbf{Y}$ can finally be written as

$$f_{\mathbf{Y}}(\mathbf{y}) = C_k |\det(\mathbf{A})|^{-1} \cdot f_{\mathbf{U}_k}(\mathbf{u}) \cdot \|(\mathbf{y} - \boldsymbol{\mu})'\mathbf{A}^{-}\|^{-(k-1)} \cdot f_R(\|(\mathbf{y} - \boldsymbol{\mu})'\mathbf{A}^{-}\|),$$

where the normalizing constant C_k is determined in (C.63).

With $t = ((\mathbf{y} - \boldsymbol{\mu})'\mathbf{A}^{-})((\mathbf{y} - \boldsymbol{\mu})'\mathbf{A}^{-})'$ and $g(t) = f_{\mathbf{U}_k}(\mathbf{u}) \cdot t^{-1/2(k-1)} \cdot f_R(t^{1/2})$, (C.56) follows, up to the normalizing constant to be derived below in (C.63). As

$$\|(\mathbf{y} - \boldsymbol{\mu})'\mathbf{A}^{-}\|^2 = [(\mathbf{y} - \boldsymbol{\mu})'\mathbf{A}^{-}(\mathbf{A}^{-})'(\mathbf{y} - \boldsymbol{\mu})]^{1/2}$$

and since by (C.52),

$$\boldsymbol{\Sigma}^{-} = \mathbf{A}^{-}(\mathbf{A}^{-})'$$

it follows that

$$f_{\mathbf{Y}}(\mathbf{y}) = \frac{C_k}{|\det(\mathbf{A})|} \cdot g((\mathbf{y} - \boldsymbol{\mu})'\boldsymbol{\Sigma}^{-}(\mathbf{y} - \boldsymbol{\mu})),$$

where $g(\cdot)$ is as in (C.56) and where $|\mathbf{A}| \equiv |\det(\mathbf{A})|$, so that (C.54) follows. For $k = n$ and if $\boldsymbol{\Sigma}$ has full rank, then

$$|\det(\mathbf{A})|^{-1} = [\det(\mathbf{A})\det(\mathbf{A}')]^{-1/2} = [\det(\boldsymbol{\Sigma})]^{-1/2} = |\boldsymbol{\Sigma}|^{-1/2}.$$

Result (C.55) follows.

Necessity of assumption (C.53):[1] Set $r := \sqrt{t}$ in (C.56), so that

$$g(r^2) = \frac{\Gamma(k/2)}{2\pi^{k/2}} \cdot r^{-(k-1)} \cdot f_R(r), \tag{C.61}$$

noting that $f_R(r)$ is an absolutely continuous density function with support $(0, \infty)$. By rearranging (C.61) and integrating,

$$1 = \int_0^\infty f_R(r)\,\mathrm{d}r = \int_0^\infty \frac{2\pi^{k/2}}{\Gamma(k/2)} \cdot r^{k-1} g(r^2)\,\mathrm{d}r. \tag{C.62}$$

Thus, the integrability condition

$$\int_0^\infty r^{k-1} g(r^2)\,\mathrm{d}r < \infty$$

is required such that $g(r^2)$ qualifies as a valid density. Set $t = r^2$, so that $\mathrm{d}t = 2r\,\mathrm{d}r$, and hence $\mathrm{d}r = \mathrm{d}t/2r$, so that (C.62) is equivalent to

$$\int_0^\infty \frac{2\pi^{k/2}}{\Gamma(k/2)} t^{1/2(k-1)} g(t) \frac{\mathrm{d}t}{2r} = \frac{\pi^{k/2}}{\Gamma(k/2)} \int_0^\infty t^{k/2-1} g(t)\,\mathrm{d}t = 1,$$

1 For the following relatively elementary statement, it is noteworthy that it does not appear in Frahm (2004) (his density function does not integrate to one because he leaves off the normalizing constant), nor in this form in Cambanis et al. (1981). Kelker (1970) just mentions the integrability condition. Fang et al. (1989) conclude the same integrability condition, but based on a quite different approach, via the Dirichlet distribution and not directly via the derived density generator. It is of course simple, once one recognizes that $f_R(r)$ is a density function and can then use (C.56).

showing (C.53). For $g(\cdot)$ to be a valid density such that $\int_0^\infty g(t)\,\mathrm{d}t = 1$, introduce the normalizing constant

$$C_k := \frac{\Gamma(k/2)}{\pi^{k/2}} \frac{1}{h_k}, \tag{C.63}$$

and (C.54) follows.

In the next step, the density of a uniformly distributed r.v. on the unit sphere, $f_{\mathbf{U}_k}(\cdot)$, as given in (C.56), is derived, and follows Anderson (2003, p. 47). Recall from Theorem C.1 that $f_{\mathbf{U}}(\cdot) = 1/\overline{S_1^{k-1}}$, $\mathrm{d}S := \mathrm{d}S_1^{k-1}$ for notational simplicity and that $\overline{S_1^{k-1}} = \int_{\mathbf{s}\in S_1^{k-1}} \mathrm{d}S(\mathbf{s})$. Therefore, it is sufficient to compute $\overline{S_1^{k-1}}$. Define the surface area of the sphere with radius r to be $\overline{S_r^{k-1}} := \int_{\mathbf{s}\in S_r^{k-1}} \mathrm{d}S(\mathbf{s})$, where $S_r^{k-1} = \{\mathbf{s} \in \mathbb{R}^k : \|\mathbf{s}\| = r > 0\}$ as in (C.60). The volume V_1^k of the unit sphere is then given by adding infinitely thin spherical shells of radius $0 < r \leqslant R$, where the parametrization R is used for the upper limit of the integral, as interest does not center on computing the volume of the sphere V_1^k in what follows; see (C.68). Hence,

$$V_1^k = \int_0^R \overline{S_r^{k-1}}(r)\,\mathrm{d}r = \int_0^R \int_{\mathbf{s}(r)\in S_r^{k-1}} \mathrm{d}S(\mathbf{s}(r))\,\mathrm{d}r \tag{C.64}$$

$$= \int_0^R \int_{\mathbf{x}(r)\in S_r^{k-1}} \mathbf{x}(r)\,\mathrm{d}r, \quad R = 1, \tag{C.65}$$

where $\mathbf{x}(r)$ is the chosen parametrization of the area element $\mathrm{d}S(\mathbf{s}(r))$. In the case of a sphere, this parametrization is particularly simple as it can be chosen to depend only on the radius r. By the (first) fundamental theorem of calculus,

$$\overline{S_1^{k-1}} = \frac{\partial(V_1^k)}{\partial(R)}, \quad R = 1. \tag{C.66}$$

With a basic coordinate transformation in the Euclidean space of dimension k from Cartesian coordinates to polar coordinates, we can choose a parametrization of $\mathbf{x}(r)$ depending only on r for $k \geqslant 3$, as

$$x_1 = r\sin\theta_1,$$

$$x_j = r\left(\prod_{l=1}^{j-1}\cos\theta_l\right)\sin\theta_j, \quad 2 \leqslant j \leqslant k-1,$$

$$x_k = r\left(\prod_{l=1}^{k-1}\cos\theta_l\right),$$

where $0 < r \leqslant 1$ and $-\pi/2 \leqslant \theta_l \leqslant \pi/2$ for $l = 1, \ldots, k-2$, and $-\pi \leqslant \theta_{k-1} \leqslant \pi$. An accessible introduction to polar coordinates is given in Apostol (1969) and Trench (2003, Sec. 6.3).

By (rotational) symmetry, we could equivalently also write sin instead of cos and vice versa in the above products. Then the absolute value of the Jacobian determinant $|\mathbf{J}_v| \equiv |\det(\mathbf{J}_v)|$ of the transformation $v : (x_1, \ldots, x_k) \mapsto (\theta_1, \ldots, \theta_{k-1}, r)$ is

$$|\mathbf{J}_v| = \left|\det\left(\frac{\partial(x_1, \ldots, x_k)}{\partial(\theta_1, \ldots, \theta_{k-1}, r)}\right)\right| = r^{k-1}\cos^{k-2}\theta_1\cos^{k-3}\theta_2 \cdots \cos\theta_{k-2}.$$

Remark In most textbooks in which this result appears, no proof is given, or it is simply claimed that this result is easy to verify by induction; see, e.g., Hassani (1999, p. 594). However, as Nguyen (2014) states, this claim is not only wrong, but also a common claim whenever the Jacobian of k-dimensional polar coordinates is stated. The main problematic issue is that in the general k-dimensional case, there is a lack of a recursive relation between the Jacobians of different orders, so that no inductive proof exists so far. The few rigorous proofs of this result can be found in Muirhead (2005, Thm. 2.13) by means of so-called exterior differential forms, and in Richter (2007, Thm. 2), and Nguyen (2014, Sec. 2.3). ■

Having this result, we then get

$$\begin{aligned} V_1^k &= \int_0^R \int_{\mathbf{x}(r)\in S_r^{k-1}} \mathbf{x}(r)\,\mathrm{d}r \\ &= \int_{-\pi/2}^{\pi/2} \cdots \int_{-\pi/2}^{\pi/2} \int_{-\pi}^{\pi} \int_0^R r^{k-1}\cos^{k-2}\theta_1 \cos^{k-3}\theta_2 \quad \cdots \quad \cos\theta_{k-2} \quad \mathrm{d}\theta_1 \quad \cdots \quad \mathrm{d}\theta_{k-1}\,\mathrm{d}r \\ &= \prod_{l=1}^{k-2} \int_{-\pi/2}^{\pi/2} (\cos\theta_l)^{k-l-1}\mathrm{d}\theta_l \int_{-\pi}^{\pi} \mathrm{d}\theta_{k-1} \int_0^R r^{k-1}\,\mathrm{d}r, \quad R = 1, \quad k \geqslant 3. \end{aligned} \tag{C.67}$$

Recall that the beta function is defined as $B(a,b) := \int_0^1 x^{a-1}(1-x)^{b-1}\,\mathrm{d}x$. Substituting $x = \cos^2\theta_l$ gives $\mathrm{d}x = 2\cos\theta_l \sin\theta_l \mathrm{d}\theta_l$, and, with $\overline{\theta}_l = \arccos(1)$ and $\underline{\theta}_l = \arccos(0)$, the integration domain is $[0, \pi/2]$, so that

$$\begin{aligned} B(a,b) &= 2\int_0^{\pi/2} (\cos^2\theta_l)^{a-1}(\sin^2\theta_l)^{b-1}\cos\theta_l \sin\theta_l \quad \mathrm{d}\theta_l \\ &= 2\int_0^{\pi/2} (\cos\theta_l)^{2a-1}(\sin\theta_l)^{2b-1} \quad \mathrm{d}\theta_l. \end{aligned}$$

With $a = (k-l)/2$, $b = 1/2$, $B(a,b) = \Gamma(a)\Gamma(b)/\Gamma(a+b)$, $\Gamma(1/2) = \sqrt{\pi}$, and by symmetry of the integral, we get

$$\int_{-\pi/2}^{\pi/2} (\cos\theta_l)^{k-l-1}\mathrm{d}\theta_l = 2\cdot\frac{\Gamma\left[\frac{1}{2}(k-l)\right]\Gamma(\frac{1}{2})}{\Gamma\left[\frac{1}{2}(k-l+1)\right]} = 2\cdot\frac{\Gamma\left[\frac{1}{2}(k-l)\right]\sqrt{\pi}}{\Gamma\left[\frac{1}{2}(k-l+1)\right]}.$$

Next,

$$\begin{aligned} &\prod_{l=1}^{k-2} \frac{\Gamma\left[\frac{1}{2}(k-l)\right]\sqrt{\pi}}{\Gamma\left[\frac{1}{2}(k-l+1)\right]} \cdot \int_{-\pi}^{\pi} \mathrm{d}\theta_{k-1} \\ &= \frac{\Gamma\left[\frac{1}{2}(k-1)\right]\sqrt{\pi}\cdot\Gamma\left[\frac{1}{2}(k-2)\right]\sqrt{\pi} \quad \cdots \quad t\Gamma\left[\frac{1}{2}(3)\right]\sqrt{\pi}\cdot\Gamma[1]\sqrt{\pi}}{\Gamma\left[\frac{1}{2}(k)\right]\cdot\Gamma\left[\frac{1}{2}(k-1)\right] \quad \cdots \quad \Gamma\left[\frac{1}{2}(3)\right]} \cdot 2\pi \\ &= \frac{\Gamma(1)\pi^{k/2-1}}{\Gamma(k/2)} \cdot 2\pi = \frac{(2\pi)^{k/2}}{\Gamma(k/2)}, \end{aligned}$$

where the last equality follows because $\Gamma(1) = 1$. Finally, we can write (C.67) as

$$V_1^k = \frac{(2\pi)^{k/2}}{\Gamma(k/2)} \int_0^R r^{k-1}\,\mathrm{d}r, \quad R = 1.$$

With (C.66), we get

$$\overline{S_1^{k-1}} = \frac{\partial(V_1^k)}{\partial(R)} = \frac{(2\pi)^{k/2}}{\Gamma(k/2)}\,\partial\left(\int_0^R r^{k-1}\,\mathrm{d}r\right)/\partial R = \frac{(2\pi)^{k/2}}{\Gamma(k/2)}\cdot 1, \quad R = 1, \tag{C.68}$$

and, therefore, $f_{\mathbf{U}}(\cdot) = 1/\overline{S_1^{k-1}}$ so that (C.56) follows. ∎

Remark In the following, we give an alternative, short, self-contained proof for the surface of a k-dimensional sphere, following Huber (1982). Note that, by (C.64), the volume V_1^k of the unit sphere is given as

$$V_1^k = \int_0^1 \overline{S_r^{k-1}(r)}\,\mathrm{d}r = \int_0^1 \int_{\mathbf{s}(r)\in S_r^{k-1}} \mathrm{d}S(\mathbf{s}(r))\,\mathrm{d}r. \tag{C.69}$$

Another way to define the volume of the unit sphere is

$$V_1^k = \int_0^1 \overline{S_r^{k-1}(r)} r^{k-1}\,\mathrm{d}r, \tag{C.70}$$

where from now on the explicit dependence on r is dropped, i.e., $\overline{S_r^{k-1}} := \overline{S_r^{k-1}(r)}$ and $\mathrm{d}S(\mathbf{s})\,\mathrm{d}r := \mathrm{d}S(\mathbf{s}(r))\,\mathrm{d}r$. Note that $r = \|\mathbf{s}\| = \left(\sum_{i=1}^k s_i^2\right)^{1/2}$ denotes the usual Euclidean norm. Hence, we can equate both expressions and get

$$\int_0^1 \int_{\mathbf{s}\in S_r^{k-1}} \mathrm{d}S(\mathbf{s})\,\mathrm{d}r = \int_0^1 \overline{S_r^{k-1}} r^{k-1}\,\mathrm{d}r. \tag{C.71}$$

Recall that, by definition, $\Gamma(a) := \int_0^\infty x^{a-1}\mathrm{e}^{-x}\,\mathrm{d}x$, $a > 0$. Our interest is for $x = b^2$, and $\mathrm{d}x = 2b\,\mathrm{d}b$, so that

$$\Gamma(a) = \int_0^\infty (b^2)^{a-1}\mathrm{e}^{-b^2} 2b\,\mathrm{d}b = 2\int_0^\infty b^{2a-1}e^{-b^2}\,\mathrm{d}b, \tag{C.72}$$

and letting $a = 1/2$ gives, by the symmetry of the integration domain,

$$\Gamma\left(\frac{1}{2}\right) = 2\int_0^\infty \mathrm{e}^{-b^2}\,\mathrm{d}b = \int_{-\infty}^\infty \mathrm{e}^{-b^2}\,\mathrm{d}b = \sqrt{\pi}. \tag{C.73}$$

Multiplying now the right-hand side of (C.71) by $\exp(-s^2)$ and integrating over the domain $(-\infty, +\infty)$ (instead of $(0, 1]$) yields

$$\int_{-\infty}^{+\infty} \left(\overline{S_r^{k-1}} r^{k-1}\right) \mathrm{e}^{-s^2}\,\mathrm{d}r = 2\int_0^{+\infty} \left(\overline{S_r^{k-1}} r^{k-1}\right) \mathrm{e}^{-s^2}\,\mathrm{d}r,$$

by symmetry of the integration domain, so that, with (C.72),

$$\int_0^{+\infty} \left(\overline{S_r^{k-1}} r^{k-1}\right) e^{-s^2}\,\mathrm{d}r = \frac{1}{2}\overline{S_r^{k-1}}\Gamma\left(\frac{1}{2}k\right). \tag{C.74}$$

Using the integration domain $(-\infty, +\infty)^k$, multiplying the left-hand side of (C.71) by $\exp(-s^2)$, and using $\mathbf{s} = (s_1, s_2, \ldots, s_k)$, $r^2 = \|\mathbf{s}\|^2 = s_1^2 + s_2^2 + \ldots + s_k^2$ gives

$$\begin{aligned}\int_{(-\infty,+\infty)^k} e^{-r^2}\,dS(\mathbf{s}) &= \int_{-\infty}^{+\infty}\cdots\int_{-\infty}^{+\infty} e^{-r^2}\prod_{i=1}^{k} dS(s_i)\\ &= \int_{-\infty}^{+\infty}\cdots\int_{-\infty}^{+\infty}\prod_{i=1}^{k}(e^{-s_i^2}\,dS(s_i)) = \prod_{i=1}^{k}\int_{-\infty}^{+\infty} e^{-s_i^2}\,dS(s_i)\\ &= \left(\int_{-\infty}^{+\infty} e^{-s_i^2}\,dS(s_i)\right)^k = \pi^{k/2},\end{aligned} \tag{C.75}$$

where the last equality follows by (C.73). Equating the last expression in (C.75) and (C.74) gives

$$\overline{S_1^{k-1}} = \frac{(2\pi)^{k/2}}{\Gamma(k/2)}, \tag{C.76}$$

so that, as above, $f_{\mathbf{U}}(\cdot) = 1/\overline{S_1^{k-1}}$ and (C.56) follows. ■

Note that the converse to Theorem C.6 holds as well: Given the density function in (C.55), it is easy to see that, with $\mathbf{X} := (\mathbf{Y} - \boldsymbol{\mu})'\mathbf{A}^{-}$, the c.f. of $\mathbf{X}$ has the functional form in Theorem C.1(b), hence $\mathbf{X}$ is spherical and, thus, $\mathbf{Y}$ is elliptical.

Example C.11 Let $g(y) = \exp(-y/2)$, so that $h_n = \int_0^\infty y^{n/2-1}e^{-y/2}\,dy = 2^{n/2}\Gamma(n/2) < \infty$ and $C_n = 2^{-n/2}\pi^{-n/2}$, giving $C_n g(\mathbf{x}'\mathbf{x}/2) = (2\pi)^{-n/2}e^{-\mathbf{x}'\mathbf{x}/2}$, which, from (8.39), is the multivariate normal distribution with zero location vector and identity variance–covariance matrix. ■

Theorem C.7 If $\mathbf{Y} \sim \mathrm{EC}_n(\boldsymbol{\mu}, \boldsymbol{\Sigma}, \varphi)$, then $\mathbf{Y}$ possesses a density generator g if and only if R has p.d.f. f_R, and

$$f_R(r) = \frac{2\pi^{n/2}}{\Gamma(n/2)} r^{n-1} g(r^2) \tag{C.77}$$

dictates the relationship between g and f_R.

Proof: This is a direct consequence of Theorem C.6. Note that, if $\mathbf{Y} \sim \mathrm{EC}_n(\boldsymbol{\mu}, \boldsymbol{\Sigma}, \varphi)$, then Theorem C.6 says that the density generator has the functional form in (C.56). Rearranging yields (C.77). This shows necessity as well as sufficiency. ■

Example C.12 ***Example C.7 cont., density generator of*** $\mathbf{X} \sim \mathrm{N}(\mathbf{0}, \mathbf{I}_n)$
Let $\mathbf{X} \sim \mathrm{N}(\mathbf{0}, \mathbf{I}_n)$ with generating variate R such that $R^2 \sim \chi_n^2$, and p.d.f. $f_{\mathbf{X}}(\mathbf{x}) = (2\pi)^{-n/2}e^{-\mathbf{x}'\mathbf{x}/2}$. Recall that the p.d.f. of χ_n^2 corresponds to

$$f_{(\chi_n^2)}(x) = \frac{x^{n/2-1}\cdot e^{-x/2}}{2^{n/2}\cdot\Gamma(n/2)}, \qquad x \geqslant 0.$$

Let $C \sim \chi_n^2$, so that $r = \sqrt{c}$. Then, a simple univariate transformation and the fact that $f_R(r)$ is absolutely continuous on $(0, \infty)$ yields

$$f_R(r) = \left|\frac{dc}{dr}\right| f_C(c) = 2r\cdot f_{(\chi_n^2)}(r^2)\mathbb{I}_{(0,\infty)} = \frac{2^{-n/2+1}}{\Gamma(n/2)} r^{n-1}e^{-r^2/2}\mathbb{I}_{(0,\infty)}(r).$$

The density generator of $\mathbf{X} \stackrel{d}{=} R \cdot \mathbf{U}_n \stackrel{d}{=} \sqrt{\chi_n^2} \cdot \mathbf{U}_n$ is then given by (C.56) so that

$$g_{\sqrt{\chi_n^2}}(t) = \frac{\Gamma(n/2))}{2\pi^{n/2}} \cdot t^{-1/2(n-1)} \cdot 2t^{1/2} \cdot f_{(\chi_n^2)}(t) = \frac{1}{(2\pi)^{n/2}} \exp(-t/2),$$

corresponding to the generator of the multivariate normal distribution. ■

Example C.13 We wish to derive the density f_R corresponding to the n-dimensional vector $\mathbf{X} \sim t_v(\boldsymbol{\mu}, \boldsymbol{\Sigma})$. Recalling (C.16), the p.d.f. of the multivariate t-distribution is

$$f_{\mathbf{X}}(\mathbf{x}; \boldsymbol{\mu}, \boldsymbol{\Sigma}, v) = \frac{\Gamma((n+v)/2)}{\Gamma(v/2)} \cdot \left(\frac{\det(\boldsymbol{\Sigma}^{-1})}{(v\pi)^n}\right)^{1/2} \cdot \left(1 + \frac{(\mathbf{x}-\boldsymbol{\mu})'\boldsymbol{\Sigma}^{-1}(\mathbf{x}-\boldsymbol{\mu})}{v}\right)^{-(n+v)/2},$$

where $v > 0$ and $\boldsymbol{\Sigma}$ is positive definite. By (C.54) and (C.56), the density generator of $\mathbf{X}$ is

$$g_{(t_n)}(t) = \frac{\Gamma((n+v)/2)}{\Gamma(v/2)} \cdot \frac{1}{(v\pi)^{n/2}} \cdot \left(1 + \frac{t}{v}\right)^{-(n+v)/2}.$$

With (C.77),

$$\begin{aligned} f_R(r) &= \frac{2\pi^{n/2}}{\Gamma(n/2)} r^{n-1} g_{(t_n)}(r^2) \\ &= \frac{2r}{n} \cdot \frac{\Gamma((n+v)/2)}{\Gamma(n/2)\cdot\Gamma(v/2)} \cdot \left(\frac{n}{v}\right)^{n/2} \cdot \left(\frac{r^2}{n}\right)^{n/2-1} \cdot \left(1 + \frac{n}{v}\frac{r^2}{n}\right)^{-(n+v)/2} \\ &= \frac{2r}{n} f_F\left(\frac{r^2}{d}\right), \end{aligned}$$

where f_F is the p.d.f. of an $F_{n,v}$ random variable. ■

Theorem C.8 Let $\mathbf{X} \sim S_n(\varphi)$ have the stochastic representation $\mathbf{X} \stackrel{d}{=} R\mathbf{U}_n$ and $\Pr(\mathbf{X} = \mathbf{0}) = 0$. Assume that $\mathbf{X}$ is partitioned as $(\mathbf{X}_1, \mathbf{X}_2)$, where $\mathbf{X}_1$ is $m \times 1$ and $\mathbf{X}_2$ is $(n-m) \times 1$, $1 \leqslant m < n$. Further, let $\mathbf{X}_1$, $\mathbf{X}_2$ have the stochastic representation $\mathbf{X}_1 \stackrel{d}{=} R_1\mathbf{U}_1$ and $\mathbf{X}_2 \stackrel{d}{=} R_2\mathbf{U}_2$ with associated distribution functions F_{R_1} and F_{R_2}.

a) $R_1 \stackrel{d}{=} BR$, where $B^2 \sim \text{Beta}(m/2, (n-m)/2)$ and is independent of R.
b) The distribution of R_1 is absolutely continuous on $(0, \infty)$ with p.d.f.

$$f_{R_1}(r) = \frac{2s^{m-1}\Gamma(n/2)}{\Gamma(m/2)\Gamma((n-m)/2)} \int_0^\infty r^{-(n-2)}(r^2 - s^2)^{(n-m)/2-1}\, dF_R(r), \quad 0 < s < \infty. \tag{C.78}$$

c) $\mathbf{X}_1$ is absolutely continuous on $(0, \infty)$ with p.d.f.

$$f_{\mathbf{X}_1}(\mathbf{x}_1) = \frac{\Gamma(n/2)}{\pi^{m/2}\Gamma((n-m)/2)} \int_{\|\mathbf{x}_1\|}^\infty r^{-(n-2)}(r^2 - \mathbf{x}_1'\mathbf{x}_1)^{(n-m)/2-1}\, dF_R(r). \tag{C.79}$$

Proof: (From Cambanis et al., 1981, and Gupta and Varga, 1993)

a) $\mathbf{X} = (\mathbf{X}_1, \mathbf{X}_2) \stackrel{d}{=} R\mathbf{U}_n \sim S_n(\varphi)$ by C.1 (**d**) $\Leftrightarrow$ (**a**), where $\mathbf{X}_1$ is of size $m \times 1$ and $\mathbf{X}_2$ is of size $(n-m) \times 1$. Then, $\mathbf{X}_1 \stackrel{d}{=} R_1\mathbf{U}_1 \sim S_m(\varphi)$. From Theorem C.2(**b**), $\mathbf{X}_1 \stackrel{d}{=} R_1\mathbf{U}_1 \stackrel{d}{=} BR\mathbf{U}_1$ where R, B, and $\mathbf{U}_1$ are independent. Thus, $\mathbf{X}_1$ has two representations and uniqueness in law of the two representations as well as uniqueness of F_R from Theorem C.1(**d**) implies $R_1 \overset{d}{=} BR$.

b) Recall that the beta density function is given by

$$f_B(x; p, q) = \frac{\Gamma(p+q)}{\Gamma(p)\Gamma(q)} x^{p-1}(1-x)^{q-1}, \quad 0 \leqslant x \leqslant 1, \quad p, q > 0.$$

From Theorem C.8(**a**), we have $R_1 \overset{d}{=} BR$, where $B^2 \sim \text{Beta}(m/2, (n-m)/2)$. Thus, $\Pr(R_1 = 0) = \Pr(R = 0) = F_R(0)$, showing that R_1 has an atom with measure $F_{R_1}(0) = 0$ at zero and so is absolutely continuous on $(0, \infty)$. Hence we can write

$$\begin{aligned}
\Pr(0 < R_1 \leqslant c) &= \Pr(0 < BR \leqslant c) = \mathbb{E}[\mathbb{I}\{0 < BR_0 \leqslant c\}] = \mathbb{E}[\ \mathbb{E}[\mathbb{I}\{0 < BR \leqslant c\} \mid R]\ ] \\
&= \int_0^\infty \Pr\left(0 < B^2 \leqslant \frac{c^2}{R^2}\right) \mathrm{d}F_R(r) \\
&= \int_0^\infty \frac{\Gamma(n/2)}{\Gamma(m/2)\Gamma((n-m)/2)} \int_0^{\min(1, c^2/r^2)} x^{m/2-1}(1-x)^{(n-m)/2-1}\,\mathrm{d}x\,\mathrm{d}F_R(r).
\end{aligned} \tag{C.80}$$

Define $x = s^2/r^2$, with $\mathrm{d}x = 2s\,\mathrm{d}s/r^2$, so that the last line of (C.80) can be written as

$$\begin{aligned}
&\int_0^\infty \frac{\Gamma(n/2)}{\Gamma(m/2)\Gamma((n-m)/2)} \int_0^{\min(r,c)} \left(\frac{2s}{r^2}\right)\left(\frac{s}{r}\right)^{m-2}\left(1-\frac{s^2}{r^2}\right)^{(n-m)/2-1} \mathrm{d}s\,\mathrm{d}F_R(r) \\
&= \frac{2s^{m-1}\Gamma(n/2)}{\Gamma(m/2)\Gamma((n-m)/2)} \int_0^\infty \int_0^{\min(r,c)} r^{-2-m+2-(n-m)+2}(r^2-s^2)^{(n-m)/2-1}\,\mathrm{d}s\,\mathrm{d}F_R(r) \\
&= \int_0^c \frac{2s^{m-1}\Gamma(n/2)}{\Gamma(m/2)\Gamma((n-m)/2)} \left(\int_s^\infty r^{-(n-2)}(r^2-s^2)^{(n-m)/2-1}\,\mathrm{d}F_R(r)\right) \mathrm{d}s,
\end{aligned} \tag{C.81}$$

where the last line in (C.81) follows by Fubini's Theorem. (In particular, the integrand $r^{-(n-2)}(r^2-s^2)^{(n-m)/2-1}$ is non-negative.) Thus, by the (first) fundamental theorem of calculus, the p.d.f. in (C.78) can be identified by the c.d.f. in the last line of (C.81).

c) From Theorem C.8(**b**) it follows that $\mathbf{X}_1$ is absolutely continuous, and, as $\mathbf{X}_1 \overset{d}{=} R_1\mathbf{U}_1$ is the stochastic representation of $\mathbf{X}_1$, R_1 has the p.d.f. in Theorem C.8(**b**). Further, from (C.77),

$$f_{R_1}(r) = \frac{2\pi^{m/2}}{\Gamma(m/2)} r^{m-1} g_m(r^2), \quad r > 0,$$

so that, with (C.78), we get (setting $s = r$)

$$g_m(r^2) = \frac{\Gamma(n/2)}{\pi^{m/2}\Gamma((n-m)/2)} \int_s^\infty r^{-(n-2)}(r^2-s^2)^{(n-m)/2-1}\,\mathrm{d}F_R(r). \tag{C.82}$$

As $\mathbf{X}_1$ has an elliptic distribution, its p.d.f. is of the form $f_{\mathbf{X}_1}(\mathbf{x}_1) = g(\mathbf{x}_1'\mathbf{x}_1)$ by Theorem C.6. Moreover, by Theorem C.2(**a**), $\mathbf{x}_1'\mathbf{x}_1 = r^2$. Hence, (C.79) follows by setting $s^2 = \mathbf{x}_1'\mathbf{x}_1$ in (C.82). ■

Theorem C.9 Let $\mathbf{X} \sim S_n(\varphi)$ with p.d.f. $f_\mathbf{X}(\mathbf{x}) = g(\mathbf{x}'\mathbf{x})$. Let $\mathbf{X}$ be partitioned as $\mathbf{X} = (\mathbf{X}_1, \mathbf{X}_2)$, where $\mathbf{X}_1$ is $m \times 1$, $1 \leqslant m < n$. Then, $\mathbf{X}_1$ is absolutely continuous and its p.d.f. is

$$f_{\mathbf{X}_1}(\mathbf{x}_1) = \frac{\pi^{(n-m)/2}}{\Gamma((n-m)/2)} \int_{\mathbf{x}_1'\mathbf{x}_1}^\infty (t - \mathbf{x}_1'\mathbf{x}_1)^{(n-m)/2-1} h(t)\,dt. \tag{C.83}$$

*Proof: (From Gupta and Varga, 1993)*Let $\mathbf{X} = R\mathbf{U}_n$ be the stochastic representation of $\mathbf{X}$ and $F_R(r)$ the c.d.f. of R. Then, from Theorem C.6, equation (C.56), the p.d.f. of R is

$$f_R(r) = \frac{2\pi^{n/2}}{\Gamma(n/2)} r^{n-1} g(r^2),$$

so that, with (C.79),

$$\begin{aligned} f_{\mathbf{X}_1}(\mathbf{x}_1) \\ &= \frac{\Gamma(n/2)}{\pi^{m/2}\Gamma((n-m)/2)} \int_{\|\mathbf{x}_1\|}^{\infty} r^{-(n-2)} (r^2 - \mathbf{x}_1'\mathbf{x}_1)^{(n-m)/2-1} \frac{2\pi^{n/2}}{\Gamma(n/2)} r^{n-1} g(r^2)\, \mathrm{d}r \\ &= \frac{2\pi^{(n-m)/2}}{\Gamma((n-m)/2)} \int_{\|\mathbf{x}_1\|}^{\infty} r(r^2 - \mathbf{x}_1'\mathbf{x}_1)^{(n-m)/2-1} g(r^2)\, \mathrm{d}r. \end{aligned}$$

Let $u = r^2$, so that, with $\mathrm{d}r = \mathrm{d}u/2r$,

$$f_{\mathbf{X}_1}(\mathbf{x}_1) = \frac{\pi^{(n-m)/2}}{\Gamma((n-m)/2)} \int_{\mathbf{x}_1'\mathbf{x}_1}^{\infty} (u - \mathbf{x}_1'\mathbf{x}_1)^{(n-m)/2-1} g(u)\, \mathrm{d}u,$$

which is (C.83). ■

The next theorem characterizes marginal densities (marginal density generators) of $\mathbf{X} \sim S_n(\varphi)$ and the fundamental relationship between them.

Theorem C.10 Let $\mathbf{X} \sim S_n(\varphi)$ with p.d.f. $f_{\mathbf{X}}(\mathbf{x}) = g(\mathbf{x}'\mathbf{x})$. Let $\mathbf{X}$ be partitioned as $\mathbf{X} = (\mathbf{X}_1, \mathbf{X}_2)$, where $\mathbf{X}_1$ is $m \times 1$. Let $\mathbf{X}_1$ have the p.d.f. $f_{\mathbf{X}_1}(\mathbf{x}_1) = g_m(\mathbf{x}_1'\mathbf{x}_1)$.

a) For $1 \leqslant m \leqslant n-2$, the marginal densities $\mathbf{X}_1$ are related by

$$g_m(u) = \pi \int_u^{\infty} g_{m+2}(t)\, \mathrm{d}t. \tag{C.84}$$

b) The marginal densities of dimension $1 \leqslant m \leqslant n-1$ are continuous and the marginal densities of dimension $1 \leqslant m \leqslant n-2$ are differentiable almost everywhere (a.e.).
c) The univariate marginal densities for $n \geqslant 2$ are non-decreasing on $(-\infty, 0)$ and non-increasing on $(0, \infty)$.
d) For $1 \leqslant m \leqslant n-2$, the (marginal) density generators are related by

$$g_{m+2}(t) = (-1/\pi) g_m'(t), \quad t > 0 \quad \text{almost everywhere (a.e.)}, \tag{C.85}$$

where $g_m'(\cdot)$ is the derivative of $g_m(\cdot)$ with respect to t.
e) Equation (C.85) allows the construction of all marginal densities knowing only the univariate marginal density.

Proof:

a) To verify $g_m(u) = \pi \int_u^{\infty} g_{m+2}(y)\, \mathrm{d}y$, set $m = n-2$, and use (C.79). This yields $g_{n-2} = \pi \int_u^{\infty} g_n(t)\, \mathrm{d}t$, and (C.84) follows for any $1 \leqslant m \leqslant n-2$.
b) Continuity is intuitively clear, and follows from differentiability. Differentiability follows from the fundamental theorem of calculus for Lebesgue integrals; see, e.g., Stein and Shakarchi (2005, Thm. 3.11).

c) This follows by setting $t := x^2$ in (C.56), where for $x \in (-\infty, 0)$, (C.56) is non-decreasing and for $x \in (0, \infty)$, (C.56) is non-increasing.
d) This follows as in Theorem C.10(**b**) and by noting that the derivative $g'(t)$ in (C.85) with respect to t is well-defined: In (C.56), the square root is uniformly continuous on $[0, \infty)$ and $f_R(t^{1/2})$ is, by assumption, continuous and non-decreasing on $[0, \infty)$ (see the proof of Theorem C.1(**d**)). Thus, $f_R'(t^{1/2})$ exists (i.e., is finite) with probability one by the Lebesgue differentiation theorem (see, e.g., Stein and Shakarchi, 2005, Cor. 3.7), hence all terms in (C.56) are differentiable with respect to t.
e) This is immediate from (C.84). ■

The following result, given as Theorem C.11, is used in the proof of the subsequent Theorem C.12.

Theorem C.11 Let X be a random variable with characteristic function φ_X. If there exists an $\epsilon > 0$ such that $|\varphi_X(t)| = 1$ for all $t \in [-\epsilon, \epsilon]$, then X is degenerate. That is, there exists a $c \in \mathbb{R}$ such that $\Pr(X = c) = 1$.

Proof: Let $t' \in \mathbb{Q} \cap [-\epsilon, \epsilon] \backslash \{0\}$. If $|\varphi_X(t')| = 1$, then there exists $\theta = \theta(t')$ such that $\theta \in [0, 2\pi)$ and $\varphi_X(t') := e^{i\theta}$. Hence,

$$\mathbb{E}[1 - e^{i(t'X-\theta)}] = 0, \quad \text{so that} \quad \mathbb{E}[\text{Re}(1 - e^{i'(t'X-\theta)})] = 0.$$

As $\text{Re}(1 - e^{i(t'X-\theta)}) \geqslant 0$, this yields

$$\begin{aligned} \Pr(e^{i(t'X-\theta)} = 1) &= \Pr(\cos(t'X - \theta) + i\sin(t'X - \theta) = 1) \\ &= \Pr(\cos(t'X - \theta) + i\cos(\pi/2 - (t'X - \theta) = 1), \end{aligned} \tag{C.86}$$

so that we need

$$X(t') \in \left\{ x : x = \frac{\theta}{t'} + \frac{2\pi z}{t'}, \quad z \in \mathbb{Z}, \quad \theta \in [0, 2\pi) \right\},$$

for (C.86) to hold. Now take $t'' \in \mathbb{Q} \cap [-\epsilon, \epsilon] \backslash \{0\}$ such that $t'' \neq t'$ on the set $\mathbb{Q} \cap [-\epsilon, \epsilon] \backslash \{0\}$. Then

$$\Pr(X(t') = X(t'')) = 1$$

by (C.86), so that

$$X \in \{X(t') \cap X(t'')\} = \{c\}$$

is at most a singleton (i.e., contains at most one element) because t' and t'' are disjoint on the set $\mathbb{Q} \cap [-\epsilon, \epsilon] \backslash \{0\}$. By a standard argument ($\mathbb{Q}$ is dense in $\mathbb{R}$, i.e., every real number can be approximated by a sequence of rational numbers), this holds for $t \in [-\epsilon, \epsilon] \backslash \{0\}$, therefore $X(t) = \{c\}$ for all $t \in \mathbb{R} \backslash \{0\}$. Hence, $\Pr(X = c) = 1$. ■

The next theorem shows that the stochastic as well as the parametric representation of non-degenerated elliptically contoured distributions are essentially unique, meaning, up to scaling with a positive constant.

Theorem C.12 Let $\mathbf{Y}$ be a random vector from a non-degenerate distribution (i.e. $\Pr(\mathbf{Y} = \mathbf{c}) \neq 1$).

a) If $\mathbf{Y} \sim \text{EC}_n(\boldsymbol{\mu}, \boldsymbol{\Sigma}, \varphi)$ and $\mathbf{Y} \sim \text{EC}_n(\boldsymbol{\mu}^*, \boldsymbol{\Sigma}^*, \varphi^*)$, there exists a constant c such that

$$\boldsymbol{\mu}^* = \boldsymbol{\mu}, \quad \boldsymbol{\Sigma}^* = c\boldsymbol{\Sigma}, \quad \varphi^*(\mathbf{t}) = \varphi(c^{-1}\mathbf{t}).$$

b) If $\mathbf{Y} \overset{d}{=} \boldsymbol{\mu} + R\mathbf{A}'\mathbf{U}_m$ and $\mathbf{Y} \overset{d}{=} \boldsymbol{\mu}^* + R^*\mathbf{A}^{*\prime}\mathbf{U}_{m^*}$, where $m^* \leqslant m$, then there exists a constant $c > 0$ such that

$$\boldsymbol{\mu}^* = \boldsymbol{\mu}, \quad \mathbf{A}^{*\prime}\mathbf{A}^* = c\mathbf{A}'\mathbf{A}, \quad R^* \overset{d}{=} \frac{RB}{\sqrt{c}},$$

where B is independent of R, $B^2 \sim \text{Beta}(m^*/2, (m - m^*)/2)$ if $m^* < m$ and $B \equiv 1$ if $m^* = m$.

Proof: (Motivated from Cambanis et al., 1981, and Gupta and Varga, 1993)

a) Uniqueness of $\boldsymbol{\mu}$: $\mathbf{Y} - \boldsymbol{\mu}$ and $\mathbf{Y} - \boldsymbol{\mu}^*$ are both symmetric around $\mathbf{0}$, so $\mathbf{Y} - \boldsymbol{\mu} \overset{d}{=} -(\mathbf{Y} - \boldsymbol{\mu})$ and $\mathbf{Y} - \boldsymbol{\mu}^* \overset{d}{=} -(\mathbf{Y} - \boldsymbol{\mu}^*)$. Therefore,

$$(\mathbf{Y} - \boldsymbol{\mu}) = \boldsymbol{\mu} - \boldsymbol{\mu}^* - (\mathbf{Y} - \boldsymbol{\mu}^*) \overset{d}{=} \boldsymbol{\mu} - \boldsymbol{\mu}^* + (\mathbf{Y} - \boldsymbol{\mu}^*) \overset{d}{=} \mathbf{Y} - (2\boldsymbol{\mu}^* - \boldsymbol{\mu}),$$

implying $\boldsymbol{\mu}^* = \boldsymbol{\mu}$.
Uniqueness (up to scaling with positive constant) of $\boldsymbol{\Sigma}$:
Write $\boldsymbol{\Sigma} = (\sigma_{ij})_{1\leqslant i,j\leqslant n}$, $\boldsymbol{\Sigma}^* = (\sigma^*_{ij})_{1\leqslant i,j\leqslant n}$ and $\boldsymbol{\mu} = (\mu_1, \dots, \mu_n)$. As $\mathbf{Y} = (Y_1, \dots, Y_n)$ is non-degenerate, at least one of its components Y_j is non-degenerate. As both representations have the same law, $\varphi_{Y_j-\mu_j}(u) = \varphi^*_{Y_j-\mu_j}(u)$, $u \in \mathbb{R}$ (see the uniqueness theorem in (C.4)), so that by Theorem C.4(**a**),

$$\exp(iu\mu_j)\varphi(\sigma_{jj}u^2) = \exp(iu\mu_j)\varphi^*(\sigma^*_{jj}u^2), \quad \sigma_{jj}, \sigma*_{jj} > 0.$$

Then, $\varphi * (\cdot) = \varphi(c^{-1}, \cdot)$ where $c = \sigma^*_{jj}/\sigma_{jj}$. Hence, the characteristic generator of $\mathbf{Y} - \boldsymbol{\mu}$ is given, again, by the same argument (uniqueness theorem in (C.4)), as $\varphi(\mathbf{t}'\boldsymbol{\Sigma}\mathbf{t}) = \varphi^*(\mathbf{t}'\boldsymbol{\Sigma}^*\mathbf{t}) = \varphi(c^{-1}\mathbf{t}'\boldsymbol{\Sigma}^*\mathbf{t})$, $\mathbf{t} \in \mathbb{R}^n$.
Now, suppose to the contrary, $\boldsymbol{\Sigma}^* \neq c\boldsymbol{\Sigma}$. We will show that $\mathbf{Y}$ is necessarily degenerate (i.e. $\Pr(\mathbf{Y} = \mathbf{c}) = 1$): For some $\mathbf{t}_0 \in \mathbb{R}^n$, we then have $\mathbf{t}_0'\boldsymbol{\Sigma}\mathbf{t}_0 \neq c\mathbf{t}_0'\boldsymbol{\Sigma}\mathbf{t}_0$. The c.f. of $\mathbf{Y}$ at $u\mathbf{t}_0$ is

$$\varphi(u(\mathbf{t}_0'\boldsymbol{\Sigma}\mathbf{t}_0)) = \varphi^*(uc(\mathbf{t}_0'\boldsymbol{\Sigma}\mathbf{t}_0)),$$

where the c.f.s of both representations need to be equal, as both representations have the same distribution by assumption; see (C.4). On the other hand, the c.f. of $\mathbf{Y}$ at $u\mathbf{t}_0$ can be expressed as $\varphi^*(u(\mathbf{t}_0'\boldsymbol{\Sigma}^*\mathbf{t}_0))$, so that we require

$$\varphi^*(uc(\mathbf{t}_0'\boldsymbol{\Sigma}\mathbf{t}_0)) = \varphi^*(u(\mathbf{t}_0'\boldsymbol{\Sigma}^*\mathbf{t}_0)). \tag{C.87}$$

Case 1: $\mathbf{t}_0'\boldsymbol{\Sigma}\mathbf{t}_0 = 0$ or $\mathbf{t}_0'\boldsymbol{\Sigma}^*\mathbf{t}_0 = 0$. From (C.87), $\varphi^*(u, \cdot) = 1$ has to hold for every $u \in \mathbb{R}$, implying by Theorem C.11 that $\mathbf{Y}$ is degenerate. But this is impossible, as $\mathbf{Y}$ is non-degenerate by assumption.

Case 2: $\mathbf{t}_0'\boldsymbol{\Sigma}\mathbf{t}_0 \neq 0$ and $\mathbf{t}_0'\boldsymbol{\Sigma}^*\mathbf{t}_0 \neq 0$. Define

$$d := c\frac{\mathbf{t}_0'\boldsymbol{\Sigma}\mathbf{t}_0}{\mathbf{t}_0'\boldsymbol{\Sigma}^*\mathbf{t}_0}.$$

Then, by the assumption $\mathbf{t}_0'\boldsymbol{\Sigma}\mathbf{t}_0 \neq c\mathbf{t}_0'\boldsymbol{\Sigma}\mathbf{t}_0$, either $d \in (0, 1)$ or $d \in (1, \infty)$ and with (C.87) we get $\varphi^*(u, \cdot) = \varphi^*(du)$. By induction, we further have the two equivalent identities

$$\varphi^*(u, \cdot) = \varphi^*(d^m u) \text{ and } \varphi^*(u, \cdot) = \varphi^*\left(\left(\frac{1}{d}\right)^m u\right), \quad m \in \mathbb{N}_+.$$

Now either $\lim_{m\to\infty} d^m = 0$ if $d \in (0,1)$ or $\lim_{m\to\infty} \left(\frac{1}{d}\right)^m = 0$ if $d \in (1,\infty)$ so that, by dominated convergence, $\varphi^*(0) = 1$, and from the (uniform) continuity of characteristic functions, we have $\varphi^*(u) = 1$ for every $u \in \mathbb{R}$. Hence, from Theorem C.11, $\mathbf{Y}$ is degenerate. By the non-degeneracy assumption of $\mathbf{Y}$, this yields a contradiction.

As such, $\mathbf{\Sigma}^* = c\mathbf{\Sigma}$.

b) By Theorem C.4(**b**),

$$\mathbf{Y} \sim \mathrm{EC}_n(\boldsymbol{\mu}, \mathbf{A}', \mathbf{A}, \varphi) \quad \text{and} \quad \mathbf{Y} \sim \mathrm{EC}_n(\boldsymbol{\mu}^*, \mathbf{A}^{*\prime}, \mathbf{A}^*, \varphi^*),$$

so that, by Theorem C.12(**a**), $\boldsymbol{\mu}^* = \boldsymbol{\mu}$, $\mathbf{A}^{*\prime}\mathbf{A}^* = c\mathbf{A}'\mathbf{A}$, $\varphi^*(\cdot) = \varphi(c^{-1}\cdot)$.

If $m^* < m$, then it follows from Theorem C.8(**a**) with $R := R_m$ and $R_1 := R_{m^*} = c^{1/2}R^*$ that $R_{m^*} \stackrel{d}{=} BR_m$, where $B = \mathbf{Y}_{m^*}/\|\mathbf{Y}_m\|$. Therefore, we have $R^* \stackrel{d}{=} c^{-1/2}BR$.

If $m^* = m$, then the uniqueness of F_R in Theorem C.1(**d**) implies $R_{m^*} \stackrel{d}{=} R_m$, and therefore $R^* \stackrel{d}{=} c^{-1/2}R$, so $B \equiv 1$. ■

The next theorem shows that the sum of independent elliptical random variables with the same dispersion matrix is elliptical. Moreover, the sum of two independent elliptical random vectors with the same dispersion matrix, which are dependent only through their radial parts, is also elliptical. This theorem is due to Hult and Lindskog (2002).

Theorem C.13

a) Let $\mathbf{Y}_i \sim \mathrm{EC}_n(\boldsymbol{\mu}_i, \mathbf{\Sigma}, \varphi_i)$. Then the sum of independent elliptically distributed random variables $\mathbf{Y}_i$, $i = 1, \ldots, n$, with identical dispersion matrices, is also elliptical.
b) Let $\mathbf{Y}_1$ and $\mathbf{Y}_2$ be two $n \times 1$ elliptically distributed random variables with respective stochastic representations

$$\mathbf{Y}_1 \stackrel{d}{=} \boldsymbol{\mu}_1 + R_1\mathbf{A}\mathbf{U}_1 \sim \mathrm{EC}_n(\boldsymbol{\mu}_1, \mathbf{\Sigma}, \varphi_1),$$

$$\mathbf{Y}_2 \stackrel{d}{=} \boldsymbol{\mu}_2 + R_2\mathbf{A}\mathbf{U}_2 \sim \mathrm{EC}_n(\boldsymbol{\mu}_2, \mathbf{\Sigma}, \varphi_2),$$

where (R_1, R_2), $\mathbf{U}_1$, and $\mathbf{U}_2$ are mutually independent, whereas R_1 and R_2 may depend on each other. Then, $\mathbf{Y}_1 + \mathbf{Y}_2 \sim \mathrm{EC}_n(\boldsymbol{\mu}_1 + \boldsymbol{\mu}_2, \mathbf{\Sigma}, \varphi)$, where $\mathbf{\Sigma} = \mathbf{A}'\mathbf{A}$. Moreover, if R_1 and R_2 are independent, then $\varphi(\mathbf{t}) = \varphi_1(\mathbf{t})\varphi_2(\mathbf{t})$.

Proof: (From Hult and Lindskog, 2002, with more details)

a) Define $\boldsymbol{\mu} := \sum_i^n \boldsymbol{\mu}_i$ and compute

$$\mathbb{E}\left[\exp\left(i\mathbf{t}'\sum_i^n(\mathbf{Y}_i - \boldsymbol{\mu}_i)\right)\right] = \prod_{i=1}^n \mathbb{E}[\exp(i\mathbf{t}'(\mathbf{Y}_i - \boldsymbol{\mu}_i))]$$

$$= \prod_i^n \varphi_{(\mathbf{Y}_i - \boldsymbol{\mu}_i)}(\mathbf{t}'\mathbf{\Sigma}\mathbf{t}) \equiv g(\mathbf{t}'\mathbf{\Sigma}\mathbf{t}),$$

which shows that the last expression has the same functional form as in Theorem C.4(**a**). Hence, the sum of independent elliptically distributed random variables is elliptically distributed.

b) Without loss of generality, assume $\boldsymbol{\mu}_1 = \mathbf{0}$, $\boldsymbol{\mu}_2 = \mathbf{0}$ (or just consider the centered random variables, $\mathbf{Y}_1 - \boldsymbol{\mu}_1$ and $\mathbf{Y}_2 - \boldsymbol{\mu}_2$) and set $\mathbf{Z}_1 := \mathbf{A}\mathbf{U}_1$ and $\mathbf{Z}_2 := \mathbf{A}\mathbf{U}_2$. Let $\varphi^{(r_1)}$ be the characteristic generator of $(R_2 \mid R_1 = r_1)\mathbf{Z}_2$, let Ω_1 be the characteristic generator of $\mathbf{Z}_1$, and let $F_{R_1}(r_1)$ be the distribution function of R_1. Then, as $\mathbf{Y}_1$ and $\mathbf{Y}_2$ are independent, $R_1\mathbf{Z}_1$ and $R_2\mathbf{Z}_2$ are independent, and, for all $\mathbf{t} \in \mathbb{R}^n$,

$$\begin{aligned}
\varphi_{R_1\mathbf{Z}_1+R_2\mathbf{Z}_2}(\mathbf{t}) &= \varphi_{R_1\mathbf{U}_1}(\mathbf{t})\varphi_{R_2\mathbf{Z}_2}(\mathbf{t}) = \mathbb{E}[\ \mathbb{E}[\varphi_{R_1\mathbf{Z}_1}(\mathbf{t})\varphi_{R_2\mathbf{Z}_2}(\mathbf{t}) \mid R_1]\ ] \\
&= \int_0^\infty \varphi_{r_1\mathbf{Z}_1}(\mathbf{t})\varphi_{(R_2|R_1=r_1)\mathbf{Z}_2}(\mathbf{t})\, \mathrm{d}F_{R_1}(r_1) \\
&= \int_0^\infty \mathbb{E}[\mathrm{e}^{i\mathbf{t}' r_1 \mathbf{Z}_1}]\mathbb{E}[\mathrm{e}^{i\mathbf{t}'(R_2|R_1=r_1)\mathbf{Z}_2}]\, \mathrm{d}F_{R_1}(r_1) \\
&= \int_0^\infty \mathbb{E}[\mathrm{e}^{i r_1^2 \mathbf{t}'(\mathbf{A}\mathbf{U}_1)'(\mathbf{A}\mathbf{U}_1)\mathbf{t}}]\mathbb{E}[\mathrm{e}^{i(R_2|R_1=r_1)^2\mathbf{t}'(\mathbf{A}\mathbf{U}_2)'(\mathbf{A}\mathbf{U}_2)\mathbf{t}}]\, \mathrm{d}F_{R_1}(r_1) \\
&= \int_0^\infty \Omega_1(r^2\mathbf{t}'\boldsymbol{\Sigma}\mathbf{t})\varphi^{(r_1)}(\mathbf{t}'\boldsymbol{\Sigma}\mathbf{t})\, \mathrm{d}F_{R_1}(r_1),
\end{aligned} \tag{C.88}$$

where the (crucial) fourth line follows from the fact that $\mathbf{U}_1$ and $\mathbf{U}_2$ are spherical random variables, and by Theorem C.1(**a**) ⇒ (**b**), i.e., the characteristic function of a spherical random variable $\varphi_{\mathbf{X}}(\mathbf{t})$ has necessarily the functional form $\varphi(\|\mathbf{t}\|^2)$. By Theorem C.4(**a**) or by Theorem C.1(**b**) ⇒ (**a**) (by setting without loss of generality $\boldsymbol{\Sigma} = \mathbf{I}_n$), it follows that the last line in (C.88) has the desired functional form of an elliptic distributed random variable. Hence, we get $\mathbf{Y}_1 + \mathbf{Y}_2 \sim \mathrm{EC}_n(\boldsymbol{\mu}_1 + \boldsymbol{\mu}_2, \boldsymbol{\Sigma}, \varphi)$, where (setting $u := \mathbf{t}'\boldsymbol{\Sigma}\mathbf{t}$)

$$\varphi(u) := \int_0^\infty \Omega_1(r^2 u)\varphi^{(r_1)}(u)\, \mathrm{d}F_{R_1}(r_1). \tag{C.89}$$

Moreover, if R_1 and R_2 are independent, then $\varphi^{(r_1)}(u) = \varphi_2(u)$ and (C.89) becomes

$$\begin{aligned}
\varphi(u) &:= \int_0^\infty \Omega_1(r^2 u)\varphi^{(r_1)}(u)\, \mathrm{d}F_{R_1}(r_1) \\
&= \varphi_2(u) \int_0^\infty \Omega_1(r^2 u)\, \mathrm{d}F_{R_1}(r_1) = \varphi_1(u)\varphi_2(u),
\end{aligned}$$

where the last equality follows by the identity in (C.35) (setting $u := \mathbf{t}'\mathbf{t}$), stated in the proof of Theorem C.1(**d**), i.e., we have $\varphi_1(u) = \int_0^\infty \Omega_1(r^2 u)\, \mathrm{d}F_{R_1}(r_1)$. ■

Remarks

a) Essentially, the sum of i.i.d. elliptical random variables is elliptical. But this does not imply that the sum is of the same type, i.e., it usually does not belong to the location-scale family of its components. This is only given for the class of multivariate sum-stable distributions; see, e.g., Embrechts et al. (2000, p. 522) and Rachev and Mittnik (2000, Sec. 7.1).
b) The property in Theorem C.13(b) is useful for time-series analysis when assuming a sequence $R_1, R_2, \ldots$ of dependent (i.e., heteroskedastic) generating (or radial) variates. See the next example. ■

Example C.14 A natural application of Theorem C.13 is in the context of multivariate time series. Let $\mathbf{Y}_t = \sigma_t \mathbf{Z}_t$, $t \in \mathbb{Z}$, where the random variables $\mathbf{Z}_t \sim \mathrm{EC}_n(\mathbf{0}, \boldsymbol{\Sigma}, \varphi_t)$ are mutually independent and independent of the non-negative (univariate) random variable σ_t for all t. The σ_t, on the other hand, are allowed to be dependent. Then, for every $t \in \mathbb{Z}$, by Theorem C.13(**b**), $\mathbf{Y}_t$ is elliptically distributed with dispersion matrix $\boldsymbol{\Sigma}$ and so are all partial sums, $S_T = \sum_{t=1}^{T} \mathbf{Y}_t$. ■

The following theorem states that a linear combination of r.v.s whose joint distribution is elliptically symmetric is also elliptically symmetric, and, thus, that the marginal distributions of **Y** are also elliptically symmetric with the same characteristic generator.

Theorem C.14 Let **Y** be elliptically symmetric with $\mathbf{Y} \sim \mathrm{EC}_n(\boldsymbol{\mu}, \boldsymbol{\Sigma}, \varphi)$. For vector $\mathbf{v} \in \mathbb{R}^m$ and real $n \times m$ matrix $\mathbf{B}$, $\mathbf{v} + \mathbf{B}'\mathbf{Y} \sim \mathrm{EC}_m(\mathbf{v} + \mathbf{B}'\boldsymbol{\mu}, \mathbf{B}'\boldsymbol{\Sigma}\mathbf{B}, \varphi)$.

Proof: We have

$$\mathbf{v} + \mathbf{B}'\mathbf{Y} \overset{d}{=} \mathbf{v} + \mathbf{B}'(\boldsymbol{\mu} + R\mathbf{A}'\mathbf{U}_k) \overset{d}{=} (\mathbf{v} + \mathbf{B}'\boldsymbol{\mu}) + R(\mathbf{A}\mathbf{B})'\mathbf{U}_k,$$

from the definition of elliptically distributed r.v.s and Theorem C.4(**b**). ■

Example C.15 The three results (C.26), (C.27), and (C.28) given above are all special cases of Theorem C.14. ■

The following theorem gives the first two moments of this class of distributions.

Theorem C.15 Let $\mathbf{Y} \sim \mathrm{EC}_n(\boldsymbol{\mu}, \boldsymbol{\Sigma}, \varphi)$ and $\mathbb{E}[R] < \infty$, where R is from the representation in (C.50). Then

$$\mathbb{E}[\mathbf{Y}] = \boldsymbol{\mu}, \qquad \mathrm{Cov}(\mathbf{Y}) = \frac{\mathbb{E}[R^2]}{\mathrm{rank}(\boldsymbol{\Sigma})}\boldsymbol{\Sigma} = -2\varphi'(0)\boldsymbol{\Sigma}, \tag{C.90}$$

$$\mathbb{E}[\mathbf{Y}\mathbf{Y}'] = \boldsymbol{\mu}\boldsymbol{\mu}' - 2\varphi'(0)\boldsymbol{\Sigma}, \tag{C.91}$$

where φ' is the first derivative of φ.

Proof: Denoting $k = \mathrm{rank}(\boldsymbol{\Sigma})$, we have $\mathbf{Y} \overset{d}{=} \boldsymbol{\mu} + R\mathbf{A}'\mathbf{U}_k$. By Theorem C.1(**e**) (i.e., $\mathbb{E}[\mathbf{U}_k] = \mathbf{0}$ and $\mathbb{V}(\mathbf{U}_k) = \mathbf{I}_k/k$), we obtain

$$\mathbb{E}[\mathbf{Y}] = \boldsymbol{\mu} + \mathbb{E}[R]\mathbf{A}'\mathbb{E}[\mathbf{U}_k] = \boldsymbol{\mu}$$

and

$$\begin{aligned}\mathrm{Cov}(\mathbf{Y}) &= \mathrm{Cov}(R\mathbf{A}'\mathbf{U}_k) = \mathbb{E}[R^2]\mathbf{A}'\ \mathrm{Cov}(\mathbf{U}_k)\mathbf{A} \\ &= \mathbb{E}[R^2]\frac{1}{k}\mathbf{A}'\mathbf{I}_k\mathbf{A} = \frac{1}{k}\mathbb{E}[R^2]\boldsymbol{\Sigma}.\end{aligned}$$

Assume without loss of generality that $\boldsymbol{\mu} = \mathbf{0}$ and $\boldsymbol{\Sigma} = \mathbf{I}_n$. Let $\mathbf{X} = (\mathbf{Y} - \boldsymbol{\mu})'\mathbf{A}^-$. The c.f. of **X** is then $\varphi_{\mathbf{X}}(\mathbf{t}) = \varphi(\mathbf{t}'\mathbf{t})$, where $\mathbf{t} = (t_1, \dots, t_n)$. Then

$$\frac{\partial \varphi_{\mathbf{X}(\mathbf{t})}}{\partial t_i} = \frac{\partial \varphi\left(\sum_{i=1}^{n} t_i^2\right)}{\partial t_i} = 2t_i\varphi'\left(\sum_{i=1}^{n} t_i^2\right).$$

Hence,

$$\frac{\partial^2 \varphi_{\mathbf{X}(\mathbf{t})}}{\partial^2 t_i} = 2\varphi'\left(\sum_{i=1}^n t_i^2\right) + 4t_i^2\varphi''\left(\sum_{i=1}^n t_i^2\right), \quad i = j,$$

and

$$\frac{\partial \varphi_{\mathbf{X}(\mathbf{t})}}{\partial t_i \partial t_j} = 4t_i t_j \varphi'\left(\sum_{i=1}^n t_i^2\right), \quad i \neq j.$$

Therefore,

$$\left.\frac{\partial^2 \varphi_{\mathbf{X}(\mathbf{t})}}{\partial^2 t_i \partial t_j}\right|_{t=0} = 2\varphi'(0), \quad i = j, \qquad \left.\frac{\partial^2 \varphi_{\mathbf{X}(\mathbf{t})}}{\partial^2 t_i \partial t_j}\right|_{t=0} = 0, \quad i \neq j.$$

From $\varphi_{\mathbf{X}}(\mathbf{t})$, it follows that $\mathrm{Cov}(\mathbf{X}) = -2\varphi'(0)$, so that, with $\mathbf{X} = (\mathbf{Y} - \boldsymbol{\mu})\mathbf{A}^-$, we have $\mathbb{E}[\mathbf{Y}\mathbf{Y}'] = \boldsymbol{\mu}\boldsymbol{\mu}' - 2\varphi'(0)\boldsymbol{\Sigma}$. Thus, $\mathrm{Cov}(\mathbf{Y}) = -2\varphi'(0)\boldsymbol{\Sigma}$. ■

Example C.16 ***(Example C.8 cont.)***
We found for $\mathbf{X} = \sqrt{k}\mathbf{Z}/S \sim t_k$ that $R_*^2/n \sim F(n, k)$, and, from (just after) Example I.9.8, $\mathbb{E}[F] = k/(k-2)$ for $F \sim F(n, k)$, so that $\mathrm{Cov}(\mathbf{X}) = \{k/(k-2)\}\mathbf{I}_n$. ■

Remarks

a) In the ANOVA setting, Butler (1986) shows that, if the assumption of spherically symmetric errors is not specifically tied to the sample size, then the standard F test must be UMPI. More specifically, this means that, if the error distribution is what is called 1-*extendible*, then the standard F test must be UMPI. An $n \times 1$ error distribution is said to be 1-extendible if it is the marginal distribution of an $(n+1) \times 1$ error distribution that is also spherically symmetric. In other words, if one more observation *were* sampled in addition to the n observations, and the spherically symmetric assumption *was* maintained with the sample of size $n+1$, then the F test must be UMPI.
Similar results from Butler (1986) show that Hotelling's T^2-test in MANOVA is UMPI if the assumption of left orthogonal invariance for the $n \times k$ error matrix is not tied to sample size n. More specifically, if the $n \times k$ matrix error distribution is k-extendible as the marginal distribution of an $(n+k) \times n$ error matrix that is also left orthogonally invariant, then Hotelling's T^2-test in MANOVA is UMPI. This extends previous results of Dawid (1977), Jensen (1981), and Kariya (1981a,b).
b) There are several ways of testing a data set for ellipticity; see Zhu and Neuhaus (2003), Bodnar and Schmid (2007), Huffer and Park (2007), Su (2012), Bianco et al. (2017), and the references therein, as well as the discussion and illustrations in McNeil et al. (2005, Sec. 3.3.5). Below in Section C.2.4, we illustrate the use of one test on financial asset returns data.
c) A confidence set for the mean of a spherically symmetric distribution based on bootstrap inference is developed by Samworth (2005).
d) Formulae for the expected shortfall (see, e.g., Section III.A.7) associated with a portfolio (weighted sums of margins) from an elliptic distribution have been derived by Landsman and Valdez (2003), Kamdem (2005), and Dobrev et al. (2017). Let $\mathbf{X} \sim S_n(g)$, where g is the density generator. Let $P = \mathbf{w}'\mathbf{X}$ be the weighted linear combination of interest. Then, with $q = \mathrm{VaR}(P, \xi)$ being the ξ-level tail quantile (where VaR is the value-at-risk) for some $0 < \xi < 1$, Dobrev et al. (2017) show that

$$\mathrm{ES}(P, \xi) = \frac{1}{\xi} \frac{\pi^{(n-1)/2}}{2\Gamma((n+1)/2)} \int_{q^2}^{\infty} (u - q^2)^{(n-1)/2} g(u) \, du.$$

For $\mathbf{X} \sim \mathrm{EC}_n(\boldsymbol{\mu}, \boldsymbol{\Sigma}, \varphi)$, the VaR quantile and the ES are just linear transforms of their corresponding elliptic values, with the location being $\mathbf{w}'\boldsymbol{\mu}$ and scale $(\mathbf{w}'\boldsymbol{\Sigma}\mathbf{w})^{1/2}$.

e) Stein's lemma (see, e.g., Section III.A.7) can be generalized to the elliptic setting; see Landsman and Nešlehová (2008).

f) Quadratic forms in elliptic random vectors have been studied. See, e.g., King (1980), Fang et al. (1989, p. 149), Díaz-García (2013), and the references therein.

g) Frahm (2004) introduces the class of generalized elliptical distributions given by

$$\mathbf{Y} \overset{d}{=} \boldsymbol{\mu} + R\mathbf{A}\mathbf{U}_k, \tag{C.92}$$

where $\mathbf{U}_k$ is, as before, a $k \times 1$ random variable distributed uniformly on S^{k-1}, R is a (not necessarily non-negative) random variable, and $\mathbf{A} \in \mathbb{R}^{d \times k}$.

In contrast to elliptical distributions, the generating variate R may become negative and even depend on the direction determined by $\mathbf{U}_k$. Hence, the dependence structure of R and $\mathbf{U}_k$ constitutes the multivariate c.d.f. of $\mathbf{Y}$. In particular, $\mathbf{X}$ does not need to be radially symmetric anymore. See Frahm (2004), Chapter 3.2 for details.

Frahm (2004) shows that the density function of general elliptic distributed random variables still has the functional form as in Theorem C.6, where the density generator is now given by two additively separable terms, i.e.,

$$g(t) := \frac{\Gamma(k/2)}{2\pi^{k/2}} \cdot t^{-1/2(k-1)} \cdot \left(f_{R|\mathbf{U}_k=-\mathbf{u}}(-t^{1/2}) + f_{R|\mathbf{U}_k=\mathbf{u}}(t^{1/2}) \right), \quad t > 0, \tag{C.93}$$

and $f_{R|\mathbf{U}_k=\mathbf{u}}$ is the conditional p.d.f. of R under $\mathbf{U}_k = \mathbf{u} \in S^{k-1}$. The proof follows along the same lines as in Theorem C.6, by noting that the transformation $h : (r, \mathbf{u}) \mapsto r\mathbf{u}$ is no longer injective (due to the domain of R) giving rise to the two additively separable terms in (C.93). Naturally, Schoenberg's key insight as stated in Theorem C.1(**d**), does not hold, as $\mathbf{Y}$ is not radial symmetric anymore. The class of generalized elliptical distribution contains the class of conditional scale distributions (see Frahm, 2004, Example 13) giving rise to asymmetric density contours and heavy tails prominent in empirical finance.

h) The class of **meta-elliptical distributions** is constructed from specified marginal distributions with a given dependence structure, where the margins can be arbitrarily chosen. The density function of a meta-elliptical distribution can be decomposed into the **density weighting function** and the product of the marginal densities. An example of their use is given in Chapter 12.

Consider the case $\mathbf{Z} \sim \mathrm{EC}_n(\mathbf{0}, \mathbf{R}, g)$, where g is a density generator and $\mathbf{R}$ is given by (12.2). In this case, all the marginal distributions of $\mathbf{Z}$ are identical, with p.d.f.

$$q_g(x) = \frac{\pi^{(n-1)/2}}{\Gamma((n-1)/2)} \int_{x^2}^{\infty} (y - x^2)^{(n-1)/2-1} g(y)\, dy \tag{C.94}$$

and c.d.f.

$$Q_g(x) = \frac{1}{2} + \frac{\pi^{(n-1)/2}}{\Gamma((n-1)/2)} \int_0^x \int_{u^2}^{\infty} (y - x^2)^{(n-1)/2-1} g(y)\, dy\, du. \tag{C.95}$$

Let $\mathbf{X} = (X_1, \ldots, X_n)'$ be a random variable with each component X_i having a given continuous density f_i and c.d.f. F_i. Let $\mathbf{Z} = (Z_1, \ldots, Z_n)' \sim \mathrm{EC}_n(\boldsymbol{\mu}, \mathbf{R}, g)$. Suppose that

$$z_i = Q_g^{-1}(F_i(x_i)), \qquad i = 1, \ldots, n, \tag{C.96}$$

where Q_g^{-1} is the inverse of Q_g. The determinant of the Jacobian of the transformation is

$$\mathrm{J}\{(z_1, \dots, z_n)' \to (x_1, \dots, x_n)'\} = \prod_{i=1}^{n} \frac{\mathrm{d}z_i}{\mathrm{d}x_i} = \prod_{i=1}^{n} \frac{f_i(x_i)}{q_g(Q_g^{-1}(F_i(x_i)))} \tag{C.97}$$

and the p.d.f. of $\mathbf{X}$ is given by

$$h(x_1, \dots, x_n) = \phi(Q_g^{-1}(F_1(x_1)), \dots, Q_g^{-1}(F_n(x_n); \mathbf{R}) \prod_{i=1}^{n} f_i(x_i), \tag{C.98}$$

where $\phi(\cdot)$ is the n-variate density weighting function

$$\phi(z_1, \dots, z_n; \mathbf{R}) = C_n |\mathbf{R}|^{-1/2} g(\mathbf{x}'\mathbf{R}^{-1}\mathbf{X}) \Big/ \prod_{i=1}^{n} q_g(z_i) \tag{C.99}$$

and C_n is the normalizing constant defined in Theorem C.6. The $n \times 1$ random vector $\mathbf{X}$ is said to have a meta-elliptical distribution if its density function is given by (C.98), and denoted $\mathbf{X} \sim \mathrm{ME}_n(\mathbf{0}, \mathbf{R}, g; F_1, \dots, F_n)$. ∎

C.2.4 Testing Ellipticity

This section was written with Christian Frey and Ludovic Mathys, who also researched and programmed the tests.

There have been several ways proposed for testing a data set for ellipticity; some of these were mentioned above in Remark (**b**). We consider here those from Schott (2002), Manzotti et al. (2002), and Huffer and Park (2007). A more recent approach such as Su (2012) builds directly on Manzotti et al. (2002), while that from Bianco et al. (2017) is an extension of Zhu and Neuhaus (2003) (which is based on the computation of the empirical characteristic function) where the assumption about existence of fourth moments for the asymptotic validity of the test statistic distribution is relaxed. Su (2012) and Bianco et al. (2017) rely on bootstrapping procedures, making their implementation computationally too expensive for a large number of dimensions and data points.

As shown in a simulation study by Huffer and Park (2007), their test (hereafter, H-P) and that of Manzotti et al. (2002) (hereafter MPQ) have higher power than the test from Schott (2002), when used for samples drawn from a multivariate generalized Laplace distribution (see Huffer and Park, 2007, Tables 1, 2, and 3 for details). As such, we investigate the performance of H-P and MPQ in more detail. Both of these tests are conceptually of the same complexity and both are computationally intractable for large dimensions d. More precisely, H-P requires generating all possible combinations of diagonal matrices with entries $+1/-1$ for a given sample, leading to 2^d diagonal matrices. Likewise, the MPQ test requires a nested computation of *spherical harmonics* (polynomials) of different degrees, where the numbers of required polynomials of a particular degree is linked to d. For large d (say, 50), this implies the nested computation of several hundreds or even thousands of polynomials, and becomes computationally intractable on a modern (at the time of writing) desktop computer.

For these reasons, the H-P test (which also enjoys accurate finite sample properties) is chosen for samples of dimension up to 10. As the implementation of this test is rather involved, we refer the interested reader to Huffer and Park (2007) for details. Program Listing C.2 gives our Matlab implementation, though, unlike most programs given throughout the book, the code does not tie in with any analytic description of the method given in the text, and is thus just a "black box" routine.

```
1  function [TestStat,Pvalue,PSerror]=ProgTestPARK(Y)
2  p=length(Y(1,:)); c=3; p0=1/(c*2^p); G=matriceO(p);
3  % Define Scaled Residual:
4  Ybar=mean(Y); N=length(Y(:,1)); S=((N-1)/N).*cov(Y); [~,q] = chol(S);
5  if q==0
6    L=chol(S,'lower'); L=L'; R=inv(L); Z=(Y-Ybar)*R'; PSerror=0;
7  else
8    Shat=nearestSPD(S); L=chol(Shat,'lower'); L=L';
9    R=inv(L); Z=(Y-Ybar)*R'; PSerror=norm(Shat-S);
10 end
11 % Define the Cell Counts
12 if p==1, Znorm=(Z').^2; else Znorm=vecnorm(Z').^2; end
13 U=zeros(2^p,c); q=zeros(1,c+1);
14 for k=1:N
15   for j=2:c+1
16     q(1,j)=quantile(Znorm,(j-1)/c);
17     for i=1:2^p
18       if (G(1:p,(i-1)*p+1:i*p)*Z(k,:).'> 0) ...
19         & (q(1,j-1)<Znorm(1,k)) & (Znorm(1,k)<=q(1,j))
20         U(i,j-1)=U(i,j-1)+1;
21       end
22     end
23   end
24 end
25 % Define statistic X^2 and Obtain p-value:
26 squareX=sum(sum((U-N*p0).^2))/(N*p0);
27 TestStat=squareX; Pvalue=LimitPvalue(TestStat,p,c);
28 end
29
30 function G=matriceO(p)
31 g = dec2bin(0:2^p-1); Q1 = zeros(2^p-1,p);
32 for i=1:size(g,1)
33   for j=1:size(g,2), Q1(i,j) = str2double(g(i,j)); end
34 end
35 Q2=Q1-1; Q=Q1+Q2; G=zeros(p,p*2^p);
36 for j=1:2^p, G(1:p,(j-1)*p+1:j*p)=diag(Q(j,:)); end
37 end
38
39 function Pvalue=LimitPvalue(TestStat,p,c)
40 N=7000; dfW0=c*(2^p-1)-p*(p+1)/2; dfW1=p; dfW2=p*(p-1)/2;
41 W0=chi2rnd(dfW0,[1,N]);
42 W1=chi2rnd(dfW1,[1,N]); W2=chi2rnd(dfW2,[1,N]); VW=0:1/c:1;
43 q0=chi2inv(VW,p);
44 A=chi2cdf(q0(1,2:c),p+1)-chi2cdf(q0(1,1:c-1),p+1);
45 B=chi2cdf(q0(1,2:c),p+2)-chi2cdf(q0(1,1:c-1),p+2);
46 astar=2*c/pi * sum(A.^2); bstar=4*c/pi^2 * sum(B.^2);
47 W=W0+(1-astar).*W1+(1-bstar).*W2; [f,x]=ecdf(W);
48 for k=1:N+1
49   if (TestStat >= x(k,1)), PV=f(k,1); end
50 end
51 Pvalue=1-PV;
52 end
53
54 function n = vecnorm(x, dim) % Already built into version R2017b
55   if nargin < 2, dim = 1; end
56   n = sqrt(sum(x.^2, dim));
57 end
```

Program Listing C.2: Program to compute the H-P test statistic from Huffer and Park (2007). Function `nearestSPD`, by John D'Errico, called in line 8, was obtained from the Matlab File Exchange. It takes as input a square real matrix, and delivers the nearest matrix (by minimizing the Frobenius norm of the difference) that is symmetric and positive definite.

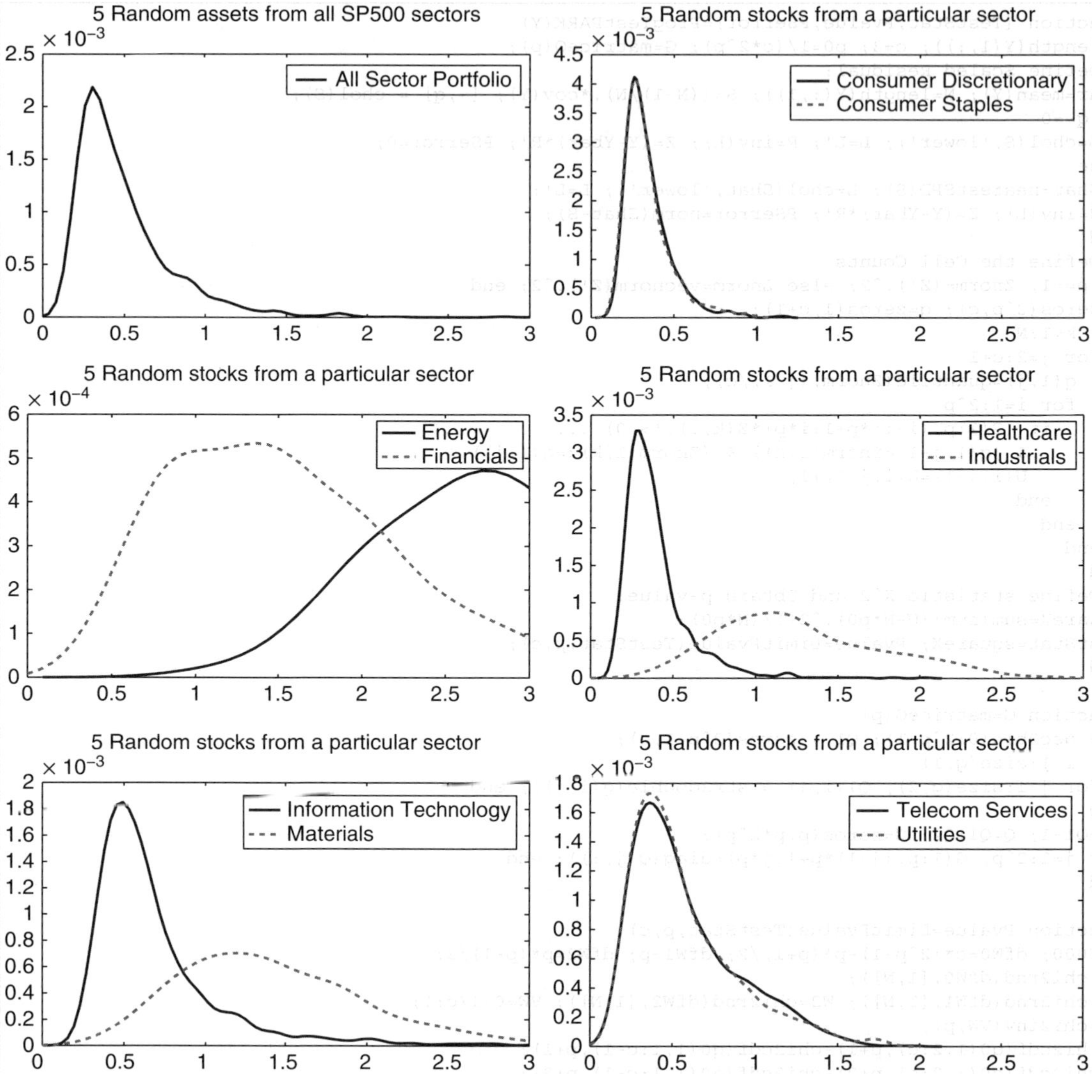

Figure C.2 Kernel density plots (truncated, so that the *x*-axis is the same in each plot) of the distribution of H-P test statistic *T* for ellipticity. The *x*-axis elements were divided by 1,000. **Top left:** Computed on the GARCH-filtered log percentage returns of $d = 5$ randomly drawn stocks out of 416 from the S&P500 index, and this done over 1,000 random draws. **The remaining plots** show the same result but when restricting the $d = 5$ stocks to be within the same of each of the 10 industry sectors that divide the stocks on the index.

As an empirical example of potential relevance, we use the Gaussian GARCH-filtered daily returns of 416 stocks listed on the S&P500 index, as initially examined in Example 12.7. With 10 years of daily data, this results in 2,592 data points. Recall that each stock belongs to one of 10 industry sectors, e.g., energy, financials, health care, utilities, etc. Due to the computational requirements of the H-P test, we restrict attention to use of $d = 5$ and $d = 10$ assets, and consider the following heuristic to assess the extent to which sets of stocks within a sector are closer to being elliptic than sets of stocks from different financial sectors.

First, $d = 5$ stocks from the available 416 are randomly drawn (without regard to industry sector) and the corresponding H-P ellipticity test statistic, T_1, is computed. This is repeated $M = 1{,}000$ times, resulting in test statistics $T_1, \dots, T_M$. The top left panel in Figure C.2 shows the resulting kernel density plot. This serves as a "null distribution" (specific to the 416 stocks, the 10 years of daily data used, and also the extent to which the Gaussian GARCH model is adequate for filtering out the time-varying scale term to result in a set of i.i.d. deviates) of the distribution of T when based on $d = 5$ stocks and the industry sector is ignored.

Second, $d = 5$ stocks from each sector, $s = 1, \dots, S$, $S = 10$, are randomly drawn for $m = 1, \dots, M$, and the corresponding ellipticity test statistics, $T_{s,1}, \dots, T_{s,M}$, are computed, $s = 1, \dots, S$. For each sector, the remaining panels in Figure C.2 show the resulting kernel density estimates. Note that the plots, starting from the second one to the bottom right, each contain two overlaid sectors (to save space). From this exercise, we see that non-ellipticity is relatively very prominent in some sectors, notably the energy and financial sectors, but not in others, such as health care and consumer staples. A similar exercise using $d = 10$ assets yields qualitatively similar results.

Appendix D

Introducing the SAS Programming Language

SAS is one of the most versatile software packages for data handling and statistical analysis. Its programming language and data structures are suited specifically for this task, and thus differ from those of more general programming languages such as C++ and Java, and their matrix-based prototyping extensions, such as Julia, Matlab, Python, R, etc. Software packages that emphasize canned, pre-written statistical methods include SAS, SPSS, and Stata, though both Matlab and R also have packages with canned statistical routines.

For development of methods, C++, Java, Matlab, Python, and R (alphabetically listed) appear at the top of many lists, with big data and machine learning practitioners tending towards Python, while for statistics, R dominates, particularly in the academic setting. For industry end-users, it appears that SAS is still the leader in medicine/clinical trials and is also popular in large organizations, including major financial institutions, as it serves as a modular and integrated computing package useful for generating, combining, and processing various (potentially large) databases.

Note that SAS is relatively rather expensive, Matlab is not cheap, while R is free. Due to its open source nature, new techniques are often available very quickly in R, though often numerous packages will exist for similar analyses, and the reliability of the code is not ensured. For decades, SAS has been the undisputed leader in the commercial space, with well-tested updates to its capabilities based on demand, as opposed to trying to include every new method in real time. R has a very large online community (as does SAS and Matlab), but no service support, while SAS has dedicated customer service support.

As a commercial product, SAS is governed by a common design unifying the data processing engine, user interfaces, procedures, and documentation, with syntax consistent across procedures. R-package developers operate independently and without a comparable design. Its diversity and speed in implementing new methods is a strong plus, but can be a drawback (and liability) for professionals in industry.

A data analyst does not have to choose to the exclusion of others, just like with human languages: Knowing more than one is often an advantage, though generally accepted advice is that being able to program is a necessary, but far from sufficient, prerequisite for success. Programming is relatively easy compared to deep acquisition of skills in statistics, distribution and probability theory, as well as subject-specific knowledge such as biostatistics, engineering, quantitative risk management, mathematical finance, or econometrics.

Linear Models and Time-Series Analysis: Regression, ANOVA, ARMA and GARCH, First Edition. Marc S. Paolella.

Some advantages in learning and using SAS include:

1) **Data handling capability.** Statistically analyzing a data set and building a model for a certain purpose is often only part of the job. Among the many steps (data acquisition, presentation of results, etc.), an important one includes reading in the data from various types of data files that might have been "uniquely" (or poorly) constructed, processing the data in various ways, such as sorting, merging with related data, and cleaning (eliminating faulty values), etc. SAS offers the user a wide variety of techniques to process data, preparing it for proper statistical analysis.
2) **Availability of many statistical and other procedures.** In statistical consulting and much applied research, the analysis of the majority of data sets requires techniques that are already available in most statistical packages. While, say, simple t-tests and basic linear models analysis can easily and quickly be conducted in a programming language such as Matlab, more advanced routines can be very time-consuming to program, not to mention the time required to test their reliability and robustness. Instead of reinventing the wheel, use of the canned routines in software such as SAS can spare many potential mistakes, and enormous amounts of time (and your employer's or client's patience).
3) **Reliability.** Although arguably less today, SAS is considered a benchmark in statistical computing. Of course, no software package is without mistakes, and journal articles occasionally appear comparing computations across software platforms and pointing out errors.
4) **Popularity.** Because so many public and private organizations use SAS, and have already developed many custom programs with it, they will be unwilling to switch to another platform.
5) **Processing speed.** SAS was traditionally known for well-coded statistical algorithms and fast execution speed, though for particular methods the gap is no doubt closing, given, for example, that the underlying vectorized operations in Matlab are efficiently written in a low-level (close to machine) language. SAS has an advantage with massive data processing, as it does not require all the data to be loaded into memory, though various workarounds for this issue do already exist for other languages, such as in Matlab with their so-called tall arrays and distributed arrays.
6) **Access to data.** SAS can interface with commercial data bases. An example of interest to researchers in empirical corporate finance, quantitative risk management, and financial econometrics is Wharton Research Data Services (WRDS), from the Wharton (business) School of the University of Pennsylvania, and includes, among others, the data from CRSP, Thomson Reuters, and OptionMetrics. Once the user has access to the WRDS, one can "Use SAS on your PC to submit jobs on the WRDS cloud or download data directly to your PC and analyze using our extensive site, programs and utilities", as stated on the WRDS webpage.

D.1 Introduction to SAS

D.1.1 Background

SAS was developed before powerful desktop computers and workstations existed and, thus, originally intended for use on mainframe computers, whereby users (typically universities, research firms, or other companies) had to pay for computer resources (memory and CPU time). This is reflected, for example, in the fact that output from a SAS job (i.e., program execution), still includes information about runtime, even in the PC version (in the LOG file). In addition, there are various options and commands that can be invoked to perform data operations in less time and/or with less memory; these are naturally somewhat obsolete in a PC environment, although still useful from time to time. Note that, with the emergence of cloud computing, this aspect might come full circle and become relevant again.

Given the existing computer technology at the time of SAS's initial development, it was not possible (perhaps not even imaginable) for interactive, real-time data exploration with easy, high-level graphics capabilities. Instead, data processing, numerical output, and simple character graphics from statistical procedures were emphasized, obtained from submitting written programs for execution.

Naturally, newer versions running on graphics-friendly devices support modern data exploration. Nevertheless, SAS's strong point is undoubtedly still its data handling capability. Before more specific software packages became available and popular, SAS's data manipulation features were rich enough to be able to serve as a database manager, spreadsheet, payroll manager, report generator and, of course, a state-of-the-art statistical software package. For this reason, we will concentrate on reading and manipulating datasets in SAS.

D.1.2 Working with SAS on a PC

A computer-savvy person with basic programming skills should be able to work through this chapter in a couple of days. To save space, not all output from the demonstration programs is shown. Thus, while helpful to just read the notes, nothing can replace brewing a pot of coffee, rolling up the sleeves, and actively working through the material.

It is useful to specify a working directory, where input files can be found and output files can be directed. This is accomplished by going to the menu **Tools**, then **Options**, then **Change Current Folder**, and entering the desired path. It is useful to have a way of putting the location of such files into the SAS code, so that a "batch" code can be run without requiring to start the SAS graphical user interface and manually use the menu structure to indicate the desired path. How this is done is shown at the beginning of Section D.3.3.

When SAS is started, several sub-windows are presented, the most useful of which for us will be LOG, OUTPUT, and PROGRAM EDITOR. A program can be typed in a PROGRAM EDITOR window, and then executed by command **Submit**, located under the menu title **Run**. This menu option only appears when the cursor is in a PROGRAM EDITOR. Another way of executing the program is to left mouse click on the icon that resembles a running person. As the program runs, SAS generates a so-called *log file*, which appears in the LOG window and provides details regarding the previously submitted program statements. Error messages are shown in red, helping to spot them. For beginners, it is worth reading the other messages (in blue) as well.

If the program was successful and output from a procedure was generated, it can be viewed in the OUTPUT window. In SAS version 9.4, output is in hypertext markup language (HTML) format. It can easily be converted to both Adobe portable document format (pdf) and rich text format (rtf); see, e.g., the commands in Listing 2.1 (Chapter 2) for how to do this.

As an example to get started:

1) (How to type in a program) In the PROGRAM EDITOR, type the following short program. Don't worry about understanding all the details of the program now—they will be explained shortly.

```
data welcome; /*  this is a comment. Note the delimiters  */
  input stage_name $ age;
  datalines; /* the old name was: cards. It can still be used */
Rodolfo 26
Marcello 24
Colline 31
Mimi 25
;
proc print;
  title 'La Boheme Quanti anni abbiamo ora?';
run;
proc means;
  var age;
run;
```

2) (How to submit a program and examine the Output) With the mouse or keyboard, **Submit** the program. If it was typed in correctly, then the LOG window will show the details about the executed program, and the OUTPUT window will show the output of the two procedures `print` and `means`.
3) (How to copy and paste a code segment) Go back into the PROGRAM EDITOR window and append to the end of the program a copy of the program segment corresponding to `proc means` (i.e., the last three lines of the above program). Edit the new `proc means` as follows:

```
proc means min max range maxdec=5;
  var age;
run;
```

4) (How to run just a piece of code) Mark the new `proc means` procedure (using the mouse). Now **Submit** the code (right mouse click, choose **Submit Selection**). This will execute just the piece of code that you marked. The LOG and OUTPUT windows will be *appended* with the new results, i.e., they keep growing until you delete them.
5) (How to clear the contents of the OUTPUT and LOG windows) With a right mouse click from somewhere in the OUTPUT window, choose **Edit** followed by **Clear All**. Do the same for the LOG window.
6) (How to save a SAS program) The code you entered in the PROGRAM EDITOR window can be saved as a text file via menu option **File**, **Save As** and specifying a path and name. If you do not provide an extension, SAS uses the default of `.sas`.
7) (How to pull a SAS program into SAS) Under MS-Windows, from the Explorer, a right mouse click on a file with extension `.sas` will result in a floating menu of options catered to SAS, one of them being **Open with SAS 9.4** (or whatever version you have). This will start the SAS GUI (the Windows Graphical User Interface) software if it is not already running and place the contents of the selected program into its editor. If the SAS GUI is already running, one can also just drag the file with the mouse into the SAS editor window, and a new editor window will be created for the file. The program can then be executed, modified, and saved, etc., using the methods already discussed. Finally, and usefully, dragging a file into the LOG or OUTPUT window will execute the code, but not instigate a new editor window.
8) (How to run SAS as a batch job) It is actually not necessary to use SAS's file editor and LOG and OUTPUT windows. From the Windows Explorer, right click the mouse on a SAS program (assuming it has the `.sas` extension) and select from the resulting floating menu the option **Batch**

Submit. The program will be executed by SAS "in the background" and will produce a corresponding `.log` file for the log output and, if at least one procedure successfully ran and generated output, a `.lst` file for the statistical output (and possibly a `.pdf` and/or `.rtf` file). This works irrespective of whether or not the SAS GUI is running.

D.1.3 Introduction to the Data Step and the Program Data Vector

It is easiest to think of a SAS data set as a matrix, with the columns representing different variables, and the number of rows representing the total number of observations. A SAS program can define and operate on numerous data sets and its ability to combine them in various ways is one of its advantages. A data set is created in SAS by specifying the keyword `data`, followed by a name consisting of letters and numbers, the first of which must be a letter; a blank space is not allowed. For instance, the statement `data erstmal` is valid, but not `data 1.mal`. In older versions of SAS, the name was restricted to at most eight characters, but this is no longer the case.

Data is input into the named data set by using the keyword `input`, followed by variable names. The variable names follow the same rules as those of the data names. The actual numbers read into the variables can either be part of the program, following the `datalines` statement (as we used above), or contained in a text file (as will be shown below). Variables can also be generated from previous variables using mathematical functions. The Program Data Vector, or PDV, is where SAS internally holds a vector of observations before they are output to the data set. The following examples are illustrated both with their output, as generated by SAS, as well as a representation of what the PDV looks like during the data step.

It is sometimes useful to designate variable names using a foreign language (obviously provided the user understands it), so that it is clearly differentiated from function names in the syntax of the programming language. We continue this below, occasionally using some simple words from German (erstmal, Leute, alle, zusammen, ein, aus, Geschlecht, Geburtstag, das Ende, drucken, Einheit, falsch, klappt, keine, Montag, Dienstag, Mittwoch, Donnerstag, Freitag).

Consider the program:

```
data Leute;
  input name $ sex age height weight;
  ratio=height/weight;
  drop height weight;
datalines;
john 0 45 101 151
mike 0 38 105 163
susan 1 50 98 142
frank 0 32 120 182
jenn 1 71 78 100
bill 0 14 43 64
mary 1 15 53 81
;
run;
proc print; run;
```

The name of the data set is `Leute` (people). The `input` statement tells SAS to input 5 variables: name, sex, age, height, and weight. Notice that `name` is followed by a dollar sign, $, as it resembles a string. This indicates that `name` is not a number, but rather a character string. The data come after the

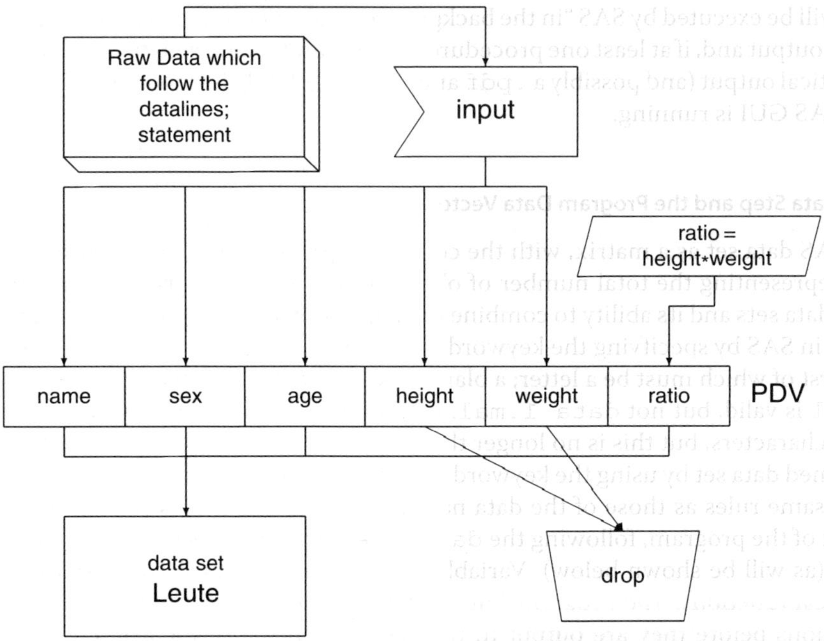

Figure D.1 PDV for input and calculation of variables.

keyword `datalines`, and the `datalines` statement comes at the end of the processing commands. The variable `ratio` is generated from the most recently input `height` and `weight` variables. The `drop` statement instructs SAS not to save the variables `height` and `weight`. We only want the ratio of these two variables, and do need to keep them in the data set. The `proc print; run;` statement at the end invokes SAS's printing procedure. It causes the contents of the most recently created data set to be printed.

To help visualize things, a flowchart-like diagram of the PDV in this case is given in Figure D.1. It serves to indicate how SAS processes *a single observation*. The first data line is read in, `ratio` is computed, and all the variables in the PDV that are not tagged with a `drop` statement are written to data set `Leute`. It is important also to envision how the data step processes *several* observations. The above depicted procedure is repeated as many times as there are observations following the `datalines` statement. In our case, there are 7. Thus, 7 "lines" or rows are written to the data set `Leute`. The output from the `print` procedure looks approximately as follows:

OBS	NAME	SEX	AGE	RATIO
1	john	0	45	0.66887
2	mike	0	38	0.64417
3	susan	1	50	0.69014
4	frank	0	32	0.65934
5	jenn	1	71	0.78000
6	bill	0	14	0.67188
7	mary	1	15	0.65432

Consider the additional code segment

```
data adults;
  set Leute;
  if age < 18 then delete;
run;
```

The `set` statement uses the previously created data set `Leute` to input observations. In this case, we want a subset of the observations from `Leute`, namely only those people at least 18 years old. We call the new data set `adults`. The PDV in this case can be represented by the diagram in Figure D.2.

There are other ways to extract this subset from data set `Leute`. Two other possibilities are

```
data adults;
  set Leute;
  if age >= 18;
  run;
```

and

```
data adults;
  set Leute(where=(age>=18));
run;
```

The first of these is logically equivalent to the original code, but in terms of easily readable (or self-documenting) code, it is less clear (if `age >= 18`, then what? And if not, then what?) The second alternative uses the `where=` statement, a feature that was added to version 6 of SAS (and revealing the age of the author). Only if the condition specified in parentheses after the `where=` statement is fulfilled is the observation allowed to enter into `adults`. With larger data sets, using the `where` statement can save execution time. It also makes for shorter and better documented programs.

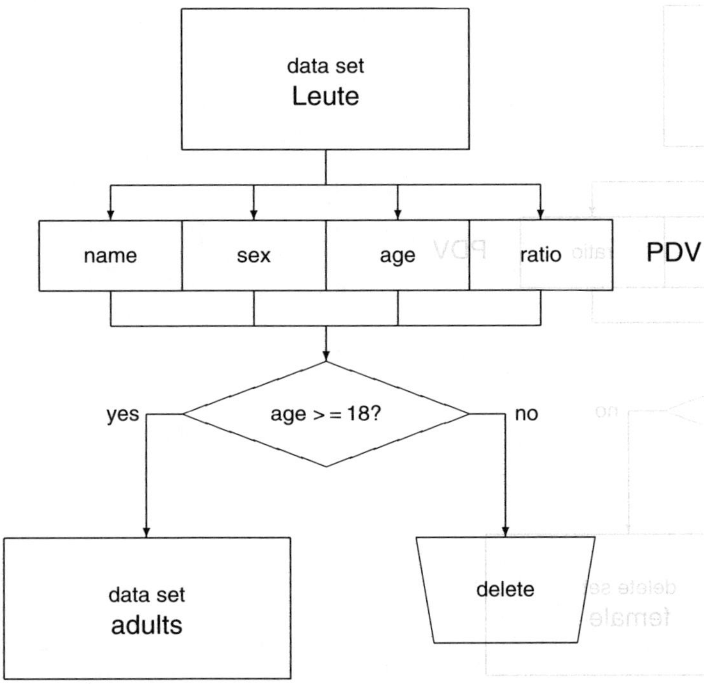

Figure D.2 PDV illustrating branching via an `if` statement.

We can create more than one data set at a time. Perhaps we want to make two separate data sets, based on the sex variable, and including both children and adults. It should be noted that although we created the data set adults from data set Leute, they are both still present in SAS. Thus, the following is used.

```
data male female;
  set leute;
  if sex=0 then output male;
  else output female;
run;
```

Notice first that the two new data sets appear on the data declaration (the first line in the program), and are not separated by commas. The output statement instructs SAS to which data set the current observation should be output. Because there is more than one data set being created in this case, we have to specify which observations go to which data set. In fact, when there is only one data set declared, like in the previous data steps, SAS implicitly inserts an output statement at the end of the data step. For instance, in the previous program, we could have explicitly written

```
data adults;
  set Leute(where=(age>=18));
  output;
run;
```

and would get exactly the same result. The PDV in this case is shown in Figure D.3.

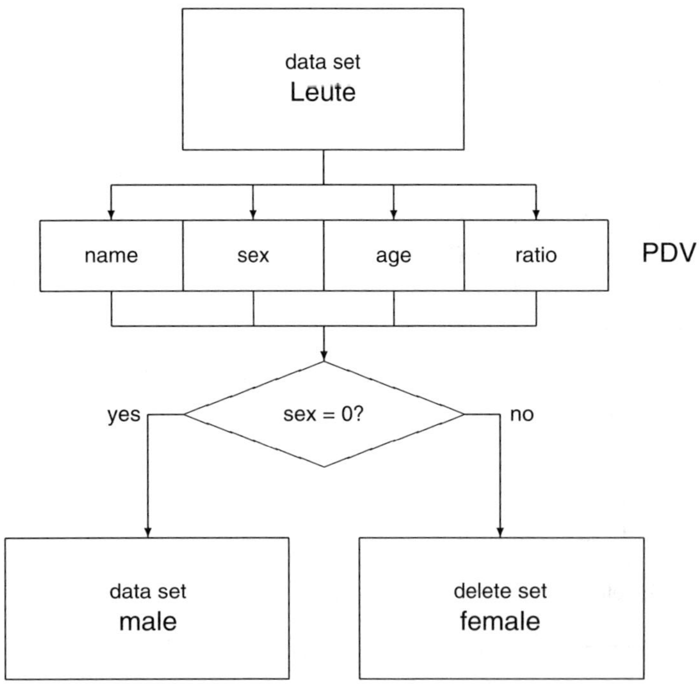

Figure D.3 PDV illustrating construction of two data sets.

The data set male just contains the observations for john, mike, frank, and bill. The data set female contains only those observations for susan, jenn, and mary.

Now imagine that two new people are being added to this study; Josh and Laura. We would create another data set with their information, and call it Leute2. Notice it is exactly the same as the data step for Leute, except that the two observations after the datalines statement are different.

```
data leute2;
  input name $ sex age height weight;
  ratio=height/weight;
  drop height weight;
datalines;
josh 0 53 130 110
laura 1 60 165 140
;
run;
```

We would like to combine all the people into one data set. To do this, we may place both data set names, Leute and Leute2, on the set statement.

```
data Alle;
  set Leute Leute2;
run;
```

SAS simply appends the two data sets together, one after another, to create the data set Alle. Using the print procedure, where this time we explicitly tell SAS which data set to print, proc print data=alle; run; we get as output,

OBS	NAME	SEX	AGE	RATIO
1	john	0	45	0.66887
2	mike	0	38	0.64417
3	susan	1	50	0.69014
4	frank	0	32	0.65934
5	jenn	1	71	0.78000
6	bill	0	14	0.67188
7	mary	1	15	0.65432
8	josh	0	53	1.18182
9	laura	1	60	1.17857

Imagine that later it is decided to ask the participants some information about how much they eat and how active they are. In particular, we ask them approximately how many calories they consume on average every day and, on a scale of 0 to 3, how active they are, where 0 indicates "absolutely lazy" and 3 means "very active in sports". With the collected data, we would type the following program, creating the new data set moreinfo. Laura unfortunately refused to answer how many calories she consumes every day, as well as how sporty she is. We therefore leave her out of this data set.

```
data moreinfo; /* notice Laura is missing! */
  input name $ calories sport;
  datalines;
susan 1250 2
jenn 3270 0
```

```
mike 2370 0
frank 1540 1
josh 1050 0
mary 5340 3
john 1040 2
bill 2080 0
;
run;
```

This is quite similar to the first program above, so we omit the PDV and the output from the `print` procedure. The goal is now to merge this data set with the data set `Alle`, where the rest of the information is contained. SAS has an appropriately named statement, `merge`, for this. However, to merge the two data sets `Alle` and `moreinfo`, they each need to be sorted by the name variable first. This is accomplished in SAS with the procedure `sort`.

```
proc sort data=moreinfo;
  by name;
run;
proc sort data=Alle;
  by name;
run;
```

Notice the `by` statement, which, somewhat obviously, indicates by which variable to sort the observations. If we were to print the data sets `moreinfo` and `Alle` now, we would see that the observations in both are sorted by name, alphabetically. Now comes the exciting part.

```
data zusammen; /* means 'together' in German */
  merge Alle moreinfo;
  by name;
run;
proc print; run;
```

The output from `proc print` looks as follows. Notice two things. First, the observations are sorted by name. Second, SAS does not give an error message or even a warning when it cannot find the information for Laura in the `moreinfo` data set. SAS simply sets those values to missing. Missing values are denoted with a period.

OBS	NAME	SEX	AGE	RATIO	CALORIES	SPORT
1	bill	0	14	0.67188	2080	0
2	frank	0	32	0.65934	1540	1
3	jenn	1	71	0.78000	3270	0
4	john	0	45	0.66887	1040	2
5	josh	0	53	1.18182	1050	0
6	laura	1	60	1.17857	.	.
7	mary	1	15	0.65432	5340	3
8	mike	0	38	0.64417	2370	0
9	susan	1	50	0.69014	1250	2

With the `merge` statement, the PDV can be thought of as the diagram in Figure D.4.

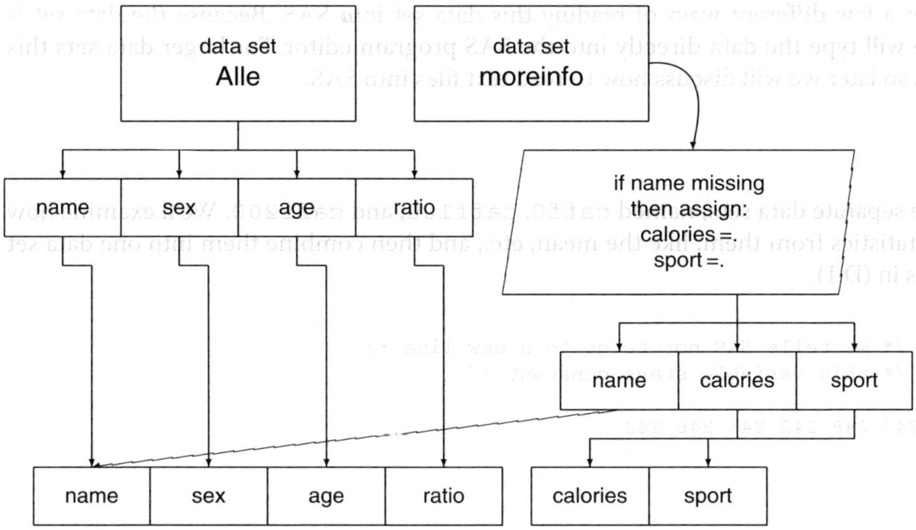

Figure D.4 PDV using `merge`.

Notice that the variable `name` is not duplicated. Because we `merged` by `name`, `name` only appears once.

D.2 Basic Data Handling

Consider the following real data set, taken from Hand et al. (1994), collected in an experiment to determine if caffeine increases ones ability to tap his/her fingers. The number of taps per minute was recorded for 30 people, 10 per each group, where the first group received no caffeine, the second group received 100 ml, and the third group 200 ml.

	Independent observations									
Caffeine										
0	242	245	244	248	247	248	242	244	246	242
100	248	246	245	247	248	250	247	246	243	244
200	246	248	250	252	248	250	246	248	245	250

Although there are several possible ways of constructing a data matrix from the above data, analysis is most easily conducted when SAS internally views the data as a matrix of two variables, with 30 observations:

Obs.	Caffeine	Taps
1	0	242
2	0	245
⋮	⋮	⋮
30	200	250

(D.1)

We now consider a few different ways of reading this data set into SAS. Because the data set is relatively small, we will type the data directly into the SAS program editor. For larger data sets this will be impractical, so later we will discuss how to read text files into SAS.

D.2.1 Method 1

We construct three separate data sets, named `caff0`, `caff100`, and `caff200`. We'll examine how to calculate basic statistics from them, like the mean, etc., and then combine them into one data set so that it appears as in (D.1).

```
data caff0;
  input taps @@; /* @@ tells SAS not to go to a new line */
  caffeine=0;    /* this variable stays constant */
  datalines;
242 245 244 248 247 248 242 244 246 242
;
data caff100;
  input taps @@;
  caffeine=100;
  datalines;
248 246 245 247 248 250 247 246 243 244
;
data caff200;
  input taps @@;
  caffeine=200;
  datalines;
246 248 250 252 248 250 246 248 245 250
;
```

The three data sets are now in memory. They would have to be combined in order to conduct, say, an F test for equality of means. Before doing so, we might be interested in printing the data and examining some simple statistics. Use the following to print the data sets just constructed:

```
proc print data=caff0;                /* indicates which data set */
  title 'No caffeine administered'; /* prints nice title */
  title2 '(Control Group)';         /* 2nd title line */
run;                                  /* ''run'' is actually not necessary */
proc print data=caff100;
  title '100 ml caffeine administered';
proc print data=caff200;
  title '200 ml caffeine administered';
run;
```

To compute various simple statistics, use the `means` procedure:

```
proc means data=caff0 maxdec=1; /* specify max # of decimal places */
  title 'No caffeine administered';
  title2 '(Control Group)';
  var taps;    /* which variable(s) to analyze? */
run;
proc means data=caff100 mean min max; /* just compute mean, min and max */
  title '100 ml caffeine administered';
  var taps;
```

```
proc means data=caff200;
  title '200 ml caffeine administered';
  var taps;
run;
```

The variable specification `var` command is not necessary. Without it, SAS uses all the variables in the data set. The output is, however, less cluttered if only those variables of interest are used.

Now we wish to combine the three data sets into one. The following accomplishes this:

```
data all;
  set caff0 caff100 caff200; /* appends them */
;
```

As before, we may use the `print` and `means` procedures.

```
proc print; /* default dataset is last one created */
  title 'All Observations';
run;
proc means maxdec=1 mean;
  by caffeine; /* data are sorted */
  var taps;
  * title 'get the MEANS for each level of caffeine';
run;
```

The `by` statement is very useful. As long as the data are sorted by the "`by` variable", we can perform essentially three `means` procedures, one for each level of caffeine. The star in front of a line (and which ends with a semicolon) serves to comment that line out, as shown above for the `title` line in the `proc means`. In this case, we comment out the `title` to illustrate the point that, if the procedure is not explicitly given a `title`, *then its output uses the title from the most recently executed procedure*, in this case, from the `print` procedure.

It might be desired to print the data grouped according to the level of caffeine. Here is one way:

```
proc print data=all noobs; /* noobs omits the observation number */
  title 'We can print BY caffeine also';
  by caffeine;
run;
```

D.2.2 Method 2

Here we directly create one data set.

```
data coffee;
  input caffeine @; /* don't go to next input line */
  do i=1 to 10;     /* start a loop */
    input taps @;   /* get a data point but stay on the line */
    output;         /* now write both vars: caffeine, taps */
  end;
  datalines;
0   242 245 244 248 247 248 242 244 246 242
100 248 246 245 247 248 250 247 246 243 244
200 246 248 250 252 248 250 246 248 245 250
;
proc print; /* just check */
  title 'Second Method for reading in data';
run;
```

The above deserves a bit more explanation. The @ sign at the end of the input line causes SAS to keep the "input pointer" on the current line until there are no more observations on that line. The do loop then reads the 10 observations on the first line, also using the @ sign to prevent SAS from going to the next line. For each observation, the command `output` is executed, and writes the variables caffeine and taps to the data set coffee. This command `output` is normally executed by SAS automatically in the data step, but for complicated data entry tasks it is a useful tool.

The difference between the single @ sign, and the double @@ sign is small, but important. Use @@ also to "hold the line", but only when complete sets of variables are followed by one another on a line. For example, if for some reason, we had typed the level of caffeine for each observation, we would use the @@ sign instead:

```
data koffee;
  input caffeine taps @@; /* several FULL obs. on each line */
  datalines;
0 242 0 245 0 244 0 248 0 247 0 248 0 242 0 244 0 246 0 242 100 248
100 246 100 245 100 247 100 248 100 250 100 247 100 246 100 243 100
244 200 246 200 248 200 250 200 252 200 248 200 250 200 246 200 248
200 245 200 250
;
proc print; /* just check again */
  title 'yet another way';
run;
```

D.2.3 Method 3

Here we'll see how to use arrays in SAS, as well as some other useful features.

```
data tapping;
  input caffeine v1-v10;     /* no @, as we read the whole line    */
  average=mean(of v1-v10); /* now we have the mean too            */
  array vv{10} v1-v10;       /* v1 is vv(1), v2 is vv(2), etc.     */
  do i=1 to 10;
    taps=vv(i);              /* we want a separate obs. for each  */
    deviate=taps-average;    /* construct a new variable          */
    output;                  /* combination of caffeine and taps  */
  end;
  drop v1-v10;               /* no need to keep these variables   */
  datalines;
0   242 245 244 248 247 248 242 244 246 242
100 248 246 245 247 248 250 247 246 243 244
200 246 248 250 252 248 250 246 248 245 250
;
proc print;
  by caffeine;
  title 'The 3rd way to read in the data set';
run;
```

Observe that the variable `average` is the mean over the levels of caffeine, and not the overall mean. To calculate the overall mean, just use `proc means` without the `by` statement, i.e.,

```
proc means data=tapping mean;
  title 'the overall mean';
  var taps;
run;
```

D.2.4 Creating Data Sets from Existing Data Sets

Using the previously created `tapping` data set, we now create three new data sets, `caf0`, `caf100`, and `caf200`, that contain the data corresponding to the caffeine level 0, 100, or 200, respectively. The `keep` statement below only takes the variables `caffeine` and `taps` from the previously created `tapping` data set. Recall we had the additional variables `average` and `deviate`, though we do not wish to use them now.

```
/* how to make three data sets from just one */
data caf0 caf100 caf200;
  set tapping (keep=caffeine taps);
  drop caffeine;
  if caffeine=0 then output caf0;
  else if caffeine=100 then output caf100;
  else if caffeine=200 then output caf200;
run;
```

Now consider making once again a single data set with all 30 observations, but this time having three separate variables, `taps0`, `taps100`, and `taps200`, within the single data set. The `rename` command will be of use here, and is of the form `rename=(oldname=newname)`. Notice that, in this case, the `set` command does work, but generates missing values. The `merge` command is what we really want to use. Understanding how the `set` and `merge` commands work is of great value.

```
/* combine them to make 3 separate vars */
data try1;
  set caf0(rename=(taps=taps0))
      caf100(rename=(taps=taps100))
      caf200(rename=(taps=taps200));
run;
proc print;
  title '3 different vars';
  title2 'but not quite what we wanted!';
run;
proc means mean min max nmiss;
  title 'the MEANS procedure ignores missing values';
run;

data try2;
  merge caf0(rename=(taps=taps0))
        caf100(rename=(taps=taps100))
        caf200(rename=(taps=taps200));
run;
proc print;
  title '3 different vars, with no Missing Values';
run;
```

D.2.5 Creating Data Sets from Procedure Output

Many procedures in SAS allow the output to be sent into a new data set. We will illustrate this idea with `proc means`, which we have seen computes such statistics as the mean, variance, minimum, maximum, etc., of a data set. It is perhaps more useful if we can merge the output of the procedure with the original data set. This is relatively easy to do in SAS, and is a common task.

If we wish to incorporate the overall, or grand, mean into the data set, we have two options, the "one shot" fast way, and the elegant, but longer way. The first way is as follows. Run `proc means` to get the overall mean, as we did at the end of Section D.2.3 above, examine the output, and then just type the mean into another data step as follows:

```
data tapping; /* notice this overwrites the old tapping */
  set tapping;
  overall=246.5;
run;
proc print;
  title 'The quick and dirty way to do this';
run;
```

To avoid having to "do it by hand", the following technique is used. We run `proc means`, but request that *its output* becomes a data file, called `tapbar`. It will be a data set with only one important variable (ignore the rest for now), and one observation, namely the mean of the 30 observations from the data set `tapping`. The option `noprint` indicates that no printed output should be generated from the procedure. The option `mean` indicates that only the mean should be computed. On the `output` line, `out=tapbar` is how we indicate the name of the new data set, and `mean=overall` indicates that we wish to output the mean and call it `overall`.

```
proc means data=tapping noprint mean;
  var taps;
  output out=tapbar mean=overall;  /* creates new dataset tapbar */
run;
proc print; /* look at the new data set */
  title 'output from MEANS procedure (overall mean)';
run;
data einheit; /* this means 'unified' in English */
  set tapbar(keep=overall in=m) tapping(keep=caffeine taps);
  retain grand;
  if m then do;
    grand=overall;
    delete;
  end;
  drop overall;
run;
proc print;
  title 'the combined data sets: data and their overall mean';
run;
```

We first notice the `keep` statements: They simplify the `einheit` data set by only allowing those variables of interest. (As an example, notice `_TYPE_` and `_FREQ_`; these are additional, sometimes useful variables that SAS generates as output from `proc means`). The special command `in=` is used to create a boolean variable (true or false), in this case we called it `m`. As data set `einheit` is being created, `m` indicates if the observation from `tapbar` entered in.

To be more specific, the `set` statement works as follows. First, all observations from `tapbar` are read in because it is the first data set listed in the `set` line. (Notice that this is an example in which order does matter). In this case, there is only one observation, the mean. Next, the 30 observations from `tapping` are read in, so that `einheit` should really have 31 observations. This first observation from `tapbar` is critical. The `if` statement tells SAS to maintain only those 30 observations, `deleting` the one observation from `tapbar`. But then how do we keep the mean? The `retain` statement tells SAS not to clear the value of the variable `grand`; it gets assigned the overall mean *from that first observation from the data set* `tapbar`, i.e., the single variable `overall`.

Yes, some practice with SAS will be necessary to understand its logic. Try removing the `retain` statement and convince yourself that it really works.

One might try to devise a simpler program to accomplish the same task. For instance, it seems like the following could work:

```
proc means data=tapping noprint mean;
  var taps;
  output out=tapbar mean=overall;
run;
data falsch;
  retain overall;
  merge tapbar(keep=overall) tapping(keep=caffeine taps);
run;
proc print; run;
```

Unfortunately, it does not. However, the following program does, and is considerably simpler than the above correct technique. This should be thought of as a "trick" because it is really not obvious why it works.

```
proc means data=tapping noprint mean;
  var taps;
  output out=tapbar mean=overall;
run;
data klappt;
  set tapping(keep=caffeine taps);
  if_N_=1 then set tapbar(keep=overall);
run;
```

Finally, we can even use the above technique for combining the group means and not just the overall mean. To do so, we would use:

```
proc sort data=tapping; /* in case it is not sorted */
  by caffeine;
run;
proc means data=tapping noprint mean;
  by caffeine;
  var taps;
  output out=grptap mean=grpmean; /* creates new data set called grptap */
run;
proc print; /* look at the new data set */
  title 'output from MEANS procedure (by caffeine)';
run;
data allinone;
  merge tapping grptap;
```

```
  by caffeine;
  keep caffeine taps grpmean;
run;
proc print;
  title 'Combined now!';
run;
```

By adding the by statement to the proc means procedure, the output contains the same variables, but now (in this case) three observations. The command merge is very useful in SAS, and combines automatically the mean for each of the three groups with the observations. Observe that, in this case, where we use the by statement with merge, it works, whereas for the overall mean, it did not.

D.3 Advanced Data Handling

D.3.1 String Input and Missing Values

To input a string, simply follow the variable name on the input line with a dollar sign. To represent a missing value, use a period. Consider the following list of authors:

```
data a;
  input name $ x1-x6;
  datalines;
Christopher  11 22 33 44 55 66
Sam          66 55 44 .  22 11
Richard      11 33 55 77 99  0
Daniel       99  .  . 33 11  0
Steven        . 11 77 33 55  .
;
proc print; run;
```

Notice that SAS understands the abbreviation x1-x6 to mean x1 x2 x3 x4 x5 x6. Missing values can appear anywhere in the data list. As long as it is surrounded by blank spaces, there will not be any confusion with a decimal point belonging to a valid number. Imagine that we wish to create a subset of this data set, including only those observations for which the entire vector contains no missing values. In other words, we want a data set containing only those observations corresponding to the names Christopher and Richard. One way is the following:

```
data b1;
  set a;
  if x1 >. & x2 >. & x3 >. & x4 >. & x5 >. & x6 >.;
run;
proc print; run;
```

Three things must be mentioned to understand how this works.

1) Internally, SAS stores a missing value as the largest (in magnitude) negative number possible. Thus, the comparison x1 >. asks if the variable x1 is greater than the value "missing". If x1 has any non-missing value (except the internal SAS code for a missing value), it will be greater than the largest negative number, and thus be true. Otherwise, if x1 is in fact missing, the comparison will be false.

2) The `if` statement checks whether the six variables `x1` through `x6` are not missing. The sign "&" stands for the logical AND mathematical operation. The OR operation is designated by the "|" sign.
3) The `if` statement has no corresponding `then` statement. SAS interprets this to mean that if the condition is true, then allow the observation into the data set, otherwise do not. We already saw this earlier. Another way of accomplishing this is to write the following:

```
data b2;
  set a;
  if x1 =. | x2 =. | x3 =. | x4 =. | x5 =. | x6 =. then delete;
run;
proc print; run;
```

That is, if `x1` is missing, **or** `x2` is missing, … , **or** `x6` is missing, `then delete` the observation, i.e., do not let it enter into data set `b2`.

Now imagine if we had 36 variables instead of six. This leads to a good illustration of the usefulness of the `array` statement introduced in Section D.2.3 above. The following accomplishes the same task as the above programs, but is not only more elegant and easier to read, but also less likely to have a mistake.

```
data c;
  set a;
  array check{6} x1-x6;
  flag=0;
  do i=1 to 6;
    if check(i)=. then flag=1;
  end;
  if flag=0;
  drop i flag; /* these are no longer needed */
run;
proc print; run;
```

The use of so-called boolean or **flag variables** is very common in all computer programming languages. Here, we initialize `flag` to zero, and set it to one if any of the variables in the array are missing. Then, we allow the observation to enter into the data set only when `flag` is zero, i.e., there are no missing values in the observation. Instead of the line `if flag=0;` we could have used the longer (but clearer) `if flag=1 then delete;`.

D.3.2 Using `set` with `first.var` and `last.var`

Consider the caffeine data set introduced earlier. Imagine we would like to construct a data set with three variables: the level of caffeine and the minimum and maximum of the 10 observations in each group. In particular, from the following data table,

	Independent observations									
Caffeine										
0	242	245	244	248	247	248	242	244	246	242
100	248	246	245	247	248	250	247	246	243	244
200	246	248	250	252	248	250	246	248	245	250

we want a data set that looks like

Caffeine	Min.	Max.
0	242	248
100	243	250
200	245	252

We have already seen most of the tools we need to address this problem. Consider the following code:

```
data tapping(keep= taps caffeine)
     extreme1(keep=grpmin grpmax caffeine);
  input caffeine v1-v10;
  grpmin=min(of v1-v10); grpmax=max(of v1-v10);
  output extreme1;
  array vv{10} v1-v10;
  do i=1 to 10;
    taps=vv(i);
    output tapping;
  end;
  drop v1-v10;
datalines;
0   242 245 244 248 247 248 242 244 246 242
100 248 246 245 247 248 250 247 246 243 244
200 246 248 250 252 248 250 246 248 245 250
;
run;
proc print data=tapping;
  by caffeine;
  title 'I''m starting to hate this data set';
  /* Observe how to get a single quote mark into the title */
run cancel;
proc print data=extreme1;
  title 'The min and max of each level of Caffeine';
  title2 'Method 1';
run;
```

The program is very similar to that in Section D.2.3. We construct two data sets at the beginning. The first is `tapping`, and is just the data set with all the observations. Data set `extreme1` contains the desired variables. Notice how the `keep` statements are used on the first line. Without them, no harm is done, but both data sets then contain superfluous variables. After the `proc print` is executed, and you are convinced that the `tapping` data set is correct and do not wish to see the output over and over again, there are at least four options:

- Delete the code corresponding to the `proc print` statement.
- Enclose the code in the comment brackets /* and */.
- "Comment out" each line by preceding it with an asterick * (each line needs to end with a semicolon).
- Use the `run cancel` option, which instructs SAS not to execute the procedure.

All four ways except the first allow the code to stay in the program; this provides clear documentation and is especially useful for longer and more advanced programs, even more so if you plan on looking at it later (and have forgotten everything in the meantime) or, worse, someone else has to look at your code.

The next method should also be familiar. We use `proc means` to generate a data set with the required variables:

```
proc sort data=tapping;
  by caffeine;
  title;
run;
proc means data=tapping noprint min max;
  by caffeine;
  var taps;
  output out=extreme2 min=grpmin max=grpmax;
run;
proc print data=extreme2;
  var caffeine grpmin grpmax;
  title 'The min and max of each level of Caffeine';
  title2 'Method 2 this time';
run;
```

The third method introduces a new data step technique: When we generate a new data set from an old one, using both the `set` and the `by` statements, say `by myvar`, SAS automatically creates two new variables, `first.myvar` and `last.myvar`, that do not get put into the data set, but can be used during the execution of the data step. The data has to be sorted by the `myvar`. Before explaining how they work, we look at an example. Because the variables are not written to the new data set, in order to see them we simply assign them to two new variables, and then use `proc print`.

```
data show;
  set tapping;
  by caffeine;
  first=first.caffeine;
  last=last.caffeine;
run;
proc print;
  title 'The first. and last. variables';
run;
```

The abbreviated out is shown in SAS Output D.1.

We see that `first.caffeine` takes on the value 1 only when the level of caffeine changes to a new level. The variable `last.caffeine` is similar, being 1 only when it is the last observation with that level of caffeine. So how might we extract the minimum and maximum using these variables? The data have to be arranged so that, for each level of caffeine, the data are sorted by taps. If we just wanted to know the first and last observation for each level of caffeine, we do not require a two-level sort, but in this case, we do. Performing a two-level sort is no more difficult (for us) than a one-level:

```
proc sort data=tapping out=tapsort;
  by caffeine taps;
run;
```

Certainly for the computer, this requires more resources, so in general this is not the recommended way to get the minimum and maximum, unless you need to sort the data anyway. The `proc sort` also allows a new data set to be created, as we have done here. Recall that the default (when the `out=`

```
OBS  CAFFEINE  TAPS  FIRST  LAST
 1   0         242   1      0
 2   0         245   0      0
                .
                .
 9   0         246   0      0
10   0         242   0      1
11   100       248   1      0
                .
                .
19   100       243   0      0
20   100       244   0      1
21   200       246   1      0
                .
                .
29   200       245   0      0
30   200       250   0      1
```

SAS Output D.1: Part of the SAS output with `first.` and `last.` variables.

option is not specified) is to rewrite the old data set. The data set `tapsort` is now sorted not only by `caffeine`, but also by `taps`, *within each level of caffeine.*

So, how do we proceed? First, the wrong way. Consider the following code, and try to tell before you run it why it will indeed work, but the data set will not be quite what you would like it to be. (Hint: at each `output`, what is the value of `grpmin` and `grpmax`?) Next, run it, and examine the output.

```
data extreme3; * NOT the correct way!;
  set tapsort;
  by caffeine;
  if first.caffeine then do;
    grpmin=taps;
    output;
  end;
  if last.caffeine then do;
    grpmax=taps;
    output;
  end;
  drop taps;
run; proc print;
  title 'NOT what we wanted!!';
run;
```

After having reflected on what went wrong above, try to determine why one way of fixing things is the following program. The key is the `retain` statement that we also met earlier.

```
data extreme3; * now it is correct;
  set tapsort;
  by caffeine;
  retain grpmin;
  if first.caffeine then grpmin=taps;
  if last.caffeine then do;
```

```
    grpmax=taps;
    output;
  end;
  drop taps;
run; proc print;
  title 'Ahh yes, the pleasures of SAS!';
run;
```

D.3.3 Reading in Text Files

Having to type in the data, or even copy/paste it from a file, is not necessary and not elegant. One can easily circumvent this using the following. The text file `elderly.asc` from Hand et al. (1994) contains the heights of 351 elderly women who participated in an osteoporosis study. We first associate the file to be read in, along with the directory path where it is located, with a name, here `ein`. Similarly, the name and directory location of an output file can be specified, as we do here with `aus`. If, as is common, one particular directory is used for a particular project, then the default directory path can be specified, as stated at the beginning of Section D.1.2.

Next, we read the file in, compute the mean, and write the mean to another file, using the `put` statement. If no `file` is specified, the `put` statement writes to the LOG file. This can be useful for debugging.

```
filename ein "u:\datasets\elderly.asc";
filename aus "u:\datasets\elderly_output.txt";
data grey;
  infile ein;
  input height @@;
run;
proc means data=grey noprint mean;
  var height;
  output out=greymean mean=themean;
run;
data _NULL_;
  set greymean;
  file aus;
  put themean=;
  put themean;
run;
proc univariate data=grey normal plot;
  var height;
run;
```

Inspect the file `elderly_output.txt` to see what the two `put` statements have done. Only one is necessary in general. We will see more uses of the `put` statement later. The data name `_NULL_` is used when we are not interested in the creation of a new data set, just (in this case) the `put` statements contained in it. This not only saves computer memory, disk space, and time, but serves also as documentation for yourself and other potential users of your program.

Finally, examine the output of `proc univariate`. The options `normal` and `plot` are not necessary, but cause `proc univariate` to calculate a test of normality statistic and plot a stem-and-leaf plot of the data, respectively.

D.3.4 Skipping over Headers

Sometimes data files have a header or titles above each column of data. For example, imagine the fictitious data file `justtest.dat` looks as follows:

```
height weight
    155      74
    182      92
    134      45
    188      53
```

To read the data into SAS, it would be quickest just to skip the first line containing the header. (More complicated SAS commands could be used to actually read the titles, see ahead). The following will work:

```
filename in "c:\justtest.dat";
data a;
  infile in;
  if _N_=1 then do;
    input;
    delete;
  end;
  else input height weight;
run;
```

The variable `_N_` is created automatically by SAS and indexes the observations as they are read in. Thus, `_N_` starts at the value 1, and we `input` without specifying any variables. We then `delete` the empty "observation". SAS then goes to the next input line, `_N_` = 2, and the rest of the file is read in. You should try the above technique by creating an artificial data set, such as the one above, and running the above program. Omit the `delete` statement to see what purpose it serves here.

D.3.5 Variable and Value Labels

In older versions of SAS, variable names were limited to eight characters, and this prevented using names that more precisely describe what the variable represents. One way to deal with this in SAS that is still useful is to accompany a name with a *variable label*. In addition, labels for actual data values are also possible, and can convey much more information than the originally coded values. These are called *value labels*, or *formats* in SAS. For example, instead of using a 1 to represent male, and 2 to represent female, it would be nice if we could print the character strings `MALE` and `FEMALE`. We begin with an example. The following description is taken from Hand et al. (1994, p. 266).

> The data come from the 1990 Pilot Surf/Health survey Study of the NSW (New South Wales) Water Board (in Sydney Australia). The first column takes values 1 or 2 according to the recruit's perception of whether (s)he is a Frequent Ocean Swimmer, the second column has values 1 or 4 according to recruit's usually chosen swimming location (1 for non-beach, 4 for beach), the third column has values 2 (aged 15–19), 3 (aged 20–25), or 4 (aged 25–29), the fourth column has values 1 (male) or 2 (female) and, finally, the fifth column has the number of self-diagnosed ear infections that were reported by the recruit.

> The objective of this study was to determine, in particular, whether beach swimmers run a greater risk of contracting ear infections than non-beach swimmers.

The data set starts like this:

```
1 1 2 1 0   2 1 2 1 0   1 4 2 1 0   2 4 2 1 0
```

At this point, we wish just to read the data set into SAS and print it with appropriate labels. Examine the following program:

```
filename in "u:\datasets\ear.asc";
proc format;
  value ocean 1='yes' 2='no';
  value beach 1='non-beach' 4='beach';
  value agegrp 2='15-19' 3='20-25' 4='25-29';
  value sex 1='male' 2='female' other='neutral?';
run;
data a;
  infile in;
  input ocean beach age sex ear @@;
  label ocean='Frequent Ocean Swimmer'
        beach='Usual Swimming Location'
        age='Age Group'
        sex='Geschlecht'
        ear='Self Diagnosed Ear Infections';
  format ocean ocean. beach beach. age agegrp. sex sex.
run;
proc print split=' ';
  title 'With nice labels';
run;
```

There are a few new things here. The `proc format` defines the value labels; it only needs to get executed once. Observe with the value `sex`, the SAS keyword `other`. This is useful for detecting outliers, typographical errors, and strange things in the data set, and should, in general, be used. The variable labels are placed in the data step, and the value labels are engaged also in the data step, but must be previously defined. *Observe that the variable name and the format name can be the same, but that need not be the case. The latter is distinguished by placing a period after its name.* Now when we use `proc print`, things look much "prettier". However, the variable labels are too long and SAS will only use them if it knows where to divide them. To help SAS do this, specify the `split` character in `proc print`. (Try it without this option to see that it works.) A sample of the output is shown in SAS Output D.2.

D.4 Generating Charts, Tables, and Graphs

> The most ubiquitous graph is the pie chart. It is a staple of the business world. Rule of Thumb: Never use a pie chart. Present a simple list of percentages, or whatever constitutes the divisions of the pie chart.
>
> (Gerald van Belle, 2008, p. 203)

```
                              With nice labels

                                                                  Self
         Frequent     Usual                                     Diagnosed
          Ocean       Swimming        Age                          Ear
  Obs    Swimmer      Location       Group     Geschlecht       Infections

   1       yes        non-beach      15-19        male               0
   2       no         non-beach      15-19        male               0
   3       yes        beach          15-19        male               0
   4       no         beach          15-19        male               0
```

SAS Output D.2: Use of `proc print` with the `split` option.

D.4.1 Simple Charting and Tables

Before beginning, it is worth emphasizing that the point of this chapter is to introduce the workings of the SAS data handling language and some of its most common statistical procedures, and not the correct analysis of data per se. As alluded to in the above quotation, the book by van Belle (2008) should be required reading for anyone who has to work with, and present, statistical data. As an example, van Belle (2008, Sec. 9.6) discusses and illustrates why bar graphs and stacked bar graphs are "a waste of ink".

We have already worked with `proc print` and `proc means`, as procedures to output the data set, and sample statistics. Another popular procedure is `proc freq`, which produces frequency tables. With the last data set still in memory, execute the following:

```
proc freq;
  tables sex age sex*age;
run;
```

Observe how the * symbol produces two-way tables (and how SAS knows that, even though the line ends with a semicolon, it is not serving as the delimiter of a comment). Notice that the variable and value labels associated with the data set are used; this considerably assists reading the output. As with most all SAS procedures, there are many possible options that can be added to this procedure; we indicate some below, while the SAS documentation, as usual, can be consulted for the full monty.

A graphical way of depicting the one-way frequency tables shown above is given next. Run the following segment of code:

```
proc chart;
  hbar age sex / discrete;
run;
```

The option `discrete` forces SAS to treat the data as discrete, which it is in this case. The default is to treat the data as continuous. Run the program without the option to see the difference. In the data description given above, the authors noted that the question of interest is whether or not beach swimmers have more ear infections than non-beach swimmers. We could attempt to answer this by an analysis of variance via `proc anova`. For now, consider a graphical approach to shed light on the question:

```
proc chart;
  vbar ear / group=beach;
run;
```

The first statement we can make is that the data are not normally distributed! As such, the inference from the usual ANOVA F test should be taken cautiously (simulation confirms that it is indeed somewhat robust; recall Section 2.4.6) but non-parametric procedures should also be invoked (these are implemented in SAS's `proc npar1way`). Either way, do the sample distributions look different? As skilled statisticians, we immediately consider the next question: Does `sex` make a difference? We will answer the question by making use of the `by` statement:

```
proc sort; by sex; run;
proc chart;
  vbar ear / group=beach;
  by sex;
run;
```

Observe how we first had to sort by the variable `sex`. Does `sex` influence your decision? Another possibility with `proc chart` is the following:

```
proc chart;
  vbar ear / group=beach subgroup=sex;
run;
```

We combine the two different `sex` graphs into one, using the letters "`m`" and "`f`" to distinguish between the two genders. Notice how SAS automatically used the value format that we specified earlier. The `chart` procedure can also make pseudo-3D charts. Consider the following, which not only produce appealing looking graphs, but conveys useful information:

```
proc chart;
  block beach / discrete type=mean sumvar=ear;
run;
```

Several of the options can be combined to produce relatively complicated (and interesting) plots. For example, the following code produces the output shown in SAS Output D.3.

```
proc chart;
  title 'Not bad for text-based graphics';
  block beach / discrete subgroup=sex group=age type=mean sumvar=ear;
run;
```

The "rules" for the `block` chart are as follows:

- The variable specified by `block` forms the x-axis (here it is `beach`).
- The optional `group=` specifies the y-axis (here it is `age`).
- The optional `subgroup=` specifies how the vertical bars are divided (in this case we used `sex`).
- The z-axis is determined by `sumvar=`. In our case we want to examine the distribution of ear infections.
- As we want the average of the ear infections in the particular category, we specify `type=mean`. Other options include `type=freq`, and `type=sum`.
- The option `discrete` is needed in our case because the values of `beach` are limited to two values. (Try without it and convince yourself.)

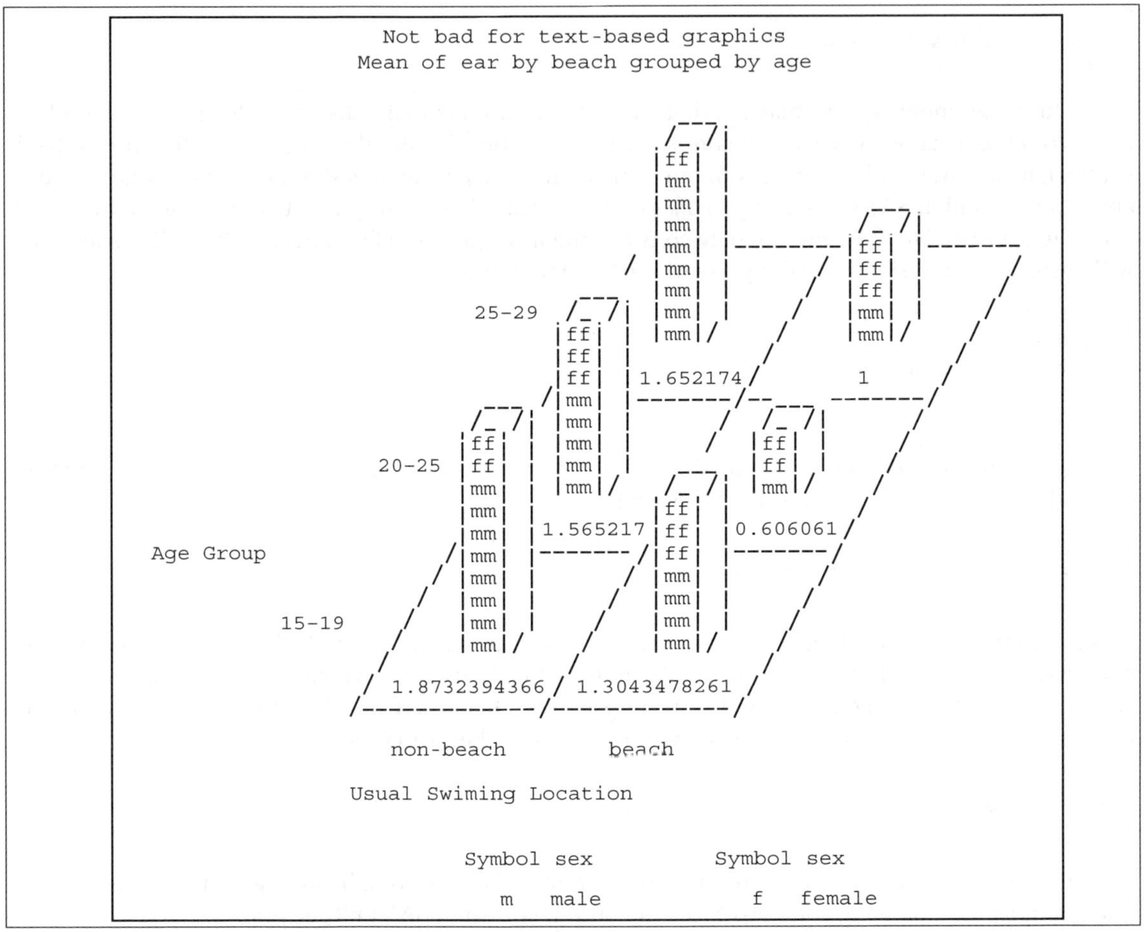

SAS Output D.3: Pseudo-graphical output from `proc chart` (converted to a simple font instead of the better looking `SAS Monospace`).

As a last method to answer the previously posed question, we could always consider using our old friend, `proc means`. However, we would like to look at the mean in four different groups, corresponding to the two levels of `beach` and the two levels of `sex`:

```
proc sort;
  by beach sex;
run;
proc means;
  var ear;
  by beach sex;
run;
```

The above procedures are all methods to condense the data somehow, either as a graph or into summary statistics. Another option when some of the variables are categorical in nature (as are `sex`, `beach`, `age`, and `ocean` in our case) is the `tabulate` procedure:

- The `class` statement specifies which variables are to be used as categorical variables.
- The `var` statement specifies which variable(s) to use in the table.
- How the variable gets used is indicated by one or more of the summary statistics that can be used in `proc means`. For example, in our case we are probably interested in not only the average number of ear infections per category (such as female ocean swimmers aged 20–25), but also the maximum and the standard deviation.
- The `table` statement dictates how the table is formed. The best way to approach the `table` statement is with trial and error. The following are three possibilities.

```
proc tabulate;
  var ear;
  title 'Separate table for each sex';
  class beach age sex;
  table sex, age*beach, ear*(mean max std);
run;
proc tabulate;
  var ear;
  title 'Everything together';
  class beach age sex;
  table sex*age*beach, ear*(mean max std);
run;
proc tabulate;
  var ear;
  title 'Still another possibility';
  class beach age sex;
  table age*beach*(mean max std), sex*ear;
run;
```

Truncated output from the last call to `proc tabulate` is shown in SAS Output D.4.

D.4.2 Date and Time Formats/Informats

SAS makes working with times and dates rather simple. SAS can store variables that contain a representation for the year, month, day, hour, minute, and second, and can manipulate them in many useful ways. For example, in the following program, assume `geburtst` is the birthday formed from the month, day, and year, as input from the `mdy` function. The `intck` function with first argument `'day'` returns the number of days between the second and third arguments, where both are date/-time variables. The function `today()` always returns the current date. Finally, the `format` statement instructs SAS to associate the `mmddyy8.` format with `geburtst`, so that when we print the variable, it appears in a familiar form.

```
data a;
  input j m d;
  geburtst=mdy(m,d,j); ntage= intck('day',geburtst,today());
  format geburtst mmddyy8.;
```

			Geschlecht	
			male	female
			Self Diagnosed Ear Infections	Self Diagnosed Ear Infections
Age Group	Usual Swimm-ing Locat-ion			
15-19	non-beach	Mean	1.79	2.50
		Max	16.00	10.00
		Std	2.67	3.82
	beach	Mean	1.15	1.52
		Max	9.00	10.00
		Std	1.89	2.64
20-25	non-beach	Mean	1.88	1.19
		Max	17.00	4.00
		Std	3.44	1.36
	beach	Mean	0.57	0.63
		Max	5.00	3.00
		Std	1.34	0.90
25-29	non-beach	Mean	2.20	0.63
		Max	10.00	2.00
		Std	2.60	0.92
	beach	Mean	0.65	1.79

SAS Output D.4: Output from `proc tabulate` (converted to a simple font instead of the better looking `SAS Monospace`).

```
  datalines;
1996 1 1 1995 12 31
;
run;
proc print; run;
```

Very useful is `intnx(a,b,c)`. It returns a date/time variable corresponding to `b` incremented by `c` periods, where the period is given by `a`. For example, `nextqtr=intnx('qtr',today(),1)` returns the date/time that is exactly one quarter of a year away from today's date.

There are many other functions, formats, and possibilities. The *SAS Users Guide: Basics* contains many examples.

D.4.3 High Resolution Graphics

D.4.3.1 The GPLOT Procedure

Although SAS offers many graphics procedures, probably the most useful is `proc gplot`, for two-dimensional graphs. The bare bones syntax is as follows:

```
proc gplot;
  plot y*x;
  title 'Yippie!';
run;
```

This generates a plot with the variable `y` on the y-axis, and the variable `x` on the x-axis. Naturally, the procedure has many other options. Consider the data set `e1.dat` from Lütkepohl (1993, App. E) giving quarterly macroeconomic data for West Germany from 1960 to 1982. The file has some header lines that describe the three columns and indicate the starting date of the data, namely 1960, first quarter. We would like to read the data in, skipping the header lines, and also create a variable in SAS that indicates the year and quarter of each observation. The following program works.

```
filename in "u:\datasets\E1.dat";
data level;
  retain period;
  format period YYQ4.;
  label income='Income'
        consume= 'Consumption'
        invest='Investment';
  infile in;
  if _N_ < 3 then do;
    input garb $;
    delete;
  end;
  else do;
    input invest income consume;
    if period=. then period=yyq(1960,1);
    else period=intnx('QTR',period,1);
  end;
  drop garb;
run;
proc print split=' ';
run;
```

For the first actual observation, period is missing, and we set it to the first value, 1960 quarter I. For further observations, we wish to use the intnx command on *the previous value* of period. This is the reason for the use of the retain statement. Without it, period will always be initially set to missing, and thus, our if statement will set it to 1960 Quarter I every time.

Just to get an idea of the range of the data, we run proc means.

```
proc means data=level min max range maxdec=0;
  var invest income consume;
run;
```

The output looks approximately as follows:

Variable	Label	Minimum	Maximum	Range
invest	Investment	179	870	691
income	Income	451	2651	2200
consume	Consumption	415	2271	1856

As income and consumption are roughly of the same scale, we could plot them on the same graph, i.e., using the same set of axes. This is quite easy to do in SAS. We would specify the plot statement as plot income*period consume*period / overlay;. The overlay option tells SAS not to generate a second graph, but rather place them on top of one another. SAS is also smart enough to set the y-axis to include both sets of variables. In other words, the y-axis would start at 415, *the minimum of consumption*, and end at 2651, *the maximum of income*. We could also overlay the plot of invest * period. However, because investment is considerably smaller than both income and consumption, SAS would be forced to choose the minimum of the y-axis to be 179, so that the plots of income and consumption would be rather small.

There is a way around this, however. Because investment shares the same x-axis, namely the variable period, we could overlay the plot of invest * period using a different scaling for the y-axis, *shown on the right side of the graph*. This is accomplished by following the above statement by: plot2 invest * period / overlay;. Notice this is not a second plot statement (and is not allowed), but rather plot2, instructing SAS to use the right side of the plot margin as a second axis.

The next problem is that all the lines are the same type and of the same color. This is changed by defining a symbol statement for each graph, and following the variable pairs to plot with "=", an equals sign, and the number of the symbol. The C= specifies the color, L= specifies the line type, and I= indicates how we would like to "connect the dots". In this case, we just wish to join the points. SAS has other options, such as polynomial smoothing, splines, etc. The symbol definitions are specified before the call to proc gplot and are then valid in any subsequent call to proc gplot or, for that matter, any high-resolution graphics procedure that makes use of them.

```
symbol1 C=blue I=join L=1;
symbol2 C=red  I=join L=2;
symbol3 C=black I=join L=20;
proc gplot;
  plot income*period=1 consume*period=2 / overlay;
  plot2 invest*period=3 / overlay;
run;
```

The next thing we need to do is to improve the axis labels. By default, SAS will use variable labels, if they are defined, and if not, just the variable name itself. As we have two variables along the left y-axis

(income and consumption), SAS just uses the first, namely income. As this is misleading, before calling `proc gplot`, add the following:

```
axis1 LABEL= (ANGLE=90 FONT=SWISS 'Income & Consumption');
axis2 LABEL=(ANGLE=90 FONT=SWISS);
```

The `axis` command is of the form `axis` *n*, where *n* is a number. The `ANGLE` statement instructs SAS to write the axis label at a 90° angle, so that it runs along the axis itself. `FONT` can be used to change which font the characters are written in. Finally, to tell SAS to actually use the axis definitions, follow the slash (where the `overlay` command is) with `VAXIS=AXIS` *n* to modify the vertical axis with the *n* th defined `axis` command, or `HAXIS=AXIS` *n* to modify the horizontal axis. In our case we would have

```
plot income*period=1 cons*period=2 / overlay VAXIS=axis1;
plot2 invest*period=3 / overlay VAXIS=axis2;
```

The last feature we discuss is how to add a legend to the graph. One defines a `legend` *n* statement, with a `SHAPE=` command to indicate what is shown. We would like a line of, say, length equivalent to four letters, with the color and type (dotted, dashed, solid, etc.) corresponding to that used in the graph. We only need to specify a length, SAS takes care of the rest. The `DOWN` command specifies how many lines are shown in a vertical direction. (The `ACROSS` command specifies the horizontal number.) The final set of graphics definitions and call to `proc gplot` look as follows:

```
symbol1 C=blue  I=join L=1;
symbol2 C=red   I=join L=2;
symbol3 C=black I=join L=20;
axis1 LABEL= (ANGLE=90 FONT=SWISS 'Income & Consumption');
axis2 LABEL=(ANGLE=90 FONT=SWISS);
legend1 SHAPE=LINE(4) DOWN=2  LABEL=(FONT=SWISS)
  POSITION=(BOTTOM LEFT INSIDE);
legend2 SHAPE=LINE(4) DOWN=1  LABEL=(FONT=SWISS)
  POSITION=(BOTTOM RIGHT INSIDE);
proc gplot;
  title 'West German Data in Billions of DM';
  plot income*period=1 consume*period=2 /
    overlay grid legend=legend1 VAXIS=axis1;
  plot2 invest * period=3 / overlay legend=legend2 VAXIS=axis2;
run;
```

Notice the `legend=` statement specifies which `legend` *n* to use. We also place a grid on the plot by adding the `grid` statement to one of the `plot` lines. The resulting graph is shown in Figure D.5.

D.4.3.2 The GCHART Procedure

This is similar to `proc chart` discussed above. Extensions to the high-resolution case include color and line fill specification, among other things. Again with the ear infection data, we had used the following to produce two vertical bar charts (histograms) next to one another (using the `group` statement), comparing beach swimmers to non-beach swimmers, dividing each bar into two segments, male and female (using the `subgroup` statement):

```
proc chart;
  vbar ear / group=beach subgroup=sex discrete;
run cancel;
```

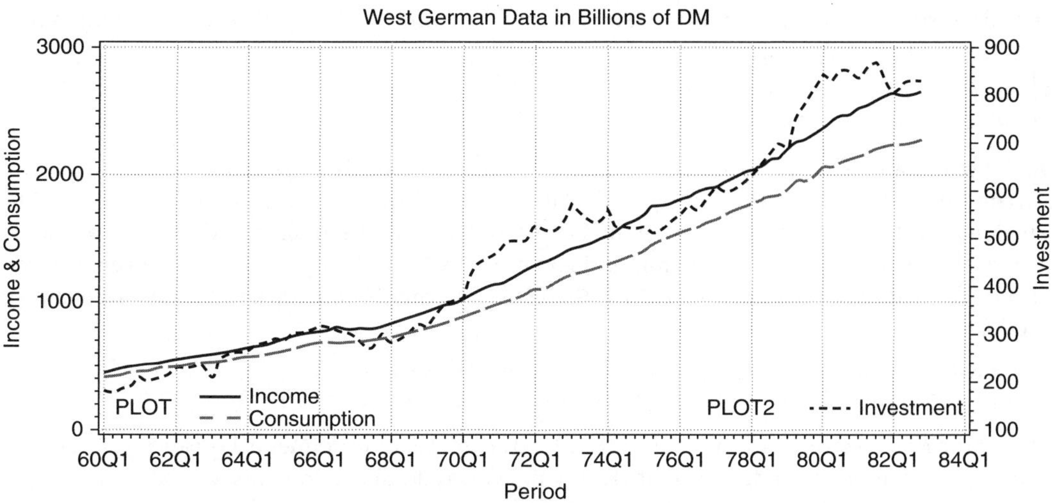

Figure D.5 Output from SAS `proc gplot` with overlaid data.

Now consider the high-resolution version. We wish to use the blue lines for the male segments and red lines for the female segments:

```
proc gchart;
  title 'High Resolution Charts';
  label ear='Infections'  beach='location';
  pattern1 C=BLUE V=L2;   pattern2 C=RED V=R4;
  vbar ear / group=beach subgroup=sex discrete;
run;
```

The `V=` option controls the appearance of the bar, in this case, `L` indicates lines in the left direction, with thickness 1. Thickness can be a number from 1 to 5. Other options are `R` for right lines, and `S` for solid fill. Because the original labels for the variables `ear` and `beach` were quite long, we shorten them somewhat, so they fit on the graph better. Figure D.6 shows the result.

We mention that there are many other useful graphical procedures in SAS; see the online help or the *SAS/Graph Users Guide* for more information.

D.4.4 Linear Regression and Time-Series Analysis

Consider the West German data that we previously plotted. Perhaps we would like to perform a regression with consumption as the dependent variable, and income and investment as independent variables. Given the nature of the data, it might be more sensible to work with first differences of the data,[1] obtained using the `dif` function. Using the `level` data set created earlier,

```
data diff;
  set level;
```

1 Excellent, technically detailed presentations of co-integration, and vector error correction models (VECM) can be found in Hamilton (1994), Hayashi (2000), and Lütkepohl (2005), while Patterson (2000a) provides a highly readable, more basic introduction.

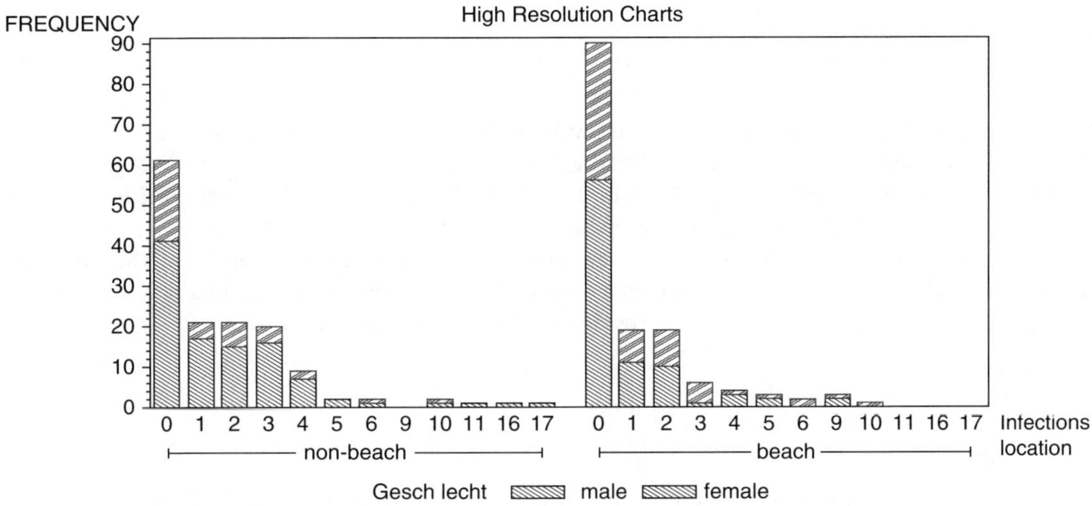

Figure D.6 Output from SAS `proc gchart`.

```
  time=_N_ - 1; inv=dif(invest); inc=dif(income); con=dif(consume);
  label inc='Income (1st diff)'
        con= 'Consumption (1st diff)'
        inv='Investment (1st diff)'
        time='time trend';
run;
```

In data set `diff`, the variable `time` is 1… 91. We subtract 1 from `_N_` when `time` is constructed because the first observation of `inv`, `inc`, and `con` will be missing (due to differencing). The following is the bare bones structure of the regression procedure, which we already encountered in Example 1.12.

```
proc reg data=diff;
  model consume = invest income;
run;
```

Suppose we want to consider several models, in particular we wish to compare the fit in levels with the fit in differences. We can specify several `model` statements under one `proc reg` call, as well as giving each one a label, so that the output is easier to identify. Also, SAS allows a data set to be generated that contains the coefficient estimates for all the models. This is accomplished using the `outest=` statement. Below we generate this, and print only some of its contents, in particular the root mean square error (RMSE) of each model. Finally, additional options can be specified on any `model` statement. There are far too many to describe here—see the SAS manual for a listing. Here we look at the correlation matrix of the coefficient estimates, `CORRB`, as well as the Durbin–Watson statistic, `DW`.

```
proc reg data=diff outest=beta;
  levels1: model consume = invest income;
  levels2: model consume = time income;
  onediff: model con = inv inc / CORRB DW;
```

```
run;
proc print data=beta;
  var _model_ _rmse_;
run;
```

Assume we decide to pursue further the last model, the one in differences, and want to plot the true value of consumption against the predicted value. To do this requires two steps. We first obtain the predicted value of the *difference of* consumption from the regression. Then we *un-difference* (or integrate) it. The first of these tasks is accomplished by creating a new data set from `proc reg` containing the predicted values. This data set, named after the `OUT=` statement, contains all the variables in the incoming data set, as well as the ones specified. Here, `P=` writes the predicted values. Other variables could also be written, such as the residuals, 95% confidence bounds, etc.

```
  proc reg data=diff;
  model con = inv inc;
  output OUT=story P=p;
run;
```

For the second task, SAS unfortunately does not have a built-in function to undifference a variable, but the following program will work. Observe that the `retain` statement is key here.

```
data story2;
  set story;
  retain p2;
  if _N_=1 then p2=consume;
  else p2=p2+p;
  label consume='Actual Consumption';
  label p2='Predicted Consumption';
run;
```

The following `gplot` statements should be familiar now; they result in Figure D.7.

```
symbol1 C=blue I=join L=1;
symbol2 C=red I=join L=2;
legend1 SHAPE=LINE(15)
        DOWN=2
        LABEL=(FONT=SWISS);
axis1 label=(ANGLE=90 "Consumption");
proc gplot data=story2;
  title 'True and Predicted Consumption';
  title2 'Using model in Differences';
  plot (p2 consume) * period / overlay grid legend=legend1 vaxis=axis1;
run;
```

Perhaps we now wish to perform the same regression analysis, but treating the error term as an autoregressive (AR) process. The `proc autoreg` is ideally suited for this. To fit the above regression model with AR(3) disturbances, we would use:

```
proc autoreg data=diff;
  model con = inv inc / nlag=3;
run;
```

To examine the generalized Durbin–Watson statistics (5.24), along with their exact p-values, use the following:

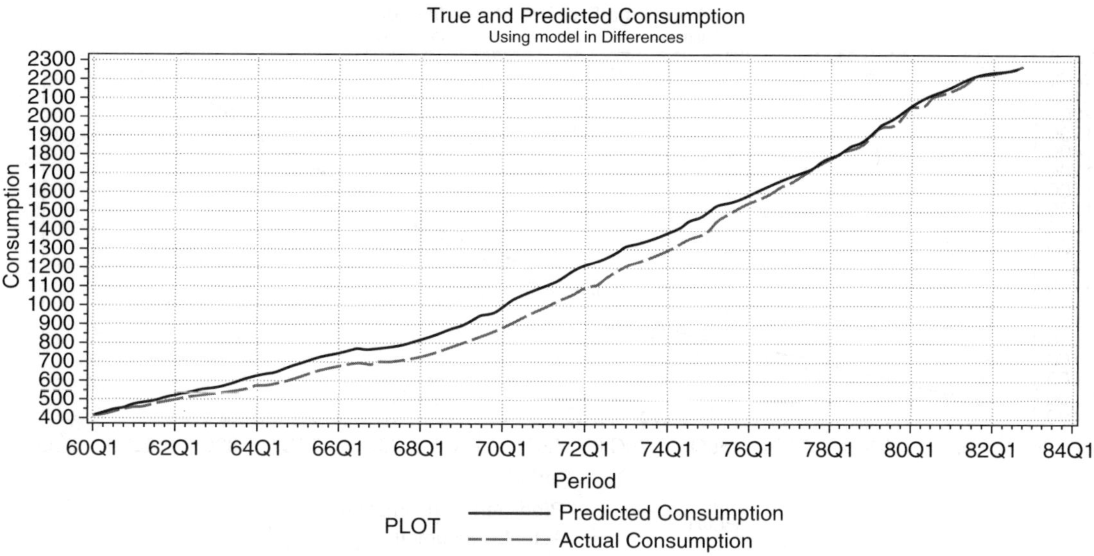

Figure D.7 Differences model for predicting consumption.

```
proc autoreg data=diff;
  model con = inv inc / DW=12 DWPROB;
run;
```

Using the `backstep` option, one could automatically pick those AR lags that are "significant" to include in the model, though, as emphasized in Chapter 9, there are better ways of model selection.

The `slstay=` option allows us to change the cutoff *p*-value determining whether an AR lag is permitted to enter the model. The default is 0.05.

```
proc autoreg data=diff;
  model con = inv inc / nlag=12 backstep slstay=0.25 method=ml;
  output out=story3 P=p3;
run;
```

Notice the `output` statement has the same form as that in the `proc reg`. We would expect that this model fits better. In fact, the `autoreg` procedure selects lags 1, 3, 6, and 7, and the RMSE improves from 10.47 to 9.35. Following the same procedure as above to generate a plot of true and predicted consumption, we get Figure D.8. Notice the fit is somewhat better.

D.5 The SAS Macro Processor

It's not daily increase but decrease—hack away the unessential!

(Bruce Lee)

D.5.1 Introduction

Macros in SAS are programs that generate SAS code to be executed. One use of this arises in the case when the program to write depends on quantities that can only be assessed at runtime.

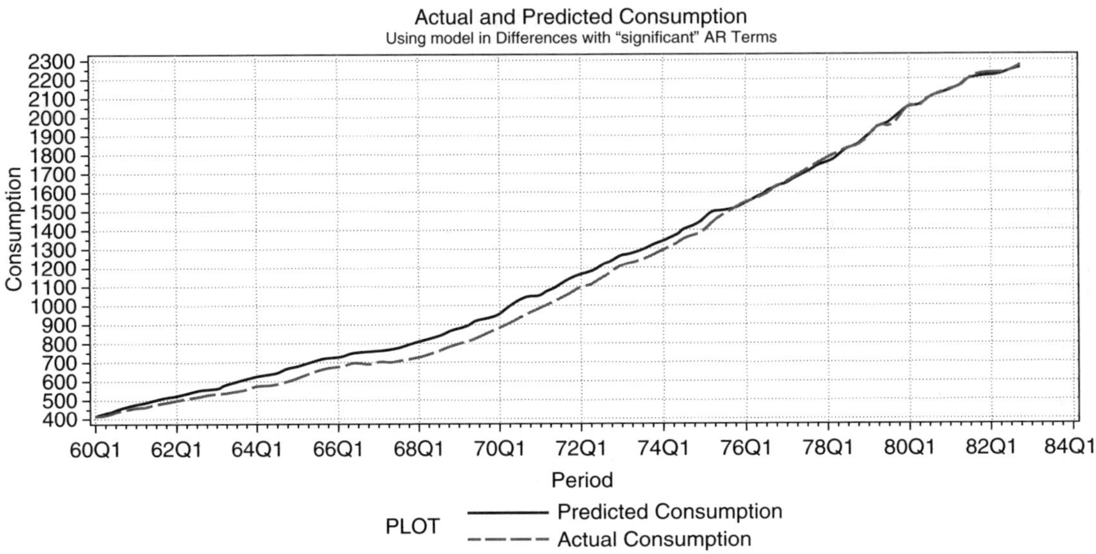

Figure D.8 Autoregressive model for predicting consumption.

Some consider the macro features of SAS to be (i) difficult, (ii) confusing, and (iii) not necessary. The first of these is true only if you are not yet comfortable with the techniques we have discussed up to this point. The second statement is sometimes true, so that some extra care and experience are indeed required when using SAS macros. Regarding the last point, there are many tasks that are either very difficult or virtually impossible to do in SAS without using macros. Many times, even if a certain task could be accomplished without macros, using them can (i) make the program much shorter, (ii) save computer memory and disk space, (iii) make the program easier to understand, and, most importantly, (iv) drastically reduce the chance of a programming error.

Another important reason is speed. For example, if a bootstrap inference is required, it is much faster to generate the large data set with the bootstrap resamples, and then use a `by` statement and use of a `do` loop via a macro to call the statistical procedure.

Instead of a general treatment, we will consider several simple, but common examples that illustrate how SAS macros can make life much easier. The more advanced features are detailed in the SAS manual dealing exclusively with the macro processor.

D.5.2 Macro Variables

A macro variable is defined with the `%let` statement and evaluated by placing an amperstand (&) before the variable. Consider the following example, where we assume that data set `a` contains at least one variable and 25 observations:

```
%let myvar=25;
proc print data=a;
  title "This data set has &myvar observations";
run;
```

The first thing to notice is that, in title statements with macro variables, we need to use the double quote mark instead of the single quote mark. With single quote marks, SAS does not parse (read) the statement for macro variables. When SAS executes the above program, it first evaluates `&myvar` before running the `proc print`, and instead "sees" the following code:

```
proc print data=a;
  title 'This data set has 25 observations';
run;
```

Of course, the whole purpose of macros is that they allow SAS code to be generated at runtime, so that the above program is not particularly useful, i.e., the variable `myvar` is fixed. In order to allow `myvar` to get defined at runtime, we need the `symput` command, which defines a macro variable during execution of a data step. Regardless of the number of observations in data set `a`, the following will work:

```
data a;
  input y x @@;
  datalines;
18 543 18 583 9 358 21 356 21 923
;
data _NULL_;
  set a end=dasEnde;
  if dasEnde then call symput('myvar',_N_);
run;
proc print data=a;
  title "This data set has &myvar observations";
run;
```

Note that, with the `end=` feature in the `set` statement, we can create a boolean variable that is always false until the last observation from the `set` is read in, in which case it is true. The variable (in this case, `dasEnde`) is not written to the data file. In the above program, when `dasEnde` is true, we `call` the `symput` function, defining the macro variable `myvar` to have the value `_N_`, which is SAS's internal counter of the number of observations. Notice that with the `symput` function, we enclose the macro variable name in single quotation marks. If you run the above, you will notice that the number of observations, in this case 5, is printed with many leading or trailing blanks. To avoid this, we can use two character manipulation functions of SAS, namely `trim` and `left`, which trim leading blanks, and left align the string, respectively. Replace the appropriate line above with the following:

```
if dasEnde then call symput('myvar',trim(left(_N_)));
```

Of course, for SAS to treat the numeric value contained in `myvar` as a character string, it must first be converted to a string. SAS does this internally for you, but does print the following to let you know:

> **NOTE: Numeric values have been converted to character values at the places given by: (Line):(Column). 108:49**

For example, the program

```
%let fname = "u:\datasets\temp.txt";
filename in &fname;
data b;
  infile in;
```

```
  input x y;
run;
```

will be interpreted by SAS as

```
filename in "u:\datasets\temp.txt";
data b;
  infile in;
  input x y;
run;
```

and the program will be executed successfully (assuming that the file exists). Assume we also want to print the filename into the title of say, `proc print`, so that, for this particular filename, we would want the following to be executed:

```
proc print;
  title 'The text file is: "u:\datasets\temp.txt" ';
run;
```

To do this with the macro variable, use the following:

```
proc print;
  title "The text file is: "&fname" ";
run;
```

We surrounded the macro variable reference `&fname` also with double quotation marks. This is in general not needed, as we saw above. In this case, however, `fname` is itself a string surrounded by double quotes, and two quotes instead of one are needed.

D.5.3 Macro Programs

Imagine you are getting tired of having to type `proc print; run;` every time you want to see the results of a data step and would like to type something shorter. We could define the following macro program:

```
%macro druck0;
  proc print data=_LAST_;
    title;
  run;
%mend druck0;
```

The data set `_LAST_` just tells SAS to use the latest data set that you created. Now, when we create a new data set and wish to print it, we can just enter `%druck0` after the data step to call the macro. Notice we don't need to follow the macro call with a semicolon because the macro itself ends in a `run;` statement.

It might be nice if we could pass it a parameter indicating to `print` using a `by` statement. However, often we won't want to use the `by` statement, so a method should be used where the default is that no `by` statement is used, only if we tell it to. Here is one way, making use of the SAS boolean operator `NE`, which means "not equal to", and noting that "keine" is German for "none":

```
%macro druck1(byvar=keine);
  proc print data=_LAST_;
    title;
```

```
    %if &byvar NE keine %then %do;
      by &byvar;
    %end;
  run;
%mend druck1;
```

Macro `druck1` takes a parameter, `byvar`, and, optionally, we have specified a default value, `keine`. If the user does not specify the value of `byvar`, then it takes on this default value, namely `keine`. In this case, it can be called as `%druck1;` (with the semicolon) or `%druck1()`, and the semicolon is not needed.

Statements that get evaluated, not generated, in the macro, are preceded by a percent (%) sign. In other words, we do not wish the macro to generate statements with `if` and `then`, but rather to actually check if `&byvar` is not equal to the value `keine`. Thus, if we call `%druck1` without any parameters, the code translates to

```
proc print data=_LAST_;
  title;
run;
```

If instead we call `%druck1(byvar=y)`, the code is

```
proc print data=_LAST_;
  title;
  by y;
run;
```

When we specify a default for the input variable in a macro, as we have done in `druck1`, we must also specify the variable name when we invoke the macro. That is why we use `%druck1(byvar=y)` and not just `%druck1(y)`. SAS will return an error message if you try this. Of course, the variable `y` must be in the previous data set, and the data set is sorted by this variable `y`. Here is an example:

```
data a;
  input y x @@;
  datalines;
18 543 18 583 9 358 21 356 21 923
;
proc sort data=a;
  by y;
run;
%druck1(byvar=y)
```

We can simplify this macro somewhat, taking advantage of SAS's somewhat forgiving syntax. The following piece of code is allowed:

```
proc print;
  by;
  title 'does this really work?';
run;
```

Here, there is no `by` variable specified with the `by` statement, but SAS does not consider this an error. Instead, it is taken to mean that SAS should print the data set *without using* a `by` variable. Try it and see. Thus, we can use the easier macro:

```
%macro druck2(byvar);
  proc print data=_LAST_;
    title;
    by &byvar;
  run;
%mend druck2;
```

If we do not pass `byvar`, and call `%druck2()` or `%druck2;`, the macro resolves to

```
proc print;
  title;
  by;
run;
```

D.5.4 A Useful Example

Suppose we wish to use SAS in batch mode to read the text file

```
1.2 11 22
2.7 33 44
3.1 55 88
4.5 77 99
```

perform a regression analysis, and write the coefficients out to a file. Because we know that `regin.txt` contains three columns with the first being the dependent variable, say y, and the next two are the independent variables, say x_1 and x_2, we could use the following program. One suggestion is to run this first without the `noprint` option in the `proc reg` statement, just to make sure things are working. Only after it is debugged should one use this.[2]

```
filename in  'u:\datasets\regin.txt';
filename out 'u:\datasets\regout.txt';
data a;
  infile in;
  input y x1 x2;
run;
proc reg data=a outest=beta noprint;
  model y = x1 x2;
run;
data _NULL_;
  file out;  set beta;
  put intercept;   put x1;   put x2;
run;
```

In the last `data` step, the `put` statements write to the file specified by the `file` statement. However, what if the number of independent variables can change? Call the number of regressors `p`.

D.5.4.1 Method 1

The first way is the following. We create the input file with the first line specifying the number of regressors. So, `regin.txt` now looks like:

2 Observe that the intercept term of the regression automatically receives the name `intercept`; in older versions that restricted the length of variable names to eight characters, it was `intercep`.

```
2
1.2 11 22
2.7 33 44
3.1 55 88
4.5 77 99
```

Our goal is to read this file twice. The first time, we just read the first number to establish the value of p. Then we read the file again, skipping the first line, but using our knowledge of p to correctly read the matrix. The first part could be accomplished by the code segment:

```
data _NULL_;
  infile in;
  if _N_=1 then do;
    input p; call symput('p',p);
  end;
  stop;
run;
```

The `stop` statement tells SAS to stop reading the input file. There is no need to continue reading it, so this saves time. There is a slightly more elegant way to do this. If we could somehow tell SAS that all we want is the first line, we would not need the `if _N_=1` statement, nor the `stop` statement. This can be accomplished as follows:

```
data _NULL_;
  infile in obs=1;
  input p; call symput('p',p);
run;
```

Here, the `obs=1` statement tells SAS precisely what we wanted. Of course, this has other uses. If we wish to test a program, we could read in just the first, say, 100 observations of a large file instead of the whole thing, and debug the program. When we are sure that it works, we would remove the `obs=` statement.

Next, we need a macro that, for a given value of p, say 4, would generate the following line: `x1 x2 x3 x4;` We could then use such a macro in the regression procedure. Here is the macro:

```
%macro xnames(name,uplim);
  %do n=1 %to &uplim; &name&n %end;
%mend xnames;
```

By calling `%xnames(x,4)`, for example, we would get the desired line. However, we will call it with the macro variable p instead, i.e., `%xnames(x,&p)`. Notice that there is no semicolon following the line `&name&n`. If there were, SAS would also insert a semicolon between each variable name, which is not what we want. Next, we need a way to generate the p `put` statements. This will work:

```
%macro varput(name,uplim);
  %do n=1 %to &uplim; put &name&'n; %end;
%mend varput;
```

Here we use a semicolon after the line `put &name&n` because we want each `put` statement to be executed separately. Putting this all together, we have

```
filename in  'u:\datasets\regin.txt';
filename out 'u:\datasets\regout.txt';
%macro xnames(name,uplim);
```

```
  %do n=1 %to &uplim; &name&n %end;
%mend xnames;
%macro varput(name,uplim);
  %do n=1 %to &uplim; put &name&n; %end;
%mend varput;
data _NULL_;
  infile in obs=1; input p; call symput('p',p);
run;
data a;
  infile in; if _N_=1 then do; input; delete; end;
  else input y %xnames(x,&p);
run;
proc reg data=a outest=beta noprint;
  model y = %xnames(x,&p);
run;
data _NULL_;
  file out; set beta; put intercept; %varput(x,&p);
run;
```

D.5.4.2 Method 2

Now assume that we either do not want to, or, for some reason, cannot write the number of regressors as the first line of the text file. What we could then do is read the first line of data and somehow figure out how many numbers are on it. Because `y` is the first variable on the line, we take `p` to be one less than this number. Once we know `p`, we can re-read the entire file.

There are a number of approaches to "parsing" the first line to determine how many numbers are there. One way would be to read the line as a character string and count the number of blank spaces. For instance, if there are three columns of numbers, then there must be a total of two blanks on the first line. This only works when the data are separated exactly by one blank space; otherwise, it gets trickier. There is a much easier way though, which works irrespective of how the numbers are spaced on the first line. Before the program is shown, a new option for the `infile` statement is described that is very useful in general.

Imagine we have a data file consisting of names, ages, and year of high school graduates. However, if the person has either not graduated yet, or never will, instead of the SAS missing character, the period, there is no entry. The text file might look like this:

```
John 23 1990
Mike 16
Susan 14
Mary 20 1992
Ed 45
```

If we were to use the following code to read this data, we would get an error message:

```
data a;
  infile people; * assume this refers to the text file above;
  input name $ age year;
run;
```

The reason it will not work is as follows. When SAS reads the entry for `Mike`, because the `year` is missing, SAS goes to the next line to find it. SAS then encounters the character string `Susan`, and everything goes wrong from there. The `missover` statement instruct SAS *not to go to the next line when something is missing*. Thus, the program

```
data a;
  infile people missover;
  input name $ age year;
run;
```

will work correctly. The default is what is called `flowover`. This means, flow over to the next line to find the data, and is exactly not what we want in this case. A third option SAS allows is `stopover`. If there is something missing, SAS stops reading and reports the mistake immediately. This is useful if you know that the data should be complete and want SAS to check.

Regarding the program we wish to construct, our strategy is as follows. Read in a large number of variables for the first line, say `v1` through `v40`, but use the `missover` option. If `p` is 3, i.e., there are 4 numbers on the line, then `v1` will be the `y` value, `v2` will be `x1`, `v3` will be `x2`, and `v4` will be `x3`. The variables `v5-v40` will all be set to missing. Thus, we only need to count the number of variables of `v2-v40` to determine the value of `p`. Of course, if there are more than 39 regressors, this method will fail, so that some "prior" knowledge about the data is required.

```
data _NULL_;
  infile in obs=1 missover; input v1-v40;
  array v{*} v1-v40;
  do i=2 to 40;
  if v(i) >. then p+1; * In SAS, p+1 is short for p=p+1;
end;
call symput('p',p);
run;
data a;
  infile in; input y %xnames(x,&p);
run;
```

In addition, we see yet another application of the `array` statement, as well as another way to increment a variable in SAS. The expression `p+1` is equivalent to `p=p+1`. We could use any number instead of 1, but for negative numbers we cannot write, for example, `count-3` to mean `count=count-3`. We could, however, write `count+(-3)`.

Much more information about macros in SAS and many examples can be found in the *SAS Guide to Macro Processing*.

D.6 Problems

Problem 4.1 You maintain a file of the names and grades of doctoral students. The information for each exam comes from a different instructor. Grades are in the Swiss format, meaning between 1.0 and 6.0, in increments of 0.25, with 6.00 being the best, 1.0 the worst, and 4.0 just passing. The file currently looks as follows.

```
Darwin    Charles  4.50  5.00
Dawkins   Richard  5.25  4.50  5.50
Fisher    Ronald   5.25  6.00
Freud     Sigmund  4.75  5.50
Mendel    Gregor   5.50  4.75
Pinker    Steven   5.75  5.00
Popper    Karl     6.00  6.00
```

This is called the master file because it contains both first and last names, and all the exam grades (and possibly other information, like student ID number, etc.). Richard had to take the third exam before anyone else. The official third exam was taken later.
You receive a text file from the instructor for the third exam with the last names (not necessarily in alphabetical order) along with the raw score (meaning, the total number of points from an exam, out of, say, 200). It looks as follows.

```
Darwin   120
Dawkins
Fisher   145
Freud    180
Mendel    90
Pinker   110
Popper   135
```

Write a program that reads in the master and exam files, merges them, and constructs a new master file. The grade (G) from the third exam is determined from the raw score (r) as $G = 1.0 + 0.25 \times \lceil 12 \times q + 8 \rfloor$, where

$$q = \frac{\max(r) - r}{\max(r) - \min(r)}$$

and $\lceil x \rfloor$ denotes the numeric rounding function, i.e., $\lceil 3.6 \rfloor = 4$.

Problem 4.2 You are a personal fitness trainer in Switzerland and have asked your client (Laura) to record information about her workout Monday through Friday during the period of November 2009 to January 2010 as follows. Each line contains the month and day, and then a sequence of numbers indicating how many repetitions she managed with the weights. For example, from the second line, which corresponds to November 16, she did three workouts, each time doing five repetitions. Here are the first five lines of the data set; the entire data set can be found in the file named `fitness.txt`.

```
11 13 5
11 16 5 5 5
11 17 6 6 7
11 18 7
11 19 7 5
```

The task is to write a program that, ideally, is more general and not dependent on this particular data set for which the maximal number of sets she accomplished on a day is five. It should generate a report containing the following:

1. A list of the data, the beginning of which might look like:

Client's Program

DATE	SETS	AVERAGE	V1	V2	V3	V4	V5
Fri, Nov 13, 2009	1	5.0	5	.	.	.	.
Mon, Nov 16, 2009	3	5.0	5	5	5	.	.
Tue, Nov 17, 2009	3	6.3	6	6	7	.	.

2. A list of the average frequency of training sessions per weekday, which should look like:

```
Average number of sets per weekday

weekday      average
-------------------------------
Montag       2.87500
Dienstag     3.37500
Mittwoch     3.28571
Donnerstag   1.60000
Freitag      1.50000
```

   For this part, the means should of course be computed using a `by` statement. Just to practice, also make a program that produces the output *without using* the `by` statement.
3. A high-resolution plot containing both number of sessions and average daily repetition number.

Hints:

1. You will need to determine a way to read in a variable number of entries per line, and a way to instruct SAS to keep only as many variables as needed.
2. The `mdy` function will be useful, as well as the `weekday` function (use the online help for details) and the `weekdate17.` format.
3. To get the mean per weekday, use `proc means` with the `output` option, and for printing the output, you will need to create a custom `format` for each day of the week.

D.7 Appendix: Solutions

1) The programs in Listing D.1 will accomplish the task.
2) The programs in Listings D.2 and D.3 will accomplish the task, and Listing D.4 shows the code that can be used if you do not wish to use the `by` statement.

```
filename masterin "u:\datasets\master.txt";
filename exam3in "u:\datasets\exam3.txt";
filename out "u:\datasets\nmaster.txt";
data master;
  infile masterin missover;
  attrib vorname length=$14 label='Given Name';
  attrib nachname length=$14 label='Family Name';
  input nachname $ vorname $ grade1-grade3;
run;
proc sort data=master; * only need this the first time;
  by nachname;
run;
data newexam;
  infile exam3in missover;
  attrib nachname length=$14 label='Family Name';
  input nachname $ raw;
run;
proc sort data=newexam;
  by nachname;
run;
proc means data=newexam(where=(raw>-1)) noprint max min;
  var raw;
  output out=extremes max=rawmax min=rawmin;
run;

data newexam;
  set newexam;
  if _N_=1 then set extremes(keep=rawmax rawmin);
  ratio = 1 - (rawmax-raw)/(rawmax-rawmin);
  grade = 1.0 + 0.25*round(12*ratio+8);
  keep nachname grade raw;
run;
data masternew;
  merge master newexam;
  by nachname;
  if grade3 <= . then grade3=grade;
run;
proc print data=masternew;
  title 'Results of 3rd test';
  var vorname grade3;
run;
data _NULL_;
  set masternew;
  file out;
  put nachname $18. vorname $12. (grade1-grade3) (5.2);
run;
```

SAS Program Listing D.1: Program for calculating grades.

```
%macro vlist(st,p);
    &st.1-&st&p
%mend vlist;
proc format;
  value myweek  2='Montag'
                3='Dienstag'
                4='Mittwoch'
                5='Donnerstag'
                6='Freitag';
run;
data a;
  infile "u:\datasets\fitness.txt" missover;
  retain themost 0;
  input monat datum v1-v40;
  if 11<=monat<=12 then jahr=2009;
  else jahr=2010;
  date=mdy(monat,datum,jahr);
  weekday=weekday(date);  * mon, tues, etc.;
  format date ddmmyy8.; * a default format;
  average=mean(of v1-v40);
  array v{*} v1-v40;
  sets=0;
  do i=1 to 40;
    if v(i) > . then sets+1;
  end;
  themost=max(themost,sets);
  call symput('themost',themost);
  drop i themost;
run;
data a;
  set a (keep=date weekday sets average %vlist(v,&themost));
  if sets > 0;
run;
```

SAS Program Listing D.2: Program to process the fitness trainer data. Continued below.

```
proc print noobs;
  title 'Client''s Program';
  format date weekdate17.;  * a nicer format;
  var date sets average %vlist(v,&themost);
run;
proc sort out=byday;
  by weekday;
run;
proc means data=byday noprint;
  var sets;
  by weekday;
  output out=mbyday mean=average;
run;
proc print data=mbyday noobs;
  title 'Average number of sets per weekday';
  format weekday myweek.;
  var weekday average;
run;

symbol1 C=blue I=join L=1;
symbol2 C=red I=join L=2;
legend1 SHAPE=LINE(5)
      DOWN=1
      LABEL=(FONT=SWISS)
      POSITION=(BOTTOM LEFT INSIDE);
legend2 SHAPE=LINE(5)
      DOWN=1
      LABEL=(FONT=SWISS)
      POSITION=(BOTTOM RIGHT INSIDE);
proc gplot data=a;
  axis1 LABEL= (ANGLE=90 FONT=SWISS
          'Average Number of "Reps"');
  axis2 LABEL=(ANGLE=90 FONT=SWISS
           'Number of "Sets"');
  title 'Client''s Progress';
  plot average*date=1 / overlay
      grid legend=legend1 VAXIS=axis1;
  plot2  sets * date=2 / overlay
      legend=legend2 VAXIS=axis2;
run;
```

SAS Program Listing D.3: Continuation of program for fitness trainer data.

```
data monday; set a(keep=sets date);
  wotag=weekday(date); if wotag=2; run;
proc means noprint mean; var sets;
  output out=tag1 mean=average; run;
data tuday;  set a(keep=sets date);
  wotag=weekday(date); if wotag=3; run;
proc means noprint mean; var sets;
  output out=tag2 mean=average; run;
data wedday; set a(keep=sets date);
  wotag=weekday(date); if wotag=4; run;
proc means noprint mean; var sets;
  output out=tag3 mean=average; run;
data thurday; set a(keep=sets date);
  wotag=weekday(date); if wotag=5; run;
proc means noprint mean; var sets;
  output out=tag4 mean=average; run;
data friday; set a(keep=sets date);
  wotag=weekday(date); if wotag=6; run;
proc means noprint mean; var sets;
  output out=tag5 mean=average; run;

data alles;
  set tag1 tag2 tag3 tag4 tag5;
  if _N_=1 then Wochentg='Montag    ';
  if _N_=2 then Wochentg='Dienstag  ';
  if _N_=3 then Wochentg='Mittwoch  ';
  if _N_=4 then Wochentg='Donnerstag';
  if _N_=5 then Wochentg='Freitag   ';
run;
proc print;
var Wochentg average; title 'Average number of sets per weekday';
run;
```

SAS Program Listing D.4: How to get the means for each weekday without using the `by` statement in `proc means`.

Bibliography

Aas, K., Haff, I., and Dimakos, X. K. (2005). Risk Estimation using the Multivariate Normal Inverse Gaussian Distribution. *Journal of Risk*, 8(2):39–60.

Aas, K. and Haff, I. H. (2006). The Generalised Hyperbolic Skew Student's t-Distribution. *Journal of Financial Econometrics*, 4(2):275–309.

Abad, P., Benito, S., and López, C. (2014). A Comprehensive Review of Value at Risk Methodologies. *Spanish Review of Financial Economics*, 12(1):15–32.

Abadir, K. M. (1998). *Explicit Distribution Theory for Simple Time Series*. John Wiley & Sons, New York.

Abadir, K. M. and Magnus, J. R. (2005). *Matrix Algebra*. Cambridge University Press, Cambridge.

Abdous, B., Genest, C., and Rémillard, B. (2005). Dependence Properties of Meta-Elliptical Distributions. In Duchesne, P. and Rémillard, B., editors, *Statistical Modeling and Analysis for Complex Data Problems*, chapter 1. Springer, New York.

Abraham, B. and Ledolter, J. (1983). *Statistical Methods for Forecasting*. John Wiley & Sons, New York.

Abraham, B. and Ledolter, J. (1984). A Note on Inverse Autocorrelations. *Biometrika*, 71:609–614.

Agresti, A. (2015). *Foundations of Linear and Generalized Linear Models*. John Wiley & Sons, Hoboken, NJ.

Aielli, G. P. (2013). Dynamic Conditional Correlation: On Properties and Estimation. *Journal of Business & Economic Statistics*, 31(3):282–299.

Aielli, G. P. and Caporin, M. (2013). Fast Clustering of GARCH Processes via Gaussian Mixture Models. *Mathematics and Computers in Simulation*, 94:205–222.

Akaike, H. (1973). Block Toeplitz Matrix Inversion. *SIAM Journal on Applied Mathematics*, 24(2):234–241.

Albuquerque, R. (2012). Skewness in Stock Returns: Reconciling the Evidence on Firm Versus Aggregate Returns. *The Review of Financial Studies*, 25(5):1630–1673.

Alexander, C. (2001). Orthogonal GARCH. *Mastering Risk*, 2:21–38.

Alexander, C. (2002). Principal component models for generating large GARCH covariance matrices. *Economic Notes*, 31(2):337–359.

Alexander, C. (2008). *Market Risk Analysis II: Practical Financial Econometrics*. John Wiley & Sons, Chichester.

Alexander, C. and Chibumba, A. (1996). Multivariate Orthogonal Factor GARCH. Working paper.

Alexander, C. and Lazar, E. (2004). The Equity Index Skew and Asymmetric Normal Mixture GARCH. ICMA Centre Discussion Papers in Finance 2004–14.

Alexander, C. and Lazar, E. (2005). Asymmetries and Volatility Regimes in the European Equity Markets. ICMA Centre Discussion Papers in Finance 2005–14.

Linear Models and Time-Series Analysis: Regression, ANOVA, ARMA and GARCH, First Edition. Marc S. Paolella.

Alexander, C. and Lazar, E. (2006). Normal Mixture GARCH(1,1): Applications to Exchange Rate Modelling. *Journal of Applied Econometrics*, 21:307–336.

Ali, M. M. (1983). A Note on Approximating the Distribution of the Durbin–Watson Statistic. *Journal of Time Series Analysis*, 4(4):217–220.

Ali, M. M. (1984). Distributions of the Sample Autocorrelations when Observations are from a Stationary Autoregressive-Moving-Average Process. *Journal of Business and Economic Statistics*, 2:271–278.

Ali, M. M. (1987). Durbin–Watson and Generalized Durbin–Watson Tests for Autocorrelations and Randomness. *Journal of Business & Economic Statistics*, 5(2):195–203.

Ali, M. M. and Sharma, S. C. (1993). Robustness to Nonnormality of the Durbin–Watson Test for Autocorrelation. *Journal of Econometrics*, 57(1–3):117–136.

Allen, D. and Satchell, S. (2014). *The Four Horsemen: Heavy-tails, Negative Skew, Volatility Clustering, Asymmetric Dependence*. The University of Sydney, Business School, Discipline of Finance. Discussion Paper 2014–004.

Allen, D. E., McAleer, M., Powell, R. J., and Singh, A. K. (2016). Down-Side Risk Metrics as Portfolio Diversification Strategies across the Global Financial Crisis. *Journal of Risk and Financial Management*, 9:1–18. Article 6.

Alvarez, L. J. and Dolado, J. J. (1994). Deriving Restricted Least Squares without a Lagrangean, Solution, 93.2.2. *Econometric Theory*, 10(2):443–445.

Amendola, A., Niglio, M., and Vitale, C. (2006). The Moments of SETARMA Models. *Statistics & Probability Letters*, 76:625–633.

Amisano, G. and Giacomini, R. (2007). Comparing Density Forecasts via Weighted Likelihood Ratio Tests. *Journal of Business & Economic Statistics*, 25(2):177–190.

Anatolyev, S. and Khrapov, S. (2015). Right on Target, or Is it? The Role of Distributional Shape in Variance Targeting. *Econometrics*, 3(3):610–632.

Andersen, R. (2008). *Modern Methods for Robust Regression*. SAGE Publications, Los Angeles.

Andersen, T. G., Bollerslev, T., Christoffersen, P. F., and Diebold, F. X. (2007). Practical Volatility and Correlation Modeling for Financial Market Risk Management. In Carey, M. and Stulz, R. M., editors, *The Risks of Financial Institutions*, chapter 11, pages 513–544. The University of Chicago Press.

Anderson, O. D. (1993). Exact General-Lag Serial Correlation Moments and Approximate Low-Lag Partial Correlation Moments for Gaussian White Noise. *Journal of Time Series Analysis*, 14(6):551–574.

Anderson, O. D. (1995). More Effective Time-Series Analysis and Forecasting. *Journal of Computational and Applied Mathematics*, 64(1–2):117–147.

Anderson, T. W. (1948). On the Theory of Testing Serial Correlation. *Skandinavisk Aktuarietidskrift*, 1948(3–4):88–116.

Anderson, T. W. (1971). *The Statistical Analysis of Time Series*. John Wiley & Sons, New York.

Anderson, T. W. (1992). The Asymptotic Distribution of Autocorrelation Coefficients. In Mardia, K. V., editor, *The Art of Statistical Science, A Tribute to G. S. Watson*. John Wiley & Sons, Chichester.

Anderson, T. W. (1994). Pattern Identification of ARMA Models. Technical Report No. 295, Stanford University, Stanford, CA.

Anderson, T. W. (2003). *An Introduction to Multivariate Statistical Analysis*. John Wiley & Sons, New York, 3rd edition.

Anderson, T. W. and Styan, G. P. H. (1982). Cochran's Theorem, Rank Additivity and Tripotent Matrices. In Kallianpur, G., Krishnaiah, P. R., and Ghosh, J. K., editors, *Statistics and Probability: Essays in Honor of C. R. Rao*, pages 1–23. North Holland, Amsterdam.

Anderson-Sprecher, R. (1994). Model Comparisons and R^2. *The American Statistician*, 48(2):113–117.

Andrews, D. W. K. (1991). Heteroskedasticity and Autocorrelation Consistent Covariance Matrix Estimation. *Econometrica*, 59:817–858.

Andrews, D. W. K. (1993). Exactly Median-Unbiased Estimation of First Order Autoregressive/Unit Root Models. *Econometrica*, 61(1):139–165.

Andrews, D. W. K. and Chen, H.-Y. (1994). Approximately Median-Unbiased Estimation of Autoregressive Models. *Journal of Business & Economic Statistics*, 12:187–204.

Andrews, D. W. K. and Guggenberger, P. (2014). A Conditional-Heteroskedasticity-Robust Confidence Interval for the Autoregressive Parameter. *The Review of Economics and Statistics*, 96(2):376–381.

Anh, V. V. (1988). On the Hildreth–Houck Estimator for Random Coefficient Regression Models. *Australian & New Zealand Journal of Statistics*, 30(2):189–195.

Ansley, C. F. and Newbold, P. (1981). On the Bias in Estimates of Forecast Mean Squared Error. *Journal of the American Statistical Association*, 76:569–578.

Apostol, T. M. (1969). *Multivariable Calculus and Linear Algebra with Applications to Differential Equations and Probability*. John Wiley & Sons, New York, 2nd edition.

Ardia, D., Bluteau, K., Boudt, K., and Catania, L. (2017a). Forecasting Performance of Markov-Switching GARCH Models: A Large-Scale Empirical Study. Available at SSRN: https://ssrn.com/abstract=2918413.

Ardia, D., Bluteau, K., Boudt, K., and Trottier, D.-A. (2017b). Markov-Switching GARCH Models in R: The MSGARCH Package. Available at SSRN: https://ssrn.com/abstract=2845809.

Arellano-Valle, R. B. and Genton, M. G. (2010). Multivariate Extended Skew-*t* Distributions and Related Families. *METRON—International Journal of Statistics*, LXVIII(3):201–234.

Arteche, J. and García-Enríquez, J. (2017). Singular Spectrum Analysis for Signal Extraction in Stochastic Volatility Models. *Econometrics and Statistics*, 1:85–98.

Ash, R. B. and Doléans-Dade, C. A. (2000). *Probability & Measure Theory*. Harcourt Academic Press, San Diego, 2nd edition.

Astatkie, T., Watts, D. G., and Watt, W. E. (1997). Nested Threshold Autoregressive (NeTAR) Models. *International Journal of Forecasting*, 13:105–116.

Azzalini, A. (1985). A Class of Distributions which Includes the Normal Ones. *Scandinavian Journal of Statistics*, 12:171–178.

Bacon, D. W. and Watts, D. G. (1971). Estimating the Transition between Two Intersecting Straight Lines. *Biometrika*, 58(3):525–534.

Badescu, A., Kulperger, R., and Lazar, E. (2008). Option Valuation with Normal Mixture GARCH Models. *Studies in Nonlinear Dynamics & Econometrics*, 12(2):5.

Baek, E. G. and Brock, W. A. (1992). A Nonparametric Test for Independence of a Multivariate Time Series. *Statistica Sinica*, 2:137–156.

Baesens, B., Rösch, D., and Scheule, H. (2016). *Credit Risk Analytics: Measurement Techniques, Applications, and Examples in SAS*. John Wiley & Sons, Hoboken, NJ.

Bai, J. (2003). Testing Parametric Conditional Distributions of Dynamic Models. *The Review of Economics and Statistics*, 85(3):531–549.

Bai, J. and Perron, P. (1998). Estimating and Testing Linear Models with Multiple Structural Changes. *Econometrica*, 66:47–78.

Bai, J. and Perron, P. (2003). Computation and Analysis of Multiple Structural Change Models. *Journal of Applied Econometrics*, 18:1–22.

Bai, X., Russell, J. R., and Tiao, G. C. (2003). Kurtosis of GARCH and Stochastic Volatility Models with Non-Normal Innovations. *Journal of Econometrics*, 114(2):349–360.

Bailey, D. H., Borwein, J. M., López de Prado, M., and Zhu, Q. J. (2014). Pseudo-Mathematics and Financial Charlatanism: The Effects of Backtest Overfitting on Out-of-Sample Performance. *Notices of the American Mathematical Society*, 61(5):458–471.

Baillie, R. T. (1996). Long Memory Processes and Fractional Integration in Economics. *Journal of Econometrics*, 73:5–59.

Baillie, R. T., Bollerslev, T., and Mikkelsen, H. O. (1996). Fractionally Integrated Generalized Autoregressive Conditional Heteroskedasticity. *Journal of Econometrics*, 74(1):3–30.

Baillie, R. T. and Chung, H. (2001). Estimation of GARCH Models from the Autocorrelations of the Squares of a Process. *Journal of Time Series Analysis*, 22:631–650.

Bali, T. G. and Engle, R. F. (2010). The Intertemporal Capital Asset Pricing Model with Dynamic Conditional Correlations. *Journal of Monetary Economics*, 57(4):377–390.

Bali, T. G., Engle, R. F., and Murray, S. (2016a). *Empirical Asset Pricing: The Cross Section of Stock Returns*. John Wiley & Sons, Hoboken, NJ.

Bali, T. G., Engle, R. F., and Tang, Y. (2016b). Dynamic Conditional Beta is Alive and Well in the Cross Section of Daily Stock Returns. *Management Science*, 68(11):3760–3779.

Baltagi, B. H. (2013). *Econometric Analysis of Panel Data*. John Wiley & Sons, New York, 5th edition.

Banulescu, D., Hansen, P. R., Huang, Z., and Matei, M. (2016). Volatility During the Financial Crisis Through the Lens of High Frequency Data: A Realized GARCH Approach.

Bao, Y. (2007). The Approximate Moments of the Least Squares Estimator for the Stationary Autoregressive Model Under a General Error Distribution. *Econometric Theory*, 23:1013–1021.

Bao, Y. and Kan, R. (2013). On the Moments of Ratios of Quadratic Forms in Normal Random Variables. *Journal of Multivariate Analysis*, 117:229–245.

Bao, Y., Lee, T.-H., and Saltoğlu, B. (2006). Evaluating Predictive Performance of Value-at-Risk Models in Emerging Markets: A Reality Check. *Journal of Forecasting*, 25(2):101–128.

Bao, Y., Lee, T.-H., and Saltoğlu, B. (2007). Comparing Density Forecast Models. *Journal of Forecasting*, 26(3):203–225.

Bao, Y. and Ullah, A. (2007). The Second-Order Bias and Mean Squared Error of Estimators in Time-Series Models. *Journal of Econometrics*, 140:650–669.

Barndorff-Nielsen, O. E. (1977). Exponentially Decreasing Distributions for the Logarithm of Particle Size. *Proceedings of the Royal Society of London A*, 353:401–419.

Barndorff-Nielsen, O. E., Blæsild, P., Jensen, J. L., and Jorgensen, B. (1982). Exponential Transformation Models. *Proceedings of the Royal Society of London A: Mathematical, Physical and Engineering Sciences*, 379(1776):41–65.

Barndorff-Nielsen, O. E., Blæsild, P., and Seshadri, V. (1992). Multivariate Distributions with Generalized Inverse Gaussian Marginals, and Associated Poisson Mixtures. *Canadian Journal of Statistics*, 20:109–120.

Barndorff-Nielsen, O. E. and Cox, D. R. (1979). Edgeworth and Saddlepoint Approximations with Statistical Applications (with discussion). *Journal of the Royal Statistical Society, Series B*, 41:279–312.

Barndorff-Nielsen, O. E. and Cox, D. R. (1989). *Asymptotic Techniques for Use in Statistics*. Chapman & Hall, London.

Barone-Adesi, G. (2016). VaR and CVaR Implied in Option Prices. *Journal of Risk and Financial Management*, 9(2):1–6.

Barone-Adesi, G., Giannopoulos, K., and Vosper, L. (1999). VaR without Correlations for Portfolios of Derivative Securities. *Journal of Futures Markets*, 19(5):583–602.

Barone-Adesi, G., Giannopoulos, K., and Vosper, L. (2002). Backtesting Derivative Portfolios with Filtered Historical Simulation (FHS). *European Financial Management*, 8(1):31–58.

Bartels, R. (1982). The Rank Version of von Neumann's Ratio Test for Randomness. *Journal of the American Statistical Association*, 77(377):40–46.

Bartels, R. (1984). The Rank von Neumann Test as a Test for Autocorrelation in Regression Models. *Communications in Statistics—Theory and Methods*, 13(20):2495–2502.

Bartels, R. (1992). On the Power Function of the Durbin–Watson Test. *Journal of Econometrics*, 51:101–112.

Bartlett, M. S. (1946). On the Theoretical Specification and Sampling Properties of Autocorrelated Time-Series. *Supplement to the Journal of the Royal Statistical Society*, 8(1):27–41.

Baum, C. F. (2006). *An Introduction to Modern Econometrics Using Stata*. Stata Press, College Station, TX.

Baumeister, C. and Peersman, G. (2013). The Role of Time-Varying Price Elasticities in Accounting for Volatility Changes in the Crude Oil Market. *Journal of Applied Econometrics*, 28(7):1087–1109.

Bauwens, L., Bos, C. S., and van Dijk, H. K. (1999). Adaptative Polar Sampling with an Application to a Bayes Measure of Value-at-Risk. Tinbergen institute discussion paper ti99-082/4, Erasmus University.

Bauwens, L., Hafner, C. M., and Rombouts, J. V. K. (2007). Multivariate Mixed Normal Conditional Heteroskedasticity. *Computational Statistics & Data Analysis*, 51(7):3551–3566.

Bauwens, L., Laurent, S., and Rombouts, J. K. V. (2006a). Multivariate GARCH Models: A Survey. *Journal of Applied Econometrics*, 21(1):79–109.

Bauwens, L., Preminger, A., and Rombouts, J. V. K. (2006b). Regime Switching GARCH Models. CORE Discussion Paper 2006/11, Center for Operations Research and Econometrics, Université Catholique de Louvain.

Bauwens, L. and Rombouts, J. V. K. (2007a). Bayesian Clustering of Many GARCH Models. *Econometric Reviews*, 26(2):365–386.

Bauwens, L. and Rombouts, J. V. K. (2007b). Bayesian Inference for the Mixed Conditional Heteroskedasticity Model. *Econometrics Journal*, 10(2):408–425.

Bauwens, L. and Storti, G. (2009). A Component GARCH Model with Time Varying Weights. *Studies in Nonlinear Dynamics & Econometrics*, 13(2):1–33.

Bechhofer, R. E. (1954). A Single-Sample Multiple Decision for Ranking Means of Normal Populations with Known Variances. *Annals of Mathematical Statistics*, 25:16–39.

Bechhofer, R. E. and Goldsman, D. M. (1989). A Comparison of the Performances of Procedures for Selecting the Normal Population Having the Largest Mean when the Variances are Known and Equal. In Gleser, L. J., Perlman, M. D., Press, S. J., and Sampson, A. R., editors, *Contributions to Probability and Statistics: Essays in Honor of Ingram Olkin*. Springer, New York.

Becker, W. E. and Kennedy, P. E. (1992). A Lesson in Least Squares and *R* Squared. *The American Statistician*, 46(4):282–283.

Beguin, J. M., Gourieroux, C., and Monfort, A. (1980). Identification of a Mixed Autoregressive-Moving Average Process: The Corner Method. In Anderson, O. D., editor, *Time Series*, pages 423–436. North-Holland, Amsterdam.

Bekaert, G. and Gray, S. F. (1998). Target Zones and Exchange Rates: An Empirical Investigation. *Journal of International Economics*, 45:1–35.

Bellini, F. and Di Bernardino, E. (2017). Risk Management With Expectiles. *European Journal of Finance*, 23(6):487–506.

Ben-Israel, A. and Greville, T. N. E. (2003). *Generalized Inverses: Theory and Applications*. Springer, 2nd edition.

Bengtsson, C. (2003). The Impact of Estimation Error on Portfolio Selection for Investors with Constant Relative Risk Aversion. Working Paper 2003:17, Department of Economics, Lund University, Lund.

Berenblut, I. I. and Webb, G. I. (1973). A New Test for Autocorrelated Errors in the Linear Regression Model. *Journal of the Royal Statistical Society, Series B*, 35(1):33–50.

Berger, J. O. (1985). *Statistical Decision Theory and Bayesian Analysis*. Springer, New York, 2nd edition.

Berkes, I. and Horváth, L. (2003). Limit Results for the Empirical Process of Squared Residuals in GARCH Models. *Stochastic Processes and their Applications*, 105:279–298.

Berkes, I., Horváth, L., and Kokoszka, P. (2003a). GARCH Processes: Structure and Estimation. *Bernoulli*, 9(2):201–227.

Berkes, I., Horváth, L., and Kokoszka, P. S. (2003b). Asymptotics for GARCH Squared Residual Correlations. *Econometric Theory*, 19(4):515–540.

Berkes, I., Horváth, L., and Kokoszka, P. S. (2004). A Weighted Goodness-of-Fit Test for GARCH(1,1) Specification. *Lithuanian Mathematics Journal*, 44:1–17.

Berkson, J. (1944). Application of the Logistic Function to Bio-Assay. *Journal of the American Statistical Association*, 39(227):357–365.

Berkson, J. (1980). Minimum Chi-Square, Not Maximum Likelihood! *Annals of Statistics*, 8(3):457–487.

Best, M. J. and Grauer, R. R. (1991). On the Sensitivity of Mean-Variance-Efficient Portfolios to Changes in Asset Means: Some Analytical and Computational Results. *Review of Financial Studies*, 4:315–342.

Best, M. J. and Grauer, R. R. (1992). The Analytics of Sensitivity Analysis for Mean-Variance Portfolio Problems. *International Review of Financial Analysis*, 1:17–37.

Bhansali, R. J. (1993). Order Selection for Linear Time Series Models: A Review. In Rao, T. S., editor, *Developments in Time Series Analysis. In honour of Maurice B. Priestley*, chapter 5. Chapman & Hall, London.

Bhargava, A. (1986). On the Theory of Testing for Unit Roots in Observed Time Series. *Review of Economic Studies*, 53:369–384.

Bianchi, M. L., Tassinari, G. L., and Fabozzi, F. J. (2016). Riding with the Four Horsemen and the Multivariate Normal Tempered Stable Model. *International Journal of Theoretical and Applied Finance*, 19(4).

Bianco, A. M., Boente, G., and Rodrigues, I. M. (2017). Conditional Tests for Elliptical Symmetry Using Robust Estimators. *Communications in Statistics—Theory and Methods*, 46(4):1744–1765.

Bickel, P. J. and Levina, E. (2008). Regularized Estimation of Large Covariance Matrices. *Annals of Statistics*, 36(1):199–227.

Billio, M. and Caporin, M. (2009). A Generalised Dynamic Conditional Correlation Model for Portfolio Risk Evaluation. *Mathematics and Computers in Simulation*, 79:2566–2578.

Billio, M., Caporin, M., and Gobbo, M. (2006). Flexible Dynamic Conditional Correlation Multivariate GARCH Models for Asset Allocation. *Applied Financial Economics Letters*, 2(2):123–130.

Binkley, J. K. and Abbott, P. C. (1987). The Fixed X Assumption in Econometrics: Can the Textbooks be Trusted? *The American Statistician*, 41(3):206–214.

Bishop, C. (2006). *Pattern Recognition and Machine Learning*. Springer, New York.

Bittner, A. C. (1974). Note on Mantell, E. H., Exact Linear Restrictions on Parameters in a Linear Regression Model. *The American Statistician*, 28(1):36.

Black, F. (1976). Studies of Stock Price Volatility Changes. *Proceedings of the 1976 Meetings of the American Statistical Association, Business and Economic Statistics Section*, pages 177–181.

Black, F. and Litterman, R. (1992). Global Portfolio Optimization. *Financial Analysts Journal*, 48:28–43.

Bloomfield, P. (1972). On the Error of Prediction of a Time Series. *Biometrika*, 59:501–507.

Bloomfield, T., Leftwich, R., and Long, J. (1977). Portfolio Strategies and Performance. *Journal of Financial Economics*, 5:201–218.

Blough, S. R. (1992). The Relationship Between Power and Level for Generic Unit Root Tests in Finite Samples. *Journal of Applied Econometrics*, 7(3):295–308.

Bluhm, C., Overbeck, L., and Wagner, C. (2010). *Introduction to Credit Risk Modeling*. Chapman & Hall/CRC, Boca Raton, 2nd edition.

Bodnar, T. and Schmid, W. (2007). Matrix Elliptical Contour Distributions versus a Stable Model: Application to Daily Stock Returns of Eight Stock Markets. In Gregoriou, G. N., editor, *Asset Allocation and International Investments*, chapter 11. Palgrave MacMillan, New York.

Bollerslev, T. (1986). Generalized Autoregressive Conditional Heteroskedasticity. *Journal of Econometrics*, 31:307–327.

Bollerslev, T. (1987). A Conditional Heteroskedastic Time Series Model for Speculative Prices and Rates of Return. *Review of Economics and Statistics*, 69:542–547.

Bollerslev, T. (1988). On the Correlation Structure for the Generalized Autoregressive Conditional Heteroskedastic Process. *Journal of Time Series Analysis*, 9(2):121–131.

Bollerslev, T. (1990). Modeling the Coherence in Short-Run Nominal Exchange Rates: A Multivariate Generalized ARCH Approach. *Review of Economics and Statistics*, 72:498–505.

Bollerslev, T. (2010). Glossary to ARCH (GARCH). In Bollerslev, T., Russell, J., and Watson, M., editors, *Volatility and Time Series Econometrics: Essays in Honor of Robert Engle*, chapter 8, pages 137–163. Oxford University Press, Oxford.

Bollerslev, T., Engle, R. F., and Nelson, D. B. (1994). ARCH Models. In Engle, R. and McFadden, D., editors, *Handbook of Econometrics*, volume 4, chapter 49. Elsevier Science B.V., Amsterdam, The Netherlands.

Bollerslev, T., Engle, R. F., and Wooldridge, J. (1988). A Capital Asset-pricing Model with Time-varying Covariances. *Journal of the Political Economy*, 96:116–131.

Bollerslev, T. and Mikkelsen, H. O. (1996). Modeling and Pricing Long Memory in Stock Market Volatility. *Journal of Econometrics*, 73(1):154–184.

Bollerslev, T. and Wooldridge, J. M. (1992). Quasi-Maximum Likelihood Estimation and Inference in Dynamic Models with Time-Varying Covariances. *Econometric Reviews*, 11(2):143–172.

Boos, D. D. and Hughes-Oliver, J. M. (1998). Applications of Basu's Theorem. *The American Statistician*, 52(3):218–221.

Bos, T. and Newbold, P. (1984). An Empirical Investigation of the Possibility of Stochastic Systematic Risk in the Market Model. *Journal of Business*, 57:35–41.

Boshnakov, G. N. (1996). Bartlett's Formula—Closed Forms and Recurrent Equations. *Annals of the Institute of Statistical Mathematics*, 48(1):49–59.

Boswijk, H. P. and van der Weide, R. (2011). Method of Moments Estimation of GOGARCH Models. *Journal of Econometrics*, 163:118–126.

Boudoukh, J., Richardson, M., and Whitelaw, R. F. (1998). The Best of Both Worlds: A Hybrid Approach to Calculating Value at Risk. *Risk*, 11(5):64–67.

Box, G. E. P. (1979). Robustness in the Strategy of Scientific Model Building. In Launer, R. L. and Wilkinson, G. N., editors, *Robustness in Statistics*, pages 201–236. Academic Press, New York.

Box, G. E. P., Jenkins, G. M., and Reinsel, G. C. (2008). *Time Series Analysis: Forecasting & Control*. John Wiley & Sons, Hoboken, NJ, 4th edition.

Box, G. E. P. and Pierce, D. A. (1970). Distribution of the Autocorrelations in Autoregressive Moving Average Time Series Models. *Journal of the American Statistical Association*, 65:1509–1526.

Bradley, B. O. and Taqqu, M. S. (2003). Financial Risk and Heavy Tails. In Rachev, S. T., editor, *Handbook of Heavy Tailed Distributions in Finance*, pages 35–103. Elsevier Science, Amsterdam.
Brandt, A. (1986). The Stochastic Equation $Y_{n+1} = A_n Y_n + B_n$ with Stationary Coefficients. *Advances in Applied Probability*, 18:211–220.
Brandt, M. W. (2010). Portfolio Choice Problems. In Aït-Sahalia, Y. and Hansen, L. P., editors, *Handbook of Financial Econometrics: Tools and Techniques*, chapter 5, pages 269–336. North-Holland, Amsterdam.
Brandt, M. W., Santa-Clara, P., and Valkanov, R. (2009). Parametric Portfolio Policies: Exploiting Characteristics in the Cross-Section of Equity Returns. *Review of Financial Studies*, 22(9):3411–3447.
Breitung, J. (2002). Nonparametric Tests for Unit Roots and Cointegration. *Journal of Econometrics*, 108:343–63.
Briggs, W. (2016). *Uncertainty: The Soul of Modeling, Probability & Statistics*. Springer, Geneva.
Britten-Jones, M. (1999). The Sampling Error in Estimates of Mean-Variance Efficient Portfolio Weights. *Journal of Finance*, 54(2):655–671.
Britten-Jones, M. and Schaefer, S. M. (1999). Non-Linear Value-at-Risk. *European Finance Review*, 2:161–187.
Brock, W. A., Dechert, W. D., Scheinkman, J. A., and LeBaron, B. (1996). A Test for Independence Based on the Correlation Dimension. *Econometric Reviews*, 15:197–235.
Brockwell, P., Liu, J., and Tweedie, R. L. (1992). On the Existence of Stationary Threshold Autoregressive Moving-Average Processes. *Journal of Time Series Analysis*, 13:95–107.
Brockwell, P. J. and Davis, R. A. (1991). *Time Series: Theory and Methods*. Springer, New York, 2nd edition.
Brockwell, P. J. and Davis, R. A. (2016). *Introduction to Time Series and Forecasting*. Springer, New York, 3rd edition.
Broda, S. and Paolella, M. S. (2007). Saddlepoint Approximations for the Doubly Noncentral t Distribution. *Computational Statistics and Data Analysis*, 51:2907–2918.
Broda, S. A., Carstensen, K., and Paolella, M. S. (2007). Bias-Adjusted Estimation in the ARX(1) Model. *Computational Statistics & Data Analysis*, 51(7):3355–3367.
Broda, S. A., Carstensen, K., and Paolella, M. S. (2009). Assessing and Improving the Performance of Nearly Efficient Unit Root Tests in Small Samples. *Econometric Reviews*, 28(5):468–494.
Broda, S. A., Haas, M., Krause, J., Paolella, M. S., and Steude, S. C. (2013). Stable Mixture GARCH Models. *Journal of Econometrics*, 172(2):292–306.
Broda, S. A., Krause, J., and Paolella, M. S. (2017). Approximating Expected Shortfall for Heavy Tailed Distributions. *Econometrics and Statistics*. in press.
Broda, S. A. and Paolella, M. S. (2009a). CHICAGO: A Fast and Accurate Method for Portfolio Risk Calculation. *Journal of Financial Econometrics*, 7(4):412–436.
Broda, S. A. and Paolella, M. S. (2009b). Evaluating the Density of Ratios of Noncentral Quadratic Forms in Normal Variables. *Computational Statistics & Data Analysis*, 53(4):1264–1270.
Broda, S. A. and Paolella, M. S. (2011). Expected Shortfall for Distributions in Finance. In Čížek, P., Härdle, W., and Rafał Weron, editors, *Statistical Tools for Finance and Insurance*, pages 57–99. Springer, Berlin.
Brooks, C. (2001). A Double-Threshold GARCH Model for the French Franc/Deutschmark Exchange Rate. *Journal of Forecasting*, 20:135–143.
Brooks, C., Burke, S. P., and Persand, G. (2001). Benchmarks and the Accuracy of GARCH Model Estimation. *International Journal of Forecasting*, 17(1):45–56.
Brooks, C. and Heravi, S. M. (1999). The Effect of (Mis-Specified) GARCH Filters on the Finite Sample Distribution of the BDS Test. *Computational Economics*, 13(2):147–162.

Brooks, R. D. (1993). Alternative Point-Optimal Tests for Regression Coefficient Stability. *Journal of Econometrics*, 57:365–376.

Brooks, R. D. (1995). The Robustness of Point Optimal Testing for Rosenberg Random Regression Coefficients. *Econometric Reviews*, 14(1):35–53.

Brooks, R. D. (1997). Using a Sequence of Point Optimal Tests to Select a Varying Coefficient Model. *Communications in Statistics—Simulation and Computation*, 26(2):671–685.

Brooks, R. D., Faff, R. W., and Lee, J. H. H. (1994). Beta Stability and Portfolio Formation. *Pacific Basin Finance Journal*, 2:463–479.

Brooks, R. D. and King, M. L. (1994). Testing Hildreth–Houck Against Return to Normalcy Random Regression Coefficients. *Journal of Quantitative Economics*, 10:33–52.

Brown, R. L., Durbin, J., and Evans, J. M. (1975). Techniques of Testing the Constancy of Regression relationships Over Time. *Journal of the Royal Statistical Society, Series B*, 37:141–192.

Brown, S. J., Hwang, I., and In, F. (2013). Why Optimal Diversification Cannot Outperform Naive Diversification: Evidence from Tail Risk Exposure. http://citeseerx.ist.psu.edu/viewdoc/summary?doi=10.1.1.295.3247.

Burch, B. D. and Iyer, H. K. (1997). Exact Confidence Intervals for a Variance Ratio (or Heritability) in a Mixed Linear Model. *Biometrics*, 53(4):1318–1333.

Burda, M. (2015). Constrained Hamiltonian Monte Carlo in BEKK GARCH with Targeting. *Journal of Time Series Econometrics*, 7(1):95–113.

Burdick, R. K. and Graybill, F. A. (1992). *Confidence Intervals on Variance Components*. Marcel Dekker, New York.

Burnham, K. P. and Anderson, D. (2002). *Model Selection and Multi-Model Inference*. Springer, New York, 2nd edition.

Buse, A. (1973). Goodness of Fit in Generalized Least Squares Estimation. *The American Statistician*, 27(3):106–108.

Busetti, F. and Taylor, A. M. R. (2004). Tests of Stationarity Against a Change in Persistence. *Journal of Econometrics*, 123:33–66.

Butler, R. J., McDonald, J. B., Nelson, R. D., and White, S. B. (1990). Robust and Partially Adaptive Estimation of Regression Models. *The Review of Economics and Statistics*, 72:321–327.

Butler, R. W. (1986). Extendibility and the Optimality of F, T^2 and Forward Variable Selection. *Scandinavian Journal of Statistics*, 13(4):257–262.

Butler, R. W. (1998). Generalized Inverse Gaussian Distributions and their Wishart Connections. *Scandinavian Journal of Statistics*, 25:69–75.

Butler, R. W. (2007). *An Introduction to Saddlepoint Methods*. Cambridge University Press, Cambridge.

Butler, R. W. and Paolella, M. S. (1998). Approximate Distributions for the Various Serial Correlograms. *Bernoulli*, 4(4):497–518.

Butler, R. W. and Paolella, M. S. (2002a). Calculating the Density and Distribution Function for the Singly and Doubly Noncentral F. *Statistics and Computing*, 12(1):9–16.

Butler, R. W. and Paolella, M. S. (2002b). Saddlepoint Approximation and Bootstrap Inference for the Satterthwaite Class of Ratios. *Journal of the American Statistical Association*, 97:836–846.

Butler, R. W. and Paolella, M. S. (2008). Uniform Saddlepoint Approximations for Ratios of Quadratic Forms. *Bernoulli*, 14(1):140–154.

Butler, R. W. and Paolella, M. S. (2017). Autoregressive Lag-Order Selection Using Conditional Saddlepoint Approximations. *Econometrics*, 5(3):1–33. Article 43.

Cai, Z. and Wang, X. (2008). Nonparametric Estimation of Conditional VaR and Expected Shortfall. *Journal of Econometrics*, 147(1):120–130.

Cambanis, S., Huang, S., and Simons, G. (1981). On the Theory of Elliptically Contoured Distributions. *Journal of Multivariate Analysis*, 11(3):368–385.

Cambanis, S., Keener, R., and Simons, G. (1983). On α-Symmetric Multivariate Distributions. *Journal of Multivariate Analysis*, 13(2):213–233.

Cameron, A. C. and Trivedi, P. K. (2005). *Microeconometrics: Methods and Applications*. Cambridge University Press, Cambridge.

Campbell, J. Y. and Mankiw, N. G. (1987). Are Output Fluctuations Transitory? *Quarterly Journal of Economics*, 102(4):857–880.

Campbell, R. A. and Kräussl, R. (2007). Revisiting the Home Bias Puzzle: Downside Equity Risk. *Journal of International Money and Finance*, 26(7):1239–1260.

Caner, M. and Hansen, B. E. (2001). Threshold Autoregression with a Unit Root. *Econometrica*, 69(6):1555–1596.

Caporale, G. M., Ntantamis, C., Pantelidis, T., and Pittis, N. (2005). The BDS Test as a Test for the Adequacy of a GARCH(1,1) Specification: A Monte Carlo Study. *Journal of Financial Econometrics*, 3(2):282–309.

Caporin, M. (2003). Identification of Long Memory in GARCH Models. *Statistical Methods and Applications*, 12(2):133–151.

Caporin, M., Kolokolov, A., and Renò, R. (2017). Systemic Co-Jumps. *Journal of Financial Economics*, 126:563–591.

Caporin, M. and McAleer, M. (2008). Scalar BEKK and Indirect DCC. *Journal of Forecasting*, 27(6):537–549.

Caporin, M. and McAleer, M. (2012). Do We Really Need Both BEKK and DCC? A Tale of Two Multivariate GARCH Models. *Journal of Economic Surveys*, 26(4):736–751.

Caporin, M. and McAleer, M. (2013). Ten Things You Should Know about the Dynamic Conditional Correlation Representation. *Econometrics*, 1(1):115–126.

Cappiello, L., Engle, R. F., and Sheppard, K. (2006). Asymmetric Dynamics in the Correlations of Global Equity and Bond Returns. *Journal of Financial Econometrics*, 4(4):537–572.

Carrodus, M. L. and Giles, D. E. A. (1992). The Exact Distribution of R^2 when the Regression Disturbances are Autocorrelated. *Economics Letters*, 38(4):375–380.

Carstensen, K. and Paolella, M. S. (2003). On Median Unbiased Inference for First Order Autoregressive Models. In Klein, I. and Mittnik, S., editors, *Contributions to Modern Econometrics: From Data Analysis to Economic Policy*. Kluwer Academic Publishers.

Casella, G. and Berger, R. L. (2002). *Statistical Inference*. Duxbury, Wadsworth, Pacific Grove, CA, 2nd edition.

Cassing, S. A. and White, K. J. (1983). An Examination of the Eigenvector Condition in the Durbin–Watson Test. *Australian and New Zealand Journal of Statistics*, 25(1):17–22.

Castellacci, G. and Siclari, M. J. (2003). The Practice of Delta-Gamma VaR: Implementing the Quadratic Portfolio Model. *European Journal of Operational Research*, 150:529–545.

Castle, J. L., Doornik, J. A., and Hendry, D. F. (2011). Evaluating Automatic Model Selection. *Journal of Time Series Econometrics*, 3(1):1–33.

Castle, J. L., Doornik, J. A., Hendry, D. F., and Pretis, F. (2015). Detecting Location Shifts during Model Selection by Step-Indicator Saturation. *Econometrics*, 3:240–264.

Castle, J. L., Hendry, D. F., and Martinez, A. B. (2017). Evaluating Forecasts, Narratives and Policy Using a Test of Invariance. *Econometrics*, 5(3):1–27. Article 39.

Cesarone, F. and Colucci, S. (2016). A Quick Tool to Forecast Value-at-Risk Using Implied and Realized Volatilities. *Journal of Risk Model Validation*, 10(4):71–101.

Chambers, M. J. (2004). Testing for Unit Roots with Flow Data and Varying Sampling Frequency. *Journal of Econometrics*, 119:1–18.

Chambers, M. J. (2013). Jackknife Estimation of Stationary Autogressive Models. *Journal of Econometrics*, 172(1):142–157.

Chambers, M. J. and Kyriacou, M. (2013). Jackknife Estimation with a Unit Root. *Statistics & Probability Letters*, 83(7):1677–1682.

Chambers, M. J. and Kyriacou, M. (2018). Jackknife Bias Reduction in the Presence of a Near-Unit Root. *Econometrics*, 6(1):11.

Chan, K.-S. (2009). *Exploration of a Nonlinear World: An Appreciation of Howell Tong's Contributions to Statistics*. World Scientific, Singapore.

Chan, K. S. and Tong, H. (1986). On Estimating Thresholds in Autoregressive Models. *Journal of Time Series Analysis*, 7:179–190.

Chan, L., Karceski, J., and Lakonishok, J. (1999). On Portfolio Optimization: Forecasting Covariances and Choosing the Risk Model. *Review of Financial Studies*, 12(5):937–974.

Chan, W.-S. (1999). A Comparison of Some of Pattern Identification Methods for Order Determination of Mixed ARMA Models. *Statistics & Probability Letters*, 42:69–79.

Chang, S. Y. and Perron, P. (2016). Inference on a Structural Break in Trend with Fractionally Integrated Errors. *Journal of Time Series Analysis*, 37(4):555–574.

Chatfield, C. (1979). Inverse Autocorrelations. *Journal of the Royal Statistical Society, Series A*, 142:363–377.

Chatfield, C. (2001). *Time-Series Forecasting*. CRC Press, London.

Chatterjee, S. and Hadi, A. S. (2012). *Regression Analysis by Example*. John Wiley & Sons, Hoboken, NJ, 5th edition.

Chavez-Demoulin, V., Embrechts, P., and Sardy, S. (2014). Extreme-quantile Tracking for Financial Time Series. *Journal of Econometrics*, 181(1):44–52.

Chen, B., Gel, Y. R., Balakrishna, N., and Abraham, B. (2011a). Computationally Efficient Bootstrap Prediction Intervals for Returns and Volatilities in ARCH and GARCH Processes. *Journal of Forecasting*, 30(1):51–71.

Chen, B. and Pearl, J. (2013). Regression and Causation: A Critical Examination of Six Econometrics Textbooks. *Real-World Economics Review*, 65:2–20.

Chen, C. W. S., Gerlach, R., and Lin, E. M. H. (2008a). Volatility Forecast Using Threshold Heteroskedastic Models of the Intra-Day Range. *Computational Statistics & Data Analysis*, 52:2990–3010.

Chen, C. W. S., Liu, F.-C., and So, M. K. P. (2008b). Heavy-Tailed Distributed Threshold Stochastic Volatility Models in Financial Time Series. *Australian & New Zealand Journal of Statistics*, 50:29–51.

Chen, C. W. S., So, M. K. P., and Gerlach, R. H. (2005). Assessing and Testing for Threshold Nonlinearity in Stock Returns. *Australian & New Zealand Journal of Statistics*, 47(4):473–488.

Chen, C. W. S., So, M. K. P., and Liu, F.-C. (2011b). A Review of Threshold Time Series Models in Finance. *Statistics and Its Interface*, 4:167–181.

Chen, H., Chong, T. T.-L., and Bai, J. (2012). Theory and Applications of TAR Model with Two Threshold Variables. *Econometric Reviews*, 31(2):142–170.

Chen, J. and Yuan, M. (2016). Efficient Portfolio Selection in a Large Market. *Journal of Financial Econometrics*, 14(3):496–524.

Chen, Q. and Giles, D. E. (2011). A Saddlepoint Approximation to the Distribution of the Half-Life Estimator in a Stationary Autoregressive Model. *Communications in Statistics: Theory and Methods*, 40(21):3903–3916.

Chen, R.-B., Guo, M., Härdle, W. K., and Huang, S.-F. (2015). COPICA-Independent Component Analysis Via Copula Techniques. *Statistics and Computing*, 25(2):273–288.

Chen, S. X. and Tang, C. Y. (2005). Nonparametric Inference of Value-at-Risk for Dependent Financial Returns. *Journal of Financial Econometrics*, 3(2):227–255.

Chen, Y. and Yu, J. (2015). Optimal Jackknife for Unit Root Models. *Statistics & Probability Letters*, 99:135–142.

Chen, Y.-T. and Kuan, C.-M. (2002). Time Irreversibility and EGARCH Effects in U.S. Stock Index Returns. *Journal of Applied Econometrics*, 17:565–578.

Cheng, X., Yu, P. L. H., and Li, W. K. (2009). On a Dynamic Mixture GARCH Model. *Journal of Forecasting*, 28:247–265.

Chester, A. D. (1984). Testing for Neglected Heterogeneity. *Econometrica*, 52:865–872.

Cheung, Y. W. and Lai, K. S. (1993). A Fractional Cointegration Analysis of Purchasing Power Parity. *Journal of Business & Economic Statistics*, 11:103–112.

Chicheportiche, R. and Bouchaud, J.-P. (2012). The Joint Distribution of Stock Returns is Not Elliptical. *International Journal of Theoretical and Applied Finance*, 15(3):1–23.

Cho, J. S. and White, H. (2011). Generalized Runs Tests for the IID Hypothesis. *Journal of Econometrics*, 162:326–344.

Choi, B.-S. (1992). *ARMA Model Identification*. Springer, New York.

Choi, I. (2015). *Almost All About Unit Roots: Foundations, Developments, and Applications*. Cambridge University Press, New York.

Chopra, V. K. and Ziemba, W. T. (1993). The Effect of Errors in Means, Variances, and Covariances on Optimal Portfolio Choice. *Journal of Portfolio Management*, 19:6–11.

Chow, G. C. (1976). A Note on the Derivation of Theil's BLUS Residuals. *Econometrica*, 44:609–610.

Christensen, R. (1987). *Plane Answers to Complex Questions*. Springer, New York.

Christensen, R. (2011). *Plane Answers to Complex Questions*. Springer, New York, 4th edition.

Christensen, R., Johnson, W., Branscum, A., and Hanson, T. E. (2011). *Bayesian Ideas and Data Analysis: An Introduction for Scientists and Statisticians*. Chapman & Hall/CRC, Boca Raton.

Christoffersen, P. (2009). Value-at-Risk Models. In Mikosch, T., Kreiss, J.-P., Davis, R. A., and Andersen, T. G., editors, *Handbook of Financial Time Series*, pages 753–766. Springer, Heidelberg.

Christoffersen, P. F. (1998). Evaluating Interval Forecasts. *International Economic Review*, 39(4):841–862.

Christoffersen, P. F. (2011). *Elements of Financial Risk Management*. Academic Press, Amsterdam, 2nd edition.

Christoffersen, P. F. and Gonçalves, S. (2005). Estimation Risk in Financial Risk Management. *Journal of Risk*, 7(3):1–28.

Chuffart, T. (2017). An Implementation of Markov Regime Switching GARCH Models in Matlab. Available at SSRN.

Chui, C. K. and Chen, G. (1999). *Kalman Filtering: With Real-Time Applications*. Springer, New York, 3rd edition.

Chung, C.-F. (1994). A Note on Calculating the Autocovariances of Fractionally Integrated ARMA Models. *Economics Letters*, 45:293–297.

Cisewski, J. and Hannig, J. (2012). Generalized Fiducial Inference for Normal Linear Mixed Models. *Annals of Statistics*, 40(4):2102–2127.

Cleveland, W. S. (1972). The Inverse Autocorrelations of a Time Series and Their Applications. *Technometrics*, 14:277–297.

Cochran, W. G. (1934). The Distribution of Quadratic Forms in a Normal System, with Applications to the Analysis of Covariance. *Mathematical Proceedings of the Cambridge Philosophical Society*, 30(2):178–191.

Cochrane, J. H. (1988). How Big is the Random Walk in GDP? *Journal of Political Economy*, 96(5):893–920.

Cochrane, J. H. (1991). A Critique of the Application of Unit Root Tests. *Journal of Economic Dynamics and Control*, 15:275–284.

Cohn, A. (1922). Über die Anzahl der Wurzeln einer Algebraischen Gleichung in einem Kreis. *Mathematische Zeitschrift*, 14:110–148.

Conrad, C. and Haag, B. (2006). Inequality Constraints in the Fractionally Integrated GARCH Model. *Journal of Financial Econometrics*, 4:413–449.

Conrad, C. and Karanasos, M. (2009). Negative Volatility Spillovers in the Unrestricted ECCC-GARCH Model. *Econometric Theory*, 26(3):838–862.

Cont, R. (2001). Empirical Properties of Asset Returns: Stylized Facts and Statistical Issues. *Quantitative Finance*, 1:223–236.

Cooley, T. F. and Prescott, E. C. (1973). An Adaptive Regression Model. *International Economic Review*, 14(2):364–371.

Corradi, V. and Swanson, N. R. (2006). Predictive Density Evaluation. In Elliott, G., Granger, C. W. J., and Timmermann, A., editors, *Handbook of Economic Forecasting*, pages 197–284. Elsevier, Amsterdam.

Corsetti, G., Pesenti, P., and Roubini, N. (1999a). Paper Tigers? A Model of the Asian Crisis. *European Economic Review*, 43(7):1211–1236.

Corsetti, G., Pesenti, P., and Roubini, N. (1999b). What Caused the Asian Currency and Financial Crisis? *Japan and the World Economy*, 11(3):305–373.

Covitz, D., Liang, N., and Suarez, G. A. (2013). The Evolution of a Financial Crisis: Collapse of the Asset-Backed Commercial Paper Market. *Journal of Finance*, 68(3):815–848.

Cox, D. R. (1983). Some Remarks on Overdispersion. *Biometrika*, 70:269–274.

Cox, D. R. and Hinkley, D. V. (1974). *Theoretical Statistics*. Chapman and Hall, London.

Cox, D. R. and Reid, N. (2004). A Note on Pseudo-Likelihood Constructed from Marginal Densities. *Biometrika*, 91(3):729–737.

Creal, D., Koopman, S. J., and Lucas, A. (2011). A Dynamic Multivariate Heavy-Tailed Model for Time-Varying Volatilities and Correlations. *Journal of Business & Economic Statistics*, 29(4):552–563.

Creal, D., Koopman, S. J., and Lucas, A. (2013). Generalized Autoregressive Score Models with Applications. *Journal of Applied Econometrics*, 28(5):777–795.

Cribari-Neto, F. (1996). On Time Series Econometrics. *The Quarterly Review of Economics and Finance*, 36(Supplement 1):37–60.

Crockett, P. W. (1985). Asymptotic Distribution of the Hildreth–Houck Estimator. *Journal of the American Statistical Association*, 80(389):202–204.

Dangl, T. and Halling, M. (2012). Predictive Regressions with Time-Varying Coefficients. *Journal of Financial Economics*, 106(1):157–181.

Daniels, H. E. (1939). The Estimation of Components of Variance. *Journal of the Royal Statistical Society Supplement*, 6(2):186–197.

Daniels, H. E. (1954). Saddlepoint Approximation in Statistics. *Annals of Mathematical Statistics*, 25:631–650.

Davidson, J. (2009). When is a Time Series I(0)? In Castle, J. L. and Shephard, N., editors, *The Methodology and Practice of Econometrics: A Festschrift in Honour of David F. Hendry*, chapter 13, pages 322–342. Oxford University Press, Oxford.

Davidson, R. and MacKinnon, J. G. (2004). *Econometric Theory and Methods*. Oxford University Press, New York.

Davies, R. B. (1977). Hypothesis Testing when a Nuisance Parameter is Present Only Under the Alternative. *Biometrika*, 64(2):247–254.

Davies, R. B. (1987). Hypothesis Testing when a Nuisance Parameter is Present Only Under the Alternatives. *Biometrika*, 74(1):33–43.

Davino, C., Furno, M., and Vistocco, D. (2014). *Quantile Regression: Theory and Applications*. John Wiley & Sons, Chichester.

Davis, R. A., Heiny, J., Mikosch, T., and Xie, X. (2016a). Extreme Value Analysis for the Sample Autocovariance Matrices of Heavy-Tailed Multivariate Time Series. *Extremes*, 19(3):517–547.

Davis, R. A., Holan, S. H., Lund, R., and Ravishanker, N., editors (2015). *Handbook of Discrete-Valued Time Series*. Chapman & Hall/CRC, Boca Raton.

Davis, R. A., Mikosch, T., and Pfaffel, O. (2016b). Asymptotic Theory for the Sample Covariance Matrix of a Heavy-Tailed Multivariate Time Series. *Stochastic Processes and their Applications*, 126(3):767–799.

Davis, T. A. (2006). *Fundamentals of Algorithms: Direct Methods for Sparse Linear Systems*. Society for Industrial and Applied Mathematics, Philadelphia.

Davison, A. C. (2003). *Statistical Models*. Cambridge University Press, Cambridge.

Davison, A. C., Hinkley, D. V., and Young, G. V. (2003). Recent Developments in Bootstrap Methodology. *Statistical Science*, 18:141–157.

Dawid, A. P. (1977). Spherical Matrix Distributions and a Multivariate Model. *Journal of the Royal Statistical Society, Series B*, 39(2):254–261.

Dawid, A. P. (1984). Statistical Theory: The Prequential Approach (with discussion). *Journal of the Royal Statistical Society, Series A*, 147:278–292.

Dawid, A. P. (1985a). Calibration-Based Empirical Probability (with discussion). *Annals of Statistics*, 13:1251–1285.

Dawid, A. P. (1985b). The Impossibility of Inductive Inference. *Journal of the American Statistical Association*, 80:340–341.

Dawid, A. P. (1986). Probability Forecasting. In Kotz, S., Johnson, N. L., and Read, C. B., editors, *Encyclopedia of Statistical Sciences*, volume 7, pages 210–218. John Wiley & Sons, New York.

de Carvalho, M. and Rua, A. (2017). Real-Time Nowcasting the US Output Gap: Singular Spectrum Analysis at Work. *International Journal of Forecasting*, 33:185–198.

De Gooijer, J. G. (1978). On the Inverse of the Autocovariance Matrix for a General Mixed Autoregressive Moving Average Process. *Statistische Hefte*, 19:114–123.

De Gooijer, J. G. (1980). Exact Moments of the Sample Autocorrelations from Series Generated by General ARIMA Processes of Order (p, d, q), $d = 0$ or 1. *Journal of Econometrics*, 14(3):365–379.

Dean, A. M. and Voss, D. T. (1999). *Design and Analysis of Experiments*. Springer, New York.

Demarta, S. and McNeil, A. J. (2005). The t Copula and Related Copulas. *International Statistical Review*, 73:111–129.

DeMiguel, V., Garlappi, L., Nogales, F. J., and Uppal, R. (2009a). A Generalized Approach to Portfolio Optimization: Improving Performance by Constraining Portfolio Norms. *Management Science*, 55(5):798–812.

DeMiguel, V., Garlappi, L., and Uppal, R. (2009b). Optimal Versus Naive Diversification: How Inefficient is the $1/N$ Portfolio Strategy? *Review of Financial Studies*, 22(5):1915–1953.

DeMiguel, V., Martin-Utrera, A., and Nogales, F. J. (2013). Size Matters: Optimal Calibration of Shrinkage Estimators for Portfolio Selection. *Journal of Banking & Finance*, 37(8):3018–3034.

DeMiguel, V., Nogales, F. J., and Uppal, R. (2014). Stock Return Serial Dependence and Out-of-Sample Portfolio Performance. *Review of Financial Studies*, 27:1031–1073.

Deng, A. and Perron, P. (2006). A Comparison of Alternative Asymptotic Frameworks to Analyse a Structural Change in a Linear Time Trend. *Econometrics Journal*, 9(3):423–447.

Dent, W. T. and Hildreth, C. (1977). Maximum Likelihood Estimation in Random Coefficient Models. *Journal of the American Statistical Association*, 72:69–72.

Devroye, L. (1986). *Non-Uniform Random Variate Generation*. Springer, New York.

Dhrymes, P. J. (2013). *Mathematics for Econometrics*. Springer, New York, 4th edition.

Diamandis, P. F., Drakos, A. A., Kouretas, G. P., and Zarangas, L. (2011). Value-at-Risk for Long and Short Trading Positions: Evidence from Developed and Emerging Equity Markets. *International Review of Financial Analysis*, 20(3):165–176.

Díaz-García, J. A. (2013). Distribution Theory of Quadratic Forms for Matrix Multivariate Elliptical Distribution. *Journal of Statistical Planning and Inference*, 143(8):1330–1342.

Dickey, D. A. and Fuller, W. A. (1979). Distribution of the Estimators for Autoregressive Time Series with a Unit Root. *Journal of the American Statistical Association*, 74:427–431.

Dickie, J. R. and Nandi, A. K. (1994). A Comparative Study of AR Order Selection Methods. *Signal Processing*, 40(2-3):239–255.

Diebold, F. X., Gunther, T. A., and Tay, A. S. (1998). Evaluating Density Forecasts with Applications to Financial Risk Management. *International Economic Review*, 39:863–883.

Diebold, F. X. and Lopez, J. A. (1996). Modeling Volatility Dynamics. In Hoover, K., editor, *Macroeconomics: Developments, Tensions and Prospects*. Kluwer, Boston.

Diebold, F. X. and Mariano, R. S. (1995). Comparing Predictive Accuracy. *Journal of Business and Economic Statistics*, 13:253–263.

Diebold, F. X. and Rudebusch, G. D. (1991). On the Power of Dickey–Fuller Tests Against Fractional Alternatives. *Economics Letters*, 35:155–160.

Diebold, F. X. and Yilmaz, K. (2015). *Financial and Macroeconomic Connectedness: A Network Approach to Measurement and Monitoring*. Oxford University Press, Oxford.

Ding, P. (2016). On the Conditional Distribution of the Multivariate t Distribution. *The American Statistician*, 70(3):293–295.

Ding, Z. (1994). *Time Series Analysis of Speculative Returns*. PhD thesis, University of California San Diego.

Ding, Z., Granger, C. W. J., and Engle, R. F. (1993). A Long Memory Property of Stock Market Returns and a New Model. *Journal of Empirical Finance*, 1:83–106.

Dobrev, D., Nesmith, T. D., and Oh, D. H. (2017). Accurate Evaluation of Expected Shortfall for Linear Portfolios with Elliptically Distributed Risk Factors. *Journal of Risk and Financial Management*, 10(1):1–14. Article 5.

Doraň, H. E. (1992). Constraining Kalman Filter and Smoothing Estimates to Satisfy Time-Varying Restrictions. *The Review of Economics and Statistics*, 74:568–572.

Doran, H. E. and Rambaldi, A. N. (1997). Applying Linear Time-Varying Constraints to Econometric Models: With an Application to Demand Systems. *Journal of Econometrics*, 79:83–95.

Dowd, K. (2005). *Measuring Market Risk*. John Wiley & Sons, New York, 2nd edition.

Dreier, I. and Kotz, S. (2002). A Note on the Characteristic Function of the t-Distribution. *Statistics & Probability Letters*, 57:221–224.

Dubbelman, C., Louter, A. S., and Abrahamse, A. P. J. (1978). On Typical Characteristics of Economic Time Series and the Relative Qualities of Five Autocorrelation Tests. *Journal of Econometrics*, 8:295–306.

Dudewicz, E. J. and Mishra, S. N. (1988). *Modern Mathematical Statistics*. John Wiley & Sons, New York.

Dufays, A. (2016). Infinite-State Markov-Switching for Dynamic Volatility. *Journal of Financial Econometrics*, 14(2):418–460.

Dufour, J.-M. and King, M. L. (1991). Optimal Invariant Tests for the Autocorrelation Coefficient in Linear Regressions with Stationary or Nonstationary AR(1) Errors. *Journal of Econometrics*, 47:115–143.

Dufour, J.-M. and Roy, R. (1985). Some Robust Exact Results on Sample Autocorrelations and Tests of Randomness. *Journal of Econometrics*, 29:257–273.

Dufrénot, G. and Jawadi, F. (2017). Special Issue: Recent Developments of Switching Models for Financial Data. *Studies in Nonlinear Dynamics & Econometrics*, 21(1):1–2.

Dungey, M., Erdemlioglu, D., Matei, M., and Yang, X. (2018). Testing for Mutually Exciting Jumps and Financial Flights in High Frequency Data. *Journal of Econometrics*, 202:18–44.

Dungey, M. and Martin, V. L. (2007). Unravelling Financial Market Linkages During Crises. *Journal of Applied Econometrics*, 22:89–119.

Duong, Q. P. (1984). On the Choice of the Order of Autoregressive Models: A Ranking and Selection Approach. *Journal of Time Series Analysis*, 5(3):145–157.

Durbin, J. (1959). Efficient Estimation of Parameters in Moving-Average Models. *Biometrika*, 46(3–4):306–316.

Durbin, J. (1987). Statistics and Statistical Sciences. *Journal of the Royal Statistical Society, Series A*, 150(3):177–191.

Durbin, J. (2000). The Foreman Lecture: the State Space Approach to Time Series Analysis and its Potential for Official Statistics (with discussion). *Australian and New Zealand Journal of Statistics*, 42(1):1–23.

Durbin, J. and Koopman, S. J. (2012). *Time Series Analysis by State Space Methods*. Oxford University Press, Oxford, 2nd edition.

Durbin, J. and Watson, G. S. (1950). Testing for Serial Correlation in Least Squares Regression. I. *Biometrika*, 37:409–428.

Durbin, J. and Watson, G. S. (1971). Testing for Serial Correlation in Least Squares Regression. III. *Biometrika*, 58:1–19.

Dziechciarz, J. (1989). Changing and Random Coefficient Models: A Survey. In Hackl, P., editor, *Statistical Analysis and Forecasting of Economic Structural Change*, pages 217–251. Springer, Berlin.

Efron, B. (1979). Bootstrap Methods: Another Look at the Jackknife. *Annals of Statistics*, 7:1–26.

Efron, B. (2003). Second Thoughts on the Bootstrap. *Statistical Science*, 18:135–140.

Efron, B. and Hastie, T. (2016). *Computer Age Statistical Inference: Algorithms, Evidence, and Data Science*. Cambridge University Press, Cambridge.

Elliott, G. and Stock, J. H. (2001). Confidence Interval for Autoregressive Coefficients Near One. *Journal of Econometrics*, 103:155–181.

Embrechts, P., Klüppelberg, C., and Mikosch, T. (2000). *Modelling Extremal Events for Insurance and Finance*. Springer, New York.

Embrechts, P., McNeil, A., and Straumann, D. (2002). Correlation and Dependency in Risk Management: Properties and Pitfalls. In Dempster, M. A. H., editor, *Risk Management: Value at Risk and Beyond*, pages 176–223. Cambridge University Press, Cambridge.

Engle, R. (1982). Autoregressive Conditional Heteroskedasticity with Estimates of the Variance of United Kingdom Inflation. *Econometrica*, 50:987–1007.

Engle, R. (2001). GARCH 101: The Use of ARCH/GARCH Models in Applied Econometrics. *Journal of Economic Perspectives*, 15(4):157–168.

Engle, R. and Kelly, B. (2012). Dynamic Equicorrelation. *Journal of Business & Economic Statistics*, 30(2):212–228.

Engle, R. F. (2002). Dynamic Conditional Correlation: A Simple Class of Multivariate Generalized Autoregressive Conditional Heteroskedasticity Models. *Journal of Business and Economic Statistics*, 20(3):339–350.

Engle, R. F. (2009). *Anticipating Correlations: A New Paradigm for Risk Management*. Princeton University Press, Princeton.

Engle, R. F. (2016). Dynamic Conditional Beta. *Journal of Financial Econometrics*, 14(4):643–667.

Engle, R. F. and Granger, C. W. J. (1987). Co-Integration and Error Correction: Representation, Estimation and Testing. *Econometrica*, 35:251–276.

Engle, R. F. and Kroner, K. F. (1995). Multivariate Simultaneous Generalized ARCH. *Econometric Theory*, 11:122–150.

Engle, R. F. and Manganelli, S. (2004). CAViaR: Conditional Autoregressive Value at Risk by Regression Quantiles. *Journal of Business & Economic Statistics*, 22(4):367–381.

Engle, R. F. and Mezrich, J. (1996). GARCH for Groups. *Risk Magazine*, 9:36–40.

Engle, R. F. and Ng, V. K. (1993). Measuring and Testing the Impact of News on Volatility. *Journal of Finance*, 48(5):1749–1778.

Engle, R. F., Ng, V. K., and Rothschild, M. (1990). Asset Pricing with a Factor-ARCH Covariance Structure: Empirical Estimates for Treasury Bills. *Journal of Econometrics*, 45:213–237.

Engle, R. F. and Sheppard, K. (2001). Theoretical and Empirical Properties of Dynamic Conditional Correlation Multivariate GARCH. NBER Working Papers 8554, National Bureau of Economic Research, Inc.

Engsted, T. and Pedersen, T. Q. (2014). Bias-Correction in Vector Autoregressive Models: A Simulation Study. *Econometrics*, 2(1):45–71.

Ericsson, N. R. (2012). Detecting Crises, Jumps, and Changes in Regime. Board of Governors of the Federal Reserve System, Washington, DC.

Etuk, E. H. (2000). On Autoregressive Model Identification. *Journal of Official Statistics*, 4(2):113–124.

Evans, M. A. and King, M. (1985). A Point Optimal Test for Heteroscedastic Disturbances. *Journal of Econometrics*, 27(2):163–178.

Evans, M. A. and King, M. (1988). A Further Class of Tests for Heteroscedasticity. *Journal of Econometrics*, 37(2):265–276.

Fabian, V. (2000). New Modifications of the Bechhofer Method. *Journal of Statistical Planning and Inference*, 91:313–322.

Fahrmeir, L., Kneib, T., Lang, S., and Marx, B. (2013). *Regression: Models, Methods and Applications.* Springer, Berlin.

Fahrmeir, L. and Tutz, G. (2001). *Multivariate Statistical Modelling Based on Generalized Linear Models.* Springer, New York, 2nd edition.

Fama, E. F. and French, K. R. (1993). Common Risk Factors in the Returns of Stocks and Bonds. *Journal of Financial Economics*, 33:3–56.

Fama, E. F. and French, K. R. (1996). Multifactor Explanations of Asset Pricing Anomalies. *Journal of Finance*, 51:55–84.

Fan, J., Fan, Y., and Lv, J. (2008). High Dimensional Covariance Matrix Estimation Using a Factor Model. *Journal of Econometrics*, 147:186–197.

Fan, J., Qi, L., and Xiu, D. (2014). Quasi-Maximum Likelihood Estimation of GARCH Models With Heavy-Tailed Likelihoods. *Journal of Business & Economic Statistics*, 32(2):178–191.

Fan, J. and Yao, Q. (2003). *Nonlinear Time Series: Nonparametric and Parametric Methods*. Springer, New York.

Fan, J., Zhang, J., and Yu, K. (2012). Vast Portfolio Selection With Gross-Exposure Constraints. *Journal of the American Statistical Association*, 107(498):592–606.

Fang, H. B., Fang, K. T., and Kotz, S. (2002). The Meta-Elliptical Distribution with Given Marginals. *Journal of Multivariate Analysis*, 82:1–16.

Fang, H. B., Fang, K. T., and Kotz, S. (2005). Corrigendum to 'The Meta-Elliptical Distribution with Given Marginals'. *Journal of Multivariate Analysis*, 94(1):222–223.

Fang, K.-T., Kotz, S., and Ng, K.-W. (1989). *Symmetric Multivariate and Related Distributions*. Chapman & Hall, London.

Farebrother, R. W. (1980). Algorithm AS 153: Pan's Procedure for the Tail Probabilities of the Durbin–Watson Statistic. *Applied Statistics*, 29(2):224–227.

Farebrother, R. W. (1984). Remark AS R52: The Distribution of a Linear Combination of Central χ^2 Random Variables: A Remark on AS 153: Pan's Procedure for the Tail Probabilities of the Durbin–Watson Statistic. *Journal of the Royal Statistical Society, Series C*, 33(3):363–366.

Farebrother, R. W. (1985). Eigenvalue-Free Methods for Computing the Distribution of a Quadratic Form in Normal Variables. *Statistische Hefte*, 26:287–302.

Farebrother, R. W. (1990). The Distribution of a Quadratic Form in Normal Variables. *Applied Statistics*, 39:294–309.

Farebrother, R. W. (1994). A Critique of Recent Methods for Computing the Distribution of the Durbin–Watson and Other Invariant Test Statistics. *Statistische Hefte*, 35:365–369.

Farkas, G. and Vicknair, K. (1996). Appropriate Tests of Racial Wage Discrimination Require Controls for Cognitive Skill: Comment on Cancio, Evans, and Maume. *American Sociological Review*, 61(4):557–560.

Fastrich, B., Paterlini, S., and Winker, P. (2015). Constructing Optimal Sparse Portfolios using Regularization Methods. *Computational Management Science*, 12(3):417–434.

Ferguson, T. S. (1967). *Mathematical Statistics: A Decision Theoretic Approach*. Academic Press, New York.

Fermanian, J.-D. (2017). Editorial: Recent Developments in Copula Models. *Econometrics*, 5(3):1–3. Article 34.

Fermanian, J.-D. and Malongo, H. (2017). On the Stationarity of Dynamic Conditional Correlation Models. *Econometric Theory*, 33(3):636–663.

Fiebig, D. G., Bartels, R., and Krämer, W. (1996). The Frisch–Waugh Theorem and Generalized Least Squares. *Econometric Reviews*, 15(4):431–443.

Figlewski, S. and Wang, X. (2000). Is the 'Leverage Effect' a Leverage Effect? mimeo, NYU Stern School of Business.

Fisher, R. A. (1925). The Influence of Rainfall on the Yield of Wheat at Rothamsted. *Philosophical Transactions of the Royal Society of London. Series B, Containing Papers of a Biological Character*, 213:89–142.

Fisher, R. A. (1938). Presidential Address by Professor R. A. Fisher. *Sankhya*, 4(1):14–17.

Fisk, P. R. (1967). Models of the Second Kind in Regression Analysis. *Journal of the Royal Statistical Society, Series B*, 29(2):266–281.

Fletcher, J. (2017). Exploring the Benefits of Using Stock Characteristics in Optimal Portfolio Strategies. *European Journal of Finance*, 23(3):192–210.

Fomby, T. B. and Guilkey, D. K. (1978). On Choosing the Optimal Level of Significance for the Durbin–Watson Test and the Bayesian Alternative. *Journal of Econometrics*, 8:203–213.

Forbes, K. and Rigobon, R. (2002). No Contagion, Only Interdependence: Measuring Stock Market Co-Movements. *Journal of Finance*, 57:2223–2261.

Forchini, G. (2000). The Density of the Sufficient Statistics for a Gaussian AR(1) Model in terms of Generalized Functions. *Statistics & Probability Letters*, 50(3):237–243.

Frahm, G. (2004). Generalized Elliptical Distributions: Theory and Applications. PhD thesis, University of Cologne.

Francioni, I. and Herzog, F. (2012). Probability-Unbiased Value-at-Risk Estimators. *Quantitative Finance*, 12(5):755–768.

Francq, C., Horváth, L., and Zakoïan, J.-M. (2011). Merits and Drawbacks of Variance Targeting in GARCH Models. *Journal of Financial Econometrics*, 9(4):619–656.

Francq, C., Horváth, L., and Zakoïan, J.-M. (2016). Variance Targeting Estimation of Multivariate GARCH Models. *Journal of Financial Econometrics*, 14(2):353–382.

Francq, C., Wintenberger, O., and Zakoïan, J.-M. (2013). GARCH Models Without Positivity Constraints: Exponential or Log GARCH? *Journal of Econometrics*, 177(1):34–46.

Francq, C. and Zakoïan, J.-M. (2004). Maximum Likelihood Estimation of Pure GARCH and ARMA-GARCH Processes. *Bernoulli*, 10(4):605–637.

Francq, C. and Zakoïan, J.-M. (2010). *GARCH Models: Structure, Statistical Inference and Financial Applications*. John Wiley & Sons Ltd, Chichester.

Francq, C. and Zakoïan, J.-M. (2016). Estimating Multivariate Volatility Models Equation by Equation. *Journal of the Royal Statistical Society, Series B*, 78(3):613–635.

Frankfurter, G. M., Phillips, H. E., and Seagle, J. P. (1971). Portfolio Selection: The Effects of Uncertain Means, Variances, and Covariances. *Journal of Financial and Quantitative Analysis*, 6(5):1251–1262.

Franses, P. H. and van Dijk, D. (2000). *Nonlinear Time Series Models in Empirical Finance*. Cambridge University Press, Cambridge.

Franzini, L. and Harvey, A. C. (1983). Testing for Deterministic Trend and Seasonal Components in Time Series Models. *Biometrika*, 70:673–682.

Freimann, K.-D. (1991). Estimating the Second Moments of Random Coefficients. In Gruber, J., editor, *Econometric Decision Models: New Methods of Modeling and Applications*, chapter 22, pages 385–403. Springer, Berlin.

Froehlich, B. R. (1973). Some Estimators for a Random Coefficient Regression Model. *Journal of the American Statistical Association*, 68(342):329–335.

Frost, P. A. and Savarino, J. E. (1986). An Empirical Bayes Approach to Efficient Portfolio Selection. *Journal of Financial and Quantitative Analysis*, 21(3):293–305.

Frühwirth-Schnatter, S. (2006). *Finite Mixture and Markov Switching Models*. Springer, New York.

Fugazza, C., Guidolin, M., and Nicodano, G. (2015). Equally Weighted vs. Long-Run Optimal Portfolios. *European Financial Management*, 21(4):742–789.

Fuller, W. A. (1996). *Introduction to Statistical Time Series*. John Wiley & Sons, New York, 2nd edition.

Gabrielsen, A., Kirchner, A., Liu, Z., and Zagaglia, P. (2015). Forecasting Value-at-Risk with Time-Varying Variance, Skewness And Kurtosis in an Exponential Weighted Moving Average Framework. *Annals of Financial Economics*, 10(1):1–29.

Galbraith, J. W. and Kisinbay, T. (2005). Content Horizons for Conditional Variance Forecasts. *International Journal of Forecasting*, 21:249–260.

Galbraith, J. W. and Zinde-Walsh, V. (1999). On the Distribution of Augmented Dickey–Fuller Statistics in Processes with Moving Average Components. *Journal of Econometrics*, 93:25–47.

Galbraith, R. F. and Galbraith, J. I. (1974). On the Inverse of some Patterned Matrices Arising in the Theory of Stationary Time Series. *Journal of Applied Probability*, 11:63–71.

Galwey, N. W. (2014). *Introduction to Mixed Modelling: Beyond Regression and Analysis of Variance*. John Wiley & Sons, Chichester, 2nd edition.

Gambacciani, M. and Paolella, M. S. (2017). Robust Normal Mixtures for Financial Portfolio Allocation. *Econometrics and Statistics*, 3:91–111.

Gao, C.-T. and Zhou, X.-H. (2016). Forecasting VaR and ES Using Dynamic Conditional Score Models and Skew Student Distribution. *Economic Modelling*, 53:216–223.

Gao, F. and Song, F. (2008). Estimation Risk in GARCH VaR and ES Estimates. *Econometric Theory*, 24:1404–1424.

Gardiner, W. P. and Gettinby, G. (1998). *Experimental Design Techniques in Statistical Practice: A Practical Software-Based Approach*. Horwood, Chichester.

Geary, R. C. (1944). Extension of a Theorem by Harald Cramér on the Frequency Distribution of the Quotient of Two Variables. *Journal of the Royal Statistical Society*, 17:56–57.

Geary, R. C. (1970). Relative Efficiency of Count of Sign Changes for Assessing Residual Autocorrelation in Least Squares Regression. *Biometrika*, 57(1):123–127.

Gelman, A., Carlin, J. B., Stern, H. S., Dunson, D. B., Vehtari, A., and Rubin, D. B. (2013). *Bayesian Data Analysis*. Chapman & Hall/CRC, Boca Raton, 3rd edition.

Genest, C., Gendron, M., and Bourdeau-Brien, M. (2009). The Advent of Copulas in Finance. *European Journal of Finance*, 15(7–8):609–618.

Gentle, J. E. (2007). *Matrix Algebra: Theory, Computations, and Applications in Statistics*. Springer, New York.

Genton, M. G., editor (2004). *Skew-Elliptical Distributions and Their Applications: A Journey Beyond Normality*. Chapman & Hall/CRC Press, Boca Raton.

Gerlach, R. and Chen, C. W. S. (2016). Bayesian Expected Shortfall Forecasting Incorporating the Intraday Range. *Journal of Financial Econometrics*, 14(1):128–158.

Gerlach, R., Lu, Z., and Huang, H. (2013). Exponentially Smoothing the Skewed Laplace Distribution for Value-at-Risk Forecasting. *Journal of Forecasting*, 32(6):534–550.

Geweke, J. and Amisano, G. (2010). Comparing and Evaluating Bayesian Prediction Distributions of Asset Returns. *International Journal of Forecasting*, 26:216–230.

Geweke, J. and Amisano, G. (2011). Hierarchical Markov Normal Mixture Models with Applications to Financial Asset Returns. *Journal of Applied Econometrics*, 26:1–29.

Ghalanos, A., Rossi, E., and Urga, G. (2015). Independent Factor Autoregressive Conditional Density Model. *Econometric Reviews*, 34(5):594–616.

Ghazal, G. A. (1994). Moments of the Ratio of Two Dependent Quadratic Forms. *Statistics & Probability Letters*, 20(4):313–319.

Ghosh, M. (2002). Basu's Theorem with Applications: A Personalistic Review. *Sankhya, Series A*, 64(3):509–531.

Giacometti, R., Bertocchi, M., Rachev, S. T., and Fabozzi, F. J. (2007). Stable Distributions in the Black-Litterman Approach to the Asset Allocation. *Quantitative Finance*, 7(4):423–433.

Giamouridis, D. (2006). Estimation Risk in Financial Risk Management: A Correction. *Journal of Risk*, 8(4):121–125.

Gill, P. E., Golub, G. H., Murray, W., and Saunders, M. A. (1974). Methods for Modifying Matrix Factorizations. *Mathematics of Computation*, 28(126):505–535.

Giot, P. and Laurent, S. (2003). Value-at-Risk for Long and Short Trading Positions. *Journal of Applied Econometrics*, 18(6):641–663.

Giot, P. and Laurent, S. (2004). Modelling Daily Value-at-Risk Using Realized Volatility and ARCH Type Models. *Journal of Empirical Finance*, 11:379–398.

Glosten, L. R., Jagannathan, R., and Runkle, D. E. (1993). On the Relation between the Expected Value and Volatility of Nominal Excess Return on Stocks. *Journal of Finance*, 48(5):1779–1801.

Gneiting, T., Balabdaoui, F., and Raftery, A. E. (2007). Probabilistic Forecasts, Calibration and Sharpness. *Journal of the Royal Statistical Society, Series B*, 69(2):243–268.

Godolphin, E. J. and Unwin, J. M. (1983). Evaluation of the Covariance Matrix for the Maximum Likelihood Estimator of a Gaussian Autoregressive Moving Average Process. *Biometrika*, 70:279–284.

Golden, R. M., Henley, S. S., White, H., and Kashner, T. M. (2016). Generalized Information Matrix Tests for Detecting Model Misspecification. *Econometrics*, 4(4). Article 46.

Golub, G. H. and Loan, C. F. V. (2012). *Matrix Computations*. John Hopkins Press, 4th edition.

Gonzalo, J. and Montesinos, R. (2002). Threshold Stochastic Unit Root Models. Manuscript, Universidad Carlos III.

Gourieroux, C. (1997). *ARCH Models and Financial Applications*. Springer, New York.

Grabchak, M. and Samorodnitsky, G. (2010). Do Financial Returns have Finite or Infinite Variance? A Paradox and an Explanation. *Quantitative Finance*, 10(8):883–893.

Grad, A. and Solomon, H. (1955). Distribution of Quadratic Forms and some Applications. *Annals of Mathematical Statistics*, 26:464–77.

Granger, C. W. J. (1980). Long Memory Relationships and the Aggregation of Dynamic Models. *Journal of Econometrics*, 14:227–238.

Granger, C. W. J. (1992). Forecasting Stock Market Prices: Lessons for Forecasters. *International Journal of Forecasting*, 8:3–13.

Granger, C. W. J. (2005). The Past and Future of Empirical Finance: Some Personal Comments. *Journal of Econometrics*, 129(1–2):35–40.

Granger, C. W. J. (2008). Non-Linear Models: Where Do We Go Next—Time Varying Parameter Models? *Studies in Nonlinear Dynamics & Econometrics*, 12(3):1–10.

Granger, C. W. J. and Ding, Z. (1995). Some Properties of Absolute Return, An Alternative Measure of Risk. *Annales D'économie et de Statistique*, 40:67–91.

Granger, C. W. J. and Joyeux, R. (1980). An Introduction to Long-Memory Time Series and Fractional Differencing. *Journal of Time Series Analysis*, 1(1):15–29.

Granger, C. W. J. and Newbold, P. (1986). *Forecasting Economic Time Series*. Academic Press, San Diego, 2nd edition.

Granger, C. W. J., Spear, S., and Ding, Z. (2000). Stylized Facts on the Temporal and Distributional Properties of Absolute Returns: An Update. In Chan, W.-S., Li, W. K., and Tong, H., editors, *Statistics and Finance: An Interface*, pages 97–120. Imperial College Press, London.

Granger, C. W. J. and Swanson, N. R. (1997). An Introduction to Stochastic Unit-Root Processes. *Journal of Econometrics*, 80:35–62.

Granger, C. W. J. and Teräsvirta, T. (1993). *Modelling Nonlinear Economic Relationships*. Oxford University Press, Oxford.

Gray, S. F. (1996). Modeling the Conditional Distribution of Interest Rates as a Regime-Switching Process. *Journal of Financial Economics*, 42:27–62.

Graybill, F. A. (1976). *Theory and Application of the Linear Model*. Duxbury Press, North Scituate, MA.

Graybill, F. A. (1983). *Matrices with Applications in Statistics*. Wadsworth, Pacific Grove, CA.

Graybill, F. A. and Iyer, H. K. (1994). *Regression Analysis: Concepts and Applications*. Duxbury, Wadsworth, Belmont, CA.

Greene, W. H. (2017). *Econometric Analysis*. Pearson, New York, 8th edition.

Greene, W. H. and Seaks, T. G. (1991). The Restricted Least Squares Estimator: A Pedagogical Note. *The Review of Economics and Statistics*, 73(3):563–567.

Greenspan, A. (1999). New Challenges for Monetary Policy. Symposisum Opening Remarks at the Federal Reserve Bank of Kansas City.

Griffiths, W. E. (1972). Estimation of Actual Response Coefficients in the Hildreth–Houck Random Coefficient Model. *Journal of the American Statistical Association*, 67(339):633–635.

Gruber, J. (2005). Religious Market Structure, Religious Participation and Outcomes: Is Religion Good for You? *Advances in Economic Analysis and Policy*, 5(1). Article 5.

Guidolin, M., Hyde, S., McMillan, D., and Ono, S. (2008). Non-Linear Predictability in Stock and Bond Returns: When and Where is it Exploitable? Working Paper 2008-010B, Federal Reserve Bank of St. Louis, Research Division.

Gupta, A. K. and Varga, T. (1993). *Elliptically Contoured Models in Statistics*. Springer Science, Dordrecht.

Haas, M. (2005). Improved Duration-Based Backtesting of Value-at-Risk. *Journal of Risk*, 8(2):17–38.

Haas, M. (2009). Value-at-Risk via Mixture Distributions Reconsidered. *Applied Mathematics and Computation*, 215(6):2103–2119.

Haas, M. (2010). Skew-Normal Mixture and Markov-Switching GARCH Processes. *Studies in Nonlinear Dynamics & Econometrics*, 14(4). Article 1.

Haas, M., Krause, J., Paolella, M. S., and Steude, S. C. (2013). Time-Varying Mixture GARCH Models and Asymmetric Volatility. *North American Journal of Economics and Finance*, 26:602–623.

Haas, M., Mittnik, S., and Mizrach, B. (2006a). Assessing Central Bank Credibility During the EMS Crises: Comparing Option and Spot Market-Based Forecasts. *Journal of Financial Stability*, 2:28–54.

Haas, M., Mittnik, S., and Paolella, M. S. (2004a). Mixed Normal Conditional Heteroskedasticity. *Journal of Financial Econometrics*, 2(2):211–250.

Haas, M., Mittnik, S., and Paolella, M. S. (2004b). A New Approach to Markov-Switching GARCH Models. *Journal of Financial Econometrics*, 2(4):493–530.

Haas, M., Mittnik, S., and Paolella, M. S. (2009). Asymmetric Multivariate Normal Mixture GARCH. *Computational Statistics & Data Analysis*, 53(6):2129–2154.

Haas, M., Mittnik, S., Paolella, M. S., and Steude, S. C. (2006b). Analyzing and Exploiting Asymmetries in the News Impact Curve. FINRISK Working Paper No. 256, Swiss National Science Foundation.

Haas, M. and Paolella, M. S. (2012). Mixture and Regime-Switching GARCH Models. In Bauwens, L., Hafner, C. M., and Laurent, S., editors, *Handbook of Volatility Models and their Applications*, chapter 3. John Wiley & Sons, Inc., Hoboken, NJ.

Haidt, J. (2006). *The Happiness Hypothesis: Finding Modern Truth in Ancient Wisdom*. Basic Books, New York.

Haldrup, N., Meitz, M., and Saikkonen, P., editors (2014). *Essays in Nonlinear Time Series Econometrics*. Oxford University Press, Oxford.

Hall, P. and Yao, Q. (2003). Inference in ARCH and GARCH Models With Heavy-tailed Errors. *Econometrica*, 71:285–317.

Halmos, P. R. (1985). *I Want to be a Mathematician: an Automathography in Three Parts*. Springer, New York.

Hamilton, J. D. (1989). A New Approach to the Economic Analysis of Nonstationary Time Series and the Business Cycle. *Econometrica*, 57:357–384.

Hamilton, J. D. (1991). A Quasi-Bayesian Approach to Estimating Parameters for Mixtures of Normal Distributions. *Journal of Business and Economic Statistics*, 9(1):21–39.

Hamilton, J. D. (1994). *Time Series Analysis*. Princeton University Press, Princeton.

Hamilton, J. D. (2008). Regime Switching Models. In Durlauf, S. N. and Blume, L. E., editors, *The New Palgrave Dictionary of Economics, Second Edition*. Palgrave Macmillan, London.

Hamouda, O. and Rowley, R. (1996). *Probability in Economics*. Routledge, London.

Hampel, F. (1996). On the Philosophical Foundations of Statistics: Bridges to Huber's Work, and Recent Results. In *Robust Statistics, Data Analysis, and Computer Intensive Methods*, pages 185–196. Springer.

Hand, D. J., Daly, F., McConway, K., Lunn, D., and Ostrowski, E. (1994). *A Handbook of Small Data Sets*. Chapman & Hall/CRC, Boca Raton, USA.

Hannan, E. J. and McDougall, A. J. (1988). Regression Procedures for ARMA Estimation. *Journal of the American Statistical Association*, 83(402):490–498.

Hansen, B. E. (1992). Testing for Parameter Instability in Linear Models. *Journal of Policy Modeling*, 14(4):517–533.

Hansen, B. E. (1994). Autoregressive Conditional Density Estimation. *International Economic Review*, 35(3):705–730.

Hansen, B. E. (1997). Inference in TAR Models. *Studies in Nonlinear Dynamics and Econometrics*, 2(1):119–131.

Hansen, B. E. (1999). Threshold Effects in Non-Dynamic Panels: Estimation, Testing and Inference. *Journal of Econometrics*, 93:345–368.

Hansen, B. E. (2000). Sample Splitting and Threshold Estimation. *Econometrica*, 68(3):575–603.

Hao, L. and Naiman, D. Q. (2007). *Quantile Regression*. SAGE Publications, Thousand Oaks, CA.

Harman, R. and Lacko, V. (2010). On Decompositional Algorithms for Uniform Sampling from n-Spheres and n-Balls. *Journal of Multivariate Analysis*, 101(10):2297–2304.

Harrell, Jr., F. E. (2015). *Regression Modeling Strategies: With Applications to Linear Models, Logistic and Ordinal Regression, and Survival Analysis*. Springer, Cham, 2nd edition.

Harrison, M. J. (1972). On Testing for Serial Correlation in Regression when the Bounds Test is Inconclusive. *Economic and Social Review*, 4(1):41–57.

Hartz, C., Mittnik, S., and Paolella, M. S. (2006). Accurate Value-at-Risk Forecasting Based on the Normal-GARCH Model. *Computational Statistics & Data Analysis*, 51(4):2295–2312.

Harvey, A. and Sucarrat, G. (2014). EGARCH Models with Fat Tails, Skewness and Leverage. *Computational Statistics & Data Analysis*, 26:320–338.

Harvey, A. C. (1993). *Time Series Models*. MIT Press, Cambridge, MA, 2nd edition.

Harvey, A. C. (2013a). *Dynamic Models for Volatility and Heavy Tails: With Applications to Financial and Economic Time Series*. Cambridge University Press, Cambridge.

Harvey, A. C. (2013b). *Dynamic Modes for Volatility and Heavy Tails: With Applications to Financial and Economic Time Series*. Cambridge University Press, Cambridge.

Harvey, A. C. and Phillips, G. D. A. (1974). A Comparison of the Power of Some Tests for Heteroskedasticity in the General Linear Model. *Journal of Econometrics*, 2:307–316.

Harvey, A. C. and Pierse, R. G. (1984). Estimating Missing Observations in Economic Time Series. *Journal of the American Statistical Association*, 79(385):125–131.

Harvey, C. R. and Liu, Y. (2016). Lucky Factors. Available at SSRN.

Harvey, C. R., Liu, Y., and Zhu, H. (2016). … and the Cross-Section of Expected Returns. *The Review of Financial Studies*, 29(1):5–68.

Harvey, C. R. and Roper, A. (1999). The Asian Bet. In Harwood, A., Litan, R. E., and Pomerleano, M., editors, *The Crisis in Emerging Financial Markets*, pages 29–115. Brookings Institutional Press.

Harvey, C. R. and Siddique, A. (1999). Autoregressive Conditional Skewness. *Journal of Financial and Quantitative Analysis*, 34(4):465–487.

Harvey, D. I., Leybourne, S. J., and Taylor, A. M. R. (2006). Modified Tests for a Change in Persistence. *Journal of Econometrics*, 134:441–469.

Harvey, D. I. and Newbold, P. (2003). The Non-Normality of some Macroeconomic Forecast Errors. *International Journal of Forecasting*, 19:635–653.

Harville, D. A. (1997). *Matrix Algebra from a Statistician's Perspective*. Springer, New York.

Hasanhodzic, J. and Lo, A. W. (2011). Blacks Leverage Effect is not Due to Leverage. Available at SSRN.

Hassani, H., Heravi, S., Brown, G., and Ayoubkhani, D. (2013a). Forecasting Before, During, and After Recession with Singular Spectrum Analysis. *Journal of Applied Statistics*, 40(10):2290–2302.

Hassani, H., Soofi, A. S., and Zhigljavsky, A. (2013b). Predicting Inflation Dynamics with Singular Spectrum Analysis. *Journal of the Royal Statistical Society, Series A*, 176(3):743–760.

Hassani, H. and Thomakos, D. (2010). A Review on Singular Spectrum Analysis for Economic and Financial Time Series. *Statistics and its Interface*, 3(3):377–397.

Hassani, S. (1999). *Mathematical Physics, A Modern Introduction to Its Foundations*. Springer, New York, 3rd edition.

Hastie, T. and Tibshirani, R. (1993). Varying-Coefficient Models (with discussion). *Journal of the Royal Statistical Society, Series B*, 55(4):757–796.

Hatanaka, M. (1996). *Time-Series-Based Econometrics: Unit Roots and Cointegration*. Oxford University Press, Oxford.

Hautsch, N., Kyj, L. M., and Malec, P. (2015). Do High-Frequency Data Improve High-Dimensional Portfolio Allocations? *Journal of Applied Econometrics*, 30(2):263–290.

Hayashi, F. (2000). *Econometrics*. Princeton University Press, Princeton.

He, C. and Teräsvirta, T. (1999a). Properties of Moments of a Family of GARCH Processes. *Journal of Econometrics*, 92(1):173–192.

He, C. and Teräsvirta, T. (1999b). Statistical Properties of the Asymmetric Power ARCH Model. In Engle, R. F. and White, H., editors, *Cointegration, Causality, and Forecasting. Festschrift in Honour of Clive W. J. Granger*, pages 462–474. Oxford University Press.

Heberle, J. and Sattarhoff, C. (2017). A Fast Algorithm for the Computation of HAC Covariance Matrix Estimators. *Econometrics*, 5(1):1–16. Article 9.

Hedayat, A. and Robson, D. S. (1970). Independent Stepwise Residuals for Testing Homoscedasticity. *Journal of the American Statistical Association*, 65:1573–1581.

Heijmans, R. (1999). When does the Expectation of a Ratio Equal the Ratio of Expectations? *Statistical Papers*, 40:107–115.

Heinen, A. and Valdesogo, A. (2012). Copula-based Volatility Models. In Bauwens, L., Hafner, C. M., and Laurent, S., editors, *Handbook of Volatility Models and their Applications*, chapter 12. John Wiley & Sons, Inc., Hoboken, NJ.

Helstrom, C. W. (1996). Calculating the Distribution of the Serial Correlation Estimator by Saddlepoint Integration. *Econometric Theory*, 12(3):458–480.

Hendry, D. F. (1980). Econometrics-Alchemy or Science? *Economica*, 47(188):387–406.

Hendry, D. F. (1995). *Dynamic Econometrics*. Oxford University Press, Oxford.

Hendry, D. F. (1999). An Econometric Analysis of US Food Expenditure, 1931–1989. In Magnus, J. R. and Morgan, M. S., editors, *Methodology and Tacit Knowledge: Two Experiments in Econometrics*, chapter 17, pages 341–361. John Wiley & Sons, Chichester.

Hendry, D. F. (2009). The Methodology of Empirical Econometric Modeling: Applied Econometrics Through the Looking-Glass. In Mills, T. C. and Patterson, K., editors, *Palgrave Handbook of Econometrics, Volume 2: Applied Econometrics*, pages 3–67. Palgrave Macmillan, London.

Hendry, D. F. and Doornik, J. A. (2014). *Empirical Model Discovery and Theory Evaluation: Automatic Selection Methods in Econometrics*. MIT Press, Cambridge, MA.

Hens, T. and Steude, S.-C. (2009). The leverage effect without leverage. *Finance Research Letters*, 6(2):83–94.

Henshaw, Jr., R. C. (1966). Testing Single-Equation Least Squares Regression Models for Autocorrelated Disturbances. *Econometrica*, 34(3):646–660. Errata, 1968, Vol. 36(3), p. 626.

Heyde, C. C. and Kou, S. G. (2004). On the Controversy over Tailweight of Distributions. *Operations Research Letters*, 32:399–408.

Hildreth, C. and Houck, J. P. (1968). Some Estimators for a Linear Model with Random Coefficients. *Journal of the American Statistical Association*, 63(322):584–595. Errata, 1969, Vol. 64(328), p. 1701.

Hill, J. B. and Renault, E. (2012). Variance Targeting for Heavy Tailed Time Series. http://citeseerx.ist.psu.edu/viewdoc/summary?doi=10.1.1.397.2318.

Hillebrand, E. (2005). Neglecting Parameter Changes in GARCH Models. *Journal of Econometrics*, 129(1–2):121–138.

Hillebrand, E. and Medeiros, M. C. (2016). Nonlinearity, Breaks, and Long-Range Dependence in Time-Series Models. *Journal of Business & Economic Statistics*, 34(1):23–41.

Hirschberg, J. G. and Slottje, D. J. (1999). The Reparameterization of Linear Models Subject to Exact Linear Restrictions. Research Paper 702, Department of Economics, University of Melbourne.

Hisamatsu, H. and Maekawa, K. (1994). The Distribution of the Durbin–Watson Statistic in Integrated and Near-Integrated Models. *Journal of Econometrics*, 61:367–382.

Hochberg, Y. and Tamhane, A. C. (1987). *Multiple Comparison Procedures*. John Wiley & Sons, New York.

Holzmann, H., Munk, A., and Gneiting, T. (2006). Identifiability of Finite Mixtures of Elliptical Distributions. *Scandinavian Journal of Statistics*, 33(4):753–763.

Horn, R. (1994). *Topics in Matrix Analysis*. Cambridge University Press, Cambridge.

Horváth, L., Kokoszka, P., and Zitikis, R. (2006). Sample and Implied Volatility in GARCH Models. *Journal of Financial Econometrics*, 4(4):617–635.

Horváth, L., Kokoszka, P. S., and Teyssiére, G. (2001). Empirical Process of the Squared Residuals of an ARCH Sequence. *Annals of Statistics*, 29(2):445–469.

Hosking, J. R. M. (1981). Fractional Differencing. *Biometrika*, 68(1):165–176.

Hotelling, H. (1933). Analysis of a complex of statistical variables into principal components. *Journal of Educational Psychology*, 24(6):417.

Hsu, J. C. (1996). *Multiple Comparisons: Theory and Methods*. CRC Press, Boca Raton.

Hubbard, J. H. and Hubbard, B. B. (2002). *Vector Calculus, Linear Algebra, and Differential Forms: A Unified Approach*. Prentice Hall, Upper Saddle River, NJ, 2nd edition.

Huber, G. (1982). Gamma Function Derivation of n-Sphere Volumes. *The American Mathematical Monthly*, 89(5):301–302.

Huber, P. J. and Ronchetti, E. M. (2009). *Robust Statistics*. John Wiley & Sons, Hoboken, NJ, 2nd edition.

Huberty, C. J. and Olejnik, S. (2006). *Applied MANOVA and Discriminant Analysis*. John Wiley & Sons, Hoboken, NJ, 2nd edition.

Huffer, F. W. and Park, C. (2007). A Test for Elliptical Symmetry. *Journal of Multivariate Analysis*, 98(2):256–281.

Hull, J. and White, A. (1998). Incorporating Volatility Updating for Value-at-Risk. *Journal of Risk*, 1(1):5–19.

Hult, H. and Lindskog, F. (2002). Multivariate Extremes, Aggregation and Dependence in Elliptical Distributions. *Advances in Applied Probability*, 336(2):587–608.

Hurst, S. (1995). The Characteristic Function of the Student t Distribution. Financial Mathematics Research Report FMRR006-95, Australian National University, Canberra. available online: http://wwwmaths.anu.edu.au/research.reports/srr/95/044/.

Ibragimov, R. and Prokhorov, A. (2017). *Heavy Tails and Copulas: Topics in Dependence Modelling in Economics and Finance*. World Scientific, Singapore.

Ilmanen, A. (2011). *Expected Returns: An Investor's Guide to Harvesting Market Rewards*. John Wiley & Sons, Chichester.

Imhof, J. P. (1961). Computing the Distribution of Quadratic Forms in Normal Variables. *Biometrika*, 48:419–26.

Ingrassia, S. and Rocci, R. (2007). Constrained Monotone EM Algorithms for Finite Mixture of Multivariate Gaussians. *Computational Statistics & Data Analysis*, 51(11):5339–5351.

Ioannidis, J. P. A. (2005). Why Most Published Research Findings Are False. *PLoS Med*, 2(8).

Ioannidis, J. P. A. (2014). Discussion: Why "An Estimate of the Science-Wise False Discovery Rate and Application to the Top Medical Literature" is False. *Biostatistics*, 15:28–36.

Jagannathan, R. and Ma, T. (2003). Risk Reduction in Large Portfolios: Why Imposing the Wrong Constraints Helps. *Journal of Finance*, 58:1651–1684.

Jenkins, G. M. and Alevi, A. S. (1981). Some Aspects of Modelling and Forecasting Multivariate Time Series. *Journal of Time Series Analysis*, 2:1–47.

Jensen, D. R. (1981). Power of Invariant Tests for Linear Hypotheses under Spherical Symmetry. *Scandinavian Journal of Statistics*, 8(3):169–174.

Jensen, M. B. and Lunde, A. (2001). The NIG-S and ARCH Model: A Fat Tailed, Stochastic, and Autoregressive Conditional Heteroscedastic Volatility Model. *Econometrics Journal*, 4:319–342.

Jeon, J. and Taylor, J. W. (2013). Using CAViaR Models with Implied Volatility for Value-at-Risk Estimation. *Journal of Forecasting*, 32(1):62–74.

Joe, H. (2015). *Dependence Modeling with Copulas*. Chapman & Hall/CRC, Boca Raton.

Johansen, S. and Nielsen, B. (2009). An Analysis of the Indicator Saturation Estimator as a Robust Regression Estimator. In Castle, J. L. and Shephard, N., editors, *The Methodology and Practice of*

Econometrics: A Festschrift in Honour of David F. Hendry, chapter 1, pages 1–36. Oxford University Press, Oxford.

Johnson, L. W. (1977). Stochastic Parameter Regression: An Annotated Bibliography. *International Statistical Review*, 45(3):257–272.

Johnson, N. L., Kotz, S., and Balakrishnan, N. (1995). *Continuous Univariate Distributions, Volumes 1 and 2*. John Wiley & Sons, New York, 2nd edition.

Jolliffe, I. T. (2002). *Principal Component Analysis*. Springer, New York, 2nd edition.

Jondeau, E. (2016). Asymmetry in Tail Dependence of Equity Portfolios. *Computational Statistics & Data Analysis*, 100:351–368.

Jondeau, E., Poon, S.-H., and Rockinger, M. (2007). *Financial Modeling Under Non-Gaussian Distributions*. Springer, New York.

Jondeau, E. and Rockinger, M. (2003). Conditional Volatility, Skewness, and Kurtosis: Existence, Persistence, and Comovements. *Journal of Economic Dynamics and Control*, 27:1699–1737.

Jondeau, E. and Rockinger, M. (2009). The Impact of News on Higher Moments. *Journal of Financial Econometrics*, 7(2):77–105.

Jondeau, E. and Rockinger, M. (2012). On the Importance of Time Variability in Higher Moments for Asset Allocation. *Journal of Financial Econometrics*, 10(1):84–123.

Jones, C. R. and Marriott, M. J. (1999). A Bayesian Analysis of Stochastic Unit Root Models. *Bayesian Statistics*, 6:785–794.

Jones, M. C. (2002). A Dependent Bivariate t Distribution with Marginals on Different Degrees of Freedom. *Statistics and Probability Letters*, 56(2):163–170.

Jones, R. H. (1980). Maximum Likelihood Fitting of ARMA Models to Time Series With Missing Observations. *Technometrics*, 22(3):389–395.

Jorion, P. (1986). Bayes–Stein Estimation for Portfolio Analysis. *Journal of Financial and Quantitative Analysis*, 21:279–292.

J.P. Morgan/Reuters (1996). RiskMetrics™ Technical Document, 4th edition. Morgan Guaranty Trust Company and Reuters Ltd, New York.

Jude, E. C. (2010). Financial Development and Growth: A Panel Smooth Regression Approach. *Journal of Economic Development*, 35(1):15–33.

Judge, G. G., Griffiths, W. E., Hill, R. C., Lütkepohl, H., and Lee, T.-C. (1985). *The Theory and Practice of Econometrics*. John Wiley & Sons, New York, 2nd edition.

Juselius, K. (2018). Editorial: Recent Developments in Cointegration. *Econometrics*, 6(1):1–5.

Kadiyala, K. R. (1970). Testing for the Independence of Regression Disturbances. *Econometrica*, 38:97–117.

Kallenberg, O. (2002). *Foundations of Modern Probability*. Springer, 2nd edition.

Kalymon, B. A. (1971). Estimation Risk in the Portfolio Selection Model. *Journal of Financial and Quantitative Analysis*, 6(1):559–582.

Kamdem, J. S. (2005). Value-at-Risk and Expected Shortfall for Linear Portfolios with Elliptically Distributed Risk Factors. *International Journal of Theoretical and Applied Finance*, 8:537–551.

Kan, R. and Wang, X. (2010). On the Distribution of the Sample Autocorrelation Coefficients. *Journal of Econometrics*, 154(2):101–121.

Kan, R. and Zhou, G. (2007). Optimal Portfolio Choice with Parameter Uncertainty. *Journal of Financial and Quantitative Analysis*, 42(3):621–656.

Kanto, A. J. (1988). Covariances Between Estimated Autocorrelations of an ARMA Process. *Economics Letters*, 26:253–258.

Kanzler, L. (1998). Very Fast and Correctly Sized Estimation of the BDS Statistic. Working Paper. Department of Economics, Oxford University.

Kapetanios, G. (2003). A Note on an Iterative Least-Squares Estimation Method for ARMA and VARMA Models. *Economic Letters*, 79(3):305–312.

Kapetanios, G. and Shin, Y. (2006). Unit Root Tests in Three-Regime SETAR Models. *Econometrics Journal*, 9(2):252–278.

Karanasos, M. (1998). A New Method for Obtaining the Autocovariance of an ARMA Model: An Exact Form Solution. *Econometric Theory*, 14:622–640. Acknowledgment of Priority and Correction Note: 2000, 16:280–282.

Karanasos, M. and Kim, J. (2006). A Re-Examination of the Asymmetric Power ARCH Model. *Journal of Empirical Finance*, 13:113–128.

Kariya, T. (1977). A Robustness Property of the Tests for Serial Correlation. *Annals of Statistics*, 5:1212–1220.

Kariya, T. (1981a). A Robustness Property of Hotelling's T^2 Test. *Annals of Statistics*, 9(1):211–214.

Kariya, T. (1981b). Robustness of Multivariate Tests. *Annals of Statistics*, 9(6):1267–1275.

Kariya, T. and Eaton, M. L. (1977). Robust Tests for Spherical Symmetry. *Annals of Statistics*, 5:206–215.

Karlsen, H. A. (1990). Existence of Moments in a Stationary Difference Equation. *Advances in Applied Probability*, 22:129–146.

Kavalieris, L., Hannan, E. J., and Salau, M. (2003). Generalized Least Squares Estimation Of ARMA Models. *Journal of Time Series Analysis*, 24(2):165–172.

Kelker, D. (1970). Distribution Theory of Spherical Distributions and a Location-Scale Parameter Generalization. *Sankhyā, Series A*, 32(4):419–430.

Kennedy, P. (2017). To Be a Genius, Think Like a 94-Year-Old. *The New York Times: Sunday Review*, April 7.

Keuzenkamp, H. A. and McAleer, M. (1997). The Complexity of Simplicity. In *11th Biennial Conference on Modelling and Simulation*, pages 553–561.

Khuri, A. I. (2010). *Linear Model Methodology*. Chapman & Hall/CRC, Boca Raton.

Khuri, A. I., Mathew, T., and Sinha, B. K. (1998). *Statistical Tests for Mixed Linear Models*. John Wiley & Sons, New York.

Khuri, A. I. and Searle, S. R. (2017). *Matrix Algebra Useful for Statistics*. John Wiley & Sons, Hoboken, NJ, 2nd edition.

Kim, D. and Perron, P. (2009). Unit Root Tests Allowing for a Break in the Trend Function at an Unknown Time Under Both the Null and Alternative Hypotheses. *Journal of Econometrics*, 148(1):1–13.

Kim, J. H. and Choi, I. (2017). Unit Roots in Economic and Financial Time Series: A Re-Evaluation at the Decision-Based Significance Levels. *Econometrics*, 5(3):1–23. Article 41.

Kim, T.-H., Leybourne, S., and Newbold, P. (2002). Unit Root Tests with a Break in Innovation Variance. *Journal of Econometrics*, 109:365–387.

King, M. L. (1980). Robust Tests for Spherical Symmetry and their Application to Least Squares Regression. *Annals of Statistics*, 8:1265–1271.

King, M. L. (1981). The Alternative Durbin–Watson Test: An Assessment of Durbin and Watson's Choice of Test Statistic. *Journal of Econometrics*, 17:51–66.

King, M. L. (1985a). A Point Optimal Test for Autoregressive Disturbances. *Journal of Econometrics*, 27:21–37.

King, M. L. (1985b). A Point Optimal Test for Moving Average Regression Disturbances. *Econometric Theory*, 1:211–222.

King, M. L. (1987a). An Alternative Test for Regression Coefficient Stability. *The Review of Economics and Statistics*, 69(2):379–381.

King, M. L. (1987b). Towards a Theory of Point Optimal Testing. *Econometric Reviews*, 6:169–218.

King, M. L. and Hillier, G. H. (1985). Locally Best Invariant Tests of the Error Covariance Matrix of the Linear Regression Model. *Journal of the Royal Statistical Society, Series B*, 47(1):98–102.

Kirchler, M. and Huber, J. (2007). Fat Tails and Volatility Clustering in Experimental Asset Markets. *Journal of Economic Dynamics & Control*, 31:1844–1874.

Kirkwood, G., Rampes, H., Tuffrey, V., Richardson, J., Pilkington, K., and Ramaratnam, S. (2005). Yoga for Anxiety: A Systemic Review of the Research Evidence. *British Journal of Sports Medicine*, 39(12):884–891.

Kitzrow, M. A. (2003). The Mental Health Needs of Today's College Students: Challenges and Recommendations. *National Association of Student Personnel Administrators (NASPA)*, 41(1):167–181.

Klaster, M. A. and Knot, K. H. W. (2002). Toward an Econometric Target Zone Model with Endogenous Devaluation Risk for a Small Open Economy. *Economic Modelling*, 19:509–529.

Klein, R. W. and Bawa, V. S. (1976). The Effect of Estimation Risk on Optimal Portfolio Choice. *Journal of Financial Economics*, 3(3):215–231.

Klotz, J. (1969). A Simple Proof of Scheffé's Multiple Comparison Theorem for Contrasts in the One-Way Layout. *American Statistician*, 23:44–45.

Koehn, U. and Thomas, D. L. (1975). On Statistics Independent of a Sufficient Statistic: Basu's Lemma. *The American Statistician*, 29(1):40–42.

Koenker, R. (2005). *Quantile Regression*. Cambridge University Press, Cambridge.

Koenker, R. and Hallock, K. F. (2001). Quantile Regression. *Journal of Economic Perspectives*, 15(4):143–156.

Koerts, J. (1967). Some Further Notes on Disturbance Estimates in Regression Analysis. *Journal of the American Statistical Association*, 62:169–183.

Koerts, J. and Abrahamse, A. P. J. (1969). *On the Theory and Application of the General Linear Model*. University Press, Rotterdam.

Kolm, P. N., Focardi, S. M., and Fabozzi, F. J. (2008). Incorporating Trading Strategies in the Black–Litterman Framework. In Fabozzi, F. J., editor, *Handbook of Finance, Volume II: Investment Management and Financial Management*, chapter 36, pages 359–367. John Wiley & Sons, New York.

Kolm, P. N., Tütüncü, R., and Fabozzi, F. J. (2014). 60 Years of Portfolio Optimization: Practical Challenges and Current Trends. *European Journal of Operational Research*, 234(2):356–371.

Konishi, S. and Kitagawa, G. (2008). *Information Criteria and Statistical Modeling*. Springer, New York.

Konstantinides, K. (1991). Threshold Bounds in SVD and a New Iterative Algorithm for Order Selection in AR Models. *IEEE Transactions on Signal Processing*, 39(5):1218–1221.

Konstantinides, K. and Yao, K. (1988). Statistical Analysis of Effective Singular Values in Matrix Rank Determination. *IEEE Transactions on Acoustics, Speech, and Signal Processing*, 36(5):757–763.

Koopman, S. J., Jungbacker, B., and Hol, E. (2005). Forecasting Daily Variability of the S&P 100 Stock Index Using Historical, Realized and Implied Volatility Measurements. *Journal of Empirical Finance*, 12:445–475.

Koreisha, S. G. and Pukkila, T. (1990). A Generalized Least Squares Approach for Estimation of Autoregressive Moving-Average Models. *Journal of Time Series Analysis*, 11:139–151.

Koreisha, S. G. and Pukkila, T. (1995). A Comparison Between Different Order-Determination Criteria for Identification of ARIMA Models. *Journal of Business & Economic Statistics*, 13(1):127–131.

Koreisha, S. G. and Yoshimoto, G. (1991). A Comparison Among Identification Procedures for Autoregressive Moving Average Models. *International Statistical Review*, 59:37–57.

Kotz, S., Kozubowski, T., and Podgorski, K. (2001). *The Laplace Distribution and Generalizations: A Revisit with Application to Communication, Economics, Engineering and Finance*. Birkhäuser, Boston.

Kotz, S., Kozubowski, T. J., and Podgórski, K. (2000). An Asymmetric Multivariate Laplace Distribution. Technical Report 367, Department of Statistics and Applied Probability, University of California at Santa Barbara.

Kotz, S. and Nadarajah, S. (2004). *Multivariate t Distributions and Their Applications*. Cambridge University Press, Cambridge.

Kozubowski, T. J. and Podgórski, K. (2001). Asymmetric Laplace Laws and Modeling Financial Data. *Mathematical and Computer Modelling*, 34:1003–1021.

Krämer, W. (1985). The Power of the Durbin–Watson Test for Regressions Without an Intercept. *Journal of Econometrics*, 28:363–370.

Krämer, W., Bartels, R., and Fiebig, D. G. (1996). Another Twist on the Equality of OLS and GLS. *Statistical Papers*, 37(3):277–281.

Krämer, W. and Zeisel, H. (1990). Finite Sample Power of Linear Regression Autocorrelation Tests. *Journal of Econometrics*, 43:363–372.

Krause, J. and Paolella, M. S. (2014). A Fast, Accurate Method for Value at Risk and Expected Shortfall. *Econometrics*, 2:98–122.

Kroner, K. F. and Ng, V. K. (1998). Modeling Asymmetric Comovements of Asset Returns. *The Review of Financial Studies*, 11(4):817–844.

Krzanowski, W. J. and Marriott, F. H. C. (1994). *Multivariate Analysis, Part 1: Distributions, Ordination and Inference*. Edward Arnold, London.

Kshirsagar, A. M. (1961). Some Extensions of the Multivariate Generalization *t* distribution and the Multivariate Generalization of the Distribution of the Regression Coefficient. *Proceedings of the Cambridge Philosophical Society*, 57:80–85.

Kuan, C.-M., Yeh, J.-H., and Hsu, Y.-C. (2009). Assessing Value at Risk with CARE, the Conditional Autoregressive Expectile Models. *Journal of Econometrics*, 150(2):261–270.

Kuester, K., Mittnik, S., and Paolella, M. S. (2006). Value-at-Risk Prediction: A Comparison of Alternative Strategies. *Journal of Financial Econometrics*, 4:53–89. Reproduced in: *The Foundations of Credit Risk Analysis*, Willi Semmler and Lucas Bernard (Eds.), chapter 14, Edward Elgar Publishing, 2007.

Kurozumi, E. (2002). Testing for Stationarity with a Break. *Journal of Econometrics*, 108:63–99.

Kwan, W., Li, W. K., and Li, G. (2012). On the Estimation and Diagnostic Checking of the ARFIMA-HYGARCH Model. *Computational Statistics & Data Analysis*, 56(11):3632–3644.

Kwiatkowski, D., Phillips, P. C. B., Schmidt, P., and Shin, Y. (1992). Testing the Null Hypothesis of Stationarity Against the Alternative of a Unit Root: How sure are we that economic time series have a unit root? *Journal of Econometrics*, 54(1–3):159–178.

LaMotte, L. R. and McWhorter, Jr., A. (1978). An Exact Test for the Presence of Random Walk Coefficients in a Linear Regression Model. *Journal of the American Statistical Association*, 73:816–820.

Lamoureux, C. G. and Lastrapes, W. D. (1990). Persistence in Variance, Structural Change, and the GARCH Model. *Journal of Business & Economic Statistics*, 8(2):225–234.

Landsman, Z. and Nešlehová, J. (2008). Stein's Lemma for Elliptical Random Vectors. *Journal of Multivariate Analysis*, 99(5):912–927.

Landsman, Z. M. and Valdez, E. A. (2003). Tail Conditional Expectations for Elliptical Distributions. *North American Actuarial Journal*, 7(4):55–71.

Lang, S. (1997). *Undergraduate Analysis*. Springer, New York, 2nd edition.

Lange, K. L., Little, R. J., and Taylor, J. M. G. (1989). Robust Statistical Modeling Using the t Distribution. *Journal of the American Statistical Association*, 84(408):881–896.

Lange, T. and Rahbek, A. (2009). An Introduction to Regime Switching Time Series Models. In Andersen, T. G., Davis, R. A., Kreiß, J.-P., and Mikosch, T., editors, *Handbook of Financial Time Series*, pages 871–887. Springer, Berlin.

Lanne, M. and Saikkonen, P. (2007). A Multivariate Generalized Orthogonal Factor GARCH Model. *Journal of Business & Economic Statistics*, 25(1):61–75.

Lansangan, J. R. G. and Barrios, E. B. (2017). Simultaneous Dimension Reduction and Variable Selection in Modeling High Dimensional Data. *Computational Statistics & Data Analysis*, 112:242–256.

Larsson, R. (1995). The Asymptotic Distribution of Some Test Statistics in Near-Integrated AR Processes. *Econometric Theory*, 11(2):306–330.

Larsson, R. (1998). Distribution Approximation of Unit Root Tests in Autoregressive Models. *Econometrics Journal*, 1(2):10–26.

Lay, D. C., Lay, S. R., and McDonald, J. J. (2015). *Linear Algebra and its Applications*. Pearson, 5th edition.

Leamer, E. E. (1978). *Specification Searches: Ad Hoc Inference with Nonexperimental Data*. John Wiley & Sons, New York.

Leamer, E. E. (1983). Let's Take the Con Out of Econometrics. *American Economic Review*, 73(1):31–43.

Leamer, E. E. (1997). Revisiting Tobin's 1950 Study of Food Expenditure. *Journal of Applied Econometrics*, 12(5):533–553.

Ledoit, O. and Wolf, M. (2004). Honey, I Shrunk the Sample Covariance Matrix. *Journal of Portfolio Management*, 30(4):110–119.

Ledoit, O. and Wolf, M. (2012). Nonlinear Shrinkage Estimation of Large-Dimensional Covariance Matrices. *Annals of Statistics*, 40:1024–1060.

Lee, J. and Khuri, A. I. (2001). Modeling the Probability of a Negative ANOVA Estimate of a Variance Component. *Calcutta Statistical Association Bulletin*, 51:31–45.

Lee, L.-F. and Griffiths, W. E. (1979). The Prior Likelihood and Best Linear Unbiased Prediction in Stochastic Coefficient Linear Models. Discussion Paper No. 79–107, January 1979.

Lee, T. and Lee, S. (2009). Normal Mixture Quasi-Maximum Likelihood Estimator for GARCH Models. *Scandinavian Journal of Statistics*, 36:157–170.

Lee, Y.-H., Hu, H.-N., and Chiou, J.-S. (2010). Jump Dynamics with Structural Breaks for Crude Oil Prices. *Energy Economics*, 32(2):343–350.

Leek, J. T. and Jager, L. R. (2017). Is Most Published Research Really False? *Annual Review of Statistics and Its Application*, 4(1):109–122.

Lehmann, E. L. (1986). *Testing Statistical Hypotheses*. John Wiley & Sons, New York, 2nd edition.

Lehmann, E. L. and Casella, G. (1998). *Theory of Point Estimation*. Springer, New York, 2nd edition.

Levina, E., Rothman, A., and Zhu, J. (2008). Sparse Estimation of Large Covariance Matrices via a Nested Lasso Penalty. *Annals of Applied Statistics*, 2(1):245–263.

Leybourne, S. J. and McCabe, B. P. M. (1994). A Consistent Test for a Unit Root. *Journal of Business and Economic Statistics*, 12:157–166.

Leybourne, S. J. and McCabe, B. P. M. (1999). Modified Stationarity Tests with Data-Dependent Model-Selection Rules. *Journal of Business and Economic Statistics*, 17:264–270.

Leybourne, S. J., McCabe, B. P. M., and Mills, T. C. (1996a). Randomized Unit Root Processes for Modelling and Forecasting Financial Time Series: Theory and Applications. *Journal of Forecasting*, 15(3):253–270.

Leybourne, S. J., McCabe, B. P. M., and Tremayne, A. R. (1996b). Can Economic Time Series be Differenced to Stationarity? *Journal of Business & Economic Statistics*, 14(4):435–446.

Li, C. W. and Li, W. K. (1996). On a Double-Threshold Autoregressive Heteroscedastic Time Series Model. *Journal of Applied Econometrics*, 11:253–274.

Lidong, E., Hannig, J., and Iyer, H. K. (2008). Fiducial Intervals for Variance Components in an Unbalanced Two-Component Normal Mixed Linear Model. *Journal of the American Statistical Association*, 103(482):854–865.

Lieberman, O. (1994a). Saddlepoint Approximation for the Distribution of a Ratio of Quadratic Forms in Normal Variables. *Journal of the American Statistical Association*, 89(427):924–928.

Lieberman, O. (1994b). Saddlepoint Approximation for the Least Squares Estimator in First-Order Autoregression. *Biometrika*, 81(4):807–11.

Lin, C.-F. J. and Teräsvirta, T. (1999). Testing Parameter Constancy in Linear Models Against Stochastic Stationary Parameters. *Journal of Econometrics*, 90(2):193–213.

Ling, S. (1999). On the Probabilistic Properties of a Double Threshold ARMA Conditional Heteroskedastic Model. *Journal of Applied Probability*, 36:688–705.

Ling, S. and McAleer, M. (2002). Necessary and Sufficient Moment Conditions for the GARCH(r, s) and Asymmetric Power GARCH(r, s) Models. *Econometric Theory*, 18(3):722–729.

Ling, S. and Tong, H. (2005). Testing for a Linear MA Model Against Threshold MA Models. *Annals of Statistics*, 33:2529–2552.

Ling, S., Tong, H., and Li, D. (2007). Ergodicity and Invertibility of Threshold Moving-Average Models. *Bernoulli*, 13:161–168.

Liu, L.-M. and Hanssens, D. M. (1981). A Bayesian Approach to Time-Varying Cross-Sectional Regression Models. *Journal of Econometrics*, 15(3):341–356.

Liu, S. and Brorsen, B. W. (1995). Maximum Likelihood Estimation of a GARCH-Stable Model. *Journal of Applied Econometrics*, 10:273–285.

Ljung, G. M. and Box, G. E. P. (1978). On a Measure of Lack of Fit in Time Series Models. *Biometrika*, 65(2):297–303.

Lo, A. W. (1991). Long-Term Memory in Stock Market Prices. *Econometrica*, 59(5):1279–1313.

Lugannani, R. and Rice, S. O. (1980). Saddlepoint Approximations for the Distribution of Sums of Independent Random Variables. *Advances in Applied Probability*, 12:475–490.

Lütkepohl, H. (1993). *Introduction to Multiple Time Series Analyis*. Springer, Berlin, 2nd edition.

Lütkepohl, H. (2005). *New Introduction to Multiple Time Series Analysis*. Springer, Berlin.

Lütkepohl, H. and Krätzig, M. (2004). *Applied Time Series Econometrics*. Cambridge University Press, Cambridge.

Lux, T. (2008). The Markov-Switching Multifractal Model of Asset Returns: GMM Estimation and Linear Forecasting of Volatility. *Journal of Business & Economic Statistics*, 26(2):194–210.

Lux, T. and Kaizoji, T. (2007). Forecasting Volatility and Volume in the Tokyo Stock Market: Long Memory, Fractality and Regime Switching. *Journal of Economic Dynamics and Control*, 31(6):1808–1843.

Lux, T. and Segnon, M. (2018). Multifractal Models in Finance: Their Origin, Properties, and Applications. In Chen, S.-H., Kaboudan, M., and Du, Y.-R., editors, *The Oxford Handbook of Computational Economics and Finance*, chapter 6. Oxford University Press.

Lux, T., Segnon, M., and Gupta, R. (2016). Forecasting Crude Oil Price Volatility and Value-at-Risk: Evidence from Historical and Recent Data. *Energy Economics*, 56:117–133.

Ma, J., Nelson, C. R., and Startz, R. (2006). Spurious Inference in the GARCH(1,1) Model When It Is Weakly Identified. *Studies in Nonlinear Dynamics and Econometrics*, 11(1). Article 1.

MacKinnon, J. G. and Smith, Jr., A. A. (1998). Approximate Bias Correction in Econometrics. *Journal of Econometrics*, 85:205–30.

MacKinnon, J. G. and White, H. (1985). Some Heteroskedasticity-Consistent Covariance Matrix Estimators with Improved Finite Sample Properties. *Journal of Econometrics*, 29:305–325.

Madan, D. B. and Seneta, E. (1990). The Variance Gamma (V.G.) Model for Share Market Returns. *Journal of Business*, 63:511–524.

Maddala, G. and Kim, I.-M. (1998). *Unit Roots, Cointegration, and Structural Change*. Cambridge University Press, Cambridge.

Magnus, J. R. (1986). The Exact Moments of a Ratio of Quadratic Forms in Normal Variables. *Annales d'Economie et de Statistique*, 4:95–109.

Magnus, J. R. (2017). *Introduction to the Theory of Econometrics*. VU University Press, Amsterdam.

Magnus, J. R. and Neudecker, H. (2007). *Matrix Differential Calculus with Applications in Statistics and Econometrics*. John Wiley & Sons, Chichester, 3rd edition.

Magnus, J. R. and Sinha, A. K. (2005). On Theil's Errors. *Econometrics Journal*, 8(1):39–54.

Mahalanobis, P. C. (1964). Professor Ronald Aylmer Fisher. *Biometrics*, 20:238–251.

Makridakis, S. and Hibon, M. (2000). The M3 Competition: Results, Conclusions and Implications. *International Journal of Forecasting*, 17:567–570.

Mancini, L. and Trojani, F. (2011). Robust Value at Risk Prediction. *Journal of Financial Econometrics*, 9(2):281–313.

Mandelbrot, B. (1963). The Variation of Certain Speculative Prices. *Journal of Business*, 36(4):394–419.

Maneesoonthorn, W., Martin, G. M., Forbes, C. S., and Grose, S. (2012). Probabilistic Forecasts of Volatility and its Risk Premia. *Journal of Econometrics*, 171(2):217–236.

Manganelli, S. (2004). Asset Allocation by Variance Sensitivity. *Journal of Financial Econometrics*, 2(3):370–389.

Mann, H. B. and Wold, A. (1943). On the Statistical Treatment of Linear Stochastic Difference Equations. *Econometrica*, 11:173–220.

Manzotti, A., Pérez, F. J., and Quiroz, A. J. (2002). A Statistic for Testing the Null Hypothesis of Elliptical Symmetry. *Journal of Multivariate Analysis*, 81(2):274–285.

Markowitz, H. (1952). Portfolio Selection. *Journal of Finance*, 7(1):77–91.

Marsh, P. W. N. (1998). Saddlepoint Approximations and Non-Central Quadratic Forms. *Econometric Theory*, 14:539–559.

Martens, M. (2001). Forecasting Daily Exchange Rate Volatility using Intraday Returns. *Journal of International Money and Finance*, 20:1–23.

Martin, R. D., Rachev, S., and Siboulet, F. (2003). Phi–alpha Optimal Portfolios and Extreme Risk Management. *Wilmott Magazine of Finance*, 6:70–83.

Martins-Filho, C., Yao, F., and Torero, M. (2016). Nonparametric Estimation of Conditional Value-at-Risk and Expected Shortfall based on Extreme Value Theory. *Econometric Theory*, 34(1):1–45.

Mathai, A. M. and Provost, S. B. (1992). *Quadratic Forms in Random Variables: Theory and Applications*. Marcel Dekker, New York.

Matilla-García, M., Marín, M. R., Dore, M., and Ojeda, R. (2014). Nonparametric Correlation Integral-Based Tests for Linear and Nonlinear Stochastic Processes. *Decisions in Economics and Finance*, 37(1):181–193.

McAleer, M., Chan, F., Hoti, S., and Lieberman, O. (2008). Generalized Autoregressive Conditional Correlation. *Econometric Theory*, 24(6):1554–1583.

McCabe, B. P. M. and Leybourne, S. J. (2000). A General Method of Testing for Random Parameter Variation in Statistical Models. In Heijmans, R. D. H., Pollock, D. S. G., and Satorra, A., editors, *Innovations in Multivariate Statistical Analysis: A Festschrift for Heinz Neudecker*, pages 75–85. Kluwer, Amsterdam.

McCleary, R. M., editor (2011). *The Oxford Handbook of the Economics of Religion*. Oxford University Press, Oxford.

McCloskey, D. N. (2000). *How to be Human: Though an Economist*. University of Michigan Press, Ann Arbor, MI.

McCulloch, J. H. (1985a). Interest-Risk Sensitive Deposit Insurance Premia: Stable ACH Estimates. *Journal of Banking and Finance*, 9:137–156.

McCulloch, J. H. (1985b). Miscellanea: On Heteroskedasticity. *Econometrica*, 53(2):483.

McDonald, J. B. (1997). Probability Distributions for Financial Models. In Maddala, G. S. and Rao, C. R., editors, *Handbook of Statistics*, volume 14. Elsevier Science.

McDonald, J. B. and Newey, W. K. (1988). Partially Adaptive Estimation of Regression Models Via the Generalized t Distribution. *Econometric Theory*, 4:428–457.

McElroy, F. W. (1967). A Necessary and Sufficient Condition that Ordinary Least-Squares Estimators be Best Linear Unbiased. *Journal of the American Statistical Association*, 62(320):1302–1304.

McLachlan, G. J. and Krishnan, T. (2008). *The EM Algorithm and Extensions*. John Wiley & Sons, Hoboken, NJ, 2nd edition.

McLachlan, G. J. and Peel, D. (2000). *Finite Mixture Models*. John Wiley & Sons, New York.

McLeod, I. (1975). Derivation of the Theoretical Autocovariance Function of Autoregressive-Moving Average Time Series. *Applied Statistics*, 24(2):255–256. Correction: 1977, 26:194.

McNeil, A. J. and Frey, R. (2000). Estimation of Tail-Related Risk Measures for Heteroscedastic Financial Time Series: An Extreme Value Approach. *Journal of Empirical Finance*, 7(3–4):271–300.

McNeil, A. J., Frey, R., and Embrechts, P. (2005). *Quantitative Risk Management: Concepts, Techniques, and Tools*. Princeton University Press, Princeton.

McNeil, A. J., Frey, R., and Embrechts, P. (2015). *Quantitative Risk Management: Concepts, Techniques, and Tools*. Princeton University Press, Princeton, revised edition.

McQuarrie, A. D. R. and Tsai, C.-L. (1998). *Regression and Time Series Model Selection*. World Scientific, River Edge, NJ.

McShane, B. B. and Gal, D. (2016). Blinding Us to the Obvious? The Effect of Statistical Training on the Evaluation of Evidence. *Management Science*, 62(6):1707–1718.

Mecklin, C. J. and Mundfrom, D. J. (2004). An Appraisal and Bibliography of Tests for Multivariate Normality. *International Statistical Review*, 72:123–138.

Medeiros, M. C. and Veiga, A. (2009). Modeling Multiple Regimes in Financial Volatility with a Flexible Coefficient GARCH(1,1) Model. *Econometric Theory*, 25:117–161.

Mencken, H. L. (1920). *Prejudices: Second Series, Volume 2*. Alfred A. Knopf, New York.

Meucci, A. (2006). Beyond Black–Litterman: Views on Non-Normal Markets. *Risk*, 19:87–92.

Mikosch, T. and Straumann, D. (2006). Stable Limits of Martingale Transforms With Application to the Estimation of GARCH Parameters. *Annals of Statistics*, 31(1):493–522.

Miller, R. (2008). *Meditions on Violence: A Comparsion of Martial Arts Training and Real World Violence*. YMAA Publication Center, Boston.

Miller, R. G. (1981). *Simultaneous Statistical Inference*. Springer, New York.

Miller, R. G. (1985). Multiple Comparisons. In Kotz, S. and Johnson, N. L., editors, *Encyclopedia of Statistical Sciences, Volume 5*. John Wiley & Sons, New York.

Miller Jr., R. G. (1997). *Beyond ANOVA: Basics of Applied Statistics*. Chapman & Hall, Boca Raton, USA.

Milliken, G. A. and Johnson, D. E. (2001). *Analysis of Messy Data Volume III: Analysis of Covariance*. Chapman & Hall/CRC, Boca Raton.

Milliken, G. A. and Johnson, D. E. (2009). *Analysis of Messy Data Volume I: Designed Experiments*. Chapman & Hall/CRC, Boca Raton, 2nd edition.

Minami, M. (2003). A Multivariate Extension of Inverse Gaussian Distribution Derived from Inverse Relationship. *Communications in Statistics—Theory and Methods*, 32:2285–2304.

Mittnik, S. (1988). Derivation of the Theoretical Autocovariance and Autocorrelation Function of Autogressive Moving Average Processes. *Communications in Statistics—Theory and Methods*, 17:3825–3831.

Mittnik, S. and Paolella, M. S. (2000). Conditional Density and Value-at-Risk Prediction of Asian Currency Exchange Rates. *Journal of Forecasting*, 19(4):313–333.

Mittnik, S. and Paolella, M. S. (2003). Prediction of Financial Downside Risk with Heavy Tailed Conditional Distributions. In Rachev, S. T., editor, *Handbook of Heavy Tailed Distributions in Finance*. Elsevier Science, Amsterdam.

Mittnik, S., Paolella, M. S., and Rachev, S. T. (2000). Diagnosing and Treating the Fat Tails in Financial Returns Data. *Journal of Empirical Finance*, 7:389–416.

Mittnik, S., Paolella, M. S., and Rachev, S. T. (2002). Stationarity of Stable Power-GARCH Processes. *Journal of Econometrics*, 106:97–107.

Montgomery, D. C. (2000). *Introduction to Statistical Quality Control*. John Wiley & Sons, New York, 4th edition.

Montgomery, D. C., Peck, E. A., and Vining, G. G. (2012). *Introduction to Linear Regression Analysis*. John Wiley & Sons, Hoboken, NJ, 5th edition.

Moosa, I. A. (2017). *Econometrics as a Con Art: Exposing the Limitations and Abuses of Econometrics*. Edward Elgar Publishing Limited, Cheltenham.

Morana, C. (2015). Semiparametric Estimation of Multivariate GARCH Models. *Open Journal of Statistics*, 5:852–858.

Morana, C. (2017). Macroeconomic and Financial Effects of Oil Price Shocks: Evidence for the Euro Area. *Economic Modelling*, 64:82–96.

Morana, C. and Sbrana, G. (2017). Temperature Anomalies, Radiative Forcing and ENSO. DEMS Working Paper no. 361.

Morimune, K. (2007). Volatility Models. *The Japanese Economic Review*, 58(1):1–23.

Morin-Wahhab, D. (1985). Moments of a Ratio of Two Quadratic Forms. *Communications in Statistics—Theory and Methods*, 14(2):499–508.

Morrison, G. W. and Pike, D. H. (1977). Kalman Filtering Applied to Statistical Forecasting. *Management Science*, 23(7):768–774.

Mosteller, F. and Tukey, J. W. (1977). *Data Analysis and Regression: A Second Course in Statistics*. Addison-Wesley, Reading, MA.

Muirhead, R. J. (2005). *Aspects of Multivariate Statistical Theory*. John Wiley & Sons, New York.

Müller, U. K. and Elliott, G. (2003). Tests for Unit Roots and the Initial Condition. *Econometrica*, 71:1269–1286.

Munkres, J. R. (1991). *Analysis on Manifolds*. Perseus Books, Cambridge, MA.

Murphy, K. P. (2012). *Machine Learning: a Probabilistic Perspective*. MIT Press, Cambridge, MA.

Nabeya, S. and Tanaka, K. (1988). Asymptotic Theory of a Test for the Constancy of Regression Coefficients Against the Random Walk Alternative. *Annals of Statistics*, 16(1):218–235.

Nadarajah, S. and Dey, D. K. (2005). Multitude of Multivariate t-Distributions. *Statistics*, 39(2):149–181.

Näf, J., Paolella, M. S., and Polak, P. (2018a). Getting out of the COMFORT Zone: The MEXI Distribution for Asset Returns. Mimeo.

Näf, J., Paolella, M. S., and Polak, P. (2018b). Heterogeneous Tail Generalized COMFORT Modeling via Cholesky Decomposition. Mimeo.

Nakamura, A. and Nakamura, M. (1978). On the Impact of the Tests for Serial Correlation Upon the Test of Significance for the Regression Coefficient. *Journal of Econometrics*, 7:199–210.

Narayan, P. K. (2006). The Behaviour of US Stock Prices: Evidence from a Threshold Autoregressive Model. *Mathematics and Computers in Simulation*, 71(2):103–108.

Neely, C. J. (1999). Target Zones and Conditional Volatility: The Role of Realignments. *Journal of Empirical Finance*, 6:177–192.

Neely, C. J. and Weller, P. A. (2002). In *Predicting Exchange Rate Volatility: Genetic Programming Versus GARCH and RiskMetrics™*, pages 43–54. The Federal Reserve Bank of St. Louis.

Nelder, J. A. (1968). Regression, Model-Building and Invariance (with discussion). *Journal of the Royal Statistical Society, Series A*, 131(3):309–329.

Nelsen, R. B. (2006). *An Introduction to Copulas*. Springer, New York, 2nd edition.

Nelson, D. (1991). Conditional Heteroskedasticity in Asset Returns: A New Approach. *Econometrica*, 59:347–370.

Nelson, D. B. (1990). Stationarity and Persistence in the GARCH(1,1) Model. *Econometric Theory*, 6:318–334.

Nelson, D. B. and Cao, C. Q. (1992). Inequality Constraints in the Univariate GARCH Model. *Journal of Business and Economic Statistics*, 10(2):229–235.

Nelson, D. B. and Foster, D. B. (1994). Asymptotic Filtering Theory For Univariate ARCH Models. *Econometrica*, 62:1–41.

Neudecker, H. (1969). Some Theorems on Matrix Differentiation with Special Reference to Kronecker Matrix Products. *Journal of the American Statistical Association*, 65:953–963.

Newbold, P., Agiakloglou, C., and Miller, J. (1993). Long-Term Inference Based on Short-Term Forecasting Models. In Rao, T. S., editor, *Developments in Time Series Analysis. In honour of Maurice B. Priestley*, chapter 2.Chapman & Hall, London.

Newey, W. K. and West, K. D. (1987). A Simple, Positive Semi-Definite, Heteroskedasticity and Autocorrelation Consistent Covariance Matrix. *Econometrica*, 55:703–708.

Newton, H. J. (2002). A Conversation with Emanuel Parzen. *Statistical Science*, 17:357–378.

Ng, S. and Perron, P. (1995). Unit Root Tests in ARMA Models with Data-dependent Methods for the Selection of the Truncation Lag. *Journal of the American Statistical Association*, 90:268–281.

Ng, S. and Perron, P. (2001). Lag Length Selection and the Construction of Unit Root Tests with good Size and Power. *Econometrica*, 69:1519–1554.

Nguyen, T. M. (2014). N-Dimensional Quasipolar Coordinates—Theory and Application. Masters thesis, University of Nevada, Las Vegas.

Nicholls, D. F. and Quinn, B. G. (1982). *Random Coefficient Autoregressive Models: An Introduction*. Springer, New York.

Nijman, T. and Sentana, E. (1996). Marginalization and Contemporaneous Aggregation in Multivariate GARCH Processes. *Journal of Econometrics*, 71:71–87.

Nikoloulopoulos, A. K., Joe, H., and Li, H. (2009). Extreme Value Properties of Multivariate *t* Copulas. *Extremes*, 12(2):129–148.

Nolan, J. P. (1999). Fitting Data and Assessing Goodness-of-fit with Stable Distributions. In *Proceedings of the Conference on Applications of Heavy Tailed Distributions in Economics, Engineering and Statistics*. American University, Washington DC.

Noureldin, D., Shephard, N., and Sheppard, K. (2014). Multivariate Rotated ARCH Models. *Journal of Econometrics*, 179:16–30.

Nyblom, J. (1989). Testing for the Constancy of Parameters Over Time. *Journal of the American Statistical Association*, 84(405):223–230.

Nyblom, J. and Mäkeläinen, T. (1983). Comparisons of Tests for the Presence of Random Walk Coefficients in a Simple Linear Model. *Journal of the American Statistical Association*, 78(384):856–864.

Pagano, M. (1973). When is an Autoregressive Scheme Stationary? *Communications in Statistics*, 1:533–544.

Palm, F. C. (1996). GARCH Models of Volatility. In Maddala, G. S. and Rao, C. R., editors, *Handbook of Statistics: Statistical Methods in Finance*, volume 14, pages 209–240. Elsevier Science.

Palm, F. C. (1997). GARCH Models of Volatility. In Maddala, G. S. and Rao, C. R., editors, *Handbook of Statistics, Volume 14*. Elsevier Science.

Paloyo, A. R. (2011). When Did We Begin to Spell "Heteroskedasticity" Correctly? Ruhr Economic Papers No. 300, Ruhr-Universität Bochum (RUB), Department of Economics, Bochum, Germany.

Pan, X., Yan, Y., Peng, X., and Liu, Q. (2016). Analysis of the Threshold Effect of Financial Development on China's Carbon Intensity. *Sustainability*, 8(3). Article 271.

Pan Jie-Jian (1964). Distributions of the Noncircular Serial Correlation Coefficients. *Shuxue Jinzhan*, 7:328–337. Translated by N. N. Chan for *Selected Translations in Mathematical Statistics and Probability, Volume* 7 (1968), 281–292.

Pankratz, A. (1983). *Forecasting with Univariate Box–Jenkins Models: Concepts and Cases*. John Wiley & Sons, New York.

Paolella, M. S. (2003). Computing Moments of Ratios of Quadratic Forms in Normal Variables. *Computational Statistics & Data Analysis*, 42(3):313–331.

Paolella, M. S. (2006). *Fundamental Probability: A Computational Approach*. John Wiley & Sons, Chichester.

Paolella, M. S. (2007). *Intermediate Probability: A Computational Approach*. John Wiley & Sons, Chichester.

Paolella, M. S. (2014). Fast Methods For Large-Scale Non-Elliptical Portfolio Optimization. *Annals of Financial Economics*, 9(2):1440001.

Paolella, M. S. (2015). Multivariate Asset Return Prediction with Mixture Models. *European Journal of Finance*, 21(13–14):1214–1252.

Paolella, M. S. (2016). Stable-GARCH Models for Financial Returns: Fast Estimation and Tests for Stability. *Econometrics*, 4(2). Article 25.

Paolella, M. S. (2017). The Univariate Collapsing Method for Portfolio Optimization. *Econometrics*, 5(2):1–33. Article 18.

Paolella, M. S. and Polak, P. (2015a). ALRIGHT: Asymmetric LaRge-Scale (I)GARCH with Hetero-Tails. *International Review of Economics and Finance*, 40:282–297.

Paolella, M. S. and Polak, P. (2015b). COMFORT: A Common Market Factor Non-Gaussian Returns Model. *Journal of Econometrics*, 187(2):593–605.

Paolella, M. S. and Polak, P. (2015c). Portfolio Selection with Active Risk Monitoring. Research paper, Swiss Finance Institute.

Paolella, M. S. and Polak, P. (2017). Density and Risk Prediction with Non-Gaussian COMFORT Models. Submitted.

Paolella, M. S., Polak, P., and Walker, P. (2018a). A Flexible Regime-Switching Model for Asset Returns. Submitted.

Paolella, M. S., Polak, P., and Walker, P. (2018b). A New Non-Gaussian Factor GARCH Model. Submitted.

Paolella, M. S. and Steude, S.-C. (2008). Risk Prediction: A DWARF-like Approach. *The Journal of Risk Model Validation*, 2(1):25–43.

Paolella, M. S. and Taschini, L. (2008). An Econometric Analysis of Emission Trading Allowances. *Journal of Banking and Finance*, 32(10):2022–2032.

Park, B. U., Mammen, E., Lee, Y. K., and Lee, E. R. (2015). Varying Coefficient Regression Models: A Review and New Developments. *International Statistical Review*, 83(1):36–64.

Park, J.-A., Choi, M.-S., and Sun-Young (2011). Quadratic GARCH Models: Introduction and Applications. *Korean Journal of Applied Statistics*, 24(1):61–69. In Korean.

Pascual, L., Romo, J., and Ruiz, E. (2006). Bootstrap Prediction for Returns and Volatilities in GARCH Models. *Computational Statistics & Data Analysis*, 50:2293–2312.

Patterson, K. (2000a). *An Introduction to Applied Econometrics: A Time Series Approach*. Palgrave Macmillan, New York.

Patterson, K. (2000b). Finite Sample Bias of the Least Squares Estimator in an AR(p) Model: Estimation, Inference, Simulation, and Examples. *Applied Economics*, 32(15):1993–2005.

Patterson, K. (2011). *Unit Root Tests in Time Series Volume 1: Key Concepts and Problems*. Palgrave Macmillan, Houndmills, Basingstoke, Hampshire.

Patterson, K. (2012). *Unit Root Tests in Time Series Volume 2: Extensions and Developments*. Palgrave Macmillan, Houndmills, Basingstoke, Hampshire.

Patton, A. (2009). Copula-Based Models for Financial Time Series. In Andersen, T. G., Davis, R. A., Kreiß, J.-P., and Mikosch, T., editors, *Handbook of Financial Time Series*, pages 767–785. Springer, Berlin.

Pearl, J. (2009). Causal Inference in Statistics: An Overview. *Statistics Surveys*, 3:96–146.

Pearson, K. (1901). On lines and planes of closest fit to systems of points in space. *The London, Edinburgh, and Dublin Philosophical Magazine and Journal of Science*, 2(11):559–572.

Pedersen, R. S. and Rahbek, A. (2014). Multivariate Variance Targeting in the BEKK-GARCH Model. *Econometrics Journal*, 17(1):24–55.

Pelletier, D. (2006). Regime Switching for Dynamic Correlations. *Journal of Econometrics*, 131(1-2):445–473.

Pelletier, D. and Wei, W. (2016). The Geometric-VaR Backtesting Method. *Journal of Financial Econometrics*, 14(4):725–745.

Peng, J.-Y. and Aston, J. A. D. (2011). The State Space Models Toolbox for MATLAB. *Journal of Statistical Software*, 41(6).

Percival, D. B. (1993). Three Curious Properties of the Sample Variance and Autocovariance for Stationary Processes with Unknown Mean. *The American Statistician*, 47:274–276.

Perri, S. (2014). The Role Of Macroeconomic Stability In The Finance Growth Nexus: Threshold Regression Approach. IPE Working Paper No. 1.

Perron, P. (1989). The Great Crash, the Oil Price Shock and the Unit Root Hypothesis. *Econometrica*, 57:1361–1401.

Perron, P. (2006). Dealing with Structural Breaks. In Mills, T. C. and Patterson, K., editors, *Palgrave Handbook of Econometrics, Volume 1: Econometric Theory*, pages 278–352. Palgrave Macmillan, Houndmills, Basingstoke, Hampshire.

Perron, P. and Zhu, X. (2005). Structural Breaks with Stochastic and Deterministic Trends. *Journal of Econometrics*, 129(1,2):65–119.

Pesaran, M. H. (2015). *Time Series and Panel Data Econometrics*. Oxford University Press, Oxford.

Pesaran, M. H. and Pick, A. (2007). Econometric Issues in the Analysis of Contagion. *Journal of Economic Dynamics & Control*, 31:1245–1277.

Phillips, G. D. A. and Harvey, A. C. (1974). A Simple Test for Serial Correlation in Regression Analysis. *Journal of the American Statistical Association*, 69:935–939.

Phillips, P. C. B. (1988). The ET Interview: Professor James Durbin. *Econometric Theory*, 4:125–157.

Phillips, P. C. B. and Perron, P. (1988). Testing for a Unit Root in Time Series Regression. *Biometrika*, 75:335–346.

Phillips, P. C. B. and Sul, D. (2003). Dynamic Panel Estimation and Homogeneity Testing Under Cross Section Dependence. *Econometrics Journal*, 6:217–259.

Phillips, P. C. B. and Yu, J. (2005). Jackknifing Bond Option Prices. *Review of Financial Studies*, 18:707–742.

Pigliucci, M. and Kaplan, J. (2006). *Making Sense of Evolution: The Conceptual Foundations of Evolutionary Biology*. University of Chicago Press, Chicago.

Pincus, S. and Kalman, R. E. (2004). Irregularity, Volatility, Risk, and Financial Market Time Series. *Proceedings of the National Academy of Sciences of the United States of America*, 101(38):13709–13714.

Pitman, E. J. G. and Williams, E. J. (1967). Cauchy-Distributed Functions of Cauchy Variates. *Annals of Mathematical Statistics*, 38(3):916–918.

Plackett, R. L. (1950). Some Theorems in Least Squares. *Biometrika*, 37:149–157.

Plackett, R. L. (1960). Models in Analysis of Variance (with discussion). *Journal of the Royal Statistical Society, Series B*, 22:195–217.

Platen, E. and Heath, D. (2006). *A Benchmark Approach to Quantitative Finance*. Springer, Berlin.

Platen, E. and Rendek, R. (2008). Empirical Evidence on Student-t Log-Returns of Diversified World Stock Indices. *Journal of Statistical Theory and Practice*, 2(2):233–251.

Podgórski, T. J. and Kozubowski, T. (2001). Asymmetric Laplace Laws and Modeling Financial Data. *Mathematical and Computer Modelling*, 34:1003–1021.

Poirier, D. J. (1995). *Intermediate Statistics and Econometrics, A Comparative Approach*. The MIT Press, Cambridge, MA. Errata: http://www.chass.utoronto.ca:8080/~ poirier.

Pollock, D. S. G. (1999). *A Handbook of Time-Series Analysis, Signal Processing and Dynamics*. Academic Press, San Diego.

Poon, S.-H. and Granger, C. (2003). Forecasting Volatility in Financial Markets: A Review. *Journal of Economic Literature*, 41(2):478–539.

Pope, N. G. (2016). How the Time of Day Affects Productivity: Evidence from School Schedules. *The Review of Economics and Statistics*, 38(1):1–11.

Pötscher, B. M. (1983). Order Estimation in ARMA Models by Lagrangian Multiplier Tests. *Annals of Statistics*, 11(3):872–885.

Pourahmadi, M. (1986). On Stationary of the Solution of a Doubly Stochastic Model. *Journal of Time Series Analysis*, 7:123–132.

Pourahmadi, M. (1988). Stationary of the solution of $X_t = A_t X_{t-1} + \epsilon_t$ and Analysis of Non-Gaussian Dependent Random Variables. *Journal of Time Series Analysis*, 9:225–239.

Pourahmadi, M. (2001). *Foundations of Time Series Analysis and Prediction Theory*. John Wiley & Sons, New York.

Pozzi, F., Di Matteo, T., and Aste, T. (2012). Exponential Smoothing Weighted Correlations. *The European Physical Journal B*, 85(6):1–21. With Erratum.

Priestley, M. B. (1981). *Spectral Analysis and Time Series, Volume I: Univariate Series*. Academic Press, San Diego.

Pritsker, M. (2006). The Hidden Dangers of Historical Simulation. *Journal of Banking & Finance*, 30(2):561–582.

Prono, T. (2016). Simple Estimators for GARCH Models. Finance and Economics Discussion Series. Washington: Board of Governors of the Federal Reserve System.

Puntanen, S. and Styan, G. P. H. (1989). The Equality of the Ordinary Least Squares Estimator and the Best Linear Unbiased Estimator (with Comments and Reply). *The American Statistician*, 43(3):153–164.

Qu, Z. and Perron, P. (2007). Estimating and Testing Structural Changes in Multivariate Regressions. *Econometrica*, 75(2):459–502.

Rachev, S. T. and Mittnik, S. (2000). *Stable Paretian Models in Finance*. John Wiley & Sons, New York.

Rachev, S. T., Mittnik, S., Fabozzi, F. J., Focardi, S. M., and Jašić, T. (2007). *Financial Econometrics: From Basics to Advanced Modeling Techniques*. John Wiley & Sons, Hoboken, NJ.

Rahman, M. S. and King, M. L. (1999). Improved Model Selection Criterion. *Communications in Statistics—Simulation and Computation*, 28:51–71.

Rao, C. R. (1965). The Theory of Least Squares when the Parameters are Stochastic and its Application to the Analysis of Growth Curves. *Biometrika*, 52(3–4):447–458.

Rao, C. R. (1968). A Note on a Previous Lemma in the Theory of Least Squares and Some Further Results. *Sankya*, 30:245–252.

Rao, C. R. (1973). *Linear Statistical Inference and Its Applications*. John Wiley & Sons, New York, 2nd edition.

Rao, C. R., Toutenburg, H., Shalabh, and Heumann, C. (2008). *Linear Models and Generalizations: Least Squares and Alternatives*. Springer, Berlin, 3rd edition.

Rao, M. J. M. (2000). Estimating Time-Varying Parameters in Linear Regression Models Using a Two-Part Decomposition of the Optimal Control Formulation. *Sankhya*, 62:433–447.

Rapach, D. and Zhou, G. (2013). Forecasting Stock Returns. In Elliott, G. and Timmermann, A., editors, *Handbook of Economic Forecasting, Volume 2*, chapter 6, pages 328–383. Elsevier, Amsterdam.

Raunig, B. (2017). On The Interpretation of Instrumental Variables in the Presence of Specification Errors: A Causal Comment. *Econometrics*, 5(3):1–6. Article 31.

Ravishanker, N. and Dey, D. K. (2002). *A First Course in Linear Model Theory*. Chapman & Hall, London.

Reams, R. (1999). Hadamard Inverses, Square Roots and Products of Almost Semidefinite Matrices. *Linear Algebra and its Applications*, 288:35–43.

Richter, W.-D. (2007). Generalized Spherical and Simplicial Coordinates. *Journal of Mathematical Analysis and Applications*, 336(2):1187–1202.

Robert, C. P. (2007). *The Bayesian Choice*. Springer, New York, 2nd edition.

Roberts, L. A. (1995). On the Existence of Moments of Ratios of Quadratic Forms. *Econometric Theory*, 11:750–774.

Rocco, M. (2014). Extreme Value Theory in Finance: A Survey. *Journal of Economic Surveys*, 28(1):82–108.

Rockafellar, R. T. and Uryasev, S. P. (2000). Optimization of Conditional Value at Risk. *Journal of Risk*, 2:21–41.

Rockinger, M. and Jondeau, E. (2002). Entropy Densities with an Application to Autoregressive Conditional Skewness and Kurtosis. *Journal of Econometrics*, 106:119–142.

Romano, J. P. and Wolf, M. (2001). Subsampling Intervals in Autoregressive Models with Linear Time Trend. *Econometrica*, 69:1283–1314.

Rombouts, J. V. K. and Stentoft, L. (2009). Bayesian Option Pricing Using Mixed Normal Heteroskedasticity Models. CREATES Research Papers 2009-07, School of Economics and Management, University of Aarhus.

Rombouts, J. V. K. and Stentoft, L. (2011). Multivariate Option Pricing with Time Varying Volatility and Correlations. *Journal of Banking & Finance*, 35(9):2267–2281.

Rosenberg, B. (1973). The Analysis of a Cross-Section of Time Series by Stochastically Convergent Parameter Regression. *Annals of Economic and Social Measurement*, 2:399–428.

Rosenkrantz, W. A. (1997). *Introduction to Probability and Statistics for Scientists and Engineers.* McGraw-Hill, New York.

Rounvinez, C. (1997). Going Greek with VaR. *Risk*, 10(2):57–65.

Roussas, G. G. (1997). *A Course in Mathematical Statistics.* Academic Press, San Diego, 2nd edition.

Rousseau, P. and Wachtel, P. (2002). Inflation Thresholds and the Finance-Growth Nexus. *Journal of International Money and Finance*, 21:777–793.

Roy, S. N. (1953). On a Heuristic Method of Test Construction and its use in Multivariate Analysis. *Annals of Mathematical Statistics*, 24(2):220–238.

Rubin, H. (1950). Note on Random Coefficients. In Koopmans, T. C., editor, *Statistical Inference in Dynamic Economic Models: Cowles Commission for Research in Economics, Monograph No. 10*, pages 419–421. John Wiley & Sons, New York.

Ruppert, D. (2004). *Statistics and Finance: An Introduction*. Springer, New York.

Russell, B. (2009). *The ABC of Relativity*. Routledge, Taylor & Francis Group, London. Originally published: A B C of Relativity, George Allen & Unwin, London, 1925.

Ruud, P. A. (2000). *An Introduction to Classical Econometric Theory*. Oxford University Press, Oxford.

Sahai, H. and Ageel, M. I. (2000). *The Analysis of Variance: Fixed, Random and Mixed Models.* Springer, New York.

Sahai, H. and Ojeda, M. M. (2004). *Analysis of Variance for Random Models Volume I: Balanced Data. Theory, Methods, Applications and Data Analysis*. Birkhäuser, Boston.

Sahai, H. and Ojeda, M. M. (2005). *Analysis of Variance for Random Models Volume II: Unbalanced Data. Theory, Methods, Applications and Data Analysis*. Birkhäuser, Boston.

Saikkonen, P. and Lütkepohl, H. (2002). Testing for a Unit Root in a Time Series with a Level Shift at Unknown Time. *Econometric Theory*, 18:313–348.

Samworth, R. (2005). Small Confidence Sets for the Mean of a Spherically Symmetric Distribution. *Journal of the Royal Statistical Society, Series B*, 67:343–361.

Santos, A. A. P. and Moura, G. V. (2014). Dynamic Factor Multivariate GARCH Model. *Computational Statistics & Data Analysis*, 76:606–617.

Santos, A. A. P., Nogales, F. J., and Ruiz, E. (2013). Comparing Univariate and Multivariate Models to Forecast Portfolio Value-at-Risk. *Journal of Financial Econometrics*, 11(2):400–441.

Sargan, J. D. and Bhargava, A. (1983). Testing Residuals from Least Squares Regression for being Generated by the Gaussian Random Walk. *Econometrica*, 51:153–174.

SAS/STAT 9.2 User's Guide (2008). SAS Institute Inc., Cary, NC, USA.

Satterthwaite, F. E. (1946). An Approximate Distribution of Estimates of Variance Components. *Biometrics Bulletin*, 2:110–114.

Sawa, T. (1972). Finite Sample Properties of the k-Class Estimator. *Econometrica*, 40(4):653–680.
Sawa, T. (1978). The Exact Moments of the Least Squares Estimator for the Autoregressive Model. *Journal of Econometrics*, 8:159–172.
Scheffé, H. (1953). A Method of Judging all Contrasts in the Analysis of Variance. *Biometrika*, 40:87–104.
Scheffé, H. (1959). *The Analysis of Variance*. John Wiley & Sons, New York.
Scherrer, A., Larrieu, N., Owezarski, P., Borgnat, P., and Abry, P. (2007). Non-Gaussian and Long Memory Statistical Characterizations for Internet Traffic with Anomalies. *IEEE Transactions on Dependable and Secure Computing*, 4(1):56–70.
Schlattmann, P. (2009). *Medical Applications of Finite Mixture Models*. Springer, Heidelberg.
Schoenberg, I. J. (1938). Metric Spaces and Completely Monotone Functions. *Annals of Mathematics, Second Series*, 39(4):811–841.
Scholz, M., Nielsen, J. P., and Sperlich, S. (2012). Nonparametric Prediction of Stock Returns Guided by Prior Knowledge. Graz economics papers, University of Graz, Department of Economics.
Scholz, M., Nielsen, J. P., and Sperlich, S. (2015). Nonparametric Prediction of Stock Returns Based on Yearly Data: The Long-Term View. *Insurance: Mathematics and Economics*, 65:143–155.
Schott, J. R. (2002). Testing for Elliptical Symmetry in Covariance-Matrix Based Analyses. *Statistics and Probability Letters*, 60(4):395–404.
Schott, J. R. (2005). *Matrix Analysis for Statistics*. John Wiley & Sons, New York, 2nd edition.
Schur, I. (1917). Über Potenzreihen, die im Innern des Einheitskreises beschränkt sind. *Journal für die reine und angewandte Mathematik*, 147:205–232.
Schwert, G. W. (1989a). Tests for Unit Roots: A Monte Carlo Investigation. *Journal of Business and Economics Statistics*, 7:147–159.
Schwert, W. G. (1989b). Why Does Stock Market Volatility Change Over Time? *Journal of Finance*, 44:1115–1153.
Searle, S. R. (1971). *Linear Models*. John Wiley & Sons, New York.
Searle, S. R. (1982). *Matrix Algebra Useful for Statistics*. John Wiley & Sons, New York.
Searle, S. R., Casella, G., and McCulloch, C. E. (1992). *Variance Components*. John Wiley & Sons, New York.
Searle, S. R. and Gruber, M. H. J. (2017). *Linear Models*. John Wiley & Sons, Hoboken, NJ, 2nd edition.
Seber, G. A. F. and Lee, A. J. (2003). *Linear Regression Analysis*. John Wiley & Sons, Hoboken, NJ, 2nd edition.
Segnon, M., Lux, T., and Gupta, R. (2017). Modeling and Forecasting the Volatility of Carbon Dioxide Emission Allowance Prices: A Review and Comparison of Modern Volatility Models. *Renewable and Sustainable Energy Reviews*, 69:692–704.
Semrl, P. (1996). On a Matrix Version of Cochran's Statistical Theorem. *Linear Algebra and its Applications*, 237–238:477–487.
Seneta, E. (2004). Fitting the Variance-Gamma Model to Financial Data. *Journal of Applied Probability*, 41, Stochastic Methods and Their Applications:177–187.
Sentana, E. (1995). Quadratic ARCH models. *The Review of Economic Studies*, 62(4):639–661.
Severini, T. A. (2005). *Elements of Distribution Theory*. Cambridge University Press, Cambridge.
Shaman, P. and Stine, R. A. (1988). The Bias of Autoregressive Coefficient Estimators. *Journal of the American Statistical Association*, 83(403):842–848.
Shao, J. and Tu, D. (1995). *The Jackknife and Bootstrap*. Springer, New York.
Shaw, W. T. and Lee, K. T. A. (2008). Bivariate Student t Distributions with Variable Marginal Degrees of Freedom and Independence. *Journal of Multivariate Analysis*, 99:1276–1287.

Shi, S. and Song, Y. (2016). Identifying Speculative Bubbles Using an Infinite Hidden Markov Model. *Journal of Financial Econometrics*, 14(1):159–184.

Shipley, B. (2016). *Cause and Correlation in Biology*. Cambridge University Press, Cambridge.

Shively, P. A. (2001). Trend-Stationary GNP: Evidence from a new exact Pointwise Most Powerful Invariant Unit Root Test. *Journal of Applied Econometrics*, 16:537–551.

Shively, P. A. (2003). The Nonlinear Dynamics of Stock Prices. *The Quarterly Review of Economics and Finance*, 43(3):505–517.

Shively, T. S. (1988a). An Analysis of Tests for Regression Coefficient Stability. *Journal of Econometrics*, 39:367–386.

Shively, T. S. (1988b). An Exact Test for a Stochastic Coefficient in a Time Series Regression Model. *Journal of Time Series Analysis*, 9(1):81–88.

Shumway, R. H. and Stoffer, D. S. (2000). *Time Series Analysis and Its Applications*. Springer, New York.

Shumway, T. (1997). The Delisting Bias in CRSP Data. *Journal of Finance*, 52(1):327–340.

Silva, E. S. and Hassani, H. (2015). On the Use of Singular Spectrum Analysis for Forecasting U.S. Trade Before, During and After the 2008 Recession. *International Economics*, 141:34–49.

Silvennoinen, A. and Teräsvirta, T. (2009). Multivariate GARCH Models. In Andersen, T. G., Davis, R. A., Kreiss, J.-P., and Mikosch, T., editors, *Handbook of Financial Time Series*, pages 201–229. Springer, Berlin.

Singh, B., Nagar, A. L., Choudhry, N. K., and Raj, B. (1976). On the Estimation of Structural Change: A Generalization of the Random Coefficients Regression Model. *International Economic Review*, 17(2):340–361.

Skovgaard, I. M. (1987). Saddlepoint Expansions for Conditional Distributions. *Journal of Applied Probability*, 24:275–287.

Slim, S., Koubaa, Y., and BenSaïda, A. (2016). Value-at-Risk Under Lévy GARCH models: Evidence from Global Stock Markets. *Journal of International Financial Markets, Institutions & Money*, 46:30–53.

Small, J. P. (1993). The Limiting Power of Point Optimal Autocorrelation Tests. *Communications in Statistics—Theory and Methods*, 22(2):3907–3916.

Smetanina, E. (2017). Real-Time GARCH. *Journal of Financial Econometrics*, 15(4):561–601.

So, M. K. P. and Choi, C. Y. (2009). A Multivariate Factor Threshold Stochastic Volatility Model. *Journal of Forecasting*, 28:712–735.

So, M. K. P. and Yip, I. W. H. (2012). Multivariate GARCH Models with Correlation Clustering. *Journal of Forecasting*, 31(5):443–468.

Sobreira, N. and Nunes, L. C. (2016). Tests for Multiple Breaks in the Trend with Stationary or Integrated Shocks. *Oxford Bulletin of Economics and Statistics*, 78(3):394–411.

Sollis, R., Newbold, P., and Leybourne, S. J. (2000). Stochastic Unit Roots Modelling of Stock Price Indices. *Applied Financial Economics*, 10(3):311–315.

Solnik, B. and Longin, F. (2001). Extreme Correlation of International Equity Markets. *Journal of Finance*, 2(LVI):649–676.

Song, D.-K., Park, H.-J., and Kim, H.-M. (2014). A Note on the Characteristic Function of Multivariate t Distribution. *Communications for Statistical Applications and Methods*, 21(1):81–91.

Sortino, F. A. and van der Meer, R. (1991). Downside Risk. *The Journal of Portfolio Management*, 17(4):27–31.

Sowell, F. (1992). Maximum Likelihood Estimation of Stationary Univariate Fractionally Integrated Time Series Models. *Journal of Econometrics*, 53(1–3):165–188.

Spiegelhalter, D. (2017). Trust in Numbers. *Journal of the Royal Statistical Society, Series A*, 180:949–965.

Srivastava, M. S. (1987). Asymptotic Distribution of Durbin-Watson Statistic. *Economics Letters*, 24(2):157–160.

Stapleton, J. H. (1995). *Linear Statistical Models*. John Wiley & Sons, New York.

Stein, E. M. and Shakarchi, R. (2005). *Real Analysis. Measure Theory, Integration and Hilbert Spaces*. Princeton University Press.

Stigler, S. M. (1981). Gauss and the Invention of Least Squares. *Annals of Statistics*, 9(3):465–474.

Stivers, A. (2018). Equity Premium Predictions with Many Predictors: A Risk-Based Explanation of the Size and Value Factors. *Journal of Empirical Finance*, 45:126–140.

Stock, J. H. (1994). Unit Roots, Structural Breaks and Trends. In Engle, R. F. and McFadden, D. L., editors, *Palgrave Handbook of Econometrics, Volume 4*, chapter 46, pages 2739–2841. Elsevier, Amsterdam.

Stock, J. H. and Watson, M. W. (1998). Median Unbiased Estimation of Coefficient Variance in a Time-Varying Parameter Model. *Journal of the American Statistical Association*, 93(441):349–358.

Stolbov, M. (2013). The Finance-Growth Nexus Revisited: From Origins to a Modern Theoretical Landscape. *Economics: The Open-Access, Open-Assessment E-Journal*, 7(2).

Stoyanov, S., Samorodnitsky, G., Rachev, S., and Ortobelli, S. (2006). Computing the Portfolio Conditional Value-at-Risk in the alpha-stable Case. *Probability and Mathematical Statistics*, 26:1–22.

Stroup, W. W. and Mulitze, D. K. (1991). Nearest Neighbor Adjusted Best Linear Unbiased Prediction. *The American Statistician*, 45(3):194–200.

Stuart, A. and Ord, J. K. (1994). *Kendall's Advanced Theory of Statistics, Volume 1, Distribution Theory*. Edward Arnold, London, 6th edition.

Stuart, A., Ord, J. K., and Arnold, S. F. (1999). *Kendall's Advanced Theory of Statistics, Volume 2A, Classical Inference and the Linear Model*. Edward Arnold, London, 6th edition.

Su, Y. (2012). Smooth Test for Elliptical Symmetry. In *2012 International Conference on Machine Learning and Cybernetics*, volume 4, pages 1279–1284.

Sucarrat, G., Pretis, F., and Reade, J. (2017). gets: General-to-Specific (GETS) Modelling and Indicator Saturation Methods. R package version 0.12. Available at: https://CRAN.R-project.org/package=gets.

Suh, S. (2016). A Combination Rule for Portfolio Selection with Transaction Costs. *International Review of Finance*, 16(3):393–420.

Sutradhar, B. C. (1986). On the Characteristic Function of Multivariate Student t-Distribution. *Canadian Journal of Statistics*, 14(4):329–337.

Swamy, P. A. V. B. (1971). *Statistical Inference in Random Coefficient Regression Models*. Springer, New York.

Swamy, P. A. V. B., Conway, R. K., and LeBlanc, M. R. (1988). The Stochastic Coefficients Approach to Econometric Modeling Part I: A Critique of Fixed Coefficients Models. *Journal of Agricultural Economics Research*, 40(2):2–10.

Swamy, P. A. V. B., Hall, S. G., Tavlas, G. S., and von zur Muehlen, P. (2017). On The Interpretation of Instrumental Variables in the Presence of Specification Errors: A Reply. *Econometrics*, 5(3):1–3. Article 32.

Swamy, P. A. V. B. and Tavlas, G. S. (1995). Random Coefficient Models: Theory and Applications. *Journal of Economic Surveys*, 9(2):165–196.

Swamy, P. A. V. B. and Tavlas, G. S. (2001). Random Coefficient Models. In Baltagi, B. H., editor, *A Companion to Theoretical Econometrics*, chapter 19, pages 410–428. Blackwell Publishing, Oxford.

Swamy, P. A. V. B., Tavlas, G. S., and Hall, S. G. (2015). On the Interpretation of Instrumental Variables in the Presence of Specification Errors. *Econometrics*, 3(1):55–64.

Tamhane, A. C. and Dunlop, D. D. (2000). *Statistics and Data Analysis: From Elementary to Intermediate.* Prentice Hall, Upper Saddle River, NJ.

Tanaka, K. (1996). *Time Series Analysis: Nonstationary and Noninvertible Distribution Theory.* John Wiley & Sons Ltd, New York.

Tanizaki, H. (2000). Bias Correction of OLSE in the Regression Model with Lagged Dependent Variables. *Journal of Computational Statistics & Data Analysis*, 34:495–511.

Tashman, A. (2010). A Regime-switching Approach to Model-based Stress Testing. *Journal of Risk Model Validation*, 3:89–101.

Tashman, A. and Frey, R. J. (2009). Modeling Risk in Arbitrage Strategies Using Finite Mixtures. *Quantitative Finance*, 9:495–503.

Tay, A. S. and Wallis, K. F. (2000). Density Forecasting: A Survey. *Journal of Forecasting*, 19(4):124–143.

Tayefi, M. and Ramanathan, T. V. (2012). An Overview of FIGARCH and Related Time Series Models. *Austrian Journal of Statistics*, 41(3):175–196.

Taylor, J. W. (2008). Estimating Value at Risk and Expected Shortfall Using Expectiles. *Journal of Financial Econometrics*, 6(2):231–252.

Taylor, J. W. and Yu, K. (2016). Using Auto-Regressive Logit Models to Forecast the Exceedance Probability for Financial Risk Management. *Journal of the Royal Statistical Society, Series A, Statistics in Society*, 179(4):1069–1092.

Taylor, S. (1986). *Modelling Financial Time Series.* John Wiley & Sons, New York.

Temme, N. M. (1982). The Uniform Asymptotic Expansion of a Class of Integrals Related to Cumulative Distribution Functions. *SIAM Journal of Mathematical Analysis*, 13:239–253.

Teräsvirta, T. (1994). Specification, Estimation, and Evaluation of Smooth Transition Autoregressive Models. *Journal of the American Statistical Association*, 89:208–218.

Teräsvirta, T. (1998). Modelling Economic Relationships with Smooth Transition Regressions. In Ullah, A. and Giles, D. E. A., editors, *Handbook of Applied Economic Statistics*, pages 507–552. Marcel Dekker, New York.

Teräsvirta, T. (2009). An Introduction to Univariate GARCH Models. In Andersen, T. G., Davis, R. A., Kreiß, J.-P., and Mikosch, T., editors, *Handbook of Financial Time Series*, pages 17–42. Springer, Berlin.

Teräsvirta, T., Tjøstheim, D., and Granger, C. W. J. (2010). *Modelling Nonlinear Economic Time Series.* Oxford University Press, Oxford.

Teräsvirta, T. and Zhao, Z. (2011). Stylized Facts of Return Series, Robust Estimates and Three Popular Models of Volatility. *Applied Financial Economics*, 21:67–94.

Theil, H. (1965). The Analysis of Disturbances in Regression Analysis. *Journal of the American Statistical Association*, 60:1067–1079.

Theil, H. (1968). A Simplification of the BLUS Procedure for Analyzing Regression Disturbances. *Journal of the American Statistical Association*, 63:242–251.

Theil, H. (1971). *Principles of Econometrics.* John Wiley & Sons, New York.

Theil, H. and Goldberger, A. S. (1961). On Pure and Mixed Statistical Estimation in Economics. *International Economic Review*, 2(1):65–78.

Theodossiou, P. (1998). Financial Data and the Skewed Generalized T Distribution. *Management Science*, 44(12):1650–1661.

Thiel, H. and Mennes, L. B. M. (1959). Multiplicative Randomness in Time Series Regression Analysis. Mimeographed Report No. 5901.

Thode, Jr., H. C. (2002). *Testing for Normality.* Marcel Dekker, New York.

Tiao, G. C. and Box, G. E. P. (1981). Modelling Multiple Time Series with Applications. *Journal of the American Statistical Association*, 76:802–816.

Tillman, J. A. (1975). The Power of the Durbin–Watson Test. *Econometrica*, 43:959–974.

Timmermann, A. (2000). Density Forecasting in Economics and Finance. *Journal of Forecasting*, 19(4):231–234.

Tjøstheim, D. (1986). Some Doubly Stochastic Time Series. *Journal of Time Series Analysis*, 7:51–72.

Tong, H. (1978). On a Threshold Model. In Chen, C. H., editor, *Pattern Recognition and Signal Processing*, pages 575–586. Sijhoff and Noordhoff, Alpen aan den Rijn.

Tong, H. (1983). *Threshold Models in Non-linear Time Series Analysis. Lecture Notes in Statistics, No. 21*. Springer, New York.

Tong, H. (1990). *Non-linear Time Series: A Dynamical System Approach*. Oxford University Press, Oxford.

Tong, H. (2007). Birth of the Threshold Time Series Model. *Statistica Sinica*, 17:8–14.

Tong, H. (2011). Threshold Models in Time Series Analysis—30 Years On (with discussion). *Statistics and Its Interface*, 4:107–136.

Tong, H. and Lim, K. S. (1980). Threshold Autoregression, Limit Cycles and Cyclical Data (with discussion). *Journal of the Royal Statistical Society, Series B*, 42(3):245–292.

Trench, W. F. (2003). *Introduction to Real Analysis*. Prentice Hall, Upper Saddle River, NJ.

Tsay, R. S. (1998). Testing and Modeling Multivariate Threshold Models. *Journal of the American Statistical Association*, 93(443):1188–1202.

Tsay, R. S. (2010). *Analysis of Financial Time Series*. John Wiley & Sons, Hoboken, NJ, 3rd edition.

Tsay, R. S. (2012). *An Introduction to Analysis of Financial Data with R*. John Wiley & Sons, Hoboken, NJ.

Tsay, R. S. (2014). *Multivariate Time Series Analysis: With R and Financial Applications*. John Wiley & Sons, Hoboken, NJ.

Tse, Y. K. and Tsui, A. K. C. (2002). A Multivariate Generalized Autoregressive Conditional Heteroscedasticity Model With Time-Varying Correlations. *Journal of Business and Economic Statistics*, 20(3):351–362.

Tunnicliffe Wilson, G. (1979). Some Efficient Computational Procedures for High Order ARMA Models. *Journal of Statistical Computation and Simulation*, 8:301–309.

Ullah, A., Srivastava, V. K., and Roy, N. (1995). Moments of the Function of Non-Normal Random Vector with Applications to Econometric Estimators and Test Statistics. *Econometric Reviews*, 14(4):459–471.

Uppuluri, V. R. R. and Carpenter, J. A. (1969). The Inverse of a Matrix Occurring in First-Order Moving-Average Models. *Sankhya, Series A*, 31(1):79–82.

van Belle, G. (2008). *Statistical Rules of Thumb*. John Wiley & Sons, Hoboken, NJ, 2nd edition.

van der Leeuw, J. (1994). The Covariance Matrix of ARMA Errors in Closed Form. *Journal of Econometrics*, 63(2):397–405.

van der Weide, R. (2002). GO-GARCH: A Multivariate Generalized Orthogonal GARCH Model. *Journal of Applied Econometrics*, 17:549–564.

van Dijk, D., Teräsvirta, T., and Franses, P. H. (2002). Smooth Transition Autoregressive Models—A Survey of Recent Developments. *Econometric Reviews*, 21:1–47.

Vandebril, R., Van Barel, M., and Mastronardi, N. (2008). *Matrix Computations and Semiseparable Matrices Volume I: Linear Systems*. The Johns Hopkins University Press, Baltimore.

Vargas, G. A. (2006). An Asymmetric Block Dynamic Conditional Correlation Multivariate GARCH Model. *The Philippine Statistician*, 55(1–2):83–102.

Vaynman, I. and Beare, B. K. (2014). Stable Limit Theory for the Variance Targeting Estimator. In Chang, Y., Fomby, T. B., and Park, J. Y., editors, *Advances in Econometrics: Essays in Honor of Peter C. B. Phillips, Volume 33*, chapter 18, pages 639–672. Emerald Group Publishing Limited, Bingley, UK.

Vecchio, A. (2003). A Bound for the Inverse of a Lower Triangular Toeplitz Matrix. *SIAM Journal on Matrix Analysis and Applications*, 24(4):1167–1174.

Vervaat, W. (1979). On a Stochastic Difference Equation and a Representation of Non-Negative Infinitely Divisible Random Variables. *Advances in Applied Probability*, 11:750–783.

Vinod, H. D. (1973). Generalization of the Durbin–Watson Statistic for Higher Order Autoregressive Processes. *Communications in Statistics*, 2:115–144.

Virbickaite, A., Ausin, M. C., and Galeano, P. (2016). A Bayesian Non-Parametric Approach to Asymmetric Dynamic Conditional Correlation Model with Application to Portfolio Selection. *Computational Statistics & Data Analysis*, 100:814–829.

Vlaar, P. J. G. and Palm, F. C. (1993). The Message in Weekly Exchange Rates in the European Monetary System: Mean Reversion, Conditional Heteroscedasticity, and Jumps. *Journal of Business & Economic Statistics*, 11(3):351–360.

von Neumann, J. (1941). Distribution of the Ratio of the Mean Square Successive Difference to the Variance. *Annals of Mathematical Statistics*, 12:367–395.

Vrontos, I. D., Dellaportas, P., and Politis, D. (2003). A Full-Factor Multivariate GARCH Model. *Econometrics Journal*, 6(2):312–334.

Wald, A. (1947). A Note on Regression Analysis. *Annals of Mathematical Statistics*, 18:586–589.

Walker, G. (1931). On Periodicity in Series of Related Terms. *Proceedings of the Royal Society of London A*, 131:518–532.

Wallis, W. A. (1980). The Statistical Research Group, 1942–1945. *Journal of the American Statistical Association*, 75(370):320–330.

Wan, A. T. K., Zou, G., and Banerjee, A. (2007). The Power of Autocorrelation Tests Near the Unit Root in Models with Possibly Mis-Specified Linear Restrictions. *Economics Letters*, 94:213–219.

Wang, C.-S. and Zhao, Z. (2016). Conditional Value-at-Risk: Semiparametric Estimation and Inference. *Journal of Econometrics*, 195(1):86–103.

Wang, M. and Li, Y. (2011). Pricing of Convertible Bond Based on GARCH Model. In Wu, D. D., editor, *Quantitative Financial Risk Management*, pages 77–86. Springer, Berlin.

Wang, Y., Wu, C., and Yang, L. (2016). Forecasting Crude Oil Market Volatility: A Markov Switching Multifractal Volatility Approach. *International Journal of Forecasting*, 32:1–9.

Watson, G. N. (1922). *A Treatise on the Theory of Bessel Functions*. Cambridge University Press, Cambridge.

Watson, M. W. (1994). Vector Autoregressions and Cointegration. In Engle, R. F. and McFadden, D. L., editors, *Palgrave Handbook of Econometrics, Volume 4*, chapter 47, pages 2843–2915. Elsevier, Amsterdam.

Watson, M. W. and Engle, R. F. (1985). Testing for Regressoin Coefficient Stability with a Stationary AR(1) Alternative. *The Review of Economics and Statistics*, 67:341–346.

Wells, C. (1996). *The Kalman Filter in Finance*. Kluwer Academic Publishing, Dordrecht.

West, B. T., Welch, K. B., and Galecki, A. T. (2015). *Linear Mixed Models: A Practical Guide Using Statistical Software*. CRC Press, Boca Raton, 2nd edition.

West, M. and Harrison, J. (1997). *Bayesian Forecasting and Dynamic Models*. Springer, New York, 2nd edition.

White, H. (1980). A Heteroskedasticity-Consistent Covariance Matrix Estimator and a Direct Test for Heteroskedasticity. *Econometrica*, 48:817–838.

White, H. (1982). Maximum Likelihood Estimation of Misspecified Models. *Econometrica*, 50(1):1–25.

White, H. (1994). *Estimation, Inference, and Specification Analysis*. Cambridge University Press, New York.

White, H., Kim, T.-H., and Manganelli, S. (2015). VAR for VaR: Measuring Tail Dependence Using Multivariate Regression Quantiles. *Journal of Econometrics*, 187(1):169–188.

Williams, J. S. (1962). A Confidence Interval for Variance Components. *Biometrika*, 49:278–281.

Winkelmann, R. (2008). *Econometric Analysis of Count Data*. Springer, Berlin, 5th edition.

Winker, P. and Maringer, D. (2009). The Convergence of Estimators Based on Heuristics: Theory and Application to a GARCH Model. *Computational Statistics*, 24(3):533–550.

Wong, C. S. and Li, W. K. (2001). On a Logistic Mixture Autoregressive Model. *Biometrika*, 88:833–846.

Wooldridge, J. M. (2009). *Introductory Econometrics: A Modern Approach*. South-Western: Cengage Learning, Mason, OH, 4th edition.

Wooldridge, J. M. (2010). *Econometric Analysis of Cross Section and Panel Data*. MIT Press, Cambridge, MA, 2nd edition.

Wright, R. (2017). *Why Buddhism is True: The Science and Philosophy of Meditation and Enlightenment*. Simon & Schuster, New York.

Wu, L., Meng, Q., and Velazquez, J. C. (2015). The Role of Multivariate Skew-Student Density in the Estimation of Stock Market Crashes. *European Journal of Finance*, 21(13–14):1144–1160.

Wu, P. and Crato, N. (1995). New Tests for Stationarity and Parity Reversion: Evidence on New Zeland Real Exchange Rates. *Empirical Economics*, 20:559–613.

Yadav, P. K., Pope, P. F., and Paudyal, K. (1994). Threshold Autoregressive Modeling in Finance: The Price Differences of Equivalent Assets. *Mathematical Finance*, 4(2):205–221.

Yajima, Y. (1985). On Estimation of Long Memory Time Series Models. *Australian Journal of Statistics*, 27(3):303–320.

Yakowitz, S. J. and Spragins, J. D. (1968). On the Identifiability of Finite Mixtures. *Annals of Mathematical Statistics*, 39(1):209–214.

Yamamoto, T. (1976). Asymptotic Mean Square Prediction Error for an Autoregressive Model with Estimated Coefficients. *Applied Statistics*, 25:123–127.

Yamamoto, Y. and Perron, P. (2013). Estimating and Testing Multiple Structural Changes in Linear Models Using Band Spectral Regressions. *Econometrics Journal*, 16(3):400–429.

Yang, F. and Leon-Gonzalez, R. (2010). Bayesian Estimation and Model Selection in the Generalized Stochastic Unit Root Model. *Studies in Nonlinear Dynamics & Econometrics*, 14(4). Article 5.

Yang, R.-C. (2010). Towards Understanding and Use of Mixed-Model Analysis of Agricultural Experiments. *Canadian Journal of Plant Science*, 90(5):605–627.

Yoon, G. (2003). A Simple Model that Generates Stylized Facts of Returns. UCSD Economics Working Paper No. 2003-04.

Young, P. C. (2011). *Recursive Estimation and Time-Series Analysis*. Springer, Berlin, 2nd edition.

Yule, G. U. (1927). On a Method for Investigating Periodicities in Disturbed Series with Special Reference to Wolfer's Sunspot Numbers. *Philosophical Transactions of the Royal Society of London A*, 226:267–298.

Zaman, A. (2002). Maximum Likelihood Estimates for the Hildreth–Houck Random Coefficients Model. *Econometrics Journal*, 5(1):237–262.

Zeileis, A. (2006). Object-Oriented Computation of Sandwich Estimators. *Journal of Statistical Software*, 16:1–16.

Zeisel, H. (1989). On the Power of the Durbin–Watson Test Under High Autocorrelation. *Communications in Statistics—Theory and Methods*, 18:3907–3916.

Zellner, A. (2001). Keep it Sophisticatedly Simple. In Zellner, A., Keuzenkamp, H. A., and McAleer, M., editors, *Simplicity, Inference and Modelling*, pages 242–262. Cambridge University Press, Cambridge.

Zhang, K. and Chan, L. (2009). Efficient Factor GARCH Models and Factor-DCC Models. *Quantitative Finance*, 9(1):71–91.

Zhigljavsky, A. (2010). Singular Spectrum Analysis for Time Series: Introduction to this Special Issue. *Statistics and Its Interface*, 3:255–258.

Zhou, T. and Chan, L. (2008). Clustered Dynamic Conditional Correlation Multivariate GARCH Model. In Song, I.-Y., Eder, J., and Nguyen, T. M., editors, *Data Warehousing and Knowledge Discovery: 10th International Conference, DaWaK 2008 Turin, Italy, September 2-5, 2008 Proceedings*, pages 206–216.

Zhu, D. and Galbraith, J. W. (2010). A Generalized Asymmetric Student-t Distribution with Application to Financial Econometrics. *Journal of Econometrics*, 157(2):297–305.

Zhu, D. and Galbraith, J. W. (2011). Modeling and Forecasting Expected Shortfall with the Generalized Asymmetric Student-t and Asymmetric Exponential Power Distributions. *Journal of Empirical Finance*, 18(4):765–778.

Zhu, L.-X. and Neuhaus, G. (2003). Conditional Tests for Elliptical Symmetry. *Journal of Multivariate Analysis*, 84(2):284–298.

Zhu, Q. J., Bailey, D. H., López de Prado, M., and Borwein, J. M. (2017). The Probability of Backtest Overfitting. *Journal of Computational Finance*, 20(4):39–69.

Zinde-Walsh, V. (1988). Some Exact Formulae for Autogressive Moving Average Processes. *Econometric Theory*, 4:384–402.

Zivot, E. (2018). Modeling Financial Time Series with R. Announced, and presumably forthcoming.

Zivot, E. and Wang, J. (2006). *Modeling Financial Time Series with S-PLUS*. Springer, New York.

Index

e

f

g

h